About the Cover

A Canada lynx (*Lynx canadensis*) with piercing green eyes seems to be looking directly at you on our cover. The image draws you in, inviting you to ask and answer questions about how life works. The Canada lynx is a visually stunning cat, with its distinctive ear tufts and long hair hanging down from its lower cheeks. Its hind limbs are longer than its front limbs, so its back slopes downward as it hunts in northern forests of North America, searching for prey. Just from looking at it you can tell it's a predator, with its sharp eyes, strong limbs, lean body, powerful muscles, and retractable claws. Predation has shaped its form and function over many generations. It has also shaped its prey—the snowshoe hare (*Lepus americanus*).

The interaction of the Canada lynx and snowshoe hare echoes and highlights the themes you will encounter throughout your AP® Biology class. *Evolution* has influenced the form and function of all organisms on Earth, and that's particularly clear when looking at predators and prey. The two have evolved in lockstep, driving each other's adaptations. Underpinning evolution is the idea of *Information Storage and Transmission,* which allows organisms to function and pass on traits from generation to generation. None of this would be possible without *Energetics*—a source of energy fueling the entire process and the ability of organisms to harness this energy to carry out their functions. And the lynx and hare don't exist in isolation, but instead are part of an ecosystem that functions as a whole. Thus, *Systems Interactions,* whether at the scale of a forest or a cell, provides a way to put all the pieces together.

While you will approach biology class by class and module by module, don't forget that these themes are inseparable in nature. The lynx on the cover is there to remind you of the four *Big Ideas of AP® Biology* and how they work together to sustain life.

Biology

for the
AP® Course

James Morris

BRANDEIS UNIVERSITY

Dominic Castignetti

LOYOLA UNIVERSITY CHICAGO (EMERITUS)

John Lepri

UNIVERSITY OF NORTH CAROLINA AT GREENSBORO

Rick Relyea

RENSSELAER POLYTECHNIC INSTITUTE

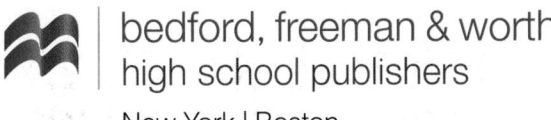
bedford, freeman & worth
high school publishers

New York | Boston

AP® is a registered trademark of the College Board, which was not involved in and does not endorse this product.

Executive Vice President, General Manager: Charles Linsmeier
Executive Program Director: Ann Heath
Executive Program Manager: Yolanda Cossio
Lead Developmental Editor: Rebecca Kohn
Executive Developmental Editor: Lisa Samols
Senior Media Editor: Kim Morté
Editorial Assistant: Meghan Kelly
Marketing Manager: Thomas Menna
Marketing Assistant: Nicollette Brady
Market Development Manager: Kelly Johnson
Senior Director, Content Management Enhancement: Tracey Kuehn
Senior Managing Editor: Lisa Kinne
Senior Content Project Manager: Martha Emry
Senior Workflow Project Manager: Paul Rohloff
Production Supervisor: Robert Cherry
Senior Media Permissions Manager: Christine Buese
Photo Researcher: Lisa Passmore, Lumina Datamatics, Inc.
Director of Digital Production: Keri deManigold
Lead Media Project Manager: Jodi Isman
Director of Design, Content Management: Diana Blume
Design Services Manager: Natasha Wolfe
Cover Designer: John Callahan
Interior Designer: Dirk Kaufman
Art Manager: Matthew McAdams
Illustrations: Troutt Visual Services, Joseph BelBruno, Imagineering Art
Printing and Binding: Transcontinental
Cover, About the Cover, pg. iii Image: Ken Canning/Getty Images
Back Cover Image: S.J. Krasemann/Getty Images

Library of Congress Control Number: 2021945885

ISBN-13: 978-1-319-11331-5
ISBN-10: 1-319-11331-1

Printed in Canada

W. H. Freeman and Company
Bedford, Freeman & Worth High School Publishers
120 Broadway
New York, NY 10271
bfwpub.com/apbio1e

To Jen Pfannerstill

All AP® Biology students need a friend. The course is stimulating and will challenge you to learn new and amazing things. Along that path, a steady, smart, knowledgeable, and concerned friend can guide and motivate you to do your best. Jen Pfannerstill was a best friend to AP® Biology, as her students and colleagues will testify. More than two decades of her fierce commitment to the AP® Biology curriculum, the exam, and biology students serve as an example of what a teacher can be. Unfortunately, we lost this dear friend too soon. We dedicate this work to her friendship and commitment to all things AP® Biology. We hope that you never feel alone in your AP® Biology journey.

Brief Contents

Contents

Unit 8: Ecology 722

About the Authors

Ben Morris

James Morris is professor of biology at Brandeis University. He teaches a wide variety of courses for majors and non-majors, including introductory biology, evolution, genetics and genomics, epigenetics, comparative vertebrate anatomy, and a first-year seminar on Darwin's *On the Origin of Species*. He is the recipient of numerous teaching awards from Brandeis and Harvard. His research focuses on the rapidly growing field of epigenetics, making use of the fruit fly *Drosophila melanogaster* as a model organism. He currently pursues this research with undergraduates in order to give them the opportunity to do genuine, laboratory-based research early in their scientific careers. Dr. Morris received a PhD in genetics from Harvard University and an MD from Harvard Medical School. He was a Junior Fellow in the Society of Fellows at Harvard University and a National Academies Education Fellow and Mentor in the Life Sciences. For the last 2 years he has served as a reader for the AP® Biology exam.

Dorothy Castignetti

Domenic Castignetti is professor emeritus of biology at Loyola University Chicago. During his 37 years at Loyola, he taught many courses, including the introductory biology two-semester course, general microbiology, microbial physiology, and biochemistry. He was awarded Loyola University's Master Teacher of the College of Arts and Sciences Award and received the first Faculty/Staff Member of the Year Award from the Student Government Association. Dr. Castignetti received an MS from Colorado State University and a PhD in microbial ecology and physiology from the University of Massachusetts Amherst. His research, conducted with undergraduate and graduate students, focused on the microbiology and biochemistry of soil microorganisms. He has been associated with the AP® Biology program since 2003 and has served as a reader, table leader, and member of the Development Committee. He was chief reader for the AP® Biology Exam for 5 years.

John Lepri

John Lepri is Professor of Biology and Faculty Fellow in the Lloyds International Honors Program at the University of North Carolina at Greensboro. He teaches courses in introductory biology, biological clocks, and animal physiology. His earlier research focused on hormones and pheromones that coordinate reproduction in mammals. Dr. Lepri earned a BS in zoology and psychology from the University of Michigan, and a PhD in zoology from North Carolina State University. His professional activities include teacher preparation programs for high school science teachers. Dr. Lepri has worked with College Board and Educational Testing Service for more than 25 years, including 5 years as chief reader for the AP® Biology Exam.

Christine Relyea

Rick Relyea is the David Darrin Senior '40 Endowed Chair in Biological Sciences and the director of the Darrin Freshwater Institute at Rensselaer Polytechnic Institute. He received a BS in environmental forest biology from the SUNY College of Environmental Science and Forestry, an MS in wildlife management from Texas Tech University, and a PhD in ecology and evolution from the University of Michigan. He has authored more than 200 scientific articles and book chapters and presented research seminars throughout the world. Dr. Relyea was a professor at the University of Pittsburgh for 15 years, where he was named the Chancellor's Distinguished Researcher and received the Tina and David Bellet Teaching Excellence Award. In 2014, he moved to Rensselaer Polytechnic Institute to direct The Jefferson Project, which is the most technologically advanced research endeavor to study freshwater lakes. Rick has a strong interest in high school education, including hosting high school science teachers who conduct research in his laboratory. He is co-author of *Environmental Science for the AP® Course,* also published by BFW publishers.

Content Advisory Board

The following individuals provided invaluable help in the development of *Biology for the AP® Course*. We thank them for giving their time and expertise to make this project a unique fit for the course.

Sean Bennett has taught AP® Biology for over 20 years. He has participated in the AP® Biology reading since 2004 and has served as a reader, table leader, and question leader. Sean has a BA in biology from Washington & Jefferson College and an MS in biology from West Virginia University. He regularly travels with students to Costa Rica and the Galapagos Islands. He currently teaches at Career Center High School in Winston-Salem, North Carolina.

Joan Carlson taught AP® Biology for 30 years and served as science department head at Bancroft School in Worcester, Massachusetts. Joan was also a grader for the AP® Biology Exam for 30 years and served on the AP® Biology Test Development Committee for 6 years. She is currently an AP® Biology New England consultant and an adjunct instructor at Fitchburg State University.

Brian Lazzaro is a Liberty Hyde Bailey Professor of Entomology and of Ecology and Evolutionary Biology at Cornell University. He has been on the faculty at Cornell since 2002 and has taught courses including Introduction to Evolutionary Biology, Ecological Genetics, Advanced Population Genetics, and Ecology and Evolution of Infectious Disease. In 2015–2018, he served on the AP® Biology Test Development Committee, which creates the annual AP® Biology Exam. Dr. Lazzaro acted as co-chair of the committee in 2015–2018 and helped to draft the current version of the AP® Biology curriculum framework. He received his BS from the University of California, Davis and his PhD from Pennsylvania State University. His research program is focused on genetic and environmental factors that alter insect resistance to microbial infection.

Paula Phillips (Teacher's Edition author) has taught AP® Biology and related biological and chemical sciences in both public and independent schools for over three decades. She has been involved with the AP® Biology reading since 1994, serving as reader, table leader, and assistant chief reader. She conducts workshops and AP® Summer Institutes nationwide. In 2016, she was a recipient of the Outstanding Biology Teacher Award from the National Association of Biology Teachers. She has been awarded scholarships to study molecular biology at both Duke University and Cold Spring Harbor Laboratories and was a grant recipient for the NYSTEM Stem Cell Research Program for Teachers at Cornell University. Paula is a graduate of North Carolina State University with a BS in forestry and environmental policy and a BS in science education and biology. She holds an MEd in science education from the University of Central Florida. She currently teaches at Lansing High School in Lansing, New York.

Katherine Smanik

James Smanik (Teacher's Edition author) has taught AP® Biology, Anatomy and Physiology, and other biological sciences for over 34 years. He has been involved with the AP® Biology reading for 28 years, serving as a reader and for 12 years as a table leader. He has presented more than 60 workshops and AP® Summer Institutes. Jim served on the AP® Biology Development Committee from 2011 to 2016. He has written and reviewed materials for Pre-AP® Biology and AP® Biology. In 2013, Jim was named Outstanding Teacher by *Cincy Magazine*. Jim graduated from the University of Cincinnati with a BS in biology and earned an MAT with an emphasis in biological science from Miami University. He currently teaches at Sycamore High School in Cincinnati, Ohio.

Robert Davidson

Mark Stephansky has over 34 years of teaching experience and 20 years as an AP® Biology instructor. In addition to teaching at the high school level, Mark is also an adjunct biology instructor at Massasoit Community College in Brockton, Massachusetts. Since 2002, Mark has served in various roles at the AP® Biology reading, including as a reader, table leader, and question leader. Mark has a BS in biology from Bridgewater State University and an MEd from Cambridge College. He is currently the Science Curriculum Director at Whitman-Hanson Regional High School in Whitman, Massachusetts.

Gordon Uno

Gordon Uno (Teacher's Edition author) is a David Ross Boyd Professor of Botany at the University of Oklahoma, Norman. He holds a BS from the University of Colorado, Boulder, and a PhD from the University of California, Berkeley. Dr. Uno teaches Introductory Plant Biology, and his research interests include plant reproductive biology and science education. He was elected president of the National Association of Biology Teachers and has also served as president of the Botanical Society of America. Dr. Uno was a member of the committee that revised the AP® Biology Course and Exam from 2010 to 2016 and served as co-chair for four years.

Jamal Hardman, Hardman Portraits

Audra Brown Ward is a College Board consultant, and a reader and table leader for the AP® Biology Exam. She has led numerous AP® Biology Summer Institutes and one-day workshops, as well as workshops at the National Association of Biology Teachers Conferences. Audra received a BS in biochemistry from Spelman College and an MS in biochemistry from Georgia Tech. She received the 2007 NABT Outstanding Biology Teacher Award for Georgia. Audra is an upper school biology teacher at the Westminster Schools of Atlanta, where she teaches AP® Biology and AP® Environmental Science.

Acknowledgments

Biology for the AP® Course is based on *How Life Works,* Third Edition, by James Morris, Daniel Hartl, Andrew Knoll, Robert Lue, Melissa Michael, Andrew Berry, Andrew Biewener, Brian Farrell, and N. Michele Holbrook, as well as assessment authors Jean Heitz, Mark Hens, Elena Lozovsky, John Merrill, Randall Phillis, and Debra Pires. The authors of this text are grateful for their permission to build on the foundation established by the *How Life Works* team.

James Morris thanks his high school biology teacher, Dr. Dorothy Andrews, who early on instilled a love of biology and encouraged him to take time to observe the natural world. Dominic Castignetti thanks his wife, Dorothy, who manages to be engaged when he inundates her with biological facts, and for her encouragement, patience, and support throughout the years. John Lepri has worked with hundreds of high school teachers of AP® Biology classes, and he sends thanks to them for their generosity in sharing teaching methods and in nurturing our young learners—you all are awesome! Rick Relyea would like to thank Christine, Isabelle, and Wyatt Relyea.

Many individuals contributed to the creation of this book. The members of our content advisory board provided thoughtful help, often daily. At BFW publishers, we wish to thank the efforts of the many individuals who have made this book possible. On the editorial side, we thank executive program director Ann Heath, program manager Yolanda Cossio, executive development editor Lisa Samols, and developmental editor Rebecca Kohn. We wish to thank Brian Lazarro and Erika Yates for help with the assessments. In art and design, we thank senior design services manager Natasha Wolfe and art manager Matthew McAdams. We thank Martha Emry, content project manager, for carefully overseeing the production process. In addition, thanks goes to Sarah Wales-McGrath for copyediting and Leslie Hoy for proofreading. Senior photo editor Christine Buese and photo researcher Lisa Passmore are responsible for the beautiful and engaging images in the book. Thanks to marketing manager Thomas Menna and market development manager Kelly Johnson for spreading the word about the book.

Reviewers

We thank the many individuals whose insights have helped shape this book.

Gavin Alexander, *Fontana High School*

Anthony J. Amato, *Palmetto High School*

Mina Basu, *Dougherty Valley High School*

Ken Batemen, *Wellesley High School*

Janet Harver Belval, *South Windsor High School*

Sean Bennett, *Career Center High School*

Mel Berg, *Waconia High School*

Adam Bernstein, *Cliffside Park High School*

Cynthia Blakeslee, *Loyola Academy*

Judy H. Bone, *Des Arc High School*

Abigail Borroto, *Londonderry High School*

Candace Bridges, *Stratford Academy*

Mary Brunson, *Lincoln-Sudbury Regional High School*

Joan Carlson, *Fitchburg State University*

Matthew J. Colwell, *Cathedral Preparatory School*

Jason Dries, *Piedmont Hills High School*

Justin S. Eadler, *Larkin High School*

Sarah J. Elacqua, *Waterville Jr./Sr. High School*

Diane Escobar, *Miami-Dade County Public Schools*

Tom Freeman, *Esperanza High School*

Lee Ferguson, *Allen High School*

JoAnn M, Gensert, *Bronx High School of Science*

Christopher Greco, *Saegertown Jr./Sr. High School*

Alexandra Hall, *Deerfield Windsor School*

Lynn Harder, *Falmouth High School*

Jonathan Harris, *Gardiner Area High School*

Dori L. Hess, *GlenOak High School*

Kathryn Hopkins, *Waco High School*

Lynda Kiesler, *Albert Lea High School*

Simone Kuklinski, *St. Michaels University School*

Brian Lazzaro, *Cornell University*

Leslie Lopez, *Round Rock High School*

Amy Magnuson, *Falmouth High School*

Jennifer McMaster Lumia, *Jamestown High School*

Heather MacDonald, *Nantucket High School*

Amy Magnuson, *Falmouth High School*

Keila Mena, *Gustine High School*

Nicole Phelps, *Peotone High School*

Paula Phillips, *Lansing High School*

Patricia Prime, *Canajoharie High School*

Jill Ronstadt, *Orange Lutheran High School*

Joe Scalice, *Sparrows Point High School*

Melissa Schumann, *Elwood Community High School*

Kevin Short, *Charleston County School of the Arts*

Stephanie Slade, *Karl G. Maeser Preparatory Academy*

Carolyn Slanetz-Chiu, *Portledge School*

James Smanik, *Sycamore High School*

Christina Snedeker, *Hanover Central High School*

Jessica Snowdon, *Londonderry High School*

Mark Stephansky, *Whitman Regional High School*

Karen Cruse Suder, *The Summit Country Day School*

Paulette L. Unger, *Christ Church Episcopal School*

Gordon Uno, *University of Oklahoma*

Greg Walls, *Yorba Linda High School*

Sherry Wantz, *Athens Drive High School*

Audra Brown Ward, *Westminster Schools of Atlanta*

Janet Wilson, *Mother McAuley High School*

Sally Winecke, *Lakeville Schools*

Allison L. Winward, *Kamiakin High School*

Andrew H Wright, *Lake Ridge Academy*

Jennifer Zinman, *Lacordaire Academy*

Third Edition College Reviewers

Danyelle Aganovic, *University of North Georgia*

Adam Aguiar, *Stockton University*

Nancy Aguilar-Roca, *University of California, Irvine*

Christine Andrews, *University of Chicago*

Jenny Archibald, *University of Kansas*

Jessica Armenta, *Austin Community College*

Hilal Arnouk, *St. Louis College of Pharmacy*

Andrea Aspbury, *Texas State University*

Ann Auman, *Pacific Lutheran University*

Felicitas Avendano, *Grand View University*

Ellen Baker, *Santa Monica College*

Gerard Beaudoin, *Trinity University*

Lauryn Benedict, *University of Northern Colorado*

Joydeep Bhattacharjee, *University of Louisiana, Monroe*

Andrew Blaustein, *Oregon State University*

Joel A. Borden, *University of South Alabama*

Andrew Bouwma, *Oregon State University*

Christoper Brown, *Georgia Gwinnett College*

Jill Buettner, *Richland College*

Sharon Bullock, *University of North Carolina, Charlotte*

Denise Carroll, *Georgia Southern University*

Dale Casamatta, *University of North Florida*

Anne Casper, *Eastern Michigan University*

Michelle Cawthorn, *Georgia Southern University*

Rebekah Chapman, *Georgia State University*

Steven Clark, *University of Michigan*

Kimberley Clawson, *Temple College*

John Cogan, *Ohio State University*

Brett Couch, *University of British Columbia*

Jennifer Cymbola, *Grand Valley State University*

Cari Deen, *University of Central Oklahoma*

Tracie Delgado, *Northwest University*

Stephanie DeVito, *University of Delaware*

Sunethra Dharmasiri, *Texas State University*

Christine Donmoyer, *Allegheny College*

Janet Duerr, *Ohio University*

Laura Eidietis, *University of Michigan*

Bert Ely, *University of South Carolina*

Miles Engell, *North Carolina State University*

Jean Everett, *College of Charleston*

David Fitch, *New York University*

Matthias Foellmer, *Adelphi University*

Caitlin Gabor, *Texas State University*

Jason Gee, *East Carolina University*

Cynthia Giffen, *University of Michigan*

Sharon Gillies, *University of the Fraser Valley*

Marcia Graves, *University of British Columbia*

Linda Green, *Georgetown University*

Kathryn Gronlund, *Lone Star College, Montgomery*

Nancy Guild, *University of Colorado, Boulder*

Susan R. Halsell, *James Madison University*

Sally Harmych, *University of Toledo*

Phillip Harris, *University of Alabama*

Joseph Harsh, *James Madison University*

Mary Haskins, *Rockhurst University*

Christiane Healey, *University of Massachusetts, Amherst*

David Hearn, *Towson University*

Susan Herrick, *University of Connecticut*

Brad Hersh, *Allegheny College*

James Hickey, *Miami University*

Sarah E. Hosch, *Oakland University*

Brian Hyatt, *Bethel University*

Joanne K. Itami, *University of Minnesota, Duluth*

Jamie Jensen, *Brigham Young University*

Michele Johnson, *Trinity University*

Seth Jones, *University of Kentucky*

Tonya Kane, *University of California, Los Angeles*

Lori Kayes, *Oregon State University*

Benedict Kolber, *Duquesne University*

Ross Koning, *Eastern Connecticut State University*

Peter Kourtev, *Central Michigan University*

Nadine Kriska, *University of Wisconsin, Whitewater*

Tim Kroft, *Auburn University at Montgomery*

Troy A. Ladine, *East Texas Baptist University*

James Langeland, *Kalamazoo College*

Neva Laurie-Berry, *Pacific Lutheran University*

Christian Levesque, *John Abbott College*

B. Larry Li, *University of California, Riverside*

Jason Locklin, *Temple College*

Dale Lockwood, *Colorado State University*

David Lonzarich, *University of Wisconsin, Eau Claire*

Kari Loomis, *University of Massachusetts, Amherst*

Patrice Ludwig, *James Madison University*

Nilo Marin, *Broward College*

Timothy McCay, *Colgate University*

Lori McGrew, *Belmont University*

Michael Meighan, *University of California, Berkeley*

James Mickle, *North Carolina State University*

Chad Montgomery, *Truman State University*

Scott Moody, *Ohio University*

Paul Moore, *Bowling Green State University*

Mark Mort, *University of Kansas*

Kimberlyn Nelson, *Pennsylvania State University*

Judy Nesmith, *University of Michigan, Dearborn*

Alexey Nikitin, *Grand Valley State University*

Matthew Nusnbaum, *Georgia State University*

Mary Olaveson, *University of Ontario Institute of Technology*

Jennifer O'Neil, *Houston Community College*

Samantha Parks, *Georgia State University*

Thomas Peavy, *California State University, Sacramento*

John Peters, *College of Charleston*

Michael Plotkin, *Mt. San Jacinto College*

Mirwais Qaderi, *Mount Saint Vincent University*

Raul Ramirez, *Oklahoma City Community College*

Amy Reber, *Georgia State University*

Trevor Rivers, *University of Kansas*

Jeremy Rose, *Oregon State University*

Caleb Rounds, *University of Massachusetts, Amherst*

Jaime L. Sabel, *University of Memphis*

Donald Sakaguchi, *Iowa State University*

Leslie Saucedo, *University of Puget Sound*

Erik Scully, *Towson University*

Miriam Segura-Totten, *University of North Georgia*

Justin Shaffer, *University of California, Irvine*

Mark Sherrard, *University of Northern Iowa*

James Shinkle, *Trinity University*

Marcia Shofner, *University of Maryland, College Park*

Karen Smith, *University of British Columbia*

William Smith, *University of Kansas*

Kay Song, *Georgia State University*

Frederick Spiegel, *University of Arkansas, Fayetteville*

Barbara Stegenga, *University of North Carolina*

Robert Steven, *University of Toledo*

Tara Stoulig, *Southeastern Louisiana University*

Mark Sturtevant, *Oakland University*

Elizabeth Sudduth, *Georgia Gwinnett College*

Brad Swanson, *Central Michigan University*

Ken Sweat, *Arizona State University, West Campus*

Jonathan Sylvester, *Georgia State University*

Annette Tavares, *University of Ontario Institute of Technology*

Casey terHorst, *California State University, Northridge*

Sharon Thoma, *University of Wisconsin, Madison*

Gail Tompkins, *Wake Technical Community College*

Ron Vanderveer, *Eastern Florida State College*

Daryle Waechter-Brulla, *University of Wisconsin, Whitewater*

Stacey Weiss, *University of Puget Sound*

Naomi L. B. Wernick, *University of Massachusetts, Lowell*

Alan White, *University of South Carolina*

Mary White, *Southeastern Louisiana University*

Frank Williams, *Langara College*

Lisa Williams, *Northern Virginia Community College*

Paul Wilson, *Trent University*

Jennifer Zettler, *Georgia Southern University, Armstrong Campus*

Hongmei Zhang, *Georgia State University*

To the Student:

Congratulations!

The decision to take AP® Biology requires curiosity, courage, and commitment. And now you are about to embark on a dramatic learning adventure. The subject is no less than how all of life works. Along the way you will undoubtedly face challenges. You will also be rewarded with a greater understanding of the world around you and a set of skills that will serve you whether or not you take additional biology courses in the future. You will emerge from this course in a better position to understand issues related to biological sciences that affect you as a person and as a citizen.

We hope that you will find this book an invaluable companion. We wrote this book to provide you with all the tools you need to succeed on the exam and gain confidence with the subject matter. We have divided the book's units into modules that are shorter than typical book chapters. This will help you pace your reading. Make sure you take the time to go over the Concept Check questions at the end of each section. Check your answers at the back of the book and then review any material you have not yet mastered. At the end of the module, Review Questions will also help you determine which concepts need more study. You will be able to practice for the exam all year with hundreds of AP® Practice Questions located at the end of modules and units, and at the end of the book you'll find a cumulative AP® Practice Exam, just like the actual exam in size and scope.

Visual analysis is an important component of the course. We have selected photos and created images that enhance your experience and bring the subject matter to life. Notice the detailed captions on figures and text discussions of figures, designed to help you follow and understand the key ideas of AP® Biology.

Pay close attention to the numerous AP® Exam Tips scattered throughout the modules. These tips, written by former chief readers for the AP® Biology Exam, will give you an inside track on content you need to know, and on key strategies for answering exam questions. Also make sure to read all the boxed material carefully. Analyzing Statistics and Data boxes walk you through critical math and statistics tools and provide additional practice. Practicing Science boxes are designed to help you develop your ability to use the critical science practice of questions and methods. Finally, make sure to read and review the four tutorials to solidify your understanding of statistics, probability, population growth, and graphing.

Good luck with your year!

James Morris   *Domenic Castignetti*   *John Lepri*   *Rick Relyea*

Get the Most from Your Book.

Your AP® Biology adventure begins here! This book has been created with YOU in mind. It is packed with features to help you learn effectively and do well on the AP® Biology Exam.

Organization, Scope, and Pacing

> Your book matches the course both in organization and scope. Module content is built around concepts specific to your course and the AP® Biology Exam.

> Content is scaffolded so each new topic is introduced at the right time so you have all the necessary foundations. **Modules** contain manageable chunks of material for you to pace yourself and review as you study each new topic.

Unit 1
Chemistry of Life

Portugal seacoast by LOOK Die Bildagentur der Fotografen GmbH/Alamy

Unit 1	0	1	2	3	4	5

Module 0

Introduction

LEARNING GOALS ▶LG 0.1 Four Big Ideas form a fundamental basis for understanding biology.
▶LG 0.2 Scientific inquiry is a deliberate way of asking and answering questions about nature.

The ears of the snowshoe hare (*Lepus americanus*) stand tall as it listens carefully and its nose twitches as it smells the air for the scent of its predator, the Canada lynx (*Lynx canadensis*). At the same time, the lynx uses its keen eyesight, its sense of smell, and its ears to locate its prey, the hare. Having spotted the hare, the lynx quietly and deliberately moves toward it, slowly closing the distance between itself and the hare. Then, suddenly, the lynx springs upon the hare and the chase for life and food begins.

The activities of predator and prey depend on a number of key processes that they both possess. For instance, both hare and lynx are strong and swift runners. Both are exquisitely adapted to sense their habitats, be aware of danger,

understanding of the **gene,** which is the unit of heredity, and the **genome,** which is all of the genetic information that an organism contains. This information is helping us come up with new ways of fighting disease in humans, other animals, and plants. It is also helping us understand how different groups of organisms have evolved and how endangered species might be saved. Many new biological frontiers have opened up in recent decades, while others are still waiting to be explored. For example, researchers are only just beginning to understand how bacteria in our digestive system affect our health and well-being. Other researchers are looking at how the temperature and acidity of seawater affect the ecology of coral reefs.

The Perfect Book for AP® Student Success

This book is your ultimate study tool.

Study and Review

LEARNING GOALS ▶ **LG 1.1** Matter and energy govern the properties of life.
▶ **LG 1.2** The atom is the fundamental unit of matter.
▶ **LG 1.3** Atoms combine to form molecules linked by chemical bonds.
▶ **LG 1.4** Carbon is the backbone of organic molecules.

The **Four Big Ideas of AP® Biology** are emphasized in every module to keep you focused on content that matters.

Module 1 Summary

LG 1.1 Matter and energy govern the properties of life.

- All organisms are made up matter and require energy to sustain life. Page 26
- Organisms obtain matter from other organisms and the environment, and matter therefore moves in a cycle, with the same atoms used and reused. Page 26

Every module is organized around two to four **learning goals**. These learning goals are revisited in module **summaries**. Use these goals to help you organize new information and to review what you have learned.

PREP FOR THE AP® EXAM

FOCUS ON THE BIG IDEAS

ENERGETICS: Look for the elements that make up all organisms, and the way in which organisms exchange these elements with the environment.

PREP FOR THE AP® EXAM

REVISIT THE BIG IDEAS

ENERGETICS: Using the content in this module, **identify** the key elements that all organisms obtain from the environment and exchange with the environment.

End of module multiple-choice **Review Questions** help you further assess your strengths and weaknesses with module content.

✓ Concept Check

3. **Describe** the components of an atom.
4. **Describe** how the rows and columns of the periodic table of the elements are organized.
5. **Identify** what the superscripts of ^{14}N and ^{15}N signify.
6. **Calculate** how many additional electrons would be needed to fill the outer energy level of C.
7. **Identify** the location of the electrons around an atom that have the most energy associated with them.

Review Questions

1. Using the periodic table shown in Figure 1.4, identify which atom is most likely to have properties similar to nitrogen.
   - (A) Oxygen (O)
   - (B) Bismuth (Bi)
   - (C) Lead (Pb)
   - (D) Carbon (C)

2. The element sodium (Na) has an atomic number of 11 and an atomic mass of 23. Calculate the number of protons and neutrons in Na.
   - (A) 11 protons and 11 neutrons
   - (B) 11 protons and 10 neutrons
   - (C) 11 protons and 12 neutrons
   - (D) 10 protons and 13 neutrons

3. Which type of bond is formed between a sodium ion and a chloride ion?
   - (A) Ionic
   - (B) Neutral
   - (C) Polar covalent bond
   - (D) Nonpolar covalent bond

4. Which type of bond is formed between two oxygen atoms?
   - (A) Ionic
   - (B) Neutral
   - (C) Polar covalent bond
   - (D) Nonpolar covalent bond

5. When two molecules undergo a chemical reaction, they
   - (A) lose atoms.
   - (B) keep the same bonding partners that they had as reactants.
   - (C) establish new bonding partners.
   - (D) incorporate atoms from the air and water as well as from the reactants.

6. Proteins, nucleic acids, and carbohydrates are all
   - (A) information molecules.
   - (B) signaling molecules.
   - (C) polymers made up of repeating subunits.
   - (D) monomers that combine to form polymers.

7. Two organic molecules that store energy for use by the cell are
   - (A) carbohydrates and nucleic acids.
   - (B) carbohydrates and lipids.
   - (C) proteins and nucleic acids.
   - (D) nucleic acids and lipids.

At the end of each section, **Concept Check** questions give you the opportunity to check your understanding of key concepts. Because you'll find the answers at the back of the book, you will know right away which concepts need extra attention.

Practice for the Exam All Year . . .

Your book is filled with many opportunities to practice AP®-style questions and prepare for the exam.

AP® Practice Opportunities at the Module Level

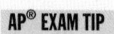

PREP FOR THE AP® EXAM

AP® EXAM TIP

Nonpolar and polar covalent bonds govern many of the properties of organic molecules. Knowing these concepts will help you understand how these molecules function, which will help you succeed on the AP® Biology Exam.

AP® Exam Tips are scattered throughout the modules to help you focus on key content you should know for the exam.

Module 14
AP® Practice Questions

PREP FOR THE AP® EXAM

At the end of each module, **AP® Practice Questions,** which are structured like the questions you'll see on the exam, help you build confidence as you sharpen your test-taking skills.

Questions 3 and 4 refer to the following information.

Glucose-6-phosphate dehydrogenase (G6PD) is an enzyme that converts a substrate, glucose-6-phosphate, to a product, 6-phosphogluconolactone. A scientist added 0.1 nanomole (nmol) of G6PD to a test tube containing 25 nmol of glucose-6-phosphate (the substrate) and measured the accumulation of 6-phosphogluconolactone (the product) over 7 hours. The data that the scientist obtained are shown in the graph.

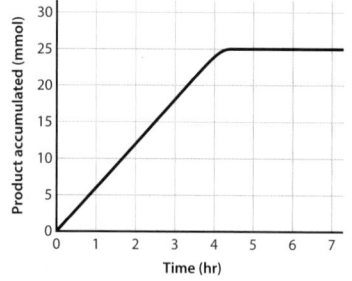

3. Based on the data provided, what is the best estimate of the rate at which G6PD converts substrate to product in this experiment?
   (A) 0.1 mmol/min
   (B) 2.0 mmol/min
   (C) 3.6 mmol/min
   (D) 5.0 mmol/min

4. Which provides the best explanation for why no further product accumulates after 5 hours?
   (A) All of the G6PD enzyme has been consumed and is no longer able to catalyze the reaction.
   (B) The reaction releases energy as heat, which causes the G6PD enzyme to become denatured.
   (C) All of the substrate has been consumed, so the reaction can no longer proceed.
   (D) The concentration of substrate becomes too high and inhibits the reaction.

Section 2: Free-Response Question

Pepsin is a digestive enzyme in animal stomachs that helps to digest, or break down, protein in the diet. A researcher tested the ability of 1 mg of pepsin to digest the protein albumin across a range of pH values. The researcher mixed six solutions, each with a different pH level, listed in the table below. Each solution was divided among 10 tubes. The researcher then added 1 mg of pepsin and 3000 mg of albumin to each tube. At the end of a 3-hour incubation period, the researcher measured the amount of albumin that had been digested in each tube. The mean amount of albumin digested at each pH and the standard error of the mean are shown in the table below.

pH	Mean albumin digested (mg)	Standard error (mg)
1	617	180
2	3000	0
3	2089	120
4	2103	170
5	1146	130
6	132	60

(a) **Draw** an appropriately labeled graph of pepsin activity across pH values, including error bars.

(b) **Identify** the pH at which pepsin activity is highest.

(c) Do you see evidence to support a claim that pepsin activity is different at pH = 3 and pH = 4? **Make a claim** and **justify** your reasoning based on the data provided.

(d) **Calculate** the percent activity of pepsin at pH = 5, relative to the maximum activity observed in the experiment. You may ignore the standard error for this calculation.

(e) **Make a claim** that would explain the observed activity of pepsin at pH = 6.

...and You Can Approach the Exam with Confidence

Regular practice helps you master the material.

AP® Practice Opportunities at the Unit and Full Book Level

Unit 1

PREP FOR THE AP® EXAM

AP® Practice Questions

Section 1: Multiple-Choice Questions
Choose the best answer for questions 1–15.

1. Using Figure 4.3 on page 67 as a reference, which amino acid would likely be found in the same cellular membrane region as the fatty acid tails of a phospholipid?
   (A) Histidine
   (B) Threonine
   (C) Leucine
   (D) Serine

2. A student is designing an experiment to test what happens to the growth of plants when they are watered with a low concentration of a sodium chloride (NaCl) solution and the presence or absence of fertilizer rich in phosphorus. The student hypothesizes that NaCl will decrease the growth of plants, no matter the type of soil present. The student plans to set up separate pots of sunflower seedlings with the treatments show in the table below.

Seedling number	Treatment
1	Low NaCl + fertilizer
2	Low NaCl + no fertilizer
3	Pure water + fertilizer
4	Pure water + no fertilizer

The seedlings will be watered daily in the student's backyard and growth will be observed and recorded over a period of 3 weeks. Which best describes the control group?

3. Which statement explains why water is a good solvent for carbohydrates, proteins, and nucleic acids, but not for lipids?
   (A) Carbohydrates, proteins, and nucleic acids are not very electronegative and will more readily bond with water than lipids.
   (B) Carbohydrates, proteins, and nucleic acids contain weaker, single bonds that water can easily break. Lipids contain stronger, double bonds that make this more difficult.
   (C) Carbohydrates, proteins, and nucleic acids have polar regions capable of forming hydrogen bonds with water. Lipids are not very polar.
   (D) Carbohydrates, proteins, and nucleic acids are much smaller macromolecules than lipids. Their size

> At the end of each unit, a comprehensive set of **Unit AP® Practice Questions** addresses content across the unit.

Cumulative AP® Biology Practice Exam

PREP FOR THE AP® EXAM

Section 1: Multiple-Choice Questions
Choose the best answer for questions 1–60.

1. Identify the main source of oxygen gas in Earth's atmosphere.
   (A) Natural geologic processes
   (B) Photosynthesis by plants, algae, and bacteria
   (C) Fermentation by microorganisms
   (D) Decay of organic material

2. Extremophiles are microorganisms that live in environments characterized by extreme temperature, acidity, salt concentrations, or pH. Many extremophiles that live in high-temperature environments without oxygen perform anaerobic cellular respiration with an electron transport chain that uses iron or sulfides as terminal electron acceptors instead of oxygen. Which of the following explains why this form of cellular respiration is observed in extreme environments without oxygen?

Use the following table to answer question 3.

First position (5' end)	Second position				Third position (3' end)
	U	C	A	G	
U	UUU Phe UUC Phe UUA Leu UUG Leu	UCU Ser UCC Ser UCA Ser UCG Ser	UAU Tyr UAC Tyr UAA Stop UAG Stop	UGU Cys UGC Cys UGA Stop UGG Trp	U C A G
C	CUU Leu CUC Leu CUA Leu CUG Leu	CCU Pro CCC Pro CCA Pro CCG Pro	CAU His CAC His CAA Gln CAG Gln	CGU Arg CGC Arg CGA Arg CGG Arg	U C A G
A	AUU Ile AUC Ile AUA Ile AUG Met/Start	ACU Thr ACC Thr ACA Thr ACG Thr	AAU Asn AAC Asn AAA Lys AAG Lys	AGU Ser AGC Ser AGA Arg AGG Arg	U C A G
G	GUU Val GUC Val GUA Val GUG Val	GCU Ala GCC Ala GCA Ala GCG Ala	GAU Asp GAC Asp GAA Glu GAG Glu	GGU Gly GGC Gly GGA Gly GGG Gly	U C A G

> A Cumulative **AP® Practice Exam** at the end of the book matches the actual AP® Biology Exam in structure and scope.

Build Essential Skills with All the Science Practices

Use these features to learn and review the statistics, math, and data analysis you need to succeed.

Skill-building with In-text Explanations and Boxes

Throughout the text, **Analyzing Statistics and Data** boxes walk through key statistical tools and techniques to help you practice and learn in context. Note the "play" icon which signifies an associated video.

Each Analyzing Statistics and Data box includes a **Your Turn** problem, which provides an opportunity for you to practice the relevant tools and gain confidence.

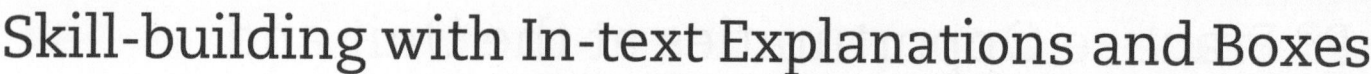

ANALYZING STATISTICS AND DATA ▶

PREP FOR THE AP® EXAM

Averages

For a detailed explanation of how to calculate mean, median, and mode, see page 20 for "Tutorial 1: Statistics." Here we review how to apply the concepts of mean, median, and mode and offer a problem for you to try.

PRACTICE THE SKILL
Below is a table listing several different ant species and the number of queens commonly found in a colony of each species. Considering the entire dataset, which measurement of the number of queens is largest: the mean, median, or mode?

Ant species	Number of queens in a small colony
Carpenter ant (*Camponotus pennsylvanicus*)	1
Red imported fire ant (*Solenopsis*	
Pavement ant (*Tetramorium caesp*	
Crazy ant (*Paratrechina longicornu*	
Pharaoh ant (*Monomorium pharac*	
Ghost ant (*Tapinoma melanoceph*	
Little black ant (*Monomorium min*	

We divide this sum by the number of values in the dataset, which in this case is 8:

$$82 \div 8 = 10.25$$

So, the mean is 10.25.

We can also do this calculation using the following equation, where $\bar{x}$ is the mean, n is the number of values in the dataset, and $\sum_{i=1}^{n} x_i$ is the sum of all of the values in the dataset:

$$\bar{x} = \frac{1}{n}\sum_{i=1}^{n} x_i$$

$$\bar{x} = \frac{1}{8}\sum_{i=1}^{8}(1 + 30 + 5 + 12 + 2 + 20 + 5 + 7)$$

$$\bar{x} = \frac{1}{8}(82) = 10.25$$

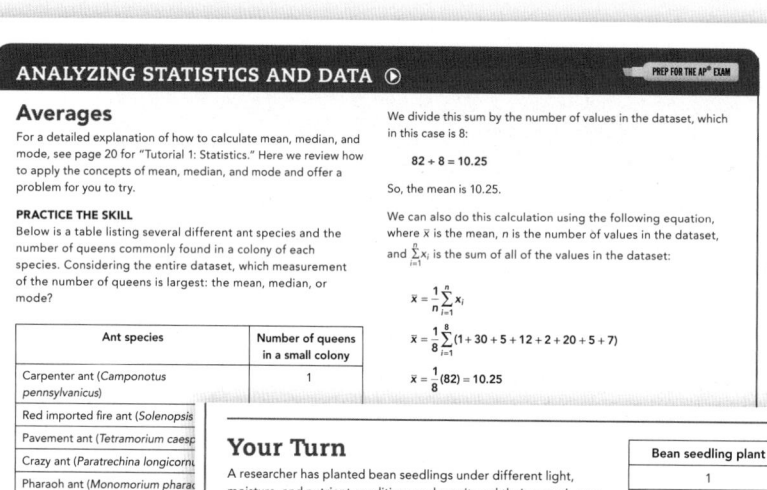

Your Turn

A researcher has planted bean seedlings under different light, moisture, and nutrient conditions and monitored their growth over several weeks. The data are recorded in the table. What are the mean, median, and mode for seedling height?

Bean seedling plant	Seedling height (cm)
1	20
2	6
3	15
4	23
5	5
	15
	21

Practicing Science boxes present and evaluate key experiments in the history of biology. Each box addresses the science practice of questions and methods with an in-depth look at research questions, experimental design, and results. Understanding questions and methods in the context of actual scientific work provides you with a better foundation for applying this science practice.

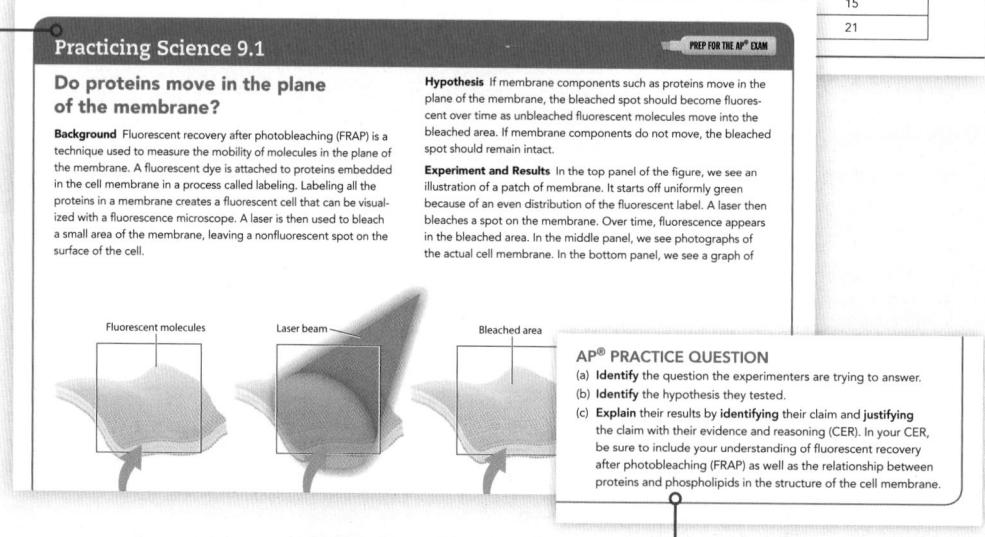

Practicing Science 9.1

PREP FOR THE AP® EXAM

Do proteins move in the plane of the membrane?

Background Fluorescent recovery after photobleaching (FRAP) is a technique used to measure the mobility of molecules in the plane of the membrane. A fluorescent dye is attached to proteins embedded in the cell membrane in a process called labeling. Labeling all the proteins in a membrane creates a fluorescent cell that can be visualized with a fluorescence microscope. A laser is then used to bleach a small area of the membrane, leaving a nonfluorescent spot on the surface of the cell.

Hypothesis If membrane components such as proteins move in the plane of the membrane, the bleached spot should become fluorescent over time as unbleached fluorescent molecules move into the bleached area. If membrane components do not move, the bleached spot should remain intact.

Experiment and Results In the top panel of the figure, we see an illustration of a patch of membrane. It starts off uniformly green because of an even distribution of the fluorescent label. A laser bleaches a spot on the membrane. Over time, fluorescence appears in the bleached area. In the middle panel, we see photographs of the actual cell membrane. In the bottom panel, we see a graph of

Fluorescent molecules   Laser beam   Bleached area

AP® PRACTICE QUESTION
(a) **Identify** the question the experimenters are trying to answer.
(b) **Identify** the hypothesis they tested.
(c) **Explain** their results by **identifying** their claim and **justifying** the claim with their evidence and reasoning (CER). In your CER, be sure to include your understanding of fluorescent recovery after photobleaching (FRAP) as well as the relationship between proteins and phospholipids in the structure of the cell membrane.

Each box concludes with a free-response-style **AP® Practice Question** that asks you to evaluate experimental design and methods or a proposed extension to the researcher's work.

Review Your Science Practice Skills Any Time

With so many review features, you can be sure to remember the essentials on exam day.

Tutorials for Extra Review of Key Statistical Tools and Data Analysis

Tutorial 1: Statistics

PREP FOR THE AP® EXAM

Scientific evidence rarely hinges on the result of a single experiment, measurement, or observation. As we discussed in Module 0, any scientific claim must be backed up by data from multiple subjects across multiple repetitions of the same experimental process or multiple observations. It is the accumulation of evidence from many independent sources, all pointing in the same direction, that lends weight to a scientific hypothesis until eventually it becomes a theory.

This means that any statement of a number—ants can carry 10 times their weight; a human cell can divide in 24 hours—is actually a statement of many numbers. How do scientists determine what number will represent all of their experimental results? And how do scientists succinctly describe all of the individual data points? Statistics is a field that helps scientists to organize, analyze, and see trends in data. In this tutorial, we will discuss several statistical tools that help us understand and interpret data. We will start with a quick look at the issue of precision in reporting calculations.

Significant Figures

When reporting any recorded data or quantitative conclusions, you should use appropriate precision. For example, if you measure something and say it is 2 meters tall, does that mean it is exactly 2 meters or perhaps 2.03 meters or even 2.10 meters? Significant figures indicate the precision of a measurement. Significant figures are numbers that carry meaning. For example, 2 meters has just one significant figure because it has only one digit. By contrast, 2.03 has three significant figures. The more significant figures, the more precise the measurement.

In general, all nonzero numbers (1, 2, 3, and so on) are significant. Zeros between other numbers are also significant. For example, as we noted, the number 2.03 has three significant figures. Leading zeros (zeros that are located to the left of another number) are not significant. The number 0.02 has only one significant figure. Trailing zeros (zeros located to the right of another number) are also not significant, except in numbers with a decimal point. For example, the number 2.10 has three significant figures.

When doing calculations, significant figures in a final answer are determined by the number that is the least precise in your dataset. However, do not round intermediate values when you perform calculations; only round your final answer. For example, let's say you are asked to calculate the area of a rectangle (*width* (*w*) × *height* (*h*)), with *w* = 4.33 feet and *h* = 2 feet. The answer is

Average: Mean, Median, and Mode

When scientists make observations or do an experiment, they gather results and collect data. The direct observations and measurements that have not been organized, processed, or interpreted are known as raw data. For example, a doctor collects all kinds of raw data about your health, such as height, weight, body temperature, breathing rate, heart rate, and blood pressure. You may give a blood or urine sample so that the concentrations of various substances can be measured. Or you may spit into a vial so that a DNA sample can be obtained to look for common genetic variants associated with disease. The data collected by physicians are all types of raw data.

After scientists collect raw data, they often process the data in some way in order to make sense of it. One way to process it involves determining the average of all of the measurements. Determining the average takes all of the individual pieces of data and provides a single, representative number. In this section, we look at several different ways to calculate the average.

Mean

The first type and most common type of average is called the *mean*. Given a set of values, the **mean** is obtained by adding all of the values in the dataset and dividing by the number of values in the dataset, which is indicated by *n*.

Example: In the dataset of nine values, 5, 9, 4, 6, 5, 1, 5, 9, 1, 9, the mean is determined by adding all of the values (5 + 9 + 4 +

> Longer **Tutorials** presented throughout the book provide comprehensive explanations of statistics, growth equations, probability, and graphing skills used throughout the course. The tutorials are a great review tool. Tutorials also point you to relevant Analyzing Statistics and Data boxes for additional practice.

Tutorial 2: Probability

PREP FOR THE AP® EXAM

Gregor Mendel's pea plant crosses provided the world with new knowledge about the presence and functioning of genes. The purebred parental generation (P_1) crosses refuted the commonly believed blending hypothesis, an idea that phenotypes or traits of offspring were the average traits of the parents, and instead revealed the existence of dominant and recessive alleles. The crosses of the offspring of those parents ($F_1 \times F_1$) contributed more evidence to the concepts of dominant and recessive alleles by producing plants with traits that were not seen in the original P_1 crosses.

These experiments laid the groundwork for more complicated genetic studies and still serve as a foundation for understanding basic hereditary patterns. Not only do these pea plant crosses represent a clear example of heredity, they also provide an excellent opportunity to understand probability and its role in predicting the outcome of events. In this tutorial, we will discuss probability and its role in Mendelian genetics.

The homozygous dominant parent's gametes and the homozygous recessive parent's gametes produce four offspring combinations. In this case, they are all heterozygous *Aa*.

The probability of a homozygous dominant offspring, *AA*, is 0; it never happens. The same can be said for a homozygous recessive offspring, *aa*. However, the probability of heterozygous offspring is 1; it always occurs.

This example replicates the findings of Mendel's initial purebred P_1 crosses. The F_1 generation inherits the *A* allele from one parent and the *a* allele from the other parent, resulting in a heterozygous genotype, *Aa*.

Your Turn

Mendel discovered that the alleles for purple flowers were dominant to white flowers in his pea plants. What is the probability of each type of offspring, in terms of genotype and phenotype, when crossing a heterozygous purple flower plant with a white flower plant?

Stunning Visuals Bring the Biological World to Life...

The book's visual program was created with you in mind.

Illustrations, Photos, and Graphs for Visual Analysis

FIGURE 1.15 Lipids

When oil, which is a lipid, is mixed in water, it forms droplets to minimize its contact with water. Oil, like all lipids, is hydrophobic.

Photo: ThomasVogel/Getty Images

> Figures and graphs integrate seamlessly with text discussions.

> Illustrations make complex structures and concepts easier to visualize.

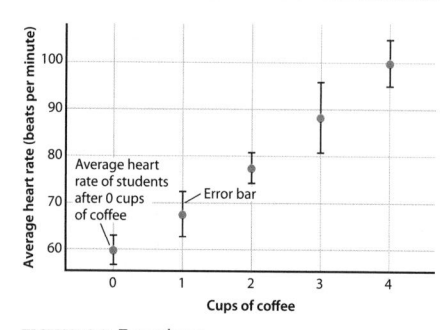

FIGURE 0.8 Error bars

An error bar represents a range of values within which the true value is likely to be. This graph shows the relationship between caffeine consumption and resting heart rate. It plots the average resting heart rate of groups of people who consumed 0, 1, 2, 3, or 4 cups of coffee, with error bars giving an indication of the uncertainty of the data.

> Data in graphs are always accompanied by a careful walk-through.

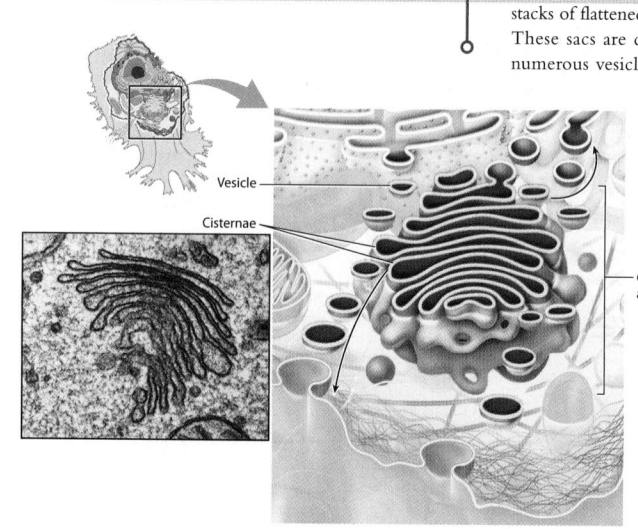

stacks of flattened membr
These sacs are called cis
numerous vesicles that tr
the ER
betwee
from t
membr
are the
which

as they
There
from th
to their
their v
tus, che
a seque
differen
An
ficatio
the Go
which
lipids
protei

FIGURE 7.5 The Golgi apparatus

The Golgi apparatus is the next stop for proteins and lipids produced in the ER. It further modifies these proteins and lipids; it sorts and targets them to other organelles, the cell membrane, or the cell exterior; and it adds sugars to proteins and lipids. Vesicles transfer molecules from the ER to the Golgi apparatus, within the Golgi apparatus, and from the Golgi apparatus to their final destinations. Photo: Biophoto Associates/Science Source

...and Help You Improve Visual Analysis Skills

Beautifully rendered works of art present key concepts in an informative and visually engaging way.

Visual Synthesis Figures

VISUAL SYNTHESIS 1.1 **THE FOUR BIG IDEAS OF AP® BIOLOGY**

We can describe four Big Ideas that connect and unite the many dimensions of biology: Evolution, Energetics, Information Storage and Transmission, and Systems Interactions. These four ideas are introduced in Unit 1 and will be visited again and again throughout the book. By the time you finish this book, you will have an understanding of how life works, from the molecular machines inside cells and the metabolic pathways that cycle carbon through the biosphere, to the process of evolution, which has shaped the living world that surrounds and includes us. The four Big Ideas are fundamental to understanding and organizing such diverse aspects of biology.

The Four Big Ideas of AP® Biology

Evolution investigates changes in the genetic makeup of a population over time. Through the process of natural selection, species become adapted to their environments. For example, ancestors of the snowshoe hare and Canada lynx evolved to have adaptations that increased their survival and reproduction. This includes the hare's ability to change coat color with the seasons.

Winter Summer

Energetics considers the processes used by cells and organisms to exchange matter and energy with their environment. For example, Canada lynx and showshoe hares acquire, store, and use the energy and matter obtained from the environment to maintain homeostasis, grow, and reproduce.

Biosphere
Ecosystem
Community
Population
Individual
Cell

Systems Interactions examines the way the components of biological systems interact with and affect one another. A system can be as small as a cell or as large as the biosphere. New properties of biological systems emerge through the interactions of organisms such as Canada lynx and snowshoe hares, and through interactions between species and their environment.

Information Storage and Transmission explores how information is stored, used, and transmitted by cells and organisms. Using information stored in DNA and through experience, the Canada lynx knows how to hunt its prey and the snowshoe hare knows how to evade its predator. These behaviors are transmitted to offspring through DNA and through the process of learning.

18

19

Special Visual Synthesis figures throughout the book present integrated concepts across each unit.

Our Digital Platform Encourages Efficient Learning

Everything you need for the course is in one place.

E-book

The interactive, mobile-ready **e-book** allows you to read and reference the text when you are online and offline. All offline highlights and notes sync when you connect to the Internet.

Homework System

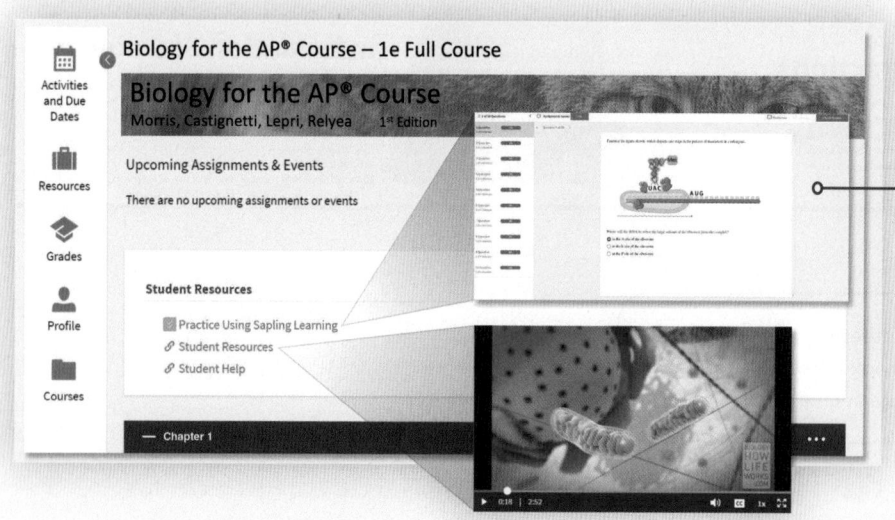

The online homework system helps you learn with targeted feedback based on common misconceptions. Students who use this system find that it significantly improves their understanding and helps them build essential foundations.

Biology

for the
AP® Course

Unit 1
Chemistry of Life

Unit 1	0	1	2	3	4	5

Module 0
Introduction

LEARNING GOALS ▶LG 0.1 **Four Big Ideas form a fundamental basis for understanding biology.**

▶LG 0.2 **Scientific inquiry is a deliberate way of asking and answering questions about nature.**

The ears of the snowshoe hare (*Lepus americanus*) stand tall as it listens carefully and its nose twitches as it smells the air for the scent of its predator, the Canada lynx (*Lynx canadensis*). At the same time, the lynx uses its keen eyesight, its sense of smell, and its ears to locate its prey, the hare. Having spotted the hare, the lynx quietly and deliberately moves toward it, slowly closing the distance between itself and the hare. Then, suddenly, the lynx springs upon the hare and the chase for life and food begins.

The activities of predator and prey depend on a number of key processes that they both possess. For instance, both hare and lynx are strong and swift runners. Both are exquisitely adapted to sense their habitats, be aware of danger, and identify potential sources of food. All of the processes employed by the hare and the lynx are part of the study of **biology,** the science of life. **Biologists,** scientists who study life, have come to understand a great deal about these and other processes. They attempt to explain why biology works as it does. That is, they explore the underlying principles and processes that shape and mold biological **organisms,** which are the living beings that display all of the properties of life. We don't know everything about how life works and there is still so much to discover, but the study of biology provides us with an organized way of understanding ourselves, other living things, and the world.

The scope of modern biology is vast and the pace at which we acquire new knowledge continues to accelerate. For example, in recent decades, we have gained unprecedented

understanding of the **gene,** which is the unit of heredity, and the **genome,** which is all of the genetic information that an organism contains. This information is helping us come up with new ways of fighting disease in humans, other animals, and plants. It is also helping us understand how different groups of organisms have evolved and how endangered species might be saved. Many new biological frontiers have opened up in recent decades, while others are still waiting to be explored. For example, researchers are only just beginning to understand how bacteria in our digestive system affect our health and well-being. Other researchers are looking at how the temperature and acidity of seawater affect the ecology of coral reefs.

By the time you finish this course, you will have an understanding of how life works. You will explore the structure and function of **cells,** which are the simplest self-reproducing unit that can exist independently. You will look at pathways that cycle carbon through the biosphere. You will understand many of the processes that have shaped the living world. You will see connections among the different ways of investigating life and gain a greater understanding of how to ask and answer scientific questions. This study will help you develop a basis for making informed decisions about issues in your life that have a direct relationship to biology.

In this module, we begin our study of biology with a first look at the four Big Ideas of AP® Biology: Evolution, Energetics, Information Storage and Transmission, and

Systems Interactions. These four Big Ideas provide a framework for understanding and examining biology. After we examine the four Big Ideas, we will consider how biologists use scientific inquiry to understand how scientists ask and answer questions about the natural world.

FOCUS ON THE BIG IDEAS

Think about ways that each of the big ideas can serve as a different lens to look at life.

0.1 Four Big Ideas form a fundamental basis for understanding biology

Let's return to the snowshoe hare and Canada lynx, which are pictured in **FIGURE 0.1**. We can view this interaction from four different lenses, each providing the perspective of one of the four Big Ideas. From the lens of *evolution*, we can ask how the ancestors of snowshoe hares and lynxes developed their keen senses and ability to run swiftly and skillfully. From the lens of *energetics*, we can explore the food sources of both animals and how they use the fuel they acquire from these sources. From the lens of information storage and transmission, we can learn how the lynx knows to hunt the hare and how the hare knows to run, as well as how this information is passed on to subsequent generations. From the lens of systems interactions, we can investigate the ecosystem dynamics that bring the hare and lynx together

or we can examine how the collection of cells that make up each individual animal communicate to enable the lynx to pursue the hare and the hare to evade the lynx.

Every unit in this text examines topics from the perspective of these four Big Ideas. As you read the modules of this book, tie in one or more of the Big Ideas to what you are learning. Doing so will help you to learn about biology and will assist you in understanding how biologists study and learn about life.

In this section, we will take a quick first look at each Big Idea. Although we present them one at a time, they are not really separate or distinct. Instead, they are inseparable in nature. To tackle biological problems—whether building an artificial cell, stopping the spread of infectious diseases like malaria or a coronavirus, feeding a growing population, or preserving endangered species—we need to understand these ideas and how they work together.

Big Idea 1: Evolution

Have you ever noticed how ants have the same body plan, but may be big or small, black or red? In fact, there are more than 10,000 different *species* of ants. A **species** is a group of interbreeding organisms that produce fertile offspring. A species is often distinct from other groups in body form, behavior, or biochemical properties. **FIGURE 0.2** shows two ant species. While there are many different species of ants and each has distinct characteristics, all ants have similar traits that enable us to recognize them as ants. Such similarities and differences are widely observed, and biologists refer to them as the unity and diversity of life. The study of **evolution,** or change over time, explains this unity and diversity of life. Evolution—Big Idea 1—is the central concept that unites all of biology, and biologists recognize it as a key principle of life.

All ants have some shared features, such as their distinctive segmented bodies and bent antennae. Ants are also diverse. For example, they vary in size and color. Some tolerate colder climates while others require warm, tropical places. Ants are not unique in displaying unity and diversity. Unity and diversity occur in all living creatures, from the

FIGURE 0.1 Canada lynx and snowshoe hare chase

The photograph shows the snowshoe hare running to escape its predator, the Canada lynx. The lynx gives chase to obtain its food, the hare. The hare runs for its life. The example of the hare and the lynx exemplifies the four Big Ideas of the AP® Biology curriculum.

Photo: Tom & Pat Leeson/Science Source

a.

b.

FIGURE 0.2 Unity and diversity in ants

All ants have common features, such as their segmented bodies and six legs. They also display considerable diversity, as evidenced by the shapes and sizes of their heads, mandibles (jaws), and abdomens (hindmost section). (a) This fire ant (*Solenopsis invicta*) is red and known for its painful sting. (b) This Costa Rican army ant (*Eciton burchellii*) has large mandibles and hunts in large groups. Photos: (a) Satrio Adi/EyeEm/Getty Images; (b) John Mason/ARDEA

in **FIGURE 0.3**. And it is why many bacteria and other disease-causing microorganisms have stopped responding to antibiotics, which is becoming a public health crisis. Life has been shaped by evolution since its origin, and the capacity for Darwinian evolution may be life's most fundamental property.

The concept of natural selection was in part inspired by observing strong competition for survival and reproduction. For example, among animals, individuals often compete for food. In a Canada lynx population, the fastest and most skilled hunters are more likely to succeed in catching hares, and therefore surviving and reproducing. Slower and less skilled Canada lynx are more likely to perish before they can reproduce. If the animals with better hunting skills reproduce more and have more viable offspring, their offspring will predominate in the next generation.

And just as natural selection has shaped predators, it has also shaped prey. Predator–prey interactions are important in every habitat on Earth. After all, every living animal can be classified as either "predator" or "prey"—and often both labels apply. The skills that a predator needs to hunt its prey and that its prey needs to escape depend in large part on

smallest single-celled organisms to the great Sequoia trees and the largest animal ever, the 150-ton blue whale.

Evolution occurs by several different mechanisms that will be discussed in detail in Unit 7. Of these, perhaps the most significant is a process first described in the nineteenth century by Charles Darwin and Alfred Russel Wallace. Both of these naturalists suggested that species change over time by a process called *natural selection*. **Natural selection** is a mechanism of evolution in which some individuals survive and reproduce more than others in a particular environment as a result of variation among individuals that can be passed on to the next generation.

As Darwin recognized, farmers have used a principle similar to natural selection for thousands of years to develop crops, such as wheat, corn, cabbage, and broccoli. It is how people around the world have developed breeds of horses, pigeons, cats, and dogs, like the two shown

FIGURE 0.3 Dog breeds

Dog breeders use a process similar to natural selection to develop many different breeds of dogs, such as this husky and terrier. Photo: © Rob Brodman 2011

a host of features of their nervous, sensory, musculoskeletal, endocrine, circulatory, and respiratory systems. In short, natural selection has left a physical imprint on both predator and prey, from nose to tail.

As we examine a wide range of biological topics throughout the book, you will notice that evolution permeates these discussions—whether we are explaining the biochemistry of cells, how organisms function and reproduce, how species interact in nature, or the remarkable biological diversity of our planet.

Big Idea 2: Energetics

All life forms require *energy* to survive, grow, move about, and reproduce. **Energy** is the ability to do work and it is absolutely essential for life. The study of **energetics** examines the properties of energy and how energy is distributed in biological, chemical, or physical processes.

Strategies to capture and use energy vary among species and depend on their evolutionary history. Plants use the energy of sunlight to produce their own food to grow, reproduce, and carry out their functions. Humans and other animals obtain energy by eating other organisms. In fact, all organisms obtain energy from just two sources—the sun or chemical compounds. Losing or reducing access to sources of energy can have damaging and sometimes fatal consequences for organisms.

Consider what happens when you eat an apple. The apple contains sugars, which store energy. By breaking down sugar, our cells harness this energy and convert it into a form that can be used to do the work of the cell. Energy from the food we eat allows us to grow, move, communicate, and do all the other things that we do. Using sugar as a source of energy to power the cell is not a strategy that is just used by human cells. It is widespread among organisms and represents another unifying characteristic of most species. This observation suggests that the ability to use sugar as a source of energy evolved early in the history of life and has been retained over time. We will delve more deeply into how organisms access and use energy in Unit 3.

Big Idea 3: Information Storage and Transmission

Big Idea 3 looks at biology from the perspective of information storage and transmission. In this context, information refers to the instructions that all cells have that in part determine what they look like and how they function. For example, how does your skin cell "know" to be a skin cell

and not a liver cell? The answer is that it contains information that is used by the cell, so it looks and acts like a skin cell. This information is stored in a cell's *DNA*. **DNA,** the abbreviation for **deoxyribonucleic acid,** is the carrier of genetic information for all organisms.

In addition to storing information, cells must also retrieve this information. In other words, they need to be able to access the information and use to it to grow and carry out their functions. Finally, cells need to be able to transmit their genetic information to the next generation. DNA is remarkable because it can store genetic information, allow this information to be retrieved and used by the cell, and transmit this information to the next generation. We will learn a lot more about DNA and its role later in this unit and throughout the book.

The transmission of genetic information from parents to their offspring enables species of organisms to maintain their identity through time. The genetic information in DNA guides the development of the offspring, ensuring that parental apple trees give rise to apple seedlings and parental geese give rise to goslings. Furthermore, variation in this genetic information allows some organisms to survive and reproduce in particular environments more than others. Through natural selection, these organisms pass on the genetic variants that account for their ability to do well in their environment. For example, if faster speed enhances survival and reproduction among hares and Canada lynx, the underlying genetic variation that accounts for this trait will likely be passed on to the next generation. Therefore, Big Idea 3—Information Storage and Transmission—is essential to the survival, growth, reproduction, and evolution of a species.

Big Idea 4: Systems Interactions

A **system** is a group of things that function together as a whole. We can consider systems at different levels of scale, from a single cell in a hare to an entire forest, which includes many interactions that occur among the organisms and nonliving materials it contains. Biologists refer to living organisms as **biotic,** and nonliving components as **abiotic.** A **biological system** is made up of both biotic and abiotic entities that interact. As a result of these interactions, biological systems show complex properties.

Big Idea 4 recognizes that biological systems exist at different levels, from the simple to the complex, and interactions among the parts of the system lead to new, emergent properties. An emergent property is a property of a system that the individual parts do not have on their own.

For example, the muscles in the leg of the hare in Figure 0.1 interact with the leg bones, tendons, and nerves to provide movement and remarkable agility. The Canada lynx, using its nervous, sensory, musculoskeletal, cardiovascular, and respiratory systems working together, is able to note the presence of the snowshoe hare, plan a strategy, hunt, chase, and possibly catch the hare.

If we examine the various components that interact with one another in an organism, we will observe the resulting emergent property, which is life. The interactions among the parts of a biological system are hallmarks of life and are found at all levels of biological organization. For example, when fresh water from a large river meets the ocean, the ocean's salty water becomes diluted with fresh water. The mixture of these waters results in a number of new environments where animals, plants, and microorganisms may live, including estuaries, like the one shown in **FIGURE 0.4**. The environment of the estuary is different from the river and ocean systems that created it and it provides a unique habitat for many different species.

From the molecular to the cellular to the organismal to the biosphere levels, biological systems are diverse and complex. These features enable biological systems to have a robustness that helps them to withstand, tolerate, and respond to the changes in the environment. Like evolution, energy, and information storage and transmission, biological systems interactions help ensure the survival and reproduction of organisms, populations, species, and—on a larger scale—life. **Visual Synthesis 1.1: The Four Big Ideas of AP® Biology,** on page 18, illustrates the connectedness of these concepts.

FIGURE 0.4 Estuaries as a biological system

Estuaries such as the Sado Estuary Natural Reserve in Portugal are formed in areas where salt water from oceans and fresh water from rivers mix. Photo: Maurício Abreu/AGE Fotostock

✓ Concept Check

1. **Describe** how a comparison of a fire ant and an army ant shows both the unity and diversity of life.

2. **Identify** why a lack of energy would result in the death of an organism.

3. **Describe** what would happen if organisms were unable to retrieve the information contained in their genes.

4. **Describe** an example of how a systems interaction allows an organism to adjust to its environment.

0.2 Scientific inquiry is a deliberate way of asking and answering questions about nature

How do we go about trying to understand the vastness and complexity of nature? **Scientific inquiry** is the process scientists use to ask questions and seek answers about the natural world in a deliberate and ordered way. Scientific inquiry is limited to investigations of the natural world. Examination of questions not in the natural world, such as questions about religion, faith, and morality, are outside the realm of scientific inquiry.

Scientific inquiry provides the opportunity to observe, investigate, and explain how natural phenomena occur.

As shown in **FIGURE 0.5**, scientific inquiry consists of three parts: exploration, investigation, and communication. We will now examine how scientists use scientific inquiry to conduct an orderly and logical investigation of the natural world and communicate their findings with others.

Making Observations and Asking Questions

For most scientists, studies of the natural world begin with exploration. In the exploration phase, scientists make observations and ask questions. **Observation** is the act of viewing the world around us. Observations allow us to ask focused questions about nature. For example, Charles Darwin initially made many observations about anatomy, embryology, domesticated plants and animals, fossils, and the distribution of organisms on the Earth. These observations led him to ask questions. Why are organisms adapted, often exquisitely, to their environment? Why do some fossil organisms resemble living ones? Why are there penguins in the Southern Hemisphere but not the Northern Hemisphere? Why are islands home to so many species that are found nowhere else in the world?

Let's say you observe a hummingbird, like the one pictured in **FIGURE 0.6,** hovering near a red flower, occasionally dipping its long beak into the bloom. What motivates this behavior? Is the bird feeding on some substance within the flower? Is it drawn to the flower by its vivid color? What benefit, if any, does the flowering plant derive from the bird?

Having formulated these questions, the next step that scientists often take is to consult the scientific literature, which is the published information about observations and experiments that others have done. In our example, the scientist would likely review those papers that focus on hummingbird feeding habits and possibly on the effects hummingbirds have on plants they visit. Having reviewed the scientific literature, the scientist should have enough

knowledge to start refining questions that would be interesting to investigate.

Regardless of where they come from or when they arise in the process, questions are the keys to scientific inquiry. Indeed, learning to ask good questions is a fundamental component of thinking like a scientist.

Formulating Hypotheses

Observations such as those about the hummingbird, the questions observations raise, and consulting the scientific literature allow us to move on to investigation. Scientists use observation and critical thinking to propose a *hypothesis*. A **hypothesis** is a tentative explanation for one or more observations, and it makes predictions that can be tested by experimentation or additional observations. A hypothesis is not just an idea or hunch. It is a working explanation that helps a researcher understand an observation and leads to a better understanding of the observation.

We might, for example, hypothesize that a hummingbird is carrying pollen from one flower to the next, facilitating reproduction in the plant. Or we might hypothesize that nectar produced by the flower provides nutrition for

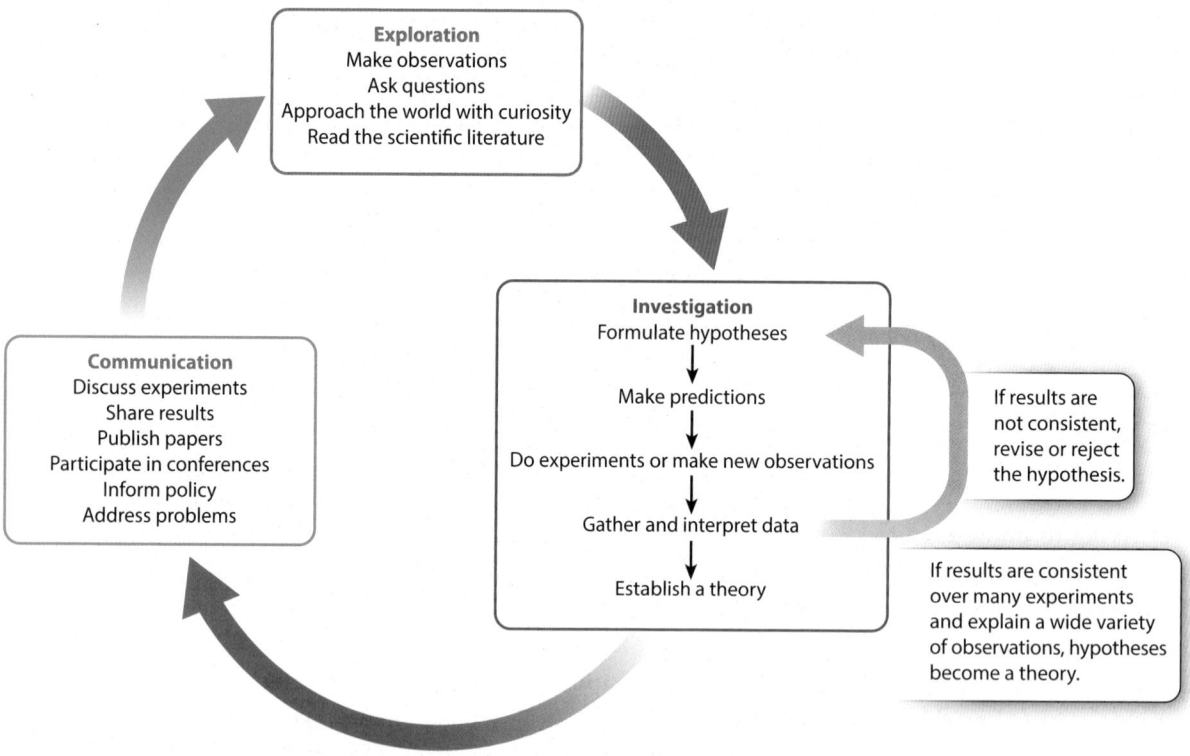

FIGURE 0.5 **Scientific Inquiry**

Scientific inquiry is the organized, deliberate process that scientists use to examine the natural world. It consists of three parts: exploration, investigation, and communication.

FIGURE 0.6 Observation

Observing a hummingbird visiting a flower may lead to a number of questions that a scientist can explore. Photo: Charles J. Smith

the hummingbird—that the hummingbird's actions reflect its need to take in food. Both hypotheses about the hummingbird's behavior provide a reasonable explanation of our observations, but they may or may not be correct. The predictions of the hypothesis lead to further observation or experimentation.

Because hypotheses make predictions, we can test them. That is, we can devise an experiment to test whether the predictions made by the hypothesis are supported by the evidence, or we can go into the field to make further observations predicted by the hypothesis.

Scientists collect data by observation or experimentation, or both. They will then analyze the data and determine if the data support the hypothesis. If the data do not support the hypothesis, the hypothesis is rejected. In this case, the researcher may generate a new hypothesis. If the data support the hypothesis, it gains support and further questions can be asked. These questions may refine or extend the hypothesis.

Returning to the hummingbird and flower, we can test the hypothesis that the bird is transporting pollen from one flowering plant to the next, enabling the plant to reproduce. Observation provides one type of test: if we catch and examine the bird just after it visits a flower, do we find pollen stuck to its beak or feathers? If so, our hypothesis is supported by the data.

The test, however, cannot prove the hypothesis. Pollen might be stuck on the bird for a different reason—perhaps it provides food for the hummingbird. However, if the birds didn't carry pollen from flower to flower, we would reject the hypothesis that they facilitate pollination. In other words, a single observation or experiment can lead

us to reject a hypothesis, or it can support the hypothesis, but it cannot prove that a hypothesis is correct. To move forward, then, we might undertake a second set of observations. Does pollen that adheres to the hummingbird rub off when the bird visits a second flower of the same species? If so, we have stronger evidence and support for our hypothesis.

We might also use observations to test a more general hypothesis about birds and flowers. Does the color red generally attract birds and thereby facilitate pollination in a wide range of flowers? To answer this question, we might catalog the pollination of many red flowers and ask whether they are pollinated mainly by birds. Or we might go the opposite direction and catalog the flowers visited by many different birds—are they more likely to be red than chance alone might predict?

Designing Controlled Experiments

Our hummingbird example used observation to test several different hypotheses, but scientists often test hypotheses through experimentation. One of the most powerful types of experiment is a *controlled experiment*. In a **controlled experiment,** the researcher sets up at least two groups to be tested; the conditions and setup of the groups are identical, except the researcher deliberately introduces a single change, or variable, in one group to see its effect.

Suppose we want to understand the relationship between caffeine consumption and the heart rate when a person is not exercising, which is known as resting heart rate. We might hypothesize that caffeine causes an increase in resting heart rate. This hypothesis could be based on our own experience and observations, or perhaps the scientific literature. To test this hypothesis, we can carry out a controlled experiment. In this case, we might have two groups of people who are similar in terms of age, gender, socioeconomic background, health, and so on. One group is given caffeine, perhaps in the form of a cup of coffee. This group is called the **test group** or **experimental group** because this group experiences the variable—it receives caffeine. In a second group, the people are not given any caffeine. This group is called the **control group** because it is not exposed to the variable.

Why is it necessary to include a control group? Imagine for a moment that there was just a test group and no control group, and the resting heart rate of the people in the test group went up after they drank a cup of coffee. In this case, you might conclude that the hypothesis is supported and caffeine increased heart rate. But how do you know that the heart rate of people in the group didn't increase on its own? How do you know the variable that was changed caused the observed effect? The control group shows what happens without caffeine so researchers can compare what happens with and without the variable, holding every other factor the same. In this way, they can determine if the variable accounts for any changes in heart rate.

In a test of this hypothesis, the action of consuming caffeine is known as the **independent variable,** the variable that is manipulated to test the hypothesis. This variable is considered "independent" because the researchers can manipulate it as they wish. The result of the experiment—resting heart rate—is known as the **dependent variable.** This variable is considered "dependent" because it is expected to vary based on the independent variable.

Scientists use controlled experiments because they are extremely powerful. By changing just one independent variable at a time, the researcher is able to determine whether that variable is important. If many independent variables were changed at once, it would be difficult, if not impossible, to draw conclusions from the experiment because the researcher would not be able to determine which variable caused the outcome.

Our experiment testing the relationship between caffeine consumption and resting heart rate is very simple. In reality, we might include more than two groups of people. For example, we could test how the amount of caffeine affects resting heart rate by giving several groups different numbers of cups of coffee. In this case, there are several test groups, each of which receives the variable of caffeine. We could also include more than one control group. One could receive nothing to drink and the other a cup of water. In both cases, the control groups do not receive caffeine. However, by providing a cup of water, we control for the potential variable of drinking. Both of these control groups are also called **negative control groups** because the expectation is that we will see no effect. We could also include a **positive control group.** This is a group that receives a treatment or variable with a known result. In our example, we could give a medicine that is known to increase heart rate to be sure that heart rate increases as expected.

If observations or experiments do not support a hypothesis, the researcher modifies or rejects the hypothesis. If observations or experiments support a hypothesis,

FIGURE 0.7 Daffodils

Daffodils, like the ones shown here, were the subject of scientific inquiry by James Kirkham Ramsbottom, who tried to track down what was causing their deaths in the early 1900s. Photo: Victoria Ambrosi/ EyeEm/Getty Images

the researcher accepts the hypothesis and then subjects it to more scrutiny by making further observations and doing additional experiments. As we saw in the hummingbird example, a hypothesis may be supported, but it is never proven because we can never know for certain whether it is true in all cases.

Let's now turn our attention to a real-world experiment that demonstrates some of the features we have been discussing. "Practicing Science 0.1: Using observation and experimentation to examine a horticultural problem" shows how a young scientist used the process of scientific inquiry to determine the cause of death in daffodils, like the ones shown in **FIGURE 0.7**, and to create an effective treatment. This study also gives us a chance to review how to use percentages when evaluating data, described in "Analyzing Statistics and Data: Percent Change" on page 10.

PREP FOR THE AP® EXAM

AP® EXAM TIP

You should know how to design a controlled experiment with a clear and precise hypothesis. The design should include experimental and control groups, and independent and dependent variables.

Using observation and experimentation to investigate a horticultural problem

Background Scientific inquiry is often called upon to address problems that arise in society and industry. In 1916, British horticulturalists were concerned with a disease that killed daffodils. Daffodils grow from bulbs, which are large underground stems that store energy and are seen in many plants, such as daffodils, tulips, and onions. However, the disease caused leaves to wither, bulbs to become discolored, and eventually death of the plant. The demise of the plants represented a substantial loss of commercial production and income to the horticultural industry. While some suspected a fungus caused the plant deaths, no one was able to determine the source of the problem.

The British Royal Horticultural Society took up the cause and assigned the problem to James Kirkham Ramsbottom. At the time he was a top student at the Royal Horticultural Society's garden in Wisley, a community near London.

Observation and Hypothesis Ramsbottom began by making observations. He examined hundreds of diseased bulbs, preparing microscopic slides and studying them closely. While he did see fungi, Ramsbottom observed that all of the diseased bulbs contained a parasitic worm, *Tylenchus devastatrix*. Ramsbottom hypothesized that the worm was the cause of the disease afflicting the plants and predicted that if he could devise a way to kill the worm without killing the bulbs, the disease would be eliminated.

Experimentation Ramsbottom launched a series of experiments where he examined a number of agents that might selectively kill the worm while keeping the plant alive. He tried chemical treatments, spraying the plants and dousing them. He experimented with both gas and formaldehyde. He settled on the use of heat. Ramsbottom immersed the bulbs for different amounts of time in hot water. The photograph shows the removable wire basket and copper boiler that permitted Ramsbottom to heat the daffodil bulbs for different periods of time. He determined that soaking them in 110°F (43°C) water for 2 to 4 hours left the bulbs intact while the parasite was eliminated. Untreated, infected daffodil bulbs failed to grow, died, and did not produce flowers. The heat-treated daffodil bulbs grew normally and produced the sought-after plant and flower. Today, the Ramsbottom heat treatment is still used in virtually the same manner as he developed it.

SOURCE
Flower Preservation, 1916. *The Scientist*, 2:64. Photo: His Majesty's Stationery Office/RHS Lindley Collections

AP® PRACTICE QUESTION

James Kirkham Ramsbottom used the process of scientific inquiry to figure out what was causing the death of daffodils. Organize the description of his experiment by identifying the following:

1. The scientific (testable) question
2. The hypothesis
3. The independent variable
4. The dependent variable
5. The experimental group
6. The control group

Analyzing and Interpreting Data

After conducting an observational or controlled experiment, the scientist has a collection of data. Data are the bedrock of science. Biologists collect, analyze, and interpret data to answer questions about the natural world. Hypotheses are supported or rejected on the basis of data, and our understanding of the world is ultimately built on a foundation of data.

What types of data will you encounter in your AP® Biology course? Data can take many forms, including observations, measurements, and facts. Data can be qualitative

Percent Change

In biology, percentages are frequently used to describe and analyze data. For example, a researcher might use percentages to describe the concentration of a solution, or to compare the numbers of each gender in a group.

Percentages describe "parts per hundred" or "parts of a whole." For example, imagine you have 10 trees in your yard and 4 of them are maple trees. You could say that $\frac{4}{10}$ or 0.4 of the trees in your yard are maple trees. However, you can also calculate the number of maple trees per 100 trees, or the percentage of maple trees. To find the percentage of maple trees, you multiply 0.4 by 100:

0.4 × 100 = 40%

In other cases, scientists might be interested in calculating percent change. This is useful to compare an initial value to a final value, which allows you to see how much something has increased or decreased. Use the following formula to calculate percent change:

$$\% \text{ change} = \frac{\text{final value} - \text{initial value}}{\text{initial value}} \times 100$$

If the final value is larger than the initial value, the percent change is a positive number, representing an increase between the values you are comparing. If the final value is smaller than the initial value, the percent change is a negative number, signifying a decrease between the values you are comparing.

PRACTICE THE SKILL

Let's look at an example of how percent change might be used. James Kirkham Ramsbottom discovered a way to eliminate parasites from daffodil bulbs by immersing them in hot water. Before he found an effective soaking time of 2 to 4 hours, he immersed 50 bulbs for 30 minutes and 50 bulbs for 1 hour. At the end of 30 minutes, 10 of the daffodil bulbs were free of parasites. After 1 hour, 25 of the bulbs were free of parasites. What was the percent change in the number of healthy, parasite-free bulbs as the immersion time increased?

To start, we must find the two values we need to calculate percent change. After 30 minutes, 10 of the 50 bulbs were free of parasites. The initial value is 10 bulbs. At the end of 1 hour, 25 bulbs were free of parasites. So, the final value is 25 bulbs. Now we can plug these values into our formula:

$$\% \text{ Change} = \frac{25 - 10}{10} \times 100$$

$$\% \text{ Change} = \frac{15}{10} \times 100$$

$$\% \text{ Change} = 1.5 \times 100$$

$$\% \text{ Change} = 150\%$$

There was a 150% increase in parasite-free bulbs as Ramsbottom changed the immersion time from 30 minutes to 1 hour.

Your Turn

The emerald ash borer is an invasive species that has destroyed ash tree populations in North America. Before the insect arrived, one forest contained 300 ash trees. A number of years after the ash borer was introduced to the area, only 60 ash trees remained. By what percent did the ash tree population decrease?

or quantitative. Qualitative data are descriptive. For example, the stem of a corn plant can be described as "short" or "tall." Quantitative data are expressed numerically. For example, a corn stem might measure 2.04 m or 2.76 m in length. Similarly, the heart rate data that we collected in our controlled experiment is an example of quantitative data. Scientists often deal with quantitative data because numbers lend themselves to statistical analysis.

Statistical analysis helps scientists interpret the data they collect. For example, when several measurements are made, they are typically not all the same. In this case, the researcher might report the average of the measurements. The researcher might also indicate the extent to which the observed measurements deviate from the average. The spread of the data can be calculated in various ways, each of which provides an indication of how tightly the data points are clustered around the average.

Statistics can also help researchers understand whether the data collected for the experimental and control groups reflect a real, or what is termed a statistically significant, difference. For example, in the caffeine experiment, let's say we observe that resting heart rate is higher in the test group compared to the control group. Is this difference the result of caffeine, in which case scientists call it a real difference? Or is the difference just due to chance?

To answer this question, researchers begin by stating a **null hypothesis,** which predicts that the intervention or treatment has no effect at all. In other words, any difference between the test and control groups is due to chance alone and nothing else is responsible for the difference between the two groups. In this case, the null hypothesis is that caffeine does not cause an increase in heart rate. They also state an **alternative hypothesis,** which predicts that the intervention or treatment has an effect, so the difference between the test and control groups is real. In this case, the alternative hypothesis is that caffeine causes an increase in resting heart rate.

A statistical test normally yields a number, called the p-value, that expresses the likelihood that an observed result could have been observed merely by chance. A p-value is a probability. If $p \leq 0.05$ (5%), there is less than or equal to a 5% chance that the observed results are the result of chance. This is a relatively small chance. In this case, it is likely that the observed results in a dataset are real and not due to chance. In other words, the null hypothesis is rejected. By contrast, if $p > 0.05$, there is greater than a 5% chance that the observed results could have been obtained by chance, so you fail to reject the null hypothesis. The phrasing "the null hypothesis has failed to be rejected" reminds us that although the results do not support the alternative hypothesis, it does not mean that the null hypothesis is correct. Rather, it simply means that the null hypothesis has not been disproven.

Finally, scientists are often interested in determining how confident they should be in their data. Uncertainty can be shown graphically as an error bar. An error bar is typically a short vertical line showing a range of values. For example, **FIGURE 0.8** shows results from our controlled experiment investigating the relationship between caffeine and resting heart rate. The data points indicate the average heart rate of the people in the groups. The vertical lines through the data points are error bars. In spite of its name, an error bar doesn't represent an error or a mistake. Instead, it shows a range of values that incorporates small differences among the individuals and perhaps even inaccuracies in the measurements.

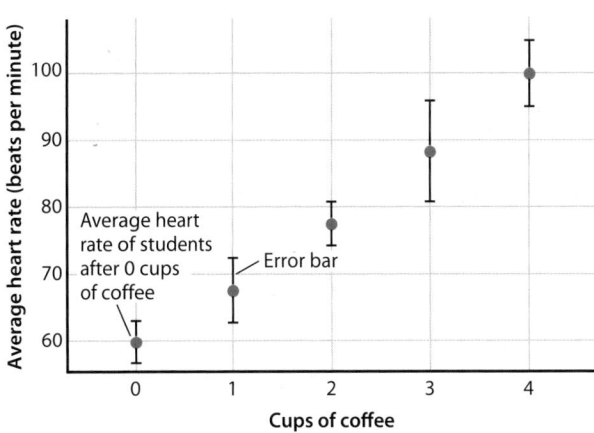

FIGURE 0.8 Error bars

An error bar represents a range of values within which the true value is likely to be. This graph shows the relationship between caffeine consumption and resting heart rate. It plots the average resting heart rate of groups of people who consumed 0, 1, 2, 3, or 4 cups of coffee, with error bars giving an indication of the uncertainty of the data.

As a result, you can think of the error bar as a way to show the uncertainty of a measurement or data point.

Throughout this course, you will have the opportunity to learn some of the techniques researchers use to evaluate data and to work with data yourself. "Tutorial 1: Statistics" on page 20 walks you through some of the statistics tools you will encounter. "Analyzing Statistics and Data: Averages" gives you a chance to practice working with these concepts.

Averages

For a detailed explanation of how to calculate mean, median, and mode, see page 20 for "Tutorial 1: Statistics." Here we review how to apply the concepts of mean, median, and mode and offer a problem for you to try.

PRACTICE THE SKILL

Below is a table listing several different ant species and the number of queens commonly found in a colony of each species. Considering the entire dataset, which measurement of the number of queens is largest: the mean, median, or mode?

Ant species	Number of queens in a small colony
Carpenter ant (*Camponotus pennsylvanicus*)	1
Red imported fire ant (*Solenopsis invicta*)	30
Pavement ant (*Tetramorium caespitum*)	5
Crazy ant (*Paratrechina longicornus*)	12
Pharaoh ant (*Monomorium pharaonis*)	2
Ghost ant (*Tapinoma melanocephalum*)	20
Little black ant (*Monomorium minimum*)	5
Argentine ant (*Linepithema humile*)	7

Data from https://www.environmentalscience.bayer.us/-/media/prfunitedstates/documents/resource-library/product-guide/ant-id-guide.ashx

To determine which number is the largest of all of the measurements, we'll have to calculate the mean, median, and mode. Let's begin by calculating the mean of the dataset. In order to do this, we add together all of the values in the dataset and divide the sum by the number of values in the dataset. The sum of the values is:

$$1 + 30 + 5 + 12 + 2 + 20 + 5 + 7 = 82$$

We divide this sum by the number of values in the dataset, which in this case is 8:

$$82 \div 8 = 10.25$$

So, the mean is 10.25.

We can also do this calculation using the following equation, where $\bar{x}$ is the mean, n is the number of values in the dataset, and $\sum_{i=1}^{n} x_i$ is the sum of all of the values in the dataset:

$$\bar{x} = \frac{1}{n} \sum_{i=1}^{n} x_i$$

$$\bar{x} = \frac{1}{8} \sum_{i=1}^{8} (1 + 30 + 5 + 12 + 2 + 20 + 5 + 7)$$

$$\bar{x} = \frac{1}{8}(82) = 10.25$$

We can find the median and mode by placing the values in numerical order:

1, 2, 5, 5, 7, 12, 20, 30

The median is the midpoint of this dataset of eight values. Because we have an even number of values, the median is the mean of the two middle values, 5 and 7.

$$5 + 7 = 12$$
$$12 \div 2 = 6$$

So, the median is 6.

We can now find the mode. The mode is the most frequent value in a dataset. If we look at the number of queens, we can see that the number 5 appears twice, more than any other value. So, the mode is 5.

The mean is 10.3, the median is 6, and the mode is 5. The largest of the three values in this case is the mean. Each of these numbers describes the dataset in a different way, which is described in "Tutorial 1: Statistics."

Your Turn

A researcher has planted bean seedlings under different light, moisture, and nutrient conditions and monitored their growth over several weeks. The data are recorded in the table. What are the mean, median, and mode for seedling height?

Bean seedling plant	Seedling height (cm)
1	20
2	6
3	15
4	23
5	5
6	15
7	21

Communicating Findings

We have seen that scientific inquiry encompasses several careful and deliberate ways of asking and answering questions about the unknown. We ask questions, make observations, collect field or laboratory samples, and design and carry out experiments to make sense of things we initially do not understand.

Another critical step in scientific inquiry, shown in Figure 0.5, is communication with other scientists and the public. Scientists publish their work in journals and present data at meetings and conferences. Sharing this information is crucial because it informs both other scientists and the public. By making the studies and data known, the results from scientific investigation are shared so that others may use the information to inform and guide future research and perhaps public action.

Scientific inquiry is typically not a linear process that proceeds in an orderly way from question to hypothesis to experiment to communication. It more accurately resembles a circle, where questions lead to experiments that lead to more questions. And there are frequent failures, false starts, and rejected hypotheses. These are all part of the process of scientific inquiry. In fact, the ability to make corrections, refine explanations, and reject hypotheses makes scientific inquiry a powerful method to understand the world around us.

Every scientist experiences failure, but good scientists learn from failed experiments, using the results to plan new ways of approaching problems. Once we obtain results that provide new understanding, we communicate what we find with other scientists and the public. Discussing and sharing ideas and results often leads to new questions, which in turn can be tested by more observations and experiments.

Establishing Theories

A hypothesis may initially be tentative. It often provides one of several possible ways to explain an observation. With repeated observation and experimentation, a good hypothesis gathers strength and researchers have more and more confidence in it. When a number of related hypotheses survive repeated testing and come to be accepted as good bases for explaining what we see in nature, scientists articulate a broader explanation that accounts for these hypotheses and the results of their tests. We call this statement a **theory,** a general explanation of the world supported by a large body of experimental evidence and observations. Examples of well-established theories include the theory of gravity, the chromosome theory, the germ theory, the cell theory, and the theory of evolution.

Scientists use the word "theory" in a very particular way. In general conversation, "theory" is often synonymous with "hypothesis," "idea," or "hunch." For example, you might say, "I've got a theory about why the car won't start." But in a scientific context, the word "theory" has a specific meaning. Scientists speak in terms of theories only if hypotheses have withstood testing to the point where they provide a general explanation for many observations and experimental results.

Just as a good hypothesis makes testable predictions, a good theory both generates hypotheses and predicts their outcomes. For example, the theory of gravity arises from a set of hypotheses you test every day by walking down the street or dropping a fork. Similarly, the theory of evolution is not just one explanation among many for the unity and diversity of life. Instead, it is a set of hypotheses that has been tested for more than a century and shown to provide an extraordinarily powerful explanation of biological observations. In fact, as we discuss throughout this book, evolution is one of the most significant theories in all of biology. It provides the most general and powerful explanation of how life works.

✓ Concept Check

5. **Describe** how a scientist turns an observation into a hypothesis and investigates that hypothesis.

6. **Describe** the differences between an experimental (test) group and a control group, and why it is important for an experiment to include both types of groups.

7. **Identify** the differences among a guess, hypothesis, and theory.

Module 0 Summary

PREP FOR THE AP® EXAM

REVISIT THE BIG IDEAS

Using the content of this module, pick your favorite organism and **describe** it in four different ways, using each of the four Big Ideas.

LG 0.1 Four Big Ideas form a fundamental basis for understanding biology.

- Evolution is the unifying principle of biology. Page 2
- Evolution explains how organisms are similar and how they are different from one another. Page 2
- Natural selection is one of the major mechanisms of evolution. Page 3
- Energy is required by all life forms in order to survive, grow, and reproduce. Page 4
- Organisms harness energy from the sun or from chemical compounds. Page 4
- DNA stores and transmits genetic information in all organisms. Page 4
- The ability to store, retrieve, and transmit genetic information is necessary for cellular function and the process of evolution. Page 4
- Cells, organisms, and the biosphere are all systems that work in an integrated and coordinated fashion to sustain life. Page 4
- Interactions of the components of a system often result in emergent or new properties that are more than a simple addition of the individual components. Page 4
- Systems interactions are diverse and occur at every level, from cells to organisms to communities to the biosphere. Page 5

LG 0.2 Scientific inquiry is a deliberate way of asking and answering questions about nature.

- Scientific inquiry involves making observations, asking questions, doing experiments or making further observations, and drawing and sharing conclusions. Page 5
- Observations are used to generate a hypothesis, a tentative explanation that makes predictions that can be tested. Page 6
- On the basis of a hypothesis, scientists design experiments and make additional observations which test the hypothesis. Page 7
- A controlled experiment involves different groups in which all the conditions are the same except for a single variable. In the test group, a variable is deliberately introduced to determine whether that variable has an effect. In the control group, the variable is not introduced. Page 8
- The independent variable in a controlled experiment is the factor that is changed by the researcher; the dependent variable is the effect or result that is being observed or measured. Page 8
- The null hypothesis predicts that an intervention or treatment in an experiment will have no effect; the alternative hypothesis predicts that the intervention or treatment will have an effect. Page 11
- Hypotheses cannot be proven, but they can be modified or rejected based on observation or experiments. Page 13
- If a hypothesis is supported by continued observation and experiments over time, it is elevated to a theory, a sound and broad explanation of some aspect of the world. Page 13

Key Terms

Biology	Energetics	Test group
Biologist	Deoxyribonucleic acid (DNA)	Experimental group
Organism	System	Control group
Gene	Biotic	Independent variable
Genome	Abiotic	Dependent variable
Cell	Biological system	Negative control group
Species	Scientific inquiry	Positive control group
Evolution	Observation	Null hypothesis
Natural selection	Hypothesis	Alternative hypothesis
Energy	Controlled experiment	Theory

Review Questions

1. Biology is the
   (A) study of components that make up organisms.
   (B) study of life.
   (C) categorization of organisms into similar groups.
   (D) study of how organisms give rise to offspring.

2. One mechanism of evolution, or change over time, is
   (A) biotic.
   (B) systems interactions.
   (C) natural selection.
   (D) DNA.

3. Many organisms use sugars as a source of energy to power the cell. This observation suggests that the pathways that break down sugars
   (A) arose relatively recently.
   (B) arose early in life's history.
   (C) are different in different species.
   (D) sugars are not necessary for life.

4. A hypothesis is
   (A) the cause of a scientific phenomenon.
   (B) the solution to a scientific problem.
   (C) a hunch or guess.
   (D) a tentative explanation of how nature functions.

5. An independent variable is
   (A) the factor that differs in the test group compared to the control group.
   (B) the order in which one measures the data collected in an experiment.
   (C) a factor that is larger or smaller than other factors in an experiment.
   (D) the result of the experiment that is being observed or measured.

6. You decide to test whether helium is light enough to cause balloons to rise. In the first group, you fill the balloons with helium. In the second group, you fill the balloons with air. What is the dependent variable in your experiment?
   (A) The presence or absence of helium
   (B) The group of balloons that receives the helium
   (C) The group of balloons that receives the air
   (D) Whether the balloons rise or not

7. A hypotheses is
   (A) supported when data contradict it.
   (B) proven when data can support it.
   (C) modified when data contradict it.
   (D) rejected when data support it.

Module 0
AP® Practice Questions

Section 1: Multiple-Choice Questions

Choose the best answer for questions 1–4.

1. Survival in some mushroom species is enhanced by the presence of gills with large surface areas. Mushrooms with larger gill surface areas produce more spores and reproduce more successfully than those with smaller gill surface areas in a particular environment. Identify which of the four Big Ideas of biology this example best describes.
   (A) Evolution
   (B) Energetics
   (C) Information Storage and Transmission
   (D) Systems Interactions

2. A student accidentally spilled a bag of rock salt on some plants while moving the bag to the garage. After some time, the student noticed that the plants in this area were turning brown and withering. The student formulated a hypothesis that the plants were dying because they do not grow well in salty soil. Among the proposed experiments below, identify the most effective way of testing this hypothesis.
   (A) Keeping a weekly record of any plant regrowth
   (B) Planting some seeds in a container of healthy soil and some in a container of salty soil and observing plant growth in each over a period of weeks
   (C) Removing the rock salt from the area, replanting grass seed, and observing any change in growth
   (D) Spreading rock salt over a larger portion of the lawn to see if the rest of the grass dies

3. A scientist is doing an experiment to see which conditions are best for bacterial growth. The scientist has set up an array of several petri dishes and will subject the bacterial cultures to various cool and warm temperatures, while comparing them to petri dishes at room temperature. Identify how the condition of temperature is used in this experiment.
   (A) As a dependent variable
   (B) As an independent variable
   (C) As a control group
   (D) As an experimental group

4. The table below shows differences in mass among males in four different species of frogs at two different life stages. A metamorph is a young individual that has nearly completed the change from tadpole to frog. Metamorphs have four legs and still have a tail. A fully grown adult frog has no tail.

	Metamorph mass (g)	Adult mass (g)
Green frog	1.1	20.2
Leopard frog	0.98	18.9
Wood frog	0.82	7.9
Gray tree frog	0.57	7.1

Which species has the largest percent change in mass from metamorph to adult?

(A) Green frog
(B) Leopard frog
(C) Wood frog
(D) Gray tree frog

Section 2: Free-Response Question

Write your answer to each part clearly. Support your answers with relevant information and examples. Where calculations are required, show your work.

An experiment was performed on a wild population of Anna hummingbirds (*Calypte anna*) to see if they preferred feeding from a specific color of sugar water. Scientists had previously observed the hummingbirds feeding from a glass feeder with a clear sugar water solution. The scientists placed five identical glass feeders side by side and filled each with a different colored sugar water solution: red, yellow, green, blue, and clear. Every 15 minutes, the scientists changed the positions of the feeders relative to one another to eliminate any position bias. The scientists recorded the color of the first sugar water solution that each hummingbird visited. The experiment was carried out for 2 days. The results of the experiment are shown in the table.

(a) **Identify** the independent and dependent variables.

(b) **Identify** the control group and the experimental group.

(c) **State** an alternative hypothesis and the corresponding null hypothesis.

(d) **Explain** what the data tell us about the hummingbirds' preference.

Average Number of Birds Approaching and Drinking from Each of Five Containers Containing Different Colored Solutions

Color	Mean counts
Red	33.9
Yellow	13.1
Green	8.4
Blue	4.3
Clear	5.3

Data from https://sora.unm.edu/sites/default/files/journals/wilson/v092n01/p0053-p0062.pdf

We can describe four Big Ideas that connect and unite the many dimensions of biology: Evolution, Energetics, Information Storage and Transmission, and Systems Interactions. These four ideas are introduced in Unit 1 and will be visited again and again throughout the book. By the time you finish this book, you will have an understanding of how life works, from the molecular

The Four Big Ideas of AP® Biology

Winter   Summer

Evolution
investigates changes in the genetic makeup of a population over time. Through the process of natural selection, species become adapted to their environments. For example, ancestors of the snowshoe hare and Canada lynx evolved to have adaptations that increased their survival and reproduction. This includes the hare's ability to change coat color with the seasons.

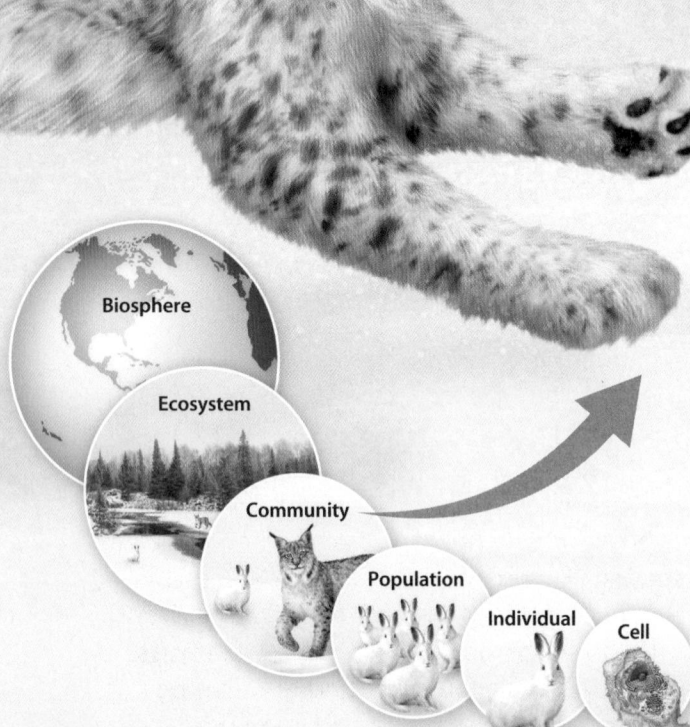

Biosphere

Ecosystem

Community

Population

Individual

Cell

Systems Interactions
examines the way the components of biological systems interact with and affect one another. A system can be as small as a cell or as large as the biosphere. New properties of biological systems emerge through the interactions of organisms such as Canada lynx and snowshoe hares, and through interactions between species and their environment.

machines inside cells and the metabolic pathways that cycle carbon through the biosphere, to the process of evolution, which has shaped the living world that surrounds and includes us. The four Big Ideas are fundamental to understanding and organizing such diverse aspects of biology.

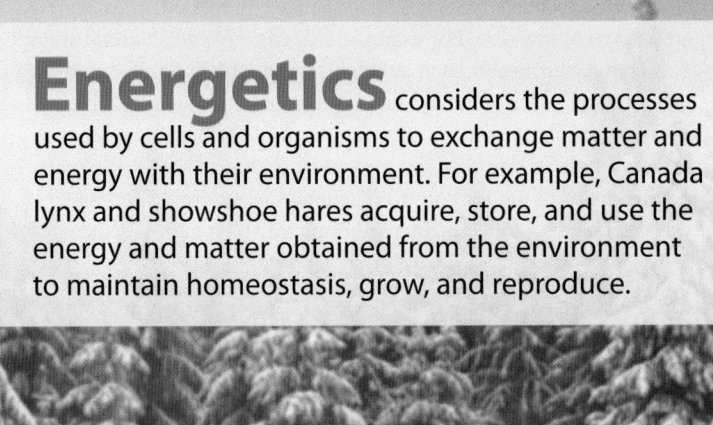

Energetics considers the processes used by cells and organisms to exchange matter and energy with their environment. For example, Canada lynx and showshoe hares acquire, store, and use the energy and matter obtained from the environment to maintain homeostasis, grow, and reproduce.

Information Storage and Transmission explores

how information is stored, used, and transmitted by cells and organisms. Using information stored in DNA and through experience, the Canada lynx knows how to hunt its prey and the snowshoe hare knows how to evade its predator. These behaviors are transmitted to offspring through DNA and through the process of learning.

Tutorial 1: Statistics

Scientific evidence rarely hinges on the result of a single experiment, measurement, or observation. As we discussed in Module 0, any scientific claim must be backed up by data from multiple subjects across multiple repetitions of the same experimental process or multiple observations. It is the accumulation of evidence from many independent sources, all pointing in the same direction, that lends weight to a scientific hypothesis until eventually it becomes a theory.

This means that any statement of a number—ants can carry 10 times their weight; a human cell can divide in 24 hours—is actually a statement of many numbers. How do scientists determine what number will represent all of their experimental results? And how do scientists succinctly describe all of the individual data points? Statistics is a field that helps scientists to organize, analyze, and see trends in data. In this tutorial, we will discuss several statistical tools that help us understand and interpret data. We will start with a quick look at the issue of precision in reporting calculations.

Significant Figures

When reporting any recorded data or quantitative conclusions, you should use appropriate precision. For example, if you measure something and say it is 2 meters tall, does that mean it is exactly 2 meters or perhaps 2.03 meters or even 2.10 meters? Significant figures indicate the precision of a measurement. Significant figures are numbers that carry meaning. For example, 2 meters has just one significant figure because it has only one digit. By contrast, 2.03 has three significant figures. The more significant figures, the more precise the measurement.

In general, all nonzero numbers (1, 2, 3, and so on) are significant. Zeros between other numbers are also significant. For example, as we noted, the number 2.03 has three significant figures. Leading zeros (zeros that are located to the left of another number) are not significant. The number 0.02 has only one significant figure. Trailing zeros (zeros located to the right of another number) are also not significant, except in numbers with a decimal point. For example, the number 2.10 has three significant figures.

When doing calculations, significant figures in a final answer are determined by the number that is the least precise in your dataset. However, do not round intermediate values when you perform calculations; only round your final answer. For example, let's say you are asked to calculate the area of a rectangle (*width* (*w*) × *height* (*h*)), with *w* = 4.33 feet and *h* = 2 feet. The answer is 9 feet. That is: 4.33 × 2 = 8.66, but the answer is rounded up to 9 because, among the numbers you used in the calculation, a height of 2 feet—with one significant figure—is the least precise. Note that if you first round 4.33 to 4 and then do the calculation, you get an answer of 4 × 2 = 8, which is incorrect.

Average: Mean, Median, and Mode

When scientists make observations or do an experiment, they gather results and collect data. The direct observations and measurements that have not been organized, processed, or interpreted are known as raw data. For example, a doctor collects all kinds of raw data about your health, such as height, weight, body temperature, breathing rate, heart rate, and blood pressure. You may give a blood or urine sample so that the concentrations of various substances can be measured. Or you may spit into a vial so that a DNA sample can be obtained to look for common genetic variants associated with disease. The data collected by physicians are all types of raw data.

After scientists collect raw data, they often process the data in some way in order to make sense of it. One way to process it involves determining the average of all of the measurements. Determining the average takes all of the individual pieces of data and provides a single, representative number. In this section, we look at several different ways to calculate the average.

Mean

The first type and most common type of average is called the *mean*. Given a set of values, the **mean** is obtained by adding all of the values in the dataset and dividing by the number of values in the dataset, which is indicated by *n*.

Example: In the dataset of nine values, 5, 9, 4, 6, 5, 1, 5, 9, 1, 9, the mean is determined by adding all of the values (5 + 9 + 4 + 6 + 5 + 1 + 5 + 9 + 1 + 9) = 54, and then dividing by the number of values (*n* = 9), or $\frac{54}{9}$ = 6.

The mean can also be calculated using the following formula:

$$\bar{x} = \frac{1}{n}\sum_{i=1}^{n} x_i$$

In this formula, $\bar{x}$ represents the mean. Σ, the Greek letter sigma, is a symbol that means "the sum of." $\sum_{i=1}^{n} x_i$ means $x_1 + x_2 + \ldots + x_n$, where x_n is the *n*th, or final, value, so it indicates that you sum all of the values. And $\frac{1}{n}$ indicates that you divide the result by the number of values.

Median

Another type of average is the *median*. The **median** is middle value of a group of values. That is, there are as many values falling above the median as below it.

The median is sometimes useful because it is less influenced than the mean to extreme values, called outliers. An outlier is a data point that is very different from all of the other data points and therefore one that falls outside the overall pattern of a group of values. One or two outliers can affect the mean much more than the median. As a result, in this case, the median might be more representative of a group of values than the mean.

To find the median of a dataset, list the values of the dataset in numerical order and identify the value in the middle of the list. If there is an even number of values, the median is the mean of the two middle values.

Example: Let's use the dataset from the previous example. We determine the median by first putting the nine values in numerical order:

1, 1, 4, 5, 5, 5, 6, 9, 9

The number in the fifth position, 5, is the median because there are four values on either side of it. Note that, in this case, the median (5) and the mean (6) are not the same.

Mode

A third type of average is the **mode,** which is the value that shows up most frequently in all the measurements.

Example: Taking the dataset 5, 9, 4, 6, 5, 1, 5, 9, 1, 9 once again, the mode is 5 because it occurs three times, which is more than any of the other values.

Your Turn

While interning at a doctor's office, you record the heights of the first 20 patients of the day:

60″, 72″, 75″, 63″, 62″, 67″ 65″, 65″, 63″, 65″,
67″, 68″, 65″, 69″, 68″, 70″, 67″, 69″, 65″, 63″

What are the mean, median, and mode of the heights of the morning's patients?

Solution

Let's begin with the mean. Start by adding the values:

60″ + 72″ + 75″ + 63″ + 62″ + 67″ + 65″ + 65″ + 63″ +
65″ + 67″ + 68″ + 65″ + 69″ + 68″ + 70″ + 67″ + 69″ +
65″ + 63″ = 1328″

Then divide the sum by the number of values, 20:

1328″ ÷ 20 = 66.4″, which is rounded to 66″ because the raw data all have just two significant figures.

Therefore, the mean height of the morning's patients is 66.4″.

To determine the median, start by writing the heights so they are in ascending order:

60″, 62″, 63″, 63″, 63″, 65″, 65″, 65″, 65″, 65″, 67″, 67″,
67″, 68″, 68″, 69″, 69″, 70″, 72″, 75″

There is an even number of values ($n = 20$), so the median is the mean of the two middle values. In a set of 20 values, the middle two are the 10th and 11th values: 65″ and 67″. The mean of these two values is:

$$\frac{(65″ + 67″)}{2} = 66″$$

So, the median height of the morning's patients is 66″.

To determine the mode, it helps to look at the values in ascending order, as we did when determining the median. Count how many instances there are of each value. Five patients have a height of 65″, more than any other height. Therefore, the mode of the heights of the morning's patients is 65″.

▶ See page 12 for "Analyzing Statistics and Data: Averages" which gives you the opportunity to apply these concepts in context.

The mean, median, and mode provide different ways to describe or characterize datasets. In a way, they are all averages, but which one you use depends on if you want to weigh all of the values equally (mean), if you want the value in the middle (median), or if you want the value that is most common (mode).

Spread of Data: Normal Distribution, Range, and Standard Deviation

The numbers that result from the multiple repetitions of each experiment involving multiple measurements or observations are typically not the same, as we discussed earlier. Instead, they may differ from one another, and on a graph will appear scattered. Experimental results may differ because conditions typically vary in some subtle way from one experiment to the next, individuals differ from one another, and the situations change slightly over the course of time. In this section, we will look at ways to visualize, calculate, and communicate the distribution of data.

Normal Distribution

The spread of all the numbers in experimental results is described as a **distribution** and can usually be graphed. Often in biology, measurements of natural phenomena form a graph pattern with a smooth, bell-shaped curve, called a **normal distribution.**

Human height is a good example. Our sample problem for mean, median, and mode used the heights of 20 patients. If you took a much larger and more representative sample of measurements and graphed them, you would end up with a normal distribution, as shown in the graph below.

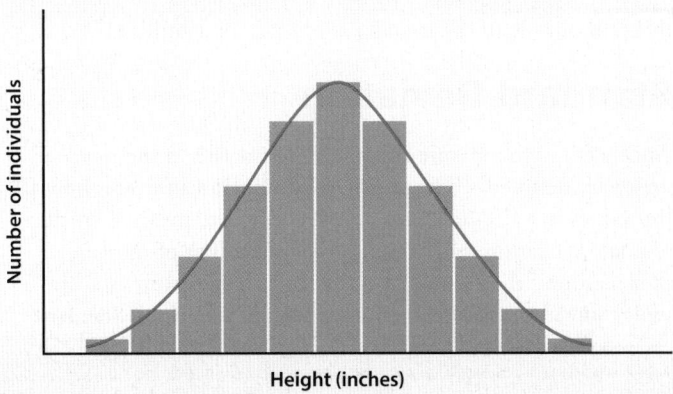

In the graph, height is plotted along the horizontal x-axis, and number of individuals is plotted along the vertical y-axis. You can see that most heights fall into the midrange along the horizontal x-axis. These people are represented by the peak in the center of the normal distribution. A few people are very short or very tall. These two extremes are found on the two ends of the distribution.

In a normal distribution, the mean, median, and mode are all the same value, shown in the graph below.

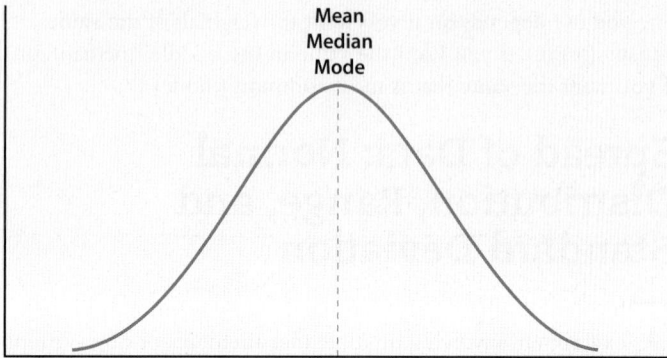

We can understand why the three values are the same by looking at where they fall on the curve. The peak of the normal distribution represents the largest number of people: the mode. In a normal distribution, the peak of the curve falls in the middle of the dataset: the median. And because there are equal numbers of values evenly distributed on either side of the median, the mean is right in the middle, too.

Not all datasets form a normal curve when graphed. In datasets that form different curves, the mean, median, and mode may not be the same.

Range

In addition to graphing the distribution of data, scientists often measure it in order to quantify how spread out the data are. One way to calculate the spread of the data is the *range*. The **range** is simply the highest value minus the lowest value. The range is useful because it describes the full spread of all of the data. However, one issue with using the range to describe the spread of data is that it is highly influenced by even a single outlier, or a data point that falls outside of the overall pattern of a distribution.

Standard Deviation

Another way to measure the spread of the data is the *standard deviation,* denoted s. The **standard deviation** indicates how far the values in a dataset as a whole are from the mean. A smaller standard deviation indicates a tighter clustering of measurements around the mean. A larger standard deviation indicates a wider spread of the measurements around the mean. Standard

deviation is influenced by outliers as well, but not nearly as much as the range.

In a normal distribution, shown below, approximately 68% of the values lie within one standard deviation on either side of the mean. In the graph below, this is the light blue area, and it is indicated by $+1s$ and $-1s$. Also in a normal distribution, 95% of the observations lie within two standard deviations on either side of the mean. In the graph below, this is the light and darker purple areas together, or $+2s$ and $-2s$.

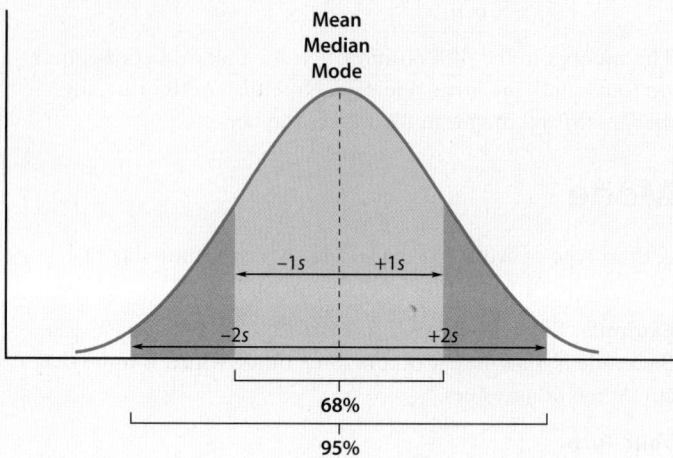

The equation for determining the standard deviation of a dataset is:

$$s = \sqrt{\frac{\Sigma(x_i - \overline{x})^2}{n - 1}}$$

In this equation, you start by calculating the difference between each individual measurement (x_i) and the mean ($\overline{x}$), noted above as ($x_i - \overline{x}$). Then you take each of the results and square it, described as ($x_i - \overline{x})^2$. Finally, you add all of these squares, divide by $n - 1$, and take the square root of the result.

Example: The dataset 1, 2, 2, 3, 3, 3, 4, 4, 5 is a normal distribution with a mean of 3. That is, $(1 + 2 + 2 + 3 + 3 + 3 + 4 + 4 + 5) = 27$, divided by 9, or $\frac{27}{9} = 3$. To find the standard deviation, we first find the differences between each value and the mean. We square the differences and add the differences together:

$$(1 - 3)^2 + (2 - 3)^2 + (2 - 3)^2 + (3 - 3)^2 + (3 - 3)^2 +$$
$$(3 - 3)^2 + (4 - 3)^2 + (4 - 3)^2 + (5 - 3)^2 = 12$$

Putting this sum into the equation above, where n is the number of values in the dataset, or 9, we get:

$$s = \sqrt{\frac{12}{9 - 1}} = \sqrt{\frac{12}{8}} = \sqrt{1.5} = 1.2$$

The standard deviation of the dataset is 1.2, which is a measure of the spread of the data on either side of the mean.

Your Turn

It's now a new day in the doctor's office in which you intern, and you want to see if this morning's patients are generally taller or shorter than yesterday's patients. You arrange the 23 patients' heights in a data table, and then graph the data.

Height (inches)	Number of patients
64	2
65	3
66	4
67	5
68	4
69	3
70	2

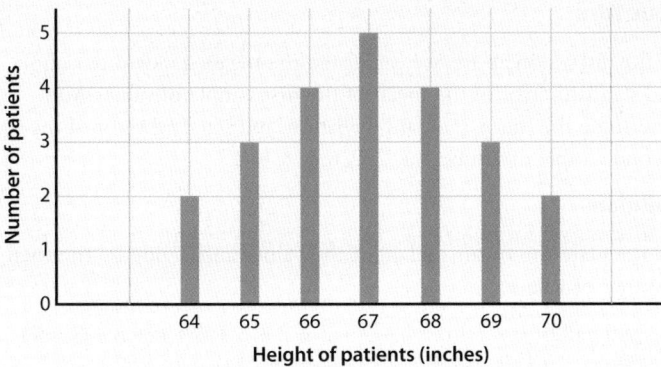

1. What is the mean, median, and mode of this dataset, and where are these points on the graph?

2. What is the standard deviation for this dataset? Draw it on the graph.

Solution

1. The mean is the sum of all of the heights, divided by the number of patients. Remember, the height in the first column of the table must be multiplied by the number of people with that height, shown in the second column of the table.

$$(2 \times 64'') + (3 \times 65'') + (4 \times 66'') + (5 \times 67'') + (4 \times 68'') + (3 \times 69'') + (2 \times 70'') = 1541''$$

Then divide the sum by the number of patients, 23.

$$1541'' \div 23 = 67''$$

The mean is 67″.

To find the median, find the middle point of the data. The data are written out in ascending order in the data table. In a sample of 23 patients, the median height is the height of the 12th patient. If you add the number of patients in the right column, the

12th patient is in the group of five patients who are 67″ tall. The median is therefore 67″.

The mode is the value that occurs most frequently. There are five patients who are 67″ Because the data fit into a normal distribution, as shown in the graph below the mean, median, and mode are all the same value.

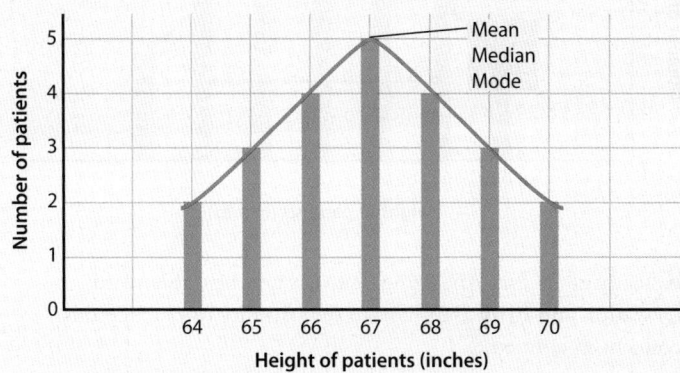

2. To use the equation to determine the standard deviation, we first need to find the difference between each value and the mean. Then we square that difference. Let's put these numbers in a table to make things easier to see.

Height (inches)	Number of patients	Difference between height (inches) and mean (67″)	Difference squared
64	2	3	9
65	3	2	4
66	4	1	1
67	5	0	0
68	4	−1	1
69	3	−2	4
70	2	−3	9

Multiple patients have the same height. Therefore, to calculate the sum of the squares of the differences, we need to look in the column that lists the number of patients at each height. We then multiply the difference squared for each height by the number of patients at each height, as follows:

$$\text{Sum} = (2 \times 9) + (3 \times 4) + (4 \times 1) + (5 \times 0) + (4 \times 1) + (3 \times 4) + (2 \times 9) = 68$$

Substituting numbers into the equation, where n = number of patients (23), we get:

$$s = \sqrt{\frac{\sum (x_i - \bar{x})^2}{n - 1}}$$

$$= \sqrt{\frac{68}{23 - 1}} = \sqrt{\frac{68}{22}} = \sqrt{3.09} = 1.76''$$

On the graph of our data shown below, the standard deviation is 1.76" in either direction of the mean, indicated by +1s and −1s.

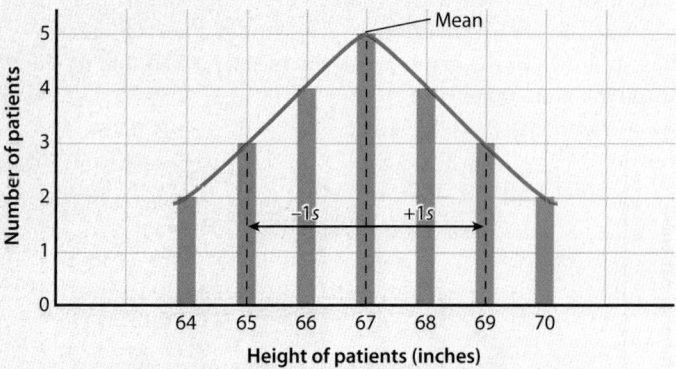

▶ See page 28 for "Analyzing Statistics and Data: Standard Deviation and Error Bars" for an opportunity to practice this concept in context.

Uncertainty in Data: Standard Error of the Mean and Error Bars

As we have seen, scientists are often interested in finding a single value that is representative of many values such as the mean, median, or mode. They also describe how spread out the values are from the average and how their data are distributed. In addition, scientists often convey how confident they are in their measurements and calculations. There are statistical tools that help scientists measure and communicate uncertainty in their data. In this section, we describe two ways to convey this information.

Standard Error of the Mean

Let's return to the sample of numbers 1, 1, 4, 5, 5, 5, 6, 9, 9. There are nine numbers in total, and these form a normal distribution. Based on this sample, the mean $\bar{x} = 5.0$ and the standard deviation $s = 2.9$.

The values of $\bar{x}$ and s are actually estimates of the true mean and standard deviation. In other words, they are calculated from a sample, not from an entire population. So, although they represent the mean and standard deviation of the sample, they may not be the true or actual mean and standard deviation of an entire population. As estimates, they will differ from one random sample to the next. Another random sample from the same distribution might yield 1, 2, 2, 2, 6, 6, 6, 8, 8. In this sample, $\bar{x} = 4.6$. A third random sample might yield 1, 2, 3, 5, 6, 6, 7, 8, 10. In this sample, $\bar{x} = 5.3$. Each of these is an estimate of the true mean. If we examined a very large number of samples, the values of $\bar{x}$ would themselves form a normal distribution.

To determine how close the estimated mean is to the true mean, we use a value called the **standard error of the mean,** which is calculated as follows:

$$SE_{\bar{x}} = \frac{s}{\sqrt{n}}$$

where $SE_{\bar{x}}$ is the standard error of the mean, s is the standard deviation, and n is the number of samples.

The estimate $\bar{x} = 5.00$ of the true mean has standard error of $\frac{s}{\sqrt{n}} = \frac{2.87}{\sqrt{9}} = 1.0$. This number tells you how close the estimate $\bar{x}$ is likely to be to the true mean. In approximately 68% of samples, $\bar{x}$ is likely to be within one standard error of the true mean. In our example with $\bar{x} = 5.00$ and $\frac{s}{\sqrt{n}} = 1.0$, in 68% of the samples, the true population mean is likely to be in the range of one standard error less than the mean (5.00 − 1.0) and one standard error more than the mean (5.00 + 1.0). This is a range of 4.0 to 6.0. This range is sometimes described in words as the mean plus or minus (±) one standard error.

Your Turn

In the discussion above, we calculated the mean, standard deviation, and standard error of the mean of the first sample of values. You determine the mean, standard deviation, and standard error of the second sample of values: 1, 2, 2, 2, 6, 6, 6, 8, 8.

Solution

To calculate the mean, add all of the values and divide by the total number of values:

Add all of the values: $1 + 2 + 2 + 2 + 6 + 6 + 6 + 8 + 8 = 41$

Divide by the total number of values: $\frac{41}{9} = 4.6$

So, $\bar{x} = 4.6$

To calculate the standard deviation, subtract the mean from each value, square the result, add up all of the squares, divide by the number of values −1, and finally take the square root:

$$s = \sqrt{\frac{62.22}{9-1}} = \sqrt{\frac{62.22}{8}} = \sqrt{7.78} = 2.79$$

To calculate the standard error of the mean, simply divide the standard deviation (s) by the square root of the number of values:

$$SE_{\bar{x}} = \frac{s}{\sqrt{n}} = \frac{2.79}{\sqrt{9}} = \frac{2.79}{3} = 0.9$$

Error Bars

In data presented graphically, you may see a short vertical line through a point. The point indicates an estimate or average of several measurements, and the vertical line is the error bar. An error bar is a vertical line on a graph that indicates a range of values within which the true value is very likely to fall. The error bar, in spite of its name, is not in fact an error. Instead, it provides a measure of the confidence that a scientist has in a particular data point. For example, in Module 0, we examined the relationship between caffeine consumption and resting heart rate. The results for each group of people were averaged and graphed in Figure 0.8. The vertical line across each point in the graph represents the error bar.

As another example, let's say we are interested in the average heights of students in four different classrooms, A–D. We can measure the heights of all of the students, take the mean for each classroom, and graph it as shown here.

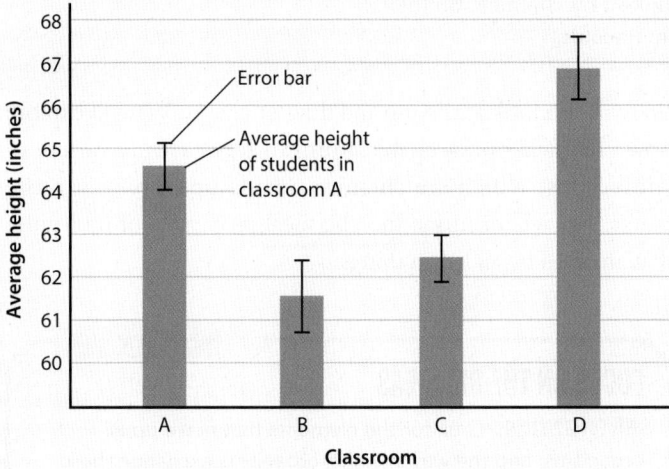

The top of each bar indicates the average height of students in each classroom, and the error bar is the short vertical line at the top of each bar.

An error bar can be calculated in different ways. It can indicate the mean plus or minus one standard deviation, plus or minus one standard error of the mean, or plus or minus two standard errors of the mean. Depending on how it is calculated, error bars will vary in length (be longer or shorter) and will convey different information (or different degrees of confidence in the data). The legend that accompanies a figure or graph typically describes how error bars are calculated, so you should always read the legend to understand what is going on in a particular graph.

In AP® Biology, error bars usually represent the mean plus or minus two standard errors of the mean. This range of values is sometimes called the 95% confidence interval because, in 95% of the samples, the confidence interval includes the true mean.

As we discussed in Module 0, scientists often use a threshold of 5% to draw conclusions about whether a result is real, or significant, or whether it is due to chance. As a result, you can look at error bars that represent 95% confidence intervals to get an indication of whether a difference between two groups may be significant or not. If the error bars of two groups do not overlap, scientists state that there is a significant difference between the two groups. If they do overlap, any observed differences may be random and due to chance.

For example, in the figure showing the heights of students in different classrooms, the error bar for classroom A does not overlap with any of the others, so it is significantly different from the others. However, the error bars for classrooms B and C do overlap, so there is probably not a significant difference between these two groups.

▶ See page 28 for "Analyzing Statistics and Data: Standard Deviation and Error Bars" for an opportunity to practice this concept in context.

Key Terms

Mean
Median
Mode
Distribution
Normal distribution
Range
Standard deviation
Standard error of the mean

Module 1

Elements of Life

LEARNING GOALS ▶LG 1.1 Matter and energy govern the properties of life.
  ▶LG 1.2 The atom is the fundamental unit of matter.
  ▶LG 1.3 Atoms combine to form molecules linked by chemical bonds.
  ▶LG 1.4 Carbon is the backbone of organic molecules.

Take a look around you. Everything you see and feel is made of matter, the material that makes up physical objects. **Matter** is anything that has mass and takes up space. Matter may be a gas, a liquid, or a solid. Even you are made of matter. To grow, reproduce, and maintain their organization, all organisms exchange matter with their environment. They also require energy. The study of biology is based on an understanding of the pathways and transformations of matter and energy. Therefore, to understand why organisms look and act as they do, we need to gain a basic understanding of matter and energy. In this module, we will examine the fundamentals of both. We will look at properties of the basic unit of matter, the **atom.**

We will also look at the use and flow of energy. We will explore how chemical bonds enable atoms to form a wide variety of **molecules,** which are chemicals made up of two or more atoms, and will examine the chemical properties of molecules that are used by all living things.

PREP FOR THE AP® EXAM

FOCUS ON THE BIG IDEAS

ENERGETICS: Look for the elements that make up all organisms, and the way in which organisms exchange these elements with the environment.

1.1 Matter and energy govern the properties of life

Because organisms are composed of matter and require energy, it follows that organisms are subject to the physical laws and principles that govern matter and energy. In this section, we will first review some of the properties of matter and energy, and how they flow through communities of living organisms.

Flow of Matter

Let's begin by following the path of matter as it flows through living systems. Imagine that we could tag a carbon atom at its moment of origin and then follow its trip through time and space. Formed in a nuclear blast furnace deep within an ancient star and then ejected into space as the star exploded in death, our atom was eventually swept up with other materials to form Earth, a small planet orbiting a newer star we now call the sun. Volcanoes introduced our carbon atom into Earth's early atmosphere as carbon dioxide (CO_2). Slowly, over millions of years, this carbon dioxide reacted with water and rocks, transferring the carbon from the air to the seafloor. Here, our atom sat for many millions of years, until

earthquakes, erosion, or other geologic activities returned it to the atmosphere as carbon dioxide once again. Slowly but surely, geologic processes on early Earth cycled carbon from atmosphere to rocks and back again. This slow movement of carbon between Earth and atmosphere continues today.

Sometime between 4 and 3.5 billion years ago, as life took hold on Earth, our carbon atom began to cycle more rapidly—much more rapidly. Microorganisms were able to convert the carbon dioxide in the environment into **organic molecules,** which are biological molecules that contain carbon. Other microorganisms broke down organic molecules and returned carbon dioxide to the environment. To this day, carbon cycles continuously from the atmosphere and oceans to organisms and back again.

This intricately linked network of geological and biological processes that shuttles carbon among rocks, soil, ocean, air, and organisms is called the carbon cycle. Why focus on carbon? The carbon cycle provides an organizing principle for understanding life on Earth. The chemistry of life is, in no small part, the chemistry of carbon because organic molecules are made up of carbon.

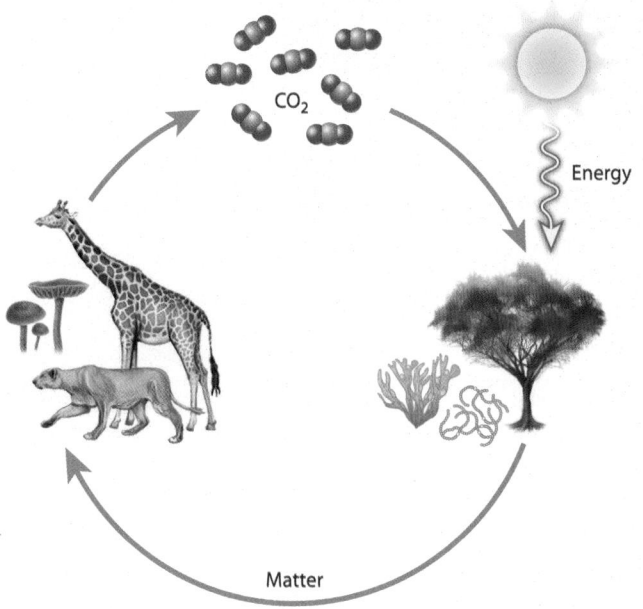

FIGURE 1.1 Flow of matter and energy

Matter, such as a carbon atom, cycles among organisms and the physical environment. Energy, like solar and chemical energy, is harnessed by organisms to do work and needs to be constantly added to the system to sustain life.

The carbon cycle also teaches that matter moves through organisms and the physical environment in a cycle, as shown in **FIGURE 1.1**. In other words, the same atoms are reused over and over. Organisms must exchange matter with other organisms and the environment to grow, reproduce, and maintain organization. For instance, when a leaf or a plant in the forest dies, it is usually consumed by the animals or microbes in the forest. These organisms use the dead leaf to build their biomass. In turn, these organisms recycle matter through the wastes they release or when they die. The released matter is used by other organisms, including plants, animals, and microbes, to build their biomass.

Flow of Energy

As organisms move carbon, they also transfer energy. Carbon and energy are closely intertwined for the simple reason that the energy sources for many organisms are the carbon-rich organic molecules in the organisms they eat or the molecules they build themselves. Unlike carbon, energy does not move in a cycle. Instead, energy must continually be harvested from the environment to sustain the community. In essentially all habitats where sunlight is available, the sun provides the entry point for energy into living systems, as

PREP FOR THE AP® EXAM

AP® EXAM TIP

You should know that matter and energy both flow through communities, but they take different paths. Matter travels in a cycle, with the same atoms moving back and forth among organisms and the physical environment. By contrast, energy is not recycled. Instead, an input of energy from the environment is constantly required to sustain cells and organisms. Be prepared to give examples of the ways that matter and energy flow through communities and therefore sustain life.

shown in Figure 1.1. Plants, algae, and certain bacteria capture energy from the sun and use it to synthesize energy-rich organic molecules. Where sunlight is absent, especially in the vast depths of the ocean, energy instead comes from chemical compounds.

Organisms transform energy. That is, they acquire energy from the environment and convert it into a chemical form that their cells can use. For instance, some of the solar energy striking a forest is captured by plants, which transform it into chemical energy in the form of sugars. Some of this energy is used by the organism to do work—such as building cellular components, moving, and reproducing. The rest of this energy is dissipated as heat and is no longer available for the organism to use. We will look more closely at the flow of energy and the cycling of matter in Units 3 and 8. "Analyzing Statistics and Data: Standard Deviation and Error Bars" gives you an opportunity to practice using data analysis skills on a question involving plants and energy capture.

PREP FOR THE AP® EXAM

AP® EXAM TIP

Make sure you understand that no overlap between error bars indicates a statistically significant difference.

✓ Concept Check

1. **Describe** how matter and energy flow through living systems.
2. **Describe** how the flow of matter depends on the flow of energy in communities.

Standard Deviation and Error Bars

For a detailed explanation of how to calculate standard deviation and standard error of the mean, and a discussion of error bars, see "Tutorial 1: Statistics" on page 20. For reference, here are the equations for standard deviation (s) and standard error of the mean ($SE_{\bar{x}}$).

$$S = \sqrt{\frac{\sum (x_i - \bar{x})^2}{n-1}} \qquad SE_{\bar{x}} = \frac{s}{\sqrt{n}}$$

PRACTICE THE SKILL

Plants capture energy from the environment and transform it into chemical energy in the form of organic molecules called sugars. They use these sugars for growth and reproduction. Sunlight is the major source of energy for plants. A lack of sunlight prevents plants from growing well. Researchers performed an experiment to demonstrate the effect of light on plant growth. The researchers grew crofton weed seedlings, which are germinated seeds, in 20 petri dishes. Ten petri dishes remained in the dark and 10 petri dishes were exposed to light. The researchers measured seedling growth in millimeters over 1 week. The data are shown in the table.

Petri Dishes	Seedling height (mm) in dark	Seedling height (mm) in light
1 and 2	12	18
3 and 4	8	22
5 and 6	15	17
7 and 8	13	23
9 and 10	6	16
11 and 12	4	18
13 and 14	13	22
15 and 16	14	12
17 and 18	5	19
19 and 20	6	17

Data from https://www.biointeractive.org/sites/default/files/media/file/2019-05/Statistics-Teacher-Guide.pdf

1. What is the mean height ($\bar{x}$) of the seedlings grown in the dark, and the mean height ($\bar{x}$) of the seedlings grown in the light?

To calculate the mean of the heights of the seedlings grown in the dark, we add all of the values in the Dark column and divide by the number of values in that dataset, which is 10:

12 mm + 8 mm + 15 mm + 13 mm + 6 mm + 4 mm + 13 mm + 14 mm + 5 mm + 6 mm = 96 mm

$$\frac{96\text{ mm}}{10} = 9.6\text{ mm}$$

We do the same calculation for the values in the Light column:

18 mm + 22 mm + 17 mm + 23 mm + 16 mm + 18 mm + 22 mm + 12 mm + 19 mm + 17 mm = 184 mm

$$\frac{184\text{ mm}}{10} = 18.4\text{ mm}$$

The mean height of the seedlings grown in the dark is 9.6 mm, and the mean height of the seedlings grown in the light is 18.4 mm.

2. Calculate the standard deviation of the heights of the seedlings grown in the dark and the light.

We know each individual measurement in the dataset (x_i), the mean ($\bar{x}$), and the sample size (n) for the dark and light treatments and can plug this information into the standard deviation formula:

Dark:

$$s = \sqrt{\frac{\sum (x_i - \bar{x})^2}{n-1}}$$

$$s = \sqrt{\dfrac{\begin{array}{c}(12-9.6\text{ mm})^2 + (8-9.6\text{ mm})^2 + (15-9.6\text{ mm})^2 + (13-9.6\text{ mm})^2 + (6-9.6\text{ mm})^2 \\ + (4-9.6\text{ mm})^2 + (13-9.6\text{ mm})^2 + (14-9.6\text{ mm})^2 + (5-9.6\text{ mm})^2 + (6-9.6\text{ mm})^2\end{array}}{10-1}}$$

$$s = \sqrt{\dfrac{158.4\text{ mm}}{9}} = 4.20\text{ mm}$$

Light:

$$s = \sqrt{\dfrac{\sum (x_i - \bar{x})^2}{n-1}}$$

$$s = \sqrt{\dfrac{\begin{array}{c}(18-18.4\text{ mm})^2 + (22-18.4\text{ mm})^2 + (17-18.4\text{ mm})^2 + (23-18.4\text{ mm})^2 + (16-18.4\text{ mm})^2 \\ + (18-18.4\text{ mm})^2 + (22-18.4\text{ mm})^2 + (12-18.4\text{ mm})^2 + (19-18.4\text{ mm})^2 + (17-18.4\text{ mm})^2\end{array}}{10-1}}$$

$$s = \sqrt{\dfrac{98.4\text{ mm}}{9}} = 3.31\text{ mm}$$

The standard deviation of the heights for seedlings grown in the dark is 4.20 mm, and the standard deviation of the heights for seedlings grown in the light is 3.31 mm.

3. Calculate the standard error of the mean of the heights of the seedlings grown in the dark and the light.

Because we have just found the standard deviation for each treatment (s) and we know the sample size (n) for each treatment as well, we can plug these into the formula for the standard error of the mean:

Dark: $SE_{\bar{x}} = \dfrac{s}{\sqrt{n}}$ $\quad SE_{\bar{x}} = \dfrac{4.20\text{ mm}}{\sqrt{10}}$ $\quad SE_{\bar{x}} = 1.33\text{ mm}$

Light: $SE_{\bar{x}} = \dfrac{s}{\sqrt{n}}$ $\quad SE_{\bar{x}} = \dfrac{3.31\text{ mm}}{\sqrt{10}}$ $\quad SE_{\bar{x}} = 1.05\text{ mm}$

The standard error of the mean of the heights for seedlings in the dark is 1.33 mm. The standard error of the mean of the heights for seedlings in the light is 1.05 mm.

4. Calculate the 95% confidence interval of the mean of the heights of the seedlings grown in the dark and the light.

The 95% confidence interval represents a range of values where the true mean is likely to be found. On a graph, this range is often portrayed with error bars. To calculate the 95% confidence interval, we will use the mean and standard error of the mean.

95% confidence interval $= \bar{x} \pm 2SE_{\bar{x}}$
Dark: 9.6 mm $\pm$ (2 $\times$ 1.33 mm) $\rightarrow$ 9.6 mm $\pm$ 2.66 mm

The 95% confidence interval of the mean for seedlings grown in the dark is 9.6 ± 2.66 mm. This can also be written as the range of values itself: 7.0–12.2 mm. On a graph, an error bar is drawn from 6.94 mm to 12.26 mm to represent the 95% confidence interval. We can be 95% sure that the confidence interval includes the true mean length of growth for seedlings grown in the dark.

Light: 18.4 mm $\pm$ (2 $\times$ 1.05 mm) $\rightarrow$ 18.4 mm $\pm$ 2.10 mm

The 95% confidence interval of the mean for seedlings grown in the light is 18.4 ± 2.10 mm. This can also be written as the range of values itself: 16.3–20.5 mm. On a graph, an error bar is drawn from 16.3 mm to 20.5 mm to represent the 95% confidence interval. We can be 95% sure that the confidence interval includes the true mean length of growth for seedlings grown in the light.

▼

Your Turn

Researchers in an aquatic biology lab have collected data on the mass of algae in a recent experiment. The data are shown in the following table:

Experimental tank	Algae mass (g)	Experimental tank	Algae mass (g)
1	0.36	9	0.66
2	0.51	10	0.31
3	0.25	11	0.22
4	0.42	12	0.29
5	0.22	13	0.33
6	0.25	14	0.32
7	0.28	15	0.48
8	0.27		

1. What is the standard deviation of this dataset?
2. If you were to graph error bars for these data, what range of values would represent a 95% confidence interval of the mean?

1.2 The atom is the fundamental unit of matter

When biologists speak of diversity, they commonly point to the 2 million or so species named and described to date, or to the 10–100 million living species thought to exist in total. Life's diversity can also be found at a very different level of observation: in the molecules within cells. Life depends critically on many essential functions, which ultimately depend on the chemical characteristics of the organic molecules that make up cells and organisms.

In spite of the diversity of molecules and functions, the chemistry of life is based on just a few types of organic molecules, which in turn are made up of just a few types of atoms. In this section, we will look at the structure of atoms and identify the major atoms that make up organic molecules.

Atomic Structure

The study of life begins with the basic unit of matter, the atom. An atom contains a dense central **nucleus.** The nucleus is made of positively charged particles called **protons** and electrically neutral particles called **neutrons.**

A third type of particle, the negatively charged **electron,** moves around the nucleus at some distance from it. For example, a carbon atom, illustrated in **FIGURE 1.2**, typically has six protons, six neutrons, and six electrons. The number of protons is known as the **atomic number.** The atomic number specifies an atom as a particular **element,** a chemical that cannot be further broken down by the methods of chemistry. For example, the atom with one proton is hydrogen (H) and the atom with six protons is carbon (C).

Each proton and neutron, by definition, has a mass of 1 atomic mass unit, whereas electrons have negligible mass. Together, protons and neutrons determine an atom's **atomic mass,** which is the total mass of the atom.

The number of neutrons in atoms of a single element can differ, which changes its mass. **Isotopes** are atoms of the same element that have different numbers of neutrons. The atomic mass is sometimes indicated as a superscript to the left of the chemical symbol. ^{12}C is the isotope of carbon with six neutrons and six protons. If we want to symbolize both the atomic number and atomic mass for an element or isotope, we indicate the atomic mass by a superscript and the atomic number by a subscript. For example, the carbon isotopes with the mass of 12 and 14 would be written as $^{12}_{6}C$ and $^{14}_{6}C$, respectively.

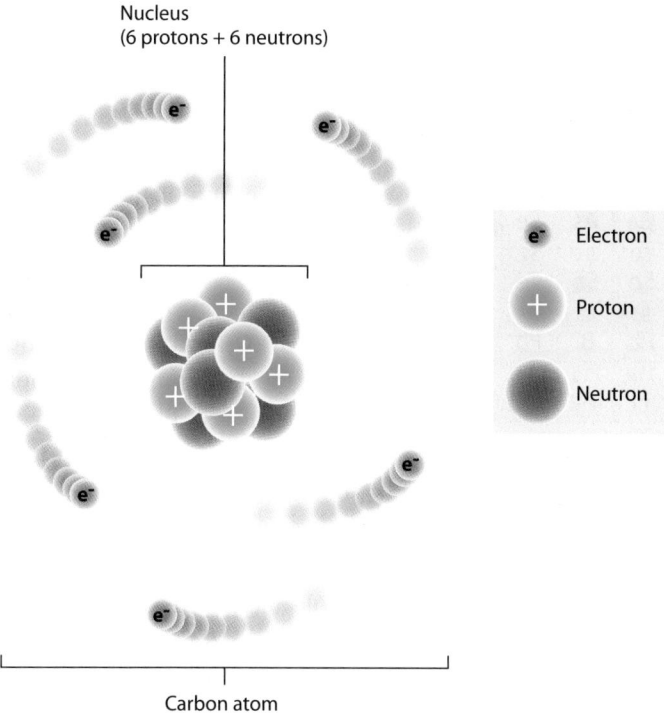

Nucleus
(6 protons + 6 neutrons)

e⁻ Electron

+ Proton

Neutron

Carbon atom

FIGURE 1.2 A carbon atom

Most carbon atoms have six protons, which are positively charged;
six neutrons, which are neutral; and six electrons, which are
negatively charged.

Typically, an atom has equal numbers of protons and
electrons. As you can see in Figure 1.2, carbon possesses
six positively charged protons and six negatively charged
electrons. The charges add up to zero, so this carbon atom is
electrically neutral.

Certain chemical processes cause an atom to either
gain or lose electrons. An atom that has lost an electron is
positively charged, and one that has gained an electron is negatively charged. Electrically charged atoms are called **ions.**
The charge of an ion is specified as a superscript to the right
of the chemical symbol. For example, H^+ indicates a hydrogen
ion that has lost an electron and is positively charged. Positively charged ions are called cations and negatively charged
atoms are anions.

Electrons

The movement of electrons from one molecule to another
is the foundation of energy transfer in many biological reactions. Therefore, understanding electron transfer is
essential to the study of cellular biochemistry.

Electrons move around the nucleus as a cloud of points
that is denser where the electron is most likely to be. The
exact path of an electron varies, but it is possible to identify a

region in space where an electron is present most of the time.
The area in space where electrons circle around the nucleus is
known as an **energy level** or **electron shell.** The innermost
energy level may contain one or two electrons. As you can see
in **FIGURE 1.3,** hydrogen has one electron in the first energy
level, which is also its only energy level. Many elements that
are important in biology, such as carbon, nitrogen, and oxygen,
contain two energy levels. In these atoms, the second energy
level may hold up to eight electrons. Figure 1.3 shows how
carbon's six electrons are placed in the energy levels: two are in
the first energy level and four are in the second energy level.

The amount of energy in a level depends on its location.
Electrons closer to the nucleus have less energy and are less
reactive than those further from the nucleus. Atoms are most
stable when their energy levels are full. When electrons fill the
energy levels closest to the nucleus, rather than ones further
away, the element becomes more stable. If an electron gains
energy and moves or "jumps" to a level further away from the
nucleus, the atom is less stable. As we will see later, molecules
are often formed when atoms share electrons to fill their outermost energy levels.

Since chemicals often react to complete their outermost
energy levels, simple diagrams like those in Figure 1.3 give
a sense of how many electrons an element must gain or lose
to have a full outermost energy level. In the case of carbon, it
must gain four electrons in the outer shell for a total of eight
electrons. Hydrogen may either gain one electron or lose an
electron to have its outermost shell complete. If an electron is
lost, a hydrogen ion (H^+) forms because it now has one more
proton than electrons. We will study hydrogen in greater
detail when we discuss cellular energetics in Unit 3.

Chemical Properties of Elements

The electrons in an element's outermost energy level are
known as the **valence electrons.** The **periodic table of**

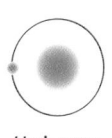

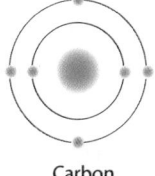

In this simplified diagram, the
electron energy levels are depicted
as circles and the electrons that
occupy them as dots. The cloud in
the center is the nucleus.

Hydrogen

Carbon

FIGURE 1.3 Energy levels for hydrogen and carbon

The hydrogen atom contains one energy level with a single electron,
while the carbon atom has two energy levels with two electrons in the
first energy level and four electrons in the second. The energy levels
are depicted as circles and the electrons as orange dots.

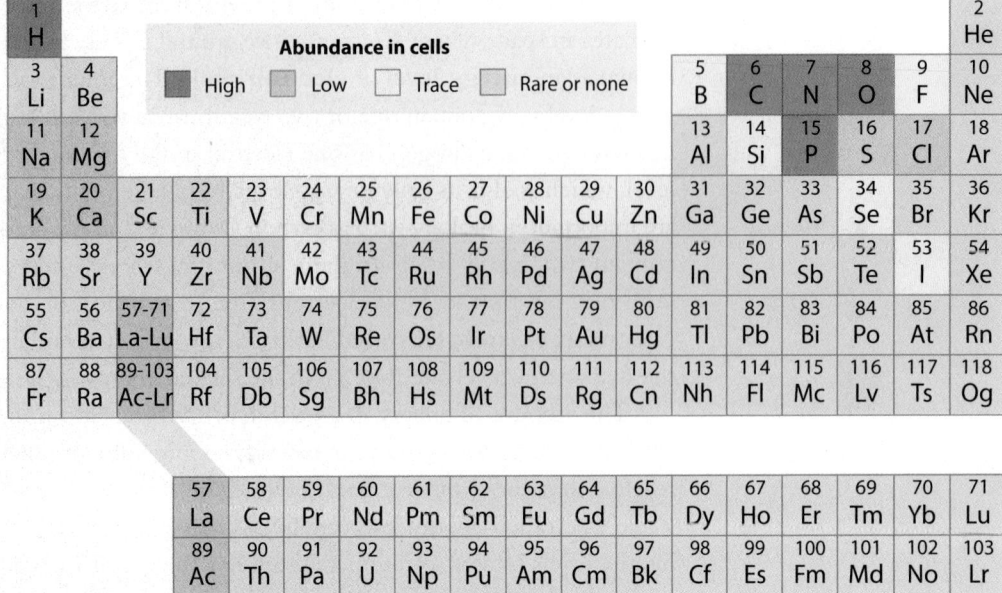

Abundance in cells

| ■ High | ■ Low | □ Trace | ■ Rare or none |

FIGURE 1.4 The periodic table of the elements

Elements are arranged by increasing number of protons, the atomic number. The atomic number is shown above the element. The elements in a column share similar chemical properties.

At the right end of the row, the energy level has a full complement of electrons.

The elements in a vertical column of the periodic table are called a group or family. Members of a group all have the same number of electrons in their outermost level. For example, carbon (C) and lead (Pb) both have four electrons in their outermost level. The number of electrons in the outermost level determines in large part how elements interact with other elements to form a diversity of molecules, as we will explore in the next section.

All living organisms are made up of atoms that can be combined to make molecules. The four elements common to every organism on the planet are carbon (C), hydrogen (H), oxygen (O), and nitrogen (N). Although organisms use elements in the first five rows of the periodic table, these most highly used elements belong to rows 1 and 2. Phosphorus, a member of the third row, is also present in large amounts in organic molecules.

the elements, shown in **FIGURE 1.4**, describes valence electrons and other properties of elements. In the periodic table, the elements are indicated by their chemical symbols and arranged in order of increasing atomic number. For example, the second row of the periodic table begins with lithium (Li), which has 3 protons and ends with neon (Ne), which has 10 protons.

For the second and third horizontal rows in the periodic table, elements in the same row have the same number of energy levels. Moving across a row, each element has one more proton and one more electron than the preceding element.

Let's take a look at the second row of elements, those that span from lithium (Li) to neon (Ne), shown in **FIGURE 1.5**. All of these elements have two energy levels. The innermost level of all of these elements is full. Only the outermost energy level has a varied number of electrons, starting with one for lithium and progressively adding one electron to the outer shell of the elements as we move from left to right across the row, ending with eight electrons in neon's outer energy level.

✓ Concept Check

3. **Describe** the components of an atom.

4. **Describe** how the rows and columns of the periodic table of the elements are organized.

5. **Identify** what the superscripts of ^{14}N and ^{15}N signify.

6. **Calculate** how many additional electrons would be needed to fill the outer energy level of C.

7. **Identify** the location of the electrons around an atom that have the most energy associated with them.

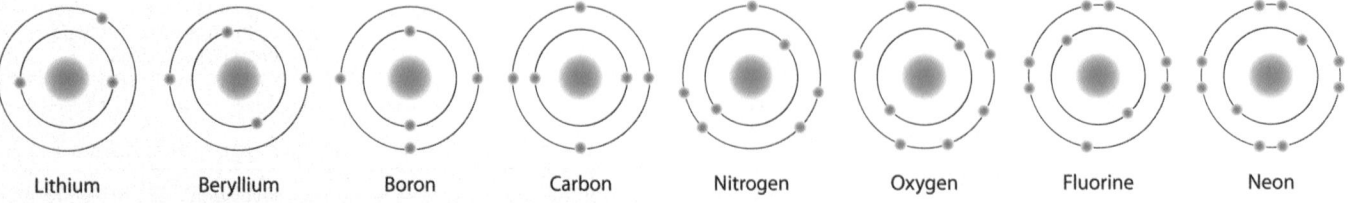

| Lithium | Beryllium | Boron | Carbon | Nitrogen | Oxygen | Fluorine | Neon |

FIGURE 1.5 Number of electrons across row 2 of the periodic table

Moving from left to right, each atom contains one more electron than the last, from lithium to neon. Neon has a full complement of eight electrons in its outer energy level.

1.3 Atoms combine to form molecules linked by chemical bonds

Atoms combine to make a great diversity of molecules in cells, which in turn leads to the diversity of life. Atoms bond with other atoms to form molecules, which are groups of two or more atoms bonded together that act as a single unit. An example of a molecule is hydrogen gas (H_2), made when two atoms of hydrogen bond, as you'll see below. Note that a chemical formula is written as the letter abbreviation for each element, followed by a subscript giving the number of that type of atom in the molecule. When molecules form, the individual atoms interact through what is called a **chemical bond,** a type of attraction between atoms that holds them together. For example, joining one atom of carbon with four atoms of hydrogen creates the compound methane (CH_4), which is also known as natural gas and is used in cooking, heating, and industry. There are several ways in which atoms can interact with one another, forming different types of chemical bonds. In this section, we will look at these chemical bonds.

Covalent Bonds

The ability of atoms to combine with other atoms is determined in large part by the electrons furthest from the nucleus, the valence electrons. When atoms combine with other atoms to form a molecule, the atoms share valence electrons with each other. Specifically, when the outermost energy levels of two atoms come in close proximity, the shells may overlap with one another so that electrons are shared between the two atoms. When a pair of electrons are shared between the two atoms, a **covalent bond** is formed.

Molecules tend to be the most stable when the two atoms forming a bond share enough electrons to completely fill the outermost energy level. Chemical stability occurs when an atom or molecule has the lowest possible energy state. Stable molecules are less likely to react because they require significant inputs of energy to change bonding partners, which is what happens during chemical reactions. Very little energy is required, for example, to get dynamite to react. A stable molecule, such as nitrogen gas, requires a large amount of energy to react.

FIGURE 1.6 shows an example of the formation of covalent bond between one carbon (C) atom and four fluorine (F) atoms. Each fluorine atom (F) has seven valence electrons. When each of these F atoms forms a covalent bond with the carbon (C) atom, the outermost shells overlap due to the

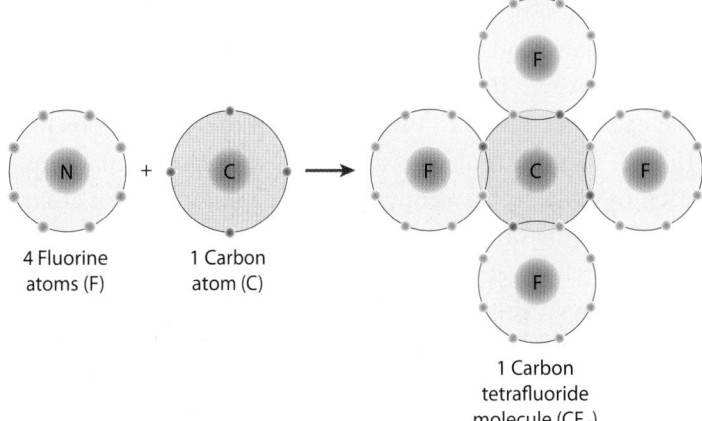

4 Fluorine atoms (F) + 1 Carbon atom (C) → 1 Carbon tetrafluoride molecule (CF_4)

FIGURE 1.6 Covalent bonds

Covalent bonds form when atoms share a pair, or pairs, of electrons in their outermost energy levels. Four individual F atoms, each containing seven electrons in its outermost energy level, form covalent bonds with a single C atom. The C atom shares four electrons, one with every F atom, while each individual F atom shares a single electron with the C atom. By sharing electrons, each F atom and the C atom fill their outer levels.

sharing of electrons, allowing each atom to fill its outer shell with electrons. The resulting compound, carbon tetrafluoride (CF_4), consists of one C atom covalently bound to four F atoms.

When two atoms share two electrons in a covalent bond, a **single bond** is formed. When two atoms share two pairs of electrons covalently, a **double bond** is formed. In practice, the word "covalent" is often omitted, as it is understood that these bonds arise through a sharing of electrons and are thus covalent. A double bond is denoted by a double line connecting the two chemical symbols for the atoms. **FIGURE 1.7** shows how double bonds are represented in ethylene, a molecule that is made of two carbon atoms and four hydrogen atoms. The two carbon atoms share four electrons with each other, creating a C=C double bond. By contrast, each carbon atom shares only two electrons with a single H atom, creating a C—H single bond.

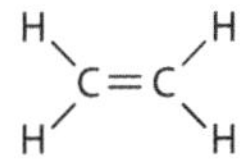

$$H_2C=CH_2$$

FIGURE 1.7 The double covalent bond of carbon atoms in ethylene

The double line represents four electrons shared between the two carbon atoms; each line denotes two shared electrons. The single lines connecting carbon and hydrogen represent single covalent bonds.

Polar Covalent Bonds

In molecules such as hydrogen (H_2) and oxygen (O_2) gases, electrons are shared equally by the atoms. In many covalent bonds, however, the electrons are not shared equally. A notable example, shown in **FIGURE 1.8**, is the water molecule (H_2O). A water molecule consists of two hydrogen atoms, each of which is covalently bound to a single oxygen atom. In a molecule of water, the region around the oxygen atom has a partial negative charge, while the area around each of the two hydrogen atoms has a partial positive charge. In the figure, charges are shown using the symbol δ^+ for a partial positive charge near the H atoms, and the δ^- symbol for a partial negative charge near the O atom.

Electrons are shared unequally because of a difference in the ability of the atoms to attract electrons, a property known as **electronegativity.** Oxygen is more electronegative than hydrogen. As a result, in a molecule of water, oxygen has a partial negative charge, while the two hydrogen atoms have a partial positive charge. When electrons are shared unequally between two atoms, the interaction is described as a **polar covalent bond.**

By contrast, a covalent bond where atoms are shared equally is sometimes referred to as a **nonpolar covalent bond.** The molecules of hydrogen gas (H_2), oxygen gas (O_2), and nitrogen gas (N_2) all have nonpolar covalent bonds. If two different kinds of atoms have similar electronegativities, then the covalent bonds between them also tend to be nonpolar because the electrons are shared equally, or nearly equally, by the atoms. Of the atoms commonly found in organic molecules, C and H frequently form nonpolar covalent bonds. For example, methane gas (CH_4) is a nonpolar compound.

Electronegativity tends to increase across a row in the periodic table. As the number of positively charged protons

across a row increases, negatively charged electrons are held more tightly to the nucleus. This principle helps to explain why O is more electronegative than N and why N is more electronegative than either C or H.

You can think of electronegativity as the "greed" of an atom for electrons. Oxygen (O) is greedier for electrons than is nitrogen (N), carbon (C), or hydrogen (H). Note, too, that carbon, hydrogen, nitrogen, and oxygen vary in their electronegativity. Carbon and nitrogen (C—N) and carbon and oxygen (C—O) each form polar covalent bonds because electrons are not shared equally between two atoms. Atoms that are closer together in electronegativity, such as carbon and hydrogen, form covalent bonds that are not polar. When a covalent bond is established between two atoms of the same type, for instance, between two H atoms or two O atoms, both of the atoms have the same degree of electronegativity and hence the electrons are shared equally.

Ionic Bonds

In a molecule of water, the difference in electronegativity between the oxygen and hydrogen atoms leads to unequal sharing of electrons. In more extreme cases, when an atom of high electronegativity is paired with an atom of low electronegativity, the difference in electronegativity may be so great that the electronegative atom "steals" the electron from its less electronegative partner. This creates an electrically charged atom, which you may recall is known as an ion. The atom with the extra electron contains more electrons than protons, which gives it a negative charge. The atom that lost the electron has a positive charge because it now has more protons than electrons. The two ions form an **ionic bond,** a chemical bond in which two ions with opposite electrical charges associate with each other because of the differences in charge. Some atoms may gain or lose more than one electron when they form an ionic bond. For instance, the calcium ion, which is often used by cells, may be either a single (Ca^+) or a double (Ca^{++}) positively charged ion.

Sodium chloride (NaCl), which is common table salt, is an example of a compound formed by the attraction of

Chemical formula $\qquad$ H_2O

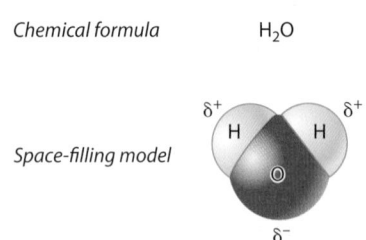

Space-filling model

FIGURE 1.8 A polar covalent bond

In a polar covalent bond, the two atoms do not share the electrons equally. In water, the shared electrons spend more time near the O atom than either of the H atoms. The result is that the O atom has a partial negative charge, written as δ^-, while the H atoms have partial positive charges, noted as δ^+.

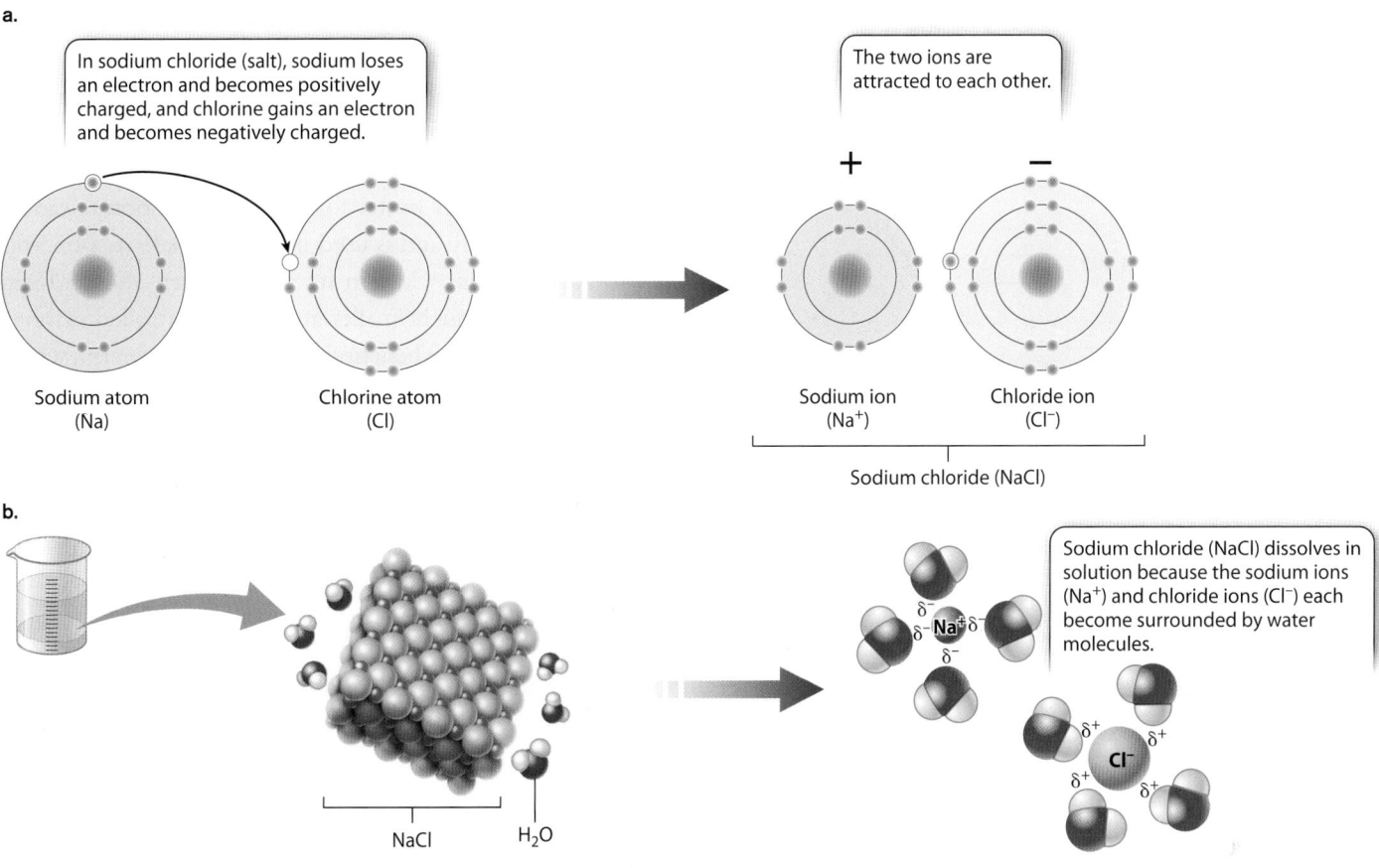

a.

In sodium chloride (salt), sodium loses an electron and becomes positively charged, and chlorine gains an electron and becomes negatively charged.

The two ions are attracted to each other.

Sodium atom
(Na)

Chlorine atom
(Cl)

+

−

Sodium ion
(Na$^+$)

Chloride ion
(Cl$^-$)

Sodium chloride (NaCl)

b.

NaCl

H$_2$O

Sodium chloride (NaCl) dissolves in solution because the sodium ions (Na$^+$) and chloride ions (Cl$^-$) each become surrounded by water molecules.

δ^- Na$^+$ δ^-

Cl$^-$

FIGURE. 1.9 An ionic bond

This figure shows how sodium chloride is formed and dissolved. (a) Sodium chloride is formed when a sodium atom gives up an electron to a chloride atom, forming Na$^+$ and Cl$^-$ ions. The two ions are then attracted to each other by their opposite charges. (b) In solution, the polarity of water causes the ions to dissociate from each other and to become surrounded by water molecules, with water's negative (oxygen) ends surrounding the Na$^+$ ions and its positive (hydrogen) ends surrounding the Cl$^-$ ions.

positive and negative ions. **FIGURE 1.9** illustrates the formation of sodium chloride. Sodium loses an electron and becomes a positively charged cation, while chloride gains an electron and becomes a negatively charged anion, as shown in Figure 1.9a. The two ions are then attracted to each other by their opposite charges. While covalent bonds are represented by lines like those shown in Figure 1.7, ionic bonds are indicated by superscripts showing charge, such as Na$^+$Cl$^-$.

Figure 1.9b illustrates what happens when sodium chloride is placed in water. The negatively charged ends of water molecules are attracted to the positively charged sodium ions, and the positively charged ends of water molecules are attracted to the negatively charged chloride ions. The ions are pulled apart in the water and become surrounded by polar water molecules as sodium chloride dissolves in the water. Chemicals that dissolve well in water tend to have polar or charged regions in the molecule.

In solution, sodium and chloride ions are completely surrounded by water molecules. If the water is then removed from this solution, ionic bonds will again form between the sodium and chloride ions. As the water evaporates, the concentrations of Na$^+$ and Cl$^-$ increase and the two ions come together to the point where they join and precipitate as salt crystals.

Chemical Reactions

The chemical bonds that link atoms in molecules can change in a **chemical reaction,** a process by which atoms or molecules are transformed into different molecules. The atoms or molecules that are changed in a chemical reaction are called **reactants.** The molecules formed from the reaction are known as **products.** In biological systems, chemical reactions provide a way to build and break down molecules for use by the cell, as well as to harness energy.

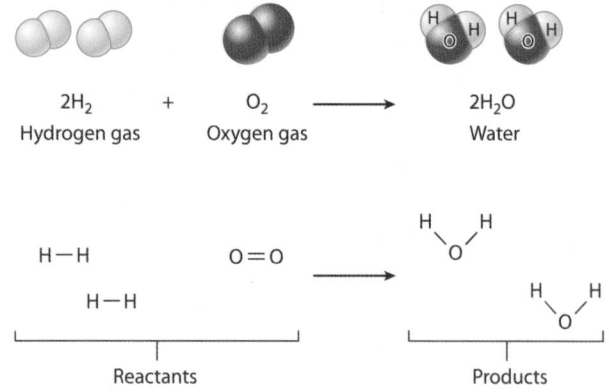

2H₂ + O₂ ⟶ 2H₂O
Hydrogen gas   Oxygen gas   Water

H—H   O=O

H—H

Reactants   Products

FIGURE 1.10 A chemical reaction

During a chemical reaction, atoms retain their identity, but their connections change as bonds are broken and new bonds are formed. In this reaction, a hydrogen molecule (H—H) reacts with one of the O atoms of an oxygen molecule (O=O), exchanging bonding partners and establishing H—O—H, also written as H_2O. Because this happens twice, two H_2O molecules are formed.

FIGURE 1.10 shows an example of a chemical reaction. In this case, two molecules of hydrogen gas ($2H_2$) and one molecule of oxygen gas (O_2) react to form two molecules of water ($2H_2O$). In this reaction, the numbers of each type of atom are conserved, meaning that the number of atoms does not change, but their arrangement in the reactants is different from the arrangement in the products. Specifically, the H—H

single bond in hydrogen gas and the O=O double bond in oxygen gas are broken. At the same time, each oxygen atom forms new covalent bonds with two hydrogen atoms, making the products of the reaction two molecules of water.

In fact, this reaction is the origin of the word "hydrogen," which literally means "water former." The reaction releases a good deal of energy and is used in some rockets as a booster in satellite launches. Certain microorganisms also perform this reaction in a much smaller and more controlled manner. These microorganisms benefit from the released energy, which they use to perform some of their cellular biochemistry.

In biological systems, chemical reactions provide a way to build and break down molecules for use by the cell, as well as to harness energy, which can be held in chemical bonds. We explore these topics in more detail in Unit 3.

✓ Concept Check

8. **Describe** the differences among covalent bonds, polar covalent bonds, nonpolar covalent bonds, and ionic bonds.

9. Use the polar property of water to **describe** the process by which water (H_2O) dissolves sodium chloride (NaCl).

10. In the reaction $3H_2 + N_2 \rightarrow 2NH_3$, **identify** the reactants and the products.

1.4 Carbon is the backbone of organic molecules

When you first learned to read, you probably began by learning the shapes and sounds of the letters of the alphabet. After that you learned letter combinations and then simple words. Our introduction to the basic chemistry of life is similar. Now that we've reviewed the basic nature of atoms and molecules, we can turn to the chemistry of life. As we mentioned earlier, the chemistry of life is based on carbon. In this section, we will examine what makes carbon well suited to its role as the chemical backbone of living things, and introduce the four major types of organic molecules.

The Chemistry of Carbon

Hydrogen and helium are by far the most abundant elements in the universe. In contrast, the solid Earth is dominated by silicon, oxygen, aluminum, iron, and calcium. In other words, Earth is not a typical sample of the universe. Similarly, the cell is not a typical sample of the solid Earth. **FIGURE 1.11**

shows the relative abundance by mass of chemical elements present in human cells after all the water has been removed. Just four elements—carbon (C), oxygen (O), hydrogen (H), and nitrogen (N)—account for approximately 90% of the total dry mass, and the most abundant element is carbon.

While other types of cells may vary somewhat, virtually all contain about the same ratios of these elements. Human life, and all life as we know it, is based on carbon. Carbon-containing molecules play such an important role in living organisms that they have a special name, as we saw earlier—they are called organic molecules. Carbon has the ability to combine with many other elements to form a wide variety of molecules, each specialized for the functions it carries out in the cell. For example, carbon-based molecules make up the structure of cells, participate in and speed up chemical reactions, and store energy for use by the cell.

Why has life evolved with carbon as its key element? Of the elements commonly observed in cells, carbon is unique in its bonding capacities. Of a carbon atom's six electrons, four are in its outermost shell and are available to form

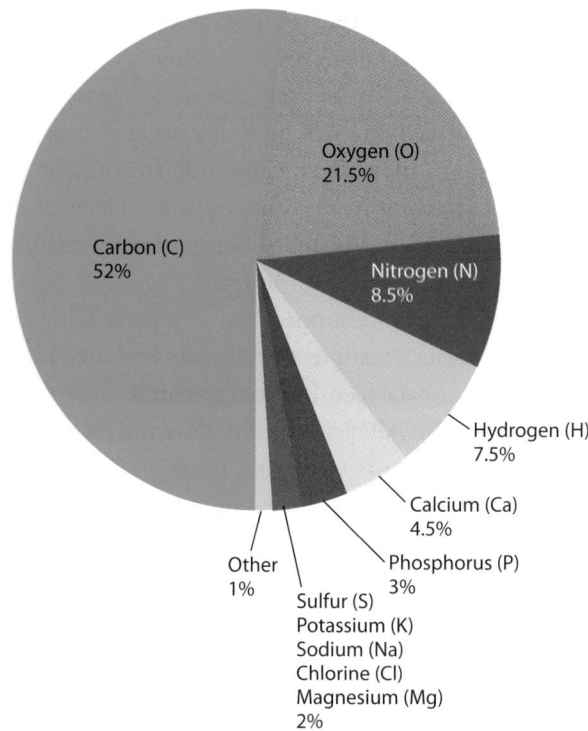

FIGURE 1.11 Approximate percentages by dry mass of chemical elements found in human cells

Carbon is the most abundant element in human cells when all of the water has been removed. Oxygen, nitrogen, and hydrogen are also relatively common. While other organisms may vary somewhat, virtually all use about the same ratios of these elements.

covalent bonds. Carbon commonly forms covalent bonds with itself, oxygen, nitrogen, and hydrogen.

FIGURE 1.12 shows methane gas (CH_4), which is formed when one atom of carbon combines with four atoms of hydrogen. Each of the four valence electrons of carbon becomes part of a covalent bond with an electron from an H atom. The bonds formed can move, or rotate, freely about their axis. In addition, the carbon atom lies at the center of a specific three-dimensional structure, called

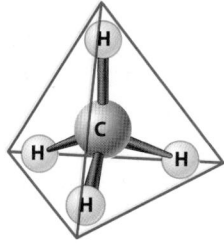

FIGURE 1.12 The shape and structure of methane

In methane gas (CH_4), a carbon atom is covalently bonded to four hydrogen atoms. The carbon atom lies at the center of a three-dimensional structure, called a tetrahedron, and the four covalent bonds with hydrogen extend the hydrogen atoms toward the four corners of this structure.

a tetrahedron, and the four covalent bonds with H extend the H atoms toward the four corners of this structure.

Because of its shape and because its single bonds rotate freely, carbon is able to make compounds in a variety of three-dimensional shapes. This ability to form many varied structures enables carbon to perform a wide variety of functions necessary to sustain and promote life.

Carbon has other special properties that contribute to its ability to form a diversity of molecules. For example, carbon atoms can link with other carbon atoms through covalent bonds, forming long chains. These chains can be branched, or two carbons at the ends of the chain or within the chain can link to form a ring. We also discussed earlier how carbon can form single and double bonds with other carbon atoms.

As a result of all of these properties, carbon-based molecules are structurally and functionally diverse. In other words, they can form an astonishing variety of molecules that can perform many different of roles in the cell. We might ask whether carbon is uniquely suited for life. Put another way, if we ever discover life on a distant planet, will it be based on carbon? Silicon, which is found just below carbon in the periodic table (see Figure 1.4), is the one other element that is both abundant on Earth and characterized by an outer shell with four valence electrons. Some scientists have speculated that silicon might therefore provide an alternative to carbon as a chemical basis for life. However, silicon readily binds with oxygen. On Earth, nearly all of the silicon atoms found in molecules are covalently bound to oxygen. Studies of Mars and meteorites show that silicon is tightly bound to oxygen throughout our solar system. As a result, the diversity of silicon-based molecules is far less than the millions of carbon-based molecules. If we ever discover life beyond Earth, very likely its chemistry will be based on carbon.

Organic Molecules

Four classes of organic molecules are of particular significance in biological systems: *proteins, nucleic acids, carbohydrates,* and *lipids.* Although they have different structures and roles, they share at least two properties. First, as we just discussed, they all contain carbon. Second, most of them are long chains, called **polymers,** built from smaller repeating subunits, called **monomers.** A polymer is like a necklace made of beads, and a monomer is a single bead, as shown in **FIGURE 1.13** on page 38. Here we take a first look at these four types of organic molecules. Later in this unit, we will examine each in greater detail.

Proteins are organic molecules that do much of the cell's work. They speed up chemical reactions and provide structural support for the cell. The white of an egg, for

FIGURE 1.13 Polymer

A polymer is a long chain of repeating subunits called monomers, similar to this necklace made up of individual beads. Photo: LionGate/Alamy Stock Photo

example, is mostly made up of proteins. Proteins are composed of subunits called **amino acids.** Returning to our necklace example, we would say that a protein is a necklace, and an amino acid is a bead.

A single cell has thousands of proteins with different functions. For example, some proteins form scaffolds that help to determine the shape of cells. Other proteins serve as chemical messengers, traveling from one cell to another to convey a message. Still other proteins accelerate the rate of chemical reactions. These various functions depend on the structure, or shape, of the protein. In fact, structure and function are closely connected. As a result, scientists can sometimes infer what a protein does by examining its shape. Furthermore, anything that disrupts the shape of a protein will often disrupt its function. The shape of a protein is determined by its sequence of amino acids. We will discuss proteins in more detail in Module 4.

Nucleic acids are responsible for encoding and transmitting genetic information. There are two types of nucleic acids. Module 0 mentioned deoxyribonucleic acid (DNA). The second nucleic acid is **ribonucleic acid (RNA).** Like proteins, nucleic acids are long polymers made up of repeating subunits, called **nucleotides.**

Nucleic acids are examples of informational molecules— that is, large molecules that carry information in the sequence, or order, of nucleotides that make them up. This molecular information is much like the information carried by the letters in an alphabet, but, in the case of nucleic acids, the information is in chemical form. DNA is the genetic material in all organisms. It is transmitted from parents to offspring, and it contains the information needed to specify the amino acid sequence of proteins. RNA has multiple functions, but one of its most important is in the synthesis of proteins. We will discuss nucleic acids in more detail in Module 5.

Many of us, when we feel tired, reach for a candy bar for a quick energy boost. The energy in a candy bar comes from sugars, which are quickly broken down to release energy. Sugars are **carbohydrates,** which are organic molecules that store energy in their chemical bonds. In addition, they are sometimes attached to proteins on the surface of cells, such as your red blood cells, and make up the external layer of the cells in plants, algae, and bacteria.

Carbohydrates are sometimes called sugars. Table sugar (sucrose) is a familiar example. Other carbohydrates include the sugars glucose, galactose, and lactose (milk sugar). Fruit, like that shown in **FIGURE 1.14**, contains a variety of sugars, including fructose. Like proteins and nucleic acids, carbohydrates are composed of repeating units of individual sugars, called monosaccharides.

As we have seen, proteins, nucleic acids, and carbohydrates all are polymers made up of smaller, repeating units. Lipids are different. Instead of being defined by a chemical structure, they share a particular property: **lipids** are organic molecules that are *hydrophobic*. **Hydrophobic** means "water fearing" and it describes nonpolar molecules that don't dissolve in water. Instead, they tend to associate with other lipids and minimize their contact with water. Think of what happens when oil, which is hydrophobic, is mixed with water. The oil forms droplets that minimize their contact with water, as pictured in **FIGURE 1.15**. By contrast, **hydrophilic** means "water loving" and it describes polar molecules that readily associate with and dissolve in water. For example, when sugar is placed in water, it dissolves as the individual sugar molecules associate with water molecules.

Because they share a property rather than a structure, lipids are chemically and functionally diverse. Their hydrophobic property allows them to be effective membranes, or barriers,

FIGURE 1.14 Carbohydrates

Fruit, such as these items at a market, contains fructose along with several other types of sugars. Sugars are types of carbohydrates.

Photo: OGphoto/Getty Images

FIGURE 1.15 Lipids

When oil, which is a lipid, is mixed in water, it forms droplets to minimize its contact with water. Oil, like all lipids, is hydrophobic.

Photo: ThomasVogel/Getty Images

between a cell's watery internal and external environments. They also include signaling molecules and familiar fats that store energy and make up part of our diet. Lipids will be discussed along with carbohydrates in more detail in Module 3.

✓ Concept Check

11. **Identify** the four most common atoms in organic molecules.

12. **Describe** how the number of valence electrons in a carbon atom is responsible for carbon's ability to form a large diversity of molecules.

13. **Identify** the four major types of organic molecules.

14. **Describe** a polymer.

15. **Describe** the property of lipids that allows them to function as a barrier between a cell's interior and external environments.

Module 1 Summary

REVISIT THE BIG IDEAS PREP FOR THE AP® EXAM

ENERGETICS: Using the content in this module, **identify** the key elements that all organisms obtain from the environment and exchange with the environment.

LG 1.1 Matter and energy govern the properties of life.

- All organisms are made up matter and require energy to sustain life. Page 26

- Organisms obtain matter from other organisms and the environment, and matter therefore moves in a cycle, with the same atoms used and reused. Page 26

- Energy from the sun or chemical compounds is used by organisms to do work but cannot be reused. Page 27

LG 1.2 The atom is the fundamental unit of matter.

- Atoms consist of protons, neutrons, and electrons. Page 30

- The atomic number is the number of protons an element contains, and it determines the identity of the element. Page 30

- The atomic mass is the number of protons and neutrons an element contains. Page 30

- Isotopes are elements with the same number of protons but different numbers of neutrons. Page 30

- Electrons occupy energy shells or levels that move around the nucleus. Page 31

- The periodic table organizes all of the elements in a way that describes their properties. Page 31

LG 1.3 Atoms combine to form molecules linked by chemical bonds.

- Valence electrons are the electrons in an atom's outermost energy shell and determine the ability of an atom to combine with other atoms to form molecules. Page 31

- Covalent bonds arise when two elements share one or more pairs of electrons. Page 33

- Covalent bonds may be either nonpolar or polar. Page 34

- Ionic bonds result from the attraction of oppositely charged ions. Page 34

- Chemical reactions involve the breaking and forming of chemical bonds, forming new molecules. Page 35

LG 1.4 Carbon is the backbone of organic molecules.

- Carbon's electron configuration allows it to form four covalent bonds. Page 36

- Carbon's bonding allows a diversity of molecules to be formed. Page 37

- The four major classes of organic molecules are proteins, nucleic acids, carbohydrates, and lipids. Page 37

- Proteins play a role in the structure of cells and can speed up the rate of chemical reactions. Page 37

- Proteins are made up of subunits called amino acids. Page 38
- Nucleic acids are information molecules and the molecules of heredity. Page 38
- Nucleic acids are made up of subunits called nucleotides. Page 38

- Carbohydrates, or sugars, store energy and make up the external layer of some types of cells. Page 38
- Carbohydrates are made up of simple sugars, called monosaccharides. Page 38
- Lipids are hydrophobic molecules that make up the cell membrane, store energy, and act as chemical messengers. Page 38

Key Terms

Matter	Energy level	Reactant
Atom	Electron shell	Product
Molecule	Valence electron	Polymer
Organic molecule	Periodic table of the elements	Monomer
Nucleus	Chemical bond	Protein
Proton	Covalent bond	Amino acid
Neutron	Single bond	Nucleic acid
Electron	Double bond	Ribonucleic acid (RNA)
Atomic number	Electronegativity	Nucleotide
Element	Polar covalent bond	Carbohydrate
Atomic mass	Nonpolar covalent bond	Lipid
Isotope	Ionic bond	Hydrophobic
Ion	Chemical reaction	Hydrophilic

Review Questions

1. Using the periodic table shown in Figure 1.4, identify which atom is most likely to have properties similar to nitrogen.
   - (A) Oxygen (O)
   - (B) Bismuth (Bi)
   - (C) Lead (Pb)
   - (D) Carbon (C)

2. The element sodium (Na) has an atomic number of 11 and an atomic mass of 23. Calculate the number of protons and neutrons in Na.
   - (A) 11 protons and 11 neutrons
   - (B) 11 protons and 10 neutrons
   - (C) 11 protons and 12 neutrons
   - (D) 10 protons and 13 neutrons

3. Which type of bond is formed between a sodium ion and a chloride ion?
   - (A) Ionic
   - (B) Neutral
   - (C) Polar covalent bond
   - (D) Nonpolar covalent bond

4. Which type of bond is formed between two oxygen atoms?
   - (A) Ionic
   - (B) Neutral
   - (C) Polar covalent bond
   - (D) Nonpolar covalent bond

5. When two molecules undergo a chemical reaction, they
   - (A) lose atoms.
   - (B) keep the same bonding partners that they had as reactants.
   - (C) establish new bonding partners.
   - (D) incorporate atoms from the air and water as well as from the reactants.

6. Proteins, nucleic acids, and carbohydrates are all
   - (A) information molecules.
   - (B) signaling molecules.
   - (C) polymers made up of repeating subunits.
   - (D) monomers that combine to form polymers.

7. Two organic molecules that store energy for use by the cell are
   - (A) carbohydrates and nucleic acids.
   - (B) carbohydrates and lipids.
   - (C) proteins and nucleic acids.
   - (D) nucleic acids and lipids.

Module 1
AP® Practice Questions

Section 1: Multiple-Choice Questions

Choose the best answer for questions 1–5.

1. Identify the main source of energy that would sustain a shallow water aquatic community.

    (A) Carbon

    (B) Sunlight

    (C) Chemical compounds

    (D) Heat

2. Which is the best example of the flow of matter through a community?

    (A) Sunlight is necessary for trees to grow → leaves wash into a cave where they are colonized by fungi and bacteria → small mites graze on the bacteria for food → insects and spiders feed on the mites.

    (B) Deep-sea volcanic activity forms vents on the ocean floor → superheated water dissolves minerals and metals → bacteria use these minerals to survive → many organisms depend on the bacteria as a food source.

    (C) The decomposition of animals from millions of years ago formed oil → oil is used by humans as a fossil fuel → burning fossil fuels releases CO_2 into the atmosphere → CO_2 is used by plants to produce O_2 → O_2 is consumed by many living animals.

    (D) A meadow of wildflowers absorbs solar energy → each flower converts this into sugars → the flowers use the sugars to grow and reproduce.

3. Fluorine (F) is a strongly electronegative element with seven valence electrons in its outermost energy level. Compared to the less electronegative sodium (Na), which has one valence electron in its outermost level, fluorine

    (A) holds electrons loosely around its nucleus.

    (B) is not as greedy to gain electrons.

    (C) will have a partial positive charge when it bonds to other elements.

    (D) is likely to become an anion.

Questions 4 and 5 refer to the chemical structure of the sugar fructose shown below.

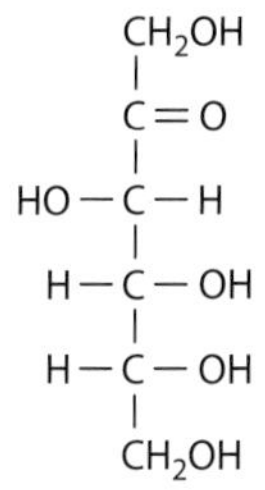

$$
\begin{array}{c}
CH_2OH \\
| \\
C=O \\
| \\
HO-C-H \\
| \\
H-C-OH \\
| \\
H-C-OH \\
| \\
CH_2OH
\end{array}
$$

4. Identify the type of bond linking the oxygen and hydrogen atoms.

    (A) Covalent

    (B) Ionic

    (C) Electronegative

    (D) Double

5. Name the organic molecule that has this structure.

    (A) Protein

    (B) Nucleic acid

    (C) Carbohydrate

    (D) Lipid

Section 2: Free-Response Question

Write your answer to each part clearly. Support your answers with relevant information and examples. Where calculations are required, show your work.

Plants use the energy from the sun to convert carbon dioxide and water into glucose and oxygen in a chemical reaction known as photosynthesis. The balanced chemical formula for this reaction is:

$$6CO_2 + 6H_2O \rightarrow C_6H_{12}O_6 + 6O_2$$

(a) **Identify** the reactants and products in this reaction.

(b) **Describe** the key difference between the monosaccharide glucose and a lipid such as vegetable oil.

(c) **Identify** the type of bond present between the carbon and oxygen atoms in carbon dioxide. **Describe** how this bond is formed.

(d) Is the role of carbon in this reaction an example of the flow of energy or the flow of matter through a system? **Justify** your answer.

A selection of North American tree species was studied in order to measure how much carbon dioxide they were capable of absorbing throughout one year. The results are given in the table below.

North American Tree Capacity for CO_2 Absorption

Type of tree	Average age of tree (years)	Average CO_2 absorption capacity (kg/tree/year) $\pm 2SE_{\bar{x}}$
Oak	5	10.02 ± 3.08
Pine	4	8.99 ± 2.34
Black walnut	4	7.26 ± 1.86
Maple	3	5.82 ± 1.79
Beech	9	18.66 ± 1.55
Ash	13	24.01 ± 4.11
Birch	10	20.42 ± 3.67

(e) **State** a hypothesis that could be explained by the data in this table.

(f) **Determine** if there is a statistically significant difference between the absorption data of black walnut trees and maple trees.

(g) **Determine** between which two tree species there is most likely to be a statistically significant difference.

Module 2
Water and Life

LEARNING GOALS ▶LG 2.1 Life depends on the properties of water.
                  ▶LG 2.2 Water is the medium of life.
                  ▶LG 2.3 Dehydration synthesis reactions build molecules, and hydrolysis reactions break them down.

All life on Earth depends on water. Indeed, life originated in water, and the availability of water strongly influences where and how different species live around the world. Many cells are surrounded by water, whether in an organism or in a habitat, such as a lake or ocean. Furthermore, water is the single most abundant molecule inside cells, so water is the medium in which the molecules of life interact. For example, in the last module, we learned about chemical reactions. All of the chemical reactions on which life depends take place in a watery, or aqueous, environment in and around cells.

Water is so important to life that in the late 1990s, the National Aeronautics and Space Administration (NASA) announced that the search for extraterrestrial life would be based on a simple strategy: follow the water. NASA's logic makes sense because Earth stands out within our solar system both for its abundance of water and for the life that water supports. So, if we want to find life elsewhere in the universe, we should start by finding water. Recently, scientists found evidence that liquid water once existed on ancient Mars. Might life have been present then too? What makes water so special as the medium of life?

In Module 1, we learned about chemical bonds that join atoms to form molecules. Water is made up of two atoms of hydrogen (H) and one atom of oxygen (O), written as H_2O. The hydrogen and oxygen atoms are joined by polar covalent bonds. The arrangement of atoms in a molecule of water, the polar covalent bonds holding the atoms together, and even how one water molecule interacts with other water molecules, all give water a set of properties that help it support life as we know it.

In this module, we will focus on the chemistry of water as a way to understand why it is essential for life. We will begin by discussing the properties of water on its own and then consider how water interacts with and influences other molecules. We will end by exploring chemical reactions that involve water. Along the way, we will see that water has many emergent properties. As we saw in Module 0, an emergent property of a system is one that is not easily predicted from the individual parts that make up the system. In the case of water, a collection of water molecules has many chemical properties that are not possessed by individual water molecules or by the hydrogen and oxygen atoms alone. These properties greatly influence, and are essential for, life.

PREP FOR THE AP® EXAM

FOCUS ON THE BIG IDEAS

SYSTEMS INTERACTIONS: Look for how the properties of water molecules affect the way they interact with each other as well as with other molecules.

2.1 Life depends on the properties of water

In Module 1, we saw that water molecules have polar covalent bonds, which are characterized by an uneven distribution of electrons. The bonds of water are polar because oxygen is much more electronegative than hydrogen. As a result, the shared electrons are more likely to be near the O atom than the H atoms, creating regions of slightly negative and positive charges. A molecule like water that has regions of positive and negative charge is called a **polar** molecule. The polar nature of water molecules gives water many unique properties, which we explore in this section.

Hydrogen Bonds

Because the oxygen and hydrogen atoms have slight charges, water molecules orient themselves to minimize the repulsion of like charges. As a result, positive charges are near negative charges. For example, because of its slight positive charge, a hydrogen atom in a water molecule tends to orient itself toward the slightly negatively charged oxygen atom in another molecule. This interaction between a hydrogen atom with a slight positive charge and an electronegative atom of

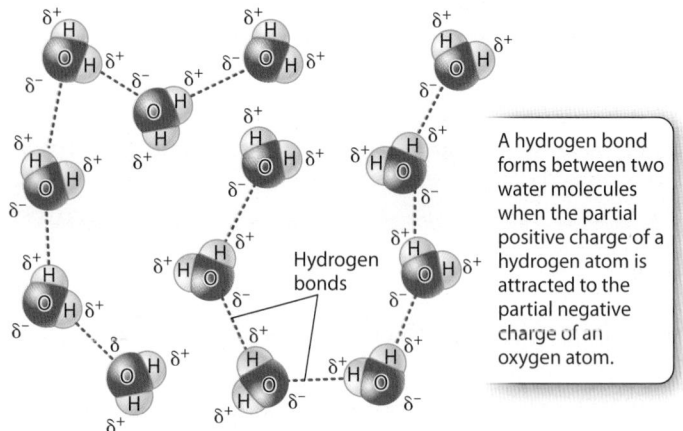

A hydrogen bond forms between two water molecules when the partial positive charge of a hydrogen atom is attracted to the partial negative charge of an oxygen atom.

Hydrogen bonds

FIGURE 2.1 Hydrogen bonds

In water, hydrogen bonds form between the partially positive hydrogen atom (δ^+) and the partially negative oxygen atom (δ^-). Hydrogen bonds are indicated by the red dashed lines. The energy and motion of molecules in liquid water causes hydrogen bonds to form and break frequently between different pairs of molecules.

another molecule is known as a **hydrogen bond.** In fact, any hydrogen atom covalently bound to an electronegative atom (such as oxygen or nitrogen) will have a slight positive charge and can form a hydrogen bond. Typically, a hydrogen bond is depicted by a dotted line, as shown in **FIGURE 2.1.**

Hydrogen bonds are much weaker than covalent bonds and ionic bonds. However, a large number of hydrogen bonds can be quite strong in total. Hydrogen bonding gives water many interesting properties, described next. In addition, the presence of many weak hydrogen bonds can help stabilize organic molecules, as in the case of nucleic acids and proteins.

Cohesion, Adhesion, and Surface Tension

Among the interesting properties that hydrogen bonding gives water is the property of **cohesion,** meaning that water molecules tend to stick to one another. Cohesion among water molecules drives the formation of drops, puddles, streams, rivers, lakes, and oceans. It also contributes to water movement in plants. As a water molecule evaporates from a leaf, a column of water is pulled from the soil, through the roots, and upward through the trunk and stems of the plant. Sometimes this water may rise as high as 100 meters above the ground in giant sequoia and coast redwood trees, which are among the tallest trees on Earth.

The cohesion of water in plant vessels is aided by the property of *adhesion.* **Adhesion** is the tendency of water to stick to materials other than water. For example, adhesion

causes droplets of dew to stick to a leaf. The cohesion and adhesion of water also allow it to stick to, and move through, transport channels in plants to the surface of leaves.

One consequence of cohesion is high **surface tension,** a measure of the difficulty of breaking the surface of a liquid. Because surface molecules do not have other molecules surrounding them on all sides, they stick to other surface molecules more strongly. This strong binding provides a film of molecules on the surface of liquid water. Water has one of the highest surface tensions of common liquids. Water's strong surface tension allows objects like leaves to float. Surface tension also makes it possible for some insects to disperse their weight over the surface of water without breaking the surface tension and being immersed in the water. These animals, such as the water strider shown in **FIGURE 2.2,** "walk on water" because of water's high surface tension.

The Structure of Water

Hydrogen bonding also accounts for the different structures of water as a solid (ice), liquid, and gas (steam), which are illustrated in **FIGURE 2.3.** The structure of ice is shown in Figure 2.3a. As you can see, most water molecules in ice are hydrogen bonded to four other water molecules and form an open crystalline lattice. As the temperature increases and ice melts to form liquid water, shown in Figure 2.3b, some of the hydrogen bonds in ice are destabilized and break. The water molecules in liquid water are able to move closer to one another than they are in ice. As a result, liquid water takes up less space, or volume, and is denser than ice.

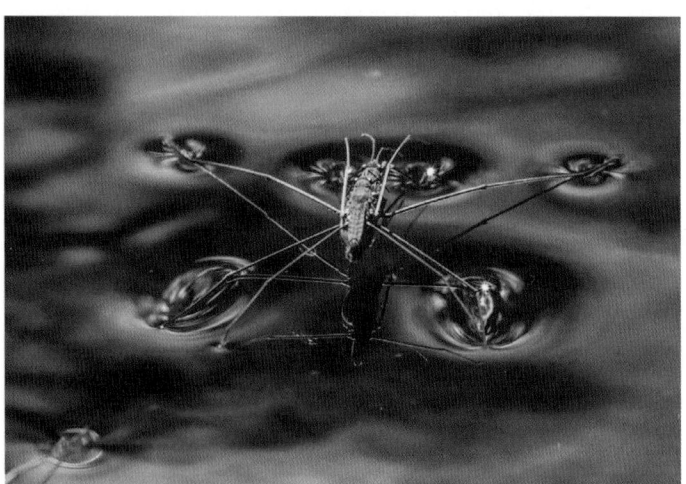

FIGURE 2.2 Surface tension of water

This water strider is capable of dispersing its weight over a large enough area that the surface tension of water prevents it from sinking.

Photo: NurPhoto/Getty Images

a. Solid water (ice)

b. Liquid water

c. Gaseous water (steam)

FIGURE 2.3 Ice, water, and steam

The structure of water is different as a solid, liquid, and gas. (a) In ice, each water molecule is hydrogen bonded to four other water molecules, resulting in a regular crystal structure. (b) In liquid water, water molecules move around, so hydrogen bonds continuously form and break apart, and water molecules are closer together than they are in ice, making liquid water more dense than ice. (c) In steam, water molecules move around even more rapidly, keeping the water molecules even further apart and preventing most hydrogen bonds from forming. Photos: (a) Ralph Lee Hopkins/Getty Images;

(b) DIRSCHERL Reinhard/hemispicture.com/Getty Images; (c) Loise Murray, Getty Images

The fact that liquid water is more dense than solid water (ice) is unusual. For most substances, it's the other way around, with the solid being denser than the liquid. This unusual property of water allows ice to float on water. Ponds and lakes freeze from the top down, and those that are deep enough do not freeze completely. In addition, surface ice acts as a temperature barrier to the water below. For example, if the air temperature falls below freezing, surface ice can insulate the deeper water below and prevent it from becoming very cold. This property helps fish and aquatic plants survive winter in the water under the layer of ice.

Gaseous water, or steam, is shown in Figure 2.3c. Unlike ice or liquid water, steam has enough heat energy and motion that the water molecules are seldom close enough to form hydrogen bonds. They move about rapidly and do not have the regular, crystalline structure of ice or the hydrogen bonds of liquid water.

Specific Heat

As we just discussed, liquid water is characterized by a large number of hydrogen bonds because each water molecule has two hydrogen atoms and a single oxygen atom that can form hydrogen bonds with other water molecules. Even though a single hydrogen bond is only about one-twentieth as strong as a covalent bond, the large number of hydrogen bonds in water give it great stability. As a consequence of this stability, changing water from a liquid to a gas requires more energy input than changing other liquids from a liquid to a gas.

Molecules are in constant motion, and this motion increases as temperature increases. When water is heated, some of the added energy is used to break apart hydrogen bonds. That energy cannot be used to cause more motion in the molecules, so the temperature increases less than it would if hydrogen bonds were not present. The numerous hydrogen bonds in water therefore make it more resistant to temperature changes than other liquids. The strength of water's hydrogen bonds helps explain why water is good for quenching fires; a substantial amount of heat is needed to break water's hydrogen bonds before the temperature of the water increases, thus cooling and eventually extinguishing the fire.

The amount of energy required to change the temperature of a substance is known as **specific heat.** Specific heat is the amount of heat required per unit mass of a substance to raise the temperature of the mass by one degree Celsius. Water's hydrogen bonding gives it an unusually high specific heat. Liquid water's specific heat is 1.000 calorie per gram. Other liquids are notably lower: ethanol (ethyl alcohol) has a specific heat that is about 0.6 times that of water.

Water's high specific heat is important for living organisms. In the cell, water resists temperature variations that would otherwise occur with the heat released by many biochemical reactions. In addition, bodies of water do not rapidly change temperature, which makes them relatively stable environments for organisms that live in water. On a global scale, oceans minimize temperature fluctuations, which helps to stabilize the temperature of Earth in a range compatible with life. "Analyzing Statistics and Data: Scientific Notation" provides an opportunity to review the use of scientific notation in the context of specific heat for two substances.

✓ Concept Check

1. **Describe** how the polarity of water enables hydrogen bonding.
2. **Identify** three properties of water that result from hydrogen bonding.

ANALYZING STATISTICS AND DATA ⊙

Scientific Notation

Scientific notation is a useful way to write very large or very small numbers. For example, the number 3,200,000,000 can be rewritten in scientific notation as 3.2×10^9, where 3.2 is the coefficient, 10 is the base, and 9 is the exponent. The coefficient is always greater than or equal to 1 and less than 10; the base is always 10; and the exponent can be either positive or negative, depending on the size of the number. Once in scientific notation, numbers can be added, subtracted, multiplied, or divided just like any other numbers.

When adding or subtracting numbers in scientific notation, the exponents must be the same. If the exponents are not the same, then one of the terms must be converted so that the exponents

▼

match. If the exponents match, then the coefficients can be added or subtracted and the exponent is kept the same. For example:

$$(3.1 \times 10^7) + (4.6 \times 10^7) = 7.7 \times 10^7$$

Significant figures for adding and subtracting numbers in scientific notation are determined by the number that is the least precise, that is, the number with the fewest decimal places. For example, when subtracting 5.22×10^3 from 8.3×10^3, the final answer is written to the tenths, as follows: 3.1×10^3.

When multiplying numbers in scientific notation, multiply the two coefficients and add the exponents:

$$(1.1 \times 10^6) \times (7.2 \times 10^2) = (1.1 \times 7.2) \times (10^{6+2}) = 7.92 \times 10^8$$
$$= 7.9 \times 10^8$$

When dividing numbers in scientific notation, divide the two coefficients and subtract the exponents:

$$\frac{6.5 \times 10^{-5}}{3.25 \times 10^{-2}} = \left(\frac{6.5}{3.25}\right) \times (10^{-5-(-2)}) = 2.0 \times 10^{-3}$$

Significant figures for multiplying and dividing numbers in scientific notation are determined by the term that has the fewest significant figures. In the multiplication example above, each term being multiplied has two significant figures; therefore, the answer is also written with two significant figures. In the division example, the numerator has two significant figures and the denominator has three significant figures. The final answer must be written with two significant figures.

Numbers that are less than 1, such as 0.0489, have negative exponents when written in scientific notation: 4.89×10^{-2}. The negative exponent means that in order to get to the original number, the decimal point in 4.89 is moved two places to the left.

Numbers that are greater than 1, such as 3,200,000,000, have positive exponents when written in scientific notation, as we have seen: 3.2×10^9. The positive exponent means that in order to get to the original number, the decimal point in 3.2 moves nine places to the right.

PRACTICE THE SKILL

1. Aluminum has a specific heat of 0.215 calories per gram. Write this number in scientific notation.

The coefficient is a number between 1 and 10. Looking at 0.215, you can see the coefficient is 2.15. All of the other options, 0.215, 21.5, and 215, are not between 1 and 10. To determine the exponent, you must know the number of places to move the decimal point. To go from 0.215 to 2.15, the decimal point is moved one place. Because 0.215 is less than 1, the exponent is negative. Therefore, the exponent is −1. The base number is 10, so 0.215 in scientific notation looks like this:

$$0.215 \text{ calories per gram} = 2.15 \times 10^{-1} \text{ calories per gram}$$

To double-check that it is written correctly, look at the exponent. The −1 indicates that the decimal in 2.15 is shifted one place to the left. This is the original number, or 0.215.

The number of significant figures in the original number is the same as when it is written in scientific notation. Both have three significant figures: 2, 1, and 5. The 0 before the decimal point is not a significant figure.

2. Gold has a specific heat of 0.0301 calories per gram. Write this number in scientific notation.

Begin by determining the coefficient, which is between 1 and 10. Looking at 0.0301, you can see the coefficient is 3.01. All of the other options, 0.0301, 0.301, 30.1, and 301, are not between 1 and 10. To go from 0.0301 to 3.01, the decimal point is shifted two places. Because 0.0301 is less than 1, the exponent is negative. Therefore, the exponent is −2. Writing this number in scientific notation looks like this:

$$0.0301 \text{ calories per gram} = 3.01 \times 10^{-2} \text{ calories per gram}$$

To double-check that it is written correctly, look at the exponent. The −2 indicates that the decimal in 3.01 is shifted two places to the left. This is the original number, or 0.0301.

Again, the number of significant figures in the original number is the same as when it is written in scientific notation. Both have three significant figures, the 3, middle 0, and 1. The zeros to the left of the 3 are not significant figures.

Your Turn

A beekeeper is looking to combine two hives. One hive contains 12,500 bees and the other contains 4.1×10^4 bees. How many bees in total will be in the combined hive? Write your answer in scientific notation with the correct number of significant figures.

2.2 Water is the medium of life

So far, we have focused on the properties of water. Water is a polar molecule that forms hydrogen bonds, which in turn lead to a host of other properties that are essential for life. Water is also the medium, or environment, in which the molecules of cells interact and undergo chemical reactions. Therefore, the chemistry of water and how it interacts with ions and molecules are critical to the health and function of a cell. In this section, we will examine how water molecules interact with other molecules in solution, and how water molecules themselves behave inside cells and organisms.

Solvent Properties of Water

Cells and organisms are mostly water. Therefore, molecules exist in a watery, or aqueous, environment, and chemical reactions essential for life take place in water. We can classify molecules, or even different regions of the same molecule, according to how they interact with water. As we saw in Module 1, hydrophilic ("water-loving") molecules are polar and readily dissolve in water, while hydrophobic ("water-fearing") molecules are nonpolar and minimize their contact with water.

Hydrophilic molecules dissolve in water because they undergo hydrogen bonding with water. Consider ammonia (NH_3), a waste product produced by many different types of organisms. **FIGURE 2.4** shows that an ammonia molecule forms hydrogen bonds with water. Note how the hydrogen (H) atom of water, which carries a partially positive charge, is attracted to the nitrogen (N) atom of ammonia,

which carries a partially negative charge. Similarly, the partial negatively charged oxygen (O) atom of water is attracted to the partial positively charged hydrogen (H) atom of ammonia. Ammonia dissolves in water if enough water molecules interact with and surround the ammonia molecules. Ammonia is toxic, so dissolving it in water helps to dilute it and is an important way that many animals excrete it.

The example of ammonia and water illustrates that water is a good **solvent** for polar or charged molecules, meaning it is capable of dissolving many substances. Similarly, organic molecules, such as proteins, nucleic acids, and carbohydrates, form hydrogen bonds with water very well. Although they may be too large to dissolve easily, they have polar or charged properties that interact with water and are important for their structure and function.

Unlike hydrophilic molecules, hydrophobic molecules are typically nonpolar. Nonpolar molecules have very few or no regions of positive and negative charges, so they arrange themselves to minimize their contact with water. Consider oil and vinegar salad dressing. The oil molecules are hydrophobic, while the vinegar molecules are hydrophilic. These properties of oil and vinegar explain why they mix together only when vigorously shaken and why they quickly separate when standing. This hydrophobic effect, in which polar molecules like water exclude nonpolar ones like oil, drives important biological processes, such as the folding of proteins and the formation of cell membranes.

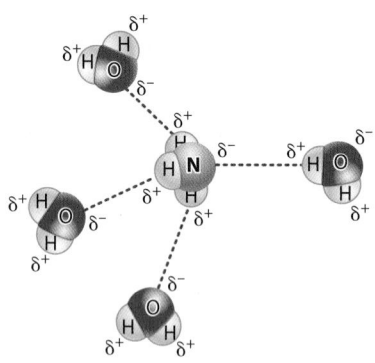

FIGURE 2.4 Water as a solvent

Ammonia is a polar molecule, with regions of partial positive and negative charge. As a result, it readily dissolves in water, which is a good solvent. Ammonia forms hydrogen bonds with water; note how both the hydrogen (H) and oxygen (O) atoms of water are attracted to the ammonia molecule.

Water Dissociation and pH

Water does not always exist as an intact molecule. In other words, it does not always remain as one oxygen atom covalently bonded to two hydrogen atoms. On occasion, it can separate, or dissociate, into two ions, a hydrogen ion (H^+) and a hydroxide ion (OH^-). This occurs when the oxygen atom of water takes the electron from one of the hydrogen atoms, as shown in **FIGURE 2.5**.

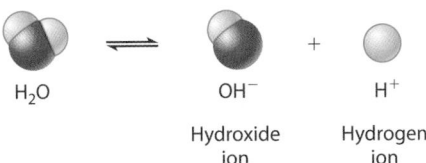

H₂O ⇌ OH⁻ + H⁺

Hydroxide ion   Hydrogen ion

FIGURE 2.5 Water dissociation

In any solution of water, a small proportion of the water molecules dissociate, forming a hydroxide ion (OH^-) and a hydrogen ion (H^+), also referred to as a proton. The forward and backward arrows indicate that the reaction proceeds in both directions.

The presence of H^+ and OH^- ions affects the characteristics of water and how it interacts with other ions and molecules. Both H^+ and OH^- ions are reactive and can have noteworthy effects on the chemistry and well-being of a cell or an organism. Excess H^+ or OH^- ions can affect the function of proteins, nucleic acids, and carbohydrates. As a result, most organisms tightly regulate the concentrations of H^+ and OH^- ions present in their cells.

In addition, the presence of H^+ and OH^- ions affects how acidic or basic a solution is. If you have ever taken a chemistry or cooking course, you have probably learned a little about *acids* and *bases*. Let's put them into a biological context. Many organic molecules such as nucleic acids, nucleotides, and amino acids have acidic and basic properties.

Acids are molecules that donate H^+ ions to solutions. As a result, when there are more H^+ ions than OH^- ions in a solution, a solution is termed acidic. Familiar examples of acids include hydrochloric acid (HCl) and carbonic acid (H_2CO_3). Lemon juice and vinegar are also acids. **Bases** are molecules that accept H^+ ions and thereby remove them from solution. When there are fewer H^+ ions than OH^- ions, the solution is basic, or alkaline. Sodium hydroxide (NaOH) is an example of a base. Common household bases include antacids and baking soda. A solution is neutral when the concentrations of H^+ and OH^- ions are equal.

The **pH scale,** shown in **FIGURE 2.6,** is a way of describing how acidic or basic a solution is; it can range from 0 to 14. A neutral solution is neither acidic nor basic and has a pH of 7. A solution with pH less than 7 is acidic. For example, lemon juice, with a pH of about 2, is a strong acid because its pH is well below 7. Apple juice, with a pH of approximately 3.5, is also an acid. A solution with a pH above 7 is basic. Seawater, with a pH of approximately 8, is basic, as is ammonia, with a pH of 11.

The pH of most cells is approximately 7 and is tightly regulated because most chemical reactions can be carried out only in a narrow pH range. Different parts of the human body contain fluids at different pH values. For instance, the human stomach often has a pH of 1.5–2.0, where the acidity starts the process of breaking down food prior to its complete digestion and absorption by the intestine. The low pH also kills many microbes and therefore provides some protection against them. By contrast, human blood is slightly basic, with a pH of approximately 7.4. This value is sometimes called physiological pH because it is the pH for the normal physiology, or function, of cells, tissues, and organs.

The pH is a measure of the concentration of H^+ ions. It is calculated using the following formula:

$$pH = -\log[H^+]$$

In words, pH equals the negative logarithm of the hydrogen ion concentration. The negative sign means that as the H^+ ion concentration increases and the solution becomes more acidic, the pH decreases. Conversely, as the H^+ ion concentration decreases and the solution becomes more basic, the pH increases.

Because the pH scale is logarithmic, a difference of one pH unit corresponds to a tenfold difference in H^+ ion concentration. Similarly, a difference of two pH units is a $10 \times 10 = 100$-fold change in the H^+ ion concentration. So, a small change in pH represents a big change in the concentration of hydrogen ions. For example, a solution at pH 5 has 10 times more H^+ ions than a solution at pH 6. A solution at pH 10 has 1000 times fewer H^+ ions than one at pH 7.

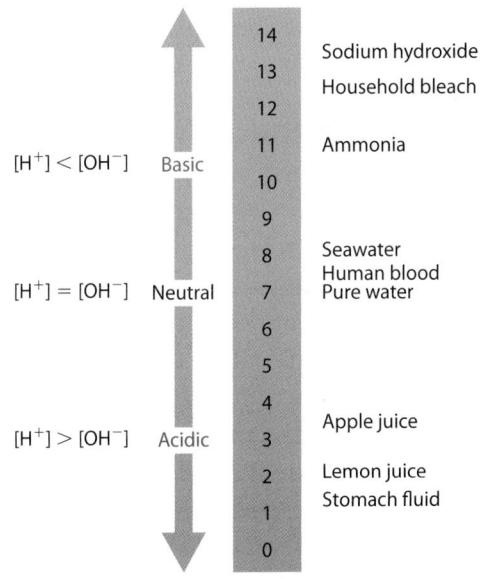

FIGURE 2.6 The pH scale

The pH scale provides a measure of the acidity of a solution. A pH of 7 is neutral. Lemon juice and stomach fluid are acidic and have pH values less than 7. Seawater, human blood, and household bleach are basic and have values greater than 7.

AP® EXAM TIP

Although you are not expected to calculate pH using the pH equation, you should be very familiar with the pH scale and understand that it is logarithmic. Be prepared to translate whole number differences in pH units to differences in H⁺ ion concentrations. For example, a difference of one pH unit translates to a tenfold difference in H⁺ ion concentration, and a difference of two pH units translates to a 100-fold difference in H⁺ ion concentration.

✓ Concept Check

3. **Describe** the process by which ammonia dissolves in water.

4. **Identify** the atoms of the water and ammonia molecules that form hydrogen bonds with one another.

5. **Describe** why the pH of a municipal water supply, at 7.2, is slightly basic.

6. **Calculate** how many times more H⁺ ions are present in a solution with a pH of 3 compared to a solution of pH 4.

2.3 Dehydration synthesis reactions build molecules, and hydrolysis reactions break them down

We have seen that water has special properties. It is the medium of the cell, and provides an environment in which other molecules function and chemical reactions take place. Water also plays a key role in chemical reactions that build and break down large molecules. Cells synthesize molecules such as proteins and lipids to build structures and to carry out functions. They break down large molecules to absorb them, for example, and to harness the energy in their chemical bonds. Building and breaking down molecules is critical to the life of a cell. In this section, we will examine two chemical reactions in which water plays an active role in building and breaking down molecules.

Dehydration Synthesis Reactions

In Module 1, we noted that many organic molecules are polymers—large molecules made up of repeating subunits. These polymers are assembled by chemical reactions in which the small subunits are joined together. The subunits are linked by covalent bonds. In some of these chemical reactions, a water molecule is released as the covalent bond is formed. This type of chemical reaction is called a **dehydration synthesis reaction** because water is removed in the process and because it synthesizes larger molecules from smaller ones. It is sometimes shortened to dehydration reaction, which emphasizes the loss of water. By repeating this process over and over again, cells are able to make the large molecules needed for a number of cellular processes.

FIGURE 2.7 illustrates a dehydration synthesis reaction. In this figure, two small molecules, one with three subunits represented by red circles, and one with one subunit, are covalently linked to form a four-unit molecule. In the

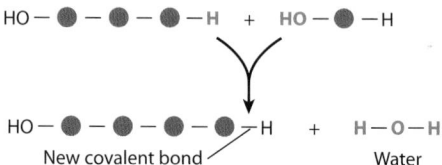

FIGURE 2.7 A dehydration synthesis reaction

A three-unit molecule, represented by the three red circles, is joined to a one-unit molecule, represented by the single red circle, to build a four-unit molecule. In the process, a water molecule is released, indicated by H—O—H.

process, a molecule of water is released; hence the name dehydration to describe this synthesis reaction.

Dehydration synthesis reactions build larger molecules from smaller subunits or add an individual subunit to an existing chain of molecules. They are commonly used to link subunits to form proteins, carbohydrates, and lipids. For example, they are used to join amino acids to form proteins and individual sugar molecules (monosaccharides) to build carbohydrates.

Some types of lipids are made up of individual components that are joined by dehydration synthesis reactions. For example, a lipid called a monoglyceride is made up of two molecules—glycerol and a fatty acid. These two molecules are covalently linked by a dehydration synthesis reaction, as shown in **FIGURE 2.8**. Note that the glycerol loses a hydrogen atom (H) and the fatty acid loses a hydroxyl (OH) group to make a water molecule.

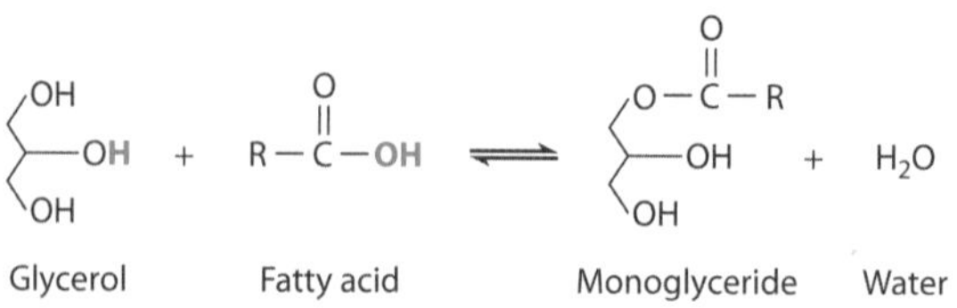

Glycerol   Fatty acid   Monoglyceride   Water

FIGURE 2.8 The dehydration synthesis of a monoglyceride

In this reaction, glycerol and a fatty acid are joined by a covalent bond to form a monoglyceride. Water (H_2O) is formed by the removal of the H atom of glycerol and the OH group of the fatty acid.

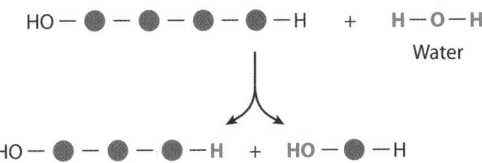

FIGURE 2.9 A hydrolysis reaction

A four-unit molecule, represented by the four red circles, is broken down by a water molecule into a three-unit molecule and a one-unit molecule. An H atom is joined to one product, and an OH group is joined to the other product.

Hydrolysis Reactions

Not only do cells synthesize polymers and other large molecules, cells also break them down. Breaking down molecules is required to harness energy held in the chemical bonds of these molecules and to recycle the building blocks of the polymers. In addition, organic molecules are often broken down into smaller units so they can be taken up and used, as occurs in fungi and bacteria. This same process occurs in the intestines of animals and allows animals to take up and use the subunits contained in large proteins, carbohydrates, and lipids.

As we saw in the case of dehydration synthesis reactions, water also plays a critical role in breaking down large molecules. **Hydrolysis reactions** are chemical reactions that break covalent bonds by adding water across the bond. As you can see in **FIGURE 2.9**, a single large molecule is broken down into two smaller molecules. In fact, "hydrolysis" means to "break using water." In the process, one of the products of the reaction gains an H atom and the other product gains an OH group, both of which came from the original water molecule. In this case, think of water as HOH, which is then split into H and OH.

An example of a hydrolysis reaction is the breakdown of a two-unit sugar molecule into two one-unit sugar molecules. This reaction is shown in **FIGURE 2.10**. Hydrolysis reactions are, in effect, the reverse of dehydration synthesis reactions and are used to break complex molecules into simpler, smaller molecules.

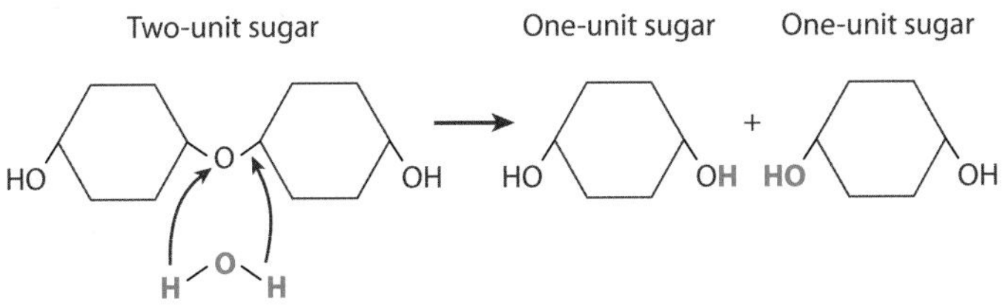

Two-unit sugar → One-unit sugar + One-unit sugar

FIGURE 2.10 The hydrolysis of a two-unit sugar

The reaction shows the hydrolysis of a two-unit sugar molecule by a molecule of water. Note how the H atom and OH atoms of the water break the indicated bonds and become part of products. In the process, a two-unit sugar is broken down into two one-unit sugars.

Throughout this module, we have explored the chemistry of water and have seen how water is essential to life on Earth. But does it fill this critical role because water is uniquely suited for life, or is it because life on Earth has adapted over time to a watery environment? We don't know the answer to this question, but probably both explanations are partly true. Chemists have proposed that under conditions of high pressure and temperature, other small molecules, including ammonia (NH_3) and some simple carbon-containing molecules, might display similar characteristics friendly to life. However, under the conditions that exist on Earth, water is the only molecule uniquely suited to life. **TABLE 2.1** summarizes the properties of water. Water is a truly remarkable substance, and life on Earth would not be possible without it.

TABLE 2.1 Special Properties of Water	
Property	**Explanation**
Polar	Regions of partial positive and negative charges
Cohesive	Hydrogen bonding between water molecules
Adhesive	Hydrogen bonding between water and other molecules
High surface tension	Extensive hydrogen bonding on the surface of liquid water
Less dense as a solid than as a liquid	Four hydrogen bonds with other water molecules in ice, creating an open crystal structure that is not present in liquid water
High specific heat	Extensive hydrogen bonding of liquid water
Good solvent	Hydrogen bonding with other polar molecules
Reactive	Participation in chemical reactions that build and break down molecules

✓ Concept Check

7. **Describe** a dehydration synthesis reaction.

8. **Describe** a hydrolysis reaction.

9. **Identify** the functions of dehydration synthesis reactions and hydrolysis reactions.

Module 2 Summary

PREP FOR THE AP® EXAM

REVISIT THE BIG IDEAS

SYSTEMS INTERACTIONS: Use the content of this module to **describe** how the polar properties of water molecules affect the way they interact with each other and with other molecules.

LG 2.1 Life depends on the properties of water.

- Water is abundant and essential for life. Page 43
- Water is a polar molecule because electrons are not shared equally between the oxygen atom and the hydrogen atoms, resulting in weak positive and negative charges. Page 43
- Water forms hydrogen bonds, which result when the hydrogen atom of one water molecule interacts with the oxygen atom of another water molecule. Page 44
- Hydrogen bonds among water molecules help explain many of water's unique properties, such as cohesion, adhesion, and surface tension. Page 44
- Water exists as a solid (ice), liquid, and gas (steam). Page 44
- Ice is less dense than liquid water, so ice floats on water, and ponds and lakes freeze from the top down. Page 44
- Water has high specific heat, so it resists rapid temperature changes. Page 46

LG 2.2 Water is the medium of life.

- Water is a good solvent. Page 48
- Hydrophilic molecules dissolve readily in water, whereas hydrophobic molecules in water tend to associate with one another, minimizing their contact with water. Page 48
- Water dissociates into H^+ and OH^- ions. Page 48
- Solutions may be acidic, neutral, or basic depending on the concentration of H^+ and OH^- ions. Page 49
- pH is a measure of the acidity of a solution. Page 49
- A pH of 7 is neutral; a pH of less than 7 is acidic; and a pH of greater than 7 is basic. Page 49

LG 2.3 Dehydration synthesis reactions build molecules, and hydrolysis reactions break them down.

- Dehydration synthesis reactions remove water and join smaller molecules to form larger molecules linked by covalent bonds. Page 50
- Hydrolysis reactions add a water molecule across a covalent bond and break large molecules into smaller ones. Page 51

Key Terms

Polar	Adhesion	Solvent	pH scale
Hydrogen bond	Surface tension	Acid	Dehydration synthesis reaction
Cohesion	Specific heat	Base	Hydrolysis reaction

Review Questions

1. Which is not a property of water?
   (A) Water has low surface tension.
   (B) Water molecules tend to stick to other water molecules.
   (C) Water dissolves polar molecules.
   (D) Water has a high specific heat.

2. Which property of water is primarily responsible for the ability of trees to draw water up from the roots to the leaves?
   (A) Neutral pH
   (B) Ability to act as a solvent
   (C) Cohesion
   (D) Polarity

3. Which statement about water is correct?
   (A) Ice is less dense than liquid water.
   (B) Ice forms on top and sinks to the bottom of lakes and rivers.
   (C) Bodies of water freeze from the bottom up.
   (D) Water molecules in ice have a disorganized, irregular arrangement.

4. Water readily dissolves compounds that are
   (A) hydrophobic.
   (B) solvent.
   (C) nonpolar.
   (D) hydrophilic.

5. Polarity is due to differences in

(A) pH between atoms.

(B) electronegativity between two atoms.

(C) the number of neutrons between two atoms.

(D) the size of two atoms.

6. Glucose is a common sugar. If two glucose molecules join by a covalent bond while at the same time releasing a water molecule, which of the following has occurred?

(A) Cohesion

(B) Adhesion

(C) A hydrolysis reaction

(D) A dehydration synthesis reaction

7. White vinegar has a pH of about 2.4, which means it is

(A) nonpolar.　　　　(C) basic.

(B) neutral.　　　　(D) acidic.

Module 2
AP® Practice Questions

PREP FOR THE AP® EXAM

Section 1: Multiple Choice Questions

Choose the best answer for questions 1–4.

1. Which explains why organic molecules like proteins act as a solute in water?

(A) Proteins have few regions of positive and negative charge, allowing them to dissolve easily in water.

(B) Proteins and water have similar crystalline lattice structures.

(C) Proteins have polar properties and can hydrogen bond with water.

(D) Proteins are very large, hydrophobic molecules.

2. Why is a lipid membrane an effective barrier between the watery interior and exterior environments of a cell?

(A) Lipids are hydrophobic and minimize their contact with water, thus helping to keep the two environments separate.

(B) The lipid membrane causes water to have high surface tension, which makes the cell impermeable.

(C) The lipid membrane easily forms hydrogen bonds with water because it is hydrophilic. This property prevents it from dissolving in water.

(D) Lipids are highly polar molecules that form strong ionic bonds with water, creating a barrier that cannot be crossed.

3. Which property of water contributes most to its ability to form hydrogen bonds with other water molecules?

(A) Water is a good solvent.

(B) Water has high surface tension.

(C) Water is a polar molecule.

(D) Water has high specific heat.

4. Students studied a lake affected by acid rain and found the pH to be 4. Using Figure 2.6 on page 49, approximately how many times more acidic is this lake water than seawater?

(A) 40 times more acidic

(B) 100 times more acidic

(C) 4000 times more acidic

(D) 10,000 times more acidic

Section 2: Free-Response Question

Write your answer to each part clearly. Support your answers with relevant information and examples. Where calculations are required, show your work.

The following reaction illustrates the breakdown of a sucrose into two other sugar molecules. The two sugars produced are commonly found in fruits, honey, and other sweet foods. When consumed and digested, they provide energy and sometimes contribute to fat production in humans.

$$C_{12}H_{22}O_{11} + H_2O \rightarrow C_6H_{12}O_6 + C_6H_{12}O_6$$
Sucrose + Water → Glucose + Fructose

(a) **Identify** the type of reaction this equation represents.

(b) **Describe** what is happening in this reaction.

(c) **Explain** why sucrose must be broken down for humans to use it.

(d) **Describe** the type of reaction that occurs when glucose and fructose are joined to form sucrose. What is the role of this type of reaction?

Module 3

Carbohydrates and Lipids

LEARNING GOALS ▶**LG 3.1** Monosaccharides are the basic units of carbohydrates.

▶**LG 3.2** Monosaccharides are joined by glycosidic bonds to make complex carbohydrates.

▶**LG 3.3** Lipids are hydrophobic molecules.

In Module 1, we introduced the four carbon-based organic molecules—proteins, nucleic acids, carbohydrates, and lipids. All of these biological macromolecules are essential to life, and each one has its own distinctive characteristics and roles. They function in an aqueous, or watery, environment, which is critical for life itself, and which we discussed in Module 2. Now we will look at each of the organic molecules in detail, starting with carbohydrates and lipids.

Carbohydrates and lipids are both energy-rich molecules and are therefore sometimes referred to as fuel molecules. They are good sources of energy because they contain many carbon–carbon and carbon–hydrogen bonds, which are rich in energy. As a result, they are familiar components of the diet for many organisms. For example, sugars are a type of carbohydrates and fats are a type of lipid.

Carbohydrates and lipids play other important roles as well. Carbohydrates provide long-term energy storage in the form of glycogen in animals and starch in plants. They also make up cell walls in plants and fungi. Carbohydrates are sometimes attached to proteins on the surface of cells.

For example, the A, B, and O blood groups result from different types of carbohydrates on the surface of red blood cells.

Lipids, as we have seen, can store energy, but they also form the barrier, or membrane, separating the inside and outside of all cells. Some lipids are chemical messengers, carrying signals from one cell the next. Familiar examples are the hormones estrogen and testosterone.

We will begin this module by focusing on carbohydrates. We will examine their basic building blocks and discuss how these building blocks are joined to form long chains and sometimes complex branching molecules. We will then turn to lipids. Lipids are chemically diverse, but they all are hydrophobic and therefore minimize interaction with water.

PREP FOR THE AP® EXAM

FOCUS ON THE BIG IDEAS

SYSTEMS INTERACTIONS: Think about how the properties of carbohydrates and lipids affect the way they interact with other molecules.

3.1 Monosaccharides are the basic units of carbohydrates

Do you like sugar on your cereal or a donut glazed with sugar? Both are everyday examples of carbohydrates. Carbohydrates are sometimes referred to by the common names of sugars or saccharides. Table sugar, or sucrose, is a familiar carbohydrate. Other examples of carbohydrates include glucose, fructose, galactose, and lactose. Although we tend to think of sugars as sweeteners, they have a number of roles in biology. Carbohydrates can provide energy reserves for cells and they form rigid structures in cell walls, as cellulose in plants and chitin in fungi.

As we discussed in Module 1, carbohydrates are polymers, which are long chains made up of repeated monomers or

subunits. For carbohydrates, the subunits are called monosaccharides, or single sugars (*mono* means "one"). Monosaccharides can join to form long chains or complex branching molecules. In this section, we will focus on the chemical formulas and structures of monosaccharides.

Monosaccharides

The simplest carbohydrates are sugars, also called saccharides. Simple sugars contain five or six carbon atoms. All six-carbon sugars have the chemical formula $C_6H_{12}O_6$. However, they differ in the configuration of the atoms. Glucose, a common sugar, and fructose, found in fruit and sweeteners, are examples. They both share the same formula ($C_6H_{12}O_6$), but differ in the arrangement of their atoms, as can be seen in **FIGURE 3.1**. Glucose and fructose are considered

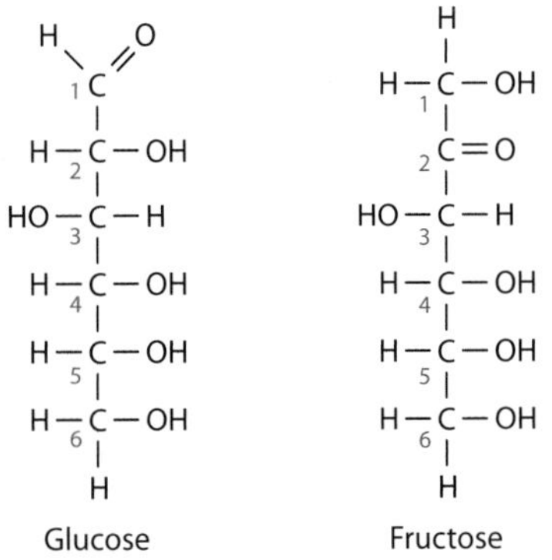

FIGURE 3.1 Glucose and fructose

Both glucose and fructose are simple sugars with the same chemical formula, $C_6H_{12}O_6$. They differ in the arrangement of their atoms, as shown here.

isomers, molecules that have the same chemical formula but differ in structure.

The chemical formula for sugars indicates that it has 6 carbon atoms, 12 hydrogen atoms, and 6 oxygen atoms. This is a ratio of 6:12:6, which can be simplified to 1:2:1. In fact, many simple sugars, or monosaccharides, have a ratio of 1 C atom to 2 H atoms to 1 O atom (1:2:1). The common sugars glucose and fructose follow this rule.

You will notice that the carbon atoms are numbered in the chemical structures in Figure 3.1. This allows us to be clear about which C atom of the sugar is being discussed. Because glucose and fructose contain six carbons, the C atoms are numbered from 1 to 6, with number 1 being at the top of the figure and 6 being at the bottom.

You will also notice that most of the C atoms in glucose and fructose are covalently bonded to a hydroxyl group (—OH). The hydroxyl group is one of a number of *functional groups* in biological molecules. **Functional groups** are groups of one or more atoms that have particular chemical properties, regardless of what they are attached to. For instance, hydroxyl groups are polar, with regions of slight negative and positive charges because of the electronegativity of the oxygen atom. As a result, the entire sugar molecule is at least somewhat polar and therefore hydrophilic.

The molecule ethane provides another example of how a functional group can influence the properties of the molecule to which it is attached. The chemical formula for ethane is C_2H_6, and it is a nonpolar gas at room temperature. If one of the hydrogen (H) atoms of ethane is replaced by a hydroxyl group (—OH), the result is ethanol, or C_2H_5OH. Ethanol is the alcohol in alcoholic beverages, such as beer and wine, and at room temperature it is a liquid that is readily soluble in water.

Glucose and fructose also contain a second functional group called a carbonyl group. Note that in Figure 3.1 the carbonyl group of glucose is the first carbon atom, while it is the second carbon atom of fructose. In both cases, as well as in all of the monosaccharides, the carbonyl group is polar and helps make the sugar a polar, hydrophilic molecule. **TABLE 3.1** lists several different types of functional groups found in carbohydrates and lipids, starting with the hydroxyl and carbonyl groups. We will add to this list as we discuss additional functional groups found in the other organic molecules.

TABLE 3.1 Functional Groups Commonly Observed in Biological Molecules, Part I

Name	Formula	Structure	Properties	Commonly found in
Hydroxyl	—OH	—O—H	Polar due to the electronegative oxygen atom, hydrophilic	Carbohydrates, proteins, nucleic acids
Carbonyl	$\diagdown C=O$	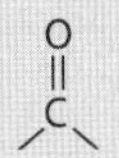 (C=O structure)	Polar due to the electronegative oxygen atom), hydrophilic	Carbohydrates, proteins
Carboxyl	—COOH	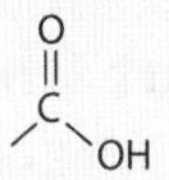 (C(=O)OH structure)	Polar, negatively charged at the pH of a cell, hydrophilic	Fatty acids, amino acids, proteins
Phosphate	$-OPO_3H_2$	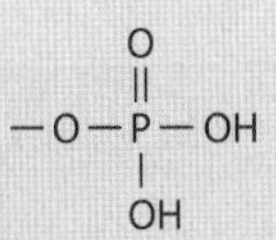 —O—P(=O)(OH)—OH	Polar, negatively charged at the pH of a cell, hydrophilic	Phospholipids, nucleic acids, ATP

Linear and Ring Structures

The simple sugars glucose and fructose, shown in Figure 3.1, are linear, meaning that they form a straight chain of C atoms. Quite often, however, monosaccharides form ring structures. Indeed, almost all of the monosaccharides form rings. This happens because the linear chain folds upon itself. **FIGURE 3.2** shows the formation of cyclic glucose. To form a ring, the carbon atom at one end of the chain forms a covalent bond with the oxygen atom of a hydroxyl group attached to another carbon atom in the same molecule. For example, cyclic glucose is formed when the oxygen atom of the hydroxyl group on carbon 5 forms a covalent bond with carbon 1, as shown in Figure 3.2. Once the ring is formed, groups attached to the carbon atoms project above or below the ring. In Figure 3.2, the double arrows between the molecules indicate that the reaction may go back and forth, indicating that the linear and ring forms exist in equilibrium. However, the ring form is strongly favored in aqueous solution such as the environment inside of a cell.

Why is it important to understand that the monosaccharides form cyclic structures? Remember that when it comes to biological function, shape is crucial. In other words, the shape of a molecule (and many other biological structures) and the function of a molecule are closely connected. The number and position of groups extending above or below the rings, as well as whether a monosaccharide is in its linear or cyclic form, may affect the properties and therefore the function of the molecule.

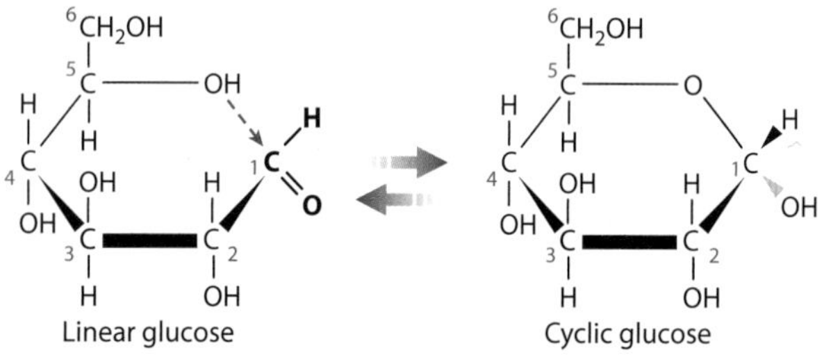

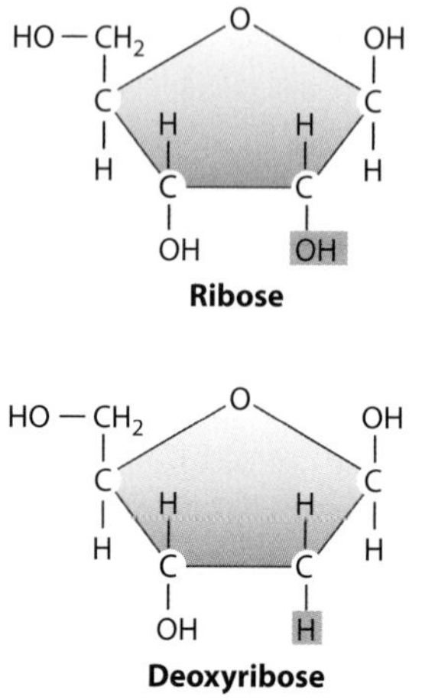

FIGURE 3.3 Ribose and deoxyribose

Ribose and deoxyribose are five-carbon sugars that are components of the nucleic acids RNA and DNA, respectively. They are shown here in their ring forms.

Six-carbon sugars are among the simplest carbohydrates. There are also five-carbon sugars. Noteworthy among the five-carbon sugars are **ribose,** which is a component of the nucleic acid RNA, and **deoxyribose,** which is a component of the nucleic acid DNA. These two sugars are shown in their ring form in **FIGURE 3.3**. Although most carbohydrates only contain the elements carbon, hydrogen, and oxygen, some contain nitrogen (N) and phosphorus (P). For example, chitin, found in the cell walls of fungi and the exoskeletons of insects, is a carbohydrate that contains nitrogen.

FIGURE 3.2 Linear and ring forms of glucose

Glucose exists in two forms: a linear form shown on the left, and a ring, or cyclic, form shown on the right. To form a ring, the hydroxyl (—OH) group attached to carbon atom 5 reacts with carbon atom 1 in linear glucose, as indicated by the red dashed arrow, to form a covalent bond. The ring is flat, or planar, so the thicker lines indicate the side of the ring that is closer to you, and the thinner lines indicate the side that is further away. The groups attached to the carbon atoms project above and below the plane of the ring.

PREP FOR THE AP® EXAM

AP® EXAM TIP

You should know that carbohydrates can take the form of monomers, chains of monomers, and branched structures.

✓ Concept Check

1. **Identify** the three elements that make up all carbohydrates.

2. **Identify** the two structures of monosaccharides.

3. **Describe** where the groups attached to the ring are located in a ring-structured monosaccharide.

3.2 Monosaccharides are joined by glycosidic bonds to make complex carbohydrates

Monosaccharides, especially the six-carbon sugars, can join to form long chains or complex branching molecules. Although they are made with the same basic subunit, carbohydrates can have different properties based on the types and arrangements of monosaccharides. We will now examine how such complex carbohydrates are formed and how their structures give rise to their properties.

Recall that the basic building blocks of carbohydrates are the simple sugars, or monosaccharides. Linking two simple sugars together by a covalent bond forms a **disaccharide** (*di* means "two"). Sucrose ($C_{12}H_{22}O_{11}$), or table sugar, is a disaccharide that combines one molecule each of glucose and fructose. Simple sugars combine in many ways to form polymers called **polysaccharides** (*poly* means "many"). Long, branched chains of monosaccharides are called **complex carbohydrates.**

Monosaccharides are attached to each other by covalent bonds called glycosidic bonds, as shown in **FIGURE 3.4**. In this figure, two molecules of glucose are connected by a covalent bond to form a single molecule. The carbon atom of one glucose molecule joins the hydroxyl group carried by a carbon atom of a different glucose molecule. In the process, a molecule of water is also produced. Because the formation of glycosidic bonds involves the loss of a water molecule, it is an example of a dehydration synthesis reaction, which we discussed in Module 2.

Carbohydrate diversity stems in part from the monosaccharides that make up carbohydrates, similar to the way that protein and nucleic acid diversity stems from the sequence of their subunits. Some carbohydrates are composed of a single type of monosaccharide, while others are a mix of different kinds of monosaccharides. Carbohydrate diversity also results from the arrangement of the monosaccharide subunits. Starch and cellulose, for example, are both composed completely of glucose molecules, but the arrangement of atoms within the glucose molecules and how they connect to each other are different in the two carbohydrates. This leads to different properties. Starch is an energy-storage molecule, while cellulose is a tough, resilient molecule that gives strength to plant cell walls and stems, like the one shown in **FIGURE 3.5**. The paper we write on, the cotton fibers in the clothes we wear, and the wood in the chairs we sit on are all composed of cellulose. Cellulose is, in fact, the most widespread organic molecule on Earth.

FIGURE 3.5 A plant stem

The stem of a plant is rigid and strong. It is primarily composed of the polysaccharide cellulose, as are paper and cotton fibers.

Photo: sukanya sitthikongsak/Getty Images

PREP FOR THE AP® EXAM

AP® EXAM TIP

When considering biological molecules, remember that structure and function are intertwined. The functions of carbohydrates, proteins, nucleic acids, and lipids are distinctive because their structures are unique and have their own set of properties.

✓ Concept Check

4. **Describe** a glycosidic bond and how it is formed.

5. **Describe** one similarity and one difference between starch and cellulose.

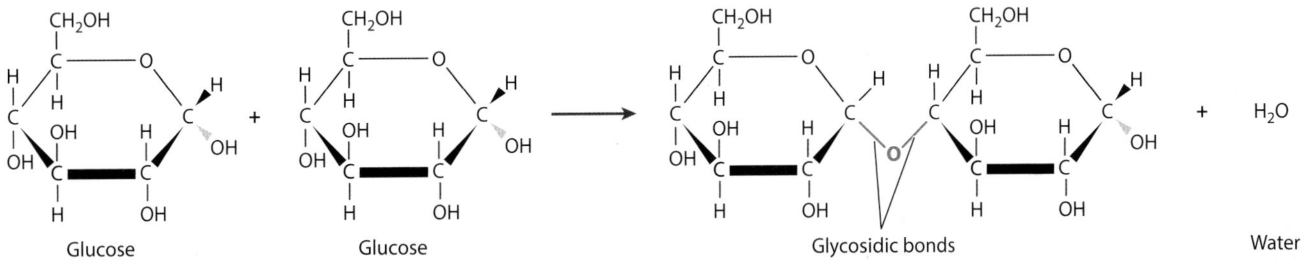

CH₂OH ... Glucose + Glucose → Glycosidic bonds + H₂O Water

FIGURE 3.4 Glycosidic bonds

Simple sugars, such as glucose, can be joined by glycosidic bonds. Glycosidic bonds are formed by a dehydration synthesis reaction in which two sugar molecules are connected, with loss of a molecule of water.

3.3 Lipids are hydrophobic molecules

Carbohydrates, as we have seen, are polymers made of repeating subunits called monosaccharides. Lipids, a second type of organic molecule, are different; they are not polymers. Instead of being defined by a particular chemical composition, lipids are more aptly defined by a chemical property: they are all hydrophobic. In other words, they do not dissolve in water and tend to aggregate, like an oil droplet in water. Lipids are thus unique when compared to the carbohydrates, proteins, and nucleic acids, which are all polymers made of monomers.

All organisms require lipids to maintain life. Like carbohydrates, they are important energy-storage molecules in cells and organisms. They include the familiar fats that make up part of our diet, such as meat fats, vegetable shortening, lard, corn oil, and canola oil. When animals have an excess of fats, they store them in cells. These cells later release the fats to the body in times of energy need. Lipids also play a critical role in cell communication, as we will see in Unit 4. Finally, they are major components of **cell membranes,** sometimes called **plasma membranes,** which are structures that define the boundary between the inside and outside of all cells. We will discuss the cell membrane in more detail in the next unit. In this section, we introduce the basic structure and function of lipids.

Triacylglycerols

Triacylglycerol is an example of a lipid that is used for energy storage. It is the major component of animal fat and vegetable oil. A triacylglycerol molecule is made up of *glycerol* joined to three *fatty acids*. **Glycerol** is a three-carbon molecule with hydroxyl (—OH) groups attached to each carbon, shown at the top of FIGURE 3.6. A **fatty acid,** shown in the middle of the figure, is a long chain of carbon atoms, called a hydrogen carbon chain, attached to a carboxyl group (—COOH) at one end.

The carboxyl group of the fatty acid is a functional group, and it is included in Table 3.1. It is able to react with the hydroxyl group of glycerol. When it reacts, the two molecules are joined by a covalent bond, with the release of a water molecule, as shown in Figure 2.8 on page 50. This is an example of a dehydration synthesis reaction, the same type of reaction that links sugar molecules to form long chains and branched structures. The bottom of Figure 3.6 shows a triacylglycerol molecule that results from the combination of a single glycerol molecule with three fatty acids.

Fatty acids differ in the length of their tails—that is, they differ in the number of carbon atoms in their hydrocarbon chain. As we discussed in the introduction, carbon–carbon and carbon–hydrogen bonds are rich sources of energy. As a result, the longer the tail, the more energy a fatty acid contains.

In addition, fatty acids differ in the number of carbon–carbon double bonds. Some fatty acids do not have any carbon–carbon double bonds, and some have one or more. Fatty acids that do not contain double bonds are described as **saturated.** An example is shown in FIGURE 3.7a. Because there are no

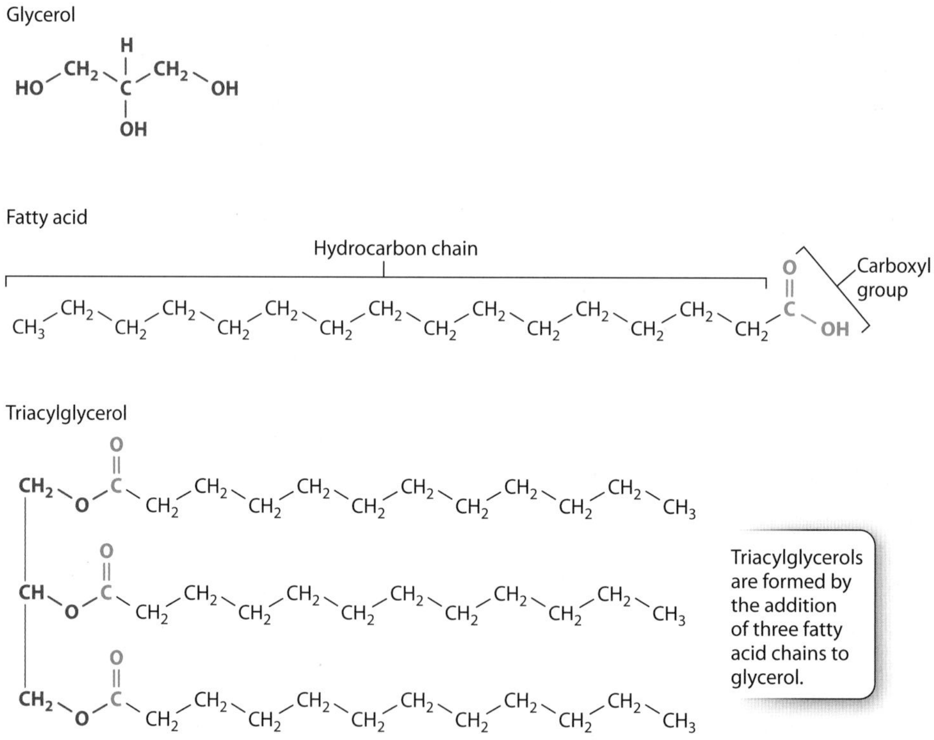

FIGURE 3.6 Triacylglycerol and its components

Triacylglycerol is a common energy-storing lipid. It is made up of one molecule of glycerol and three fatty acids. The resulting molecule, shown here, is synthesized by a dehydration synthesis reaction and is hydrophobic, like all lipids.

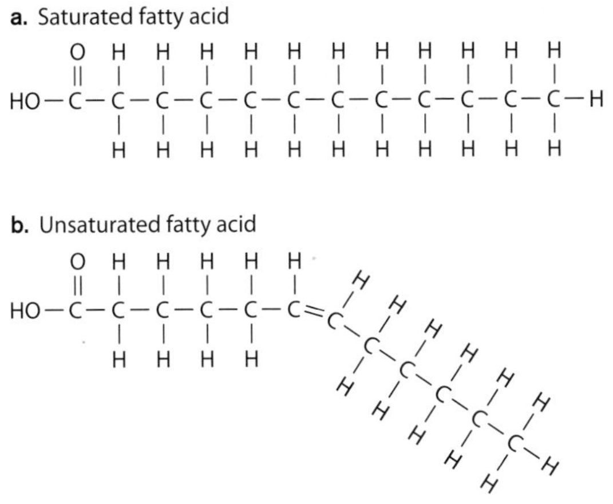

a. Saturated fatty acid

b. Unsaturated fatty acid

FIGURE 3.7 Saturated and unsaturated fatty acids

Fatty acids differ in the length of their tails and the number and position of carbon–carbon double bonds. (a) This saturated fatty acid tail has only carbon–carbon and carbon–hydrogen single covalent bonds. Saturated fatty acids have a linear, or straight, structure. (b) This unsaturated fatty acid tail has a carbon–carbon double bond between two of the carbon atoms. A double bond at this position puts a kink in the tail, resulting in a bent structure.

double bonds, the maximum number of hydrogen atoms is attached to each carbon atom, so all of the carbon atoms are said to be "saturated" with hydrogen atoms. Fatty acids that contain carbon–carbon double bonds are **unsaturated.** An example is shown in Figure 3.7b. They are described as unsaturated because at least two of the carbon atoms are not attached to two hydrogen atoms. The chains of saturated fatty acids are straight, while the chains of unsaturated fatty acids have a kink or bend at each double bond, as shown in Figure 3.7.

Triacylglycerols can contain different types of fatty acids attached to the glycerol backbone. The chains of fatty acids do not contain polar covalent bonds like those in a water molecule. Instead, their electrons are distributed uniformly over the whole molecule. As a result, triacylglycerols are all hydrophobic and, therefore, form oil droplets inside cells, such as fat cells. Triacylglycerols are an efficient form of energy storage because, by excluding water molecules, a large number can be packed into a small volume and because they contain many carbon–carbon and carbon–hydrogen bonds.

Although fatty acid molecules are nonpolar and uncharged, the constant motion of electrons leads to regions of partial positive and partial negative charges, illustrated in **FIGURE 3.8.** These partial charges, indicated by δ^+ and δ^- in the figure, either attract or repel electrons in neighboring molecules, setting up areas of positive and negative charges

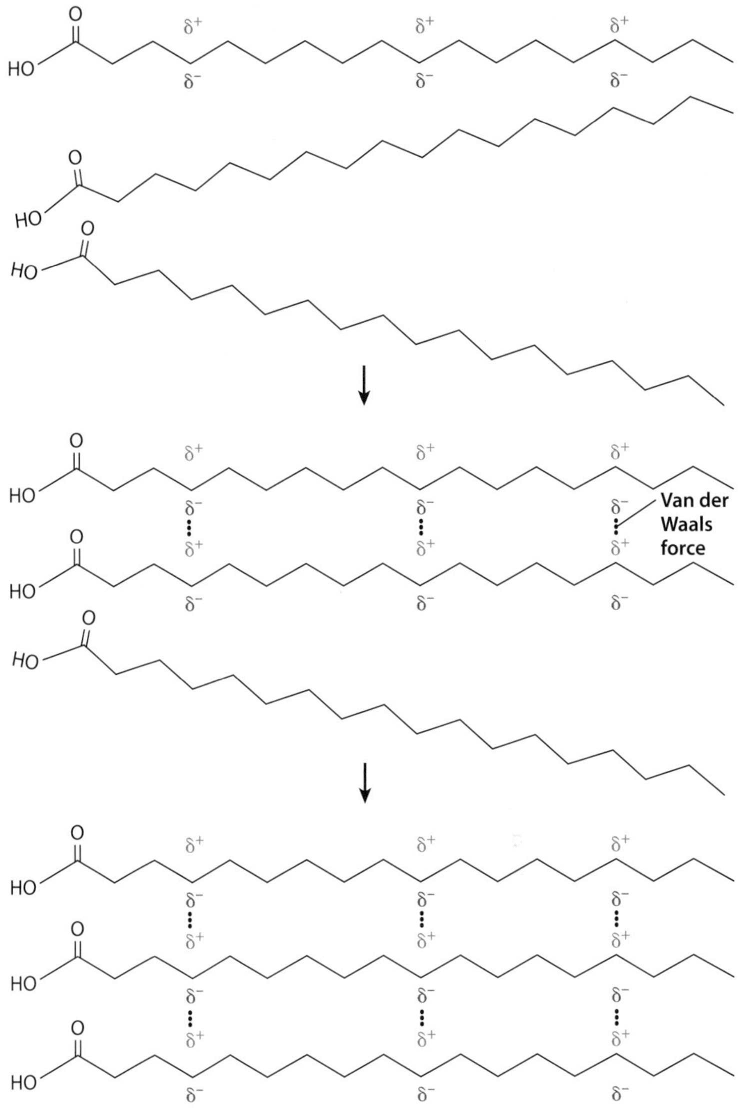

FIGURE 3.8 Van der Waals forces

The motion of electrons leads to transient regions of partial positive (δ^+) and partial negative (δ^-) charges. In turn, these lead to partial charges in neighboring molecules, resulting in weak attractions by opposite charges, indicated by the dotted lines.

in those molecules as well. The temporarily polarized molecules weakly associate with one another because of the attraction of opposite charges. An interaction of temporarily polarized molecules because of the attraction of opposite charges is known as **van der Waals forces.** The van der Waals forces come into play only when atoms are very close to one another, and they are weaker than hydrogen bonds. Even so, many van der Waals forces acting together help to stabilize molecules.

Because of van der Waals forces, the melting points of fatty acids depend on their length and level of saturation. As the length of the hydrocarbon chains increases, the

Unit Conversion

Unit conversions are used frequently in biology. Conversions can make it easier to compare multiple values or to perform calculations. Whether converting from millimeters to meters or gallons to liters, it is important to pay attention to the units in a question and understand how and when you may need to convert them.

PRACTICE THE SKILL

A student needs to fill several 8-liter tanks to do an aquatic experiment. The hose will fill the tanks at a rate of 500 milliliters per second (mL/s). Convert this rate into liters per second (L/s). How many seconds will it take to fill one tank?

The metric system is full of opportunities for relatively straightforward unit conversions. Remember that the different prefixes before standard measurements like meter or liter signify different magnitudes of value. In this example, we want to convert from mL/s to L/s. The prefix "milli" has a value of 10^{-3}. This can also be written as

$$\frac{1}{10^3} \text{ or } \frac{1}{1000}$$

It is often helpful to do unit conversions with fractions. Because of the prefix "milli," we know that there are 1000 mL in 1 L. With

this in mind, our first step is to convert mL into L. We can ask: if 1000 mL = 1 L, how many L does 500 mL equal? This question can also be written and solved with fractions:

$$500 \text{ mL} \times \frac{1 \text{ L}}{1000 \text{ mL}} = 0.5 \text{ L}$$

The fractions are set up so that the mL of our original number, 500, and the mL of the conversion rate, 1000, cancel and leave us with our new units, L. Now we know that 500 mL = 0.5 L. The hose fills a tank at a rate of 500 mL/s, which is the same as filling at a rate of 0.5 L/s.

Because we have converted to the correct units for the rate of water flowing from the hose, we can answer the second part of the question. If it takes 1 second to fill 0.5 L, how many seconds will it take to fill 8 L? This question can be solved with the same setup as the above conversion from mL to L:

$$8 \text{ L} \times \frac{1 \text{ s}}{0.5 \text{ L}} = 16 \text{ s}$$

The L of our original number and the L of the conversion rate cancel once again and leave us with our answer in seconds. It will take 16 seconds to fill one 8-L tank.

Your Turn

An agar solution must be cooled to 55°C in order to be poured safely into petri dishes by a student. The student's thermometer measures Fahrenheit. What temperature must the agar be cooled to in Fahrenheit? The conversion from Fahrenheit to Centigrade is $T(°F) = T(°C) \times \frac{9}{5} + 32$.

number of van der Waals interactions between the chains also increases. The melting temperature increases because more energy is needed to break the greater number of van der Waals interactions.

By contrast, kinks introduced by double bonds reduce the tightness of the molecular packing and, therefore, the number of van der Waals interactions. As a result, the melting temperature is lower. Thus, an unsaturated fatty acid has a lower melting point than a saturated fatty acid of the same length. Animal fats such as butter are composed of triacylglycerols with saturated fatty acids and are solid at room temperature, whereas plant and fish oils are composed of triacylglycerols with unsaturated fatty acids and are liquid at room temperature.

If you have ever cooked, you have probably used fats such as butter and cooking oil. You might have had to convert units from a recipe into other units. Being able to do unit

conversions is also a necessary skill for conducting scientific research. "Analyzing Statistics and Data: Unit Conversion" provides an opportunity to practice unit conversions.

Steroids and Phospholipids

Steroids are a second type of lipid. A common steroid is cholesterol, which is found in animal products and in mammalian cell membranes. Cholesterol is also the starting point by which animals, such as mammals, make other steroid molecules, including hormones such as estrogens and testosterone. Like other steroids, cholesterol has a core composed of 20 carbon atoms bonded to form four fused rings. The structure is shown in **FIGURE 3.9**. Because of all of the carbon–carbon and carbon–hydrogen bonds, it is hydrophobic. Notice that the structure of cholesterol is very different from the structure of triacylglycerol. However, they are both considered lipids because they are both hydrophobic.

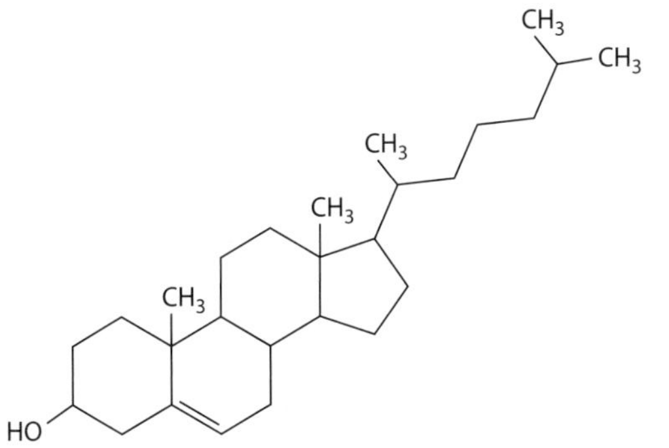

FIGURE 3.9 Cholesterol

The chemical structure of cholesterol is shown here. Although not shown, there is a carbon atom at each place two or more lines meet. Cholesterol is a component of some membranes surrounding cells and is the starting molecule for building certain types of chemical messengers called hormones.

Phospholipids are a third type of lipid. They are a major component of the cell membrane that surrounds every cell, shown in **FIGURE 3.10**. Like triacylglycerol, phospholipids have a glycerol backbone attached to fatty acids, shown in the inset of Figure 3.10. However, in the case of phospholipids, glycerol is attached to two fatty acids, not three. These are called the phospholipid "tails," and like all lipids, they are hydrophobic. In addition, glycerol is attached to a chemical structure that includes a phosphate group ($-OPO_3H_2$) and forms the "head" of the phospholipid. A phosphate group is a functional group that is hydrophilic (see Table 3.1 on page 55). As a result, phospholipids have both hydrophilic and hydrophobic regions in the same molecule: the head of the phospholipid is hydrophilic, while the tails are hydrophobic.

Two layers of phospholipids make the up the cell membrane, as shown in Figure 3.10. This structure allows it to create an effective barrier between the inside and outside of the cell. The hydrophobic tails point inward, away from water, creating a hydrophobic space that acts as a barrier. The hydrophilic heads point outward, interacting with the watery environment inside and outside of the cell. We will explore the structure and function of cell membranes in more detail in Module 9.

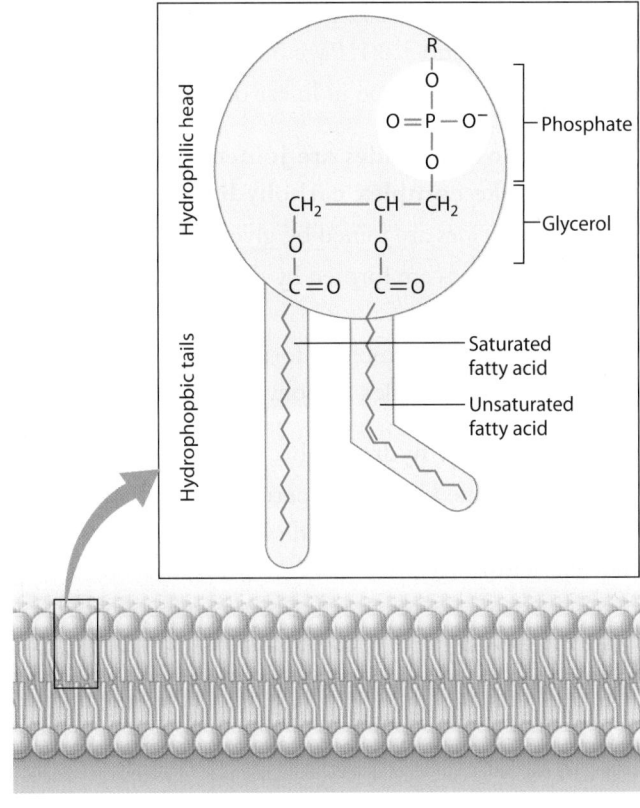

FIGURE 3.10 Phospholipids

Phospholipids are a major component of the membranes surrounding cells. They have hydrophilic heads, which point outward toward water, and hydrophobic tails, which point inward away from water.

✓ Concept Check

6. **Identify** a chemical property shared by all lipids.

7. **Identify** the components of triacylglycerol.

8. **Describe** the effect that unsaturation has on a fatty acid.

9. **Identify** three types of lipids and describe a role for each one.

Module 3 Summary

PREP FOR THE AP® EXAM

REVISIT THE BIG IDEAS

SYSTEMS INTERACTIONS: Using the material in this module, **describe** the properties of carbohydrates and lipids and **explain** how these properties affect the way they function.

LG 3.1 Monosaccharides are the basic units of carbohydrates.

- Carbohydrates are sources of energy, as well as components of the cell wall of plants and fungi. Page 54
- Carbohydrates are polymers built from subunits called monosaccharides. Page 54

- Monosaccharides are composed of C, H, and O atoms, usually in the ratio 1:2:1. Page 55
- Monosaccharides can be in linear or ring forms. Page 56

LG 3.2 Monosaccharides are joined by glycosidic bonds to make complex carbohydrates.
- Monosaccharides are joined by glycosidic bonds. Page 57
- Glycosidic bonds are formed by dehydration synthesis reactions, with loss of a water molecule. Page 57
- Monosaccharides assemble to form disaccharides or longer polymers called complex carbohydrates. Page 57

LG 3.3 Lipids are hydrophobic molecules.
- Lipids are defined by the property of being hydrophobic; they are not polymers like carbohydrates, proteins, and nucleic acids. Page 58

- Lipids include fats, oils, steroids, and phospholipids. Page 58
- Triacylglycerols store energy and are made up of glycerol and fatty acids. Page 58
- Triacylglycerols are synthesized by a dehydration synthesis reaction. Page 58
- Fatty acids consist of a linear hydrocarbon chain of variable length with a carboxyl group at one end. Page 58
- Fatty acids are either saturated (no carbon–carbon double bonds) or unsaturated (one or more carbon carbon double bonds). Page 58
- The tight packing of fatty acids in lipids is the result of van der Waals forces, a type of weak, noncovalent interaction. Page 60
- Steroids and phospholipids are two other types of lipids. Page 60

Key Terms

Functional group	Cell membrane	Unsaturated
Ribose	Plasma membrane	Van der Waal force
Deoxyribose	Triacylglycerol	Steroid
Disaccharide	Glycerol	Phospholipid
Polysaccharide	Fatty acid	
Complex carbohydrate	Saturated	

Review Questions

1. The building blocks of carbohydrates are
   (A) disaccharides.
   (B) polysaccharides.
   (C) monosaccharides.
   (D) lipids.

2. When a glycosidic bond is formed, which is true?
   (A) Multiple monosaccharide units are released.
   (B) Two monosaccharide units are released, and a water molecule is consumed.
   (C) A monosaccharide unit is joined to either another monosaccharide or to a chain of monosaccharides, and a water molecule is released.
   (D) Two monosaccharide units are joined, and a water molecule is consumed.

3. The groups of atoms that give a carbon molecule distinct characteristics are called
   (A) operational groups.
   (B) hydrophilic groups.
   (C) electronegative groups.
   (D) functional groups.

4. A triacylglyceride is made from
   (A) three glycerol molecules and one fatty acid molecule.
   (B) three glycerol molecules and three fatty acid molecules.
   (C) one glycerol molecule and one fatty acid molecule.
   (D) one glycerol molecule and three fatty acid molecules.

5. Fatty acid molecules contain
   (A) a carboxyl group and a hydrocarbon tail.
   (B) a carboxyl group and three hydrocarbon tails.
   (C) three carboxyl groups and a hydrocarbon tail.
   (D) only hydrocarbon tails.

6. Which is not a function of lipids?
   (A) A source of energy
   (B) Genetic information storage
   (C) Component of the cell membrane
   (D) Type of signaling molecule

7. Lipids have which property/properties?

(A) Hydrophobic and nonpolar

(B) Water soluble and polar, but uncharged

(C) Form double helical structures when dissolved in water

(D) Have primary, secondary, tertiary, and quaternary structures

---

Module 3
AP® Practice Questions

Section 1: Multiple-Choice Questions

Choose the best answer for questions 1–5.

1. How do hydroxyl groups affect the properties of a simple sugar?

(A) They decrease the polarity of simple sugars.

(B) They increase the number of ionic bonds formed between simple sugars.

(C) They decrease the electronegativity of simple sugars.

(D) They increase the solubility of simple sugars.

2. The carbohydrates cellulose and starch are functionally different due to the

(A) specific monosaccharides used to build them.

(B) arrangement of their monosaccharides.

(C) functional groups contained within their monosaccharides.

(D) reaction used to join their monosaccharides.

3. Unlike saturated fatty acids, unsaturated fatty acids

(A) have a high melting point.

(B) have a kinked structure.

(C) have the maximum number of hydrogen atoms attached to carbon atoms.

(D) do not contain carbon–carbon double bonds.

4. Which is the correct ranking from strongest to weakest?

(A) covalent bonds > van der Waals forces > hydrogen bonds

(B) van der Waals forces > hydrogen bonds > covalent bonds

(C) covalent bonds > hydrogen bonds > van der Waals forces

(D) hydrogen bonds > van der Waals forces > covalent bonds

5. Which is true of monosaccharides and phospholipids?

(A) They contain polar, hydrophilic functional groups.

(B) They are built from long linear chains of fatty acids.

(C) They are highly insoluble.

(D) They are each long chains joined with glycosidic bonds.

Section 2: Free-Response Question

Write your answer to each part clearly. Support your answers with relevant information and examples. Where calculations are required, show your work.

Pictured below are two organic molecules, phosphatidic acid and glycogen. One of these is a carbohydrate and the other is a lipid. The primary role of phosphatidic acid is to act as a chemical messenger and signal between cells. It contains phosphate and carboxyl functional groups. Glycogen is an important molecule for energy storage in animals. It contains several hydroxyl functional groups.

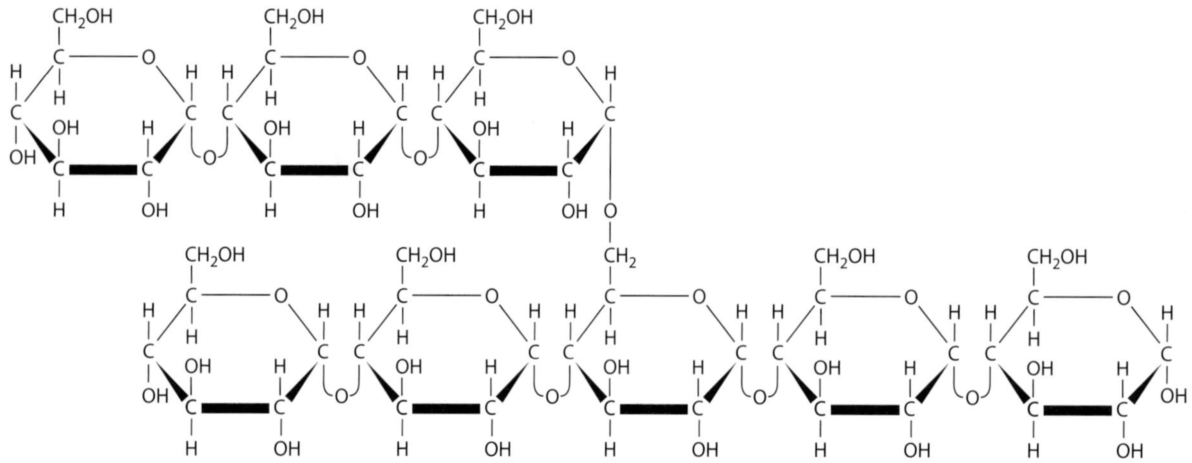

(a) Based on the structures, functional groups, and cellular functions described, **identify** the specific type of organic macromolecules represented by phosphatidic acid and glycogen.

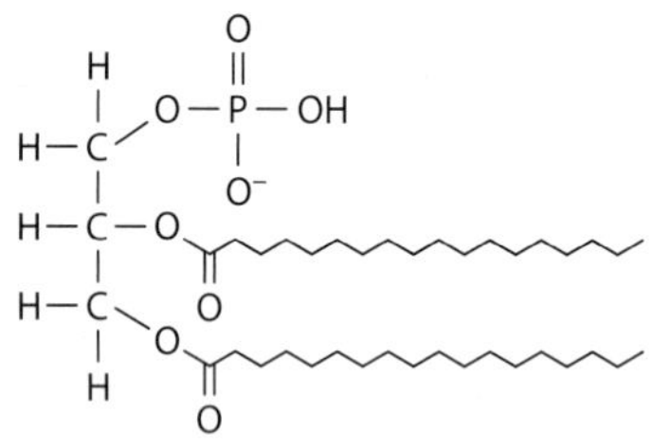

(b) **Identify** the names of the main building blocks of phosphatidic acid and glycogen.

(c) **Describe** the specific mechanism and bonds formed through dehydration synthesis reactions in the building of both lipids and complex carbohydrates.

(d) Each molecule contains different functional groups. **Describe** how the functional groups of each molecule affect the solubility of the molecule.

Phosphatidic Acid

Glycogen

Module 4

Proteins

LEARNING GOALS ▶**LG 4.1** Amino acids are the building blocks of proteins.

▶**LG 4.2** Peptide bonds link successive amino acids to form proteins.

▶**LG 4.3** Proteins are folded into three-dimensional shapes that determine their functions.

In the last module, we discussed two organic molecules—carbohydrates and lipids. In this module, we will continue our discussion of organic molecules with proteins. Hardly anything happens in the life of a cell that does not require proteins. They are extremely versatile, and able to carry out many cellular functions. For example, some proteins form stiff filaments that help define the cell's shape. Others span the membrane surrounding cells and form channels through which ions and small molecules move. Many others accelerate the rate of the thousands of chemical reactions needed to sustain life. Still others act as molecular signals that enable cells to coordinate their internal activities or to communicate with other cells.

Want to see some proteins? Look at the white of an egg. Apart from the 90% or so that is water, most of what you see is protein. The predominant type of protein in egg white is ovalbumin. Easy to obtain in large quantities, ovalbumin was one of the first proteins studied through scientific inquiry.

Ovalbumin consists largely of four elements: carbon, hydrogen, oxygen, and nitrogen, like all proteins. Some proteins also contain a small amount of sulfur. Ovalbumin is a very large molecule, made up of thousands of carbon atoms. Proteins such as ovalbumin can be so large because—as we discussed in Module 1—they are polymers, molecules made up of repeated subunits. Recall that the repeated units of proteins are called amino acids.

In this module, we will look closely at proteins. We will begin by examining their amino acid building blocks. Then we will see how these amino acids are joined together to form long chains. Finally, we will see how these chains of amino acids are folded into three-dimensional shapes, which are closely connected to the functions of the proteins. Indeed, the tight relationship between structure and function is a recurring theme in biology.

PREP FOR THE AP® EXAM

FOCUS ON THE BIG IDEAS

SYSTEMS INTERACTIONS: Think about how the various properties of amino acids and their interaction with each other and with water affect the final structure and function of proteins.

4.1 Amino acids are the building blocks of proteins

Proteins do much of the work of the cell and help to form its structure. As we noted, proteins are a major component of the white of an egg, like the one in **FIGURE 4.1**. Proteins are made up of subunits called amino acids. If you think of a protein as being like a word in the English language, then the amino acids are like letters. In fact, there are nearly as many amino acids in proteins as there are letters in the alphabet, and the order of both amino acids and letters is important. For example, the word PROTEIN has the same letters as POINTER, but the two words have completely different meanings. Similarly, the exact order of amino acids in a protein determines the protein's shape and function.

FIGURE 4.1 Proteins

The white of an egg contains high quantities of the protein ovalbumin.

Photo: Valentyn Volkov/Alamy Stock Photo

In this section, we will examine the structure and properties of amino acids.

Amino Acid Structure

Amino acids are the subunits, or monomers, that make up proteins. The general structure of an amino acid is shown in **FIGURE 4.2a**. It consists of a carbon atom, called the α (alpha) carbon. This central carbon atom is connected by covalent bonds to four chemical groups: an amino group ($-NH_2$, shown in dark blue), a carboxyl group ($-COOH$, shown in brown), a hydrogen atom ($-H$, shown in black), and an R group or side chain (shown in green).

The R groups of the amino acids differ from one amino acid to the next. They are what make the "letters" of the amino acid "alphabet" distinct from one another. Just as letters differ in their shapes and sounds—vowels like E, I, and O and consonants like B, P, and T—amino acids differ in their chemical and physical properties. For example, the R group of the amino acid alanine is $-CH_3$, while the R group of serine is $-CH_2OH$. In most amino acids, the α carbon is covalently linked to four different groups. Glycine is the exception, since its R group is a hydrogen (H) atom.

In the environment of a cell, where the pH is approximately 7.4, the amino and carboxyl groups are ionized, or charged, as shown in Figure 4.2b. The amino group gains a proton (H^+), changing from $-NH_2$ to become $-NH_3^+$, and the carboxyl group loses a proton, changing from $-COOH$ to become $-COO^-$.

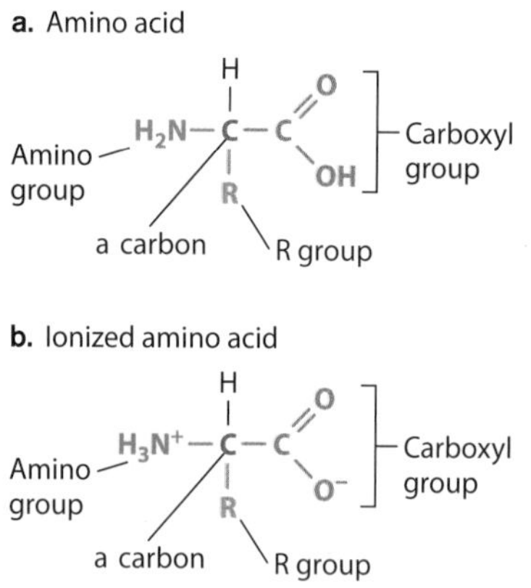

a. Amino acid

b. Ionized amino acid

FIGURE 4.2 Amino acid structure

This figure shows the general structure of an amino acid, the monomers that make up proteins. (a) An amino acid contains four groups attached to a central, or alpha (α), carbon atom. The four groups are an amino group in dark blue, a carboxyl group in brown, a hydrogen atom in black, and an R group in green. (b) At the pH commonly found in cells, amino acids are ionized and contain NH_3^+ and COO— groups.

The Twenty Common Amino Acids

Twenty different amino acids make up proteins. These are listed in **FIGURE 4.3**. The names of the amino acids all have standard three-letter abbreviations and one-letter abbreviations. For example, the amino acid alanine is also indicated by Ala and A in the figure.

The R groups are shown in green in Figure 4.3. They are chemically diverse and are grouped according to their properties, specifically whether they are hydrophobic, are hydrophilic, or have special characteristics that affect a protein's structure. These properties strongly influence how a protein folds and hence affect the shape and function of the protein.

Hydrophobic amino acids do not readily interact with water. Most hydrophobic amino acids have nonpolar R groups. Because water molecules in the cell form hydrogen bonds with each other instead of with the hydrophobic R groups, the hydrophobic R groups aggregate with each other. The tendency for hydrophobic molecules to interact with each other instead of with water is the same tendency that leads to the formation of oil droplets in water. It is also the reason most hydrophobic amino acids tend to be buried in the interior of folded proteins, where they do not interact with water.

Hydrophilic amino acids are divided into three groups in Figure 4.3: polar, basic, and acidic. Amino acids with polar R groups have one end that is slightly more negatively charged than the other end. As we saw in Module 1, polar molecules are hydrophilic, and they tend to form hydrogen bonds with each other or with water molecules.

The R groups of the basic and acidic amino acids are typically charged and, therefore, are strongly polar. At the pH of a cell, the R groups of the basic amino acids gain a proton (H^+) and become positively charged, whereas those of the acidic amino acids lose a proton (H^+) and become negatively charged. Because the R groups of these amino acids are charged, they are usually located on the outside surface of the folded molecule, where they interact with water and other polar molecules. The R groups can form ionic bonds with each other and with other charged molecules. Such bonds form when a negatively charged group or molecule attracts a positively charged group or molecule.

This ability to bind another molecule of opposite charge is an important way in which proteins associate with each other or with other molecules such as DNA.

Three amino acids—glycine, proline, and cysteine—have special properties. Their R groups are shown in Figure 4.3. The properties of these amino acids are noteworthy because they affect the structure of the protein in specific ways. Glycine's R group is simply a hydrogen (H) atom, which

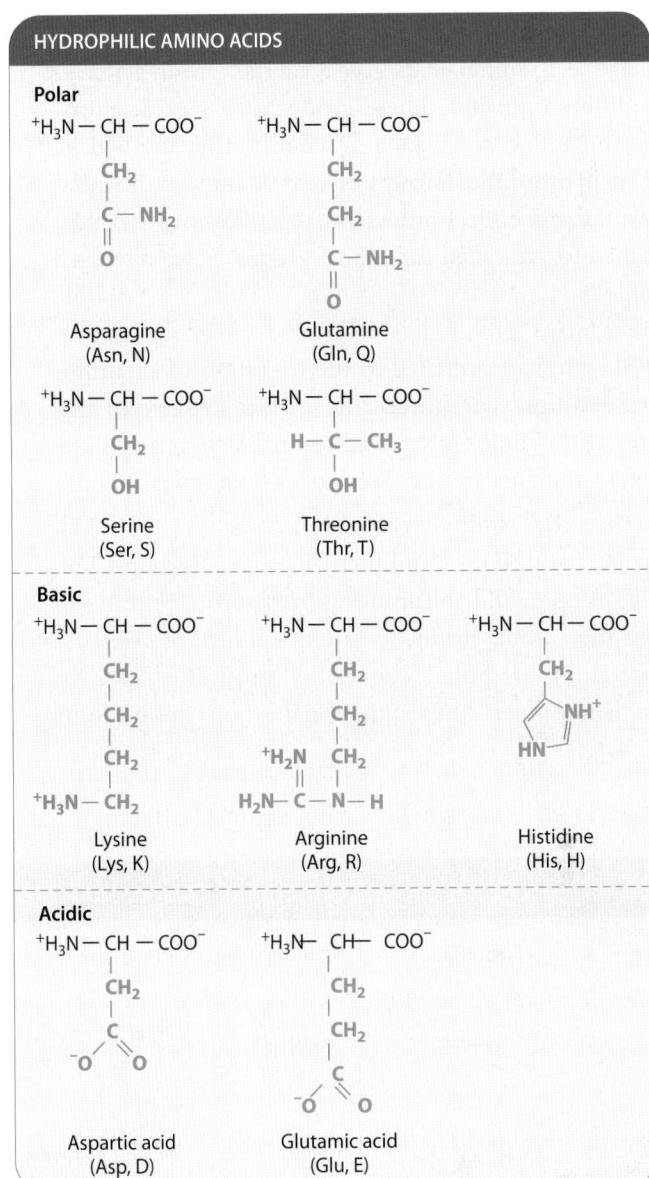

FIGURE 4.3 Structures of the 20 amino acids commonly found in proteins

The amino acids are divided into three groups: hydrophobic, hydrophilic, and special. The hydrophilic amino acids are further divided into polar, basic, and acidic. All of these groups are based on the chemical properties of their R groups, shown in green. Amino acids have names, three-letter abbreviations, and one-letter abbreviations, as indicated below each chemical structure.

is nonpolar. In addition, the small size of glycine's R group also allows for rotation around the C—N bond, since its R group does not get in the way of the R groups of neighboring amino acids. Thus, glycine increases the flexibility of the polypeptide backbone, which can be important in the folding of the protein. Proline is the only amino acid that has a carbon chain attached to its amino group, so it introduces kinks in the structure of proteins. Cysteine affects protein folding through its sulfhydryl (—SH) group. When two cysteines are close together, they can form an S—S disulfide bond, which is a type of covalent bond that can connect different parts of the same protein or even different proteins.

PREP FOR THE AP® EXAM

AP® EXAM TIP

You do not need to memorize the structures of individual amino acids, but you should be familiar with the components of an amino acid and understand that the R group conveys an amino acid's unique properties.

Functional Groups

As we have seen, amino acids have a central carbon attached to four different groups. These groups give specific properties to the amino acids, such as whether they are hydrophobic or hydrophilic. Groups of one or more atoms that have particular chemical properties, regardless of what they are attached to, are called functional groups, as we discussed in Module 3.

The amino group ($-NH_2$) and carboxyl group ($-COOH$) that are attached to the α carbon of an amino acid are examples of functional groups. Many of the R groups also contain functional groups, which give each amino acid their particular properties. Examples include hydroxyl ($-OH$), sulfhydryl ($-SH$), carbonyl $\left(\begin{array}{c} \diagdown \\ \diagup \end{array} C=O \right)$, amide ($-C(=O)NH-$), and methyl ($-CH_3$) groups. The nitrogen, oxygen, and sulfur atoms are more electronegative than the carbon atoms, so functional groups containing these atoms are polar. The methyl group ($-CH_3$), by contrast, is nonpolar. These functional groups are listed in **TABLE 4.1**.

Because many functional groups are polar, molecules that contain these groups—molecules that would otherwise be nonpolar—become polar and soluble in the cell's aqueous environment. Moreover, because many functional groups are polar, they are reactive. Notice in the following sections that the reactions joining simpler molecules into polymers usually take place between functional groups. Functional groups are not only present in proteins; they are also present in other organic molecules, such as carbohydrates, lipids, and nucleic acids. We will introduce additional functional groups as we encounter them.

✓ Concept Check

1. **Explain** what gives a particular amino acid its unique properties.

2. **Identify** the components that are common to all amino acids.

3. **Identify** the three main groups of amino acids.

TABLE 4.1 Functional Groups Commonly Observed in Biological Molecules, Part 2

Name	Formula	Structure	Properties	Commonly found in
Amino	$-NH_2$		Polar, positively charged at the pH of a cell, behaves as a base, hydrophilic	Amino acids, proteins
Carboxyl	$-COOH$		Polar, negatively charged at the pH of a cell, behaves as an acid, hydrophilic	Fatty acids, amino acids, proteins
Hydroxyl	$-OH$	$-O-H$	Polar, hydrophilic	Carbohydrates, proteins, nucleic acids
Sufhydryl	$-SH$	$-S-H$	Polar, forms S—S disulfide bonds	The amino acid cysteine, proteins
Carbonyl	$\diagdown C=O$		Polar, hydrophilic	Carbohydrates, proteins
Amide	$-C(=O)NH$		Polar, hydrophilic	Proteins
Methyl	$-CH_3$		Nonpolar	Amino acids, proteins, nucleic acids

4.2 Peptide bonds link successive amino acids to form proteins

As we have seen, cellular proteins are composed of combinations of 20 different amino acids, each of which can be classified according to the chemical properties of its R group. The particular sequence, or order, in which amino acids are present in a protein determines how it folds into its three-dimensional structure. The three-dimensional structure, in turn, determines the protein's function. Our next step is to explore how amino acids are linked to one another to form long chains, and then discover some of the properties of these linked amino acids.

Peptide Bond

Amino acids are linked in a chain to form a protein. The carbon atom in the carboxyl group of one amino acid is joined to the nitrogen atom in the amino group of the next. The resulting covalent linkage is called a **peptide bond.** The formation of a peptide bond is shown in **FIGURE 4.4.** As with glycosidic bonds, the formation of a peptide bond involves the loss of a water molecule and is therefore an example of a dehydration synthesis reaction, which we introduced in Module 2. That is, to form a C—N bond, the carbon atom of the carboxyl group releases a hydroxyl group (—OH) and the nitrogen atom of the amino group releases a hydrogen atom (H). This hydroxyl group and hydrogen atom then combine to form a water molecule (H_2O). The

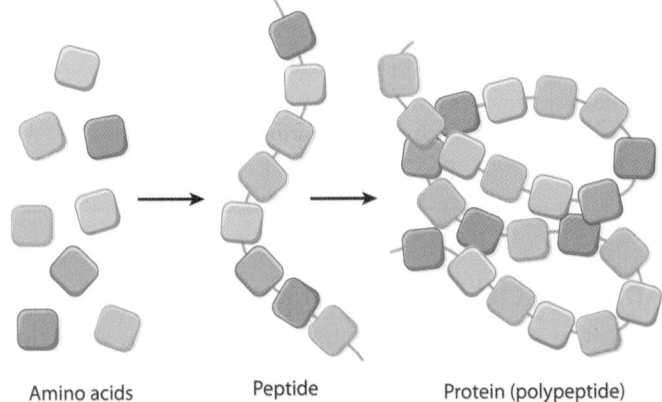

FIGURE 4.5 Amino acids, peptides, and proteins

Amino acids are the building blocks of proteins. In the figure on the left, each color represents a different amino acid. In the middle figure, the short string of linked amino acids is called a peptide. Long peptides are called polypeptides. In figure on the right, the polypeptide is folded into its final shape and is called a protein.

loss of a water molecule also occurs when subunits link to form other polymers such as carbohydrates.

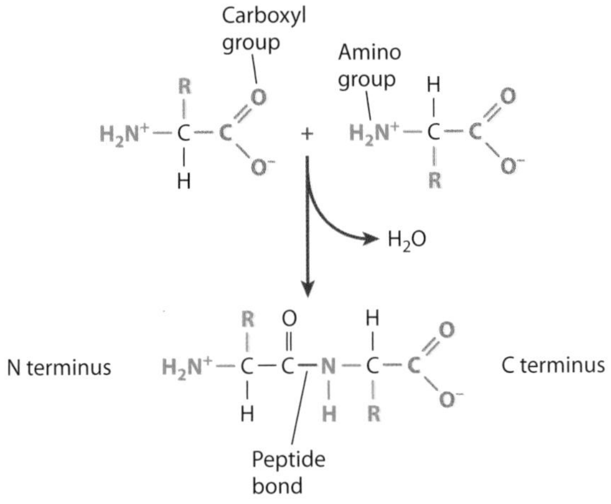

In Figure 4.4, the resulting chain of amino acids includes just two amino acids, and the peptide bond is indicated in red. A peptide bond is made every time an amino acid is added to a growing chain of a protein. The C=O group on one side of the peptide bond is known as a carbonyl group, and the N—H group on the other is an amide group. These are both examples of functional groups, as we discussed earlier, and are included in Table 4.1.

Polymers ranging from as few as two amino acids to many hundreds linked together share a chemical feature common to individual amino acids: namely, the ends are chemically distinct from each other. One end, shown at the left in Figure 4.4, has a free amino group; this is the amino end, or N terminus, of the molecule. The other end has a free carboxyl group; it is the carboxyl end, or C terminus, of the molecule. As a result, proteins show directionality.

Because successive amino acids are linked by peptide bonds, a short polymer of amino acids is called a **peptide,** and a long polymer of amino acids is called a **polypeptide** (*poly* means "many"). The term "protein" is often used as a synonym for "polypeptide," especially when the polypeptide chain has folded into a stable, three-dimensional conformation. The relationship among these various terms is illustrated in **FIGURE 4.5.** Typical proteins produced in cells consist of a few hundred amino acids. In human cells, the shortest proteins are approximately 100 amino acids in length; the longest is the muscle protein titin, with 34,350 amino acids.

FIGURE 4.4 Formation of a peptide bond

A peptide bond is formed between the carboxyl group of one amino acid and the amino group of another by a dehydration synthesis reaction, in which water is released. The peptide bond is shown in red. Proteins have directionality, with an N terminus and a C terminus as shown.

AP® EXAM TIP

You should understand that a dehydration synthesis reaction removes a water molecule when linking two monomers, such as in the formation of proteins, carbohydrates, and triacylglycerols.

Amino Acid Sequence

Up to this point, we have considered the amino acids that make up a protein. The sequence, or order, of the amino acids in a protein is its **primary structure.** FIGURE 4.6 shows the primary structure of a protein made up of 25 amino acids. The primary structure of a protein ultimately determines how a protein folds.

Proteins have remarkably diverse functions in the cell, ranging from serving as structural elements to communicating with the external environment to accelerating the rate of chemical reactions. The ability of a protein to carry out its function depends on its three-dimensional shape. When fully folded, some proteins contain pockets with positively or negatively charged side chains at just the right positions to trap small molecules; others have surfaces that can bind another protein or a sequence of nucleotides in DNA or RNA; some form rigid filaments for structural support; and still others keep their hydrophobic side chains away from water molecules by inserting into the membrane of the cell.

The sequence of amino acids in a protein—its primary structure—is usually represented by a series of three-letter or one-letter abbreviations for the amino acids, such as Thr–Ile–Pro–Ser. By convention, the amino acids in a protein

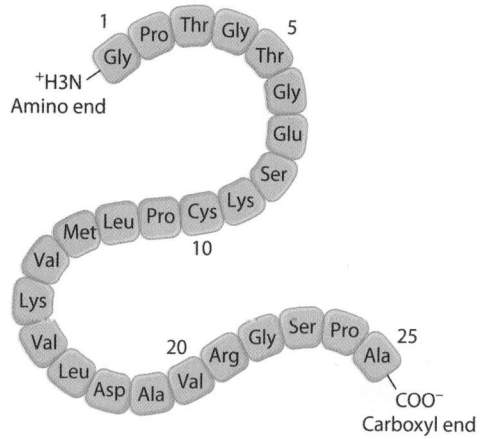

FIGURE 4.6 Primary structure

The primary structure of a protein is the sequence of its amino acids from its amino end to its carboxyl end. For this protein with 25 amino acids, the protein begins with the amino end glycine and ends with the carboxyl end alanine.

are listed in order from left to right, starting at the amino end and proceeding to the carboxyl end. The amino and the carboxyl ends are different, so the order matters. Just as TIPS is not the same word as SPIT, the sequence Thr–Ile–Pro–Ser is not the same as Ser–Pro–Ile–Thr.

✓ Concept Check

4. **Identify** the type of bond that connects amino acids to form a protein.
5. **Identify** the two atoms that form a covalent bond in a peptide bond.
6. **Describe** the directionality present in a peptide.

4.3 Proteins are folded into three-dimensional shapes that determine their functions

In the last section, we discussed the sequence of amino acids that make up a protein, or its primary structure. The primary structure defines the backbone of the protein and determines how a protein folds, which in turn determines how it functions. The primary structure, however, is just the first of several levels of protein structure. In this section, we will introduce the other levels of protein structure.

Secondary Structure

Interactions between stretches of amino acids in a protein form **secondary structures.** Hydrogen bonds can form between the carbonyl group in one peptide bond and the amide group in another, thus allowing different regions of the polypeptide chain to fold. There are two types of secondary structures that are commonly found in many proteins: the *alpha helix* and the *beta sheet*. Both of these secondary structures are stabilized by hydrogen bonding along the polypeptide backbone.

In an **alpha helix** (α helix), like the one shown in FIGURE 4.7, the polypeptide backbone is twisted, forming a

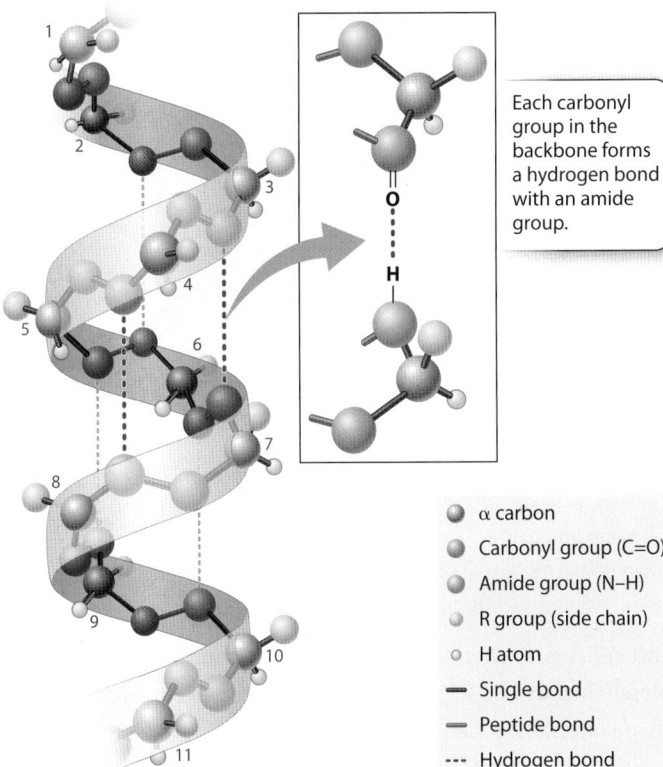

Each carbonyl group in the backbone forms a hydrogen bond with an amide group.

- ● α carbon
- ● Carbonyl group (C=O)
- ● Amide group (N–H)
- ● R group (side chain)
- ○ H atom
- — Single bond
- — Peptide bond
- --- Hydrogen bond

FIGURE 4.7 An alpha helix

An alpha helix is a secondary structure of a protein in which the backbone of the protein is twisted as shown. It is stabilized by hydrogen bonds, indicated by the red dashed lines. Hydrogen bonds form between carbonyl and amide groups.

spiral or helix. The helix is stabilized by hydrogen bonds that form between each amino acid's carbonyl group (C=O) and an amide group (N—H), as indicated by the dashed lines in Figure 4.7.

Another secondary structure is a **beta sheet** (β sheet), depicted in **FIGURE 4.8**. In a beta sheet, the polypeptide forms a pleated sheet that is stabilized by hydrogen bonds between carbonyl groups in one chain and amide groups in the other chain across the way, as shown by dashed lines. Broad arrows are typically used to show beta sheets, where the direction of the arrow runs from the amino end of the polypeptide segment to the carboxyl end. In Figure 4.8, the arrows run in opposite directions, and the polypeptide chains are said to be antiparallel. Beta sheets can also be formed by hydrogen bonding between polypeptide chains that are parallel (pointing in the same direction).

Tertiary Structure

The **tertiary** (third) **structure** of a protein is the three-dimensional shape of a single polypeptide chain. The three-dimensional shapes of proteins can be illustrated in different ways, as shown in **FIGURE 4.9** on page 72. The ball-and-stick model in Figure 4.9a draws attention to the atoms in the amino acid chain. The ribbon model in Figure 4.9b emphasizes secondary structures, with alpha helices depicted as twisted ribbons and beta sheets as broad arrows. Finally, the space-filling model in Figure 4.9c shows the overall shape of the folded protein.

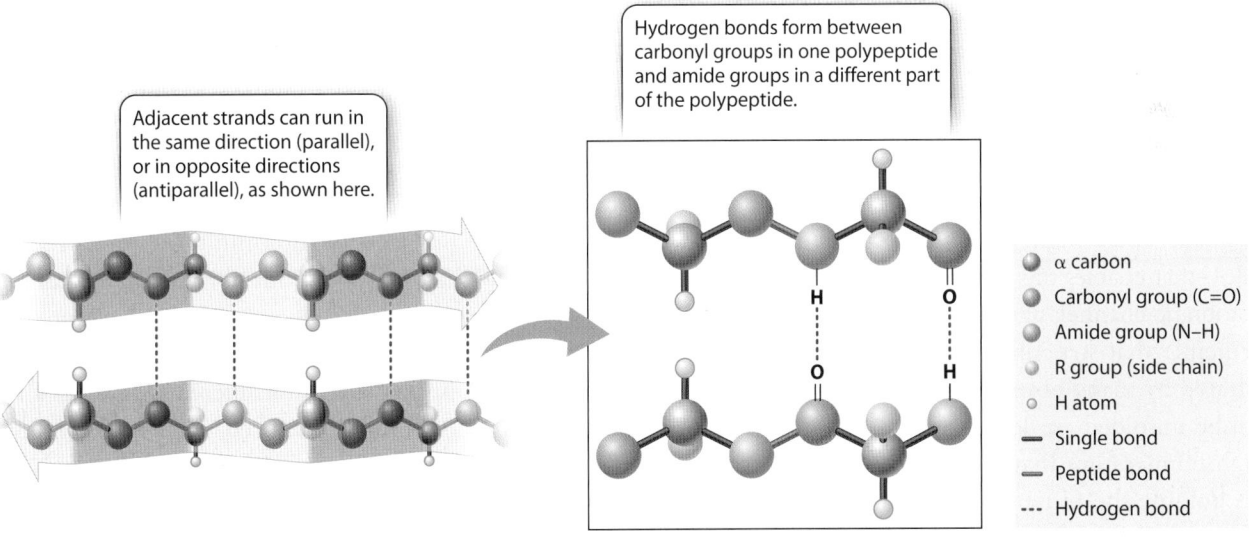

Adjacent strands can run in the same direction (parallel), or in opposite directions (antiparallel), as shown here.

Hydrogen bonds form between carbonyl groups in one polypeptide and amide groups in a different part of the polypeptide.

- ● α carbon
- ● Carbonyl group (C=O)
- ● Amide group (N–H)
- ● R group (side chain)
- ○ H atom
- — Single bond
- — Peptide bond
- --- Hydrogen bond

FIGURE 4.8 A beta sheet

A beta sheet is a secondary structure that forms when hydrogen bonds form two adjacent chains of amino acids. Each sheet is pleated and runs in the same or opposite direction. The direction is indicated by the arrows. Hydrogen bonds between amide groups and carbonyl groups on neighboring strands stabilize the structure.

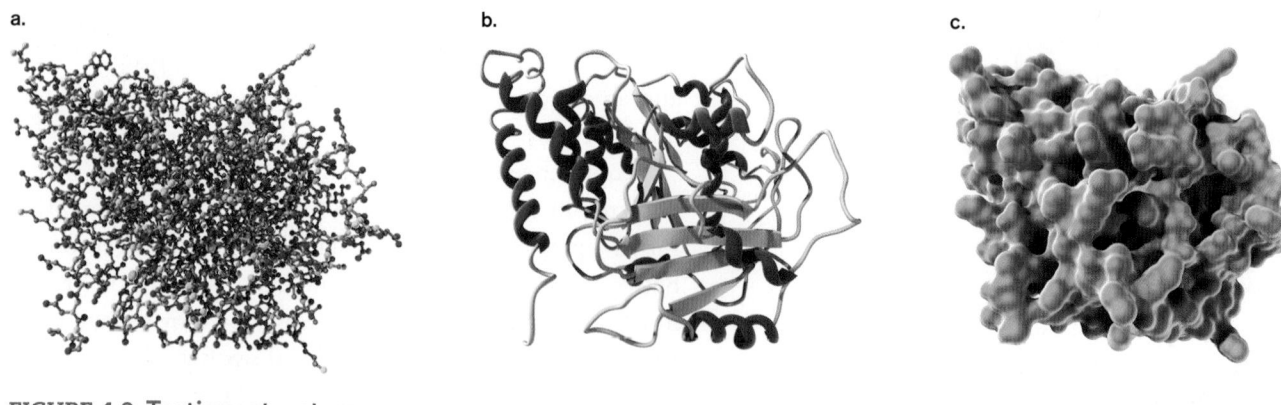

FIGURE 4.9 Tertiary structure

The tertiary structure is the overall shape of a protein. In this figure, the tertiary structure of the same protein is depicted in three different ways. (a) This model is known as a ball-and-stick representation and allows you to see the individual atoms. (b) This ribbon model uses twisted ribbons to represent alpha helices and wide arrows to represent beta sheets, allowing you to see the secondary structures that contribute to the tertiary structure. (c) This space-filling model of tertiary structure shows the overall shape and contours of the folded protein.

Tertiary structure is defined largely by interactions between the amino acid R groups. The spatial distribution of hydrophilic and hydrophobic R groups along the molecule, and different types of chemical bonds, such as ionic and hydrogen bonds, that form between R groups all contribute to the tertiary structure. The amino acids whose R groups form bonds with each other may be far apart in the polypeptide chain, but can end up near each other in the folded protein. Hence, the tertiary structure usually includes loops or turns in the backbone that allow these R groups to sit near each other in space and enable bonds to form.

The primary structure determines both the secondary and tertiary structures. In other words, the primary structure is responsible for how a protein folds. Furthermore, tertiary structure determines function because it is the three-dimensional shape of the molecule—the contours and distribution of charges on the outside of the molecule and the presence of pockets that might bind with smaller molecules on the inside—that enables the protein to serve as structural support, membrane channel, signaling molecule, and so on.

The principle that structure determines function can be demonstrated by many observations. For example, most proteins can be unfolded, or denatured, by chemical treatment or high temperature that disrupts the hydrogen and ionic bonds holding the tertiary structure together. When denatured, the proteins lose their activity. Similarly, proteins with an incorrect amino acid often do not fold properly and therefore do not function properly or are inactive.

One experimental demonstration of the close connection between protein structure and function was based on studies of a protein called ribonuclease A. In 1961, Christian Anfinsen and colleagues chemically denatured ribonuclease A and completely inactivated it. Most proteins are not able to recover their shape after being denatured. For example, the white of an egg will never go back to being clear and runny after being heated and denatured. However, using careful methods, the researchers were able to remove the denaturing agents and found that the protein recovered its activity. For these experiments, Anfinsen was awarded a Nobel Prize in 1972. This experiment is described in more detail in "Practicing Science 4.1: What is the relationship between protein structure and function?" on page 74.

Quaternary Structure

Although many proteins are complete and fully functional as a single polypeptide chain with a tertiary structure, some proteins are composed of two or more polypeptide chains or subunits. The resulting ensemble is the **quaternary** (fourth) **structure.** The function of such multi-subunit proteins depends on the quaternary structure formed by the combination of the various tertiary structures that make it up.

In a protein with quaternary structure, the polypeptide subunits may be either identical or different. **FIGURE 4.10a** shows an example of a protein produced by the human immunodeficiency virus (HIV) that consists of two identical polypeptide subunits. By contrast, many proteins are composed of different subunits. One example is hemoglobin, which binds oxygen in red blood cells and is shown in Figure 4.10b.

Quaternary structures often have emergent properties that the individual subunits that make it up do not have on their own. For example, the ability of hemoglobin to pick up oxygen and deliver it to tissues depends on its four subunits,

a.

This protein consists of two identical polypeptide subunits, shown in light green and dark green.

b.

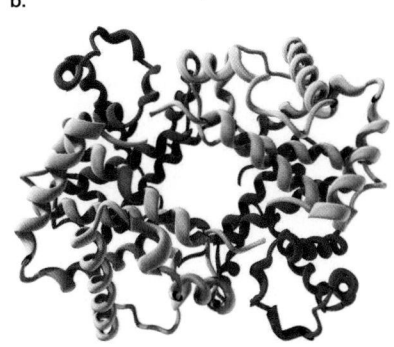

Hemoglobin is made up of four subunits: two copies of the polypeptide depicted in magenta and two copies of the polypeptide depicted in blue.

FIGURE 4.10 Quaternary structure

Some proteins are made up of more than one polypeptide chain, or subunit. All of the subunits as a group are the quaternary structure. (a) This HIV protein is made from the combination of two identical polypeptide subunits. (b) The protein hemoglobin is made from two pairs of different subunits, or four subunits in total.

which would not be possible with any one of the subunits alone. As a result, quaternary structures exemplify the idea that "the sum is greater than its parts." The four levels of protein structure are summarized in **FIGURE 4.11**.

✓ Concept Check

7. **Identify** two types of secondary structure in proteins.

8. **Describe** what holds the secondary structures of a protein in place.

9. **Describe** the tertiary structure of a protein.

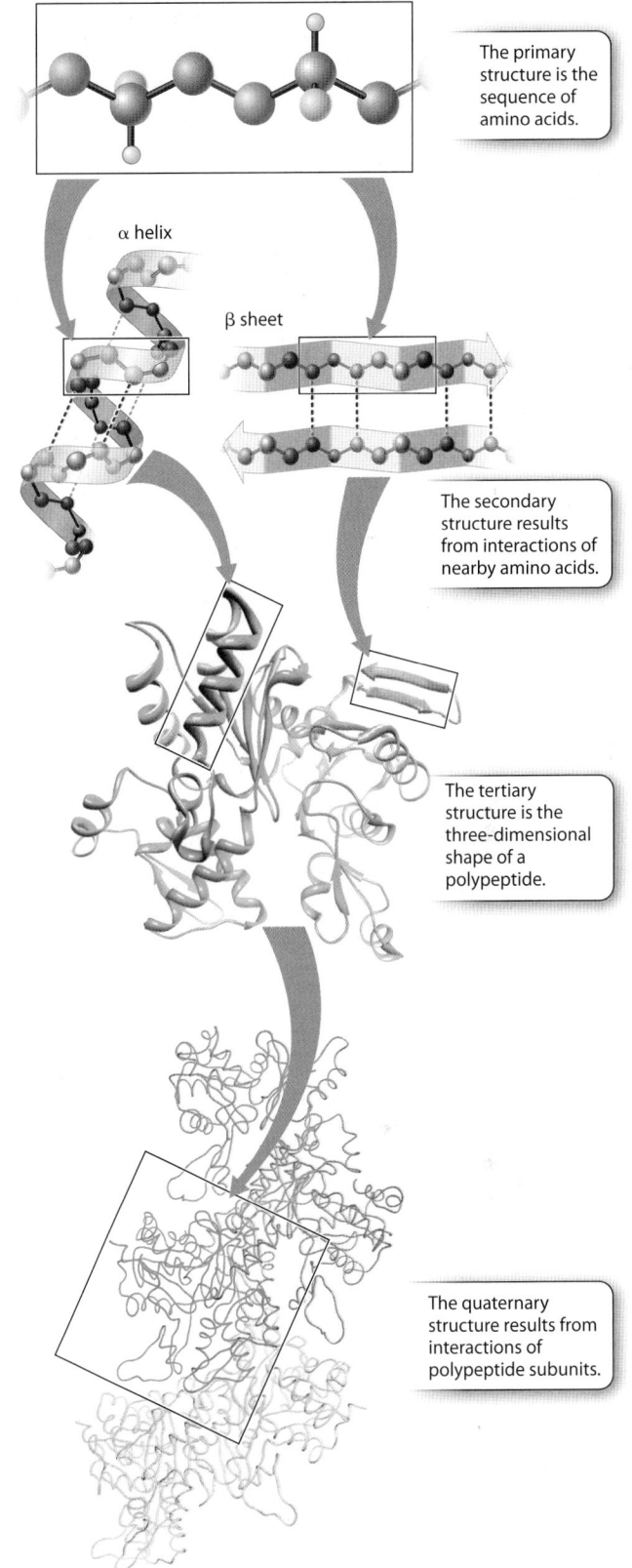

The primary structure is the sequence of amino acids.

α helix

β sheet

The secondary structure results from interactions of nearby amino acids.

The tertiary structure is the three-dimensional shape of a polypeptide.

The quaternary structure results from interactions of polypeptide subunits.

FIGURE 4.11 Levels of protein structure

The folding of a polypeptide or protein is dependent on its primary structure, which ultimately governs its shape and function, and gives rise to the protein's secondary, tertiary, and—if present—quaternary structures.

Practicing Science 4.1

What is the relationship between protein structure and function?

Background Proteins come in a wide variety of shapes and carry out many different functions. From this observation, scientists hypothesized that the shape of a protein and its function are closely connected.

Experiment Christian B. Anfinsen and his colleagues at the National Institutes of Health conducted experiments to examine the relationship between the structure and function of proteins. They focused on a protein called ribonuclease A, a small protein of 124 amino acids. Anfinsen and colleagues measured the activity of ribonuclease A after treating it with various combinations of two chemicals, 2-mercaptoethanol (ME) and urea. In ribonuclease A, as in many other proteins, the sulfhydryl groups (—SH) of pairs of cysteines form a covalent disulfide bond (—S—S—) that links the cysteine side chains. These disulfide bonds break in the presence of ME. Urea forms stronger hydrogen bonds than do water molecules, and for this reason urea disrupts the hydrogen bonds that help stabilize a polypeptide chain.

Results When the researchers treated the purified ribonuclease A with a mixture of urea and ME, the enzyme completely lost activity, shown in part a of the figure. When the ME was removed and the protein was treated with urea only, the protein was misfolded and also inactive, shown in part b. However, when the urea was removed and only a trace of ME was retained, the protein regained activity, shown in part c.

Interpretation Treating the protein with urea and ME results in the loss of all disulfide bonds and complete unfolding of the protein. As a result, the protein loses its activity. When the ME is removed but urea is retained, the protein folds in such a way that the disulfide bonds form incorrectly between the cysteine amino acids, shown in part b of the figure. The improper disulfide bonds result in loss of activity. When the urea is removed and only a trace of ME is retained, the trace ME allows the incorrect disulfide bonds to break and re-form with the proper partner, so that the protein folds into the correct structure, regaining its activity, shown in part c of the figure.

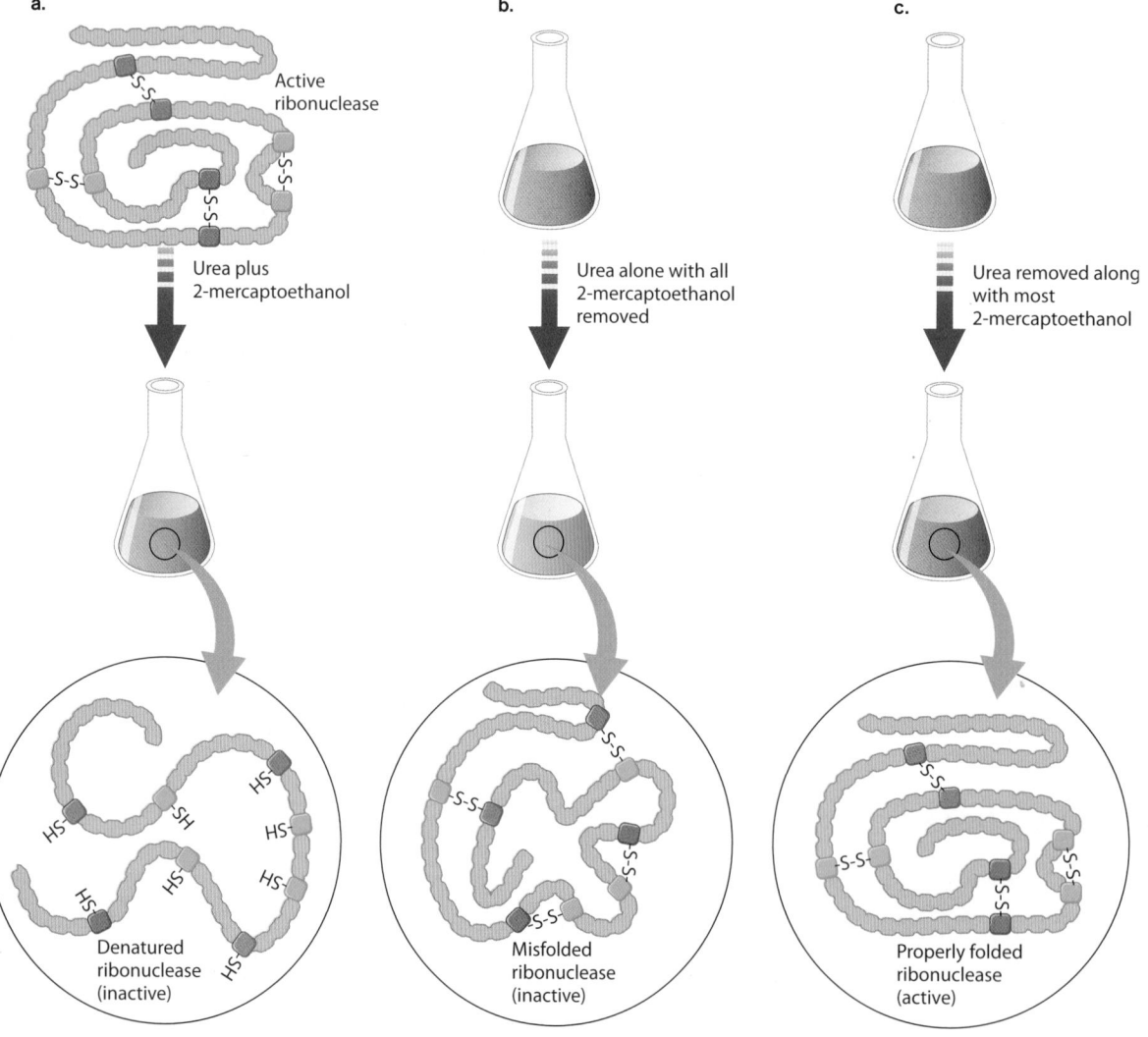

Conclusion These findings support the hypothesis that the shape of a protein is important for the function of the protein.

Follow-up work Many diseases, including Alzheimer's disease, Parkinson's disease, and mad cow disease (bovine spongiform encephalopathy), result from collections, or aggregates, of improperly folded proteins.

SOURCE
Anfinsen, C. B., Haber, E., Sela, M., and White, F. H., Jr. 1961. "The Kinetics of Formation of Native Ribonuclease during Oxidation of the Reduced Polypeptide Chain." *Proceedings of the National Academy of Sciences U.S.A.* 47:1309–1314.

AP® PRACTICE QUESTION

Polyphenol oxidase is a protein inside the cells of apples. It seeps out when an apple is cut open or bruised. Polyphenol oxidase causes the production of brown pigments when it is in the presence of oxygen and other compounds, so this is what we see when an apple turns brown after it is cut open and exposed to the air for a while. Polyphenol oxidase has an optimum pH of 5.5. However, soaking apples in lemon juice, which is a strong acid with a pH of approximately 2, prevents cut apples from turning brown when exposed to the air. Based on your understanding of the relationship between protein structure and function, **explain** how lemon juice prevents the browning of exposed apples.

Module 4 Summary

PREP FOR THE AP® EXAM

REVISIT THE BIG IDEAS

SYSTEMS INTERACTIONS: Describe the various properties of amino acids. **Explain** how these properties affect amino acid interaction with other amino acids and water and how these interactions affect the final structure and function of proteins.

LG 4.1 Amino acids are the building blocks of proteins.

- Proteins are linear polymers of amino acids that form three-dimensional structures with specific functions. Page 65
- An amino acid consists of an α carbon connected by covalent bonds to an amino group, a carboxyl group, a hydrogen atom, and a side chain or R group. Page 66
- The 20 common amino acids differ in their R groups. Page 66
- Amino acids are categorized by the chemical properties of their R groups—hydrophobic, hydrophilic (polar, acidic, basic), and special. Page 66

LG 4.2 Peptide bonds link successive amino acids to form proteins.

- Amino acids are connected by peptide bonds to form proteins. Page 69

- The formation of a peptide bond is a dehydration reaction where the amino group of an amino acid is covalently bonded to the carboxyl acid group of another amino acid. In the process, a water molecule is released. Page 69
- The primary structure of a protein is its amino acid sequence. This structure determines how a protein folds, which in turn determines how it functions. Page 70
- Proteins have directionality: the beginning of a protein is the amino, or N, terminus and the end is the carboxy, or C, terminus. Page 70

LG 4.3 Proteins are folded into three-dimensional shapes that determine their functions.

- The primary structure of a protein determines the other three structural properties of a peptide or protein. Page 70
- The secondary structure of a protein results from hydrogen bonds between nearby amino acids. Examples include the α helix and β sheet. Page 70
- Tertiary structure is the three-dimensional shape of a peptide. Page 71
- Quaternary structure is the structure that forms when two or more polypeptide subunits interact to form a protein. Page 72
- Emergent properties that are not present in their individual units arise in quaternary proteins. Page 72

Key Terms

Peptide bond
Peptide
Polypeptide

Primary structure
Secondary structure
Alpha helix (α helix)

Beta sheet (β sheet)
Tertiary structure
Quaternary structure

Review Questions

1. Which is not a component of an amino acid?
   (A) α carbon
   (B) Amino group
   (C) Phosphate group
   (D) R group

2. The fully folded structure of a functional protein composed of a single polypeptide chain is referred to as the
   (A) primary structure.
   (B) secondary structure.
   (C) tertiary structure.
   (D) quaternary structure.

3. The bond between two amino acids in the primary structure of a protein is referred to as a(n)
   (A) hydrogen bond.
   (B) peptide bond.
   (C) ionic bond.
   (D) hydrophilic bond.

4. Secondary structures are stabilized by which type of interaction?
   (A) Hydrophobic clustering
   (B) Disulfide bonds
   (C) Ionic bonds
   (D) Hydrogen bonds

5. Which statement is true regarding a basic amino acid?
   (A) The hydrophilic R group of a basic amino acid will be located in the interior of a protein.
   (B) The R group of a basic amino acid is positively charged.
   (C) The R group of a basic amino acid would only be able to form covalent bonds with other molecules.
   (D) A basic amino acid is hydrophobic.

6. The bond that forms between amino acids
   (A) connects amino groups and carboxyl groups of adjacent amino acids.
   (B) connects hydrogen atoms of adjacent amino acids.
   (C) connects amino groups and R groups of adjacent amino acids.
   (D) connects water molecules and carboxyl groups of adjacent amino acids.

Module 4
AP® Practice Questions

Section 1: Multiple-Choice Questions
Choose the best answer for questions 1–5.

1. The primary structure of a protein is determined by the
   (A) R group properties.
   (B) number of peptide bonds present.
   (C) order of amino acids.
   (D) number of attached functional groups.

2. Looking at Figure 4.3 on page 67, what characteristic do all hydrophobic amino acids share?
   (A) They only contain carbon.
   (B) They have nonpolar R groups.
   (C) They readily interact with water.
   (D) They are acidic.

3. Which statement is true of both proteins and carbohydrates?
   (A) Both are formed by dehydration synthesis reactions.
   (B) Both are relatively short in length.
   (C) Both contain the C:H:O ratio of 1:2:1.
   (D) Both types of macromolecules are nonpolar and hydrophobic.

4. Lys, Gln, and Asp are the three-letter abbreviations for the amino acids lysine, glutamine, and aspartic acid. How many different peptide combinations could be made from these three amino acids? (For example, Lys–Gln–Asp is one peptide combination.)
   (A) 6
   (B) 9
   (C) 27
   (D) 36

5. Which protein structure introduces alpha helices and beta sheets?
   (A) Primary structure
   (B) Secondary structure
   (C) Tertiary structure
   (D) Quaternary structure

Section 2: Free-Response Question

Write your answer to each part clearly. Support your answers with relevant information and examples. Where calculations are required, show your work.

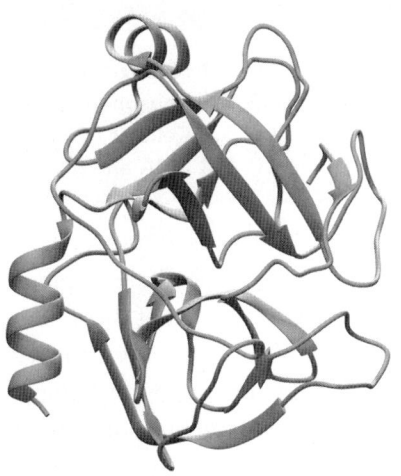

The amino acid structure at the top in the figure is serine in its ionized form. Serine is one of nine hydrophilic and polar amino acids. The structure beneath it is trypsin, a serine-containing protein that is commonly found in the digestive system.

(a) For each of the following, **identify** the atoms and their location relative to the central carbon atom that makes up the designated groups of serine.

1. Amino

2. Carboxyl

3. R group

(b) **Describe** how one serine molecule is capable of joining to another serine molecule. What type of bond would form to link these two molecules?

(c) **Identify** the type of protein structure trypsin displays. **Describe** the bonds involved in producing the conformation seen in trypsin.

(d) **Describe** three different environmental factors that can affect noncovalent interactions holding proteins together and how these factors cause changes in structure.

Module 5
Nucleic Acids

LEARNING GOALS ▶LG 5.1 Nucleotides are the building blocks of nucleic acids.
▶LG 5.2 Phosphodiester bonds join nucleotides to form nucleic acids.
▶LG 5.3 Cellular DNA takes the form of a double helix.
▶LG 5.4 DNA and RNA have similarities and differences.

In the last few modules, we introduced three of the four organic molecules—carbohydrates, lipids, and proteins. We have seen that the structures of these biological macromolecules are closely related to their functions. Proteins, discussed in Module 4, are a good example. Composed of long, linear strings of 20 different kinds of amino acids in various combinations, each protein folds into a specific three-dimensional shape due to chemical interactions between the amino acids along the chain. The three-dimensional structure of the protein determines its functional properties and enables the protein to carry out its job in the cell.

Another notable example of the close relationship between structure and function can be seen in the molecule deoxyribonucleic acid (DNA), which we briefly introduced in Module 0. DNA molecules from all cells and organisms have a very similar three-dimensional structure, reflecting their shared ancestry. This structure allows it to carry out its two major functions, which we briefly introduced in Module 0. First, DNA stores genetic information. This information is encoded in the sequence of subunits along its length. Some of the information in DNA encodes proteins that provide structure and do much of the work of the cell. Second, DNA transmits genetic information to other molecules and from one generation to the next.

In this module, we will focus on DNA. As we did with carbohydrates and proteins, we will begin by discussing its subunits. Then we will examine how these subunits are joined to form nucleic acids, and how information is stored in DNA. We will then discuss the structure of DNA and how the structure gives clues to its functions. Finally, we will end with a comparison of DNA and its close molecular relative—ribonucleic acid, or RNA.

PREP FOR THE AP® EXAM

FOCUS ON THE BIG IDEAS

INFORMATION STORAGE AND TRANSMISSION: Look for the properties of nucleotides and nucleic acids that allow them to store genetic information and to pass it accurately from cell to cell and parent to offspring.

5.1 Nucleotides are the building blocks of nucleic acids

Just as proteins are polymers made up of amino acids and carbohydrates are built from simple sugars, nucleic acids such as DNA and RNA are polymers of nucleotides. In this section, we will examine the structure of nucleotides, the building blocks of nucleic acids.

Nucleotides consist of three basic components: a *base,* a 5-carbon sugar, and a phosphate group. These three components are shown in **FIGURE 5.1**. Each component plays an important role in the overall structure of DNA. Let's consider each one in turn.

The first component is a **nitrogenous base,** which is a cyclic molecule that contains nitrogen, carbon, hydrogen, and oxygen. In DNA, there are four different bases, which are shown in **FIGURE 5.2**. Two of the bases are double-ring structures known as **purines;** these are the bases

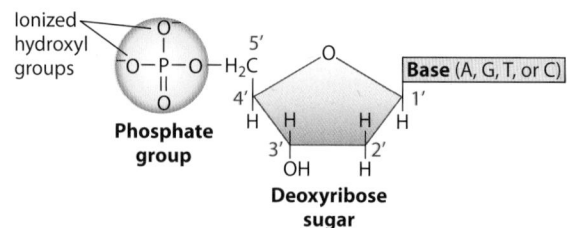

FIGURE 5.1 DNA nucleotide structure

Nucleotides consist of a nitrogenous base, a 5-carbon sugar, and a phosphate group. This figure uses the convention of not labeling the carbon atoms in the ring structure. They are understood to be at the intersection where two lines meet.

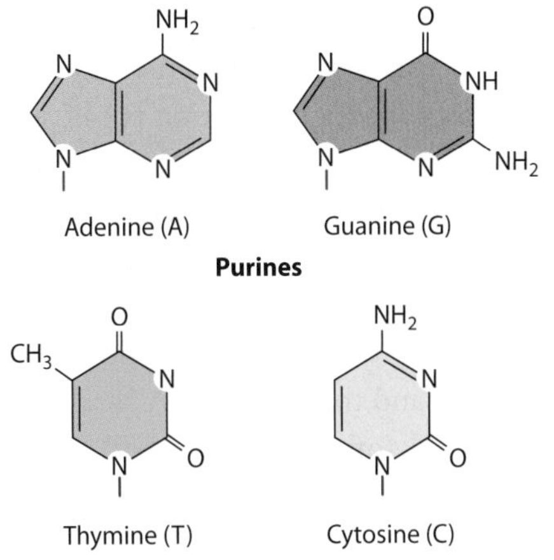

FIGURE 5.2 Bases found in DNA

The bases commonly found in DNA include the purines adenine (A) and guanine (G), and the pyrimidines thymine (T) and cytosine (C). Purines have two rings, and pyrimidines have one ring. In nucleic acids, each base is attached to the sugar by the bond indicated in red.

adenine (A) and guanine (G), and they are shown across the top of the figure. The other two bases are single-ring structures known as **pyrimidines;** these are the bases **thymine (T)** and **cytosine (C),** shown across the bottom of the figure. Just as the order of amino acids provides the information carried in proteins, so, too, does the sequence of these bases determine the information in a DNA molecule.

Attached to each base is a 5-carbon sugar. In the nucleotide illustrated in Figure 5.1, the 5-carbon sugar is indicated by the pentagon, in which four of the five points represent the position of a carbon atom. The sugar in DNA is deoxyribose. By convention, the carbons in the sugar are numbered with primes—$1'$, $2'$, $3'$, and so on—to distinguish them from carbons in the base—1, 2, 3, and so

on. In deoxyribose, the $2'$ carbon has a H atom, and the $3'$ carbon has a hydroxyl group.

The third component of a nucleotide is a phosphate group, shown in Figure 5.1. A phosphate group consists of a central phosphorus (P) atom covalently bound to four oxygen atoms. Recall from Module 3 that a phosphate group is a functional group with the properties of being polar and negatively charged. Note in Figure 5.1 that the phosphate group is attached to the $5'$ carbon and it has negative charges on two of its oxygen atoms. These charges are present because at cellular pH (around 7), the free hydroxyl groups attached to the phosphorus atom are ionized by the loss of a proton and, therefore, are negatively charged. These negative charges make DNA a mild acid, which you will recall from Module 2 is a molecule that tends to lose protons to the aqueous environment.

AP® EXAM TIP

Make sure not to confuse amino acids and nucleic acids. Remember that the subunits of nucleic acids are nucleotides, the subunits of proteins are amino acids, and the subunits of carbohydrates are simple sugars.

✓ Concept Check

1. **Identify** the purines and pyrimidines in DNA.
2. With respect to chemical structure, **describe** the difference between a purine and a pyrimidine.
3. **Identify** the sugar in DNA.

5.2 Phosphodiester bonds join nucleotides to form nucleic acids

DNA is usually a very large molecule and each strand of DNA consists of many nucleotides linked one to the next. We have seen that monosaccharides are joined by glycosidic bonds to make disaccharides and polysaccharides. Amino acids are joined by peptide bonds to make proteins. Nucleotides are joined by phosphodiester bonds, which we discuss in this section.

The chemical linkages between nucleotides in DNA are shown in **FIGURE 5.3** on page 80. As in earlier figures, the

carbon atoms in the sugar deoxyribose are not written but are present at each of the points of the pentagon. The characteristic covalent bond that connects one nucleotide to the next is indicated by the vertical lines that connect the $3'$ carbon of one nucleotide to the $5'$ carbon of the next nucleotide in line through the $5'$-phosphate group. This C—O—P—O—C linkage, consisting of a series of covalent bonds, is known collectively as a phosphodiester bond. In DNA, it is a relatively stable bond that can withstand stresses such as heat and substantial changes in pH that would break weaker bonds.

The succession of phosphodiester bonds traces the backbone of the DNA strand. In other words, the 5-carbon

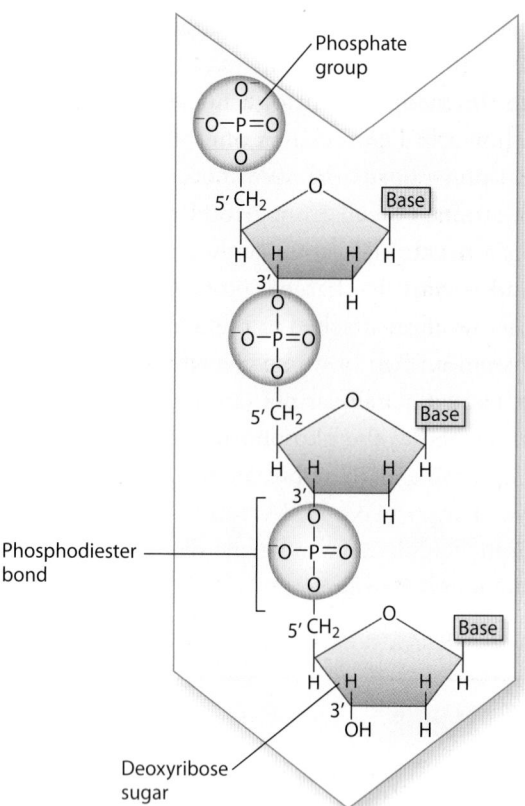

FIGURE 5.3 Phosphodiester bonds

Phosphodiester bonds link successive nucleotides, forming the backbone of the DNA strand.

sugars and phosphate groups form the backbone of the molecule, with each sugar being linked to the phosphate group of the neighboring nucleotide by phosphodiester bonds. Each strand of DNA consists of an enormous number of nucleotides linked one to the next.

The phosphodiester linkages in a DNA strand give it **directionality,** sometimes referred to as **polarity,** which means that one end differs from the other. In **FIGURE 5.4,** the nucleotide at the top has a free 5′ phosphate, and is known as the **5′ end** of the molecule. The nucleotide at the bottom has a free 3′ hydroxyl and is known as the **3′ end.** The DNA strand in Figure 5.4 has the sequence of bases AGCT from top to bottom, but because of strand directionality we need to specify which end is which. For this strand of DNA, we could say that the base sequence is 5′–AGCT–3′. When a base sequence is stated without specifying the 5′ end, by convention the end at the left is the 5′ end. Therefore, we can also say the sequence in Figure 5.4 is AGCT, which means 5′–AGCT–3′.

Phosphodiester bonds, as we have seen, link successive nucleotides. Therefore, both DNA and RNA can grow, or

polymerize, by the formation of new phosphodiester bonds. These bonds are formed when a nucleotide with three phosphate groups, called a nucleotide triphosphate, joins a growing chain. This process is called DNA synthesis, and it is shown in **FIGURE 5.5.** A nucleotide triphosphate reacts with an existing DNA molecule to extend the molecule by adding the new nucleotide. As you can see in the figure, the incoming nucleotide triphosphate is added to the 3′ OH group of the DNA and therefore DNA synthesis takes place only in the 5′ to 3′ (often written as 5′ → 3′) direction. Only one of the three phosphate groups of the incoming nucleotide triphosphate is used to make the sugar–phosphate

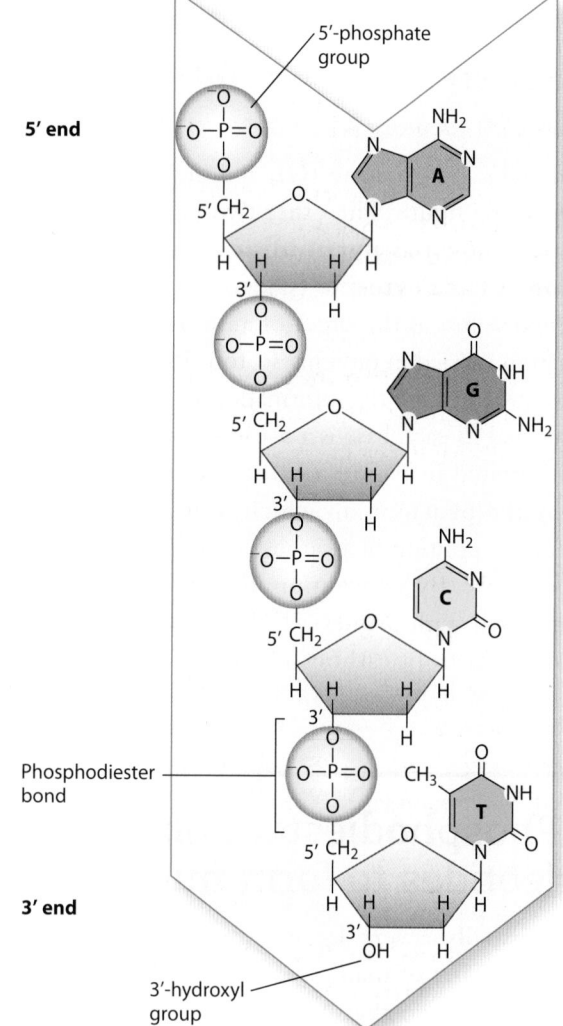

FIGURE 5.4 DNA directionality

The two ends of DNA are different from each other, which reflects its directionality or polarity. The directionality of this DNA molecule is indicated by the light blue arrow shading. One end, called the 5′ end, has a free 5′-phosphate group. The other end, called the 3′ end, has a free 3′-hydroxyl group. The sequence of this DNA strand is 5′-AGCT-3′.

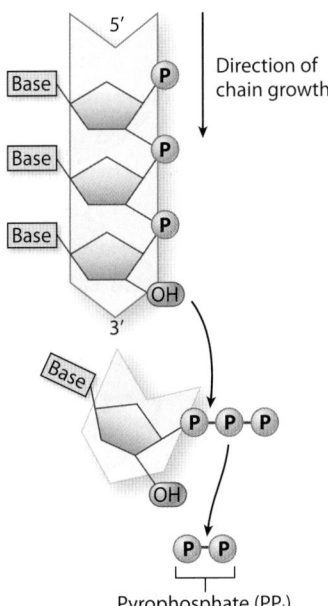

FIGURE 5.5 DNA synthesis

The synthesis of DNA occurs as an incoming nucleotide triphosphate is added to an existing DNA chain. As the incoming nucleotide is added, it is joined by the first phosphate group of its triphosphate. Two phosphates, called pyrophosphate, are released in the reaction. New nucleotides are added to the 3'-hydroxyl group, so chain growth is always in the 5'-to-3' direction.

backbone. The other two are released as pyrophosphate, or two phosphate groups attached to each other and abbreviated PP_i, as shown in the figure.

Nucleic acids are examples of informational molecules—that is, large molecules that carry information in the sequence of nucleotides that make them up. The genetic information in DNA is contained in the sequence, or order, in which successive nucleotides occur along the molecule. The bases attached to the sugar give each nucleotide its chemical identity. Successive nucleotides along a DNA strand can occur in any order, so a long molecule could contain any of an immense number of possible nucleotide sequences. This is one reason why DNA is an efficient carrier of genetic information.

AP® EXAM TIP

You should know what is meant by a 5' → 3' direction, and understand that it applies to nucleic acids, such as DNA and RNA, not other organic molecules, such as proteins and carbohydrates.

✓ Concept Check

4. **Identify** the bond that links successive nucleotides in DNA.

5. **Describe** how genetic information is stored in a molecule of DNA.

5.3 Cellular DNA takes the form of a double helix

Up to now, we have discussed the building blocks of DNA and how these building blocks connect to one another to form a long chain. In 1953, using key data and insights obtained by Rosalind Franklin, James Watson and Francis Crick of Cambridge University announced a description of the three-dimensional structure of DNA. In the cell and even in a laboratory solution, DNA consists of two chains wound around each other to form what is known as a **double helix** because each strand takes the shape of a helix. This discovery marked a turning point in modern molecular biology, as the structure of DNA revealed a great deal about its function. Let's examine this shape in more detail and how it was determined.

Double Helix

With the knowledge of the chemical makeup of the nucleotides and their linkages in a DNA strand, Watson and Crick set out to build a molecular model of the structure of DNA. To do this, they combined three critical pieces of information. The first consisted of results from X-ray crystallography of DNA carried out by Rosalind Franklin and Maurice Wilkins, also at Cambridge University. X-ray crystallography is a technique in which X-rays are passed through crystals of a substance and the pattern that results on the X-ray film provides information about the structure of the substance. Franklin's results, in particular, suggested that DNA had some sort of helical structure with a simple repeating structure all along its length.

The second piece of information consisted of results of experiments carried out by Erwin Chargaff, a biochemist at Columbia University. Chargaff had shown that DNA from many organisms has a characteristic feature: the number of molecules of the nucleotide adenine (A) always equals the number of molecules of thymine (T), and the number of molecules of guanine (G) always equals the number of molecules of cytosine (C).

The third piece of information came from Jerry Donohue and John Griffith, colleagues of Watson and Crick at Cambridge University. They determined that if the bases were to pair in some way, the most likely way would be that A paired with T and that G paired with C.

An accurate model of DNA had to account for the results of all of these pieces of information. Watson and Crick went to work, using sheet metal cutouts of the bases and wire ties for the sugar–phosphate backbone. After many false starts, they finally found a structure that worked: a double-helical structure with the backbones on the outside, the bases pointing inward, and A paired with T and G paired with C.

The two scientists realized immediately that they had made one of the most important discoveries in all of biology. That day, February 28, 1953, they lunched at the Eagle, a pub across the street from their laboratory, where Crick loudly pronounced, "We have discovered the secret of life." The Eagle is still there in Cambridge, England, and on its wall is a commemorative plaque marking the table where the two ate. Watson and Crick published the structure of DNA in 1953. Knowing the structure of DNA opened the door to understanding how genetic information is stored, faithfully replicated, and able to direct the synthesis of other macromolecules.

The structure of DNA is shown in **FIGURE 5.6**. Figure 5.6a is a space-filling model, in which each atom is represented as a color-coded sphere. The big surprise of the structure is that it consists of two DNA strands, each wrapped around the other in the form of a double helix coiling to the right, with the sugar–phosphate backbones winding around the outside of the molecule and the bases pointing inward. Many of us are familiar with the iconic shape of DNA—it shows up on everything from T-shirts to coffee mugs. The elegant shape of the twisting strands relies on the structure of the nucleotides that make it up.

The individual DNA strands in the double helix are **antiparallel,** which means that they run in opposite directions. That is, the 3′ end of one strand is opposite the 5′ end of the other. In Figure 5.6a, the strand that starts at the bottom left and coils upward begins with the 3′ end and terminates at the top with the 5′ end. Its partner strand begins with its 5′ end at the bottom and terminates with the 3′ end at the top.

Figure 5.6b shows a different depiction of double-stranded DNA, called a ribbon model. In this model, the sugar–phosphate backbones wind around the outside with the bases paired between the strands. The ribbon model of the structure closely resembles a spiral staircase, with

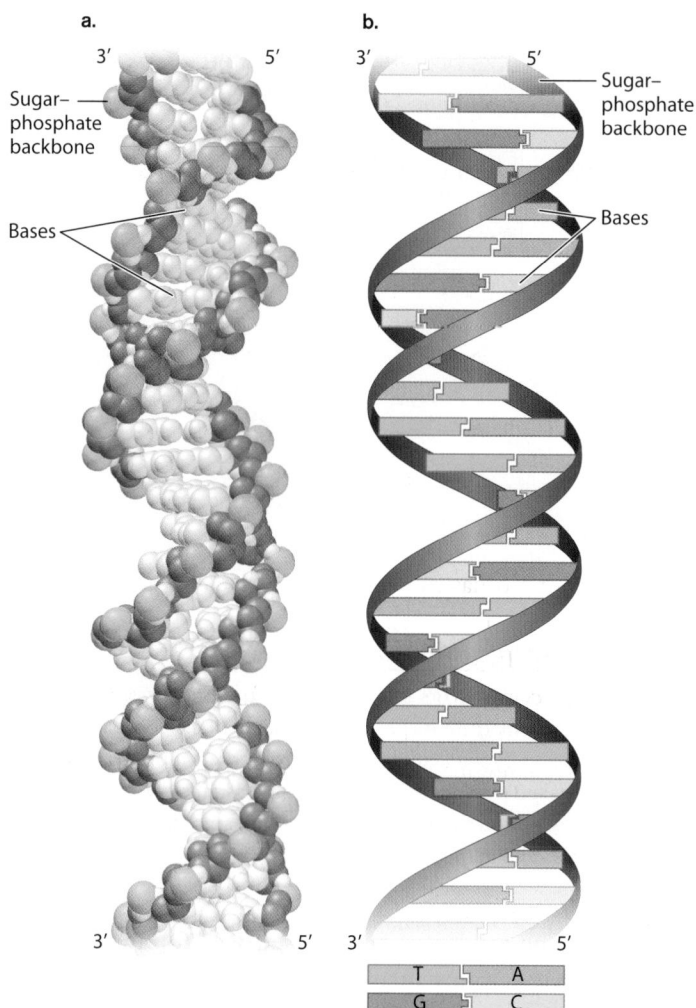

FIGURE 5.6 DNA structure

The DNA double helix is shown here in two ways. (a) In the space-filling model, the atoms are shown as solid spheres. (b) In the ribbon model, the backbones appear as ribbons. In both cases, we see how the sugar–phosphate backbones twist around each other, with the bases pointed inward. The two strands run in opposite directions and are described as antiparallel.

the backbones forming the banisters and the base pairs the steps. If the amount of DNA in a human egg or sperm—or 3 billion base pairs in total—were scaled to the size of a real spiral staircase, it would reach from Earth to the moon.

Base Pairing

As shown in Figure 5.6b, an A in one strand pairs only with a T in the other strand, and G pairs only with C. Each base pair contains a purine and a pyrimidine. The pairing of A with T and of G with C nicely explains Chargaff's observations, now called Chargaff's rule. This precise pairing maintains the structure of the double helix. If the base pairing instead occurred between two purines, the backbones would bulge, and if the pairing occurred between two pyrimidines, the backbones would narrow. In both cases, there would be excessive strain on the covalent bonds in the sugar–phosphate backbone. The pairing of one purine (with one ring) and one pyrimidine (with two rings) preserves the distance between the backbones along the length of the entire molecule.

Because they form specific pairs, the bases A and T are said to be **complementary,** as are the bases G and C. The formation of only A–T and G–C base pairs means that the paired strands in a double-stranded DNA molecule have different base sequences. The strands are paired like this:

5′–ATGC–3′

3′–TACG–5′

where one strand has the base A, the other strand across the way has the base T. Likewise, where one strand has a G, the other has a C. In other words, the paired strands are not identical but complementary. Because of the A–T and G–C base pairing, knowing the base sequence in one strand tells you the base sequence in its partner strand.

Why is it that A pairs only with T, and G only with C? **FIGURE 5.7** illustrates the answer. The specificity of base pairing is brought about by hydrogen bonds that form between A and T, which have two hydrogen bonds, and between G and C, which have three hydrogen bonds. A hydrogen bond in DNA is formed when an electronegative atom (O or N) in one base shares a hydrogen atom (H) with another electronegative atom in the base across the way. Hydrogen bonds are relatively weak bonds, and can be disrupted by high pH or heat. However, added together, millions of these weak bonds along the molecule contribute to the stability of the DNA double helix.

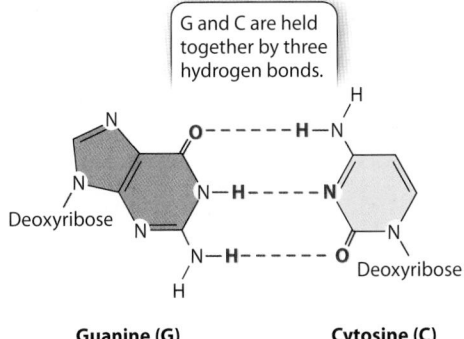

Adenine (A)   **Thymine (T)**

Guanine (G)   **Cytosine (C)**

FIGURE 5.7 Base pairing

In DNA, the base A pairs with T, and G pairs with C. This pairing results from hydrogen bonds between the two bases. An A–T base pair has two hydrogen bonds, and a G–C base pair has three hydrogen bonds.

✓ Concept Check

6. DNA is shaped like a double helix that resembles a spiral staircase. **Describe** which parts of the molecule make up the banisters and which parts make up the steps of the staircase.

7. **Identify** which nucleotides pair with each other in DNA.

5.4 DNA and RNA have similarities and differences

In this module, we have focused on the nucleic acid DNA. Like DNA, RNA is a nucleic acid. As a result, it has many similarities with DNA. At the same time, there are important differences between the two molecules, which influence both their structure and function. In this section, we will take a look at the similarities of DNA and RNA, as well as the differences that distinguish DNA from RNA.

DNA and RNA are both nucleic acids. They are polymers made up of repeating subunits, called nucleotides, which are joined to each other by phosphodiester bonds. Like DNA,

each RNA strand has directionality, or polarity, determined by which end of the chain carries the 5′ phosphate group and which end carries the 3′ hydroxyl group (—OH).

A number of important differences distinguish RNA from DNA, however. First, the sugar in DNA is deoxyribose, while the sugar in RNA is ribose, shown in **FIGURE 5.8**. The sugars differ in that ribose has a hydroxyl (—OH) group on the second carbon (designated the 2′ carbon), whereas deoxyribose has a hydrogen atom at this position (hence, *deoxy*ribose, which means "minus an oxygen"). These groups are highlighted in pink in the figure. Hydroxyl groups are reactive functional groups, so the additional hydroxyl group on ribose in part explains why RNA is a less stable molecule than DNA.

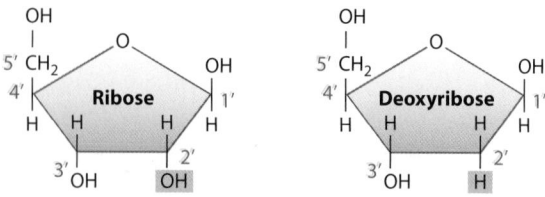

FIGURE 5.8 Ribose and deoxyribose

RNA contains the sugar ribose, and DNA contains the sugar deoxyribose. Ribose has a hydroxyl group (—OH), whereas deoxyribose has a hydrogen atom (—H) at the 2′ carbon.

Like DNA, RNA has four bases, with two purines and two pyrimidines. However, RNA has the base **uracil (U)** instead of the base thymine (T), shown in **FIGURE 5.9**. As a result, DNA contains the bases A, T, G, and C, whereas RNA contains the bases A, U, G, and C. The groups that participate in hydrogen bonding are identical, so that uracil pairs with adenine in RNA (U–A) just as thymine pairs with adenine in DNA (T–A).

Third, RNA molecules are usually much shorter than DNA molecules. A typical RNA molecule consists of a few thousand nucleotides, whereas a typical DNA molecule consists of millions or tens of millions of nucleotides.

Finally, most RNA molecules in the cell are single stranded, whereas DNA molecules, as we saw, are double stranded. Although RNA molecules are single stranded, they can fold into complex, three-dimensional structures containing one or more double-stranded regions where the RNA pairs with itself. Folded RNA structures can have a three-dimensional complexity rivaling that of proteins, and some can even accelerate chemical reactions, just as some proteins do.

The differences in structure between DNA and RNA result in different functions. DNA, as we have seen, stores genetic information and transmits it to offspring. The genetic information in DNA is contained in the sequence of A's, T's, G's, and C's along its length. DNA specifies the sequence of amino acid subunits of which each protein is composed, and this sequence in turn determines the three-dimensional

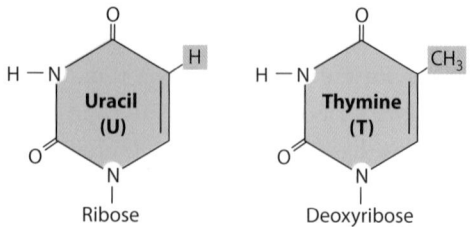

FIGURE 5.9 Uracil and thymine

RNA contains the base uracil (U), whereas DNA contains the base thymine (T). Uracil has a hydrogen atom (—H), and thymine has a methyl group (—CH₃) in the position highlighted in the figure.

FIGURE 5.10 The central dogma of molecular biology

The central dogma defines the flow of information in all organisms from DNA to RNA to protein.

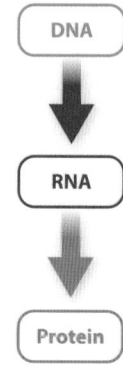

structure of the protein, its chemical properties, and its biological activities, as we discussed in Module 4. To specify the amino acid sequence of proteins, DNA acts through RNA. In other words, one of RNA's many roles in the cell is to act as an intermediate between DNA and protein during protein synthesis. The flow of information in a cell from DNA to RNA to proteins is known as the **central dogma** of molecular biology. **FIGURE 5.10** illustrates the central dogma. **TABLE 5.1** summarizes the differences between DNA and RNA. We will explore the functions of DNA and RNA in detail in Unit 6.

TABLE 5.1 Comparison of DNA and RNA

	DNA	RNA
Nucleic acid	Yes	Yes
Composed of nucleotides	Yes	Yes
Nucleotides linked by phosphodiester bonds	Yes	Yes
Polarity	Yes	Yes
Bases	A, T, G, C	A, U, G, C
Sugar	Deoxyribose	Ribose
Length	Typically long	Typically short
Structure	Typically double stranded	Typically single stranded
Function	Storage and transmission of genetic information	Many roles, including protein synthesis

✓ Concept Check

8. **Identify** three characteristics that DNA and RNA share.

9. **Identify** three characteristics that differ between DNA and RNA.

Module 5 Summary

PREP FOR THE AP® EXAM

REVISIT THE BIG IDEAS

INFORMATION STORAGE AND TRANSMISSION: Describe the properties of nucleotides and nucleic acids that allow them to store genetic information and to pass it accurately from cell to cell and parent to offspring.

LG 5.1 Nucleotides are the building blocks of nucleic acids.

- Nucleotides assemble to form nucleic acids, which store and transmit genetic information. Page 78
- Nucleotides are composed of a nitrogen-containing base, a 5-carbon sugar, and a phosphate group. Page 78
- The four bases of DNA are adenine (A), guanine (G), cytosine (C), and thymine (T). Page 79
- Adenine and guanine are purines, which are bases with a double-ring structure. Page 79
- Cytosine and thymine are pyrimidines, which are bases with a single-ring structure. Page 79
- Nucleotides in DNA incorporate the sugar deoxyribose. Page 79

LG 5.2 Phosphodiester bonds join nucleotides to form nucleic acids.

- Successive nucleotides are linked by phosphodiester bonds to form a linear strand of DNA. Page 79

- DNA strands have polarity, with a 5′-phosphate group at one end and a 3′-hydroxyl group at the other end. Page 79
- Nucleotide triphosphates react with a strand of DNA to make the polymer one nucleotide longer. In the process, the phosphodiester bond of the DNA backbone is formed. Page 80
- Genetic information is stored in the sequence, or order, of the nucleotides in nucleic acids. Page 81

LG 5.3 Cellular DNA takes the form of a double helix.

- DNA forms a double helix where the two strands run in antiparallel directions to one another. Page 82
- In a DNA double helix, A pairs with T, and G pairs with C. Page 83

LG 5.4 DNA and RNA have similarities and differences.

- RNA, like DNA, is a polymer of nucleotides linked by phosphodiester bonds. Page 83
- Unlike DNA, RNA incorporates the sugar ribose instead of deoxyribose and the base uracil instead of thymine. Page 84
- DNA is typically double stranded, while RNA is typically single stranded. Page 84

Key Terms

Nitrogenous base
Purine
Adenine (A)
Guanine (G)
Pyrimidine
Thymine (T)

Cytosine (C)
Directionality
Polarity
5′ end
3′ end
Double helix

Antiparallel
Complementary
Uracil (U)
Central dogma

Review Questions

1. Which is not a component of a nucleotide?
   (A) A 5-carbon sugar
   (B) A phosphate group
   (C) A nitrogen-containing base
   (D) A carboxyl group

2. In a deoxyribonucleotide, which chemical group is found at the 2′ carbon of the sugar component?
   (A) An oxygen atom
   (B) A hydroxyl group
   (C) A hydrogen atom
   (D) One or more phosphate groups

3. The two strands in a double helix of DNA are
   (A) parallel.
   (B) complementary.
   (C) held together by covalent bonds.
   (D) held together by ionic bonds.

4. The linkage between nucleotides is referred to as a(n)
   (A) ionic bond.
   (B) peptide bond.
   (C) phosphodiester bond.
   (D) glycosidic bond.

5. Which represents a standard base pair interaction?
   (A) G–C
   (B) G–T
   (C) C–A
   (D) A–G

6. An organism's genome is analyzed and found to contain 18% thymine. What percentage of that organism's DNA is cytosine?
   (A) 18%
   (B) 32%
   (C) 36%
   (D) 82%

7. Ribose differs from deoxyribose in that a
   (A) ribose has an extra hydroxyl group.
   (B) ribose is missing a hydroxyl group.
   (C) ribose has an extra phosphate group.
   (D) ribose is missing a phosphate group.

8. The base uracil pairs with
   (A) adenine.
   (B) thymine.
   (C) guanine.
   (D) cytosine.

Module 5
AP® Practice Questions

Section 1: Multiple-Choice Questions

Choose the best answer for questions 1–4.

1. Which is true of nucleic acids and proteins?
   (A) The subunits of nucleic acids are nucleotides, and the subunits of proteins are sugars.
   (B) Nucleic acids and proteins are formed when their subunits join together with phosphodiester bonds.
   (C) Dehydration synthesis reactions are necessary to produce nucleic acids, while hydrolysis reactions are necessary to produce proteins.
   (D) Nucleic acids carry genetic information, while proteins provide structure and do much of the work of the cell.

2. Which is an example of a nucleotide?
   (A) Adenine–ribose–a carboxyl group
   (B) Guanine–deoxyribose–a phosphate group
   (C) Uracil–deoxyribose–a phosphate group
   (D) Thymine–ribose–a carboxyl group

3. One strand of DNA has the sequence 5′-GTGCA-3′. What is the 5′ → 3′ sequence of the complementary DNA strand?
   (A) CACGT
   (B) TGCAC
   (C) UGCAC
   (D) CACGU

4. Which is a shared structural trait of DNA and RNA?
   (A) 5′ → 3′ directionality
   (B) A hydroxyl group at the 2′ carbon
   (C) Base pairing of adenine and thymine
   (D) The double helical shape

Section 2: Free-Response Question

Write your answer to each part clearly. Support your answers with relevant information and examples. Where calculations are required, show your work.

A researcher was successful in performing a DNA extraction. A short sequence of this DNA was isolated for further study. The figure shows the basic structure of DNA with four pairs of nucleotides.

(a) **Identify** the location of the 3′ and 5′ ends of each strand of DNA and **describe** why you chose those places.

(b) Write out the order of the bases of the unlabeled DNA strand with 5′ → 3′ directionality. Briefly **describe** why you chose these bases.

(c) **Describe** the process by which this DNA strand could be lengthened.

(d) **Describe** how the figure would look different if it were RNA instead of DNA.

(e) **Explain** the process and bonds involved in the ability of RNA to fold into complex three-dimensional shapes, similar to those of tertiary proteins.

Unit 1

AP® Practice Questions

Section 1: Multiple-Choice Questions

Choose the best answer for questions 1–15.

1. Using Figure 4.3 on page 67 as a reference, which amino acid would likely be found in the same cellular membrane region as the fatty acid tails of a phospholipid?
    (A) Histidine
    (B) Threonine
    (C) Leucine
    (D) Serine

2. A student is designing an experiment to test what happens to the growth of plants when they are watered with a low concentration of a sodium chloride (NaCl) solution and the presence or absence of fertilizer rich in phosphorus. The student hypothesizes that NaCl will decrease the growth of plants, no matter the type of soil present. The student plans to set up separate pots of sunflower seedlings with the treatments shown in the table below.

Seedling number	Treatment
1	Low NaCl + fertilizer
2	Low NaCl + no fertilizer
3	Pure water + fertilizer
4	Pure water + no fertilizer

The seedlings will be watered daily in the student's backyard and growth will be observed and recorded over a period of 3 weeks. Which best describes the control group?
    (A) Placing a pot of sunflower seedlings in a protected environment indoors
    (B) Watering sunflower seedlings with pure water and no fertilizer present
    (C) Using other flower seedlings in addition to sunflowers
    (D) Adding both fertilizer and the NaCl solution to each seedling

3. Which statement explains why water is a good solvent for carbohydrates, proteins, and nucleic acids, but not for lipids?
    (A) Carbohydrates, proteins, and nucleic acids are not very electronegative and will more readily bond with water than lipids.
    (B) Carbohydrates, proteins, and nucleic acids contain weaker, single bonds that water can easily break. Lipids contain stronger, double bonds that make this more difficult.
    (C) Carbohydrates, proteins, and nucleic acids have polar regions capable of forming hydrogen bonds with water. Lipids are not very polar.
    (D) Carbohydrates, proteins, and nucleic acids are much smaller macromolecules than lipids. Their size allows water molecules to bond more easily with their atoms than with those of lipids.

4. The interactions between R groups of amino acids contribute greatly to the overall tertiary structure of a protein. Which type of bond or interaction is commonly found between these R groups?
    (A) van der Waals interactions
    (B) Nonpolar covalent bonds
    (C) Hydrogen bonds
    (D) Polar covalent bonds

5. A triacylglycerol with three unsaturated fatty acid chains is likely to have which characteristic compared to a triacylglycerol with three saturated fatty acid chains?
    (A) Fewer van der Waals interactions between hydrocarbon chains
    (B) A higher melting point and thus a solid structure at room temperature
    (C) Straight, tight hydrocarbon chains
    (D) Increased polarity due to phosphate groups

6. When the body temperature of certain animals becomes too high, they will often sweat to cool down. Which property of water helps to account for the ability of sweat to cool the body?
   (A) The adhesion of water molecules
   (B) The high specific heat of water
   (C) The ability of water to act as a solvent
   (D) The high surface tension of water

7. A small piece of potato is placed into a sucrose solution. A student measured the initial mass to be 10 g. The next day, the student removed the potato from the sucrose solution and measured the final mass as 7 g. What is the percent change in mass of the potato?
   (A) −30%
   (B) +30%
   (C) −43%
   (D) +43%

8. Which best explains the type of bond between monomers and its role in the function of the macromolecule it forms?
   (A) The glycosidic bonds that join amino acids together determine the overall structure of the molecule, which determines how the protein will function.
   (B) The dehydration synthesis reaction that occurs between lipid monomers forms only single bonds, giving lipids their tightly packed structure.
   (C) The covalent bonds in carbohydrates contribute to the folding of complex tertiary carbohydrate structures that lead to their role as energy sources.
   (D) The combination of phosphodiester bonds and hydrogen bonds in nucleic acids allows for a stable structure that is easily unzipped to transfer genetic information.

9. In a phosphorus-limited environment, the production of which macromolecule would be affected the most?
   (A) Carbohydrates
   (B) Nucleic acids
   (C) Proteins
   (D) Lipids

10. The oxygen atom in the carboxyl group of a protein becomes part of a water molecule during the formation of a peptide bond. That water molecule may then go on to take part in the hydrolysis of a polysaccharide. The path this oxygen atom takes is an example of the flow of
    (A) matter.
    (B) information.
    (C) energy.
    (D) gas.

11. Depending on the arrangement of atoms in the ring structure of a monosaccharide, it may have which trait compared to its linear structure?
    (A) Different functional groups
    (B) Different bonds joining monosaccharides
    (C) A different C:H:O ratio
    (D) A different function

Use the following information to answer questions 12 and 13.

The bright coloration of beaks is an important indicator of body condition and overall health in birds. The intensity of the color is heavily based on a type of pigment called carotenoids. Carotenoids can be found in many of the foods that birds consume. Researchers manipulated the diets of male zebra finches with a combination of good- or poor-quality food and the presence or absence of carotenoids. The color of the beak was then measured, with high chroma signifying a bright color. The results are shown in the figure below.

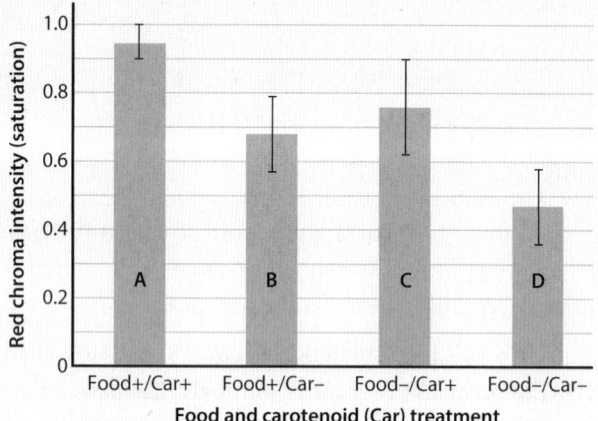

Effect of food and carotenoid intake on beak coloration. Data from https://www.nature.com/articles/srep23546

12. Which could be a null hypothesis for this experiment?

(A) Good-quality food and the presence of carotenoids will result in a bright red beak.

(B) Only birds with diets rich in carotenoids will have bright beaks. Food quality does not matter.

(C) Beak color intensity will not be influenced by food quality or the presence of carotenoids.

(D) A diet of poor quality and low in carotenoids yields dull coloration.

13. Based on the error bars in the graph, which two treatments are significantly different from each other?

(A) A and C   (C) B and C

(B) A and D   (D) B and D

14. The algae population of a lake in the Midwest United States has increased dramatically within the past few years. A group of researchers has hypothesized that this is due to an abundance of phosphorus in the water caused by fertilizers used on nearby lawns. Which is the best way to test this hypothesis?

(A) Set up multiple tanks of water and add different concentrations of the fertilizer rich in phosphorus to the tanks. Observe and measure the algal growth.

(B) Continue observing the algae growth in the lake. Compare its growth to areas around the lake that do not have lawns.

(C) Take water samples from the lake to quantify the exact amount of phosphorus in each area. Use these data to add phosphorus back into the lake and observe the growth.

(D) Remove the algae from the lake and use it to start an algal culture in a lab setting. Place different amounts of algae in clean water and see if it survives.

15. The amino acid glutamine is ionized in a neutral cellular environment of pH 7.4. As the pH becomes more acidic, glutamine becomes a positive ion. What is the relationship between H^+ and ^-OH ions in solution in an environment with a pH of 5.4?

(A) There would be 2 times fewer H^+ ions and 2 times more ^-OH ions in solution.

(B) There would be 10 times fewer H^+ ions and 10 times more ^-OH ions in solution.

(C) There would be 20 times more H^+ ions and 20 times fewer ^-OH ions in solution.

(D) There would be 100 times more H^+ ions and 100 times fewer ^-OH ions in solution.

Section 2: Free-Response Questions

Write your answer to each part clearly. Support your answers with relevant information and examples. Where calculations are required, show your work.

1. Metformin (Met) is a diabetes treatment drug that, in some cases, can reduce the rate of cancer development and mortality. In people with diabetes, blood-glucose levels are elevated. Researchers studied the effects that metformin has on groups of mice that were fed two different diets, a standard chow diet (STD) and a high-fat–sucrose diet (HFS). The STD mice represented normal and healthy individuals, while the HFS mice represented people with diabetes. Four-week-old mice were injected with tumor cells and then divided into four treatment categories: a standard chow diet (STD), a standard chow diet plus metformin (STD + Met), a high-fat–sucrose diet (HFS), and a high-fat–sucrose diet plus metformin. Tumor growth was monitored over a period of 25 days and the results are shown in the graph.

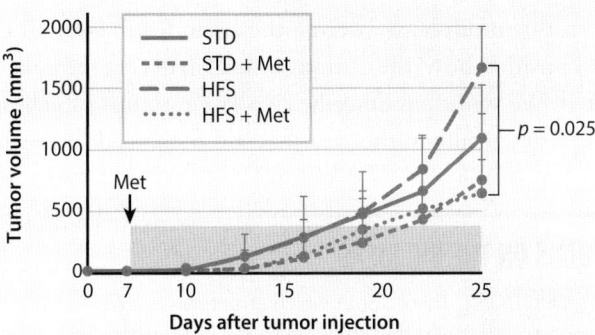

Data from https://www.nature.com/articles/s41598-020-71946-32

(a) Diabetes is tightly linked to diet; however, certain cases can also be attributed to genetics. **Name** the class of macromolecule and a specific macromolecule that would be most helpful in determining a genetic predisposition to diabetes.

(b) **Determine** if there is a statistically significant difference between the HFS and HFS + Met treatments after 25 days of tumor growth. **Justify** your answer.

(c) **Describe** the trends we would see for each treatment if the experiment were extended for another 10 days.

(d) **Describe** a possible hypothesis for this experiment.

2. The chemical structure shown below is cellulose. It is made up of repeating glucose molecules.

(a) **Identify** the type of bond joining the glucose molecules together.

(b) Cellulose is rigid and gives structure to plant stems and cell walls. Starch is also made of repeating glucose molecules, but its function is for energy storage. **Describe** how the chemical structures of these molecules are a major factor in their functional difference within living systems.

(c) **Describe** the sequence of chemical events which leads to building the polymer cellulose from glucose monomers.

(d) **Explain** the difference between the sequence of chemical events to build cellulose with the chemical events necessary to break down cellulose. Use arrows to **represent** where bonds would be broken within the cellulose molecule.

Unit 2
Cell Structure and Function

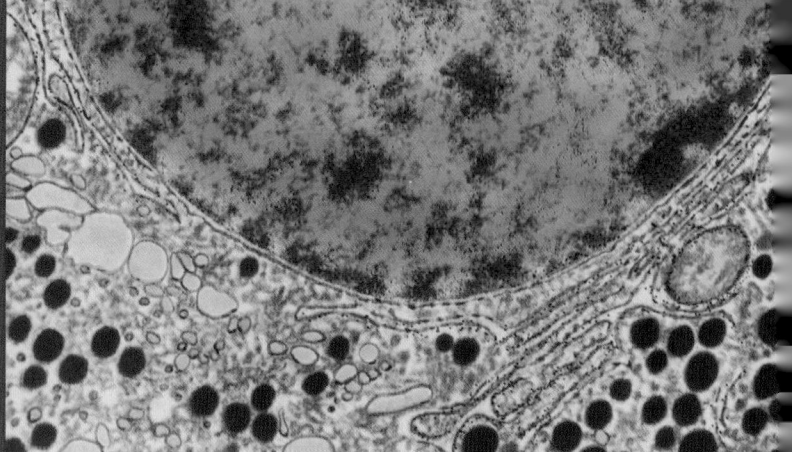

Cell producing growth hormone by Quest/Science Source

Module 6
An Introduction to the Cell

LEARNING GOALS ▶ **LG 6.1** The cell is the fundamental unit of life.
▶ **LG 6.2** All cells maintain homeostasis, store and transmit information, and transfer energy.
▶ **LG 6.3** The structure and function of cells are closely related.
▶ **LG 6.4** Prokaryotes and eukaryotes differ in internal organization.

❝ **With the discovery of the cell,** biologists found their atom." So stated François Jacob, the French biologist who shared the Nobel Prize in Physiology or Medicine in 1965. Just as the atom is the smallest, most basic unit of matter, the cell is the smallest, most basic unit of living organisms. All organisms, from single-celled algae to multicellular organisms like humans, are made up of cells. Therefore, essential properties of life, including growth, reproduction, movement, and evolution, must be understood in terms of cell structure and function.

In this module, we define the basic features of all cells and consider how the function and structure of cells are related. We will then describe two large groups of cells that differ dramatically in the way they are organized internally.

PREP FOR THE AP® EXAM

FOCUS ON THE BIG IDEAS

EVOLUTION: Think about which structures are so fundamental to life that they are found in all organisms.

6.1 The cell is the fundamental unit of life

The cell is the simplest structure that exists as an independent unit of life. Every living organism is either a single cell— unicellular—or a group of a few to many cells— multicellular. **FIGURE 6.1** shows examples of both unicellular and multicellular organisms. Most bacteria, yeasts, and the tiny algae that float in oceans and ponds spend their lives as single cells. In contrast, plants and animals contain billions to trillions of cells that function in a coordinated fashion.

FIGURE 6.2 shows examples of different types of cells. Most cells are tiny, with dimensions that are too small for the naked eye to see. The cells that make up the layers of

your skin, shown in Figure 6.2a, average about 30 micrometers (μm), or 0.03 mm, in diameter, which means that nearly 20 would fit in a row across the period at the end of this sentence. Many bacteria are less than a micrometer long. Other cells can be quite large. Some nerve cells in humans, like those pictured in Figure 6.2b, extend slender projections that communicate with other nerve cells. Still other nerve cells, like the ones that run from your spinal cord to your foot, have projections that extend as long as a meter. The cannonball-size egg of an ostrich in Figure 6.2c is, remarkably, a single giant cell.

Cells were first seen in the mid-1600s, when the English scientist Robert Hooke built a microscope that he used to observe thin sections of dried cork tissue derived from plants.

a.
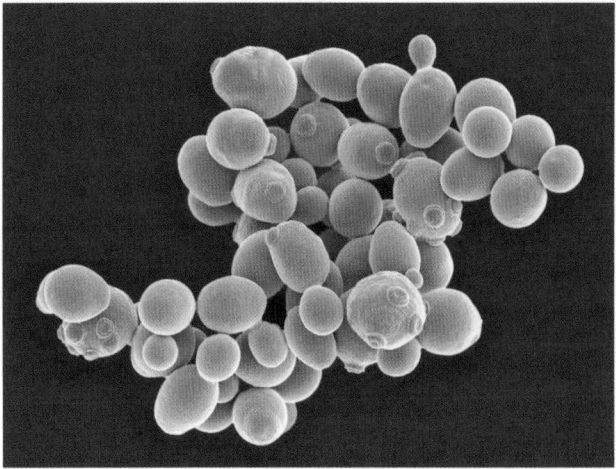

b.

FIGURE 6.1 Unicellular and multicellular organisms

All living organisms are made up of cells. (a) Brewer's yeast (*Saccharomyces cerevisiae*) are single-celled organisms. (b) These cheetahs (*Acinonyx jubatus*) in Kenya are multicelled organisms. Photos: (a) Steve Gschmeissner/Science Source; (b) Sven-Olof Lindblad/Science Source

In these sections, Hooke observed arrays of small cavities and named them cells. **FIGURE 6.3**, on page 94, shows what Hooke observed. Figure 6.3a is a drawing by Hooke, and Figure 6.3b shows a modern micrograph of the same tissue that Hooke studied. These and other observations contributed to the idea that cells are the fundamental unit of life. Later in the seventeenth century, the Dutch microbiologist Anton van Leeuwenhoek greatly improved the magnifying power of microscope lenses, enabling him to see and describe many different unicellular organisms, including bacteria, protists, and algae. Today, modern microscopy provides unprecedented detail of the structure of internal cell structures, and a deep understanding of the life of cells.

Cells differ in size and shape, but they share many features. This similarity in the microscopic organization of all living organisms led to the one of the pillars of modern biology: the *cell theory*, which was developed in the middle of the nineteenth century. Based on the work and ideas of Matthias Schleiden, Theodor Schwann, Rudolf Virchow, and others, the **cell theory** makes three interrelated observations:

- All organisms are made up of cells.
- The cell is the fundamental unit of life.
- Cells come from preexisting cells.

We have already explored the first idea, that all organisms are made up

a.
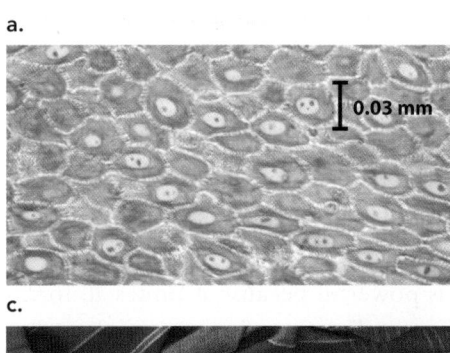
0.03 mm

c.

b.
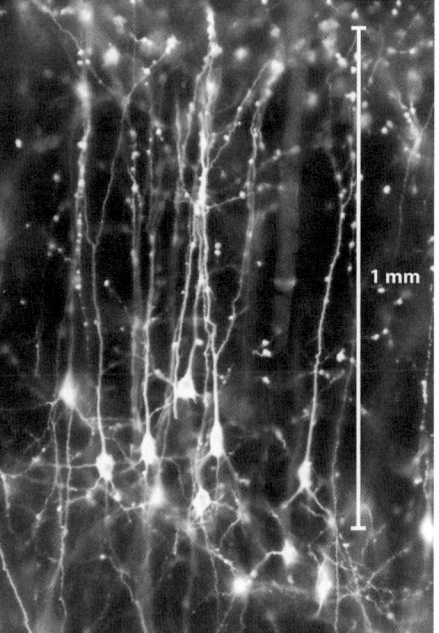
1 mm

FIGURE 6.2 Cell diversity

Cells vary greatly in size and shape. (a) Skin cells average about 30 micrometers (μm), or 0.03 mm, in diameter, and are relatively flat. (b) Nerve cells, like these from the cerebral cortex, can be short or long, and have projections that communicate with other nerve cells. (c) An ostrich egg weighs as much as 5 pounds but is still a single oval-shaped cell. Photos: (a) Biophoto Associates/Science Source; (b) Dr. Jonathan Clarke/Wellcome Images; (c) netsuthep/Shutterstock

a.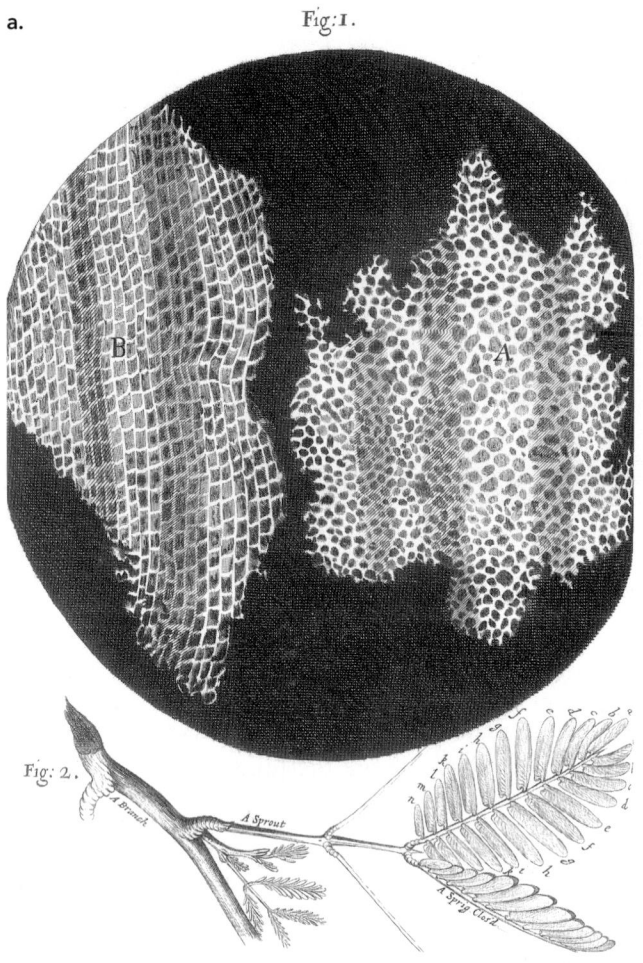

b.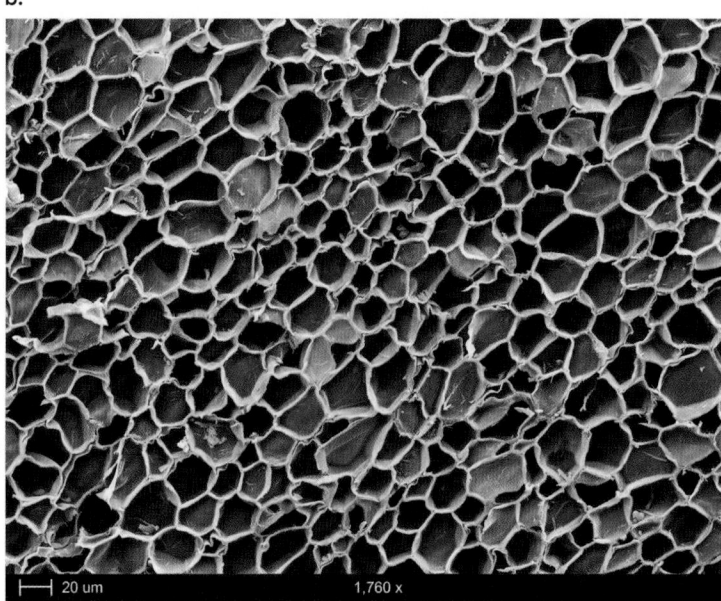

FIGURE 6.3 The first observation of cells

Robert Hooke used a simple microscope to observe small chambers in a sample of cork tissue that he described as "cells." (a) This page from Hooke's notebook shows a drawing of his observations. (b) This modern micrograph shows the same type of tissue that Hooke observed. Photos: (a) Science Museum/SSPL/The Image Works; (b) Ted Kinsman/Science Source

of cells. Some organisms are unicellular, spending their lives as a single cell. Other organisms are multicellular, made up of hundreds, thousands, millions, or even trillions of cells. In multicellular organisms, cells are specialized to carry out different functions. For example, in humans, skin cells provide protection from the outside world; skeletal muscle cells help you move about; liver cells process and help to break down the food you eat; and nerve cells process information and help to control and coordinate the functions of your various organs.

The second observation states that the cell is the fundamental unit of life. When we say the "fundamental unit" of life, we mean that the cell is the simplest entity we can define as living. Life is difficult to define, but it has certain features, such as the ability to reproduce, respond to the environment, harness energy from the environment, and evolve over time. Cells also have these features. Anything smaller or simpler, like a membrane or a molecule, does not have all of these features, so it is not alive. In other words, the cell is the smallest, most basic unit of life, and there is no life as we know it without cells.

The third observation notes that cells arise from preexisting cells through the process of cell division. When a single parent cell divides, it produces daughter cells. We will address the question of where the first cell came from in Unit 7.

The cell theory is powerful because it unites all forms of life. In spite of the great diversity of organisms, the cell is a common entity that is shared among the millions of different species on Earth. The cell theory also provides a way for us to study and understand life. In other words, we can study life's basic processes, such as growth and reproduction, at the level of the cell. If we understand how these processes work at the cellular level, we can extend what we learn to all forms of life.

✓ Concept Check

1. **Identify** the three claims of the cell theory.
2. **Describe** what is meant by the fundamental unit of life.

6.2 All cells maintain homeostasis, store and transmit information, and transfer energy

Cells such as bacteria, yeasts, skin cells, nerve cells, and eggs might seem very different from one another, but all are organized along broadly similar lines. They all have a discrete boundary that separates the interior of the cell from its external environment and maintains the inside in a way that is compatible with life; they all contain information in molecular form that they can use and pass on to other cells; and they all have the ability to harness energy and materials from the environment to carry out their functions. In this section, we describe these three essential features of all cells.

The Cell Membrane and Homeostasis

The first feature of all cells is a cell membrane that separates the living material within the cell from the nonliving environment around it. The cell membrane was introduced briefly in Module 3. FIGURE 6.4 shows a photograph of a hamster cell taken with an electron microscope, with a

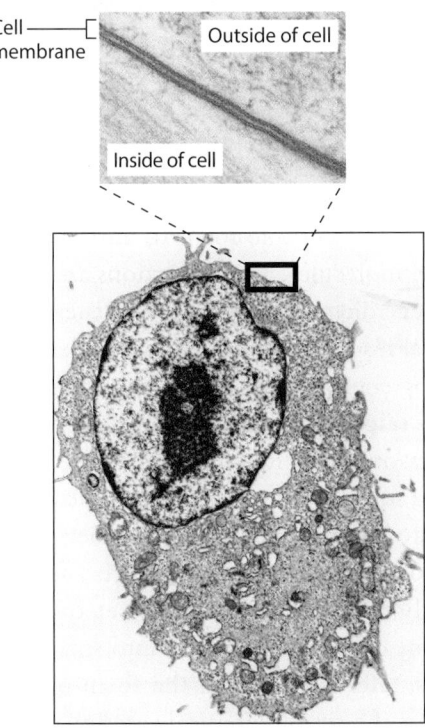

FIGURE 6.4 The cell membrane

The cell membrane, also known as the plasma membrane, surrounds all cells and controls the exchange of material with the environment. Photos: (top) Don W. Fawcett/Science Source; (bottom) Dr. Gopal Murti/SPL/Science Source

close-up view of the cell membrane. This boundary between inside and outside does not mean that cells are closed systems independent of the environment. On the contrary, there is an active and dynamic interplay between cells and their surroundings that plays out at the cell membrane. All cells must continually acquire and exchange ions and the building blocks required to build macromolecules from their surroundings. Cells also release waste products into the environment. As we will see in Module 10, the cell membrane controls the movement of materials into and out of the cell.

The environment outside the cell is constantly changing. In contrast, the internal environment of a cell is much more stable. In fact, the inside of a cell operates within a narrow window of conditions, such as a particular pH range and salt concentration. These conditions are important for chemical reactions, protein folding, and other cellular functions, which occur efficiently only within a narrow range of conditions. The cell membrane actively maintains the conditions inside a cell so they remain compatible with life.

The active maintenance of stable internal conditions is called **homeostasis.** Homeostasis is a critical feature of cells and of life itself. Homeostasis is important not just for individual cells but also for the body as a whole in multicellular organisms. Many physiological parameters are maintained in a narrow range, including temperature, heart rate, blood pressure, blood sugar, blood pH, and water content. Some scientists even extend the concept of homeostasis to ecosystems, where a stable equilibrium is achieved through many independent but interacting elements.

Although the conditions within a cell or organism are relatively stable, homeostasis is not a passive process and the internal environment is not stagnant. Instead, homeostasis is an active process. That is, a cell or organism uses energy to maintain homeostasis. Furthermore, homeostasis requires a continual interplay between the inside and outside of the cell. Maintaining steady and stable conditions takes work in the face of changing environmental conditions.

Information

A second essential feature of a cell is its ability to store, use, and transmit information. All cells have a stable archive or library of information that encodes and helps determine their physical features. Just as the construction and maintenance of a house depends upon a blueprint or plan that defines the walls, plumbing, and electrical wiring, organisms have an accessible and reliable archive of information that helps determine their structure and function. This information is not only stored and used, but also passed on to daughter cells. To reproduce, cells must be able to copy their archive of information rapidly and accurately. In all organisms, the information archive is deoxyribonucleic acid, or DNA, described in Unit 1.

As illustrated in **FIGURE 6.5**, DNA is a double-stranded helix, with each strand made up of varying sequences of four different kinds of molecules connected end to end. The arrangement of these molecular subunits, or nucleotides, makes DNA special; in essence, these nucleotides provide a four-letter alphabet that encodes cellular information.

The information stored in DNA directs the formation of proteins, the key structural and functional molecules that do the work of the cell. Virtually every aspect of the cell's existence—its internal architecture, its shape, its ability to move, and its various chemical reactions—depends on proteins. To make proteins, the information in DNA is first used to guide the synthesis of a closely related molecule called ribonucleic acid, or RNA. The information in RNA is then used to direct the synthesis of a protein. Protein synthesis requires many components, but a key one is the *ribosome*. A **ribosome** is a complex structure that is the site where the protein is assembled. In essence, the ribosome is the place that translates the language of RNA, which is a sequence of nucleotides, into the language of proteins, which is a sequence of amino acids. Ribosomes consist of a small subunit and a large subunit, each composed of 1 to

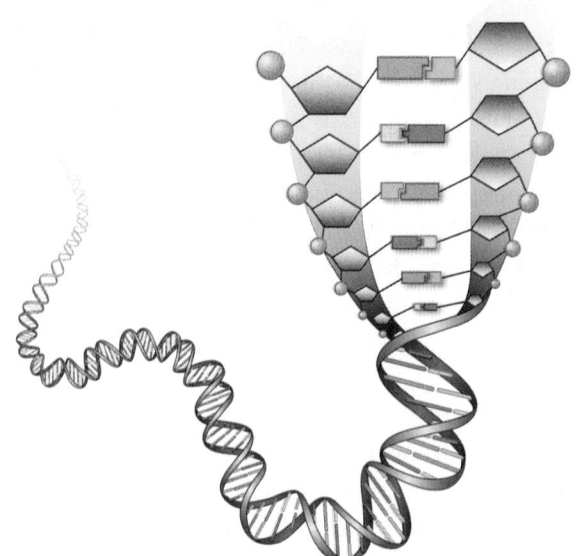

FIGURE 6.5 DNA structure

A molecule of DNA is a double helix made up of varying sequences of four different subunits called nucleotides.

3 types of ribosomal RNAs and 20 to 50 types of ribosomal proteins.

The pathway from DNA to RNA to protein is known as the central dogma of molecular biology, as we introduced briefly in Module 5. The central dogma describes the basic flow of information in all cells. While there are a few exceptions to the central dogma, it remains a key concept in molecular biology and will be revisited in more detail in Unit 6.

In addition to storing information and using this information to direct the synthesis of proteins, DNA has one more important feature: it is easily copied, or replicated. This feature allows genetic information to be passed from cell to cell or from an organism to its progeny. The ability of DNA to be faithfully copied and transmitted is the molecular basis for inheritance, as information in the sequence of nucleotides is stably and reliably passed from generation to generation.

Metabolism

A third essential feature of cells is the ability to transfer energy from the environment. When you eat, you take in molecules that store energy in their chemical bonds. By breaking down these molecules, your cells harness this energy and convert it into a form that can be used to do the work of the cell. Energy from the food you eat allows you to grow, move, communicate, and do all the other things that you do.

Organisms obtain energy from just two sources—the sun and chemical compounds. The term **metabolism** describes the entire set of chemical reactions by which cells transfer energy from one form to another and build and break down molecules. These reactions are required to sustain life. Regardless of their source of energy, all organisms use chemical reactions to break down molecules, in the process releasing energy that is stored in a chemical form called **adenosine triphosphate,** or **ATP.** This molecule enables cells to carry out many functions, including growth, division, and moving substances into and out of the cell. ATP will be discussed later in this unit and in more detail in Unit 3.

Metabolism is divided into two branches: catabolism and anabolism. Catabolism is the set of chemical reactions that break down molecules into smaller units and, in the process, release energy in the form of ATP and heat. Anabolism is the set of chemical reactions that build molecules from smaller units and require an input of energy, usually in the form of ATP. For example, during digestion, carbohydrates we eat are broken down, or catabolized, into sugars; fats into fatty acids and glycerol; and proteins into amino acids. These initial products can be broken down further to release energy stored in their chemical bonds. By contrast, the synthesis of macromolecules such as carbohydrates and proteins is anabolic.

Many metabolic reactions are highly conserved between organisms, meaning the same reactions are found in many different organisms. This observation suggests that the reactions evolved early in the history of life, were passed on to a great diversity of organisms, and have been maintained for billions of years because of their fundamental importance to cellular biochemistry.

✓ Concept Check

3. **Describe** the role that the cell membrane plays in the maintenance of homeostasis.

4. **Identify** the molecule that stores and transmits information in a cell.

5. **Describe** the hand-off of information from molecule to molecule according to the central dogma of molecular biology.

6. **Describe** what is meant by metabolism.

6.3 The structure and function of cells are closely related

So much in biology depends on shape. Take your hand, for example. You can pick up a pin, text on a smartphone, or touch your pinky to your thumb. These activities are made possible by the coordinated movement of dozens of bones, muscles, nerves, and blood vessels that give your hand its shape. The functional abilities of your hand emerge from its structure.

A close connection between structure and function exists at all levels of scale in biology, from molecules to cells to tissues to organs. Proteins are a good example. Composed of long, linear strings of amino acids in various combinations, each protein folds into a specific three-dimensional shape. Recall that the amino acid sequence is called the primary structure. Primary structure determines how a protein folds into its final shape, or tertiary structure. Amino acids connect with each other by covalent peptide bonds, and many types of interactions, including hydrogen, ionic, and covalent bonds, are involved in protein folding. The three-dimensional structure of the protein determines its functional properties and enables the protein to carry out its job in the cell. A single change in the amino acid sequence of a protein can change its shape, which in turn can disrupt the function of the protein. This observation underscores the close connection between structure and function.

The double helix of DNA provides another good example of how structure and function come together. The structure of the molecule suggests how genetic information is stored in DNA—in the linear order or sequence of the nucleotides. In addition, the double-helical structure of DNA suggests how it is able to make copies of itself: each strand of the double helix serves as a template, or model, for a new daughter strand. This was first noted by James Watson and Francis Crick, who described the structure of DNA using key data from Rosalind Franklin. Even to those seeing the double-helix model for the first time, the structure of the double helix suggested a mechanism by which DNA is copied. This process is critical because it enables DNA to pass genetic information from cell to cell and from parent to offspring.

The close relationship between structure and function can also be seen at the level of cells. **FIGURE 6.6**, on page 98, shows different types of cells, all with different shapes. Figure 6.6a shows red blood cells. Each one has a distinctive biconcave shape, in which both sides of the cell curve inward toward the cell interior. This unusual shape allows the cell to be flexible so it can pass through narrow blood vessels with diameters smaller than that of the red blood cell itself. In addition, its shape gives it a relatively high surface area compared to its volume, which helps it to pick up and release oxygen throughout the body.

Spherical cells in the liver, shown in Figure 6.6b, build and break down macromolecules. These cells look and function very differently from the long, slender muscle cells shown in Figure 6.6c; muscle cells contract to exert force. A neuron, shown in Figure 6.6d, has long and extensively branched projections that communicate with other cells and organs. It is structurally and functionally quite distinct from the cell shown in Figure 6.6e, a cell that absorbs nutrients and is located in the lining of the intestine.

Within many cells are smaller compartments, each specialized for different functions. As we discuss in the next module, the shapes and abundance of these compartments give clues to their various functions.

a. Red blood cells

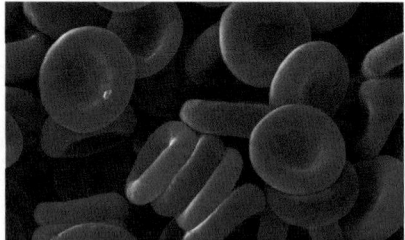

b. Liver cells

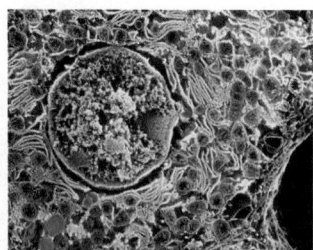

c. Muscle cells

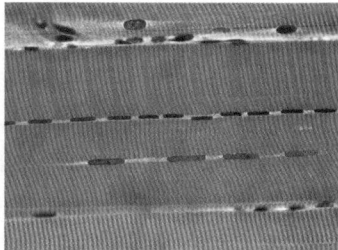

d. Motor neuron

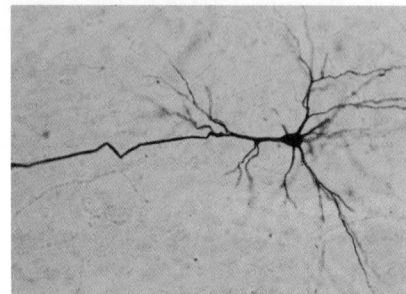

e. Intestinal cell

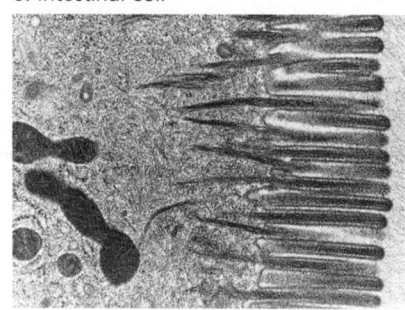

FIGURE 6.6 Diverse cell types

Cells differ in shape and are well adapted for their various functions. (a) Red blood cells have a biconcave shape that maximizes their surface area for gas exchange and allows them to deform readily. (b) Liver cells are spherical and involved in metabolism. (c) Muscle cells are long and specialized for contraction. (d) Neurons, like this motor neuron, have long, slender projections that allow them to communicate with other cells. (e) Intestinal cells have comblike projections that increase their surface absorption.

Photos: (a) Cheryl Power/Science Source; (b) Professors Pietro M. Motta & Tomonori Naguro/Science Source; (c) Innerspace Imaging/Science Source; (d) Biophoto Associates/Science Source; (e) Don W. Fawcett/Science Source

---

PREP FOR THE AP® EXAM

AP® EXAM TIP

Note that descriptions in biology often include structure and function. Make sure you are able to explain the relationship between structure and function and give examples.

✓ Concept Check

7. **Justify** the claim that there is a close relationship between structure and function.

8. Use an example to **describe** how the shape of a cell can help it perform its function.

6.4 Prokaryotes and eukaryotes differ in internal organization

We have seen that all cells have a cell membrane and contain genetic material in the form of DNA. In some cells, this genetic material is housed in a membrane-bound space called the **nucleus.** Like the cell membrane, the nuclear membrane selectively controls movement of molecules into and out of it. As a result, the nucleus occupies a discrete space within the cell, separate from the space outside the nucleus, called the **cytoplasm.** Cells can be divided into two broad classes based on whether they have a nucleus. Cells without a nucleus are **prokaryotic,** and cells with a nucleus are called **eukaryotic.** In this section, we compare and contrast these two major groups of cells.

Prokaryotes

The first cells that emerged about 4 billion years ago were prokaryotic. Most prokaryotes live as single-celled organisms, but some have simple multicellular forms. Prokaryotes include two groups, or **domains,** of organisms—**Bacteria** and **Archaea.** Bacteria and Archaea are mostly single-celled microorganisms that lack a nucleus and are therefore classified as prokaryotes.

Bacteria are familiar organisms, found today nearly everywhere that life can persist. Some bacteria live in peaceful coexistence with humans, inhabiting our gut and aiding digestion. Others cause disease—salmonellosis, tuberculosis, and cholera are familiar examples of such bacterial diseases.

Many archaea tolerate environmental extremes, such as heat and acidity. In fact, for many environmental conditions, archaea define the known limits of life. For example, some archaea exhibit the ability to grow and reproduce at temperatures of 80°C or more. These organisms do not simply tolerate high temperatures; they require them for survival. At present, the world record for high-temperature growth is held by *Methanopyrus kandleri,* an archaeon that grows at 122°C among hydrothermal vents on the seafloor beneath the Gulf of California. Another group of archaea illustrates the limits of tolerance to another environmental condition—salinity, or saltiness. In many parts of the globe where humans mine coal or metal ores, abandoned mines fill up with very acidic water. The pH of these waters is commonly in the range of 1 to 2 and can actually fall to near 0. To us, acid mine drainage is an environmental catastrophe, but a number of archaea thrive in it, as shown in **FIGURE 6.7**.

Environmental studies have yielded another surprise discovery: the immense volumes of water beneath the surface waters of the oceans contain vast numbers of archaea. In the deep ocean, archaea may be nearly as abundant as bacteria in the oceans. The success of prokaryotes evolutionarily and ecologically depends in part on their small size, their ability to reproduce rapidly, and their ability to obtain energy and nutrients from diverse sources.

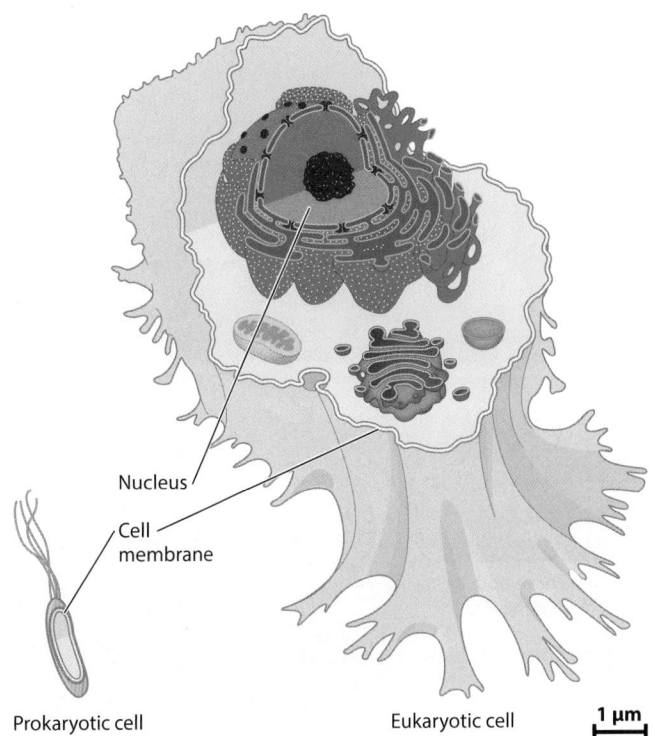

Nucleus

Cell membrane

Prokaryotic cell   Eukaryotic cell   1 μm

FIGURE 6.8 A prokaryotic and eukaryotic cell

Prokaryotic cells lack a nucleus and are very small. Eukaryotic cells have a nucleus and are typically much larger than prokaryotic cells.

FIGURE 6.7 Archaea

Archaea from the headwaters of the Rio Tinto, in southwestern Spain, thrive in these highly acidic waters, pH 1–2. The red color comes from iron sulfate ions that form in highly acidic water. Photo: Andrew Knoll, Harvard University

Although the absence of a nucleus is a defining feature of prokaryotes, other features also stand out. For example, prokaryotes are small, typically just 1–2 micrometers (a micrometer is $\frac{1}{1,000,000}$ of a meter) in diameter or smaller. By contrast, eukaryotic cells are commonly much larger, on the order of 10 times larger in diameter and 1000 times larger in volume. This dramatic difference in size between prokaryotic cells and eukaryotic cells can be seen in **FIGURE 6.8**. As we will see in Module 8, the small size of prokaryotic cells helps them absorb nutrients from the environment.

Several key structures of prokaryotes are illustrated in **FIGURE 6.9** on page 100. The DNA of prokaryotes is concentrated in a discrete region of the cell interior known as the nucleoid. In most prokaryotes, it is organized as one circular molecule of DNA arranged in many loops. Prokaryotes also have a cell wall surrounding the plasma membrane, which helps to maintain their shape. The bacterial cell wall is made of peptidoglycan, a complex polymer of sugars and amino acids. Some bacteria have thick walls made up of multiple peptidoglycan layers, while others have thin walls surrounded by an outer layer of lipids. Some bacteria and archaea have flagella (singular, flagellum), structures that extend from their surface and allow them to move.

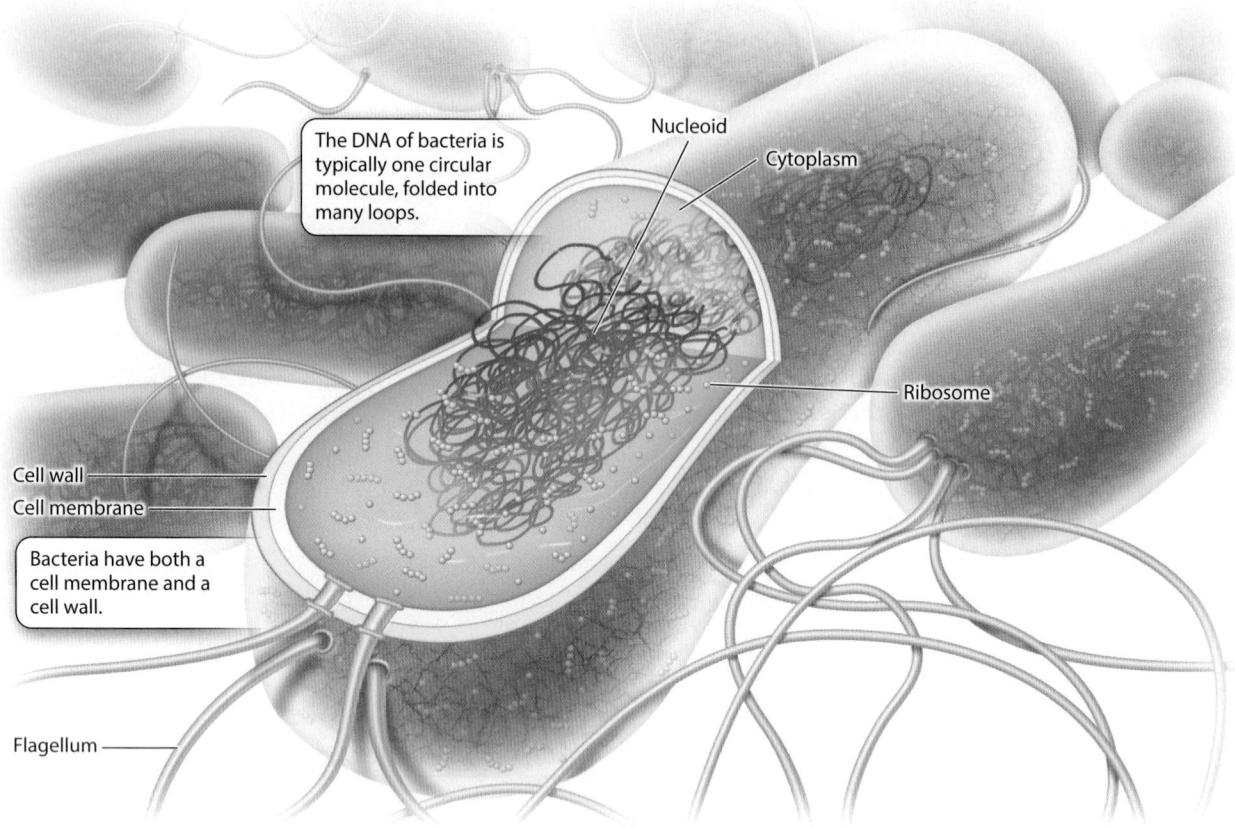

The DNA of bacteria is typically one circular molecule, folded into many loops.

Nucleoid

Cytoplasm

Ribosome

Cell wall

Cell membrane

Bacteria have both a cell membrane and a cell wall.

Flagellum

FIGURE 6.9 A bacterial cell

The DNA of bacteria is located in the nucleoid. Bacteria have a cell wall outside the cell membrane. Some bacteria also have one or more flagella.

In addition to the large circle of DNA, bacteria and archaea often contain much smaller circular molecules of DNA known as plasmids. These plasmids usually carry just a few genes. Plasmids are commonly transferred between cells through the action of threadlike structures known as pili (singular, pilus), which extend from one cell to another when they transfer plasmids, as shown in **FIGURE 6.10**. Genes that may confer advantages for bacteria living in certain conditions are often located on plasmids and then are able to spread rapidly through a population of bacteria. For example, they might help bacteria survive and reproduce in particular environments.

Because of their small size and simple cell organization, bacteria and archaea were long dismissed as primitive organisms, distinguished mostly by the eukaryotic features they lack: they have no nucleus and no internal compartments. This point of view turns out to be more than a little misleading. Bacteria and archaea are the diverse and remarkably successful products of nearly 4 billion years of evolution. Today, prokaryotic cells outnumber eukaryotic cells by several orders of magnitude. Even in your own body, bacteria

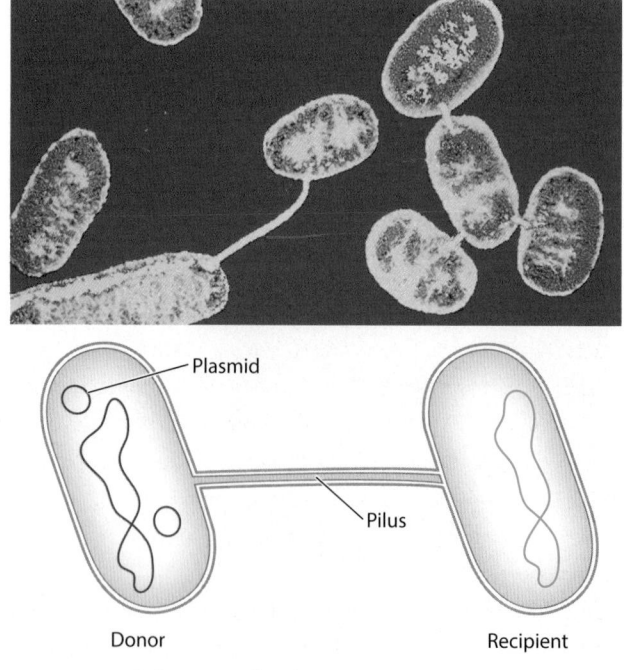

Plasmid

Pilus

Donor

Recipient

FIGURE 6.10 A bacterial pilus

A pilus is used to transfer plasmids from one bacterium to another.

Photo: Dr. Linda M. Stannard, University of Cape Town/Science Source equal or outnumber human cells. They also are metabolically diverse, able to harness energy and transfer materials in many different ways. Bacteria and archaea are present almost everywhere on Earth. What they lack in complexity of cell structure, these tiny cells more than make up for in their dazzling metabolic diversity. Bacteria and archaea underpin the efficient operation of ecosystems on our planet and life would not be possible without them.

Although Bacteria and Archaea are grouped together as prokaryotes, there are important differences between these two domains of organisms. Some of the differences relate to the structure of the cell wall; others relate to the details of how DNA is used to synthesize RNA. Finally, they differ in their relationship to eukaryotes: from an evolutionary perspective, archaea and eukaryotes are more closely related to each other than either is to bacteria.

Eukaryotes

Eukaryotes evolved much later than prokaryotes, roughly 2 billion years ago, from prokaryotic ancestors. They make up a third domain of life called **Eukarya.** Eukaryotes include familiar groups such as animals, plants, and fungi, along with a wide diversity of mostly single-celled microorganisms called protists. Eukaryotic organisms exist as single cells like yeasts or as multicellular organisms like plants and animals. In multicellular organisms, cells may specialize to perform different functions. For example, in animals, muscle cells contract; red blood cells carry oxygen to tissues; and skin cells provide an external barrier.

Eukaryotes are defined by the presence of a nucleus, which houses the vast majority of the cell's DNA. In eukaryotes, the DNA is not circular, as it is in most prokaryotes, but instead is organized as multiple linear molecules. The nuclear membrane allows for more complex regulation of gene expression than is possible in prokaryotic cells. In eukaryotes, the processes of synthesizing RNA from DNA, and the process of synthesizing proteins from RNA, are separated in both space and time: DNA is used to direct the synthesis of RNA in the nucleus first, and RNA is used to direct the synthesis of proteins later in the cytoplasm. Each of these steps can therefore be regulated separately. By contrast, in prokaryotes, these two processes occur immediately one after the next in the same space of the cell.

Eukaryotes have an extensive internal array of membranes, which are for the most part absent in prokaryotes. These membranes define compartments, called **organelles,** that divide the cell contents into smaller spaces specialized for different functions. While the entire content of a cell

outside the nucleus is the cytoplasm, the jelly-like material outside of the nucleus and organelles is called the **cytosol.**

There are other differences between prokaryotes and eukaryotes. For example, RNA and protein synthesis differs in detail between prokaryotes and eukaryotes, and eukaryotic ribosomes are larger than prokaryotic ribosomes. In addition, there are differences in the types of lipids that make up their cell membranes. Finally, whereas some bacteria have flagella, many eukaryotic cells have cilia. Cilia are rodlike structures that extend from the surface of cells. There are two types of cilia: those that don't move, called nonmotile cilia, and those that move, called motile cilia. Nonmotile cilia are very common and are found on many eukaryotic cells. In these cells, they often serve a sensory function, taking in environmental signals and conveying them to the cell interior. For example, nonmotile cilia are found in specialized cells in the nose and eyes.

Motile cilia propel the movement of cells or fluid surrounding the cells. These are shown in **FIGURE 6.11**. They can be found in single-celled eukaryotes, such

a. *Chlamydomonas*

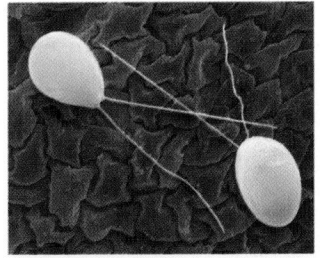

b. *Paramecium*

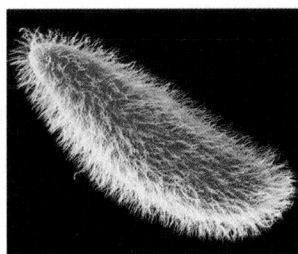

c. Sperm cells

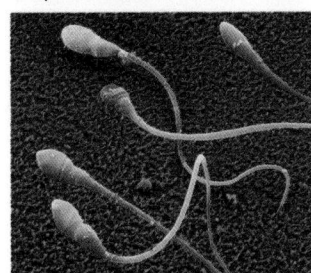

d. Airway epithelial cells

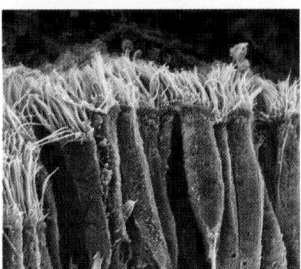

FIGURE 6.11 Motile cilia

Motile cilia move cells or allow cells to propel substances. (a) The green alga *Chlamydomonas* moves by the motion of two long cilia. (b) The protist *Paramecium* is covered with short cilia that beat in a coordinated fashion, allowing the organism to move through its environment. (c) Sperm cells swim by movement of one long cilium, which is sometimes called a flagellum. (d) Epithelial cells lining the upper respiratory tract have cilia that sweep fluid with debris and pathogens out of the airway. Photos: (a) Andrew Syred/Science Source; (b) SPL/ Science Source; (c) Juergen Berger/Science Source; (d) Eye of Science/Science Source

as the green alga *Chlamydomonas* (Figure 6.11a) and the protist *Paramecium* (Figure 6.11b), where they propel the organisms through the water. They are also found in sperm cells (Figure 6.11c). Long, elaborate cilia present in *Chlamydomonas* and sperm cells are sometimes called flagella, but they are similar in structure to the shorter cilia and are unrelated to the flagella found on bacterial cells, so most biologists now prefer the term "cilia." Cilia are also found in the upper airway of humans and other mammals (Figure 6.11d), where they beat to move mucus and debris out of the lungs.

PREP FOR THE AP® EXAM

AP® EXAM TIP

You should know the similarities and differences between the structures of prokaryotic cells and eukaryotic cells. Make sure you can explain the roles and functions of the cell membrane, nucleus, and ribosomes.

✓ Concept Check

9. **Describe** three major differences between prokaryotic cells and eukaryotic cells.

10. **Identify** two major groups of prokaryotes.

Module 6 Summary

PREP FOR THE AP® EXAM

REVISIT THE BIG IDEAS

EVOLUTION: Using the content of this module, **describe** the structure of cells and the processes that take place in all living organisms.

LG 6.1 The cell is the fundamental unit of life.

- The cell is the simplest biological entity that can exist independently. Page 92
- The cell theory states that organisms consist of cells; cells are the basic unit of life; and cells come from preexisting cells. Page 93

LG 6.2 All cells maintain homeostasis, store and transmit information, and transfer energy.

- The cell membrane is the boundary that separates the cell from its environment. Page 95
- The cell membrane maintains homeostasis for the cell. Page 95
- Information in a cell is stored in the form of the nucleic acid DNA. Page 95
- The central dogma describes the usual flow of information in a cell, from DNA to RNA to protein. Page 96
- Metabolism is the set of chemical reactions in cells that build and break down macromolecules and harness energy. Page 96

LG 6.3 The structure and function of cells are closely related.

- There is a close and intimate relationship between the shape of biological structures and their function. Page 97
- The close relationship between form and function can be seen at all levels of scale, including molecules, cells, and organs. Page 97

LG 6.4 Prokaryotes and eukaryotes differ in internal organization.

- Prokaryotic cells lack a nucleus and extensive internal membrane-enclosed compartments. Page 98
- Prokaryotes include bacteria and archaea. Page 98
- Prokaryotic cells are typically much smaller than eukaryotic cells. Page 98
- The three domains of life are Bacteria, Archaea, and Eukarya. Page 101
- Eukaryotes evolved much later than prokaryotes in the history of life. Page 101
- Eukaryotes include animals, plants, fungi, and protists. Page 101
- Eukaryotic cells have a nucleus and other internal compartments called organelles. Page 101

Key Terms

Cell theory	Nucleus	Bacteria
Homeostasis	Cytoplasm	Archaea
Ribosome	Prokaryotic	Eukarya
Metabolism	Eukaryotic	Organelle
Adenosine triphosphate (ATP)	Domain	Cytosol

Review Questions

1. Which statement is correct?
   (A) Eukaryotic cells have a nucleus and prokaryotic cells do not.
   (B) Prokaryotic cells have a nucleus and eukaryotic cells do not.
   (C) Neither eukaryotic cells nor prokaryotic cells have a nucleus.
   (D) Both eukaryotic cells and prokaryotic cells have a nucleus.

2. In which domains do cells store their genetic information in a nucleus?
   (A) Eukarya and Archaea
   (B) Eukarya
   (C) Archaea and Bacteria
   (D) Archaea

3. All cells have
   (A) a nucleus.
   (B) genetic information.
   (C) internal compartments.
   (D) a cell wall.

4. The genetic material of a prokaryotic cell is located in its
   (A) nucleolus.
   (B) nucleus.
   (C) nucleoid.
   (D) nucleosome.

5. Which statement is true regarding cytosol?
   (A) Cytosol includes organelles and the nucleus.
   (B) Cytosol includes the nucleus but excludes organelles.
   (C) Cytosol includes organelles but excludes the nucleus.
   (D) Cytosol excludes the nucleus and organelles.

6. Which do a bacterial cell, a plant cell, and an animal cell have in common?
   (A) Nucleus
   (B) Cell wall
   (C) Vacuole
   (D) Cytoplasm

7. The term "homeostasis" refers to the ability of the cell to
   (A) transfer energy from the environment.
   (B) seek out an optimal environment.
   (C) store and use information.
   (D) control and maintain its internal environment.

Module 6
AP® Practice Questions

PREP FOR THE AP® EXAM

Section 1: Multiple-Choice Questions
Choose the best answer for questions 1–5.

1. A student describes a newly discovered species of organism. The student notes that the organism is comprised of hundreds of microscopic cells. The student also suggests that the new cells of the organism form from materials that are in the lab environment outside the cells. Which statement suggests that the student's observation was incorrect?
   (A) Organisms are comprised of cells.
   (B) Cells come only from preexisting cells.
   (C) The basic unit of life is the cell.
   (D) Cells are made from proteins, nucleic acids, lipids, and carbohydrates.

2. Some viruses use RNA as their genetic information molecule. When this kind of virus infects a cell, it uses its RNA as the information molecule from which DNA is built. How does information storage in these viruses differ from information storage in all cells?

(A) It doesn't: in cells, information is stored in RNA.

(B) In cells, information is stored in DNA.

(C) In cells, information is stored in proteins.

(D) In cells, RNA is stored in DNA.

3. Some organisms use only very simple inorganic compounds, such as H_2, CO_2, NH_4^+, and PO_4^{2-}, to make all the macromolecules they require to maintain homeostasis, grow, and reproduce. Which is the most likely explanation of how these organisms make the much larger molecules required of life, such as sugars, fats, and proteins?

(A) These organisms use the simple inorganic molecules to make the subunits and then use the subunits to make the macromolecules.

(B) These organisms use the simple inorganic molecules to make only the subunits, as they do not require macromolecules.

(C) These organisms require neither anabolism nor catabolism.

(D) These organisms do not require catabolism, only anabolism, as they use inorganic molecules in their metabolism.

4. Many fungi and bacteria break down proteins that they acquire from the environment. They break down the proteins into amino acids and other molecules, and then use some of the molecules resulting from breaking down proteins to build their own cellular materials. Using what you know about catabolism and anabolism, identify which is also necessary for fungi and bacteria to perform anabolism using the amino acids and molecules from protein breakdown.

(A) Additional macromolecules generated by catabolism

(B) Additional subunits generated by anabolism

(C) ATP generated during catabolism

(D) ATP generated during anabolism

5. A newly discovered organism has been isolated from ocean water. It has a cell membrane, a nucleus, ribosomes, and organelles. It has a diameter of 10 micrometers and requires salt water to grow. Given these data, identify the organism's most likely classification.

(A) The organism is an animal.

(B) The organism is a plant.

(C) The organism is a bacterium.

(D) The organism is a eukaryote.

Section 2: Free-Response Question

Write your answer to each part clearly. Support your answers with relevant information and examples. Where calculations are required, show your work.

Students analyze water samples from a local pond. They observe two types of cells, which they call species A and species B. Species A is a single-cell microorganism that is approximately 100 µm in length and 40 µm in width. The students observe that species A is surrounded by a cell membrane that contains cilia. Species A also has a nucleus and a number of organelles, including a rough endoplasmic reticulum, ribosomes, mitochondria, and food and contractile vacuoles. Species B is also a microorganism and has a length of about 4 µm and a width of 1.5 µm. Its genetic material is not bordered by a membrane and its cytoplasm contains ribosomes. Species B has a cell wall and a single flagellum extending from the cell wall.

(a) **Identify** the domains to which species A and species B most likely belong.

(b) **Justify** the assignment of these microorganisms to their domains.

Module 7

Subcellular Compartments of Eukaryotes

LEARNING GOALS ▶**LG 7.1** The endomembrane system compartmentalizes the eukaryotic cell.
▶**LG 7.2** Mitochondria and chloroplasts harness energy for use by the cell.
▶**LG 7.3** The cytoskeleton and cell wall help to maintain cell shape.

In the last module, we focused on the cell, the fundamental unit of life. All living things are made up of cells, and there is no life without cells. We discussed the universal features of all cells. All cells have a cell membrane that regulates the movement of substances into and out of the cell and maintains the cell's interior compatible with the various functions required for life, a process called homeostasis. All cells have genetic material in the form of DNA. The information in DNA is used to guide the synthesis of RNA. In turn, the information in RNA guides the synthesis of proteins on structures called ribosomes. Finally, all cells carry out metabolic reactions that build and break down molecules and transfer energy from the environment for use by the cell.

We also introduced two broad classes of cells—prokaryotes and eukaryotes. Prokaryotes do not have a nucleus or extensive internal membranes, whereas eukaryotes do have a nucleus and extensive internal membranes. In this module, we will examine eukaryotic cells more closely, paying particular attention to how the internal membranes of eukaryotes carve out different spaces within cells that are specialized for different functions. As we saw in the discussion of cells, the shapes of these subcellular compartments and their functions are closely related to each other. Finally, we will look at how the cytoskeleton helps to maintain the shape of cells, organize compartments within the cell, and provide tracks for the movement of organelles and other substances.

> **PREP FOR THE AP® EXAM**
>
> **FOCUS ON THE BIG IDEAS**
>
> **SYSTEMS INTERACTIONS:** Look for examples of how living systems are organized in a hierarchy of structural levels that interact.

7.1 The endomembrane system compartmentalizes the eukaryotic cell

Eukaryotes are defined by the presence of a nucleus, which sets them apart from prokaryotes. In addition, eukaryotes have an extensive array of membranes inside the cell. The total *surface area* of these internal membranes is about tenfold greater than that of the cell membrane. **Surface area** is the total amount of area of the outer surface of an object. This observation highlights the large amount of internal membrane area in a eukaryotic cell and the extent to which it is divided into internal compartments.

The internal membranes define the subcellular compartments called organelles. Each organelle carries out a specific function. Eukaryotic cells can be compared to a large organization with many different departments. Each department has a specific function and internal organization that

contribute to the work of the organization. Eukaryotic cells have well-defined compartments that carry out different functions that are important to the life of the cell. In this section, we will discuss the form and function of these subcellular compartments.

The Endomembrane System

Many of the organelles inside cells are not isolated entities, but instead are connected with one another. The membranes of these organelles are either physically connected by membrane bridges or temporarily connected by **vesicles,** small membrane-enclosed sacs that transport substances within a cell or from the interior to the exterior of the cell. These vesicles form by budding from an organelle. They take with them a piece of the membrane and internal contents of the organelle from which they derive. The vesicles then fuse with another organelle or the cell membrane to re-form a continuous membrane and unload their contents.

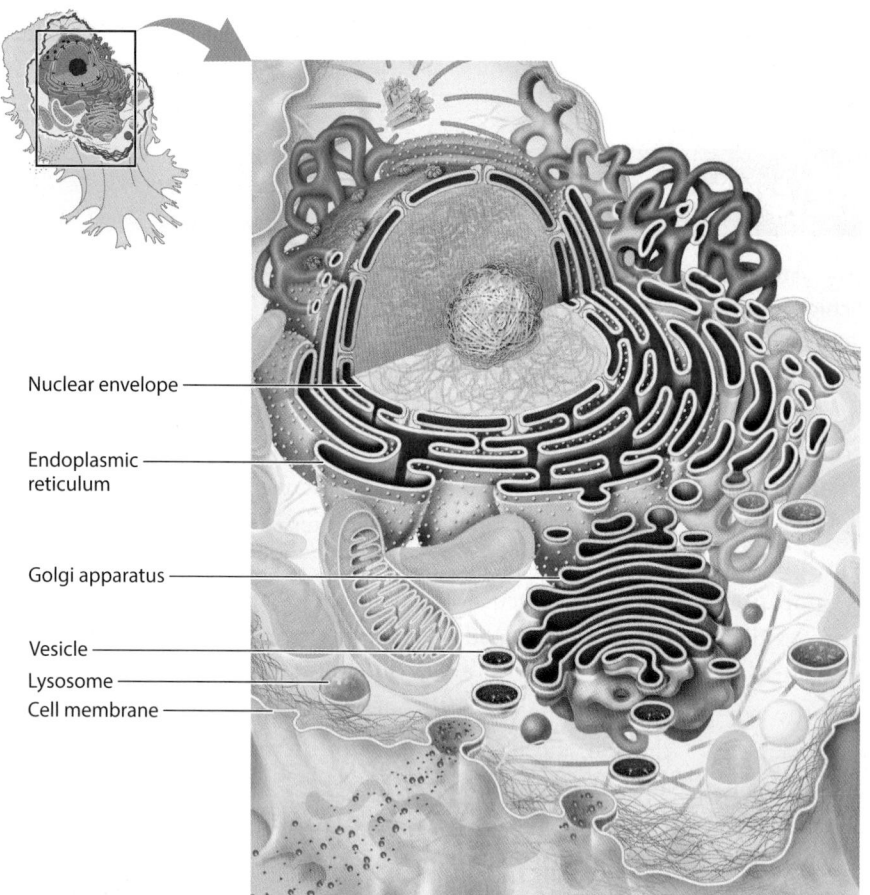

FIGURE 7.1 The endomembrane system

Nuclear envelope

Endoplasmic reticulum

Golgi apparatus

Vesicle

Lysosome

Cell membrane

The endomembrane system is a series of interconnected membrane-bound compartments in eukaryotic cells. It includes the nuclear envelope, endoplasmic reticulum, Golgi apparatus, vesicles, lysosomes, and cell membrane.

In total, the interconnected membranes of the cell make up the **endomembrane system.** The endomembrane system includes the cell membrane, *nuclear envelope, endoplasmic reticulum, Golgi apparatus, lysosomes,* and the vesicles that move between them. These organelles are shown in **FIGURE 7.1** and discussed below. In plants, the endomembrane system is actually continuous between cells through intercellular connections. Extensive internal membranes are not common in prokaryotic cells. However, photosynthetic bacteria have internal membranes that are specialized for harnessing light energy.

Because many types of molecules are unable to cross cell membranes on their own, the endomembrane system divides the interior of a cell into two distinct spaces, one inside the compartments defined by these membranes and one outside these compartments. As we saw in the previous module, the space outside the organelles is called the cytosol. Substances within a compartment are therefore in a different physical space from those in the cytosol, separated by membranes of the endomembrane system. This physical separation allows specific functions to take place within the spaces defined by the membranes. In spite of forming an interconnected system, the various compartments have unique properties and maintain distinct identities.

Nucleus

The innermost organelle of the endomembrane system is the nucleus, which—as we saw in Module 6—stores DNA, the genetic material that encodes the information for all the activities and structures of the cell. The **nuclear envelope,** illustrated in **FIGURE 7.2,** defines the boundary of the nucleus. It actually consists of two membranes, an inner membrane and an outer membrane. Each of these membranes consists of a lipid bilayer.

The inner and outer membranes of the nuclear envelope are perforated by openings called nuclear pores, which can be seen in Figure 7.2. Nuclear pores are large protein complexes with an inner passageway that regulate which molecules move into and out of the nucleus. They are essential for

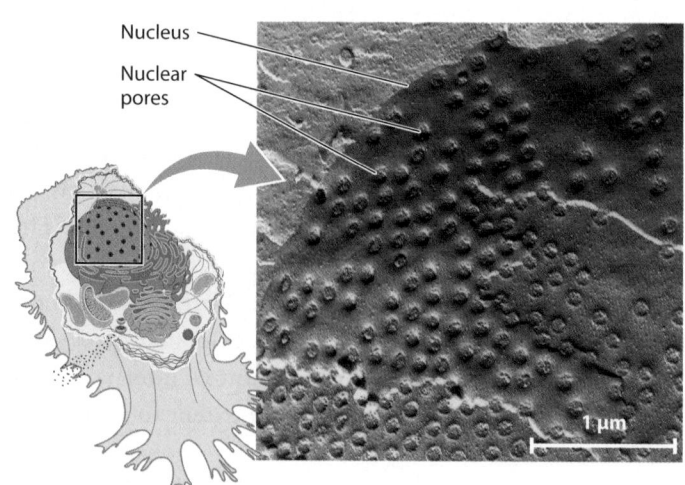

Nucleus

Nuclear pores

1 μm

FIGURE 7.2 A surface view of the nuclear envelope

The nucleus is surrounded by a double membrane and houses the cell's DNA. The nuclear envelope is perforated by membrane protein openings called nuclear pores. Photo: Don W. Fawcett/Science Source

communication between the nucleus and the rest of the cell. For example, some proteins that are synthesized in the cytosol, such as proteins that turn genes on and off, move through nuclear pores to enter the nucleus, where they control how and when genetic information is expressed. In addition, the transfer of information encoded by DNA depends on the movement of RNA out of the nucleus through these pores.

Endoplasmic Reticulum

The outer membrane of the nuclear envelope is physically continuous with the **endoplasmic reticulum (ER).** The ER is an organelle that is involved in the production of proteins and lipids. It is shown in **FIGURE 7.3**. In contrast to the nuclear envelope, the ER is bounded by a single membrane. It produces and transports many of the proteins and lipids used inside and outside the cell. The proteins can be transported by vesicles to the cell membrane, other organelles of the endomembrane system, or the cell exterior. The ER is also the site of production of most of the lipids that make up the various cell membranes. The ER is a conspicuous feature of many eukaryotic cells, accounting in some cases for as much as half of the total amount of membrane.

Unlike the nucleus, which is a single spherical structure in the cell, the ER consists of a complex network of interconnected tubules and flattened sacs. Its interior is continuous throughout and is called the lumen. As shown in Figure 7.3, the ER has an almost mazelike appearance when sliced and viewed in cross section. Its membrane is extensively convoluted, allowing a large amount of membrane surface area to fit within the cell.

When viewed through an electron microscope, ER membranes have two different appearances, which are shown in **FIGURE 7.4** on page 108. Some look rough because they are studded with ribosomes; this portion of the ER is referred to as rough endoplasmic reticulum and is shown in Figure 7.4a. Recall from Module 6 that ribosomes are the site of protein synthesis, where the information in RNA is used to guide the synthesis of proteins. Ribosomes can float freely in the cytosol or become associated with the ER membrane. Proteins synthesized by free ribosomes end up in the cytosol and then are directed to their final destinations. Proteins synthesized by the rough ER can stay associated with cell membranes as transmembrane proteins or be transported to the interior of organelles or out of the cell. Cells that secrete large quantities of protein have extensive rough ER, including cells of the pancreas that produce the hormone insulin and cells of the gut that secrete digestive *enzymes.* **Enzymes** are proteins that accelerate the rate of a chemical reaction. All cells have at least some rough ER for the production of transmembrane and organelle proteins.

A small amount of ER membrane in most cells appears smooth because it lacks ribosomes; this portion of the ER is called smooth endoplasmic reticulum and is shown in Figure 7.4b. Smooth ER is the site of fatty acid and phospholipid synthesis. Thus, this type of ER predominates in cells specialized for the production of lipids.

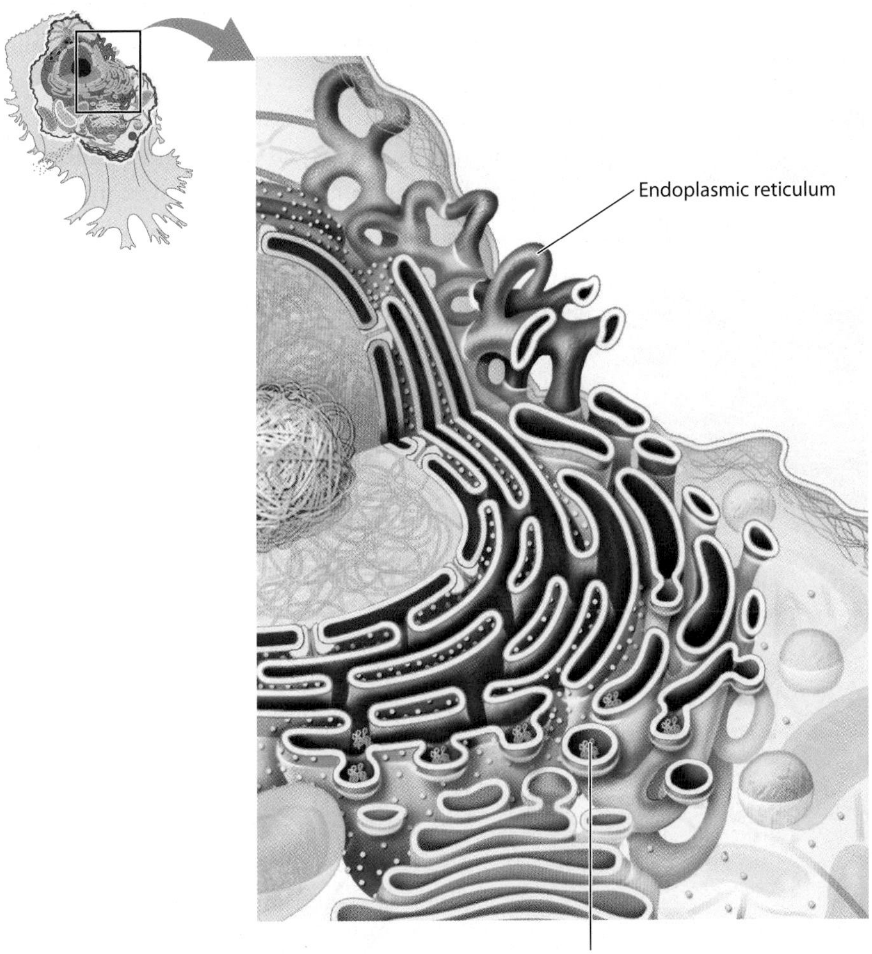

Endoplasmic reticulum

Protein

FIGURE 7.3 The endoplasmic reticulum (ER)

The ER is a major site for protein and lipid synthesis. It consists of interconnected membranes that are continuous with each other and the outer membrane of the nuclear envelope. The membrane is highly convoluted, providing a large amount of surface area.

a. Rough endoplasmic reticulum

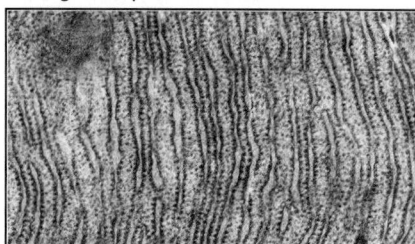

b. Smooth endoplasmic reticulum

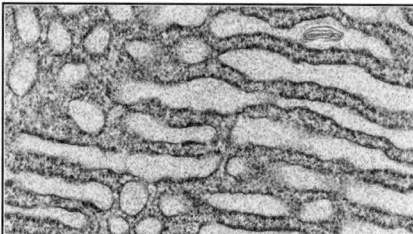

FIGURE 7.4 Rough and smooth endoplasmic reticulum

These photos show examples of rough and smooth endoplasmic reticulum. (a) Rough ER has a rough appearance because it is studded with ribosomes. Many proteins are synthesized by ribosomes associated with the rough ER. (b) Smooth ER appears smooth because it does not have ribosomes. It is the primary site of lipid synthesis, such as steroid hormones. Photos: (a) Biophoto Associates/Science Source; (b) David M. Phillips/Science Source

For example, cells that synthesize steroid hormones, which are lipids, have well-developed smooth ER that produces large quantities of cholesterol, an important lipid and precursor of other products. Enzymes within the smooth ER convert cholesterol into steroid hormones. Smooth ER also contains enzymes that help to detoxify certain drugs and harmful products of metabolism.

FIGURE 7.5 The Golgi apparatus

The Golgi apparatus is the next stop for proteins and lipids produced in the ER. It further modifies these proteins and lipids; it sorts and targets them to other organelles, the cell membrane, or the cell exterior; and it adds sugars to proteins and lipids. Vesicles transfer molecules from the ER to the Golgi apparatus, within the Golgi apparatus, and from the Golgi apparatus to their final destinations. Photo: Biophoto Associates/Science Source

Vesicle

Cisternae

Golgi apparatus

Golgi Apparatus

The **Golgi apparatus** modifies and sorts proteins and lipids produced by the endoplasmic reticulum. Although it is not physically continuous with the ER, the Golgi apparatus is often the next stop for vesicles that bud off the ER. These vesicles carry lipids and proteins, either within the vesicle interior or embedded in their membranes. The movement of these vesicles from the ER to the Golgi apparatus and then to the rest of the cell is part of a pathway in which lipids and proteins are sequentially modified and delivered to their final destinations.

The Golgi apparatus has three primary roles. First, it further modifies proteins and lipids produced by the ER. Second, it acts as a sorting station as these proteins and lipids move to their final destinations. Third, it is a major site where carbohydrates are added to proteins and lipids.

Under the microscope, the Golgi apparatus appears to be stacks of flattened membrane sacs, as shown in **FIGURE 7.5**. These sacs are called cisternae. They are surrounded by numerous vesicles that transport proteins and lipids from the ER to the Golgi apparatus, and then between the various cisternae, and finally from the Golgi apparatus to the cell membrane or other organelles. Vesicles are therefore the primary means by which proteins and lipids move through the Golgi apparatus to their final destinations.

Proteins and lipids are chemically modified by enzymes as they pass through the Golgi apparatus. There is a general movement of vesicles from the ER through the Golgi apparatus to their final destinations. As they make their way through the Golgi apparatus, chemical modifications take place in a sequence of steps, each performed in a different region of the Golgi apparatus.

An example of a chemical modification that occurs predominantly in the Golgi apparatus is glycosylation, in which sugars are covalently linked to lipids or proteins. As these lipids and proteins move through the Golgi apparatus, they encounter different enzymes in each region that add or trim sugars. Proteins with added sugars are important components of the eukaryotic cell surface. The sugars can protect the protein on the cell surface from digestive enzymes outside of the cell by blocking access to the peptide chain. The distinctive shapes that sugars contribute to proteins and lipids on the cell surface also allow them to be recognized specifically by other cells in the external environment. For example, human blood types (A, B, AB, and O) are defined by the particular sugars that are linked to proteins and lipids on the surface of red blood cells.

While traffic usually travels from the ER to the Golgi apparatus, a small amount of traffic moves in the reverse direction, from the Golgi apparatus to the ER. This reverse pathway is important to retrieve proteins in the ER or Golgi that were accidentally moved forward, and to recycle membrane components. As a result of the reverse pathways, the synthesized compounds can be recycled and used in other pathways.

Lysosomes

The ability of the Golgi apparatus to sort and dispatch proteins to particular destinations is dramatically illustrated by *lysosomes*. **Lysosomes,** specialized vesicles derived from the Golgi apparatus, degrade damaged or unneeded macromolecules. As a result, lysosomes play a key role in intracellular digestion and in recycling organic compounds. They also are involved in programmed cell death, in which a cell degrades in an orderly and controlled fashion. Lysosomes are shown in **FIGURE 7.6**. They contain a variety of enzymes that break down macromolecules such as nucleic acids, lipids, and complex carbohydrates. These enzymes are packaged into lysosomes by the Golgi apparatus. The Golgi apparatus also packages into vesicles macromolecules that are destined for degradation. These vesicles then fuse with lysosomes and deliver their contents to the lysosome interior.

The inside of a lysosome is much more acidic than the inside of a cell. The pH, a measure of acidity, inside a lysosome is around 5, while the pH of the cytosol is around 7. As we saw in Module 2, the lower the pH, the more acidic a solution is. The function of an enzyme, like all proteins, depends on its shape, and in many cases the shape of a protein depends on the pH of the solution where it acts. Lysosomal enzymes cannot function in the cytosol, and many of a cell's proteins would unfold and degrade if the entire cell were at a pH of 5, like the inside of a lysosome. By restricting the activity of particular enzymes to the lysosome, the cell protects proteins and organelles in the cytosol from degradation. The different environments inside and outside lysosomes underscore the importance of having separate compartments within the cell bounded by membranes.

The formation of lysosomes also illustrates the ability of the Golgi apparatus to target key proteins to specific destinations in the cell. The enzymes inside the lysosomes are synthesized in the rough ER, sorted in the Golgi apparatus, and then packaged into lysosomes. In addition, the Golgi apparatus sorts and delivers specialized proteins that become embedded in lysosomal membranes. These include transmembrane pumps that keep the internal environment at an acidic pH of about 5, the optimal pH for the activity of the enzymes inside lysosomes. Other proteins in lysosomal membranes transport the broken-down products of macromolecules, such as amino acids and simple sugars, across the membrane to the cytosol so they can be recycled by the cell.

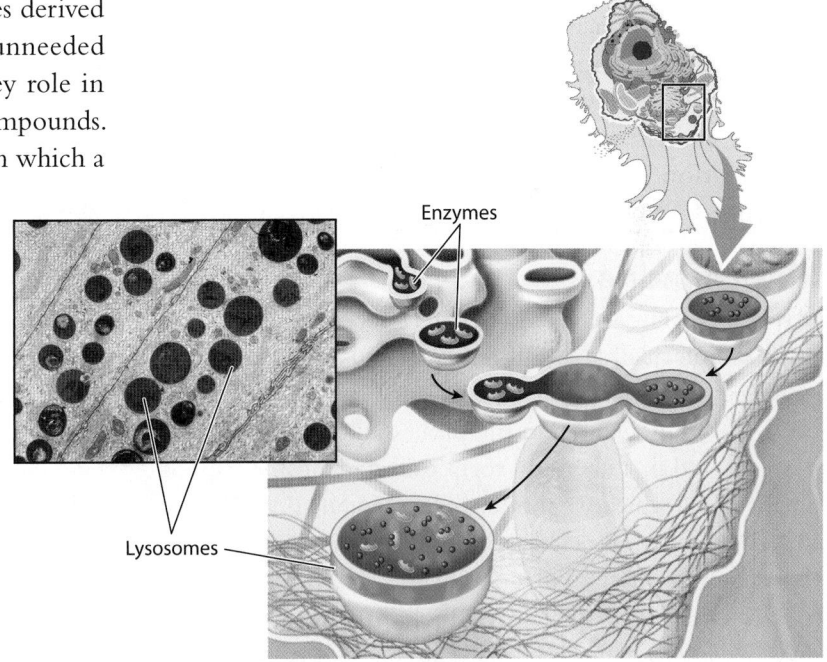

FIGURE 7.6 Lysosomes

Lysosomes are vesicles that degrade macromolecules, recycle key molecules, and play a role in programmed cell death. The Golgi apparatus packages enzymes that break down molecules into lysosomes. The Golgi apparatus also sorts macromolecules destined for degradation into vesicles that fuse with lysosomes.

Photo: Don W. Fawcett/Getty Images

AP® EXAM TIP

You should be able to predict a likely function of a particular cell by looking at the relative amounts of organelles in it. For example, a cell with a lot of rough ER probably secretes proteins.

✓ Concept Check

1. **Identify** the names and functions of the organelles that make up the endomembrane system.

2. **Describe** one difference between the environment inside an organelle of the endomembrane system and the environment outside it. **Describe** the significance of this difference.

7.2 Mitochondria and chloroplasts harness energy for use by the cell

The membranes of two organelles, *mitochondria* and *chloroplasts,* are not part of the endomembrane system. Both of these organelles are specialized to harness energy for the cell. Interestingly, they are able to grow and multiply independently of the other membrane-bound compartments, and they contain their own DNA, separate from the DNA in the nucleus of the cell. The similarities between the DNA of these organelles and the DNA of certain bacteria have led scientists to conclude that these organelles originated as bacteria that were captured and encapsulated by another cell and, over time, evolved to their current function. We will discuss the structure and function of mitochondria and chloroplasts in this section.

Mitochondria

Mitochondria are organelles that harness energy from organic molecules, such as carbohydrates. Regardless of their source of energy, all organisms use chemical reactions to break down molecules. In the process, they release energy that is stored in a chemical form called adenosine triphosphate (ATP), which we introduced in Module 6. ATP drives the many chemical reactions in the cell, and is often called the universal energy currency of the cell. It enables cells to carry out all sorts of work, including cell growth and division, and moving substances into and out of the cell. As a result, mitochondria provide eukaryotic cells with most of their usable energy.

Mitochondria are the site of a metabolic process called **cellular respiration,** a series of chemical reactions in which organic molecules such as carbohydrates are broken down, and the energy stored in organic molecules is

converted to ATP. In this process, oxygen is consumed and carbon dioxide is released. Does this process sound familiar? It also describes your own breathing, or respiration. The oxygen that you take in with each breath is used by mitochondria to synthesize ATP, and carbon dioxide is released as a waste product.

Mitochondria are rod-shaped organelles with two membranes: an outer membrane and a highly convoluted inner membrane whose folds project into the interior, as seen in **FIGURE 7.7**. These membranes define two spaces: the space between the inner and outer membranes is called the intermembrane space, and the space enclosed by the inner membrane is called the **mitochondrial matrix.** These spaces are shown in **FIGURE 7.8**.

The process of cellular respiration does not take place in a single reaction, but instead occurs in a series of steps. As a result, the energy held in the chemical bonds of carbohydrates and other organic molecules is not released all at once, but can be transferred to the chemical bonds of ATP. These steps will be described in detail in Unit 3.

Each of these steps occurs in a different cellular space. Some steps occur outside of the mitochondria in the cytosol; others take place in the mitochondrial matrix; and still others take place along the inner mitochondrial membrane. This last step, in particular, produces most of the ATP in cellular respiration. The presence of folds of the inner mitochondrial membrane increases the surface area available for the biochemical machinery that synthesizes ATP. The more folds there are, the more surface is available and the more ATP is synthesized. This is another example where structure and function come together.

Chloroplasts

Mitochondria are present in nearly all eukaryotic cells, including both plant and animal cells, providing

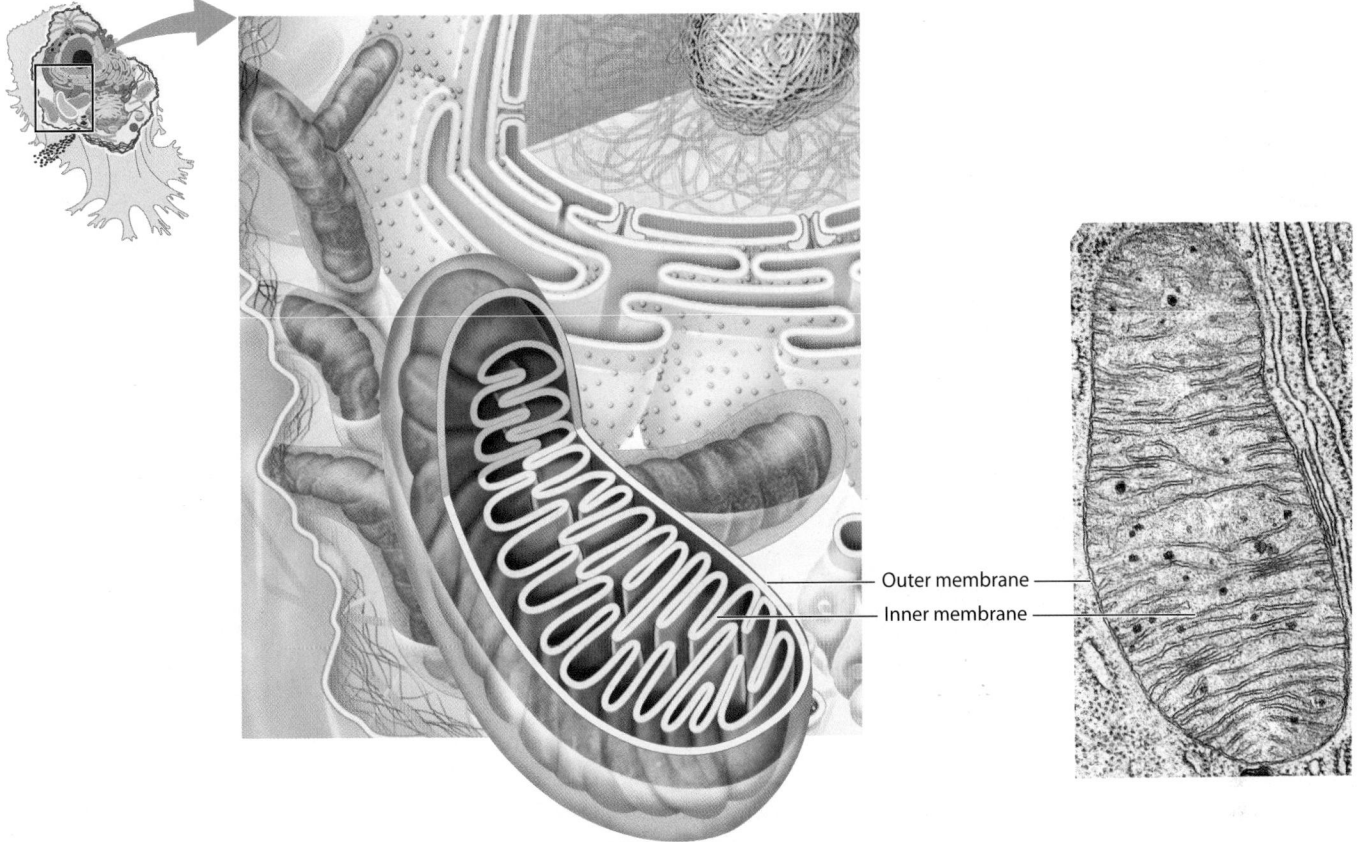

FIGURE 7.7 Mitochondria

Mitochondria have a double membrane, consisting of an outer and inner membrane. Mitochondria synthesize most of the cell's ATP. Photo: Keith R. Porter/Science Source

these cells with life-sustaining ATP. In addition, plant cells and green algae have organelles called *chloroplasts*. **Chloroplasts** are organelles that capture the energy of sunlight to synthesize simple sugars. This process of capturing sunlight to synthesize simple sugars is called **photosynthesis.** During photosynthesis, carbon dioxide is consumed, and oxygen is released as a waste product. We will explore the process of photosynthesis in detail in Unit 3.

Like the nucleus and mitochondria, chloroplasts are enclosed by a double membrane consisting of an outer and inner membrane, shown in **FIGURE 7.9**. Inside chloroplasts is a structure called the **thylakoid.** Thylakoids resemble flattened sacs. In fact, the word "thylakoid" is derived from *thylakois,* the Greek word for "sac." These sacs in turn are grouped into structures called **grana** (singular, granum) that look like stacks of pancakes. Grana are connected to one another by membrane bridges in such a way that the thylakoid membrane encloses a single interconnected compartment.

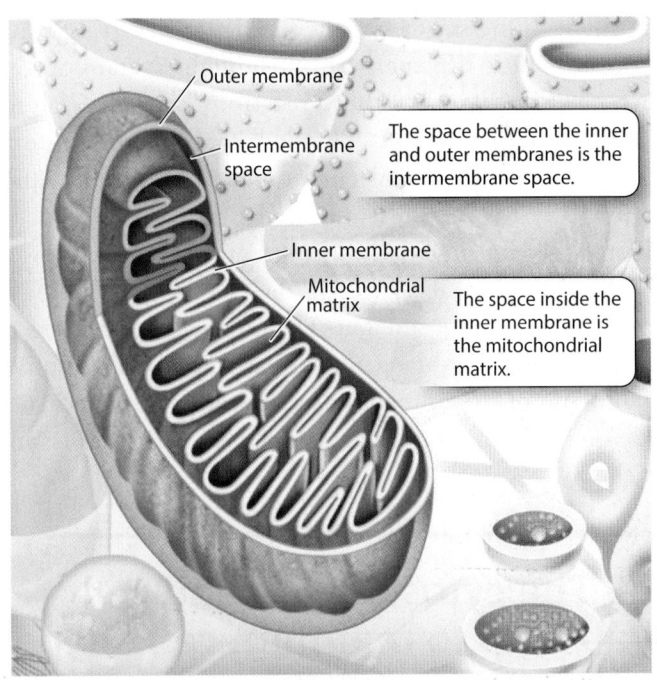

FIGURE 7.8 Mitochondrial membranes and spaces

The inner and outer membranes of mitochondria create two spaces, the intermembrane space and the mitochondrial matrix.

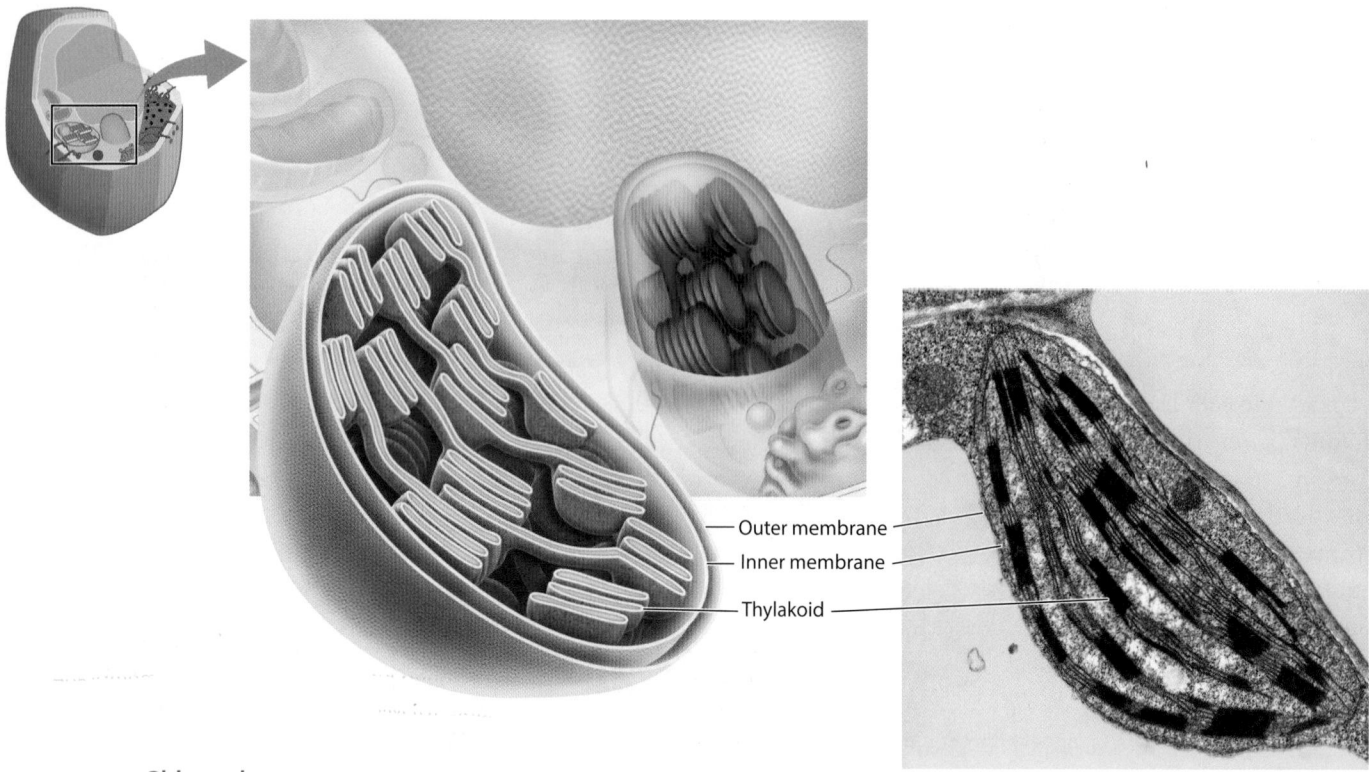

FIGURE 7.9 Chloroplasts

Chloroplasts have a double membrane, consisting of an outer and inner membrane. They also have internal structures called thylakoids, which appear like flattened sacs. They capture energy from sunlight and use it to synthesize sugars. Photo: Dr. Jeremy Burgess/Science Source

Like cellular respiration, photosynthesis consists of many steps which take place in different cellular compartments. One set of steps is the light-harvesting reactions, in which the energy from sunlight is transformed into chemical energy. These reactions take place along the thylakoid membrane. The thylakoid membrane contains light-collecting molecules called pigments, of which **chlorophyll** is the most important. The green color of chlorophyll explains why so many plants have green leaves. Chlorophyll plays a key role in the chloroplast's ability to capture energy from sunlight. The interconnected stacks of the thylakoid membrane are similar to the folds of inner membrane of the mitochondria: in both cases, the structure increases the surface area available for the harnessing of energy.

✓ Concept Check

3. **Identify** the types of cells that have mitochondria.

4. **Describe** three similarities of mitochondria and chloroplasts.

7.3 The cytoskeleton and cell wall help to maintain cell shape

The shape of a cell depends on the cell membrane, as we discussed, as well as a system of protein filaments that are collectively called the **cytoskeleton.** Just as the bones of vertebrate skeletons provide internal support for the body, the cytoskeleton provides internal support for cells and tracks within cells for the transport of vesicles and other organelles. The cytoskeleton, which helps to determine cell shape, is a universal and ancient feature of all cells. In some cells, cytoskeletal elements perform other functions as well, allowing cells to change shape, move about, and transport substances within the cell. Some, but not all, cells also have a cell wall outside of the cell membrane. Like the cytoskeleton, the cell wall provides structural support for the cell. In this section, we take a closer look at the structure and function of the cytoskeleton and cell wall.

Cytoskeleton

All eukaryotic cells have at least two cytoskeletal elements, microtubules and microfilaments, shown in **FIGURE 7.10**. These cytoskeletal elements are long chains, or polymers, made up of protein subunits. They provide structural support, and enable cells to change shape, move about, and transport substances.

Microfilaments are present in various locations in the cytoplasm. As can be seen in Figure 7.10a, they are extensively branched in the area of the cytoplasm just beneath the cell membrane, where they help to reinforce the cell membrane and organize proteins associated with it.

Microfilaments also play important roles in maintaining the shape of a cell, such as the biconcave shape of red blood cells discussed in Module 6 (see Figure 6.6a on page 98). In addition, long bundles of microfilaments form a band that extends around the circumference of epithelial cells. Microfilaments also maintain the shape of epithelial cells such as those lining the small intestine that are specialized to absorb nutrients in the gut (see Figure 6.6e).

Microtubules are hollow tubelike structures pictured in Figure 7.10b. Microtubules help maintain cell shape and internal structure. In animal cells, these structures radiate outward from a microtubule organizing center to the cell periphery. This spokelike arrangement of microtubules helps cells withstand compression. Many organelles are tethered to microtubules, so microtubules also help to guide the arrangement of organelles in the cell.

Microfilaments and microtubules are found in all eukaryotic cells, and their structure and function have remained relatively unchanged throughout the course of evolution. Not long ago, it was believed that cytoskeletal proteins were present only in eukaryotic cells. Recently, a number of studies have shown that many prokaryotes also have a system of proteins similar in structure to the cytoskeletal elements of eukaryotic cells. At least one of these prokaryotic cytoskeleton-like proteins is

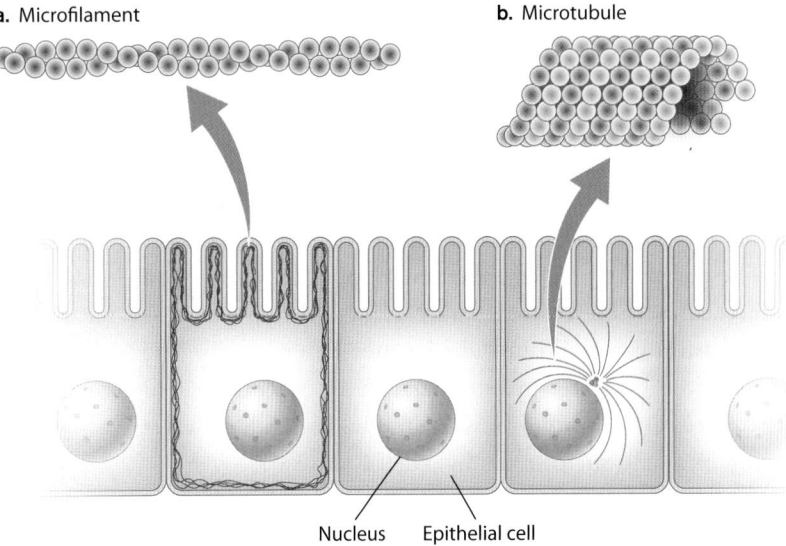

a. Microfilament **b.** Microtubule

Nucleus Epithelial cell

FIGURE 7.10 Cytoskeleton

Eukaryotic cells contain at least two cytoskeletal elements. (a) Microfilaments are thin, helical protein fibers and can be found just underneath the cell membrane, where they help to maintain cell shape. (b) Microtubules are hollow tubes and radiate outward from the cell center. They support the cell and provide tracks for vesicles and other organelles.

expressed in the mitochondria and chloroplasts of eukaryotic cells. The presence of this protein in these organelles lends further support to the theory that mitochondria and chloroplasts were once independent prokaryotic cells.

Cell Wall

Some cells have an additional structure, called the cell wall, outside the cell membrane. **FIGURE 7.11** shows the cell

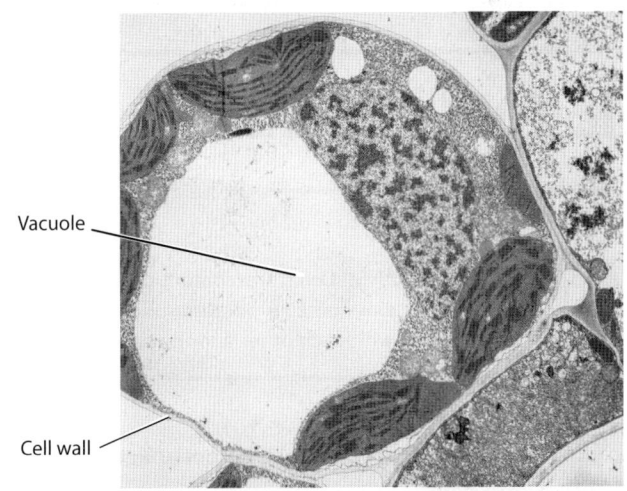

Vacuole

Cell wall

FIGURE 7.11 The plant cell wall and vacuoles

The plant cell wall is a rigid structure that maintains the shape of the cell. The pressure exerted by water absorbed by vacuoles also provides support. Photo: Biophoto Associates/Getty Images

wall in a typical plant cell. The cell wall is present not only in plants but also algae, fungi, and bacteria. It is absent in animal cells. The cell wall plays a critical role in the maintenance of cell size and shape. When Hooke looked at cork through his microscope and saw images similar to those pictured in Figure 6.3b on page 94, what he saw was not living cells but rather the remains of cell walls.

The cell wall provides structural support and protection for the cell. Because the cell wall is rigid and resists expansion, pressure can build up when water enters a cell. The force exerted by water pressing against an object is called **turgor pressure.** Turgor pressure builds as a result of water moving into cells surrounded by a cell wall.

The pressure exerted on the cell wall by water inside the cell provides structural support for many organisms that is similar to the support provided by skeletons in animals. In addition, plant and fungal cells have another structure, called the **vacuole,** that absorbs water and contributes to turgor pressure. Figure 7.11 shows a vacuole in a plant cell. Its function helps to explain why plants wilt when dehydrated: the loss of water from the vacuoles reduces turgor pressure, so the cells can no longer maintain their shape within the cell wall. Plant vacuoles have many other functions and are often the most conspicuous feature of plant cells. In addition to water, they can also store nutrients, ions, and wastes. The large size of vacuoles also explains in part why plant cells are typically larger than animal cells.

The cell wall is made up of many different components, including carbohydrates and proteins. The specific components differ depending on the organism. The main component of the plant cell wall is the polysaccharide cellulose. The paper we write on, the cotton fibers in the clothes we wear, and the wood in the chair we sit on are all composed of cellulose as well as other polysaccharides. Cellulose is, in fact, the most abundant and widespread organic molecule on Earth. Many types of algae have cell walls made up of cellulose, as in plants, but others have cell walls made of silicon or calcium carbonate. Most fungi have cell walls made of chitin, another polymer of sugars. In bacteria, the cell wall is made up primarily of peptidoglycan, a polymer of amino acids and sugars.

Visual Synthesis 2.1: A Typical Animal Cell and a Typical Plant Cell, on pages 118–119 summarizes the major organelles and structural features of a typical animal and plant cell. These figures do not show specific types of cells, but instead composite cells that highlight key features of animal and plant cells. The functions of the organelles and other cellular structures are described in the textboxes. Although animal and plant cells have many differences, they share a cell membrane, mitochondria, an endomembrane system, and a cytoskeleton. Cell walls, chloroplasts, and vacuoles are present in plant cells but absent in animal cells.

✓ Concept Check

5. **Identify** a shared function of the cytoskeleton and cell wall.

6. **State** which types of organisms have cells with cell walls.

Module 7 Summary

REVISIT THE BIG IDEAS

PREP FOR THE AP® EXAM

SYSTEMS INTERACTIONS: Using the content of this module, **support the claim** that cells possess subcellular structures that interact with each other and the environment.

LG 7.1 The endomembrane system compartmentalizes the eukaryotic cell.

- The endomembrane system is an interconnected system of membranes that defines spaces in the cell, synthesizes important molecules, and traffics and sorts these molecules into and out of the cell. Page 106

- The endomembrane system includes the nuclear envelope, endoplasmic reticulum, Golgi apparatus, lysosomes, vesicles, and cell membrane. Page 106

- The nucleus, which is enclosed by a double membrane called the nuclear envelope, houses the genetic material. Page 106

- The endoplasmic reticulum is continuous with the outer membrane of the nuclear envelope and manufactures proteins and lipids. Page 107

- The Golgi apparatus communicates with the endoplasmic reticulum by vesicles; it receives proteins and lipids from the endoplasmic reticulum and directs them to their final destinations. Page 108
- Lysosomes break down macromolecules such as proteins into simpler compounds that can be used by the cell. Page 109

LG 7.2 Mitochondria and chloroplasts harness energy for use by the cell.

- Mitochondria and chloroplasts likely evolved from free-living prokaryotes and have their own DNA. Page 110
- Mitochondria transfer energy from chemical compounds for use by nearly all cells. Page 110
- Cellular respiration takes place in the cytosol; in the mitochondrial matrix; and along the inner membrane of the mitochondria, which is highly folded. Page 110

- Chloroplasts harness the energy of sunlight to make ATP and build sugars in plant and algal cells. Page 111
- Photosynthesis takes place in the cytosol, chloroplast, and thylakoid membrane. Page 111

LG 7.3 The cytoskeleton and cell wall help to maintain cell shape.

- The cytoskeleton is present in all cells. Page 112
- In eukaryotes, the cytoskeleton consists of microfilament and microtubules that help maintain cell shape and structure. Page 113
- The cell wall of plants is made of cellulose and helps to maintain cell structure. Page 113
- The cell wall is present in plants, algae, fungi, archaea, and bacteria. Page 113
- Vacuoles are conspicuous organelles in plant cells and help to maintain turgor pressure. Page 114

Key Terms

Surface area	Enzyme	Mitochondrial matrix	Chlorophyll
Vesicle	Golgi apparatus	Chloroplast	Cytoskeleton
Endomembrane system	Lysosome	Photosynthesis	Turgor pressure
Nuclear envelope	Mitochondrium	Thylakoid	Vacuole
Endoplasmic reticulum (ER)	Cellular respiration	Grana	

Review Questions

1. In which region of the cell can protein synthesis occur in eukaryotes?

    (A) Rough endoplasmic reticulum

    (B) Nucleus

    (C) Golgi apparatus

    (D) Lysosomes

2. The process of synthesizing simple sugars using sunlight as an energy source is referred to as

    (A) cellular respiration.

    (B) transcription.

    (C) translation.

    (D) photosynthesis.

3. Which is part of the endomembrane system of a eukaryotic cell?

    (A) Ribosomes

    (B) Golgi apparatus

    (C) Chloroplasts

    (D) Mitochondria

4. Where are ribosomes found inside a cell?

    (A) Attached to the Golgi apparatus

    (B) Attached to the smooth endoplasmic reticulum

    (C) On the rough endoplasmic reticulum and in the cytosol

    (D) In the nucleus

5. Which organelle is not part of the endomembrane system?

   (A) Lysosome

   (B) Chloroplast

   (C) Nucleus

   (D) Vesicle

6. Which statement about mitochondria is true?

   (A) Mitochondria generate ATP.

   (B) Mitochondria are subdivided into compartments known as thylakoids.

   (C) Mitochondria are rich in pigments.

   (D) Mitochondria are part of the endomembrane system.

7. Chloroplasts and mitochondria are most closely related to certain

   (A) bacteria.

   (B) fungi.

   (C) single-celled eukaryotes.

   (D) algae.

Module 7
AP® Practice Questions

PREP FOR THE AP® EXAM

Section 1: Multiple-Choice Questions

Choose the best answer for questions 1–6.

1. Scientists discover a new organism. The organism is single-celled and contains a nucleus, endoplasmic reticulum, Golgi apparatus, vesicles, and lysosomes. The organism also contains a double-membraned organelle with no thylakoids or light-harvesting pigments, as well as another double-membraned organelle that does contain thylakoids and light-harvesting pigments. The new organism is most appropriately classified as

   (A) a photosynthetic eukaryotic microbe.

   (B) a photosynthetic bacterium.

   (C) a photosynthetic archaeon.

   (D) a single-celled eukaryote similar to an amoeba.

2. A new photosynthetic marine algal species is discovered. Scientists observe that the organelles of its endomembrane system and nonendomembrane system are similar to those found in photosynthetic algae. Scientists decide to examine the alga's metabolism in the dark. Which would most likely yield a positive result?

   (A) Measuring and collecting data on the amount of glucose produced by the alga

   (B) Measuring and collecting data on the amount of carbon dioxide produced by the alga

   (C) Measuring and collecting data on the activity of vacuoles

   (D) Measuring and collecting data is not possible, as photosynthetic organisms do not have metabolic activity unless given a light source.

Use the following information for questions 3 and 4.

A scientist conducts an experiment with a species of yeast, which is a eukaryotic microbe, called *Saccharomyces cerevisiae*. An external chemical named compound X is applied to the yeast, which binds to DNA and causes the yeast to synthesize a specific protein. The scientist gives compound X to the *S. cerevisiae*. At the same time, the scientist also gives these *S. cerevisiae* cells a second chemical, named Blockem, which blocks nuclear pores. A second group of *S. cerevisiae* cells serves as a control group and receives only compound X.

The scientist later found that the protein was synthesized in *S. cerevisiae* cells that received only compound X. In contrast, *S. cerevisiae* cells that receive both compound X and Blockem do not produce the protein.

3. Which best explains the results observed?

   (A) Blockem destroyed both the DNA and the RNA in the nucleus.

   (B) Blockem destroyed the RNA and the proteins in the nucleus.

   (C) Blockem prevented the DNA in the nucleus from reaching the cytoplasm.

   (D) Blockem prevented compound X from entering the nucleus and, as a result, RNA was not made and did not enter the cytoplasm.

4. The same experiment is performed on the bacterium *Escherichia coli*. The scientist gives the experimental group of *E. coli* cells compound X and Blockem, and gives only compound X to the control group. The scientist notes that both groups of *E. coli* cells make the protein. The scientist then creates a second control group of *E. coli* cells that receives neither compound X nor Blockem, and a third control group that receives only Blockem. Neither the second nor third control group produces the protein. Which of the following best explains the experimental results observed?

(A) Bacterial ribosomes do not use RNA to make proteins.

(B) Bacteria do not contain a nuclear membrane that separates their DNA from the cytoplasm.

(C) Bacteria use both compound X and Blockem as stimulants to make the protein.

(D) Unlike *S. cerevisiae*, *E. coli* makes the protein whether or not compound X is present.

5. Researchers isolate and study a cell from a mammal. The cell has an extensive rough endoplasmic reticulum but little smooth endoplasmic reticulum. Which describes the likely function of this cell?

(A) The cell is engaged in extensive lipid metabolism.

(B) The cell is engaged in extensive detoxification of harmful chemicals.

(C) The cell is engaged in abundant protein synthesis.

(D) The cell is engaged in extensive processing of proteins synthesized by the cell.

6. In humans, fatigue is defined as feeling tired and having little energy. Medical researchers have noted that patients suffering from fatigue sometimes have impairments in cellular respiration. Cellular respiration is a series of chemical reactions, with each step mediated by specific proteins. If one or more of the proteins that participate in cellular respiration is dysfunctional, cellular respiration does not occur in the usual manner, resulting in fatigue. Which best describes why people suffering from impaired cellular respiration suffer from fatigue?

(A) Impaired respiration results in too little oxygen being in their blood.

(B) Impaired respiration results in too much oxygen being in their blood.

(C) Impaired respiration results in too little ATP being produced.

(D) Impaired respiration results in too much ATP being produced.

Section 2: Free-Response Question

Write your answer to each part clearly. Support your answers with relevant information and examples. Where calculations are required, show your work.

The protein elastase is made by the pancreas and is normally secreted by pancreatic cells into serum, a component of blood. The protein is synthesized on the rough endoplasmic reticulum (rough ER) of pancreatic cells. Some of the carbohydrates found on completed elastase molecules are added to the protein in the rough ER.

While isolating and examining elastase from the serum of an individual, researchers note that the protein is not fully glycosylated, which means that the normal, complete set of carbohydrates is not present. Rather, only the carbohydrates attached by the rough ER are present on the enzyme.

(a) **Describe** how the gene for elastase in pancreatic cells normally results in the protein being secreted into the serum.

(b) **Make a claim** about why the elastase of the individual studied lacks the normal, complete set of carbohydrates.

VISUAL SYNTHESIS 2.1 A TYPICAL ANIMAL CELL AND A TYPICAL PLANT CELL

Animal and plant cells have many features and organelles in common. Animal cells have organelles of the endomembrane system, including a nucleus, endoplasmic reticulum, Golgi apparatus, vesicles, and cell membrane. They also have mitochondria and a cytoskeleton. In addition, plant cells have chloroplasts, vacuoles, and a cell wall.

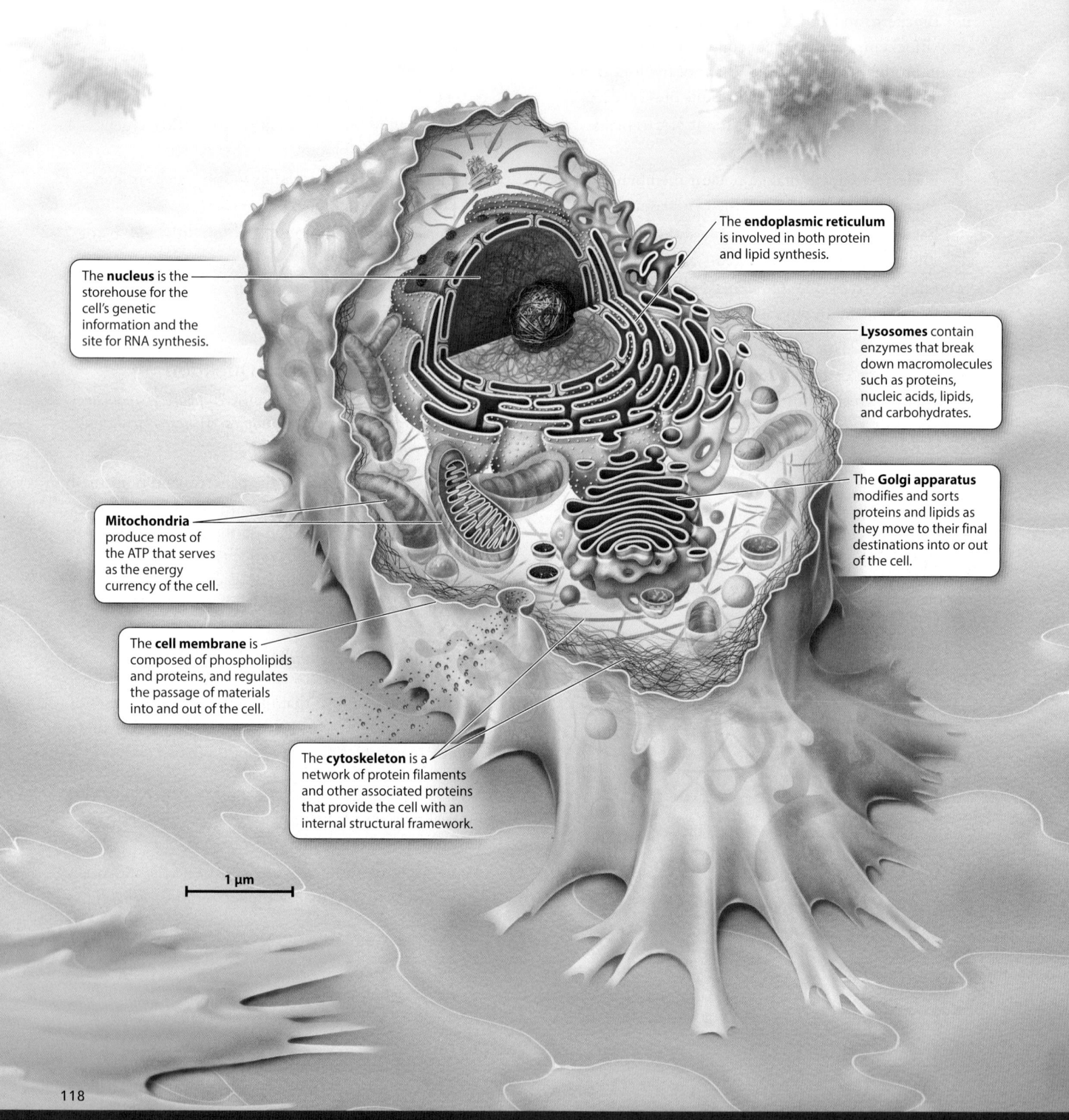

The **nucleus** is the storehouse for the cell's genetic information and the site for RNA synthesis.

The **endoplasmic reticulum** is involved in both protein and lipid synthesis.

Lysosomes contain enzymes that break down macromolecules such as proteins, nucleic acids, lipids, and carbohydrates.

Mitochondria produce most of the ATP that serves as the energy currency of the cell.

The **Golgi apparatus** modifies and sorts proteins and lipids as they move to their final destinations into or out of the cell.

The **cell membrane** is composed of phospholipids and proteins, and regulates the passage of materials into and out of the cell.

The **cytoskeleton** is a network of protein filaments and other associated proteins that provide the cell with an internal structural framework.

1 μm

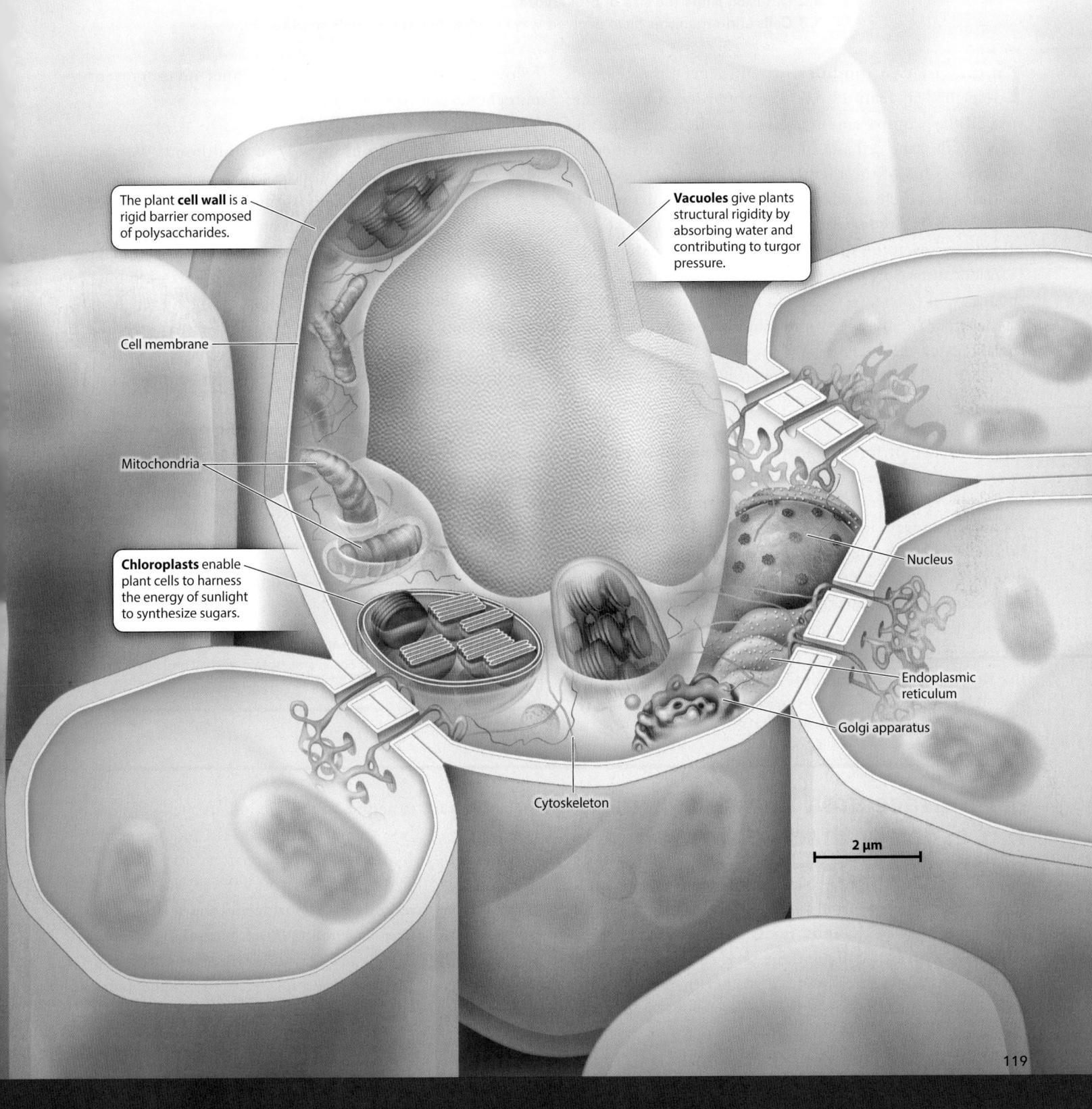

The plant **cell wall** is a rigid barrier composed of polysaccharides.

Vacuoles give plants structural rigidity by absorbing water and contributing to turgor pressure.

Cell membrane

Mitochondria

Chloroplasts enable plant cells to harness the energy of sunlight to synthesize sugars.

Nucleus

Endoplasmic reticulum

Golgi apparatus

Cytoskeleton

2 μm

Module 8

Cell and Organism Size

LEARNING GOALS ▶**LG 8.1** Surface area increases more slowly than volume as an object gets larger.
▶**LG 8.2** Diffusion limits the size of prokaryotes.
▶**LG 8.3** Cells and organisms have evolved ways to circumvent the limits of diffusion.

In the last two modules, we introduced the cell and described its features. We also discussed some of the subcellular compartments, or organelles, that define spaces within the cell and that are specialized for different functions. One of the themes that emerged in these two modules is the close relationship between structure and function. This close relationship can be seen in the shapes of cells and even the shapes of the organelles within cells. Consider surface area, the total amount of area of the outer surface of an object. As we saw in the last module, the folds of the inner membrane of mitochondria and the stacked thylakoids of chloroplasts both provide a large amount of surface area. The large surface area enables cells to carry out their functions more effectively and provide more energy for the cell to use.

The shape or form of a structure depends in part on its size. This may seem counterintuitive at first, but it turns out that the size is an important determinant of shape. Consider the relationship of surface area and **volume,** where volume is the total amount of space an object occupies. As an object gets larger, both its surface area and its volume increase, but the volume increases much more quickly than the surface area. The mismatch between the increase in surface area and volume has implications for the shapes of many biological structures, from organelles to cells to organs and even to organisms. As you will learn below, it helps to explain why the inner membrane of mitochondria is so highly folded.

In biology, we see differences in size as organisms develop and grow from infancy to adulthood. We also observe that organisms themselves differ tremendously in size. The smallest free-living organisms are bacteria, specifically mycoplasma. Mycoplasma are approximately 200 nanometers long, or 200 billionths of a meter. The largest organisms to have ever existed on Earth are blue whales. At about 30 meters long, they are larger, even, than the dinosaurs were. This vast difference in size matters for many functions—how the organisms move about, how they obtain energy and nutrients, how they gain and lose heat, and so on.

In this module, we will look at how the size of an object influences its shape. We will begin with simple geometric objects, such as cubes and spheres, and apply what we learn from these examples to biological structures, such as cells and organisms.

> PREP FOR THE AP® EXAM
>
> **FOCUS ON THE BIG IDEAS**
>
> **ENERGETICS:** Look for examples of cellular organization that affect how cells and organisms obtain energy and nutrients.

8.1 Surface area increases more slowly than volume as an object gets larger

In Module 6, we compared and contrasted prokaryotes and eukaryotes. One of the most notable differences between these two types of cells is their size. Prokaryotic cells, like bacteria, are generally much smaller than eukaryotic cells, like a skin cell. This difference in size has important implications for structure. In this section, we will consider how surface area and volume change as cells and organisms get larger, and why this relationship matters for how they look and how they function. As you will see, the concept of ratios is very useful when considering the relationship between surface area and volume. Before looking at how size affects shape in cells and organisms, "Analyzing Statistics and Data: Ratios" will help you to review the mathematics behind ratios.

Ratios

A ratio is used to compare one quantity to another quantity. For example, if there are 3 sunflowers and 7 tulips in a garden, the ratio of sunflowers to tulips can be expressed as:

$$3{:}7 \text{ or } \frac{3}{7}$$

The ratio of tulips to sunflowers is the reverse:

$$7{:}3 \text{ or } \frac{7}{3}$$

Ratios can also be expressed as one quantity compared to the total quantity. In our flower example, the ratio of sunflowers to all flowers is:

$$3{:}10 \text{ or } \frac{3}{10}$$

A percentage is also a type of ratio. Let's convert our ratio of $\frac{3}{10}$ into a percent by multiplying by 100.

$$\left(\frac{3}{10}\right) \times 100 = 30\%$$

When simplifying ratios, divide each number by the same value. The ratio of 4:8 can be simplified by dividing each number by 4, resulting in a ratio of 1:2.

PRACTICE THE SKILL

We can practice using ratios by comparing the number of atoms in molecules of water and salt. What is the ratio of atoms in 5 molecules of water (H_2O) to 5 molecules of oxygen gas (O_2)?

First, we determine the number of atoms in 1 molecule of each of these molecules. The chemical formula of water is H_2O. It contains 2 atoms of hydrogen and 1 atom of oxygen for a total of 3 atoms in 1 molecule of water. So, in 5 molecules of water, there are $5 \times 3 = 15$ atoms. The chemical formula for oxygen gas is O_2. It contains 2 atoms of oxygen in 1 molecule of oxygen gas. In 5 molecules of oxygen gas, there are $5 \times 2 = 10$ atoms.

The ratio of atoms in 5 molecules of water to 5 molecules of oxygen gas is therefore 15:10 or $\frac{15}{10}$. We can simplify this ratio to 3:2 or $\frac{3}{2}$.

Your Turn

A scientific journal reported a ratio of 5:2 hermit crabs to gulls in a tidal community. Based on this ratio, how many hermit crabs would you estimate to be on the beach if 28 gulls were spotted?

Surface Area and Volume

Any three-dimensional object, including a cell, organelle, or organism, can be described by the amount of area of its outer surface, or surface area, and the amount space it occupies, or its volume. For all shapes, surface area describes a flat, two-dimensional structure, so is calculated mathematically by multiplying the lengths of two sides. Volume describes a three-dimensional structure, so is calculated by multiplying the lengths of three sides. Biological structures tend to be irregularly shaped. Therefore, to describe the surface area and volume of biological structures, we start by considering standard geometric shapes such as cubes, spheres, and cylinders.

Let's begin with a cube, shown in **FIGURE 8.1**. Each face of a cube is a square. The area of a square is

OBJECT		SURFACE AREA	VOLUME	VARIABLES
Cube 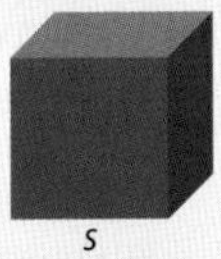		$SA = 6s^2$	$V = s^3$	s = side length
Rectangular solid 		$SA = 2lh + 2lw + 2wh$	$V = lwh$	l = length w = width h = height
Sphere 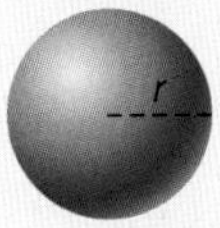		$SA = 4\pi r^2$	$V = \frac{4}{3}\pi r^3$	r = radius
Cylinder 		$SA = 2\pi r^2 + 2\pi rh$	$V = \pi r^2 h$	r = radius h = height

FIGURE 8.1 Surface area and volume

The surface area and volume of various three-dimensional objects can be calculated using the indicated formulas.

the length of a side $(s) \times$ the length of a side (s), or s^2. A cube has six faces, so the total surface area is $6 \times s^2$. The volume of a cube is the area of one face, or s^2, multiplied by the length of a side, or s^3.

In summary:

Surface area of a cube $= 6 \times (\text{length of a side})^2 = 6s^2$

Volume of a cube $= \text{length of a side}^3 = s^3$

Notice that the surface area of a cube is calculated by squaring a length of a side $(s \times s)$, whereas the volume is calculated by cubing the length of a side $(s \times s \times s)$. This is a key observation: areas are proportional to a squared factor, while volumes are proportional to a cubed factor. In fact, this observation holds for all three-dimensional objects. For example, consider a sphere, shown in Figure 8.1. If the radius of the sphere is r, then

Surface area of a sphere $= 4\pi r^2$

Volume of a sphere $= \dfrac{4}{3}\pi r^3$

These formulas are different from those used to calculate the surface area and volume of a cube, but again we see that surface area is described by a square $(r \times r, \text{ or } r^2)$, whereas volume is described by a cube $(r \times r \times r, \text{ or } r^3)$.

Figure 8.1 also includes formulas for the surface area and volume of a rectangular solid and a cylinder. In each case, we see that surface area is determined by multiplying two lengths, and volume is determined by multiplying three lengths.

Surface Area-to-Volume Ratio

The formulas for the various objects in Figure 8.1 have implications for what happens when an object gets larger. As an object gets larger, both the surface area and the volume increase, but the volume increases faster than the surface area increases. Or, conversely, the surface area does not increase as quickly as the volume does. The same is true when comparing a small object to a large object of the same shape. The large object has much more volume relative to its surface area than the small object does. Biologists consider the relationship between surface area and volume when determining what happens as organisms grow or when comparing small and large organisms; some functions depend on surface areas and others depend on volumes, but the two have to work together.

We can describe the amount of surface area and volume for a given object by the ratio of surface area to volume. The surface area-to-volume ratio is equal to the surface area divided by the volume, as follows:

Surface area-to-volume ratio $= \text{surface area}: \text{volume}$

$$= \dfrac{SA}{V}$$

2   4

FIGURE 8.2 A small and big cube

The small cube has sides of length 2, and the big cube has sides of length 4. The surface area-to-volume ratio of the big cube, 1.5, is smaller than the surface area-to-volume of ratio of the small cube, 3.

Let's compare the surface area-to-volume ratio of a small and big cube. As shown in **FIGURE 8.2**, the small cube has sides of length $= s = 2$, and the big cube has sides of length $= s = 4$. Recall that all four sides of a cube are the same length. For the small cube,

$SA = 6 \times (s \times s) = 6 \times (2 \times 2) = 24$

$V = s \times s \times s = 8$

$\dfrac{SA}{V} = \dfrac{24}{8} = \dfrac{3}{1} = 3$

For the big cube,

$SA = 6 \times (s \times s) = 6 \times (4 \times 4) = 96$

$\text{Volume} = s \times s \times s = 4 \times 4 \times 4 = 64$

$\dfrac{SA}{V} = \dfrac{96}{64} = \dfrac{6}{4} = 1.5$

We see that the surface area of the big cube is larger than the surface area of the small cube. We also see that the volume of the big cube is larger than the volume of the small cube. However, the surface area-to-volume ratio of the big cube (1.5) is actually smaller than the surface area-to-volume ratio of the small cube (3). In other words, the big cube has less surface area relative to its volume than the small cube has.

The same is true of biological structures, like a grape and a grapefruit. A grapefruit is bigger than a grape, and it has more surface area and a bigger volume than a grape. However, because volume increases more quickly than surface area as an object gets bigger, the grapefruit has less surface area relative to its volume than a grape has, or a smaller surface area to volume ratio. Conversely, a grape has more surface area relative to its volume than a grapefruit has.

> **PREP FOR THE AP® EXAM**
>
> **AP® EXAM TIP**
>
> Questions that involve calculating the ratio of surface area-to-volume frequently appear on the AP® Biology Exam. Make sure you are comfortable using the relevant formulas and understand the significance of the ratio of surface area-to-volume.

Scaling

The mismatch between how quickly surface area increases relative to volume has many implications for biology. For example, ants look the way they do in part because of their size. If an ant suddenly became 10 times larger than it usually is, the strength of its legs would increase 100 times because strength is proportional to the cross-sectional area of the legs. However, the weight would increase 1000 times because weight is proportional to volume. As a result, its weight would quickly outstrip the strength of its legs, and it would come crashing down. Therefore, a human-sized ant would not be able support its body.

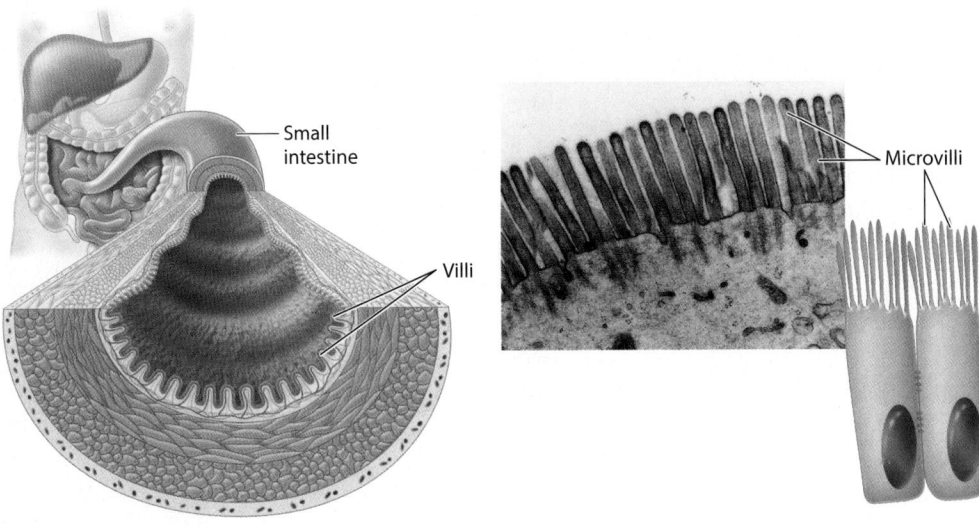

FIGURE 8.3 Intestinal villi and microvilli

Villi and microvilli are folds in the small intestine. These structures increase the surface area of the gut, allowing larger organisms to absorb sufficient nutrients to support the volume of the body.

Photo: Biophoto Associates/Science Source

Now let's consider an ant-sized human. An ant-sized human would have more surface area to volume compared with a normal human, which would create problems. For example, we generate heat through the many metabolic reactions occurring in the numerous cells throughout our body, which is a volume. However, we lose heat through our skin, which is a surface. With more surface area relative to volume, an ant-sized human would lose too much heat, unless that human constantly ate and moved around to generate enough heat to keep up with the increased loss.

The problem with imagining a human-sized ant or an ant-sized human is that they increase in size but keep their existing overall shape. This kind of change is called isometry, meaning same measure. By contrast, increases in size are often accompanied by changes in shape. This kind of change is called allometry, meaning different measure. For example, in the 1600s, Galileo Galilei noticed that the bones of larger animals are not scaled-up versions of the bones of smaller animals. Instead, in many cases, to support greater weight, the bones of larger animals are disproportionately wider than those of smaller animals. The British biologist J. B. S. Haldane wrote, "For every type of animal there is a most convenient size, and a large change in size inevitably carries with it a change of form."

We have seen that as any three-dimensional object gets bigger, both surface area and volume increase, but surface area increases more slowly than does volume. Therefore, shape changes often accompany size differences to balance the two.

For example, nutrients are absorbed through the lining of the gut, which is one large surface area. However, these nutrients must support the entire body, which represents a volume. Larger organisms, therefore, have adaptations of the lining of the gut to increase its surface area. These adaptations include folds of the lining of the gut, called villi, and projections of the surface of the intestinal cells, called microvilli. As shown in **FIGURE 8.3**, these structures greatly increase the surface area available to absorb nutrients. The increased surface area of the lining of the gut enables larger organisms to absorb enough nutrients to support the increased volume.

Just like the lining of the gut, the inner membrane of mitochondria is highly folded. These folds increase the surface area of the inner membrane, which is the site of ATP production in mitochondria. ATP is produced across a membrane, so it is limited by the amount of surface area available, but it provides energy for an entire cell, which is a volume. The folds of the inner membrane therefore are an adaptation that allows enough surface area for ATP synthesis to match the energy needs of the volume of the cell.

✓ Concept Check

1. **Describe** what happens to surface area and volume as an object gets bigger.

2. **Compare** the surface area to volume ratio of a eukaryotic cell and a prokaryotic cell.

3. **Calculate** the surface area to volume ratio of a cube with sides of length 5.0 cm.

8.2 Diffusion limits the size of prokaryotes

As we have seen, the relationship between surface area and volume influences many aspects of biology, from the structure of the lining of the gut to the shape of the inner membrane of mitochondria. It also influences the size of cells.

Prokaryotic cells are smaller than eukaryotic cells. Let's focus on one group of prokaryotes—bacteria. Most bacterial cells are tiny: the smallest are only 200–300 nanometers (nm) in diameter, and relatively few are more than 1–2 micrometers (μm) long. Several bacterial cells are shown in **FIGURE 8.4**. You can see in the figure that they are tiny spheres, rods, and spirals. Recall that a small object has more surface area relative to its volume than a large object. Because bacteria are so small, they have a large amount of surface area compared to volume. Most bacteria take in nutrients and other substances they need through the surface area of their cell membrane. Their small size means that they have a lot of surface area for the movement of substances across their cell membranes to support the volume of the interior of the cell.

Let's consider how some substances enter cells. All molecules are in constant motion in their environments. If you could watch the movement of any particular molecule in air or water, you would see that its motion is random, sometimes going in one direction and sometimes in another. When there are differences in concentration in the distribution of a molecule, molecules tend to move from a region with a higher concentration of the same molecule to a region with a lower concentration of the molecule. In other words, random motion leads to a net movement of molecules from regions of higher to lower concentration, a process called **diffusion.** Net movement stops only when the two regions achieve equal concentrations of the molecule, although random motion continues.

Bacteria rely on diffusion to take in the molecules they require and remove wastes. For example, photosynthetic bacteria gain the carbon dioxide they need by the diffusion of carbon dioxide from the environment into the cell across their cell membrane. Similarly, respiring bacteria take in small organic molecules and oxygen by diffusion.

a. Spheres

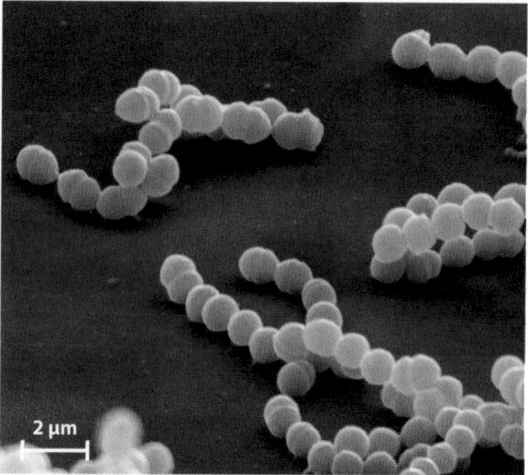

b. Rods

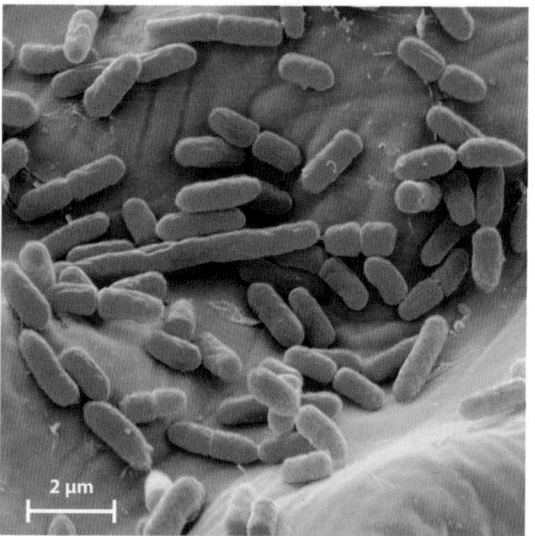

c. Spirals

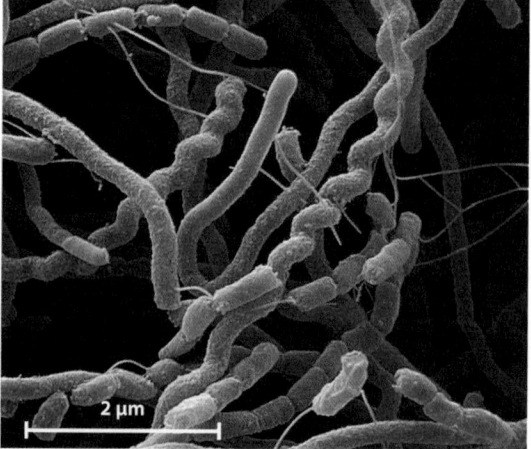

FIGURE 8.4 Cell size and shape in bacteria

Bacteria are very small, typically 1–2 micrometers (μm) across. As cell size increases, it becomes harder to supply the cell with materials needed for growth. They also come in different shapes: (a) *Streptococcus* form strings of spheres. (b) *E. coli* are shaped like rods. (c) *Streptomyces* are helical bacteria. Photos: (a) Eye of Science/Science Source; (b) Scimat Scimat/Getty Images; (c) David Scharf/Science Source

Diffusion is fast over short distances. However, as distances become greater, diffusion takes a long time and is therefore not an effective way to transport nutrients and other key molecules. The average time for a glucose molecule to diffuse across a 10-μm cell is only 1 second and to diffuse across a 100-μm cell is 100 seconds. But to diffuse 1 mm would take, on average, 16 minutes, while to diffuse across 1 m would take almost 32 years!

Diffusion helps to explain why bacterial cells tend to be small. The interior parts of a small cell are closer to the surrounding environment than are those of a larger cell. As a consequence, slowly diffusing molecules do not have to travel as far to reach every part of a small cell's interior. Because bacterial cells rely on diffusion, most are small enough for molecules to diffuse into the cell's interior.

The ratio of surface area to volume also helps to explain the small size of bacteria. A small cell, as we have seen, has more surface area relative to its volume than a large cell. The surface area of a spherical cell—the area available for taking up molecules from the environment—increases as the square of the radius. However, the cell's volume—the amount of cytoplasm that is supported by diffusion—increases as the cube of the radius. Therefore, as cell size increases, there is less surface area relative to volume and diffusion takes too long to distribute the materials needed for growth.

We can learn even more about how diffusion works by considering an exceptionally large bacterium. The largest known bacterium, *Thiomargarita namibiensis,* exceeds 100 μm in length and lives in oxygen-poor sediments off the coast of southwestern Africa. It is shown in **FIGURE 8.5**. Its total volume is approximately 100 million times larger than that of a typical bacterial cell. But, in one sense, *T. namibiensis* cheats: 98% of its volume is taken up by a large vacuole, so the cytoplasm is restricted to a thin film around the cell's periphery.

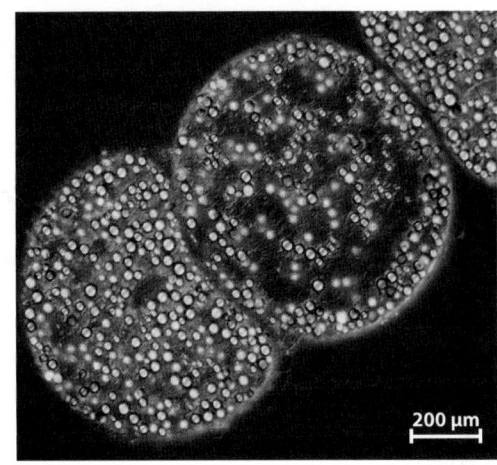

FIGURE 8.5 *Thiomargarita namibiensis,* **the largest known bacterial cell**

A vacuole takes up most of the volume of the cell, so the active cytoplasm is only a few micrometers thick. Thus, the distance through which nutrients move by diffusion is only a few micrometers, as in other bacteria. The small spheroidal granules consist of sulfur, formed by the metabolism of hydrogen sulfide in their environment. Photo: Courtesy Heide Schulz-Vogt, Leibniz Institute for Baltic Sea Research, Department of Biological Oceanography

Despite its size, the distance through which nutrients move by diffusion is only a few micrometers, as in many other bacteria. "Analyzing Statistics and Data: Surface Area-to-Volume Ratio" reviews how to calculate surface area to volume ratios and provides an opportunity to practice a problem.

✓ Concept Check

4. **Describe** what is meant by diffusion.

5. **Describe** the relationship between diffusion and the small size of bacteria.

8.3 Cells and organisms have evolved ways to circumvent the limits of diffusion

Eukaryotic cells are larger than prokaryotic cells. As we discussed, the larger size of eukaryotes means that compared to prokaryotic cells, they have less surface area relative to their volume. We have already noted one adaptation that makes larger size viable: internal membranes. Eukaryotic cells have an extensive array of internal membranes and compartments. Many of these membranes are highly folded or convoluted, such as the ER, Golgi apparatus, inner mitochondrial membrane, and thylakoid. Highly folded membranes provide a large amount of surface area to serve the volume of the cell. This adaptation allows for an increase in cell size in eukaryotic cells compared to prokaryotic cells.

What about multicellular organisms? Multicellular organisms are even larger than unicellular organisms. As a result, the problem of how to obtain nutrients and remove wastes is even more challenging because they cannot rely simply on diffusion. In this section, we discuss how large, multicellular organisms use both diffusion and another mechanism, called bulk flow, to obtain the nutrients they need for their functions.

Surface Area-to-Volume Ratio

As a cell, organ, or entire organism increases in size, its surface area and volume also increase. However, surface area increases more slowly than volume. This means that the surface area-to-volume ratio (SA:V) decreases as cell size increases. Small cells have high SA:V, which is especially useful for nutrient uptake or waste removal by diffusion. Folding of membranes increases surface area without greatly increasing volume and leads to a closer match between surface area and volume.

PRACTICE THE SKILL

Microvilli demonstrate the importance of the SA:V ratio. As noted, microvilli are small cellular projections of the membrane of an intestinal cell. They greatly increase the surface area for absorption or secretion, while minimizing an increase in volume. Let's compare the surface area, volume, and SA:V ratios of a section of a smooth small intestine without microvilli to a smooth small intestine tissue with 100 microvilli. We will assume that the smooth tissue is a thin rectangular solid with $l = 8$ mm, $w = 4$ mm, and $h = 2$ mm.

The surface area for this rectangular solid is:

$$SA = 2lh + 2lw + 2wh$$
$$SA = 2(8 \text{ mm}) \times (2 \text{ mm}) + 2(8 \text{ mm}) \times (4 \text{ mm}) + 2(4 \text{ mm}) \times (2 \text{ mm})$$
$$SA = 112 \text{ mm}^2$$

The volume for this rectangular solid is:

$$V = lwh$$
$$V = (8 \text{ mm}) \times (4 \text{ mm}) \times (2 \text{ mm})$$
$$V = 64 \text{ mm}^3$$

To calculate the surface area to volume ratio, we simply divide surface area by volume:

$$\text{Surface area to volume ratio} = \frac{SA}{V} = \frac{112}{64} = 1.75,$$

or approximately 1.8

Now let's calculate the quantities for 100 microvilli. Microvilli are roughly cylindrical in shape and have $h = 1$ μm and $r = 0.045$ μm. First, we will calculate the surface area and volume for a single microvillus.

The surface area for a cylinder is:

$$SA = 2\pi rh + 2\pi r^2$$
$$SA = 2\pi(0.045 \text{ μm}) \times (1 \text{ μm}) + 2\pi(0.045 \text{ μm})^2$$
$$SA = 0.3 \text{ μm}^2$$

For 100 microvilli, we multiply this surface area by 100, as follows: $0.3 \text{ μm}^2 \times 100 = 30 \text{ μm}^2$.

The volume for a cylinder is:

$$V = \pi r^2 h$$
$$V = \pi(0.045 \text{ μm})^2(1 \text{ μm})$$
$$V = 0.006 \text{ μm}^3$$

To calculate the volume for 100 microvilli, multiply by 100: $0.006 \text{ μm}^3 \times 100 = 0.6 \text{ μm}^3$.

Now we will calculate the surface area to volume ratio of 100 microvilli:

$$\text{Surface area-to-volume ratio} = \frac{SA}{V} = \frac{30}{0.6} = 50$$

Comparing the two ratios, we can see that the 100 microvilli have a much greater SA:V ratio than the smooth small intestine tissue (50 > 1.8). The presence of microvilli results in a vast increase in surface area, but not volume. Microvilli enhance the ability of the small intestine to absorb nutrients quickly and efficiently by diffusion.

Your Turn

Calculate the surface area and volume of a cylindrical cell with a height of 3.0 μm and a radius of 2.0 μm. Is this cell more or less efficient at absorbing nutrients by diffusion than a spherical cell with a surface area to volume ratio of 1.1 μm^{-1}? Why?

Diffusion in Multicellular Organisms

We have seen that diffusion is net movement from areas of higher concentration to areas of lower concentration due to random motion. Diffusion is effective only over small distances. Because diffusion supplies key molecules for metabolism, it exerts a strong constraint on the size, shape, and function of cells and organisms.

The need for oxygen provides a good example. Most eukaryotes require oxygen for respiration. If a cell or tissue must rely on diffusion for its oxygen supply, it cannot be far away from the source of oxygen in the environment. Aquatic organisms obtain oxygen from the water. In shallow water that is in direct contact with the atmosphere, animals that rely on diffusion for oxygen cannot exceed thicknesses of approximately 1 mm to 1 cm because active cells must remain close to the water. Of course, a quick swim along the seacoast will reveal many animals that are much larger. And, obviously, you are larger. How do you and other large animals get enough oxygen to all of your cells?

Sponges, such as those shown in **FIGURE 8.6a**, can reach overall dimensions of a meter or more, but they actually consist of only a few types of cell that line a dense network of pores and canals. As a result of this organization, their cells remain in close contact with circulating seawater. In essence, sponges can grow large without placing active cells at any great distance from their environment.

Similarly, sea jellies (also called jellyfish), shown in Figure 8.6b, confine active cells to thin tissues that line the inner and outer surfaces of the body. Essentially, a large flat surface is folded up to produce a three-dimensional structure. The sea jelly's bell-shaped body is often thicker than the active tissue, but its massive interior is filled by the mesoglea, the sea jelly's "jelly." The mesoglea provides structural support but does not require very much oxygen. Sponges and sea jellies are able to reach a large size while relying on diffusion to take in oxygen because their active cells and tissues are near to, or in direct contact with, the environment and they provide a large amount of surface area for oxygen absorption.

Similarly, humans and other animals have organs that provide a large amount of surface area for oxygen absorption. The lung is a prime example of diffusion in action. The lung consists of airways that branch extensively until they reach the terminal air sacs called alveoli. The two human lungs together consist of about 600 million alveoli, which provide a very large amount of surface area, about the size of a tennis court. At the same time, the walls of the alveoli are exceedingly thin, which allows oxygen to diffuse readily from the air spaces to the blood. The combination of large surface area and thin walls allow diffusion to be an effective means of transporting oxygen from the air to the blood.

a.

b.

FIGURE 8.6 Sponges and sea jellies

Sponges and sea jellies are both large, multicellular organisms that rely on diffusion. (a) Sponges like this azure vase sponge (*Callyspongia plicifera*) can attain a large size because the many pores and canals in their bodies ensure that all cells are in close proximity to the environment. (b) Sea jellies, such as this purple-striped jellyfish (*Chrysaora colorata*), also have thin layers of active tissue, but their familiar bell can be relatively thick because it is packed with inert molecules that make up the mesoglea, or "jelly." Photos: (a) Andrew J. Martinez/SeaPics.com; (b) D. R. Schrichte/SeaPics.com

Bulk Flow in Multicellular Organisms

Diffusion provides an efficient way for oxygen to move from the lung to the bloodstream, but how does it get from the lungs to our brains or toes? And how does it get into the lungs in the first place? The distances are far too great for diffusion to be effective. To get from our lungs to other parts of the body, oxygen travels by a mechanism called bulk flow. Bulk flow is the movement of a fluid driven by pressure differences. In the case of oxygen, after it diffuses through the alveolar wall and enters the bloodstream, it is carried throughout the body by molecules in red blood cells that bind oxygen, called hemoglobin. This movement is driven by the pumping action of your heart, which drives the movement of blood and oxygen along with it. In short, we circumvent the limits of diffusion by actively pumping oxygen-rich blood through our bodies, driven by pressure differences created by a beating heart.

In humans and other vertebrate animals, the active pumping of blood by the heart through blood vessels supplies oxygen to tissues that may be more than a meter distant from the lungs, as illustrated in **FIGURE 8.7a**. Most invertebrate animals lack well-defined blood vessels but pump fluids into the body cavity, where they circulate freely. In addition to oxygen, bulk flow also distributes nutrients throughout the body. Intestinal cells absorb organic molecules, but it is bulk flow that distributes these molecules over large distances through the bloodstream. Signaling molecules such as hormones also move rapidly through the body by means of bulk flow in the blood.

Humans and other mammals also rely on bulk flow to get air from outside of the lung to the inside of the lung. Mammals have a muscle, called the diaphragm, to bring in and expel air. When the diaphragm contracts, it creates a pressure difference between the inside and outside of the lung, so that air rushes in. When the diaphragm relaxes, the pressure difference is reversed, and air moves out. In this way, bulk flow brings oxygen-rich air over much greater distances than would be possible by diffusion. As a result, we use bulk flow driven by the diaphragm to get oxygen into the lungs, diffusion across the thin lung tissue to move oxygen from the inside of the lung to the bloodstream, and bulk flow driven by the heart to transport it around the body.

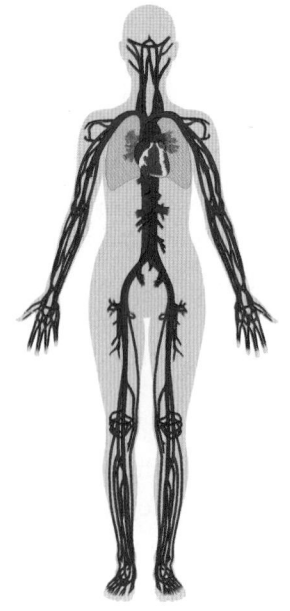

FIGURE 8.7 Bulk flow

Bulk flow is the movement of substances driven by pressure differences. It allows substances to be transported long distances that would not be possible by diffusion. (a) Transport in the circulatory system of animals is driven by the beating heart. (b) Transport in the vascular system of plants is driven by evaporation of water from leaves. In both cases, bulk flow allows organisms to get around the size limits of diffusion.

Plants also rely on bulk flow. A redwood tree transports water upward from its roots to leaves that may be 100 m above the soil. Plants that rely on diffusion to transport water are only a few millimeters tall. How, then, do most plants move water? Bulk flow operates in plants through a system of vascular channels powered by the evaporation of water from leaf surfaces, as illustrated in Figure 8.7b. The evaporation of water creates a pressure difference between the base and top of a tree that helps to pull water up through the tree.

In general, when some cells within an organism are buried within tissues, far from the external environment, bulk flow is required to supply those cells with the molecules needed for metabolism. Indeed, without a mechanism like bulk flow, plants and animals could not have achieved the range of size, shape, and function that is familiar to us.

✓ Concept Check

6. **Identify** two factors that increase the rate of diffusion.

7. **Describe** two ways that large, multicellular organisms circumvent the limitations of diffusion.

Module 8 Summary

PREP FOR THE AP® EXAM

REVISIT THE BIG IDEAS

ENERGETICS: Using the content of this module, explain how the form of organelles, cells, organs, and organisms allows them to function effectively.

LG 8.1 Surface area increases more slowly than volume as an object gets larger.

- Surface area is a measure of the total surface of a three-dimensional object. Page 120
- Volume is a measure of the total space occupied by a three-dimensional object. Page 120
- Surface area to volume ratio equals the surface area divided by the volume. Page 122
- The surface area to volume ratio gets smaller as objects get larger. Page 122
- Inner membrane folds of mitochondria and villi and microvilli of the intestine both increase surface area relative to volume. Page 123

LG 8.2 Diffusion limits the size of prokaryotes.

- Diffusion is net movement of molecules from regions of higher concentration to regions of lower concentration due to random motion. Page 124

- Diffusion is effective only over short distances. Page 125
- Diffusion places limits on the size of cells and multicellular organisms. Page 125

LG 8.3 Cells and organisms have evolved ways to circumvent the limits of diffusion.

- Some multicellular organisms have active cells along the periphery close to the source of nutrients and other key molecules. Page 127
- Some multicellular organisms have organs, such as the lungs, with large surface areas and thin walls that allow for diffusion. Page 127
- Bulk flow is a process in which fluids move by pressure differences at rates much greater than is possible by diffusion. Page 128
- Bulk flow allows multicellular organisms to nourish cells located far from the external environment, thereby circumventing the constraints imposed by diffusion. Page 128

Key Terms

Volume
Diffusion

Review Questions

1. As an object gets larger
   (A) the surface area decreases and the volume decreases.
   (B) the surface area increases and the volume decreases.
   (C) the surface area decreases and the volume increases.
   (D) the surface area increases and the volume increases.

2. As an object gets larger
   (A) the surface area-to-volume ratio increases.
   (B) the surface area-to-volume ratio decreases.
   (C) the surface area-to-volume ratio stays the same.
   (D) the surface area-to-volume ratio cannot be determined.

3. As an object gets larger
   (A) the volume increases more quickly than the surface area.
   (B) the volume increases more slowly than the surface area.
   (C) the volume increases at the same rate as the surface area.
   (D) the volume increases, but the surface area decreases.

4. Diffusion is defined as

    (A) the net movement of molecules from areas of higher concentration to areas of lower concentration due to random motion.

    (B) the net movement of molecules from areas of lower concentration to areas of higher concentration due to random motion.

    (C) the highly ordered movement of molecules from areas of higher concentration to areas of lower concentration.

    (D) the highly ordered movement of molecules from areas of lower concentration to areas of higher concentration.

5. Which is not a mechanism that organisms use to circumvent the limits imposed by diffusion?

    (A) Placing active cells deep in the interior of an organism

    (B) Using vascular tissue to transport materials throughout the plant body

    (C) Using a circulatory system to distribute materials throughout the animal body

    (D) Placing inert structures along the exterior surface

6. Which statement about diffusion is false?

    (A) Diffusion is effective only over small distances.

    (B) Diffusion limits the size of prokaryotic cells.

    (C) Diffusion supplies key molecules for metabolism.

    (D) Diffusion transports materials actively, powered by ATP.

7. Which statement about bulk flow is false?

    (A) Bulk flow is the movement of molecules throughout an organism's body driven by pressure differences.

    (B) Bulk flow is found in unicellular organisms but not multicellular organisms.

    (C) Bulk flow moves materials over greater distances than is possible with diffusion.

    (D) Bulk flow allows animals and plants to grow larger than a few millimeters.

Module 8
AP® Practice Questions

Section 1: Multiple-Choice Questions

Choose the best answer for questions 1–4.

1. Some bacteria have shapes that very nearly approximate a sphere. If we assume that these bacteria are completely spherical in shape, what is the ratio of surface area to volume for these cells?

    (A) $\dfrac{0.3}{r}$

    (B) 3

    (C) $\dfrac{3}{r}$

    (D) $3r$

2. Bacterium A and bacterium B are both spherical bacteria. The radius of bacterium A is 1 μm, and the radius of bacterium B is 3 μm. How much greater is the surface area to volume ratio of bacterium A than that of bacterium B?

    (A) It is 3 times greater.

    (B) There is no difference; they both have the same surface area to volume ratio.

    (C) It is actually smaller and is $\frac{1}{3}$ (one-third) as much as the surface area to volume ratio of bacterium B.

    (D) The answer cannot be calculated from the data given.

3. A researcher compares two different eukaryotic cells, cell 1 and cell 2. Both cells have the same volume and number of mitochondria, but cell 2 has 8 times the number of folds in its inner mitochondrial membrane as does cell 1. Which best explains the difference between the functions of the two cells?

    (A) Cell 1 produces more ATP than cell 2.

    (B) Cell 1 produces less ATP than cell 2.

    (C) Cell 1 and cell 2 produce the same amount of ATP.

    (D) Cell 1 and cell 2 do not produce ATP.

4. Human lungs normally contain about 600 million alveoli. A patient is analyzed and is severely lacking in the total number of alveoli, having only about 300 million alveoli. Which of the following is a likely consequence of having so few alveoli?

   (A) Bulk flow will only transport oxygen half as far from the lungs.

   (B) The patient will have lower than normal blood oxygen levels.

   (C) Oxygen will not diffuse from the alveoli into the blood.

   (D) The patient's body will be unable to take up nutrients from the blood.

Section 2: Free-Response Question

Write your answer to each part clearly. Support your answers with relevant information and examples. Where calculations are required, show your work.

Penguins live in cold environments and have a cylinder-like body shape with short and thin cylinder-like legs. Working with their local zoo, AP® Biology students take measurements of a penguin's body. For the main body, they measure its height to be 90 cm and its radius to be 25 cm. Each leg has a height of 18 cm and a radius of 5 cm.

(a) **Calculate** the surface area-to-volume ratio of the penguin's main body and of one of its legs if both the main body and the legs are considered to be cylinders.

(b) Consider the surface area-to-volume ratios that you calculated for the penguin's main body and one of its legs. **Predict** which is more likely to gain or lose heat. Justify your answer.

(c) In addition to having a thick coat of feathers, penguins keep warm in their cold environment because of the way their blood vessels are organized. Vessels carrying warm blood from the heart to the feet run alongside vessels carrying cooled blood back from the feet to the heart. As the warm blood and cool blood move past each other in opposite directions, some of the heat from the warm blood flows into the cooler blood. The result is that the cooled blood is already somewhat warm by the time it gets back to the heart. **Describe** why this paired arrangement of blood vessels is more necessary in the penguin's feet than in the penguin's body.

Module 9

Cell Membranes

LEARNING GOALS ▶**LG 9.1** Cell membranes are composed of two layers of lipids.
▶**LG 9.2** Cell membranes are dynamic.
▶**LG 9.3** Proteins and carbohydrates associate with cell membranes.

In Module 6, we learned that all cells have a cell membrane, also called a plasma membrane. It is present in prokaryotes and eukaryotes, and in unicellular organisms and multicellular organisms. It is a universal feature of cells. The cell membrane encloses the space of the cell; without it, the cellular contents would disperse, resulting in the loss of organization that is necessary for life itself.

However, the cell membrane is not simply a uniform barrier between the inside and outside of a cell. Instead, it is made up of a mix of components, including lipids, proteins, and carbohydrates, that work together so that cell membranes are able to maintain homeostasis, or a stable internal environment inside of cells or organisms compatible with life. An organism's external environment is constantly changing—it may be hot or cold, wet or dry, windy or still. By contrast, the inside of a cell or organism is stable. The internal environment is kept within a very narrow range of conditions, such as temperature, ion concentration, and pH. These stable internal conditions are required for chemical reactions that sustain life, the proper folding of proteins, and other biological processes.

Homeostasis plays out at the cell membrane that encloses and defines a cell. The cell membrane actively controls what enters and what exits the cell, and in this way keeps the internal environment remarkably stable. In addition, cells require nutrients and other substances to grow and carry out their functions and generate wastes that must be removed. Movement into and out of the cells is controlled by the cell membrane and its components.

In Module 7, we learned that eukaryotic cells have an extensive array of internal membranes known as the endomembrane system. These membranes define compartments in the cell called organelles that carry out specific functions. Movement into and out of these compartments is also carefully regulated by membranes.

In this module, we will examine the structure of cell membranes to gain a better understanding of how they define spaces and carry out their functions. We will examine the structure and function of each of the membrane components, including lipids, proteins, and carbohydrates. We will also learn that the cell membrane is not a static structure. The dynamic nature of a cell membrane allows it to serve as an effective gatekeeper between the inside and outside of the cell.

PREP FOR THE AP® EXAM

FOCUS ON THE BIG IDEAS

ENERGETICS: Consider how the structure of a cell's membrane influences its ability to take in nutrients and other forms of matter from the environment.

9.1 Cell membranes are composed of two layers of lipids

Cells are physically defined by their cell membranes. Membranes, shown in **FIGURE 9.1**, separate cells from their external environment and define compartments within eukaryotic cells that allow them to carry out their diverse functions. Lipids, which were introduced in Unit 1, are the main component of cell membranes. Lipids have particular properties that allow them to form a barrier in an aqueous, or watery, environment. In this section, we will discuss the role of the lipid component of cell membranes.

The major types of lipids found in cell membranes are phospholipids, which were introduced in Module 3.

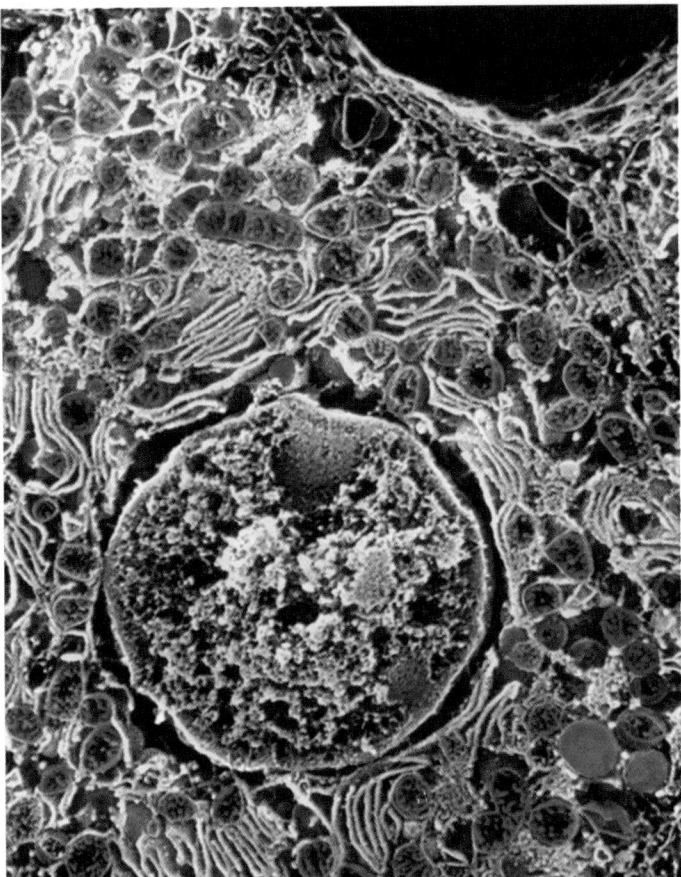

FIGURE 9.1 Cell membranes

All cells have membranes that define and enclose them. Eukaryotic cells, like the liver cell shown here, also have extensive internal membranes that enclose various compartments, or organelles.

Photo: Pietro M. Motta & Tomonori Naguro/Science Source

Most phospholipids are made up of a glycerol backbone attached to a phosphate group and two fatty acids, as shown in **FIGURE 9.2.** The phosphate head group is hydrophilic. It is polar and can form hydrogen bonds with water. By contrast, the two fatty acid tails are hydrophobic. They are nonpolar and do not form hydrogen bonds with water. Molecules such as phospholipids that have both hydrophilic and hydrophobic regions in a single molecule are termed amphipathic.

In an aqueous environment, amphipathic molecules like phospholipids behave in a particular way. Namely, they spontaneously arrange themselves into various structures in which the polar head groups are positioned on the outside and interact with water, and the nonpolar tail groups come together on the inside and interact with other fatty acid regions away from water. This arrangement results from the tendency of polar molecules like water to exclude nonpolar molecules or nonpolar groups of molecules.

The shape of the structure that forms is determined in part by the bulkiness of the head group relative to the hydrophobic tails. Several of these shapes are illustrated in **FIGURE 9.3.** For example, lipids with bulky head groups and a single fatty acid tail are wedge-shaped and pack into a spherical structure called a micelle, shown in Figure 9.3a. By contrast, lipids with less bulky head groups and two hydrophobic tails are roughly rectangular and form a *lipid bilayer,* shown in Figure 9.3b. A **lipid bilayer** is a structure formed of two layers of lipids in which the hydrophilic head groups are on the outside surfaces of the bilayer and therefore interact with water, and the hydrophobic tails are sandwiched in between, isolated from contact with water. All membranes found in cells are lipid bilayers, including the membrane enclosing the cell and internal membranes, such as the nuclear envelope and other membranes of the endomembrane system.

When phospholipids are placed in water at neutral pH, which is approximately 7, they spontaneously form bilayers that take on the shape of a liposome, shown in Figure 9.3c. Liposomes are spherical structures with an inner and outer space, resembling a cell. This structure forms on its own because its outer surface is polar and readily interacts with water, while the nonpolar fatty acid tails are sandwiched in between and therefore kept away from water. Liposomes form spontaneously, without a net input of energy. They are also self-healing when their structure is disrupted. For example, small tears in the membrane are rapidly sealed by the rearrangement of the lipids surrounding the damaged region.

In this way, liposomes create an enclosed, spherical space. In addition, the structure, with polar groups on the externally facing surfaces and nonpolar groups buried in the interior, allows the membrane to interact with the watery environment that is present in and out of the cell. Finally, the hydrophobic interior of the membrane enables it to act as a barrier to polar molecules, which cannot readily pass through the membrane on their own.

The ability of phospholipids to form bilayers and liposomes spontaneously in a watery environment is not only important for modern cells but also may have played a critical and important role in the origin of the first cell. We return to this topic in Unit 7.

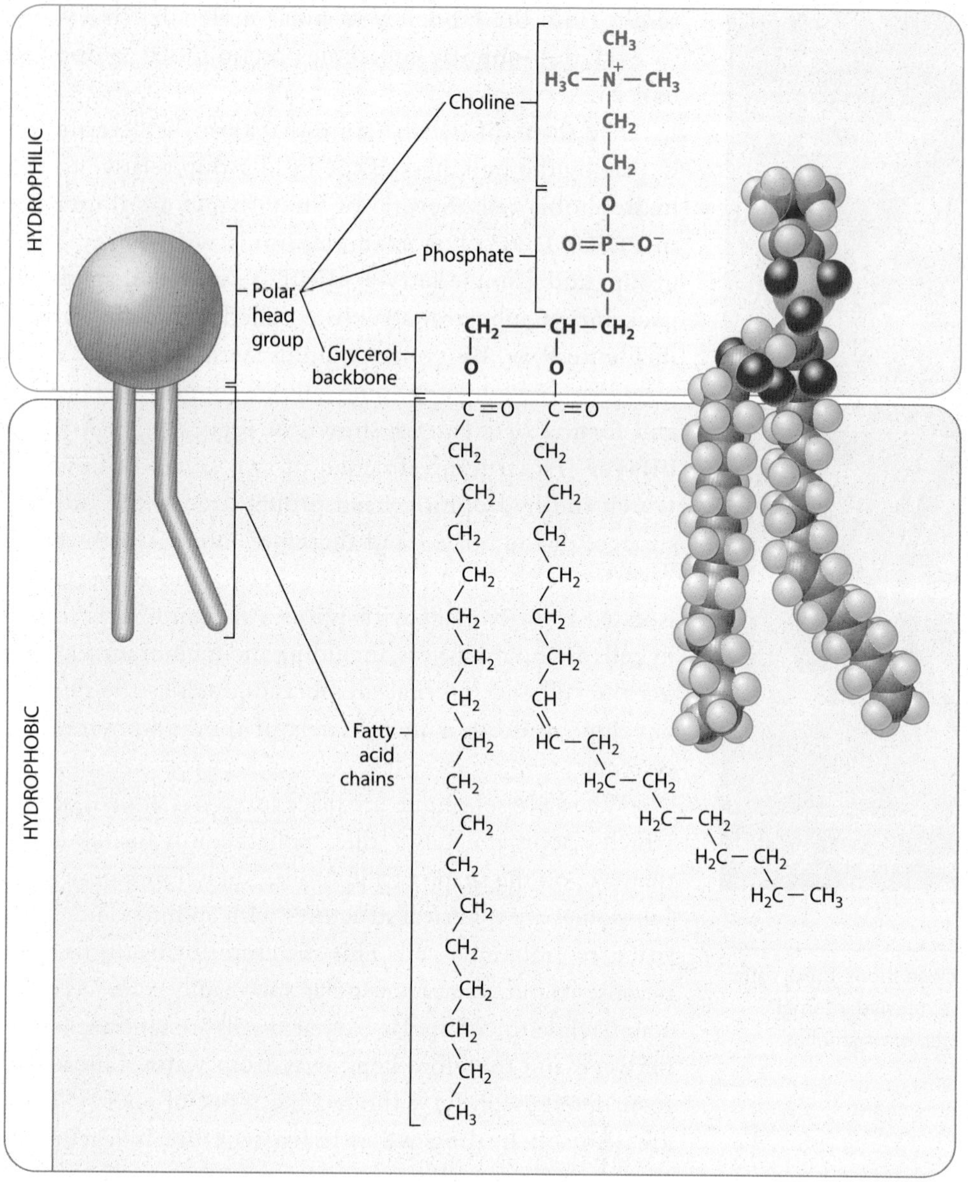

Choline

$$CH_3$$
$$H_3C-\overset{+}{N}-CH_3$$
$$CH_2$$
$$CH_2$$
$$O$$

Phosphate

$$O=P-O^-$$
$$O$$

Polar head group

Glycerol backbone

$$CH_2—CH—CH_2$$
$$O \qquad O$$
$$C=O \quad C=O$$
$$CH_2 \quad CH_2$$
$$CH_2 \quad CH_2$$
$$CH_2 \quad CH_2$$
$$CH_2 \quad CH_2$$
$$CH_2 \quad CH_2$$
$$CH_2 \quad CH_2$$
$$CH_2 \quad CH$$
$$CH_2 \quad HC—CH_2$$
$$CH_2 \quad H_2C—CH_2$$
$$CH_2 \qquad H_2C—CH_2$$
$$CH_2 \qquad H_2C—CH_2$$
$$CH_2 \qquad\qquad H_2C—CH_3$$
$$CH_2$$
$$CH_2$$
$$CH_2$$
$$CH_3$$

Fatty acid chains

FIGURE 9.2 Phospholipid structure

Phospholipids are the major component of cell membranes. They are made up of glycerol attached to a phosphate group and two fatty acid tails. The head group is hydrophilic, and the fatty acid tails are hydrophobic. Phospholipids can be represented in a variety of ways. The model at the left highlights the hydrophilic head group and hydrophobic fatty acid tails. The model in the middle shows the chemical structure. The model at the right reveals the three-dimensional shape.

a. Micelle

b. Bilayer

c. Liposome

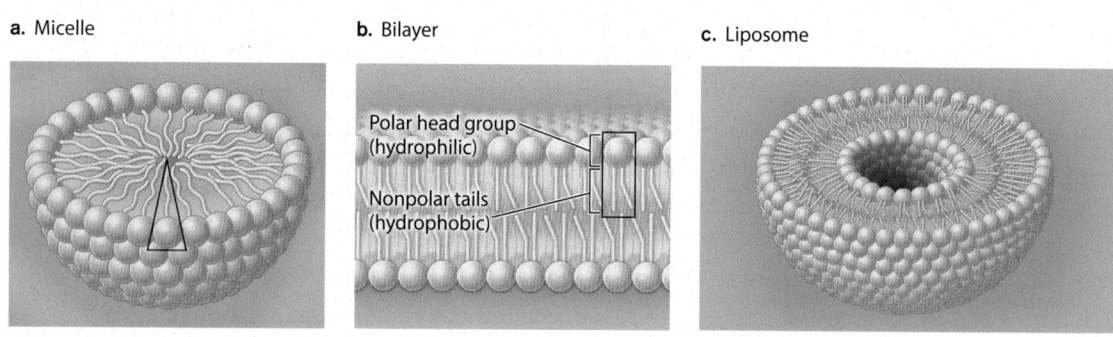

Polar head group (hydrophilic)

Nonpolar tails (hydrophobic)

FIGURE 9.3 Different phospholipid structures

Phospholipids can form different structures when placed in an aqueous solution.
(a) Phospholipids with bulky head groups and one fatty acid tail form micelles. (b) Phospholipids with less bulky head groups and two fatty acid tails form bilayers. (c) Liposomes are spherical structures with a lipid bilayer forming a barrier between inside and out.

✓ **Concept Check**

1. Amphipathic molecules have both hydrophilic and hydrophobic regions. **Describe** how amphipathic molecules orient themselves with respect to water.

2. **Describe** the structures that amphipathic molecules form in an aqueous (watery) environment.

9.2 Cell membranes are dynamic

Membranes serve as a barrier for the cell and internal organelles, but they are not static structures. Instead, membranes are dynamic, with lateral movement of lipids and other components in the membrane. The dynamic nature of membranes is the reason for their versatility and allows them to carry out their functions. For example, it allows vesicles to break off and rejoin easily, which is important for the continuity of the endomembrane system, as we saw in Module 7. It also allows some cells to change shape, move about, and even engulf particles and other cells. In this section, we will look more closely at lipid bilayers to understand how their structure allows them to be dynamic.

Lipids that make up cell membranes have long fatty acid tails that are oriented inward, away from the watery environment inside and outside of the cell. These tails freely associate with one another because of extensive van der Waals forces, which are weak intermolecular interactions, as discussed in Unit 1. Because they are weak, they are easily broken, allowing phospholipids to move about, with their tails interacting first with one fatty acid tail and then another. As a result, phospholipids are able to move laterally within the plane of the membrane. This movement can be surprisingly rapid. For example, a single phospholipid molecule can move across the entire length of a bacterial cell in less than one second. Lipids can also rotate rapidly around their vertical axis, so that they stay in the same place but spin around. Finally, individual fatty acid tails are able to flex, or bend. As a result, membranes are dynamic: they are continually moving, forming, and re-forming during the lifetime of a cell.

Because membrane lipids are able to move, or float, in the plane of the membrane, the membrane is said to be fluid. The degree of membrane fluidity depends on the composition of the membrane. For example, the length of the fatty acid tails influences the degree of fluidity, as shown in **FIGURE 9.4**. The longer the fatty acid tails, the less fluid the membrane. This is because long fatty acid tails have more surface area to participate in van der Waals interactions with one another than short fatty acid tails. These

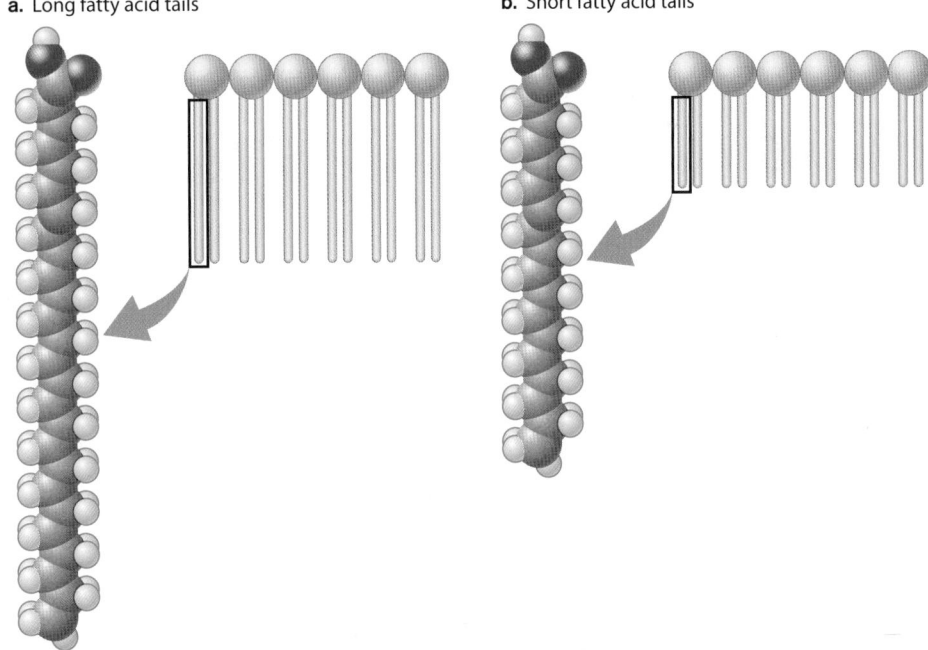

a. Long fatty acid tails

b. Short fatty acid tails

FIGURE 9.4 Long and short fatty acid tails

Longer fatty acid tails have more surface area to participate in interactions with other fatty acid tails than shorter fatty acid tails do. Therefore, membranes with phospholipids having long fatty acid tails are less fluid than those with short fatty acid tails.

interactions are weak but attractive, holding fatty acid tails in place. Therefore, the more interactions, the more stable and less fluid the membrane.

In addition to the length of the fatty acid tails, membrane fluidity is also affected by the number of carbon–carbon bonds in the fatty acid tails. As you can see in **FIGURE 9.5a**, saturated fatty acids do not have double bonds. They are straight and therefore able to pack tightly, held together by van der Waals interactions. These interactions are relatively weak, but their strength depends on distance: the closer the atoms or molecules, the stronger the interaction. As a result, membranes with saturated fatty acids are more stable and less fluid. Unsaturated fatty acids, like the ones shown in Figure 9.4b, have one or more double bonds. These double bonds introduce kinks or bends in the fatty acid tails. Therefore, the fatty acid tails cannot pack as tightly as saturated fatty acid tails, so have weaker van der Waals interactions. As a result, membranes with unsaturated fatty acids are less stable and more fluid.

The effect of saturation on membrane fluidity is easily observed in the kitchen. Like cell membranes, many fats and oils are made up of fatty acids. Most animal fats like butter are solid at room temperature because they contain saturated fatty acids and are therefore less fluid. By contrast, plant fats like canola oil tend to be liquid at room temperature because they contain unsaturated fatty acids and are therefore more fluid. Like plant oils, fish oils tend to be unsaturated and liquid at room temperature.

In addition to phospholipids, cell membranes often contain other types of lipids, which also influence membrane fluidity. For example, cholesterol is a major component of animal cell membranes, representing approximately 30% by mass of the membrane lipids. **FIGURE 9.6** shows the chemical structure of cholesterol. Like phospholipids, cholesterol is amphipathic, with both hydrophilic and hydrophobic groups present in the same molecule. The hydrophilic region is simply a hydroxyl group (—OH) and the hydrophobic region consists of four planar carbon rings with an attached hydrocarbon

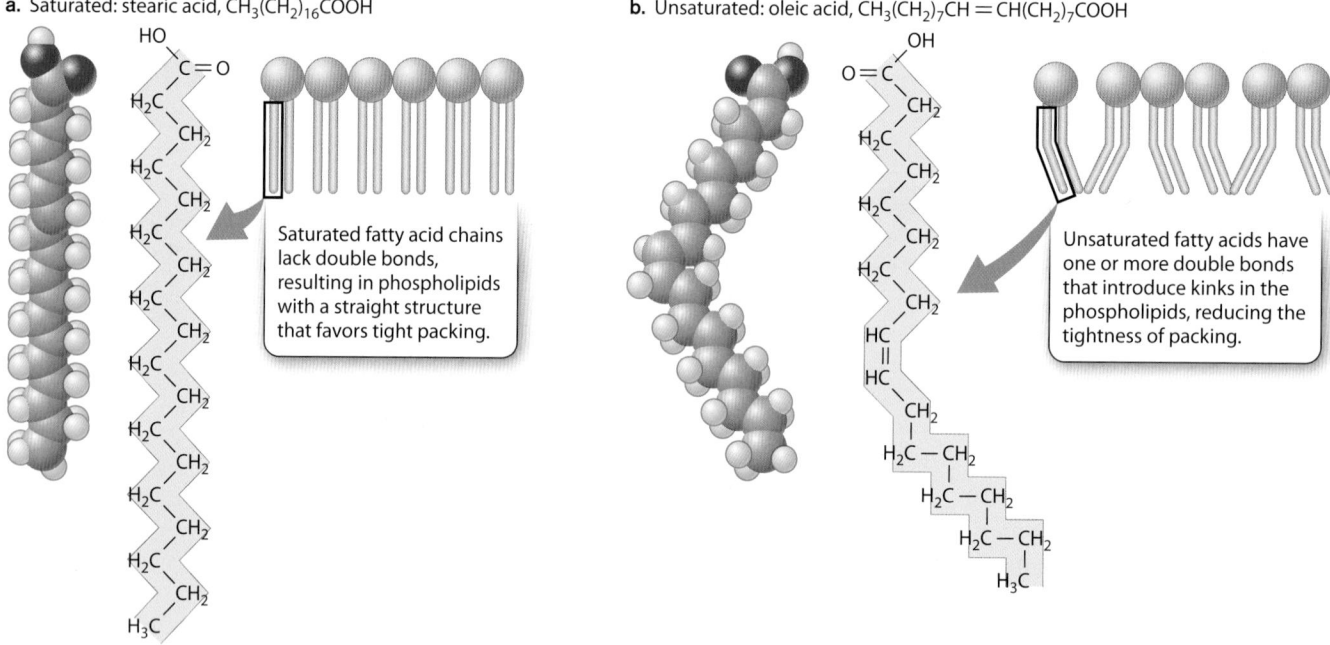

a. Saturated: stearic acid, $CH_3(CH_2)_{16}COOH$

Saturated fatty acid chains lack double bonds, resulting in phospholipids with a straight structure that favors tight packing.

b. Unsaturated: oleic acid, $CH_3(CH_2)_7CH=CH(CH_2)_7COOH$

Unsaturated fatty acids have one or more double bonds that introduce kinks in the phospholipids, reducing the tightness of packing.

FIGURE 9.5 Saturated and unsaturated fatty acid tails

The degree of saturation of fatty acids affects membrane fluidity. (a) Saturated fatty acids do not have double bonds and are straight. They therefore pack tightly with one another and are less fluid. (b) Unsaturated fatty acids have one or more double bonds. A double bond causes the tail to bend. As result, phospholipids with unsaturated fatty acid tails do not pack tightly and are more fluid.

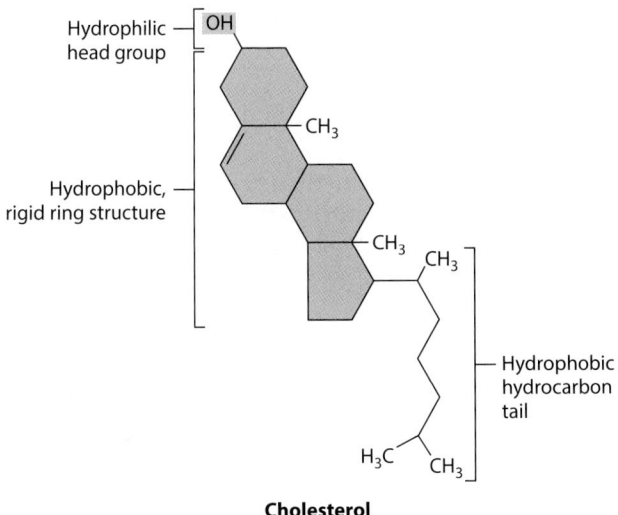

Hydrophilic head group — OH

— CH₃

Hydrophobic, rigid ring structure —

— CH₃

CH₃

— Hydrophobic hydrocarbon tail

H₃C   CH₃

Cholesterol

FIGURE 9.6 Cholesterol

Cholesterol has both hydrophilic and hydrophobic regions. The hydrophilic region is the hydroxyl group (—OH), highlighted in blue, and the hydrophobic region includes the orange rings and hydrocarbon tail.

chain. This structure allows a cholesterol molecule to insert itself into the lipid bilayer, as shown in **FIGURE 9.7**. In this figure, you can see that the head group interacts with the hydrophilic head group of phospholipids, while the ring structure participates in van der Waals interactions with the fatty acid chains.

The effect of cholesterol on membrane fluidity depends on temperature. In the absence of cholesterol in the membrane, higher temperatures increase membrane fluidity and lower temperatures decrease membrane fluidity. Cholesterol in the membrane has the opposite effect on membrane fluidity, so the two effects tend to cancel each other out. At higher temperatures, cholesterol decreases membrane fluidity, making it more stable. In this case, the rigid ring structure of cholesterol interacts with the phospholipid fatty acid tails, thereby reducing the mobility of the phospholipids. At lower temperatures, cholesterol increases membrane fluidity. In this case, cholesterol prevents phospholipids from packing tightly with other phospholipids.

Thus, cholesterol helps maintain a consistent state of membrane fluidity by preventing dramatic transitions in fluidity as the temperature changes. The effect of cholesterol on membrane fluidity in the face of changing temperatures is a form of homeostasis. The fluidity of the membrane is maintained at a particular level even as the outside temperature

fluctuates. As a result, membranes are able to stay fluid and carry out functions such as vesicle budding and fusing or cell movement, even in the face of temperature changes outside of the cell.

For many decades, it was thought that the various types of lipids found in the membrane were randomly distributed throughout the bilayer. More recent studies show that specific types of lipids sometimes assemble into defined patches called lipid rafts. Cholesterol and other membrane components also appear to accumulate in some of these regions. Thus, cell membranes are not always a uniform fluid bilayer, but instead can contain regions with discrete components.

Although lipids are free to move in the plane of the membrane, the spontaneous transfer of a lipid between layers of the bilayer, known as lipid flip-flop, is very rare.

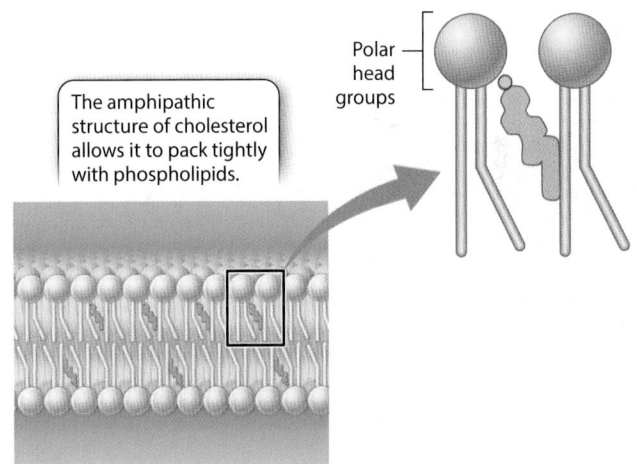

The amphipathic structure of cholesterol allows it to pack tightly with phospholipids.

Polar head groups

FIGURE 9.7 Cholesterol in the lipid bilayer

Cholesterol has a hydrophilic region (indicated in blue) and hydrophobic region (indicated in orange). As a result, it is able to embed itself in the lipid bilayer, affecting the fluidity of the membrane. At higher temperatures, cholesterol makes the membrane less fluid, and at low temperatures, cholesterol makes the membrane more fluid.

This is not surprising because flip-flop requires the hydrophilic head group of a phospholipid molecule to pass through the hydrophobic interior of the cell membrane. As a result, little exchange of components occurs between the two layers of the membrane, which in turn allows the two layers to differ in composition. In fact, in many cell membranes, different types of lipids are present primarily in one layer or the other.

9.3 Proteins and carbohydrates associate with cell membranes

Up to this point, we have considered two types of lipids that make up cell membranes: phospholipids and cholesterol. However, membranes are made up not only of lipids. Proteins are present as well. For example, proteins represent as much as 50% by mass of the membrane of a red blood cell. Proteins embedded in the cell membrane have many functions. Some transport molecules into and out of the cell. Other proteins, such as those in the inner membrane of mitochondria and the thylakoid

membrane of chloroplasts, pass electrons along the membrane in the process of harnessing energy for use by the cell. In addition to proteins, carbohydrates are sometimes present in cell membranes. In this section, we discuss the types of proteins and carbohydrates that are found in cell membranes.

Many different types of proteins are associated with membranes, and these membrane proteins serve different functions, which are shown in **FIGURE 9.8**. Some act as **transport proteins,** moving ions or other molecules across the membrane. Other membrane proteins act as **receptor proteins,** which allow the cell to receive signals

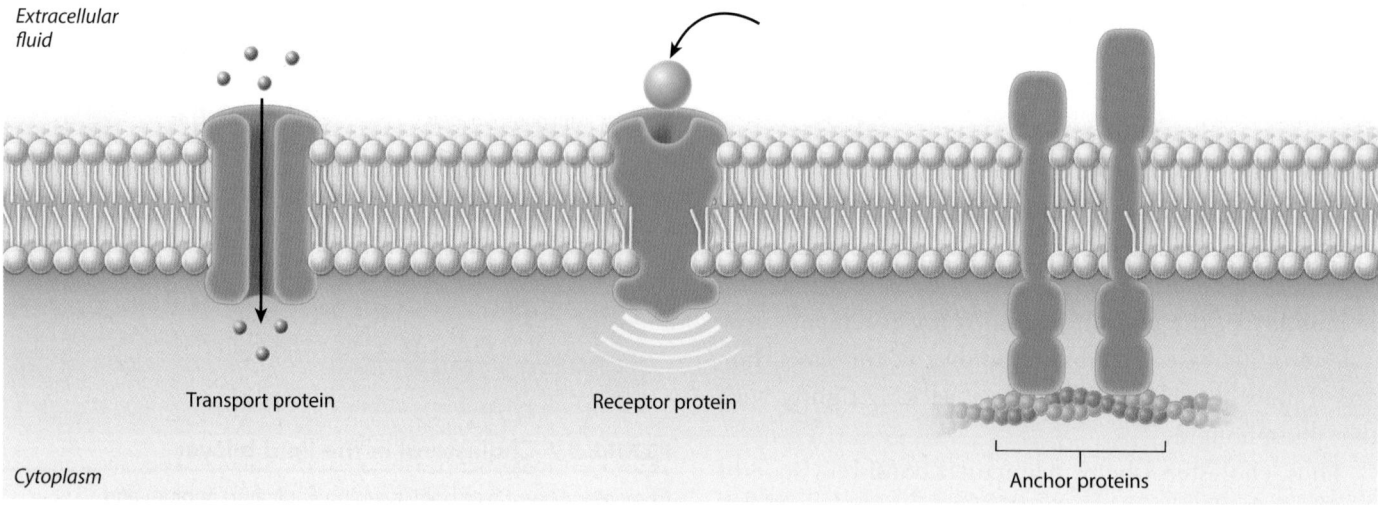

FIGURE 9.8 Membrane proteins

Many types of proteins are associated with cell membranes. These proteins act as transporters that move molecules from one side of the membrane to the other, receptors that receive and transmit signals, and anchors that connect with molecules inside the cell.

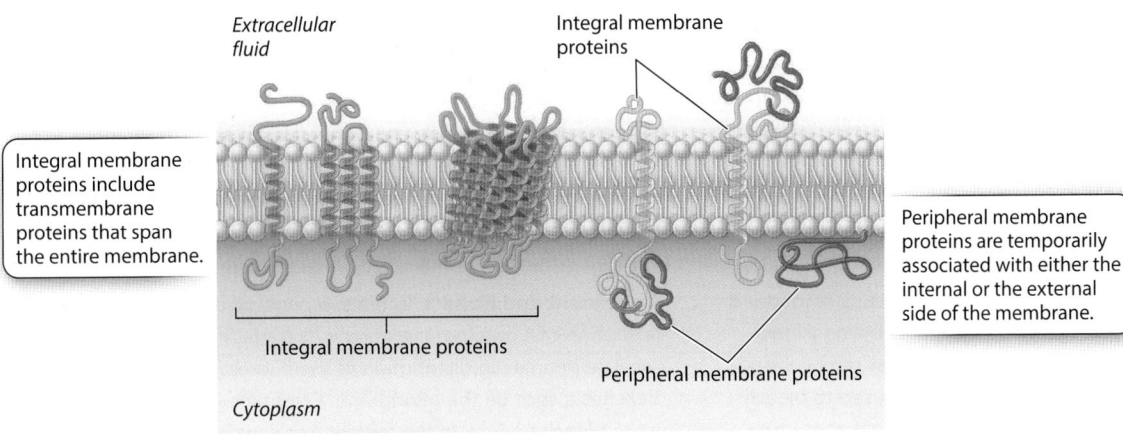

Extracellular fluid

Integral membrane proteins

Integral membrane proteins include transmembrane proteins that span the entire membrane.

Integral membrane proteins

Cytoplasm

Peripheral membrane proteins are temporarily associated with either the internal or the external side of the membrane.

Peripheral membrane proteins

FIGURE 9.9 Integral and peripheral membrane proteins

Integral membrane proteins are permanently associated with the membrane. Peripheral membrane proteins are temporarily associated with one of the two lipid bilayers or with an integral membrane protein.

from the environment. Still others speed up chemical reactions. Finally, there are proteins that serve as anchors that attach to other proteins and help to maintain cell structure and shape.

The various membrane proteins can be classified into two groups, depending on how they associate with the membrane. These two groups are shown in **FIGURE 9.9**. **Integral membrane proteins** are highlighted on the left of Figure 9.9 and are permanently associated with cell membranes. They cannot be separated from the membrane experimentally without destroying the membrane itself. **Peripheral membrane proteins** are highlighted on the right of Figure 9.9 and are temporarily associated with the lipid bilayer or with integral membrane proteins through weak noncovalent interactions. They are easily separated from the membrane while leaving the structure of the membrane intact.

Most integral membrane proteins are **transmembrane proteins** that span the entire lipid bilayer, as shown in Figure 9.9. These proteins are composed of three regions. There are two hydrophilic regions, one protruding from each face of the membrane in contact with the aqueous environment inside and outside of the cell. There is also one hydrophobic region that spans the hydrophobic interior of the membrane. The hydrophobic region holds the protein in the membrane, resulting in stability of the protein within the membrane. The structure of transmembrane proteins also allows separate functions at each end. For example, the hydrophilic region on the external side

of the membrane can act as a receptor that binds a signal, a topic explored in detail in Unit 4. The hydrophilic region on the internal side of the membrane can interact with other proteins in the cytoplasm of the cell to pass along the message.

Peripheral membrane proteins may be associated with either the internal or external side of the membrane, as can be seen in Figure 9.9. These proteins interact either with the polar heads of lipids or with integral membrane proteins by weak noncovalent interactions such as hydrogen bonds. Peripheral membrane proteins are only transiently associated with the membrane and can play a role in transmitting information received from external signals. Other peripheral membrane proteins limit the ability of transmembrane proteins to move within the membrane and assist proteins in clustering in lipid rafts, which we described in the previous section.

Proteins, like lipids, are free to move in the membrane. The mobility of proteins in the cell membrane can be demonstrated using an elegant experimental technique called fluorescence recovery after photobleaching (FRAP), which is shown in "Practicing Science 9.1: Do proteins move in the plane of the membrane?" In this technique, membrane proteins are labeled with a fluorescent tag so they are visible. A laser is used to bleach an area of the membrane, making it nonfluorescent. Over time, the bleached area becomes fluorescent again, suggesting that fluorescent proteins that were not bleached are able to move into the bleached area.

Do proteins move in the plane of the membrane?

Background Fluorescent recovery after photobleaching (FRAP) is a technique used to measure the mobility of molecules in the plane of the membrane. A fluorescent dye is attached to proteins embedded in the cell membrane in a process called labeling. Labeling all the proteins in a membrane creates a fluorescent cell that can be visualized with a fluorescence microscope. A laser is then used to bleach a small area of the membrane, leaving a nonfluorescent spot on the surface of the cell.

Hypothesis If membrane components such as proteins move in the plane of the membrane, the bleached spot should become fluorescent over time as unbleached fluorescent molecules move into the bleached area. If membrane components do not move, the bleached spot should remain intact.

Experiment and Results In the top panel of the figure, we see an illustration of a patch of membrane. It starts off uniformly green because of an even distribution of the fluorescent label. A laser then bleaches a spot on the membrane. Over time, fluorescence appears in the bleached area. In the middle panel, we see photographs of the actual cell membrane. In the bottom panel, we see a graph of

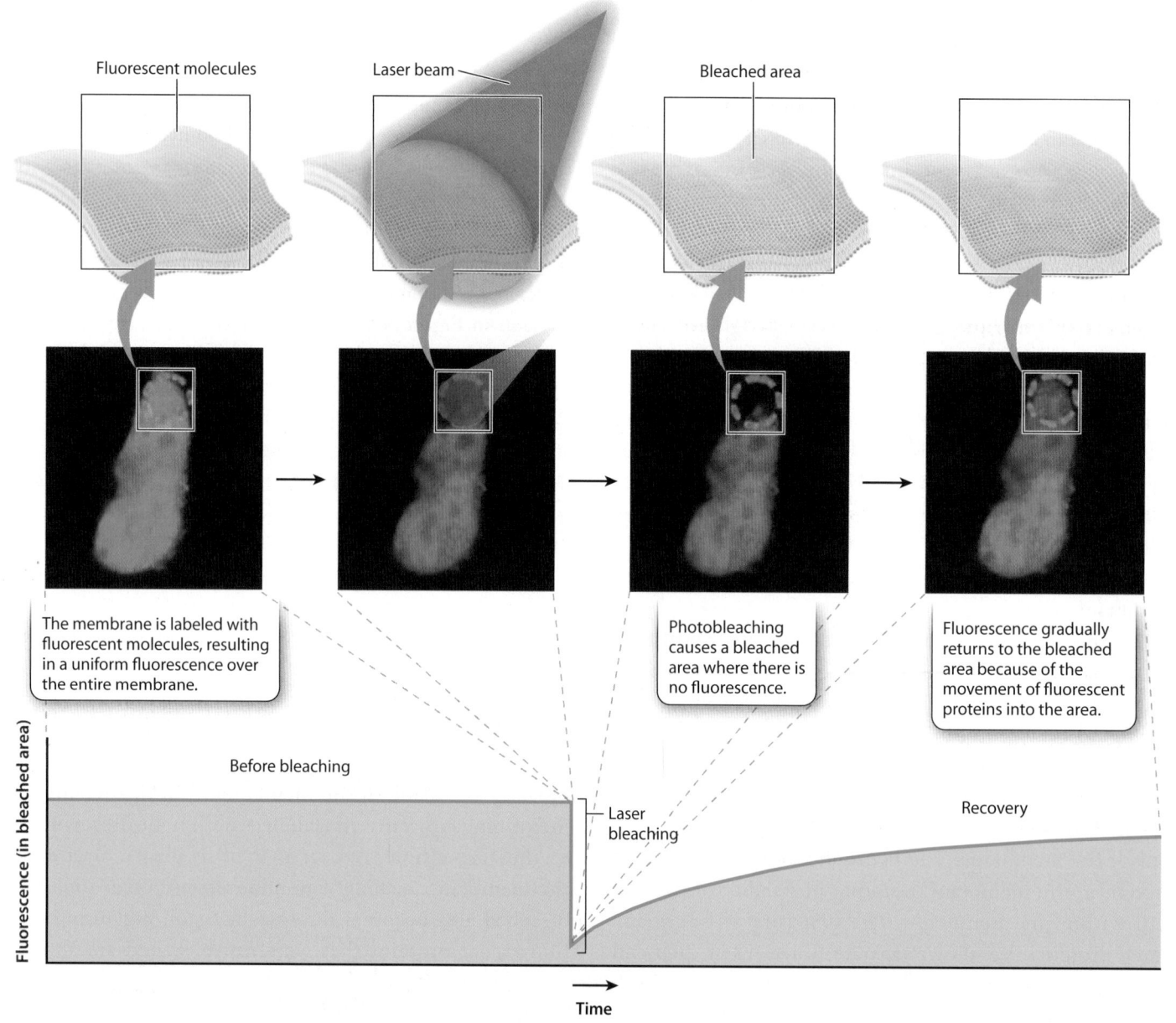

Fluorescent molecules

Laser beam

Bleached area

The membrane is labeled with fluorescent molecules, resulting in a uniform fluorescence over the entire membrane.

Photobleaching causes a bleached area where there is no fluorescence.

Fluorescence gradually returns to the bleached area because of the movement of fluorescent proteins into the area.

Before bleaching

Laser bleaching

Recovery

Fluorescence (in bleached area)

Time

the amount of fluorescence in the membrane over time. Before bleaching, the level of fluorescence is high. When the laser is applied, the level of fluorescence decreases rapidly. Over time, the level of fluorescence gradually increases, telling us that fluorescent proteins that were not bleached moved into the bleached area.

Conclusion The gradual recovery of fluorescence in the bleached area indicates that proteins move in the plane of the membrane.

SOURCE
Peters, R., et al. 1974. "A Microfluorimetric Study of Translational Diffusion in Erythrocyte Membranes." *Biochimica et Biophysica Acta* 367:282–294. Photo: FRAP of cytoplasmic EGFP in living HeLa cells, performed using an UltraVIEW spinning disk confocal system (PerkinElmer Inc.). HeLa cells were transfected with pEGFP-C1 (Clontech Laboratories, Inc.) using GeneJuice transfection reagent (Novagen). A region of interest in the cytoplasm was photobleached using the UltraVIEW photokinesis unit, and the recovery of fluorescence in this region was observed (fluorescence recovery after photobleaching (FRAP) using the UltraVIEW PhotoKinesis accessory, PerkinElmer Technical Note).

AP® PRACTICE QUESTION
(a) **Identify** the question the experimenters are trying to answer.
(b) **Identify** the hypothesis they tested.
(c) **Explain** their results by **identifying** their claim and **justifying** the claim with their evidence and reasoning (CER). In your CER, be sure to include your understanding of fluorescent recovery after photobleaching (FRAP) as well as the relationship between proteins and phospholipids in the structure of the cell membrane.

In addition to proteins, carbohydrates can be found in cell membranes. Carbohydrates in cell membranes are usually attached to other membrane components by covalent bonds. For example, they can be attached to lipids. A carbohydrate covalently attached to a lipid is called a **glycolipid.** Carbohydrates can also be attached to proteins. A carbohydrate covalently linked to a protein is known as a **glycoprotein.** Like other membrane macromolecules, glycolipids and glycoproteins are free to move about in the plane of the membrane.

The idea that lipids, proteins, and carbohydrates are all present in cell membranes, and that they are able to move in the plane of the membrane, led American biologists S. Jonathan Singer and Garth Nicolson to propose the *fluid mosaic model* in 1972. According to the **fluid mosaic model,** the lipid bilayer is a structure within which molecules move laterally—meaning it is fluid—and is a mixture—meaning it is a mosaic—of various components.

The structure of cell membranes helps to explain their function. Membranes are selective barriers that control the movement of molecules between the inside and outside of the cell or organelle. In other words, some molecules pass freely across cell membranes; others can pass only under certain conditions; and still others are unable to pass through at all. Selective permeability results from the combination of macromolecules that make up cell membranes. In this way, they maintain homeostasis for the cell.

✓ Concept Check

5. **Describe** two ways in which proteins associate with membranes.

6. **Predict** what would happen in the FRAP experiment if proteins did not move in the plane of a membrane.

Module 9 Summary

REVIST THE BIG IDEAS

PREP FOR THE AP® EXAM

ENERGETICS: Using the content of this module, **describe** the structure of the cell membrane and how its structure influences its ability to take in nutrients and other forms of matter from the environment.

LG 9.1 Cell membranes are composed of two layers of lipids.

- Cell membranes are made up of phospholipids and form a bilayer. Page 132

- Phospholipids have both hydrophilic and hydrophobic regions. Page 133
- Phospholipids spontaneously form structures such as micelles and bilayers when placed in an aqueous environment. Page 133

LG 9.2 Cell membranes are dynamic.

- Membranes are fluid, meaning that membrane macromolecules are able to move laterally in the plane of the membrane. Page 135

- Membrane fluidity is influenced by the length of fatty acid chains, the presence of carbon–carbon double bonds in fatty acid chains, and the amount of cholesterol. Page 135

LG 9.3 Proteins and carbohydrates associate with cell membranes.

- Many membranes contain proteins as well as lipids. Page 138
- Proteins that span the membrane are transmembrane proteins. Page 139

- Proteins that are temporarily associated with one or the other layer of the lipid bilayer are peripheral proteins. Page 139
- Proteins embedded in the cell membrane can act as transporters, receptors, enzymes, or anchors. Page 139
- The fluid mosaic model of cell membranes states that membranes are dynamic and made up of several components, including lipids, proteins, and carbohydrates. Page 141

Key Terms

Lipid bilayer	Integral membrane protein	Glycolipid
Transport protein	Peripheral membrane protein	Glycoprotein
Receptor protein	Transmembrane protein	Fluid mosaic model

Review Questions

1. What type of molecule is not associated with a cell membrane?
   (A) Carbohydrate
   (B) Lipid
   (C) Nucleic acid
   (D) Protein

2. Exposure of phospholipids to water results in the spontaneous formation of
   (A) lipid bilayers.
   (B) triacylglycerols.
   (C) steroids.
   (D) polypeptides.

3. The interior region of a phospholipid bilayer is
   (A) hydrophilic.
   (B) hydrophobic.
   (C) polar.
   (D) hydrophilic and polar.

4. A protein that is temporarily associated with a cell membrane is a(n)
   (A) transmembrane protein.
   (B) integral membrane protein.
   (C) peripheral membrane protein.
   (D) receptor protein.

5. An amphipathic molecule is one that
   (A) contains both hydrophilic and hydrophobic regions.
   (B) contains only hydrophobic regions.
   (C) makes up a carbohydrate.
   (D) can fully dissolve in water.

6. Some lipid rafts are characterized by an accumulation of cholesterol. What does this mean for the fluidity of the raft domain?
   (A) These lipid rafts are more fluid than the surrounding membrane at normal temperatures.
   (B) These lipid rafts are less fluid than the surrounding membrane at normal temperatures.
   (C) These lipid rafts are equally fluid as the surrounding membrane, regardless of temperature.
   (D) These lipid rafts are less fluid than the surrounding membrane at low temperatures.

7. Which lipid composition option has the least membrane fluidity?
   (A) Phospholipids with long-chain, saturated fatty acids
   (B) Phospholipids with long-chain, unsaturated fatty acids
   (C) Phospholipids with short-chain, saturated fatty acids
   (D) Phospholipids with short-chain, unsaturated fatty acids

Module 9
AP® Practice Questions

Section 1: Multiple-Choice Questions

Choose the best answer for questions 1–5.

1. Consider the structure of a micelle, which is depicted in Figure 9.3a. Which is true of micelles?

   (A) The tails are repelled by hydrophobic, nonpolar molecules and are attracted to hydrophilic, polar molecules.

   (B) The heads are repelled by hydrophilic, polar molecules and are attracted to hydrophobic, nonpolar molecules.

   (C) The tails are repelled by water and attracted to one another.

   (D) The interior of a micelle contains both hydrophobic and hydrophilic components.

2. Liposomes are vesicles enclosed by a phospholipid bilayer that forms in aqueous solutions. Interior compartments of liposomes are also aqueous. Which is a feature common to both liposomes and the cell membrane of a living organism?

   (A) Both liposomes and cell membranes contain electron transport proteins.

   (B) Both liposomes and cell membranes perform endocytosis and exocytosis.

   (C) Both liposomes and cell membranes have the ability to change their fatty acid content.

   (D) Both liposomes and cell membranes separate the interior content from the external environment.

3. The sugar glucose, $C_6H_{12}O_6$, is a polar molecule with 5 hydroxyl groups and 1 aldehyde group. Mammalian cells have a number of types of transporter proteins that move glucose across the membrane into the cells. Which best explains why mammalian cells contain glucose transporter proteins?

   (A) Glucose is a polar, hydrophilic molecule that would otherwise be repelled by the cell's hydrophobic cytoplasm.

   (B) Glucose is a polar, hydrophilic molecule that would otherwise be repelled by the cell's hydrophilic cytoplasm.

   (C) Glucose is a polar, hydrophilic molecule that would otherwise be repelled by the membrane's hydrophobic membrane core.

   (D) Glucose is a polar, hydrophilic molecule that would otherwise be repelled by the membrane's hydrophilic membrane core.

Questions 4 and 5 refer to the graph of fluorescence versus time in the "Practicing Science 9.1: Do proteins move in the plane of the membrane?" on page 140.

4. Researchers labeled membrane proteins with fluorescent molecules, and then bleached a section of the cell membrane. The researchers measured the membrane's fluorescence both before and after bleaching, as indicated on the graph. Why is the level of fluorescence on the membrane so low immediately after bleaching?

   (A) The laser tears the membrane and the cell leaks out its fluorescent proteins.

   (B) The fluorescent proteins in the bleached part of the membrane are pushed to other parts of the membrane by the laser.

   (C) The phospholipids of the membrane absorb the laser energy and release it by fluorescing at a different wavelength of light.

   (D) The fluorescent molecules in the part of the membrane that was bleached are destroyed.

5. Which best explains why the amount of fluorescence at the end of the recovery period approximates that amount of fluorescence prior to bleaching?

(A) The fluorescently labeled proteins in the bleached part of the membrane regained their ability to fluoresce.

(B) Fluorescently labeled proteins from other parts of the membrane moved into the bleached area of the membrane.

(C) The cell made new proteins and inserted them into the bleached part of the membrane.

(D) The cell divided and what was recorded was the joint fluorescence of the two new daughter cells.

Section 2: Free-Response Question

Write your answer to each part clearly. Support your answers with relevant information and examples. Where calculations are required, show your work.

An early hypothesis of the cell membrane was termed the sandwich model, which is shown on the top of the figure below. It incorporated data showing that both phospholipids and proteins are present in the cell membrane. A later model, one that is widely accepted today, is the fluid mosaic model, shown on the bottom. Both models show a cell membrane that is composed of a phospholipid bilayer and proteins. They also show that the membrane can act as a barrier.

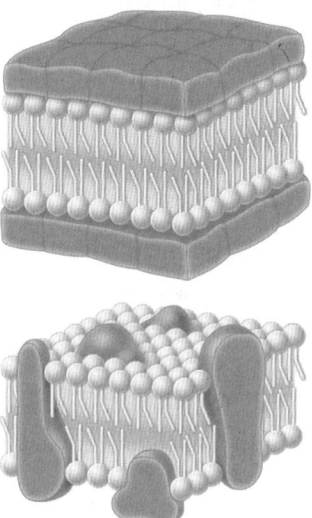

(a) **Describe** two differences between the models.

(b) Many hydrophilic molecules cross the membrane into a cell, even though hydrophilic molecules are repelled by the hydrophobic core of the lipid membrane. **Predict** which of the two models would be more likely to allow a hydrophilic molecule to enter the cell. **Justify** your answer.

Module 10
Membrane Transport

LEARNING GOALS ▶**LG 10.1** Passive transport involves diffusion.
▶**LG 10.2** Active transport requires energy.
▶**LG 10.3** Endocytosis and exocytosis move large molecules into and out of a cell.

In the last module, we explored the structure of cell membranes. We learned that membranes are made up of phospholipids arranged as a bilayer. This bilayer has a hydrophilic exterior and hydrophobic interior. The hydrophilic exterior interacts with the watery environment inside and outside of the cell, while the hydrophobic interior creates a barrier to the free movement of many substances into and out of the cell. The membrane also consists of cholesterol and other kinds of lipids. Finally, there are proteins embedded in the membrane. Some of these proteins span the entire membrane, from inside to outside, while others are associated with the outer or inner face of the bilayer.

The structure of the membrane helps it to perform one of its key functions: the cell membrane controls what enters and exits the cell, or what enters and exits the organelles within cells. It maintains homeostasis within cells by keeping the inside of the cell under a stable set of conditions that are compatible with life itself. The membrane is selectively permeable, which means that some molecules pass freely while the movement of other molecules is regulated. The mixture of proteins and lipids of a cell membrane makes it selectively permeable and allows it to maintain homeostasis.

In addition to being made up of several different kinds of molecules, membranes are fluid, with movement of molecules in the plane of the membrane. The degree of fluidity is influenced by the types of lipids that make it up. This fluidity allows membranes to break and re-form, which is important for movement of substances through the endomembrane system by vesicles, which bud off from and join different organelles, as well as how some substances enter and exit cells.

In the previous module, we introduced the fluid mosaic model of cell membranes. Cells are fluid, or dynamic, and made up of a mosaic, or mixture, of different components. In this module, we will examine how these two features of membranes—their fluidity and mosaic nature—control the movement of various molecules into and out of cells. We will see how the proteins in the cell membrane and the fluidity of the membrane allow cells to take in molecules they need for growth, metabolism, and other key functions, as well as remove wastes and actively maintain a stable set of interior conditions.

PREP FOR THE AP® EXAM

FOCUS ON THE BIG IDEAS

ENERGETICS: Focusing on cell membrane structure, look for mechanisms that cells use to capture matter and energy from the environment.

10.1 Passive transport involves diffusion

The active maintenance of a stable environment within cells, known as homeostasis, is a critical attribute of cells and of life itself. The cell membrane helps to maintain cellular homeostasis because it is selectively permeable. This means that it lets some molecules in and out freely; it lets others in and out only under certain conditions; and it prevents still other molecules from passing through at all.

The cell membrane is able to act as a selective barrier because of the combination of lipids and embedded proteins of which it is composed. In general, polar, charged, and large molecules are unable to cross the membrane on their own, while nonpolar, uncharged, and small molecules are able to cross the membrane on their own. The hydrophobic nonpolar interior of the lipid bilayer prevents charged ions and polar molecules from moving across it easily. Furthermore, many macromolecules such as proteins and polysaccharides are too large to cross the cell membrane on their own. By contrast, small gases such as oxygen, carbon dioxide, and nitrogen, and nonpolar molecules such as lipids can readily move across the lipid bilayer.

Small uncharged polar molecules, such as water, are able to move through the lipid bilayer to a limited extent. However, transport proteins in the membrane can greatly facilitate the movement of molecules, including water, ions, and nutrients, that cannot easily cross the lipid bilayer on their own. In this section, we will examine how some molecules move across cell membranes in a way that does not require an input of energy by the cell.

Simple Diffusion

Molecules are always moving in their environments. For example, molecules in water at room temperature move around at about 500 m/sec, which means that they can move only about 3 molecular diameters before they run into another molecule, leading to about 5 trillion collisions per second. The frequency of these collisions has important consequences for chemical reactions. Chemical reactions depend on the interaction of molecules and cannot occur unless molecules bump into each other.

Net movement of molecules can occur from one region to another when there is a **concentration gradient** in the way the molecules are distributed, meaning that there are areas of higher and lower concentrations. **FIGURE 10.1**

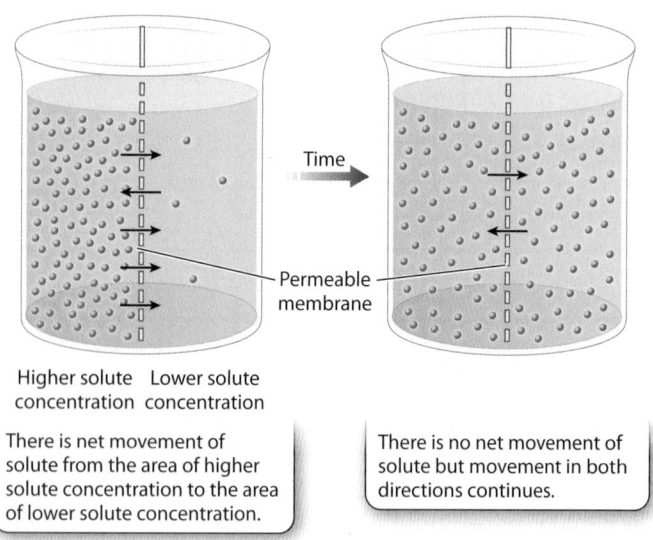

Higher solute concentration   Lower solute concentration

There is net movement of solute from the area of higher solute concentration to the area of lower solute concentration.

There is no net movement of solute but movement in both directions continues.

FIGURE 10.1 Diffusion

Diffusion is the net movement of molecules from areas of high to low concentration due to random motion of molecules. In this case, molecules move from left to right through a permeable membrane, which allows passage of all molecules. Diffusion continues until the concentration is the same on both sides of the membrane.

shows such a concentration gradient. As you can see, there is initially a high concentration of a molecule on the left and low concentration on the right, with a membrane that is freely permeable to all molecules in between. In this case, net movement occurs from the area of higher concentration to the area of lower concentration. The net movement of molecules from areas of higher to lower concentration due to the random movement of molecules is called diffusion, which was introduced in Module 8. When there is no longer a concentration gradient, net movement stops but movement of molecules in both directions continues, a condition known as dynamic equilibrium.

Passive transport occurs when molecules move across a membrane by diffusion. These molecules move as a result of differences in concentration between the inside and outside of the cell. Some molecules diffuse directly through the cell membrane, a process called simple diffusion. Oxygen and carbon dioxide, for example, move into and out of the cell in this way because they are small and uncharged. Certain hydrophobic molecules, such as steroids and other lipids, are also able to diffuse directly through the cell membrane, which is not surprising because the lipid bilayer is also hydrophobic.

When molecules move by simple diffusion, the cell does not need to expend any energy to take in or remove substances. Instead, these substances enter and exit the cell simply as a result of concentration differences. Simple diffusion, like all forms of diffusion, only works from high to low concentration, never in the reverse direction. It allows cells to take in nutrients that are in higher concentration outside the cell than inside the cell. It also allows cells to export wastes that are in higher concentration inside the cell than outside the cell.

Facilitated Diffusion

In addition to simple diffusion, passive transport also works when molecules move by diffusion across the cell membrane through transport proteins. As we saw in the previous module, a transport protein is a transmembrane protein that spans the cell membrane and provides a route for substances to enter and exit the cell. **Facilitated diffusion** is diffusion across a cell membrane through a transport protein. In both forms of diffusion, molecules move from areas of high concentration to low concentration. In the case of facilitated diffusion, molecules move through a transport protein in the cell membrane, while in the case of simple diffusion, molecules move directly through the lipid bilayer.

Two types of transport proteins may be used in facilitated diffusion, both of which are shown in **FIGURE 10.2**. The first type is a **channel protein,** which provides an opening between the inside and outside of the cell through which certain molecules can pass, depending on their shape and charge. Some channel proteins are gated, which means that they open in response to some sort of signal, which may be chemical or electrical. The second type is a **carrier protein,** which binds to and then transports specific molecules across the cell membrane. Carrier proteins exist in two conformations: one that is open to one side of the cell, and another that is open to the other side of the cell. When a molecule binds to a carrier protein, the shape of the carrier protein changes in a way that allows the molecule to move across the lipid bilayer, as shown on the right in Figure 10.2.

We will discuss these transport proteins in more detail in Unit 4.

The identity and abundance of transport proteins vary among cell types, reflecting the specific functions of different cells. For example, many cells have water channels called **aquaporins** that allow water to enter or exit the cell by facilitated diffusion. Cells in the intestine have transport proteins that specialize in the uptake of glucose from the gut into cells lining the gut. Nerve cells rely on channel proteins to generate electrical signals. These channel proteins allow the passage of charged ions, such as sodium ions (Na^+) and potassium ions (K^+). As a result of the movement of these ions, the inside of the cell can have a different amount of electrical charge than the outside of the cell. The charge difference between the inside and outside of a cell due to differences in charged ions is known as the cell's membrane potential. At rest, there are more negative ions inside the cell, so the inside is negatively charged relative to the outside. The key to producing an electrical signal in a nerve cell is the movement of positively and negatively charged ions across the cell membrane through channel proteins. This movement creates a change in membrane potential that constitutes the electrical signal.

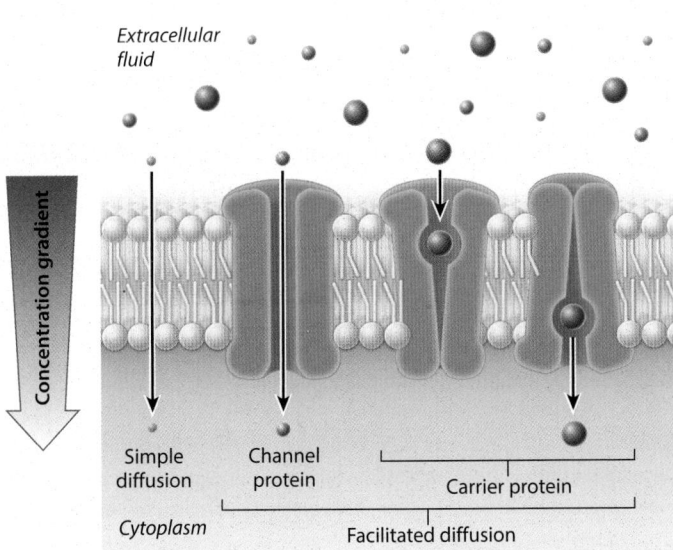

FIGURE 10.2 Simple diffusion and facilitated diffusion

Both simple diffusion and facilitated diffusion result in net movement of molecules down a concentration gradient, from a region of higher concentration (in this case, outside the cell) to a region of lower concentration (in this case, inside the cell). In simple diffusion, molecules move directly through the cell membrane. In facilitated diffusion, molecules move through a transport protein, such as a channel protein or a carrier protein.

PREP FOR THE AP® EXAM

AP® EXAM TIP

Make sure you understand the function of aquaporins. Although water molecules are small enough to make their way through the cell membrane by simple diffusion to a very limited extent, cells often need to move water faster than simple diffusion permits. Aquaporins provide a channel that allows rapid movement of water by facilitated diffusion.

✓ Concept Check

1. **Describe** how molecules move by the process of diffusion.

2. **Describe** the roles of lipids and proteins in the selective permeability of membranes.

3. **Identify** two molecules that move across the cell membrane by simple diffusion and **describe** the properties that allow them to cross the cell membrane.

4. **Identify** two molecules that move across the cell membrane by facilitated diffusion.

10.2 Active transport requires energy

Passive transport, as we have seen, only works in one direction: from areas of higher concentration to areas of lower concentration. However, many molecules that cells need to take in are not highly concentrated outside of the cell. Cells have to move these molecules from areas of lower concentration outside the cell to areas of higher concentration inside the cell. The "uphill" movement of substances against a concentration gradient is called **active transport.**

Active transport requires an input of cellular energy. In other words, the cell has to expend energy to move these molecules against their concentration gradients. In **primary active transport,** cells use ATP directly to move molecules across the cell membrane. In **secondary active transport,** cells use ATP indirectly to move molecules across the cell membrane. Either way, cells use a great deal of energy moving molecules across their cell membranes and therefore keep the environment inside the cell different from the environment outside the cell.

Primary Active Transport

As we have noted, active transport requires cellular energy to move a substance across the cell membrane against its concentration gradient from an area of lower concentration to an

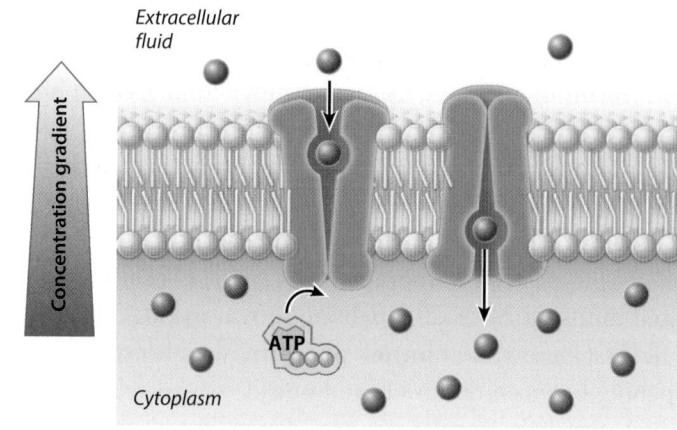

FIGURE 10.3 Primary active transport

In primary active transport, the energy of ATP is used directly to move a molecule across the cell membrane against its concentration gradient. In this case, the energy of ATP is transferred to a transport protein, which changes its shape and moves a molecule across the membrane.

area of higher concentration. In primary active transport, this energy comes from the breakdown of adenosine triphosphate (ATP), as shown in **FIGURE 10.3**. As a result, the transport protein changes its shape and pumps a molecule uphill against its concentration gradient across the cell membrane.

A good example of primary active transport is provided by the sodium–potassium pump. illustrated in **FIGURE 10.4**. Sodium ions (Na^+) are kept at concentrations much lower

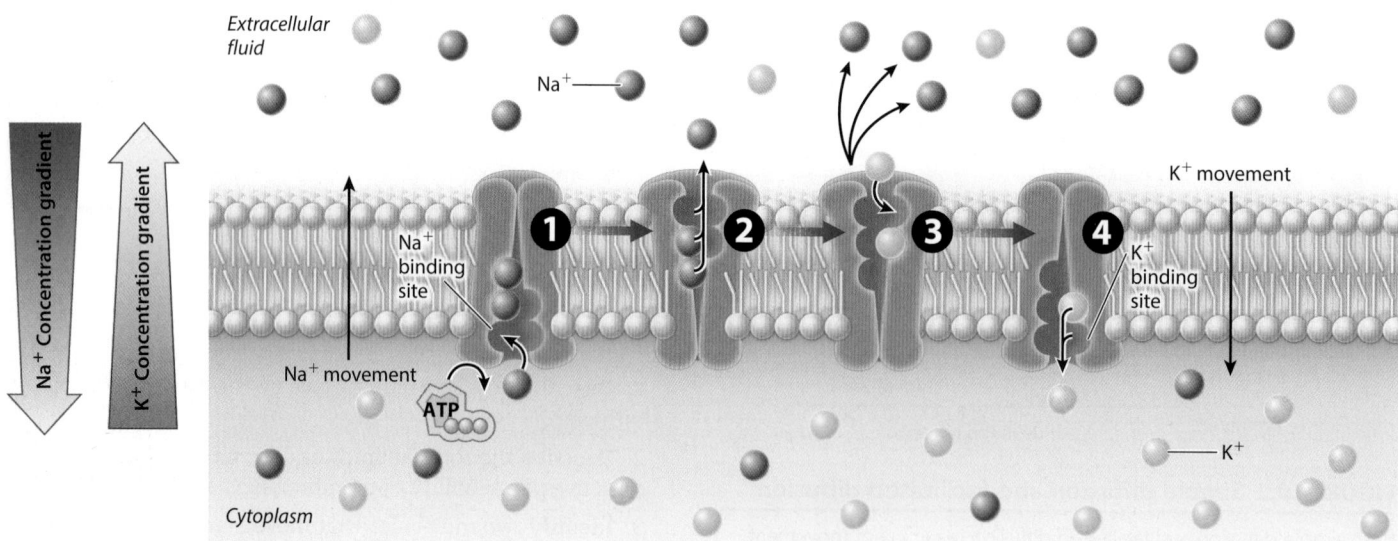

FIGURE 10.4 The sodium–potassium pump

The sodium–potassium pump is a transmembrane protein that uses the energy stored in ATP to move sodium and potassium ions against their concentration gradients. In step 1, a sodium ion binds to the pump. In step 2, three sodium ions are pumped out of the cell. In step 3, a potassium ion binds to the pump. In step 4, two potassium ions are pumped into the cell.

inside the cell than outside the cell; potassium ions (K⁺) are kept at concentrations much higher inside the cell than outside the cell. Therefore, both sodium ions and potassium ions have to be moved against their concentration gradients. The sodium–potassium pump is a transmembrane protein that actively moves sodium ions out of the cell and potassium ions into the cell, using the energy stored in the chemical bonds of ATP.

Notice in Figure 10.4 that the sodium ions and potassium ions move in opposite directions; one molecule moves into the cell and a different molecule moves out of the cell. Transport proteins that work in this way are referred to as antiporters. Other transport proteins move two molecules in the same direction; they may move two molecules into the cell, or two molecules out of the cell. These transport proteins are referred to as symporters.

Secondary Active Transport

As we just discussed, cells use energy to move a molecule against its concentration gradient. Primary active transport uses the chemical energy in ATP directly to move molecules across the membrane against their concentration gradients. Active transport can also work in another way. Because charged ions cannot cross the lipid bilayer on their own, many cells use a transport protein to pump ions across the cell membrane. As a result, the concentration of the ion builds up on one side of the membrane. This concentration gradient stores energy that can be harnessed to drive the movement of other molecules across the membrane against their concentration gradient. As ions move from areas of higher to lower concentration, the other molecule moves from areas of lower to higher concentration. Because the movement of the coupled molecule is driven by the movement of ions and not by ATP directly, this form of active transport is called secondary active transport.

For example, some cells actively pump protons (H⁺)

across the cell membrane. These pumps are transport proteins that use the energy of ATP to move protons from one side of the membrane to the other. This is step 1 in **FIGURE 10.5**. As a result of the action of the pump, the concentration of protons builds up on one side of the membrane. In step 2 of Figure 10.5, we see that the concentration of protons is higher on one side of the membrane and lower on the other side.

In other words, the pump generates a concentration gradient, also called a chemical gradient because the entity forming the gradient is a chemical. Concentration differences favor the movement of protons back to the other side of the membrane. By blocking the movement of protons to the other side, the lipid bilayer creates a store of energy, just as a dam or battery does.

In addition to the chemical gradient, another force favors the movement of protons back across the membrane: a difference in charge. Because protons carry a positive charge, the side of the membrane with more protons is more positive than the other side, also shown in step 2 of Figure 10.5. This difference in charge is called an electrical gradient.

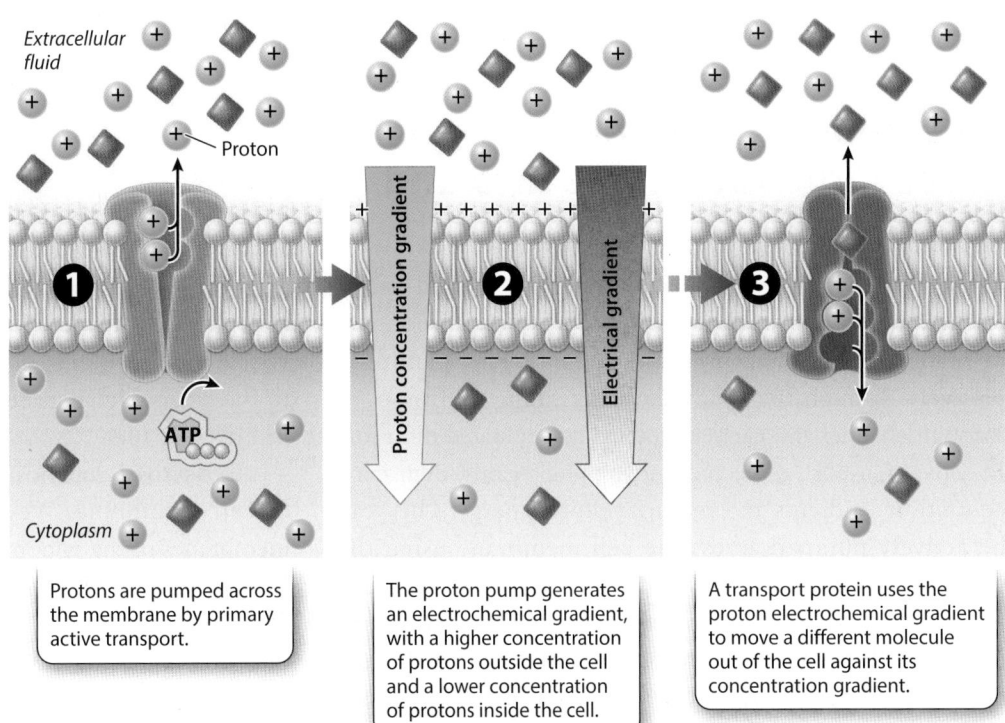

1 Protons are pumped across the membrane by primary active transport.

2 The proton pump generates an electrochemical gradient, with a higher concentration of protons outside the cell and a lower concentration of protons inside the cell.

3 A transport protein uses the proton electrochemical gradient to move a different molecule out of the cell against its concentration gradient.

FIGURE 10.5 Secondary active transport

In secondary active transport, the movement of molecules against their concentration gradients is driven by the energy stored in an electrochemical gradient. In step 1, protons are pumped across a membrane using the chemical energy of ATP. This movement generates an electrochemical gradient, which is shown in step 2. In step 3, protons move down the electrochemical gradient, which drives the movement of another molecule against its concentration gradient from a region of low concentration to high concentration.

Protons (and other ions) move from areas of like charge to areas of unlike charge, driven by an electrical gradient. A gradient that has both chemical and charge components is known as an electrochemical gradient.

If protons are then allowed to pass through the cell membrane by a transport protein, they will move down their electrochemical gradient toward the region of lower proton concentration. These transport proteins can use the movement of protons down their gradient to drive the movement of other molecules against their concentration gradient, shown in step 3 of Figure 10.5.

The movement of protons is from regions of higher to lower concentration, while the movement of the other molecules is from regions of lower to higher concentration. The movement of the other molecules is driven by the movement of protons and not by ATP directly and therefore is called secondary active transport. Secondary active transport uses the stored energy of an electrochemical gradient to drive the movement of molecules; by contrast, primary active transport uses the chemical energy of ATP directly.

The use of an electrochemical gradient as a temporary energy source is a common and widespread cellular strategy.

For example, cells use a sodium electrochemical gradient generated by the sodium–potassium pump to transport glucose and amino acids. In addition, cells use a proton electrochemical gradient to move other molecules and, as we discuss later in this unit and in Unit 3, to synthesize ATP.

✓ **Concept Check**

5. **Describe** two differences between passive and active transport.

6. **Describe** one similarity and one difference between primary active transport and secondary active transport.

10.3 Endocytosis and exocytosis move large molecules into and out of a cell

As we have seen, molecules can enter and exit a cell in different ways. In passive transport, they rely on diffusion and move down their concentration gradients. They can move passively through the lipid bilayer by simple diffusion, or through channel and carrier proteins by facilitated diffusion. In active transport, molecules move uphill against their concentration gradients and require cellular energy. They can be actively pumped across the cell membrane using the chemical energy stored in ATP by primary active transport. Alternatively, they can move uphill across the cell membrane driven by the movement of an ion down an electrochemical gradient by secondary active transport. How they are transported across the cell membrane depends on their concentration, size, charge, and polarity. When a molecule crosses the cell membrane by any of these routes, it ends up in the cytoplasm of the cell.

Some molecules are exported out of or imported into a cell by an entirely different process that does not rely on passive or active transport, but instead uses vesicles. Vesicles were introduced in Module 7; they are small spherical organelles that travel between organelles of the endomembrane system in eukaryotic cells. These vesicles bud off from and fuse with organelles and also with the cell membrane. When a vesicle fuses with the cell membrane, its contents are released outside of the cell. This process, in which a vesicle fuses with the cell membrane and releases its contents to the extracellular space, is called **exocytosis.** It is shown in **FIGURE 10.6.**

Exocytosis depends on the fluid and dynamic nature of cell membranes. Vesicles form by budding off from membranes of the endomembrane system, and they become continuous with other membranes by fusing with them. They would not be able to bud off from and fuse with membranes if cell membranes were rigid or static structures.

Exocytosis is used to remove cytoplasmic waste formed within the cell. The waste is packaged in vesicles, which then fuse with the cell membrane so that the waste is released outside of the cell. In addition, nerve cells use exocytosis to send molecular signals called neurotransmitters to adjacent nerve cells. Neurotransmitters are packaged into vesicles by one nerve cell and then released by exocytosis. They then travel in the extracellular space, where they bind to receptors

Exocytosis

Endocytosis

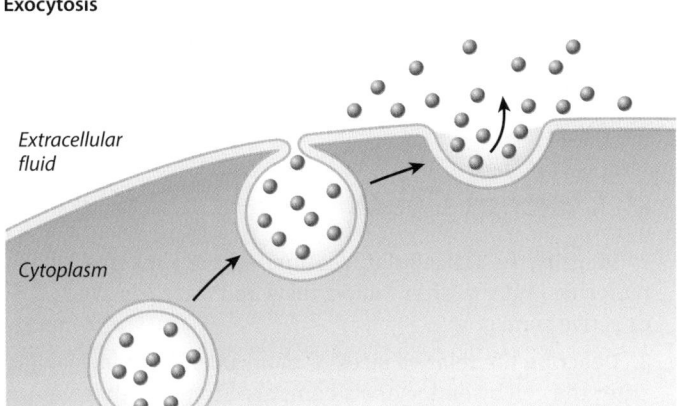
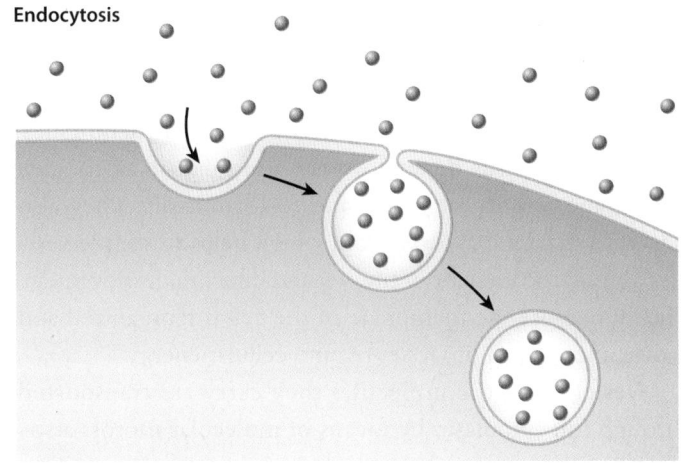

Extracellular fluid

Cytoplasm

FIGURE 10.6 Exocytosis and endocytosis

Exocytosis moves substances out of the cell and endocytosis brings substances into a cell. Both processes use vesicles to transport molecules and other substances into and out of the cell.

on the surface of an adjacent nerve cell. In this way, signals are passed from one nerve cell to the next. Finally, cells use exocytosis to deliver proteins to the cell membrane. These proteins are embedded in the membrane of the rough endoplasmic reticulum (ER) as they are synthesized. As vesicles bud off from the ER, they carry with them any proteins associated with the membrane. When the vesicles fuse with the cell membrane surrounding the cell, these proteins become part of the cell membrane.

This process also works in reverse, as shown in Figure 10.6. In a process called **endocytosis,** a vesicle buds off from the cell membrane toward the cell interior, or invaginates, enclosing material from outside the cell and bringing it into the cell. In some cases, a cell can ingest large particles, microorganisms, or even dead cells. In this case, the process is called phagocytosis ("cellular eating"). When packaged in this way, the particles can be transported into the cell interior for digestion by lysosomes. **FIGURE 10.7** shows an amoeba engulfing a yeast cell by phagocytosis.

Together, exocytosis and endocytosis provide a way to move material into and out of cells without passing through the cell membrane at all. As a result, molecules end up in

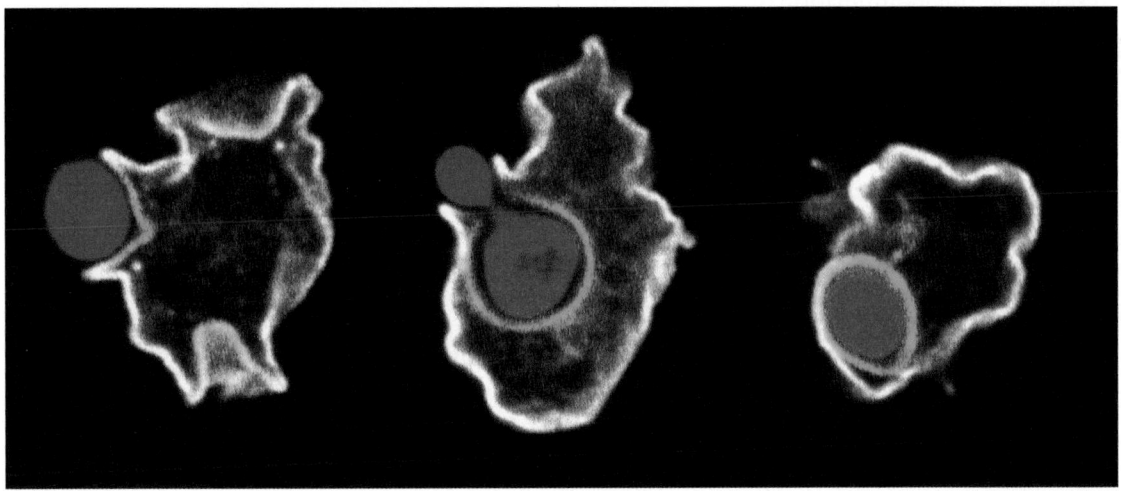

FIGURE 10.7 Phagocytosis

The flexibility of the eukaryotic cytoskeleton and endomembrane system enables eukaryotic cells to engulf food particles, including other cells. This sequence shows an amoeba engulfing a yeast cell. Note how the cell membrane (green) folds inward locally to form a vesicle around the red food particle. Photo: Valentina Mercanti et al. 2006. "Selective Membrane Exclusion in Phagocytic and Macropinocytic cups." *Journal of Cell Science* 119:4079–4087. doi:10 1242/jcs.03190

or out of vesicles, not in the cytosol, as is the case for passive and active transport. Exocytosis and endocytosis are unique to eukaryotes and do not take place in prokaryotes, which you will recall do not have extensive internal membranes. The two processes depend on the fluid nature of the cell membrane, which can break and re-form easily. They also rely on a dynamic cytoskeleton, which helps to shuttle vesicles around the interior of the cell and direct their movement. Finally, because the membrane of the cell is reorganized and moves about, both processes require cellular energy.

Vesicles and the molecules they carry are transported through the cytoplasm by means of molecular motors associated with the cytoskeleton. In this way, both nutrients and signaling molecules move through the cell at speeds much greater than diffusion allows. A major consequence of this accelerated pace of transport is that eukaryotic cells are much larger than most bacteria, as we discussed in Module 8.

✓ Concept Check

7. **Identify** the subcellular locations where a macromolecule ends up after endocytosis and after passive or active transport.

8. **Describe** the relative sizes of molecules that can enter the cell by endocytosis compared to those that can enter by passive or active transport.

9. **Describe** the difference between endocytosis and exocytosis.

Module 10 Summary

REVISIT THE BIG IDEAS
PREP FOR THE AP® EXAM

ENERGETICS: Focusing on cell membrane structure, use the content of this module to **describe** how cells capture nutrients of different sizes, polarity, charge, and concentration.

LG 10.1 Passive transport involves diffusion.

- Selective permeability results from the combination of lipids and proteins that make up cell membranes. Page 145

- Passive transport is the movement of molecules by diffusion. Page 146

- In passive transport, there is net movement of molecules from regions of higher concentration to regions of lower concentration. Page 146

- Passive transport can occur by the diffusion of molecules directly through the cell membrane (simple diffusion) or be aided by transport proteins (facilitated diffusion). Page 146

LG 10.2 Active transport requires energy.

- Active transport moves molecules from regions of lower concentration to regions of higher concentration and uses cellular energy. Page 148

- Primary active transport uses the chemical energy stored in ATP. Page 148

- Secondary active transport uses the energy stored in an electrochemical gradient. Page 149

LG 10.3 Endocytosis and exocytosis move large molecules into and out of a cell.

- Endocytosis is a process in which a vesicle is formed by invagination of the cell membrane, taking in molecules from the extracellular space to the cell interior. Page 150

- Phagocytosis is a form of endocytosis in which microorganisms, dead cells, or large particles are engulfed by the cell and eventually digested in lysosomes. Page 151

- Exocytosis is a process in which a vesicle fuses with the cell membrane, releasing its contents to the extracellular space. Page 152

Key Terms

Concentration gradient
Passive transport
Facilitated diffusion
Channel protein

Carrier protein
Aquaporin
Active transport
Primary active transport

Secondary active transport
Exocytosis
Endocytosis

Review Questions

1. The process of a vesicle fusing with the cell membrane and depositing its contents into the extracellular space is referred to as
    (A) active transport.
    (B) endocytosis.
    (C) budding.
    (D) exocytosis.

2. The net movement of molecules from regions of high to low concentration due to random motion of molecules is referred to as
    (A) transport.
    (B) osmosis.
    (C) diffusion.
    (D) secondary active transport.

3. Which is not required for net movement of a substance to occur by facilitated diffusion?
    (A) A concentration gradient
    (B) ATP
    (C) A cell membrane
    (D) A transport protein

4. Which molecules do not easily diffuse directly through a cell membrane?
    (A) Oxygen gas
    (B) Carbon dioxide
    (C) Polar molecules
    (D) Nonpolar molecules

5. The process of diffusion requires
    (A) a cell membrane.
    (B) ATP.
    (C) the random movement of molecules.
    (D) transport proteins.

6. Which is an example of secondary active transport?
    (A) The use of an electrochemical gradient of one type of molecule to move a second type of molecule
    (B) The movement of potassium ions in the opposite direction as sodium ions
    (C) The use of more than one type of transport protein for the movement of a molecule
    (D) The use of a chemical gradient to generate an electrical gradient

7. How is the energy stored in a molecule of ATP used by the sodium–potassium pump?
    (A) It is used to transport potassium ions out of the cell.
    (B) It is used to transport sodium ions into the cell.
    (C) It is used to alter the shape of the pump protein.
    (D) It is used to randomly move sodium and potassium ions by diffusion.

Module 10
AP® Practice Questions

Section 1: Multiple-Choice Questions
Choose the best answer for questions 1–5.

1. The cell membrane is a fluid, or dynamic, structure that is a mosaic of phospholipids and proteins. The composition of the membrane predicts which chemicals will freely pass through the membrane?
    (A) Small, nonpolar molecules such as oxygen gas (O_2) or nitrogen gas (N_2)
    (B) Large, negatively charged molecules such as fatty acids
    (C) Small, charged ions such as chloride (Cl^-) or potassium (K^+) ions
    (D) Highly polar molecules such as glucose or ribose

Questions 2 and 3 refer to the figure below.

Time 1   Time 2

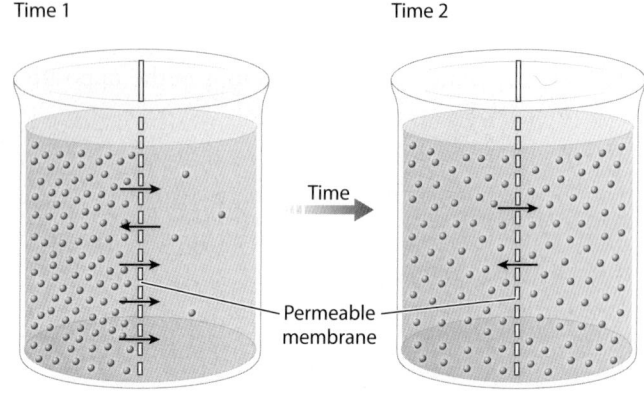

Time

Permeable
membrane

2. The figure depicts a solute in solution at an initial time, time 1, and at a later time, time 2. The solute present in both beakers is able to freely move across the permeable membrane. Which beaker is at equilibrium with respect to the movement of the solute?

   (A) The beaker at time 1

   (B) The beaker at time 2

   (C) Both beakers

   (D) Neither beaker

3. The figure models the movement of small molecules across a barrier such as a cell membrane, but it is an extremely simplified model of a cell membrane. Imagine the membrane in the beaker is a functioning cell membrane impermeable to the solute, with the solution on the left side representing the extracellular environment and the solution on the right representing the cell's cytoplasm. How would the membrane keep the concentration of the solute on the left high while maintaining a low concentration of the solute on the right?

   (A) Use passive transport of the solute molecule

   (B) Use facilitated transport of the solute molecule

   (C) Use passive transport of the solute molecule while increasing the fatty acid content of the membrane

   (D) Use active transport of the solute molecule

4. In primary active transport, transport proteins in the membrane use ATP to move a chemical into a cell. One such chemical is glucose, a sugar used by many cells to power their metabolic reactions, especially when these cells have an active metabolism and require high glucose concentrations to maintain their metabolism. Predict what would happen to glucose concentrations in such cells if the cells ran out of ATP.

   (A) Glucose concentrations would increase and then decline.

   (B) Glucose concentrations would stay the same.

   (C) Glucose concentrations would decline.

   (D) Glucose concentrations would increase.

5. A eukaryotic cell is prevented from using exocytosis. Which is the most likely consequence?

   (A) The cell will be unable to remove wastes.

   (B) The cell will be unable to take up materials from its environment.

   (C) The cell will evolve to acquire antiport systems.

   (D) There will be little, if any, consequence to the cell.

Section 2: Free-Response Question

Write your answer to each part clearly. Support your answers with relevant information and examples. Where calculations are required, show your work.

In animals, oxygen from the air enters the lungs, where it diffuses into the bloodstream. Blood then travels in vessels, carrying oxygen to various organs, such as the heart, liver, and muscles. Under normal circumstances, the blood flows continuously past the cells that make up these organs, and the oxygen diffuses from the blood into the cells, where it is consumed in metabolic reactions.

(a) **Explain** why the oxygen diffuses from the blood into the cells that make up organs such as the heart, liver, and muscles.

(b) **Predict** what would happen to the concentration gradient and movement of oxygen over time if blood flow in the vessels were to cease, but metabolic activity consuming oxygen in the cells of the various organs continued.

Module 11

Water Movement: Osmosis, Tonicity, and Osmoregulation

LEARNING GOALS ▶**LG 11.1** Osmosis governs the movement of water across cell membranes.

▶**LG 11.2** Water potential combines all of the factors that influence water movement.

▶**LG 11.3** Osmoregulation is a form of homeostasis.

In the last module, we focused on the movement of substances across cell membranes. These substances can be ions such as sodium (Na^+) and potassium (K^+), or molecules such as lipids and glucose. Different substances move across cell membranes in different ways, including passive transport, which is driven by diffusion, and active transport, which requires an input of energy. Other molecules enter and exit the cell in bulk by endocytosis and exocytosis. In these cases, molecules don't actually cross the cell membrane, but are initially located in vesicles or released into the space outside the cell. The movement of ions and molecules across cell membranes is critical for homeostasis, so that the inside of the cell actively maintains conditions that are compatible with life. Movement of ions and molecules across cell membranes also allows cells to take in nutrients required for cell function and growth, and remove waste products.

Up until this point, we have focused our attention on the movement of ions and molecules dissolved in water. In this module, we take a different approach and look at how water moves into and out of a cell across cell membranes. The movement of water is critical because it determines how much water is inside a cell or organism, which is known as water balance. Water balance itself is a form of homeostasis. The amount of water in a cell also determines in part the concentration of ions and other solvents. These concentrations are often maintained in a narrow range in the cell. Finally, the amount of water in a cell, in some cases, determines cell size and shape. As we discussed earlier, the size and shape of a cell are essential for its function.

This module explores the movement of water into and out of cells. We will begin by discussing how water crosses cell membranes and then consider how a difference in the concentration of ions or molecules on either side of a cell membrane leads to the net movement of water. The movement of water is influenced by several factors, which we will discuss. Finally, we will consider how water movement affects water balance and homeostasis in organisms.

PREP FOR THE AP® EXAM

FOCUS ON THE BIG IDEAS

ENERGETICS: Look for examples of how an organism utilizes energy and matter to manage the movement of water.

11.1 Osmosis governs the movement of water across cell membranes

As we discussed in Module 2, water is the medium of life. Life originated and diversified in a watery environment. Consequently, the chemical functions of cells and organisms crucially depend on water and the amount of water inside a cell. Cells are mostly water, and the biochemical processes needed to sustain life require that cells maintain their water content within a narrow limit. A healthy person can live for several weeks without food, but cannot survive for more than a few days without water. Water balance is critical to normal physiological function, not just in humans, but in all organisms. In this section, we consider how water enters and exits cells.

Water molecules, like all molecules, are in constant, random motion, as shown in **FIGURE 11.1** on page 156. As a result, water moves by diffusion from regions of higher water concentration to regions of lower water concentration. Differences in water concentration often occur because substances such as ions, amino acids, and sugars are dissolved

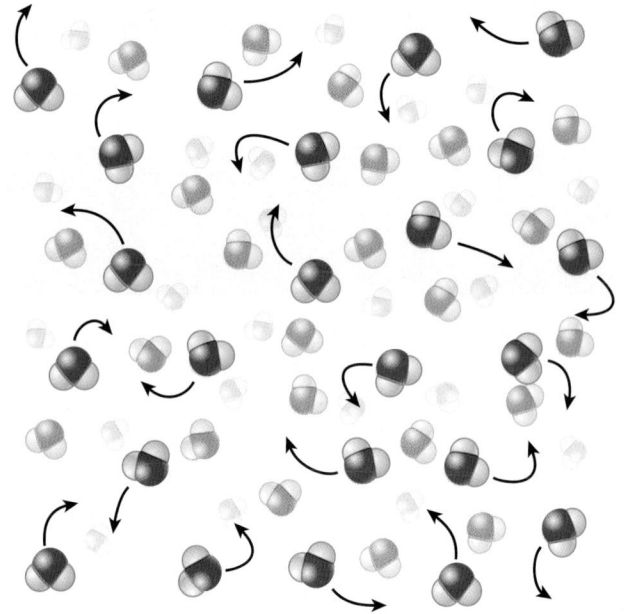

FIGURE 11.1 Random motion

Water molecules are in constant, random motion in many environments.

in water. Dissolved substances are collectively called **solutes.** Solutes are dissolved in a liquid, known as the solvent, to make a solution. In biological systems, the solvent is typically water. The concentration of a solute in a solution is known as **molarity,** or molar concentration. It is expressed as the number of moles of a solute per liter of the solution (mol/L), and abbreviated M.

Sometimes a membrane separates two solutions. When that membrane allows water or solutes to diffuse through it freely, it is said to be permeable. When it blocks the diffusion of water and solutes entirely, it is described as impermeable. When it allows the movement of some molecules but not others, it is selectively permeable.

As we saw in earlier modules, the cell membrane is a lipid bilayer with embedded proteins, and so it is selectively permeable. Ions and charged molecules, for example, are unable to cross the lipid bilayer on their own. Lipids, by contrast, can easily diffuse across the lipid bilayer. Water is a polar molecule, so it has difficulty crossing the hydrophobic interior of the cell membrane. However, because water is a small molecule, it is able to cross the membrane to a very limited extent on its own by simple diffusion. This rate is too slow for cells to maintain water balance. In most cell membranes, channel proteins called aquaporins are present (see Module 10). Aquaporins allow water to diffuse freely across the cell membrane much more quickly than would be possible by simple diffusion alone.

Water moves by diffusion across a cell membrane by following its concentration gradient. As in all forms of diffusion, water moves from an area where the concentration of water is high to an area where the concentration of water is low. Another way to think about the movement of water is to consider the solute concentration. Where the solute concentration is low, water concentration is high; and where the solute concentration is high, the water solution is low. Therefore, water moves by diffusion across a cell membrane from regions of low solute concentration to high solute concentration. Whether considered in terms of water concentration or solute concentration, the direction of net water movement is the same.

The movement of water across a selectively permeable membrane in response to a difference in solute concentration is known as **osmosis.** When there is a difference in the concentration of solutes on two sides of a selectively permeable membrane, water moves across the membrane from the side with a lower concentration of solutes toward the side with a larger concentration of solutes.

Consider the situation illustrated in **FIGURE 11.2**, which shows a difference in solute concentration on the two sides of a selectively permeable membrane. The membrane is freely permeable to water but impermeable to the solute. Water moves by osmosis from regions of higher water concentration (where the solute concentration is lower, side A in the figure) to regions of lower water concentration (where the solute concentration is higher, side B in the figure).

Osmosis can seem mysterious because it appears as if solutes on one side of the membrane influence the movement

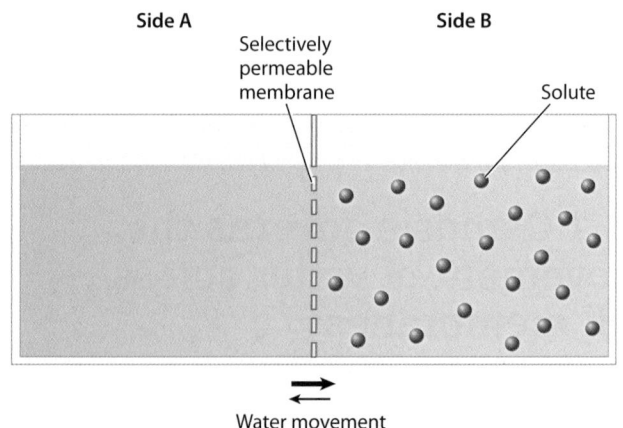

FIGURE 11.2 Osmosis

Osmosis is the net movement of water across a selectively permeable membrane from an area of lower solute concentration to an area of higher solute concentration. In this case, water moves from left to right across the selectively permeable membrane.

of water on the other side of the membrane and therefore exert some sort of force at a distance. To gain an intuitive understanding of the process, let's think about what is happening at a molecular level. Without any solutes, water molecules diffuse equally in both directions across a selectively permeable membrane. Now let's consider what happens if we add a solute, such as glucose, to one side of the membrane, as shown in **FIGURE 11.3**. If we add glucose to side B, the rate at which water molecules diffuse from side A to side B remains unchanged. But on side B, the glucose solute molecules will move about and bounce off the membrane, colliding with water molecules. The collisions slow the movement of water from side B to side A. As a result, the net, or overall, movement of water molecules will be from side A, which has no solutes, to side B, where solutes are present.

Whether water moves by osmosis into or out of a cell depends on the solute concentration inside the cell relative to the solute concentration outside the cell. Similarly, whether water moves by osmosis into or out of an animal depends on the solute concentration inside an animal relative to its environment, as well as the permeability to water of the sheets of cells that make up the skin and other surfaces exposed to the environment.

Osmosis occurs even if there are solutes on both sides of a selectively permeable membrane. The net movement of water will be from the side that has a lower concentration of solutes toward the side that has a higher concentration of solutes. When the concentration of solutes is the same on the two sides of the membrane, net movement of water molecules by osmosis stops, although water molecules continue to move in both directions.

In general, the identity of the solute does not matter. Water moves across selectively permeable membranes in response to concentration differences of any solute. That is, the solute can be an ion, such as sodium or potassium, or a dissolved molecule, such as glucose, an amino acid, or

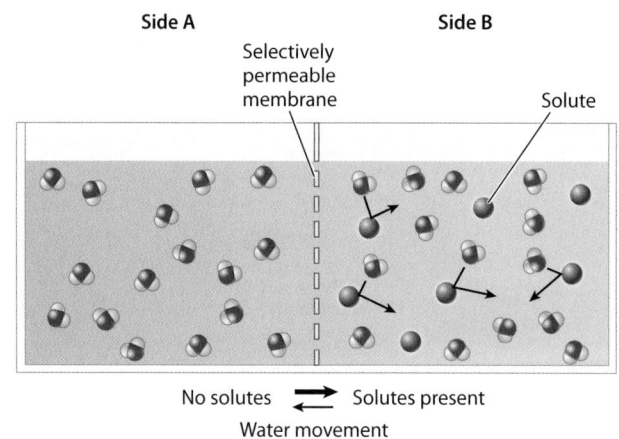

FIGURE 11.3 A molecular view of osmosis

A container is separated by a selectively permeable membrane. Solutes are added to side B. Water molecules move easily from side A to side B. On side B, water molecules collide with solute molecules, so move more slowly from side B to side A. Therefore, net movement of water molecules is from side A to side B.

even waste, as long as the membrane is impermeable to it. Furthermore, if several different solutes are dissolved in a single solution, the total concentration of all of the solutes determines the direction of water movement. This is because the concentration of water is determined by considering all of the dissolved solutes.

> ### ✓ Concept Check
>
> 1. **Identify** one solute that easily crosses cell membranes, and one solute that does not.
> 2. **Describe** why water, even though it is polar, is able to cross cell membranes easily by facilitated diffusion.
> 3. **Describe** the direction of water movement by osmosis across a selectively permeable membrane in terms of solute concentration and in terms of water concentration.

11.2 Water potential combines all of the factors that influence water movement

Water movement in cells is largely driven by osmosis, but there are other forces that affect water movement as well. The cell wall can exert pressure on water, as can gravity. Scientists use the term **water potential** to describe all of the chemical and physical forces that affect the movement

of water, such as osmosis, pressure, and gravity. Water potential helps us to understand whether water moves into a cell, or even up a tall tree. In this section, we consider some of the major forces that influence how water moves in cells and organisms.

Osmotic Pressure

The tendency of water to move from one solution into another by osmosis is called **osmotic pressure.** The higher

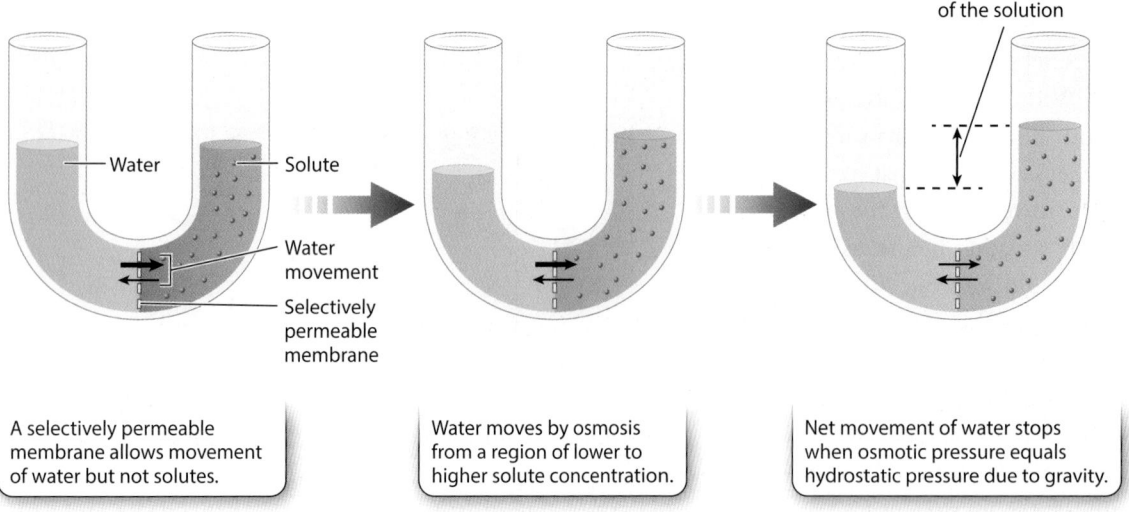

Water Solute

Water movement

Selectively permeable membrane

A selectively permeable membrane allows movement of water but not solutes.

Water moves by osmosis from a region of lower to higher solute concentration.

Osmotic pressure of the solution

Net movement of water stops when osmotic pressure equals hydrostatic pressure due to gravity.

FIGURE 11.4 Osmotic and hydrostatic pressure

A U-shaped tube with a selectively permeable membrane is filled with water. Solutes are added to one side, as shown. Water moves by osmosis into the side with the solutes until the pressure exerted by gravity, called hydrostatic pressure, is balanced by osmotic pressure. The height of the water column provides one way to measure osmotic pressure.

the solute concentration, the higher the osmotic pressure of that solution. In other words, there is a greater tendency for water to move into that solution by osmosis. Therefore, osmotic pressure is a way of describing the tendency of a solution to draw water in by osmosis.

We can measure osmotic pressure in two ways. The first is shown in **FIGURE 11.4**, which shows a U-shaped tube and two solutions separated by a selectively permeable membrane. If a solute is added to one side of the tube, water molecules will move from the side without the solute to the side with the solute. As a result, the level of the solution will rise against the force of gravity. The pressure that gravity exerts on the solution is called the hydrostatic pressure. When the water level stops rising, hydrostatic pressure equals osmotic pressure, as shown on the right in the figure. At this point, the net movement of water stops and equilibrium is reached. The height of the water column provides a measure of osmotic pressure. At equilibrium, water molecules still move across the membrane in both directions, but there is no longer any net movement of water molecules.

A second way to measure osmotic pressure is shown in **FIGURE 11.5**. In this case, instead of letting the water level rise on one side of a selectively permeable membrane, we can apply a force to prevent the water level from rising. The amount of external force that needs to be applied to prevent water from moving into the solution with the higher solute concentration is equal to the osmotic pressure.

Tonicity

The term **tonicity** is used to describe osmotic pressure and the direction of water movement. "Tonicity" literally means "strength" and it describes how strongly water is pulled into one solution compared to another. Note that tonicity is a relative term, so it only makes sense as a comparison of one solution to another. A solution with higher tonicity

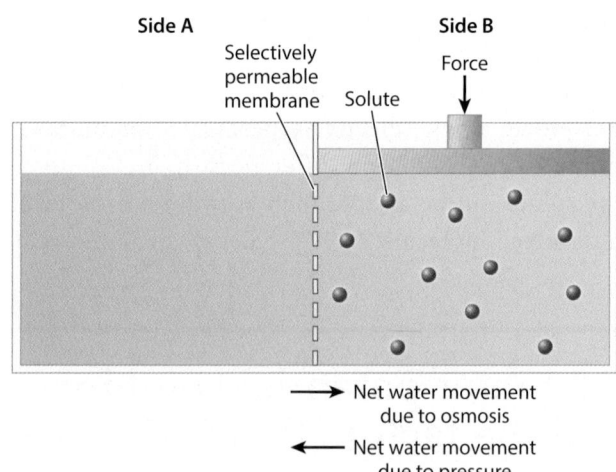

Side A   Side B

Selectively permeable membrane   Solute   Force

⟶ Net water movement due to osmosis

⟵ Net water movement due to pressure

FIGURE 11.5 Osmotic pressure and an external force

Osmotic pressure can be measured by the force required to stop osmosis. In this figure, water would move into the side with more solutes, but this movement is prevented by the externally applied force. The amount of force needed to stop the net movement of water by osmosis provides another way to measure osmotic pressure.

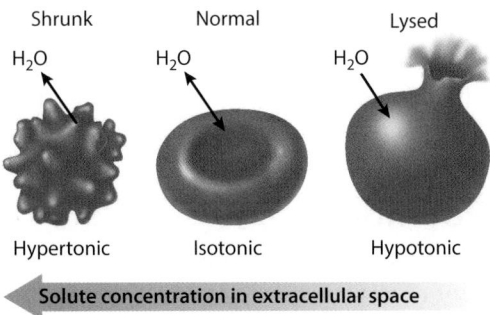

Shrunk   Normal   Lysed

H_2O H_2O H_2O

Hypertonic   Isotonic   Hypotonic

Solute concentration in extracellular space

FIGURE 11.6 Changes in red blood cell shape due to osmosis

Red blood cells shrink, swell, or burst because of net water movement driven by differences in solute concentration between the inside and the outside of the cell.

has a higher solute concentration than another solution and draws water into it by osmosis. A **hypertonic** solution is one with a higher solute concentration than another solution. A **hypotonic** solution is one with a lower concentration than another solution. Finally, an **isotonic** solution has the same concentration as another solution.

The concept of tonicity can be illustrated by placing a red blood cell in three different solutions, as shown in **FIGURE 11.6**. When a red blood cell is placed in a solution that is hypertonic relative to the inside of the cell, water leaves the cell by osmosis and the cell shrinks. By contrast, when a red blood cell is placed in a solution that is hypotonic relative to the inside the cell, water moves into the cell by osmosis and the cell lyses, or bursts. Animal cells generally solve the problem of water movement in part by keeping the intracellular fluid isotonic with the extracellular fluid. Many animal cells use the active transport of ions to maintain equal concentrations of solutes inside and outside the cell, as discussed in Module 10.

PREP FOR THE AP® EXAM

AP® EXAM TIP

Make sure you can describe the characteristics of a hypotonic, isotonic, or hypertonic solution in relation to the solute concentration of an original solution.

Some single-celled eukaryotes, or protists, have another strategy to avoid cell lysis resulting from water movement. The paramecium shown in **FIGURE 11.7** is one example. These single-celled organisms live in freshwater habitats, where the extracellular environment is hypotonic relative

to the cell's interior. As a result, they face the risk of bursting from water moving in by osmosis, just like the red blood cell in Figure 11.6. However, these organisms contain a vacuole that solves this problem. **Contractile vacuoles** are organelles that take up excess water from inside the cell and then, by contraction, expel it into the external environment.

Turgor Pressure

In addition to osmotic pressure, there are other forces that affect water movement. One of these is turgor pressure. In Module 7, we introduced turgor pressure and defined it as the force exerted by water pressing against an object. In cells, turgor pressure develops as a result of water moving

a.

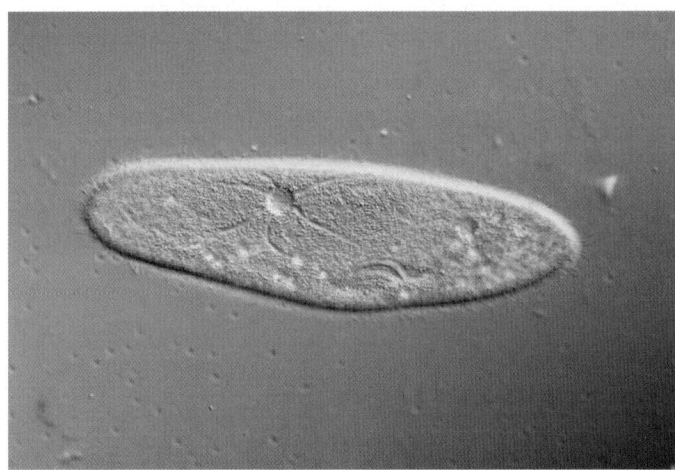

b.

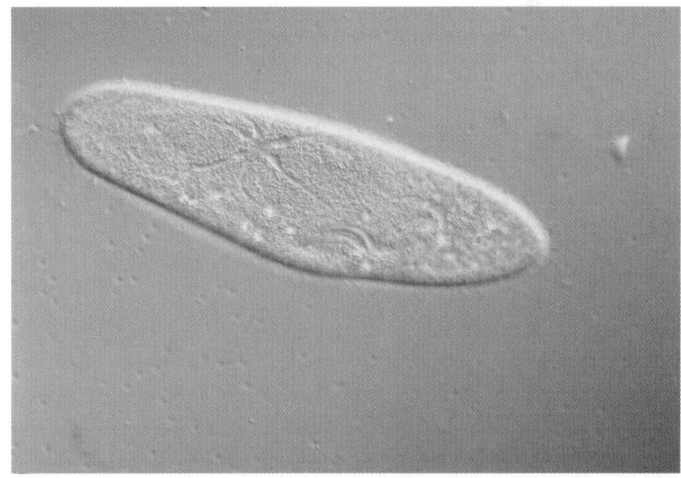

FIGURE 11.7 Contractile vacuole

A paramecium is a single-celled eukaryote, or protist, with a contractile vacuole. (a) In this image, the contractile vacuole in the paramecium is full. (b) In this image, the contractile vacuole has emptied. Photos: Michael Abbey/Science Source

by osmosis into cells surrounded by a cell wall. Cells contain high concentrations of solutes. When in contact with fresh water, which has much lower concentrations of solutes, a cell will swell as water moves into the cell by osmosis. Turgor pressure develops because the cell wall resists being stretched and pushes back on the cell interior, just as a balloon pushes back on the air inside it. Water will continue to enter the cell until the turgor pressure increases to a level sufficient to stop osmosis. Many organisms have cell walls, including bacteria, fungi, many protists, most algae, and all plants. When these organisms live in terrestrial or freshwater environments where solute concentrations are low, their cells have high turgor pressures due to the movement of water into cells by osmosis.

The effect of turgor pressure on a plant can be seen in **FIGURE 11.8**. When a plant cell is placed in a hypotonic solution, water moves into a cell and pushes against the cell wall, causing high turgor pressure, as shown on the right. This turgor pressure contributes to the mechanical support of plant tissues and we describe the plant as turgid. By contrast, when a plant cell is placed in a hypertonic

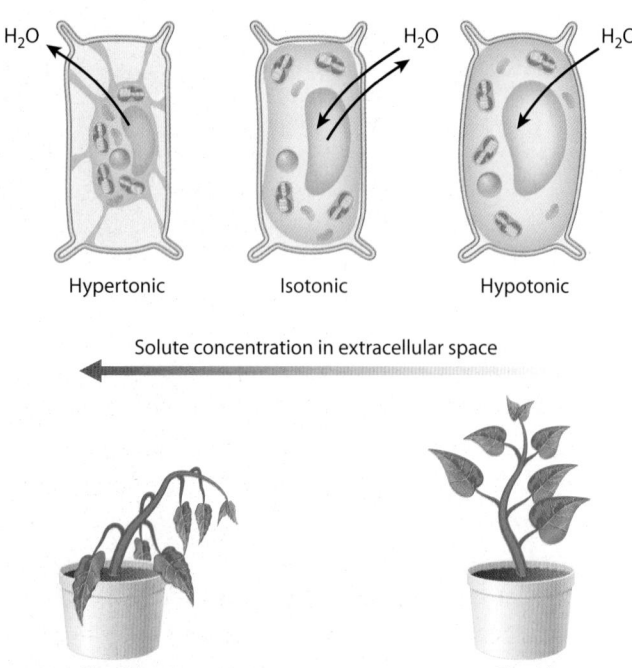

FIGURE 11.8 **Turgor pressure**

Turgor pressure results when water enters a plant cell by osmosis and the cell wall pushes back, as shown on the top right. As a result, the plant on the bottom is upright and described as turgid. In the absence of turgor pressure, the cell shrinks, as shown on the top left and the plant visibly wilts, as shown on the bottom left. When the solute concentration is the same inside and outside of the cell, as shown in the middle panel, water moves equally in both directions.

solution, water moves out of the cell by osmosis, resulting in low turgor pressure, as shown on the left. In this case, plants visibly deform, or wilt, and we describe the plant as flaccid.

Water Potential

We have discussed how the presence of solutes can lead to the movement of water across a selectively permeable membrane and generate osmotic pressure. We also know from our daily life that pressure can cause water to move, for example, when water gushes from a garden hose. This is similar to the pressure exerted by cell walls, or turgor pressure. Gravity can also affect the movement of water, as when water flows downhill. In large organisms such as trees, these various factors also come into play and together influence how water moves. In this section, we will combine all of these factors into a single formula. This formula tells us in which direction water moves, such as into or out of a cell.

Water potential is a measure of all of the factors that influence the movement of water. Specifically, it describes the potential energy (or stored energy) of water compared to a reference, such as pure water with no solutes at ground level. It allows us to determine how water moves due to solutes, turgor pressure, gravity, and any other force that acts on it. Water moves from higher to lower water potential, in the same way that water flows downhill.

Water potential is abbreviated by the Greek symbol psi, which is written as Ψ. We can add up the individual forces acting on water movement. Here, we will just focus on two forces, as follows:

$$\Psi = \Psi_p + \Psi_s$$

where:

Ψ is the water potential, measured in units of pressure expressed as bars

Ψ_p is the **pressure potential,** or the effect of pressure on the movement of water, and

Ψ_s is the **solute potential,** or the effect of solutes on the movement of water.

Let's consider each variable in turn. In order to add these different variables, they need to be expressed on the same scale. Therefore, we compare each one to a reference, which is typically pure water with no solutes at ground level. We designate this reference as having a $\Psi = 0$. The pressure potential Ψ_p then is the pressure of the water measured relative to our reference of $\Psi = 0$. Because turgor pressure exerts a force on the movement of water, the higher the turgor pressure, the higher the pressure potential. In turn, the higher the pressure potential, the higher the water potential.

Note that an open container has a pressure potential Ψ_p of 0. In this case, the water potential simply equals the solute potential, as follows:

$$\Psi = \Psi_p + \Psi_s$$
$$\Psi = 0 + \Psi_s$$
$$\Psi = \Psi_s$$

The solute potential Ψ_s measures the effect of solutes in a solution on water movement. It is calculated as follows:

$$\Psi_s = -iCRT$$

where:

  i = the ionization constant, which equals 1 for a substance that does not ionize in water

  C = the solute concentration (molar concentration, or molarity)

  R = the pressure constant $\left(0.0831 \dfrac{\text{L} \cdot \text{bars}}{\text{mol} \cdot \text{K}}\right)$, and

  T = the temperature in Kelvin (°C + 273).

The reference solution, or pure water, has a solute potential of 0 because it does not contain any solutes. As solutes are added, the solute potential Ψ_s becomes more negative, as indicated by the negative sign in the formula. This is because water flows from higher water potential (pure water) to lower water potential (solution), so the solute potential of a solution must be less than 0.

Recall, however, that adding solutes increases the osmotic pressure. Therefore, the more concentrated a solution is, the higher the osmotic pressure, but the lower (more negative) the solute potential and the more likely it draws water across the membrane by osmosis. Therefore, the solute potential, Ψ_s, is the negative of the osmotic pressure.

We can use these different forces to figure out the direction that water moves. For example, in Figure 11.7, water moves into a red blood cell placed in a hypotonic solution because the water potential is higher outside the cell compared to inside the cell. The low water potential inside the cell results from the presence of solutes, or a low (more negative) solute potential inside the cell compared to the outside of the cell.

Similarly, in Figure 11.8, the flaccid cell on the left might have a pressure potential Ψ_p of 0 due to the loss of turgor pressure, and a solute potential Ψ_s of −0.3 bars due to the presence of solutes inside of the leaf cell. As a result, its water potential Ψ is 0 + −0.3 = −0.3. If the cell is placed in pure water with a water potential of 0, water will flow from outside of the cell to inside the cell because water moves from a region of higher water potential to one with lower water potential. "Analyzing Statistics and Data: Water Potential and Solute Potential" on page 162 shows you how to use these concepts in context and provides a problem for practice.

We can also apply the concept of water potential to understand what causes water to flow up a tall tree, as shown in **FIGURE 11.9**. During the day, water is lost by the leaves. As water is lost, the turgor pressure of the leaf cells decreases, similar to the cell on the left in Figure 11.8. As turgor pressure decreases, the pressure potential Ψ_p decreases. In addition, as water is lost, the solute concentration inside the cell increases, which increases the osmotic pressure and therefore decreases the solute potential Ψ_s. As a result, the water potential Ψ is low at the top of the tree, approximately −2.00 bars, as indicated in the figure. By contrast, water is drawn in by the roots at the base of the tree. As a result, the pressure potential resulting from turgor pressure and the solute potential resulting from solutes are both higher than at the top of tree. The water potential Ψ at the base of the tree is approximately −0.10 bar, as shown in the figure.

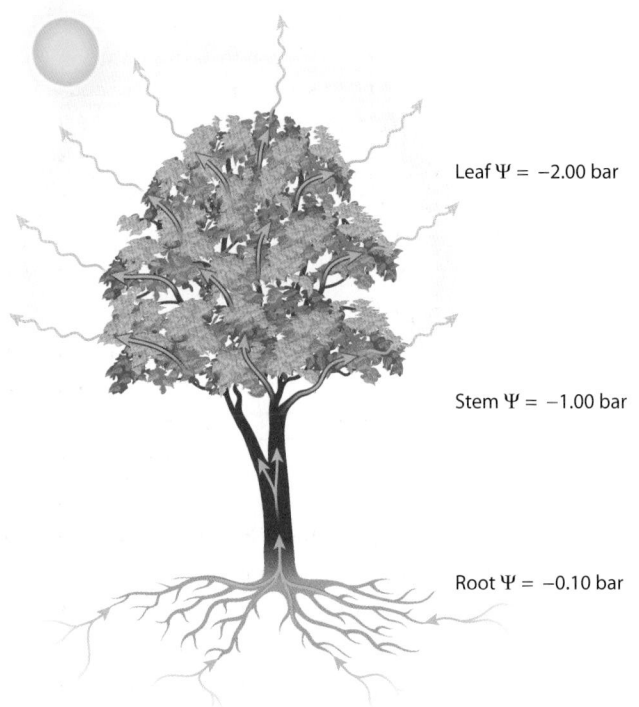

Leaf Ψ = −2.00 bar

Stem Ψ = −1.00 bar

Root Ψ = −0.10 bar

FIGURE 11.9 Water movement up a tree

During the day, when water evaporates from leaves, the water potential Ψ is lower at the top of the tree than at the bottom of the tree, so water moves up the tree.

Water and Solute Potential

Water potential (Ψ) is a measurement of all of the chemical and physical forces that affect water movement. This exercise focuses on the forces of pressure potential (Ψ_p) and solute potential (Ψ_s). Remember that water moves from areas of high Ψ to low Ψ. Knowing this will allow us to determine which direction water will flow between cells and the environment.

The equation for water potential is:

$$\Psi = \Psi_p + \Psi_s$$

When the pressure potential is zero, as it would be in an open system, the water potential Ψ is equal to the solute potential Ψ_S. To calculate solute potential, we use the following equation:

$$\Psi_s = -iCRT$$

In this equation, i is the ionization constant, C is the molar concentration, R is the pressure constant $\left(0.0831\dfrac{L \bullet bars}{mole \bullet K}\right)$, and T is the temperature in Kelvin (°C + 273).

PRACTICE THE SKILL

1. A cell has a solute potential of –2 bars. When placed in pure water, water will move into the cell, since water moves across a membrane from high water potential to low water potential. Equilibrium is reached when the water potentials are equal. What would the pressure potential be at equilibrium?

   At equilibrium, there is no net water movement between the cell and its environment. In pure water, water potential is equal to 0. We can enter our known variables and solve for Ψ_p.

   $$\Psi = \Psi_p + \Psi_s$$
   $$0 = \Psi_p + (-2\text{ bars})$$
   $$0 = \Psi_p - 2\text{ bars}$$
   $$\Psi_p = 2\text{ bars}$$

A pressure potential of 2 bars is needed to keep the cell in equilibrium with the solution, resulting in no net movement of water and a water potential of 0.

2. A student is asked if a cell with a solute potential of –0.5 bar would lose or gain water when placed in a sucrose solution. Calculate the solute potential for a 0.1 M sucrose solution (ionization constant = 1) at 23°C to determine the direction of water movement into or out of the cell.

   The ionization constant for a sucrose solution is 1 because it does not ionize in water. The temperature in Celsius can be converted to Kelvin with a simple calculation:

   $$T = 23°C + 273$$
   $$T = 296\text{ K}$$

We can now substitute all of our values into the solute potential equation.

$$\Psi_s = -iCRT$$
$$\Psi_s = -(1) \times (0.1\text{ M}) \times \left(0.0831\dfrac{L \bullet bars}{mol \bullet K}\right) \times (296\text{ K})$$
$$\Psi_s = -2.5\text{ bars}$$

In this example, the water potential is equal to the solute potential. Water will move from higher water potential, –0.5 bar, to lower water potential, –2.5 bars. So, the cell will lose water when placed in this sucrose solution.

Your Turn

Consider three cells, with various water potential values shown in the diagram. Calculate the Ψ of plant cell b if $\Psi_p = 2$ and $\Psi_s = -3$. Then describe the direction that water would move between the three cells.

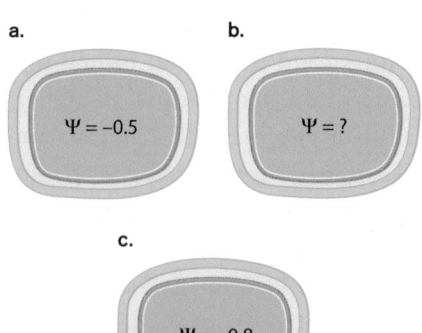

Notice that the water potential is higher at the base of the tree in Figure 11.9 than at the top of the tree. Because water moves from regions of higher water potential to lower water potential, water flows upward from the base to the top of the tree.

PREP FOR THE AP® EXAM

AP® EXAM TIP

You should be very familiar with the equations for water potential and the solute potential of a solution and know how to apply them.

As we saw in Module 8, bulk flow is driven by differences in pressure. The direction of flow is always from higher to lower pressure. In animals, a mechanical pump—the heart—raises the pressure of the blood, causing it to flow from the heart through your arteries and veins. In plants, water loss in leaves lowers the pressure of the water at the top of the tree, causing water to flow from the soil to the leaves, as we just discussed. Plants face a challenge owing to this process: to transport water from the soil to the leaves at rates equal to the water loss from leaves, the pressures in the leaf xylem must be very low, typically well below zero.

"Practicing Science 11.1: Do plants generate negative pressures?" on page 164 describes an experiment which showed that the pressures generated in plants are actually negative and stronger than what a vacuum produces.

Water stands out among liquids for its ability to sustain large negative pressures. In Module 2, we noted that hydrogen bonds form between water molecules. These bonds allow tensile (pulling) forces to be transmitted from one water molecule to another. Hydrogen bonds between water molecules create a continuous stream of water as water moves from the soil to the roots to stems and finally to leaves.

✓ Concept Check

4. **Describe** what happens with regard to the movement of water when an animal cell is placed in a hypotonic solution.

5. **Describe** what happens with regard to the movement of water when an animal cell is placed in a hypertonic solution.

6. **Identify** three factors that influence the movement of water.

11.3 Osmoregulation is a form of homeostasis

Animals can survive in a wide range of environments by adopting different means of *osmoregulation*. **Osmoregulation** is the regulation of osmotic pressure inside cells and organisms. It can be thought of as the regulation of water content, which keeps internal fluids from becoming too concentrated (high osmotic pressure) or too dilute (low osmotic pressure).

Osmoregulation is a form of homeostasis, as it maintains water balance inside cells and organisms. High osmotic pressure inside a cell can lead to damage, sometimes even causing the cell to burst, as we have seen. Low osmotic pressure can lead to dehydration. Although cells and tissues can tolerate some degree of dehydration, excessive dehydration impairs a cell's metabolic function. Many chemical reactions depend on the presence of water, such as hydrolysis reactions that break down proteins and nucleic acids into individual subunits, as discussed in Unit 1. In this section, we will consider two general strategies that cells and organisms use to maintain water balance.

Osmoconformers

Some animals keep their internal fluids at the same osmotic pressure as the surrounding environment. These animals are called osmoconformers. Because the solute concentrations inside and outside their bodies are similar, osmoconformers do not spend a lot of energy regulating osmotic pressure. However, they must adapt to the solute concentration of their external environment. Osmoconformers tend to live in environments that have stable solute concentrations, such as seawater, because maintaining stable internal solute concentrations is easier in stable environments compared to fluctuating environments.

Although osmoconformers generally match the overall concentration of solutes within their tissues with the overall concentration of solutes in their external environment, they often expend energy to regulate the concentrations of particular ions, such as sodium, potassium, and chloride, and particular molecules, such as amino acids and glucose. For example, whereas the intracellular space of nearly all animals has a relatively high concentration of potassium ions and a low concentration of sodium ions, the extracellular space

Do plants generate negative pressures?

Background The idea that water is pulled through the plant by negative pressures in leaves was first suggested in 1895. However, without a way to measure these pressures, there was no way to know how large they are.

Hypothesis In 1912, German physiologist Otto Renner hypothesized that leaves are able to exert stronger suctions than pressures generated with a vacuum pump.

Experiment Renner placed water in a U-shaped container with a reservoir at one end. In one trial, shown on the left in the figure, he dipped the cut tip of a branch of a plant into the reservoir. He then measured the rate at which water flowed from the reservoir into the branch. In a second trial, shown on the right in the figure, he cut off a short segment of the branch and attached a vacuum pump in place of the plant. He then measured the rate at which water flowed from the reservoir into the vacuum pump. Finally, he compared the flow rate through the branch when attached to the plant with the flow rate through the branch generated by the vacuum pump. The flow rates are indicated by the height of the bars in the histogram in the figure.

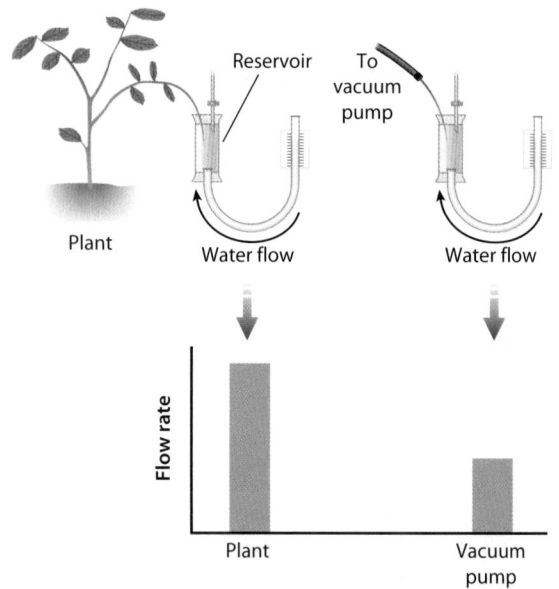

Results Renner found that the flow rates generated by plants were two to nine times greater than the flow rates generated by the vacuum pump.

Conclusion A plant pulls water through a branch faster than a vacuum pump. Therefore, the pulling force generated by a plant must be greater than that of a vacuum pump. Because vacuum pumps create suction by reducing the pressure in the air, a vacuum pump can reduce the pressure only to zero. For the leaves to pull even harder, the pressures must be less than zero. Thus, plants generate negative pressures.

The maximum height that one can lift water using a vacuum pump is 10 m (33 feet). The forces that pull water through plants have no comparable limit because the hydrogen bonds between water molecules mean that the vascular system can sustain very large negative pressures. This is why plants can pull water from dry soils and why they can grow to more than 100 m, the height of a 30-story building.

Follow-Up Work In 1965, Per Fredrick Scholander developed a pressurization technique that uses reverse osmosis to measure the magnitude of the negative pressures exerted by leaves of intact plants. Using this technique, he demonstrated that the leaves at the top of a Douglas fir tree 100 m tall exert pressures that are even more negative than those exerted by leaves closer to the ground.

SOURCES
Renner, O. 1925. "Zum Nachweis negativer Drucke im Gefässwasser bewurzeler Holzgewachse." *Flora* 118/119:402–408; Scholander, P. F. et al. 1965. "Sap Pressure in Vascular Plants." *Science* 148:339–346.

AP® PRACTICE QUESTION

After learning about Renner's experiment, a student predicts that the flow rate in a plant is nine times greater than the flow rate due to a vacuum tube.

(a) **Describe** the mathematical steps Renner used to calculate the flow rate of the water.

(b) **Calculate** the flow rate of water from the U-tube to the vacuum pump in mL/s if the volume of water in the U-tube decreased 1 mL in 10 s.

(c) **Calculate** the expected flow rate in mL/s in a plant.

---

has concentrations that are just the opposite: low in potassium and high in sodium. This means that these cells actively pump sodium out and potassium in.

Most marine invertebrates, such as sea stars, mussels, lobsters, and scallops, are osmoconformers. **FIGURE 11.10** shows sea stars, which are osmoconformers. These organisms maintain high concentrations of sodium and chloride in their cells to achieve an overall solute concentration close to that of seawater. Consequently, they have few specializations for osmoregulation beyond the need to regulate specific internal ion concentrations relative to their environment. Some marine vertebrates, such as hagfish and lampreys, are also osmoconformers that maintain high internal concentrations of particular electrolytes.

FIGURE 11.10 An osmoconformer

Sea stars, like these *Linckia laevigata* in Indonesia, are marine invertebrates and are osmoconformers, matching their internal solute concentration to the salt water around them. Photo: Azure Computer & Photo Services/Animals Animals

Other osmoconformers, including sharks, rays, and coelacanths, match seawater's solute concentration by maintaining a high internal concentration of a compound called urea. Urea is a waste product of protein metabolism that many animals excrete, thereby ensuring it does not build up to high concentrations. By retaining this solute, these marine vertebrates are able to achieve osmotic equilibrium with the surrounding seawater.

Osmoregulators

The second way in which animals achieve water and electrolyte homeostasis is by maintaining an internal solute concentration that is different from that of the

FIGURE 11.11 An osmoregulator

Bony fish like these char are osmoregulators. Their internal solute concentration is different from that in the salt water around them.
Photo: JOEL SARTORE/National Geographic Creative

environment. Animals that rely on this mechanism are called osmoregulators. Osmoregulators expend considerable energy pumping ions across cell membranes to regulate the movement of water into or out of their bodies. Some freshwater and marine fishes that are osmoregulators are estimated to expend 50% of their resting metabolic energy on osmoregulation. However, the ability to regulate osmotic pressure allows osmoregulators to live in diverse environments—in salt water, in fresh water, and on land.

FIGURE 11.11 shows an example of an osmoregulator: teleosts, also called bony fishes. All freshwater and terrestrial animals, including humans, are osmoregulators. Marine bony fishes maintain concentrations of electrolytes much lower than those in the surrounding seawater, whereas freshwater fishes and amphibians maintain concentrations that are higher than those in the surrounding fresh water.

✓ Concept Check

7. **Describe** what is meant by osmoregulation.

8. **Describe** the difference between osmoconformers and osmoregulators.

9. **Identify** one animal that is an osmoconformer and one animal that is an osmoregulator.

Module 11 Summary

REVISIT THE BIG IDEAS

ENERGETICS: Use the content of this module to **explain** how organisms utilize energy and matter to manage the movement of water.

LG 11.1 Osmosis governs the movement of water across cell membranes.

• Cell membranes act as selectively permeable membranes, allowing the passage of water but restricting or controlling the movement of many solutes. Page 155

• Water moves across selectively permeable membranes by osmosis from a region of lower solute concentration to a region of higher solute concentration. Page 156

LG 11.2 Water potential combines all of the factors that influence water movement.

• Osmotic pressure is a measure of the tendency for water to move by osmosis across a selectively permeable membrane into a solution with higher solute concentration. Page 157

• Turgor pressure develops when water moves into a cell by osmosis and the cell wall resists being stretched and pushes back on the cytoplasm. Page 159

• Water potential is a parameter that combines all of the various factors that influence the movement of water, such as the effects of solutes and pressure. Page 160

• Water moves from areas with higher water potential to areas with lower water potential, just as water flows downhill. Page 161

LG 11.3 Osmoregulation is a form of homeostasis.

• All animals regulate the water level within their cells. Page 163

• Osmoregulation is the regulation of osmotic pressure inside cells or organisms. Page 163

• Osmoconformers maintain an internal solute concentration similar to that of the environment. Page 163

• Osmoregulators have an internal solute concentration different from that of the environment. Page 165

Key Terms

Solute	Tonicity	Pressure potential
Molarity	Hypertonic	Solute potential
Osmosis	Hypotonic	Osmoregulation
Water potential	Isotonic	
Osmotic pressure	Contractile vacuole	

Review Questions

1. Water moves
   (A) from regions of high water concentration to regions of low water concentration.
   (B) from regions of low water concentration to regions of high water concentration.
   (C) from regions of high solute concentration to regions of low solute concentration.
   (D) from regions of low water concentration to regions of low solute concentration.

2. Turgor pressure is the result of
   (A) the activity of the contractile vacuole.
   (B) water accumulation within a cell with a cell wall.
   (C) the rigidity of the cell membrane.
   (D) active transport.

3. Animals that match their internal osmotic pressures to the osmotic pressures of their external environments are called
   (A) osmoregulators.
   (B) osmosis.
   (C) osmoconformers.
   (D) diffusion regulators.

4. Osmoregulation is defined as
   (A) the control of osmotic pressure through regulation of water and solute levels.
   (B) the movement of water from an area of lower solute concentration to an area of higher solute concentration.
   (C) the difference in solute concentration on the two sides of a selectively permeable membrane.
   (D) the movement of water through aquaporins.

5. If the fluid inside of a cell has a solute concentration of 5%, what would happen if the cell is placed in a solution with a solute concentration of 10%?
   (A) Water molecules would enter the cell by osmosis.
   (B) Water molecules would leave the cell by osmosis.
   (C) Water molecules would cross the membrane in both directions, but there would be no net movement of water molecules.
   (D) Solute molecules would leave the cell by osmosis.

6. If solution X has a 2% concentration of NaCl and solution Y has an 8% concentration of NaCl, which of the following statements is correct?
   (A) Solution X has a higher osmotic pressure than solution Y.
   (B) Solution X has a lower concentration of water than solution Y.
   (C) Solution X is hypotonic relative to solution Y.
   (D) Solution X is hypertonic relative to solution Y.

7. Two solutions of water and dissolved potassium and glucose are separated by a selectively permeable membrane that only permits the passage of water. If the two solutions have the same total solute concentration, but solution 1 has a higher concentration of potassium and a lower concentration of glucose than solution 2, which of the following statements is correct?
   (A) There will be no net movement of water molecules between the solutions.
   (B) There will be net water movement from solution 1 to solution 2.
   (C) There will be net water movement from solution 2 to solution 1.
   (D) There will be net glucose movement from solution 2 to solution 1.

Module 11
AP® Practice Questions

Section 1: Multiple-Choice Questions

Choose the best answer for questions 1–5.

1. A 4-liter container of 100% (pure) water sits on a lab bench. A student carefully places a sugar cube on the bottom right side of the container. In which direction will the net movement of water occur until the sugar cube is dissolved?
   (A) The net movement of water will be away from the sugar cube.
   (B) The net movement of water will be toward the sugar cube.
   (C) There will be no net movement of water.
   (D) The net movement of water will first be away from the sugar cube but when half the cube is dissolved, the net movement of water will be toward the sugar cube.

Question 2 refers to the following figure.

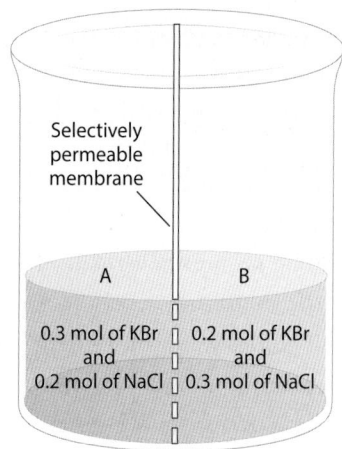

Question 3 refers to the following figure.

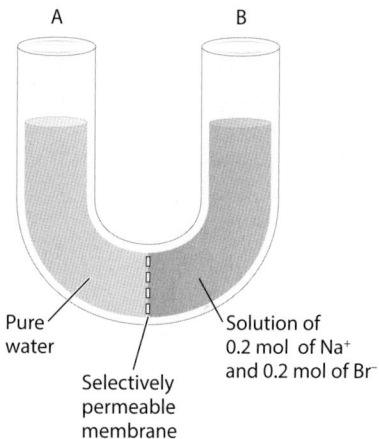

2. A beaker of water is separated into two equal volumes by a selectively permeable membrane. The membrane is permeable to water but not to the ions Na^+, K^+, Cl^-, or Br^-. Into side A of the beaker 0.3 mol of KBr and 0.2 mol of NaCl are dissolved, while into side B 0.2 mol of KBr and 0.3 mol of NaCl are dissolved. In which direction will the net movement of water occur?

    (A) The net movement of water will occur from side A to side B.

    (B) The net movement of water will occur from side B to side A.

    (C) There will be no net movement of water from one side to the other.

    (D) Once the KBr is fully dissolved, the net movement will be from side A to side B.

3. An experiment is conducted in which a selectively permeable membrane separates two sides of a U-tube. Side A contains pure water and Side B is an aqueous solution of 0.2 mol of Na^+ and 0.2 mol of Br^-. The membrane is permeable to water and Na^+ but not to Br^-. When equilibrium is reached and there is no net movement of water across the selectively permeable membrane, which side will have a lower column of fluid?

    (A) Side A

    (B) Side B

    (C) Both sides will be the same height.

    (D) When equilibrium is reached for the Na^+ concentration, side B will be lower but when equilibrium for the Br^- concentration is reached, side A will be lower.

4. Solutions that are 0.85% to 0.90% sodium chloride are isotonic to body fluids. If red blood cells (RBCs) are placed into a solution of 0.10% sodium chloride, predict what will happen to the RBCs.

(A) They will gain fluid and swell.

(B) They will lose fluid and burst.

(C) They will lose fluid and shrink.

(D) They will remain the same and neither gain nor lose fluid.

5. The protist *Paramecium* commonly lives in aquatic environments where the external solution is hypotonic to the cell. In experiments, researchers determined that when the external environment was a standard saline solution (0.9% sodium chloride) and hypotonic to the *Paramecium*, the solute concentration of the *Paramecium*'s cytosol differed from the solute concentration of its contractile vacuole. The cytosol of the *Paramecium* had a solute concentration that was 1.5 times higher than the solute concentration of the fluid of the *Paramecium*'s contractile vacuole. Which of the following describes the movement of water relative to the *Paramecium*?

(A) Water moves from the contractile vacuole to the cytosol to the external environment.

(B) Water moves from the cytosol to the external environment to the contractile vacuole.

(C) Water moves only from the cytosol to the contractile vacuole.

(D) Water moves from the external environment to the cytosol to the contractile vacuole.

Section 2: Free-Response Question

Write your answer to each part clearly. Support your answers with relevant information and examples. Where calculations are required, show your work.

A group of students in an AP® Biology laboratory make a 2.0 M (molar) solution of sucrose at a room temperature of 23°C.

(a) **Calculate** the Ψ_s of the sucrose solution. The ionization constant for sucrose is 1.

(b) **Calculate** the water potential if the pressure potential is 0. Note that the water potential of pure water in an open container is 0.

(c) The students fill a dialysis bag that is permeable to water but not sucrose with their 2.0 M sucrose solution, and place the bag into a beaker containing pure water. **Predict** whether water will move into or out of the dialysis bag. **Justify** your answer. Assume a temperature of 23°C.

(d) The students place a second dialysis bag containing the 2.0 M solution of sucrose into a 3.0 M solution of sucrose. Calculate the water potential of the solution in the dialysis bag and the water potential of the solution outside the dialysis bag. The ionization constant for sucrose is 1, the pressure potential is 0, and the temperature is 23°C. **Predict** whether water will move into or out of the dialysis bag. **Justify** your answer.

Module 12

Origin of Compartmentalization and the Eukaryotic Cell

LEARNING GOALS ▶**LG 12.1**   The organization of the eukaryotic cell helps explain eukaryotic diversity.

                  ▶**LG 12.2**   Chloroplasts and mitochondria originated by endosymbiosis.

In the last six modules, we have discussed the structure and function of cells. All cells on the planet, including all prokaryotic and eukaryotic cells, have many features in common, such as a cell membrane, an archive of information in the form of DNA that they can use to build proteins, and the ability to harness energy from the environment. All cells regulate what substances enter and exit the cell across their cell membrane, including water, ions, nutrients, and so on. Because cells are able to regulate the substances that move across their membrane, all cells maintain homeostasis.

Some features are not shared by all cells but are specific to particular cell types. For example, prokaryotic cells are small, do not have a nucleus, and do not have extensive internal membranes. They are metabolically diverse, meaning that they are able to harness carbon and energy from their environments in many different ways. Eukaryotic cells are generally much larger, have a nucleus, and have extensive internal membranes that define compartments within cells. They are not as metabolically diverse as prokaryotes and are capable of obtaining carbon and energy in just a limited number of

ways. Carbon is the backbone of all of the macromolecules, and energy is required to do the work of the cell, such as grow, divide, move about, and change shape.

Why, then, have eukaryotes been so successful? The source of eukaryotic success lies in their remarkable capacity to form cells of diverse shapes and sizes, and to produce multicellular structures in which multiple cell types specialize and act together. These innovations opened up a range of possibilities for structure and function not found among prokaryotes. Eukaryotic diversity includes cells that can engulf other cells, as well as organisms as large as redwood trees and as complex as humans. In this module, we will review some of the key features of eukaryotic cells and then discuss how they originated and evolved over time.

PREP FOR THE AP® EXAM

FOCUS ON THE BIG IDEAS

EVOLUTION: Consider how the origin of compartmentalization in cells increased the diversity of life, and how these changes increased the ability of organisms to survive and reproduce.

12.1 The organization of the eukaryotic cell helps explain eukaryotic diversity

Eukaryotes are defined by the presence of a nucleus. However, many features distinguish them from prokaryotes. One key feature of eukaryotic cells is an extensive network of internal membranes that compartmentalizes the cell. These compartments provide spaces in the cell with suitable environments for specific processes and reactions to take place. For example, the rough endoplasmic reticulum is one of the sites of protein synthesis; the Golgi apparatus processes and sorts proteins and lipids; lysosomes break down and recycle proteins

and other macromolecules; and mitochondria harness energy from carbohydrates and other organic molecules. In this section, we will review these and other features of eukaryotes that account in part for their evolutionary success.

Dynamic Cytoskeleton and Membranes

All cells require a mechanism to maintain spatial order in the cytoplasm. As we saw in Module 6, prokaryotes such as bacteria and archaea rely primarily on walls that support the cell from the outside, along with a relatively rigid framework of proteins within the cytoplasm.

Some eukaryotic cells have cell walls, but many others, including the cells in your body, do not. Eukaryotes rely mainly on an internal scaffolding of proteins, the cytoskeleton, to organize the cell. This cytoskeleton, which is found in all eukaryotic cells, differs in one key property from the protein framework of bacteria: it can be remodeled quickly, enabling cells to change shape.

Dynamic cytoskeletons require dynamic membranes that enable the cell to continue functioning even as it changes shape. As we have seen, eukaryotes maintain within their cells a remarkably dynamic network of membranes called the endomembrane system. This network includes the nuclear envelope, an assembly of membranes that runs through the cytoplasm called the endoplasmic reticulum (ER) and Golgi apparatus, and a cell membrane that surrounds the cytoplasm, as we discussed in Module 7 and saw in Figure 7.1 on page 106. All membranes of the endomembrane system are interconnected, either directly or indirectly by the movement of membrane-bound vesicles. They are also capable of changing shape rapidly.

In fact, the different membranes are interchangeable in the sense that material originally added to the endoplasmic reticulum may in time be transferred to the cell membrane or nuclear envelope by the movement of vesicles or continuity between organelles. Biologists like to say that the membranes of eukaryotic cells are in dynamic continuity.

By contrast, some membranes within a eukaryotic cell are more stable than those of the endomembrane system. These include the membranes of the organelles responsible for energy transfer in a cell—mitochondria and chloroplasts. Recall from Module 7 that mitochondria are the site of many of the steps of cellular respiration, and chloroplasts are the site of photosynthesis. The inner membrane of the mitochondria and the thylakoid membrane of chloroplasts both include membrane-embedded proteins that require a stable membrane for these organelles to function.

In combination, the dynamic cytoskeleton and membranes provide eukaryotes with possibilities for movement that are not available to prokaryotes. For example, amoebas extend fingerlike projections that pull the rest of the cell forward. The cytoskeleton and membrane system also enable eukaryotic cells to engulf molecules or particles, including other cells, in a process called endocytosis, discussed in Module 10. Prokaryotic cells cannot perform this activity. Phagocytosis is a specific form of endocytosis in which eukaryotic cells surround food particles and package them in vesicles that bud off from the cell membrane. When packaged in this way, the particles can be transported into the cell interior for digestion. For example, **FIGURE 12.1** shows a eukaryotic cell engulfing a bacterial cell.

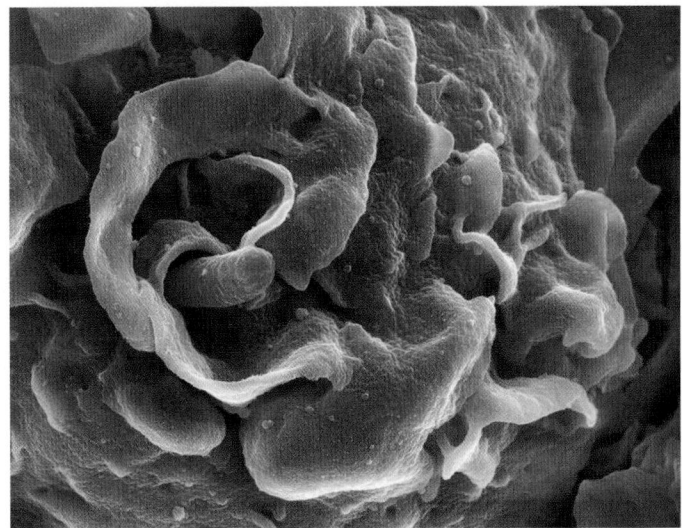

FIGURE 12.1 Phagocytosis

This photomicrograph shows a eukaryotic cell called a macrophage engulfing a rod-shaped bacterium called *Shigella* (orange) in the process of phagocytosis. *Shigella* are a group of bacteria that infect humans, causing mild to severe bloody diarrhea. Macrophages provide one line of defense against *Shigella*. Photo: SPL/Science Source

Energy Metabolism

Prokaryotic cells, including bacteria and archaea, can only take in molecules by passive and active transport. Although they are limited in the ways they take in molecules, they can use and metabolize these molecules in a diverse set of ways to obtain the carbon and energy they need to function.

Compared to prokaryotic organisms, eukaryotes are fairly limited in the ways they use and transfer carbon and energy. Moreover, many of the metabolic processes that power eukaryotic cells take place in specific organelles. Cellular respiration occurs in mitochondria, and photosynthesis takes place in chloroplasts, as discussed in Module 7.

Although eukaryotic cells and organisms are not metabolically diverse, they can take in molecules and food in a diverse set of ways. In addition to passive and active transport, many eukaryotic cells engulf food particles and package them inside vesicles, which are then transported into the cytoplasm. These vesicles may fuse with lysosomes, which have enzymes that break down the particles into molecules, some of which are processed by the mitochondria. As a result of having sites of intracellular digestion, many single-celled eukaryotes are able to feed on bacteria or other eukaryotic cells. Multicellularity opened up even more possibilities for obtaining food. Animals, for example, are able to ingest large foodstuffs, including other animals and plants. As a consequence of all of these ways of obtaining food, eukaryotes can exploit sources of food that are not readily available to prokaryotes. This ability opened up the great new ecological possibility of predation, in which one organism, a predator, eats another, the prey. The evolution of predation vastly increased the complexity of interactions among organisms.

The structural flexibility of eukaryotic cells also allows photosynthetic eukaryotes to interact with their environment to a much larger extent than photosynthetic bacteria can. Unicellular algae, which are eukaryotes, can move effectively through surface waters vertically as well as horizontally and therefore can seek and exploit local patches of nutrients in their environments.

Plants have evolved multicellular bodies with many different cell types that work together. For example, leaves have cells that harness the energy of sunlight to build carbohydrates; cells in the stem have channels that transport these carbohydrates to other parts of the plant and transport water up from the ground; cells in the root are specialized for absorbing water and nutrients from the soil. In other words, specialized cell types working together can provide leaves high in the canopy of trees with water and nutrients from the soil in which the plants are rooted. Thus supplied, leaves can capture sunlight many meters above the ground, giving plants a tremendous advantage on land.

✓ Concept Check

1. **Describe** two key features of eukaryotic cells that distinguish them from prokaryotic cells.
2. **Identify** two ways of obtaining nutrients that are seen in eukaryotes, but not in prokaryotes.

12.2 Chloroplasts and mitochondria originated by endosymbiosis

Eukaryotes have two organelles that are involved in energy metabolism. Chloroplasts are present in certain algae and plants, and mitochondria can be found in most eukaryotic cells and all multicellular eukaryotes. Chloroplasts are the site of photosynthesis: they use the energy of sunlight to build energy-rich carbohydrates. Mitochondria are the site of cellular respiration: they harness the energy stored in carbohydrates and other organic molecules and transfer it to a form—ATP—that they can use easily to carry out cellular reactions. Remarkably, these two organelles were once free-living bacteria that were engulfed by an ancestral cell. Instead of being digested, however, they were maintained in the cell interior, and over evolutionary time became the chloroplasts and mitochondria that we see today in eukaryotic cells. In this section, we will discuss how these organelles originated. In Unit 3, we will discuss the processes of cellular respiration and photosynthesis in detail.

Chloroplast and Mitochondrial Origins

The chloroplasts found in plant cells closely resemble certain photosynthetic bacteria, specifically cyanobacteria. The molecular workings of photosynthesis are nearly identical in the two, and internal membranes organize the photosynthetic machinery in similar ways. The Russian botanist Konstantin Sergeevich Merezhkovsky recognized this similarity more than a century ago. He also realized that corals and some other organisms harbor algae within their tissues that aid the growth of their host. A symbiont is an organism that lives in closely evolved association with another species, like the algae in corals shown in **FIGURE 12.2**. This close association between two species is called a **symbiosis.**

Putting these two observations together, Merezhkovsky came up with a radical hypothesis. Chloroplasts, he argued,

FIGURE 12.2 Symbiosis

Many corals harbor algae in their tissues, and so represent an example of a symbiosis, which is a close association between two species.

Photo: Takashi Images/Shutterstock

originated as symbiotic cyanobacteria that through time became permanently incorporated into their hosts. Such a symbiosis, in which one partner lives within the other, is called an **endosymbiosis.** Merezhkovsky's hypothesis of chloroplast origin by endosymbiosis was difficult to test with the tools available in the early twentieth century, and his idea was dismissed, more neglected than disproved, by most biologists.

In 1967, American biologist Lynn Margulis resurrected the endosymbiotic hypothesis. She supported her arguments with new types of data made possible by the then-emerging techniques of cell and molecular biology. Transmission electron microscopy revealed that the structural similarities between chloroplasts and cyanobacteria extend to the microscopic level, such as in the organization of the photosynthetic membranes, as shown in the photographs on page 174 in "Practicing Science 12.1: What is the evolutionary origin of chloroplasts?"

It also became clear that the chloroplasts present in algal and plant cells are separated from the cytoplasm that surrounds them by two membranes. This arrangement would be expected if a cyanobacterial cell had been engulfed by a eukaryotic cell. The inner membrane corresponds to the cell membrane of the cyanobacterium, and the outer membrane is part of the engulfing cell's membrane system, as shown in **FIGURE 12.3.** In addition, the biochemistry of photosynthesis is essentially the same in cyanobacteria and in chloroplasts of eukaryotic cells.

Such observations kindled renewed interest in the endosymbiotic hypothesis, but the decisive tests were made possible by another, and unexpected, discovery. It turns out that chloroplasts have their own DNA, which is organized as a single circular molecule, like that of bacteria. Because of this, researchers were able to use the tools of molecular sequence comparison to study chloroplast genes. The sequences of nucleotides in chloroplast genes closely match those of cyanobacterial genes, as described in the Experiment section of "Practicing Science 12.1." This finding provides strong support for the hypothesis of Merezhkovsky and Margulis.

The microscopic images and the molecular data support the idea that chloroplasts are indeed the descendants of symbiotic cyanobacteria that lived within eukaryotic cells. Eukaryotes acquired the ability to perform photosynthesis because they engulfed and then retained cyanobacterial cells.

Like chloroplasts, mitochondria closely resemble free-living bacteria in organization and biochemistry. Also, like chloroplasts, mitochondria have a small, circular genome like that of bacteria. Finally, the genome of mitochondria is made up of DNA that confirms their close relationship to a type of bacteria—in this case, proteobacteria. In fact, the DNA sequence of mitochondria is very similar to the DNA sequence in proteobacteria. Like chloroplasts, then, mitochondria, present in your cells and indeed most eukaryotic cells, likely originated as endosymbiotic bacteria.

PREP FOR THE AP® EXAM

AP® EXAM TIP

You should be familiar with the concept of endosymbiosis and be able to cite evidence that both the mitochondrion and the chloroplast are descendants of free-living prokaryotic organisms.

Other Symbioses

The evolution of the eukaryotic cell is marked by intimate associations between formerly free-living organisms. We might view this kind of arrangement as unusual, but symbioses between eukaryotic cells and bacteria are all around us.

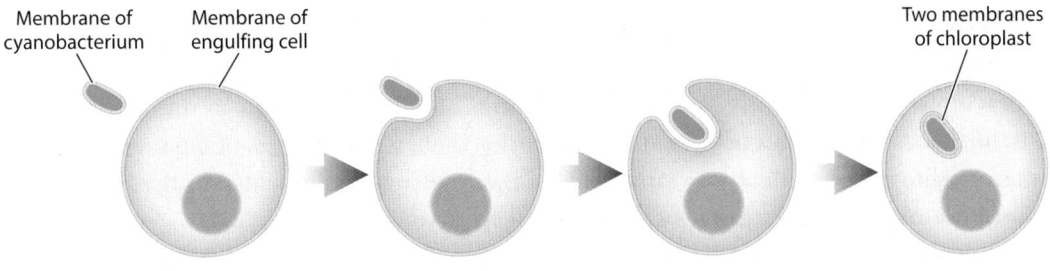

Membrane of cyanobacterium Membrane of engulfing cell Two membranes of chloroplast

FIGURE 12.3 Chloroplast membranes

Chloroplasts have two membranes, an inner and an outer membrane. The inner membrane evolved from the membrane of the cyanobacterium, and the outer membrane evolved from the membrane of the engulfing cell.

What is the evolutionary origin of chloroplasts?

Background Transmission electron microscopy in the photos shows that cyanobacteria (top photo), red algal chloroplasts (middle photo), and plant chloroplasts (bottom photo) have very similar internal membranes. These structures are recognizable in the images as parallel laminae within cells or chloroplasts, and they are involved in photosynthesis. Other structural and biochemical features also indicate a strong similarity between cyanobacteria and the chloroplasts found in photosynthetic eukaryotes.

Hypothesis The similarities between cyanobacteria and chloroplasts reflect descent from a common ancestor: chloroplasts evolved from cyanobacteria living as endosymbionts within a eukaryotic cell.

Experiment Like cyanobacteria, chloroplasts have DNA organized in a single circular chromosome. Biologists sequenced genes in chloroplasts and compared the sequences to those in the nuclei of various eukaryotes and in bacteria, including cyanobacteria. These studies showed that genes in chloroplasts are very similar to genes in cyanobacteria.

Conclusion Molecular and electron microscope data support the hypothesis that chloroplasts originated as endosymbiotic cyanobacteria.

SOURCE
Giovannoni, S. J. et al. 1988."Evolutionary Relationships among Cyanobacteria and Green Chloroplasts." *Journal of Bacteriology* 170:3584–3592. Photos (top to bottom): Dr. Kari Lounatmaa/Science Source; Biology Pics/Getty Images; Dr. Jeremy Burgess/Science Source.

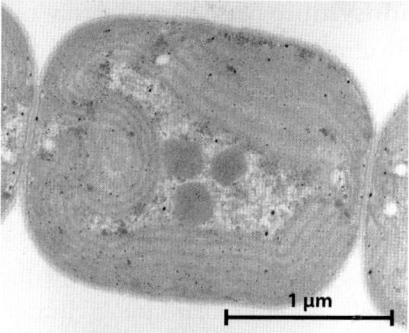

1 µm

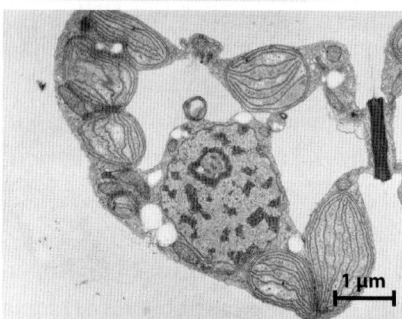

1 µm

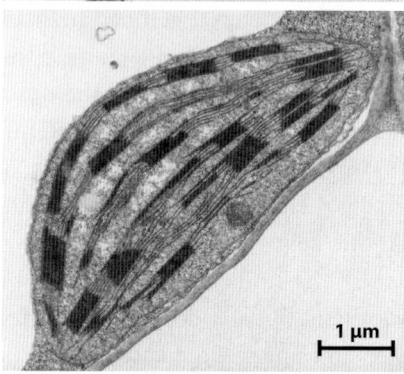

1 µm

AP® PRACTICE QUESTION

(a) **Identify** the following:
1. The question the experiment is trying to answer
2. The hypothesis tested
3. The independent variable
4. The dependent variable

(b) The summary of the studies did not include a control group. **Describe** an appropriate control group.

(c) **Predict** the results you would expect from your control group.

(d) **Identify** other questions that could be investigated based on the results of this experiment.

(e) Explain the results of the experiment by **identifying** their claim and **justifying** the claim with their evidence and reasoning (CER).

For example, an unexpected symbiosis was recently discovered in the Santa Barbara Basin, a local depression in the seafloor off the coast of southern California. Sediments accumulating in this basin contain little oxygen but large amounts of hydrogen sulfide (H_2S) generated by bacteria within the sediments. These bacteria can live in the absence of oxygen and are therefore called anaerobic. We might predict that eukaryotic cells would be uncommon in these sediments. Not only is oxygen scarce, but also sulfide is toxic because it inhibits cellular respiration in mitochondria. It came as a surprise, then, when samples of sediment from the Santa Barbara Basin were found to contain large populations of single-celled eukaryotes.

Some of these cells thrive by supporting populations of symbiotic bacteria on or within their cells. In one case, rod-shaped bacteria cover the surface of their eukaryotic host, as shown in **FIGURE 12.4**. Research has shown that the bacteria metabolize hydrogen sulfide in the local environment, thereby protecting their host from this toxic chemical. Scientists hypothesize that the bacteria benefit from this association as well because they get a free ride as their host moves through the sediments.

The diversity of these eukaryotic–bacterial symbioses is remarkable, yet poorly studied. The evidence gathered to date shows that single-celled eukaryotes have evolved numerous symbiotic relationships with bacteria. Typically, the bacteria feed and protect the eukaryotes, enabling them to colonize habitats where most eukaryotes cannot live. Clearly, then, the types of symbioses that led to mitochondria and chloroplasts on the early Earth were not rare events, but basic associations between cells that continue to evolve today.

Symbioses between a host and a photosynthetic partner are also common in eukaryotes. For example, as already mentioned, reef corals harbor photosynthetic algae that live symbiotically within their tissues. The coral provides an environment and some molecules for photosynthesis, and the algae provide carbohydrates produced by photosynthesis. Some corals have even lost the capacity to capture food from surrounding waters, relying on their photosynthetic partners for their food. Giant clams of the genus *Tridacna,* found in

FIGURE 12.5 Cow–bacterial symbiosis

Cows eat grass, but they can't digest it without bacteria that live in a chamber of their stomach. Photo: Alan Hopps/Getty Images

tropical Pacific waters, also obtain some or even most of their nutrition from symbiotic algae that live within their tissues.

In these cases, the symbiosis is very close, with one organism living inside the cells or tissues of another. Looser symbiotic associations are also common. Cows, for example, eat grass and other plants, as shown in **FIGURE 12.5**. However, they are unable to digest it on their own. Instead, they harbor billions of bacteria that live in a specialized digestive chamber of their stomach. And even humans harbor a host of bacteria called the human microbiome. Scientists are only just beginning to understand the role of our bacterial partners in health and disease.

Eukaryotic Cell Origins

If, as the evidence strongly suggests, chloroplasts originated as cyanobacteria and mitochondria originated as proteobacteria, where does the rest of the eukaryotic cell come from? Many biologists, including American biologists Carl Woese and George Fox, propose that archaea-like microorganisms played a key role in the origin of the eukaryotic cell. This would make eukaryotes more closely related to archaea than eukaryotes are related to bacteria.

Two different hypotheses have been proposed to explain the origin of the eukaryotic cell. In hypothesis 1, pictured in **FIGURE 12.6**, an archaeal cell first evolved into a eukaryotic cell. This cell had a nucleus, a cytoskeleton, and an endomembrane system, but only limited ability to derive energy from organic molecules. Then this early eukaryotic cell engulfed a proteobacterium that evolved over time to become the mitochondria present in nearly all eukaryotic cells.

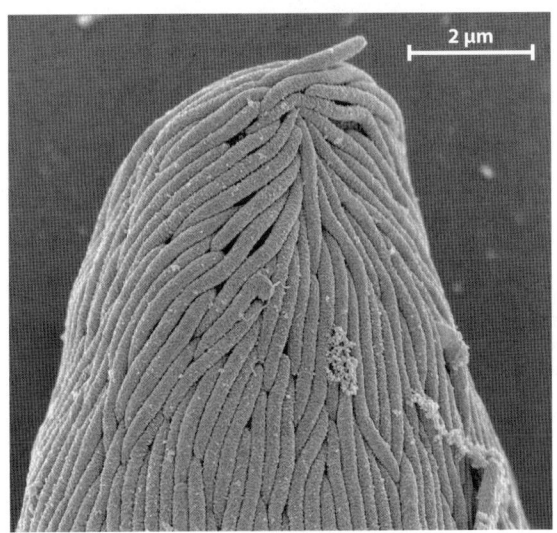

FIGURE 12.4 A modern bacteria–eukaryotic cell symbiosis

A scanning electron microscope image shows bacterial cells on the surface of a single-celled eukaryote found in oxygen-depleted sediments in the Santa Barbara Basin, off the coast of California. The bacteria are hypothesized to metabolize the H_2S in this environment, thereby protecting the eukaryotes that they enclose. Photo: Reprinted by permission from Macmillan Publishers Ltd: Edgcomb, V. P., S. A. Breglia, N. Yubuki, D. Beaudoin, D. J. Patterson, B. S. Leander, and J. M. Bernhard. 2011. "Identity of Epibiotic Bacteria on Symbiontid Euglenozoans in O_2-Depleted Marine Sediments: Evidence for Symbiont and Host Co-evolution," *ISME Journal* 5:231–243. Copyright 2010. Courtesy of Naoji Yubuki.

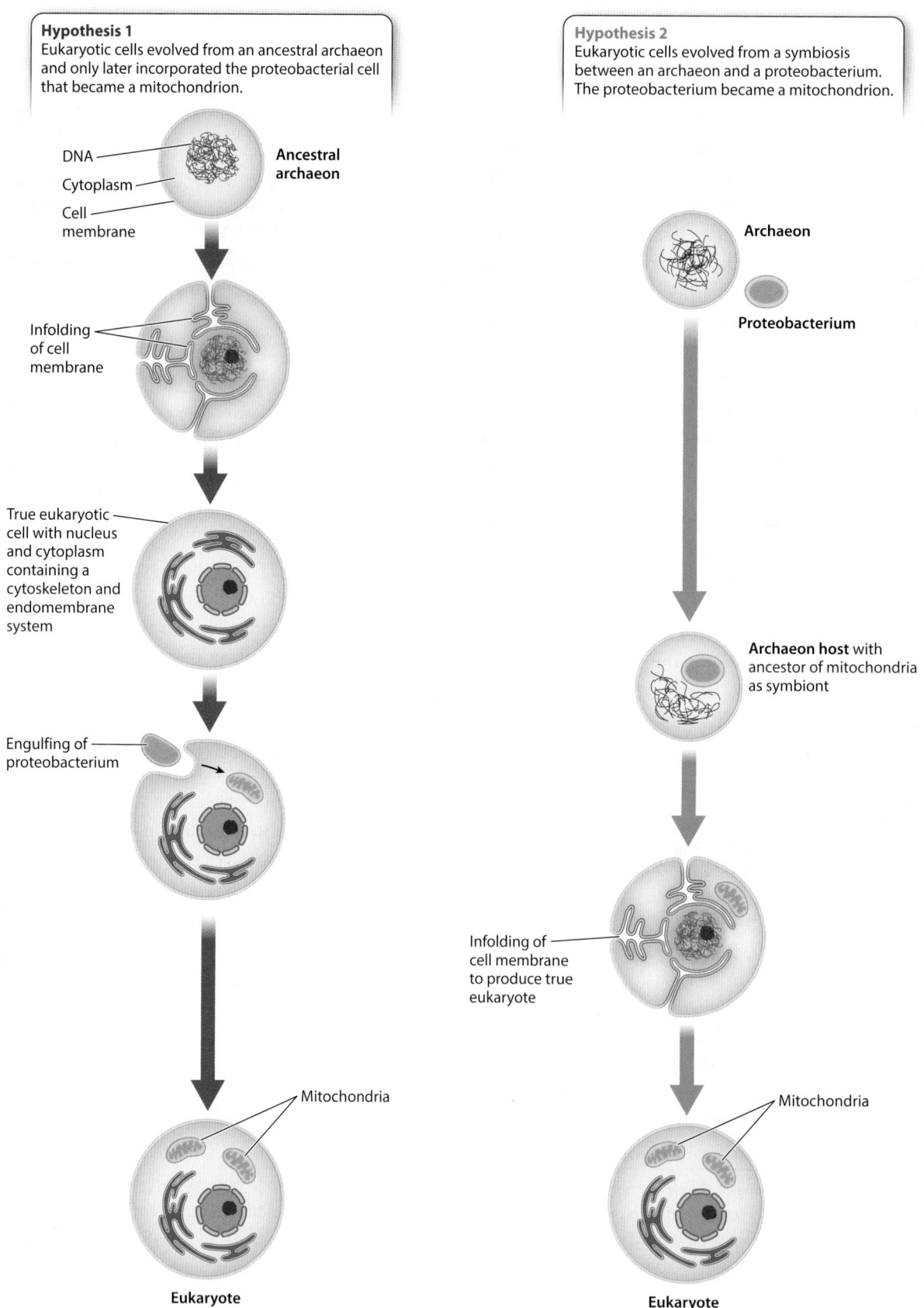

Hypothesis 1
Eukaryotic cells evolved from an ancestral archaeon and only later incorporated the proteobacterial cell that became a mitochondrion.

DNA
Cytoplasm
Cell membrane

Ancestral archaeon

Infolding of cell membrane

True eukaryotic cell with nucleus and cytoplasm containing a cytoskeleton and endomembrane system

Engulfing of proteobacterium

Mitochondria

Eukaryote

Hypothesis 2
Eukaryotic cells evolved from a symbiosis between an archaeon and a proteobacterium. The proteobacterium became a mitochondrion.

Archaeon

Proteobacterium

Archaeon host with ancestor of mitochondria as symbiont

Infolding of cell membrane to produce true eukaryote

Mitochondria

Eukaryote

FIGURE 12.6 Two hypotheses for the origin of the eukaryotic cell

The origin of the eukaryotic cell remains a mystery. In hypothesis 1, an archaeal cell evolves into a eukaryotic cell, which engulfs a proteobacterium. In hypothesis 2, an archaeal cell engulfs a proteobacterium, and the resulting cell evolves into a eukaryotic cell.

In other words, an archaeal cell had already undergone massive changes in cell biology before it engulfed an endosymbiotic proteobacterium.

A different sequence of events is hypothesis 2 in Figure 12.6. According to this hypothesis, it was an archaeal cell, not a eukaryotic cell, that engulfed a proteobacterium. In other words, no eukaryotic cell existed before the mitochondria emerged. The eukaryotic cell as a whole began as a symbiotic association between an archaeon and a proteobacterium. The proteobacterium, over time, became the mitochondria.

Biologists continue to debate these two alternative hypotheses. Neither fully explains the origin and evolution of the nucleus, the eukaryotic cytoskeleton, or a cytoplasm subdivided into compartments by ever-changing membranes. There is no consensus on the origins of eukaryotes: it remains one of biology's deepest unanswered questions, awaiting novel observations by a new generation of biologists with new technologies. Once a dynamic cytoskeleton became coupled to a flexible membrane system, however, the evolutionary possibilities of eukaryotic form were established.

PREP FOR THE AP® EXAM

AP® EXAM TIP

You should be able to describe the two hypotheses of the origin of eukaryotes. Your description should include how eukaryotes have developed their current forms through the incorporation of a protobacterium that became the mitochondria and a cyanobacterium that became the chloroplasts.

✓ Concept Check

3. **Describe** how chloroplasts originated.

4. **Describe** how mitochondria originated.

5. **Identify** the prokaryotic group that scientists think is most closely related to eukaryotes.

6. **Describe** two hypotheses for the origin of eukaryotic cells.

Module 12 Summary

PREP FOR THE AP® EXAM

REVISIT THE BIG IDEAS

EVOLUTION: Using the content in this module, **describe** how the origin of compartmentalization in cells increased diversity of life, and how these changes increased the ability of organisms to survive and reproduce.

LG 12.1 The organization of the eukaryotic cell helps explain eukaryotic diversity.

- Eukaryotic cells are defined by the presence of a nucleus, but features such as a dynamic cytoskeleton and membrane system explain their evolutionary success. Page 170

- Eukaryotic cells have a network of proteins inside the cell that allow them to change shape, move, and transfer substances into and out of the cell. Page 171

- Eukaryotic cells have dynamic membranes that provide new possibilities for movement and feeding. Page 171

- Eukaryotic cells compartmentalize their machinery for energy metabolism into mitochondria and chloroplasts. Page 171

LG 12.2 Chloroplasts and mitochondria originated by endosymbiosis.

- The endosymbiotic hypothesis proposes that the chloroplasts and mitochondria of eukaryotic cells were originally free-living bacteria. Page 172

- The endosymbiotic hypothesis is based on physical, biochemical, and genetic similarities between chloroplasts and cyanobacteria, and between mitochondria and proteobacteria. Page 173

- Chloroplasts and mitochondria have their own genomes. Page 173

- Symbioses are common in nature. Page 173

- Eukaryotes are more closely related to Archaea than they are to Bacteria. Page 175

- The ancestor of the modern eukaryotic cell was either a primitive eukaryote descended from archaeal ancestors or an archaeon that engulfed a bacterium. Page 175

Key Terms

Symbiosis
Endosymbiosis

Review Questions

1. All of the membranes of the eukaryotic cell are in dynamic continuity except for membranes in the

   (A) mitochondria and chloroplasts.

   (B) endoplasmic reticulum.

   (C) Golgi apparatus.

   (D) nuclear membrane.

2. Which best describes a symbiotic relationship between two organisms?

   (A) Two organisms live in the same habitat.

   (B) Two organisms live in the same habitat at the same time.

   (C) Two organisms live in close, physical association with each other where one organism benefits to the detriment of the other.

   (D) Two organisms live in close, long-term, physical association with each other.

3. Which is a key feature of eukaryotic cells that is absent in prokaryotic cells?

   (A) Eukaryotic cells have a cell membrane, which allows them to transfer materials into and out of the cell.

   (B) Eukaryotic cells have ribosomes, which allow them to make more complex proteins.

   (C) Eukaryotic cells have a flexible cytoskeleton, which allows them to change shape and gives them the ability to engulf other materials.

   (D) Eukaryotic cells have genetic material, which allows them to store and transmit information.

4. Which structure has the greatest membrane stability?

   (A) Cell membrane

   (B) Nuclear membrane

   (C) Golgi apparatus

   (D) Mitochondria

5. Evidence supports the hypothesis that chloroplasts evolved from

   (A) plants.

   (B) mitochondria.

   (C) cyanobacteria.

   (D) proteobacteria.

6. Evidence supports the hypothesis that mitochondria evolved from

   (A) amoeba.

   (B) chloroplasts.

   (C) cyanobacteria.

   (D) proteobacteria.

7. Chloroplasts and mitochondria in eukaryotes are thought to have originated by

   (A) endosymbiosis.

   (B) parasitism.

   (C) ectosymbiosis.

   (D) membrane infolding.

Module 12
AP® Practice Questions

Section 1: Multiple-Choice Questions

Choose the best answer for questions 1–4.

1. Dynamic continuity of the eukaryotic endomembrane membrane system refers to the fact that membrane components initially found in one organelle may at a different time be located in another organelle of the endomembrane system. Which explains how dynamic continuity of the endomembrane system occurs?

   (A) RNA from the nucleus passes through the endoplasmic reticulum and the Golgi apparatus, carrying the membranes with it.

   (B) Phospholipids regularly dissolve into their component parts and then are reassembled in new organelles.

   (C) Membrane components move from one organelle to another by vesicles.

   (D) Membrane proteins can be synthesized on any organelle.

2. Which best explains why prokaryotic cells, including bacteria and archaea, can only take in molecules by passive and active transport?

(A) Other modes of uptake require many genes, and prokaryotic cells are too small to contain the required number of genes.

(B) Other modes of uptake require very large quantities of ATP, and prokaryotic cells cannot make these amounts of ATP.

(C) Molecules that are too big to be taken in by passive and active transport are too big to be metabolized by prokaryotes.

(D) Prokaryotes have rigid cell walls that cannot open or bend to allow molecules to enter by endocytosis.

3. Which statement provides evidence for the endosymbiont theory of the origin of the chloroplast?

(A) Modern eukaryotic cells contain a nucleus with a membrane.

(B) Both chloroplasts and plant cells contain external and internal membranes.

(C) The steps of photosynthesis are similar in eukaryotic photosynthetic organisms and cyanobacteria.

(D) Both the chloroplast and photosynthetic eukaryotes contain a cytosol.

4. Figure 12.6 on page 176 offers two hypotheses of the origin of the eukaryotic cell. Which best describes the difference between the two?

(A) In hypothesis 1, an archaean engulfed a proteobacterium but in hypothesis 2, a proteobacterium engulfed an archaean.

(B) In hypothesis 1, cellular organelles developed before the uptake of the proteobacterial endosymbiont but in hypothesis 2, the organelles developed afterward.

(C) In hypothesis 1, the original archaean had a metabolism that used oxygen but in hypothesis 2, the original archaean did not use oxygen.

(D) In hypothesis 1, other eukaryotic cells had already begun to develop when the proteobacterium was engulfed but in hypothesis 2, no other eukaryotic cells had begun to develop.

Section 2: Free-Response Question

Write your answer to each part clearly. Support your answers with relevant information and examples. Where calculations are required, show your work.

Cows eat grass, which contains a large amount of cellulose, a polymer made up of subunits of glucose. Cows, however, cannot digest cellulose on their own. The cellulose they consume is broken down by bacteria and archaea living in the cow's digestive tract. The stomach of a cow has multiple chambers, one of which is called a rumen. A rumen is in many respects a fermentation vat where bacteria and archaea break down molecules such as cellulose without using oxygen. The bacteria and archaea convert the cellulose into nutrients that are accessible to themselves as well as to the cow. The process of breaking down cellulose by the bacteria and archaea also leads to the production of carbon dioxide and methane. These gases are belched or emitted by flatulence from the cows into the air. By emitting these gases to the atmosphere, the cow ensures that the waste products of cellulose degradation are removed, and further metabolism of the grass can continue.

(a) **Describe** the symbiosis that occurs in the cow's rumen between the cow and the bacteria and archaea. **Explain** the benefits to both the cow and the prokaryotes.

(b) The archaean *Methanobacterium formicium* is one of a number of archaea responsible for making methane in the rumen. These archaea metabolize some of the molecules produced from the breakdown of cellulose to gain the energy they need to grow. Methane-producing archaea cannot continue their metabolism without making methane. Similarly, the reactions that cause cellulose breakdown by both the bacteria and other archaea in the rumen cannot continue if methane, as one of the two ultimate products (carbon dioxide and methane), is not made.

A fellow student proposes an experiment whereby the *M. formicium* of the rumen are inhibited by a chemical that prevents them from making methane. **Predict** the effect of such an inhibitor on the cow and its symbionts in the rumen. In your prediction, make sure to consider the ability of the cow and the archaea to gain the nutrients they require. **Justify** your answer.

Unit 2
AP® Practice Questions

Section 1: Multiple-Choice Questions

Choose the best answer for questions 1–14.

1. An AP® Biology teacher gives a student a saliva sample from an unidentified animal. The student isolates a few cells from the saliva and examines the cells under a microscope. In one of the cells, the student observes an organelle that is able to replicate itself independently of the other organelles and contains its own DNA. Identify which organelle the student observed.

   (A) The chloroplast

   (B) The Golgi apparatus

   (C) The mitochondrion

   (D) The nucleus

2. Scientists discover a new eukaryotic organism and note that it contains a variety of organelles. They do not recognize one of the organelles but observe that it contains its own DNA, has a double-membrane system, has the ability to harness energy, and likely evolved from a bacterial ancestor. Which organelles are most similar to this unidentified organelle?

   (A) The rough endoplasmic reticulum and the smooth endoplasmic reticulum

   (B) The vacuole and the cilium

   (C) The mitochondrion and the chloroplast

   (D) The Golgi apparatus and the lysosome

Questions 3 and 4 refer to the following information.

Epithelial cells are cells that line the surfaces of a tissue or an organ. Simple cuboidal epithelia are a type of epithelial cell that lines the surfaces of ovaries and parts of the thyroid gland and eye. As the name suggests, these cells are arranged in a single layer (simple) and are square in shape (cuboidal).

3. A simple cuboidal epithelial cell with a side length of 10 µm divides into two daughter cells, each with a side length of 5 µm. What are the volumes of the mother cell and the daughter cells?

   (A) The mother cell has a volume of 200 μm^3, and each daughter cell has a volume of 250 μm^3.

   (B) The mother cell has a volume of 100 μm^3, and each daughter cell has a volume of 25 μm^3.

   (C) The mother cell has a volume of 1000 μm^3, and each daughter cell has a volume of 125 μm^3.

   (D) The mother cell has a volume of 10,000 μm^4, and each daughter cell has a volume of 625 μm^4.

4. A simple cuboidal epithelial cell with a side length of 10 µm divides into two daughter cells, each with a side length of 5 µm. Which cell, the mother cell or daughter cell, has the greater surface area-to-volume (SA:V) ratio, and by how much is it greater?

   (A) The mother cell has the greater SA:V ratio; it is 4 times greater than the daughter cell's SA:V.

   (B) The mother cell has the greater SA:V ratio; it is 2 times greater than the daughter cell's SA:V.

   (C) The daughter cell has the greater SA:V ratio; it is 4 times greater than the mother cell's SA:V.

   (D) The daughter cell has the greater SA:V ratio; it is 2 times greater than the mother cell's SA:V.

5. Most bacteria are small organisms with small volumes. Because of their size, bacteria can sometimes obtain nutrients by diffusion alone. In addition, bacteria contain numerous active transporters that move nutrients into the bacteria. Which explains the need for active transporters in bacteria?

(A) Diffusion is too slow to work effectively to bring in all of the nutrients that bacteria need.

(B) Diffusion can only bring nutrients into the bacteria across the membrane if their concentration is higher inside the bacteria than outside.

(C) Diffusion can bring large nonpolar molecules, like glucose, into the bacteria across the membrane, but cannot move ions across the membrane.

(D) Diffusion can bring nutrients into the bacteria across the membrane only if their concentration is higher outside the bacteria than inside.

Questions 6 and 7 refer to the following information.

Brewer's yeast (*Saccharomyces cerevisiae*) is a eukaryote that is approximately spherical in shape and can obtain most of its nutrients through diffusion across its cell membrane. A researcher sorts yeast cells into four groups by volume, with group 1 cells having the smallest volume and group 4 cells having the largest volume. She then provides the yeast cells with the amino acid glutamine, which supports growth and reproduction. Glutamine can diffuse across the cell membrane. The same number of yeast cells are present in each group. Data collected for this experiment are listed in the table.

Group	1	2	3	4
Surface area (μm^2)	12.6	50.3	201	452
Volume (μm^3)	4.19	33.5	268	905
Rate of glutamine absorption ($\mu mol/min$)	75	59	45	30

6. What is the average rate of glutamine uptake for all of the groups?

(A) 59 $\mu mol/min$

(B) 55 $\mu mol/min$

(C) 52 $\mu mol/min$

(D) 49 $\mu mol/min$

7. Which best explains the results of the experiment?

(A) The larger cells need less glutamine than do the smaller cells.

(B) The larger cells have a greater need for glutamine than do the smaller cells.

(C) Larger cells have more plasma membrane, which slows down the rate of glutamine uptake.

(D) As the surface area-to-volume ratio decreases, so does the rate of glutamine absorption.

8. Scientists have observed species of bacteria with membranes unlike those found in other bacteria. Each layer of the lipid bilayer in these bacteria has a distinct composition. One layer is composed of the same kind of phospholipid found in most cell membranes, with a glycerol backbone attached to two fatty acids and a phosphate group. The other layer, however, is composed of fatty acids attached to a sugar instead of to a glycerol and phosphate group. The two layers remain separate over time. Which is the most likely reason that the layers remain distinct from one another?

(A) The flip-flopping of components between layers is a rare event.

(B) The hydrocarbon chains of the fatty acids in both layers are not hydrophobic.

(C) There is no lateral movement of the membrane components in either layer.

(D) The hydrocarbon chains of the fatty acids in both layers are hydrophilic.

9. Substance X normally moves into a cell by facilitated diffusion. An AP® Biology student uses a microinjection procedure to inject enough substance X into a cell so that its concentration inside the cell is eight times its concentration outside of the cell. Predict the direction that substance X will move the instant after it is injected into the cell.

(A) Substance X will move into the cell until its concentration is ten times as high inside the cell as it is outside the cell.

(B) The movement of substance X in any direction will cease.

(C) Substance X will move into the cell from the outside environment.

(D) Substance X will move out of the cell into the outside environment.

10. Solutions of 0.85% to 0.90% sodium chloride are isotonic to body fluids. Predict what will happen to red blood cells that are placed into a solution of 10.00% sodium chloride.

(A) They will gain fluid and swell.

(B) They will gain fluid, swell, and burst.

(C) They will lose fluid and shrink.

(D) They will remain the same and neither gain nor lose fluid.

11. Osmoconformers keep their internal fluids at the same osmotic pressure as the surrounding environment. Some osmoconformers, however, have internal concentrations of ions and molecules (for example, K^+, Na^+, Cl^-, amino acids, glucose) that are not equal to what is in their external environments. How do these osmoconformers maintain these internal concentrations of these ions and molecules despite different concentrations in the environment?

(A) They use active transport processes.

(B) They use passive transport processes.

(C) They use facilitated transport processes.

(D) They use vesicle mechanisms.

12. German physiologist Otto Renner conducted an experiment to determine if plants could generate negative pressures, as described in "Practicing Science 11.1: Do plants generate negative pressures?" He measured the flow rates of water from the U-tube when a plant was immersed in the water and when a vacuum pump was attached to the water. The results of the experiment are summarized in the figure on page 164. Which is supported by the data presented in the graph in the figure?

(A) The efficiency of the vacuum pump was approximately 2.0 to 2.5 times that of the plant.

(B) The efficiency of the plant was approximately 2.0 to 2.5 times that of the vacuum pump.

(C) Water flow from the U-tube was equal for both the plant and the vacuum pump.

(D) Water flow from the U-tube was in an opposite direction for the plant compared to the vacuum pump.

Questions 13 and 14 refer to the following information.

Approximately 86% of the fatty acids in the cell membranes of coconut tree cells are saturated, which is a higher percentage of saturated fatty acids than the average plant. Soybean cell membranes, however, have a lower-than-average percentage of fatty acids in their membranes, with about 16% saturated fatty acids.

13. Predict what will happen to the membrane fluidity of both plants when the environmental temperature is lowered from room temperature (around 23°C) to 5°C.

(A) The membranes of the coconut tree cells will become more fluid, but the membranes of the soybean plant cells will become more rigid.

(B) The membranes of the coconut tree cells will become more rigid, but the membranes of the soybean plant cells will become more fluid.

(C) The membranes of both plants' cells will become more rigid.

(D) The membranes of both plants' cells will become more fluid.

14. Coconut trees can only grow in environments where the temperature is 22°C or higher. Soybean plants, however, can grow at considerably lower temperatures. How might the difference in saturated fatty acid content between the two plants account for their different tolerances for low temperatures?

(A) Membranes with high amounts of saturated fatty acids become rigid at lower temperatures, preventing essential cell functions like vesicle budding.

(B) Membranes with high amounts of saturated fatty acids become fluid at lower temperatures, allowing transporter proteins to dissociate.

(C) Membranes with high amounts of saturated fatty acids become rigid at lower temperatures, preventing large molecules from diffusing through channels.

(D) Membranes with high amounts of saturated fatty acids become fluid at lower temperatures, enabling endosymbiosis of additional chloroplasts.

Section 2: Free-Response Questions

Write your answer to each part clearly. Support your answers with relevant information and examples. Where calculations are required, show your work.

1. *Lactococcus lactis* is a spherical bacterium commonly found in products such as yogurt. *L. lactis* ferments dairy products like yogurt by absorbing the sugar lactose through its membrane and breaking the lactose into its component sugars, glucose and galactose. Researchers grew *L. lactis* and grouped the cells by volume in order to study the rate of breakdown of lactose. The same number of *L. lactis* cells are present in each group. Data collected for this experiment are listed in the table.

Group	1	2	3	4
Radius (μm)	0.3	0.5	0.7	1.0
Surface area (μm²)	1.13	3.14	6.16	12.6
Volume (μm³)	0.11	0.53	1.44	4.19
Rate of lactose breakdown (μmol/min)	9.0	7.1	4.9	3.2

(a) Construct a graph using the data in the table to show the rate of galactose breakdown of the groups.

(b) **Determine** the surface area-to-volume ratio of the four *L. lactis* groups. **Identify** the group with the greatest surface area-to-volume ratio and **explain** what this largest ratio means with respect to galactose breakdown.

2. In plants, potassium is involved in the movement of water, other nutrients, and carbohydrates. It is also used to activate enzymes and is involved in protein, starch, and ATP production. Sulfate serves as a source of the element sulfur, which is an essential component of the amino acids methionine and cysteine. In order to make a solution of potassium sulfate (K_2SO_4) to be used in plant growth experiments, an AP® Biology teacher enters the laboratory and makes a 0.20 M (molar) solution of K_2SO_4 at a temperature of 27°C.

(a) **Calculate** the Ψ_s of the potassium sulfate solution. The ionization constant for K_2SO_4 is 3.

(b) Calculate the water potential if the pressure potential is 4.0.

(c) The AP® Biology teacher places the K_2SO_4 solution into a dialysis bag that is permeable to water but not K_2SO_4. The teacher then places the filled bag into a beaker containing pure water. **Predict** whether water will move into or out of the dialysis bag. **Justify** your answer. Assume a temperature of 27°C and that the Ψ_p is 0.

(d) The teacher fills more of the 0.20 molar solution of K_2SO_4 into a new dialysis bag that is permeable to water but not to K_2SO_4 or sucrose. The teacher then places the filled bag into a beaker containing a 2.0 molar solution of sucrose. **Calculate** the water potentials of the solution inside the dialysis bag and the solution in the beaker. Assume the pressure potential is 0 and the temperature is 27°C. **Predict** whether water will move into or out of the dialysis bag. **Justify** your answer.

Unit 3
Cellular Energetics

Photo of a large wave by Purestock/Getty Images

Unit 3	13	14	15	16	17	18	19

Module 13
Cellular Energy

LEARNING GOALS ▶**LG 13.1** Kinetic and potential energy are two forms of energy.
▶**LG 13.2** The laws of thermodynamics govern energy flow in biological systems.
▶**LG 13.3** Chemical reactions transform molecules and transfer energy in cells.

In the last unit, we focused on the structure and function of cells. Cells come in many different shapes and sizes, but they are all able to harness energy from the environment. This unit focuses on what energy is and how cells use it to carry out their functions.

As we discussed in Unit 1, energy is the ability to do work. Cells need energy to do all kinds of work. They grow and divide. They move, change shape, pump ions in and out, and transport vesicles. They also synthesize macromolecules such as nucleic acids, proteins, carbohydrates, and lipids. These activities are all considered work, and they all require energy. Cells are also highly organized, with cell membranes and genetic material interacting as a system in a precise manner so that a cell can carry out its functions. Maintaining this high level of organization also requires a sustained input of energy.

We are all familiar with different sources of energy, including the sun and wind, as well as fossil fuels such as oil and natural gas. We have learned to harness the energy from these sources and convert it to other forms, such as electricity, to provide needed power to our homes, schools, cars, planes, towns, and cities. In a similar way, cells harness energy from the environment and convert it to a form that allows them to do the work necessary to sustain life.

Cells harness energy from just two sources: the sun and chemical compounds. You learned about these chemical compounds in Unit 1, including carbohydrates, lipids, and proteins. The food we eat, for example, contains energy. Although the source of energy may differ among cells, all cells transfer energy to a form that can be easily used to drive cellular processes.

In this module, we consider energy in the context of cells. We will introduce two forms of energy, and then consider what happens when energy is transferred from one form to another, and from one molecule to another. Finally, we will examine chemical reactions. Chemical reactions are processes by which molecules are transformed into other molecules by the breaking and forming of chemical bonds. They are also the way in which energy is transferred in cells.

PREP FOR THE AP® EXAM

FOCUS ON THE BIG IDEAS

ENERGETICS: Look for the many ways that cells make use of energy and note the strategies cells utilize to capture, convert, and store energy.

13.1 Kinetic and potential energy are two forms of energy

Energy is important in biological systems because it is needed to do work. A cell is doing work when it carries out the processes we discussed in earlier units, such as synthesizing macromolecules, pumping substances across the cell membrane, and moving vesicles between its various compartments. A cell also does work to maintain the high level of order required for it to function as an integrated system. In this section, we will consider two forms of energy. **Kinetic energy** is the energy of motion. **Potential energy** is stored energy. We will also look at a form of potential energy known as **chemical energy,** which is held in the chemical bonds between pairs of atoms in a molecule. Both kinetic energy and potential energy are relevant to biological systems, from a carbohydrate molecule to a transport protein in a cell membrane to a swimming fish.

Kinetic and Potential Energy

Kinetic energy is the energy of motion. It is perhaps the most familiar form of energy. A moving object, such as a ball bouncing down a set of stairs, possesses kinetic energy, as shown in the middle panel of **FIGURE 13.1**. Kinetic energy is associated with any kind of movement, such as a flowing river; contracting muscle; or running cheetah, like the one pictured in **FIGURE 13.2**. Light, electricity, and thermal

FIGURE 13.2 Kinetic energy

Kinetic energy is the energy of motion, as illustrated by this running cheetah. Photo: GP232/Getty Images

energy (perceived as heat) are also forms of kinetic energy because light is associated with the movement of photons, electricity with the movement of electrons, and thermal energy with the movement of molecules.

Energy is not always associated with movement. An immobile object can still possess a form of energy called potential energy, or stored energy. Potential energy depends on the structure of the object or its position relative to a field, such as a gravitational, an electrical, or a magnetic field. This stored energy is then released when the object changes its structure or position. For example, the potential energy of a ball is higher at the top of a flight of stairs than at the bottom, as seen in the left and right panels of Figure 13.1. If the ball were not blocked by the floor, the force of gravity would move it from the position of higher potential energy at the top of the stairs to the position of lower potential energy at the bottom of the stairs.

A dam in a river, like the one shown in **FIGURE 13.3** on page 186, also stores potential energy. Although the water is not moving, we can "see" the stored energy if the dam is released and water flows through it. Similarly, a cell membrane can act like a dam and stores potential energy. For example, as we saw in Module 10, an electrochemical gradient of protons (H^+) across a membrane is a form of potential energy. Given a pathway through the membrane, such as a channel protein, the protons move down their concentration and electrical gradients from a region of higher potential energy to a region of lower potential energy. This energy can be harnessed to move another molecule against its concentration gradient in secondary active transport.

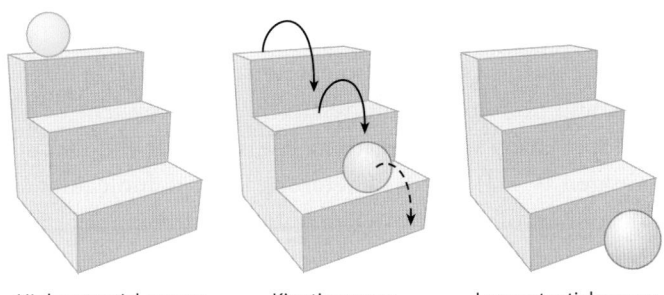

High potential energy   Kinetic energy   Low potential energy

FIGURE 13.1 Kinetic and potential energy

Kinetic energy is energy of motion, and it is shown in the middle panel by the ball moving down the stairs. Potential energy is stored energy. The ball at the top of the stairs on the left has high potential energy because of its position. The high amount of potential energy is indicated by the bright yellow color. Some of the potential energy of a ball at the top of a set of stairs is transformed to kinetic energy as the ball rolls down the stairs, leaving the ball with low potential energy at the bottom of the stairs, indicated by the lighter shade of yellow.

FIGURE 13.3 Potential energy

Potential energy is stored energy. This dam in Austria blocks the flow of water and therefore stores energy. Photo: rusm/Getty Images

Energy can be converted from one form to another. When a dam breaks, the potential energy is converted to the kinetic energy of rushing water. Similarly, as shown in Figure 13.1, the ball at the top of the stairs has a certain amount of potential energy because of its position. As it rolls down the stairs, this potential energy is converted to kinetic energy associated with movement of the ball and the molecules in the surrounding air. When the ball reaches the bottom of the stairs, the remaining energy is stored as potential energy. So, in this example, potential energy is converted to kinetic energy of the ball and air. Conversely, it takes an input of energy to move the ball back to the top of the stairs, and this input of energy is stored as potential energy.

Chemical Energy

We obtain the energy we need from the food we eat, which contains chemical energy, a form of potential energy held in the chemical bonds between pairs of atoms in a molecule. Recall from Unit 1 that a covalent bond results when two atoms share electrons. Covalent bonds are relatively strong because the shared electrons in a molecule are in a more stable configuration than if they were not shared.

Although all covalent bonds are strong and stable, some are stronger and more stable than others. The stronger the bond, the less chemical energy the molecule contains. This may seem counterintuitive at first, but it is similar to the observation that a ball at the bottom of a flight stairs is more stable and has less potential energy than a ball at the top of the stairs.

A strong covalent bond has a very stable configuration of electrons. The stability of the bond means that it does not require a lot of energy to remain intact and therefore does not contain very much chemical energy, similar to the potential energy of the ball at the bottom of a flight of stairs in Figure 13.1. Furthermore, these bonds are strong, or hard to break, because sharing electrons between the two atoms is much more stable than if the two atoms do not share electrons. Examples of molecules with strong covalent bonds that contain relatively little chemical energy are carbon dioxide (CO_2) and water (H_2O).

Other covalent bonds are weaker than those in carbon dioxide and water. The weaker the bond, the more chemical energy the molecule contains. The arrangement of electrons in these bonds is less stable than that in strong covalent bonds. As a result, weak covalent bonds require a lot of energy to stay intact and contain a lot of chemical energy, similar to the potential energy of the ball at the top of the flight of stairs in Figure 13.1. These bonds are weak, or easily broken, because sharing electrons in these molecules is only somewhat more stable than if the two atoms did not share electrons. Organic molecules such as carbohydrates, lipids, and proteins contain relatively weak covalent bonds, including many carbon–carbon (C–C) bonds and carbon–hydrogen (C–H) bonds. Therefore, organic molecules are rich sources of chemical energy and are sometimes called fuel molecules.

ATP

The chemical energy carried in carbohydrates, lipids, and proteins is harnessed by cells to do work. But cells do not use this energy all at once. Instead, they break down these molecules in a series of chemical reactions that make up a *metabolic pathway*. A **metabolic pathway** is a series of chemical reactions in which molecules are broken down or built up. The reactions that cells use to break down the carbohydrate glucose is an example of a metabolic pathway.

In a metabolic pathway, the product of one reaction is the reactant for the next. As large molecules with lots of chemical energy are broken down into smaller ones with less chemical energy, cells package the energy that is released into a chemical form that is readily accessible to the cell. The most common form is adenosine triphosphate, or ATP, which was introduced in Module 6. The chemical energy in the bonds of ATP is used in turn to drive many cellular processes, such as muscle contraction, cell movement, and membrane pumps. In this way, ATP serves as a go-between, acting as an intermediary between fuel molecules that store a large amount of potential energy in their bonds and the activities of the cell that require an input of energy. The use of ATP as a ready, accessible form of chemical energy in nearly all cells reflects its use in cells that evolved early in Earth's long history. In other words, the earliest cells on Earth likely used ATP, and this feature of cells has been retained over evolutionary time.

The chemical structure of ATP is shown in **FIGURE 13.4a.** The core of ATP is made up of the base adenine and the 5-carbon sugar ribose. To complete the molecule, ribose is attached to triphosphate, or three phosphate groups. The accessible chemical energy of ATP is held in the bonds connecting the phosphate groups. At the pH typical of a cell, which is approximately pH 7, these phosphate groups are negatively charged. Because like charges repel one another, these phosphate groups tend to repel each other and are thus unstable. The chemical bonds connecting the phosphate groups therefore contain a lot of chemical energy to keep the phosphate groups connected. This energy is released when new, more stable bonds are formed that contain less chemical energy. The released energy in turn can be associated, or coupled, with chemical reactions that require energy, and therefore harnessed to power the work of the cell.

ATP has two chemical relatives. One is **adenosine diphosphate,** or **ADP,** which has two phosphate groups attached to the sugar ribose and is shown in Figure 13.4b. ADP contains less chemical energy in its bonds than ATP because it has fewer phosphate groups to repel each other. Another chemical relative is adenosine monophosphate, or AMP, with just a single phosphate group attached to the sugar ribose and shown in Figure 13.4c. AMP has even less chemical energy than ADP because it has just one phosphate group and it is not repelled by another group.

ATP is often called the universal "currency" of cellular energy because it provides energy in a form that

a. ATP

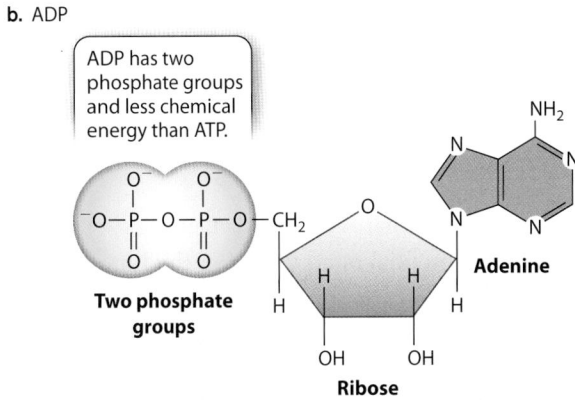

b. ADP

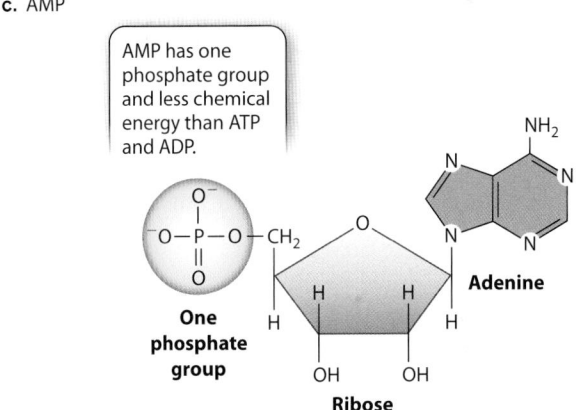

c. AMP

FIGURE 13.4 ATP and its chemical relatives

Adenine triphosphate (ATP) and its chemical relatives, adenine diphosphate (ADP) and adenine monophosphate (AMP), hold chemical energy in their bonds. (a) ATP is made up of the sugar ribose, the base adenine, and three phosphate groups. The bonds linking the phosphate groups have a large amount of chemical energy, which can be harnessed to do the work of the cell. (b) ADP has two phosphate groups and therefore less chemical energy than ATP. (c) AMP has one phosphate group and even less chemical energy than ATP and ADP.

nearly all cells can readily use to perform work. Although currency provides a useful analogy for the role that ATP plays, there is an important distinction between actual currency, such as a dollar bill or a bitcoin, and

ATP. A dollar bill or a bitcoin represents a certain value but does not have any value in itself. By contrast, ATP does not represent energy; it actually contains energy in its chemical bonds. Nevertheless, the analogy is useful because ATP, like currency, engages in a broad range of energy "transactions" in the cell.

✓ Concept Check

1. **Describe** two forms of energy and provide an example of each form.

2. **Describe** the relationship between the strength of a covalent bond and the amount of chemical energy it contains.

3. **Describe** the structure of ATP, indicating the bonds that contain a high amount of chemical energy.

13.2 The laws of thermodynamics govern energy flow in biological systems

Plants and other organisms convert energy from the sun into chemical potential energy held in the bonds of organic molecules, such as carbohydrates. Chemical reactions can transfer this chemical energy between different molecules. Some of this energy is transferred to the chemical bonds of ATP, which cells use to do work. In fact, we can think of living organisms as energy transformers. They acquire energy from the environment and transform it into a chemical form that cells can use.

In all these energy transfers, energy is subject to the laws of thermodynamics. In fact, all processes in a cell, from diffusion to osmosis to the pumping of ions to cell movement, are subject to these laws. There are four laws of thermodynamics, but the first and second laws are particularly relevant to biological processes and are discussed in this section.

The First Law of Thermodynamics

The **first law of thermodynamics** is the law of conservation of energy and it states that the universe contains a constant amount of energy. Therefore, energy is neither created nor destroyed. New energy is never formed, and existing energy is never lost. Instead, energy is converted from one form to another. For example, kinetic energy can change to potential energy and vice versa, but the total amount of energy always remains the same, as illustrated in **FIGURE 13.5**.

In the simple example of a ball rolling down a set of stairs, the potential energy stored in the ball at the top of the stairs is converted to kinetic energy associated with the movement of the ball and the surrounding air as the ball collides with molecules in the air. The total amount of energy in this transformation is constant—that is, the total potential

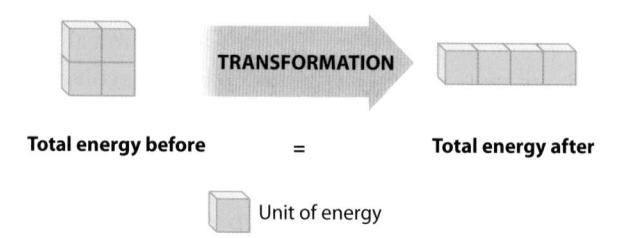

FIGURE 13.5 The first law of thermodynamics

The first law of thermodynamics is the law of conservation of energy. When energy is transformed from one form to another, the total amount of energy remains the same, as indicated by the four yellow squares representing a certain amount of energy that exists both before and after transformation.

energy of the ball at the top of the stairs equals the amount of kinetic energy associated with the movement of the ball and air molecules plus the potential energy of the ball at the bottom of the stairs.

Because energy is not created or destroyed, chemical reactions do not create or use up energy. Instead, they transfer energy between different types of molecules and the environment. Energy is ultimately not produced or consumed; it just changes its form. The total amount of energy in the universe is constant.

The Second Law of Thermodynamics

We have seen that energy changes its form, but the total amount of energy remains constant. However, in going from one form of energy to another, the energy available to do work decreases. In fact, the amount of energy available to do work decreases every time energy changes form. This is because energy transformations are never 100% efficient. The loss of energy to do work is like a universal price to pay in transforming energy from one form to another.

The energy that is not available to do work takes the form of an increase in disorder, or **entropy,** which is

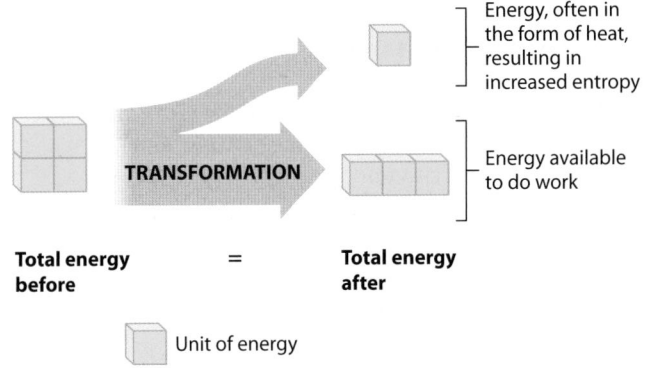

Total energy = Total energy
before   after

☐ Unit of energy

FIGURE 13.6 The second law of thermodynamics

The second law of thermodynamics states that energy transformations are always associated with an increase in entropy and therefore a loss of energy to do work. Because the amount of disorder increases when energy is transformed from one form to another, some of the total energy is not available to do work.

illustrated in **FIGURE 13.6**. This principle is summarized by the **second law of thermodynamics,** which states that the transformation of energy is associated with an increase in disorder of the universe. For example, when kinetic energy is changed into potential energy, the amount of disorder always increases.

The degree of disorder is just one way to describe entropy. Another way to think about entropy is to consider the number of possible positions and motions a molecule can take on in a given system. As entropy increases, so does the number of positions and motions available to the molecule. Consider the expansion of a gas after the lid is removed from its container. Given more space, the molecules of the gas are less constrained and able to move about more freely. They have more positions available to them and move about at a larger range of speeds, so they have more entropy.

In chemical reactions, most of the entropy increase occurs through the transformation of various forms of energy into thermal energy, which we experience as heat. Thermal energy is a type of kinetic energy corresponding to the random motion of molecules and results in a given temperature. The higher the temperature, the more rapidly the molecules move and the higher the disorder. With every energy transfer, some of the starting energy is released as thermal energy, or heat, which manifests as disorder.

Let's consider a biological example that illustrates both the first and second laws of thermodynamics—the contraction of a muscle. Muscle contraction is a form of kinetic energy associated with the shortening of muscle cells. The contraction of a muscle is powered by chemical potential

energy in ATP. Some of this potential energy is transferred into the kinetic energy of movement, and the rest is dissipated as thermal energy, which is why your muscles warm as you exercise. Consistent with the first law of thermodynamics, the amount of chemical potential energy expended is equal to the amount of kinetic energy plus the amount of thermal energy. Consistent with the second law of thermodynamics, thermal energy flowing as heat is a necessary by-product of the transformation of chemical potential energy into kinetic energy.

Creating and Maintaining Order

Because entropy increases every time energy changes form, creating order in a system requires energy. Think about a box full of red and blue marbles distributed more or less randomly; if you want to line up all the red ones or blue ones in a row, you have to do work—that is, you have to add energy. In this case, adding energy increases the order of the system. That is, adding energy decreases disorder in the system.

Cells are also highly organized. As with lining up marbles in a row, a constant input of energy is needed to maintain this organization. Given the tendency toward greater disorder, the high level of organization of even a single cell would appear to violate the second law of thermodynamics. But it does not. The key is that a cell is not an isolated system and therefore cannot be considered on its own; it exists in an environment. As a result, we need to take into account the whole system—the cell and the environment that surrounds it. Only some of the energy harnessed by cells is used to do work; the rest is dissipated as heat. Heat is a form of energy, so the total amount of energy is conserved, as dictated by the first law. In addition, heat corresponds to the motion of small molecules—the greater the heat, the greater the motion, and the greater the motion, the greater the degree of disorder. Therefore, the release of heat as organisms harness energy means that the total entropy for the combination of the cell and its surroundings increases, in keeping with the second law. In short, a local decrease in entropy is always accompanied by an even higher increase in the entropy of the surroundings.

A single cell or a multicellular organism requires a constant input of energy to maintain its high degree of function and organization. This input of energy comes either from the sun or from the energy stored in chemical compounds. This energy allows molecules to be built and other work to be carried out, but it also leads to greater entropy in the surrounding environment.

✓ Concept Check

4. **State** the first two laws of thermodynamics.

5. **Describe** how the first two laws of thermodynamics apply to chemical reactions.

13.3 Chemical reactions transform molecules and transfer energy in cells

Chemical reactions are central to life. Living organisms build and break down molecules, pump ions across membranes, move, and in general function largely through chemical reactions. Chemical reactions transform molecules into other molecules by breaking and forming bonds. When we write out a chemical reaction, we can see which bonds break and which bonds form. However, it is not as obvious that chemical reactions also transfer energy in cells. In other words, cells harness and use energy largely through chemical reactions. For example, organisms break down organic molecules, such as glucose, and store energy in the bonds of ATP. ATP then powers chemical reactions that sustain life. In this section, we will discuss chemical reactions and then consider how the laws of thermodynamics apply to chemical reactions. We will also discuss how ATP drives many chemical reactions in cells.

Chemical Reactions and Metabolic Pathways

Chemical reactions were introduced in Unit 1. A chemical reaction is the process by which molecules, called reactants, are transformed into other molecules, called products. During a chemical reaction, atoms keep their identity, but the atoms that share bonds change. For example, carbon dioxide (CO_2) and water (H_2O) can react to produce carbonic acid (H_2CO_3), as shown in **FIGURE 13.7**. This reaction is common in nature. For example, it occurs when carbon dioxide in the air enters ocean waters, and it explains why the oceans are becoming more acidic as carbon dioxide levels in the atmosphere increase. This reaction also takes place in your blood. When cells break down glucose, they produce carbon dioxide. In animals, this carbon dioxide diffuses out

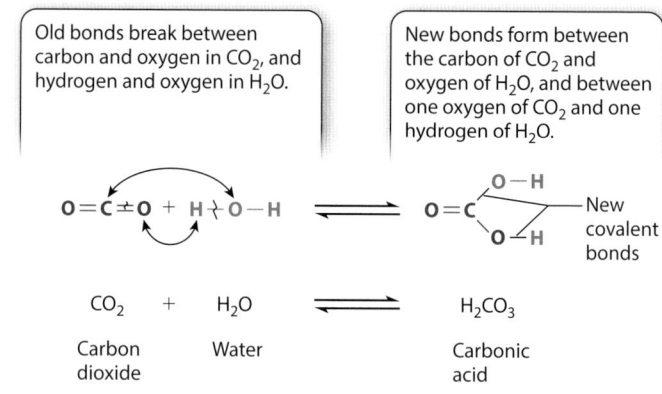

FIGURE 13.7 A chemical reaction

In a chemical reaction, old bonds break and new bonds form, while atoms maintain their identity. In this reaction, carbon dioxide and water are the reactants, and carbonic acid is the product. The two arrows indicate that the reaction can go in both directions.

of cells and into the blood. Rather than remaining dissolved in solution, most of the carbon dioxide is converted into carbonic acid.

Many chemical reactions in cells are readily reversible: the products can react to form the reactants. For example, carbon dioxide and water react to form carbonic acid, and carbonic acid can break apart to produce carbon dioxide and water. This reverse reaction also occurs in the blood, allowing carbon dioxide to be removed from the lungs or gills. The reversibility of the reaction is indicated by a double arrow in the figure.

The way the reaction is written defines forward and reverse reactions: a forward reaction proceeds from left to right and the reactants are located on the left side of the arrow; a reverse reaction proceeds from right to left and the reactants are located on the right side of the arrow. The direction of a reaction can be influenced by the concentrations of reactants and products. For example, increasing the concentration of the reactants or decreasing the concentration of the products favors the forward reaction. In the case of the reaction shown in Figure 13.7, the forward reaction is favored in tissues, where the concentration of carbon dioxide is high because it is produced as a by-product of metabolism. By contrast, the reverse reaction is favored in lungs, where the concentration of carbon dioxide is low because it is exhaled.

Now imagine a string of chemical reactions, one after the next, in which the products of one reaction are the reactants in the next, forming a metabolic pathway. Because the products of these reactions are quickly consumed by the next reaction, each reaction will tend to move forward. This is true of many metabolic pathways in a cell.

Gibbs Free Energy

We have seen that cells use chemical reactions to perform much of their work. The amount of energy available to do work is called **Gibbs free energy (G).** In a chemical reaction, we can compare the free energy of the reactants and products to determine whether the reaction releases energy that is available to do work. The difference between two values is denoted by the Greek letter delta (Δ). Therefore, in a chemical reaction, ΔG is the free energy of the products minus the free energy of the reactants, as written mathematically here:

$$\Delta G = G_{products} - G_{reactants}$$

If the products of a reaction have more free energy than the reactants, then ΔG is positive and a net input of energy is required to drive the reaction forward, which is illustrated in the graph shown in **FIGURE 13.8a.** Reactions with a positive ΔG that require an input of energy and are not spontaneous are called **endergonic.** By contrast, if the products of a reaction have less free energy than the reactants, ΔG is negative and energy is released and available to do work, illustrated in the graph shown in Figure 13.8b. Reactions with a negative ΔG that release energy and proceed spontaneously are called **exergonic.** The term "spontaneous" does not imply instantaneous or even rapid. "Spontaneous" in this context means that a reaction releases energy; "non-spontaneous" means that a reaction requires a sustained input of energy.

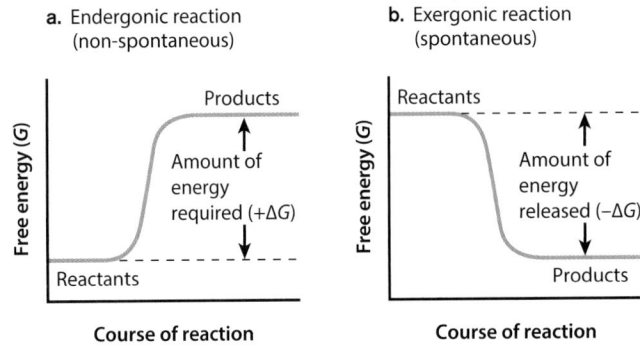

a. Endergonic reaction (non-spontaneous)

b. Exergonic reaction (spontaneous)

FIGURE 13.8 Endergonic and exergonic reactions

Chemical reactions require or release energy. (a) In endergonic reactions, the reactants have less free energy than the products. As a result, they require an input of energy and have a positive ΔG. (b) In exergonic reactions, the reactants have more free energy than the products. As a result, they release energy and have a negative ΔG.

Let's apply these concepts to a specific chemical reaction. Earlier, we introduced ATP, the molecule that drives many cellular processes using the chemical energy in its chemical bonds. ATP is able to carry out this function because it is broken down in cells, releasing energy that is used to drive other reactions that require an input of energy.

ATP is broken down by reacting with water to form ADP and inorganic phosphate, P_i (HPO_4^{2-}), as shown here and in **FIGURE 13.9** on page 192:

$$ATP + H_2O \rightarrow ADP + P_i$$

This is an example of a hydrolysis reaction, which was introduced in Unit 1. In a hydrolysis reaction, a water molecule is split into a proton (H^+) and a hydroxyl group (OH^-). Hydrolysis reactions often break down polymers into their subunits, and in the process one product gains a proton and the other gains a hydroxyl group. The reaction of ATP with water is an exergonic reaction because there is less free energy in the products than in the reactants. It has a negative value of ΔG.

The release of free energy during ATP hydrolysis comes from breaking weaker bonds with more chemical energy in the reactants and forming more stable bonds with less chemical energy in the products. The release of free energy can then drive chemical reactions and other processes that require a net input of energy, as we discuss next.

Energetic Coupling

If the conversion of reactant A into product B is spontaneous, the reverse reaction converting reactant B into product A is not. The ΔG's for the forward and reverse reactions have the same absolute values but opposite signs. However, in living

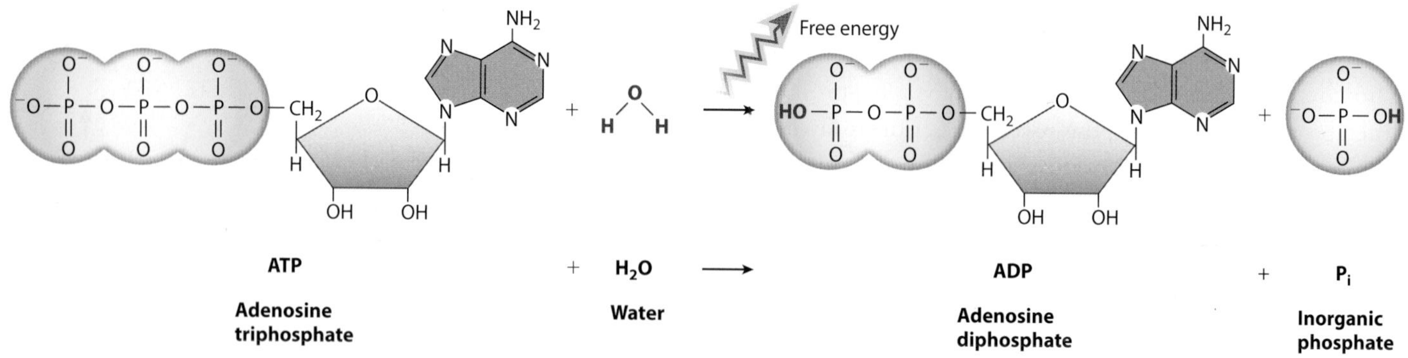

FIGURE 13.9 ATP hydrolysis

ATP hydrolysis occurs when ATP reacts with water to produce ADP and inorganic phosphate (P_i). This reaction is exergonic, and it releases free energy, as indicated by the yellow arrow, which is then available to do the work of the cell.

organisms, not all chemical reactions are spontaneous, but they nevertheless occur. Many of these reactions are driven by **energetic coupling,** a process in which a spontaneous reaction (negative ΔG) drives a non-spontaneous reaction (positive ΔG). It requires that the sum of the ΔG's of the two reactions be negative. As a result, if the sum of the ΔG's is positive or 0, there is not enough energy for the two reactions to occur. In addition, the two reactions must occur together. In some cases, this coupling can be achieved if the two reactions share a molecule in common.

For example, ATP hydrolysis can be used to drive a non-spontaneous reaction, as shown in **FIGURE 13.10**. The first reaction shows ATP hydrolysis, in which ATP reacts with water to form ADP and inorganic phosphate (P_i). Recall that this reaction is spontaneous with a negative ΔG. The second reaction is the addition of a phosphate group (P_i) to glucose to produce glucose 6-phosphate and water. This reaction is non-spontaneous with a positive ΔG. However, the sum of the ΔG's for the two reactions is negative, so the two reactions are able to proceed if they occur together. In addition, the phosphate group (P_i) released in the first reaction is transferred to glucose in the second reaction.

This shared phosphate group links the two reactions, so they occur together, as shown in the third coupled reaction.

Following ATP hydrolysis, the cell needs to replenish its ATP so that it can carry out additional chemical reactions. The synthesis of ATP from ADP and P_i is an endergonic reaction with a positive ΔG, requiring an input of energy. In some cases, exergonic reactions can drive the synthesis of ATP by energetic coupling. The sum of the ΔG's of the two reactions is negative and the reactions share a phosphate group, allowing the two reactions to proceed together.

Metabolic Pathways

Chemical reactions in the cell break down organic molecules such as glucose, and, in the process, harness energy. The term "metabolism" encompasses the entire set of chemical reactions that convert molecules into other molecules and transfer energy in living organisms. These chemical reactions occur all the time in your cells and the cells of all living organisms. Many of these reactions are linked, in that the products of one are the reactants of the next, forming long pathways and intersecting networks. A metabolic pathway is a series of chemical reactions that build or break down molecules in cells.

ATP	+ H_2O	$\longrightarrow$	ADP	+ P_i	$\Delta G_1 = -7.3$ kcal/mol	**Exergonic reaction**
Glucose	+ P_i	$\longrightarrow$	Glucose 6-phosphate	+ H_2O	$\Delta G_2 = +3.3$ kcal/mol	**Endergonic reaction**
Glucose	+ ATP	$\longrightarrow$	Glucose 6-phosphate	+ ADP	$\Delta G = -4$ kcal/mol	**Coupled reaction**

FIGURE 13.10 Energetic coupling

Energetic coupling occurs when an exergonic reaction drives an endergonic reaction. In this case, the hydrolysis of ATP drives the formation of glucose 6-phosphate from glucose. The coupled reaction, shown at the bottom, proceeds because ΔG is negative for the two reactions and the P_i is shared between the two reactions.

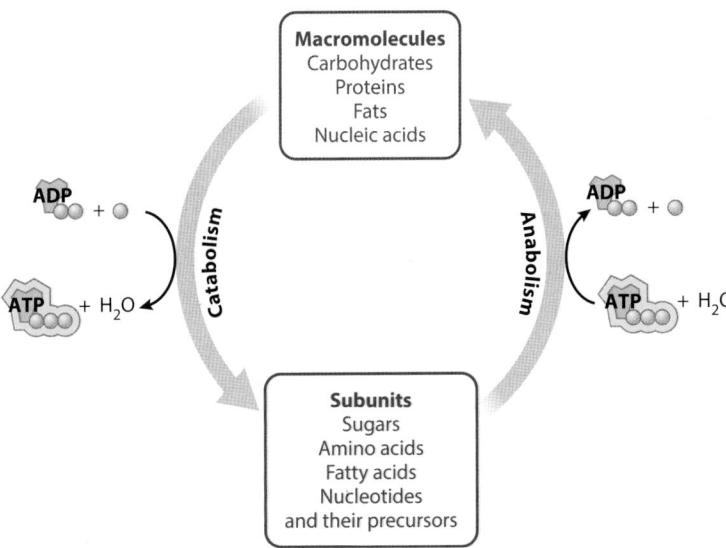

FIGURE 13.11 Catabolism and anabolism

Catabolism describes the set of reactions that breaks down macromolecules into smaller units. Anabolism is the set of reactions that builds macromolecules from smaller units. The energy released during the breakdown of molecules in catabolism can be used to synthesize ATP from ADP and inorganic phosphate (P_i). ATP is hydrolyzed, or broken down to ADP and P_i, to synthesize molecules in anabolism. The yellow highlight around ATP indicates that it carries more chemical energy than ADP, and the yellow spheres denote phosphate groups.

Metabolism is divided into two branches: *catabolism* and *anabolism*. **Catabolism** is the set of chemical reactions that breaks down molecules into smaller units and, in some cases, produces ATP from ADP and P_i. **Anabolism** is the set of chemical reactions that builds molecules from smaller units

and requires an input of energy, usually in the form of ATP, which is hydrolyzed to ADP and P_i. **FIGURE 13.11** illustrates these two complementary processes. For example, carbohydrates can be broken down, or catabolized, into sugars; fats into fatty acids and glycerol; and proteins into amino acids. These products can be broken down even further into their precursors, such as carbon dioxide and water, to release energy stored in their chemical bonds. In the process, ATP is produced from ADP and P_i. By contrast, the synthesis of macromolecules such as carbohydrates and proteins is anabolic. We discussed how these macromolecules are synthesized from their subunits in Unit 2 by dehydration synthesis reactions. The synthesis of large molecules from smaller ones requires an input of energy, such as from hydrolysis of ATP.

Many metabolic reactions are highly conserved between organisms, meaning the same reactions are found in many different organisms. This observation suggests that the reactions evolved early in the history of life and have been maintained for billions of years because of their fundamental importance to cellular energetics.

✓ Concept Check

6. **Describe** how to determine whether there is energy available to do work in a chemical reaction.

7. **Describe** the similarities and differences between catabolism and anabolism.

Module 13 Summary

PREP FOR THE AP® EXAM

REVISIT THE BIG IDEAS

ENERGETICS: Using the content of this module, **provide evidence** to support the following statement: organisms require energy to maintain life functions, and that energy is captured, converted, and stored through a combination of metabolic pathways.

LG 13.1 Kinetic and potential energy are two forms of energy.

- Kinetic energy is energy associated with movement. Page 185

- Potential energy is stored energy; it depends on the structure of an object or its position relative to its surroundings. Page 185

- Chemical energy is a form of potential energy held in the bonds of molecules. Page 186

- The bonds linking phosphate groups in ATP have high potential energy that can be harnessed for use by the cell. Page 187

LG 13.2 The laws of thermodynamics govern energy flow in biological systems.

- The first law of thermodynamics states that energy cannot be created or destroyed. Page 188

- The second law of thermodynamics states that there is an increase in entropy in the universe over time. Page 188

- Entropy is the degree of disorder or a measure of the number of positions and motions a molecule can take. Page 188

LG 13.3 Chemical reactions transform molecules and transfer energy in cells.

- Chemical reactions involve the breaking and forming of bonds. Page 190

- In a chemical reaction, atoms themselves do not change; the bonds among them change as they form new molecules. Page 190

- The direction of a chemical reaction is influenced by the concentration of reactants and products. Page 190

- Gibbs free energy (G) is the amount of energy available to do work. Page 191

- Exergonic reactions are spontaneous ($\Delta G < 0$) and release energy. Page 191

- Endergonic reactions are non-spontaneous ($\Delta G > 0$) and require energy. Page 191

- The hydrolysis of ATP is an exergonic reaction that drives many endergonic reactions in a cell. Page 191

- In living systems, non-spontaneous reactions are often coupled with spontaneous reactions. Page 192

Key Terms

Kinetic energy
Potential energy
Chemical energy
Metabolic pathway
Adenosine diphosphate (ADP)

First law of thermodynamics
Entropy
Second law of thermodynamics
Gibbs free energy (G)
Exergonic

Endergonic
Energetic coupling
Catabolism
Anabolism

Review Questions

1. The entire set of chemical reactions that sustain life is called
   (A) catabolism.
   (B) anabolism.
   (C) metabolism.
   (D) hydrolysis.

2. Large molecules are broken down and ATP is produced in the process of
   (A) anabolism.
   (B) catabolism.
   (C) diffusion.
   (D) active transport.

3. The food in your lunch is broken down into smaller molecules in your gastrointestinal tract. These chemical reactions are categorized as
   (A) anabolism.
   (B) catabolism.
   (C) diffusion.
   (D) active transport.

4. The ATP molecule contains
   (A) glucose.
   (B) glutamic acid.
   (C) ribose.
   (D) deoxyribose.

5. Nearly all cells use ATP as their primary energy source. This is evidence that
   (A) life continues to evolve and use novel energy sources.
   (B) ATP evolved recently (in the past 100 years) as a potential source of energy.
   (C) the use of ATP by cells began long ago and has been conserved over time.
   (D) only eukaryotic cells can produce ATP.

6. In chemical reactions, most of the entropy increase occurs as
   (A) heat.
   (B) chemical energy.
   (C) ATP production.
   (D) glycolysis.

7. The lid of a container containing a gas is removed. As a consequence, the molecules of the gas have more space available to them to spread out. For the gas released from the container, which increases?
   (A) Temperature
   (B) Pressure
   (C) Entropy
   (D) Energy

Module 13
AP® Practice Questions

Section 1: Multiple-Choice Questions

Choose the best answer for questions 1–3.

1. The second law of thermodynamics states that transformation of energy results in increased disorder within any closed system over time. Yet biological cells are extremely ordered and organized. Which best explains why cells do not violate the second law of thermodynamics?

   (A) The organization of cells is coupled with catabolism and decay of other biological material.

   (B) Cells rely primarily on chemical reactions that release energy and therefore decrease disorder.

   (C) Cells are not closed systems and obtain energy from the sun and chemical sources.

   (D) Cells maintain a negative net energy.

2. Cells use a process called energetic coupling, in which chemical reactions that release energy are paired with chemical reactions that require energy. One common reaction that releases energy is the hydrolysis of ATP to produce ADP. Which of the following is the primary use of energetic coupling in cells?

   (A) To convert energy from kinetic energy to chemical energy

   (B) To drive chemical reactions that would otherwise not occur in cells

   (C) To prevent accumulation of excess ATP within cells

   (D) To reduce the temperature of the cell

3. The hydrolysis of ATP to form ADP releases approximately 30 kJ of energy per mole of ATP. The fungus *Arachniotus citrinus* hydrolyzes starch. Hydrolysis of starch by *A. citrinus* glucoamylase requires approximately 18 kJ of energy per mole of starch. How many moles of starch can *A. citrinus* hydrolyze using 3.0 moles of ATP?

   (A) 0.6   (C) 1.8

   (B) 1.7   (D) 5.0

Section 2: Free-Response Question

Write your answer to each part clearly. Support your answers with relevant information and examples. Where calculations are required, show your work.

The human nervous system is extremely metabolically active and has high energy demands. On average, the nervous system requires approximately 0.67 mole (120 grams) of glucose every day in order to maintain maximum function, which is approximately 60% of the glucose requirement for the entire body. The energy contained in 0.67 mole of glucose is approximately 420 kilocalories.

(a) Nerve cells produce 32 moles of ATP for every mole of glucose. **Calculate** how many mole of ATP the nervous system releases from glucose every day.

(b) ATP provides chemical energy for nerve cells to function. Hydrolysis of ATP releases 7.3 kilocalories per mole. Assuming the nervous system uses all of the ATP it generates from glucose every day, **calculate** how many kilocalories of chemical energy it uses.

(c) **Make a claim** to explain any observed difference between the amount of energy the nervous system releases from the metabolism of a daily supply of glucose, and the amount of energy from ATP that the nervous system uses daily.

(d) The nervous system uses approximately 60% of the body's daily glucose requirement, but only approximately 20% of the body's overall energy requirement. **Identify** the source of the additional energy required by the body.

Module 14

Enzymes

LEARNING GOALS ▶**LG 14.1** Enzymes increase the rates of chemical reactions.
▶**LG 14.2** Enzymes bind reactants at active sites, which are highly specific.
▶**LG 14.3** Enzymes are influenced by the environment, activators, and inhibitors.

In the last module, we introduced two forms of energy—kinetic and potential energy. Kinetic energy is energy of motion and potential energy is stored energy. Molecules hold a form of potential energy, called chemical energy, in the bonds that hold atoms together. Cells in turn can transform molecules into other molecules by breaking bonds and forming new ones in chemical reactions. As bonds break and new bonds are formed, energy is also used or released. Cells are energy transformers: they transfer energy from the environment, including the energy in sunlight and the energy in chemical compounds, and make use of it to perform their functions. In all of these transformations, chemical reactions are subject to the laws of thermodynamics, which describe what happens to energy as it is passed from molecule to molecule.

Countless chemical reactions are occurring in cells all of the time. Often, the product of one reaction is a reactant for the next reaction, forming long metabolic pathways that in turn connect with other pathways. Taken together, all of these chemical reactions constitute metabolism. Some of these pathways are anabolic, building larger molecules from smaller ones, and some are catabolic, breaking down larger molecules into smaller ones. Many of these reactions are spontaneous. However, as we explained in the last unit,

a spontaneous reaction is not necessarily a fast reaction. For example, glucose is readily broken down by a series of chemical reactions in cells all of the time, but the rates of these reactions on their own are close to zero. In cells, the rates of reactions are often increased by **enzymes,** which are proteins that accelerate the rate of a chemical reaction. Enzymes speed up the rate of reactions, so they occur at rates that are fast enough for cells to carry out their functions.

By accelerating the rates of reactions, enzymes help to determine which reactions, out of all of the possible reactions, occur at specific times and places in cells. In this module, we will discuss how enzymes increase the rate of chemical reactions and how this ability gives them a central role in metabolism. We will end with a discussion of factors that influence the activity of enzymes and therefore affect the life of a cell.

PREP FOR THE AP® EXAM

FOCUS ON THE BIG IDEAS

ENERGETICS: Focus on the structural and functional relationship between enzymes and substrates; pay particular attention to how enzyme-mediated reactions are influenced by environmental factors.

14.1 Enzymes increase the rates of chemical reactions

In a chemical reaction, reactants are converted to products. In the last unit, we focused on the spontaneity and direction of chemical reactions. A spontaneous reaction releases free energy because the products have less free energy than the reactants. A non-spontaneous reaction consumes free energy because the products have more free energy than the reactants. We also saw how the direction of a chemical reaction can be influenced by the concentration of reactants

and products. Increasing the concentration of reactants and decreasing the concentration of products favors the forward reaction. In this section, we turn our attention to the rate of chemical reactions. Many chemical reactions that occur commonly in cells would be imperceptibly slow if not for the presence of enzymes. We will focus on how enzymes increase the rate of chemical reactions.

Catalysis

The rate of a chemical reaction is defined as the amount of product formed or reactant consumed per unit of time.

Catalysts are substances that increase the rate of chemical reactions.

Catalysts have certain properties in common. As just mentioned, they all increase the rate of chemical reactions. In addition, catalysts participate in a chemical reaction, interacting with reactants and products, but they emerge from the reaction unchanged. In other words, catalysts are not consumed in the reaction, so they are available to catalyze one reaction after the next. In addition, catalysts don't change the equilibrium of a reaction. Chemical equilibrium is the point in which the concentrations of reactants and products are not changing because the forward and reverse reactions proceed at the same rate. Catalysts increase the rate of the forward reaction and reverse reaction to the same extent, so the equilibrium of the reaction is not changed. Finally, catalysts tend to be specific, participating in some—but not all—reactions.

There are several different types of catalysts. In cells, enzymes serve as biological catalysts. Enzymes are proteins that increase the rate of chemical reactions in cells. For instance, you have enzymes in your digestive system that help to break down the food you eat, as shown in **FIGURE 14.1**. Like all catalysts, enzymes participate in the reaction but are not consumed, and are specific for certain chemical reactions. Because enzymes are proteins, the shapes of enzymes determine their functions, and anything that changes the shapes of enzymes will alter their functions.

For example, if an enzyme is denatured, or unfolded, it completely loses its function. Occasionally, enzymes can refold after being denatured, regaining their function. In addition to enzymes, chemical elements, such as iron and nickel, can serve as catalysts. They often speed up chemical reactions that take place in the lab. Certain RNA molecules also act as catalysts. These are called ribozymes. Therefore, catalysts include some proteins called enzymes, certain chemical elements, and some RNA molecules. So, all enzymes are catalysts, but not all catalysts are enzymes.

Enzymes catalyze many thousands of chemical reactions that occur in cells. For example, we can return to the chemical reaction we introduced in Module 13, in which carbon dioxide reacts with water to produce carbonic acid, as follows:

$$CO_2 + H_2O \rightleftharpoons H_2CO_3$$
Carbon dioxide   Water   Carbonic acid

This reaction occurs in your bloodstream and is catalyzed by an enzyme called carbonic anhydrase. In the absence of the enzyme, the reaction proceeds very slowly. In the presence of the enzyme, it is about 10^7 times faster. The enzyme does not affect the concentrations of the reactants and products when the reaction reaches equilibrium. Nor does the enzyme affect the direction of the reaction, which is determined in part by the relative concentrations of the reactants and products. In muscles, where the concentration of carbon dioxide is high because it is produced as a by-product of metabolism, the forward reaction is favored. In the lungs, where the concentration of carbon dioxide is low because it is exhaled, the reverse reaction is favored. The enzyme allows the reaction to proceed at rates that help organisms function effectively. Evolution has favored organisms that have carbonic anhydrase because its presence allows them to carry out the reaction at rates which are needed for efficient control of carbon dioxide and carbonic acid. "Analyzing Statistics and Data: Rates" on page 198 demonstrates how to use the equation for calculating rates and provides a practice problem.

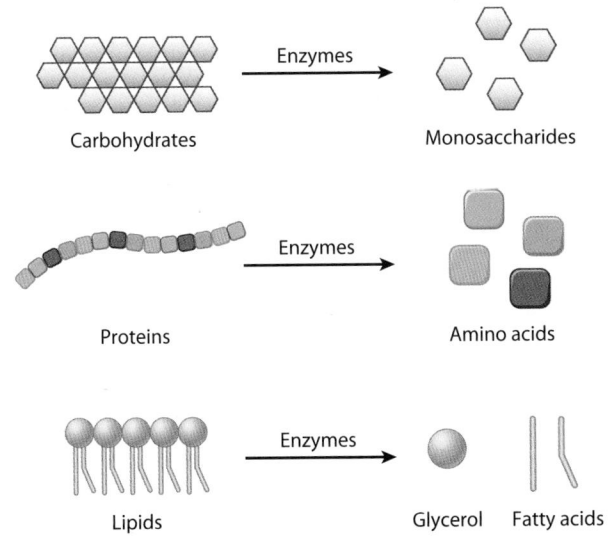

Carbohydrates → Enzymes → Monosaccharides

Proteins → Enzymes → Amino acids

Lipids → Enzymes → Glycerol   Fatty acids

FIGURE 14.1 Enzymes

Enzymes in your digestive tract help to break down macromolecules, such as carbohydrates, proteins, and lipids, into their subunits, including monosaccharides, amino acids, and glycerol and fatty acids.

Rates

Similar to a ratio, a rate is a comparison between two numbers. The numbers typically have different units or represent different quantities of something, with one related to time. For example, a bee may visit 100 flowers every 2 hours before returning to the hive to deposit the nectar it collected. Expressed as a rate, this is 100 flowers per 2 hours. To determine how many flowers the bee visited per hour, we divide each number by 2. This yields 50 flowers per hour.

In chemical reactions, a rate is defined as the amount of product formed or the reactant consumed per unit of time. The following equation is used to determine a rate, where dY is the change in a quantity called Y and dt is the change in time:

$$\text{Rate} = \frac{dY}{dt}$$

PRACTICE THE SKILL

Amylase is an enzyme that breaks down starches into sugars. It is present in human saliva and helps begin the digestion process. A researcher measured the effect amylase has on starch found in potatoes. The researcher added amylase to a concentrated starch solution and measured the amount of glucose produced every 15 seconds. The table below shows the data.

Time (seconds)	Total amount of glucose produced (mg) at each time point
0	0
15	0.04
30	0.09
45	0.21
60	0.33

What is the rate of this enzymatic reaction in mg/s from 15 to 60 seconds?

All of the information needed to use the rate equation is provided in the table. The change in sugar produced between 15 and 60 seconds is dY, and the change in time between 15 and 60 seconds is dt.

$$\text{Rate} = \frac{dY}{dt}$$

$$\text{Rate} = \frac{(0.33 \text{ mg} - 0.04 \text{ mg})}{(60 \text{ s} - 15 \text{ s})}$$

$$\text{Rate} = 0.006 \text{ mg/s}$$

This reaction progressed at a rate of 0.006 mg/s.

Your Turn

A researcher decided to compare the rates of a reaction containing the enzyme carbonic anhydrase in room temperature conditions (20°C) versus elevated temperature conditions (30°C). The elevated temperature was moderate enough not to cause protein denaturation, or unfolding. Given the information about each of the reactions in the table below, calculate the rate of reaction for each temperature between 1 and 4 minutes. Explain why the rates do or do not differ.

Time (minutes)	Total amount of carbonic acid released at 20°C (mL) at each time point	Total amount of carbonic acid released at 30°C (mL) at each time point
0	0	0
1	5	8
2	10	15
3	17	27
4	22	34
5	30	34

Activation Energy

In Module 13, we saw that exergonic reactions release free energy and endergonic reactions require free energy. Nevertheless, all chemical reactions require an initial input of energy to proceed, even exergonic reactions that release energy. For an exergonic reaction, the energy released is more than the initial input of energy, so there is a net release of energy.

An input of energy is needed for all chemical reactions because they all proceed through a state that is highly unstable and has a large amount of free energy. As a chemical reaction proceeds, existing chemical bonds break and new bonds form. For an extremely brief time, a compound is formed in which the old bonds are breaking, and the new ones are forming. This unstable intermediate state between reactants and products is known as the **transition state.**

In all chemical reactions, reactants adopt at least one transition state before they convert into products. The graph in **FIGURE 14.2** represents the free energy levels of the reactant, transition state, and product. This reaction is spontaneous because the free energy of the reactant is higher than the free energy of the product and ΔG is negative. However, the highest free energy value corresponds to the transition state.

To reach the transition state, the reactant must absorb energy from its surroundings, which is represented as the uphill portion of the curve in Figure 14.2. In other words, the reactants must gain some energy if they are to react, regardless of whether the ΔG of the reaction is positive or negative. This energy needed to reach the transition state is called the **activation energy (E_A).** You can think of the activation energy as an energy barrier that the reactants have to overcome if the reaction is to proceed. Once the transition state is reached, the reaction continues, products are formed, and energy is released into the surroundings, which is the downhill portion of the curve in Figure 14.2.

There is an inverse correlation between the rate of a reaction and the height of the energy barrier: the lower the activation energy, the faster the reaction, and the higher the activation energy, the slower the reaction. In chemical reactions that take place in the laboratory, heat is a common source of energy used to overcome the activation energy. In most organisms, an increase in heat would disrupt

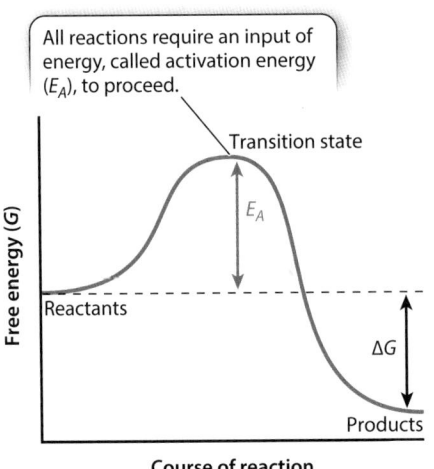

FIGURE 14.2 Transition state

All chemical reactions proceed through a transition state, which has a high amount of free energy. The difference between the free energy of the transition state and the free energy of the reactants is the activation energy (E_A). The activation energy is a barrier that all reactions have to overcome to proceed. ΔG is the difference in free energy between the reactants and products. In this case, ΔG is negative, so the reaction is exergonic.

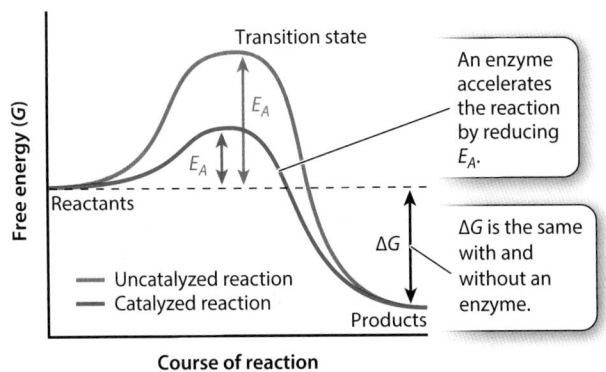

FIGURE 14.3 Enzyme function

The blue line represents the path of an uncatalyzed reaction, one that occurs in the absence of an enzyme. The red line represents the path of a catalyzed reaction, one that occurs in the presence of an enzyme. Enzymes increase the rate of chemical reactions by reducing the activation energy (E_A).

the function of many macromolecules, including proteins. Instead, cells make use of enzymes to accelerate chemical reactions. Enzymes do not act by supplying heat. Instead, they reduce the activation energy by stabilizing the transition state and decreasing its free energy. The red curve in **FIGURE 14.3** shows the path of an enzyme-catalyzed reaction. Note that the activation energy is lower in the presence of the enzyme (red) compared to the activation energy in the absence of the enzyme (blue). As the activation energy decreases, the rate of the reaction increases.

Although an enzyme reduces the activation energy, the difference in free energy between reactants and products (ΔG) does not change. In other words, an enzyme changes the path of the reaction between reactants and products, but not the starting or end point, as can be seen by following the blue and red lines in Figure 14.3. Note that they start and end in the same place but take a different route between these points. Similarly, consider the breakdown of glucose into carbon dioxide and water. The ΔG of the reaction is the same whether it proceeds by combustion in a lab or by the action of multiple enzymes in a metabolic pathway in a cell.

✓ Concept Check

1. **Describe** three characteristics of the role of enzymes in chemical reactions.

2. **Describe** how enzymes increase the rate of chemical reactions.

14.2 Enzymes bind reactants at active sites, which are highly specific

Like all catalysts, enzymes work by reducing the activation energy between reactants and products in a chemical reaction. In this way, they provide a different chemical route that makes it more likely for the reaction to proceed. This alternative pathway requires a smaller input of energy than if the enzyme were not present and so the reaction proceeds more quickly. In the absence of enzymes, most—if not all—chemical reactions that occur regularly in the cell would not proceed, and life would grind to a halt. Enzymes perform their functions by binding to reactants and forming a complex. This complex stabilizes the transition state and therefore reduces its free energy, reducing the activation energy of the reaction and allowing it to proceed more quickly.

In this section, we will focus on the formation of this complex. Although it forms for just a brief time, it is key to understanding why the activation energy of the reaction is reduced and therefore why the reaction proceeds much more rapidly in the presence of an enzyme. We will also look at the sites where enzymes and reactants bind, which will help us understand why enzymes are highly specific.

Enzyme–Substrate Complex

As catalysts, enzymes participate in a chemical reaction but are not consumed in the process. They emerge from a chemical reaction unchanged, ready to catalyze the same reaction again. Enzymes accomplish this chemical feat by forming a short-lived complex with the reactants and products.

In a chemical reaction, the reactant is often referred to as the **substrate.** In such a reaction, the substrate (S) is converted to a product (P):

$$S \rightleftharpoons P$$

In the presence of an enzyme (E), the substrate first forms a complex with the enzyme (enzyme–substrate, or ES). While still part of the complex, the substrate is converted to product (enzyme–product, or EP). Finally, the complex dissociates, releasing the enzyme and product. Therefore, a reaction catalyzed by an enzyme can be described as follows:

$$E + S \rightleftharpoons ES \rightleftharpoons EP \rightleftharpoons E + P$$

FIGURE 14.4 illustrates the steps of an enzyme-catalyzed reaction. Figure 14.4a shows the steps for a catabolic reaction, in which one molecule is broken up into two, and Figure 14.4b

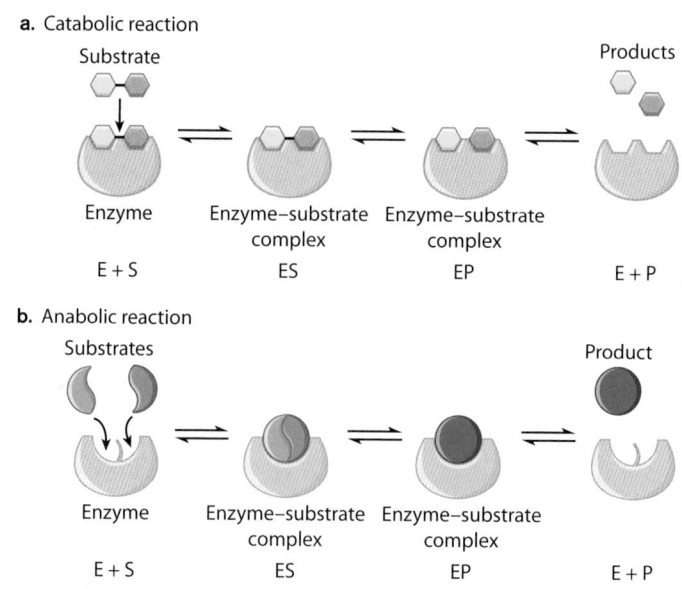

a. Catabolic reaction

Substrate / Products

Enzyme / Enzyme–substrate complex / Enzyme–substrate complex

E + S / ES / EP / E + P

b. Anabolic reaction

Substrates / Product

Enzyme / Enzyme–substrate complex / Enzyme–substrate complex

E + S / ES / EP / E + P

FIGURE 14.4 Steps of an enzyme-catalyzed reaction

In a reaction catalyzed by an enzyme, the enzyme (E) binds to the substrate (S). They form a complex (ES) in which the substrate is converted to product (EP). The enzyme (E) and product (P) then dissociate, and the enzyme emerges from the reaction unchanged, ready to catalyze another reaction. (a) In a catabolic reaction, a large molecule is broken down into two smaller ones. (b) In an anabolic reaction, two small molecules are joined to form a larger one.

shows the steps for an anabolic reaction, in which two molecules are joined into one. In both cases, the formation of an enzyme–susbstrate complex is critical for accelerating the rate of a chemical reaction. Specific interactions between the enzyme and substrate in the complex stabilize the transition state and decrease the activation energy. As a result, the reaction proceeds more quickly than it would without the enzyme.

The formation of a complex between an enzyme and a substrate can be demonstrated experimentally, as illustrated in "Practicing Science 14.1: Do enzymes form complexes with substrates?" In this experiment, a beaker is separated by a selectively permeable membrane. The membrane is permeable to the substrate (S), but not the enzyme (E). The enzyme can bind the substrate but cannot convert it to a product because of the specific structure of the substrate. The substrate is added to one side of the membrane and the enzyme to the other. Radioactivity is used to follow the substrate as it diffuses across the membrane. Over time, the substrate diffuses to the side with the enzyme and accumulates to a greater degree on that side compared to when the enzyme is not present because it forms a complex with the enzyme. By forming a complex, the bound substrate is removed from the chemical equilibrium.

Do enzymes form complexes with substrates?

Background The idea that enzymes form complexes with substrates to catalyze a chemical reaction was first proposed in 1888 by the Swedish chemist Svante Arrhenius. In the 1930s, American chemist Kurt Stern performed one of the earliest experiments that supported Arrhenius's idea. Stern studied an enzyme called catalase, which is very abundant in animal and plant tissues. In the conclusion of his paper, he wrote, "It remains to be seen to which extent the findings of this study apply to enzyme action in general."

To illustrate that Stern's findings do indeed apply to enzymes in general, another experiment that analyzes a different enzyme and technique are described here. The enzyme β-galactosidase catalyzes the cleavage (splitting) of a bond in its substrate β-galactoside, which includes the sugar lactose. However, it binds but cannot split the bond in a similar molecule called β-thiogalactoside.

Method As shown in the figure, a container is separated into two compartments by a selectively permeable membrane. The membrane is permeable to β-thiogalactoside, but it is not permeable to the enzyme β-galactosidase.

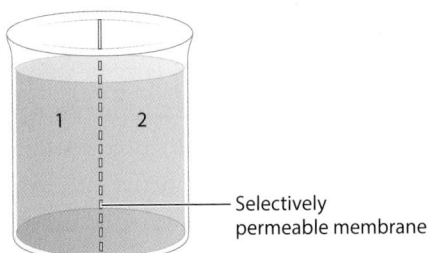

Selectively permeable membrane

Experiment 1 and Results Radioactively labeled β-thiogalactoside (the substrate, S) is added to compartment 1 and the movement of S is followed by measuring the level of radioactivity in the two compartments. Radioactivity provides a way to tag molecules so they can be visualized and followed. As you can see in the illustration, over time, the level of radioactivity becomes the same in the two compartments as the substrate diffuses from compartment 1 to compartment 2.

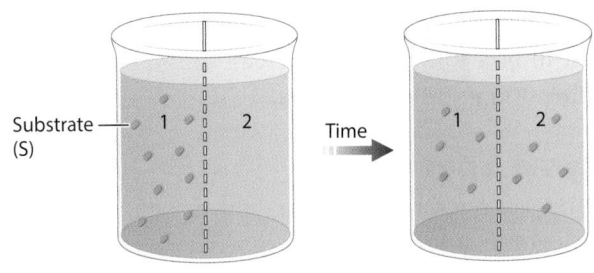

Substrate (S)   Time

Experiment 2 and Results Radioactively labeled β-thiogalactoside (S) is added to compartment 1, the enzyme β-galactosidase (E) is added to compartment 2, and the movement of S is followed by measuring the level of radioactivity in the two compartments. As shown in the figure, over time, the level of radioactivity becomes greater in compartment 2 than in compartment 1.

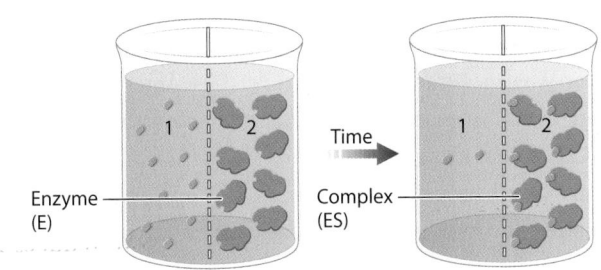

Enzyme (E)   Time   Complex (ES)

Interpretation The substrate β-thiogalactoside (S) diffuses from compartment 1 to compartment 2, forms a complex (ES) with the enzyme β-galactosidase (E), and is not released because, in this case, the enzyme (E) does not catalyze the conversion of substrate (S) to product (P). The level of radioactivity is greater in compartment 2 compared to the level in compartment 1 because E and S form a complex (ES) and the radioactive substrate is not released from the complex to diffuse back across the membrane.

SOURCES
Adapted from Doherty, D. G., and F. Vaslow. 1952. "Thermodynamic Study of an Enzyme–Substrate Complex of Chymotrypsin." *J. Am. Chem. Soc.* 74: 931–936. Stern, K. G. 1936. "On the Mechanism of Enzyme Action: A Study of the Decomposition of Monoethyl Hydrogen Peroxide by Catalase and of an Intermediate Enzyme–Substrate Compound." *J. Biol. Chem.* 114:473–494.

AP® PRACTICE QUESTION

(a) **Identify** the following:
   1. The question the experimenters are trying to answer
   2. The hypothesis they tested
   3. The independent and dependent variables utilized
(b) **Describe** the role of the control.
(c) Explain the results by **identifying** the researchers' claim and **justifying** the claim with evidence and reasoning (CER). In your CER, be sure to include your understanding of the experimental set-up.

Active Site of the Enzyme

The formation of a complex is critical for accelerating the rate of a chemical reaction. Complexes form because enzymes bind substrates at a region of the enzyme called the *active site*. The **active site** is the part of the enzyme that binds substrate and catalyzes its conversion to the product, as shown in **FIGURE 14.5**. In the active site, the enzyme and substrate interact with each other. They may form transient, or very temporary, covalent bonds. More commonly, the enzyme and substrate form noncovalent interactions, such as ionic bonds between positive and negative charges, hydrogen bonds, and van der Waals interactions. Together, these interactions stabilize the transition state and decrease the activation energy.

Recall that enzymes are proteins, which were discussed in Unit 1. All proteins adopt three-dimensional shapes and the shape of a protein is linked to its function. This shape in turn is determined by the sequence of amino acids that make up the protein. The sequence determines how a protein folds, which determines its shape, which in turn determines its function. In other words, the form and function of proteins are intimately related.

Enzymes, like all proteins, fold into specific three-dimensional shapes. As they fold, they form the active site, and the shape of the active site is in turn critical for its ability to bind substrate and catalyze a chemical reaction. The active site is made up of amino acids. Specific amino acids need to be in precise positions in order to bind the substrate.

If there is more than one substrate for a reaction, the active site may position the substrates in a specific way so that they are able to react with each other. Within the active site, the reactive chemical groups of the substrates are aligned and their motion relative to each other is restrained. As a

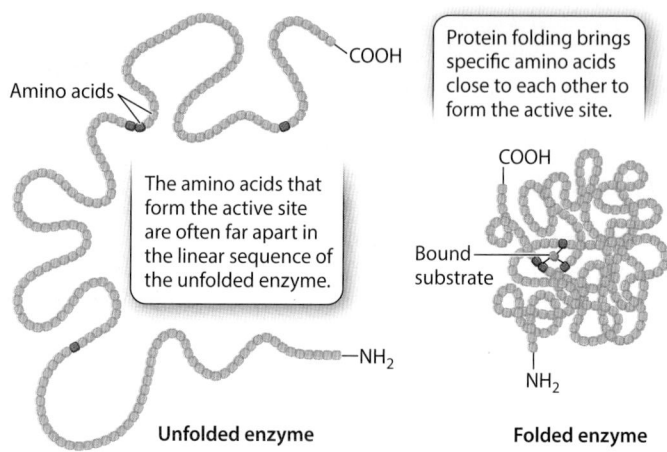

FIGURE 14.6 Enzyme folding

The amino acids that form the active site may be far apart in the sequence of amino acids that make up the protein, as shown in the primary structure on the left. They are brought close together by protein folding, as shown in tertiary structure on the right. As a result, enzymes are often large molecules even though the active site is small enough to allow the catalytic amino acids to be positioned in a precise manner.

result, they are more likely to come into contact with each other and react to form product.

The size of the active site is extremely small compared with the size of the enzyme. Nevertheless, enzymes are often large molecules. The large size of enzymes allows the amino acids that make up the active site to be positioned so they can participate in the chemical reaction. Furthermore, of the many amino acids that form the active site, only a few contribute to catalysis. Each of these amino acids has to occupy a very specific spatial position to align with the complementary reactive group on the substrate. If the few essential amino acids of the active site were part of a short peptide, the alignment of chemical groups between the amino acids of the active site and the substrate would be difficult or even impossible because the length of the bonds and the bond angles in the peptide would constrain its three-dimensional structure. In many cases, the catalytic amino acids are spaced far apart in the primary structure of the enzyme, but brought close together by protein folding, as shown in **FIGURE 14.6**. In other words, the large size of many enzymes is required at least in part to bring the catalytic amino acids into very specific positions in the active site of the folded enzyme.

Enzyme Specificity

Enzymes are remarkably specific both for the substrate and the reaction that is catalyzed. In general, enzymes recognize either a unique substrate or a class of substrates that share common chemical structures. In addition, enzymes typically

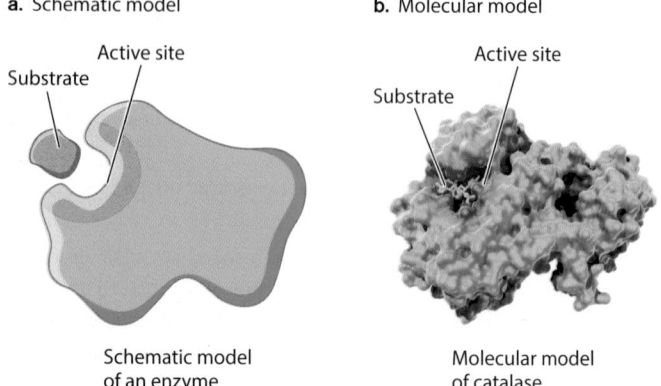

a. Schematic model   **b.** Molecular model

Schematic model of an enzyme

Molecular model of catalase

FIGURE 14.5 Active site of an enzyme

A substrate binds and is converted to product in the active site. (a) The enzyme is shown as a schematic model. (b) The enzyme is shown as a molecular model.

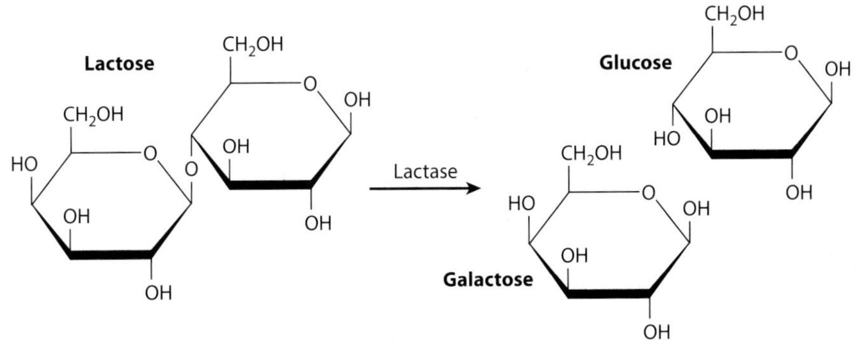

FIGURE 14.7 Enzyme specificity

The enzyme lactase is specific for its substrate, lactose. Lactase breaks lactose into its two constituent sugars, glucose and galactose, in the gut.

catalyze only one reaction or a very limited number of reactions. The specificity of an enzyme can be attributed to the structure of its active site. The active site interacts only with substrates having a precise three-dimensional structure.

For example, the enzyme lactase is specific for the milk sugar lactose. Lactose is a disaccharide made up of glucose and galactose. Lactase binds lactose in its active site, breaking down into its two sugars, as illustrated in **FIGURE 14.7**. Lactase is produced in the gut, where it breaks down lactose, allowing the individual sugars to be absorbed by the bloodstream. Mammals such as humans produce lactase in infancy to help them digest breast milk. After weaning, many individuals stop producing lactase, which leads to symptoms of lactose intolerance after drinking milk or eating a bowl of ice cream. Lactase acts on lactose, but not other disaccharides, so it is specific for lactose. The specificity of lactase results from the structure of its active site.

The binding of a substrate to the active site of an enzyme can be modeled in two different ways, shown in **FIGURE 14.8**. This first model is described as lock and key. In this model, the substrate fits into the active site in the same way that a key fits into a lock. This model emphasizes the specificity of the interaction. Just as keys are specific for certain locks, enzymes are specific for certain substrates.

The second model is described as induced fit. In this model, the binding of the substrate to the active site actually modifies the shape of the active site. In some cases, the shape of the substrate is also influenced by binding the active site. The substrate and active site require a close fit to form a complex, but the fit becomes even closer as the two interact. Essentially, this model suggests that the substrate and active site are not rigid structures, but instead are able to mold to some degree to each other, like a handshake or the way a hand fits into a glove. These two models help us to visualize and understand how enzymes form a complex and therefore stabilize the transition state.

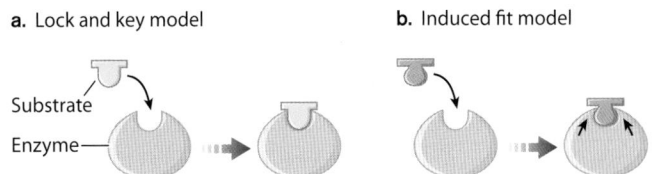

FIGURE 14.8 Two models of enzyme–substrate interaction

Enzymes and substrates interact in a specific manner, and two models describe this interaction. (a) In the lock and key model, the substrate and enzyme fit together like a key fits into a lock. (b) In the induced fit model, the enzyme molds itself to some extent to fit the substrate, similar to the way that a glove molds itself to fit your hand.

✓ Concept Check

3. **Describe** how enzymes reduce the activation energy of a chemical reaction and thereby increase its rate.

4. **Describe** how protein folding leads to enzyme specificity.

14.3 Enzymes are influenced by the environment, activators, and inhibitors

Up to this point, we have focused on how an individual enzyme affects a chemical reaction.

Taking a broader view, cells are essentially chemical factories, with a myriad of chemical reactions occurring all of the time. Most of these chemical reactions are catalyzed by enzymes. When molecules are built or broken down, the reactions are catalyzed by enzymes. When a molecule is pumped across a cell membrane, this is an enzyme-catalyzed reaction. When DNA is copied in the process of cell division, this is an enzyme-catalyzed reaction. When the information in DNA is used to build RNA, or the information in RNA is used to build a protein, these are enzyme-catalyzed reactions. When you eat, the food is broken down before being absorbed into the bloodstream by enzymes (see Figure 14.1). In short, enzymes are the master controllers of the chemical reactions that essentially define the function of cells.

Enzymes, like all proteins, can be influenced by environmental conditions, such as temperature and pH. They can also be influenced by binding interactions with other molecules called activators and inhibitors. In this section, we will focus on factors that affect the activities of enzymes, which in turn affect the metabolic activity of a cell.

Enzyme Activity and Temperature

Chemical reactions occur when molecules interact, and these interactions depend on collisions between molecules. As temperature increases, there is an increase in thermal energy, which in turn increases the motion of molecules. So, molecules move around more rapidly, increasing the chance they will collide and therefore interact with one another.

The activity of enzymes is also sensitive to temperature. Enzymes have an optimal temperature, which is the temperature at which the enzyme is most active. This is the peak of the graph in **FIGURE 14.9**. At low temperatures, enzyme activity is typically low as molecules move slowly, or have lower kinetic energy, reducing the probability that reactants come together with one another and interact with enzymes. As the temperature increases, molecules move around more rapidly, or have higher kinetic energy, which increases their chance of interacting with one another. As we have seen, enzyme-catalyzed reactions depend on reactants interacting with enzymes to form a complex.

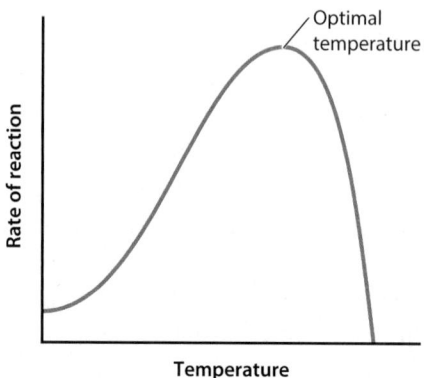

FIGURE 14.9 Effect of temperature on enzyme activity

Enzyme activity tends to increase with increasing temperature, as indicated by the uphill portion of the graph. At high temperature, enzymes unfold, or denature, and lose their function, as indicated by the downhill portion of the graph.

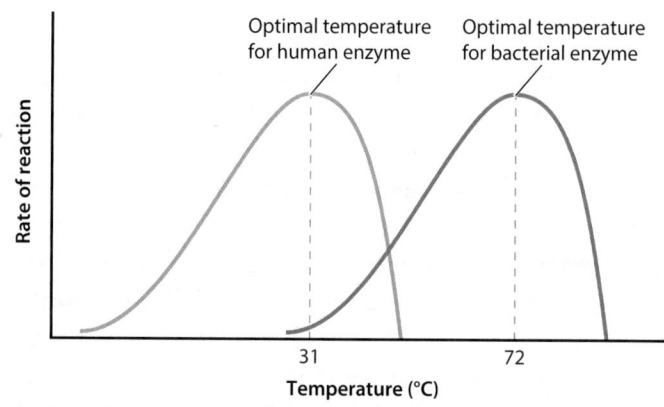

FIGURE 14.10 Optimal temperatures for two enzymes

Different enzymes have different optimal temperatures for their function. Enzymes that copy DNA in humans and other mammals work best around 31°C, while enzymes that copy DNA in certain bacteria that thrive at high temperatures work best around 72°C.

Temperature tends to increase the activity of enzymes up to a point, represented by the peak of the graph in Figure 14.9. At very high temperatures, enzymes unfold, or **denature.** Enzymes are proteins, so their function is influenced by the same set of conditions that influence other proteins. Because the ability of an enzyme to function depends on its shape, a denatured enzyme loses its ability to catalyze a chemical reaction.

The amino acid sequence of a protein, called its primary structure, determines how it folds and therefore its overall shape, as described in Unit 1. For most proteins, the folding process takes place within milliseconds as the molecule is synthesized. Some proteins fold more slowly, however, and for these molecules folding is a dangerous business. For example, unfolded proteins can aggregate or misfold. In both cases, they lose their function. Cells have evolved proteins called chaperones that help protect slow-folding or denatured proteins until they can attain their proper three-dimensional structure.

Different enzymes work best at different temperatures. At one extreme, there are bacteria called thermophiles, which thrive at high temperatures. These bacteria have evolved specialized enzymes that function at higher temperatures than their human counterparts, as shown in the graph in **FIGURE 14.10**. Such high temperatures would denature most enzymes. As we will see in Unit 6, we have learned to take advantage of these enzymes for techniques in molecular biology that require high temperature.

Enzyme Activity and pH

The pH of an environment also influences the activity of enzymes. Recall from Module 2 that pH is a measure of the hydrogen ion (H⁺) concentration. The lower the pH, the higher the H⁺ concentration and the more acidic a solution is. The higher the pH, the lower the H⁺ concentration and the more basic a solution is.

The pH also influences enzyme activity. All enzymes have an optimal pH for their function, indicated by the peaks of the graph in **FIGURE 14.11**. The optimal pH is the one in which the enzyme is most active. Most enzymes, such as trypsin, function at a pH of about 7 to 8, shown on the right of the graph, but there are some exceptions. For example, digestive enzymes in your stomach, such as pepsin, have evolved to act at a very low pH, in the range of 1 to 3, shown on the left of the graph. The acidic environment of the stomach kills ingested microbes and facilitates digestion. Most enzymes, by contrast, are denatured in acidic environments. That is, they are unfolded and thus made inactive.

Similarly, as we saw in Unit 2, enzymes that are packaged in lysosomes, which break down macromolecules in a cell, work most effectively at about pH 5. These enzymes cannot function in the normal cellular environment, which has a pH of about 7. Many of a cell's enzymes would unfold and not function properly if the entire cell were at the pH of the inside of a lysosome.

The pH of an environment affects the activity of enzymes in two ways. First, the pH can affect how the enzyme folds. This is because pH affects charges of the amino acids that make up the enzyme and these in turn can affect how the amino acids interact as they fold into a three-dimensional shape. The shape of the enzyme affects the shape of the active site.

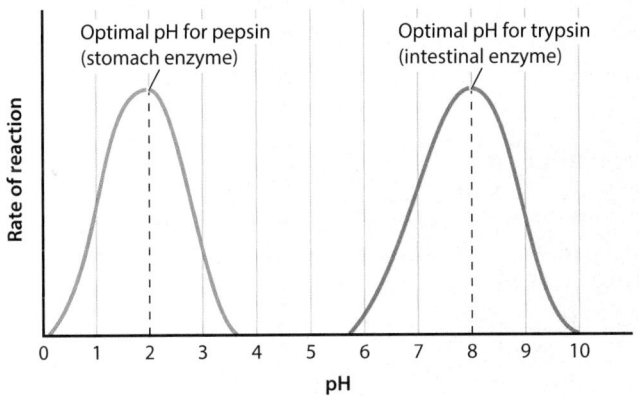

FIGURE 14.11 Effect of pH on enzyme activity

Enzymes typically have an optimal pH for their function, which is represented here by the peaks of the graph. At higher and lower pH, the activity of the enzyme is reduced. The optimal pH is different for different enzymes. The optimal pH for pepsin, a stomach enzyme, is 1 to 2, while the optimal pH for trypsin, an intestinal enzyme, is 7 to 8.

Second, the pH can affect the charges of the amino acids that make up the active site of the enzyme. At low pH, with a high concentration of protons (H⁺), protons bind to functional groups, leading to an overall increase in positive charge. By contrast, at high pH, with a low concentration of protons, protons dissociate from functional groups, leading to an overall increase in negative charge. The charges of the amino acids can then influence how well it binds the substrate, as hydrogen bonds are often involved in interactions of enzymes and substrates. So, pH influences both the charge and shape of the active site of the enzyme.

In short, the pH can affect the overall shape of the enzyme, the shape of the active site of the enzyme, and the various charges present in the active site. As a result, enzymes typically have an optimal pH for full activity.

Enzyme Activators and Inhibitors

The activity of enzymes can be influenced by *activators* and *inhibitors*. **Activators** increase the activity of enzymes, while **inhibitors** decrease the activity of enzymes. Enzyme inhibitors are quite common. They are synthesized naturally by many plants as a defense against herbivores. Similarly, pesticides and herbicides often target enzymes to inactivate them. Many drugs used in medicine are enzyme inhibitors. For example, many antibiotics used to treat bacterial infections are enzyme inhibitors, including penicillin, which is an inhibitor of an enzyme that some bacteria use to build their cell walls. By inhibiting the function of this enzyme, penicillin acts to kill the bacteria. Some drugs used to treat cancer, or uncontrolled cell division, are enzyme inhibitors, such as methotrexate. Poisons like cyanide are also enzyme inhibitors. Given the importance of chemical reactions and the role of enzymes in metabolism, it is not surprising that enzyme inhibitors have such widespread applications.

There are two classes of inhibitors. Irreversible inhibitors usually form covalent bonds with enzymes and irreversibly inactivate them. Reversible inhibitors form weak bonds with enzymes and therefore easily dissociate from them.

Inhibitors can act in many different ways, two of which are shown in **FIGURE 14.12**. Figure 14.12a shows a typical enzyme-catalyzed reaction, in which an enzyme binds a substrate in its active site, where it converts substrate to product. A competitive inhibitor, as its name suggests, competes with the substrate for binding to the active site. Competitive inhibitors are typically similar in structure to the substrate and therefore are able to bind to the active site of the enzyme, as illustrated in Figure 14.12b. Binding of the inhibitor prevents the binding of the substrate. This type of

a. An enzyme-catalyzed reaction

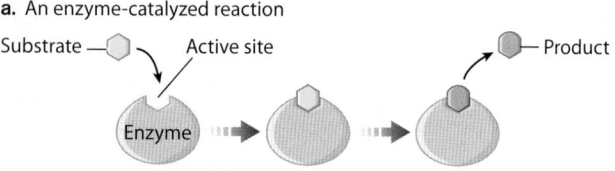

b. Competitive inhibition

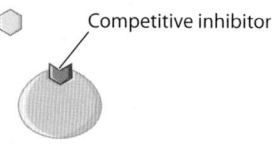

c. Non-competitive inhibition

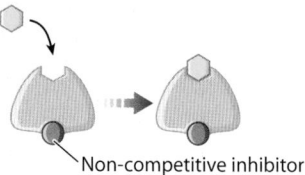

FIGURE 14.12 Competitive and non-competitive inhibition

Inhibitors reduce the activity of an enzyme and therefore decrease the rate of enzyme-catalyzed reactions. (a) In an enzyme-catalyzed reaction, the enzyme binds the substrate, which is converted to a product. (b) Competitive inhibitors bind to the active site of the enzyme and compete with the substrate for binding. (c) Non-competitive inhibitors bind to a site that is different from the active site. They do not block substrate from binding, but they change the shape of the enzyme and block the conversion of substrate to product.

inhibitor can be overcome by increasing the concentration of substrate because they both compete for binding to the same site.

Non-competitive inhibitors do not compete with the substrate for binding to the active site. They often bind to a site other than the active site of the enzyme, called the **allosteric site,** but still inhibit the activity of the enzyme. In the case illustrated in Figure 14.12c, binding of the inhibitor changes the shape and therefore the activity of the enzyme, so that it is still able to bind substrate, but it cannot convert it to product. This type of inhibitor usually has a structure very different from that of the substrate. Furthermore, this type of inhibitor cannot be overcome by increasing the concentration of the substrate.

An inhibitor that binds to a site other than the active site is known as an allosteric inhibitor. An example of an allosteric inhibitor can be seen in the synthesis of isoleucine from threonine, a metabolic pathway found in some bacteria. This conversion requires five reactions, each catalyzed by a

different enzyme and pictured in **FIGURE 14.13**. Once the bacterium has enough isoleucine for its needs, it would be a waste of energy to continue synthesizing the amino acid. To shut down the pathway once it is no longer needed, the cell relies on a non-competitive enzyme inhibitor. The inhibitor is isoleucine, the final product of the five reactions. Isoleucine binds to the first enzyme in the pathway, threonine dehydratase, at a site distinct from the active site. The binding of isoleucine changes the shape of the enzyme and in this way inhibits its function.

The isoleucine pathway provides an example of a common mechanism used widely in organisms to maintain homeostasis—that is, to actively maintain stable conditions or steady levels of a substance. In this case, the final product of a reaction inhibits the first step of the reaction. As a result, a relatively steady amount of product is maintained.

Enzymes that are regulated by molecules that bind at sites other than their active sites are called allosteric enzymes. The activity of allosteric enzymes can be influenced by both inhibitors and activators. They play a key role in the regulation of metabolic pathways. In the next two modules, we examine more closely two key metabolic processes, photosynthesis and cellular respiration. Both of these processes require many chemical reactions acting in a coordinated

FIGURE 14.13 Regulation of threonine dehydratase, an allosteric enzyme

Threonine is converted to isoleucine by a series of chemical reactions. The first reaction is catalyzed by the enzyme threonine dehydratase. This enzyme is inhibited by the final product, isoleucine, which binds to a site that is different from the active site.

Threonine → Threonine dehydratase → P1 → P2 → P3 → P4 → Isoleucine

Allosteric inhibitor

fashion. Allosteric enzymes catalyze key reactions in these and other metabolic pathways. These enzymes are usually found at or near the start of a metabolic pathway or at the crossroads between two metabolic pathways. Allosteric enzymes are one way that the cell coordinates the activity of multiple metabolic pathways.

✓ **Concept Check**

5. **Describe** how temperature affects the activity of an enzyme.

6. **Describe** how activators and inhibitors affect enzyme function.

Module 14 Summary

PREP FOR THE AP® EXAM

REVISIT THE BIG IDEAS

ENERGETICS: Use the content of this module to describe the structural relationship between enzymes and substrates. Using an example from the text, describe how the functional relationship between enzyme and substrate may change if the enzyme is exposed to extreme high or low temperatures.

LG 14.1 Enzymes increase the rates of chemical reactions.

- Enzymes are proteins that act as biological catalysts. Page 197

- Enzymes reduce the free energy level of the transition state between reactants and products, thereby reducing the energy input, or activation energy, required for a chemical reaction to proceed. Page 199

LG 14.2 Enzymes bind reactants at active sites, which are highly specific.

- During catalysis, the substrate and product form a complex with the enzyme. Page 200

- Transient covalent bonds, weak noncovalent interactions, or both stabilize the complex. Page 202

- The size of the active site of an enzyme is small compared to the size of the enzyme as a whole. Page 202

- The active site amino acids occupy a very specific spatial arrangement. Page 202

- An enzyme is highly specific for its substrate and for the types of reaction it catalyzes. Page 202

LG 14.3 Enzymes are influenced by the environment, activators, and inhibitors.

- Increasing temperature increases the activity of enzymes up to a point, as molecules move about more rapidly. Page 204

- High temperature can unfold, or denature, enzymes, reducing their activity. Page 204

- Enzymes function optimally in a narrow range of pH. Page 205

- Activators increase the activity of enzymes. Page 205

- Inhibitors reduce the activity of enzymes. Page 205

- Allosteric enzymes bind activators and inhibitors at sites other than the active site, resulting in a change in their shape and activity. Page 206

- Allosteric enzymes are often found at or near the start of a metabolic pathway or at the crossroads of multiple pathways. Page 206

Key Terms

Enzyme	Substrate	Inhibitor
Catalyst	Active site	Allosteric site
Transition state	Denature	
Activation energy (E_A)	Activator	

Review Questions

1. The energy of activation of a reaction is:
   (A) the net change in free energy.
   (B) the difference in energy between substrate and product.
   (C) the energy input needed to reach the transition state.
   (D) the difference in energy between the transition state and the product.

2. In a reaction, enzymes change the
   (A) ΔG.
   (B) laws of thermodynamics.
   (C) equilibrium of the reaction.
   (D) activation energy.

3. Enzymes are typically
   (A) highly specific.
   (B) most active at $0°C$.
   (C) most active at $65°C$.
   (D) polysaccharides.

4. Allosteric inhibitors of an enzyme bind to
   (A) the substrate.
   (B) the active site of the enzyme.
   (C) a site on the enzyme that is not the active site.
   (D) a site on the substrate that is not the active site.

5. Which statement is true regarding enzymes?
   (A) Enzymes are relatively unselective and can recognize several substrates.
   (B) Enzymes only form complexes with reactants and never with products.
   (C) Enzymes can either decrease or increase the rate of a chemical reaction.
   (D) Enzymes catalyze chemical reactions.

6. In a given reaction, which has the highest amount of free energy?
   (A) Products
   (B) Reactants
   (C) The transition state when the reaction is catalyzed by an enzyme
   (D) The transition state when the reaction is not catalyzed by an enzyme

7. Most enzymes work best at
   (A) very high temperatures.
   (B) a narrow range of pH.
   (C) very low temperatures.
   (D) a wide range of pH.

Module 14
AP® Practice Questions

Section 1: Multiple-Choice Questions
Choose the best answer for questions 1–4.

1. Which statement best describes the role of enzymes in a cell?
   (A) Enzymes supply energy to drive chemical reactions.
   (B) Enzymes drive chemical reactions in the direction opposite to which they would normally go.
   (C) Enzymes stabilize the products of chemical reactions.
   (D) Enzymes make chemical reactions happen more quickly.

Use the following schematic to answer question 2.

$$S \xrightarrow{E} P$$

Enzyme E catalyzes the conversion of substrate S to product P as shown in the schematic above.

2. A scientist discovered a molecule, I, that reduces the rate at which E converts S to P. The scientist added both E and I to test tubes that contained different amounts of S. As a control, the scientist added only E to a set of test tubes that contained the same amounts of S. The scientist measured the amount of P produced after 4 hours and obtained the graphs below.

a. Experimental treatment: E and I added

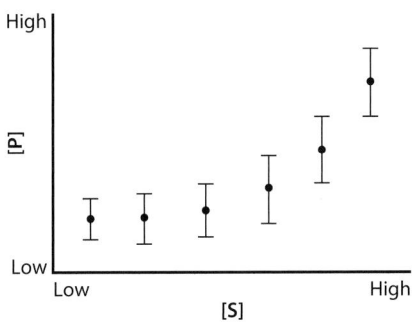

b. Control: only E added

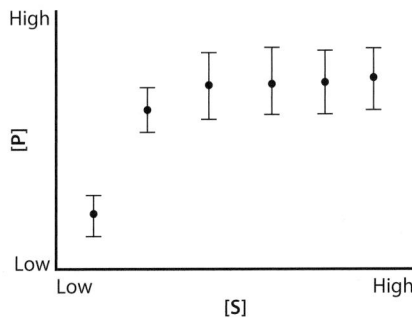

Which statement best describes I?

(A) I probably binds irreversibly to E.

(B) I probably binds to the active site of E.

(C) I probably increases the stability of E.

(D) I probably binds to P.

Questions 3 and 4 refer to the following information.

Glucose-6-phosphate dehydrogenase (G6PD) is an enzyme that converts a substrate, glucose-6-phosphate, to a product, 6-phosphogluconolactone. A scientist added 0.1 nanomole (nmol) of G6PD to a test tube containing 25 nmol of glucose-6-phosphate (the substrate) and measured the accumulation of 6-phosphogluconolactone (the product) over 7 hours. The data that the scientist obtained are shown in the graph.

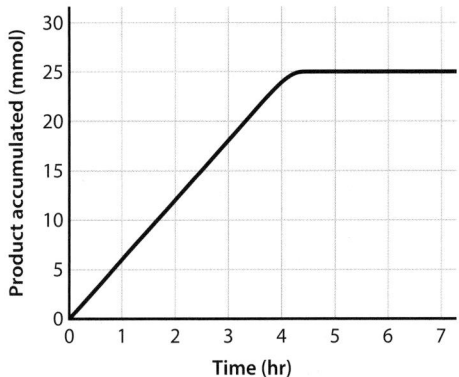

3. Based on the data provided, what is the best estimate of the rate at which G6PD converts substrate to product in this experiment?

(A) 0.1 mmol/min

(B) 2.0 mmol/min

(C) 3.6 mmol/min

(D) 5.0 mmol/min

4. Which provides the best explanation for why no further product accumulates after 5 hours?

(A) All of the G6PD enzyme has been consumed and is no longer able to catalyze the reaction.

(B) The reaction releases energy as heat, which causes the G6PD enzyme to become denatured.

(C) All of the substrate has been consumed, so the reaction can no longer proceed.

(D) The concentration of substrate becomes too high and inhibits the reaction.

Section 2: Free-Response Question

Pepsin is a digestive enzyme in animal stomachs that helps to digest, or break down, protein in the diet. A researcher tested the ability of 1 mg of pepsin to digest the protein albumin across a range of pH values. The researcher mixed six solutions, each with a different pH level, listed in the table below. Each solution was divided among 10 tubes. The researcher then added 1 mg of pepsin and 3000 mg of albumin to each tube. At the end of a 3-hour incubation period, the researcher measured the amount of albumin that had been digested in each tube. The mean amount of albumin digested at each pH and the standard error of the mean are shown in the table below.

pH	Mean albumin digested (mg)	Standard error (mg)
1	617	180
2	3000	0
3	2089	120
4	2103	170
5	1146	130
6	132	60

(a) **Draw** an appropriately labeled graph of pepsin activity across pH values, including error bars.

(b) **Identify** the pH at which pepsin activity is highest.

(c) Do you see evidence to support a claim that pepsin activity is different at pH = 3 and pH = 4? **Make a claim** and **justify** your reasoning based on the data provided.

(d) **Calculate** the percent activity of pepsin at pH = 5, relative to the maximum activity observed in the experiment. You may ignore the standard error for this calculation.

(e) **Make a claim** that would explain the observed activity of pepsin at pH = 6.

Module 15

Photosynthesis I: Overview

LEARNING GOALS ▶**LG 15.1** Photosynthesis is the major entry point for energy into biological systems.
▶**LG 15.2** Photosynthesis is a redox reaction.
▶**LG 15.3** Photosynthesis takes place in two main stages.

We have examined what energy is and how it is transformed in cells. Cells build molecules and break them down using chemical reactions. In the process, they capture and release energy. Chemical reactions form long metabolic pathways, in which the products of one reaction are the reactants for the next. These reactions are regulated by enzymes, protein catalysts that increase the rate of reactions. By influencing key steps in metabolic pathways, enzymes ultimately control which reactions occur in a cell out of the many that could potentially occur.

An essential feature of cells is the ability to harness energy from the environment. All cells obtain energy from just two sources—the sun and chemical compounds. In this module, we will focus on how some cells transform energy from sunlight into chemical energy by photosynthesis. Photosynthesis is one of the major anabolic reactions in cells. During photosynthesis, the energy of sunlight is transformed into chemical energy in the bonds of carbohydrates. Recall that anabolic reactions build molecules and require energy, whereas catabolic reactions break down molecules and release energy.

Photosynthesis involves many reactions taking place in different parts of the cell. We will discuss these reactions in this module and the next. In this module, we will provide an overview of photosynthesis. We will start with the overall chemical formula for photosynthesis, focusing on the major reactants and products. We will then consider what types of organisms carry out photosynthesis and where they live. Finally, we will discuss the two major stages of the process, the first in which light energy is captured into chemical forms and the second in which this chemical energy is used to build carbohydrates.

In the next module, we will build on the foundation we set up in this module and discuss the biochemical details of how light energy is captured and how carbohydrates are synthesized. By the end of these two modules, you will have a solid understanding of the biochemistry, cell biology, and ecology of this critical anabolic pathway.

PREP FOR THE AP® EXAM

FOCUS ON THE BIG IDEAS

ENERGETICS: Focus on the biochemical reactions of photosynthesis, being sure to note at which stage the reactants enter the pathway and at which stage the products are released.

15.1 Photosynthesis is the major entry point for energy into biological systems

Walk through a forest and you will be struck, literally if you aren't careful, by the substantial nature of trees. Where does the material to construct these massive organisms come from? Because trees grow upward from a firm base in the ground, a reasonable first guess is the soil. In the first recorded experiment on this question, the Flemish chemist and physiologist Jan Baptist van Helmont (1580–1644) found that the 200 pounds of dry soil into which he had planted a small willow tree decreased by only 2 ounces over a 5-year period. During this same period, the tree gained 164 pounds. Van Helmont concluded that water must be responsible for the tree's growth. He was, in fact, half right: a tree is roughly half liquid water. However, what he missed completely is that the other half of his tree had been created almost entirely out of thin air.

The process that allowed Van Helmont's tree to increase in mass using molecules pulled from the air is called photosynthesis. Photosynthesis is a biochemical process for building carbohydrates using energy from sunlight and carbon dioxide (CO_2) taken from the air. Specifically, carbon dioxide (CO_2)

and water (H_2O) react to form the carbohydrate glucose ($C_6H_{12}O_6$), with oxygen (O_2) given off as a by-product. We can write the overall formula for photosynthesis as follows:

$$Energy + 6CO_2 + 6H_2O \rightarrow C_6H_{12}O_6 + 6O_2$$

Carbon   Water   Glucose   Oxygen
dioxide

The carbohydrates that are produced are used both as a means of storing energy in chemical bonds which can later be broken down for use by the cell, and as starting points for the synthesis of other molecules.

Photosynthesis is the major entry point for energy into cells, from the energy in sunlight to the energy in chemical bonds. It is the source of all of the food we eat, both directly when we consume plant material, and indirectly when we consume meat. It is also the source of much of the fuel we use for heating and transportation. Fossil fuels are the legacy of ancient photosynthesis: oil has its origin in the bodies of marine algae and the organisms that grazed on them, whereas coal represents the geologic remains of land plants. Thus, photosynthesis plays an essential and major role for life on Earth.

Photosynthesis is also the source of all the oxygen that we breathe. In fact, the Earth's atmosphere contained little to no oxygen gas before the evolution of photosynthesis. The first organisms to evolve the biochemical machinery for photosynthesis in which oxygen is produced as a by-product were a group of photosynthetic bacteria called cyanobacteria.

Photosynthesis occurs almost everywhere. We are most familiar with photosynthesis that takes place on land by plants. Grasses, shrubs, and trees, like the one shown in **FIGURE 15.1**, are all examples of photosynthetic organisms.

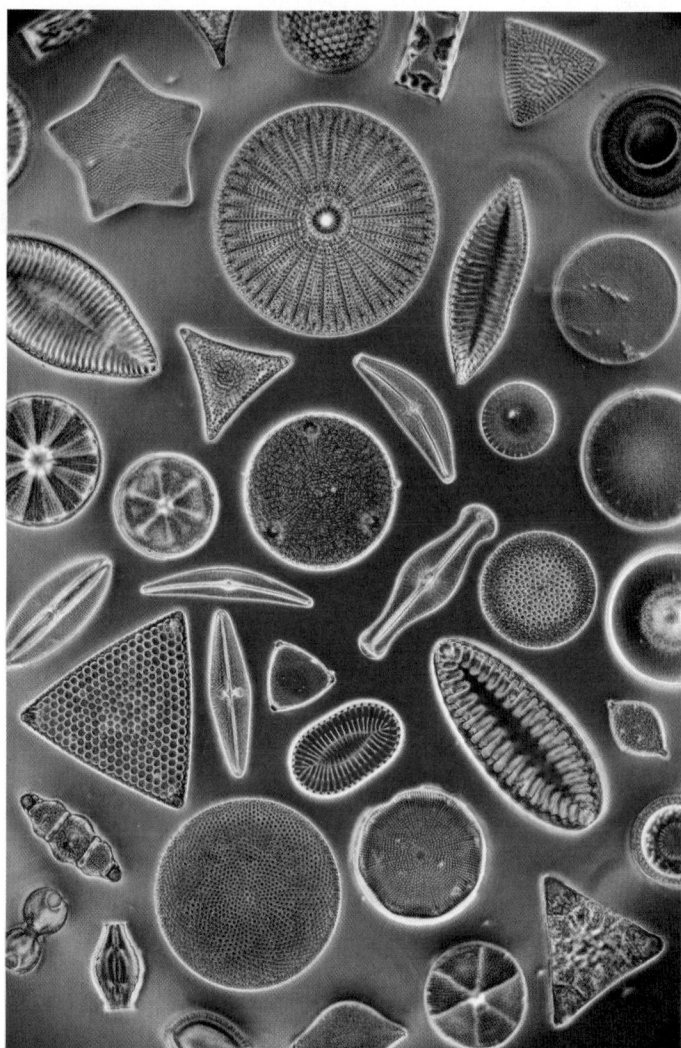

FIGURE 15.2 Photosynthesis by phytoplankton in the ocean

The majority of photosynthesis in the ocean is carried out by phytoplankton, which includes prokaryotes and unicellular eukaryotes.
Photo: Scenics & Science/Alamy Stock Photo

FIGURE 15.1 Photosynthesis by plants on land

On land, photosynthesis is carried out by plants, including trees, shrubs, and grasses. Photo: ooyoo/Getty Images

However, photosynthesis is carried out in the ocean as well as on land. Approximately half of global photosynthesis is carried out on land, with the remaining half taking place in the ocean. The majority of photosynthesis in marine environments is carried out by a diverse group of organisms collectively called phytoplankton, shown in **FIGURE 15.2**. Some of these are prokaryotes, such as cyanobacteria, and some are unicellular eukaryotes.

Photosynthesis is also carried out in the oceans by multicellular algae, which are informally called seaweeds and

FIGURE 15.3 Photosynthesis by multicellular algae in the ocean

Some photosynthesis in the ocean is carried out by multicellular algae, which are informally called seaweeds. Photo: Pete Niesen/Shutterstock

a.

b.

are pictured in **FIGURE 15.3**. In summary, photosynthesis occurs widely across the planet—on land and in the sea, by prokaryotes and eukaryotes, and by unicellular and multicellular organisms.

Photosynthesis takes place almost everywhere sunlight is available to serve as a source of energy. In the ocean, photosynthesis occurs in the surface layer extending to about 100 m deep, called the photic zone, through which enough solar energy penetrates to enable photosynthesis. On land, photosynthesis occurs most readily in environments that are both moist and warm. Tropical rain forests have high photosynthetic productivity, as do grasslands and forests in the temperate zone.

In addition, photosynthetic organisms have evolved adaptations that allow them to tolerate a wide range of environmental conditions, like those illustrated in **FIGURE 15.4**. In very dry regions, a combination of photosynthetic bacteria (a prokaryote) and unicellular algae (a eukaryote) forms an easily disturbed layer on the surface of the soil known as desert crust. At the other extreme, unicellular algae can grow on the surfaces of glaciers, causing the snow to appear red.

✓ Concept Check

1. **Describe** the overall chemical reaction for photosynthesis in words and chemical formulas.

2. **Identify** three different types of organisms that carry out photosynthesis.

FIGURE 15.4 Photosynthesis by bacteria and algae

Photosynthesis is carried out by bacteria and algae, sometimes in extreme environments. (a) This desert crust in the Colorado Plateau is formed by photosynthetic bacteria and algae. (b) The surface of this permanent snowpack in Colorado is red because of photosynthetic algae that thrive in cold conditions. Photos: (a) NPS Photo by Neal Herbert; (b) Shattil & Rozinski/Naturepl.com

15.2 Photosynthesis is a redox reaction

Carbohydrates are synthesized from carbon dioxide (CO_2) and water (H_2O) molecules during photosynthesis. Carbohydrates have more energy stored in their chemical bonds than is contained in the bonds of CO_2 or H_2O molecules. Therefore, to build carbohydrates using CO_2 and H_2O requires an input of energy. This energy comes from sunlight.

The process by which energy from sunlight is incorporated into chemical bonds and then transferred from one molecule to the next involves two types of chemical reactions. **Reduction** reactions are reactions in which a molecule gains both electrons and energy. **Oxidation** reactions are reactions in which a molecule loses electrons and releases energy.

The gain and loss of electrons always occur together: electrons are transferred from one molecule to another so that one molecule gains electrons and one molecule loses those same electrons. The molecule that gains electrons is reduced, and the molecule that loses electrons is oxidized. Although it may sound counterintuitive that the molecule that gains electrons is reduced, the term "reduction" was first coined because some reduction reactions involve a loss of an oxygen atom. At that time, scientists believed that reduction was the loss of an oxygen atom and oxidation was the gain of an oxygen atom. Eventually, scientists came to understand that the key to these reactions was the transfer of electrons, not oxygen. However, the names remain. Because reduction and oxidation reactions always occur together, they are known as oxidation—reduction reactions, or redox reactions for short. They are an example of coupled reactions, which we discussed in Module 13.

Let's take a simple example by considering a molecule called an electron carrier. Electron carriers do just that—they gain and lose electrons and therefore transfer electrons and energy in cells. One electron carrier is **nicotinamide adenine dinucleotide phosphate (NADPH).** NADPH exists in two forms—an oxidized form ($NADP^+$) and a reduced form (NADPH). $NADP^+$ can accept electrons and be converted to its reduced form, NADPH, as follows:

$$NADP^+ + 2e^- + H^+ \rightarrow NADPH$$

NADPH can also release electrons and be converted to its oxidized form $NADP^+$ as follows:

$$NADPH \rightarrow NADP^+ + 2e^- + H^+$$

In redox reactions involving organic molecules such as NADPH, the gain (or loss) of electrons is often accompanied by the gain (or loss) of protons (H^+). Therefore, you can easily recognize reduced molecules by an increase in H covalent bonds, and the corresponding oxidized molecules by a decrease in H covalent bonds: NADPH is the reduced form, and $NADP^+$ is the oxidized form. Furthermore, as electron carriers gain and lose electrons, they also gain and lose energy: NADPH has more chemical energy in its bonds than $NADP^+$. Therefore, this electron carrier can pick up electrons and energy, and release electrons and energy in a cell.

The reaction for photosynthesis can itself be understood as a redox reaction, even though it consists of many steps and electrons do not show up in the overall formula. In photosynthesis, water (H_2O) is oxidized to oxygen (O_2), and at the same time, carbon dioxide (CO_2) is reduced to the carbohydrate glucose ($C_6H_{12}O_6$):

$$\text{Energy} + 6\,CO_2 + 6\,H_2O \longrightarrow C_6H_{12}O_6 + 6\,O_2$$

When considering the overall reaction, we will not see the complete gain or loss of electrons between molecules as we saw in the case of the electron carrier NADPH. Instead, we will see that electrons are partially gained or lost. Let's first consider the reduction reaction, where carbon dioxide is converted to glucose. In carbon dioxide, the oxygen atom is more electronegative than the carbon atom, so the electrons are more likely to be found near the oxygen atom, as shown in **FIGURE 15.5**. The red shading in the figure shows the distribution of electrons. Note that the intensity of the red color is greater near the oxygen atom compared to the carbon atom, indicating that electrons spend more time near oxygen than they do around carbon. In glucose, there are many C—C and C—H covalent bonds, in which electrons are shared about equally between the two atoms, as indicated by the even red shading around the two carbon atoms. As a result, carbon has partially gained electrons and is reduced when carbon dioxide is converted to glucose.

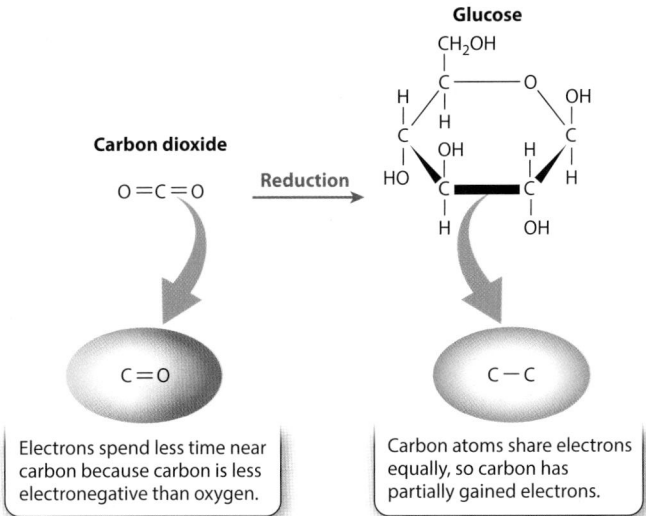

FIGURE 15.5 Reduction

Reduction involves the gain of electrons. When carbon dioxide is converted to glucose, carbon dioxide is reduced because the carbon atom in carbon dioxide partially gains electrons. Note that the red shading indicates electron density. The shading is lighter around the carbon atom in carbon dioxide than the shading around the carbon atom in glucose.

Now let's consider the oxidation reaction, where water is converted to oxygen. In water, the electrons that are shared between hydrogen and oxygen are more likely to be found near oxygen because oxygen is more electronegative than hydrogen, as shown by the red shading in **FIGURE 15.6**. In oxygen gas, electrons are shared equally between two oxygen atoms. As a result, oxygen has partially lost electrons and is oxidized when water is converted to oxygen.

During photosynthesis, lower-energy CO_2 molecules are reduced to form higher-energy carbohydrate molecules. We mentioned earlier that energy for this process comes from sunlight. Where do the electrons used to reduce CO_2 come from? In photosynthesis carried out by plants, many algae, and some bacteria, water is the source of the electrons. Because water is the electron donor, oxygen gas is produced as a by-product of photosynthesis.

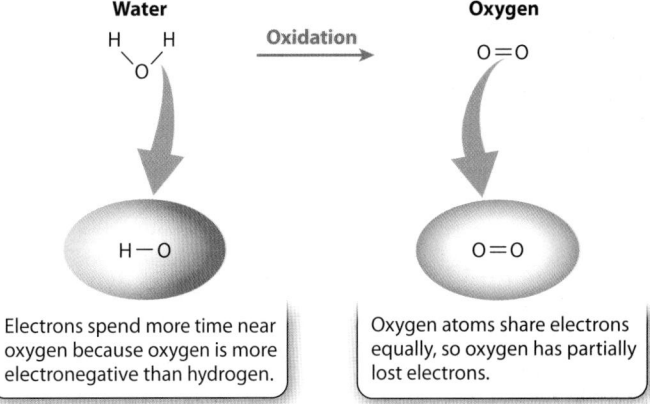

FIGURE 15.6 Oxidation

Oxidation involves the loss of electrons. When water is converted to oxygen, water is oxidized because the oxygen atom in the water molecule partially loses electrons. Note that there is more red shading, indicating electron density, around the oxygen atom in water compared to the oxygen atom in oxygen.

"Practicing Science 15.1: Does the oxygen released by photosynthesis come from H_2O or CO_2?" on page 216 describes an experiment that was designed to figure out whether water or carbon dioxide is the source of the oxygen released during photosynthesis. In this experiment, isotopes of oxygen were used. As we saw in Unit 1, isotopes are elements that have different numbers of neutrons and therefore can be distinguished on the basis of their mass. When a heavy form of oxygen was incorporated into water, this heavy form was found in the oxygen gas that was produced in photosynthesis. When a heavy form of oxygen was incorporated into carbon dioxide, this heavy form was not found in the oxygen gas that was produced in photosynthesis. Therefore, the oxygen produced in photosynthesis comes from water and not from carbon dioxide.

PREP FOR THE AP® EXAM

AP® EXAM TIP

You should be familiar with the overall equation that describes photosynthesis. Make sure you understand that water is oxidized to oxygen gas, and carbon dioxide is reduced to become a carbohydrate, typically noted as $C_6H_{12}O_6$, or glucose.

✓ Concept Check

3. **Identify** the molecule that is oxidized in photosynthesis.

4. **Identify** the molecule that is reduced in photosynthesis.

Does the oxygen gas released by photosynthesis come from H_2O or CO_2?

Background The reactants in photosynthesis are water (H_2O) and carbon dioxide (CO_2). Both of these molecules contain oxygen atoms (O), so it is unclear from the formula alone which molecule is the source of oxygen gas that is produced in photosynthesis. How did scientists determine that the source of oxygen is water and not carbon dioxide?

Method Researchers used isotopes to identify the source of oxygen in photosynthesis. Isotopes are the same elements that differ in the number of neutrons. Most of the oxygen in the atmosphere is ^{16}O, an isotope containing 8 protons and 8 neutrons. A small amount (0.2%) is ^{18}O, an isotope with 8 protons and 10 neutrons. The relative abundance of molecules containing ^{16}O versus ^{18}O can be measured using an instrument called a mass spectrometer.

Experiment As shown in the figure, water and carbon dioxide containing a high percentage of ^{18}O were prepared, and separate solutions of the algae *Chlorella* were exposed to each one. The algae were then allowed to carry out photosynthesis for 2 hours. Samples were taken at the start (initial) and end (final) of the experiment, and the amount of $^{18}O_2$ produced was measured.

Initial

This test tube has $H_2{}^{18}O$ and CO_2.

This test tube has H_2O and $C^{18}O_2$.

1 Researchers placed *Chlorella* into two test tubes.

Final

2 The tubes were exposed to sunshine for two hours–enough time for photosynthesis to occur.

3 The percentage of $^{18}O_2$ in each tube was measured.

Results The percentage of $^{18}O_2$ produced by photosynthesis in the algae increased when water contained a high percentage of ^{18}O, as shown in the graph on the left. By contrast, the percentage of $^{18}O_2$ did not increase when carbon dioxide contained a high percentage of ^{18}O, as shown in the graph on the right.

Conclusion These findings indicate that the oxygen gas produced during photosynthesis comes from water and not carbon dioxide.

Follow-up Work Scientists often use isotopes as a way to follow particular atoms through chemical reactions. Like oxygen, carbon also has several isotopes. The amount of carbon dioxide has been increasing in the atmosphere over at least the last 100 years. Where does this carbon dioxide come from? Scientists have measured the relative abundance of carbon isotopes to answer this question. They found that the increased carbon dioxide in the atmosphere comes from the burning of fossil fuels and not from other potential sources, such as volcanic activity.

SOURCE
Adapted from Ruben, S., M. Randall, M. Kamen, and J. L. Hyde. 1941. "Heavy Oxygen (O^{18}) as a Tracer in the Study of Photosynthesis." *Journal of the American Chemical Society* 63:877–879. Photo: Sinclair Stammers/ Science Source

AP® PRACTICE QUESTION

(a) **Describe** the following:
   1. The question being investigated by the experimenters
   2. The method used by the experimenters to answer the question
   3. A possible hypothesis that was utilized in this experiment

(b) Based on the graphs shown in the figure, **identify** the independent and dependent variables in this experiment.

(c) **Explain** the results of the experiment by **identifying** the researchers' claim and **justifying** the claim with the evidence and the reasoning researchers used (CER).

(d) Carbon dioxide, like oxygen, exists with several isotopes of carbon. They are $^{12}CO_2$, $^{13}CO_2$, and $^{14}CO_2$. In experiments related to the follow-up work done on sources of carbon dioxide, scientists used carbon isotopes to determine ingredients in the foods typically found in our grocery stores. In corn plants, $^{13}CO_2$ reacts more readily with water than any of the other carbon dioxide isotopes do. The use of $^{13}CO_2$ by corn in photosynthesis is unique among plants. Using claims, evidence, and reasoning (CER), **explain** how this information could be used to determine if the foods we purchase from grocery stores possess ingredients that originate from the process of photosynthesis in corn plants.

15.3 Photosynthesis takes place in two main stages

Up to this point, we have considered the overall reaction for photosynthesis. This reaction occurs in two main parts, or stages. In this section, we will provide an overview of these stages, reserving the biochemical details until Module 16. In this way, we can focus on the major inputs and outputs here, and then in the next module consider how these transformations actually occur.

The first stage consists of the **light reactions,** in which energy from sunlight is converted into chemical energy. This chemical energy is held in the bonds of ATP and NADPH, which temporarily store the energy for use in the second stage. The second stage is the **Calvin cycle,** in which this energy is used to convert an inorganic molecule (carbon dioxide) to an organic molecule (a carbohydrate). These two stages are pictured in **FIGURE 15.7.** The carbohydrates that are synthesized in photosynthesis are considered fuel molecules because they store energy. A bag of sugar on your kitchen shelf represents a large and stable source of energy. This sugar was produced by photosynthesis, and the energy that you extract when you eat it came from sunlight. Carbohydrates can also be used as a starting point for the synthesis of other organic molecules, such as proteins and lipids.

The light reactions begin with the absorption of sunlight by pigments. Pigments are molecules that are able to absorb light energy. The major pigment for photosynthesis is chlorophyll. The absorbed sunlight provides the energy that drives the entire process. Specifically, it initiates a series of redox reactions that are collectively called the **photosynthetic electron transport chain.** The photosynthetic electron transport chain is a series of redox reactions in which light energy absorbed by chlorophyll is used to power the movement of electrons.

In green plants, as well as some algae and bacteria, the electrons come from water, forming oxygen gas, and the final electron acceptor is $NADP^+$, forming NADPH. In addition, as electrons are transferred, so is energy. This energy is not only used to reduce $NADP^+$ to NADPH but also used to synthesize ATP. Collectively these steps are sometimes referred to as the light reactions or the light-dependent reactions. The light reactions are a series of chemical reactions that begin with absorption of light energy by chlorophyll, the movement of electrons along the photosynthetic electron transport chain, and the synthesis of NADPH and ATP. The major inputs of the light reactions are sunlight, $NADP^+$,

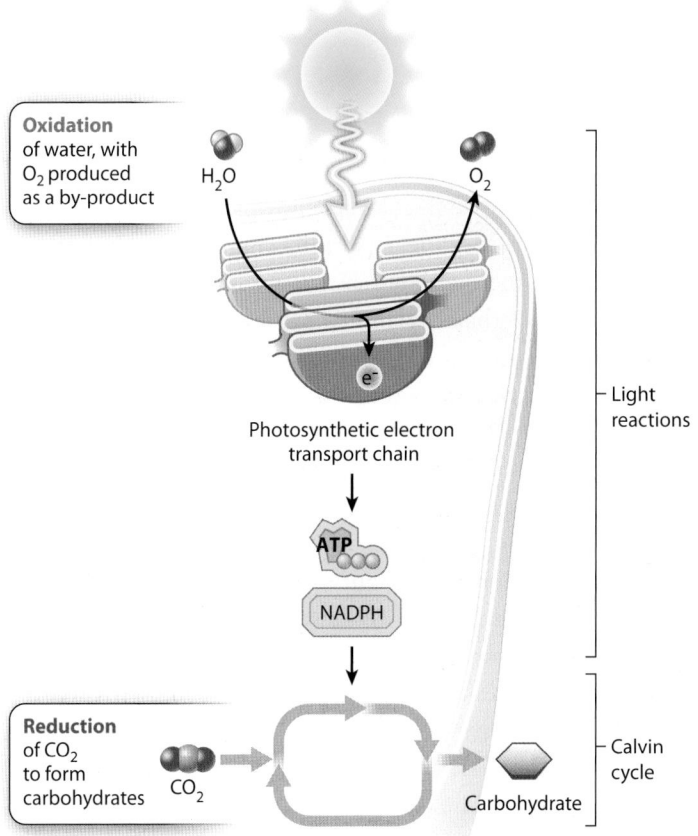

FIGURE 15.7 Two stages of photosynthesis

Photosynthesis occurs in two stages. In the first stage, light capture, energy from sunlight is converted into chemical energy in the bonds of ATP and NADPH. In the second stage, carbon fixation, this energy is used to convert carbon dioxide into carbohydrates.

ADP, and water, and the major outputs are NADPH, ATP, and oxygen gas.

The chemical energy that is held in the bonds of NADPH and ATP is then used to build carbohydrates from carbon dioxide. In other words, the light reactions and Calvin cycle are linked by the transfer of energy in the form of NADPH and ATP. The Calvin cycle occurs in many organisms, including plants, algae, and photosynthetic bacteria. In the Calvin cycle, carbohydrates are produced using CO_2 as the carbon source and energy supplied by NADPH and ATP from the light reactions. The major inputs of the Calvin cycle are NADPH, ATP, and carbon dioxide, and the major outputs are carbohydrates, as well as $NADP^+$ and ADP.

The Calvin cycle does not use sunlight directly and, for this reason, this pathway is sometimes referred to as the light-independent or even the dark reactions of photosynthesis. However, this pathway cannot operate without the energy input provided by a steady supply of NADPH and ATP produced in the light reactions. Thus, in a photosynthetic cell, the Calvin cycle occurs only in the light, when the products of the light reactions are produced.

In photosynthetic eukaryotes, both the light reactions and Calvin cycle take place in chloroplasts, which were introduced and described in Module 7. Within chloroplasts, there is a structure called the thylakoid, shown in **FIGURE 15.8**. Recall from Module 7 that thylakoids are highly folded structures that are made up of stacks of pancake-like grana. The photosynthetic electron transport chain consists of a series of protein complexes that are embedded in the thylakoid membrane. The complex and intricate folding of the thylakoid membrane greatly increases its surface area, allowing for the production of high levels of NADPH and ATP within these organelles during the light reactions.

Grana are connected to one another by membrane bridges in such a way that the thylakoid encloses an interconnected compartment called the thylakoid space. This space forms a single compartment, although this may not be apparent at first glance. The region surrounding the thylakoid is called the **stroma.** The Calvin cycle takes place in the stroma.

Recall that bacteria do not contain organelles like eukaryotes have, but some bacteria are capable of carrying out photosynthesis. In these bacteria, the photosynthetic electron transport chain is located in membranes within the cytoplasm or, in some cases, directly in the outer cell membrane, while carbon fixation takes place in the cytoplasm. In this and the next module, we focus on photosynthesis as carried out by eukaryotic cells, but remarkable diversity exists in the way it is carried out in bacteria.

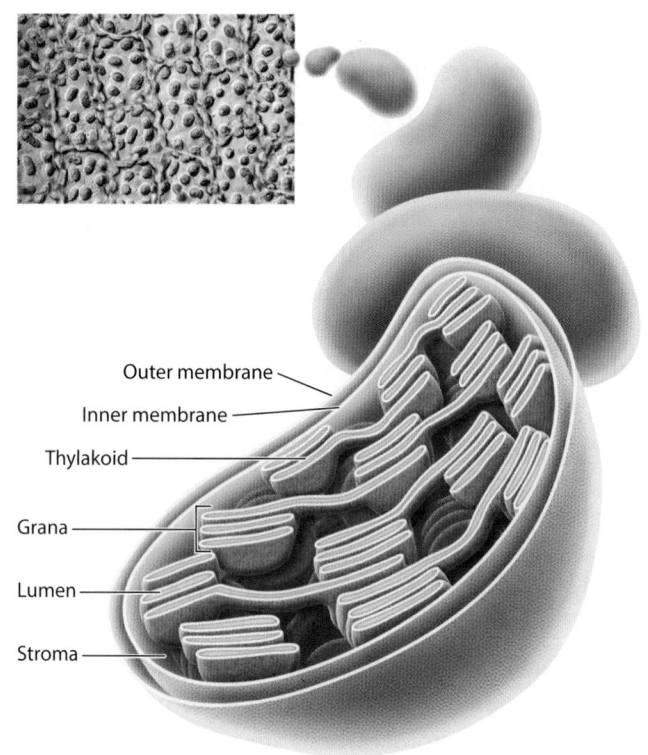

Outer membrane
Inner membrane
Thylakoid
Grana
Lumen
Stroma

FIGURE 15.8 Chloroplast

Chloroplasts contain highly folded structures called thylakoids. The photosynthetic electron transport chain is located in the thylakoid membrane. Source: Biophoto Associates/Science Source

✓ Concept Check

5. **Identify** the main products of the light reactions of photosynthesis.

6. **Identify** the main products of the Calvin cycle.

Module 15 Summary

PREP FOR THE AP® EXAM

REVISIT THE BIG IDEAS

ENERGETICS: Using the overall chemical reaction of photosynthesis described in this module, **identify** the stage in the reaction pathway that each reactant enters and the stage in the reaction pathway where each product is released.

LG 15.1 Photosynthesis is the major entry point for energy into biological systems.

- Photosynthesis is a set of biochemical reactions in which carbon dioxide and water react to form carbohydrates and oxygen gas. Page 212

- Photosynthesis is an anabolic process. Page 212

- Photosynthesis uses the energy of sunlight to synthesize carbohydrates, which are the major output of the process. Page 212

LG 15.2 Photosynthesis is a redox reaction.

- Oxidation–reduction, or redox reactions, transfer electrons and energy. Page 214

- Reduction reactions are chemical reactions in which molecules gain electrons and energy. Page 214

- Oxidation reactions are chemical reactions in which molecules lose electrons and energy. Page 214

- Oxidation and reduction reactions always occur together in a set of coupled reactions. Page 214

- In photosynthesis, water is oxidized, releasing oxygen, and carbon dioxide is reduced, forming carbohydrates. Page 215

LG 15.3 Photosynthesis takes place in two main stages.

- Photosynthesis consists of two sets of reactions. Page 217

- The light reactions begin with the absorption of light energy by the pigment chlorophyll. Page 217

- Chlorophyll hands off electrons to the photosynthetic electron transport chain, which consists of a series of protein complexes located in the thylakoid membrane. Page 217

- The photosynthetic electron transport chain receives electrons from water, which is oxidized to oxygen gas, and donates them to $NADP^+$, which is reduced to NADPH. Page 217

- The energy released by the photosynthetic electron transport chain is also used to synthesize ATP. Page 217

- The Calvin cycle is a set of reactions in which carbon dioxide is reduced to form carbohydrates, using the ATP and NADPH synthesized in the light reactions. Page 217

- The Calvin cycle takes place in the stroma of the chloroplast. Page 218

Key Terms

Reduction

Oxidation

Nicotinamide adenine dinucleotide phosphate (NADPH)

Photosynthetic electron transport chain

Light reaction

Calvin cycle

Stroma

Review Questions

1. Photosynthesis is the pathway used to synthesize carbohydrates from

   (A) sunlight.

   (B) carbon dioxide.

   (C) water.

   (D) sunlight, carbon dioxide, and water.

2. In plants and algae, what is the source of the electrons needed for photosynthesis?

   (A) O_2

   (B) CO_2

   (C) H_2O

   (D) NADPH

3. In plants and algae, which is a by-product of photosynthesis?

   (A) O_2

   (B) CO_2

   (C) H_2O

   (D) NADPH

4. During photosynthesis in plants and algae, what molecule is oxidized?

   (A) O_2

   (B) CO_2

   (C) H_2O

   (D) $C_6H_{12}O_6$

5. During photosynthesis in plants and algae, what molecule is reduced?

   (A) O_2

   (B) CO_2

   (C) H_2O

   (D) $C_6H_{12}O_6$

6. Where is the photosynthetic electron transport chain located in plant cells?

   (A) In the thylakoid space of the chloroplast

   (B) In the outer membrane of the chloroplast

   (C) In the stroma of the chloroplast

   (D) In the thylakoid membranes of the chloroplast

7. In the chloroplasts of plant cells, the synthesis of carbohydrates takes place in the
   (A) thylakoid space of the chloroplast.
   (B) outer membrane of the chloroplast.
   (C) stroma of the chloroplast.
   (D) thylakoid membranes of the chloroplast.

Module 15
AP® Practice Questions

PREP FOR THE AP® EXAM

Section 1: Multiple-Choice Questions

Choose the best answer for questions 1–3.

1. Identify the source from which plants obtain the molecules that make up most of their biomass.
   (A) Air
   (B) Soil
   (C) Oxygen gas
   (D) Light

2. Which statement correctly describes the production of carbohydrates by photosynthesis?
   (A) Photosynthetic production of carbohydrates produces energy.
   (B) Photosynthetic production of carbohydrates requires energy input.
   (C) Photosynthetic production of carbohydrates converts potential energy to kinetic energy.
   (D) Photosynthetic production of carbohydrates converts chemical energy to kinetic energy.

3. Which statement correctly explains the role of oxidation–reduction reactions in the transfer of energy?
   (A) When a molecule gains an electron, it is reduced because it goes to a lower energy state.
   (B) A molecule becomes oxidized when it gains an electron and energy from O_2.
   (C) Reduction and oxidation reactions are coupled so they always occur together.
   (D) Reduction reactions result in loss of energy and therefore lower entropy within a cell.

Section 2: Free-Response Question

Write your answer to each part clearly. Support your answers with relevant information and examples. Where calculations are required, show your work.

Chloridazon is an herbicide that can be used to prevent growth of many types of weeds. A scientist wanted to determine how chloridazon prevents weed growth. The scientist grew seedlings of a weed called black nightshade. The scientist applied chloridazon to half of the seedlings and left the other half untreated. The scientist then measured the rate of O_2 production and the abundance of NADPH in chloroplasts that were isolated from the black nightshade leaves. The scientist repeated the experiment both with the chloroplasts exposed to light and with the chloroplasts kept in the dark. The scientist obtained the data shown in the figure below.

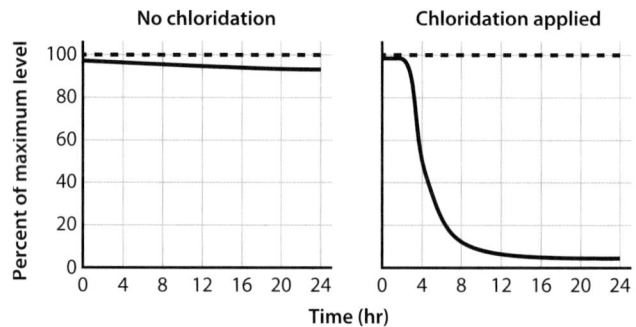

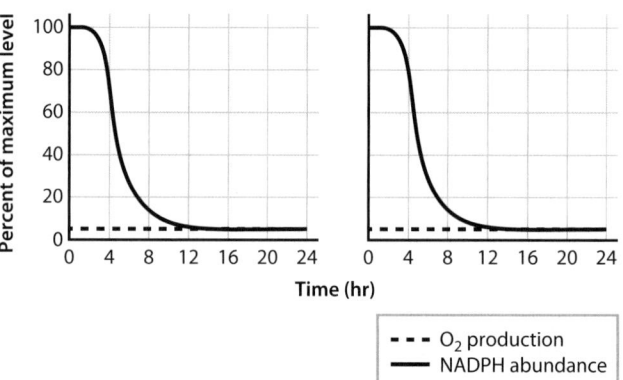

(a) **Identify** the treatment that has the highest level of NADPH at 12 hours.

(b) **Identify** the source of O_2 that is produced.

(c) Based on the data provided, **identify** the photosynthetic process that is probably disrupted by chloridazon.

(d) **Make a claim** that would explain why NADPH levels decline after chloridazon is applied.

Module 16

Photosynthesis II: Biochemistry

LEARNING GOALS ▶**LG 16.1** The light reactions use sunlight to produce ATP and NADPH.
▶**LG 16.2** The Calvin cycle uses ATP and NADPH to build carbohydrates.

In the last module, we took a broad view of photosynthesis. We learned that photosynthesis is the major entry point for energy into cells. Specifically, it is a series of biochemical reactions that transforms the energy from sunlight into the energy in chemical bonds. In photosynthesis, carbon dioxide and water react to form sugar molecules that the cell uses to make glucose, with oxygen given off as a by-product. Glucose has a large amount of chemical energy in its bonds. As a result, it can be thought of as a fuel molecule that stores chemical energy, which can be harnessed to power the work of the cell. In addition, glucose is the starting point for the synthesis of other molecules that cells use to carry out their functions. The oxygen that is produced as a by-product of photosynthesis is the source of all of the oxygen that we and many other organisms breathe. Photosynthesis is therefore critical for life on Earth as we know it.

We also learned that the overall reaction for photosynthesis is a redox reaction. Redox reactions are coupled reduction and oxidation reactions. In reduction reactions, electrons and energy are gained, and in oxidation reactions, electrons and energy are lost. In this module, we will see that photosynthesis takes place in many steps or reactions, and we will examine these reactions in more detail.

The reactions that make up photosynthesis can be grouped into two major stages: the light reactions and the Calvin cycle. We begin by examining the light reactions. These reactions use the energy of sunlight to produce ATP and NADPH, both of which carry chemical energy in their bonds. These molecules, in turn, are used to synthesize carbohydrates in the Calvin cycle, which we will also discuss.

PREP FOR THE AP® EXAM

FOCUS ON THE BIG IDEAS

ENERGETICS: Look for the connection between the light reactions, which yield the energy-rich molecules ATP and NADPH, and the Calvin cycle, which requires energy to synthesize carbohydrates.

16.1 The light reactions use sunlight to produce ATP and NADPH

As we saw in Figure 15.7 on page 217, photosynthesis occurs in two stages: the light reactions and the Calvin cycle. In the light reactions, energy from sunlight is transformed into chemical energy in the bonds of ATP and NADPH. In this section, we will examine how pigments absorb light energy, how this light energy drives the flow of electrons through the photosynthetic electron transport chain, and how the flow of electrons in turn leads to the formation of both NADPH and ATP.

Light Capture by Chlorophyll

To understand how energy from sunlight is captured and stored by photosynthesis, we need to know a little about light. The sun, like all stars, produces a broad spectrum of electromagnetic radiation, ranging from gamma rays to radio waves. Each point along the electromagnetic spectrum has a different energy level and a corresponding wavelength. **Visible light** is the portion of the electromagnetic spectrum apparent to our eyes, as shown in **FIGURE 16.1** on page 222. The wavelengths of visible light range from 400 to 700 nanometers (nm), which includes the range of wavelengths used in photosynthesis. Approximately 40% of the sun's energy that reaches Earth's surface is in this range.

Pigments are molecules that absorb some wavelengths of visible light. Pigments look colored because they reflect light enriched in the wavelengths that they do not absorb. The graph in **FIGURE 16.2** on page 222 shows an absorption spectrum, which indicates the extent to which wavelengths of visible light are absorbed by pigments in an intact leaf. As you can see in the graph, leaves are good at absorbing

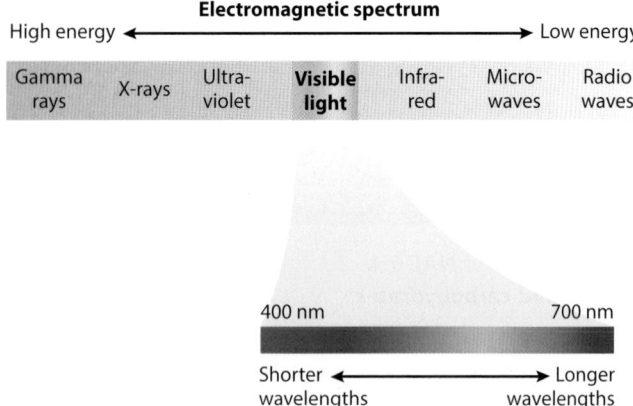

FIGURE 16.1 The electromagnetic spectrum

Visible light is only one part of the full electromagnetic spectrum, with wavelengths from 400 to 700 nm.

blue and red wavelengths, which are at the two ends of the visible light spectrum. However, they are poor at absorbing green wavelengths, which fall in the middle of the visible light spectrum. Leaves therefore appear green because they don't absorb green light, but instead reflect it back to us.

Chlorophyll, introduced in Module 7, is the major photosynthetic pigment. It is therefore the major entry point for light energy in photosynthesis. It consists of a large, light-absorbing head that contains a magnesium atom at its center and a long hydrocarbon tail. Chlorophyll is bound to integral membrane proteins in the thylakoid membrane of the chloroplast in eukaryotic cells. There are actually two types of chlorophyll molecules present in green plants, called

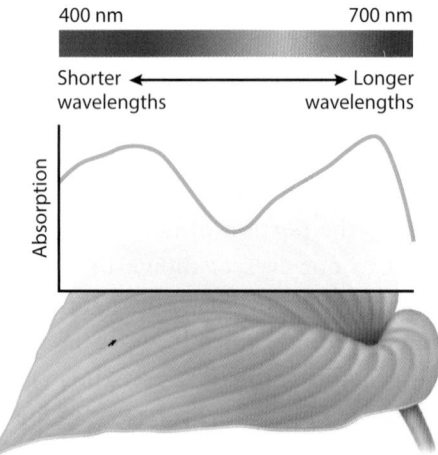

FIGURE 16.2 Light absorbed by leaves

The graph shows the extent to which wavelengths of visible light are absorbed by pigments in an intact leaf. Leaves efficiently absorb light across most of the visible spectrum. Note, however, that they absorb green wavelengths poorly, which explains why most leaves appear green.

chlorophyll *a* and *b*. They absorb slightly different wavelengths of light, as shown in FIGURE 16.3.

Chloroplasts also contain pigments other than chlorophyll, called accessory pigments. The most notable are the orange-yellow carotenoids, which absorb wavelengths of visible light that are poorly absorbed by chlorophyll, such as green, as shown in Figure 16.3. The presence of these accessory pigments allows photosynthetic cells to absorb a broader range of wavelengths than would be possible with chlorophyll alone. In some cases, they also provide protection against excess light energy, which we will explain below.

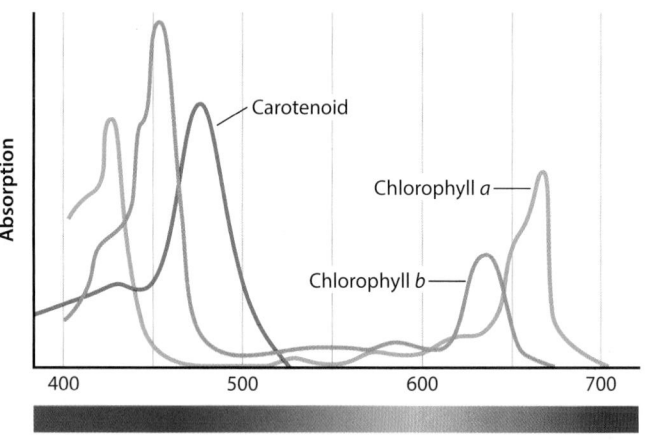

FIGURE 16.3 Absorption spectra of individual pigments

Leaves contain different pigments, each of which absorbs different wavelengths of light. This graph shows the absorption spectra of two types of chlorophyll, called chlorophyll *a* and *b*, and an accessory pigment called carotenoid.

Antenna Chlorophyll and Reaction Centers

When chlorophyll absorbs visible light, one of its electrons is excited, or elevated to a higher energy level. For chlorophyll molecules that have been extracted from chloroplasts in the laboratory, this absorbed light energy is rapidly released, allowing the electron to return to its initial, or ground, energy level. Most of the energy is converted into heat; a small amount is reemitted as light, as shown in FIGURE 16.4.

By contrast, chlorophyll within an intact chloroplast in a cell does not convert the energy to heat and light when electrons return to their ground level. Instead, it transfers energy to an adjacent chlorophyll molecule. FIGURE 16.5 shows this energy transfer. The top row shows a chlorophyll molecule absorbing light energy, raising its electron from a ground state to an excited state. Instead of emitting heat and light as it returns to the ground state, as in Figure 16.4, it transfers the

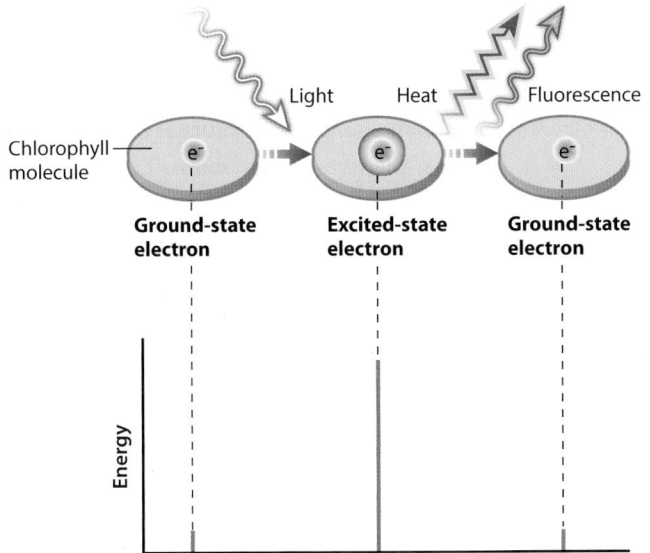

FIGURE 16.4 Absorption of light energy by chlorophyll in the lab

Absorption of visible light by an isolated chlorophyll molecule in solution in the lab results in an elevation of the energy level of an electron so that it is excited. This energy is converted to heat and fluorescence (light) when the electron returns to its ground, or initial, level.

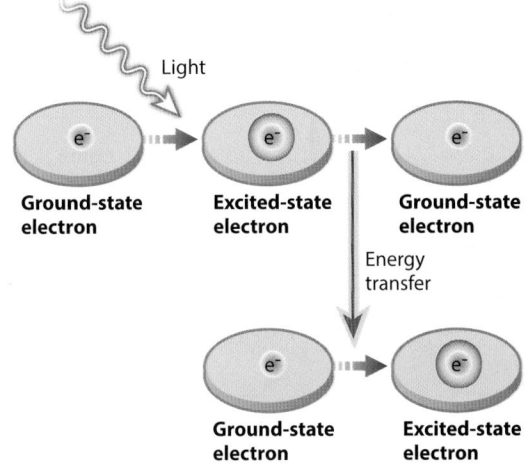

FIGURE 16.5 Absorption of light energy by chlorophyll in a plant cell

When light energy is absorbed by a chlorophyll molecule (top row), it excites an electron. The energy of this excited-state electron is transferred to a neighboring chlorophyll molecule (bottom row), so the electron of the neighboring chlorophyll molecule goes from a ground (initial) state to an excited (high-energy) state.

energy to a neighboring chlorophyll molecule. Energy is still released as the excited electron returns to its ground level, but the release of this energy raises the energy level of an electron in an adjacent chlorophyll. This mode of energy transfer is extremely efficient. In other words, very little energy is lost as heat, which allows energy that is initially absorbed from sunlight to be transferred from one chlorophyll to another and then to another. This chain of chlorophyll molecules is called antenna chlorophyll because it takes in energy and passes it along, just like an antenna.

Energy is transferred along antenna chlorophyll until it reaches the *reaction center*. The **reaction center** includes a pair of chlorophyll molecules that accept and lose electrons. When excited, the reaction center transfers an electron to an adjacent molecule that acts as an electron acceptor, shown in **FIGURE 16.6a**. When the transfer takes place, the reaction center is oxidized and the adjacent electron–acceptor molecule is reduced. Once the reaction center has lost an electron, it can no longer absorb light or contribute additional electrons. Thus, another electron must be delivered to take its place, as shown in Figure 16.6b. These electrons come from water (H_2O).

In many ways, water is an ideal source of electrons for photosynthesis. Water is so abundant within cells that it is always available to serve as an electron donor. In addition, O_2, the by-product of pulling electrons from water, diffuses readily away.

The Photosynthetic Electron Transport Chain

Up to this point, we have followed the energy from sunlight to antenna chlorophylls to reaction centers. The transfer of an

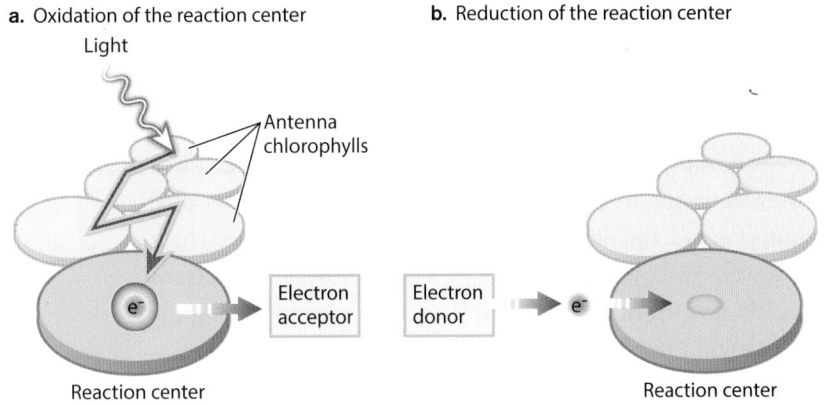

FIGURE 16.6 The reaction center

The reaction center includes chlorophyll molecules that accept and donate electrons. (a) Antenna chlorophylls deliver absorbed light energy to the reaction center, allowing an electron to be transferred to an electron acceptor. (b) After the reaction center has lost an electron, it must gain an electron from an electron donor before it can absorb additional light energy.

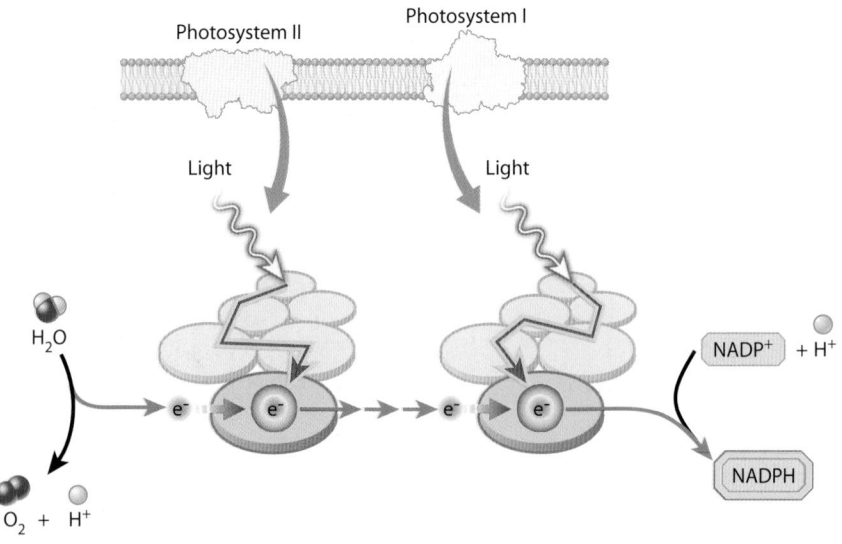

FIGURE 16.7 Two photosystems

Photosystems II and I are arranged in a series along the thylakoid membrane. Photosystem II accepts electrons from water, with oxygen given off as a by-product. Electrons are then passed to photosystem I, which transfers them to $NADP^+$, forming NADPH.

electron from the reaction center to an electron acceptor is the first in a series of redox reactions carried out by the photosynthetic electron transport chain, introduced in the previous module. In these redox reactions, light energy absorbed by chlorophyll is used to power the movement of electrons. This movement in turn leads to the production of NADPH and ATP. In green plants, some algae, and cyanobacteria, the electrons come from water and the final electron acceptor is $NADP^+$, forming NADPH.

The photosynthetic electron transport chain consists of protein–pigment complexes embedded in the thylakoid membrane in chloroplasts in eukaryotic cells. These protein–pigment complexes, referred to as **photosystems,** are the structural and functional units that absorb light energy and use it to drive electron transport. There are two photosystems, arranged in a series along the membrane. The two photosystems are called photosystem I and photosystem II, based on the order in which they were discovered and characterized. However, electrons actually flow from photosystem II to photosystem I along the electron transport chain. **FIGURE 16.7** shows these two photosystems in the thylakoid membrane. The energy captured by photosystem II allows electrons to be pulled from water, as shown on the left in the figure; the energy captured by photosystem I allows electrons to be transferred to $NADP^+$ to form NADPH, as shown on the right in the figure.

If you follow the flow of electrons from water through both photosystems and on to $NADP^+$, you can see a large increase in energy as the electrons pass through each of the

two photosystems, shown in the graph in **FIGURE 16.8**. This increase in energy comes from sunlight. You can also see an overall decrease in energy that occurs as electrons move between the two photosystems. The decrease in energy occurs as electrons are passed from one molecule to the next, and indicates that these are exergonic reactions, which we described in Module 13. Because the overall energy trajectory has an up-down-up configuration resembling a Z on its side, the photosynthetic electron transport chain is sometimes referred to as a Z scheme.

ATP Synthesis

Up to this point, we have followed the movement of electrons along the photosynthetic electron chain from photosystem II to photosystem I. We have seen how water donates electrons to one end of the photosynthetic electron transport chain. The enzyme that pulls electrons from water (H_2O), releasing both protons (H^+) and oxygen (O_2), is located in photosystem II. We have also seen how $NADP^+$ accepts electrons at the other end

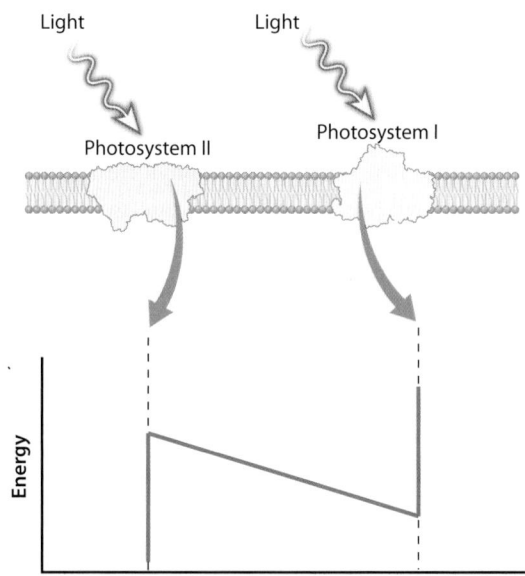

FIGURE 16.8 The Z scheme

The use of water as an electron donor requires input of light energy at two places in the photosynthetic electron transport chain. Absorption of light energy by photosystem II allows electrons pulled from water to enter the photosynthetic electron transport chain. A second input of light energy by photosystem I produces electron donor molecules capable of reducing $NADP^+$ to NADPH. As a result, energy levels increase, decrease, and increase again, resembling the letter Z on its side.

in photosystem I, forming NADPH. However, to build carbohydrates in the Calvin cycle, cells require the energy not only in NADPH but also in ATP. In this section, we will focus on how electron transport leads to the synthesis of ATP.

As electrons move along the photosynthetic electron transport chain, protons (H^+) accumulate inside the thylakoid, as shown in **FIGURE 16.9**. Two features of the photosynthetic electron transport chain are responsible for the buildup of protons in the thylakoid space. First, the oxidation of water (H_2O) by photosystem II releases protons (H^+) and oxygen (O_2) directly into the thylakoid space. Second, as electrons are transferred along the photosynthetic electron transport chain, protons are pumped from the outside (stroma) to the inside (thylakoid space) of the thylakoid. Together, these mechanisms are quite powerful.

Like all cell membranes, the thylakoid membrane is selectively permeable: protons cannot passively diffuse across this membrane, and the movement of other molecules is controlled by transport and channel proteins, as discussed in Module 10. We have just seen that the movement of electrons through the electron transport chain is coupled to the accumulation of protons in the thylakoid space. The result is a proton gradient, a difference in proton concentration across the thylakoid membrane. When the photosynthetic electron transport chain is operating at full capacity, the concentration of protons in the thylakoid space can be approximately

10,000 times greater than their concentration in the stroma. As a result, the thylakoid space can become quite acidic, with a pH of about 4, compared to a pH of about 8 in the stroma. Recall that the pH scale is logarithmic, so a 4-point difference in pH translates to a 10^4 (or 10,000)-fold difference in H^+ concentration.

The proton gradient has two components: a chemical gradient that results from the difference in concentration, and an electrical gradient that results from the difference in charge between the two sides of the membrane, as discussed in Module 10. To reflect these two components, the proton gradient is also called an electrochemical gradient.

The proton gradient is a source of potential energy, as discussed in Module 13. It stores energy in much the same way that a dam in a river does. There is a tendency for protons to diffuse from the thylakoid space to the stroma, driven by a difference in concentration and charge on the two sides of the thylakoid membrane. However, this movement is blocked by the thylakoid membrane. The gradient therefore stores potential energy.

In a river, water can be blocked by a dam, storing potential energy. If there is an opening in the dam, water will flow through the dam, and potential energy will be converted to kinetic energy. Similarly, the thylakoid membrane acts like a dam, preventing the flow of protons. An opening, or channel, is provided by an enzyme called **ATP synthase.** ATP synthase is a transmembrane protein that provides a channel

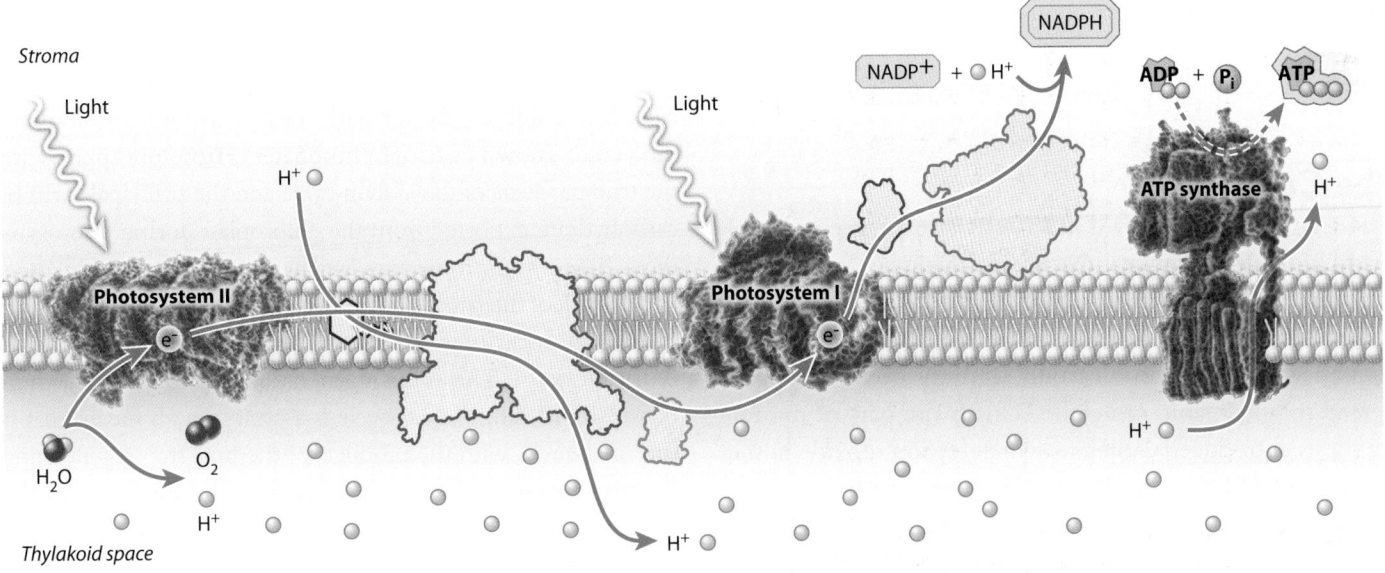

FIGURE 16.9 Electron transport, proton accumulation, and ATP synthesis

As electrons are passed along the photosynthetic electron transport chain, protons (H^+) accumulate in the thylakoid space due to the oxidation of water and the pumping of protons. These protons then pass through an enzyme called ATP synthase, which uses the energy stored in the proton gradient to synthesize ATP from ADP and P_i.

for protons to move down their electrochemical gradient from the thylakoid space to stroma, as shown in Figure 16.9. As they move down the gradient, ATP synthase harnesses the kinetic energy of proton movement, and uses it to synthesize ATP from ADP and inorganic phosphate (P_i). The process in which a phosphate is added to a molecule is called phosphorylation. ATP synthase uses the energy from sunlight to phosphorylate ADP and synthesize ATP, a process known as *photophosphorylation*. **Photophosphorylation** is the process by which phosphorylation is powered by the energy of sunlight.

In summary, the photosynthetic electron transport chain leads to the generation of a proton gradient. This gradient stores potential energy. As protons move through ATP synthase, potential energy is converted to kinetic energy. The kinetic energy in turn is harnessed to synthesize ATP. ATP holds chemical energy, a form of potential energy, in its chemical bonds. In turn, ATP, along with NADPH, is used by the cell to synthesize carbohydrates, which we will discuss next.

PREP FOR THE AP® EXAM

AP® EXAM TIP

Make sure you are able to describe how the movement of electrons in the photosynthetic electron transport chain pumps protons across the thylakoid membrane, and how the protons diffuse back through ATP synthase, which provides the energy needed to synthesize ATP.

✓ Concept Check

1. **Describe** the how antenna chlorophylls differ from reaction center chlorophylls.

2. **Describe** the role of the two photosystems in the photosynthetic electron transport chain.

3. **Describe** how the energy from sunlight is used to produce ATP.

4. **Identify** the products of the light reactions of photosynthesis.

16.2 The Calvin cycle uses ATP and NADPH to build carbohydrates

We have explored how light energy is used to synthesize ATP and NADPH. The chemical energy these molecules carry in their bonds is used in the Calvin cycle. We will now look at the Calvin cycle, which is a series of enzymatic reactions that synthesize carbohydrates from carbon dioxide (CO_2).

The Calvin Cycle

The Calvin cycle is shown in **FIGURE 16.10**. It consists of many individual chemical reactions, which can be grouped into three steps. In the first step, CO_2 is added to a 5-carbon molecule to make a 6-carbon molecule. This is the fixation step, in which inorganic carbon (carbon dioxide) is converted to an organic molecule. This is the part of photosynthesis, introduced earlier, in which carbon dioxide in the air is converted to an organic molecule in a cell.

This step is catalyzed by an enzyme called ribulose bisphosphate carboxylase oxygenase, or rubisco for short. The long name comes from the fact that enzymes that add carbon dioxide to another molecule are carboxylases, and the 5-carbon molecule to which carbon dioxide is added is ribulose 1,5-bisphophate (or RuBP). Rubisco is one of the most abundant enzymes on Earth. For this reaction to occur, CO_2 and RuBP must diffuse into the active site of the enzyme. Once the active site is occupied, the addition of CO_2 to RuBP proceeds spontaneously in the sense that no addition of energy is required. The resulting 6-carbon molecule immediately breaks into two 3-carbon molecules.

In the second step of the Calvin cycle, the 3-carbon molecules are converted into carbohydrates. This is the reduction step (step 2) shown in Figure 16.10. This step requires both ATP and NADPH produced in the light reactions discussed earlier. It results in the formation of 3-carbon carbohydrate molecules known as triose phosphates. Triose phosphates are the true products of the Calvin cycle and the principal form of carbohydrate exported from the chloroplast during photosynthesis. These sugar molecules are later converted into glucose.

Not all of the triose phosphates, however, are exported from the chloroplasts. Some are used to produce the 5-carbon sugar, RuBP, of step 1. This is the regeneration step (step 3) shown in Figure 16.10. The Calvin cycle is a cycle, which means that it runs in a circle, with the products of the final step regenerating the reactants of the first step. Therefore, the third step of the Calvin cycle involves regenerating the 5-carbon sugar from the 3-carbon triose phosphates. ATP is required for this step.

As we mentioned in Module 15, the Calvin cycle does not use sunlight directly. However, it cannot operate without NADPH and ATP produced by the light reactions. These are used in steps 2 and 3 of the Calvin cycle. As a result, the Calvin cycle occurs only in the light.

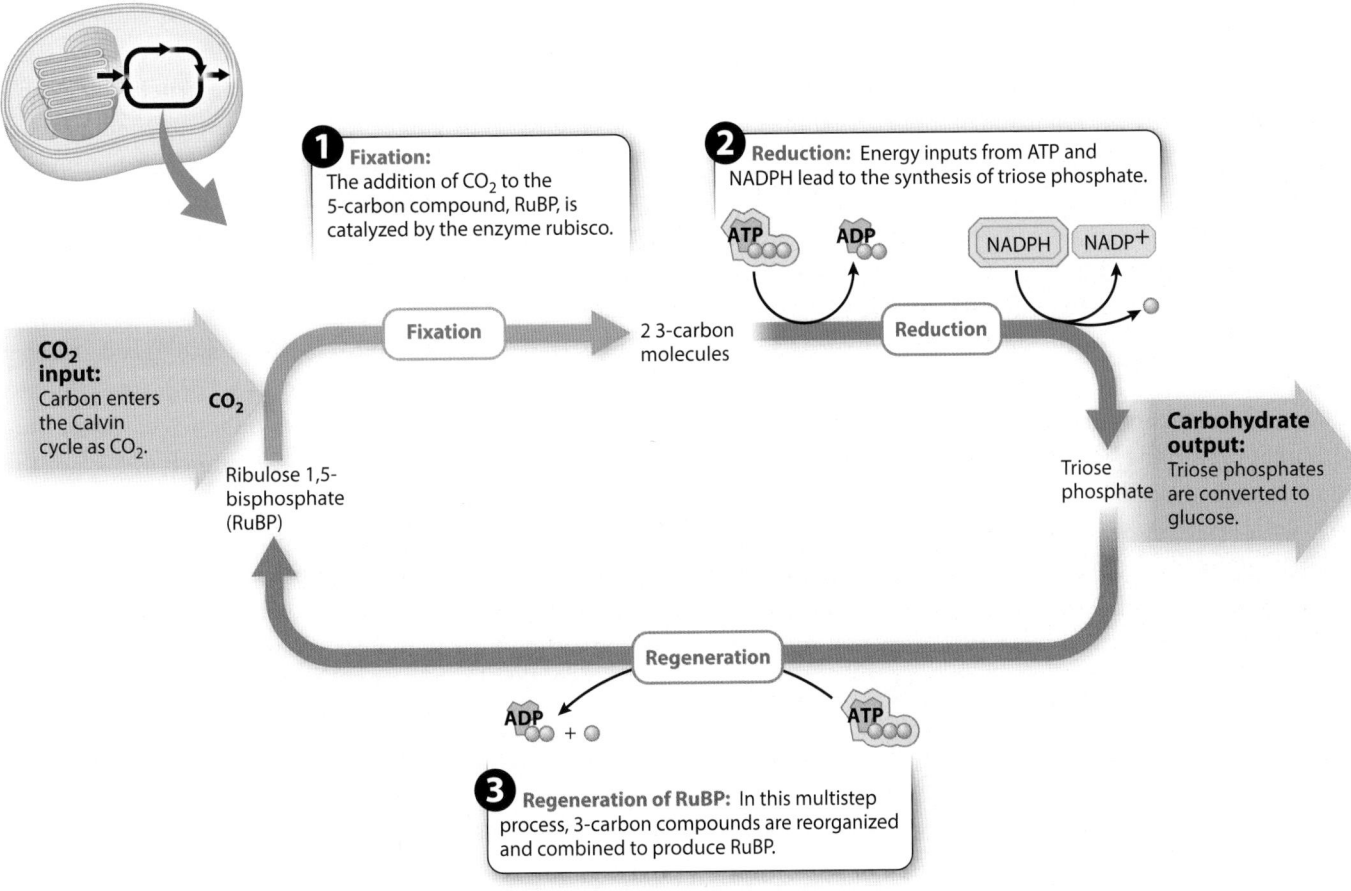

1 **Fixation:** The addition of CO_2 to the 5-carbon compound, RuBP, is catalyzed by the enzyme rubisco.

2 **Reduction:** Energy inputs from ATP and NADPH lead to the synthesis of triose phosphate.

CO₂ input: Carbon enters the Calvin cycle as CO_2.

CO_2

Fixation

2 3-carbon molecules

ATP ADP

NADPH NADP⁺

Reduction

Ribulose 1,5-bisphosphate (RuBP)

Triose phosphate

Carbohydrate output: Triose phosphates are converted to glucose.

Regeneration

ADP + ATP

3 **Regeneration of RuBP:** In this multistep process, 3-carbon compounds are reorganized and combined to produce RuBP.

FIGURE 16.10 The Calvin cycle

The Calvin cycle uses ATP and NADPH produced in the light reactions to incorporate carbon dioxide into carbohydrates. It takes place in three steps: fixation, reduction, and regeneration. In the first step, carbon dioxide is added to a 5-carbon sugar. In the second step, energy inputs from ATP and NADPH lead to the synthesis of triose phosphates, which are carbohydrates. In the third step, the 5-carbon sugar is regenerated.

The Calvin cycle is capable of producing more carbohydrates than the cell needs or, in a multicellular organism, more than the cell is able to export. If carbohydrates accumulated in the cell, they would cause water to enter the cell by osmosis, perhaps damaging the cell. Instead, excess carbohydrates are converted to starch, a storage form of carbohydrates, shown as the yellow granules in **FIGURE 16.11**. This reaction is a dehydration synthesis reaction, which we discussed in Module 2. Because starch molecules are stored in granules that are not water soluble, they provide a means of carbohydrate storage that does not lead to osmosis. Starch formed during the day provides photosynthetic cells with a source of carbohydrates that they can use during the night.

Rubisco

The enzyme rubisco, as we just discussed, catalyzes the first set of reactions in the Calvin cycle. In this step, rubisco uses CO_2 as a substrate and adds it to a 5-carbon molecule in a process called carbon fixation. However, rubisco can actually use

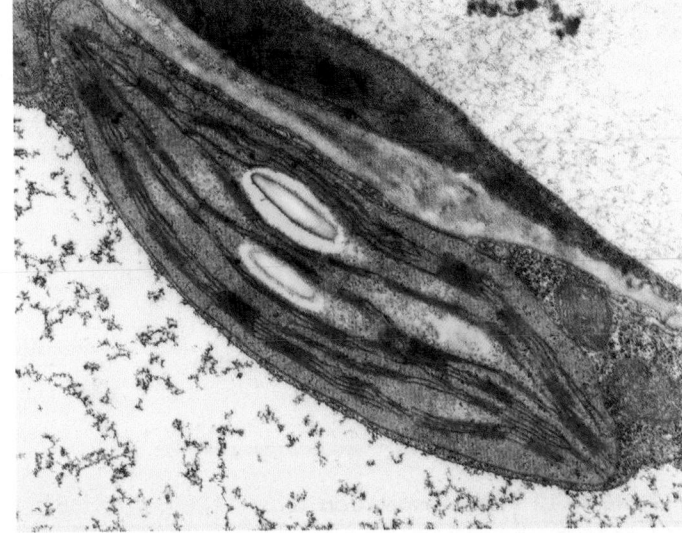

FIGURE 16.11 A chloroplast containing starch granules

Carbohydrates can be stored temporarily as starch. Starch is stored is granules, which are shown in yellow in this broad bean plant.

Source: Biophoto Associates/Science Source

both CO_2 and O_2 as substrates. If O_2 instead of CO_2 diffuses into the active site of rubisco, the reaction can still proceed, although O_2 is added to the 5-carbon molecule in place of CO_2. This explains why the full name of rubisco, ribulose bisphosphate carboxylase oxygenase, includes "oxygenase." An oxygenase is an enzyme that adds oxygen to another molecule.

When rubisco adds O_2 instead of CO_2 to the 5-carbon molecule, it results in the release of CO_2. Carbon dioxide is also released when you breathe during cellular respiration. For this reason, the process in which rubisco adds O_2 to the 5-carbon molecule is referred to as photorespiration. **FIGURE 16.12** compares photosynthesis (on the right) and photorespiration (on the left). In photosynthesis, ATP is produced in the light reactions and carbon dioxide is consumed in the Calvin cycle. In photorespiration, by contrast, ATP is consumed and carbon dioxide is produced. In fact, photorespiration represents a net energy drain. Not only does it consume ATP, but also it results in the oxidation and loss, in the form of CO_2, of carbon atoms that had already been incorporated and reduced by the Calvin cycle.

The Calvin cycle originated long before the accumulation of oxygen in Earth's atmosphere. Without oxygen, carbon dioxide was the only substrate present in the air. Still, why would photorespiration persist in the face of what must

be strong evolutionary pressure to reduce or eliminate the unwanted reaction with oxygen? The difficulty is that CO_2 and O_2 are similar in size and chemical structure. For this reason, rubisco can achieve selectivity only by binding more tightly in the transition state of the reaction. If it binds more tightly, the reaction will slow down. So, the price of high selectivity for CO_2 over O_2 is speed. In other words, the better rubisco is at binding to CO_2 over O_2, the slower its catalytic rate.

This trade-off between selectivity and speed is a key constraint for photosynthetic organisms. For plants, rubisco's low catalytic rate means that photosynthetic cells must produce huge amounts of this enzyme. As much as 50% of the total protein within a leaf is rubisco, and it is estimated to be the most abundant protein on Earth. At the same time, because O_2 is approximately 500 times more abundant in the atmosphere than CO_2 is, as much as one-quarter of the reduced carbon formed in photosynthesis can be lost through photorespiration.

Excess Light Energy

The light reactions and Calvin cycle work together during photosynthesis. For example, the light reactions synthesize NADPH and ATP for use by the Calvin cycle. In turn, the

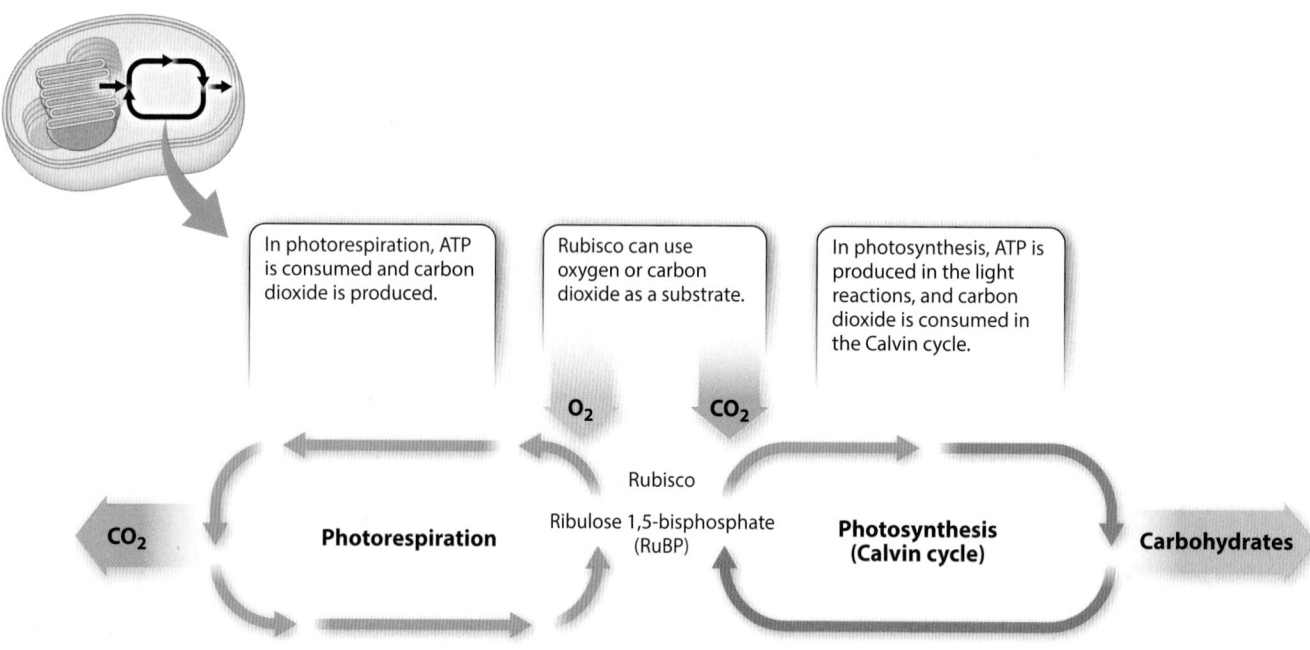

FIGURE 16.12 Photorespiration

Photorespiration (left) occurs when rubisco uses oxygen rather than carbon dioxide as a substrate. In this process, ATP is consumed, and carbon dioxide is produced. By contrast, in photosynthesis (right), ATP is produced in the light reactions, and carbon dioxide is consumed in the Calvin cycle.

Calvin cycle produces $NADP^+$ and ADP, which are used by the light reactions. These reactions therefore need to occur at similar rates.

If there is an imbalance between the two, with the light reactions proceeding more quickly than the Calvin cycle, $NADP^+$ will be in short supply. Under these circumstances, the electron transport chain backs up and leaves electrons and energy with no place to go. This greatly increases the probability of creating reactive forms of oxygen known collectively as reactive oxygen species. These highly reactive molecules can cause substantial damage to the cell. They cause damage by indiscriminately oxidizing lipids, proteins, and nucleic acids.

$NADP^+$ is returned to the photosynthetic electron transport chain by the Calvin cycle's use of NADPH. Therefore, any factor that causes the rate of NADPH use to fall behind the rate of light-driven electron transport can potentially lead to damage. Such an imbalance is likely to occur, for example, in the middle of the day when light intensity is highest. As a result, excess light energy is an everyday event for photosynthetic cells.

The rate at which the Calvin cycle can make use of NADPH is also influenced by a number of factors that are independent of light intensity. For example, at cold temperatures, the enzymes of the Calvin cycle function more slowly, but temperature has little impact on the absorption of light energy. On a cold, sunny day, more light energy is absorbed than can be used by the Calvin cycle.

Photosynthetic organisms employ two major lines of defense to avoid the stresses that occur when the Calvin cycle cannot keep up with the light reactions. First among these are chemicals that detoxify reactive oxygen species. Ascorbate (vitamin C), beta-carotene, and other antioxidants are able to neutralize reactive oxygen species. These compounds exist in high concentration in chloroplasts. Some of these antioxidant molecules are brightly colored, like the red pigments found in algae that live on snow shown in Figure 15.4b

on page 213. The presence of antioxidant compounds is one of the many reasons eating green, leafy vegetables is good for your health.

A second line of defense is to prevent reactive oxygen species from forming in the first place. Xanthophylls are yellow-orange pigments that slow the formation of reactive oxygen species by reducing excess light energy. These pigments accept absorbed light energy directly from chlorophyll, and then convert this energy to heat. Photosynthetic organisms that live in extreme environments often appear brown or yellow because they contain high levels of xanthophyll pigments, as seen in Figure 15.4a. Plants that lack xanthophylls grow poorly when exposed to moderate light levels and die in full sunlight.

PREP FOR THE AP® EXAM

AP® EXAM TIP

Make sure you know that photosynthesis is comprised of two parts: the light reactions and the Calvin cycle. You should be able to explain how the light reactions provide the energy in the form of NADPH and ATP needed for the Calvin cycle, which uses carbon dioxide from the air to synthesize carbohydrates.

✓ Concept Check

5. **Identify** the major inputs and outputs of the Calvin cycle.
6. **Describe** the three steps of the Calvin cycle.
7. **Describe** the role of rubisco in the Calvin cycle.
8. **Identify** why rubisco has a low catalytic rate.
9. **Describe** two ways plants limit the formation and effects of reactive oxygen species.

Module 16 Summary

PREP FOR THE AP® EXAM

REVISIT THE BIG IDEAS

ENERGETICS: Using the content in this module, **describe** how energy captured in the light reactions powers the synthesis of carbohydrates in the Calvin cycle.

LG 16.1 The light reactions use sunlight to produce ATP and NADPH.

- Chlorophyll absorbs visible light. Page 221
- Antenna chlorophyll molecules transfer absorbed light energy to the reaction center. Page 222

- Reaction centers transfer electrons to an electron acceptor, thus initiating the photosynthetic electron transport chain. Page 223
- The photosynthetic electron transport chain consists of a series of electron transfer or redox reactions. Water is the electron donor and $NADP^+$ is the final electron acceptor. Page 224
- The photosynthetic electron transport chain consists of two photosystems in series, photosystem II and photosystem I. Page 224
- Photosystem II pulls electrons from water, resulting in the production of oxygen, and photosystem I reduces $NADP^+$, producing NADPH. Page 224
- The movement of electrons through the electron transport chain is coupled with the transfer of protons from the stroma to the thylakoid space. Page 225
- The buildup of protons in the thylakoid space drives the production of ATP by ATP synthase. Page 225

LG 16.2 The Calvin cycle uses ATP and NADPH to build carbohydrates.
- The first step of the Calvin cycle is the addition of CO_2 to a 5-carbon sugar. Page 226
- The first step of the Calvin cycle is catalyzed by the enzyme rubisco, considered the most abundant protein on Earth. Page 226

- The second step of the Calvin cycle uses ATP and NADPH produced by the light reactions and produces triose phosphate molecules, a type of carbohydrate. Page 226
- Some of these triose phosphates are exported from the chloroplast for use by the cell, and some are used in the third step to regenerate the 5-carbon sugars, allowing the cycle to continue. Page 226
- Starch formation provides chloroplasts with a way of storing carbohydrates that will not cause water to enter the cell by osmosis. Page 226
- Rubisco can bind with oxygen as well as with carbon dioxide. When it reacts with oxygen, energy is lost and CO_2 is released. Page 228
- Rubisco has evolved to favor carbon dioxide over oxygen, but the cost of this selectivity is reduced speed. Page 228
- An imbalance between the rate at which NADPH is produced by the photosynthetic electron transport chain and the rate at which NADPH is used by the Calvin cycle can lead to the formation of reactive oxygen species. Page 229
- Protection from excess light energy includes antioxidant molecules that neutralize reactive oxygen species and xanthophyll pigments that dissipate excess light energy as heat. Page 229

Key Terms

Visible light	Photosystem	Photophosphorylation
Reaction center	ATP synthase	

Review Questions

1. The Z scheme refers to
   (A) the changes in the energy level of electron donors along the photosynthetic electron transport chain.
   (B) the use of H_2O as an electron donor.
   (C) proton transport along the photosynthetic electron transport chain.
   (D) proton pumping along the photosynthetic electron transport chain.

2. In the light reactions of photosynthesis, $NADP^+$ is
   (A) oxidized.
   (B) reduced.
   (C) phosphorylated.
   (D) synthesized.

3. Which products of the light reactions of photosynthesis are required by reactions of the Calvin cycle?
   (A) ATP and oxygen
   (B) ATP and NADPH
   (C) $NADP^+$ and oxygen
   (D) Water and NADPH

4. Chloroplast ATP synthases are powered by the flow of protons from the
   (A) thylakoid space to the stroma.
   (B) stroma to the intermembrane space.
   (C) intermembrane space to the cytoplasm.
   (D) stroma to the thylakoid space.

5. Which is a characteristic of rubisco?
   (A) It is a fast catalyst.
   (B) It uses just one substrate.
   (C) It functions in the light reactions of photosynthesis.
   (D) It is one of the most abundant enzymes on Earth.

6. Photorespiration
   (A) produces carbon dioxide and consumes ATP.
   (B) consumes carbon dioxide and produces ATP.
   (C) produces carbon dioxide and ATP.
   (D) consumes carbon dioxide and ATP.

7. Which statement is true regarding reactive oxygen species?
   (A) Reactive oxygen species form when the Calvin cycle runs faster than the light reactions.
   (B) Reactive oxygen species protect the cell from damage.
   (C) Reactive oxygen species form when there is not very much light energy.
   (D) Reactive oxygen species can be neutralized by antioxidants.

Module 16
AP® Practice Questions

PREP FOR THE AP® EXAM

Section 1: Multiple-Choice Questions

Choose the best answer for questions 1–4.

1. Which statement best explains the role of light energy in the Calvin cycle?
   (A) Light energy enables production of NADPH and ATP that are required for the Calvin cycle.
   (B) Light provides energy used by the Calvin cycle to produce NADPH and ATP.
   (C) Light absorbed by pigments in the Calvin cycle provides energy to establish a proton gradient.
   (D) Light provides energy directly to the Calvin cycle in order to produce sugar from CO_2 and H_2O.

Questions 2–4 use the following information:

The diagram below illustrates the energy associated with electrons flowing through photosystem II (PS II) and photosystem I (PS I). This diagram is sometimes called a Z scheme because it resembles the letter Z turned on its side. Four key steps in the process are labeled 1–4 in the figure.

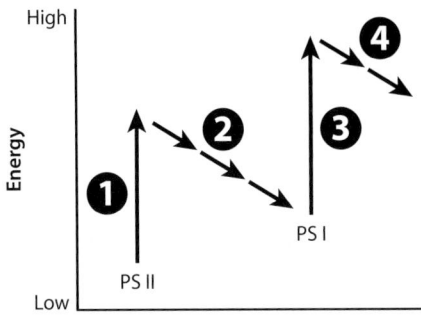

2. The level of energy increases dramatically at step 3. Identify the source of the energy at this step.
   (A) Cleavage of water
   (B) Light
   (C) ATP
   (D) NADPH

3. During two steps in the light reactions of photosynthesis, protons are pumped across the thylakoid membrane. Identify the numbers that represent these two steps in the diagram.
   (A) 1 and 3
   (B) 2 and 4
   (C) 1 and 2
   (D) 3 and 4

4. Identify the step at which $NADP^+$ is reduced to become NADPH.
   (A) 1
   (B) 2
   (C) 3
   (D) 4

Section 2: Free-Response Question

Write your answer to each part clearly. Support your answers with relevant information and examples. Where calculations are required, show your work.

The graph in Figure 16.3 on page 222 was created with data from a technique called spectrophotometry. Spectrophotometry measures the amount of light that is absorbed by an object or a chemical over a range of wavelengths. To get the graph in Figure 16.3, researchers used a biochemical technique to extract pigments from a plant leaf and subjected each of the three pigments that they found to spectrophotometry. The technique revealed that each of the three pigments found in plant leaves—chlorophyll *a* and *b*, and carotenoid—absorbs a unique set of wavelengths of visible light.

(a) **Describe** why carotenoids appear orange in a leaf, but chlorophylls appear green.

(b) Assuming all three of the pigment types are present in roughly equal proportion in the surface of the leaf, **draw** the predicted light absorbance of an intact leaf.

(c) **Explain** why a leaf requires more than one pigment for photosynthesis. Use a claim, evidence, reasoning format in your answer.

(d) A scientist proposes that because they are green, plants should grow best under green light. **Support** or **refute** this claim. Provide your reasoning

Module 17

Cellular Respiration I: Overview

LEARNING GOALS ▶**LG 17.1** Cellular respiration harnesses energy of carbohydrates and other fuel molecules.
▶**LG 17.2** Cellular respiration takes place in four stages.
▶**LG 17.3** Cellular respiration can make use of diverse organic molecules.

All organisms transfer energy from the environment and then harness it to do work. We saw in Module 13 that energy is needed for many different functions, including the synthesis of macromolecules, cell movement and division, muscle contraction, and growth and development. Organic molecules such as carbohydrates, lipids, and proteins are good sources of chemical energy. As we saw with photosynthesis, some organisms, such as plants, synthesize these molecules from inorganic molecules like carbon dioxide. They harness the energy of sunlight to produce ATP and NADPH, and then use these molecules to build carbohydrates. Other organisms, such as humans and other animals, consume organic molecules in their diet.

Regardless of how they obtain organic molecules such as carbohydrates, nearly all organisms break them down in the process of *cellular respiration*. **Cellular respiration** is a series of chemical reactions that convert the chemical energy in fuel molecules into the chemical energy of adenosine triphosphate (ATP). While both organic molecules and ATP store chemical energy in their bonds, ATP is a ready, available source of energy that cells use to carry out their functions. ATP hydrolysis yields small, manageable amounts of free energy that cells use for many chemical reactions. Therefore, you can think of carbohydrates as a storage form of chemical energy, and ATP as a usable form of chemical energy.

It is tempting to think that organic molecules are converted into energy in the process of cellular respiration, but this is not the case. Recall from Module 13 that the first law of thermodynamics states that energy cannot be created or destroyed. Rather than creating energy, cellular respiration converts energy from one form to another. Specifically, it converts the chemical energy stored in carbohydrates and other fuel molecules into the chemical energy stored in ATP. Similarly, as we saw in Modules 15 and 16, photosynthesis converts light energy into the chemical energy of carbohydrates.

It is also easy to forget that organisms other than animals, such as plants, use cellular respiration. If plants use sunlight as a source of energy, why would they need cellular respiration? Plants use the energy of sunlight to make carbohydrates. Plants then break down these carbohydrates in the process of cellular respiration to produce ATP for use by the cell, just like animals.

We will begin our study of cellular respiration with an overview. We will focus on the overall reaction, how ATP is synthesized, the four main stages of cellular respiration, and the diverse set of starting molecules that can funnel into this critical metabolic pathway. This overview will lay the foundation for the next module, where we will examine each of the stages of cellular respiration in more detail.

PREP FOR THE AP® EXAM

FOCUS ON THE BIG IDEAS

ENERGETICS: Focus on the biochemical reactions of cellular respiration, being sure to note at which stage the reactants enter the pathway and at which stage the products are released.

17.1 Cellular respiration harnesses energy of carbohydrates and other fuel molecules

Cellular respiration is the major metabolic pathway by which you, and nearly all organisms, break down carbohydrates and other organic molecules in order to sustain life. It is a series of catabolic reactions that convert the energy stored in organic molecules, such as glucose, into the energy stored in ATP. ATP then can be used to power many of the chemical reactions that occur all of the time in cells. Carbon dioxide is produced as a waste product. Cellular respiration that occurs in the presence of oxygen is known as aerobic respiration. Cellular respiration that

FIGURE 17.1 Organisms that use cellular respiration

Nearly all organisms carry out cellular respiration, including animals such as elephants and plants such as spring wildflowers. Photo sources: (left) Steve Allen/Getty Images; (right) Kerstin Hinze/naturepl.com/NaturePL

occurs in the absence of oxygen is known as anaerobic respiration. Most organisms that you are familiar with carry out aerobic respiration, such as the ones shown in **FIGURE 17.1**. In this section, we will focus on the overall reaction for aerobic cellular respiration, and we will discuss the two different ways that ATP is produced.

Overall Reaction

Cellular respiration and photosynthesis are complementary biochemical processes. Recall that in photosynthesis, carbon dioxide and oxygen react to form glucose and oxygen, as we discussed in Modules 15 and 16. In cellular respiration, glucose and oxygen react to form carbon dioxide and water. The overall reaction therefore can be described as follows:

$$C_6H_{12}O_6 \; + \; 6O_2 \; \rightarrow \; 6CO_2 \; + \; 6H_2O \; + \; Energy$$

Glucose   Oxygen   Carbon   Water

dioxide

In Module 13, we saw that molecules such as glucose have a large amount of potential energy in their chemical bonds. By contrast, molecules such as carbon dioxide and water have less potential energy in their bonds. Cellular respiration releases a large amount of free energy—energy available to do work—because the sum of the potential energy in all of the chemical bonds of the reactants (glucose and oxygen) is higher than that of the products (carbon dioxide and water). The maximum amount of free energy released during cellular respiration is −686 kcal per mole of glucose. Recall from Module 13 that when $\Delta G < 0$, energy

is released. Much of this energy is transferred into to the chemical bonds of ATP.

This overall reaction is a reduction–oxidation, or redox, reaction. Recall from Module 15 that oxidation is the loss of electrons, and reduction is the gain of electrons. This loss and gain always happen in a single coupled reaction in which electrons are transferred from one molecule to another. One molecule loses electrons, and one molecule gains those same electrons. As molecules lose and gain electrons, they also lose and gain energy.

In cellular respiration, glucose is oxidized to carbon dioxide, and, at the same time, oxygen is reduced to water:

Oxidation

$$C_6H_{12}O_6 \; + \; 6O_2 \longrightarrow 6CO_2 \; + \; 6H_2O \; + \; Energy$$

Reduction

Let's first focus on the oxidation reaction, in which glucose is converted to carbon dioxide, shown in **FIGURE 17.2**. In glucose, there are many C–C and C–H covalent bonds, in which electrons are shared about equally between the two atoms, as illustrated by the uniform red shading between the

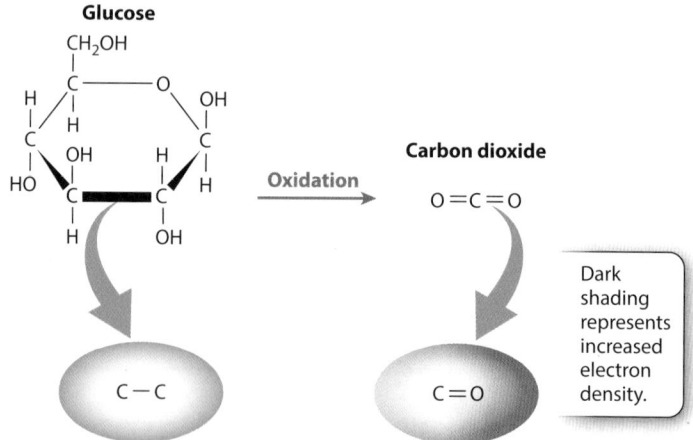

FIGURE 17.2 Oxidation

In glucose, carbon atoms share electrons equally, as indicated by the uniform red shading. By contrast, in carbon dioxide, electrons spend more time around oxygen than carbon, as indicated by the dark red shading around oxygen. Therefore, as glucose is converted to carbon dioxide, carbon loses electrons and is oxidized.

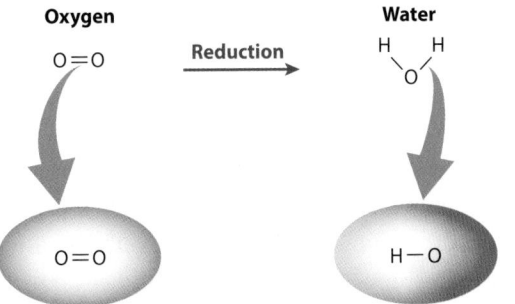

Oxygen   **Water**

O=O   Reduction →   H   H
                           O

O=O   H—O

FIGURE 17.3 Reduction

In oxygen, electrons are shared equally, as indicated by the uniform red shading. In water, electrons spend more time around oxygen than hydrogen, as indicated by the dark red shading around oxygen. Therefore, as oxygen is converted to water, oxygen gains electrons and is reduced.

carbon atoms in the figure. By contrast, in carbon dioxide, electrons are not shared equally. The oxygen atoms are more electronegative than the carbon atom, so the electrons are more likely to be found near the oxygen atoms, as illustrated by the darker red shading around oxygen compared to carbon. As a result, carbon loses electrons to oxygen and is oxidized in the reaction. The carbon atoms in glucose are oxidized because they go from sharing electrons equally in the carbon–carbon bonds to losing electrons in the carbon–oxygen bonds of the carbon dioxide molecule.

Let's now focus on the reduction reaction, in which oxygen gas is converted to water, shown in **FIGURE 17.3**. In oxygen gas, electrons are shared equally between two oxygen atoms, indicated by the uniform red shading between the oxygen atoms. In water, the electrons that are shared between hydrogen and oxygen are more likely to be found near oxygen, indicated by the darker red shading, because oxygen is more electronegative than hydrogen. As a result, oxygen gains electrons and is reduced in the reaction. The oxygen atoms go from sharing electrons equally to gaining electrons when water is formed.

Cellular respiration does not take place in one single reaction in a cell. Instead, it takes place in a long series of reactions, many of which are redox reactions. In the next module, we will follow the individual reactions of cellular respiration. Recognizing that many of these are redox reactions will help you follow the flow of electrons and energy in these reactions.

ATP Production

The main function of cellular respiration is to synthesize ATP for use by the cell. Let's now focus on how it is produced. During cellular respiration, ATP is produced in two different ways. In the first, an organic molecule transfers a phosphate group directly to ADP, as we discussed in Module 13. This type of reaction is shown on the left in **FIGURE 17.4**. In this case, a single enzyme carries out two coupled reactions: the hydrolysis of an organic molecule to yield a phosphate group, and the addition of that phosphate group to ADP. The hydrolysis reaction releases enough free energy to drive the synthesis of ATP. In this mechanism of generating ATP, a phosphate group is transferred to ADP from an enzyme substrate. As a result, the reaction is called **substrate-level phosphorylation.**

Substrate-level phosphorylation produces only a small amount of the total ATP generated in cellular respiration, about 12% if the fuel molecule is glucose. The remaining 88% is produced by a different process, shown on the right in Figure 17.4. In this case, the chemical energy of organic molecules is transferred first to electron carriers. The role of these electron carriers is exactly what their name suggests— they carry electrons from one set of reactions to another. Because they carry electrons, they also carry energy. In the last module, we saw that NADPH is the electron carrier used in photosynthesis. In cellular respiration, the electron carriers are nicotinamide adenine dinucleotide (NADH) and flavin adenine dinucleotide ($FADH_2$).

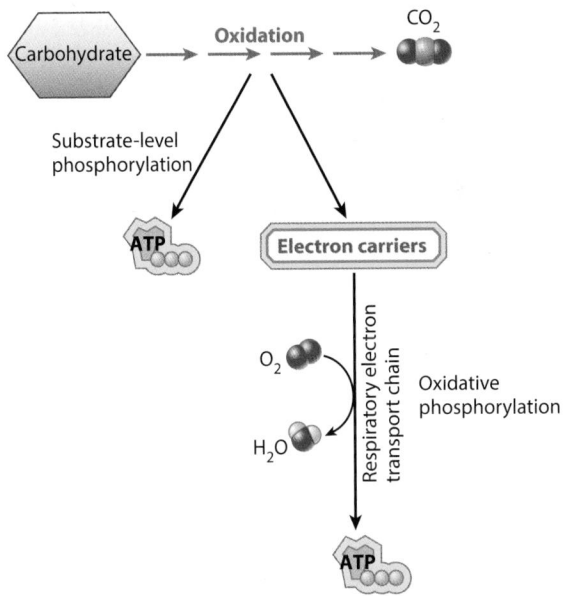

FIGURE 17.4 ATP production in cellular respiration

In most organisms, cellular respiration consumes carbohydrates and oxygen, and produces carbon dioxide and water. In this process, ATP is produced in two ways: substrate-level phosphorylation, shown on the left, and oxidative phosphorylation, shown on the right. Most of the ATP in cellular respiration comes from oxidative phosphorylation.

These electron carriers exist in two forms—an oxidized form (NAD^+ and FAD) and a reduced form (NADH and $FADH_2$). As glucose is oxidized step by step in a series of redox reactions, electrons are gained by these electron carriers, as follows:

$$NAD^+ + 2e^- + H^+ \rightarrow NADH$$
$$FAD + 2e^- + 2H^+ \rightarrow FADH_2$$

In these reactions, NAD^+ and FAD accept electrons and energy, and are converted to their reduced forms, NADH and $FADH_2$. In their reduced forms, NADH and $FADH_2$ transport electrons and energy to the respiratory electron transport chain. In this way, electron carriers act as shuttles, transferring electrons and energy derived from the oxidation of fuel molecules such as glucose to the respiratory electron transport chain.

The respiratory electron transport chain is a series of membrane-associated proteins in the inner membrane of mitochondria that transfers electrons to a final electron acceptor. As with the photosynthetic electron transport chain, electron movement drives the pumping of protons across the membrane, and the resulting electrochemical gradient in turn is used to produce ATP. Because oxygen is the final electron acceptor, this mechanism of synthesizing ATP is called **oxidative phosphorylation.** When oxygen gains an electron, water is produced.

The respiratory electron transport chain is similar to the photosynthetic electron transport chain we discussed in Modules 15 and 16. In cellular respiration, the electron transport chain harnesses energy from fuel molecules such as glucose, and in photosynthesis it harnesses energy from sunlight. In cellular respiration, the electron transport chain is located in the inner membrane of mitochondria, and in photosynthesis it is located in the thylakoid membrane. In both cases, the high surface area of the folded membranes allows for the harnessing of a great deal of energy.

Cellular respiration is widespread in nature and is carried out by many different organisms, including both plants and animals. Although photosynthetic organisms, such as plants, use the energy of sunlight to make their own carbohydrates, they also require a constant supply of ATP because it is the cell's ready and usable form of energy. ATP is produced within chloroplasts during photosynthesis, as we saw in Modules 15 and 16, but only carbohydrates—not ATP—are exported from chloroplasts to the cytosol. Therefore, cellular respiration is required to synthesize ATP for use by the cell. Furthermore, plants require cellular respiration when it is dark at night, when there is no light energy to make ATP. As a result, cells that have chloroplasts, where photosynthesis occurs, also contain mitochondria, where cellular respiration occurs.

✓ Concept Check

1. **Identify** the reactants and products of cellular respiration.

2. **Identify** the molecule that is oxidized and the molecule that is reduced in the overall reaction for cellular respiration.

3. For each of the following pairs of molecules, **identify** the member of the pair that is reduced and the member that is oxidized, and which one has more chemical energy and which has less chemical energy: NAD^+/NADH; FAD/$FADH_2$; CO_2/$C_6H_{12}O_6$.

4. **Describe** two different ways ATP is generated in cellular respiration.

17.2 Cellular respiration takes place in four stages

In cellular respiration, glucose is broken down in the presence of oxygen to produce carbon dioxide and water. Up to this point, we have focused on the overall reaction of cellular respiration and the chemistry of redox reactions. Let's now divide the entire process into several stages. These stages take place in different regions of the cell and they all have different inputs, or reactants, and outputs, or products.

Each stage of cellular respiration consists of a series of reactions, many of which are redox reactions. Therefore, glucose is not oxidized all at once to carbon dioxide. Instead, it is oxidized slowly and in a controlled manner, allowing for energy to be harnessed. Some of this energy is used to synthesize ATP directly by substrate-level phosphorylation, and some of it is stored temporarily in reduced electron carriers and then used to generate ATP by oxidative phosphorylation.

The four stages are illustrated in **FIGURE 17.5**. In stage 1, called **glycolysis,** glucose is partially broken down to produce pyruvate. This stage takes place in the cytoplasm. In this stage, energy is transferred to ATP by substrate-level phosphorylation and reduced electron carriers.

In stage 2, called pyruvate oxidation, pyruvate is oxidized to another molecule called acetyl-coenzyme A (acetyl-CoA). This stage takes place in mitochondria in eukaryotes. It produces reduced electron carriers, and releases carbon dioxide.

Acetyl-CoA enters stage 3, the **Krebs cycle,** also called the citric acid cycle. This stage also takes place in mitochondria in eukaryotes. In this series of redox reactions, the acetyl group is completely oxidized to carbon dioxide, and free energy is transferred to ATP by substrate-level phosphorylation and reduced electron carriers. The amount of energy transferred to ATP and reduced electron carriers in this stage is nearly twice that transferred by stages 1 and 2 combined.

Stage 4 is oxidative phosphorylation. In this series of reactions, reduced electron carriers generated in stages 1–3 donate electrons to the respiratory electron transport chain. The electron transport chain is made up of proteins and small molecules associated with the inner mitochondrial membrane. Oxidative phosphorylation generates a large amount of ATP.

We can connect these four stages to your everyday experience. The oxygen you take in with each breath is the final electron acceptor in stage 4, oxidative phosphorylation. The carbon dioxide you breathe out is produced in stages 2 and 3, pyruvate oxidation and the Krebs cycle. The food you eat, including glucose, provides the fuel for the process, entering in stage 1, glycolysis. Finally, the ATP that is produced in stages 1, 3, and 4 allows your cells to work and function.

FIGURE 17.6 shows the change of free energy (ΔG) at each step in the catabolism of glucose. Recall from Module 13 that if the products of a reaction have less free energy than the reactants, ΔG is negative, and energy is available to do work (see Figure 13.8b on page 191). These are exergonic reactions. Exergonic reactions are shown in

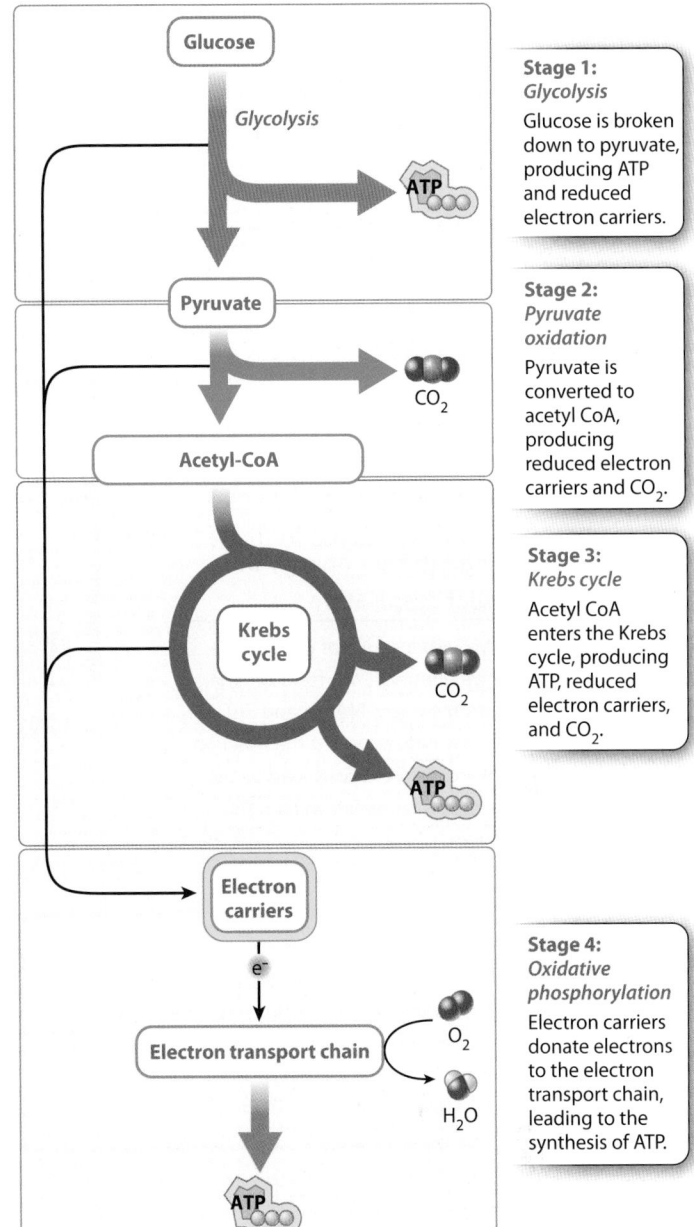

FIGURE 17.5 The four stages of cellular respiration

Cellular respiration consists of glycolysis, pyruvate oxidation, the Krebs cycle, and oxidative phosphorylation. Stages 1–3 produce reduced electron carriers, indicated by the yellow highlighting. They pass on their electrons to the respiratory electron transport chain, which leads to the synthesis of ATP by oxidative phosphorylation.

Figure 17.6 on page 238 by a decrease in the line in the graph. Note that these exergonic reactions lead to the production of ATP and reduced electron carriers. As you can see, the individual reactions allow the initial chemical energy present in a molecule of glucose to be "packaged" into molecules of ATP and the reduced electron carriers NADH and FADH$_2$.

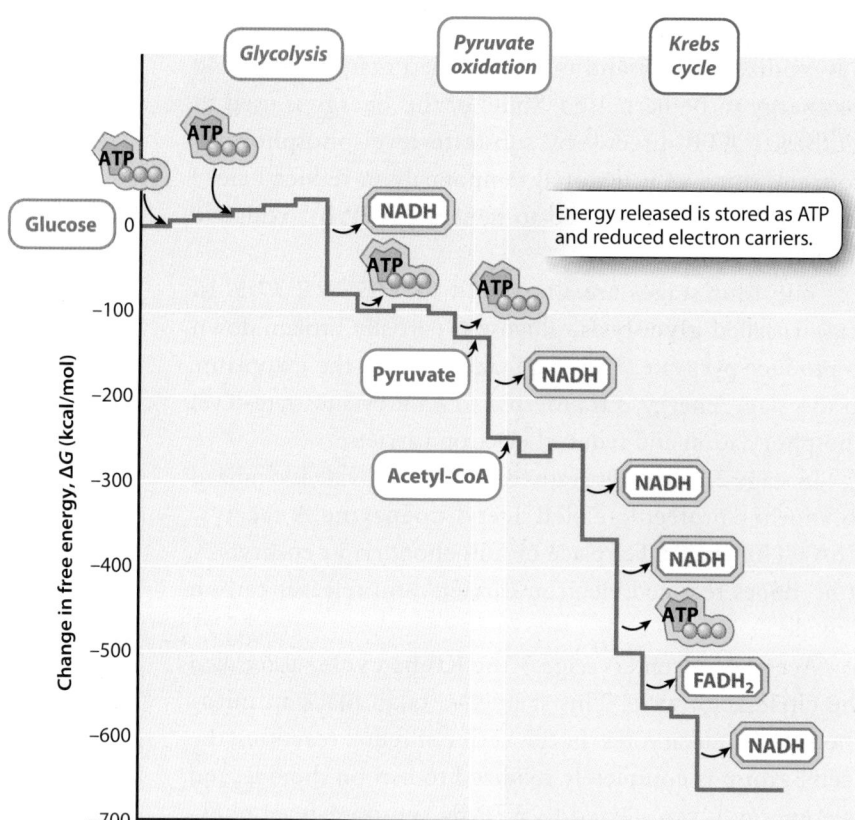

FIGURE 17.6 Change in free energy in each step of cellular respiration

Glucose is oxidized through a series of chemical reactions, releasing free energy in the form of ATP and the reduced electron carriers NADH and FADH$_2$. A decrease in free energy indicates that the reaction is exergonic. These exergonic reactions lead to the production of ATP and reduced electron carriers.

PREP FOR THE AP® EXAM

AP® EXAM TIP

Remember that the large energy release that occurs when glucose is oxidized to carbon dioxide and water is captured in discrete, small steps. In this manner, the cell is able to harness the energy of glucose.

✓ Concept Check

5. **Identify** the four major stages of cellular respiration.

6. **Describe** the four stages of cellular respiration, including the starting and ending molecules.

17.3 Cellular respiration can make use of diverse organic molecules

In this and the next module, we focus on glucose as the starting molecule for cellular respiration. However, many different types of fuel molecules can enter the pathways of cellular respiration. In this section, we discuss other types of fuel molecules and how they enter the process.

Storage Forms of Glucose

Glucose is a useful fuel molecule for organisms, but it is not always broken down immediately. Excess glucose can be stored in cells and then mobilized when necessary. Glucose is stored in two major forms: as glycogen in animals, described in Module 3, and as starch in plants. These molecules are both large branched polymers of glucose and are shown in **FIGURE 17.7**.

Carbohydrates in the diet are broken down into glucose and other simple sugars and circulate in the blood. The level of glucose in the blood is actively maintained at relatively constant levels and therefore represents an example of homeostasis. When the blood glucose level is high, as it is after a meal, glucose molecules that are not needed are linked to form glycogen in liver and muscle cells. Glycogen stored in muscle cells provides the substrate for the catabolic processes described in this module, thereby generating ATP as needed for repeated muscle contractions. By contrast, the liver does not store glycogen primarily for its own use but is a glycogen storehouse for the whole body. Glucose molecules located at the end of glycogen chains can be cleaved one by one, released, and then broken down during cellular respiration to produce usable energy for the cell in the form of ATP.

Other Fuel Molecules

The carbohydrates in your diet are digested to produce a variety of sugars, which were described in Unit 1. Some of these are disaccharides, such as maltose, lactose, and sucrose, with two sugar units; others are monosaccharides, such as fructose, mannose, and galactose, with one sugar unit, as shown in **FIGURE 17.8**. The disaccharides are hydrolyzed into monosaccharides, which are transported into cells.

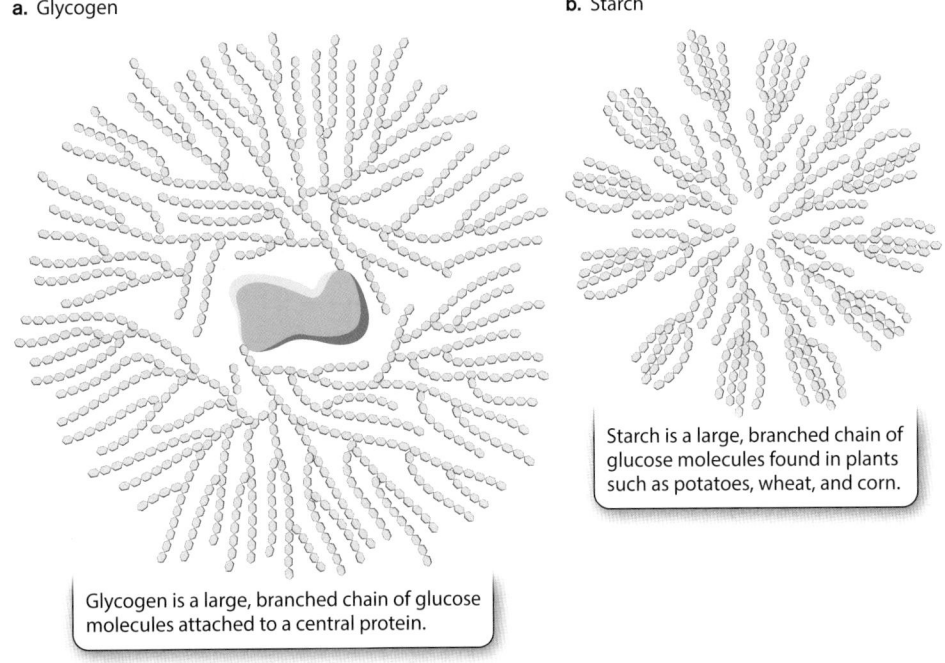

a. Glycogen   **b.** Starch

Starch is a large, branched chain of glucose molecules found in plants such as potatoes, wheat, and corn.

Glycogen is a large, branched chain of glucose molecules attached to a central protein.

FIGURE 17.7 Storage forms of glucose

Both glycogen and starch are large branched polymers of glucose. (a) Glycogen is a storage form of glucose in animal cells. (b) Starch is a storage form of glucose in plant cells. Each small circle in the images represents a single glucose molecule monomer that is incorporated into the polymers.

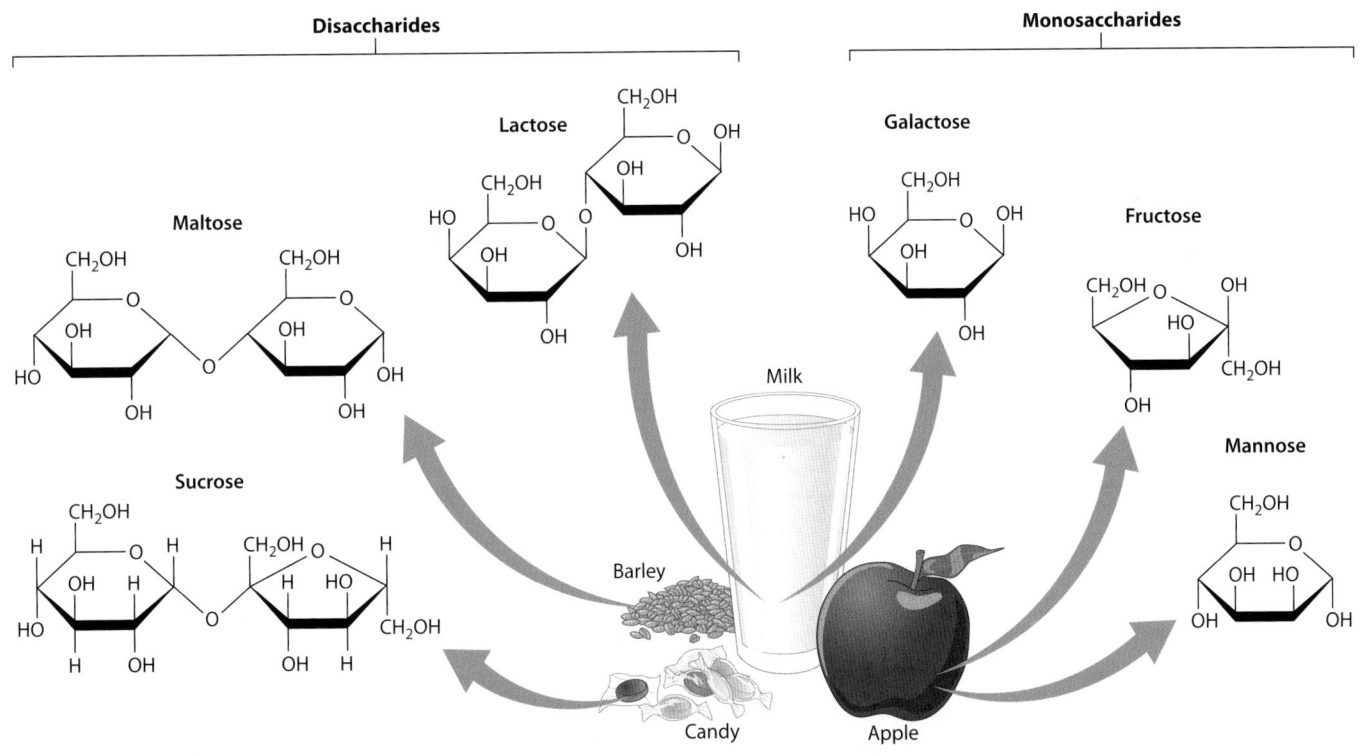

FIGURE 17.8 Common sugars in the diet

Common sugars in the human diet include disaccharides, with two sugar units, and monosaccharides, with one sugar unit.

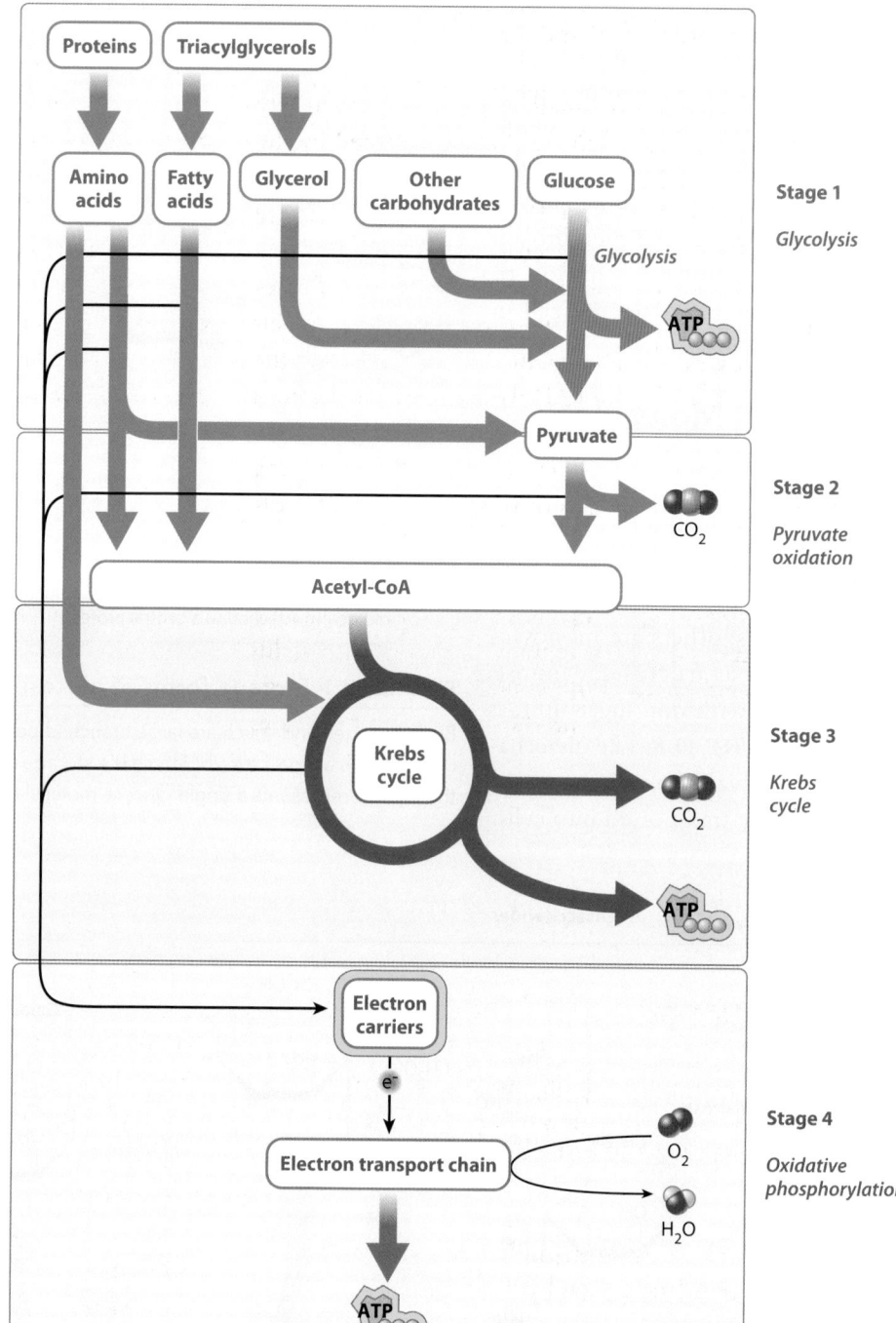

FIGURE 17.9 Other fuel molecules

Sugars other than glucose, fats such as triacylglycerol, and proteins are all fuel molecules that are broken down and enter the cellular respiration pathway at different places. Some enter glycolysis; others are converted to acetyl-CoA; and still others enter the Krebs cycle.

Glucose molecules released during digestion in the gut are absorbed into the bloodstream and then are taken up by cells, where they directly enter cellular respiration. Other sugars, too, enter cellular respiration, although not as glucose. Instead, they are converted into intermediates that enter the process later in the pathway. **FIGURE 17.9** shows where in the cellular respiration pathway they enter. Note that carbohydrates and glycerol enter glycolysis; fatty acids and some amino acids are converted to acetyl-CoA; and other amino acids enter the Krebs cycle.

We know from common experience that lipids are also a good source of energy. Butter, oils, ice cream, and the like all contain lipids and are high in calories, which are units of energy. From the chemical structure of lipids, we can also infer that they are a good source of energy. Recall from Unit 1 that a type of fat called triacylglycerol is composed of three fatty acid molecules bound to a glycerol backbone. (See Figure 3.6, page 58.) These fatty acid molecules are rich in carbon–carbon and carbon–hydrogen bonds, which, as we saw earlier, carry a large amount of chemical potential energy.

Following a meal, the small intestine very quickly absorbs triacylglycerols, which are then transported by the bloodstream and either consumed or stored in fat (adipose) tissue. As indicated in Figure 17.9, triacylglycerols are broken down inside cells to glycerol and fatty acids. Then the fatty acids themselves are shortened by a series of reactions that sequentially remove two carbon units from their ends. This process does not generate ATP but produces the electron carriers NADH and $FADH_2$ that provide electrons for the synthesis of ATP by oxidative phosphorylation. In addition, the product of the reaction is acetyl-CoA, which feeds the Krebs cycle and leads to the production of additional reduced electron carriers.

The oxidation of fatty acids produces a large amount of ATP. For example, the complete oxidation of a molecule of palmitic acid, a fatty acid containing 16 carbons, yields about 106 molecules of ATP. By contrast, the complete oxidation of a glucose molecule produces about 32 molecules

of ATP. Fatty acids therefore are a useful and efficient source of energy, but they cannot be used by all tissues of the body. Notably, the brain and red blood cells depend primarily on glucose for energy.

Proteins, like fatty acids, are also a source of chemical energy that can be broken down, if necessary, to power the cell. Proteins are typically first broken down to amino acids, which enter at various points in glycolysis (stage 1), pyruvate oxidation (stage 2), and the Krebs cycle (stage 3), as indicated in Figure 17.9. Proteins, however, are an energy source of last resort as they serve structural and other roles important for cells.

✓ Concept Check

7. **Identify** three fuel molecules that can be broken down by cellular respiration.

8. **Identify** the storage forms of glucose in plants and animals.

Module 17 Summary

PREP FOR THE AP® EXAM

REVISIT THE BIG IDEAS

ENERGETICS: Using the overall chemical reaction of cellular respiration as a guide, **describe** the stage in the reaction pathway that each reactant enters and the stage in the reaction pathway at which each product is released.

LG 17.1 Cellular respiration harnesses energy of carbohydrates and other fuel molecules.

- During cellular respiration, organic molecules such as glucose are broken down in the presence of oxygen to produce carbon dioxide and water. Page 233
- Cellular respiration releases energy because the potential energy of the reactants is greater than that of the products. Page 234
- The overall reaction of cellular respiration is an oxidation–reduction reaction, in which glucose is oxidized to carbon dioxide and oxygen is reduced to water. Page 234
- In oxidation–reduction reactions, electrons are transferred from one molecule to another. Page 234

- Oxidation is the loss of electrons, and reduction is the gain of electrons. Page 234
- The electron carriers NADH and $FADH_2$ transfer electrons to the respiratory electron transport chain, which harnesses the energy of these electrons to generate ATP. Page 235
- ATP is generated in two ways during cellular respiration: substrate-level phosphorylation and oxidative phosphorylation. Page 236
- Cellular respiration is carried out by many organisms, including animals and plants. Page 236

LG 17.2 Cellular respiration takes place in four stages.

- Cellular respiration is a four-stage process. Page 237
- The first stage is glycolysis, in which glucose is broken down to pyruvate. This stage produces ATP and reduced electron carriers. Page 237
- The second stage is pyruvate oxidation, in which acetyl-CoA, reduced electron carriers, and carbon dioxide are produced. Page 237

- The third stage is the Krebs cycle, which generates ATP, reduced electron carriers, and carbon dioxide. Page 237
- The fourth stage is oxidative phosphorylation, in which electrons are transferred along an electron transport chain to oxygen, producing water, and ATP is synthesized. Page 237
- In eukaryotes, stage 1 takes place in the cytoplasm, and stages 2–4 take place in mitochondria. Page 237

LG 17.3 Cellular respiration can make use of diverse organic molecules.

- Excess glucose molecules are linked and stored in polymers called glycogen in animals and starch in plants. Page 238
- Other monosaccharides derived from the digestion of dietary carbohydrates are converted into intermediates of glycolysis. Page 239
- Fatty acids contained in triacylglycerols are an important form of energy storage in cells. Page 240
- Proteins can also be broken down when other fuel molecules are depleted. Page 241

Key Terms

Cellular respiration
Substrate-level phosphorylation

Oxidative phosphorylation
Glycolysis

Krebs cycle

Review Questions

1. Which is not a product of cellular respiration?
   (A) Carbon dioxide
   (B) Water
   (C) Oxygen gas
   (D) ATP

2. Cellular respiration is a series of
   (A) catabolic reactions.
   (B) anabolic reactions.
   (C) glycolytic reactions.
   (D) phosphorylation reactions.

3. Which best describes ATP production during cellular respiration?
   (A) A small amount of ATP is produced by substrate-level phosphorylation; most is produced by oxidative phosphorylation.
   (B) A small amount of ATP is produced by oxidative phosphorylation; most is produced by substrate-level phosphorylation.
   (C) An equal amount of ATP is produced by oxidative phosphorylation and substrate-level phosphorylation.
   (D) Cellular respiration does not produce ATP.

4. In cellular respiration, oxygen
   (A) gains electrons and is reduced.
   (B) loses electrons and is oxidized.
   (C) gains electrons and is oxidized.
   (D) loses electrons and is reduced.

5. In cellular respiration, glucose is
   (A) oxidized.
   (B) reduced.
   (C) deoxygenated.
   (D) phosphorylated.

6. Which statement is true regarding a molecule that is oxidized?
   (A) It is an electron acceptor.
   (B) It is usually oxygen.
   (C) It loses electrons.
   (D) It gains electrons.

7. Excess glucose is stored in large branched molecules of
   (A) starch in bacteria.
   (B) starch in animals.
   (C) glycogen in plants.
   (D) glycogen in animals.

Module 17
AP® Practice Questions

PREP FOR THE AP® EXAM

Section 1: Multiple-Choice Questions

Choose the best answer for questions 1–4.

1. Which of the following processes most directly generates the greatest amount of ATP?

   (A) Glycolysis

   (B) Fermentation

   (C) The Krebs cycle

   (D) The electron transport chain

2. Which best describes the role of the electron transport chain in the generation of ATP during respiration?

   (A) It generates a proton gradient to power ATP production.

   (B) It provides the electrons that are required for ATP hydrolysis.

   (C) It reduces ADP to become ATP.

   (D) It generates NAD^+ to drive glycolysis.

Questions 3 and 4 refer to Figure 17.5 on page 237, and Figure 17.6 on page 238, a schematic overview of the four stages of cellular respiration.

3. Identify the molecule that contains the most chemical energy.

   (A) Glucose

   (B) Pyruvate

   (C) Acetyl-CoA

   (D) H_2O

4. A drug called PS10 blocks the conversion of pyruvate to acetyl-CoA in mice. Predict which molecule would be most likely to accumulate in the tissues of a mouse that is given PS10.

   (A) Glucose

   (B) Pyruvate

   (C) CO_2

   (D) Acetyl-CoA

Section 2: Free-Response Question

Write your answer to each part clearly. Support your answers with relevant information and examples. Where calculations are required, show your work.

A researcher was investigating the biological properties of two strains of the bacterium *Escherichia coli* (*E. coli*). The two strains were designated strain A and strain B. Glucose and lactose are sugars that some bacteria are able to metabolize for energy. The researcher grew each strain in liquid growth medium that contained glucose as the only sugar, lactose as the only sugar, or no sugar at all. No further sugar was added after the bacteria strains began to grow. The researcher measured the number of cells in each culture tube over a 24-hour period and obtained the data shown in the figure.

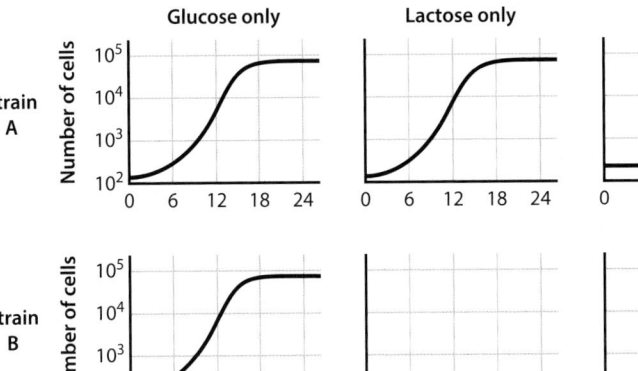

(a) **Identify** the dependent variable in this experiment.

(b) **Identify** the negative control in this experiment.

(c) **Determine** approximately how many cells the researcher added to each culture tube at the beginning of the experiment.

(d) **Make a claim** to explain the difference between the graphs for strain A and strain B bacteria that are grown in liquid growth medium containing only lactose. Provide reasoning to support your claim.

(e) **Make a claim** to explain why the number of strain A cells is approximately the same at 18 and 24 hours when glucose is provided.

Module 18

Cellular Respiration II: Biochemistry

LEARNING GOALS ▶**LG 18.1** Glycolysis, pyruvate oxidation, and the Krebs cycle break down organic molecules.
▶**LG 18.2** Oxidative phosphorylation involves an electron transport chain and ATP synthesis.
▶**LG 18.3** Fermentation and cellular respiration meet the energy needs of a cell.

In Module 14, we saw that catabolism describes the set of chemical reactions that breaks down molecules into smaller units. In the process, these reactions convert chemical energy that can be stored in molecules of ATP. Anabolism, by contrast, is the set of chemical reactions that builds molecules from smaller units. Anabolic reactions require an input of energy, usually in the form of ATP. Cellular respiration is one of the major sets of catabolic reactions in a cell, as we learned in the previous module. During cellular respiration, fuel molecules such as glucose, lipids, and proteins are catabolized into smaller units, releasing the energy stored in their chemical bonds to produce ATP. In turn, ATP powers the work of the cell.

In Module 17, we learned that cellular respiration is the process in which glucose reacts with oxygen to produce carbon dioxide and water. In this process, some of the energy held in the bonds of the original glucose molecule is used to synthesize ATP, as shown in Figure 17.4 on page 235. This overall reaction helps us focus on the starting reactants, final products, and release of energy. However, it misses the many intermediate steps that occur in the breakdown of glucose. Tossing a match into gasoline releases a tremendous amount of energy in the form of an explosion, but this energy is not used to do work. Instead, it is released as light and heat.

Similarly, if all the energy stored in glucose were released all at once in a single reaction, most of it would be released as heat and the cell would not be able to harness it to do work. In cellular respiration, energy is released gradually in a series of chemical reactions. This gradual release allows the energy to be harnessed by the cell and used to form ATP.

We also learned in Module 17 that cellular respiration can be divided into four main stages: glycolysis, pyruvate oxidation, the Krebs cycle, and oxidative phosphorylation, as illustrated in Figure 17.5 on page 237. In this module, we will look at each of these stages in more detail. We will start with glucose and follow it as it is oxidized step by step in each stage. In this way, we will learn how and where ATP is generated. Then we will consider other metabolic pathways, such as those that occur when oxygen is not present and those that are used by exercising muscle.

PREP FOR THE AP® EXAM

FOCUS ON THE BIG IDEAS

ENERGETICS: Focus on the movement of electrons and energy during cellular respiration, starting with glucose and oxygen gas, and ending with carbon dioxide and water.

18.1 Glycolysis, pyruvate oxidation, and the Krebs cycle break down organic molecules

As we discussed in the previous module, cellular respiration is divided into four stages. In this section, we will focus on the first three stages, which are glycolysis, pyruvate oxidation, and the Krebs cycle. These three stages fully oxidize glucose into carbon dioxide. In the process, they produce a small amount of ATP by substrate-level phosphorylation and reduced electron carriers. These reduced electron carriers generate much more ATP in stage 4.

Glycolysis

Glucose is a common fuel molecule in animals, plants, and microbes. It is the starting molecule for glycolysis, which results in the partial oxidation of glucose and the synthesis of a relatively small amount of both ATP and reduced electron carriers. The ATP can be used in cellular reactions and the electron carriers support more ATP production, as you will see. Glycolysis literally means "splitting sugar," an apt name because glucose (*glyco-*), a 6-carbon sugar, is split (*-lysis*) into two pyruvate molecules, which are 3-carbon molecules. Glycolysis is anaerobic because oxygen is not consumed. It evolved very early in the evolution of life, when oxygen was not present in Earth's atmosphere, and was retained over time. As a result, it occurs in nearly all living organisms and is probably the most widespread metabolic pathway. In this section, we will look more closely at this first stage of cellular respiration.

Glycolysis begins with a 6-carbon molecule of glucose and produces two 3-carbon molecules of pyruvate and a net total of 2 molecules of ATP and 2 molecules of the electron carrier NADH. ATP is produced directly by substrate-level phosphorylation in some of the reactions of glycolysis.

Glycolysis is a series of 10 chemical reactions that take place in the cytoplasm of cells. These reactions can be divided into three phases, shown in **FIGURE 18.1**. Phase 1 is the preparatory phase, which prepares glucose for the next

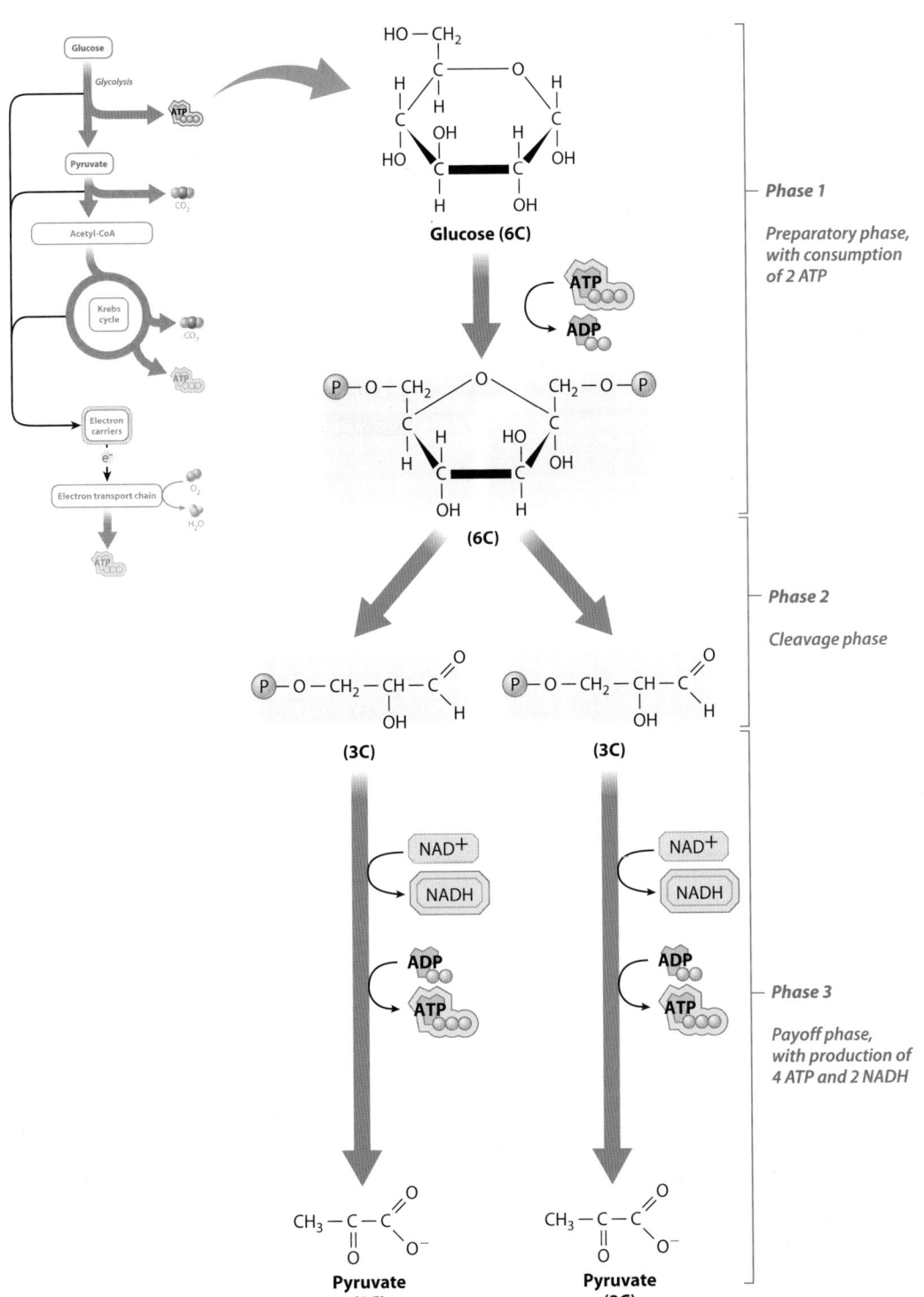

Phase 1

Preparatory phase, with consumption of 2 ATP

Phase 2

Cleavage phase

Phase 3

Payoff phase, with production of 4 ATP and 2 NADH

FIGURE 18.1 Glycolysis

During glycolysis, glucose, with six carbons (6C), is partially oxidized to pyruvate, with three carbons (3C). It occurs in three phases: preparatory, cleavage, and payoff phases. The process results in the net production of 2 molecules of ATP and 2 molecules of NADH. The yellow highlighting around ATP and NADH indicates that they carry chemical energy.

two phases by the addition of 2 phosphate groups to glucose. This phase requires an input of energy. To supply that energy and provide the phosphate groups, 2 molecules of ATP are hydrolyzed per molecule of glucose. In other words, the first phase of glycolysis is an endergonic process, as we discussed in Module 13.

The phosphorylation of glucose has two important consequences. First, it confines the product of the reaction to the inside of the cell. Although glucose enters and exits cells through specific membrane transporters, phosphorylated glucose is trapped inside the cell. Second, the presence of two negatively charged phosphate groups near each other repels and therefore destabilizes the molecule so that it can be broken apart in the second phase of glycolysis.

The next phase in Figure 18.1, phase 2, is the cleavage phase. In this phase, the 6-carbon (6C) molecule is split into two 3-carbon (3C) molecules. For each molecule of glucose entering glycolysis, two 3-carbon molecules enter the third phase of glycolysis.

Phase 3 of glycolysis is the payoff phase because ATP and the reduced electron carrier NADH are produced. Later, NADH will contribute to the synthesis of ATP during oxidative phosphorylation. This phase ends with the production of two molecules of pyruvate.

In summary, glycolysis begins with a single molecule of glucose (six carbons) and produces 2 molecules of pyruvate (three carbons each). These reactions yield 4 molecules of ATP and 2 molecules of NADH. However, 2 ATP molecules are consumed during the initial phase of glycolysis, resulting in a net gain of 2 ATP molecules and 2 molecules of NADH.

Pyruvate Oxidation

Glycolysis is a common and widespread metabolic pathway, but it does not produce very much ATP. The end product, pyruvate, still contains a good deal of chemical potential energy in its bonds. In the presence of oxygen, pyruvate can be further oxidized to release more energy, first to acetyl-CoA and then even further in the series of reactions of the Krebs cycle. In this section, we examine pyruvate oxidation, a key step that links glycolysis to the Krebs cycle.

In eukaryotes, pyruvate oxidation is the first step that takes place inside mitochondria. Mitochondria were introduced in Module 8 and are illustrated in **FIGURE 18.2**. Recall that they are rod-shaped organelles with a double membrane. The inner membrane has folds that project inward. These membranes define two spaces: the

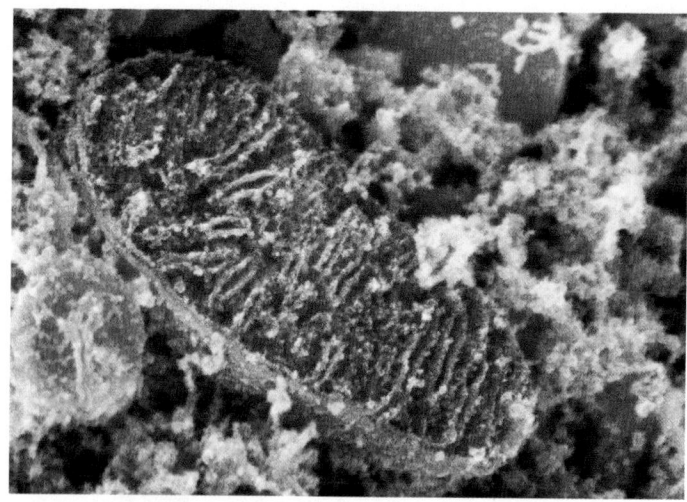

FIGURE 18.2 Mitochondria

In eukaryotes, pyruvate oxidation takes place inside mitochondria. Mitochondria are rod-shaped organelles with two membranes. There is an outer membrane and a highly folded inner membrane. The space between the inner and outer membranes is the intermembrane space, and the space inside the inner membrane is the mitochondrial matrix. Pyruvate oxidation and the Krebs cycle both take place in the mitochondrial matrix. Photo: Dr. David Furness, Keele University/ Science Source

space between the inner and outer membranes is called the intermembrane space, and the space enclosed by the inner membrane is called the mitochondrial matrix.

Pyruvate produced by glycolysis moves from the cytoplasm into the mitochondrial matrix, where it is converted into a molecule called acetyl-CoA. This reaction is shown in **FIGURE 18.3**. First, part of the pyruvate molecule is oxidized and splits off to form carbon dioxide, the most oxidized, and therefore the least energetic, form of carbon. Carbon dioxide is released as a waste product of the reaction. The electrons lost by pyruvate are gained by NAD^+, which is reduced to NADH. The remaining part of the pyruvate molecule, an acetyl group ($COCH_3$), still contains a large amount of potential energy. It is transferred to coenzyme A (CoA), a molecule that carries the acetyl group to the next set of reactions.

Overall, the synthesis of 1 molecule of acetyl-CoA from pyruvate results in the formation of 1 molecule of carbon dioxide and 1 molecule of NADH. Recall, however, that a single 6-carbon molecule of glucose forms two 3-carbon molecules of pyruvate during glycolysis. Therefore, 2 molecules of carbon dioxide, 2 molecules of NADH, and 2 molecules of acetyl-CoA are produced from a single glucose

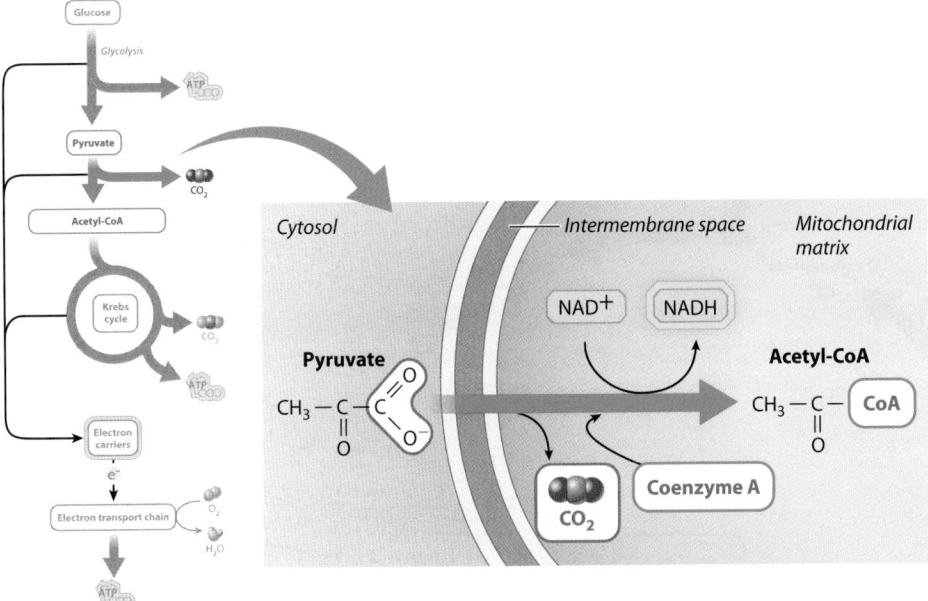

FIGURE 18.3 Pyruvate oxidation

Pyruvate is oxidized in the mitochondrial matrix, forming acetyl-CoA. In the process, NADH and carbon dioxide are produced. Acetyl-CoA is the first substrate in the Krebs cycle.

molecule in this stage of cellular respiration. Acetyl–CoA is the substrate, or reactant, of the first step in the Krebs cycle.

The Krebs Cycle

During the Krebs cycle, the oxidation of glucose is completed. Specifically, the acetyl group of acetyl-CoA is fully oxidized to carbon dioxide, and the chemical energy is transferred to ATP by substrate-level phosphorylation and to the electron carriers NADH and $FADH_2$. In this way, the Krebs cycle supplies electrons to the electron transport chain, leading to the production of much more energy in the form of ATP than is obtained by glycolysis alone. In this section, we look closely at the Krebs cycle.

Like the synthesis of acetyl-CoA, the Krebs cycle takes place in the mitochondrial matrix. It is composed of eight reactions and is called a cycle because the starting molecule is regenerated at the end, just like the Calvin cycle in photosynthesis. The cycle is illustrated in **FIGURE 18.4** on page 248.

In the first reaction, the 2-carbon acetyl group of acetyl-CoA is transferred to the 4-carbon molecule oxaloacetate to form the 6-carbon molecule citric acid. This reaction gives the Krebs cycle the alternative name citric acid cycle. Citric acid is then oxidized in a series of reactions. The last reaction of the cycle regenerates the 4-carbon molecule oxaloacetate, which can join a new 2-carbon acetyl

group to form the 6-carbon molecule citric acid and allow the cycle to continue.

Because the first reaction creates a molecule with six carbons (citric acid) and the last reaction regenerates a 4-carbon molecule (oxaloacetate), two carbons are eliminated during the cycle. These carbons are released as 2 molecules of carbon dioxide. Along with the release of carbon dioxide from pyruvate during pyruvate oxidation, these reactions are the sources of carbon dioxide released during cellular respiration and therefore the sources of the carbon dioxide that we exhale when we breathe. Four redox reactions, including the two that release carbon dioxide, produce the reduced electron carriers NADH and $FADH_2$. In this way, energy released in the oxidation reactions is transferred to a large quantity of reduced electron carriers: 3 molecules of NADH and 1 molecule of $FADH_2$ per turn of the cycle. These electron carriers donate electrons to the respiratory electron transport chain.

A single substrate-level phosphorylation reaction eventually generates a molecule of ATP from ADP. Overall, 2 molecules of acetyl-CoA produced from a single molecule of glucose yield 2 molecules of ATP, 6 molecules of NADH, and 2 molecules of $FADH_2$ in the Krebs cycle.

Let's summarize the first three stages of cellular respiration. Glycolysis starts with a 6-carbon molecule, glucose,

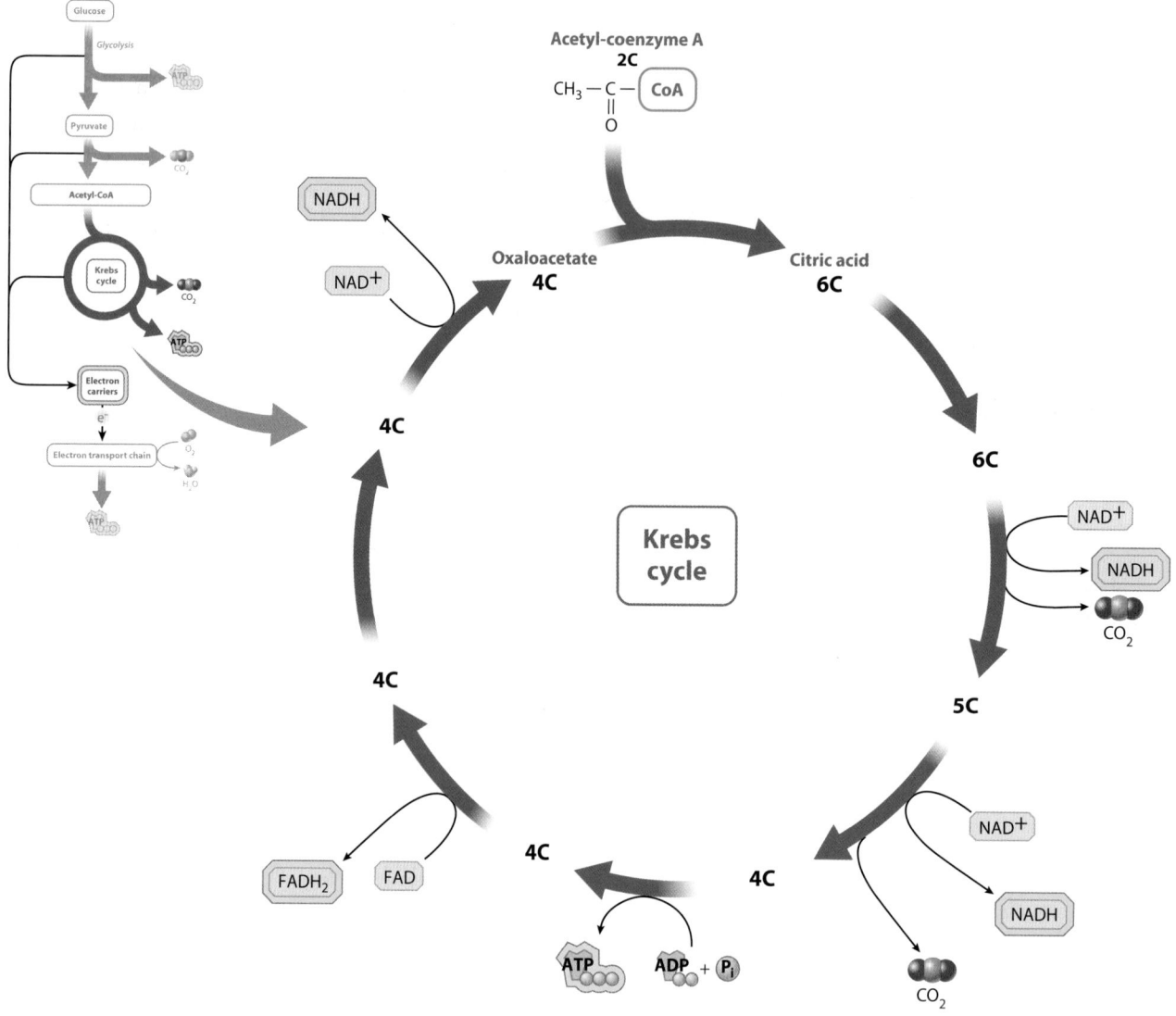

FIGURE 18.4 The Krebs cycle

In the Krebs cycle, the starting molecule and ending molecule are the same—oxaloacetate with four carbons (4C). Oxaloacetate joins with acetyl-CoA (2C) to make a 6-carbon (6C) molecule. Oxaloacetate is regenerated at the end. During the Krebs cycle, the acetyl group of acetyl-CoA is completely oxidized. The cycle results in the net production of 1 molecule of ATP, 3 molecules of NADH, and 1 molecule of FADH$_2$. The yellow highlighting is a reminder that the molecules carry chemical energy.

and generates two 3-carbon molecules, pyruvate. In the process, ATP and reduced electron carriers are produced. Pyruvate oxidation removes a carbon from each of the two 3-carbon molecules, producing carbon dioxide as a waste product, as well as more reduced electron carriers. The Krebs cycle completes the oxidation process, producing ATP, reduced electron carriers, and carbon dioxide. The ATP generated in these three stages can be used by the cell. The reduced electron carriers hand their electrons off to the respiratory electron transport chain, which synthesizes much more ATP by oxidative phosphorylation, as we will discuss next.

✓ Concept Check

1. **Identify** the major inputs and outputs of glycolysis.

2. **Identify** which molecules contain some of the chemical energy held in the original glucose molecule at the end of glycolysis, but before the subsequent stages in cellular respiration.

3. **Identify** which molecules contain the energy held in the original glucose molecule at the end of pyruvate oxidation, before the subsequent stages of cellular respiration.

4. **Identify** the molecules that contain the energy held in the original glucose molecule at the end of the Krebs cycle, but before the subsequent stages of cellular respiration.

18.2 Oxidative phosphorylation involves an electron transport chain and ATP synthesis

The complete oxidation of glucose during the first three stages of cellular respiration results in the production of two kinds of reduced electron carriers: NADH and $FADH_2$. We are now going to see how the energy stored in these electron carriers is used to synthesize ATP.

These reduced electron carriers hand off their electrons to the respiratory electron transport chain. Recall that the respiratory electron transport chain is made up of a series of redox reactions that occur in the inner mitochondrial membrane. Electrons are passed along the chain until they reach the final electron acceptor, oxygen, which is reduced to water. The energy released by these redox reactions is not converted directly into the chemical energy of ATP, however. Instead, the passage of electrons is coupled to the transfer of protons (H^+) across the inner mitochondrial membrane, creating a concentration and charge gradient. The electrochemical gradient provides a source of potential energy that is then used to drive the synthesis of ATP. This process is similar to what occurs in the photosynthetic electron transport chain described in Modules 15 and 16. In this section, we will explore the electron transport chain, the role of the proton gradient, and the synthesis of ATP. Keep in mind that the function of cellular respiration is to produce ATP for use by the cell. ATP is also produced in photosynthesis, but the main function of photosynthesis is to produce carbohydrates.

The Respiratory Electron Transport Chain

The respiratory electron transport chain is similar to the photosynthetic electron transport chain that we discussed in Modules 15 and 16. As you can see in **FIGURE 18.5**, it is made up of four large protein complexes known as complexes I to IV that are embedded in the inner mitochondrial membrane. The inner mitochondrial membrane contains one of the highest concentrations of proteins found in eukaryotic membranes.

Electrons donated by NADH and $FADH_2$ are transported along the series of protein complexes shown in the figure. Electrons enter the electron transport chain at either complex I or II. Electrons donated by NADH enter through complex I, and electrons donated by $FADH_2$ enter through complex II. These electrons are transported to complex III and then through

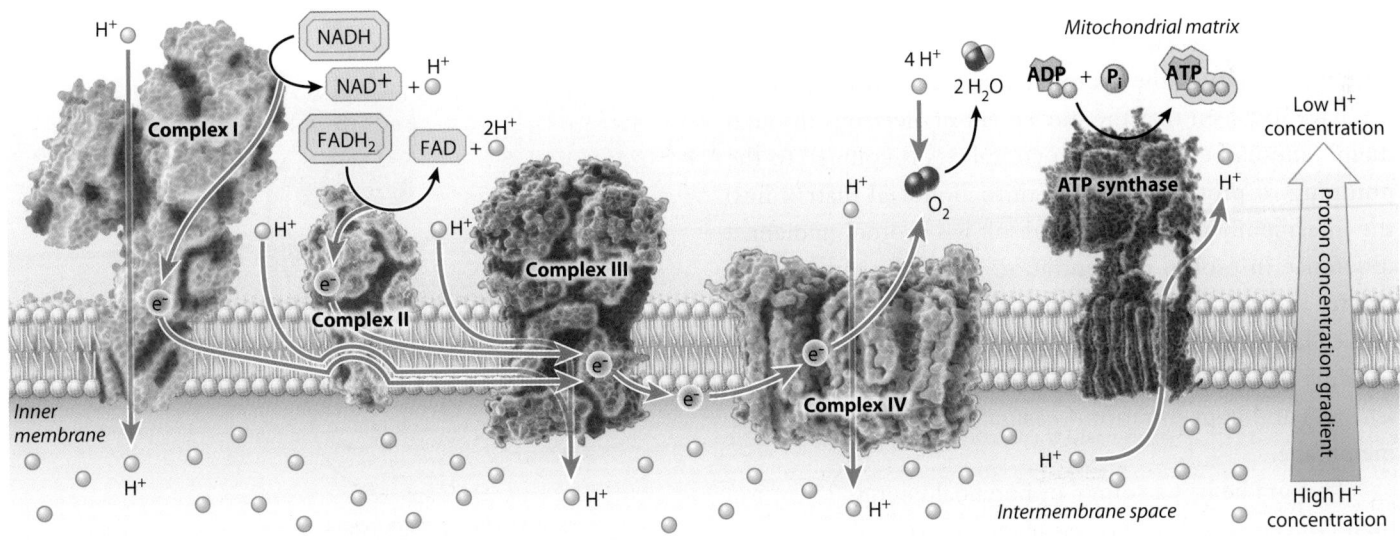

FIGURE 18.5 The respiratory electron transport chain

The respiratory electron transport chain consists of four complexes (I to IV) in the inner mitochondrial membrane. Electrons flow from electron carriers to oxygen, the final electron acceptor. The proton gradient formed from the electron transport chain has potential energy that is used to synthesize ATP.

complex IV. As NADH and $FADH_2$ hand off their electrons, they are oxidized, as shown follows:

$$NADH \rightarrow NAD^+ + 2e^- + H^+$$
$$FADH_2 \rightarrow FAD + 2e^- + 2H^+$$

NAD^+ and FAD can then accept electrons from the breakdown of organic molecules in stages 1 through 3, allowing the process of cellular respiration to continue.

Within each protein complex of the electron transport chain, electrons are passed from electron donors to electron acceptors. Each donor and acceptor pair is a redox couple, consisting of an oxidized and a reduced form of a molecule. The electron transport chain contains many of these redox couples. When oxygen accepts electrons at the end of the electron transport chain in the presence of protons (H^+), it is reduced to form water.

The electron transfer steps within complexes are each associated with the release of free energy. Some of this free energy is used to reduce the next electron acceptor in the chain, but some of it is used to pump protons (H^+) across the inner mitochondrial membrane, from the mitochondrial matrix to the intermembrane space, as shown in Figure 18.5. Thus, the transfer of electrons through the electron transport chain is coupled with the pumping of protons. The result is an accumulation of protons in the intermembrane space.

Proton Gradient

Like all membranes, the inner mitochondrial membrane is selectively permeable: protons cannot passively diffuse across this membrane because they are positively charged, and so cannot pass through the hydrophobic core of the lipid bilayer. We have just seen that the movement of electrons through membrane-embedded protein complexes is coupled to the pumping of protons from the mitochondrial matrix into the intermembrane space. The result is a proton gradient, a difference in proton concentration across the inner membrane, shown in Figure 18.5. The proton gradient is also called an electrochemical gradient, as we discussed in Module 16, because there is both a difference in charge (*electro-*) and concentration (*chemical*) of protons across the membrane.

This gradient is a source of potential energy. There is a high concentration of protons in the intermembrane space and a low concentration of protons in the mitochondrial matrix. As a result, there is a tendency for protons to diffuse back to the mitochondrial matrix, driven by a difference in concentration and charge on the two sides of the membrane. This movement, however, is blocked by the membrane. The gradient stores potential energy because if a pathway is opened

through the membrane, the resulting movement of the protons through the membrane can be used to perform work.

In sum, the oxidation of the electron carriers NADH and $FADH_2$ leads to the generation of a proton electrochemical gradient. This gradient is a source of potential energy used to synthesize ATP.

ATP Synthesis

In 1961, British biochemist Peter Mitchell proposed a hypothesis to explain how the energy stored in the proton electrochemical gradient is used to synthesize ATP. In 1978, he was awarded the Nobel Prize in Chemistry for work that fundamentally changed the way we understand how a cell harnesses energy.

According to Mitchell's hypothesis, the gradient of protons provides the driving force that is converted into chemical energy stored in ATP. First, for the potential energy of the proton gradient to be released, there must be an opening in the membrane for the protons to flow through. Mitchell suggested that protons in the intermembrane space diffuse down their electrical and concentration gradients through a transmembrane protein channel into the mitochondrial matrix. Second, the movement of protons through the channel must be coupled with the synthesis of ATP. This coupling is made possible by ATP synthase, the same enzyme that produces ATP in photosynthesis. Proton flow through the channel makes it possible for the enzyme to synthesize ATP, which is illustrated in **FIGURE 18.6**.

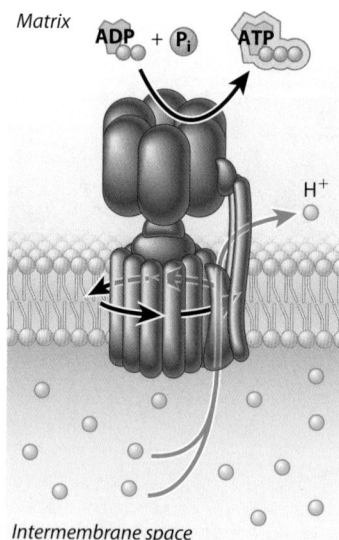

FIGURE 18.6 ATP synthase

ATP synthase drives the synthesis of ATP. Protons move down their electrochemical gradient from high concentration to low concentration through a channel in ATP synthase, causing it to rotate and leading to the formation of ATP from ADP and P_i by means of an electrochemical proton gradient.

Proton flow through the channel causes it to rotate, converting the energy of the proton gradient into mechanical rotational energy, a form of kinetic energy. The rotation in turn causes a conformational (shape) change in the enzyme that allows it to catalyze the synthesis of ATP from ADP and P_i. In this way, energy stored in the proton gradient is eventually converted into the chemical energy of ATP.

In this process, protons move from an area of high concentration to an area of low concentration across a selectively permeable membrane. The process in which protons move across the membrane and their movement is coupled to the synthesis of ATP is sometimes called **chemiosmosis** because it is similar to the movement of water across a selectively permeable membrane during osmosis. Direct experimental evidence for Mitchell's idea, called the chemiosmotic hypothesis, did not come for over a decade. One of the key experiments that provided support for his idea is illustrated in "Practicing Science 18.1: Can a proton gradient drive the synthesis of ATP?" In this experiment, the researchers created a vesicle consisting of a membrane, a proton pump activated by light, and ATP synthase. In the presence of light, the pump turned on, generating a proton gradient and leading to the production of ATP.

Practicing Science 18.1

PREP FOR THE AP® EXAM

Can a proton gradient drive the synthesis of ATP?

Background Peter Mitchell's hypothesis that a proton gradient can drive the synthesis of ATP was met with skepticism because he proposed the idea before experimental evidence existed to support it. In the 1970s, biochemist Efraim Racker and his collaborator Walther Stoeckenius tested the hypothesis.

Experiment Racker and Stoeckenius built an artificial system consisting of a collection of vesicle-shaped membranes, each with ATP synthase and a bacterial proton pump that was activated by light. They measured the concentration of protons (pH) in the external medium and the amount of ATP produced in the presence and absence of light.

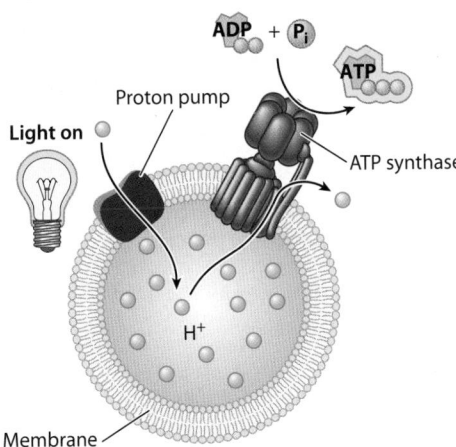

Results In the presence of light, the pH of the external medium increased, as shown in the graph. Notice that the pH increased rapidly when the light was turned on. An increase in pH indicates a decrease in concentration of protons as a result of movement of protons from the outside to the inside of the vesicles. This result demonstrates that the proton pump was activated and the vesicles were taking up protons. In the dark, the pH rapidly decreased.

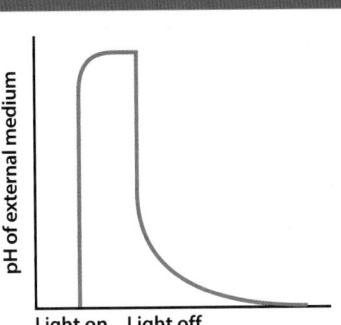

Furthermore, ATP was generated in the light, but not in the dark, as indicated in the table below.

Condition	Relative level of ATP
Light	594
Dark	23

Interpretation In the presence of light, the proton pump was activated, and protons were pumped to one side of the membrane, leading to an increase in pH and the formation of a proton gradient. The proton gradient in turn powered the synthesis of ATP by ATP synthase. As a result, high levels of ATP were produced in the light, but not in the dark.

Conclusion A membrane, proton gradient, and ATP synthase are sufficient to synthesize ATP. This result provided an important piece of experimental evidence for Mitchell's hypothesis.

SOURCES
Mitchell, P. 1961. "Coupling of Phosphorylation to Electron and Hydrogen Transfer by a Chemiosmotic Type of Mechanism." *Nature* 191:144–148; Racker, E., and W. Stoeckenius. 1974. "Reconstitution of Purple Membrane Vesicles Catalyzing Light-Driven Proton Uptake and Adenosine Triphosphate Formation." *Journal of Biological Chemistry* 249:662–663.

AP® PRACTICE QUESTION

(a) **Identify** the following:
   1. The question the experimenters are trying to answer
   2. The hypothesis they tested
   3. The independent variable
   4. The dependent variable

(b) The graph shows the change in the amount of hydrogen ions (H⁺) pumped from the outside of the vesicles to the inside as the concentration of proton pumps increases. The proton pumps are called purple protein because the purple membrane from *Halobacterium halobium* was incorporated into the phospholipid vesicles used in the experiment. **Justify** the following claims with evidence and reasoning:
   1. Claim: The purple protein are proton pumps.
   2. Claim: A graph showing ATP production would also increase in direct proportion to the increase in purple protein.

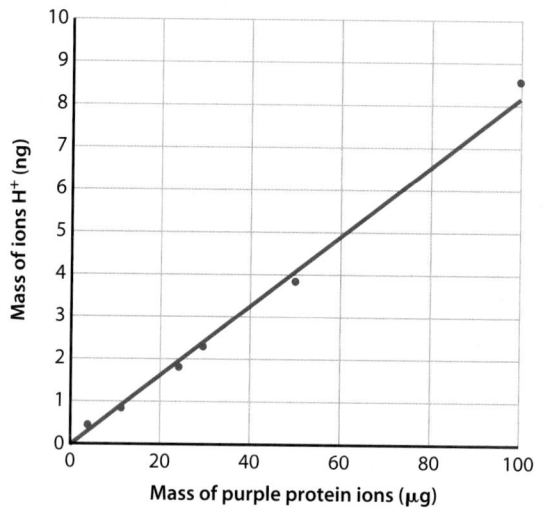

Total ATP Production

At this point, we have followed the oxidation of glucose through stages 1 through 4. The role of cellular respiration is to produce ATP for use by the cell. Approximately 2.5 molecules of ATP are produced for each NADH that donates electrons to the chain and 1.5 molecules of ATP are produced for each $FADH_2$. Overall, the complete oxidation of glucose yields about 32 molecules of ATP from all four stages of cellular respiration, shown in **TABLE 18.1**. Some of the energy held in the bonds of glucose is now available in the form of ATP that can be readily used by cells.

The energy needed to form one mole of ATP from ADP and P_i is approximately 7.3 kcal. Thus, cellular respiration harnesses at least $32 \times 7.3 = 233.6$ kcal of energy in ATP for every mole of glucose that is broken down in the presence of oxygen. About 34% of the total energy released by aerobic respiration is harnessed in the form of ATP ($233.6 \div 686 = 0.34 = 34\%$), with the remainder of

TABLE 18.1 Approximate Total ATP Yield in Cellular Respiration

Pathway	Substrate-level phosphorylation	Oxidative phosphorylation	Total ATP
Glycolysis (Glucose → 2 Pyruvate)	2 ATP	2 NADH = 5 ATP	7
Pyruvate oxidation (2 Pyruvate → 2 Acetyl-CoA)	0 ATP	2 NADH = 5 ATP	5
Krebs cycle (2 turns, 1 for each acetyl-CoA)	2 ATP	6 NADH = 15 ATP 2 $FADH_2$ = 3 ATP	20
Total	4 ATP	28 ATP	32

the energy given off as heat. This degree of efficiency compares favorably with the efficiency of a gasoline engine, which also burns fuel and which has an efficiency of approximately 25%.

We can appreciate how the proton gradient stores energy by seeing what happens when the gradient is dissipated. Uncoupling agents are proteins spanning the inner mitochondrial membrane that allow protons to pass through the membrane and bypass the ATP synthase channel. Uncoupling agents decrease the proton gradient and therefore decrease levels of ATP. The energy of the proton gradient is not used for oxidative phosphorylation but instead is dissipated as heat. Uncoupling agents are found naturally in certain tissues, such as fat, for heat generation. They can also act as poisons.

Let's follow the flow of energy in cellular respiration, which is illustrated in its full form in **FIGURE 18.7**. All four stages are shown, with their inputs and outputs, as well as where each of the stages takes place. We began with glucose and noted that it holds chemical potential energy in its covalent bonds. This energy is released in a series of reactions and captured in chemical form. Some of these reactions generate ATP directly by substrate-level phosphorylation. Others are redox reactions that transfer energy to the electron carriers NADH and FADH$_2$.

These reduced electron carriers donate electrons to the electron transport chain. Electrons are passed along a series of protein complexes embedded in the inner mitochondrial membrane to oxygen, the final electron acceptor. The movement of electrons along the respiratory electron transport chain is coupled to the pumping of protons across the inner membrane of the mitochondria. As a result, the energy of the reduced electron carriers is transformed into energy stored in a proton electrochemical gradient. ATP synthase then converts the energy of the proton gradient into rotational energy, which drives the synthesis of ATP by oxidative phosphorylation. The cell now has a useable form of energy that it can use in many ways to perform work.

Photosynthesis, which we discussed in Modules 15 and 16, and cellular respiration are complementary metabolic processes. Cellular respiration breaks down carbohydrates in the presence of oxygen, producing carbon dioxide and water. Photosynthesis synthesizes carbohydrates from carbon dioxide and water, and produces oxygen as a by-product. These two metabolic processes are summarized in **Visual Synthesis 3.1: Photosynthesis and Cellular Respiration** on page 260.

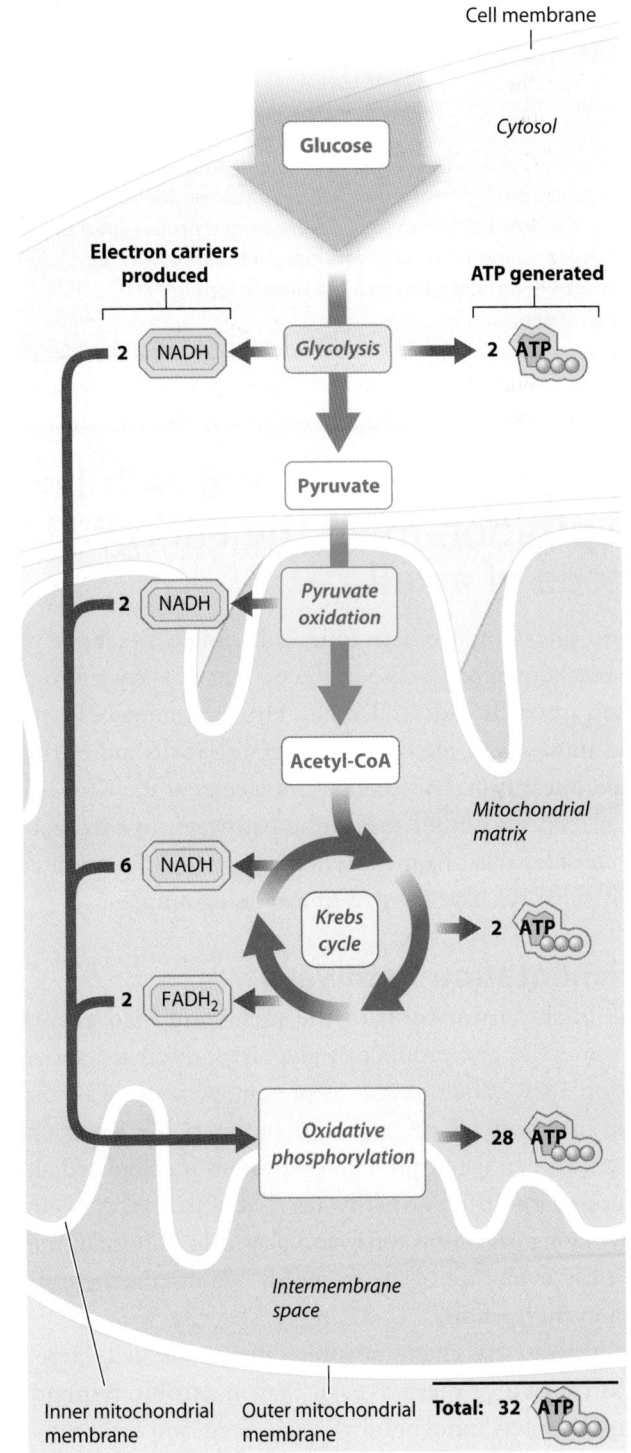

FIGURE 18.7 Total ATP production in cellular respiration

Cellular respiration consists of four stages. Glycolysis takes place in the cytosol; pyruvate oxidation and the Krebs cycle occur in the mitochondrial matrix; and oxidative phosphorylation occurs along the inner mitochondrial membrane. ATP is produced by substrate-level phosphorylation in glycolysis and the Krebs cycle, and by oxidative phosphorylation. A single glucose molecule yields a total of 32 molecules of ATP.

AP® EXAM TIP

Remember that during oxidative phosphorylation, electron transport and ATP formation are coupled, but they are not the same. The electron transport system pumps protons from the mitochondrial matrix to the intermembrane space, creating a proton gradient. The gradient in turn powers the synthesis of ATP as protons diffuse back into the matrix through ATP synthase.

✓ Concept Check

5. Animals breathe in air that contains more oxygen gas than the air they breathe out. **Identify** the process in which the oxygen gas is consumed.

6. **Describe** how the movement of electrons along the electron transport chain leads to the generation of a proton gradient.

7. **Describe** how a proton gradient is used to generate ATP.

18.3 Fermentation and cellular respiration meet the energy needs of a cell

Up to this point, we have followed a single metabolic path: the breakdown of glucose in the presence of oxygen to produce carbon dioxide and water. However, metabolic pathways more often resemble intersecting roads rather than a single, linear path. For example, molecules of the Krebs cycle often feed into other metabolic pathways. In this section, we consider what happens when oxygen is not present, and apply what we have learned to exercising muscles.

Fermentation Pathways

One of the major forks in the metabolic road occurs at pyruvate, the end product of glycolysis, discussed above in section 18.1. When oxygen is present, pyruvate is incorporated into acetyl-CoA, which then enters the Krebs cycle. When oxygen is not present, pyruvate is metabolized along a number of different pathways. These pathways occur in many living organisms today and played an important role in the early evolution of life on Earth. We discuss these pathways in this section.

Pyruvate has many possible fates in the cell. Here we focus on two of them. Recall that in aerobic respiration, pyruvate enters mitochondria, is oxidized, and feeds into the Krebs cycle. These processes produce reduced electron carriers that donate electrons to the respiratory electron transport chain, where oxygen is the final electron acceptor. If oxygen is not present, then the movement of electrons stops. NADH and $FADH_2$ cannot donate their electrons to the electron transport chain, so NAD^+ and FAD are not regenerated. Therefore, the Krebs cycle and the production of acetyl-CoA stop, and the concentration of pyruvate builds up in the cytoplasm.

In this case, pyruvate can be broken down by **fermentation**, a process that extracts energy from fuel molecules without using oxygen or an electron transport chain. Fermentation is accomplished through a wide variety of metabolic pathways. These pathways are important for anaerobic organisms, which live without oxygen, as well as for yeast and some other organisms that favor fermentation over oxidative phosphorylation, even in the presence of oxygen. In addition, aerobic organisms sometimes use fermentation when oxygen cannot be delivered quickly enough to meet the cell's metabolic needs, as in exercising muscle.

Recall that during glycolysis, glucose is oxidized to form pyruvate, and NAD^+ is reduced to form NADH. For glycolysis to continue, NADH must be oxidized back to NAD^+. Without NAD^+, glycolysis would grind to a halt. In the presence of oxygen, NAD^+ is regenerated when NADH donates its electrons to the electron transport chain. In the absence of oxygen during fermentation, NAD^+ is regenerated when pyruvate or a derivative of pyruvate is reduced in one of the fermentation pathways.

One fermentation pathway is lactic acid fermentation, which is shown in **FIGURE 18.8**. During lactic acid fermentation, electrons from NADH are transferred to pyruvate to produce lactic acid and NAD^+. The overall chemical reaction is written as follows:

$$\text{Glucose} + 2\text{ ADP} + 2\text{ P}_i \rightarrow 2\text{ Lactic acid} + 2\text{ ATP} + 2H_2O$$

Lactic acid fermentation is carried out by some bacteria and in animals, particularly in muscle cells, which is discussed below. We also use lactic acid fermentation to make yogurt, sauerkraut, kimchi, and pickles, to name a few applications.

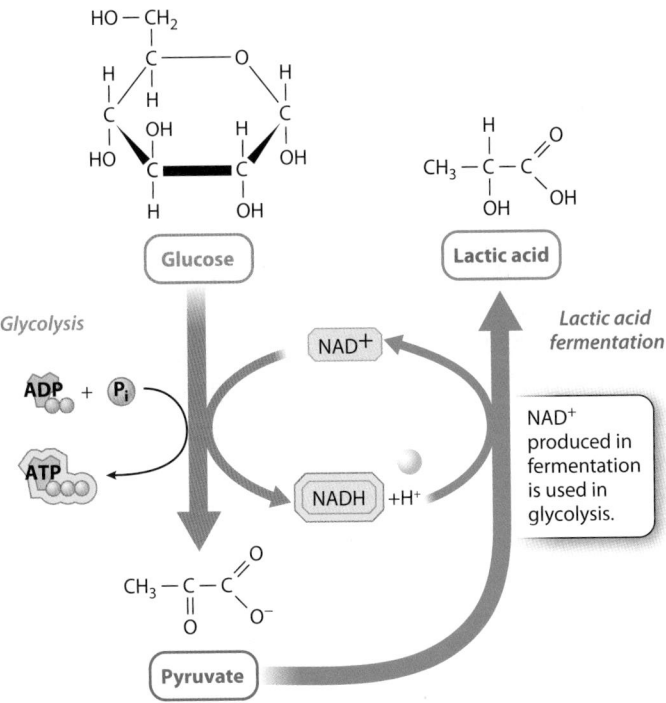

FIGURE 18.8 Lactic acid fermentation

ATP is generated by glycolysis in both aerobic and anaerobic conditions. In lactic acid fermentation, pyruvate is converted to lactic acid, and NAD⁺ is regenerated so that glycolysis can continue.

A second fermentation pathway is ethanol fermentation, shown in **FIGURE 18.9**. It occurs in plants and fungi. During ethanol fermentation, pyruvate releases carbon dioxide to form acetaldehyde, and electrons from NADH are transferred to acetaldehyde to produce ethanol and NAD⁺. The overall chemical reaction is written as follows:

$$\text{Glucose} + 2\,\text{ADP} + 2\,\text{P}_i \rightarrow 2\,\text{Ethanol} + 2\text{CO}_2 \\ + 2\,\text{ATP} + 2\text{H}_2\text{O}$$

Ethanol fermentation is the basis of the production of alcoholic beverages, such as wine and beer, and bread making, shown in **FIGURE 18.10** on page 256. In wine production, for example, sugars from grapes are fermented by yeasts to produce ethanol, an alcohol. In bread making, sugars in flour are fermented by yeasts to produce ethanol, which is removed in baking, and carbon dioxide, which causes the bread to rise.

In both fermentation pathways, NADH is oxidized to NAD⁺. However, NADH and NAD⁺ do not appear in the overall chemical equations because there is no net production or loss of either molecule. NAD⁺ molecules that are reduced during glycolysis are oxidized when lactic acid or ethanol is formed.

The breakdown of a molecule of glucose by fermentation yields only 2 molecules of ATP. The energetic gain is relatively small compared with the yield of aerobic respiration because the end products, lactic acid and ethanol, are not fully oxidized and still contain a large amount of chemical energy in their bonds. Organisms that produce ATP by fermentation therefore must consume a large quantity of fuel molecules to power the cell.

Exercise

In the last several modules, we considered what energy is and how it is harnessed by cells. Let's apply the concepts we discussed to a familiar example: exercise. Exercise, such as running, walking, and swimming, is a form of kinetic energy that is powered by ATP in muscle cells. Where does this ATP come from? Exercise requires several types of fuel molecules and the coordination of metabolic pathways.

Muscle cells, like all cells, do not contain a lot of ATP, and stored ATP is depleted by exercise in a matter of

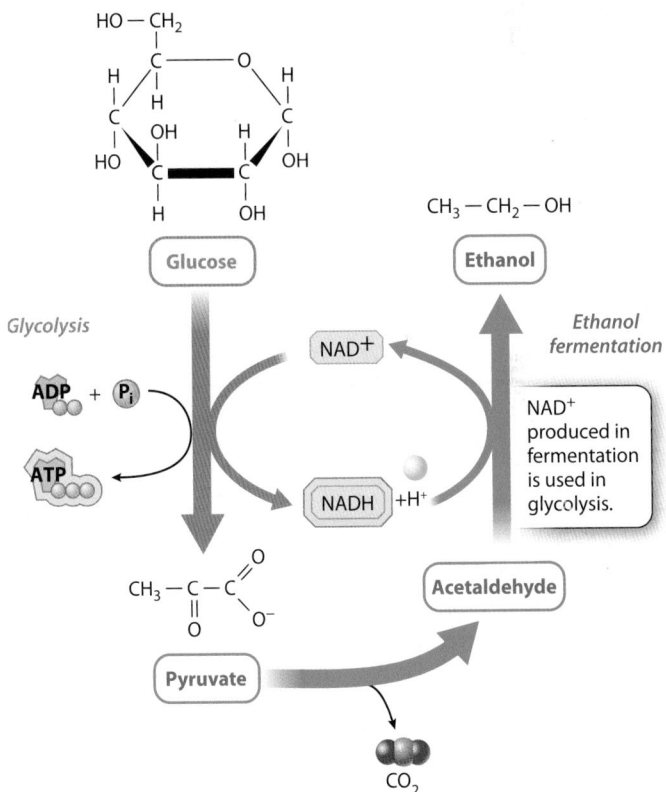

FIGURE 18.9 Ethanol fermentation

In ethanol fermentation, pyruvate is converted in two steps to ethanol. NAD⁺ is regenerated, as in other fermentation pathways. Carbon dioxide is produced as a waste product.

 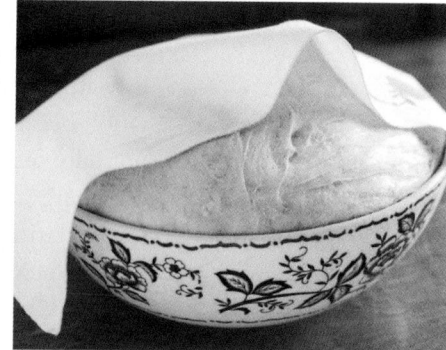

FIGURE 18.10 Wine production and bread making

The production of both wine and bread make use of ethanol fermentation.

Photos: (left) Morsa Images/Getty Images; (right) Alice Day/Alamy Stock Photo

seconds. As a result, muscle cells rely on fuel molecules to generate ATP. For a short sprint or a burst of activity, muscle can convert stored glycogen to glucose, and then break down glucose anaerobically by lactic acid fermentation. This pathway is rapid, but it does not generate very much ATP.

For longer, more sustained exercise, other metabolic pathways come into play. Muscle cells contain many mitochondria, which produce ATP by aerobic respiration. The energy yield of aerobic respiration is much greater than that of fermentation, but the process is slower. This slower production of ATP by aerobic respiration in part explains why runners cannot maintain the pace of a sprint for longer runs.

For even longer exercise, liver glycogen supplements muscle glycogen: the liver releases glucose into the blood that is taken up by muscle cells and oxidized to produce ATP. In addition, fatty acids are released from adipose tissue and taken up by muscle cells, where they are broken down. Using fatty acids as a source of energy yields even more ATP than does the complete oxidation of glucose—remember, the more C—H bonds, the more energy in a molecule—but the process is again slower. Storage forms of energy molecules, such as fatty acids and glycogen, contain large reservoirs of energy because they have many C—C and C—H bonds, but are slow to mobilize. Thus, exercise takes coordination between different cells, tissues, and metabolic pathways to ensure adequate ATP to meet the needs of working muscle.

✓ Concept Check

8. Bread making involves ethanol fermentation and typically uses yeast, sugar, flour, and water. **Describe** why yeast and sugar are used.

9. **Describe** two different metabolic pathways that pyruvate can enter.

10. **Describe** how muscle tissue generates ATP during short-term and long-term exercise.

Module 18 Summary

PREP FOR THE AP® EXAM

REVISIT THE BIG IDEAS

ENERGETICS: Using the content in the module, **describe** how the electron transport is related to the production of ATP during the metabolic pathways of cellular respiration.

LG 18.1 Glycolysis, pyruvate oxidation, and the Krebs cycle break down organic molecules.

- Glycolysis is a series of reactions in which glucose is oxidized to pyruvate. Page 244
- Glycolysis takes place in the cytoplasm. Page 245
- Glycolysis consists of preparatory, cleavage, and payoff phases. Page 245

- For each molecule of glucose broken down during glycolysis, a net gain of 2 molecules of ATP and 2 molecules of NADH are produced. Page 246
- The synthesis of ATP in glycolysis results from the direct transfer of a phosphate group from a substrate to ADP, a process called substrate-level phosphorylation. Page 246
- Pyruvate oxidation and the Krebs cycle occur in the mitochondrial matrix. Page 246
- The oxidation of pyruvate to acetyl-CoA results in the production of 1 molecule of NADH and 1 molecule of carbon dioxide. Page 246
- The acetyl group of acetyl-CoA is completely oxidized in the Krebs cycle. Page 247

- A complete turn of the Krebs cycle results in the production of 1 molecule of ATP, 3 molecules of NADH, and 1 molecule of $FADH_2$. Page 248

LG 18.2 Oxidative phosphorylation involves an electron transport chain and ATP synthesis.

- The reduced electron carriers NADH and $FADH_2$ donate electrons to the respiratory electron transport chain. Page 249
- The electron transport chain is made up of four complexes in the inner mitochondrial membrane. Page 249
- The transfer of electrons along the electron transport chain is coupled with the pumping of protons across the inner mitochondrial membrane into the intermembrane space. Page 250
- The buildup of protons in the intermembrane space results in a proton electrochemical gradient, which stores potential energy. Page 250
- The movement of protons back into the mitochondrial matrix through ATP synthase is coupled with the production of ATP. Page 253

LG 18.3 Fermentation and cellular respiration meet the energy needs of a cell.

- Glycolysis and fermentation are ancient biochemical pathways and were likely used in the common ancestor of all organisms living today. Page 254
- Pyruvate, the end product of glycolysis, is processed differently in the presence and the absence of oxygen. Page 254
- In the absence of oxygen, pyruvate enters one of several fermentation pathways in which NADH is oxidized to NAD^+. Page 254
- In lactic acid fermentation, pyruvate is reduced to lactic acid. Page 254
- In ethanol fermentation, pyruvate is converted to acetaldehyde, which is reduced to ethanol, and carbon dioxide is released. Page 255
- The ATP in muscle cells used to power exercise is generated by lactic acid fermentation and aerobic respiration and uses both carbohydrates and fatty acids as fuel molecules. Page 255

Key Terms

Chemiosmosis
Fermentation

Review Questions

1. At the end of glycolysis, the carbon molecules originally found in the starting glucose molecule are in the form of
   (A) 1 pyruvate molecule.
   (B) 2 pyruvate molecules.
   (C) 2 ATP molecules.
   (D) 2 NADH molecules.

2. In eukaryotes, pyruvate oxidation takes place in the
   (A) cytoplasm.
   (B) chloroplast.
   (C) intermembrane space of mitochondria.
   (D) mitochondrial matrix.

3. During the Krebs cycle
   (A) fuel molecules are completely reduced.
   (B) ATP is synthesized by substrate-level phosphorylation.
   (C) high-energy electrons are removed from electron carriers.
   (D) ATP is synthesized by oxidative phosphorylation.

4. The final electron acceptor of the electron transport chain is
   (A) water.   (C) ATP synthase.
   (B) oxygen.   (D) NAD^+.

5. For the potential energy of a proton gradient to be converted to the chemical energy of ATP, the movement of protons down their electrochemical gradient must be coupled with ATP synthesis. This coupling is made possible by

(A) ATP.

(B) a proton gradient.

(C) redox couples.

(D) ATP synthase.

6. Fermentation takes place in the

(A) cytoplasm.

(B) chloroplast.

(C) intermembrane space of mitochondria.

(D) mitochondrial matrix.

7. During lactic acid fermentation, pyruvate is

(A) oxidized.

(B) reduced.

(C) phosphorylated.

(D) carboxylated.

Module 18
AP® Practice Questions

PREP FOR THE AP® EXAM

Section 1: Multiple-Choice Questions

Choose the best answer for questions 1–5.

1. Identify the molecule that is the terminal electron acceptor at the end of the electron transport chain in aerobic cellular respiration.

(A) FAD

(B) NAD^+

(C) H_2O

(D) O_2

Questions 2 and 3 refer to the figure below, which illustrates the steps in glycolysis when all enzymes in the process are functioning normally.

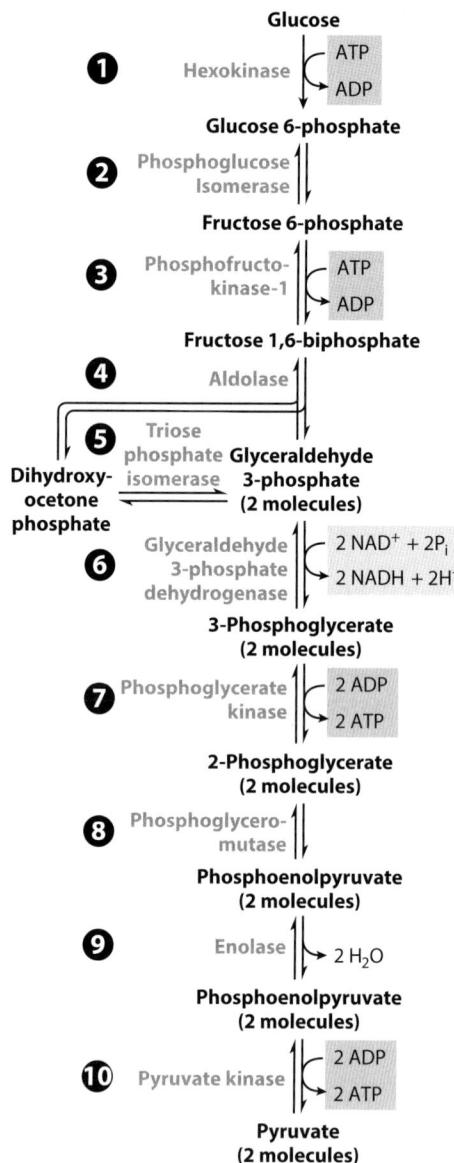

2. Which step in the process illustrates substrate-level phosphorylation?

(A) Step 3   (C) Step 7

(B) Step 6   (D) Step 9

3. In the pathway shown, the enzyme enolase catalyzes step 9. If a cell lacked functional enolase enzymes, what would be the net amount of ATP produced from each molecule of glucose?

(A) Loss of one ATP

(B) No net change in ATP

(C) Loss of two ATP

(D) Gain of four ATP

Use the following information to answer questions 4 and 5.

As shown in the figure below, mitochondria have two membranes, an inner membrane and an outer membrane. The space between the two membranes is called the intermembrane space. The inner membrane is folded into structures called cristae. The space inside the inner membrane is called the matrix.

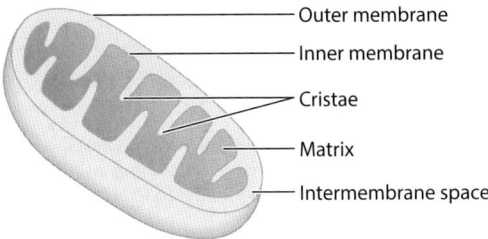

4. Which statement best explains the role of cristae in a mitochondrion?

    (A) The cristae increase the surface area of the inner membrane relative to the volume of the intermembrane space in order to establish a larger proton gradient.

    (B) The cristae allow a larger inner membrane in order to increase the volume of the matrix and allow more efficient accumulation of $FADH_2$.

    (C) The cristae increase the surface area of the outer membrane in order to more efficiently import glucose into the mitochondrion.

    (D) The cristae allow the matrix to be larger than the intermembrane space in order to promote more efficient glycolysis in the mitochondrion.

5. Which part of a functioning mitochondrion has the lowest pH?

    (A) Intermembrane space

    (B) Inner membrane

    (C) Cristae

    (D) Matrix

Section 2: Free-Response Question

Write your answer to each part clearly. Support your answers with relevant information and examples. Where calculations are required, show your work.

Thermogenin is a protein found in the mitochondria of brown adipose (fat) cells in mammals. It allows animals to regulate their body temperature by generating heat through metabolic means. Thermogenin is a transport protein. When an animal's body temperature is low, thermogenin moves protons across the membrane in which it is embedded. The energy of the movement of the protons, which normally powers ATP synthesis by ATP synthase, is instead released as heat.

(a) **Predict** the effect of a drug that inhibits thermogenin function in an animal.

(b) **Identify** where in the mitochondrion thermogenin is most likely to be found.

(c) A scientist claims that thermogenin dissipates proton gradients in the mitochondria of brown adipose tissue. **Support or refute the claim.** Provide your reasoning.

(d) Brown adipose tissue is more abundant in juvenile mammals, especially small babies, than it is in larger adults, even though juveniles and adults typically have similar body temperatures. **Make a claim** that supports this observation.

VISUAL SYNTHESIS 3.1 PHOTOSYNTHESIS AND CELLULAR RESPIRATION

Photosynthesis builds carbohydrates from carbon dioxide and water using the energy of sunlight. Cellular respiration breaks down carbohydrates, supplying ATP for use by the cell.

Photosynthesis
Energy captured from sunlight is stored in carbohydrates.

Photosynthesis requires light energy, CO_2, and H_2O, and produces carbohydrates and O_2. The *light reactions* use energy from sunlight to drive the movement of electrons from water to $NADP^+$ along a photosynthetic electron transport chain, producing NADPH and ATP.

Carbon dioxide is reduced by NADPH and, using the energy provided by ATP, is incorporated into a carbohydrate molecule in the *Calvin cycle*.

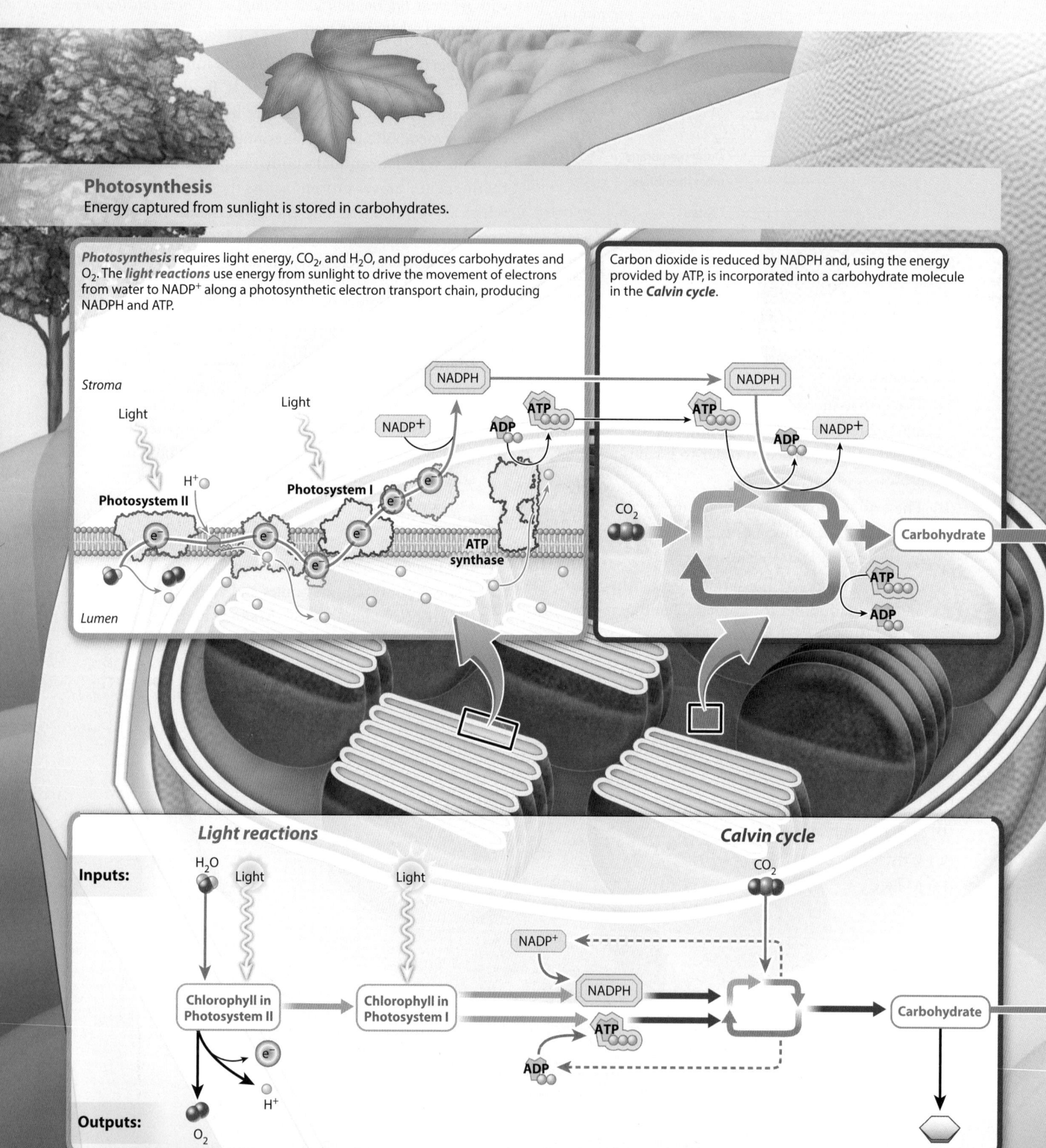

Cellular respiration

Energy released from the oxidation of carbohydrates is used to make ATP.

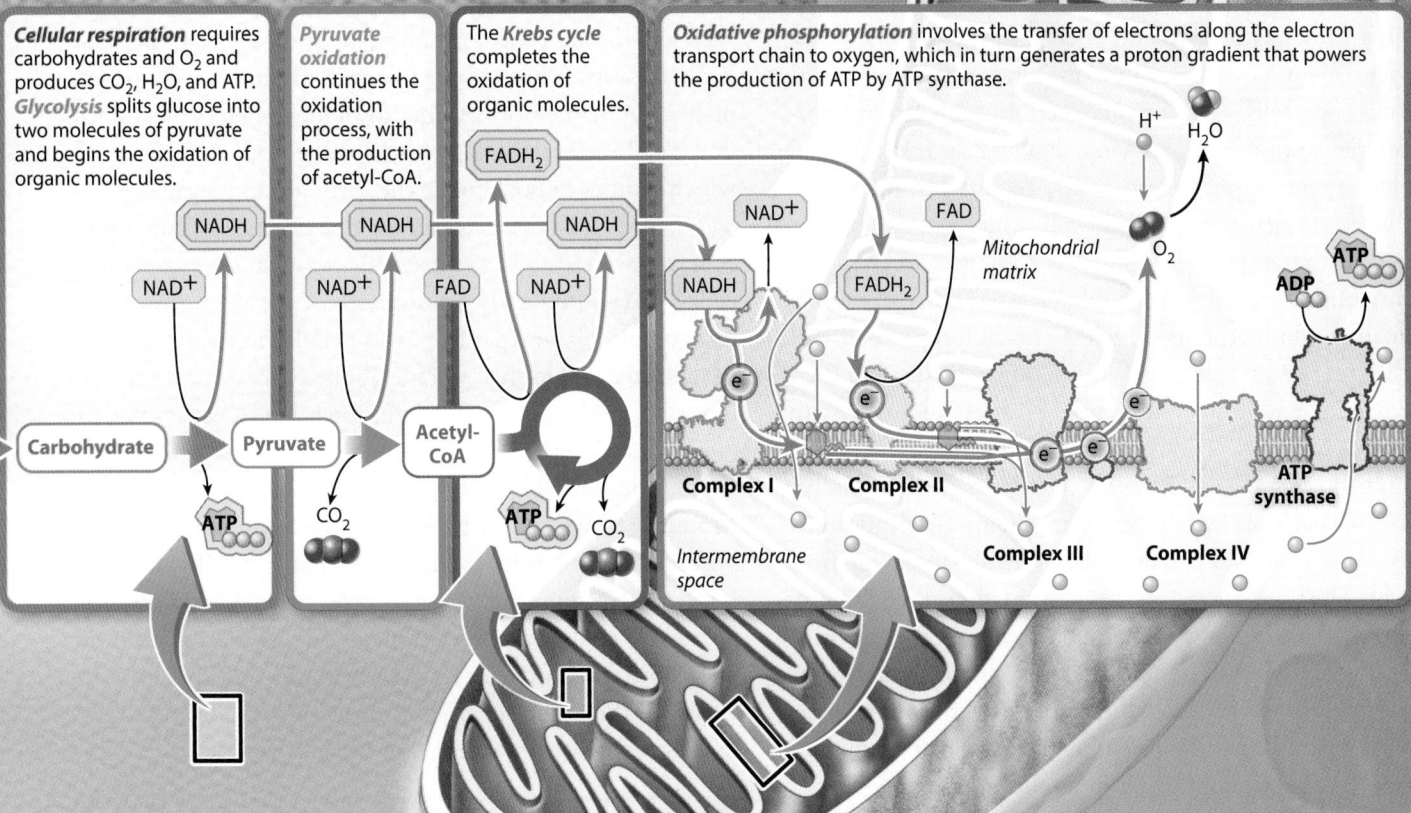

Cellular respiration requires carbohydrates and O_2 and produces CO_2, H_2O, and ATP. *Glycolysis* splits glucose into two molecules of pyruvate and begins the oxidation of organic molecules.

Pyruvate oxidation continues the oxidation process, with the production of acetyl-CoA.

The *Krebs cycle* completes the oxidation of organic molecules.

Oxidative phosphorylation involves the transfer of electrons along the electron transport chain to oxygen, which in turn generates a proton gradient that powers the production of ATP by ATP synthase.

NADH
NAD+

NADH
NAD+   FAD

FADH₂

NADH
NAD+

Carbohydrate → Pyruvate → Acetyl-CoA

ATP

CO_2

ATP CO_2

NAD+   FAD H^+ H_2O

Mitochondrial matrix

NADH   FADH₂ O_2 ADP   ATP

e⁻   e⁻   e⁻   e⁻   e⁻

Complex I   Complex II

Complex III   Complex IV   ATP synthase

Intermembrane space

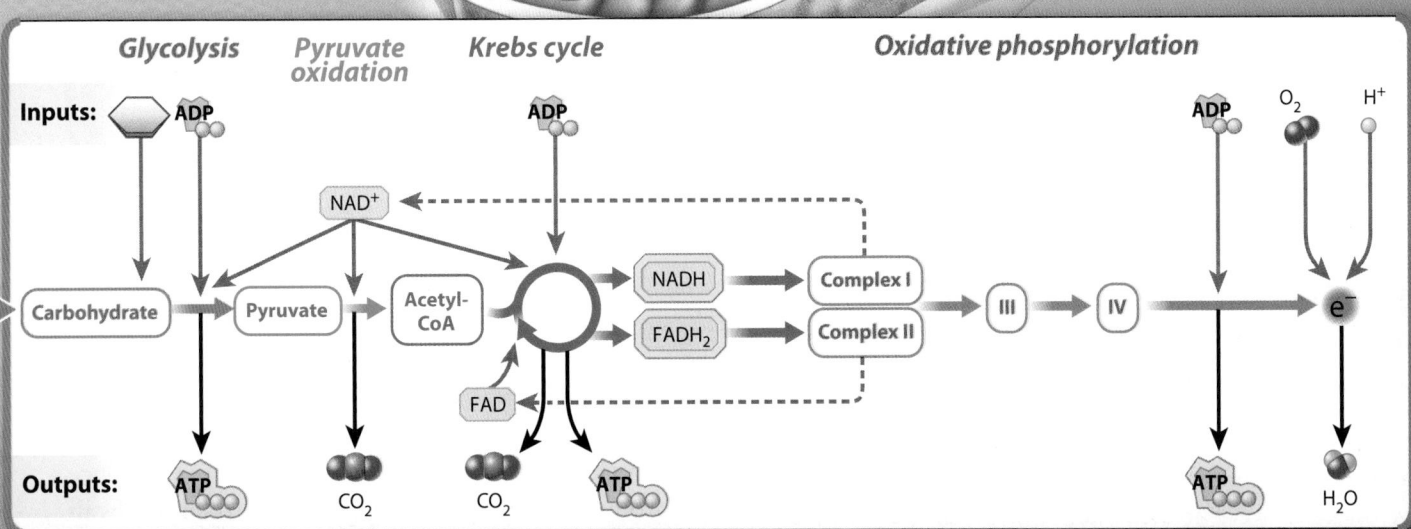

Glycolysis   *Pyruvate oxidation*   *Krebs cycle*   *Oxidative phosphorylation*

Inputs: ADP   ADP   ADP O_2 H^+

NAD+

Carbohydrate → Pyruvate → Acetyl-CoA

NADH → Complex I
FADH₂ → Complex II

III → IV → e⁻

FAD

Outputs: ATP CO_2 CO_2 ATP   ATP H_2O

Module 19

Metabolism, the Environment, and Evolutionary Fitness

LEARNING GOALS ▶**LG 19.1** The earliest cells used anaerobic metabolism to survive and reproduce.
▶**LG 19.2** Photosynthesis transformed life on Earth.
▶**LG 19.3** Metabolic variation affects survival and reproduction.

This unit has focused on energy. Energy comes in two forms—potential and kinetic—and is used by cells to do work. For cells, work involves actions such as synthesizing biological molecules, moving, dividing, and transporting substances in and out. Energy is also required to maintain a high level of organization of the cell, which we explored in Unit 2. In the absence of a sustained input of energy, cells and organisms would not be able to stay highly organized and maintain homeostasis, which is critical for life itself.

Cells, like everything else, are subject to the laws of thermodynamics, which describe how energy behaves in a system. For cells, chemical reactions not only transform reactants into products but also transfer energy. Cells obtain their energy from just two sources: the sun and chemical compounds. Plants, some algae, and certain bacteria transform energy from sunlight. They use the set of chemical reactions called photosynthesis to convert light energy into chemical energy held in the bonds of carbohydrates. Plants, animals, fungi, and many other organisms break down these energy-rich carbohydrates in another set of chemical reactions called cellular respiration, packaging the energy into ATP to do the work of the cell.

Photosynthesis and cellular respiration are complementary metabolic processes. In photosynthesis, carbon dioxide and water react to form carbohydrates and oxygen, and in cellular respiration, carbohydrates and oxygen react to form carbon dioxide and water. If we follow carbon through these transformations, we see that it cycles from the air, into organisms, and back into the air. When organisms die and decay, carbon is buried in the earth. And, over vast periods of time, ancient organic matter can be transformed into fossil fuels, such as oil and natural gas. Together, these processes make up the carbon cycle, which shuttles carbon among the earth, air, and organisms.

The carbon cycle is one of the engines of life. However, it hasn't always operated the way we see it operating today. In fact, the early atmosphere of the young Earth did not contain oxygen gas to power cellular respiration. In this module, we will look at the metabolic processes we explored in earlier modules from an evolutionary perspective. We will start by considering how the earliest cells met their energy needs, using one of the fermentation pathways described in Module 18. We will then explore the evolution of photosynthesis and its impact on life on Earth, setting the stage for aerobic respiration. We will also consider how variation in the molecules involved in metabolic processes can influence how efficiently they are carried out, and therefore the **fitness**—the ability of organisms to survive and reproduce in a particular environment—of organisms.

PREP FOR THE AP® EXAM

FOCUS ON THE BIG IDEAS

EVOLUTION: Focus on the connection between variation in the types of molecules found in the cells of an organism and the ability of that organism to survive and reproduce in changing environments.

19.1 The earliest cells used anaerobic metabolism to survive and reproduce

All cells require energy to carry out their functions and maintain their internal organization, and the first cells to evolve on Earth were no exception. These cells used simple chemical reactions to harness energy from macromolecules in the environment. In this section, we will discuss how the earliest cells on Earth met their energy requirements.

Earth formed approximately 4.6 billion years ago. For the first half billion years or so, Earth was molten, so there were no rocks and no evidence of life. But we see evidence of life from the time rocks solidified. Microscopic fossils document diverse bacterial communities that stretch back more than 3.5 billion years. Larger fossils include layered sedimentary structures called stromatolites that were produced by microbial communities on shallow seafloors and lake bottoms. **FIGURE 19.1** shows stromatolites in Shark Bay, Australia. Chemical signatures of microbial life further support the view that life originated at least 3.5 billion years ago. We don't know exactly when life originated because as we go further back in time, we run out of rocks before we run out of biological evidence.

Scientists aren't entirely clear how life emerged from nonlife. It is possible that early chemical reactions produced the first organic molecules, or that a meteor seeded Earth with its first organic molecules. Either way, multiple lines of evidence suggest that all species on Earth today can trace their ancestry back to a single living organism. We explore how life originated in Unit 7.

The earliest forms of life were prokaryotic: single-celled organisms without a nucleus or other internal compartments. The first 2 billion years of life on Earth, almost half of the history of Earth, belonged to these prokaryotic organisms, including bacteria and archaea. It was a time of great diversification. Although the bacteria and archaea remained unicellular, they evolved many different metabolic pathways.

Ancient rocks tell us something else: for its first billion or more years, life existed in a world without oxygen gas. Because

FIGURE 19.1 Stromatolites

Stromatolites are microbial communities that build the domed structures shown in this photograph of modern-day stromatolites taken at Shark Bay, Australia. Stromatolites are among the earliest records of life on Earth. Photo: Andrew Knoll, Harvard University

the early atmosphere contained little to no oxygen gas when life first evolved, the earliest organisms probably used one of the fermentation pathways described in Module 18 to generate ATP, carry out their work, survive, and reproduce. As we learned, fermentation does not require oxygen, occurs in the cytoplasm, and does not make use of proteins embedded in organelle membranes to transfer electrons and energy along an electron transport chain. During fermentation, glucose is only partially oxidized, so just some of the energy held in its chemical bonds is transferred to ATP. Nearly all organisms are capable of partially breaking down glucose, suggesting that this metabolic pathway evolved very early in the history of life and has been maintained over evolutionary time.

Fermentation begins with glycolysis, the first of the four stages of cellular respiration. As we discussed in earlier modules, cellular respiration leads to the full oxidation of glucose and the transfer of a large amount of energy from the chemical bonds of glucose to the chemical bonds of ATP. Cellular respiration makes use of a respiratory electron transport chain composed of protein complexes embedded in a membrane. It transfers electrons from one complex to the next and, in the process, pumps protons across the membrane. The resulting proton gradient powers the synthesis of a large amount of ATP.

Like fermentation, the electron transport chain can operate in the absence of oxygen, but in this case molecules other than oxygen gas, such as sulfate and nitrate, are the final electron acceptors. When oxygen gas is the final electron acceptor, we call the process aerobic cellular respiration. When sulfate, nitrate, or anything other than oxygen is the final electron acceptor, we call the process anaerobic cellular respiration. Anaerobic cellular respiration likely evolved after fermentation, as a way of harnessing additional energy from glucose in the absence of oxygen. It still occurs in some present-day bacteria. The electron transport chain in these bacteria is located in the cell membrane, not in the inner mitochondrial membrane, as it is in eukaryotes.

How might such a system have evolved? One hypothesis is shown in **FIGURE 19.2** on page 264. The environment of early Earth was likely acidic. To maintain a neutral pH of about 7, early prokaryotes evolved protein pumps to drive protons (H^+) out of the cell. Some of these pumps might have used the energy of ATP directly to pump protons. Others might have used the energy from electron transport to pump protons, in the same way that the respiratory and photosynthetic electron transport chains do today. These two proton pumps are illustrated in Figure 19.2a.

At some point, proton pumps powered by electron transport might have generated a large enough electrochemical gradient that the protons could pass back through the ATP-driven pumps, running them in reverse to synthesize ATP, as shown in Figure 19.2b. In this way, early forms of cellular respiration, operating in the absence of oxygen, could generate ATP to power the work of the cell.

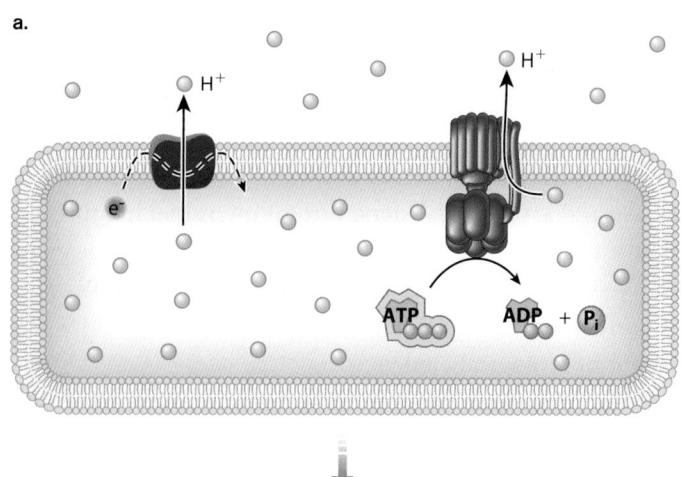

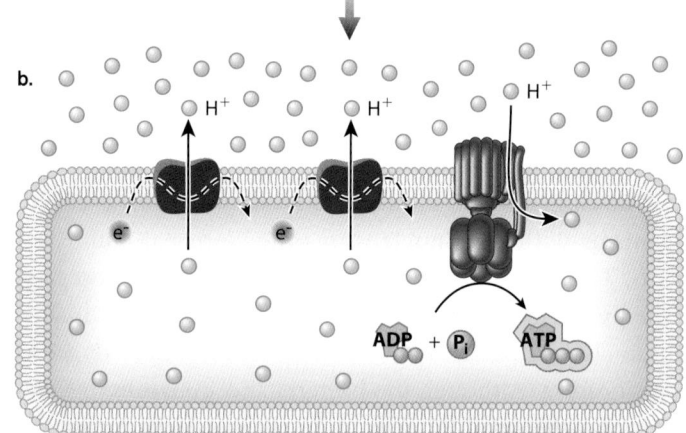

> ## ✓ Concept Check
>
> 1. **Describe** the early atmosphere of Earth.
> 2. **Describe** how the earliest cells met their energy requirements.

FIGURE 19.2 Evolution of the electron transport chain

How the electron transport chain evolved is uncertain, but one hypothesis is shown here. (a) Early cells evolved mechanisms to pump protons (H^+) out of the cell, including pumps powered by electron transport (left) and pumps powered by ATP (right). Those cells that transported protons out of the cell reproduced more successfully than those that did not. (b) Over time, electron transport–powered pumps might have generated such a large proton gradient that the ATP-driven pump could be run in reverse, synthesizing rather than using ATP.

19.2 Photosynthesis transformed life on Earth

The evolution of photosynthesis had a profound impact on the history of life on Earth. Not only did photosynthesis provide organisms with a new source of energy—the sun—but it also released oxygen into the atmosphere for the first time. In this section, we consider hypotheses for the evolution of the photosynthetic pathways that are the dominant entry point for energy into the biosphere today.

As we discussed in the last section, early Earth had little to no oxygen gas. Chemical evidence in ancient sedimentary rocks tells us that oxygen gas first began to accumulate in the atmosphere and oceans about 2.4 billion years ago, almost halfway across the timeline of the history of Earth. At this time, one group of bacteria—cyanobacteria—evolved the ability to carry out the form of photosynthesis that produces oxygen. They used the energy from sunlight to build sugars and produced, as a by-product, oxygen, as described in Modules 15 and 16. Present-day cyanobacteria, descendants of these early photosynthetic bacteria, are shown in **FIGURE 19.3**.

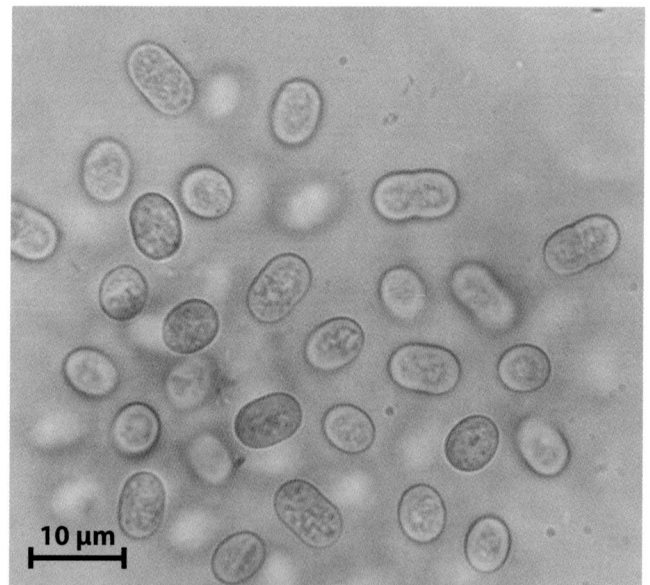

FIGURE 19.3 Cyanobacteria

Cyanobacteria were the first organisms to evolve the biochemical machinery to carry out the form of photosynthesis in which water donates electrons and oxygen is produced as a by-product and released into the ocean and atmosphere. Pictured here are modern cyanobacteria in the genus *Aphanothece*. Photo: Dr. Ralf Wagner

Cyanobacteria evolved a photosynthetic electron transport chain that used water as an electron donor. This form of photosynthesis is called oxygenic photosynthesis because it generates oxygen. Other forms of photosynthesis already existed at the time. These forms used molecules other than water as the electron donor, and therefore did not produce oxygen as a by-product of the reaction.

The ability to use water as an electron donor in photosynthesis had two major impacts on life on Earth. First, photosynthesis could now occur in any environment with both sunlight and sufficient water for cells to survive. Second, because water releases oxygen as it donates electrons, oxygen began to accumulate in Earth's atmosphere. Over time, oxygen levels in the atmosphere increased as it accumulated from production by these tiny bacteria.

No other group evolved this ability in the history of life on Earth. It evolved just once. Organisms such as plants and algae, which are also capable of oxygenic photosynthesis, can do so only because they incorporated these bacteria by endosymbiosis, as discussed in Module 12.

Even at relatively low concentrations, oxygen gas opened up new possibilities for biological diversification, including the evolution of the eukaryotic cell, with its membrane-bounded nucleus and compartmentalized organelles for energy metabolism. Photosynthesis is hypothesized to have gained a foothold among eukaryotes when a free-living cyanobacterium took up residence inside a eukaryotic cell. **FIGURE 19.4** shows this series of events. Over time, the cyanobacterium lost its ability to survive outside its host cell and evolved into the chloroplast. The outer membrane of the chloroplast is thought to have originated from the cell membrane of the ancestral eukaryotic cell, which surrounded the ancestral cyanobacterium as it became incorporated into the cytoplasm of the eukaryotic cell. The inner chloroplast membrane is thought to correspond to the cell membrane of the ancestral free-living cyanobacterium. The thylakoid membrane then corresponds to the internal photosynthetic membrane found in cyanobacteria. Finally, the stroma corresponds to the cytoplasm of the ancestral cyanobacterium.

The evolution of oxygenic photosynthesis changed the composition of Earth's atmosphere. For some organisms, the increase in oxygen gas proved deadly. For example, as we saw in Module 16, photosynthesis can produce reactive oxygen species, which are toxic to the cell. In fact, the accumulation of oxygen gas in the atmosphere is often called the Oxygen Catastrophe or Crisis to describe the massive loss of life that occurred. However, for other organisms, this

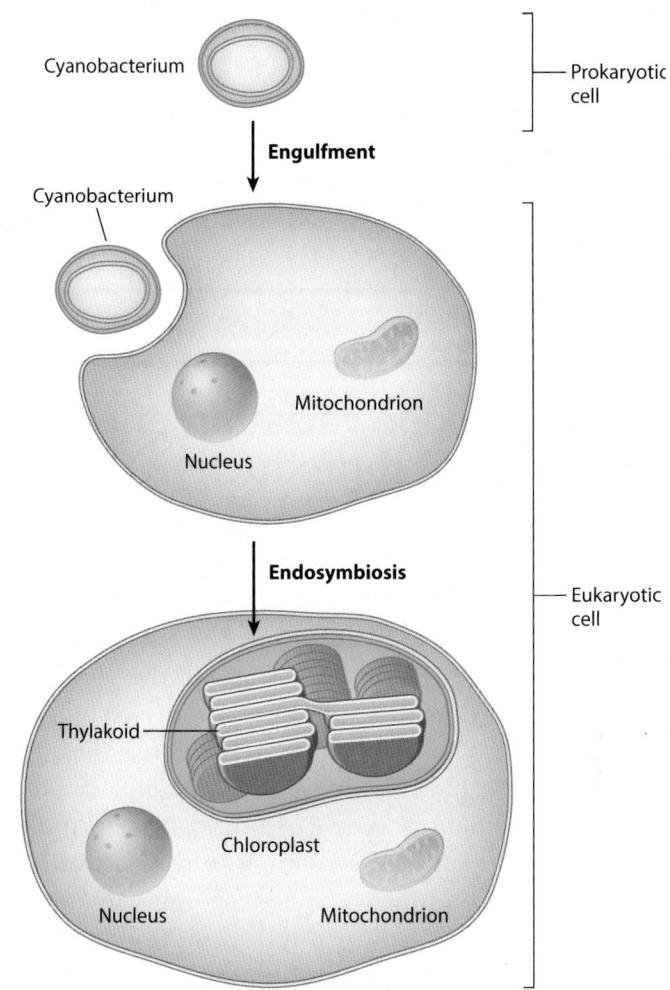

FIGURE 19.4 Endosymbiosis

An ancestral eukaryotic cell engulfed a cyanobacterium, which over time evolved into the chloroplast. Chloroplasts replicate within the host cell and are transmitted from one generation of the host cell to the next.

dramatic change to Earth's atmosphere led to the evolution of new ways to extract energy from organic molecules such as glucose. Aerobic cellular respiration, in which oxygen serves as the final electron acceptor in the respiratory electron transport chain, releases much more energy in the form of ATP than does anaerobic metabolism, such as fermentation and anaerobic cellular respiration.

The evolution of metabolic pathways illustrates that evolution often works in a step-wise fashion, building on what is already present. In this case, aerobic respiration picked up where anaerobic respiration left off, making it possible to harness more energy from organic molecules to power the work of the cell. In turn, these new metabolic pathways increased the ability of some organisms to survive and reproduce, and therefore increased their evolutionary fitness.

✓ Concept Check

3. **Identify** the organism that was the first to evolve the biochemical machinery of oxygenic photosynthesis.

4. **Describe** how eukaryotes evolved the ability to carry out photosynthesis.

19.3 Metabolic variation affects survival and reproduction

The metabolic processes of photosynthesis and cellular respiration are carried out by many species and drive the biological carbon cycle. As we have seen, these metabolic processes make use of many different molecules and cellular structures. Small changes in these components can affect how these processes are carried out, and therefore the survival and reproduction of organisms. The ability of an organism to survive and reproduce in a particular environment is known as fitness, and it strongly influences the course of evolution. In this section, we will focus on two of these components—photosynthetic pigments such as chlorophyll, and the oxygen-carrying molecule hemoglobin—to illustrate how small changes in key molecules can affect the fitness of individuals.

Photosynthetic Pigments

In Modules 15 and 16, we discussed the major photosynthetic pigment, chlorophyll. Chlorophyll absorbs the energy of sunlight and transfers it to other components of the photosystem. Pigments like chlorophyll absorb certain wavelengths of light and reflect others. Chlorophyll, for example, absorbs blue and yellow wavelengths of light, and reflects green, so appears green. There are several different forms of chlorophyll and each absorbs a slightly different wavelength of light. In this way, they are well adapted for different environments with different wavelengths of available light.

Chlorophyll is not the only pigment found in plants and other photosynthetic organisms. There are also other pigments that absorb different wavelengths of light and have diverse roles. One such pigment is anthocyanin. The color it reflects depends on pH, and can appear red, blue, or purple. Carotenoids are typically red, orange, and yellow pigments. There are two groups of carotenoids, carotenes and xanthophylls. Carotenes give carrots their orange color. Xanthophylls are yellow-orange pigments. During autumn, as leaves die, these various pigments become visible, producing the brilliant fall colors many of us enjoy each year.

As a group, these accessory pigments serve two basic roles: they assist in harvesting light energy for the cell, and also help protect cells from excess light energy. For example, xanthophylls accept absorbed light energy directly from chlorophyll, and then convert this energy to heat. In this way, they reduce excess light energy, which can lead to the formation of potentially dangerous reactive oxygen species that can damage the cell.

Photosynthetic organisms that live in extreme environments often appear brown or yellow because they contain high levels of xanthophyll pigments, as seen in **FIGURE 19.5**. As an indication of the importance of

FIGURE 19.5 Accessory photosynthetic pigments

This hot spring in Yellowstone National Park is yellow colored due to photosynthetic bacteria that use the pigment xanthophyll. Photo: f11photo/Shutterstock

xanthophylls, plants that lack them grow poorly when exposed to moderate light levels and die in full sunlight because they are not protected from the full range of damaging light energy. In this way, the presence or absence of specific pigments can affect how well particular organisms survive in different environments. The fact that plants have a variety of pigments in different amounts means that when the light availability increases or decreases, some individuals will be able to survive and reproduce more successfully than others.

Hemoglobin

Variation in molecules involved in cellular respiration can also affect survival and reproduction. We breathe in oxygen to power cellular respiration. The oxygen we inhale by our lungs is carried in the blood so it can be delivered to cells, such as those of an exercising muscle. In the blood, oxygen is bound to a molecule called hemoglobin in red blood cells. Hemoglobin is an ancient molecule that has diverse roles related to oxygen binding and transport. It has been found in organisms from all kingdoms of life. Both vertebrates and invertebrates have hemoglobin that binds and transports oxygen in the circulatory fluid. This observation indicates that the common ancestor of invertebrate and vertebrate animals evolved a shared form of hemoglobin.

Hemoglobin is also present in organisms other than animals. Plant hemoglobins bind and transport O_2 within the cell. In soybeans, for example, hemoglobin binds O_2 in root cells to keep concentrations low enough to avoid inhibiting enzymes important for the functioning of the cells. In bacteria, archaea, and fungi, forms of hemoglobin evolved as the final electron acceptor in cellular respiration. These hemoglobins may have also served a role in scavenging O_2 to avoid damage due to oxidative stress. Earlier, we saw an example of oxidative stress when oxygen accumulated in the atmosphere for the first time, leading to the extinction of organisms that were adapted to an oxygen-free environment. So, we see that a single molecule can evolve over time to take on diverse roles important for the function of organisms, which is a common theme in evolution.

In vertebrates, hemoglobin exists in large concentrations within red blood cells. It consists of four globin subunits, shown in **FIGURE 19.6**. Each of these four units includes a heme group with an iron atom, which reversibly binds one O_2 molecule. The reversible binding allows hemoglobin to pick up oxygen in the lungs and deliver oxygen to tissues, where it is used to power cellular

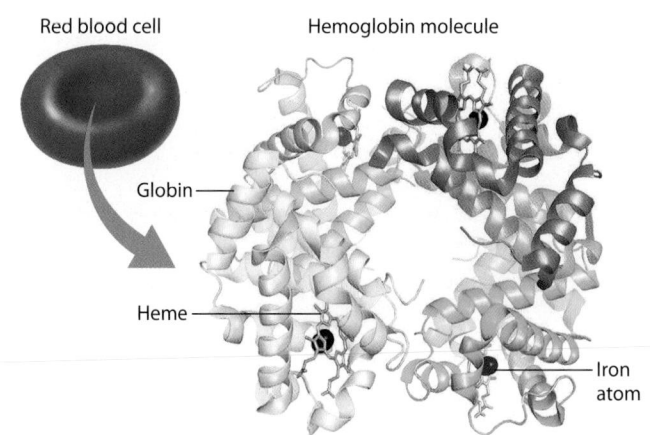

Red blood cell   Hemoglobin molecule

Globin

Heme

Iron atom

FIGURE 19.6 Hemoglobin

Hemoglobin consists of four subunits, each of which has an iron-containing heme group that binds one oxygen molecule. Variants of hemoglobin have different affinities for oxygen, allowing organisms to live in a wide range of environments, and cells within a single organism to work together.

respiration. Therefore, hemoglobin efficiently binds to, transports, and delivers oxygen to tissues.

There are different forms, or variants, of hemoglobin, which differ slightly in their structure and how tightly they bind oxygen. These different forms allow organisms to carry out their functions, survive, and reproduce. For example, in mammals, the mother's respiratory and cardiovascular systems provide the developing fetus with O_2 and remove CO_2, as well as supply nutrients and remove wastes. Maternal and fetal blood do not mix, but rather exchange gases across a structure called the placenta. Yet the O_2 concentration gradient at this exchange point does not strongly favor the movement of O_2 from the mother's blood to the fetal blood. Fetal mammals solve the problem of extracting O_2 from their mother's hemoglobin by expressing a form of hemoglobin, called fetal hemoglobin, that has a higher affinity for O_2 than does their mother's hemoglobin. As a result, the fetus is supplied with sufficient O_2. At birth, when the newborn begins to breathe on its own, there is a shift to expressing the adult form of hemoglobin, maintaining this form throughout life.

Many animals live at high altitude. For example, some birds, such as the Rüppell's vulture (*Gyps rueppellii*) shown in **FIGURE 19.7**, have been observed flying over Mt. Everest, the highest peak in the world at an elevation of approximately 29,000 feet (or 9000 m), even though air at this altitude has less than one-third the O_2 content of air at sea level. Similarly, llamas inhabit the Andes at altitudes up to 5000 m even though the air has about one-half the

FIGURE 19.7 Rüppell's vulture

This bird has a form of hemoglobin that binds oxygen with high affinity. As a result, it can fly at very high altitudes where most other birds cannot fly, which opens up new environments. Photo: Copyright by hellboy2503/ Jörg David/Getty Images

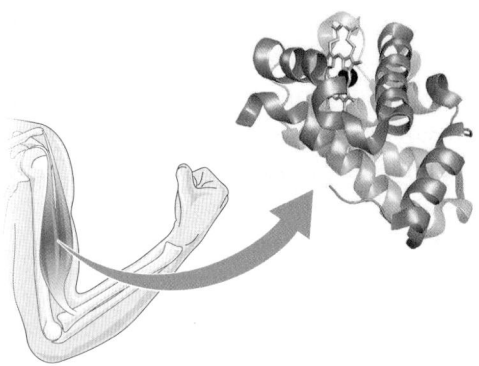

FIGURE 19.8 Myoglobin

Myoglobin is found in muscle, where it binds oxygen. It consists of a just a single subunit with one iron-containing heme group.

a.

b.

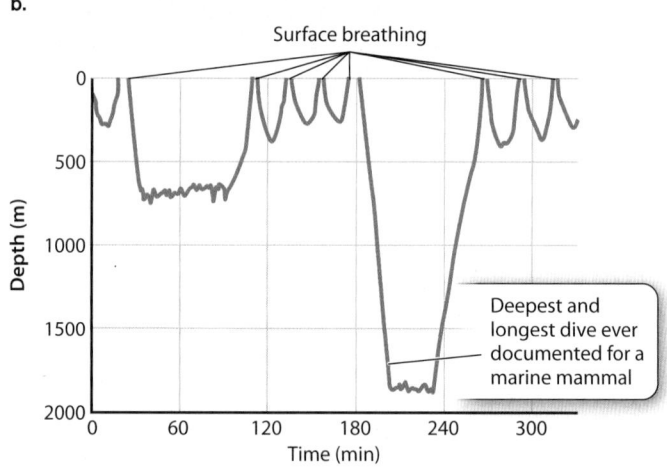

FIGURE 19.9 Long-lasting dives in whales

Myoglobin binds to and stores oxygen, allowing whales and other marine mammals to hold their breath for long periods. (a) Diving marine mammals, such as this beaked whale (*Ziphius cavirostris*), have large amounts of myoglobin within their muscles, which allows them to live in the water. (b) The graph plots the dives of a beaked whale fitted with a tracking device. Data from Figure 3A in P. L. Tyack, M. Johnson, N. Aguilar de Soto, A. Sturlese, and P. T. Madsen. 2006. "Extreme Diving Behaviour of Beaked Whale Species Known to Strand in Conjunction with Use of Military Sonars." *Journal of Experimental Biology* 209:4238–4253. Photo: Bill Curtsinger/Getty Images

O_2 content of air at sea level. How do these animals obtain enough O_2? Their hemoglobin has a higher affinity for O_2 and binds O_2 more readily than does the hemoglobin of animals living at sea level.

A close molecular relative of hemoglobin is myoglobin, shown in **FIGURE 19.8**. Myoglobin is a specialized O_2 carrier within the cells of vertebrate muscles. In contrast to hemoglobin, myoglobin consists of a single subunit instead of four subunits. This subunit contains a single heme group. Myoglobin has a higher affinity for O_2 than hemoglobin does, and it binds O_2 more tightly. As a result, hemoglobin in the blood releases O_2 to exercising muscles.

Muscle cells that depend mainly on aerobic respiration to produce ATP store large amounts of myoglobin. The myoglobin in these cells can release O_2 quickly at the onset of activity, before the respiratory and circulatory systems have had time to increase the supply of O_2. Diving marine mammals, such as whales and seals, have large amounts of myoglobin within their muscles. The myoglobin loads up with O_2 when the animals breathe at the surface before beginning a dive. The O_2 bound to the myoglobin is then used to supply ATP during the dive, when the animal cannot breathe. These marine mammals can stay under water for 30 minutes or longer and can dive to considerable depths, as shown in **FIGURE 19.9**.

As with photosynthetic pigments, we see that different forms of hemoglobin allow organisms to survive and reproduce in diverse environments. Variation in hemoglobin allows oxygen to be transported efficiently to respiring cells, allowing them to meet their energy needs and carry out their functions.

✓ **Concept Check**

5. **Describe** two roles of accessory pigments.

6. **Describe** how the various globin molecules, such as adult hemoglobin, fetal hemoglobin, the hemoglobin of animals that live at high altitude, and myoglobin, are similar and how they are different from one another, and what benefit these differences confer.

Module 19 Summary

PREP FOR THE AP® EXAM

REVISIT THE BIG IDEAS

EVOLUTION: Using an example from the module, provide reasoning to **justify** the following claim: Organisms with variation in the molecules of their cells have a greater likelihood of surviving and reproducing in changing environments.

LG 19.1 The earliest cells used anaerobic metabolism to survive and reproduce.

- Earth is approximately 4.6 billion years old. Page 262
- First life evolved at least 3.5 billion years ago. Page 263
- The early atmosphere of the young Earth had little or no oxygen gas. Page 263
- Fermentation is an ancient biochemical pathway that does not require oxygen; it was likely used in the common ancestor of all species living today. Page 263
- Anaerobic cellular respiration makes use of final electron acceptors other than oxygen gas. Page 263

LG 19.2 Photosynthesis transformed life on Earth.

- The evolution of photosynthesis had a profound impact on life on Earth as oxygen gas accumulated in the atmosphere for the first time. Page 264
- The ability to use water as an electron donor in photosynthesis, with oxygen gas produced as a by-product, first evolved in cyanobacteria approximately 2.4 billion years ago. Page 265

- Photosynthesis in eukaryotes evolved by endosymbiosis, in which a cyanobacterium was engulfed by another cell and over time became the chloroplast. Page 265
- All of the oxygen in Earth's atmosphere results from photosynthesis by photosynthetic bacteria, algae, and plants. Page 265

LG 19.3 Metabolic variation affects survival and reproduction.

- The ability to survive and reproduce in a particular environment is called fitness. Page 266
- Variation in molecules involved in metabolic processes can affect fitness. Page 266
- There are several different forms of chlorophyll, which absorb slightly different wavelengths of light. Page 266
- Pigments other than chlorophyll assist in light harvesting and provide protection against excess light energy. Page 266
- Hemoglobin is an ancient molecule. Page 266
- Red blood cells contain hemoglobin with iron-containing heme groups that reversibly bind and release oxygen gas. Page 267
- Fetal hemoglobin expressed by the mammalian fetus has a higher binding affinity for oxygen than adult hemoglobin does, allowing for oxygen uptake from the mother's blood. Page 267
- Myoglobin binds and stores oxygen in muscle cells, increasing the delivery of oxygen to muscle mitochondria for activity in general and for diving in marine mammals. Page 268

Key Term

Fitness

Review Questions

1. Which statement is true regarding Earth's earliest organisms?
   (A) These organisms likely carried out cellular respiration.
   (B) These organisms carried out fermentation because early Earth's atmosphere was rich in oxygen.
   (C) These organisms likely used oxygen as an electron acceptor.
   (D) These organisms likely carried out fermentation and may have had rudimentary proton pumps.

2. The process in which one cell takes up residence in another cell and loses its ability to survive on its own is called
   (A) endosymbiosis.
   (B) parasitism.
   (C) endocytosis.
   (D) exocytosis.

3. Which was the first group of organisms able to use water as an electron donor in photosynthesis and therefore produce oxygen as a by-product?
   (A) Land plants
   (B) Aquatic plants
   (C) Algae
   (D) Cyanobacteria

4. Xanthophylls are
   (A) part of the respiratory electron transport chain.
   (B) the first organism to evolve the biochemical machinery for photosynthesis.
   (C) accessory pigments in photosynthesis.
   (D) a symbiosis between a photosynthetic organism and a host.

5. What is the main benefit of having hemoglobin in the blood?
   (A) It keeps oxygen separated from carbon dioxide within the blood.
   (B) It increases the amount of oxygen that can be carried in the blood.
   (C) It increases the solubility of carbon dioxide in the blood.
   (D) It is an important source of iron for body cells.

6. How is a fetus able to extract O_2 from its mother's blood?
   (A) The maternal blood mixes with the fetal blood in the placenta to provide O_2.
   (B) Fetal hemoglobin has a lower affinity for O_2 than maternal hemoglobin does.
   (C) Fetal blood has a higher concentration of hemoglobin than maternal blood does.
   (D) Fetal hemoglobin has a higher affinity for O_2 than maternal hemoglobin does.

7. Which statement about myoglobin is true?
   (A) It transports O_2 in the bloodstream.
   (B) It is found within muscle tissue.
   (C) It is not found in marine mammals.
   (D) It irreversibly binds to O_2.

Module 19
AP® Practice Questions

Section 1: Multiple-Choice Questions

Choose the best answer for questions 1–4.

1. Hemoglobin is an iron-containing molecule that, in animals, transports O_2 in blood. Hemoglobin is also found in plants and microbes, where it regulates the concentration of free oxygen within cells. Different forms of hemoglobin have different affinities for oxygen. Which of the following provides the best evidence that hemoglobin evolved early in the history of life?
   (A) Hemoglobin is found in almost all forms of life.
   (B) Hemoglobin is extremely important for health and fitness.
   (C) Some animals use different forms of hemoglobin at different stages of their lives.
   (D) Hemoglobin has been modified in diverse organisms over evolutionary time.

2. The early atmosphere on Earth had little or no oxygen gas, but O_2 became a major component of the atmosphere following the evolution of widespread photosynthesis. Which of the following is a consequence of the change in Earth's environment?

(A) The presence of O_2 in the atmosphere enabled photosynthesis and allowed plants to become a dominant form of life on Earth.

(B) The presence of O_2 in the atmosphere allowed primitive bacteria to obtain energy from carbohydrates through anaerobic fermentation.

(C) The presence of O_2 in the atmosphere established a symbiotic relationship between prokaryotes and eukaryotes.

(D) The presence of O_2 in the atmosphere resulted in the extinction of organisms that could not tolerate a high-oxygen environment.

Use the following information to answer questions 3 and 4.

The graph below indicates the percentage of Earth's atmosphere that was composed of oxygen gas over its history. The time that key prokaryotic and eukaryotic groups arose is indicated at the top of the graph.

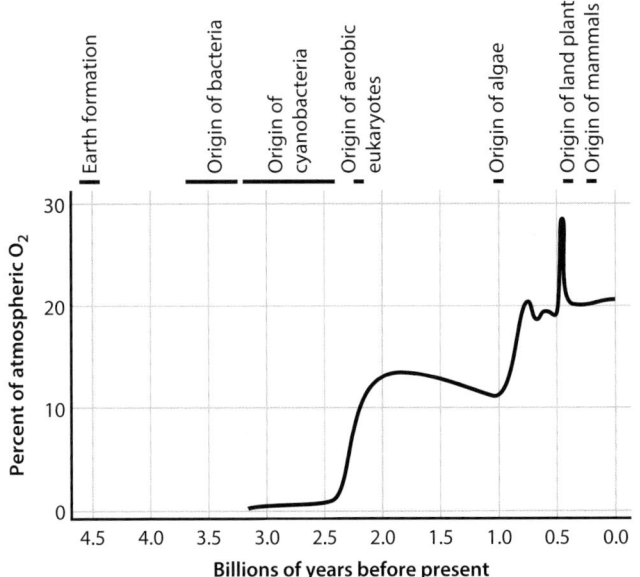

3. Based on the data in the figure, identify the approximate time at which Earth's atmosphere had the highest abundance of O_2 gas.

(A) More than 2.3 billion years before present

(B) 0.7 billion years before present

(C) 0.4 billion years before present

(D) Less than 0.2 billion years before present

4. Based on the data in the figure, identify the organisms that had the largest effect on the abundance of O_2 in Earth's atmosphere.

(A) Bacteria

(B) Cyanobacteria

(C) Eukaryotes

(D) Land plants

Section 2: Free-Response Question

Write your answer to each part clearly. Support your answers with relevant information and examples. Where calculations are required, show your work.

Scientists can make inferences about the deep history of life on Earth by combining observations made about presently living organisms with evidence from the fossil record. Scientists have used this process of reasoning to make inferences about metabolism in the earliest forms of life on Earth.

(a) **Identify** the group of organisms that first evolved the form of photosynthesis that produces oxygen.

(b) **Explain** the process by which eukaryotes evolved the ability to perform photosynthesis.

(c) A scientist claimed that this process provided an evolutionary advantage to two different organisms. Provide reasoning to **justify** that claim.

(d) Scientists have suggested that fermentation may have been one of the earliest metabolic pathways to evolve on Earth. Provide two pieces of evidence to **support this claim.**

Unit 3

AP® Practice Questions

Section 1: Multiple-Choice Questions

Choose the best answer for questions 1–15.

1. Many chemical reactions that take place inside a cell require an input of energy. Which best describes how these reactions can occur?

    (A) The reactions utilize heat from the body as a source of energy.

    (B) The reactions are catalyzed by enzymes that supply energy.

    (C) The reactions utilize energy released from hydrolysis of ATP.

    (D) The reactions capture energy that is released when NAD^+ is reduced.

2. Which diagram best describes the relationship between H_2O, CO_2, and O_2 as driven by photosynthesis and cellular respiration?

    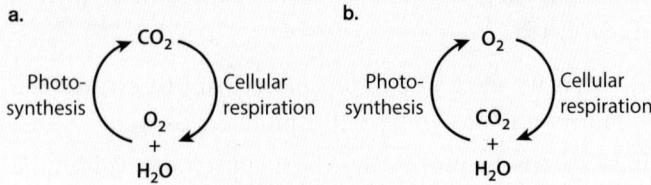

    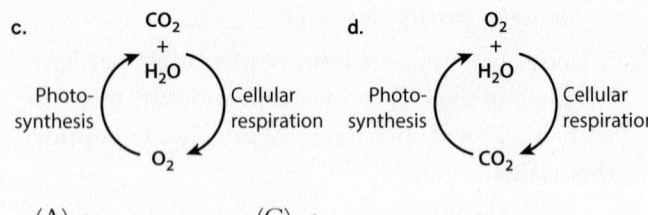

    (A) a (C) c

    (B) b (D) d

3. Which statement accurately describes a similarity between photosynthesis and cellular respiration?

    (A) Both generate ATP using energy from an electrochemical proton gradient.

    (B) Both use H_2O as a source of electrons.

    (C) Both produce NADPH.

    (D) Both increase the entropy of the cell.

4. Which molecule is reduced in both glycolysis and the Krebs cycle?

    (A) Acetyl-CoA (C) Pyruvate

    (B) CO_2 (D) NAD^+

5. Seaweeds such as kelp are not plants. They are photosynthetic algae that can live 10 to 30 meters underwater. Long wavelength light, which produces orange and red colors, does not penetrate deeper than 10 meters underwater. Short wavelength light, which produces green and blue colors, may penetrate as deep as 40 meters underwater. Kelp and other underwater algae are often orange or red in color. Which provides the best explanation for why these algae are not green like land plants?

    (A) Photosynthetic algae like kelp only appear brown, orange, or red when they are not in water; they appear green in their natural aquatic environment.

    (B) Green color comes from chlorophyll, which is a photosynthetic pigment found only in plants.

    (C) Dull colors like brown, orange, and red contrast with the ocean floor to make the kelp visible to herbivores.

    (D) Algal pigments have adapted to absorb energy from shorter wavelength green and blue light so the algae appear orange or red under full-spectrum visible light.

6. In a study to evaluate new agricultural products, a researcher treated plants with an experimental synthetic chemical to see if it would kill the plants. The researcher observed that after the chemical treatment, the plants continued to consume water and produce oxygen, but they stopped producing ATP and NADPH. The plants subsequently died as a consequence of having insufficient ATP. Based on the data provided, which is the most likely description of the chemical's activity?

(A) The chemical probably inhibits the photosynthetic electron transport chain.

(B) The chemical probably inhibits the Calvin cycle.

(C) The chemical probably inhibits the Krebs cycle.

(D) The chemical probably inhibits the respiratory electron transport chain.

7. Identify the stage in cellular respiration in which CO_2 is produced.

(A) Glycolysis

(B) The Calvin cycle

(C) The Krebs cycle

(D) The respiratory electron transport chain

Questions 8 and 9 refer to the following information.

A researcher grew a tomato plant in a sealed container. The research measured the amount of CO_2 and O_2 in the air inside the container over a 48-hour period. The researcher provided the plant with light during hours 0–12 and 24–36, and kept the plant in the dark during hours 12–24 and 36–48. The data that the researcher obtained over the course of the experiment are shown in the figure. Periods of light are indicated with open bars, and periods of dark are indicated with filled bars.

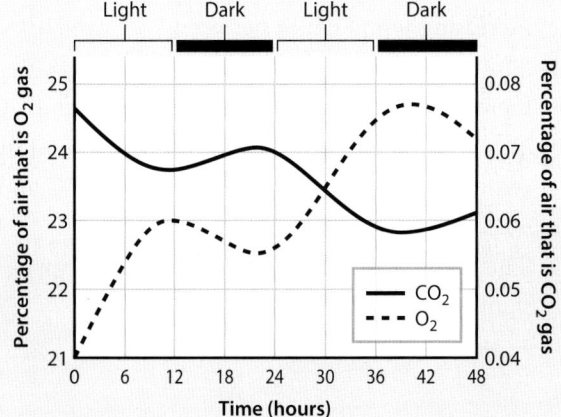

8. Identify the time at which CO_2 percentage in the air was highest.

(A) 0 hr   (C) 36 hr

(B) 24 hr   (D) 48 hr

9. Which provides the best explanation for the change in O_2 and CO_2 levels between 12 and 24 hours?

(A) In the absence of light, the photosynthetic electron transport chain runs backward, consuming O_2 and producing CO_2.

(B) In the absence of light, the plant relies on cellular respiration to harness energy from glucose, consuming O_2 and producing CO_2.

(C) In the absence of light, the plant shifts to the light-independent Calvin cycle, consuming O_2 and producing CO_2.

(D) In the absence of light, the plant stops producing ATP, consuming CO_2 and producing O_2.

Use the following information to answer questions 10 and 11.

A hypothetical metabolic pathway is shown below. In this pathway, a substrate is converted to a product by a series of four enzymes. The rate at which each enzyme performs its reaction is shown below the arrow for that enzyme.

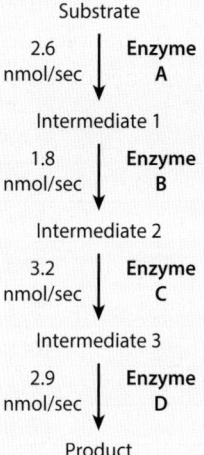

10. Predict which molecule would accumulate to the highest level in the presence of a chemical that inhibits enzyme B.

(A) Intermediate 1

(B) Intermediate 2

(C) Intermediate 3

(D) Product

11. If a second metabolic pathway, independent from the one diagrammed, produced additional intermediate 2 that could be used in the illustrated pathway, which best predicts the effect on production of product?

(A) The amount of product would increase because the extra intermediate 2 could serve as a substrate for enzyme 3.

(B) The amount of product would remain the same because the pathway is already operating at its maximum rate.

(C) The amount of product would decrease because the extra intermediate 2 would inhibit enzyme B.

(D) The amount of product would decrease because intermediate 2 would be diverted for use in a different biological process.

Use the following information to answer questions 12–15.

Rubisco is an enzyme that functions in the Calvin cycle, sometimes called the light-independent reactions. Rubisco catalyzes the attachment of CO_2 to a molecule with five carbon atoms, resulting in a molecule with six carbon atoms that is further metabolized in the cycle to produce a sugar. The incorporation of CO_2 into an organic molecule is called carbon fixation. A researcher was interested in the effect of temperature on carbon fixation and measured the rate at which carbon was consumed in the presence of rubisco and the 5-carbon molecule to which rubisco attaches CO_2. The researcher obtained the data shown in the table.

Temperature (°C)	CO_2 Consumed $\pm$ 2 SE_x (μmol/sec)
25	91±9
30	96±5
35	93±8
40	104±7
45	71±7
50	59±4

12. In the figure below, which graph best illustrates the data in the table?

(A) a   (C) c

(B) b   (D) d

a.
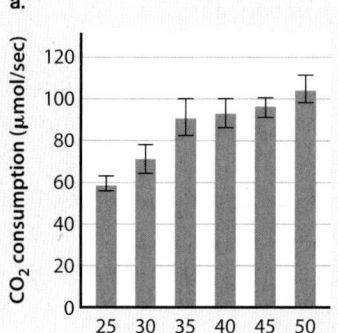

b.
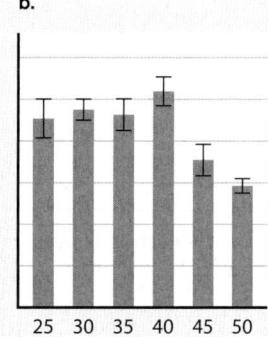

c.
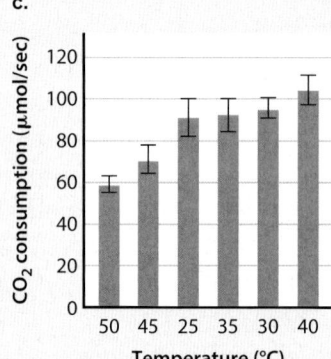

d.
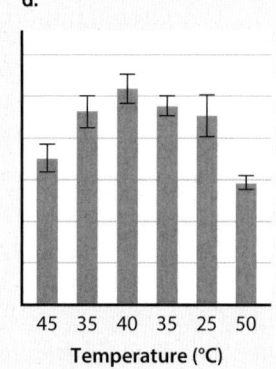

13. Identify the dependent variable in the study.

(A) Temperature

(B) Amount of the 5-carbon molecule

(C) Amount of rubisco

(D) CO_2 consumption

14. Based on the data provided, which best describes the effect of temperature on carbon fixation in the range of 25°C to 40°C?

(A) The amount of carbon fixation increases with increasing temperature between 25°C and 40°C.

(B) The amount of carbon fixation decreases with increasing temperature between 25°C and 40°C.

(C) The amount of carbon fixation varies randomly with increasing temperature between 25°C and 40°C.

(D) The amount of carbon fixation does not vary significantly between 25°C and 40°C.

15. Which best explains the rate of carbon fixation observed at 50°C?

(A) The high temperature causes the reaction to proceed very quickly.

(B) The high temperature causes some of the rubisco to become denatured and become inactive.

(C) The high temperature causes some of the CO_2 to remain as a gas and be unavailable for fixation.

(D) The high temperature begins to cause the 6-carbon molecule to decay.

Section 2: Free-Response Questions

Write your answer to each part clearly. Support your answers with relevant information and examples. Where calculations are required, show your work.

1. Enzyme Z converts substrate S to product P. A scientist combined enzyme Z and substrate S in a reaction tube and measured the accumulation of product P at three different temperatures. The data obtained are shown in the figure.

Amount of Product P Generated by Enzyme Z at Three Different Temperatures

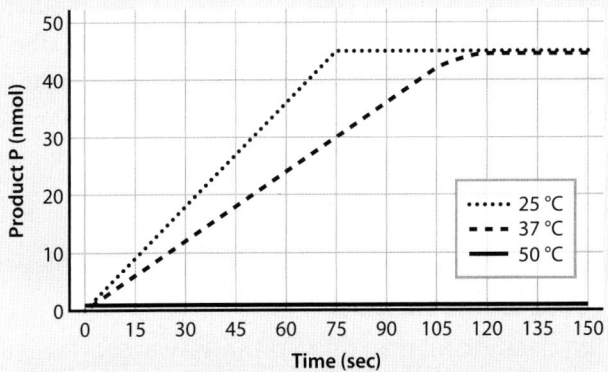

(a) At 25°C, **identify** how much more product P was there at 75 seconds than there was at 45 seconds.

(b) **Calculate** the rate of the reaction between seconds 30 and 60 at 37°C. Show all steps in the calculation.

(c) At 25°C, which time point showed the slowest rate of reaction: 15 sec, 60 sec, 90 sec, or 135 sec?

(d) **Make a claim** to explain the observed rate of reaction at 50°C.

(e) The scientist hypothesizes that there is an optimal pH where the activity of enzyme E is highest. **Propose** an experiment to test this hypothesis. **Identify** the dependent variable in the experiment you propose.

2. Acetyl-CoA reacts with oxaloacetate to generate citric acid and CoA-SH in the Krebs cycle. This reaction is catalyzed by the enzyme citrate synthase. Acetyl-CoA can also go down a different metabolic pathway, in which it is converted to malonyl-CoA by an enzyme called acetyl-CoA carboxylase. Malonyl-CoA is subsequently used for lipid biosynthesis. A simplified diagram of the Krebs cycle and steps leading to lipid biosynthesis is given in the figure below. The figure shows steps leading to lipid biosynthesis, with intermediate molecules named in boxes and the names of important enzymes written next to arrows.

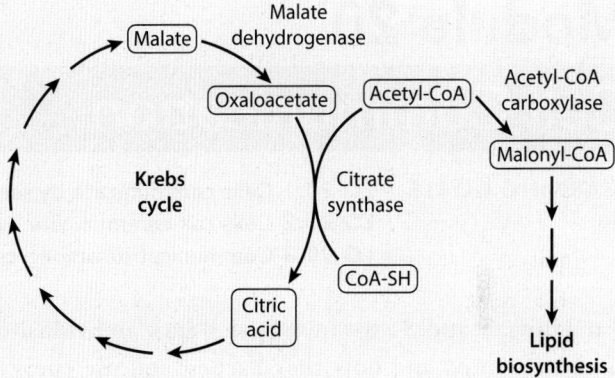

A scientist developed a drug, called compound X, that can inhibit citrate synthase. When compound X is provided to an animal in a low dose, citrate synthase activity is reduced but not completely eliminated. However, when compound X is provided at a high dose, citrate synthase activity is completely blocked.

(a) Malate dehydrogenase is the enzyme that converts malate to oxaloacetate. **Identify** an organelle of a eukaryotic cell that is likely to have a high concentration of malate dehydrogenase. **Identify** the compartment within this organelle where the concentration of malate dehydrogenase is likely to be highest.

(b) A scientist gave a low dose of compound X to a mouse, so citrate synthase activity was reduced but not eliminated. Based on the information provided, **predict** the consequence of this low dose of the drug for lipid biosynthesis in the mouse. **Provide reasoning** to support your prediction.

(c) A scientist gave a high dose of compound X to a mouse, completely blocking citrate synthase activity. Based on the information provided, **predict** the consequence for the mouse of the high dose of this drug. **Provide reasoning** to support your prediction.

Unit 4
Cell Communication and Cell Cycle

Fluorescent micrograph of a human epithelial cell dividing by Dr. Torsten Wittman/Science Source

Module 20
Cell Communication

LEARNING GOALS ▶**LG 20.1** Cells communicate by sending and receiving chemical messengers.
▶**LG 20.2** Cells can communicate over various distances.
▶**LG 20.3** Communication among cells, tissues, and organs underlies homeostasis.

In Units 2 and 3, we investigated how individual cells are structured and how they harness and use energy. In this unit, we will zoom out from individual cells to groups of cells, examining how cells communicate with each other. We will show how bacteria in a **population**—that is, all the individuals of a given species that live and reproduce in a particular geographical area—can communicate with each other to spur processes that support them in changing environments. We will also see that cells of multicellular organisms send chemical messages to one another to coordinate the activities of different organs and tissues. There are substantial benefits to fast and effective cell communication, from coordinating cell activities to enabling multicellular organisms to function.

We begin this module by introducing the basic steps of cellular communication. We will see that communication can take place between cells that are near one another and even between very distant cells. We will then describe how physical connections between some cells coordinate their activities. Finally, we will examine how cell communication allows multicellular organisms to maintain a stable set of internal conditions, a process called homeostasis.

PREP FOR THE AP® EXAM

FOCUS ON THE BIG IDEAS

INFORMATION STORAGE AND TRANSMISSION: Focus on the structures of cells and the mechanisms that make it possible for cells to communicate with other cells.

20.1 Cells communicate by sending and receiving chemical messengers

Cell communication makes multicellular life possible. Your body, in fact, depends on communication among the estimated 30 trillion cells that make it up. This communication helps to coordinate cell activities so the cells of an organism function together as a whole. Cell communication also takes place among unicellular organisms in response to the environment. In this section, we will consider the general principles of cell communication, including how cells send and receive signals and how a cell responds after it receives a signal. These mechanisms first evolved in unicellular organisms as a way to sense and respond to their environment, and also as a way to influence the behavior and activity of other cells. These basic principles apply to all cells, both prokaryotic and eukaryotic, and to all organisms, both unicellular and multicellular. Finally, we will examine some specifics of prokaryotic and eukaryotic signaling.

An Overview of Cell Communication

Birds do an amazing job of rearing their hungry, rapidly growing offspring. The communication that takes place between a hungry chick and its parent is comparable to the communication that takes place between two different cells. The chick's hunger leads the chick to express begging behaviors, chirping and gaping with its beak wide open when the parent is nearby. Chirping and gaping are very noticeable signals, as seen in **FIGURE 20.1**. The parent sees and hears the chick's begging behaviors, and these signals stimulate the parent to respond by delivering food to the chick. In this simple communication system, hunger is the stimulus that starts the communication. The chick sends signals—begging behaviors—which are received by a parent and cause the parent to alter its behavior and feed the chick. The change in the parent's behavior is a response to the signal. This communication helps the baby bird grow.

Communication, whether between birds or between cells, has four sequential steps: a stimulus, the release of a signal, signal reception, and response. In our bird example, the two birds in the communication chain have different roles. When a hungry chick begs an adult for food, the chick is the signaling source. When the parent observes the begging behavior, the parent is the target of the signal. When the parent provides food in response to the begging behavior, the parent is the responder.

FIGURE 20.1 Communication in birds

Hunger stimulates a chick to beg for food, sending a signal that it is hungry. The parent receives the signal and responds by sharing food. By analogy, cells release signals that are received by other cells and cause the cells to respond. Photo: Panther Media GmbH/Alamy Stock Photo

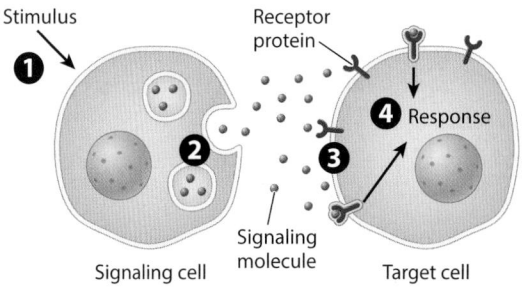

FIGURE 20.2 Cell communication

In step 1, signaling cells receive a stimulus. In step 2, the cell produces and releases signaling molecules. In step 3, signaling molecules bind to receptor proteins on a target cell. In step 4, the target cell responds.

Note that the parent is both the target of the chick's signal and the responder. Similarly, when cells communicate, one cell is the signaling source and the cell that receives the signal and responds is both the target and responder.

FIGURE 20.2 illustrates the four steps of cell communication. The cell on the left is the signaling cell, which is a cell that responds to a stimulus—for example, a chemical—by releasing signals to communicate with other cells. In step 1 in the figure, the signaling cell receives a stimulus. In step 2, the signaling cell makes and releases **signaling molecules,** chemicals released by cells that alter the activities of other cells. In step 3, the released signals bind to the target cell's receptor proteins, which are proteins on a cell's surface or in a cell's interior that bind specific signaling molecules. The cell that receives the signal in step 3 is called the **target cell.** Signaling molecules are also called **ligands** because they bind to receptor proteins on or in a target cell. ("Ligand" comes from *ligare,* which means "to bind.") The interaction between the signaling molecule and the receptor protein initiates step 4, the target cell's response. For convenience, signaling molecules are often simply called signals, and the receptor proteins are simply called receptors.

The binding of a signaling molecule to the receptor protein is the key step in communication between cells; it allows a message to be received by a target cell. The target cell's receptor proteins and the signaling molecules must be specifically matched to one another. In Figure 20.2 this specificity is represented by the way that the pink signaling molecules fit into, or bind to, the crescent-shaped receptor proteins on the cell membrane of the target cell. The shapes and chemical properties of the signaling molecules and receptor proteins complement, or match, one another. This precise matchup is known as a ligand–receptor interaction. In this type of interaction, a ligand can only bind to a specific receptor protein that has a shape and distribution of charges that complement the ligand exactly. Ligand–receptor interactions ensure that the signal is received only by the appropriate cell. The interaction between a ligand and its receptor involves relatively weak, noncovalent bonds. The weak bonds allow the termination of the communication when the time is right, as the signal departs from the receptor.

Prokaryotic Signaling

Let's consider an example of cell communication in bacteria. It may be hard to imagine why it would be beneficial for single-celled organisms to communicate with one another. Though each bacterium is an individual organism, populations of bacteria often have the greatest effect when working in coordination. For example, some types of bacteria in the ocean produce light using biochemical reactions like those in fireflies, a process called bioluminescence. Bioluminescence has many different functions in a wide range of species, such as attraction of mates and defense against predators. In bacteria, it is thought to be a way to monitor population density. One of the most astonishing things to see in the ocean at night is the coordinated bioluminescence of bacteria, shown in **FIGURE 20.3**. This bioluminescence is the result of a coordinated effort among many individual cells.

For a bacterium to display bioluminescence in ocean waters, it must first produce the molecules needed for bioluminescence. However, bacteria only produce bioluminescent chemicals when the bacteria are present at high population densities. This is an example of a communication pathway called **quorum sensing,** which is communication among

FIGURE 20.3 Bacterial bioluminescence

Populations of bioluminescent bacteria live in many parts of the ocean, such as the coast of California. When present in high enough numbers, the bacteria signal each other to glow, which is visible as blue light.

Photo: Kevin Key/Slworking/Getty Images

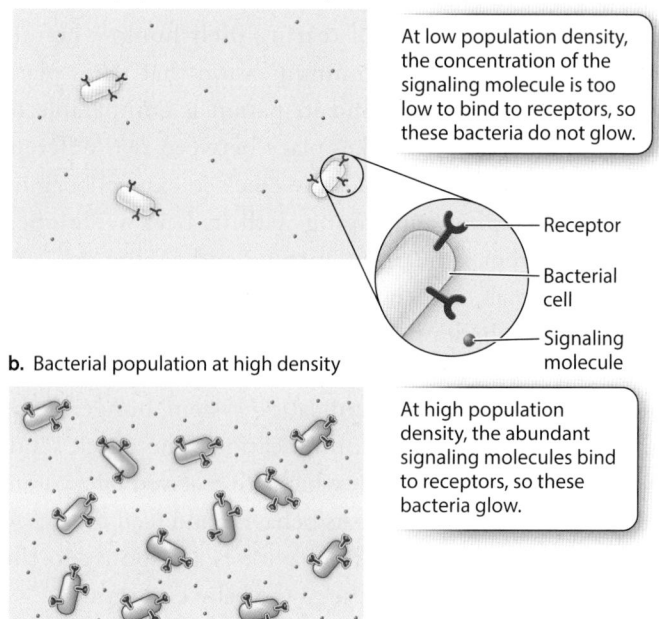

a. Bacterial population at low density

At low population density, the concentration of the signaling molecule is too low to bind to receptors, so these bacteria do not glow.

— Receptor
— Bacterial cell
— Signaling molecule

b. Bacterial population at high density

At high population density, the abundant signaling molecules bind to receptors, so these bacteria glow.

FIGURE 20.4 Communication by quorum sensing among bacterial cells

Quorum sensing in bacteria controls the production of molecules used for bioluminescence. (a) At low density of bacteria, there is not enough of the signaling molecule for target bacteria to bind and respond to the signal. (b) At high density of bacteria, there is enough signaling molecule to bind receptor proteins, leading to bioluminescence.

bacteria that leads to a response only when there are a high number of bacteria present in a given area. The bacteria must be present in high population densities to turn on a communication pathway. For human social and political organizations, the term "quorum" refers to the minimum number of members of a group that must be present in order to validate a change in the group's rules.

Bioluminescent bacteria produce a signaling molecule that leads to the production of bioluminescent chemicals, as shown in **FIGURE 20.4**. At low population densities, there is very little signal in the environment because there are very few bacteria, as shown in Figure 20.4a. The few bacteria present at low population densities produce so little signal collectively that the signaling molecules are unlikely to bind to the bacteria's receptors. When the population density of the bacteria increases, the signal becomes more abundant in the water, as shown in Figure 20.4b. When the signal is abundant, the receptors on the bacteria bind to this signaling molecule. The bacteria with receptors that bind the signaling molecule respond by synthesizing the chemicals needed for bioluminescence. The stage is now set for a light show in the ocean at night, in which agitation or disturbance of the water triggers the soft glow.

Bacteria use quorum sensing to control and coordinate many other types of responses, including a pathway that helps some bacteria adapt to their environments. For example, quorum sensing can initiate a pathway in which bacteria take up DNA that other bacteria have released into their environment. In this way, bacteria can express new traits encoded by the DNA, helping them survive and reproduce. The rate at which bacteria take up these pieces of DNA increases as their population density increases. At low population density, there is not enough signal being released to stimulate responses. However, as population density increases, the amount of signal increases. Soon there is enough of the signal in the environment to bind to receptors on the bacteria. The bacteria respond by increasing their rate of DNA uptake.

In these cases, each bacterial cell is both a signaling cell and a target cell because each cell makes the signaling molecule and its corresponding receptor protein. Nevertheless, it illustrates the general idea that cells are able to communicate by sending a signaling molecule that binds to a receptor protein on or in a target cell. Indeed, the ability of cells to communicate is universal among prokaryotes and eukaryotes. The four steps of cell communication occur in very much the same way in many examples of cellular communication.

Eukaryotic Signaling

The importance of cell signaling in a multicellular eukaryote can be seen by considering what happens when you are suddenly startled or scared. You likely experience a strange feeling in the pit of your stomach and your heart beats faster. Collectively, these changes to your body induced in an emergency are known as the fight-or-flight response. Cell communication is responsible for the fight-or-flight response. Many different organs and tissues are involved in the communication chain that quickly coordinates your response, as shown in **FIGURE 20.5**. In response to being startled, the nervous system directs the release of a signaling molecule called adrenaline from the adrenal glands, located on top of the kidneys. The adrenal glands release adrenaline into the blood. Adrenaline in the blood circulates throughout the body and acts on cells in many different organs and tissues, such as the heart, lungs, and stomach. All of the target cells have the adrenaline receptor proteins that allow them to receive the adrenaline-signaling molecule. The different target cells react differently to adrenaline. For example, when receptor proteins on the cells of your heart bind adrenaline, the receptor proteins change the activity of heart cells so that the heart pumps more strongly, moving more blood and

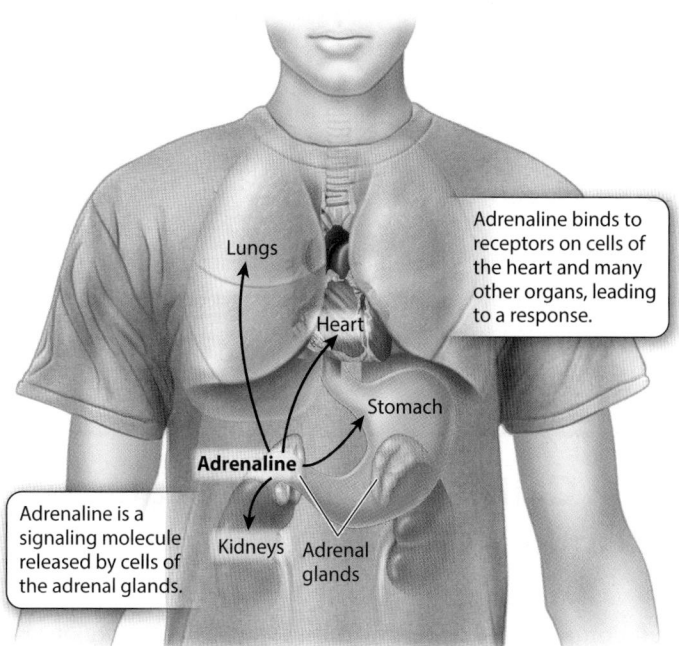

FIGURE 20.5 Communication among cells in a multicellular organism

The signaling molecule, adrenaline, circulates in the blood to reach target cells in many organs. Each organ has a different response to adrenaline. Working in coordination, these responses are known as the body's fight-or-flight response.

delivering more oxygen to the body. When adrenaline binds to receptors on lung cells, the lung cells respond by relaxing, allowing more air intake.

The example of the fight-or-flight response illustrates the four steps of cellular communication: stimulus, signal release, signal reception, and response. The stimulus is associated with being startled. The signaling cells are adrenaline-producing cells in the adrenal glands. The cells release the signaling molecule, adrenaline. Adrenaline binds to receptor proteins located on the surface of target cells, such as those of the heart, which responds by delivering greater blood flow. The full set of the body's responses to adrenaline allows you to cope with whatever startled you in the first place, preparing you to respond in an appropriate manner.

Adrenaline travels to its target cells through the blood. Although it has the potential to interact with many different cells, it only interacts with cells that have a specific receptor protein to which it can bind. It is the presence of the receptor protein that determines which cells are target cells and therefore which cells respond to adrenaline. Adrenaline is an example of a **hormone,** a type of signaling molecule that travels in the circulatory system and affects distant cells. However, not all signaling molecules travel such long distances.

AP® EXAM TIP

Writing out sequential steps in a signaling system demonstrates that you understand how the system works in a cause-and-effect manner that includes signaling cells and target cells. Be sure that you include complete sentences with detailed descriptions of each step: stimulus, signal release, signal reception, and response.

✓ Concept Check

1. **Describe** each of the four sequential steps of communication between a signaling cell and a target cell, as depicted in Figure 20.2.

2. **Identify** each of the four steps of cell communication in quorum sensing by bacteria.

20.2 Cells can communicate over various distances

As we have seen, in unicellular organisms such as bacteria, cell communication occurs between individual organisms. In multicellular organisms, cell communication occurs between cells within the same organism. The same principles apply in both instances, although some important differences exist. In multicellular organisms, the distance between communicating cells varies considerably. When the two cells are far apart, the signaling molecule is transported by the circulatory system, as we have seen in the example of adrenaline. When they are close, the signaling molecule simply moves by diffusion. In addition, many cells in multicellular organisms are physically attached to one another, in which case the signaling molecule is not released from the signaling cell at all. In this section, we will explore communication over long and short distances, as well as communication between cells that are physically attached to one another.

Long-Distance Signaling

Signaling molecules released by a cell may have to travel great distances to reach receptor cells in the body. In such a case, they are often carried in the circulatory system. Signaling by hormones that travel through the circulatory system is called endocrine signaling. Endocrine signaling is illustrated in **FIGURE 20.6a**. In the figure, signaling molecules are released from the cell on the left into a blood vessel, represented by the pink tube. The signaling molecules travel through the bloodstream until they reach the target cell on the right. Coordination of activities across cells, tissues, and organs of a multicellular organism is typically directed by endocrine signaling.

a. Endocrine signaling

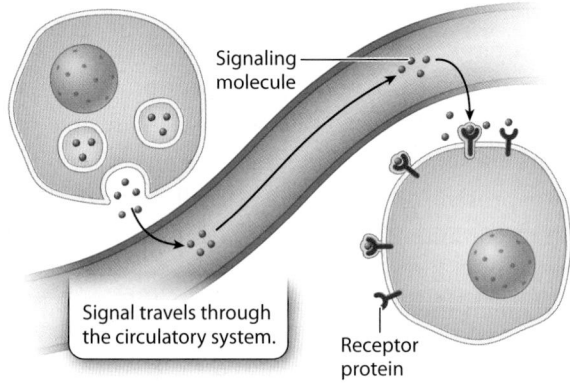

b. Paracrine signaling

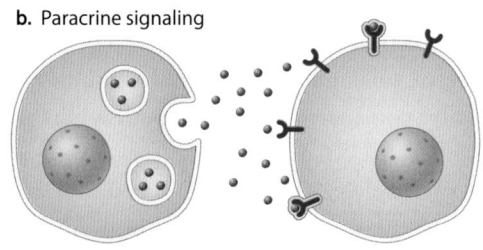

c. Autocrine signaling

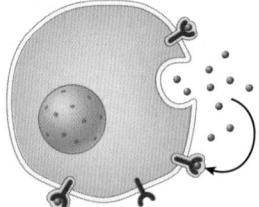

FIGURE 20.6 Signaling distances

Cell communication can be classified according to the distance between the signaling cells and the responding cells. (a) Endocrine signals are transported in the circulatory system to their target cells in another part of the body. (b) Paracrine signals move by diffusion and stimulate neighboring cells. (c) Autocrine signals are released and received by the same cell.

Adrenaline provides a good example of endocrine signaling. As we discussed earlier, adrenaline is a hormone produced in the adrenal glands, and then carried by the bloodstream to target cells that may be far from the signaling cells. Other examples of hormones include the mammalian steroid hormones estradiol (an estrogen) and testosterone. These hormones are primarily produced by the ovaries and the testes, respectively; travel through the bloodstream; and interact with target cells in various tissues throughout the body. An increased amount of these hormones in the blood causes physiological changes associated with puberty.

Short-Distance Signaling

Signaling can also occur between two cells that are close to each other. In this case, movement of the signaling molecule through the circulatory system is not needed. Instead, the signaling molecule can simply move by diffusion between the two cells. Recall from Unit 2 that diffusion occurs when there is a concentration difference between two regions, with movement from a region of high concentration to a region of low concentration. This form of signaling is called paracrine signaling. Paracrine signaling is shown in Figure 20.6b. This figure closely resembles our overview of cell communication in Figure 20.2, with a signaling cell releasing signaling molecules that travel a short distance to receptor proteins on a target cell.

Growth factors are an example of signaling molecules that often act over short distances. These are small, water-soluble molecules that, as their name suggests, cause the target cell to grow, divide, or differentiate. Scientists have identified many different growth factors; in most cases, their effects are confined to neighboring cells. During the development of embryos, growth factors secreted by cells work over short distances to influence the kind of cells their neighboring cells will become. For example, in developing vertebrates, paracrine signaling by the growth factor Sonic Hedgehog (named after a video game character) ensures that the motor neurons in in the spinal cord are in the proper locations, that the bones of the vertebral column form correctly, and that the thumb and pinky fingers are on the correct sides of the hands.

Paracrine signaling also plays an essential role in the nervous system, where one neuron signals another neuron across a short distance. Similarly, motor neurons communicate with muscles by paracrine signaling. The nerve ending releases a chemical signal onto the muscle, and when the signal binds to receptors on the muscle cells the interaction leads to a muscle contraction.

Some cells actually communicate or "talk" to themselves. In autocrine signaling, the signaling cell and the target cell are the same. As shown in Figure 20.6c, the cells that make and release the signaling molecule have receptor proteins for the signaling molecule. Autocrine signaling occurs in bacterial quorum sensing, as we saw earlier. It is also important to multicellular organisms during embryonic development. For example, once a cell differentiates into a specialized cell type, autocrine signaling is sometimes used to maintain this developmental decision. In addition, autocrine signaling is used by some cancer cells to promote cell division.

Autocrine, paracrine, and endocrine signaling draw our attention to the distance between the signaling cell and the target cell. However, in all three cases, the steps of cell communication are the same: a stimulus leads to the release of a signaling molecule, which binds with a receptor protein in or on a target cell, leading to a response.

Contact-Dependent Signaling

Sometimes a cell communicates with another cell through direct contact, without diffusion or circulation of the signaling molecule. This form of signaling requires that the two communicating cells be in physical contact with each other, and is therefore called contact-dependent signaling. This form of cell signaling occurs in two ways. In the first way, a transmembrane protein on the surface of one cell acts as the signaling molecule, and a transmembrane protein on the surface of an adjacent cell acts as the receptor, as shown in **FIGURE 20.7**. As a result, the signaling molecule is not released from the cell, but instead remains associated with the cell membrane.

As an example, let's consider the development of the central nervous system of vertebrate animals. In the brain,

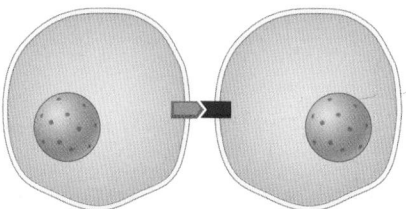

FIGURE 20.7 Contact-dependent signaling

Contact-dependent signaling relies on physical contact between two cells. A transmembrane protein on one cell acts as the signal, and a different transmembrane protein on an adjacent cell acts as the receptor. In this case, the signaling molecule does not travel in a circulatory system or diffuse away from the signaling cell.

there are neurons that transmit information in the form of electrical signals, and glial cells that nourish and insulate the neurons. Both the neurons and glial cells start out as similar cells in the embryo, but some of these cells become neurons and others become glial cells. During development, the amount of a transmembrane protein called Delta dramatically increases on the surface of some of these undifferentiated cells. These cells will become neurons. After developing into neuron cells, the Delta proteins on each new neuron bind to different transmembrane proteins called Notch on the surface of adjacent, undifferentiated cells. In this case, the signaling cell is the cell with elevated levels of Delta. Delta is the signaling molecule and Notch is its receptor. Cells with Notch receptors bound to Delta transmembrane proteins on neighboring cells become glial cells.

Contact-dependent signaling can also work in a completely different way. Some cells have passages or channels that connect the interior of one cell to the interior of an adjacent cell. The passages between cells allow ions and molecules inside one cell to move directly into the interior of another cell. In contrast to other forms of signaling we have discussed, there is no receptor protein in this form of cell communication.

Many kinds of animal cells communicate with adjacent cells through passages that connect them. Recall from Module 9 that the cell membrane is a lipid bilayer studded with proteins, some of which are channels that span the membrane and connect the inside and outside of the cell. In animal cells, neighboring cells can send signals to one another through gap junctions, shown in **FIGURE 20.8**, which are transmembrane protein channels between two neighboring cells. Adjacent cells with gap junctions can closely coordinate activities without having to release any signals into the extracellular space. The heart is an example of an organ that uses gap junctions to coordinate the activities of neighboring cells. Gap junctions between cells of the heart allow ions that control the heart's contraction to rapidly move from one cell to the next. This rapid movement of ions produces the coordinated contraction of the heart muscle that efficiently pumps blood out of the heart.

Adjacent cells in land plants and many algae also have passages that connect adjacent cells, allowing them to communicate. Recall from Module 7, however, that plant cells have an additional layer that animal cells do not: plant cells have a rigid cell wall. In order to send signal molecules from the interior of one plant cell to the interior of the neighboring plant cell, there must be openings that cross through the membranes and the cell walls of both cells. These openings are called **plasmodesmata** (singular, plasmodesma). **FIGURE 20.9** shows a cross-section of two adjacent plant cells connected by plasmodesmata. Like gap junctions,

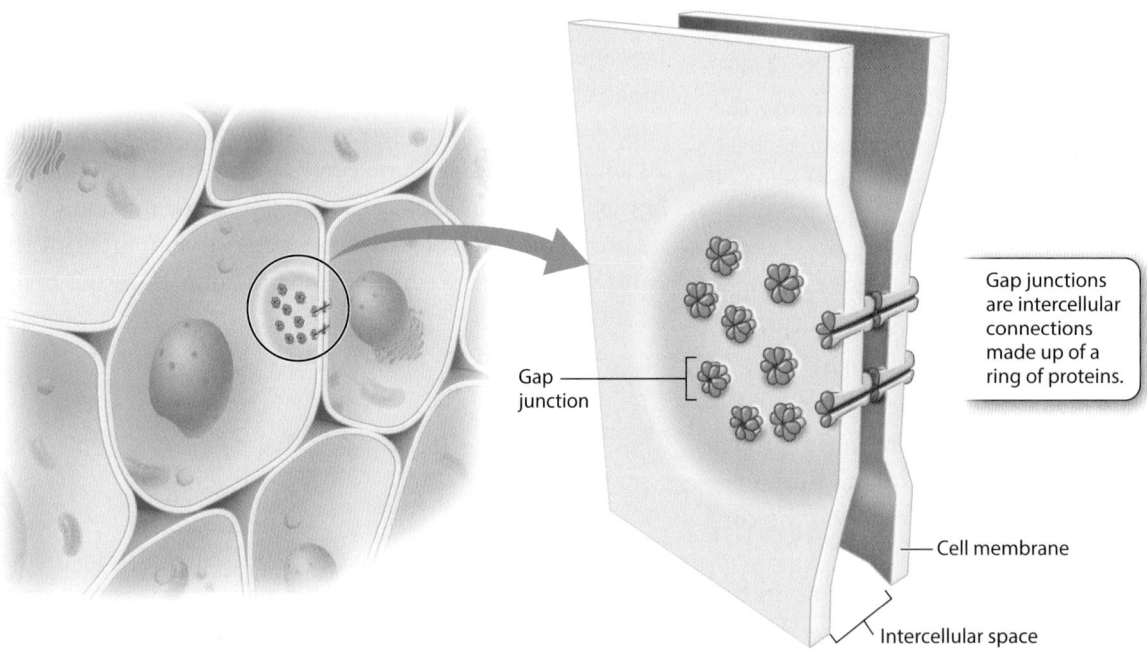

Gap junction

Gap junctions are intercellular connections made up of a ring of proteins.

Cell membrane

Intercellular space

FIGURE 20.8 Gap junctions

Gap junctions are protein channels embedded in the membranes of two neighboring cells. Signaling molecules and ions can travel directly from one cell to the next through gap junctions, allowing neighboring cells to communicate rapidly with one another directly.

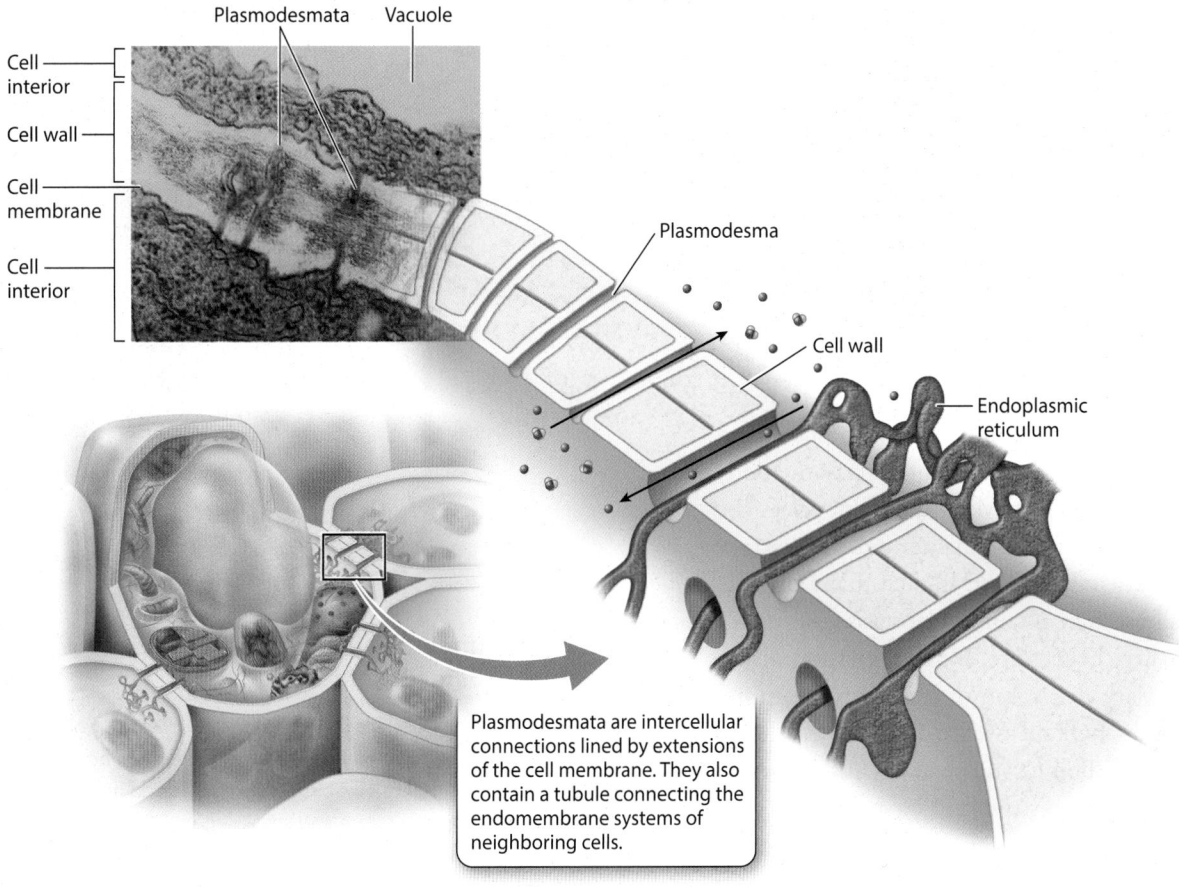

Cell interior

Plasmodesmata   Vacuole

Cell wall

Cell membrane

Cell interior

Plasmodesma

Cell wall

Endoplasmic reticulum

Plasmodesmata are intercellular connections lined by extensions of the cell membrane. They also contain a tubule connecting the endomembrane systems of neighboring cells.

FIGURE 20.9 Plasmodesmata

Plasmodesmata are passages connecting one plant cell to another. Signaling molecules and ions can pass through the plasmodesmata, allowing direct communication between neighboring cells. In addition, larger substances, such as organelles, can pass through plasmodesmata. Photo: Biophoto Associates/Science Source

plasmodesmata allow cells to exchange ions and signaling molecules, but the similarity ends there. In plasmodesmata, the cell membranes of the two connected cells are actually continuous, which is not the case in gap junctions. In addition, the size of the opening in plasmodesmata is considerably larger than that in gap junctions. The figure shows many substances crossing from one cell to the next through the plasmodesmata. Signaling molecules, thin strands of cytoplasm, and organelles all can flow through plasmodesmata. Plasmodesmata connect one cell to the next, creating a continuous space all the way across a leaf, for example.

Some fungi also have passages between cells that are similar to plasmodesmata, allowing direct cell–cell communication. Ancient eukaryotic organisms have not been found to have anything like plasmodesmata or gap junctions, suggesting that channels connecting neighboring cells appeared later in life's history and were an important step toward the evolution of organisms with greater complexity.

As you can see from these various examples, the same fundamental principles are at work when signaling guides a developing embryo, allows neurons to communicate with other neurons or muscles, triggers DNA uptake by bacterial cells, or allows your body to respond to stress. All of these forms of communication are based on signaling molecules that are sent from a signaling cell to a target cell. These signaling molecules are the language of cellular communication.

✓ **Concept Check**

3. **Describe** endocrine, paracrine, and autocrine signaling, including what differentiates them from one another.

4. **Describe** the similarities between the gap junctions of animals and the plasmodesmata of plants.

20.3 Communication among cells, tissues, and organs underlies homeostasis

Cell communication is critical for homeostasis in multicellular organisms. In Module 9, we saw that cellular homeostasis is the ability of a cell to regulate and stabilize its internal environment. Here we extend the concept of cellular homeostasis to an entire organism. On this scale, homeostasis is the set of mechanisms that allow the whole organism to regulate and stabilize its internal environment. Homeostasis is an active process that relies on cells in different parts of the body communicating with one another. Many difficulties can arise if a stable internal state cannot be maintained in the face of a changing external environment. In fact, many diseases develop from the failure to maintain homeostasis.

Body temperature, heart rate, blood pressure, blood sugar, and blood pH are just some of the many examples of parameters that must be kept within a narrow range for an organism to thrive. When circumstances cause a departure from this range, organisms respond in specific ways to restore steady conditions. Homeostasis in this context refers to the ongoing maintenance of steady conditions in the face of forces that would change these conditions.

Cell communication is critical for maintaining homeostasis because cells of tissues and organs only "know" to actively restore conditions if a change in conditions is communicated to them. This is our familiar four-step communication process: stimulus, release of signal, signal reception, and response. In homeostasis, the stimulus is a departure from normal conditions and the response is the set of actions that restore normal conditions. Homeostasis depends on signals and responses that act in a loop, so that when normal conditions are again achieved, cells stop sending out signals. Stopping the production of signals pauses the responses by target cells. We will develop this idea in the following section, starting with a simple example of temperature regulation by a heating system and then discussing temperature regulation by an organism.

Temperature Regulation by a Heating System

The way an organism's body maintains homeostasis through cell communication is similar to the way that a heating system maintains a comfortable temperature in a room. **FIGURE 20.10** shows the loop of signals and responses that allows a thermostat and a heater to keep a room at a stable

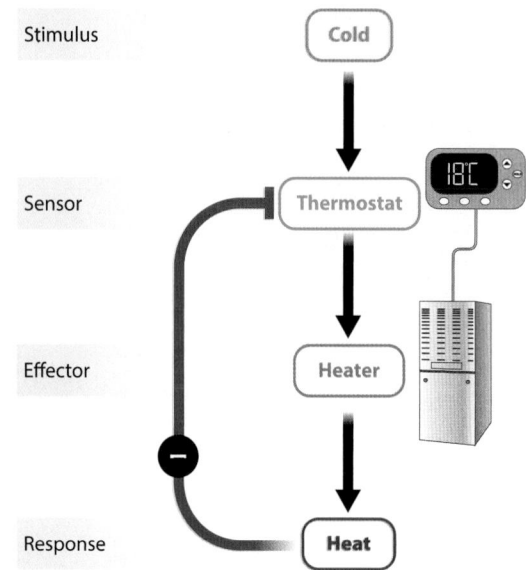

FIGURE 20.10 Temperature regulation by negative feedback in a home heating system

Cold temperatures in a house are sensed by the thermostat, which signals the heater to turn on and heat the house. When the house's temperature returns to its set point because of the added heat, the thermostat senses the temperature and stops sending signals to the heater. The heat acts as negative feedback in this homeostatic loop.

temperature. When cold weather arrives, the thermostat is stimulated by the colder temperature. The thermostat sends a signal that turns on the heater. The heater responds to the signal by producing heat. Again, note the familiar steps of the stimulus, release and reception of signal, and response. In homeostasis, however, the players have different names. Let's look at this system as part of a homeostasis feedback loop that produces a stable temperature.

For a heating system, the cool temperature is the stimulus, which is detected by the thermostat. The thermostat is known as the **sensor** because it senses the cool air. In general, a sensor is a component of the system that detects when conditions have moved away from a particular level. This level is called the **set point,** which in this example is the temperature at which the thermostat is set. In response to the cool air, the thermostat sends a signal to the heater. In this example, the signal is an electrical one. The heater is known as the effector. An effector is the component in the system that restores normal conditions following a disturbance in those conditions. The heater's response to the electrical signal from the thermostat is to heat the air in the room. After the room has warmed up, the thermostat detects that the temperature is back at the set point and turns off the signal to the heater.

Note in this example that the response (heat) opposes the initial stimulus (cold). The heat feeds back on the thermostat, which stops signaling the heater. This is an example of *negative feedback*. In **negative feedback,** a stimulus acts on a sensor that signals an effector, which produces a response that opposes the initial stimulus and therefore turns off the signal. In this way, a stable temperature, or set point, is achieved. No additional heat is produced by the heater until the temperature drops once again below the set point.

While negative feedback maintains a set point, *positive feedback* does the opposite. **Positive feedback** describes a pattern in which the output or response of a system increases the activity of the system. Once initiated, this pattern is self-reinforcing and tends to snowball, moving the system away from a set point. Positive feedback functions in childbirth, where a signaling molecule leads to contractions of the uterus, which in turn causes release of more of the signaling molecule, leading to further contractions.

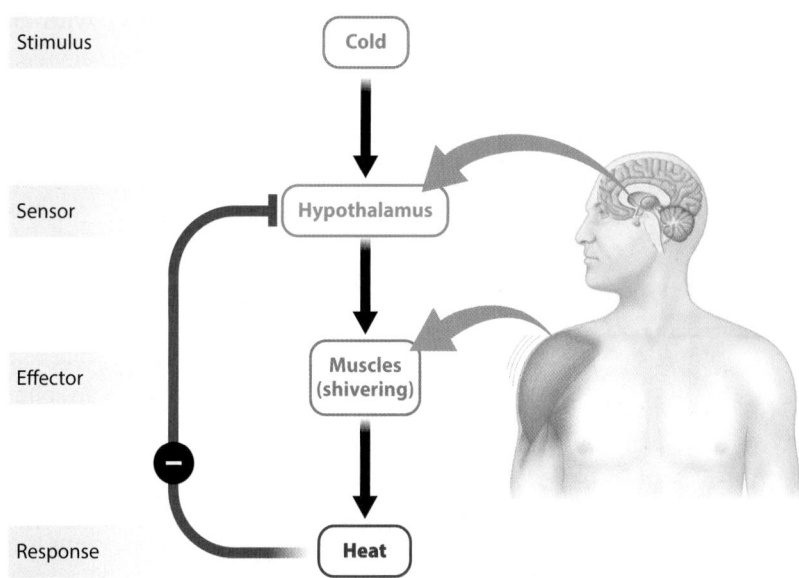

FIGURE 20.11 Temperature regulation by negative feedback in a human

A lower body temperature is sensed by the hypothalamus, which signals muscle cells to start a shivering response to generate heat. The heat produced by shivering warms the body back to its normal temperature. When the hypothalamus senses the return to normal temperature, it stops sending signals to the muscles. The heat produced by shivering is the negative feedback.

Homeostasis in Biology

All parts of a home heating system work together to maintain the temperature set by the thermostat, and a stable temperature is maintained by negative feedback. This is similar to how some organisms maintain a stable body temperature. For example, your body uses the feedback loop shown in **FIGURE 20.11.** Cool temperature is the stimulus detected by the sensor, which is the temperature-regulating part of the brain, located in the hypothalamus. When the hypthalamus receives the stimulus, it sends out signals to many parts of your body. Some of these signals travel to skeletal muscles, which start to shiver. During shivering, muscles undergo very rapid cycles of contraction and relaxation, producing heat that warms the body. The shivering muscles therefore act as effectors for restoring the body's normal temperature.

Signals are also sent to blood vessels, which respond by constricting and therefore reducing blood flow to the body's surface in order to minimize heat loss to the outside environment. The combined responses of shivering and reduced blood flow at the body's surface work together to restore the normal body temperature. When the normal body temperature is reached, there is no longer any stimulus saying that the body is cold. Without the stimulus of cold, the signaling and the responses are terminated, establishing a negative feedback pattern.

Homeostatic regulation relies on negative feedback to maintain a set point. The ability to maintain a constant body temperature is known as thermoregulation, and body temperature is just one of many parameters that is actively maintained at relatively stable levels by the body. Another example is the concentration of the sugar molecule glucose in the bloodstream, diagrammed in **FIGURE 20.12.** In healthy individuals, the amount of glucose in the blood is maintained at relatively constant levels. After a meal the level of glucose in the blood increases. Increased glucose is a stimulus that is sensed by cells in the pancreas. In response to the stimulus, the pancreas cells release insulin. Insulin is a hormone-signaling molecule that causes many target cells in the body to increase uptake of glucose from the blood. As these target cells around the body respond by taking up glucose out of the blood, the glucose level in the blood decreases back to the set point. At the set point, cells of the pancreas are no longer stimulated to produce insulin. In this way, blood glucose is maintained at a relatively constant level.

People with a condition called diabetes are unable to maintain constant levels of glucose in their blood. Their blood glucose levels can rise dramatically after a meal as a result of defects in the insulin feedback loop. Diabetes

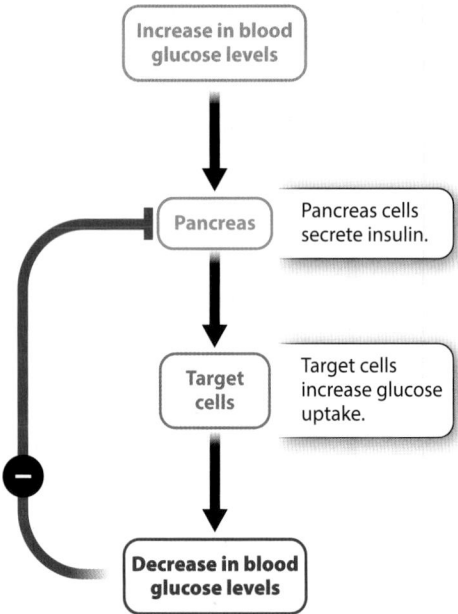

FIGURE 20.12 Blood glucose regulation by negative feedback

Eating a meal increases levels of glucose in the blood. Cells of the pancreas sense the high levels of glucose and secrete the hormone insulin. Insulin travels in the circulatory system to target cells, which increase their glucose uptake. Glucose uptake by cells reduces the amount of glucose in the blood back to normal levels. The pancreas senses the return to normal glucose levels and stops releasing insulin.

mellitus comes in two forms, known as type 1 diabetes and type 2 diabetes. In type 1 diabetes, the pancreas does not produce insulin and in type 2 diabetes, insulin is produced but cells do not respond to it as they normally do. People

with diabetes can monitor their blood glucose level with a device. In a way, this device supplements the sensor functions of the pancreas. Patients can then use the information from the device to make dietary changes or administer insulin to help restore normal glucose levels.

Homeostasis is a fundamental and universal attribute of living organisms, not only protecting cells and organisms from dangerous conditions but maintaining steady conditions in which the organism can function. There are many different variables that are actively maintained by homeostasis, but they all work similarly. In each case, a disturbance is detected by a sensor, which sends a signal to an effector that initiates a response that will bring the system back to a set point. Cell communication allows components of the system to work together in an intergrated fashion.

✓ Concept Check

5. A student came home on a winter night and the house was cold. **Identify** the sensor and effector in the house's heating system, as shown in Figure 20.10, and **predict** which element might be malfunctioning.

6. **Describe** how body temperature is maintained in a human exposed to cold temperature. In your answer, be sure to include the terms "stimulus," "sensor," "effector," and "response," and describe the role of negative feedback.

7. **Identify** the stimulus, sensor, signal, effector, and response in the regulation of glucose levels in the blood.

Module 20 Summary

REVISIT THE BIG IDEAS

PREP FOR THE AP® EXAM

INFORMATION STORAGE AND TRANSMISSION: Using the content of this module, **describe** the structures cells use to send and receive information and how the various components interact to produce a response.

LG 20.1 Cells communicate by sending and receiving chemical messengers.

- Cell communication consists of four sequential steps: stimulus, signal release, signal reception, and response. Page 277

- Cells communicate by sending and receiving chemical messengers called signaling molecules, which are also called ligands. Page 277

- Ligands bind to specific receptor proteins when there is complementarity between molecular shape and distribution of charge. Page 277

- A signaling molecule affects only those cells that have the complementary receptor protein. Page 277

- Quorum sensing in prokaryotes occurs in a density-dependent manner, leading to various responses, such as bioluminescence or DNA uptake. Page 278

- Secretion of the hormone adrenaline achieves cell communication in the fight-or-flight response. Page 279

LG 20.2 Cells can communicate over various distances.

- Endocrine signaling takes place over long distances and relies on the circulatory system for transport of signaling molecules. Page 280
- Paracrine signaling takes place over short distances between neighboring cells and relies on diffusion. Page 280
- Autocrine signaling occurs when a cell signals itself. Page 281
- Some forms of cell communication depend on direct contact between two cells. Page 281
- Gap junctions in animals and plasmodesmata in plants are passages that connect two cells and allow direct communication. Page 282

LG 20.3 Communication among cells, tissues, and organs underlies homeostasis.

- Communication between cells allows coordination of many of an organism's functions, including maintaining homeostasis. Page 284
- Homeostasis describes the dynamic processes used by living organisms to maintain steady internal conditions. Page 284
- Sensors are components in a homeostatic system that are stimulated when a variable is altered from its set point. Page 284
- Effectors are components in a homestatic system that help restore the set point. Page 284
- Homeostasis is often maintained by negative feedback. Page 285
- Diabetes, which results in high blood glucose levels, is an example of a disease the results from a problem of homeostasis. Page 286

Key Terms

Population	Quorum sensing	Set point
Signaling molecule	Hormone	Negative feedback
Target cell	Plasmodesmata	Positive feedback
Ligand	Sensor	

Review Questions

1. In order to receive signaling molecules, a target cell must have
   (A) blood circulation.
   (B) a nucleus.
   (C) mitochondria.
   (D) receptor proteins.

2. In an emergency, a signal from the adrenal glands
   (A) is converted to ATP to provide energy.
   (B) binds to receptors found in many locations.
   (C) affects all cells whether or not receptors are present.
   (D) produces a relaxed sensation to calm activity in the brain.

3. In bacterial responses regulated by quorum sensing, an individual bacterium will respond to a signal if
   (A) there are enough bacteria in the area producing enough signal to bind receptor proteins.
   (B) there are enough bacteria in the area that they are stimulated to produce receptors for the signal.
   (C) there are bacteria in direct contact with one another so signals can be exchanged through gap junctions.
   (D) there are bacteria in different organs of a host organism, so signals can be exchanged through blood vessels.

4. Receptor proteins for signaling molecules
   (A) will bind only specific signaling molecules.
   (B) will bind only to amino acids but not to proteins.
   (C) determine the set point of the cell.
   (D) are involved in energy production but not in homeostasis.

5. Homeostasis describes an organism's capacity for

(A) achieving a one-time goal, such as childbirth.

(B) actively regulating its internal conditions.

(C) alternating between levels above and below a set point.

(D) locating and migrating to the most supportive habitat.

6. In discussions of cellular homeostasis, the "set point" refers to

(A) the lowest temperature at which the cell can remain alive.

(B) the maximum population density at which transfer of DNA between cells can occur.

(C) the steady conditions for the cell to function.

(D) the highest temperature at which the cell can remain alive.

7. Channels that allow ions to move between adjacent heart cells are called

(A) plasmodesmata.   (C) desmosomes

(B) receptor proteins.   (D) gap junctions.

Module 20
AP® Practice Questions

⬛ PREP FOR THE AP® EXAM

Section 1: Multiple-Choice Questions

Choose the best answer for questions 1–4.

1. A multicellular organism has a signaling system in which cells of tissue A send a signal to cells of tissue B, stimulating a response from tissue B. In one of these organisms, the cells of tissue A make and send the signal, but there is no response from tissue B. Which best describes why this organism's tissue B cells do not respond to tissue A's signal?

(A) Tissue B's cells produce too many of the receptor proteins, causing the cells of tissue B to bind to one another rather than to tissue A's signal.

(B) Tissue B's cells have faulty receptor proteins that do not bind tissue A's signal.

(C) Tissue A's signal binds to not just the receptors on the cells of tissue B but also the receptors on cells of at least three other tissues in the organism.

(D) Tissue A's cells are incapable of responding to the stimulus that normally results in the cells of tissue A producing the signaling molecule.

2. A new signaling molecule is discovered in the hypothalamus of the brain. The newly discovered molecule is produced and released by cells of the hypothalamus, and travels through the blood to its target organs, the kidneys and the liver. The newly discovered hypothalamus signaling molecule is best described as

(A) a hormone.

(B) a paracrine signal.

(C) an autocrine signal.

(D) a positive feedback signal.

3. In response to high concentrations of glucose in the blood, cells of the pancreas release the signaling molecule insulin, which binds to receptors on the surface of target cells. The binding of insulin to its receptor sets in motion the taking up of glucose from the blood, as diagrammed in Figure 20.12 on page 286. In individuals with type 2 diabetes, cells do not take up glucose from the blood in response to insulin. One hypothesis for why type 2 diabetes occurs is that insulin binds to the receptor normally, but the signal is not sent into the cell. If this hypothesis is correct, which identifies the step where the malfunction first occurs?

(A) Stimulus

(B) Sensor

(C) Effector

(D) Response

Question 4 refers to the following figure.

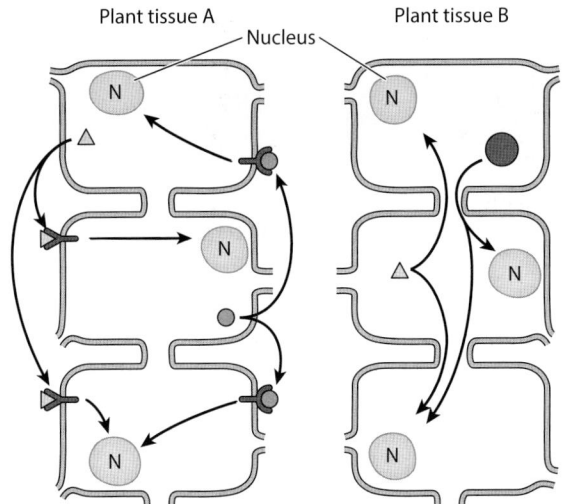

Plant tissue A   Plant tissue B
          Nucleus

4. The figure depicts connected plant cells of two different tissues. In both tissues, signal molecules move from signaling cells to their target cells, as shown by the arrows. In both plant tissues, the triangles and dark circles are the signaling molecules. The circles labeled with an N represent the nucleus of each cell. Determine which statement is correct.

(A) Tissue A cells communicate with one another through cell-surface receptor proteins, and tissue B cells communicate with one another through cell-surface receptor proteins.

(B) Tissue A cells communicate with each other through cell-surface receptors, and tissue B cells do not communicate with each other.

(C) Tissue A cells communicate with each other through the cell-surface receptor proteins, while tissue B cells communicate with each other through plasmodesmata.

(D) Tissue A cells communicate with each other through plasmodesmata, and tissue B cells communicate with each other through cell-surface receptor proteins.

Section 2: Free-Response Question

Write your answer to each part clearly. Support your answers with relevant information and examples. Where calculations are required, show your work.

In the human fight-or-flight response, a signal of danger, such as a scream, stimulates the brain to direct the cells of the adrenal glands to secrete the hormone adrenaline. Adrenaline travels around the body through the blood and binds to adrenaline receptors on the cells of a number of organs and tissues. In response to adrenaline, heart rate and blood pressure increase, which in turn increases the blood flow to the brain, muscles, and legs. The brain, muscles, and legs can then work harder. Adrenaline also signals the lungs to increase the rate of breathing so that tissues have abundant oxygen with which to generate ATP through cellular respiration.

(a) **Identify** the stimulus, signaling cells, target cells, and response(s) in the example of the fight-or-flight response in humans.

(b) **Predict** the effect to a person's fight-or-flight response if the person's lungs lack adrenaline receptors. **Justify** your answer.

Module 21

Signal Transduction

LEARNING GOALS ▶**LG 21.1** Signaling molecules bind to specific receptor proteins.
▶**LG 21.2** Hydrophilic, polar signaling molecules can produce rapid biochemical responses.
▶**LG 21.3** Hydrophobic, nonpolar signaling molecules can turn genes on or off.

Chemical signaling helps cells to coordinate their activities with one another, allowing different parts of a multicellular organism to support each other. In Module 20, we saw that cells work together because they communicate: one cell produces a signal and another cell, the target cell, receives and responds to the signal. We explored the many routes that a signal from one cell can take to get to another cell, whether the second cell is adjacent or in a completely different part of the body. Once a target cell receives a signal, how is that signal converted into a response? The short answer is **signal transduction,** the process by which a signal from outside the cell leads to a response inside the cell. In this module, we will explore signal transduction.

Recall our analogy for communication in Module 20, in which a baby bird communicates its hunger to a parent by begging. The chick is the signaling partner, and its gaping and chirping behaviors are the signals it releases. The signals from the hungry chick are seen and heard by the parent bird, the receiving partner in this communication. The parent then responds to the chick's signals by feeding it. The gaping and chirping signals undergo a conversion or transduction from sounds and sights to trigger a pathway of behavioral responses that lead the parent to feed the chick. The step in which the signals are converted to the parent's behavioral response is like signal transduction in cellular communication.

In this module, we will investigate four distinct signal transduction pathways set into motion by receptor activation. These pathways can be categorized by whether they bind signals outside or inside the cell. The first three types of signal transduction produce rapid and dramatic responses and the fourth type produces changes that start slowly but are sustained over long periods. As we explore how signals are transduced to responses, we will continue to emphasize the role of cell signaling in maintaining homeostasis.

PREP FOR THE AP® EXAM

FOCUS ON THE BIG IDEAS

SYSTEMS INTERACTIONS: Consider the relationship between the structure of molecules and their ability to transmit information from one cell to another.

21.1 Signaling molecules bind to specific receptor proteins

Cell communication occurs in both unicellular and multicellular organisms, helping to coordinate the functions of groups of cells so they are able to work together. Regardless of which cells are involved, there is a consistent pattern of steps involved in cell communication: signal binding and receptor activation, a cellular response, and termination of the response. We will look at many different signal transduction pathways in this module, and we will again see a consistent pattern of steps across all signal transduction pathways: signal binding and receptor activation, response, and termination. **FIGURE 21.1**

outlines these basic steps. First, a signaling molecule binds to a receptor. The receptor changes its shape, and this conformational change is what activates the receptor. In the figure, the yellow highlight around the receptor bound to the signaling molecule indicates that the receptor is activated. The activated receptor then triggers a pathway that leads to the cell's response. The response is ended, or terminated, a short time after it begins. In this section we will look at how the binding of a signaling molecule leads to an activated receptor.

Signal–Receptor Binding

A cell can be bombarded with many different types of signaling molecules. However, a cell will respond to a signaling

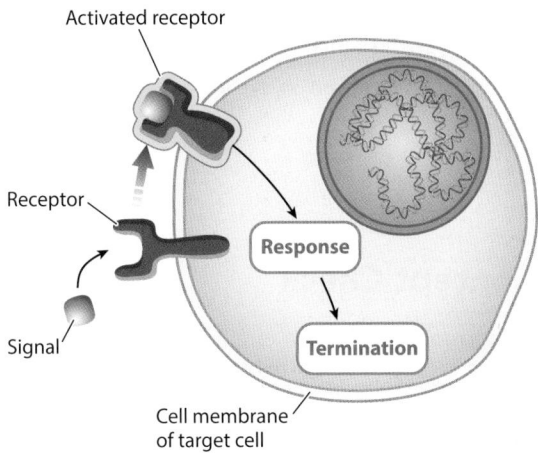

FIGURE 21.1 Signal transduction

All signal transduction pathways go through the same basic steps: the receptor is activated when it binds the signal; the information in the signal is relayed within the cell; the cell responds to the signal; and the response to the signal is terminated.

molecule only if the cell has receptor proteins that match up with the signaling molecule and a binding interaction occurs. In the chemical interaction between signaling molecules and receptor proteins there is precise matching between the shapes and other characteristics of signals and receptors. This matching allows a chemical interaction to take place between the signaling molecule and receptor protein. When that interaction takes place, the receptor protein changes shape and becomes activated, setting in motion the pathway that leads to the target cell's responses.

We first saw two molecules complementing each other and binding in Figure 14.4 on page 200, which demonstrates that enzymes are specific to complementary reaction substrates. In cell communication, a similar match needs to occur: receptor proteins must match the signaling molecules in molecular structure, shape, and charge distribution in order for the receptor and signal to fit together so that they can bind to one another. The shape-matching part of the signal and receptor interaction is shown in **FIGURE 21.2**. On the left side of the figure, the circular ligand fits into the semicircular binding site on the receptor protein embedded in the membrane. The circular ligands can therefore initiate responses in the cell. On the right side of the figure, the triangle-shaped ligands do not fit the receptor protein and therefore the triangle-shaped ligands do not alter the cell's activity.

In addition to the ligand and receptor having shapes that match, it is also likely that regions of the ligand and the receptor have opposite charges that attract one another. In cell signaling, the ligand–receptor complex is the signaling molecule and its receptor protein bonded together by non-covalent molecular interactions. The ligand–receptor complex, held together by weak attractions between opposite charges, can be easily uncoupled to end the communication. This reversibility of the signal–receptor interaction prevents most signals from causing permanent or irreversible changes in the target cell's activity.

Tissue-Specific Responses

A signaling molecule can lead to different responses in different cells and tissues. In other words, one signal can trigger different tissue-specific responses. A signal of stress, adrenaline, for example, can have different effects in the heart and lungs. Recall from Module 20 that adrenaline is a hormone that is released into the blood during emergencies and during exercise. Only cells with receptor proteins that bind adrenaline will respond to adrenaline. Receptor proteins that bind adrenaline are found all around the body, including target cells in the heart, lungs, and many other organs. When adrenaline binds its receptor on heart cells, it causes the heart rate to increase. When adrenaline binds its receptor on lung cells, it causes airways to widen. The presence of the adrenaline-binding receptor protein allows cells to respond to adrenaline, but the type

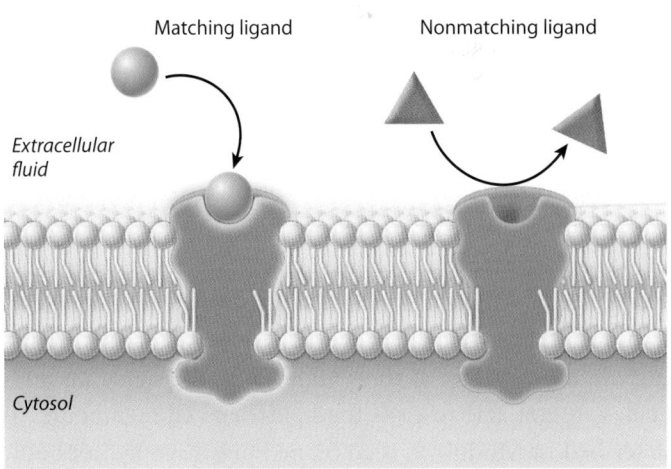

FIGURE 21.2 Specific binding of ligand and receptor protein

A cell is a target for only the signaling molecules that are complementary to receptor molecules on the cell. When a complementary ligand binds to a receptor, the receptor changes conformation and is activated. The activated receptor can then initiate a signal transduction pathway.

of cell determines the nature of the response. Many tissues, including the muscles that move your arms and legs, do not respond directly to adrenaline because they do not have adrenaline receptor proteins.

Insulin, the hormone that helps regulate glucose levels in the blood, is another example of one signal producing different effects in different cells. Insulin is a ligand that binds to its receptor proteins on muscle, fat, and liver cells, but these cell types do not all have the same response to binding insulin. The activated insulin–receptor complex in muscle and fat cells stimulates an increase in the number of glucose transporters in the cell membranes. More glucose transporters allow more glucose to move from the blood into insulin-stimulated muscle and fat cells. However, when the liver's insulin receptor proteins bind to insulin, the activated receptor triggers a pathway that increases the rate at which the cell assembles glycogen from glucose, as described in Module 17.

✓ Concept Check

1. The hormone adrenaline circulates in the blood and so has access to all parts of an animal's body. **Describe** why some cells in the body are affected by adrenaline and others are not.

2. **Describe** two events that cause an inactive receptor protein to become activated.

21.2 Hydrophilic, polar signaling molecules can produce rapid biochemical responses

Now that we have discussed that binding of signals to receptors leads to receptor activation, we will sample the diverse ways that activated receptors influence cells and cause responses. There are two broad groups of signaling molecules: hydrophilic, polar signaling molecules, and hydrophobic, nonpolar signaling molecules. The two groups are distinguished by the water solubility of the signaling molecules. In this section, we will focus on hydrophilic signaling molecules. In the next section, we will study hydrophobic signals.

In Module 1, you learned that water is a polar solvent, and that polar molecules readily dissolve in water. Hydrophilic signaling molecules are molecules that include regions of polarity. Since extracellular fluids are more than 95% water, the hydrophilic signals readily dissolve and are easily transported in extracellular fluids. When hydrophilic signaling molecules arrive at their target cells, the signaling molecules are not able to cross the lipid-rich cell membrane. As described in Module 9, membrane lipids have hydrophobic tails in the middle of the membranes, and these tails repel hydrophilic molecules. The receptor proteins for hydrophilic signals are on the surface of cells, where they are able to interact with the signals.

FIGURE 21.3 shows a polar, hydrophilic signaling molecule interacting with a receptor protein on the surface of a target cell. Note that the receptor spans the membrane, with

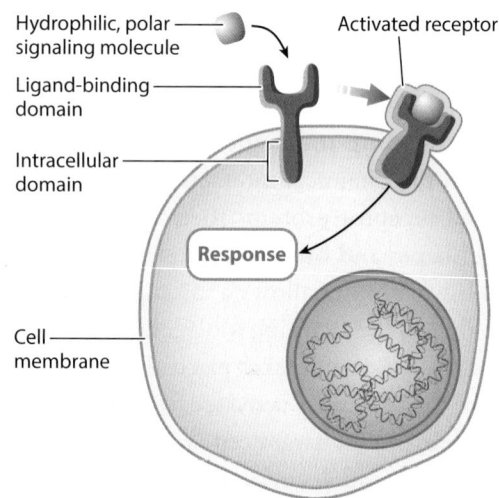

FIGURE 21.3 Signal transduction of hydrophilic signaling molecules

Hydrophilic signaling molecules cannot pass through the cell membrane to enter the cell, and so bind to ligand-binding domains of membrane-bound receptor proteins. The activated receptor then initiates a signal transduction pathway, leading to a cellular response.

one part, or domain, projecting out of the membrane and one domain to the inside of the membrane. The portion of the receptor that is outside the cell is the ligand-binding domain, and the portion that is inside the cell is the intracellular domain. Once the signaling molecule binds to the ligand-binding domain of the receptor, the receptor, including its intracellular domain, is switched to its activated form. The activated receptor turns on a pathway that triggers cellular responses.

The receptor proteins that bind the hydrophilic signaling molecules are organized into three groups: *G protein-coupled receptors, receptor-protein kinases,* and *ligand-gated ion channels,* each of which is defined below. Although each transduction mechanism is different, there are three unifying facts that hold true for all three groups that bind hydrophilic signaling molecules. First, all of the signals in this category are water soluble. Second, the receptor proteins for water-soluble signals are located on the target cell's plasma membrane as embedded proteins. Third, the responses by the target cells are usually rapid but short-lived.

G Protein-Coupled Receptors

Imagine that you are sitting in a classroom, and you smell gas. Alarmed, you tell your two closest neighbors. They tell their two closest neighbors. Now seven of you have been alerted to the leak. The message spreads throughout the class, with each student telling another two students that they smell gas. Before long, there are enough alarmed students for your teacher to hear the message, evacuate the room, and seek assistance. It took several relayed whispers, but the outcome was an **amplification,** or increase in volume, of your original message, and this generated a response from your teacher. A similar chain of events occur in cells with *G protein-coupled receptors.* A **G protein-coupled receptor** is a type of receptor protein that, when activated, uses a molecule called a G protein as part of the signal transduction pathway. In G protein-coupled receptor pathways, the external signal is the first message to the target cell that it will be changing its activity. Activation of this receptor leads to the production of a very large amount of a **second messenger,** which is an intermediate signaling molecule, amplifying the response inside the cell. In the gas leak analogy, the gas smell that alerts you to the leak is the external signal, you are the receptor, and all of your classmates are the second messengers.

Almost every eukaryotic organism has cells that communicate through signaling pathways that include G protein-coupled receptors. **FIGURE 21.4** shows the cascade of events that take place when G protein-coupled receptors are activated. The receptor is shown as a membrane-embedded protein that loops back and forth through the cell membrane, crossing the membrane seven times. In the top panel of Figure 21.4, before the extracellular signal is bound, the receptor is a short distance away from an inactive G protein on the inside of the cell membrane. Activation of the receptor by binding the signal allows the receptor to bind to the G protein. In response, the G protein releases a molecule of GDP and replaces it with GTP, activating it. The activated

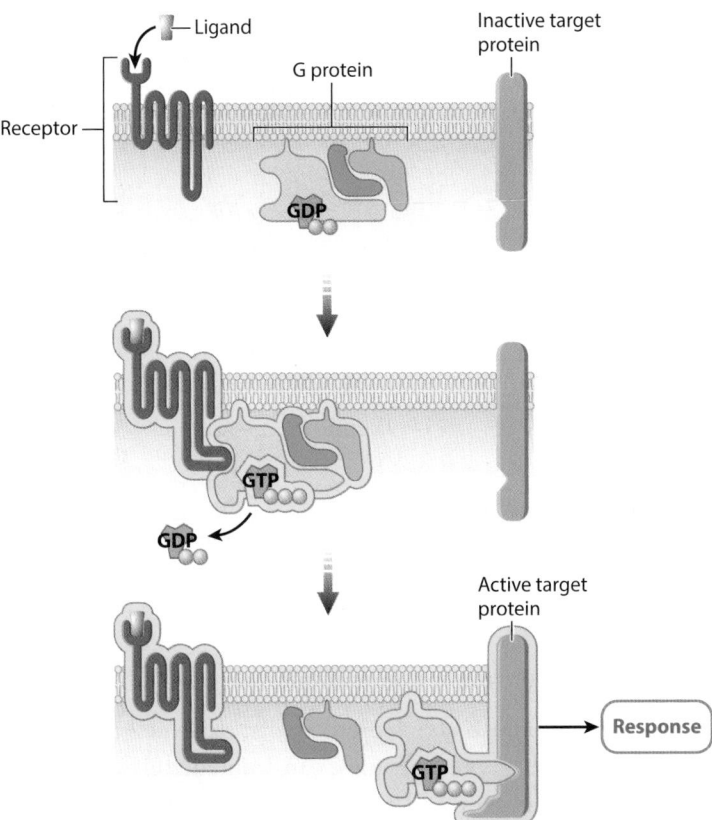

FIGURE 21.4 G protein-coupled receptor activation

When a signal binds to the extracellular part of the receptor protein (top), the G protein binds to the signal–receptor complex inside the cell (middle). As a result of binding to the complex, the G protein's GDP is exchanged for GTP. The G protein then binds to and activates a target protein (bottom). The active target protein produces intracellular events, leading to a cellular response.

G protein separates into two different subunits, and the GTP-bound subunit then binds to and activates a target protein, as seen in the bottom panel of Figure 21.4. The activated target protein can then trigger the internal pathway that will change the activity of the target cell.

There are several possible target proteins in G protein-coupled pathways, such as one called adenylyl cyclase, which is shown in **FIGURE 21.5** on page 294. When the G protein activates adenylyl cyclase, the active adenylyl cyclase converts ATP to a derivative called **cyclic adenosine monophosphate (cAMP)**. The molecule cAMP is a second messenger. In its role as second messenger, cAMP binds to and activates proteins called protein *kinases.* A **kinase** is an enzyme that, when activated, catalyzes the transfer of phosphate groups from ATP to proteins, as described in Unit 3. Adding a phosphate group to a protein changes its shape and alters it function. Kinases activated by cAMP convert

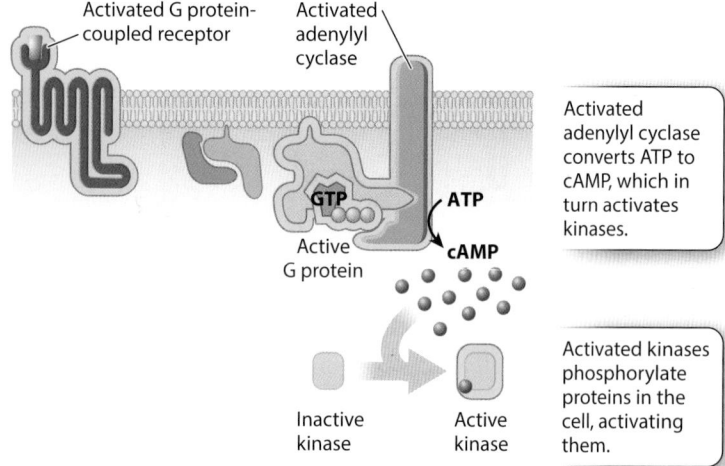

FIGURE 21.5 Protein kinase activation by second messengers

Adenylyl cyclase is a target protein for active G proteins. Activated adenylyl cyclase converts ATP into the second messenger cAMP, which can activate protein kinases.

inactive proteins to active proteins that drive cell responses. The kinases, like all enzymes, can immediately serve the same function over and over, so that many reactions take place through the chain of events following the cell binding a signal, producing a very large cellular response.

The effect of the hormone adrenaline on heart cells is an example of a small amount of signal being amplified by the signal transduction pathway. As shown in **FIGURE 21.6**, a single activated adrenaline receptor can lead to the activation of multiple adenylyl cyclase enzymes, which each produce many cAMP molecules. Each cAMP molecule activates a protein kinase, and each kinase can activate multiple target proteins. The result is that many target proteins are activated after one signal binds its receptor. We call this a **signaling cascade,** where a small amount of signal is amplified to produce a large cellular response. Elsewhere in the body, adrenaline activates other fight-or-flight responses through similar transduction pathways.

Cellular responses to the activation of G protein-coupled receptors all tend to be rapid and short-lived, meaning that the response to the signal ends quickly. This rapid response is a characterstic of all hydrophilic signaling molecules. The fast on/off cycling of G protein-coupled receptors permits an organism to quickly respond to changes in their environment. In the example of adrenaline, the heart rate increases rapidly in response to stress. In addition, when the emergency is over or exercise is completed, the heart rate quickly returns to resting levels.

Termination of the signaling pathway begins when adrenaline dissociates from its receptor, as shown in **FIGURE 21.7** on page 296. The top panel of the figure shows active signaling with adrenaline bound to the receptor. Soon after adrenaline detaches from the receptor, shown in the bottom panel, the GTP that was attached to the G protein is dephosphorylated to become GDP. The conversion from GTP to GDP inactivates the G protein. The inactive G protein dissociates from the adenylyl cyclase, switching adenylyl cyclase to the inactive state, so it stops producing more molecules of cAMP. Enzymes in the cell constantly convert cAMP into adenosine monophosphate (AMP). The AMP molecules cannot activate kinases, and inactived kinases cannot activate any more downstream target proteins. In addition, an enzyme called phosphatase removes the phosphate groups from activated kinases, so they revert to the inactive state. This reversal of cell responses happens quickly when the signal is gone.

G protein-coupled receptor pathways illustrate the hallmarks of a cell's response to a hydrophilic signal: the signal binds and activates the receptor, beginning a signal transduction pathway that quickly leads to a cellular response. This response is quick and lasts only as long as the signal is bound to the receptor. The presence of a second messenger, in this case, cAMP, allows the action of an external signal to be converted and amplified to intracellular signals that control cell responses. Following the removal of the external signal, the process quickly terminates. However, it is poised to be activated again in response to a signal. Because most forms of life use this pathway, this was likely one of the first signaling pathways to appear in primitive eukaryotes.

Receptor-Protein Kinases

The second group of cell-surface receptor proteins is a **receptor-protein kinase,** which is a two-in-one protein: it has both a signal-binding extracellular domain and a kinase intracellular domain. The binding of an external signal to the ligand-binding domain of the receptor-protein kinase rapidly activates the kinase domain, and the kinase's activity triggers a signal transduction pathway. The activated kinases are familiar: recall that in G protein-coupled receptor pathways, kinases activated other enzymes by adding a phosphate group, changing the shape and the functions of the enzymes. The activated kinase part of receptor-protein kinases transfers a phosphate group from ATP to a protein and thus changes the shape of the protein to drive the responses of target cells.

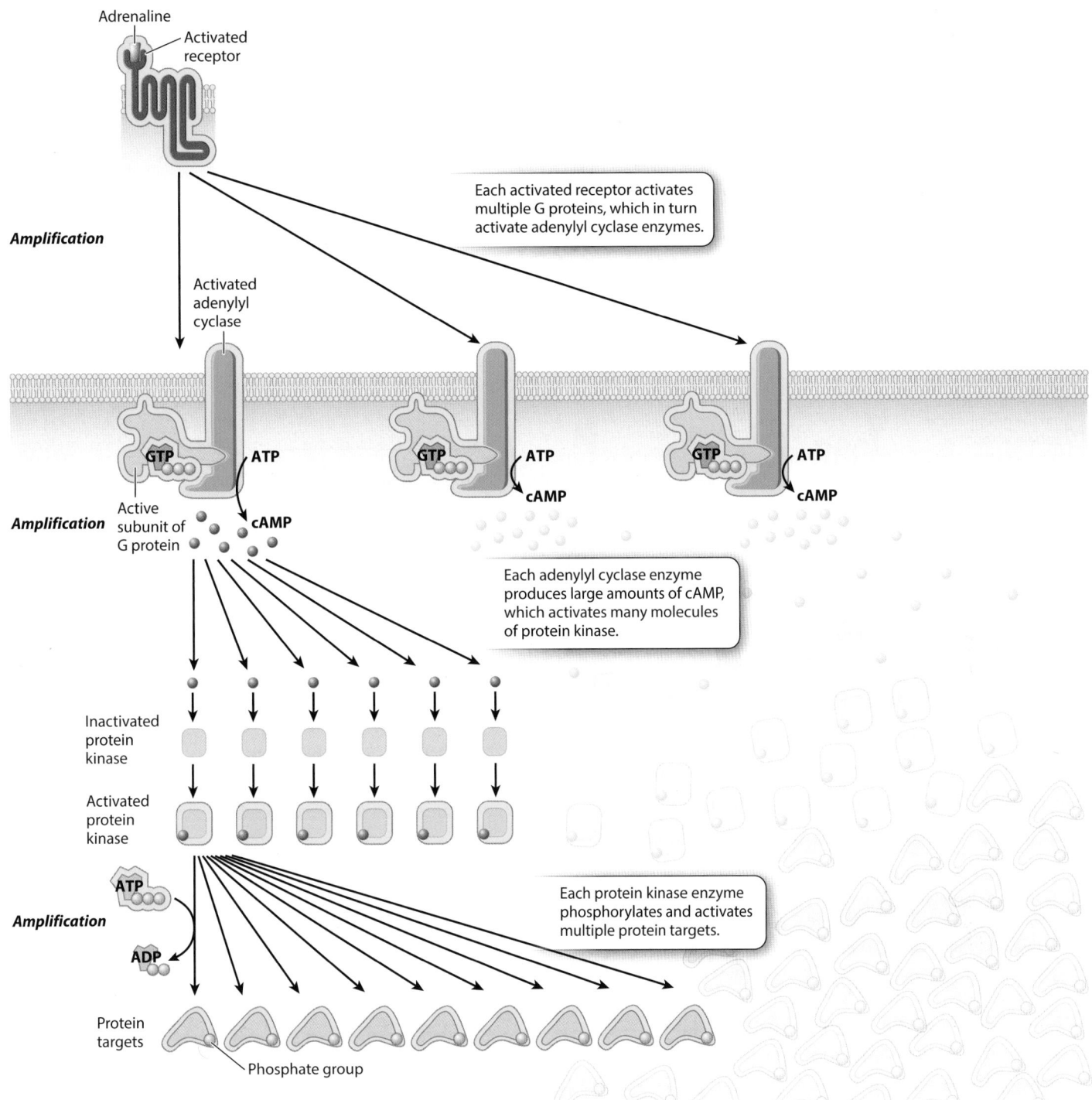

Amplification

Adrenaline
Activated receptor

Each activated receptor activates multiple G proteins, which in turn activate adenylyl cyclase enzymes.

Activated adenylyl cyclase

Amplification

GTP   ATP
Active subunit of G protein   cAMP

GTP   ATP
cAMP

GTP   ATP
cAMP

Each adenylyl cyclase enzyme produces large amounts of cAMP, which activates many molecules of protein kinase.

Inactivated protein kinase

Activated protein kinase

Amplification

ATP

ADP

Each protein kinase enzyme phosphorylates and activates multiple protein targets.

Protein targets

Phosphate group

FIGURE 21.6 A signaling cascade

A single signal can be amplified inside the cell, resulting in a large response. One activated adrenaline receptor leads to multiple activated adenylyl cyclase enzymes. Adenylyl cyclase enzymes generate many second messenger cAMP molecules. Each molecule of cAMP binds to and activates many kinases, which then activate many more protein targets. The activated proteins lead to a cellular response.

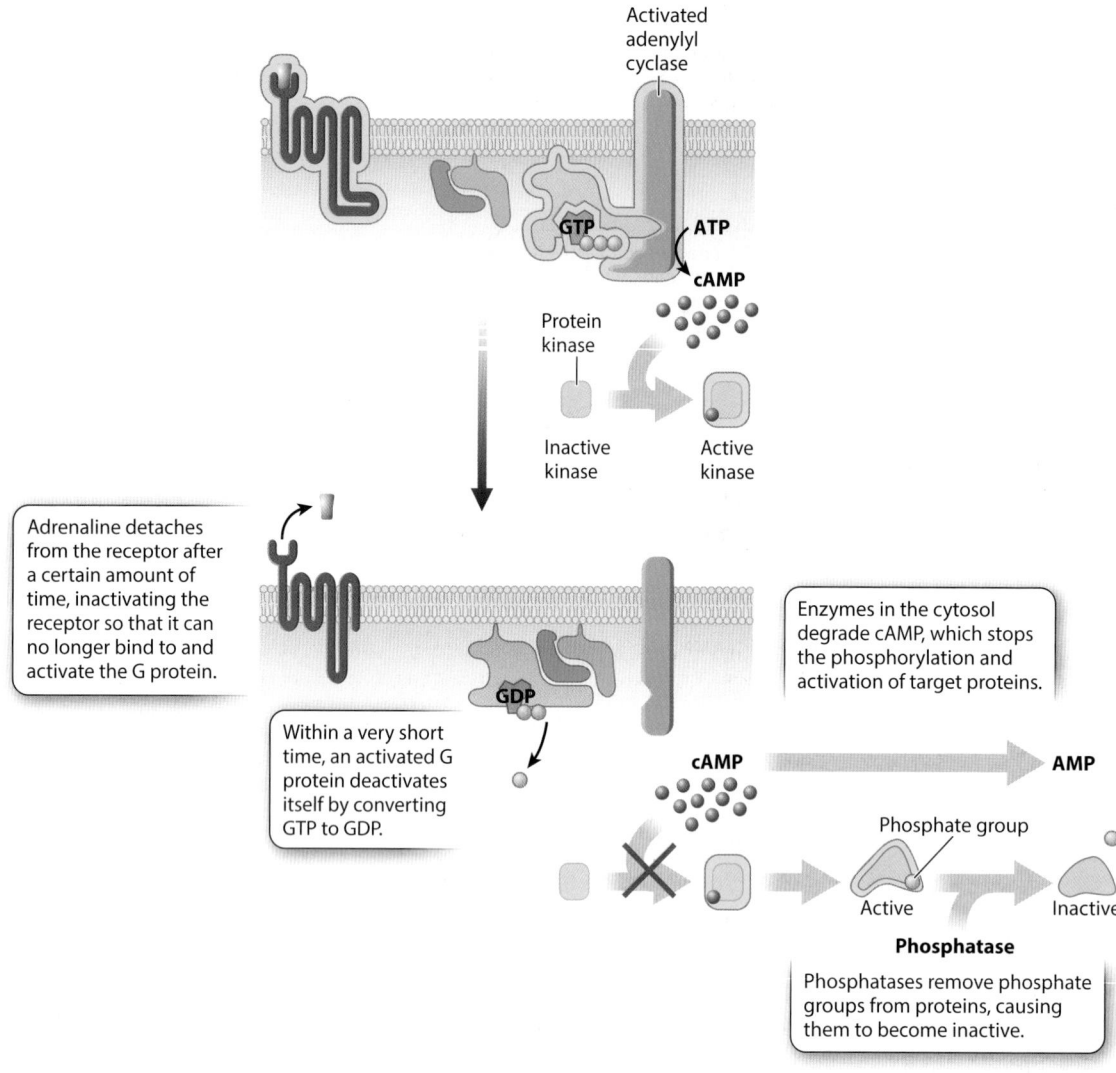

Activated
adenylyl
cyclase

GTP **ATP**

cAMP

Protein
kinase

Inactive
kinase

Active
kinase

Adrenaline detaches
from the receptor after
a certain amount of
time, inactivating the
receptor so that it can
no longer bind to and
activate the G protein.

Enzymes in the cytosol
degrade cAMP, which stops
the phosphorylation and
activation of target proteins.

GDP

Within a very short
time, an activated G
protein deactivates
itself by converting
GTP to GDP.

cAMP **AMP**

Phosphate group

Active Inactive

Phosphatase

Phosphatases remove phosphate
groups from proteins, causing
them to become inactive.

FIGURE 21.7 Termination of G protein-coupled signaling

The release of adrenaline from the receptor protein causes the conversion of GTP to GDP on
the G protein. The inactive G protein then separates from adenylyl cyclase, inactivating
adenylyl cyclase and halting cAMP production. The ongoing breakdown of cAMP terminates
the activation of the kinases. The inactivated kinases no longer activate target proteins.
Ongoing phosphatase activity removes phosphates from active target proteins, and
nonphosphorylated proteins no longer contribute to responses.

The dual functions of receptor-protein kinases are shown
in **FIGURE 21.8**. The receptor-protein kinase is made up of a
ligand-binding domain where extracellular signals bind and an
intracellular domain that serves as a kinase, as identified in the
first panel of the figure. In the second panel of Figure 21.8,
the arrival and binding of the external signal brings together
the two subunits of the receptor kinase, and as a result, the
domain containing the kinase becomes active. The kinases
then cause phosphate groups to be added to the intracellular
domain of the receptor protein, as shown in the third panel of
Figure 21.8. The addition of a phosphate group to the receptor

protein changes the receptor protein's shape and activity. The
change in shape allows the activated receptor protein to interact with proteins called cytoplasmic-signaling proteins, shown
in the final panel of Figure 21.8. The cytoplasmic-signaling
proteins then trigger the rest of the signal transduction pathway that triggers the responses of the target cells.

Cells that respond to hydrophilic signals by means of
receptor-protein kinases do so in varied times and places. For
example, receptor-protein kinases have important roles during
the development of embryos and in the insulin-signaling
pathway described earlier.

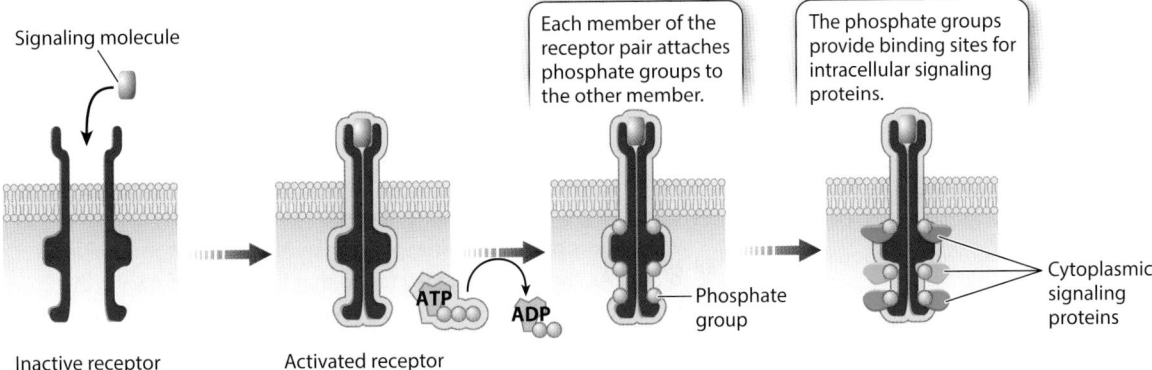

FIGURE 21.8 Receptor-protein kinase pathway

A signaling molecule binding to the ligand-binding domain of a receptor-protein kinase pathway brings the two halves of the receptor together. The conformation change activates the receptor's kinase activity, which allows cytoplasmic signaling proteins to bind and begin a relay pathway.

Receptor-protein kinase pathways also play a role in controlling blood loss after a skin injury. The last time you got a small cut, it likely bled for a minute or two, and then the bleeding stopped and the healing started. A receptor-protein kinase signaling pathway is responsible for the body responding to the injury and stopping the bleeding. In this signaling system, fragments of cells in the blood called platelets release a signaling molecule called platelet-derived growth factor (PDGF) at the injury site. PDGF binds to PDGF-specific receptor-protein kinases on cells at the site of a wound. The PDGF receptor-protein kinases becomes activated after binding PDGF. Phosphate groups then bind to the receptor-protein kinase at several sites in its intracellular domain. The addition of these phosphate groups provides places on the activated receptor-protein kinase where other proteins in the cytoplasm bind and become active in the pathway.

As we saw in the G protein-coupled receptor pathways, a small quantity of an extracellular signal causes many proteins to become active, thus amplifying the signal. In the body's response to an injury, some of the proteins activated by the receptor-protein kinase response turn on genes that are needed for cell division so that your cut can heal. "Practicing Science 21.1: Where do growth factors come from?" describes a pathway where receptor kinases mediate cell division after an injury.

Receptor-protein kinase signaling is terminated by the same basic mechanisms that are involved in G protein-coupled receptor pathways. The removal of the signaling molecule from the receptor-protein kinase terminates the kinase activity, and the response pathway stops as other enzymes remove the phosphate groups from proteins activated by kinase. In short, signaling through receptor kinases follows the same basic sequence of events that we saw in signaling by G protein-coupled receptor proteins, including receptor activation, signal transduction, cellular response, and termination.

Practicing Science 21.1

PREP FOR THE AP® EXAM

Where do growth factors come from?

Background Biomedical researchers perform many informative experiments using a model system in which human and nonhuman cells are grown in culture, a solution of purified cells growing in a flask. Cell culture systems allow researchers to conduct carefully controlled experiments where cell interactions do not complicate the analysis, and where pain and suffering are kept to a minimum. However, cells grown in culture survive and grow well only under certain conditions. For a long time, researchers hypothesized that certain substances are required for cells to grow in culture, but the identity and source of these substances remained a mystery. A key insight came from the observation that chicken cells grow much better if they are cultured in blood serum rather than in blood plasma. Blood serum is the liquid component of blood that is collected after blood has been allowed to clot. Blood plasma is also the liquid component of blood, but it is collected from blood that has not clotted. Platelets in the blood cause blood to clot. Serum contains proteins involved in clotting, while plasma does not. In the 1970s, American biologists Nancy Kohler and Allan Lipton were interested in identifying the chemical factor in blood serum that allows cells to survive in culture.

Hypothesis Kohler and Lipton knew that clotting depends on the release of substances from platelets. Because cells grow better in serum collected from clotted blood than in plasma collected from unclotted blood, they hypothesized that a growth-promoting factor was released by platelets during clotting.

Experiment 1 and Results The researchers first confirmed earlier observations using mouse cells called fibroblasts that are easily grown in culture. They grew two sets of fibroblasts in small plastic dishes, illustrated in the figure below. They added serum to one of the dishes and plasma to the other. They counted the number of cells in both culture dishes over time to determine the rate of cell division. Fibroblasts grown in serum divided more rapidly than those grown in plasma, as shown In the graph below.

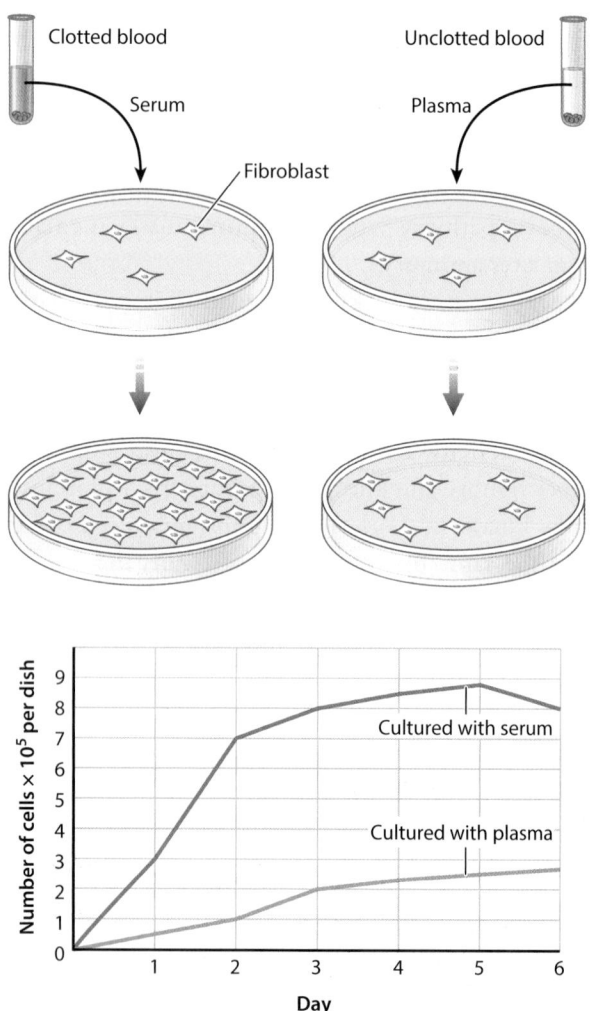

Experiment 2 and Results To see if a growth-promoting factor is released from platelets, the researchers prepared a solution of proteins made from purified platelets, and added these proteins to cultured fibroblasts. Cells cultured with plasma were used for comparison. The researchers then counted the number of cells over time, comparing counts in cultures that received serum from clotted blood with counts of cells in cultures that received plasma from unclotted blood. As shown below, they found that fibroblasts grown in the presence of the platelet proteins found in serum divided more rapidly than fibroblasts grown in plasma, which lacks platelet proteins.

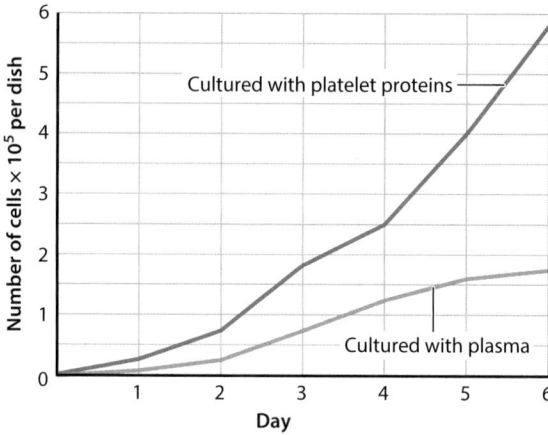

Conclusion Kohler and Lipton concluded that the growth-promoting factor is a protein that is released by platelets upon clot formation, and that is normally present in serum but absent in plasma.

Follow-up Work Over the next few years, Kohler, Lipton, and other researchers purified and characterized the growth factor we know today as platelet-derived growth factor, or PDGF.

SOURCE
Kohler, N., and A. Lipton. 1974. "Platelets as a Source of Fibroblast Growth-Promoting Activity." *Experimental Cell Research* 87:297–301.

AP® PRACTICE QUESTION

(a) **Identify** the following:
  1. The question the experimenters are trying to answer
  2. The hypothesis they tested
  3. The independent variable for both experiments
  4. The dependent variable for both experiments

(b) Explain the results of both experiments by answering the following:
  1. **Identify** the claim (conclusion) made by the researchers for both experiments.
  2. **Justify** the claim with evidence from the experiments and your own reasoning.

Ligand-gated Ion Channels

The third group of hydrophilic signaling mechanisms is the **ligand-gated channel,** which is a receptor protein that includes a membrane-spanning channel that opens or closes (is gated) in response to binding a signaling molecule (ligand). The opening and closing of the channels alters the rate of ion movements across the cell membrane. Ion movement across cell membranes is an important part of many cell types' responses to signals. Imagine a complex behavior, like juggling three oranges. Ligand-gated ion channels, with their very rapid cycling patterns of response, underly the precise contractions of individual muscles needed for juggling. Ligand-gated ion channels also play a role in our nervous system's processing of visual information, allowing the complex movements and coordination between your eyes and hands that allow you to juggle the oranges. Muscles and nerves are examples of excitable cells, cells that are dependent on the very rapid and transient movements of ions into and out of the cell to stimulate the electrical and chemical process of muscle excitation and relaxation.

FIGURE 21.9 illustrates how ligand-gated ion channels respond to signaling molecules to alter the flow of ions into and out of the cell. On the left side of the figure, no ligand is bound to the receptor protein, and so the ion channel is closed. As we saw in Unit 2, ions, because they are charged, are not able to cross the hydrophobic membrane unless an open channel is available. On the right side of the figure, the ligand is bound to the receptor and this binding alters the shape of the ion channel so that it is open. The channel is only open as long as the ligand is bound, so the cell's response is quickly terminated when the ligand is released.

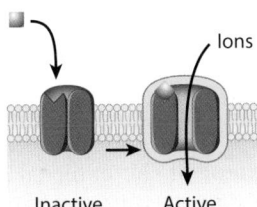

Inactive   Active

FIGURE 21.9 Ligand-gated channel

Binding a ligand changes the shape of the ligand-gated ion channel, opening a channel through which ions can flow across the cell membrane. Until it binds the ligand, the receptor is in its inactive, closed conformation. Binding the ligand activates the receptor, opening the channel.

Ligand-gated ion channels have very specific shapes and charges so each type of ion channel only allows a specific ion to move through to cross the membrane. One type of ligand-gated ion channel is a potassium ion channel, another is a chloride ion channel, and one is a sodium ion channel.

Muscle contractions result from the actions of ligand-gated ion channels. Muscle cells have ligand-gated sodium channels that bind the signaling molecule acetylcholine. When the sodium channel binds acetylcholine, sodium ions flow into the cell, changing the concentration of ions inside the cell and triggering a pathway that leads to a muscle contraction. The sodium ion channel closes as soon as acetylcholine is released from the receptor, immediately terminating the cell's response. In excitable cells such as muscles, rapid responses can be essential for survival. For example, the rapid movement of sodium ions is the basis for the reflex of quickly removing your arm from a hot stove. Acetylcholine from the nerves controlling muscles acts as a ligand to bind to acetylcholine receptors on muscles. This binding alters the shape of the receptor protein and this produces a small passageway through the protein, just right for sodium ions to move through and enter the muscle. The rapid sodium entry leads to rapid muscle contraction. Acetylcholine is quickly degraded after binding, producing a short-lived but dramatic response in this transduction mechanism.

PREP FOR THE AP® EXAM

AP® EXAM TIP

Rapid and accurate signaling and cellular communication are essential functions in healthy organisms. You should be able to describe the cause-and-effect processes in each of the three groups of hydrophilic signaling molecules: G protein-coupled receptor proteins, receptor-protein kinases, and ligand-gated ion channels. Be sure you can identify the characteristics that set them apart from one another.

✓ Concept Check

3. **Describe** how a very small amount of the signal adrenaline causes a very large cellular response.

4. **Identify** the meaning of the name receptor-protein kinase and describe the dual function activites of these proteins.

5. **Describe** how the signaling molecule acetylcholine causes the entry of sodium ions into a muscle cell.

21.3 Hydrophobic, nonpolar signaling molecules can turn genes on or off

We now look at the hydrophobic signaling molecules, which are nonpolar lipids that do not readily dissolve in water. Although less soluble in water, the hydrophobic signaling molecules are readily soluble in lipids. Lipid-soluble molecules can diffuse into and out of cells through the lipid-rich cell membrane. In contrast, as we saw, hydrophilic signaling molecules do not cross cell membranes or enter cells. Because the hydrophobic signals easily cross the membranes and enter cells, the receptor proteins for these signaling molecules are located inside the cells. Receptors for hydrophobic signals are free-floating in the cytosol and in the nucleus.

These internal receptor proteins, when activated, cause changes in **gene expression,** turning some genes on and turning other genes off. Because the response to hydrophobic signaling molecules is a change in gene expression, rather than activation of enzymes and other proteins, the response to hydrophobic signaling molecules takes time to get started, but the responses are sustained over time. Nevertheless, cell communication by hydrophobic signaling has the same essential ingredients for signal transduction as communication by hydrophilic signaling: a signal is released from a signaling cell, the signal binds to its receptor on a target cell, the receptor becomes activated, and the activated receptor triggers a signal transduction pathway that causes a cellular response.

In Module 20, we saw that hormones are signaling molecules that travel to target cells through the blood. Steroid hormones are hydrophobic signaling molecules. Steroid hormones are lipids derived from the molecule cholesterol, which was described in Module 3. Once in the bloodstream, the steroids freely diffuse through the cell membrane of any cell they encounter, even entering cells that do not have the appropriate receptor proteins and so are not target cells. Though present in very low concentrations, these small and nonpolar signals are nontheless powerful influences in the cell. Steroid hormones serve essential communication roles in many homeostatic processes and in reproduction.

After diffusing into a target cell, a steroid binds to its intracellular receptors, as shown in **FIGURE 21.10**. The activated ligand–receptor complex then moves into the nucleus

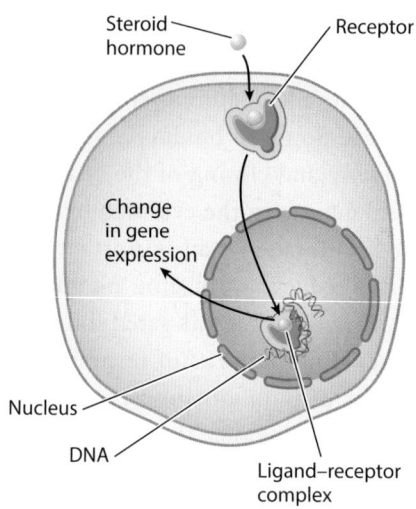

FIGURE 21.10 Hydrophobic signaling molecule and intracellular receptor

Because steroid hormones are hydrophobic, they can cross the cell membrane. After diffusing into the cell, the hormone binds an intracellular receptor protein, which activates the receptor. The activated receptor binds to DNA to turn genes on or off.

of the target cell, where it interacts with the DNA and turns certain genes on or off to either produce proteins or stop production of proteins.

Proteins are the products of gene expression, as you learned in Module 6. The presence of certain proteins in a cell means that the cell has responded to a steroid hormone's stimulus to make those proteins. For example, testosterone produced in a male fetus reaches a large number of target cells with testosterone receptors. As the fetus grows and develops, the male fetus's body responds to testosterone by turning on genes to synthesize proteins that produce the features that distinguish male babies from female babies anatomically. These proteins determine the pattern of development of the external genitals and guide the formation of the ducts and passageways that will eventually deliver sperm when the male reaches reproductive maturity.

In summary, the hydrophobic signals diffuse into their target cells, where they bind to and activate receptor proteins. The activated ligand–receptor complexes move into the cell's nucleus, where they control the expression of specific genes and therefore the synthesis of specific proteins in the responding cells. Responses to hydrophobic signals are slow to start but are usually sustained. In general,

TABLE 21.1 Signal Transduction Pathways

Signaling molecule	Cell response	Receptor protein	Example
Hydrophilic	Rapid, short-lived	G protein-coupled receptor	Adrenaline-signaling pathway, resulting in increased heart rate
		Receptor-protein kinase	PDGF-signaling pathway in the clotting response to injury
		Ligand-gated ion channel	Acetylcholine signaling, leading to the opening of sodium ion channels in muscle cells
Hydrophobic	Slow, long-lived	Intracellular receptor protein	Steroid hormones leading to changes in gene expression

all of the steps leading to the synthesis of proteins take some time, but the synthesis of the new proteins continues for a while once it is started, so the responses are sustained. When the hydrophobic signaling molecules dissociate from their receptor proteins, the response will slowly begin to wane.

In this module, we saw four types of signal transduction pathways that follow the same basic steps: receptor activation, signal transduction pathway, response, and termination. **TABLE 21.1** summarizes the similarities and differences among these four pathways.

✓ Concept Check

6. **Describe** the chemical characteristics of the two broad groups of signaling molecules and explain how these characteristics give clues to the location of receptor proteins.

7. **Describe** why receptors for hydrophobic signaling molecules function inside target cells rather than on the surface of the target cell.

8. Many hydrophilic signaling molecules cause rapid responses because they trigger enzyme activation. **Describe** why slower responses are seen in systems that depend on hydrophobic signals.

Module 21 Summary

PREP FOR THE AP® EXAM

REVISIT THE BIG IDEAS

SYSTEMS INTERACTIONS: Explain how the structures and properties of molecules facilitate the transmission of information from one cell to another.

LG 21.1 Signaling molecules bind to specific receptor proteins.

- Signal transduction is the process by which external signals cause internal responses in target cells. Page 290

- A close match between the shapes and chemical characteristics of signaling molecules and receptor proteins allows them to interact with and bind to each other. Page 290

- A cell will respond to a given signaling molecule only if the cell has the matching receptor proteins to allow this interaction. Page 291

- The noncovalent binding interactions between signaling molecules and receptor proteins form a ligand–receptor complex that produces responses in target cells. Page 291

- Different target cells can respond in different ways to the same signal. Page 291

LG 21.2 Hydrophilic, polar signaling molecules can produce rapid biochemical responses

- Hydrophilic signaling molecules are not able to cross cell membranes and so their receptor proteins are on the surface of target cells. Page 292

- G protein-coupled receptors, upon binding signals, change conformation and use membrane-bound proteins called G proteins to trigger a signal transduction cascade in which second messengers drive the response mechanisms in target cells. Page 293

- Signals can be amplified by activating multiple second messengers, which cause cascades of kinase activation. Page 294
- The second messenger cAMP is deactivated to terminate responses initiated by G protein-coupled receptors. Page 294
- Receptor-protein kinases have a signal-binding function and an enzyme function that is activated upon binding the signal. Page 294
- Responses initiated by receptor-protein kinases are terminated by the release of the signal, which terminates the kinase enzyme's activity. Page 296
- Ligand-gated ion channels are proteins that have both signal-binding and ion channels that can open or close to alter ion movements across the membrane. Page 299
- The rapidity of responses by ligand-gated channels allows rapid communication in neurons and reflex actions by muscles. Page 299

LG 21.3 Hydrophobic, nonpolar signaling molecules can turn genes on or off.
- Hydrophobic, nonpolar signals are soluble in the lipid-rich membranes of cells and readily enter cells. Page 300
- Target cells that respond to hydrophobic signals have receptor proteins inside the cell. Page 300
- The ligand–receptor complex formed when a hydrophobic signal binds to its receptors interacts with genes in the nucleus of the responding cell, altering the synthesis of proteins in the response pathway. Page 300
- Steroid hormones, synthesized from a substrate of cholesterol, are examples of hydrophobic signals. Page 300
- When the hydrophobic ligands dissociate from the receptors, the target cells' responses slowly start to wane. Page 301

Key Terms

Signal transduction
Amplification
G protein-coupled receptor protein
Second messenger

Cyclic adenosine monophosphate (cAMP)
Kinase
Signaling cascade

Receptor-protein kinase
Ligand-gated channel
Gene expression

Review Questions

1. Insulin is a hydrophilic hormone that reduces the level of the sugar glucose in the blood after a meal. How does insulin affect levels of glucose in the blood?
   (A) Insulin is transported from the blood directly into muscle cells, carrying glucose molecules with it.
   (B) Insulin moves to the nucleus of cells and rapidly turns on genes after binding to intracellular receptor proteins.
   (C) Insulin binds to transmembrane receptor proteins to activate signal transduction pathways that promote glucose uptake.
   (D) Insulin stimulates the liver to synthesize glucose and release it into the blood.

2. The molecular interaction of a signaling molecule and its receptor protein
   (A) forms an irreversible bond, sharing electrons between signal and receptor.
   (B) has high specificity in the match between signal and receptor.
   (C) is based on a covalent bond forming between signal and receptor.
   (D) does not affect the conformation of the signal or the receptor.

3. Signaling molecules that are proteins are likely to
   (A) bind to cytosolic proteins that act as receptors.
   (B) bind to cell-surface, membrane-embedded proteins that act as receptors.
   (C) diffuse readily across the cell membranes of target cells.
   (D) bind to segments of DNA and RNA that act as receptors.

4. During stress, adrenaline is a hydrophilic signal that
   (A) is transported into the target cell.
   (B) binds to and activates adenylyl cyclase.
   (C) replaces GDP on the G protein.
   (D) binds to a cell-surface receptor on the cell.

5. The acetylcholine-gated sodium channel is an example of
   (A) a G protein-coupled receptor.
   (B) a receptor kinase.
   (C) a ligand-gated ion channel.
   (D) a transcription factor.

6. Why are responses to hydrophobic signals slow and sustained, relative to responses to hydrophilic signals?
   (A) Hydrophobic signals cause a change in gene activity, which is a process that cannot be turned off quickly.
   (B) Hydrophobic signals are amplified, so removing a single signal does not end the activity of the many proteins activated by the signal.
   (C) Hydrophobic signals cannot dissociate from their receptors because they are inside the cell.
   (D) Hydrophobic signals open ion channels in their receptors, and it takes time to rebalance the charge across the cell membrane after the signal dissociates.

7. In the nucleus of cells responding to signals, the hormone–receptor complexes that turn genes on or off are associated with signaling molecules that are
   (A) proteins.
   (B) gases.
   (C) kinases.
   (D) hydrophobic.

Module 21
AP® Practice Questions

Section 1: Multiple-Choice Questions

Choose the best answer for questions 1–4.

1. In humans, cells take in glucose in response to insulin, a hormone and signaling molecule. One type of diabetes results when the body's immune system destroys the insulin-producing cells in the pancreas. The result is that glucose is not taken up by cells as normally occurs. Which is the first component of the signal transduction pathway that prevents the normal cellular response of taking up glucose in this type of diabetes?
   (A) Lack of activation of the receptor on target cells
   (B) Lack of function of enzymes in the signal transduction pathway in the cell
   (C) The cell's response to the signal
   (D) Lack of a termination signal

2. Figure 21.6 on page 295 shows the activation of a G protein-coupled receptor pathway in a heart cell. If adenylyl cyclase were unable to bind an activated G protein, which of the following would be inactive?
   (A) Only the adenylyl cyclase
   (B) The adenylyl cyclase and the kinases
   (C) The adenylyl cyclase, the kinases, and the target proteins
   (D) Only the target proteins

3. In Figure 21.7 on page 296, if the enzymes that remove phosphate groups from the kinases were inactivated in the cell, which would be still be active after adrenaline is no longer present?
   (A) The G protein-coupled receptor
   (B) The adenylyl cyclase
   (C) Cyclic AMP (cAMP)
   (D) The kinases and the target proteins

4. Cells that are targets for ligands have not one, but many, receptor proteins in their membranes. When a ligand dissociates from a G protein-coupled receptor on a target cell, the signal transduction pathway quickly turns off, starting with the inactivation of the receptor protein once the ligand leaves. After a short interval of time, the G protein will also become inactive by converting its GTP to GDP, as shown in Figure 21.7 on page 296. The G protein then returns to its initial position at the inner side of the membrane, which allows it to be activated at another time by an activated G protein-coupled receptor. If the G protein-coupled receptor is inactive, the G protein will remain inactive as well. Which explains why it is necessary for target cells to have multiple receptor proteins and multiple G proteins?

(A) G proteins spontaneously deactivate by converting GTP to GDP, so having multiple G proteins means a cell always has activated G proteins that can respond to activated receptors.

(B) Having multiple receptor proteins and G proteins means that the cell can have a response to the signal even if not every receptor is activated.

(C) Once a ligand binds a receptor, the receptor is permanently activated, so having multiple receptors means the cell can continue to respond to new ligands.

(D) Signals can be amplified only if multiple receptors are activated by multiple ligands.

Section 2: Free-Response Question

Write your answer to each part clearly. Support your answers with relevant information and examples. Where calculations are required, show your work.

Cholera is a debilitating disease. It results in massive diarrhea and dehydration that may lead to shock and death. The cholera toxin interferes with the normal function of a transmembrane protein called the cystic fibrosis transmembrane conductance regulator (CFTR). CFTR spans the membrane, where it acts as channel that moves chloride ions (Cl^-) and water out of the cell and into the intestinal cavity. Under normal circumstances, CFTR pumps ions and water out of the cell only in response to a signal transduction pathway that is stimulated by binding a signal.

The cholera toxin circumvents the normal signal transduction pathway that activates CFTR. The toxin is an A–B toxin, meaning it has A and B subunits. The B subunits bind to the cell membrane and release the A subunit into the cell, where it makes its way into the cytoplasm and activates the G proteins in such a way that they can no longer convert GTP to GDP, resulting in permanent activation of the G proteins. Permanent activation of the G proteins in turn causes cyclic AMP (cAMP) to rise to levels of up to 100 times what they are normally. The excess cAMP activates an abnormally high number of CFTR proteins, and the cell pumps a large amount of ions and water into the intestinal cavity, resulting in diarrhea and dehydration.

(a) A signal transduction pathway includes the following steps: signal binding, receptor activation, signal transduction pathway, response, and termination. **Identify** which of these steps the cholera toxin directly affects.

(b) **Identify** which steps of the signal transduction pathway the cholera toxin bypasses and describe why bypassing these steps results in an abnormal response from the cell.

Module 22

Changes in Signal Transduction Pathways

LEARNING GOALS ▶**LG 22.1** Receptor agonists activate signal transduction pathways and receptor antagonists inhibit them.
▶**LG 22.2** Genetic mutations can disrupt signal transduction pathways.

Up to this point, we have learned about the steps required for cells to communicate with one another. Cell communication is extremely precise: the signaling molecule must be complementary and specific to the receptor protein. After the signal and receptor bind, a signal transduction pathway is triggered, leading to a response in the target cell. We have seen that cell communication is not only beneficial—but in fact vital—to organisms, coordinating the actions of millions of cells in a multicellular organism. However, the many steps of these signal-and-response systems can also be points of vulnerability. A change in any of the molecules or proteins that play a role in the signal transduction pathway can affect everything downstream of the change.

This module begins with disruptions to the first step of signaling: the binding of the signal molecule to its receptor protein on the target cell. Changes to, or replacements for, the signaling molecule can interrupt proper cell communication. Chemicals in foods and drugs can activate or inhibit receptor proteins and influence cell responses. In the case of medicines used to treat disease, the cellular responses may be beneficial, but in the case of certain chemicals like poisons, the cellular responses may be harmful to the organism. Following that discussion, we will see that **mutations,** which are heritable changes in genetic material, can produce altered receptor proteins that also interfere with cell communication. Disruption of cell signaling can disrupt homeostasis and result in poor health for the organism. Although there are many steps of the signaling process during which alterations can have effects, we will limit our focus to the signal and the receptor in this module.

PREP FOR THE AP® EXAM

FOCUS ON THE BIG IDEAS

INFORMATION STORAGE AND TRANSMISSION: Think about how changes in the molecules participating in signal transmission affect the transmission of information.

22.1 Receptor agonists activate signal transduction pathways and receptor antagonists inhibit them

Many chemicals in food, drugs, and the environment are similar enough to signaling molecules produced by cells that they act as ligands and can fit into the ligand-binding site of the receptor protein. The binding of these ligands to a receptor protein can lead to abnormal or unexpected responses by target cells. In general, we can describe ligands produced by the body as endogenous ligands (*endo* means "inside") and ones from outside the body as exogenous ligands (*exo* means "outside").

Exogenous ligands can affect signaling pathways in different ways. Some mimic an endogenous ligand and therefore produce a typical cellular response. Others over-stimulate signal transduction pathways, resulting in a stronger or a longer response. In both of these cases, the ligand acts as an *agonist*. An **agonist** is any ligand that binds to a receptor and causes a response. Finally, some ligands inhibit the signal transduction pathway. In this case, the ligand is acting as an *antagonist*. An **antagonist** is any ligand that binds to a receptor and inhibits a response. In this section, we will examine the effects of agonists and antagonists of signaling pathways.

Agonists of G Protein-Coupled Receptors

FIGURE 22.1 on page 306 compares the effects of two agonists of the same receptor. In the top panel of Figure 22.1, a signaling molecule binds to its receptor, initiating a signal transduction pathway that leads to a typical cellular response. As shown in the bottom panel of the figure, an exogenous

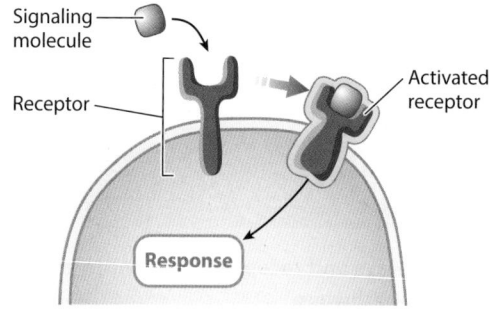

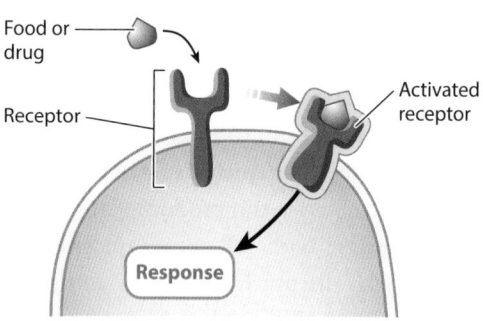

FIGURE 22.1 Agonists

An agonist binds to a receptor and leads to a response. Signaling molecules bind to receptor proteins and cause a cellular response, as shown in the top panel. Chemicals, foods, and drugs can bind to receptor proteins that would otherwise bind to signaling molecules, as shown in the bottom panel. They might bind more tightly, leading to a longer or stronger response. Alternatively, they might appear at the wrong time or under conditions that differ from those driving the arrival of the endogenous signal.

therefore bind to intracellular receptors, as discussed in the previous module. Examples include steroids found in the environment, such as in birth control pills or muscle-building anabolic steroids, which can affect reproduction and development.

Nature has produced an abundance of chemicals that act as agonists of receptor proteins in signal transduction pathways. Pharmaceutical companies have learned to mimic and synthesize some of these compounds. The drug albuterol, used to open airways in the lungs, is an example. Recall from Module 20 that during exercise or an emergency, the body releases adrenaline, a hormone that increases the heart rate and widens the airways of the lungs. As a result, more oxygen enters the bloodstream and is circulated around the body to help the organism respond to the exercise or emergency. For hundreds of years practitioners of traditional medicine used an extract of the leaves of the ephedra plant, shown in **FIGURE 22.2**, to treat breathing difficulties. Pharmaceutical researchers identified several chemicals in ephedra leaves that act as agonists at adrenaline receptors in the lungs. Researchers isolated one that works reliably and effectively. With a few chemical modifications, the drug known as albuterol was developed.

The receptor for adrenaline is a G protein-coupled receptor, which we described in Module 21. The drug albuterol is an agonist of the adrenaline receptors in

ligand, such as a chemical in a food or drug, might bind the receptor more tightly than the signaling molecule does. As a result, the exogenous ligand stays attached to the receptor protein for a longer period of time. With prolonged binding, this ligand stimulates a response that is longer lasting, and thus greater than, the typical cellular response.

An exogenous ligand may also bind as tightly to the receptor as an endogenous signaling molecule does, staying bound to the receptor only as long as the endogenous signaling molecule would. However, signaling molecules are carefully regulated and are released from the signaling cell at the right times and levels. An exogenous ligand may be present at the wrong times or at higher levels, leading to altered responses. More ligands at a target cell mean more activated ligand–receptor complexes, which lead to more of a response from the cell. Although Figure 22.1 shows a hydrophilic signaling molecule, exogenous ligands can also be hydrophobic signaling molecules and

FIGURE 22.2 Ephedra plant

The leaves of the ephedra plant have been used for centuries in traditional medicine to treat asthma and other breathing difficulties. The leaves contain a chemical that is an agonist in adrenaline pathways, causing the airways to open wider and make breathing easier.

Photo: Horst Mahr/Getty Images

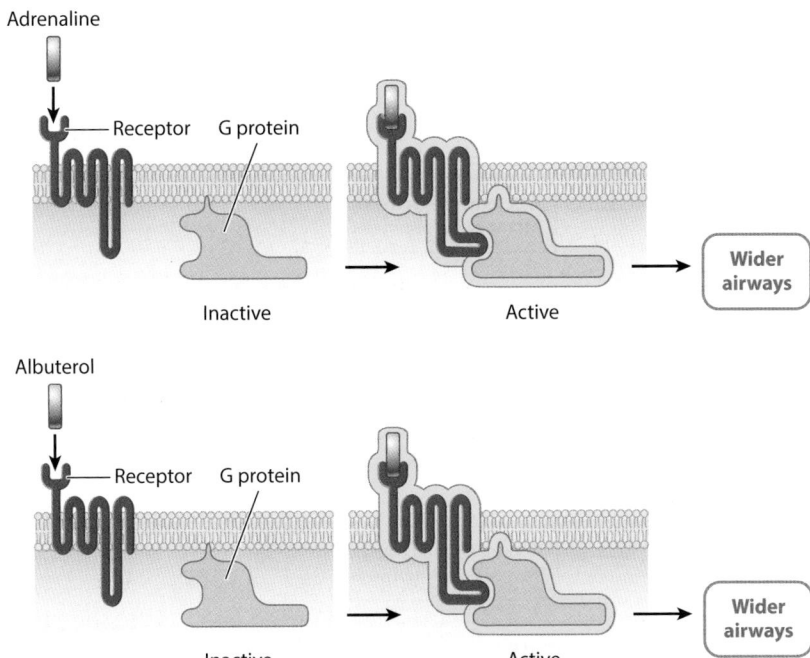

FIGURE 22.3 Albuterol as an agonist of adrenaline receptors

Adrenaline normally binds to the complementary adrenaline receptor. Albuterol is a drug that is an agonist at adrenaline receptors; it binds to the receptor and stimulates the signal transduction pathway.

the airways, as shown in **FIGURE 22.3**. Like adrenaline, it causes the airways in the lungs to widen, making it easier to breathe. The chemical structures of adrenaline and albuterol, one made by animals and the other made by plants, are quite similar, making them both matches for the adrenaline receptor. When either ligand binds to the receptor, the receptor activates, triggering the signal transduction pathway that opens up the airways to facilitate breathing.

Antagonists of G Protein-Coupled Receptors

Antagonists, as we have seen, bind to receptor proteins and block their function. **FIGURE 22.4** compares the action of a signaling molecule and antagonist. On the top is the signaling molecule bound to the receptor protein, activating the signal transduction pathway and eliciting a response from the cell. On the bottom is an antagonist bound to the receptor protein. The antagonist binds to the receptor protein, but it does not activate it. As a result, there is no cellular response. In addition, when antagonists are in abundance, they can block the signaling molecule from binding the receptor protein, so cellular responses can be effectively suppressed.

The caffeine found in drinks like coffee, tea, soda, and energy drinks provides an example of how we commonly put antagonists to use. Caffeine causes many physiological changes in our bodies, one of which is alertness. This alertness is a result of altered cell signaling. Caffeine is an antagonist at G protein-coupled receptors that normally bind the signaling molecule adenosine, as shown in **FIGURE 22.5** on page 308. When adenosine binds to its receptor on target cells, the activated adenosine receptor activates G proteins, starting the signal transduction pathway that results in drowsiness, as shown in the top panel of the figure. The shape of caffeine is similar enough to adenosine that caffeine can fit into and bind to the adenosine receptor. But, unlike adenosine, caffeine does not cause the receptor to change shape and the receptor remains inactive. The responses that cause drowsiness are therefore suppressed.

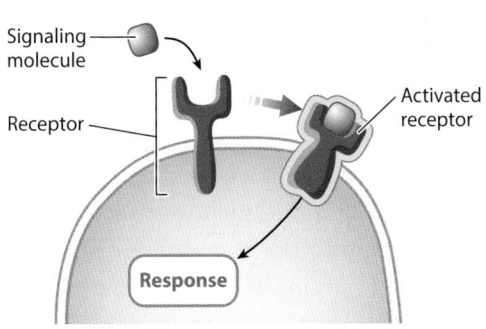

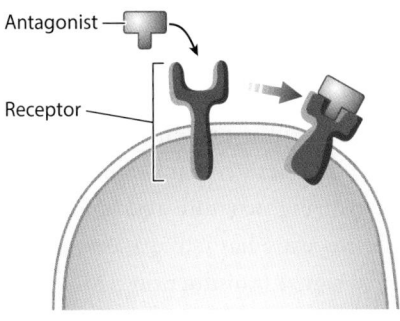

FIGURE 22.4 Antagonists

In the top panel, a signaling molecule binds its complementary receptor protein, setting in motion the signal transduction pathway that leads to a typical cellular response. As shown in the bottom panel, antagonists of the same receptor proteins bind the receptor protein in such a way that it blocks the ligand-binding site but does not activate the receptor. Blocking the ligand-binding site means that no other ligand can bind the receptor, and so the cell's response to the signaling molecule is suppressed.

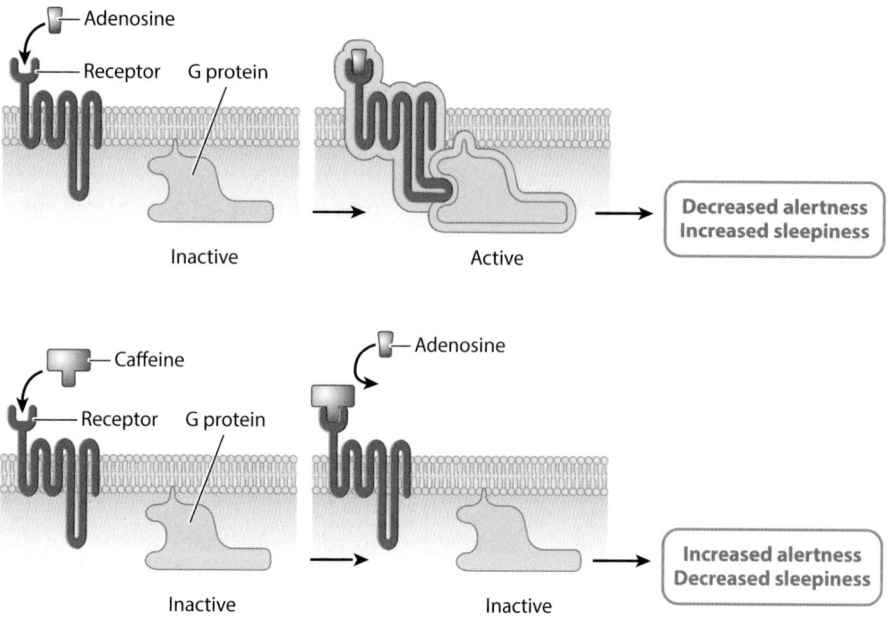

FIGURE 22.5 Caffeine, an antagonist of adenosine receptors

Adenosine normally binds to the adenosine receptor, activating the signal transduction pathway that leads to responses that cause sleepiness. Caffeine is an antagonist at adenosine receptors, blocking the ligand-binding site of the adenosine receptor and suppressing the signal transduction pathway, thus promoting alertness.

In addition to keeping the adenosine receptor in its inactive state, caffeine binding to the adenosine receptor blocks adenosine from binding to the receptor. Normally, adenosine is present at high levels in the human brain at bedtime, and the responses to adenosine binding to receptors in the brain make people feel drowsy and help them fall asleep.

FIGURE 22.6 shows the chemical structures of caffeine and adenosine. Both caffeine and adenosine have a two-ring structure. As a result, it is not surprising that they are able to bind to the same receptor protein. However, adenosine has an additional ring structure that caffeine does not have. Adenosine's additional ring structure and other structural differences from caffeine are what make adenosine activate its receptors, leading to signal transduction pathways associated with drowsiness. Although caffeine binds to the adenosine receptor, caffeine does not activate the receptor protein. The receptor remains inactive when bound to caffeine, preventing cellular responses that lead to drowsiness.

Caffeine and albuterol both come from plants, but bind to receptor proteins in humans and therefore have physiological effects. Why would a compound in a plant bind to a human protein? Plants have evolved many different kinds of chemical defenses to protect themselves from plant-eating mammals. Because these chemicals have evolved to interact with our various organ systems, we are sometimes able to benefit from them in small doses or by modifying them in specific ways.

Many of the drugs we use in medicine in fact come from plants and other organisms. Aspirin, used to treat headaches, comes from the bark of a shrub of the genus *Spiraea,* where aspirin gets its name; taxol, used to treat breast, ovarian, and lung cancers, comes from the Pacific yew tree; and digitalis, used to treat congestive heart failure, comes from foxglove, a common flower in English gardens. There are countless other examples.

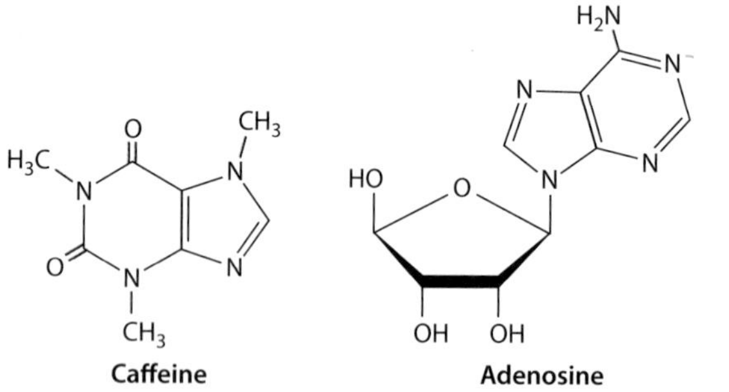

FIGURE 22.6 Adenosine and caffeine

Adenosine and caffeine have similar chemical structures that allow them to bind to the adenosine receptor in the brain. Adenosine binds to the receptor, activating the receptor and stimulating the sleep pathway. Caffeine binds to the receptor but does not activate it, so acts an antagonist of the adenosine receptor.

Many drugs act as agonists and antagonists of receptor proteins involved in cell signaling. These act on all of the types of receptors discussed in Module 21, including G protein-coupled receptors, receptor-protein kinases, ligand-gated ion channels, and intracellular receptors that bind hydrophobic signaling molecules. Other drugs alter cell signaling by interacting with proteins further along in the signal transduction pathways.

✓ Concept Check

1. **Describe** the effects of agonists and antagonists on the responses of target cells.

2. **Describe** how albuterol stimulates the same lung responses as adrenaline.

3. **Describe** how caffeine blocks the effects of adenosine in the brain.

22.2 Genetic mutations can disrupt signal transduction pathways

We have seen that drugs and foods can act as exogenous ligands that affect cell communication. It is also possible that changes from within cells can interrupt, block, change, or enhance signal transduction pathways. The sequence of bases in DNA determines the structure of proteins, including signaling molecules, receptor proteins, and all the proteins that play a role in signal transduction pathways. A mutation, or change in the DNA sequence, can lead to a change in the amino acid sequence of a protein, which in turn can affect its shape and function. As a result, the mutant protein that is produced might no longer function in cell communication. Today, we know of numerous types of inherited disorders that are the result of genetic errors in making signaling molecules. We also know of additional types of disorders that are the result of genetic errors in making functional receptor proteins.

Mutations can occur at any step in a signal transduction pathway. The effect of the mutation depends in part on the nature of the mutation and in part on where it occurs in the pathway. For example, if it occurs early, everything downstream of the mutant protein can be affected. **FIGURE 22.7** shows the points of cell communication that can be disrupted by a mutant protein. Mutations in proteins that act as signaling molecules or receptor proteins, play a role in signal transduction pathways, or participate in termination of the signal can all disrupt cell communication.

Leptin is a protein signal molecule that is produced by fat cells known as adipocytes.

Leptin receptor proteins are found on target cells in many parts of the body. In the brain, leptin helps to regulate appetite. Leptin acts as a signal that reflects the number of fat cells in an individual's body. Leptin reduces the intensity of hunger, so animals with lots of fat cells, and therefore a lot of leptin, exhibit less interest in food and eating, as shown in **FIGURE 22.8** on page 310.

Leptin has been closely studied in humans because it may play a role in obesity, a significant public health problem in many populations. Researchers have found mutations in the leptin gene of lab mice that lead to leptin deficiency. As a result, leptin is not present to bind with its receptors on target cells in the brain. These lab mice become obese very early in life. In the absence of leptin and a subsequent cellular response, appetite is increased even in the presence of abundant fat cells. The leptin-signaling pathway can be disrupted by mutations in the leptin-signaling molecule or the leptin receptor protein. In both cases, the brain stimulates appetite due to the absence of a leptin signal.

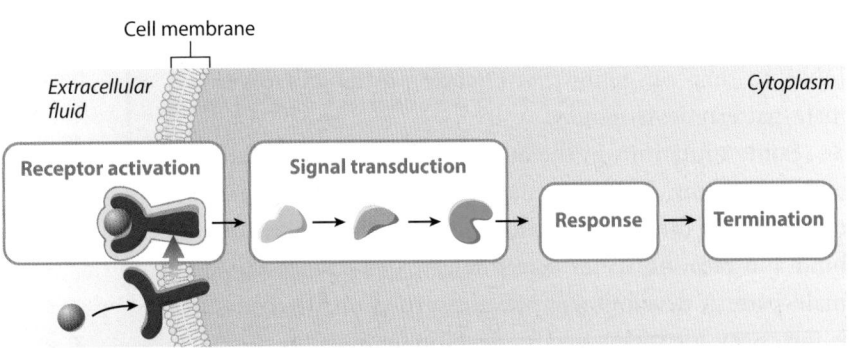

FIGURE 22.7 Altered proteins in cell communication

Mutations in DNA can lead to mutant proteins. Mutations in proteins involved at any of the four points in the cell communication pathway shown can disrupt cell communication. Mutant proteins can inhibit ligand binding or the activation of any of the second messenger or target proteins, altering the cell's response to the signal.

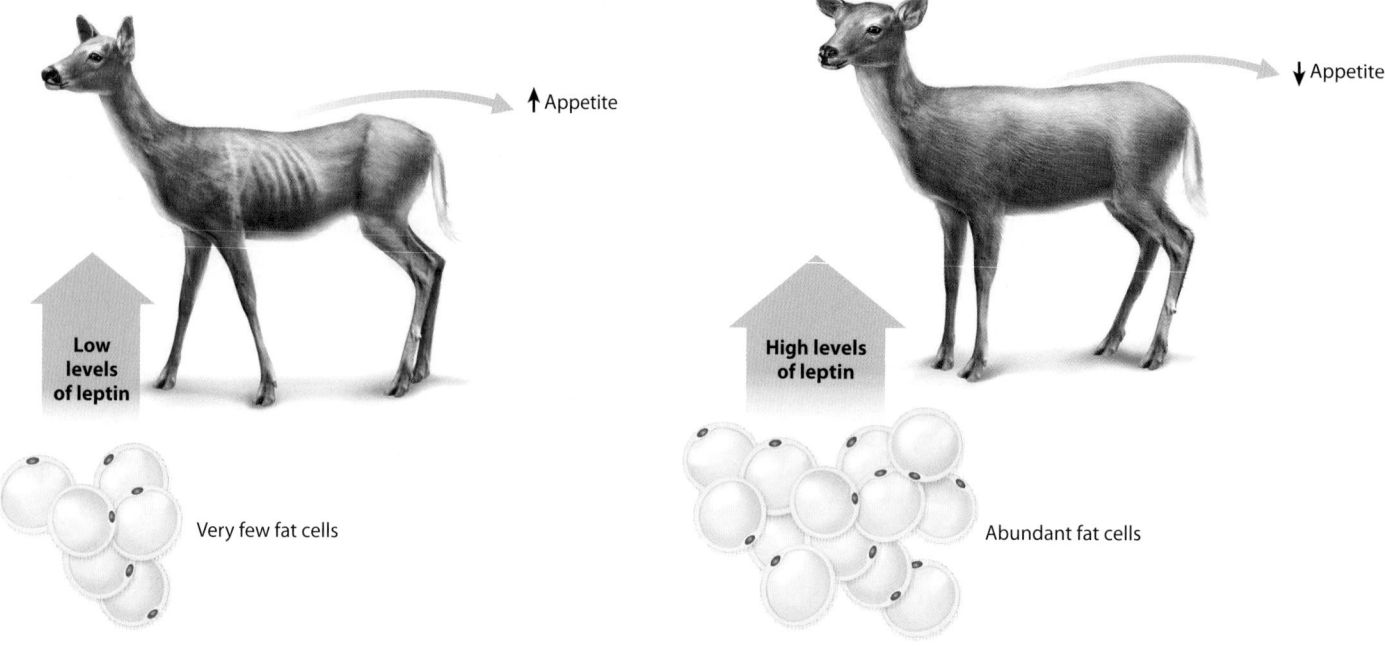

Low levels of leptin

↑Appetite

Very few fat cells

High levels of leptin

↓ Appetite

Abundant fat cells

FIGURE 22.8 Leptin signaling

Leptin, a protein made by fat cells, is a signaling molecule. Leptin travels in the blood to bind to receptor proteins in the brain. Low levels of leptin stimulate appetite, as shown on the left. High levels of leptin, resulting from an abundance of fat cells, reduce appetite, as shown on the right.

Another example of a change in a signaling pathway resulting from a genetic mutation focuses on the receptor for the steroid hormone testosterone. Testosterone is a hydrophobic signaling molecule produced by the testes before and after birth. As we saw in Module 21, testosterone is part of the signaling pathway that directs male development in embryos and fetuses. Testosterone binds receptor proteins called androgen receptor proteins on target cells. The androgen receptor, like all receptors for hydrophobic signals, is an intracellular receptor rather than a membrane-bound receptor. Once activated by binding testosterone, the activated receptor interacts with the cell's DNA to turn on genes that encode proteins involved in male-pattern development.

Some mutations in the androgen receptor prevent testosterone from binding its receptor. Because the target cells without functional androgen receptor proteins cannot bind and respond to testosterone, the genes necessary for male-pattern development are not turned on. Testosterone is necessary for male-specific development; in the absence of testosterone, embryos follow female-pattern anatomical development. As a result, genetic males without androgen receptors appear female. This example demonstrates the consequences of disrupting the androgen receptor proteins. Up to 1 in 20,000 male births are accompanied by genetic changes that disrupt androgen receptor functions.

All of these examples highlight the importance of cell signaling in the functioning of organisms. Cell signaling is very precise, so that any alteration in the pathway can lead to physiological changes, whether in the heart, lungs, brain, or any other system. Changes can come from outside the body in the form of foods or drugs, or from within, in the form of mutations. Either way, they can have dramatic effects on the function of an organism.

PREP FOR THE AP® EXAM

AP® EXAM TIP

When you are asked to discuss the biochemical interactions between two molecules in a biological setting, recognize that either molecule might be responsible when the interaction is disrupted. For example, in cell signaling, there might be a mutation that disrupts the signaling molecule or there might be a mutation in the receptor proteins or any of the proteins in the signal transduction pathway.

✓ Concept Check

4. **Describe** two different ways leptin signaling might become disrupted at the genetic level and the consequences of the disruption.

5. **Describe** why a cell might not respond to an abundance of signaling molecules.

Module 22 Summary

PREP FOR THE AP® EXAM

REVISIT THE BIG IDEAS

INFORMATION STORAGE AND TRANSMISSION: Using the information in this module, **identify** one example of a mutation altering information transmission and one example of a chemical altering information transmission. **Describe** the effect of the changes for each example.

LG 22.1 Receptor agonists activate signal transduction pathways and receptor antagonists inhibit them.

- Food and drugs can be sources of chemicals that act as exogenous ligands in cell communication. Page 305
- Exogenous ligands can disrupt the usual activity of a signal transduction pathway, thereby altering the responses of target cells. Page 305
- An agonist stimulates a response in a target cell. Page 305
- Albuterol is a drug that acts as an agonist of adrenaline receptors to support breathing in lung diseases. Page 306

An antagonist prevents a response in a target cell, even when the normal ligand is present. Page 307

- Compounds in coffee, especially caffeine, bind to and block adenosine receptors, blocking drowsiness responses by acting as antagonists to adenosine receptors. Page 307

LG 22.2 Genetic mutations can disrupt signal transduction pathways.

- A mutation is a change in genetic material. Page 309
- Mutations in the DNA encoding any of the proteins involved in signal transduction can interfere with the signal transduction pathway, affecting the cell's responses. Page 309
- Failure to produce functional leptin, or to produce functional leptin receptors, can lead to increased appetite and potentially obesity. Page 309
- Nonfunctioning androgen receptors that fail to bind testosterone lead to female external anatomy. Page 310

Key Terms

Mutation   Agonist   Antagonist

Review Questions

1. An exogenous ligand
   (A) may be present in food, drugs, or the environment.
   (B) is produced by the body.
   (C) acts as a receptor protein.
   (D) binds and inactivates endogenous ligands.

2. What is the effect of caffeine binding to receptors in the brain?
   (A) Caffeine is an agonist in the heart but an antagonist in the lungs.
   (B) Caffeine is an antagonist in the heart but an agonist in the lungs.
   (C) Caffeine is an agonist to adenosine receptors in the brain, enhancing adenosine responses.
   (D) Caffeine is an antagonist to adenosine receptors in the brain, blocking adenosine responses.

3. The drug albuterol is an agonist to adrenaline signaling because it
   (A) inhibits the synthesis of adrenaline receptors.
   (B) evokes responses similar to those induced by adrenaline.
   (C) reduces the breakdown of adrenaline molecules.
   (D) decreases the synthesis of adenosine molecules.

4. In mice, appetite is stimulated when
   (A) mice have an abundance of fat calls.
   (B) mice have a mutant form of leptin that fails to bind to leptin receptors.
   (C) abnormally sensitive leptin receptors are synthesized in the brain.
   (D) fat cells deliver high levels of leptin to the brain.

5. A mutant, nonfunctional androgen receptor is the result of
   (A) the failure to synthesize testosterone.
   (B) the lack of the ability to synthesize any proteins.
   (C) changes in the DNA coding for the receptor.
   (D) changes in the DNA coding for testosterone.

6. Albuterol is an agonist of the hydrophilic signaling molecule adrenaline, so its receptor at a target cell is
   (A) in the nucleus.
   (B) on the cell membrane.
   (C) in the cytosol.
   (D) on the DNA.

Module 22
AP® Practice Questions

PREP FOR THE AP® EXAM

Section 1: Multiple-Choice Questions

Choose the best answer for questions 1–3.

1. A group of scientists have isolated a protein from a plant that they hypothesize is an agonist to a particular receptor on an animal cell. The scientists grow a group of target cells from the animal in a solution that mimics the typical environment of the cells. They then add the plant protein and measure the response of the animal cells. What outcome would support the hypothesis?
   (A) There would be a cellular response.
   (B) There would be no cellular response.
   (C) The signal transduction pathway would not progress past the activation of the receptor.
   (D) The signal transduction pathway would be bypassed, and the response terminated.

2. Caffeine is an antagonist of the adenosine-signaling pathway that promotes drowsiness and sleep. A group of scientists is searching for a molecule that increases alertness more quickly than caffeine does. The scientists conduct an experiment in which they purify a compound that is structurally similar to caffeine and derived from a plant related to the coffee plant. When the scientists use the newly isolated compound to make a drink in which the compound's concentration is similar to the concentration of caffeine in a cup of coffee, they observe that people taking the drink do not become more alert. In fact, the test subjects become notably less alert and fall asleep easily. Which best explains the actions of this newly discovered molecule?
   (A) The new molecule is an antagonist, not an agonist, of the adenosine receptor.
   (B) The new molecule is an agonist, not an antagonist, of the adenosine receptor.
   (C) The new molecule goes into the cell and directly activates the relay part of the signaling pathway that results in drowsiness and sleep.
   (D) The new molecule binds adenosine and prevents it from binding the adenosine receptor, thus turning on the signaling pathway that results in drowsiness and sleep.

3. Insulin is a protein hormone that results in cells, such as liver and muscle cells, taking up glucose from the blood. Researchers discover a family of cows whose insulin receptor contains an extra three amino acids compared to the insulin receptor of most cows. The remainder of the insulin-signaling pathway is examined and is found to be normal. The researchers measure the amount of glucose taken in by liver cells from the cows with the altered insulin receptor, and liver cells from typical cows with the normal insulin receptor. The researchers observe that the liver cells from the cows with the altered insulin receptor take up glucose at only 10% of the rate of the liver cells from cows with normal insulin receptors. Which best explains the results of these experiments?

(A) The altered insulin receptors bind insulin to a much greater extent than do the normal insulin receptors.

(B) The altered insulin receptors bind insulin to a much lesser extent than do the normal insulin receptors.

(C) The altered insulin receptors are enzymes that destroy insulin, which thus destroys the ligand of the insulin receptor in these cows.

(D) The altered insulin receptors are enzymes that destroy glucose, which thus destroys the ligand of the insulin receptor in these cows.

Section 2: Free-Response Question

Write your answer to each part clearly. Support your answers with relevant information and examples. Where calculations are required, show your work.

A team of cell biologists developed two groups of human cells that can be grown, or cultured, in flasks containing a liquid medium with all the materials needed for cell growth. The first group of cells have receptor proteins for adrenaline in their membranes. In these cells, the receptors are engineered to fluoresce when the receptor binds adrenaline. The second group of cells have androgen receptors. The androgen receptors are engineered to fluoresce when bound by their ligand, testosterone.

The scientists then conduct an experiment. Both groups of cells are treated by adding enzymes called proteases to the cultures of cells. Proteases are enzymes that hydrolyze proteins, breaking down the proteins into amino acids so that they can no longer function. Under the conditions of the experiment, the proteases can only function in the culture medium, digesting proteins that are outside of the cells and proteins in the cell membrane which are exposed to the culture medium.

After the protease treatment, the scientists wash the proteases out of the cultures. They then add adrenaline to the first group of cells and testosterone to the second group of cells. The scientists observe that the first group of cells no longer emit the fluorescent signal when adrenaline is added. The second group of cells, in contrast, fluoresce when testosterone is added to the culture.

(a) **Describe** why the second group of cells emit the fluorescent signal after the protease treatment, while the first group of cells do not.

(b) **Predict** the effect of the protease treatment on the cell signaling pathways associated with the androgen receptor and the adrenaline receptor. **Justify** your answer.

Module 23

Feedback in Cell Communication

LEARNING GOALS ▶**LG 23.1** Negative feedback helps to maintain homeostasis.
　　　　　　　　▶**LG 23.2** Positive feedback is self-reinforcing and amplifies responses.

In the previous modules, we saw that cell communication helps to maintain steady conditions in the body, a state known as homeostasis. In order to thrive in a dynamically changing world, cells must coordinate their responses to changes in the environment. They must remain within a livable range of temperature, pH, salt concentration, and many other conditions. Cell communication begins when a stimulus, such as a change in the environment, causes a signaling cell to release signaling molecules, as described in Module 20. The signaling molecules then travel and bind to receptor proteins on target cells, starting a signal transduction pathway that causes responses by target cells. In this module, we will learn more about the conditions that lead to cell communication and how communication is precisely regulated through negative and positive feedback.

Negative feedback was introduced in Module 20 in the discussion of the regulation of body temperature. By more closely examining negative feedback in this module, you will gain further insights on how cell communication underlies homeostasis. You will see that negative feedback responds to a change or disturbance in a system, like a

change in temperature, by returning the system to a set point. In this way, it maintains relatively steady conditions. While negative feedback is important in many signaling pathways, positive feedback is used in others. In contrast to negative feedback, positive feedback moves a system away from a set point. In positive feedback, a small change or disturbance in a system is amplified over time. Positive feedback is therefore important in situations where homeostasis is not maintained. In this module we will see that cell communication helps organisms maintain a steady internal state. It also helps organisms move away from these steady conditions from time to time when needed. We begin with negative feedback and then follow that discussion with a look at positive feedback.

PREP FOR THE AP® EXAM

FOCUS ON THE BIG IDEAS

SYSTEMS INTERACTIONS: Look for examples of feedback mechanisms and how they contribute to cell and organism responses.

23.1 Negative feedback helps to maintain homeostasis

In Unit 2, we looked at how cells maintain homeostasis. The environment outside cells may change, but the environment inside cells remains relatively constant. This consistency is critical for many functions that support life. Chemical reactions and protein folding, for example, are carried out efficiently only within a narrow range of conditions, such as pH range or salt concentration. For cells, the selectively permeable membrane actively maintains intracellular conditions compatible with life.

Homeostasis is also maintained for the body as a whole. Many physiological parameters are maintained in a narrow

range throughout the body, including temperature, heart rate, blood pressure, blood sugar, blood pH, and other ion concentrations. Homeostasis is therefore critical for the life of cells and organisms. Multicellular organisms maintain homeostasis through cell signaling pathways that use negative feedback to maintain a set point. In this section, we will look closely at how these signaling pathways work and provide several examples.

FIGURE 23.1 shows a signaling system that maintains homeostasis by negative feedback. In this system, a departure from steady conditions provides the stimulus. In response to this stimulus, signaling cells release signaling molecules. The signaling molecules travel to effector cells, bind to receptor proteins, and trigger signal transduction pathways that cause

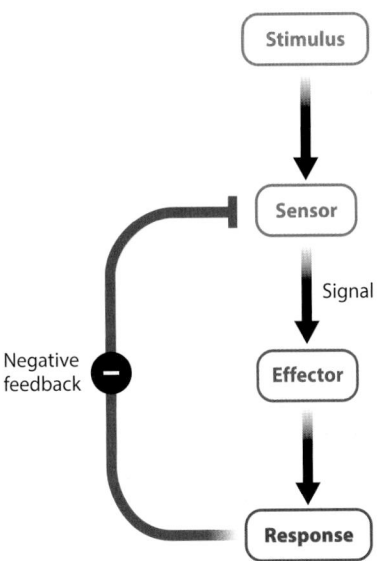

FIGURE 23.1 Negative feedback

Negative feedback maintains homeostasis. After an initial stimulus causes the release of a signal from a sensor, the effector elicits a response. The response in turn feeds back on the sensor, turning off further production of the signal. The result is that the system maintains steady conditions over time.

responses. These responses restore steady conditions. In negative feedback, the response counters the stimulus, so signaling cells stop releasing signaling molecules when the initial conditions are restored. In other words, the restoration of steady conditions has a negative effect on the further release of signals, also shown in Figure 23.1.

Antarctica, with average temperatures ranging from minus 10°C to minus 40°C, offers many examples of organisms that use negative feedback to maintain steady conditions. For example, emperor penguins (*Aptenodytes forsteri*) are well adapted to the harsh conditions of Antarctica. They have tissues and communication pathways that allow them to maintain a steady, warm body temperature. Emperor penguins respond to the stimulus of extreme cold in many ways, one of which is to huddle in groups, as shown in **FIGURE 23.2**. Temperature-sensitive cells sense a drop of temperature and rapidly communicate with the parts of the penguin's brain that direct specific behavioral responses. This signaling system directs the penguin to respond by joining a huddle for warmth.

Penguins huddle up when they are too cold, but how do they prevent their body temperature from getting too hot? Negative feedback is used for that as well. The huddle of penguins is often so tight that a penguin in the middle can become overheated. The penguin's body detects the heat buildup and directs behaviors that will resolve the problem—the penguin finds its way out of the huddle so it can cool off. Outside the huddle, a solitary penguin can radiate excess heat to the environment and cool itself back down to its set temperature. As the body temperature returns to its set point, the penguin resumes its normal activities outside the huddle.

To respond to both cold and heat, and therefore maintain a steady temperature, penguins use negative feedback. Each of these feedback loops consist of stimuli, signaling cells, signaling molecules, receptor proteins, target cells, and responses. Cell communication directs the appropriate behavioral responses in both directions, too cold and too hot, thus regulating temperature for the challenges that are encountered.

Let's consider another example of homeostasis maintained by negative feedback. In humans and other animals, the glucose level in the blood is remarkably stable. It is maintained in a very narrow range in spite of the fact that several times a day, after eating, glucose and other nutrients are absorbed from the digestive system and move into the blood. By contrast, between meals, glucose is removed from the blood and used by cells as a source of energy or stored for later use. Even though glucose is added to and removed

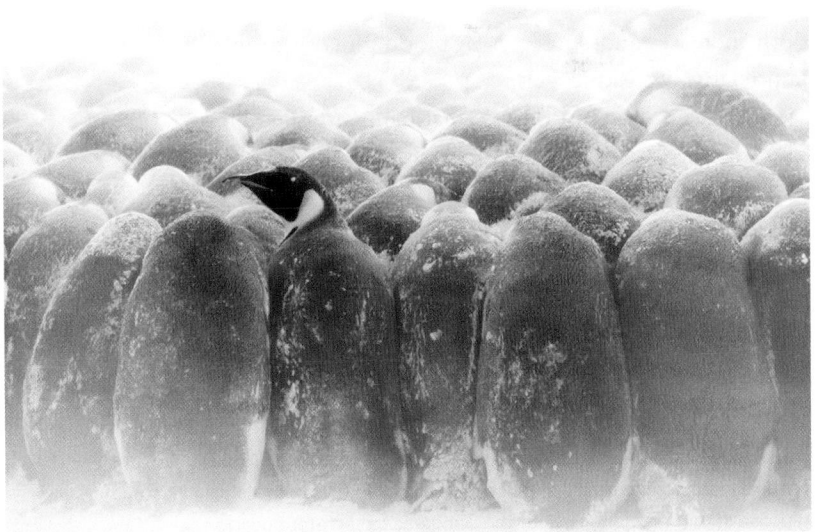

FIGURE 23.2 Emperor penguin huddling

Penguins huddle when their bodies receive signals from the brain that their temperature has fallen below their set level. They leave the huddle when their bodies receive signals from the brain that their body temperature is above their set level.

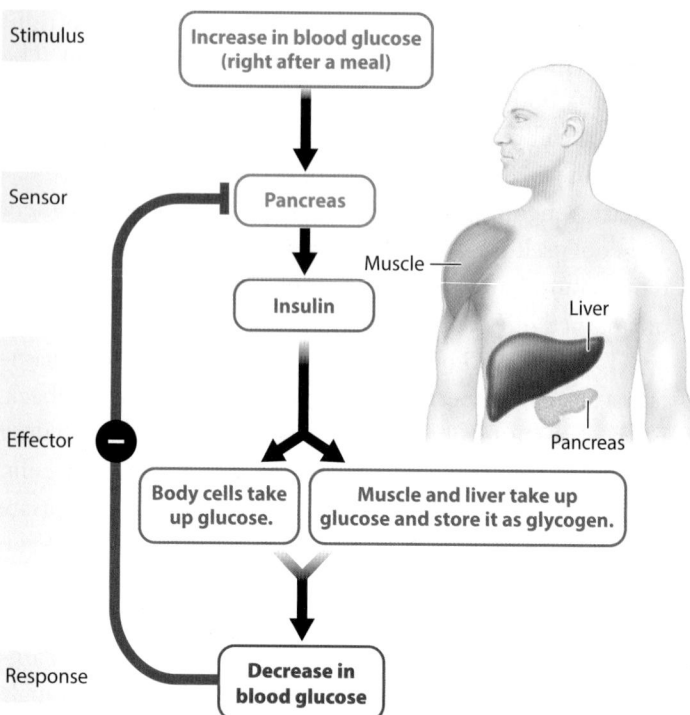

Stimulus — Increase in blood glucose (right after a meal)

Sensor — Pancreas

Muscle

Insulin

Liver

Pancreas

Effector — ⊖

Body cells take up glucose.

Muscle and liver take up glucose and store it as glycogen.

Response — Decrease in blood glucose

FIGURE 23.3 Negative feedback and insulin secretion

Eating increases the level of glucose in the blood. An increase in blood glucose stimulates beta cells in the pancreas to secrete insulin. Insulin binds to its receptors on many cells, which respond by transporting glucose out of the blood and into the cells. As a result, the level of glucose in the blood decreases, reducing the stimulus for the pancreas to secrete insulin. As a result, blood glucose levels are maintained at steady levels by negative feedback.

from the blood, the blood glucose level does not change very much. Negative feedback is used to respond to both increases and decreases in blood glucose to maintain relatively steady levels.

Let's start by considering what happens when blood glucose levels rise after a meal. **FIGURE 23.3** shows the negative feedback loop stimulated by an increase in blood glucose. The elevated levels of glucose in the blood are sensed by cells in the pancreas called beta cells, which respond to the stimulus by secreting the hormone insulin.

FIGURE 23.4 Negative feedback and glucagon secretion

Fasting leads to low levels of glucose in the blood. Low levels of glucose stimulate alpha cells of the pancreas to secrete glucagon. Glucagon binds to its receptors on muscle and liver cells, which respond by breaking down glycogen into glucose and releasing glucose into the blood. As a result, the concentration of glucose in the blood increases, reducing the stimulus for the pancreas to secrete glucagon. Blood glucose levels are maintained at steady levels by negative feedback.

Many cells in the body are targets for insulin, including muscle, fat, and liver cells. When insulin receptors on some target cells bind insulin, those cells respond to the signal by incorporating more glucose transporters into their cell membranes. Glucose transporters bring glucose into the cells, pulling glucose out of the blood and lowering the amount of glucose in the blood. Inside cells, glucose is broken down, driving ATP production. In muscle and liver cells, glucose is stored as a molecule of multiple glucose monomers bound together, called glycogen, as described in Unit 1. Glucose is taken up by cells and then either broken down or stored, lowering the level of glucose in the blood until the set point for blood glucose is once again attained. When there is no longer a stimulus of high blood glucose, the beta cells of the pancreas stop releasing insulin.

Let's now consider how the body responds to a decrease in blood glucose levels. Between meals, or when the body is fasting or starving, blood glucose levels fall. In this case, different cells in the pancreas, called alpha cells, act as sensors that detect low levels of glucose, initiating the negative feedback loop shown in **FIGURE 23.4**. In response to low blood glucose, the alpha cells secrete the hormone glucagon. Glucagon in the blood travels throughout the body and binds to its receptors on many target cells. Muscle and liver cells respond by breaking down glycogen, a process that frees glucose from storage

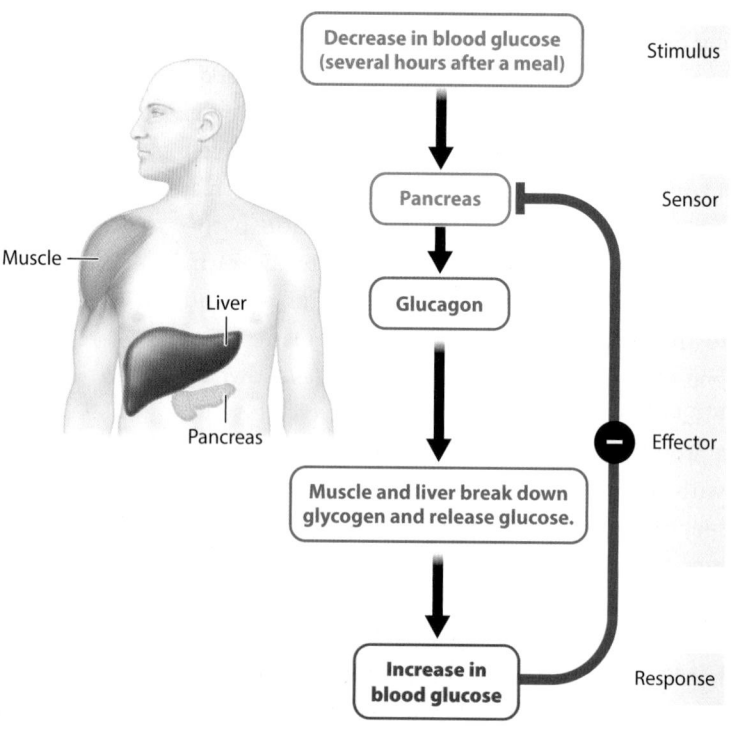

Stimulus — Decrease in blood glucose (several hours after a meal)

Muscle

Liver

Pancreas

Sensor — Pancreas

Glucagon

Effector — ⊖

Muscle and liver break down glycogen and release glucose.

Response — Increase in blood glucose

and releases it into the blood. The released glucose elevates the level of glucose in the blood back to the set point. The increase in blood glucose levels inhibits the further secretion of glucagon.

Insulin and glucagon have opposite effects on blood glucose levels, and so are considered counter-regulatory hormones. In other words, each hormone counters the action of the other. They work together to maintain homeostasis for the level of glucose in the blood. Because both hormones are subject to negative feedback, each only operates until the set point for blood glucose is reached again.

✓ **Concept Check**

1. **Describe** the pathway that drives an emperor penguin to exit from the center of a thermoregulatory huddle.

2. **Describe** a cell's responses to binding insulin, including insulin's effects on blood glucose levels and the negative feedback effects on the secretion of additional insulin.

3. **Describe** what is meant by the statement that insulin and glucagon are counter-regulatory hormones for the regulation of glucose levels in the blood.

23.2 Positive feedback is self-reinforcing and amplifies responses

In negative feedback loops, the response to the signal turns off the system. The opposite occurs in positive feedback loops. In positive feedback loops, signals become amplified after a signaling system has been initiated. Rather than maintaining homeostasis, positive feedback loops work toward a peak or climax. Positive feedback loops are self-reinforcing, which means they "snowball." As a snowball rolls down a hill, it picks up snow. This causes it to roll faster, which causes it to pick up more snow, which in turn causes it to roll even faster, and so on. The process continues until something "breaks" the system, which in the snowball analogy is the bottom of the hill. In this section, we will provide several examples where positive feedback loops come into play, and how they are stopped once they are started.

Positive feedback starts out the same way as negative feedback, as shown in **FIGURE 23.5**. An initial stimulus causes signaling cells to release signaling molecules. The signaling molecules travel to target cells and bind to receptor proteins, setting in motion signal transduction pathways and cellular responses. However, rather than the cellular responses terminating the stimulus as in negative feedback, in positive feedback the cellular responses further stimulate signal release. Instead of reaching a set point, processes mediated by positive feedback amplify over time, leading to stronger and stronger responses.

Childbirth provides an example of a process that uses positive feedback. In response to hormonal changes that occur at the end of pregnancy, the uterus, the organ where the fetus has been developing, starts contracting. The

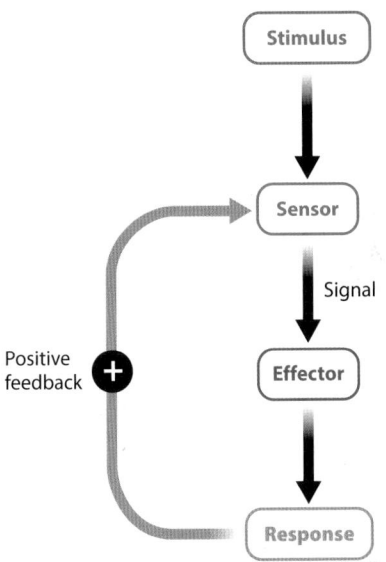

FIGURE 23.5 Positive feedback

In positive feedback, responses are amplified. A stimulus acts on a sensor to release a signal. The signal acts on an effector to cause a response. The response in turn stimulates more signal release from the sensor, amplifying the overall response to the initial stimulus over time.

beginning of childbirth is marked by much stronger contractions of the uterus, called labor, and additional physiological events required for delivery take place. **FIGURE 23.6** on page 318 shows the positive feedback loop that occurs to stimulate uterine contractions. During labor, the contractions of the uterus push the baby's head downward, stretching the birth canal. This pressure stimulates sensor cells in the birth canal, leading a part of the brain to secrete the hormone oxytocin. Oxytocin binds to its receptors on cells of the uterus, and the uterus cells respond by contracting more strongly. The increased contraction strength pushes the baby's head against

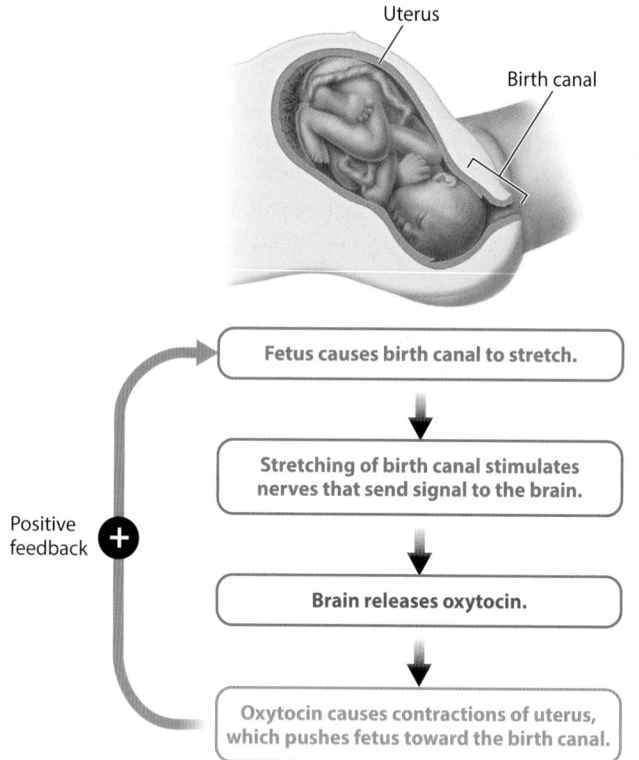

Uterus

Birth canal

Positive feedback **+**

Fetus causes birth canal to stretch.

↓

Stretching of birth canal stimulates nerves that send signal to the brain.

↓

Brain releases oxytocin.

↓

Oxytocin causes contractions of uterus, which pushes fetus toward the birth canal.

FIGURE 23.6 Positive feedback during labor and delivery

Oxytocin is a hormone that binds to its receptors on uterine muscle and promotes uterine contraction. Oxytocin secretion is initiated by nerves stimulated by the pushing of the baby's head against the birth canal. The uterine contraction response leads to more pushing of the baby's head against the birth canal, causing more oxytocin release and more uterine contractions until delivery is complete.

the birth canal more forcefully, stimulating the release of more oxytocin and further increasing the strength of uterine contractions. This positive feedback loop continues until delivery of the baby. In cases where labor is not progressing quickly enough, doctors can administer synthetic oxytocin-like drugs, which act as agonists to the oxytocin receptors. The drugs mimic the actions of oxytocin, promoting labor and delivery.

Oxytocin also plays an additional role in mammals after birth. Oxytocin stimulates the release of milk from the mammary glands in the mother's breasts. When a baby suckles from the mother's breast, sensor cells are stimulated, leading the brain to again increase oxytocin secretion. Oxytocin binds target cells in the mammary gland, which respond by increasing the contraction of the gland, forcing milk out of the breast through the ducts leading to the nipple, where the baby takes its nourishment by suckling. Suckling in turn causes ever more oxytocin release, and ever more release of milk from the breast, rapidly delivering food to the baby. Positive feedback speeds milk delivery and

quiets a hungry baby. For a nonhuman mammal, this urgent quieting by feeding might reduce predator attention to the newborn and mother.

Positive feedback also occurs in plants, leading to fruit ripening. Imagine biting into an unripe peach. Its skin is tough, its flesh is hard, and its taste is bitter. Ripening occurs after the seeds in the fruit have reached maturity. At the earliest stages of ripening, the peaches on a tree produce small amounts of the gas ethylene. Ethylene produced by the ripening fruit travels into cells and binds to intracellular receptors. The cells respond with biochemical changes that cause further ripening, including increases in sweetness and juiciness, as well as color changes. Cells also respond to ethylene by increasing their production of ethylene. This sets off a positive feedback loop, where a little bit of ethylene produced by one peach stimulates ripening, leading to the production of more ethylene, which further stimulates ripening, leading to the production of even more ripening and so on. Ripening culminates in the fruit falling off the tree. The fruit continues to produce ethylene even after it is no longer on the tree, and any fruit nearby with ethylene receptors will ripen in response. Caught in the positive feedback loop for too long, a ripe peach will continue to ripen until it rots.

In each of these three examples of positive feedback, an initial stimulus leads to a response, which is amplified over time, so the response gets stronger and stronger. In addition, each of these responses continues until something stops them: a baby is delivered, milk is released, and fruit ripens and falls. In this way, the organism achieves a coordinated and timely overall response.

✓ Concept Check

4. **Describe** the role of positive feedback in childbirth.

5. **Draw** a diagram of a positive feedback loop that illustrates the sequence of events of a suckling baby setting off increased oxytocin secretion in its mother.

6. **Describe** why ethylene's effects on peach ripening is an example of positive feedback.

Module 23 Summary

PREP FOR THE AP® EXAM

REVISIT THE BIG IDEAS

SYSTEMS INTERACTIONS: Provide one example of negative feedback and one example of positive feedback and **explain** how their responses to stimuli contribute to a particular response.

LG 23.1 Negative feedback helps to maintain homeostasis.

- Negative feedback is used to maintain steady conditions inside of cells and organisms, a condition known as homeostasis. Page 314
- Emperor penguins respond to an increase and decrease in temperature by behavioral changes, which stop when their baseline temperature is restored. Page 315
- An increase in the level of glucose in the blood, typically after eating a meal, stimulates the secretion of insulin from beta cells of the pancreas. Page 316
- Insulin binding to its target cells causes the target cells to increase their uptake of glucose for use or storage, reducing the level of glucose in the blood. Page 316
- When insulin decreases blood glucose levels, there is no longer a stimulus to release insulin. Page 316
- A decrease in the concentration of glucose in the blood, typically after several hours of fasting, stimulates the secretion of glucagon from pancreatic alpha cells. Page 316

- Glucagon binding to its target cells, especially in the liver, causes the release of glucose, thereby increasing the glucose level in the blood. Page 316
- When glucagon increases blood glucose levels, there is no longer a stimulus to release glucagon. Page 316
- Insulin and glucagon are counter-regulatory hormones due to their opposite effects on levels of glucose in the blood. Page 317

LG 23.2 Positive feedback is self-reinforcing and amplifies responses.

- Positive feedback leads to stronger and stronger responses and does not maintain homeostasis. Page 317
- The stretching of the birth canal due to a baby's head pressing against it as a result of a uterine contraction stimulates oxytocin secretion, which further stimulates uterine contraction in a positive feedback loop. Page 318
- The physical pressure of a mammalian infant suckling stimulates oxytocin secretion in the mother, leading to milk release, along with more vigorous suckling and greater release of oxytocin. Page 318
- Ripening fruit gives off the signal ethylene gas which stimulates accelerated ripening and increased release of ethylene gas, promoting ripening in other nearby fruits. Page 318

Review Questions

1. Pushing into a huddle of penguins when cold and wiggling out of the huddle when overheated is an example of
   - (A) social conflict that disrupts thermoregulatory homeostasis.
   - (B) inaccurate sensing of body temperature.
   - (C) the conflict between hunger and warmth.
   - (D) behavioral responses that maintain temperature homeostasis.

2. Negative feedback refers to
   - (A) continuous stimulation of every step in a pathway leading to a response.
   - (B) inhibition of only the final step in a pathway by the response.
   - (C) signaling that restores steady conditions.
   - (D) continuous accelerating stimulation of the final step in a response pathway.

3. Insulin and glucagon are considered counter-regulatory hormones because
   (A) the glands that make these hormones are on opposite ends of the body.
   (B) the secretion of insulin and glucagon occurs at the same time.
   (C) insulin and glucagon have opposite effects on blood glucose levels.
   (D) one is synthesized in the liver and one is catabolized in the liver.

4. The synthesis and secretion of oxytocin, a hormone secreted in nursing mothers
   (A) is part of a positive feedback loop that accelerates milk release to a suckling baby.
   (B) shifts to a negative feedback loop when the baby is done sucking.
   (C) is very high when mammary glands are filled with milk, even when the baby is not present.
   (D) is slow and steady throughout the months or years of breastfeeding.

5. Ethylene in plants works by positive feedback, as shown by the fact that
   (A) ethylene inhibits further synthesis and release of ethylene.
   (B) ethylene synthesis and release inhibit fruit ripening.
   (C) ethylene synthesis and release start out at high levels and consistently reduce over time as fruit ripening progresses.
   (D) ethylene stimulates further synthesis and release of ethylene.

Module 23
AP® Practice Questions

Section 1: Multiple-Choice Questions
Choose the best answer for questions 1–4.
Use the following figure to answer question 1.

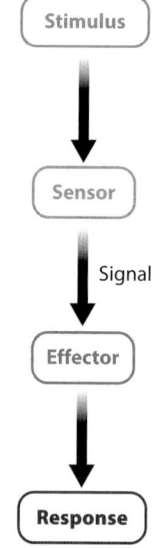

1. The figure shows a hypothetical stimulus in a response from a tissue. What element should be added to show a negative feedback?
   (A) A means of producing more stimulus
   (B) A means of responding to the stimulus
   (C) A means of amplifying the signal
   (D) A means of turning off the process

2. A patient has muscle and liver cells that bind the hormone glucagon but do not break down glycogen and release glucose to the blood. Predict what will happen when this patient's blood glucose levels decrease below the set point.
   (A) Even though the liver cells bind glucagon, no glucose will be released into the blood and so blood glucose levels will remain below the set point.
   (B) Because the liver cells can bind glucagon, blood glucose levels will remain below the set point.
   (C) Glucagon secretion from the alpha cells of the pancreas will not occur and so the patient's blood glucose level will decrease at a higher rate.
   (D) Glucagon secretion from the alpha cells of the pancreas will occur but glucagon will not bind to its receptors on the muscle and liver cells.

3. During childbirth, pressure on the birth canal stimulates the release of the hormone oxytocin, which causes uterine contractions, further stimulating oxytocin release. Which would eliminate the positive feedback in this pathway?

   (A) Stretching the birth canal to the head circumference of the baby

   (B) Increasing oxytocin secretion in the mother's brain

   (C) Completing the mother's labor and delivery of the baby

   (D) Stimulating more frequent uterine contractions

4. Fruit that is shipped great distances is often picked and shipped well before it is ripe. A student conducts an experiment with 20 unripe pears. The student divides the pears into two groups of 10. Each group is put into a paper bag. Both bags, A and B, are identical and both are made of paper that is thin enough for air to pass through easily. Bag A contains just 10 pears. Bag B contains 10 pears and an open cup of activated carbon. Activated carbon is a substance that is very good at absorbing gases such as acetaldehyde, acetylene, butylene, and ethylene. Predict which bag of pears will ripen first.

   (A) Because there is no ethylene in paper bag B to ripen the pears, the pears in paper bag A will ripen first due to the positive feedback caused by ethylene produced by the pears.

   (B) Because there is no ethylene in bag B, the pears in bag B will ripen first since ethylene normally turns off the feedback loop that ripens fruit.

   (C) The pears in bag B will ripen first as the activated carbon will absorb the ethylene produced by the pears and thus stop the negative feedback that ethylene causes in ripening fruit.

   (D) The pears in bag B will ripen first because in the absence of ethylene, oxygen, and nitrogen in air ripens the fruit through a positive feedback mechanism.

Section 2: Free-Response Question

Write your answer to each part clearly. Support your answers with relevant information and examples. Where calculations are required, show your work.

In humans, lactation, or secretion of breast milk, is a positive feedback process. Lactation occurs when an infant suckles on the mother's nipple, stimulating sensor cells and causing the brain to release the hormone oxytocin. Oxytocin, in turn, stimulates the release of breast milk from the mammary glands. Continued suckling leads to further stimulation, the secretion of the hormones, and the secretion of breast milk by the mother. As long as the child continues to suckle, the system remains active.

(a) There are two components of the lactation pathway which are the response part of the positive feedback loop. **Identify** these two components and **describe** how they result in the positive loop becoming, and remaining, active.

(b) **Predict** what would occur if a nursing mother's brain constantly secreted oxytocin. **Justify** your answer.

Module 24

The Cell Cycle

LEARNING GOALS ▶**LG 24.1** Cell division produces cells needed for growth, maintenance, and reproduction.
▶**LG 24.2** The division of eukaryotic cells occurs in phases that make up the cell cycle.
▶**LG 24.3** During mitotic cell division, a parental cell produces two daughter cells.

In the last few modules, we have focused on cell communication. We saw that signaling cells release signaling molecules that bind to receptor proteins on or in target cells, which in turn leads to a cellular response. The target cell may respond by changing shape, turning on or off the expression of genes, or even signaling other cells. Another way the target cell might respond is by dividing in two. **Cell division** is the process by which cells make more cells.

Cells come from preexisting cells. This is one of the fundamental principles of biology and a key component of the cell theory, which was introduced in Unit 2. Multicellular organisms begin life as a single cell, and then cell division produces the millions, billions, or trillions of cells that make up the fully developed organism. Even after a multicellular organism has achieved its adult size, cell division continues. In plants, cell division is essential for continued growth. In many animals, cell division replaces worn-out blood cells, skin cells, and cells that line much of the digestive tract. If you fall and scrape your knee, the cells at the site of the wound begin dividing to replace the damaged cells and heal the scrape.

Cell division is a regulated process. In other words, it only occurs at specific times, places, and environments.

It is often regulated by chemical signals like the signaling molecules we have been studying in this unit. If cell division occurs when or where it is not supposed to occur, the result in some cases is cancer, which is uncontrolled cell division.

In this module, we will focus on cell division. We will begin by looking at how prokaryotic cells and eukaryotic cells divide. Although cell division is universal, the process of cell division is different in these two groups of cells. We will then examine eukaryotic cell division in more detail, starting with the various steps or phases the cell goes through from the time it is produced to the time it divides. We will end by describing the steps of cell division in eukaryotic cells. We will see how cell division ensures that the cell's genetic information is passed from parent cell to daughter cells.

PREP FOR THE AP® EXAM

FOCUS ON THE BIG IDEAS

INFORMATION STORAGE AND TRANSMISSION: Think about the relationship between the structures of molecules and their ability to transmit information from one cell to another.

24.1 Cell division produces cells needed for growth, maintenance, and reproduction

Cell division is the process by which a single cell produces two daughter cells. In order for a cell to divide successfully, it must be large enough to divide in two and contribute sufficient nuclear and cytoplasmic components to each daughter cell. Before the cell divides, therefore, the cell grows and key cellular components are duplicated. All cells go through a period of preparation followed by division, but the details

are different in prokaryotes and eukaryotes. In this section, we will explore the different ways in which prokaryotic and eukaryotic cells divide.

Binary Fission in Prokaryotes

Prokaryotes include bacteria and archaea, and were the first forms of life on Earth. Today, these organisms are of great interest to biologists because of their central roles in health and disease, the environment, and biotechnology. When environmental conditions are unfavorable, cell growth is slow and cell division is infrequent. However, when conditions

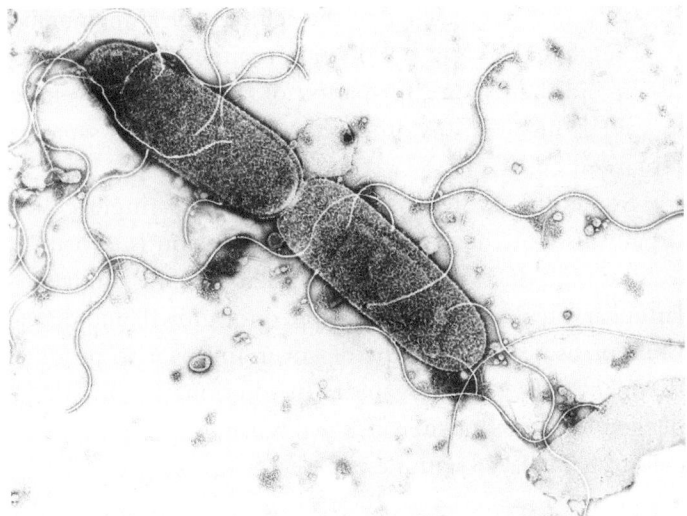

FIGURE 24.1 Binary fission in prokaryotes

Bacteria and other prokaryotes divide by binary fission. This photo shows *Escherichia coli* in the process of binary fission, with one bacterial cell becoming two bacterial cells. *E. coli* is commonly found in the human gut. Photo: Hazel Appleton, Health Protection Agency Centre for Infections/Science Source

are favorable, cell growth and division can be rapid, quickly increasing population numbers. Bacteria and other prokaryotes divide by binary fission, shown in **FIGURE 24.1**. In this process, one prokaryotic cell divides to produce the two daughter cells. *Binary* refers to the number two, the number of daughter cells that result from binary fission. *Fission* refers to splitting in two, and in this case a single cell divides into two cells.

FIGURE 24.2 illustrates the steps of binary fission in the bacteria *Escherichia coli*. In step 1, the prokaryote has grown large enough to begin the process of cell division. In step 2, the cell copies, or replicates, its DNA so that each daughter cell receives one copy. In step 3, the circular DNA is attached by proteins to the inside of the cell membrane. After it is copied, there are two DNA molecules, each attached to the cell membrane at a different site. As step 4 begins, the two attachment sites are initially close together, but as the cell grows, they move apart. In step 5, when the cell is about twice its original size and the DNA molecules are well separated, a constriction forms at the midpoint of the cell. Eventually, a new cell membrane and cell wall are synthesized at the site of the constriction, dividing the single cell into two (step 6). The result is two daughter cells, each having the same DNA as the parent cell.

Amazingly, the entire cycle of prokaryote cell division is completed in less than 20 minutes for some species. Each daughter cell then becomes a parent cell and divides, and the

cycle continues. Each parent cell has the ability to produce two daughter cells, so this means that the number of cells present in a location can increase rapidly.

Cell division in prokaryotes is a form of **asexual reproduction,** which is the reproduction of organisms in which an offspring inherits its DNA from a single parent. Prokaryotes live their entire lives as single cells. As a result, prokaryotic cell division is identical to prokaryotic reproduction, the production of a new generation. Cell division in prokaryotes also provides insight into the division of mitochondria and chloroplasts, which are eukaryotic organelles that resemble prokaryotes, as you saw in Module 7. Like prokaryotic cells, mitochondria and chloroplasts divide by binary fission, providing

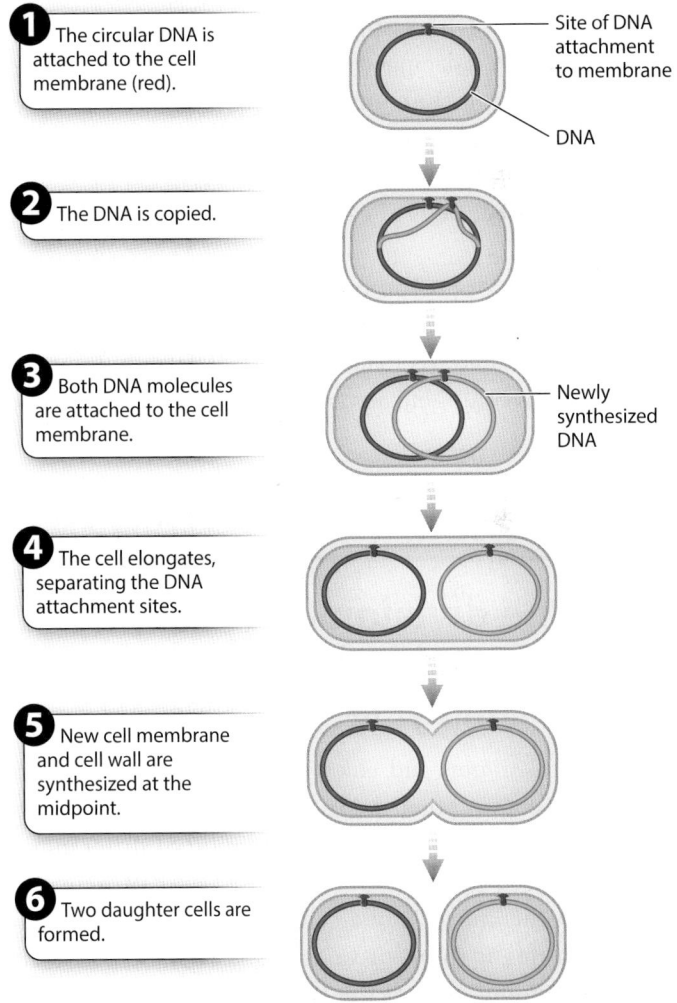

1 The circular DNA is attached to the cell membrane (red).

Site of DNA attachment to membrane

DNA

2 The DNA is copied.

3 Both DNA molecules are attached to the cell membrane.

Newly synthesized DNA

4 The cell elongates, separating the DNA attachment sites.

5 New cell membrane and cell wall are synthesized at the midpoint.

6 Two daughter cells are formed.

FIGURE 24.2 Steps of binary fission

Binary fission begins with the duplication of DNA. As DNA synthesis progresses, the cell elongates, separating the two copies of DNA. Finally, new cell membrane and cell wall are synthesized, producing two daughter cells.

another piece of evidence supporting the endosymbiont theory of eukaryotic cell origin, which was described in Module 12.

Mitotic Cell Division in Eukaryotes

The basic steps of binary fission that we just saw—copying DNA, movement of the two DNA molecules to daughter cells, and division of one cell into two—occur in all forms of cell division. However, cell division in eukaryotes is more complicated than cell division in prokaryotes. When eukaryotic cells divide, they first divide the nucleus by **mitosis.** Following this, the cytoplasm is divided into two daughter cells by **cytokinesis.** Together, mitosis and cytokinesis are called mitotic cell division.

Both prokaryotes and eukaryotes contain cellular structures known as **chromosomes** that consist of a single DNA molecule with associated proteins. The DNA of a prokaryotic cell is arranged in a single, relatively small, circular chromosome. In comparison, the DNA of a eukaryotic cell is typically much larger and is organized into one or more linear chromosomes. Each of these chromosomes must be copied, or replicated, and then separated into daughter cells.

The DNA of prokaryotes is attached to the inside of the cell membrane, allowing the copied DNA to be separated into daughter cells simply by cell growth. By comparison, the DNA of eukaryotes is located in the nucleus of the cell and is not attached to the cell or nuclear membrane. As a result, eukaryotic cell division requires first the breakdown and then the re-formation of the nuclear envelope, as well as mechanisms other than cell growth to separate replicated DNA.

Some eukaryotes, like prokaryotes, are single-celled organisms. For these organisms, cell division is a form of asexual reproduction, as it is for prokaryotes. In multicellular eukaryotes, cell division is needed for growth and development, as a single cell divides over and over to produce the many cells of the organism. In addition, cell division is needed for tissue maintenance and repair, allowing damaged cells to be replaced with new ones.

✓ Concept Check

1. **Describe** what is needed before a cell is able to divide.
2. **Describe** the steps of binary fission.
3. **Identify** the process by which eukaryotic cells divide.

24.2 The division of eukaryotic cells occurs in phases that make up the cell cycle

In both prokaryotes and eukaryotes, cells prepare for cell division by copying their DNA and other key components, as well as growing in size. This duplication of material followed by cell division is achieved in a series of steps that constitutes the life cycle of a cell. When you think of a life cycle, you might think of various stages beginning with birth and ending with death. In the case of a cell, the life cycle begins and ends with cell division. This period, from the production of a new cell to the time it divides, is known as the **cell cycle.** The cell cycle describes the life cycle of a cell. In this section, we will discuss the cell cycle of eukaryotic cells.

M Phase and Interphase

The eukaryotic cell cycle has two major phases, **M phase** and **interphase,** which are shown in **FIGURE 24.3**. During M phase, the parent cell divides into two daughter cells. M phase is so named because it includes mitosis, but actually consists of two different events: (1) mitosis, the division of the nucleus, and (2) cytokinesis, the division of the cell itself into two separate cells. Typically, these two processes go hand in hand, with cytokinesis beginning even before mitosis is complete.

The second stage of the cell cycle, called interphase, is the time between two successive M phases. For many years, it was thought that the relatively long period of interphase is uneventful. Today, we know that the cell makes many preparations for division during this stage. These preparations include copying DNA and cell growth. The DNA

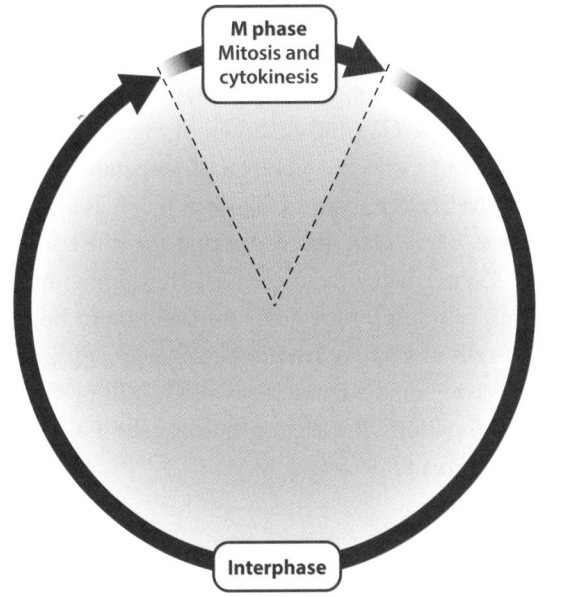

FIGURE 24.3 The eukaryotic cell cycle

The eukaryotic cell cycle consists of interphase, in which the cell prepares to divide, and M phase, in which the cell divides.

in the nucleus is copied so that each daughter cell receives genetic material. The cell then increases in size so that each daughter cell receives sufficient amounts of cytoplasmic and membrane components to allow it to survive on its own.

Interphase: S Phase and Gap Phases

Interphase can be divided into three phases, as shown in **FIGURE 24.4**. During **S phase**, DNA molecules are copied. This phase is called S phase because copying DNA involves the synthesis of DNA. In eukaryotes, DNA is present in the form of linear chromosomes. In order for cell division to proceed normally, every chromosome in the parent cell must be duplicated so that each daughter cell receives a full set of chromosomes. This duplication occurs during S phase. Even though the DNA in each chromosome duplicates, the two identical copies, called **sister chromatids,** do not separate. They stay side by side, physically held together at a constriction called the **centromere,** as shown in **FIGURE 24.5** on page 326. For example, human cells have 23 pairs of chromosomes, or 46 chromosomes in total. At the beginning of mitosis, the nucleus of a human cell still contains 46 chromosomes, but each chromosome is a pair of identical sister chromatids linked together at the centromere.

In most cells, S phase does not immediately precede or follow mitosis but is separated from it by two gap phases: **G_1 phase** between the end of M phase and the start of

S phase, and **G_2 phase** between the end of S phase and the start of M phase. Historically, the letter "G" was an abbreviation for "gap." Initially, scientists could not distinguish any physical changes in the appearance of the cells during G_1 and G_2 when viewed through a microscope, and so assumed there was a gap in cell activity at these stages. Later, better microscopes and biochemical assessments revealed that many essential processes occur during both "gap" phases, despite the name. For example, during G_1, the cell prepares to copy the DNA, and during G_2, both the size and protein content of the cell increase. Thus, G_1 is a time of preparation for S-phase DNA synthesis, and G_2 is a time of preparation for M-phase mitosis and cytokinesis. Some describe this phase as the growth phase.

How long a cell takes to pass through the cell cycle depends on the type of cell and the organism's stage of development. Most actively dividing cells in your body take about 24 hours to complete the cell cycle. Cells in your skin and intestine that require frequent replenishing usually take about 12 hours. A unicellular eukaryote such as yeast can complete the cell cycle in just 90 minutes. The embryonic cells of some frog species complete the cell cycle even faster. Early cell divisions divide the cytoplasm of the large frog

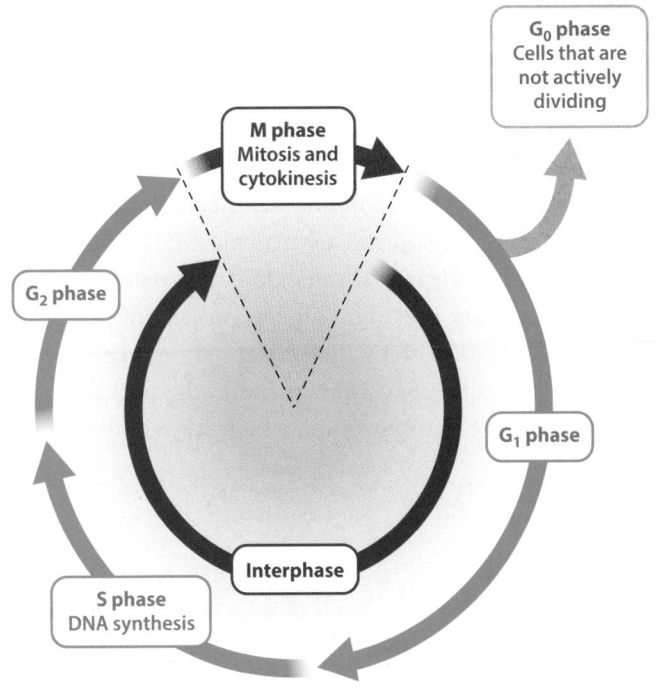

FIGURE 24.4 Interphase

Interphase consists of three phases: G_1 phase, S phase, and G_2 phase. During G_1, the cell prepares for DNA replication. In S phase, DNA synthesis takes place, duplicating each chromosome. In G_2, the cell prepares for M phase.

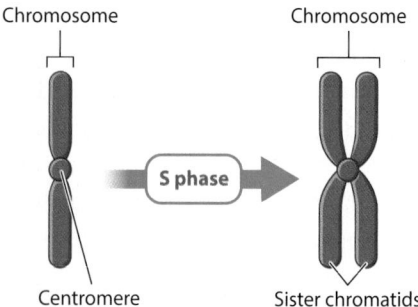

Chromosome   Chromosome

S phase

Centromere   Sister chromatids

FIGURE 24.5 DNA synthesis and duplicated chromosomes

In eukaryotes, DNA is organized as a linear chromosome with a constriction called a centromere, shown on the left. In S phase, the cell's DNA replicates. Each chromosome in the cell replicates in such a way that the copy remains attached to the original at the centromere, as shown on the right. The duplicated DNA molecules are called sister chromatids. Each pair of sister chromatids is a single chromosome.

egg cell into many smaller cells and so no growth period is needed between cell divisions. Consequently, there are virtually no G_1 and G_2 phases, and as little as 30 minutes pass between cell divisions.

Not all the cells in your body actively divide. Instead, many cells pause in the cell cycle somewhere between M phase and S phase for periods ranging from days to more than a year. This period is the **G_0 phase,** shown in Figure 24.4. It is distinguished from G_1 by the absence of preparations for DNA synthesis. Liver cells remain in G_0 for as much as a year. Other cells such as nerve cells and those that form the lens of the eye enter G_0 permanently; these cells are nondividing. Thus, many brain cells lost to disease or damage cannot be replaced. Although cells in G_0 have exited the cell cycle, they are active in other ways—in particular, cells in G_0 still perform their specialized functions. For example, liver cells in G_0 still carry out metabolism and detoxification.

✓ **Concept Check**

4. **Identify** the two main phases of the cell cycle.

5. **Describe** what happens during S phase of the cell cycle.

24.3 During mitotic cell division, a parental cell produces two daughter cells

Mitotic cell division (mitosis followed by cytokinesis) is the basis for asexual reproduction in single-celled eukaryotes, and the means by which a multicellular eukaryote's cells, tissues, and organs develop and are maintained. During mitosis and cytokinesis, the parental cell's DNA is copied and passed on to two daughter cells. This process is continuous, but it is divided into separate steps. These steps are marked by dramatic changes in the cytoskeleton and in the packaging and movement of the chromosomes, which we will describe in this section.

Mitosis

Mitosis is the division of the nucleus of a eukaryotic cell. It takes place in five steps, each of which is easily identified by events that can be observed under a microscope. When you look in a microscope at a cell in interphase, you cannot distinguish individual chromosomes because they are long and thin. Though the chromosomes in interphase

often appear as long thin threads, each chromosome is actually a highly organized complex of DNA, RNA, and proteins called **chromatin.** As the cell moves from G_2 phase to the start of mitosis, the chromosomes condense and become visible in the nucleus. The first stage of mitosis is known as **prophase** and is characterized by the appearance of visible chromosomes, as shown in step 1 of **FIGURE 24.6.**

Outside the nucleus, in the cytosol, the cell begins to assemble the **mitotic spindle.** The mitotic spindle is a group of fibers made up of microtubules, which are part of the cytoskeleton introduced in Unit 2. The mitotic spindle pulls the chromosomes to opposite ends of the dividing cell. In essence, the mitotic spindle serves as the guide wires for chromosome movement during cell division. In animal cells, the mitotic spindle radiates outward from a structure called the **centrosome.** During S phase, the centrosome duplicates and each one migrates around the nucleus. The two centrosomes halt at opposite poles of the cell at the start of prophase. The final locations of the centrosomes define the two ends of the cell that will eventually be separated into two daughter cells.

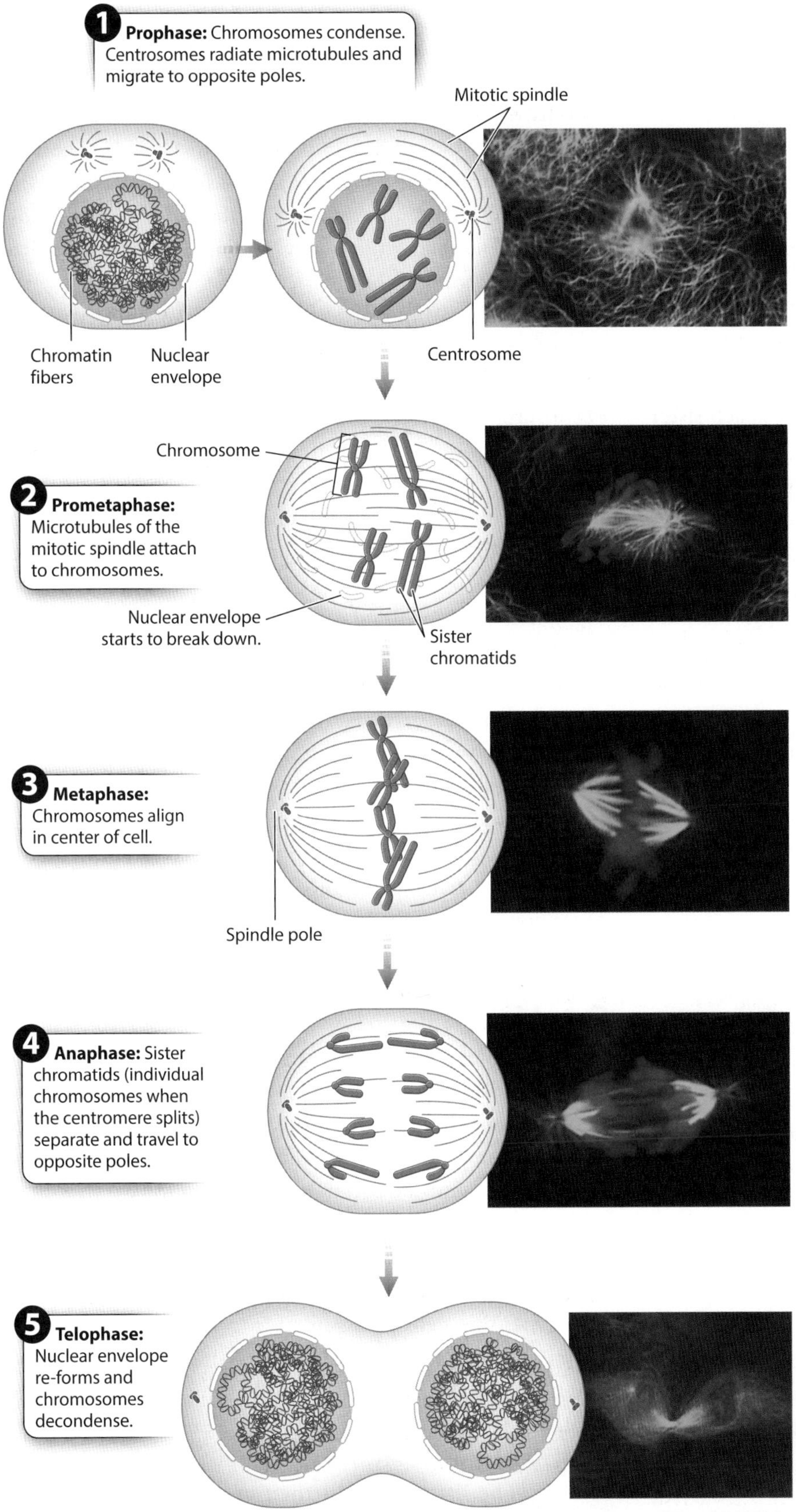

1 **Prophase:** Chromosomes condense. Centrosomes radiate microtubules and migrate to opposite poles.

Mitotic spindle

Chromatin fibers Nuclear envelope Centrosome

Chromosome

2 **Prometaphase:** Microtubules of the mitotic spindle attach to chromosomes.

Nuclear envelope starts to break down. Sister chromatids

3 **Metaphase:** Chromosomes align in center of cell.

Spindle pole

4 **Anaphase:** Sister chromatids (individual chromosomes when the centromere splits) separate and travel to opposite poles.

5 **Telophase:** Nuclear envelope re-forms and chromosomes decondense.

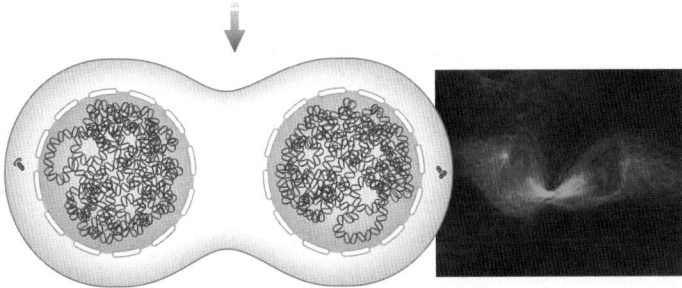

FIGURE 24.6 Steps of mitosis

In prophase, the chromosomes condense and become visible. In prometaphase, spindle fibers attach to the centromeres of the sister chromatids. In metaphase, the chromosomes align across the middle of the cell. In anaphase, the sister chromatids separate from each another and move toward the opposite ends of the cell. In telophase, two new nuclear membranes form. Photos: Jennifer Waters/Science Source

In the next stage of mitosis, called **prometaphase,** the nuclear envelope breaks down and the mitotic spindle attaches to the chromosomes, as shown in step 2 of Figure 24.6. The mitotic spindles radiating from the centrosomes grow and shrink as they explore the space of the cell. When the spindle encounters a chromosome, it attaches to the chromosome at the centromere. Associated with the centromere of each chromosome are two protein complexes called **kinetochores,** one located on each side of the constriction, as shown in **FIGURE 24.7.** Each kinetochore is associated with one of the two sister chromatids and forms the site of attachment for a single mitotic spindle. This arrangement ensures that each sister chromatid is attached to a mitotic spindle radiating from one of the poles of the cell. The symmetrical tethering of each chromosome to the two poles of the cell is essential for proper chromosome movement into daughter cells.

Once each chromosome is attached to the mitotic spindles from both poles of the cell, the mitotic spindles lengthen or shorten to move the chromosomes to the middle of the cell. There the chromosomes are lined up in a single plane that is roughly the same distance from each pole of the cell. This stage of mitosis, when the chromosomes are aligned in the middle of the dividing cell, is called **metaphase,** which

is shown in step 3 of Figure 24.6. It is one of the most visually distinctive stages under the microscope.

In the next stage of mitosis, called **anaphase,** the sister chromatids separate, as shown in step 4 of Figure 24.6. The centromere holding a pair of sister chromatids together splits, allowing the two sister chromatids to separate from each other. The spindles attached to the kinetochores gradually shorten, pulling the newly separated chromosomes to the opposite poles of the cell.

After separation, each chromatid is considered to be a full-fledged chromosome. All of the resulting chromosomes are then equally separated into two daughter cells. For example, during S phase in a human cell, each of the 46 chromosomes is duplicated to yield 46 pairs of sister chromatids. Then, when the chromatids are separated at anaphase, an identical set of 46 chromosomes arrives at each spindle pole, the complete genetic material for one of the daughter cells.

When a complete set of chromosomes arrives at a pole, the chromosomes have entered the area that will form the cytosol of a new daughter cell. This event marks the beginning of **telophase.** During this stage, the cell prepares for its division into two new cells, shown in step 5 of Figure 24.6. The mitotic spindle breaks down and disappears, and a nuclear envelope re-forms around each set of chromosomes, creating two new nuclei. As the nuclei become increasingly distinct in the cell, the chromosomes contained within them decondense, becoming less visible under the microscope. This stage marks the end of mitosis.

Cytokinesis

At the end of mitosis, the single dividing cell has two fully formed nuclei, each with identical genetic material. Cytokinesis is the division of the cytoplasm to produce the two daughter cells. Usually, as mitosis is nearing its end, cytokinesis begins and the parent cell divides into two daughter cells, as shown in **FIGURE 24.8.** In animal cells, this stage begins when a structure called the contractile ring forms against the inner face of the cell membrane in the center of the dividing cell. As if pulled by a drawstring, the ring contracts, pinching the cytoplasm of the cell and dividing it in two. This process is superficially similar to what occurs in binary fission, although different proteins are involved. Cytokinesis results in two daughter cells, each with its own nucleus. The daughter cells then enter G_1 phase and are ready to start the process again.

For the most part, mitosis is similar in animal and in plant cells, but cytokinesis is different, as can be seen in

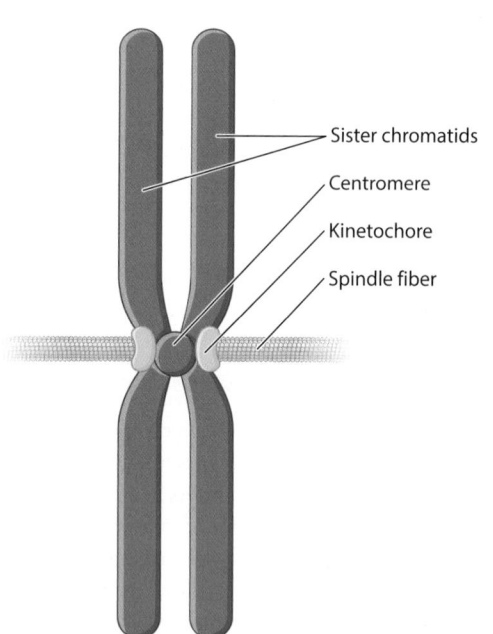

Sister chromatids

Centromere

Kinetochore

Spindle fiber

FIGURE 24.7 Kinetochores

Kinetochores are sites of spindle attachment. There is one kinetochore on each side of the centromere. These connections allow the mitotic spindle to guide chromosome movement during mitosis.

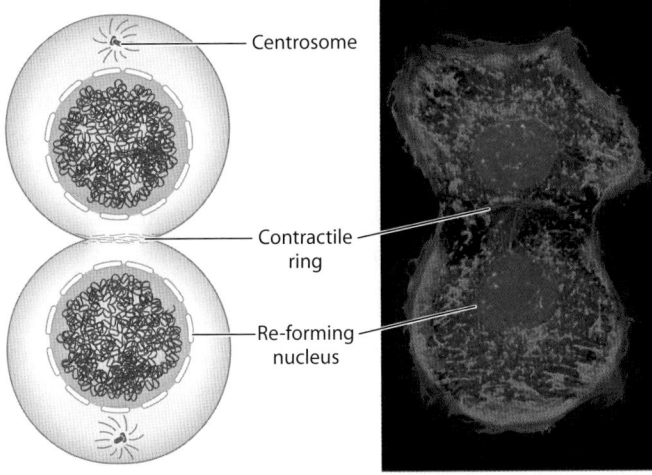

a. Animal cell cytokinesis

Centrosome

Contractile ring

Re-forming nucleus

b. Plant cell cytokinesis

Cell wall

Re-forming nucleus

FIGURE 24.8 Cytokinesis in animal and plant cells

The successful duplication of chromosomes and their migration to opposite poles sets the stage for completing the cell cycle by cytokinesis. (a) During animal cell cytokinesis, membranes pinch off in the middle to completely separate the daughter cells. (b) In addition to forming a membrane between the dividing cells, plant cells also synthesize a new length of cell wall in order to separate the two daughter cells.

Photos: (a) Dr. Paul Andrews, University of Dundee/Science Source; (b) Carolina Biological/Medical Images

Figure 24.8. Because plant cells have a cell wall, the cell divides in two by constructing a new cell wall. During telophase, dividing plant cells form a structure that guides vesicles containing cell wall components to the middle of the cell. During late anaphase and telophase, these vesicles fuse to form a new cell wall in the middle of the dividing cell. Once this developing cell wall is large enough, it fuses with the original cell wall at the perimeter of the cell. Cytokinesis is then complete and the plant cell has divided into two daughter cells.

"Visual Synthesis 4.1: Cell Division and Cell Communication" on page 334 illustrates the concepts of mitotic cell division and cellular communication that we have been exploring over the last few modules.

✓ **Concept Check**

6. **Describe** the relationships between centrosomes, mitotic spindles, and kinetochores, and state when they are active.

7. **Describe** the relationship between sister chromatids and chromosomes.

8. **Identify** the phase of the cell cycle in which the centrosome is duplicated and **describe** what would happen to a cell that failed to duplicate this organelle.

9. **Describe** how cytokinesis differs in animal cells versus plant cells.

Module 24 Summary

LG 24.1 Cell division produces cells needed for growth, maintenance, and reproduction.

- During cell division, a single parental cell gives rise to two daughter cells. Page 322

- Before a cell divides, it increases in size, copies its DNA, and has adequate resources to support two daughter cells. Page 322

- Prokaryotic cell division is called binary fission. Page 322

- Eukaryotic cells divide by mitotic cell division. Page 324

- Mitotic cell division consists of mitosis, or nuclear division, and cytokinesis, or cytoplasmic division. Page 324

- Binary fission and mitotic cell division are the basis of asexual reproduction. Page 324

LG 24.2 The division of eukaryotic cells occurs in phases that make up the cell cycle.

- The eukaryotic cell cycle consists of M phase, when the cell divides, and interphase, when the cell prepares for division. Page 324

- Interphase is divided into G_1, S, and G_2 phases. Page 325

- G_1 and G_2 phases are gap phases, when cells prepare for S phase and M phase, respectively. Page 325

- G_0 phase occurs between M and S phases and is a time that cells temporarily or permanently leave the cell cycle. Page 325

- In S phase, the cell's DNA is copied or synthesized, with the two copies of each chromosome connected at the centromere as sister chromatids. Page 325

LG 24.3 During mitotic cell division, a parental cell produces two daughter cells.

- Mitosis consists of five steps. Page 326

- In prophase, chromosomes condense and become visible under a microscope. Page 326

- Prophase also includes the assembly of the mitotic spindle that will be needed later to move the chromosomes. Page 326

- In prometaphase, the mitotic spindle attaches to chromosomes. Page 328

- In metaphase, the chromosomes line up in the middle of the cell. Page 328

- In anaphase, the sister chromatids separate, and the chromosomes move toward the opposite poles of the cell. Page 328

- In telophase, new nuclear membranes form and chromosomes decondense. Page 328

- Mitosis is followed by cytokinesis, the complete separation of daughter cells with new plasma membranes. Page 328

- Cytokinesis in plant cells requires synthesis of a new cell wall, which is not present in animal cells. Page 328

Key Terms

Cell division	M phase	G_2 phase	Prometaphase
Asexual reproduction	Interphase	G_0 phase	Kinetochores
Mitosis	S phase	Chromatin	Metaphase
Cytokinesis	Sister chromatid	Prophase	Anaphase
Chromosome	Centromere	Mitotic spindle	Telophase
Cell cycle	G_1 phase	Centrosome	

Review Questions

1. Which statement about binary fission is true?
   - (A) Binary fission takes place during M phase of the cell cycle.
   - (B) Binary fission refers to nuclear division, but not cytoplasmic division.
   - (C) Binary fission includes duplication of DNA but not division of the cell.
   - (D) Binary fission is the process by which prokaryotic cells, mitochondria, and chloroplasts divide.

2. Which event of the eukaryotic cell cycle takes place in interphase?
   - (A) Chromosome movement to opposite poles of the cell
   - (B) DNA synthesis
   - (C) Formation of cell walls in plant cells
   - (D) Attachment of the mitotic spindle to centromeres

3. Eukaryotic cells that never divide
   - (A) remain in S phase throughout their lives.
   - (B) are produced by a process other than mitotic cell division.
   - (C) remain in G_0 phase throughout their lives.
   - (D) move from G_1 to G_2 without DNA synthesis occurring.

4. The S phase of the eukaryotic cell cycle
   - (A) is the phase when the cell's DNA is copied.
   - (B) is the phase when new nuclear membranes are synthesized.
   - (C) is also called a gap phase.
   - (D) is the phase in which the new part of the cell wall is synthesized.

5. The mitotic spindle connects the kinetochores to
   - (A) chromatids.
   - (B) centromeres.
   - (C) chromatin masses.
   - (D) mitotic spindles.

6. Giraffes have 62 chromosomes. Immediately following the S phase of the cell cycle in a giraffe cell, there will be
   - (A) 31 centromeres.
   - (B) 62 centromeres.
   - (C) 124 centromeres.
   - (D) 248 centromeres.

Module 24
AP® Practice Questions

Section 1: Multiple-Choice Questions

Choose the best answer for questions 1–4.

1. Prokaryotic cells reproduce by the process of binary fission. The increase in the population size of bacteria undergoing binary fission may be expressed mathematically as 2^n, where n is the number of generations. If bacterial cells of a particular species can reproduce every 30 minutes, a single bacterium of the species reproducing at this rate would result in how many cells after 3 hours?
   - (A) 12
   - (B) 32
   - (C) 64
   - (D) 128

2. Tubulin is a protein that is the major component of microtubules, which make up the mitotic spindle. In mitosis, microtubules extend from the centrosomes at the poles of the dividing cell to the centromeres of each sister chromatid pair, moving the chromosomes and eventually separating the sister chromatids from one another. A group of students conduct an experiment on onion cells that they have identified as being in prophase. After adding the drug colchicine to the onion cells, the students observe that the cells do not advance to prometaphase, metaphase, anaphase, telophase, and cytokinesis. The students claim that colchicine is an inhibitor of cell division that interacts with tubulin. Which best explains the claim and experimental results?

(A) Colchicine prevents tubulin from assisting in the duplication of DNA necessary to make sister chromatids.

(B) Colchicine binds to tubulin in the nuclear membrane and prevents the membrane from dissolving prior to mitosis.

(C) Colchicine binds tubulin, resulting in the inhibition of microtubule formation.

(D) Colchicine binds to tubulin and prevents it from organizing the DNA into the sister chromatids needed for mitosis.

Question 3 refers to the following figure:

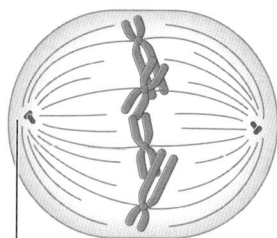

Spindle pole

3. The figure depicts a cell in metaphase. The cell contains 4 individual chromosomes. If one of the sister chromatids fails to separate during anaphase, how many chromosomes would the two resulting daughter cells contain?

(A) One cell would have 2 chromosomes and one cell would have 6 chromosomes.

(B) Both cells would have 4 chromosomes.

(C) One cell would have 3 chromosomes and one cell would have 5 chromosomes.

(D) One cell would have 3.5 chromosomes and one cell would have 4.5 chromosomes.

4. A liver cell completes the cell cycle and gives rise to two daughter cells, A and B. Daughter cell A then goes through the cell cycle itself, giving rise to two more cells. Daughter cell B, however, does not pass G_1. Which prediction of what will happen to daughter cell B is the most likely?

(A) The cell will revert to mitosis and go through the process for a second time.

(B) The cell will skip S phase and "jump" to G_2.

(C) The cell will enter G_0.

(D) The cell will divide into two daughter cells.

Section 2: Free-Response Question

Write your answer to each part clearly. Support your answers with relevant information and examples. Where calculations are required, show your work.

Using a microscope, a student observes two different cells that are undergoing cell division. The student's drawing of each cell is shown below.

a.

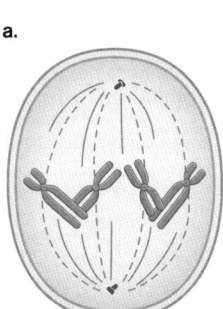

b.

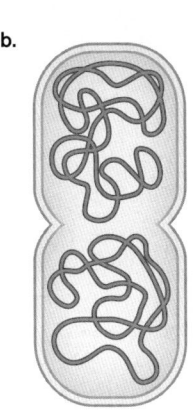

(a) **Identify** the phase of the cell cycle shown in figure a. **Justify** your answer using the features shown.

(b) **Make a claim** about whether the cell in figure a is a bacterial cell or a eukaryotic cell. Provide evidence to **support your claim.**

(c) The student shows the cell in figure b to another student. The first student claims that it is a eukaryotic cell in cytokinesis, and the second student claims that it is a bacterial cell that has almost completed binary fission. **Make a claim** about which type of cell the drawing shows. Provide evidence to **support your claim.**

Cells in tissues like skin divide by mitotic cell division, a tightly regulated process that results in two copies of the parent cell. Cells communicate with one another by molecular signals to coordinate the functions of tissues, organs, and organ systems within an organism.

Cell division

Mitotic cell division is the process by which a single cell divides into two cells that, except for rare mutations, are genetically identical to the parent cell.

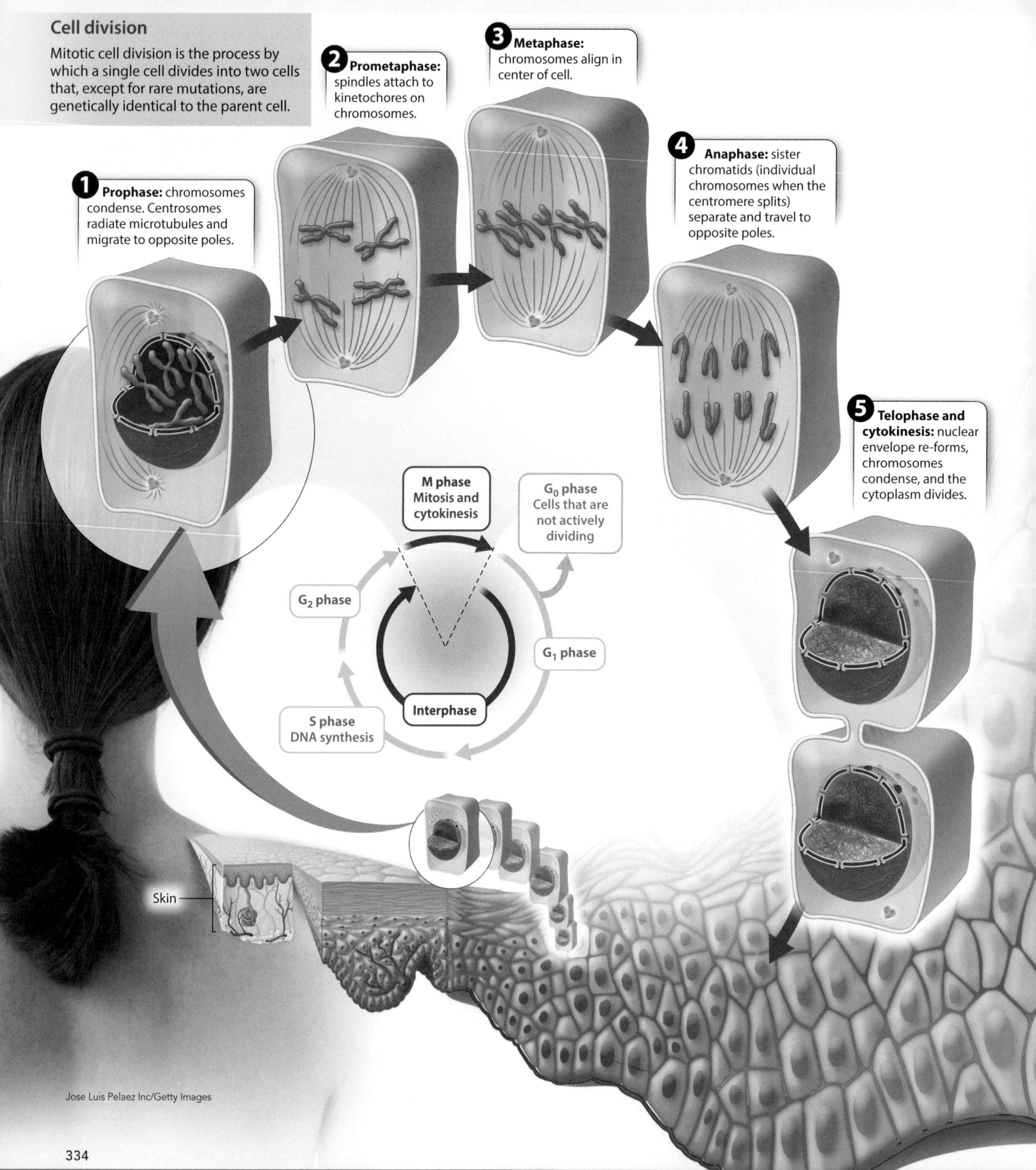

1 Prophase: chromosomes condense. Centrosomes radiate microtubules and migrate to opposite poles.

2 Prometaphase: spindles attach to kinetochores on chromosomes.

3 Metaphase: chromosomes align in center of cell.

4 Anaphase: sister chromatids (individual chromosomes when the centromere splits) separate and travel to opposite poles.

5 Telophase and cytokinesis: nuclear envelope re-forms, chromosomes condense, and the cytoplasm divides.

M phase
Mitosis and cytokinesis

G₀ phase
Cells that are not actively dividing

G₂ phase

G₁ phase

S phase
DNA synthesis

Interphase

Skin

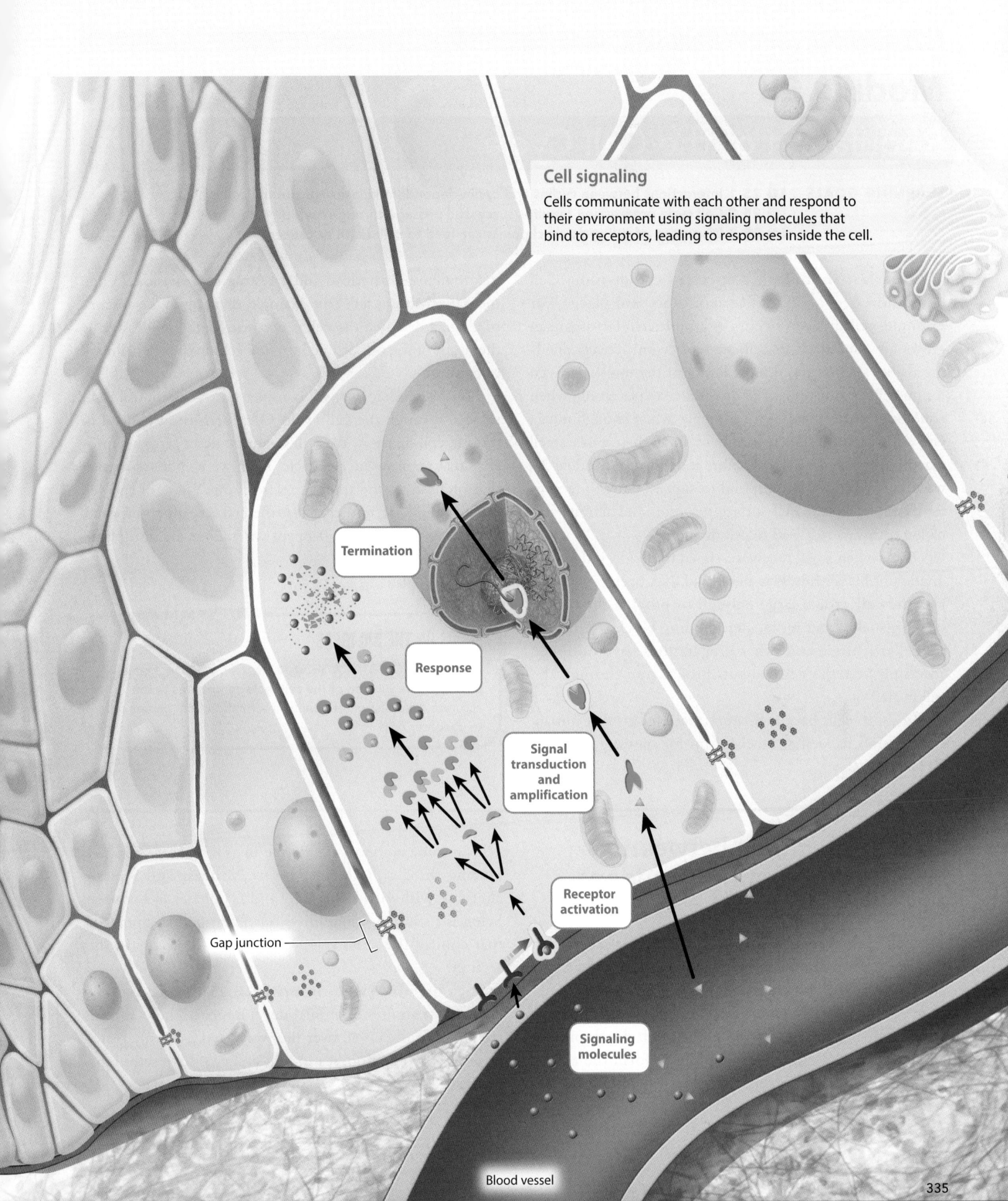

Cell signaling

Cells communicate with each other and respond to their environment using signaling molecules that bind to receptors, leading to responses inside the cell.

Termination

Response

Signal transduction and amplification

Receptor activation

Gap junction

Signaling molecules

Blood vessel

335

Module 25

Regulation of the Cell Cycle

LEARNING GOALS ▶**LG 25.1** Interactions between cyclins and cyclin-dependent kinases control the cell cycle.
    ▶**LG 25.2** Cell cycle progression requires successful passage through multiple checkpoints.
    ▶**LG 25.3** Disruption of cell cycle checkpoints can lead to cell death or cancer.

In the last module, we discussed cell division. Cell division occurs only at certain times and places. For example, cell division occurs as a multicellular organism grows, a wound heals, or cells are replaced in actively dividing tissues such as the skin or lining of the intestine. Even for unicellular organisms, cell division takes place only when conditions are favorable—for example, when enough nutrients are present in the environment. In all these cases, a cell may have to receive a signal before it divides. In Module 20, we discussed how cells respond to signals. Growth factors, for example, bind to cell-surface receptors and activate signaling pathways that lead to cell division.

Even when a cell receives a signal to divide, it does not divide until it is ready. Has all of the DNA been replicated? Has the cell grown large enough to provide the necessary organelles and molecular building blocks to daughter cells? If these and other preparations have not been accomplished, the cell halts its progression through the cell cycle.

In short, cells have regulatory mechanisms that initiate cell division, as well as mechanisms for spotting faulty or incomplete preparations and arresting cell division. When these mechanisms fail—for example, dividing in the absence of a signal or when the cell is not ready—the result may be the death of the cell or uncontrolled cell division, a hallmark of **cancer.**

In this module, we will consider how cells control their passage through the cell cycle. Our focus is on control of mitotic cell division. We will begin by describing key proteins that regulate the cell cycle. Then we will discuss signals that orchestrate the transitions from one phase of the cell cycle to the next, ensuring that the cell is ready to proceed. Finally, we will see what happens when there are disruptions in the cell cycle.

PREP FOR THE AP® EXAM

FOCUS ON THE BIG IDEAS

INFORMATION STORAGE AND TRANSMISSION: Focus on how the regulation of the cell cycle contributes to the accurate transmission of genetic information from parent cell to daughter cells.

25.1 Interactions between cyclins and cyclin-dependent kinases control the cell cycle

The cell cycle is regulated so that cells divide only when they are ready and only at the right time and in the right place. One of the first clues about how it is regulated came from studies of early embryos. The cell cycle is perhaps at its most impressive during embryonic development, when a fertilized egg divides rapidly to form the tissues and organs of an organism. As noted in Module 24, the cells in some embryos can divide in as little as 30 minutes. During these rapid cell divisions, M phase and S phase alternate with virtually no G_1 and G_2 phases in between. Scientists studying these divisions identified key proteins that control the cell cycle, which we will discuss in this section.

Sea urchins, shown in **FIGURE 25.1**, were an early favorite for studies of animal development. These animals are found in oceans all over the world and are easily reared in the lab. In addition, sea urchin embryos have semitransparent cells that are easy to study with a microscope.

FIGURE 25.1 A sea urchin

Sea urchins, like this *Sphaerechinus granularis*, have been used as a model organism to study the cell cycle and cell division. Photo: Jose B. Ruiz/naturepl.com

In the 1980s, British biochemist Tim Hunt and his colleagues extracted proteins from embryonic cells as they rapidly divided. They discovered that the members of a family of proteins were present in large amounts in some phases of the cell cycle and were much less abundant in other phases. Hunt named the members of this protein family **cyclins.** Although he named them cyclins because of his fondness for cycling, the name stuck because it also describes their cyclic appearance and disappearance as cells divide. There are several different types of cyclins in eukaryotic cells.

The graph in **FIGURE 25.2** shows the cyclical increase and decrease in the amounts of cyclins in interphase and mitosis of two successive cell divisions in sea urchin embryos. As shown in the graph, the concentration of the cyclins increases during interphase and peaks in mitosis, then decreases before increasing again as interphase transitions to mitosis. This pattern indicates that there is an interval of cyclin synthesis followed by an interval of cyclin degradation.

In addition to the appearance and disappearance of cyclins, the researchers made a second observation. They noticed that several enzymes that were present at constant concentrations throughout the cell cycle cycled between active and inactive forms as the cells divided. These enzymes are kinases. As discussed in Module 21, kinases phosphorylate target proteins, changing their shapes and functions. The timing of kinase activity is delayed slightly relative to the appearance of the cyclical proteins.

These and many other observations led to the following view of cell cycle control. The levels of proteins called cyclins rise and fall with each turn of the cell cycle. These proteins bind to kinases and activate them. These kinases are called **cyclin-dependent kinases** (**CDKs**) because their activity depends on binding cyclins. CDKs are always present within the cell but are active only when bound to a specific cyclin. The kinase activity of the cyclin–CDK complexes allows the cell cycle to progress.

The cyclins and CDKs do not remain bound together and active for long. **FIGURE 25.3** on page 338 shows how CDKs cycle from active to inactive and back to active. At the top of the figure, CDK binds a cyclin protein, forming an active cyclin–CDK complex. The active cyclin–CDK complex then binds a target protein and transfers a phosphate group from a molecule of ATP to the target protein to activate the target protein. Once the cyclin–CDK complex activates a target protein, the cyclin dissociates from

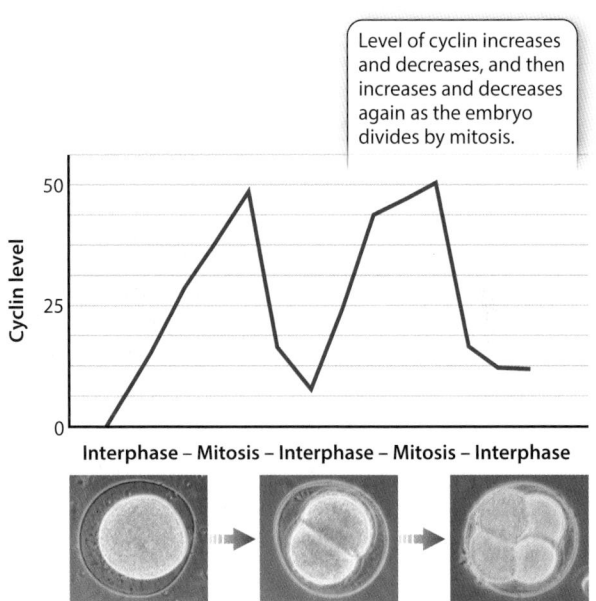

Level of cyclin increases and decreases, and then increases and decreases again as the embryo divides by mitosis.

Interphase – Mitosis – Interphase – Mitosis – Interphase

FIGURE 25.2 Cyclins

Researchers found that the amounts of the protein cyclin increase and decrease in coordination with the cell cycle phases. Two divisions of a sea urchin embryo are shown here. In the photos, one round of cell division produces two cells and a second round produces four. Cyclin levels are lowest at interphase and highest during mitosis, indicating the importance of cyclin for regulating the phases of the cell cycle.

Photos: Biology Pics/Science Source

the complex and degrades. Without its cyclin partner, the inactive CDK no longer transfers a phosphate group to the target protein, and other proteins in the cell continuously remove phosphate groups from target proteins. The result is a short-lived signal.

There are several types of cyclins that function at different times in the cell cycle. These cell cycle proteins, including both cyclins and CDKs, are widely conserved across eukaryotes, reflecting their fundamental role in controlling cell cycle progression. They have been extensively studied not just in sea urchins, but also in yeast, mice, and humans.

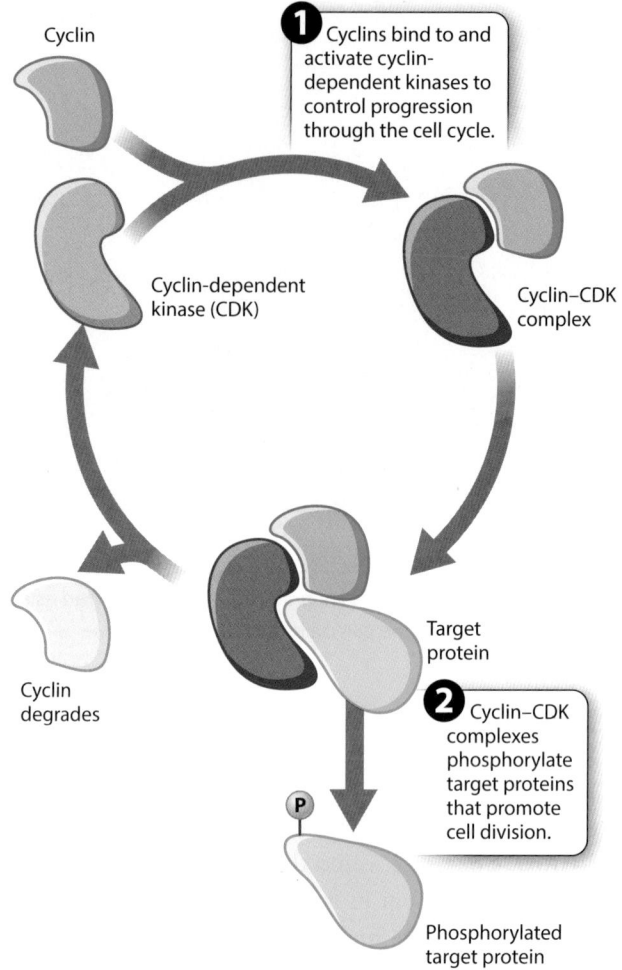

✓ Concept Check

1. **Describe** why the name cyclin is a fitting name for this family of proteins.

2. **Define** the term "CDK" and **describe** how this term relates to the functions of these proteins.

FIGURE 25.3 Cyclins and cyclin-dependent kinases (CDKs)

When a cyclin binds to a CDK, the kinase function of the CDK is activated. Kinases are enzymes that transfer a phosphate group to target proteins, shown at the bottom of the diagram. The phosphorylated target proteins are required for key steps in the cell cycle. Cyclins then degrade and dissociate from CDKs, terminating the kinase activity of the CDK. The result is a short-lived signal that moves the cell forward in the cell cycle.

25.2 Cell cycle progression requires successful passage through multiple checkpoints

Cyclins and CDKs not only allow a cell to progress through the cell cycle but also give the cell opportunities to halt the cell cycle should something go wrong. For example, the preparations for the next stage of the cell cycle may be incomplete or there may be some kind of damage. In these cases, there are mechanisms that block the cyclin–CDK activity required for the next step, pausing the cell cycle until preparations are complete or the damage is repaired. Each of these mechanisms is called a **checkpoint.** There are many cell cycle checkpoints that monitor different points of the cell cycle. In this section, we will discuss three of the major ones.

In order to successfully divide, a cell must complete many tasks at the right time. The DNA must be fully and accurately copied in S phase. Cell organelles must be produced in numbers sufficient to supply both of the daughter cells that will be produced. Membrane synthesis must occur. Coordination and regulation are required to make smooth transitions between phases. Once it is underway, the cell cycle normally proceeds to the end of cytokinesis, when one cell divides into two cells. Researchers have discovered that dividing cells must pass through several checkpoints, during which the cell ensures that everything is ready for each phase to occur.

FIGURE 25.4 shows three major checkpoints. The DNA damage checkpoint is at the transition from G_1 to S phase. It monitors for damaged DNA before the DNA is copied in S phase. DNA can be damaged by environmental factors such as ultraviolet radiation or chemical agents. Typically, damage takes the form of double-stranded breaks in the DNA. If the cell progresses through mitosis with DNA damage, daughter cells might inherit the damage, or the chromosomes might not segregate normally. The checkpoint late in G_1 delays progression through the cell cycle by interfering with the formation of a cyclin–CDK complex until DNA damage is repaired.

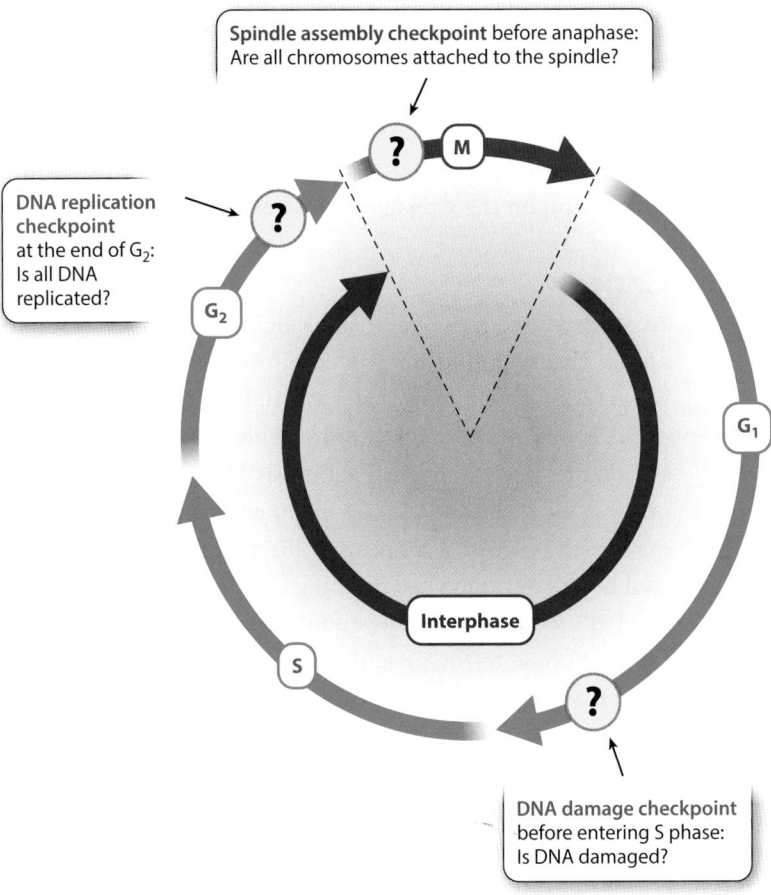

Spindle assembly checkpoint before anaphase:
Are all chromosomes attached to the spindle?

DNA replication checkpoint at the end of G₂:
Is all DNA replicated?

M

G₂

G₁

Interphase

S

DNA damage checkpoint before entering S phase:
Is DNA damaged?

FIGURE 25.4 Cell cycle checkpoints

At multiple points during the cell cycle, the cell ensures that it is ready for the next phase of cell division. At the end of G₁, the cell checks that the DNA is not damaged before it moves on to S phase, when the DNA is copied. At the end of G₂ before entering mitosis, the cell ensures that its DNA has been completely copied in preparation for nuclear division. During mitosis, at the end of metaphase and before anaphase, the cell checks to make sure each pair of sister chromatids are connected to spindle fibers so that the sister chromatids can be separated from one another in anaphase.

A second major checkpoint occurs at the transition between G₂ and M phase. This checkpoint relates to the process of duplicating a DNA molecule, known **DNA replication.** This process, which we will explore in Unit 6, allows genetic information to be passed from cell to cell and from an organism to its progeny. This checkpoint is called the DNA replication checkpoint because it monitors the cell to be sure that the parental DNA has been completely copied before the cell moves on to mitosis. The presence of uncopied or partially copied DNA will cause the cell cycle to pause here until the process of copying DNA is complete. If a cell enters mitosis without fully copying its DNA, the daughter cells will not inherit a full complement of genetic material.

A third checkpoint, called the spindle assembly checkpoint, occurs during metaphase of mitosis, before the cell moves into anaphase when the sister chromatids separate and move to opposite ends of the dividing cell. At this checkpoint, the cell ensures that its chromosomes are connected to the mitotic spindle that will move them to opposite poles in anaphase.

✓ Concept Check

3. **Identify** three major cell cycle checkpoints.
4. **Identify** when in the cell cycle each checkpoint occurs.

25.3 Disruption of cell cycle checkpoints can lead to cell death or cancer

As we discussed, cell cycle checkpoints do not just prevent damaged cells from duplicating, they give cells several opportunities to pause the cell cycle when something goes wrong. If the preparations for the next phase of the cell cycle are incomplete or there is damage to DNA or other cell components, the cycle is halted. In these cases, cell cycle checkpoint mechanisms block the formation of the cyclin–CDK pairs whose activity is required before moving to the next step. These mechanisms pause the cell cycle until preparations are complete or the damage is repaired. When the damage is too extensive to repair, the cell may be induced to undergo a set process that leads to its death. In other cases, the cell proceeds past the checkpoint even when it is not ready, which can lead to uncontrolled cell division, or cancer. In this section, we will consider what happens when checkpoints fail, starting with cell death and ending with cancer.

Cell Death

When the cell cycle's checkpoints are functional, either a healthy cell is duplicated or a damaged cell is prevented from going through the cell cycle. Cells that are too damaged to be fixed when stopped at a checkpoint are sometimes programmed to die. There are two different routes by which cells die. A cell that is damaged or lacks a blood supply will die by necrosis, in which case the contents of the cell leak out, possibly damaging neighboring cells. By contrast, some cells are programmed to die in a process called *apoptosis*. During **apoptosis,** cells die in an orderly way with a distinct set of cellular changes. Apoptosis occurs as a normal part of development to remove unneeded cells or to sculpt parts of the body, such as the spaces between your fingers. Apoptosis is also known as programmed cell death because the cell death pathway is genetically programmed, or directed, by the cell.

A cell that does not meet the requirements to pass through a checkpoint in the cell cycle will attempt to make DNA repairs or synthesize components, as needed, for the next phase. Successful repairs and synthesis of components will allow the checkpoint to be met, and the cell cycle proceeds. However, if the repair efforts are not successful, the cell may undergo apoptosis, producing enzymes that disassemble the cell components.

Overexposure to the ultraviolet radiation in sunlight, for example, can damage DNA in skin cells, which in turn may activate enzymes that cause the death of the cell. As the enzymes break down the cell components, the cell shrinks, DNA is fragmented, and the remnants are removed by the immune system. Apoptosis is irreversible: once the program to produce enzymes that disassemble the cell has started, it cannot be stopped. Like cell cycle checkpoints, apoptosis is a safeguard against proliferating dangerous cells. Cells with irreparable errors undergo apoptosis rather than dividing and passing on the damage to the next generation of cells.

Cancer

As we have seen, the cell has several safeguards in place to ensure that cell division proceeds smoothly. One safeguard is the set of checkpoints that act at different times in the cell cycle. If one of these checkpoints fails, the cell has another safeguard—it may undergo apoptosis. If a cell bypasses both of these safeguards, it may divide uncontrollably, and cancer may result. Cancer is a group of diseases that are characterized by uncontrolled cell division.

Let's focus on the checkpoint that occurs at the end of G_1 in response to the presence of damaged DNA to see how cancer can develop. DNA can be damaged by exposure to radiation or to chemicals that cause cancer, called carcinogens. DNA damage often includes breaks in the linear strands of the DNA, and these breaks in the DNA begin the signaling process that stops the cell cycle dead in its tracks.

When DNA is damaged by radiation, a protein kinase is activated that phosphorylates a protein called p53. FIGURE 25.5 shows how p53 normally works in the regulation of the G_1 checkpoint after DNA damage. The green oval represents the protein p53. In a healthy cell in the leftmost

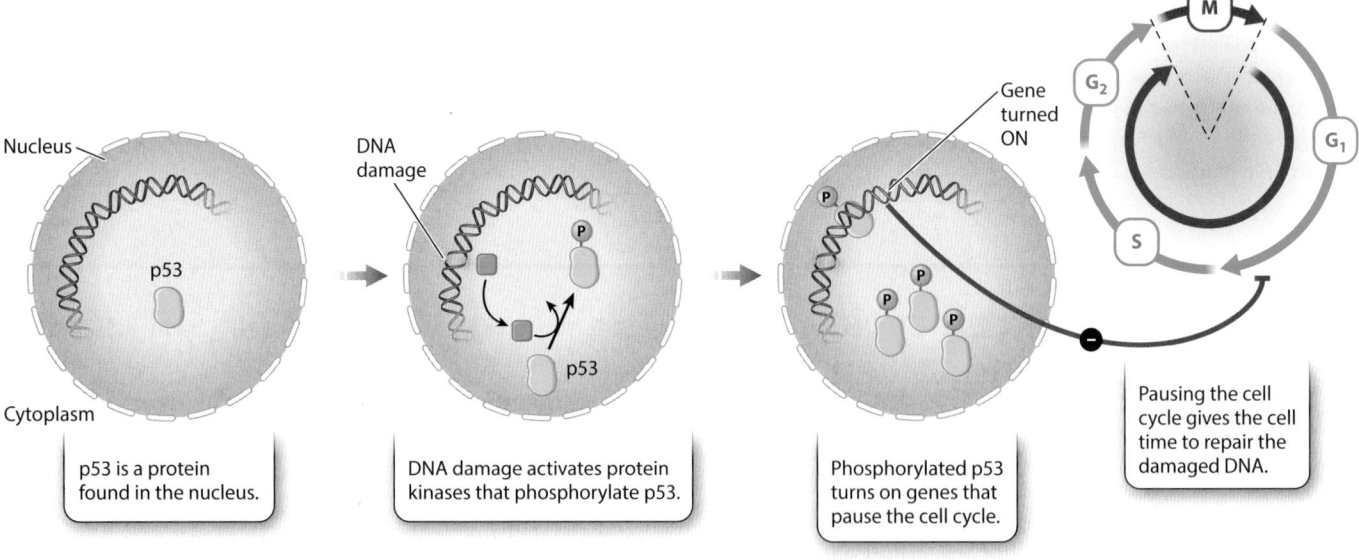

FIGURE 25.5 Response to DNA damage

The protein p53 is normally found in a cell's nucleus. Should DNA damage occur, kinases that add a phosphate group to p53 are activated. Activated p53 proteins lead to the production of proteins that arrest the cell cycle at the end of the G_1 phase by blocking the G_1 checkpoint.

diagram of the figure, the p53 protein is present without any phosphates and in very small amounts. When no phosphate groups are bound to p53, it degrades quickly. When DNA is damaged, however, a specific protein kinase is activated in the nucleus. This kinase adds phosphates to p53, as shown in the middle diagram of Figure 25.5. The activated p53 binds to DNA, where it turns on several genes, as shown in the rightmost diagram of the figure. One of these genes guides the production of a protein that blocks the cyclin and its CDK from forming the complex needed for the transition from G_1 phase to S phase. Active p53 arrests the cell at the G_1 transition, giving the cell time to repair the damaged DNA.

The p53 protein is a good example of a protein involved in halting the cell cycle when the cell is not ready to divide. Because of its role in protecting the genome from accumulating DNA damage, p53 is sometimes called the guardian of the genome.

When the p53 protein is mutated or its function is inhibited, the cell can divide before the DNA damage is repaired. Such a cell continues to divide in the presence of damaged DNA, leading to the accumulation of mutations that promote cell division. The p53 protein is mutated in many types of human cancer, highlighting its critical role in regulating the cell cycle.

FIGURE 25.6 illustrates how a single mutation that impairs the function of p53 can snowball into malignant cancer, which is cancer that can invade nearby tissues and spread throughout the body. An initial mutation that affects the p53 gene eliminates the safeguard against dividing cells with damaged DNA. All cells descended from this first damaged cell also lack functional p53. A second error in a daughter cell's DNA might result in a protein that speeds up the cell cycle, allowing a tumor to form. It might take a while, even years, before that second mistake occurs. After another interval of time, a third error in the DNA might generate a protein that enhances tumor growth. A fourth error results in the movement of the fast-growing cells to other parts of the body, known as metastatic cancer. Fortunately, not all cancers become metastatic, and it is the goal of preventative cancer screening to find and treat cancers early.

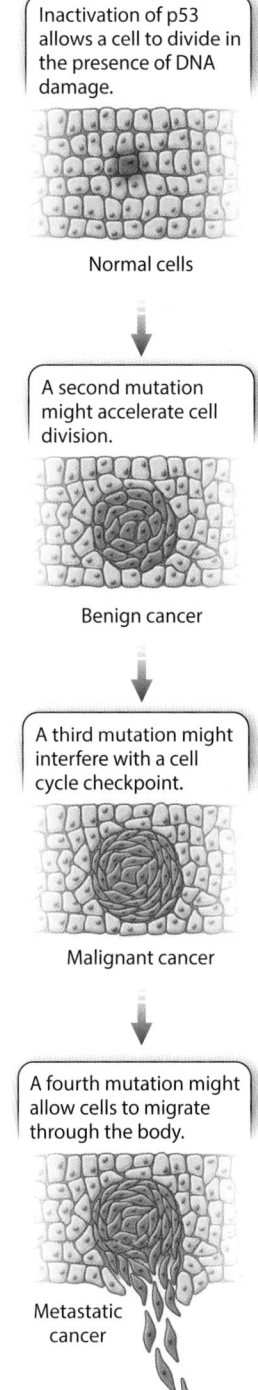

FIGURE 25.6 Cancer progression

Cancers require multiple mutations that affect cell division, growth, and mobility. An initial mutation in the gene for p53 allows the cell to continue through the cell cycle and divide in the presence of DNA damage. Now there are two cells with mutant p53 genes. Additional mutations in daughter cells speed up cell division, producing even more mutant cells with more chances for mutations to occur, until the mass of cells becomes a tumor that can then spread, or metastasize.

✓ Concept Check

5. **Describe** how a pause in the cell cycle can benefit a cell.

6. **Describe** cellular conditions in which you would expect to find p53 protein that has had phosphates added.

7. **Describe** how multiple errors in DNA can accumulate in a cell and drive cancer.

Module 25 Summary

REVISIT THE BIG IDEAS

INFORMATION STORAGE AND TRANSMISSION: Explain how the regulation of the eukaryotic cell cycle contributes to the accurate transmission of genetic information from parent cell to daughter cells.

LG 25.1 Interactions between cyclins and cyclin-dependent kinases control the cell cycle.

- Levels of proteins called cyclins increase and decrease during the cell cycle. Page 336
- Cyclins form complexes with cyclin-dependent kinases (CDKs), activating the CDKs to phosphorylate target proteins that promote cell division. Page 337
- Cyclin–CDK complexes are activated by signals that promote cell division. Page 337
- Different cyclin–CDK complexes control progression through the cell cycle at key steps, including G_1/S phase, S phase, and M phase. Page 337

LG 25.2 Cell cycle progression requires successful passage through multiple checkpoints.

- The cell has checkpoints that halt progression through the cell cycle if something is not right. Page 338

- The DNA damage checkpoint in G_1 monitors for damaged DNA. Page 338
- The DNA replication checkpoint in G_2 ensures that DNA has been fully copied. Page 339
- The spindle assembly checkpoint in M phase ensures that the spindles are attached to the chromosomes. Page 339

LG 25.3 Disruption of cell cycle checkpoints can lead to cell death or cancer.

- Cells with DNA damage may undergo a genetically programmed cell death called apoptosis, thereby reducing the risk of developing cancer, or uncontrolled cell division. Page 339
- Damage to DNA can result in the addition of phosphate groups to p53 proteins, bringing the cell cycle to a halt in G_1 phase. Page 340
- Activated p53 proteins turn on genes that pause the cell cycle. Page 341
- Some cancers have mutations in p53, rendering it non-functional. Page 341
- Cancer results from multiple mutations that promote cell division and remove the normal checks on cell division. Page 341

Key Terms

Cancer	Cyclin-dependent kinase (CDK)	DNA replication
Cyclin	Checkpoint	Apoptosis

Review Questions

1. Referring to Figure 25.2, which statement best describes measurements of amounts of cyclins in the sea urchin studies?
   (A) Cyclin levels increased only during the first, not the second, cell division.
   (B) Four peaks of cyclin production were evident.
   (C) Cyclin abundance reached its lowest levels during mitosis.
   (D) Cyclin abundance peaked during M phase.

2. Which response is most likely to occur when a CDK is activated by its partner cyclin?
   (A) An increase in concentration of free phosphate ions in the cytosol
   (B) A slowdown in the cell cycle progress
   (C) A cell cycle arrest at the G_1 phase
   (D) An increase in the amount of phosphorylated target proteins

3. What controls the enzyme function of the cyclin-dependent kinases?

   (A) Binding to phosphorylated proteins needed for the cell cycle

   (B) Binding with their cyclin partners

   (C) Binding with DNA

   (D) Binding to receptor proteins in the cell membrane

4. Once synthesized, an individual cyclin

   (A) will decline in abundance after its checkpoint function has passed.

   (B) continues to increase in concentration until mitosis is halted.

   (C) remains at stable concentrations across the rest of the cell cycle.

   (D) is secreted from one cell to alter the cell cycle in neighboring cells.

5. A cell that is unable to synthesize the cyclin needed for the G_2 checkpoint is likely to

   (A) destroy one copy of duplicated DNA and return to the G_0 phase.

   (B) complete mitosis more quickly than most cells.

   (C) undergo programmed cell death.

   (D) enter cytokinesis without having duplicated any of its DNA.

6. The abnormally early appearance of a cyclin needed for the S checkpoint could result in

   (A) the persistence of the nuclear membrane throughout mitosis.

   (B) the copying of damaged DNA.

   (C) the abnormal production of four copies of DNA prior to M phase.

   (D) the presence of fully assembled mitotic spindles in S phase.

7. Cancers that spread by metastasis

   (A) are usually the result of apoptosis.

   (B) usually include many different mutations in DNA.

   (C) are present only at their site of origin.

   (D) are highly sensitive to cell cycle checkpoints.

Module 25
AP® Practice Questions

Section 1: Multiple-Choice Questions

Choose the best answer for questions 1–4.

1. A human liver cell reaches the end of G_1 but fails to move into S phase. Researchers treat the cell with chemicals that stain the DNA to make it visible. The stains indicate that the liver cell's DNA is undamaged. Which is a possible explanation for why the cell cannot enter S phase?

   (A) The spindle fibers are unable to attach to the centromeres of the chromosomes.

   (B) The undamaged DNA triggers the start of apoptosis.

   (C) The cell's p53 proteins cannot be activated.

   (D) The cyclin–CDK complex for G_1 is unable to form.

2. Receptor proteins in cell signaling and cyclin-dependent kinases (CDKs), in the regulation of the cell cycle, function similarly in which way?

   (A) Each kind of receptor protein and CDK can bind a variety of ligands and cyclins.

   (B) Both receptor proteins and CDKs are in their active form until they bind a ligand or cyclin.

   (C) Both receptor proteins and CDKs change shape and activate upon binding a ligand or cyclin.

   (D) Receptor proteins and CDKs are only found in cell membranes.

3. There are multiple types of cyclins, each of which binds its CDK during a different phase of the cell cycle. The graph shows the point in the cell cycle that each cyclin increases and decreases in abundance. Using the data in the graph, determine what percentage of cyclin types are necessary to move the cell from S phase to G_2.

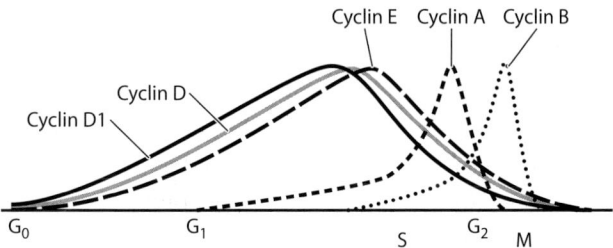

(A) 20%

(B) 40%

(C) 60%

(D) 80%

4. Researchers find a mutation that arose in a cell that results in its p53 proteins being constantly phosphorylated. Predict what will occur when the cell goes through the cell cycle now that its p53 proteins are always phosphorylated.

(A) The cell will not progress beyond G_1.

(B) The cell will move from G_1 to S but will not pass into G_2.

(C) The cell will undergo apoptosis.

(D) The cell will bypass G_1, S, and G_2 and enter into mitosis.

Section 2: Free-Response Question

Write your answer to each part clearly. Support your answers with relevant information and examples. Where calculations are required, show your work.

In humans, a gene called *BRCA1* encodes a protein that plays a crucial role in DNA repair. Scientists are continuing to research the precise mechanism by which the BRCA1 protein repairs damaged DNA, but it is thought that the BRCA1 protein is involved in fixing breaks in the phosphodiester bonds of the backbones of DNA molecules.

(a) *BRCA1* stands for *breast cancer susceptibility gene 1*. Mutations in *BRCA1* are risk factors for the development of breast and ovarian cancers. Despite the risk of disease when *BRCA1* is mutated, the *BRCA1* gene is beneficial to dividing cells. **Describe** how *BRCA1* aids in the completion of the cell cycle.

(b) **Make and support a claim** about how mutations in *BRCA1* could result in the development of breast and ovarian cancers.

Unit 4

AP® Practice Questions

Section 1: Multiple-Choice Questions

Choose the best answer for questions 1–14.

1. Which is the best prediction of what will happen if a ligand permanently binds to and activates its receptor on a target cell?
   (A) The signaling cell will stop secreting the signaling molecule.
   (B) The target cell's receptors will inactivate, ending the signal transduction pathway.
   (C) The target cell will activate p53 and apoptosis.
   (D) The target cell will constantly respond.

Use the following illustration to answer question 2.

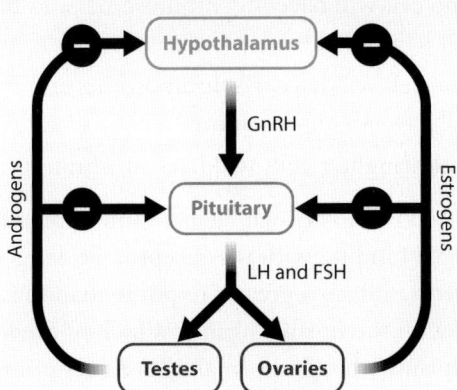

2. In mammals, the family of hormones called estrogens and androgens controls the development of female and male characteristics during fetal development and during puberty. The production and release of estrogens and androgens is controlled by a feedback loop involving the hypothalamus and anterior pituitary gland, and the ovaries in females and testes in males. The hypothalamus, in response to hormones involved in sexual maturation and the menstrual cycle, secretes a hormone called gonadotropin-releasing hormone (GnRH). GnRH binds to receptors on target cells in the anterior pituitary gland. In response to GnRH, the target cells release hormones called follicle-stimulating hormone (FSH) and luteinizing hormone (LH), which are part of a family of hormones called gonadotropins. The gonadotropins in turn bind to ovaries, which release estrogens, and testes, which release androgens. Based on the diagram of this pathway, which best explains the control of GnRH secretion?
   (A) When sufficient gonadotropins are present, GnRH secretion increases due to negative feedback by the hypothalamus.
   (B) When sufficient gonadotropins are present, GnRH secretion decreases due to positive feedback by the hypothalamus.
   (C) When sufficient estrogen is present, GnRH secretion decreases due to negative feedback by the high estrogen concentrations.
   (D) When sufficient estrogen is present, GnRH secretion increases due to positive feedback by the high estrogen concentrations.

3. In humans, insulin binds to its receptor protein on the surface of target cells in various organs of the body. The target cells respond to insulin by taking up glucose from the blood. Some researchers hypothesize that in individuals with type 2 diabetes, insulin binds to the receptor, but the signal is not sent into the cell. If this hypothesis about type 2 diabetes is correct, which cell signaling component is deficient in the disease?

(A) Stimulus

(B) Signal release

(C) Signal reception

(D) Response

4. A liver cell begins the process of cell division but is unable to pass the DNA damage checkpoint. After attempts to correct the DNA damage that stopped the cell from passing the checkpoint, the cell fails to pass the checkpoint. Which will most likely occur next?

(A) The cell will necrotize.

(B) The cell will undergo apoptosis.

(C) The cell will return to G_1 and attempt to undergo cell division.

(D) The cell will bypass the S checkpoint and begin G_2.

Questions 5 and 6 refer to the following material.

In vertebrate animals, blood pressure is controlled by a negative feedback loop. Low blood pressure stimulates the kidneys to release an enzyme that leads to the production of angiotensin. Angiotensin causes arteries to constrict and eventually causes the kidneys to absorb water into the blood, which increases the volume of blood. Artery constriction and the increase of blood volume result in an increase in blood pressure, counteracting the low blood pressure.

5. In this pathway, which component is the sensor?

(A) Low blood pressure

(B) The kidneys

(C) Angiotensin

(D) The arteries

6. Which will result from the release of angiotensin?

(A) Continuous release of angiotensin

(B) Increase in blood pressure past the set point

(C) Negative feedback on the kidneys

(D) Maintenance of low blood pressure

7. Scientists develop a molecule that chemically mimics a signaling molecule of a signal transduction pathway. The new molecule differs from the signaling molecule in that it binds to the receptor but does not cause a conformational change in the receptor. Predict the effect of the newly developed chemical on the signal transduction pathway when the chemical is bound to the receptor.

(A) The pathway will exhibit 50% activation because the receptor while bound does not change its conformation.

(B) The pathway will be unaffected and will function as it normally does.

(C) The pathway will be inactivated.

(D) The pathway will be constantly activated.

8. Human cells have 46 chromosomes. If a cell undergoing mitosis fails to separate one of its sister chromatid pairs, how many chromosomes will each daughter cell have?

(A) One cell will have 46 chromosomes and the other will have 44 chromosomes.

(B) One cell will have 46 chromosomes and the other will have 48 chromosomes.

(C) One cell will have 47 chromosomes and the other will have 45 chromosomes.

(D) Both daughter cells will have 46 chromosomes.

9. Compared to the normal ligand, some agonists have prolonged binding with the receptor. Prolonged binding often results in a greater response than that which occurs with the normal ligand. Which of the following best explains why the normal ligand does not bind as well as the agonist?

(A) The normal ligand has fewer noncovalent interactions with the receptor than does the agonist.

(B) The normal ligand has more noncovalent interactions with the receptor than does the agonist.

(C) The normal ligand has more covalent bonds with the receptor than does the agonist.

(D) The normal ligand has fewer covalent bonds with the receptor than does the agonist.

Use the following information to answer questions 10–14.

Glucose level in the blood in humans is tightly regulated. The hormone insulin influences one of the mechanisms used by the body to regulate glucose uptake by cells. Researchers investigated the effects of insulin and glucose concentrations on glucose uptake from the blood in six healthy adults. The researchers prepared the subjects for the start of the experiment by purposely creating conditions in which the subjects had minimal or elevated concentrations of insulin, and normal or elevated concentrations of glucose in their blood. The subjects were separated into four groups: subjects with normal glucose levels and minimal insulin; subjects with elevated glucose levels and minimal insulin; subjects with normal glucose levels and elevated insulin; and subjects with elevated glucose levels and elevated insulin. After creating these four conditions in the subjects, the researchers measured the amount of glucose uptake in the subjects' whole body and in their muscles in particular. The results of these investigations are listed in the table below. Values are means ±2 times the standard errors of the mean ($\pm 2\ SE\bar{x}$).

	Conditions in subjects		Measurements taken	
	Blood glucose concentration (mg/dL)	Insulin concentration (µUnits/mL)	Glucose uptake in whole body (mg/min)	Glucose uptake in muscle (mg/min)
Group 1	88 ± 4 (normal glucose)	<4 (minimal insulin)	128 ± 12	17 ± 6
Group 2	201 ± 16 (elevated glucose)	<4 (minimal insulin)	213 ± 36	81 ± 24
Group 3	81 ± 2 (normal glucose)	80 ± 6 (elevated insulin)	536 ± 98	409 ± 244
Group 4	219 ± 18 (elevated glucose)	69 ± 6 (elevated insulin)	1218 ± 304	1395 ± 562

Data from: Baron, A. D., G. Brechtel, P. Wallace, and S. V. Edelman. 1988. "Rates and Tissue Sites of Non-Insulin- and Insulin-Mediated Glucose Uptake in Humans." *American Journal of Physiology* 255(6, Part 1):E769–E774.

10. The researchers hypothesized that glucose uptake by cells can increase in the presence of only minimal (<4 µUnits/mL) amounts of insulin if high concentrations of glucose are in the blood. Using the data in the table, where normal blood glucose concentration is 88 ± 4 mg/dL and elevated blood glucose concentration is 207 ± 16 mg/dL, which best describes the null hypothesis?

(A) When minimal amounts of insulin are present, increasing blood glucose concentrations from 88 ± 4 mg/dL to 207 ± 16 mg/dL will increase the uptake of glucose from the blood in the whole body.

(B) When minimal amounts of insulin are present, increasing blood glucose concentrations from 88 ± 4 mg/dL to 207 ± 16 mg/dL will not increase the uptake of glucose from the blood in the whole body.

(C) When minimal amounts of insulin are present, increasing blood glucose concentrations from 88 ± 4 mg/dL to 207 ± 16 mg/dL will not increase the uptake of glucose from the blood in muscles.

(D) When minimal amounts of insulin are present, increasing blood glucose concentrations from 88 ± 4 mg/dL to 207 ± 16 mg/dL will increase the uptake of glucose from the blood in muscles.

11. Among the subjects with minimal insulin, is the uptake of glucose by the whole body significantly different between subjects with elevated blood glucose concentrations (207 ± 16 mg/dL) and subjects with normal blood glucose concentrations (88 ± 4 mg/dL)?

(A) Yes, glucose uptake by the whole body is significantly different under these two conditions.

(B) No, glucose uptake by the whole body is not significantly different under these two conditions.

(C) There is insufficient data in the table to determine if there are significant differences in glucose uptake under these two conditions.

(D) It is impossible to determine if these means are different because the number of samples used to calculate the mean values is not stated.

12. Which describes a pattern in the data?

(A) Glucose uptake in the whole body is highest at elevated glucose concentrations.

(B) Glucose uptake in the whole body is highest at normal glucose concentrations.

(C) Glucose uptake in the whole body is lowest at elevated insulin concentrations.

(D) Glucose uptake in the whole body is highest at minimal insulin concentrations.

13. Using the data in the table, identify the conditions that resulted in the greatest uptake of glucose from the blood.

(A) Minimal insulin concentration in the presence of elevated glucose concentration resulted in the greatest uptake of glucose from the blood.

(B) Elevated insulin concentration in the presence of elevated glucose concentration resulted in the greatest uptake of glucose from the blood.

(C) Elevated insulin concentration in the presence of normal glucose concentration resulted in the greatest uptake of glucose from the blood.

(D) Minimal insulin concentration in the presence of normal glucose concentration resulted in the greatest uptake of glucose from the blood.

14. When glucose concentrations are elevated, what is the percentage change in mean glucose uptake by the whole body between the group with minimal insulin concentrations and the group with elevated insulin concentrations?

(A) 713%

(B) 613%

(C) 14%

(D) 7%

Section 2: Free-Response Questions

Write your answer to each part clearly. Support your answers with relevant information and examples. Where calculations are required, show your work.

1. Refer to the experiment described in Multiple-Choice Questions 10–14, as well as the data presented in the table on page 347.

   (a) **Create** a graph, including error bars, of glucose uptake by the muscles under the four conditions studied in the experiment.

   (b) A student claims that in the presence of minimal insulin, the muscle cells of the subjects with elevated glucose concentrations took up significantly more glucose than the muscle cells of the subjects with normal glucose concentrations. **Calculate** the ratio of glucose uptake in subjects with elevated glucose to glucose uptake in subjects with normal glucose concentrations. Use statistically valid data and analysis to either support or refute the student's claim.

   (c) **Determine** whether significantly more glucose was taken up by the muscle cells of subjects with elevated blood glucose in the presence of elevated insulin than the muscle cells of subjects with elevated blood glucose and minimal insulin. Citing the significance of the error bars of the graph, justify your answer.

   (d) **Make a claim** about the effect of insulin on the uptake of glucose by the muscles. Use data to support your claim.

2. When a cell's DNA is exposed to DNA-damaging agents or events, the resulting DNA damage is defined as genotoxic stress. Cells respond to genotoxic stress by either pausing the cell cycle and initiating mechanisms that repair damaged DNA, or, if the damage is too severe to fix, by triggering apoptosis. The p53 protein is vital to cells' responses to genotoxic stress, playing a role in both DNA repair and apoptosis. The figure below illustrates that the number of phosphate groups on p53 determines whether the cell pauses the cell cycle to repair its DNA, or triggers apoptosis.

   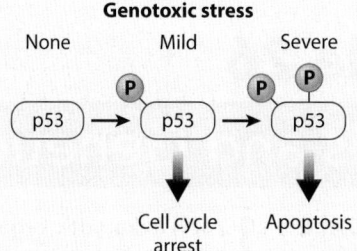

   (a) p53 may be unphosphorylated, phosphorylated partially with one phosphate group, or phosphorylated more completely, with two phosphate groups. **Explain** the role of unphosphorylated p53 in the cell.

   (b) Based on the model, **describe** what happens when mild genotoxic stress results in one phosphate group binding to p53 and what effects it has on the cell.

   (c) In the figure, when p53 has two phosphate groups attached to it, apoptosis results. **Describe** why this result occurs in a cell undergoing severe genotoxic stress.

Unit 5
Heredity

Group of young kittens with different phenotypes by Edoma/Getty Images

Module 26
Meiosis and Genetic Diversity

LEARNING GOALS ▸**LG 26.1** Eukaryotic organisms use sexual or asexual reproduction.
▸**LG 26.2** Meiosis produces four genetically unique daughter cells that are haploid.
▸**LG 26.3** Meiosis is the basis of sexual reproduction.

In the previous unit, we discussed mitotic cell division in eukaryotic cells. After chromosomes are duplicated during S phase of interphase, chromosomes are pulled apart as a parent cell divides into two new daughter cells during mitotic cell division. When this happens, each daughter cell ends up with the same number of chromosomes as the parent cell. Mitotic cell division is the way in which multicellular organisms grow, develop, and repair damaged tissues. Another challenge faced by organisms is how to reproduce. In this module, we will examine how plants, animals, and other eukaryotes use cell division for the purpose of reproduction. We will examine each step of this process to understand how a single cell containing two sets of chromosomes produces four daughter cells that each contain a single set of chromosomes. Based on this understanding, we will also explore how the process produces genetic variation.

> **PREP FOR THE AP® EXAM**
>
> **FOCUS ON THE BIG IDEAS**
>
> **INFORMATION STORAGE AND TRANSMISSION:** Look for events in the meiotic process that lead to variation among gametes.

26.1 Eukaryotic organisms use sexual or asexual reproduction

Reproduction happens when offspring are produced by inheriting copies of chromosomes from their parents. Each chromosome has many different genes, which are physical locations on chromosomes that code for proteins and influence particular traits. When we think about offspring inheriting genes, we can consider the **genotype,** which is the set of genes that an organism carries, and the **phenotype,** which is the set of traits that the organism expresses as a result of the genes it carries, such as hair color or eye color.

Reproduction by organisms can be either sexual or asexual. As we saw in Module 24, asexual reproduction occurs when an offspring inherits its DNA from a single parent. We saw an example of this in Unit 4 when we discussed binary fission in bacteria. One way that asexual reproduction can occur in plants and animals is when offspring are produced by nonsexual tissues of a parent. For example, many people obtain new houseplants by clipping a stem with a few leaves off an existing houseplant and then putting the stem into a glass of water. After several weeks, the stem grows roots, as shown in **FIGURE 26.1a**, and it can be put into a pot of soil.

a.

b.

FIGURE 26.1 Asexual reproduction

In asexual reproduction, offspring inherit their DNA from a single parent. (a) Asexual reproduction can occur when an offspring is produced from the nonsexual tissues of a parent, such as a clipping of a plant that grows new roots when the stem is cut from a parent plant. (b) Asexual reproduction can also occur when a single female produces embryos without any fertilization by another individual, which occurs in all-female species, such as this Tremblay's salamander. Photos: (a) Araddara/Shutterstock; (b) Breck P. Kent/AGE Fotostock

This new plant is genetically identical to the parent plant because it was produced from nonsexual structures. Asexual reproduction also occurs in some species where a parent creates an embryo without fertilization by another individual. In animals, species that reproduce asexually are typically all female. For example, the Tremblay's salamander (*Ambystoma tremblayi*), shown in Figure 26.1b, is an all-female salamander species that produces only female embryos.

Sexual reproduction occurs when an offspring inherits its DNA from two parents. This is the most common type of reproduction in plants and animals, with males and females combining their DNA to make offspring. Sexual reproduction is found in most amphibians, reptiles, birds, mammals, plants, and fungi. It is also a more energetically expensive strategy. For example, most sexually reproducing plants build colorful flowers that produce nectar to attract insects that pollinate flowers. Similarly, sexually reproducing animals commonly have elaborate courtship displays to attract mates. For example, some male mammals such as elk do not eat for weeks while they focus their attention on courting females. Similarly, many species of birds grow colorful feathers and spend weeks building elaborate nests and producing mating calls to try to attract a female to breed. Given these high costs of sexual

reproduction, why is this reproductive strategy so pervasive in plants and animals?

One of the major benefits of sexual reproduction is the creation of offspring with increased **genetic variation,** which is the range of different genotypes found among individuals. Given that environments typically vary over space and time, different genotypes produce a variety of phenotypes that may be favored in different environments. As a result, parents that produce offspring with greater genetic variation are more likely to have at least some offspring with phenotypes that are well suited to a wider range of different environments they might encounter in the future. Asexual reproduction produces little to no genetic variation in the offspring. Sexual reproduction can produce a large amount of genetic variation because the offspring receive different combinations of genes from two parents.

A fundamental challenge to sexual reproduction is that two parents need to combine their chromosomes without doubling the total number of chromosomes in each generation of offspring. Most plant and animal species are **diploid,** which means that their cells possess two sets of chromosomes (denoted as $2n$). For instance, humans are diploid because we have 23 pairs of chromosomes. When a cell contains a single set of chromosomes it is called

haploid (denoted as 1*n*). As you saw in Unit 4, mitosis starts with a diploid parent cell and produces diploid daughter cells. Thus, the fundamental challenge for sexual reproduction is how a diploid parent can produce haploid cells so that a haploid cell from one parent can combine with a haploid cell of another parent to produce a diploid offspring.

The solution to this challenge is a different type of cell division known as *meiosis*. In **meiosis**, cell division causes a diploid parental cell to divide in two stages to produce four haploid daughter cells. The four daughter cells are genetically different from each other and genetically different from the parent cell. In the next section, we will explore each step in the process of meiosis.

✓ **Concept Check**

1. **Describe** the relationship between genes and chromosomes.
2. **Identify** the difference between asexual and sexual reproduction.
3. **Describe** how sexual reproduction leads to greater genetic variation.

26.2 Meiosis produces four genetically unique daughter cells that are haploid

As we discussed in the previous section, mitotic cell division involves one cell division that produces two daughter cells that are diploid and genetically identical, while meiotic cell division involves two cell divisions that produce four daughter cells that are haploid and genetically unique. In this section, we will examine how DNA is arranged into chromosomes and then walk through the detailed steps of how these chromosomes are separated through the process of meiosis.

Homologous Chromosomes

As we have discussed, eukaryotic chromosomes are the structures that contain DNA. Each chromosome represents one DNA molecule, which contains two interwoven chains of nucleotides. Chromosomes are categorized as either *sex chromosomes* or *autosomes*. **Sex chromosomes** are chromosomes that determine whether an individual is male or female. In contrast, **autosomes** are all of the other chromosomes that are not sex chromosomes. Because most plants and animals are diploid, individuals have two sets of sex chromosomes and autosomes; one set came from the individual's mother and the other set came from the father. For example, the 46 chromosomes in human DNA include 22 pairs of autosomes and one pair of sex chromosomes. A pair of chromosomes that is similar in size and shape, and carry the same genes are known as **homologous chromosomes.** Thus, humans have 22 pairs of homologous autosomes. The sex chromosomes of female mammals are both *X* chromosomes, so they are also homologous chromosomes. However, the sex chromosomes of male mammals include one *X* and one *Y* chromosome, which are very different in size and different in the genes they contain, so they are not considered homologous chromosomes.

The distinctive sizes and shapes of each pair of chromosomes are visible when they are condensed during the early stages of mitosis or meiosis. When we display the condensed pairs of chromosomes visually, it is known as a **karyotype.** You can see an example of a human karyotype in **FIGURE 26.2**, which illustrates the 22 pairs of autosomes and one pair of sex chromosomes. Note that the chromosomes in the photograph did not originally line up so neatly on their own. Karyotypes are based on microscope images of chromosomes. The images have been edited to align each chromosome with its appropriate homologous match. Each of the 23 pairs of human chromosomes has been assigned a number and each chromosome has its own characteristic set of genes that it passes along. Note that the karyotype is from a female individual, so it has two *X* chromosomes.

An Overview of Meiosis

As noted, meiosis is a process that involves two stages of cell division to move from a diploid parent cell to four haploid

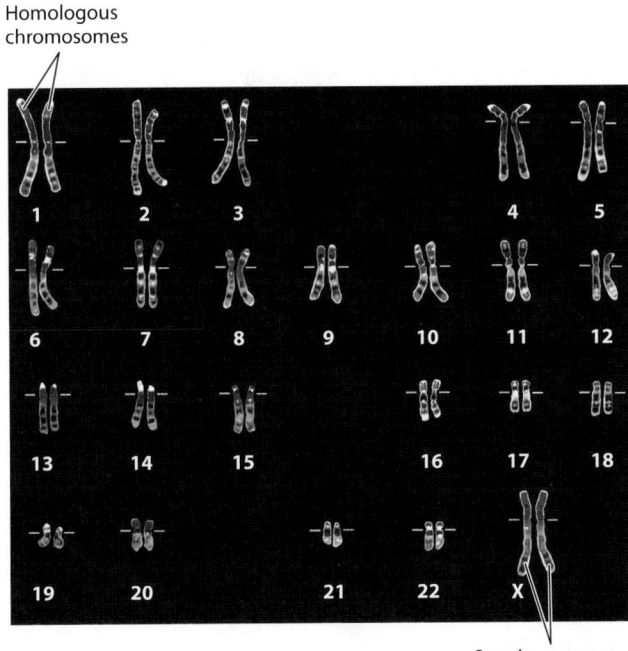

Homologous
chromosomes

Sex chromosomes

FIGURE 26.2 A human karyotype

When the chromosomes are condensed, we can identify
homologous pairs based on their size and shape. This image shows
the 23 pairs of chromosomes in a human, which includes 22 pairs
of autosomes and one pair of sex chromosomes. Photo: ISM/Sovereign/
Medical Images

daughter cells. **FIGURE 26.3** presents an overview of meiosis.
As you can see in the top half of the figure, **meiosis I** is
the first cell division in which we begin with chromosomes that have already been duplicated prior to the start
of meiosis I. The homologous chromosome pairs then line
up along a central line in the cell. The homologous chromosomes (and nonhomologous sex chromosomes in male
mammals) are then pulled to opposite sides of the parent
cell. Once the cell divides its cytoplasm by the process of
cytokinesis, the result is two daughter cells that each contains half as many chromosomes as the parent cell. Thus,
these daughter cells are haploid.

This is followed by **meiosis II,** which is the second cell
division shown in the bottom half of Figure 26.3. During
meiosis II, the chromosomes once again line up along a
central line in the cell. However, this time it is the sister
chromatids that are pulled to opposite sides of the parent
cell. This process closely resembles what happens in mitosis.

FIGURE 26.3 An overview of meiosis

Meiosis involves two distinct cell divisions, known as meiosis I and
meiosis II.

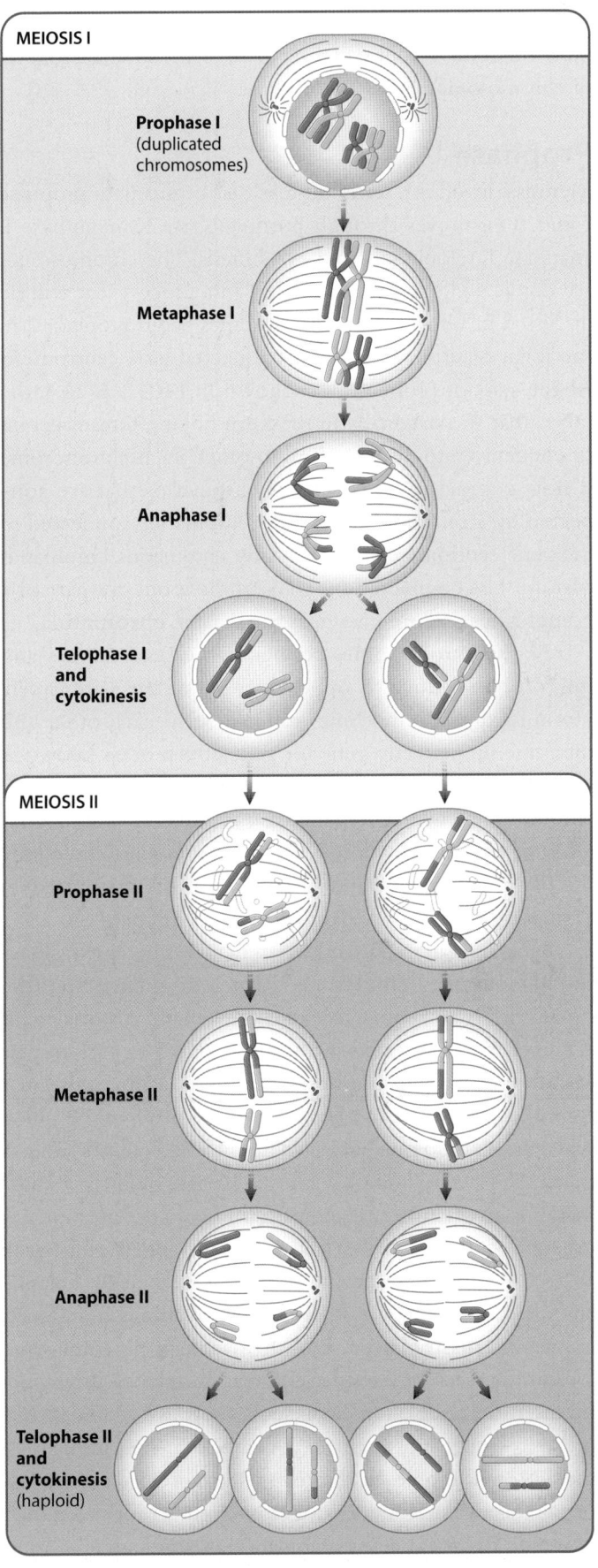

MEIOSIS I

Prophase I
(duplicated
chromosomes)

Metaphase I

Anaphase I

Telophase I
and
cytokinesis

MEIOSIS II

Prophase II

Metaphase II

Anaphase II

Telophase II
and
cytokinesis
(haploid)

Once the cell divides its cytoplasm by the process of cytokinesis, the result is four daughter cells, each with a haploid set of chromosomes.

Prophase I

Meiosis I involves a series of steps that begins with prophase I and then moves through prometaphase I, metaphase I, anaphase I, telophase I, and cytokinesis. The chromosomes are duplicated prior to prophase I in a process that will be discussed in Unit 6. The chromosomes begin prophase I in the form of sister chromatids connected by a centromere. At the start of prophase I, as shown in **FIGURE 26.4**, the DNA that was originally in the form of long threads begins to condense into distinct chromosomes. Each chromosome is now comprised of two sister chromatids that are connected by a centromere. Whereas chromatids connected to the same centromere are called sister chromatids, chromatids that are not connected by a centromere but are part of a homologous pair are known as **non-sister chromatids.**

As the chromosomes condense, the centrosomes and microtubules move to opposite poles of the parent cell. Homologous chromosomes move next to each other and then line up perfectly, gene for gene, in a process known as synapsis. The exception to this synapsis rule is the two sex chromosomes, which in mammals can either be homologous (*XX*) or nonhomologous (*XY*). If they are nonhomologous, the two sex chromosomes achieve synapsis by joining only at their tips, where they have a few genes in common.

Synapsis is an important step in meiosis I because it allows the homologous chromosomes, which originated from the organism's two parents, to exchange genetic material with each other, as illustrated in **FIGURE 26.5**. For a given pair of homologous chromosomes, the chromosome inherited from the individual's mother is called the maternal homolog while the chromosome inherited from the individual's father is called the paternal homolog. During synapsis, non-sister chromatids from the maternal and paternal homologs overlap each other in a cross-like pattern called a chiasma (plural, chiasmata). When this happens, each chromatid can break at the chiasma and then connect to the other non-sister chromatid, a process known as **crossing over.** Crossing over happens at the same location on each of the sister chromatids, so there is typically no insertion or deletion of nucleotides—the building blocks of DNA—after crossing over is completed. This means that the DNA chain on each chromatid does not experience a loss or gain of any nucleotides. Given that crossing over causes each homolog to have a new combination of genes coming from both parents, we say that there has been **recombination** of

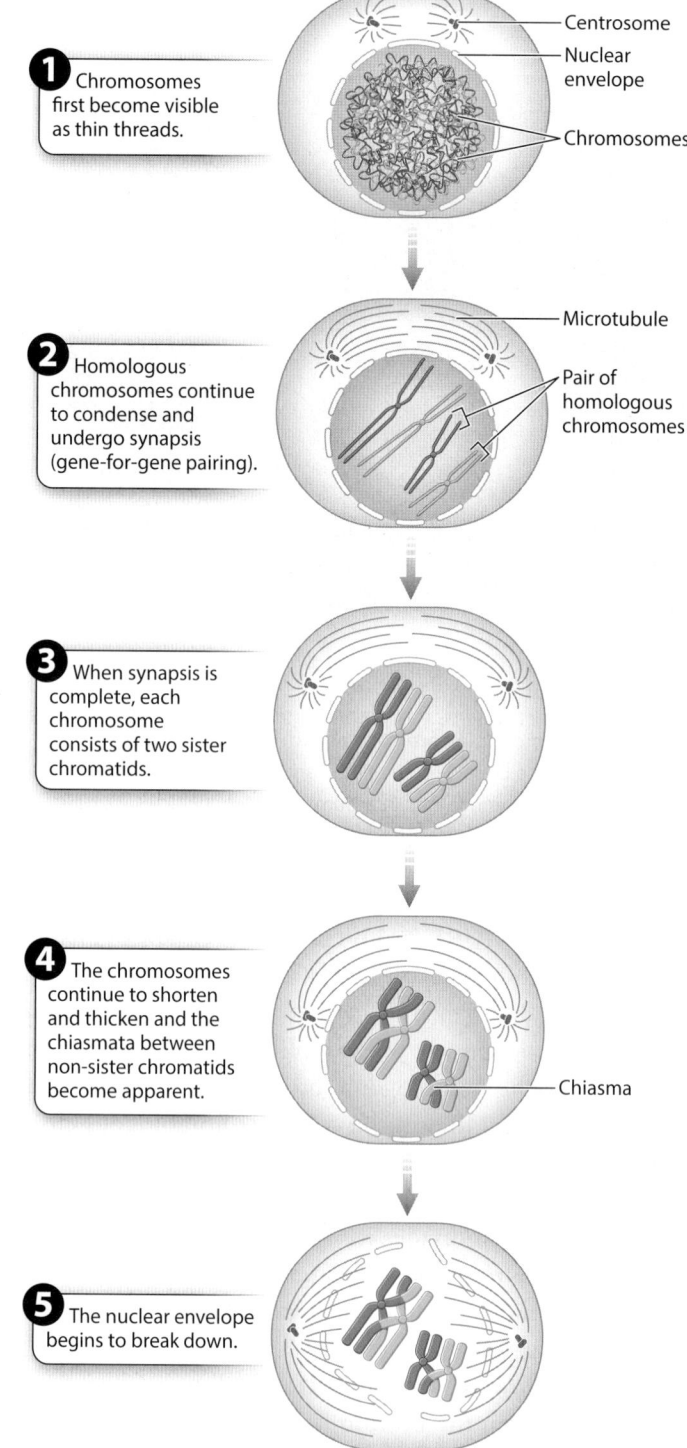

1 Chromosomes first become visible as thin threads.

Centrosome

Nuclear envelope

Chromosomes

2 Homologous chromosomes continue to condense and undergo synapsis (gene-for-gene pairing).

Microtubule

Pair of homologous chromosomes

3 When synapsis is complete, each chromosome consists of two sister chromatids.

4 The chromosomes continue to shorten and thicken and the chiasmata between non-sister chromatids become apparent.

Chiasma

5 The nuclear envelope begins to break down.

FIGURE 26.4 Prophase I

During prophase I, the duplicated chromosomes condense. As the centromeres migrate to opposite poles of the cells with microtubules attached, the homologous chromosomes, each comprised of two sister chromatids, move next to each other in a process called synapsis. During synapsis, homologous chromosomes can cross over at sites known as chiasmata and pieces of chromosomes can be exchanged. As the chromosomes continue to shorten and thicken, the nuclear envelop begins to break down.

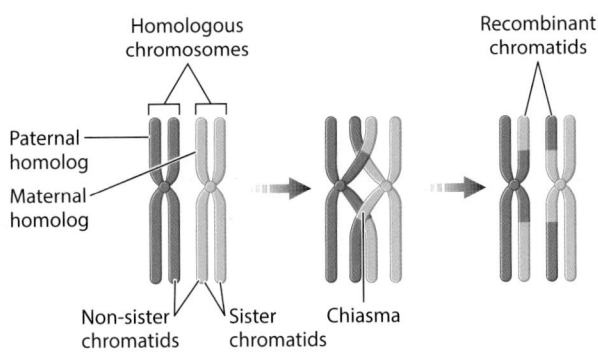

FIGURE 26.5 Crossing over

During prophase I, homologous chromosomes are in close proximity. The maternal homolog and paternal homolog can overlap at points known as chiasmata. When this happens, genetic material can be exchanged between non-sister chromatids in a process known as crossing over. After crossing over has occurred, each chromatid has a new combination of genes, so they are now referred to as recombinant chromatids.

the genes. Chromatids that have experienced crossing over are subsequently referred to as recombinant chromatids.

Crossing over can bring new combinations of genes together in the haploid cells that will ultimately become the next generation of offspring. As a result, the next generation of offspring can have greater genetic diversity and therefore express a greater range of different phenotypes. As we have noted, if the two sex chromosomes are nonhomologous, they only have a few genes in common which line up during synapsis. At this location, the sex chromosomes can also form a chiasma and experience crossing over, but this is rare and such an event typically leads to sexual development disorders. The number of chiasmata, and therefore the amount of crossing over, varies among species.

The Subsequent Steps of Meiosis I

Prophase I is the first stage of meiosis I and it is followed by several subsequent stages that cause the first cell division, as illustrated in **FIGURE 26.6**. At the end of prophase I, the nuclear envelope begins to break down and the meiotic spindles form on opposite sides of the parental cell. As noted in Module 24, these spindles are made up of microtubules that emerge from opposite poles in the vicinity of the migrating centrosomes.

The next step is prometaphase I, which can be seen as the second step in Figure 26.6. During prometaphase I, the nuclear envelope is completely broken down and the meiotic spindles attach to the kinetochores of each chromosome. Similar to what we saw in mitosis, the meiotic spindles move the chromosomes to the middle of the cell.

Next is metaphase I, which is step 3 in Figure 26.6. During metaphase I, the meiotic spindles cause the homologous pairs of

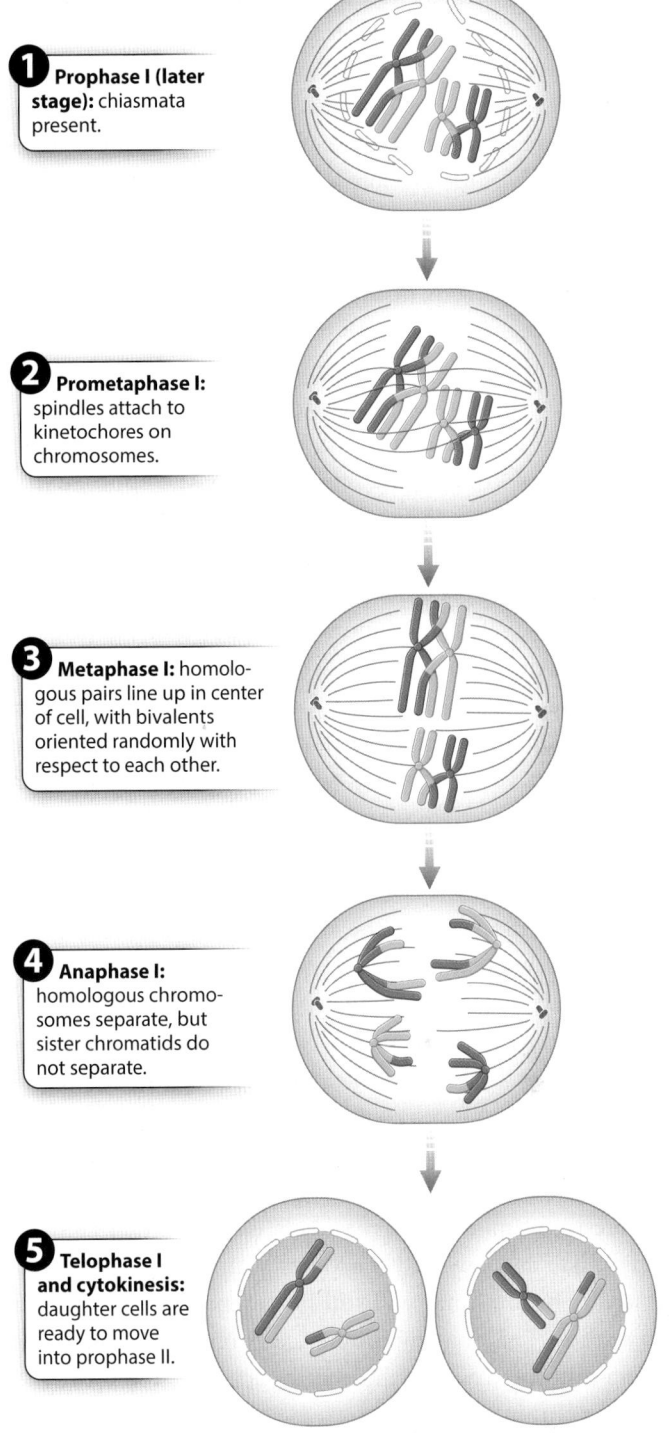

1 **Prophase I (later stage):** chiasmata present.

2 **Prometaphase I:** spindles attach to kinetochores on chromosomes.

3 **Metaphase I:** homologous pairs line up in center of cell, with bivalents oriented randomly with respect to each other.

4 **Anaphase I:** homologous chromosomes separate, but sister chromatids do not separate.

5 **Telophase I and cytokinesis:** daughter cells are ready to move into prophase II.

FIGURE 26.6 Meiosis I

Meiosis I begins with a diploid parent cell. As chromosomes condense, homologous chromosomes are moved to be next to each other in a synapsis that allows crossing over. The nuclear envelope then breaks down, and the meiotic spindles move the homologous chromosomes to line up in the middle of the cell. The meiotic spindles pull the homologous chromosomes apart and then the parent cell divides in half, thereby creating two haploid daughter cells.

chromosomes to line up along an imaginary line down the middle of the cell, which is the metaphase plate. When comparing the steps in mitosis versus meiosis, you might notice that during metaphase of mitosis, each chromosome straddles the metaphase plate. However, during metaphase I of meiosis, each *pair* of chromosomes straddles the metaphase plate. As you will see, this difference in the orientation of the chromosomes in meiosis I is what ultimately causes the daughter cells to become haploid.

Note that there is no specific orientation of the chromosomes that originally came from each parent. For a given pair of homologous chromosomes, the maternal chromosome may be on the left side of the metaphase plate while the paternal chromosome may be on the right side. For another pair of homologous chromosomes, the opposite orientation may occur. As a result, when the pairs of homologous chromosomes get pulled apart in the next stage, each daughter cell will have a mix of maternal and paternal chromosomes. The random segregation of the homologous chromosomes means that the maternal and paternal chromosomes received by each daughter cell is also random. Therefore, each haploid cell that combines during sexual reproduction, the **gametes,** will likely carry a mixture of maternal and paternal genes from the original parent cell. This mixing allows new combinations of genes to appear in each cell, which can allow new phenotypes to be expressed. We will talk more about the impacts of increased genetic variation in the next module.

The next step is anaphase I, which is step 4 in Figure 26.6. During anaphase I, each pair of homologous chromosomes is pulled apart by the meiotic spindles and each chromosome in the pair goes to opposite poles. Because the spindle fibers from one side of the cell are attached to both kinetochores of a given chromosome, the chromosome's centromere is not split apart. As a result, each daughter cell will contain half as many chromosomes at the end of meiosis I, but each chromosome will still be comprised of two sister chromatids. This is a key difference compared to anaphase during mitosis, in which the centromeres are split apart and the sister chromatids separate.

The final step is telophase I and cytokinesis, which is shown as step 5 in Figure 26.6. During this step, the chromosomes become a bit less condensed, the cell cytoplasm divides to create two daughter cells by cytokinesis, and a nuclear envelope briefly reappears around the chromosomes. Each daughter cell now contains half the number of chromosomes, so each cell is haploid. However, because of crossing over and the random alignment of maternal and paternal chromosomes during metaphase I, the two daughter cells are not genetically identical. At this point, the cells are ready to undergo meiosis II.

Meiosis II

In meiosis II, the two haploid daughter cells produced by meiosis I further divide into four haploid daughter cells, as shown in **FIGURE 26.7**. The steps in meiosis II are similar to the steps in mitosis, in that sister chromatids separate. In meiosis II, however, there are half as many chromosomes to split in half, since the two daughter cells begin meiosis II as haploid cells.

The process begins with prophase II, in which the chromosomes condense, the nuclear envelope breaks down, and the meiotic spindles begin to form. This is very similar to prophase I. In prometaphase II, the meiotic spindles from each side of the cell attach to the kinetochores on opposite sides of each chromosome. This is different from prometaphase I, where meiotic spindles attached to the kinetochores on both sides of each chromosome. This difference is what allows the centromere to break apart in anaphase II and allows sister chromatids to be pulled in opposite directions. During metaphase II, the chromosomes are moved by the meiotic spindles to line up along the metaphase plate in the middle of the cell. In anaphase II, the sister chromatids are pulled apart by the meiotic spindles. Finally, in telophase II, the cytoplasm of the two haploid cells from meiosis I is divided by cytokinesis and the nuclear envelope returns to produce four haploid cells.

Comparing Mitosis and Meiosis

The processes of mitosis and meiosis have a number of similarities and differences. As illustrated in **FIGURE 26.8** on page 358, mitosis involves a single cell division, whereas meiosis involves two cell divisions. In addition, mitosis begins and ends with diploid cells and there is no crossing over between homologous chromosomes. In fact, homologous chromosomes do not even closely associate with each other in mitosis.

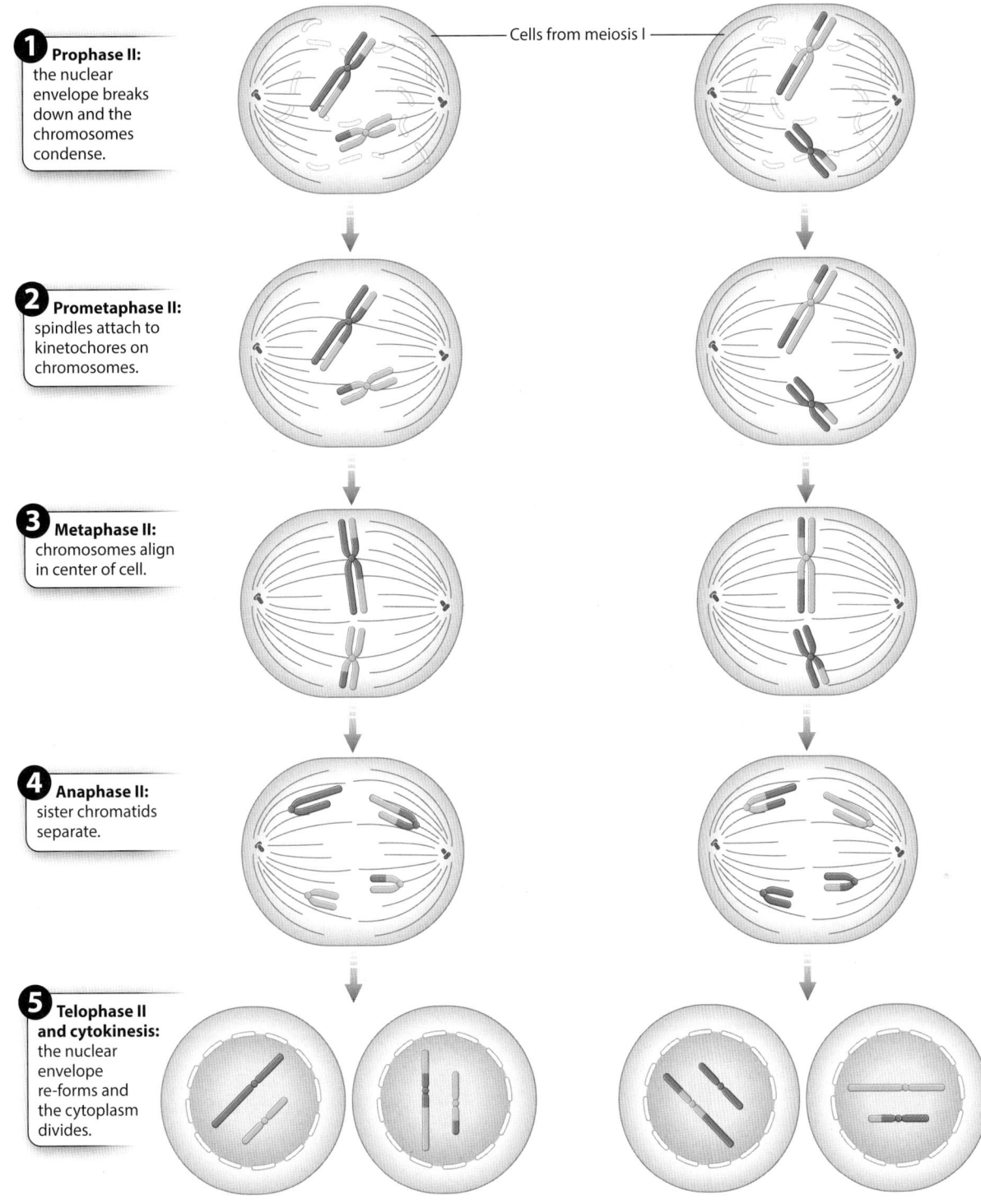

1 **Prophase II:** the nuclear envelope breaks down and the chromosomes condense.

2 **Prometaphase II:** spindles attach to kinetochores on chromosomes.

3 **Metaphase II:** chromosomes align in center of cell.

4 **Anaphase II:** sister chromatids separate.

5 **Telophase II and cytokinesis:** the nuclear envelope re-forms and the cytoplasm divides.

Cells from meiosis I

FIGURE 26.7 Meiosis II

Meiosis II begins with the haploid daughter cells produced during meiosis I. During prophase II, the cells experience a breakdown of their nuclear envelope and the chromosomes condense. During prometaphase II, the meiotic spindles attach to the kinetochores on opposite sides of each chromosome's centromere. In metaphase II, the meiotic spindles move the chromosomes to align along the metaphase plate. During anaphase II, the meiotic spindles pull the sister chromatids to opposite poles of the cell. In the final step, the cytoplasm divides by cytokinesis and the nuclear envelope forms again. The result is four genetically distinct haploid daughter cells.

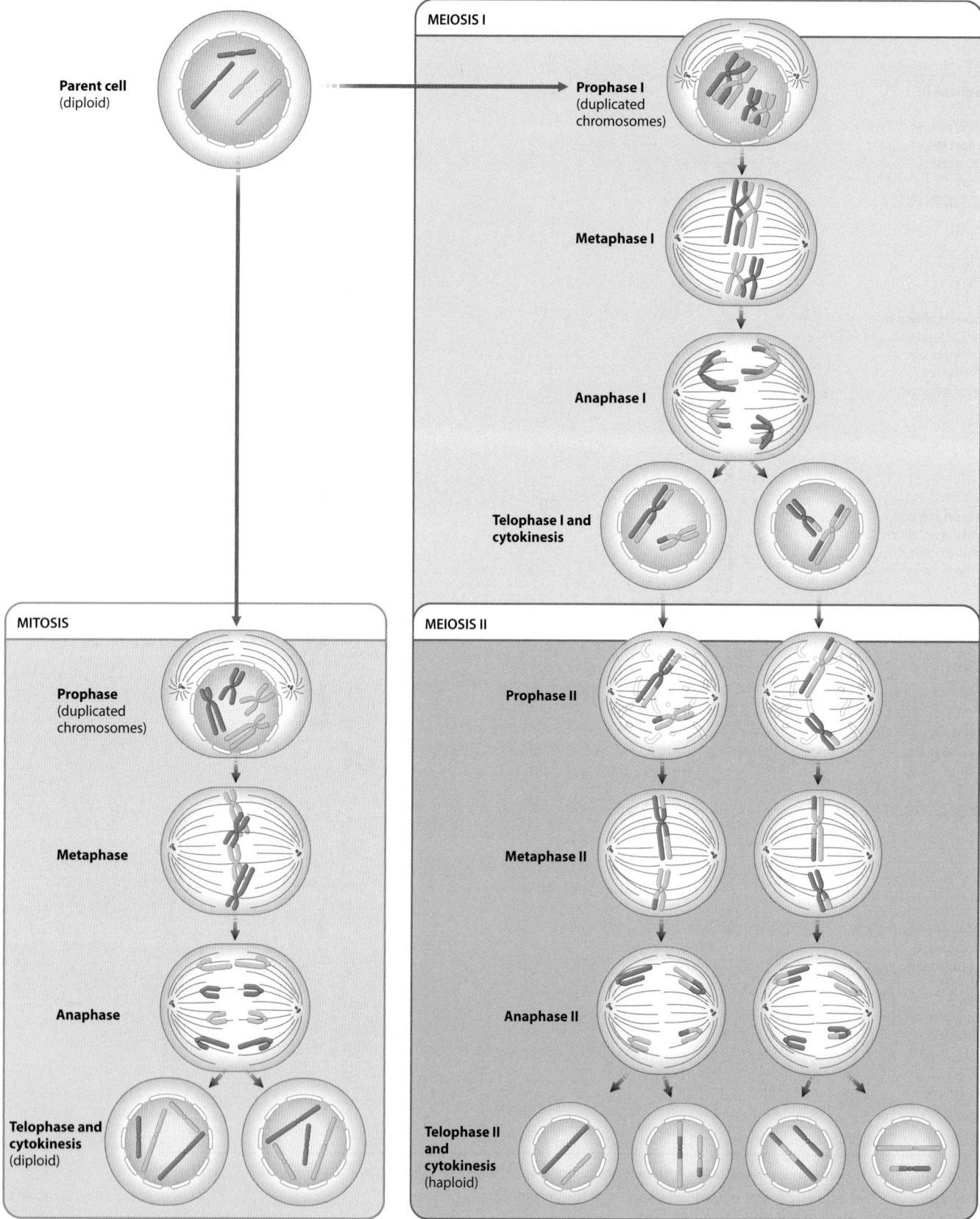

FIGURE 26.8 Comparing mitosis and meiosis

The processes of mitosis and meiosis have many similarities, but they also have some key differences. In mitosis, sister chromatids are pulled apart, producing two daughter cells that are diploid and genetically identical. In meiosis I, DNA is also duplicated, but the process begins with the synapsis of homologous chromosomes and crossing over. The homologous chromosomes are then separated to produce two haploid daughter cells. In mitosis and meiosis II, the sister chromatids are pulled apart. While mitosis produces two daughter cells that are genetically identical, meiosis produces four daughter cells that are genetically unique.

In contrast, meiosis begins with diploid cells and ends with haploid cells. In meiosis homologous chromosomes are so closely associated that they experience crossing over, which creates new genetic combinations in the daughter cells. Moreover, the random positions of maternal and paternal homologs along the metaphase plate causes a mixing of maternal and paternal chromosomes in each of the daughter cells. This provides even more variation in the genes carried by each daughter cell compared to the parent cell. As noted earlier, these new combinations can produce changes in the phenotypes in the offspring, which can be very beneficial when the offspring may experience a range of different environmental conditions. There are also many similarities between mitosis and meiosis II, including the chromosomes aligning along the metaphase plate and the sister chromatids being pulled apart to opposite poles of the cell. **TABLE 26.1** provides an additional comparison of mitosis and meiosis.

AP® EXAM TIP

You should understand the difference between mitosis and meiosis and be able to describe each process.

✓ Concept Check

4. **Describe** how meiosis I causes a mixing of the maternal and paternal genes.

5. **Identify** the two processes that cause increased genetic variation during meiosis.

6. **Describe** the fundamental differences in the final outcome of cells that experience mitosis versus those that experience meiosis.

TABLE 26.1 Comparison of Mitosis and Meiosis

	Mitosis	Meiosis
Function	Asexual reproduction in unicellular eukaryotes Development in multicellular eukaryotes Tissue regeneration and repair in multicellular eukaryotes	Sexual reproduction Production of gametes and spores
Organisms	All eukaryotes	Most eukaryotes
Number of rounds of DNA synthesis	1	1
Number of cell divisions	1	2
Number of daughter cells	2	4
Chromosome complement of daughter cell compared to parent cell	Same	Half
Pairing of homologous chromosomes	No	Meiosis I: yes Meiosis II: no
Crossing over	No	Meiosis I: yes Meiosis II: no
Separation of homologous chromosomes	No	Meiosis I: yes Meiosis II: no
Centromere splitting	Yes	Meiosis I: no Meiosis II: yes
Separation of sister chromatids	Yes	Meiosis I: no Meiosis II: yes

26.3 Meiosis is the basis of sexual reproduction

As we have discussed, the process of meiosis results in diploid parental cells becoming haploid daughter cells. This is an important process for passing on genes to the next generation by sexual reproduction because it allows an individual to combine its own chromosomes with the chromosomes of another individual, without increasing the total number of chromosomes in the next generation. As the haploid cells produced by meiosis mature in animals, they become gametes, which are the male or female cells that combine during sexual reproduction. When two gametes fuse, they become a diploid **zygote.** When the gametes fuse, they become a diploid zygote. In this section, we will discuss how meiosis produces four genetically unique daughter cells, and how it can also produce daughter cells of very different sizes.

Sex difference in the division of the cytoplasm

In many eukaryotic organisms the size of cells produced by meiosis can be very different. In animals, for example, females produce relatively large eggs while males produce small sperm. These size differences are the result of how the cytoplasm is divided during meiosis, as shown in **FIGURE 26.9a**. In female animals, both daughter cells have a haploid number of chromosomes at the end of meiosis I, but one of those daughter cells receives much more cytoplasm, so it is a much larger cell. During meiosis II, the smaller cell gives rise to two new small cells, while the larger cell gives rise to one large cell and one small cell. By the end of meiosis, the female has produced a single large haploid cell, called an oocyte, and three small haploid cells, which are called polar bodies. In most animals, the oocyte will subsequently develop into the female egg while the polar bodies typically disintegrate within a day.

In contrast to what happens in female animals, meiosis in male animals divides up the cytoplasm equally. As a result, the male's four haploid daughter cells are of similar size, as you can see in Figure 26.9b. As these cells further develop into sperm cells, most of the cytoplasm is eliminated, which allows the sperm cells to be very small compared to the egg cell. The tiny sperm cells contain a head that carries the DNA inside the nucleus and a long tail to help it swim toward the egg cell.

Sexual Reproduction and Genetic Diversity

As mentioned earlier in this module, the process of meiosis is essential for sexual reproduction because it allows individuals to produce haploid cells that can be fertilized by another haploid cell, thereby forming a diploid zygote. In humans, for example, cells not involved in reproduction are diploid and contain 46 chromosomes; 23 of them come from the mother and the other 23 come from the father. As a result of meiosis, human gametes are haploid so they contain just one set of 23 chromosomes, including one of the sex chromosomes. When the egg is fertilized by a sperm, the resulting zygote has the full set of 46 chromosomes.

The process of meiosis to create haploid gametes followed by the fertilization of one individual's gamete by another individual's gamete can create a great deal of genetic diversity. In meiosis I, the crossing over of homologous chromosomes during synapsis causes genetic material to be exchanged between the maternal and paternal chromosomes. In addition, random segregation of the homologous chromosomes into two daughter cells in meiosis I means that daughter cells are genetically unique from each other. Finally, fertilization can bring even more new genetic combinations together by pooling the genetics of two individuals. All of these new genetic combinations increase the likelihood that at least some of the offspring will have genotypes and phenotypes that are well suited to current and changing environmental conditions compared to species that use asexual reproduction. For example, when an oak tree makes thousands of offspring in the form of acorns each summer and fall, each offspring carries a unique genotype as a result of meiosis that produced the male gamete, meiosis that produced the female egg, and the coming together of the male and female genes when the fertilized embryo was produced. The thousands of acorns containing different genotypes ensures that at least some of the offspring will be well suited to the environmental conditions that are present when the acorns germinate the following spring.

✓ Concept Check

7. **Describe** how meiosis leads to daughter cells of two different sizes in female mammals.

8. **Describe** how fertilization contributes to genetic variation.

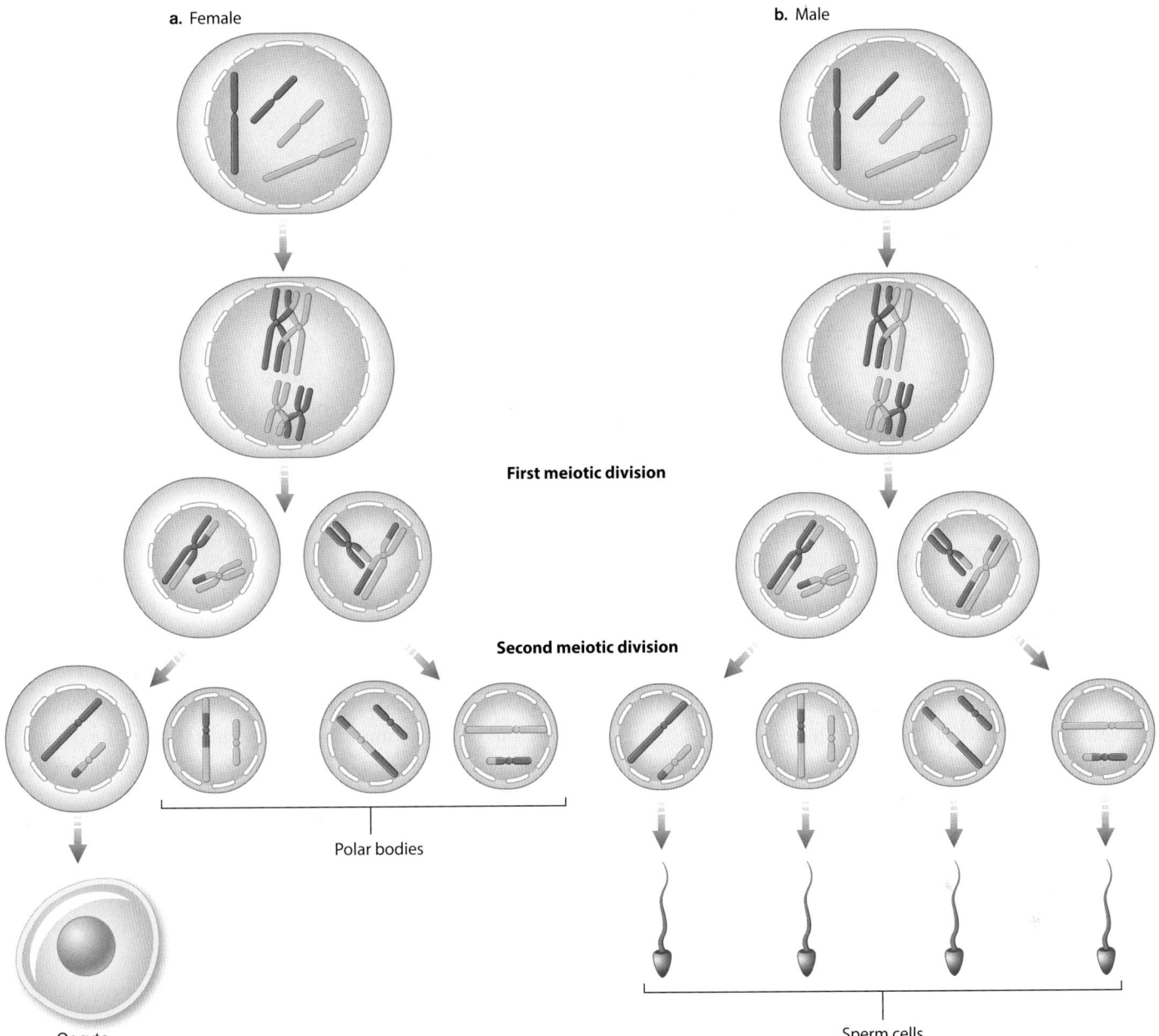

a. Female
b. Male

First meiotic division

Second meiotic division

Polar bodies

Oocyte

Sperm cells

FIGURE 26.9 Unequal allocation of cytoplasm during meiosis in female and male animals

Female animals produce haploid cells of unequal sizes, while male animals produce haploid cells of equal sizes. (a) In female mammals, there is unequal division of the cytoplasm during both meiosis I and meiosis II. As a result, there is a single large haploid cell, known as the oocyte, and three small haploid cells, known as the polar bodies. (b) In male mammals, there is equal division of the cytoplasm during both meiosis I and meiosis II. As a result, the four daughter cells are of equal size.

Module 26 Summary

REVISIT THE BIG IDEAS

INFORMATION STORAGE AND TRANSMISSION: Using the content of this module, **describe** two events that lead to variation among gametes during the process of meiosis.

LG 26.1 Eukaryotic organisms use sexual or asexual reproduction.

- Each chromosome has many different genes, which are physical locations on chromosomes that code for particular traits. Page 350

- A genotype is the set of genes that an organism carries, whereas the phenotype is the set of traits that the organism expresses, such as hair color or eye color. Page 350

- Reproduction by organisms can be sexual or asexual. Page 351

- Asexual reproduction occurs when an offspring inherits its DNA from a single parent, such as reproduction from nonsexual tissues of a plant, or when a parent can create an embryo without fertilization by another individual. Page 351

- Sexual reproduction occurs when an offspring inherits its DNA from two parents; this is the most common type of reproduction in plants and animals. Page 351

- Sexual reproduction is more costly than asexual reproduction, but it provides more genetic variation in the offspring, which is beneficial when environments vary over space and time. Page 351

- Most plant and animal species are diploid, which means that their cells possess two sets of chromosomes (denoted as $2n$). Page 351

- The fundamental challenge for sexual reproduction is how a diploid parent can produce haploid cells, so that a haploid cell from one parent can combine with a haploid cell of another parent to produce a diploid offspring. Page 352

- In meiosis, cell division causes a diploid parental cell to divide in two stages to produce four haploid daughter cells. Page 352

LG 26.2 Meiosis produces four genetically unique daughter cells that are haploid.

- Chromosomes can be categorized as either sex chromosomes or autosomes. Page 352

- Two chromosomes that are similar in size and shape, and carry the same genes are known as homologous chromosomes. Page 352

- A karyotype is a visual display of condensed pairs of chromosomes. Page 352

- Meiosis is a process that involves two stages of cell division to move from a diploid parent cell to four haploid daughter cells. Page 353

- Meiosis I produces two daughter cells, each containing half as many chromosomes as the parent cell. Page 353

- Synapsis, which occurs in in meiosis I, allows the homologous chromosomes, which originated from the organism's two parents, to exchange genetic material with each other by crossing over. Page 354

- Meiosis II produces four daughter cells that contain a haploid set of chromosomes. Page 356

LG 26.3 Meiosis is the basis of sexual reproduction.

- In animals, meiosis produces gametes that can be very different in size as a result of a difference in how much cytoplasm is allocated to male and female gametes. Page 360

- Meiosis in female animals produces a single large oocyte and three small polar bodies. Page 360

- Meiosis in male animals produces four similarly sized sperm. Page 360

- The process of meiosis to create haploid gametes followed by the fertilization of one individual's gamete by another individual's gamete can create a great deal of genetic diversity. Page 360

Key Terms

Genotype	Meiosis	Meiosis II
Phenotype	Sex chromosomes	Non-sister chromatids
Sexual reproduction	Autosomes	Crossing over
Genetic variation	Homologous chromosomes	Recombination
Diploid	Karyotype	Gametes
Haploid	Meiosis I	Zygote

Review Questions

1. Which step does not occur in meiosis I?

    (A) Sister chromatids are pulled apart by meiotic spindles.

    (B) Cytoplasm divides.

    (C) Chromatids cross over during prophase I.

    (D) Homologous chromosomes align in the center of the cell.

2. Which statement is true about meiosis II?

    (A) Diploid cells divide to become haploid cells.

    (B) Homologous chromosomes experience the process of synapse.

    (C) Meiotic spindles attach to kinetochores and cause sister chromatids to separate.

    (D) Homologous chromosomes separate to opposite side of the parental cell.

3. Which process does not lead to new genetic combinations during meiosis?

    (A) Crossing over

    (B) Random segregation of homologous chromosomes

    (C) Recombination

    (D) Fertilization

4. If the diploid number of chromosomes in a dog is 78, how many chromosomes are present in a single cell during prophase I, prophase II, telophase II, and in a zygote, respectively?

    (A) 39, 39, 78, 78

    (B) 39, 39, 78, 156

    (C) 78, 39, 156, 78

    (D) 78, 39, 39, 78

5. Which of the following processes is not unique to meiosis compared to mitosis?

    (A) Homologous chromosomes undergoing synapse

    (B) Spindles attaching to opposite sides of a centromere

    (C) Production of haploid cells

    (D) Crossing over

Module 26
AP® Practice Questions

Section 1: Multiple-Choice Questions

Choose the best answer for questions 1–4.

1. Which statement best describes the difference between sexual and asexual reproduction?

    (A) Sexual reproduction can generate substantial genetic diversity in offspring, whereas asexual reproduction generates offspring that are generally identical to the parents.

    (B) Sexual reproduction depends on mitosis to generate gametes, whereas asexual reproduction can occur through cell division.

    (C) Sexual reproduction produces offspring that are well-suited to their original environment, while asexual reproduction produces offspring that may be well-suited to a variety of environments.

    (D) Sexual reproduction allows for rapid population growth, whereas asexual reproduction results in slow population growth.

Use the following illustrations of a diploid parental cell and daughter cells to answer question 2.

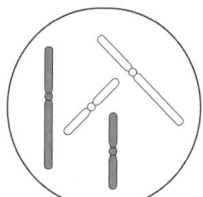

Parental cell
2n = 4

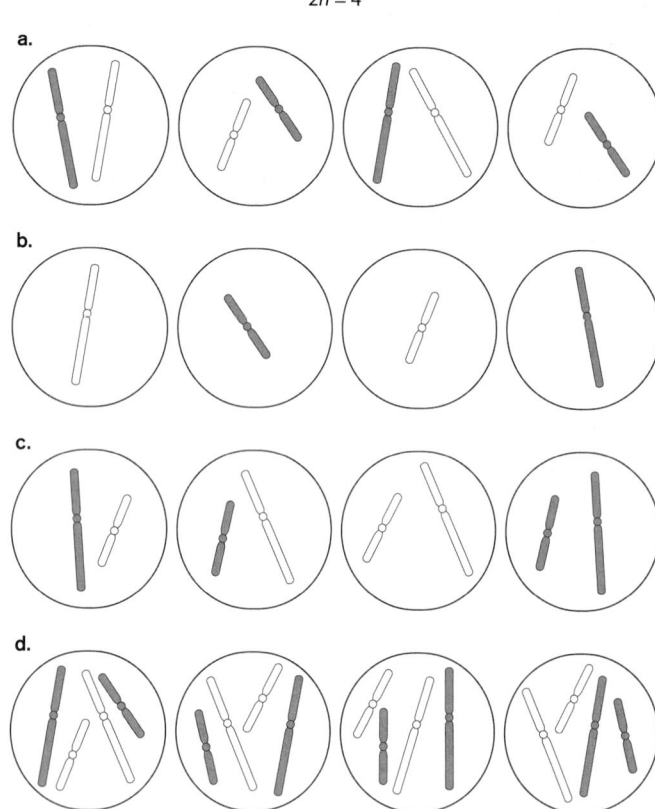

a.

b.

c.

d.

2. Which accurately illustrates a possible outcome of meiosis in an organism that has a diploid chromosome number equal to 4?

   (A) a   (C) c

   (B) b   (D) d

Use the following figure to answer question 3.

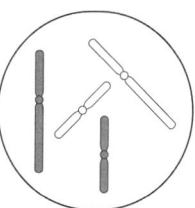

Parental cell
2n = 4

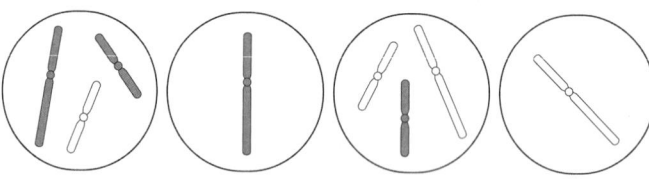

Gametes produced

3. Occasional errors in meiosis result in gametes that contain either too many or too few chromosomes. Embryos that are generated from these gametes typically do not survive. The figure illustrates gametes arising from one possible error in meiosis. Identify the most likely step at which the error in meiosis occurred.

   (A) Anaphase I

   (B) Telophase I

   (C) Metaphase II

   (D) Anaphase II

4. The painted turtle (*Chrysemys picta*) is a diploid species with 50 chromosomes in its nonreproductive cells. Assuming no mutations, what is the total number of chromosomes present in a single cell after prophase I, after prophase II, after telophase II, and after fertilization, respectively?

(A) 50, 100, 50, 50

(B) 100, 50, 25, 50

(C) 50, 25, 25, 50

(D) 100, 100, 50, 100

Section 2: Free-Response Question

Write your answer to each part clearly. Support your answers with relevant information and examples. Where calculations are required, show your work.

A scientist claims that sexual reproduction increases genetic variation that could be favored in different environments. Answer the following questions related to this claim.

(a) **Describe** the difference between the genotype and the phenotype of an organism.

(b) **Identify** two ways in which sexual reproduction increases genetic diversity within species.

(c) **Provide reasoning** to explain why increased genetic diversity might increase the ability of a species to survive in a changing environment.

Module 27

Mendelian Genetics

LEARNING GOALS ▶**LG 27.1** Organisms are linked by lines of descent due to common ancestry.

▶**LG 27.2** The segregation of homologous chromosomes during meiosis is the basis for the segregation of alleles.

▶**LG 27.3** Independent assortment during meiosis causes independent assortment of genes on different chromosomes.

▶**LG 27.4** We can visualize inheritance patterns using pedigrees.

You have probably noticed that you share some of the physical characteristics of your parents and siblings, such as hair color, eye color, and perhaps even height. One reason for these shared traits is that you have inherited genes from your parents and therefore have many genes in common with them. Similarly, your parents inherited their genes from their parents. Going back even more generations in a family tree, you might still notice some features that resemble your own because you are carrying copies of your ancestor's genes. In fact, every organism inherits genes from its ancestors. How does the inheritance of these genes determine the phenotypes that are produced by an individual? The answers are not only interesting but also relevant in a number of applications. For example, they help us understand how to breed species for different desirable characteristics, including domesticated animals and varieties of crop plants. As we will see, the earliest insights on the inheritance of genes came from the studies conducted by a nineteenth-century monk in Europe who studied pea plants. These early studies set the stage to inform our later knowledge of meiosis and the different combinations of genes that can be produced during sexual reproduction.

In Module 26, we discussed the process of meiosis and we learned how chromosomes are divided when an organism is making haploid gametes for reproduction. In this module, we will see how the events in meiosis explain patterns of gene inheritance from parents to their offspring. We will begin by examining the concept of common ancestry and how early research on plant breeding identified the way that traits are inherited from parents to offspring. We will then discuss how the process of meiosis results in chromosomes segregating and independently sorting. Finally, we will see how we can examine family trees to infer patterns of inheritance.

PREP FOR THE AP® EXAM

FOCUS ON THE BIG IDEAS

INFORMATION STORAGE AND TRANSMISSION: Look for connections between the laws that govern the movement of chromosomes in the meiotic process and the patterns of gene inheritance.

27.1 Organisms are linked by lines of descent due to common ancestry

In this section, we examine the evidence that many distantly related modern species have a common ancestor and we discuss how nineteenth-century plant-breeding experiments first discovered how offspring inherit traits from their parents.

Evidence of Common Ancestry

While it can seem daunting to think about a common ancestor for all living organisms, there are many examples

of a common ancestor within particular groups. For example, we know from research on DNA that all breeds of dogs have been produced by the domestication of wolves. For dog breeds developed over the past 200 years, we even have a paper trail of how these breeds were produced, based on dog-breeding records. Similarly, we know that the original corn plant was a very short species of grass that, over time, was bred by humans to be much taller with much larger ears of corn. In these examples, researchers have been able to trace common ancestry. However, what is the evidence that all species on Earth have a common ancestor?

To answer this question, we need to look at molecules and processes that are fundamental to all life and are found in every modern species. Such molecules and processes are said to be conserved, in that they are still retained in modern species. Conserved molecules and processes would be evidence that they were originally present in a common ancestor. In Unit 2, for example, we discussed the fact that all organisms are composed of one or more cells that are separated from their environment by a cell membrane. In addition, all organisms contain ribosomes, which synthesize proteins. We also learned that many essential metabolic pathways—including cellular respiration, photosynthesis, cell signaling, and cell division—are found across a wide range of distantly related organisms. When we examine the genetic material, we find that DNA is the genetic material for all cells. Moreover, the structure of these molecules and how they function within a cell is similar across all species. The fact that all of these features are common across a wide diversity of species suggests that these features were likely inherited from an ancient common ancestor.

Mendel's Crosses of Pea Plants

Our understanding that common ancestors gave rise to new varieties of plants and animals has long motivated humans to breed plants and animals to enhance the food that we eat and the work that we do. For example, Native Americans bred corn plants that over time yielded taller plants and bigger ears. People living in Eurasia bred horses with temperaments suitable for riding and pulling carts. Throughout human history people understood that characteristics found in adult organisms could be passed on to their offspring. People did not know about the existence of chromosomes that carry genes, so they bred pairs of animals or plants that exhibited a desirable phenotype, such as a large corn plant or a calm horse. It was commonly believed that the traits found in offspring represented an average of the phenotypes found in the two parents, which was called the blending hypothesis. This long-standing idea was challenged in the 1800s when a monk named Gregor Mendel, shown in **FIGURE 27.1**, began conducting breeding experiments on garden peas. Mendel's experiments focused on the externally visible characteristics of individuals, which we refer to as phenotypes or traits, such as seed color or seed shape. His observations led to the discovery of how these different phenotypes are determined by inherited genotypes.

Mendel lived in a monastery in what is now the Czech Republic in Europe, where he grew different varieties of garden peas that he obtained from local seed suppliers. As illustrated in **FIGURE 27.2** on page 368, each pea variety had different traits, including differences in flower color, seed

FIGURE 27.1 Gregor Mendel's pea plants

Gregor Mendel conducted experiments on pea plants in which he examined the inheritance of different phenotypes, including flower color, seed color, and seed shape. He discovered that when two parents possess different phenotypes, the phenotypes of offspring are not an average of the two parents. Photos: (left) Juliette Wade/Getty Images; (right) Authenticated News/Getty Images

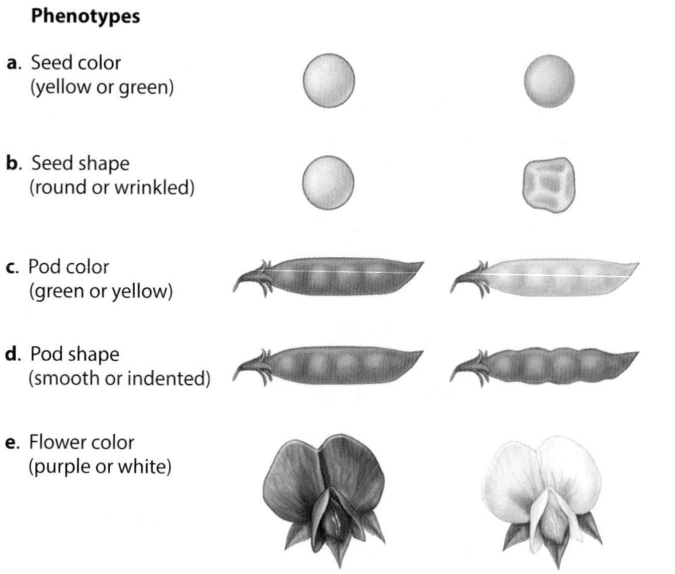

Phenotypes

a. Seed color
(yellow or green)

b. Seed shape
(round or wrinkled)

c. Pod color
(green or yellow)

d. Pod shape
(smooth or indented)

e. Flower color
(purple or white)

FIGURE 27.2 Pea plant phenotypes

Mendel focused on just a few phenotypes in the pea plants, including flower color, pod color, pod shape, pea color, and pea shape.

color, seed shape (round or wrinkled), pod color, and pod shape (smooth or indented). He started with these different purebred varieties (also known as true-breeding varieties), which consistently produced only one type of flower, pod, and pea seed, generation after generation. Mendel decided to see what would happen when he used the pollen from one purebred pea variety to fertilize the egg of a different purebred pea variety. What would their offspring look like? When we breed two individuals, we call it a cross. Mendel's goal was to examine the phenotypes of the offspring when crossing two different parental purebred plants and determine if there were repeatable patterns. He intended to use his data to explore possible hypotheses about how traits are inherited in sexually reproducing organisms.

To cross two purebred plants, Mendel began by opening an immature flower from one purebred variety and removing the male flower parts that make pollen, known as anthers, so that the flower could not fertilize itself. Once he did this, the flower only had its female part remaining, which consists of the ovary that contains the ovule and egg cells and a projection known as a stigma, which receives the pollen that fertilizes the eggs. After creating a flower with only female parts remaining, Mendel clipped the anthers from a different pea variety and applied pollen from those anthers to the stigma of the first flower. Finally, he placed a cloth bag around the female flower to make sure that no other pollen could arrive and fertilize the flower. After a few days, the fertilized egg cells began to develop into individual pea seeds within a pea pod.

Mendel's Crosses of Two Purebred Plants

Mendel began with crosses in which the two purebred pea varieties only differed in one trait, such as the color of their pea seeds. In one such experiment, he crossed one purebred variety that produced yellow seeds with another purebred variety that produced green seeds. Since these two varieties are the parents in the cross, we refer to them as the parental, or P_1, generation. He then examined the color of the peas that were produced in the next generation, which we refer to as the first filial, or F_1, generation. In this experiment, Mendel observed that a cross of yellow and green purebred parents produced offspring that had 100% yellow pea seeds, as depicted in **FIGURE 27.3**. The same outcome was found when he bred purebred varieties that differed in flower color, pea shape, pod color, and pod shape; only one of the two

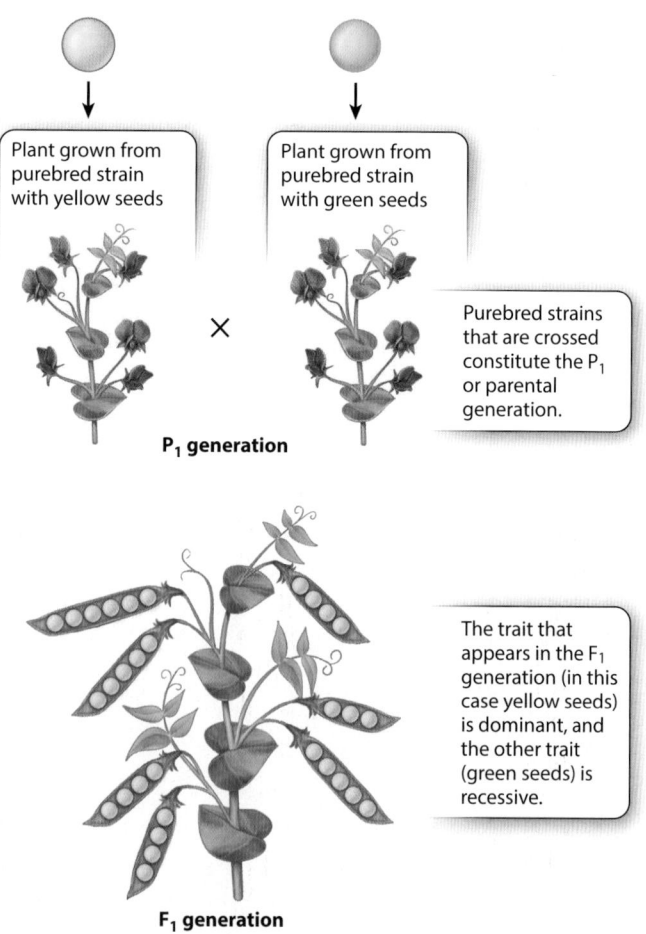

Plant grown from purebred strain with yellow seeds

Plant grown from purebred strain with green seeds

Purebred strains that are crossed constitute the P_1 or parental generation.

P_1 generation

The trait that appears in the F_1 generation (in this case yellow seeds) is dominant, and the other trait (green seeds) is recessive.

F_1 generation

FIGURE 27.3 Crossing different P_1 varieties to create the F_1 generation

When Mendel conducted a hybrid cross between two parents (P_1) that were purebred for yellow or green pea seeds, he discovered that the F_1 offspring all exhibited the yellow seed phenotype.

phenotypes was observed in the offspring, and it was evident in 100% of the offspring. When two different purebred varieties are crossed and only one phenotype is produced, we say that the exhibited phenotype is **dominant.** In contrast, the phenotype that does not appear is called **recessive.** In Mendel's experiment on pea color, we would say that the yellow pea color is dominant and that the green pea color is recessive. This observation refuted the long-held hypothesis that the phenotypes of offspring were the average of the traits of the parents due to blending inheritance.

When Mendel conducted his research in the 1800s, the scientific world did not yet know about inheritance and genes. Mendel hypothesized that offspring were inheriting what he called factors from their parents. We now know that genes, which are specific locations in a chromosome, can have different versions of a DNA sequence, which we call **alleles.** Mendel's observations of pea plant crosses revealed that each purebred parental variety carried a pair of alleles that were either dominant or recessive. When the cross was made, the offspring possessed a dominant allele from one parent and a recessive allele from the other parent. As a result, offspring exhibited the phenotype of the dominant allele—yellow pea seeds in the example above.

AP® EXAM TIP

Be sure you can differentiate between a gene and an allele. Using these terms precisely will earn points on the exam.

✓ Concept Check

1. **Identify** the cell structures and molecules that are common to all species and therefore suggest a common ancestor.

2. **Identify** the cell processes that are common to a wide range of distantly related species and therefore suggest a common ancestor.

3. **Describe** Mendel's evidence that refuted the blending hypothesis.

4. **Describe** the data that would be needed from a cross of two purebred pea varieties that would provide evidence for one allele being dominant.

5. **Identify** why it was useful that Mendel started with purebred pea plants in the P_1 generation.

27.2 The segregation of homologous chromosomes during meiosis is the basis for the segregation of alleles

The crosses between purebred P_1 pea varieties demonstrated to Mendel that some phenotypes were dominant and others were recessive in the F_1 generation. Given that Mendel's inherited "factors" seemed to determine the phenotypes of the F_1 generation, he wondered what phenotypes would appear if he were to cross F_1 individuals. Would the offspring continue to exhibit a single dominant phenotype, or would some individuals exhibit the recessive phenotype that was last seen in the parental (P_1) plants? If recessive phenotypes did reappear, then it would suggest that the two alleles inherited from the two purebred parents could segregate from each other when the plants were crossed, as opposed to being inherited together as a pair. In this section, we will examine how crossing hybrid F_1 offspring led Mendel to discover that pairs of alleles possessed by a parent segregate from each other when gametes are produced and used to create offspring.

Discovering the Law of Segregation

After crossing the purebred parents to produce hybrid F_1 offspring, Mendel decided to cross the hybrid offspring with themselves. A cross between individuals that are hybrids for a single gene, meaning they possess two different alleles, is a **monohybrid cross.** You can see an example of a monohybrid cross using Mendel's pea plants in **FIGURE 27.4** on page 370. At the top of this figure, we see the two purebred parents; one carries two alleles for yellow seeds while the other carries two alleles for green seeds. As we now know, when the purebred parents are crossed, we obtain F_1 hybrid offspring that carry one dominant allele (yellow) and one recessive allele (green). As a result, all of the F_1 offspring exhibit yellow seeds, as you can see in the middle of the figure. Mendel then crossed the F_1 plants by allowing crossing all of the F_1 offspring. The next generation produced is called the second filial, or F_2, generation. When he did this, he discovered that the recessive traits, which were absent in the F_1 generation, re-appeared in the F_2 generation, as you can see in the bottom of the figure. Moreover, he discovered a consistent pattern: the dominant phenotype appeared in about 75% of the F_2 generation, while the recessive phenotype appeared

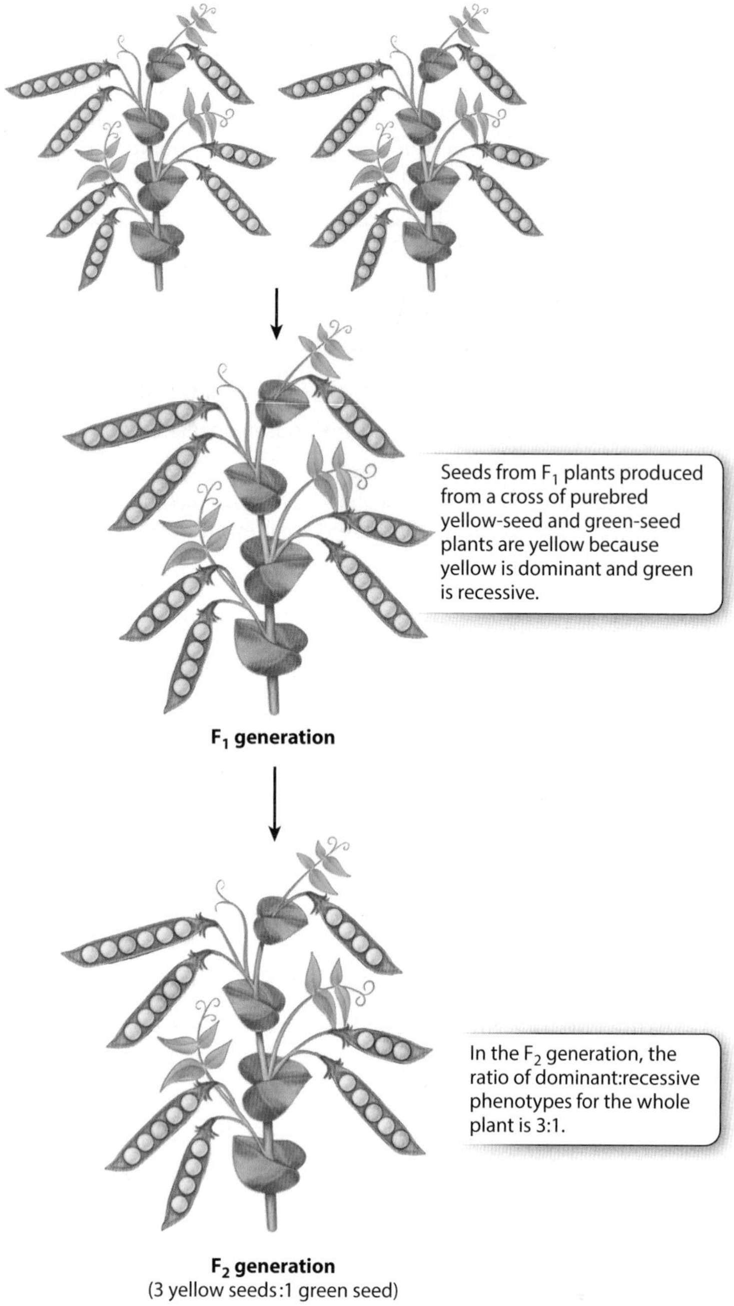

Seeds from F₁ plants produced from a cross of purebred yellow-seed and green-seed plants are yellow because yellow is dominant and green is recessive.

F₁ generation

In the F₂ generation, the ratio of dominant:recessive phenotypes for the whole plant is 3:1.

F₂ generation
(3 yellow seeds:1 green seed)

FIGURE 27.4 Crossing F₁ hybrids to produce the F₂ generation

When Mendel created the hybrid F₁ generation of pea plants by crossing two different purebred parents, he found that all of the F₁ offspring exhibited yellow seeds. However, when he crossed the hybrid F₁ offspring using a monohybrid cross, he discovered that the F₂ offspring expressed both the dominant (yellow) and recessive (green) phenotypes. Moreover, the two phenotypes were present in a 3:1 ratio of yellow and green seeds.

TABLE 27.1 Observed F₂ Ratios of Dominant and Recessive Phenotypes in Mendel's Pea Plant Experiments That Used a Monohybrid Cross

Trait	Dominant phenotype	Recessive phenotype	Ratio
Seed color	6022	2001	3.01:1
Seed shape	5474	1850	2.96:1
Pod color	428	152	2.82:1
Pod shape	882	299	2.95:1
Flower color	705	224	3.15:1

The underlying reason for Mendel's discovery of a 3:1 ratio of dominant and recessive traits for a single gene in the F₂ generation is known as the **law of segregation,** which reflects the fact that half of the daughter cells receive the maternal allele for a given trait while the other half of the daughter cells receive the paternal allele for the trait. Although Mendel was not aware of chromosomes and alleles at the time of his research, we now know the maternal and paternal homologs segregate during meiosis, as we discussed in the previous module.

Using our knowledge of meiosis, we can say that a purebred parent possesses two identical alleles for a given trait, a condition that we call **homozygous** (*homo* = same, *zygous* = referring to alleles present in the fertilized egg or zygote). For example, a purebred plant that produces yellow seeds is homozygous because it carries two yellow-seed alleles. As a result of being homozygous, all daughter cells produced by meiosis and which are destined to become gametes will have the same single allele. We denote an individual that is homozygous for the dominant alleles as having an *AA* genotype while an individual that is homozygous for the recessive alleles has an *aa* genotype. Each letter represents the two alleles that the individual carries. Thus, the purebred pea plant with yellow seeds is denoted as *AA* since it carries two dominant alleles, whereas the purebred pea plant with green seeds is denoted as *aa* since it carries two recessive alleles.

When two different purebred parents are used in a monohybrid cross, each offspring receives a dominant allele from one parent and a recessive allele from the other parent. When this happens, we say that the offspring

in about 25% of the offspring. Thus, there was a consistent 3:1 ratio of dominant and recessive phenotypes in the F₂ generation. You can see Mendel's original data for five different traits in **TABLE 27.1**.

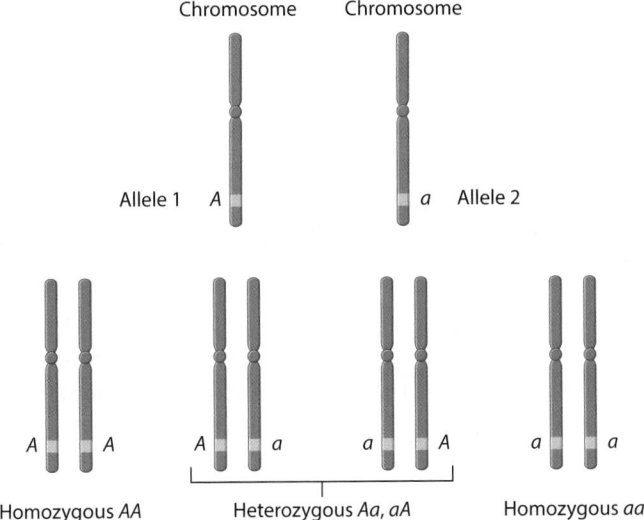

FIGURE 27.5 Homozygous and heterozygous allele combinations

On the top row of the figure, we see two pea plants represented by a single chromosome with a gene of interest shown as a light blue band. At this gene location, a pea plant can possess either a dominant allele (*A*) or a recessive allele (*a*). On the bottom row, we see all of the possible allele combinations in a group of pea plants possessing both alleles. When breeding occurs, the offspring can be homozygous dominant (*AA*), heterozygous (*Aa* or *aA*), or homozygous recessive (*aa*).

are **heterozygous,** because they possess two nonidentical alleles for a given trait (*hetero* = different). We denote an individual that is heterozygous as having an *Aa* genotype, since the individual has one dominant allele (*A*) and one recessive allele (*a*).

We can understand the relationship of genes and allele combinations in **FIGURE 27.5**. At the top of the figure, we see a single chromosome for each of two individuals. At a single location along each chromosome, we see there is a gene that is indicated as a light blue band. At this gene, the individual on the left carries a dominant allele (*A*) while the individual on the right carries a recessive allele (*a*). On the bottom of the figure, we see the different combinations of alleles that are possible when we conduct a monohybrid cross. Allele combination *AA* is homozygous dominant, allele combinations *Aa* and *aA* are both heterozygous, and allele combination *aa* is homozygous recessive.

With an understanding that individuals may be homozygous or heterozygous, we can now investigate why Mendel's F₂ generation produced a 3:1 ratio of dominant and recessive phenotypes. In **FIGURE 27.6**, you can see the purebred cross at the top of the figure that produces an F₁ generation in the middle of the figure, which are all heterozygous. The F₁ generation is then used in a monohybrid cross to produce

the F₂ generation. Notice that the gametes of each F₁ individual have an equal chance of containing either the *A* allele or the *a* allele. As a result, there are four ways in which the two alleles of one F₁ individual can combine at random with the two alleles of another F₁ individual: *AA*, *Aa*, *aA*, and *aa*.

One way to graphically show these four combinations is through the use of a Punnett square, which is named after its inventor, the British geneticist Reginald Punnett. The Punnett square is shown in the bottom of Figure 27.6. We begin by drawing a square box that has four compartments. At the top of the box we list the two possible alleles that the first parent can produce as a result of meiosis. On the left side of the box, we list the two possible alleles of the second parent. Finally, within each compartment of the box, we combine the alleles from the appropriate row and column. In doing so, we see that a monohybrid cross produces a genotypic ratio of 1:2:1. This translates into a phenotypic ratio of 3:1, which reflects the fact that the *AA* and *Aa* genotype

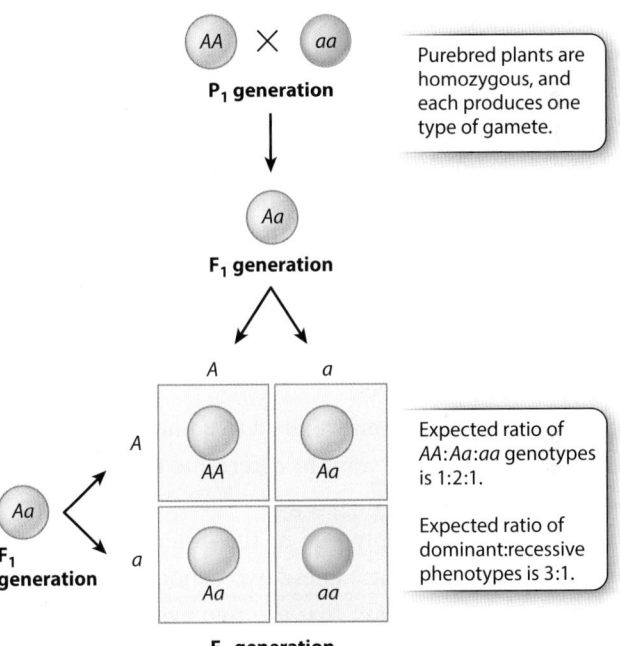

FIGURE 27.6 How the law of segregation creates the F₁ and F₂ generations

When the two different purebred P₁ individuals produce gametes, each individual has two identical alleles that segregate during meiosis. When the gametes experience fertilization, the F₁ offspring are all heterozygous (*Aa*). When the F₁ individuals are allowed to self-fertilize, each individual's pair of alleles once again segregates during meiosis, such that half of the gametes contain a dominant allele (*A*) while the other half contain a recessive allele (*a*). The two alleles for each individual can produce four combinations, two of which are heterozygotes (*Aa*). Thus, while the ratio of the genotypes is 1:2:1, the ratio of the dominant: recessive phenotypes is 3:1.

both produce the same phenotype, which in this case is a yellow pea seed. This is the same 3:1 ratio of phenotypes that Mendel observed.

Connecting the Law of Segregation to Laws of Probability

The Punnett square in the last figure shows different ratios of genotypes and phenotypes produced during a mating event. These ratios are based on the likelihood of various alleles being present in the male and female gametes. The likelihood of an event happening is its **probability.** Probability ranges from 0 to 1, which can also be written as 0% to 100%. When the probability of an event happening is 0, it will never happen; when the probability of an event happening is 1, it will always happen. As an example, we can revisit Figure 27.6 on page 371. When we cross the purebred parents, the probability that we will obtain an F_1 offspring with the aa or AA genotype is 0. The probability that we will obtain an offspring with the Aa genotype is 1. In contrast, if we were to cross an Aa individual with another Aa individual, as we do in the monohybrid cross in the figure, the probability that we will obtain an offspring with the aa genotype is $\frac{1}{4}$, which is equivalent to 25%. When we consider the results of Mendel's pea experiments in Table 27.1 on page 370, we can see that when there is a large sample size, the observed ratios of dominant and recessive phenotypes are very close to the predicted probabilities of 75% and 25%, which is equivalent to a 3:1 expected ratio of phenotypes. We can also see these probabilities appear when we work with Punnett squares to predict the different ratios of genotypes and phenotypes.

There are times when we need to combine probabilities. For example, we might want to determine the probability of the $Aa \times Aa$ cross producing genotypes that are either AA or Aa, rather than simply one or the other. The key words here are "either" and "or." These two genotypes are two distinct outcomes, so we say that the two events are mutually exclusive. When we wish to know the probability of either one event or another happening, we apply the **addition rule of probability,** which states that the probability of two mutually exclusive events happening is the sum of their individual probabilities. This rule can be simplified as:

If A and B are mutually exclusive, then:
$P(\text{A or B}) = P(\text{A}) + P(\text{B})$

For example, we know from our Punnett square in Figure 27.6 that the $Aa \times Aa$ cross has a $\frac{1}{4}$ probability of producing the AA genotype. From the same Punnett square, we can see that the $Aa \times Aa$ cross has a $\frac{1}{2}$ probability of producing the Aa genotype. Therefore, the probability of

producing genotypes that are either AA or Aa is $\frac{1}{4} + \frac{1}{2} = \frac{3}{4}$, which is equivalent to 75%.

Mutually exclusive events do not happen at the same time. In other situations, we need to know the probability of two independent events occurring at the same time. Independent events are those in which the probability of one event occurring has no influence on the probability of the other event happening. For example, if you flipped two coins, the probability of obtaining a "heads" on one coin has no effect on the probability obtaining a "heads" on the other coin. To know the probability of two independent events both happening, we use the **multiplication rule of probability,** which states that the probability of two independent events both happening is the product of their individual probabilities. This rule can be simplified as:

If A and B are independent, then:
$P(\text{A and B}) = P(\text{A}) \times P(\text{B})$

For example, consider an $Aa \times Aa$ cross for pea color, in which green peas are recessive (a) and yellow peas are dominant (A). For a pea pod with four seeds, we might want to know the probability that the first seed (nearest the stem) is green and that the seeds in the second, third, and fourth positions in the pod are yellow. In the case of pea seeds, the fertilization of each seed and the genotype it has is independent of the other seeds in the pod. The key word here is "and." In this case, we multiply the probability that the first seed will be green $\left(\frac{1}{4}\right)$ by the probability that the second seed will be yellow $\left(\frac{3}{4}\right)$, the third seed will be yellow $\left(\frac{3}{4}\right)$, and the fourth seed will be yellow $\left(\frac{3}{4}\right)$. When we multiply these four probabilities, we obtain $\frac{1}{4} \times \frac{3}{4} \times \frac{3}{4} \times \frac{3}{4} = \frac{27}{256}$, which is a probability of 10.54%. This means the probability that the first seed in the pea pod is green and that the seeds in the second, third, and fourth positions are yellow is about 11%.

All of these probabilities are based on the law of segregation of alleles during meiosis. This means we can predict the genotypic and phenotypic outcomes that will occur, as we have done with Punnett squares. It also means that when we know the genotypic or phenotypic outcomes, we may be able to infer the genotypes of the parents that were mated. For more practice on the addition rule of probability and the multiplication rule of probability, see "Analyzing Statistics and Data: Calculating Probability."

In considering Mendel's peas, it is important to note that these examples illustrate Mendel's laws of inheritance because they involve single genes with only two alleles. However, most traits are much more complex because they are determined by many genes and each of the genes can have a large number of alleles. In some cases, there can be hundreds of different alleles for a given gene.

Calculating Probability

Probability is the likelihood of an event or observation when two or more outcomes can occur. It can be used to predict the results of simple everyday things, such as the likelihood of a flipped coin coming up heads or tails, but it is also useful in biological contexts such as predicting the outcomes of experiments and mating events. Probability is particularly useful for understanding Mendel's pea plant crosses.

The probability for any event can range from 0 to 1, or 0% to 100%. If an event has a probability of 0, then it will never happen, while an event with a probability of 1 will always happen. There are two basic rules to remember when confronting a question on probability: the addition rule and the multiplication rule.

The addition rule states that if two events are mutually exclusive, such that both events cannot occur at the same time, the probability of one or the other happening is the sum of their individual probabilities. The key words to look out for in a problem asking you to use the addition rule are "either" and "or." The formula for the addition rule states:

If event A and event B are mutually exclusive, then:
$P(\text{A or B}) = P(\text{A}) + P(\text{B})$

The multiplication rule states that the probability of two independent events both happening is the product of their individual probabilities. The key word to look out for in a problem asking you to use the multiplication rule is "and." The formula for the multiplication rule states:

If event A and event B are independent, then:
$P(\text{A and B}) = P(\text{A}) \times P(\text{B})$

PRACTICE THE SKILL

With this background, we can use the laws of probability to understand the genotypes and phenotypes resulting from plant and animal crosses. For example, a heterozygous cross between two brown mice, *Aa*, yields a 3:1 phenotypic ratio of brown mice to white mice. What is the probability of producing either homozygous brown offspring or homozygous white offspring? If there are 60 offspring, how many offspring would be homozygous brown or homozygous white?

To begin, it is helpful to visualize monohybrid crosses with Punnett squares. We know that both heterozygous parents have the genotype *Aa* and exhibit the brown phenotype. With this information we create a Punnett square that looks like this:

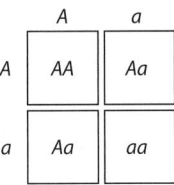

The gametes of one heterozygous parent are to the left of the square and the gametes of the other heterozygous parent are on the top of the square. Combining these gives us the potential offspring that could result from this cross.

The question tells us that there is a 3:1 phenotypic ratio of brown to white mice. This means that the *AA* and *Aa* offspring in the square will be brown and the *aa* offspring will be white. Now, we can answer the probability portion of the question.

Only $\frac{1}{4}$, or 25%, of the offspring are expected to exhibit the *AA* genotype that codes for brown fur. The same can be said for the *aa* genotype that codes for white fur; the probability is $\frac{1}{4}$, or 25%.

The question asked for the probability of having either the *AA* brown offspring or the *aa* white offspring. The "or" word tells us that we should use the addition rule to find the final probability for this cross. To double-check that this is the right rule to use, we can determine if these events are mutually exclusive. If one mouse is *AA* brown, it cannot be *aa* white, so these are indeed mutually exclusive events; a mouse cannot be both brown and white.

The final step for this question is to add the probabilities together:

AA brown mice + aa white mice
$$\frac{1}{4} + \frac{1}{4} = \frac{1}{2}$$

The probability for producing either homozygous brown or homozygous white offspring is $\frac{1}{2}$, or 50%. In a cross that produced 60 offspring, this would account for 30 of the mice: $60 \times \frac{1}{2} = 30$.

Your Turn

In his studies on pea plants, Mendel found that smooth pods are dominant to indented pods, green pods are dominant to yellow pods, and yellow seeds are dominant to green seeds. In a cross between two *Aa* individuals, what is the probability of having offspring with a smooth and yellow pod and either four green or four yellow seeds?

Chi-square Calculations for Statistically Comparing Expected and Observed Phenotypes

When examining the phenotypic data from crosses such as Mendel's pea plants, we can determine the ratios of the various observed phenotypes. At the same time, we can have predicted ratios of phenotypes based on our hypotheses about how many genes and alleles are involved. For example, in a monohybrid cross involving a single gene and only two alleles, we can state a null hypothesis that the phenotypes should follow a 3:1 ratio, as we have seen in Figure 27.6. However, the ratio of observed phenotypes from a monohybrid cross rarely produces phenotypes in an exact 3:1 ratio, as we can see from Mendel's data in Table 27.1. The question is, how different does the observed ratio have to be for us to reject the hypothesis of a 3:1 ratio? As we discussed in Unit 1, we can use statistical tests to objectively determine whether we can reject or fail to reject a hypothesis. To do this, we conventionally use a threshold value of $P = 0.05$, which means that we will only incorrectly reject the null hypothesis 5% of the time.

When our data consist of counts in different categories, such as the number of individuals exhibiting different phenotypes, we can test whether the observed data are significantly different from a set of expected phenotypic counts using a *chi-square test*. A **chi-square (χ^2) test** is a statistical test that determines whether the number of observed events in different categories differs from a number of expected events. That is, it helps determine whether a difference between the number of expected and observed counts is large enough to be considered statistically significant so that we can reject the null hypothesis or small enough to be due to random effects so that we fail to reject the null hypothesis.

As an example, consider the color of pea pods from Mendel's experiment in Table 27.1 in which the green pods are dominant. From a monohybrid cross, we expect to observe a 3:1 ratio of green pods:yellow pods. However, Mendel actually observed 428 green pods and 152 yellow pods, which is a ratio of 2.82:1. We can use a chi-square test to decide whether this observed ratio of 2.82:1 is significantly different from the expected ratio of 3:1. This will allow us to decide whether to reject or fail to reject the hypothesis that the phenotypes are determined by a single gene and two alleles.

The first step in conducting a chi-square analysis is determining the values for the observed and expected phenotype. Determining the observed values is straightforward; we can

obtain the numbers from Mendel's data in Table 27.1. Determining the expected values takes a bit more work. Since the hypothesis is predicting a 3:1 ratio of green:yellow pods, we need to take Mendel's total number of pea pods in his experiments ($428 + 153 = 581$) and then calculate how many of the 581 pods would be green and yellow if they followed a 3:1 ratio. Given that our expected ratio is 75% green pods and 25% yellow pods (a 3:1 ratio), our expected number of green pods is $581 \times 75\% = 436$. Similarly, our expected number of yellow pods is $581 \times 25\% = 145$. Note that we do not use the predicted percentages for our expected values; we need to multiply these percentages by the total number of observed counts. We can now put these two expected values into the table below.

	Observed	Expected
Green pods	428	436
Yellow pods	153	145
Total pods	581	581

To calculate the chi-square value (χ^2), we use the following equation that includes our observed (o) and expected (e) values for each pod color. The summation sign (Σ) tells us that we do the calculation for the first phenotype (green pods) and then add it to the calculation for the second phenotype (yellow pods). When making your calculations, you should not round the intermediate values.

$$\chi^2 = \sum \frac{(o - e)^2}{e}$$

$$\chi^2 = \frac{(428 - 436)^2}{436} + \frac{(153 - 145)^2}{145}$$

$$\chi^2 = \frac{(428 - 436)^2}{436} + \frac{(153 - 145)^2}{145}$$

$$\chi^2 = 0.147 + 0.441 = 0.59$$

We now have to compare our calculated chi-square value to a table of values shown in **TABLE 27.2** to determine if our observed distribution is significantly different from the expected distribution. The first step in that process is determining the degrees of freedom for the chi-square test, which is always one less than the number of categories that we are using. In the current example, we have two categories (yellow pods and green pods), so:

Degrees of freedom = (number of observed categories − 1) = (2 − 1) = 1

Using the chi-square table, we can compare our calculated chi-square value (0.59) to the critical chi-square value

TABLE 27.2 Chi-square (χ^2) Table

P-value	Degrees of freedom							
0.05	3.84	5.99	7.81	9.49	11.07	12.59	14.07	15.51
0.01	6.63	9.21	11.34	13.28	15.09	16.81	18.48	20.09

when we have 1 degree of freedom and we use a threshold P-value = 0.05. In this case, the critical chi-square value is 3.84. Since our calculated value is less than the critical value, we fail to reject the null hypothesis. Therefore, we conclude that any difference between the observed and expected results may have occurred simply by chance.

While this is a relatively simple example using two categories (yellow versus green pods), chi-square tests can be used when there are more than two phenotypes. For example, later in this module we will discuss the more complex situation of crossing individuals that differ in two genes, which can produce four different combinations of phenotypes. For more practice working with chi-square tests, see "Analyzing Statistics and Data: Chi-square Analysis."

ANALYZING STATISTICS AND DATA ▶

Chi-square Analysis

When working with plant and animal crosses to understand Mendelian genetics, Punnett squares provide us with expected ratios of each genotype or phenotype. When we actually conduct the crosses, such as in the pea plant experiments of Mendel, we typically observe that the number of each genotype or phenotype resembles the predicted ratios, but the observed ratios are not exactly the same as the predicted ratios. What we want to know is how different does the observed ratio have to be for us to reject the hypothesized expected ratio?

To statistically compare the observed versus the expected genotypes or phenotypes of an experiment, we use the chi-square test. This test helps scientists determine if differences between observed and expected data are significant or due to chance. Significance is determined by a 5% threshold, which means that if we calculate a chi-square value that exceeds this threshold, we reject the hypothesis that the observed ratios of genotypes or phenotypes are the same as the expected ratios. On the critical chi-square table (Table 27.2), this is represented by a P-value of 0.05. The closer the observed and expected data are, the smaller the calculated chi-square value χ^2 will be and the more likely it is that we will accept the hypothesis.

There are several steps to calculating the chi-square test and we will walk through the steps more thoroughly in the practice section below. We begin by having both the observed and expected values to plug into the chi-square formula:

$$\chi^2 = \sum \frac{(o - e)^2}{e}$$

Using this calculated chi-square value, we then compare it to the critical chi-square values in Table 27.2 that use the threshold P-value of 0.05. Comparing these two numbers allows us to interpret whether differences between the observed and expected data are significantly different, allowing us to either reject or not reject the hypothesis provided by the Punnett square.

If the calculated chi-square value is greater than the critical chi-square value, then the difference between the two groups of data is large and the observed counts of genotypes or phenotypes are significantly different from the expected counts.

If the calculated chi-square value is less than the critical chi-square value, then the difference between the two groups of data is small and those differences in data may have been due to chance. We would fail to reject our null hypothesis.

PRACTICE THE SKILL

Mendel's pea plant experiments offer a good opportunity to practice chi-square analysis. In the table below are the results of Mendel's F_1 crosses. They show a 2.96:1 ratio of the dominant to

▼

recessive phenotype for seed shape. For this kind of heterozygous cross ($Aa \times Aa$) we would normally expect a 3:1 ratio. Are the observed and expected results significantly different? We can use chi-square analysis to find out.

Trait	Dominant phenotype (round seed)	Recessive phenotype (wrinkled seed)	Ratio
Seed shape	5474	1850	2.96:1

To begin, we need both the observed and expected data for the seed shapes. The pea plant crosses have provided us with the observed numbers of each phenotype and their ratio. However, we must calculate the expected numbers. To do this, we will use the expected ratio for this cross, 3:1, and the total number of seeds, which is 7324. The 3:1 ratio can also be written as 75%:25%.

Expected number of round seeds: 7324 × 0.75 = 5493
Expected number of wrinkled seeds: 7324 × 0.25 = 1831

With these data, we can fill out a table that includes the observed and expected number of round versus wrinkled seeds.

	Observed	Expected
Round seeds	5474	5493
Wrinkled seeds	1850	1831
Total seeds	7324	7324

With these data, we can apply the chi-square formula. Remember to sum the observed and expected data for the round seeds with the observed and expected data for the wrinkled seeds.

$$\chi^2 = \sum \frac{(o - e)^2}{e}$$
$$\chi^2 = \frac{(5474 - 5493)^2}{5493} + \frac{(1850 - 1831)^2}{1831}$$
$$\chi^2 = 0.066 + 0.197$$
$$\chi^2 = 0.26$$

This is our calculated chi-square value. We must now compare this value to the critical chi-square values in Table 27.2 to conclude if the observed and expected data are significantly different.

In order to use Table 27.2, we have to find our degrees of freedom for this problem. We determine the degrees of freedom by subtracting 1 from the number of categories in the experiment. In this case, we have 2 categories, round seeds and wrinkled seeds, so our degree of freedom is 1.

To read Table 27.2 and get our results, we first find our degrees of freedom, 1, in the correct column and then read from the 0.05 P-value row. This P-value represents our 5% threshold of significance. The critical chi-square value read from this table is 3.84.

Our calculated chi-square value (0.27) is much less than the critical chi-square value of 3.84. Therefore, we can conclude that the difference between the observed data and the expected data is not significantly different from the prediction of a 3:1 ratio of phenotypes, so we fail to reject our null hypothesis.

Your Turn

Chi-square tests can be used to compare observed and expected values from a wide variety of different experiments. For example, scientists studying behavior decided to set up a choice experiment to determine if a species of ant have a preference for fresh or crushed fruit. Two small covered circular chambers were connected by a narrow channel. Thirty ants were placed in the channel between the chambers and were free to move between the two chambers. The scientists put a fresh raspberry in one chamber and a crushed raspberry in the other. They observed the movement of the ants between chambers and recorded their positions every minute for 20 minutes. The number of ants in each location after 1, 10, and 20 minutes are shown in the table.

Time (minutes)	Position in chamber		
	Side with fresh raspberry	Middle channel	Side with crushed raspberry
1	4	19	7
10	7	8	15
20	8	3	19

1. What is the null hypothesis for this experiment?
2. Using chi-square analysis for the data collected after 20 minutes, determine if the null hypothesis will be rejected or fail to be rejected.

Mendelian Segregation and Genetic Variation

Mendel's experiments produced substantial insights about heredity. His experiments with peas showed that phenotypes are not simply a blending of the two parents' phenotypes, as earlier researchers had hypothesized. Instead, each parent has two sets of alleles that segregate when gametes are formed. As a result, recessive alleles can remain in the population and not appear as phenotypes for many generations, but then reappear as recessive phenotypes in later generations.

This means that populations can continue to possess genetic variation for multiple generations. As we have noted, such genetic variation can be very important when there is variation in the environment such that different environments favor different genotypes and phenotypes. We will discuss the importance of genetic variation for evolution in much more detail in Unit 7.

Incomplete Dominance and Codominance

While Mendel was fortunate to have chosen traits in pea plants that are determined by a single gene, with one allele being dominant over another, the patterns of inheritance for most traits do not follow Mendel's expected ratios. Evidence of this is found when we observe phenotypic ratios that do not closely align with Mendel's predicted ratios. Although most traits are determined by multiple genes, even traits that are determined by a single gene can sometimes exhibit incomplete dominance, in which the heterozygous genotype expresses a phenotype that is intermediate to the homozygous dominant and homozygous recessive genotypes.

We can observe incomplete dominance in a species of flower known as the snapdragon (*Antirrhinum majus*). The homozygous genotypes can have red ($C^R C^R$) or white flowers ($C^W C^W$), while the heterozygous genotypes have pink flowers ($C^R C^W$). In the case of incomplete dominance, we no longer use uppercase and lowercase letters to indicate the alleles because there are no dominant or recessive alleles. Instead, we use all capital letters with superscripts that indicate the allele. Incomplete dominance using snapdragons is illustrated in **FIGURE 27.7**, in which a red parent and a white parent are crossed to produce an all-pink F₁ generation. When the F₁ individuals are allowed to self-fertilize, the F₂ generation consists of $\frac{1}{4}$ red flowers, $\frac{1}{2}$ pink flowers, and $\frac{1}{4}$ white flowers. As a result of incomplete dominance, the 1:2:1 genotypic ratio produces a 1:2:1 phenotypic ratio.

A different pattern is codominance, in which both alleles are expressed in the heterozygote, rather than one or the other. A good example of codominance is the human phenotypes for blood groups. At the gene for blood group, a person can carry one of three alleles: *A, B,* or *O.* These alleles code for a group of proteins known as antigens, which the body's immune system recognizes. Allele *A* results in the production of *A* antigens, allele *B* results in the production of *B* antigens, and allele *O* results in the production of no antigens. If a person inherits the *A* allele from one parent and the *B* allele

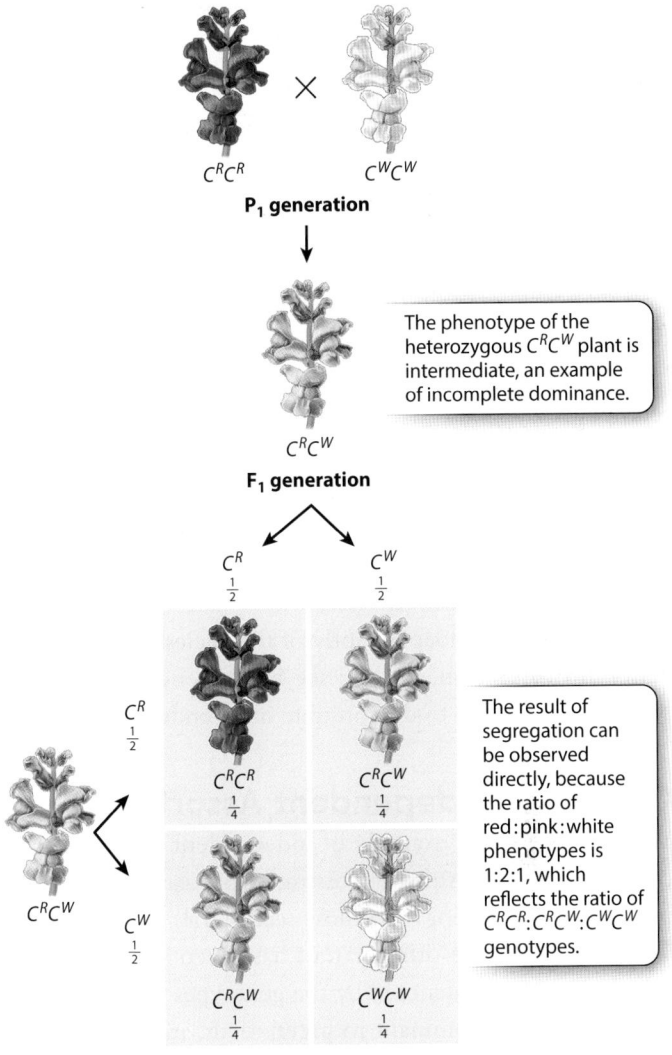

The phenotype of the heterozygous $C^R C^W$ plant is intermediate, an example of incomplete dominance.

The result of segregation can be observed directly, because the ratio of red:pink:white phenotypes is 1:2:1, which reflects the ratio of $C^R C^R$:$C^R C^W$:$C^W C^W$ genotypes.

FIGURE 27.7 Incomplete dominance in snapdragons

When a red parent is crossed with a white parent, all individuals of the F₁ generation are heterozygous and produce pink flowers. When individuals of the F₁ generation are self-fertilized, we achieve $\frac{1}{4}$ red flowers, $\frac{1}{2}$ pink flowers, and $\frac{1}{4}$ white flowers.

from the other parent, then that person produces both antigens. Because the *A* and *B* both cause the production of both antigens, they are considered codominant alleles.

✓ Concept Check

6. **Describe** how meiosis explains the law of segregation.

7. **Describe** how the law of segregation preserves genetic variation.

8. **Describe** how incomplete dominance contradicts the idea of dominant and recessive alleles.

27.3 Independent assortment during meiosis causes independent assortment of genes on different chromosomes

We have looked at genotypes and phenotypes that result from two alleles and a single gene. However, Mendel took his research further by examining what happened when he crossed individuals that differed in two traits—such as seed color and seed shape—that are determined by alleles at two different genes. Fortunately for Mendel, the traits he examined were determined by genes that were located in different chromosomes rather than two genes in the same chromosome. This allowed him to discover the **law of independent assortment,** which states that the alleles for one gene sort independently of the alleles for a different gene. In this section, we will see how Mendel conducted crosses in pea plants to demonstrate independent assortment.

Observing Independent Assortment

We can observe an example of independent assortment by considering the result of crossing two purebred parental pea plants: one parent has yellow seeds (*AA*) that are wrinkled (*bb*), while the other parent has green seeds (*aa*) that are round (*BB*). As denoted by the genotypes in parentheses, yellow seeds are dominant to green seeds, and round seeds are dominant to wrinkled seeds. As shown in **FIGURE 27.8**, the F₁ generation is comprised entirely of individuals that are heterozygous for both traits (*Aa Bb*), so their phenotypes are yellow, round seeds. Mendel then crossed these F₁ hybrids. Because all individuals were hybrids for two traits that are coded by two different genes, it is known as a **dihybrid cross.** In these experiments, Mendel discovered that the F₂ generation includes four different combinations for seed phenotypes: yellow and round, green and round, yellow and wrinkled, and green and wrinkled. Moreover, the numbers of individuals expressing each phenotypic combination were not equal, as you can see in **TABLE 27.3**. In fact, the ratios of the four phenotypes came close to a ratio of 9:3:3:1.

This ratio of dominant and recessive phenotypes intrigued Mendel, since this is the expected ratio of phenotypes when two traits are independently assorted. How did Mendel derive this expected ratio of 9:3:3:1? He used the multiplication rule of probability by assuming that the gene that determines one trait (pea color) was independent of the gene that determines the other trait (pea shape). Thus, he could use the probabilities from a monohybrid cross (3:1) for each of the two traits.

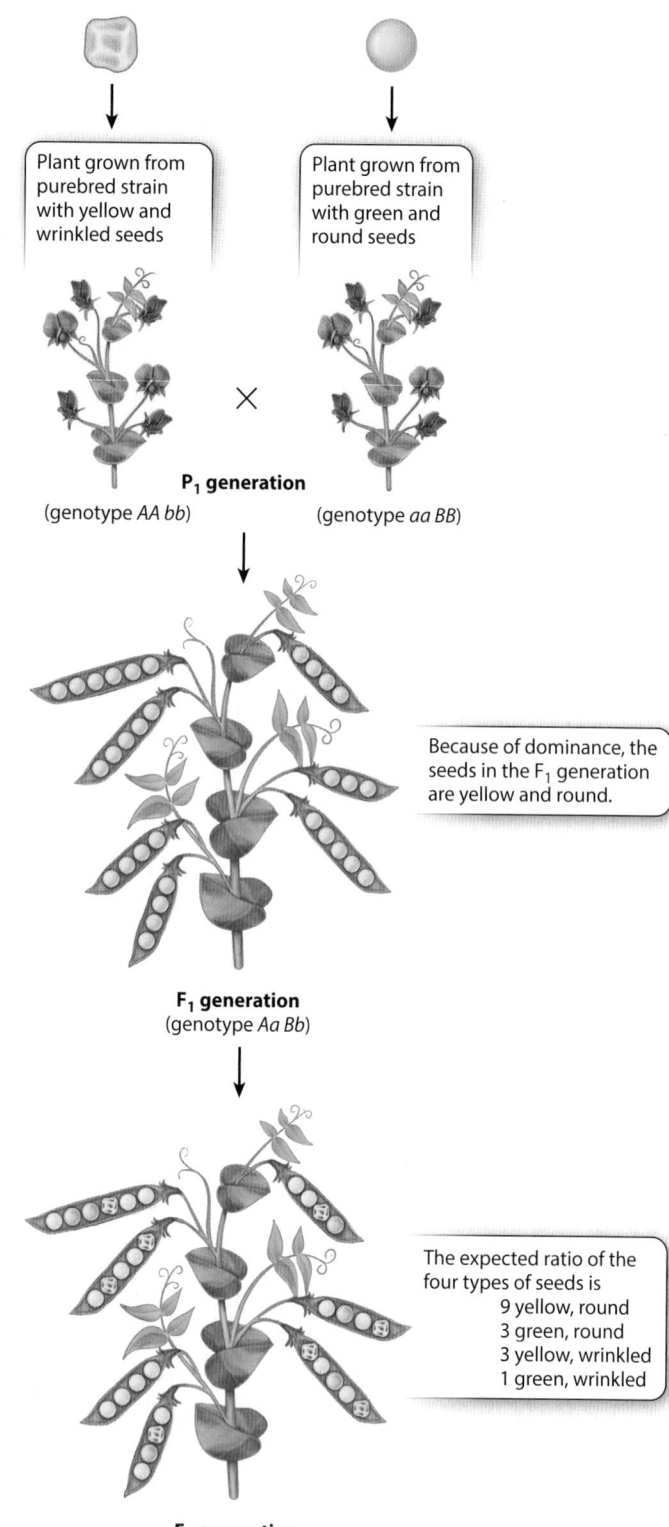

Plant grown from purebred strain with yellow and wrinkled seeds

Plant grown from purebred strain with green and round seeds

P₁ generation

(genotype *AA bb*)   (genotype *aa BB*)

Because of dominance, the seeds in the F₁ generation are yellow and round.

F₁ generation
(genotype *Aa Bb*)

The expected ratio of the four types of seeds is
9 yellow, round
3 green, round
3 yellow, wrinkled
1 green, wrinkled

F₂ generation

FIGURE 27.8 Crossing pea plants that differ in two traits (a dihybrid cross)

Starting with two purebred parents that differ in their homozygous genotypes, the F₁ generation is heterozygous for both traits (*Aa Bb*). When individuals from the F₁ generation were allowed to self-fertilize, the F₂ generation exhibited four different combinations of the two traits, with the phenotypic ratio approximating 9:3:3:1.

TABLE 27.3	Number of Mendel's F_2 Individuals Exhibiting Different Combinations of Seed Color and Seed Shape	
Phenotype	**Number of individuals**	**Phenotypic ratio**
Yellow, round	367	9.9
Green, round	122	3.3
Yellow, wrinkled	113	3.1
Green, wrinkled	37	1

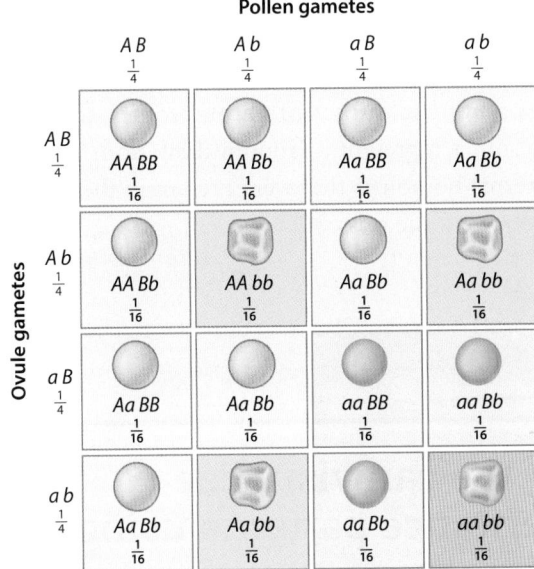

FIGURE 27.9 Punnett square for two independently assorting traits

When pea plants are heterozygous for two traits such as pea color and pea shape, a Punnett square can show how the four possible gametes from one individual can combine with those of another individual.

Given that the F_1 generation was heterozygous for both traits, the F_2 generation has a $\frac{3}{4}$ probability of being yellow and a $\frac{1}{4}$ probability of being green. It also has a $\frac{3}{4}$ probability of being round and a $\frac{1}{4}$ probability of being wrinkled. Thus, we can multiply these different probabilities to predict the ratios of the four phenotypic combinations:

$$\text{yellow and round} = \frac{3}{4} \times \frac{3}{4} = \frac{9}{16}$$

$$\text{green and round} = \frac{1}{4} \times \frac{3}{4} = \frac{3}{16}$$

$$\text{yellow and wrinkled} = \frac{3}{4} \times \frac{1}{4} = \frac{3}{16}$$

$$\text{green and wrinkled} = \frac{1}{4} \times \frac{1}{4} = \frac{1}{16}$$

As you can see, these four probabilities are in a ratio of 9:3:3:1. Another way to view this outcome is by using a large Punnett square that lists the four possible gametes of each doubly heterozygous F_1 pea plant. As illustrated in **FIGURE 27.9**, the four possible gametes for one individual are represented by pea pollen and the four possible gametes for the other individual are represented by the pea ovule. This 4 × 4 Punnett square results in 16 combinations of F_2 genotypes, which produces an expected phenotypic ratio of 9:3:3:1.

The underlying reason for this ratio is that during meiosis, each pair of maternal and paternal homologs move into daughter cells independently of every other pair of chromosomes. This happens because the homologous pairs line up randomly during metaphase I and are then pulled apart during anaphase I, as shown in **FIGURE 27.10**. In the cell on the left side of the figure, the A and B alleles go into one daughter cell while the a and b alleles go into the other daughter cell. Equally likely, however, is the scenario on the

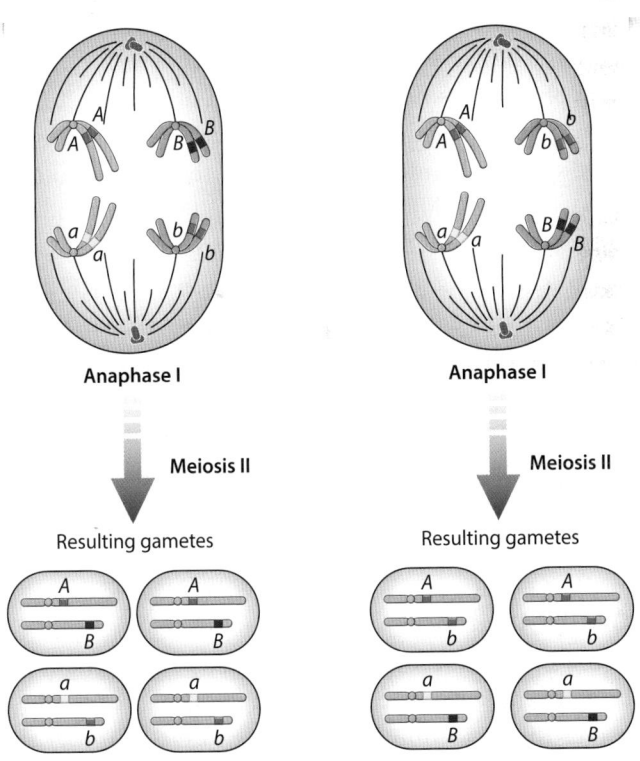

FIGURE 27.10 Independent assortment of genes during meiosis

During meiosis, the maternal and paternal homologs sort independently. As a result, the pattern of allele sorting in the cell on the left is as likely as the pattern of allele sorting in the cell on the right. The result is that all possible genotype combinations occur in the gametes.

right side of the figure, in which the *A* and *b* alleles go into one daughter cell while the *a* and *B* alleles go into the other daughter cell. As a result, while an individual daughter cell receives a full haploid set of chromosomes, the particular chromosomes represent a random draw from the maternal and paternal homologs that were present in the parental cell that underwent meiosis.

✓ **Concept Check**

9. **Describe** Mendel's evidence for the law of independent assortment.

10. **Describe** how meiosis is the underlying mechanism for the law of independent assortment.

11. **Identify** the ratios of dominant and recessive phenotypes for two traits in a dihybrid cross that involves genes in different chromosomes.

27.4 We can visualize inheritance patterns using pedigrees

In some species, such as humans, we do not do controlled breeding experiments to determine patterns of inheritance of dominant and recessive alleles. Mating events in humans are typically not random and the number of offspring is not large. This makes it difficult to compare observed and expected phenotypes using a chi-square test. One way that researchers have worked to understand patterns of inheritance in such species is to use a **pedigree,** which is a visual map of phenotypes that uses ancestral relationships. To help identify dominant and recessive traits, pedigrees typically use unique symbols to identify males and females and unique colors to identify different phenotypes. From these identifications, we can infer the existence of different genotypes. You can see an example of a pedigree in **FIGURE 27.11**. Here, males and females are given different shapes and the individuals with a given phenotype are indicated by red. In this example, the figure indicates the phenotypes of the parents and their offspring, their offspring's mates, and their grand-offspring. Based on the phenotypes of these individuals, we can infer the genotypes of the individuals and whether the phenotypes are dominant or recessive.

Dominant Traits

When a trait is dominant, it appears in every generation, although not necessarily in every individual. For example, consider the condition brachydactyly, a rare disorder that appears as a shortening of the fingers caused by a mutation that interferes with bone growth. The trait for bone growth is controlled by two autosomal alleles of a single gene. A recessive allele produces a phenotype of normal finger length when present in two copies and is common

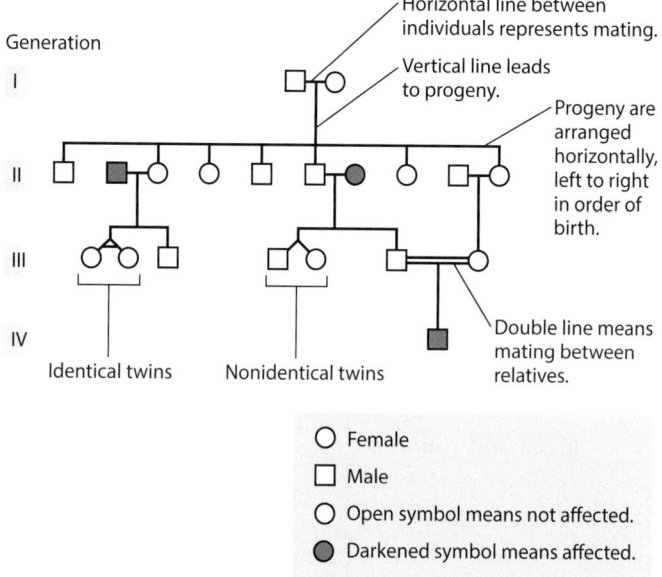

FIGURE 27.11 A pedigree

Using unique symbols and colors, researchers can map males and females and the existence of dominant and recessive phenotypes.

in human populations. In contrast, a dominant allele produces a phenotype with shorter fingers and is rare in human populations. If both parents do not have the brachydactyly trait, then their genotypes would both be homozygous recessive (*aa*) and they could not have a child with the trait. However, if a child does have short fingers, then at least one of the child's parents must possess the dominant allele.

FIGURE 27.12 shows a pedigree that maps the presence of brachydactyly in a family. When we examine the appearance of this disorder throughout the pedigree, we notice several patterns. Matings that produce offspring with the disorder have only one affected parent, which suggests that the disorder is caused by a dominant allele. In addition, affected individuals are equally likely to be females or males,

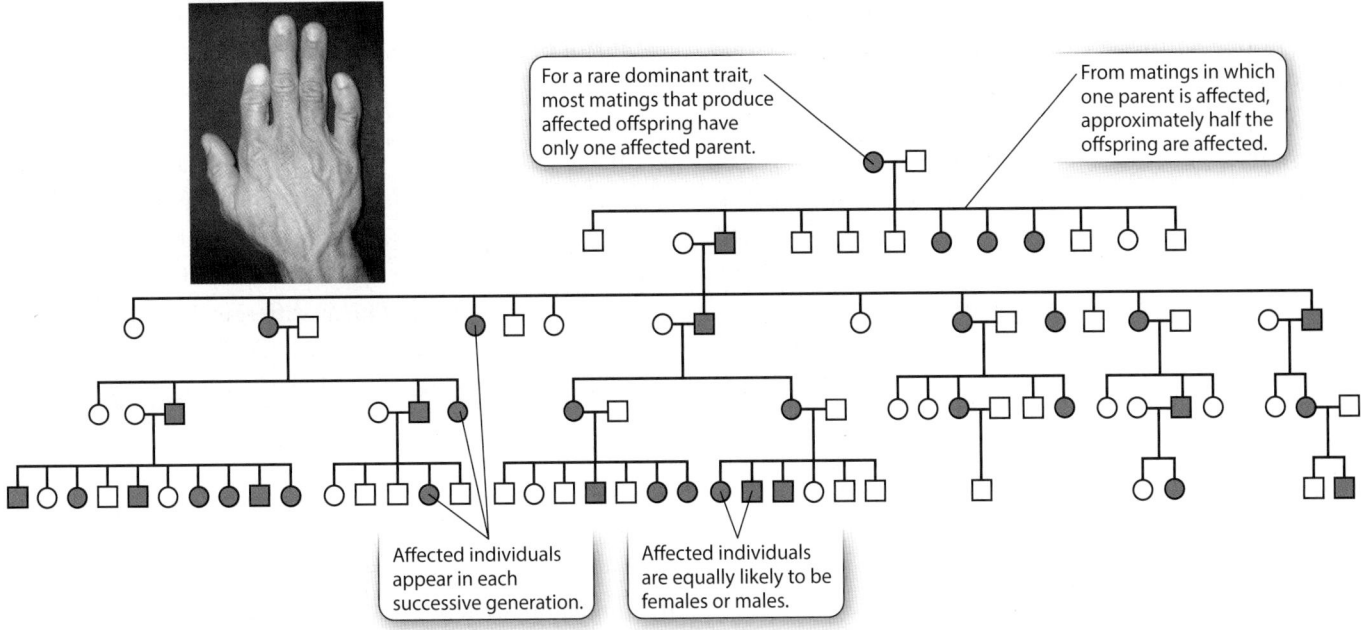

FIGURE 27.12 A pedigree of the rare dominant gene for the brachydactyly phenotype

Since the condition is rare, it is likely that any individuals exhibiting the phenotype will be heterozygous (*Aa*). If so, then the parent not exhibiting the phenotype will be homozygous recessive (*aa*) and only approximately half of the offspring will be heterozygous (*Aa*) and exhibit the phenotype. Photo: Stefan Mundlos, Institute for Medical Genetics, Charité, Berlin, Germany

which suggests the gene is located on an autosome and not on one of the sex chromosomes. Because the allele for the bone disorder is rare, if a parent expresses the phenotype, that parent is most likely heterozygous (*Aa*) for the trait rather than homozygous (*AA*). Among matings in which one parent has the phenotype (*Aa*) and the other does not (*aa*), there is a 50% probability of the offspring expressing the trait, which we would expect given a 50% probability of the offspring being heterozygous (*Aa*) and exhibit the disorder, and a 50% probability the offspring would be homozygous recessive (*aa*) and not exhibit the disorder. To see this effect with a sufficiently large sample size, we need to examine all of the offspring within a given generation. As you can see, the patterns of inheritance in a pedigree tells us a great deal about the type of allele that is being inherited.

Recessive Traits

When we examine pedigrees for autosomal recessive alleles, we see a very different pattern from that of autosomal dominant alleles. For example, the trait for blue eyes is a relatively rare recessive trait, found in 8% to 10% of people worldwide, that can appear in pedigrees such as the one illustrated in **FIGURE 27.13**. In this pedigree, double lines indicate mating between first cousins. As you can see, recessive traits

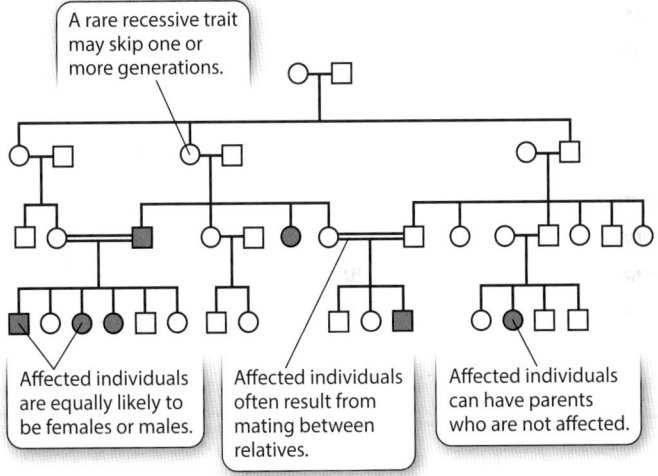

FIGURE 27.13 A pedigree of the rare recessive gene for blue eyes

Since the recessive allele is relatively rare, the blue-eyed phenotype (*aa*) only occasionally appears. It appears when two heterozygous individuals (*Aa*) mate. The probability of expression increases when close relatives, such as cousins (shown as double horizontal lines), mate because they are both more likely to be carrying the recessive allele.

show a number of distinct patterns when mapped onto a pedigree. Affected individuals, those who display the phenotype, are equally likely to be females or males, which provides evidence that the gene is located in an autosome and not in one of the sex chromosomes. In addition, the trait can skip one or more generations, which suggests that the allele is recessive. Affected individuals may have unaffected parents, which is also consistent with the idea that the blue eye allele is recessive. For a recessive trait that is sufficiently rare, almost all affected individuals have unaffected parents. Finally, affected individuals often result from mating between relatives—typically first cousins—since the relatives are more likely to both be heterozygous for the trait. As you can see in these examples, while we cannot conduct controlled crosses in humans to examine patterns of inheritance as Mendel did with pea plants, we can use pedigrees to infer whether alleles are dominant or recessive and whether the alleles are found in an autosome or a sex chromosome.

Incomplete Penetrance and Variable Expressivity

Two major factors can complicate the patterns in pedigrees, even when we are examining a trait that involves a single gene. The first factor is incomplete penetrance, which occurs when some individuals who have the genotype that codes for a particular phenotype fail to exhibit that phenotype. Whenever less than 100% of individuals carrying the genotype express the expected phenotype, we consider it to be incompletely penetrant. For example, type 2 diabetes is associated with certain mutations in the gene *TCF7L2* that affect insulin secretion and glucose production. However, some people who carry the mutation do not develop type 2 diabetes because they make healthy lifestyle choices, including eating more fruits and vegetables and getting regular exercise. As a result, some people who are at risk of getting diabetes because they carry the mutant allele do not exhibit the phenotype, so there is incomplete penetrance.

A second factor that complicates the patterns observed in pedigrees is the problem of variable expressivity, which happens when the phenotype is expressed by all individuals who carry the allele, but the degree of expression differs among individuals. This is different from incomplete penetrance because all individuals exhibit the trait, but to different degrees. This variation in expressivity is often due to environmental factors that vary among individuals. For example, there is a mutant allele that causes a reduction in lung elasticity and emphysema in humans. In people that smoke tobacco, the symptoms can be much more severe compared to nonsmokers. These complications of incomplete penetrance and variable expressivity can make it difficult to infer the inheritance of dominant and recessive alleles on a pedigree since the genotypes do not always express the expected phenotypes. We will discuss the interaction of genes and the environment in more detail in subsequent modules.

✓ Concept Check

12. **Identify** the conditions under which pedigrees are a useful way to understand allele inheritance.

13. **Describe** the patterns expected in a pedigree for a rare dominant allele.

14. **Describe** how incomplete penetrance can complicate interpretations of inheritance when using pedigrees.

Module 27 Summary

PREP FOR THE AP® EXAM

REVISIT THE BIG IDEAS

INFORMATION STORAGE AND TRANSMISSION: Using the content of this module, **describe** one example of an inheritance pattern that illustrates Mendel's law of segregation and one example that illustrates Mendel's law of independent assortment.

LG 27.1 Organisms are linked by lines of descent due to common ancestry.

- All organisms are composed of one or more cells that are separated from their environment by a plasma membrane, they share core metabolic pathways, and their genetic code is found in the molecules of DNA and RNA. Page 366

- Common traits suggest that all organisms have inherited their genetics from a common ancestor. Page 367

- People once thought that the traits found in offspring represented an average of the phenotypes found in the two parents, which was called the blending hypothesis. Page 367
- Mendel's experiments on garden peas used crosses of purebred pea varieties and demonstrated that the offspring inherited dominant and recessive "factors," which we now know are dominant and recessive alleles of a given gene. Page 368

LG 27.2 The segregation of homologous chromosomes during meiosis is the basis for the segregation of alleles.

- When Mendel crossed the heterozygous F_1 generation using a monohybrid cross, he discovered that the recessive traits that were not appearing in the F_1 generation reappeared in the F_2 generation. Page 368
- Mendel also discovered that the ratio of dominant: recessive phenotypes was 3:1, which illustrated the law of segregation. Page 370
- The 3:1 ratio reflects the phenotypes exhibited by homozygous dominant, heterozygous, and homozygous recessive genotypes. Page 370
- The law of segregation can be illustrated by a Punnett square in which we can examine all combinations of alleles contributed by two individuals that are crossed. Page 371
- The Punnett square is tied to the laws of probability, including the addition rule and multiplication rule of probability. Page 372
- The predicted phenotypic ratios can be tested against the observed ratios by using a chi-square test. Page 374
- The law of segregation maintains genetic variation in a population because the recessive alleles are not lost from the population; their phenotype is just not frequently exhibited because they are recessive. Page 376
- Incomplete dominance occurs when heterozygous individuals express intermediate phenotypes. Page 377
- In codominance, both alleles are expressed in the heterozygote, such as the *ABO* alleles for blood type in humans. Page 377

LG 25.3 Independent assortment during meiosis causes independent assortment of genes on different chromosomes.

- The law of independent assortment states that the segregation of one set of alleles for one gene sorts independently of a second set of alleles for a different gene. Page 378
- When the F_1 generation consists of doubly heterozygous individuals (*Aa Bb*) and they are mated in a dihybrid cross, the F_2 generation produces phenotypes in a ratio of 9:3:3:1. Page 378
- The 9:3:3:1 ratio is predictable from the multiplication rule of probability by assuming that the gene that determines one trait is independent of the gene that determines the other trait. Page 379
- The underlying reason for the 9:3:3:1 ratio is that during meiosis, each pair of maternal and paternal homologous chromosomes sorts independently of every other pair of chromosomes. Page 379

LG 25.4 We can visualize inheritance patterns using pedigrees.

- In some species, such as humans, it is not possible to do controlled breeding experiments to determine patterns of inheritance of dominant and recessive alleles. Page 380
- Pedigrees are visual maps of phenotypes using ancestral relationships. Page 380
- Rare dominant traits typically appear in every generation in a pedigree; when one parent has the phenotype, half of the offspring are expected to exhibit the phenotype. Page 380
- Rare recessive traits may not appear in the parents, but then reappear in the offspring. Page 381
- Individuals expressing rare recessive traits often result from mating between relatives, since the relatives are more likely to both be heterozygous for the trait. Page 382
- Incomplete penetrance and variable expressivity can complicate our interpretation of inheritance patterns in pedigrees. Page 382

Key Terms

Dominant	Homozygous	Chi-square (χ^2) test
Recessive	Heterozygous	Law of independent assortment
Allele	Probability	Dihybrid cross
Monohybrid cross	Addition rule of probability	Pedigree
Law of segregation	Multiplication rule of probability	

Review Questions

1. In regard to genotypes and phenotypes, which statement is true?

   (A) Genotypes indicate the alleles that an organism carries while phenotypes are the traits that organisms express.

   (B) Phenotypes indicate the alleles that an organism carries while genotypes are the traits that organisms express.

   (C) Phenotypes only exhibit dominant or recessive phenotypes.

   (D) Phenotypes always express complete penetrance.

2. Which statement about the law of segregation is true?

   (A) It helps explain why offspring exhibit the average phenotype of their parents.

   (B) It helps explain how genetic variation is maintained over generations.

   (C) It is the result of mitosis.

   (D) It helps explain how different traits sort independently.

3. What is a likely outcome of self-fertilizing an F_1 generation of heterozygotes that was produced by crossing purebred parents that are homozygous for a different allele?

   (A) The F_2 generation will be made up entirely of heterozygous individuals.

   (B) The F_2 generation will be comprised entirely of dominant phenotypes.

   (C) The F_2 generation will be comprised of $\frac{3}{4}$ dominant phenotypes and $\frac{1}{4}$ recessive phenotypes.

   (D) The F_2 generation will be comprised of $\frac{1}{2}$ dominant phenotypes and $\frac{1}{2}$ recessive phenotypes.

4. A cross begins with purebred parents ($AA\ BB \times aa\ bb$) and then the F_1 generation is allowed to self-fertilize. Which statement is false?

   (A) The F_1 generation will be comprised entirely of individuals that are heterozygous for both traits.

   (B) The F_1 generation will have a phenotypic ratio of 9:3:3:1.

   (C) The F_2 generation will be comprised of individuals that are both heterozygous and homozygous for both traits.

   (D) The F_2 generation will have a phenotypic ratio of 9:3:3:1.

5. Which statement is true about inferring inheritance patterns from pedigrees?

   (A) Incomplete penetrance can complicate the interpretation of inheritance.

   (B) Pedigrees are not useful for examining trait inheritance in humans.

   (C) Interpretation of inheritance is unaffected by variable expressivity.

   (D) Rare recessive alleles appear in every generation.

Module 27
AP® Practice Questions

Section 1: Multiple-Choice Questions

Choose the best answer for questions 1–5.

1. Superoxide dismutase is a gene found in the azure damselfly (*Coenagrion puella*). Two alleles of superoxide dismutase are designated as SOD^1 and SOD^2. If a male damselfly with genotype SOD^1/SOD^1 mates with a female damselfly with genotype SOD^1/SOD^2, which of the following correctly predicts the expected genotype frequencies of their offspring?

   (A) $25\%SOD^1/SOD^1, 50\%SOD^1/SOD^2, 25\%SOD^2/SOD^2$

   (B) $0\%SOD^1/SOD^1, 50\%SOD^1/SOD^2, 50\%SOD^2/SOD^2$

   (C) $100\%SOD^1/SOD^1, 0\%SOD^1/SOD^2, 0\%SOD^2/SOD^2$

   (D) $50\%SOD^1/SOD^1, 50\%SOD^1/SOD^2, 0\%SOD^2/SOD^2$

2. Natural populations of garter snake (*Thamnophis sirtalis*) are polymorphic in body color and pattern. Most garter snakes are black with a yellow stripe, but some are entirely black. Whether a snake is striped or solid black is determined by a single gene. Solid black is recessive to striped. If a black female mates with a purebred striped male, what frequency of their offspring are expected to be black?

   (A) 0%

   (B) 25%

   (C) 50%

   (D) 100%

3. Ear length in goats is a genetic trait determined by a single gene. Goats with a genotype *EE* have long ears, goats with genotype *Ee* have intermediate-length ears, and goats with genotype *ee* have short ears. The natural presence or absence of horns is also determined by a single gene in goats. Goats with an *HH* or an *Hh* genotype have horns, and goats with an *hh* genotype are born without horns. The genes for ear length and horn presence are present in separate chromosomes. What proportion of the offspring from a cross between a male goat with genotype *EeHh* and a female goat with genotype *eeHh* are expected to have short ears and no horns?

(A) 0.125

(B) 0.25

(C) 0.50

(D) 0.75

Use the following material to answer questions 4 and 5:

Wall-eye is a trait in swamp buffalo (*Bubalis bubalis*) in which the iris of the eye is very light in color instead of being the typical brown. The wall-eye trait in swamp buffalo is determined by a single gene, and the allele for wall-eye is dominant to the allele for brown eyes. The figure shows a pedigree for a lineage of swamp buffalo in which the wall-eye phenotype is present.

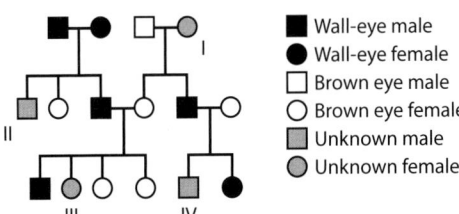

■ Wall-eye male
● Wall-eye female
□ Brown eye male
○ Brown eye female
▨ Unknown male
◑ Unknown female

4. If the wall-eye allele is designated as *A* and the brown eye allele is designated as *a*, what is the genotype of the brother of individual II?

(A) *AA*

(B) *Aa*

(C) Either *AA* or *Aa*

(D) *aa*

5. Which of the individuals indicated as having an unknown phenotype is most likely to be wall-eye?

(A) I

(B) II

(C) III

(D) IV

Section 2: Free-Response Question

Write your answer to each part clearly. Support your answers with relevant information and examples. Where calculations are required, show your work.

Petal color in snapdragons (*Antirrhinum majus*) is hypothesized to be determined by a single gene with two incompletely dominant alleles. Plants with an *RR* genotype have red petals. Plants with an *Rr* genotype have pink petals. Plants with an *rr* genotype have white petals. A researcher set up a cross between two plants with pink petals and collected 184 seeds. The researcher then planted the 184 seeds and recorded the petal color of the plants that grew. The data from this experiment is shown in the table.

Petal color	Number of progeny
Red	37
Pink	103
White	44

(a) **Determine** the proportion of the progeny plants that have an *RR* genotype.

(b) Using the hypothesis that petal color is determined by a single gene with incompletely dominant alleles, **calculate** the number of progeny plants that are expected to have pink petals.

(c) Use the data provided to **determine** the equation for a χ^2 test.

(d) Based on the data provided, do you find evidence to reject the hypothesis that petal color is determined by a single gene with incompletely dominant alleles? Provide reasoning to **justify** your response.

Tutorial 2: Probability

Gregor Mendel's pea plant crosses provided the world with new knowledge about the presence and functioning of genes. The purebred parental generation (P_1) crosses refuted the commonly believed blending hypothesis, an idea that phenotypes or traits of offspring were the average traits of the parents, and instead revealed the existence of dominant and recessive alleles. The crosses of the offspring of those parents ($F_1 \times F_1$) contributed more evidence to the concepts of dominant and recessive alleles by producing plants with traits that were not seen in the original P_1 crosses.

These experiments laid the groundwork for more complicated genetic studies and still serve as a foundation for understanding basic hereditary patterns. Not only do these pea plant crosses represent a clear example of heredity, they also provide an excellent opportunity to understand probability and its role in predicting the outcome of events. In this tutorial, we will discuss probability and its role in Mendelian genetics.

What Is Probability?

Probability is the likelihood that some event, observation, process, or experiment that can have two or more outcomes will occur. The probability for each of these outcomes can be calculated and will range from 0 to 1. If the probability of an event is 0, then that event will never happen. This can also be written as having a 0% chance of happening. Alternatively, if the probability of an event is 1, or 100%, then it will always happen.

To show this simply, let's use the example of rolling a 6-sided die once. There are 6 possible outcomes of rolling this die. It could land on a 1, 2, 3, 4, 5, or 6. Because we are only rolling the die once, there is an equal chance for each number to appear. The die has a 1 in 6 chance of landing on the 1, 2, 3, 4, 5, or 6. In terms of probability, each number has a $\frac{1}{6}$ or 0.167 chance of occurring.

Example: We can reexamine basic probability through Mendel's ideas as well. For example, let's say we are crossing a homozygous dominant pea plant, *AA*, with a homozygous recessive pea plant, *aa*. What would the probability be for each possible offspring genotype, *AA, Aa,* and *aa*? The probabilities for this cross can be deduced from the parental genotypes in a mating event and the principle of segregation, which states thawt alleles separate into different gametes. Punnett squares are a useful tool to visualize these crosses.

The Punnett square for this cross between a homozygous dominant pea plant, *AA*, and a homozygous recessive pea plant, *aa*, looks like this:

	a	*a*
A	*Aa*	*Aa*
A	*Aa*	*Aa*

The homozygous dominant parent's gametes and the homozygous recessive parent's gametes produce four offspring combinations. In this case, they are all heterozygous *Aa*.

The probability of a homozygous dominant offspring, *AA,* is 0; it never happens. The same can be said for a homozygous recessive offspring, *aa*. However, the probability of heterozygous offspring is 1; it always occurs.

This example replicates the findings of Mendel's initial purebred P_1 crosses. The F_1 generation inherits the *A* allele from one parent and the *a* allele from the other parent, resulting in a heterozygous genotype, *Aa*.

Your Turn

Mendel discovered that the alleles for purple flowers were dominant to white flowers in his pea plants. What is the probability of each type of offspring, in terms of genotype and phenotype, when crossing a heterozygous purple flower plant with a white flower plant?

Solution

To begin, let's assign the proper genotypes to the pea plant parents. The genotype is the set of genes that an organism carries. In this case, the combination of alleles *A* and *a* make up the genotype for an individual. These alleles can make three potential genotypes, *AA, Aa,* and *aa*. Each genotype influences an individual's phenotype, or the traits that an organism expresses. Here, the phenotype is the purple or white color of the plant. A plant with the genotype *AA* or *Aa* appears purple because of the genes it possesses, while a plant with the genotype *aa* exhibits a white phenotype.

For our question, the heterozygous purple flower would have the genotype *Aa*. The white flower must possess two recessive alleles, so it has the genotype *aa*. Now we can set up our Punnett square to help us visualize this cross:

	a	*a*
A	*Aa*	*Aa*
a	*aa*	*aa*

Based on this Punnett square, we predict that half of the offspring will be heterozygous and half will be homozygous recessive. In other words, the probability of observing the *Aa* genotype and the purple phenotype is 50%. The probability of observing the *aa* genotype and a white phenotype is also 50%. The probability of observing the *AA* genotype is 0%.

In some cases, it may be necessary to combine the probabilities of two or more possible outcomes of a cross or an event. When this happens, two rules of probability will be of help.

The Addition Rule

When we want to know the probability that one event or another will happen, we use the addition rule of probability. The addition rule states that the probability of two mutually exclusive events happening is the sum of their individual probabilities. Mutually exclusive events are those that cannot occur simultaneously in one individual. The probability of rolling either a 4 or a 5 on a die is an example of mutually exclusive events. You cannot roll both numbers simultaneously, so the addition rule would be used to calculate that probability. This rule can be simplified as:

If A and B are mutually exclusive, then:
$P(\text{A or B}) = P(\text{A}) + P(\text{B})$

Example: Let's use another of Mendel's discoveries to walk through this rule. Round seeds are known to be dominant to wrinkled seeds. What would the probability be of producing either round AA or wrinkled aa offspring in a cross of two round-seeded Aa parent plants? The Punnett square for this cross is shown below:

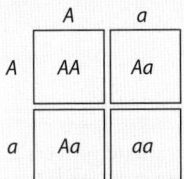

The probability of producing offspring with round seeds and the genotype AA is $\frac{1}{4}$. The probability of producing offspring with wrinkled seeds and the genotype aa is also $\frac{1}{4}$. If one seed is round, it cannot also be wrinkled, so we can conclude that these two events are mutually exclusive and the addition rule will apply to find the total probability for this cross. The probability of producing offspring that are either homozygous dominant or homozygous recessive is $\frac{1}{4} + \frac{1}{4} = \frac{1}{2}$. Thus, there is a 50% probability of observing either homozygous round or homozygous wrinkled seeds when the two plants are crossed.

The Multiplication Rule

Mutually exclusive events do not happen at the same time. Independent events are those in which the probability of one event occurring has no bearing on the probability of the other. For example, when rolling a single die multiple times or rolling two dice simultaneously, the outcome of one roll does not affect the outcome of the other. Rolling a 2 and then a 6 would be considered independent events. Obtaining a 2 on the first roll has no effect on the probability of obtaining a 6 for the second roll. So, what if we want to know the probability of two independent events occurring? In this situation, we would use the multiplication rule.

The multiplication rule states that the probability of two independent events both happening is the product of their individual probabilities. In contrast to the conditions where we would use the addition rule, these outcomes can occur simultaneously. While the words "either, or" are a sign of when to use the addition rule, the word "and" is key to knowing when to use the multiplication rule. Because fertilization events are independent of each other, this rule is often used in determining the probability of successive offspring of a cross. This rule can be simplified as:

If A and B are independent, then:
$P(\text{A and B}) = P(\text{A}) \times P(\text{B})$

Example: Continuing with the $Aa \times Aa$ seed cross, if pea plants typically produce pods with 5 seeds in them, let us find the probability of a pod with seeds that alternate in shape, such that the first seed is round, the second is wrinkled, the third is round, the fourth is wrinkled, and the fifth is round. Because the fertilization of each seed is an independent event, the shape of one seed does not influence the shape of the next seed. Therefore, we can use the multiplication rule.

Round is the dominant seed phenotype, represented by the genotypes AA and Aa, and thus has a probability of $\frac{3}{4}$. A wrinkled seed, aa, has a probability of $\frac{1}{4}$. We would then multiply the probabilities of the alternating seed shapes:

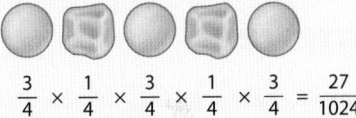

$$\frac{3}{4} \times \frac{1}{4} \times \frac{3}{4} \times \frac{1}{4} \times \frac{3}{4} = \frac{27}{1024}$$

The probability of having this sequence of seeds is $\frac{27}{1024}$, or 2.6%.

Your Turn

Green pea pods are dominant to yellow pea pods, and yellow seeds are dominant to green seeds. In an $Aa \times Aa$ cross, what is the probability of a plant producing either a green or a yellow pod and each pod containing 4 seeds in the following order: yellow, yellow, green, and green?

Solution

First, let's find the probability of a green pod with 4 seeds in the order described. The probability of a green pod is $\frac{3}{4}$. A pod containing 2 yellow seeds followed by 2 green seeds would make use of the multiplication rule because these events are independent of each other—the color of one seed does not influence the color of the next seed. Therefore, the seed colors would have a probability of $\frac{3}{4} \times \frac{3}{4} \times \frac{1}{4} \times \frac{1}{4} = \frac{9}{256}$. Because the pod color and the seed color are occurring simultaneously, we can also multiply their probabilities to get $\frac{3}{4} \times \frac{9}{256} = \frac{27}{1024}$.

We can follow the same process for the yellow pod with 2 yellow and 2 green seeds. The probability of a yellow pod is $\frac{1}{4}$ and the seed colors would have the same probability as those in the green pod, $\frac{9}{256}$. Multiplying these two terms together yields $\frac{1}{4} \times \frac{9}{256} = \frac{9}{1024}$.

Now that we have the individual probabilities sorted out, we can figure out how to combine them. The question asks for the probability of either the green pod and colored seeds or the yellow pod and colored seeds; this is a mutually exclusive event and so the addition rule of probability should be used to find the final answer.

Green pod with 2 yellow and 2 green seeds + yellow pod with 2 yellow and 2 green seeds:

$$\frac{27}{1024} + \frac{9}{1024} = \frac{36}{1024} = \frac{18}{512} = \frac{9}{256}$$

There is a $\frac{9}{256}$, or 3.5%, chance of having a green pod or a yellow pod and each pod containing 4 seeds in the following order: yellow, yellow, green, and green.

▶ See page 373 for "Analyzing Statistics and Data: Calculating Probability," which provides an opportunity to practice these concepts in context.

Probability and its accompanying rules can be confusing. Punnett squares are a great first step to visualizing the genetic combination of a mating event. Remember to seek out the key phrases of each rule to determine which to use in the problem you are solving. We can predict the outcomes of many biological events when following the above guidelines.

Chi-square Test

Probability is used for predicting the results of certain events and outcomes. There are also analyses known as chi-square tests that statistically compare the observed and expected phenotypes in an experiment. Scientists frequently use chi-square tests to determine whether differences in data are significant. Chi-square tests help determine whether observed outcomes differ significantly from expected outcomes that are based on a hypothesis.

Example: The table below shows Mendel's flower color data for his pea plants. If the observed phenotypic ratio of this monohybrid cross is 3.15:1, is that statistically different from the expected ratio of 3:1, or is it similar enough to allow us to fail to reject the hypothesis that the plants will produce an expected ratio of 3:1?

Trait	Dominant phenotype (purple flowers)	Recessive phenotype (white flowers)	Ratio
Flowers	705	224	3.15:1

To answer this and other questions about statistical differences between observed and expected groups, we must follow a few steps.

First, determine the values for the observed and expected phenotypes. The observed value is normally given in a table, such as the one above. It will be helpful to organize the observed and expected data in a table of your own.

	Observed	Expected
Purple flowers	705	
White flowers	224	
Total flowers	929	929

The second step is to find the expected values. Here, we will need to do some basic calculations to fill in the rest of our table. We can do this by applying our expected ratio, 3:1 or 75% to 25%, to the total number of flowers, 929.

Expected purple flowers: $929 \times 0.75 = 697$
Expected white flowers: $929 \times 0.25 = 232$

We can fill in these values to complete our table.

	Observed	Expected
Purple flowers	705	697
White flowers	224	232
Total flowers	929	929

Now that we have both the observed and expected values, we can begin performing the chi-square test. The formula for this chi-square is as follows:

$$\chi^2 = \sum \frac{(o - e)^2}{e}$$

where:

o = observed result
e = expected result
Σ = sum of all

The third step in the chi-square process is to use this formula. The "sum of all" means that we must add the observed and expected calculation for the purple flower phenotype to the observed and expected calculation for the white flower phenotype. Entering these values yields:

$$\chi^2 = \frac{(705 - 697)^2}{697} + \frac{(224 - 232)^2}{232}$$
$$\chi^2 = 0.092 + 0.276$$
$$\chi^2 = 0.37$$

Our chi-square value is 0.37.

Now we must compare this value to a table of critical chi-square values given in Table 27.2 on page 375 to conclude if the observed results differ significantly from the expected results.

If the calculated chi-square value is less than the critical chi-square value found in Table 27.2, then we can conclude that any difference between the observed and expected results may have been due to chance and we cannot reject the null hypothesis. This means that the differences between the observed and expected counts were not very large.

A calculated chi-square value that is larger than the critical chi-square value would mean that the difference between the observed and expected groups was large. The observed counts are significantly different from the expected counts and we must reject the null hypothesis.

To use Table 27.2, we must begin our fourth step and determine the degrees of freedom for our question. The degrees of freedom can be found by taking the number of categories in the data and subtracting one. For our problem, there are 2 categories, purple flowers and white flowers. $2 - 1 = 1$, so our degree of freedom is 1.

We now have all of the information necessary to read Table 27.2. The first number to look for is the degrees of freedom, 1. From here, we will look at the P-value for 0.05. Remember, this is our 5% threshold for considering if one group is significantly different from another. The critical chi-square value we get from the table is 3.84. Comparing our calculated chi-square value of 0.37 to the table value of 3.84, we see that our calculated value is much less ($0.37 < 3.84$).

The final step in this process is to interpret our findings and draw a conclusion about whether the observed flower phenotypes differ significantly from the expected flower phenotypes. Because our calculated chi-square value is less than the critical value, we fail to reject the null hypothesis. Any difference between the observed and expected results may have been due to chance. Thus, our hypothesis that the trait is inherited as a single gene with a dominant and recessive allele cannot be rejected.

Your Turn

The data shown in the table below were collected after mating the heterozygous F_1 generation for mice where brown fur is dominant to white fur. What is the chi-square value for these data? Interpret the results in terms of the null hypothesis.

	Observed	Expected
Brown mice	125	
White mice	27	
Total	152	152

Solution

To start, we must first calculate the expected number of mice for this mating. Since it was an $Aa \times Aa$ cross, we can conclude that our expected ratio would be 3:1 brown mice to white mice. Applying this ratio to our total number of mice gives us:

Expected brown mice: $152 \times 0.75 = 114$
Expected white mice: $152 \times 0.25 = 38$

With these variables found, we can now use the chi-square formula to find our calculated chi-square value.

$$\chi^2 = \frac{(125 - 114)^2}{114} + \frac{(27 - 38)^2}{38}$$
$$\chi^2 = 1.061 + 3.184$$
$$\chi^2 = 4.25$$

The next step is to calculate our degrees of freedom for this experiment. Our categories this time are brown mice and white mice; therefore, our degrees of freedom are $2 - 1 = 1$.

Taking all of this information to Table 27.2, we see that with 1 degree of freedom and a P-value of 0.05, the critical chi-square value is 3.84. Our calculated chi-square value is greater than this value ($4.25 > 3.84$), so we reject our null hypothesis. The differences between the observed and expected values are significantly different.

▶ See page 375 for "Analyzing Statistics and Data: Chi-square Analysis," which provides an opportunity to practice these concepts in context.

Like probability, chi-square tests take some practice to get comfortable with. Questions can range from relatively simple setups with only two categories, like those above, to complex dihybrid-cross scenarios. No matter the complexity, as long as the proper steps are followed, chi-square tests are a useful tool that provides a way to determine statistical significance when our data consist of observed and expected counts.

Module 28

Non-Mendelian Genetics

LEARNING GOALS ▸**LG 28.1** Epistasis is caused by multiple genes that interact to determine a phenotype.
▸**LG 28.2** Sex-linked traits are determined by genes in the sex chromosomes.
▸**LG 28.3** Genes linked in the same chromosome do not sort independently unless recombination occurs.
▸**LG 28.4** Mitochondrial and chloroplast genes are also inherited.

In the previous module, we discussed the pea plant experiments of Gregor Mendel. In those experiments, Mendel discovered that the flower, pod, and seed traits he examined were determined by dominant and recessive alleles for a single gene, which was located in a single chromosome. He further described how the frequencies of these alleles were determined by the law of segregation and the principle of independent assortment. Mendel did not know that he was very fortunate to have selected traits that were determined in a very simple and straightforward manner. The pea plant traits he worked on are determined by a single gene with only two alleles. It turns out that few phenotypes follow such simple patterns. Most phenotypes are determined in much more complex ways. Thus, we refer to these more complex inheritance patterns as non–Mendelian genetics.

In this module, we will consider some of the more complex ways in which phenotypes can be determined by genes. We will discuss how multiple genes can work together to determine phenotypes. We will then examine the novel situation of genes that only exist in one chromosome, such as the *X* or *Y* sex chromosomes. We will also discuss situations in which two genes are located in close proximity in a single chromosome. Finally, we will examine the ways in which mitochondrial and chloroplast genes are inherited.

PREP FOR THE AP® EXAM

FOCUS ON THE BIG IDEAS

INFORMATION STORAGE AND TRANSMISSION: Look for examples of inheritance patterns that illustrate deviation from the expected ratios predicted by Mendelian patterns of inheritance.

28.1 Epistasis is caused by multiple genes that interact to determine a phenotype

In Module 27, we discussed how Mendel discovered the law of independent assortment for two genes that code for different traits. For example, in Mendel's pea plants, the genes that coded for one trait such as seed color were independent from the genes that coded for a second trait such as seed shape. As a result, his dihybrid crosses produced a phenotypic ratio of 9:3:3:1. However, more recent experiments have found that phenotypic ratios can diverge from this expected ratio. One way this can happen is when one gene affects the phenotype that is coded by another gene. For example, one gene may code for an enzyme that is needed in the biochemical pathway that is coded by a second gene. As a result, the first gene can prevent the second gene from producing

a particular phenotype, or it might cause the second gene to produce a very different phenotype. When one gene modifies the expression of another gene, we call it epistasis.

An excellent example of epistasis can be found in the feather color of White Wyandotte and White Leghorn chickens, which you can see in **FIGURE 28.1**. Their phenotype (white feathers) is caused by two distinct genetic mechanisms. To understand the two genetic mechanisms, we first need to know that two different genes determine feather color in these chickens and that each gene has two possible alleles. The first gene codes for the production of a dark pigment color. The *C* allele is dominant and codes for the production of a dark pigment. The *c* allele is recessive and codes for the production of no pigment, which results in white feathers. The second gene codes for a protein that determines whether pigment production is inhibited. Thus, even if a chicken has the dominant *C* allele for producing

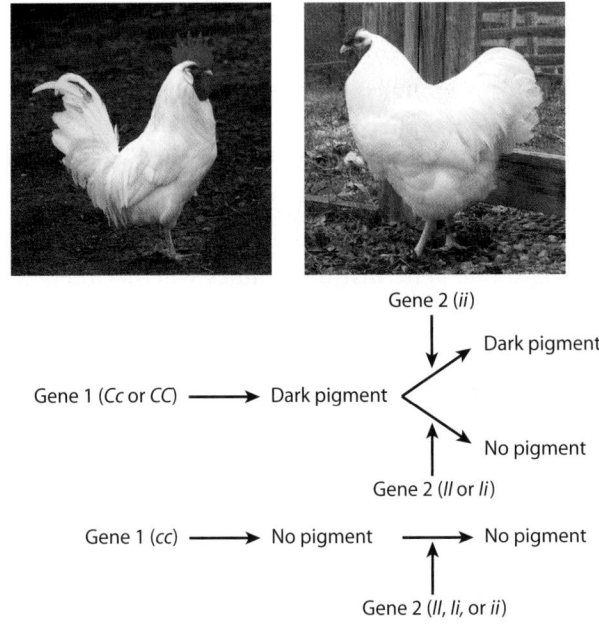

Gene 2 (*ii*)

Gene 1 (*Cc* or *CC*) ⟶ Dark pigment

Dark pigment

No pigment

Gene 2 (*II* or *Ii*)

Gene 1 (*cc*) ⟶ No pigment ⟶ No pigment

Gene 2 (*II, Ii,* or *ii*)

FIGURE 28.1 Epistasis in chickens

The white feathers of the White Leghorn (left) and the White Wyandotte (right) are the result of epistasis. As shown in the genetic pathway, the first gene codes for the production of a dark pigment color. The second gene codes for inhibiting or not inhibiting any pigment production. Photos: (left) Andrea Mangoni/Shutterstock; (right) Dr. Don Monke

the dark pigment, the second gene can determine whether the chicken has white or pigmented feathers. The allele *I* is dominant and codes for a protein that inhibits pigment production. The allele *i* is recessive and codes for a protein that does not inhibit pigment production. If the first gene codes for the production of a pigment with allele *C*, the second gene can inhibit that pigment production with allele *I* and the chicken will have white feathers.

With this information about the two genes, we can understand the two distinct genetic reasons that two chicken breeds both have white feathers. As illustrated in **FIGURE 28.2**, the White Wyandotte is white because it carries two recessive alleles for pigment production (*cc*). It also carries alleles for not inhibiting pigment production (*ii*), but this makes no difference since the first gene does not produce any pigments. In contrast, the White Leghorn is white because it carries the genotype *CC II*. This genotype means that it carries the alleles to produce the dark pigment (*CC*), but it also carries the alleles that block pigment production (*II*). Because of the alleles that block pigment color, the production of pigment cannot happen and the White Leghorn makes white feathers.

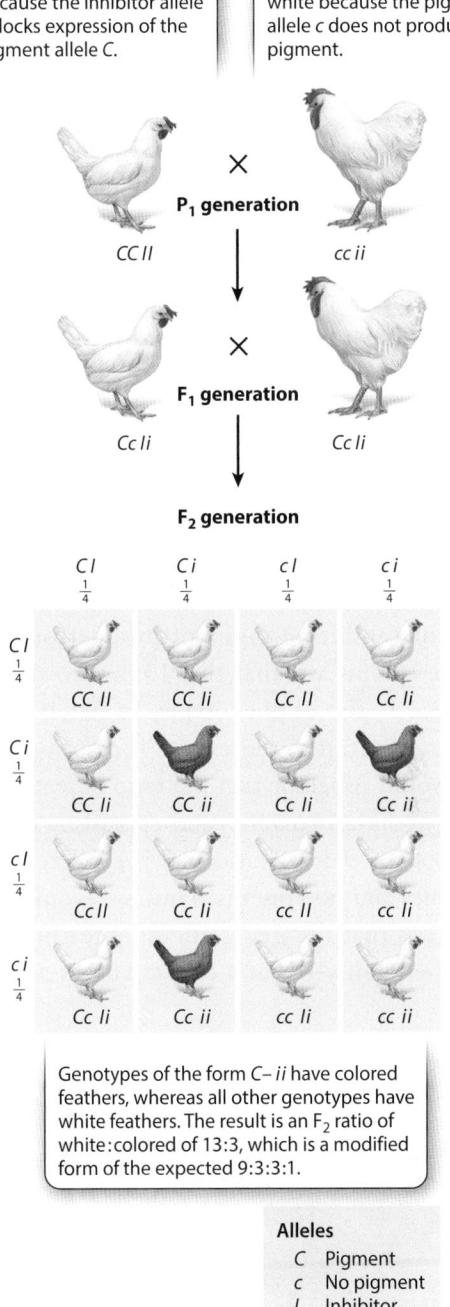

The White Leghorn is white because the inhibitor allele *I* blocks expression of the pigment allele *C*.

The White Wyandotte is white because the pigment allele *c* does not produce pigment.

P₁ generation

CC II × *cc ii*

F₁ generation

Cc Ii × *Cc Ii*

F₂ generation

Genotypes of the form *C– ii* have colored feathers, whereas all other genotypes have white feathers. The result is an F₂ ratio of white:colored of 13:3, which is a modified form of the expected 9:3:3:1.

Alleles

C Pigment
c No pigment
I Inhibitor
i No inhibitor

FIGURE 28.2 Crossing purebred chickens to observe epistasis

The White Leghorn breed of chicken has white feathers because the alleles that code for pigment production (*CC*) are influenced by the alleles of another gene that code for making a protein that inhibits pigment production (*II*). The White Wyandotte breed has white feathers because it lacks the alleles that code for pigment production (*cc*). Crossing the two breeds results in an F₁ generation that is doubly heterozygous, which means it would make the pigment (*Cc*), but pigment production is inhibited (*Ii*). When the dihybrid F₁ generation is crossed, it results in an F₂ phenotypic ratio of 13 white (nonpigmented) chickens for every 3 brown (pigmented) chickens.

What would the offspring phenotypes look like if we crossed these two pure breeds of white-feathered chickens? As shown in Figure 28.2, the F$_1$ generation would consist entirely of individuals that are heterozygous for both traits (*Cc Ii*). Based on this genotype, the chickens have the dominant allele for pigment production (*C*), but they also have the dominant allele for inhibiting pigment production (*I*). As a result, all of the F$_1$ chickens have white feathers.

It gets more interesting when we conduct a dihybrid cross of the F$_1$ generation to produce the F$_2$ generation, as shown in the Punnett square. In this case, the F$_2$ generation has a phenotypic ratio of 13 white chickens and 3 dark-pigmented chickens. All of the dark F$_2$ chickens have genotypes to produce the pigment (either *CC* or *Cc*) and they are homozygous recessive for the allele that inhibits the pigment (*ii*). In contrast, the white F$_2$ chickens are white either because they carry the homozygous recessive alleles for pigment production (*cc*) or because they carry at least one allele for inhibiting pigment production (*II* or *Ii*). It is because of epistasis that this 13:3 ratio differs from Mendel's 9:3:3:1 ratio of phenotypes. In fact, if we were to apply a chi-square test to these data, as described in Module 27, we would conclude that the ratio of feather phenotypes when we have two genes and epistasis is significantly different from the 9:3:3:1 ratio that is expected when we have two genes and no epistasis. Thus, we would reject the null hypothesis that the genes affecting feather color are acting independently of each other. This is an excellent example of epistasis because one gene blocks the phenotype produced by another gene.

While many examples of epistasis involve one gene completely blocking the phenotype produced by another gene, as we saw in chickens, in other cases one gene can subtly modify the effects of another gene. An interesting example of this phenomenon can be found in the coat colors of horses. Horses have one gene that determines whether their coats are black or red. We know that this color difference is determined by two alleles. The *E* allele is dominant and codes for a black coat while the *e* allele is recessive and codes for a red coat. However, there is a second gene that can modify the red coat color to different degrees, depending on which alleles are present for the second gene. This second gene exhibits incomplete dominance, so the alleles are written in a different way, as we saw Module 27. The second gene can have either a C^{CR} allele, which can modify coat color, or a *C* allele, which does not modify coat color. If the first gene codes for a red coat (*ee*) and the other gene has two recessive alleles (*CC*), then the coat color is not modified and the horse has a reddish coat that we refer to as chestnut. However, if the first gene codes for a red coat (*ee*) and the second gene has one modifying allele ($C^{CR}C$), then the horse has a light brown coat, with a white tail and mane, coloring that we refer to as palomino. Finally, if the first gene codes for a red coat (*ee*) and the second gene has two modifying alleles ($C^{CR}C^{CR}$), then the horse has a very light coat that we refer to as cremello. You can see these three different horse phenotypes in **FIGURE 28.3**.

FIGURE 28.3 Epistasis in horse coat color

The coat color of horses can be black (*EE* or *Ee*) or red (*ee*). When the gene for coat color codes for a red coat, it can be modified by a second gene that lightens the coat color. (a) If the second gene has two *CC* alleles, the coat color is not modified, so it remains a reddish brown color that we call chestnut. (b) If the second gene has one C^{CR} allele, the horse has a lighter brown coat, with a white tail and mane. We refer to this color as palomino. (c) If the second gene has two C^{CR} alleles, the horse has a very light, whitish coat that we call cremello.

Photos: (a) Alexia Khruscheva/Shutterstock; (b) Gary Chalker/Getty Images; (c) arthorse/Alamy Stock Photo

As you can appreciate from these examples, epistasis provides us with a very different perspective of how genotypes determine phenotypes. As a reminder, Mendel's pea plant crosses did not involve epistasis, so the phenotypes he observed were the product of genes acting independently on the pea phenotypes. However, we now know that epistasis is quite common in a wide range of species.

✓ Concept Check

1. **Describe** the reason epistasis causes a modification of Mendel's phenotypic ratio of 9:3:3:1.

2. **Describe** the reason that the doubly homozygous recessive F_2 chickens (*cc ii*) have the same phenotype as chickens that are homozygous recessive for producing pigments (*cc*) but carry a dominant allele for pigment inhibition (*II* or *Ii*).

3. **Describe** how the palomino horse coat color represents an example of epistasis.

28.2 Sex-linked traits are determined by genes in the sex chromosomes

When we think about the genes that code for most traits, we commonly think about the genes that are in the many different autosomes. However, there are also some very important genes located in the sex chromosomes. As we discussed in Module 26, mammals have two different-sized chromosomes that are designated as *X* and *Y*. Individuals with the *XX* genotype exhibit the female phenotype, while individuals with the *XY* genotype exhibit the male phenotype. Genetic sex differences also occur in many other species of animals. For example, in birds the sex chromosomes are named *Z* and *W*. In contrast to mammals, the heterozygous genotype (*ZW*) produces females while the homozygous phenotype (*ZZ*) produces males. In some insect species, genetic sex determination can be even more extreme. In bees, for instance, fertilized eggs are diploid and develop into females, while unfertilized eggs remain haploid and develop into males.

Sex chromosomes that determine the sex of the animals also contain genes that code for other traits. In this section, we will explore some of these genes and the phenotypes they control. As we will see, when the genes only exist in one of the two sex chromosomes, the pattern of genetic inheritance is substantially different from what we saw in the pea plant experiments of Mendel.

Genetically Determined Sex Based on Chromosomal Differences

In Module 26, we mentioned that while homologous chromosomes have all of their genes in common, the *X* and *Y* chromosomes have only a few genes in common. You can see an image of the two sex chromosomes in **FIGURE 28.4**. In the lower portion of the figure, we can see a photo of the

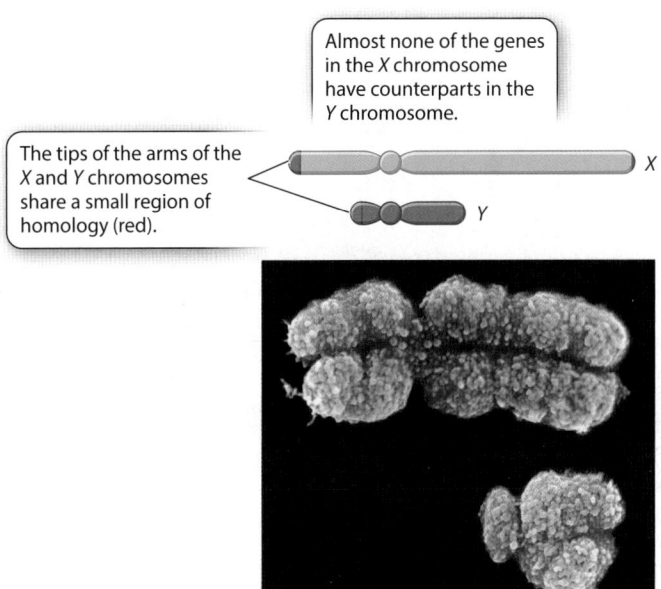

Almost none of the genes in the *X* chromosome have counterparts in the *Y* chromosome.

The tips of the arms of the *X* and *Y* chromosomes share a small region of homology (red).

FIGURE 28.4 *X* and *Y* chromosomes in humans

The photo shows the larger *X* chromosome positioned above the much smaller *Y* chromosome, immediately after chromosome duplication has occurred. The two sex chromosomes have a few genes in common on their ends, as shown in red on the top schematic image of the two sex chromosomes. This allows the two sex chromosomes to form a chiasma during prophase I of meiosis. The remainder of the two chromosomes has unique genes, and those in the *X* chromosome do not have homologous counterparts in the *Y* chromosome. Photo: Science Photo Library/Science Source

large *X* chromosome positioned above the much smaller *Y* chromosome. Above the photo, we can see a diagram version of the two sex chromosomes, with the red color indicating the locations of the few genes that the *X* and *Y* chromosomes share in common.

In humans, the larger *X* chromosome has about 900 genes, whereas the much smaller *Y* chromosome only has about 50 genes. Thus, a very large number of genes (about 850) exist only in the *X* chromosome. When genes are

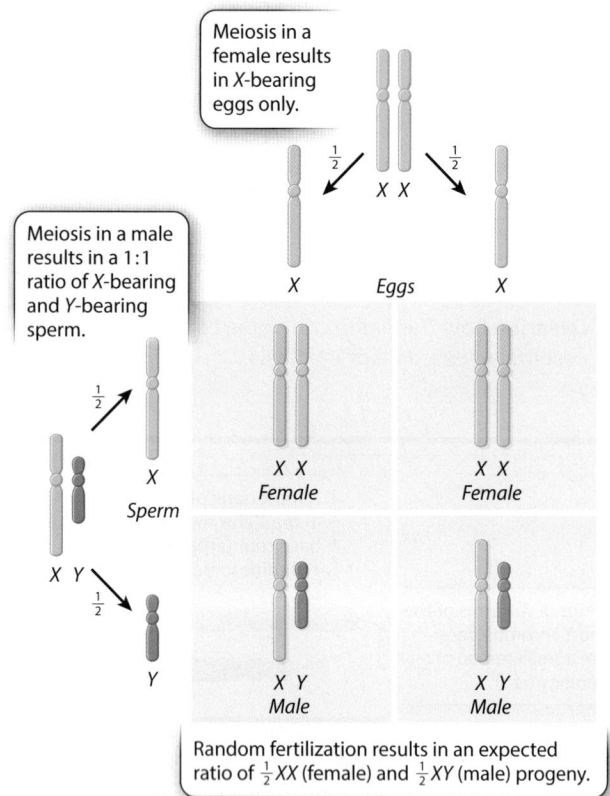

FIGURE 28.5 A Punnett square of sex chromosomes

A mating between a female mammal (*XX*) and a male mammal (*XY*) results in a 50% probability of producing daughters (*XX*) and a 50% probability of producing sons (*XY*).

located in only one of the two sex chromosomes, we call them **sex-linked genes.** More specifically, genes located in only the *X* chromosome are called *X*-linked genes, while genes located in only the *Y* chromosome are called *Y*-linked genes. These sex-linked genes produce sex-linked traits, which are interesting to us because they have very different patterns of inheritance compared to genes in autosomes.

The inheritance of sex chromosomes in mammals produces a very interesting pattern. As individuals produce gametes by meiosis, a female will produce eggs that either have the *X* chromosome she inherited from her mother or the *X* chromosome she inherited from her father. In contrast, a male will produce sperm that has either the *X* chromosome he inherited from his mother or the *Y* chromosome he inherited from his father. As shown in **FIGURE 28.5**, we can consider these different gametes in terms of a Punnett square. When we examine the four possible mating combinations, we can see that matings result in a 50% probability of producing daughters (*XX*) and a 50% probability of producing sons (*XY*). Thus, matings result in a 1:1 sex ratio of males and females in the next generation.

Inheritance of Genes in the X Chromosome

The first discovery of *X*-linked genes was in a population of fruit flies. The flies typically have red eyes, but the researchers discovered a male fly with a mutation that caused him to have white eyes. Presuming that the mutation was recessive, the researchers crossed the mutant male with red-eyed females. The F_1 generation all had red eyes, which we would expect for a recessive allele. However, when they crossed individuals from the F_1 generation, they were surprised by what they saw in the F_2 generation. The researchers did not observe the typical 3:1 ratio of dominant and recessive phenotypes in both male and female flies, as we would have predicted from Mendel's studies of pea plants. Instead, all of the females were red-eyed while half of the males were red-eyed and the other half were white-eyed, as shown in **FIGURE 28.6**.

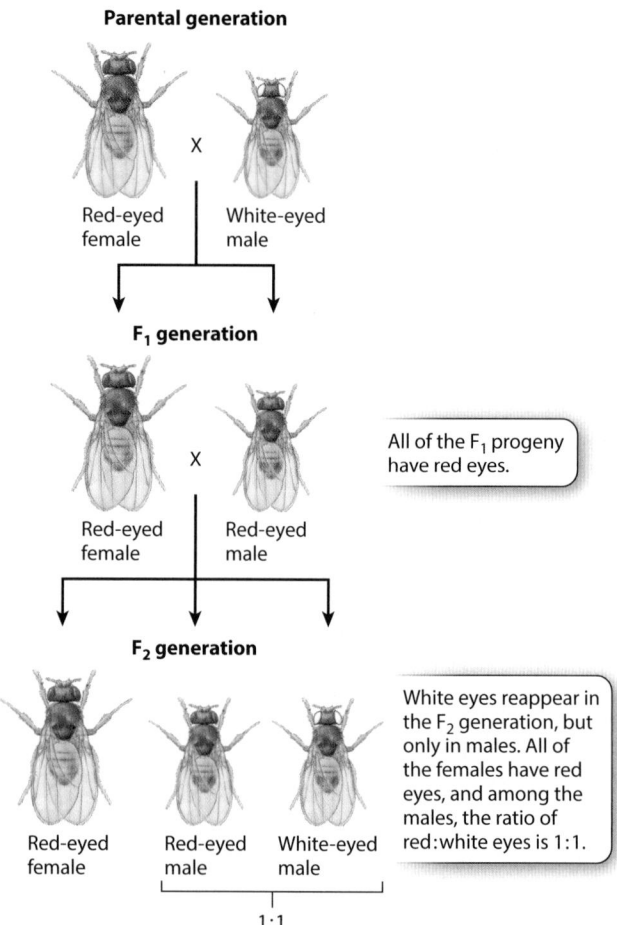

FIGURE 28.6 Fruit fly crosses with an X-linked trait

In fruit flies, white eyes are recessive to red eyes, so when a red-eyed female is crossed with a white-eyed male, all of the F_1 generation has red eyes. When the F_1 generation is crossed, the F_2 generation of females all have red eyes. However, half of the F_2 males have red eyes, while the other half of the males have white eyes. This unusual outcome occurs because the recessive white-eye allele is an *X*-linked trait.

Clearly, something different was happening in the inheritance of eye color in the fruit flies.

Why did the researchers observe such an unusual ratio of phenotypes that differed between the sexes? It turns out that eye color is determined by a gene that is located in the X chromosome of the fruit fly. Because there is no homologous gene on the Y chromosome, the gene that determines eye color in fruit flies is an X-linked gene. Females carry two X chromosomes, so they exhibit the recessive white-eye phenotype only if they have two X chromosomes that both carry the recessive allele. In contrast, because males carry one X and one Y chromosome, they exhibit the recessive white-eye phenotype whenever they have just one copy of the recessive allele.

You can see these outcomes in the form of the two Punnett squares in **FIGURE 28.7**. In part a, we see that when a homozygous female mates with a male who carries the recessive allele, all of the offspring have either one or two dominant alleles. Because none of the males has the recessive allele in his X chromosome and none of the females has both X chromosomes with the recessive allele, all of the offspring express red eyes. In part b, we see a contrasting situation. When the female carries one recessive allele and the

male does not, all of the females have red eyes because they have one X chromosome with the dominant allele for red eyes. Among the males, half of them have red eyes because they lack the recessive allele in their single X chromosome, while the other half have white eyes because they possess the recessive allele in their single X chromosome.

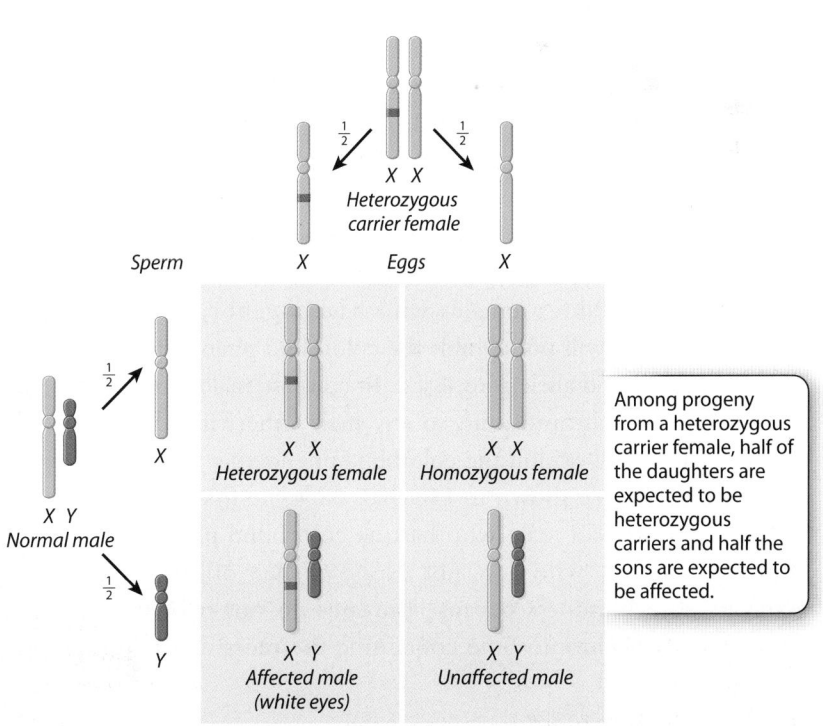

FIGURE 28.7 A Punnett square for X-linked traits

We can use a Punnett square to determine the different genotypes and phenotypes for X-linked genes that code for X-linked traits. (a) When a female that is homozygous for red eyes mates with a male carrying a recessive white-eye allele in his X chromosome, the resulting offspring consist of females that are heterozygous for the red-eye allele and males that have a single red-eye allele, so all of the offspring express the red-eye phenotype. (b) When a heterozygous female is mated with a male carrying a dominant red-eye allele in his X chromosome, the resulting female offspring are either heterozygous for red eyes or homozygous for red eyes. Thus, all of the female offspring have red eyes. In contrast, the resulting male offspring consist of individuals that carry the recessive allele or the dominant allele in their single X chromosome. Thus, half of the male offspring have white eyes, while the other half have red eyes.

These Punnett squares highlight two important patterns about the inheritance of X-linked genes. First, males transmit only their X chromosome to their daughters and only their Y chromosomes to their sons. Therefore, if a male has a recessive allele in his X chromosome, his daughters will inherit it. Second, sons inherit their X chromosome from their mothers. Therefore, if the mother is heterozygous, the male offspring have a 50% probability of inheriting the recessive allele and a 50% probability of expressing the recessive phenotype.

Collectively, this means that when genes are X-linked, we can predict the pattern of inherited genotypes and the phenotypes that will be produced. It also means we can infer that a gene is X-linked when we see specific phenotypic ratios emerging from a hybrid cross. Mendel demonstrated a typical 3:1 ratio of dominant and recessive phenotypes that emerge from an F_1 hybrid cross for autosomal traits. An F_1 hybrid for X-linked traits produces a very different ratio of phenotypes that also differs dramatically between the sexes. An F_1 hybrid for X-linked genes produces F_2 females that all express the dominant phenotype and F_2 males that express 50% dominant phenotypes and 50% recessive phenotypes. Thus, by examining the ratios of different phenotypes, we can infer whether a trait is coded for by genes in the autosomes or genes in the X chromosome.

We can also examine patterns of inheritance for X-linked genes using pedigrees of human families, which we discussed in Module 27. For example, there is a recessive mutation for red-green colorblindness in humans that is exhibited by about 10% of males. Individuals expressing this phenotype cannot tell the difference between red and green colors. For example, they are unable to see the number 5 when comprised of green dots against a background of red, orange, and yellow dots in **FIGURE 28.8**.

We can denote the dominant, common vision gene as X^+ and the recessive colorblind allele as X^C. In this case, the superscript "+" refers to the common or "wild-type" allele while the superscript "C" refers to a mutant, less common allele. Thus, females who have a genotype of X^+X^+ or X^+X^C will not exhibit the colorblind phenotype, since the colorblind allele is recessive. In contrast, males only have a single X chromosome, so any male inheriting the recessive allele will exhibit the colorblind phenotype.

In the pedigree shown in **FIGURE 28.9**, the first generation consists of a male who has the colorblind phenotype and a female who does not. None of their offspring exhibits colorblindness because the sons do not inherit their father's X chromosome containing the recessive X^C

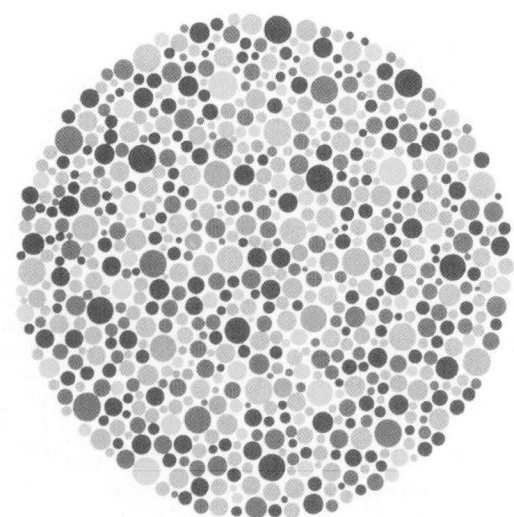

FIGURE 28.8   A colorblindness test for humans

Individuals who have red-green colorblindness have a hard time discerning certain colors, such as the green dots the make up the number 5 in a field of red, orange, and yellow dots. Photo: Dorling Kindersley/Getty Images

allele. Moreover, if the mother in the first generation does not carry the recessive allele, all of the daughters would be heterozygous because the X chromosome they receive from their father does carry the recessive allele. Across the entire the third generation, about half the sons have the colorblind phenotype, while none of the daughters does, which is similar to what we observed in the fruit flies carrying the recessive allele for white eyes. Thus, the patterns of inheritance in the human pedigree strongly suggests that the recessive allele for colorblindness is an X-linked gene. Many other traits are also determined by X-linked genes, including more than 100 disorders, among them hemophilia, which causes a person's blood to not clot properly.

Inheritance of Genes in the Y Chromosome

As we have mentioned, there are a large number of genes in the X chromosome in humans and other mammals, but

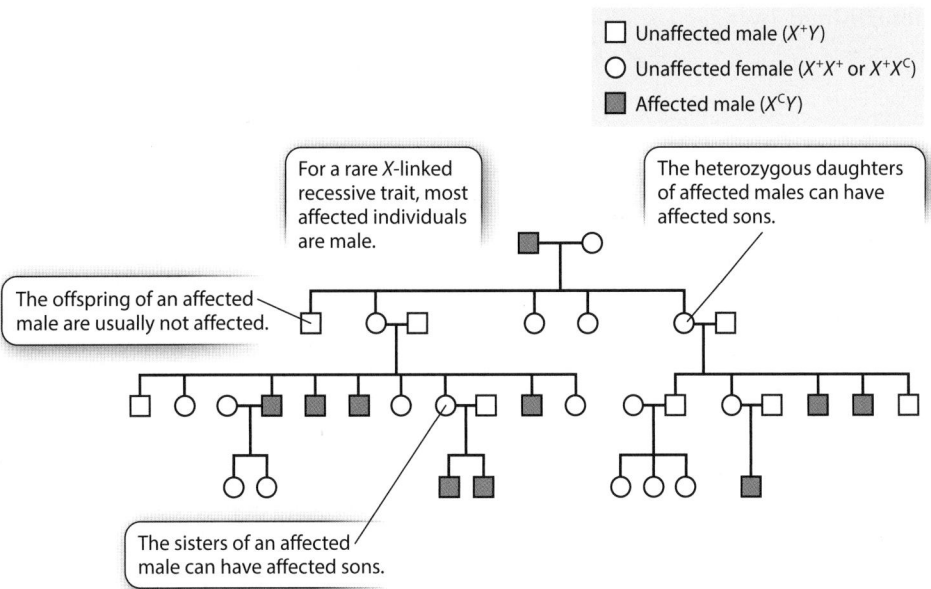

FIGURE 28.9 A pedigree for human colorblindness, an X-linked trait

Colorblindness results from a mutation in the X chromosome that is expressed only in females that carry two copies or in males that carry one copy. In a pedigree of humans, when a color-blind father mates with a mother who we presume lacks the mutation, none of the offspring is colorblind because the daughters are heterozygous and the sons received their X chromosome from their mother, who lacks the mutation. In the grandchildren, none of the granddaughters is colorblind because each granddaughter carries only (at most) one recessive allele from their mother. However, half of the grandsons inherit the recessive allele from their heterozygous mother and are colorblind, while the other half of the grandsons inherit the dominant allele from their mother and are not colorblind.

there are not many genes in the smaller *Y* chromosome. One important gene in the *Y* chromosome is called *SRY,* which stands for "sex-determining region in the *Y* chromosome." This gene triggers male development in embryos so that an *XY* geno-type leads to a male phenotype. Other genes are also important to male fertility, and muta-tions in these genes can result in impaired fer-tility, including low sperm counts.

For such *Y*-linked traits, we can see that they have a unique pattern of inheritance that is very different from *X*-linked genes and auto-somal genes, as illustrated in **FIGURE 28.10**. In this pedigree, males that carry a mutant allele for fertility are shown in red. As you can see, fathers can transmit the mutant allele only to their sons and never to their daughters. Moreover, females can never inherit the mutant allele nor can they transmit the allele to the next generation. Finally, all sons of

affected fathers will also be affected because the mutant allele is transmitted in the *Y* chromosome. Understand-ing these patterns allows us to determine when a pattern

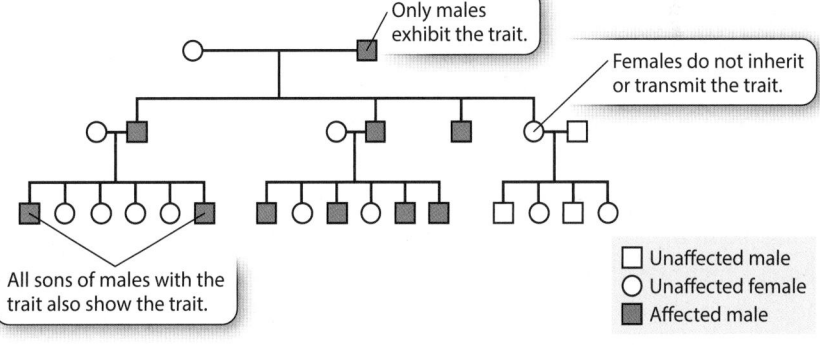

FIGURE 28.10 A pedigree for a Y-linked gene

When a father carries a mutant allele in the *Y* chromosome, shown in red, he passes it to all of his sons and grandsons, but never to his daughters because daughters do not inherit a *Y* chromosome. In addition, mothers can never inherit or transmit the mutant allele, since they do not possess a *Y* chromosome.

of inheritance suggests that a trait is determined by a Y-linked gene.

✓ Concept Check

4. Use a Punnett square to **illustrate** how the existence of sex chromosomes leads to a 1:1 sex ratio of sons and daughters.

5. **Describe** the pattern of inheritance for X-linked genes.

6. **Describe** the pattern of inheritance for Y-linked genes.

28.3 Genes linked in the same chromosome do not sort independently unless recombination occurs

When we discussed Mendel's pea plant experiments, we highlighted how his crosses led him to discover the law of independent assortment. This law states that the alleles for one gene sort independently of a second gene. This law fit Mendel's data very nicely because, as we now know, the phenotypes he examined were coded by genes that are located in different chromosomes. However, we also know that thousands of genes exist in a single chromosome. What happens to inheritance patterns when two genes are located in close proximity in the same chromosome? In this section, we will discuss how such genes have different patterns of inheritance and how these patterns can change when chromosomes experience crossing over during meiosis.

Linkage of Nearby Genes on the Same Chromosome

In the previous section, we talked about genes that are located in either the X or Y chromosome and we referred to them as sex-linked genes. Genes can also be linked to one another by being physically located near each other in the same chromosome, which we refer to as **linked genes.** Because of their close physical proximity, linked genes segregate together in the same chromosome during meiosis, so they do not sort independently of each other. Thus, the expectation is that the inheritance pattern for the allele at one gene in a given chromosome will match the inheritance pattern for an allele of a nearby gene in the same chromosome.

For example, in humans the genes for eye color and hair color are linked, which is why we commonly see people who have brown hair and brown eyes or people with blond hair and blue eyes. While these pairs of phenotypes are common, we also know that there can be exceptions, such as some people with brown hair and blue eyes. How does this happen if the alleles of the two genes are linked?

To understand how alleles of linked genes can occasionally become unlinked from each other, we have to return to our discussion of meiosis in Module 26. As you likely recall, during prophase I, pairs of homologous chromosomes line up on either side of the metaphase plate. The two homologs line up perfectly, gene for gene, in a process known as synapsis. Synapsis allows an opportunity for the two homologs to overlap each other in a cross-like pattern called a chiasma. At the point where the two homologs cross, they exchange genetic material in a process we refer to as crossing over. Crossing over allows each chromosome to swap genes with the other. When the location of the crossing over event occurs at a point that is physically located between two linked genes in a single chromosome, the allele of one gene is no longer physically linked to the allele of the other gene.

We can visualize the process and consequences of crossing over by examining the pair of homologous chromosomes in **FIGURE 28.11.** As we can see in the top half of the figure, the two chromosomes pair up during prophase I, form a chiasma, and the blue and red chromosomes exchange a section of their chromosomes. While the blue chromosome began with the genotype *AB* on both chromatids and the red chromosome began with the genotype *ab* on both chromatids, after crossing over one chromatid on each of the chromosomes now contains a different allele. After crossing over, one blue chromatid still has an *AB* genotype, but the other now has an *Ab* genotype. Similarly, one red chromatid still has an *ab* genotype, but the other now has an *aB* genotype. When meiosis is complete, shown on the right side of the figure, we have two chromosomes that have new combinations of alleles. When crossing over produces new combinations of alleles in a chromosome, we call it **recombination.** We also refer to the chromosomes that

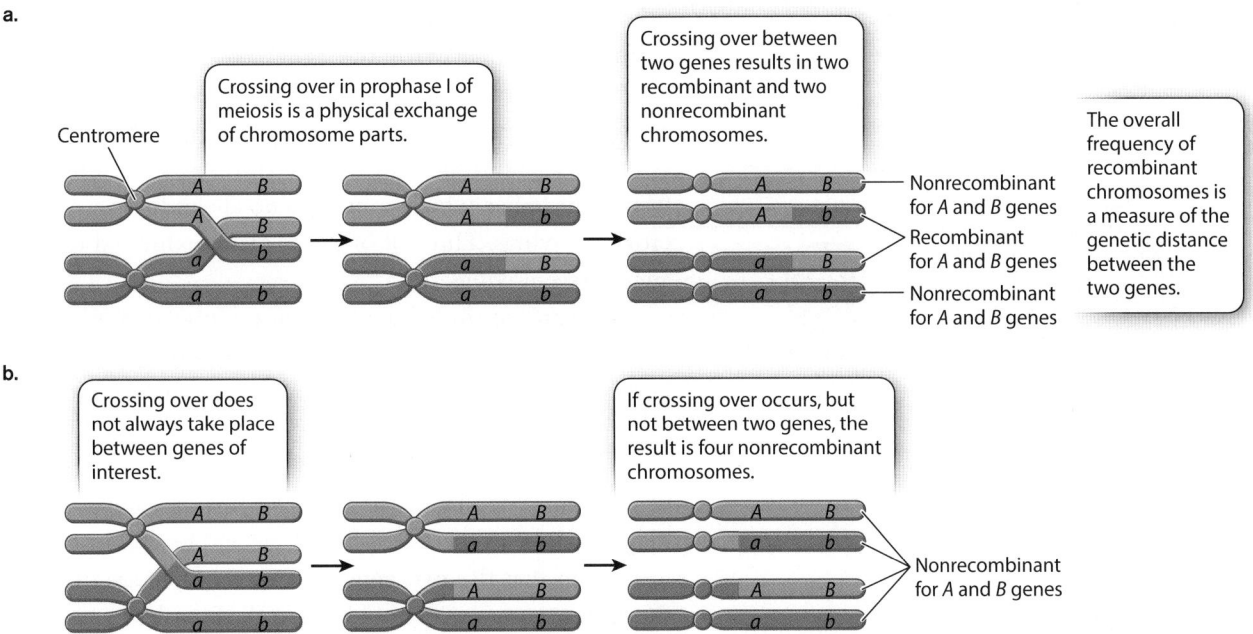

a.

Centromere

Crossing over in prophase I of meiosis is a physical exchange of chromosome parts.

Crossing over between two genes results in two recombinant and two nonrecombinant chromosomes.

The overall frequency of recombinant chromosomes is a measure of the genetic distance between the two genes.

Nonrecombinant for *A* and *B* genes

Recombinant for *A* and *B* genes

Nonrecombinant for *A* and *B* genes

b.

Crossing over does not always take place between genes of interest.

If crossing over occurs, but not between two genes, the result is four nonrecombinant chromosomes.

Nonrecombinant for *A* and *B* genes

FIGURE 28.11 Crossing over between linked genes

Linked genes are those genes that are physically located near each other in a chromosome. As a result, the alleles at linked genes are inherited together rather than sorting independently. However, when the pairs of homologous chromosomes come together during meiosis I, synapsis occurs. During synapsis, the homologs can form a chiasma and exchange portions of their chromosomes. (a) When this crossing over occurs at a position between two genes of interest, two of the four daughter cells will contain a chromosome that has a new combination of alleles, known as recombinant chromosomes, while the other two do not. (b) When this crossing over occurs at a position that is not between two genes of interest, none of the four daughter cells produces new combinations of alleles.

experienced the recombination as recombinant chromosomes. In contrast, chromosomes that did not experience the recombination are called nonrecombinant chromosomes. As you can see in the bottom half of Figure 28.11, it is important to keep in mind that sometimes crossing over occurs without causing any recombination of alleles between two genes, even though there is recombination at the level of the entire chromosome. This can happen whenever the crossing over point occurs at a location that is not between the two genes of interest. In this case, pieces of the chromosomes are still exchanged, but the alleles of the two nearby genes continue to be linked and not recombined.

An Example of Linked Genes and Recombination

With our understanding of linked genes and the possibility of genetic recombination due to crossing over during meiosis, let's look at a real example in an experiment using fruit flies with linked genes. In the fruit fly, researchers commonly look for linked genes that are located near each other in the *X* chromosome, which means they are both linked genes and X-linked genes. As you will see, when we work with linked genes in the *X* chromosome, we can easily see the outcome of linkage and recombination in the males,

since the *Y* chromosome in males has so few homologous genes that could affect the phenotype.

Fruit flies have a gene that codes for eye color and a gene that codes for wing veins. For eye color, there is a gene (*w*) that affects eye color. The wild type for fruit flies is a dominant allele that codes for red eyes (w^+), while a recessive allele codes for white eyes (w^-). For wing veins, there is a gene (*cv*) that codes for crossveins in the fly wings. The wild type for fruit flies is a dominant allele that codes for the presence of crossveins in the wing (cv^+), while a recessive allele codes for the absence of crossveins (cv^-). You can see these four phenotypes in **FIGURE 28.12** on page 400. The two genes that code for these two traits are physically located close together in the *X* chromosome, so we hypothesize that they sort together rather than sort independently.

To test this hypothesis, researchers started with a female fly that was homozygous for the dominant red eyes (w^+w^+) and homozygous for dominant crossvein presence (cv^+cv^+). They crossed this female with a male with recessive alleles for each trait in its *X* chromosome and, of course, no alleles for these two traits in its *Y* chromosome. You can see this parental generation at the top of Figure 28.12. The outcome of this cross was an F_1 generation in which the female offspring inherited one *X* chromosome from their

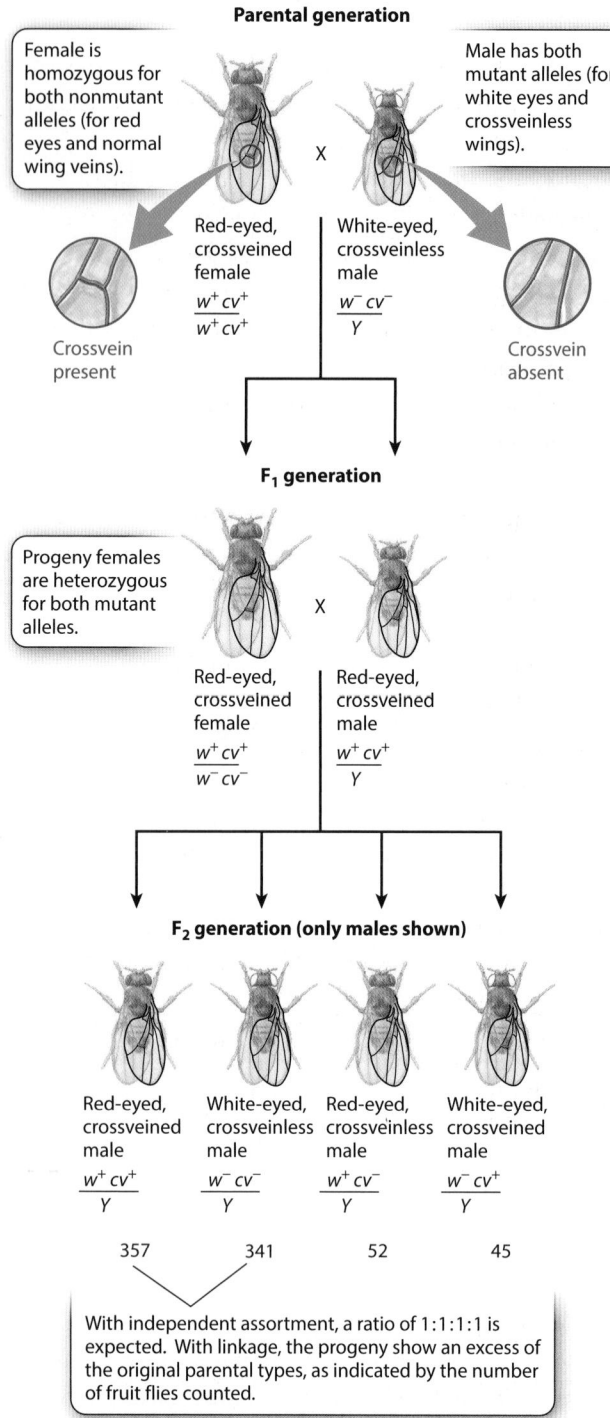

Parental generation

Female is homozygous for both nonmutant alleles (for red eyes and normal wing veins).

X

Male has both mutant alleles (for white eyes and crossveinless wings).

Red-eyed, crossveined female
$\dfrac{w^+ cv^+}{w^+ cv^+}$

White-eyed, crossveinless male
$\dfrac{w^- cv^-}{Y}$

Crossvein present

Crossvein absent

F₁ generation

Progeny females are heterozygous for both mutant alleles.

X

Red-eyed, crossveined female
$\dfrac{w^+ cv^+}{w^- cv^-}$

Red-eyed, crossveined male
$\dfrac{w^+ cv^+}{Y}$

F₂ generation (only males shown)

Red-eyed, crossveined male
$\dfrac{w^+ cv^+}{Y}$
357

White-eyed, crossveinless male
$\dfrac{w^- cv^-}{Y}$
341

Red-eyed, crossveinless male
$\dfrac{w^+ cv^-}{Y}$
52

White-eyed, crossveined male
$\dfrac{w^- cv^+}{Y}$
45

With independent assortment, a ratio of 1:1:1:1 is expected. With linkage, the progeny show an excess of the original parental types, as indicated by the number of fruit flies counted.

FIGURE 28.12 Linked genes in fruit flies

Researchers crossed fruit flies that have linked genes for eye color and crossvein wings. Because these two genes are located in the X chromosome, the F₁ generation has heterozygous females that all have red eyes and crossvein wings. Males have only one X chromosome, so they also all have red eyes and crossvein wings. In the F₂ generation, most males have either red eyes and crossvein wings or white eyes and no crossveins, which is expected because the two genes are linked. However, a small number of males have one dominant allele and one recessive allele, which is evidence that the X chromosome that they received previously experienced crossing over, causing the alleles in the two linked genes to become unlinked.

mother with the dominant alleles and one X chromosome from their father with the recessive alleles. Thus, the F₁ females all exhibited red eyes and crossvein wings. The F₁ males inherited the X chromosome from their mother, containing the dominant alleles, and the Y chromosome from their father, which contained no alleles for eye color or crossvein wings. Thus, all of the F₁ males exhibited red eyes and crossvein wings.

We can see the effects of genetic linkage and recombination when the F₁ generation is crossed to produce an F₂ generation, as shown in Figure 28.12. Given that the males inherit only a single X chromosome, the males will exhibit both the recessive and dominant phenotypes because the phenotypes are coded by the single alleles that they inherit. Thus, they can show us the effects of linkage and recombination.

Given that the two genes in our example are linked in the X chromosome, we would expect the F₂ males to receive an X chromosome that has either two dominant alleles ($w^+ cv^+$) or two recessive alleles ($w^- cv^-$). Thus, if there is no recombination, we should see two types of males: half of the males should have red eyes and crossveins ($w^+ cv^+$) and the other half should have white eyes and no crossveins ($w^- cv^-$). However, when the researchers examined the phenotypes of the F₂ males, they observed four phenotypic combinations rather than two. At the bottom of Figure 28.12, you can see that most of the males had red eyes and crossveins or white eyes and no crossveins, which is what the researchers expected, given that the two genes are linked. However, there were also a small number of males with red eyes and no crossveins and another small group of males with white eyes and crossveins. The only way that these last two groups could exist is if the linked genes experienced crossover during meiosis, which would unlink the two dominant alleles and unlink the two recessive alleles in the X chromosome.

This example demonstrates that linked genes do not sort independently as the unlinked genes in Mendel's peas did, so the pattern of inheritance involves the two alleles being inherited together, unless recombination unlinks them.

✓ Concept Check

7. **Describe** what is meant by linked genes.

8. **Describe** why linked genes do not fit Mendel's law of independent assortment.

9. **Describe** how crossing over can increase the number of phenotypic combinations that we observe in an X chromosome.

28.4 Mitochondrial and chloroplast genes are also inherited

Back in Module 12, we discussed how two important cell organelles in eukaryotes—mitochondria and chloroplasts—had their evolutionary origin as free-living bacteria. Approximately 1.4 billion years ago, an ancient eukaryotic cell engulfed a bacterium that became the modern-day mitochondria. Approximately 0.6 to 1.5 billion years ago, the ancestral cell of algae and plants engulfed a cyanobacteria that became the modern-day chloroplast. As you may recall, one important piece of evidence that these two organelles originated by endosymbiosis is that they possess their own genes. Moreover, the DNA sequences in chloroplasts are closely related to the DNA sequences found in modern-day cyanobacteria. Given that mitochondria and chloroplasts contain genes, in this section we examine how these genes are inherited.

Patterns of Inheritance in Mitochondria and Chloroplasts

How the cytoplasmic DNA in mitochondria and chloroplasts gets transmitted from one generation to the next varies widely among species. In animals and plants, the mitochondria typically exhibit **maternal inheritance** of the organelles, in which the organelles are transferred from the mother to the offspring, which is a very different pattern of inheritance compared to the autosomal genes examined by Mendel. This makes sense if we think back to how mammalian gametes are formed during meiosis, which we examined in Module 26. When a female mammalian cell undergoes meiosis to form gametes, the products are one large egg cell and three small polar bodies. As you can see in **FIGURE 28.13**, the egg cell has all of the necessary organelles, including the mitochondria. In contrast, male mammals make sperm cells that are very small by eliminating most of the cytoplasm, including most of the typical cell organelles except for a nucleus, centrioles, some mitochondria to provide power for the swimming tail, and a vacuole known as the acrosome that contains enzymes that help the sperm enter the egg cell. During fertilization to produce a zygote, the entire sperm cell fuses with the egg cell and therefore the sperm's mitochondria enter the egg cell. Shortly after fertilization, however, the sperm's mitochondria are degraded, so the inheritance of male mitochondria in the zygote does not persist very long. As a result, by the time a mammal is born, all of the mitochondria it has are maternally inherited.

While the vast majority of species exhibit maternal inheritance of mitochondria and chloroplasts, there are a few exceptions to this pattern. For example, species such as the giant redwood (*Sequoiadendron giganteum*) exhibit **paternal inheritance** of the organelles, in which the organelles transfer from the father to the offspring. Much less common is biparental inheritance, in which the organelles are inherited from both parents. We can see biparental inheritance in some species of ferns. While such exceptions exist, most species inherit their mitochondrial and chloroplast genes from their mothers.

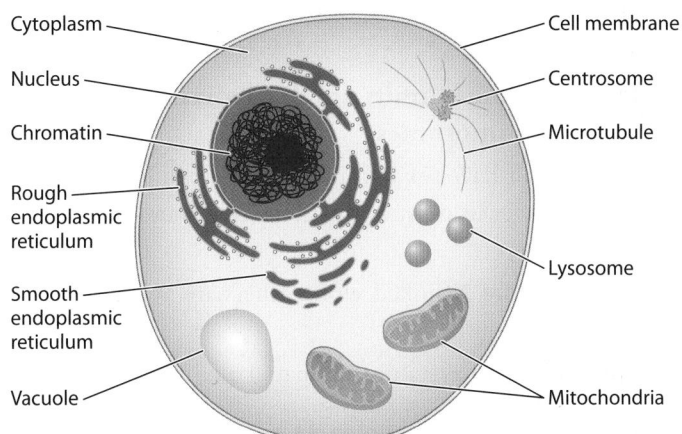

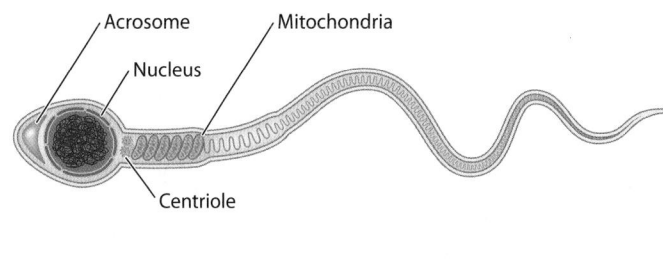

FIGURE 28.13   The organelles of egg and sperm cells

Mammalian egg cells (left) contain a full set of organelles, while sperm cells (right) contain only a nucleus, centrioles, mitochondria for powering the swimming tail, and an acrosome that contains enzymes that help the sperm fuse with the egg cell.

Inheritance of Mitochondrial Diseases

Given that mitochondria are maternally inherited in mammals, what are the patterns of genetic inheritance for mutations that occur in the mitochondria? Mutations in mitochondria cause a number of disorders including heart disease, liver disease, kidney disease, and muscle weakness. For example, there is a disease known as MERFF syndrome (the acronym stands for "myoclonic epilepsy with ragged red fibers"), which causes muscle weakness from insufficient production of ATP by the mitochondria. The pedigree in **FIGURE 28.14** shows many individuals who carry genes for the MERFF disorder. The pedigree demonstrates several patterns of mitochondrial inheritance. Because mothers provide mitochondria to both sons and daughters, both sexes can inherit the mutant allele. Moreover, all of the offspring of an affected mother will exhibit the affected phenotype. Finally, because male mammals do not pass on their mitochondria to their offspring, males that have mitochondria with the mutant allele cannot pass these recessive mitochondria alleles to any of their offspring. Once again, this pattern of inheriting an allele in mitochondria is completely different from the patterns of inheritance that we saw in Mendel's pea plants because Mendel focused on genes located in the nuclear DNA in chromosomes.

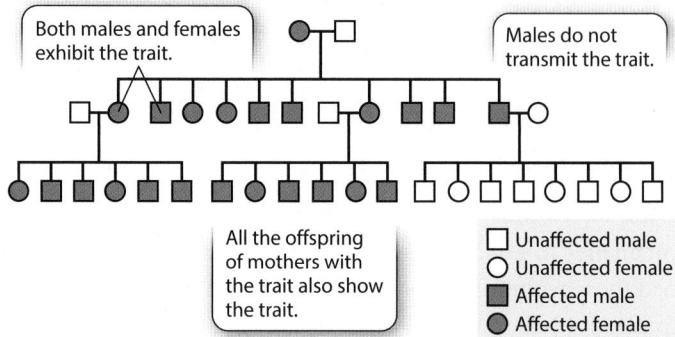

FIGURE 28.14 A pedigree for a human disease caused by a mutant allele in mitochondria

Given that only females pass on the mitochondria in mammals, all offspring of an affected female will exhibit the disorder. However, none of the offspring of an affected male will inherit the disorder.

✓ Concept Check

10. **Describe** the reason that mammals receive their mitochondria from their mothers.

11. Based on a pedigree, **describe** the patterns of inheritance that you would expect to see for a mitochondrial disorder caused by a mutant mitochondria gene.

Module 28 Summary

PREP FOR THE AP® EXAM

REVISIT THE BIG IDEAS

INFORMATION STORAGE AND TRANSMISSION: Using the content of this module, **identify** one example that illustrates a deviation from Mendel's prediction of the inheritance of traits, and **provide reasoning** for the deviation.

LG 28.1 Epistasis is caused by multiple genes that interact to determine a phenotype.

- Epistasis occurs when one gene modifies the phenotype produced by another gene. Page 390

- Epistasis is demonstrated in chicken feather colors, where one gene determines whether a protein is produced to make a pigment and another gene determines whether pigment production by the first gene will be inhibited. Page 391

- While some examples of epistasis involve one gene completely blocking the phenotype produced by another gene, in many cases one gene can subtly modify the effects of another gene. Page 392

LG 28.2 Sex-linked traits are determined by genes in the sex chromosomes.

- Most animals have sex chromosomes that differ in size and they differ in most of the genes that they carry. Page 393

- Some genes are X-linked while others are Y-linked. Page 394

- X-linked genes have a distinct pattern of inheritance in which males only inherit their X chromosome from their mothers and only transmit their X chromosome to their daughters. Page 395

- If a mother is heterozygous, the offspring have a 50% probability of inheriting a recessive allele; since sons receive only one X chromosome, they have a 50% probability of expressing the recessive phenotype. Page 396
- Y-linked genes also have a distinct pattern of inheritance in which fathers can transmit only Y-linked genes to their sons and never to their daughters. Page 397
- While mothers can never inherit the Y-linked genes nor can they transmit the allele to the next generation, all sons of affected fathers will be affected. Page 397

LG 28.3 Genes linked in the same chromosome do not sort independently unless recombination occurs.

- Genes can be linked to one another by being physically located near each other in the same chromosome, which we refer to as linked genes. Page 398
- Linked genes do not sort independently as those studied by Mendel did, so the pattern of inheritance involves the two alleles being inherited together, unless recombination unlinks them. Page 398

- Alleles that occupy linked genes can occasionally become unlinked by crossing over during meiosis. Page 398
- When crossing over produces new combinations of alleles in a chromosome, we call it recombination. Page 398

LG 28.4 Mitochondrial and chloroplast genes are also inherited.

- Two important cell organelles in eukaryotic cells—mitochondria and chloroplasts—had their evolutionary origin as free-living bacteria. Page 401
- In many species, including mammals, the mitochondria and chloroplasts are maternally inherited. Page 401
- Given that mitochondria are maternally inherited in most plants and animals, mothers pass down any mitochondrial disorders to both sons and daughters, but fathers cannot pass down mitochondrial disorders to any of their offspring. Page 402

Key Terms

Sex-linked genes
Linked genes

Recombination
Maternal inheritance

Paternal inheritance

Review Questions

1. A Punnett square on a dihybrid cross that involves epistasis will likely show
   (A) a phenotypic ratio that differs from 9:3:3:1.
   (B) a phenotypic ratio of 3:1.
   (C) maternal inheritance of the recessive allele.
   (D) paternal inheritance of the recessive allele.

2. If researchers crossed one breed of chicken with a *CC ii* genotype with another chicken having a *cc II* genotype, what will be the phenotypes of the offspring? Use Figure 28.1 on page 391.
   (A) All offspring will have dark feathers.
   (B) All offspring will have white feathers.
   (C) Half of the offspring will have dark feathers.
   (D) Three-quarters of the offspring will have white feathers.

3. Which statement is true about X-linked genes?
   (A) There is a homologous gene in the Y chromosome.
   (B) Females express the recessive trait when they carry one recessive allele.
   (C) Males inherit their X chromosome from their fathers.
   (D) Males express the recessive trait when they carry one recessive allele.

4. Which statement is true about Y-linked genes?
   (A) There is a homologous gene in the X chromosome.
   (B) In mammals, there are many fewer Y-linked genes than X-linked genes.
   (C) Males inherit their Y chromosome from their mothers.
   (D) Only half of all sons of affected fathers will also be affected.

5. Which statement is true regarding the inheritance of mitochondrial genes in mammals?

(A) They are maternally inherited.

(B) Mutant genes cannot be passed from mothers to daughters.

(C) Mutant genes are passed from fathers to daughters.

(D) Mutant genes cannot be passed from mothers to sons.

Module 28
AP® Practice Questions

Section 1: Multiple-Choice Questions

Choose the best answer for questions 1–5.

Use the figure below to answer question 1.

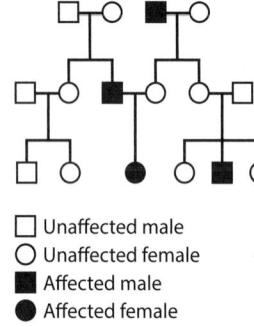

☐ Unaffected male
○ Unaffected female
■ Affected male
● Affected female

1. The pedigree illustrates a family in which some members exhibit an unknown disorder. Based on the information shown, identify the most likely mode of inheritance of the disorder.

(A) *X*-linked dominant

(B) *X*-linked recessive

(C) *Y*-linked dominant

(D) Mitochondrial

2. Leber's hereditary optic neuropathy (LHON) is an inherited disease where the retina of the eye degenerates, resulting in blindness. Clinical scientists gathered family data from 1,163 people who are affected with LHON and from 60,038 people who are not affected with LHON. Only adults who have biological children were included in the study. Each person was asked whether they were affected with LHON, whether either parent was affected with LHON, and whether any of their children were affected with LHON. The number of people who answered "Yes" to each question are shown in the table.

Sex	Mother affected with LHON	Father affected with LHON	Children affected with LHON	Total number of people surveyed
Male affected with LHON	568	0	0	568
Female affected with LHON	595	0	595	595
Male not affected with LHON	0	1	0	29,419
Female not affected with LHON	0	1	0	30,619

Based on the data provided, identify the chromosome that carries the LHON trait.

(A) An autosome

(B) The *X* chromosome

(C) The *Y* chromosome

(D) The mitochondrial chromosome

Questions 3–5 refer to the following material.

Corn seeds are contained in kernels. An autosomal gene influences the color of the kernels. The dominant *C* allele causes kernels to be yellow, whereas plants that are homozygous for the recessive *c* allele have white kernels. A different gene influences the shape of the kernel. Plants carrying the dominant *S* allele have smooth kernels, whereas plants that are homozygous for the recessive *s* alleles have wrinkled kernels. A corn breeder established a cross using pollen from a male plant that grew from a wrinkled white kernel to fertilize a female plant that grew from a smooth yellow kernel. The breeder determined the kernel phenotypes from offspring of this cross and obtained the following data.

Offspring kernel phenotype	Number of offspring with the phenotype
Smooth, yellow	43
Wrinkled, white	46
Smooth, white	873
Wrinkled, yellow	848

3. Which statement best explains the inheritance pattern observed in the cross?

   (A) Kernel color and shape are epistatic traits.

   (B) The genes controlling kernel color and shape are in the same biochemical pathway.

   (C) The genes determining kernel color and shape are physically close to each other on a chromosome.

   (D) Kernel color and kernel shape are determined by the same genes.

4. Based on the data provided, identify the genotype of the female plant.

   (A) *CS/CS*   (C) *Cs/cS*

   (B) *CS/cs*   (D) *cs/cs*

5. If the two traits are sorting independent of each other, how many of the offspring from this cross would be expected to have smooth, white kernels?

   (A) 25   (C) 861

   (B) 453   (D) 905

Section 2: Free-Response Question

Write your answer to each part clearly. Support your answers with relevant information and examples. Where calculations are required, show your work.

Labrador dogs have three coat colors recognized by breeders. Black Labradors have dark black fur. Chocolate Labradors have brown fur. Yellow Labradors have very light tan fur. In Labradors, fur color is determined by two genes, *Ext* and *TyrP1*. A recessive mutation in the *Ext* gene (*Ext⁻*) prevents the production of black pigment. A single *Ext⁺* allele is sufficient for wild-type *Ext* activity, which produces black pigment. Yellow Labradors have the genotype *Ext⁻/Ext⁻* and therefore do not produce black pigment. The production of black pigment can be blocked by recessive mutation in the *TyrP1* gene (*TyrP1⁻*) even in the presence of wild-type *Ext* alleles. Brown Labradors have at least one wild-type *Ext* allele and a *TyrP1⁻/TyrP1⁻* genotype. A single *TyrP1⁺* allele is sufficient to allow the production of black pigment.

A breeder crossed a male Labrador that is heterozygous for both genes (*Ext⁺/Ext⁻*; *TyrP1⁺/TyrP1⁻*) with a female Labrador with the same genotype. The *Ext* and *TyrP1* genes are inherited independently.

(a) **Identify** the phenotype of the female (mother) dog in the cross.

(b) **Calculate** the proportion of the offspring from this cross that are expected to have a *TyrP1⁻/TyrP1⁻* genotype.

(c) **Calculate** the proportion of the offspring from this cross that are expected to be chocolate Labradors.

Module 29

Environmental Effects on Phenotypes

LEARNING GOALS ▶**LG 29.1** Phenotypes are the product of an organism's genes and its environment.
▶**LG 29.2** Genes and the environment affect human health and disease.

Earlier in this unit, we discussed how traits are inherited when they are controlled by a single gene, such as Mendel's pea plants. We then examined more complex gene interactions, such as epistasis, and the unusual patterns of inheritance caused by sex-linked genes and linked genes that experience recombination. In all of our discussions, we have only talked about how two alleles at one or two genes determine the phenotype that is produced. Most phenotypes, however, are influenced by a large number of genes rather than one or two. Moreover, an organism's genes are not the only determinant of its phenotype. The environment can also have a major impact on phenotypes. In this module, we will talk about how living in different environments can

affect the phenotypes that we see in nature. As we will see, genes and environments interact to affect the phenotypes of all organisms, including the domesticated plants and animals that we rely on for food. The interactions of genes and the environment even affect the prevalence of some human diseases.

PREP FOR THE AP® EXAM

FOCUS ON THE BIG IDEAS

SYSTEMS INTERACTIONS: Focus on how the transmission of information encoded in the genotype interacts with environmental factors to produce varying phenotypes.

29.1 Phenotypes are the product of an organism's genes and its environment

When we look at plants and animals of a given species, including humans, we notice a tremendous amount of variation in their appearance. For example, in a crowd of adult people, there is a great deal of variation in height and weight. Similarly, if we look at a field of corn, as shown in **FIGURE 29.1**, we notice that some plants are taller than others, depending on where in the field the corn is growing. While all of the corn seeds are the same genotype and are planted at the same time, the individual corn plants grow to different heights because of differences in available sunlight, soil nutrients, and water. The ability of a single genotype to produce different phenotypes in different environments is known as **phenotypic plasticity.** The term "plasticity" denotes that the phenotype is flexible, similar to a flexible piece of plastic. Such plasticity can include changes in behavior, morphology, and physiology. In this section, we will examine several examples of phenotypic plasticity in nature and in domesticated animals and crops. We will also investigate the costs and benefits that

FIGURE 29.1 Variation in corn height

When planting corn, farmers typically plant individuals of a single genotype, yet the corn grows to different heights depending on the environmental conditions of available sun, nutrients, and water in the field. Photo: Bruce Leighty/Getty Images

a. Tadpoles raised with and without predators

b. Tadpole activity with and without predator

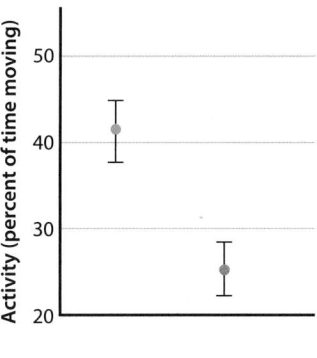

c. Tail depth with and without predator

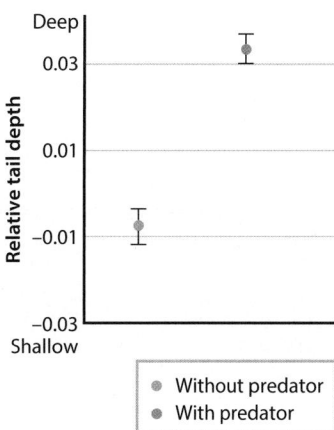

FIGURE 29.2 Phenotypic plasticity in gray treefrog tadpoles

When gray treefrog tadpoles are raised with and without predators, they develop different phenotypes. (a) The upper tadpole was raised in the absence of predators and developed a smaller tail with a drab color. The lower tadpole was raised in the presence of predator smells and developed a much larger and more colorful tail, which allows it to escape predators better. (b) When researchers monitor how much the tadpoles move, they find that tadpoles living with predators are less active. (c) Tadpoles living with predators also develop deeper tails. Data are means ±1 SE. Data from N. Schoeppner and R. Relyea. (2005). Damage, Digestion, and Defense: The Roles of Alarm Cues and Kairomones for Inducing Prey Defenses. *Ecology Letters* 8:505–512. Photo: John I. Hammond

favor the evolution of plastic phenotypes rather than non-plastic phenotypes.

Phenotypic Plasticity in Nature

Most of the phenotypic plasticity that exists in nature represents evolved responses to the different environments in which plants and animals live. For instance, when gray treefrogs (*Hyla versicolor*) lay their eggs in a pond and the eggs hatch into tadpoles, the pond might contain a large number of tadpole predators or it might be predator free. Fortunately, tadpoles can smell predators in the water. When tadpoles of the gray treefrog smell a predator in the water, their behavioral phenotype changes in ways that reduce their risk of being captured as prey. They become less active and hide more.

In addition, over the period of a few days, the tadpoles can change their morphological phenotype by altering their shape and color, as you can see in **FIGURE 29.2**. In this photo, the two tadpoles are siblings, so they have similar genotypes, yet the upper tadpole was raised for several weeks in the absence of predator odors, while the lower tadpole was raised in the presence of predator odors. As a result of living in the environment with predator odors, the lower tadpole developed a much larger and more colorful tail and a smaller body.

In Figures 29.2b and 29.2c, you can see data from experiments in which researchers observed that tadpoles living with predators reduced their swimming activity and displayed an increased tail fin depth. This large tail allows the tadpoles to escape predators better. The reason for the red color is a mystery; it may encourage predators to strike at the tail, which can be regrown, rather than striking at the body, which would be fatal. Thus, the tadpole's genes allow it to build two different phenotypes, depending on whether its mother lays her eggs in a predator-filled or predator-free pond. As we will discover in "Practicing Science 29.1: How do we test for fitness trade-offs in environmentally induced traits?" on page 409, these predator-induced phenotypes have evolved due to fitness trade-offs.

While many species alter their phenotypes in response to predators, others alter their phenotypes in response to changes in the physical environment. For example, the snowshoe hare lives in Canada and the northern United States. During the summer, the hare grows a brown coat. As fall approaches, however, the brown coat is replaced with a white coat, as illustrated in **FIGURE 29.3** on page 408. This change in coat color is induced by the days getting shorter and the nights getting longer, which is a reliable cue that winter is coming. Given where the hare lives, it is typically cold and snowy in the winter. Thus, a white coat helps the

FIGURE 29.3 Phenotypic plasticity in snowshoe hares

During the summer, the snowshoe hare grows a brown coat, which helps it blend in among the brown stems of shrubs and trees. In the winter, induced by shorter day length, the hare grows a white coat, which helps it blend into the snowy background. Photos: (left) Joe Austin Photography/Alamy; (right) FotoRequest/Alamy Stock Photo

hare blend into the snowy background much better, so it is less likely to be seen by predators. In contrast, when there is no snow in the summer, the hare's brown coat is better camouflage in the forest against the brown stems of shrubs and trees. Once again, a single genotype can produce multiple phenotypes, with each phenotype providing a benefit within a specific environment.

Reptiles have an amazing form of phenotypic plasticity, whereby temperature can determine the sex of their offspring. While the sex of most animals is determined by the sex chromosomes, the sex of many species of turtles, lizards, and alligators is determined by the temperature at which the eggs are incubated. In most alligators and lizards, embryos that are incubated at lower temperatures develop into females while those incubated at higher temperature develop into males. In many species of turtles, such as the loggerhead sea turtle (*Caretta caretta*) shown in **FIGURE 29.4**, the opposite pattern occurs. These sea turtles dig a hole in the sand on the ocean shoreline and then deposit their eggs. The warmer eggs are near the surface and develop into females while the eggs that are deeper in the nest are cooler and develop into males. Because the sex of the offspring is determined by the environment, we call this environmental sex determination. Environmental sex determination gives species the opportunity to favor having more male offspring or more female offspring, depending on which is rarer in the

FIGURE 29.4 Environmental sex determination

When this loggerhead sea turtle lays her eggs on a beach, the eggs at the bottom of the nest incubate at a cooler temperature and are more likely to develop into male offspring. Eggs at the top of the nest incubate at a warmer temperature and are more likely to develop into female offspring. Photo: A & J Visage/Alamy Stock Photo

population. If females happen to be rarer in the population and males are common, many males will not be able to find a mate and have offspring. In this situation, if a parent makes more female offspring, these offspring are much more likely to have mates and their own offspring.

How do we test for fitness trade-offs in environmentally induced traits?

Background As we have seen, tadpoles develop large tails and small bodies when predators are present, but they develop small tails and large bodies when predators are absent. Early experiments showed the benefit of growing a larger tail; tadpoles could escape predators better. Given the benefit of having a larger tail, why don't tadpoles always grow a large tail? When predators are absent and there is less danger of being killed, tadpoles live in higher numbers, which causes increased competition for the algae that they eat. Given these observations, researchers wondered whether tadpoles with larger tails and smaller bodies experience a performance trade-off in an environment without predators. Perhaps the smaller body, which contains a smaller mouth and smaller digestive system, causes the large-tailed tadpoles to grow more slowly. Growing more slowly results in adult frogs that take longer to reach sexual maturity and female frogs that produce fewer offspring. In short, predator-induced traits can improve tadpole survival, but they may come at the cost of lower fitness as an adult frog.

Hypothesis Tadpoles initially raised in a predator environment will subsequently grow more slowly than tadpoles initially raised in a no-predator environment.

Experiment The researchers examined this question in two stages. In the first stage, they raised groups of 75 wood frog tadpoles (*Rana sylvatica*) in large tanks of water that simulated wetlands. Each tank contained either no predators or caged predators (larval dragonflies) that emit chemicals. Tadpoles can detect these chemicals in the water and they respond by altering their phenotypes. Within each of these two environments, the researchers created low and high competition environments by adding 0 or 75 tadpoles of a competing species (leopard frogs; *R. pipiens*). The wood frogs grew in these inducing environments for 3 weeks.

In the second stage of the experiment, the researchers removed a sample of tadpoles from each tank and then raised them in new tanks of water to determine how well each phenotype grew in the absence of any predators or leopard frog competitors. After 10 days, they weighed the tadpoles and determined the relative growth rate of each tadpole, which is calculated as the tadpole's final mass divided by its mass at the start of the second stage of the experiment. Thus, a relative growth rate of 1.0 would indicate that the tadpole's final mass was the same as its initial mass.

Results As you can see in the figure, the researchers discovered that tadpoles initially raised with predators experienced slower growth compared to tadpoles living without predators. In addition, tadpoles that were initially raised with leopard frog competitors experienced faster growth in the second stage of the experiment compared to tadpoles raised without competitors.

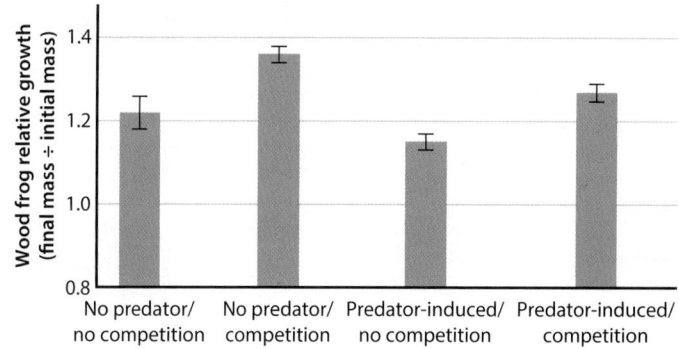

Data are means ±1 SE.

Conclusion The researchers discovered that while tadpoles with larger tails and smaller bodies can escape predators more easily, this benefit comes at the cost of slower growth. In addition, when the presence of competitors induces smaller tails and larger bodies, tadpoles experience greater growth. These changes in growth among the different phenotypes are driven by changes in the size of the tadpole's body, including changes in their mouth size and intestine length. Having a larger mouth allows the tadpoles to consume more algae, and having a longer intestine allows the tadpoles to absorb more energy from the algae that they consume. Predator-induced tadpoles have relatively smaller bodies, so their smaller mouths consume less algae and their shorter intestines are less efficient at absorbing energy from the algae. In contrast, competitor-induced tadpoles have relatively larger bodies, so their larger mouths consume more algae, and their longer intestines are more efficient at absorbing energy from the algae.

Follow-up Work The discovery that predator and competitor environments induce opposite phenotypes led to the discovery that tadpoles can fine-tune their phenotypes, with genes being turned on and off in response to a wide range of predator species and competitors. Tadpoles produce a range of phenotypes that balance the need to avoid predation, while growing as rapidly as possible to improve fitness.

SOURCE
Relyea, R. A. 2002. "Competitor-Induced Plasticity in Tadpoles: Consequences, Cues, and Connections to Predator-Induced Plasticity." *Ecological Monographs* 72:523–540.

AP® PRACTICE QUESTION

(a) **Identify** the following:
  1. The question the experimenters are trying to answer
  2. Two hypotheses tested
  3. The independent variable
  4. The dependent variable
  5. The control group

(b) **Explain** the conclusion using a claims, evidence, reasoning format (CER).

FIGURE 29.5 Phenotypic plasticity in jewelweed

When the plants experience increased shade, they grow taller, which allows them to obtain more sunlight. In crowded patches of jewelweed such as in this photo, the most crowded plants extend their stems to be taller. Photo: AdamLongSculpture/Getty Images

Phenotypic plasticity is also common in plants. When insects chew on plants, some plants start growing spines or hairs, or release toxic chemicals to deter the insects from attacking them again in the future. Plants can also respond to reductions in sunlight when neighboring plants begin to shade them. For example, a plant known as jewelweed (*Impatiens capensis*), shown in **FIGURE 29.5**, has a striking ability to rapidly grow taller when it is shaded by neighboring plants. The jewelweed leaves can detect that they are being shaded, and then the plant responds by growing taller. As a result, in very crowded areas of jewelweed, we often see the plants in the middle are the tallest because they are experiencing the densest shade, which causes them to produce a taller phenotype.

We can see examples of phenotypic plasticity in domesticated farm animals, such as cattle, sheep, pigs, and chickens. For instance, the amount of feed and the nutritional quality of the feed that is given to farm animals has a large effect on how large they can grow. If they are not given sufficient feed when they are young, the animals can be stunted at a smaller body size. These stunted individuals can never achieve a full adult body size, even if they are given abundant feed later in life. In contrast to our earlier examples, these changes

in phenotype do not improve evolutionary fitness because there is no benefit to having stunted growth.

Animal nutrition also affects a wide range of other phenotypes, including egg production in chickens and milk production in cows. As you can see, phenotypic plasticity is pervasive in both wild and domesticated plants and animals. In each of these cases, the environmental conditions induce phenotypic changes in the organism by activating certain genes and inactivating others. Thus, although the environment can induce phenotypic changes, it can do so only because of the genes present in the organism. We will have much more to say about genes being activated to cause phenotypic changes in Unit 6.

The Interaction of Genes and Environments in Determining Phenotypes

The interaction of genes and the environment can occur in a diversity of ways; some genotypes exhibit no response to environmental change while others exhibit a large response to environmental change. A good example of this observation can be found in different strains of corn seed, which represent distinct genotypes. If we grow strain 1 in soils containing different amounts of nitrogen, as shown in red in **FIGURE 29.6**, we see that there is only a small increase in the amount of grain produced. In contrast, if we grow strain 2

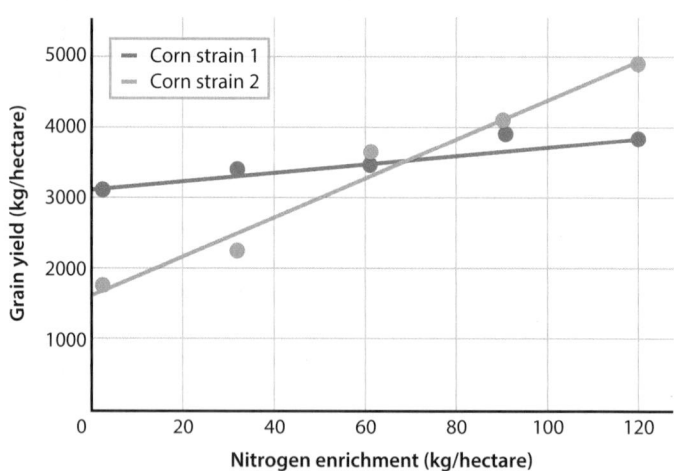

FIGURE 29.6 Different corn strains in a range of nitrogen environments

The graph shows that the responses of two corn strains to soil nutrients differs. When strain 1 was grown in soils with increased amounts of nitrogen, it displayed a small increase in grain produced. However, when strain 2 was grown in the same soils, it exhibited a large increase in grain produced.

in the same soils, as shown in blue, we see that the grain yield increases dramatically. Thus, the two distinct genotypes respond very differently to the same range of environmental conditions. Knowing how different strains of domesticated crops such as corn respond to higher nitrogen in the soil becomes very important when farmers need to determine how much nitrogen to add for increased grain production. As you can see in the corn example, the phenotype of an organism is the combined outcome of both the genes and the environment.

The Evolution of Phenotypically Plastic Traits and Phenotypic Trade-offs

Many environmentally induced changes in phenotypes represent evolved strategies that increase the organism's performance in its environment. Some phenotypes are well suited for one environment, but poorly suited for another environment. We saw an example of this pattern in the tadpoles that were responding to predator smells in the water. In the presence of predators, tadpoles with larger tails escape predators better than tadpoles with smaller tails. As we discussed earlier, in a pond that does not have predators, building a large tail comes at the cost of building a smaller body. A smaller body includes a smaller mouth and a smaller digestive system, so having a smaller body means the tadpoles cannot eat as much and therefore they will not grow as fast. Tadpoles with large tails and small bodies perform better in a predator environment, but tadpoles with small tails and large bodies perform better in a no-predator environment. In short, there is a fitness trade-off between which phenotype performs better in two different environments.

A striking example of the trade-offs that occur with different phenotypes can be seen in the intestines of a snake known as the Burmese python (*Python bivittatus*). The python commonly eats a large rodent only once a month, but the rodent can be 25% of the python's body weight. Within 24 hours of eating the rodent, the snake enlarges the diameter and length of its intestine, as shown in **FIGURE 29.7**. These intestinal changes result from an increase in the size of its cells and an increase in the rate of

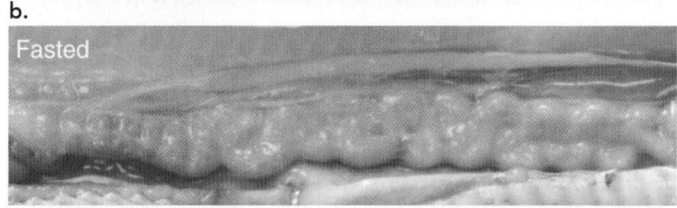

Fasted

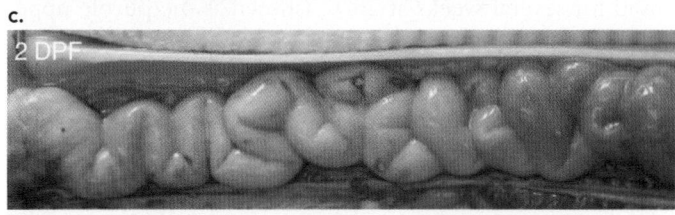

2 DPF

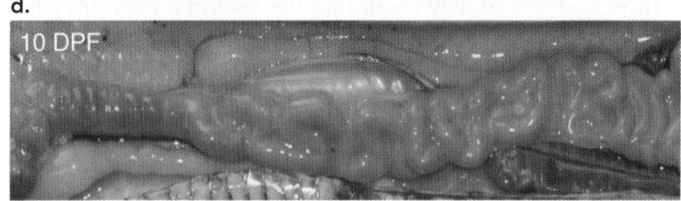

10 DPF

FIGURE 29.7 Phenotypic plasticity in a python's intestines

Burmese pythons exhibit tremendous plasticity in their intestines. (a) Burmese pythons eat large rodents, but they may eat only one large rodent per month. (b) When the python has not eaten for several weeks, its intestine is relatively small. (c) When it consumes a large rodent, the intestine enlarges by increasing its diameter and doubling in length for improved digestive ability. Note that "DPF" means "days postfeeding." (d) Ten days after consuming the rodent, digestion is complete and the intestine once again shrinks down to a much smaller size. Photos: (a) Bryan Rourke, California State University; (b through d) Reproduced/adapted with permission from Secor S M J Exp Biol 2008;211:3767-3774 © 2008 by The Company of Biologists Ltd. Permission conveyed through Copyright Clearance Center, Inc.

cell division. These changes help the python efficiently digest the rodent. At the same time, the snake rapidly increases the size of its heart, liver, and kidneys. Ten days later, digestion is complete and an enlarged intestine is no longer useful. In turn, the intestine's length and diameter dramatically shrink.

These changes allow the snake to carry much less weight during the weeks spent moving around between meals and also reduce the amount of energy needed to sustain the enlarged organs. Thus, the python benefits from having an enlarged intestine when it has consumed a large rodent, and it benefits from having a small intestine when it is not digesting.

Another interesting example of phenotypic plasticity that is shaped by performance trade-offs can be found in the swimming performance of fish at different temperatures. For example, goldfish (*Carassius auratus*) can experience a wide range of water temperatures throughout the seasons and can adjust their physiology accordingly. In one experiment, goldfish were raised for several weeks in 5°C water or 25°C water and then tested for how fast they could swim at different water temperatures. As you can see in **FIGURE 29.8**, fish raised for several weeks at 5°C (shown as a blue, lower curve) swam most quickly at lower water temperatures and their performance declined as the temperature rose above 15°C. In contrast, fish raised for several weeks at 25°C (shown as the purple upper curve) swam most quickly at higher water temperatures and their performance declined as the temperature dropped below 25°C. Goldfish make this adjustment in physiology by altering the enzymes that they produce; some enzymes perform well at low temperature while others perform well at high temperatures. You may recall that we discussed how enzymes are influenced by environmental conditions in Module 14. As we have seen in each of these examples, phenotypic plasticity has evolved due to performance trade-offs when species find themselves living in different environments.

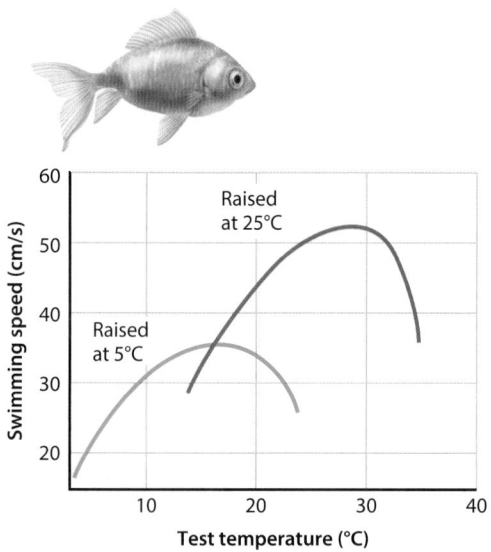

FIGURE 29.8 Phenotypic trade-offs in goldfish swimming when raised at cold and warm temperature

Researchers raised goldfish in water at 5°C or 25°C for several weeks and then assessed their swimming performance across a range of different temperatures. The fish raised under the colder temperature swam most quickly at lower water temperatures while the fish raised under the higher temperature swam most quickly at higher water temperatures. Data from F. E. J. Fry and J. S. Hart. 1948. "Cruising Speed of Goldfish in Relation to Water Temperature." *Journal of the Fisheries Board of Canada* 7:169–174.

PREP FOR THE AP® EXAM

AP® EXAM TIP

Note that phenotypic plasticity is the result of an evolutionary process. It causes a species to experience altered gene expression in ways that change phenotypes and increase the chances of survival and reproduction under differing environmental circumstances.

✓ Concept Check

1. **Identify** the two factors that determine an organism's phenotype.

2. **Evaluate** the claim that all phenotypes are determined only by the genetics of an organism.

3. **Describe** how phenotypic trade-offs favor the evolution of phenotypic plasticity.

29.2 Genes and the environment affect human health and disease

The interaction of genes and the environment is not limited to wild and domesticated plants and animals. We see the same phenomenon in humans. In this section, we will highlight a few examples of genes and environments affecting human health and also discuss how researchers have studied human twins to better understand the contributions of genes and the environment.

Common Diseases as the Product of Multiple Genes and Exposures to Different Environments

While some phenotypes are determined by a single gene, the vast majority of phenotypes are determined by a large number of genes. For example, the genetic contribution to human height is determined by more than 600 genes. Of course, human height is also determined by the environment, particularly the amount of food available to a person early in life. The same can be said for most human diseases. While certain diseases are indeed caused by a mutation at a single gene, most diseases are determined by a large number of genes within the context of specific environmental exposures during a person's life.

The cholesterol level in people's bodies is an excellent example of the interaction of genes and the environment. Researchers have identified nearly 100 genes that contribute to a person's cholesterol level. Some genes have larger effects while others have smaller effects. At the same time, we know that a person's diet can also affect cholesterol levels. For example, a diet that is high in fat, carbohydrates, and cholesterol increases a person's cholesterol more than a diet that is high in fruits and vegetables. An individual's cholesterol level is therefore also affected by the environment, in the form of the person's diet. A person's cholesterol level is the product of both genotype and environment. As a result, some people with a genotype that favors high cholesterol can have a difficult time lowering the cholesterol in their bodies even when they modify their diet by reducing their consumption of animal products. At the same time, people with a genotype that favors low cholesterol can consume a cholesterol-rich diet, yet not experience high cholesterol levels in their bodies.

Emphysema is another good example of genes and the environment contributing to human diseases. Emphysema is a disease of the lungs in which the alveoli become inflamed and damaged, often as a result of smoking tobacco products. Lifelong smokers can develop emphysema or related lung diseases. In addition, people exposed to secondhand smoke or air pollution such as wood smoke can develop emphysema. Thus, the environmental exposure to smoke plays a major role in determining whether a person will experience lung inflammation and damage. However, not everyone who is exposed to smoke will develop emphysema. Moreover, we are learning that genetics can also have an effect; researchers have identified a number of genes that make people more likely to develop the disease if they smoke. Once again, the likelihood of having this lung disease is the product of genes and environment.

Twin Studies and the Effects of Genes and Environment

A useful approach to examine the role of genes versus the environment on human diseases is examining twins. Twins can arise from one fertilized egg (zygote) that divides into two genetically identical embryos. These twins are known as monozygotic twins or identical twins. Twins can also arise when two eggs are fertilized by two different sperm. These twins are known as dizygotic twins, nonidentical twins, or fraternal twins. You can see an example of these two types of twins in **FIGURE 29.9**. Studying phenotypes in twins is particularly useful because identical twins carry identical genotypes, whereas nonidentical twins carry different genotypes. Thus, any differences in phenotypes between identical twins cannot be due to differences in their genetics; it can only be due to differences in their environments. In contrast, differences in phenotypes between nonidentical twins can be due to both differences in their genotypes and differences in their environments.

a.

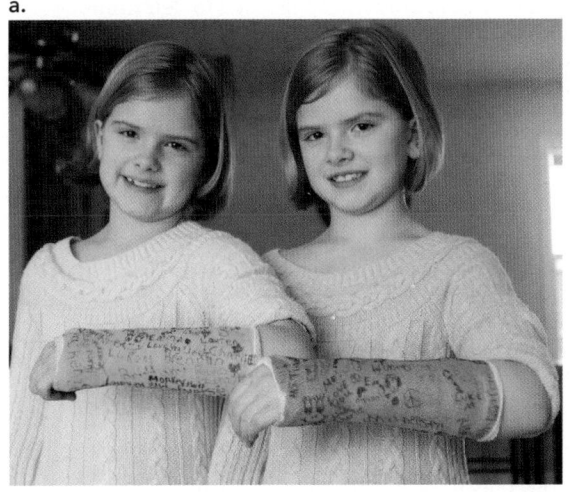

b.

FIGURE 29.9 Identical and nonidentical twins

Twins can arise in two different ways. (a) Identical twins, also known as monozygotic twins, are produced when a fertilized egg (a zygote) divides into two genetically identical embryos. (b) Nonidentical twins, also known as dizygotic twins, are produced when two different eggs are fertilized by two different sperm, resulting in twins that have different genotypes. Photos: (a) Craig Gaffiels/AP Images; (b) Wendy Connett/Getty Images

TABLE 29.1 Comparing Disease Phenotypes in Identical and Nonidentical Twins

Disease	Percentage of cases in which the identical twin expressed the disease	Percentage of cases in which the nonidentical twin expressed the disease
Type I diabetes	65	12
Type 2 diabetes	77	25
Graves' disease	32	4
Obesity	73	30
Growth hormone deficiency	60	15

Data from K. V. Hari Kumar and K. D. Modi. 2014. "Twins and Endocrinology." *Indian Journal of Endocrinology and Metabolism* 18:S48–52.

In studying human diseases, we can ask whether identical or nonidentical twins are more likely to have a disease. To do this, we identify sets of twins in which at least one twin has a given disease and then we determine the percentage of cases in which the other twin also has the disease. If the disease has a strong genetic contribution, we would expect that a higher percentage of identical twins both have the disease compared to nonidentical twins. Examples of twin comparisons for different diseases can be found in **TABLE 29.1**. In this table, we can see that if one twin has type 1 diabetes, type 2 diabetes, Graves' disease, obesity, or growth hormone deficiency, there is a higher probability that the identical twin also has the disease, and a lower probability that a nonidentical twin has the disease. This observation suggests that genetics plays an important role in determining the likelihood of experiencing these four diseases. In contrast, experiencing measles or an acute infection leading to death have only small differences in the percentages between identical and nonidentical twins. This observation suggests that genetics plays only a small role in determining the likelihood of experiencing measles or an acute infection. Instead, the likelihood of a person experiencing measles and acute infections has much more to do with whether they are exposed to the viruses and bacteria that cause these diseases and whether they have received a vaccine or antibiotic to combat these diseases.

In **"Visual Synthesis 5.1: The Combined Effects of Genes and the Environment in Determining Phenotypes"** on page 418, you can see an overview of the roles that genes and the environment play in determining phenotypes, including the phenotypes associated with diseases. For example, we can see that a mutation in the *BRCA1* gene can cause an increase in the risk of breast and ovarian cancers. While such mutations can increase the risk of having cancer, we know that a person's environment can play a role as well. As a result, we can have two individuals carry the same mutation, but one individual develops cancer and the other individual does not. The difference can be in the environments of the two individuals, including the food they eat, pollutants in the water they drink, and pollutants in the air they breathe.

✓ Concept Check

4. **Evaluate** the claim that a person's cholesterol level is determined solely by diet.

5. **Describe** how twin studies can be used to determine whether genetics plays a large role in determining a disease phenotype.

Module 29 Summary

REVIST THE BIG IDEAS

PREP FOR THE AP® EXAM

SYSTEMS INTERACTIONS: Using an example from the module, **explain** how two individuals with identical genotypes may exhibit variation in phenotypes.

LG 29.1 Phenotypes are the product of an organism's genes and its environment.

- Most phenotypes are influenced by a large number of genes. Page 406

- The environment can have a major impact on phenotypes. Page 406

- The ability of a single genotype to produce different phenotypes in different environments is known as phenotypic plasticity. Page 406

- Phenotypic plasticity is pervasive in nature. Page 407

- The effects of genes and the environment can interact in a variety of ways, with some genotypes exhibiting no response to environmental change while others exhibit a large response to environmental change. Page 410

- Many phenotypically plastic traits have evolved due to phenotypic trade-offs. Page 411

LG 29.2 Genes and the environment affect human health and disease.

- Genes and environments affect human health and disease. Page 412
- Most diseases are determined by a large number of genes along with a substantial contribution from the environments to which a person is exposed. Page 412

- A useful approach to examine the role of genetics versus the environment on human diseases is examining twins. Page 413
- Any differences in phenotypes between identical twins cannot be due to differences in their genetics; it can only be due to differences in their environments. Page 414
- Any differences in phenotypes between nonidentical twins can be due to differences in the genotypes and differences in their environments. Page 414

Key Term

Phenotypic plasticity

Review Questions

1. In a field of cotton plants where the farmer has planted a single genotype, you see differences in plant height. Which statement does this evidence support?
   (A) The differences in height are due to differences in genotypes.
   (B) The differences in height are due to differences in environment.
   (C) The differences in height are due to differences in genotypes and environment.
   (D) The differences in height are due to neither genotypes nor environment.

2. The arctic fox grows a brown coat in the summer and a white coat in the winter. This is an example of
   (A) environmental sex determination.
   (B) phenotypic plasticity.
   (C) incomplete dominance.
   (D) dominant alleles.

3. Why do phenotypic trade-offs favor the evolution of plastic phenotypes?
   (A) A single phenotype performs well in all environments.
   (B) A single phenotype performs poorly in all environments.
   (C) A phenotype that performs well in one environment also performs well in other environments.
   (D) A phenotype that performs well in one environment performs poorly in other environments.

4. Which statement about human diseases is not true?
   (A) Some diseases are caused by a mutation at a single gene.
   (B) Most diseases have a genetic contribution from a large number of genes.
   (C) Most diseases are affected by both genetics and the environment.
   (D) Most diseases are not affected by the environment.

5. Why are twin studies useful in understanding human diseases?
   (A) Differences in disease susceptibility in identical twins are only due to environmental differences.
   (B) Differences in disease susceptibility in nonidentical twins are only due to environmental differences.
   (C) Differences in disease susceptibility in identical twins are only due to genetic differences.
   (D) Differences in disease susceptibility in nonidentical twins are only due to genetic differences.

Module 29
AP® Practice Questions

PREP FOR THE AP® EXAM

Section 1: Multiple-Choice Questions

Choose the best answer for questions 1–5.

1. Which statement best explains why traits such as human height are highly variable within populations?

   (A) Traits such as height are determined by allelic variation in a large number of genes, and also are influenced by the environment.

   (B) Traits such as height are determined by a very large number of alleles in a small number of genes, and are also influenced by the environment.

   (C) Traits such as height are primarily determined by the environment, and are only slightly influenced by genotype.

   (D) Traits such as height are entirely determined by the environment, and do not have a genetic basis.

2. A study of human twins in London evaluated the frequency of dyslexia in monozygotic and dizygotic twins. In monozygotic twins, if one twin was diagnosed with dyslexia, the other twin had a 37% chance of also being dyslexic. In dizygotic twins, if one twin was diagnosed with dyslexia, the other twin had a 21% chance of also being dyslexic. Which statement explains why researchers studied both monozygotic and dizygotic twins rather than monozygotic twins only?

   (A) Dizygotic twins do not share alleles and therefore reveal environmental influences on the chance of having dyslexia.

   (B) Dizygotic twins are genetically identical, so an increased chance of having dyslexia indicates a genetic basis for the trait.

   (C) Dizygotic twins are less closely genetically related than monozygotic twins, so a lower chance of having dyslexia in dizygotic twins indicates a genetic basis for the trait.

   (D) Dizygotic twins experience different environments while monozygotic twins do not, so the chance of having dyslexia in dizygotic twins reveals environmental influences.

Use the following information to answer questions 3–5.

Norwegian white sheep (NWS) and Spel sheep (Spel) are two breeds of sheep that are commonly reared in Norway. Body mass at weaning, which is the time when baby sheep stop drinking their mother's milk, is an important measure for sheep farmers. Scientists measured body mass at weaning in NWS and Spel from 40 flocks of sheep distributed around Norway in 1989 and 1990. The table shows the mean body mass at weaning for the two breeds. The weather was very different between the two years. Daily temperature averaged 1.9°C colder in 1990 and there was 4% less precipitation in 1990. Thus, the 2 years can be considered different environments.

Year	NWS mass (in kg)	Spel mass (in kg)
1989	44.0	40.8
1990	44.7	42.9

3. On average, how much larger at weaning were sheep from the Spel breed in 1990 than in 1989?

   (A) 0.7 kg   (C) 1.8 kg

   (B) 1.1 kg   (D) 2.1 kg

4. Which graph best illustrates the data?

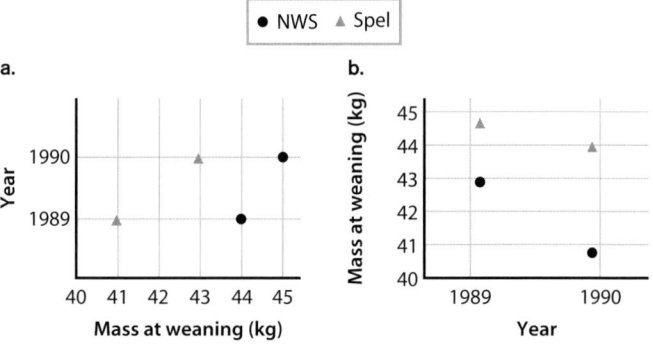

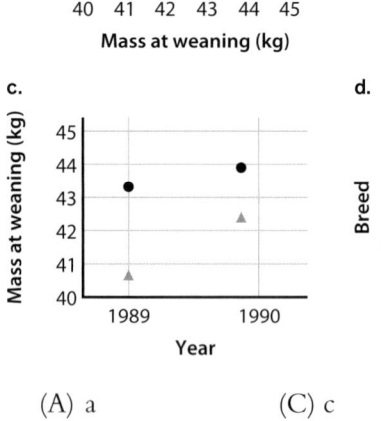

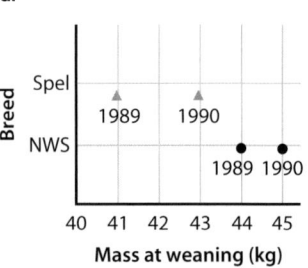

   (A) a   (C) c

   (B) b   (D) d

5. Based on the data provided, which statement best describes how genotype and environment interact to determine the mass at weaning in Norwegian sheep?

(A) Environmental conditions do not influence mass at weaning in the NWS breed because NWS always have higher mass at weaning than Spel.

(B) Mass at weaning is more strongly affected by environmental conditions in the Spel breed than in the NWS breed.

(C) The difference between the NWS breed and the Spel breed in mass at weaning is mostly due to environmental effects.

(D) Temperature has a larger effect on mass at weaning than rainfall does in the Spel breed.

Section 2: Free-Response Question

Write your answer to each part clearly. Support your answers with relevant information and examples. Where calculations are required, show your work.

Indole is a molecule produced by plants in response to herbivory by insects. Indole is one of several molecules that regulates chemical defenses against insect herbivores. A researcher investigated production of indole by two genotypes of rice plant after fall armyworm caterpillars (*Spodoptera frugiperda*) fed upon them. Indole production by each rice strain was measured for 1 hour after the caterpillars were allowed to eat varying percentages of a rice leaf. The figure shows the mean value of indole that the researcher obtained for each treatment. Error bars are calculated as ±1 standard deviation on either side of the mean.

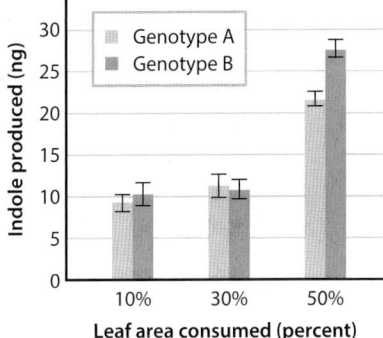

(a) **Identify** the dependent variable in the study.

(b) **Identify** the conditions that induce the highest level of indole production by genotype B.

(c) **Calculate** the approximate difference in indole production by genotype A when 50% of the leaf area is consumed compared to when 30% of the leaf area is consumed.

(d) The researcher claimed that the two rice genotypes differ in indole production only after 50% of the leaf area has been consumed. **Provide reasoning** to support or refute the researcher's claim.

(e) **Make a claim** to explain why rice plants do not produce indole when 0% of the leaf area has been consumed. Considering the fitness of the plants, provide reasoning to justify your claim.

In the case of some cancers, researchers have identified genetic mutations that increase the risk of a particular type of cancer. However, not all individuals with the mutation will experience cancer. One reason is because the

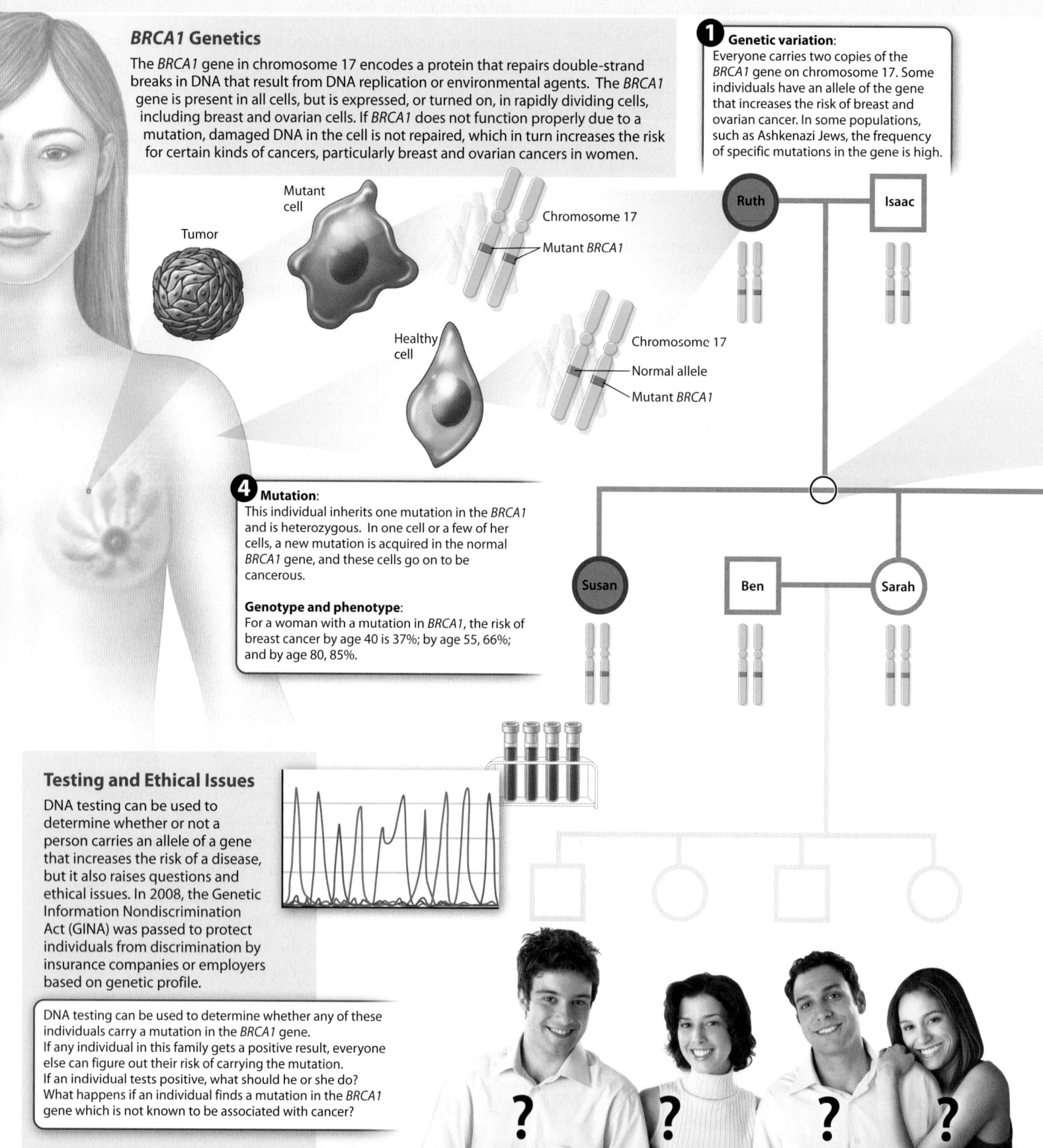

BRCA1 Genetics

The *BRCA1* gene in chromosome 17 encodes a protein that repairs double-strand breaks in DNA that result from DNA replication or environmental agents. The *BRCA1* gene is present in all cells, but is expressed, or turned on, in rapidly dividing cells, including breast and ovarian cells. If *BRCA1* does not function properly due to a mutation, damaged DNA in the cell is not repaired, which in turn increases the risk for certain kinds of cancers, particularly breast and ovarian cancers in women.

Mutant cell

Tumor

Chromosome 17

Mutant *BRCA1*

Healthy cell

Chromosome 17

Normal allele

Mutant *BRCA1*

1 Genetic variation:
Everyone carries two copies of the *BRCA1* gene on chromosome 17. Some individuals have an allele of the gene that increases the risk of breast and ovarian cancer. In some populations, such as Ashkenazi Jews, the frequency of specific mutations in the gene is high.

Ruth

Isaac

4 Mutation:
This individual inherits one mutation in the *BRCA1* and is heterozygous. In one cell or a few of her cells, a new mutation is acquired in the normal *BRCA1* gene, and these cells go on to be cancerous.

Genotype and phenotype:
For a woman with a mutation in *BRCA1*, the risk of breast cancer by age 40 is 37%; by age 55, 66%; and by age 80, 85%.

Susan

Ben

Sarah

Testing and Ethical Issues

DNA testing can be used to determine whether or not a person carries an allele of a gene that increases the risk of a disease, but it also raises questions and ethical issues. In 2008, the Genetic Information Nondiscrimination Act (GINA) was passed to protect individuals from discrimination by insurance companies or employers based on genetic profile.

DNA testing can be used to determine whether any of these individuals carry a mutation in the *BRCA1* gene.
If any individual in this family gets a positive result, everyone else can figure out their risk of carrying the mutation.
If an individual tests positive, what should he or she do?
What happens if an individual finds a mutation in the *BRCA1* gene which is not known to be associated with cancer?

? ? ? ?

environments to which the person is exposed can also play a role in determining whether the person will come down with the disease.

Photos: (left to right) Michael Poehlman/Getty Images, LWA/Larry William/Getty Images, Digital Vision/Getty Images, Fabrice Lerouge/AGE fotostock, Piotr Marcinsk/Dreamstime.com

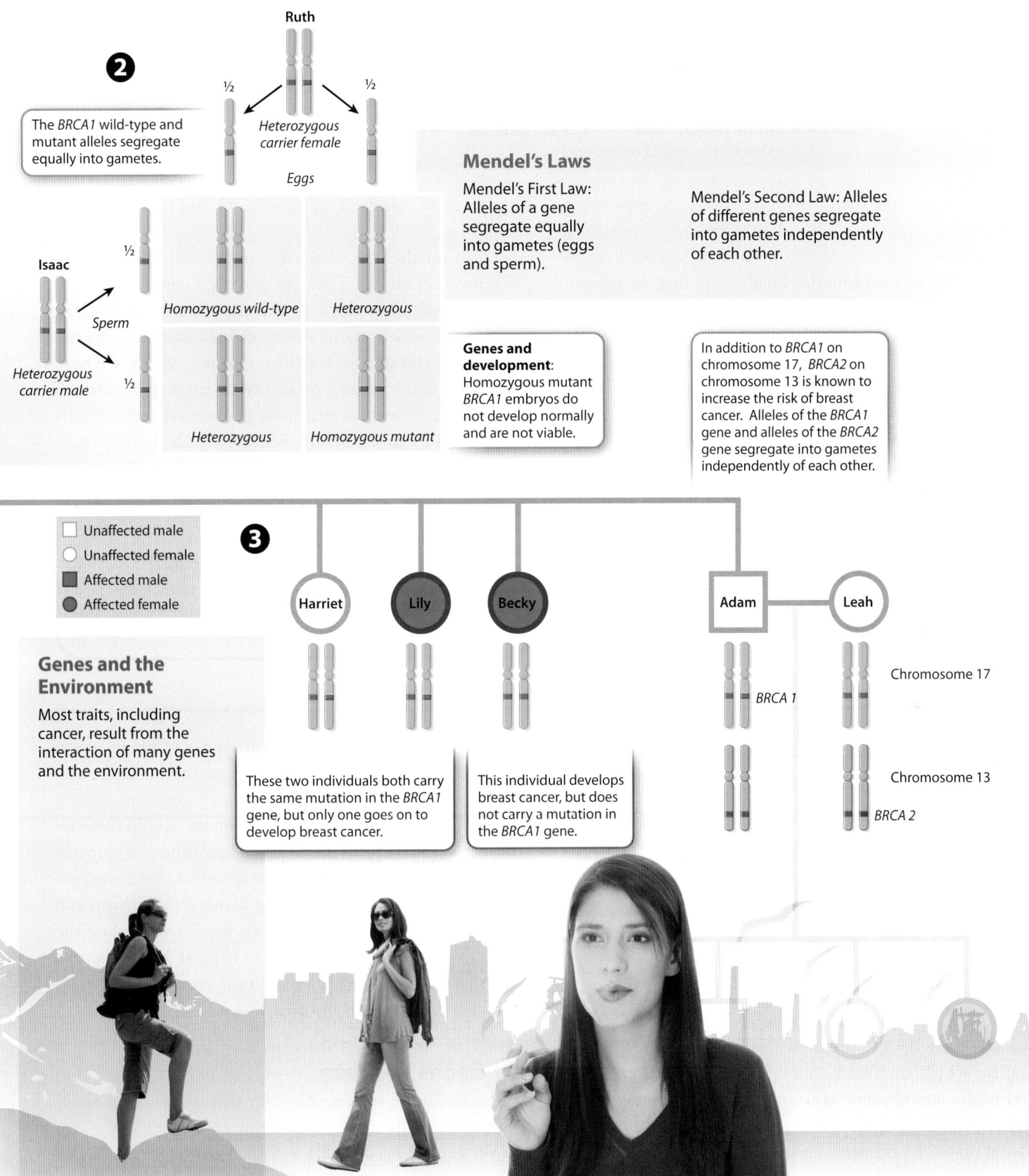

2

The BRCA1 wild-type and mutant alleles segregate equally into gametes.

Ruth

½   ½

Heterozygous carrier female

Eggs

Isaac

½

Sperm

½

Heterozygous carrier male

Homozygous wild-type   *Heterozygous*

Heterozygous   *Homozygous mutant*

Mendel's Laws

Mendel's First Law: Alleles of a gene segregate equally into gametes (eggs and sperm).

Mendel's Second Law: Alleles of different genes segregate into gametes independently of each other.

Genes and development: Homozygous mutant BRCA1 embryos do not develop normally and are not viable.

In addition to BRCA1 on chromosome 17, BRCA2 on chromosome 13 is known to increase the risk of breast cancer. Alleles of the BRCA1 gene and alleles of the BRCA2 gene segregate into gametes independently of each other.

3

☐ Unaffected male
○ Unaffected female
■ Affected male
● Affected female

Harriet   **Lily**   **Becky**   **Adam**   **Leah**

Genes and the Environment

Most traits, including cancer, result from the interaction of many genes and the environment.

These two individuals both carry the same mutation in the BRCA1 gene, but only one goes on to develop breast cancer.

This individual develops breast cancer, but does not carry a mutation in the BRCA1 gene.

BRCA 1

Chromosome 17

BRCA 2

Chromosome 13

419

Module 30
Chromosomal Inheritance

LEARNING GOALS ▶**LG 30.1** Chromosomal inheritance can result in mutations that increase genetic diversity but can also cause genetic disorders.

▶**LG 30.2** Sexual reproduction can increase genetic diversity by creating offspring with extra or missing chromosomes.

Earlier in this unit, we introduced the process of meiosis, whereby diploid species that experience sexual reproduction produce haploid gametes. We learned about the law of segregation, which tells us that the pairs of homologous chromosomes separate from each other during meiosis I. We then introduced the law of independent assortment, which tells us that the way in which two alleles segregate in one chromosome has no effect on how alleles segregate in another chromosome. As a result, a single gamete can contain some chromosomes that originally came from the individual's mother and other chromosomes that originally came from the individual's father. We also noted that segregation and independent assortment generate a considerable amount of genetic variation among the gametes. Additional genetic variation is generated in zygotes that are produced when sperm cells fertilize egg cells by bringing together new combinations of alleles that collectively influence the phenotype. In short, the processes associated with sexual reproduction can create a substantial amount of genetic variation.

However, these processes are not the only way that sexual reproduction can generate genetic variation through chromosomal inheritance. In this module, we will discuss how genetic variation can also be created by mutations and by having extra or missing chromosomes. While we have previously emphasized our ability to predict patterns of inheritance, we will see that these phenomena can complicate our ability to predict the genotypes and phenotypes of offspring based on the genotypes of their parents.

PREP FOR THE AP® EXAM

FOCUS ON THE BIG IDEAS

INFORMATION STORAGE AND TRANSMISSION: Focus on the relationship between the movement of chromosomes in meiosis and how inherited mutations and extra chromosomes generate genetic variation.

30.1 Chromosomal inheritance can result in mutations that increase genetic diversity but can also cause genetic disorders

You likely recall that DNA is comprised of a long string of nitrogenous bases that are arranged in a specific sequence to code for a specific order of amino acids. You might also recall that the ordering of these amino acids is critically important to building the appropriate proteins. As discussed briefly in Module 22, a mutation is a change in DNA sequence that can alter the order of the amino acids that comprise the protein for which the DNA codes. DNA mutations can happen when chromosomes are copied during the processes of replication. If mutations occur prior to meiosis, they can be passed on to the offspring of sexually reproducing species.

While mutations in genes can sometimes result in a new beneficial allele that can improve fitness and be favored by evolution over time, most mutations in genes are harmful because they lead to an altered protein, which includes enzymes, and that protein can no longer function properly. As we will see, these harmful mutations can be passed on to generations of offspring during sexual reproduction and cause a variety of disorders. In Unit 6, we will explore the mechanisms that allow mutations to happen. In this section, we will discuss a number of harmful mutations at single genes that cause genetic disorders.

Tay-Sachs Disease

Tay-Sachs disease is a rare disorder that is most commonly diagnosed in infancy. It occurs due to a mutation in the *HEXA* gene that codes for a critical enzyme in the brain

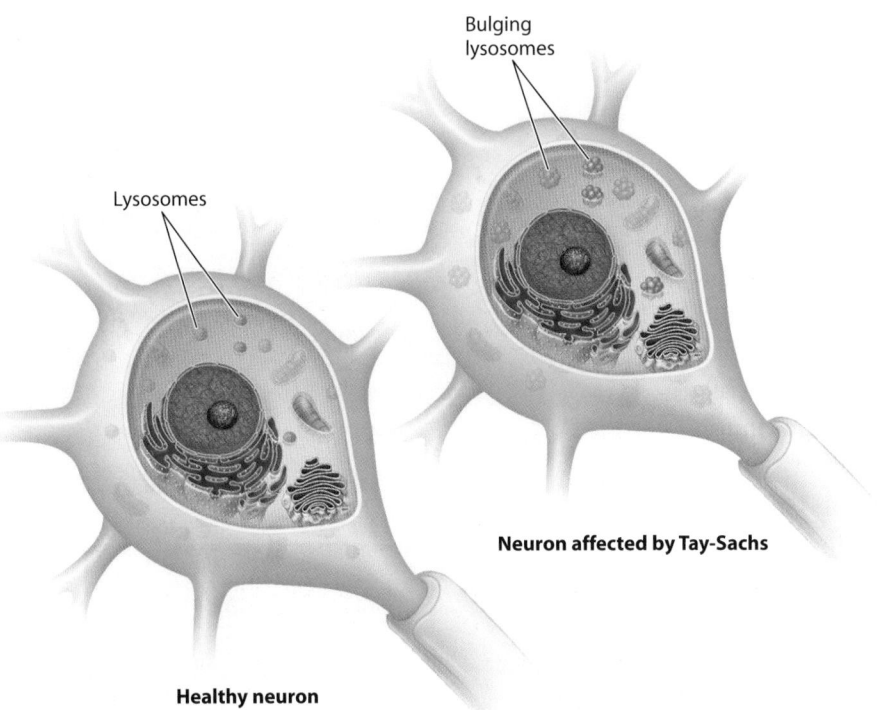

Bulging
lysosomes

Lysosomes

Neuron affected by Tay-Sachs

Healthy neuron

FIGURE 30.1 Tay-Sachs disease

A single mutation in the *HEXA* gene causes an enzyme (known as beta-hexosaminidase A) to no longer break down fatty substances (known as GM2 gangliosides) in the lysosomes of nerve cells. In the nerve cell of a person that is not homozygous recessive for the allele (top image), the fatty substances are broken down and the lysosomes remain relatively small. In the nerve cell of a person that is homozygous recessive for the allele (bottom image), the fatty substances accumulate, causing the GM2 gangliosides to reach toxic levels, which leads to the destruction of the neurons and the symptoms of Tay-Sachs disease.

and spinal cord, which is known as beta-hexosaminidase A. The enzyme is present in the lysosomes of the nerve cells, where they break down fatty substances known as GM2 ganglioside. As shown in **FIGURE 30.1**, when there is a mutation, the lysosomes can no longer break down GM2 ganglioside, so it accumulates to toxic levels and destroys the neurons, which causes the symptoms of Tay-Sachs disease. Infants with this disease slowly lose their physical and mental abilities and rarely live beyond 4 years old. The mutation is recessive, so only infants that carry two copies of the recessive allele express the disease. Thus, a child can have Tay-Sachs disease only if both parents are heterozygous carriers of the recessive allele.

Huntington's Disease

Huntington's disease is another rare human disorder caused by a mutation in a single gene that mostly affects adults in their 30s and 40s. Huntington's disease is caused by a breakdown of brain cells that increasingly impairs a person's muscle control, coordination, and ability to think clearly. It appears when there is a mutation in the *HTT* gene in which there are extra repeats of the *CAG* trinucleotide, which help code for a protein known as huntingtin. Most people have 10 to 35 copies of the *CAG* trinucleotide, but people with Huntington's disease can have 40 to 120 copies. While the function of the protein remains unclear, we know that it plays a role in nerve cells and that the more repeats of *CAG* present, the more extreme the symptoms are in an individual. Unlike Tay-Sachs disease, which is caused by a recessive mutant allele, Huntington's disease results from a dominant mutant allele. As a result, a person needs only one parent to carry the allele to potentially inherit the allele and experience the disease.

Phenylketonuria

Phenylketonuria (PKU) is another disorder caused by a mutation in a single gene, but in this case more than 400 different mutant alleles of the *PAH* gene have been identified that can cause the disease. These mutations cause the gene to produce an impaired form of the enzyme phenylalanine hydroxylase, which is a key enzyme for the metabolic pathway that breaks down phenylalanine. Phenylalanine is an amino acid found in eggs, meat, and the milk of mammals, including the breast milk that infants consume. When children are born with PKU, they slowly accumulate more phenylalanine, since the impaired form of the enzyme cannot properly break phenylalanine down fast enough. High concentrations of accumulated phenylalanine become toxic to tissues throughout the body, and nerve cells are especially sensitive. As a result, increasing levels of phenylalanine impair brain development and cause intellectual disability as well as behavioral problems and seizures. The alleles that cause PKU are recessive, so a child needs to get a copy of a recessive allele from each parent to exhibit the disease. If infants are diagnosed early, PKU can be treated with medication and a modified diet to ensure proper brain development.

Crooked-tail Syndrome in Cattle

Genetic mutations also occur in nonhuman animals. For example, there is a strain of cattle, known as Belgian Blue cattle, that have been bred for incredibly large muscles to the point that the phenotype is known as double muscling. This increase in muscle size, which you can see in **FIGURE 30.2**, is caused by genes that cause calf embryos to build twice as many muscle fibers. This phenotype allows farmers to grow cattle with more muscle, which translates into more meat to sell. However, unknown to the original breeders of Belgian Blue cattle, the small number of cows that were used to start the breeding program for larger muscles also contained individuals that carried a mutation in the *MRC2* gene, which causes a disorder known as crooked-tail syndrome. Thus, when the cows were bred for the double-muscle phenotype, the cattle were inadvertently also bred for the crooked tail mutation, which is currently carried by about 25% of all individuals. Crooked-tail syndrome causes the tail to grow at an odd angle compared to a typical cow, as you can see in the figure. In addition, individuals with this genetic mutation exhibit retarded growth and abnormal legs, which is not desirable for farmers trying to grow large cattle. With genetic information identifying which individuals carry the mutation and controlled breeding of specific cows and bulls, cattle breeders are gradually reducing the number of calves that carry the mutation.

Variegated Leaves in Plants

Plants can also experience disorders from single-gene mutations. For example, some plants have leaves that are a mix

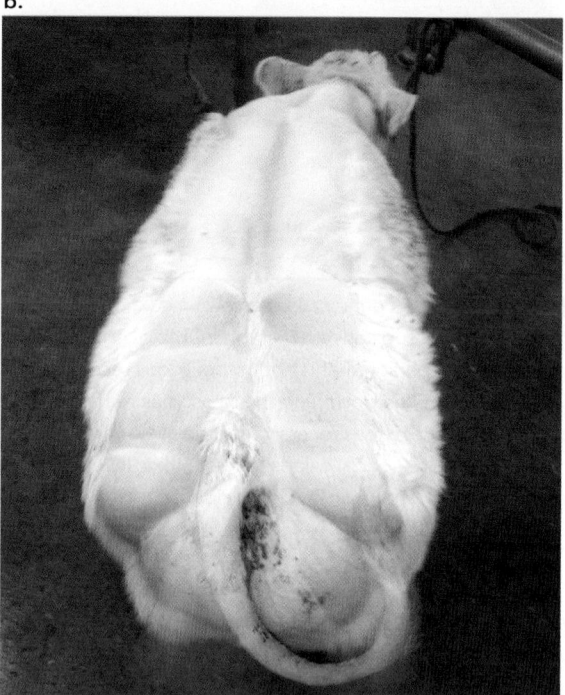

FIGURE 30.2 Genetic mutations in cattle

When cattle breeders bred for larger cattle, they inadvertently bred for a harmful mutation. (a) Cattle breeders bred Belgian Blue for a double-muscle phenotype. (b) However, about 25% of cattle in the breeding program carried a mutation in the *MRC2* gene, which causes the animal to develop a crooked tail. When the offspring carry two copies of the recessive allele, they develop a crooked tail, retarded growth, and abnormal legs, which is not desirable for farmers trying to grow large cattle. Photos: (a) David Bagnall/Alamy Stock Photo; (b) Arnaud Sartelet

of green and white colors, which we call variegated leaves, as shown in **FIGURE 30.3**. These two-colored leaves can be induced by disease-causing organisms or a lack of nutrients, but they can also be caused by genetic mutations. The genetic mutations can occur in the nuclear DNA that inhibits the production of the green chlorophyll in the

FIGURE 30.3 Variegated leaves in the evening primrose

Variegated leaves have a mix of green and white tissues, which represents leaf cells with functioning and nonfunctioning chloroplasts. When variegated leaves are caused by mutations, the mutations can be in nuclear DNA, mitochondrial DNA, or chloroplast DNA. These mutations interfere with the proper functioning of the chloroplasts to make the green chlorophyll that gives leaves their green color.

Photo: Courtesy of TERRA NOVA® Nurseries, Inc., www.terranovanurseries.com

chloroplasts. It can also occur when a mutation in the mitochondrial DNA affects the proper functioning of the chloroplast. Finally, some plants exhibit variegated leaves when a mutation in a nuclear gene causes deletions or insertions of the chloroplast DNA. This prevents the chloroplast from functioning properly, resulting in white leaf tissues. While many house and garden plants are bred for the mutation that causes leaf variegation, such plants are at a disadvantage in nature since they cannot photosynthesize as much as plants that lack the mutation.

As you can see in all of these examples, most mutations in genes are harmful to the normal performance of organisms because they impair the proteins that are coded for by DNA.

✓ Concept Check

1. **Describe** how a single-gene mutation causes Tay-Sachs disease.
2. **Describe** the difference in the inheritance pattern of Tay-Sachs disease and Huntington's disease.
3. **Describe** why most mutations are harmful.

30.2 Sexual reproduction can increase genetic diversity by creating offspring with extra or missing chromosomes

While mutations to single genes can cause important changes in in the genotypes and phenotypes of organisms, even bigger mutations can occur when entire chromosomes are missing, or extra chromosomes are present. Every species has a specific number of chromosomes; for example, humans have 22 pairs of autosomes plus a pair of sex chromosomes for a total of 46 chromosomes. In contrast, horses have 64 chromosomes and corn has 20. We can see the distinctive size and shapes of each homologous pair of chromosomes when they are condensed. As discussed in Module 26, when we display the pairs of chromosomes visually, we call it a karyotype.

Missing or extra chromosomes can be a consequence of errors that occur during mitosis and meiosis. Although we do not commonly observe this phenomenon, it happens with some frequency. However, we rarely observe individuals with extra or missing chromosomes because this outcome is typically lethal. In some cases, an individual is able to survive, but with substantial effects on its phenotype. In this section, we will discuss the causes of extra or missing chromosomes and how this phenomenon creates an increased diversity in genotypes and phenotypes in sexually reproducing organisms.

Extra or Missing Chromosomes and Nondisjunction

Nondisjunction is the failure of chromosomes to separate during cell division. This can be a failure of two sister chromatids to separate during mitosis, a failure of the homologous chromosomes to separate during meiosis I, or a failure of the sister chromatids to separate in meiosis II. In each of these cases, one daughter cell has an extra chromosome, and another daughter is missing a chromosome. When nondisjunction happens during mitosis, it results in a lineage of cell descendants with an abnormal number of chromosomes. While having an abnormal number of chromosomes is often lethal, this is not always the case. In human cancer cells, for example, some tumor cells can have as many as 100 chromosomes instead of the normal 46 chromosomes.

FIGURE 30.4 shows the process of disjunction during meiosis. This figure shows a simplified situation with only one pair of homologous chromosomes. In meiosis I, shown in Figure 30.4a, the homologous pair of chromosomes are separated from each other and then meiosis II the sister chromatids being separated. The result of this process is that each gamete has an identical number of chromosomes, which in this simplified example is just one chromosome.

Nondisjunction can happen early or late in meiosis. As shown in Figure 30.4b, during the first division of meiosis I, nondisjunction results in a pair of homologous chromosomes not being split evenly between the two daughter cells. After meiosis I, one cell keeps both homologous chromosomes and the other cell has none. The subsequent step of meiosis II operates as usual by pulling apart the sister chromatids, which results in two daughter cells having an extra chromosome and the other two daughter cells missing a chromosome.

When nondisjunction happens during meiosis II, as shown in Figure 30.4c, we observe a different outcome. In this case, meiosis I operates as usual and produces two daughter cells, each containing an equal number of homologous chromosomes. During meiosis II, however, the sister chromatids can fail to be evenly separated. When nondisjunction happens in just one of the two daughter cells produced by

meiosis I, this results in two daughter cells having the correct number of chromosomes, one daughter cell having an extra chromosome, and one daughter cell missing a chromosome.

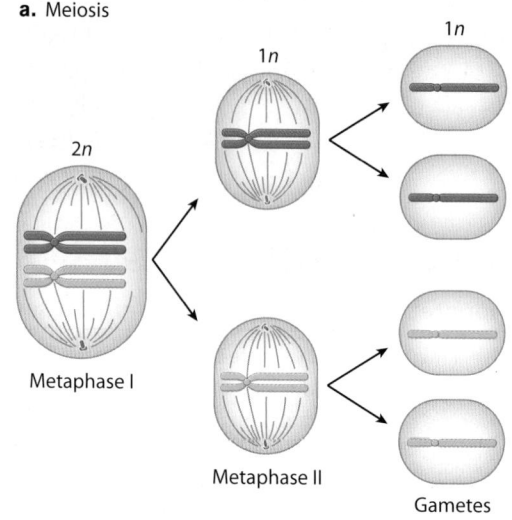

a. Meiosis

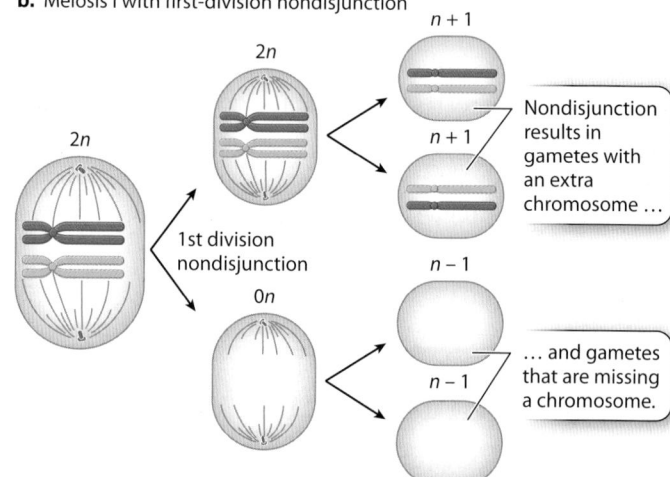

b. Meiosis I with first-division nondisjunction

Nondisjunction results in gametes with an extra chromosome …

… and gametes that are missing a chromosome.

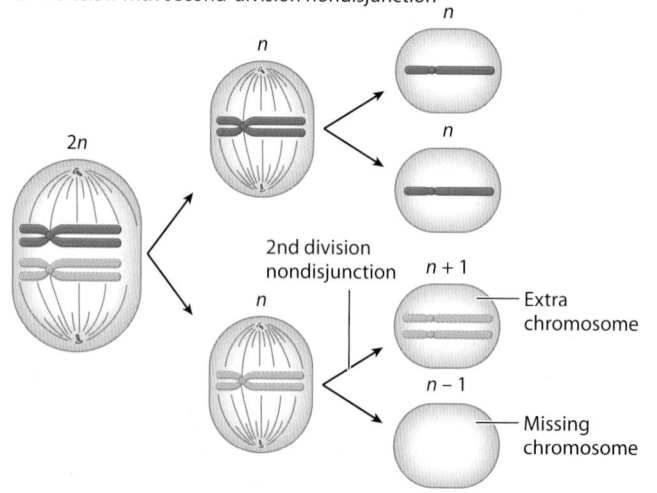

c. Meiosis II with second-division nondisjunction

Extra chromosome

Missing chromosome

FIGURE 30.4 Nondisjunction

Nondisjunction during meiosis can result in extra or missing chromosomes. (a) Typically, the homologous pairs of chromosomes segregate equally during meiosis I and the sister chromatids segregate equally during meiosis II. (b) When nondisjunction occurs in meiosis I, a homologous chromosome pair fails to segregate equally. As a result, half of the daughter cells have an extra chromosome and the other half of the daughter cells are missing a chromosome. (c) When nondisjunction occurs in meiosis II, sister chromatids fail to segregate equally. When nondisjunction happens in just one of the two daughter cells after meiosis I, it produces two daughter cells that have a complete complement of chromosomes, one daughter cell that has an extra chromosome, and one daughter cell that has a missing chromosome.

Human Disorders Caused by Nondisjunction

About 1 out of 300 people (0.3%) are born with an extra chromosome or missing chromosome due to nondisjunction. This can include an entire chromosome or just a portion of a chromosome. In such cases, the gametes of one parent had an extra or missing chromosome in the sperm or egg cell. Having a third chromosome is frequently lethal in human embryos, so we do not typically see examples in living humans. However, there are a few exceptions for specific chromosomes.

One of the best-known examples of a disorder caused by an extra autosome in humans is trisomy 21, also known as **Down syndrome,** in which a person has a third chromosome 21 (or a portion of an extra chromosome 21). We can see an example of trisomy 21 by looking at a karyotype of an individual with this condition. In **FIGURE 30.5**, you can see that this female has two copies of each chromosome, including two *X* chromosomes, but three copies of chromosome 21. Given that a person with trisomy 21 possesses extra genes in the third chromosome 21, this causes a wide range of phenotypic changes. Affected people tend to be shorter, have low muscle tone, and commonly have heart defects. Effects also include mild to moderate impaired intellectual development, but with special education and support individuals can acquire basic communication, self-help, and social skills.

In humans, the rate of trisomy 21 is relatively rare, with only 1 in 750 children possessing the extra chromosome. However, this risk increases dramatically with the age of the mother when the child is conceived. In mothers who are 45–50 years old, the risk of a child with trisomy 21 is 50 times higher than in mothers who are 18–20 years old.

Trisomy 21 is not the only example of humans having an extra chromosome. For example, humans can also produce offspring with trisomy 13 and trisomy 18, but the physical and mental effects on the phenotype are so severe that the offspring typically do not survive their first year.

Extra or Missing Sex Chromosomes in Humans

While having an extra or missing autosome is often lethal or very harmful, having an extra or missing *X* or *Y* sex chromosome often has fewer effects. As we have discussed in previous modules, sex determination varies a good amount among different groups of animals. In most mammals, the

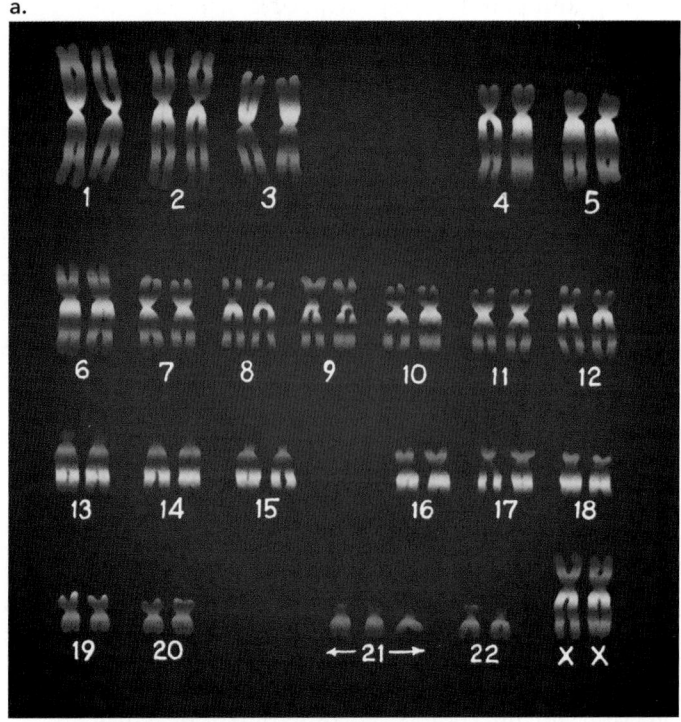

a.

b.

FIGURE 30.5 Trisomy 21

Trisomy 21 is one of the best-known examples of a disorder caused by an extra autosome in humans. (a) An individual with trisomy 21 has a third copy of chromosome 21 (or a portion of a third chromosome 21). (b) Trisomy 21 causes a number of phenotypic changes, including a tendency to be shorter, low muscle tone, heart defects, and intellectual disability. Photos: (a) Bisphoto Associates/Science Source; (b) Denis Kuvaev/Shutterstock

sex chromosomes are *XX* for females and *XY* for males. When it comes to the development of the embryos, the *Y* chromosome causes the embryo to become male. Without the *Y* chromosome, the embryo becomes female.

In some cases, an added sex chromosome has no discernible effect on a person's phenotype. For example, if we were to compare a group of people who are *XX* to a group of people who are *XXX*, we would see the same range of female phenotypes. Similarly, if we were to compare a group of people who are *XY* to a group of people who are *XYY*, we would see the same range of male phenotypes. In both cases, the individuals with the extra chromosome have the same physical attributes and mental abilities. In fact, most people who are *XXX* or *XYY* do not even know they have an extra sex chromosome unless they are tested for some other reason.

We also understand why individuals who are *XXX* or *XYY* do not show any effects of the extra sex chromosome. In the case of an *XYY* individual, there are very few genes on the small *Y* chromosome, and there is just one gene that makes an embryo male. Thus, there are not a lot of extra genes in an *XYY* individual to cause developmental problems. A different mechanism causes *XXX* individuals to have the same phenotypes as *XX* individuals. In female mammals, only one *X* chromosome is active and any additional *X* chromosomes are inactivated. Thus, regardless of whether an individual has two or three *X* chromosomes, only one is active, so the extra set of genes is not expressed, and the phenotype is therefore not altered.

In contrast to these two situations, we do see larger phenotypic effects on individuals who are *XXY*. Individuals who are *XXY* have many male characteristics as a result of the *Y* chromosome, but they make less testosterone, so they exhibit a number of distinctive traits that are collectively known as Klinefelter syndrome. As you can see in **FIGURE 30.6a**, these traits include small testes, a tall stature, a high-pitched voice even after puberty, and some breast development. Males who are *XXY* also do not produce sperm, so they are infertile, and about half of the individuals experience mental impairment.

Another possible alteration in the sex chromosomes are females who have only one *X* chromosome, which is associated with a number of phenotypic changes known as **Turner syndrome.** As shown in Figure 30.6b, individuals with Turner syndrome generally have a short stature, a web of skin between the neck and shoulders, and no sexual maturation. As a result, they have poor breast development, underdeveloped gonads, and no menstruation. Unlike individuals with Klinefelter syndrome, individuals with Turner syndrome have unimpaired mental abilities.

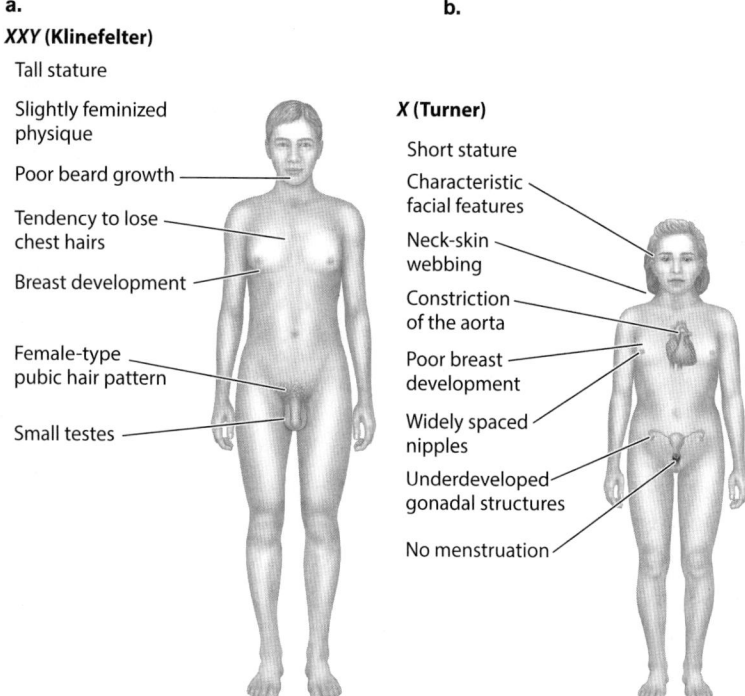

FIGURE 30.6 Phenotypic effects of have an extra or missing sex chromosome

Individuals who have an extra sex chromosome or are missing a sex chromosome exhibit distinctive traits. (a) An individual with *XXY* sex chromosomes exhibits Klinefelter syndrome, which includes male genitalia, small testes, some breast enlargement, poor beard growth, and a tall stature. (b) An individual with just one *X* chromosome exhibits Turner syndrome, which includes underdeveloped female gonads, poor breast development, a web of skin between the neck and shoulders, and a short stature.

Extra or Missing Sex Chromosomes in Plants

As we have seen in humans, having an extra chromosome or a missing chromosome causes major problems in animals, with most individuals not being viable. In contrast, having an extra chromosome or a missing chromosome is quite common in plants and the plants continue to be viable. It is not currently understood why the outcome is so fundamentally different for animals versus plants. It is clear that while the plants can remain viable, the change in chromosome number can cause phenotypic changes that range from small to large, including changes in stem diameter and leaf size.

For example, in **FIGURE 30.7a**, you can see the typical phenotype of a plant known as mouse-ear cress (*Arabidopsis thaliana*) when it is diploid and has no extra chromosomes. In contrast, the same plant has much wider leaves when it

a.

b.

c.

FIGURE 30.7 Phenotypic effects of having extra chromosomes in the mouse-ear cress

When individuals have an extra chromosome, their phenotypes can change dramatically. (a) A diploid plant with no extra chromosomes has normal leaves. (b) A plant of the same species with an extra chromosome 2 develops shorter and wider leaves. (c) A plant of the same species with an extra chromosome 3 and chromosome 5 develops very long, skinny leaves.

Photos: E. Han Tan/University of Maine

has an extra chromosome 2, as shown in Figure 30.7b. However, it develops long and narrow leaves when it has an extra chromosome 3 and chromosome 5, which you can see in Figure 30.8c. In these examples, all three plants are viable because plants are better at surviving and reproducing with an extra chromosome compared to animals. Moreover, the extra chromosomes of the plants create new genetic diversity that can cause major changes in the plant's phenotype.

✓ **Concept Check**

4. **Describe** the process of nondisjunction.

5. **Describe** how nondisjunction can cause trisomy 21.

6. **Describe** why *XXX* females have similar phenotypes to *XX* females.

Module 30 Summary

PREP FOR THE AP® EXAM

REVISIT THE BIG IDEAS

INFORMATION STORAGE AND TRANSMISSION: Using the content of this module, **describe** how changes in chromosome number or structure may occur and how those changes can influence an organism's overall health.

LG 30.1 Chromosomal inheritance can result in mutations that increase genetic diversity but can also cause genetic disorders

- Tay-Sachs disease is caused by a single-gene mutation that is recessive and causes impaired nerve function in infants. Page 420

- Huntington's disease is caused by a dominant single-gene mutation that causes physical and mental impairment of adults in their 30s and 40s. Page 421

- Phenylketonuria (PKU) is a single-gene mutation with more than 400 mutant alleles that prevent the normal breakdown of phenylalanine in the diet of infants, leading to impaired brain development. Page 421

- Single-gene mutations also occur in nonhuman animals, such as the crooked-tail syndrome observed in some breeds of cattle. Page 422

- Single-gene mutations also occur in plants, such as those plants that have variegated leaves, resulting from some chloroplasts being inhibited from producing chlorophyll. Page 422

30.2 Sexual reproduction can increase genetic diversity by creating offspring with extra or missing chromosomes

- While mutations can cause important changes in in the genotypes and phenotypes of organisms, even bigger impacts can occur when entire chromosomes are missing or extra copies are present. Page 423
- Missing or extra chromosomes can occur during mitosis and meiosis due to nondisjunction. Page 423
- Nondisjunction is the failure of a chromosome to separate during anaphase of cell division, which causes one daughter cell to have an extra chromosome and another daughter cell to have a missing chromosome. Page 424
- During meiosis, nondisjunction can occur in either meiosis I or meiosis II. Page 424
- One of the best-known examples of a disorder caused by an extra autosome in humans is trisomy 21, also known as Down syndrome, in which a person has three copies of chromosome 21. Page 425
- Most cases of extra or missing chromosomes in animals cause such extreme physical and mental effects that the offspring do not survive. Page 425
- While having extra or missing autosomes is often lethal or very harmful, having an extra or a missing X or Y sex chromosome is often much less harmful. Page 426
- People with XXX or XYY sex chromosomes do not exhibit altered phenotypes. Page 426
- Individuals who are XXY have many male characteristics as a result of the Y chromosome, but they also exhibit a number of other distinctive traits that are collectively known as Klinefelter syndrome. Page 426
- Females who have only one X chromosome exhibit a number of phenotypic changes known as Turner syndrome. Page 426

Key Terms

Nondisjunction   Down syndrome   Turner syndrome

Review Questions

1. Tay-Sachs disease
   (A) affects adults.
   (B) results from a recessive single-gene mutation.
   (C) affects teenagers.
   (D) prevents the breakdown of phenylalanine.

2. Phenylketonuria (PKU) is a mutation that
   (A) has hundreds of mutant alleles.
   (B) appears in plants.
   (C) causes the production of an impaired form of an enzyme.
   (D) is caused by an extra chromosome.

3. Which statement is true about single-gene mutations?
   (A) They only occur in humans.
   (B) They only occur in animals.
   (C) They only occur in plants.
   (D) They occur in all organisms.

4. Which is an accurate description of the effects of extra or missing chromosomes?
   (A) They cause less extreme phenotypic impacts than single-gene mutations.
   (B) They cause large changes when they involve XXX individuals.
   (C) They are often lethal in animals.
   (D) They are often lethal in plants.

5. Which statement is true about nondisjunction?
   (A) It can happen during meiosis and mitosis.
   (B) It can only happen during mitosis.
   (C) It can only happen during meiosis.
   (D) It does not happen during mitosis or meiosis.

Module 30
AP® Practice Questions

Section 1: Multiple-Choice Questions

Choose the best answer for questions 1–3.

1. Which statement correctly describes a mechanism by which a diploid parent could give rise to an offspring that has one or more extra chromosomes?

   (A) Failure of sister chromatids to separate during meiosis I

   (B) Failure of sister chromatids to separate during mitosis I

   (C) Failure of sister chromatids to separate during meiosis II

   (D) Failure of sister chromatids to separate during mitosis II

Use the following figure to answer question 2.

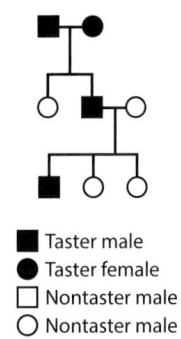

■ Taster male
● Taster female
□ Nontaster male
○ Nontaster male

2. The chemical phenylthiocarbamide (PTC) has a very bitter taste to some people but has no taste at all to other people. Researchers found a specific gene that encodes a receptor protein on taste buds that detects bitter flavors. One allele of the gene results in a receptor that can bind PTC, and another allele results in a receptor that does not respond to PTC. A researcher mapped the ability to taste PTC onto a pedigree and obtained the data shown in the figure. Based on this pedigree, identify the most likely mode of inheritance of PTC tasting.

   (A) Autosomal recessive

   (B) Autosomal dominant

   (C) Mitochondrial

   (D) X-linked dominant

3. Grass carp are large plant-eating fish that are naturally diploid and are native to eastern Asia. Grass carp are frequently used in North America to control weed growth in residential and community ponds. However, when they are introduced to new environments, grass carp have the potential to outcompete and overtake the native fish populations. Therefore, most U.S. states have regulations requiring that only artificially generated grass carp with three sets of chromosomes may be used for aquatic weed control. Which statement best explains why grass carp with three sets of chromosomes are allowed in the United States but diploid grass carp are not?

   (A) Grass carp with three sets of chromosomes have shorter life spans than diploid grass carp and therefore are less likely to establish in natural environments.

   (B) Grass carp with three sets of chromosomes are unable to produce gametes with balanced numbers of chromosomes and therefore are sterile.

   (C) Grass carp with three sets of chromosomes have larger bodies than diploid carp because of their larger genome and therefore are more effective at controlling vegetation.

   (D) Grass carp with three sets of chromosomes are native to North America and therefore cannot be harmful to pond communities.

Section 2: Free-Response Question

Write your answer to each part clearly. Support your answers with relevant information and examples. Where calculations are required, show your work.

Phenylketonuria (PKU) is an autosomal recessive genetic disorder in mammals. PKU is caused by a mutation in the enzyme phenylalanine hydroxylase, which is responsible for converting phenylalanine to tyrosine. Individuals with a mutated enzyme accumulate toxic high levels of phenylalanine in the blood. A team of scientists developed an experimental therapy for PKU in mice. The scientists injected the therapeutic compound into mice with the PKU mutation and then measured the concentration of phenylalanine in the blood for 24 weeks. The data they obtained are shown in the figure. The dashed line indicates the threshold for phenylalanine toxicity. Error bars are calculated as ±1 standard deviation on either side of the mean.

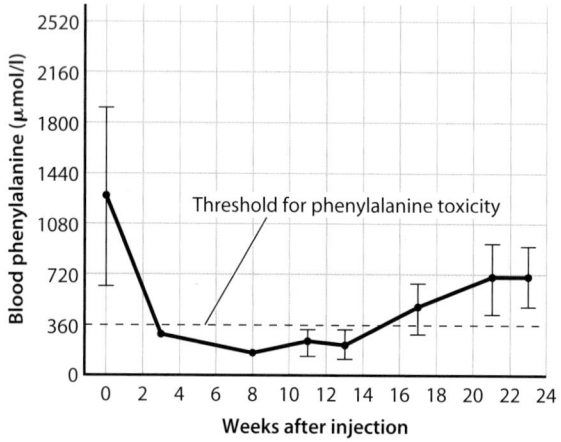

Data from A. Rebuffat, C.O. Harding, Z. Ding, and B. Thöny. (2009, November). "Comparison of Adeno-Associated Virus Pseudotype 1, 2, and 8 Vectors Administered by Intramuscular Injection in the Treatment of Murine Phenylketonuria." *Human Gene Therapy* 21.

(a) **Identify** the time point at which phenylalanine levels in the blood are the highest.

(b) Assuming the scientists want to prevent phenylalanine in the blood from reaching toxic levels while providing the therapy as infrequently as possible, **identify** the time at which mice should be reinjected with the experimental compound.

(c) **Calculate** the average percent reduction in blood phenylalanine level in mice at 3 weeks after the injection.

(d) Phenylalanine is an essential amino acid for mice, meaning mice do not have the capacity to synthesize phenylalanine and must acquire it from food. **Predict** the health consequences for a mouse with PKU that is provided with a diet that has no phenylalanine in it. Provide reasoning to **support** your prediction.

Unit 5

AP® Practice Questions

Section 1: Multiple-Choice Questions

Choose the best answer for questions 1–15.

1. Which is the ultimate source of genetic variation for every domain of life on the planet?
   (A) Recombination between homologous chromosomes
   (B) Independent assortment of alleles
   (C) Random fertilization of gametes
   (D) Genetic mutation

2. Which best describes the cell divisions in meiosis I and meiosis II?
   (A) The first division reduces the chromosome number from diploid to haploid, and the second division separates homologous chromosomes.
   (B) The first division reduces the chromosome number from diploid to haploid, and the second division separates sister chromatids.
   (C) The first division separates the homologous chromosomes, and the second division reduces the chromosome number from diploid to haploid.
   (D) The first division separates sister chromatids, and the second division reduces the chromosome number from diploid to haploid.

Use the following figure to answer question 3.

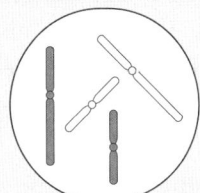

Parental cell
2n = 4

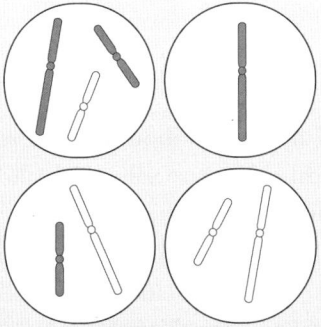

Gametes produced

3. Occasional errors in meiosis result in gametes that contain either too many or too few chromosomes. In animals, embryos that are generated from these gametes typically do not survive. The figure illustrates gametes arising from one possible error in meiosis. Identify the most likely step at which the error in meiosis occurred.
   (A) Anaphase I
   (B) Telophase I
   (C) Metaphase II
   (D) Anaphase II

4. Although most wolves have ears that stand upright, a few have floppy ears. Fur color in wolves can be gray or black. A cross between two wolves with black fur and upright ears produced offspring with the following four phenotypes: black fur and upright ears, gray fur and upright ears, black fur and floppy ears, gray fur and floppy ears. Which principle in genetics best explains the observed offspring phenotypes?
   (A) Independent assortment
   (B) Incomplete dominance
   (C) Mitotic recombination
   (D) Blending inheritance

Use the following figure to answer question 5.

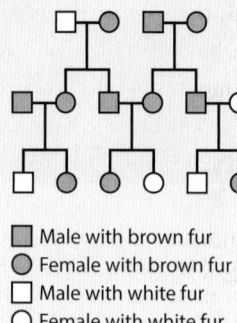

■ Male with brown fur
● Female with brown fur
□ Male with white fur
○ Female with white fur

5. Guinea pigs are small rodents. Female guinea pigs carry two X chromosomes and male guinea pigs carry one X and one Y. Fur color in guinea pigs can be determined by a single gene. The figure shows the pedigree of a family of guinea pigs in which some individuals have white fur and others have brown fur. Based on the pedigree, identify the inheritance pattern of white fur.

(A) Autosomal dominant

(B) Autosomal recessive

(C) X-linked recessive

(D) Y-linked recessive

Use the following figure to answer question 6.

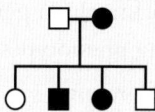

■ Male with dominant phenotype
● Female with dominant phenotype

6. The pedigree shown traces a dominant allele from the parent generation to the F_1 generation. An investigator claims that the dominant allele being traced on this pedigree is found on a chromosome in the nucleus rather than a mitochondrial chromosome. Which provides the best justification for the claim?

(A) All inherited chromosomes are found in the nucleus.

(B) Alleles found on mitochondrial chromosomes would be present only in the females in the pedigree.

(C) The allele being traced would be present in all the offspring in this pedigree if it were on a mitochondrial chromosome.

(D) Dominant alleles of genes found on nuclear chromosomes are more likely to be observed in females.

7. In pea plants, the genes determining pea color and pea shape are not linked. If the probability that a gamete carries an allele for yellow pea color is $\frac{1}{2}$, and the probability that a gamete carries an allele for round pea shape is $\frac{1}{2}$, then what is the probability that a gamete carries alleles for both yellow pea color and round pea shape?

(A) $\dfrac{1}{2} + \dfrac{1}{2}$

(C) $\dfrac{1}{2} \div \dfrac{1}{2}$

(B) $\dfrac{1}{2} \times \dfrac{1}{2}$

(D) $\left(\dfrac{1}{2} + \dfrac{1}{2}\right)^2$

8. A female chicken with gray feathers mates with a male chicken that also has gray feathers. The offspring produced from this cross are 15 chicks with gray feathers, 6 chicks with black feathers, and 8 chicks with white feathers. Based on these data, and assuming chicken feather color is controlled by a single gene, predict the phenotype frequencies expected from a mating between a male with gray feathers and a female with black feathers.

(A) 50% gray feathers and 50% black feathers

(B) 75% gray feathers and 25% black feathers

(C) 33% gray feathers, 33% black feathers, and 33% white feathers

(D) 25% gray feathers, 50% black feathers, and 25% white feathers

9. Ear length is a genetically determined trait in rabbits. A scientist crossed two rabbits that had intermediate ear length and obtained offspring that had either long, short, or intermediate length ears. The scientist performed a chi-squared test using the observed offspring phenotypes and a null hypothesis that ear length is determined by a single gene with two codominant alleles. The scientist obtained a critical chi-square value of 5.991 with 2 degrees of freedom. Using a P-value of 0.05, which statement provides the best interpretation of the scientist's data?

(A) The null hypothesis should be rejected because the chi-squared test statistic is greater than the critical chi-square value.

(B) The null hypothesis should be rejected because the chi-squared test statistic is larger than the critical chi-square value.

(C) The null hypothesis should not be rejected because the critical chi-square value is smaller than the chi-squared test statistic.

(D) The null hypothesis should not be rejected because the chi-squared test statistic is smaller than the critical chi-square value.

10. Snowshoe hares exhibit seasonal camouflage to help them avoid predation. In the spring and summer months, the hares produce a pigment that causes their fur to be brown. However, in the winter, the hares stop producing the pigment, and their fur is white, which enables them to blend into snowy landscapes. In the absence of predation, a snowshoe hare may live up to 5 years and can display both fur colors during its life. Which statement best explains the seasonal change in fur color in snowshoe hares?

(A) Warm summer temperatures cause pigment production to become an incompletely dominant trait.

(B) Warm summer temperatures cause a different phenotype to be exhibited than do cold winter temperatures.

(C) Cold winter temperatures cause only the recessive allele for the pigment gene to be passed to offspring.

(D) Cold winter temperatures prevent the alleles for the pigment gene from segregating during meiosis.

Use the following information for questions 11 and 12.

The Punnett square below illustrates possible offspring genotypes from a cross between two cats that have the genotype *Mm*. Cats with an *mm* genotype have white hair, cats with an *MM* genotype have black hair, and cats with an *Mm* genotype have gray hair.

Female

	M	m
M	MM	Mm
m	Mm	mm

Male

11. Which genetic principle is best illustrated by the cross?

(A) Independent assortment

(B) Incomplete dominance

(C) Incomplete penetrance

(D) Phenotypic plasticity

12. If four kittens are born from this cross, how many of them would be expected to have black hair?

(A) 1   (C) 3

(B) 2   (D) 4

Use the following information to answer questions 13–15.

Blue peafowl are birds that are native to the Indian subcontinent but are famous worldwide because the males display enormous showy tail feathers during the breeding season. Like all birds, peafowl have a *ZW* sex determination system, where males carry two *Z* chromosomes and females are heterozygous for *Z* and *W* chromosomes. Both male peafowl (peacocks) and female peafowl (peahens) typically have bright blue neck feathers during the breeding season. However, a recessive *Z*-linked mutation in the *cameo* gene results in brown feathers instead of the wild type blue. An independent autosomal gene, *myostatin,* regulates muscle growth in peafowl. A recessive mutation in the *myostatin* gene results in extreme muscle growth and large body size. A peacock with brown neck feathers that was heterozygous at the *myostatin* gene was crossed with a peahen that had blue neck feathers and large body size.

13. What percent of the offspring from this cross are expected to have large body size?

(A) 25%   (C) 75%

(B) 50%   (D) 100%

14. What percent of the offspring from this cross are expected to be males with brown necks?

(A) 0 %   (C) 50%

(B) 25%   (D) 100%

15. What percent of the offspring from this cross are expected to have large body size and blue neck feathers?

(A) 6.25%   (C) 25.0%

(B) 18.75%   (D) 50.0%

Section 2: Free-Response Questions

Write your answer to each part clearly. Support your answers with relevant information and examples. Where calculations are required, show your work.

1. In rice plants, the *GS3* gene encodes a protein that regulates the length of rice grains. The wild-type allele, $GS3^+$, is dominant to two mutant alleles, GSA^B and $GS3^C$. DNA sequencing indicates that the $GS3^C$ allele does not encode a functional protein. Plants that are homozygous for each of the three alleles vary significantly in the size of their rice grains when grown at 24°C, as shown in Table 1. Grain length is a trait that is also sensitive to environmental conditions. On average, rice plants grown at 24°C have grains that are approximately 1 mm longer than plants grown at 28°C, as shown in Table 2.

TABLE 1 Average Rice Grain Length from Three Genotypes of Rice Plant Grown at 24°C

Genotype	Grain length (mm)	Standard error
$GS3^+/GS3^+$	5.72	0.45
$GS3^B/GS3^B$	7.02	0.50
$GS3^C/GS3^C$	7.35	0.22

TABLE 2 Average Rice Grain Length from Three Genotypes of Rice Plant Grown at 28°C

Genotype	Grain length (mm)	Standard error
$GS3^+/GS3^+$	5.42	0.50
$GS3^B/GS3^B$	6.02	0.37
$GS3^C/GS3^C$	6.75	0.21

(a) Using the data provided, draw a histogram that illustrates the grain length for each of the three homozygous genotypes when grown at 24°C and at 28°C.

(b) Long rice grains have the highest value in food markets. **Identify** the genotype and growth temperature that will yield the most valuable rice.

(c) A breeder crosses a rice strain with the genotype $GS3^+/GS3^+$ to a strain that is $GS3^C/GS3^C$. **Predict** the average grain length of the resulting progeny if the plants are grown at 28°C.

(d) A scientist makes a claim that the *GS3* gene encodes a protein that suppresses the growth of rice grains. **Provide reasoning** to support or refute this claim.

(e) A scientist makes a claim that the $GS3^B$ allele encodes a protein that has some activity but is not fully functional. **Provide evidence** to support this claim.

2. In *Drosophila melanogaster*, the gene *garnet* (*g*) affects eye color, and the gene *scalloped* (*sd*) affects wing shape. Both of these genes are on the *X* chromosome. *D. melanogaster* have an *XY* sex determination system, where individuals carrying two *X* chromosomes are female. A recessive mutation in garnet (g^-) results in a darkened eye color compared to the wild-type red color produced by the g^+ allele. A recessive mutation in scalloped (sd^-) results in a misshapen wing edge compared to the rounded wing produced by the wild-type sd^+ allele. A scientist conducted dozens of crosses between females that were heterozygous at both genes with males who displayed the mutant phenotypes at both genes. The scientist obtained the data shown in the table.

Offspring phenotypes	Number of offspring
Red eyes, rounded wings	53,819
Dark eyes, misshapen wings	31,570
Dark eyes, rounded wings	4233
Red eyes, misshapen wings	2483

(a) **Identify** the genotype of the male parent in this cross.

(b) **Predict** the proportion of offspring from this cross that are expected to have dark eyes.

(c) **Make a claim** to explain why offspring that display either dark eyes or scalloped wings, but not both, are so much less common than offspring that display both mutant phenotypes. **Provide reasoning** to support your claim.

Unit 6
Gene Expression and Regulation

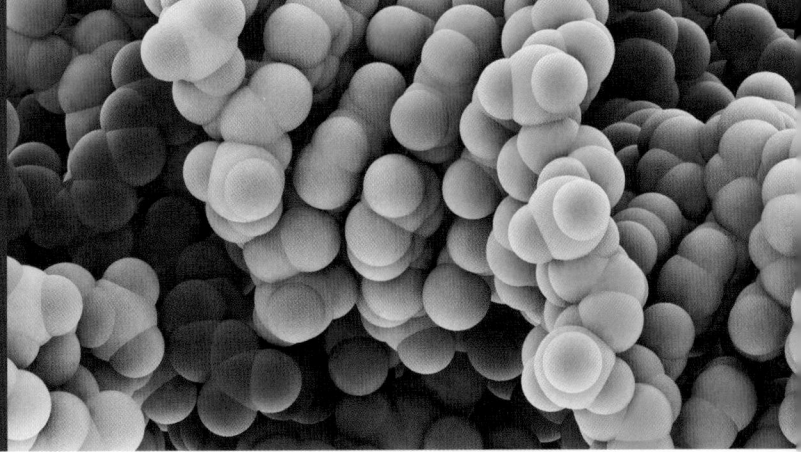

Computer module of protein molecules by LAGUNA DESIGN/Getty Images

Unit 6		31	32	33	34	35	36	37	38	39

Module 31
DNA and RNA Structure and Function

LEARNING GOALS ▶LG 31.1 Nucleic acids encode genetic information in their nucleotide sequence.
▶LG 31.2 Genetic information typically flows from DNA to RNA to protein.
▶LG 31.3 Chromosomes are typically circular in prokaryotes and linear in eukaryotes.

One of the four Big Ideas in biology, introduced in the first unit, is Information Storage and Transmission. Let's consider each aspect of this Big Idea, starting with Information Storage. All cells store information that they use to maintain their identity, carry out their functions, and sustain life. As we discussed in Module 5, the information storage molecule for all cells is the nucleic acid DNA. DNA is a long polymer made up of repeated subunits called nucleotides. Cells store information in the sequence, or order, of the nucleotides in DNA.

Now, let's consider Information Transmission. Cells pass information along, or transmit it, to other cells in the process of cell division. For example, in Unit 4, we discussed the cell cycle, which describes the life of a cell. During the cell cycle, a cell grows, makes copies of its chromosomes, and divides in two by mitotic cell division. In the process, the parent cell transmits genetic information to the two daughter cells. Similarly, in Unit 5, we discussed inheritance, the process by which traits are passed from one generation to the next. In sexually reproducing organisms, inheritance involves meiotic cell division, which produces gametes, followed by fertilization. As in the cell cycle, genetic information is transmitted to the next generation in this process.

In this unit, we will look more closely at how cells transmit and use the information encoded in DNA. The key point, one that you will see throughout this unit, is that a DNA molecule can hand off the information that it stores to other molecules. For example, when a cell divides and replicates its chromosomes, a single DNA molecule is copied in two, so information is transmitted from one DNA molecule to another DNA molecule. Similarly, when a cell uses the information stored in DNA, it first passes this information from DNA to RNA. The process by which the information encoded in DNA is passed to other molecules is at the heart of how cells transmit and use this information.

In this module, we will start by reviewing the structure of DNA and RNA, first introduced in Module 5. We will also provide an overview of the functions of DNA and RNA. Finally, we will discuss how information is stored, transmitted, and used in prokaryotes and eukaryotes, with a focus on how chromosomes are organized in the two groups.

PREP FOR THE AP® EXAM

FOCUS ON THE BIG IDEAS

INFORMATION STORAGE AND TRANSMISSION: Focus on how the structure of DNA and RNA ensure the accurate storage of and transmission of genetic information.

31.1 Nucleic acids encode genetic information in their nucleotide sequence

In Unit 1, we introduced the two types of nucleic acids, deoxyribonucleic acid (DNA) and ribonucleic acid (RNA). Let's review their structures here so we are ready to learn more about their functions. Nucleic acids are examples of informational molecules—that is, large molecules that carry information in the sequence of nucleotides that make them up. This molecular information is much like the information carried by the letters in an alphabet, but in the case of nucleic acids, the information is in chemical form.

DNA is the genetic material in all cells and organisms. It is transmitted from cell to cell, and from parents to offspring. It also contains the information needed to specify the amino acid sequence of all the proteins synthesized in an organism. RNA has multiple functions; it is a key player in protein synthesis and the regulation of **gene expression.**

DNA and RNA are long molecules consisting of nucleotides bonded covalently one to the next. Nucleotides, in turn, are composed of three components: a 5-carbon sugar, a nitrogenous base, and a phosphate group, as shown in **FIGURE 31.1**.

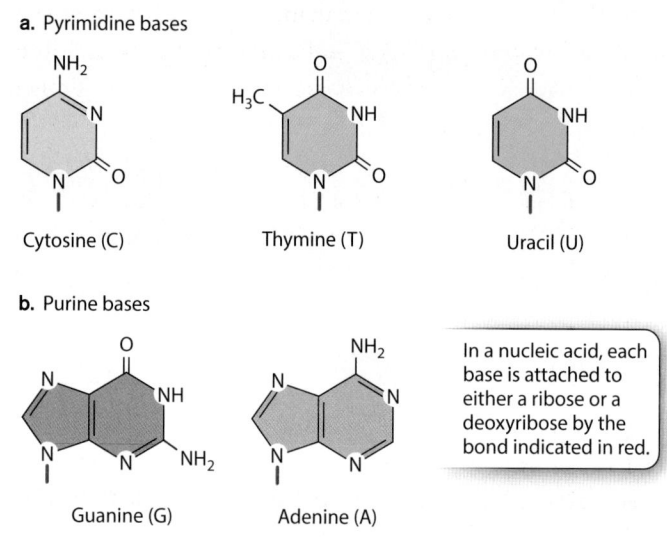

a. Pyrimidine bases

Cytosine (C)   Thymine (T)   Uracil (U)

b. Purine bases

Guanine (G)   Adenine (A)

In a nucleic acid, each base is attached to either a ribose or a deoxyribose by the bond indicated in red.

FIGURE 31.2 Bases found in DNA and RNA

There are two types of bases. (a) Pyrimidines have a single-ring structure, and (b) purines have a double-ring structure.

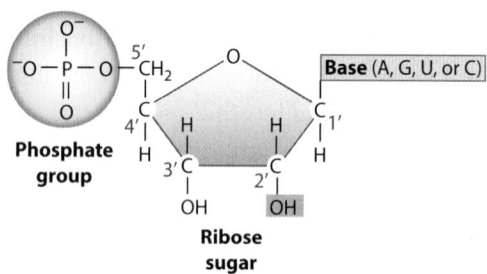

Phosphate group

Ribose sugar

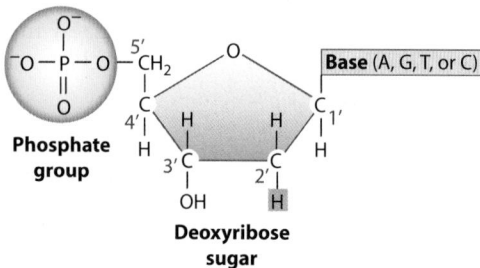

Phosphate group

Deoxyribose sugar

FIGURE 31.1 Nucleotides

Nucleotides are the building blocks of nucleic acids. The upper image is a nucleotide that contains ribose in RNA. The lower image is a nucleotide that contains deoxyribose in DNA. The two sugars differ by the presence of a hydroxyl (—OH) group in ribose and an H atom in deoxyribose at the highlighted position.

The sugar in RNA is ribose, and the sugar in DNA is deoxyribose. The sugars differ in that ribose has a hydroxyl (—OH) group on the second carbon (designated the 2′ carbon), whereas deoxyribose has a hydrogen atom at this position (hence, *deoxyribose*). The difference between the two sugars is highlighted in Figure 31.1.

The bases are built from nitrogen-containing rings. In Figure 5.2 on page 79, we introduced the bases found in DNA. In **FIGURE 31.2**, we show bases found in both DNA and RNA. There are two types of bases, pyrimidines and purines. The pyrimidine bases, shown in part a, have a single ring and include cytosine (C), thymine (T), and uracil (U). The purine bases, shown in part b, have a double-ring structure and include guanine (G) and adenine (A). DNA contains the bases A, T, G, and C, whereas RNA contains the bases A, U, G, and C. In DNA and RNA, each adjacent pair of nucleotides is connected by a phosphodiester bond, which forms when a phosphate group in one nucleotide is covalently joined to the sugar unit in another nucleotide to form the sugar-phosphate backbone.

DNA and RNA have different structures in cells. DNA usually consists of two strands of nucleotides twisted around each other in the form of a double helix, like the ones shown in **FIGURE 31.3**. The sugar-phosphate backbones of the strands wrap like a ribbon around the outside of the double helix, and the bases point inward. The bases form specific purine–pyrimidine pairs that are complementary: where one strand carries an A, the other carries a T; and where one strand carries a G, the other carries a C. **FIGURE 31.4** illustrates these base pairs. Base pairing results from hydrogen bonding between the bases, indicated by the dashed lines in the figure. Note that A–T base pairs have two hydrogen bonds, while G–C base pairs have three hydrogen bonds. As a result, G–C base pairs are stronger than A–T base pairs. By contrast, RNA is typically much shorter than DNA and single stranded. However, some RNAs do exhibit base pairing in regions of the molecule. In the case of RNA, A pairs with U, and G pairs with C.

Cells store information in DNA. The genetic information in DNA is contained in the sequence, or order, in which nucleotides occur along the molecule. Successive nucleotides can occur in any order, so a long molecule can contain a very large number of possible nucleotide sequences. One of the most important features of DNA is that there is no restriction on the sequence of bases along a DNA strand. Any A, for example, can be followed by another A or by T, C, or G. With any of four possible bases at each nucleotide site, the information-carrying capacity of a DNA molecule is unimaginable. The number of possible base sequences of a DNA molecule only 133 nucleotides in length is equal to the estimated number of electrons, protons, and neutrons in the entire universe! This is how DNA can carry the genetic information for so many different species of organisms, and

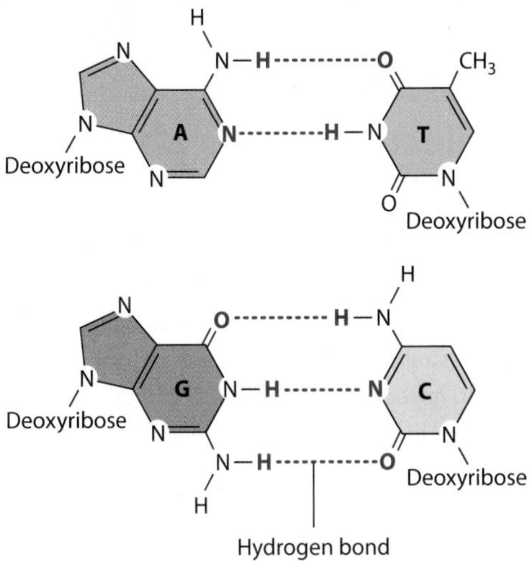

FIGURE 31.4 Base pairing

The bases are complementary: A is always paired with T, and G is always paired with C. Base pairing results from hydrogen bonds, indicated by the dashed lines.

how variation in DNA sequence even within a single species can underlie genetic differences among individuals.

Today, the role of DNA as the molecule that stores and transmits genetic information from generation to generation is well known. However, this was not always the case. Before about 1950, any poll of biologists would have shown overwhelming support for the idea that proteins are life's information molecules. Compared with the seemingly monotonous, featureless structure of DNA, the three-dimensional structures of proteins are highly diverse. Proteins carry out most of the essential activities in a cell, so it seemed logical to assume that they would play a key role in heredity too.

The first hint that DNA is the molecule that stores and transmits genetic information came in 1928, when Frederick Griffith conducted studies on the transmission of genetic information and showed that molecules in extracts from bacteria could transmit genetic information from one bacterial cell to another. Almost 20 years later, experiments carried out by Oswald Avery, Colin MacLeod, and Maclyn McCarty showed that this information-carrying macromolecule is DNA, rather than RNA or protein, as described on page 438 in "Practicing Science 31.1: Which molecule carries genetic information?"

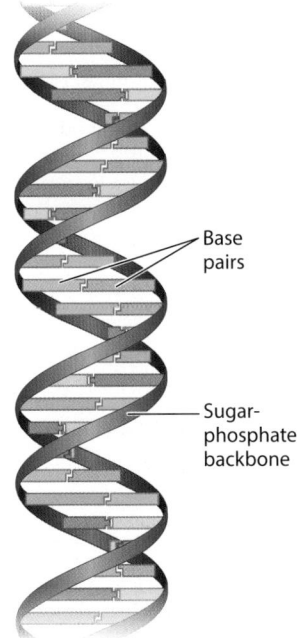

FIGURE 31.3 The structure of DNA

DNA most commonly occurs in the form of a double helix, with the sugar and phosphate groups forming the backbone and the bases oriented inward.

✓ Concept Check

1. **Identify** the four bases of DNA and the four bases of RNA.

2. **Describe** how genetic information is encoded in a molecule of DNA.

Practicing Science 31.1

Which molecule carries genetic information?

Background In the 1920s, it was not known which molecule carries genetic information. Frederick Griffith studied different strains of bacteria. One strain was virulent and caused pneumonia when injected into mice; another strain was nonvirulent and did not cause pneumonia when injected into mice. Griffith discovered that this trait—virulence— could be transferred between the two strains of bacteria. This result showed that genetic traits can be transferred from one organism to another, but it did not indicate which molecule carried genetic information. In the early 1940s, researchers Oswald Avery, Colin MacLeod, and Maclyn McCarty set out to identify the molecule that is responsible for transforming nonvirulent bacteria into virulent bacteria.

Experiment Avery, MacLeod, and McCarty killed virulent bacterial cells with heat, and then purified the remains of the dead virulent cells to make a solution. They found that this solution of dead virulent cells transformed nonvirulent bacteria into virulent ones. To identify what caused the transformation, they treated the solution with three different enzymes. Each enzyme (DNase, RNase, and protease) destroyed one of the three types of molecules (DNA, RNA, and protein) that they believed might be responsible for transformation. The experiment is shown in the figure below.

Virulent bacteria extract

DNA, RNA, and protein are extracted from heat-killed virulent bacteria and purified to make a solution.

Solution only / Solution + RNase / Solution + protease / Solution + DNase

Nonvirulent bacteria / Nonvirulent bacteria / Nonvirulent bacteria / Nonvirulent bacteria

The enzymes DNase, RNase, and protease, which degrade DNA, RNA, and protein, respectively, are added to the solution.

Virulent and nonvirulent bacteria / Virulent and nonvirulent bacteria / Virulent and nonvirulent bacteria / Nonvirulent bacteria only

The control solution of killed but untreated virulent bacteria transforms nonvirulent bacteria into virulent bacteria. So do the solutions treated with RNase and protease.

The solution treated with DNase does not transform nonvirulent bacteria into virulent bacteria.

Hypothesis The researchers hypothesized that transformation would not occur if the molecule responsible for transformation was destroyed by the enzymes.

Results When the researchers treated the solution of dead virulent cells with enzymes that destroy RNA (RNase) or protein (protease), the solution was still able to transform nonvirulent bacteria into virulent cells. In contrast, when the solution was treated with the enzyme that destroys DNA (DNase), the extract was not able to transform nonvirulent bacteria into virulent cells.

Conclusion DNA is the molecule responsible for transforming nonvirulent bacteria into virulent bacteria. This experiment provided a key piece of evidence for the idea that DNA is the genetic material. These experiments were followed up by Alfred Hershey and Martha Chase, as well as countless others, who have all confirmed that DNA is the molecule that carries genetic information.

SOURCE
Avery, O., C. MacLeod, and M. McCarty. 1944. "Studies on the Chemical Nature of the Substance Inducing Transformation of Pneumococcal Types." *Journal of Experimental Medicine* 79:137–158.

AP® PRACTICE QUESTION

Identify the following:

1. The question the experimenters are trying to answer
2. The hypothesis they tested
3. The experimental group
4. The control group
5. The independent variable
6. The dependent variable
7. The conclusion, using a claim, evidence, reasoning format (CER)

31.2 Genetic information typically flows from DNA to RNA to protein

We have seen how nucleic acids store genetic information. DNA stores genetic information in the order of nucleotides along its length, with each strand made up of varying sequences of four different kinds of nucleotides connected end to end. In this section, we will focus on how that information is transmitted from cell to cell, and how the information is used by cells to carry out their functions. This section will provide an overview of key processes involving DNA and RNA. We will return to each of these processes in more detail in subsequent modules.

DNA Replication

One of the overarching themes of biology is that the functional unit of life is the cell. This theme rests, in turn, on the fundamental concept that all cells come from preexisting cells. In Units 4 and 5, we considered the mechanics of cell division and how it is regulated. Prokaryotic cells divide by binary fission, whereas eukaryotic cells divide by mitosis and cytokinesis. These processes ensure that cellular reproduction results in daughter cells that are like the parental cell. In other words, in cell division, like begets like.

The molecular basis for the resemblance between parental cells and daughter cells is that a double-stranded DNA molecule in the parental cell duplicates and gives rise to two double-stranded daughter DNA molecules. The process of duplicating a DNA molecule is called DNA replication. DNA replication allows genetic information to be passed from cell to cell and from an organism to its progeny. It is therefore the molecular basis for inheritance. Replication occurs in virtually the same way in all organisms, reflecting that it evolved very early in life's history.

An organism's DNA can be stably and reliably passed from generation to generation in large part because of its double-stranded helical structure. **FIGURE 31.5** shows a DNA molecule in the process of replication. On the left is the original, or parental, DNA molecule, and on the right are two new, or daughter, DNA molecules. Note how the two strands of the parental DNA molecule separate from each other, like a zipper in the process of unzipping, as the hydrogen bonds between the bases are broken. When they are separated, each parental strand serves as a template, or a model, for a daughter DNA strand. The parental strand

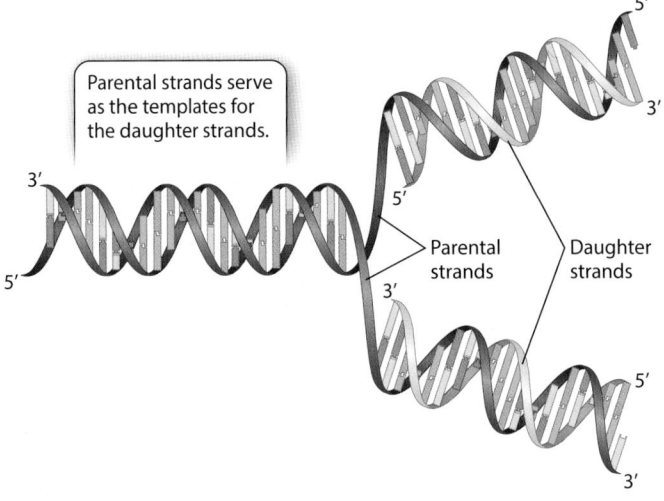

FIGURE 31.5 DNA replication

The structure of the double helix gave an important clue to how DNA replicates. Each strand of the parental DNA molecule serves as a template, or model, for a daughter strand. In the process of DNA replication, one DNA molecule gives rise to two DNA molecules.

dictates the sequence of the daughter strand because of base pairing, with A pairing with T, and G pairing with C. As a result of DNA replication, a single parental DNA molecule produces two daughter DNA molecules.

Replication is necessarily precise and accurate because mistakes introduced into the cell's information archive may be lethal to the cell. That said, errors in DNA can and do occur occasionally during the process of replication. Environmental insults, such as radiation and carcinogens, can damage DNA as well. Such changes to the DNA sequence are known as mutations, a topic that was introduced in Unit 5. Most mutations are neutral or harmless, and do not affect the survival or reproduction of a cell or organism. Some are harmful, spelling death for the cell or organism. And a precious few are beneficial, leading to the variations that underlie the diversity of life. DNA replication will be discussed in detail in the next module.

Information Flow

In addition to storing and transmitting genetic information, cells use the information encoded in DNA to survive and carry out their diverse functions. For example, cells use genetic information when they produce proteins. Proteins, as we have seen, do much of the work of the cell. They act as enzymes to accelerate the rate of chemical reactions, serve as channels and transporters in the cell membrane, and provide structural support for the cell, to name just a few of their many functions. The amino acid sequence of proteins, or its primary structure, determines how the protein folds and functions. The amino acid sequence ultimately is determined by the genetic information in DNA. In other words, the information encoded in DNA directs the formation of proteins.

How does the information stored in DNA direct the synthesis of proteins? As we learned in Units 1 and 2, the central dogma of molecular biology describes the pathway of information from DNA to RNA to protein. It is illustrated again in **FIGURE 31.6**, with additional detail showing the processes by which each step occurs. First, the information in DNA is used to build a molecule of RNA. The process by which RNA is synthesized from DNA is called **transcription.** In everyday usage, the term "transcription" is used when spoken language is written down. Biologists use the same term because it emphasizes that the information is copied from DNA to RNA using the same language—that of nucleotides. Transcription will be discussed in detail in Module 33.

After transcription, specialized molecular structures in the cell "read" the RNA molecule to determine which amino acids to use to build a protein. The process in which a protein is synthesized from RNA is called **translation.** In everyday usage, the term "translation" is used when one language is converted into a different language. Biologists similarly use this term to indicate a change of languages, from nucleotides that make up nucleic acids to amino acids that make up proteins. Translation will be discussed in detail in Module 34.

The central dogma describes the basic flow of information in a cell and, while there are exceptions, it constitutes a fundamental principle in biology. As proteins are ultimately encoded by DNA, we can define specific stretches of DNA

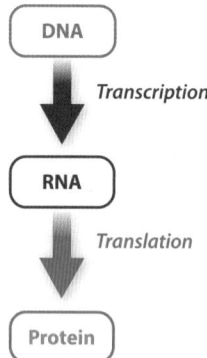

FIGURE 31.6 The central dogma of molecular biology

The usual flow of information in a cell is from DNA to RNA to protein. DNA is transcribed to RNA, which in turn is translated to protein.

according to the proteins that they encode. For example, a segment of DNA encodes the protein actin that makes up part of the cytoskeleton, which was introduced in Unit 2. Other segments of DNA encode enzymes that replicate DNA, protein channels in the cell membrane, and proteins that bind to oxygen in red blood cells. Each of these segments of DNA is called a gene, which is a DNA sequence that corresponds to a specific product, such as a protein.

Through the years, some exceptions to the usual flow from DNA to RNA to protein have been discovered, most notably in viruses, which will be discussed in Module 39. For example, genetic information can flow in the reverse direction, from RNA to DNA, in the human immunodeficiency virus (HIV), which causes acquired immune deficiency syndrome (AIDS). It can also flow from RNA to RNA, which occurs during replication of the influenza virus, which causes the flu, or in the virus SARS CoV-2, which causes COVID-19. Nevertheless, in most cases and in all cells, the flow of information is from DNA to RNA to protein.

The process by which a gene is transcribed and translated is called gene expression, as noted in Unit 4. When a gene is transcribed and translated to produce a protein, it is said to be turned on or expressed. When it is not transcribed and translated, it is turned off or not expressed. Gene expression is regulated, meaning that it does not occur at all times in all cells, even though nearly all cells in an individual contain the same DNA. The processes that control whether gene expression occurs at a given time, in a given cell, or at what level are collectively called **gene regulation.** You can think of gene regulation as the where, when, and how much of gene expression.

Some genes, called housekeeping genes, are expressed almost all of the time by almost all cells, like those involved in metabolism. By contrast, other genes are expressed, or turned on, only at certain times and places, and not expressed, or turned off, at other times and places. In multicellular organisms, for instance, cells are specialized for certain functions, and these different functions depend on which genes are on and which genes are off in specific cells. For example, muscle cells express genes that encode for proteins involved in muscle contraction, but these genes are not expressed in skin cells or liver cells. Similarly, during development of a multicellular organism, the expression of genes may be required at certain times but not at others. In this case, the timing of gene expression is carefully regulated. The timing of gene expression is critical for development from fertilized egg to adult, as we will see in Module 37. Even for single-celled organisms like bacteria, certain genes are expressed only in response to environmental signals, such as the availability of nutrients.

As we have seen, DNA stores and transmits genetic information, and RNA acts as an intermediary between DNA and proteins when genes are expressed. RNA has other roles as well. In some cases, a stretch of DNA is transcribed to RNA, but that RNA is not translated to proteins. These small RNAs are called noncoding RNAs because they don't code for proteins. Nevertheless, noncoding RNAs perform important cellular functions. Some are involved in the process of translation, which will be discussed in Module 34. Some have catalytic activity and therefore act as enzymes, similar to proteins. Still others participate in the regulation of gene expression.

PREP FOR THE AP® EXAM

AP® EXAM TIP

Make sure you understand the terms "replication," "transcription," and "translation." These terms come up frequently on the AP® Biology Exam.

✓ Concept Check

3. **Describe** the typical flow of information in a cell.

4. **Describe** the relationship between gene expression and gene regulation.

31.3 Chromosomes are typically circular in prokaryotes and linear in eukaryotes

The flow of genetic information from DNA to RNA to protein applies to both prokaryotes and eukaryotes, but the differences in cell structure between the two groups mean the details of the processes differ. In this section, we will highlight some of the major differences.

Perhaps the clearest difference between eukaryotic and prokaryotic cells is the presence or absence of a nucleus. Recall from Unit 2 that eukaryotes have a nucleus. Therefore, in eukaryotes, transcription and translation are spatially separated from each other, with transcription occurring in the nucleus and translation in the cytoplasm. This also means that the two processes occur one after the next. The separation of transcription and translation in space and time in eukaryotic cells allows each step to be regulated separately.

This degree of regulation is not possible in prokaryotic cells, which do not have a nucleus. In prokaryotes, transcription and translation both occur in the cytoplasm. Because no nuclear membrane separates DNA from the surrounding cytoplasm, transcription and translation occur in the same space and time. In spite of this and other differences in the details of transcription and translation between prokaryotes and eukaryotes, the processes are sufficiently similar that they must have evolved early in the history of life.

Another difference between prokaryotic cells and eukaryotic cells is the way DNA is organized into chromosomes. Recall from Unit 4 that chromosomes are structures in cells that are made up of a single molecule of DNA with associated

proteins. Chromosomes replicate and are passed from cell to cell in the process of cell division, so are the basis of inheritance.

Chromosomes provide a way to package long DNA molecules into the small space of a cell. For example, if the single DNA molecule of the chromosome of the intestinal bacterium *Escherichia coli* were fully extended, its length would be 200 times greater than the diameter of the cell itself. The fully extended length of DNA in human chromosome 1, our longest chromosome, would be 10,000 times greater than the diameter of the average human cell. Consequently, an enormous length of DNA must be packaged into a form that will fit inside the cell while still allowing the DNA to replicate and carry out its coding functions.

The mechanism of packaging differs substantially in prokaryotes, such as bacteria, and eukaryotes, such as humans. In prokaryotes, the cell's DNA is usually present in a single circular chromosome. In addition, the DNA double helix makes fewer turns in going around the circle than it would need in order for every base in one strand to pair with its partner base in the other strand. As a result, there is strain on the DNA molecule, which is relieved by the formation of supercoils, in which the DNA molecule coils on itself, like the rubber band shown in **FIGURE 31.7** on page 442. Supercoiling allows all the base pairs to form. You can make a supercoil by stretching and twirling the ends of a rubber band, then relaxing the stress slightly to allow the twisted part to form coils around itself, like the one in the figure.

In bacteria, the supercoils of DNA form a structure with multiple loops called a nucleoid. **FIGURE 31.8** on page 442 shows a bacterial nucleoid. The loops are bound together by proteins, which are represented by the red circles in the

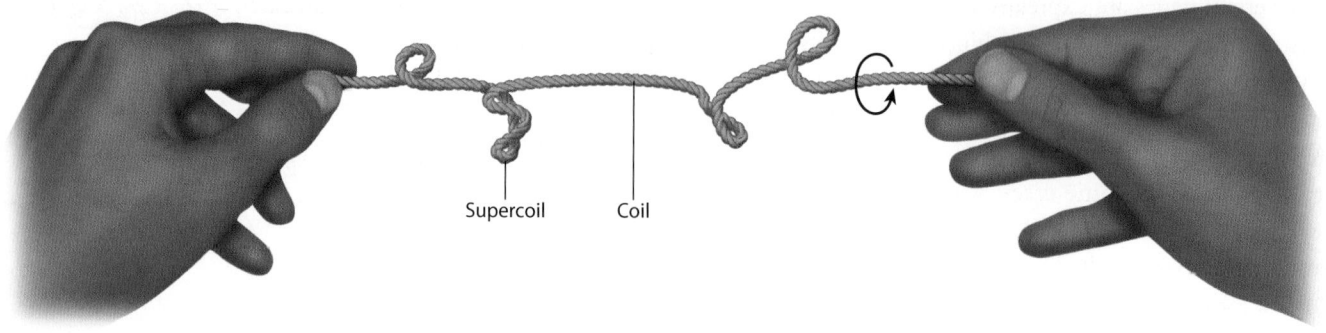

FIGURE 31.7 Supercoils

A highly twisted rubber band forms coils of coils, called supercoils. Supercoils also form in a circular DNA molecule in order for every base to pair with its partner and relieve the strain on the molecule.

bottom image of the figure. The protein binding that forms the loops as well as the supercoiling of the DNA compress the molecule into a compact volume.

In addition to a single circular chromosome, many bacteria carry additional DNA in the form of **plasmids,** small circles of DNA that replicate independently of the cell's circular chromosome. In general, plasmid DNA is not essential for the cell's survival, but it may contain genes that help survival or reproduction under specific environmental conditions.

In contrast to the single circular chromosome of prokaryotes, most eukaryotes usually have multiple linear chromosomes. **FIGURE 31.9** shows the 23 pairs of human

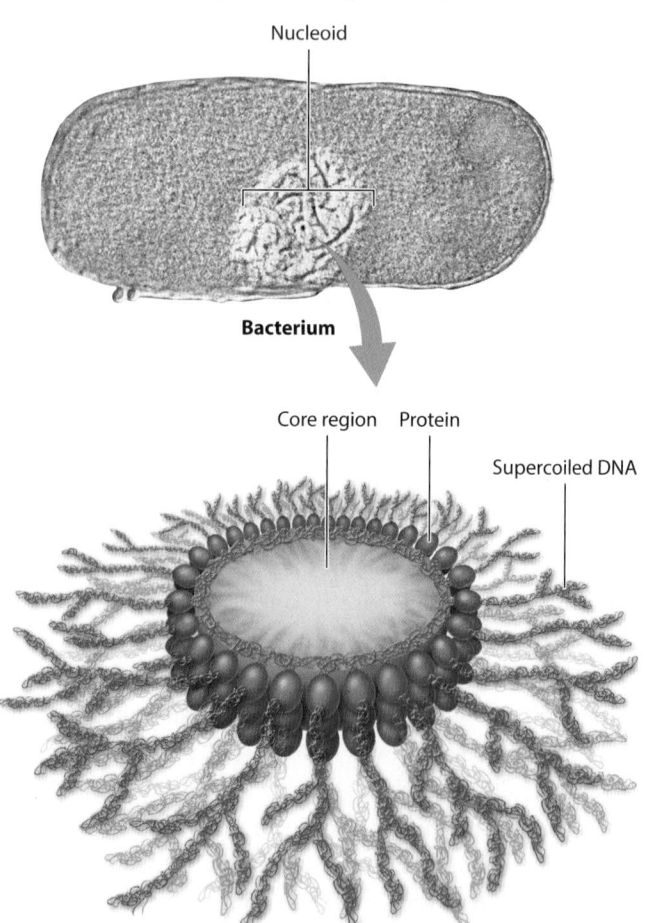

FIGURE 31.8 Bacterial nucleoid

The circular bacterial chromosome twists on itself to form supercoils, which are anchored by proteins, indicated by the red circles in the lower image. Photo: Dr. Klaus Boller/Science Source

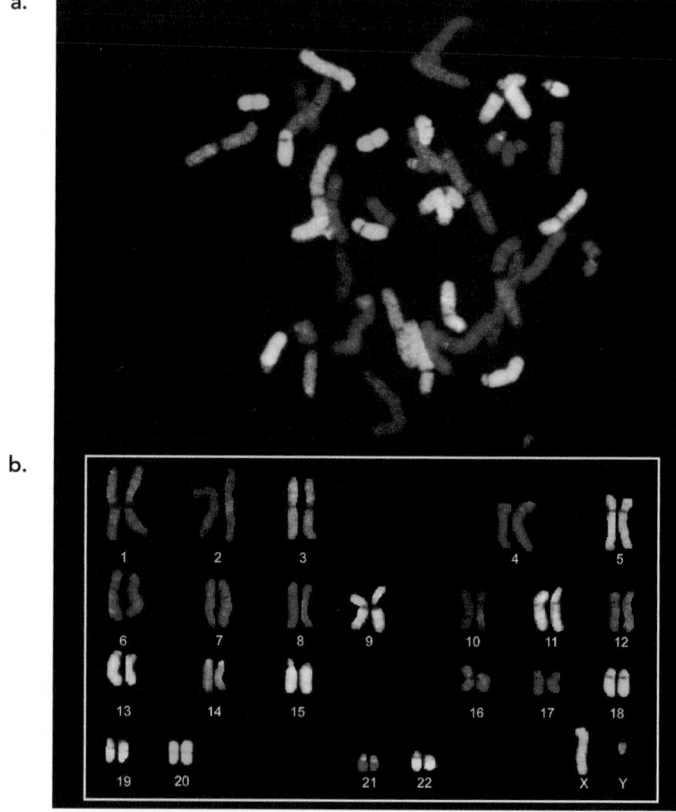

FIGURE 31.9 A chromosome paint of chromosomes from a male human

Humans have 23 pairs of chromosomes, or 46 chromosomes in total. (a) Condensed chromosomes are "painted" with fluorescent dyes so that each chromosome pair is a different color. (b) Chromosomes are arranged from largest (chromosome 1) to smallest (chromosome 22), followed by the X and Y sex chromosomes. Photo: NHGRI, www.genome.gov

chromosomes. Fluorescent dyes are used to "paint" each chromosome a different color. The chromosomes are numbered from 1 to 22 from biggest to smallest, in addition to the *X* and *Y* sex chromosomes. In addition, the DNA in eukaryotic chromosomes is packaged very differently from the way it is organized in prokaryotes. Eukaryotic DNA is packaged with proteins to form a DNA–protein complex called chromatin.

There are several levels of chromosome packaging, which are all shown in **FIGURE 31.10**. First, eukaryotic DNA winds around histone proteins. Histone proteins are found in all

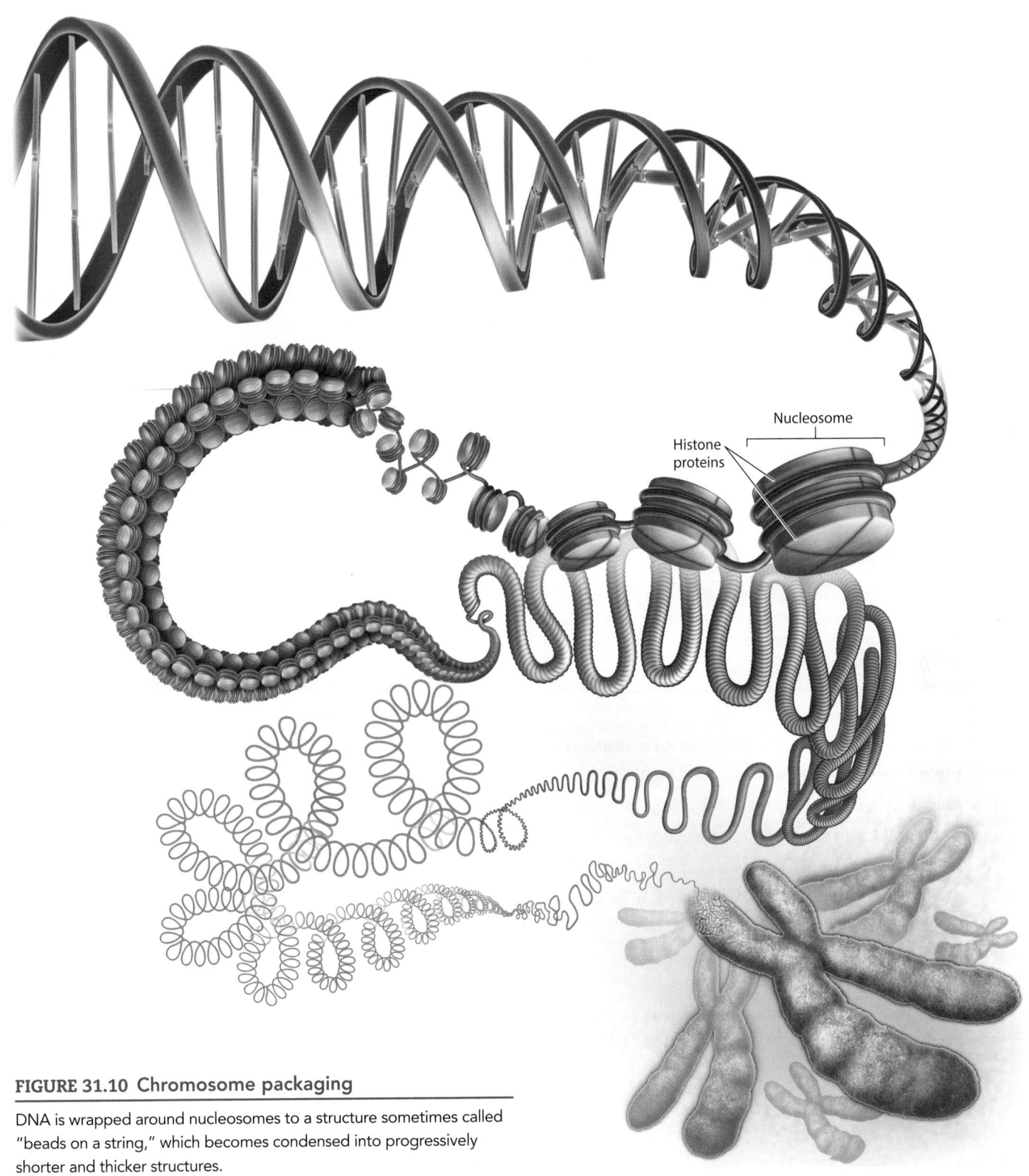

Nucleosome

Histone
proteins

FIGURE 31.10 Chromosome packaging

DNA is wrapped around nucleosomes to a structure sometimes called "beads on a string," which becomes condensed into progressively shorter and thicker structures.

eukaryotes, and they interact with any double-stranded DNA, no matter what its sequence is. Histone proteins are evolutionarily conserved, which means that they are very similar in sequence from one organism to the next, and histone proteins from one eukaryote can associate with DNA from another. The histone proteins are organized as a nucleosome, which is made up of eight histone proteins. The histone proteins are rich in the amino acids lysine and arginine, whose positive charges are attracted to the negative charges of the phosphates along the backbone of each DNA strand.

This first level of packaging of the DNA is sometimes referred to as "beads on a string," where the nucleosomes are the beads and the DNA is the string, as can be seen toward the top of Figure 31.10. The next level of packaging occurs when the chromatin is more tightly coiled. As the chromosomes in the nucleus condense in preparation for cell division, each chromosome becomes progressively shorter and thicker. The progressive packaging constitutes chromosome condensation, an active, energy-consuming process.

Underlying all of this is a supporting protein structure called the chromosome scaffold. Despite intriguing similarities between the nucleoid model in prokaryotes and the chromosome scaffold model in eukaryotes, the structures evolved independently and make use of different types of proteins. Furthermore, the size of the eukaryotic chromosome is vastly greater than the size of the bacterial nucleoid. To appreciate the difference in scale, keep in mind that the volume of a fully condensed human chromosome is five times larger than the volume of a bacterial cell.

✓ Concept Check

5. **Describe** the structure of chromosomes in prokaryotes and eukaryotes.

6. **Describe** where chromosomes are located in prokaryotic and eukaryotic cells.

Module 31 Summary

REVISIT THE BIG IDEAS   PREP FOR THE AP® EXAM

INFORMATION STORAGE AND TRANSMISSION: Using the content of this module, **describe** how the structure of DNA and RNA ensure the accurate storage of and transmission of genetic information.

LG 31.1 Nucleic acids encode genetic information in their nucleotide sequence.

- Nucleic acids include DNA and RNA. Page 436
- DNA stores and transmits genetic information in all cells. Page 436
- Nucleic acids are long polymers made up of subunits called nucleotides. Page 436
- Nucleotides are composed of a 5-carbon sugar, a nitrogenous base, and a phosphate group. Page 436
- Nucleotides in DNA incorporate the sugar deoxyribose, and nucleotides in RNA incorporate the sugar ribose. Page 436

- The bases are pyrimidines (cytosine, thymine, and uracil) and purines (guanine and adenine). Page 436
- Nucleic acids store information in the sequence of nucleotides. Page 436

LG 31.2 Genetic information typically flows from DNA to RNA to protein.

- DNA structure suggests a mechanism for replication in which each parental strand serves as a template for a daughter strand. Page 439
- The central dogma of molecular biology states that the usual flow of genetic information is from DNA to RNA to protein. Page 440
- DNA is transcribed to RNA, and RNA is translated to protein. Page 440
- Gene expression is the process by which a gene produces a protein. Page 440
- Gene regulation is the control of gene expression. Page 440

LG 31.3 Chromosomes are typically circular in prokaryotes and linear in eukaryotes.

- The orderly packaging of DNA into a chromosome allows it to carry out its functions and fit inside the cell. Page 441

- Prokaryotes, such as bacteria, typically have a single circular chromosome, which is packaged in a structure called a nucleoid. Page 441

- Eukaryotes typically have multiple linear chromosomes, which are located in the nucleus of the cell. Page 441

- DNA in eukaryotes is wound around groups of histone proteins called nucleosomes to form a thin structure, which in turn coils to form higher-order structures. Page 443

Key Terms

Gene expression
Transcription

Translation
Gene regulation

Plasmid

Review Questions

1. Ribose differs from deoxyribose in that
   (A) ribose has an extra hydroxyl group.
   (B) ribose is missing a hydroxyl group.
   (C) ribose has an extra phosphate group.
   (D) ribose is missing a phosphate group.

2. Which of the following is a component of a nucleotide?
   (A) A five-carbon sugar
   (B) A six-carbon sugar
   (C) A carboxyl group
   (D) ATP

3. The central dogma of molecular biology states that information flows from
   (A) DNA directly to protein.
   (B) protein to RNA to DNA.
   (C) DNA to RNA to protein.
   (D) RNA to protein to DNA.

4. Translation is the process by which
   (A) proteins are synthesized from DNA.
   (B) RNA is synthesized from protein.
   (C) proteins are synthesized from RNA.
   (D) RNA is synthesized from DNA.

5. Nucleoids are supercoils of DNA forming a structure with multiple loops located in
   (A) plants.
   (B) insects.
   (C) yeast.
   (D) bacteria.

6. Eukaryotic chromosomes are typically
   (A) single circular structures.
   (B) multiple circular structures.
   (C) single linear structures.
   (D) multiple linear structures.

7. The level of DNA packaging brought about by the formation of nucleosomes looks like
   (A) beads on a string.
   (B) 30-nm chromatin fibers
   (C) 1400-nm chromatin fibers
   (D) scaffolds

Module 31
AP® Practice Questions

PREP FOR THE AP® EXAM

Section 1: Multiple-Choice Questions

Choose the best answer for questions 1–4.

Use the following figure to answer question 1.

```
a. ACAGTTGCA        b. ACAGTTGCA
   | | | | | | | | |    | | | | | | | | |
   ACAGTTGCA           TGTCAACGT

c. ACAGTTGCA        d. ACAGTTGCA
 ~ | | | | | | | | |    | | | | | | | | |
   GTGACCATG           CACTGGTAC
```

1. Which correctly illustrates DNA base pairing?

   (A) a

   (B) b

   (C) c

   (D) d

2. A scientist studying a previously undiscovered virus determined that its genetic material had the following composition.

   adenosine = 16%

   guanine = 37%

   thymine = 0%

   cytosine = 24%

   uracil = 23%

   Which correctly describes the viral genome?

   (A) The viral genome is single-stranded DNA.

   (B) The viral genome is double-stranded DNA.

   (C) The viral genome is single-stranded RNA.

   (D) The viral genome is double-stranded RNA.

3. Identify the double-stranded DNA molecule that would require the greatest energy input to separate into single strands.

   (A) One that is 30% A + T

   (B) One that is 40% A + T

   (C) One that is 50% G + C

   (D) One that is 60% G + C

4. Identify the DNA molecule that is most likely to be linear.

   (A) A mitochondrial genome

   (B) A plasmid

   (C) A bacterial chromosome

   (D) A mammalian sex chromosome

Section 2: Free-Response Question

Write your answer to each part clearly. Support your answers with relevant information and examples. Where calculations are required, show your work.

DNA has been described as the biological molecule that is most central to life.

(a) **Describe** the role of DNA in the production of proteins in a cell.

(b) **Describe** the structure of DNA and explain how that structure allows for DNA replication.

(c) **Calculate** the number of distinct sequences that are possible from a piece of DNA that is 10 nucleotides long.

446   **UNIT 6**   GENE EXPRESSION AND REGULATION

Module 32

DNA Replication

LEARNING GOALS ▶**LG 32.1** DNA replication is the basis of inheritance and occurs semiconservatively.
▶**LG 32.3** During DNA replication, one strand is synthesized continuously and the other discontinuously.
▶**LG 32.3** DNA polymerase is self-correcting because of its proofreading function.

DNA is the genetic material for all cells and organisms. To serve as the genetic material, DNA has three key properties. First, it stores information in the same way that a library stores books or the Internet stores digital information. As we saw in Modules 5 and 31, DNA is a long, linear molecule made up of repeated subunits called nucleotides. DNA stores genetic information in the order of the A's, C's, T's, and G's along its length. Second, the information in DNA is used by cells to maintain their identity and carry out their functions. This is possible because the information in DNA is passed along to RNA, which is passed along to proteins. Proteins, in turn, do much of the work of the cell. Third, DNA is able to make copies of itself in the process of DNA replication. When DNA is replicated, genetic information is transmitted from one DNA molecule to another.

The ability of DNA to store genetic information, direct the synthesis of other macromolecules in the cell, and make copies of itself allows it to serve as the genetic material. How can one molecule do all of this? The answer rests in part on its structure. DNA is a molecule in which structure and function come together.

In this module, we will focus on the process of DNA replication—how DNA makes copies of itself. We will start by illustrating how the structure of DNA suggests a mechanism for how it replicates. We will then look at the process in detail, focusing on the many enzymes that are involved and the mechanism by which the two DNA strands of the double helix are copied. We will end by considering how it is able to make copies of itself with few, if any, errors.

PREP FOR THE AP® EXAM

FOCUS ON THE BIG IDEA

INFORMATION STORAGE AND TRANSMISSION: Focus on how the mechanism of DNA replication ensures the accurate transmission of genetic information from parent to offspring and contributes to genetic variation.

32.1 DNA replication is the basis of inheritance and occurs semiconservatively

DNA serves as the genetic material because it is able to specify exact copies of itself, a process known as DNA replication. When a cell divides, DNA is first replicated, so that each of the daughter cells receives genetic information from a parent cell. Faithful replication enables DNA to pass genetic information from cell to cell and from parent to offspring. As a result, DNA replication is the molecular basis for inheritance. In this section, we will discuss how DNA is replicated.

Base Pairing

You may recall from Module 5 that DNA in cells is a double helix, with two strands wound around each other. The two strands of DNA coil in such a way that a purine base (A or G) in one strand is paired with a pyrimidine base (T or C, respectively) in the other strand. In addition, the two strands are antiparallel, meaning that they run in opposite directions: one strand runs in the 5'-to-3' direction and the other runs in the 3'-to-5' direction. When James Watson and Francis Crick, with critical data from Rosalind Franklin, published their paper describing the structure of DNA in 1953, they also noted that "It has not escaped our notice that the specific pairing we have

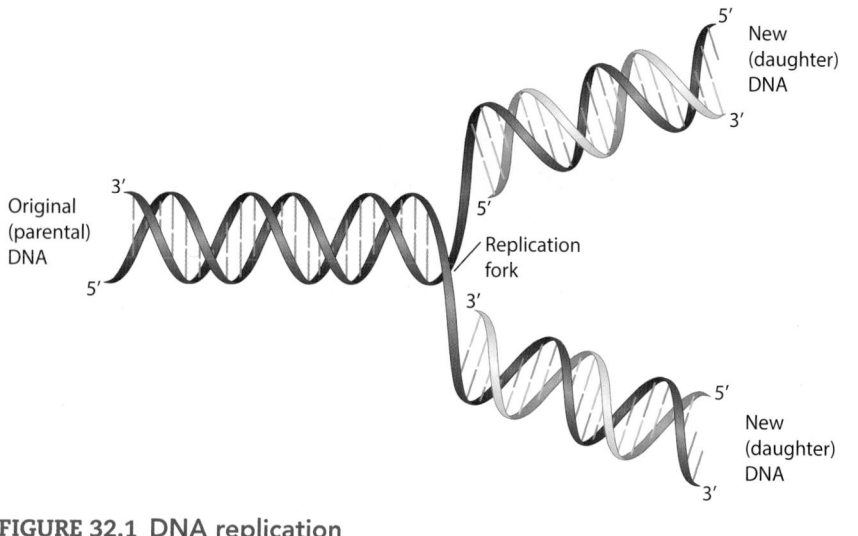

FIGURE 32.1 DNA replication

During DNA replication, the two strands of the original (parental) DNA molecule separate at the replication fork, where DNA synthesis takes place. As a result, one parental DNA molecule gives rise to two daughter DNA molecules.

postulated immediately suggests a copying mechanism for the genetic material." In other words, the structure of DNA immediately suggested how it makes copies of itself.

A diagram of DNA replication is shown in **FIGURE 32.1.** The two strands of a parental DNA double helix unwind and separate into single strands at a site called the **replication fork.** The replication fork is like a fork in the road, with a single road splitting into two roads going off in different directions, as can be seen in the figure.

As the two strands separate, each of the parental strands serves as a template, or model, for the synthesis of

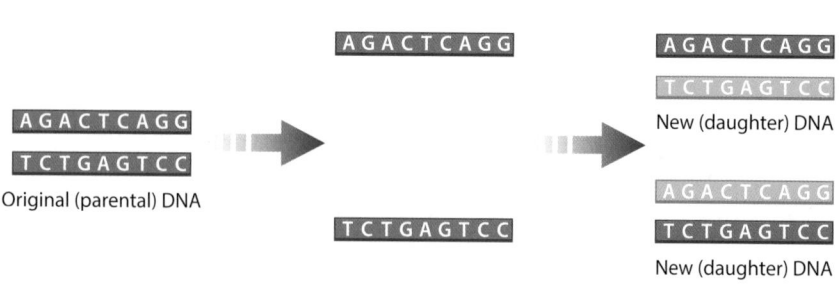

FIGURE 32.2 Base pairing

During DNA replication, each parental strand serves as a template for the synthesis of a daughter strand following the base-pairing rules in which A pairs with T, and G pairs with C. As a result, the two daughter DNA molecules have the same sequence as the parental DNA molecule.

a complementary daughter strand. The sequence of bases along the parental strand determines the sequence of bases along the daughter strand because wherever one strand carries an A, the other must carry a T, and wherever one carries a G, the other must carry a C, as can be seen in **FIGURE 32.2.** As a result, each of the daughter DNA molecules is identical in sequence to the parental DNA molecule, except possibly for rare errors called mutations.

Semiconservative Replication

When the process of DNA replication is complete, there are two DNA molecules. There are two possible mechanisms to explain how DNA replicates. In one mechanism, each new DNA molecule consists of one strand that was originally part of the parental molecule and one newly synthesized strand. This mechanism is called **semiconservative replication.** An alternative mechanism, known as conservative replication, proposes that the original DNA molecule remains intact and the daughter DNA molecule is completely new. Which mechanism is correct? We now know that DNA replicates semiconservatively. Note in Figures 32.1 and 32.2 that the daughter DNA molecules each contain one parental strand and one daughter strand.

Evidence for semiconservative replication came from a famous experiment carried out by American molecular biologists Matthew S. Meselson and Franklin W. Stahl. They reasoned that if there were a way to distinguish newly synthesized daughter DNA strands ("new strands") from previously synthesized parental strands ("old strands"), the products of replication could be observed and the mechanism of replication determined. This experiment is described in "Practicing Science 32.1: How is DNA replicated?" It has been called "the most beautiful experiment in biology" because it so elegantly demonstrates the process of generating hypotheses, making predictions, and performing an experiment to test the hypotheses.

How is DNA replicated?

Background The discovery of the structure of DNA in 1953 suggested a mechanism by which DNA is replicated. Experimental evidence of how DNA is replicated came from research by American molecular biologists Matthew Meselson and Franklin Stahl in 1958.

Hypothesis DNA replicates in a semiconservative manner, meaning that each new DNA molecule consists of one parental strand and one newly synthesized strand.

Alternative Hypothesis DNA replicates in a conservative manner, meaning that after replication, one DNA molecule consists of two parental strands and the other DNA molecule consists of two newly synthesized strands.

Method Meselson and Stahl distinguished parental strands ("old strands") from daughter strands ("new strands") using two isotopes of nitrogen atoms. Recall from Unit 1 that an isotope has the same number of protons as a particular element but a different number of neutrons, and therefore a different atomic mass. The researchers used two isotopes of nitrogen: ^{15}N with an extra neutron, which is a heavy form of nitrogen, and ^{14}N, a lighter and more typical form of nitrogen. Old strands were labeled with the heavy form of nitrogen (^{15}N), and new strands were labeled with the light form of nitrogen (^{14}N).

Experiment The researchers first grew bacterial cells on a medium containing only the light ^{14}N form of nitrogen. These cells incorporated ^{14}N into the DNA bases. The researchers could not observe the DNA directly, but they were able to measure the density of the DNA by spinning it in a high-speed centrifuge in tubes containing a solution of cesium chloride. They found that the DNA has a density of 1.708 gm/cm^3. This is the density of a DNA molecule containing two light ^{14}N strands, indicated by light/light in the figure below.

Light/Light Heavy/Heavy

Density (gm/cm^3) 1.708 1.722

They then grew bacterial cells on medium containing only the heavy ^{15}N form of nitrogen for several generations. As the cells grew, ^{15}N was incorporated into the DNA bases. Again, they measured the density of the DNA and found it has a density of 1.722 gm/cm^3. This is the density of a DNA molecule containing two heavy ^{15}N strands, indicated by heavy/heavy in the figure.

Finally, they took bacterial cells that had been growing for several generations in heavy ^{15}N nitrogen, and then transferred the cells into medium containing only light ^{14}N nitrogen. After one round of DNA replication in light ^{14}N nitrogen, cell division was halted, DNA was extracted, and the density of DNA was measured.

Prediction According to the hypothesis of semiconservative replication, daughter DNA molecules should be made up of one parental (old) and one daughter (new) strand. Therefore, after one round of DNA replication, the DNA molecules should consist of one heavy ^{15}N strand (old) and one light ^{14}N strand (new). According to the hypothesis of conservative replication, one daughter DNA molecule should be made up of two parental (old) strands, and one daughter DNA molecule should be made up of two daughter (new) strands. Therefore, after one round of DNA replication, half of the DNA in the cells should be composed of two heavy ^{15}N strands (old), and half should be composed of two light ^{14}N strands (new).

Results After one round of replication, indicated by 1st generation in the figure below, the researchers found that the DNA concentrated in a single band with a density of 1.715 gm/cm^3. This is shown in the test tube. This density is intermediate between the density of DNA with two light ^{14}N strands and the density of DNA with two heavy ^{15}N strands. It is the density of a DNA molecule containing one heavy ^{15}N strand and one light ^{14}N strand, indicated by the light/heavy label and coloring in the figure.

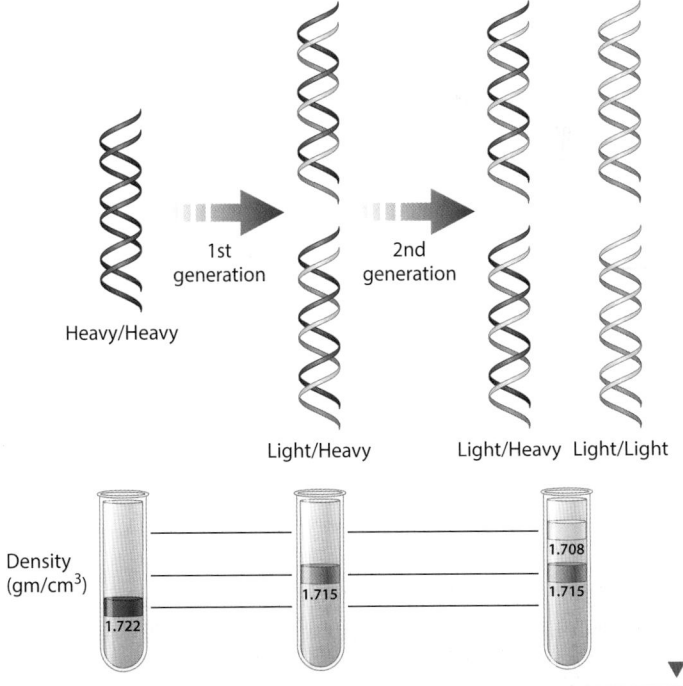

Heavy/Heavy   1st generation   2nd generation   Light/Heavy Light/Heavy Light/Light

Density (gm/cm^3)   1.722   1.715   1.708 1.715

Conclusion The data support the first hypothesis that DNA replicates semiconservatively. According to this model, daughter DNA molecules are made up of one parental (old) and one daughter (new) strand. As a result, the density of a daughter DNA molecule should be intermediate between fully heavy and fully light DNA, which is what the researchers observed. If DNA replicates conservatively, the researchers would have observed two bands after one round of DNA replication. One band would correspond to DNA with two heavy strands, and one band would correspond to DNA with two light strands. This is not what the researchers observed, so they rejected the alternative hypothesis of conservative replication.

Follow-up Work The researchers took this experiment one step further. They asked what happens after two rounds of DNA replication, indicated by 2nd generation in the figure on page 449. Their model of semiconservative replication predicts one daughter molecule should consist of two light strands, and one should consist of one light strand and one heavy strand. As a result, they should observe two bands after spinning DNA in the centrifuge. One of these bands corresponds to DNA with two light strands (light/light). The other band corresponds to DNA with one light strand and one heavy strand (light/heavy). This is precisely what they observed, providing even more evidence for the model of semiconservative replication.

SOURCE
Meselson, M., and F. W. Stahl. 1958."The Replication of DNA in *Escherichia coli.*" *PNAS* 44:671–82.

AP® PRACTICE QUESTION

(a) **Identify** the following:
   1. The question the researchers are trying to answer
   2. The hypothesis they tested
   3. The alternative hypothesis
   4. The densities of the three possible products of DNA replication in the experiment: $^{14}N/^{14}N$ (light/light), $^{14}N/^{15}N$ (light/heavy), $^{15}N/^{15}N$ (heavy/heavy)
   5. The independent variable
   6. The dependent variable

(b) **Explain** the conclusion using a claim, evidence, reasoning (CER) format.

(c) Suppose Meselson and Stahl had done their experiment the other way around, starting with cells fully labeled with ^{14}N light DNA and then transferring them to medium containing only ^{15}N heavy DNA. **Predict** the density of DNA molecule after one round and two rounds of replication.

PREP FOR THE AP® EXAM

AP® EXAM TIP

Make sure you can differentiate and describe the processes of replication, transcription, and translation.

The Meselson–Stahl experiment demonstrated that semiconservative replication occurs in bacteria. This left open the possibility that DNA replication in eukaryotes might be different. A few years after the Meselson–Stahl experiment, methods for labeling DNA with fluorescent nucleotides were developed. These methods allowed researchers to visualize entire strands of eukaryotic DNA and follow each strand through replication.

FIGURE 32.3 Eukaryotic DNA replication

Further evidence that DNA replication is semiconservative came from observing the uptake of fluorescent nucleotides into chromosomal DNA. After two rounds of DNA replication in a labeled medium, one daughter molecule (top) is half-labeled, and the other daughter molecule (bottom) is fully labeled. Photo: Daniel Hartl

FIGURE 32.3 shows a human chromosome with unlabeled DNA that subsequently underwent two rounds of replication in medium containing a fluorescent

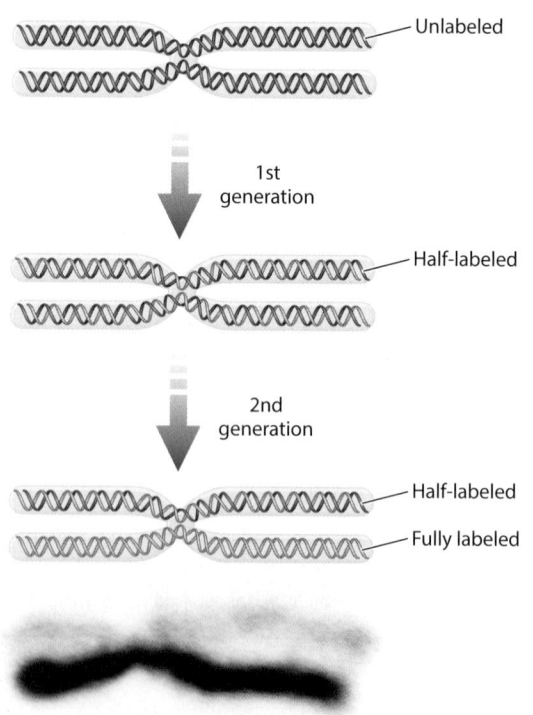

Unlabeled

1st generation

Half-labeled

2nd generation

Half-labeled

Fully labeled

nucleotide. The chromosome was photographed at metaphase of the second round of mitosis, after chromosome replication but before the separation of the chromatids into the daughter cells. Notice that one chromatid contains DNA with one labeled strand and one unlabeled strand, which fluoresces faintly (light); the other chromatid contains two strands of labeled DNA, which fluoresces strongly (dark). This result is conceptually the same as what was seen by Meselson and Stahl after two rounds of replication and exactly as predicted by the semiconservative replication mechanism.

AP® EXAM TIP

Make sure you can use illustrations such as Figure 32.3 to interpret and make conclusions about the data presented.

✓ Concept Check

1. **Describe** how base pairing ensures that parental strands and daughter strands have the same DNA sequence.
2. **Identify** the mechanism by which DNA replicates.

32.2 During DNA replication, one strand is synthesized continuously and the other discontinuously

We have discussed how parental DNA molecules determine the sequence of daughter DNA molecules through base pairing. We have also described semiconservative replication, in which each newly synthesized DNA molecule consists of a parental strand and a daughter strand. In this section, we will look at the copying mechanism in more detail. Specifically, we will consider the enzymes involved in the process of DNA replication and look at how each of the two daughter strands is synthesized.

DNA Replication Enzymes

As we have seen, DNA replication begins at a replication fork, where the parental strands separate. Many enzymes are involved in DNA replication, some of which are shown in **FIGURE 32.4** on page 452. An enzyme called **helicase** separates the strands of the parental double helix by breaking hydrogen bonds holding the base pairs together. Then single-strand binding protein binds to these single-stranded regions to prevent the parental strands from coming back together. Another enzyme called **topoisomerase** works upstream from the replication fork to relieve the stress that results from unwinding the double helix at the replication fork. Topoisomerases are a family of enzymes that wind or unwind DNA to help relieve stress that occurs during both replication and transcription.

DNA is replicated by an enzyme called **DNA polymerase,** also shown in Figure 32.4. This enzyme is a component of a large protein complex that makes copies of DNA and therefore carries out DNA replication. DNA polymerases exist in all organisms and are highly conserved, meaning that they vary little from one species to another because they carry out an essential function.

DNA polymerase has several properties, two of which are especially important for understanding the details of DNA replication. First, it can only attach one nucleotide to another nucleotide. In other words, it can only elongate the end of an existing piece of DNA or RNA and cannot lay down the first nucleotide of a newly synthesized strand on its own. As a result, each new DNA strand must begin with a short stretch of RNA that serves as a **primer,** or starter, for DNA synthesis, shown in red in Figure 32.4. The primer is made by an enzyme called RNA primase, which synthesizes a short piece of RNA complementary to the DNA parental strand. Once the RNA primer has been synthesized, DNA polymerase extends it, adding successive DNA nucleotides to the 3′ end of the growing strand.

The second key property of DNA polymerase is that it can only add nucleotides to the 3′ end of another nucleotide, as was discussed in Module 5. Recall that DNA has directionality, or polarity, with a 5′ end and a 3′ end, as can be seen in **FIGURE 32.5** on page 452. At the 5′ end is a phosphate, and at the 3′ end is a hydroxyl group. DNA polymerase can add a nucleotide to the 3′ end, but not the 5′ end. DNA synthesis occurs when the 3′ hydroxyl group attacks the phosphate group of an incoming nucleotide triphosphate, as shown in the figure. As a result, DNA synthesis, or polymerization, occurs only in the 5′-to-3′ direction. As the incoming nucleotide triphosphate is added to the growing DNA strand, one of the nucleotide's high-energy phosphate bonds is broken, providing the energy for the reaction. The outermost two phosphates are released in the process.

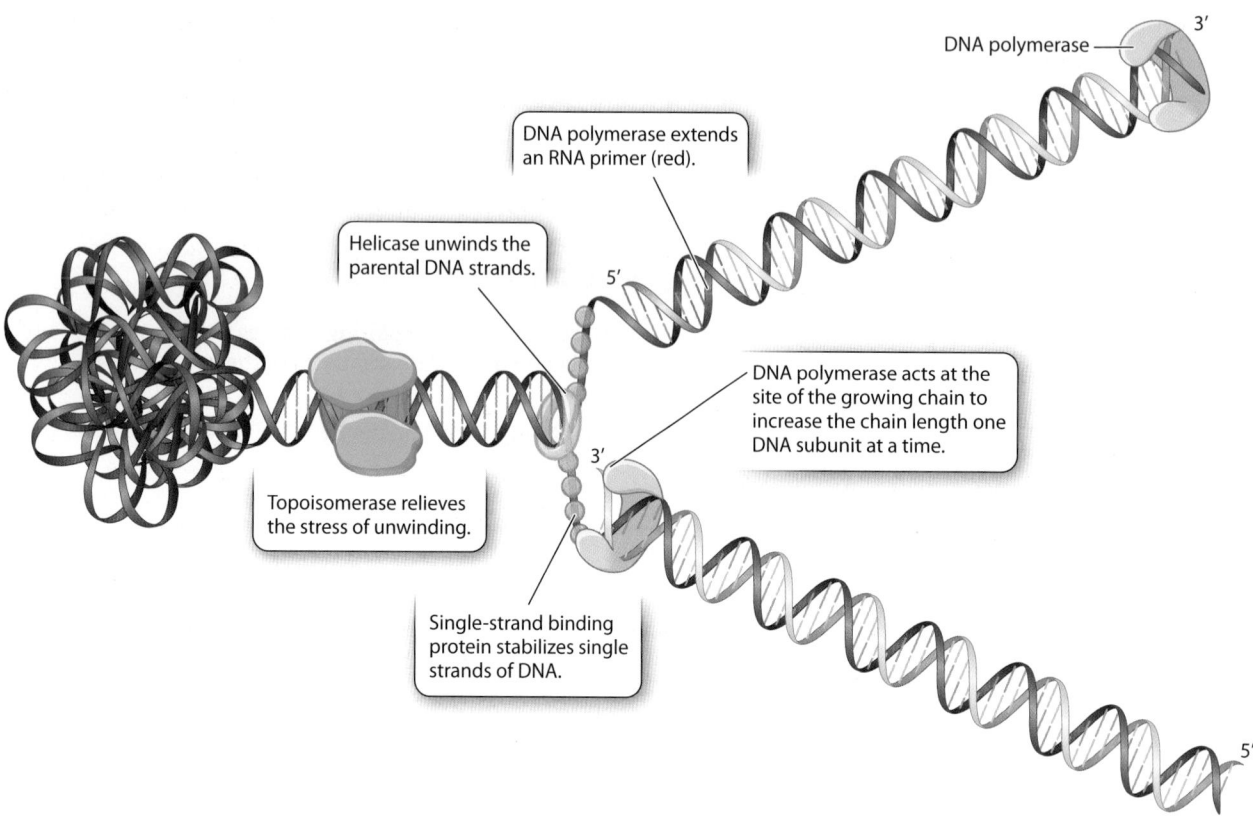

FIGURE 32.4 Some enzymes involved in DNA replication

Many enzymes play roles in DNA replication. Helicase unwinds the parental strands at the replication fork. Single-stranded binding protein helps to stabilize single strands of DNA. Topoisomerase relieves the stress that results from unwinding the double helix. DNA polymerase synthesizes new DNA strands.

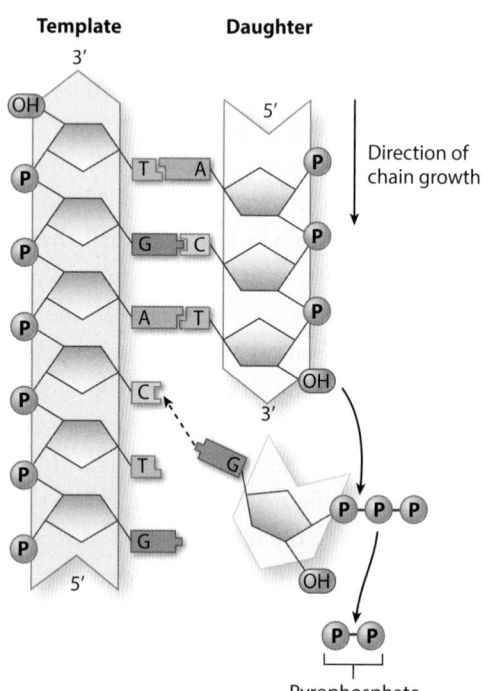

FIGURE 32.5 DNA synthesis

A new DNA strand is synthesized by the addition of a new nucleotide to the 3′ end of a growing DNA strand. The 3′ OH group of the growing strand attacks the high-energy phosphate bond of the incoming nucleotide, with release of a pyrophosphate molecule. As a result, DNA grows in a 5′-to-3′ direction.

Leading and Lagging Strands

Because the two DNA strands in a double helix are antiparallel and a new DNA strand can be synthesized only at the 3′ end, the two daughter strands are synthesized in quite different ways. This is illustrated in **FIGURE 32.6**. In this figure, one of the parental strands (the dark blue strand on the bottom) has a left-to-right 5′-to-3′ orientation, whereas the other parental strand (the dark blue strand on the top) runs in the opposite direction, with a left-to-right 3′-to-5′ orientation.

Let's focus on the bottom dark blue strand first. The replication fork is moving to the left in the figure, as can be seen by the three panels. As the replication fork moves along, it creates a region of single-stranded DNA. An RNA primer is first laid down, as shown in red in the figure. Then DNA polymerase can take over. Note that the light blue daughter strand that is being synthesized using the bottom strand as a template has its 3′ end pointed toward the replication fork. As a result, as the parental double helix unwinds, nucleotides can be added onto the 3′ end, and this daughter strand can be synthesized as one long, continuous polymer. This daughter strand is called the **leading strand.**

Now let's turn to the top dark blue strand in Figure 32.6. This strand has a left-to-right 3′-to-5′ orientation, which means the daughter strand has a left-to-right 5′-to-3′ orientation. As a result, the 5′ end of the daughter strand is pointed toward the replication fork, as can be seen in the figure. However, the daughter strand cannot grow in that direction because DNA polymerase can add nucleotides only to the 3′ end of a growing strand. To enable synthesis of the top daughter strand, the replication fork moving to the left first creates a single-stranded region of parental DNA ranging in length from a few hundred to a few thousand nucleotides. An RNA primer is laid down, shown in red. Synthesis of the new strand occurs at its 3′ end as usual, which means that the daughter strand grows in a direction away from the replication fork, not toward it. The result is that the top daughter strand in Figure 32.6 is synthesized in relatively short, discontinuous pieces called Okazaki fragments. In the figure, two fragments are shown. As the parental double helix unwinds, a new piece is initiated at intervals, and each new piece is elongated at its 3′ end until it reaches the piece in front of it. This daughter strand is called the **lagging strand.**

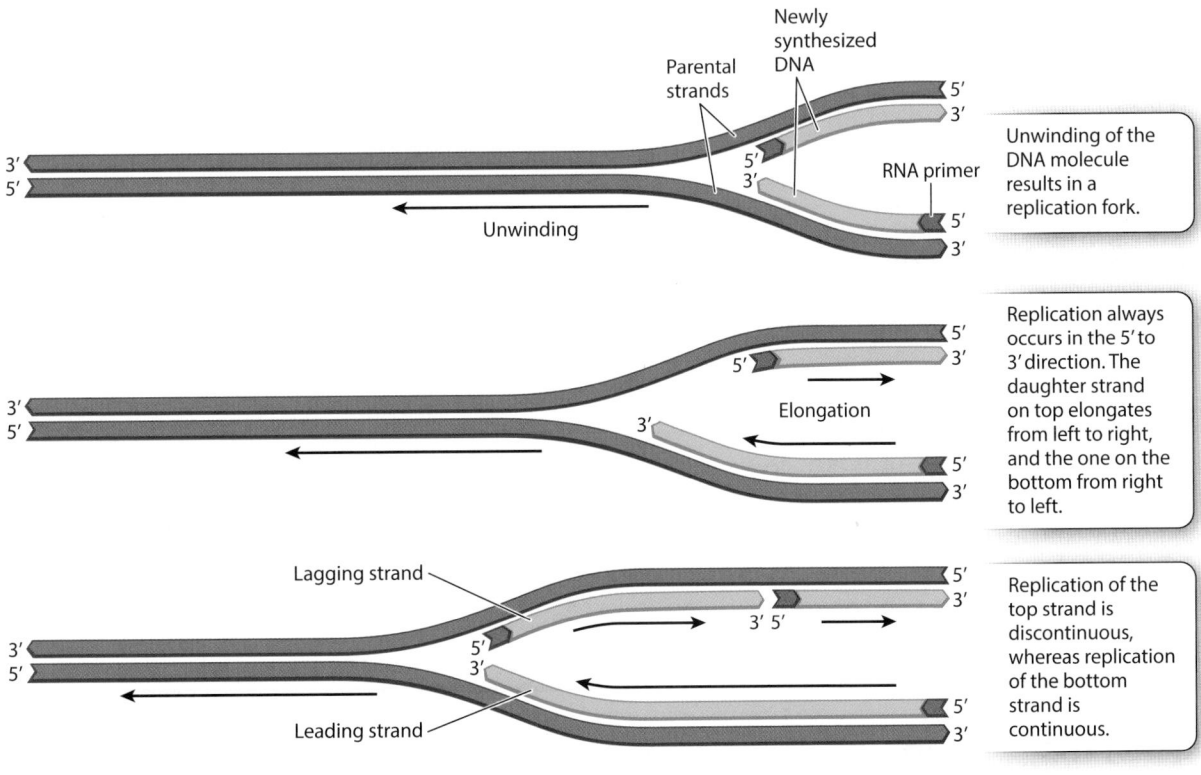

FIGURE 32.6 Leading and lagging strands

Because the parental DNA molecule is made of two antiparallel strands and DNA polymerase can add nucleotides to only the 3′ end, the two daughter strands are synthesized differently. One daughter strand is synthesized continuously and is called the leading strand. The other daughter strand is synthesized discontinuously and is called the lagging strand.

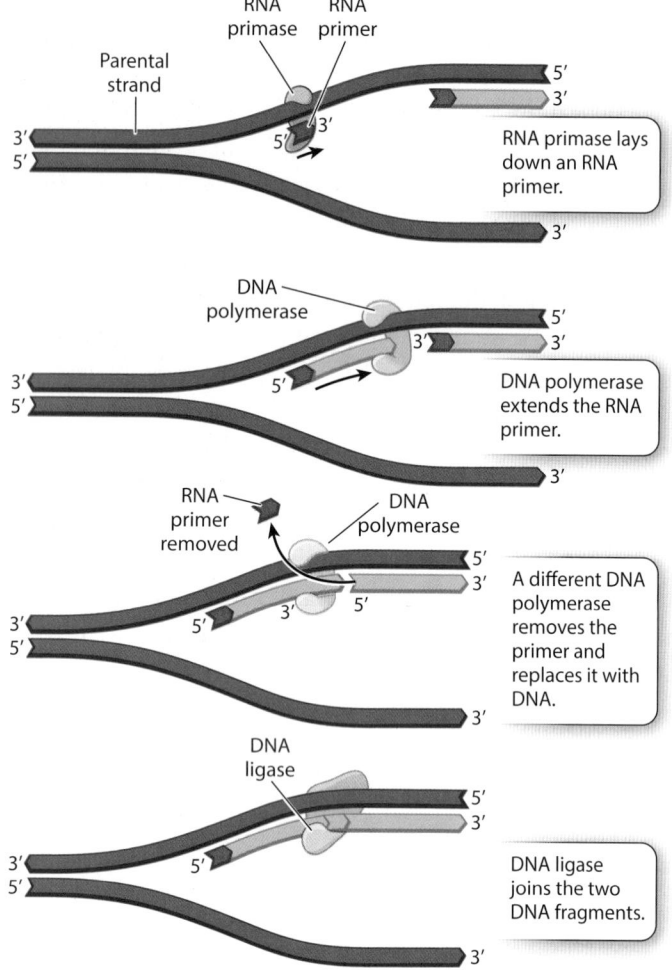

FIGURE 32.7 RNA primers

DNA polymerase requires a short stretch of RNA to begin DNA synthesis. RNA primase therefore synthesizes a short stretch of RNA. Then DNA polymerase begins synthesis until it runs into another RNA primer, which is removed and replaced with DNA. DNA ligase then connects the two DNA fragments.

Labels in figure (top to bottom):

- Parental strand
- RNA primase
- RNA primer
- RNA primase lays down an RNA primer.
- DNA polymerase
- DNA polymerase extends the RNA primer.
- RNA primer removed
- DNA polymerase
- A different DNA polymerase removes the primer and replaces it with DNA.
- DNA ligase
- DNA ligase joins the two DNA fragments.

For both the leading and lagging strands, each new DNA strand begins with a short stretch of RNA, as discussed earlier. These are shown in red in Figure 32.6. Because DNA polymerase extends an RNA primer, all new DNA strands have a short stretch of RNA at their 5′ end. For the lagging strand, there are many such primers, one

for each of the discontinuous (Okazaki) fragments of newly synthesized DNA. As each of these fragments is elongated by DNA polymerase, it grows toward the primer of the fragment in front of it, as shown in **FIGURE 32.7**. When the growing fragment comes into contact with the primer of the fragment synthesized earlier, a different DNA polymerase takes over, removing the earlier RNA primer and extending the growing fragment with DNA nucleotides to fill the space left by its removal. When the replacement is completed, the adjacent fragments are joined, or ligated, by an enzyme called **DNA ligase.**

The presence of leading and lagging strands during DNA replication is a consequence of the antiparallel nature of the two strands in a DNA double helix and the fact that DNA polymerase can synthesize DNA only in the 5′-to-3′ direction. Nevertheless, the copying mechanism is the same on both strands: after the two strands of the parental DNA separate, each serves as a template for the synthesis of a daughter strand according to the base-pairing rules of A–T and G–C, with each successive nucleotide being added to the 3′ end of the growing strand.

PREP FOR THE AP® EXAM

AP® EXAM TIP

Make sure you can describe "leading strand" and "lagging strand" and can differentiate them.

✓ Concept Check

3. **Describe** the functions of helicase, topoisomerase, RNA primase, DNA polymerase, and DNA ligase during DNA replication.

4. **Describe** the orientation of the DNA strands and the direction of DNA synthesis.

5. **Describe** one similarity between the two resulting daughter strands during DNA replication.

6. **Describe** one difference in the way the two daughter strands are synthesized at a replication fork during DNA replication.

32.3 DNA polymerase is self-correcting because of its proofreading function

We have seen that during DNA replication, a single parental molecule of DNA produces two daughter molecules. These two daughter molecules are identical, or nearly identical, to each other and the parental DNA molecule. In this section, we will discuss how DNA is able to be copied so faithfully.

Reproducing the sequence of nucleotides as precisely as possible is important because mistakes that go unrepaired may be harmful to the cell or organism. An unrepaired error in DNA replication results in a mutation, which is a change in the genetic information in DNA. For example, a mutation in DNA causes the genetic difference between virulent and nonvirulent bacteria, as discussed in Practicing Science 31.1 on page 438.

Most DNA polymerases are capable of **proofreading,** which is a process in which a DNA polymerase can immediately correct its own errors. When each new nucleotide comes into line in preparation for attachment to the growing DNA strand, the nucleotide is temporarily held in place by hydrogen bonds that form between the base in the new nucleotide and the base across the way in the parental strand, as shown in Figure 32.3. The strand being synthesized and the parental strand therefore have complementary bases— A paired with T, or G paired with C.

However, on rare occasions, an incorrect nucleotide is attached to the new DNA strand, as shown in **FIGURE 32.8.** When this happens, DNA polymerase can correct the error because it detects mispairing between the template and the most recently added nucleotide. After it detects mispairing between a base in the parental strand and a newly added base in the daughter strand, it removes the incorrect nucleotide and inserts the correct one in its place.

Mutations resulting from errors in nucleotide incorporation still occur, but proofreading reduces their number. In the bacterium *Escherichia coli,* for example, about 99% of the incorrect nucleotides that are incorporated during replication are removed and repaired by the proofreading function of DNA polymerase. Those that slip past proofreading and other repair systems lead to mutations, which are then copied and passed on to daughter cells. Some of these mutations may be harmful, but others are neutral and a rare few may be beneficial. These mutations are the ultimate source of genetic variation that we see among individuals of the same species and among species. They are essential in the process of evolution because, as we will see in Unit 7, they allow populations of organisms to change through time and adapt to their environment.

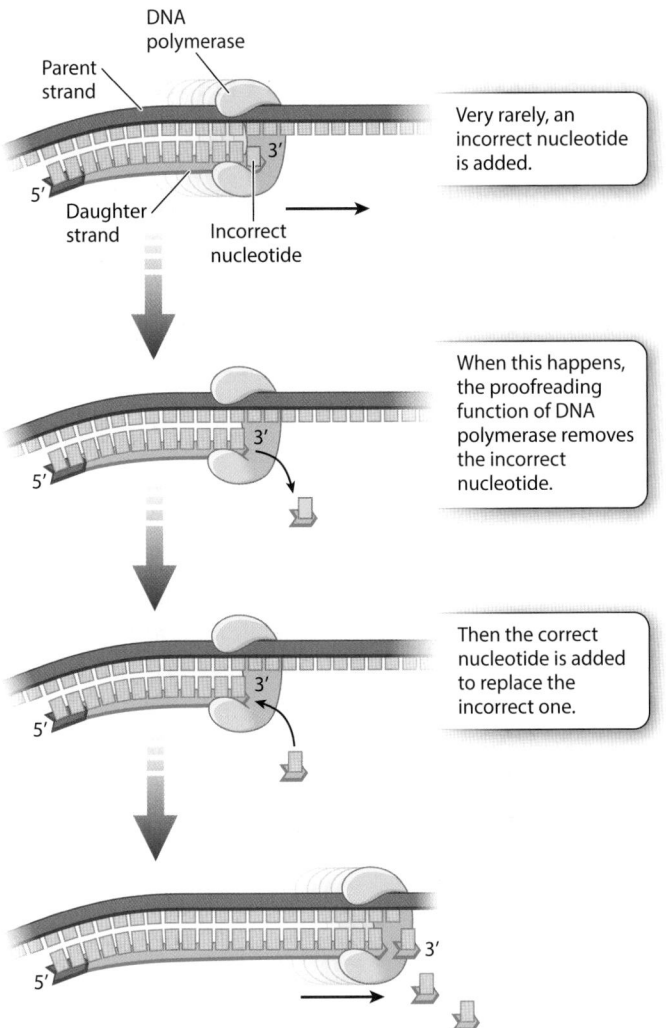

FIGURE 32.8 Proofreading

DNA polymerase has a proofreading function, allowing it to identify and replace an incorrect nucleotide during DNA replication.

PREP FOR THE AP® EXAM

AP® EXAM TIP

You should know the roles and functions of the enzymes DNA polymerase, DNA ligase, topoisomerase, and helicase.

✓ Concept Check

7. **Describe** why it is important that DNA replication occurs accurately.

8. **Identify** the function of DNA polymerase that allows it to correct its own mistakes.

Module 32 Summary

PREP FOR THE AP® EXAM

REVISIT THE BIG IDEA

INFORMATION STORAGE AND TRANSMISSION: Describe how the mechanism of DNA replication ensures the accurate transmission of genetic information from parent to offspring and contributes to genetic variation.

LG 32.1 DNA replication is the basis of inheritance and occurs semiconservatively.

- In DNA replication, a single parental molecule of DNA produces two daughter molecules. Page 448

- DNA replication involves the separation of the two strands of the double helix at a replication fork and the use of these strands as templates to direct the synthesis of new strands. Page 448

- DNA replication is semiconservative, meaning that each daughter DNA molecule consists of a newly synthesized strand and a strand that was present in the parental DNA molecule. Page 448

LG 32.3 During DNA replication, one strand is synthesized continuously and the other discontinuously.

- Nucleotides are added to the 3′ end of the growing strand, so DNA synthesis occurs in a 5′-to-3′ direction. Page 451

- DNA polymerase requires RNA primers for DNA synthesis. Page 451

- Other enzymes required for DNA replication include helicase, topoisomerase, and DNA ligase. Page 451

- One new strand is synthesized continuously and is called the leading strand. Page 453

- The other strand is synthesized in small pieces and is called the lagging strand. Page 453

LG 32.3 DNA polymerase is self-correcting because of its proofreading function.

- DNA polymerase can correct its own mistakes by detecting a pairing mismatch between a template base and an incorrect new base. Page 455

- Unrepaired errors in DNA replication result in mutations. Page 455

Key Terms

Replication fork	DNA polymerase	DNA ligase
Semiconservative replication	Primer	Proofreading
Helicase	Leading strand	
Topoisomerase	Lagging strand	

Review Questions

1. The enzyme that catalyzes the addition of new nucleotides to a growing DNA strand is
   (A) DNA polymerase.
   (B) helicase.
   (C) topoisomerase.
   (D) DNA ligase.

2. Which statement about the strands of a newly replicated DNA molecule is correct?
   (A) Both strands are made up of newly assembled nucleotides.
   (B) One strand is new and the other is from the original molecule.
   (C) Both strands contain some nucleotides from the original molecule.
   (D) The sugar-phosphate chains are from the original molecule and new bases are inserted between them.

3. What is the result of DNA ligase's action?

(A) DNA is broken up at specific sites.

(B) DNA fragments are joined together.

(C) DNA translation occurs.

(D) DNA transcription occurs.

4. Short DNA fragments are found on which strand of DNA?

(A) The leading strand

(B) The lagging strand

(C) The parental strand

(D) Both the leading and lagging strands

5. A new nucleotide can be added to only which end of a growing DNA strand?

(A) 5′

(B) 3′

(C) 2′

(D) 1′

6. DNA always grows in which direction?

(A) 5′-to-5′

(B) 5′-to-3′

(C) 3′-to-5′

(D) 3′-to-5′

7. Which enzyme is primarily responsible for proofreading a growing DNA strand?

(A) DNA polymerase

(B) Helicase

(C) Topoisomerase II

(D) DNA ligase

Module 32
AP® Practice Questions

Section 1: Multiple-Choice Questions

Choose the best answer for questions 1–4.

1. A scientist grows a population of bacteria in the presence of a chemical that labels the strands of DNA. While grown in the presence of the chemical, all DNA molecules become permanently labeled on both strands. The scientist then removes the labeling chemical and allows the bacterial cells to undergo two rounds of cell division. The DNA that was synthesized during these rounds of cell division was unlabeled. Which statement correctly describes the population of bacterial cells after the second round of cell division?

(A) After the second round, 100% of cells will have one strand of DNA labeled.

(B) After the second round, 50% of cells will have one strand of DNA labeled, and 50% of cells will have none of their DNA labeled.

(C) After the second round, 25% of the cells will have both strands of DNA labeled, and 75% of cells will have none of their DNA labeled.

(D) After the second round, 25% of the cells will have one strand of DNA labeled, and 75% of cells will have none of their DNA labeled.

Use the following figure to answer question 2.

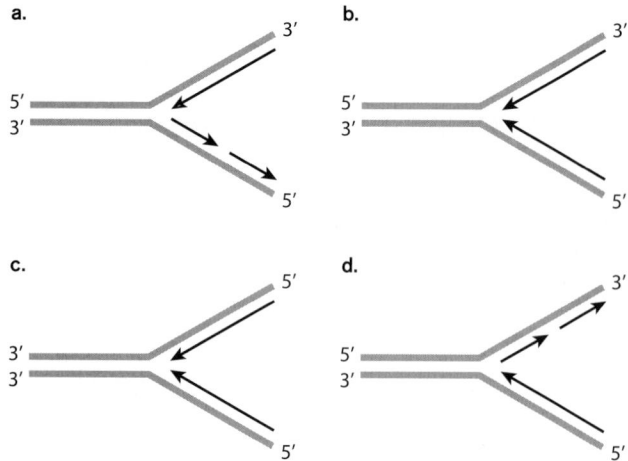

2. Which correctly illustrates the process of DNA replication? Arrowheads indicate the direction in which a newly synthesized DNA strand is growing.

   (A) a

   (B) b

   (C) c

   (D) d

3. Which of the following correctly describes the role of topoisomerase in DNA replication?

   (A) Topoisomerase relaxes supercoiling in front of the replication fork.

   (B) Topoisomerase unwinds and separates DNA strands to enable replication.

   (C) Topoisomerase produces RNA primers that can initiate DNA synthesis.

   (D) Topoisomerase incorporates complementary nucleotides into a growing DNA strand.

4. Which correctly describes the proofreading activity of DNA polymerase?

   (A) DNA polymerase removes RNA primers so they can be replaced with DNA during DNA replication.

   (B) DNA polymerase proofreads in a 5′-to-3′ direction, incorporating complementary nucleotides.

   (C) Proofreading by DNA polymerase reduces the amount of genetic diversity in a population.

   (D) Proofreading by DNA polymerase separates DNA strands to allow DNA replication.

Section 2: Free-Response Question

Write your answer to each part clearly. Support your answers with relevant information and examples. Where calculations are required, show your work.

The human female diploid genome size is 3.2×10^9 base pairs (bp). The male genome is slightly smaller because the Y chromosome is smaller than the X chromosome. DNA polymerase makes an error once in approximately every 10^8 nucleotides incorporated.

(a) Using the information provided, **calculate** the number of mutations that DNA polymerase creates every time a diploid female genome is replicated.

(b) A scientist makes a claim that the vast majority of DNA replication errors are not passed from parents to offspring. **Provide reasoning** to support or refute the scientist's claim.

(c) **Describe** why mutations in DNA that encodes protein sequences are often harmful.

Module 33

Transcription and RNA Processing

LEARNING GOALS ▶**LG 33.1** During transcription, DNA is used as a template to guide the synthesis of RNA.
▶**LG 33.2** RNA polymerase adds successive nucleotides to the 3′ end of the transcript.
▶**LG 33.3** The primary transcript is processed to make mature mRNA.

We have seen that DNA stores information in the sequence of nucleotides that make it up, and it transmits information to daughter DNA molecules in the process of DNA replication. DNA replication is the molecular basis of inheritance; it allows genetic information to be transmitted from cell to cell and generation to generation through the process of cell division.

Both of these functions of DNA—information storage and transmission—can be seen in the structure of DNA. DNA is a long polymer composed of repeated units called nucleotides. In cells, DNA takes the form of a double helix, with two strands wound around each other in an antiparallel fashion. Information is encoded in the order of nucleotides along the length of a DNA strand, and information is transmitted when the double helix separates and each parental strand serves as a template for a complementary daughter strand.

The information in DNA is also used by cells to produce proteins. Specifically, the sequence of nucleotides in DNA is used to specify the sequence of nucleotides in RNA, which in turn is used to specify the sequence of amino acids in proteins. Biologists say that information "flows" from DNA to RNA to protein, a process called the central dogma of molecular biology.

In this module, we will look closely at the first step of this process of information flow—how DNA hands off genetic information to RNA. This process is known as transcription and it is a universal feature of all cells. It is the first step of gene expression that allows cells to retrieve and use the information encoded in DNA. The RNA that is produced then undergoes a series of modifications, which we will discuss as well. These modifications are necessary for it to be able to produce proteins, which allow cells to function.

PREP FOR THE AP® EXAM

FOCUS ON THE BIG IDEAS

INFORMATION STORAGE AND TRANSMISSION: Focus on how the processes of transcription and RNA processing ensure the accurate transmission of genetic information for the production of proteins.

33.1 During transcription, DNA is used as a template to guide the synthesis of RNA

Although the double helix structure of DNA gave important clues about how DNA stores and transmits information, it left open many questions about how the genetic information in DNA is retrieved to control cellular processes. In 1953, when the double helix was discovered, almost nothing was known about these processes. Within a few years, however, an accumulating body of evidence began to show that DNA carries the genetic information for proteins, and that proteins are synthesized on ribosomes. But in eukaryotes, DNA is located in the nucleus and ribosomes are located in the cytoplasm. Thus, there must be an intermediary molecule that cells use to transfer the genetic information from the DNA in the nucleus to ribosomes in the cytoplasm. That molecule is RNA. In this section, we will discuss how the information in DNA is transferred to RNA in the process of transcription. We will also introduce the sequences in DNA that mark the beginning and end of individual transcripts.

Transcription

Transcription is the process by which RNA is synthesized from a DNA template. Conceptually, the process is similar to DNA replication, except that a molecule of RNA is

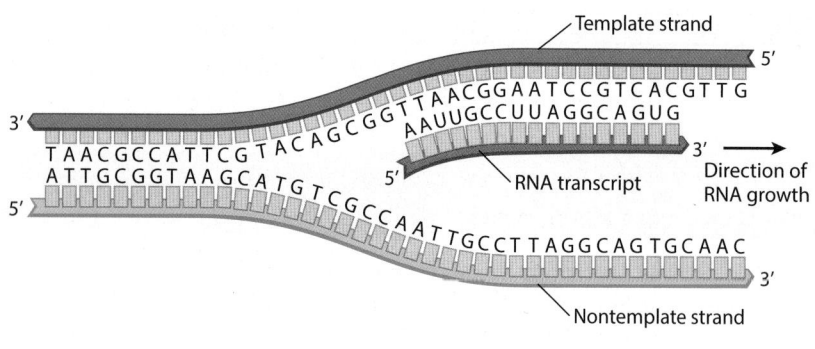

Template strand

5′

3′

T A A C G C C A T T C G T A C A G C G G T T A A C G G A A T C C G T C A C G T T G
A T T G C G G T A A G C A T G T C G C C A A U U G C C U U A G G C A G U G

5′

3′

RNA transcript

Direction of
RNA growth

A T T G C C T T A G G C A G T G C A A C

5′

3′

Nontemplate strand

FIGURE 33.1 Transcription

The DNA double helix unwinds for transcription. For each gene, only one strand, the template strand, is transcribed. The template strand and the RNA transcript run in opposite directions, and RNA is synthesized in the 5′-to-3′ direction.

synthesized. As a region of the DNA double helix unwinds, one strand, known as a **template strand,** is used as a model for the synthesis of an RNA transcript, as shown in **FIGURE 33.1.** The newly synthesized RNA is complementary to the DNA template according to the base-pairing rules. However, because RNA contains U (uracil) instead of T (thymine), an A (adenine) in the DNA template pairs with a U (uracil) in RNA. The RNA molecule is synthesized by an enzyme called **RNA polymerase,** similar to the way in which DNA polymerase synthesizes DNA during DNA replication. Like DNA polymerase, RNA polymerase acts by adding successive nucleotides to the 3′ end of the newly synthesized RNA, also called the growing transcript.

The direction of growth of the RNA transcript is the opposite of the direction that the DNA template is read. All nucleic acids are synthesized by addition of nucleotides to the 3′ end of the strand. That is, they grow in a 5′-to-3′ direction. Just like the two strands of DNA in a double helix, the DNA template and the RNA strand transcribed from it are antiparallel, meaning the DNA template runs in the opposite direction from the RNA, as shown in Figure 33.1. As the RNA transcript is synthesized in the 5′-to-3′ direction, the DNA template is read in the opposite direction, which is 3′-to-5′.

As mentioned above, the DNA strand that is transcribed is called the template strand because it serves as a template, or model, for RNA synthesis. Note in Figure 33.1 that this strand is oriented in the opposite, or reverse, direction from the RNA transcript. Also note that the

template strand is the complement of the RNA transcript. In other words, where the template strand has a C, the RNA strand has a G, and so on. Because the RNA transcript runs in the opposite direction and is the complement of the template strand, the RNA transcript is described as the reverse complement of the template strand.

Furthermore, the sequence of the RNA transcript is the same as the DNA strand (with U's in place of T's) that is not used as the template for transcription, which is called the **nontemplate strand.** For example, in Figure 33.1, the sequence of nontemplate strands is 5′-AATTGC . . . -3′ and the sequence of the RNA transcript is 5′-AAUUGC . . . -3′. Like the RNA transcript, the nontemplate strand is the reverse complement of the template strand. For this reason, the nontemplate strand is also called the coding strand, sense strand, or plus strand. The template strand, by contrast, is sometimes called the noncoding strand, antisense strand, or minus strand.

Transcription takes place in three stages. The first stage is initiation, in which RNA polymerase and other proteins bind double-stranded DNA, the DNA strands are separated, and transcription of the template strand begins. The second stage is elongation, in which successive nucleotides are added to the 3′ end of the growing RNA transcript as the RNA polymerase proceeds along the template strand. The third stage is termination, in which the RNA polymerase encounters a sequence in the template strand that causes transcription to stop and the RNA transcript to be released.

Promoters and Terminators

Transcription initiates and terminates at specific places along a DNA molecule. A long DNA molecule typically contains thousands of genes, most of them coding for proteins or RNA molecules with specialized functions, and hence thousands of different transcripts are produced. For example, the DNA molecule in the bacterium *Escherichia coli* has about 4.6 million base pairs and produces RNA transcripts for about 4300 proteins.

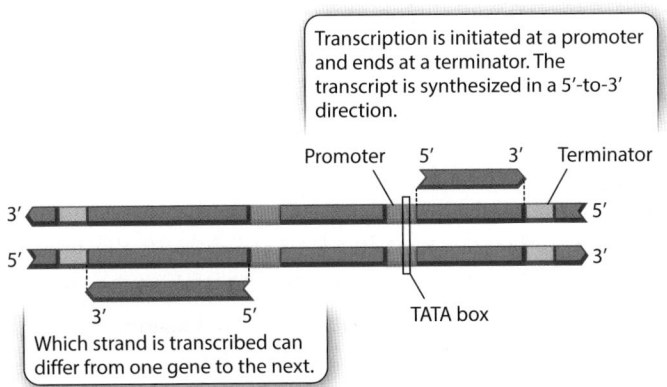

Transcription is initiated at a promoter and ends at a terminator. The transcript is synthesized in a 5'-to-3' direction.

Which strand is transcribed can differ from one gene to the next.

FIGURE 33.2 Promoters and terminators

Promoters are sequences where transcription initiates, and terminators are sequences where transcription terminates. A common sequence in eukaryotic promoters is called the TATA box. A DNA molecule usually contains many genes that are transcribed individually and at different times, often from opposite strands.

A typical map of a small part of a long eukaryotic DNA molecule is shown in **FIGURE 33.2**. Each green segment indicates the position where transcription is initiated, and each purple segment indicates the position where it ends. The green segments in Figure 33.2 are **promoters,** regions of typically a few hundred base pairs where RNA polymerase and associated proteins bind to the DNA molecule and initiate transcription. Although the term "promoter" refers to a region in double-stranded DNA, transcription is initiated on only one strand, as discussed above. In many promoters found in eukaryotes and archaea, the promoter includes the sequence 5'-TATAAA-3' or something similar, which is known as a TATA box because of the sequence 5'-TATA-3', as indicated in Figure 33.2. Transcription initiation takes place about 25 nucleotides downstream of the TATA box, and transcription elongation occurs as the RNA polymerase moves along the template strand in the 3'-to-5' direction.

Transcription continues until the RNA polymerase encounters a sequence known as a terminator, indicated by purple in Figure 33.2. Transcription stops at the terminator, and the transcript is released.

A long DNA molecule contains the genetic information for hundreds or thousands of genes. For any one gene, only one DNA strand is transcribed. However, different genes in the same double-stranded DNA molecule can be transcribed from opposite strands. In other words, there is not one fixed template strand for a double-stranded DNA molecule; the template strand can be different for different genes. Which strand is transcribed depends on the orientation of the promoter. When the promoters are in opposite orientation, as they are in Figure 33.2, transcription occurs in opposite directions because, as noted earlier, transcription always begins downstream of the promoter. Furthermore, transcription can proceed only by successive addition of nucleotides to the 3' end of the transcript.

Transcription does not take place all the time from all promoters, but rather is a regulated process. For housekeeping genes, whose products are needed at all times in all cells, transcription takes place continually. Most genes, however, are transcribed only at certain times, under certain conditions, or in certain cell types. In the bacterium *E. coli,* for example, the genes that encode proteins needed to utilize the sugar lactose (milk sugar) are transcribed only when lactose is present in the environment. For such genes, regulation of transcription often depends on whether the RNA polymerase and associated proteins are able to bind with the promoter.

PREP FOR THE AP® EXAM

AP® EXAM TIP

Make sure you understand the potentially confusing terms "template," "noncoding," "antisense," and "minus strand," which all describe the same DNA strand, the one that is used as a template during the synthesis of RNA. Similarly, remember that the terms "nontemplate," "coding," "sense," and "plus strand" describe the complementary DNA strand, which is not transcribed.

✓ Concept Check

1. **Describe** the usual flow of genetic information in a cell.

2. A segment of the template strand of a double-stranded DNA molecule has the sequence 5'-ACTTTCAGCGAT-3'. **Predict** the sequence of an RNA molecule synthesized from this DNA template.

33.2 RNA polymerase adds successive nucleotides to the 3′ end of the transcript

As we have seen, RNA polymerase is the enzyme responsible for synthesizing RNA from a DNA template. RNA polymerase is a remarkable molecular machine that carries out many functions. It separates the DNA strands, allows an RNA–DNA hybrid molecule to form, elongates the transcript nucleotide by nucleotide, releases the finished transcript, and restores the original DNA double helix. It can add thousands of nucleotides to a transcript before dissociating from the DNA template. In this section, we will look at how RNA polymerase works in more detail.

After transcription initiation takes place, successive nucleotides are added to grow the transcript in the process of transcription elongation. Transcription elongation takes place in a sort of bubble shown in **FIGURE 33.3**. In this figure, you can see that the strands of the DNA molecule are separated from each other, allowing transcription to occur from the template strand. One end of the RNA transcript is paired with the template strand, forming a DNA–RNA hybrid molecule. The other end of the RNA transcript is single stranded and grows as RNA polymerase moves along the DNA template. The RNA transcript is synthesized in a 5′-to-3′ direction, so RNA polymerase moves along the DNA template in a 3′-to-5′ direction, as shown in the figure.

RNA synthesis is also called RNA polymerization because a long molecule, or polymer, is made. Details of the RNA polymerization reaction are shown in **FIGURE 33.4**. The incoming nucleotide, with three phosphates, is shown at the bottom right. It is accepted by the RNA polymerase only if it undergoes proper base pairing with the base in the template DNA strand. In Figure 33.4, there is a proper match because U from the incoming nucleotide pairs with A of the DNA template strand.

At this point, the RNA polymerase orients the 3′ end of the growing strand so that the oxygen in the hydroxyl (—OH) group can attack the innermost phosphate of the triphosphate of the incoming nucleotide. This hydroxyl group is necessary for the reaction. If it is replaced with a hydrogen atom, for example, polymerization will not occur. The bond connecting the innermost phosphate to the next phosphate is a high-energy phosphate bond. Breaking this bond provides the energy to drive the reaction. The result is a phosphodiester bond attaching the incoming nucleotide to the 3′ end of the growing chain, which creates the sugar-phosphate backbone of the RNA molecule. The term "high-energy" here refers to the amount of energy released when the phosphate bond is broken; this energy can be used to drive chemical reactions.

The synthesis, or polymerization, reaction releases a phosphate–phosphate group, shown at the lower right in Figure 33.4. The next nucleotide that complements the template is brought into line. In this way, transcription occurs

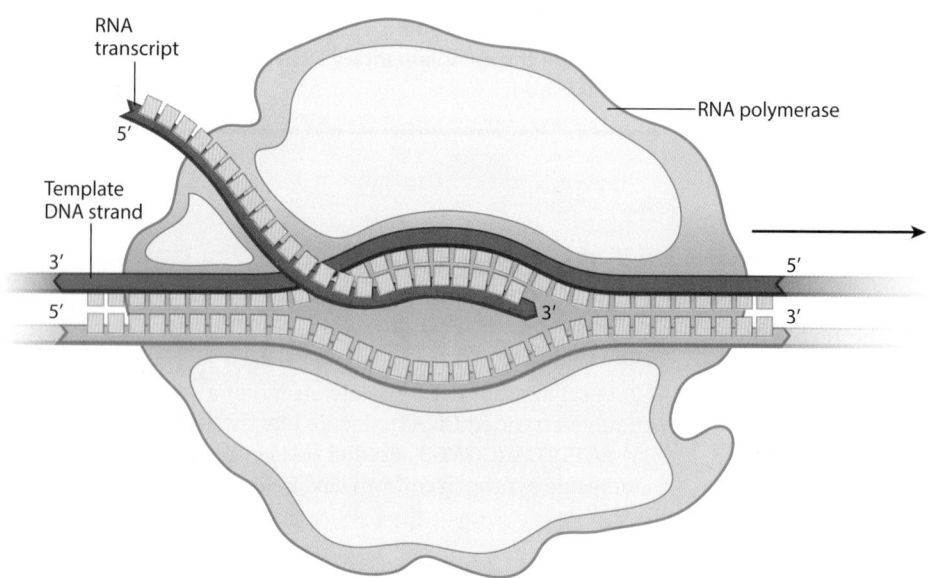

FIGURE 33.3 Transcription elongation

RNA polymerase is the enzyme that synthesizes RNA from a DNA template. Within the RNA polymerase, the two strands of DNA separate during transcription initiation, and the growing RNA strand forms a hybrid DNA–RNA molecule during transcription elongation.

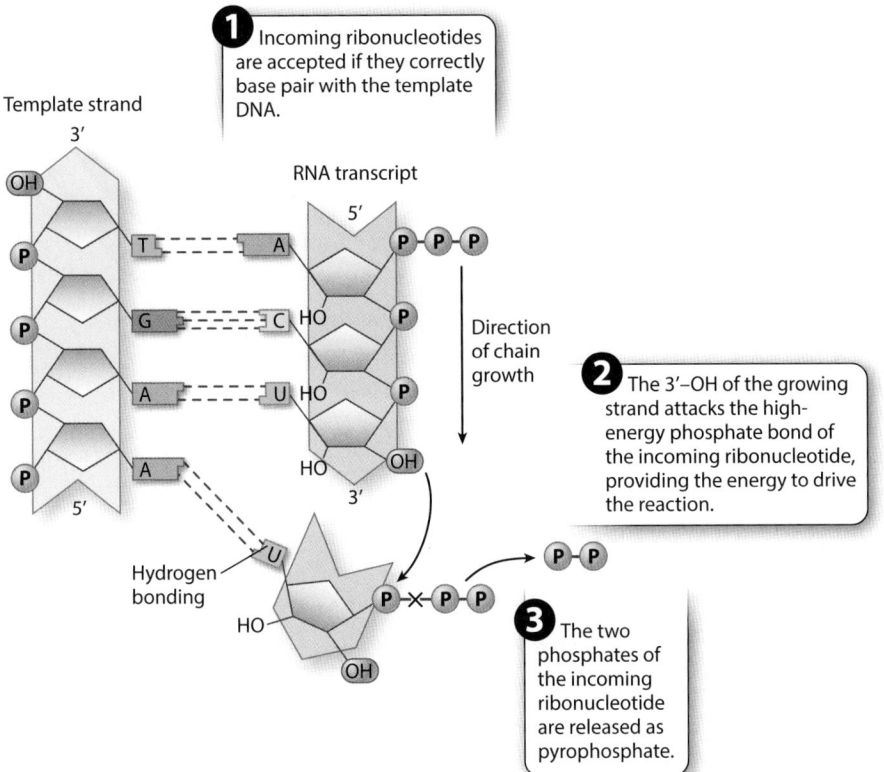

① Incoming ribonucleotides are accepted if they correctly base pair with the template DNA.

Template strand

3′

RNA transcript

5′

Direction of chain growth

② The 3′–OH of the growing strand attacks the high-energy phosphate bond of the incoming ribonucleotide, providing the energy to drive the reaction.

③ The two phosphates of the incoming ribonucleotide are released as pyrophosphate.

Hydrogen bonding

FIGURE 33.4 RNA synthesis

RNA is synthesized, or elongates, by the addition of nucleotides to the 3′ end of the growing chain. The 3′-OH group attacks an incoming nucleotide triphosphate to make a phosphodiester bond, with release of two phosphates.

quickly and efficiently. It is also fairly accurate, with fewer than 1 incorrect nucleotide incorporated per 10,000 nucleotides. However, RNA polymerase does not have a proofreading function like DNA polymerase does. As a result, the error rate of RNA polymerase is much higher than that of DNA polymerase. This difference makes sense evolutionarily, as an error during transcription will only affect a transcript in a cell, while an error in a DNA molecule will be passed from cell to cell and generation to generation in the process of cell division.

✓ **Concept Check**

3. **Identify** the enzyme that synthesizes RNA from DNA.

4. **Describe** the consequence for a growing RNA transcript if a nucleotide with a 3′-H atom is incorporated rather than a 3′-OH group.

33.3 The primary transcript is processed to make mature mRNA

The RNA transcript that comes off the template DNA strand is known as the **primary transcript,** and it contains the complement of every base that was transcribed from the DNA template strand. For protein-coding genes, this means that the primary transcript includes the information needed to direct the ribosome to produce the protein corresponding to the gene. The RNA molecule that combines with the ribosome to direct protein synthesis is known as **messenger RNA (mRNA)** because it carries the genetic "message" from the DNA to the ribosome. Because mRNA is able to bind to ribosomes, it is sometimes called mature mRNA. In some cases, the primary transcript undergoes a series of chemical modifications that convert the primary transcript into mRNA, a process known as **RNA processing.** As we will see in this section, there is a major difference between prokaryotes and eukaryotes in how the primary transcript is processed to become mRNA. We will also see that some transcripts are not processed at all, but rather have functions of their own.

Prokaryotes

In prokaryotes, RNA processing does not occur. As a result, the RNA that is transcribed from DNA is mRNA and is able to bind ribosomes, as shown on the left in **FIGURE 33.5**. Even as the 3′ end of the transcript is still being synthesized, ribosomes bind to the 5′ end of the transcript and begin the process of protein synthesis, illustrated in **FIGURE 33.6**. Recall from Module 31 that the process of synthesizing proteins from mRNA is called translation. This intimate connection between transcription and translation can take place because prokaryotes have no nucleus to spatially separate transcription from translation. Thus, the two processes occur together in the same place and at the same time.

Primary transcripts for protein-coding genes in prokaryotes have another feature not shared with those in eukaryotes: they often contain the genetic information for the synthesis of two or more different proteins. Usually these proteins code for successive steps in the biochemical reactions that produce small molecules needed for growth, or for successive steps needed to break down a small molecule used for nutrients or energy.

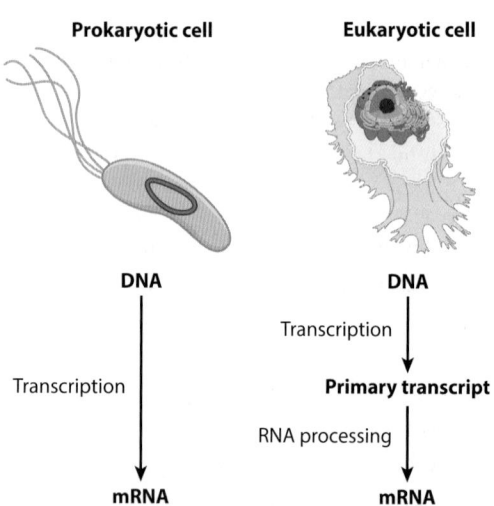

FIGURE 33.5 Production of mRNA in prokaryotes and eukaryotes

In prokaryotes, the RNA that is transcribed from DNA can bind ribosomes and be translated, so it is called mRNA. In eukaryotes, the RNA that is transcribed from DNA is called the primary transcript. The primary transcript is processed to become mRNA, which can bind to ribosomes and be translated.

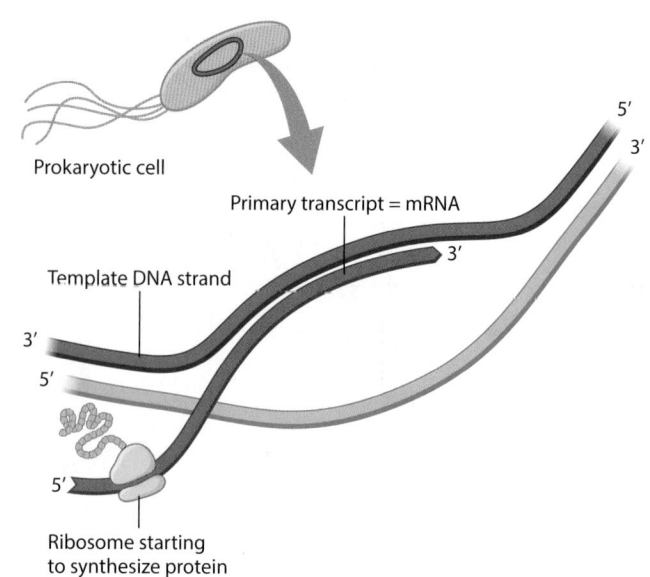

FIGURE 33.6 Transcription and translation in prokaryotes

In prokaryotes, the primary transcript is not processed, so it can serve as mRNA immediately. The mRNA associates with a ribosome, as shown, and begins the process of translation in which a protein is synthesized. As a result, transcription and translation occur together, with translation beginning even before transcription is complete.

Eukaryotes

In eukaryotes, the nuclear envelope is a barrier between the processes of transcription and translation. Transcription takes place in the nucleus, and translation takes place in the cytoplasm. Before the primary transcript is exported out of the nucleus, it undergoes a series of chemical modifications to become mRNA, which can then be translated by the ribosome, as shown on the right in Figure 33.5.

RNA processing consists of three main types of chemical modifications, all of which are illustrated in **FIGURE 33.7**. First, the 5′ end of the primary transcript is modified by the addition of a special nucleotide attached in an unusual linkage. This addition, which is called the **GTP cap** or 5′ cap, consists of a modified nucleotide called 7-methylguanosine. An enzyme attaches the modified nucleotide to the 5′end of the primary transcript. In a typical linkage between two nucleotides, the phosphodiester bond forms between the 3′-OH group of one nucleotide and the 5′-carbon of the next nucleotide, but here the 5′-carbon links with the 5′-carbon of the two nucleotides. The GTP cap is essential for translation because in eukaryotes, the ribosome recognizes an mRNA by its GTP cap. Without the cap, the ribosome would not attach to the mRNA and translation would not occur.

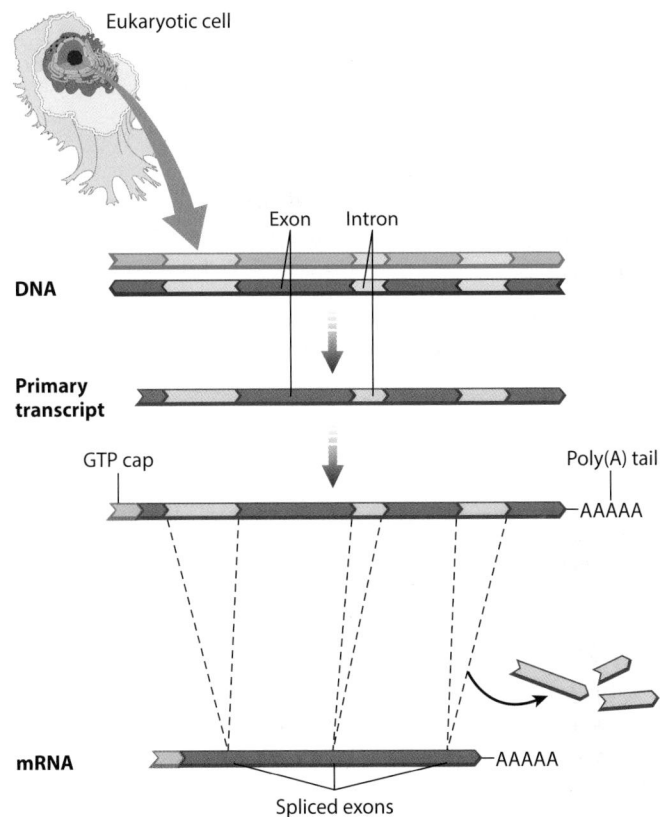

Eukaryotic cell

Exon   Intron

DNA

Primary transcript

GTP cap   Poly(A) tail

—AAAAA

mRNA   —AAAAA

Spliced exons

FIGURE 33.7 RNA processing in eukaryotes

The primary transcript in eukaryotes is modified by the addition of a GTP cap on the 5′ end; the addition of a poly(A) tail to the 3′ end; and the joining, or splicing, of exons, and removal of introns, which leaves only exons in mRNA.

The second modification is the addition of a string of about 250 consecutive adenine nucleotides to the 3′ end, forming a **poly(A) tail,** also shown in Figure 33.7. This tail plays an important role in the export of mRNA into the cytoplasm. In addition, both the GTP cap and poly(A) tail help to stabilize the RNA transcript. Single-stranded nucleic acids can be unstable and are even susceptible to enzymes that break them down. In eukaryotes, the GTP cap and poly(A) tail protect the two ends of the transcript and increase the stability of the RNA transcript until it is translated in the cytoplasm.

The third modification involves removal of segments of the primary transcript. Not every stretch of the RNA transcript ends up being translated into protein. Transcripts in eukaryotes often contain regions called **exons,** which are expressed as proteins, and regions called **introns,** which are interspersed with exons and not expressed as proteins. Introns are not expressed because they are removed from the primary transcript, shown in Figure 33.7. The process by which exons are stitched together and introns are removed is known as **RNA splicing.**

The presence of introns has important implications for gene expression. Approximately 90% of all human genes contain at least one intron. Although most genes contain 6 to 9 introns, the largest number is 147 introns, found in a muscle gene. Most introns are just a few thousand nucleotides in length, but roughly 10% are longer than 10,000 nucleotides. The presence of multiple introns in most genes allows for a process known as **alternative splicing,** in which primary transcripts from the same gene can be spliced in different ways to yield different mRNAs and, therefore, different protein products, shown in **FIGURE 33.8** on page 466. In this figure, the primary transcript has 4 exons labeled E1, E2, E3, and E4. Alternative splicing produces two different mRNAs, one with exons E1, E2, and E4 and the other with exons E1, E2, and E3. The result is two different proteins with slightly different functions.

More than 80% of human genes are alternatively spliced. In most cases, the alternatively spliced forms differ in whether a particular exon is or is not removed from the primary transcript along with its flanking introns. Alternative splicing allows the same transcript to be processed in diverse ways to produce mRNA molecules with different combinations of exons coding for different proteins. As a result, many different proteins can be made from the same gene and primary transcript. During evolution, alternative splicing allowed for the mixing and matching of different exons, producing different proteins with new or varied functions, all from a single gene.

PREP FOR THE AP® EXAM

AP® EXAM TIP

When explaining how the primary transcript is processed in eukaryotes to become mRNA, make sure you can identify the functions of the GTP cap and poly(A) tail. Your description should also include the details of how introns are removed from the primary transcript, and how RNA splicing allows multiple proteins to be synthesized from the same primary transcript.

Noncoding RNAs

Not all primary transcripts are processed into mRNA. Some RNA transcripts are not translated into proteins, but instead have functions of their own. These are called noncoding RNAs, and were briefly introduced in Module 31.

There are many different types of noncoding RNAs. For example, **ribosomal RNA (rRNA)** makes up the bulk of ribosomes and is essential in translation. In eukaryotic cells, the genes and transcripts for rRNA are concentrated in the nucleolus, a distinct, dense, non–membrane-bound spherical structure observed within the nucleus. **Transfer RNA (tRNA)** carries individual amino acids for use in translation.

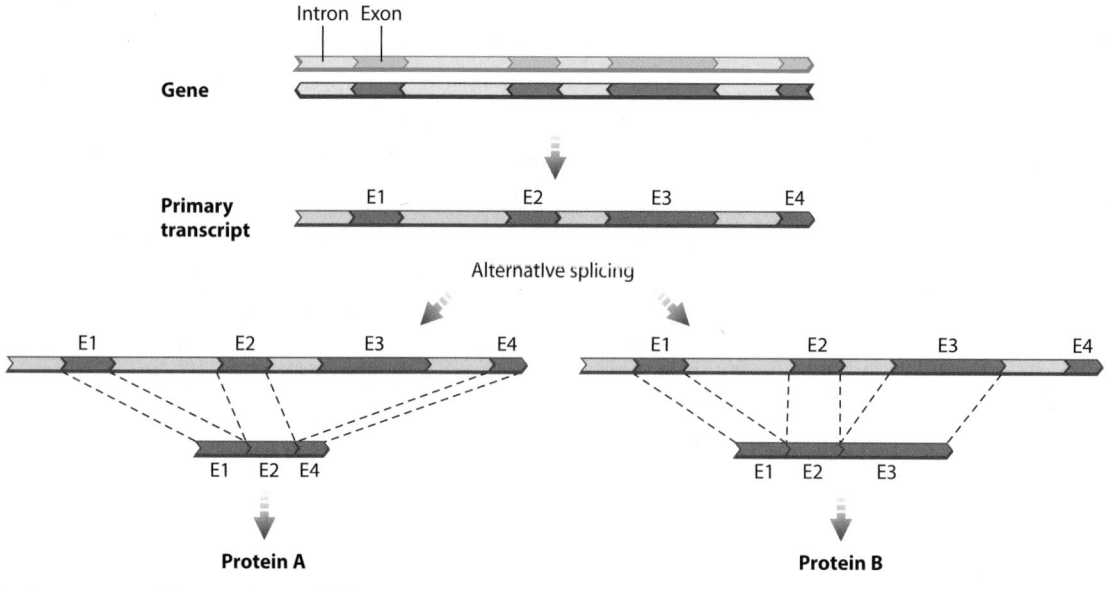

Intron Exon

Gene

Primary transcript E1 E2 E3 E4

Alternative splicing

E1 E2 E3 E4   E1 E2 E3 E4

E1 E2 E4   E1 E2 E3

Protein A   **Protein B**

FIGURE 33.8 Alternative splicing

A single primary transcript can be spliced in different ways to produce different proteins. E1, E2, E3, and E4 are exons present in the primary transcript. One mRNA is spliced to bring exons E1, E2, and E4 together, which is translated to protein A. Another mRNA is spliced in a different way from the same primary transcript to bring exons E1, E2, and E3 together, which is translated to protein B. In this way, one primary transcript can produce different proteins.

There are also many types of small RNA molecules that play important roles in regulating the expression of genes.

By far, the most abundant transcripts in mammalian cells are those for ribosomal RNA and transfer RNA. In a typical mammalian cell, approximately 80% of the RNA consists of ribosomal RNA, and another approximately 10% consists of transfer RNA. Why are these types of RNA so abundant? The answer is that they are needed in large amounts to synthesize the proteins encoded in the messenger RNA, which is the subject of the next module.

✓ Concept Check

5 . **Identify** three types of modifications that occur during RNA processing in eukaryotic cells.

6. **Describe** the function of the GTP cap and poly(A) tail.

7. **Describe** what occurs during RNA splicing and its significance.

8. **Identify** three types of noncoding RNAs.

Module 33 Summary

REVISIT THE BIG IDEAS

INFORMATION STORAGE AND TRANSMISSION: Describe how transcription followed by the processing of RNA helps ensure transmission of genetic information from DNA to ribosomes.

LG 33.1 During transcription, DNA is used as a template to guide the synthesis of RNA.

- RNA is synthesized from one of the two strands of DNA, called the template strand. Page 460

- RNA synthesis starts at a promoter and ends at a terminator in the DNA template. Page 460

LG 33.2 RNA polymerase adds successive nucleotides to the 3′ end of the transcript.

- RNA is synthesized by RNA polymerase in a 5′-to-3′ direction. Page 462
- RNA polymerase separates the two DNA strands, allows an RNA–DNA hybrid molecule to form, elongates the transcript, releases the transcript, and restores the DNA molecule. Page 462

LG 33.3 The primary transcript is processed to make mature mRNA.

- In prokaryotes, the primary transcript is immediately translated into protein. Page 464

- In eukaryotes, transcription and translation are separated in time and space, with transcription occurring in the nucleus and then translation in the cytoplasm. Page 464
- In eukaryotes, there are three major types of modification to the primary transcript—the addition of a GTP cap, the addition of a poly(A) tail, and RNA splicing. Page 465
- Alternative splicing is a process in which primary transcripts from the same gene are spliced in different ways to yield different protein products. Page 465
- Some RNAs, called noncoding RNAs, do not code for proteins, but instead have functions of their own. Page 465

Key Terms

Template strand	Messenger RNA (mRNA)	Intron
RNA polymerase	RNA processing	RNA splicing
Nontemplate strand	GTP cap	Alternative splicing
Promoter	Poly(A) tail	Ribosomal RNA (rRNA)
Primary transcript	Exon	Transfer RNA (tRNA)

Review Questions

1. An RNA transcript is synthesized in which direction?
   - (A) N terminus to C terminus
   - (B) C terminus to N terminus
   - (C) 5′-to-3′
   - (D) 3′-to-5′

2. Transcription starts at a
   - (A) promoter.
   - (B) GTP cap.
   - (C) terminator.
   - (D) 3′ end.

3. Transcription ends at a
   - (A) promoter.
   - (B) 3′ cap.
   - (C) terminator.
   - (D) 5′ end.

4. A transcribed region of DNA has a 5′-to-3′ sequence GCGAC. The 5′-to-3′ sequence of an RNA transcribed from this DNA is
   - (A) 5′-GCGAC-3′
   - (B) 5′-CAGCG-3′
   - (C) 5′-CGCUG-3′
   - (D) 5′-GUCGC-3′

5. Alternative splicing means that
   - (A) alternating introns are removed.
   - (B) some transcripts are spliced while others are not.
   - (C) some transcripts are spliced correctly and others incorrectly.
   - (D) different spliced forms contain different combinations of exons.

6. In eukaryotes, messenger RNA consists of the
   (A) primary RNA transcript.
   (B) primary RNA transcript with cap added.
   (C) primary RNA transcript with cap added and introns removed.
   (D) primary RNA transcript with cap added, introns removed, and poly(A) tail added.

7. Which type of RNA carries information for making a protein?
   (A) Transfer RNA
   (B) Messenger RNA
   (C) Small noncoding RNA
   (D) Ribosomal RNA

Module 33
AP® Practice Questions

PREP FOR THE AP® EXAM

Section 1: Multiple-Choice Questions

Choose the best answer for questions 1–4.

1. Which correctly describes the flow of information in a eukaryotic cell?
   (A) tRNA → ribosome → protein
   (B) DNA → tRNA → protein
   (C) Protein → DNA → mRNA
   (D) DNA → mRNA → protein

2. A portion of the template strand of a DNA molecule is illustrated below. Identify the RNA sequence that would be transcribed from this DNA sequence:

   5′-GAG CTG AGT CAG GGA-3′

   (A) 5′-CUC GAC UCA GUC CCU-3′
   (B) 5′-UCC CUG ACU CAG CUC-3′
   (C) 3′-CUC GAC UCA GUC CCU-5′
   (D) 3′-UCC CUG ACU CAG CUC-5′

3. The *white* gene of the fruit fly *Drosophila melanogaster* is 5868 base pairs (bp) long. However, the *white* mRNA is 2352 bases. Which of the following best describes why an mRNA transcript can be shorter than the gene from which it was transcribed?
   (A) RNA transcripts are single stranded and therefore smaller than DNA genes.
   (B) The nontemplate strand of DNA contains additional sequences not found on the template strand.
   (C) The gene contains segments of sequence that have been removed from the corresponding mRNA.
   (D) RNA nucleotides are smaller than DNA nucleotides.

4. Which statement explains why prokaryotes can begin translation of an RNA molecule while it is still being transcribed, but eukaryotes cannot?
   (A) Eukaryotic RNA cannot be translated unless its introns have been removed, but prokaryotic RNA is spliced as translation occurs.
   (B) Prokaryotes add a GTP cap to mRNA as soon as transcription begins, but eukaryotes add the GTP cap only after transcription is complete.
   (C) Ribosomes and DNA are found in the same part of the cell in prokaryotes but are in different parts of the cell in eukaryotes.
   (D) Eukaryotes rely on a poly(A) tail to initiate translation but prokaryotes do not use poly(A) tails for translation.

Section 2: Free-Response Question

Write your answer to each part clearly. Support your answers with relevant information and examples. Where calculations are required, show your work.

A team of scientists is studying a gene in crickets that has four exons. The primary transcript can be alternatively spliced to produce three different mRNAs. A schematic of the gene, showing the sizes of the four exons, is shown in part a below. The exons are represented by rectangles. The size of each version is shown in the table. The relative abundance of the three mRNAs changes as the cricket develops, beginning when the egg that will become the cricket is laid, as shown in part b.

a.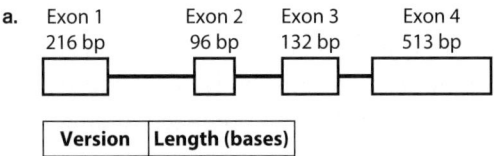

Version	Length (bases)
mRNA 1	861
mRNA 2	825
mRNA 3	729

b.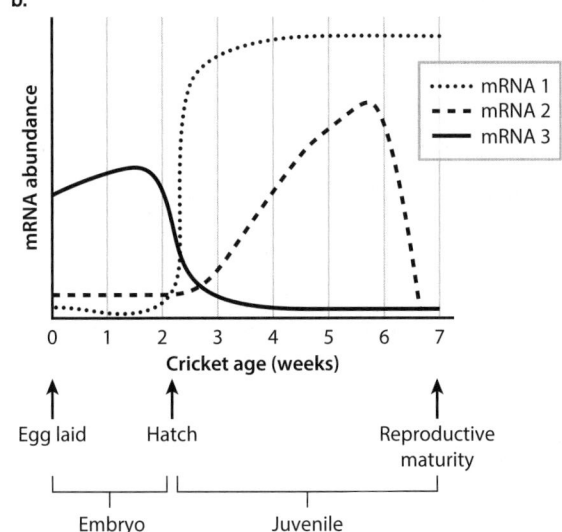

(a) **Identify** the exons that are included in mRNA 2.

(b) **Identify** the version of mRNA that is most abundant during the egg stage.

(c) **Describe** how alternative splicing allows multiple different forms of a protein to be produced from the same gene.

(d) A scientist claims that high expression of mRNA 2 is required for crickets to successfully reproduce. Use the data provided to **support or refute this claim.**

Module 34

Translation

LEARNING GOALS ▶**LG 34.1** The sequence of bases in mRNA specifies the order of amino acids in a protein.
▶**LG 34.2** The genetic code specifies codons in mRNA and their corresponding amino acids.
▶**LG 34.3** Translation consists of initiation, elongation, and termination.

In the last module, we discussed how information in DNA is transmitted to RNA in the process of transcription. Specifically, the sequence of bases along part of a DNA strand is used as a template for the synthesis of the complementary sequence of bases in a molecule of RNA. This is just the first step in the flow of information in cells. The information in RNA is then transmitted to protein in the process of translation. During translation, the sequence of bases in an RNA molecule known as messenger RNA (mRNA) is used to specify the sequence of amino acids in a newly synthesized protein.

Recall from Module 4 that the three-dimensional structure of a protein determines what it can do and how it works. The immense diversity in the tertiary and quaternary structures among proteins explains their wide range of functions in cellular processes. Yet it is the sequence of amino acids along a polypeptide chain—a protein's primary structure—that governs how the molecule folds into a stable three-dimensional configuration. This sequence is ultimately specified by the sequence of nucleotides in DNA.

Information in a cell flows from DNA to RNA to protein according to the central dogma of molecular biology.

In this module, we will focus on the second step of this process—how information flows from mRNA to protein, which takes place during translation. We will begin by describing how the sequence of nucleotides in mRNA is translated to the sequence of amino acids in proteins. Then we will examine the relationship between the sequence of nucleotides in mRNA and the sequence of amino acids in proteins. Finally, we will examine each of the steps in translation. By the end of this module, we will have followed the transfer of information from DNA to RNA to protein, and therefore have a better understanding of how cells retrieve and use the information encoded in DNA.

PREP FOR THE AP® EXAM

FOCUS ON THE BIG IDEAS

INFORMATION STORAGE AND TRANSMISSION: Focus on how the process of translation allows the information in mRNA to be transferred to proteins, which in turn influences the phenotype of a cell or organism.

34.1 The sequence of bases in mRNA specifies the order of amino acids in a protein

Translation is the process by which a molecule of mRNA is used to guide the synthesis of a protein. Specifically, specialized molecular structures within the cell read the mRNA molecule to determine which amino acids to use to build a protein, as shown in **FIGURE 34.1**. Note that the mRNA is along the bottom of the figure and the newly synthesized protein is along the top. Between the mRNA and the protein are two key cellular components that translate information from mRNA to protein. These components are the ribosome and transfer RNA (tRNA). In this section, we will

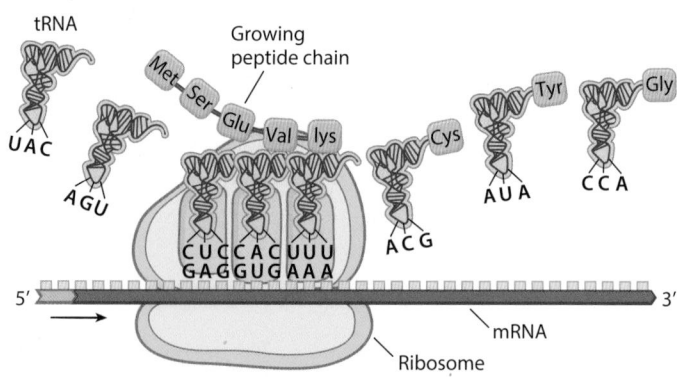

FIGURE 34.1 An overview of translation

During translation, an mRNA molecule is read by ribosomes and tRNAs to direct the synthesis of a protein. The order of nucleotides in the mRNA determines the order of amino acids in the protein.

look at the roles of ribosome and transfer RNAs (tRNAs) in translation.

Ribosomes

As we learned in Module 6, ribosomes are the site of protein synthesis in all cells. Ribosomes are structures made up of **ribosomal RNA (rRNA)** and protein. They organize the process of translation by binding to and "reading" the sequence of nucleotides in mRNA. They then catalyze the reaction that joins the sequence of amino acids to make a protein chain. In prokaryotes, translation occurs as soon as the mRNA comes off the DNA template while transcription continues. In eukaryotes, the processes of transcription and translation are physically separated: transcription takes place in the nucleus, and translation takes place in the cytoplasm. In both eukaryotes and prokaryotes, the ribosome consists of a small subunit and a large subunit, shown in **FIGURE 34.2**. Each of these subunits is composed of 1 to 3 types of ribosomal RNAs and 20 to 50 types of ribosomal proteins. Eukaryotic ribosomes are larger than prokaryotic ribosomes.

One of the key roles of the ribosome is to ensure that when the mRNA is in place on the ribosome the sequence in the mRNA coding for amino acids is read in successive, non-overlapping groups of three nucleotides, much as you would read the following sentence:

THEBIGBOYSAWTHEMANRUN

Each non-overlapping group of three adjacent nucleotides (like THE or BIG or BOY in our sentence analogy) constitutes a **codon**. A codon in the mRNA codes for a single amino acid in the polypeptide chain.

In this example, it is clear that the sentence begins with THE. However, in a long linear mRNA molecule, the ribosome could begin at any nucleotide. As an analogy, if we knew that the letters THE were the start of the phrase, then we would know immediately how to read

ZWTHEBIGBOYSAWTHEMANRUN

However, without knowing where to start reading this string of letters, we could find three ways to break the sentence into three-letter words:

ZWT HEB IGB OYS AWT HEM ANR UN

Z WTH EBI GBO YSA WTH EMA NRU N

ZW THE BIG BOY SAW THE MAN RUN

The different ways of parsing the string of letters into three-letter words are known as reading frames. The protein-coding sequence in an mRNA consists of its sequence of bases, and just as in our sentence analogy, it can be translated into the correct protein only if it is translated in the proper reading frame.

Transfer RNA

In addition to the ribosome, tRNA plays a key role in translation, shown in Figure 34.1. The tRNA molecules translate each codon in the mRNA into one amino acid. They are able to carry out this function because one end of the tRNA binds to the mRNA, and the other end of the tRNA is attached to an amino acid. The ribosome holds in place all of the various components required for translation and catalyzes the reaction that connects one amino acid to the next, and the tRNAs act as bridges from the mRNA sequence to the protein sequence.

A transfer RNA is a small RNA molecule consisting of 70 to 90 nucleotides. As you can see in **FIGURE 34.3a**, it has a characteristic structure in which some of the bases pair with each other to form a cloverleaf. The letters in the figure indicate the bases common to all tRNA molecules. The actual structure, though, is more like that in Figure 34.3b. Three bases at the bottom of the structure make up the **anticodon**, which consists of three nucleotides that undergo base pairing with the corresponding codon in mRNA.

FIGURE 34.4 on page 472 shows the base pairing between the mRNA and the tRNA. The first (5′) base in the codon in mRNA pairs with the last (3′) base in the anticodon because, as noted in Module 33, nucleic acid strands that undergo base pairing must be antiparallel. The other end of the tRNA carries a specific amino acid that corresponds to that codon. In the figure, the tRNA molecule carries the amino acid Met (methionine). In short, the tRNA molecule brings the correct amino acid to the place specified by the codon on the mRNA.

The amino acids are attached to tRNAs by enzymes called aminoacyl tRNA synthetases. These enzymes connect

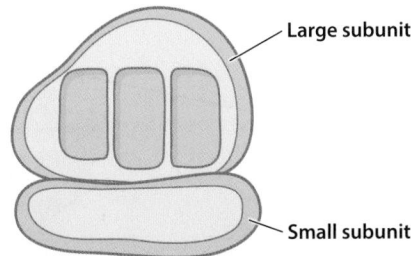

FIGURE 34.2 Ribosome

This is a simplified diagram of a ribosome in eukaryotes and prokaryotes. Ribosomes have a large subunit and a small subunit. The large subunit binds tRNA molecules.

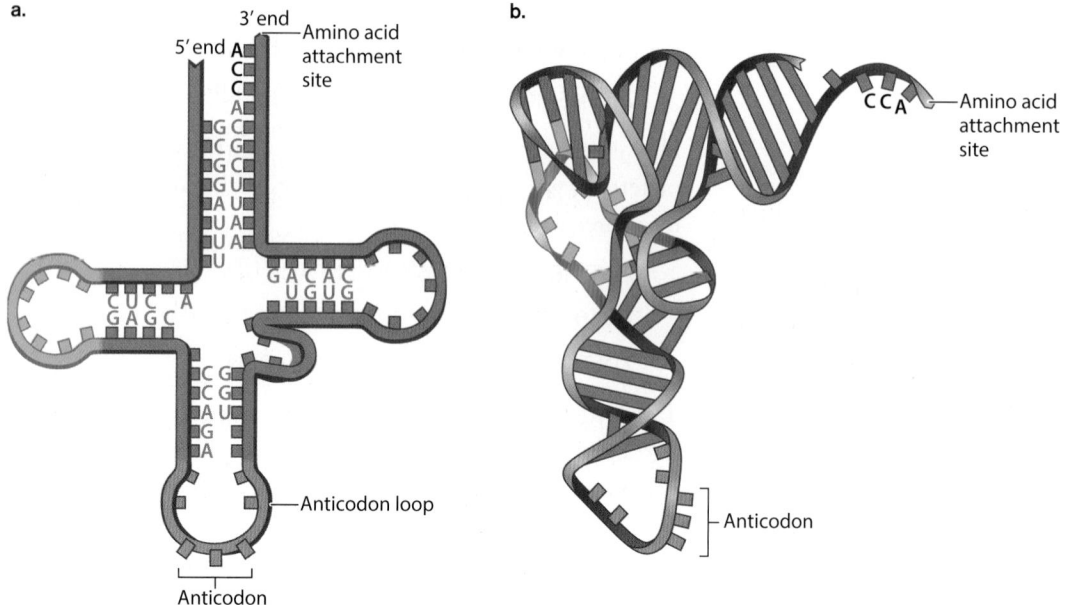

FIGURE 34.3 Transfer RNA (tRNA)

The structure of a tRNA is depicted in two ways. (a) This cloverleaf configuration shows the specific base pairing between different regions of the molecule. (b) The three-dimensional form is more realistic. All tRNAs have an anticodon at one end that pairs with the codon, and a sequence at the other end that binds a specific amino acid.

specific amino acids to specific tRNA molecules, as shown in **FIGURE 34.5**. Therefore, these enzymes are directly responsible for actually translating the codon sequence in a nucleic acid to a specific amino acid in a polypeptide chain. Most organisms have one aminoacyl tRNA synthetase for each amino acid. So, with 20 different amino acids, there are 20 different aminoacyl tRNA synthetases.

✓ Concept Check

1. **Describe** the role of ribosomes in translation.

2. **Describe** the role of tRNAs in translation.

3. **Describe** the relationship between a codon in mRNA and an amino acid in a protein.

4. **Describe** the relationship between a codon in mRNA and an anticodon in tRNA.

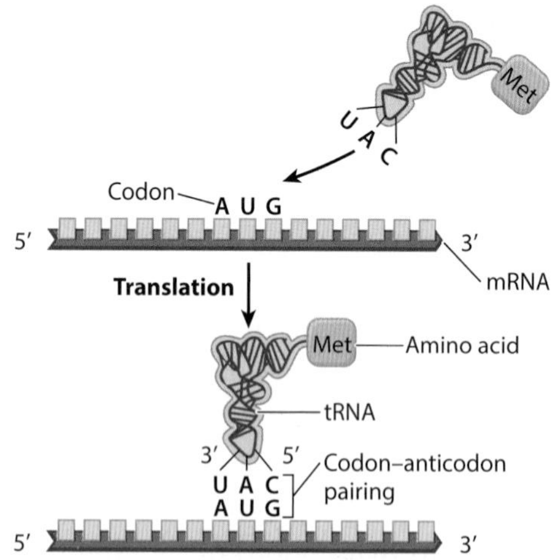

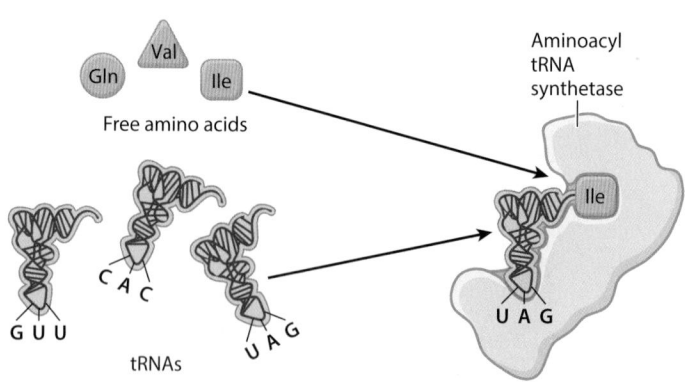

FIGURE 34.4 Codon–anticodon base pairing

Base pairing between the codon in mRNA and the anticodon in tRNA brings in the amino acid methionine (Met). The first base in the codon (5′ end) pairs with the last base (3′ end) in the anticodon.

FIGURE 34.5 Aminoacyl tRNA synthetase

Aminoacyl tRNA synthetases attach specific amino acids to specific tRNAs. In this way, they are responsible for translating the codon sequence in an mRNA into an amino acid sequence in a protein.

34.2 The genetic code specifies codons in mRNA and their corresponding amino acids

As we just discussed, three nucleotides in mRNA correspond to a specific amino acid in a protein. The relationship between these three nucleotides in a codon, and the amino acid that it specifies, is called the **genetic code.** In this section, we will introduce the genetic code and what it tells us about the relationships of all living organisms.

Genetic Code

The genetic code is shown in **FIGURE 34.6**. It has 61 codons that specify 20 amino acids. For example, from the table, you can see that the codon UUU corresponds to the amino acid phenylalanine (abbreviated Phe). Because there are more codons than amino acids, many amino acids are specified by more than one codon. For example, both UUU and UUC correspond to the amino acid phenylalanine. As a result, the genetic code is described as redundant. This code is sometimes called the "standard" genetic code because, while it is used by almost all cells, some minor differences are found in a few organisms as well as in mitochondria.

The codon at which translation begins is called the initiation codon, which is the THE in the example in the last section. It is coded by AUG, which specifies the amino acid methionine (Met). The polypeptide is synthesized from the amino end to the carboxyl end, so Met forms the amino end of any polypeptide being synthesized. For proteins with a methionine at their amino ends, the starting methionine is retained in the finished protein; for proteins without a methionine at their amino ends, the methionine is cleaved off by an enzyme after synthesis is complete. The AUG codon also specifies the incorporation of methionine at internal sites within the polypeptide chain.

In Figure 34.6, you can also see that three codons— UAA, UAG, and UGA—do not specify amino acids. These codons are called stop codons, which are like punctuation marks in a sentence. They are sometimes called termination codons or nonsense codons. Stop codons do not code for any amino acid but signal where translation terminates and the protein is released from the ribosome.

The standard genetic code was deciphered in the 1960s by a combination of techniques, but among the most ingenious were the chemical methods for making synthetic RNAs of known sequence developed by American biochemist Har Gobind Khorana and his colleagues. This experiment is illustrated on page 474 in "Practicing Science 34.1: How was the genetic code deciphered?"

First position (5′ end)	Second position				Third position (3′ end)
	U	C	A	G	
U	UUU Phe UUC Phe UUA Leu UUG Leu	UCU Ser UCC Ser UCA Ser UCG Ser	UAU Tyr UAC Tyr **UAA Stop** **UAG Stop**	UGU Cys UGC Cys **UGA Stop** UGG Trp	U C A G
C	CUU Leu CUC Leu CUA Leu CUG Leu	CCU Pro CCC Pro CCA Pro CCG Pro	CAU His CAC His CAA Gln CAG Gln	CGU Arg CGC Arg CGA Arg CGG Arg	U C A G
A	AUU Ile AUC Ile AUA Ile **AUG Met/Start**	ACU Thr ACC Thr ACA Thr ACG Thr	AAU Asn AAC Asn AAA Lys AAG Lys	AGU Ser AGC Ser AGA Arg AGG Arg	U C A G
G	GUU Val GUC Val GUA Val GUG Val	GCU Ala GCC Ala GCA Ala GCG Ala	GAU Asp GAC Asp GAA Glu GAG Glu	GGU Gly GGC Gly GGA Gly GGG Gly	U C A G

FIGURE 34.6 The genetic code

The genetic code specifies the relationship between three nucleotides in mRNA, called codons, and the amino acids in proteins that the codons specify.

How was the genetic code deciphered?

Background The genetic code is the correspondence between three-letter nucleotide codons in RNA and amino acids in a protein. American biochemist Har Gobind Khorana performed key experiments that helped to crack the code. For this work, he shared the Nobel Prize in Physiology or Medicine in 1968 with Robert W. Holley and Marshall E. Nirenberg.

Method Khorana and his group made RNAs of known sequence. They then added these synthetic RNAs to a solution containing all of the other components needed for translation. By adjusting the experimental conditions, the researchers could get the ribosome to initiate synthesis with any codon, even if that codon was not AUG.

Experiment 1 and Results When a synthetic poly(U), which is an mRNA molecule made up entirely of the base U (UUU . . .), was used as the mRNA, the resulting polypeptide was polyphenylalanine (Phe–Phe–Phe . . .), as shown below.

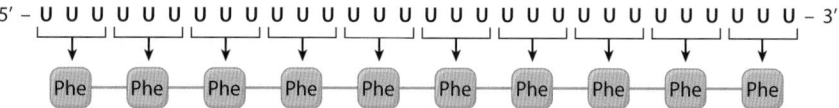

Conclusion The codon UUU corresponds to Phe. The poly(U) mRNA can be translated in three possible reading frames, depending on which U is the 5′ end of the start codon. In each of those reading frames, all the codons are UUU.

Experiment 2 and Results When a synthetic mRNA with alternating U and C was used, the resulting polypeptide had alternating serine (Ser) and leucine (Leu), as shown below.

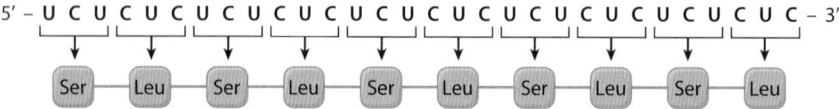

Conclusion Here again there are three reading frames, but each has alternating UCU and CUC codons. The researchers could not deduce from this result whether UCU corresponds to Ser and CUC to Leu, or the other way around; the correct assignment, shown in the figure, came from experiments using other synthetic mRNA molecules.

Experiment 3 and Results When a synthetic mRNA with repeating UCA was used, three different polypeptides were produced— polyserine (Ser), polyhistidine (His), and polyisoleucine (Ile)—because of the three different reading frames, as shown below.

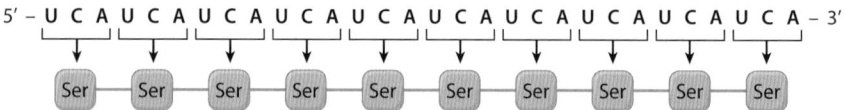

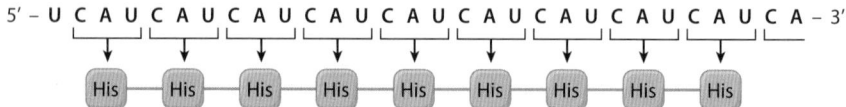

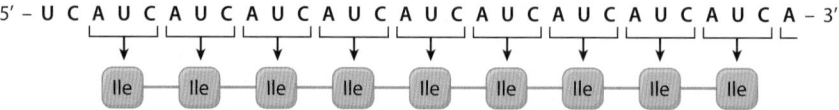

Conclusion The results do not reveal which of the three reading frames corresponds to which amino acid. The correct assignment, shown in the figure, was sorted out by studies of other synthetic polymers.

SOURCE
Khorana, H. G. 1972. "Nucleic Acid Synthesis in the Study of the Genetic Code." In *Nobel Lectures, Physiology or Medicine 1963–1970.* Amsterdam: Elsevier.

AP® PRACTICE QUESTION

(a) **Identify** the following:
    1. The question being investigated   3. The independent variable
    2. The hypothesis for each of the three experiments   4. The dependent variable

(b) **Describe** the conclusion for each experiment using a claim, evidence, reasoning format (CER).

AP® EXAM TIP

While you do not need to memorize the genetic code, you should be able to correctly identify which amino acids will go into a peptide based on a given mRNA sequence.

Common Ancestry

The genetic code is nearly universal. In other words, it is used, with a few rare exceptions, by all living organisms. As a result, the genetic code used by the bacterium *Escherichia coli* (*E. coli*) is the same genetic code used by humans. In both *E. coli* and humans, it is a triplet code, with three nucleotides specifying a particular amino acid; there are four nucleotides and 20 amino acids; and the relationship between codons and amino acids is the same, as indicated in Figure 34.6.

The near universality of the genetic code is one of many pieces of evidence that all living organisms evolved from a single common ancestor. There are other pieces of evidence as well, including fossils, traits of organisms, details of development, types of proteins and carbohydrates found in living organisms, and so on. But the fact that the genetic code is so widely shared across so many different types of organisms provides compelling evidence for a single common ancestor.

It is also worth noting that the genetic code has strong patterns. For example, we noted earlier that the genetic code is redundant. This redundancy results almost exclusively from the third codon position. For example, UUU and UUC both correspond to phenylalanine, as noted earlier. Furthermore, when one amino acid is specified by two codons, they differ either in whether the third position is a U or C, which are pyrimidines, or in whether the third position is an A or a G, which are purines. Again, the UUU and UUC codons for phenylalanine illustrate this point because the third position is a U or C. Finally, when one amino acid is specified by four codons, the identity of the third codon position does not matter; it could be U, C, A, or G, as can be seen for the four codons that specify the amino acid serine in Figure 34.6.

✓ Concept Check

5. **Identify** the three reading frames from a synthetic mRNA with the repeating sequence 5'-UUUGGGUUUGGGUUUGGG-3'.

6. Using the genetic code in Figure 34.6, **identify** the amino acid specified by UUU and the amino acid specified by GGG.

34.3 Translation consists of initiation, elongation, and termination

At this point, we have discussed the roles of ribosomes and tRNA in translation. We have also introduced the genetic code, showing how codons and amino acids relate to one another. Now we are in a position to look at how translation is actually carried out. Translation can be divided into three processes: *initiation, elongation,* and *termination*. During **translation initiation,** the initiator AUG codon is recognized and Met is established as the first amino acid in the new polypeptide chain. Next, during **translation elongation,** successive amino acids are added one by one to the growing chain. Finally, during **translation termination,** the addition of amino acids stops and the completed polypeptide chain is released from the ribosome. In this section, we will discuss these three steps in more detail.

Translation Initiation

We have seen that translation initiation begins with the start codon AUG. Figure 34.4 on page 472 shows how the codon AUG specifies the amino acid methionine (Met) by base pairing with the anticodon of a tRNA. As can also be seen in the figure, the AUG codon that initiated translation is preceded by a region in the mRNA that is not translated. The position of the initiator AUG codon in the mRNA establishes the reading frame that determines how the downstream codons (those following the AUG) are to be read.

Initiation of translation is shown in **FIGURE 34.7** on page 476. It requires a number of protein initiation factors, shown in purple, that bind to the mRNA. Some initiation factors bind to the small subunit of the ribosome, and other initiation factors bind to a tRNA with Met, shown in Figure 34.7. The initiation complex then moves along the mRNA until it encounters the first AUG codon. The position of this AUG establishes the translational reading frame. When the first

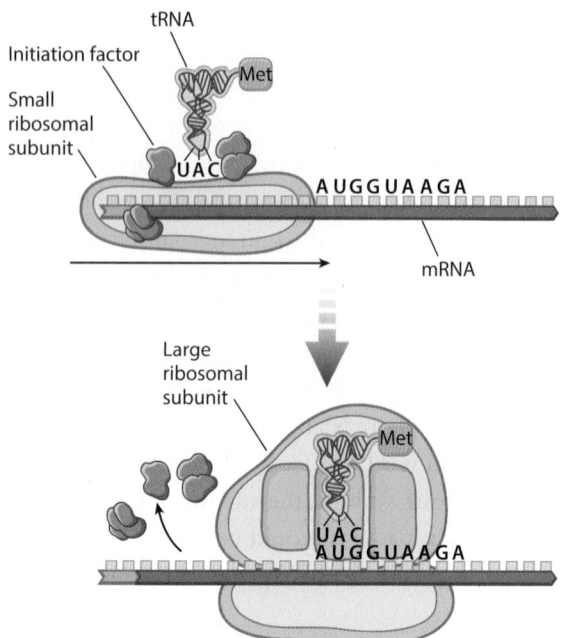

FIGURE 34.7 Translation initiation

Translation begins when initiation factors bind the small ribosomal subunit, mRNA, and tRNA carrying methionine. This complex moves along the mRNA until it encounters an AUG start codon. When the AUG start codon is reached, the large ribosomal subunit joins and the initiation factors are released.

AUG codon in mRNA is encountered, the large ribosomal subunit joins the complex. Then the initiation factors are released and the next tRNA is ready to join the ribosome.

Translation Elongation

During translation elongation, amino acids are added to the growing polypeptide chain. The process is illustrated in **FIGURE 34.8**. In the top panel of the figure, the tRNA attached to Met is bound to the ribosome at one site, and the next tRNA in line binds at a site adjacent to the first one. Once the next tRNA is in place, a dehydration synthesis reaction takes place in which a peptide bond is formed

FIGURE 34.8 Translation elongation

Translation elongation is the process by which amino acids are added one-by-one to the growing polypeptide chain. In the upper panel, a tRNA carrying an amino acid binds to the ribosome. In the second panel, a peptide bond is formed between the first two amino acids. In the third panel, the ribosome moves down one codon, the first tRNA is released from the ribosome, and a new tRNA binds to the ribosome. In the bottom panel, a peptide bond is formed between the next two amino acids.

between the two amino acids, shown in the second panel of Figure 34.8. An RNA in the large subunit is the catalyst for this reaction. In this way, a short polypeptide is formed. The ribosome then shifts one codon to the right, as shown in the third panel of Figure 34.8. With this move, the tRNA that originally was attached to Met is released from the ribosome. The movement of the ribosome provides a new site for the next tRNA in line to come into place.

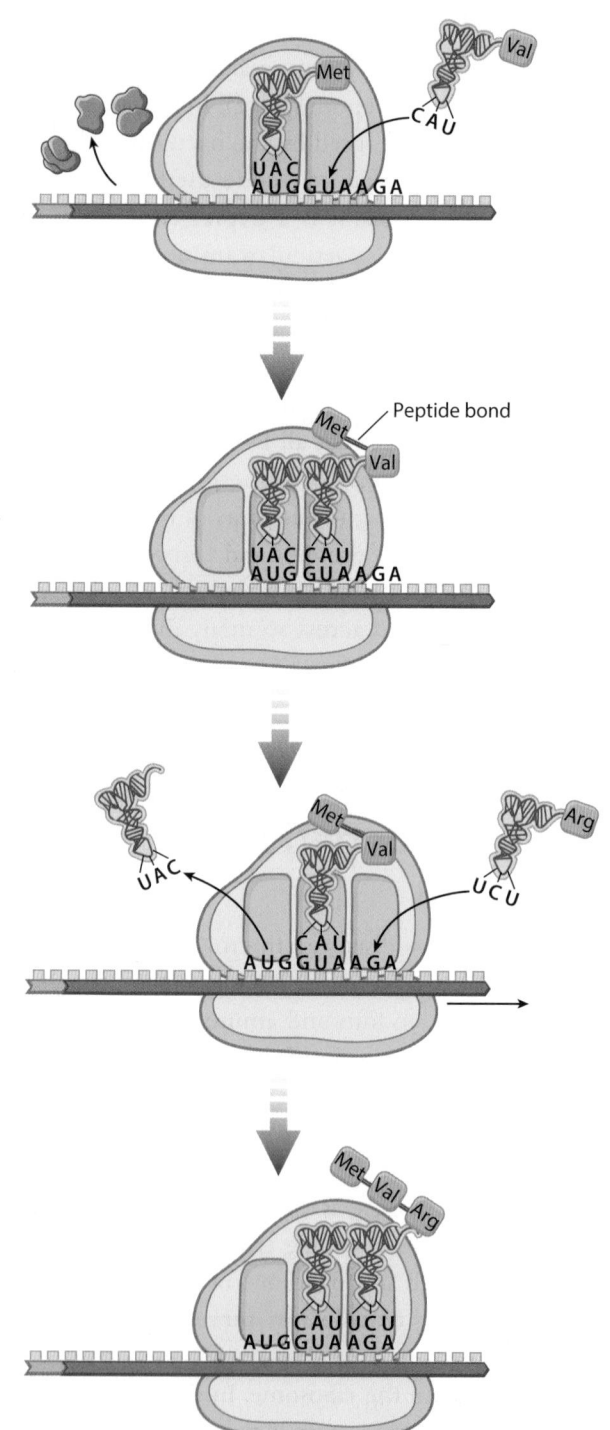

The steps shown in Figure 34.8—the binding of a new tRNA to the ribosome, the formation of a peptide bond between the amino acids, and the release of a tRNA—are then repeated over and over again in the process of elongation. Once the initial Met creates the amino end of a new polypeptide chain, the downstream codons are read one by one in non-overlapping groups of three bases (codons). At each step, the ribosome binds to a tRNA with an anticodon that can base pair with the codon. Then the amino acid on that tRNA is attached to the growing chain to become the new carboxyl end of the polypeptide chain.

As a result, the polypeptide chain grows in length through the addition of successive amino acids. Ribosome movement along the mRNA and formation of the peptide bonds require energy, which is obtained with the help of proteins called elongation factors. Elongation factors are bound to GTP molecules and break their high-energy bonds to provide energy for the elongation of the polypeptide.

Translation Termination

The elongation process shown in Figure 34.8 continues until the ribosome encounters one of the stop codons (UAA, UAG, or UGA). These codons signal termination of polypeptide synthesis. Termination takes place because the stop codons do not have corresponding tRNA molecules. Rather, when the ribosome encounters a stop codon, a protein release factor binds to the ribosome, as shown in **FIGURE 34.9**. The release factor causes the bond connecting the polypeptide to the tRNA to break, creating the carboxyl terminus of the polypeptide and completing the chain. Once the finished polypeptide is released, the small and large ribosomal subunits disassociate both from mRNA and from each other.

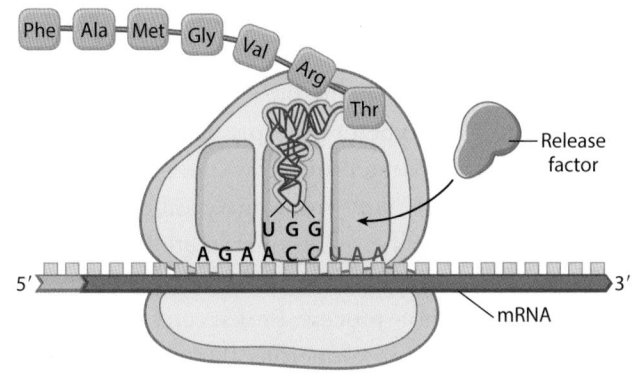

FIGURE 34.9 Translational termination

When the ribosome encounters a termination codon (UAA, UAG, or UGA, shown in red), a release factor binds the ribosome. Binding triggers release of the polypeptide chain from the final tRNA and dissociation of the ribosomal subunits.

Translation in Prokaryotes and Eukaryotes

Although elongation and termination are very similar in prokaryotes and in eukaryotes, translation initiation differs between the two. This difference has implications for the number of proteins that a single mRNA molecule can produce and is illustrated in **FIGURE 34.10**. In eukaryotes, the initiation complex forms at the GTP cap and scans along the mRNA until the first AUG is encountered, as shown in Figure 34.10a. In prokaryotes, the mRNA molecules have no GTP cap, as discussed in the previous module. Instead, the initiation complex is formed at one or more internal sequences present in the mRNA, as shown in Figure 34.10b. This internal sequence is followed by an AUG codon farther downstream that serves as the initiation codon for translation.

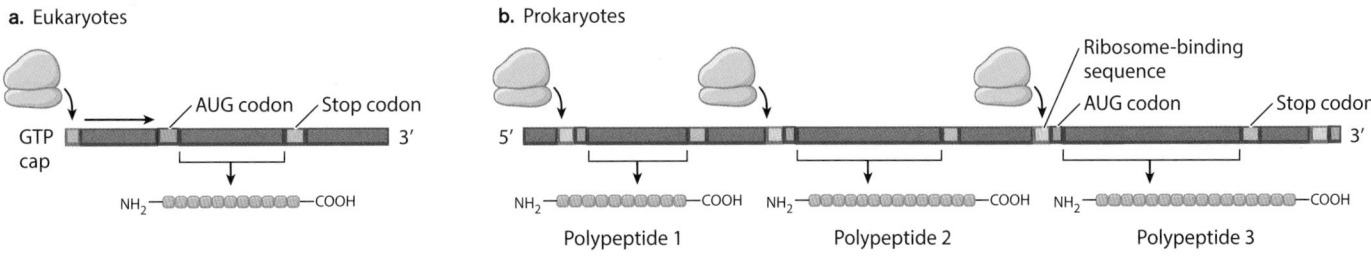

FIGURE 34.10 Translation in eukaryotes and prokaryotes

Translation is different in eukaryotes and prokaryotes. (a) In eukaryotes, the ribosome binds the GTP cap, with translation beginning when the first AUG codon is reached. As a result, each mRNA codes for a single polypeptide. (b) In prokaryotes, the ribosome binds at specific internal sequences, with translation beginning at the next downstream AUG codon. This organization, called an operon, allows a single mRNA to code for multiple polypeptides.

The ability to initiate translation internally allows prokaryotic mRNAs to code for more than one protein. For example, in Figure 34.10b, the mRNA codes for three different proteins. This type of gene organization, in which a group of genes are located in a row along the DNA and transcribed as a single unit from one promoter, is known as an **operon.** The genes in an operon are typically functionally related, which means that their products work together in a common process. Prokaryotes have many of their genes organized into operons. This organization has the advantage that it allows all the protein products to be expressed together whenever they are needed. Typically, the genes organized into operons are those whose products are needed either for successive steps in the synthesis of an essential small molecule, such as an amino acid, or for successive steps in the breakdown of a source of energy, such as the sugar lactose, which we will discuss in more detail in the next module.

Origin and Evolution of Translation

During transcription and translation, proteins and nucleic acids work together to convert the information stored in DNA into proteins. If we think about how such a system might have originated, however, we immediately confront a chicken-and-egg problem: cells require nucleic acids to make proteins, but proteins are required to make nucleic acids. Which came first?

Like DNA, RNA is able to store genetic information in its sequence of nucleotides. Like proteins, some RNA molecules can also act as catalysts, accelerating the rate of chemical reactions. Because of these features of RNA, scientists have suggested that early in the evolution of life on Earth, RNA carried out many of the essential functions that sustain life. Sometime later in evolutionary history, then, proteins were added to the mix. No one fully understands how

they were incorporated, but researchers are looking closely at tRNA, the molecule involved in the "translating" step of translation.

In modern cells, tRNA shuttles amino acids to the ribosome, but an innovative hypothesis suggests that in early life tRNA-like molecules might have served a different function. This hypothesis holds that the early precursors of the ribosome were RNA molecules that facilitated the replication of other RNAs, not proteins. Precursors to tRNA would have shuttled nucleotides to growing RNA strands. Researchers hypothesize that tRNAs bound to amino acids may have acted as simple catalysts, facilitating more accurate RNA synthesis. Through time, amino acids brought into close proximity in the process of building RNA molecules might have bound together to form polypeptide chains. From there, evolution might have favored the formation of polypeptides that enhanced replication of RNA molecules, bringing proteins into the chemistry of life.

All of the steps of gene expression, including transcription, RNA processing, and translation, are illustrated in **"Visual Synthesis 6.1: Gene Expression"** on page 482.

✓ Concept Check

7. **Describe** how the first amino acid is brought to the ribosome during translation initiation.

8. **Describe** how the second amino acid joins the growing polypeptide chain.

9. **Describe** what takes place during translation elongation.

10. **Describe** what takes place during translation termination.

Module 34 Summary

REVISIT THE BIG IDEAS

INFORMATION STORAGE AND TRANSMISSION: Explain how translation transmits information from mRNA to protein, resulting in gene expression.

LG 34.1 The sequence of bases in mRNA specifies the order of amino acids in a protein.

- Translation is the process by which the sequence of bases in messenger RNA specifies the order of successive amino acids in a newly synthesized protein. Page 470

- Translation requires many cellular components, including ribosomes and tRNAs. Page 470

- Ribosomes are composed of a small and a large subunit, each consisting of RNA and protein; the large subunit contains three tRNA-binding sites that play different roles in translation. Page 471
- An mRNA transcript of a gene has three possible reading frames composed of three-nucleotide codons. Page 471
- tRNAs have an anticodon that base pairs with the codon in the mRNA and carries a specific amino acid. Page 471
- Aminoacyl tRNA synthetases attach specific amino acids to tRNAs. Page 471

LG 34.2 The genetic code specifies codons in mRNA and their corresponding amino acids.

- The genetic code defines the relationship between the three-letter codons of nucleic acids and their corresponding amino acids. Page 473
- The genetic code was deciphered using synthetic RNA molecules. Page 473

- The genetic code is redundant, in that many amino acids are specified by more than one codon. Page 475

LG 34.3 Translation consists of initiation, elongation, and termination.

- In initiation, the initiator AUG codon is recognized and Met is established as the first amino acid. Page 475
- In elongation, successive amino acids are added one by one to the growing polypeptide chain. Page 476
- In termination, the addition of amino acids stops and the completed polypeptide chain is released from the ribosome. Page 477
- In eukaryotes, an mRNA encodes for a single polypeptide, whereas in prokaryotes, an mRNA can code for several polypeptides. Page 477
- Early in the evolution of life, RNA might have played both the roles of information molecule and chemical catalyst. Page 478

Key Terms

Codon
Anticodon
Genetic code

Translation initiation
Translation elongation

Translation termination
Operon

Review Questions

1. Which is specific to the synthesis of only one type of protein?
   (A) A ribosome
   (B) An aminoacyl tRNA synthetase
   (C) A messenger RNA
   (D) A transfer RNA

2. How many nucleotides make up a codon?
   (A) 1
   (B) 2
   (C) 3
   (D) 4

3. How many different types of aminoacyl tRNA synthetases are there?
   (A) 2
   (B) 4
   (C) 20
   (D) 64

4. The codon used to initiate protein synthesis is
   (A) AAA.
   (B) AUG.
   (C) GGG.
   (D) GUA.

5. A group of functionally related genes transcribed as a single transcriptional unit under the control of a single promoter is referred to as a(n)
   (A) operator.
   (B) mRNA.
   (C) tRNA.
   (D) operon.

6. Which type of protein interacts directly with a stop codon?

(A) Initiation factor

(B) Elongation factor

(C) Release factor

(D) Transcription factor

7. Which processes occur in the cytoplasm of eukaryotic cells?

(A) DNA replication and transcription

(B) Transcription and translation

(C) Translation only

(D) DNA replication only

Module 34
AP® Practice Questions

Section 1: Multiple-Choice Questions

Choose the best answer for questions 1–4.

1. Which of the following correctly describes the function of tRNA molecules?

(A) tRNA molecules transport mRNA molecules to the ribosome for translation.

(B) tRNA molecules transform mRNA into protein at the ribosome.

(C) tRNA molecules transfer amino acids to a growing polypeptide chain at the ribosome.

(D) tRNA molecules transmit the protein sequence to the ribosome.

2. The following DNA sequence represents a portion of the template strand of a protein-coding gene:

5'...GTG AGA TCG AGT CAG GCA...3'

Identify the change to the DNA sequence that is likely to have the largest effect on the function of the encoded protein. The alteration to the DNA is underlined in each response. You may refer to Figure 34.6 to see the genetic code.

(A) 5'...GTG AGA TCG AG<u>C</u> AGT CAG GCA...3'

(B) 5'...GTG AGA TC<u>A</u> AGT CAG GCA...3'

(C) 5'...GTG AGA <u>TG</u> AGT CAG GCA...3'

(D) 5'...GTG AGA TCG AGT C<u>GG</u> GCA...3'

3. The following DNA sequence represents a portion of the template strand of a protein-coding gene.

5'...GTG AGA TCG AGT CAG GCA...3'

Identify the change to the DNA sequence that is likely to have no effect on the function of the encoded protein. The alteration to the DNA is underlined in each response. You may refer to Figure 34.6 to see a table of the genetic code.

(A) 5'...GTG AGA TCG AG<u>C</u> AGT CAG GCA...3'

(B) 5'...GTG AGA TC<u>A</u> AGT CAG GCA...3'

(C) 5'...GTG AGA <u>TG</u> AGT CAG GCA...3'

(D) 5'...GTG AGA TCG AGT C<u>GG</u> GCA...3'

4. Which of the following attributes of the genetic code provides evidence that life on Earth evolved from a common ancestor?

(A) The genetic code is redundant, with multiple nucleotide sequences coding for the same amino acid.

(B) The genetic code is defined by three-nucleotide codons.

(C) Mutations in the genetic code can lead to amino acid changes in proteins.

(D) All living organisms have a virtually identical genetic code.

Section 2: Free-Response Question

Write your answer to each part clearly. Support your answers with relevant information and examples. Where calculations are required, show your work.

Arabidopsis thaliana is a small, diploid plant in the mustard family. Like all plants, *A. thaliana* has structures on its leaves called stomata that open and close to allow exchange of gasses with the environment. The *HT1* gene of *A. thaliana* encodes a protein that controls opening of leaf stomata in response to CO_2 concentrations. In wild-type plants, the stomata open when CO_2 concentrations are low and close when they are high. The protein-coding portion of the *HT1* mRNA is 1170 bases long. In a recessive mutant allele of the *HT1* gene, called $ht1^1$, the 796th nucleotide is mutated, causing lysine—instead of arginine—to be incorporated into the protein. The stomata of plants that are homozygous for the $ht1^1$ allele are always closed, even when CO_2 concentrations are low.

(a) Using the information provided, **calculate** the length in amino acids of the wild-type HT1 protein.

(b) In the $ht1^1$ allele, nucleotides 795 through 797 are AGA, which, according to the genetic code shown in Figure 34.6, codes for arginine. Knowing that the $ht1^1$ allele incorporates an arginine instead of the wild-type lysine, **predict** the 796th nucleotide in a wild-type *HT1* allele.

(c) **Describe** how a plant that is heterozygous for an $ht1^1$ allele and a wild-type *HT1* allele can have the same phenotype as a plant that is homozygous for wild-type *HT1* alleles.

(d) An *A. thaliana* plant that is homozygous for wild-type *HT1* alleles is crossed with a plant that is homozygous for the $ht1^1$ allele. Using the information provided, **predict** the proportion of the progeny from this cross that will have stomata that are unresponsive to CO_2 concentration. **Provide reasoning** to support your prediction.

VISUAL SYNTHESIS 6.1 GENE EXPRESSION

Gene expression is the process by which a gene is turned on to produce a product. It begins with transcription, in which DNA is used as a template to synthesize a complementary RNA molecule. Transcription is typically followed by RNA processing, in which introns are removed and the ends of the RNA molecule are modified to produce mRNA. The final step is translation, in which the

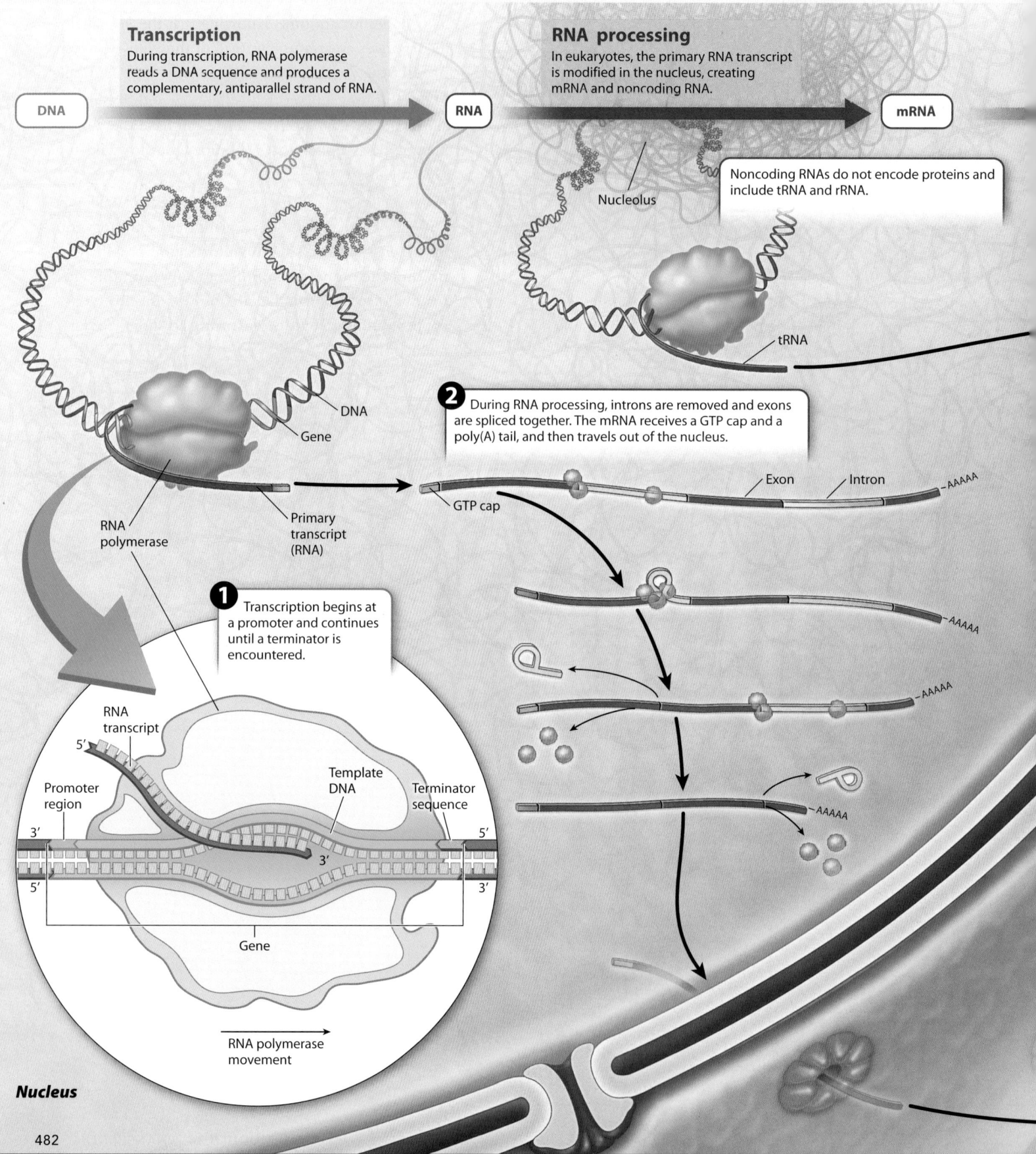

Transcription
During transcription, RNA polymerase reads a DNA sequence and produces a complementary, antiparallel strand of RNA.

DNA → RNA

RNA processing
In eukaryotes, the primary RNA transcript is modified in the nucleus, creating mRNA and noncoding RNA.

RNA → mRNA

Nucleolus

Noncoding RNAs do not encode proteins and include tRNA and rRNA.

tRNA

DNA

Gene

2 During RNA processing, introns are removed and exons are spliced together. The mRNA receives a GTP cap and a poly(A) tail, and then travels out of the nucleus.

Exon   Intron   ~AAAAA

RNA polymerase

Primary transcript (RNA)

GTP cap

~AAAAA

~AAAAA

~AAAAA

1 Transcription begins at a promoter and continues until a terminator is encountered.

RNA transcript

5'

Promoter region

Template DNA

Terminator sequence

3'   5'

3'

5'   3'

Gene

RNA polymerase movement

Nucleus

482

mRNA is read by a ribosome to produce a protein. Proteins play many roles in the cell. They provide structural support, act as enzymes that facilitate chemical reactions, and are involved in cell signaling.

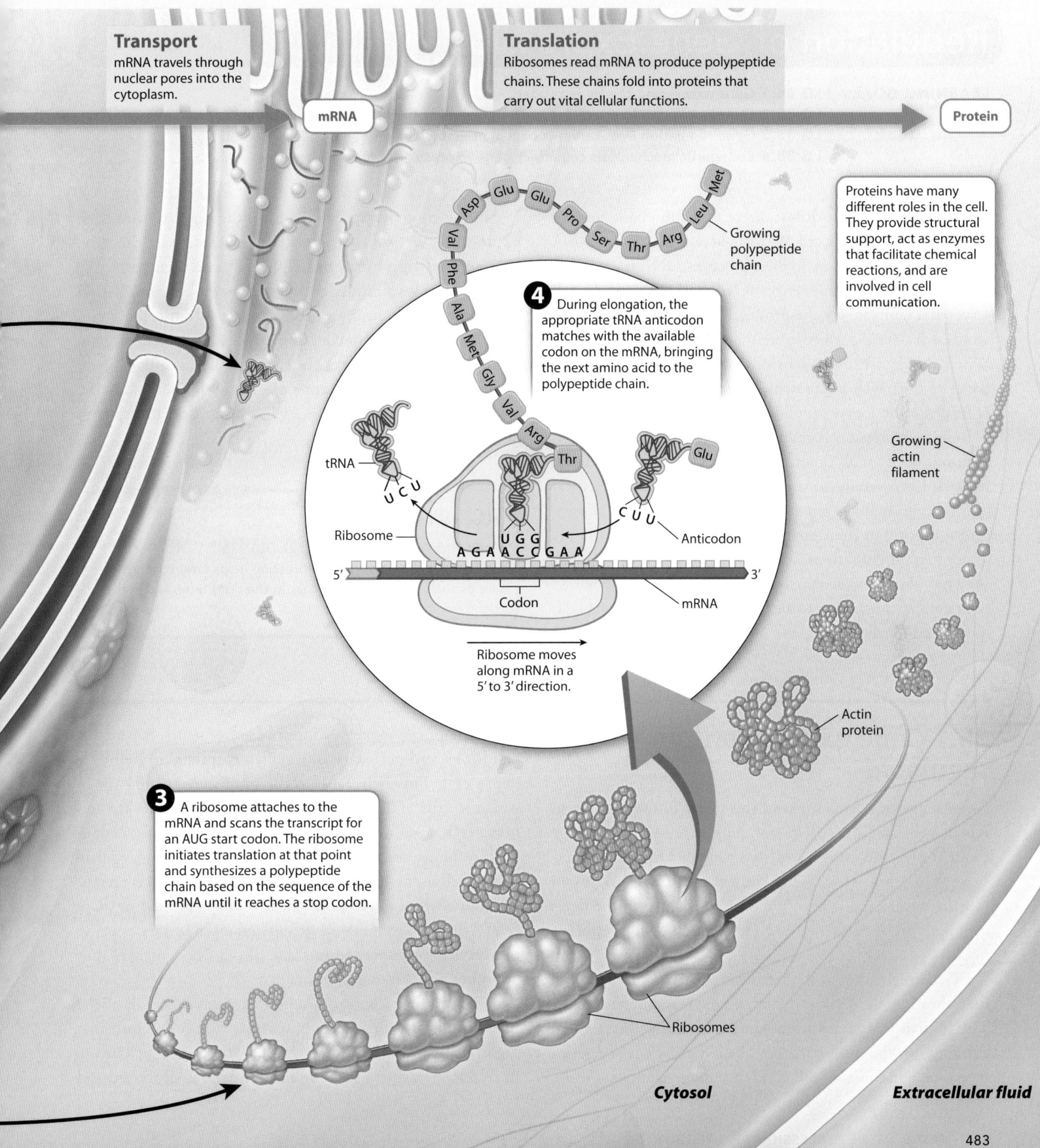

Transport
mRNA travels through nuclear pores into the cytoplasm.

mRNA

Translation
Ribosomes read mRNA to produce polypeptide chains. These chains fold into proteins that carry out vital cellular functions.

Protein

Proteins have many different roles in the cell. They provide structural support, act as enzymes that facilitate chemical reactions, and are involved in cell communication.

Growing polypeptide chain

4 During elongation, the appropriate tRNA anticodon matches with the available codon on the mRNA, bringing the next amino acid to the polypeptide chain.

Growing actin filament

tRNA

Ribosome

Anticodon

mRNA

Codon

Ribosome moves along mRNA in a 5′ to 3′ direction.

Actin protein

3 A ribosome attaches to the mRNA and scans the transcript for an AUG start codon. The ribosome initiates translation at that point and synthesizes a polypeptide chain based on the sequence of the mRNA until it reaches a stop codon.

Ribosomes

Cytosol

Extracellular fluid

Module 35

Regulation of Gene Expression

LEARNING GOALS ▶**LG 35.1** Gene expression can be regulated.
▶**LG 35.2** Gene regulation in prokaryotes often occurs at the level of transcription.
▶**LG 35.3** Gene regulation in eukaryotes occurs at many levels.
▶**LG 35.4** Epigenetic mechanisms can affect gene expression.

In the last few modules, we outlined the basic steps of information flow in a cell. We focused on transcription, RNA processing, and translation. In these processes, DNA is a template for the production of messenger RNA (mRNA), which itself serves as a template for the production of a protein. In other words, what we described were the steps of gene expression. When a gene is expressed, or turned on, it produces a functional product, such as a protein. When a gene is not expressed, or turned off, it does not produce a functional product.

Gene expression is regulated. That is, gene expression does not occur at all times, in all cells, and in all conditions. Instead it is carefully controlled. Gene regulation is critical for life itself. It allows single-celled organisms such as bacteria to respond to their environment, expressing genes only under certain conditions, such as when particular nutrients are available or molecular signals are received. It also allows multicellular organisms to develop, function, and have cells specialized for different tasks.

In this module, we will focus on the regulation of gene expression. We will begin with an overview of the process, emphasizing broad principles of gene regulation. We will follow this introduction with a look at gene regulation in prokaryotes, focusing on how bacteria transcribe genes only when they are needed. Finally, we will turn to gene regulation in eukaryotes. Eukaryotes can regulate gene expression at many different steps in the process, allowing for very fine control of the level, timing, and location of gene expression.

PREP FOR THE AP® EXAM

FOCUS ON THE BIG IDEAS

INFORMATION STORAGE AND TRANSMISSION: Focus on how regulation of gene expression helps ensure genes are expressed in the right place, at the right time, and in the right amount.

35.1 Gene expression can be regulated

Gene regulation encompasses the many different ways in which cells are able to control gene expression. It can be thought of as the *where, when,* and *how much* of gene expression. Where (in which cells) are genes turned on? When (during development or in response to changes in the environment) are they turned on? How much gene product is made?

Gene regulation in multicellular eukaryotes leads to cell specialization: different types of cell express different genes. The human body contains about 200 major cell types, and although for the most part they share the same set of genes, they look and function differently from one another because each type of cell expresses different sets of genes. A pancreas cell looks and acts very differently from a skin cell or a nerve cell. This is because some genes are turned on in pancreas cells, and different genes are turned on in skin and nerve cells, as shown in **FIGURE 35.1**. For example, certain cells in the pancreas express insulin, which helps to regulate sugar levels in the blood. The gene encoding insulin is present in skin cells and nerve cells, but it is not expressed in these cells. This is an example of gene regulation. In this case, the insulin gene is expressed in some places (the pancreas), but not other places (the skin or brain).

Even in the pancreas, insulin is not expressed all of the time. Insulin is expressed only when it receives a signal that sugar levels in the blood are high, as described in Unit 4. In response to this signal, the insulin gene is expressed and

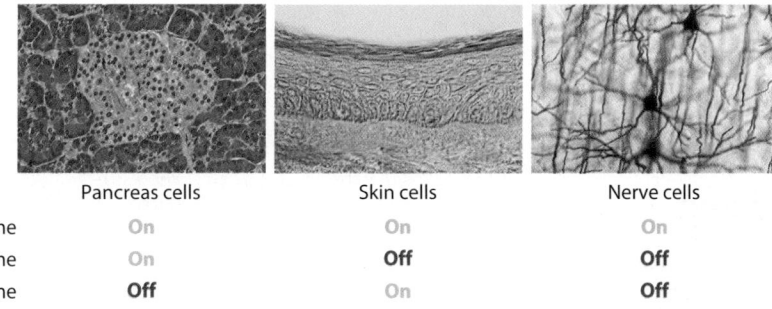

	Pancreas cells	Skin cells	Nerve cells
Housekeeping gene	On	On	On
Insulin gene	On	**Off**	**Off**
Keratin gene	**Off**	On	**Off**

FIGURE 35.1 Gene regulation

Gene regulation is the process by which genes are turned on and off. Housekeeping genes are turned on, or expressed, by all cells at all times. However, different cells in multicellular organisms express different genes, which gives cells such as pancreas, skin, and nerve cells their different forms and functions. For example, cells in the pancreas express the hormone insulin, which helps to regulate blood glucose levels. The gene encoding insulin is expressed in pancreas cells, but not in skin or nerve cells. On the other hand, skin cells express the protein keratin, which covers the skin, but pancreas and nerve cells don't. Photos: Ed Reschke/Getty Images

insulin is produced and released into the bloodstream, where it helps to bring sugar levels back to normal levels. This is also an example of gene regulation. In this case, the insulin gene is expressed only at certain times (when blood sugar levels increase) and not at other times (when blood sugar levels decrease). This example illustrates one of the roles of gene regulation—genes are typically expressed for the most part only when they are required. In this way, cells and organisms do not expend energy and resources when they are not needed.

Another important principle of gene regulation is that it can occur at almost any step in the path from DNA to mRNA to protein—at the level of the chromosome itself; during transcription or translation; and, perhaps surprisingly, even after the protein is made. Each of these steps of gene expression may be subject to regulation. In other words, each successive event that takes place in the expression of a gene is a potential control point for gene expression. Scientists often use the term "level" to describe each of these steps where gene expression can be controlled, so that gene expression can be regulated at different levels. Having

many different points of control allows organisms to finely tune the amount, timing, and location of proteins that are produced.

The regulation of gene expression occurs in unicellular and multicellular organisms, as well as prokaryotes and eukaryotes. All life, from the simplest to the most complex, requires gene regulation. Gene regulation relies on similar processes in all organisms because all life shares a common ancestor. Nevertheless, certain features of eukaryotes—the way that DNA is packaged into chromosomes, RNA processing, and the separation in space of transcription and translation—provide additional levels of gene regulation in eukaryotes that are not possible in prokaryotes.

✓ Concept Check

1. **Describe** the difference between gene expression and gene regulation.
2. **Identify** three different levels of gene regulation.

35.2 Gene regulation in prokaryotes often occurs at the level of transcription

Because gene regulation in prokaryotes is simpler than gene regulation in eukaryotes, prokaryotes have served as model organisms for our understanding of how genes are turned on

and off. In prokaryotes, DNA is packaged into chromosomes differently from eukaryotes. In addition, mRNA is not processed after transcription, and transcription and translation are not separated by a nuclear envelope. In prokaryotes, gene expression involves transcription of the gene into mRNA and translation of the mRNA into protein. Each of these levels of gene expression is subject to regulation. In this section, we will focus on gene regulation at the level of transcription, where gene expression is commonly controlled in prokaryotes. Transcriptional control allows prokaryotes to respond quickly to changes in the environment.

Positive and Negative Regulation

In both eukaryotes and prokaryotes, transcription can be positively or negatively regulated. In positive regulation, a **positive regulatory molecule** (usually a protein) binds to the DNA at a site near the gene in order for transcription to take place. In negative regulation, a **negative regulatory molecule** (again, usually a protein) binds to the DNA at a site near the gene and prevents transcription.

FIGURE 35.2 illustrates positive regulation. The main players are DNA, RNA polymerase, and a regulatory protein called an activator. The DNA contains two binding sites: one for the activator and one for RNA polymerase. When the activator is bound to its site in DNA, RNA polymerase binds to its site, called the promoter, and transcription takes place, as shown in Figure 35.2a. When the activator is not present or is not able to bind with the DNA, RNA polymerase cannot bind to the promoter and transcription does not occur, as shown in Figure 35.2b.

Sometimes, the activator protein combines with a small molecule in the cell and undergoes a change in shape that alters its binding affinity for DNA. For example, the activator on its own might not be able to bind with DNA, but when the activator is bound to the small molecule, the activator undergoes a shape change that allows it to bind with the DNA. This change

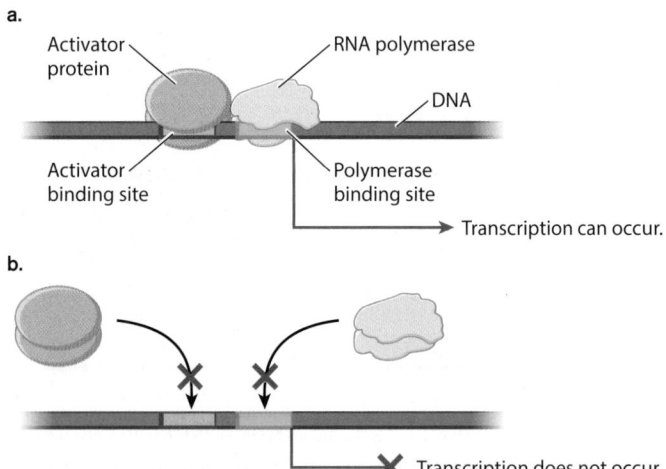

FIGURE 35.2 Positive regulation

In positive regulation of gene expression, an activator is required to turn on transcription of a gene. (a) When the activator is present and binds to DNA, transcription occurs. (b) When the activator is not present or does not bind to DNA, transcription does not occur.

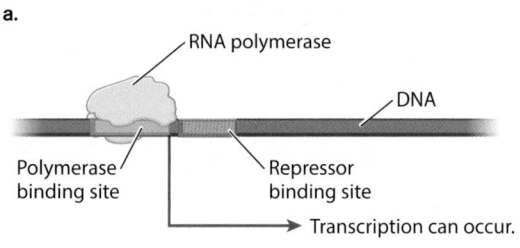

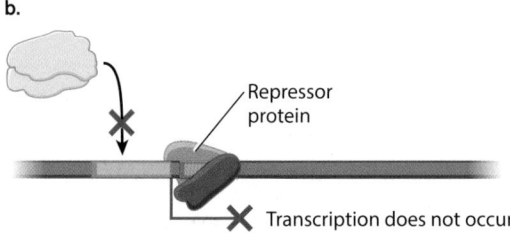

FIGURE 35.3 Negative regulation

In negative regulation of gene expression, a gene is transcribed, but is turned off by the binding of a repressor to DNA. (a) In the absence of a repressor, RNA polymerase binds to DNA and transcription occurs. (b) In the presence of a repressor, RNA polymerase does not bind to DNA and transcription does not occur.

in shape is an example of an allosteric effect. An allosteric effect occurs when a molecule binds a protein, changing the shape and altering the activity of the protein. In Unit 3, for example, we described an allosteric inhibitor that binds an enzyme at a site different from the active site, changing the shape and activity of the enzyme. For some activators, we see a similar phenomenon, in which a small molecule binds to and changes the shape of the activator, affecting its activity. Specifically, the binding of a small molecule to the activator allows it to bind to DNA, resulting in transcription of the gene.

FIGURE 35.3 illustrates negative regulation. In this case, RNA polymerase can bind to DNA on its own and transcription can take place, as shown in Figure 35.3a. However, a negative regulatory protein called a repressor can also bind DNA and turn off transcription, as shown in Figure 35.3b. As with positive regulation, the ability of the repressor to bind DNA is often determined by an allosteric interaction with a small molecule. For example, a small molecule might interact with the repressor and prevent it from binding DNA. This type of small molecule is called an inducer. In the presence of an inducer, the repressor cannot bind to DNA and therefore gene expression is turned on.

The *lac* Operon

Bacteria such as *Escherichia coli* contain many different genes. Some of these, like the genes involved in glycolysis, are expressed all of the time. Other genes, such as those involved in the breakdown of the sugar lactose, are only expressed when they are needed. Lactose is a disaccharide made up of glucose and galactose. An enzyme called β-(beta-)galactosidase cleaves lactose, producing glucose and galactose. Both molecules can then be broken down further and used as a source of carbon and energy, as discussed in Unit 3.

The regulation of genes involved in the breakdown of the sugar lactose in *E. coli* was extensively studied in the 1960s by François Jacob and Jacques Monod. Since then, it has been used as a model for transcriptional gene regulation. Jacob and Monod began their research with an interesting observation: active β-galactosidase enzyme is observed in cells in the presence of lactose, but not in the absence of lactose. In other words, the expression of the gene is regulated—it is only turned on when it is needed by the cell to breakdown lactose. Otherwise, it is not expressed. In this way, *E. coli* does not waste energy producing an enzyme that it does not need.

To understand how the genes for lactose utilization are regulated, you need to know the main players, which are shown in **FIGURE 35.4**. Let's start with the three genes on the right side of the figure:

- *lacZ* is the gene for β-galactosidase which cleaves the lactose molecule into its glucose and galactose subunits.

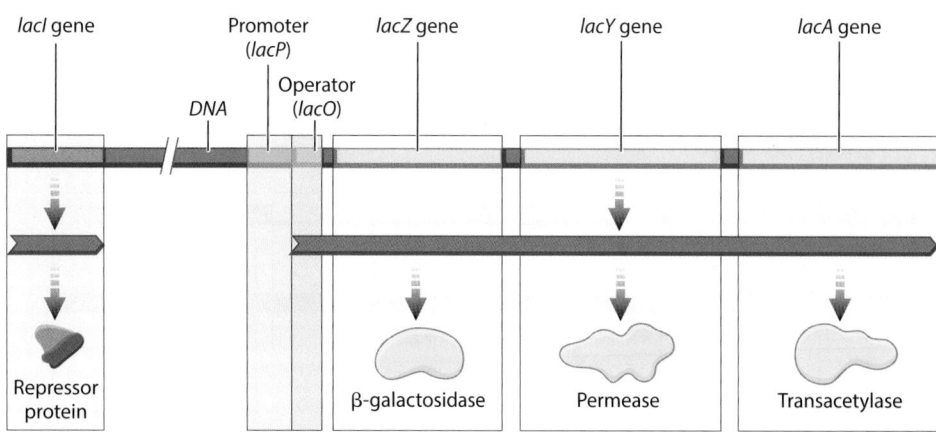

FIGURE 35.4 The *lactose* operon

The *lac* operon includes genes that encode β-galactosidase (*lacY*), a permease (*lacZ*), and a transacetylase (*lacA*). It also consists of a promoter (*lacP*) that binds RNA polymerase and an operator (*lacO*) that binds a repressor protein, which is encoded by *lacI*.

- *lacY* is the gene for permease, which transports lactose from the external medium into the cell.
- A third gene, called *lacA*, is the gene for transacetylase and is not required for the breakdown of lactose.

These three genes are transcribed together when RNA polymerase binds to its promoter, *lacP*, and initiates transcription. This gene organization, in which a set of genes is transcribed together, is called an operon, as discussed in the previous module. Operons are common in bacteria. An advantage of this organization is that all of the genes required to use lactose are expressed together at the same time.

Regulation of the three genes is controlled by the product of another gene, called *lacI*, shown in purple in Figure 35.4. This gene encodes a repressor protein. The repressor protein binds to a regulatory DNA sequence called an operator, *lacO*. The lactose operon is negatively regulated by the repressor protein. That is, the three genes are always expressed unless the operon is turned off by the repressor.

FIGURE 35.5 shows what the operon looks like in the presence of the repressor. The repressor binds to the operator, blocking RNA polymerase from binding to DNA, and transcription does not take place. This is how the operon operates in the absence of lactose. As a result, when lactose is not present, the genes that encode proteins required for the breakdown of lactose are not transcribed.

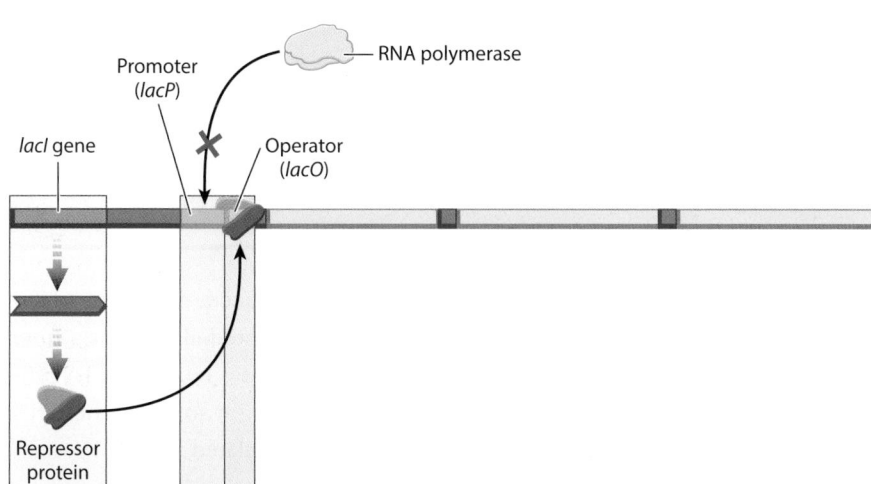

FIGURE 35.5 The *lactose* operon in the absence of lactose

In the absence of lactose, the repressor protein binds to the operator and prevents RNA polymerase from binding to the promoter. As a result, transcription does not occur.

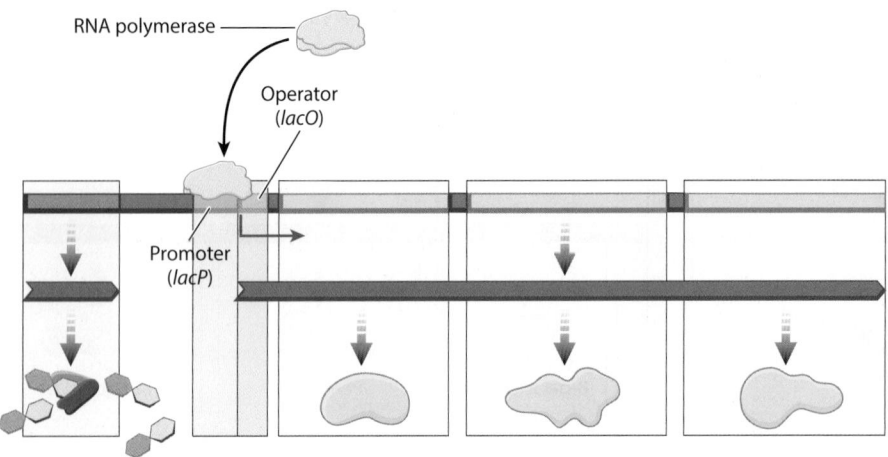

FIGURE 35.6 The *lactose* operon in the presence of lactose

In the presence of lactose, the repressor protein is unable to bind to the operator. As a result, RNA polymerase is able to bind to the promoter and transcription occurs.

When lactose is present, the repressor is unable to bind to the operator. As a result, the RNA polymerase is able to bind to DNA and transcription occurs, as shown in **FIGURE 35.6**. In other words, lactose acts as an inducer of the lactose operon because it prevents binding of the repressor. The inducer is not actually lactose itself, but rather an isomer of lactose called allolactose. Lactose is always accompanied by a small amount of allolactose in the cell, and so the lactose operon is induced in the presence of lactose.

The binding of the inducer to the repressor results in a change in shape of the repressor that blocks the repressor from binding to the operator. When the repressor is not bound to DNA, permease and β-galactosidase are expressed. Permease allows lactose to be transported into the cell, and β-galactosidase cleaves the molecules to allow the subunits to be used as a source of energy and carbon. In this way, the genes for lactose utilization are only expressed when lactose is actually present in the environment.

The function of the various components of the *lac* operon was determined in part by using mutations that disrupt their functions. For example, bacteria that contain mutations that eliminate the function of the *lacZ* or *lacY* gene cannot use lactose as a source of energy. Without permease from *lacY*, lactose cannot enter the cell, and without β-galactosidase from *lacZ*, lactose cannot be broken down. As another example, consider a mutation in the *lacI* repressor gene that encodes a mutant form of the repressor that is able to bind to the operator but is not able to bind to the inducer. In this case, the repressor will bind to the operator even in the presence of the inducer. This will lead to a cell that is not able to produce permease and β-galactosidase at all, whether lactose is present or absent. In other words, with such a mutation, the lactose operon will not be inducible.

✓ **Concept Check**

3. Both an activator and an inducer turn on transcription. **Describe** how they are different from each other.

4. **Describe** how the binding of a small molecule can change the ability of a protein to bind to a specific DNA sequence.

35.3 Gene regulation in eukaryotes occurs at many levels

Gene regulation in eukaryotes is more complex than gene regulation in prokaryotes. Eukaryotes can control the expression of genes at many different levels. There are many different processes that can be regulated to control the expression of genes in eukaryotes, including transcription, RNA processing, RNA stability, and translation. Regulation can even occur after a protein is produced. The presence of many different steps that can be regulated allows for a finer control of gene expression in eukaryotes compared to prokaryotes. In this section, we will highlight some of the steps, or levels, where gene expression is regulated in eukaryotes.

Transcriptional Regulation

Transcriptional regulation in eukaryotic cells is similar to transcriptional regulation in prokaryotic cells, but it requires the coordinated action of many proteins that interact with one another and with regulatory sequences near the gene. Initially, a group of proteins called general **transcription factors** bind to the promoter. Once bound to the promoter, the transcription factors bind RNA polymerase, allowing transcription to proceed. In this way, they act as positive regulatory molecules.

The binding of general transcription factors is controlled by proteins called regulatory transcription factors. Some regulatory transcription factors have two binding sites, one of which binds with a particular regulatory DNA sequence in or near a gene known as an enhancer and the other of which binds to one or more general transcription factors, as shown in **FIGURE 35.7**. The general transcription factors then bind to RNA polymerase, and transcription can begin. Transcription does not occur if the regulatory transcription factors fail to bring the general transcription factors to the gene.

Hundreds of different regulatory transcription factors control the transcription of thousands of genes. Some bind with enhancers and stimulate transcription; others bind with DNA sequences known as silencers and repress transcription. Enhancers and silencers are often in or near the genes they regulate, but, in some cases, they may be many thousands of nucleotides distant from the genes. A typical gene may be regulated by multiple enhancers and silencers, each with one or more regulatory transcription factors that can bind with it. Transcription takes place only when the proper combination of regulatory transcription factors is present in the same cell, as shown in Figure 35.7. Transcription of a gene with multiple silencers and enhancers depends on the presence of a particular combination of regulatory transcription factors, so this type of regulation is called combinatorial control.

RNA Processing

A great deal happens in the nucleus after transcription takes place. The initial transcript, called the primary transcript, undergoes several modifications, which are collectively called RNA processing, discussed in Module 33. RNA processing includes the addition of a GTP cap to the 5′ end and a string of tens to hundreds of adenosine nucleotides to the 3′ end to form the poly(A) tail. These modifications are necessary for the mRNA molecule to be transported to the cytoplasm and recognized by the ribosome. The poly(A) tail also helps to determine how long the mRNA will persist in the cytoplasm before being degraded. Therefore, RNA processing is an important point where gene regulation can occur.

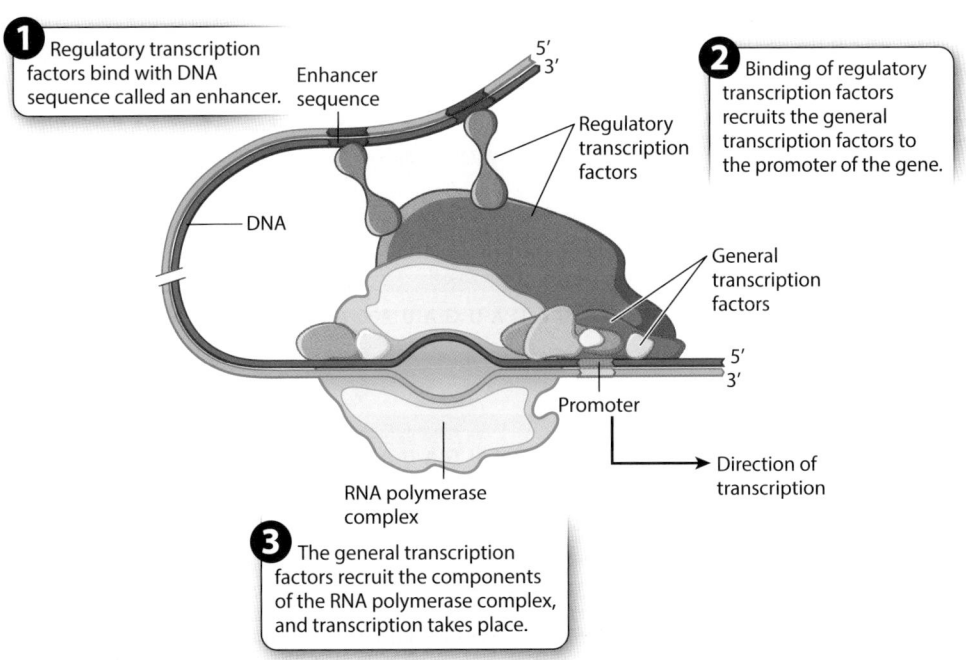

1 Regulatory transcription factors bind with DNA sequence called an enhancer.

Enhancer sequence

Regulatory transcription factors

5′
3′

2 Binding of regulatory transcription factors recruits the general transcription factors to the promoter of the gene.

DNA

General transcription factors

5′
3′

Promoter

Direction of transcription

RNA polymerase complex

3 The general transcription factors recruit the components of the RNA polymerase complex, and transcription takes place.

FIGURE 35.7 Transcriptional regulation in eukaryotes

Transcriptional regulation in eukaryotes requires the coordinated action of many different proteins called transcription factors. Regulatory transcription factors bind to regulatory sequences, such as enhancers, that can be far from the gene. These regulatory transcription factors then interact with general transcription factors, which bind to the promoter and help to recruit RNA polymerase so that transcription can occur.

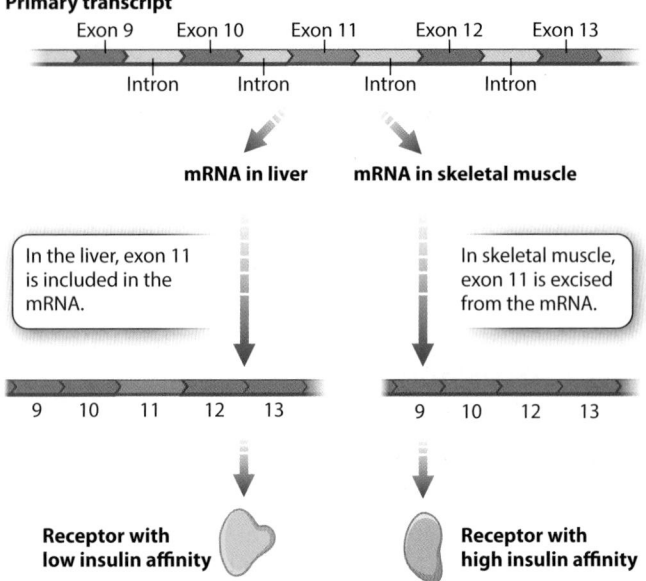

FIGURE 35.8 Alternative splicing of an insulin-receptor transcript

Alternative splicing generates different processed mRNAs and different proteins from the same primary transcript. In liver cells, exon 11 is retained, resulting in a receptor with a low affinity for insulin. In muscle cells, exon 11 is spliced out, resulting in a receptor with a high affinity for insulin.

In eukaryotes, the primary transcript of many protein-coding genes is far longer than the mRNA ultimately used in protein synthesis. The long primary transcript consists of regions that are retained and expressed in the mRNA, called exons, and others that are interspersed and removed, called introns. The exons are joined and the introns are removed during RNA splicing, also described in Module 33, forming the processed mRNA.

RNA splicing provides another opportunity for regulating gene expression because the same primary transcript can be spliced in different ways to yield different proteins in a process called alternative splicing. The alternatively spliced forms may be produced in the same cells or in different types of cell. Alternative splicing accounts in part for the observation that humans produce many more proteins than our total number of genes. By some estimates, over 90% of human genes undergo alternative splicing.

FIGURE 35.8 shows the primary transcript of a gene encoding an insulin receptor found in humans and other mammals. The insulin receptor binds insulin and allows the cell to take up glucose, an important source of energy. During RNA splicing in liver cells, exon 11 is included in the mRNA, and the insulin receptor produced from this mRNA has low affinity for insulin. In contrast, in skeletal

muscle cells, exon 11 is spliced out of the primary transcript. The resulting protein has a high affinity for insulin. The different forms of the receptor are important: the higher affinity of muscle cells for insulin enables them to absorb enough glucose to fulfill their energy needs.

Some mRNAs are substrates for enzymes that modify particular bases in the RNA, thus changing their sequences and what they code for. This process is known as RNA editing. One type of editing enzyme removes the amino group (—NH₂) from adenosine. Another enzyme removes the amino group from cytosine. In the human genome, hundreds if not thousands of transcripts undergo RNA editing. In many cases, not all copies of the transcript are edited, and some copies may be edited more extensively than others.

Transcripts from the same gene may undergo different editing in different cell types. An example is shown in **FIGURE 35.9**. The mRNA fragments show part of the coding sequence for apolipoprotein B, a protein that circulates in the blood. RNA editing does not occur in liver cells. The unedited mRNA is shown in Figure 35.9a. It is translated into

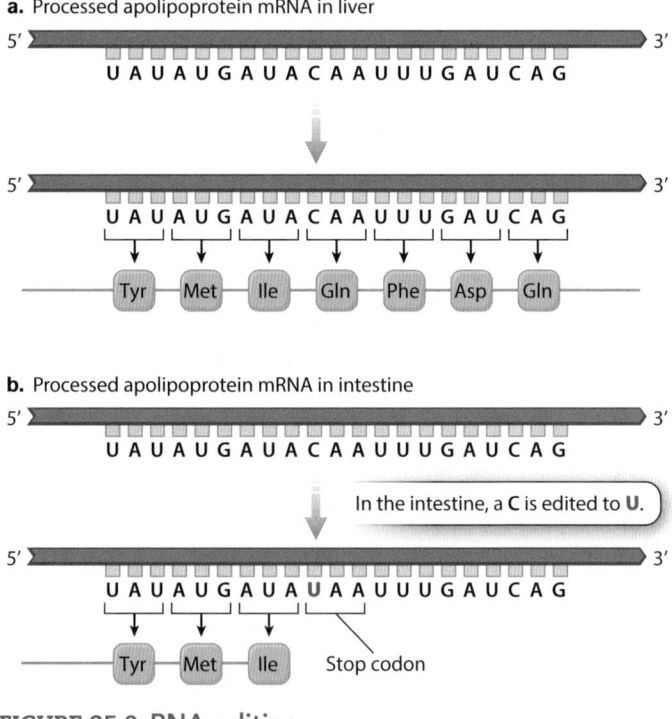

FIGURE 35.9 RNA editing

The human apolipoprotein B transcript is not edited in liver cells, but is edited in intestine cells. The result is that the same gene can produce two different proteins with different functions. (a) In the liver, the unedited mRNA encodes a protein that transports cholesterol in the blood. (b) In the intestine, the edited mRNA encodes a much shorter protein due to the introduction of a stop codon. This protein functions to absorb lipids from foods.

a protein that transports cholesterol in the blood. In contrast, RNA editing occurs in the intestine. The cytosine nucleotide in codon 2153 is edited to uracil, as shown in Figure 35.9b. The edited codon is UAA, which is a stop codon. Translation therefore terminates at this point, releasing a protein only about half as long as the liver form. This shorter form of the protein helps the cells of the intestine absorb lipids from the foods we eat. The result is that transcripts from the same gene can produce multiple types of proteins.

RNA Stability and Translation Regulation

In eukaryotes, a processed mRNA exits the nucleus before the translation step of gene expression can occur. The mRNA migrates to the cytoplasm through one of a few thousand nuclear pores, which are large protein complexes that span both layers of the nuclear envelope and regulate the flow of macromolecules into and out of the nucleus. Once the mRNA is in the cytoplasm, there are multiple opportunities for gene regulation at the levels of mRNA stability, translation, and protein activity.

The stability of mRNA can be regulated by small regulatory RNA molecules. Regulatory RNA molecules are a type of noncoding RNA that do not code for proteins, but have functions of their own. The existence of these molecules is among the most exciting recent discoveries in gene regulation. They are being extensively studied by biologists and drug researchers because their small size allows easy synthesis in the laboratory, and researchers can design their sequences to target transcripts of interest.

Two important types of small regulatory RNAs are known as siRNA (small interfering RNA) and miRNA (microRNA). These RNAs can be incorporated into a larger protein complex, which is targeted to specific mRNA molecules, resulting in degradation of RNA transcripts or inhibition of translation. Regulation by small regulatory RNAs is widespread in eukaryotes. Human chromosomes are thought to encode about 1000 microRNAs, each of which can target tens or hundreds of mRNA molecules. Half or more of human proteins may have their synthesis regulated in part by such small regulatory RNAs.

Translation of mRNA into protein provides another level of control of gene expression. **FIGURE 35.10** shows the structure of an mRNA molecule in a eukaryotic cell and highlights some of the features that help regulate its translation. Not all mRNA molecules have all the features shown, but almost all mRNA molecules have a GTP cap, a 5′ untranslated region (5′ UTR), a 3′ untranslated region (3′ UTR), and a poly(A) tail.

The 5′ UTR and 3′ UTR may contain regions that bind with regulatory proteins. These RNA-binding proteins help control mRNA translation and degradation. The UTRs may also contain binding sites for small regulatory RNAs. During development, some RNA-binding proteins interact with molecular motors that transport the mRNA to particular regions of the cell. In other cases, the proteins are only in particular locations in the cell and repress translation of the mRNAs that are transported there and to which they bind. By either transport or repression, these proteins cause the mRNA to be translated only in certain places in the cell.

Post-Translational Modification

Once translation is completed, the resulting protein can alter the phenotype of the cell or organism by affecting metabolism, signaling, gene expression, or cell structure. After translation, proteins are modified in multiple ways that regulate their structure and function. Collectively, these processes are called post-translational modification. Regulation at this level is essential because some proteins are downright dangerous. For example, proteases, such as the digestive enzyme trypsin, must be kept inactive until they are secreted out of the cell. If they were not kept inactive, their activity would kill the cell. These types of protein are often controlled by being translated in inactive forms that are made active by modification after secretion.

Folding and acquiring stability are key control points for some proteins. Although many proteins fold properly as they come off the ribosome, others require help from proteins called chaperones that act as folding facilitators. Correct folding is important because improperly folded proteins may bind together to form aggregates that are destructive to cell

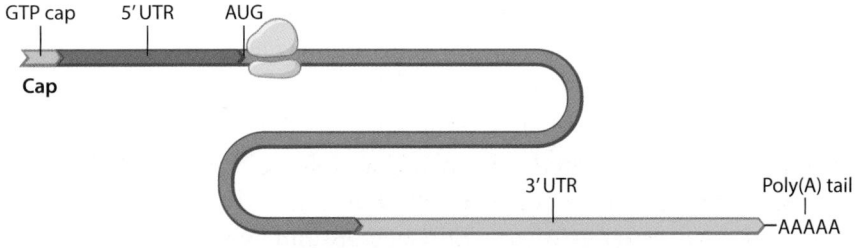

FIGURE 35.10 Features of mRNA that affect gene expression

Translation of mRNA can be affected by the GTP cap, 5′ untranslated region (5′ UTR), 3′ untranslated region (3′ UTR), and poly(A) tail.

function. Many diseases are associated with such aggregates, including Alzheimer's disease, Huntington's disease, and the neurodegenerative disorder known as Creutzfeldt-Jakob disease, which is similar to mad cow disease.

Post-translational modification also helps regulate protein activity. Many proteins are modified by the addition of one or more sugar molecules to the side chains of some amino acids. This modification can alter the protein's folding and stability or target the molecule to particular cellular compartments, such as the endoplasmic reticulum, Golgi apparatus, or lysosome. In other cases, functional groups such as a phosphate, a methyl, or an acetyl group, are added to proteins, changing their shape and function. Addition of a phosphate group to the side chains of amino acids such as serine, threonine, or tyrosine is a key regulator of protein activity. Introduction of the negatively charged phosphate group alters the conformation of the protein, in some cases switching it from an inactive state to an active state and in other cases the reverse. Because the function of a protein molecule results from its shape and charge, a change in protein conformation affects protein function.

✓ Concept Check

5. **Describe** how one gene can code for more than one protein.

6. **Describe** how small regulatory RNAs differ from mRNAs.

7. **Identify** three ways in which gene expression can be influenced after mRNA is processed and leaves the nucleus.

35.4 Epigenetic mechanisms can affect gene expression

Even before transcription takes place, there is an additional level of gene regulation in eukaryotes, at the chromosome itself. The manner in which DNA is packaged in the nucleus in eukaryotes provides an important opportunity for regulating gene expression. Regulation at this level of gene expression is determined in part by whether the proteins necessary for transcription can gain access to the genes they transcribe. In this final section, we will introduce how DNA packaging in chromosomes affects gene expression.

DNA in eukaryotes is packaged as chromatin, a complex of DNA, RNA, and proteins that gives chromosomes their structure, as described in Module 31. When chromatin is in its coiled state, the DNA is not accessible to the proteins that carry out transcription. The chromatin must loosen to allow space for transcriptional proteins and enzymes to work. This is accomplished through chromatin remodeling, in which the nucleosomes are repositioned to expose different stretches of DNA to the environment of the nucleus.

One way in which chromatin is remodeled is by chemical modification of the histones around which DNA is wound, as shown in **FIGURE 35.11**. Modification

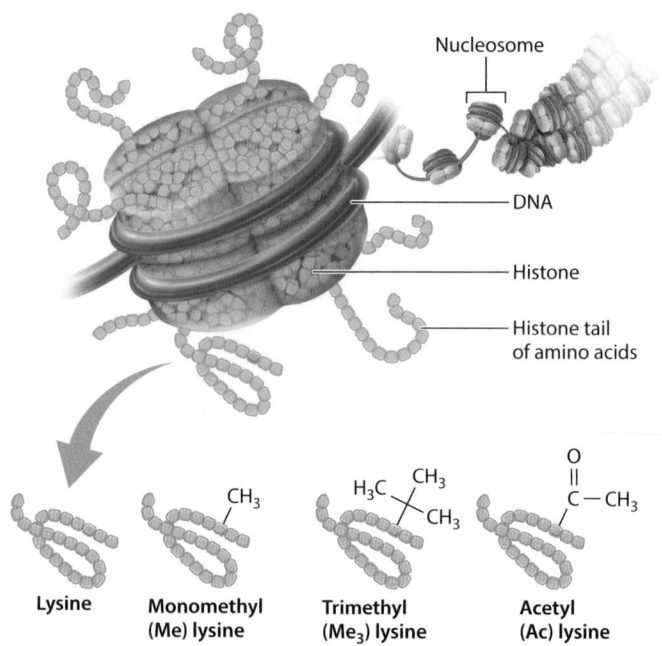

FIGURE 35.11 Histone modifications

The amino acid lysine in the histone tails of nucleosomes can be modified. Typical modifications include addition of a methyl group (Me), addition of three methyl groups (Me₃), and acetylation (Ac). All of these types of post-translational modifications change the shape of the tail and can affect gene expression.

usually occurs on histone tails, which are strings of amino acids that protrude from the histone proteins in the nucleosome. Individual amino acids in the tails can be post-translationally modified by the addition or later removal of different chemical groups, including methyl groups ($-CH_3$) and acetyl groups ($-COCH_3$). Some of these modifications tend to activate transcription and others to repress transcription. Modification of histones takes place at key times in development to ensure that the proper genes are turned on or off, as well as in response to environmental cues.

In many eukaryotic organisms, gene expression is also affected by chemical modification of certain bases in the DNA, the most common of which is the addition of a methyl group to the base cytosine. DNA methylation leads to changes in chromatin structure, histone modification, and nucleosome positioning that restrict the access of transcription factors to promoters. Heavy cytosine methylation is usually associated with transcriptional repression of a nearby gene. The methylation state of cytosines can change over time or in response to environmental signals, providing a way to turn genes on or off.

Together, the modification of cytosine bases, changes to histones, and alterations in chromatin structure are often termed **epigenetic**, from the Greek *epi* ("over and above") and "genetic" ("inheritance"). That is, epigenetic mechanisms of gene regulation typically involve changes not to the DNA sequence itself but to the manner in which the DNA is packaged. Epigenetic modifications can, in some cases, affect gene expression. They can be inherited through mitotic cell divisions, just as genes are, but are often reversible and responsive to changes in the environment. Furthermore, epigenetic modifications present in sperm or eggs are sometimes transmitted from parent to offspring, meaning that these chemical modifications can be inherited.

For example, in humans and other mammals, about 100 genes are repressed by chemical modifications such as DNA methylation in the egg and sperm in a sex-specific manner. This sex-specific silencing of gene expression is known as imprinting. For some genes, the gene inherited from the mother is imprinted and therefore silenced, so only the gene inherited from the father is expressed. For other imprinted genes, the gene inherited from the father is imprinted and therefore silenced, so only the gene inherited from the mother is expressed.

Some of the genes that are imprinted affect the growth rate of the embryo. For example, in matings between lions and tigers that take place in captivity, the offspring of a male tiger and a female lion (a "tigon") is about the same size as its parents, whereas the offspring of a male lion and a female tiger (a "liger") is a giant cat—the largest of any big cat that exists, as shown in **FIGURE 35.12**. Much of the size difference between a tigon and liger is thought to result from whether imprinted genes for rapid embryonic growth are inherited from the mother or the father, and therefore expressed or not expressed.

FIGURE 35.12 A tigon and liger

A tigon (top) is the result of a mating between a male tiger and a female lion, whereas a liger (bottom) is the result of a mating between a male lion and a female tiger. A liger is much bigger than a tigon in part because of imprinting, in which certain genes are silenced if they are inherited from the male parent or female parent. Many imprinted genes control the growth rate of embryos, and therefore affect adult size. Photos: (top) Zigmund Leszczynski/AGE Fotostock; (bottom) YuriyKot/Shutterstock

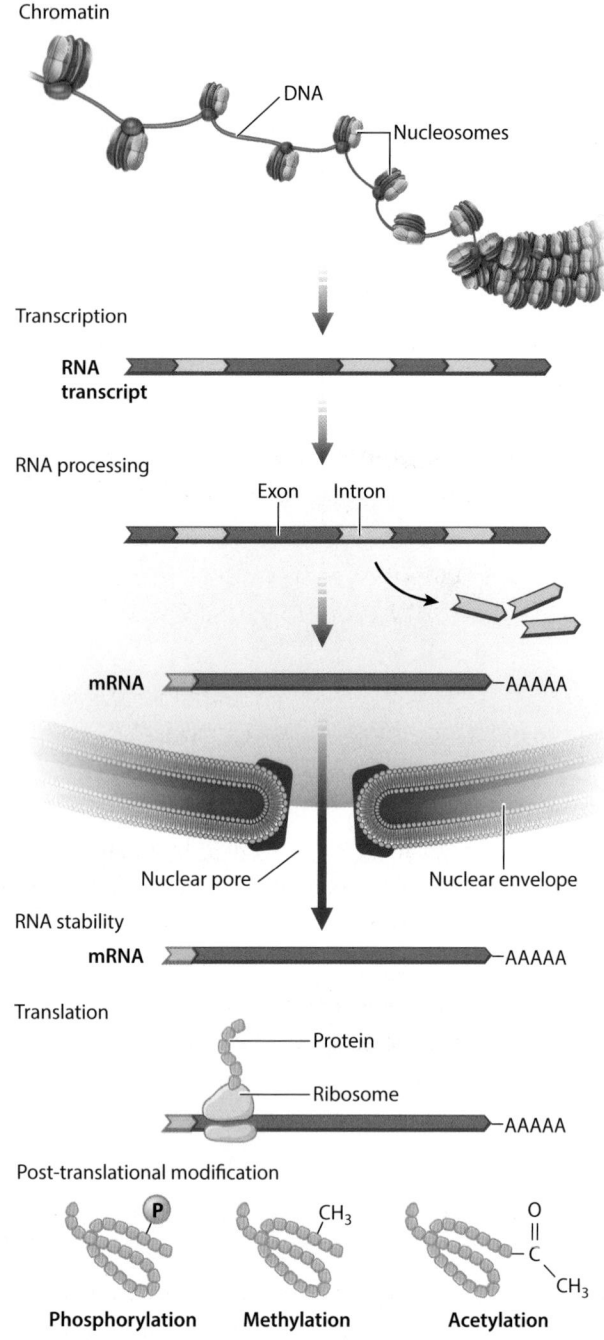

Chromatin

DNA

Nucleosomes

Transcription

RNA transcript

RNA processing

Exon   Intron

mRNA ——AAAAA

Nuclear pore   Nuclear envelope

RNA stability

mRNA ——AAAAA

Translation

Protein

Ribosome

——AAAAA

Post-translational modification

P   CH₃   O
                             ‖
                             C
                                CH₃

Phosphorylation   Methylation   Acetylation

FIGURE 35.13 Levels of gene regulation in eukaryotes

Gene regulation in eukaryotes can occur at many different levels, including chromatin, transcription, RNA processing, RNA stability, translation, and after the protein is produced. These steps allow for fine control of when, where, and how much protein is produced.

The various steps, or levels, of gene regulation are summarized in **FIGURE 35.13**. If you examine Figure 35.13 as a whole and consider the DNA sequence shown as your personal genome, you might be led to believe that genes dictate everything, and that biology is destiny. However, if you focus on the lower levels of regulation in Figure 35.13, a different picture emerges. The picture is different because much of the regulation that occurs after transcription is determined by the physiological state of your cells, which, in turn, is strongly influenced by your behaviors and lifestyle choices, as also described in Module 29. For example, your cells can synthesize 12 of the amino acids in proteins, but if any of these is present in sufficient amounts in your diet, it is absorbed during digestion and not synthesized. The amino acid you ingest blocks the amino acid–synthesizing pathway through feedback effects.

Similarly, it has been shown that dietary intake of fats and cholesterol affects not only the activity of enzymes directly involved in the metabolism of fats and cholesterol but also the levels of transcription of the genes encoding these enzymes by affecting the activity of their regulatory transcription factors. Similarly, lifestyles that combine balanced diets with exercise and stress relief have been shown to increase transcription of genes whose products prevent cellular dysfunction and decrease transcription of genes whose products promote disease. The central message of Figure 35.13 is that the regulation of gene expression occurs through a hierarchy of regulatory mechanisms acting at different levels (and usually at multiple levels) from DNA to protein.

✓ Concept Check

8. **Describe** how chromatin can be modified and how this modification affects gene expression.

9. **Describe** how DNA bases can be modified and how this modification affects gene expression.

Module 35 Summary

LG 35.1 Gene expression can be regulated.

- Gene expression involves the turning on or turning off of a gene. Page 484

- Gene regulation determines where, when, and how much gene product is made. Page 484

- Gene regulation occurs at many levels, or steps, in the process of gene expression. Page 485

LG 35.2 Gene regulation in prokaryotes often occurs at the level of transcription.

- Transcriptional regulation controls whether transcription of a gene occurs. Page 486

- Positive regulation occurs when a gene is usually off and is turned on in response to the binding to DNA of a regulatory protein called an activator. Page 486

- Negative regulation occurs when a gene is usually on and is turned off in response to the binding to DNA of a regulatory protein called a repressor. Page 486

- Jacob and Monod demonstrated how production of the enzymes encoded in the lactose operon in *E. coli* occurs only in the presence of lactose in the culture medium. Page 487

- When lactose is added to culture of bacteria, the genes for enzymes that facilitate the uptake of lactose (permease) and cleavage of lactose (β-galactosidase) are expressed. Page 488

- The lactose operon is negatively regulated by the repressor protein, which binds to a DNA sequence known as the operator when lactose is absent and prevents transcription of permease and β-galactosidase. Page 488

- When lactose is added to the medium, it induces an allosteric change in the repressor protein, preventing it from binding to the operator. In this way, lactose acts as an inducer of the lactose operon. Page 488

LG 35.3 Gene regulation in eukaryotes occurs at many levels.

- Transcription can be regulated by regulatory transcription factors that bind to specific DNA sequences known as enhancers and silencers that can be near, in, or far from genes. Page 489

- Further levels of regulation after a gene is transcribed to mRNA include RNA processing, splicing, and editing. Page 489

- After an mRNA is exported to the cytoplasm, gene expression in eukaryotes can be regulated at the level of mRNA stability, translation, and post translational modification of proteins. Page 491

- Translational regulation is determined by many features of an mRNA molecule, including the 5′ and 3′ untranslated region (UTR), the GTP cap, and the poly(A) tail. Page 491

- Post translational modification comes into play after a protein is synthesized and includes chemical modification of side groups of amino acids, affecting the structure and activity of a protein. Page 491

LG 35.4 Epigenetic mechanisms can affect gene expression.

- Regulation at the level of chromatin involves chemical modifications of DNA and histones that make a gene accessible or inaccessible to the transcriptional machinery. Page 492

Key Terms

Positive regulatory molecule   Transcription factor   Epigenetic
Negative regulatory molecule

Review Questions

1. In prokaryotes, inducers are small molecules that
   (A) bind to activators and promote transcription.
   (B) bind to activators and inhibit transcription.
   (C) bind to repressors and promote transcription.
   (D) bind to repressors and inhibit transcription.

2. For the lactose operon, lactose acts as a(n):
   (A) inducer.
   (B) operator.
   (C) activator.
   (D) repressor.

3. The process by which a single primary RNA transcript is used to make multiple proteins is called
   (A) combinatorial control.
   (B) alternative splicing.
   (C) regulatory splicing.
   (D) translational control.

4. One regulatory step in the process of gene expression and synthesis of proteins is the actual modification of proteins themselves, which is called
   (A) alternative splicing.
   (B) transcriptional modification.
   (C) post translational modification.
   (D) post protein modification.

5. Sometimes an activator protein interacts with a small molecule in the cell and undergoes a change in shape that alters its binding to DNA. This change in shape is an example of a(n)
   (A) allometric effect.
   (B) allosteric effect.
   (C) van der Waals effect.
   (D) hydrophobic effect.

6. Histone modification
   (A) is fixed; once a histone is modified, it stays that way and the genes with which it is associated are turned on or off permanently.
   (B) is fixed, but this has no effect on whether genes are expressed.
   (C) is random—sometimes the lysines are modified and sometimes they're not, but the state is independent of the environment or cell type.
   (D) can change over time in response to environmental cues, allowing genes to be turned on or off as needed.

Module 35
AP® Practice Questions

PREP FOR THE AP® EXAM

Section 1: Multiple-Choice Questions

Choose the best answer for questions 1–4.

1. There are several mechanisms for coordinating expression of multiple genes with related functions. Which mechanism is seen in prokaryotes but not in eukaryotes?
   (A) Transcription of multiple genes may be controlled by the same set of transcription factors.
   (B) Regulatory RNAs may control coordinated translation of mRNAs.
   (C) Multiple genes may be linked in an operon that has a single promoter.
   (D) Introns are removed from functionally related RNA transcripts together, allowing simultaneous translation of their proteins.

2. Which of the following statements correctly describes epigenetic modifications?
   (A) Epigenetic modifications do not change DNA sequence but can be inherited from parent to offspring.
   (B) Epigenetic modifications are changes to DNA sequence that alter gene expression.
   (C) Epigenetic modifications impact the rate at which mRNAs are translated.
   (D) Epigenetic modifications are changes to mRNA sequence after transcription is complete.

Use the following information to answer questions 3 and 4.

A scientist was studying the effect of a chemical on mammalian cells grown in culture. The scientist applied the chemical to growing cells and observed a rapid increase in the abundance of a particular protein, called protein A, in the nucleus of the cells. The scientist also observed an increase in the expression of a specific gene, called *gene X,* one hour later.

3. Which statement best explains the data observed by the scientist?
   (A) The chemical edited the mRNA transcribed from *gene X.*
   (B) *Gene X* encodes protein A.
   (C) Protein A activates expression of *gene X.*
   (D) The chemical activates the DNA in the nucleus of the cells.

4. Which would be most appropriate as a negative control in this experiment?
   (A) Cells that do not have the chemical added to them
   (B) Cells that have *gene X* removed from their DNA
   (C) Cells that always have protein A in their nucleus
   (D) Cells that always express *gene X* at a high level

Section 2: Free-Response Question

Write your answer to each part clearly. Support your answers with relevant information and examples. Where calculations are required, show your work.

E. coli bacteria are able to acquire the amino acid tryptophan from their environment. However, if the amount of tryptophan in the environment is low, *E. coli* can synthesize tryptophan using proteins encoded by genes in the *Trp* operon. Expression of the genes in the *Trp* operon increases when tryptophan levels in the cell are low and decreases when cellular tryptophan levels are high. The five genes in the *Trp* operon are *trpA, trpB, trpC, trpD,* and *trpE.*

TrpR is a protein that binds tryptophan in the cytosol. When TrpR is bound to tryptophan, it also binds DNA of the regulatory region upstream of the *Trp* operon as shown in the figure below.

A researcher grew *E. coli* in growth medium in three different batches. The researcher supplemented one batch with 0.1 mM (millimolar) tryptophan, the second batch with 10 mM tryptophan, and the third batch with 20 mM tryptophan. The researcher then measured expression of *TrpE* under each growth condition and obtained the data shown in the graph. Error bars are ± 1 standard error.

(a) **Describe** the role of the TrpR protein in regulating expression of the *Trp* operon.

(b) **Make a claim** to explain why *E. coli* do not continuously synthesize tryptophan.

(c) **Identify** the concentration of tryptophan in the medium that results in the highest expression of the *TrpE* gene.

(d) **Determine** how many times greater the concentration of tryptophan is in the medium supplemented with 10 mM tryptophan than it is in the medium supplemented with 0.1 mM tryptophan.

(e) Based on the data provided, **predict** whether a strain of *E. coli* that carried a mutation eliminating function of the *trpB* would be able to grow in culture medium that contained no tryptophan. **Provide reasoning** to support your prediction.

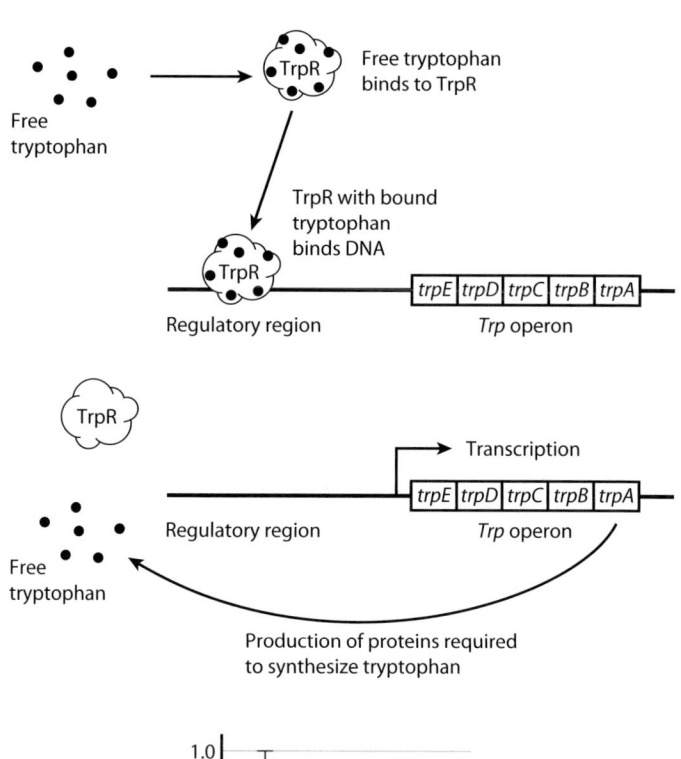

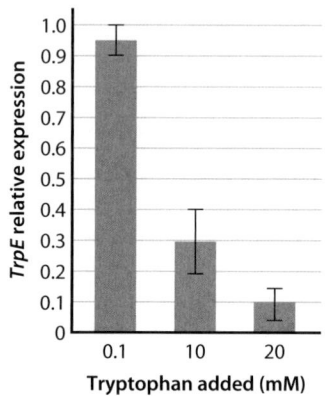

Production of proteins required to synthesize tryptophan

Module 36

Cell Specialization and Development

LEARNING GOALS ▶LG 36.1 In multicellular organisms, cells become increasingly specialized during development.
▶LG 36.2 Some transcription factors act as master regulators of development.
▶LG 36.3 Transcription factors may work in combination to turn a gene on or off.

In the last few modules, we looked at gene expression and regulation. When a gene is expressed, it is turned on, so that DNA is transcribed to mRNA, and mRNA is translated to protein. When a gene is not expressed, it is turned off and no functional product is made. We also noted that genes are not simply on or off. In a multicellular organism, for example, they may be on in some cells, but not in others. During development, they may be on at some times, but not at other times. Some genes may be expressed at low levels in one time or place, but at high levels at another time or place. These are examples of gene regulation, which is the set of processes that control gene expression so that genes are only turned on at the right times, in the right cells, and at the right levels.

We also discussed how gene expression can be regulated at many different steps, from the chromosome, to transcription, to mRNA stability and processing, to translation, and even after the final protein is produced. These different points of control, or levels of gene regulation, provide eukaryotic cells the ability to fine tune gene expression.

One of the best examples of the importance of gene regulation is the process by which cells specialize during development. Altogether, the human body contains about 200 different types of cell, all of which derive from a single cell, the zygote. Some cells derived from the zygote

become muscle cells, others nerve cells, and still others liver cells. All of these cells have almost exactly the same genome: they differ not in what genes they have but instead in when and where these genes are expressed or repressed. In other words, these cell types differ as a result of gene regulation, which was discussed in Module 35.

In this module, we will focus on development as a way to illustrate the importance of the regulation of gene expression. We will see that as cells differentiate along one pathway they progressively lose their ability to differentiate along other pathways. We will then focus on eye development and floral development as examples to show that some key molecular mechanisms of development are used over and over again in different organisms. However, they have evolved to yield many differences in shape and form that underlie the diversity of life on Earth.

PREP FOR THE AP® EXAM

FOCUS ON THE BIG IDEA

INFORMATION STORAGE AND TRANSMISSION: Look at how information for cell specialization and development is transmitted from DNA to proteins at the correct time and in the correct order.

36.1 In multicellular organisms, cells become increasingly specialized during development

In multicellular organisms, gene regulation underlies development, the process in which a fertilized egg undergoes multiple rounds of cell division to become an embryo with specialized tissues and organs. During development, cells undergo changes in gene expression as genes are turned on

and off at specific times and places. Gene regulation causes cells to become progressively more specialized, a process known as **differentiation,** which we will focus on in this section.

In all sexually reproducing organisms, the fertilized egg, shown at the top in **FIGURE 36.1** on page 500, is special because it can give rise to a complete organism. The fertilized egg is said to be totipotent (from the Latin *toti* meaning "all"). At each successive stage in development, cells lose the

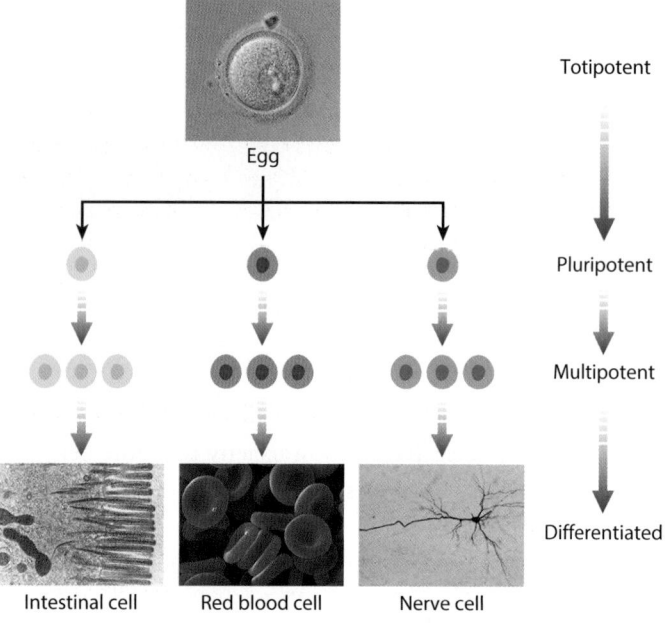

FIGURE 36.1 Differentiation

As cells develop, they become increasingly specialized. The egg is a totipotent cell which can give rise to a complete organism. Pluripotent cells can give rise to any type of cell, but not a complete organism. Multipotent cells can form a limited number of cell types. Differentiated cells, like intestinal cells, red blood cells, and nerve cells, are specialized for different functions. Photos: (top) Claude Cartier/Science Source; (bottom left) Don W. Fawcett/Science Source; (bottom middle) Cheryl Power/Science Source; (bottom right) Biophoto Associates/Science Source

potential to develop into any kind of cell and become more restricted in the types of cells they can become. For example, after many cell divisions, cells of the developing embryo become pluripotent (from the Latin *pluri* meaning "more") because they can give rise to any cell of the body. However, they cannot give rise to an entire organism on their own, as a totipotent cell can. Cells further along in development are multipotent (from the Latin *multi* meaning "many"); these cells can form a limited number of types of cells. Totipotent, pluripotent, and multipotent cells are all stem cells, which are cells that are capable of differentiating into different cell types.

Why do differentiating cells increasingly lose their developmental potential? One hypothesis focuses on gene regulation, discussed in the previous module. When cells become committed to a particular developmental pathway, genes that are no longer needed are turned off, or repressed, and are difficult to turn on again. This hypothesis has been supported by many experiments. Another hypothesis is genome reduction. As cells become differentiated, they delete DNA for genes they no longer need. This hypothesis turns out to be incorrect.

These two hypotheses were tested by experiments in which differentiated cells were placed in an environment that reprograms them to mimic earlier states. If loss of developmental potential is due to gene regulation, then differentiated cells could be reprogrammed to become stem cells again because the genes are present but turned off. If loss of developmental potential is due to genome reduction, then differentiated cells could not be reprogrammed to become stem cells because the genes for alternate developmental routes are deleted and no longer present.

British developmental biologist John Gurdon carried out such experiments in the early 1960s with the African clawed frog *Xenopus laevis*. This work is described in "Practicing Science 36.1: How do stem cells lose their ability to differentiate into any cell type?" Gurdon used a procedure called nuclear transfer, in which the nucleus of one cell is placed in the cytoplasm of another cell in which the nucleus has been inactivated or removed. He used a hollow glass needle to insert the nucleus of a differentiated cell (in this case, an intestinal cell) into an egg with an inactivated nucleus. He found that at least some nuclei from intestinal cells can be reprogrammed to develop into normal tadpoles. These findings support the first hypothesis: all of the same genes are present in intestinal cells as in early embryonic cells, but some of the genes are turned off, or repressed, during development.

FIGURE 36.2 summarizes the results of many nuclear transfer experiments in mammals and amphibians.

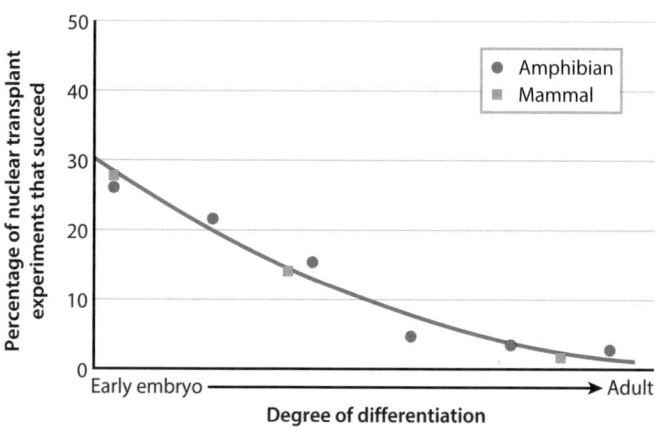

FIGURE 36.2 Results of nuclear transfer experiments

Cells that are further along in development are more difficult to reprogram, as shown by the declining success of nuclear transfer experiments as cells are more differentiated. Data from: J. B. Gurdon and D. A. Melton. 2008. "Nuclear Reprogramming in Cells." *Science* 322:1811–1815.

How do stem cells lose their ability to differentiate into any cell type?

Background During differentiation, cells become progressively more specialized and their developmental paths more restricted over time. Early studies left the mechanism of differentiation unclear.

Hypotheses According to one hypothesis, differentiation occurs when patterns of gene expression in the cell change as it develops, and genes that are no longer needed are repressed. An alternative hypothesis suggests that differentiation occurs as a result of genome reduction, in which genes that are not needed are deleted.

Experiment John Gurdon carried out experiments in the African clawed frog to test these hypotheses. He transferred nuclei from differentiated intestinal cells into unfertilized eggs whose nuclei had been inactivated with ultraviolet light. If differentiation is due to changes in gene expression, then the differentiated nucleus might be able to be reprogrammed into an earlier state when placed in an egg, and then develop into a tadpole. If differentiation is the result of loss of genes, then differentiation is irreversible, and development cannot proceed.

Results As you can see in the figure, the experiment was carried out 726 times. In 716 cases, development did not occur; in 10 cases, development proceeded normally.

Follow-up Work Gurdon's work was controversial. Some critics argued that the successful experiments resulted from a small number of undifferentiated cells present in the lining of the intestine. Others accepted the conclusion but expressed misgivings about possible applications to humans. Later experiments that succeeded in cloning mammals from fully differentiated cells confirmed Gurdon's original conclusion.

SOURCE
Gurdon, J. B. 1962. "The Developmental Capacity of Nuclei Taken from Intestinal Epithelium Cells of Feeding Tadpoles." *Journal of Embryology & Experimental Morphology* 10:622–640.

AP® PRACTICE QUESTION

a. **Identify** the following:
   1. The question being investigated by the experimenters
   2. The hypotheses they tested
   3. The experimental group
   4. The control group (one is not specified; propose one)
   5. The independent variable
   6. The dependent variable

b. **Explain** the results of the experiment by **identifying** the researchers' claim and **justifying** the claim with the evidence and the reasoning researchers used (CER).

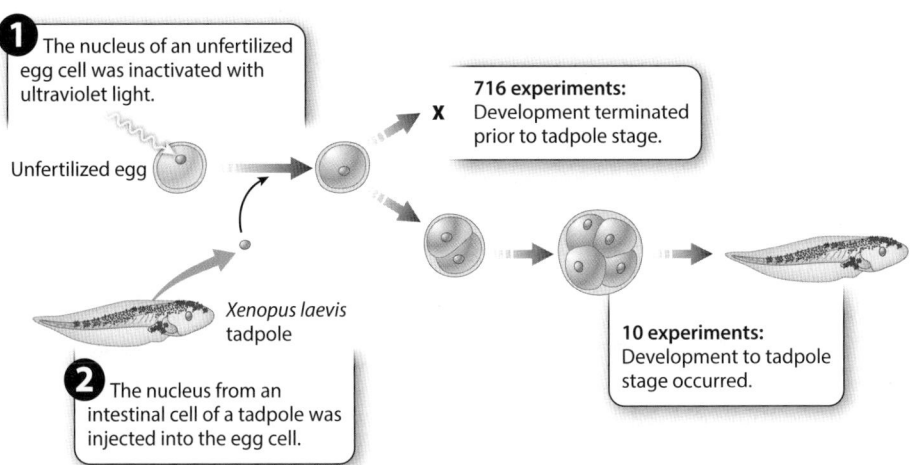

1 The nucleus of an unfertilized egg cell was inactivated with ultraviolet light.

Unfertilized egg

Xenopus laevis tadpole

2 The nucleus from an intestinal cell of a tadpole was injected into the egg cell.

X **716 experiments:** Development terminated prior to tadpole stage.

10 experiments: Development to tadpole stage occurred.

Conclusion Although the experiment succeeded in only 10 of 726 attempts, it demonstrated that the nucleus of a differentiated cell in the cytoplasm of the unfertilized egg was able to support complete development of an animal. This result allowed Gurdon to reject the hypothesis that differentiation occurs by the loss of genes. The first hypothesis—that cells become differentiated as a result of changes in gene expression—was supported. However, because of the small number of successes in reprogramming, additional experiments were needed to validate the conclusions.

FIGURE 36.3 Clones

Dolly is a genetic clone of her mother and was produced by nuclear transfer. Photo: JOHN CHADWICK/AP Images

The percentage of reprogramming experiments that fail increases as cells differentiate. The best chance of success is to use stem cell nuclei from early development. However, even some experiments using nuclei from fully differentiated cells have been successful.

When nuclear transfer succeeds, the result is a clone—an individual that carries an exact copy of the genome of another individual. In this case, the new individual shares the same genome as that of the individual from which the donor nucleus was obtained. The first mammalian clone was a lamb called Dolly, shown in **FIGURE 36.3**. Born in 1996, Dolly was produced from the transfer of the nucleus of a cell in the mammary gland of a sheep to an egg cell with no nucleus, and was the only successful birth among 277 nuclear transfers. Successful cloning in sheep soon led to cloning in cattle, pigs, and goats. The first household pet to be cloned was a kitten named CopyCat, born in 2001 and derived from the nucleus of a differentiated ovarian cell.

Stem cells play a prominent role in regenerative medicine, which aims to use the natural processes of cell growth and development to replace diseased or damaged tissues. Stem cells are already used in bone marrow transplantation and may someday be used to treat Parkinson's disease, Alzheimer's disease, heart failure, certain types of diabetes, severe burns and wounds, and spinal cord injury.

A major breakthrough in regenerative medicine occurred in 2006 when Japanese scientists demonstrated that adult differentiated cells can be reprogrammed to become stem cells by activation of just a handful of genes, most of them encoding transcription factors or chromatin proteins. As we saw in Module 35, transcription factors bind to DNA and can activate gene expression. So, when the researchers activated key genes, the proteins they produced turned on other genes that were turned off during development, returning the cell to an earlier developmental state. This kind of reprogramming opens the door to personalized stem cell therapies. The goal is to create stem cells derived from the adult cells of the individual patient. Because these cells contain the patient's own genome, problems with tissue rejection are minimized or eliminated. Researchers hope that someday soon, your own cells containing your personal genome could be reprogrammed to restore cells or organs damaged by disease or accident.

✓ Concept Check

1. The terms "totipotent," "pluripotent," and "multipotent" derive from Latin roots: *toti* (all), *pluri* (more), and *multi* (many). **Describe** how these terms accurately convey the developmental potential of each type of cell.

2. **Describe** why a cell from an early embryo has more developmental potential than a cell from a later embryo.

3. **Describe** how an individual's own cells might be used in stem cell therapy.

36.2 Some transcription factors act as master regulators of development

Transcription factors play key roles during development. Recall from Module 35 that transcription factors are proteins that bind to sequences near genes, such as promoters, and turn gene expression on or off. Many transcription factors that are important in development are the same or very similar in many different species, even in two species that are distantly related to each other. This observation suggests that the transcription factor was present in the last common ancestor of all of these species, and then maintained for long periods as the species evolved over time, which is known as being **evolutionarily conserved.** An impressive example of evolutionary conservation is found in the development of animal eyes. We will focus on eye development in this section as a way to illustrate the importance of transcription factors in development.

Animal eyes show amazing diversity, as illustrated in **FIGURE 36.4.** Among the simplest eyes are those of the planarian flatworm, shown in Figure 36.4a, which are only pit-shaped cells containing light-sensitive photoreceptors. The planarian eye has no lens to focus the light, but the animal can perceive differences in light intensity. Even such a simple eye provides benefits to the organism, allowing it to differentiate light from dark. The incorporation of a spherical

a. Planarian

b. Sea jelly

c. Squid

d. Human

e. House fly

f. Trilobite

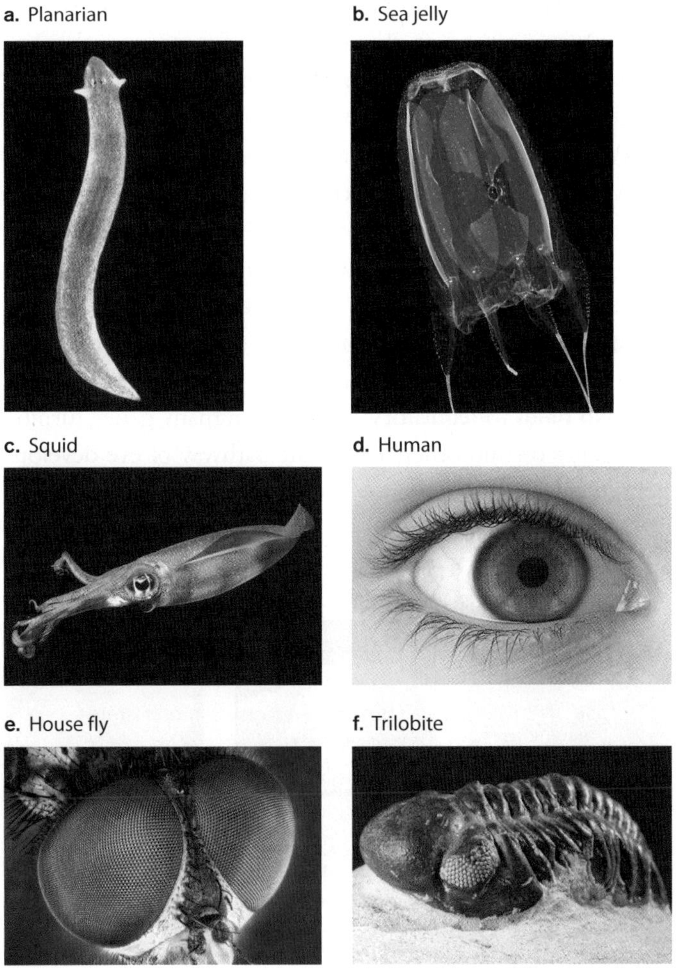

FIGURE 36.4 Diversity of animal eyes

Animals have many different types of eyes. (a) Planaria have an eye cup. (b) Some sea jellies have a lens to focus light. Squids (c) and humans (d) both have single-lens eyes. House flies (e) and extinct trilobites, shown here as a fossil (f) have compound eyes made up of many small lenses. Sources: (a) David M. Dennis/AGE Fotostock; (b) Masa Ushioda/SeaPics.com; (c) Reinhard Dirscherl/AGE Fotostock; (d) Andrey Armyagov/iStockphoto; (e) Julian Brooks/AGE Fotostock; (f) Walter Geiersperger/AGE Fotostock

lens, as in some sea jellies, shown in Figure 36.4b, improves the image, allowing the organism to detect prey and avoid predators.

Among the most complex eyes are the camera-type eyes of the squid, shown in Figure 36.4c, and humans, shown in Figure 36.4d, which have a single lens to focus light onto a light-sensitive tissue, the retina. Though the single-lens eyes of squids and vertebrates are similar in external appearance, they are vastly different in their development, anatomy, and physiology. Some organisms, such as the house fly, shown in Figure 36.4e, and other insects, have compound eyes. Compound eyes consist of hundreds of small lenses arranged on a convex surface and pointing in slightly different directions. This arrangement of lenses allows a wide viewing angle and detection of rapid movement.

In the history of life, eyes are very ancient. By the time of the extraordinary diversification of animals 542 million years ago, known as the Cambrian explosion, organisms such as trilobites, shown in Figure 36.4f, already had well-formed compound eyes. These differed greatly from the eyes in modern insects, however. For example, trilobite eyes had hard mineral lenses composed of calcium carbonate (the world's first safety glasses, so to speak). Altogether, about 96% of living animal species have true eyes that produce an image, as opposed to simply being able to detect differences in light intensity.

The wide diversity of structure and function of animal eyes suggests that eyes may have evolved independently in different organisms. However, more recent evidence suggests an alternative hypothesis—that they evolved once, very early in evolution, and subsequently diverged over time. One piece of evidence that supports the hypothesis of a single origin for light perception is the observation that the light-sensitive molecule in all light-detecting cells is the same, a derivative of vitamin A in association with a protein called opsin. The presence of the same light-sensitive molecule in diverse eyes suggests that it may have been present in the common ancestor of all animals with eyes and retained over time.

Another piece of evidence in favor of a single origin of eyes comes from studies of eye development. Researchers identified a gene called *eyeless* in the fruit fly *Drosophila*. As its name implies, the phenotype of *eyeless* mutants is abnormal eye development, which is shown in **FIGURE 36.5a** on page 504. When the protein product of the *eyeless* gene was identified, it was found to be a transcription factor called Pax6. (By convention, in *Drosophila*, gene names are italicized, but protein names are not.) Mutant forms of a *Pax6*

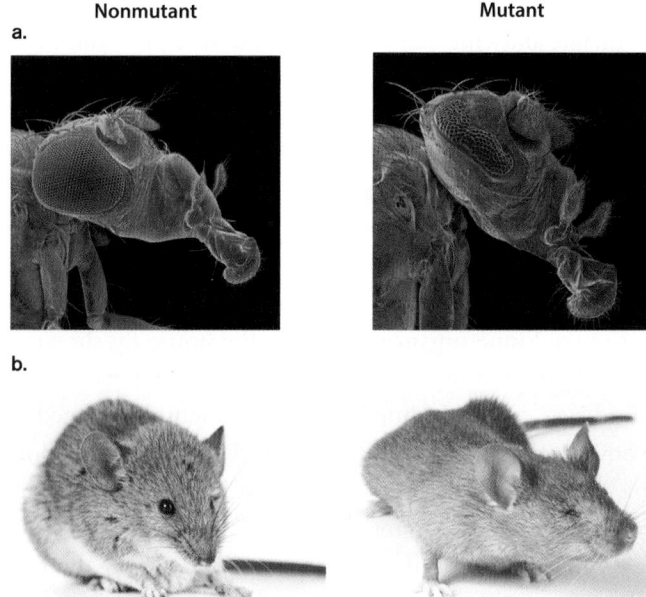

a.　Nonmutant　　　　　Mutant

b.

FIGURE 36.5 Effect of *Pax6* mutations on eye development

Pax6 is important for eye development in fruit flies and mice. (a) Fruit flies typically have round, compound eyes. Mutations in *Pax6* lead to small or absent eyes. (b) Similar eye phenotypes are observed in mice without mutations in *Pax6* and mice with mutations in *Pax6*. Sources: (a) David Scharf/Science Source; (b) left: INSADCO Photography/Alamy; right: Jennifer L. Torrance, Photographer, The Jackson Laboratory

gene were already known to cause small eyes in the mouse, shown in Figure 36.5b, and absence of the iris, known as aniridia, in humans.

The similarity in phenotypes of *Pax6* mutants, along with the observation that Pax6 is evolutionarily conserved in *Drosophila* and mouse, led Swiss developmental biologist Walter Gehring to hypothesize that Pax6 might be a master regulator of eye development. In other words, he hypothesized that Pax6 binds to regulatory sequences in many genes, turning them on or off. These genes are part of a developmental program that leads to eye development. Like all hypotheses, this one makes a prediction that can be tested: it suggests that Pax6 can induce the development of an eye in any tissue in which it is expressed.

The expression of *Pax6* is regulated—it is expressed in the eye, but not elsewhere in the body of the fruit fly. To test the hypothesis, Gehring and collaborators genetically engineered fruit flies that produce the Pax6 transcription factor in the antenna, where it is not normally expressed. They found that the antenna developed into a miniature compound eye. A normal antenna is shown in **FIGURE 36.6a**, and an antenna that expresses Pax6 and

produces an eye is shown in Figure 36.6b. The researchers also engineered fruit flies which produced eyes on the legs, wings, and other tissues, which the *New York Times* publicized in an article headlined "With New Fly, Science Outdoes Hollywood."

Gehring and his group then went one step further. They took the *Pax6* gene from mice and expressed it in fruit flies to see whether the mouse *Pax6* gene is similar enough to the fruit fly version of the gene that it could induce eye development in the fruit fly. Specifically, they created fruit flies that expressed the *Pax6* gene from mice in the fruit fly antenna. The mouse gene induced a miniature compound eye in the fly, shown in Figure 36.6c.

The ability of mouse *Pax6* to make an eye in fruit flies suggests that mouse and fruit fly *Pax6* are not only similar in DNA sequence but also similar in function, and act as a master switch that can turn on a developmental program that leads to the formation of an eye. Transcription factors like Pax6 interact with their target genes by binding to short DNA sequences that are adjacent to the gene. These regulatory sequences help determine whether the adjacent DNA is transcribed. When bound to regulatory sequences, some transcription factors act as activators that lead to transcription of the target gene, and others act as repressors that prevent transcription of the target gene.

Pax6 binds to regulatory sequences in many genes, turning some genes on and others off. In the pathway of eye development, these target genes are considered to be downstream of

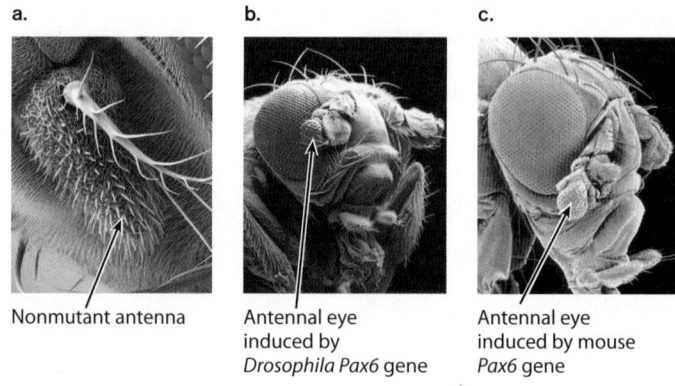

a. Nonmutant antenna

b. Antennal eye induced by *Drosophila Pax6* gene

c. Antennal eye induced by mouse *Pax6* gene

FIGURE 36.6 *Pax6*, a master regulator controlling eye development

Pax6 leads to eye development when it is expressed in tissues where it is normally not expressed. (a) A typical fruit fly antenna does not form eye tissue. (b) Eye tissue is formed on the antenna when fruit fly *Pax6* is expressed in the antenna. (c) Eye tissue is also formed on the antenna when mouse *Pax6* is expressed in the antenna. Photos: (a) Cheryl Power/Science Source; (b) Eye of Science/Science Source; (c) Prof. Walter Gehring/Science Source

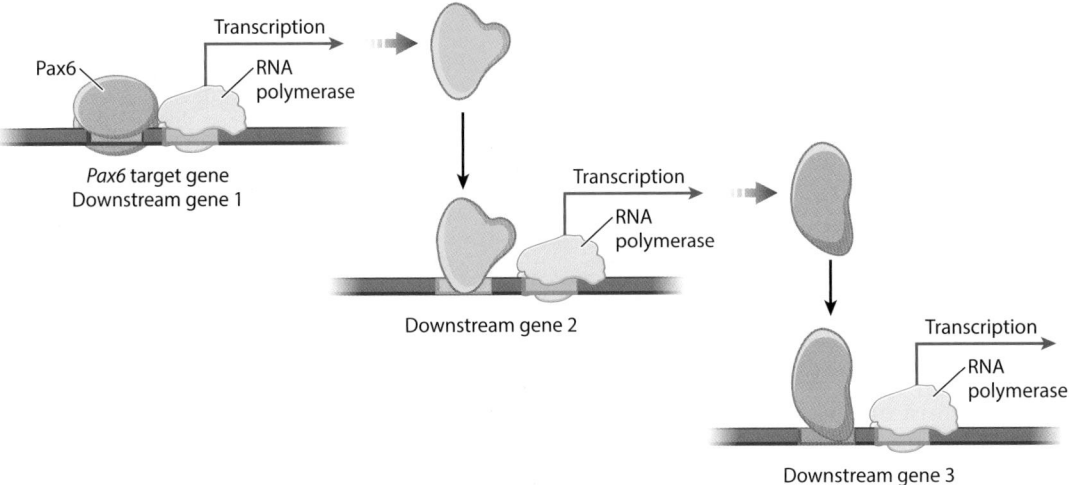

FIGURE 36.7 Downstream genes

Pax6 binds to DNA sequences, where it turns on transcription of target genes, also called downstream genes. In turn, the protein products of these downstream genes turn on genes that are further downstream. In this figure, only one downstream gene for each protein is shown.

Pax6, as shown in **FIGURE 36.7**. The products of these downstream genes, in turn, affect the expression of further downstream genes. The total number of genes that are needed for eye development is estimated at about 2000. So, we see that the activation of one gene can turn on an entire developmental pathway, which is why *Pax6* is considered a master regulator.

One hypothesis of how *Pax6* became a master regulator for eye development in a wide range of organisms, yet produces a diversity of eyes, is that it evolved early in the history of life as a transcription factor that was able to bind to and regulate genes involved in early eye development. Over time, different genes in different organisms acquired new Pax6-binding sequences by mutation, and if these were beneficial they persisted. Therefore, the downstream genes that are targets of Pax6 are different in different organisms. In this way, a single master regulator turns on the pathway for eye development in many different organisms, but the eyes that are produced are quite diverse.

AP® EXAM TIP

Remember the role of transcription factors and their importance to regulating genes so development progresses appropriately.

✓ Concept Check

4. *Pax6* is considered a master regulator of eye development. **Describe** what the gene encodes.

5. **Describe** what it means when scientists say that a gene is evolutionarily conserved.

36.3 Transcription factors may work in combination to turn a gene on or off

In the last section, we focused on a single transcription factor, Pax6, and how it affects the expression of genes involved in eye development. Similarly, in Module 35, we discussed the *lac* operon, where a single repressor, when bound to DNA, blocks transcription of genes involved in lactose use. In both of these cases, we focused on one regulatory molecule and how it affects gene expression. However, many genes have binding sites for more than one regulatory molecule. Some of these molecules are activators of transcription and others repressors of transcription. Therefore, whether a gene is turned on or off often depends on the combination of transcription factors that are present in a cell and bound to DNA. It also depends on the relative balance of transcription factors that are activators and those that are repressors. Regulation of gene transcription according to the mix, or combination, of transcription factors in the cell is known as combinatorial control. Combinatorial control of transcription is a general principle often seen in multicellular organisms at many stages of development. Here, we discuss flower development as an example of this general principle of gene regulation.

The plant *Arabidopsis thaliana* is a model organism for developmental studies and presents a clear example of combinatorial control. As in all plants, *Arabidopsis* has regions of undifferentiated cells where growth of shoots, roots, and flowers takes place. Meristem cells in plants are similar to stem cells in animals because they consist of cells that can differentiate into different structures. In floral meristems of *Arabidopsis,* the flowers develop from a pattern of four concentric circles of cells, or whorls, each of which differentiates into a distinct type of floral structure.

From the outermost whorl (whorl 1) of cells to the innermost whorl (whorl 4), the floral organs are formed as shown in **FIGURE 36.8**. Note the following:

- Cells in whorl 1 form the green sepals, which are modified leaves forming a protective sheath around the petals in the flower bud.
- Cells in whorl 2 form the petals.
- Cells in whorl 3 form the stamens, which are the male sexual structures in which pollen is produced. The small granules on the stamens in Figure 36.8 are pollen grains.

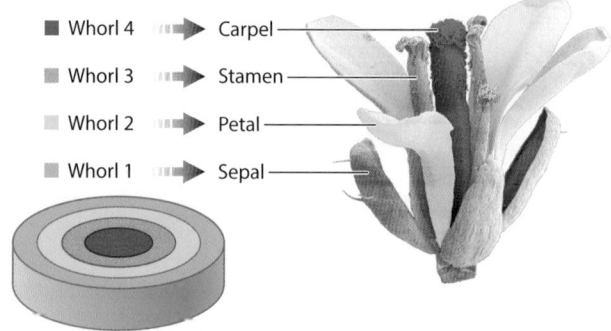

FIGURE 36.8 Flower development in *Arabidopsis*

Cells in the floral meristem are organized in a pattern of four concentric circles, or whorls, shown on the left. These whorls give rise to different parts of the flower, shown on the right. Photo: Juergen Berger/Max Planck Institute for Developmental Biology, Tuebingen, Germany

- Cells in whorl 4 form the carpels, which are the female reproductive structures that receive pollen and contain the ovaries.

The genetic control of flower development in *Arabidopsis* was discovered by the analysis of mutant plants, which is a common method in genetics. The plant mutants fell into three classes showing characteristic floral abnormalities, which are shown in **FIGURE 36.9**. A nonmutant (wild-type) flower is shown in Figure 36.9a. One set of mutants affects whorls 1 and 2 (Figure 36.9b); another affects whorls 2 and 3 (Figure 36.9c); and a third affects whorls 3 and 4 (Figure 36.9d).

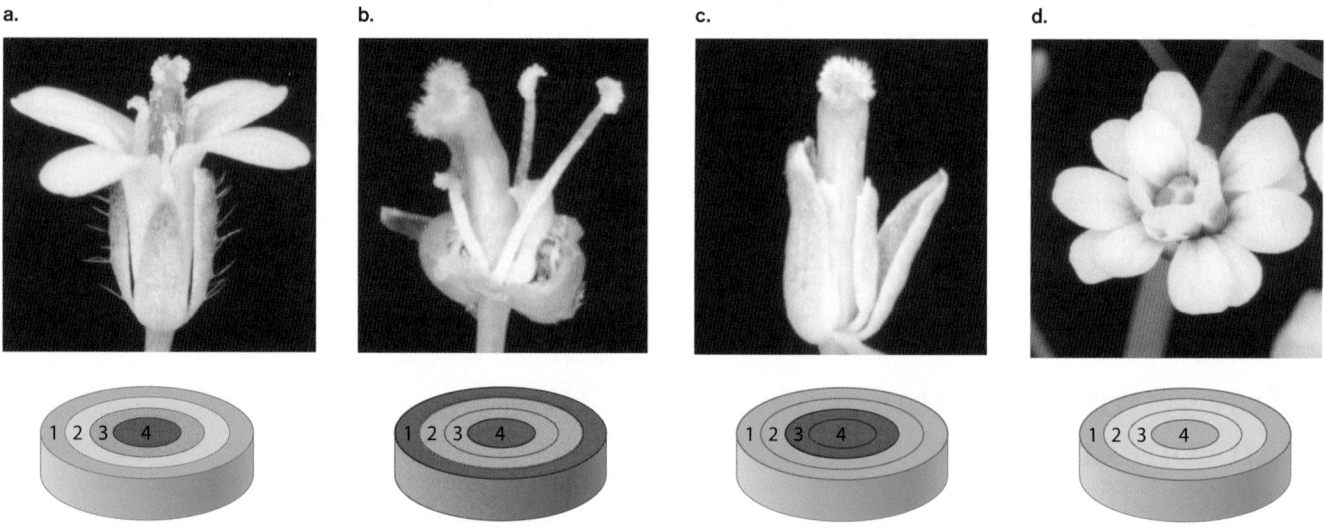

FIGURE 36.9 Phenotypes of the normal flower and floral mutants of *Arabidopsis*

A normal flower is shown in part a. Various mutant flowers are shown in parts b–d.

Photos: P. John Bowman

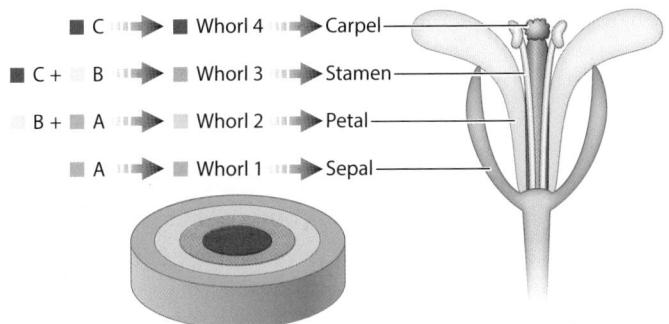

FIGURE 36.10 The ABC model for flower development

The combination of A, B, and C activities in each whorl determines which flower organs develop from that whorl. For example, activity A leads to the development of sepals, activities A and B lead to the development of petals, and so on.

These observations led researchers to a model of floral development called the ABC model. The model invokes three activities called A, B, and C. These activities represent the function of one or more different protein products of different genes that are hypothesized to be present in cells of each whorl as shown in **FIGURE 36.10**. In the course of floral differentiation, the cells of each whorl become different from those in the other whorls, and different combinations of genes are activated or repressed in each whorl. Activity A is present in whorls 1 and 2, activity B in whorls 2 and 3, and activity C in whorls 3 and 4. Furthermore, activity A alone results in the formation of sepals, A and B together result in petals, B and C together result in stamens, and C alone results in carpels.

Like any good hypothesis, the ABC model makes specific predictions. Mutants lacking activity A will have defects in whorls 1 and 2; mutants lacking activity B will have defects in whorls 2 and 3; and mutants lacking activity C will have defects in whorls 3 and 4, just like those shown in Figure 36.9. The identification of the products of these genes demonstrates combinatorial control and explains the mutant phenotypes. All of the genes encode transcription factors. The transcription factors bind to regulatory sequences of genes that encode proteins that are necessary for each whorl's development, similar to what we saw earlier for Pax6 and eye development.

Although other transcription factors associated with other activities that also contribute to flower development were discovered later, the original ABC model is still valid and serves as an elegant example of combinatorial control. As in the case of the *Pax6* gene in the development of animals' eyes, evolution of the regulatory sequences in downstream genes has resulted in a wide variety of floral shapes and forms in different plant groups, some of which are shown in **FIGURE 36.11**.

✓ Concept Check

6. **Describe** combinatorial control.

7. A gene has regulatory sequences that bind transcription factors, some of which act as activators and others as repressors. However, only activators are present in the cell and bound to DNA. **Describe** whether the gene is transcribed.

a.
b.
c.
d.

FIGURE 36.11 Flower diversity

Flowers, like eyes, show remarkable diversity in part as a result of differences in gene expression. (a) Colorado blue columbine (*Aquilegia caerulea*); (b) slipper orchid (*Paphiopedilum holdenii*); (c) ginger flower (*Smithatris supraneanae*); (d) garden rose (hybrid). Photos: (a) ljh images/Shutterstock; (b) Korkiat/Getty Images; (c) Nonn Panitvong; (d) Maria Mosolova/AGE Fotostock

Module 36 Summary

PREP FOR THE AP® EXAM

REVISIT THE BIG IDEAS

INFORMATION STORAGE AND TRANSMISSION: Describe how *Pax6* exemplifies the mechanism for cell specialization and development used in cells to ensure the transfer of information from DNA to proteins at the proper time and in the correct order.

LG 36.1 In multicellular organisms, cells become increasing specialized during development.

- In the development of humans and other animals, cells become progressively more restricted in their possible pathways of cellular differentiation. Page 500

- The fertilized egg is totipotent, which means it can give rise to a complete organism. Page 500

- At each successive stage in development, cells lose developmental potential as they differentiate. Page 500

- Pluripotent cells can give rise to any type of cell, but not a complete organism. Page 500

- Multipotent cells can form only a limited number of specialized cell types. Page 500

- Stem cells play a prominent role in regenerative medicine, in which stem cells—in some cases, reprogrammed cells from the patient's own body—are used to replace diseased or damaged tissues. Page 502

LG 36.2 Some transcription factors act as master regulators of development.

- Many proteins that are important in development are similar in sequence from one organism to the next. Such proteins are evolutionarily conserved. Page 503

- Although animals exhibit an enormous diversity in eyes, the observation that the proteins involved in light perception are evolutionarily conserved suggests that the ability to perceive light may have evolved once, early in the evolution of animals. Page 503

- The Pax6 transcription factor is a master regulator of eye development. Page 504

- Pax6 activates a pathway for eye development in many different organisms. Page 504

LG 36.3 Transcription factors may work in combination to turn a gene on or off.

- Combinatorial control is a developmental mechanism in which cellular differentiation depends on the particular combination of transcription factors present in a cell. Page 505

- By analyzing mutants that affect flower development in the plant *Arabidopsis,* researchers were able to determine the genes involved in normal flower development. Page 505

- The ABC model of flower development invokes three activities (A, B, and C) present in circular regions (whorls) of the developing flower, with the specific combination of factors determining the developmental pathway in each whorl. Page 507

Key Terms

Differentiation
Evolutionarily conserved

Review Questions

1. Differentiation refers to the process by which
   (A) fertilized eggs undergo multiple rounds of cell division to become embryos.
   (B) changes in gene expression allow cells to produce the correct proteins at the correct time.
   (C) cells become progressively more specialized during development.
   (D) gene regulation determines which proteins are produced in a given cell during development.

2. The *Pax6* gene acts as a master regulator of eye development. If a *Pax6* gene from a mouse were engineered to be expressed in the antenna of *Drosophila*, what would you expect to observe?
   (A) A compound eye develops on the antenna.
   (B) A mouse eye develops on the antenna.
   (C) A compound eye develops on the head.
   (D) A mouse eye develops on the head.

3. Using the natural processes of cell growth and development to replace diseased or damaged tissue is called
   (A) cloning.
   (B) gene regulation.
   (C) genetic engineering.
   (D) regenerative medicine.

4. A cell in the lining the human gut is very different in structure and function from a white blood cell. How would you describe the genetic basis for this difference?
   (A) These different cell types express different sets of genes, although their genomes are identical.
   (B) These different cell types contain different sets of genes as a result of modifications during development.
   (C) Genes are gradually lost as these cell types differentiate into specialized tissues.
   (D) The genes evolve as the body develops the specialized tissues needed for full development.

5. In the ABC model of floral development, A, B, and C stand for the activities of different:
   (A) transcription factors.
   (B) enzymes.
   (C) RNA molecules
   (D) carbohydrates.

6. The fertilized egg is totipotent, which means
   (A) it only contains genetic material from the female.
   (B) it forms the membranes that surround and support the developing embryo.
   (C) it can give rise to a complete organism.
   (D) it can be removed and survive in cell culture.

Module 36
AP® Practice Questions

PREP FOR THE AP® EXAM

Section 1: Multiple-Choice Questions
Choose the best answer for questions 1–4.

1. GABA receptors are a class of cell-surface receptors found in cells of the mammalian brain. GABA receptors are not present in a different mammalian organ, the kidney. Which of the following best explains why GABA receptors are found in the brain but not in the kidney?
   (A) The DNA encoding the gene for GABA receptors is deleted from kidney cells during their development.
   (B) The mRNA encoding GABA receptors is translated in the brain but is not recognized by ribosomes in the kidney.
   (C) The cells of the brain produce proteins that drive expression of GABA receptors but the kidney does not produce those proteins.
   (D) GABA receptors have an important function in the kidney but are not required in the brain.

2. The expression level of *gene X* was measured under four different experimental conditions. The abundance of four different transcription factors was measured under the same four conditions. The data obtained are shown in the table, where "+++" indicates high gene expression or protein abundance and "+" indicates low gene expression or protein abundance.

	Abundance of four different transcription factors				Expression of *gene X*
	TF 1	TF 2	TF 3	TF 4	
Condition 1	+++	++	++	+	+++
Condition 2	+	+++	++	+++	+
Condition 3	+	++	++	+++	+
Condition 4	+++	++++	++	+	+++

Identify the transcription factor that is most likely to be a negative regulator of *gene X*.

(A) TF 1

(B) TF 2

(C) TF 3

(D) TF 4

Use the following information to answer questions 3 and 4.

The fruit fly *Drosophila melanogaster* can fight off pathogenic bacterial infections by producing antimicrobial peptides. In the absence of pathogenic bacteria, the pathway that leads to the production of the antimicrobial peptides is repressed by a negative regulatory protein called cactus. When no pathogenic bacteria are present, cactus binds a protein called DIF in the cytoplasm of *D. melanogaster* immune cells. If pathogenic bacteria are present, the immune cells activate a signaling cascade that results in cactus becoming phosphorylated and releasing DIF. This process is illustrated in the figure. In part a, cactus is bound to DIF in the cytoplasm when no pathogen is present. In part b, recognition of a pathogen activates a signaling cascade that results in the phosphorylation of cactus and release of DIF.

a.

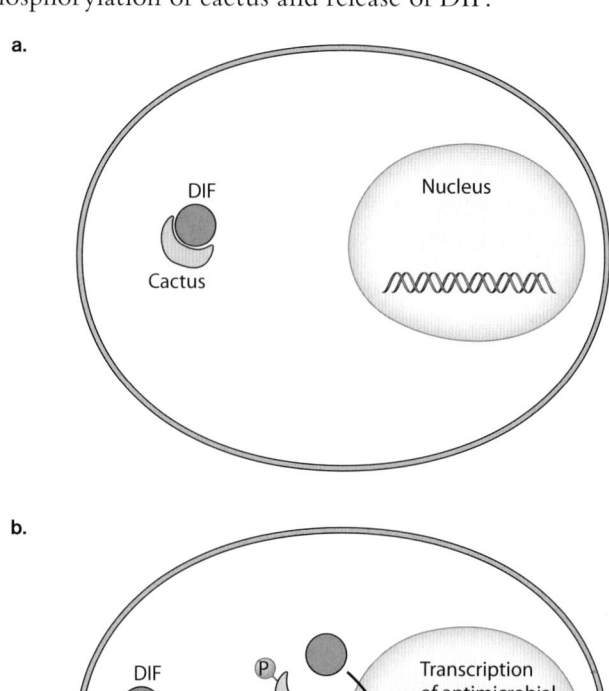

b.

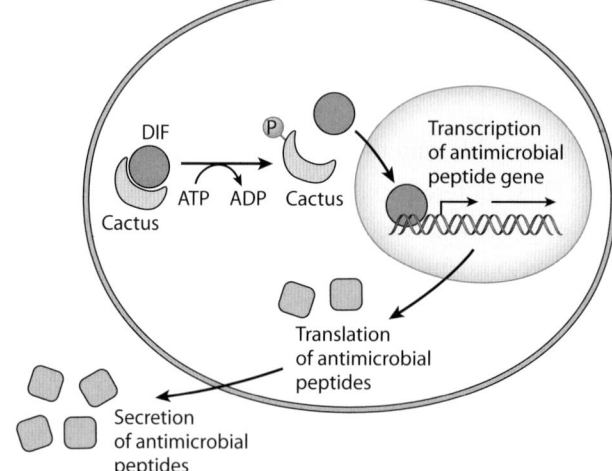

3. Which best describes the role of DIF in the *D. melanogaster* immune response?
   (A) DIF activates expression of genes encoding antimicrobial peptides that are secreted from the cell to kill the pathogen.
   (B) DIF protein is cleaved to form antimicrobial peptides that are secreted from the cell to kill the pathogen.
   (C) DIF activates a signaling cascade that drives expression of antimicrobial peptide genes to kill the pathogen.
   (D) Translation of DIF protein is activated in the nucleus, and DIF is then secreted from the cell to kill the pathogen.

4. Predict the phenotype of a *D. melanogaster* that carries a mutation that prevents Cactus from being phosphorylated.
   (A) The *D. melanogaster* cells will not activate the signaling cascade.
   (B) The *D. melanogaster* cells will be unable to express antimicrobial peptides in response to infection.
   (C) The *D. melanogaster* cells will always have high levels of DIF protein in the nucleus.
   (D) The *D. melanogaster* cells will be unable to express the gene encoding Cactus.

Section 2: Free-Response Question

Write your answer to each part clearly. Support your answers with relevant information and examples. Where calculations are required, show your work.

The backbone of a vertebrate animal is made up of a series of individual bones known as vertebrae. These vertebrae have different shapes and sizes in different parts of the spine. A mouse has seven vertebrae in the neck, or cervical, region of the backbone, which are numbered C1–C7. Vertebrae C3, C4, and C5 are nearly identical to each other, but C1, C2, C6, and C7 have different shapes and sizes.

A scientist measured the expression of six *Hox* genes in the developing vertebrae of mice and observed the pattern shown in the table, where "+" means the gene is expressed and "−" means the gene is not expressed.

	C1	C2	C3	C4	C5	C6	C7
Hox 1.6	+	+	+	+	+	−	−
Hox 1.5	+	+	+	+	+	+	+
Hox 1.4	−	+	+	+	+	+	+
Hox 1.3	−	−	+	+	+	+	+
Hox 3.4	−	−	−	−	−	+	+
Hox 3.3	−	−	−	−	−	−	+

(a) Using the information provided, **identify** the *Hox* genes that must be expressed in order to develop a vertebra the same size and shape as C6.

(b) **Predict** the effect that blocking expression of *Hox 1.3* would have on the size and shape of a vertebra in the C4 position.

(c) Some mice carry a recessive mutation that destroys the function of *Hox 1.4*. **Predict** the phenotypic consequences for C1 vertebra in a mouse that is homozygous for this mutation. **Provide reasoning** to support your prediction.

Module 37

Mutations

LEARNING GOALS ▶**LG 37.1** Mutations can affect one or a few bases or entire chromosomes.
▶**LG 37.2** The genotype is the genetic makeup, and the phenotype is an observable characteristic.
▶**LG 37.3** Mutations can be inherited, occur spontaneously, or be induced by mutagens.

Up to this point, we have looked at the process of gene expression, in which a gene is transcribed to messenger RNA, which is then translated to protein. We have also examined how this process is regulated, so that a gene is turned on or off in particular cells, at particular times, and at particular levels. As we saw in the last module, the regulation of gene expression is critical for the development of an organism from a single cell to a complex, multicellular individual with different cells specialized for different tasks.

Throughout, we have treated the DNA sequence of a gene as a single sequence of nucleotides. However, in reality, the sequence of a particular gene may be different in different individuals of the same species. Indeed, the sequence of nucleotides in the entire genome is certainly different from one individual to the next. These differences are known as mutations, which we discussed in Units 4 and 5. Any heritable change in the nucleotide sequence of DNA is a mutation. A change might be a substitution of one nucleotide for another, a small deletion or insertion of a few nucleotides, or a larger alteration affecting a segment of a chromosome. By "heritable," we mean that the mutation persists through DNA replication and is transmitted during cell division, such as meiotic and mitotic cell division in eukaryotes and binary fission in prokaryotes.

Mutations in DNA can be transmitted to mRNA and then to proteins, where they in turn may affect the form and function of an organism, or its phenotype, which was discussed in Unit 5. This is not surprising because proteins make up cells and do much of the work of the cell. In the last module, for example, we saw that a mutation in the *Pax6* gene can lead to loss of eyes in flies and mice. In this module, we will look closely at mutations, focusing on different types of mutations, their effects, and how they occur.

FOCUS ON THE BIG IDEAS

PREP FOR THE AP® EXAM

INFORMATION STORAGE AND TRANSMISSION: Focus on how changes in genotypes relate to changes in phenotypes, and how changes in phenotypes can in turn affect the survival, growth, and reproduction of organisms.

37.1 Mutations can affect one or a few bases or entire chromosomes

The central dogma of molecular biology tells us that information in a cell typically flows from DNA to RNA to protein. We have followed the processes—transcription and translation—that allow this information flow to occur. We can now focus on how a change in the information in DNA, a mutation, can be transmitted to mRNA and then to a protein, and what effect such a change might have on an organism. In Module 30, for example, we saw that mutations in the *HEXA* gene lead to Tay-Sachs disease and mutations in the *PAH* gene results in phenylketonuria.

In this section, we will discuss different types of mutations in DNA and the effects they have on mRNA and protein, and therefore on the organism.

Point Mutations

The simplest type of mutation is a change in a single nucleotide, known as a point mutation or nucleotide substitution. For example, the base thymine (T) might be replaced with the base cytosine (C). However, because DNA is double stranded, a point mutation involves replacing one base pair (in this example, T–A) with another base pair (in this example, C–G). Point mutations are quite common. For example, any two human genomes are likely to differ from each other at about 3 million nucleotide sites out of a total of 3 billion sites, or about one difference per thousand nucleotides across the genome. In other words, we differ at the level of DNA from one person to the next, but only in a very small fraction of nucleotides.

The effect of a point mutation depends in part on where in the genome it occurs. Consider a mutation that does not occur in a gene. Such a mutation is unlikely to have any observable effect on the organism because it is not transcribed to mRNA and then translated to protein. In many multicellular eukaryotes, including humans, the vast majority of DNA in the genome does not code for protein or for RNAs with known functions. This observation helps to explain why many point mutations have no detectable effects on the organism: they likely occur in regions of the genome that do not encode for proteins.

By contrast, point mutations that occur in coding sequences of genes do have predictable consequences for an organism because these are transcribed to mRNA and translated to protein. **FIGURE 37.1** shows an example in the human DNA sequence coding for β-(beta-)globin, a subunit of hemoglobin, which carries oxygen in red blood cells. Figure 37.1a shows a small part of the nonmutant DNA sequence of β-globin. When this sequence is transcribed and translated, it results in incorporation of the amino acids Pro–Glu–Glu in the β-globin polypeptide.

Figure 37.1b shows an example of a neutral, or harmless, nucleotide substitution in this coding sequence. In this mutation, an A–T base pair (shown in red) is substituted for the nonmutant G–C base pair. When transcribed into mRNA, the mutation changes the nonmutant GAG codon into the mutant GAA codon. But GAG and GAA both code for the same amino acid, glutamic acid (Glu), as you can see from looking at the genetic code (see Figure 34.6 on page 473). In other words, they are synonymous codons, which are codons that specify the same amino acid. Therefore, the resulting amino acid sequences are the same: Pro–Glu–Glu. Such mutations are called silent or synonymous mutations. Synonymous codons typically differ at their third position (the 3′ end of the codon). A quick look at the genetic code shows that most amino acids can be specified by more than one codon, and that in most cases these synonymous codons differ in the identity of the nucleotide at the third position.

A point mutation in a coding sequence can sometimes change the amino acid sequence of the resulting protein. Figure 37.1c shows such an example, in which a T–A base pair (shown in red) is substituted for the nonmutant A–T base pair. The result is a change in the mRNA from GAG, which specifies Glu (glutamic acid), to GUG, which specifies Val (valine). The resulting protein therefore contains the amino acids Pro–Val–Glu instead of Pro–Glu–Glu. This single amino acid change affects the ability of hemoglobin to function normally. In this case, the nucleotide substitution results in an amino acid replacement. Point mutations that cause amino acid replacements are called missense or nonsynonymous mutations.

FIGURE 37.2 on page 514 shows a third way that a point mutation can affect a protein. A mutation can occur that changes a codon that specifies an amino acid to a stop codon. Recall that a stop codon terminates translation. Due to the presence of a stop codon in place of a codon that specifies an amino acid, such a mutation produces a shortened protein. This type of mutation is called a nonsense mutation.

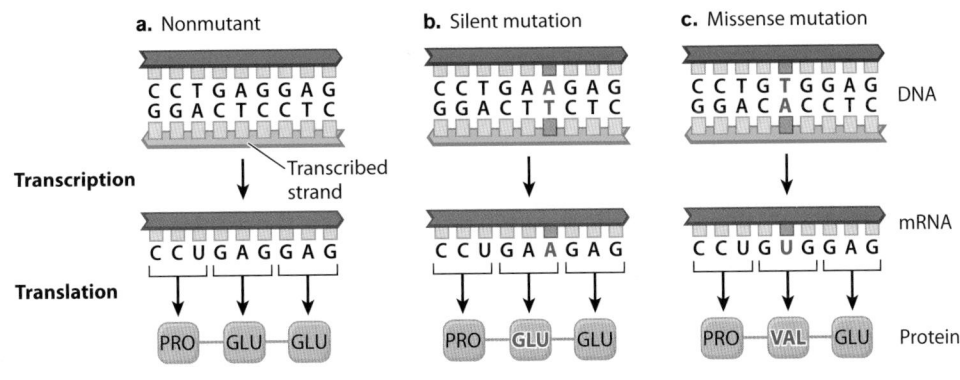

FIGURE 37.1 Silent and missense mutations

Point mutations in genes can be silent or missense. (a) A small portion of the nonmutant sequence of the β-globin gene leads to the synthesis of a protein with amino acids Pro–Glu–Glu. (b) Silent mutations do not change the amino acid sequence of the resulting protein, so it remains the same. (c) Missense mutations change the amino acid sequence, in this case to Pro–Val–Glu.

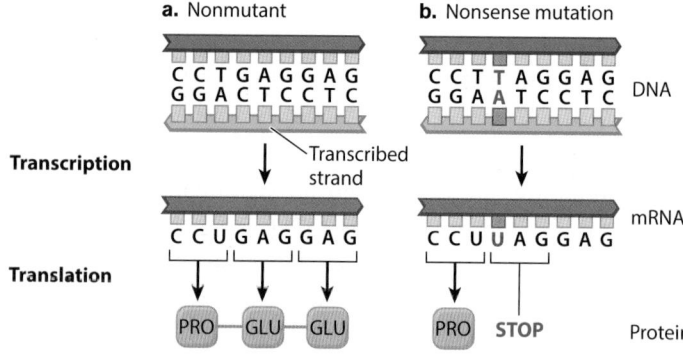

a. Nonmutant

```
C C T G A G G A G
G G A C T C C T C
```
Transcribed strand

Transcription ↓

```
C C U G A G G A G
```
mRNA

Translation ↓

PRO — GLU — GLU

b. Nonsense mutation

```
C C T T A G G A G
G G A A T C C T C
```
DNA

Transcription ↓

```
C C U U A G G A G
```
mRNA

Translation ↓

PRO   STOP

Protein

FIGURE 37.2 A nonsense mutation

A point mutation that creates a stop codon is called a nonsense mutation. (a) A small portion of the nonmutant sequence of the β-globin gene leads to the synthesis of a protein with amino acids Pro–Glu–Glu. (b) A nonsense mutation changes an amino acid to a stop codon, resulting in a shortened and unstable protein, in this case ending in the amino acid Pro.

In Figure 37.2, the mutation creates a UAG codon in the mRNA. Because UAG is a stop codon, the resulting polypeptide terminates after Pro. Nonsense mutations nearly always have harmful effects. Polypeptides that are truncated are typically nonfunctional, unstable, and quickly broken down by the cell.

Small Insertions and Deletions

Another relatively common type of mutation is the deletion or insertion of a small number of nucleotides. In noncoding DNA, which is a region of DNA that does not code for RNAs or proteins, such mutations have little or no effect, similar to what we described for point mutations. In protein-coding regions, the effects of deletions or insertions depend on their size. A small deletion or insertion that is an exact multiple of three nucleotides, such as three, six, nine, and so on, results in a polypeptide with fewer (in the case of a deletion) or more (in the case of an insertion) amino acids as there are codons deleted or inserted. For example, a deletion of three nucleotides eliminates one amino acid, and an insertion of six nucleotides adds two amino acids.

The effects of a deletion of three nucleotides can be seen in the disease cystic fibrosis. This disease is characterized by the production of abnormal secretions in the lungs, liver, and pancreas and other glands. Patients with cystic fibrosis have an accumulation of thick, sticky mucus in their lungs, which often leads to respiratory complications, including recurrent bacterial infections. With proper medical care, including

regular physical therapy to clear the lungs, antibiotics, pancreatic enzyme supplements, and good nutrition, the average life expectancy of a person with cystic fibrosis is currently about 35 to 40 years, and continues to rise.

The mutations responsible for cystic fibrosis occur in the gene encoding the cystic fibrosis transmembrane conductance regulator (CFTR). The CFTR protein is a chloride channel, which acts as a transporter to pump chloride ions out of the cell. Mutations in the *CFTR* gene disrupt the function of the chloride channel, interfering with the usual flow of chloride ions. Because water follows ions by osmosis, the result is a buildup of thick mucus in the lungs, where the *CFTR* gene is expressed. Many different mutations can contribute to cystic fibrosis, including a mutation known as *Δ508* (*delta 508*), which is a deletion of three nucleotides that eliminates a phenylalanine normally present at position 508 in the protein, shown in **FIGURE 37.3**. The

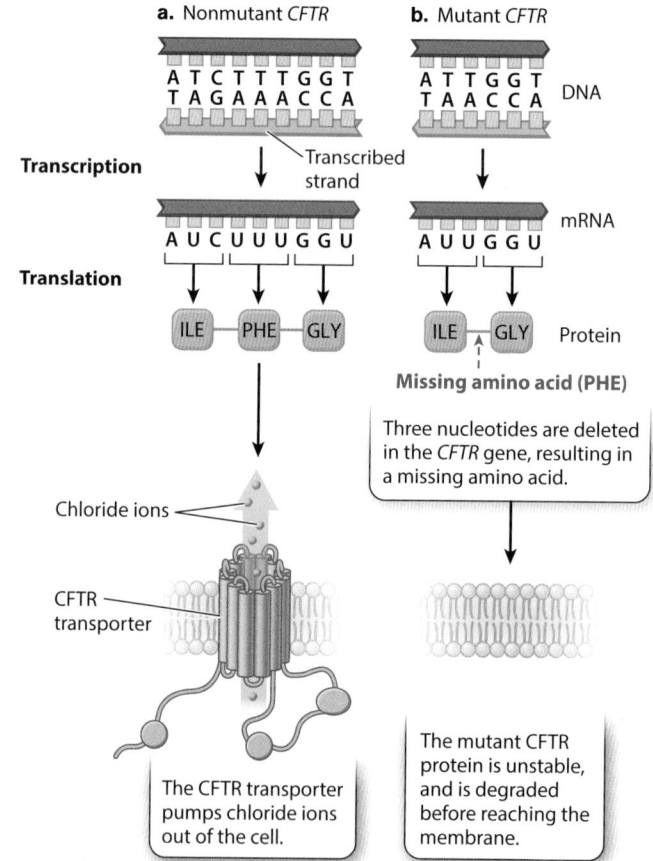

a. Nonmutant *CFTR*

```
A T C T T T G G T
T A G A A A C C A
```
Transcribed strand

Transcription ↓

```
A U C U U U G G U
```

Translation ↓

ILE — PHE — GLY

b. Mutant *CFTR*

```
A T T G G T
T A A C C A
```
DNA

Transcription ↓

```
A U U G G U
```
mRNA

Translation ↓

ILE — GLY

Protein

Missing amino acid (PHE)

Three nucleotides are deleted in the *CFTR* gene, resulting in a missing amino acid.

Chloride ions

CFTR transporter

The CFTR transporter pumps chloride ions out of the cell.

The mutant CFTR protein is unstable, and is degraded before reaching the membrane.

FIGURE 37.3 A small deletion

A deletion of three nucleotides results in a protein with a missing amino acid. In this case, the loss of the amino acid Phe in CFTR, the cystic fibrosis transmembrane conductance regulator, leads to the loss of a membrane transport protein and causes the disease cystic fibrosis.

missing amino acid results in a CFTR protein that does not fold properly and is degraded before it reaches the membrane. Researchers have created drugs that stabilize the Δ508 mutant protein and improve its function.

Small deletions or insertions that are not exact multiples of three can cause major changes in the amino acid sequence of the corresponding proteins because they do not insert or delete entire codons. The effect of such a mutation can be appreciated by seeing how the deletion of a single letter makes a perfectly sensible sentence of three-letter words unintelligible. Consider this sentence:

THE BIG DOG SAW THE CAT EAT THE BUG

If the E in THE is deleted, the new reading frame for three-letter words is as follows:

THB IGD OGS AWT HEC ATE ATT HEB UG

Similarly, an insertion of a single nucleotide causes a one-nucleotide shift in the reading frame of the mRNA, and it changes all codons following the site of insertion. Frameshift mutations are mutations in which an insertion or deletion of nucleotides in a number that is not a multiple of three causes a shift in the reading frame of the mRNA, changing all following codons. Because frameshift mutations so profoundly alter the amino acid sequence, the mutant protein does not fold properly into its tertiary structure and becomes nonfunctional.

FIGURE 37.4 shows the consequences of a frameshift mutation in the β-globin gene. The nonmutant sequence in Figure 37.4a corresponds to amino acids 5–10. The frameshift mutation in Figure 37.4b is caused by the insertion of a C–G base pair. In turn, the mRNA transcript of the DNA also has a single-base insertion. When this mRNA is translated, the one-nucleotide shift in the reading frame results in an amino acid sequence that bears no resemblance to the original protein. All amino acids downstream of the site of insertion are changed, resulting in loss of protein function.

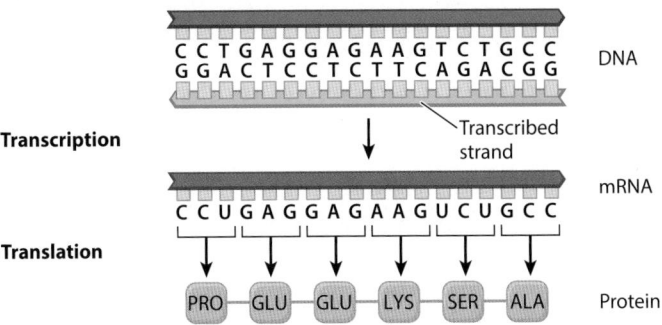

a. Nonmutant

b. Frameshift mutation (insertion)

An insertion or deletion that is not an exact multiple of three nucleotides changes the reading frame of translation.

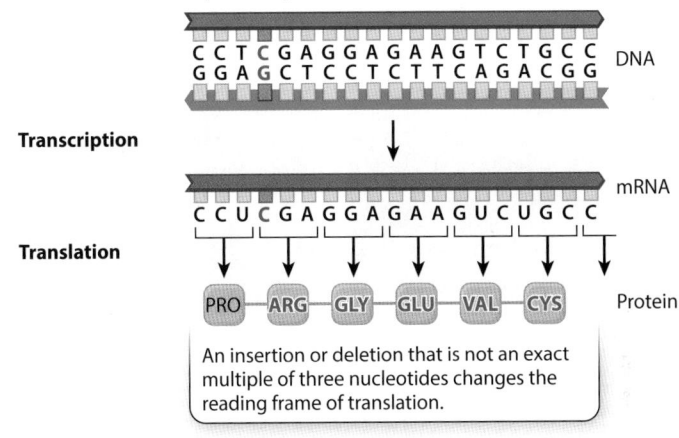

FIGURE 37.4 A frameshift mutation

A frameshift mutation results from the insertion or deletion of nucleotides that are not multiples of three. (a) The nonmutant sequence of a gene leads a protein with amino acids Pro–Glu–Glu–Lys–Ser–Ala. (b) The insertion of just one nucleotide results in a frameshift mutation, which changes the translational reading frame, resulting in a protein with amino acids Pro–Arg–Gly–Glu–Val–Cys.

Transposable Elements

An important source of new mutations in many organisms is the insertion of movable DNA sequences into or near a gene. Such movable DNA sequences are called transposable elements or transposons. Transposable elements can move or "jump" around the genome, leaving one place and inserting into another, in a process called **transposition**. The genomes of virtually all organisms contain several types of transposable elements, each present in multiple copies per genome.

Transposable elements were discovered by American geneticist Barbara McClintock in the 1940s. She studied corn (maize) because genetic changes that affect pigment formation can be observed directly in the kernels, which are normally purple. Since McClintock's work, biologists have discovered

What causes sectoring in corn kernels?

Background In the late 1940s, Barbara McClintock studied corn (*Zea mays*). Nonmutant corn, shown in the photo below, has purple kernels, resulting from synthesis of purple anthocyanin pigment in all cells of the kernel. Mutant cells produce no purple pigment, resulting in yellow kernels. McClintock noticed that, in some mutant corn, speckles or streaks of purple pigment could be seen in many otherwise yellow kernels, as shown in the photo.

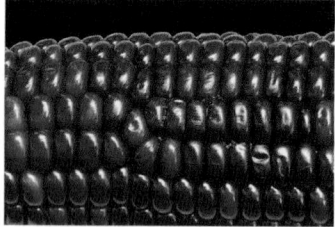

Nonmutant

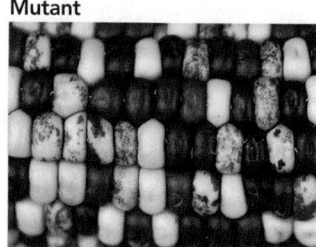

Mutant

Hypothesis McClintock hypothesized that the yellow mutation resulted from insertion of a transposable element that had jumped into a site near or in the anthocyanin pigment gene and disrupted its function. She attributed the speckles of purple color in the yellow kernels to cells in which the transposable element had jumped out again, restoring the function of the anthocyanin gene.

Experiment and Results By a series of genetic crosses, McClintock observed that cells in mutant yellow kernels reverted to normal purple, resulting in purple speckles in an otherwise yellow kernel. From this observation, she inferred that a transposable element had jumped out of the anthocyanin gene in these cells, restoring its function, shown in the figure. She also observed that in cells in which the original purple color was restored, there were mutations elsewhere in the genome. From this observation, she inferred that after jumping out of the anthocyanin gene, the transposable element had become integrated elsewhere in the genome, where it disrupted the function of a different gene.

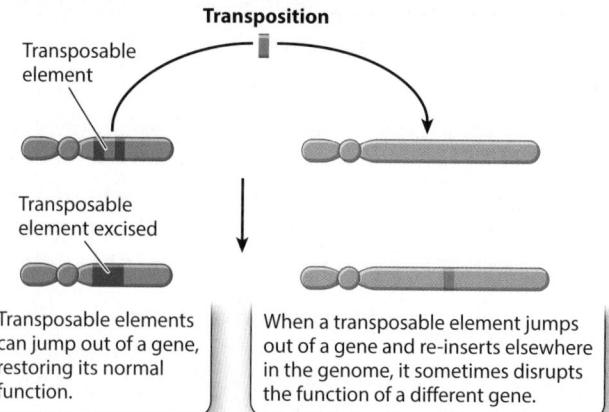

Transposition

Transposable element

Transposable element excised

| Transposable elements can jump out of a gene, restoring its normal function. | When a transposable element jumps out of a gene and re-inserts elsewhere in the genome, it sometimes disrupts the function of a different gene. |

Conclusion McClintock demonstrated the presence of a transposable element that could jump from one place to another in the genome. When it jumped into a gene, it disrupted the gene's function. When it jumped out, the function of the gene was restored.

Follow-up Work McClintock won the Nobel Prize in Physiology or Medicine in 1983. Much additional work has shown that many different types of transposable elements exist and that they are common among organisms.

SOURCE
McClintock, B. 1950. "The Origin and Behavior of Mutable Loci in Maize." *Proceedings of the National Academy of Sciences of the USA* 36:344–355. Photos: (Nonmutant) photo journey/Shutterstock; (Mutant) Robert Martienssen, Cold Spring Harbor Laboratory

AP® PRACTICE QUESTION

(a) **Identify** the following:

1. The question being investigated
2. The hypotheses tested
3. The independent variable
4. The dependent variable

(b) **Explain** the results of the experiment by **identifying** the researcher's claim and **justifying** the claim with the evidence and the reasoning researchers used (CER).

---

many different types of transposable elements, most ranging in length from a few hundred to a few thousand base pairs. When such a large piece of DNA inserts into a gene, it can interfere with transcription, cause errors in mRNA processing, or disrupt the reading frame. The result in the case of maize is that the cell is unable to produce pigment, so the kernels will be yellow, as described in "Practicing Science 37.1: What causes sectoring in corn kernels?"

Chromosomal Mutations

Whereas most mutations involve only one or a few nucleotides, some affect larger regions extending over hundreds of thousands or millions of nucleotides and have effects on chromosome structure that are often large enough to be visible through an ordinary light microscope.

Chromosomal mutations can delete or duplicate regions of a chromosome containing several or many genes, and the resulting change in gene copy number also changes the amount of the products of these genes in the cell. Chromosomal mutations can also alter the order of genes along a chromosome, or interchange the arms of nonhomologous chromosomes. While these types of chromosomal mutations do not change gene copy number, they do affect chromosome pairing and segregation in meiosis.

As we discussed in Module 30, errors during mitosis and meiosis can lead to cells with extra or missing chromosomes. Specifically, a pair of chromosomes fail to separate during anaphase of cell division, known as nondisjunction. The result is that one daughter cell receives an extra copy of one chromosome, and the other daughter cell receives no copy of that chromosome. Nondisjunction occurs quite commonly in humans. It is infrequently observed, however, because the result is typically lethal. Nevertheless, a few major chromosomal differences are found in the general population. Some are common enough that their effects are familiar. Others are less common, and a few have no major effects at all.

Perhaps the most familiar of the conditions resulting from nondisjunction is Down syndrome, which results from the presence of an extra copy of chromosome 21. Down syndrome is also known as trisomy 21 because affected individuals have three copies of chromosome 21. The presence of an extra chromosome 21 leads to a set of traits that, taken together, constitute a syndrome. These traits result from an increase in the number of genes and other elements present in chromosome 21. For most genes, there is a direct relation between the number of copies of a gene, called the gene dosage, and the level of expression of the gene, or the amount of protein produced. Therefore, an increase in the number of genes due to the presence of an extra chromosome 21 leads to an increased amount of protein, which in turn results in the traits we associate with Down syndrome.

Two other trisomies of autosomes (chromosomes other than the X and Y chromosomes) are sometimes found in live births. These are trisomy 13 and trisomy 18. Both are much rarer than Down syndrome and the effects are more severe. In both cases, the developmental abnormalities are so profound that newborns with these conditions usually do not survive the first year of life. Other trisomies of autosomes also occur, but they are not compatible with life. In each of these cases, the presence of an extra chromosome results in an increase in expression of the genes on that chromosome, resulting in severe developmental abnormalities.

Children born with an abnormal number of autosomes are rare, but extra or missing sex chromosomes (X or Y) are relatively common among live births. As we saw in Module 30, some of these do not show any effects, but two do have effects. One of these is Klinefelter syndrome, with two X chromosomes and one Y chromosome. Such individuals are male because of the presence of the Y chromosome, but they do not produce sperm and hence are sterile. The other is Turner syndrome, in which individuals have just one X chromosome instead of two. Like individuals with Klinefelter syndrome, individuals with Turner syndrome are usually sterile.

In all of these cases, just one chromosome is extra or missing. More extreme mutations result in changes to the number of complete sets of chromosomes. "Ploidy" refers to the number of sets of chromosomes. Humans, for example, are diploid because they have two sets of 23 chromosomes, or 46 chromosomes in total. **Polyploidy** describes more than two sets of chromosomes. Polyploidy is typically lethal in humans and other animals. However, it is common in plants. The polyploid bread wheat *Triticum aestivum,* for example, has six sets of seven chromosomes. Some ferns take polyploidy to an extreme: one species has 84 copies of a set of 15 chromosomes, or 1260 chromosomes altogether.

Polyploidy has played an important role in plant evolution. **Triploidy** is the condition of having three sets of chromosomes in the genome. While triploids are often sterile, other polyploids sometimes show increased vigor. Many agricultural crops are polyploid, including wheat, potatoes, olives, bananas, sugarcane, and coffee. Among flowering plants, it is estimated that 30% to 80% of existing species have polyploidy in their evolutionary histories, either because of the duplication of the complete set of chromosomes in a single species or because of hybridization, or crossing, between related species followed by duplication of the chromosome sets in the hybrid.

✓ Concept Check

1. **Describe** the effects of missense, silent, and non-sense mutations on a protein.

2. **Describe** the effects of an insertion or deletion of three nucleotides on a protein.

37.2 The genotype is the genetic makeup, and the phenotype is an observable characteristic

We can describe mutations in two different ways. The first is to look at the change at the level of DNA. From this perspective, we can consider whether the mutation is a point mutation, a small insertion or deletion, or a larger mutation affecting an entire chromosome. A second way of describing mutations is to look at the effect of a mutation on the traits of a cell or an organism. In this section, we will focus on the relationship between changes in DNA and changes in traits.

As we discussed in Unit 5, the genotype is the genetic makeup of a cell or organism. By genetic makeup, we mean the specific DNA sequence at a particular gene or even across the entire genome. For example, we can describe different DNA sequences of the β-globin gene, as in Figure 37.1. The different forms of any gene are called alleles, and they correspond to different DNA sequences in the genes. An individual's phenotype is the individual's observable characteristics or traits, such as height, weight, eye color, and so forth. For example, three copies of chromosome 21 (a genotype) causes Down syndrome (a phenotype).

The relationship between genotype and phenotype is not always straightforward. For example, when mutations occur in noncoding regions of the genome, they typically don't have any observable effect. In this case, a change in genotype does not lead to any change in phenotype. Similarly, some mutations in genes, such as silent mutations, do not lead to any change in proteins and therefore have no effect on the organism. By contrast, some mutations in genes can have dramatic effects on an organism, such as missense, nonsense, and frameshift mutations. Extra or missing chromosomes almost always lead to observable effects, as we saw in the cases of Down syndrome, Klinefelter syndrome, and Turner syndrome. In these cases, a change in genotype leads to a change in phenotype.

In diploid organisms, the relationship between genotype and phenotype also depends on whether the mutation is homozygous or heterozygous. Recall that an individual who inherits the same allele of a gene from each parent is homozygous. By contrast, an individual who inherits a different allele of a gene from each parent is heterozygous. Consider, for example, pea color in Mendel's pea plants. In Unit 5, we discussed how pea color (yellow or green) results from variation at a single gene. Two alleles of this gene are the dominant *A* (yellow) allele and the recessive *a* (green) allele.

AA homozygotes and *Aa* heterozygotes produce yellow peas, whereas *aa* homozygotes produce green peas. Therefore, the *a* allele causes green peas only when homozygous, not when heterozygous.

The molecular basis for these observations is the fact that only one copy of the *A* gene is needed to produce yellow seeds. The yellow color in pea seeds results from an enzyme that breaks down green chlorophyll, allowing yellow pigments to show through. Green seeds result from a mutation in this gene that inactivates the enzyme, so that the green chlorophyll is retained. In an *Aa* heterozygote, the seeds are yellow because the single *A* allele produces enough enzyme to break down the chlorophyll to yield a yellow seed. In an *aa* homozygote, by contrast, both copies of the enzyme are defective, so the green pigment is retained and can be seen.

Mutations can affect not only observable characteristics of an organism but also an organism's ability to survive and reproduce in a particular environment. Viewed in this way, mutations can be harmful, neutral, or even beneficial, as introduced in Module 30. In humans, for example, there are many forms, or alleles, of genes that increase the susceptibility to particular diseases, such as coronary artery disease, Alzheimer's disease, and others. These are considered harmful mutations. Other mutations have no effect at all on an organism and are considered neutral. Many mutations are neutral because they occur in noncoding DNA. Neutral mutations are especially likely to occur in organisms with large genomes and abundant noncoding DNA, such as humans. Some mutations are actually beneficial, increasing the chance of survival and reproduction. For example, there are mutations in humans that provide protection against acquired immune deficiency syndrome (AIDS), which is caused by the human immunodeficiency virus (HIV).

Whether a mutation is harmful, neutral, or beneficial often depends on factors other than the mutation itself. For example, some mutations can be harmful in some environments, but are neutral or even beneficial in others. Let's return to the example of mutations in β-globin. **FIGURE 37.5** shows two alleles of the β-globin gene: *A* and *S*. The *A* allele has a GAG codon in the position indicated in the figure. This codon translates to glutamic acid (Glu) in the resulting polypeptide. The *S* allele contains mutations that change the Glu found in the *A* allele to valine (Val). Although only one amino acid of the β-globin protein is affected, this change can have a dramatic effect on the function of the protein, since the amino acid sequence determines how a protein folds, and protein folding in turn determines the protein's function.

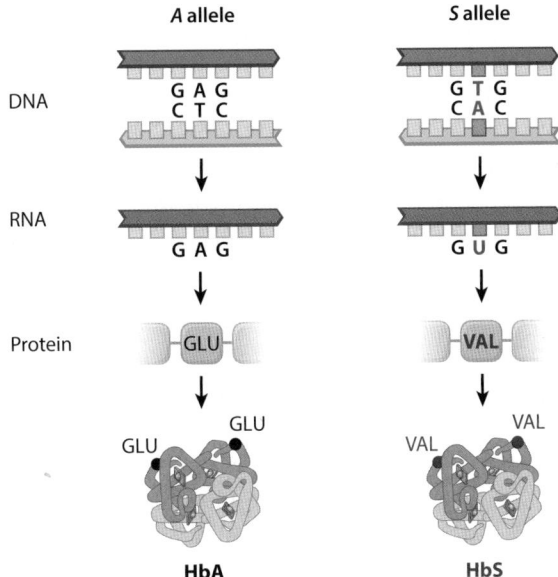

FIGURE 37.5 Two alleles of the gene encoding β-globin, a subunit of hemoglobin

The *A* allele leads to a protein with the amino acid Glu at the indicated position. The *S* allele has a point mutation that leads to a protein with Val instead of Glu. The *S* allele causes sickle-cell anemia when present in two copies as a homozygote, but provides protection against malaria when present in a single copy as a heterozygote.

What is the effect of the *S* mutation? Is it beneficial, harmful, or neutral? The short answer is, it depends. Homozygous *SS* individuals have sickle-cell anemia. In this condition, the hemoglobin molecules tend to bind together, or polymerize, when exposed to lower than normal levels of oxygen. The polymerization of hemoglobin causes the cell to change from its normal oval shape into the shape of a half-moon, or sickle. In this form, the red blood cell is unable to carry the normal amount of oxygen. In addition, the sickled cells can block small capillaries, interrupting the blood supply to vital tissues and organs and resulting in severe pain. In the absence of proper medical care, patients with sickle-cell anemia usually die before adulthood.

Heterozygous *AS* individuals have only a mild anemia called the sickle-cell trait. Furthermore, in Africa, where malaria is widespread, being a heterozygote is actually beneficial because it provides some protection against malaria. *AA* individuals don't have anemia, but also don't have any protection against malaria.

This example illustrates two important principles about the connection between genotype and phenotype. First, the effect of a mutation often depends on whether the mutation is homozygous or heterozygous, as we also saw in the example of color in peas. In areas with malaria, the *S* allele is harmful as a homozygous genotype, but is beneficial as a heterozygous genotype. Second, the effect of a mutation may depend on the environment. The *S* allele, when heterozygous, is beneficial only in malaria-prone regions, where it offers protection from the disease that outweighs its other effects. In areas without malaria, it is harmful.

✓ Concept Check

3. SS individuals have sickle-cell anemia. **Identify** the genotype and phenotype and **explain** the connection between the two.

4. With regard to mutations, **describe** what is meant by the terms "harmful," "beneficial," and "neutral."

5. **Describe** why it is difficult to state whether the S allele of β-globin is harmful, beneficial, or neutral.

6. The human genome is large and is mostly made up of noncoding regions. **Predict** whether most mutations in humans are harmful, beneficial, or neutral.

37.3 Mutations can be inherited, occur spontaneously, or be induced by mutagens

Up to this point, we have discussed different types of mutations and their effects. If we look at any population of individuals, such as humans, we will observe that they have differences in their DNA sequences; this is known as genetic variation. Differences among individuals' DNA, or genotypes, can lead to differences among the individuals' mRNA and proteins, which affect the molecular functions of the cell and ultimately can lead to physical differences that we can observe, or phenotypes. Genetic differences among apples produce varieties whose mature fruits differ in taste and color, such as the green Granny Smith, the yellow Golden Delicious, and the scarlet Red Delicious. In this final section, we will discuss where genetic variation comes from.

The genetic differences that we see among individuals are ultimately the result of mutations. Some mutations result from mistakes that occur during DNA replication. In Module 32 we mentioned that DNA polymerase, the enzyme that replicates DNA, has a proofreading function. This means that it corrects mistakes as it proceeds. Some mistakes, however, escape the proofreading function and therefore become mutations. These mutations are sometimes called spontaneous mutations.

Other mutations result from damage that occurs to DNA. The damage may result from reactive molecules produced in the normal course of metabolism; chemicals in the environment such as cigarette smoke; or radiation of various types, including X-rays and ultraviolet light. **Mutagens** are agents that increase the probability of mutation. The presence of a mutagen can increase the probability of mutation by a factor of 100 or more. Cells have mechanisms to repair DNA damage, but damage that goes unrepaired results in mutations. These mutations are sometimes called induced mutations. Another source of mutations, discussed earlier, are transposable elements. Most genomes contain these DNA sequences, which can "jump" from one position to another in the genome.

If we look at any present-day population of individuals, they will differ genetically one to the next. Most of these mutations have occurred in the past and have been passed along from one generation to the next. These mutations are inherited. A much smaller number are new, occurring in the lifetime of an individual, either spontaneous or induced. Although mutations are the ultimate source of all of the genetic variation we see in any present-day population of individuals, recombination by crossing over followed by segregation of homologous chromosomes during meiotic cell division shuffles existing mutations to create new combinations. Recombination and mutations both create new alleles and are therefore considered sources of genetic variation.

In eukaryotic organisms, genes generally pass from parent to offspring, which is sometimes referred to as vertical gene transfer. Bacteria likewise inherit most of their genes from parental cells by binary fission, discussed in Module 24. However, they can also obtain new genes from other individuals by **horizontal gene transfer,** sometimes called horizontal acquisition of genes. Horizontal gene transfer occurs when a cell receives DNA from another cell that is not its parent cell. Horizontal gene transfer is a major source of genetic diversity in bacteria, in addition to mutation.

Horizontal gene transfer can occur in various ways. For example, a bacterial cell can directly transfer DNA to another bacterial cell, shown in **FIGURE 37.6**. In the figure, thin strands of membrane-bound cytoplasm called pili connect one bacterial cell to another bacterial cell. After joining to a second cell, the pilus contracts, drawing the two cells close together. A pore-like opening develops where the two cells are in close contact, providing a route for the direct cell-to-cell transfer of DNA. This process is called **conjugation.** Conjugation provides a way to transfer plasmids from one cell to another. Recall from Module 6 that plasmids are small circular DNA molecules that are typically present in prokaryotes. They can carry a small number of genes. Conjugation, then, provides a way to spread novel genes on plasmids throughout a population very quickly.

For example, antibiotics are used to treat infections caused by bacteria. Bacteria can develop resistance to antibiotics, which means the bacteria are not killed by the antibiotics, but instead continue to survive and grow. Resistance often occurs when bacteria acquire mutations in genes carried on plasmids. One bacterial cell with a gene that provides resistance to antibiotics can quickly spread it to other bacterial cells by conjugation. Genes that confer resistance to antibiotics are a well-studied example of horizontal gene transfer by conjugation.

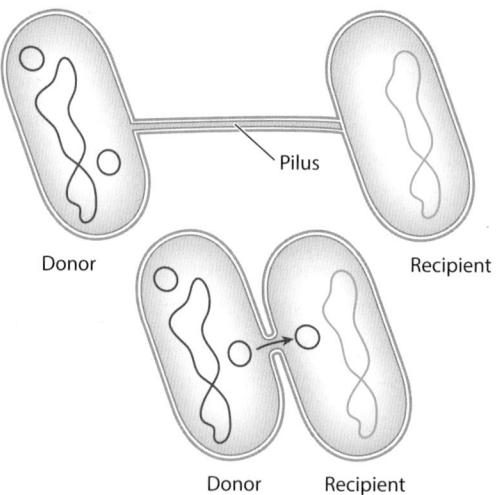

FIGURE 37.6 Conjugation

Conjugation is a form of horizontal gene transfer, or the transfer of genetic material between two cells that are not parent and offspring. DNA shown in red originates from the donor cell. DNA shown in blue is that of the recipient cell. In conjugation, DNA (usually a plasmid) from a donor cell is transferred directly to an adjacent recipient cell by a structure called a pilus.

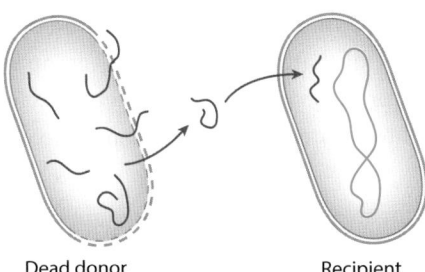
Dead donor   Recipient

FIGURE 37.7 Transformation

Transformation is a form of horizontal gene transfer in which DNA released into the environment by dead cells is taken up by a recipient cell. DNA shown in red originates from the dead cell, which acts as the donor; DNA shown in blue is that of the recipient cell, which takes up the DNA from the environment.

Genes can also be transferred from one cell to another without any direct bridge between cells. DNA released to the environment by cell breakdown can be taken up by other cells in a process called **transformation,** which is shown in **FIGURE 37.7**. Transformation was discovered when experiments showed that harmless strains of the bacteria causing pneumonia could be transformed into virulent strains by exposure to media containing dead cells of disease-causing strains, as described in "Practicing Science 31.1: Which molecule carries genetic information?" on page 438. Scientists reasoned that transformation occurred because the living bacteria took up some substance from the dead cells. The "transforming substance" was later shown to be DNA. Today, biologists commonly use transformation in the laboratory to introduce genes into cells.

Viruses provide a third mechanism of horizontal gene transfer. Viruses are small, infectious agents that infect all kinds of cells. Viruses that infect bacterial cells sometimes integrate their DNA into the host's DNA. This viral DNA persists within the cells as they grow and divide. Before the virus leaves the cell to infect others, the viral DNA removes itself from the bacterial genome and is packaged in a protein capsule. The complete virus particle can then be released into the environment. This excision is not always precise, and sometimes genetic material from the bacterial host is incorporated into the virus. Viruses released from their bacterial host cell go on to infect others, bringing bacterial host genes with them. Horizontal gene transfer by means of viruses is called **transduction** and it is illustrated in **FIGURE 37.8**. It is common in nature and is also widely used in the laboratory to introduce novel genes into bacteria for medical research.

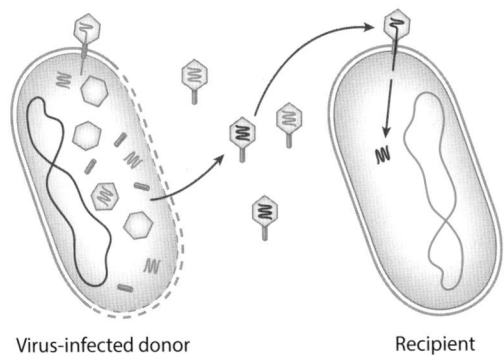
Virus-infected donor   Recipient

FIGURE 37.8 Transduction

Transduction is a form of horizontal gene transfer in which DNA is transferred from a donor to a recipient cell by a virus. Here, a virus infects a bacterial cell. Following infection, new viruses are released, taking with them some bacterial host DNA, shown in red. When these viruses infect another cell, DNA is sometimes also transferred to the recipient cell.

✓ Concept Check

7. **Describe** the effect of a mutagen and provide two examples.
8. **Describe** three forms of horizontal gene transfer.

Module 37 Summary

PREP FOR THE AP® EXAM

REVISIT THE BIG IDEAS

INFORMATION STORAGE AND TRANSMISSION: Use the content of this module to **identify** a phenotypic change that results from DNA mutations or chromosomal mutations. **Describe** how the change in the genome of a species affects its survival, growth, or reproduction and therefore its long-term survival and evolution as a species.

LG 37.1 Mutations can affect one or a few bases or entire chromosomes.

• A point mutation, or nucleotide substitution, is a change of one base for another. Page 512

• The effect of a point mutation depends on where it occurs. Page 513

- If a point mutation arises in noncoding DNA, it will likely have little or no effect on an organism. Page 513

- If a point mutation arises in a protein-coding gene, it can result in no change in the amino acid sequence (silent mutation), a change in the amino acid sequence (missense mutation), or the introduction of a stop codon (nonsense mutation). Page 513

- Small insertions (or deletions) in DNA add (or remove) one base or a few contiguous bases. Page 514

- An insertion or deletion of nucleotides that are not multiples of three in a protein-coding gene results in a frameshift mutation, in which all the codons downstream of the insertion or deletion are changed. Page 515

- Transposable elements are DNA sequences that can jump from one place in a genome to another. They can affect the expression of a gene if they are inserted into or near a gene. Page 515

- Chromosomal mutations involve large regions of one or more chromosomes. Page 516

- Trisomy 21 (Down syndrome) is usually characterized by three copies of chromosome 21. Page 517

- Extra or missing sex chromosomes (X or Y) result in syndromes such as Klinefelter syndrome and Turner syndrome. Page 517

- Ploidy refers to the number of complete sets of chromosomes; diploidy describes two sets; triploidy describes three sets; and polyploidy describes more than two sets, such as three, four, and so on. Page 517

LG 37.2 The genotype is the genetic makeup, and the phenotype is an observable characteristic.

- The genotype of an organism is its genetic makeup. Page 518

- The phenotype of an organism is any observable characteristic of that organism, such as its appearance, physiology, or behavior. The phenotype results from a complex interplay of the genotype and the environment. Page 518

- Different forms (sequences) of genes are called alleles. Page 518

- A person's genotype can be homozygous, with two copies of the same allele, or heterozygous, with two different alleles. Page 518

- Mutations can be neutral (no effect on survival or reproduction), harmful (associated with decreased survival or reproduction), or beneficial (associated with increased survival or reproduction). Page 518

LG 37.3 Mutations can be inherited, occur spontaneously, or be induced by mutagens.

- Genetic variation results from mutation and recombination. Page 520

- Mutations occur as a result of errors in DNA replication or are induced by mutagens. Page 520

- DNA polymerase has a proofreading function and most DNA damage is repaired by the cell, so mutations are very rare. Page 520

- In bacteria, DNA can be transferred from cell to cell by horizontal gene transfer, which increases genetic variation. Page 521

Key Terms

Transposition	Mutagen	Transformation
Polyploidy	Horizontal gene transfer	Transduction
Triploidy	Conjugation	

Review Questions

1. A point mutation that causes no change in the amino acid sequence of a protein is called a
   (A) silent mutation.
   (B) missense mutation.
   (C) nonsense mutation.
   (D) stop mutation.

2. A point mutation that causes an amino acid replacement is called a
   (A) silent mutation.
   (B) missense mutation.
   (C) nonsense mutation.
   (D) stop mutation.

3. Movable DNA sequences are called
   (A) frameshifts.
   (B) transposable elements.
   (C) transition elements.
   (D) centromeres.

4. Which statement about mutations is true?
   (A) A mutation may leave the amino acid sequence of a protein unchanged.
   (B) A mutation will always result in a change in phenotype.
   (C) A mutation will always be corrected.
   (D) A mutation will always be passed onto offspring.

5. A nonsense mutation
   (A) changes the identity of one amino acid in a polypeptide chain.
   (B) shifts the reading frame of a messenger RNA.
   (C) has no effect on a protein.
   (D) changes a codon for an amino acid into a codon for chain termination.

6. Imagine a gene in which the sequence that is transcribed into a GAG codon, which codes for glutamic acid, is mutated to GUG, which codes for valine. What type of mutation is this?
   (A) Silent
   (B) Missense
   (C) Nonsense
   (D) Frameshift

7. Which enzyme is responsible for proofreading during DNA replication?
   (A) DNA ligase
   (B) DNA polymerase
   (C) AP endonuclease
   (D) RNA polymerase

Module 37 AP® Practice Questions

Section 1: Multiple-Choice Questions

Choose the best answer for questions 1–4.

1. In mice, the transcription factor FOSB activates transcription of a gene called *scd*. The *scd* gene encodes an enzyme called stearoyl co-A desaturase, which promotes cell division. Predict the phenotypic effect on mice that are homozygous for a mutation in *scd* that prevents FOSB from binding.
   (A) There will be no effect on the mouse phenotype because the mutation does not affect the protein sequences of stearoyl co-A desaturase or FOSB.
   (B) The mice will have a reduced rate of cell division because they will have lower expression of the *scd* gene.
   (C) The mice will have an increased rate of cell division because they will have higher levels of the stearoyl co-A desaturase protein.
   (D) The mice will have an increased rate of cell division because FOSB will be unable to bind to the DNA.

2. Transposable elements are segments of DNA that can jump from spot to spot within the genome. Many transposable elements insert randomly into the genome, and they may insert in or near genes. Which is the most likely consequence of transposable elements transposing randomly in somatic cells?

(A) Random insertion of transposable elements in the genomes of somatic cells will decrease genetic diversity in the population.

(B) Genetic variation created by transposable elements may be inherited from parents and passed on to offspring.

(C) Different cells of the body may have different genome sequences due to transposable elements inserting themselves in different positions in different cells' genomes.

(D) Some organisms may confer improved function to their offspring if the transposable element inserts near a beneficial gene.

Questions 3 and 4 refer to the following information.

In many bacterial species, some strains are pathogenic, which means that they cause disease in a host organism, while other strains are not. The ability to cause infections is often determined by sets of genes called virulence genes, which are often encoded on plasmids. Strains that contain the plasmid bearing the virulence genes are pathogenic bacteria, and strains that lack the plasmid are nonpathogenic bacteria. Virulence genes are essential for infection but may even be costly when bacteria grow in noninfection conditions.

A researcher cultured two strains of the bacterium *Pseudomonas aeruginosa* together in the same petri dish: a pathogenic strain and a related, nonpathogenic strain of the same species. When the researcher separated the two strains from each other, the researcher found that a subset of cells from the formerly nonpathogenic strain had become pathogenic. The researcher also noticed that the pathogenic strain became nonpathogenic after culturing it in the laboratory in the absence of the nonpathogenic bacteria for hundreds of generations.

3. Which statement provides the best explanation for how cells of the formerly nonpathogenic strain could have become pathogenic?

(A) A pathogenic cell may have transferred virulence genes to a nonpathogenic cell by conjugation.

(B) Some nonpathogenic cells may have acquired the proteins necessary for virulence from the culture medium.

(C) The nonpathogenic strain induced a mutation in response to being cultured with pathogenic bacteria.

(D) The nonpathogenic strain may have activated expression of virulence genes in its genome.

4. Which statement is the best explanation for how the formerly pathogenic strain became nonpathogenic after hundreds of generations in culture?

(A) The pathogenic strain may have acquired DNA from the nonpathogenic strain that eliminated virulence.

(B) A viral infection in the pathogenic strain may have eliminated the virulence genes required for being pathogenic.

(C) The strain may have acquired nonpathogenic DNA from the environment by transformation.

(D) Natural selection could have favored mutations that eliminated the virulence genes when the strain was grown in culture.

Section 2: Free-Response Question

Write your answer to each part clearly. Support your answers with relevant information and examples. Where calculations are required, show your work.

A researcher hypothesized that a chemical called compound X might be mutagenic. To test this hypothesis, the researcher designed an experiment with a strain of the bacterium *Salmonella typhimurium* that is unable to synthesize the amino acid histidine. The researcher exposed *S. typhimurium* to compound X in five different doses. The researcher then spread the exposed cells on a solid growth medium. The growth medium did not contain enough histidine to support the growth of the strain of *S. typhimurium* being studied. However, mutant *S. typhimurium* that have gained the ability to synthesize histidine were able to grow on the plate. The number of colonies that grew on the medium therefore provides an estimate of the number of mutations that compound X induced. The data that the researcher obtained from using compound X at five doses are shown in the table.

Dose of compound X (mg/mL)	Number of *S. typhimurium* cells exposed	Number of colonies observed on medium
0	10^6	3
5	10^6	15
10	10^6	32
20	10^6	59
40	10^6	124
80	10^6	0

(a) **Describe** the purpose of the treatment in which the bacteria were exposed to 0 mg/mL of compound X.

(b) **Calculate** the rate of mutations that enable histidine synthesis at 20 mg/mL.

(c) Based on the data provided, **make a claim** about whether there is evidence that compound X is mutagenic. **Provide reasoning** to support your claim.

(d) **Make a claim** to explain the number of colonies observed after the *S. typhimurium* was exposed to 80 mg/mL of compound X.

(e) **Draw** a histogram that shows the number of mutant *S. typhimurium* colonies observed as a function of exposure to compound X.

Module 38

Biotechnology

LEARNING GOALS ▶**LG 38.1** The polymerase chain reaction selectively amplifies regions of DNA.
▶**LG 38.2** Gel electrophoresis separates DNA molecules by size.
▶**LG 38.3** DNA sequencing determines the order of nucleotides in a DNA strand.
▶**LG 38.4** Restriction enzymes cut DNA at particular short sequences.
▶**LG 38.5** Genetic engineering allows researchers to alter DNA sequences in living organisms.

In this unit, we have seen how cells store, transmit, and use the genetic information in DNA. In the process of DNA replication, cells faithfully make copies of DNA, allowing them to transmit genetic information from parent cell to daughter cell. In the processes of transcription and translation, cells express the information in DNA, allowing them to make proteins that form the structure of cells and do much of the work of the cell. Cells use many different processes to regulate the expression of genes, so that genes are expressed at the right time, place, and level. Finally, different cells, even within a single multicellular organism, often have different DNA sequences, or genotypes, due to mutations. In some cases, these mutations can affect protein structure and function, and ultimately the visible characteristics, or phenotype, of the organism.

An understanding of the structure and function of DNA has allowed biologists to understand some of life's central processes and also to create tools to study how life works. Biologists often need to isolate, identify, and determine the nucleotide sequence of particular DNA fragments. Such procedures have many uses. For example, they can determine whether a genetic risk factor for diabetes has been inherited or if blood at a crime scene matches that of a suspect. They can identify a variety of wheat that carries a gene for insect resistance, or determine if two species are closely or distantly related.

Many of the experimental procedures for the isolation, identification, and sequencing of DNA are based on knowledge of the structure and physical properties of DNA. Others make use of the principles of DNA replication. In this module, we will examine techniques for the amplification, separation, sequencing, and combining of DNA fragments, as well as their practical applications.

PREP FOR THE AP® EXAM

FOCUS ON THE BIG IDEAS

INFORMATION STORAGE AND TRANSMISSION: Focus on how technological tools have allowed humans to affect the transmission of information within and between species.

38.1 The polymerase chain reaction selectively amplifies regions of DNA

DNA in the nuclei of your cells is present in two copies per cell, except for DNA in the X and Y chromosomes in males, which are each present in only one copy per cell. In the laboratory, it is very difficult to manipulate or visualize a sample containing just one or two copies of a DNA molecule. Instead, researchers typically work with many identical copies of the DNA molecule they are interested in studying. A common method for making copies of a piece of DNA is the **polymerase chain reaction (PCR)**. PCR is a technique that allows a targeted region of a DNA molecule to be replicated many times, or amplified, into as many copies as desired. Starting with just a single DNA molecule, a researcher can generate millions of copies of the DNA molecule. In this section, we will discuss how PCR is performed and why it is used.

Because the PCR reaction is essentially a DNA synthesis reaction, it requires the same basic components used by the cell to replicate its DNA. In this case, the procedure takes place in a small plastic tube containing a solution that includes four components:

1. DNA. At least one molecule of double-stranded DNA containing the region to be amplified serves as the template that will be replicated over and over, or amplified.

2. DNA polymerase. The enzyme DNA polymerase is used to replicate the DNA.

3. All four nucleotides with different bases. The bases A, T, G, and C are used as building blocks for the synthesis of new DNA strands.

4. Two primers. Two short sequences of single-stranded DNA are required for the DNA polymerase to start synthesis. Recall from Module 32 that DNA polymerase cannot begin synthesis on its own, but requires a short primer. During DNA replication, the primer is a short strand of RNA. In PCR, DNA primers are commonly used. Their base sequences are complementary to the ends of the region of DNA to be amplified. In other words, the primers flank the specific region of DNA to be amplified. The 3′ end of each primer is oriented toward the region to be amplified so that, when DNA polymerase extends the primer, it creates a new DNA strand complementary to the targeted region.

PCR creates new DNA fragments in three steps that are repeated over and over again in a cycle. These steps are illustrated in **FIGURE 38.1**. The first step, denaturation, involves heating the solution in a plastic tube to a temperature just short of boiling so that the individual DNA strands of the template separate, or denature, as a result of the breaking of hydrogen bonds between the complementary bases. The second step, annealing, begins as the solution is cooled. The two primers bind, or anneal, to their complementary sequence on the DNA. In the final step, the solution is heated to the optimal temperature for DNA polymerase and the polymerase elongates, or extends, each primer. These steps make a single copy of the DNA.

After sufficient time to allow new DNA synthesis, the solution is heated again, and the cycle of denaturation, annealing, and extension is repeated over and over, as indicated in **FIGURE 38.2** on page 528, usually for 25–35 cycles. In each cycle, the number of copies of the DNA is doubled. The first round of PCR makes 2 copies, the next 4, the next 8, then 16, 32, 64, 128, 256, 512, 1024, and so forth. The doubling in the number of copies of the DNA fragment is the basis for the term "chain reaction" in PCR.

Although PCR is elegant in its simplicity, DNA polymerase from many species, including humans, irreversibly loses both structure and function at the high temperature required to separate the DNA strands. At each cycle, you would have to open the tube and add fresh DNA polymerase. This is possible, and, in fact, it was how PCR was performed when the technique was first developed. However, the procedure is time consuming and tedious.

To solve this problem, we now use DNA polymerase enzymes that are heat stable and so do not denature during the denaturation step. Such DNA polymerases are found in the bacterial species *Thermus aquaticus,* which lives at the near–boiling point of water in natural hot springs, including those at Yellowstone National Park. This polymerase, called *Taq* polymerase, does not denature at the high temperature of the denaturation step, but instead remains active. With this heat-stable DNA polymerase, the entire procedure can be carried out in a fully automated machine. The time of each cycle, temperatures, number of cycles, and other variables

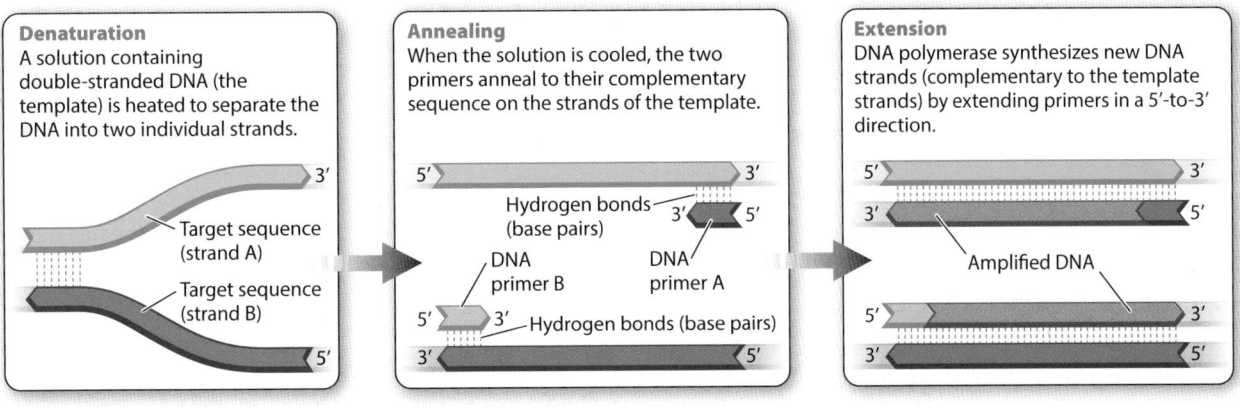

FIGURE 38.1 The polymerase chain reaction (PCR)

PCR consists of three steps. The first step is denaturation, in which the solution is heated, causing DNA strands to separate from each other. The second step is annealing, in which the solution is cooled, allowing primers to bind to their complementary sequences. The third step is extension, in which DNA polymerase synthesizes new strands. These three steps are repeated over and over again, resulting in amplification of a region of DNA.

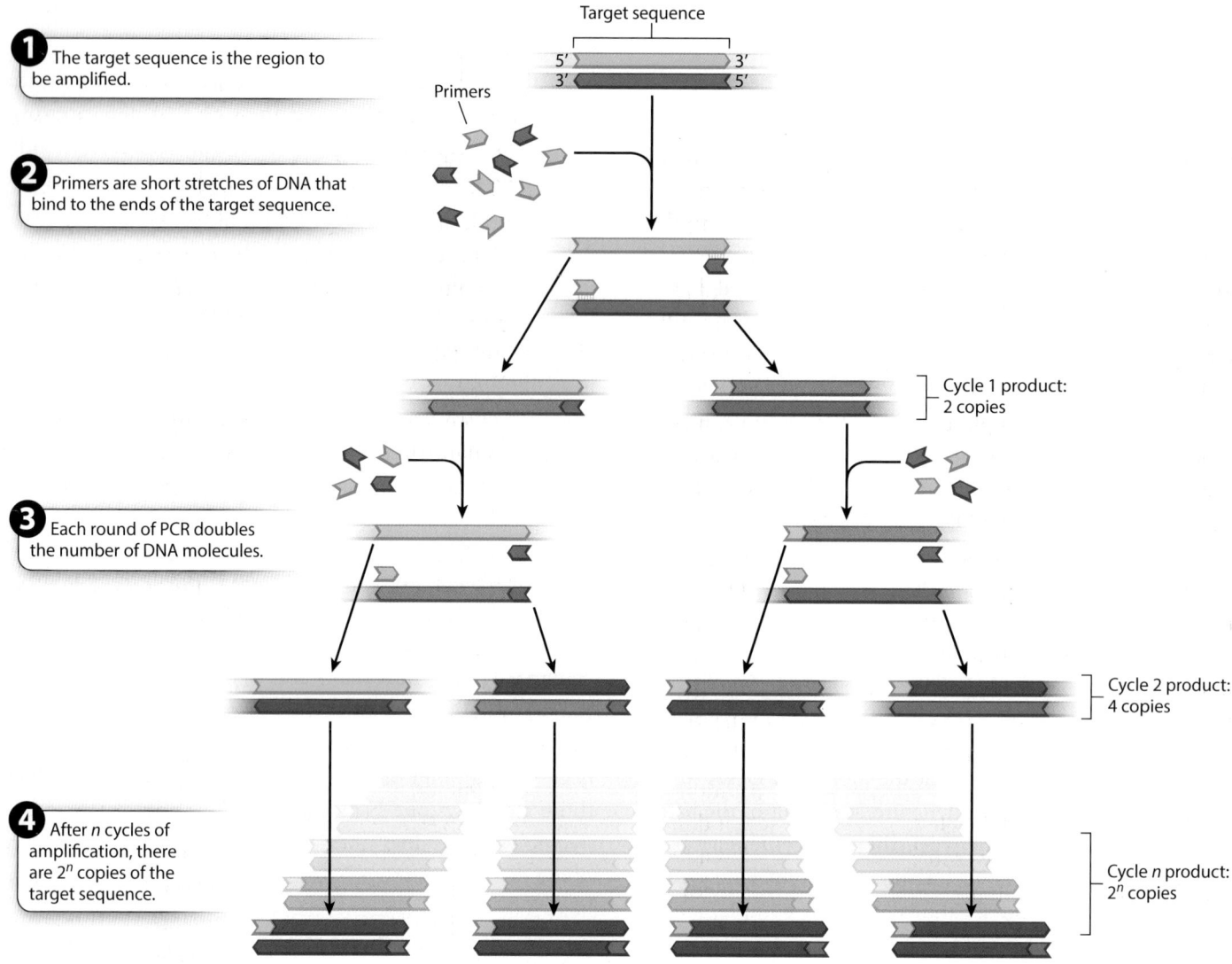

1 The target sequence is the region to be amplified.

Target sequence

Primers

2 Primers are short stretches of DNA that bind to the ends of the target sequence.

Cycle 1 product: 2 copies

3 Each round of PCR doubles the number of DNA molecules.

Cycle 2 product: 4 copies

4 After n cycles of amplification, there are 2^n copies of the target sequence.

Cycle n product: 2^n copies

FIGURE 38.2 PCR amplification

PCR results in amplification of the DNA sequence flanked by the two primers. Each round of replication doubles the number of molecules that have the same sequence as the target DNA. After n cycles of amplification, there are 2^n copies of the target DNA. For example, when $n = 30$, there are 2^{30} or approximately 1,000,000,000 copies.

can all be programmed. The fact that DNA polymerase from bacteria that live in hot springs can be used to amplify DNA from any organism reminds us again of the conserved function and evolution of this enzyme early in the history of life.

PCR is both selective and highly sensitive, so it is used to amplify and detect small quantities of nucleic acids, such as the virus that causes COVID-19 in nasal swabs or the human immunodeficiency virus (HIV) in blood bank supplies. PCR can even be used to study DNA samples as minuscule as those left by a smoker's lips on a cigarette butt dropped at the scene of a crime or in a fingerprint smudge

on a coffee cup. It can also be used to identify species based on known, conserved DNA sequences. In all of these applications, the starting sample can be as small as a single molecule of DNA.

✓ Concept Check

1. **Describe** what PCR does.
2. **Describe** the three steps of PCR.
3. **Identify** two uses of PCR.

38.2 Gel electrophoresis separates DNA molecules by size

PCR amplification does not always work as you might expect. Sometimes the primers have the wrong sequence and thus fail to anneal properly; sometimes they anneal to multiple sites and several different fragments are amplified. To determine whether PCR has yielded the expected product, a researcher must determine the size of the amplified DNA molecules. The size of a DNA fragment is measured in base pairs (bp). Usually, the researcher knows what the size of the correctly amplified fragment should be, making it possible to compare the expected size to the actual size. In this section, we will describe one technique that is used to determine the size of any DNA fragment.

One way to determine the size of a DNA fragment is by **gel electrophoresis,** a procedure in which DNA fragments are separated by size as they migrate through a porous substance in response to an electrical field. This procedure is illustrated in **FIGURE 38.3**. DNA samples are first inserted into slots or wells near the edge of a rectangular slab of porous material that resembles solidified agar (the "gel"). The porous nature of the gel is due to its composition of a tangle of polymers. This tangle of polymers makes it difficult for large molecules to pass through. The gel is then inserted into an apparatus and immersed in a solution that allows an electric current to be passed through it, from negative at the top of the gel where the DNA samples are initially placed, to positive at the bottom of the gel. This electrical field is what causes the DNA to move through the gel. DNA is negatively charged because of the ionized (negative) phosphate groups along the DNA backbone. Therefore, DNA moves toward the positive pole of the electric field, which is toward the bottom of the gel, over time. The electric current is applied for a certain amount of time and then stopped, allowing the DNA to move in the gel but not run off the end of the gel.

The DNA molecules move according to their size. Small fragments pass through the pores of the gel more readily than large fragments, and so in a given amount of time, small fragments move a greater distance in the gel than large fragments. The rate of migration is dependent only on size, not on sequence of nucleotides, and so all fragments of a given size move together at the same rate and form a discrete band, or stripe. Such bands can be made visible by dyes that bind to DNA and fluoresce under ultraviolet light, shown in **FIGURE 38.4** on page 530. A solution consisting of DNA fragments of known sizes is usually placed in one of the wells, resulting in a series of bands, called a ladder, which can be used for size comparison.

Consider a PCR experiment in which the expected region to be amplified is 300 bp. To determine if the PCR experiment worked as expected, a sample of the product is checked by gel electrophoresis. This sample is loaded into the well of the gel, a current is applied for a short period of time, and DNA fragments migrate to positions in the gel that correspond to their size. If a single band of 300 bp is seen on the gel, then the experiment worked as expected. Sometimes, however, the reaction might yield no bands, a single band of the incorrect size, or multiple bands. The researcher would then need to go back and investigate why the experiment did not work as expected.

Gel electrophoresis can separate DNA fragments produced by any means, not only PCR. Genomic DNA can be cut with certain enzymes and the resulting fragments separated by gel electrophoresis. Gel electrophoresis can also be used to separate RNA and protein molecules. RNA, like DNA, moves according to its size. The rate at which the proteins move from one end of a

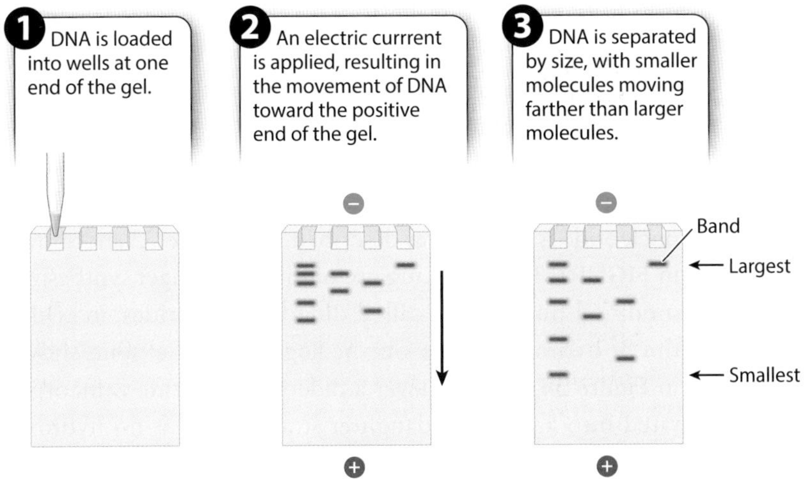

① DNA is loaded into wells at one end of the gel.

② An electric current is applied, resulting in the movement of DNA toward the positive end of the gel.

③ DNA is separated by size, with smaller molecules moving farther than larger molecules.

Band
← Largest
← Smallest

FIGURE 38.3 Gel electrophoresis

A gel electrophoresis apparatus consists of a plastic tray, gel, and solution. DNA molecules separate by size: larger molecules move more slowly than smaller ones because larger molecules take longer to work their way through the pores in the gel. Therefore, after a given amount of time, small molecules will move farther than large molecules. A band in the gel represents DNA of a particular size.

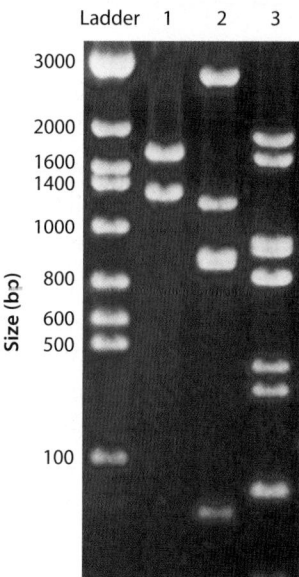

FIGURE 38.4 A gel

In gel electrophoresis, DNA bands can be visualized after staining with a dye that fluoresces under ultraviolet light. A ladder with DNA fragments of known lengths is used for size comparison. Photo: Guy Tear/Wellcome Images

gel to the other, however, is determined by both their charge and their size. Early studies of protein gel electrophoresis focused on enzymes that catalyze reactions that can be induced to produce a dye when the substrate for the enzyme is added. When researchers add the substrate, they can see the locations of the proteins in the gel. Some of these studies focused on different forms of the same enzyme, called allozymes. Protein gel electrophoresis was used to identify and study these allozymes.

✓ Concept Check

4. **Describe** how the properties of DNA determine the direction that DNA fragments move through a gel.

5. **Describe** how the size of DNA fragments affects their movement through a gel.

38.3 DNA sequencing determines the order of nucleotides in a DNA strand

One of the high points of modern biology has been determining the complete nucleotide sequence of the DNA in a large number of species. Scientists have sequenced the genomes of humans, model organisms like mice and fruit flies, farm animals like horses and cows, crops like wheat and rice, and thousands of bacteria and viruses. DNA sequencing is a technique that is used to figure out the order of nucleotides in DNA. Sequencing the human genome and many others has helped us to see in unprecedented detail how species are related to one another. It has also helped us understand human health and disease. For example, scientists can sequence the DNA from a healthy cell and a cancer cell in the same individual. By comparing the two sequences, scientists can figure out what mutations occurred in the cancer cell. In this section, we will describe one procedure that is used to sequence DNA.

Techniques used to sequence DNA make use of the principles of DNA replication. Consider a solution containing identical single-stranded molecules of DNA, each being used as the template for the synthesis of a complementary daughter strand that originates at a short primer sequence, as in PCR. The challenge is to determine the nucleotide sequence of the template strand. A brilliant answer to this problem was developed by the English geneticist Frederick Sanger.

Recall that a free 3′ hydroxyl group is essential for each step in DNA synthesis because that is where the incoming nucleotide is attached, as described in Module 32 and shown in **FIGURE 38.5a**. Making use of this fact, Sanger synthesized modified nucleotides called dideoxynucleotides, in which the 3′ hydroxyl group on the sugar ring is absent, shown in Figure 38.5b. Whenever a dideoxynucleotide is incorporated into a growing daughter strand, there is no hydroxyl group to attack the incoming nucleotide, and strand growth is stopped dead in its tracks. For this reason, a dideoxynucleotide is known as a chain terminator. By including a small amount of each of the chain terminators in a tube along with larger quantities of all four normal nucleotides, a DNA primer, a DNA template, and DNA polymerase, Sanger was able to produce a series of short daughter strands, each

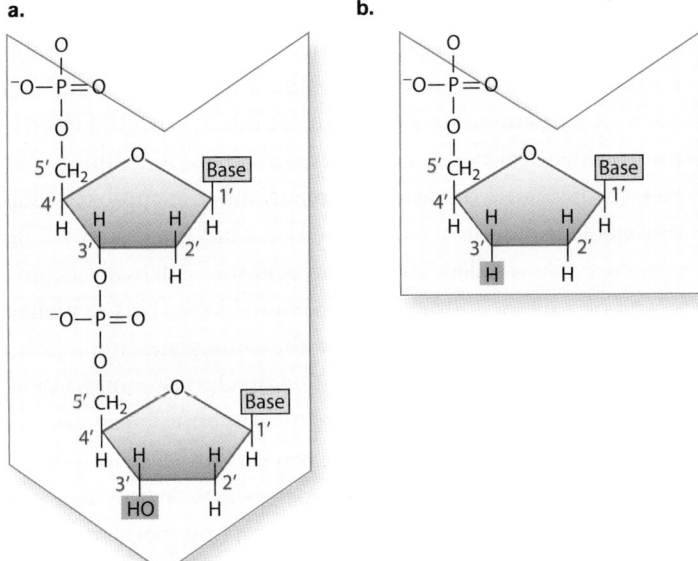

a.

b.

FIGURE 38.5 A deoxynucleotide and a dideoxynucleotide

Sanger sequencing makes use of two kinds of nucleotides. (a) A deoxynucleotide has a hydroxyl (—OH) group on the 3′ carbon, allowing this end to be elongated. (b) A dideoxynucleotide lacks the 3′ hydroxyl group and instead has a 3′-H atom. It cannot be elongated because there is no hydroxyl group to attack an incoming nucleotide triphosphate. As a result, incorporation of a dideoxynucleotide prevents strand elongation.

terminating at the site at which a dideoxynucleotide was incorporated.

FIGURE 38.6 shows how the daughter strands help us to determine the DNA sequence by the procedure now called Sanger sequencing. In a tube containing dideoxy-C and all the other elements required for many rounds of DNA replication, a strand of DNA is synthesized complementary to the template until, when it reaches a G in the template strand, it incorporates a C. If it incorporates a dideoxy-C, DNA synthesis terminates. However, only a small fraction of the C nucleotides in the sequencing reaction is in the dideoxy form, so only a fraction of the daughter strands incorporates

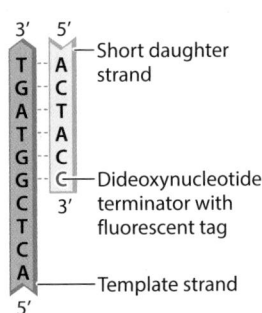

FIGURE 38.6 Chain termination

The incorporation of an incoming dideoxynucleotide, in this case dideoxy-C, stops the elongation of a new strand.

a dideoxy-C at that point, resulting in termination. The rest of the strands incorporate a normal C and continue synthesis. Most of these will be stopped at some point farther along the line when a G is reached again. Similarly, in a reaction containing dideoxy-A, DNA fragments will be produced whose sizes correspond to the positions of the T's, and likewise for dideoxy-T and dideoxy-G.

Each of the four dideoxynucleotides is chemically color coded with a different fluorescent dye, as indicated by the different colors of A, C, T, and G in **FIGURE 38.7**. After DNA synthesis is complete, the daughter strands are separated by size with gel electrophoresis. The smallest daughter molecules migrate most quickly and, therefore, are the first to reach the bottom of the gel, followed by the others in order of increasing size. A fluorescence detector at the bottom of the gel "reads" the colors of the fragments as they exit the gel. What the scientist sees is a trace (or graph) of the fluorescence intensities, such as the one shown in Figure 38.7. The differently colored peaks, from left to right, represent the order of fluorescently tagged DNA fragments emerging from the gel. Thus, a trace showing peaks colored green-purple-red-green-purple-purple-blue-green-blue-red corresponds to a daughter strand having the sequence 5′-ACTACCGAGT-3′.

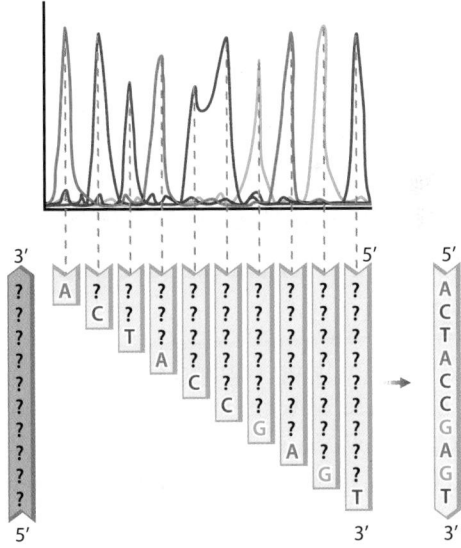

FIGURE 38.7 Sanger sequencing

Dideoxynucleotides terminate strands at different points in the template sequence. Separation of the short daughter strands by size shows where each terminator was incorporated and hence the identity of the corresponding nucleotide in the template strand. A fluorescent scanner "reads" the sequence, as indicated by the colored peaks at the top. From this information, the sequence can be deduced, as shown on the right.

The human genome and many others were sequenced by Sanger sequencing. This technique works well and is still the gold standard for accuracy, but it takes time and is expensive for large genomes like the human genome. Since it was first put into practice, Sanger sequencing has undergone many improvements. Together, these methods have increased the speed and decreased the cost of DNA sequencing considerably.

However, the ability to sequence everyone's genome, including yours, will require new technologies to further reduce the cost and increase the speed of sequencing. Collectively, these are often called next-generation sequencing technologies. The first human genome sequence, completed in 2003 at a cost of approximately $2.7 billion, stimulated great interest in large-scale sequencing and the development of devices that increased scale and decreased cost. As in the development of computer hardware, emphasis was on making the sequencing devices smaller while increasing their capacity through automation.

The goal of technology development was captured in the catchphrase "the $1000 genome," a largely symbolic target indicating a reduction in sequencing cost. For all practical purposes, the $1000-genome target has been achieved, but $1000 is still too costly to make genome sequencing a routine diagnostic procedure. The technologies are evolving and the costs decreasing, and so it likely that in the coming years, your own personal genome could be sequenced quickly and cheaply.

Why sequence more genomes? And if the human genome is sequenced, why sequence yours? There is really no such thing as *the* human genome, any more than there is *the* fruit fly genome or *the* mouse genome. With the exception of identical twins, every person's genome is unique. In fact, the genomes of any two humans differ at approximately one nucleotide out of every one thousand nucleotides. The sequence that is called "the human genome" is actually a composite of sequences from different individuals. This sequence is useful because most of us share the same genes and regulatory regions, organized the same way on chromosomes. However, knowledge of your own personal genome can also be valuable. For example, it can help you predict whether you are at risk of developing certain genetic diseases or how you will respond to medicines. There may come a time, perhaps within your lifetime, when personal genome sequencing becomes part of routine medical testing. Information about a patient's genome will bring benefits, but it will also raise ethical concerns and pose risks to confidentiality and insurability.

✓ Concept Check

6. In addition to dideoxynucleotides, **identify** the components needed for Sanger sequencing.

7. You have determined that the newly synthesized strand of DNA in your sequencing reaction has the sequence 5'-ACCGAGG-3'. **Identify** the sequence of the template (complementary) strand.

38.4 Restriction enzymes cut DNA at particular short sequences

In addition to amplifying segments of DNA, researchers often also make use of techniques that cut DNA at specific sites. Cutting DNA molecules allows whole genomes to be broken into smaller pieces for further analysis, such as DNA sequencing. Because techniques for cutting DNA depend on specific DNA sequences, it is also a way to determine whether specific sequences are present in a segment of DNA. Finally, cutting DNA allows pieces from the same or different organisms to be brought together in recombinant DNA technology, which is discussed below.

The method for cutting DNA makes use of enzymes that recognize specific, short nucleotide sequences in double-stranded DNA and cut the DNA at these sites. These enzymes are known as restriction enzymes. There are about 1000 different kinds of restriction enzymes isolated from bacteria and other microorganisms, but they can cut DNA from any organism whatsoever. The recognition sequences that the enzymes cut, called restriction sites, are typically four or six base pairs long. Most restriction enzymes cut double-stranded DNA at or near the restriction site. For example, the enzyme *Eco*RI recognizes the following restriction site:

$$\downarrow$$
5'–GAATTC–3'
3'–CTTAAG–5'
$$\uparrow$$

Wherever the enzyme finds this site in a DNA molecule, it cuts each strand at the position indicated by the

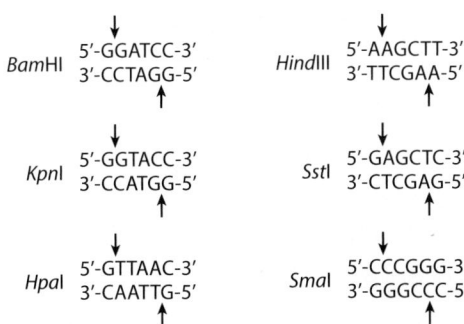

BamHI	5′-GGATCC-3′ 3′-CCTAGG-5′		HindIII	5′-AAGCTT-3′ 3′-TTCGAA-5′
KpnI	5′-GGTACC-3′ 3′-CCATGG-5′		SstI	5′-GAGCTC-3′ 3′-CTCGAG-5′
HpaI	5′-GTTAAC-3′ 3′-CAATTG-5′		SmaI	5′-CCCGGG-3′ 3′-GGGCCC-5′

FIGURE 38.8 Examples of restriction enzymes

Restriction enzymes recognize and cut DNA at specific sequences. Some examples are shown here. In some cases, such as *Bam*HI, they make a jagged cut, leaving a single-stranded overhang. In other cases, such as *Hpa*I, they cut in the middle of the sequence, leaving a blunt end.

vertical arrows. Note that the *Eco*RI restriction site is symmetrical: reading from the 5′ end to the 3′ end, the sequence of the top strand is the same as the sequence of the bottom strand. This kind of symmetry is palindromic (it reads the same in both directions) and is typical of restriction sites.

Note also that the site of cleavage is not in the center of the recognition sequence. The cleaved double-stranded molecules, therefore, each terminate in a short single-stranded overhang. In this case, the overhang is at the 5′ end, as shown below:

$$
\begin{array}{ll}
5'-G & -3' \\
3'-CTTAA & -5'
\end{array}
+
\begin{array}{ll}
5'- & AATTC-3' \\
3'- & G-5'
\end{array}
$$

FIGURE 38.8 shows more examples of restriction enzymes. Some cut their restriction site to produce a 5′ overhang, others produce a 3′ overhang, and still others cut in the middle of their restriction site and leave blunt ends with no overhang. The standard symbols for restriction enzymes include both italic and roman letters. The italic letters stand for the species from which the enzyme is derived (*E. coli* in the case of *Eco*RI, *Bacillus amyloliquefaciens* in the case of *Bam*HI), and the Roman letters designate the particular restriction enzyme isolated from that species.

An example of using a restriction enzyme to cut up plasmid DNA is shown in **FIGURE 38.9**. The restriction enzyme cuts the circular DNA into fragments of specific sizes, shown in Figure 38.8a. The individual fragments can

a.

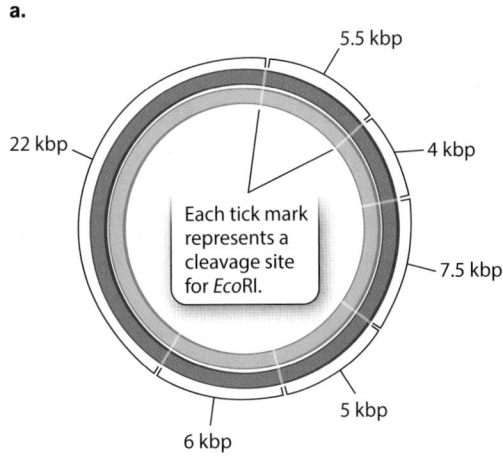

b.

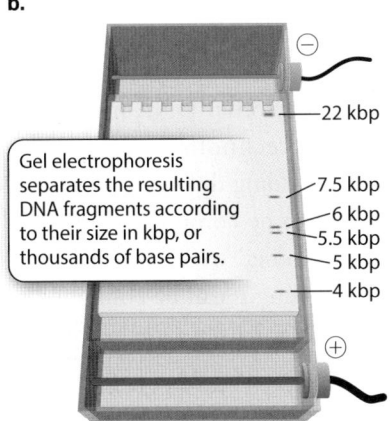

FIGURE 38.9 Cutting by restriction enzymes

Restriction enzymes can be used to cut a circular DNA molecule into short, linear fragments. (a) A circular DNA molecule, or plasmid, is cut by the restriction enzyme *Eco*RI at the sites indicated. Cutting with *Eco*RI produces DNA fragments of specific sizes, indicated in kilo base pairs (kbp), or 1000 base pairs. (b) These fragments can be separated and visualized on a gel.

be visualized in a gel, as shown in Figure 38.8b. These fragments can then be extracted from the gel for further analysis or manipulation. For instance, the extracted fragments can be sequenced or joined to other fragments, which we will discuss next.

✓ Concept Check

8. **Describe** what a restriction enzyme does.

9. **Identify** a technique that is used to determine the sizes of DNA fragments produced by restriction enzymes.

38.5 Genetic engineering allows researchers to alter DNA sequences in living organisms

As methods to manipulate DNA fragments developed, so did the ability to isolate genes from one species and introduce them into another. This type of genetic engineering is called recombinant DNA technology because it literally recombines DNA molecules from two or more different sources into a single molecule. This technology is possible because the DNA of all organisms differs only in sequence and not in chemical or physical structure. When DNA fragments from different sources are combined into a single molecule and incorporated into a cell, they are replicated and transcribed just like any other DNA molecule.

Recombinant DNA technology can combine DNA from any two sources, including different species. DNA from one species of bacteria can be combined with another, or a human gene can be combined with bacterial DNA, or the DNA from a plant and a fungus can be combined into a single molecule. These new sequences may be unlike any found in nature, raising questions about their possible effects on human health and the environment.

This section will discuss one of the basic methods for producing recombinant DNA, some of the important applications of recombinant DNA technology such as genetically modified organisms, and a new method that can be used to change the DNA sequence of any organism into any other desired sequence.

DNA Transformation

The first application of recombinant DNA technology was the introduction of foreign DNA fragments into the cells of bacteria in the early 1970s. The method is simple and straightforward, and remains one of the mainstays of modern molecular research. It can be used to generate a large quantity of a protein for study or therapeutic use.

The method is illustrated in **FIGURE 38.10**. It requires a fragment of double-stranded DNA that serves as the donor.

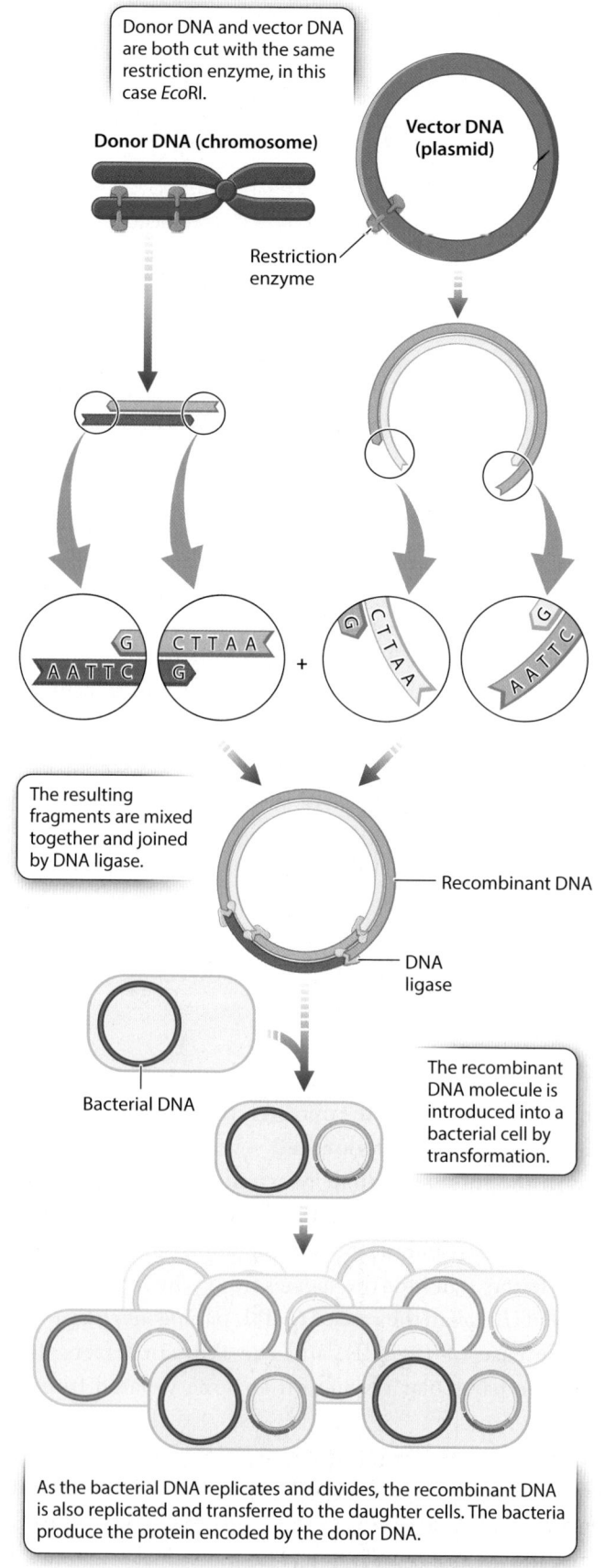

FIGURE 38.10 Recombinant DNA

Recombinant DNA combines DNA from two or more sources. In this case, human donor DNA is inserted in a bacterial plasmid. The recombinant plasmid is then transformed into bacteria, where it replicates and expresses the protein encoded by the donor DNA.

The donor fragment may be a protein-coding gene, a regulatory part of a gene, or any DNA segment of interest. If you are interested in generating bacteria that can produce human insulin, you would use the coding region of the human insulin gene as your donor DNA molecule.

Once a molecule of donor DNA has been isolated, another requirement for recombinant DNA is a vector sequence into which the donor fragment is inserted. The vector is the carrier of the donor fragment, and it must have the ability to be maintained in bacterial cells. A frequently used vector is a bacterial plasmid, a small circular molecule of DNA found in certain bacteria that can replicate when the bacterial genomic DNA replicates and be transmitted to the daughter bacterial cells when the parental cell divides. Many plasmids have been modified by genetic engineering to make them suitable for use as vectors in recombinant DNA technology. The donor and vector DNA molecules are both cut with restriction enzymes, and then joined by an enzyme called DNA ligase. The joining of the donor DNA to the vector creates the recombinant DNA molecule.

The next step in the procedure is transformation, in which the recombinant DNA is mixed with bacteria that have been chemically coaxed into a physiological state in which they take up DNA from outside the cell. Recall from Module 37 that transformation is a form of horizontal gene transfer. Having taken up the recombinant DNA, the bacterial cells are transferred into growth medium, where they reproduce. Because the vector part of the recombinant DNA molecule contains all the DNA sequences needed for its replication and partition into the daughter cells, the recombinant DNA multiplies as the bacterial cell multiplies. As a result, the bacteria expresses the protein encoded by the donor DNA. For example, if the donor DNA encodes human insulin, the bacterial cells will produce human insulin, which can then be used for medical purposes. This is much less expensive and simpler than manufacturing insulin from scratch or collecting large quantities of insulin from animals, such as pigs.

Genetically Modified Organisms (GMOs)

Applications of recombinant DNA have gone far beyond genetically engineered bacteria. Using methods that are conceptually similar to those described for bacteria but differing in details, scientists have been able to produce varieties of genetically engineered viruses and bacteria, laboratory organisms, agricultural crops, and domesticated

FIGURE 38.11 Genetically modified rice

Rice has been genetically modified in many different ways to make it more nourishing, produce higher yields, and resist insect pests. Non–genetically modified white rice is shown on the left, and genetically modified golden rice, which produces more vitamin A, is shown on the right. Photo: REUTERS/Alamy Stock Photo

animals. These are called **genetically modified organisms (GMOs)** or transgenic organisms.

One example of a genetically modified plant is shown in **FIGURE 38.11**. The photo shows rice that has been engineered to produce a higher content of vitamin A. Additional examples include sheep that produce a human protein in their milk that is used to treat emphysema, chickens that produce eggs containing human antibodies to help fight harmful bacteria, and salmon with increased growth hormone for rapid growth. Plants such as corn, canola, cotton, and many others have been engineered to resist insect pests. Additional engineered plants include tomatoes with delayed fruit softening, potatoes with waxy starch, and sugarcane with increased sugar content. To model disease, researchers have used recombinant DNA to produce organisms such as laboratory mice that have been engineered to develop heart disease and diabetes. By studying these organisms, researchers can better understand human diseases and begin to find new treatments for them.

Transgenic laboratory organisms are indispensable in the study of gene function and regulation and for the identification of genetic risk factors for disease. In crop plants and domesticated animals, GMOs promise enhanced resistance to disease, faster growth and higher yields, more efficient utilization of fertilizer or nutrients, and improved taste and quality. However, there are concerns about unexpected effects on human health or the environment and the increasing power and influence of agribusiness conglomerates.

There are also ethical objections to tampering with the genetic makeup of animals and plants. Nevertheless, more than 250 million acres of GMO crops are grown annually in more than 20 countries. The majority of this acreage is in the United States and South America. Resistance to the use of GMOs in Europe remains strong and vocal.

DNA Editing

Recombinant DNA technology combines existing DNA from two or more different sources. Its usefulness in research, medicine, and agriculture, however, is limited by the fact that it can make use only of existing DNA sequences. Therefore, scientists have also developed many different techniques to alter the nucleotide sequence of almost any gene in a deliberate, targeted fashion. In essence, these techniques allow researchers to "rewrite" the nucleotide sequence so that specific mutations can be introduced into genes to better understand their function, or mutant versions of genes can be corrected to restore normal function. Collectively, these techniques are known as **DNA editing.**

One of the newest and most exciting ways to edit DNA is CRISPR, which was discovered in an unexpected way. Researchers noted that about half of all species of Bacteria and most species of Archaea contain similar small segments of DNA of about 20–50 base pairs derived from viruses. Their function was a mystery until it was discovered that they play a role in bacterial defense against viruses. When a bacterium is infected by a virus for the first time, it makes a copy of part of the viral genome and incorporates it into its genome. On subsequent infection by the same virus, the DNA copy of the viral genome is transcribed to RNA that combines with a protein called Cas9 that cuts DNA. The RNA serves as a guide to identify target DNA in the virus by complementary base pairing. The Cas9 protein then cuts the target DNA, thus ending the viral threat, as shown in **FIGURE 38.12**. In this way, bacteria "remember" past infections and defend themselves from reinfection by the same virus. CRISPR is an acronym for "clustered regularly interspaced short palindromic repeats," which describes the organization of the viral DNA segments in the bacterial genome.

Scientists have learned to use CRISPR to alter the nucleotide sequence of almost any gene in any kind of cell.

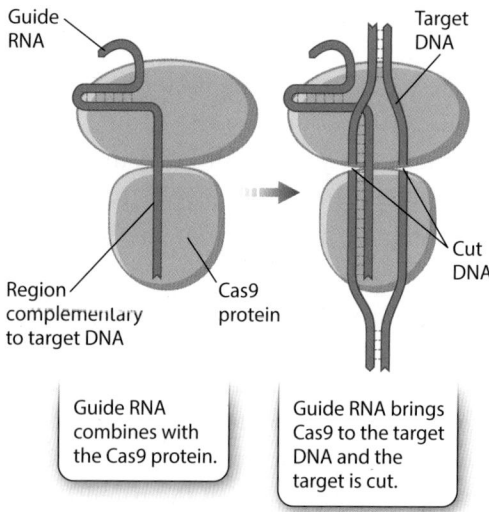

FIGURE 38.12 CRISPR

CRISPR is a mechanism of bacterial defense against viruses that is now used to edit DNA sequences. It uses a guide RNA that binds to target DNA and a protein called Cas9 that cuts DNA. The RNA binds to its target DNA by complementary base pairing, and then Cas9 cuts the target DNA.

DNA editing by CRISPR is technically straightforward and highly efficient. The method has generated great interest because of its potential to correct genetic disorders of the blood, immune system, or other tissues and organs. CRISPR technology is so powerful that it can be used to change existing genes in virtually any cell type from any type of organism. In principle, the method could also be used to manipulate human eggs and sperm, thereby affecting future generations. The uses and potential risks of human gene editing prompted a 2015 meeting of experts from around the world to discuss the science, ethics, and governance of the research, which resulted in a statement urging caution in the use of the technology.

✓ Concept Check

10. **Describe** a recombinant DNA molecule.

11. **Describe** how recombinant DNA techniques can be used to express a mammalian gene in bacteria.

Module 38 Summary

PREP FOR THE AP® EXAM

REVISIT THE BIG IDEAS

INFORMATION STORAGE AND TRANSMISSION: Using the content of this module, **identify** one biotechnology tool and **describe** how it allows humans to affect the transformation of genetic information within and between species.

LG 38.1 The polymerase chain reaction selectively amplifies regions of DNA.

- Techniques for manipulating DNA follow from the basics of DNA structure and replication. Page 526
- The polymerase chain reaction (PCR) is a technique for amplifying a segment of DNA. Page 527
- PCR requires a DNA template, DNA polymerase, the four nucleotides, and two primers. Page 527
- PCR is a repeated cycle of denaturation, annealing, and extension. Page 529

LG 38.2 Gel electrophoresis separates DNA molecules by size.

- Gel electrophoresis allows DNA fragments to be separated according to size. Page 529
- Small DNA fragments migrate farther than big fragments in a gel. Page 529
- Electrophoresis can also be used to separate RNA and protein molecules. Page 529

LG 38.3 DNA sequencing determines the order of nucleotides in a DNA strand.

- In Sanger sequencing of DNA, dideoxynucleotide chain terminators are used to stop the DNA synthesis reaction and produce a series of short DNA fragments from which the DNA sequence can be determined. Page 530

- New DNA sequencing technologies are being developed to increase the speed and decrease the cost of sequencing, perhaps making it possible to sequence everyone's personal genomes. Page 532

LG 38.4 Restriction enzymes cut DNA at particular short sequences.

- Restriction enzymes cut DNA at specific recognition sequences. Page 532
- The recognition sequences for restriction enzymes are called restriction sites. Page 532
- Cutting with restriction enzymes leaves single-stranded overhangs or blunt ends. Page 533

LG 38.5 Genetic engineering allows researchers to alter DNA sequences in living organisms.

- A recombinant DNA molecule can be made by cutting DNA from two organisms and then using DNA ligase to join them. Page 534
- Recombinant DNA is the basis for genetically modified organisms (GMOs), which offer both potential benefits and risks. Page 535
- Almost any DNA sequence in an organism can be altered by means of a form of DNA editing called CRISPR. Page 536
- CRISPR uses modified forms of molecules found in bacteria and archaea. Page 536

Key Terms

Polymerase chain reaction (PCR)
Gel electrophoresis

Genetically modified organism
(GMO)

DNA editing

Review Questions

1. The technique that involves isolating genes from one species and introducing them into another is called
   (A) Sanger sequencing.
   (B) recombinant DNA technology.
   (C) DNA hybridization.
   (D) gel electrophoresis.

2. The name of the technique used to amplify specific sequences of DNA is
   (A) gel electrophoresis.
   (B) Southern blot.
   (C) PCR.
   (D) DNA hybridization.

3. In gel electrophoresis, DNA fragments migrate toward
   (A) the negative pole.
   (B) the positive pole.
   (C) both the negative and positive poles.
   (D) neither the negative nor the positive pole.

4. You run a PCR reaction for five cycles starting with a single DNA molecule. Theoretically, how many copies of your sequence would you now have?
   (A) 8
   (B) 16
   (C) 24
   (D) 32

5. What is the benefit of using *Taq* polymerase in PCR?
   (A) Because it is taken from bacteria that live in high temperatures, it doesn't have a proofreading function.
   (B) Because it is taken from bacteria, this enzyme works much more efficiently than other types of DNA polymerase.
   (C) Because it is taken from bacteria that live in high temperatures, it stays active during the denaturation steps of the reaction
   (D) Because it is taken from bacteria, it makes fewer mistakes.

6. Which of the following is an example of a transgenic organism?
   (A) A human infected with the virus hepatitis C who expresses viral proteins
   (B) A patient who has received a kidney transplant from a close relative
   (C) A fish that "glows in the dark" by expressing fluorescent jellyfish proteins
   (D) A patient who has received a heart valve from a pig

Module 38
AP® Practice Questions

Section 1: Multiple-Choice Questions
Choose the best answer for questions 1–4.

1. Which of the following is an application in a clinical health setting where DNA sequencing could be applied?
   (A) Determining whether the specific mutation that causes a health disorder is inherited
   (B) Determining whether a gene that may be related to a disease is being expressed in an individual
   (C) Determining whether a specific cell type is producing a particular protein
   (D) Determining whether an enzyme that could be linked to a disease is active

2. Restriction enzymes cut double-stranded DNA at specific short nucleotide sequences. A particular piece of DNA is 6216 bp long. The restriction enzyme *XbaI* cuts this piece of DNA at nucleotide positions 2316, 3142, and 4513. A scientist incubated the DNA with *XbaI* until it was completely digested, then ran the digested DNA on a gel. Which of the gels in the figure correctly represents the expected banding pattern? A labeled size ladder is shown on each gel to indicate fragment sizes.

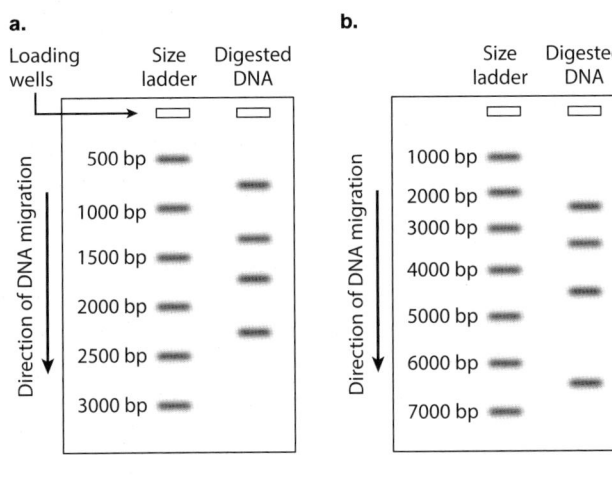

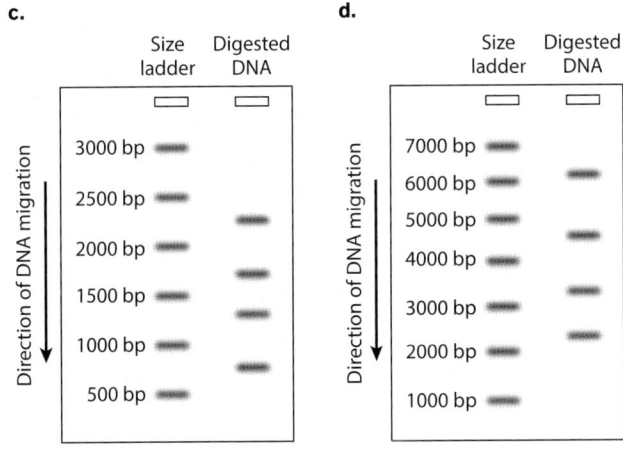

(A) a   (C) c

(B) b   (D) d

Question 3 refers to the following information.

Forensic analysis often relies on very small amounts of DNA recovered from crime scenes. The recovered DNA can be compared to the DNA of people suspected to be involved in the crime. Forensic investigators examined the DNA recovered from a hair that was found at a crime scene and obtained the banding pattern shown on the gel below. They additionally examined the banding patterns of DNA from four potential suspects, which are also shown on the gel.

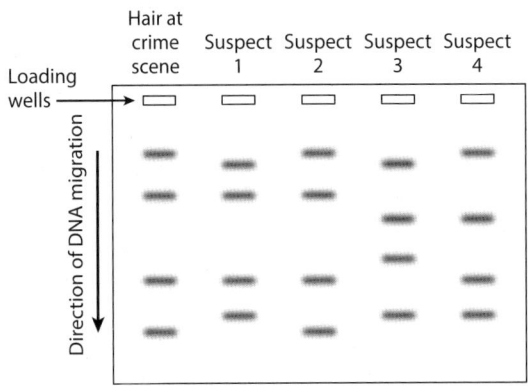

3. Which step must be completed before running the gel?
   (A) Polymerase chain reaction (PCR) to copy the DNA found at the crime scene
   (B) Bacterial transformation to express the DNA found at the crime scene
   (C) Translation of the DNA at the crime scene to identify the encoded proteins
   (D) DNA editing to purify the DNA found at the crime scene

Section 2: Free-Response Question

Write your answer to each part clearly. Support your answers with relevant information and examples. Where calculations are required, show your work.

Variation in the DNA sequences of genes encoding proteins can result in different individuals in a population carrying versions of the protein that have different amino acid sequences. The proteins can be isolated from tissue samples and separated with gel electrophoresis. The migration patterns of the isolated proteins reflect the genotype of the individual from which the proteins were isolated.

A researcher isolated a specific protein from blood samples taken from a mother bird and the six chicks in her nest. The researcher separated the proteins on a gel and obtained the pattern shown in the figure. The gene encoding the protein is encoded by a single gene that is on an autosomal chromosome.

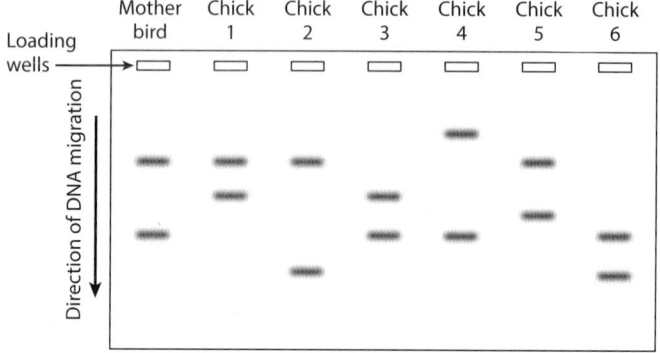

(a) **Identify** two properties of proteins that cause them to migrate to different positions on gels.

(b) **Predict** the number of bands that would be observed on the gel if proteins were isolated from a bird that is homozygous for one allele of the gene encoding the protein.

(c) The researcher claims that the gel shown in the figure indicates that some birds in the nest may have different fathers. Provide evidence to **justify** whether you support or refute this claim.

Module 39

Viruses

LEARNING GOALS ▶LG 39.1 Viruses are diverse, but all contain genetic material and a protein coat.
▶LG 39.2 Viruses can lyse a cell or integrate into the host genome.
▶LG 39.3 Some viruses cause human disease.
▶LG 39.4 Viruses are everywhere.

Throughout this unit, we have focused on how DNA stores and transmits genetic information, and how this information is used by the cell. Stretches of DNA called genes can be turned on and off by the process of gene expression, and the regulation of gene expression allows for cell specialization during the development of a multicellular organism. Our understanding of the structure and function of DNA has allowed us to manipulate it in the lab, and many of the techniques we use in the lab are actually borrowed from nature. For example, both PCR and DNA sequencing make use of basic principles of DNA replication. Similarly, the gene editing technique called CRISPR was adapted from bacteria, which use the technique to defend themselves against viruses. In this module, we will focus on viruses, which are diverse, abundant, and found just about everywhere.

We tend to think of viruses as agents of disease. And, in fact, they do cause diseases in humans and other organisms. The novel coronavirus (SARS-CoV-2), flu (influenza virus), and human immunodeficiency virus (HIV) are familiar examples of viruses that cause human diseases. But viruses play other roles as well. For example, viruses can transfer genetic material from one cell to another by transduction, a form of horizontal gene transfer, as discussed in Module 37. Horizontal gene transfer has played a major role in the evolution of bacteria and archaea. Molecular biologists have learned to use viral transduction as a technique to deliver genes into cells. In addition, the study of viruses has helped us understand how gene expression is regulated.

We will begin by describing the anatomy of a virus—its basic structure and components. Then we will examine the life cycles of viruses and what viruses have taught us about gene regulation. Finally, we will turn to several viruses that cause human diseases and the role that viruses play in ecosystems.

PREP FOR THE AP® EXAM

FOCUS ON THE BIG IDEAS

SYSTEMS INTERACTIONS: Focus on virus–host cell interactions over the short term and over many generations.

39.1 Viruses are diverse, but all contain genetic material and a protein coat

A **virus** is a small agent that infects cells. Viruses are extremely tiny and relatively simple in structure. Most viruses have only a few genes inside a protein coat. Viruses cannot reproduce on their own, but instead require a cell to reproduce and make more viruses. A cell in which viral reproduction occurs is called a host cell. When a virus infects a host cell, it uses the cellular machinery to make copies of itself. These new viruses, in turn, go on to infect other cells. In this section, we will focus on the basic structure of viruses.

The Anatomy of a Virus

All viruses contain a genome made up of DNA or RNA. In this way, they are different from cells, where the genome is always DNA. The genome is packaged inside a protein coat called a capsid. Some, but not all, viruses also have an outer phospholipid envelope that surrounds the genome and capsid, as shown in **FIGURE 39.1**. The envelope comes from the host cell where the virus reproduced.

A virus infects a cell by binding to the cell's surface, inserting its genetic material into the cell, and hijacking the cell's molecular machinery to produce more viruses. A virus must infect a cell to reproduce, because viruses use the cell's biochemical machinery to replicate, transcribe, and translate their genome. The infected cell may produce more viruses,

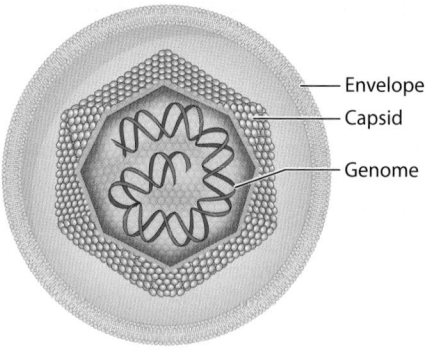

Envelope
Capsid
Genome

FIGURE 39.1 The anatomy of a virus

All viruses have a genome and a protein capsid. Some also have a phospholipid envelope, like the one shown here.

leading to lysis, or breakage, of the cell, and the new viruses can then infect more cells. In some cases, such as with the human immunodeficiency virus (HIV), the genetic material of the virus becomes integrated into the DNA of the host cell.

In Unit 2, we noted that the cell is the smallest living entity, and the fundamental unit of life. This concept is central to the cell theory. Viruses are typically much smaller than cells, so why aren't viruses considered the smallest living entity? Recall that all cells have three essential features—the capacity to store and transmit information, a membrane that selectively controls movement of substances into and out of the cell, and the ability to harness energy from the environment. Do viruses share these features?

Viruses contain genetic information that is stored and transmitted, just as cells do. However, by themselves, viruses cannot read and use the information contained in their genetic material. To replicate their genome and synthesize proteins, they require a host cell. In addition, viruses cannot regulate the passage of substances across their protein coats or phospholipid envelopes in the same way that cells do. Likewise, they cannot harness energy from the environment without the help of a host cell. For all these reasons, most scientists do not consider viruses to be living. They exist at the border between life and nonlife. That said, viruses exhibit one of the most fundamental properties of living systems: they evolve.

Viral Size and Shape

Viruses come in a wide variety of sizes. All viruses are microscopic, with some being hardly larger than a ribosome, 25–30 nm in diameter. Roughly speaking, the average size of a virus, relative to that of the host cell it infects, may be compared to the size of an average person relative to that of a commercial airliner.

Most viral genomes range in size from 3000 base pairs (bp) to 300,000 bp and contain a small number of genes. For example, the genome of influenza A virus (one of the four types of flu viruses) contains seven genes, and the genome of HIV contains nine genes. However, a few viruses contain much larger genomes. So-called giant viruses, which infect a wide range of single-celled eukaryotes, have double-stranded DNA genomes of 300,000 to 1,000,000 bp or more. *Pandoravirus salinus,* found in coastal sediments off the coast of Chile, has the largest genome on record. Its 2,500,000 bp genome codes for approximately 2500 proteins, including genes for nearly every component needed for translation except ribosomes, as well as genes for sugar, lipid, and amino acid metabolism that are not found in other viruses.

Viruses also come in different shapes. Some of these are shown in **FIGURE 39.2**. T4 virus, shown in Figure 39.2a, has a complex structure that includes a head composed of protein surrounding a molecule of double-stranded DNA, a tail, and tail fibers. It infects cells of the bacterium *Escherichia coli.* Viruses that infect bacterial cells are called bacteriophages, which literally means "bacteria eaters." When infecting a host cell, the T4 tail fibers attach to proteins on the surface of the bacterium, and the DNA and some proteins are injected into the cell through the tail.

Most viruses are not so structurally complex. Consider the tobacco mosaic virus, which infects plants. It has a helical shape that is formed by the arrangement of protein subunits entwined with a molecule of single-stranded RNA. Its shape is shown in Figure 39.2b. Tobacco mosaic virus causes brown spots in tobacco leaves, pictured in **FIGURE 39.3**, and was the first virus to be discovered. Its presence was revealed in experiments showing that the infectious agent causing discoloration of tobacco leaves was so small that it could pass through the pores of filters that could trap even the smallest bacterial cells.

Some viruses have an icosahedral shape. An icosahedron has 20 identical triangular faces. This shape is formed from protein subunits that come together at their edges to form a capsid. The example in Figure 39.2c is an adenovirus, a common cause of upper respiratory infections in humans.

Many viruses that infect eukaryotic cells, such as influenza, HIV, and coronaviruses, are surrounded by an envelope composed of a phospholipid bilayer with embedded proteins that recognize and attach to host cell receptor proteins. Figure 39.2d shows a coronavirus.

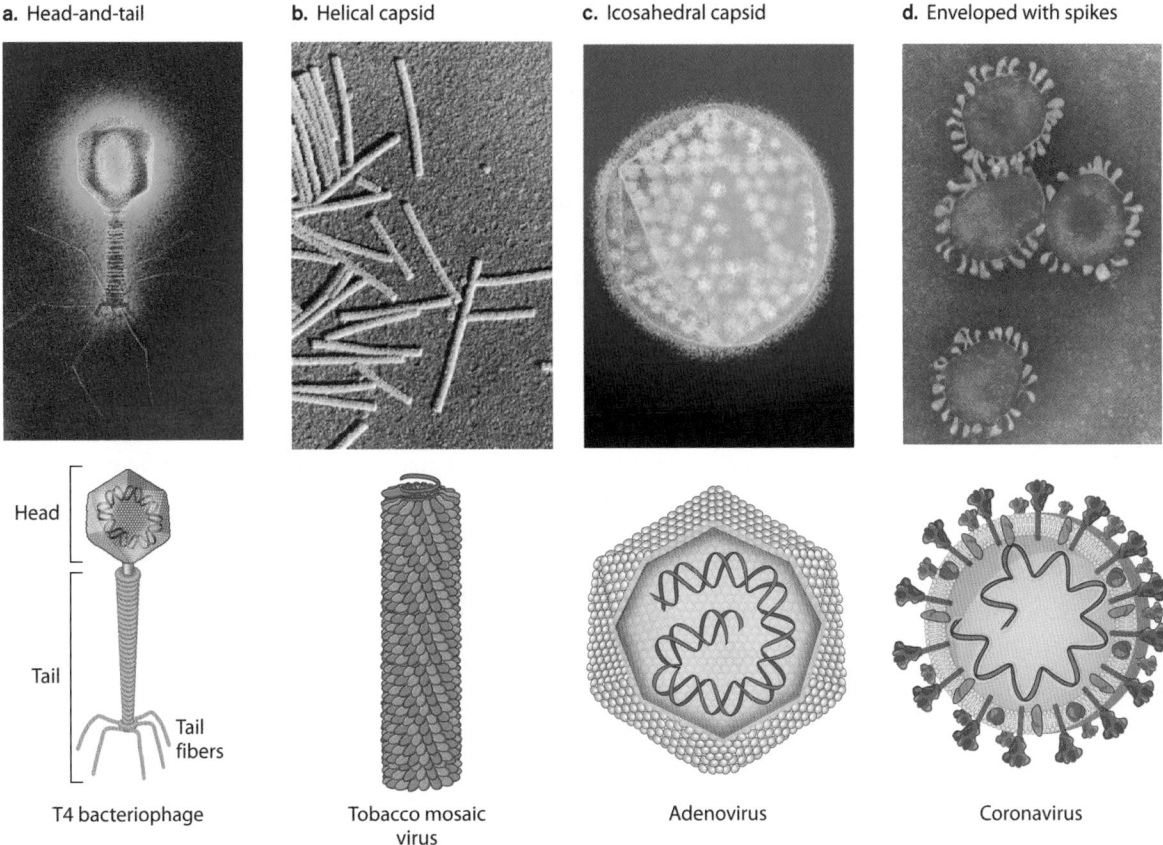

a. Head-and-tail   **b.** Helical capsid   **c.** Icosahedral capsid   **d.** Enveloped with spikes

Head

Tail

Tail
fibers

T4 bacteriophage   Tobacco mosaic
virus   Adenovirus   Coronavirus

FIGURE 39.2 Viral shapes

Viruses come in a variety of shapes. (a) This bacteriophage displays a head-and-tail structure.
(b) The tobacco mosaic virus has a helical capsid. (c) Adenoviruses have an icosahedral capsid,
which is a polyhedron with 20 faces. (d) An envelope with protein spikes is characteristic of
SARS-CoV-2 and other coronaviruses. Photos: (a) Department of Microbiology, Biozentrum, University of Basel/Science
Source; (b) Biology Pics/Science Source; (c) BSIP/Science Source; (d) BSIP/Getty Images

FIGURE 39.3 Tobacco mosaic virus

This plant is infected by tobacco mosaic virus, as indicated by the
brown, discolored areas. Source: Courtesy H. D. Shew; Reproduced, by permission,
from Shew, H. D., and Lucas, G. B., eds. 1991. *Compendium of Tobacco Diseases.* American
Phytopathological Society, St. Paul, MN.

Viral Classification

As we mentioned, all viruses have a nucleic acid genome. In
this way, they resemble cells. However, there are important
differences between viral genomes and cellular genomes. For
example, viral genomes are typically very small and compact.
In addition, the genomes of viruses are far more diverse than
the genomes of cellular organisms, which all have genomes
of double-stranded DNA. Some viral genomes are composed of RNA and others of DNA. Some viral genomes are
single stranded, others are double stranded, and still others
have both single- and double-stranded regions. Some viral
genomes are circular, while others are made up of a single
piece of DNA or multiple linear pieces of DNA.

    Unlike forms of cellular life, viruses do not seem to
share a single common ancestor. Different types of virus may
have evolved independently more than once. As a result,
viruses are no more related to bacteria than they are to
mammals, for example. In fact, classification of viruses based

on evolutionary relatedness is not possible, as it is for organisms. Instead, classification of viruses is based on the type of genome they have and how the genome is replicated. This classification scheme known as the Baltimore system (named after David Baltimore, who devised it) divides viruses into seven major groups.

DNA viruses synthesize mRNA using RNA polymerase, the same enzyme that cells use. DNA viruses include some important agents of disease, including adenoviruses, herpes simplex virus, and the African swine fever virus that since 2018 has killed half the domestic pigs in China.

Other viruses have single-stranded RNA genomes that replicate by means of an RNA-dependent RNA polymerase. This enzyme uses viral RNA as a template to make more RNA molecules. In other words, in these viruses, information flows from RNA to RNA to proteins. This contrasts with cells, in which genetic information flows from DNA to RNA to protein. Examples include SARS-CoV-2, which is the virus that causes the respiratory disease COVID-19; influenza, which causes hundreds of thousands of deaths each year and occasionally causes worldwide outbreaks that can kill millions; and the highly infectious Ebola viruses, which cause deadly hemorrhagic fever.

Some RNA viruses, such as HIV, reverse the usual flow of information. Recall that information in a cell usually flows from DNA to RNA. In these viruses, information instead flows from RNA to DNA. They have an enzyme called **reverse transcriptase,** which uses single-stranded RNA as a template to synthesize a complementary DNA strand. Viruses whose replication makes use of reverse transcriptase are called retroviruses. (*Retro* means "backward.") This capability is so unusual and was so unexpected that many molecular biologists at first doubted whether such an enzyme could exist. Soon after the enzyme was purified and its properties verified, its discoverers, Howard Temin and David Baltimore, were awarded the Nobel Prize in Physiology or Medicine in 1975, which they shared with Renato Dulbecco.

✓ Concept Check

1. **Identify** two structural features that all viruses share.

2. **Describe** four shapes of viruses.

3. **Describe** why most scientists do not consider viruses to be living organisms.

39.2 Viruses can lyse a cell or integrate into the host genome

Viruses have unique reproductive cycles that have helped us to understand basic biological processes, such as gene expression and regulation. In this section, we will introduce the reproductive cycle of viruses and what we have learned from them.

Viral Host Range

All known cells and organisms, including bacteria, archaea, and eukaryotes, are susceptible to viral infections. As discussed earlier, viruses reproduce within the cells of organisms, called host cells. Some viruses kill the host cell; others do not. Although viruses can infect all types of organisms, a given virus can infect only certain species or certain types of cell. For example, smallpox infects only a single species: humans. In cases like this, we say that the virus has a narrow host range.

For other viruses, the host range is broad. For example, rabies infects many different types of mammals, including humans, dogs, and racoons. Similarly, coronaviruses as a group infect many different kinds of animals, including humans, bats, cows, and camels. Tobacco mosaic virus, a plant virus, infects more than 100 different species of plants. No matter how broad the host range, however, plant viruses cannot infect bacteria or animals, and bacterial viruses cannot infect plants or animals.

Host specificity results from the way that viruses gain entry into cells. Proteins on the surface of the capsid or envelope of a virus bind to proteins on the surface of host cells. These proteins interact in a specific lock and key manner, similar to a substrate binding an enzyme (Unit 3) or a ligand binding its receptor (Unit 4), so viruses can infect only cells that include the host protein on the surface. For example, a protein on the surface of the SARS-CoV-2 virus binds to a protein called ACE2 on the surface of lung and some other cells. Thus, SARS-CoV-2 infects these cells but not others. Because viruses can only infect cells that have specific proteins on their surface, it is the interaction of viral and host surface proteins that determines host range. If many species have the specific cell-surface protein, the host range is broad; if only one or a few species have the specific cell-surface protein, the host range is narrow.

Lytic and Lysogenic Pathways

Viruses are vital research tools in studying certain forms of gene regulation, especially at the level of transcription. A good example of transcriptional regulation is found in certain viruses that infect bacteria. Bacterial cells are susceptible to infection by bacteriophages, or phages for short. One type of phage, called bacteriophage λ (lambda), infects cells of *E. coli*. On infection, the virus injects its linear DNA genome into the bacterial cell, and almost immediately the ends of the DNA molecule join to form a circle. Phage λ can follow one of two possible pathways after it infects a bacterial cell. These two pathways are illustrated in **FIGURE 39.4**.

The usual outcome of viral infection by phage is the lytic pathway, shown on the left side of Figure 39.4. In this pathway, the virus hijacks and uses the cellular machinery to replicate the viral genome and produce viral proteins. After about an hour, the infected cell undergoes lysis and bursts open to release a hundred or more phages that are capable of infecting other bacterial cells.

The alternative to the lytic pathway is the lysogenic pathway, shown on the right side of Figure 39.4. In this pathway, the bacteriophage DNA becomes integrated into the bacterial DNA. Lysogeny often takes place in cells growing in good conditions with abundant nutrients or in cells infected by large numbers of phages at once. When the bacteriophage DNA is integrated, the only bacteriophage gene transcribed and translated is one that represses the transcription of other phage genes that prevent entry into the lytic pathway. The bacteriophage DNA is replicated along with the bacterial DNA and transmitted to the bacterial progeny when the cell divides. Under stress, such as exposure to ultraviolet light or poor nutrients, the viral DNA is freed, initiating the lytic pathway.

Whether a phage follows a lytic or lysogenic pathway depends on many factors, such as the availability of nutrients in the environment. These conditions are detected by the phage, leading to the turning on and off of key genes involved in the two pathways. In other words, at the molecular level, which pathway is activated is determined by the positive and

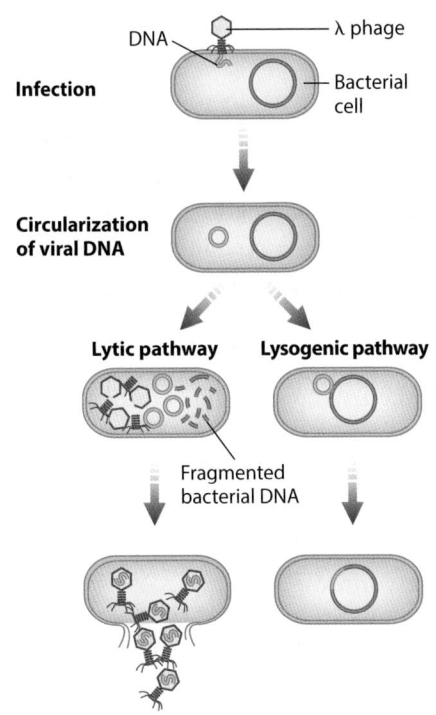

FIGURE 39.4 Lytic and lysogenic pathways

On infection of a bacterial cell, bacteriophage λ can enter either the lytic pathway (left) or the lysogenic pathway (right). In the lytic cycle, the phage replicates in the bacterial cell, leading the cell to burst, or lyse. In the lysogenic cycle, the phage integrates its DNA into the genome of the bacterial cell.

negative regulatory effects on transcription of a small number of bacteriophage proteins produced soon after infection. Therefore, the two pathways provide a good example of, and have served as, a model of transcriptional gene regulation.

✓ Concept Check

4. **Describe** what determines the types of cell that a virus can infect.

5. **Describe** the difference between the lytic and lysogenic pathways.

39.3 Some viruses cause human disease

Although viruses play diverse roles, viruses that cause human disease receive the most attention. They differ in the way they spread from person to person: some are transmitted in air; others in water; still others in blood or other body fluids; and yet others by insects, including mosquitoes and ticks or other vectors. Viruses also differ in the severity of the diseases they cause. In this section, we will discuss how viruses cause disease and how we use vaccines and other public health measures to prevent disease.

We will also highlight a few viruses that cause diseases in humans.

Viral Diseases and Vaccines

The pathogenicity of a virus or other infectious agent is its ability to cause disease: a pathogenic virus causes disease, whereas a nonpathogenic one does not. Virulence refers to the degree of severity of the disease caused. Flu is an example of a virus that shows a wide range of virulence. Scientists think that about 20% to 30% of infections proceed without symptoms. When symptoms appear, they are usually mild and include fever, headaches, muscle pain, and shortness of breath. However, flu can also cause severe disease and even death. In the 1918 flu pandemic, approximately 50 million people died worldwide, more than twice as many as had been killed in World War I.

The virulence of a virus relates in part to the virus itself, but often also reflects how strongly our immune system reacts to it. The primary response of the immune system includes the production of cytokines. Cytokines are chemical messengers that activate other components of the immune system. These cytokines, which are produced in abundance, lead to many of the symptoms commonly associated with the flu and other viruses, such as fever, muscle aches, and chills.

The primary response also involves the activation of immune cells that help protect us from the virus. Some of these cells are memory cells that circulate in the blood long after the body has rid itself of the virus. Then, on re-exposure to the same virus, these memory cells mount a secondary immune response, which is faster and stronger than the primary immune response. We have learned to take advantage of immunological memory through the use of vaccines. Vaccines trigger a primary response and the production of memory cells, but do not lead to disease. By creating a store of memory cells ready to attack their pathogenic targets, vaccines protect us upon subsequent exposure to those pathogens.

Vaccines can take many forms. Some are weakened forms of a pathogen that do not cause disease, but are able to trigger an immune response. They are sometimes called live vaccines. Another type of vaccine, called inactivated vaccines, use killed or destroyed pathogens. There are also vaccines that include molecules produced by pathogens, such as surface proteins or carbohydrates. New types of vaccines continue to be developed. For example, scientists have developed nucleic acid or mRNA vaccines. In this case, the nucleic acid encodes for a protein made by the pathogen. Some of the vaccines against SARS-CoV-2 are mRNA vaccines. Regardless of what form they take, vaccines are among the most effective public health measures ever developed. They protect us from a wide range of infections that used to cause widespread illness and even death.

The Flu Virus

The influenza virus, or flu virus, causes symptoms of fever, chills, sore throat, cough, weakness, fatigue, and muscle pain. This virus is notable for causing seasonal outbreaks, called epidemics. Four times in the past 100 years, the flu virus spread more widely than usual, resulting in worldwide outbreaks and millions of deaths: the Spanish flu of 1918, the Asian flu of 1957–1958, the Hong Kong flu of 1968–1969, and the swine flu of 2009. These are called pandemics, which are worldwide outbreaks affecting many different countries. Pandemics have occurred repeatedly over the course of human history. Some, like the Black Death and Great Plagues, were caused by bacteria, but many were caused by viruses, including smallpox, influenza, HIV, and SARS-CoV-2, as shown in **FIGURE 39.5**.

The influenza virus is an enveloped virus with an RNA genome. It is spread in small water droplets in the air, which are often released through sneezing, coughing, or talking. Regular handwashing and flu vaccination are two basic public health measures that can limit the spread of the virus and protect vulnerable individuals, such as elderly individuals and people with weakened immune systems. Soap is particularly effective because it disrupts the phospholipid envelope surrounding the virus, thereby destroying the virus.

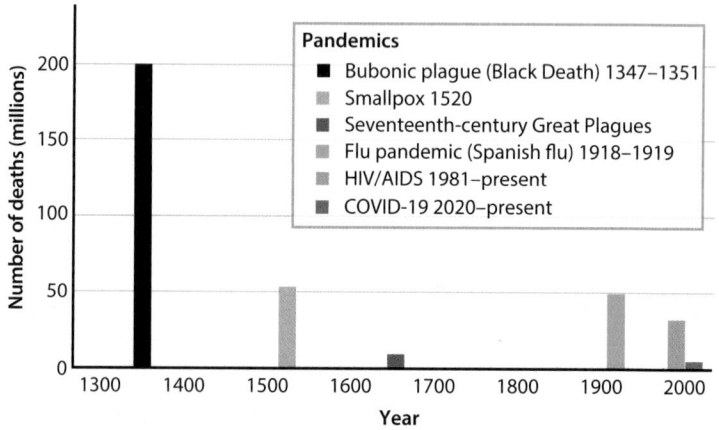

FIGURE 39.5 Pandemics in history

Worldwide outbreaks of infectious diseases, called pandemics, have occurred throughout human history. This graph shows some of the major pandemics. Smallpox, flu, AIDS, and COVID-19 are caused by viruses, while the Black Death and Great Plagues were caused by bacteria.

The flu virus, like all viruses, is subject to evolution and natural selection. Indeed, it can change quickly over time. Viruses like the flu have a high mutation rate, which is typical of RNA viruses, making them moving targets for our immune system and vaccines. For example, proteins on the virus surface can change from season to season. In such a case, memory cells produced from earlier infections will not recognize the new form of the virus. This explains in part why a new flu vaccine is needed every year.

In addition, the flu virus is capable of undergoing more significant changes. The flu genome consists of eight linear RNA strands. If a single cell is infected with two or more different flu strains at the same time, the RNA strands can re-assort to generate a completely novel strain. For example, swine flu (also called H1N1) has RNA segments from human, pig, and bird flu viruses. The swine flu spread worldwide and caused a pandemic in 2009.

HIV

HIV is an RNA enveloped virus, like influenza. It can be transmitted sexually between partners, from mother to fetus during pregnancy or birth, and by blood in contaminated needles or blood products. HIV infects and kills specific cells of its host's immune system. The resulting compromised state of the immune system leaves the body vulnerable to infections by other pathogens that would normally be warded off. Because patients with HIV become immunocompromised over time, the disease that results from HIV infection is called acquired immunodeficiency syndrome, or AIDS.

As discussed earlier, HIV is a retrovirus, meaning that it uses the enzyme reverse transcriptase to synthesize DNA from an RNA template. The DNA can then become integrated into the host cell, where it uses the host cell machinery to make more viruses. The mutation rate of HIV is very high because of the high error rate associated with reverse transcriptase. Thus, HIV has very high genetic variability, meaning that the genome sequence is likely to differ from one virus to the next. A single patient infected with HIV may actually harbor many variants of HIV, helping the virus evade the immune system and evolve resistance to antiviral treatments. Furthermore, if a single cell is infected by two different HIV variants, they can undergo recombination, resulting in viral progeny that are composites of the variants. Recombination is an integral part of the reproductive cycle of HIV, and it makes the virus an even more elusive target for the immune system—and for the development of a vaccine.

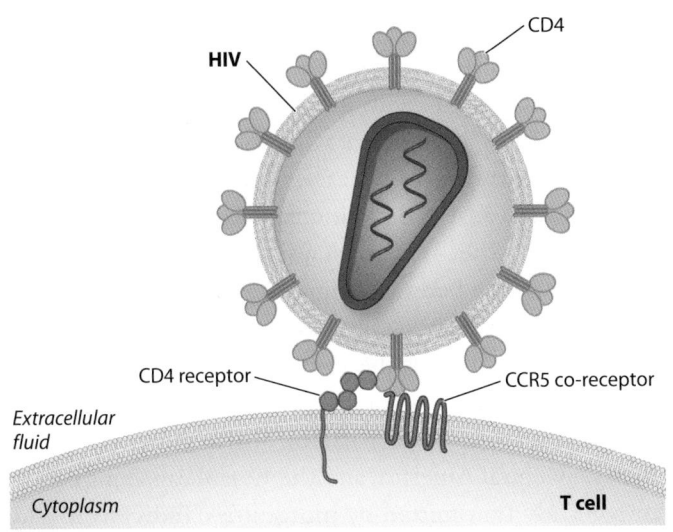

FIGURE 39.6 HIV infection

The first step in viral infection involves a specific interaction between a viral protein and protein on the surface of the host cell. This interaction allows the virus to bind to and then enter the cell. In the case of HIV, shown here, a protein on the surface of the virus called CD4 binds to a CD4 receptor protein and a CCR5 co-receptor on the surface of a T cell.

HIV virus gains entry into certain immune system cells, called T cells, through interaction of a viral surface protein with a T cell receptor protein called CD4, shown in **FIGURE 39.6**. This interaction is necessary but not sufficient for viral entry. Viral entry also requires interaction with a co-receptor on the T cell, called CCR5.

As we discussed in Module 37, some mutations are beneficial. A beneficial effect of a particular mutation in the *CCR5* gene was discovered in studies focusing on HIV-infected patients whose infection had not progressed to full-blown AIDS after 10 years or more. The protective allele is denoted as *Δ32* (*delta 32*) because the mutation is a deletion that removes 32 base pairs in the coding sequence of the *CCR5* gene. The mutant protein that results from this deletion is completely inactive. The *Δ32* allele has quite a pronounced effect. In individuals with the homozygous *Δ32/Δ32* genotype, HIV progression to AIDS is rarely observed. This mutation even provides some protective effect in individuals with heterozygous *Δ32* genotypes, in whom progression to AIDS is delayed by an average of about 2 years after infection compared to individuals without the mutation.

Emerging Diseases

Emerging diseases are those infectious diseases that have appeared recently or spread rapidly. Although such diseases are caused by a variety of pathogens, many result from viruses. For example, Ebola viruses caused two outbreaks in the Democratic Republic of the Congo and South Sudan in 1976. The most recent and largest epidemic occurred in 2013–2016 in West Africa, where several countries were affected. Even today, new cases continue to be reported. Ebola viruses cause a disease characterized by fever, sore throat, and diarrhea. They can also lead to bleeding inside and outside of the body, leading to a very high rate of death in infected individuals.

Another example is the Zika virus. It was first isolated in 1947 in Uganda, but was recently found in Brazil. From there, it spread to other parts of North America, South America, Central America, and the Caribbean. This pathogen is mostly transmitted by mosquitoes. Individuals who contract the Zika virus generally exhibit mild symptoms, but the virus can also be transmitted from a mother to her fetus, causing severe birth defects in the offspring.

A third example is the novel coronavirus called SARS-CoV-2, which caused a global pandemic in 2020–2021, as shown by the mostly empty streets of New York City in **FIGURE 39.7**. In this virus, a phospholipid envelope surrounds a capsid protein–RNA complex. This envelope has several different types of proteins associated with it, called spike, envelope, and membrane proteins. The spike-studded phospholipid envelope gives coronavirus its name: *corona* means "crown," which describes its appearance under the microscope, shown in **FIGURE 39.8**.

SARS-CoV-2 infects cells in a way that is typical of RNA viruses. The viral spike protein binds to proteins on the surface of certain human cells, such as those in the

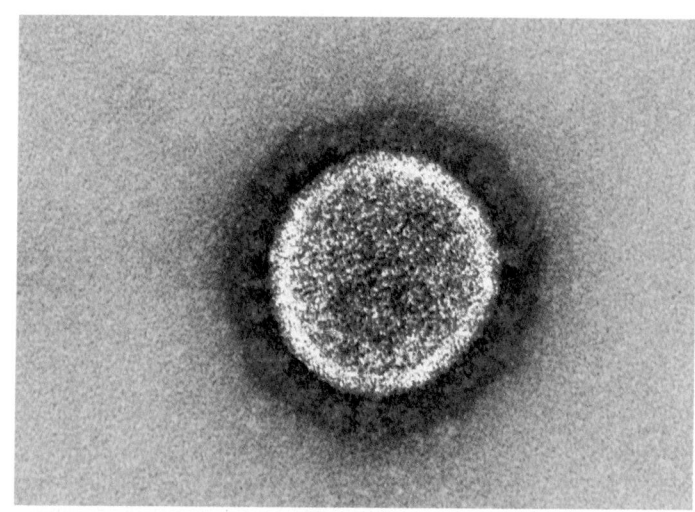

FIGURE 39.8 SARS-CoV-2

SARS-CoV-2 causes COVID-19, a potentially lethal respiratory infection. The virus has proteins on its lipid envelope, giving it a spiked appearance under the microscope, resembling a crown, or *corona*.

Photo: Science Source/Science Source

respiratory tract. Only cells with surface proteins capable of binding to the viral spike protein can be infected. Consequently, the virus is able to infect only certain cells, such as cells in the nose, mouth, and lungs.

After it latches on, the virus enters the cell and releases viral RNA. In the cytoplasm, the viral RNA is translated to make RNA-dependent RNA polymerase. As discussed earlier, this enzyme synthesizes RNA from an RNA template. Some of these newly produced RNAs are then translated to make viral proteins. Next, the RNA and protein components are packaged into new virus particles that bud from the host cell; the particles wrap themselves up in a piece of cell membrane before exiting the host cell, allowing the newly assembled viruses to infect other cells.

Emerging diseases, such as COVID-19, result from ongoing evolution in our time. For example, viruses that typically cause mild disease may evolve to become more virulent, and vice versa. SARS-CoV-2 is one of several different types of coronaviruses that infect humans. Coronaviruses usually cause mild respiratory symptoms and are often grouped into the informal category of the common cold. In recent years, however, we have seen the emergence of more virulent coronaviruses, such as those that caused outbreaks of severe acute respiratory syndrome (SARS) in 2002 and Middle East respiratory syndrome (MERS) in 2012. Similarly, SARS-CoV-2 continues to evolve—with more transmissible variants, for example, being selectively favored—as

FIGURE 39.7 The 2020 coronavirus pandemic

A novel coronavirus caused unprecedented medical, economic, and social upheaval in 2020, as shown by this largely empty street in Times Square, New York City. Photo: Anadolu Agency/Getty Images

we attempt to slow its spread with vaccines and other public health measures.

Viruses may also evolve in such a way that they change their host specificity. In other words, viruses that infect one species may evolve the ability to infect a new species, or a virus with a narrow host range may evolve to have a broader host range. The host range of a virus can change as viral surface proteins evolve over time to interact with new host cell proteins. SARS-CoV-2 most likely came from bats, but may have passed through another animal before being able to infect humans.

Viruses and other pathogens that cause emerging diseases not only are subject to evolution and natural selection but also are sensitive to the ecological landscape. Human movement and population density, interactions with domesticated and wild animals, ecosystem disruption, and climate change can set the stage for or exacerbate the spread of emerging diseases. As a result, COVID-19 and other emerging diseases will continue to pose challenges for twenty-first-century public health and medicine.

PREP FOR THE AP® EXAM

AP® EXAM TIP

Viruses with high mutation rates, such as HIV, give rise to much genetic population diversity. This in turn provides for a large number of mutations, some of which may be favored by natural selection.

✓ Concept Check

6. **Identify** two human diseases and the viruses that cause them.

7. **Describe** the difference between pathogenicity and virulence.

39.4 Viruses are everywhere

The numbers and diversity of viruses are staggering. Every day, huge numbers present in sea spray or soil dust are swept up into the lowest region of the atmosphere, where the winds can transport them thousands of kilometers before they fall back to Earth. The number of airborne viral particles that rain down on Earth each day is estimated as one hundred thousand million billion (10^{20}). Such observations have convinced many biologists that it is time not only to think of viruses as agents of disease but also to appreciate their vital roles in the functioning of ecosystems around the world. In fact, on a global scale, relatively few viruses cause disease, but many participate in ecosystem health. In this final section, we will discuss some of the roles that viruses play in the ocean and on land.

Viruses in the Ocean

For many years, biologists thought of the oceans as a vast pool of water inhabited by animals and algae. Ongoing surveys of marine life now show that bacteria dominate the oceans in terms of biomass, and viruses account for the largest numbers of individuals, as shown by the pie charts in **FIGURE 39.9**. It has been estimated that approximately 10^{30} viruses reside in the sea—that's equivalent to the number of stars in the universe multiplied by 10 million. There are more viruses in a liter of seawater than there are humans on Earth. Furthermore, an estimated 10^{24} marine viruses infect their hosts every second. Genomic surveys of the oceans have identified several hundred thousand different types of viruses, most of them distinct from any of the well-characterized virus families. And, as you might imagine, because of the overwhelming numbers, the accounting remains incomplete.

In the ocean, the carbon cycle begins with photosynthesis, mostly by phytoplankton—algae and cyanobacteria

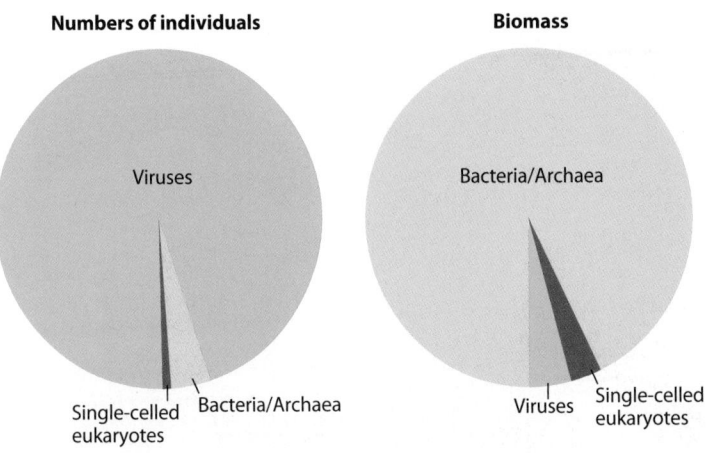

Numbers of individuals

Viruses

Single-celled eukaryotes

Bacteria/Archaea

Biomass

Bacteria/Archaea

Viruses

Single-celled eukaryotes

FIGURE 39.9 Viruses in the oceans

These pie charts show the relative biomass and abundances of prokaryotes, single-celled eukaryotes, and viruses in the world's oceans. Viruses are tiny but remarkably abundant.

floating in sunlit waters. The phytoplankton get eaten by small grazers; these, in turn, are consumed by larger organisms; and so on to the top carnivores. This food-chain model provides an important foundation for understanding marine ecology, but it is incomplete. Biologists estimate that 20% to 40% of all phytoplankton in the surface ocean, and an equal proportion of all bacteria in the oceans as a whole, are lysed by viruses every day. This process decreases the flow of energy and nutrients upward through the system, while sustaining the growth of bacteria.

Viruses lyse many cells and return their nutrients to the water. Occasionally, however, viruses play a more dramatic role in controlling phytoplankton populations. Seasonally, when abundant nutrients become available, single species of phytoplankton expand rapidly to form huge populations called blooms. Blooms of the alga *Emiliani huxleyi* are so extensive that they can be seen from space, as shown in **FIGURE 39.10**. As the algal population achieves high population density, viruses spread rapidly, lysing cells and thereby driving the phytoplankton populations down.

Viruses influence their photosynthetic hosts in another way. Analysis of the genomes of viruses that infect cyanobacteria in the open ocean reveal that many of these viruses contain genes that play a role in photosynthesis. It is thought that expression of these genes within the host cell increases the energy available for viral replication. As our understanding of viruses in the natural world increases, it is becoming clear that many viruses contain genes related to their host's metabolism. That is, the viruses are not simply entering their hosts but are altering the host's biology in ways that benefit the viruses.

Viruses on Land

Assessing viral abundance and diversity in soils presents difficult technical challenges, but recent advances suggest that what is true of the oceans is also true for the land. Viruses are everywhere, and they contribute to the structuring and function of land and freshwater communities. A gram of soil typically contains 10^7–10^9 virus particles, with the abundance being highest in relatively wet environments with limited seasonal freezing, much like the organisms that viruses infect. The diversity of soil viruses is high, but essentially unquantified. Soils also contain diverse bacteria, fungi, protists, and small animals such as nematodes. Viruses influence their hosts' population sizes and in turn their interactions with other species.

Viruses also infect plants (see Figure 39.3). Many plant viruses spread from plant to plant by hitchhiking on insects. Some viruses have the potential to inflict great damage on crop plants, but others actually improve the health of crops. For example, several viruses have been shown to increase crop resistance to drought or temperature extremes. And, as they do in the oceans, viruses can limit the populations of fungi, bacteria, and other viruses that infect or consume plant tissues. Just as viruses arrest blooms in the oceans, so they can reduce outbreaks of harmful organisms on land. For example, gypsy moths and tent caterpillars can cause immense damage to leaves, as shown in **FIGURE 39.11**. As their population densities increase, however, rapid spread of viruses that infect gypsy moths and tent caterpillars can arrest the outbreak of the insects.

Viruses play a key role in ecology and provide invaluable tools and models for our understanding of basic biological processes, such as gene regulation. "**VISUAL SYNTHESIS 6.2: VIRUSES**" on page 554 summarizes our discussion of viral structure, diversity, replication, host range, and effects on organisms.

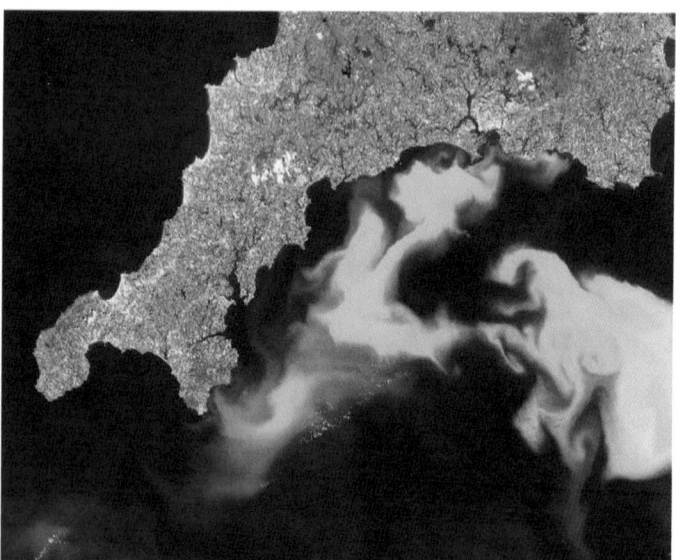

FIGURE 39.10 An algal bloom

A satellite image captures an algal bloom off the southwestern coast of England visible as light blue streamers in the otherwise dark blue ocean. Photo: National Oceanography Centre, UK

✓ Concept Check

8. **Identify** the types of organisms that viruses are able to infect.

9. **Describe** one way that viruses affect phytoplankton in the oceans.

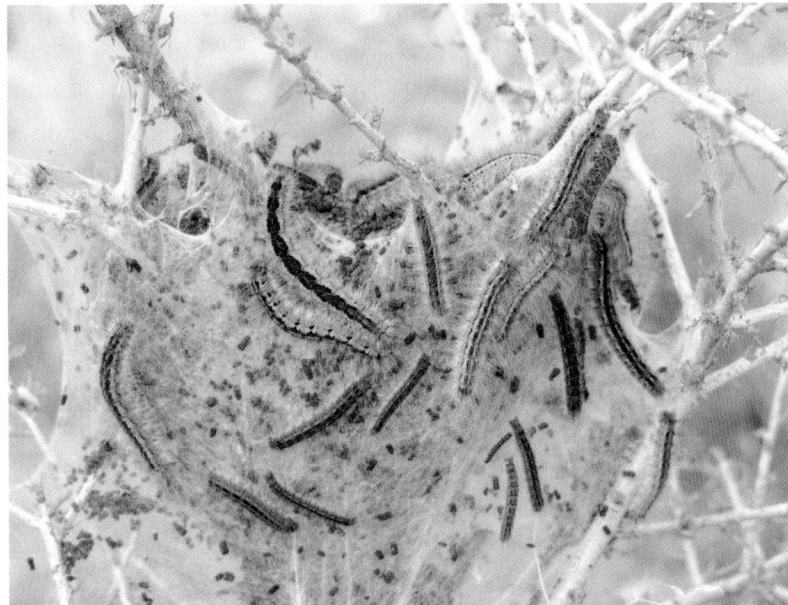

FIGURE 39.11 Viruses in pest control

Outbreaks of (a) gypsy moths and (b) tent caterpillars can do immense damage to plants. Viruses sometimes play a role in reducing infestations. Photos: (a) Scott Conner/Alamy Stock Photo; (b) Jeremy Christensen/Getty Images

Module 39 Summary

REVISIT THE BIG IDEAS

SYSTEMS INTERACTIONS: Describe characteristics of the virus–host cell interaction that result in viruses changing over time and possibly becoming capable of infecting other hosts.

LG 39.1 Viruses are diverse, but all contain genetic material and a protein coat.

- A virus is an infectious agent. Page 541
- All viruses have a nucleic acid genome and a protein capsid; some also have a phospholipid envelope. Page 541
- A virus infects a cell by binding to the cell's surface, inserting its genetic material into the cell, and using the cellular machinery to produce more viruses. Page 541
- Viruses cannot replicate on their own, so they generally are not considered to be living. Page 542
- Viruses occur in a wide variety of sizes, but even the largest is barely visible by a light microscope. Page 542

- Viruses show a diversity of shapes, including head-and-tail, helical, icosahedral, and enveloped with spikes. Page 542
- Viruses can be classified by the Baltimore system, which defines seven groups on the basis of type of genome and the way they replicate. Page 544

LG 39.2 Viruses can lyse a cell or integrate into the host genome.

- Viruses can infect all types of organisms, but a given virus can infect only specific types of cells and specific species. Page 544
- The host range of a virus can be narrow or broad and is determined by interactions of viral molecules with cell-surface molecules. Page 544
- The lytic and lysogenic pathways of bacteriophage λ have been studied as a model of gene regulation. Page 545
- When bacteriophage λ infects *E. coli,* it can lyse the cell (lytic pathway), or its DNA can become integrated into the bacterial genome (lysogenic pathway). Page 545

LG 39.3 Some viruses cause human disease.

- The pathogenicity of a virus relates to whether it causes disease, while the virulence of a virus describes the severity of disease. Page 546
- Influenza is an RNA virus that cause the flu. Page 546
- Influenza is a moving target for the immune system and vaccines. Page 546
- HIV causes acquired immunodeficiency syndrome (AIDS). Page 547
- HIV is a retrovirus that produces reverse transcriptase, an enzyme that copies the viral RNA genome into DNA, thereby reversing the usual flow of information in a cell. Page 547
- SARS-CoV-2 is an RNA virus that causes a disease called COVID-19. Page 548
- SARS-CoV-2 uses an RNA-dependent RNA polymerase to replicate its genome. Page 548

- Emerging diseases are new, rapidly spreading infectious diseases, and many are caused by viruses, such as HIV, Ebola, Zika, and SARS-CoV-2. Page 548
- Emerging diseases may cause more severe symptoms, spread more easily, or infect new species. Page 549

LG 39.4 Viruses are everywhere.

- Viruses are the most abundant biological entities on Earth, infecting essentially all species of bacteria, archaea, and eukaryotes. Page 549
- Viruses in the oceans lyse bacteria and phytoplankton, influencing the movement of carbon, energy, and nutrients throughout the sea. Page 550
- Viruses can control blooms of photosynthetic organisms in water and outbreaks of grazers on land. Page 550

Key Terms

Virus

Reverse transcriptase

Emerging disease

Review Questions

1. Viruses that infect bacteria are called
   (A) bacteriophages.
   (B) macrophages.
   (C) prokaryophages.
   (D) adenoviruses.

2. The concept of reverse transcription was at first controversial because it
   (A) suggested that RNA could sometimes function as an enzyme.
   (B) implied that DNA could be translated into protein.
   (C) showed that viruses could function with an RNA genome.
   (D) contradicted the "central dogma" that DNA makes RNA.

3. A cell in which viral reproduction occurs is called a
   (A) host cell.   (C) sex cell.
   (B) daughter cell.   (D) somatic cell.

4. A virus surrounded by a protein coat is known as a
   (A) reverse transcriptase.   (C) capsid.
   (B) surface protein.   (D) protease.

5. A virus infects only some types of cells because of specific interactions between
   (A) lipids on the surface of the virus and lipids on the cell.
   (B) proteins on the surface of the virus and proteins on the cell.
   (C) sugars on the surface of the virus and sugars on the cell.
   (D) nucleic acids on the surface of the virus and nucleic acids on the cell.

6. According to the Baltimore system, the major groups of viruses are classified according to
   (A) their capsids.
   (B) their host organisms.
   (C) their ability to cause disease.
   (D) their genomes.

7. The enzyme reverse transcriptase synthesizes
   (A) DNA from RNA.
   (B) RNA from DNA.
   (C) proteins from RNA
   (D) proteins from DNA.

Module 39
AP® Practice Questions

Section 1: Multiple-Choice Questions

Choose the best answer for questions 1–4.

1. Researchers discovered a novel virus and determined that its genome contained 20% adenine, 23% uracil, and 30% cytosine. How much of the genome is expected to be composed of thymine?

   (A) 0%   (C) 27%

   (B) 20%   (D) 30%

2. Identify the trait that all viruses share with living cells.

   (A) A genome encoded by nucleic acids

   (B) An external membrane made of phospholipids

   (C) The ability to synthesize proteins

   (D) An RNA to DNA flow of information

3. Which statement correctly describes viruses?

   (A) Viruses are rare in the environment because they cannot survive for very long outside of cells.

   (B) Viruses are highly specific and can only infect the host species in which they first evolved.

   (C) Viral epidemics cannot occur in populations if the virus is not able to be transmitted from one host to another by direct contact.

   (D) The virulence of a pathogenic virus is partly determined by the host immune response.

4. Retroviruses such as HIV have RNA genomes that are reverse transcribed to create DNA templates. The DNA templates can then be expressed to produce viral proteins. The enzyme that performs reverse transcription, called reverse transcriptase, has a much higher error rate than DNA polymerase. Which statement correctly explains a consequence of the high mutation rate of reverse transcriptase?

   (A) The high mutation rate increases the probability that the virus will integrate into the host genome.

   (B) Novel virus variants may arise within an individual host, making them moving targets for the immune system.

   (C) The high rate of mutation results in many nonfunctional virus particles and causes the virus to eventually go extinct.

   (D) Reverse transcriptase errors increase the rate of recombination between different elements of the viral genome.

Section 2: Free-Response Question

Write your answer to each part clearly. Support your answers with relevant information and examples. Where calculations are required, show your work.

Bacteriophages are viruses that infect bacteria. Many bacteriophages can follow two different pathways in their reproductive cycles. During the lysogenic pathway, the bacteriophage DNA is integrated into the bacterial chromosome and most genes of the bacteriophage are not expressed. During the lytic pathway, bacteriophage genes are expressed and the infected bacterial cell produces all of the proteins required to make new infectious bacteriophage particles. The bacterial cell in the lytic pathway eventually bursts, releasing the bacteriophage particles into the environment, where they are able to infect new bacterial cells.

A researcher grew a culture of bacteria infected with bacteriophage, recording the number of living bacterial cells and the number of bacteriophage genome copies. After 24 hours of growing the infected bacteria culture, the researcher then altered the conditions in the culture to induce the bacteriophage to enter the lytic pathway. The researcher counted the number of live bacterial cells and number of bacteriophage genome copies for another 24 hours. The data that the researcher obtained are shown in the table.

	Observation time (hour)	Number of viable bacterial cells	Number of bacteriophage genome copies
Lysogenic phase			
	0	5.22×10^2	5.21×10^2
	12	4.18×10^5	4.17×10^5
	24	3.78×10^6	3.78×10^6
Lytic phase induced			
	36	6.20×10^3	6.84×10^8
	48	8.35×10^1	4.38×10^9

(a) **Describe** how the genome of bacteriophage is replicated during the lysogenic pathway.

(b) **Calculate** the ratio of bacteriophage to bacterial cells at 36 hours.

(c) **Calculate** the percent increase in bacteriophage number at 36 hours compared to 24 hours.

(d) **Make a claim** to explain the percent increase in bacteriophage number between 24 and 36 hours, as well as the decrease in the number of bacterial cells between 24 and 36 hours.

VISUAL SYNTHESIS 6.2 VIRUSES

Viruses are small infectious agents that exist at the border between life and nonlife. They have RNA and DNA genomes, and come in different shapes and sizes. They require a host cell to reproduce. Following infection of a

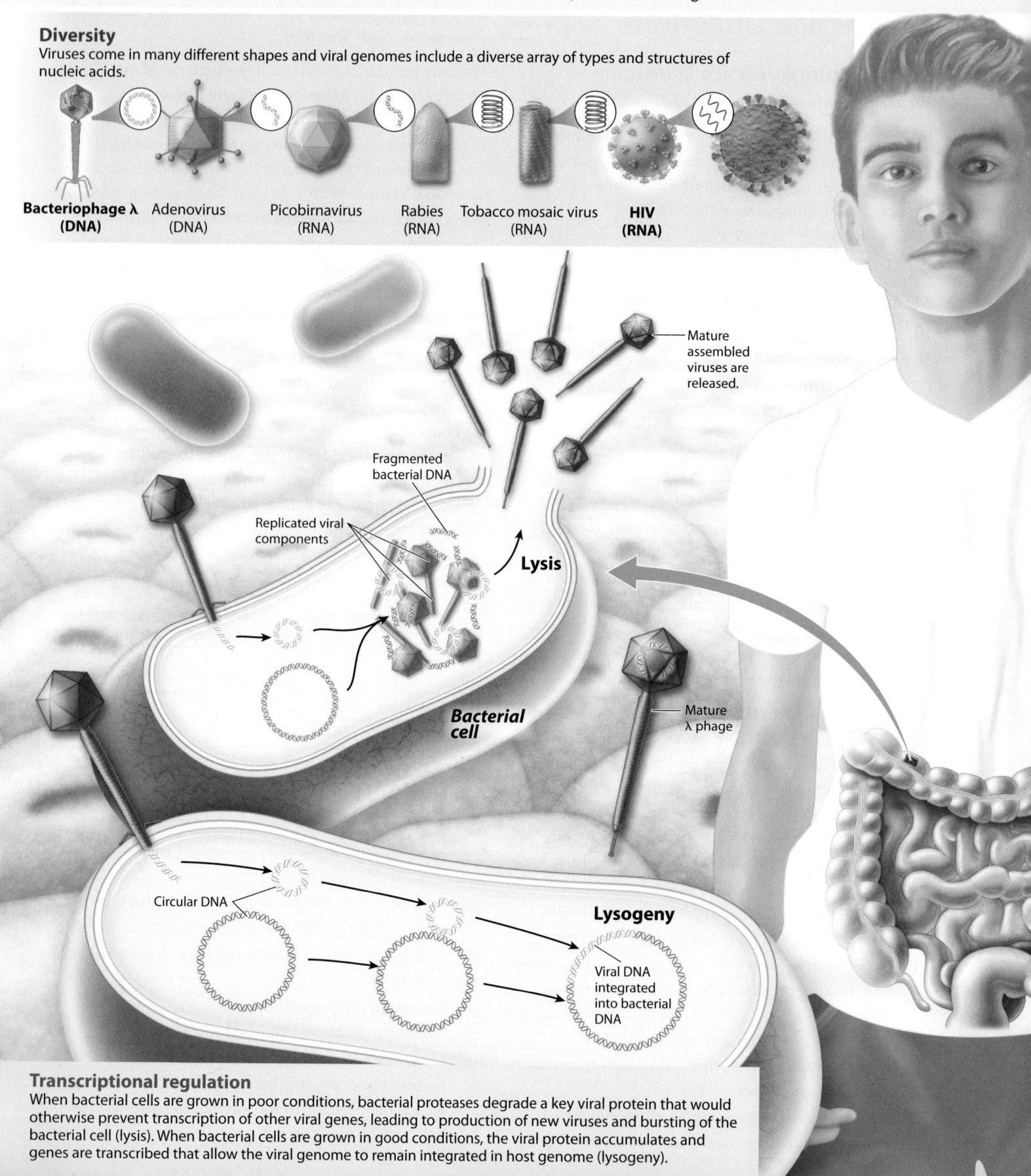

Diversity
Viruses come in many different shapes and viral genomes include a diverse array of types and structures of nucleic acids.

Bacteriophage λ (DNA)

Adenovirus (DNA)

Picobirnavirus (RNA)

Rabies (RNA)

Tobacco mosaic virus (RNA)

HIV (RNA)

Mature assembled viruses are released.

Fragmented bacterial DNA

Replicated viral components

Lysis

Bacterial cell

Mature λ phage

Circular DNA

Lysogeny

Viral DNA integrated into bacterial DNA

Transcriptional regulation
When bacterial cells are grown in poor conditions, bacterial proteases degrade a key viral protein that would otherwise prevent transcription of other viral genes, leading to production of new viruses and bursting of the bacterial cell (lysis). When bacterial cells are grown in good conditions, the viral protein accumulates and genes are transcribed that allow the viral genome to remain integrated in host genome (lysogeny).

cell, they may integrate into the genome or replicate until the cell bursts or releases new viruses capable of infecting more cells.

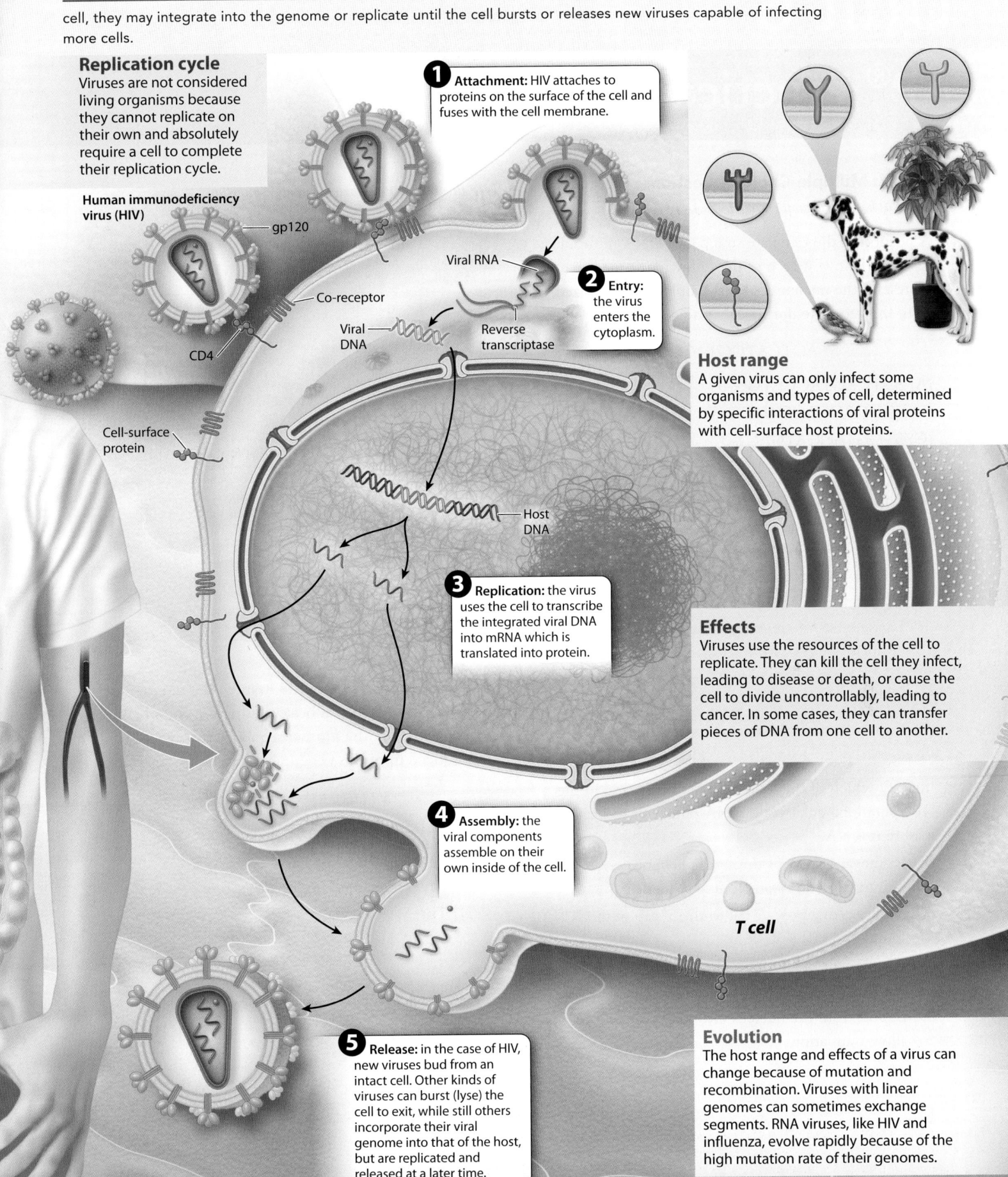

Replication cycle
Viruses are not considered living organisms because they cannot replicate on their own and absolutely require a cell to complete their replication cycle.

Human immunodeficiency virus (HIV)

gp120

Co-receptor

Viral DNA

CD4

Cell-surface protein

1 Attachment: HIV attaches to proteins on the surface of the cell and fuses with the cell membrane.

Viral RNA

2 Entry: the virus enters the cytoplasm.

Reverse transcriptase

Host DNA

3 Replication: the virus uses the cell to transcribe the integrated viral DNA into mRNA which is translated into protein.

4 Assembly: the viral components assemble on their own inside of the cell.

T cell

5 Release: in the case of HIV, new viruses bud from an intact cell. Other kinds of viruses can burst (lyse) the cell to exit, while still others incorporate their viral genome into that of the host, but are replicated and released at a later time.

Host range
A given virus can only infect some organisms and types of cell, determined by specific interactions of viral proteins with cell-surface host proteins.

Effects
Viruses use the resources of the cell to replicate. They can kill the cell they infect, leading to disease or death, or cause the cell to divide uncontrollably, leading to cancer. In some cases, they can transfer pieces of DNA from one cell to another.

Evolution
The host range and effects of a virus can change because of mutation and recombination. Viruses with linear genomes can sometimes exchange segments. RNA viruses, like HIV and influenza, evolve rapidly because of the high mutation rate of their genomes.

Unit 6

AP® Practice Questions

Section 1: Multiple-Choice Questions

Choose the best answer for questions 1–13.

1. A portion of the template strand of a DNA molecule is shown below. Which of the following sequences represents the nucleic acid that would be synthesized from this template during DNA replication?

   5′-AGTCATGGTACC-3′

   (A) 5′-TCAGTACCATGG-3′
   (B) 5′-UCAGUACCAUGG-3′
   (C) 5′-GGTACCATGACT-3′
   (D) 5′-GGUACCAUGACU-3′

2. Which statement correctly describes the role of the enzyme ligase during DNA replication?

   (A) Ligase relaxes the DNA supercoil in front of the replication fork to allow the strands of DNA to separate.
   (B) Ligase binds the duplicated strand of DNA to the template strand with hydrogen bonds.
   (C) Ligase initiates DNA duplication from RNA primers.
   (D) Ligase connects the fragments of duplicated DNA on the lagging strand.

3. Identify which of the following steps is critical in the processing of a eukaryotic RNA transcript to form a mature mRNA.

   (A) Removal of introns through splicing to create an uninterrupted protein coding sequence
   (B) Addition of a poly-A tail to the 5′ cap of the RNA strand
   (C) Attachment of ribosomes to produce the protein encoded by the RNA
   (D) Export from the nucleus into the cytoplasm to allow translation

Use the following chart to answer questions 4 and 5.

First position (5′ end)	Second position				Third position (3′ end)
	U	C	A	G	
U	UUU Phe UUC Phe UUA Leu UUG Leu	UCU Ser UCC Ser UCA Ser UCG Ser	UAU Tyr UAC Tyr UAA Stop UAG Stop	UGU Cys UGC Cys UGA Stop UGG Trp	U C A G
C	CUU Leu CUC Leu CUA Leu CUG Leu	CCU Pro CCC Pro CCA Pro CCG Pro	CAU His CAC His CAA Gln CAG Gln	CGU Arg CGC Arg CGA Arg CGG Arg	U C A G
A	AUU Ile AUC Ile AUA Ile AUG Met/Start	ACU Thr ACC Thr ACA Thr ACG Thr	AAU Asn AAC Asn AAA Lys AAG Lys	AGU Ser AGC Ser AGA Arg AGG Arg	U C A G
G	GUU Val GUC Val GUA Val GUG Val	GCU Ala GCC Ala GCA Ala GCG Ala	GAU Asp GAC Asp GAA Glu GAG Glu	GGU Gly GGC Gly GGA Gly GGG Gly	U C A G

4. A fragment of the template strand of a protein-coding gene is shown below. Using the codon table above, identify the protein sequence that would be produced from this gene fragment.

   5′-TGT CAG GTT ACC-3′

   (A) Thr–Val–Gln–Cys
   (B) Cys–Gln–Val–Thr
   (C) Gly–Asn–Leu–Thr
   (D) Thr–Val–Leu–Trp

5. A fragment of the coding strand of a DNA sequence is shown below. Using the codon table, predict the effect that mutating the underlined T to G would have on the function of the encoded protein.

5'-GTA CC<u>T</u> GTA TAC-3'

(A) The mutation might alter protein function because there would be a change in the amino acid sequence.

(B) The mutation would destroy the protein function because it would introduce a premature stop codon.

(C) The mutation would probably have no effect on protein function because it would not change the encoded amino acid.

(D) The mutation would destroy the protein function because it would introduce a frameshift.

6. A scientist determined that a strain of the bacterium *Yersinia pestis* is resistant to the antibiotic rifampicin. The scientist also determined that strain of a related bacterial species, *Y. pseudotuberculosis*, is susceptible to rifampicin. The scientist cultured equal amounts of the two species together in a single flask without rifampicin for 24 hours, then grew the mixed bacteria on agar plates on a solid medium containing rifampicin. The scientist then isolated many individual colonies of each species of bacteria. The scientist discovered that a small number of *Y. pseudotuberculosis* colonies had become resistant to rifampicin. Which provides the best explanation for this observation?

(A) When the two strains were cultured together, some of the *Y. pseudotuberculosis* acquired DNA that provides resistance to rifampicin from the *Y. pestis*.

(B) When the two strains were cultured together, some of the *Y. pseudotuberculosis* internalized rifampicin-resistance proteins that were produced by the *Y. pestis*.

(C) When the two strains were cultured together, the *Y. pestis* produced signals that stimulated some of the *Y. pseudotuberculosis* to express rifampicin-resistance genes.

(D) When the two strains were cultured together, the presence of rifampicin stimulated *Y. pseudotuberculosis* to produce resistance mutations.

Use the following figure to answer question 7.

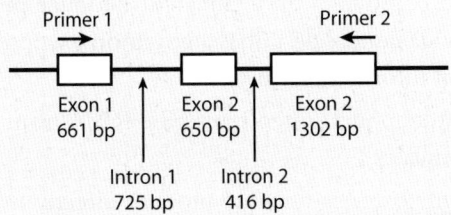

7. The figure shows a model of a particular gene. The gene has three exons and two introns. The illustration is labeled with the length, in base pairs (bp), of each exon and intron. The positions of two PCR primers are also shown. A researcher obtained mRNA from this gene and used an enzyme called reverse transcriptase to produce a single-stranded DNA molecule using the mRNA as a template. The researcher then used this single-stranded DNA as template in a PCR reaction. The researcher performed gel electrophoresis using the product of the PCR reaction and a size standard that contained DNA fragments of known sizes. In the figure below, which correctly illustrates the expected migration pattern?

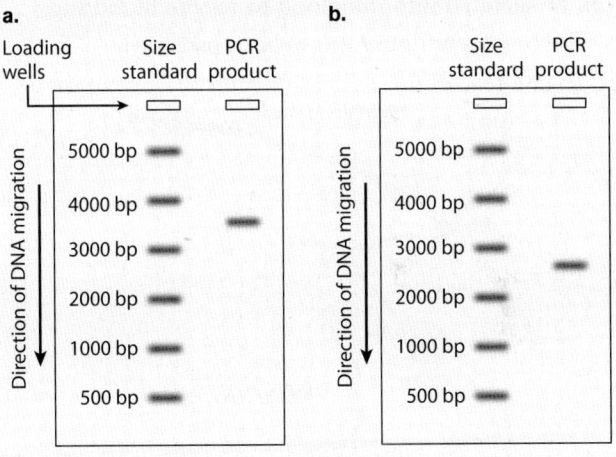

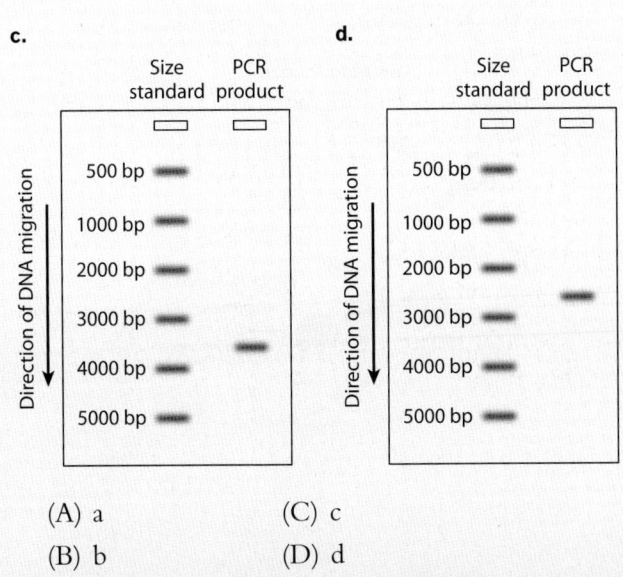

(A) a

(B) b

(C) c

(D) d

8. In a cell, DNA provides the template to produce RNA, and RNA provides the template to produce protein. Which accurately describes a common exception to this order of information flow?

(A) Mitochondria often transcribe RNA from protein templates.

(B) Some bacteria translate proteins directly from DNA sequence.

(C) Retroviruses transcribe DNA from RNA templates.

(D) Some proteins in plant cells can duplicate themselves without using nucleic acids.

Questions 9–11 refer to the following information.

Insulin is a hormone in animals that regulates blood sugar levels. Insulin is produced when blood sugar is high, for example, after eating. When bound to target cells, insulin leads to the activation of a protein called Akt that inhibits another protein named FOXO. When blood sugar is low, genes required for gluconeogenesis are expressed. Gluconeogenesis converts stored carbohydrates into glucose that can be secreted into the blood to restore blood sugar levels. The following figure shows both pathways.

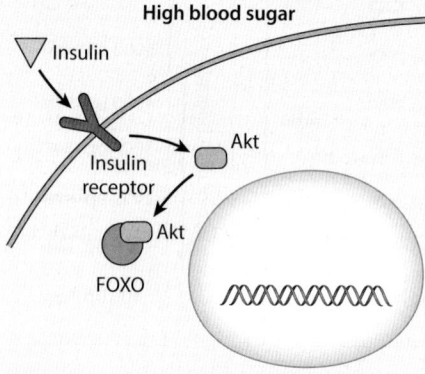

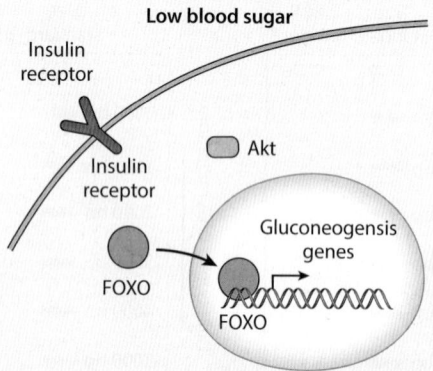

9. Which best describes the role of FOXO in the low blood sugar pathway?

(A) FOXO binds to DNA and activates gene expression.

(B) FOXO binds to a receptor and activates gene expression.

(C) FOXO binds to DNA and represses gene expression.

(D) FOXO binds to a receptor and represses gene expression.

10. Based on the information provided, predict the effect that eating a large meal high in carbohydrates would have on expression of gluconeogenesis genes.

(A) Expression of gluconeogenesis genes would be high because Akt would be active.

(B) Expression of gluconeogenesis genes would be high because Akt would be inactive.

(C) Expression of gluconeogenesis genes would be low because Akt would be active.

(D) Expression of gluconeogenesis genes would be low because Akt would be inactive.

11. Based on the information provided, predict the effect of a mutation that eliminated FOXO function on blood sugar levels in an animal that has not eaten recently.

(A) Blood sugar level would be high because gluconeogenesis genes would always be expressed.

(B) Blood sugar level would be low because gluconeogenesis genes would always be expressed.

(C) Blood sugar level would be high because gluconeogenesis genes would never be expressed.

(D) Blood sugar level would be low because gluconeogenesis genes would never be expressed.

Questions 12 and 13 refer to the following information.

Bacteriophages are viruses that infect bacteria. One type of phage uses a specific protein receptor on the surface of the bacterium *Salmonella enterocolitica* to gain entry into the cell. Some *S. enterocolitica* carry a mutation in the gene that encodes the protein receptor. This mutation prevents phages from entering the bacterial cell.

A scientist started two bacterial cultures that had an equal mix of wild-type and mutant bacteria. The scientist added phages to one of the two cultures, and then measured the abundance of each bacterial type over 18 hours. The scientist obtained the data shown in the figure.

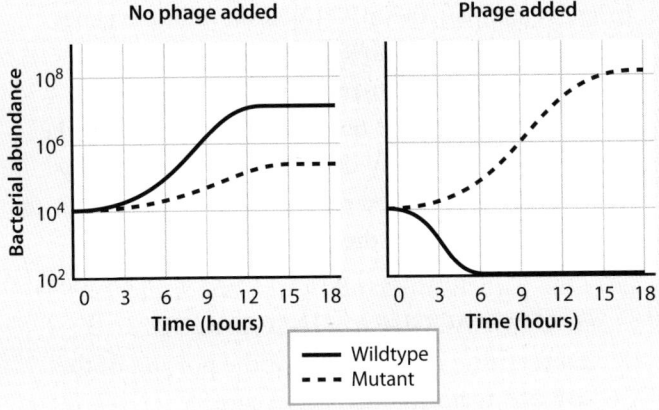

12. Which statement best explains the observed data?
    (A) The bacteria express the mutation only when phages are present.
    (B) The presence of phages regulates the expression of bacterial genes necessary for growth.
    (C) The mutation is harmful in the absence of phages but is beneficial when phages are present.
    (D) The resistant strain is unable to grow in the absence of phages.

13. In the treatment where no phages were added, what is the approximate ratio of wild-type cells to mutant cells at 18 hours?
    (A) 2:1
    (B) 7:5
    (C) 100:1
    (D) 107:105

14. A scientist identified two genes that are required for cell wall synthesis in the bacterium *Staphylococcus aureus*. The scientist hypothesized that the two genes might be encoded in a single operon. Which of the following experiments would provide the best test of the scientist's hypothesis?
    (A) The scientist could determine whether the proteins encoded by both genes are required at the same time for cell wall synthesis. If the proteins encoded by both genes are required at the same time, that would provide evidence that the genes are encoded in an operon. However, if only one gene is required at a time, the genes cannot be in the same operon.
    (B) The scientist could determine whether the proteins encoded by the two genes bind to one another. If the proteins encoded by the two genes bind to one another, that would provide evidence that the genes are encoded in an operon. However, if the proteins do not bind to one another, they cannot be encoded in the same operon.
    (C) The scientist could measure the amount of mRNA transcribed from each gene. If the two genes have identical expression levels, that would provide evidence that the genes are encoded in an operon. However, if the genes have different expression levels, they cannot be encoded in the same operon.
    (D) The scientist could measure the amount of protein in the cell that is encoded by each of the two genes. If the two proteins have equal abundance, that would provide evidence that the genes are encoded in an operon. However, if the proteins are at different levels in the cell, they cannot be encoded in the same operon.

Section 2: Free-Response Questions

Write your answer to each part clearly. Support your answers with relevant information and examples. Where calculations are required, show your work.

1. A scientist is interested in using genetically modified yeast cells to produce large quantities of viral protein that can be used to make vaccine. The scientist inserted a gene from the hepatitis B virus into the yeast *Saccharomyces cerevisiae*. The yeast produces the hepatitis B protein when the scientist adds a chemical that activates expression of the gene, and the protein can then be purified and used in vaccines. The scientist wanted to test the efficiency of two different chemicals (chemical W and chemical Z) for activating production of the hepatitis B protein. The scientist suspended each chemical in water and added the chemicals to separate yeast cultures. The scientist added water with no chemical to a third yeast culture. The scientist measured the rate (in mg/s) of hepatitis B protein production in all three cultures and obtained the data shown in the table (mean and standard error of the mean).

Time after addition of chemical (min)	Chemical W	Chemical Z	Water only
0	4 mg/s (SE = 2 mg/s)	6 mg/s (SE = 2 mg/s)	5 mg/s (SE = 2 mg/s)
30	32 mg/s (SE = 6 mg/s)	28 mg/s (SE = 5 mg/s)	6 mg/s (SE = 2 mg/s)
60	71 mg/s (SE = 7 mg/s)	38 mg/s (SE = 4 mg/s)	4 mg/s (SE = 2 mg/s)
90	63 mg/s (SE = 6 mg/s)	80 mg/s (SE = 5 mg/s)	4 mg/s (SE = 2 mg/s)
120	58 mg/s (SE = 6 mg/s)	110 mg/s (SE = 6 mg/s)	6 mg/s (SE = 2 mg/s)
150	50 mg/s (SE = 6 mg/s)	102 mg/s (SE = 5 mg/s)	4 mg/s (SE = 2 mg/s)

(a) **Draw** a graph that represents the data shown in the table.

(b) **Describe** the role in the experiment of the treatment where water is added with no chemical.

(c) **Calculate** the difference between the maximum rate for production of the hepatitis B protein in the culture treated with chemical Z and the maximum rate for production of the hepatitis B protein in the culture treated with chemical W.

(d) **Determine** whether there is likely a significant difference in hepatitis B protein at 30 minutes after addition of chemical W compared to 30 minutes of chemical Z. Provide reasoning to support your claim.

2. microRNAs are small, short single-stranded RNA molecules that regulate the amount of protein produced from an mRNA molecule. microRNAs are usually around 22 nucleotides long and they bind to their target mRNA, creating a short stretch of double-stranded RNA. This region of double-stranded RNA indicates to the cell that the mRNA should be degraded, and the cell destroys the mRNA transcript instead of translating it into protein. Thus, microRNAs regulate protein production after transcription is complete.

Multiple genes whose proteins participate in the same processes may be regulated by the same microRNA. In such examples, the mRNAs from multiple genes all have a sequence that is recognized by the same microRNA, so the transcripts of all of these genes are degraded when the microRNA is expressed. The sequence recognized by the microRNA is typically found in an untranslated region of the mRNA after the stop codon but before the 3′ end of the mRNA.

(a) One particular microRNA has the sequence 5′-UGGAAUGUAAAGAAGUAUGUAU-3′. **Determine** the sequence of the portion of the mRNA that this microRNA targets.

(b) **Make a claim** to explain the benefit to the cell of using a single microRNA sequence to prevent translation of the mRNAs from multiple different genes.

(c) Scientists have proposed that eukaryotic cells' use of microRNAs to degrade mRNAs evolved from a related mechanism to degrade genetic material from viruses infecting the cell. **Predict** whether the sequence of the viral mRNAs that were targeted by the ancestral microRNA gene regulation mechanism were the same or different from cellular mRNAs that were essential for the host cell's function. **Provide reasoning** to support your prediction.

Unit 7
Evolution and Natural Selection

Photo of an Española cactus finch (*Geospiza conirostris*), one of Darwin's finches, by Zoonar GmbH/Alamy

Module 40

Introduction to Evolution and Natural Selection

LEARNING GOALS ▶**LG 40.1** Evolution explains both the diversity and unity of life.
▶**LG 40.2** Variation in populations provides the raw material for evolution.
▶**LG 40.3** Natural selection results in adaptations.

In the last two units, we looked at how traits are passed down in families from one generation to the next. We examined how both genes and the environment shape the traits that we see in all organisms. We discussed gene expression—whether a gene is "on" or "off"—and how this process can be regulated as organisms develop and respond to their environment. We considered how mutations lead to changes in the DNA sequence.

In this unit, we take a broader view of biology, moving from individual organisms to entire populations, and from brief periods of time to longer periods of time. For example, we can consider what happens to mutations in a population of organisms over time. Do they spread throughout a population so that everyone has them? Do they disappear from the population over time? Evolution is a change in the genetic makeup of a population over time, and it has shaped life on Earth since life first originated. In Module 0, we introduced four Big Ideas that together help us understand the natural world. The first, and perhaps the most important, is Big Idea 1: Evolution. Evolution is so sweeping and significant because it can explain many aspects of biology, from the shape of molecules to the traits of organisms to the interactions among species and even to the functioning of whole ecosystems. One of the major mechanisms by which evolution occurs is natural selection. Natural selection leads not only to changes over time but to **adaptations,** which are traits of organisms that closely fit their environments. In this module, we will explore both evolution and natural selection in more detail and show how evolution can be demonstrated in the laboratory.

PREP FOR THE AP® EXAM

FOCUS ON THE BIG IDEAS

EVOLUTION: Think about how populations evolve over time to become better adapted to their environment.

40.1 Evolution explains both the diversity and unity of life

There are millions of species on Earth. These species differ from one another in all kinds of ways. At the same time, species have many traits in common. These two aspects of the natural world—diversity and unity—can be explained by evolution. In the process of evolution, species evolve from common ancestors, as well as diverge over time. In this section, we will examine both diversity and unity. We will also visualize the process of evolution as a branching tree, with a trunk representing a single common ancestor, and the tips of the branches representing the diversity of life on Earth.

Diversity and Unity of Life

One of the most notable features of the natural world is the diversity of life. In **FIGURE 40.1**, you can see that organisms differ from each other in all kinds of ways. They come in different sizes, colors, and shapes. Some are very large, such as the blue whale (*Balaenoptera musculus*) shown in the figure, the largest species ever to have lived on Earth; others are very small, such as bacteria and other microscopic organisms. Organisms also differ in how they move about: some walk on land; others swim in the water; still others fly, such as the butterfly in the figure; and some don't move at all, like Douglas fir trees (*Pseudotsuga menziesii*). Some organisms harness energy from the sun, whereas others obtain energy by eating other organisms. There are organisms that reproduce sexually and others that reproduce asexually. In short, organisms are incredibly diverse.

Scientists often quantify or measure organisms, processes, and phenomena as a way to better understand them. If we want to quantify biological diversity, or **biodiversity,** how might we do it? One approach is to count the number of species alive today. While this seems like a relatively simple task, in fact scientists don't know the total number of species. We do know the number of species that have been described and given a name. That number is approximately 2 million. However, it's clear that this number represents only a fraction of the number of species on Earth. Based on the number of known species, we can estimate the total number of species, which is approximately 10 million. In other words, there is a tremendous number of species on Earth, and we know only a fraction of them. Furthermore, we have named and described virtually all mammals, but only a tiny fraction of insects and bacteria.

Given that we know only a small fraction of the total number of species, it should not come as a surprise that we find new species all the time. Here are just a few new species we have discovered recently out of the hundreds and thousands of new species scientists have recently found. A tiny frog measuring just 1 centimeter was found in Borneo in 2010, shown in **FIGURE 40.2**; a new species of dolphin was discovered off the coast of Australia in 2011; a new species

FIGURE 40.1 Diversity of life

Organisms differ in size, shape, color, how they move about, and even how they reproduce. Clockwise from top left: blue whales are the largest species ever to have lived on Earth; evergreen trees such as these Douglas firs grow in a forest; bacteria are microscopic and reproduce asexually; caterpillars undergo metamorphosis to become butterflies, which are capable of flight. Photos (clockwise from left): Charles J. Smith; petekarici/Getty Images; STEVE GSCHMEISSNER/SCIENCE PHOTO LIBRARY/ Getty Images; Ralph A. Clevenger/Getty Images

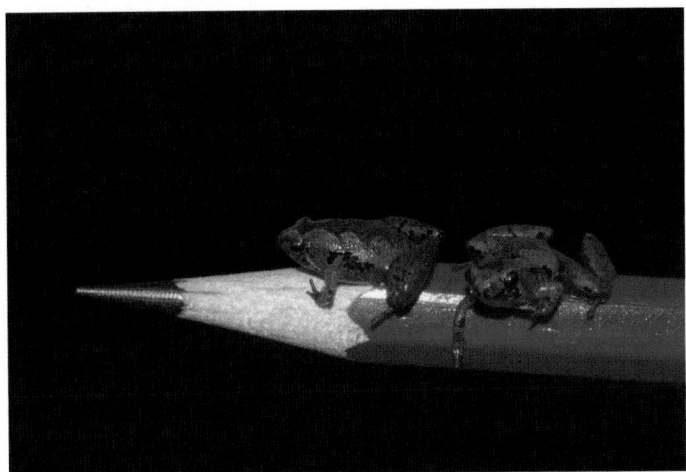

FIGURE 40.2 New species

New species are discovered all the time, indicating that we have named and described only a small fraction of the total number of species alive today. This miniature frog (*Microhyla nepenthicola*) was found living in puddles in pitcher plants in Borneo in 2010. Photo: Indraneil Das/Alamy Stock Photo

of monkey was found in 2012; a new species of beetle was discovered by biology students in the bustling city of Manila in 2013; a new deep-sea animal that defies easy classification was discovered in 2014; new dwarf dragons were found in the Andes in 2014; an Appalachian lichen named for the singer Dolly Parton was unearthed in 2015; an entirely new species of whale was found in 2016; a new species of gibbon named Skywalker from *Star Wars* and a crab named after Harry Potter were discovered in 2017; a new species of giant salamander was found in Florida in 2018; a sea slug that looks like seaweed was found in 2019; and an iridescent snake was discovered in Vietnam in 2020.

New species can be found almost everywhere, in rainforests and coral reefs to be sure, but also in such unlikely places as Central Park in New York City. A new species of lacewing (a kind of insect) was even discovered on, of all places, the Internet! A Malaysian photographer took a picture of the lacewing and posted it online. The photograph was seen by an entomologist (a scientist who studies insects), who suspected and later confirmed that it was indeed a new species.

Just as one of the clear features of the natural world is the diversity of organisms, another feature is completely the opposite—organisms have many traits in common, which we can describe as shared traits. Consider, for example, vertebrates, or animals like us that have a skull, known as the cranium, and vertebrae, which are the small bones that make up the backbone. Although these animals can look different from each other superficially, they all share a common body plan, which is illustrated in **FIGURE 40.3**.

As embryos, vertebrates look more alike. And if we dive down to the molecular level, they appear even more similar. In fact, all organisms, from bacteria to whales, share many traits, including the same genetic material in the form of DNA, the same amino acids that assemble to make proteins, similar machinery to build these proteins, and so on. Furthermore, if we look at how organisms harness energy from the environment, the widespread use of fermentation reactions, glycolysis, the citric acid cycle, and electron transport chains by numerous organisms attests to the unity of life.

How can we make sense of diversity and unity, which seem to be two different ways to describe the natural world? How can organisms exhibit so many differences, but also have so many features in common? Evolution provides a simple way to explain these two seemingly contradictory aspects of the natural world. Shared traits were present in a common ancestor and maintained over time, and traits that set them apart often arise over time after they diverged from a common ancestor.

Tree of Life

The process of evolution can be modeled as a branching tree, with a common trunk and branches that split off from this trunk. Such a tree is called an evolutionary or **phylogenetic tree.** The shared features sometimes imply inheritance from a common ancestor. For example, the fundamental features

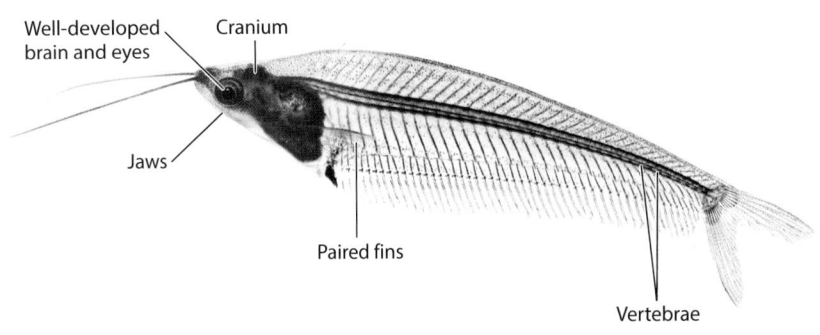

Well-developed brain and eyes

Cranium

Jaws

Paired fins

Vertebrae

FIGURE 40.3 Vertebrate body plan

Organisms often look different from one another, but also have many traits in common. Most vertebrates, for example, have a cranium, well-developed brain and eyes, jaws, and so on, or modified versions of these structures, as shown in this photograph of a fish. Photo: Stockbroker xtra/AGE Fotostock

shared by all organisms reflect inheritance from a common ancestor that lived at least 3.5 billion years ago. These features include the ability to store and transmit information, a cell membrane to keep in the internal environment of the cell compatible with life, and the ability to harness energy from the environment. The differences that characterize the many branches on the tree of life have formed through the continuing action of evolution since the time of this common ancestor. These differences reflect what Charles Darwin called descent with modification—evolutionary changes that have accumulated over time since two groups split from each other.

We can start with the base of a tree and move forward in time, as we just did. Alternatively, we can start from the tips of the branches and work backward. This approach, in fact, is how evolutionary trees are built. For example, let's start with humans and successively add more species to an evolutionary tree. Closely related species often resemble each other more closely than they do more distantly related species. You know this to be true from common experience. All of us recognize the similarity between a chimpanzee's face and body and our own, as shown in FIGURE 40.4a. Biologists have long known that humans share more features with chimpanzees than we do with any other species.

Humans and chimpanzees, in turn, share more features with gorillas than they do with any other species. And humans, chimpanzees, and gorillas share more features with orangutans than they do with any other species. And so on. We can continue to include more species, successively adding monkeys, lemurs, and other primates, to build a set of evolutionary relationships that can be depicted as a tree, as shown in Figure 40.4b. In this tree, time runs from left to

right, as you can see from the arrow below the tree. The tips of the branches represent different groups of organisms, such as humans, chimpanzees, gorillas, and orangutans. **Nodes** are points where one species splits into two, so they represent the most recent common ancestor of the groups at the tips of those lines. Finally, the **root** is the base of the tree, representing the common ancestor of all of the groups.

We can continue to add other mammals and then other vertebrate animals, in the process generating a pattern of evolutionary relationships that forms a larger tree, with the primates confined to one branch. Using comparisons of DNA sequences among species, we can generate still larger trees, ones that include plants as well as animals and the full diversity of microscopic organisms. Biologists call the full set of evolutionary relationships among all organisms the tree of life and it is shown in FIGURE 40.5.

This tree has three major branches—Bacteria (blue), Archaea (red), and Eukaryotes (green). Recall from Unit 2 that Bacteria and Archaea are microscopic organisms that lack a nucleus and are therefore described as prokaryotes (*pro* meaning "before" and *karyon* meaning "nucleus"). Eukaryotes include organisms whose cells have a nucleus (*eu* meaning "true" and *karyon* meaning "nucleus"). Although eukaryotes include large multicellular organisms such as humans, most branches on the tree of life consist of microorganisms. Plants and animals, so conspicuous in our daily life, make up only two branches on the eukaryotic branch of the tree. The last common ancestor of all living organisms forms the root or base of the tree, shown in black in the figure.

Evolution is the unifying theory of biology. It explains why organisms look the way they do and how organisms are related to one another. It explains the remarkable diversity of life on

a.

b.

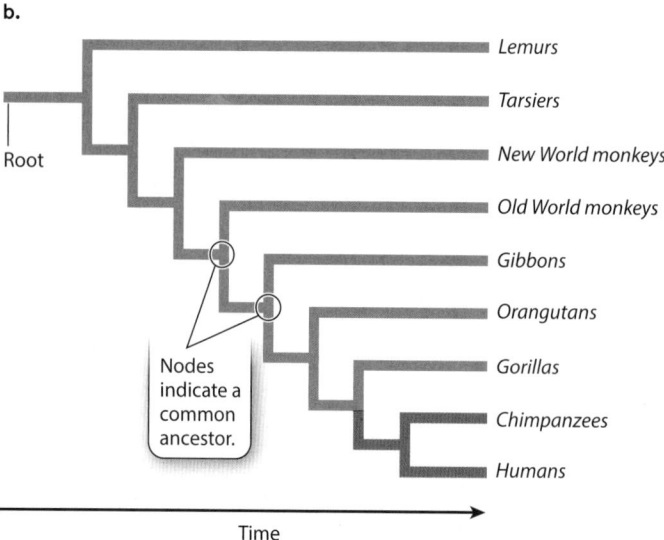

Root

Nodes indicate a common ancestor.

Lemurs
Tarsiers
New World monkeys
Old World monkeys
Gibbons
Orangutans
Gorillas
Chimpanzees
Humans

Time

FIGURE 40.4 Evolutionary relationships among primates

Evolutionary relationships can be depicted as a tree, shown here. (a) Humans share many features with chimpanzees, suggesting they are very closely related, as can be seen by looking at the primatologist Jane Goodall and the chimpanzee. (b) Humans and chimpanzees, in turn, share more features with gorillas than they do with other species, and so on. Treelike patterns of nested similarities are the result of evolution. Photo: BELA SZANDELSZKY/ AP Images

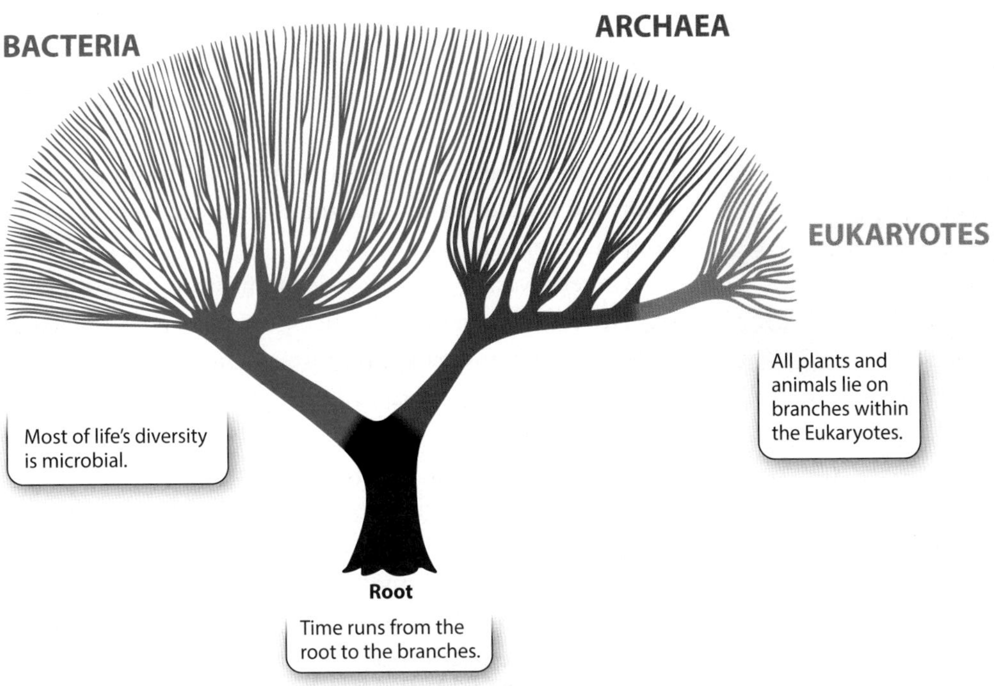

BACTERIA

ARCHAEA

EUKARYOTES

Most of life's diversity is microbial.

All plants and animals lie on branches within the Eukaryotes.

Root

Time runs from the root to the branches.

FIGURE 40.5 The tree of life

This evolutionary tree of all organisms has three major branches, or groups: Bacteria, Archaea, and Eukaryotes.

Information from Sprang, A., and T. G. Ettema. 2016. "The Tree of Life Comes of Age." *Nature Microbiology* 1:16056.

this planet, as well as the traits that organisms share. This is why Theodosius Dobzhansky, a geneticist, famously wrote, "Nothing in biology makes sense except in the light of evolution."

✓ Concept Check

1. **Describe** three ways in which species differ from one another.

2. **Describe** three ways in which species are similar.

3. **Describe** how evolution accounts for both the unity and diversity of life.

40.2 Variation in populations provides the raw material for evolution

As we have seen, organisms often differ from one another. Let's take a specific example—apples, shown in **FIGURE 40.6** on page 566. Strolling through a local market, you come across a bin full of crisp apples, pick one up, and take a bite. The apples in the bin from which you made your choice didn't all look alike. Had you picked your apple in an orchard, you would have seen that all apples, even those from the same tree, looked different—some smaller, some greener, some misshapen, a few damaged by worms. Such variation is so commonplace that we scarcely pay attention to it. Variation is observed among individuals in virtually every species of

organism. Variation that can be inherited provides the raw material on which evolution acts.

The causes of variation among individuals within a species are usually grouped into two broad categories. Variation among individuals is sometimes due to differences in the environment; this is called environmental variation. Among apples on the same tree, some may have good exposure to sunlight; some may be hidden in the shade; some were lucky enough to escape the female codling moth, whose egg develops into a caterpillar that eats its way into the fruit. These are all examples of environmental variation. Environmental variation is not passed on to the next generation, so does not play a role in evolutionary change over time.

The other cause of variation among individuals is differences in the genetic material that is transmitted from parents

FIGURE 40.6 Variation among apples

Apples come in many different varieties, which are very different from one another. Even within a particular variety, there is some variation. Variation among species, varieties, and individuals results from both environmental and genetic factors. Photo: BruceBlock/Getty Images

to offspring; this is known as genetic variation. Differences among DNA in individuals can lead to differences among RNA and proteins, which affect the functions of the cell and can lead to observable physical differences, as we discussed in Units 5 and 6. Genetic differences among apples produce varieties whose mature fruits differ in taste and color, such as the green Granny Smith, the yellow Golden Delicious, and the scarlet Red Delicious. Genetic variation is inherited, so plays an important role in evolution.

But even on a single tree, each apple contains seeds that are genetically distinct because the apple tree is a sexual organism, as we discussed in Unit 5. Bees carry pollen from the flowers of one tree and deposit it in some of the flowers of another, enabling the sperm from pollen grains to fertilize egg cells within that single flower. All the seeds on

an apple tree contain shared genes from one parent, the tree on which they developed. But they contain distinct sets of genes contributed by sperm transported from pollen from other trees. In all sexual organisms, fertilization produces unique combinations of genes, which explains in part why sisters and brothers with the same parents can be so different from one another.

Genetic variation ultimately stems from mutations. As we have seen, a mutation is a change in the genetic material (DNA). Mutations arise either from random errors during DNA replication, as described in Module 32, or from environmental factors such as ultraviolet (UV) radiation, which can damage DNA. If these mutations are not corrected, they are passed on to the next generation. To give a human example, lung cancer can result from an environmental insult, such as cigarette smoking, or from a genetic susceptibility inherited from the parents.

In nature, mutations that harm survival, growth, and reproduction do not persist in populations, but instead tend to disappear after a handful of generations. Those that are neither harmful nor beneficial can persist for hundreds or thousands of generations. And those that are beneficial to survival, growth, and reproduction can gradually become incorporated into the genetic makeup of every individual in the species. That is how evolution works: the genetic makeup of a population changes over time. Without variation, evolution would not be possible. Therefore, variation is the starting point for understanding the process of evolution.

✓ Concept Check

4. **Identify** two causes of variation among different organisms.
5. **Identify** the source of genetic variation.

40.3 Natural selection results in adaptations

Species are often well adapted to their environment. Desert plants resist drying out and sharks have powerful jaws and fins that make them effective predators. These are examples of adaptations—the close fit between an organism and its environment. Charles Darwin developed the theory of evolution by natural selection to explain not only how populations change over time but also how they become

exquisitely adapted to their environment. In this section, we will provide an overview of natural selection, and describe how it results in both evolution and adaptations.

Natural Selection

Evolution is change over time. Specifically, evolution is a change in the genetic makeup of a population from one generation to the next. There are many different mechanisms by which evolution occurs. One of these is natural selection. Natural selection is a mechanism of evolution that leads to adaptations.

The main principles of evolution by natural selection are straightforward. When there is variation within a population of organisms, and when that variation can be passed from one generation to the next, the variants best suited for growth and reproduction in a given environment will contribute disproportionately to the next generation. The result is that populations change over time; that is, they evolve. In addition, they become more fit over time; that is, they become better adapted to their environment.

As Darwin recognized, farmers have used this principle for thousands of years to select crops with high yield or improved resistance to drought and disease. It is also how people around the world have developed the many breeds of dog, shown in **FIGURE 40.7**. Similarly, we have used selection to breed cows, chickens, horses, and other livestock for work and food. In these cases, humans control the evolutionary process.

We can see evolution by natural selection in the development of antibiotic resistance among bacteria and other disease-causing microorganisms. As certain species of bacteria evolve to resist all known antibiotics, the emergence of antibiotic-resistant strains is making some diseases difficult to treat. This is becoming a global public health crisis.

Antibiotic-resistant strains develop when there is variation among individual bacteria. Most bacteria are antibiotic sensitive and die when exposed to an antibiotic targeted to them. However, a few may be antibiotic resistant. These bacteria survive and pass resistance on to daughter cells; that is, the resistance is inherited. In the presence of antibiotics, the antibiotic-resistant bacteria survive and are able to reproduce more successfully than the antibiotic-sensitive bacteria. As a result, the population evolves over time, from being mostly antibiotic sensitive to mostly antibiotic resistant. In addition, the population becomes adapted to an environment with antibiotics.

In fact, all life has been shaped by evolution since its origin, and the capacity for Darwinian evolution may be life's most fundamental property.

Demonstrating Evolution in the Laboratory

Many people think that evolution occurs slowly and is difficult to see. For the most part, this is correct. However, it sometimes occurs quickly and we can directly see the changes over time. In other words, we can capture evolution by looking at natural selection in action. One way to accomplish this goal is through laboratory experiments.

Bacteria are ideal for these experiments because they reproduce rapidly and can form populations with millions of individuals. Large population size means that mutations are likely to form in nearly every generation, even though the probability that any individual cell will acquire a mutation is small. In contrast to bacteria, think about trying evolutionary experiments on elephants!

One such experiment is illustrated on page 568 in "Practicing Science 40.1: Can evolution be demonstrated in the laboratory?" Microbiologist Richard Lenski grew populations of the common intestinal bacterium *Escherichia coli* in liquid medium containing small amounts of the sugar glucose as the only source of food. Lenski and his colleagues hypothesized that any bacterium with a mutation that increased its ability to use glucose would grow and reproduce at a faster rate than other bacteria in the population.

FIGURE 40.7 Dog breeds

Selection over many centuries has resulted in remarkable variation among dogs. Charles Darwin called this selection under domestication and noted that it resembles selection that occurs in nature. Photo: Igor Mojzes/Alamy Stock Photo

Can evolution be demonstrated in the laboratory?

Background *Escherichia coli* is a common bacterium found in your gut and in the intestines of many organisms. It is also used in the laboratory as a model organism to study all kinds of biological processes, including evolution. In 1988, the biologist Richard Lenski asked whether *E. coli* grown generation after generation on a medium containing limiting amounts of glucose as food would evolve in ways that improved their ability to metabolize glucose.

Hypothesis Any bacterium with a random mutation that increases its ability to take up and utilize glucose will reproduce at a faster rate than other bacteria in the population. Over time, such a mutant will increase in frequency relative to other types of bacteria, thereby demonstrating evolution by natural selection in a bacterial population.

Experiment Cells of *E. coli* can be frozen at –80 °C, which keeps them in a sort of suspended animation in which the cells survive without any biological processes taking place. At the beginning of the experiment, Lenski and his students froze samples of the starting bacteria in the flask labeled "Ancestral" in the figure below. Then they grew the bacteria with limiting amounts of glucose, taking samples of the bacterial populations every 500 generations in flasks labeled "Descendant" in the figure. They mixed the contents of the two flasks and grew them together in the presence of glucose. They then compared the rate of growth of the ancestral bacteria with that of the descendant bacteria samples taken at later time points. The color of the media in the figure provides a visual measure of the number of different bacteria, with white as an indicator of the ancestral bacteria and red as an indicator of the descendant bacteria.

At the first time point (*t* = 0), the pink color of the flask contents shows that there are similar numbers of ancestral and descendant bacteria. The ratio is 1:1.

At the second time point (*t* = 1), the redder color of flask contents indicates that there are greater numbers of descendant bacteria and fewer ancestral bacteria. The ratio is no longer 1:1.

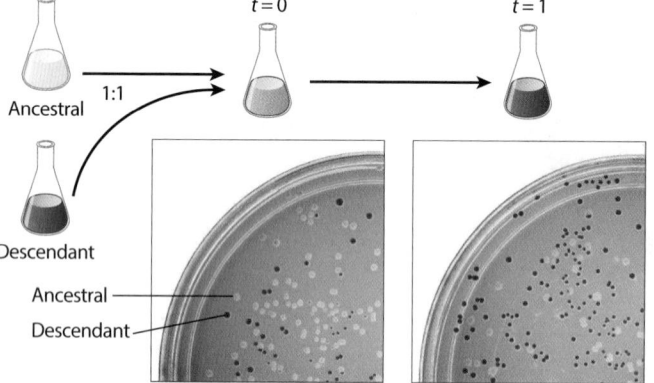

Results At early time points, there are about equal numbers of ancestral and descendant bacteria, as indicated by the pink color of the media in the flask on the left side of the figure. However, as the experiment progressed, there were more descendant bacteria compared to ancestral bacteria when the two were grown together, as indicated by the red color on the right side of the figure.

The growth rate of the two populations of bacteria can be compared in graph form as well. The graph below shows the growth rate of the descendant bacteria relative to the growth rate of the ancestral bacteria. Note that the line on the graph increases over time. This observation indicates that the bacteria from later time points grew increasingly more rapidly than the ancestral cells when the two populations were grown together on the glucose medium.

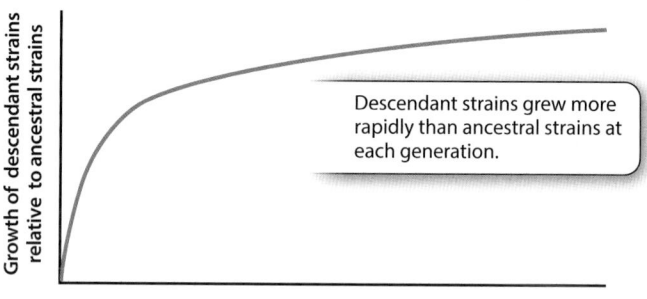

Descendant strains grew more rapidly than ancestral strains at each generation.

Conclusion The hypothesis was supported. This experiment demonstrates that the *E. coli* bacteria evolved an improved ability to metabolize glucose over time. In other words, evolution occurred in the population. In addition, the bacteria became adapted to living in an environment in which glucose was limiting.

Follow-up Work As of 2017, the experiment had continued for more than 60,000 generations! Using modern biotechnology, Lenski and colleagues have been able to sequence the full genomes of the bacteria to identify the specific mutations that allow the bacteria to metabolize glucose more efficiently and follow what happens to these variants through time.

SOURCES
(experimental diagram) Elena, S. F., and R. E. Lenski. 2003. "Evolution Experiments with Microorganisms." *Nature Review Genetics* 4:457–469; (graph) Lenski, R. E. 2017. "Experimental Evolution and the Dynamics of Adaptation and Genome Evolution in Microbial Populations." *ISME Journal* 11:2181–2194. Photos: Neerja Hajela, Michigan State University

AP® PRACTICE QUESTION

a. **Identify** the following:
   1. The question the experimenters are trying to answer
   2. The hypothesis they tested
   3. The independent variable
   4. The dependent variable
   5. The control group

b. **Explain** the conclusion using a claims, evidence, reasoning format (CER).

In this experiment, then, they asked: did bacteria from later generations grow and reproduce more rapidly than those of earlier generations—that is, did evolution occur?

In fact, the bacteria did evolve an improved ability to use glucose, as demonstrated by the results shown in Practicing Science. This experiment both illustrates how experiments can be used to test hypotheses and shows evolution in action. Furthermore, follow-up studies of the bacterial DNA identified differences in genetic makeup that resulted in the improved ability of *E. coli* to use glucose. Many experiments of this type have been carried out, applying scientific inquiry to demonstrate how bacteria adapt through mutation and natural selection to any number of environments.

Experiments in laboratory evolution have an immensely important practical side. They allow biologists to develop new and beneficial strains of microorganisms that, for example, remove toxins from lakes and rivers. In addition, they show how some of our worst pathogens develop resistance to drugs designed to eliminate them. They also help us understand how evolution by natural selection works and demonstrate how populations become adapted to their environment over time.

AP® EXAM TIP

You should know that for a population to evolve by natural selection, some of the members of the population must possess a favorable trait and be able to pass the selected trait on to subsequent generations of the population.

✓ Concept Check

6. **Describe** the difference between evolution and natural selection.

7. The use of antibiotics has led, over time, to antibiotic resistance among bacteria. **Identify** the process that leads to the evolution of antibiotic resistance and **describe** the process.

Module 40 Summary

REVISIT THE BIG IDEAS

EVOLUTION: Using the mechanism of natural selection discussed in this module, **describe** how populations can evolve to become better adapted to their environments over time.

LG 40.1 Evolution explains both the diversity and unity of life.

- Species differ from one another in many different ways. Page 562
- Species also share many traits. Page 562
- Evolution can be visualized as a branching tree, with a common root (ancestor) and living organisms at the tips of the branches. Page 563

LG 40.2 Variation in populations provides the raw material for evolution.

- Variation among species and individuals within a species results from differences in the environment, or environmental variation, as well as differences in genetic material, or genetic variation. Page 565
- Genetic variation stems from mutations, or changes in the genetic material. Page 566

LG 40.3 Natural selection results in adaptations.

- An adaptation is a close fit between a trait of an organism and the environment where the organism lives. Page 566
- Natural selection is a mechanism of evolution that leads to adaptations. Page 566
- Evolution is change in a population's genetic makeup over time. Page 567

Key Terms

Adaptation
Biodiversity

Phylogenetic tree
Node

Root

Review Questions

1. On an evolutionary tree, a branch point, or node, represents
   (A) time.
   (B) present-day species.
   (C) a common ancestor.
   (D) a trait.

2. Which of the following domains have cells with a nucleus?
   (A) Eukarya and Archaea
   (B) Eukarya
   (C) Archaea and Bacteria
   (D) Archaea

3. How many major branches are there on the evolutionary tree of life?
   (A) One
   (B) Two
   (C) Three
   (D) Six

4. The majority of branches on the tree of life represent which of the following types of organisms?
   (A) Mammals
   (B) Microorganisms
   (C) Plants
   (D) Humans

5. Which does the base, or root, of an evolutionary tree represent?
   (A) The common ancestor of all organisms included in the tree
   (B) The common ancestor of a subset of the organisms included in the tree
   (C) An organism that is unrelated to all other organisms included in the tree
   (D) An organism that is the common ancestor for organisms not included in the tree

6. Which statement is true regarding the initial work of Richard Lenski and his colleagues?
   (A) They demonstrated that populations of bacteria can evolve in culture.
   (B) They demonstrated that an individual bacterium can evolve in culture.
   (C) They demonstrated that bacteria from earlier and later generations use glucose similarly.
   (D) They demonstrated that mutations in DNA are responsible for the evolution of bacteria in culture.

7. Why are bacteria a useful model in which to study evolution?
   (A) Bacteria can produce large populations in a short amount of time.
   (B) Bacteria have a long generation time.
   (C) Mutations do not occur in bacterial populations.
   (D) Bacteria are difficult to grow in the lab.

Module 40
AP® Practice Questions

PREP FOR THE AP® EXAM

Section 1: Multiple-Choice Questions
Choose the best answer for questions 1–4.

1. Identify which provides evidence that all life on Earth evolved from a common ancestor.
   (A) The use of specialized organelles to carry out cellular functions
   (B) The use of the same set of amino acids to build proteins
   (C) The diversity of metabolic pathways used by organisms in different environments
   (D) The presence of a nucleus in eukaryotic cells

2. Which illustrates an evolutionary process in action?

(A) The fur on an individual snowshoe hare turns from brown to white in the winter.

(B) Bacterial cells activate specific metabolic pathways in response to the availability of a nutrient.

(C) Within a single species of bacteria, only cells that are genetically resistant to an environmental toxin survive and reproduce when the toxin is present.

(D) Recessive traits "skip" generations, reappearing in the offspring of two individuals who carry the recessive genes for the trait.

3. Which is required for long-term evolution by natural selection?

(A) Phenotypes must be influenced by the environment.

(B) Environmental conditions must rapidly change.

(C) Offspring must inherit phenotypes from their parents.

(D) Mutation must create new genetic variation every generation.

4. Cats such as lions, tigers, and even house cats are extremely capable hunters. A key trait that enables cats to hunt successfully is their sharp claws, which can be contrasted to the blunt claws of dogs or the flat fingernails of humans. Which statement best explains the evolution of sharp claws in the cat family?

(A) The ancestors of modern cats evolved sharp claws because they needed them to hunt and survive.

(B) The ancestors of cats hunted prey that also had sharp claws, so the cats evolved sharp claws in response.

(C) The ancestors of cats that had sharper claws were more successful at hunting and had more offspring.

(D) The ancestors of cats that had sharper claws had lower evolutionary fitness and were eliminated by natural selection.

Section 2: Free-Response Question

Write your answer to each part clearly. Support your answers with relevant information and examples. Where calculations are required, show your work.

Natural selection is one of the primary mechanisms by which biological evolution occurs and is central to the generation of Earth's biological diversity.

(a) **Identify** the ultimate source of all variation on which evolution acts.

(b) Both genotype and environment can generate phenotypic variation among individuals in a population. A scientist claims that evolution by natural selection occurs when phenotypic variation in a trait is determined by genetic variation in a population, but variation in a trait due to differences in environment cannot result in evolution. **Provide reasoning** to support or refute the claim.

Module 41

Natural and Artificial Selection

LEARNING GOALS ▶**LG 41.1** Natural selection brings about adaptations and increases fitness.
▶**LG 41.2** Natural selection acts on phenotypes.
▶**LG 41.3** Artificial selection has produced many different breeds of plants and animals.
▶**LG 41.4** Sexual selection increases an individual's ability to find and attract mates.

In Modules 0 and 40, we introduced the concept of evolution, or change over time. Specifically, evolution is a change in the genetic makeup of a population from generation to generation. As we discussed in Unit 5, the genotype of an organism can affect its phenotype, or outward appearance. Phenotypic variation in a population is important because, in some cases, the phenotype affects the ability of an organism to survive or reproduce in a particular environment. Those that survive and reproduce more than others will pass on their traits to the next generation, and the population will change over time. Charles Darwin recognized that this process, which he called natural selection, provides a mechanism of evolutionary change over time. Natural selection not only leads to changes in populations over time but also brings about adaptations, the close fit between an organism and its environment. And just as populations change over time, so does the environment. So, evolution is an ongoing process that began with the first cell and continues today.

In this module, we will take a closer look at natural selection and how it brings about adaptations. We will also look at different types of selection and the effect they have on phenotypes of organisms that make up a population. Even before Darwin proposed his theory of evolution by natural selection, humans understood that they could modify breeds of animals and plants through selective breeding, which we will also discuss. Finally, we will look at a form of selection that is responsible for many of the bright colors and displays among birds and other animals.

FOCUS ON THE BIG IDEAS

PREP FOR THE AP® EXAM

EVOLUTION: Think about the how a diverse gene pool affects species survival when there is a change in environmental conditions.

41.1 Natural selection brings about adaptations and increases fitness

A conspicuous aspect of the natural world is that organisms are well adapted to the environment in which they live. Charles Darwin proposed his theory of evolution by natural selection to explain how these adaptations come about. He suggested that organisms change over time, and that this change is not random. Instead, organisms that are well suited to a particular environment leave more offspring than those that are less suited to that environment. The result is that populations change and become adapted to their environment over time. In this section, we will describe many examples of adaptations, and then look closely at how Darwin proposed they come about.

Adaptations

Organisms are often exquisitely adapted to their environments. Think of an eagle, pictured in **FIGURE 41.1.** Everything about this animal tells you it's a predator—the powerful wings; strong beak; keen sense of vision; and sharp, hooked claws called talons. In fact, if you didn't know that an eagle is a predator, you could probably figure it out just by looking at the photograph. All of these traits are examples of adaptations.

Adaptations are commonplace in nature. Your eyes are well adapted for seeing; your hands for grasping; your legs for walking. Desert plants are adapted for a dry, hot environment: they resist drying out, or desiccation. Other plants are adapted for growing in full sunlight or deep shade. The burs of plants are adapted to catch on to the fur of a passing animal, helping to disperse their seeds. Sharks have muscular

FIGURE 41.1 Adaptations

Eagles have many adaptations for hunting, such as a strong wings and powerful claws. Photo: Feng Wei Photography/Getty Images

fins, sharp teeth, powerful jaws, a keen sense of smell, and large eyes facing forward. Some bacteria are well adapted to live in what we call extreme environments, such as high salt, low pH, or high temperatures that are lethal to many other types of organisms.

For thousands of years, people have been struck by these and other adaptations. The fit is sometimes so close and so wonderful that many people have asked why organisms are so well adapted. How did they get this way? How did the woodpecker get its powerful chisel of a bill? And the hummingbird its long delicate bill for probing nectar in flowers? Charles Darwin, pictured in **FIGURE 41.2**, came up with an elegant answer to these questions. First, he suggested that species are not unchanging, but rather have evolved over time. Second, he proposed a mechanism, natural selection, that brings about adaptations. Natural selection is a brilliant solution to the central problem of biology: how organisms come to fit so well in their environments. Darwin showed how a simple mechanism leads to an extraordinary range of adaptations.

How did Darwin come up with his revolutionary idea? He famously sailed around the world as a naturalist on board the HMS *Beagle* in the 1830s. During this voyage, he saw species that he had never seen before on Pacific islands like the Galápagos and in rainforests in South America and Southeast Asia. He was also struck by fossil organisms—the remains of species that lived long ago but are now extinct. He collected beetle specimens and spent many years studying barnacles. He raised pigeons and bred them, seeing if he could favor certain traits over others.

For 20 years after first conceiving the essence of his theory, Darwin collected supporting evidence for that theory. In 1858, he was finally spurred to publish his ideas by a letter he received from another naturalist named Alfred Russel Wallace, who is pictured in **FIGURE 41.3** on page 574. By a remarkable coincidence, Wallace had independently come up with the theory of evolution by natural selection. Aware that Darwin was interested in the problem but having no idea that Darwin was working on the same theory, Wallace wrote to Darwin to ask what Darwin thought of his idea.

Suddenly, Darwin was confronted with the prospect of losing his claim on the theory that he had been nurturing for 20 years. But all was not lost. Darwin's colleagues arranged for the publication of a joint paper by Wallace and Darwin in 1858. Then, one year later in 1859, Darwin published *On the Origin of Species*. It was this work that brought both evolution and natural selection, its underlying mechanism of adaptation, to public attention.

FIGURE 41.2 Charles Darwin

This portrait shows Darwin just after he returned from his five-year voyage around the world as a naturalist on board the HMS *Beagle*.
Photo: © Historic England/Bridgeman Images

FIGURE 41.3 Alfred Russel Wallace

This portrait shows Wallace collecting and describing specimens in the tropics, where he spent many years. Photo: Historic England/Bridgeman Images

Darwin's Postulates

Darwin's theory of evolution by natural selection is a simple idea with profound implications. It rests on certain observations or claims that are sometimes called Darwin's postulates. Although he didn't use that term, he did describe them in the introduction of *On the Origin of Species.*

The first postulate is that the members of a species vary from one another, a point we highlighted in the previous module. This may seem obvious to you. A walk down any street reveals how variable our species is: skin color and hair color, for example, are different from person to person, as can be seen in **FIGURE 41.4**. However, until the publication of *On the Origin of Species,* naturalists tended to view all the variation we see in humans and other species as biologically unimportant. Darwin drew our attention to the variation that we see all around us and suggested that it is key to understanding how species evolve.

Note that when we talk about variation in appearance, we are referring to variation in phenotype. As we saw in Unit 5, phenotype is an observable trait, such as human height or butterfly wing color. Two factors contribute to phenotype: an individual's genotype (genetic makeup) and the environment in which the individual lives. For example, human height is influenced by thousands of genes, as well as by nutrition during fetal development and childhood. Phenotypic variation, or variation in appearance, is the result of genetic variation and environmental variation.

The second postulate, already noted in Unit 5, is that some phenotypic variation is heritable, or passed on to the next generation. We know this from common experience—tall parents tend to have tall children, for example. And humans have used this knowledge to select traits in breeds of farm animals and crops for thousands of years. The phrase "like begets like" nicely captures this well-known and well-accepted observation.

The third postulate is that in nature, organisms frequently compete for resources. Populations tend to grow rapidly, but the resources on which they depend, such as food, water, and living space, do not. Some individuals are more successful in this competition because of certain traits; as a result of the variation among individuals, some are more likely to survive and reproduce than others.

Both Darwin and Wallace came across this idea when reading the writings of a British clergyman, Thomas Robert Malthus. In his *An Essay on the Principle of Population,* Malthus pointed out that populations have the potential to increase in size extremely rapidly. For example, imagine a sexually reproducing group of organisms in which each pair has four offspring, two males and two females. If reproduction continues, so that each pair has four offspring, the population will double in size every generation. Starting with a single pair, the population will have grown to more than a million by the 20th generation. In reality, this rapid expansion of populations does not typically occur. In fact, many population sizes are often stable from generation to generation, largely because the resources upon which populations depend are

FIGURE 41.4 Human variation

Darwin's first postulate states that individuals of a species look different from one another. Photo: Ryan McVay/Getty Images

limited. In each generation, many individuals fail to survive or reproduce. There are not enough resources to go around, so individuals in a population often compete with one another, and some variants are more successful than others.

The fourth postulate is that variations that increase survival and reproduction will be passed on to individuals in the next generation. Individuals with variations that are best suited to the environment will tend to survive more and leave more offspring, so that the particular beneficial trait or variant will be more common in the next generation. For example, a desert plant, such as the cactus pictured in **FIGURE 41.5**, that is a little better at minimizing water loss will pass this trait to its offspring. Offspring with this trait will be more likely to survive and reproduce in the desert climate. The next generation will therefore have a higher proportion of cacti with this beneficial trait. Note that the genetic makeup of the population will now be different from the previous, ancestral population. Darwin used the term "natural selection" for the filtering process that acts in favor of beneficial traits and against harmful traits. Specifically, natural selection is the differential reproductive success of genetic variants.

The claims of Darwin's four postulates are powerful because they are simple and testable. In addition, they lead to two conclusions. The first is that populations will change over time, or evolve, because in each generation the variants that leave more offspring will be overrepresented and the variants that leave fewer offspring will be underrepresented. The second is that organisms will become adapted to their environment over time because those that do survive and reproduce are not random, but those best suited to the environment in which they live.

Fitness

Competitive advantage is a function of how well an organism is adapted to its environment. An organism that is better adapted to its environment is more fit. Fitness is a measure of the ability of an individual to survive and reproduce in a particular environment. More specifically, it is the extent to which the individual's genotype is represented in the next generation.

A desert plant is more efficient at minimizing water loss than other plants and is therefore better adapted to the desert environment. We say that the one plant's fitness is higher than another plant's fitness if one is more likely to survive or leaves more offspring than another plant. One type of plant could leave more offspring because it survives longer, which gives it more opportunities to reproduce, or because it has some other reproductive advantage, such as the ability to produce more seeds. Note that fitness is relative, meaning that it only makes sense to talk about the fitness of one organism in relation to the fitness of another. Fitness describes how well an individual survives and reproduces compared to other individuals of the same species.

The words "fit" and "fitness" therefore have different meanings in evolutionary biology than they do in everyday language. In common usage, being fit has to do with working

FIGURE 41.5 A desert plant adapted for a dry climate

This saguaro cactus is well adapted to the desert because it minimizes water loss. For example, it has spines rather than leaves to reduce the surface area available for evaporation. Photo: Charles Harker/Getty Images

out and being healthy. In evolution, however, the meaning is quite different: being fit is the ability of an individual to survive and reproduce in the environment in which it lives.

Assuming that a trait—for example, the ability to minimize water loss—remains advantageous in a particular environment, natural selection acts over generations to increase the overall fitness of a population. A plant newly arrived in a desert may be poorly adapted to its environment. Over time, however, traits that minimize water loss increase under natural selection, resulting in a population of plants that is better adapted to the desert environment. So, although natural selection acts on individuals, it is populations, not individuals, that evolve over time.

Such changes in populations often take a long time. Darwin, borrowing from the geologists of his day, recognized that time was a critical ingredient of his theory. Geologists had put forward a view of Earth's history that argued large geological changes—like the carving of the Grand Canyon—can be explained by simple, day-to-day processes operating over vast timescales of thousands, millions, or even billions of years. Darwin applied this view to biology. He recognized that small changes can add up to major changes over long time periods. What might seem to be a trivial change over the short term can, over the long term, result in substantial differences among populations, eventually leading to the origin of new species and the remarkable diversity of living organisms we see today.

Role of the Environment

A key point about fitness is that it depends on the environment in which an organism lives. A trait that is well suited to one environment may not be well suited to another. This is particularly important in the face of a changing environment. As an environment changes, an organism that was more fit may become less fit, and the other way around. For example, a desert plant that is well adapted to the dry, arid environment of a desert will be less well adapted if it is moved to a wet, humid environment. Changes in the environment might be natural, such as the changes in a forest ecosystem over time, or they might be caused by human activity, such as the introduction of a non-native species to an area or the use of pesticides and herbicides.

Biologists use the term **selective pressure** to describe the full set of environmental conditions, both biotic (biological factors such as other organisms) and abiotic (physical factors such as air, soil, and water), that allow some organisms to survive and reproduce better than others. In a constant environment, selective pressures tend to be stable. In the face of an environmental disturbance, whether natural or caused by humans, selective pressures can change. The term is useful because it draws attention to the role of the environment in influencing how natural selection acts on a population. Note that natural selection is not a forward-looking process and cannot anticipate environmental conditions that might occur in the future. Natural selection acts on populations with respect to current environmental conditions. In a similar way, natural selection does not have a goal or endpoint in mind.

Fitness depends on the environment. Therefore, to understand the fitness of different organisms we need to know something about the environment in which they live. Let's consider two types of bacteria, and ask the following question: which is more fit, antibiotic-sensitive bacteria or antibiotic-resistant bacteria? While it's tempting to think that antibiotic-resistant bacteria are more fit because, after all, they are able survive and reproduce in the presence of antibiotics, the fact is that we don't know which is more fit because we didn't describe the environment in which the bacteria are living.

In the presence of antibiotics, we can answer the question: the antibiotic-resistant bacteria are more fit because they will survive better and reproduce more quickly than the antibiotic-sensitive bacteria. The antibiotic-resistant bacteria will pass on this trait to their offspring, and the population will evolve antibiotic resistance over time.

What about in the absence of antibiotics? In this case, you might think that the two strains of bacteria are equally fit. After all, without antibiotics in the environment, it might seem that it doesn't matter whether bacteria are antibiotic sensitive or resistant. But, in this case, antibiotic-sensitive bacteria typically outcompete antibiotic-resistant bacteria because it takes energy to be antibiotic resistant, and this energy can't be used for other essential functions, like survival and reproduction.

✓ Concept Check

1. **Describe** three adaptations displayed by organisms.

2. **Identify** Darwin's four postulates.

3. **Describe** fitness.

4. **Describe** one example of how you might measure fitness of individuals in a population.

41.2 Natural selection acts on phenotypes

When we consider interactions of organisms and their environments, we are focusing on the traits of organisms—the shape of a bird's beak, or drought tolerance in a desert plant, or the streamlined body of a fish. These are all examples of phenotypes, the outward appearance of an organism. Natural selection acts on phenotypes. In a particular environment, certain traits are beneficial and individuals with those traits tend to survive and reproduce more than others. These traits then become more common in succeeding generations until they spread to the entire population. Other traits may be harmful and therefore individuals with these traits leave fewer offspring, making these traits less common in future generations. In this section, we will focus on the effect of natural selection on phenotypes of organisms.

Phenotypic Change

The effect that natural selection has on phenotypes is well illustrated when different species are placed in similar environments. In this situation, they tend to converge on similar phenotypes. For example, sharks have a streamlined shape, which is an adaptation for swimming in water. Marine mammals, such as whales and dolphins, evolved from four-legged, land-based animals that returned to the water. Over time, they became adapted to moving around in the water, so ended up with a similar streamlined shape. **FIGURE 41.6** shows the uncanny resemblance between a shark and dolphin, even though many of their traits evolved independently of each other. The similarities between sharks and dolphins is an example of **convergent evolution,** a process by which two different species evolve similar traits in response to similar selective pressures, in this case, moving efficiently in an aquatic environment.

Phenotype, as we have seen, depends in part on the underlying genotype. Weight, for example, depends on the genetic makeup of an individual, as well as such factors as diet and exercise. Even though natural selection acts on phenotypes, it is genotypes that are passed on to the next generation. In this way, the genetic makeup of a population changes over time.

The genetic variation that gives rise to phenotypic differences is ultimately the result of mutations. Although we all have the same set of genes, we have different versions or forms of those genes. As we saw in Unit 5, alleles are different forms of a particular gene. The different alleles of a gene are the result of mutations. When we talk about a beneficial mutation, we mean that it results in a phenotype that is beneficial to an

FIGURE 41.6 Convergent evolution

Dolphins (top) and sharks (bottom) closely resemble each other because they are both adapted to swim in the water. The process by which two species independently develop similar traits under similar selective pressures is called convergent evolution. Photos: (top) Chase Dekker Wild-Life Images/Getty Images; (bottom) ullstein bild/Getty Images

individual in a given environment. The individual with this trait will tend to survive and reproduce more, and therefore that phenotype and the underlying genetic mutation will be more represented in the next generation. Conversely, a harmful mutation confers a phenotype that results in an individual with a reduced ability to survive or reproduce. This individual is therefore less fit in this environment.

Natural selection increases the frequency of beneficial alleles, resulting in adaptation. To start with, a new beneficial allele will exist as a single copy in a single individual. Subsequently, under the influence of natural selection, the beneficial allele will increase in frequency over time and may replace all the other alleles in the population. In some cases, natural selection increases the frequency of a beneficial allele until the population is fixed for the allele, which means all the individuals in the population have the allele. By contrast, a harmful allele will become less common over time, until it is eliminated from the population.

In short, natural selection results in the frequency of alleles changing from generation to generation according to the allele's impact on the survival and reproduction of individuals. New mutations that are harmful and eliminated by natural selection have no long-term evolutionary impact. Neutral mutations do not affect survival and reproduction and therefore do not affect fitness and are not acted on by natural selection. In contrast, mutations that are beneficial become more common and result in adaptation to the environment over time.

Stabilizing, Directional, and Disruptive Selection

Because natural selection acts on phenotypes, we can follow the effects of natural selection by looking at changes over time in a particular trait of an organism. For example, we might track the evolution of height in a population, despite not knowing the specifics of the genetic basis of height differences. When we look at natural selection from this perspective, we see three patterns: stabilizing, directional, and disruptive.

Stabilizing selection maintains the status quo and acts against extremes. Human birth weight is a good example. The trait of human birth weight is affected by a number of factors, including many fetal genes and the environment. If a baby is very small, its chances of survival at birth are low. However, if it is very large, there may be complications during delivery that endanger both mother and baby. The optimal birth weight is intermediate, somewhere between these two extremes. In this case, natural selection acts against the extremes.

In **FIGURE 41.7**, we see a graph of human birth weight. Birth weight is plotted along the *x*-axis, from 0 to 5 kilograms (kg). Percent of population is indicated along the left-side *y*-axis. The blue bell-shaped curve, with its peak around 3 kg, indicates that most newborns weigh around 3 kg, with fewer weighing less (1–2 kg) or more (4–5 kg). The figure also plots mortality, indicated by the red dots and measured along the right-side *y*-axis. As you can see, mortality is higher for low-birth-weight and high-birth-weight babies, but lower for babies of intermediate weight.

The vast majority of natural selection is stabilizing; harmful mutations that cause a departure from an intermediate phenotype are selected against. These are the basis for adaptations and tend to occur in environments that are relatively stable.

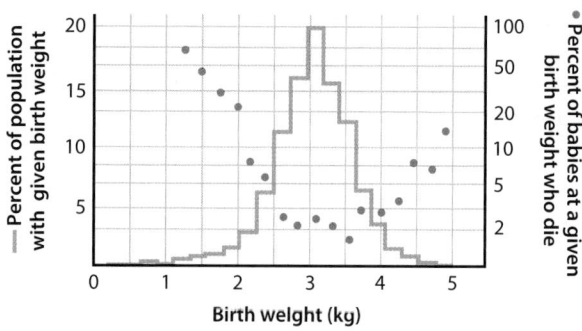

FIGURE 41.7 Stabilizing selection

Stabilizing selection acts against extreme phenotypes and favors intermediates phenotypes. Birth weight provides an example. Mortality, indicated by the red dots, is higher for very small babies and very large babies compared to intermediate-sized babies. The result is that most babies are born at an intermediate size, indicated by the blue line. Although modern medicine has improved the survival rate of low-birth-weight babies since this study was done, the overall trend that we see in the graph still holds. Data from Cavalli-Sforza, L. L. and W. F. Bodmer. 1971. *The Genetics of Human Populations*. San Francisco, CA: W.H. Freeman, p. 613.

While stabilizing selection keeps a trait the same over time, directional selection leads to a change in a trait in a population over time. The color forms of the peppered moth (*Biston betularia*) provide a classic example of directional selection. The moth comes in two distinct forms: a light one, *typica,* and a dark one, *carbonaria*. The difference is determined by two alleles at a single gene, with the *carbonaria* allele dominant to the *typica* one. The *carbonaria* form is absent from collections of moths from the United Kingdom until around 1850, when its frequency started to increase in urban areas. By about 1900, *typica* had disappeared from these populations.

This striking shift is attributable to pollution caused by industry. Soot darkens trees and sulphur dioxide in smoke kills the lichens against which the *typica* form is camouflaged. With industrialization, *typica* went from being camouflaged, and therefore protected against bird predation, to readily visible; during this same time, *carbonaria* went from being easily seen to camouflaged, as shown in **FIGURE 41.8a**. Selection therefore strongly favored the *carbonaria* allele. With anti-pollution legislation in the 1950s, however, lichen returned to trees in urban areas, and the direction of selection flipped, with *typica* now favored over *carbonaria*. Today, the *carbonaria* allele is once again extremely rare. Over the past 170 years, directional selection on these moth populations has operated strongly at different times in two different directions.

Figure 41.8b shows what happened in graph form. In the graph, time is plotted on the *x*-axis and the frequency

a.

FIGURE 41.8 Directional selection

Directional selection results in a change in phenotype over time, as illustrated by the peppered moth. (a) The two color forms of the moth are shown against two different backgrounds. On a tree with lichen shown on the left, the light form (*typica*) is camouflaged and the dark form (*carbonaria*) is visible. Industrial pollution kills lichen and darkens trees, shown on the right, making the light form visible and the dark form camouflaged. (b) The graph plots the frequency of the *carbonaria* form of the moth over time in Manchester, an industrial town in north England. The frequency increased as trees became darker, providing an example of directional selection. The frequency decreased as the trees became lighter, providing another example of directional selection. Data from Hof, A., P. Campagne, D. Rigden, et al. 2016. "The industrial melanism mutation in British peppered moths is a transposable element," Extended Data Figure 4: The Rise and Fall of *Carbonaria* in the Manchester Area. *Nature* 534:102–105. Photos: (left) Bill Coster/Alamy Stock; (right) Breck P. Kent/AGE Fotostock

b.

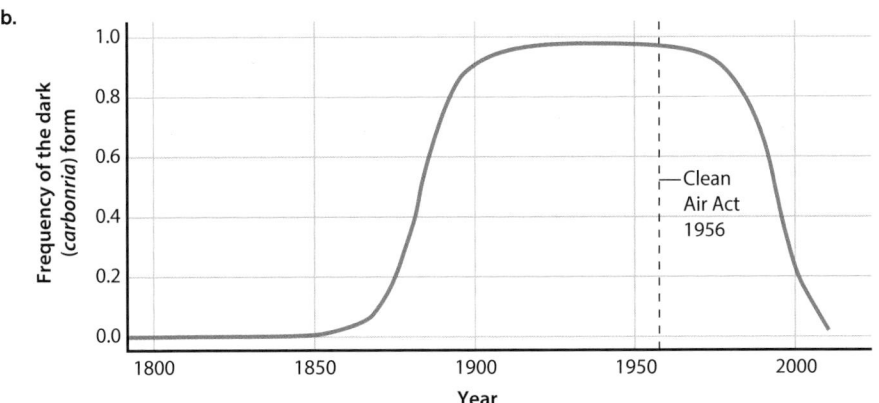

of the *carbonaria* form is plotted on the *y*-axis. As you can see, the frequency of *carbonaria* increased from 1850 to 1900, from 0 to almost 1, when it leveled off. It then decreased again from about 1950 to the present. Directional selection tends to occur when the environment changes, such as increasing or decreasing pollution in the case of the peppered moth. As the environment changes, selective pressure pushes certain traits in organisms in one direction or another, such as toward darker forms or lighter forms of the moth over time.

A third mode of selection, known as disruptive selection, operates in favor of extremes and against intermediate phenotypes. Apple maggot flies of North America (*Rhagoletis pomonella*), shown in **FIGURE 41.9a**, provide an example. The larvae of these flies originally fed on the fruit of hawthorn trees that bloom and produce fruit late in the summer.

FIGURE 41.9 Disruptive selection

Disruptive selection operates in favor of extremes. (a) An apple maggot fly feeds on an apple. (b) Before the introduction of apple trees, all apple maggot fly life cycles were coordinated with hawthorn trees. Disruptive selection and differences in the timing of fruiting have created two peaks in the distribution—one for apple-specializing flies and one for hawthorn-specializing flies. Each population is coordinated with the fruiting time of its host tree species. Photo: Rob Oakleaf, Michigan State University, Tree Fruit Entomology Lab; Data from Filchak et al. 2000. "Natural Selection and sympatric divergence in the apple maggot Rhagoletis pomonella," *Nature* 407:739–742.

a.

b.

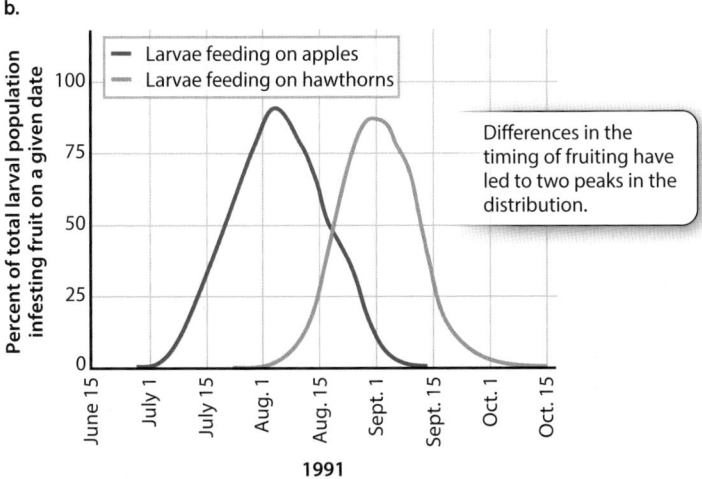

However, since the introduction of apples from Europe about 150 years ago, some apple maggot flies have evolved to eat apples that flower and produce fruit earlier in the summer. Disruptive selection has led to the evolution of two genetically distinct groups of flies, one that feeds on apples earlier and the other on hawthorn fruit later. Disruptive selection acts against intermediates between the two groups, which would miss the peaks of both the apple and hawthorn seasons.

Figure 41.9b illustrates the process of disruptive selection. The x-axis shows time, from mid-June to mid-October. The y-axis shows percent of larvae feeding on the two trees. Purple indicates apple trees and blue indicates hawthorn trees. Notice the two peaks, which show that larvae feed on apple trees earlier in the summer and on hawthorn trees later in the summer. For these flies, there is selection on the genes that influence the timing of larval feeding, in favor of feeding early or late, but not at intermediate times when neither tree bears fruit. As we will discuss in Module 47, this pattern of selection can lead to the evolution of new species, as the two populations become more and more different from one another over time.

✓ **Concept Check**

5. Birds and bats both have wings but evolved them independently of each other. **Identify** the term that describes the acquisition of similar structures or traits by two different groups independently of one another.

6. **Describe** how natural selection affects the frequency of beneficial and harmful alleles in a population over time.

7. **Describe** stabilizing selection and provide an example.

8. **Describe** directional selection and provide an example.

9. **Describe** disruptive selection and provide an example.

41.3 Artificial selection has produced many different breeds of plants and animals

Selection has been practiced by humans since at least the dawn of agriculture. We have bred many different types of livestock and crops, shaping them over time so they exhibit traits we find useful. For example, we have selected for corn with bigger kernels, cows that produce more milk, chickens that lay larger eggs, horses that run more quickly, wheat that produces a lot of grain, and so on. In each case, the breeder notices variation in a trait—such as running speed in horses—and then only allows the fastest runners to mate and have offspring. This process, played out over many generations, has led to faster and faster horses. The process by which humans select for traits is known as **artificial selection.**

Many people are familiar with different breeds of dogs. All of these have been shaped by humans over thousands of years, making them strikingly different from one another and from their common ancestor. The different breeds of dogs demonstrate the power of artificial selection. Darwin himself was interested in this process but didn't breed dogs or cats. Instead, he chose to work with pigeons. Pigeon breeding was common in the 1800s, and breeders have selected for all kinds of traits—larger tail feathers, fancier plumage, and different color patterns, to name a few. Today, there are hundreds of different breeds of pigeons, some of which are shown in **FIGURE 41.10**. Yet, as Darwin points out in *The Origin,* all these different breeds have a common ancestor—the rock pigeon, pictured in Figure 41.10d.

As Darwin himself noted, artificial selection is analogous to natural selection, but the competitive element is removed. Successful phenotypes are selected by the breeder, not through competition. Because it can be carefully controlled by the breeder, artificial selection is astonishingly efficient at generating change in populations. When practiced over many generations, artificial selection can create a population in which the selected phenotype is quite different from that of the starting population.

Although there are many similarities between artificial selection and natural selection, there is an important difference: in artificial selection, the breeder typically has some sort of goal in mind, such as breeding faster racehorses. In natural selection, however, there is no goal or endpoint. Those that survive and reproduce are simply better adapted to the current environment.

a.

b.

c.

d.

FIGURE 41.10 Artificial selection

Artificial selection has produced a remarkable variety and number of pigeon breeds, including the (a) Pygmy Pouter, (b) Old Dutch Capuchine, and (c) Blue Grizzle Frillback. All are descended from the (d) Rock Pigeon. Photos: (a)–(c) Richard Bailey/Getty Images; (d) Jose A. Bernat Bacete/ Getty Images

"Practicing Science 41.1: How far can artificial selection be taken?" shows the result of artificial selection for oil content in kernels of corn. In this experiment, researchers selected for corn that has high oil content, shown in red in the graph, and separately selected for corn that has low oil content, shown in blue in the graph on page 582. Over time, the line of corn that was selected for high oil content steadily increased in oil content, while the line that was selected for low oil content steadily decreased in oil content.

✓ Concept Check

10. **Identify** the process by which humans have produced breeds of dogs and cats.

11. **Describe** two ways in which artificial selection and natural selection are similar.

12. **Describe** two ways in which artificial selection and natural selection are different.

Practicing Science 41.1

PREP FOR THE AP® EXAM

How far can artificial selection be taken?

Background From the 1890s to the present day, an experiment at the University of Illinois has attempted to manipulate the properties of corn. This experiment has become one of the longest-running biological experiments in history.

Hypothesis Researchers hypothesized that there is a limit to the extent to which a population can respond to continued directional selection.

Experiment Corn was artificially selected for either high oil content or low oil content. Every generation, researchers bred together just the plants that produced corn with the highest oil content and did the same for the plants that produced corn with the lowest oil content. Each generation, kernels showed a range of oil levels, but only the 12 kernels with the highest or the lowest oil content were used for the next generation.

Results The graph below shows the results of the experiment. The percent oil content is indicated on the y-axis, and the time in generations is indicated on the x-axis. In the line selected for high oil

▼

content, shown in red, the percentage of oil more than quadrupled, from about 5% to more than 20%. In the line selected for low oil content, shown in blue, the oil content fell so close to zero that it could no longer be measured accurately, and the selection was terminated. Both selected lines are completely outside the range of any phenotype observed at the beginning of the experiment.

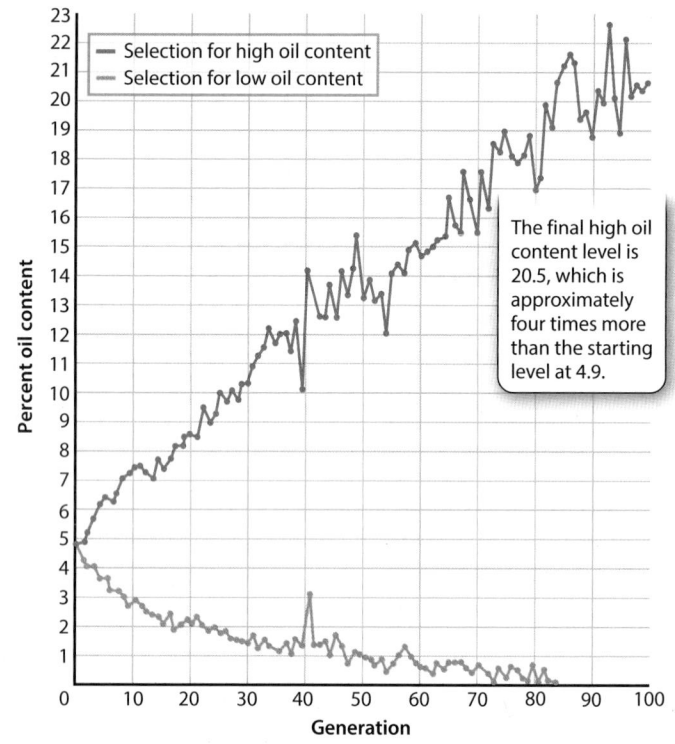

The final high oil content level is 20.5, which is approximately four times more than the starting level at 4.9.

Follow-up Work Genetic analysis of the selected lines indicates that the differences in oil content are due to the effects of at least 50 genes.

SOURCE
Moose, S. P., J. W. Dudley, and T. R. Rocheford. 2004. "Maize Selection Passes the Century Mark: A Unique Resource for 21st Century Genomics." *Trends in Plant Science* 9:358–364.

AP® PRACTICE QUESTION

a. **Identify** the following:
   1. The question the experimenters are trying to answer
   2. The hypotheses tested
   3. The independent variable
   4. The dependent variable

b. **Describe** the conclusion using a claim, evidence, reasoning format (CER).

41.4 Sexual selection increases an individual's ability to find and attract mates

Initially, Darwin was puzzled by certain traits of organisms, such as the tail of a peacock or the bright, showy colors of many birds. These traits seem to reduce an individual's chances of survival. How then did they evolve over time? In a letter dated a few months after the publication of *The Origin,* Darwin wrote, "The sight of a feather in a peacock's tail, whenever I gaze at it, makes me sick!" The tail is energetically expensive to produce; it is an advertisement to potential predators; and it is a hindrance to any attempt to escape a predator.

In his 1871 book, *The Descent of Man, and Selection in Relation to Sex,* Darwin proposed a solution to this problem. Natural selection indeed acts to reduce the size and showiness of the peacock's tail, but another form of selection, *sexual selection,* acts in the opposite direction. **Sexual selection** is a form of selection that promotes traits that increase an individual's ability to find and attract mates. That is, large, showy tails in male peacocks increase their chances of attracting females, so that trait is passed on to the next generation.

Darwin recognized that this process can play out in two different ways. In one form of sexual selection, members of one sex, usually the males, compete with one another for access to the other sex, usually the females. Because competition typically occurs among males, it is in males that we often see physical traits such as large size, horns, and other elaborate weaponry. Larger, more powerful males tend to win more fights, hold larger territories, and have access to more females. When sexual selection focuses on interactions between individuals of one sex, it is called intrasexual

a.

b.

FIGURE 41.11 Sexual selection

There are two types of sexual selection. (a) Intrasexual selection often involves competition between males, as in this battle between two male elk. (b) Intersexual selection often involves bright colors and displays by males to attract females, as shown by these Japanese red-crowned cranes, one male and one female. Photos (a) Kelly Funk/Getty Images; (b) Steven Kaufman/Getty Images

selection (*intra* means "within"). One example of intrasexual selection can be seen in the horns of male elks (*Cervus canadensis*) shown in **FIGURE 41.11a**.

Darwin also recognized a second form of sexual selection. In this case, males typically do not fight with one another, but instead compete for the attention of females with bright colors, songs, or other advertisement displays. This form of selection is called intersexual selection because it focuses on interactions between males and females (*inter* means "between"). Intersexual selection can be seen in Japanese red-crowned cranes (*Grus japonensis*), like those shown in Figure 41.11b where a male is doing an elaborate courtship dance to attract a female.

The peacock's tail is also thought to be the product of intersexual selection: its evolution has been driven by a female

preference for ever-showier tails in males. In the absence of sexual selection, natural selection would act to minimize the size of the peacock's tail. Presumably, the peacock tails we see today are a compromise, a trade-off between the conflicting demands of attracting a mate and survival.

✓ Concept Check

13. Sexual selection tends to cause bigger size, more elaborate weaponry, or brighter colors in males. **Identify** the pattern of selection that is observed (stabilizing, directional, or disruptive).

14. **Describe** the difference between intrasexual selection and intersexual selection.

Module 41 Summary

PREP FOR THE AP® EXAM

REVISIT THE BIG IDEAS

EVOLUTION: Using an example from the module, **explain** how a diverse gene pool improves species survival when there is a significant environmental change.

LG 41.1 Natural selection brings about adaptations and increases fitness.

- An adaptation is the close fit between an organism and the environment. Page 572

- Natural selection is the differential reproductive success of genetic variants and was independently conceived by Charles Darwin and Alfred Russel Wallace. Page 573

- Natural selection explains how adaptations come about. Page 573

- Natural selection rests on four postulates: (1) individuals of a species are variable; (2) variation is at least in part heritable; (3) there is competition so that some individuals survive and reproduce more than others; and (4) those variants that survive and reproduce more leave more offspring. Page 574
- Fitness is the ability to survive and reproduce in a particular environment. Page 575

LG 41.2 Natural selection acts on phenotypes.

- Natural selection acts on individuals with particular traits, or phenotypes. Page 577
- As a result of natural selection, the genetic makeup of populations changes over time. Page 577
- Stabilizing selection favors intermediate phenotypes and acts against extreme phenotypes, thereby maintaining the status quo. Page 578
- Directional selection favors one extreme, so the phenotype shifts in one direction. Page 578
- Disruptive selection favors both extremes and acts against intermediate phenotypes. Over time, it can lead to the evolution of two species from one. Page 579

LG 41.3 Artificial selection has produced many different breeds of plants and animals.

- Artificial selection is the process by which humans govern the selection process. Page 580
- Artificial selection has produced many of the common breeds of farm animals, crops, cats, dogs, and pigeons. Page 580
- Artificial selection is a form of directional selection. Page 580

LG 41.4 Sexual selection increases an individual's ability to find and attract mates.

- Sexual selection involves the evolution of traits that increase an individual's access to members of the opposite sex. Page 582
- In intrasexual selection, individuals of the same sex compete with one another, resulting in traits such as large size and horns in males. Page 583
- In intersexual selection, individuals of one sex advertise to individuals of the opposite sex, resulting in traits such as colorful plumage in male birds. Page 583

Key Terms

Selective pressure   Artificial selection   Sexual selection
Convergent evolution

Review Questions

1. Which of the following may be a characteristic of a beneficial mutation?
   (A) It facilitates successful reproduction.
   (B) It decreases lifetime reproductive output.
   (C) It has no effect on phenotype.
   (D) It decreases an organism's fitness.

2. Which of the following is a source of genetic variation?
   (A) Adaptation
   (B) Natural selection
   (C) Convergent evolution
   (D) Mutation

3. Which of the following is considered a beneficial mutation?
   (A) A mutation that makes an individual more visible to predators
   (B) A mutation that increases an organism's ability to find food
   (C) A mutation that decreases the offspring's chance of survival
   (D) A mutation that changes hair color in humans

4. Which of the following is an example of stabilizing selection?
   (A) Selection for average birth weight in humans
   (B) Selection for antibiotic resistance in bacteria
   (C) A decrease in the number of birds that have intermediate-sized beaks in an environment with large or small seeds
   (D) Breeding dogs from wolves

5. What is a key difference between natural selection and artificial selection?
   (A) Natural selection results in evolution; artificial selection does not.
   (B) Artificial selection is driven by humans.
   (C) Natural selection may select traits that are detrimental.
   (D) Natural selection is controlled.

6. Which form of selection works on the ability to attract mates, even if the traits selected for reduce the organism's chance of survival?
   (A) Sexual
   (B) Asexual
   (C) Disruptive
   (D) Stabilizing

Module 41
AP® Practice Questions

Section 1: Multiple-Choice Questions
Choose the best answer for questions 1–4.

1. Which provides the most direct estimate of evolutionary fitness?
   (A) The birth weight of a human baby
   (B) The volume of water lost by a desert plant over the course of a day
   (C) The swimming speed of a great white shark
   (D) The number of offspring produced by a seed-eating finch

2. Which figure best illustrates disruptive selection?

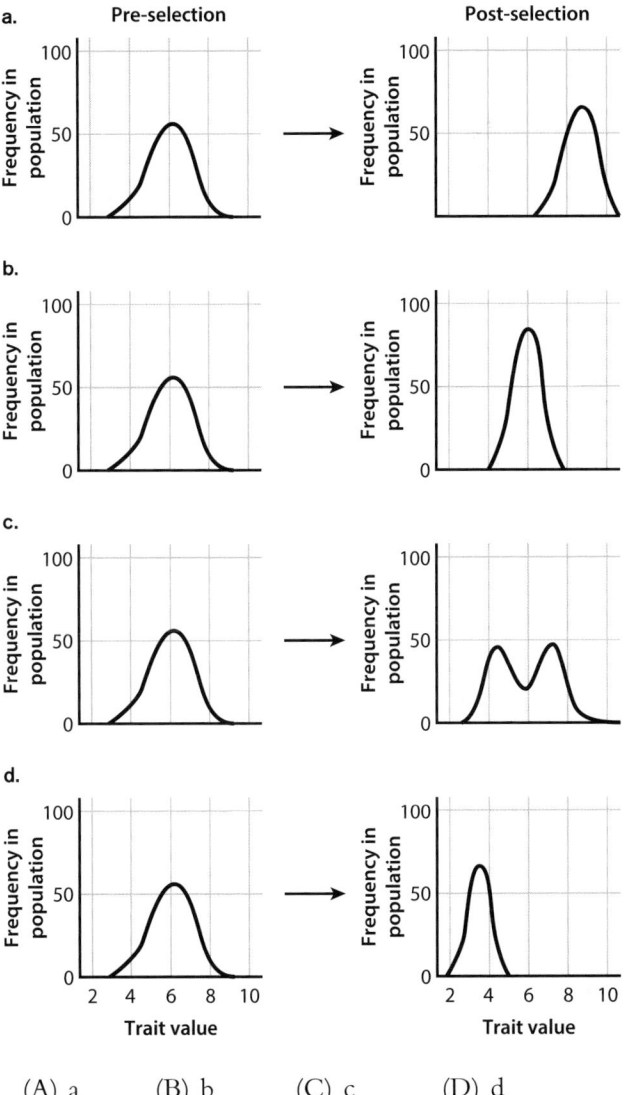

(A) a   (B) b   (C) c   (D) d

Questions 3 and 4 refer to the following information:

Cardinals are small songbirds that live in the northeastern United States. Male cardinals have bright red feathers, which makes them extremely visible during breeding season. Female cardinals have duller gray or brown feathers. Males with the brightest red color are most attractive to females and have the highest reproductive success. The red color of male cardinals is generated from carotenoid pigments obtained from the diet. The male cardinal produces enzymes that convert carotenoid pigments in the diet to the bright red pigment displayed in the feathers.

A scientist captured 16 male cardinals from a wild population and used a tool called a spectrophotometer to measure the percentage of pure red in the feathers. The scientist also measured the percentage of pure red in a sample of 10 male cardinals reared in captivity. The data obtained are shown in the histogram below. Error bars are calculated as ±1 standard deviation on either side of the mean.

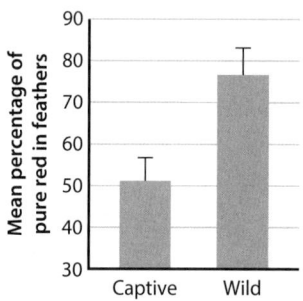

Data from McGraw, K. J., Hill, G. E. (2001). "Carotenoid access and intraspecific variation in plumage pigmentation in male American Goldfinches (*Carduelis tristis*) and Northern Cardinals (*Cardinalis cardinalis*)," *Functional Ecology*, 20 December, 2001 vol 15, issue 6, 732–739.

3. What is the difference in mean percentage of pure red color in feathers between the cardinals captured in the wild versus those reared in captivity?

   (A) 76.7%

   (B) 25.3%

   (C) 51.4%

   (D) 65.1%

4. Lutein is a pigment that is commonly found in plant seeds. Male cardinals produce an enzyme that converts lutein to α-(alpha-)doradexanthin, which is one of the four red pigments in the feathers of red cardinals. Predict the impact of a nonsense mutation in the gene encoding this enzyme on the fitness of a male cardinal.

   (A) There would be no effect on fitness on the male cardinal because the enzyme is not essential for life.

   (B) The male cardinal would have higher fitness because lutein is toxic and it would accumulate in the feathers.

   (C) The male cardinal would have lower fitness because it would be less attractive to female cardinals.

   (D) The male cardinal would have lower fitness because it would be less visible to predators.

Section 2: Free-Response Question

Write your answer to each part clearly. Support your answers with relevant information and examples. Where calculations are required, show your work.

Sexual selection often results in exaggerated traits in male animals such as large and flashy tails in peacocks, large antlers in elk, and conspicuous courtship displays. The traits are often most attractive to females when they are most extreme.

(a) **Identify** two reasons natural selection may act against sexually selected traits.

(b) **Make a claim** about whether it is possible for a trait to be under both intersexual selection and intrasexual selection at the same time. **Justify** your response.

Module 42

Population Genetics

LEARNING GOALS ▶**LG 42.1** Variation is a notable feature of the natural world.
        ▶**LG 42.2** Mutation and recombination are the two sources of genetic variation.
        ▶**LG 42.3** Genetic variation is measured by allele frequencies.

In the last module, we focused on Charles Darwin's theory of evolution by natural selection. Natural selection is a mechanism of evolution that leads to adaptations—the remarkable fit that organisms often have with the environment in which they live. Darwin made four claims about the natural world, which are known as Darwin's postulates. The first one is that individuals of a species are variable.

Before Darwin, these variations were largely ignored and naturalists focused instead on ideal forms or types that defined a group. Since Darwin, however, we have learned that a species consists of individuals that vary one to the next. In our own species, we can observe that people may be tall, short, dark-haired, fair-haired, and so on. In fact, variation is an essential ingredient of Darwin's theory because the mechanism he proposed for evolution, natural selection,

depends on the differential survival and reproduction of variants.

Darwin changed how we view variation. Before Darwin, variation was irrelevant, something to be ignored. After Darwin, it was recognized as the key to the evolutionary process. In this module, we will focus on variation—how to describe it, where it comes from, and how to measure it. When we have a way to measure genetic variation, we can follow it over time. If it changes, then we know that evolution is occurring.

PREP FOR THE AP® EXAM

FOCUS ON THE BIG IDEAS

EVOLUTION: Focus on types and sources of genetic variation among individuals in a population.

42.1 Variation is a notable feature of the natural world

Darwin started off with the observation that individuals of a species differ from one another, whether looking at pigeons, Galápagos finches, or even humans. In other words, variation is abundant in nature. Observable differences among individuals of a species are the result of differences in their genetic material (DNA) as well as differences in the environment. We will explore these differences in this section.

Phenotypic Variation

Variation is a major feature of the natural world. We humans are particularly good at noticing variation among individuals of our own species. We differ in height, weight, facial features, skin color, eye color, hair color, hair texture, and many other ways.

As we saw in Unit 5, the observable traits of an organism are called the phenotype. The trait may be visible, as in characteristics such as weight or height, or it may be observed in the development, physiology, or behavior of a cell or an organism. For example, color blindness and lactose intolerance are phenotypes, even though you can't "see" these traits by simple observation. They are considered phenotypes because they are both traits of organisms.

A phenotype results in part from an organism's genotype, its genetic makeup. For example, some individuals have a genotype with a mutation that affects expression of an enzyme that normally breaks down the sugar lactose. This genotype can lead to the phenotype of lactose intolerance. In addition, the environment commonly plays an important role in phenotype. For example, weight is affected by diet, and skin color is affected by exposure to sunlight. As a result, it is most accurate to say that a phenotype reflects an interaction between the genotype and the environment.

Genotypic Variation

As we just discussed, the genetic makeup of a cell or organism constitutes its genotype. A population that has variation in many different genes consists of organisms with many different genotypes. Genotypic or genetic variation refers to differences in the DNA sequence among individuals of a species. Genetic variation accounts in part for the physical differences we see among individuals, such as differences in hair color, eye color, and height. And, on a much larger scale, genetic differences among species has led to the diversity of organisms on this planet, from bacteria to blue whales.

The relationship between phenotype and genotype is typically not straightforward. Humans are phenotypically very variable—that is, many visible traits differ among humans. However, in spite of having a high degree of phenotypic variation, humans actually rank low in terms of overall genetic variation compared to other species. Any two randomly selected humans differ from each other, on average, by one DNA base per thousand bases, which means that they are 99.9% genetically identical. By way of comparison, Adélie penguins (*Pygoscelis adeliae*), shown in **FIGURE 42.1**, are two to three times more genetically variable than humans (2–3 bases per thousand), even though they are one of the most phenotypically uniform species on the planet. And fruit flies (*Drosophila melanogaster*) are 10 times more genetically variable than humans (10 bases per thousand).

Genetic variation arises from mutations. Any heritable change in the nucleotide sequence of DNA is a mutation. By heritable, we mean that the mutation persists through DNA replication and is transmitted from parent cell to daughter cell through cell division. A change might be a substitution of one base for another, a small deletion or insertion of a few nucleotides, or a larger alteration affecting a segment of a chromosome, as described in Units 5 and 6. Most mistakes in DNA replication or chemical damage to DNA are immediately repaired by specialized enzymes in the cell. Some, however, escape these repair mechanisms and are transmitted from parent cell to daughter cell.

A mutation occurs in an individual cell at a particular point in time. Through evolution, the proportion of individuals in a population carrying this mutation may increase or decrease. Therefore, if we look at any present-day population of organisms, such as the human population, we will find that it harbors many genetic differences, all the result of mutations that occurred sometime in the past.

We tend to think of mutations as something negative or harmful because in everyday language the term "mutant" suggests something abnormal. In reality, because mutations result in genetic variation among individuals, we are all "mutants" that differ from one another in our DNA. Although it is true that many mutations are harmful, other mutations have no or negligible effects on the organism, and a precious few are beneficial. Without mutations, evolution would not be possible. Mutations generate the occasional favorable variants that allow populations to evolve and organisms to become adapted to their environment over time, as we discussed in Module 41.

PREP FOR THE AP® EXAM

AP® EXAM TIP

Remember that genetic variation is necessary for evolution by natural selection, but it is not the only factor. Genetic variation results in phenotypic variation which in turn can lead to differential survival and reproduction of individuals in a particular environment. As a result, the genetic makeup of the population changes from one generation to the next.

✓ Concept Check

1. **Describe** the difference between a phenotype and a genotype.

2. **Identify** two factors that result in phenotypic variation among individuals of a species.

3. **Describe** why a mutation may be harmful, be beneficial, or have no impact on the survival or reproduction of an individual.

FIGURE 42.1 Genetic diversity in Adélie penguins

Adélie penguins look very similar to one another but are more genetically diverse than humans. Photo: Ralph Lee Hopkins/Getty Images

42.2 Mutation and recombination are the two sources of genetic variation

Genetic variation refers to any difference in the DNA sequence among individuals of a species. If we focus our attention on genes, the regions of DNA that encode for proteins, we find that individuals often have different gene sequences as well. The different forms or variants of a gene are alleles, and they correspond to different DNA sequences in the genes. In this section, we will discuss the two sources of genetic variation—mutation and recombination—and then characterize mutations.

Sources of Genetic Variation

Genetic variation that we observe in any present-day population of organisms has two sources. First, as we have seen, mutation generates new variation in DNA sequences. Second, recombination shuffles existing variation to make new combinations. Recall from Module 26 that, in sexually reproducing organisms, a form of cell division called meiosis produces gametes, such as eggs and sperm. Meiosis generates daughter cells with half the number of chromosomes as the parent cell. Then, when fertilization occurs, the combination of genetic material from the egg and the sperm produces a new organism with the same number of chromosomes as the parents. During meiosis, chromosomes exchange DNA segments through recombination. This process shuffles mutations to create new combinations.

Because mutation and recombination lead to the formation of new alleles, both processes increase genetic variation. These two processes are shown in **FIGURE 42.2**. In generation 1, there is one allele. In generation 2, mutation of T to C creates a new allele (allele 2), so now there are two alleles in the population. In generation 3, mutation of T to G creates another new allele (allele 3), or three alleles in the population. In generation 4, recombination between allele 1 and allele 3 creates a new allele (allele 4). Over four generations, we see new alleles arising by both mutation and recombination.

Although mutation and recombination can both create new genetic sequences and therefore increase genetic variation, all genetic variation ultimately comes from mutations. Recombination only increases genetic variation if the genetic sequences are different because of mutations that occurred in the past, as Figure 42.2 demonstrates. As a result,

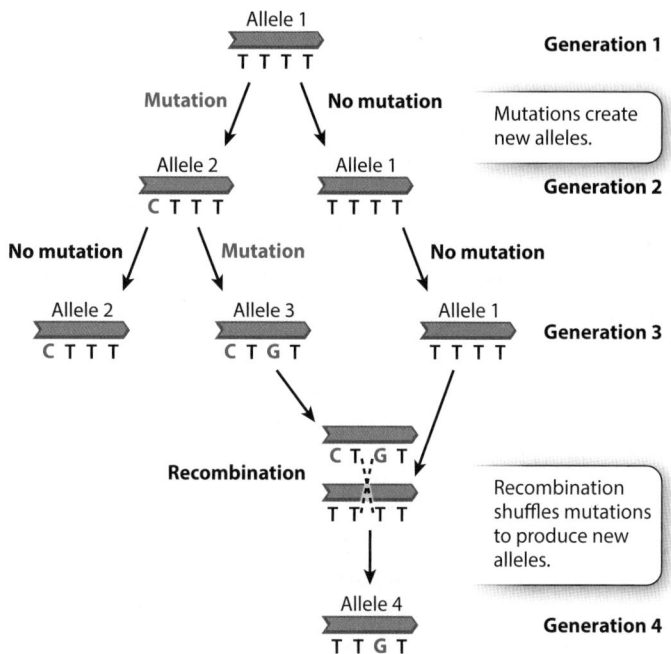

FIGURE 42.2 Mutation and recombination

The formation of new alleles occurs by mutation in generations 1–3 and recombination in generation 4.

we say that mutations are the ultimate source of genetic variation, even though both processes contribute to genetic variation in present-day populations.

Mutation Frequency

Mutations arise from mistakes in DNA replication or from unrepaired damage to DNA. Sources of the damage may include reactive molecules produced in the normal course of metabolism; chemicals in the environment; or radiation of various types, including X-rays and ultraviolet light. Most genomes also contain DNA sequences that can "jump" from one position to another in the genome. Insertion of these sequences into or near genes is also a source of mutation.

The rate at which mutations occur differs among species. **FIGURE 42.3** on page 590 compares the rates of newly arising mutations in a given nucleotide pair in a single round of replication in several viruses and several types of organisms. As seen in the graph, the mutation rates range across almost eight orders of magnitude, from about 10^{-3} mutations per nucleotide per replication in some viruses to about 10^{-11} mutations per nucleotide per replication in humans. This range reflects differences in the accuracy of DNA replication and the efficiency of DNA repair mechanisms.

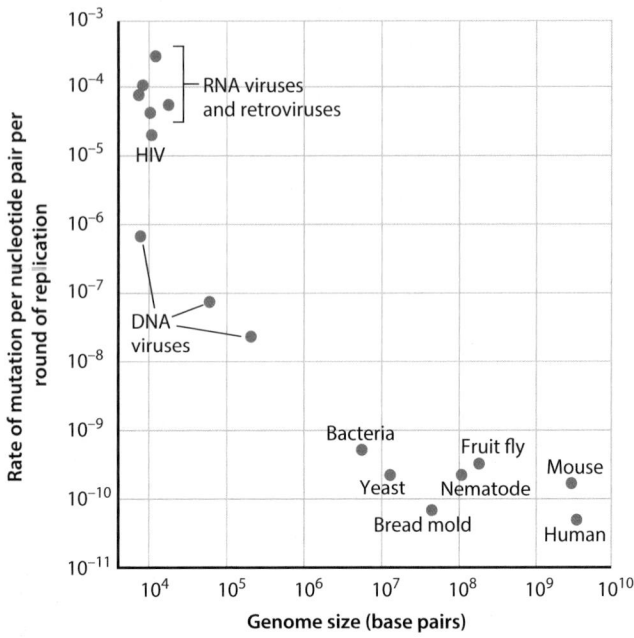

FIGURE 42.3 Mutation rate

The rate of mutation for a specific nucleotide in one round of DNA replication is very low, and it varies in different organisms. Data from Drake, J. W., B. Charlesworth, and J. F. Crow. 1998. "Rates of Spontaneous Mutation." *Genetics* 148:1667–1686.

Mutations are rare events when considered from the point of view of a particular base (nucleotide). In other words, if we consider a particular nucleotide and ask what's the chance that it will undergo a mutation in a single round of replication, it's very small. While the rate of mutation per nucleotide per replication in most organisms is low, the rate of mutation across an entire genome for *any* nucleotide is higher, especially for organisms with large genomes. Therefore, although the probability of a change in a specific nucleotide is very low, the probability of a mutation anywhere in the genome is much higher.

The Random Nature of Mutations

Mutations occur randomly. What exactly does that mean? Consider the following experiment: an antibiotic is added to bacterial cells that are growing and dividing. Most of the cells are killed, but a few survivors continue to grow and divide. These survivors contain mutations that confer resistance to the antibiotic. That is, the survivors with the mutations are not killed by the antibiotic. This simple observation raises two alternative hypotheses. First, the antibiotic-resistant mutations might have already been present before the antibiotic was added. Second, the antibiotic-resistant mutations might arise in response to the application of the antibiotic.

These alternative hypotheses have profound implications for all of biology because they suggest two very different ways in which mutations occur. The first suggests that mutations occur without regard to the needs of an organism. According to this hypothesis, the presence of the antibiotic in the experiment with bacterial cells does not direct or induce antibiotic resistance in the cells. Instead, it kills off all but the small number of preexisting antibiotic-resistant mutants, which flourish. The second hypothesis suggests that some sort of feedback occurs between the needs of an organism and the process of mutation, as if the environment directs specific mutations that are beneficial to the organism.

We now know from many experiments in many species that the first hypothesis is correct—mutations are random and do not occur out of need. One of these experiments was carried out by Joshua and Esther Lederberg in 1952. This experiment is summarized below and described in more detail in "Practicing Science 42.1: Do mutations occur randomly, or are they directed by the environment?"

In this experiment, bacterial cells were grown and formed colonies on agar plates in the absence of antibiotic. Then, using a technique called replica plating, the Lederbergs transferred these colonies to new plates containing antibiotic. Only bacteria that were resistant to the antibiotic grew on the new plates. Because replica plating preserved the physical arrangement of the colonies, the Lederbergs were able to go back to the original plate and identify the colony that produced the antibiotic-resistant colony on the replica plate. From that original colony, they were then able to isolate a pure culture of antibiotic-resistant bacteria.

In the experiments, isolating antibiotic-resistant bacteria from the original plate showed that the mutation for antibiotic resistance existed before the bacteria were ever exposed to antibiotic. This result supported the hypothesis that mutations occur randomly without regard to the needs of the organisms. The environment does not cause specific mutations, but instead selects for them. The principle that the Lederbergs demonstrated is true of all organisms so far examined.

Types of Mutations

Some features of mutations are particularly important for evolution. For example, we can consider whether mutations are passed on to offspring. Somatic mutations occur in the body's tissues in nonreproductive cells. A somatic mutation affects only the cells descended from the cell in which the mutation originally arose, so it affects only that one individual and is not passed off to offspring.

Do mutations occur randomly, or are they directed by the environment?

Background Researchers have long observed that beneficial mutations tend to persist in environments where they are beneficial to an organism. For example, in the presence of antibiotics, bacterial populations evolve antibiotic resistance; in the presence of insecticides, insect populations evolve insecticide resistance.

Two Hypotheses These kinds of observations lead to two hypotheses about how a mutation, such as one that confers antibiotic resistance in bacteria, might arise. The first suggests that mutations occur randomly in bacterial populations and over time become more common in the population in the presence of an antibiotic that kills those bacteria that do not have the mutation. In other words, they occur randomly and independent of the needs of an organism. The second hypothesis suggests that the environment, in this case the application of antibiotic, induces mutations that confer antibiotic resistance.

Method To distinguish between these two hypotheses, Joshua and Esther Lederberg developed a technique called replica plating. In this technique, bacteria are grown on plates of a solid gel-like medium containing agar and nutrients, where they form colonies of cells. The cells in any one colony result from the division of a single original cell, so all individual bacteria in a single colony are genetically identical except for rare mutations that occur in the course of growth and division. Then a disk of sterilized velvet is pressed onto the plate, shown in the figure to the right. The velvet disk picks up cells from each colony and preserves their geometrical positions on the plate. The velvet disk is pressed onto the surface of a fresh plate, creating a replica of the original plate, hence the name replica plating. The velvet transfers to the new plate some of the cells that originate from the colony on the first agar plate in their original positions. This is key

because it allows the researcher to expose the colonies on the replica plate to some sort of treatment, in this case, antibiotic, while also being able to go back and examine the exact same colonies on the original plate that were never exposed to antibiotic. In essence, the two plates create a "before" and "after" scenario of the same bacterial colonies.

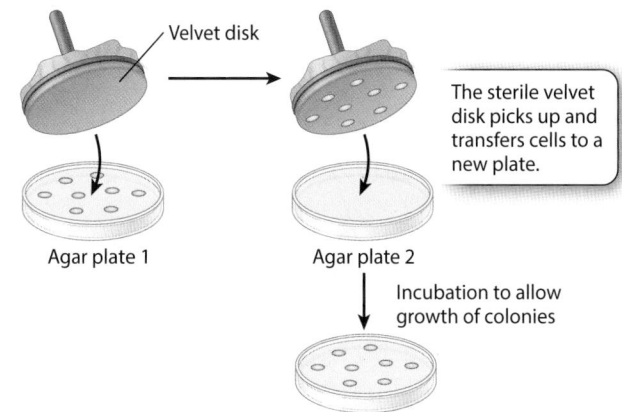

The sterile velvet disk picks up and transfers cells to a new plate.

Agar plate 1   Agar plate 2

Incubation to allow growth of colonies

Experiment and Results First, the Lederbergs grew bacterial colonies on a master plate containing medium without antibiotic. Then, by replica plating, they transferred some cells from each colony to two plates, shown in the figure below. One plate contained medium without antibiotic. This was the nonselective plate and served as the control; all cells were able to grow and form colonies on it. The second plate contained medium with antibiotic. This was the selective plate and served as the experimental treatment: only antibiotic-resistant cells multiplied and formed colonies on this plate because it contained antibiotic.

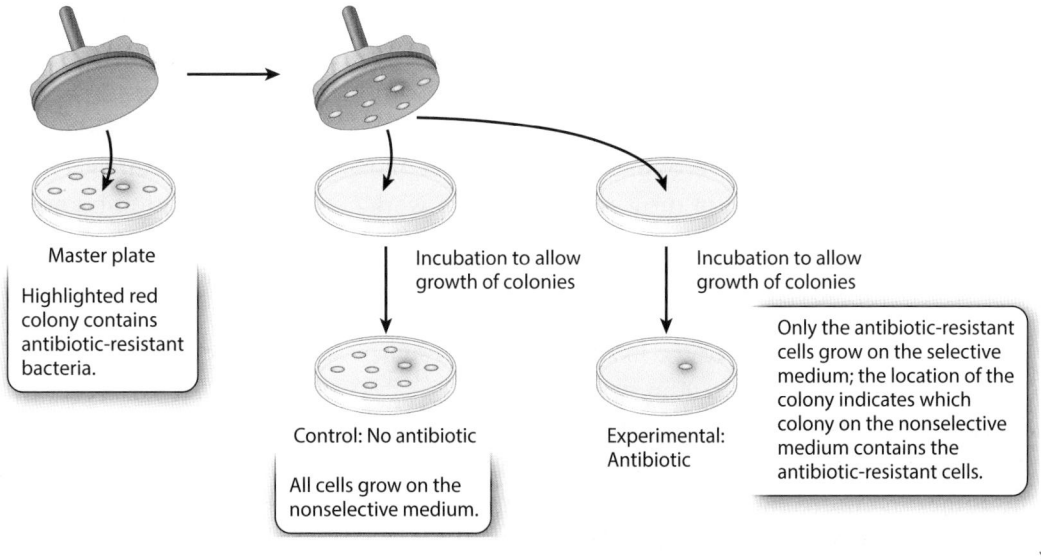

Master plate

Highlighted red colony contains antibiotic-resistant bacteria.

Incubation to allow growth of colonies

Incubation to allow growth of colonies

Control: No antibiotic

Experimental: Antibiotic

Only the antibiotic-resistant cells grow on the selective medium; the location of the colony indicates which colony on the nonselective medium contains the antibiotic-resistant cells.

All cells grow on the nonselective medium.

Because replica plating preserves the arrangement of the colonies, the location of an antibiotic-resistant colony that grew on the selective medium revealed the location of its parental colony on the control plate. This allowed the researchers to go back to the master plate and find the colony that grew on the selective plate before it was exposed to the antibiotic. Therefore, in the final step of the experiment, shown below, the Lederbergs went back to the parental colony, which had not been exposed to the antibiotic, and grew a culture of bacteria from this colony. This is a pure culture of antibiotic-resistant bacteria that were present on the original plate before the bacteria ever encountered antibiotics.

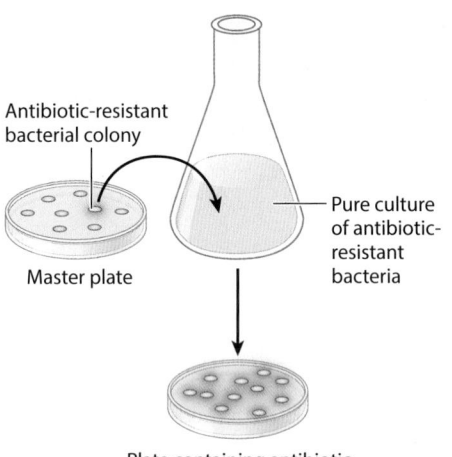

Antibiotic-resistant bacterial colony

Master plate

Pure culture of antibiotic-resistant bacteria

Plate containing antibiotic

Conclusion The Lederbergs' experiment showed that antibiotic-resistant mutants arise before exposure to antibiotics, not as a result of it. Antibiotic-resistant bacteria were isolated in these experiments even though the parental cells on nonselective medium never came into contact with the antibiotic.

Follow-up Work These results have been extended to other types of mutation and other organisms, demonstrating that mutations are random and are not directed or induced by the environment.

SOURCE
Lederberg, J., and E. M. Lederberg. 1952. "Replica Plating and Indirect Selection of Bacterial Mutants." *Journal of Bacteriology* 63:399–406.

AP® PRACTICE QUESTION

a. **Identify** the following:
   1. The question being investigated by the experimenters
   2. The hypotheses they tested
   3. The experimental group
   4. The control group
   5. The independent variable
   6. The dependent variable
b. **Explain** the results of the experiment by **identifying** the researchers' claim and **justifying** the claim with the evidence and the reasoning researchers used (CER).

There are also germ-line mutations, which occur in the reproductive cells, such as eggs and sperm. Because germ-line mutations are passed to the next generation, they are of particular interest from an evolutionary standpoint.

We can also consider the effects that mutations have on an organism. Recall from Unit 6 that DNA is transcribed to mRNA, which is translated to proteins. Therefore, a change or mutation in DNA can be transferred to mRNA and then to a protein. Proteins, in turn, do much of the work of the cell and make up the structure of cells and organisms. Therefore, mutations often exert their effects on an organism by changing the structure and hence the function of proteins. In other words, a change in a protein can, in some cases, result in a change in a trait in an organism.

A mutation can have many effects on an organism, but here we will focus on the influence a mutation has on the ability of an organism to survive and reproduce in a particular environment. Viewed in this way, mutations can be neutral, meaning that they have little or no effect on the survival and reproduction of an organism. Every human is born with about 60 new mutations, but, because most of the genome consists of noncoding DNA, most of these are neutral. In contrast, a large proportion of mutations that occur in genes or protein-coding regions have a harmful effect on an organism and are known as deleterious mutations. These decrease a carrier's chances of survival and reproduction. Rarely, a mutation occurs that has a beneficial effect. A mutation is known as an advantageous mutation if it improves a carrier's chances of survival and reproduction. As we have seen, through natural selection, advantageous mutations can increase in frequency in a population until eventually they are carried by every member of a species. These mutations result in a species that is adapted to its environment—better able to survive and reproduce in that environment.

Finally, we can consider the types of mutations that contribute to present-day genetic variation among individuals in a population. Different types of mutations are common in

a. Point mutation

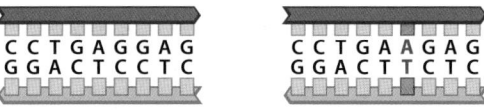

b. Duplication and deletion

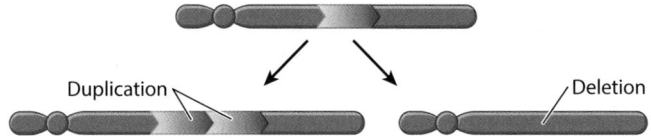

c. Movement of a transposable element

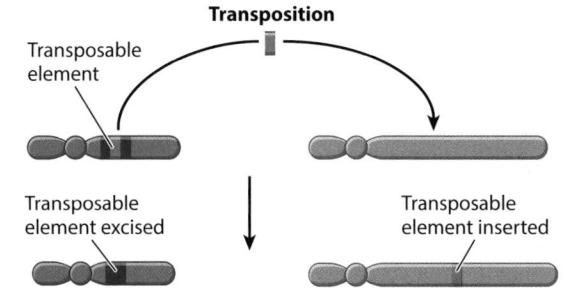

FIGURE 42.4 Common mutations in populations

Mutations are the ultimate source of genetic variation among individuals in a population. Common types of mutations we see in present-day populations include point mutations (a), duplications and deletions (b), and insertions of a transposable element (c).

present-day populations of many species, including humans. These mutations occurred in the past, but then were retained and spread so that many individuals in a population share them.

FIGURE 42.4 reviews three major types of genetic variation, which we introduced in Unit 6. A mutation in which one base pair is replaced by a different base pair is called a point mutation. A point mutation in which a G–C base pair is replaced by an A–T base pair is shown in Figure 42.4a. Other relatively common types of mutations are duplications, in which a segment of DNA is repeated, and deletions, in which a segment of DNA is removed, which is shown in Figure 42.4b. If these mutations include one or more genes, the resulting change in gene copy number changes the amount of the protein products of these genes in the cell. Another type of mutation is the insertion of movable DNA sequences into or near a gene. Such a "jumping" DNA sequence is called a transposable element and is shown in Figure 42.4c. The genomes of virtually all organisms contain several types of transposable elements, and each is present in multiple copies per genome, so they are quite common.

✓ Concept Check

4. **Identify** the two sources of all present-day genetic variation among organisms.

5. **Identify** the ultimate source of all present-day genetic variation.

6. **Describe** what it means when scientists say that mutations are random.

7. **Identify** and **describe** three types of mutations.

42.3 Genetic variation is measured by allele frequencies

Most species harbor abundant genetic variation, such as the types of mutations shown in Figure 42.4. What factors determine the amount of variation in a species? Why are humans genetically less variable than fruit flies and Adélie penguins? What factors affect the distribution of particular variations? The field of population genetics addresses these and other questions about patterns of variation. In this section, we will look at how scientists measure genetic variation and follow genetic variation in populations over time.

Populations

Every individual of a given species is different. For example, they have different combinations of alleles. All of the alleles present in all of the individuals in a species is called the gene pool for that species. The human gene pool includes alleles that cause differences in skin color, hair type, eye color, and so on. Each one of us has a different set of those alleles—alleles that cause brown hair and brown eyes, for example, or alleles that cause black hair and blue eyes— drawn from that gene pool.

Individuals of a species share alleles through mating. Species represent individuals that are able to mate with each other and have fertile offspring. Therefore, a species consists of all the individuals that are capable, through reproduction, of sharing alleles with one another. A population is a subset of the individuals of a particular species that lives in one area. In other words, a population is an interbreeding group of organisms of the same species living in the same geographical area.

Measuring Genetic Variation

To understand patterns of genetic variation in populations, we require information about allele frequencies. Allele frequencies provide a measure of genetic variation in a population. The allele frequency of an allele x is the number of allele x's present in the population divided by the total number of alleles of that gene. In other words, allele frequency is the proportion, among all the alleles of a gene in a population, that consists of a particular allele.

Consider, for example, pea color in Mendel's pea plants, which we discussed in Unit 5. Pea color (yellow or green) is determined by variation at a single gene. Two alleles of this gene include the dominant A (yellow) allele and the recessive a (green) allele. AA homozygotes and Aa heterozygotes produce yellow peas, whereas aa homozygotes produce green peas. Imagine a population in which every pea plant produces green peas, meaning that only one allele, a, is present. In this population, the allele frequency of a is 100%, and the allele frequency of A is 0%. When a population exhibits only one allele at a particular gene, we say that the population is fixed for that allele. In other words, the frequency of the allele is 1, and no other alleles for that gene exist in the population.

The genotype frequency of a population is the proportion of each genotype at a particular gene or set of genes. Consider another population of 100 pea plants with genotype frequencies of 50% aa, 25% Aa, and 25% AA. The genotype numbers for this population are 50 green pea homozygotes (aa), 25 yellow pea heterozygotes (Aa), and 25 yellow pea homozygotes (AA).

What is the allele frequency of a in this population? First, we find the total number of a alleles in the population and divide by the total number of alleles in the population. Each of the 50 aa homozygotes has two a alleles and each of the 25 heterozygotes has one a allele. There are no a alleles in AA homozygotes. The total number of a alleles is therefore $(2 \times 50) + 25 = 125$. Because each pea plant is diploid, meaning that it has two alleles, the total number of alleles in the population of 100 pea plants is 200. To determine the allele frequency of a, we divide the number of a alleles, 125, by the total number of alleles in the population, 200:

$$\frac{125}{200} = 0.625 = 62.5\%$$

Because we are dealing with only two alleles in this example, we can calculate the allele frequency of A by subtracting the allele frequency of a from 100%:

$$100\% - 62.5\% = 37.5\%$$

The allele frequencies of A and a provide a measure of genetic variation at one gene in a given population. It would be a simple matter to measure genetic variation in a population if we could always use observable traits, as we did with the pea plant example. By simply counting the individuals that displayed variant forms of a trait, we were able to determine a measure of the variation of that trait's gene. However, this approach works only rarely for two important reasons. The first reason is that many traits are encoded by a large number of genes. A good example is human height, which is influenced by variation in thousands of genes. In this and other similar cases, it is difficult, if not impossible, to make direct inferences from a phenotype to the underlying genotype. Even traits that seem to have a simple set of phenotypes often prove to have a complicated genetic basis. For instance, human skin color is determined by at least six different genes. The second reason that we often can't use phenotype as a measure of genotype is that the phenotype is a product of both the genotype and the environment, as we discussed earlier.

Until the 1960s, only one workable solution was available: limit population genetics to the study of phenotypes that are encoded by a single gene, like color in pea plants. As these phenotypes are relatively rare, the number of genes that population geneticists could study was extremely small. Another example is certain markings in invertebrates. For example, the coloring of the two-spot ladybug *Adalia bipunctata* shown in **FIGURE 42.5** is controlled by a single gene.

Genetic variation became much easier to measure in the 1960s with the application of gel electrophoresis, described in Unit 6. Gel electrophoresis separates segments of DNA according to their size. Before DNA technologies were developed, the same basic process was applied to proteins to separate them according to their electrical charge and their size. In gel electrophoresis, the proteins being studied migrate through a gel when an electrical charge is applied. The rate at which the proteins move from one end of the gel to the other is determined by their charge and their size.

Early studies of protein electrophoresis focused on enzymes that catalyze reactions that produce a dye when the substrate for the enzyme is added. If researchers add the substrate, they can see the locations of the proteins in the gel. The bands in the gel provide a visual picture of genetic variation in the population, revealing which alleles are present and what their frequencies are.

FIGURE 42.5 A single-gene trait

A genetic difference in color in the two-spot ladybug, *Adalia bipunctata*, results from variation in a single gene. Photos: (left) © Biopix: G Drange; (right) Howard Marsh/Shutterstock

"Practicing Science 42.2: How did gel electrophoresis allow us to detect genetic variation?" shows this sort of experiment. In this experiment, proteins from fruit flies are run on a gel. Each well of the gel is loaded with proteins from one individual. When an electrical current is applied, the proteins are separated by size and charge. The gel is then stained in such a way that a particular enzyme, called Adh, is visualized. The resulting bands give a picture of the genetic variation of the *Adh* gene in a population of eight fruit flies.

Practicing Science 42.2

How did gel electrophoresis allow us to detect genetic variation?

Background The introduction of protein gel electrophoresis in 1966 gave researchers the opportunity to identify differences in amino acid sequence in proteins both among individuals and, in the case of heterozygotes, within individuals. In gel electrophoresis, proteins with different amino acid sequences move at different rates through a gel in an electric field. Often, a single amino acid difference is enough to affect the mobility of a protein in a gel.

Method Researchers load material derived from crude tissue—in this case, the ground-up whole body of a fruit fly known as *Drosophila melanogaster*—on a gel. Each well of the gel (shown at the top of the gel) contains material from one fruit fly. Then the researchers turn on an electric current, placing the negative charge at the sample end and the positive charge at the other end of the gel. Proteins migrate toward the positive charge. The rate at which any given protein migrates depends on its size and its charge, both of which may be affected by its amino acid sequence.

To visualize the protein at the end of the gel run, researchers use a biochemical indicator that produces a stain when the protein of interest is active. In the experiment shown in the figure to the right, researchers used a biochemical indicator that detects the activity of a protein called alcohol dehydrogenase (Adh). A series of bands on the gel shows where the Adh proteins in each lane migrated.

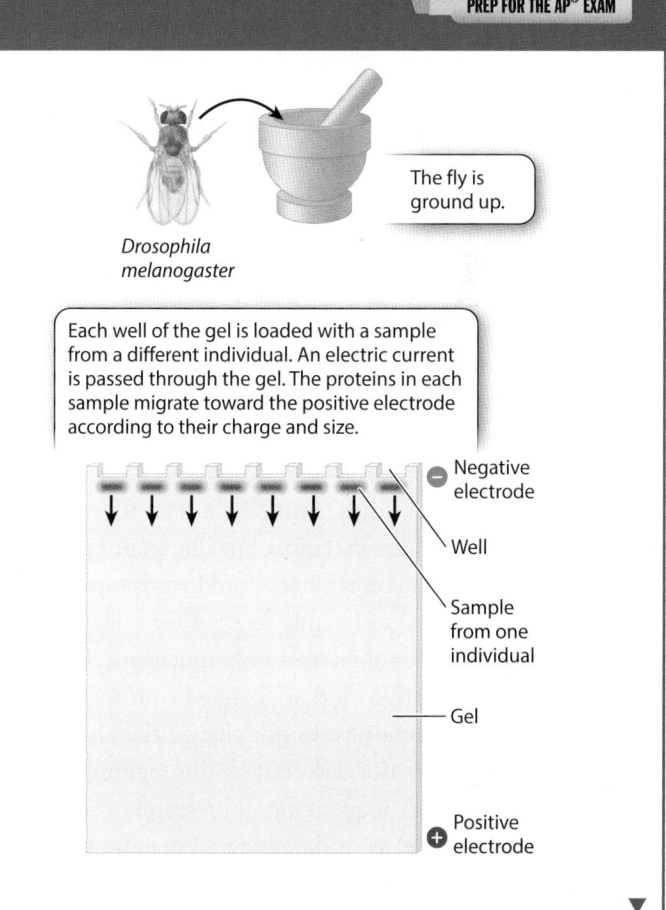

Drosophila melanogaster

The fly is ground up.

Each well of the gel is loaded with a sample from a different individual. An electric current is passed through the gel. The proteins in each sample migrate toward the positive electrode according to their charge and size.

— Negative electrode

— Well

— Sample from one individual

— Gel

— Positive electrode

Results The *Adh* gene has two common alleles, which are distinguished by a single amino acid difference that changes the charge of the protein. One allele, *Fast* (*F*), runs more quickly than the other, *Slow* (*S*). In the figure below, the gel shows that four individuals are *SS* homozygotes, two are *FF* homozygotes, and two are *FS* heterozygotes. Note that the *FS* heterozygotes do not stain as strongly on the gel because each band has half the intensity of the single band in the homozygotes.

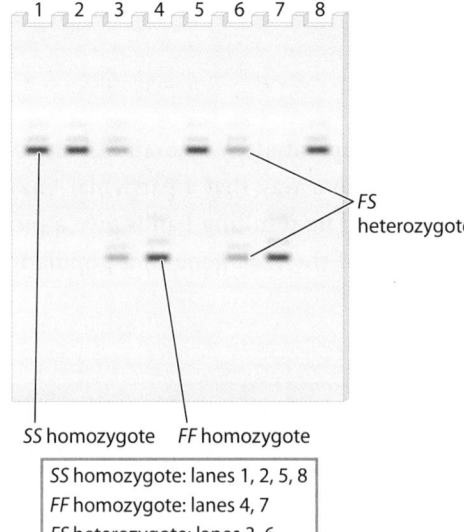

The gel is stained with a biochemical agent that produces color in the presence of Adh enzyme.

1 2 3 4 5 6 7 8

FS heterozygote

SS homozygote   *FF* homozygote

SS homozygote: lanes 1, 2, 5, 8
FF homozygote: lanes 4, 7
FS heterozygote: lanes 3, 6

We can measure the allele frequencies by counting the alleles. Each homozygote has two of the same allele, and each heterozygote has one of each.

Total number of alleles in the population of 8 individuals = $8 \times 2 = 16$

Number of *S* alleles in the population = $2 \times$ (number of *SS* homozygotes) + number of *FS* heterozygotes = $(2 \times 4) + 2 = 8 + 2 = 10$

Frequency of $S = \dfrac{10}{16} = \dfrac{5}{8}$

Number of *F* alleles in the population = $2 \times$ (number of *FF* homozygotes) + number of *FS* heterozygotes = $(2 \times 2) + 2 = 4 + 2 = 6$

Frequency of $F = \dfrac{6}{16} = \dfrac{3}{8}$

Note that the allele frequency of *S* and the allele frequency of *F* add to 1.

Conclusion The experiment provides a profile of genetic variation at this gene for these eight individual fruit flies. Population genetics involves comparing data such as these with data collected from other populations to determine the mechanisms shaping patterns of genetic variation.

Follow-up Work This technique is seldom used now because it is easy to recover much more detailed genetic information about genetic variation from DNA sequencing.

SOURCE
Lewontin, R. C., and J. L. Hubby. 1966. "A Molecular Approach to the Study of Genic Heterozygosity in Natural Populations: II. Amount of Variation and Degree of Heterozygosity in Natural Populations of *Drosophila pseudoobscura*." *Genetics* 54:595–609.

AP® PRACTICE QUESTION

a. **Identify** the following:
   1. The question the experimenters are trying to answer
   2. The hypothesis they tested
   3. The independent variable
   4. The dependent variable
b. **Explain** the results of the experiment by **identifying** the researchers' claim and **justifying** the claim with the evidence and the reasoning researchers used (CER).

Protein gel electrophoresis was a leap forward in our ability to detect genetic variation, but this technique had significant limitations. Researchers could study only enzymes because they needed to be able to stain for enzyme activity. The technique also could detect only mutations that resulted in amino acid substitutions that changed a protein's mobility in the gel. Some mutations do not change the charge or size of the protein, so are not detected by this technique.

Only with DNA sequencing did researchers finally have an unambiguous means of detecting all genetic variation in a stretch of DNA, whether in genes or not. The variations studied by modern population geneticists are differences in DNA sequence, such as an A rather than a G at a specified nucleotide position in a particular gene.

Calculating allele frequencies with DNA sequencing involves collecting a population sample and counting the number of occurrences of a given mutation. For example, researchers can look even more closely at the example of the *Drosophila Adh* gene identified in Practicing Science 42.2 to focus not on the amino acid difference between the *Fast* and *Slow* phenotypes, but on the A or G nucleotide difference underlying the two phenotypes.

If researchers sequence the *Adh* gene from 50 individual flies, they then have 100 gene sequences from these individuals because fruit flies are diploid and carry two copies of the *Adh* gene. If they find that 70 sequences have an A and 30 have a G at the position in question, the allele frequency of A is $\frac{70}{100} = 0.7$ and the allele frequency of G is 0.3. In general, in a sample of *n* diploid individuals, the allele frequency is the number of occurrences of that allele divided by twice the number of individuals in the sample.

The ability to determine allele frequencies gives scientists the ability to measure genetic variation. With this information in hand, they can then follow genetic variation over time. If they find that it changes, they can infer that evolution is occurring.

PREP FOR THE AP® EXAM

AP® EXAM TIP

Make sure you know how to calculate allele frequencies from genotype frequencies and genotype numbers; past exams have required this skill.

✓ **Concept Check**

8. **Describe** how population geneticists measure genetic variation.

9. **Describe** allele frequency.

10. **Calculate** the allele frequency if there are 25 copies of allele A in a total of 100 copies of alleles of that gene.

Module 42 Summary

PREP FOR THE AP® EXAM

REVISIT THE BIG IDEAS

EVOLUTION: Describe two processes that can contribute to genetic variation in a population.

LG 42.1 Variation is a notable feature of the natural world.

- Phenotypic variation is differences in the observable traits of a cell or organism. Page 587
- Phenotypic variation results from genetic variation and environmental variation. Page 587
- Genotypic or genetic variation is differences at the level of DNA sequences. Page 588

LG 42.2 Mutation and recombination are the two sources of genetic variation.

- Different forms (sequences) of genes are called alleles. Page 589
- Mutation and recombination are the two sources of genetic variation, but all genetic variation ultimately comes from mutation. Page 589
- Mutations are rare for a given base (nucleotide), but common across an entire genome. Page 590

- Mutations are random with respect to the needs of an organism. Page 590
- Mutations can be either somatic (in body tissues) or germ line (in gametes); germ-line mutations are the only ones that can be passed on to the next generation. Page 592
- Mutations can be neutral, deleterious, or advantageous. Page 592
- A point mutation is a change of one base for another. Page 593
- Small insertions (or deletions) in DNA add (or remove) one base or a few bases. Page 593
- A duplication is a region of the chromosome that is present two times, and a deletion is the loss of part of a chromosome. Page 593
- Transposable elements are DNA sequences that can jump from one place to another. Page 593

LG 42.3 Genetic variation is measured by allele frequencies.

- Patterns of genetic variation are described by allele frequencies. Page 593
- An allele frequency is the number of occurrences of a particular allele divided by the total number of all alleles of that gene in a population. Page 594

- In the past, population geneticists relied on observable traits determined by single genes and protein gel electrophoresis to measure genetic variation. Page 594

- DNA sequencing is now the standard technique for measuring genetic variation in a population of individuals. Page 597

Review Questions

1. All the alleles present in all individuals in a species are referred to as
   (A) allele frequency.
   (B) genotype frequency.
   (C) genotype.
   (D) gene pool.

2. Genetic variation has two sources: mutation and
   (A) recombination.
   (B) natural selection.
   (C) nonrandom mating.
   (D) cell division.

3. What kind of mutation has no effect on an organism's ability to survive or reproduce?
   (A) A deleterious mutation
   (B) A neutral mutation
   (C) An advantageous mutation
   (D) An adapted mutation

4. A population refers to a group of organisms that are in the same geographic area and that
   (A) are genetically related.
   (B) have the similar phenotype.
   (C) cannot interbreed.
   (D) are the same species.

5. Genetic variation in a population refers to
   (A) multiple individuals breeding together.
   (B) multiple alleles within a gene pool.
   (C) multiple phenotypes with the same genotype.
   (D) multiple genes within a genome.

6. Using visual observation of phenotypes to identify allele frequencies rarely works because
   (A) most alleles are dominant and will not produce a visible phenotype.
   (B) phenotype is not due to the alleles.
   (C) multiple genes may control the phenotype.
   (D) visual phenotypes are rare.

7. If a gene has two alleles, and allele A has a frequency of 83%, allele a has a frequency of:
   (A) 17%.
   (B) 117%.
   (C) 41.5%.
   (D) 7%.

Module 42
AP® Practice Questions

PREP FOR THE AP® EXAM

Section 1: Multiple-Choice Questions
Choose the best answer for questions 1–3.

1. The rate of mutation in human germ-line cells has been estimated to be approximately 1.3×10^{-8} per base pair per generation. The human genome size, consisting of a diploid complement of chromosomes, is approximately 6.4×10^9 base pairs. Which is the most likely number of new mutations present in a human baby but not present in either parent?
   (A) 166.4
   (B) 83.2
   (C) 41.6
   (D) 24.6

Use the following information to answer questions 2 and 3.

Threespine sticklebacks are small fish that sometimes have bony plates along their sides, which protect them from the bite of predators. Genetic variation in the *Eda* gene of threespine sticklebacks determines the amount of the bony plating. Fish that are homozygous for the Eda^{218} allele are completely plated, fish that are homozygous for the Eda^{158} allele are unplated, and fish that are heterozygous for the two alleles are partially plated. The figure shows data that researchers collected in a population of sticklebacks over several different years in a single place.

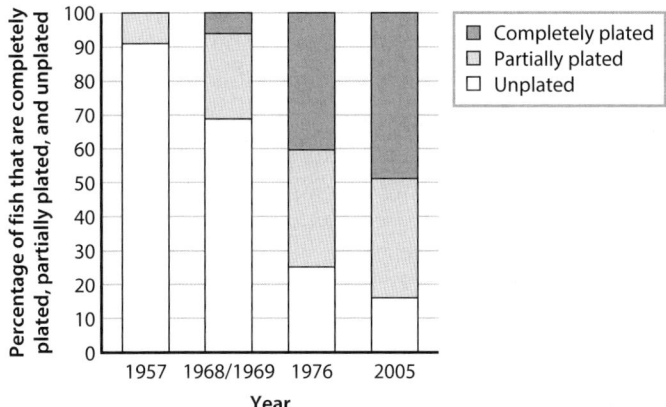

Data from Kitano, J., et al. 2008. "Reverse Evolution of Armor Plates in the Threespine Stickleback." *Cell* 18(10):769–774.

2. What percentage of threespine sticklebacks sampled in 1976 were completely plated?

    (A) 25%

    (B) 40%

    (C) 50%

    (D) 60%

3. What is the difference in the frequency of the of the Eda^{158} allele between 1957 and 2005?

    (A) 49%

    (B) 55%

    (C) 62%

    (D) 75%

Section 2: Free-Response Question

Write your answer to each part clearly. Support your answers with relevant information and examples. Where calculations are required, show your work.

Atlantic cod is a species of fish. Variation in the hemoglobin protein of cod can be observed with gel electrophoresis. Protein encoded by the Hbl^1 allele of the *hemoglobin* gene moves more quickly through a gel than the protein encoded by the Hbl^2 allele of the gene. A scientist collected 15 cod from the ocean near the east coast of Norway and 15 cod from the ocean near the west coast of Norway, purified the hemoglobin protein, and separated the hemoglobin proteins on a gel. Each well of the gel was loaded with the hemoglobin proteins of a single fish. The figure shows the data that were obtained.

East Coast of Norway

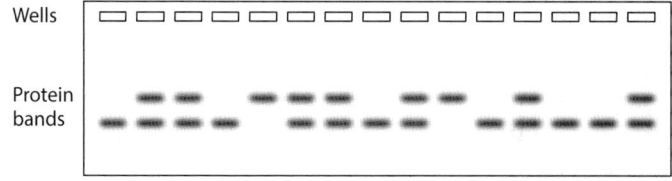

West Coast of Norway

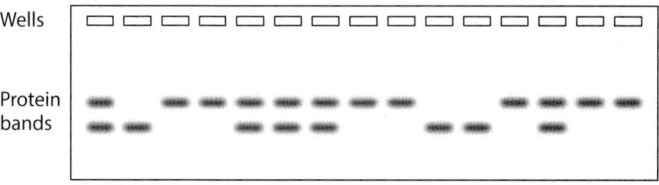

(a) **Identify** the sampling site with a higher frequency of heterozygotes for the *hemoglobin* gene.

(b) **Calculate** the allele frequency of the Hbl^1 allele in the sample from the west coast of Norway.

Module 43

Hardy–Weinberg Equilibrium

LEARNING GOALS ▶**LG 43.1** Evolution is a change in the genetic makeup of a population over time.

▶**LG 43.2** There are five mechanisms of evolution.

▶**LG 43.3** The Hardy–Weinberg equilibrium occurs when the genetic makeup of a population does not change over time.

In the last module, we focused on variation. Individuals of a species are different from one another. These observable differences result from both genetic and environmental variation. Genetic variation in turn results from mutation and recombination. We can measure genetic variation in a population by determining allele frequencies—the number of a specific allele out of the total number of alleles of a gene in the entire population.

When we have allele frequencies, we can then follow them over time. In other words, we can sample a population at one time and come back after several generations and sample it again. What will we find? We might find that allele frequencies have changed. This is evolution—the change in the genetic makeup of a population over time. Allele frequencies can change for a variety of reasons. These reasons are associated with different mechanisms of evolution. We have already discussed natural selection in Module 40. Natural selection is one of five mechanisms of evolution. In this module, we will take a closer look at the relationship between genetic change and evolution. We will consider the five mechanisms by which the genetic makeup of a population changes over time. Finally, we will look at conditions in which allele frequencies do not change.

PREP FOR THE AP® EXAM

FOCUS ON THE BIG IDEAS

EVOLUTION: Focus on the mechanisms or processes that can lead to changes in the genetic makeup of populations over time.

43.1 Evolution is a change in the genetic makeup of a population over time

Measuring and following changes in genetic variation over time is key to understanding the genetic basis of evolution. Genetic variation can be measured in two ways. One way is by determining the allele frequency of a particular gene. Recall that the allele frequency is the number of occurrences of a particular allele divided by the total number of occurrences of all the alleles of a gene in the population. We can also measure genetic variation by determining genotype frequency, which is the number of occurrences of a particular genotype divided by the total number of occurrences of all the genotypes for a gene in a population.

Let's consider a simple example. Red blood cells in your body, shown in **FIGURE 43.1**, have molecules on their surfaces that determine your blood type. One group of molecules goes by the name ABO blood group. The ABO blood group is the result of variation in a single gene, called the *ABO* gene. This gene has three alleles—*A, B,* and *O.* The allele frequency of allele *A* is the number of *A* alleles in the population divided by the total number of alleles of the *ABO* gene. For example, if there are 25 copies of *A* in a population, and the total number of alleles *A, B,* and *O* is 100, then the frequency of allele *A* is 25 divided by 100, or $\frac{25}{100} = \frac{1}{4} = 0.25$. This provides one way to measure genetic variation in the ABO blood group.

Although there are three alleles of the *ABO* gene in the population, an individual can only have at most two different alleles because humans are diploid, meaning that they have two copies of each chromosome with one allele on each chromosome. So, there are six possible genotypes—*AA, AO, BB, BO, AB,* and *OO.* To calculate the genotype frequency,

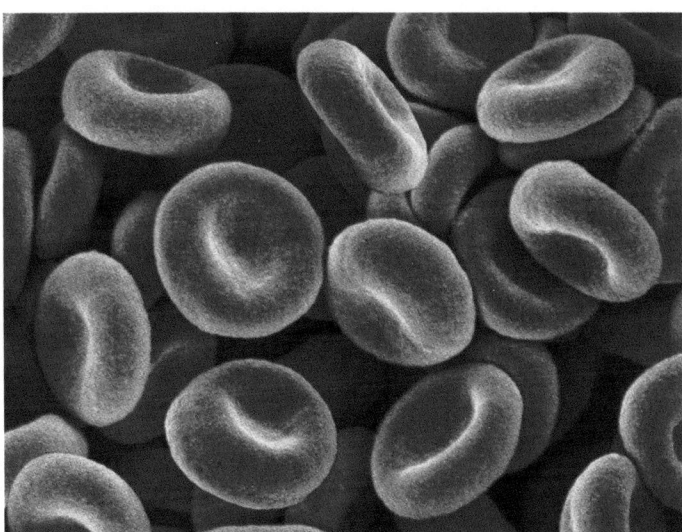

FIGURE 43.1 Red blood cells

Red blood cells have molecules on their surface that determine blood type. Photo: Micro Discovery/Getty Images

we divide the number of people with one of the genotypes by the total number of people. So, if 10 people are *AA* in a population of 100 people, then the genotype frequency of *AA* is 10 divided by 100, or $\frac{10}{100} = \frac{1}{100} = 0.1$.

The phenotype is blood type. Blood type A consists of people who have the genotypes *AA* or *AO;* blood type B consists of people who have the genotypes *BB* or *BO;* blood type AB is genotype *AB;* and blood type O consists of people who have the genotype *OO*. So, for the *ABO* gene, there are three alleles, six genotypes, and four phenotypes. **TABLE 43.1** summarizes the ABO blood group.

When we have a measure of genetic variation, we can follow it over time. At the genetic level, evolution is a change in the frequency of an allele or a genotype from one generation to the next. For example, if there are 200 copies of an allele that causes blue eye color in a population in generation 1, and there are 300 copies of that allele in a population of the same size in generation 2, evolution has occurred.

Evolution is a change in the genetic makeup of a population over time. In principle, evolution may occur when allele frequencies change, genotype frequencies change, or both allele and genotype frequencies change. For instance, in the example of blood types, we might see a change in the *A, B,* and *O* allele frequencies over time, which is evolution. Because genotypes are made up of two alleles (*AA, AO, BB, BO, AB, OO*), a change in allele frequency will also cause a change in genotype frequencies. Furthermore, even if the *A, B,* and *O* allele frequencies stay the same from one generation to the next, the frequencies of the different genotypes may change. This would be evolution without allele frequency change.

Note that this definition of evolution makes it clear that populations rather than individuals evolve. In other words, if we consider a population of individuals at one time and compare it to a population at a later time, we may see a change in allele frequencies, genotype frequencies, or both. Any of these changes in the genetic makeup of a population represents evolution. Individuals may also change over time, such as the changes that you experience as you grow up and develop, but this type of change is not considered evolution.

The definition of evolution does not specify a mechanism for this change. As we will see, several mechanisms can cause allele or genotype frequencies to change. Regardless of which mechanisms are involved, any change in allele frequencies, genotype frequencies, or both constitutes evolution.

TABLE 43.1 The ABO Blood Group

Phenotype	Genotype
A	*AA* or *BO*
B	*BB* or *BO*
AB	*AB*
O	*OO*

✓ Concept Check

1. **Describe** the genetic basis for evolution.

2. A particular gene has two alleles denoted C and c. **Identify** all the possible genotypes.

3. Assume there are 30 C alleles and 20 c alleles at a particular gene in a population. **Calculate** the frequency of C in the population.

43.2 There are five mechanisms of evolution

We have seen that evolution is a change in allele frequency, genotype frequency, or both in a population over time. What causes these frequencies to change from one generation to the next? There are five ways that they can change, and therefore five mechanisms of evolution. Natural selection is one of them, but not the only one. However, natural selection is the only mechanism that causes both evolution and adaptation. In this section, we will examine the five mechanisms of evolution, beginning with natural selection.

Natural Selection

In Modules 40 and 41, we discussed evolution by natural selection. In natural selection, some individuals survive and reproduce more than others. For example, in a population of birds, some may have lighter-colored feathers than others, as shown in **FIGURE 43.2**. If the lighter birds survive and reproduce more than others in a particular environment, the population will consist of more lighter-colored individuals over time.

Natural selection causes allele and genotype frequencies to change over time. Let's consider a gene with two alleles,

FIGURE 43.3 A camouflaged katydid

This katydid looks like a dead leaf, providing camouflage and protection against predators. Adaptations such as this one evolve by natural selection. Photo: Atelopus/Getty Images

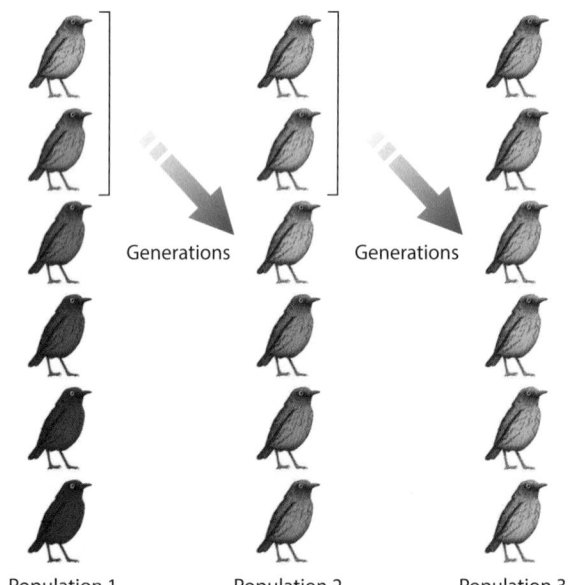

Generations   Generations

Population 1   Population 2   Population 3

FIGURE 43.2 Natural selection

In this population of birds, there is variation in color, with some having lighter feathers and some having darker feathers. The lighter birds survive and reproduce more than the darker birds in a particular environment. As a result, the population of birds becomes lighter over time.

A and *a*. Now let's imagine that the alleles affect the ability of individuals to survive. For example, the allele *a* could be a recessive lethal mutation. In other words, all individuals of genotype *aa* die. In every generation, then, there is a selective elimination of *a* alleles, meaning that the frequency of *a* will gradually decline and the frequency of *A* will correspondingly increase over the generations.

As we discussed earlier, the individuals carrying these alleles have different levels of fitness, the ability to survive and reproduce in a particular environment. In this case, *AA* and *Aa* individuals survive and reproduce well, but *aa* individuals don't. Because there is differential success of these alleles based on differences in the fitness of individuals, we call this mechanism of evolution natural selection.

Natural selection is evolution's major driving force, preferentially passing on to each successive generation those mutations that best fit organisms to their environments. As a result, they lead to impressive adaptations, like the camouflage of the katydid shown in **FIGURE 43.3**. This katydid is almost indistinguishable from a dead leaf, providing protection against predators. However, natural selection is not the only evolutionary mechanism. Other mechanisms can cause allele and genotype frequencies to change as well. Unlike natural selection, however, they do not lead to a population being better adapted to their environment or to a population being more fit.

Genetic Drift

Allele and genotype frequencies can also change over time simply as a result of chance. **Genetic drift** is the random change in allele frequencies from generation to generation. By "random," we mean that frequencies go up or down simply by chance. Let's go back to our example of birds with light and dark feathers. Let's imagine that a sudden storm eliminates much of a bird population, leaving by chance birds with an intermediate feather color, as shown in **FIGURE 43.4**. The survivors will breed and the population will end up being an intermediate color. The survival of the intermediate-colored birds had nothing to do with their color; they survived simply by chance.

In the last section, we considered what happens to beneficial and harmful mutations under the influence of natural selection. What about neutral mutations that are neither beneficial nor harmful? Natural selection, by definition, does not affect what happens to neutral mutations. Consider a neutral mutation, *m,* which has no effect on survival and reproduction (fitness). At first, it is present in just a single individual. What happens if that individual fails to reproduce for reasons unrelated to *m?* In this case, *m* is lost from the population, but not by natural selection because *m* is neutral and natural selection has selected neither for nor against *m.*

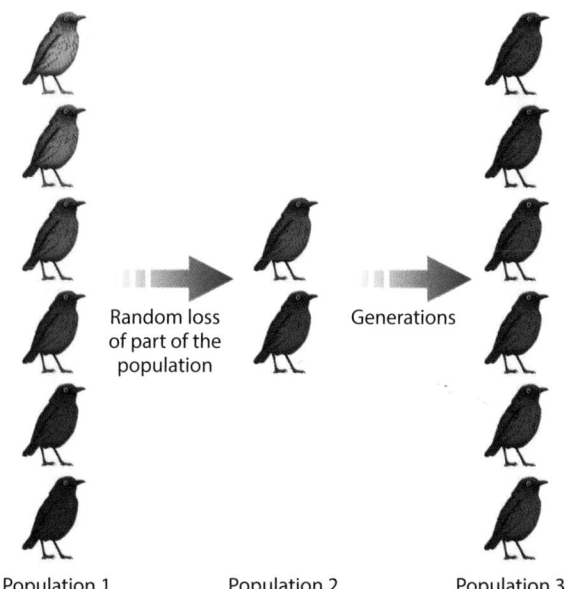

FIGURE 43.4 Genetic drift

Genetic drift occurs when allele and genotype frequencies change by chance. In this example, a sudden storm eliminates much of the population, leaving by chance intermediate-colored birds. These birds go on to breed, resulting in a population of intermediate-colored birds.

Rather, the loss of *m* is due to chance, so it represents a case of genetic drift. Alternatively, the *m*-bearing individual might by chance leave many offspring, again for reasons unrelated to *m.* In this case, the frequency of *m* increases due to genetic drift.

While natural selection only acts on beneficial and harmful alleles, but not neutral alleles, genetic drift acts on all alleles regardless of their effects on fitness because it is a random process. Like natural selection, genetic drift leads to allele frequency changes and, therefore, evolution. Unlike natural selection, however, genetic drift does not lead to adaptations because the alleles whose frequencies are changing do not affect an individual's ability to survive or reproduce.

Now let's consider two situations in which we commonly see genetic drift in action. Consider a rare allele, *A,* with a frequency of $\frac{1}{1000}$. Imagine that habitat destruction reduces the population to just one pair of individuals, and one of those, by chance, is a heterozygote carrying *A.* The frequency of *A* in this new population is $\frac{1}{4} = 0.25$ because each individual has two alleles, giving a total of four alleles. So, we see that there was a large change in allele frequency following the habitat destruction event, from $\frac{1}{1000}$ to $\frac{1}{4}$.

When a large population is reduced to just a few individuals, we say the population experiences a **bottleneck,** as in the example of the birds shown in Figure 43.4. Genetic drift often results from population bottlenecks. A population of Galápagos tortoises (*Chelonoidis*) probably went through such a bottleneck about 100,000 years ago when a volcanic eruption eliminated most of the tortoises' habitat. One of these tortoises is shown in **FIGURE 43.5** on page 604. As a result, allele frequencies likely changed, just as in the example above.

Genetic drift also commonly occurs by a *founder effect.* A **founder effect** occurs when a few individuals from a larger population establish a new population. Such events occur, for example, when a small number of individuals arrive on an island from a nearby continent and colonize it, as illustrated in **FIGURE 43.6** on page 604. We also see founder effects in the human population when small groups of people leave one area and set up a population elsewhere. Once again, allele frequencies are randomly changed compared to the parent population.

With both a bottleneck effect and a founder effect, the allele frequency of the new population is different from the original population due to genetic drift. In addition, there is typically a large loss of genetic variation that occurs in both processes. The new, smaller population typically contains

FIGURE 43.5 A Galápagos tortoise

A population of Galápagos tortoises underwent a population bottleneck after a volcanic eruption, resulting in genetic drift and reducing genetic variation. Photo: James Morris

just a subset of the alleles present in the original population. As a result, the original population and the new population may end up being quite different from each other genetically.

Genetic drift occurs in all finite populations, but its impact depends on population size. Genetic drift acts

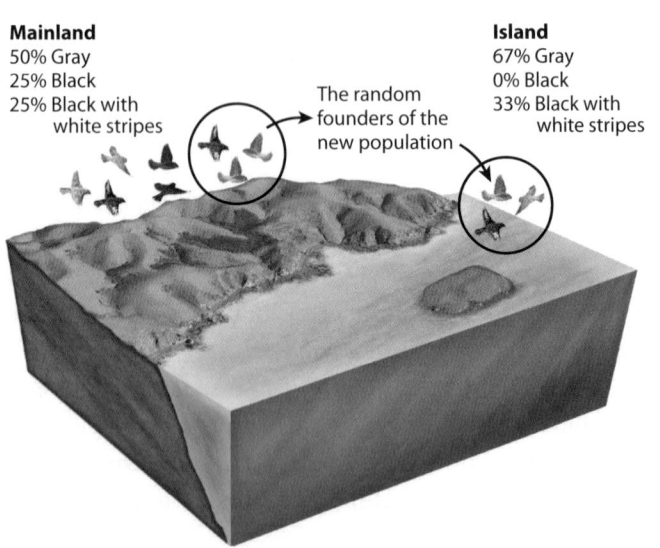

Mainland
50% Gray
25% Black
25% Black with
  white stripes

The random founders of the new population

Island
67% Gray
0% Black
33% Black with
  white stripes

FIGURE 43.6 Founder effect

A founder effect occurs when a few, random individuals of a larger population colonize a new area, such as the island shown here. The result is that the new population and parent population have different allele frequencies.

strongly in small populations and weakly in large populations. The reason is that random effects have more of an impact on small populations than on large ones. If you flip a coin 1000 times, you expect approximately 50% heads and 50% tails. However, if you flip a coin four times, you wouldn't be surprised if you got three heads and one tail, which is 75% heads and 25% tails. In other words, in small samples, chance plays a bigger role than in large ones. The same is true of genetic drift: it is likely to be much more significant in small populations than in large ones.

FIGURE 43.7 uses four computer simulations to show the effect of population size on allele frequencies subject to genetic drift. In Figure 43.7a, the population size is very small, with just 4 individuals. At the start of the simulation, the frequency of a particular allele is 0.5. The change in allele frequency that is seen in each generation is just the result of chance. After a few generations, the frequency increases to 1.0, meaning that the allele is fixed in the population and all members of the population have it. The frequency of the allele changes very quickly because the population is so small.

In Figure 43.7b, the same simulation is run six times, indicated by the blue lines. Note that the blue lines are all different, meaning that genetic drift follows different paths each time. This is because genetic drift is random. Note too that sometimes the allele is fixed (frequency = 1) and sometimes it is lost (frequency = 0). This is another feature of genetic drift—alleles can be fixed or lost randomly.

Figure 43.7c and Figure 43.7d show larger populations of 40 and 400 individuals, respectively. In each figure, the simulation is again run six times. We see the same patterns in these two graphs—populations take random paths, and alleles are fixed or lost—but the process is slower in these larger populations than in the smaller population shown in Figure 43.7b. So, we see that genetic drift acts strongly in small populations and less strongly in large populations.

Migration

Natural selection and genetic drift are two mechanisms of evolution. A third mechanism is *migration*. **Migration** is the movement of individuals from one population to another, as shown in **FIGURE 43.8**. In this example, there is a population of light-colored birds and a separate, nearby population of dark-colored birds of the same species. If some dark-colored birds migrate into the territory of the of the light-colored bird population, the genetic makeup of the population of light-colored birds will change.

a. Population size = 4

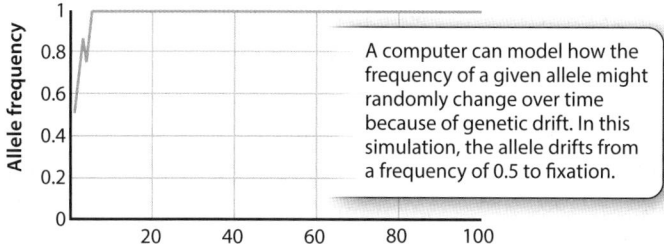

A computer can model how the frequency of a given allele might randomly change over time because of genetic drift. In this simulation, the allele drifts from a frequency of 0.5 to fixation.

b. Population size = 4

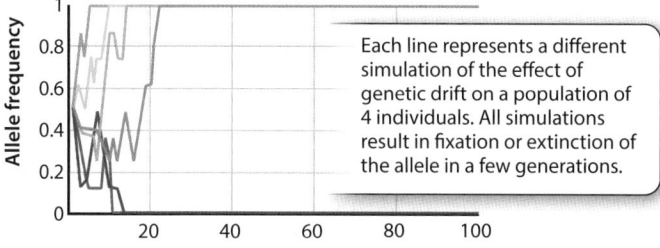

Each line represents a different simulation of the effect of genetic drift on a population of 4 individuals. All simulations result in fixation or extinction of the allele in a few generations.

c. Population size = 40

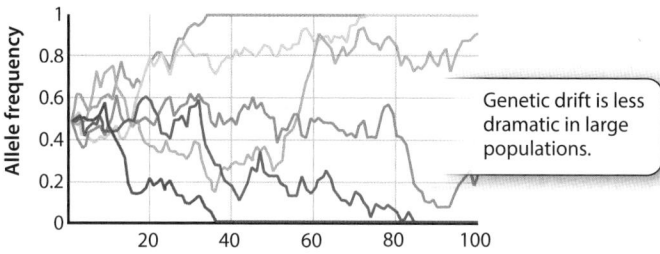

Genetic drift is less dramatic in large populations.

d. Population size = 400

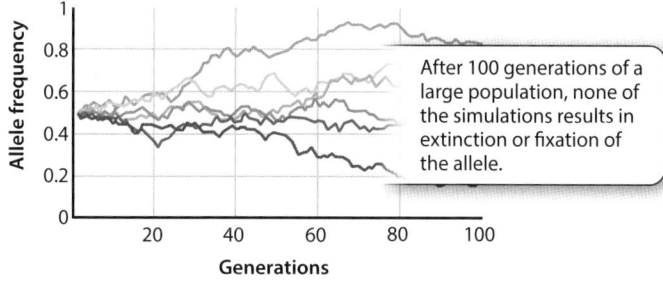

After 100 generations of a large population, none of the simulations results in extinction or fixation of the allele.

Generations

FIGURE 43.7 The effect of population size on genetic drift

Four computer simulations are shown. In each case, the starting allele frequency is 0.5, but the size of the population varies from 4 (a and b) to 40 (c) to 400 (d). In all simulations, the allele is fixed (frequency = 1) or lost (frequency = 0), but the process is more rapid in the smaller populations compared to the larger ones.

Migration causes **gene flow,** the movement of alleles from one population to another. To see how migration and gene flow can lead to changes in allele frequencies, let's consider two populations. The first population contains only A alleles and the second contains a mix of A and a

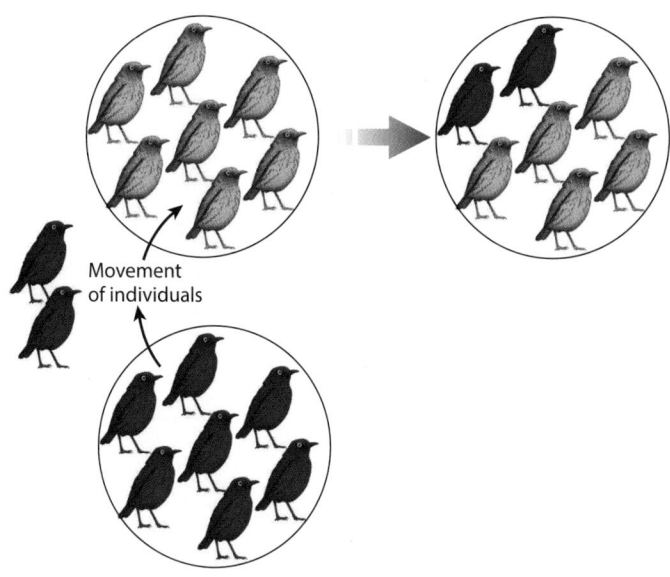

Movement of individuals

FIGURE 43.8 Migration

Migration is the movement of individuals between populations. Consider two populations of birds all of one species—one population with light-colored birds, and another population with dark-colored birds. Some of the dark-colored birds move into the population of light-colored birds, which in turn changes the genetic makeup of the population of light-colored birds.

alleles. If there is a sudden influx of individuals from the second population into the first, the frequency of A in the first population will decline in proportion to the number of immigrants as a alleles enter the population. In other words, we see a change of allele frequency in the first population due to the migration of individuals and genes into that population.

Migration typically makes populations more similar to each other, reducing the genetic differences between them. Consider two isolated island populations of rabbits, one brown and the other black. Now imagine that the isolation breaks down—a bridge is built between the islands—and migration occurs between the two populations. Over time, black alleles enter the brown population, and vice versa, and the allele frequencies of the two populations gradually become the same.

Migration is sometimes used in conservation efforts as a way to increase genetic variation in endangered or threatened species, such as the Arctic fox (*Vulpus lagopus*) shown in **FIGURE 43.9** on page 606. In the late nineteenth century, hunting significantly reduced the population of Arctic foxes in Scandinavia. This is an example of a population bottleneck, which reduced both the size and genetic variation of the population of foxes. As part of an effort to restore the

FIGURE 43.9 An Arctic fox

A population of Artic foxes was reduced to low levels in Scandinavia due to hunting. As one part of an effort to restore the population, foxes were moved from a captive breeding program into the threatened population, thereby restoring the health and genetic variation of the population. Migration between populations is a mechanism of evolution because it changes allele frequencies.

Photo: Daniel Parent/Getty Images

population, foxes from a captive breeding program were introduced into the area. This assisted migration increased the size and genetic variation of the population and helped the population to recover.

Mutation

Mutation is a fourth mechanism of evolutionary change. As we have seen, mutation is the source of new alleles and the raw material on which the other mechanisms of evolution act. Without mutation, there would be no genetic variation and no evolution. In addition, mutation itself is a mechanism of evolution. Going back to our example of a population of birds, a mutation that causes yellow wing bands might occur in an individual, as shown in **FIGURE 43.10**. This new mutation in the population introduces genetic variation and changes the genetic makeup of the population.

From a genetic perspective, if *A* alleles mutate into *a* alleles, then we will see changes in the allele frequencies over the generations. However, because mutation is so rare, it has a very small effect on changing allele frequencies on the timescales studied by population geneticists. Mutation has a significant effect on populations in the long term but does

not typically have a large effect on an entire population in the short term.

Nonrandom Mating

A fifth way that a population can evolve is by mating nonrandomly. In our population of birds that vary in color, if light birds tend to mate with light birds, and dark birds tend to mate with dark birds, they are mating nonrandomly with respect to feather color, as shown in **FIGURE 43.11**.

In random mating, individuals select mates without regard for genotype. In **nonrandom mating**, by contrast, individuals preferentially choose mates according to their genotypes. The result is that certain genotypes increase and others decrease in frequency. For example, in random mating, an *AA* homozygote, when offered a choice of mate from among *AA, Aa,* or *aa* individuals, chooses at random. In contrast, in nonrandom mating, *AA* homozygotes might preferentially mate with other *AA* homozygotes, and *aa* homozygotes might preferentially mate with other *aa* homozygotes.

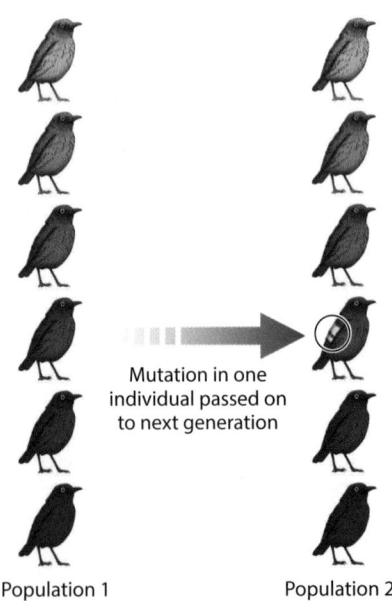

Mutation in one individual passed on to next generation

Population 1   Population 2

FIGURE 43.10 Mutation

Mutation is both the source of genetic variation and a mechanism of evolution. In this example, a mutation occurs in one individual that results in yellow bands on the wings.

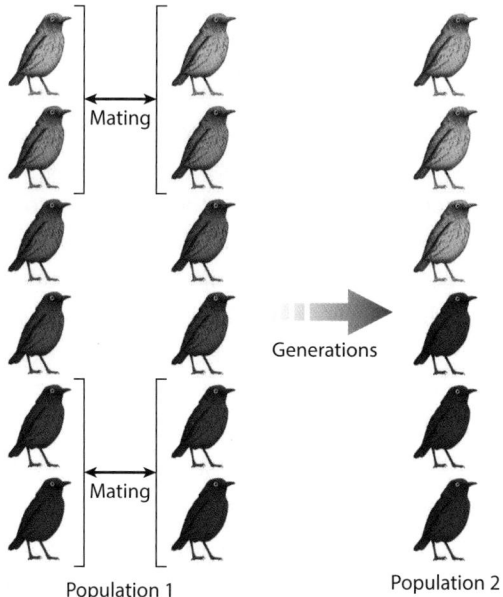

FIGURE 43.11 Nonrandom mating

Nonrandom mating occurs when individuals do not mate randomly. In this example, light-colored birds mate with each other, and dark-colored birds mate with each other.

Nonrandom mating rearranges alleles already in a population's gene pool. As a result, genotype frequencies change. Unlike migration or mutation, nonrandom mating does not add new alleles to the population. As a result, allele frequencies do not change. In this way, nonrandom mating is unique among the mechanisms of evolution in that it affects genotype frequencies from generation to generation, but does not affect allele frequencies. As we saw, the other four mechanisms change allele frequencies (as well as genotype frequencies).

An example of nonrandom mating that we see in some populations is inbreeding. Inbreeding occurs when there is mating between close relatives. Consider a rare allele b that has a frequency of 0.001 in a population. In random mating, the expected frequency of a bb homozygote is $0.001 \times 0.001 = 0.000001$.

What happens to the frequency of bb homozygotes when there is inbreeding? Suppose a male with one b allele has two offspring—one male and one female. The probability that the male offspring inherited the allele from his father is 0.5, and the same is true for the female offspring. Now suppose these offspring mate with each other. The probability of a bb homozygote from this mating is the probability each sibling inherited a copy from their father (0.5×0.5) multiplied by the probability that their offspring inherited b from both of them (0.5×0.5), or a total probability of $(0.5)^4 = 0.0625$.

Clearly, 0.0625 is a considerably higher frequency than 0.000001. So, we see that the frequency of homozygotes increases in nonrandom mating compared with random mating. As the frequency of homozygotes increases, the frequency of heterozygotes decreases because genotypes can only be heterozygous or homozygous. This is significant from an evolutionary point of view because rare recessive mutations that are present in the population have a higher probability of being homozygous and therefore expressed when inbreeding occurs. As some rare recessive mutations are harmful (for example, they may cause disease when homozygous), they can in turn affect the fitness of the population.

✓ Concept Check

4. **Identify** the five mechanisms of evolution.

5. **Identify** the mechanism of evolution that leads to adaptations.

6. **Describe** the process by which adaptations evolve.

7. **Identify** and **describe** the mechanism of evolution that causes genotype frequencies to change, but not allele frequencies.

43.3 The Hardy–Weinberg equilibrium occurs when the genetic makeup of a population does not change over time

Allele and genotype frequencies change over time only if specific mechanisms act on the population. These mechanisms are natural selection, genetic drift, migration, mutation, and nonrandom mating. If these mechanisms are not acting, then allele and genotype frequencies don't change. This principle, which was demonstrated independently in 1908 by both the English mathematician G. H. Hardy and the German physician Wilhelm Weinberg, has become known as the **Hardy–Weinberg equilibrium.** In this section, we will describe the conditions in which allele and genotype frequencies do not change. We will also see how the Hardy–Weinberg equilibrium provides a useful null hypothesis for studies about evolution.

The Five Conditions of the Hardy–Weinberg Equilibrium

In essence, the Hardy–Weinberg equilibrium describes the situation in which evolution does not occur. When none of the five mechanisms of evolution are acting, evolution does not occur. The conditions or assumptions of the Hardy–Weinberg equilibrium are therefore the following:

1. There is no difference in survival and reproductive success among individuals. In other words, selection is not operating.
2. The population is infinitely large so the role of chance or genetic drift is eliminated.
3. Populations are not added to or subtracted from by migration.
4. There is no mutation.
5. Individuals mate at random.

According to the Hardy–Weinberg equilibrium, the allele and genotype frequencies of a population don't change when these five conditions are met. If allele and genotype frequencies do not change, we can conclude that evolutionary mechanisms are not acting in the population we are studying. In many ways, then, the Hardy–Weinberg equilibrium provides a useful baseline or starting point for evolutionary studies. We can begin by testing whether a population is at Hardy–Weinberg equilibrium. If it is at Hardy–Weinberg equilibrium, then the population is not evolving. If it is not at Hardy–Weinberg equilibrium, then we can conclude that one or more of the conditions are not met and evolutionary mechanisms are at work. This is just a starting point, as the next step would be to figure out which of the five evolutionary mechanisms is acting on the population under study.

Relationship between Allele and Genotype Frequencies

As we have seen, when the five conditions of the Hardy–Weinberg equilibrium are met, the population is not evolving. Furthermore, under these conditions, we are able to calculate genotype frequencies from allele frequencies. Let's return to the example of Mendel's peas that we looked at in Unit 5. In this example, there are two alleles, *A* and *a,* that determine pea color. In a population of pea plants, we find 70 *A* alleles and 30 *a* alleles. What are the genotype frequencies? In other words, how many *AA* homozygotes, *Aa* heterozygotes, and *aa* homozygotes are there in the population? If the population is not in Hardy–Weinberg equilibrium, it is not possible to figure out genotype frequencies from allele frequencies. However,

if the population is in Hardy–Weinberg equilibrium, we can determine genotype frequencies from allele frequencies.

Recall that random mating is a condition that must be met if a population is in Hardy–Weinberg equilibrium. Random mating is the equivalent of putting all the population's gametes into a single pot and drawing out pairs of them at random to form zygotes. This is the same principle we saw in action in the discussion of independent assortment in Module 27. If there are 70 *A* alleles and 30 *a* alleles and mating is random, what is the probability of picking an *AA* homozygote? That is, what is the probability of picking an *A* allele followed by another *A* allele? The probability of picking an *A* allele is the same as the *A* allele's frequency in the population. Therefore, the probability of picking the first *A* is $\frac{70}{100} = 0.7$. What is the probability of picking the second *A*? Also 0.7. The probability of picking an *A* followed by another *A* is the product of the two probabilities: $0.7 \times 0.7 = 0.49$. In other words, there is a 49% probability of picking an *AA* homozygote. That means that the frequency of an *AA* genotype is 0.49.

We can follow the same steps to determine the genotype frequency for the *aa* genotype. The probability of picking the first *a* is $\frac{30}{100} = 0.3$. The probability of picking a second *a* allele is also 0.3. The probability of picking an *a* followed by another *a* is the product of the two probabilities: $0.3 \times 0.3 = 0.09$. So, there is a 9% probability of picking an *aa* homozygote. That means the frequency of an *aa* genotype is 0.09

What about the genotype frequency of the heterozygote, *Aa?* If mating is random and a population is in Hardy–Weinberg equilibrium, what is the probability of an *Aa* heterozygote? We compute this value by determining the probability of drawing *A* followed by *a,* or *a* followed by *A.* In other words, there are two different ways in which we can generate the heterozygote. Its frequency is therefore $(0.7 \times 0.3) + (0.3 \times 0.7) = 0.42$, so there is a 42% probability of picking an *Aa* heterozygote. Note that the frequencies of the three genotypes (*AA, Aa, aa*) sum to 1 because there are only three genotypes in the entire population.

We can generalize these calculations by substituting letters for the numbers. Let the allele frequency of one allele, *A,* equal *p,* and the frequency of the other allele, *a,* equal *q.* Because there are no other alleles at this gene, the allele frequencies must add to 1 (or 100%). Therefore, we can write the following equation:

$$p + q = 1$$

We can also determine genotype frequencies given a set of allele frequencies. If the frequency of A is *p,* then the frequency of the AA genotype is $p \times p$ or p^2. If the frequency of a is q, then the frequency of the aa genotype is $q \times q$ or q^2. Finally, the frequency of the Aa genotype is $2pq$. These relationships are summarized here:

Frequency of $AA = p^2$

Frequency of $Aa = 2pq$

Frequency of $aa = q^2$

Because there are no other genotypes in the population, the genotype frequencies must also add to 1 (or 100%). Therefore, we can write the following equation:

$$p^2 + 2pq + q^2 = 1$$

These relationships are summarized in **TABLE 43.2**.

The relationship between allele and genotype frequencies when a population is in Hardy–Weinberg equilibrium can also be shown visually as shown in **FIGURE 43.12**. In this figure, the allele frequencies in the population are shown along the side and across the top. The genotype frequencies are indicated by the area of the squares. So, we see that the area of the square AA is $p \times p$ or p^2; the area of the square aa is $q \times q$ or q^2; and the area of the square Aa is pq, but because there are two Aa squares, the total area is $2pq$.

Not only does the Hardy–Weinberg relation predict genotype frequency from allele frequencies, but it works in reverse too: genotype frequencies can be used to determine allele frequencies. Knowing the genotype frequency of aa, for example, permits us to calculate allele frequencies. Let's return to our pea example. In that case, we found that 9% of the population has genotype aa. Therefore,

$$p^2 = 0.09$$

$$p = \sqrt{p^2} = \sqrt{0.09} = 0.3$$

And, because

$$p + q = 1$$

then

$$p = 1 - q = 1 - 0.3 = 0.7$$

We can also illustrate the relationship between allele and genotype frequencies by the graph shown in **FIGURE 43.13** on page 610. In this graph, allele frequencies are shown on the x-axis along the bottom. Genotype frequencies are plotted on the y-axis; and the three colored lines represent the frequency of the three genotypes AA, Aa, and aa for all allele frequencies.

For example, if no a alleles are present, then $q = 0$. This is the far right side of the x-axis. At this point along the

TABLE 43.2 Relationship Between Allele and Genotype Frequencies Under the Hardy–Weinberg Equilibrium

Alleles	A	a	
Allele frequencies	p	q	
Genotypes	AA	Aa	aa
Genotype frequencies	p^2	$2pq$	q^2

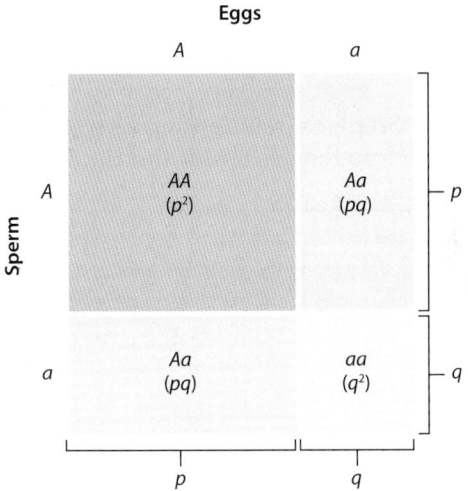

FIGURE 43.12 Visual relationship between allele and genotype frequencies under the Hardy–Weinberg equilibrium

This figure shows the frequency of A and a by the length along the side or top, and the frequency of the three possible genotypes by the area of the squares.

graph, we can see that $p = 1$. This makes sense, as $p + q = 1$. To find the genotype frequencies, we look at the colored lines in the graph along the line where $p = 1$ and $q = 0$ (far right side of the graph). Along this line, we see that the blue line representing the frequency of AA is at 1, and the red line representing the frequency of aa and the purple line representing the frequency of Aa are both at 0. In other words, when there are no a alleles, the frequency of AA is 1, the frequency of Aa is 0, and the frequency of $aa = 0$.

The graph in Figure 43.13 works in reverse as well: we can use it to determine allele frequencies given a set of genotype frequencies. For example, the purple line represents the frequency of heterozygotes Aa. When the purple line is

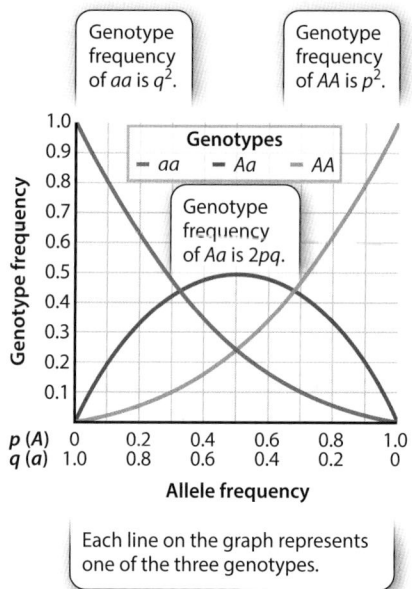

Each line on the graph represents one of the three genotypes.

FIGURE 43.13 Graphical relationship between allele and genotype frequencies under the Hardy–Weinberg equilibrium

This graph shows the frequency of A and a along the horizontal x-axis, the genotype frequency along the vertical y-axis, and the resulting three genotype frequencies for all allele frequencies by the colored lines.

0.5, we can read the corresponding place along the x-axis to determine the allele frequencies. At this place, we find that the allele frequencies for p and q are both 0.5.

Note that these relationships hold only if the Hardy–Weinberg conditions are met. If not, then we cannot figure out genotype frequencies from allele frequencies. "Analyzing Statistics and Data: The Hardy–Weinberg Equilibrium" reviews the math and provides a practice problem.

The Hardy–Weinberg Equilibrium as a Null Hypothesis

Recall the definition of evolution: a change in the genetic makeup (allele or genotype frequency) from one generation to the next. Given this definition, it might seem odd that we have spent so much time discussing factors necessary for allele frequencies to stay the same. Why study the Hardy–Weinberg equilibrium at all?

The Hardy–Weinberg equilibrium not only provides a means of converting between allele and genotype frequencies but also serves as a null hypothesis for evolutionary studies. We discussed null hypotheses in Module 0. In science, we start by making observations of some aspect of the

natural world. Observations often lead to questions. Scientists then propose a hypothesis, or a tentative answer to the question or explanation of what they observe. Hypotheses are useful because they allow us to make predictions that can be tested by experiments or additional observations.

Often, it is also useful to also propose a null hypothesis, which is a hypothesis that states that nothing is occurring. For example, if the researcher performs a treatment or changes conditions in some way, the null hypothesis states that the treatment or change has no effect at all on whatever is being studied. In evolutionary biology, the Hardy–Weinberg equilibrium can serve as a null hypothesis. It states that the five mechanisms of evolution are not acting on the population. In essence, it proposes that nothing interesting is happening in a population from an evolutionary viewpoint.

By contrast, when we find a population whose allele or genotype frequencies are not in Hardy–Weinberg equilibrium, we can infer that evolution has occurred in that population. In other words, we can reject the null hypothesis that evolution is not occurring, which in turn supports the alternative hypothesis that evolution is occurring. With further study of the population, we can consider, for the gene in question, whether the population is subject to one of the five mechanisms of evolution: natural selection, genetic drift, migration, mutation, or nonrandom mating. Thus, the Hardy–Weinberg equilibrium gives us a baseline from which to explore the evolutionary processes affecting populations.

AP® EXAM TIP

You should be familiar with the Hardy–Weinberg equilibrium and its use. Make sure you are able to use it to test whether a population is undergoing evolution. This may be done by noting if allele and genotype frequencies of a population change from time 1 to time 2, or by determining whether the Hardy–Weinberg equilibrium conditions are met.

✓ Concept Check

8. **Describe** how you would determine genotype frequencies given two alleles at frequencies p and q for a population at Hardy–Weinberg equilibrium.

9. **Describe** what it means if a population is not in Hardy–Weinberg equilibrium.

The Hardy–Weinberg Equilibrium

The Hardy–Weinberg equilibrium is a method used to compare allele and genotype frequencies to determine if the process of evolution has or has not occurred. There are five conditions that must be in place in order for allele frequencies not to change over time, and evolution not to occur: selection is not operating, there is no genetic drift, there is no migration, there is no mutation, and there is random mating. If these conditions are met, then the population being examined is at Hardy–Weinberg equilibrium. If the conditions are not met, then the population is not at Hardy–Weinberg equilibrium. When populations are at Hardy–Weinberg equilibrium, it is possible to calculate genotype frequencies from allele frequencies, and allele frequencies from genotype frequencies.

The equations used to determine if a population is at Hardy–Weinberg equilibrium are:

$$p + q = 1 \quad \text{and} \quad p^2 + 2pq + q^2 = 1$$

where:

p is the dominant allele frequency A

q is the recessive allele frequency a

p^2 is the genotype frequency AA

$2pq$ is the genotype frequency Aa, and

q^2 is the genotype frequency aa

PRACTICE THE SKILL

A small population of 50 rabbits was found on an island a wcentury ago. Scientists discovered the rabbits had two alleles for ear type, A for straight ears and a for floppy ears. Imagine that we know that the allele frequency of A was 0.6. The scientists have monitored the island throughout the years as the population of rabbits increased to 10,000 and have noted that there are 1600 floppy-eared aa rabbits in the present day. With this information, determine the allele and genotype frequencies for each time period and interpret whether the frequencies of the alleles A and a changed over this time period.

Our goal is to compare all the allele and genotype frequencies. We have the frequency for the dominant allele, A, and can determine the allele frequency of the recessive allele, a, using the following equation:

$$p + q = 1$$
$$0.6 + q = 1$$
$$q = 0.4$$

We can use the above values to calculate the genotype frequencies, assuming the population was at Hardy–Weinberg equilibrium.

AA homozygotes: $p^2 = (0.6)^2 = 0.36$

Aa heterozygotes: $2pq = 2 \times 0.6 \times 0.4 = 0.48$

aa homozygotes: $q^2 = (0.4)^2 = 0.16$

Now that we have all the frequencies for the first time period, we can calculate the frequencies for the present time period. To begin, we'll calculate the genotype frequency q^2 for the floppy-eared aa homozygous rabbits.

$$q^2 = \text{frequency of } aa = \text{number of } aa \text{ individuals/}$$
$$\text{total number of individuals} = \frac{1600}{10,000} = 0.16$$

With this value, we can now solve for the allele frequency of q and then p, assuming the present-day population is at Hardy–Weinberg equilibrium.

$$\text{If } q^2 = 0.16 \quad \text{then} \quad q = \sqrt{q^2} = 0.4$$
$$p + q = 1$$
$$p + 0.4 = 1$$
$$p = 0.6$$

Calculating the genotype frequencies yields the following:

AA homozygotes: $p^2 = (0.6)^2 = 0.36$

Aa heterozygotes: $2pq = 2 \times 0.6 \times 0.4 = 0.48$

Looking at each set of allele and genotype frequencies, we can see that neither has changed over the past century, even with the explosion of the rabbit population's size. Because the allele and genotype frequencies have remained the same, we can conclude that evolution has not occurred.

	p (A)	q (a)	p^2 (AA)	$2pq$ (Aa)	q^2 (aa)
100 years ago	0.6	0.4	0.36	0.48	0.16
Present day	0.6	0.4	0.36	0.48	0.16

Your Turn

A rare neuromuscular disease is known to be caused by a recessive mutation. The recessive homozygous genotype occurs at a frequency of 1 in 5,000,000. Calculate how many people would be carriers of this recessive allele in an area with a population of 10,000 people, assuming Hardy–Weinberg conditions are met.

Module 43 Summary

PREP FOR THE AP® EXAM

REVISIT THE BIG IDEAS

EVOLUTION: Choose one of the five mechanisms of evolution discussed in the module and **describe** how that mechanism can lead to a change in allele or genotype frequencies in a population over time or both.

LG 43.1 Evolution is a change in the genetic makeup of a population over time.

- Genetic variation can be measured by allele frequency and genotype frequency. Page 600
- Evolution is a change in the frequency of alleles or genotypes in a population over time. Page 600
- Populations, not individuals, evolve. Page 601

LG 43.2 There are five mechanisms of evolution.

- Natural selection leads to a decrease in frequency of a harmful allele and an increase in frequency of a beneficial allele. Page 602
- Natural selection is the only mechanism of evolution that results in adaptations. Page 602
- Genetic drift is a random change in allele frequency change. Page 603
- Genetic drift acts more strongly in small populations than in large ones. Page 603
- Two forms of genetic drift are a population bottleneck and founder effect. Page 603

- Migration involves the movement of individuals and alleles between populations and tends to have a homogenizing effect among populations. Page 604
- Mutation is the ultimate source of variation, but it also can change allele frequencies on its own. Page 606
- Nonrandom mating occurs when individuals choose mates based on their genotypes. Page 606
- Inbreeding is a form of nonrandom mating. Page 607
- Nonrandom mating, such as inbreeding, changes genotype frequencies, but does not change allele frequencies. Page 607

LG 43.3 The Hardy–Weinberg equilibrium occurs when the genetic makeup of a population does not change over time.

- When certain conditions are met, allele and genotype frequencies do not change, a state called the Hardy–Weinberg equilibrium. Page 607
- The five conditions of the Hardy–Weinberg equilibrium are that the population experiences no selection, no chance events due to small population size, no migration, no mutation, and random mating. Page 608
- The Hardy–Weinberg equilibrium allows allele frequencies and genotype frequencies to be calculated from each other. Page 609
- By determining whether a population is in Hardy–Weinberg equilibrium, we can determine whether evolution is occurring in a population. Page 610

Key Terms

Genetic drift
Bottleneck
Founder effect

Migration
Gene flow

Nonrandom mating
Hardy–Weinberg equilibrium

Review Questions

1. At the genetic level, evolution is
   (A) a change in fitness over time.
   (B) a change in the number of individuals in a population over time.
   (C) a change in the frequency of an allele or genotype over time.
   (D) an increase in fitness over time.

2. If there is inbreeding in a population, which would you expect to observe with respect to allele frequencies and genotype frequencies?
   (A) Allele frequencies would change, and genotype frequencies would change.
   (B) Allele frequencies would change, and genotype frequencies would not change.
   (C) Allele frequencies would not change, and genotype frequencies would change.
   (D) Allele frequencies would not change, and genotype frequencies would not change.

3. A gene in a diploid species has five different alleles. In any one population, the allele frequencies of all five alleles must add up to
   (A) one.
   (B) two.
   (C) five.
   (D) ten.

4. If there are 100 individuals in a population and 20 are homozygous for *B*, 60 are heterozygous, and 20 are homozygous for *b*, what is the allele frequency of *B*?
   (A) 0.20
   (B) 0.40
   (C) 0.50
   (D) 0.80

5. If a gene has two alleles, and allele *A* has a frequency of 83%, then allele *a* has a frequency of
   (A) 17%.
   (B) 117%.
   (C) 41.5%.
   (D) 7%.

6. What does the term *2pq* represent in the Hardy–Weinberg relation?
   (A) The frequency of homozygous dominant individuals
   (B) The frequency of heterozygotes
   (C) The frequency of homozygous recessive individuals
   (D) The frequency of deleterious mutations

7. The only mechanism of evolution that leads to adaptations is
   (A) natural selection.
   (B) genetic drift.
   (C) migration.
   (D) nonrandom mating.

Module 43
AP® Practice Questions

PREP FOR THE AP® EXAM

Section 1: Multiple-Choice Questions

Choose the best answer for questions 1–4.

1. Which would tend to increase genetic similarity between two geographically separated populations of the same species?
   (A) Genetic bottlenecks
   (B) Mutation
   (C) Migration
   (D) Genetic drift

2. Scientists observed a population of lizards living on a small, forested island in the Pacific Ocean for many consecutive years. All the lizards observed were the same light-green color, matching many of the plants in the forest. In the middle of the observation period, a volcanic eruption covered much of the island with black volcanic rock. In the years following the eruption, the scientists began to observe that lizards living on the volcanic rock tended to be darker in color than lizards living in the surrounding forest, which retained their light-green coloring. Which provides the best explanation for the observation?

(A) A random mutation that results in darker color increased in frequency after the eruption.

(B) The lizards needed to adapt to the volcanic rock, so they mutated a gene that controls color.

(C) Light-green lizards migrated from another island and colonized the forest.

(D) A genetic bottleneck increased the frequency of a mutation that controls color.

Use the following information to answer questions 3 and 4.

In 1983, a new housing development was constructed in a forested region that provides habitat for timber rattle-snakes. The construction of the buildings and associated major roads resulted in a small patch of forest becoming physically disconnected from the remaining forest. Heavily trafficked roads and highways create barriers to movement for timber rattlesnakes because the snakes cross roads slowly and therefore are frequently killed by cars.

Every 5 years between 1985 and 2020, a researcher measured the frequency of an allele that is unrelated to fitness in timber rattlesnakes. The researcher took samples both from the isolated small patch of forest and from the larger forest that the small patch was previously connected to. These allele frequencies were compared to the frequency of the mutant allele measured in the forest in 1980, before construction of the housing development. The data are plotted in the figure below.

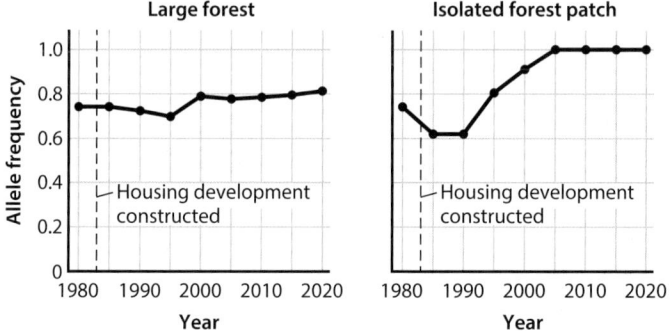

3. Which statement is best supported by the data?

(A) The isolated forest patch population is in Hardy–Weinberg equilibrium.

(B) Migration between the large forest and the isolated forest patch maintains allele frequencies at a constant level in the large forest.

(C) In the isolated forest patch, new alleles arose for the gene whose allele frequencies were measured.

(D) The isolated forest patch has a smaller population of timber rattlesnakes than the large forest does.

4. Which would be most likely to occur if the housing development is removed and the isolated patch becomes reconnected to the larger forest?

(A) Allele frequencies in the area that is now the isolated patch would become similar to allele frequencies in the forest.

(B) Timber rattlesnakes from the area that is now the isolated patch would migrate and replace the snakes in the forest.

(C) Timber rattlesnakes in the area that is now the isolated patch would go extinct.

(D) The allele frequencies in the whole population would become the average of the allele frequencies in the patch and the allele frequencies in the forest.

Section 2: Free-Response Question

Write your answer to each part clearly. Support your answers with relevant information and examples. Where calculations are required, show your work.

Yellow monkeyflower (*Mumulus guttatus*) is an annual plant that is sometimes able to grow in soil contaminated with copper, including soil near copper mines. A scientist hypothesized that the GST^{Cu} allele of the gene encoding glutathione S-transferase (GST) might cause resistance to environmental copper, but that plants homozygous for GST^+ alleles would not grow well in the presence of copper in the soil.

The scientist planted populations of *M. guttatus* in a laboratory greenhouse under two growth treatments. In treatment 1, a population was grown in soil contaminated with a level of copper similar to what is observed near copper mines. In treatment 2, the population was grown in soil without excess copper. The populations in both treatments were started with 50 GST^+/GST^+ plants, 100 GST^+/GST^{Cu} plants, and 50 GST^{Cu}/GST^{Cu} plants. Every generation, the scientist allowed the plants to cross randomly within each treatment population. For each treatment, the scientist collected the seeds from the flowers in the population and planted 100 new plants under the same treatment conditions. The scientist measured the frequency of the GST^+ and GST^{Cu} alleles every generation for five generations and obtained the data shown in the table.

	Treatment 1 (copper contamination)		Treatment 2 (no excess copper)	
Generation	Frequency of GST^+ allele	Frequency of GST^{Cu} allele	Frequency of GST^+ allele	Frequency of GST^{Cu} allele
1	0.50	0.50	0.50	0.50
2	0.42	0.58	0.44	0.56
3	0.37	0.63	0.41	0.59
4	0.31	0.69	0.43	0.57
5	0.28	0.72	0.38	0.62

(a) On a single graph, **plot** the frequencies of the GST^+ and GST^{Cu} alleles in each generation for both treatments.

(b) **Identify** the null hypothesis and the control in this experiment.

(c) **Determine** whether the data obtained from the experiment do or do not support the scientist's hypothesis.

(d) **Make a claim** to explain the data obtained from treatment 2. Propose an additional experiment that would test your claim.

Module 44

Evidence of Common Ancestry and Evolution

LEARNING GOALS ▶**LG 44.1** Homologous traits provide evidence for common ancestry.
▶**LG 44.2** Vestigial structures provide a window on the past.
▶**LG 44.3** Fossils provide direct evidence of past life.
▶**LG 44.4** The distribution of organisms on Earth sheds light on evolutionary history.

In the last few modules, we have discussed evolution, which is a change in the genetic makeup of a population over time. We have seen how to measure genetic variation by calculating allele and genotype frequencies. We can follow these frequencies over time to see if they change or stay the same. We have also discussed the five mechanisms that bring about evolutionary change from one generation to the next. These mechanisms are natural selection, genetic drift, migration, mutation, and nonrandom mating. All these mechanisms lead to a change in the allele frequency, genotype frequency, or both in populations over time. However, only natural selection leads to adaptations, the fit between an organism and its environment.

Up to this point, we have focused on the genetic basis of evolution and used mathematical models to show how it occurs. Can we see any evidence of evolution? The answer is a resounding yes. In this module, we will see that evidence of evolution is literally all around us—in the traits of organisms, in the fossil record, and even in the distribution of species on Earth. Each one of these teaches us something about the evolutionary history of life on Earth, and therefore provides strong evidence for evolution.

PREP FOR THE AP® EXAM

FOCUS ON THE BIG IDEAS

EVOLUTION: Look for examples of structural, functional, and molecular data that provide evidence of evolution.

44.1 Homologous traits provide evidence for common ancestry

Different species sometimes have traits in common. As we saw in Module 41, convergent evolution can account for some of these similarities; the trait may have evolved independently in the two groups as an adaptation to similar environments. Alternatively, some of these similarities reflect common ancestry; the trait may have been present in the common ancestor of the two groups and retained over time. Traits that are similar because of convergent evolution are analogous. Traits that are similar because of descent from a common ancestor are homologous, or a homology among descendant species. Homologous traits provide evidence for evolution. They show us what common ancestors might have looked like and help us to group species based on their evolutionary history.

In this section, we will look at various traits among organisms that are shared because of descent from a common ancestor. We will examine many different types of traits, including anatomical, cellular, and molecular traits.

Anatomical Traits

Let's consider some anatomical traits that were present in a common ancestor and retained over time. These anatomical traits are considered **morphological homologies** because they are similarities based on shared ancestry (homologies) and they are structural (morphological). Mammals, birds, and reptiles all have a water-tight egg called an amniotic egg. This trait is so important that mammals, birds, and reptiles are collectively called amniotes. The amniotic egg allows these groups to reproduce in dry environments, away from water. Why do these different groups all share this trait? Multiple lines of evidence suggest that it was present in the common ancestor of these groups and retained over time, so that all descendants of this ancestor now have the trait.

We can visualize what happened using a phylogenetic tree like the ones we discussed in Module 40. In **FIGURE 44.1,**

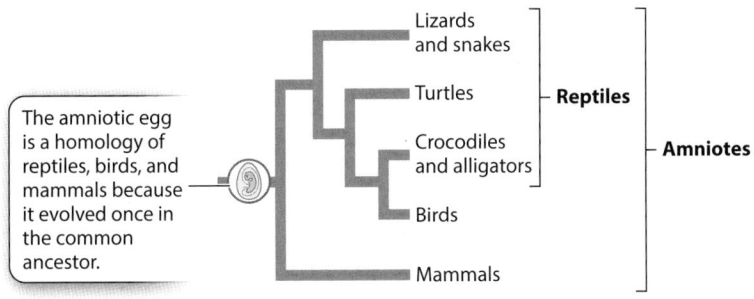

FIGURE 44.1 Descent from a common ancestor

The amniotic egg is a homologous trait shared by reptiles, birds, and mammals. It evolved once in the common ancestor of these animals and was retained over time.

time runs from left to right, and living organisms are placed at the tips of the trees. Note that the amniotic egg evolved at the base, or root, of the tree. All of the descendant species have the trait because it was present in their common ancestor.

Let's consider another example—our four limbs. This trait evolved in the first vertebrates to leave the water and move about on land. Amphibians such as frogs and salamanders, reptiles such as turtles and lizards, birds, and mammals all show remarkable anatomical similarities in the structure of these limbs. In fact, this structure is so important that these groups are collectively called tretrapods, which means "four-footed animals." This basic structure has been retained in all the descendants of this organism. It has also been modified over time, so that some animals walk on four legs, while others walk on two legs. Still others—such as birds and bats—fly; in these cases, two of the four limbs have evolved into wings, while the remaining two are legs. In marine mammals, the forelimbs have been modified into fins. In spite of their superficial differences, these structures are all variations on the same theme. The fact that they all share the same anatomical structure provides evidence that they evolved from a common ancestor with this structure.

We can illustrate the evolution of four limbs using an evolutionary tree, as shown in **FIGURE 44.2**. In this tree, time runs from left to right. The trait "walking legs" evolved at the base of the tree, on the left. All of the descendant species, including amphibians, reptiles, birds, and mammals have this trait or a modification of this trait.

Cellular Traits

When we think of traits, we often think of anatomical structures. However, we can also consider cellular structures. Humans, other animals, plants, fungi,

and many other groups or organisms all have cells with a nucleus. Hence, we are part of a group called eukaryotes, from the Greek *eu* meaning "true" and *karyon* meaning "nucleus." The cells of these organisms share traits in addition to a nucleus, including a cell membrane; mitochondria; and membrane-bound organelles such as the endoplasmic reticulum, Golgi apparatus, and vesicles, as shown in **FIGURE 44.3** on page 618. Genetically, they all have one or more linear chromosomes and genes that contain exons, which are expressed, and introns, which are intervening sequences that are spliced out and therefore not expressed. Multiple lines of evidence demonstrate that these traits were present in the first eukaryotic cell and so are homologous traits in all eukaryotes.

Finally, we can consider traits shared by all living organisms, including eukaryotes and prokaryotes, such as bacteria and archaea. These universal traits include the ability to grow and reproduce, evolve, respond to the environment, maintain homeostasis, and harness energy. These were certainly present in the universal common ancestor of all living things and, collectively, provide a way to think about the definition of life itself.

Some of these universal cellular traits are structural, such as a cell membrane enclosing the cell, cytoplasm making up the cell interior, genetic material in the form of DNA, and ribosomes that synthesize proteins from RNA. Some are processes, such as metabolic reactions that produce ATP, the ability to maintain homeostasis, a genetic code in which

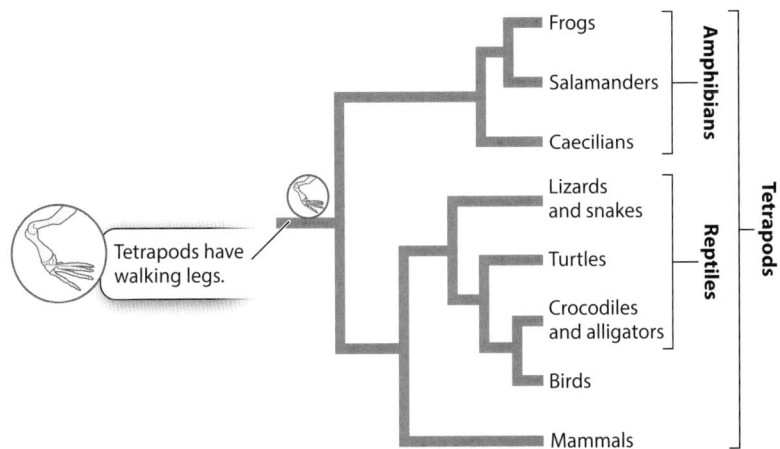

FIGURE 44.2 An evolutionary tree showing the origin of walking legs

The trait "walking legs" is a homology shared by amphibians, reptiles, birds, and mammals. It evolved once in the common ancestor of these animals and was retained over time.

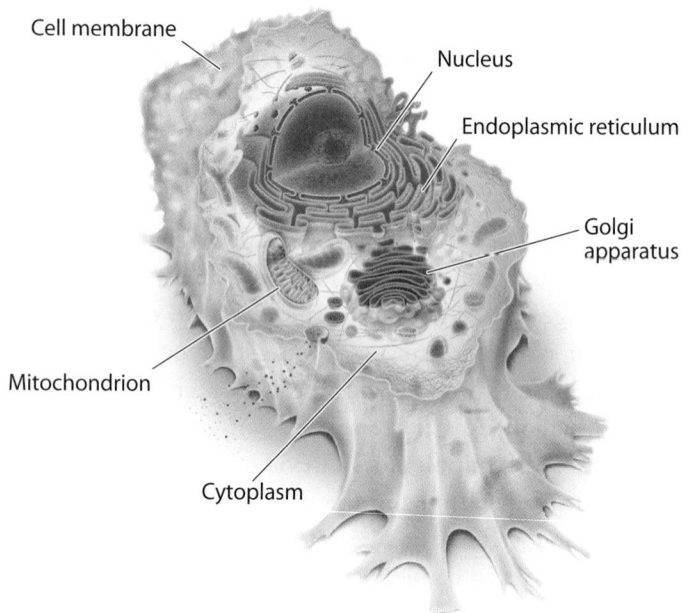

Cell membrane

Nucleus

Endoplasmic reticulum

Golgi apparatus

Mitochondrion

Cytoplasm

FIGURE 44.3 Shared cellular traits in eukaryotes

Eukaryotic cells have a nucleus, mitochondria, and membrane-bound organelles such as the endoplasmic reticulum and Golgi apparatus, all of which were present in their common ancestor.

three nucleotides specify a single amino acid during translation, and the capacity for evolution over time. These structural and functional traits are conserved in all organisms and provide evidence for the relatedness of organisms across all three domains of life.

Molecular Traits

Up to this point, we have considered anatomical traits, such as the amniotic egg and walking legs, as well as cellular structures, such as the cell membrane, nucleus, and ribosomes. We can also consider molecular traits. The nucleotide sequence of DNA and RNA, and the amino acid sequence of proteins are molecular traits that can be compared between two species. These too can provide evidence for evolution and common ancestry.

The extent of genetic difference, or genetic divergence, between two species is a function of the time they have been genetically isolated from each other. The longer they have been apart, the greater the opportunity for mutations to occur in each population and for those mutations to increase in frequency through the population over time. This correlation between the time for which two species have been evolutionarily separated and the

amount of genetic divergence between them is known as the **molecular clock.**

The slowest molecular clock on record belongs to the histone genes, which encode the proteins around which DNA is wrapped to form chromatin (see Module 31). These proteins are exceptionally similar in all organisms; only two amino acids in a chain of about 100 are different in plant and animal histones. Plants and animals last shared a common ancestor more than 1 billion years ago. Despite this lengthy span of time, the histones have hardly changed at all. In other words, the histone genes are evolutionarily conserved in distantly related species. Almost any amino acid change fatally disrupts the histone protein, preventing it from carrying out its proper function. Natural selection against these variants has thus been extremely effective in eliminating just about every amino acid–changing histone mutation over time. As a consequence, the histone molecular clock is breathtakingly slow.

The histone gene is a homologous trait among all eukaryotic cells, present in the common ancestor and retained, essentially unchanged, in the descendant species. There are many examples of shared genes between groups that were present in a common ancestor and thus homologous in these groups. Many of these genes change more rapidly than the histone genes, but they retain enough similarity that they are unmistakably homologous, just like the leg of a lizard, arm of a human, wing of a bird, and flipper of a whale.

Some examples of homologous genes include *Pax6* in vertebrates and related genes in other organisms. Recall from Module 36 that *Pax6* encodes a transcription factor that controls eye development in a wide range of species. This gene evolved at least 500 million years ago and has been conserved over time. This means that it has changed very little and has retained its function over the course of evolution. Similarly, hemoglobin, which binds oxygen in your red blood cells, is encoded by an ancient gene that is present in many different species, including animals, plants, and fungi.

✔ Concept Check

1. **Describe** a homologous trait.

2. **Identify** two examples of homologous traits.

3. **Identify** three homologous traits shared by all eukaryotes.

44.2 Vestigial structures provide a window on the past

In addition to homologous traits shared by different species, we can also look at *vestigial structures*. **Vestigial structures** are traits that were once useful and functional but are no longer and so tend to be reduced in size or function. The presence of a trait in a reduced or nonfunctional structure in an organism suggests that it is a vestigial structure. These traits give us a window on the past.

Consider the hip bone in whales. Whales and other marine mammals evolved from land-based mammals that moved about on four legs. When they returned to the water, the legs were no longer useful and might have even been detrimental to moving about in the water. Over time, hind legs that were reduced in size were likely selected for and therefore became more common in the population. Now, they are no longer present at all. However, the hip bone to which the limbs attach is small but still present, providing one piece of evidence that these animals once walked on land with hind legs, like other mammals.

A vestigial hip bone of a whale is shown in **FIGURE 44.4**. Note that vestigial structures are examples of morphological homologies. The vestigial hip bone of a whale is homologous to the larger and more functional hip bone of other vertebrate animals that move about on land.

One of the most famous examples of a vestigial structure is the appendix, a short appendage or blind sac of the gastrointestinal tract in humans and other mammals.

Although the appendix has no clear function in humans, it may play a minor role in the immune system. The appendix can become infected when partially digested material enters or blocks it, resulting in appendicitis. Unless the infected appendix is removed, it can burst, releasing gut contents into the abdominal cavity. Such a rupture can be life threatening.

Why then do we have an appendix? Charles Darwin put forth a hypothesis explaining how the appendix might have evolved to take on its current form in humans. He suggested that in the ancestors of modern humans, this part of the gut harbored bacteria that helped to break down plant material such as cellulose. However, as their diet became less reliant on plant material, this part of the gut no longer served a function. As a result, selection on its form and function relaxed and it became smaller and lost its function over time. Today, what remains of this structure is the appendix.

Other vestigial structures, such as small eyes in cave-dwelling animals and rudimentary wings in flightless birds and insects, provide an illuminating window into the past and strong evidence for evolution.

Just like an anatomical structure, genes that encode proteins can also lose function over time. A pseudogene is a gene that was once functional but is no longer. In other words, the gene used to produce a protein product, but no longer does, so it is not expressed. Pseudogenes are the genetic equivalent of the whale pelvis or the human appendix, a genetic remnant. Because all mutations in a pseudogene are by definition neutral—there is no function for a mutation to disrupt, so a mutation is neither harmful nor beneficial—we expect to see a pseudogene's molecular clock tick at a very fast rate. In the histone genes, virtually all mutations are selected against, slowing the rate of evolution. In pseudogenes, no such selection against harmful variants is operating.

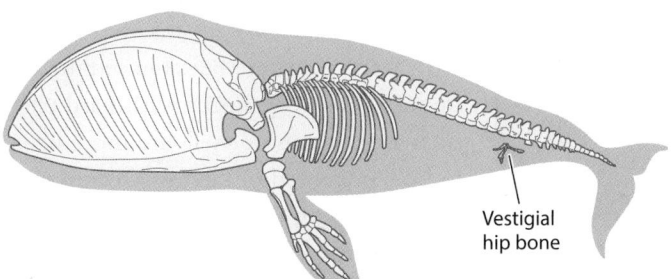

FIGURE 44.4 Vestigial structure

The hip bone in a whale is a good example of a vestigial structure, one that was once functional, but now is much reduced in size.

Vestigial hip bone

✓ Concept Check

4. **Describe** what a vestigial structure is.
5. **Identify** two examples of vestigial structures.

44.3 Fossils provide direct evidence of past life

The anatomical, cellular, and molecular traits of living organisms are useful ways to learn about evolutionary history, but the information is indirect. Direct evidence from the past comes from *fossils*. **Fossils** are the remains of once-living organisms. Fossils provide unique information about the past that we would otherwise not know. However, the fossil record is an incomplete record of the past because it preserves only a small fraction of all the organisms that have ever lived. In this section, we will discuss how fossils form, consider the unique information fossils provide, and learn about how fossils are dated.

The Fossil Record

Fossilization is the process by which fossils form. Fossilization begins with burial. For example, a clam that dies on the seafloor may be quickly covered by sand, or a leaf that falls to the forest floor may be covered in mud during a flood event. Through time, accumulating sediments from sand, mud, or soil harden into sedimentary rocks, such as those exposed so dramatically in the walls of the Grand Canyon, shown in **FIGURE 44.5**. If the remains of an organism are not buried, they are eventually recycled by biological and physical processes as discussed in Module 1, and no fossil forms.

The fossil record should not be thought of as a complete encyclopedia of everything that ever walked, crawled, or swam across our planet's surface. In general, the fossil record of marine life is more complete than that for land-dwelling creatures because sediments in marine habitats are more likely to accumulate and become rock. Because of the way fossils form, organisms such as trees and elks living high in the Rocky Mountains have a low probability of fossilization, whereas clams and corals on the shallow seafloor are commonly buried and become fossils.

Biological factors also limit the fossil record. Most fossils preserve just the hard parts of organisms—those features that resist decay after death. For animals, this usually means hard skeletons. Clams and snails that have shells also have excellent fossil records. More than 80% of the clam species found today along California's coast also occur as fossils in sediments deposited during the past million years. In contrast, nematodes, tiny worms that may be the most abundant animals on Earth, have no hard skeletons and are not well represented in the fossil record. The wood and pollen of plants, which are made in part of decay-resistant organic compounds, enter the fossil record far more commonly than do flowers. And, among unicellular organisms, the skeleton-forming diatoms, foraminifera, and radiolarians photographed in **FIGURE 44.6** have exceptionally good fossil records, whereas unicellular organisms without hard parts, like amoebas, are underrepresented.

Together, then, the environment and properties of organisms influence the probability that an ancient species will be represented in the fossil record. Organisms that live in environments where burial in sediment is likely are more often preserved than organisms that live in environments where burial in sediment is unlikely. And organisms with hard parts such as skeletons or shells are more likely to be represented in the fossil record than organisms with only soft parts.

In some cases, organisms that lack hard parts can also leave a fossil record. Some animals leave tracks and trails as they move about or burrow into sediments. These fossils of trails, rather than body parts, are called trace fossils. Trace fossils range from dinosaur tracks

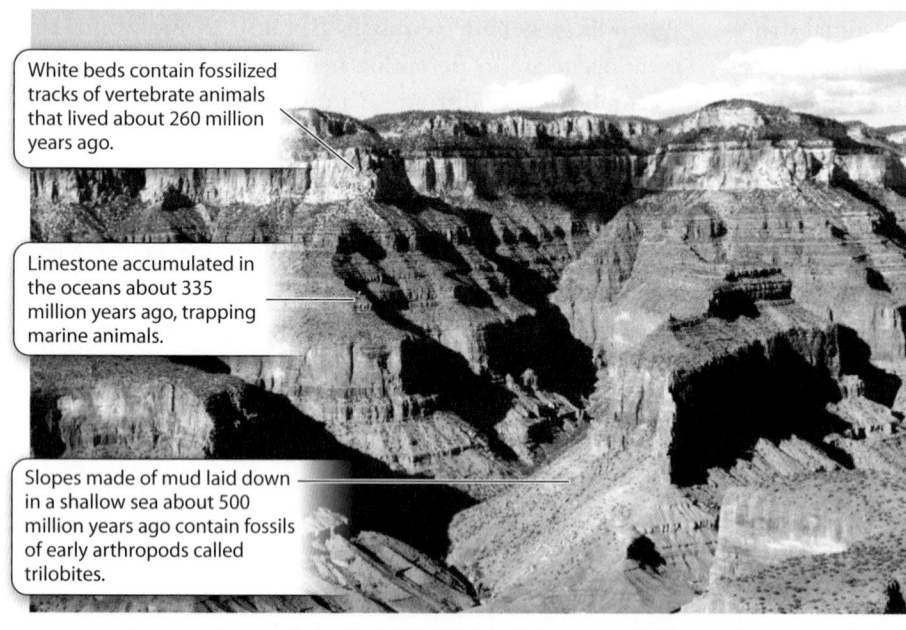

White beds contain fossilized tracks of vertebrate animals that lived about 260 million years ago.

Limestone accumulated in the oceans about 335 million years ago, trapping marine animals.

Slopes made of mud laid down in a shallow sea about 500 million years ago contain fossils of early arthropods called trilobites.

FIGURE 44.5 A record of Earth's history at the Grand Canyon

Erosion has exposed layers of sedimentary rock that record Earth's history. Photo: National Park Service

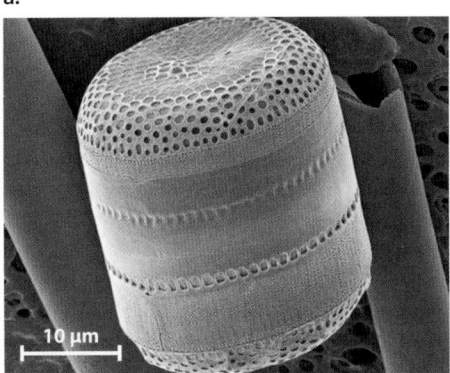

a.

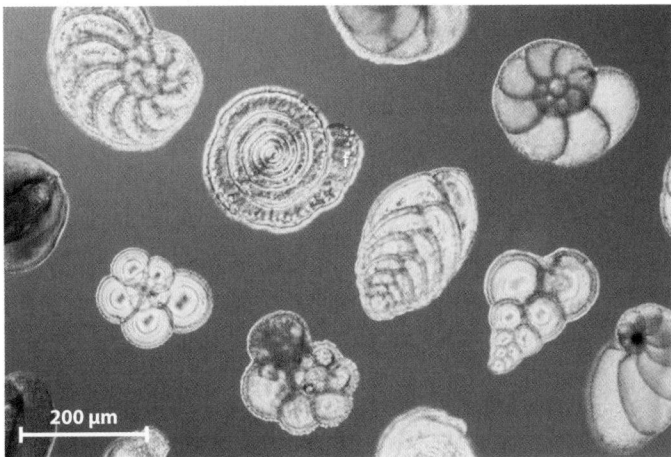

b.

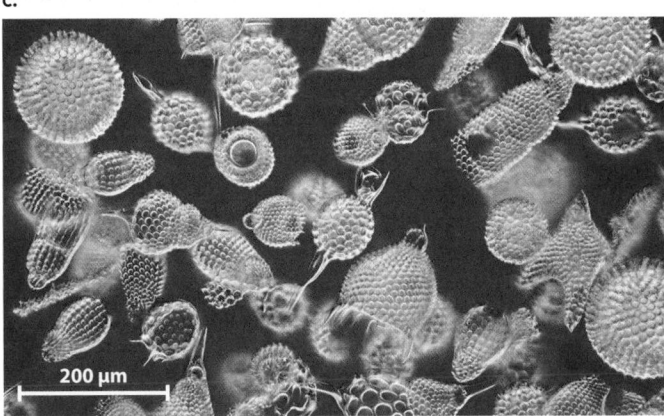

c.

FIGURE 44.6 Unicellular organisms with skeletons

Diatoms (a), foraminifera (b), and radiolarians (c) all have hard skeletons, which preserve well as fossils. Photos: (a) Gert Hansen, SCCAP, University of Copenhagen; (b) Astrid & Hanns-Frieder Michler/Science Source; (c) M. I. Walker/Science Source

like those shown in **FIGURE 44.7**, to the feeding trails of snails and trilobites. In essence, they preserve a record of both anatomy and behavior.

Organisms can also leave behind molecular fossils. Most biological molecules decay quickly after death. Proteins and DNA, for example, generally break down before they can be preserved, although, remarkably, the genome of an early human species known as Neanderthal has been pieced together from DNA in 40,000-year-old bones. Other molecules, especially lipids like cholesterol, are more resistant to decomposition. Sterols, bacterial lipids, and some pigment molecules can accumulate in sedimentary rocks. Scientists can study molecular fossils for organisms such as bacteria and single-celled eukaryotes, which rarely form conventional fossils.

Sometimes unusual conditions preserve fossils of unexpected quality, including animals without shells, delicate flowers or mushrooms, fragile seaweeds or bacteria, and even the embryos of plants and animals. For example, during the Cambrian Period (541–485 million years ago), a sedimentary rock formation called the Burgess Shale accumulated on a relatively deep seafloor covering what is now British Columbia. Waters just above the basin floor contained little or no oxygen. Thus, when mud swept into the basin,

FIGURE 44.7 Trace fossils

These footprints in 150-million-year-old rocks record both the structure and the behavior of the dinosaurs that made them. Photo: José Antonio Hernaiz/AGE Fotostock

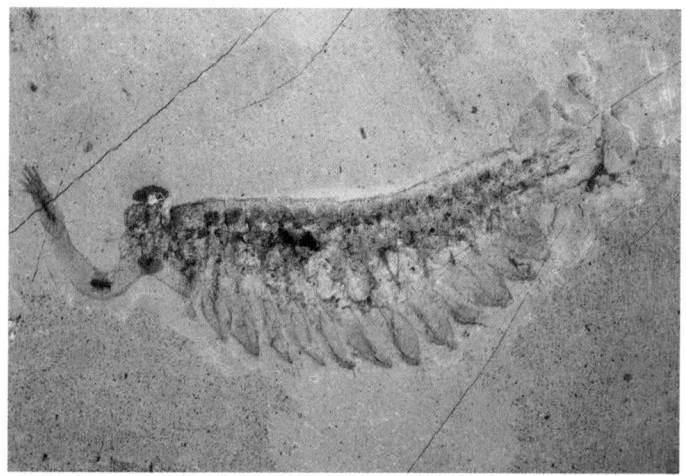

FIGURE 44.8 A Burgess Shale fossil

This fossil is *Opabinia regalis*, an extinct early relative of the arthropods, from the 505-million-year-old Burgess Shale. Photo: © Smithsonian Institution—National Museum of Natural History (USNM 57683) Jean-Bernard Caron

entombed animals were sealed off from scavengers, disruptive burrowing activity, and even bacterial decay. For this reason, the Burgess Shale preserves remarkable fossils of marine life over 500 million years ago during the initial diversification of animals, one of which is shown in **FIGURE 44.8**.

In general, the fossil record preserves some aspects of biological history well and others poorly. Fossils provide a good sense of how the forms, functions, and diversity of skeletonized animals have changed over the past 500 million years. The same is true for land plants and unicellular organisms that form mineralized skeletons. Collectively, these fossils shed light on major patterns of evolution and diversity change through time; their geographic distributions record the movements of continents over millions of years; and the radiations and extinctions they document show how life responds to environmental change, both gradual and catastrophic.

Unique Information from Fossils

Although the fossil record is incomplete, there is information we can learn from fossils that we would not otherwise know or even be able to guess. Not only do fossils record past life, but they also provide our only direct record of extinct species. The evolutionary tree shown in Figures 44.1 and 44.2 contain a great deal of information, but they don't give a hint about dinosaurs. Fossils demonstrate that dinosaurs once roamed Earth.

Fossils also give us direct evidence of transitional forms, sometimes called missing links, between modern groups of organisms. In 1861, just two years after publication of *On the Origin of Species,* German quarry workers discovered a remarkable fossil that remains one of paleontology's most famous examples of a form that is transitional between two groups. *Archaeopteryx lithographica,* shown in **FIGURE 44.9**, lived 150 million years ago. Its skeleton shares many characters with a group of small, agile dinosaurs, but several features—its pelvis; its braincase; and, especially, its winglike forearms—are distinctly birdlike. Spectacularly, the fossils preserve evidence of feathers. *Archaeopteryx* clearly suggests a close relationship between birds and dinosaurs. In fact, today we know that modern birds are descended from a group of dinosaurs and are therefore considered living dinosaurs.

Tiktaalik roseae is another famous fossil of a transitional form, in this case between fish and amphibians. The fossil is about 370 million years old and it documents the colonization of land by vertebrates. All land vertebrates, from amphibians to mammals, are descended from a common ancestral fish. *Tiktaalik* had fins, gills, and scales like other fish of its day, but its skull was flattened, more like that of a crocodile than a fish, and it had a functional neck and ribs that could support its body—features today found only in tetrapods. Along with many other fossils, *Tiktaalik* captures a key moment in the evolutionary transition from water to land. In "Practicing Science 44.1: Do fossils bridge the evolutionary gap between fish and tetrapod vertebrates?" we describe *Tiktaalik* in more detail and point out structures that indicate that it is a transitional form between fish and the first land-based, walking vertebrates.

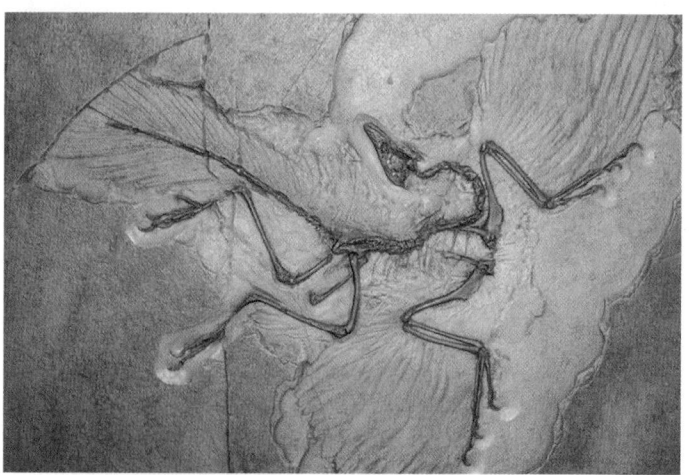

FIGURE 44.9 *Archaeopteryx*

This fossil has features of dinosaurs and modern birds, including feathers, indicating that birds and dinosaurs are closely related.

Photo: Jason Edwards/Getty Images

Do fossils bridge the evolutionary gap between fish and tetrapod vertebrates?

Background Anatomical and molecular traits of living animals indicate that fish are the closest relatives of four-legged land vertebrates.

Hypothesis Land vertebrates evolved from fish by modification of the skeleton and internal organs that made it possible for them to live on land.

Observation Fossil skeletons 390 to 360 million years old show a mix of features seen in living fishes and amphibians. Older fossils have fins, fishlike heads, and gills, whereas younger fossils have weight-bearing legs, skulls with jaws able to grab prey, and ribs that help ventilate lungs. Paleontologists predicted that key intermediate fossils would be preserved in 380- to 370-million-year-old rocks.

In 2004, Edward Daeschler, Neil Shubin, and Farish Jenkins discovered the remarkable fossil *Tiktaalik* in rocks laid down by a meandering stream of just the right age. *Tiktaalik* had fins, scales, and gills like fishes, but an amphibian-like skull and a true neck (which fishes lack). Its limbs preserve wrist bones and fingers broadly similar to those in tetrapod vertebrates. Even so, as the reconstruction shows, its forelimbs were not completely adapted for movement on land.

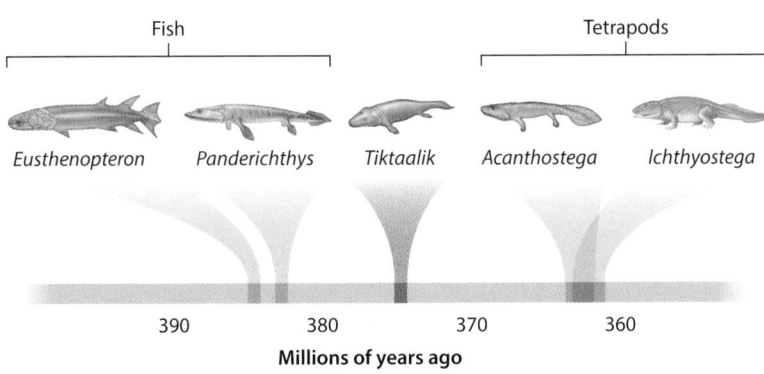

Fish | Tetrapods

Eusthenopteron *Panderichthys* *Tiktaalik* *Acanthostega* *Ichthyostega*

390   380   370   360
Millions of years ago

Conclusion Fossils provide evidence that tetrapod vertebrates evolved from fish by the developmental modification of limbs, skulls, and other features.

Follow-up Work Research into the genetics of vertebrate development shows that the limbs of fish and amphibians are shaped by similar patterns of gene expression, providing further support for the connection between the two groups.

SOURCE
Daeschler, E. B., N. H. Shubin, and F. A. Jenkins Jr. 2006. "A Devonian Tetrapod-like Fish and the Evolution of the Tetrapod Body Plan." *Nature* 440:757–763. Photo: Neil Shubin Lab

AP® PRACTICE QUESTION

In newspaper articles, *Tiktaalik* was proclaimed as the missing link between fish and tetrapods. The figure to the left correctly shows *Tiktaalik* having come into existence "in between" fish and tetrapods. This conclusion is supported by the fossil record, which shows early fish fossils in older rocks than those in which *Tiktaalik* were found, and tetrapod fossils in younger rocks than those in which *Tiktaalik* were found. However, the term "missing link" and the chronological representation of the fossils can lead to the misconception that modern fish gave rise to *Tiktaalik*, which subsequentially gave rise to modern tetrapods and that there is an unbroken chain that connects modern forms.

(a) Explain the evolutionary relationship *Tiktaalik* has with fish and tetrapods, and the reason "missing link" may lead to misunderstandings about evolution.

(b) Place the following statements in the correct order that they most likely occurred:
- The group possessing the flattened head and the true neck split into two groups, with one of them continuing to evolve into *Tiktaalik*.
- A common ancestor to all three groups existed, possessing fishlike traits.
- The second group from the second split continued to evolve into the present-day tetrapods.
- The second group developed a flattened head and a true neck.
- The common ancestor split into two groups, one of which continued to evolve into the present-day fishes.

Fossils also place evolutionary events within the context of Earth's dynamic environmental history. Again, dinosaurs illustrate the point. Geologic evidence from several continents suggests that a large meteorite triggered drastic changes in the global environment 66 million years ago, leading to the extinction of dinosaurs (other than birds). In fact, five times in the past, large environmental disturbances sharply decreased Earth's biological diversity. These events, called mass extinctions, have played a major role in shaping the course of evolution, and we know about them from the fossil record.

Dating Fossils

How do we know the age of a fossil? Beginning in the nineteenth century, geologists recognized that the types of fossils found in layers of rock change in a regular fashion from the bottom of a sedimentary rock formation to its top. As more of Earth's surface was mapped and studied, it became clear that certain fossils always occur in layers that lie beneath (and so are older than) layers that contain other species. From these patterns, geologists concluded that fossils mark time in Earth's history. At first, geologists could not explain why fossils changed from one bed to the next, but after Darwin the reason became apparent: fossils record the evolution of life on Earth.

The layers of fossils in sedimentary rocks tell us that some rocks are older than others, but they cannot by themselves provide an absolute age. This information became possible with the discovery of radioactive decay. Radioactive decay makes use of isotopes, variants of an element that differ from one another in terms of the number of neutrons they contain. Many isotopes are unstable and spontaneously break down, ejecting neutrons, electrons, or other particles to form other, more stable isotopes. In the laboratory, scientists can measure how quickly unstable isotopes decay. Then, by measuring the amounts of the unstable isotope and its stable daughter isotope inside a mineral, they can determine when the mineral formed.

Scientists commonly use the radioactive decay of the isotope **carbon-14,** or **^{14}C,** to date wood and bone, a process called **radiometric dating.** ^{14}C is an isotope of carbon with eight neutrons and six protons instead of the more stable (and more common) configuration of six neutrons and six protons.

As shown in **FIGURE 44.10,** cosmic rays continually generate neutrons that collide with nitrogen in the atmosphere to form ^{14}C, much of which is incorporated

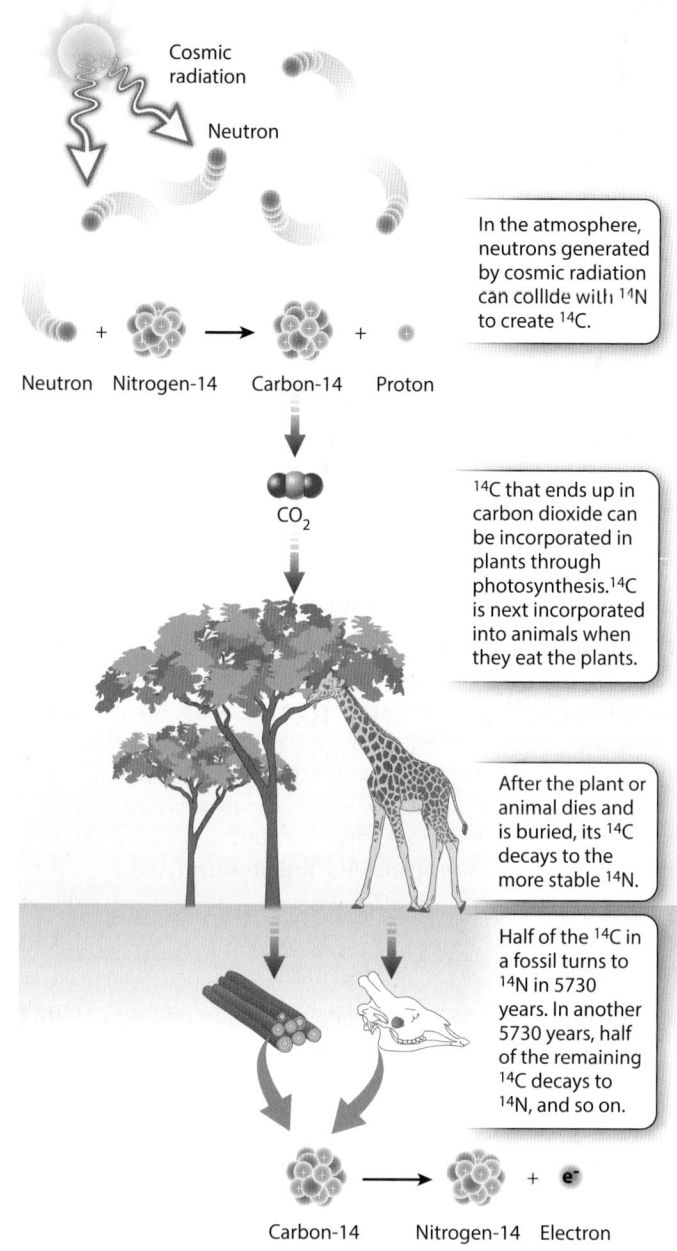

In the atmosphere, neutrons generated by cosmic radiation can collide with ^{14}N to create ^{14}C.

^{14}C that ends up in carbon dioxide can be incorporated in plants through photosynthesis. ^{14}C is next incorporated into animals when they eat the plants.

After the plant or animal dies and is buried, its ^{14}C decays to the more stable ^{14}N.

Half of the ^{14}C in a fossil turns to ^{14}N in 5730 years. In another 5730 years, half of the remaining ^{14}C decays to ^{14}N, and so on.

Carbon-14   Nitrogen-14   Electron

FIGURE 44.10 Carbon dating

Scientists can determine the age of relatively young materials such as wood and bone from the amount of ^{14}C they contain.

into atmospheric carbon dioxide (CO_2). Through photosynthesis, carbon dioxide that contains ^{14}C is incorporated into wood, and animals incorporate small amounts of ^{14}C into their tissues when they eat plant material. The unstable ^{14}C in these tissues begins to break down immediately, losing an electron to form ^{14}N, a stable isotope of nitrogen. Once the organism dies, ^{14}C is no longer added, so the clock of radioactive carbon starts to tick.

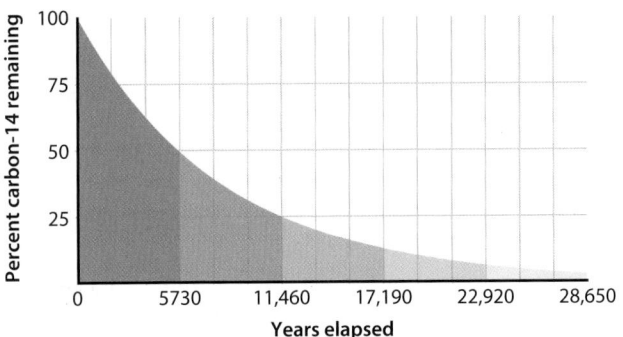

FIGURE 44.11 Half-life

The half-life of a substance is the time it takes for an amount of a substance to reach half its original amount. In the case of ^{14}C, the half-life is 5730 years, so half of it remains after every 5730 years.

Laboratory measurements indicate that half of the ^{14}C in a given sample will decay to nitrogen in 5730 years. The **half-life** is the time it takes for an amount of a substance to reach half its original value. Radioactive half-life is the time it takes for half of the atoms in a given sample of a substance to decay.

The graph in **FIGURE 44.11** shows the decay of ^{14}C. After 5730 years, 50% of the starting amount remains. After another 5730 years, or 11,460 years in total, half of 50%, or 25%, remains. After another 5730 years, or 17,190 years in total, half of 25%, or 12.5%, remains, and so on.

All isotopes of elements have a unique half-life, which vary in length depending on their stability. Armed with this information, scientists can measure the amount of ^{14}C in an archaeological sample and, by comparing it to the amount of ^{14}C in a sample of known age, determine the age of the sample. The known age of a sample can be based on annual rings in trees, for example, or yearly growth of coral skeletons.

Because its half-life is so short by geologic standards, ^{14}C is useful only in dating material younger than 50,000 to 60,000 years. Beyond that point, too little ^{14}C is left in the material to measure accurately. Older geologic materials are commonly dated using the radioactive decay of uranium (U) to lead (Pb): ^{238}U, which may be incorporated in trace amounts into the minerals of volcanic rocks, breaks down to ^{206}Pb with a half-life of 4.47 billion years; ^{235}U decays to ^{207}Pb with a half-life of 704 million years. Calibration of the geologic timescale is based mostly on the ages of volcanic ash found in sedimentary rocks that contain key fossils, as well as volcanic rocks that intrude into, and so are younger

than, layers of rock containing fossils. "Analyzing Statistics and Data: Radiometric Dating" on page 626 reviews how to calculate problems involving half-lives and provides a problem for you to try.

The sedimentary rocks that contain fossils also preserve, encrypted in their physical features and chemical composition, information about the environment in which they formed. Sandstone beds, for example, may have rippled surfaces, like the ripples produced by currents that we see today in the sand of a seashore or lake margin. This finding suggests the presence of water. Pyrite (FeS_2), or fool's gold, forms when H_2S generated by anaerobic bacteria reacts with iron. Because these conditions generally occur where oxygen is absent, the presence of pyrite in ancient sedimentary rocks can indicate that the environment was depleted of oxygen at the time that the rock formed.

In summary, fossils are a very powerful way to learn about our evolutionary past and provide strong evidence for evolutionary change over time. Although the fossil record is incomplete, what we have provides direct evidence of organisms that lived in the past. In addition, radioactive decay of isotopes in rock formations where fossils are found allows us to date fossils and place them along Earth's long timeline. Fossils can be thought of as snapshots of evolutionary history, giving us stunning pictures of past life.

PREP FOR THE AP® EXAM

AP® EXAM TIP

You should be familiar with how scientists use the age of sedimentary rocks and radioactive decay to help determine the age of ancient organisms or fossils. You should be able to explain these methods and use examples to describe the evolution and relatedness of species.

✓ Concept Check

6. **Describe** conditions in which fossilization is likely.

7. **Describe** the technique that is used to date fossils.

Radiometric Dating

Fossils are formed when an organism or plant is buried in sediment. Over time and as more layers of sediment are added, the buried material is preserved and replaced by minerals, forming a fossil. The age of a fossil can often be determined using radiometric dating and half-lives. A half-life is the amount of time it takes for a radioactive isotope to decay to the point that half of it remains in a sample. Figure 44.11 shows this explanation graphically.

Scientists use several different radioactive elements to date materials depending on their age. The radioactive decay of carbon-14 is used for organic materials that are 40,000 to 50,000 years old. The half-life of ^{14}C is 5730 years. For older rocks or fossils, scientists can use the radioactive decay of uranium-235, potassium-40, or uranium-238.

PRACTICE THE SKILL

An earthquake in California caused a shift in the layers of a sedimentary rock formation. This shift exposed several young fossil specimens. Scientists wish to determine the age of these fossils using the radioactive decay rate of carbon-14. They have measured the amount of ^{14}C in one fossil to be approximately 0.0313 as much as what was in the original atmosphere. Use this information to calculate how many half-lives have occurred and the estimated age of the fossil.

To begin, we know that the ^{14}C in the fossil is only a fraction of that found in the original atmosphere, 0.0313 as much. We can use this number to figure out how many ^{14}C half-lives have taken place.

One half-life can also be expressed as the fraction $\frac{1}{2}$, the decimal 0.5, or the percentage 50%. Two half-lives would then be $\frac{1}{2} \times \frac{1}{2}$ or $\frac{1}{4}$, 0.25, and 25%. We can continue this way to see how many half-lives it takes to get to the value 0.0313.

3 half-lives: $\frac{1}{2} \times \frac{1}{2} \times \frac{1}{2} = \frac{1}{8} = 0.125 = 12.5\%$

4 half-lives: $\frac{1}{2} \times \frac{1}{2} \times \frac{1}{2} \times \frac{1}{2} = \frac{1}{16} = 0.0625 = 6.25\%$

5 half-lives: $\frac{1}{2} \times \frac{1}{2} \times \frac{1}{2} \times \frac{1}{2} \times \frac{1}{2} = \frac{1}{32} = 0.0313 = 3.13\%$

Given the amount of ^{14}C left in the fossil the scientists discovered is 0.0313, we can see that 5 half-lives have occurred.

Now we can calculate how many years this is to figure out the age of the fossil. ^{14}C has a half-life of 5730 years. Each 5730-year period reduces the original amount of ^{14}C by 50%. Because we know that 5 half-lives have occurred, we can multiply 5730 years by 5 half-lives to determine the age of the fossil:

$$5 \text{ half-lives} \times \frac{5730 \text{ years}}{1 \text{ half-life}} = 28,650 \text{ years}$$

Using radiometric dating, the fossil is about 28,650 years old.

Your Turn

A new isotope has been discovered to be useful in the radiometric dating of rocks and fossils. It has a half-life of 1.5 million years.

1. Draw a graph showing the decay of this isotope over the course of 7.5 million years. Include the time elapsed on the x-axis and the percentage of isotope remaining on the y-axis.

2. Indicate how many total half-lives have occurred at each half-life.

3. Calculate how many grams of the isotope would remain in a fossil that is 6 million years old if it originally contained 0.58 gram of the isotope.

44.4 The distribution of organisms on Earth sheds light on evolutionary history

Traits of living organisms and fossils provide evidence of evolution and give us a window on what common ancestors likely looked like. A third piece of evidence comes from what might at first seem like an unlikely source—the distribution of organisms on the face of Earth, what scientists call biogeography.

Species are typically adapted to their environment, but that doesn't mean that species live in all the places they possibly can. Penguins, for example, live in the southern hemisphere, but not the northern hemisphere, even though there are similar environments in both places. Similarly, kangaroos and other marsupial mammals live in Australia, but not Africa, in spite of similar habitats on both continents.

Why do species live in some places and not in others? Why do they only live in a subset of the places to which they are adapted? The answer is based on their evolutionary history: species originate in one place and then they disperse. But they can only spread out until they reach a barrier to migration. For penguins, the barrier is the warm weather toward the equator, effectively blocking them from making their way northward. For kangaroos, the barrier is the Indian Ocean because kangaroos can't swim.

In other words, we only need two pieces of information to figure out the distribution of all species on Earth: where a species first originated, and how it is able to migrate or disperse from this point of origin. Therefore, the distribution of organisms gives us information about the evolutionary past.

Islands provide an interesting case study of biogeography. They are often far from the nearest mainland, and therefore may be difficult to get to. In addition, many of them, such as the Galápagos Islands and Hawaiian Islands, are volcanic in origin. As a result, when they formed, there were no organisms present. All the organisms on these islands had to migrate there in some way. They may have migrated on their own, by flying or swimming, or they may have hitched a ride in some way, on a floating log or with the help of humans, for example. Regardless of how they got there, when they arrived, there were many open habitats available to them—habitats that were likely not open to them on the mainland. These open, new habitats provided selective pressures that resulted in the evolution of many new species, all descended from the original first species to arrive.

Species that evolved in this way are often found only in one place and not in others, and are therefore considered endemic. Islands have many examples of endemic species. Darwin's finches on the Galápagos are an example: they are found there but nowhere else in the world. Other examples of endemic species include lemurs in Madagascar, Tasmanian devils on the island of Tasmania, and Komodo dragons on Komodo and a few surrounding islands, as shown in **FIGURE 44.12**.

FIGURE 44.12 Endemic species

Endemic species are found in one place and nowhere else. (a) Darwin's finches such as the large ground finch (*Geospiza magnirostris*) are found in the Galápagos Islands; (b) lemurs (*Lemur catta*) are found in Madagascar; (c) Tasmanian devils (*Sarcophilus harrisii*) are found in Tasmania; and (d) Komodo dragons (*Varanus komodoensis*) on found on Komodo and nearby islands. Photos: (a) Anton Sorokin/Alamy Stock Photo; (b) Floridapfe from S.Korea Kim in cherl/Getty Images; (c) renelo/Getty Images; (d) Afriandi/Getty Images

We can learn another lesson from biogeography. Because species don't typically live in all the places they could, if they are moved to a new, suitable habitat, they sometimes spread rapidly due to the absence of predators and other checks on population growth. Non-native species that become established in new ecosystems are termed **invasive species.** Freed from the natural constraints on their population growth, these species can expand dramatically when introduced into new areas, sometimes with negative consequences for native species and ecosystems.

Species can sometimes be transported long distances by natural processes. For example, the Gulf Stream and North Atlantic Drift occasionally carry coconuts from the Caribbean to the coast of Ireland. Migratory birds can transport microorganisms across continents. Recently, however, humans have dramatically increased dispersal of organisms, resulting in many examples of invasive species. Ships fill their ballast tanks with seawater in Indonesia and empty it into Los Angeles Harbor, carrying organisms across the Pacific Ocean. Insects in fruit from South America wind up on a dock in London. Humans have become major agents of dispersal, adding new complexity to twenty-first-century ecology.

Some of the invasive species that have become established in their new surroundings do little beyond increasing community diversity, but others put strong pressure on native species. Two highly disruptive invasive species are shown in **FIGURE 44.13**. The zebra mussel, shown in Figure 44.13a, provides an example. It is a European clam that hitchhiked to the Great Lakes in the ballast water of cargo ships. Like many invasive species, it has multiplied dramatically in its new habitat, displacing native species and clogging intake pipes to power plants. Nearly 500 introduced species contribute disproportionately to crops lost each year to competition from weeds. And in Australia, more than 10% of all indigenous mammal species have become extinct in the past 200 years, largely because European colonists introduced cats and red foxes into their new home.

Invasive species have particularly devastating effects on islands because island species have commonly evolved with relatively few competitors or predators. Bones in Hawaiian lava caves record more than 40 bird species that existed 1000 years ago but are now extinct, eliminated by the rats and pigs introduced by Polynesian colonists, as well as by habitat destruction. Introduced predators have had a similar impact on the island of Guam. Brown tree snakes, shown in Figure 44.13b, introduced from Australasia since

a.

b.

FIGURE 44.13 Invasive species

Invasive species are non-native species that become established in a new ecosystem and spread rapidly, often to the detriment of native species. They illustrate that species live in only a subset of the habitats that they could. (a) The zebra mussel (*Dreissena polymorpha*), introduced into the Great Lakes from Europe in 1985, disrupts lake ecosystems, threatening native biological diversity. (b) The brown tree snake (*Boiga irregularis*), introduced into Guam from Asia and Australia shortly after World War II, has decimated indigenous bird populations across the island. AfriPhotos: (a) Dr. David J. Jude; School of Natural Resources and Environment, University of Michigan; (b) John Mitchell/Science Source

World War II have reduced native bird and reptile diversity by 75%.

Such introductions contribute to the changing diversity and ecology of twenty-first-century landscapes, but it may be the spread of modern diseases that should concern us most. In a world filled with airplanes and oceangoing ships, bacteria and viruses spread much more rapidly than they did in our pre-industrial past. The SARS-CoV-2 virus, which causes the disease COVID-19, is a recent and tragic example. The first diagnosed case was identified in December 2019 (hence the 19 in the name of the disease), and in just a few months became a global pandemic.

✓ **Concept Check**

8. **Describe** two factors that determine the distribution of organisms around the world.

9. **Describe** invasive species.

Module 44 Summary

PREP FOR THE AP® EXAM

REVISIT THE BIG IDEAS

EVOLUTION: Using examples from the module, **describe** how molecular data can provide evidence of evolution.

LG 44.1 Homologous traits provide evidence for common ancestry.

- Traits that are shared by two or more species because of common ancestry are homologous. Page 616
- The amniotic egg evolved in the common ancestor of reptiles, birds, and mammals. Page 617
- Walking legs evolved in the common ancestor of tetrapods, including amphibians, reptiles, birds, and mammals. Page 617
- The nucleus, mitochondria, membrane-bound organelles, linear chromosomes, and genes containing exons and introns evolved in the common ancestor of eukaryotes. Page 617
- Homologous traits can be anatomical (morphological), cellular, or molecular. Page 618
- The nucleotide sequence of DNA and RNA, and the amino acid sequence of proteins provide evidence common ancestry. Page 618

LG 44.2 Vestigial structures provide a window on the past.

- A vestigial structure is a structure that has lost its original function over evolutionary time and is now much reduced in size and/or function. Page 619
- Examples of vestigial structures include the whale hip bone, the human appendix, and wings in flightless birds. Page 619

- Genes can also lose function over time and become pseudogenes; because pseudogenes have no function, they mutate rapidly and can be used as a measure of evolutionary relatedness between species. Page 619

LG 44.3 Fossils provide direct evidence of past life.

- Fossils are the remains of organisms preserved in sedimentary rocks. Page 620
- The fossil record is incomplete because fossilization requires burial in sediment, and fossils typically preserve only the hard parts of organisms, such as skeletons and shells. Page 620
- Fossils provide unique information, such as species that are now extinct, transitional forms between modern-day species, and clues about environmental history. Page 622
- Rock layers can give relative dates of fossils, with older ones deeper than younger ones. Page 624
- Radioactive decay of unstable isotopes of elements provides a means of dating rocks. Page 624

LG 44.4 The distribution of organisms on Earth sheds light on evolutionary history.

- Biogeography is the study of the distribution of species on the Earth. Page 626
- Biogeography provides information about the evolutionary history of organisms, specifically where they originated and how they were able to disperse. Page 627
- Islands provide unique habitat and have high numbers of endemic species, which are found in one place and nowhere else. Page 627
- Invasive species are non-native species that become established in new habitats, where they can disrupt ecosystems. Page 628

Key Terms

Morphological homology

Molecular clock

Vestigial structure

Fossil

Carbon-14 (^{14}C)

Radiometric dating

Half-life

Invasive species

Review Questions

1. Traits that are similar in two species as a result of common ancestry are referred to as
   (A) homologous.
   (B) analogous.
   (C) convergent.
   (D) characteristics.

2. The layers of rock give information about relative dates; absolute dates weren't established until the discovery of
   (A) radioactive decay.
   (B) atoms.
   (C) sedimentary rocks.
   (D) fossils.

3. How do fossils provide evidence of evolutionary history?
   (A) Fossils provide an accurate account of the number of species that exist at any particular time.
   (B) DNA from older fossils allows us to construct complete molecular phylogenies.
   (C) Fossils provide a record of extinct species.
   (D) Fossils provide full information on extinct species.

4. What is the first event in the fossilization process?
   (A) Erosion of soils around the organism
   (B) Burial of the organism by sediments
   (C) Conversion of organic tissue to minerals (rock)
   (D) Hardening of sediments into rock.

5. Which is most likely to fossilize?
   (A) A crayfish that died in its muddy burrow by a creek
   (B) A jellyfish that died and sunk to the bottom of the ocean
   (C) A lizard that died on a mountainside
   (D) A leaf that fell from a tree and landed on the forest floor

6. The half-life of carbon-14 is 5730 years. If a sample contained 100% carbon-14 and 0% nitrogen-14 at a point in time 17,190 years ago, then what percentage of carbon-14 would it contain today?
   (A) 12.5%   (C) 50%
   (B) 25%   (D) 75%

7. The discovery of *Tiktaalik roseae* was significant for which of the following reasons?
   (A) It provided evidence that terrestrial (land-based) vertebrates are descended from fish.
   (B) It provided evidence that birds and dinosaurs are closely related.
   (C) It was the first fossil to be found with soft parts preserved.
   (D) It provided evidence that the continents once formed a supercontinent.

Module 44
AP® Practice Questions

Section 1: Multiple-Choice Questions

Choose the best answer for questions 1–4.

1. Which is a potential consequence of fossils being more likely to form in some environments than others, and some organic materials being more likely to fossilize than others?
   (A) The numbers and types of species that once lived on land could be underestimated.
   (B) Fossils of marine organisms could be more reliable measurements of relatedness than molecular data could.
   (C) Organisms without a skeleton or a hard shell could be overrepresented in the fossil record.
   (D) Fossils of extinct organisms could lead to incorrect interpretations of data from transitional forms.

2. Which is likely to evolve with the slowest molecular clock?

(A) A pseudogene

(B) A gene for a vestigial structure

(C) An intron of a conserved eukaryotic gene

(D) The exons of a conserved eukaryotic gene

Questions 3 and 4 refer to the following information.

Scientists examined a cliff composed of four layers of sedimentary rock. In the surface layer, the scientists found the fossil stem and petals of a flowering plant. In the second layer, the next layer down from the surface, they found a fossilized dragonfly. In the third layer, the scientists found the fossilized bones of a fish. In the fourth and lowest layer, the scientists found a fossilized shell. Radiometric dating was used to determine that the fourth layer of sedimentary rock is 500 million years old and the second layer of sedimentary rock is 250 million years old.

3. Which places the fossils in the correct order, from oldest to youngest?

(A) Shell, fish, dragonfly, flower

(B) Shell, dragonfly, fish, flower

(C) Flower, fish, dragonfly, shell

(D) Flower, dragonfly, fish, shell

4. Which provides the best estimate of the age of the fish?

(A) Older than 500 million years

(B) Between 500 and 250 million years old

(C) 125–250 million years old

(D) Younger than 125 million years old

Section 2: Free-Response Question

Write your answer to each part clearly. Support your answers with relevant information and examples. Where calculations are required, show your work.

The rate at which radioactive isotopes decay can be used to determine the age of fossils. A scientist studying fossil remains of a wooly mammoth determined that the fossil tissue contained 5.5 mg of the isotope ^{14}C. Based on the total mass of the fossil and the abundance of ^{14}C found in living organisms, the scientist estimated that the fossil material originally contained 44 mg of ^{14}C when the mammoth was alive. The half-life of ^{14}C is 5730 years.

(a) **Calculate** the age of the fossil.

(b) **Describe** whether ^{14}C can be used to determine the age of fossils that are older than approximately 50,000 years.

(c) Scientists use an isotope of uranium, ^{235}U, to test the age of ancient rock sediments. The half-life of ^{235}U is 704 million years. However, uranium is not found at high enough levels in plant and animal tissues to be directly measured in fossils. **Describe** how scientists can use the decay of ^{235}U to estimate the age of biological fossils.

Module 45

Continuing Evolution

LEARNING GOALS ▶LG 45.1 Evolution is an ongoing process.
▶LG 45.2 Pathogens evolve resistance to antibiotics and cause new diseases.
▶LG 45.3 Climate change leads to adaptation, migration, and extinction.

In the last few modules, we have defined evolution, described the mechanisms by which it occurs, and provided evidence for it. From these discussions, it might seem that evolution is something that occurred in the past. Indeed, the traits of living organisms, the features of fossils, and the distribution of organisms on Earth help us reconstruct the past. From these pieces of data, we can understand the evolutionary history of organisms and understand the paths they took to their current forms.

Just as evolution has shaped life on Earth from its inception, it remains an ongoing process and continues to shape life on this planet. All the mechanisms of evolution—natural selection, genetic drift, mutation, migration, and nonrandom mating—continue to operate in modern-day populations. Furthermore, as environments change because of human activities—building cities and roads; transforming forests and grasslands to farms and fields; using antibiotics, pesticides, and herbicides; and burning fossil fuels—the selective pressures on organisms are changing as well. In essence, humans are agents of evolution. While humans have played a role in shaping the environment for tens of thousands of years, the human footprint has become bigger and the pace of change has increased rapidly over the last century. We can measure and record these changes in populations in real time.

In this module, we will discuss evolution as an ongoing, continuing, and contemporary force, shaping organisms and populations today as it has done in the past. Although evolution happens in all populations over time, we see it most dramatically when the environment is altered. Therefore, in this module we will primarily focus on examples in which organisms respond to a change in the environment, such as the movement from land to water during the evolution of whales, a recent drought on the Galápagos Islands, the widespread use of antibiotics, and climate change.

PREP FOR THE AP® EXAM

FOCUS ON THE BIG IDEAS

EVOLUTION: Focus on how changes in the environment affect the ongoing evolution of populations.

45.1 Evolution is an ongoing process

As we discussed in earlier modules, many species are remarkably adapted to their environment. This close fit is the result of natural selection, in which organisms that survive and reproduce best in a particular environment leave more offspring, and these offspring inherit the variations that allowed their parents to do so well in that environment. In a stable environment, we see the results of past selective pressure and the adaptations of organisms well suited to the environment. As an environment changes, populations will change under the influence of natural selection. This is an example of directional selection, as we discussed in Module 41.

Life continues to evolve in a changing environment, now as it has done in the past. In this section, we will look at fossil and genomic evidence of recent evolutionary change, and then look at two examples of evolution that are happening today.

Continuous Change in the Fossil Record

As we discussed in Module 44, fossils provide some of the best evidence for evolutionary change. Fossils provide a direct link to the past. One of the most dramatic and best documented cases of evolutionary change in the fossil record is the evolution of whales from land-based mammals.

Today, there are two groups of whales, both shown in **FIGURE 45.1**—baleen whales such as humpback whales

a.

b.

FIGURE 45.1. Two groups of whales

There are two major groups of whales—baleen and toothed whales. (a) The humpback whale is a baleen whale and a member of the Mysticetes group of whales. (b) The killer whale is a toothed whale and a member of the Odontocetes group of whales. Photos: (a) M Swiet Productions/ Getty Images; (b) Nature Picture Library/Alamy Stock Photo

(*Megaptera novaeangliae*), and toothed whales such as killer whales (*Orcinus orca*). Both of these groups of whales are mammals. They have the defining characteristics of all mammals, including mammary glands, hair, a four-chambered heart, three middle ear bones, and warm bloodedness, among others.

Whales have descended from land-based mammals over the last 50–60 million years, which is relatively recently in evolutionary history. During this time, their features changed dramatically as they adapted to life in the water. For example, the hindlimbs became shorter and shorter over time, until they were lost completely in modern whales. The forelimbs were modified from legs to fins. The tail evolved to become shorter and more powerful, as it is used to propel

whales and their recent ancestors through the water. The eyes shifted to the side of the head and the nasal opening moved to the top of the head, both of which are adaptations for living in the water.

These anatomical changes are all documented in the fossil record, as we can see in **FIGURE 45.2**. In this figure, we see a timeline across the bottom, and reconstructions of some of the major fossils showing the changes that occurred as whales moved from land to water. These adaptations evolved by natural selection: in the water, individuals with smaller hindlimbs, a tail, and other changes in body form would have survived better and left more

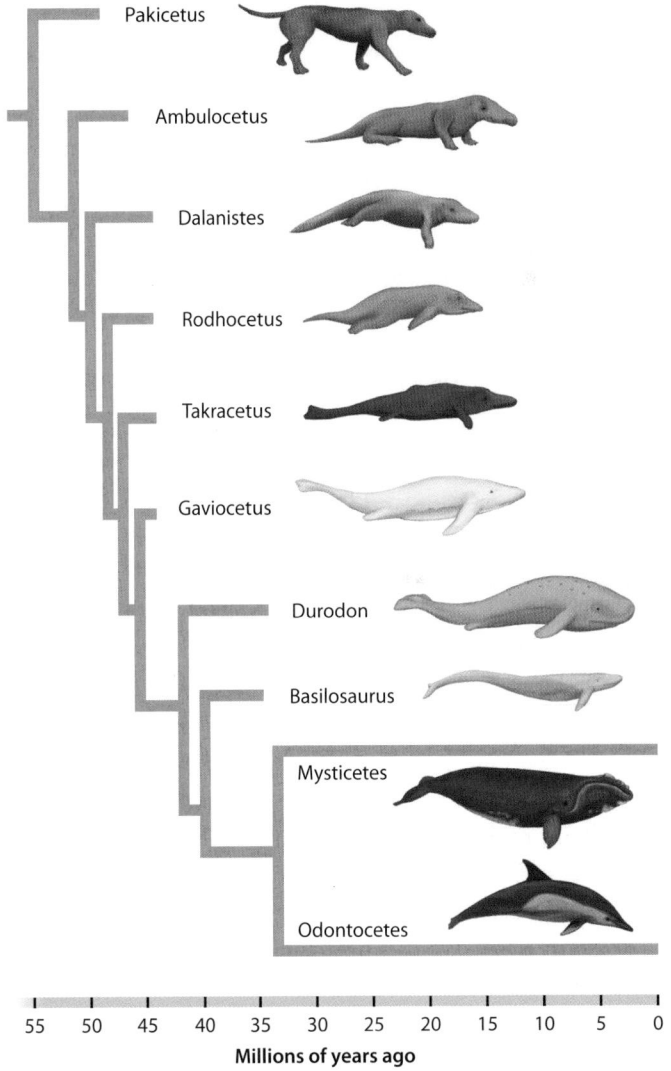

FIGURE 45.2 Whale evolution

The fossil record has revealed continuous change in the anatomy of whales as they transitioned from land to water. Adapted from: Zimmer, C. 2001. *Evolution: The Triumph of an Idea.* New York: Harper Collins.

offspring than other individuals, leading to a change in the population over time.

Whales are mammals but look like sharks and other fish. This is a dramatic example of convergent evolution, in which similar traits evolve independently in two groups because of similar selective pressures. It demonstrates how the selective pressure of living in the water can shape distantly related organisms to look very similar to one another. Recall from Module 44 that similarities that result from convergent evolution are analogous, like the tail of fishes and the tail of whales.

The evolution of whales is not the only example of well-documented, continuous change in the fossil record. We have may other examples, including the evolution of amphibians from fish about 380 million years ago, birds from their dinosaur relatives about 150 million years ago, horses over the last 60 million years, and humans and their extinct relatives over the last 6 million years. The fossil record also reveals species that have not changed very much over time, such as sharks and horseshoe crabs, which are sometimes called living fossils. In these cases, any changes in their outward forms are selected against, providing an example of stabilizing selection.

Genomic Changes over Time

The changes that we see in the fossil record over time as the ancestors of modern whales moved from the land to the water are anatomical. Physiological changes occurred alongside anatomical ones as whales adapted to life in the water. For example, whales are able to hold their breath for long periods and have to deal with the high pressure of the deep sea and the high salt environment of the oceans. Note that the environment doesn't cause changes in anatomy or physiology. Instead, variation in anatomical and physiological traits are present in the population, and the environment acts as a selective force, favoring those variations that increase survival and reproduction in that environment. Anatomical and physiological changes in turn require changes in genes that control development, physiology, and anatomy.

In some cases, we can find ancient DNA that provides evidence of genomic change over time. The oldest DNA that has been found and sequenced comes from the tooth of a mammoth that lived approximately 1 million years ago. While we can learn a lot from ancient DNA, we do not have DNA that stretches back 60 million years to record the evolution of whales from their land-based ancestors. To understand changes that occurred in the genomes of whales over the last 60 million years, we can instead take a comparative

approach. In other words, we can look at genomes of modern whales and compare them with the genomes of other mammals to see if we find changes in key genes that allowed whales to adapt to a watery environment.

One study looked at myoglobin, which is the protein in your muscle that binds oxygen and gives muscle (and meat) its red color. The more myoglobin there is in muscle, the redder it is. In whales, the muscle is almost black as a result of the high concentration of myoglobin. More myoglobin means more oxygen in muscle tissue. The high concentration of myoglobin in muscle is one adaptation that allows whales to store oxygen for so long.

Proteins tend to aggregate at high concentration, which can interfere with their function. Whale myoglobin, however, has been modified over time. Compared with myoglobin from other mammals, whale myoglobin has a higher negative charge on its surface, allowing the proteins to repel one another and not clump together, even though they are so densely packed. The high negative charge on the surface of the protein in turn is a result of changes in the myoglobin gene. Small changes in the myoglobin gene led to changes in mRNA and protein, leading to an altered protein that can be present in high concentrations without aggregating, thereby allowing whales to store more oxygen in their tissues.

Myoglobin is encoded by a single gene. Can we see examples of larger changes in genomes over time? The genomes of thousands of species have been sequenced, allowing researchers to look at sequence differences, gene order along a chromosome, and number of chromosomes in different species. The whale genome has been compared to other mammalian genomes. One of the major differences is the inactivation of 85 genes in whales as they transitioned to life in the water. To understand why and how these genes were silenced over time, scientists are studying how they are turned off and their function.

In addition to comparing whale genomes to genomes of land-based mammals, we can compare other genomes. The human genome and chimpanzee genome are remarkably similar because the two species are very closely related. The human genome and the mouse genome are much more different because there has been more time since they shared a common ancestor. The genomes of humans and mice are roughly the same size with the same number of genes. However, humans have 23 chromosomes, while mice have 20 chromosomes; the centromere is centrally located in most human chromosomes, but it is located toward the end in mice chromosomes; and there are some areas in which the order of genes has been shuffled in humans compared to mice. Scientists continue to study how these genomic changes lead to changes in the anatomy, physiology, behavior, and other traits of humans and mice.

Evolution in Our Time

The evolution of whales from their land-based ancestors is recent, but still occurred in the evolutionary past. Can we see any examples of evolution in animals happening today?

For many years, Peter and Rosemary Grant have studied in exquisite detail the populations of Darwin's finches on a tiny island in the Galápagos, Daphne Major. Because the island is so small, they have been able both to track every bird in the population and to monitor the resources, such as seeds for food, available to the birds. The island typically has a dry climate, with just 130 mm of rain falling during the wet season. Occasional years are much drier, with just 24 mm of rainfall, and in 1977 there was a severe drought. This had a dramatic impact on the island's vegetation, markedly reducing the amount of food available to the birds.

As a result, the population of the medium ground finch (*Geospiza fortis*) crashed during this time. Curiously, the size of seeds available as food shifted significantly over the course of the drought because plants that produce large seeds survived better than ones producing small seeds. Birds with small beaks had trouble breaking open these large, hard seeds, meaning that birds with bigger beaks were at an advantage.

Beak size is a trait that the Grants, following Darwin, have studied intensively. They had already established that it is highly heritable, meaning that most of the variation seen in beak size from bird to bird is attributable to genetic differences (rather than differences in nutrition, for example). They did this by comparing the beak size of offspring to the beak sizes of their parents and finding a strong positive correlation between the two. Put simply, two large-beaked parents tended to produce large-beaked offspring, and two small-beaked parents tended to produce small-beaked offspring.

Because of the differential survival of small- versus large-beaked birds in the drought, the average beak size of the survivors was larger than that of the pre-drought population. In the next generation, the offspring of the drought survivors had larger beaks than the pre-drought birds. **FIGURE 45.3** shows what happened in graph form. Beak size is plotted on the *x*-axis and number of finches on the *y*-axis. The distribution of beak sizes in 1976 is shown in Figure 45.3a, and the distribution of beak sizes in 1978 is shown in Figure 45.3b. We can see a clear difference in the distribution of beak size before and after the drought. In 1978, after the drought, the distribution has shifted to the right. The peak of the beak distribution in each graph reflects the most common beak size. The most common beak size in 1976 is 8.8 cm, whereas the most common beak size in 1978 is 9.8 cm, which is a full centimeter larger in just 2 years. The increase in beak size over this 2-year period is an example of directional selection, as discussed in Module 41.

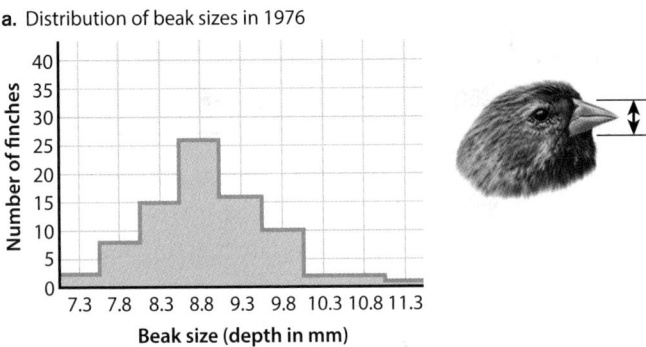

a. Distribution of beak sizes in 1976

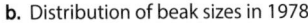

b. Distribution of beak sizes in 1978

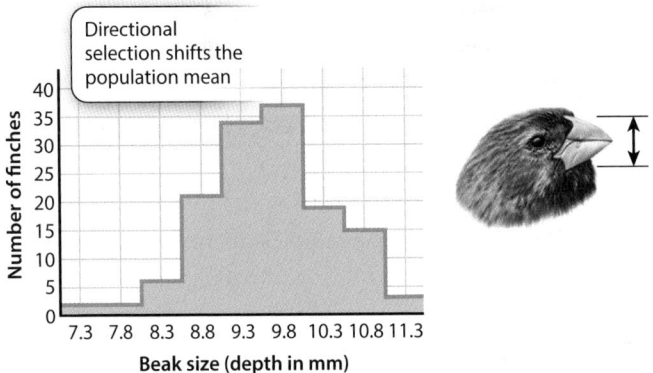

FIGURE 45.3 Evolution of larger beaks in response to a drought

Directional selection in a population of Galápagos finches caused a shift toward larger beaks as a result of a drought in 1977. The average beak size in 1976 distribution (a) is about 8.8 cm, and the average beak size in the 1978 distribution (b) is about 9.9 cm. The drought led to the evolution of larger beaks from 1976 to 1978. Data from Freeman & Herron *Evolutionary Analysis* 2004.

What is remarkable about this observation is that the impact of a single bout of severe weather has had, in effect, an instantaneous evolutionary effect on the population. We typically think of natural selection as a long, slow process, but, sometimes, it can happen extremely quickly.

A second interesting example comes from Nebraska, where another husband-and-wife team, Charles and Mary Brown, studied cliff swallows, shown in **FIGURE 45.4**. These birds usually nest on rock ledges and cliffs, as their name suggests, but more recently have started to build nests under bridges as highways have expanded across their range. This location puts them at risk of being hit by passing cars as they travel under the bridge. In effect, bridges are providing a new environment for the cliff swallows, and cars a new selective pressure.

The researchers studied road kill—birds that were hit by cars. They noticed that the number of birds hit by cars declined over time, suggesting that the birds were better able to avoid getting hit by cars over time. They also noticed that the wing length of birds that were killed was longer than that of the general population, suggesting that birds with longer wings tended to get hit by cars more than birds with shorter wings. Putting the two observations together, they reasoned that the birds evolved shorter wings over time as an adaptation to living under bridges. The shorter wings made them better able to maneuver and avoid being hit by oncoming cars.

These two examples demonstrate that, by doing careful measurements, it is possible to see the effects of evolution over just a few generations. And they both provide examples of evolution occurring in our time.

FIGURE 45.4 Cliff swallows

Cliff swallows are social birds that form large colonies. They typically build their nests underneath rock ledges in mountains and canyons, and more recently on buildings and under bridges, as shown here.

Photo: Richard Mittleman/Alamy Stock Photo

✓ Concept Check

1. **Describe** three anatomical changes that occurred as the ancestors of modern whales moved from the land to the water.

2. **Explain** what the Grants learned about evolution from the 1977 drought in the Galápagos.

45.2 Pathogens evolve resistance to antibiotics and cause new diseases

One of the most dramatic examples of continuing evolution comes from bacteria and other pathogens that cause human diseases. These are evolving in two ways, which we will discuss in this section. First, the widespread use of antibiotics is leading to the evolution of antibiotic resistance. Second, new pathogens are emerging that we have not been previously exposed to. Both of these are dramatic demonstrations of evolution in action, and they are also fast becoming a public health crisis.

Antibiotic Resistance

From the Black Death of the fourteenth century to the flu outbreak of 1918, infectious diseases have played a major role in human history. Today, we have unprecedented opportunities to control or even eradicate some of these infectious diseases through improved public health measures, vaccines, and antibiotics. At the same time, the use of antibiotics and similar drugs inevitably leads to the evolution of antibiotic resistance.

The strong selective pressures placed on pathogens by antibiotics pose a key challenge for twenty-first-century medicine. Tuberculosis provides an example. Tuberculosis is very common, affecting an estimated one-third of the world's population and causing 1.8 million deaths annually, mostly in sub-Saharan Africa. Symptoms are absent in most cases. Approximately 10% of cases, however, are active. Active cases are characterized by chronic cough, fever, night sweats, and weight loss.

Tuberculosis is caused by the bacterium *Mycobacterium tuberculosis*. Antibiotic treatment is effective against

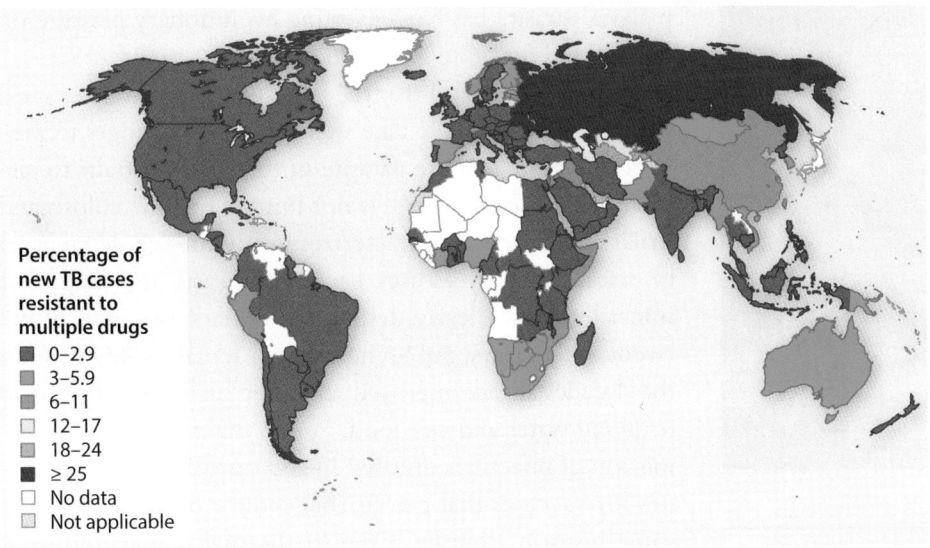

FIGURE 45.5 Distribution of extremely drug-resistant tuberculosis around the world

Percentage of new TB cases resistant to multiple drugs
- 0–2.9
- 3–5.9
- 6–11
- 12–17
- 18–24
- ≥ 25
- No data
- Not applicable

Recently, 8000 patients with extremely drug-resistant tuberculosis were reported.

Data from https://apps.who.int/iris/bitstream/handle/10665/329368/9789241565714-eng.pdf?ua=1

tuberculosis, but antibiotic resistance is on the rise. In 2018, approximately half a million new cases of antibiotic-resistant tuberculosis were reported worldwide. A small percentage of these infections has been characterized as "extremely drug resistant," meaning that the pathogens resist both standard treatments and alternative antibiotics developed over the past 25 years. **FIGURE 45.5** shows the distribution of extremely drug-resistant tuberculosis around the world. In effect, extremely drug-resistant tuberculosis is untreatable.

The evolution of antibiotic resistance is an example of natural selection in action, as we discussed in Modules 40 and 41. There is genetic variation among the bacteria that cause tuberculosis in terms of their sensitivity to antibiotics: the majority of bacteria are sensitive, but some are resistant. When we use antibiotics, we select for those that are antibiotic resistant. That is, the antibiotic-resistant bacteria survive and reproduce more than the antibiotic-sensitive ones. As a result, the population changes from being mostly antibiotic sensitive to mostly antibiotic resistant.

Malaria provides another example of natural selection. Malaria is a disease that causes fevers, chills, headaches, muscle pain, and sometimes even death. It is caused by a single-celled eukaryote, or protist, of the genus *Plasmodium*. There are five species of *Plasmodium* that infect humans. One of them, *P. falciparum*, is particularly dangerous, accounting for the vast majority of malaria fatalities. The malaria parasite is transmitted to humans by the bite of a female *Anopheles*

mosquito, shown in **FIGURE 45.6**. In any given year, malaria infects an estimated 500 million people, causing 1 million or more deaths. Infection is particularly severe in parts of Africa, where the direct and indirect effects of malaria account for almost half of all childhood mortality.

For centuries, humans have fought the infection with quinine, a chemical found in the bark of the South American *Cinchona* tree. In the 1940s, scientists developed a more sophisticated drug based on quinine, called chloroquine. Chloroquine was effective, inexpensive, and easy to administer. As a result, it was widely used. Unfortunately, in the late 1950s, the malaria parasite began showing signs of resistance to chloroquine. Within 20 years, the resistance had spread to Africa, so that today the once-potent medication is no longer effective against most strains of the parasite.

Since that time, many new antimalarial drugs have been developed. Over time, their usefulness has declined as resistance to each of these drugs has evolved in populations of *Plasmodium*. For example, in 2015, Chinese scientist Tu Youyou won the Nobel Prize in Medicine or Physiology for her work in identifying artemisinin, the active

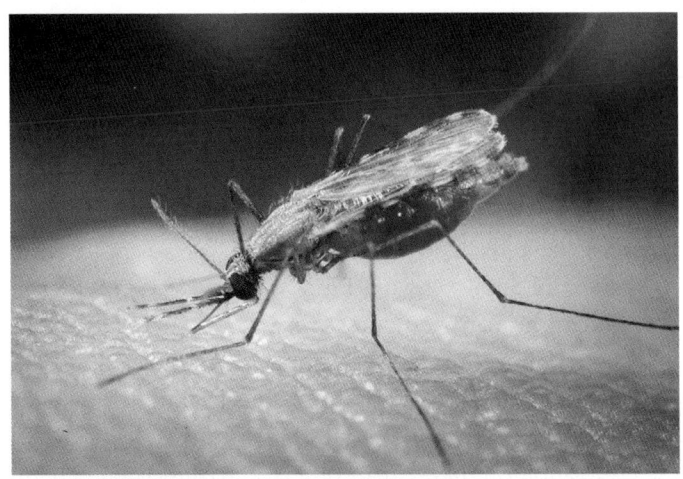

FIGURE 45.6 Malaria

The malaria parasite, a single-celled eukaryote, is transmitted by the bite of a female *Anopheles* mosquito, shown here. Photo: CDC/James Gathany/ Science Source

FIGURE 45.7 Sweet wormwood

Sweet wormwood (*Artemisia annua*) is the source of artemisinin, which has recently been used to treat infection by the malaria parasite. Photo: Kanjana Wattanakungchai/Shutterstock

antimalarial component in sweet wormwood (*Artemisia annua*). This a plant, shown in **FIGURE 45.7**, has long been used in Chinese traditional medicine. Today, artemisinin and its derivatives are key to our attempts to combat malaria, but even here, within a few years of their first being deployed against malaria, resistance has emerged.

A similar process plays out when we treat insects with insecticides, pests with pesticides, and plants with herbicides. In each case, we are selecting for resistance among these groups of organisms over time. In the case of malaria, which is transmitted by mosquitoes, insecticide-soaked bed nets have proved a highly effective means of protection against malaria in affected areas, reducing transmission by as much as 90%. Natural selection, however, has begun to chip away at success in controlling the disease: mosquito strains have begun to emerge that are resistant to some insecticides.

Not only are humans affecting the malaria parasite; the malaria parasite has also affected humans. The battle between malaria and humankind has raged through the ages. Scientists have recovered DNA of the malaria parasite from the bodies of 3500-year-old Egyptian mummies—evidence that ancient humans were infected with the malaria parasite. The close connection between humans, mosquitoes, and the malaria parasite almost certainly extends back much further.

In fact, people who come from regions where malaria is endemic are more likely than others to have certain genetic changes compared with people who come from malaria-free regions. For example, as we discussed in Module 37, the *S* allele as a heterozygote provides some protection from infection by malaria. This observation indicates that the malaria parasite has been exerting evolutionary pressure on humankind for quite some time.

This back-and-forth between antibiotics and pathogens—where we take steps or evolve changes to prevent infection, and the parasite in turn evolves traits to circumvent these defenses—is not limited to tuberculosis and malaria, but instead characterizes nearly the whole range of diseases caused by viruses, bacteria, protists, and fungi that infect humans. Clearly, despite the remarkable gains of the twentieth century, public health will remain a key issue in the decades ahead. Improved sanitation, in the form of clean drinking water and safe food, can do much to reduce transmission of infectious diseases. In addition, the search continues for vaccines that can further reduce or even eliminate some historic scourges. Even so, the ongoing evolution of drug resistance requires the continuing efforts of scientists to stay one step ahead of pathogens.

PREP FOR THE AP® EXAM

AP® EXAM TIP

Remember that evidence of evolution can come from the present as well as from the past. In the present, we see examples such as antibiotic resistance in bacteria, the size of finch beaks, and variation in gene sequences. From the past, we can compare the traits of fossil organisms with those of living organisms, and compare ancient and modern DNA.

Emerging Diseases

In addition to resistance, new strains of viruses, bacteria, and other pathogens continue to evolve. As we saw in Module 39, an emerging disease is an infectious disease that has appeared recently or spreads rapidly. A recent example is SARS-CoV-2, the virus that causes the disease COVID-19 and caused a worldwide pandemic in 2020 and 2021, which is discussed in Module 39. Other examples include the human immunodeficiency virus (HIV), Zika virus, certain strains of the flu virus, and Ebola. Emerging diseases like these and others evolve in such a way that they are able to spread more easily, cause more severe disease, or even infect new species of organisms.

For example, pathogens that typically cause mild disease may evolve to become more virulent. This has occurred multiple times in the flu (influenza) virus. Influenza is an RNA virus that infects mammals and birds. It is notable for causing seasonal outbreaks. However, four times in the last century, most notably in 1918, the virus spread more widely than usual, infected more people, and caused more severe disease. These changes result at least in part from mutations that occur in the virus and then spread through the population of viruses. Some of these mutations lead to new traits, such as the ability to spread more quickly from person to person, infect a host cell more easily, or influence the immune response of the host.

Pathogens may also change their host specificity over time. That is, pathogens that infect one group of organisms may change in such a way that they are able to infect a new group of organisms. Host specificity results from the way that viruses enter cells, as discussed in Module 39. Proteins on the surface of the virus bind to proteins on the surface of host cells. These proteins interact in a specific manner, and it is this interaction that determines the types of cells and organisms that a virus can infect.

The host range of a virus can change as viral surface proteins evolve over time to interact with new host cell proteins. For example, avian, or bird, flu is a type of influenza virus that was once restricted primarily to birds but now can infect humans. The first reported case in humans was in 1996, and the disease has since spread widely. Similarly, HIV once infected only nonhuman primates such as chimpanzees, but its host range expanded in the twentieth century to include humans. The SARS-CoV-2 virus is thought to come from bats, but recently evolved in such a way that it now infects humans. Today, we see SARS-CoV-2 evolve with the emergence of new variants. Scientists are studying how these variants differ from one another in terms of ease of spread and severity of disease.

These examples illustrate how ongoing evolution in viruses can lead to changes in their host range, causing new, emerging diseases.

✓ Concept Check

3. **Describe** how bacteria and other pathogens evolve resistance to antibiotics over time.

4. **Describe** two kinds of evolutionary changes that we see in pathogens.

45.3 Climate change leads to adaptation, migration, and extinction

One of the most dramatic ways the environment is changing today is the warming of the planet, leading to global changes in the climate and oceans. Species are responding to the changing climate in different ways. We tend to think about natural selection in terms of populations adapting to a constant or slowly changing environment. In the twenty-first century, however, the environment will probably be a rapidly moving target, so genotypes that conferred high fitness in the past may not prove advantageous in the future. In this section, we will briefly introduce the topic of climate change and then discuss three ways that populations are responding: some are adapting; some are migrating; and some are going extinct.

Climate Change

Every year, humans are adding significant amounts of carbon dioxide (CO_2) to the atmosphere. We do this primarily by burning fossil fuels and setting fire to rainforests, as shown in **FIGURE 45.8** on page 640. Clearing forests for agriculture generates still more CO_2. When preparing land for agriculture, the existing vegetation is generally burned off, converting much of the carbon in biomass and soil to CO_2.

CO_2 is a greenhouse gas, which is a gas in Earth's atmosphere that helps to warm the planet and is discussed more fully in Module 57. It acts like the glass panes of a greenhouse, which allow sunlight to enter but prevent heat from leaving. CO_2 is just one of several greenhouse gases in the atmosphere: water vapor is another, and methane a third. Without these gases absorbing and trapping heat, average surface temperatures would fall below freezing and life would not be possible. However, because CO_2 levels are increasing rapidly, its greenhouse effect is also increasing. Likewise, methane levels are rising rapidly, in large part because of increasing food production. The thawing of permafrost at high latitudes releases additional methane that was trapped in frozen soils when the ice formed long ago.

As CO_2 levels in the atmosphere are increasing, the surface temperature of Earth is also rising. For the past century, scientists, sailors, and citizen scientists have monitored temperature at weather stations around the world. More recently, satellites have enabled us to measure temperature in places as remote as the high Arctic and the middle of the ocean. The results are clear: in most parts of the world, the surface temperature of Earth is increasing.

FIGURE 45.8 Human-generated sources of carbon dioxide

Every year, humans add carbon dioxide to the atmosphere, predominantly by (a) burning fossil fuels and (b) burning rainforests, as seen in the Amazon. Photos: (a) Tim McCaig/Getty Images; (b) Stiockbyte/Getty Images

Effects of Changing Climate

What will be the consequences of climate change? Some areas will benefit. For example, temperature increase in New England and Scandinavia will mean longer growing seasons. Other regions will suffer. As precipitation patterns change, many places will become drier, including already water-limited areas of the southwestern United States. A number of climate models predict that some of the strongest declines in rainfall will occur in regions that currently produce much of the corn and wheat that feed the world. Already, farmers in southeastern Australia have experienced the worst droughts in a century, and with them unprecedented damage from brushfires. Climate models also predict an increased frequency of strong hurricanes along the western margin of the Atlantic Ocean. Scientists will be watching carefully to

see whether such extreme events increase in frequency in the coming years.

Plants illustrate the complicated ways in which organisms respond to the new selective pressure brought on by environmental change. Meticulous notes taken during the 1840s by Henry David Thoreau provide an inventory of plant species in woodlands in Concord, Massachusetts. The graph in **FIGURE 45.9a** shows that mean annual temperature in this area has increased by about 2.5°C over the last 150 years. The graph

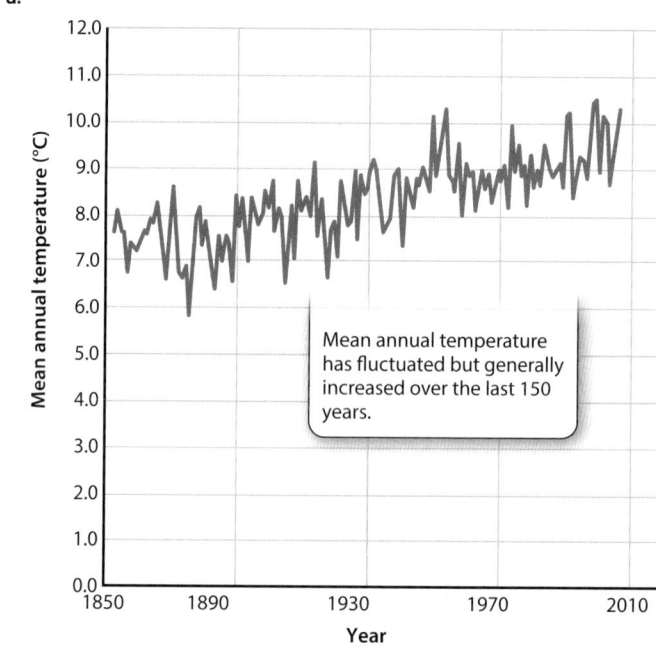

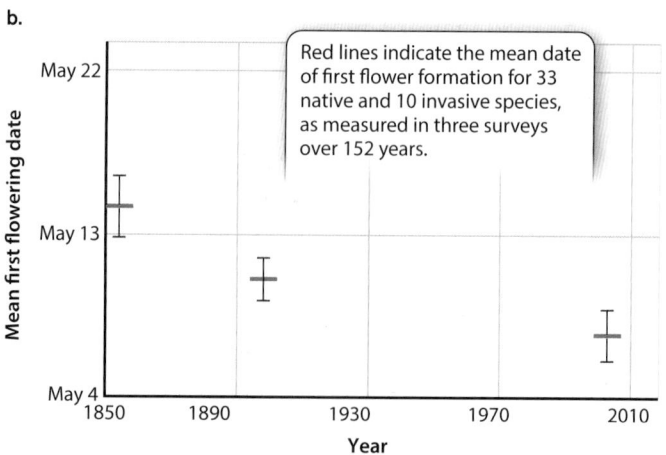

FIGURE 45.9 Plant response to warming temperatures

Observations over the last 150 years show the effects of warming temperatures in Concord, Massachusetts. (a) Regional temperatures have become warmer in the woods of Concord, Massachusetts. (b) As temperatures have become warmer, the time of first flowering has become earlier. Error bars are calculated as ±1 standard deviation on either side of the mean. Data from Miller-Rushing, A. J., and R. B. Primack. 2008. "Global Warming and Flowering Times in Thoreau's Concord: A Community Perspective." *Ecology* 89(2):332–341. doi: 10.1890/07-0068.1

in Figure 45.9b shows that many species of plants flower a week earlier than they did in the nineteenth century. Earlier onsets of leaves, flowers, and fruit have also been documented throughout the temperate zones of North America, Europe, and Asia.

Another response that plants show to elevated levels of CO_2 is increased growth. Where nitrogen in the soil is abundant, increased CO_2 levels tend to stimulate growth. However, in most terrestrial environments, soil nitrogen levels are low, limiting any growth response to elevated CO_2. Moreover, increased drought stress and more frequent outbreaks of insect pests and wildfires may counteract any capacity for accelerated growth.

Plants grown in air with CO_2 levels like those predicted for the end of this century show other changes as well: the amount of nitrogen in plant tissues commonly declines, as does the proportion of resources devoted to reproduction. This difference doesn't reflect genetic change, but rather an individual plant's physiological response to altered environmental conditions. But if environmental change persists for a long time, the varying abilities of different plants to respond physiologically will result in changing allele frequencies, as some variants survive and reproduce better than others. That is, plant populations may evolve by natural selection in response to climate change so they are better adapted to a warmer world.

Adaptation is one response to climate change, but, given the pace at which atmospheric composition and climate are changing, many populations may not have time to adapt, a process that typically takes place slowly over many generations. Another possible response to climate change is migration. Fossils deposited as Earth's climate warmed at the end of the last ice age show that many plant species dramatically changed their geographic distributions in response to changing climate. Trees don't move, but they can migrate by dispersing seeds to new areas.

Migration is a potentially important response to environmental changes projected for the twenty-first century, but it requires a continuous route to get from one place to another, and it can be a challenge to find these direct paths on continents broken up by agricultural lands and cities. For this reason, some biologists advocate "assisted migration": the deliberate transplantation of plant populations from existing habitats to new ones more favorable to growth. To make sure that assisted migration does more good than harm, we need to do further research to

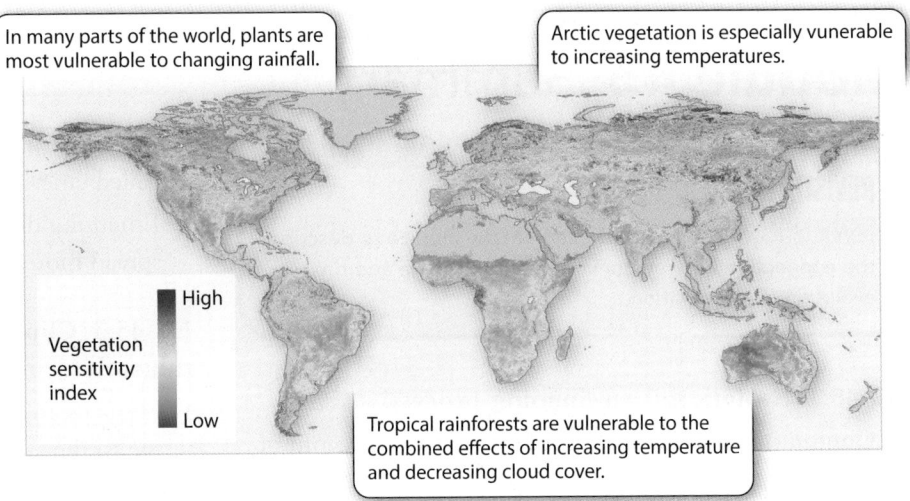

In many parts of the world, plants are most vulnerable to changing rainfall.

Arctic vegetation is especially vunerable to increasing temperatures.

Tropical rainforests are vulnerable to the combined effects of increasing temperature and decreasing cloud cover.

Vegetation sensitivity index — High / Low

FIGURE 45.10 Estimated sensitivity of global vegetation to twenty-first-century climate change

Red and orange indicate high vulnerability of vegetation, and green and yellow indicate low vulnerability of vegetation. Data from Seddon, A. W. R., M. Macias-Fauria, P. R. Long, D. Benz, and K. J. Wilis. 2016. "Sensitivity of Global Terrestrial Ecosystems to Climate Variability." *Nature* 531:229–232.

understand how plants grow and compete in new ecological situations and under different environmental conditions.

In addition to adaptation and migration, species may also go extinct if the environmental change is so extreme that there are no variants in the population capable of surviving and reproducing. A number of species have declined in abundance or even disappeared since Thoreau's time, and these tend to be the plants least able to change their flowering time. If they are unable to adapt to warmer temperatures, they may flower when there are no pollinators around, for example. This observation highlights the fact that different plant species respond in distinct ways to changing climate, which favors some populations over others.

FIGURE 45.10 shows an estimate of the sensitivity of plants around the world to projected twenty-first-century climate. Areas in red are especially vulnerable: plants at high latitudes are particularly sensitive to warming, whereas tropical rain forests are more vulnerable to predicted decreases in cloud cover. In other areas, plants are most sensitive to predicted changes in water availability. Species in these regions could face diminished population size, decreased growth rates, or even extinction.

✓ Concept Check

5. **Describe** how a greenhouse gas warms the planet.

6. **Identify** three greenhouse gases.

7. **Describe** three ways that species may respond to global changes in climate.

Module 45 Summary

PREP FOR THE AP® EXAM

REVISIT THE BIG IDEAS

EVOLUTION: Use one example from the module to **describe** the connection between environmental change and the evolution of a population.

LG 45.1 Evolution is an ongoing process.

- Continuous evolution is recorded in the fossil record of whales over the last 50–60 million years. Page 632
- Whales evolved from land-based mammals and adapted to life in the water. Page 633
- Continuous evolution is also apparent by comparing genomes of different species. Page 634
- Evolution is occurring today and is visible in Darwin's finches and cliff swallows. Page 635

LG 45.2 Pathogens evolve resistance to antibiotics and cause new diseases.

- The use of antibiotics is one of the success stories of twentieth-century medicine, but has led to antibiotic resistance, a major challenge for twenty-first-century biologists and physicians. Page 636
- Insecticide, pesticide, and herbicide resistance is also common as organisms evolve by natural selection. Page 637

- Evolution has led to new, rapidly spreading diseases, called emerging diseases. Page 638
- Emerging diseases may cause more severe symptoms, spread more easily, or infect new species. Page 639

LG 45.3 Climate change leads to adaptation, migration, and extinction.

- Humans are adding significant amounts of carbon dioxide to the atmosphere, primarily through the burning of fossil fuels and the clearing of rainforests. Page 639
- Carbon dioxide is a greenhouse gas, trapping heat and thereby keeping the surface of Earth warm. Page 639
- Other greenhouse gases include methane and water vapor. Page 639
- Greenhouse gases play an important role in maintaining temperatures on Earth that are compatible with life, but increasing levels of greenhouse gases are resulting in higher surface temperatures. Page 639
- Climate change has already affected the distribution of species and the composition of communities, and will likely continue throughout the twenty-first century. Page 641

Review Questions

1. All of the following changes occurred in the evolution of whales from land-based mammals except
   (A) loss of hindlimbs.
   (B) evolution of fins from forelimbs.
   (C) development of gills.
   (D) evolution of a streamlined shape with a powerful tail.

2. The human genome is most similar to the
   (A) mouse genome.
   (B) chimpanzee genome.
   (C) whale genome.
   (D) HIV genome.

3. Emerging diseases
   (A) spread slowly.
   (B) are ancient diseases.
   (C) do not cause illness or death.
   (D) have only recently infected humans.

4. In addition to carbon dioxide and methane, which is also considered to be a greenhouse gas?
   (A) Oxygen
   (B) Carbon monoxide
   (C) Ozone
   (D) Water vapor

5. In response to global climate change, it is likely that some plant species will survive while others will go extinct. Which does not explain this pattern?

(A) Greenhouse gases are toxic at current levels, so increasing levels of these gases will cause some plants to go extinct.

(B) Plants differ in their abilities to respond physiologically to persistent environmental change, and some will survive increased temperatures.

(C) Temperatures may increase more rapidly than some species can adapt, which will lead to their extinction.

(D) Some plants are able to migrate more effectively than others and can move into areas with more favorable conditions for survival.

6. Using records made by Henry David Thoreau in the 1840s, scientists have documented what pattern of change in flowering plants in Concord, Massachusetts?

(A) As temperatures have warmed, flowering occurs earlier in many species.

(B) As temperatures have warmed, flowering occurs later in many species.

(C) As temperatures have become more variable, flowering occurs earlier in many species.

(D) As temperatures have become more variable, flowering occurs later in many species.

Module 45
AP® Practice Questions

Section 1: Multiple-Choice Questions

Choose the best answer for questions 1–4.

1. Some organisms, such as crocodiles and horseshoe crabs, have shown very little evolution in their morphology, or the shape and structure of their bodies, over hundreds of millions of years. These organisms are sometimes called living fossils. Which explains why these "living fossils" remain morphologically unchanged?

(A) Mutations that change body structure do not occur in these species.

(B) There is no natural selective pressure for these species to evolve a different morphology.

(C) These species are incapable of evolution.

(D) Morphological evolution in these species is not heritable.

Use the following information to answer questions 2 and 3.

The cerulean-capped manakin (*Lepidothrix coeruleocapilla*) is a tropical bird that lives in forests in the Andes mountains in South America. Global climate change has resulted in increased average temperature in the Andes, and, as a consequence, change in the ranges of the plants and animals that live there. The abundance of *L. coeruleocapilla* was measured at different elevations on a mountain in Peru in 1985, and then again on the same mountain in 2017. The number of *L. coeruleocapilla* observed at each elevation in each year is given in the table below.

Sampling elevation (m)	450	650	850	1050	1350
Number of birds observed in 1985	0	0	1	24	22
Number of birds observed in 2017	0	0	0	10	29

Use the following figure to answer question 2.

a.
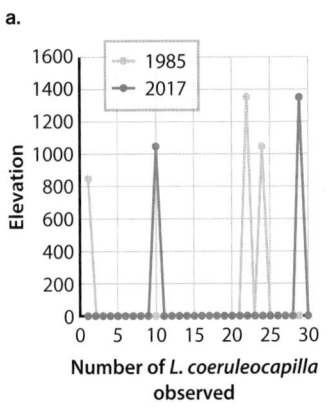

b.
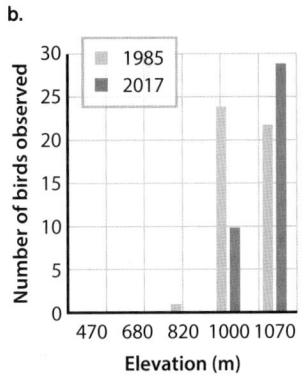

c.
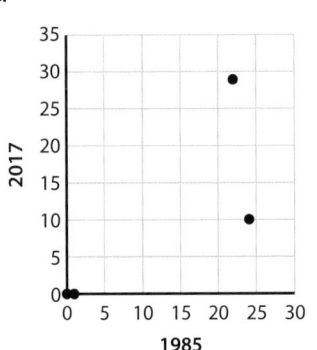

d.
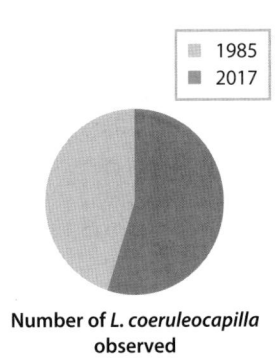

2. Which plot best illustrates the data?

(A) a

(B) b

(C) c

(D) d

3. Based on the data provided, which is the best prediction of what would happen to the *L. coeruleocapilla* population on this mountain if climate change were to continue on its present trajectory?

   (A) The birds would adapt to the higher temperatures.

   (B) The birds would be forced to migrate to a new mountain.

   (C) The birds would expand their range into lower elevations.

   (D) The birds would be at risk of extinction from the mountain if the highest elevations became too warm.

4. Pyrethroids are a class of chemical insecticides that are used to kill pest insects on farms. One use of pyrethroids is to control the number of house flies on dairy farms. However, after many years of use, pyrethroids have become less effective at reducing the number of house flies on farms. Which of the following statements is most likely to explain this observation?

   (A) The house flies needed to resist the pyrethroid so they developed a mutation that prevents them from being killed.

   (B) House flies carrying a mutation that protects them from being killed by pyrethroids are able to survive and reproduce more than house flies that don't carry this mutation.

   (C) Each house fly evolved resistance to pyrethroids over their life span.

   (D) After many years of exposure, the house flies have learned to avoid environments contaminated with pyrethroids.

Section 2: Free-Response Question

Write your answer to each part clearly. Support your answers with relevant information and examples. Where calculations are required, show your work.

An industrial factory built on the shore of a river pumps in water from the river to maintain a low temperature, and then pumps the water back out into the river. The factory was built in 1950. In 1970, environmental scientists collected aquatic snails from a site just downstream of the spot where the factory released water back into the river. The environmental scientists collected snails from the same spot again in 1994. In both years, the scientists also collected samples of the same species of snail from a site in the river upstream of the factory. Additionally, the scientists were able to secure museum samples of the snail that were collected from both locations in 1921, before the factory was built.

The *Tol1* gene encodes a protein that is involved in stress responses. There are two known alleles of the *Tol1* gene: *T* and *t*. The scientists determined the genotype at the *Tol1* gene of all snail specimens and obtained the data shown in the table.

Location	Year	Genotype (percent)		
		TT	Tt	tt
Upstream of factory	1921	0	11	89
Upstream of factory	1970	0	9	91
Upstream of factory	1994	1	11	88
Downstream of factory	1921	0	10	90
Downstream of factory	1970	16	52	32
Downstream of factory	1994	84	16	0

(a) **Construct** a histogram to illustrate the frequencies of all three genotypes in all three years at both locations.

(b) **Describe** the role of sampling the site upstream of the factory in the experiment.

(c) **Make a claim** to explain the difference in genotype frequencies between the two locations in 1994.

(d) **Propose a new experiment** to test the hypothesis that is described in part c above. **Describe** the results that would be expected from this experiment if the hypothesis is true.

Module 46

Phylogeny

LEARNING GOALS ▶**LG 46.1** The history of life on Earth can be described as a branching tree.
▶**LG 46.2** Shared traits are present in more than one species.
▶**LG 46.3** Evolutionary trees can be built in several ways.
▶**LG 46.4** Phylogenetic trees can help solve practical problems.

Life originated on Earth more than 3.5 billion years ago. Today, an estimated 10 million species inhabit the planet. Short of inventing a time machine, how can we reconstruct those 3.5 billion years of evolutionary history in order to understand the extraordinary series of events that have ultimately resulted in the biological diversity we see around us today?

One way to explore evolutionary history is to look at fossils, which we discussed in Module 44. Fossils provide direct evidence of past life. The fossil record provides insights that we would otherwise not know or be aware of, such as information about extinct organisms and mass extinctions. At the same time, the fossil record is incomplete because some organisms are more likely to be preserved than others.

Another way to understand evolutionary history is to look at traits of living organisms. As noted in Module 40, humans and chimpanzees are more similar to each other than they are to other primates. Humans, chimpanzees, and other primates in turn are more similar to one another than any one of them is to a mouse. And humans, chimpanzees, other primates, and mice are more similar to one another than any of them is to a catfish. This pattern of nested

similarity was recognized more than 200 years ago and used by the Swedish naturalist Carolus Linnaeus to classify all life on Earth. A century later, Charles Darwin recognized this pattern as the expected outcome of a process of "descent with modification," or evolution.

The process of evolution can be visualized as an evolutionary tree. Evolutionary trees provide a visual model of the past and show us how all the species alive today are related to one another. In this module, we will discuss how we can use the traits of living organisms to work backward and build an evolutionary tree that describes what happened over time. We will discuss both how to read a tree and how to build a tree given a set of species. Finally, we will put evolutionary trees to practical use and learn how they can be used to solve problems.

PREP FOR THE AP® EXAM

FOCUS ON THE BIG IDEAS

EVOLUTION: Think about what is possible for scientists to learn about the evolutionary history of a species and how species are related from a phylogenetic tree.

46.1 The history of life on Earth can be described as a branching tree

Darwin recognized that the species he observed were the modified descendants of earlier species. Distinct populations of an ancestral species separate and diverge through time, again and again, giving rise to multiple descendant species through the evolutionary processes we explored in earlier modules. The result is the pattern of nested similarities observed in nature. The evolutionary history showing the relatedness of a group of organisms is called a **phylogeny.**

A phylogeny is much like the genealogy that records our own family histories. We introduced phylogenetic trees in Module 40. In this section, we will look at them more closely and describe how scientists use them.

Phylogenetic Trees

As you learned in Module 40, a phylogenetic tree is a visual representation of the evolutionary history of organisms over time. Phylogenetic trees can also be called phylogenies, evolutionary trees, or simply trees. The tree in **FIGURE 46.1** on page 646 is a phylogenetic tree of chimpanzees, humans, and cows. In this tree, time runs from left to right, from the

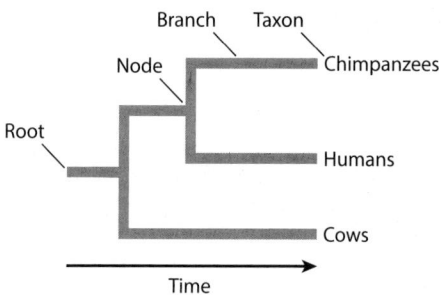

FIGURE 46.1 A phylogenetic tree of chimpanzees, humans, cows, and their ancestors

In this tree, time runs from left to right, from the root to the tips of the branches. Nodes represent common ancestors.

root of the tree to the tips of the branches. The root, which is at the base of the tree, represents the common ancestor of all the groups represented in the tree, and the tips of the branches represent the descendants of that ancestor. As you move from the root of a phylogenetic tree to the tips, you are moving forward in time.

The groups of interest are listed at the tips of the branches. These groups, called taxa (singular, taxon), can be any group of organisms, including individuals, species, or even domains. In Figure 46.1, the three taxa are chimpanzees, humans, and cows. The point on a phylogenetic tree where a branch splits is a node. A node represents the common ancestor from which the descendant species diverged, and therefore also the time when one species split into two.

A phylogenetic tree shows us the evolutionary history of groups of organisms, and it also illustrates how they are related to one another. The more recent the common ancestor—which is indicated by a node—the more closely related two groups are to each other. **FIGURE 46.2** shows

that humans and chimpanzees share a more recent common ancestor (about 7 million years ago) than humans and cows (about 95 million years ago). This means that humans and chimpanzees are more closely related to each other than either is to cows. Groups that are more closely related to each other than to any other group on the tree are sister groups. For example, in Figure 46.2, chimpanzees and humans are sister groups, indicated in red.

Nodes in a phylogenetic tree can be rotated without changing the evolutionary relationships of the taxa. In the tree of humans, chimpanzees, and cows, for example, humans can be placed above chimpanzees simply by rotating the node where the chimpanzee and human lines diverge, as shown in **FIGURE 46.3**. Evolutionary relatedness is determined by following nodes from the root to the tips of the tree, rather than by reading the order of the tips on the page. In other words, what matters is the branching order over time: the line that leads to cows branches off the line that leads to chimpanzees and humans earlier than the chimpanzee line and human line branch off from each other. What's not important is the order of the taxa on the page (from top to bottom or bottom to top in this tree), as this arrangement can be changed by rotating nodes.

In addition to rotating nodes, trees can be drawn in different styles. Some of these styles are shown in **FIGURE 46.4**. In 46.4a, time runs from left to right, but in Figure 46.4b, time runs from right to left. In both Figures 46.4c and 46.4d, time runs from bottom to top, but the lines are horizontal or vertical in Figure 46.4c, while they are angled in Figure 46.4d. You can tell that these four trees are equivalent by following the branching order over time. In each case, if you start from the root of the tree, the first node, or branch point, separates the line that leads to cows from the line that leads to chimpanzees and humans, and the second

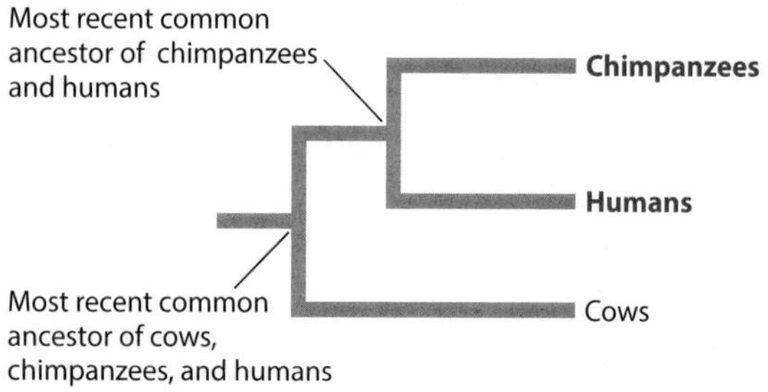

FIGURE 46.2 Close and distant relationships on a phylogenetic tree

Humans and chimpanzees are sister groups, indicated in red; they are more closely related to each other than either is to cows because they share a more recent common ancestor with each other than they share with cows.

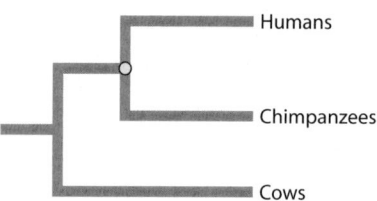

FIGURE 46.3 Rotating nodes

Nodes can be rotated without changing evolutionary relationships. The node representing the common ancestor of humans and chimpanzees has been rotated in this tree compared to that in Figure 46.2. Notice that the branching order over time is the same here as in Figure 46.2, but the order of the taxa is different along the tips of the branches.

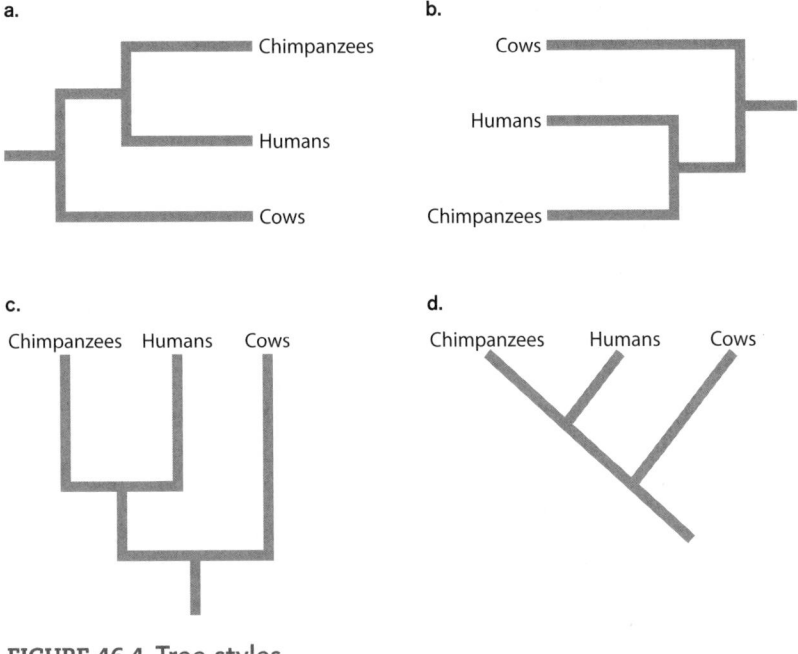

FIGURE 46.4 Tree styles

All four trees show the same evolutionary relationships and history. In (a), time runs from left to right; in (b), time runs from right to left. In (c) and (d), time runs from bottom to top, but the angles of the branch points are different.

Figure 46.5, just focusing on turtles, crocodiles and alligators, and birds. Each tree is made by rotating one of the nodes of the previous tree. Although the trees look different, they are all the same because they show the same branching order over time and therefore the same evolutionary relationships. For example, in all four trees, the closest living relatives of birds are crocodiles and alligators because the two groups share a more recent common ancestor than they do with any other group in the tree.

A phylogenetic tree does not in any way imply that more recently evolved groups are more "advanced" than groups that arose earlier. A modern lungfish, for example, is not more primitive or "less evolved" than an alligator, even though its group branches off the branch of the vertebrate tree in Figure 46.5 from an earlier node than the alligator group does. After all, both species are the products of the same interval of evolution since their divergence from a common ancestor.

node separates the line that leads to chimpanzees from the line that leads to humans.

FIGURE 46.5 shows a phylogenetic tree that includes more vertebrate animals. The informal name at the end of each branch represents a group of organisms, many of them familiar. We sometimes find it useful to refer to groups of species this way, for example, "frogs," rather than name all the individual species of frogs or list the characteristics they have in common. Remember, though, that such named groups represent a number of species. If, for example, we could zoom in on the branch labeled "Frogs," we would see that it consists of many smaller branches, each representing a distinct species of frog, either living or extinct.

The tree in Figure 46.5 provides information about evolutionary relationships among vertebrates. For example, it suggests that the closest living relatives of birds are crocodiles and alligators because the two groups share a more recent common ancestor than they do with any other group. The tree also proposes that the closest relatives of all tetrapod (four-legged) vertebrates are lungfish, which are fish with lobed limbs and the ability to breathe air.

Nodes in this tree can be rotated without changing the evolutionary relationships, just like in all trees. **FIGURE 46.6** on page 648 shows four trees from the larger tree in

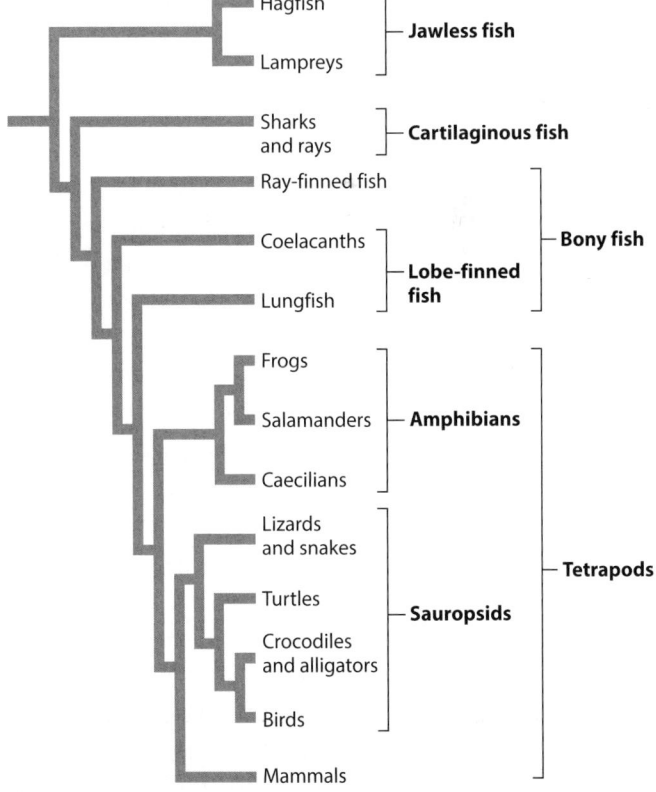

FIGURE 46.5 A phylogenetic tree of vertebrate animals

This tree shows the evolutionary history and relationships among different groups of vertebrates.

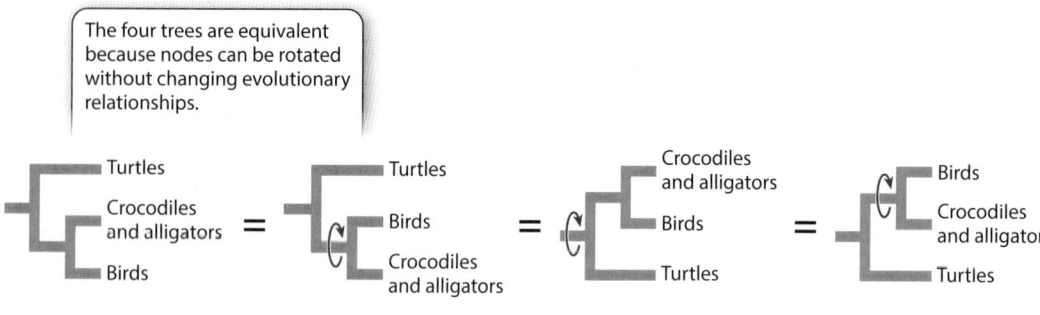

The four trees are equivalent because nodes can be rotated without changing evolutionary relationships.

Turtles	Turtles	Crocodiles and alligators	Birds
Crocodiles and alligators	Birds	Birds	Crocodiles and alligators
Birds	Crocodiles and alligators	Turtles	Turtles

FIGURE 46.6 Equivalent trees

Rotating nodes does not change the evolutionary relationships of the groups. These four trees are equivalent.

Phylogenetic Trees as Hypotheses

A phylogenetic tree is a hypothesis about the evolutionary history, or phylogeny, of a species. Phylogenetic trees are hypotheses because they represent the best model, or explanation, of the relatedness of organisms on the basis of existing data. As is the case with all hypotheses, as more information is collected that supports a given tree, confidence that the tree accurately represents evolutionary relationships grows. And, like all hypotheses, if new information is not consistent with a given tree, the tree is revised.

For example, the tree in Figure 46.5 was built from careful analyses of the morphological features, such as the presence or absence of a jaw or walking legs, and molecular attributes of the species or other groups under study. In essence, a tree is a hypothesis about the order of branching events in evolution, and that hypothesis can be tested by gathering more information about anatomical and molecular traits.

What happens if different sources of data produce trees showing different evolutionary relationships? How do scientists select one of many trees? There are several ways to try to resolve conflicting evolutionary trees. The principle of parsimony, for example, states that the simplest explanation is the most likely. Applied to phylogenetic trees, we assume that the phylogenetic tree requiring the fewest evolutionary steps or changes is likely the most accurate. Over time, most features of organisms stay the same: we have the same number of ears as our ancestors, the same number of fingers, and so on.

In some cases, there is no way to determine which evolutionary hypothesis is most likely. In these cases, all trees are considered equally likely until more data are collected to support one tree over another. Today, DNA sequencing is providing a wealth of molecular data that can be used to build and evaluate evolutionary trees. Molecular data on its own is not more accurate than morphological data, but the sheer quantity of molecular data means that scientists have high confidence in the trees that they construct to show the evolutionary history and relatedness of groups of organisms.

Classification

The nested pattern of similarities among species has been recognized by naturalists for centuries and is the basis for our modern classification system, called the Linnaean classification system. **FIGURE 46.7** shows how the classification system works for dogs. Note that as you move up from the level of species, each group includes more and more organisms. For example, closely related species are grouped into a genus (plural, genera). Closely related genera in turn belong to a larger, more inclusive branch of the tree, known as a family. Closely related families form an order, orders form a class, classes form a phylum (plural, phyla), and phyla form a kingdom, with each more inclusive group occupying a successively larger branch on the tree. Biologists today refer to the three largest branches of the entire tree of life as domains, which were introduced in Unit 2. The three domains are Eukarya, Bacteria, and Archaea.

Although the Linnaean classification system was developed before Darwin proposed his theory of evolution, it reflects, in essence, the evolutionary process over time. We can see this relationship by superimposing the classification system on an evolutionary tree of life, as shown in

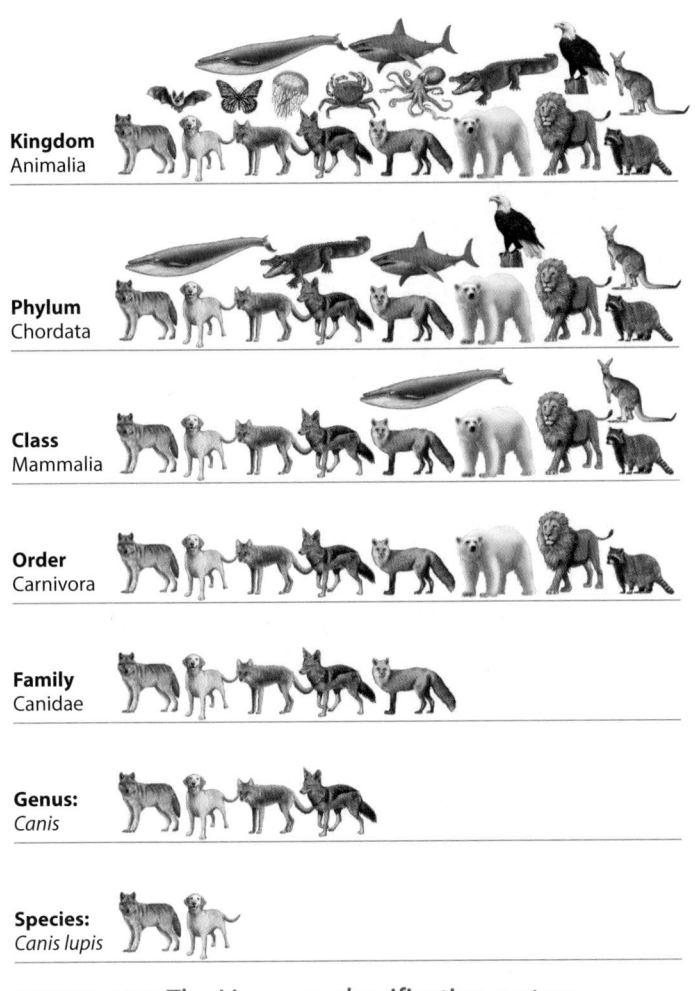

Kingdom
Animalia

Phylum
Chordata

Class
Mammalia

Order
Carnivora

Family
Canidae

Genus:
Canis

Species:
Canis lupis

FIGURE 46.7 The Linnaean classification system

The levels are more and more inclusive as you move up in the classification system.

FIGURE 46.8 Relationship between the Linnaean classification system and the evolutionary process

Classification reflects our understanding of phylogenetic relationships, and the groupings reflect the order of branching.

FIGURE 46.8. In this tree, we can see that closely related species belong to the same genus; closely related genera belong to the same family; and so on.

Cladograms

Up to this point, we have used the word "group" to mean all the species at some taxonomic level, such as a genus or family. A different way to group organisms is by reference to their common ancestor. A clade, also called a monophyletic group, is a group of organisms that includes a common ancestor and all of its descendants. For example, consider the group labeled Amphibians in Figure 46.5. Amphibians are monophyletic because all of the groups classified as amphibians share a common ancestor not shared by any other taxa.

In contrast, consider the group of animals traditionally recognized as reptiles in Figure 46.5 on page 647, which includes turtles, snakes, lizards, crocodiles, and alligators. Traditionally, birds were not considered reptiles, although they share a common ancestor with the included animals. The traditional group reptiles is paraphyletic, meaning that it includes some, but not all, of the descendants of a common ancestor. However, if we make one large group that includes birds, turtles, snakes, lizards, crocodiles, and alligators, that group is monophyletic because it includes a common ancestor and all of its descendants. This group (birds + traditional reptiles) is called sauropsids.

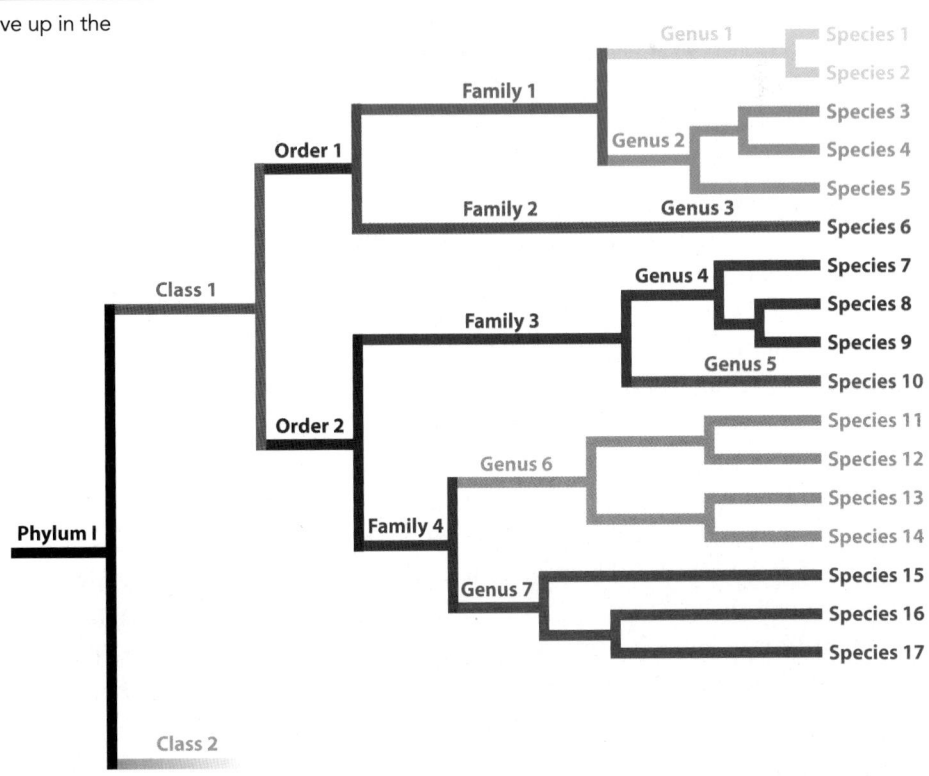

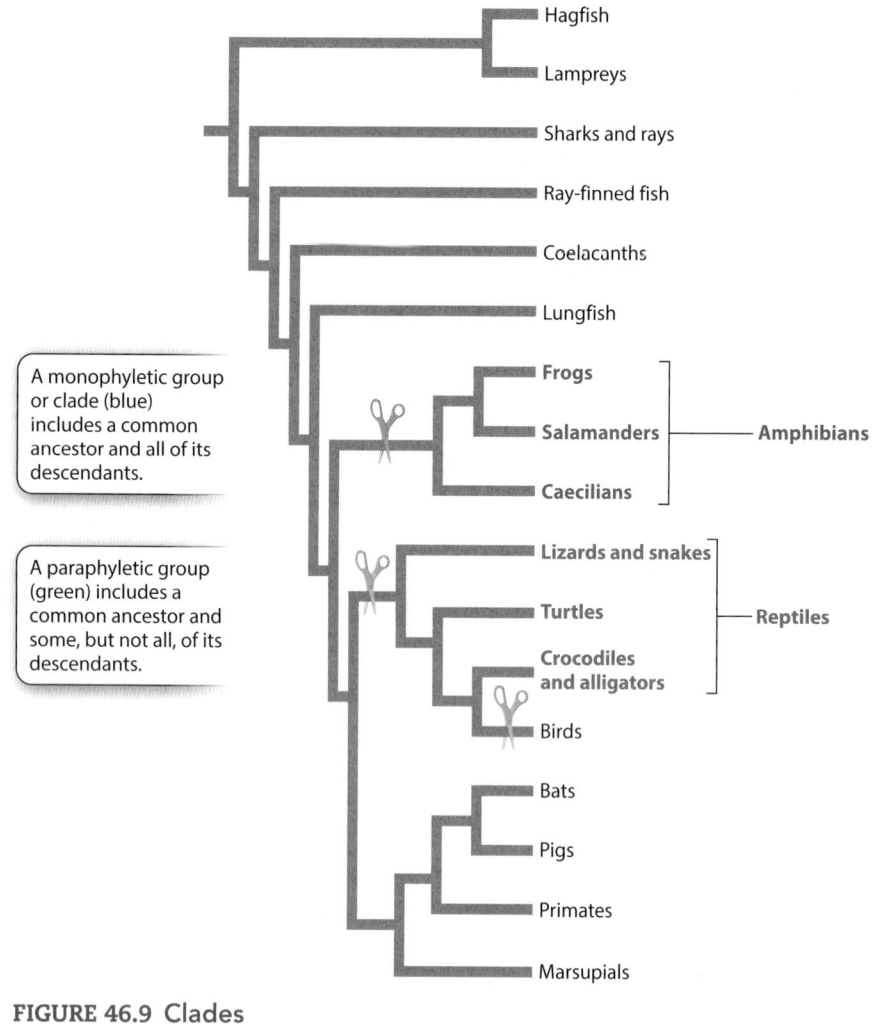

A monophyletic group or clade (blue) includes a common ancestor and all of its descendants.

A paraphyletic group (green) includes a common ancestor and some, but not all, of its descendants.

FIGURE 46.9 Clades

Only clades, or monophyletic groups, reflect evolutionary relationships because only they include all the descendants of a common ancestor.

There is a simple way to distinguish between monophyletic and paraphyletic groups, which is illustrated in **FIGURE 46.9**. If you need to make only one cut to separate a group from the rest of the phylogenetic tree, the group is monophyletic, as shown in blue in the figure. The group amphibians is a clade because only one cut separates it from the tree. If you need to make two cuts to separate the group from the rest of the tree, the group is paraphyletic, as shown in green in Figure 46.8. The group reptiles requires two cuts to separate it from the tree because birds are traditionally not considered to be reptiles. Therefore, the traditional group reptiles is not a clade.

Identifying clades is a major goal of phylogenetics because they include all descendants of a common ancestor and only the descendants of that common ancestor.

As a consequence, clades alone show the evolutionary path a given group has taken since its origin. Omitting some members of a group, as in the case of reptiles and other paraphyletic groups, can provide a misleading sense of evolutionary history. By referencing clades, we effectively convey our knowledge of their evolutionary history.

A **cladogram** is a type of evolutionary tree based on clades. Some biologists use the terms "phylogenetic tree" and "cladogram" interchangeably. Some biologists, however, make a distinction between the two types of trees. In a phylogenetic tree, the lengths of the branches may be drawn proportional to the amount of time that elapsed, whereas in a cladogram, the branch lengths are arbitrary.

The amount of time represented by the length of a branch can be determined in some cases by finding a fossil that represents the common ancestor of two groups. Alternatively, scientists can use the number of DNA sequence differences between two groups to estimate the time that they have been separated from each other. The more time that has elapsed since two groups have separated, the more opportunities for different mutations to occur in the two groups and for them to spread throughout the populations, so the DNA sequences will be more different in the two groups. The less time, the more similar the DNA sequences of the two groups will be. As we saw in Module 44, using DNA sequence differences between two groups to measure the time since they shared a common ancestor is known as a molecular clock.

✓ Concept Check

1. **Identify** the parts of a phylogenetic tree.
2. **Explain** why you should not read a phylogenetic tree by looking at the order of taxa along the tips.
3. **Describe** a clade.

46.2 Shared traits are present in more than one species

Phylogenetic trees are built by comparing traits, or *characters*, of different groups of organisms. **Characters** are the anatomical, physiological, or molecular features that make up organisms. In general, characters have several observed conditions, called character states.

In the simplest case, a character can be present or absent. For example, lungs are present in tetrapods and lungfish, but absent in other vertebrate animals. Alternatively, there may be multiple character states for a given trait. Petals are a character of flowers, for example, and each observed arrangement—petals arranged in a helical pattern, petals arranged in a whorl, or petals fused into a tube—can be considered a state of the character of petal arrangement.

Characters sometimes present in two or more groups of organisms. In this case, they are considered **shared characters.** Shared characters arise for one of two reasons. The character may have been present in the common ancestor of the two groups and retained over time. In other words, the character is shared because of common ancestry. Alternatively, the character may have evolved independently in the two groups as an adaptation to similar environments. In other words, the character is shared because of convergent evolution, which is discussed in Module 41. As we saw in Module 44, characters that are similar because of common ancestry are homologous, or a homology among different species. Characters that are similar due to convergent evolution are analogous, or an analogy among different species.

Consider two examples. Mammals and birds both produce amniotic eggs, which are water-tight eggs that allow reproduction on land. Amniotic eggs occur only in groups descended from the common ancestor at the node connecting the mammal and sauropsid branches of the tree shown in **FIGURE 46.10**. Thus, we reason that birds and mammals each inherited this character from a common ancestor in which the amniotic egg first evolved. The other sauropsids inherited this trait from the same common ancestor.

Now let's consider a second example. Wings are a character shared by birds and bats. Much evidence supports the view that wings in these two groups do not reflect descent from a common, winged ancestor; instead, wings evolved independently in the two groups as adaptations to their environments. Notice in Figure 46.10 that wings are shown twice, indicating that they evolved separately and

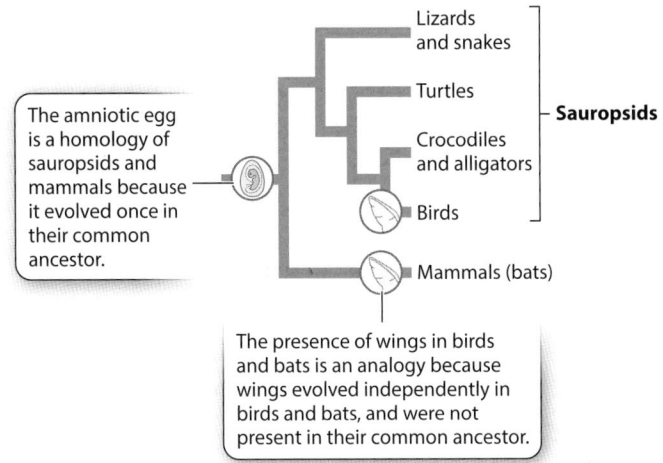

FIGURE 46.10 Homology and analogy

A homology is a similarity that results from shared ancestry, such as the amniotic egg in sauropsids and mammals. An analogy is a similarity that results from convergent evolution, such as wings in birds and bats.

independently in birds and bats. They are the result of convergent evolution.

We know of many examples of convergent evolution less dramatic than wings. In some cases, we understand the genetic basis of the convergence. For example, some organisms use echolocation to locate objects in their surroundings by emitting sounds and interpreting the patterns of echoes. Echolocation has evolved in both bats and dolphins, but not in most other mammals. Echolocation requires a protein called prestin. Prestin is in the hair cells of mammalian ears and is involved in hearing ultrasonic frequencies. Bats and dolphins independently evolved similar changes in their prestin genes, apparently convergent adaptations for echolocation. Similarly, unrelated fishes that live in freezing water at the poles, Arctic and Antarctic, have evolved similar glycoproteins that act as molecular "antifreeze," preventing the formation of ice in their tissues. They evolved this trait independently, so it's an example of an analogy in these fishes.

A phylogenetic tree can help you see the difference between homologous and analogous traits. If a trait appears once in the tree and then is maintained over time, it's a homology among the descendant species. If a trait appears more than once, it evolved multiple times and is an analogy among the descendant species. Both types of traits can be seen in Figure 46.10.

How do scientists determine whether a shared trait among different species is homologous or analogous?

Typically, they weigh evidence that suggests where other traits place the two organisms on a phylogenetic tree, look at where on the organisms the trait occurs, and examine the anatomical or genetic details of how the trait is constructed. Wings in birds and bats are similar in morphological position—both are modified forelimbs. However, wings in the two differ in details of construction. For example, the bat wing is supported by long fingers, while the bird wing is formed by the entire forelimb. Moreover, all other traits of birds and bats place them at the tips of different lineages, with many nonwinged species between them and their most recent common ancestor. Collectively, these pieces of information suggest that bats and birds did not retain wings from a common ancestor, but rather evolved wings independently.

✓ Concept Check

4. **Describe** why two species might share the same characteristic or trait.

5. **Describe** similarities and differences between homologous traits and analogous traits.

46.3 Evolutionary trees can be built in several ways

Up to this point, we have focused on how to read a phylogenetic tree. But how do we infer evolutionary history from a group of organisms? That is, how do we actually build a phylogenetic tree? Biologists use traits or characteristics of organisms to figure out their relationships. These traits can be morphological, such as legs, wings, feathers, and so on. Traits can also be developmental, for example, three germ layers or the amniotic egg. Finally, scientists can use molecular traits, such as specific nucleic acid or protein sequences.

In this section, we will discuss two different ways that evolutionary trees can be built given the traits of living organisms. The first approach uses homologies and the second approach uses overall similarity.

Shared, Derived Characters

Similarities among organisms are particularly important for building trees because they sometimes suggest shared ancestry. However, a key principle of constructing trees is that only some similarities can actually tell us about evolutionary relationships among species. Other similarities can be misleading. We have already seen that homologous traits reflect inheritance from a common ancestor and therefore are useful for building trees. Analogous traits arise by convergent evolution, so are not useful for building trees.

In addition, it turns out that only some homologies are useful for building trees. For example, character states that are unique to a given species or clade can't tell us anything about its sister group. Because they evolved after the divergence of the group from its sister group, they can be used to characterize a group but not to relate it to other groups. Consider wings again: we know that wings evolved in an ancestor of bats, but they do not occur in any other mammal. Because no other mammal has wings, this character cannot tell us which mammalian group is the sister group to bats.

Similarly, homologies formed in the common ancestor of the entire group and therefore present in all its descendants do not help to identify sister-group relationships among the descendants of that common ancestor. All bats have wings, so wings by themselves do not help us to sort out evolutionary relationships among different bat species. To build phylogenetic trees, we need homologies that are shared by some, but not all, of the members of the group under consideration.

A **derived character** state is an evolutionary innovation, such as the change from five toes to a single toe—the hoof—in the ancestor of horses and donkeys. When this kind of novelty arises in the common ancestor of two taxa, it is shared by both and is known as a synapomorphy. The hoof is a synapomorphy, or shared derived character, defining horses and donkeys as sister groups.

The opposable thumb is a derived trait found in both humans and chimpanzees. This shared trait indicates that humans and chimpanzees have a more recent common ancestor than either group has with cows, which don't have opposable thumbs. The opposable thumb was an evolutionary innovation in the lineage that eventually produced humans and chimpanzees and is therefore shared by both humans and chimpanzees.

FIGURE 46.11 illustrates the major shared derived characters that have helped us construct the phylogeny of vertebrates. For example, the lung is a character present in lungfish and tetrapods, but absent in other vertebrates. The presence of lungs provides one piece of evidence that lungfish are the sister group of tetrapods.

Building a phylogenetic tree on the basis of shared derived characters is called cladistics. FIGURE 46.12a on page 654 uses cladistics to build a phylogenetic tree of horses, humans, and lizards. Humans and horses are both mammals with many shared derived traits, such as mammary glands, hair, and three middle ear bones. Humans and horses are

therefore more closely related to each other than either is to lizards. Put another way, humans and horses share a more recent common ancestor than either shares with lizards.

We can also use blood temperature to build a tree that includes humans, horses, and lizards. Humans and horses are warm-blooded, that is, they maintain a relatively stable body temperature; lizards are cold-blooded, that is, they have a body temperature that changes with the environment. Warm bloodedness is a shared derived trait; it is shared by horses and humans because it arose in a mammal from which both are descended. Based on their common physiological trait, humans and horses are placed together as they are in

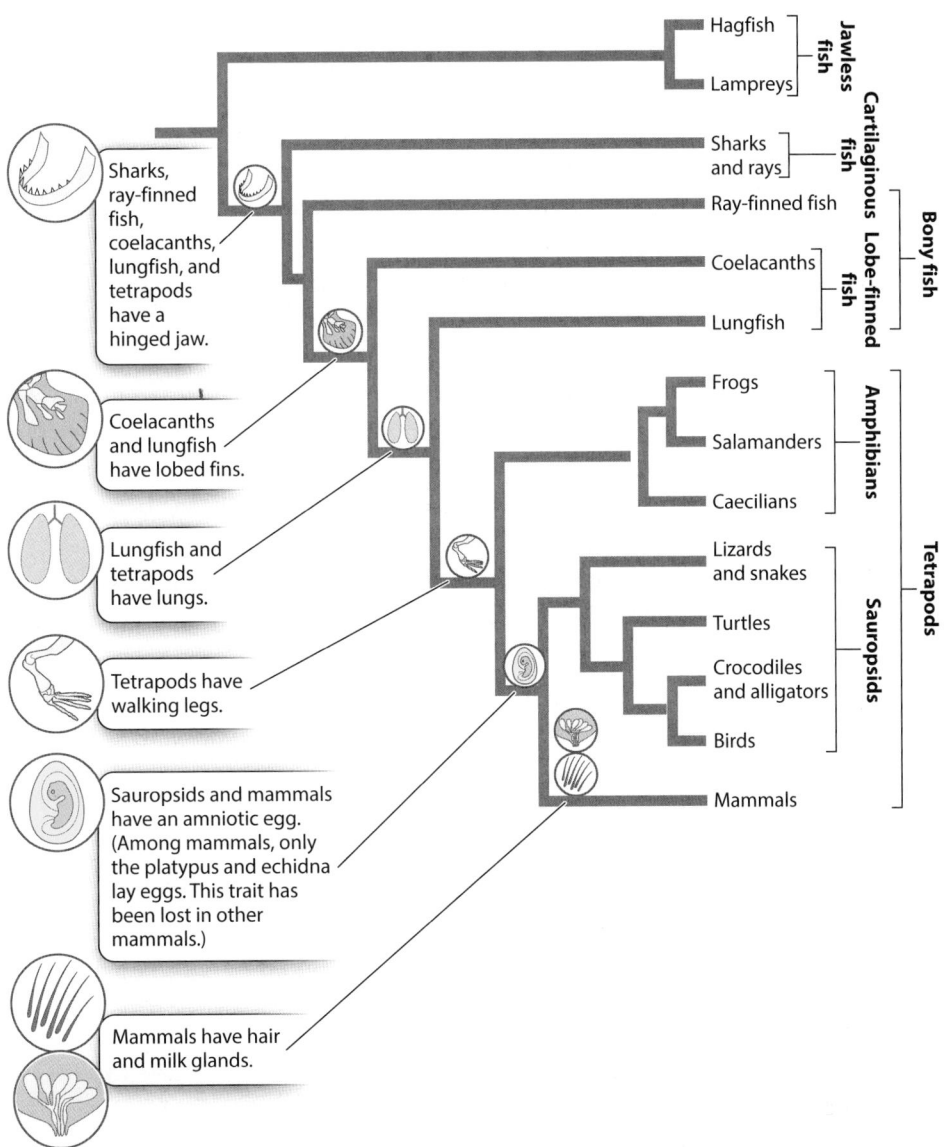

FIGURE 46.11 Shared derived characters of vertebrate animals

Homologies that are present in some, but not all, members of a group are called shared derived characters and they can be used to construct phylogenetic trees.

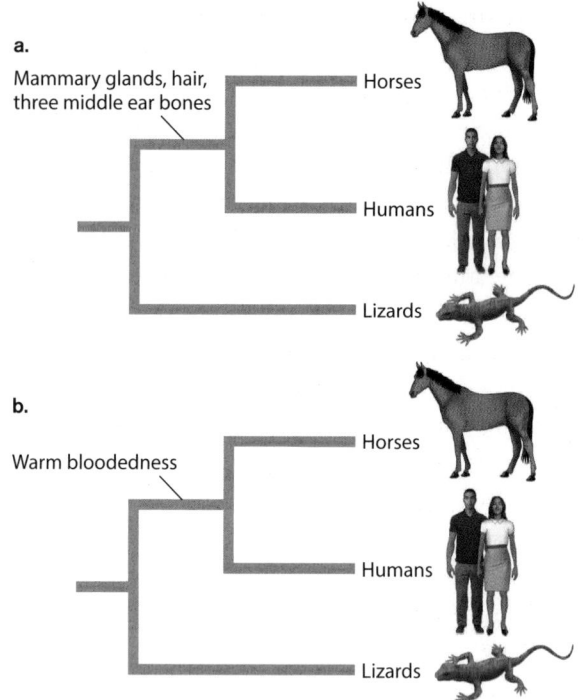

a.

Mammary glands, hair, three middle ear bones

Horses

Humans

Lizards

b.

Warm bloodedness

Horses

Humans

Lizards

FIGURE 46.12 Phylogenetic trees of horses, humans, and lizards

These two phylogenetic trees were built using shared derived traits. (a) This tree was constructed using the shared derived traits of mammary glands, hair, and three middle-ear bones. (b) This tree was constructed using the shared derived trait of warm bloodedness.

Figure 46.12b, which is the same tree as in Figure 46.12a. Using shared derived characters for phylogenetic reconstruction is the foundation of the cladistics approach.

Warm bloodedness is a more recent trait in evolutionary history that is derived from the older, ancestral condition of cold bloodedness. In a cladistics-based approach, scientists determine whether a particular trait is ancestral or derived, and then use shared derived traits when building a tree. Consider again the number of fingers in humans and horses. How do scientists know that five fingers is ancestral and one finger (the hoof) is derived, and not the other way around? One way to answer this question is by looking in the fossil record. The phylogenetic tree shown in **FIGURE 46.13** illustrates a human, a horse, and *Proterogyrinus,* a four-legged vertebrate (tetrapod) that is a common ancestor of horses and humans. Because the common ancestor of horses and humans had five fingers, we infer that five fingers is ancestral and the single hoof of horses is derived.

What if there are no available fossils? How do scientists determine whether a trait is ancestral or derived? In this case, they can make an *outgroup* comparison. An **outgroup** is a

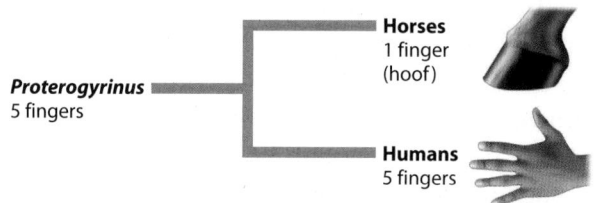

Proterogyrinus
5 fingers

Horses
1 finger
(hoof)

Humans
5 fingers

FIGURE 46.13 The evolution of hooves and hands from a five-fingered tetrapod

Because the fossil organism *Proterogyrinus* had five fingers, we infer that five fingers is the ancestral state and one finger (the hoof) is the derived state.

taxon whose common ancestor with the group of interest, known as the ingroup, is older than the common ancestor of the group of interest. Outgroup comparison is based on the idea that traits are more likely to stay the same than they are to change. For this reason, scientists expect an outgroup to have preserved the ancestral state in most instances.

If the group of interest includes horses and humans, an outgroup can be lizards, as shown in **FIGURE 46.14**. Lizards can serve as an outgroup for horses and humans because the common ancestor of lizards, humans, and horses, indicated by the node CA1 on the tree, is older than the common ancestor of humans and horses, indicated by the node CA2 on the tree. Because the outgroup—the lizard—has five fingers, you can conclude that having five fingers is the ancestral state and having one finger (a hoof) is the derived state. Note that this is the same answer at which we arrived by looking at the fossil record. In science, when two different approaches or datasets give you the same answer, scientists have more confidence in the result.

So far, we have seen how morphological traits, such as the number of fingers, and physiological traits, such as blood

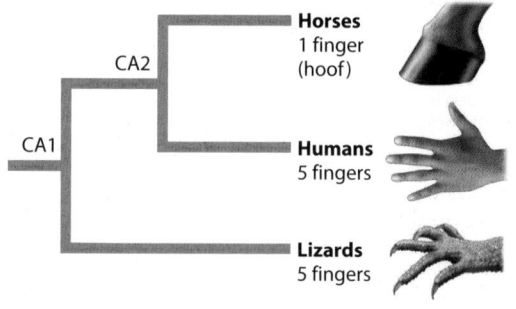

CA2

CA1

Horses
1 finger
(hoof)

Humans
5 fingers

Lizards
5 fingers

FIGURE 46.14 Outgroup comparison

The lizard is considered an outgroup with respect to humans and horses because the common ancestor of lizards and humans/horses (CA1) is older than the common ancestor of humans and horses (CA2).

temperature, are used to build evolutionary trees. Today, biologists typically use molecular data to build trees. The data are different, but the logic is the same. The sequence of amino acids of a protein can be used for this purpose, as can the nucleotides along a strand of DNA.

From genealogy to phylogeny, tracing mutations in DNA or RNA sequences has revolutionized the reconstruction of historical genetic connections. Whether we are tracing the paternity of the children of Sally Hemings, the mother of several of Thomas Jefferson's children; identifying the origin of a recent cholera epidemic in Haiti; or placing whales in a phylogenetic tree of mammals, molecular data are a rich source of phylogenetic insight.

Molecular data do not provide an intrinsically better record of history than anatomical data; molecular data simply provide more information because there are more characters that can vary among the species. A sequence of DNA with hundreds or thousands of nucleotides can represent that many characters, as opposed to the tens of characters usually visible in morphological studies. For microbes and viruses, for example, very little morphology is available, so molecular information is critical for building trees. Once a gene or other stretch of DNA or RNA is identified that seems likely to vary among the species to be studied, sequences are obtained and aligned to identify homologous nucleotide sites. Analyses of this kind commonly involve comparisons of sequences of about 1000 nucleotides from one or more genes.

Increasingly, though, the availability of whole-genome sequences is changing the way we do molecular phylogenetics. With this approach, rather than comparing the sequences of a few genes, we compare the sequences of entire genomes. The process of using molecular data is conceptually similar to the process described earlier for morphological data. Through comparison to an outgroup, we can identify shared derived molecular characters, whether DNA nucleotides or amino acids in proteins, and generate the phylogeny on the basis of synapomorphies as before.

Distance Methods

We have seen that phylogenetic trees can be built using shared derived traits. An alternative approach to building phylogenetic trees is based on overall similarity. With this approach, the assumption is that the descendants of a recent common ancestor will have had relatively little time to evolve differences, whereas the descendants of an ancient common ancestor have had a lot of time to do so. Thus, the extent of similarity indicates how recently two groups shared a common ancestor.

Biologists measure similarity using some measure of distance between the two species, such as morphological (anatomical) similarity or the number of base pair differences between DNA sequences. In other words, according to this distance-based approach, the more similar two species appear or the more similar their DNA sequences are, the more closely related they are.

This distance-based approach works well in some cases. For example, humans and chimpanzees appear more similar to each other than humans appear to cows, so we conclude that humans and chimpanzees share a more recent common ancestor than humans share with cows. Humans and chimpanzees have had less time to evolve differences and therefore appear similar to each other. Humans and cows have had more time to evolve differences and therefore look quite different from one another.

Note that this distance-based approach works for molecular analysis of humans, chimpanzees, and cows, too: for a given gene, there are fewer base pair differences between humans and chimpanzees than there are between humans and cows.

FIGURE 46.15 on page 656 shows how distance methods work, using great apes and molecular traits. In the top part of the figure, we see DNA sequences for humans, chimpanzees, gorillas, and gibbons. The bases in black are the same in all four species; the bases in red are variable. We then count the number of base differences between each pair of species. For example, the human and chimp sequences differ at two nucleotides, and the human and gorilla sequences differ at four nucleotides. We continue to compare the DNA sequences of all pairs of species, and then record the number of sequence differences in the table below the sequences. Those with the fewest differences are considered to be most closely related; those with a greater number of differences are considered to be more distantly related. Therefore, we build the tree starting with humans and chimps because they have the fewest differences. Then we add gorillas, and so on. In the end, we have a phylogenetic tree that includes all four species on the right side of the figure.

Underpinning this approach is the assumption that the rate of evolution, or change, is more or less constant in different groups. If evolutionary rates vary, then we can't use this approach because it can produce an incorrect tree. Consider the tree in **FIGURE 46.16a** on page 656. As you can see in the tree, sharks and rays are more closely related to each other than either of them is to the other major group of living fishes, called bony fish. If you were to use distance measures to build a tree, you could focus on morphological

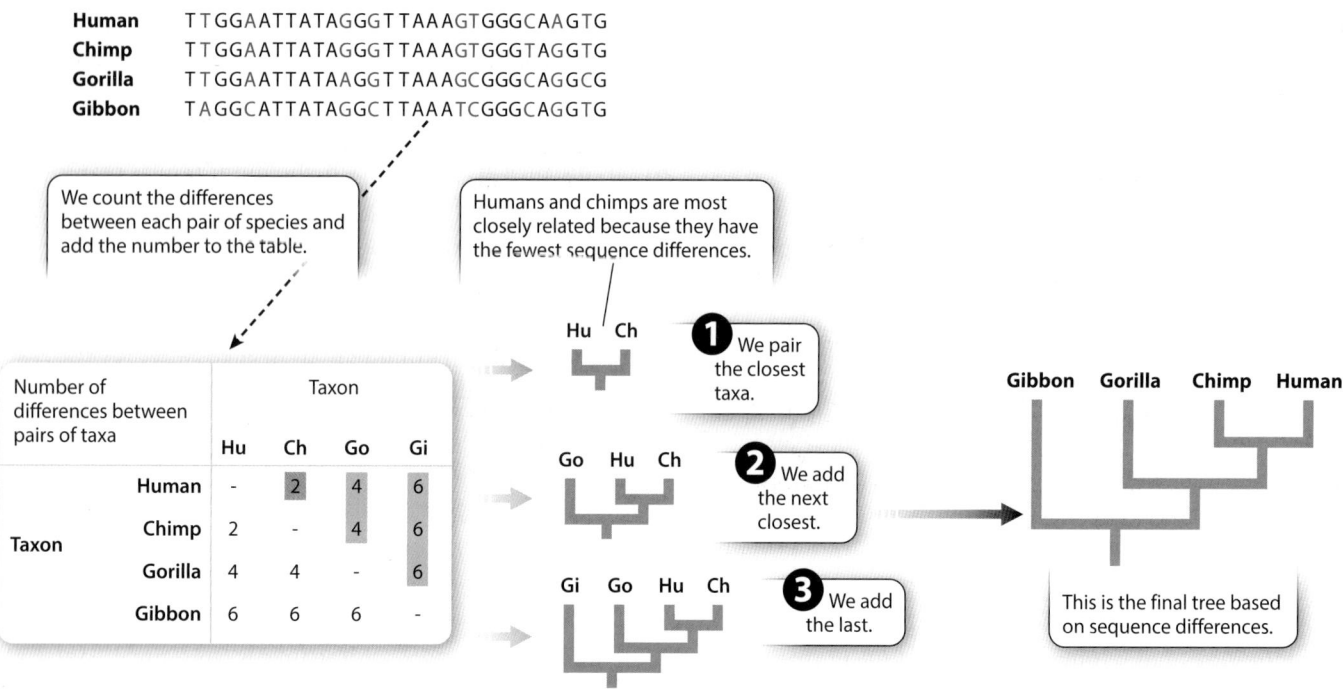

Human	T T G G A A T T A T A G G G T T A A A G T G G G C A A G T G
Chimp	T T G G A A T T A T A G G G T T A A A G T G G G T A G G T G
Gorilla	T T G G A A T T A T A A G G T T A A A G C G G G C A G G C G
Gibbon	T A G G C A T T A T A G G C T T A A A T C G G G C A G G T G

We count the differences between each pair of species and add the number to the table.

Humans and chimps are most closely related because they have the fewest sequence differences.

1 We pair the closest taxa.

2 We add the next closest.

3 We add the last.

This is the final tree based on sequence differences.

Number of differences between pairs of taxa		Taxon			
		Hu	Ch	Go	Gi
Taxon	Human	-	2	4	6
	Chimp	2	-	4	6
	Gorilla	4	4	-	6
	Gibbon	6	6	6	-

FIGURE 46.15 Building a phylogenetic tree using distance methods

Sequence data can be used to reconstruct phylogeny on the basis of distance, where distance is measured in the number of sequence differences between species.

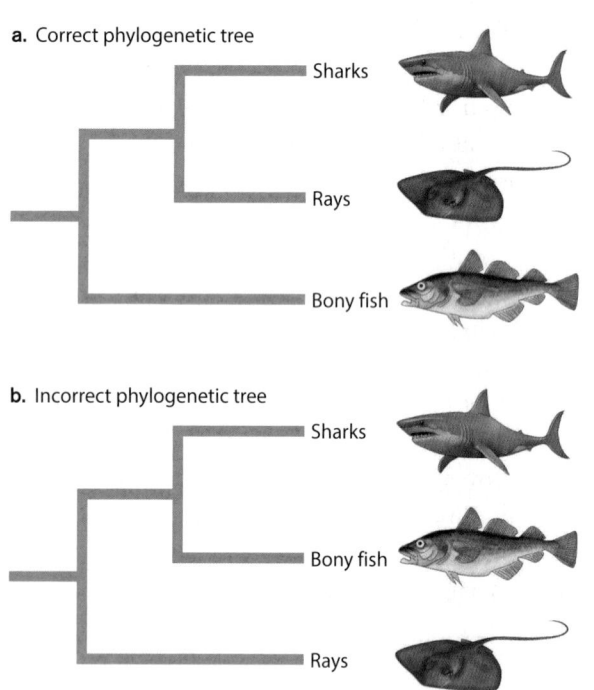

a. Correct phylogenetic tree

Sharks

Rays

Bony fish

b. Incorrect phylogenetic tree

Sharks

Bony fish

Rays

FIGURE 46.16 Correct and incorrect phylogenetic trees of sharks, rays, and bony fish

Distance measures sometimes produce an incorrect tree. (a) This tree correctly indicates that sharks are more closely related to rays than sharks are to bony fish. Note that sharks and rays share a more recent common ancestor than sharks and bony fish do. (b) This incorrect tree assumes that sharks are more closely related to bony fish than sharks are to rays because sharks look very similar to bony fish but very different from rays.

similarity among sharks, rays, and bony fish. Because sharks and bony fish have similar body forms, you would conclude, incorrectly, that sharks and bony fish have a more recent common ancestor and are therefore more closely related than sharks and rays, as shown in Figure 46.16b.

This is a case in which distance measures have resulted in the incorrect tree. What went wrong? The distance-based approach assumes that evolution proceeds at equal or constant rates in different groups over time. However, in this example, new morphological traits appeared relatively quickly in the line that led to rays, indicated by the red lines along the branches of the tree in **FIGURE 46.17**, and much more slowly in the lines that led to sharks and bony fish. As a result, rays look very different from sharks, even though rays and sharks are actually closely related to each other.

In other words, the mistake stems from grouping sharks and bony fish together on the basis of neither of them having changed very much over time. Rays, in contrast, have changed a great deal. So even though sharks and bony fish look more similar to each other than either of them looks like rays, sharks are actually more closely related to rays than they are to bony fish. In this case, looks are deceiving.

This type of problem is often less of an issue with molecular data than it is with morphological data. As a result, distance-based measures are commonly used to build phylogenetic trees with data obtained from DNA sequencing.

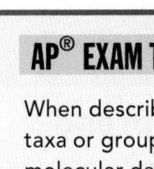

New morphological trait — Sharks

Rays

Bony fish

FIGURE 46.17 Varying rates of evolution in different groups

Many new morphological traits, indicated by the red lines, appeared in the line that led to rays, and fewer appeared in the lines that led to sharks and rays, which explains why a distance approach does not work in this case.

AP® EXAM TIP

When describing an evolutionary relationship between taxa or groups, be sure to use morphological, fossil, and molecular data whenever possible.

✓ Concept Check

6. **Describe** two different ways to build evolutionary trees.

7. **Describe** why molecular traits are useful when building trees.

46.4 Phylogenetic trees can help solve practical problems

The sequence of changes on a tree from its root to its tips documents evolutionary changes that have accumulated through time. Trees suggest how species are related to one another, and which traits came first and which followed later. Building phylogenetic trees, therefore, reveals a great deal about evolutionary history.

It can have practical uses as well. For example, oomycetes, which are the microorganisms responsible for potato blight and other important diseases of food crops, were long thought to be fungi because they look like some fungal species. The discovery, using molecular characters, that oomycetes belong to a very different group of eukaryotic organisms has opened new possibilities for understanding and controlling these plant pathogens.

Similarly, in 2006, researchers used DNA sequences to identify the Malaysian parent population of a species of butterfly called lime swallowtails that have become an invasive species in the Dominican Republic, pinpointing the source populations from which natural predators of this pest can be sought.

Phylogenetics was also employed to solve a famous case in which a dentist in Florida was accused of infecting his patients with human immunodeficiency virus (HIV). HIV nucleotide sequences evolve so rapidly that biologists can build phylogenetic trees that trace the spread of specific strains from one individual to the next. Phylogenetic study of HIV present in samples from several infected patients, the HIV-positive dentist, and other individuals provided evidence that the dentist had, indeed, infected his patients, as described on page 658 in "Practicing Science 46.1: Did an HIV-positive dentist spread the AIDS virus to his patients?"

Similarly, phylogenetic studies of influenza virus, which causes the flu, and SARS-CoV-2, which causes COVID-19, have been used to determine where strains and variants originate and their subsequent movements among geographic regions and individual patients. Today, there is a growing effort to use specific DNA sequences as a kind of fingerprint or barcode for tracking biological material. These efforts rely on the DNA sequencing techniques described in Module 38. Such information could quickly identify samples of shipments of meat as being from endangered species or track newly emerging pests. The Consortium for the Barcode of Life has already accumulated species-specific DNA barcodes for more than 100,000 species. Phylogenetic evidence provides a powerful tool for evolutionary analysis and is useful across timescales ranging from months to the entire history of life, from the rise of pandemics to the origins of metabolic diversity.

✓ Concept Check

8. **Identify** the type of trait (morphological, developmental, or molecular) that is used to track HIV or influenza samples and build trees.

9. **Describe** one practical use of building evolutionary trees.

Did an HIV-positive dentist spread the AIDS virus to his patients?

Background In the late 1980s, several patients of a Florida dentist contracted acquired immunodeficiency syndrome (AIDS), the disease caused by HIV infection. Molecular analysis showed that the dentist was HIV-positive.

Hypothesis It was hypothesized that the patients acquired HIV during dental procedures carried out by the infected dentist.

Method Researchers obtained two HIV samples each, denoted 1 and 2 in the figure below, from several people. Samples were obtained from the dentist, denoted Dentist 1 and Dentist 2; several of his patients, denoted Patients A through G; and other HIV-positive individuals chosen at random from the local population, denoted as LP followed by numbers to indicate different people. In addition, a strain of HIV from Africa (HIVELI) was included in the analysis as an outgroup.

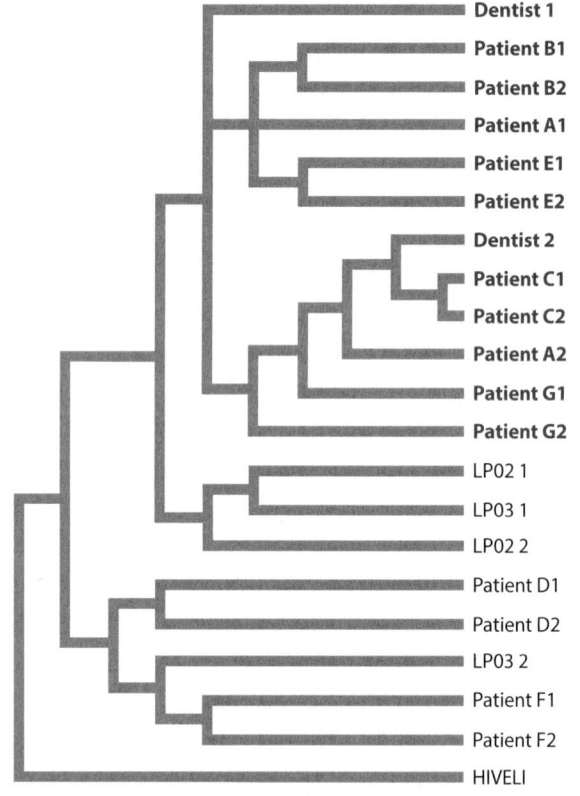

Results Biologists constructed a phylogeny based on the nucleotide sequence of a rapidly evolving gene in the genome of HIV. Because the gene evolves so quickly, its mutations preserve a record of evolutionary relatedness on a very fine scale. The HIV samples taken from some of the infected patients—patients A, B, C, E, and G—were more closely related to the dentist's HIV than they were to samples taken from other infected individuals. Some patients' sequences, however, were different enough from the dentist's that they were not on the same part of the tree, suggesting that these patients, D and F, had acquired their HIV infections from other sources.

Conclusion HIV phylogeny makes it highly likely that the dentist infected several of his patients. The details of how the patients were infected remain unknown, but the safety practices now rigidly observed in dentistry make it unlikely that such a tragedy could occur again.

Follow-up Work Phylogenies based on molecular sequence characters are now routinely used to study the origin and spread of infectious diseases, such as swine flu and Ebola.

SOURCE
Hillis, D. M., J. P. Huelsenbeck, and C. W. Cunningham. 1994. "Application and Accuracy of Molecular Phylogenies." *Science* 264:671–677.

AP® PRACTICE QUESTION

In Unit 6, you learned that the processing of genetic information is not perfect and results in genetic variation. One method utilized to construct a phylogenetic tree like the one shown is to score the degree of similarity between sequences of nucleotides. **Explain** the evolution of the HIV virus based on the placement of patients D-1, C-1, and C-2 relative to Dentist 2 in terms of their corresponding nucleotide sequences by answering the following:

(a) **Make a claim** identifying which of the three patients were infected by Dentist-2.

(b) **Justify** your claim with evidence based on the placement of each patient in the tree.

(c) **Provide reasoning** for your claim based on the evolution of nucleotide sequences over time.

Module 46 Summary

PREP FOR THE AP® EXAM

REVISIT THE BIG IDEAS

EVOLUTION: Use the content of this module to **describe** two types of evidence that can be used to determine placement of a species on a phylogenetic tree.

LG 46.1 The history of life on Earth can be described as a branching tree.

- The nested pattern of similarities seen among organisms is a result of descent with modification and can be represented as a phylogenetic tree. Page 645
- A node is a branching point on a phylogenetic tree; it represents a common ancestor and it can be rotated without changing evolutionary relationships. Page 646
- Sister groups are more closely related to one another than they are to any other group. Page 646
- Evolutionary relatedness is determined by following the sequence of nodes on a phylogenetic tree over time; more closely related groups share more recent common ancestors than more distantly related groups. Page 646
- A phylogenetic tree is a hypothesis or model of the evolutionary relationships among organisms. Page 648
- Organisms are classified into species, genus, family, order, class, phylum, kingdom, and domain. Page 648
- A monophyletic group, or clade, includes a common ancestor and all of its descendants. Page 649
- A paraphyletic group includes a common ancestor and some, but not all, of its descendants. Page 649

- A cladogram is a phylogenetic tree that shows clades, or monophyletic groups. Page 650

LG 46.2 Shared traits are present in more than one species.

- Shared character states are sometimes used to build phylogenetic trees. Page 651
- Homologies are similarities based on common ancestry. Page 651
- Analogies are similarities based on convergent evolutions. Page 651

LG 46.3 Evolutionary trees can be built in several ways.

- A phylogenetic tree can be built on the basis of shared derived characters. Page 652
- Molecular data, such as DNA sequences, provide a wealth of characters that complement other types of information in building phylogenetic trees. Page 655
- A phylogenetic tree can also be built by looking at overall similarity in form or molecular sequences. Page 655

LG 46.4 Phylogenetic trees can help solve practical problems.

- Phylogenetic trees can be used to understand evolutionary relationships of organisms. Page 657
- They can also be used to solve practical problems, such as how viruses evolve over time. Page 657

Key Terms

Phylogeny	Character	Derived character
Cladogram	Shared character	Outgroup

Review Questions

1. The evolutionary history of a group of organisms is called a
   (A) phylogeny.
   (B) taxonomy.
   (C) morphology.
   (D) fossil.

2. A group that includes a single common ancestor and all its descendants is
   (A) monophyletic.
   (B) paraphyletic.
   (C) polyphyletic.
   (D) phyletic.

3. Which statement is true regarding phylogenetic trees?

   (A) Phylogenetic trees can be considered physical representations of hypotheses that seek to establish the evolutionary relationships between different organisms.

   (B) Given the sheer number of species on Earth, it is impossible to create a phylogenetic tree encompassing all of these organisms.

   (C) Phylogenetic trees are constructed based solely on the morphological characteristics of species, not molecular characteristics.

   (D) Within a phylogenetic tree, the order of groups located at the tree tips determines evolutionary relationships.

4. The nodes on a phylogenetic tree represent

   (A) sister groups.

   (B) descendant lineages.

   (C) common ancestors.

   (D) present-day groups.

5. Two taxa that are more closely related to each other than to any other taxon are called

   (A) sister groups.

   (B) paraphyletic groups.

   (C) homologous groups.

   (D) analogous groups.

6. If two organisms are in the same class, then which of the following must also be true?

   (A) They are in the same phylum.

   (B) They are in the same family.

   (C) They are in the same genus.

   (D) They are in the same order.

7. Traits that are similar in two species as a result of common ancestry are referred to as

   (A) convergent.

   (B) analogous.

   (C) homologous.

   (D) derived.

Module 46
AP® Practice Questions

Section 1: Multiple-Choice Questions

Choose the best answer for questions 1–4.

Use the following figure to answer question 1.

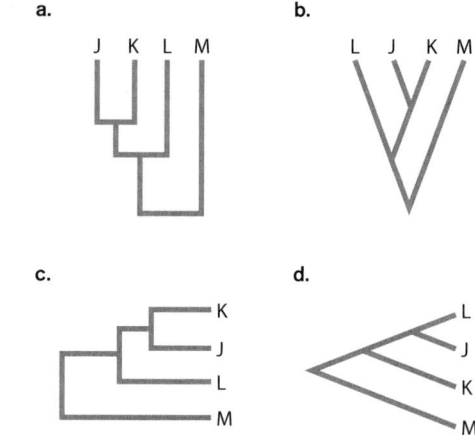

1. Which tree indicates an evolutionary relationship that is different from the others?

   (A) a

   (B) b

   (C) c

   (D) d

Use the following figure to answer question 2.

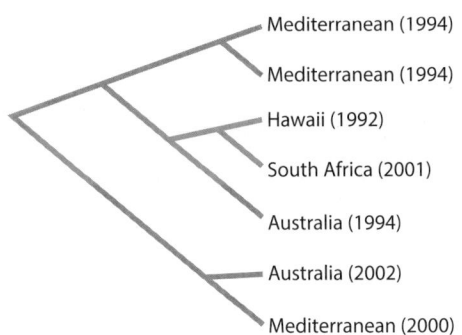

Mediterranean (1994)
Mediterranean (1994)
Hawaii (1992)
South Africa (2001)
Australia (1994)
Australia (2002)
Mediterranean (2000)

2. *Ceratitis capitata* is a species of fruit fly that lives in the Mediterranean region of the world. Multiple times in its history, *C. capitata* has been accidentally transported to other parts of the world with shipments of produce. The imported *C. capitata* establish new, reproductively isolated populations in each new location. A team of scientists used DNA sequences to study the evolutionary relationships among *C. capitata* captured around the world between 1992 and 2002, and constructed the tree shown. Based on the tree, what is most likely the number of times that Australia was invaded by the Mediterranean population of *C. capitata*?

(A) 1   (C) 3

(B) 2   (D) 4

Questions 3 and 4 refer to the following figure, which illustrates the evolutionary relationships among seven species, labeled J–P.

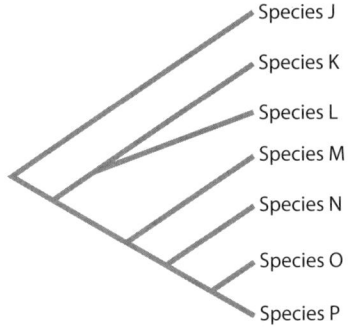

Species J
Species K
Species L
Species M
Species N
Species O
Species P

3. Which species is most closely related to species M?

(A) Species J   (C) Species L

(B) Species K   (D) Species P

4. Which species is the outgroup on the phylogenetic tree?

(A) Species J   (C) Species M

(B) Species L   (D) Species P

Section 2: Free-Response Question

Write your answer to each part clearly. Support your answers with relevant information and examples. Where calculations are required, show your work.

The figure illustrates the evolutionary relationships between major plant groups. Key traits are indicated with letters on the phylogeny.

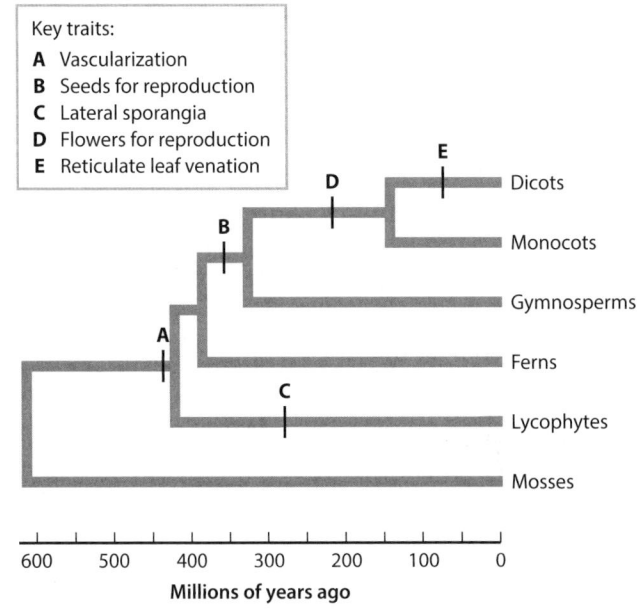

Key traits:
A Vascularization
B Seeds for reproduction
C Lateral sporangia
D Flowers for reproduction
E Reticulate leaf venation

Dicots
Monocots
Gymnosperms
Ferns
Lycophytes
Mosses

600 500 400 300 200 100 0
Millions of years ago

(a) **Identify** the outgroup relative to all other organisms in the tree.

(b) **Describe** which node represents the common ancestor between lycophytes and monocots.

(c) **Determine** how long ago gymnosperms and dicots shared a common ancestor.

(d) **Identify** which trait is shared by monocots, dicots, and gymnosperms but is absent from all other plants.

(e) **Identify** which plant group is vascular and produces flowers and seeds for reproduction, but does not have reticulate leaf venation.

Module 47

Speciation

LEARNING GOALS ▶**LG 47.1** Species are reproductively isolated from other species.

▶**LG 47.2** Reproductive isolation is caused by barriers to reproduction before or after fertilization.

▶**LG 47.3** New species arise by the process of speciation.

In the last module, we discussed phylogenetic trees. These trees provide a visual model or representation of the evolutionary history of different groups of organisms. They also illustrate how organisms are related to one another: species that share a more recent common ancestor are more closely related to each other than those that share a more ancient common ancestor.

Phylogenetic trees have a root at the base; branches that proceed from the past to the present; and tips of the branches, where we place groups of organisms, called taxa. The branches split into two at points called nodes. In the last module, we described these nodes as representing common ancestors. That is, if you move forward in time from a node, you can find the various organisms that descended from this common ancestor. Nodes split into two branches, so they also represent events where one species gives rise to two separate species. In this module, we will take a closer look at these places of divergence.

We will begin by examining the concept of species. In the Linnaean classification scheme introduced in Module 46, a species is the smallest, least inclusive group of organisms. However, this is not a very precise definition, so we will discuss in more detail what it means to be a member of a species. Then we will consider why species are distinct from one another and what prevents them from merging into one large group of organisms. Finally, we will discuss the process by which one species becomes two species, the point on a phylogenetic tree where one branch splits into two.

> **PREP FOR THE AP® EXAM**
>
> ### FOCUS ON THE BIG IDEAS
>
> **EVOLUTION:** Think about the types of barriers that can isolate populations and prevent reproduction and gene flow between them.

47.1 Species are reproductively isolated from other species

The definition of "species" has been a long-standing subject of debate in biology. In *On the Origin of Species,* Darwin wrote, "No one definition has as yet satisfied all naturalists; yet every naturalist knows vaguely what he means when he speaks of a species." The difficulty of defining species has come to be called the species problem.

Here is the problem in a nutshell. The key point of the Darwinian revolution is that species are not fixed and unchanging. Therefore, a species must be fluid and capable of changing in a way that gives rise through evolution to new species. Yet how can we establish a fixed definition of the concept of species if a species can change over time, sometimes enough to create something entirely different—a new species? In this section, we will look at different ways to think about the concept of species.

The Biological Species Concept

We can plainly see the diversity of organisms all around us. When we call a group of organisms a species, is it a separate biological entity or are we using the term as just a convenient way to organize the natural world? To test whether the concept of species reflects reality, we can start by examining characteristics of different living organisms we see in the natural world. **FIGURE 47.1** shows a graph that plots antenna length and wing length of three different types of butterflies that appear to be different. Note that the dots, which represent individual organisms, fall into non-overlapping clusters. The fact that the clusters are distinct implies that each cluster is a biologically distinct species.

While the data points in Figure 47.1 cluster into three distinct groups, we also see distances between dots within each cluster. These distances reflect variations among individuals within a species. Similarly, humans are more similar

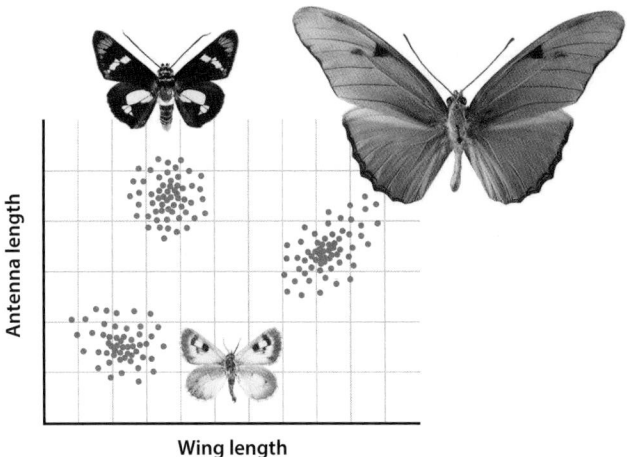

FIGURE 47.1 Comparing characteristics of species

When we graph two characteristics, such as antenna length and wing length, for different species of butterflies, we find that they fall into discrete clusters, which suggests that species are real biological groupings. Photos: Museum of Comparative Zoology, Harvard University

to one another than to our most humanlike relative, the chimpanzee, and also, within the group, humans are highly variable. On a plot such as the one in Figure 47.1, humans would form a messy cluster, but that cluster would not overlap with the chimpanzee cluster.

Species, then, are real biological entities, not just a way to group organisms. The definition of "species" continues to be debated to this day. The most widely used and generally accepted definition of a species is known as the *biological species concept*. The **biological species concept (BSC)** was described by the evolutionary biologist Ernst Mayr as follows:

> *Species are groups of actually or potentially interbreeding populations that are reproductively isolated from other such groups.*

Let's look at this definition closely. At its heart is the idea of reproductive compatibility. Members of the same species are capable of producing offspring, whereas members of different species are incapable of producing offspring. Therefore, members of different species are reproductively isolated from one another.

As Mayr and many others realized, however, reproductive compatibility entails more than just the ability to produce offspring. The offspring must be fertile and, therefore, capable of passing their genes on to their own offspring. For example, although a horse and a donkey—which are two different species—can mate to produce a mule, the mule is infertile. If a hybrid offspring is infertile, then it represents a genetic dead end.

Note, too, the part of Mayr's definition that indicates species are groups that can potentially interbreed. Consider an Asian elephant living on the island of Sri Lanka and one living in nearby India, pictured in **FIGURE 47.2**. These two elephants are considered members of the same species, *Elephas maximus,* even though Sri Lankan and Indian Asian elephants never have a chance to mate with each other in nature because they are geographically separated. If they were placed in the same region, they could interbreed and produce fertile offspring. In other words, they can potentially interbreed, even if they don't actually interbreed because of their geographic separation.

Therefore, whether two individuals are members of the same species is a reflection of whether they are able to exchange genetic material by producing fertile offspring. Consider a new advantageous mutation that appears initially in a single individual. That individual and its offspring that inherit the mutation will have a competitive advantage over other members of the population—they will be more fit. As a result, the mutation will increase in frequency until it reaches 100%, or become fixed in the population. The mutation spreads within the species but, with some rare exceptions, cannot spread beyond it. Therefore, a species represents a closed gene pool, in which gene variants (alleles) are shared among members of that species, but usually not with members of other species.

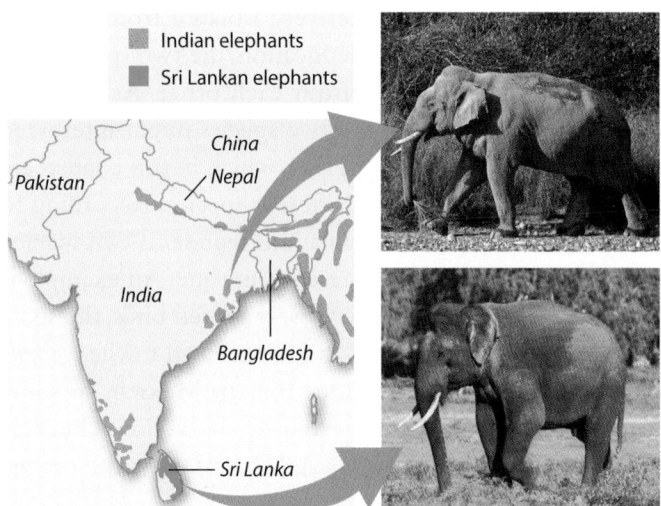

FIGURE 47.2 Reproductive compatibility in Asian elephants

Asian elephants in India and Sri Lanka are considered to be the same species, even though they will never actually interbreed with each other because they don't live in the same place. However, they could potentially interbreed if they were moved to the same geographical area. Photos: (top) Louise Morgan/Getty Images, (bottom) mike smith/AGE Fotostock

Shortcomings of the BSC

The BSC is the most widely accepted way to define species. Nevertheless, the BSC has shortcomings. First, it can be difficult to apply. Imagine you are on a field expedition in a rainforest and you find two insects that look reasonably alike. Are they members of the same species? To use the BSC, you would need to test whether they are capable of producing fertile offspring. In practice, you probably will not have the time and resources to perform such a test. Even if you do attempt such a test, it is possible that members of the same species will not reproduce when you place them together in a laboratory setting or in your field camp. The conditions may be too unnatural for the insects to behave in a normal way. Finally, if they do mate, you may not have time to determine whether their offspring are fertile. Thus, we consider the BSC to be a valid framework for thinking about species, but one that is difficult to apply to real-world settings.

In addition to the difficulty of putting the BSC into practice, a second problem is that it cannot be applied to all organisms. For example, because it is based on reproduction, or the sexual exchange of genetic information, the BSC does not apply to species that reproduce asexually, such as bacteria. Furthermore, because it depends on reproduction, the BSC cannot be applied to species that are extinct and known only through the fossil record, such as different groups of trilobites and dinosaurs.

Finally, populations of one species may vary in the extent to which they are reproductively isolated from another, closely related species. In one location, the two species may be reproductively isolated from each other. At the same time, in another location, the two species might interbreed to produce fertile hybrid offspring, which in this context are offspring produced by two different species. According to the BSC, the first pair of populations represents two species, whereas the second pair represents just one. An example is shown in **FIGURE 47.3**. Two closely related birds, the spotted towhee (*Pipilo maculatus*) and the collared towhee (*Pipilo ocai*), co-occur in various locations in Mexico. In some places they interbreed, indicated by the yellow checkers in the map; in others they do not, indicated by the orange checkers in the map.

This kind of incomplete reproductive isolation seems to be especially common in plants. For example, many different species of willow, oak, and dandelion are still capable of exchanging genes with other species in their genera, even though we call them separate species because they maintain their different appearances in spite of interbreeding. According to the BSC, these different forms should be

Pipilo maculatus   *Pipilo ocai*   Hybrid

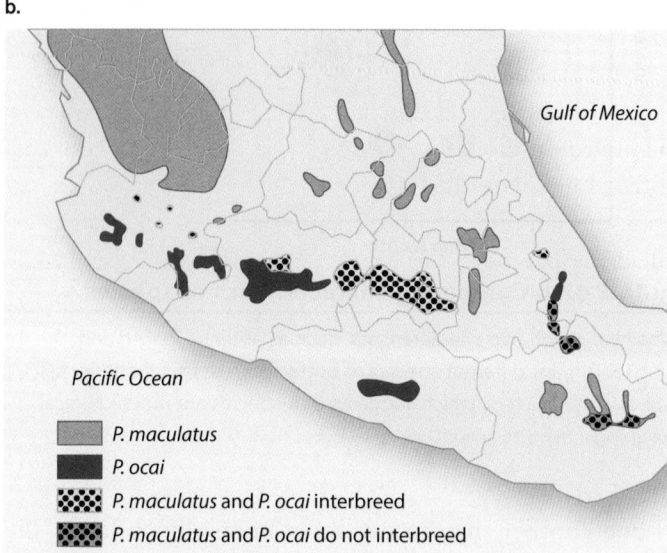

- ▢ *P. maculatus*
- ◼ *P. ocai*
- ▦ *P. maculatus* and *P. ocai* interbreed
- ▨ *P. maculatus* and *P. ocai* do not interbreed

FIGURE 47.3 Hybridization in towhee species

Bird species called towhees interbreed in some places but not in others. (a) The two species of closely related birds, *Pipilo maculatus* and *Pipilo ocai*, produce hybrid offspring. (b) The two species overlap in places in their distributions in central Mexico. In some locations, shown in yellow checkers, they hybridize; in other locations, shown in orange checkers, they do not. Data from Futuyma, D. J. 2009. *Evolution*, 2nd ed. Sunderland, MA: Sinauer Associates, Fig. 18.7, p. 484.

considered one large species because they are able to reproduce and produce fertile offspring. However, because the species maintain their distinct appearances, we call them distinct species.

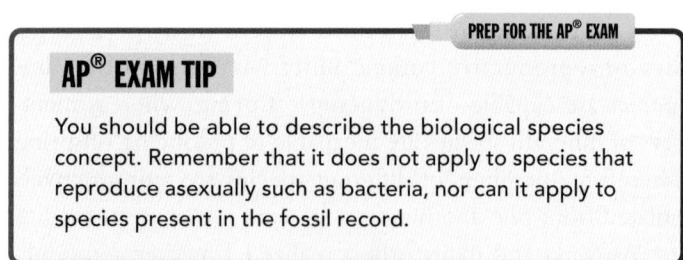

PREP FOR THE AP® EXAM

AP® EXAM TIP

You should be able to describe the biological species concept. Remember that it does not apply to species that reproduce asexually such as bacteria, nor can it apply to species present in the fossil record.

Other Species Concepts

Scientists have addressed the shortcomings of the BSC by modifying it and suggesting many alternative definitions of species. In general, these efforts highlight how difficult it is to make all species fit easily into one definition. The natural world truly defies neat categorization!

One alternative way of categorizing species is called the morphospecies concept, which is used by biologists on a day-to-day basis. The morphospecies concept holds that members of the same species usually look alike. For example, the shape, size, and coloration of the bald eagle (*Haliaeetus leucocephalus*) make it reasonably straightforward to determine whether a bird observed in the wild is a member of the bald eagle species.

Today, the morphospecies concept has been extended to the molecular level. As we discussed in Module 46, members of the same species usually have similar DNA sequences that are distinct from those of other species. A remarkable project called the Barcode of Life has established a database linking DNA sequences to species. Scientists can use it to identify species from DNA alone by sequencing the relevant segment of DNA and finding the matching sequence and therefore the species in the Barcode of Life database. Through the use of this database scientists can determine those biological clusters at the molecular level, like the morphological clusters shown in Figure 47.1 in which members of a species fall close to one another within a single, discrete group.

Although the morphospecies concept is useful and generally applicable, it is not without its own set of issues. Members of a species may not always look alike. Some species show different phenotypes. For example, some species of birds display color differences. Sometimes, males of a species may look different from females—think of the showy peacock, which differs dramatically from the relatively drab peahen. Young can also be very different from old: caterpillars that mature into butterflies are a striking example.

We have seen that members of the same species can look quite different from one another. But it is also true that members of different species can look quite similar. For example, some species of butterfly appear similar but have chromosomal differences that can be observed only with the aid of a microscope, as illustrated in **FIGURE 47.4**. These chromosomal differences prevent the formation of hybrid offspring, so those butterflies meet the BSC definition—they are different species because they do not interbreed—but they do not meet the morphospecies definition because they look so much alike.

Biologists sometimes describe species according to their ecological *niche*. An ecological **niche** is a complete description of the role the species plays in its environment and of its abiotic and biotic requirements. We will discuss the concept of niche in more detail in Unit 8. Niche can sometimes be relevant to determining species because two species can not coexist in the same location if their niches are too similar.

This microscopic view of stained chromosomes reveals that the three butterflies have different numbers of chromosomes.

FIGURE 47.4 Three species of *Agrodiaetus* butterflies

These butterflies look very similar to each other but have different numbers of chromosomes, so they are unable to interbreed, and are therefore different species. Photos: Vladimir Lukhtanov, Zoological Institute of Russian Academy of Sciences, St. Petersburg, Russia

If two very similar species attempt to exist in the same niche, competition for resources inevitably leads to the extinction of one species. This observation has given rise to the ecological species concept, the idea that there is a one-to-one correspondence between a species and its niche.

We can apply this concept to bacteria, which don't reproduce sexually and which therefore can't be separated into different species based on the BSC. Instead, we can determine whether different bacteria are distinct species on the basis of differences or similarities in their ecological requirements. If two types of bacteria have very different nutritional needs, for example, we can infer on ecological grounds that they are separate species.

Another species concept is the phylogenetic species concept, which emphasizes that members of a species all share a common ancestry and a common fate, like a single branch on an evolutionary tree. After all, species rather than individuals originate and become extinct. The phylogenetic species concept requires that all members of a species be descended from a single common ancestor. It can be useful when thinking about asexual species such as bacteria and archaea. However, in other cases, its usefulness is limited. For example, all mammals descend from a common ancestor, but include many species, not one.

It is worth bearing these ecological and evolutionary considerations in mind when thinking about species. Despite the shortcomings of the BSC and the usefulness of alternative ideas, the BSC remains the most constructive way to think about species and still the most widely used and accepted among biologists. In particular, by focusing on reproductive isolation—the inability of different species to produce viable, fertile offspring—the BSC gives us a means of studying and understanding how two populations, originally members of the same species, become distinct species.

47.2 Reproductive isolation is caused by barriers to reproduction before or after fertilization

At the heart of the biological species concept is reproductive isolation, in which two different species cannot interbreed with each other to produce viable, fertile offspring. Factors that cause reproductive isolation are generally divided into two categories, depending on when they act relative to the zygote, which is a fertilized egg. **Prezygotic** factors act before the fertilization of an egg and prevent fertilization from taking place. **Postzygotic** factors come into play after fertilization and prevent the fertilized egg from developing into a fertile individual. In this section, we will provide examples of both prezygotic and postzygotic factors that serve to reproductively isolate different species.

Prezygotic Factors

Most species are reproductively isolated by prezygotic factors, which can take many forms. Plants and animals can be isolated in space, called geographic isolation or ecological isolation. This type of isolation can be dramatic, like two species on either side of a canyon. It can also be subtle. For example, the two Japanese species of ladybug beetle shown in **FIGURE 47.5** can be found living side by side in the same field, but they feed and mate on different plants. Because their life cycles are intimately associated with their host plants, these two species never breed with each other. This ecological separation is what leads to their prezygotic isolation.

Both plants and animals may also be prezygotically isolated in time, called temporal isolation. For example, closely related plant species may flower at different times of the year, so there is no chance that the pollen of one species will come into contact with the flowers of the other. Similarly, members of a nocturnal animal species are unlikely to encounter members of closely related species that are active only during the day.

a.

b.

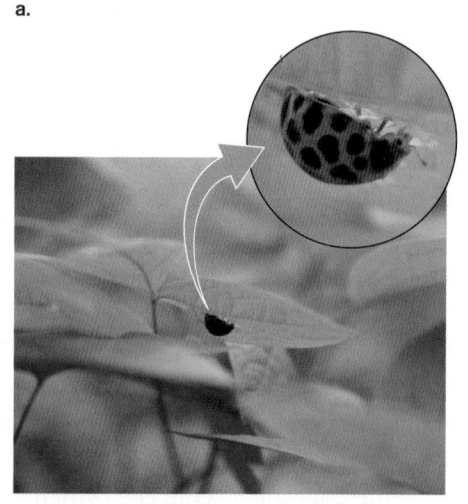 

FIGURE 47.5 Ecological isolation

The ladybugs (a) *Henosepilachna yasutomii* and (b) *H. niponica* are reproductively isolated from each other because they feed and mate on different host plants. Photos: Courtesy Dr. Haruo Katakura

Among animals, species often show behavioral isolation, meaning that individuals mate only with other individuals based on specific courtship rituals, songs, or other behaviors. Birds of paradise in the rainforests of Southeast Asia, for example, have elaborate courtship dances that involve very specific and ritualized displays, songs, and dances. Closely related birds can have very different displays that only work within members of a species, but fail to provoke even the faintest reproductive impulse in a different species. In this case, the prezygotic reproductive isolation is behavioral.

Behavior does not play a role in plants, but prezygotic factors can still be important in their reproductive isolation. Prezygotic isolation in plants can take the form of incompatibility between the incoming pollen and the receiving flower, such that fertilization fails to take place. We see similar forms of isolation between members of marine animals, such as abalone (a sea snail), which simply discharge their gametes into the water. In these cases, membrane-associated proteins on the surface of sperm interact specifically with membrane-associated proteins on the surface of eggs of the same species, but not with those of different species. These specific interactions ensure that a sperm from one abalone species, *Haliotis rufescens*, fertilizes only an egg of its own species and not an egg from *H. corrugata*, a closely related species. Incompatibility between the gametes of two different species is called gametic isolation.

In some animals, especially insects, incompatibility arises earlier in the reproductive process. As an example, the genitalia of males of the fruit fly *Drosophila melanogaster* are configured in such a way that they fit only with the genitalia of females of the same species. Thus, attempts by males of *D. melanogaster* to mate with females of another species of fruit fly, *D. virilis*, are prevented by mechanical isolation.

Postzygotic Factors

Postzygotic factors involve mechanisms that come into play after fertilization of the egg. Typically, they involve some kind of genetic incompatibility between the two organisms. One example, which we saw earlier in Figure 47.4, is the case of two organisms with different numbers of chromosomes.

In some instances, the effect can be extreme. For example, the zygote may fail to develop after fertilization because the two parental genomes are sufficiently different to prevent normal development. In this case, the hybrid offspring is not viable.

In others, the effect is less obvious. Some matings between different species produce perfectly viable adults, as in the case of the horse–donkey hybrid, the mule. As we have seen, though, all is not well with the mule from an evolutionary perspective. Because horses and donkeys have different chromosome numbers, proper homologous chromosome pairing in meiosis in the mule cannot take place, resulting in infertility of the mule. In this case, the hybrid offspring is viable, but sterile.

These examples of prezygotic and postzygotic isolating mechanisms are summarized in **TABLE 47.1**. Prezygotic factors are often considered to be more efficient than postzygotic factors because organisms do not waste the time, energy, and resources in forming and at least partially

TABLE 47.1 Prezygotic and Postzygotic Isolating Factors

Isolating factor	Example	Explanation
Prezygotic	Geographic	Individuals are separated in space.
	Ecological	Individuals are separated by habitat or other ecological specialization.
	Temporal	Individuals are reproductively active at different times.
	Behavioral	Individuals only mate with other individuals based on specific courtship rituals, songs, or displays.
	Gametic	Gametes are incompatible and do not fuse to form a fertilized egg.
	Mechanical	Individuals are unable to mate.
Postzygotic	Hybrid inviability	Embryo forms but does not fully develop.
	Hybrid sterility	Offspring are produced but are sterile.

supporting an embryo when prezygotic isolating factors are in place. Nevertheless, both prezygotic and postzygotic factors present barriers to gene flow. And when there is no gene flow between groups of organisms, they are considered to be different species according to the BSC.

✓ Concept Check

3. **Identify** two prezygotic factors.
4. **Identify** two postzygotic factors.

47.3 New species arise by the process of speciation

Recognizing that species are groups of individuals that are reproductively isolated from other such groups, we are now in a position to recast the key question addressed in this module. That is, instead of asking, "How do new species arise?" we can ask, "How does reproductive isolation arise between populations?" In this section we will examine **speciation,** the process by which new species are produced. We will focus on two types of speciation, one that involves a geographic barrier that separates one population into two populations, and one that does not involve a geographic barrier. We will end with a discussion about rates of speciation in different groups of organisms.

Allopatric Speciation

The key to speciation is the fundamental evolutionary process of genetic divergence between isolated populations. **Divergent evolution** occurs when two populations of organisms become genetically and physically different from each other. **FIGURE 47.6** illustrates this process. If a single population is split into two populations that are isolated from each other, different mutations appear by chance in the two populations. Like all mutations, these are subject to genetic drift, natural selection, or both. Over time, these mechanisms lead to the genetic divergence of the two populations. Two separate populations that are initially identical will, over a long time, gradually become distinct as different mutations occur and spread through each of the populations. At some stage in the course of evolutionary divergence, changes occur in one population that lead to its members becoming reproductively isolated from members of the other population. It is this process that results in speciation.

Speciation—the development of reproductive isolation between populations—is, therefore,

typically a by-product of the genetic divergence of separated populations. Because this process requires genetic isolation between the diverging populations and because geography is the easiest way to ensure physical and therefore genetic isolation, many models of speciation focus on geography.

The process usually begins with the creation of **allopatric** (literally, "different place") populations, which are geographically separated from each other. Clearly, physical separation does not immediately cause reproductive isolation. If a single population is split in two by a geographic barrier for just a few generations and then individuals from each population are able to interbreed

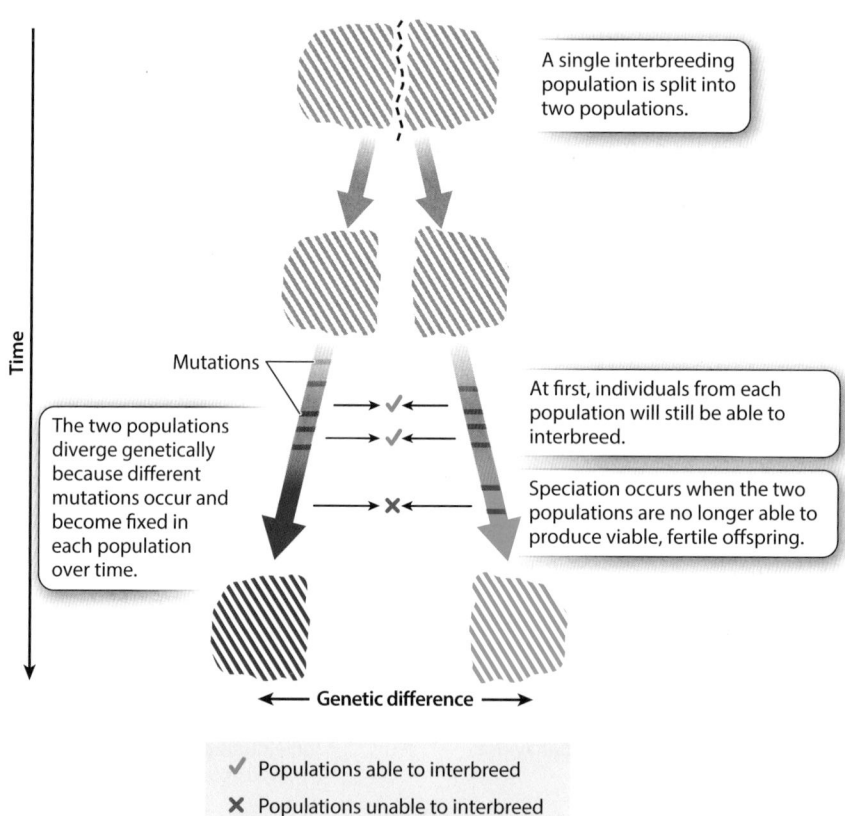

Time

A single interbreeding population is split into two populations.

Mutations

The two populations diverge genetically because different mutations occur and become fixed in each population over time.

At first, individuals from each population will still be able to interbreed.

Speciation occurs when the two populations are no longer able to produce viable, fertile offspring.

← Genetic difference →

✓ Populations able to interbreed
✗ Populations unable to interbreed

FIGURE 47.6 Speciation, the process by which one species becomes two species

Speciation occurs when two populations, A and B, genetically diverge until they become reproductively isolated from each other. At first, there is not enough genetic divergence and so they are able to interbreed, as indicated by the green check marks. Later, the populations have diverged genetically to such an extent that they can no longer interbreed, as indicated by the red x.

again, they will still be capable of producing fertile offspring. In this case, for speciation to occur, geographic separation is not enough; the process also requires enough time for different mutations to occur, increase in frequency, and eventually become fixed in the two separated populations so that the two populations become reproductively isolated from each other.

There are two ways in which populations may become allopatric. The first is by dispersal, in which some individuals colonize a distant place, such as an island, far from the main source population. The second is by vicariance, in which a geographic barrier arises within a single population, separating it into two or more isolated populations.

Let's consider each way in more detail. For dispersal, a few individuals from a mainland population, the central population of a species, disperse to a new location remote from the original population. This new population then evolves separately from the original population. The change in location may be an intentional act of dispersal, such as young mammals migrating away from where they were raised, or it could be an accident brought about by, for example, an unusual storm that blows migrating birds off their normal route. The result is a distant, isolated island population.

"Island" in this case may refer to a true island—such as Hawaii—or may simply refer to a patch of habitat on the mainland that is appropriate for the species but is geographically remote from the mainland population's habitat area. For a species adapted to life on mountaintops, a new island might be another previously uninhabited mountaintop. For a rainforest tree species, that new island might be a patch of lowland forest on the far side of a range of mountains that separates it from its mainland forest population.

Evolutionary change may occur rapidly in the island population for two reasons. First, the island population is often small and, as we discussed in Module 43, genetic drift is more pronounced in smaller populations than in larger ones. Second, the environment may differ between the mainland and the island, and the process of natural selection can drive differences between the two populations as they adapt to their different environments. Together, these two mechanisms cause genetic divergence of the island population from the mainland one, ultimately leading to speciation.

FIGURE 47.7 provides a glimpse of this process in action. Studies of the kingfisher *Tanysiptera galatea* on the island of New Guinea and other nearby islands in the Pacific

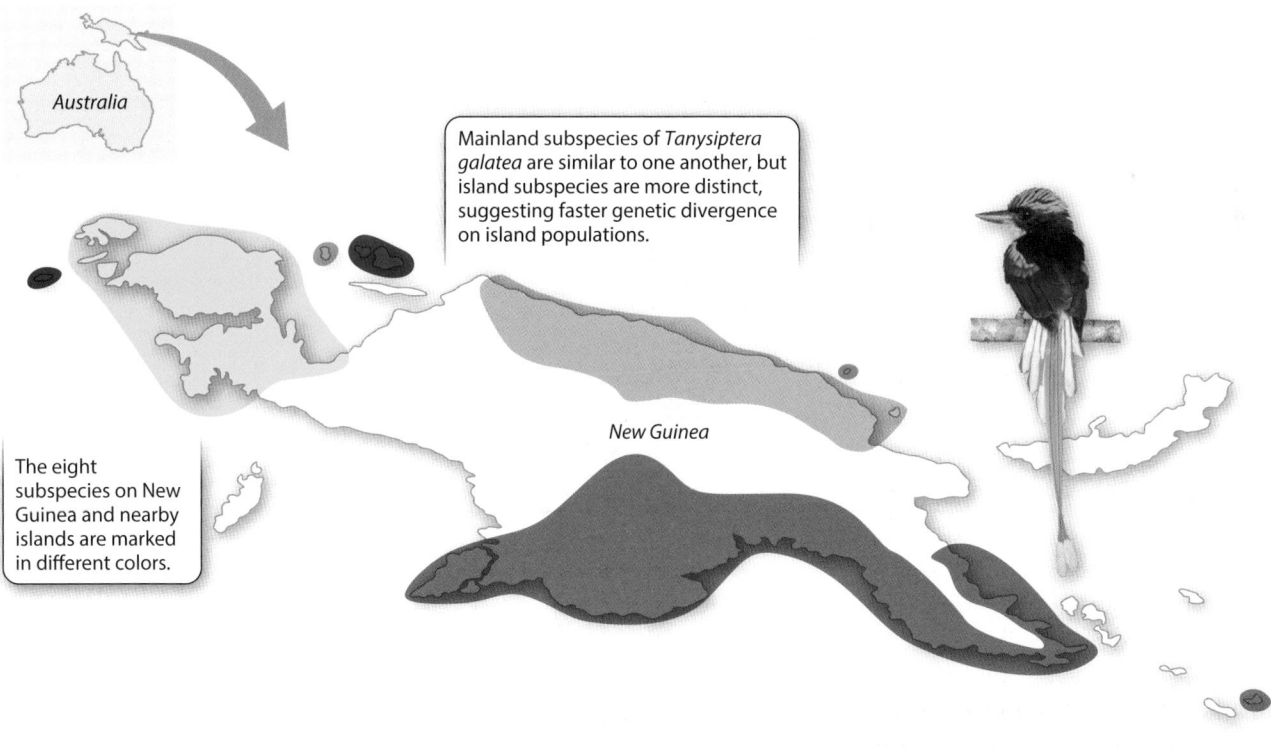

Australia

Mainland subspecies of *Tanysiptera galatea* are similar to one another, but island subspecies are more distinct, suggesting faster genetic divergence on island populations.

New Guinea

The eight subspecies on New Guinea and nearby islands are marked in different colors.

FIGURE 47.7 Allopatric speciation by dispersal

There are eight forms of kingfishers on New Guinea and the surrounding islands near Australia, indicated by the eight different colors. The three on New Guinea are very similar to one another, but the five on the nearby islands are quite distinct from one another, suggesting that speciation is underway. Data from Futuyma, D. J. 2009. *Evolution*, 2nd ed. Sunderland, MA: Sinauer Associates, Fig. 18.7, p. 484. Photo: C.H. Greenewalt/VIREO

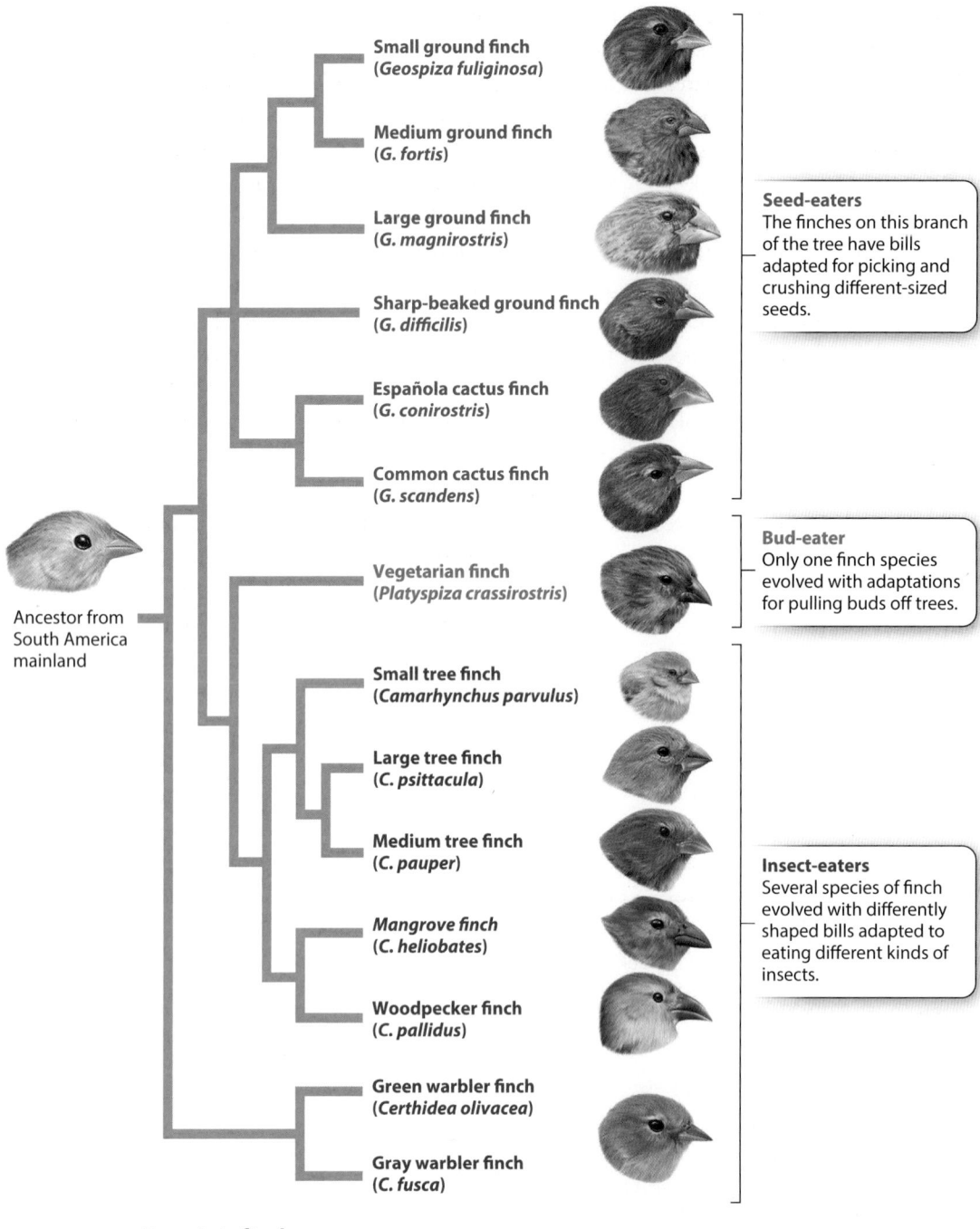

Small ground finch
(*Geospiza fuliginosa*)

Medium ground finch
(*G. fortis*)

Large ground finch
(*G. magnirostris*)

Sharp-beaked ground finch
(*G. difficilis*)

Española cactus finch
(*G. conirostris*)

Common cactus finch
(*G. scandens*)

Seed-eaters
The finches on this branch of the tree have bills adapted for picking and crushing different-sized seeds.

Vegetarian finch
(*Platyspiza crassirostris*)

Bud-eater
Only one finch species evolved with adaptations for pulling buds off trees.

Ancestor from South America mainland

Small tree finch
(*Camarhynchus parvulus*)

Large tree finch
(*C. psittacula*)

Medium tree finch
(*C. pauper*)

Mangrove finch
(*C. heliobates*)

Woodpecker finch
(*C. pallidus*)

Insect-eaters
Several species of finch evolved with differently shaped bills adapted to eating different kinds of insects.

Green warbler finch
(*Certhidea olivacea*)

Gray warbler finch
(*C. fusca*)

FIGURE 47.8 Darwin's finches

Today, 18 different species of Darwin's finches are recognized. Here we show the 14 major groups, all descended from a common ancestor from the South American mainland. Data from: Grant, P. R., and B. R. Grant. 2008. *How and Why Species Multiply*. Princeton, NJ: Princeton University Press, Fig. 2.1, p. 15.

near Australia show the process under way. There are eight recognized forms of *T. galatea*. Three forms are found on mainland New Guinea where they exist in large populations separated by mountain ranges. Five forms live in small populations on small nearby islands. The mainland forms are still quite similar to one another, but the island forms are much more distinct, suggesting that genetic divergence is occurring rapidly in the small island populations. If we wait long enough, these different groups of birds will probably diverge into new species.

FIGURE 47.8 shows the best-known example of dispersal, the finches of the Galápagos Islands, made famous by

Charles Darwin. We know that the oldest of the Galápagos Islands were formed 4–5 million years ago by volcanic action. Early in the history of the Galápagos, individuals of a small South American finch species arrived there. Conditions on the Galápagos are very different from those on the South American continent, where the mainland population of this ancestral finch lived, so the isolated island population evolved to become distinct from its mainland ancestor and eventually became a new species.

The finches subsequently dispersed among the other islands of the Galápagos, which led to further speciation. Today, there are at least 15 different species of finches on the Galápagos, as shown in the phylogenetic tree in Figure 47.8. Furthermore, the number of finch species is correlated with the number of islands in the archipelago. As the number of islands increased over time, so did the number of finch species, as shown in the graph in **FIGURE 47.9**. This is a clear indication of the importance of geographic separation (allopatry) in speciation: the availability of islands provided opportunities for populations to become isolated from one another, allowing speciation. The result was the evolution of many species of finches, collectively known today as Darwin's finches.

Darwin's finches provide a clear and ongoing example of the evolution of a single population through dispersal. In contrast, vicariance involves a geographic barrier arising and splitting one population into two or more populations. For example, when sea levels rose at the end of the most recent ice age, new islands formed along coastlines as the low-lying land around them was flooded. The populations on those new islands suddenly found themselves isolated from other populations of their species. Note that the individuals in this case did not disperse to these new islands, as was the case in the Galápagos. Instead, the individuals were already there, and the islands formed around them by the rising sea level. This kind of island formation is called a vicariance event.

Regardless of how the allopatric populations came about—whether through dispersal or vicariance—the outcome is the same: the two separated populations diverge genetically until speciation occurs. However, allopatric speciation by vicariance is often easier to study than allopatric speciation by dispersal because we can date the time at which the populations were separated based on the date when the barrier arose. One such event whose history is well known is the formation of the Isthmus of Panama between Central and South America, described on page 672 in "Practicing Science 47.1: Can a new geographic barrier cause speciation?" This event took place approximately 3.5 million years ago. In its wake, populations of marine organisms in the western Caribbean and eastern Pacific that had formerly been able to interbreed freely were separated from each other. The result was the formation of many distinct species, with each one's closest relative being on the other side of the isthmus.

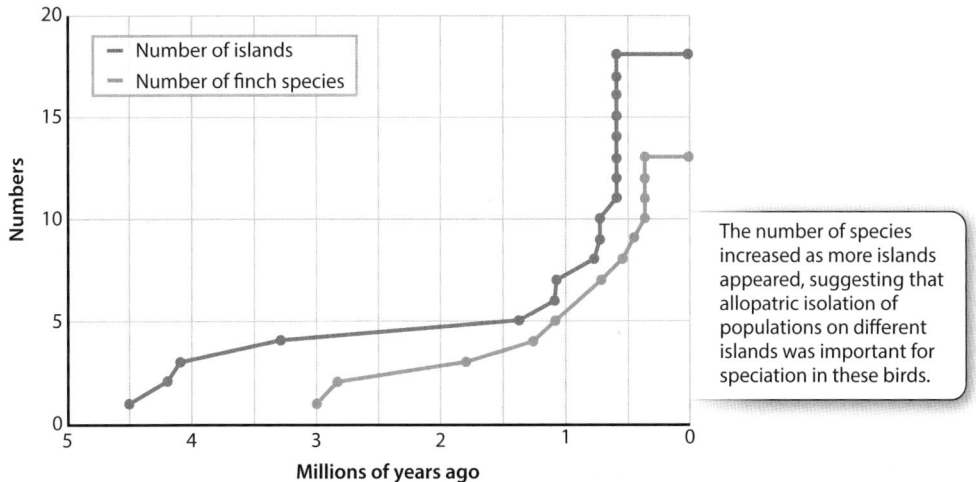

FIGURE 47.9 Graph of number of islands and number of species over time

As the number of islands in the Galápagos increased, so did the number of species of finches. This observation supports the idea that geographic separation is a key factor in promoting reproductive isolation, and therefore speciation. Data from Grant, P. R., and B. R. Grant. 2008. *How and Why Species Multiply*. Princeton, NJ: Princeton University Press, Fig. 2.4, p. 23.

Can a new geographic barrier cause speciation?

Background More than 3.5 million years ago, the Isthmus of Panama connecting North and South America was not completely formed, as shown in the map. An isthmus is a narrow stretch of land connecting two larger areas of land. Several marine corridors remained open, indicated by the black arrows. These corridors allowed interbreeding between marine populations in the Caribbean Sea and the eastern Pacific Ocean. Subsequently, about 3.5 million years ago, the marine corridors were completely closed, separating the Caribbean and eastern Pacific populations and preventing gene flow between them, shown in the map.

> Interbreeding between eastern Pacific and Caribbean populations was possible via the corridors that existed before the final formation of the Isthmus of Panama.

3.5 million years ago

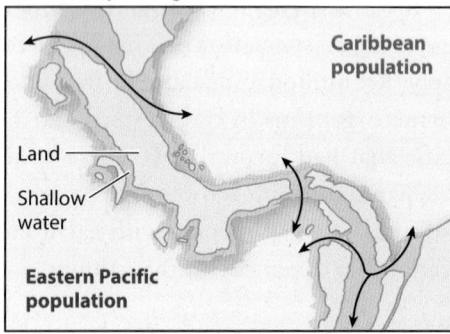

> Interbreeding between eastern Pacific and Caribbean populations is no longer possible because of the geographic barrier.

Today

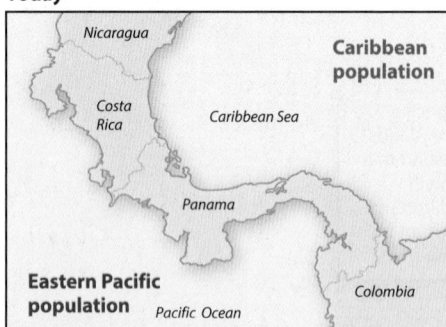

Hypothesis In 1993, American marine biologist Nancy Knowlton and her colleagues hypothesized that patterns of speciation would reflect the fact that direct marine connections between the Pacific and the Caribbean were closed. Specifically, they reasoned that each ancestor species from the time before the formation of the isthmus would

have split into two daughter species, one in the Caribbean and the other in the Pacific. They predicted that the closest relative of each current Pacific species would be a Caribbean species, and that the closest relative of each current Caribbean species would be a Pacific species.

Experiment One study focused on 17 species of snapping shrimp in the genus *Alpheus*, a group that is distributed on both sides of the isthmus. The first step was to sequence the same segment of DNA from each species. The next step was to compare those sequences in order to reconstruct the phylogenetic relationships among species.

Results The phylogenic tree that the researchers created from their data is shown in the figure. It reveals that the closest relative of each species is one from the other side of the isthmus. For example, the closest relative of Pacific 6 is Caribbean 6. These two species share a common ancestor, indicated by the yellow dot, that is more recent than either shares with any other species. This paired pattern of relatedness holds for many of the species studied and results from the geographic separation of the species by vicariance and subsequent speciation.

> The closing of marine corridors resulted over time in the speciation of *Alpheus* into eastern Pacific and Caribbean species.

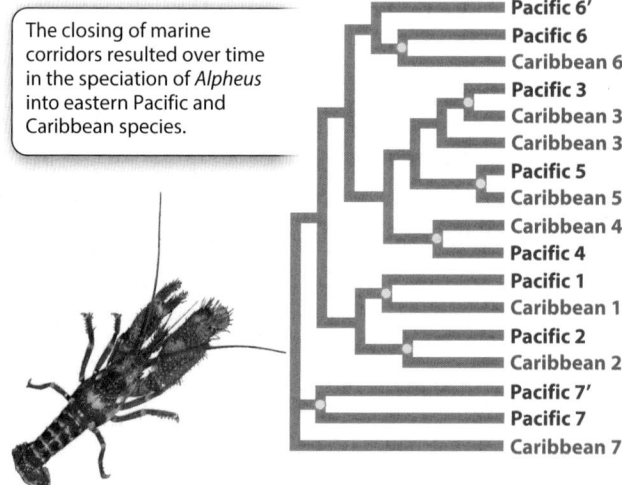

Conclusion The consistent Pacific–Caribbean sister species pairings strongly support the hypothesis that the vicariance caused by the formation of the Isthmus of Panama has driven speciation in *Alpheus*. Each Pacific–Caribbean pairing is derived from a single ancestral species whose continuous distribution between the Caribbean and eastern Pacific was disrupted by the formation of the isthmus. Here, we see striking evidence of the role of vicariance in multiple speciation events.

Follow-up Work Speciation is about more than just genetic differences between isolated populations, as shown here. Knowlton and colleagues also tested the different species for reproductive isolation and found that there were high levels of isolation between Caribbean–Pacific pairs. Because we know that the Isthmus of Panama formed 3.5 million years ago, we know that speciation in these shrimp played out within this timeframe.

SOURCE
Knowlton, N., et al. 1993. "Divergence in Proteins, Mitochondrial DNA, and Reproductive Compatibility Across the Isthmus of Panama." *Science* 260(5114):1629–1632. Photo: Dr. Arthur Anker, NUS, Singapore.

AP® PRACTICE QUESTION

(a) **Identify** the following:
   1. The question being investigated
   2. The hypotheses tested
   3. The experimental group
   4. The control group: against what were the results compared?
   5. The independent variable
   6. The dependent variable

(b) **Explain** the results of the experiment by **identifying** the researchers' claim and **justifying** the claim with the evidence and the reasoning the researchers used (CER).

Sympatric Speciation

Can speciation occur without geographical separation of populations? The answer to this question is "yes," although evolutionary biologists are still exploring how common this phenomenon is. Recall our discussion of Figure 47.6, which shows how separated populations inevitably diverge genetically over time. If a mutation arises in population A after it has separated from population B, that mutation will be present only in population A. The mutation may eventually become fixed in population A—its frequency will be 100% and all members of the population have it. This can happen either through natural selection, if the mutation is advantageous, or through genetic drift, if it is neutral. Once the mutation is fixed in population A, it represents a genetic difference between populations A and B. Over time, repeated independent fixations of different mutations in the two populations will cause the populations to diverge genetically and become reproductively isolated.

Now imagine that populations A and B are not completely separated, and there is some gene flow between them. The mutation that arose in population A can, in principle, appear in population B as members of the two populations interbreed. Gene flow prevents the genetic divergence of populations. If there is gene flow, a pair of populations may change over time, but they do so together. How, then, can speciation occur if gene flow exists? The term we use to describe populations that are in the same geographic location is **sympatric** (literally, "same place"). So, we can rephrase the question as follows: how can speciation occur sympatrically?

For speciation to occur sympatrically, natural selection must act strongly to counteract the effect of gene flow. Consider two sympatric populations of finch like birds, represented in the graph in **FIGURE 47.10**. One population begins to specialize in feeding on small seeds and the other in feeding on large seeds. If the two populations freely interbreed, no genetic differences between the two occur, and speciation does not take place.

Now suppose that the offspring produced by the pairing of a big-seed specialist with a small-seed specialist is an individual best adapted to eat medium seeds, and there are no medium-sized seeds available in the environment. Natural selection acts against the hybrids, which starve to death

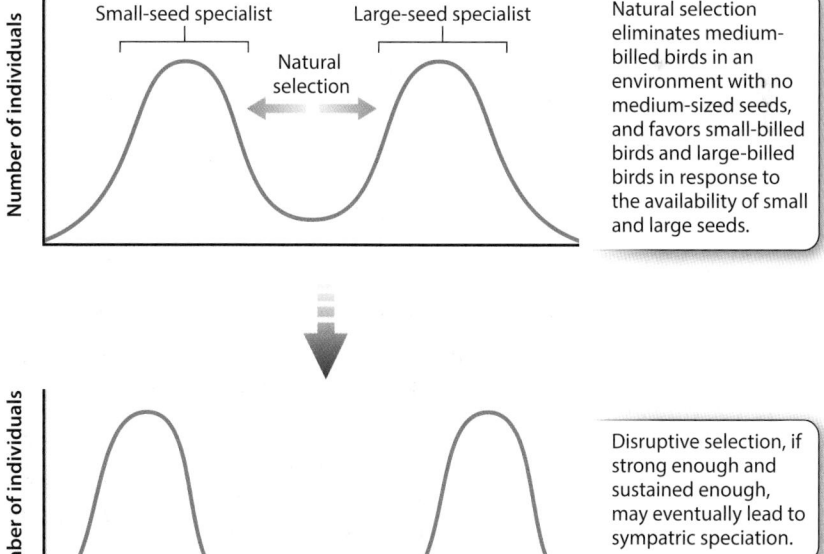

FIGURE 47.10 Sympatric speciation by disruptive selection

Natural selection eliminates individuals in the middle of the spectrum, in this case birds with medium bills, because of the absence of medium-sized seeds. It favors the extremes, birds with small and large bills, because of the presence of small and large seeds.

because there are no medium-sized seeds for them to eat and they are not well adapted to compete with the big- or small-seed specialists. Natural selection, in effect, eliminates the products of gene flow. So, although gene flow is occurring, it does not affect the divergence of the two populations because the hybrid individuals do not survive to reproduce. As discussed in Module 41, this form of natural selection, which operates against the middle of a spectrum of variation, is called disruptive selection.

It turns out to be difficult to find evidence of sympatric speciation in nature. We might find two very closely related species in the same location and argue that they arose through sympatric speciation. But there is an alternative explanation; one species could have arisen elsewhere by allopatric speciation and subsequently moved into the environment of the other species.

However, recent studies of plants on an isolated island have provided strong, if not definite, evidence of sympatric speciation. Lord Howe Island is a tiny island off the coast of Australia. Two species of palm tree that are found only on this island are each other's closest relatives. Because the island is so small, there is little chance that the two species could be geographically separated from each other, meaning that the species are and were sympatric. Moreover, because of the distance between Lord Howe Island and Australia or other islands, it is extremely likely that the palms evolved and speciated from a common ancestor on the island and were not transported there from Australia.

The two palm tree species on Lord Howe Island and other evidence show that sympatric speciation can occur. We must recognize, however, that we do not yet know just how much of all speciation is sympatric and how much is allopatric.

Speciation Rates

Speciation is typically a gradual process. If we try to cross members of two populations that have genetically diverged but not yet diverged far enough for full reproductive isolation to have arisen, we may find that the populations are partially reproductively isolated. In other words, they are not truly separate species, but the genetic differences between them are extensive enough that the hybrid offspring they produce have reduced fertility or viability compared to offspring produced by crosses between individuals within each population.

Because genetic divergence is typically gradual, we often find allopatric populations that have yet to evolve even partial reproductive isolation but that have accumulated a few

population-specific traits. This genetic distinctness is sometimes recognized by taxonomists, who call each geographic form a subspecies and add a further designation after its species name. For example, Sri Lankan Asian elephants, subspecies *Elephas maximus maximus,* are generally larger and darker than their Indian counterparts, subspecies *Elephas maximus indicus,* both shown in Figure 47.2 on page 663.

Occasionally, speciation can occur rapidly. Let's return to the example of the Galápagos finches. Consider the finch ancestor immigrants arriving on the Galápagos. Many ecological opportunities were open and available to them. Until the arrival of the colonizing immigrant birds, there were no birds on the islands to eat the plant seeds, or to eat the insects on the plants, and so on. Suppose that the finch ancestors fed specifically on medium-sized seeds on the South American mainland; that is, they were medium-seed specialists, with bills that were the right size for handling medium-sized seeds. On the mainland, they were constrained to that size of seed because any attempt to eat larger or smaller seeds brought them into conflict with other species—a large-seed specialist and a small-seed specialist—that already used these resources.

When the medium-billed immigrants first arrived on the Galápagos, however, no such competition would have existed. Therefore, natural selection could have promoted the formation of new species of small- and large-seed specialists from the original medium-seed-eating ancestral stock. Selection would have favored the large and small extremes of the bill-size spectrum because these individuals could take advantage of the abundance of unused resources, consisting of small and large seeds. In addition, some finches would have specialized in other food sources, like insects. In the end, one species of finch gave rise to about 15 species of finch in a relatively short period of time.

The Galápagos finches and their frenzy of speciation illustrate the idea of **adaptive radiation,** a bout of unusually rapid evolutionary diversification in which natural selection accelerates the rates of both speciation and adaptation. Adaptive radiation occurs when an environment offers many ecological opportunities for exploitation.

In some cases, speciation can occur even more rapidly, in a single generation. Typically, cases of such rapid speciation are caused by hybridization between two species in which the offspring are reproductively isolated from both parents. For example, hybridization in the past between two sunflower species, *Helianthus annuus* and *H. petiolaris* (the ancestor of the cultivated sunflower), has apparently given rise to three new sunflower species, one of which, *H. anomalus,* is

Helianthus annuus

Helianthus petiolaris

×

↓

Helianthus anomalus

FIGURE 47.11 Speciation by hybridization in sunflowers

Helianthus anomalus (bottom) is the product of hybridization between *H. annuus* (top left) and *H. petiolaris* (top right). Photos: (top left) Gary A. Monroe, hosted by the USDA-NRCS PLANTS Database; (top right) Jason Rick; (bottom) Gerald J. Seiler, USDA-ARS

shown in **FIGURE 47.11**. *H. petiolaris* and *H. annuus* have probably formed innumerable hybrids in nature, virtually all of them not viable. However, a few hybrids survived to yield these three daughter species. Each of these new species acquired a different mix of parental chromosomes. It is this species-specific chromosome complement that makes all three distinct and reproductively isolated from the parent species and from one another.

In many cases of hybridization, chromosome numbers are different in the offspring compared to the parents. Two diploid parent species with 5 pairs of chromosomes, for a total of 10 chromosomes each, may produce a hybrid offspring with double the number of chromosomes. That is, the hybrid inherits a full paired set of chromosomes from each parental species for a total of 20. In this case, the hybrid has 4 genomes rather than the diploid number of 2. We call such a double diploid a tetraploid. Tetraploidy is a type of polyploidy, where organisms have more than two sets of chromosomes, as described in Module 37. In general, animals cannot sustain this kind of expansion in chromosome number, but plants often can. As a result, the

formation of new species through polyploidy has been relatively common in plants.

Do we ever see rapid speciation without changes in chromosome number? Analysis of Darwin's finches in the Galápagos Islands suggests that hybridization between closely related species may occasionally result in the evolution of a new species, as shown in **FIGURE 47.12** on page 676. Detailed study of a population of finches on one small island, Daphne Major, allowed scientists to identify a hybridization event between a resident *Geospiza fortis* female and an incoming male *G. conirostris* from another island, Española. The offspring of this pairing were able to breed to produce a population that is reproductively isolated from Daphne Major's resident *G. fortis*.

The bill size of the new species is different from those of other finch species, so the new species can specialize on food resources not readily accessed by other species. Remarkably, this new species, informally called Big Bird, evolved over just two generations, so is another great example of evolution in our time, discussed in Module 44. Undoubtedly this was an unusual event, but, given the vast periods of time over which

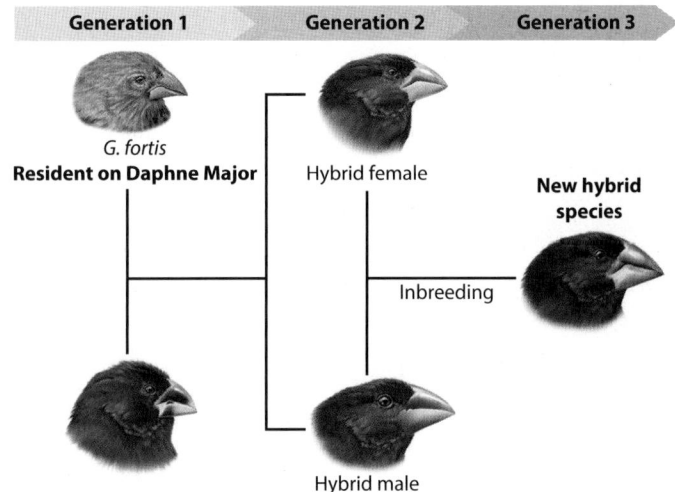

| Generation 1 | Generation 2 | Generation 3 |

G. fortis
Resident on Daphne Major   Hybrid female

New hybrid species

Inbreeding

Hybrid male

FIGURE 47.12 Speciation by hybridization in Darwin's finches

A new species of Darwin's finch derived from a mating between members of two closely related finch (*Geospiza*) species. Data from Sangeet Lamichhaney, modified from Lamichhaney et al. 2018. "Rapid Hybrid Speciation in Darwin's Finches." *Science* 359:214–218.

the evolutionary process unspools, perhaps rare events such as this one play an important role in the generation of new species.

Given the different rates for speciation, from slow to rapid, there are different models to explain change over time. In one extreme, called **gradualism,** species change slowly over time, adapting to environments that are changing slowly. At the other extreme is **punctuated equilibrium,** a model in which there are periods of relatively little change, followed by periods of rapid diversification. In other words, periods of stability, or equilibrium, are interrupted or punctuated by periods of rapid change in the number of species. Rapid changes often occur when the environment changes suddenly or when species find themselves in novel environments. In both cases, we see evolution occurring in response to environmental change; they differ in the rate and pattern of change of the environment.

"Visual Synthesis 7.1: Speciation" on page 680 summarizes the evolutionary mechanisms that lead to speciation. Speciation is caused by the accumulation of genetic differences between populations. Mutation is therefore a key component of the process, but so, too, is the fixation process. Recall from Module 42 that fixation occurs when the frequency of a mutation or allele reaches 1, or 100%, meaning that all of the individuals in a population have it. For example, a mutation can be fixed by selection if it is advantageous, or by genetic drift if it is neutral. Multiple evolutionary mechanisms acting over long periods on each population produce the differences that accumulate between diverging populations.

✓ Concept Check

5. **Describe** the difference between allopatric and sympatric speciation.

6. **Describe** the difference between allopatric speciation by dispersal and allopatric speciation by vicariance.

7. **Describe** why there are so many examples of adaptive radiation on volcanic islands such as the Galápagos.

Module 47 Summary

REVIST THE BIG IDEAS

PREP FOR THE AP® EXAM

EVOLUTION: Using the content of this module, **describe** one type of barrier that might lead to reproductive isolation between populations.

LG 47.1 Species are reproductively isolated from other species.

- The biological species concept (BSC) states that species are groups of actually or potentially interbreeding populations that are reproductively isolated from other such groups. Page 663

- The BSC does not apply to asexual or extinct organisms. Page 664

- Hybridization among closely related species further demonstrates that the BSC is not a comprehensive definition of species. Page 664

- Other species concepts include the morphospecies concept, ecological species concept, and phylogenetic species concept. Page 665

- The BSC is especially useful because it emphasizes reproductive isolation. Page 666

LG 47.2 Reproductive isolation is caused by barriers to reproduction before or after fertilization.

- Reproductive barriers can be prezygotic, occurring before egg fertilization, or postzygotic, occurring after egg fertilization. Page 666
- Prezygotic isolation may be geographic, ecological, temporal, behavioral, gametic, or mechanical. Page 667
- In postzygotic isolation, mating occurs but genetic incompatibilities prevent the development of viable, fertile offspring. Page 667

LG 47.3 New species arise by the process of speciation.

- Speciation underlies the diversity of life on Earth. Page 668
- Speciation is typically a by-product of genetic divergence that occurs as a result of the fixation of different mutations in two populations that are not regularly exchanging genes. Page 668

- Most speciation is thought to be allopatric, involving two geographically separated populations. Page 668
- If divergence continues long enough, chance differences will arise that result in reproductive barriers between the two populations. Page 669
- Geographic separation may be caused by dispersal, resulting in the establishment of a new and distant population, or by vicariance, in which the range of a species is split by a change in the environment. Page 669
- Speciation may be sympatric, meaning that there is no geographic separation between the diverging populations. Page 673
- Speciation is typically gradual, but can also occur rapidly or even instantaneously. Page 674
- Adaptive radiation, in which speciation occurs rapidly to generate a variety of ecologically diverse forms, is best documented on oceanic islands following the arrival of a single ancestral species. Page 676

Key Terms

Biological species concept (BSC)	Speciation	Adaptive radiation
Niche	Divergent evolution	Gradualism
Prezygotic	Allopatric	Punctuated equilibrium
Postzygotic	Sympatric	

Review Questions

1. The biological species concept states that species are reproductively isolated from other species and are actually or potentially
   (A) interbreeding.
   (B) cohabitating.
   (C) naturally selected.
   (D) physically similar.

2. When you use a field guide to identify a species by its appearance, you are applying the
   (A) ecological species concept.
   (B) biological species concept.
   (C) morphospecies concept.
   (D) evolutionary species concept.

3. Speciation requires that two populations become
   (A) geographically separated.
   (B) reproductively isolated.
   (C) temporally separated.
   (D) behaviorally separated.

4. Populations that are geographically separated from one another are called
   (A) allopatric.
   (B) sympatric.
   (C) patric.
   (D) hybrids.

5. Which is the process by which two populations of the same species living in the same habitat diverge into separate species?

(A) Allopatric speciation

(B) Sympatric speciation

(C) Adaptive radiation

(D) Co-speciation

6. It is difficult to apply the biological species concept to bacteria because

(A) bacteria are haploid.

(B) bacteria reproduce asexually.

(C) bacteria lack morphological variation.

(D) bacteria are found everywhere.

7. The biological species concept is least useful for

(A) identifying plant species.

(B) identifying animal species.

(C) identifying species represented by fossils.

(D) identifying species in marine environments.

Module 47
AP® Practice Questions

Section 1: Multiple-Choice Questions

Choose the best answer for questions 1–4.

1. Which is an example of postzygotic isolation?

(A) Members of two species do not recognize each other's courtship displays.

(B) Members of two species attempt to mate but are physically unable to do so.

(C) Members of two species mate but the sperm is unable to enter the egg.

(D) Members of two species mate but their offspring are sterile.

2. Which scenario would most likely result in allopatric speciation?

(A) A rapid change in the environment imposes new selective pressures on a species.

(B) A major environmental event eliminates the dominant species from an ecosystem, allowing previously rare species to flourish.

(C) A major geologic event separates a formerly continuous species range into two isolated ranges.

(D) A long period of environmental stability promotes gradual evolution in a species.

Use the following information to answer questions 3 and 4.

The caterpillars of the European corn borer (ECB) moth do tremendous damage to cornfields in North America. There are two common types of ECB, both common across the entire species range. As shown in the graph, the univoltine type produces one generation per year with adult moths emerging in late June and early July. The bivoltine type produces two generations per year with adult moths emerging in early June and early August. The adult moths mate and lay eggs to form the next generation. In the laboratory, univoltine and bivoltine moths are capable of mating and producing viable and fertile offspring.

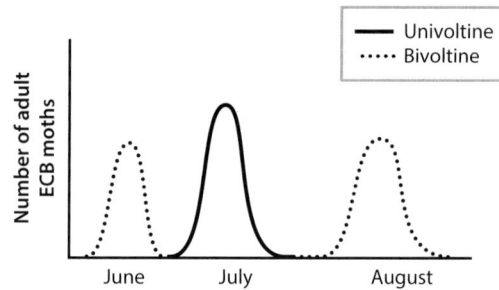

3. Based on the data provided, which statement most accurately describes this system?

(A) The two moths are different species because the adult moths never encounter each other and therefore cannot interbreed.

(B) The two moths are the same species because they interbreed and produce fertile offspring in the laboratory.

(C) The two forms are different species because they are developmentally and morphologically distinct.

(D) The two forms are the same species because they reach similar numbers in the natural environment.

4. Chemical insecticides are used to control ECB outbreaks in agricultural settings. Which would be most likely to occur if a mutation conferring resistance to insecticide appeared in a univoltine individual?

(A) Natural selection would increase the frequency of the resistance mutation in univoltine ECB only.

(B) Natural selection would increase the frequency of the resistance mutation in bivoltine ECB only.

(C) Natural selection would increase the frequency of the resistance mutation in both univoltine and bivoltine ECB.

(D) Natural selection would increase the frequency of bivoltine ECB in the population.

Section 2: Free-Response Question

Write your answer to each part clearly. Support your answers with relevant information and examples. Where calculations are required, show your work.

The island of Plaza Sur is home to two groups of iguanas. The marine iguana (*Amblyrhynchus cristatus*) has a solid gray or black color and lives near the water's edge. The marine iguana is a strong swimmer whose food source is underwater seaweed. The land iguana (*Conolophus subcristatus*) has a brown or gray back and yellow or orange legs and underside. Land iguanas remain on land and primarily eat cactus. Occasionally marine iguanas and land iguanas mate. The offspring from these matings have dark bodies with yellow stripes, are capable of swimming, and eat both cactus and seaweed. The offspring from matings between marine iguanas and land iguanas are sterile.

(a) Based on the data provided, should marine iguanas and land iguanas be considered separate species? **Justify** your claim with reasoning.

(b) Imagine that a new mutation occurs in a marine iguana. The new mutation causes the marine iguana to become more attracted to the mating displays of other marine iguanas and less attracted to the mating displays of land iguanas. **Predict** whether this mutation will be advantageous, disadvantageous, or selectively neutral. **Support your claim** with reasoning.

(c) Imagine that a new mutation occurs in a land iguana. The new mutation causes the land iguana to become more attracted to the mating displays of marine iguanas. **Predict** whether this mutation will be advantageous, disadvantageous, or selectively neutral. **Support your claim** with reasoning.

VISUAL SYNTHESIS 7.1 SPECIATION

Speciation is the process by which one species becomes two species. In this example, a single population of fish, the parent population, is physically separated into two populations, daughter populations A and B, by a geographic barrier. Different mutations occur and then are fixed in the two daughter populations by processes such as genetic drift and natural selection.

Genetic Drift

Neutral mutations increase and decrease in frequency in the population because of the random effects of generation-to-generation sampling error.

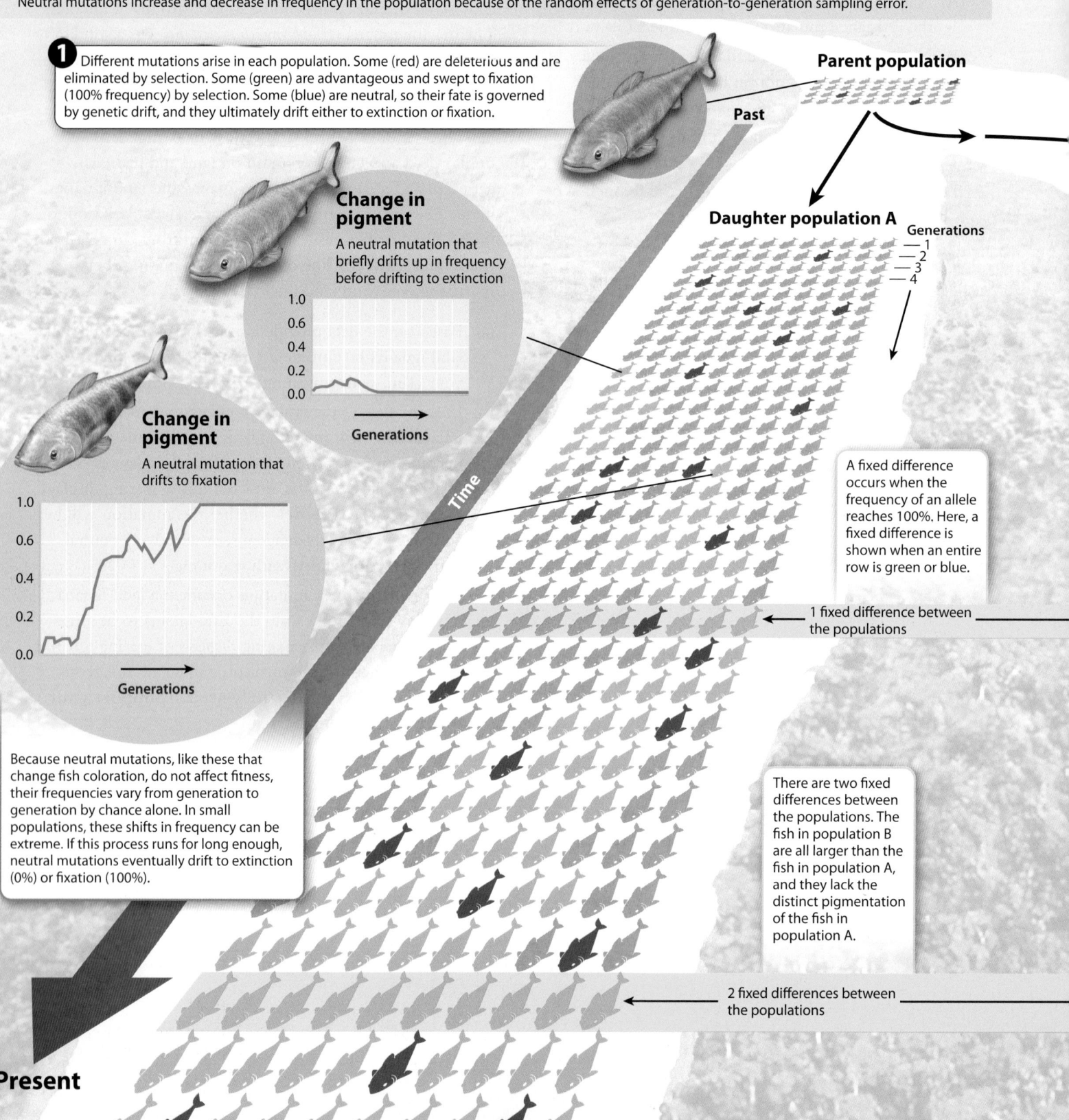

1 Different mutations arise in each population. Some (red) are deleterious and are eliminated by selection. Some (green) are advantageous and swept to fixation (100% frequency) by selection. Some (blue) are neutral, so their fate is governed by genetic drift, and they ultimately drift either to extinction or fixation.

Parent population

Past

Change in pigment

A neutral mutation that briefly drifts up in frequency before drifting to extinction

Generations

Daughter population A

Generations
— 1
— 2
— 3
— 4

Time

Change in pigment

A neutral mutation that drifts to fixation

Generations

A fixed difference occurs when the frequency of an allele reaches 100%. Here, a fixed difference is shown when an entire row is green or blue.

1 fixed difference between the populations

Because neutral mutations, like these that change fish coloration, do not affect fitness, their frequencies vary from generation to generation by chance alone. In small populations, these shifts in frequency can be extreme. If this process runs for long enough, neutral mutations eventually drift to extinction (0%) or fixation (100%).

There are two fixed differences between the populations. The fish in population B are all larger than the fish in population A, and they lack the distinct pigmentation of the fish in population A.

2 fixed differences between the populations

Present

Over time, the two populations become so genetically different from each other that they can no longer interbreed with each other and produce fertile offspring. At this point, speciation is complete.

Natural Selection

Mutations that increase the fitness of individuals become more common, and those that decrease the fitness of individuals become less common over time.

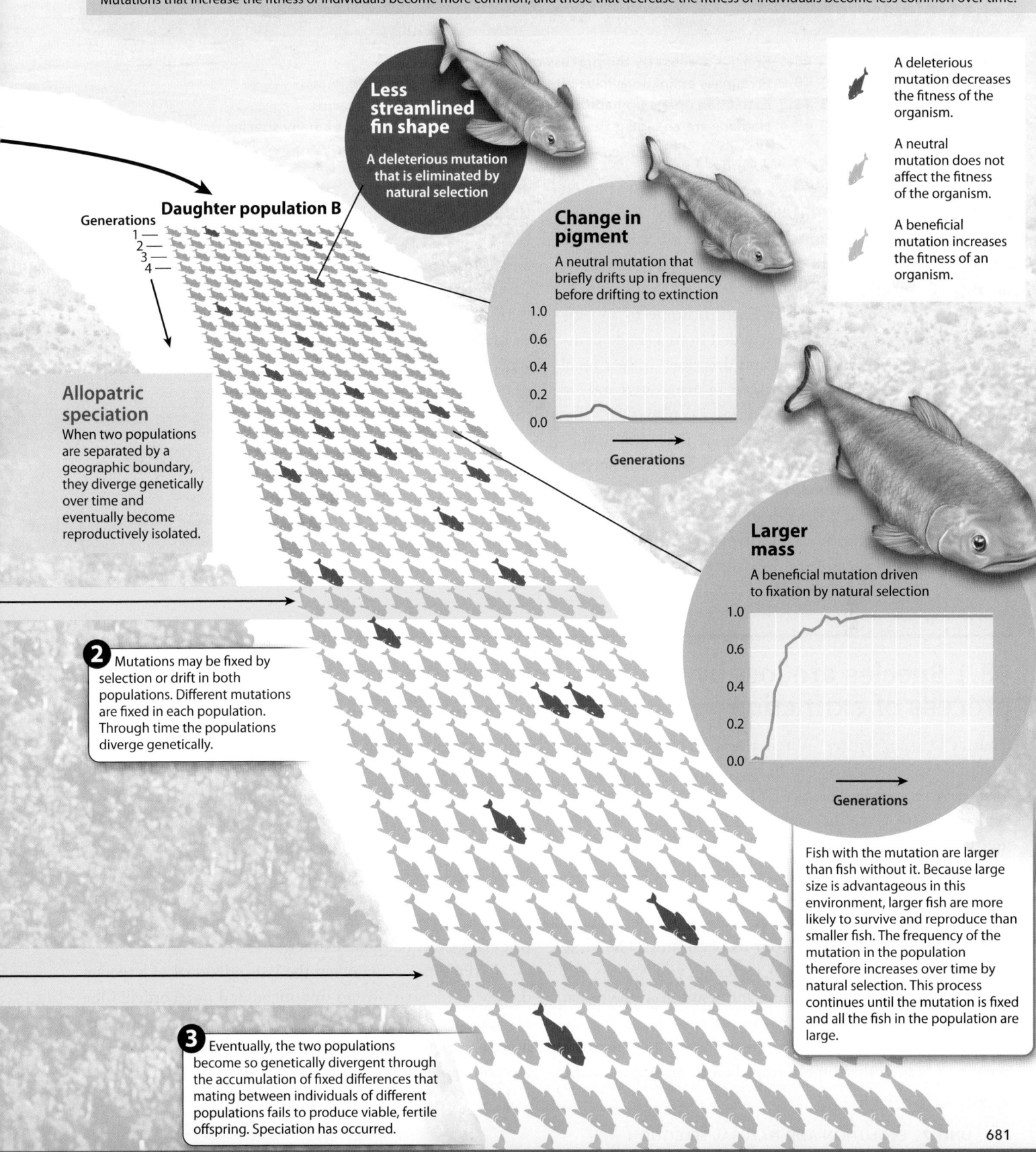

Less streamlined fin shape

A deleterious mutation that is eliminated by natural selection

Daughter population B

Generations
1 —
2 —
3 —
4 —

Allopatric speciation

When two populations are separated by a geographic boundary, they diverge genetically over time and eventually become reproductively isolated.

A deleterious mutation decreases the fitness of the organism.

A neutral mutation does not affect the fitness of the organism.

A beneficial mutation increases the fitness of an organism.

Change in pigment

A neutral mutation that briefly drifts up in frequency before drifting to extinction

1.0
0.6
0.4
0.2
0.0

Generations

Larger mass

A beneficial mutation driven to fixation by natural selection

1.0
0.6
0.4
0.2
0.0

Generations

2 Mutations may be fixed by selection or drift in both populations. Different mutations are fixed in each population. Through time the populations diverge genetically.

Fish with the mutation are larger than fish without it. Because large size is advantageous in this environment, larger fish are more likely to survive and reproduce than smaller fish. The frequency of the mutation in the population therefore increases over time by natural selection. This process continues until the mutation is fixed and all the fish in the population are large.

3 Eventually, the two populations become so genetically divergent through the accumulation of fixed differences that mating between individuals of different populations fails to produce viable, fertile offspring. Speciation has occurred.

Module 48

Extinction

LEARNING GOALS ▶**LG 48.1** Species are lost by the process of extinction.
                    ▶**LG 48.2** Five mass extinctions have altered the course of evolution.
                    ▶**LG 48.3** Extinction opens up habitats that can be filled by other species.
                    ▶**LG 48.4** Humans are changing the environment, leading to the loss of many species.

In the last module, we discussed speciation, the process by which one species becomes two species. On a phylogenetic tree, we represent speciation as a branch point, where one line splits into two. This process of branching, played out over time, has led to the remarkable diversity of species we see today. Currently, Earth is home to an estimated 10 million species, all descended from a single common ancestor. We have named and described only a small fraction of these species, which differ in many ways and are adapted to a multitude of habitats across the planet.

At the same time, scientists estimate that 99% of species that have ever lived no longer exist. So, just as speciation is common, so is **extinction,** the loss of a group of organisms. Extinction is the flip side of speciation: whereas speciation is the engine of biodiversity, extinction represents the loss of species and biodiversity. Both speciation and extinction have occurred repeatedly over the history of life on Earth. At any point of time, even today, the number of species present reflects the rate of both speciation and extinction.

In this module, we will focus on the process of extinction—what it is, when it has occurred in the history of life, the open habitats it often leaves in its wake, and the staggering rates of extinction we are seeing today as a result of human activity.

> PREP FOR THE AP® EXAM
>
> **FOCUS ON THE BIG IDEAS**
>
> **EVOLUTION:** Focus on the relationship between changing environmental conditions and the risk of extinction.

48.1 Species are lost by the process of extinction

Charles Darwin wrote, "When a species has once disappeared from the face of the Earth . . . the same identical form never reappears." He wasn't being sentimental or nostalgic when he wrote this sentence. Instead, he was describing the process of extinction in clear terms. Every species is the result of its unique history. Therefore, when a species goes extinct, that group of organisms is lost forever. It cannot be recreated because it is impossible to go back in time and replay the path it took from its beginning. The same can be said about any individual, including you. Species and individuals exist at a specific time and result from their own unique genetic and evolutionary histories.

On a phylogenetic tree, we can represent an extinction event as a branch that simply stops before making it to the present day. **FIGURE 48.1** shows a phylogenetic tree of birds and their extinct relatives, the dinosaurs. Note that many branches stop before making it to the present day. These are species that have gone extinct.

Extinction is exceedingly common in the history of life on Earth. Some species go extinct as a result of environmental change because they are unable to adapt quickly enough to survive. Some go extinct as a result of predation, hunting, or overharvesting. And some species simply change and adapt, so that, over time, the descendants of the original species are so different from the original that we say that the original species has gone extinct.

Extinction occurs when the last individual of a species dies. Most extinctions go unnoticed and unrecorded. One exception was the death of Martha, the very last passenger pigeon (*Ectopistes migratorius*), who died in the Cincinnati Zoo on September 1, 1914. She is pictured in **FIGURE 48.2**.

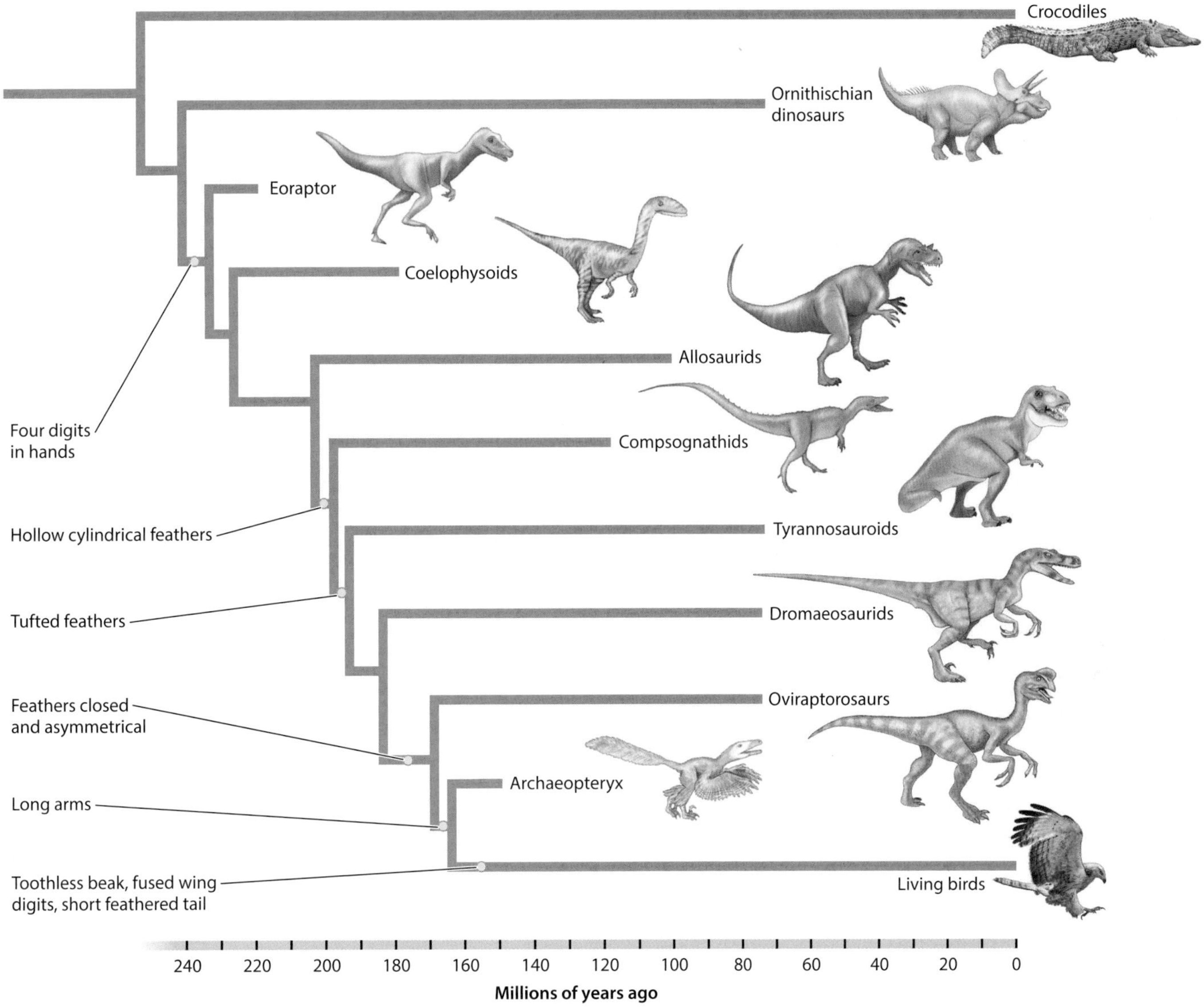

Crocodiles

Ornithischian
dinosaurs

Eoraptor

Coelophysoids

Four digits
in hands

Allosaurids

Compsognathids

Hollow cylindrical feathers

Tyrannosauroids

Tufted feathers

Dromaeosaurids

Feathers closed
and asymmetrical

Oviraptorosaurs

Archaeopteryx

Long arms

Toothless beak, fused wing
digits, short feathered tail

Living birds

240 220 200 180 160 140 120 100 80 60 40 20 0

Millions of years ago

FIGURE 48.1 Birds, crocodiles, and their extinct relatives

This phylogenetic tree shows the evolutionary relationships of crocodiles, extinct dinosaurs, and living birds. Notice that the branches that lead to different species of dinosaurs do not extend to the present. These are extinct groups of dinosaurs. Data from Zimmer, C. 2009. *The Tangled Bank.* Greenwood Village, CO: Roberts and Company.

Although extinction occurs when the last surviving member of a species dies, species are often doomed to extinction well before then; once their populations fall to very low levels, the species is often not able to recover.

Extinction helps us to understand a key feature of any ecosystem, biodiversity. Biodiversity, or biological diversity,

FIGURE 48.2 Martha, the last passenger pigeon

This is a photograph of Martha, who lived in the Cincinnati Zoo and was the last surviving member of her species. She died on September 1, 1914. The 100th anniversary of the extinction of her entire species was marked by many exhibits and articles.

Photo: Abbus Archive Images/Alamy Stock Photo

often refers to the number of species, but more broadly it includes the variety of genetic sequences among individuals of a species; the different cell types, metabolisms, and life histories among species; or even groups of organisms in different habitats around the world. We can also consider biodiversity for a particular place, such as a single tropical rainforest, or the entire planet.

A common way to measure biodiversity is to count the number of species present in a particular area. We don't know how many species exist on our planet, although the number certainly runs into the millions and may exceed 10 million. Whether local or global, patterns of biodiversity reflect the processes of both speciation and extinction. The balance of the two processes determines the total number of species. That is, the number of species at any point in time is the result of rates of speciation and extinction that have occurred in the past. So, a group of organisms, like a genus with many species, might have experienced high rates of speciation, low rates of extinction, or both.

When the number of species increases over time, we can infer that that rate of speciation exceeds the rate of extinction. When the number of species decreases over time, we can infer that the rate of extinction exceeds the rate of speciation. The two processes together help us to understand levels of biodiversity.

> ## ✓ Concept Check
>
> 1. **Describe** extinction.
>
> 2. **Describe** how to represent extinction on a phylogenetic tree.
>
> 3. **Identify** the two processes that determine the total number of species, which is one measure of biodiversity.

48.2 Five mass extinctions have altered the course of evolution

Beginning in the 1970s, American paleontologist Jack Sepkoski scoured the paleontological literature, recording the first and last appearances in the fossil record for every genus of marine animal he could find. The results of this monumental effort are shown in **FIGURE 48.3**. Sepkoski's diagram shows that animal diversity has increased over the past 500 million years. In fact, the biological diversity of animals in today's oceans may well be higher than it has ever been in the past. At the same time, it is clear that animal evolution is not simply a history of continual accumulation. Instead, it is punctuated by periods of extinction.

Five times during the past 500 million years or so, animal diversity in the oceans has dropped both rapidly and substantially; waves of extinctions have also occurred on land. These are known as mass extinctions, which are periods of rapid and substantial loss of biodiversity. These events eliminated ecologically important taxa, that is, those that either played a key role in the functioning of an ecosystem or were dominant in terms of numbers of organisms. As a result, these extinctions provided evolutionary opportunities for the survivors.

The best-known mass extinction occurred 66 million years ago, at the end of the Cretaceous Period, indicated by the rightmost arrow in Figure 48.3. On land, dinosaurs disappeared abruptly, following approximately 150 million years of dominance in terrestrial ecosystems. In the oceans, mollusks called ammonites, which had long been abundant

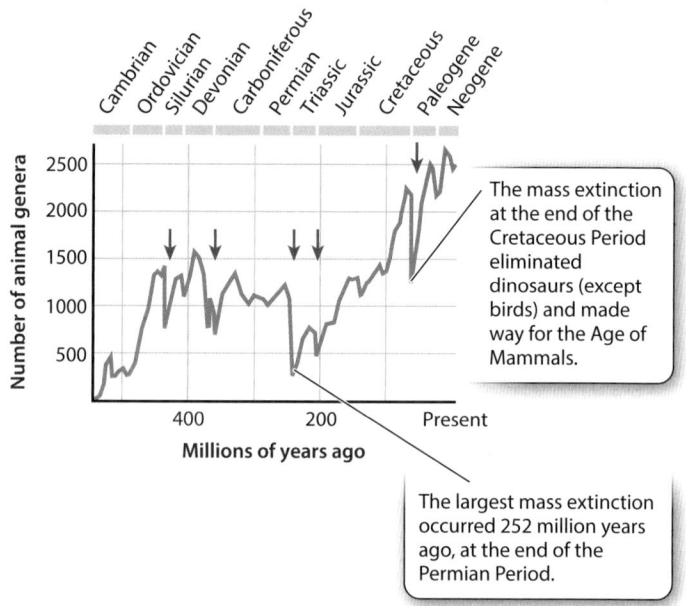

FIGURE 48.3 Mass extinctions

Five times in the past 500 million years, Earth experienced a sudden and large loss of groups of organisms, called a mass extinction. Each significant drop in the graph represents a mass extinction, indicated by an arrow. Time periods are shown across the top. Data from *Sepkoski's Online Genus Database.* http://strata.geology.wisc.edu/jack/

predators, became extinct, and most skeleton-forming microorganisms in the oceans disappeared as well. A large body of geologic evidence supports the hypothesis that this biological catastrophe was caused by the impact of a giant meteorite, as discussed in detail in "Practicing Science 48.1: What caused the extinction of the dinosaurs?"

What caused the extinction of the dinosaurs?

Background Dinosaurs dominated Earth for approximately 150 million years, but became extinct approximately 66 million years ago.

Observation Rock layers accumulate over time, with older layers deeper than younger layers. These layers can be dated using radioactive decay of isotopes, as described in Module 44. Iridium, a type of metal, was found in layers of rock that correspond to the time of the dinosaur extinction. Iridium does not originate on Earth, but is common in space, particularly in meteorites. The iridium layer can be seen in the photograph below as the white band. (The knife is included for size comparison.) The graph next to the photo shows the peak of iridium in rock layers dated to the end of the Cretaceous Period.

Iridium, rare in rocks on Earth but common in meteorites, was discovered in rock layers corresponding to the time of extinction.

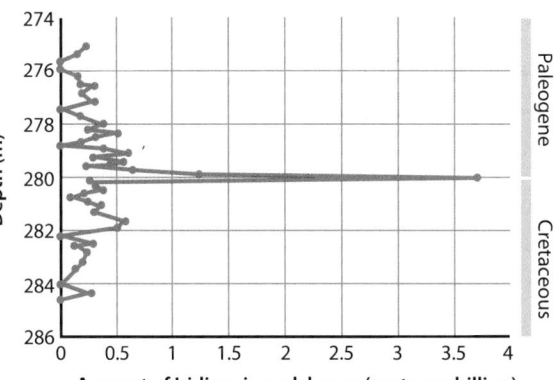

Hypothesis Geologist Walter Alvarez and his colleagues hypothesized that a large meteorite struck Earth. The meteorite was so large and hit Earth with such force that it devastated the area of impact, and also sent soot and other debris into the atmosphere, reducing sunlight. Lack of sunlight disrupted plant communities, which in turn had a domino-like effect on other organisms. Whole communities and ecosystems on land and in the sea were devastated, which led to the extinction of the dinosaurs and many other species.

Predictions A hypothesis makes predictions that can be tested by observation and experiments. To support the hypothesis, independent evidence of a meteor impact should be found in rock layers corresponding to the time of the extinction and be rare or absent in older and younger layers.

Further Observations A form of quartz called shocked quartz was also found in layers that correspond to the time of the extinction. Shocked quartz, shown in the photo below, only forms at very high temperature and pressure, which is consistent with a meteorite impact. In addition, a large crater of the right age was found off the coast of present-day Mexico, as shown on the map and in the photo below.

Quartz crystals that form only at high temperature and pressure—conditions met by giant meteors as they crash into Earth—occur abundantly in rock layers dated to the time of the extinction.

By 1990, geologists located a crater (image on the right) of just the right age and size off the coast of the Yucatán Peninsula of Mexico (image on the left).

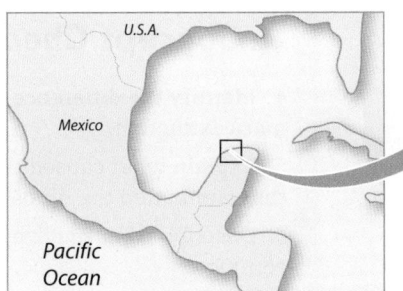

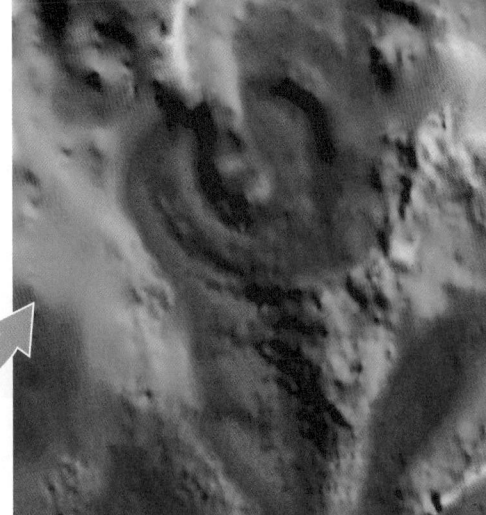

Conclusion A giant meteor struck Earth 66 million years ago, causing the extinction of many species, including the dinosaurs.

Follow-up Work Researchers have documented other mass extinctions, but the event that eliminated the dinosaurs appears to be the only one associated with a meteorite impact. Researchers are also studying other factors that contributed to the mass extinction event, such as massive volcanic activity in what is now India and a long period of global climate change both before and after the meteorite impact.

SOURCES
Alvarez, W. 1998. *T. rex and the Crater of Doom*. New York: Vintage Press. Photos: (top) Denver Museum of Nature and Science; (bottom left) Dr. David Kring/Science Source; (bottom right) Image courtesy of V. L. Sharpton/Lunar and Planetary Institute.

AP® PRACTICE QUESTION

Identify the following:

1. The question being investigated by the experimenters
2. The hypothesis
3. The evidence that supports a meteor striking Earth at approximately the same time that the dinosaurs went extinct
4. The conclusion

Before dinosaurs ever walked on Earth, there were at least four other mass extinction events. The largest of all of these occurred 252 million years ago, at the end of the Permian Period, when environmental catastrophe eliminated half of all groups of organisms called families in the oceans and approximately 80% of all genera. Estimates of marine species losses run as high as 90%. This mass extinction was so extreme that it is sometimes called the Great Dying.

Geologists have hypothesized that this large mass extinction occurred in the aftermath of massive volcanic eruptions. At the end of the Permian Period about 250 million years ago, most continents were gathered into the supercontinent Pangaea, shown in **FIGURE 48.4**, and a huge ocean covered more than half of Earth. Levels of oxygen in the deep waters of this ocean were low because of sluggish ocean circulation and warm ocean temperatures.

Then a massive outpouring of ash and lava—a million times larger than any volcanic eruption experienced by humans—erupted across what is now Siberia in Russia.

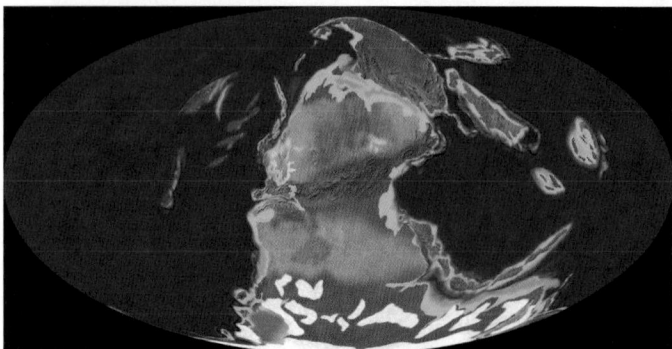

FIGURE 48.4 The supercontinent Pangaea

Pangaea is a supercontinent that formed around the time of Permian mass extinction 252 million years ago. Photo: Pangea from © Global Paleogeography and Tectonics © 2016 Colorado Plateau Geosystems Inc.

Enormous emissions of carbon dioxide and methane from the volcanoes caused global warming, so even less oxygen reached the deep oceans. As some of the increased carbon dioxide in the atmosphere was absorbed by the oceans, the pH of the seawater decreased. Because a decrease in pH corresponds to an increase in acidity, this phenomenon is known as ocean acidification. Ocean acidification is harmful to marine life, including corals and algae that use calcium carbonate to make their skeletons.

The three-way insult of lack of oxygen, global warming, and ocean acidification caused the extinction of many species in the seas and on land. Seascapes that had been dominated for 200 million years by corals and shelled invertebrates called brachiopods disappeared. On land, roughly a third of large animals went extinct and most of the trees disappeared.

As both the Cretaceous and Permian events illustrate, extinction rates can be extremely rapid during times of global ecological stress and disruption. Mass extinction events also teach us the important lesson that current levels of biodiversity—both in terms of number of species and what species are present—reflect the interplay through time of evolution, natural selection, and speciation, as well as rare massive perturbations to ecosystems. The fossil record provides a silent witness to our planet's long and complex evolutionary history.

✓ Concept Check

4. **Identify** the difference between extinction and mass extinction.

5. **Explain** what caused the mass extinction event that eliminated the dinosaurs.

6. **Describe** what the five mass extinctions had in common.

48.3 Extinction opens up habitats that can be filled by other species

In Module 47, we discussed periods of rapid diversification that are commonly seen on islands where there is little or no competition or predation. These bouts of speciation are called adaptive radiation and result from the large number of open habitats that often exist on islands before the arrival of colonists.

What happens on islands is a microcosm of what occurs following a mass extinction. A mass extinction eliminates many dominant species, opening habitats that were formerly occupied. So, for the survivors, mass extinctions provide ecological opportunities on a grand scale. As ecosystems recover from a mass extinction, they come to be dominated by new groups descended from survivors of the extinction. In other words, periods of extinction of some groups of organisms are often followed by periods of adaptive radiation of other groups of organisms.

For example, by eliminating dinosaurs and much of the biological diversity that had built up over millions of years on land, the mass extinction at the end of the Cretaceous Period had the effect of generating new evolutionary possibilities for those animals that survived the event, including mammals. During the Mesozoic Era, dinosaurs were dominant both in terms of numbers and importance in land and ocean ecosystems. As a result, this time period in Earth history is sometimes called the Age of Dinosaurs.

Mammals were present during this period as well but were relatively small and not ecologically dominant. The first fossils that clearly show evidence of hair, a key mammalian trait, appeared approximately 210 million years ago, during the Triassic Period. The earliest mammals, the monotremes, lay eggs like birds. Monotremes still living today include the platypus and spiny anteater. The first mammals that gave birth to live young appeared in the Jurassic Period, about 160 million years ago. These animals gave rise to marsupial and placental mammals. The young of marsupial mammals are born at an early stage of development and today include kangaroos, koalas, and opossums. The placental mammals are named for an organ called the placenta, which provides nutrition for the embryo and enables offspring to be larger and more independent when born. Placental mammals include carnivores, such as cats, dogs, and weasels; primates, including monkeys, apes, and humans; and hooved mammals, which include cattle, pigs, and deer.

Although mammals evolved and diversified during the Age of Dinosaurs, their numbers and ecological importance did not increase until after the end-Cretaceous extinction. In other words, the extinction of the dinosaurs provided many open ecological opportunities for mammals, so they were well poised to become ecologically dominant in the next time period, the Cenozoic Era, which is sometimes called the Age of Mammals.

Similarly, following the Great Dying at the end of the Permian Period, new groups took the place of old ones. Bivalve and gastropod mollusks diversified in the absence of their former competitors; new groups of arthropods radiated, including the ancestors of the crabs and shrimps we see today. Surviving sea anemones evolved a new ability to make skeletons of calcium carbonate, resulting in the corals that build reefs. In short, this mass extinction reset the course of evolution, much as it did at the end of the Cretaceous Period. So, when you stroll through a zoo or snorkel above a coral reef, you are not simply seeing the products of natural selection played out over Earth history. Instead, you are encountering the descendants of Earth's biological survivors.

✓ Concept Check

7. **Identify** the process by which speciation happens relatively quickly following a mass extinction or in any situation where there are lots of available habitats.

8. Dinosaurs once dominated many ecosystems. **Identify** the group of organisms that became dominant after the dinosaurs went extinct.

48.4 Humans are changing the environment, leading to the loss of many species

As we have seen, high rates of extinction often occur during periods of ecological stress. While species are often able to adapt to a slowly changing environment, sudden changes pose a particular challenge. In this case, there may be no variants present that can survive and reproduce in the new environment. Therefore, if the environment changes too quickly and species are unable to adapt, they may go extinct. We saw examples of severe ecological stress in the aftermath of a meteorite striking Earth at the end of the Cretaceous Period and following massive volcanism at the end of the Permian

Period. Today, humans are changing the environment on a global scale and in a relatively short amount of time, especially when we keep in mind the long arc of the history of life on this planet. Our actions are affecting, directly or indirectly, other species. By changing the environment, we are changing the selective landscape in which organisms live and evolve.

Our effect on the environment is partly the result of our sheer numbers. More humans are alive today than ever before, and our numbers are climbing rapidly. A world that had approximately 500 million inhabitants in the year 1500 now supports more than 7.8 billion people. By the time you retire, our numbers will probably exceed 9 billion. You can see this remarkable growth of the human population in **FIGURE 48.5**. Although our numbers have grown exponentially, the population growth rate is slowing down and our numbers are expected to level off around 2100.

As our numbers have grown, we have seen remarkable successes, particularly in the areas of public health and medicine. We have built great cities and created art, literature, and culture. We have developed sophisticated technologies that allow for rapid transportation, communication, and computing. At the same time, we use increasing amounts of energy, take up more and more space, and consume ever more food. Along the way, we are also destroying habitats, overharvesting many species, polluting ecosystems, introducing invasive species, and changing the composition of the atmosphere and oceans by burning fossil fuels. Increasingly, scientists refer to the modern time as the Anthropocene, from the Greek *anthropos,* meaning "human." This name emphasizes the dominant impact of humans on the present-day Earth.

As our population has increased, some species have expanded along with us. Some of these species are shown in **FIGURE 48.6**. Our pursuit of agriculture has sharply increased the abundance and distribution of animals and plants that we cultivate, such as corn, wheat, sheep, and cows. At the same time, we have inadvertently helped other species to expand—the crowded and not always clean environments of cities provide excellent habitats for cockroaches and rats.

Other species are in decline, their populations reduced by hunting and fishing, changes in land use, and other human activities. In fact, the numbers of species we are losing and the rapidity of this loss have forced scientists to consider whether we are in the midst of a sixth mass extinction, reminiscent in size and scale of the other five mass extinctions. However, this one differs from the other five in at least two important ways. First, it is occurring much more rapidly than the others. Second, current threats come not from meteorites or massive volcanism as in past mass extinctions, but rather from humans. This also means there is reason to hope, as we can do something about it if we choose to.

Several species that are endangered or extinct as a result of human activities are shown in **FIGURE 48.7**. At one time, large flightless birds called dodos (*Raphus cucullatus*) on the Indian Ocean island of Mauritius were plentiful. Within a century of Europeans arriving on the island, the dodo was extinct. Other species such as the Bali tiger (*Panthera tigris balica*) have gone extinct in recent decades. Still others are imperiled by human activities. Victimized by habitat destruction and poaching, the last surviving male of the northern white rhinoceros (*Ceratotherium simum cottoni*) died in 2018.

A 2016 census of population size for more than 3700 species of mammals, birds, fishes, and amphibians found that the total number of vertebrate animals in nature has declined by 60% over the past 50 years, prompting concern about mass extinction comparable to the mass extinctions of the

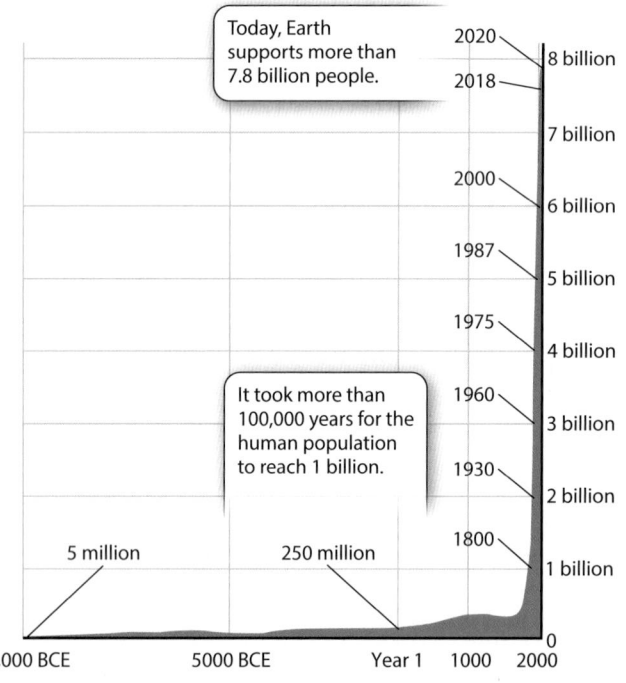

FIGURE 48.5 Growth of the human population over time

The human population has grown exponentially over time. Currently, humans number over 7.8 billion people.

FIGURE 48.6 Species that have benefited from human activity

Species that have benefited from human activity include (a) corn, (b) rats, and (c) cockroaches.

Photos: (a) Fred Dimmick/iStockphoto; (b) Arndt Sven-Erik/age footstock; (c) Nigel Cattlin/Alamy

FIGURE 48.7 Species that are extinct as a result of human activity

Through hunting, habitat destruction, and other activities, humans have caused many species to go extinct, including (a) the dodo, (b) the Bali tiger, and (c) the northern white rhinoceros.

Photos: (a) Photo Researchers/Science Source; (b) Look and Learn/Bridgeman Images; (c) James Warwick/Science Source

past. Furthermore, as shown in **FIGURE 48.8** on page 690 many species face the threat of extinction. Among mammals, for example, about a quarter of all species are vulnerable, endangered, or critically endangered, meaning that they are unlikely to survive without human intervention.

The reasons for loss of species are complex and various. Some are simply overhunted, prized for their hides or tusks or as sources of food. Some, like the dodo, have evolved in the absence of predators and so are vulnerable to human hunting and introduced species. Still others, like amphibians,

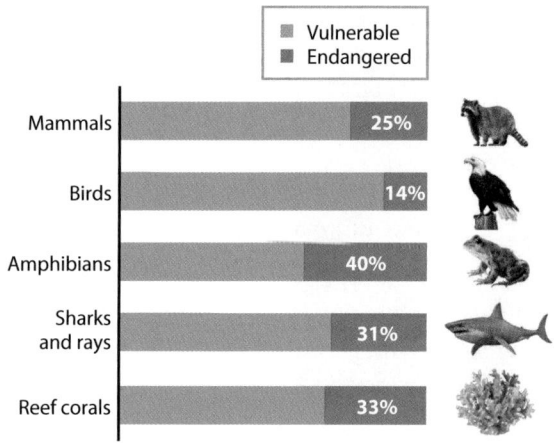

Legend:
- Vulnerable
- Endangered

Mammals 25%
Birds 14%
Amphibians 40%
Sharks and rays 31%
Reef corals 33%

Figure 48.8 Biodiversity at risk

Many mammal, bird, amphibian, fish, and coral species are at risk of extinction. These have been classified as vulnerable (likely to become endangered in the absence of intervention), endangered (likely to become extinct in the absence of intervention), or critically endangered (very high risk of extinction). Data from IUCN Red List of Threatened Species.

are affected by multiple factors. Amphibian habitat is being destroyed in many parts of the world, especially old-growth forests that provide damp, food-rich environments in abundance. Amphibians also take in gases through their skin, making them vulnerable to pesticides and other toxins. Recently, infection has also been implicated in amphibian decline. The fungus *Batrachochytrium dendrobatidis* (*Bd*) infects the skin of frogs and impairs many of the skin's functions.

The renowned biologist E. O. Wilson has estimated that Earth is losing 0.25% of its species annually—that is, 5000 to 25,000 species per year— depending on how many species actually exist. Even where ecosystems remain physically intact, pollution can erode populations. For example, recent studies in Europe show that levels of certain pesticides in soil

and water accurately predict patterns of population decline in bees and birds.

Without question, there will be fewer species when the twenty-first century ends than when it began. Lost species mean lost opportunities for the discovery of novel compounds for medical research. Diminished biodiversity may make ecological communities less productive and less resilient to fires, hurricanes, or other environmental events. There is a pressing need to feed the world, but there are also many good reasons to conserve the biodiversity that has evolved over 4 billion years. Balancing these two imperatives will entail creative approaches to both science and policy, including thoughtful ways of using land that serve both people and conservation.

In the words of the distinguished biologist Paul Ehrlich, "The fate of biological diversity for the next 10 million years will almost certainly be determined during the next 50–100 years by the activities of a single species." In this way, there is reason to be optimistic, as we have the unique ability to make choices that will ensure levels of biodiversity for a healthy planet.

> **PREP FOR THE AP® EXAM**
>
> **AP® Exam Tip**
>
> When answering questions about extinction or changing environments, remember to consider the role of human activity.

✓ Concept Check

9. **Describe** three ways in which humans are causing biodiversity loss.

10. **Identify** one species that has gone extinct as a result of human activities.

Module 48 Summary

> **PREP FOR THE AP® EXAM**
>
> **REVISIT THE BIG IDEAS**
>
> **EVOLUTION:** Choose one extinct species discussed in the module and **describe** the series of environmental changes that led to the extinction event.

LG 48.1 Species are lost by the process of extinction.

- Extinction is the loss of a group of organisms, such as a species. Page 682

- Extinction is represented on a phylogenetic tree by a branch that does not extend to the present. Page 682

- Levels of biodiversity are determined by rates of speciation and rates of extinction. Page 684

LG 48.2 Five mass extinctions have altered the course of evolution.

- This history of life on Earth is characterized by five mass extinctions, large and rapid losses of biodiversity. Page 684

- Mass extinctions have changed the course of evolution by eliminating some groups and providing opportunities for other groups to diversify. Page 684
- A meteorite struck Earth 66 million years ago, leading to environmental havoc and extinction of the dinosaurs, among other organisms. Page 685
- The largest documented mass extinction in the history of Earth occurred at the end of the Permian Period 252 million years ago. Page 686

LG 48.3 Extinction opens up habitats that can be filled by other species.

- Mass extinctions are often followed by adaptive radiation among the survivors due to the availability of many open habitats. Page 686

- When the dinosaurs went extinct at the end of the Cretaceous Period, habitats were opened up that were filled in part by mammals. Page 687

LG 48.4 Humans are changing the environment, leading to the loss of many species.

- Humans have altered the landscape in dramatic ways, leading to the expansion of some species, such as corn and wheat, and the decrease and extinction of others, such as the dodo and Bali tiger. Page 688
- Some scientists believe that we are currently experiencing extinction at a level and rate similar to other mass extinctions in Earth's history. Page 688
- Current biodiversity loss is due to human activities, such as habitat destruction, pollution, overharvesting, invasive species, and climate change. Page 690

Key Term

Extinction

Review Questions

1. Dinosaurs went extinct at the end of which period?
   (A) Cretaceous
   (B) Devonian
   (C) Permian
   (D) Triassic

2. During the Permian extinction, what percentage of species in the ocean disappeared?
   (A) 90%
   (B) 25%
   (C) 50%
   (D) 75%

3. The largest documented mass extinction on Earth occurred
   (A) 2.5 million years ago.
   (B) 25 million years ago.
   (C) 65 million years ago.
   (D) 252 million years ago.

4. One cause of biodiversity loss is the introduction of non-native species, or
   (A) convergent species.
   (B) invasive species.
   (C) natural species.
   (D) endemic species.

5. Which statement regarding biodiversity is true?
   (A) Earth is gaining 2.5% of its species annually.
   (B) There will be fewer species at the end of the twenty-first century than there are now.
   (C) Loss of biodiversity may result in more productive and resilient communities.
   (D) Habitat degradation is not a threat to biodiversity.

6. If the rate of extinction exceeds the rate of speciation, which will occur?
   (A) The number of species will increase.
   (B) The number of species will decrease.
   (C) The number of species will stay the same.
   (D) The number of species will fluctuate up and down.

Module 48
AP® Practice Questions

Section 1: Multiple-Choice Questions

Choose the best answer for questions 1–4.

1. Which of the following is a common cause of mass extinction?
   (A) Adaptive radiation by previously rare species leads to increased competition and extinction.
   (B) Gradual change in the environment leads to extinction of species that are unable to adapt.
   (C) Rapid environmental change causes ecosystems around the world to collapse.
   (D) Global populations become too large and require more environmental resources than are available.

2. The number of species in an environment is frequently used as a measure of biodiversity. A scientist studying ferns in the fossil record over a defined geological period estimated that there were approximately 335 species at the beginning of the period. The scientist estimated that 42 of these species went extinct before the end of the period. The scientist also observed 15 speciation events where a single ancestral species gave rise to two descendant species. What is the best estimate of the number of fern species present at the end of the geological period?
   (A) 293
   (B) 308
   (C) 323
   (D) 392

3. Jack Sepkoski was a scientist who used the fossil record from marine environments to document five mass extinctions over the past 500 million years. Which of the following is the most likely reason Sepkoski focused his research on marine animals?
   (A) Fossils are more likely to form in marine environments than in terrestrial environments.
   (B) The biodiversity in terrestrial environments is too low for identification of mass extinctions.
   (C) Terrestrial organisms do not experience mass extinctions.
   (D) Marine environments are more variable than terrestrial environments.

4. Approximately 66 million years ago, a massive meteor impacted Earth, resulting in the extinction of most dinosaurs and an adaptive radiation of mammals. Which statement most accurately describes this event and its consequences?
   (A) The meteor impact changed the temperature of Earth, which gave mammals a competitive advantage over dinosaurs.
   (B) Plant life on Earth changed dramatically as a result of the meteor impact, generating ecosystems that could support mammals but not dinosaurs.
   (C) The meteor impact changed the chemical composition of Earth's atmosphere, and the new atmosphere was toxic to dinosaurs but not mammals.
   (D) The meteor impacted caused a major loss of dinosaur biodiversity, which created open ecological habitats that mammals exploited.

Section 2: Free-Response Question

Write your answer to each part clearly. Support your answers with relevant information and examples. Where calculations are required, show your work.

White-tailed deer are currently extremely common in the eastern United States, but their populations have historically varied in size. The figure shows the number of white-tailed deer in Missouri from 1600 to 2020. Four important changes in human activity are indicated on the graph.

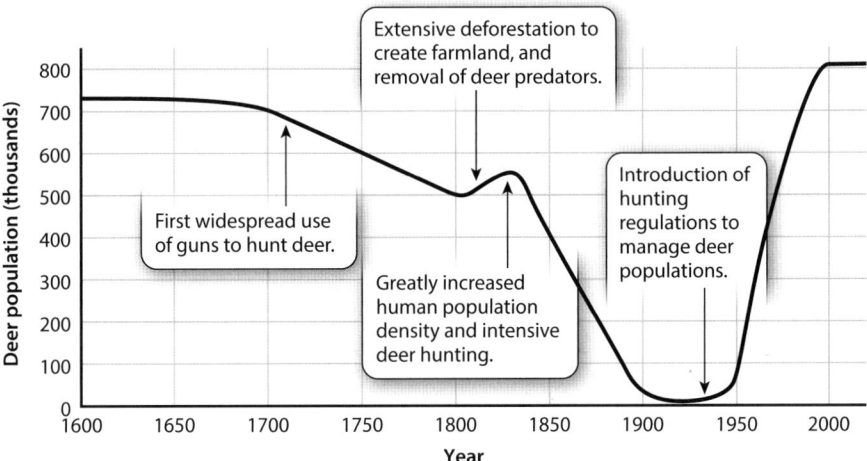

Population size of white-tailed deer in Missouri

(a) **Identify** which of the four human factors had the greatest impact on the white-tailed deer population in Missouri.

(b) Wolves were hunted to extinction in Missouri between 1800 and 1950. Based on the data provided, and assuming all other factors remain unchanged after 2022, **predict** the effect that re-introducing wolves into Missouri would have on the white-tailed deer population.

(c) A scientist claimed that conversion of forest to farmland creates improved habitat for white-tailed deer. Use the data provided to **support or refute** the scientist's claim.

Module 49

Variation in Populations

LEARNING GOALS ▶**LG 49.1** Populations with low levels of genetic variation are at risk of extinction.

▶**LG 49.2** Populations with high levels of genetic variation are often resilient to environmental change.

▶**LG 49.3** Variants that are well adapted to one environment may not be well adapted to another environment.

In the last module, we discussed the process of extinction, in which a group of organisms is lost. Extinction occurs when the last member of a group, such as a species, dies. However, well before this occurs, a species that is reduced in numbers is at risk of further decline and perhaps extinction. Small populations are at greater risk of decline and extinction because they often contain very little genetic variation. Populations with little genetic variation have a limited ability to adapt to a changing environment. In contrast, populations with considerable genetic variation tend to be resilient and can often adapt to a changing environment.

Genetic variation is fundamental to the ability of a species to adapt because of the role it plays in the evolutionary process. We have seen that genetic variation is the raw material on which evolution acts. It's the starting place for change over time. In fact, variation among members of a species is Darwin's first postulate, as we discussed in Module 41. If there is no variation, then individuals don't differ in fitness and natural selection can't operate. Without variation, there would be no evolution. And if a species can't change and adapt, it is vulnerable to changing environmental conditions.

The relationship between genetic variation and the ability to adapt also affects the outcome of conservation efforts. Successful conservation programs do not focus only on saving individual members of a species. A sanctuary or zoo with few individuals of a particular species is often not enough to sustain that species over time. Rather, where there are enough individuals to have a genetically diverse and therefore robust population, conservation of habitat is more effective.

In this module, we will discuss what happens to populations with low levels of genetic variation and those with high levels of genetic variation. We will also consider what happens to a population that is well adapted to a particular environment when the environment changes.

PREP FOR THE AP® EXAM

FOCUS ON THE BIG IDEAS

SYSTEMS INTERACTIONS: Look for examples of how a diverse gene pool increases the likelihood of survival of the species in the face of changing environmental conditions.

49.1 Populations with low levels of genetic variation are at risk of extinction

When we think about natural selection, we often focus on the variants that are best suited to a particular environment. These are the individuals that survive and reproduce, so that, over time, a population of individuals becomes exquisitely adapted to a particular environment.

However, it's important to realize that an environment is not fixed. Instead, it changes all of the time. The environment does not just include factors such as the weather, soil, and other physical aspects of a particular area, collectively called abiotic factors. The environment also includes all the organisms in the area, called biotic factors. These organisms are subject to natural selection and other evolutionary mechanisms, and often change over time. Therefore, it's best to think of many environments as constantly in flux and changing. Sometimes, environmental change can be major, such as in the aftermath of a meteorite striking Earth, or it can be more subtle, such as changes in rainfall patterns or the numbers of predators. Either way, the environment in which organisms live is constantly changing.

Populations that harbor considerable genetic variation have many different variants available. Some of these may be well adapted to the current environment, and therefore have high fitness, and others less so, and therefore have low fitness. But those that are less well adapted may be better adapted as the environment shifts. Conversely, those populations with little or no genetic variation are vulnerable to environmental change.

Let's take the example of crops, which are bred in such a way that they often have little genetic variation. We eat surprisingly few of the more than 400,000 species of plants on Earth today. Most of the plants we eat come from species that are grown in cultivation, and many of these cultivated species have, as a consequence of artificial selection, lost the ability to survive and reproduce on their own. Of the approximately 200 such cultivated species, 12 account for more than 80% of human caloric intake. Just three—corn, wheat, and rice—make up more than two-thirds of the food we eat.

Crop breeders select for varieties that grow well under cultivation and are uniform so are easy to harvest. However, one consequence of selective breeding of crop species is low genetic diversity. Because pathogens and pests are always evolving, low genetic diversity of crop species creates substantial risks. For example, before 1970, the fungal pathogen *Bipolaris maydis* destroyed less than 1% of the U.S. corn crop annually. In 1970, however, it destroyed more than 15% of the corn crop in a single year. The severity of the epidemic was due to the genetic uniformity of the corn varieties planted at that time. In natural plant populations, genetic variation for pathogen resistance makes it highly unlikely that a newly evolved pathogen strain will be able to infect every plant.

Bananas (*Musa acuminata*) are similarly vulnerable. You may have noticed that while your grocery store or supermarket has several types of apples for sale, there is usually only one type of banana. In fact, it's possible that every banana you have ever eaten has been the same variety. Now the popular yellow fruit may be in trouble.

Years ago, the Western world favored a banana variety known as the Gros Michel. In the first part of the twentieth century, a devastating fungal infection called Panama disease wiped out Gros Michel plantations around the world. In the 1950s, banana growers turned instead to the Cavendish, a variety that displayed some natural resistance to Panama disease. Half a century later, the Cavendish banana, shown in **FIGURE 49.1**, remains the top-selling banana in supermarkets in North America and beyond.

Yet the reign of the Cavendish may be coming to an end. For years, banana growers have battled black sigatoka,

FIGURE 49.1 Cavendish bananas

Cavendish bananas are a cultivated crop and the typical bananas that you find in a supermarket. They are in danger because of a lack of genetic variation, which is common among cultivated crops.

Photo: Jacek Sopotnicki/Alamy Stock Photo

a fungal infection that can cause losses of 50% or more of the banana yield in infected regions. Making matters worse, Panama disease is once again posing a threat. A strain of the disease has emerged against which the Cavendish plants have no resistance. The disease has spread across Asia and Australia, and experts fear it's only a matter of time before it reaches prime banana-growing regions in Latin America.

For those who enjoy bananas sliced into their breakfast cereal, the loss of bananas would be disappointing. For the millions of people in tropical countries who depend on bananas and related plantains for daily sustenance, the destruction of those crops would be much more serious.

Cultivated bananas are vulnerable to infection in a way that wild bananas are not. The bananas that we eat are sterile; they do not produce seeds. This means they must be propagated vegetatively, a form of asexual reproduction, by replanting cuttings taken from parent plants. As a result, banana plantations contain virtually no genetic diversity. As Panama disease and black sigatoka are showing, that lack of diversity can have disastrous effects.

Another fungus is threatening global wheat production. Wheat, shown in **FIGURE 49.2**, is a staple crop. Throughout much of human history, wheat and other crops have been vulnerable to infection by the fungus *Puccinia graminis*. *P. graminis* first infects wheat leaves, and then extends within the plant's tissues to fuel its own growth. Rust-colored pustules, shown in **FIGURE 49.3**, erupt along the stem, giving the infection the name stem rust. These pustules release a large number of spores, after which the plant withers and dies. Crop losses from *P. graminis* can be devastating.

During World War II, the plant biologist Norman Borlaug traveled to Mexico to help combat crop losses due to rust. Borlaug was eventually able to breed resistant wheat varieties. Much of the resistance shown by his varieties could be attributed to variation in a small number of genes. Wheat rusts were apparently vanquished by Borlaug's work.

FIGURE 49.3 Fungal wheat rust

The rust coloration on the stems of these wheat plants is the result of infection by a fungus, a major threat to the world's wheat crops.

Photo: Yue Jin, USDA-ARS, Cereal Disease Laboratory

FIGURE 49.2 Wheat, a vulnerable staple

Wheat crops harbor little genetic variation and are therefore susceptible to pathogens, such as fungal wheat rust. Photo: Nigel Cattlin/ Science Source

However, in 2010, a new strain of wheat rust fungus emerged in the Middle East. In its first year, the disease wiped out as much as half of Syria's wheat crop. It has since spread to other countries in the region. Meanwhile, crop scientists have warned that Ug99, a particularly damaging strain of *P. graminis*, could devastate global wheat crops. This strain first appeared in the east African country of Uganda, for which it is named.

Like corn farms and banana plantations, most modern wheat farms consist largely of monocultures, single crop species grown over a large area. In a given field of wheat, the plants are often genetically similar to one another. Growing only a single crop at a time makes it much easier to mechanize planting and harvesting. Unfortunately, it also means that a pathogen that can overcome a plant's natural defenses has the potential to wipe out an entire field.

At the time that Ug99 was discovered, it was capable of defeating all known resistance genes in wheat. These genes provide protection against rusts. In the absence of variation in these and other genes, wheat is susceptible to fungal infection. Given that wheat accounts for 20% of daily global calorie consumption by humans and that 90% of cultivated wheat has little or no resistance to Ug99, the appearance of this resistant fungus is cause for major concern.

The map in **FIGURE 49.4** shows areas where Ug99 is currently found. It is currently devastating wheat production in Kenya and has been found in Yemen, Iran, South Africa, and Egypt. As of 2019, it was present in 13 countries. In addition, Ug99 shows every sign of continuing to spread. The spread of Ug99 is not surprising given that a single

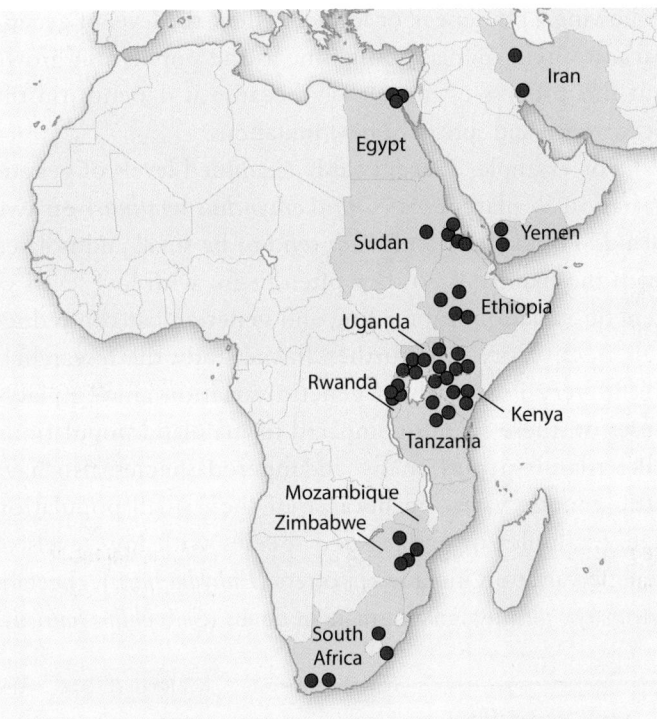

FIGURE 49.4 The spread of Ug99

Ug99 is a highly damaging stem rust of wheat and a major threat to global food production. It originated in Uganda and has spread to places indicated by the red circles. Source: https://rusttracker.cimmyt.org/?page_id=22

hectare of infected wheat produces upward of 1 billion spores. As you read this, Ug99 is poised to reach the major wheat-growing regions of Turkey and South Asia. Even more rapid spread is possible if spores accidentally lodge on cargo or airline passengers.

Just before his death in 2009, Norman Borlaug urged the world to take this threat seriously. Today, plant breeders are actively searching ancestral wheat varieties for genetic sources of resistance to Ug99, while other scientists try to understand how plants defend themselves against rusts and what makes Ug99 so virulent.

Nicolai Vavilov, a Russian botanist and geneticist, was one of the first to recognize the importance of safeguarding the genetic diversity of crop species and their wild relatives. In the early twentieth century, he mounted a series of expeditions to collect seeds from around the globe. Vavilov's observations of cultivated plants and their relatives led him to hypothesize that plants had been domesticated in specific locations. He called these regions centers of origin and they are shown in **FIGURE 49.5**. He believed that they coincided with the centers of diversity for both crop species and their wild relatives. From these locations, plant domestication spread through human migration and commerce.

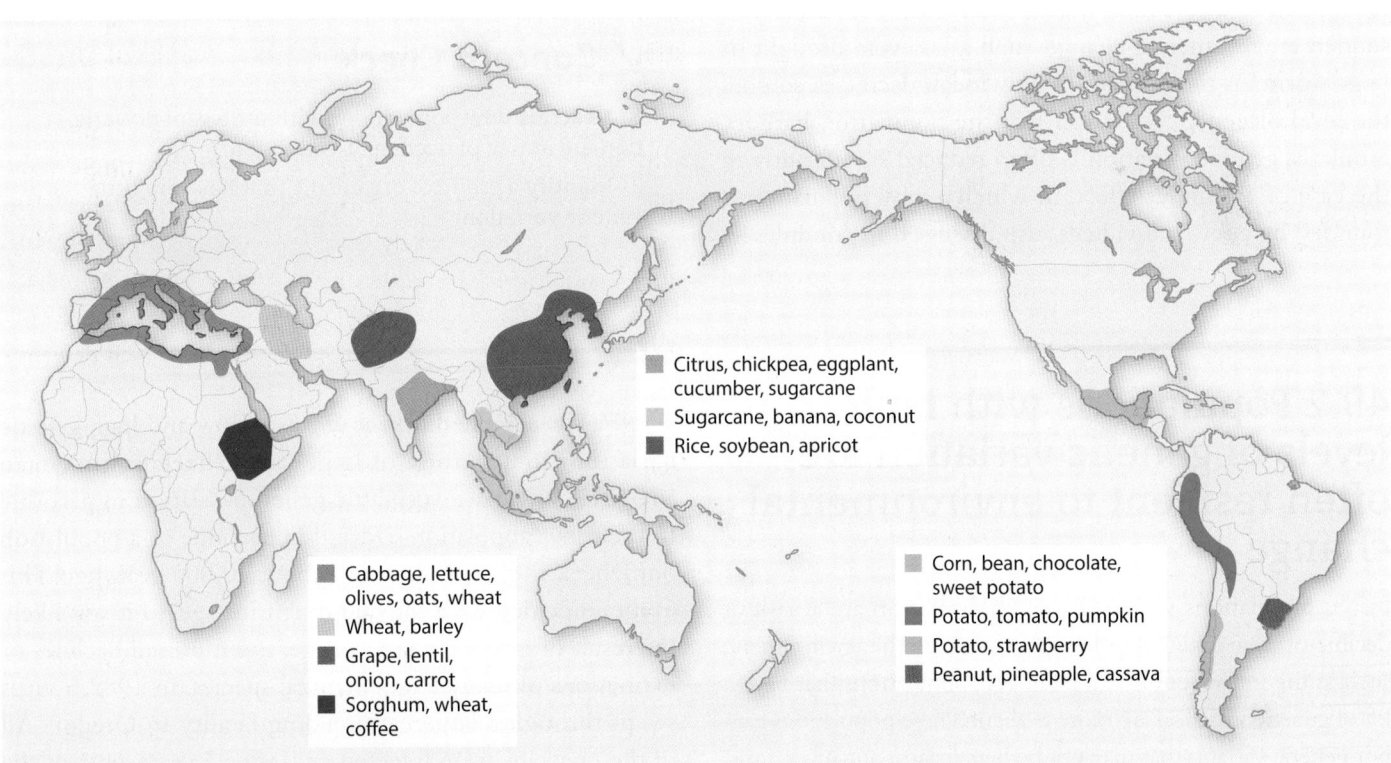

FIGURE 49.5 Centers of origin for crops

The colored areas indicate where different crop species were first domesticated. They represent places where we might be able to find natural populations of our cultivated crops with high levels of genetic variation.

In World War II, during the siege of Leningrad, scientists at the Vavilov Institute sought to protect what was then the world's largest collection of seeds. The dedication of these scientists illustrates the priceless nature of the genetic diversity on which our food supply rests.

Today, seed banks help to preserve the genetic diversity of crop species and their wild relatives. One of the most significant seed banks is Svalbard Global Seed Vault, which was started in 2008 and currently holds approximately 6000 species of plants. However, seed banks can store only a fraction of the genetic diversity present in nature. To help make up the difference, some have suggested establishing protected areas that coincide with Vavilov's centers of origin. The resurgence of *P. graminis* and other fungi as serious threats to world food production underscores the importance of safeguarding the genetic diversity of agricultural species so that the evolutionary battles with pathogenic species may be waged successfully.

The examples of corn, bananas, and wheat all focus on crops that have low levels of genetic variation because they have been bred this way by humans. In natural populations, we often see little genetic variation in populations that have undergone a population bottleneck. Recall from Module 43 that a population bottleneck occurs when a large population is reduced to a few individuals, which can happen after a sudden environmental change, such as a severe drought or large storm. As the number of individuals decreases, so does the level of genetic variation that the population harbors. Similarly, genetic variation is often reduced significantly in the case of a founder effect, in which a new population is founded by a few individuals, also discussed in Module 43.

Following a bottleneck or founder effect, the level of genetic variation will increase over time as the population grows, but this process occurs slowly because it depends on the occurrence and spread of new mutations.

For example, a recent study examined levels of genetic variation in platypuses (*Ornithorhynchus anatinus*) on two islands. Populations on islands tend to be small and isolated from the mainland, so they often begin with low levels of genetic variation. Inbreeding and genetic drift can reduce genetic diversity even further. In the study, the researchers found very low levels of genetic variation among platypuses on these islands compared to mainland populations. Like island species, many endangered species also have little genetic variation because of their small population size. Examples of endangered species with low levels of genetic variation include sea otters (*Enhydra lutris*), cheetahs (*Acinonyx jubatus*), and Tasmanian devils (*Sarcophilus harrisii*).

✓ Concept Check

1. **Describe** why populations with little genetic variation are at risk of decline or extinction.

2. **Identify** a group of organisms that has very little genetic variation.

49.2 Populations with high levels of genetic variation are often resilient to environmental change

Just as populations with little genetic variation are at risk of decline or even extinction due to a change in the environment, such as the introduction of a pathogen, populations that have a lot of genetic variation are more resilient. These populations harbor genetic variants that may not be beneficial at any given time, but may be beneficial if the environment changes. Most natural populations have high levels of genetic variation as a result of mutation and recombination that have occurred in the past.

We can see the different effects of low and high genetic variation from a study of large cats. Cheetahs, shown in FIGURE 49.6, have very little genetic variation in part due to their low population size and in part due to a population bottleneck they went through about 10,000 years ago. This event coincided with the end of last ice age, so it was likely the result of global climate change, and it caused declines or extinctions of many large mammal species. In 1982, a virus swept through a large cat breeding facility in Oregon. All of the cheetahs were infected and, after 3 years, 60% of the cheetahs died. By contrast, lions in the same facility, with much higher genetic variation, did not become ill. Presumably, the lions had gene variants (alleles) that provided

FIGURE 49.6 Low levels of genetic variation in cheetahs

Cheetahs have very little genetic diversity because of a population bottleneck that occurred in the past and their small population size today. Photo: Anna Omelchenko/Dreamstime.com

protection against the virus and such variants were absent among the cheetahs.

Bacteria provide another example. In Module 41, we discussed antibiotic-sensitive and antibiotic-resistant bacteria. Imagine a population of genetically identical bacteria, all of which are sensitive to an antibiotic. In this case, application of an antibiotic represents a change in the environment. Due to their genetic uniformity, all the bacteria will be wiped out by the antibiotic.

Now imagine a population in which some bacteria are sensitive and some bacteria are resistant to an antibiotic. Because resistance to antibiotics has a genetic basis, the bacteria are genetically different from one another. In this case, application of the antibiotic will not kill all the bacteria. There will be differential reproductive success, with those bacteria that are antibiotic resistant surviving and reproducing better than the bacteria that are antibiotic sensitive. Over time, the population will evolve so that all the bacteria are resistant to the antibiotic. This is natural selection in action.

If a second antibiotic is now applied to the bacteria, one that works differently from the first, the population will again be vulnerable. However, the more genetic variation the population of bacteria has, the more likely it will harbor a genetic variant that causes it to be resistant to an antibiotic that we use. The presence of this variant will allow the population of bacteria to evolve resistance over time when we apply antibiotics. Essentially, we are selecting for antibiotic-resistant bacteria.

In fact, this is the history of the use of antibiotics. Every time we use a new antibiotic, it is effective, but only for a short time, as the population of bacteria evolve resistance as a result of preexisting mutations that were already present in the population. These mutations are then selected by our use of antibiotics. Today, antibiotic resistance is fast becoming a public health crisis, as bacteria are being found that are resistant to all known antibiotics. This is the result of applying antibiotics to a population of bacteria that have a lot of genetic variation.

In addition to allowing a population to adapt if the environment changes, genetic variation might allow members of a species to live in a range of different environmental conditions that are already present. For example, consider an allele that is favored by natural selection in a dry environment and a different allele that is favored by natural selection in a wet environment. A species with both alleles will be well adapted to live in both environmental conditions, whereas a species with one or the other would be limited to just one condition.

This form of natural selection which acts to maintain two or more alleles of a given gene in a population is called balancing selection. This is different from earlier examples where we saw beneficial alleles increase in frequency in the population or harmful alleles decrease in frequency under the influence of natural selection. In balancing selection, by contrast, both alleles are maintained in the population at some intermediate frequency.

✓ Concept Check

3. **Explain** why populations with a lot of genetic variation are resilient in the face of a changing environment.

4. **Identify** a group of organisms that has a lot of genetic variation.

49.3 Variants that are well adapted to one environment may not be well adapted in another environment

As we discussed in Module 41, the fitness of a particular variant depends in part on the variant and in part on the environment in which it is found. As a result, a genetic variant that is beneficial in one environment may not be beneficial or in fact may be harmful in a different environment. As an example, let's consider malaria. Malaria is a disease that causes intermittent fevers, headaches, and sometimes death. It is caused by a parasite that is transmitted by mosquitoes. The malaria parasite spends part of its life cycle in human red blood cells. Within these cells is a molecule called hemoglobin, which carries oxygen and affects the structure of red blood cells.

Hemoglobin is made up of four subunits, including β-(beta-)globin. Mutations in β-globin can have a negative effect on the malaria parasite. For example, consider the *A* and *S* alleles of the β-globin gene, discussed in Module 37. The *A* allele has a GAG codon that translates to glutamic acid (Glu) in the resulting protein. The *S* allele is a mutation that changes glutamic acid (Glu) to valine (Val). Although only one amino acid of the β-globin protein is affected, this change has a dramatic effect on the form and function of the protein, since the amino acid sequence determines how a protein folds, and protein folding in turn determines the protein's function, as discussed in Module 4.

For the *A* and *S* alleles, there are two homozygous genotypes, *AA* and *SS,* and one heterozygous genotype, *AS*. In *AA* individuals, a normal form of hemoglobin is produced. As a result, the red blood cells are round and fully functional, as shown on the left in **FIGURE 49.7**. In *SS* individuals, the hemoglobin that is produced causes the red blood cell to distort into a half-moon or sickle shape, shown on the right in Figure 49.7. In this form, the red blood cell is unable to carry the normal amount of oxygen. In addition, the sickled cells can block small capillaries, interrupting the blood supply to vital tissues and organs, which causes debilitating, painful, and sometimes fatal episodes. This condition is called sickle-cell anemia. *AS* individuals have mild anemia, called sickle-cell trait. Sickle-cell trait is less severe than sickle-cell anemia, but, like all forms of anemia, it reduces the ability of red blood cells to carry oxygen to tissues and organs, and so reduces the health of the individual.

These three genotypes also differ in their susceptibility to malaria. *AA* individuals are healthy, but are susceptible to infection by malaria. *AS* individuals have mild anemia, but have some protection against infection by malaria. *SS* individuals also have protection against infection by malaria, but have severe anemia. In areas where malaria is prevalent, *AS* individuals have higher fitness than either of the homozygotes because they only have mild disease and are protected from malaria. As a result, natural selection acts to maintain both the *A* and *S* alleles in the population. This form of natural selection is called heterozygote advantage. It is a form of balancing selection because, like the example we discussed in the last section, both alleles remain in the population at intermediate frequencies.

In areas where there is no malaria, however, this balance is shifted. While *AS* is favored by heterozygote advantage in areas with malaria and is therefore the most fit genotype, it is not favored in malaria-free regions. In this case, *AA* is favored and is the most fit genotype. The *AS* and *SS* genotypes have lower fitness in areas without malaria than the *AA* genotype because they both cause anemia, although to different degrees, and there is no advantage to carrying the *S* allele when malaria is not present. So, a genetic variant that is favored in one area is not in another.

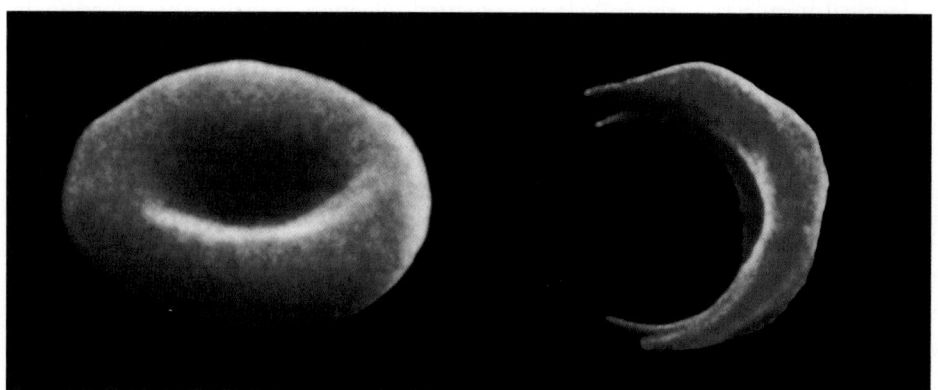

Figure 49.7 Healthy and sickle-shaped red blood cells

A healthy red blood cell is shown on the left, and a sickle-shaped red blood cell is shown on the right. Red blood cells contain hemoglobin, which carries oxygen to tissues and organs. The sickle-shaped red blood cell results from a single mutation in the gene encoding a subunit of hemoglobin and results in the disease called sickle-cell anemia.

Photo: Mary Martin/Science Source

Many people whose ancestors came from regions where malaria was prevalent but now live in a malaria-free environment still carry the S allele, even though the allele is no longer beneficial to them in their current environment. If natural selection were to run its course among these people, the S allele would gradually be eliminated because it causes anemia as a heterozygote and homozygote. This is a slow process, however, and many more people will continue to suffer from sickle-cell anemia before it is eliminated from the population. Fortunately, medical interventions are available that can help those affected with the disease.

A second example of a genetic variant being beneficial in one environment but not in another is antibiotic resistance in bacteria. A bacterium that has resistance to a particular antibiotic will likely not have resistance to a different type of antibiotic. The mutation that provides resistance is beneficial in one environment, but neutral or even harmful in a different environment. Both examples remind us that being well adapted, or having high fitness, depends on the environment, as we discussed in Module 41.

✓ Concept Check

5. **Describe** why AS individuals are well adapted to regions where there is malaria.

6. **Describe** why AA individuals are well adapted to regions without malaria.

7. **Describe** why a variant that is beneficial in one environment may not be beneficial in another.

Module 49 Summary

> **PREP FOR THE AP® EXAM**
>
> ## REVISIT THE BIG IDEAS
>
> **SYSTEMS INTERACTIONS:** Using one example from the module, **provide evidence** that a diverse gene pool increases the likelihood that a species will survive in the face of a changing environment.

LG 49.1 Populations with low levels of genetic variation are at risk of extinction.

- Populations that are genetically uniform are vulnerable to a change in the environment. Page 694

- Many crop species, such as corn and wheat, harbor very little genetic variation and so are vulnerable to new pathogens. Page 696

LG 49.2 Populations with high levels of genetic variation are often resilient to environmental change.

- Populations of individuals that are genetically different have a better chance of withstanding changing environment. Page 698

- Antibiotic resistance among bacteria is a public health crisis because some bacteria are genetically variable and have evolved resistance to all known antibiotics. Page 699

LG 49.3 Variants that are well adapted to one environment may not be well adapted to another environment.

- Variants that are well suited to one environment may not be well suited to a different environment. Page 700

- The S allele of β-globin provides protection against malaria in AS individuals, so it is beneficial as a heterozygote, but only in malaria-prone regions. Page 700

Review Questions

1. Which population is likely to be most resilient to a changing environment?
   (A) A population with very little genetic variation
   (B) A population in which individuals are genetically identical
   (C) A population with a lot of genetic variation
   (D) A population that is genetically uniform

2. Variation is important for evolution and the ability of organisms to adapt to an environment because
   (A) it increases the likelihood that the population will go extinct.
   (B) it is the raw material on which evolution acts.
   (C) it is the source of harmful mutations.
   (D) it makes individuals of a population all the same.

3. Ug99 is a
   (A) fungus that infects wheat.
   (B) bacterium that infects bananas.
   (C) protist that infects humans.
   (D) virus that infects the malaria parasite.

4. Most of the crops we eat are
   (A) genetically diverse.
   (B) genetically uniform.
   (C) made up of individuals that are genetically different from one another.
   (D) relatively resistant to new pathogens.

5. In areas where malaria is present, which genotype is the most fit?
   (A) *AA*
   (B) *AS*
   (C) *SS*
   (D) *AAA*

6. In areas where malaria is absent, which genotype is the most fit?
   (A) *AA*
   (B) *AS*
   (C) *SS*
   (D) *AAA*

7. Genetic variation is ultimately the result of
   (A) natural selection.
   (B) genetic drift.
   (C) migration.
   (D) mutation.

Module 49
AP® Practice Questions

PREP FOR THE AP® EXAM

Section 1: Multiple-Choice Questions
Choose the best answer for questions 1–3.

1. Which of the following describes a reason why small populations have a higher risk of extinction?
   (A) Genetic drift is a weaker force in small populations.
   (B) Individuals in small populations often choose not to reproduce with close relatives.
   (C) Small populations are more vulnerable to high levels of predation.
   (D) Small populations lack genetic variation that enables adaptation to changing environments.

2. Which of the following statements about environmental variation is most accurate?
   (A) Different natural selective pressures in different environments can maintain genetic variation in populations.
   (B) Environmental change cannot increase genetic variation in populations because natural selection acts on individuals.
   (C) Environmental change usually increases genetic variation because natural selection drives novel mutations to fixation.
   (D) Environmental change leads to extinction because genetic diversity prevents adaptation.

3. Which of the following statements about heterozygote advantage is most accurate?
   (A) Heterozygote advantage is most common in haploid organisms like bacteria.
   (B) Heterozygote advantage can only occur when one of the alleles is dominant.
   (C) Heterozygote advantage occurs when a heterozygote has higher fitness than all homozygotes.
   (D) Heterozygote advantage often leads to the formation of new species.

Section 2: Free-Response Question

Write your answer to each part clearly. Support your answers with relevant information and examples. Where calculations are required, show your work.

Common frogs (*Rana temporaria*) are amphibians. Because they can live both on land and in water, adult frogs can move between different bodies of water. Adult frogs lay their eggs in water, and when the eggs hatch, the tadpoles remain in the water until they undergo metamorphosis to become adult frogs. Thus, the aquatic environment imposes a strong evolutionary pressure on the frogs.

A researcher hypothesized that some genotypes of *R. temporaria* might have higher fitness than others in different aquatic environments. To test this hypothesis, the researcher collected frog eggs from a pond with high levels of salt, or salinity. The researcher also collected eggs from a pond with low salinity. The frogs in these two ponds are known to be genetically different from each other.

The researcher transferred half of the eggs from each collection into a high-salinity aquarium in the lab, and the other half into a low-salinity aquarium in the lab. After the eggs hatched, the researcher measured the proportion of tadpoles from each collection that survived to metamorphosis in each aquarium. The data that the researcher obtained are illustrated in the figure.

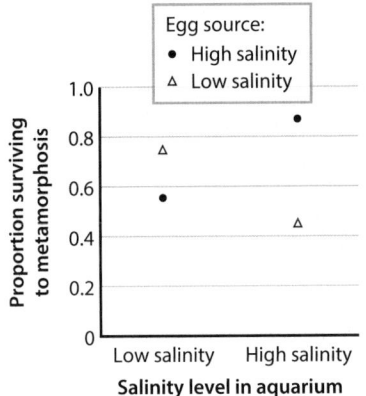

(a) **Identify** the source of the eggs that have the highest survival to metamorphosis in the high-salinity aquarium.

(b) The researcher makes a claim that the egg genotypes in each pond are adapted to their native salinity environment. **Provide reasoning** to support or refute this claim.

(c) Environmental factors like salinity can be highly variable over geography and over time. **Explain** why genetic diversity can help populations survive environmental change and avoid extinction.

Module 50

Origin of Life on Earth

LEARNING GOALS ▶**LG 50.1** The building blocks of organic molecules can be synthesized from inorganic precursors.
▶**LG 50.2** The formation of organic molecules set the stage for the first cell.
▶**LG 50.3** Life originated on Earth at least 3.5 billion years ago.
▶**LG 50.4** The history of life was mostly unicellular, with familiar organisms appearing later.

In the preceding 10 modules, we have explored the process of evolution. We have described how life changes over time, and discussed mechanisms by which this change occurs. We have seen how one of these mechanisms—natural selection—can account for the remarkable adaptations all around us. The evolution of life on Earth began with the very first cell—the first form of life—and continues today.

Where did life come from in the first place? This is a fascinating question and is the subject of much research. The question is also intimately connected with the search for life on other planets, such as Mars. If we understand the conditions in which life arises, perhaps we will have a better chance of finding it elsewhere in our solar system or the universe.

It's important to emphasize that the process by which life evolves is not the same as the process by which life originates. These are distinct and separate processes. Life evolves by the five mechanisms we discussed in Module 43—natural selection, genetic drift, migration, mutation, and nonrandom mating. We have a good understanding of these processes and can explain them in detail. By contrast, the process by which life arises from nonliving materials—called abiogenesis—does not occur by the evolutionary mechanisms we have explored and is not as well understood.

To understand how life first arose, let's work backward. As we saw in Unit 2, the smallest living entity is the cell. This is why the cell is considered the fundamental unit of life. Anything smaller is not alive. So, the question of how life originated becomes: how did the first cell arise? Cells have certain features, such as the ability to store and transmit information, maintain homeostasis, and harness energy from the environment, so we can ask: how did these functions emerge? Finally, cells are made up of organic molecules, such as proteins, nucleic acids, carbohydrates, and lipids, so we can ask: how were these molecules first made?

In this module, we will consider how life originated on this planet by working forward through these steps—How were organic molecules synthesized on the young Earth? How did they perform functions essential for life? And how did they come together to build the first cell? We will end with a brief overview of what happened after that time, which comprises the history of life on Earth.

PREP FOR THE AP® EXAM

FOCUS ON THE BIG IDEAS

SYSTEMS INTERACTIONS: Look for connections between the chemical building blocks available on the early Earth and the organic molecules necessary for life.

50.1 The building blocks of organic molecules can be synthesized from inorganic precursors

Estimates of the age of Earth indicate that it formed about 4.6 billion years ago. Analyses of ancient rocks, dated to within 0.3 billion years of Earth's formation, suggest that the early Earth was markedly different from Earth today. For instance, the young Earth was very hot and had a lot of volcanic activity. Liquid water is thought to have been first present anywhere from shortly after the planet was formed to 4.3 billion years ago. This is relevant to life's origin because water is essential to life as we know it. In fact, where we find water, we often find life. In this section, we will look at the environment of the young Earth and describe the types of molecules that were present.

The Atmosphere of Ancient Earth

Geological evidence suggests that a number of molecules existed early in Earth's history. These included relatively large amounts of the gases carbon dioxide (CO_2) and nitrogen (N_2), with smaller amounts of carbon monoxide (CO), hydrogen (H_2), hydrogen sulfide (H_2S), ammonia (NH_3), and methane (CH_4).

A notable absence was oxygen gas (O_2), which today makes up approximately 21% of Earth's atmosphere. Because oxygen is now present in our atmosphere, we have what is sometimes called an oxidative atmosphere. This description refers to oxygen's highly reactive properties that allow it to react with a large number of compounds. For example, when iron is exposed to oxygen and a little moisture, it quickly oxidizes and becomes rust, a familiar site on many cars as shown in **FIGURE 50.1a**. Similarly, hydrogen gas (H_2) explodes when it reacts with oxygen, releasing large quantities of energy, as dramatically demonstrated when the *Hindenburg*, filled with hydrogen gas, burst into flames as shown in Figure 50.1b.

In contrast to the oxidative atmosphere that we are familiar with, the atmosphere of ancient Earth was what is called a reducing atmosphere because of the lack of oxygen gas. The absence of oxygen gas was significant because of oxygen's reactive properties. Its presence in the early Earth's atmosphere would likely have destroyed any newly made molecules that had been synthesized from the other chemicals that were present. Therefore, any chemicals formed before the presence of oxygen gas would have been far more stable. The stability of the newly formed, ancient molecules could then lead to further reactions that could synthesize the molecules necessary for life.

The Building Blocks of Organic Molecules

As we saw in Unit 2, the molecules of life, or organic molecules, include proteins, nucleic acids, carbohydrates, and lipids. As we discussed, these molecules are made up of simpler molecules joined together. Proteins are made up of amino acids; nucleic acids are made up of nucleotides; carbohydrates are made up of simpler sugars; and some lipids are made up of glycerol and fatty acids. To understand how these organic molecules may have emerged on the early Earth, we first need to understand how these building blocks formed. Thus, if we want to understand how proteins formed, we should begin with the synthesis of amino acids, and if we are interested in nucleic acids, we should focus on nucleotides.

Research into the origins of life was catapulted into the experimental age in 1953 with an elegant experiment carried out by Stanley Miller, then a graduate student at the University of Chicago in the laboratory of Nobel laureate Harold Urey. This experiment is outlined on page 706 in "Practicing Science 50.1: Could the building blocks of organic molecules have been generated on the early Earth?" Miller started with water vapor, methane (CH_4), ammonia (NH_3), and hydrogen gas (H_2)—all of which are thought to have been present in the early atmosphere. He put these gases into a sealed flask and then passed a spark through the mixture. On the primitive Earth, lightning or volcanic heat might have supplied the energy needed to drive chemical reactions, and the spark was meant to simulate these effects. Analysis of the contents of the flask showed that several amino acids were generated.

a.

b.

Figure 50.1 The reactivity of oxygen

Oxygen reacts with a large number of chemicals. (a) Rust forms on metals like cars when oxygen and water react with iron. (b) The *Hindenburg* was an airship called a zeppelin filled with hydrogen gas to keep it afloat. It burst into flames in 1937 when oxygen in the air reacted with hydrogen gas as it attempted to dock. Photos: (a) Akintevs/Shutterstock; (b) Bettmann/Getty Images

Could the building blocks of organic molecules have been generated on the early Earth?

Background In the 1950s, it was widely believed that Earth's early atmosphere was rich in water vapor, methane, ammonia, and hydrogen gas, with no free oxygen.

Experiment Stanley Miller built an apparatus, illustrated in the figure, designed to simulate Earth's early atmosphere. He began by heating liquid water to produce water vapor (H_2O). This water vapor then mixed with other gases thought to be present on the early Earth, including methane (CH_4), ammonia (NH_3), and hydrogen gas (H_2). To simulate lightning, Miller passed a spark through the mixture of gases. The gas mixture was then cooled, allowing liquid water to form again and dissolve any molecules that were synthesized. These were collected at the bottom of the apparatus and obtained through a sampling valve for analysis.

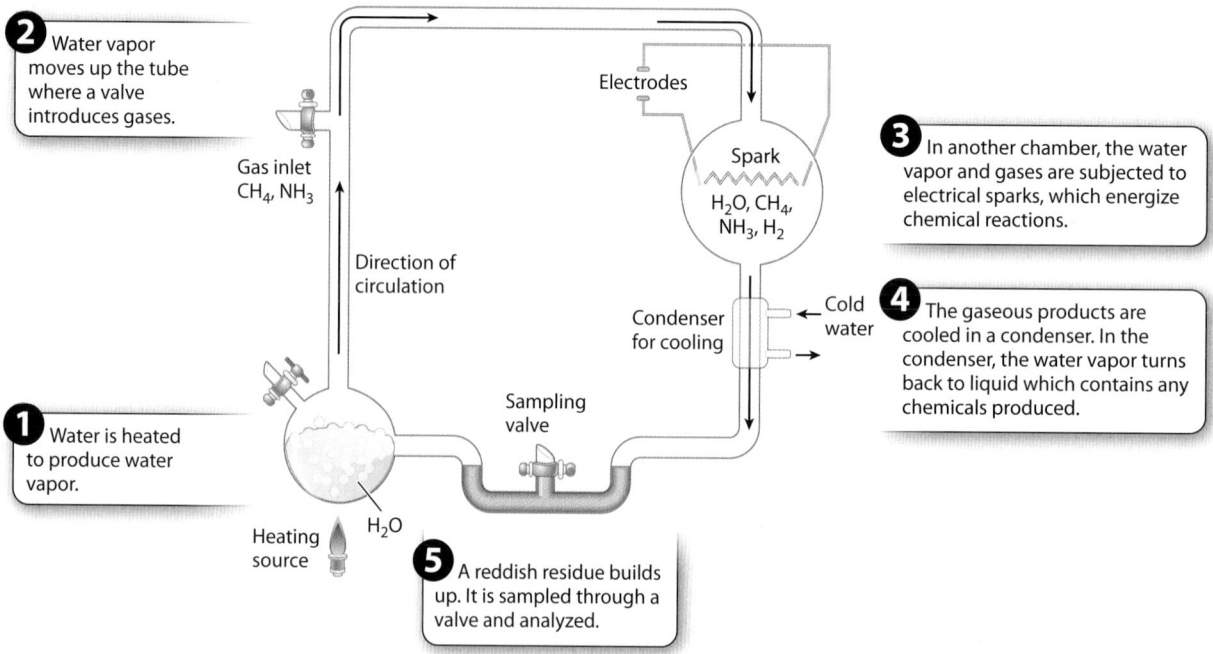

2 Water vapor moves up the tube where a valve introduces gases.

Gas inlet CH_4, NH_3

Direction of circulation

1 Water is heated to produce water vapor.

Heating source

H_2O

Electrodes

Spark
H_2O, CH_4, NH_3, H_2

3 In another chamber, the water vapor and gases are subjected to electrical sparks, which energize chemical reactions.

Condenser for cooling

Cold water

4 The gaseous products are cooled in a condenser. In the condenser, the water vapor turns back to liquid which contains any chemicals produced.

Sampling valve

5 A reddish residue builds up. It is sampled through a valve and analyzed.

Results As the experiment proceeded, reddish material accumulated at the bottom of the apparatus. Analysis showed that this material included a number of amino acids, the building blocks that make up proteins, which are the key structural and functional molecules that do much of the work of the cell.

Conclusion Amino acids can be generated in conditions that mimic those of the early Earth.

Follow-up Work Recent analysis of the original extracts, saved by Miller, shows that the experiment produced about 20 different amino acids, not all of them found in organisms.

SOURCES:
Miller, S. L. 1953. "Production of Amino Acids under Possible Primitive Earth Conditions." *Science* 117:528–529; Johnson, A. P., et al. 2008. "The Miller Volcanic Spark Discharge Experiment." *Science* 322:404.

AP® PRACTICE QUESTION

a. **Identify** the following:
   1. The question being investigated by the experimenters
   2. The hypothesis they tested
   3. The independent variable
   4. The dependent variable
   5. Future testable questions that arise in light of the results of this experiment

b. **Explain** the results of the experiment by **identifying** the researchers' claim and **justifying** the claim with the evidence and the reasoning researchers used (CER).

Recall that amino acids are the basic units from which proteins are made. Their synthesis in the laboratory under conditions that mimic the early Earth opens the possibility that amino acids could have been made on ancient Earth in a similar way. We now know that amino acids can be used by today's cells to make not only proteins but also the nucleotides required for DNA and RNA synthesis. While such a pathway was unlikely to have existed at the time, the synthesis of amino acids from basic inorganic molecules, as demonstrated by Miller and Urey's experiment, supports the concept that the basic units of organic molecules could have been formed on ancient Earth prior to the existence of living organisms.

Miller and others conducted many variations on his original experiment, all with similar results. Later experiments showed that other chemical reactions can generate nucleotides, simple sugars, and lipids. For example, in 2009, John Sutherland and his colleagues synthesized nucleotides from their sugar, base, and phosphate components.

Another striking observation comes from examining meteorites. Meteorites provide chemical samples of the early solar system. Analyses of certain meteorites have found a variety of different molecules, including amino acids. Finding these molecules on meteorites brings up the possibility that the building blocks of organic molecules were synthesized in space and brought to Earth by a meteorite.

Whether formed in space or on Earth, it's clear that the building blocks of organic molecules can be built by chemical reactions from even simpler chemicals such as water, H_2, CH_4, CO_2, N_2, and NH_3 in the absence of life. Furthermore, these observations provide evidence that the building blocks of organic molecules can be synthesized under conditions thought to resemble those at the time of the young Earth.

✓ Concept Check

1. **Describe** the atmosphere of the early Earth.
2. **Describe** the significance of the Miller–Urey experiment.

50.2 The formation of organic molecules set the stage for the first cell

We have discussed some of the evidence that life's building blocks can be generated under conditions likely to have been present on the early Earth. In this section, we will consider how these subunits can be linked to form larger molecules, such as proteins, nucleic acids, and carbohydrates. These molecules are called polymers, which are large molecules made up of smaller subunits repeated over and over.

Synthesis of Organic Molecules

Today we know that cells synthesize many polymers such as proteins using dehydration synthesis reactions, as we discussed in Module 4. Careful experiments have shown how polymers could have formed in the conditions of the early Earth. Researchers have demonstrated that dry mixtures of amino acids, when heated sufficiently, will join and form larger molecules, effectively demonstrating that proteins can be synthesized from their basic units.

Another important observation that helps us understand how subunits could have assembled into polymers on the young Earth comes from clay minerals, pictured in **FIGURE 50.2** on page 708. Clay minerals form from volcanic rocks and are able to bind nucleotides on their surfaces in such a way that nucleotides are placed near one another. As a result, the nucleotides are able to form bonds and form chains or single strands of nucleic acids like RNA.

Following up on the observation that a single strand of nucleic acid can be built on clay minerals, British biochemist Leslie Orgel wondered whether a single strand of nucleic acid, however it may have been formed, could serve as a template for synthesis of a complementary strand. In a classic experiment, he placed a short nucleic acid sequence into a tube and then added individual nucleotides that had been chemically modified to join, or polymerize, if brought close together. The added nucleotides did polymerize, and they

FIGURE 50.2 Surface for the spontaneous polymerization of nucleotides

The image shows stacked sheet like crystals of clay from Australia. Each layer is about 8 microns wide. Clays, such as the one shown here, provide a surface on which individual nucleotides can join, or polymerize, to form nucleic acids. Photo: Ray L. Frost, Professor of Chemistry, Queensland University of Technology, Australia

did so in a way to form a sequence complementary to the nucleic acid already present, as shown in **FIGURE 50.3**. In the first panel, we see the single-stranded nucleic acid and added nucleotides. In the second panel, we see the growing

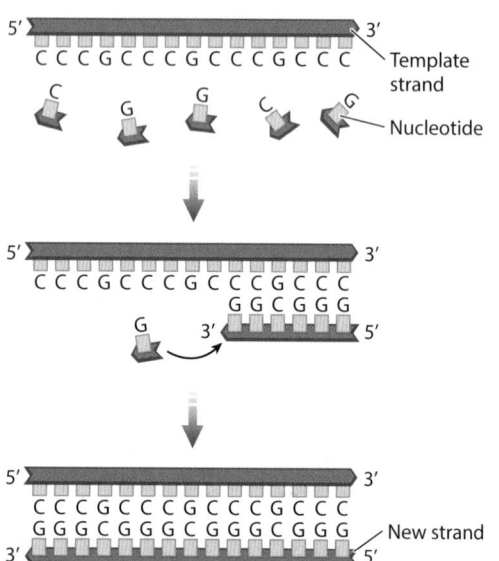

FIGURE 50.3 Spontaneous polymerization of nucleotides

Addition of a short RNA molecule (the template strand) to a flask containing nucleotides results in the formation of a complementary RNA strand.

strand that is complementary to the template strand. In the third panel, we see both the template strand and the newly synthesized daughter strand.

RNA World

For life to arise on Earth, a critical step was the synthesis of a molecule that could store information and pass it on through the process of replication. In this way, a cell would be able to reproduce and pass traits from parental to daughter generations. Of all the molecules present in cells today, only nucleic acids have the ability to be replicated and passed to daughter cells.

Today, DNA serves the function of storing and transmitting information present in cells, as we discuss in Module 32. RNA, like DNA, is made up of nucleotides, can store information in its sequence of nucleotides, and is able to make copies of itself by serving as a template for a daughter strand. In addition, RNA has one function that DNA does not have: some RNA molecules actually act as enzymes that facilitate chemical reactions. Such RNA molecules are called ribozymes and act as catalysts in biochemical reactions.

Ingenious experiments carried out by Jack W. Szostak and collaborators showed how RNA could have evolved the ability to catalyze a simple reaction. These researchers synthesized a strand of RNA in the laboratory and then replicated it many times to produce a large population of identical RNA molecules. Next, they exposed the RNA to a chemical that induced random changes in the identity of some of the nucleotides in these molecules. These random changes were mutations that created a population of diverse RNA molecules, much in the way that mutation builds genetic variation in cells.

Next, the researchers placed all these RNA molecules into a tube, and those RNA variants that successfully catalyzed a simple reaction—cleaving a strand of RNA, for example, or joining two strands together—were isolated, and the cycle was repeated. In each round of the experiment, the RNA molecules that functioned best were retained, replicated, subjected to treatments that induced additional mutations, and then tested for the ability to catalyze the same reaction. With each generation, the RNA catalyzed the reaction more efficiently, and after only a few dozen rounds of the procedure, very efficient RNA catalysts had evolved.

Experiments such as these suggest that RNA molecules can evolve over time and act as catalysts. This type of experimental evidence has led many scientists to conclude that RNA, with its dual functions of information storage and catalysis, was a key molecule in the very first forms of life.

Because RNA has properties of both DNA (information storage and transmission) and proteins (enzymes), many scientists think that RNA, not DNA, was the original information-storage molecule in the earliest forms of life on Earth. This idea is known as the **RNA world hypothesis.** The RNA world hypothesis is supported by other evidence as well. Notably, as we discussed in earlier modules, RNA is involved in key cellular processes, including DNA replication (Module 32), transcription (Module 33), and translation (Module 34). Many scientists think that this involvement is a vestige of a time when RNA played a more central role in life's fundamental processes.

If RNA played a key role on ancient Earth, why do cells now use DNA for information storage and proteins to carry out many other cellular processes? RNA is much less stable than DNA, and proteins are more versatile catalysts, so a reasonable hypothesis is that life evolved from an RNA-based world to one in which DNA, RNA, and proteins are specialized for different functions.

✓ Concept Check

3. **Describe** the importance of experiments that showed nucleic acids and proteins can be made by the polymerization of their respective building blocks.

4. **Describe** the significance of experiments that demonstrated certain RNAs have catalytic activities.

50.3 Life originated on Earth at least 3.5 billion years ago

Up to this point, we have discussed how organic molecules formed on the early Earth and we have paid special attention to RNA. As we saw, scientists have gathered evidence suggesting that RNA, rather than DNA, stored information in early cells. Very likely, RNA performed other functions in early cells as well, catalyzing chemical reactions much as proteins do today. However, all life as we know it is cellular. Biologists express this idea as the cell theory. The cell theory states that the cell is the basic unit of life and that all organisms are made up of cells. Today, all cells come from pre-existing cells, which brings up an important question: how and when did the first cell arise on ancient Earth?

This is one of the biggest questions in biology. Chemical evidence from 3.5-billion-year-old rocks in Australia indicates that biologically based carbon and sulfur cycles existed at the time those rocks formed. Because living organisms drive these cycles, life must have been established then, and therefore originated even earlier. Evidence from other continents appears to push the origin of life back to 3.8 billion years or even earlier.

All cells require an archive of information, a membrane to separate the inside of the cell from its surroundings, and the ability to harness energy from the environment. We have already discussed the earliest information-storing molecule. In this section, we will focus on how simple membranes and metabolisms might have originated.

Early Cell Membranes

All cells, whether found as single-celled bacteria or by the trillions in trees or humans, are individually encased in a cell membrane, or plasma membrane. Membranes are made up of lipids, specifically phospholipids, among other molecules. Research shows that phospholipids possess properties that may have led them to form spontaneously on the early Earth. As we discussed in Unit 2, phospholipids are able to form micelles, liposomes, and lipid bilayers when placed in water. If the conditions are appropriate and certain phospholipids are present, bilayer structures form spontaneously and without the action of an enzyme. For example, if phospholipids are added to a test tube of water at neutral pH, they spontaneously form spherical bilayer structures called liposomes that surround a central space. A liposome is shown in **FIGURE 50.4** on page 710. This observation is significant, as it suggests that cell-like structures might have formed spontaneously on the early Earth.

Such processes may have been at work in the early evolution of life on Earth. Experiments have shown that liposomes can form, break, and re-form in environments such as tidal flats that are repeatedly dried and flooded with water. Liposomes can even grow, incorporating more and more lipids from the environment, capturing nucleic acids and other molecules in their interiors. Note that a liposome is a sphere, with a fluid-filled space in its interior. Thus, when liposomes form, they may capture molecules present in solution. A critical step in the formation of a cell might have occurred when a liposome captured a self-replicating RNA.

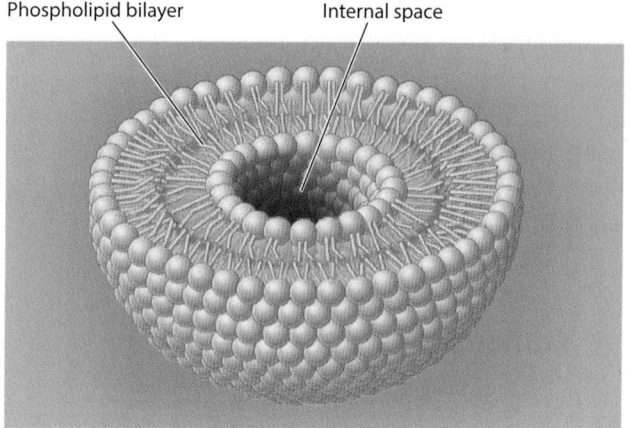

Phospholipid bilayer   Internal space

FIGURE 50.4 A liposome

Phospholipids can form liposomes when placed in water. Liposomes have a central fluid-filled space and so may have been the precursor of the first cell.

At first, the membranes were probably quite simple—straightforward but leaky barriers that kept the contents of early cells separated from the outside world. Over time, as chance variations arose, those membranes that provided a better barrier were favored by natural selection. Moreover, proteins became embedded in membranes, providing gates or channels that regulated the transport of ions and small molecules into and out of the cell. In this way, early organisms gained the capacity to regulate cell interiors—what biologists call homeostasis. Homeostasis, as we saw in Unit 2, is the active maintenance of a steady internal environment of cells and organisms, even when the external environment changes.

Given that Earth existed for approximately 800 hundred million years before geological evidence suggests that life arose, it is possible that self-replicating RNAs could have been encapsulated by liposomes numerous times and that a prototype "first cell" was made. Evidence thus suggests that membranes formed originally by straightforward physical processes and that their composition, complexity, and functions evolved over time to yield what exists today.

Early Cell Metabolisms

In addition to storing information and maintaining homeostasis, a third essential characteristic of living things is the ability to harness energy from the environment. Here, too, a series of natural chemical processes might have led to cells that could achieve this feat. Simple reactions that produced carbon-containing molecules as by-products would have enabled more complex reactions down the road. Ultimately,

that collection of reactions—combined with an archive of information and enclosed in some kind of primitive membrane—evolved into individual units that could grow, reproduce, and evolve.

Metabolism is the set of all the chemical reactions a cell requires and the capturing of the energy necessary to perform those reactions, as we discussed in Module 13. An ancient cell, even one with a primitive metabolism, would have a definite advantage compared to those cells that lacked a metabolism. While it is not known for sure, we can speculate on how a metabolism may have been gained by an ancient cell.

One of the simplest types of metabolism known today is fermentation. Recall from Unit 3 that fermentation is a process that extracts energy from organic molecules but does not require oxygen, so is anaerobic. There are many different fermentation reactions, all of which begin with glycolysis. Glycolysis breaks down a molecule of glucose into two molecules of pyruvate, generating ATP and NADH. Fermentation reactions then regenerate NAD^+ from NADH and allow the process to continue. Notably, these reactions do not require oxygen gas (O_2) and thus could have occurred on ancient Earth as a way for early cells to generate ATP.

A second type of metabolism is anaerobic respiration, which is respiration that occurs without oxygen. Respiration reactions involve an electron transport chain in which electrons are removed from one molecule and delivered to another molecule, an electron acceptor, that is not derived from the original molecule. In the process, energy is released and ATP is synthesized for use by the cell. Anaerobic respiration probably evolved later than the fermentation reactions, but the two metabolic processes were likely important for the function of early cells on the young Earth.

A third type of early metabolism is photosynthesis. In photosynthesis, light energy is used to excite electrons and move them from one compound to another, in the process releasing energy. In Unit 3, we discussed a form of photosynthesis in which oxygen is produced as a by-product, called oxygenic photosynthesis. Earlier forms of photosynthesis did not use water as a source of electrons and so did not produce oxygen gas. As with fermentation and anaerobic respiration, the energy released by photosynthesis can be captured, resulting in the production of ATP. Some bacteria today carry out photosynthesis that doesn't use water (H_2O) as an electron donor; instead, they use other molecules, such as hydrogen sulfide (H_2S). This form of photosynthesis, called anoxygenic photosynthesis, may also have evolved relatively early in the history of life on Earth.

Once ancient cells acquired a metabolism, the path was set for cells to become more independent and self-sufficient. Metabolism would allow cells to exercise greater control over their interactions with their environments and to gain complexity that further supported their survival and reproduction. While the earliest cells were very likely to be primitive compared to today's existing cells, they would have captured the essential qualities of having a self-directed, reproducing heredity material; the ability to separate themselves from their environment; and at least a rudimentary metabolism that provided a source of usable energy.

✓ Concept Check

5. **Identify** three essential functions of all cells.

6. **Describe** the significance of the observation that phospholipids can spontaneously assemble themselves into various structures, such as liposomes.

7. **Identify** one metabolic reaction that might have been present in the earliest cells.

50.4 The history of life was mostly unicellular, with familiar organisms appearing later

We have seen how organic molecules might have formed on the young Earth, and how they might have come together to form the first cell capable of storing information, maintaining homeostasis, and harnessing energy. It is difficult to say whether life originated once and then continued to evolve over time, or whether it emerged and died off multiple times before it took hold. Either way, multiple lines of evidence support the idea that all organisms alive today descended from a single common ancestor. Life was certainly established 3.5 billion years ago, likely earlier, as shown in the timeline of Earth and life in **FIGURE 50.5** on page 712.

What did this universal common ancestor look like? It was prokaryotic: a single-celled organism without a nucleus. Modern descendants of these early life forms include two of the three great domains of life—Bacteria and Archaea. Note that the bacteria and archaea that are alive today are not the same as the ones that were present on the young Earth. Like all organisms, they have continued to evolve since their origin.

The first 2 billion years of life on Earth—almost half of the time life has existed on the planet—were dominated by these prokaryotic organisms, as can be seen from the timeline in Figure 50.5. It was a time of great diversification. The bacteria and archaea remained unicellular, but they evolved many different ways to gather materials and harness energy from the environment. Today, these prokaryotes cycle elements in diverse ways, and life as we know it would therefore not be possible without them.

One group of bacteria—cyanobacteria—evolved the ability to carry out oxygenic photosynthesis. As we saw in

Unit 3, oxygenic photosynthesis is the process in which energy from sunlight is used to build sugars from carbon dioxide and water. This form of photosynthesis is similar to earlier forms of photosynthesis, but by using water as the source of electrons, it produces oxygen gas as a by-product. So, for the first time, oxygen gas started to accumulate in the atmosphere. This event is indicated on the timeline in Figure 50.5 as the "First evidence of an oxygen-rich environment." As we discussed earlier in this module, O_2 is a very reactive molecule. Therefore, although we take it for granted today and can't live without it, at the time, the accumulation of O_2 in the atmosphere represented a major crisis for life on Earth, earning the name oxygen catastrophe as many forms of life, adapted for an oxygen-free atmosphere, went extinct.

No other group evolved this ability—ever—in the history of life on Earth. It evolved just once in cyanobacteria. Organisms alive today, like plants and algae, that are also capable of oxygenic photosynthesis can only do so because they incorporated, in one way or another, these bacteria. For example, organelles called chloroplasts, where photosynthesis takes place in plants, were once free-living cyanobacteria, as we discussed in Module 12.

The next key event on the timeline of life in Figure 50.5 is the evolution of the third domain of life—the Eukaryotes—with cells that have a nucleus, like our own. This event occurred about 2 billion year ago, over half of the way along the timeline. These cells are the ancestors of our cells, but the organisms, like the bacteria and archaea that preceded them, were single-celled.

It took another billion years, about 1 billion years ago, before cells started getting together to form first simple and then more complex associations. Multicellularity didn't evolve once, but several times independently in different

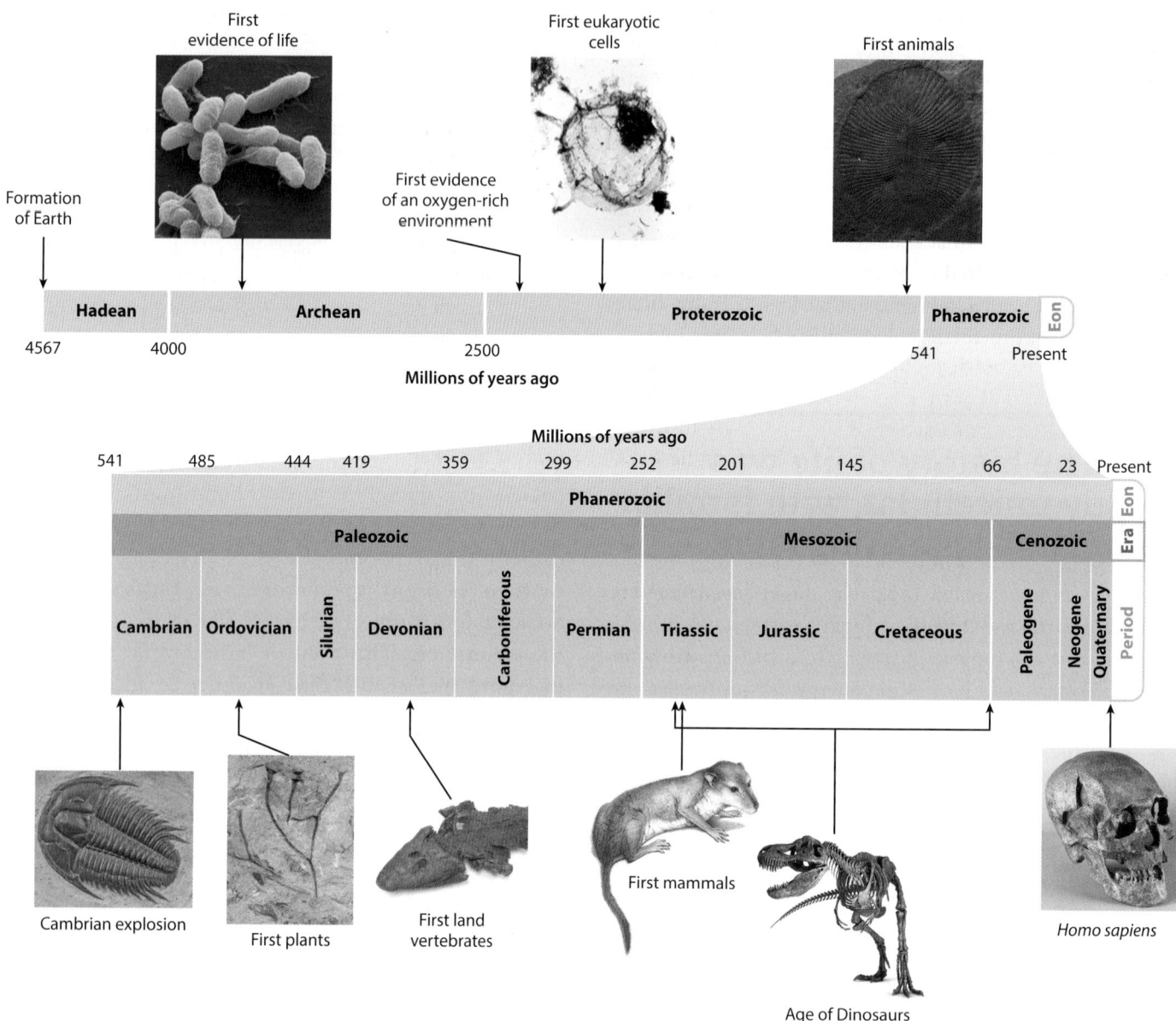

FIGURE 50.5 A timeline of the history of Earth and life

The Earth is 4.6 billion years old. Life originated at least 3.5 billion years ago. Other major events in the history of life are also shown on the timeline. Note that animals, plants, and other familiar organisms evolved relatively late, with modern humans evolving only very recently and occupying just a tiny slice of time. Photos: (top, left to right) Eye of Science/Science Source; Andrew Knoll, Harvard University; Andrew Knoll, Harvard University; (bottom, left to right) Scott Orr/iStockphoto.com; Hans Steur, The Netherlands; T. Daeschler/VIREO; Reconstruction illustration of Hadrocodium wui. Reconstruction artwork: Mark A. Klingler/Carnegie Museum of Natural History. From the cover of *Science* Vol. 292, no. 5521, 25 May 2001. Reprinted with permission from AAAS; dimair/Shutterstock; DEA/G. Cigolini/Getty Images.

groups of organisms. For example, it evolved one time in the line of organisms that led to modern-day plants, and it evolved separately in a different line of organisms that led to modern-day animals. Put another way, the common ancestor of plants and animals was unicellular, not multicellular.

About 541 million years ago, or roughly 90% of the way along Earth's timeline, there was a sudden appearance in the fossil record of many marine animals. This relatively rapid diversification of animals is called the Cambrian explosion, and it is marked on the timeline in Figure 50.5. We know about it from a remarkable set of fossils collected from what is now British Columbia in Canada and called the Burgess Shale, as we discussed in Module 44. It preserves a record of early animal marine life.

Up until this time, life existed entirely in the oceans. The movement onto land presented many challenges for organisms that were adapted for aquatic life. Fungi and plants likely moved onto land first, followed by animals. The invasion of land required adaptation to a dry environment. The first land vertebrates were amphibians; reptiles, birds, and mammals evolved later, well after life was well established on land. And what about us? Modern humans evolved just 200,000–300,000 years ago in Africa. In other words, all humans alive today can trace their ancestry back to Africa not that long ago. If the 4.6-billion-year history of the Earth played out over 60 seconds, we would only get a very tiny fraction of a second of time.

In short, the history of life on Earth is one with humble beginnings that eventually gave rise to the abundant diversity of life we see around us. This arc is described memorably by Charles Darwin in the final paragraph of *On the Origin of Species*:

It is interesting to contemplate an entangled bank, clothed with many plants of many kinds, with birds singing on the bushes, with various insects flitting about, and with worms crawling through the damp earth, and to reflect that these elaborately constructed forms, so different from each other, and dependent on each other in so complex a manner, have all been produced by laws acting around us . . . There is grandeur in this view of life, with its several powers, having been originally breathed into a few forms or into one; and that, whilst this planet has gone cycling on according to the fixed law of gravity, from so simple a beginning endless forms most beautiful and most wonderful have been, and are being, evolved.

✓ Concept Check

8. **Describe** the first cell.

9. **Describe** where oxygen in the atmosphere comes from and when in the history of life on Earth it first accumulated.

10. **Describe** the significance of the Cambrian explosion.

Module 50 Summary

PREP FOR THE AP® EXAM

REVISIT THE BIG IDEAS

SYSTEMS INTERACTIONS: Use one scientific theory from the module to **describe** how the chemical building blocks available on the early Earth might have been utilized to synthesize the components of the earliest cells.

LG 50.1 The building blocks of organic molecules can be synthesized from inorganic precursors.

- The age of the Earth is about 4.6 billion years. Page 704
- The early Earth had the inorganic molecules carbon dioxide, nitrogen gas, carbon monoxide, hydrogen gas, hydrogen sulfide, ammonia, and methane. Page 705
- Oxygen gas (O_2) was absent during the early periods of Earth's history. Page 705
- Experiments indicate that the building blocks of organic molecules can be generated from inorganic molecules under conditions that mimic those of ancient Earth. Page 707

- The building blocks of organic molecules are also found on meteorites. Page 707

LG 50.2 The formation of organic molecules set the stage for the first cell.

- Experiments suggest that polymerization of amino acids to form proteins and nucleotides to make nucleic acids could have occurred without cells being present. Page 707
- Some RNA molecules have catalytic activity and can therefore serve both as the template and the catalyst to make additional RNA molecules. Page 708
- The proposal that RNA was the original self-replicating, hereditary material is known as the RNA world hypothesis. Page 709

LG 50.3 Life originated on Earth at least 3.5 billion years ago.

- The earliest record of life is about 3.8 billion years ago, and it was certainly in place 3.5 billion years ago. Page 709
- Early cells required a way to store and transmit information, a membrane to separate inside from outside, and the ability to harness energy. Page 709
- The formation of liposomes from phospholipids could have been the first cell membrane. Page 709
- This liposome might have encapsulated a self-replicating RNA molecule. Page 709
- Metabolism is the full set of biochemical reactions that are required to sustain life. Page 710
- Fermentation, anaerobic respiration, and photosynthesis, all of which produce ATP without the involvement of O_2, may have been metabolic reactions present in early cells. Page 710

LG 50.4 The history of life was mostly unicellular, with familiar organisms appearing later.

- The first cells were prokaryotic. Page 711
- Cyanobacteria evolved the ability to carry out oxygenic photosynthesis about 2.5 billion years ago, and oxygen gas accumulated in the atmosphere for the first time. Page 711
- Eukaryotic cells, which have a nucleus and other internal membranes, evolved around 2 billion years ago. Page 711
- Multicellularity evolved several times independently about 1 billion years ago. Page 711
- The Cambrian explosion was a great diversification of marine animals that occurred 541 million years ago. Page 711
- Humans appeared very recently, approximately 200,000–300,000 years ago in Africa. Page 712

Key Term

RNA world hypothesis

Review Questions

1. Which chemical was not present in the ancient atmosphere of Earth?
   (A) Oxygen gas
   (B) Nitrogen gas
   (C) Ammonia
   (D) Carbon dioxide

2. Which statement does not support the RNA world hypothesis?
   (A) RNA can catalyze reactions.
   (B) RNA can make copies of itself.
   (C) RNA molecules can evolve to become more efficient.
   (D) RNA is unstable and degrades quickly.

3. Which is the fundamental unit of life?
   (A) A self-replicating molecule
   (B) A liposome
   (C) A cell
   (D) An organism

4. Based on the geological record, how long do scientists estimate it took for the first cells to appear on Earth?
   (A) 8 million years
   (B) 80 million years
   (C) 800 million years
   (D) 8000 million (8 billion) years

5. What was the significance of liposomes in the evolution of life on Earth?
   (A) Liposomes provided energy sources to the self-replicating RNA they contained.
   (B) Liposomes may have encapsulated a self-replicating RNA and provided protection from the environment.
   (C) Liposomes were the sites of protein synthesis in the first cells.
   (D) The phospholipids that make up liposomes were able to self-replicate, which led to the formation of cells.

6. Which best describes the role of oxygen in the making of early organic molecules?

   (A) The lack of oxygen on the early Earth was necessary as organic molecules do not contain oxygen atoms.

   (B) The lack of oxygen gas on the early Earth provided an environment in which the building blocks of organic molecules were stable.

   (C) The presence of oxygen on the early Earth was necessary to act as a catalyst in the formation of the building blocks of organic molecules.

   (D) The presence of oxygen on the early Earth prevented predators from consuming the organic molecules present.

7. Where did oxygen gas in the atmosphere first come from?

   (A) It was present in the atmosphere from the time of the formation of Earth.

   (B) It slowly seeped out of cracks in Earth and accumulated over time.

   (C) It was produced by lush rainforests.

   (D) It was produced by cyanobacteria.

2. Which of the following is most likely to have been a metabolic process found in early forms of life on Earth?

   (A) Aerobic respiration

   (B) Krebs cycle

   (C) Fermentation

   (D) Calvin cycle

3. Which of the following most accurately describes the possible role of spontaneous liposome formation in the early evolution of life on Earth?

   (A) Liposomes could have served as primitive membranes for the Krebs cycle.

   (B) Liposomes could have provided mechanical structures that would bring enzymes into contact with their substrates.

   (C) Liposomes could have protected catalytically active RNAs from the external environment.

   (D) Liposomes could have acted as essential intermediates in the synthesis of DNA.

Module 50
AP® Practice Questions

PREP FOR THE AP® EXAM

Section 1: Multiple-Choice Questions

Choose the best answer for questions 1–3.

1. Which of the following is an attribute of the early Earth atmosphere that might have contributed to spontaneous accumulation of organic molecules?

   (A) Nitrogen gas was absent from the early Earth atmosphere, which could have enabled spontaneous formation of amino acids.

   (B) Oxygen gas was absent from the early Earth atmosphere, which could have lowered the rate at which organic molecules were degraded.

   (C) Nitrogen gas was present in the early Earth atmosphere, which could have enabled spontaneous formation of DNA.

   (D) Oxygen gas was present in the early Earth atmosphere, which could have increased the rate at which organic molecules were synthesized.

Section 2: Free-Response Question

Write your answer to each part clearly. Support your answers with relevant information and examples. Where calculations are required, show your work.

Hydrogen cyanide and ammonia are chemicals that were abundant in the early Earth atmosphere. Scientists wanted to determine whether these chemicals could spontaneously react to form molecules essential for life under early Earth conditions. The scientists mixed a limiting amount of hydrogen cyanide with a large excess of ammonia in a reaction flask and allowed the chemicals to react at 70°C (158°F) for 8 days. The scientists measured the amounts of adenine and hydrogen cyanide in the flask on days 0, 1, 3, 5, and 8 and obtained the data shown in the figure.

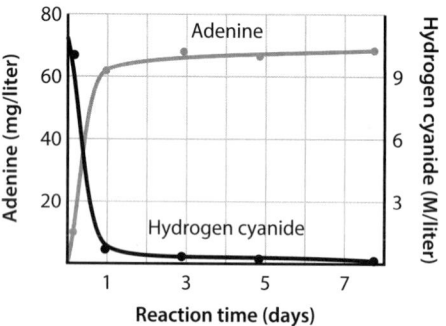

(a) **Identify** two biologically important molecules that incorporate adenine.

(b) **Predict** what would have happened to the level of adenine if additional hydrogen cyanide had been added to the flask at day 4. **Provide reasoning** to support your prediction.

(c) **Predict** what would have happened to the level of adenine if additional ammonia had been added to the flask at day 4. **Provide reasoning** to support your prediction.

(d) **Explain** how the data from this experiment could be used to support the hypothesis that life on Earth originated from nonliving chemicals.

Unit 7

AP® Practice Questions

Section 1: Multiple-Choice Questions

Choose the best answer for questions 1–15.

1. Which of the following provides the most complete definition of evolution?
   (A) Any genetic change in a population over time
   (B) An increase in the frequency of alleles that give adaptive benefits
   (C) Creation of new genetic variation through mutation
   (D) Phenotypic change in a species due to natural selection

2. Which of the following statements about fitness is most accurate?
   (A) The fitness of an individual is always determined by its genotype.
   (B) Fitness is irrelevant to evolution when there is competition among individuals.
   (C) Individuals with the same phenotype could have different fitnesses in different environments.
   (D) Individuals with high fitness are not affected by genetic drift.

3. Nematodes are microscopic, soft-bodied worms that do not fossilize. In the absence of fossils, which of the following is a method that can be used to determine how nematode species are related to each other?
   (A) The phylogenetic relationship of known predators of nematodes can be used to infer relatedness of the nematodes.
   (B) Fossilized burrows or other traces that ancient nematodes left in the environment can be used to reconstruct their relationships.
   (C) The DNA sequence of currently living nematodes can be used to infer common ancestry and relatedness.
   (D) Identification of extinct nematodes can illustrate the relatedness of living nematodes.

Use the following figure to answer question 4.

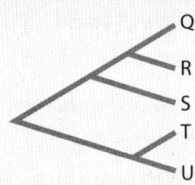

4. Which tree represents the same evolutionary relationships as the tree shown in the figure?

a.

b.
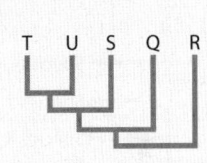

c.
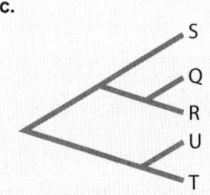

d.
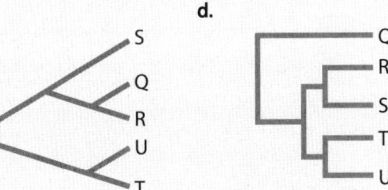

(A) a
(B) b
(C) c
(D) d

Use the following character matrix to answer question 5.

	Trait 1	Trait 2	Trait 3	Trait 4	Trait 5
Species J	+	+	–	–	+
Species K	–	–	+	+	–
Species L	+	+	–	–	+
Species M	–	+	–	–	–
Species N	–	–	+	–	–

5. Which tree in the figure correctly represents the phylogenetic relationship that would be inferred from the character matrix?

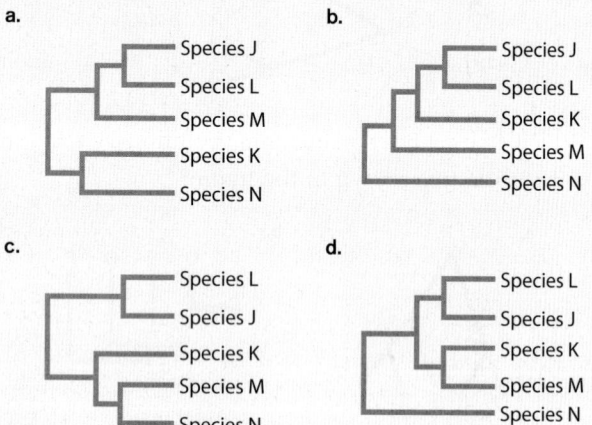

a.
- Species J
- Species L
- Species M
- Species K
- Species N

b.
- Species J
- Species L
- Species K
- Species M
- Species N

c.
- Species L
- Species J
- Species K
- Species M
- Species N

d.
- Species L
- Species J
- Species K
- Species M
- Species N

(A) a

(B) b

(C) c

(D) d

6. Which of the following modes of natural selection is most likely to increase variability within a population?

(A) Directional selection

(B) Disruptive selection

(C) Random mating selection

(D) Stabilizing (or purifying) selection

Use the following information to answer questions 7 and 8.

A student grew tomatoes in a backyard garden for 10 consecutive years. Different sections of the garden retain water differently in the soil and have different exposure to sunlight. In the first year, the student started the garden from a packet of genetically true-breeding seeds purchased from a local store. Every year, the student counted how many tomatoes were produced by each plant in the garden. The student collected seeds from the plant that produced the most fruit and used those seeds to plant the garden the following year. The student plotted the average number of tomatoes per plant each year and obtained the following graph. Error bars are calculated as ±1 standard deviation on either side of the mean.

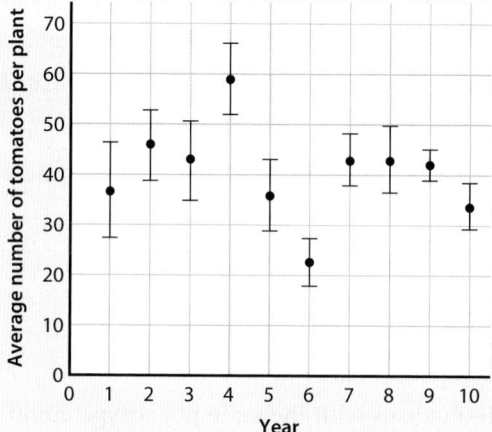

7. What is the most likely reason the average number of tomatoes is not greater in the tenth year than it is in the first?

(A) It is not possible to increase the number of tomatoes by artificial selection.

(B) The tomato plants became inbred and unhealthy during the 10-year period.

(C) There is no heritable variation for tomato number.

(D) New negative mutations arose and eliminated the effects of artificial selection.

8. What is the mostly likely explanation for the large number of tomatoes harvested in year 4?

(A) The environmental conditions were very good for tomato production in year 4.

(B) Artificial selection reached peak efficiency in year 4.

(C) A beneficial mutation went to fixation in year 4.

(D) Genetic variation was highest in year 4.

9. Why are molecular clocks valuable for interpreting phylogenetic trees?

(A) They can be used to demonstrate divergence from species that have not been sampled.

(B) They can be used to measure differences in protein or DNA sequences used to construct a tree.

(C) They can be used to estimate the age of the common ancestor of two species.

(D) They can be used to determine the age of fossils.

Use the following information to answer questions 10 and 11.

White campion (*Silene latifolia*) is a flowering plant whose species range is distributed throughout Europe, western Asia, and northern Africa. Wild populations of *S. latifolia* are sometimes infected by a fungal pathogen called anther smut (*Microbotryum lychnidid-dioiciae*). A team of researchers collected seeds from a population of white campion in Germany and another population in Hungary. The two populations are more than 1000 kilometers apart and the researchers demonstrated that they are genetically distinct. The researchers additionally collected spores of anther smut in Germany and Hungary and demonstrated that they also are genetically distinct.

In the laboratory, the researchers grew white campion from the seeds collected at each location. They then inoculated each plant with anther smut spores either collected from the same population or collected from the opposite population and measured the proportion of plants that became infected. The data that the researchers obtained are shown in the table.

Proportion of White Campion Plants That Became Infected by Anther Smut Fungus

		Origin of anther smut	
		Germany	Hungary
Origin of white campion	Germany	0.45	0.91
	Hungary	0.79	0.68

10. Which graph in the figure most accurately represents the data shown in the table?

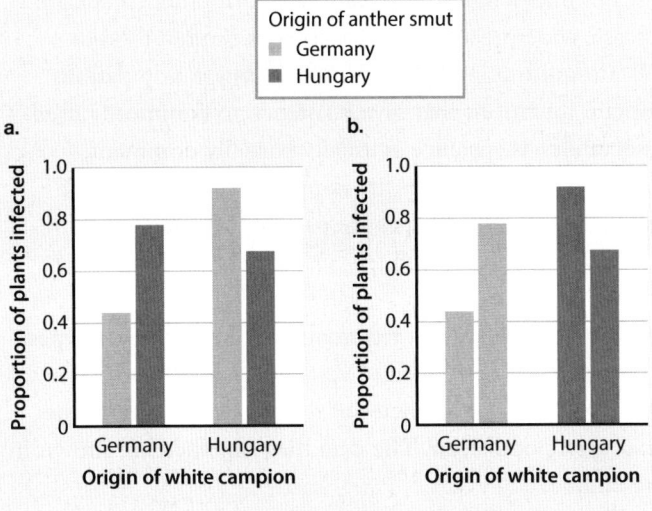

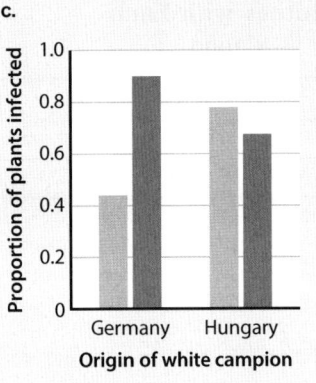

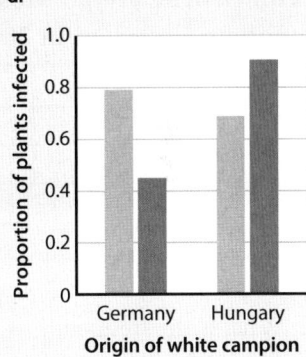

(A) a   (C) c

(B) b   (D) d

11. Which of the following statements provides the most likely explanation for the data shown in the table?

(A) The anther smut is highly adapted to the plant genotypes it encounters most.

(B) The white campion and the anther smut do not coevolve because the populations are geographically far apart.

(C) Genetic diversity is higher in anther smut than it is in white campion.

(D) The white campion evolves resistance to local anther smut genotypes.

Use the following information to answer questions 12–14.

Aedes aegypti mosquitoes transmit viruses that cause human diseases like dengue and yellow fever. In order to control disease, widespread use of insecticides is often used to kill mosquito populations. Some *A. aegypti* populations contain mutations that give resistance to commonly used insecticides. Resistance is usually partially dominant, such that *RR* genotypes are completely resistant to the insecticide, "±" genotypes are partially resistant, and *SS* genotypes are completely susceptible. A team of researchers collected *A. aegypti* from Veracruz, Mexico, in 2007 and measured the frequencies of resistant genotypes. The researchers returned to Veracruz in 2009 and again determined the frequencies of resistant genotypes in the *A. aegypti* population. The data they obtained are shown in the table.

Number of *A. aegypti* Mosquitoes with Each Genotype Collected in 2007 and 2009

	RR	RS	SS
2007	35	45	58
2009	66	51	21

Data from https://bioone.org/journals/Journal-of-the-American-Mosquito-Control-Association/volume-27/issue-4/11-6149.1/Update-On-the-Frequency-of-Ile1016-Mutation-In-Voltage-Gated/10.2987/11-6149.1.short

12. What is the frequency of the *R* allele in 2007?
    (A) 0.17
    (B) 0.25
    (C) 0.42
    (D) 0.5

13. What is the difference in the frequency of the *SS* genotype between 2007 and 2009?
    (A) 0.13
    (B) 0.22
    (C) 0.25
    (D) 0.27

14. What would be the expected number of *RS* heterozygotes if the population were in Hardy–Weinberg equilibrium in 2009?
    (A) 37
    (B) 49
    (C) 51
    (D) 62

15. If researchers used a statistical test to determine whether the *A. aegypti* population is in Hardy–Weinberg equilibrium in 2009, would they see any evidence that the assumptions of Hardy–Weinberg equilibrium were violated in 2009?
    (A) Yes, because the observed genotype frequencies do not match the expected genotype frequencies
    (B) No, because the departure from Hardy–Weinberg expectations is not statistically significant
    (C) Yes, because a chi-squared test gives a *P*-value smaller than 0.5
    (D) Yes, because natural selection changed the frequency of the *R* allele between 2007 and 2009

Section 2: Free-Response Questions

Write your answer to each part clearly. Support your answers with relevant information and examples. Where calculations are required, show your work.

1. Caribbean islands are home to diverse species of lizards in the genus *Anolis*. Distinct *Anolis* species occupy different habitats on each island. Three common habitats for *Anolis* on each island are treetops, thin tree branches (twigs), and the ground. *Anolis* species on every island show similar morphological adaptations to these environments, as shown in the table. Using DNA sequence to determine the phylogenetic relationships among *Anolis* species results in the tree shown in the figure. *Leiocephalus barahonensis* is a distant relative to *Anolis*.

Species name	Island	Habitat	Lizard color	Lizard size	Adaptation
A. valencienni	Jamaica	Twigs	Brown	Small	Short legs and long toes
A. grahami	Jamaica	Treetops	Green	Medium	Large toe pads, sticky feet
A. lineotopus	Jamaica	Ground	Brown	Large	Fast running
A. evermanni	Puerto Rico	Treetops	Green	Medium	Large toe pads, sticky feet
A. cristatellus	Puerto Rico	Ground	Brown	Large	Fast running
A. angusticeps	Cuba	Twigs	Brown	Small	Short legs and long toes
A. porcatus	Cuba	Treetops	Green	Medium	Large toe pads, sticky feet

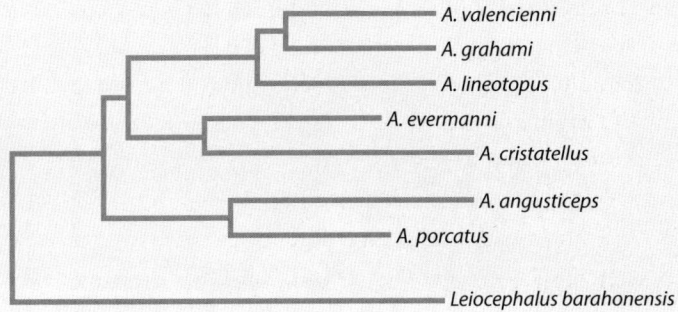

A. valencienni
A. grahami
A. lineotopus
A. evermanni
A. cristatellus
A. angusticeps
A. porcatus
Leiocephalus barahonensis

(a) **Identify** the role of *Leiocephalus barahonensis* in the figure.

(b) No competing species of lizard are believed to have lived on any of the islands before they were colonized by *Anolis*. The colonization of each island by *Anolis* has been called an adaptive radiation. **Define** adaptive radiation and **describe** how adaptive radiations result in phenotypic diversity.

(c) Is the adaptive radiation of *Anolis* lizards an example of evolutionary gradualism or punctuated equilibrium? **Provide reasoning** to explain your response.

(d) **Identify** a trait that you can claim evolved convergently in two species of *Anolis*. Provide evidence to **support** your claim.

2. The European chiffchaff (*Phylloscopus collybita abientinus*) and the Siberian chiffchaff (*Phylloscopus tristis*) are two species of songbird. The species formed during the Pleistocene ice age when the geographic range of their ancestral species was fragmented by glaciers. The currently known species diverged from each other during the ice age, but have recently reestablished contact with each other in a small geographic range (known as a hybrid zone) since the glaciers melted at the end of the ice age. European chiffchaff and Siberian chiffchaff have evolved differences in their mating songs, and each species prefers to mate with members of its own species. However, occasionally members of opposite species will mate with each other in the hybrid zone. The offspring of these matings have lower fitness than either pure species.

Scientists collected DNA samples from birds in the hybrid zone. They also collected DNA from birds in the native range of each pure species. The data from a representative genetic polymorphism are shown in the table.

Genotypes Obtained from Chiffchaff Birds in Three Geographic Regions

	European chiffchaff range	Hybrid zone	Siberian chiffchaff range
AA	40	10	0
Aa	0	4	0
aa	0	26	40

(a) **Identify** the speciation process that led to the separation of European chiffchaff and Siberian chiffchaff.

(b) Based on the information provided, is there prezygotic isolation between European chiffchaff and Siberian chiffchaff? Is there postzygotic isolation between European chiffchaff and Siberian chiffchaff? **Justify** your responses.

(c) Based on the data provided in the table, is the population in the hybrid zone in Hardy–Weinberg equilibrium? Use a statistical test to **evaluate** the null hypothesis.

(d) A scientist makes a claim that European chiffchaff females would have increased fitness from a mutation that made them more willing to mate with Siberian chiffchaff males because they would have more reproductive opportunity. **Provide reasoning** to support or refute the scientist's claim.

Unit 8
Ecology

Photo of a coral reef, Ari Atoll, Maldives by Franco Banfi/Getty Images

Module 51
Responses to the Environment

LEARNING GOALS
▶**LG 51.1** Ecological systems are organized in a hierarchy that starts with individuals.
▶**LG 51.2** Individuals adjust their physiology and behavior in response to environmental changes.
▶**LG 51.3** Individuals use communication to improve fitness.
▶**LG 51.4** Individual responses to the environment are vital to natural selection.

As we have proceeded through the units of this book, we have examined a wide range of topics including molecules, cells, cell cycles, genes, and gene expression. We have also examined the processes of heredity and evolution. With this understanding, we are now prepared to move to higher levels of complexity, ranging from individual organisms all way up to the entire biosphere of Earth. This is the realm of ecology, which examines how the adaptations of species help them survive and reproduce in the context of interactions with other species and the nonliving world. We will begin this journey by examining the different levels of ecology. We will then focus on the physiological and behavioral responses of individuals that help them improve their fitness. We will then discuss how individuals use communication to further improve their survival and reproduction, and how innate and learned behaviors represent critical adaptations that have been favored by natural selection.

> **PREP FOR THE AP® EXAM**
>
> **FOCUS ON THE BIG IDEAS**
>
> **EVOLUTION:** Look for relationships between the way an organism responds to its environment and its ability to survive and achieve reproductive success.

51.1 Ecological systems are organized in a hierarchy that starts with individuals

When we examine the ecology of the natural world, we find it helpful to categorize the world into a hierarchy of levels: individuals, populations, communities, ecosystems, and the biosphere. You can see these different levels in **FIGURE 51.1**. As we will see in this section, moving from individual organisms to the entire biosphere, each level is a subset of the next higher level, such that they exist in a hierarchy.

Individuals

The individual is the smallest and most fundamental unit in the ecological hierarchy. While there are lower biological levels including organs, cells, and molecules, the individual is the first level that can live on its own in the environment. At the individual level, we focus on examining the adaptations that allow organisms to optimize their survival and reproduction. For example, you might recall the story of the Canada lynx and snowshoe hare at the beginning of the book. At the individual level, biologists focus on the physiological, morphological, and behavioral adaptations that help

Population
Population dynamics—
the unit of evolution

Community
Interactions
among species

Ecosystem
Flow of energy
and matter

Individual
Survival and reproduction—
the unit of natural selection

Biosphere
Global processes

FIGURE 51.1 The hierarchy of ecological systems

At the lowest level, an individual, ecologists focus on the adaptations that improve survival and reproduction. The next level is a population, which is a group of individuals that experience changes in abundance and is the focus of evolutionary changes. The third level is a community, which is a collection of species in an area that interact with each other. The next higher level is an ecosystem, which combines the living and nonliving world with a focus on how energy and matter move throughout the system. Finally, we have the biosphere, which includes all ecosystems on Earth.

the lynx evolve to be a better predator and the hare evolve to avoid being killed by the lynx.

Populations

A population consists of all individuals belonging to the same species within a given area. The area may include every individual of the species, such as an entire continent, or it may be delineated to be much smaller, such as the number of squirrels of a given species within Pennsylvania or the number of catfish in a pond.

One challenge in defining a population is the ability to first define a species. As discussed in Module 47, biologists have historically defined a species as a group of individuals that naturally breed and produce fertile offspring. Over time, we have realized that some species do not fit this definition. For example, some species of salamanders consist of only female individuals that—without any breeding— produce only daughters. In addition, we now know that bacteria exchange genetic material with other very distantly

related bacteria. Thus, there is no single definition of species that works for all organisms. However, the historic definition continues to work well for many species of plants and animals.

> **PREP FOR THE AP® EXAM**
>
> **AP® EXAM TIP**
>
> Remember that evolution works at the level of populations, not individuals; populations evolve, and individuals within a population are selected by evolution.

Communities

Moving up the ecological hierarchy, a community is a collection of populations in a given area that interact with each other. These interactions include predators eating prey, pollinators fertilizing flowers, and plants competing with each other for nutrients. Similar to a population, the boundaries of a community can vary a great deal. For example, we may want to examine the community of species living in a city park or at the top of a mountain.

Ecosystems

Ecosystems are comprised of multiple communities as well as the nonliving physical and chemical environments, including sunlight, water, and temperature. For instance, we might examine the Great Lakes ecosystem that includes eight midwestern U.S. states and the province of Ontario, or the Greater Yellowstone ecosystem, which spans parts of Wyoming, Montana, and Idaho. At the ecosystem level, we are primarily interested in the movement of energy and elements, such as carbon, nitrogen, and phosphorus, which make up a large proportion of the elements needed for living organisms.

Biosphere

The highest ecological level is the biosphere, which includes all ecosystems on Earth. At this level, we focus on the movement of matter and energy by air currents, water currents, and the long-distance migration of organisms. Within the biosphere, nutrients cycle within and between the ecosystems, whereas energy is primarily provided by the sun and some fraction of the energy leaves Earth and is lost to space.

The Importance of Physical Principles

For each level of ecology, it is helpful to remember that the level is governed by physical principles. One important physical principle is the law of conservation of matter, discussed in Unit 3. As you will recall, the law of conservation of matter states that matter is not created or destroyed, but it can change form. We will see an example of this later in this unit when we examine how important elements such as carbon and nitrogen cycle through ecosystems in different forms. The second physical principle is the law of conservation of energy, which states that energy cannot be created or destroyed, but it can be converted into different forms. Taken together, these physical principles tell us that we should be able to track the movement of elements and energy at each level in the hierarchy. We will see many examples of that principle throughout this unit.

✓ Concept Check

1. **Describe** why populations are considered a subset of communities in the ecological hierarchy.
2. **Describe** the levels of communities and ecosystems.
3. **Describe** how physical principles help us track the movement of matter and energy in ecological systems.

51.2 Individuals adjust their physiology and behavior in response to environmental changes

At the individual level, ecologists focus on the adaptations that individuals possess as a result of natural selection. As we saw in the discussion of phenotypic plasticity in Module 29, a wide variety of organisms can detect changes in their environment and respond by altering the phenotypes, including changes in their physiology and behavior. In this section, we will examine some other fascinating examples of how organisms make these adjustments in response to changes in their environment. These changes include individuals orienting toward or away from environmental stimuli, individuals navigating over long distances, and individuals possessing a biological clock. In each case, we will see how possessing physiological and behavioral responses improves individual

performance and increases their fitness, suggesting that these responses are the product of natural selection.

Orientation Toward and Away From Environmental Stimuli

Organisms can respond to the environment by moving toward favorable stimuli and moving away from unfavorable stimuli. For example, bacteria and protozoa that experience water that is too warm or too salty increase their movement speed and start turning in random directions. This type of random movement is known as kinesis. Once they encounter more favorable conditions, they reduce their speed and reduce their turning rate. In other organisms, the movement is not random but in a specific direction, such as a gazelle walking toward a field of grass or running away from a predator. This movement in a specific direction is known as taxis (pronounced *tak-ses*). In all of these cases, moving toward favorable stimuli and away from unfavorable stimuli

results in the organism experiencing more favorable conditions, resulting in greater evolutionary fitness.

A common type of taxis is the movement of plants in response to the direction of sunlight. If you have ever grown houseplants near a window, you may have noticed that over a time span of several weeks the plant grows toward the window where the sunlight is the strongest. This process, known as phototropism, occurs because plants possess hormones in the tips of their stems, as shown in FIGURE 51.2. For phototropism, an important group of plant hormones is called auxins. If the stem tip receives sunlight from above, so that both sides of the stem tip are in the sun, the stem tip sends auxins down both sides of the stem and the plant grows straight up toward the light. However, if the stem tip receives sunlight on only one side, it sends auxins down the shady side of the stem. As auxins build up in concentration, they cause the cells of the stem to elongate on the shady side of the stem, which causes the stem to slowly bend toward the light. As a result, the plant can bend toward the light and increase the rate of photosynthesis to make more sugars that are essential for the plant's growth and reproduction.

Plants can also respond to day length in ways that help the plant properly time when to flower and make seeds to ensure optimal fitness. The number of sunlit hours of the day, known as the photoperiod, can have different effects on different species of plants. Some species of plants, such as potatoes and spinach, only flower in the summer when days are the longest, so we refer to them as long-day plants. Other plants, such as the poinsettia and Christmas cactus, only flower in the winter when days are the shortest, so we refer to them as short-day plants. Finally, some plants such as tomatoes and corn, are not affected by day length, so we refer to them as day-neutral plants. The plant can determine day length using photoreceptors in its leaves that cause an accumulation of a protein. When night arrives, the protein is degraded. In the case of long-day plants, as the days get longer from spring to summer, the amount of protein produced each day becomes greater until it reaches a threshold amount that triggers the plant buds to make flowers. These adaptations to day length help ensure that plant species reproduce at the time of year that will give them the best conditions for survival and reproduction, including opportunities for pollination, the appropriate temperature and amount of sunlight, and the necessary water availability to make their seeds.

Photoperiods are also very important to animals. For example, migrating birds in the Northern Hemisphere use photoperiods to know when to migrate north and start breeding, and when to fly south as winter approaches.

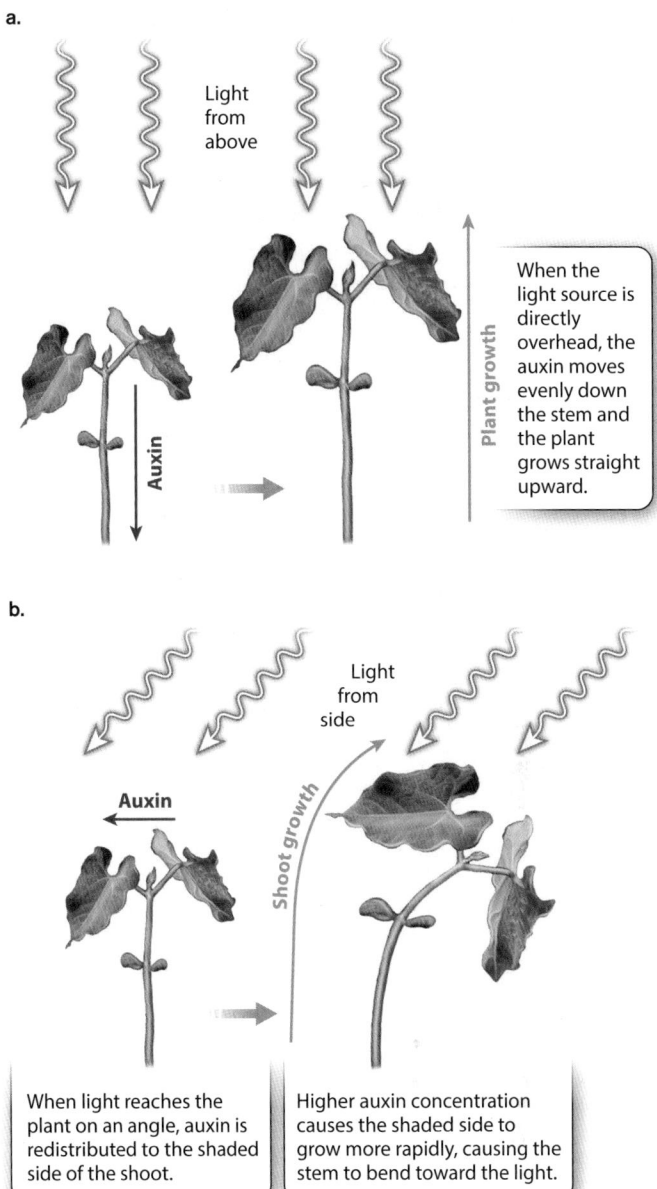

a.

Light from above

When the light source is directly overhead, the auxin moves evenly down the stem and the plant grows straight upward.

Auxin

Plant growth

b.

Light from side

Auxin

Shoot growth

When light reaches the plant on an angle, auxin is redistributed to the shaded side of the shoot.

Higher auxin concentration causes the shaded side to grow more rapidly, causing the stem to bend toward the light.

FIGURE 51.2 Phototropism in plants

Plants can respond to the direction of incoming sunlight by bending their stems toward the source of the light. (a) When the light is straight above the plant, the plant sends a hormone, known as auxin, down both sides of the stem and the stem does not bend. (b) When the light strikes the plant from one side, the plant sends auxin to the shady side of the stem. This causes cells on the shady side to elongate, which causes the stem to bend toward the source of the light.

Mammals also rely on photoperiods to stimulate the production of hormones that lead to breeding behavior. For example, white-tailed deer breed in the fall and then give birth in the late spring when there is abundant food available for their fawns. In all these cases, responding to the photoperiod helps plants and animals time their behavior and physiology to increase their fitness.

Navigation

Some species need to orient themselves for very long distances, and they have evolved methods of navigation. A classic example of navigation occurs in the homing pigeon, which can travel hundreds of kilometers away from home and then needs to find its way back. The bird has evolved to use a combination of cues, including being able to navigate based on star patterns at night and having the ability to detect magnetic north, much like a compass. In fact, when researchers attached a magnet to the heads of homing pigeons, the birds could no longer navigate home because it interfered with their internal compass. Such experiments have confirmed that the homing pigeon possesses an internal compass in its head that helps it to navigate.

Biological Clocks

Animals also respond to the presence of daylight and darkness. For instance, the retinas in the eyes of vertebrates sense daylight and darkness and they communicate this information to the pineal gland in the brain. Some species of fish and reptiles have specialized photosensitive cells on the skull between the eyes that detect a light versus dark environment, as you can see in **FIGURE 51.3**. In both situations, when the animal experiences darkness, the pineal gland releases a hormone known as melatonin. What is particularly fascinating is that nocturnal animals (those active at night) and diurnal animals (those active during the day) respond differently to increases in melatonin. In nocturnal animals, increased melatonin causes the animals to become active. In diurnal animals, however, increased melatonin causes the animals to sleep.

These natural cycles of daylight and darkness help set a circadian clock, which is an approximately 24-hour clock in organisms that regulates a wide range of activities including when to sleep, when to eat, and when to release specific hormones. Circadian clocks are common in all groups of organisms including animals, plants, fungi, and bacteria. The clocks work by having a group of "clock genes" that turn on and off in feedback loops on a 24-hour cycle. In fact, many

FIGURE 51.3 A third eye

Some species of lizards and fish, such as this tuatara (*Sphenodon punctatus*), have a region of photosensitive cells located between the two eyes that are used to detect light. Photo: Linda Kennedy/Alamy Stock Photo

species, including humans, continue to experience the same circadian clock of activities if they are put in an environment having 24 hours of darkness. By knowing what time of day it is, organisms can adjust their physiology and behavior to improve their performance and increase their fitness. For example, some species of algae can migrate up to shallow water at sunrise to obtain the more abundant sunlight and then migrate down to deeper water at sunset to obtain more abundant nutrients.

✓ Concept Check

4. **Describe** why organisms have evolved behavioral and physiological responses to changes in their environment.

5. **Describe** how migrating birds could use photoperiods to know when to fly south.

6. **Identify** two types of movement in response to an environmental stimulus.

51.3 Individuals use communication to improve fitness

When individuals determine the best behavioral response to their environment, they improve their performance, which improves their evolutionary fitness. Individuals can gain a substantial amount of information about the best response from others around them. This can include individuals of the same species or different species, such as prey responding to predators. In considering communication, we can think about a sender of information and a receiver of information. We can then define communication as a transfer of information between two individuals. In this section, we will explore the many different ways that animals communicate including visual, auditory, chemical, tactile, and electrical signals.

Visual Communication

Visual signals are signals that can be seen by another individual. For example, in the northern cardinal (*Cardinalis cardinalis*), a common bird in North America, the male cardinal has bright red feathers while the female has feathers that are quite drab, as shown in **FIGURE 51.4**. The male's bright red feathers are a visual signal to females that the bird is a male. The male's red color is also more intense when he consumes a high-quality diet. Thus, intensely red feathers can serve as a signal to females that the male is a high-quality mate. At the same time, the drab feathers of the female cardinal are a signal to the male that the bird is a potential mating partner. Other visual signals include antlers on male deer, brightly colored flowers to attract pollinating bees, and the elaborate courtship dances that some bird species perform to attract a mate.

Auditory Communication

Auditory signals can also communicate information. Some auditory signals communicate an impending danger, such as the rattling tail of a rattlesnake when it feels threatened. Auditory signals can also be used to attract mates, such as the songs of breeding birds and frogs. Animals also use calls

FIGURE 51.4 Visual signals

In the northern cardinal, the male on the left has bright red feathers that signal the cardinal is a male and that he has been eating a high-quality diet. The female on the right is much drabber, which helps to communicate that she is a female. Photo: Bonnie Taylor Barry/Shutterstock

to warn other members of their species when a predator is approaching. For example, the vervet monkey (*Chlorocebus pygerythrus*) gives unique warning calls depending on the situation. It uses one call to communicate to its relatives that a predator such as a hawk is approaching from the sky and a different call if a predator such as a leopard or snake is approaching from the ground. Such warning signals can induce a wide range of responses, such as hiding or the forming of aggregations, including herds of mammals, schools of fish, and flocks of birds.

Chemical Communication

Chemical signals are also important in communication. Some chemicals are used to attract mates. By releasing chemicals into the air, an animal can advertise over a large area that it is ready to mate. For example, female dogs and cats release chemicals that communicate that they are in heat and ready to mate. Chemicals can also be dropped along a food trail by insects such as ants to help communicate the path from the ant nest to a source of food. Chemicals can also be used to warn others about the presence of predators. For instance, many species of fish have specialized skin cells that release a chemical into the water whenever the fish is attacked by a predator. When other fish smell the chemical in the water, they may hide and become less active to reduce their risk of being detected by the predator. Finally, chemicals can be used to mark the boundaries of animal territories to warn other members of a species to stay away from another individual's territory or risk being attacked by the territory owner.

Many plants also use chemical communication. For example, consider the goldenrod plants that are found in abandoned farm fields throughout the eastern United States. When these plants are attacked by a beetle that chews on its leaves, the leaves release a chemical into the air that neighboring plants can detect. When they detect the chemical in the air, the neighboring plants begin producing anti-herbivore chemicals in their leaves to defend themselves against being eaten by the beetles if they arrive.

Other species of flowering plants emit odors that help attract pollinators in addition to any visual cues, such as flower color. In some cases, as illustrated by the "corpse flower" in **FIGURE 51.5** on page 728, the smell is not at all pleasant. This plant, officially known as the titan arum (*Amorphophallus titanum*), is pollinated by flies that feed on dead animals. To attract these flies to pollinate the plant, the flower releases a chemical that smells like a rotting corpse.

Tactile and Electrical Communication

Perhaps less obvious are tactile and electrical signals in communication. In tactile communication, organisms respond to a stimulus through the sense of touch. For example, the vibrations in a spider web, caused by a struggling prey, communicate to the spider that is standing elsewhere on the web that a meal has just arrived. In amphibians, species such as the white-lipped frog (*Leptodactylus albilabris*) make a breeding call while their body is pressed against the ground and this causes a vibration through the ground that attracts female frogs. Similarly, some insects that feed on plant leaves can send vibrations through the leaves to communicate with other insects of the same species. Electrical communication is less common, although there are species of fish, such as the brown ghost knifefish (*Apteronotus leptorhynchus*), that send weak electrical signals through the water to communicate with each other, including information on the sex of the fish that is sending out the signal.

FIGURE 51.5 Chemical signals in plants

The titan arum, also known as the corpse flower, releases chemicals that smell like a decomposing animal. This odor attracts flies that it needs for pollination. When not pollinating the flower, the flies feed on dead animals. Photo: Paul Marcus/Shutterstock

✓ Concept Check

7. **Describe** why animals responding to day length would not be an example of communication.

8. **Describe** how warning calls could improve the fitness of the individual that makes the call.

9. **Describe** how the predator-specific warning calls of the vervet monkey are more beneficial to other members of its species than a more generalized predator call.

51.4 Individual responses to the environment are vital to natural selection

We have already seen examples of physiological and behavioral responses that include avoiding predators and herbivores, finding food, locating a mate, and attracting pollinators. Given that these responses can have major impacts on an individual's survival and reproduction, you can appreciate that responses to environmental changes and responses to communication have important consequences for fitness. In this section, we continue to explore behavioral responses to the environment by examining how both innate and learned behaviors can be favored by natural selection and the importance of cooperative behaviors on improving the fitness of individuals.

Innate Behaviors

As we saw in Module 29, all phenotypes are the product of the genes that an individual carries and the environments that an individual experiences. When it comes to understanding any phenotype, such as a specific behavior, biologists have come to appreciate that there are **innate behaviors,** also known as instinctive behaviors, which means that the behavior is performed without any previous experience. Innate behavior is largely determined by genes, with a much smaller effect of the individual's environment. In contrast, there are **learned behaviors,** which means while the genes

FIGURE 51.6 Innate courtship behavior

In the Raggiana bird-of-paradise (*Paradisaea raggiana*), the male (top) exhibits courtship behaviors that are innate, including puffing up his tail feathers and outstretching his wings in an attempt to convince the female (bottom) to breed. Photo: BRUCE BEEHLER/AGE Fotostock

are necessary to perform the behavior, the individual's experience in its environment plays a large role in determining the behavior.

As you might expect, innate behaviors are very similar among individuals of the same species. In many bird species, for example, there are elaborate courtship displays that males perform to impress females in an effort to convince the females to mate, as shown in **FIGURE 51.6**. These displays commonly involve the male puffing up his feathers to appear large, flapping his wings in a particular way, and walking or flying around the female to get her attention. Courtship behaviors such as these are innate. In some birds, researchers have discovered that a male held in isolation his whole life is still capable of producing a perfect courtship display for females. Providing a proper courtship dance at the first mating opportunity would be of great value to the male bird since it would increase his probability of reproducing and increasing his fitness.

Learned Behaviors

Learned behaviors are also common in a wide variety of animals, where the learning leads to improved fitness. One form of learning is through the process of habituation, in

which animals learn over time to ignore a stimulus. For example, if you place a scarecrow in a garden, the crows are initially scared away by the human-like shape. After a few days, however, the crows learn that the scarecrow is not an actual human and poses no real risk, so they begin to fly back into the garden, as shown in **FIGURE 51.7**. This learning process allows the crows to capitalize on a food source.

Another form of learning is operant conditioning, in which an animal is either rewarded or punished after a behavior. Rats, for example, are unable to vomit, so they are very hesitant to try new foods. When they find a new food, they sample a very small bite and then wait to see if it makes them sick. If the new food does make them sick, then they have been careful to only have consumed a small amount. If it does not, they will eat more of the new food the next time they find it. By using this learning process, rats minimize harm from bad foods while being open to learning about new foods that could improve their growth, survival, and reproduction.

A third fascinating type of learning is imitation, which occurs when individuals learn a behavior from watching other individuals perform. For example, in years past, people in Britain had milk delivered to their doorsteps in the form of glass bottles with foil lids. At some point, a bird learned that it could peck at the lid, make an opening in the foil,

FIGURE 51.7 Habituation

Habituation is a learned behavior. Scarecrows are designed to scare away crows and other birds by appearing to be a human. Over time, however, the birds learn that the scarecrow is not a real human, so they return to the area to feed on the crops. Photo: Juniors Bildarchiv GmbH/ Alamy Stock Photo

and access the cream at the top of the milk jar, shown in **FIGURE 51.8**. By observing this single bird's behavior and imitating it, the behavior spread rapidly across several species of birds in Britain. The ability to learn by imitation provided an entirely new food source for the birds, demonstrating that imitation has fitness advantages.

Another important behavior is imprinting, which commonly occurs when animals are young, and they learn what their parents look like. In a series of classic experiments, researchers discovered that bird offspring that quickly leave their nest after hatching fix on the first animal they see as a parent. This makes sense from an evolutionary perspective because the vast majority of the time the first animal a

FIGURE 51.9 Learning by imprinting

In a classic experiment, researcher Konrad Lorenz made himself the first animal that would be seen by hatchling ducks. Throughout their life, the ducks would follow him as if he were their mother.

FIGURE 51.8 Learning by imitation

In Britain, milk used to be delivered with a foil cap. Researchers discovered that once one bird learned to pierce the foil to consume the cream at the top of the milk bottle, such as this English blue tit (*Cyanistes caeruleus*), other birds observed this behavior and learned to imitate it.

hatchling bird sees is indeed its parent, and knowing who to follow is critical for the hatchling to be fed and protected. However, the researchers discovered that if a human is the first animal seen by the hatchlings, the hatchlings follow the human everywhere as if it is the parent. **FIGURE 51.9** shows a particularly well-known example. In fact, the hatchlings continued to follow their human "parent" even when they were later presented with their real mother. Thus, there are many ways for animals to learn and it might be tempting to think that only larger animals, such as birds and mammals, are intelligent enough to learn. However, as you will see in Practicing Science 51.1, "To what extent are insects capable of learning?" even small insects are capable of learning, and doing so has tremendous fitness benefits.

Practicing Science 51.1

PREP FOR THE AP® EXAM

To what extent are insects capable of learning?

Background European digger wasps (*Philanthus triangulum*) live in burrows in the sand and they hunt honeybees to feed their developing young. After mating, each female wasp digs a long burrow with a few chambers at the end where she lays her eggs. She then forages for honeybees that she brings back to these chambers for her larvae to eat. The wasp faces a navigational challenge: having captured her prey, how can she find her way back to her nest? An early researcher of animal behavior, Niko Tinbergen, noticed that wasps lingered

briefly near their nest before heading off to hunt. He hypothesized that they were learning local landmarks associated with the nest.

Hypothesis Wasps learn visual cues around their nests to help locate the nest upon their return.

Method Tinbergen recognized that a good test of the learning abilities of an insect should take place in its natural environment. He devised an elegant demonstration of the way in which female wasps learn landmarks for navigation. He placed a ring of pine cones around the nest of a wasp, as shown in the first figure below. Once she had left to hunt, he shifted them to a new location away from the nest

▼

entrance, as shown in the second figure. For other nests, he placed a pine cone ring around the nest of a wasp and the ring was not shifted away after the wasp left to hunt. If visual cues are key to the wasp's ability to locate the nest, all of the wasps should return to the pine cone rings. In contrast, if cues are, for example, olfactory, the wasp should return directly to the nest rather than the pine cone rings.

Results Female digger wasps carried out a brief landmark-learning flight on departure from the nest. When the landmarks were moved, the females returned to the wrong location.

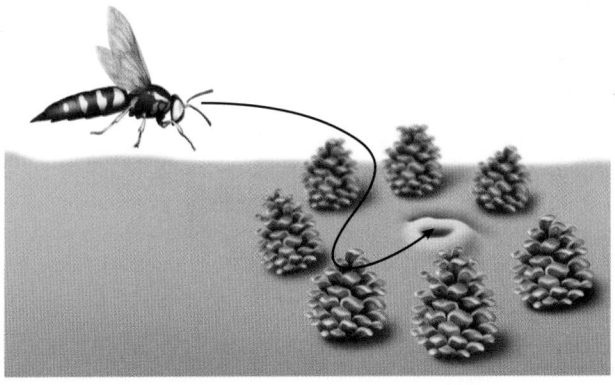

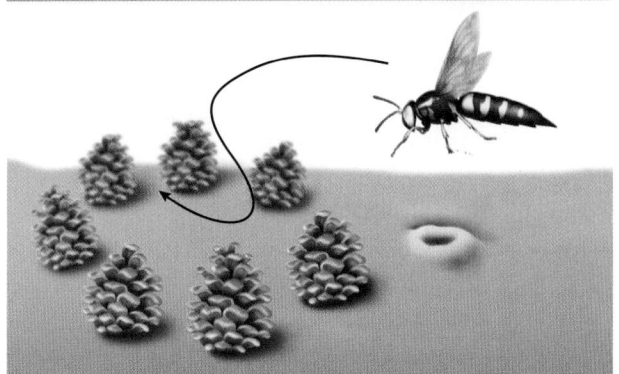

Conclusion Female digger wasps learn and use visual landmarks, such as the experimental ring of pine cones, as cues to the nest location.

SOURCE
Tinbergen, N. 1958. *Curious Naturalists.* New York: Basic Books.

AP® PRACTICE QUESTION

a. **Identify** the following:
   1. The question being investigated by the experimenters
   2. The hypothesis they tested
   3. The experimental group
   4. The control group
   5. The independent variable
   6. The dependent variable

b. **Explain** the results of the experiment by **identifying** the researchers' claim and **justifying** the claim with the evidence and the reasoning researchers used (CER).

Cooperative Behaviors

Some animal species are very social with one another and show high levels of **cooperative behavior,** in which members of the same species exhibit behaviors that improve another individual's fitness. Being social with other members of a species has a number of large fitness advantages. For example, spending time in groups results in more eyes searching for food and looking out for predators. It also allows prey to use group defense against predators, such as when muskox (*Ovibos moschatus*) form an outward facing circle of adults to protect the vulnerable calves, as shown in **FIGURE 51.10**. Among predators, such as wolves and lions, living in groups allows them to hunt down and subdue much larger prey than a single predator could do alone. Despite these advantages, living in groups also has some disadvantages, including having to share food with others, being more detectable as a group to predators, and being susceptible to the rapid spread of disease due to living close together. When the benefits of cooperative behavior outweigh the costs, a species should be favored by evolution to exhibit cooperative behavior.

Some species take cooperative behavior to the extreme by living in groups with very large numbers, with only one or a few individuals doing the breeding. For example, many species of bees, wasps, and ants live in groups ranging from hundreds to tens of thousands. In a hive of honeybees, there is typically one queen who lays all the eggs. She lays thousands of eggs that are diploid and develop into daughters, which are also known as the workers. These workers build the honeycomb, fly out to flowers to collect nectar and pollen, convert

FIGURE 51.10 Benefits of cooperative social behavior

When a muskox herd is threatened by predators, the adults form a circle with their horns facing outward. The vulnerable young muskox are placed in a protective position in the middle of the circle. Photo: All Canada Photos/Alamy Stock Photo

the nectar into honey back at the hive, and raise the next generation of offspring. Only occasionally does the queen bee lay haploid eggs, which develop into sons. The sons are not workers, but instead leave the hive to fertilize other queens, thereby increasing the fitness of the queen bee. Such highly social groups have evolved because of the large benefit of working together and the low probability of any single individual being able to survive and reproduce on its own.

✓ **Concept Check**

10. **Describe** how to test whether mice can habituate to a plastic model of a hawk.

11. **Describe** the difference between innate versus learned behaviors.

12. **Describe** two benefits of cooperative behavior.

Module 51 Summary

PREP FOR THE AP® EXAM

REVISIT THE BIG IDEAS

EVOLUTION: Using the content of this module, select one example of an organism and **describe** the relationship between a specific response to the environment and its fitness.

LG 51.1 Ecological systems are organized in a hierarchy that starts with individuals.

- The individual is the smallest and most fundamental unit in the ecological hierarchy. Page 722

- At the individual level, we focus on examining the adaptations that allow organisms to optimize their survival and reproduction. Page 722

- A population consists of all individuals belonging to the same species within a given area. Page 723

- A community is a collection of populations in a given area that interact with each other. Page 723

- Ecosystems are comprised of multiple communities as well as the nonliving physical and chemical environments, including sunlight, water, and temperature. Page 724

- The highest ecological level is the biosphere, which includes all ecosystems on Earth. Page 724

- At all levels of ecology, it is helpful to remember that each level is governed by physical principles, including the laws of conservation of matter and conservation of energy. Page 724

LG 51.2 Individuals adjust their physiology and behavior in response to environmental changes.

- Individuals can orient toward and away from environmental stimuli. Page 724

- Random movement toward or away from environmental stimuli is known as kinesis while directional movement is known as taxis. Page 724

- Phototropism causes plants to grow toward sources of sunlight. Page 725

- Plants can respond to day length, also known as the photoperiod, in ways that helps the plant properly time when to flower and make seeds to ensure optimal fitness. Page 725

- Animals can respond to photoperiods to know when to migrate and when to breed. Page 725

- Some species that need to orient for very long distances have evolved methods of navigation. Page 726

- Natural cycles of daylight and darkness help set a circadian clock, which is an approximately 24-hour clock in organisms that regulates a wide range of activities including when to sleep, when to eat, and when to release specific hormones. Page 726

LG 51.3 Individuals use communication to improve fitness.

- Communication is a transfer of information between two individuals in which the sender attempts to manipulate the behavior of the receiver. Page 726

- Species can communicate using visual, auditory, chemical, tactile, and electrical signals. Page 727

LG 51.4 Individual responses to the environment are vital to natural selection.

- Responses include avoiding predators and herbivores, finding food, locating a mate, and attracting pollinators. These responses can have major impacts on an individual's survival and reproduction. Page 728

- Some behaviors are innate, also known as instinctive, which means that the behavior is performed without any previous experience. Page 728

- Other behaviors are learned, which means that while the genes are necessary to perform the behavior, the individual's experience in its environment plays a large role in determining the behavior. Page 728

- Examples of learned behaviors include habituation, operant conditioning, imitation, and imprinting. Page 729

- Being social with other members of the same species has a number of large fitness advantages, including finding food, identifying predators, group defense, and prey capture. Page 731

- Some species take social behavior to the extreme by living in groups with very large numbers, with only one or a few individuals doing the breeding. Page 731

Key Terms

Innate behaviors   Learned behaviors   Cooperative behaviors

Review Questions

1. Which is the correct order of hierarchy for the level of ecology, from the simplest to the most complex?
   - (A) Individual, population, community, ecosystem, biosphere
   - (B) Individual, community, population, ecosystem, biosphere
   - (C) Population, individual, community, ecosystem, biosphere
   - (D) Population, individual, ecosystem, community, biosphere

2. Which statement about responses to environmental stimuli is true?
   - (A) Short-day plants flower when there is a long photoperiod.
   - (B) Plants experiencing phototropism move toward sunlight.
   - (C) Navigating birds lack an internal compass, so they rely on stars.
   - (D) A circadian clock operates during daylight hours.

3. Which statement about communication is false?

   (A) Brightly colored flowers communicate a source of nectar to pollinators.

   (B) Rattlesnakes use auditory communication when threatened.

   (C) Animals mark their territories using chemical communication.

   (D) Plants are unable to use chemical communication.

4. Which statement is true about innate or learned behaviors?

   (A) Innate behaviors are performed better over time.

   (B) Imprinting on a parent is an example of an innate behavior.

   (C) Imitation is an example of a learned behavior.

   (D) Courtship displays in birds are commonly learned behaviors.

5. Which statement about social behavior is false?

   (A) The fitness costs of social living outweigh the fitness benefits.

   (B) Benefits include the ability of predators to capture and subdue large prey.

   (C) Costs include having to share food with other members of the same species.

   (D) Some species are highly social, living in groups that number in the thousands.

Module 51
AP® Practice Questions

PREP FOR THE AP® EXAM

Section 1: Multiple-Choice Questions

Choose the best answer for questions 1–4.

1. Scientists observed that a population of bacteria growing in a flask of liquid medium do not distribute evenly throughout the medium if there is a magnet on one side of the flask. Instead, the bacteria tend to cluster close to the magnet. Which best describes the behavior exhibited by the bacteria?

   (A) The bacteria are responding to their biological clocks.

   (B) The bacteria are communicating to coordinate their movement.

   (C) The bacteria are exhibiting taxis.

   (D) The bacteria are exhibiting kinesis.

Questions 2 and 3 refer to the following information.

Students conduct an experiment with a population of bacteria named *Alcaligenes faecalis*. These bacteria move by using a tail-like structure called a flagellum. The students place two groups of bacteria into two beakers containing the same medium. The students place a capillary tube containing water into the first beaker, and they place a capillary tube filled with glucose dissolved in water into the second tube. Both the water and the glucose can diffuse out of the capillary tubes and into the medium in the beakers. The students observe the movements of the bacteria in both beakers. After a period of time, the students record the location of the bacteria in the beakers, which are represented as the white dots in the figure.

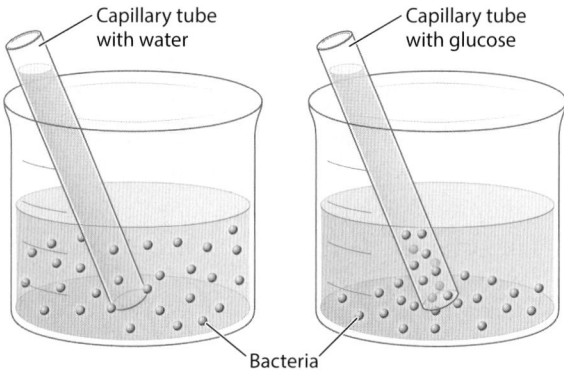

2. Identify the independent variable in the experiment.

   (A) The presence of glucose

   (B) The presence of *Alcaligenes faecalis*

   (C) The presence of the water-containing capillary tube

   (D) The presence of the medium into which *Alcaligenes faecalis* are placed

3. Which best describes the movement of the bacteria in the beaker with glucose?

   (A) Taxis

   (B) Kinesis

   (C) Navigation

   (D) Directed movement

4. Rattlesnakes rattle their tails to ward off potential predators. The rattling sound communicates to other animals that the rattlesnake is poisonous and will retaliate if attacked. In recent decades, ecologists in the western United States have observed an increase in the percentage of the rattlesnake population that is unable to make any rattling sound. Researchers also observed an increase in humans finding and killing rattlesnakes in the same region during the same time. Which best explains why the rattlesnake population is increasingly unable to make a rattling sound?

(A) Rattlesnakes that rattle their tails have an increased fitness because they are less likely to be found by animals that are hunting them.

(B) Rattlesnakes that do not rattle their tails have a decreased fitness because they cannot scare off humans that are hunting them.

(C) Rattlesnakes that do not rattle their tails have an increased fitness because they are less likely to be heard by humans hunting them.

(D) Rattlesnakes that rattle their tails have an increased fitness because they scare off humans that are hunting them.

Section 2: Free-Response Question

Write your answer to each part clearly. Support your answers with relevant information and examples. Where calculations are required, show your work.

A group of scientists conducted an experiment to study how a type of fish called the stickleback fish behave in response to seeing brine shrimp, which are a food source for the fish. The scientists placed 35 shrimp in a clear glass tube and then put the tube into the middle of an aquarium containing one stickleback fish. The glass tube prevented the shrimp from being eaten by the fish. The data presented in the figure were collected from observing how many times, over a period of 10 minutes, the fish bit at the glass tube in an attempt to eat the shrimp. The number of attempts each fish made were averaged and plotted on the graph. Error bars are calculated as ±1 standard error on either side of the mean.

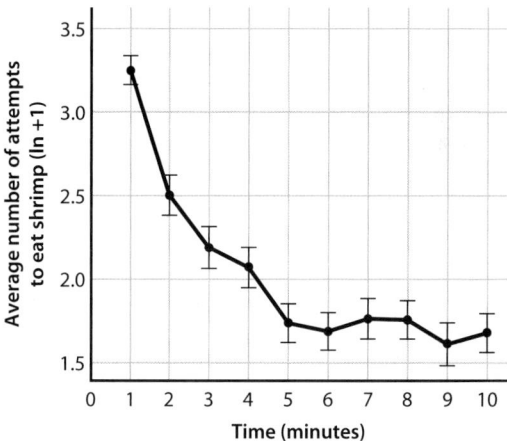

Data from Bell, A. M., and H. V. S. Peeke. 2012. Individual variation in habituation: behavior over time toward different stimuli in threespine sticklebacks (*Gasterosteus aculeatus*). *Behaviour* 149:1339–1365.

(a) **Identify** the type of behavior, including whether the behavior is innate or learned, that the fish displayed over the 10 minutes of the experiment. **Justify** your answer.

(b) **Describe** how this kind of behavior benefits the stickleback fish.

(c) In a separate fish tank that is used to maintain the stickleback colony and not subjected to the experimental conditions, the researcher feeds the fish every day by putting food into the water at the rear of the tank. The researcher notices that as soon as the lid of the tank is lifted, the fish move to the rear of the tank. One day, the caretaker feeds the fish by lifting the lid and putting the food into the front of the tank. **Predict** how the fish will respond. **Provide reasoning** for your prediction.

Module 52

Energy Flow through Ecosystems

LEARNING GOALS ▶**LG 52.1** There are different strategies for acquiring and using energy.
▶**LG 52.2** Energy is captured and moved through ecosystems.
▶**LG 52.3** Matter cycles in ecosystems.

In the previous module, we introduced the field of ecology, which studies a wide range of spheres from individuals, to populations, to communities, and the biosphere. We then focused on the physiological and behavioral responses that allow species to live in specific environmental conditions. We also discussed the different forms of communication individuals use and how natural selection has favored both innate and learned behaviors. With that background, we are now ready to focus on how energy flows through individuals, communities, and ecosystems.

In this module, we will explore how species have evolved to acquire and use energy to fuel their survival, growth, and reproduction. We will examine the ways in which species use energy to affect their body temperature and how energy needs change with body size. We will also look at how energy availability changes through the seasons,

which affects the reproductive strategies of species. We will then consider how energy moves from those species that capture energy to species that obtain their energy by consuming other species. We will also consider the efficiency of moving energy among different groups of organisms to understand how the need for energy limits the number of individuals of different species that can be supported in an ecosystem. Finally, we will examine how elements cycle in ecosystems, using the cycling of carbon as a key example.

> PREP FOR THE AP® EXAM
>
> ## FOCUS ON THE BIG IDEAS
>
> **ENERGETICS:** Think about the strategies that organisms employ to capture energy from their environment, and look for examples of strategies that are conserved among related organisms.

52.1 There are different strategies for acquiring and using energy

As we saw in the previous module, individuals need energy for growth, reproduction, and survival. In this section, we will examine the two strategies that have evolved to regulate energy acquisition and metabolic rate: having a constant body temperature that is internally regulated or having a body temperature that is more strongly affected by the external environment. We will then examine how energy availability and metabolic rate affect growth, survival, and reproductive strategies.

Using Energy to Regulate Body Temperature and Metabolism

As you may recall from Unit 2, cells maintain stable conditions in a process we call homeostasis. Homeostasis is typically achieved through negative feedback loops using

cell signaling molecules that bind to receptors on target cells. Once the cell achieves its set point, the signal is turned off to avoid overcorrecting, which helps the cell maintain steady conditions. Homeostasis happens not only at the level of the cell but also at the level of the entire organism. Maintaining homeostasis allows organisms to grow and reproduce, which both require a particular range of steady internal conditions. Here we will examine how homeostasis operates at the level of the whole organism, with a focus on how organisms control their body temperature.

The process of an organism controlling its body temperature is known as thermoregulation. In thinking about thermoregulation, we can make a distinction between organisms with two different mechanisms, *endotherms* and *ectotherms*. The body temperature of organisms known as **endotherms** is regulated internally. Well-known endotherms include mammals and birds, which use negative feedback loops to maintain body temperatures that are typically warmer than their external environment. As described in Unit 4, a negative feedback loop works like a thermostat in a house. When the thermostat detects that the house is colder than desired, it sends a signal to the home's heater and the heater makes more heat. Once the house warms up to the desired temperature, the thermostat stops sending a signal to the home's heater, so the heater stops making heat. In a similar way, when the body of a bird or mammal becomes colder than some internal set point, an organ in the brain known as the hypothalamus sends a signal to make muscles begin shivering to generate heat. When the body becomes warm enough, the hypothalamus stops sending the signal. (See Figures 20.10 on page 284 and 20.11 on page 285.)

Ectotherms are organisms that typically generate relatively small amounts of heat, so their body temperatures are strongly impacted by external conditions. Most plants are ectotherms, although some have fascinating physiological adaptations to internally generate heat. For example, a plant known as skunk cabbage (*Symplocarpus foetidus*) is one of the first plants to emerge in early spring in eastern North America, as illustrated in **FIGURE 52.1**. Even when there is some snow on the ground, the skunk cabbage uses its mitochondria to generate large amounts of metabolic heat to increase its internal temperature, up to 10°C higher than the air temperature. While generating this heat requires a large amount of energy, the plant benefits by emerging early. It experiences more rapid growth of its flowers, attracts the flies that pollinate it, and protects itself from freezing air temperatures that would kill most plants. Thus, the skunk cabbage's heat generation provides major fitness benefits by allowing it to emerge and make flowers before any competing species of plants can emerge and attract pollinators.

For animals that are ectotherms, it is tempting to think that their internal temperature simply reflects the temperature of their external environment. However, the reality is far more fascinating because many species of ectothermic animals can adjust the temperature of their bodies by altering their behavior. For example, scorpions living in a hot desert can avoid lethal overheating by seeking shade. Alternatively, when the air is cool and ectothermic animals want to increase their body temperature, they can bask in the sun to make their body temperature increase. This basking behavior improves an animal's ability to move, capture food, and digest its food. As a result, this improved performance increases its evolutionary fitness. Such behavioral thermoregulation is very common in species of ectotherms including reptiles, amphibians, and insects.

"Visual Synthesis 8.1: Homeostasis and Thermoregulation" on page 748 provides an overview

FIGURE 52.1 Skunk cabbage

Skunk cabbage species are an unusual group of plants which can internally generate heat. Using their mitochondria to generate metabolic heat, the plants can elevate their temperature to be 10°C warmer than the surrounding air temperature and melt their way up through the snow. Photo: Ray Coleman/Science Source

of homeostasis and thermoregulation. The rock squirrel (*Otospermophilus variegatus*), an endotherm, can curl up in a burrow during cold nights to reduce heat loss, and contract its muscles rapidly (shiver) to generate heat. It also uses a mechanism known as vasoconstriction, which narrows arteries in its arms and legs. This constriction reduces the amount of blood in the limbs to reduce heat loss. As noted earlier, the hypothalamus acts as the body's thermostat and controls the animal's physiological responses, such as shivering and vasoconstriction.

The motor cortex of the brain controls behavior, such as burrowing behavior and activity level. When the rock squirrel experiences cold days, it can increase its activity and generate more heat from its muscles. Should it get too hot in the middle of the day, the squirrel alters its physiology by panting, moving very little, and using vasodilation, which increases the flow of blood in the veins and arteries in its limbs to get rid of excess heat. The squirrel also changes its behavior by seeking shade and licking its forelimbs so that the evaporating saliva removes heat from its body.

In contrast, ectotherms such as the zebra-tailed lizard (*Callisaurus draconoides*) rely a great deal on behavioral responses to achieve a desired body temperature. During cold nights, the lizard's hypothalamus triggers the motor cortex to induce burrowing behavior. During cool mornings, the lizard comes out of the burrow, uses vasodilation to move more blood out to its limbs to gain heat from warm rocks, increases its activity to generate muscle heat, and basks in the sun to gain heat. Should the daytime temperature get too hot, the lizard can adjust by decreasing its activity, seeking shade, and lying on cool rocks and dirt to lower its body temperature. As you may have noticed, these responses to temperature changes are all examples of stimulus and response, which we examined in our discussion of cell communication in Unit 4. Through these activities, the lizard can have a warmer body temperature, which improves its ability to move, catch prey, and digest its prey—all with the effect of improving its ability to grow and reproduce.

Ecological Strategies in Response to Energy Availability and Energy Needs

When it comes to the evolution of different ecological strategies for growth and reproduction, the issue of energy availability becomes an important factor. This is because a net gain in energy allows an individual to experience growth and reproduction, which can lead to an increase in the size of a population. In contrast, a net loss of energy results in lost growth and reproduction. If a net loss of energy occurs over extended periods, it can also lead to the death of an individual and a decline in the population.

The energy needs of individual organisms differ depending on whether they are endotherms or ectotherms. We can compare the energy needs of different organisms by measuring their **metabolic rate,** which is the number of calories that an organism burns over time while at rest. Given that endotherms maintain a constant body temperature that is typically higher than their outside environment, endotherms have a higher metabolic rate than ectotherms. Thus, endotherms have higher energy requirements to feed their higher metabolic rate. This has implications in terms of the ecology and evolution of these species. Given that endotherms maintain a constant body temperature, they can remain active over a wide range of different environments,

from the Arctic to hot deserts. However, they need to eat frequently. In contrast, ectotherms are more limited in the habitats in which they can live, but they have much lower energy requirements. As a result, ectotherms can often go for long periods without eating. Some snakes, for example, only eat once a month.

Energy requirements also vary with body size. When researchers have measured the metabolic rate of different species, they discovered that larger species have a higher metabolic rate than smaller species, as you can see in **FIGURE 52.2a**. This makes sense in that an elephant requires more energy each day than a mouse, so the elephant needs

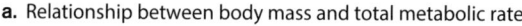

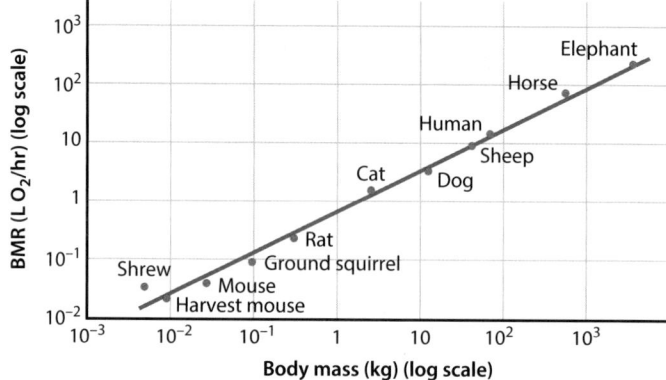

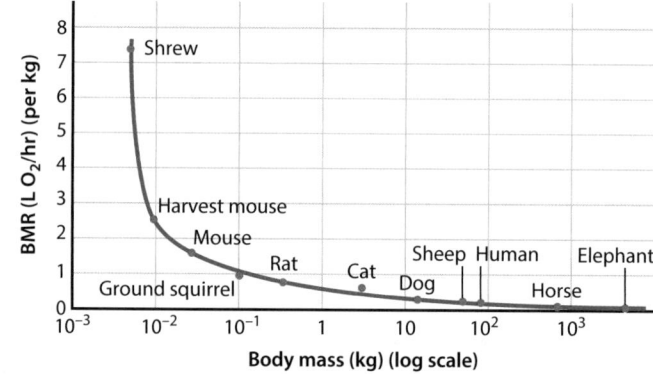

FIGURE 52.2 Metabolic rates of mammals and body size

We can measure metabolic rate by measuring the amount of oxygen gas consumed by different animals when at rest. (a) Larger mammals have a higher metabolic rate than smaller mammals. (b) When we examine metabolic rate on a per-kilogram basis, we see that larger mammals have a lower metabolic rate per kilogram of body mass. Larger mammals have a small surface area relative to their volume, so they experience less heat loss to the environment. As a result, large mammals need to generate less heat to maintain their body temperature. Data from Brody, S. 1945. *Bioenergetics and Growth.* New York: Reinhold Publishing Corporation.

to eat a lot more food every day to provide this energy. This also means that a given area of land can support a much larger population of mice than elephants. Finally, this means if more food becomes available in the environment, the populations of elephants and mice can both increase.

Another way of looking at metabolic rates among mammals of different sizes is to examine the relationship on a per kilogram basis, as shown in Figure 52.2b. In this case, we see that the largest mammals have the lowest metabolic rate on a per-kilogram basis. As you will recall from Unit 2, as an organism increases in size, its volume increases rapidly while its surface area increases slowly. As a result, larger bodies have a smaller surface area relative to their volume than smaller bodies, so they lose heat to the environment more slowly and therefore they need less energy on a per kilogram basis. Thus, while larger animals require more total food, they require less food per kilogram of mass.

Energy Availability and Reproductive Strategies

Energy availability also determines the reproductive strategies of organisms. We can see examples of this by looking at the seasonal nature of reproduction. For example, in North America, many species of plants flower in the spring when rising temperatures warm the air and soil, and more hours of sunlight are available so that the plants can generate sugars through photosynthesis. Other species of plants have evolved to initially put all their energy into their growth throughout the spring and summer and then flower in the fall, once they have accumulated sufficient energy reserves for making flowers and fruits.

Some plants take even longer to reproduce. For example, the common mullein (*Verbascum thapsus*) lives in open fields and requires 2 years to gain enough energy to flower. In the first year, it only grows into a short plant with many large leaves, which is illustrated in **FIGURE 52.3a**. As the plant photosynthesizes during its first year, it stores much of its energy in its root. In the second year, the mullein continues to photosynthesize and then uses its large amount of stored energy from 2 years of photosynthesis to grow a tall flowering stem that is more than 1 m high, which you can see in Figure 52.3b.

Large plants such as trees typically require many years of growth before they have accumulated enough energy to start reproducing. This also allows the trees to put their energy into growing taller to rise above the height of many other competing plants before allocating any energy to making flowers and seeds. Thus, different species of plants have evolved different strategies to gain evolutionary fitness.

a.

b.

FIGURE 52.3 Timing reproduction to available energy

The common mullein grows in fields around the world, and it requires 2 years to flower. (a) In the first year, the mullein is a short plant that photosynthesizes and stores much of its surplus energy in its roots. (b) In the second year, the mullein grows much taller and uses the energy it has gained over 2 years to produce a flower stalk that is more than 1 m tall. Photos: (a) Karin Jaehne/Shutterstock; (b) Scott T. Smith/Getty Images

Animals also time their reproduction to energy availability. For instance, large mammals breed in the fall so that their offspring are born in the spring, when food is very abundant. This means that the offspring have plenty of food to begin their life, but it also means that the offspring have the entire summer to grow and store fat before the onset of winter, when food is much less abundant. During the winter, they must either live off their fat reserves and the scarce food that they can find or have the energy to migrate to warmer climates, such as geese, which fly south every fall and then fly back north in the spring. As you can see, the different ecological strategies of how organisms live their lives is strongly affected by energy availability and energy use.

52.2 Energy is captured and moved through ecosystems

All organisms need energy to survive, grow, and reproduce. Some obtain energy from the sun and from inorganic molecules, while others obtain energy from consuming other organisms. In this section, we will examine how energy is captured and how it is moved through different species that live in an ecosystem.

Energy Capture

There are multiple ways that organisms have evolved to capture the energy that they need. In Unit 3, we discussed how some organisms use photosynthesis to capture solar energy and convert carbon dioxide and water into sugars. An alternative strategy, especially in places that have no sunlight available, is to obtain energy from chemical compounds to build sugars, which we call **chemosynthesis.** For example, at the bottom of the deep ocean there are thermal vents that continuously release hydrogen sulfide gas (H_2S). At these depths, there is no sunlight, so specialized bacteria have evolved to use the energy in the chemical bonds of the hydrogen sulfide to power the conversion of carbon dioxide and water into sugars, as shown in **FIGURE 52.4**. We can see an example of this in the specialized species of bacteria that live in tube worms that live near sea vents. Regardless of the process used to capture this energy, the energy stored in these sugars can be extracted through the process of cellular respiration, as we noted in Unit 3.

Trophic levels and Food Chains

Ecologists categorize organisms according to how they obtain energy. Organisms that obtain energy by photosynthesis or

FIGURE 52.4 Chemosynthesis

In the deepest part of the ocean, vents on the ocean floor release hydrogen sulfide gas. Specialized bacteria have evolved to obtain the energy in hydrogen sulfide bonds to power the conversion of carbon dioxide and water into sugar. (a) Vents on the ocean floor releasing hydrogen sulfide gas. (b) Tube worms, which contain chemosynthetic bacteria, live near vents that release hydrogen sulfide gas. Photos: (a) Ralph White/Getty Images; (b) NOAA Okeanos Explorer Program, Galapagos Rift Expedition 2011

chemosynthesis are known as **autotrophs,** or producers. Autotrophs include plants and algae. Organisms that obtain energy by eating other organisms are called **heterotrophs,** or consumers. Heterotrophs acquire energy from consuming the organic compounds that are synthesized by other organisms. They do this through the process of catabolism in which they obtain energy by breaking down the carbohydrates, lipids, and proteins of other organisms. Heterotrophs are further categorized depending on which types of organisms they consume. Primary consumers are herbivores that eat the plants or algae, such as zebras that eat grasses on the plains of Africa. Secondary consumers are carnivores that eat primary consumers, such as the lions that eat the zebras. Tertiary consumers are carnivores that eat secondary consumers, such as bald eagles that eat fish, which eat tiny herbivores.

Autotrophs and heterotrophs can be organized into successive levels, which we call **trophic levels.** Based on these trophic levels, we can draw food chains that depict the direction of energy and matter flow between trophic levels. You can see an example of trophic levels and food chains in **FIGURE 52.5,** which begins with primary producers at the base and is followed by primary, secondary, and tertiary consumers. Given

that larger rectangles in the pyramid indicate a greater amount of biomass, we can see that most of the biomass resides in the producers with less biomass in each higher level of the consumers. In some communities there are only three trophic levels. For example, in the African plains, lions eat gazelles and gazelles eat grasses, so we only have three trophic levels. In other communities, there can be four trophic levels, such as the food chain in the figure in which grass is eaten by mice, which are eaten by snakes, which are eaten by hawks.

Now that we understand that energy is transferred up through multiple trophic levels, we can ask how efficiently this energy moves from one trophic level to another. For example, of the 1000 kg of biomass present in the producers, what percentage is transferred to the primary consumers and secondary consumers? We can answer this question by examining the amount of biomass present in each trophic level. In this example, only 10% of the producer mass is transferred to the primary consumers. This low efficiency occurs because not all producers are consumable, such as plants with thorns which deter herbivores. Moreover, not all consumed producers are fully digestible. In addition, the transformation of energy is never 100%, as noted in our discussion of the second law of thermodynamics (see Module 13).

In the same way, we see in our example that only 10% of the mass from the primary consumers is transferred to the secondary consumers. This happens because some primary consumers have evolved defenses against being consumed, those primary consumers that are consumed are not fully digestible, and a substantial portion of the energy consumed by secondary consumers is used for maintaining their bodies and generating heat. The remaining energy can be used for growth and reproduction. Collectively, this means that of all the energy available in the producer level, only 1% makes it to the secondary consumer level. While a 10% ecological efficiency is a helpful rule of thumb, actual measurements of ecological efficiency generally range from 5% to 20%.

Based on our knowledge of ecological efficiency, we can also see that it places a limit on how many trophic levels can exist in a community. Having more energy available in the producer level and having higher ecological efficiencies would result in more energy available in the secondary consumers, which would allow a larger population of secondary consumers. As the number of secondary consumers increases, it can become high enough to support a population of tertiary consumers. For example, imagine a field of plants that can support a population of 20 white-tailed deer. If we fertilize the field to grow more plants, the field will then also be able to feed more deer. With more deer present, there may

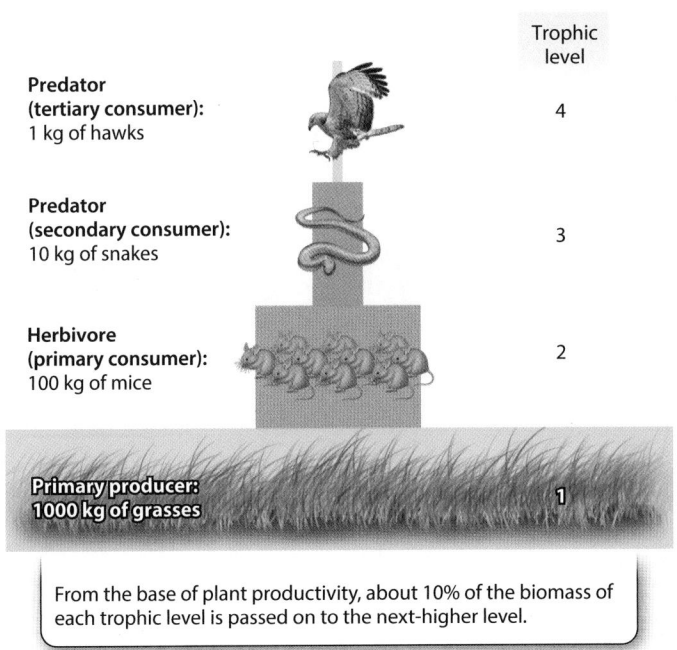

Predator
(tertiary consumer):
1 kg of hawks

Predator
(secondary consumer):
10 kg of snakes

Herbivore
(primary consumer):
100 kg of mice

Primary producer:
1000 kg of grasses

Trophic level

4

3

2

1

From the base of plant productivity, about 10% of the biomass of each trophic level is passed on to the next-higher level.

FIGURE 52.5 Trophic levels

Based on how organisms obtain their energy, we can order autotrophs and the various types of heterotrophs into successive levels that represent the flow of energy in an ecosystem into simplified food chains. Some food chains, such as those that include lions, zebras, and grasses, only have three trophic levels. Other food chains, such as a lake containing algae, zooplankton, fish, and bald eagles, can have four trophic levels.

now be enough deer to support a population of deer predators, such as wolves or mountain lions. In the same way, anything that reduces the abundance and growth of the producers will reduce the abundance of the higher trophic levels and may cause a loss of the highest trophic level because there is not enough energy available at that level.

Food Webs

While categorizing species into a small number of trophic levels helps us see the big picture of how energy flows, real communities generally have a large number of species and many of them do not fit neatly into the four trophic categories shown in Figure 52.5. Instead, we commonly find that species cannot be lined up in a simple linear way. In addition, we can have species that are omnivores, which feed on both producers and other consumers. Finally, our trophic levels and simple food chains did not include those consumers of dead organic matter. These consumers include scavengers such as condors that eat dead animals, detritivores such as earthworms that break down dead organic and waste products into smaller particles, and decomposers such as fungi and bacteria that convert the dead organic matter into molecules and elements that can be recycled by the producers.

To represent a community of interacting species in a more realistic way, we use food webs, which are depictions of how energy and matter flow among a large number of species. The food web in **FIGURE 52.6** shows species interacting in adjacent aquatic and terrestrial habitats. Note that

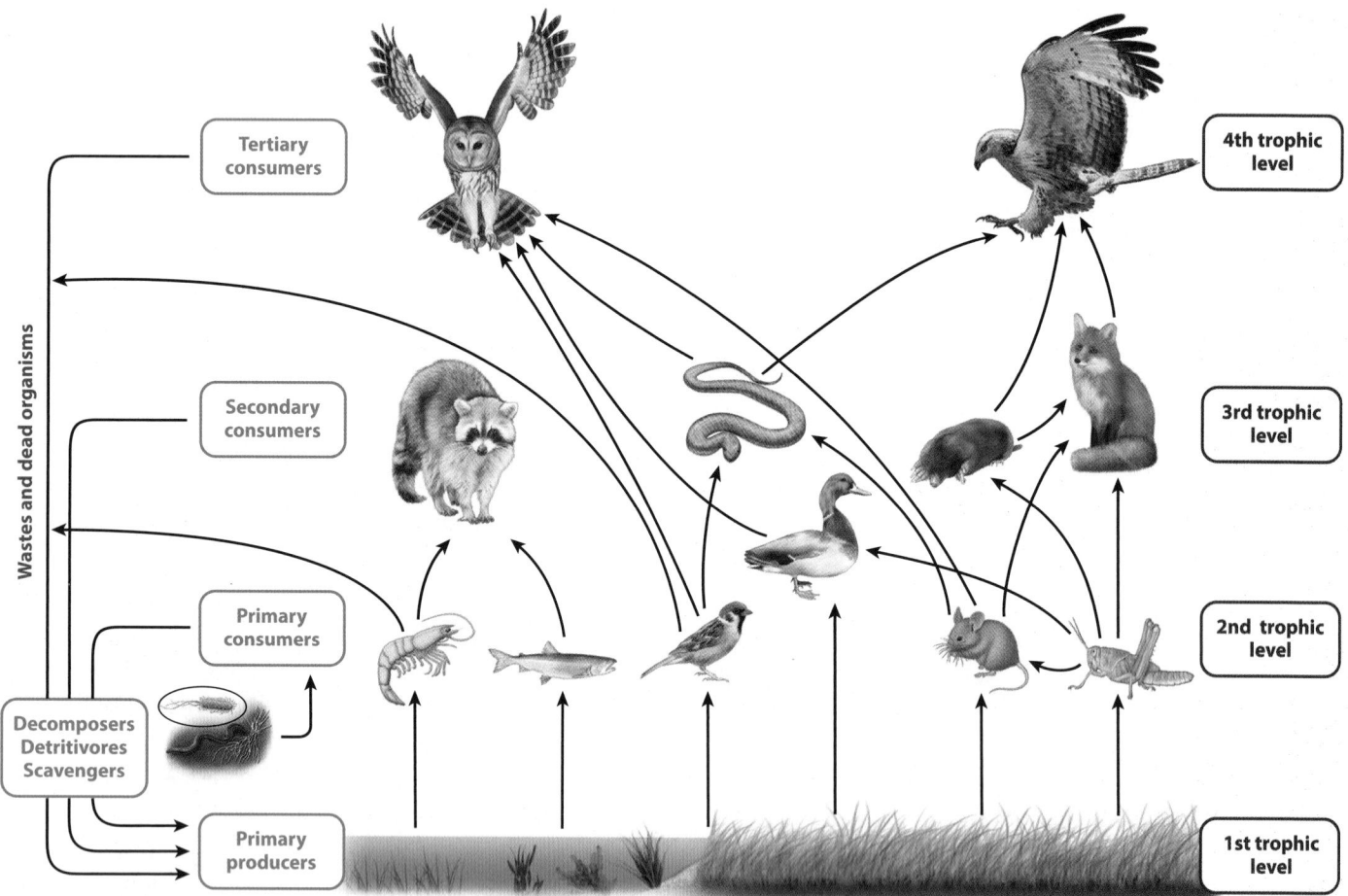

FIGURE 52.6 A simplified food web

In contrast to food chains, food webs show many more species interacting and the ways in which a given species might consume multiple trophic levels rather than only a single trophic level below, as depicted in food chains. Arrows indicate the direction of flowing energy and matter. A complete food web would show many more species and many more arrows depicting the flow of energy and matter.

the arrows indicate the direction of energy movement. In this figure, you can see a large number of species interacting, with some species feeding on multiple trophic levels. You can also see scavengers, detritivores, and decomposers feeding on the waste products and dead organisms.

✓ **Concept Check**

4. **Identify** four trophic levels.
5. **Describe** the difference between a food chain and a food web.
6. **Identify** why the amount of energy in successive trophic levels declines.

52.3 Matter cycles in ecosystems

Most of the energy that moves through communities and ecosystems originates from the sun and is captured through photosynthesis, although a small fraction originates from chemosynthesis. As this energy moves through the food web, much of it is converted into heat and this heat is ultimately lost to space. Thus, energy comes into an ecosystem, moves through the ecosystem, and eventually leaves the ecosystem.

In contrast, matter cycles within an ecosystem. This matter includes many important elements that cycle through the air, water, soils, and organisms, such as nitrogen, phosphorus, sulfur, and carbon. Given that these elements move around in a cycle, we can think about pools and processes. The pools are places in which an element resides whereas the processes are the ways in which an element moves from one pool to another. For example, we can find a substantial amount of nitrogen in plants that die each fall, so the plants represent a pool of nitrogen. The nitrogen can then be released to the soil through plant decomposition, which represents a process. A prime example of element cycling can be found by tracking the movement of carbon and seeing how human activities are having major impacts on both carbon pools and processes.

The Carbon Cycle

As you may recall from previous modules, carbon is an incredibly important element that is found in a wide variety of molecules on Earth including the molecules that comprise cell membranes, proteins, sugars, and the ATP molecule used as an accessible source of cellular energy. In fact, carbon makes up about 20% of an organism's dry mass, which is the mass of the organism's body after removing the water. In the carbon cycle, illustrated in **FIGURE 52.7** on page 744, we see that carbon moves between pools in the air, water, and land using several processes.

The two most familiar processes in the carbon cycle are photosynthesis and cellular respiration, which we discussed

in Unit 3. Organisms that photosynthesize take CO_2 out of the air and water and convert it into sugars that can subsequently provide the energy for growth and reproduction. This carbon can be moved through the food web to the consumers, scavengers, detritivores, and decomposers. At the same time, all species carry out cellular respiration, which converts organic molecules such as sugars back into CO_2. In addition, decomposers break down dead organisms to obtain energy and release CO_2 into the air and water.

Carbon also moves between pools using the processes of exchange, sedimentation, and burial. Exchange is the process of CO_2 moving between bodies of water and the atmosphere; the amount leaving the ocean is approximately equal to the amount that enters the ocean. Sedimentation occurs when dissolved CO_2 in the water combines with calcium ions to form calcium carbonate ($CaCO_3$). Calcium carbonate is not very soluble in water, so the compound precipitates out of solution, landing on the ocean bottom. Over time, this calcium carbonate accumulates into large layers that are known as limestone or dolomite. Over millions of years, this process has produced a massive pool of carbon as ocean sediments. In some cases, dead organic matter is not completely decomposed in elements before the organic matter gets buried. In this case, the organic matter can become fossilized and a portion of it can be transformed into fossil fuels. The rate of carbon removal from oceans by burial is offset by approximately the same amount of carbon that is released by the weathering of limestone rocks and by volcanic eruptions.

The final two processes in the carbon cycle are extraction and combustion. Extraction is the removal of buried carbon, including coal, oil, and natural gas. Extraction by itself does not convert carbon into another form. It simply moves the buried fossil fuels from deep underground to the surface where it can be used. The combustion of fossil fuels is the process that actually transforms the fossil fuels from fossilized carbon sources into CO_2, much like the process of cellular respiration. Combustion also occurs with volcanoes and forest fires,

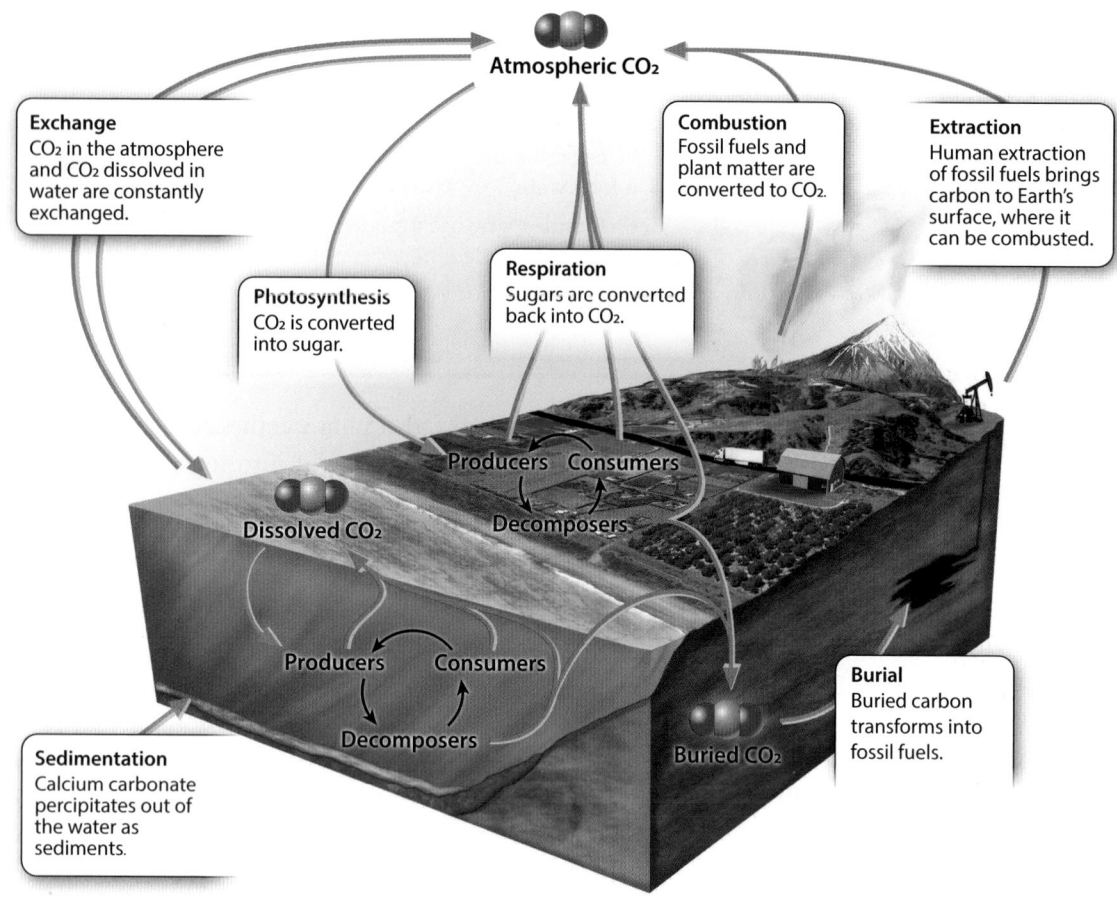

Exchange
CO_2 in the atmosphere and CO_2 dissolved in water are constantly exchanged.

Combustion
Fossil fuels and plant matter are converted to CO_2.

Extraction
Human extraction of fossil fuels brings carbon to Earth's surface, where it can be combusted.

Photosynthesis
CO_2 is converted into sugar.

Respiration
Sugars are converted back into CO_2.

Atmospheric CO_2

Producers   Consumers
Decomposers

Dissolved CO_2

Producers   Consumers
Decomposers

Burial
Buried carbon transforms into fossil fuels.

Buried CO_2

Sedimentation
Calcium carbonate precipitates out of the water as sediments.

FIGURE 52.7 The carbon cycle

There are seven important processes that move carbon among the many pools where it resides. Photosynthesis by plants and algae takes up CO_2 while cellular respiration and decomposition cause the release of CO_2. Exchange of CO_2 between waterbodies and the air occurs at nearly equal rates. Some CO_2 that is dissolved in the ocean combines with calcium to form $CaCO_3$, which precipitates out into an ocean sediment. Some organic matter that is not fully decomposed can become buried and get converted into fossil fuels. Humans have used extraction to move these fossil fuels to Earth's surface and then have used combustion to power homes, businesses, and automobiles.

both of which produce substantial amounts of CO_2. **"Visual Synthesis 8.2: The Movement of Carbon Through Ecosystems"** on page 750 provides an additional illustration of the terrestrial and marine carbon cycles.

Human Impacts on the Carbon Cycle

With our understanding of the carbon cycle, we can now examine how human activities have disrupted the carbon cycle. As we have just seen, without human impacts, the amount of carbon moved among pools by one process is approximately offset by some other process. Human activities have altered that natural balance. For example, the

large-scale combustion of fossil fuels by humans is a relatively recent activity, growing rapidly only during the past two centuries since the Industrial Revolution.

Today, humans contribute a large amount of CO_2 into the atmosphere from burning fossil fuels to heat their houses and businesses, power their cars, and provide the energy for industry and agriculture. In addition, logging large forests, often accompanied by burning the logged areas, has converted a large amount of carbon that was locked up in the bodies of the trees into CO_2 that was sent into the atmosphere. We will have much more to say about the consequences of increasing CO_2 in the atmosphere in Module 57.

Module 52 Summary

LG 52.1 There are different strategies for acquiring and using energy.

- The process of an organism controlling its body temperature is known as thermoregulation. Page 736

- We can distinguish between organisms with body temperatures that are regulated internally, known as endotherms, and organisms with body temperatures that are regulated by external conditions, known as ectotherms. Page 737

- Most plants are ectotherms, although some have fascinating physiological adaptations to internally generate heat. Page 737

- Many species of ectothermic animals can adjust the temperature of their bodies by altering their behavior. Page 737

- A net gain in energy allows an individual to experience growth and reproduction, which can lead to an increase in the size of a population. Page 738

- A net loss of energy results in lost growth, reproduction, and survival, which can lead to population declines. Page 738

- Endotherms have a higher metabolic rate than ectotherms, which means that endotherms have higher energy requirements to feed their higher metabolic rate. Page 738

- Larger species have a higher metabolic rate than smaller species, but larger species have a lower metabolic rate on a per–unit–mass basis than smaller ones. Page 739

- Energy availability also determines the reproductive strategies of organisms, including the seasons for breeding. Page 739

LG 52.2 Energy is captured and moved through ecosystems.

- Ecologists categorize organisms according to how they obtain energy. Autotrophs, or producers, obtain energy by chemosynthesis or photosynthesis. Heterotrophs, or consumers, obtain energy by eating other organisms. Page 740

- We can arrange the different producers and consumers into successive levels, which we call trophic levels. Page 741

- Based on these trophic levels, we can draw food chains that depict the direction of energy and matter flow between trophic levels. Page 741

- To represent a community of interacting species in a more realistic way, we use food webs, which are depictions of how energy and matter flow among a large number of species. Page 741

- A trophic pyramid depicts the amount of energy present in each trophic level. Page 741

- We can determine the percentage of available energy that makes its way from one trophic level to the next, which is called ecological efficiency. Page 742

LG 52.3 Matter cycles in ecosystems.

- Matter moves through communities and ecosystems in a cycle. Page 743

- Pools are places in which an element resides, whereas processes are the ways in which an element moves from one pool to another. Page 743

- The processes that move carbon in a cycle are photo-synthesis, cellular respiration, exchange, sedimentation, burial, extraction, and combustion. Page 743

- Today, human activity releases a large amount of CO_2 into the atmosphere, including burning fossil fuels for heat, to power cars, and to provide energy for industry and agriculture. Page 744

Key Terms

Endotherm

Ectotherm

Metabolic rate

Chemosynthesis

Autotroph

Heterotroph

Trophic level

Review Questions

1. Which statement is false about endotherms and ectotherms?
   (A) Ectotherms maintain a body temperature that matches their environment.
   (B) Endotherms maintain a constant body temperature.
   (C) Ectotherms can vary their body temperature using thermoregulatory behaviors.
   (D) Endotherms require more energy than ectotherms.

2. Which statement is true about endotherms versus ectotherms?
   (A) Ectotherms need to consume more food than similar-sized endotherms.
   (B) Endotherms can occupy a wider range of habitats than ectotherms.
   (C) Large endotherms require less energy than small ectotherms.
   (D) Endotherms can go for long periods of time without eating.

3. What is the relationship between mammal size and metabolic rate?
   (A) Larger mammals have a lower metabolic rate than smaller mammals.
   (B) Smaller mammals have a higher metabolic rate than larger mammals.
   (C) Larger mammals have a lower per-kg metabolic rate than smaller mammals.
   (D) Smaller mammals have a lower per-kg metabolic rate than larger mammals.

4. Which statement is true about how energy availability affects reproductive strategies of organisms?
   (A) Animals can reproduce at any time of the year.
   (B) Larger plants such as trees commonly reproduce in their first year of life.
   (C) Plant reproduction is not affected by energy availability.
   (D) Animals reproduce when abundant food is available.

5. Which statement is false about the carbon cycle?
   (A) Human combustion of fossil fuels has had no effect on the carbon cycle.
   (B) Sedimentation brings carbon to the bottom of the ocean.
   (C) Photosynthesis and respiration are approximately equal in their movement of carbon.
   (D) Exchange is the movement of carbon between the water and the air.

Module 52
AP® Practice Questions

PREP FOR THE AP® EXAM

Section 1: Multiple-Choice Questions

Choose the best answer for questions 1–4.

1. Carbon is a key element necessary for life. If all decomposers were eliminated from the Earth, which situation would be likely to happen?
   (A) Life would be unaffected as there are many other processes in the carbon cycle that would compensate for the lack of decomposers.
   (B) Life would eventually stop because all Earth's carbon would be trapped in dead materials.
   (C) Life would slow down but continue as the function of the decomposers would be taken over by the respiring organisms.
   (D) Life would continue unaffected as the rest of the carbon cycle would increase its rate.

Questions 2 and 3 refer to the following information.

The body temperature data in the table were collected from two different species in the same habitat over the course of the same 15 hours of the same day. Numbers in parentheses are ±2 standard errors of the mean.

Time of day	Outside temperature (°C)	Body temperature species 1 (°C)	Body temperature species 2 (°C)
6 a.m.	19	22 (±1)	41.3 (±0.2)
9 a.m.	27	28 (±2)	42.0 (±0.3)
12 p.m.	33	32 (±1)	41.9 (±0.4)
3 p.m.	32	35 (±1)	41.8 (±0.3)
6 p.m.	25	22 (±2)	40.9 (±0.2)
9 p.m.	20	19 (±1)	41.5 (±0.2)

2. Determine the average body temperature of species 1 over the course of this experiment.

   (A) 25°C   (C) 26°C

   (B) 24°C   (D) 28°C

3. Which conclusion is supported by the data in the table?

   (A) Species 1 regulates its body temperature.

   (B) Species 2 regulates its body temperature.

   (C) Species 1 had its lowest body temperature at the coldest time of the day.

   (D) Species 2 had its highest body temperature at the coldest time of the day.

Questions 4 and 5 refer to the following information.

Blue jays are a species of bird that lives the eastern and central United States. Students in an AP® Biology class observed a population of blue jays near their school eating acorns, berries, and seeds.

4. Ecologists often follow the movement of energy from one trophic level to another. If we consider the energy of the acorns, berries, and seeds to be 100%, what portion of that energy would be expected to be in the biomass of the blue jays that consume these food sources?

   (A) 100%   (C) 10%

   (B) 50%   (D) 1%

5. The students observe that hawks prey upon the blue jays living near their school. What portion of the energy of the acorns, berries, and seeds would be expected to be in the biomass of the hawks that consume the blue jays?

   (A) 100%   (C) 10%

   (B) 50%   (D) 1%

Section 2: Free-Response Question

Write your answer to each part clearly. Support your answers with relevant information and examples. Where calculations are required, show your work.

Researchers studying a grassland habitat observe interactions among four animal species: hawk, fox, mouse, and grasshopper. They record their observations of the interactions in the food web shown below.

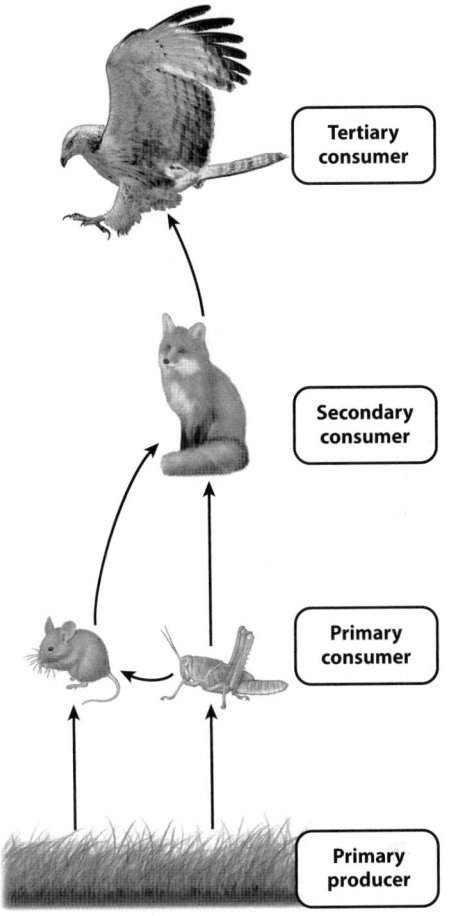

(a) In the habitat being studied, the total mass of the grass is 8000 kg. **Calculate** the total mass of the hawk population in the habitat.

(b) At some point during the observation period, an herbicide is sprayed on the grass. The herbicide kills 85% of the grass and prevents the surviving grass from growing and producing seeds. **Predict** what will happen to the hawk populations in the terrestrial habitat. **Justify** your prediction.

(c) Shortly after the herbicide kills 85% of the native grass, a quick-growing, herbicide-tolerant grass species is planted in this ecosystem. **Predict** what effect this grass species will have on the ecosystem and **justify** your answer.

VISUAL SYNTHESIS 8.1 HOMEOSTASIS AND THERMOREGULATION

Endotherms, such as the rock squirrel, use a variety of physiological and behavioral adjustments to maintain a stable body temperature as they experience cold nights, cool days, and hot days. In ectotherms, such as the zebra-tailed lizard, body

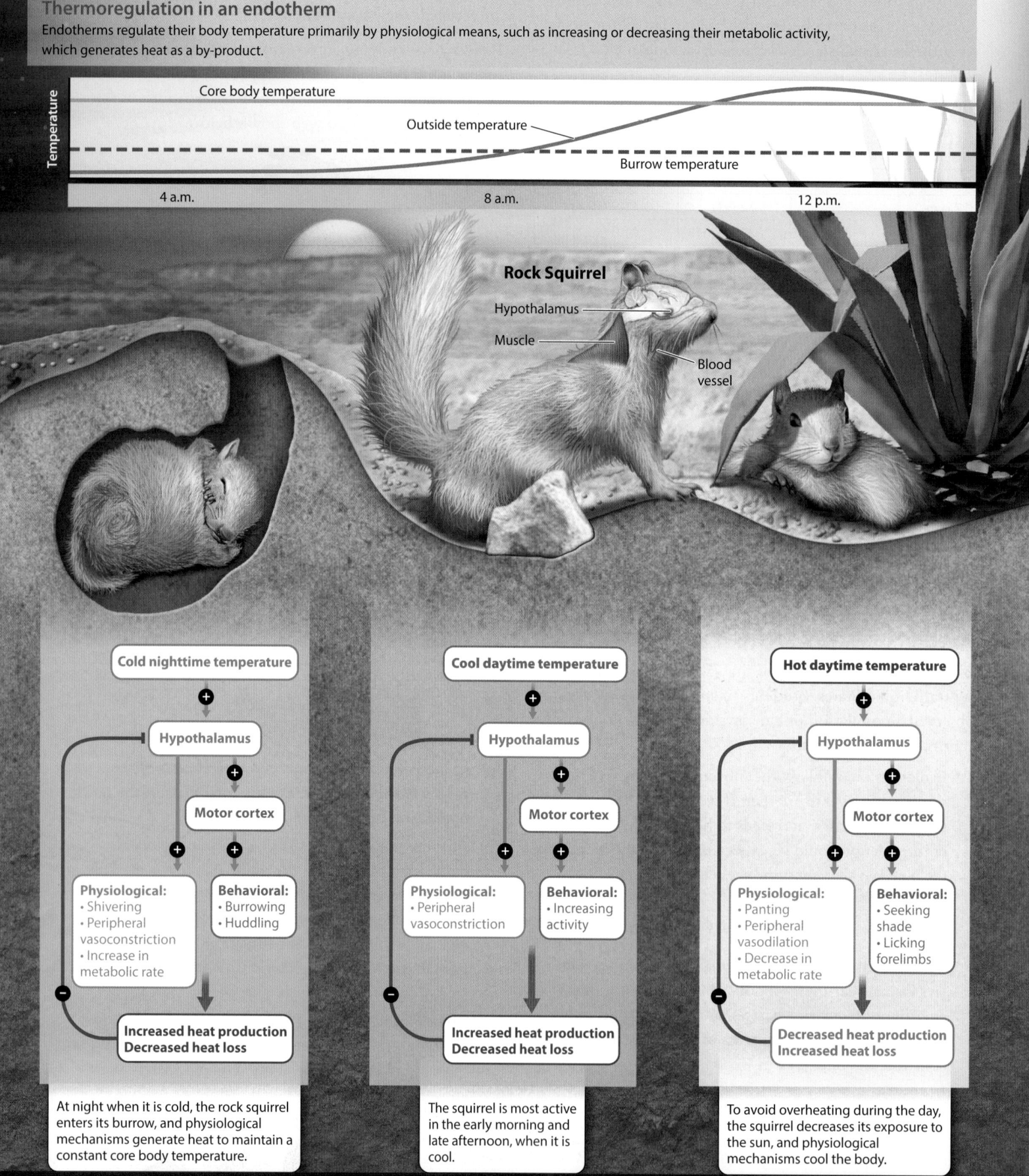

Thermoregulation in an endotherm

Endotherms regulate their body temperature primarily by physiological means, such as increasing or decreasing their metabolic activity, which generates heat as a by-product.

Temperature

Core body temperature

Outside temperature

Burrow temperature

4 a.m.   8 a.m.   12 p.m.

Rock Squirrel

Hypothalamus
Muscle
Blood vessel

Cold nighttime temperature
→ (+)
Hypothalamus
→ (+)
Motor cortex
→ (+) (+)

Physiological:
• Shivering
• Peripheral vasoconstriction
• Increase in metabolic rate

Behavioral:
• Burrowing
• Huddling

(−)

**Increased heat production
Decreased heat loss**

At night when it is cold, the rock squirrel enters its burrow, and physiological mechanisms generate heat to maintain a constant core body temperature.

Cool daytime temperature
→ (+)
Hypothalamus
→ (+)
Motor cortex
→ (+) (+)

Physiological:
• Peripheral vasoconstriction

Behavioral:
• Increasing activity

(−)

**Increased heat production
Decreased heat loss**

The squirrel is most active in the early morning and late afternoon, when it is cool.

Hot daytime temperature
→ (+)
Hypothalamus
→ (+)
Motor cortex
→ (+) (+)

Physiological:
• Panting
• Peripheral vasodilation
• Decrease in metabolic rate

Behavioral:
• Seeking shade
• Licking forelimbs

(−)

**Decreased heat production
Increased heat loss**

To avoid overheating during the day, the squirrel decreases its exposure to the sun, and physiological mechanisms cool the body.

temperature is much more variable and is determined by the air temperature of the environment as well as behavioral and physiological changes to maintain a small range of temperatures.

Thermoregulation in an ectotherm

Ectotherms gain heat from or lose heat to their environment, and so their primary means of regulating their body temperature is behavioral, increasing or decreasing their exposure to the sun and hot surfaces.

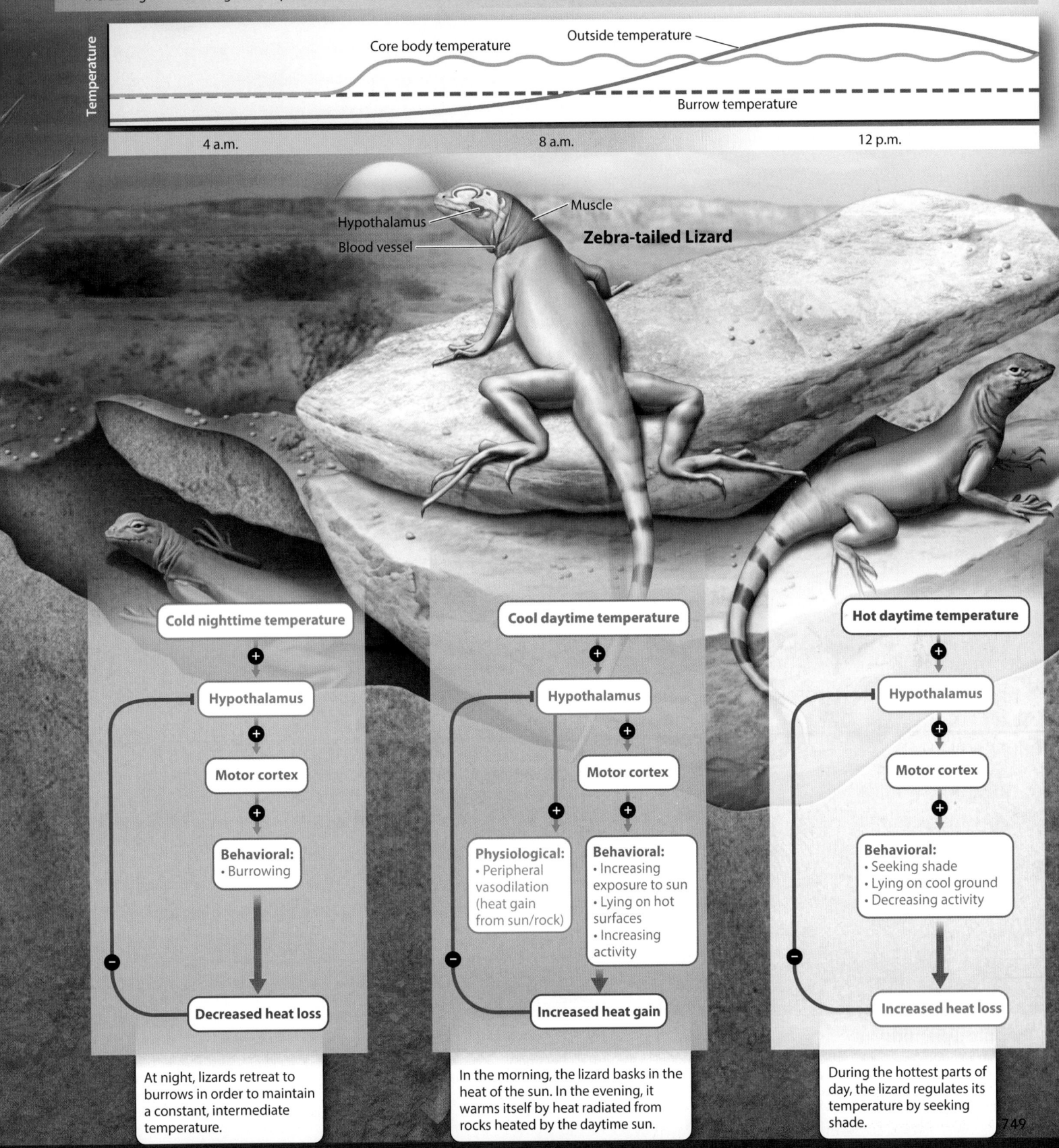

Temperature

Core body temperature

Outside temperature

Burrow temperature

4 a.m.   8 a.m.   12 p.m.

Hypothalamus
Blood vessel
Muscle

Zebra-tailed Lizard

Cold nighttime temperature
+
Hypothalamus
+
Motor cortex
+
Behavioral:
• Burrowing
−
Decreased heat loss

At night, lizards retreat to burrows in order to maintain a constant, intermediate temperature.

Cool daytime temperature
+
Hypothalamus
+
Motor cortex
+   +
Physiological:
• Peripheral vasodilation (heat gain from sun/rock)
Behavioral:
• Increasing exposure to sun
• Lying on hot surfaces
• Increasing activity
−
Increased heat gain

In the morning, the lizard basks in the heat of the sun. In the evening, it warms itself by heat radiated from rocks heated by the daytime sun.

Hot daytime temperature
+
Hypothalamus
+
Motor cortex
+
Behavioral:
• Seeking shade
• Lying on cool ground
• Decreasing activity
−
Increased heat loss

During the hottest parts of day, the lizard regulates its temperature by seeking shade.

749

In both terrestrial and marine ecosystems, producers take up carbon dioxide for photosynthesis to produce sugars that are used to fuel the growth and reproduction of the producers. This carbon then moves through the food web. All organisms produce

Terrestrial carbon cycle

Carbon (C) enters the cycle through photosynthesis and is transferred through food webs by consumers. Decomposers complete the return of carbon to the atmosphere as CO_2.

CO_2

Photosynthesis

Aerobic respiration

Tree

Adult moth

Moth larva

Solenodon

Aerobic respiration
Anaerobic respiration
Fermentation

Decomposers

Weathering

PO_4^{3-}

Primary producers:
Autotrophs that fix CO_2 into organic matter

Photosynthesis:

$$6\,CO_2 + 6\,H_2O \longrightarrow C_6H_{12}O_6 + 6\,O_2$$

Primary consumers:
Heterotrophs that feed on primary producers

Aerobic respiration:

$$C_6H_{12}O_6 + 6\,O_2 \longrightarrow 6\,CO_2 + 6\,H_2O$$

Secondary consumers:
Heterotrophs that feed on other consumers

Decomposers:
Heterotrophs that return CO_2 and nutrients to environment
• Fungi
• Bacteria
• Amoeba
• Protists

Trophic pyramids

As each trophic group obtains carbon, a portion gets transferred to the next higher trophic group through consumption. For this reason, only a portion of the energy taken in is available to consumers further up the food web.

carbon dioxide during respiration. Carbon dioxide is also exchanged between terrestrial and aquatic environments at the air–water interface.

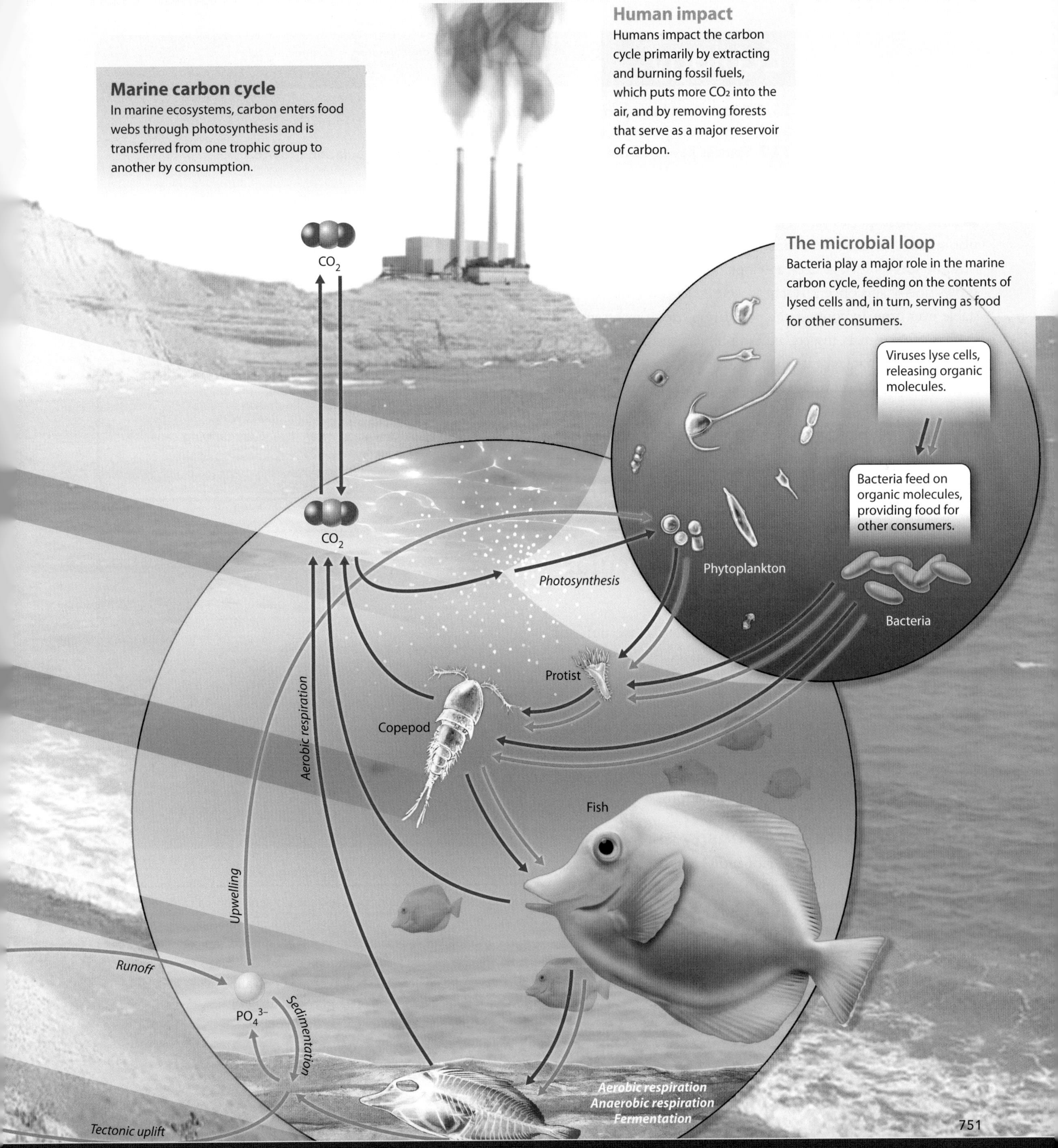

Marine carbon cycle

In marine ecosystems, carbon enters food webs through photosynthesis and is transferred from one trophic group to another by consumption.

Human impact

Humans impact the carbon cycle primarily by extracting and burning fossil fuels, which puts more CO_2 into the air, and by removing forests that serve as a major reservoir of carbon.

The microbial loop

Bacteria play a major role in the marine carbon cycle, feeding on the contents of lysed cells and, in turn, serving as food for other consumers.

Viruses lyse cells, releasing organic molecules.

Bacteria feed on organic molecules, providing food for other consumers.

CO_2

CO_2

Photosynthesis

Phytoplankton

Bacteria

Protist

Aerobic respiration

Copepod

Fish

Upwelling

Runoff

$PO_4{}^{3-}$

Sedimentation

Aerobic respiration
Anaerobic respiration
Fermentation

Tectonic uplift

Module 53

Population Ecology

LEARNING GOALS ▶**LG 53.1** Populations have particular characteristics.
▶**LG 53.2** Population growth is determined by births and deaths.
▶**LG 53.3** Species have evolved adaptations that are related to population growth.

In the field of ecology, one of the most interesting observations is that species vary a great deal in how much their population size fluctuates over time. We see that the population size of some species such as deer, antelope, and other large mammals grows slowly and once it reaches a certain level it remains fairly stable over time. In contrast, the population size of many smaller species of mammals, insects, and amphibians grows rapidly and experiences high rates of death early in life, and therefore exhibits large fluctuations over time. Understanding the factors that drive these changes in population size is the science of population ecology. Population ecology provides us with insights into the factors that promote population growth as well as factors that limit population growth. These factors are relevant to all organisms including humans. In this module, we will explore several characteristics of populations that can affect population growth. We will then examine a mathematical model of population growth, which allows us to predict the future growth of populations. With this understanding, we will examine some adaptations of species that are related to how their populations grow.

PREP FOR THE AP® EXAM

FOCUS ON THE BIG IDEAS

SYSTEMS INTERACTIONS: Think about what can happen to the size of a population when it reaches carrying capacity. Consider the energy and matter requirements of the population.

53.1 Populations have particular characteristics

In our introduction to ecology in Module 51, we defined a population as all individuals of a species living within a given area. The area may be very large, such as the number of deer on an entire continent, or it may be delineated to be much smaller, such as the number of mountain lions in Montana or the number of frogs in a pond. In this section, we will examine several characteristics of populations that affect the size and growth rate of populations.

Population Size and Density

Population size is the number of individuals in a given area. For example, we could count the number of barn owls in your town or the number of dandelions in your yard. When a population size is very large or the area is expansive, it can be difficult or impossible to determine the population size for a species. In such cases, ecologists have devised ways of estimating population size. This is discussed in "Practicing Science 53.1: How many butterflies are in a given population?" As shown in **FIGURE 53.1**, one of the techniques scientists use to estimate the size of a population is called mark-and-recapture.

FIGURE 53.1 Mark-and-recapture

Biologists can estimate population sizes by collecting a sample of a populations, marking the collected individuals, and then releasing them back into the larger population. In this example, a sample of monarch butterflies in Florida was collected and each was given a unique mark on the wing. A second collection of the population made a few days later, which included both marked and unmarked individuals, help biologists estimate the size of the entire population. Photo: Doug Wechsler/ Nature Picture Library

Practicing Science 53.1

How many butterflies are in a given population?

Background Biologists estimate population sizes of many animals using a technique known as mark-and-recapture, which involves capturing and marking a sample of individuals, releasing them back into the wild, and then capturing another sample of individuals a few days later. This technique was used to estimate the population size of monarch butterflies (*Danaus plexippus*) in Florida.

Method Biologists captured the butterflies in a field, marked the butterflies with an identification sticker on the wing, and then released them. The number of marked butterflies was recorded. The next day, a second sample of monarch butterflies was captured, and the number of marked and unmarked butterflies was recorded. The marked butterflies that were captured again on the second day are called recaptures.

Using this technique, researchers estimated the butterfly population size (*N*) in several steps. First, they determined the total number of marked butterflies and unmarked butterflies caught on the second day (*C*). They divided *C* by the number of recaptures (*R*). Finally, they multiplied $\frac{C}{R}$ by the number of butterflies marked on the first day (*M*), as follows:

$$N = \left(\frac{C}{R}\right) \times M$$

Results Let's say that 100 butterflies are captured, marked, and released on the first day. On the second day, 120 butterflies are captured, 30 of which are recaptures that have marks. The rest are unmarked. Using the above equation:

$$N = \left(\frac{C}{R}\right) \times M$$

$$N = \left(\frac{120}{30}\right) \times 100$$

$$N = (4) \times 100$$

$$N = 400$$

We conclude that the population size is 400 butterflies. As you can see, in our second sample of 100 butterflies, 25% were marked. Since we originally marked 100 butterflies, the entire population must be four times greater than the number of marked butterflies.

Related Methods This mark-and-recapture technique is extremely useful in estimating population sizes of organisms ranging from polar bears to possums. Many animals have distinctive color markings, such as the blotches on humpback whale tailfins and skin-spotting patterns on frogs; these animals can be recorded without having to be marked. This technique assumes that the population size is approximately the same on the second day as it was on the first day. Biologists also assume that the captured and released animals mix with others in the population and do not intentionally avoid recapture on the second day.

SOURCE
Knight, A., L. P. Brower, and E. H. Williams. 1999. "Spring Remigration of the Monarch Butterfly *Danaus plexippus* (Lepidoptera: Nymphalidae) in North-Central Florida: Estimating Population Parameters Using Mark–Recapture." *Biological Journal of the Linnean Society* 68:531–556.

AP® PRACTICE QUESTION

Some high school students read that planting milkweed on the fields behind their school would increase the size of the monarch butterfly population because monarchs lay their eggs on the plant and the caterpillars feed on the leaves. The local middle school, about 5 miles away, has a field of about the same size, but without any additional milkweeds being planted. Design an experiment, using the mark-and-recapture method described, to determine if planting 100 milkweed plants affects the number of monarch visits to the field over the period of 1 year.

a. In your experimental design, **identify** the following:
   1. The question being investigated
   2. The hypothesis
   3. The experimental group
   4. The control group
   5. The independent variable
   6. The dependent variable
b. **Describe** how you could utilize standard error of the mean to help you analyze the data.

Population density is the number of individuals in a given area divided by the size of the area. By calculating density, we can determine whether a population is densely crowded or relatively scarce compared to other populations. For example, the density of coyotes in some regions of Texas is 12 individuals per square kilometer, while in other regions it is 1 individual per square kilometer. Knowing a population's density can be useful if we wish to know whether a species has enough food in an area or if it will outstrip its food supply. It can also be useful when state fish and wildlife agencies are setting annual limits on how many birds and mammals can be harvested in different regions of a state by hunters and how many fish can be harvested from different lakes and streams by anglers.

Population Range and Distribution

The geographic range of a population is the area over which a population is spread. The geographic range is determined by where there are favorable conditions for a species to live

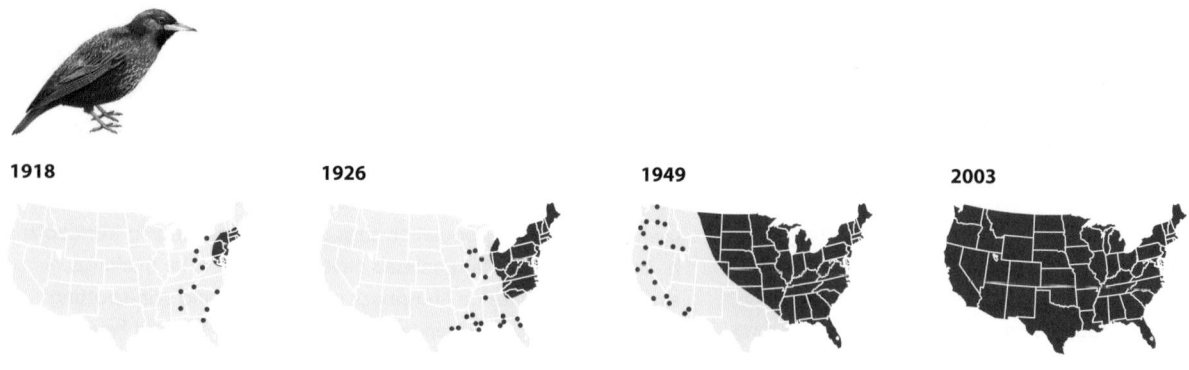

1918 **1926** **1949** **2003**

FIGURE 53.2 Geographic range

When the starling was introduced to the United States in 1890 and 1891, it began as a population of 160 birds near New York City. Over the subsequent decades, the population rapidly grew and expanded its geographic range, which is now all 48 contiguous states. Data from Kessel, B. 1953. "Distribution and Migration of the European Starling in North America." *Condor* 55: 49–67; U.S. Geological Survey. 2003. https://www.mbr-pwrc.usgs.gov/bbs/htm03/ra2003_red/ra04930.htm

and whether individuals can get there. For example, many species of birds and mammals living in South America may be able to also live in parts of Africa, but the Atlantic Ocean prevents them from dispersing to Africa, so their geographic range remains restricted to South America.

In some cases, new species are accidentally introduced to a continent where they have never lived before. It can be helpful to know what abiotic and biotic conditions the species requires so that we can estimate where the introduced species is likely to spread. For example, the European starling is a bird that was first introduced to the United States in 1890 and 1891 at a location near New York City. The original released population was 160 birds, but in less than 100 years the population's geographic range included the entire 48 contiguous states, as illustrated

in **FIGURE 53.2**, and it numbered more than 7 million starlings.

Population distribution is a measure of how clumped individuals are within their geographic range. The distribution of individuals can be random, uniform, or clumped. In a random distribution, there is no pattern in where individuals are with respect to other individuals, such as the trees shown in **FIGURE 53.3a**. In contrast, a uniform distribution occurs when all individuals are evenly spaced apart from each other, such as the nesting seabirds that evenly space out their nests shown in Figure 53.3b. A clumped distribution occurs when individuals aggregate together, often in response to clumped distributions of food or when there are benefits of social behavior. Examples include schools of fish, flocks of birds, and the clumped distributions of meerkats (*Suricata suricatta*) shown in Figure 53.3c.

a.

b.

c.

FIGURE 53.3 Population distributions

Individuals in a population can be distributed in three different ways. (a) Many species, such as these trees in the northeastern United States, show a random distribution. (b) Species of seabirds, such as these Australian gannets (*Morus serrator*), establish small territories, which results in each bird's nest being evenly spaced from other nests. (c) Some species, such as these meerkats, have clumped distributions, which can reflect clumped food sources or benefits of being social, including watching out for predators. Photos: (a) Scott Smith/Getty Images; (b) Weimann, Peter/Animals Animals/Earth Scenes; (c) P. Burghardt/Shutterstock

Population Sex Ratio and Age Structure

The population sex ratio is the ratio of males to females. In most species, this ratio is roughly 50:50, which means there is an equal number of males and females. In some species, such as fig wasps, this ratio can be 1 male for every 20 females. Uneven sex ratios can have a large impact on the number of offspring that can be produced by a population, which means that it can also affect how rapidly a population can grow.

The population age structure describes the number of individuals in different age categories. As we will see in the next section in which we discuss mathematical models of population growth, we commonly assume that all individuals are capable of breeding. In real populations, however, there are young individuals that have not yet achieved sexual maturity; middle-aged individuals that are in their breeding prime; and old individuals that are too old to breed, or breed less. As you might imagine, the number of individuals in these different age categories can have a large effect on the number of offspring produced by a population each year.

✓ Concept Check

1. **Describe** how age structure can affect population growth.

2. **Identify** the number of squirrels in a population if an initial sample of 20 squirrels is marked and released, and the next day another 20 squirrels are captured and half of them are marked.

3. **Describe** the difference between the geographic range and distribution of a population.

53.2 Population growth is determined by births and deaths

Changes in population size have a wide range of implications. In humans, for example, we often want to predict future population sizes to determine whether the world can continue producing enough food for a growing population. Similarly, ecologists often want to understand the increases and decreases in plant and animal populations to help protect them from going extinct. In this section, we will examine how ecologists have created mathematical models of population growth.

Birth and Death Rates

At the most fundamental level, the growth or decline of a population is determined by the number of births versus the number of deaths. If births exceed deaths, the population will increase. If deaths exceed births, the population will decrease. The movements of individuals into a population—immigration—and out of a population—emigration—can also be important factors in population change, as summarized in **FIGURE 53.4**.

To understand population growth, scientists use a mathematical model. To keep things simple, we will assume that we do not have individuals immigrating into the population or emigrating out of the population. We begin by defining the population size as N. We can then define the change in population size per unit time as $\frac{dN}{dt}$, where dN indicates a change in population size and dt indicates a change in time. Thus, $\frac{dN}{dt}$

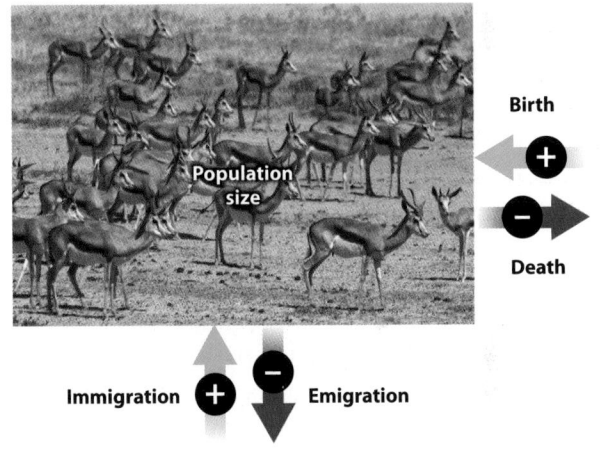

FIGURE 53.4 Changes in population size

A population can increase as a result of births and immigration of individuals into the populations. It can decrease as a result of deaths and emigration of individuals out of the population. Photo: Fernando Quevedo de Oliveira/Alamy Stock Photo

indicates the rate of change in population size over time. To determine the rate of change in population size, we can identify population sizes at different points in time. For example, if we define N_0 as the population size at time 0 and N_1 as the population size at time 1, the following equation shows how a population changes in size from time 0 to time 1:

$$\frac{dN}{dt} = N_1 - N_0$$

When we measure births and deaths over a particular time, such as births and deaths per day, we are measuring birth and death rates. Given that the change in population size is fundamentally the result of birth rate (B) and death rate (D), we can rewrite the above equation as:

$$\frac{dN}{dt} = B - D$$

where:

  dt = change in time
  B = birth rate
  D = death rate
  N = population size

For example, imagine that we wanted to know the change in population size from one year to the next. If the birth rate that year is 20 individuals per year and the death rate is 10 individuals per year, then the change in population size in that year is:

$$\frac{dN}{dt} = 20 - 10$$

$$\frac{dN}{dt} = 10 \text{ individuals per year}$$

Per Capita Growth Rate

It can be difficult to determine whether an increase in population size is substantial or trivial unless we know the size of the initial population. If the population starts with 10 individuals and it grows by 10 individuals in a year, then the population has increased by 100% and is therefore growing rapidly. However, if the population starts with 1000 individuals and it grows by 10 individuals in a year, then the population has only increased

by 1%. To give us a sense of how much a population is growing, we therefore need to assess the growth of the population relative to the initial size of the population. We can do this by calculating the **per capita growth rate (r),** which is the rate of population change divided by the size of the population:

$$r = \frac{dN}{dt} \div N$$

The per capita growth rate allows us to assess how rapidly a population is expected to increase or decrease. When r is positive, the population will increase; when it is large in magnitude, the population will increase quite rapidly. For instance, in our earlier example, we had a population with $\frac{dN}{dt} = 10$ and a population size of 10. In this case:

$$r = \frac{dN}{dt} \div N$$

$$r = 10 \div 10$$

$$r = 1$$

However, if $\frac{dN}{dt} = 10$ and the population size is 1000, then r has a much smaller positive value:

$$r = \frac{dN}{dt} \div N$$

$$r = 10 \div 1000$$

$$r = 0.01$$

In contrast to these examples, r can also be a negative value. When r is negative, the population will decrease because the birth rate is exceeded by the death rate. Finally, when $r = 0$, the population is stable because the birth rate exactly offsets the death rate. In this case, the population does not change in size over time.

"Analyzing Statistics and Data: Population Growth" reviews the use of the population growth equation and the per capita growth rate and provides a practice problem for you to try.

ANALYZING STATISTICS AND DATA ▶

Population Growth

The growth of a population can be challenging to track due to a number of natural processes that happen in an environment, such as immigration, emigration, and limited amounts of food. The population growth model exists to simplify the conditions under which scientists try to predict growth from one time point to another by assuming that there is no immigration or emigration and that food is never limiting.

The most important factor in determining population growth is the number of births and deaths occurring in that population. When births

outnumber deaths, the population increases in size; when deaths outnumber births, the population decreases in size. If births equal deaths, then the population size does not change. As we have already seen, the basic population growth formula is $\frac{dN}{dt} = B - D$.

The per capita growth rate (r) formula is also useful to know in the context of population growth. It tells us how quickly a population is increasing or decreasing.

$$r = \frac{dN}{dt} \div N \quad \text{or} \quad r = \frac{B - D}{N}$$

PRACTICE THE SKILL

A group of scientists is closely monitoring an endangered hawk population. Over the course of 1 year, the team has noted that in the population of 150 adult individuals, 120 eggs hatched, 98 of those offspring did not survive, and 13 adult hawks died. Calculate the change in population size over this year and the per capita growth rate.

The question has provided us with the birth and death rates to use in the population growth formula. The birth rate is 120 individuals per year and the death rate is 98 + 13 = 111 individuals per year. We can plug these numbers into the growth formula to find the change in population size over this 1-year period:

$$\frac{dN}{dt} = B - D$$

$$\frac{dN}{dt} = 120 - 111$$

$$\frac{dN}{dt} = 9$$

The hawk population grew by 9 individuals that year.

To answer the second part of the question, we can use the per capita growth formula. We know the population size and have just solved for population change, so these two numbers can be entered into the equation:

$$r = \frac{dN}{dt} \div N$$

$$r = 9 \div 150$$

$$r = 0.06$$

Because r is positive, the birth rate of the hawk population exceeds the death rate and the population increased by 6%. In this case, r is relatively small in magnitude, meaning that the hawk population is increasing by only a few individuals per year. If r had been negative, the opposite would have been true. The death rate would exceed the birth rate and the population would be decreasing at a rapid rate.

Your Turn

The graph at right depicts the population size of Japan from 1900 to 2005. Calculate the approximate population growth rate for the years 1920, 1960, 2000, and 2005. Describe what has happened to the population growth of Japan throughout this century.

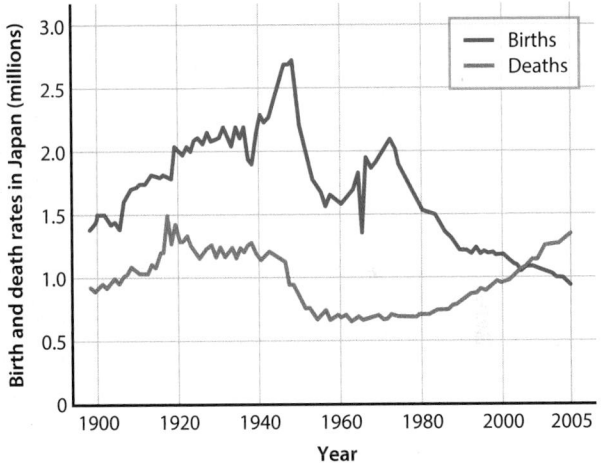

Data from https://theatlas.com/charts/B10_pHGg7

The Exponential Growth Model

Now that we understand how birth rates and death rates affect the growth of a population, we can examine a population growth model that predicts population growth into the future. Ecologists have made many different variations of population models, but we will begin with a model that makes a number of assumptions to simplify the model so we can see some important relationships. For this model, we will assume that the population experiences no immigration or emigration. We will also assume that births and deaths occur continuously throughout the year, such as we see in humans and bacteria. However, many plants and animals have distinct

breeding seasons, and we could alter our model to accommodate this reality. Finally, while populations in nature commonly have constraints on their growth, such as a limited amount of food or a limited number of nesting sites, in this first model we are going to assume that the population has no constraints on its growth. Under these assumptions, a population initially grows slowly but as the population increases it begins to experience very rapid growth because at each time step there are more and more individuals that can reproduce, as shown in **FIGURE 53.5** on page 758. As you can see in this figure, the increasing growth rate produces a J-shaped curve. While an assumption of growth without any constraints is not

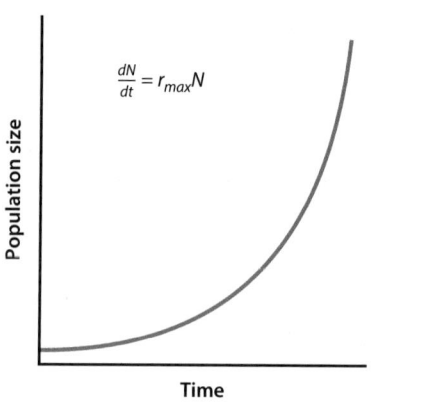

$$\frac{dN}{dt} = r_{max}N$$

FIGURE 53.5 Exponential growth model

In the exponential growth model, the population increases in size exponentially over time since at each time step there are more individuals that can reproduce.

realistic, it allows us to begin with a simple model that we will later build on to include constraints on growth.

This J-shaped population growth curve is produced by a mathematical model based on population size and population growth rate. Note that earlier we used r to indicate per capita growth rate, but here we use r_{max} to indicate the **maximum per capita growth rate** of a population, which is the growth rate when the population is living without any constraints. Under this assumption, the population grows rapidly over time and we call it the **exponential growth model:**

$$\frac{dN}{dt} = r_{max}N$$

where:

$$\frac{dN}{dt} = \text{change in population size per unit time}$$

$r_{max} = $ maximum per capita growth rate of the population
  $N = $ population size

In this model, we can predict the change in population size per unit time based on the current population size and the maximum per capita growth rate. Because we assume no constraints on growth, the population will grow faster whenever it has an increase in the maximum per capita growth rate (r_{max}) or a larger number of individuals that are reproducing (N).

Based on this equation, we can calculate the rate of population increase at any given population size. For example, if $r_{max} = 0.4$ and $N = 10$, then:

$$\frac{dN}{dt} = r_{max}N$$

$$\frac{dN}{dt} = 0.4 \times 10$$

$$\frac{dN}{dt} = 4$$

This means that when the population has 10 individuals, we will add 4 more individuals during the next time step. If the population is much larger, such as 80 individuals, the rate of population increase will be eight times larger, adding 32 individuals, which causes the J-shaped curve to rise much more quickly.

$$\frac{dN}{dt} = r_{max}N$$

$$\frac{dN}{dt} = 0.4 \times 80$$

$$\frac{dN}{dt} = 32$$

Despite its many simplifying assumptions, the exponential growth model is an excellent starting point for understanding how populations grow over time. There are real populations in nature that exhibit exponential growth, at least when they begin at low population sizes. However, as we noted earlier, no population can experience indefinite growth, because the growing populations become constrained by a limited food supply or some other constraint. In the next module, we will examine limits on population growth and how we modify our models to include such limits.

"Analyzing Statistics and Data: Exponential Growth" provides a review of the exponential growth model and a problem for you to try.

✓ Concept Check

4. **Describe** how birth rate and death rate determines population growth rate.

5. **Distinguish** the difference that a low versus high maximum per capita growth rate has on population growth.

6. **Describe** why the exponential growth model produces a J-shaped curve of population growth.

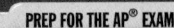
Exponential Growth

The exponential growth model is commonly used by scientists to predict the future growth or decline of a population. Exponential growth is represented by a J-shaped curve. Growth begins slowly, but as more and more individuals reach reproductive maturity, the population will rapidly increase. This curve is shown by the graph in Figure 53.5. As we have seen, the formula for exponential population growth is $\frac{dN}{dt} = r_{max}N$.

The exponential growth model assumes that there are no constraints acting on the population, such as limited food or space, so the population is free to experience unlimited and rapid growth. The model also assumes that the population experiences no emigration or immigration, and that births and deaths occur throughout the year.

The maximum per capita growth rate (r_{max}) is equal to $\frac{B-D}{N}$. Remember that B is the birth rate, D is the death rate, and N is the population size. While r may change depending on the variation in the birth rate or the death rate, r_{max} is constant even if the population size becomes very large because the population is not limited by resources.

PRACTICE THE SKILL

A group of scientists is investigating what would happen to a population of rabbits if it experienced exponential growth with different maximum per capita growth rates over 6 months. The initial population of rabbits consisted of only 2 individuals.

Using the exponential growth rate formula, fill in the missing data in the two tables below, one where $r_{max} = 0.5$ and one where $r_{max} = 1.5$. Graph each set of data and describe how different r_{max} values affect the population.

$r_{max} = 0.5$	
Month	Population size
1	2
2	3
3	5
4	7
5	
6	

$r_{max} = 1.5$	
Month	Population size
1	2
2	
3	12
4	
5	75
6	187

To use the exponential growth equation, we must have two of three variables. The question provides the number of rabbits in the population, N, and the r_{max} values. We can enter each of these into the growth equation to solve for the increase in population size and then fill in the tables.

To find the size of the population at 5 months, we must first find the exponential growth rate at 4 months:

$$r_{max}\ 0.5: \frac{dN}{dt} = r_{max}N$$

$$\frac{dN}{dt} = 0.5 \times 7$$

$$\frac{dN}{dt} = 3.5$$

The exponential growth rate when there are 7 rabbits in the population and r_{max} is 0.5 equals 3.5. This means that 3.5 rabbits, which we will round down to 3 rabbits, will be added to the population in the next time step of 1 month.

Population at 5 months = population at 4 months + 3 rabbits

Population at 5 months = 7 + 3 = 10 rabbits

Now that we know there are 10 rabbits after 5 months, we can use this number to solve for the exponential growth rate at 5 months:

$$r_{max}\ 0.5: \frac{dN}{dt} = r_{max}N$$

$$\frac{dN}{dt} = 0.5 \times 10$$

$$\frac{dN}{dt} = 5$$

The exponential growth rate when there are 10 rabbits in the population and r_{max} is 0.5 equals 5. This means that 5 rabbits will be added to the population in the next time step of 1 month.

Population at 6 months = population at 5 months + 5 rabbits

Population at 6 months = 10 + 5 = 15 rabbits

Now we can move on to completing the table for $r_{max} = 1.5$. Again, to find the size of the population at 2 months, we must first find the exponential growth rate at 1 month:

$$r_{max}\ 1.5: \frac{dN}{dt} = r_{max}N$$

$$\frac{dN}{dt} = 1.5 \times 2$$

$$\frac{dN}{dt} = 3$$

When the rabbit population has 2 individuals and a max per capita growth rate of 1.5, then the population will add 3 individuals at the next time step of 1 month.

Population at 2 months = population at 1 month + 3

Population at 2 months = 2 + 3 = 5 rabbits

▼

For the last empty cell in the table, we will calculate the exponential growth rate at 3 months to find the population size at 4 months:

$$r_{max} 1.5: \frac{dN}{dt} = r_{max}N$$

$$\frac{dN}{dt} = 1.5 \times 12$$

$$\frac{dN}{dt} = 18$$

When the rabbit population has 12 individuals and a max per capita growth rate of 1.5, the population will add 18 individuals at the next time step of 1 month.

Population at 4 months = population at 3 months + 18

Population at 4 months = 12 + 18 = 30 rabbits

We have completed the second table and can graph the data to show the exponential growth rate of each rabbit population.

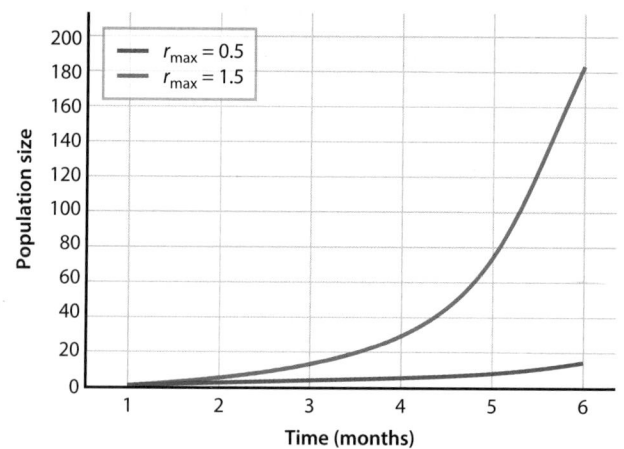

The larger r_{max} value has drastically increased how quickly the rabbit population is growing. The number of individuals added to the population at each time step is much greater with an r_{max} value of 1.5 than with an r_{max} value of 0.5. A smaller r_{max} means a slower rate of growth, resulting in fewer individuals being added after each period of time passes. A larger r_{max} means a faster rate of growth, resulting in more individuals being added after each period of time passes, and a steeper curve.

Your Turn

The protozoa *Paramecium caudatum* is known for being used in population growth studies. This species can exhibit exponential growth if laboratory conditions are favorable. *P. caudatum* has a max per capita growth rate of 0.94. Calculate the exponential growth rate of *P. caudatum* at N of 50; 2500; and 10,000. Describe the trend that you observe.

53.3 Species have evolved adaptations that are related to population growth

As we have seen, different species have different population growth rates, which are the result of evolution. Some species have evolved very high reproductive rates, which can be beneficial in species that face a risk of dying early in life. Other species have evolved relatively low reproductive rates, which reflects their low risk of dying early in life. In this section, we will examine how different species face different survival probabilities over time and how this has shaped the evolution of their maximum per capita growth rates (r_{max}).

Survivorship Curves

When we examine the probability of species surviving over their lifetime, they generally follow one of three patterns, as shown in **FIGURE 53.6**. For a Type I survivorship curve, individuals experience high survival throughout most of their life and then their survival declines sharply later in life.

Examples of Type I species include humans and other large mammals including deer, buffalo, and elephants. For a Type II survivorship curve, individuals experience steadily declining survival throughout their entire life. Species following this pattern include birds and small mammals such as squirrels. For a Type III survivorship curve, individuals experience a sharp drop in survival early in life and then a slower decline throughout the rest of their life, with few individuals reaching adulthood. Examples of such species include mosquitoes, amphibians, and many small species of plants such as dandelions.

K- and r-selected Species

The evolved differences in survivorship over time among different species are associated with differences in evolved reproductive strategies, as shown in **FIGURE 53.7**. For example, the exponential model assumes no constraints on growth. However, most species experience a limit on how many individuals can be supported in a given area, which we define as the environment's **carrying capacity,** and denote as K.

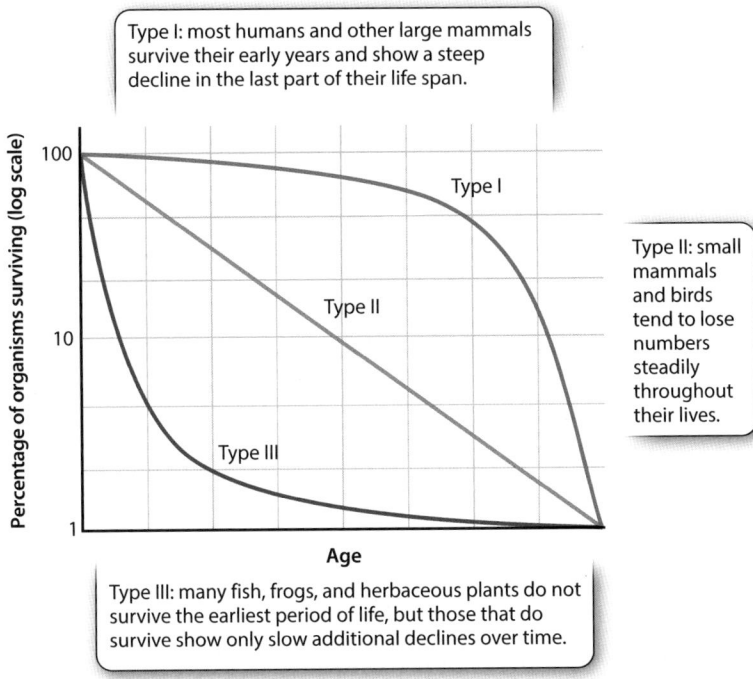

Type I: most humans and other large mammals survive their early years and show a steep decline in the last part of their life span.

Type I

Type II: small mammals and birds tend to lose numbers steadily throughout their lives.

Type II

Type III

Type III: many fish, frogs, and herbaceous plants do not survive the earliest period of life, but those that do survive show only slow additional declines over time.

FIGURE 53.6 Survivorship curves

Species with Type I survivorship curves experience high survival throughout most of their life and then low survival late in life. Species with Type II survivorship experience steadily declining survival throughout their entire life. Species with Type III survivorship experience low survival early in life and then slower decline throughout the rest of their life.

Over evolutionary time, some species have evolved to have relatively slow reproduction and remain near the environment's carrying capacity, so we refer to them as *K*-selected species. These species—including chimps, elephants, horses, and humans—commonly exhibit a Type I survivorship curve.

Given that *K*-selected species are those with populations that tend to stay near the environment's carrying capacity, they generally have low maximum per capita growth rates. These low growth rates are often found in large mammals that take many years to reach sexual maturity and do not produce many offspring each year. For example, an elephant is a *K*-selected species because it has a low maximum per capita growth rate as a result of taking 13 years to become reproductively mature, giving birth only once every 2–4 years, and giving birth to only one offspring each time. Such slow reproduction means that elephants have a low maximum per capita rate of growth, so they cannot rapidly increase in population size. Instead, their population remains fairly stable near the environment's carrying capacity.

Species that have evolved rapid reproduction that results in large fluctuations in population size over time are known as *r*-selected species. These species commonly exhibit a Type III survivorship curve, with a sharp decline in survivorship early in life. For example, in many species of frogs, one female will lay hundreds of eggs that soon hatch into tadpoles. The vast majority of these tadpoles, perhaps as much as 95%, never survive to metamorphose into a frog that will eventually become a breeding adult. In response to such a low probability of surviving to adulthood, these species have evolved high reproductive rates. One result of these high maximum per capita growth rates is that *r*-selected species experience rapid population increases. If their population exceeds the environment's carrying capacity, the population runs out of food and

a.

b.

FIGURE 53.7 Comparing *K*- versus *r*-selected species

(a) The chimpanzee takes about 10 years to achieve maturity and then has a single offspring every 3–6 years. (b) In contrast, frogs mature in 1–2 years and lay hundreds of eggs every spring.

Photos: (a) © Andre Gilden/natureinstock.com/ardea.com; (b) Chris Mansfield/Getty Images

sharply declines. Thus, *r*-selected species exhibit large variations in population sizes over time. Many species of insects, amphibians, small weedy plants, and small mammals are *r*-selected species. These species typically reproduce early in life, reproduce multiple times in rapid succession, and produce a high number of offspring each time they reproduce. For example, the house mouse (*Mus musculus*) achieves reproductive maturity at 6 weeks of age and then continues to breed again every 5 weeks, producing a dozen offspring with each breeding event.

✓ **Concept Check**

7. **Describe** the Type I and Type III survivorship curves.

8. **Describe** the characteristics of a *K*-selected species and include an example.

9. **Describe** the characteristics of an *r*-selected species and include an example.

Module 53 Summary

PREP FOR THE AP® EXAM

REVISIT THE BIG IDEAS

SYSTEMS INTERACTIONS: Using the content of this module, describe why populations do not remain at carrying capacity permanently. Be sure to connect your response to the energy and matter requirements of the population.

LG 53.1 Populations have particular characteristics.

- Population size is the number of individuals in a given area. Page 752

- Population density is the number of individuals in a given area divided by the size of the area. Page 753

- The geographic range of a population is the area over which a population is spread. Page 754

- Population distribution is a measure of how clumped individuals are within their geographic range. Page 754

- The population sex ratio is the ratio of males to females. Page 755

- The population age structure describes the number of individuals in different age categories. Page 755

LG 53.2 Population growth is determined by births and deaths.

- The growth or decline of a population is determined by the number of births versus the number of deaths in the population. Page 755

- We can write an equation for how a population changes in size from time 0 to time 1 as: $\frac{dN}{dt} = N_0 - N_1$. Page 755

- We can rewrite the above equation as: $\frac{dN}{dt} = B - D$. Page 756

- To give us a sense of how much a population is growing, we can calculate the per capita growth rate: $r = \frac{dN}{dt} \div N$. Page 756

- When *r* is positive, the population will increase; when it is large in magnitude, the population will increase quite rapidly. Page 756

- When *r* is negative, the population will decrease because the birth rate is exceeded by the death rate. Page 756

- For the exponential growth model, we assume that the population experiences no immigration or emigration, that births and deaths occur continuously throughout the year, and that the population has no constraints on its growth. Page 757

- Under these assumptions, the population grows rapidly over time. We call the model of this growth the exponential growth model: $\frac{dN}{dt} = r_{max}N$. Page 758

- In the exponential growth model, we can predict the change in population size per unit time based on the current population size and the maximum per capita growth rate. Page 758

- We can also use the exponential growth model to determine the rate of change in the population anywhere along the curve of population growth over time: $\frac{dN}{dt} = r_{max}N$. Page 758

LG 53.3 Species have evolved adaptations that are related to population growth.

- When we examine the probability of species surviving over their lifetime, they generally follow one of three patterns: Type I, Type II, or Type III. Page 760

- The evolved differences in survivorship over time among different species are associated with differences in evolved reproductive strategies. Page 761

- Over evolutionary time, some species have evolved to have relatively slow reproduction and remain near the environment's carrying capacity, so we refer to them as K-selected species. Page 761

- Other species have evolved to have rapid reproduction that results in large fluctuations in population size over time, so we refer to them as r-selected species. Page 761

Key Terms

Per capita growth rate (r)

Maximum per capita growth rate of a population (r_{max})

Exponential growth model
Carrying capacity (K)

Review Questions

1. Which statement is true about population characteristics?
   (A) Geographic range is the number of individuals divided by the area.
   (B) Density is the area over which a population is spread.
   (C) Distribution is the number of males and females in an area.
   (D) Age structure is the number of individuals in different age categories.

2. Which equation for population growth rate is incorrect?
   (A) $\dfrac{dN}{dt} = N_0 - N_1$
   (B) $\dfrac{dN}{dt} = B - D$
   (C) $\dfrac{dN}{dt} = r_{max} N$
   (D) $\dfrac{dN}{dt} = D - B$

3. Which is an assumption of the exponential growth model?
   (A) Population growth depends on birth rates and death rates.
   (B) Emigration and immigration occur in the population.
   (C) Population growth is constrained by limited food.
   (D) Population breeding happens once each year.

4. Using the exponential growth model, how much will a population increase in 1 year if we begin with 20 individuals and $r_{max} = 0.5$?
   (A) 20
   (B) 10
   (C) 30
   (D) 40

5. Which strategy is correctly paired with an evolved characteristic?
   (A) K-selected: slow reproduction
   (B) r-selected: long time to reproductive maturity
   (C) K-selected: fluctuating population sizes
   (D) r-selected: few offspring per reproductive event

Module 53
AP® Practice Questions

Section 1: Multiple-Choice Questions

Choose the best answer for questions 1–3.

Use the following information to answer question 1.

A bird-watching club collects data about the population densities of two bird species living on an abandoned farm. The first group of bird watchers collects data on the population density of red-winged blackbirds living in field 1 of the farm, while the second group collects data on the population density of robins living in field 2 of the farm. The data collected are listed in the table.

Species	Area (acres)	Number of individuals
Red-winged blackbirds (field 1)	1.7	13
Robins (field 2)	2.5	26

1. Which bird species has a higher population density?
   (A) Red-winged blackbirds
   (B) Robins
   (C) The population densities of both bird species are the same.
   (D) The population densities of the bird species cannot be determined from the data provided.

2. Wildlife biologists have reported that female wolves seem to have increased their rates of reproduction in a Rocky Mountain forest. Which of the following calculations would determine whether the wolf population in this forest is growing at a significantly higher rate than average? Assume that immigration and emigration in the wolf population is negligible.

(A) Determining the wolves' population distribution in the forest

(B) Determining how many female wolves are of reproductive age in the forest

(C) Determining the wolves' population range in the forest

(D) Determining the per capita growth rate of the wolves in the forest

3. Population A of the grey squirrel consists of 55 individuals. After 1 year, population A consists of 58 squirrels. Population B of grey squirrels has 206 individuals and after 1 year its population consists of 222 squirrels. Which population has the greater per capita growth rate?

(A) Population A

(B) Population B

(C) Both populations have the same per capita growth rate.

(D) The per capita growth rate cannot be calculated from the data given.

Section 2: Free-Response Question

Write your answer to each part clearly. Support your answers with relevant information and examples. Where calculations are required, show your work.

Ecologists conduct an experiment on the rocky coast of the Atlantic Ocean after a hurricane has removed all the organisms that were attached to the rocks. The ecologists use two identical stretches of the rocky coast to relocate two marine mussel species. Species 1 is introduced to rocky coast 1, and species 2 is introduced to rocky coast 2. The ecologists seed each coast with 10 male and 10 female mussels of the appropriate species. After each generation produces offspring, the ecologists record the number of individuals of each mussel species on each stretch of coast. They collected the data shown in the table.

Generation number	Individuals of species 1	Individuals of species 2
0	20	20
1	34	33
2	58	55
3	98	90
4	167	148
5	284	245
6	483	404
7	821	666
8	1395	1099
9	2372	1813
10	4032	2991
11	6854	4936
12	11,652	8144
13	19,809	13,438
14	33,676	22,172

(a) **Draw** a graph that shows the number of generations of the mussels versus the number of individuals in each generation.

(b) **Identify** what type of growth model is illustrated by the data of the rocky coast mussel experiment. **Justify** your answer.

(c) **Calculate** the number of individuals that will be in generation 15 of species 1.

Module 54

Effect of Density of Populations

LEARNING GOALS ▶**LG 54.1** Population growth can be limited by density-independent and density-dependent factors.
▶**LG 54.2** The logistic growth model includes density-dependent population growth.

In the previous module, we introduced a mathematical model of how populations grow, known as the exponential growth model. We noted that this model made a number of simplifying assumptions so that we could focus on the core idea that populations can grow exponentially over time, and that this growth is faster in species that possess a high maximum rate of per capita growth (r_{max}). The exponential growth model assumes that populations can grow without any limits, which we know is not realistic. In this module, we focus on the factors that place limits on population growth and then modify our mathematical model to incorporate these factors. This will allow us to build a much more realistic model of how populations increase and decrease in nature.

PREP FOR THE AP® EXAM

FOCUS ON THE BIG IDEAS

SYSTEMS INTERACTIONS: Focus on specific environmental reasons populations cannot grow indefinitely.

54.1 Population growth can be limited by density-independent and density-dependent factors

The growth of populations can be limited by a number of factors. Some factors can reduce the size of a population regardless of its original size. Other factors have a stronger limiting effect as the population approaches the carrying capacity of the environment. In this section, we will look at these two categories of factors that limit population growth.

Density-independent Factors

When population growth is reduced regardless of its density, we say it is limited by **density-independent factors.** Common density-independent factors include droughts, extended periods of cold weather, floods, fires, tornadoes, and hurricanes. When these conditions occur, we typically see a drop in birth rates and an increase in death rates, which together cause the population to decline. These declines typically happen quickly, such as when a hurricane strikes an island and immediately kills many of its plants and animals. Similarly, a tornado can quickly rip through a forest and destroy trees, as you can see in **FIGURE 54.1**. The probability of a given tree being destroyed by a tornado is not affected by the number of other trees around it. With

FIGURE 54.1 Density-independent population control

Natural disasters such as fires, hurricanes, and tornadoes can substantially reduce populations of plants and animals regardless of the population's density. Photo: NurPhoto/Getty Images

density-independent factors such as a hurricane or a tornado, the magnitude of a population drop is not related to the size of the population. Both large and small populations are equally likely to experience large declines, which means the impacts of these factors are density independent.

A classic study of how density-independent factors affect population growth was conducted in Australia on an insect known as the apple thrip (*Thrips imaginis*). The apple thrip

a.

b.

FIGURE 54.2 Apple thrips

Apple thrips cause substantial damage on roses and many fruit trees. (a) Thrips suck the sap from leaves, flowers, and fruits. (b) One symptom of thrip damage appears as patches on the fruit. Photos: (a) Tomasz Klejdysz/Shutterstock; (b) Erik Knutzen/rootsimple.com

can be a major pest on fruit trees and rose bushes because it sucks the sap out of leaves, flowers, and fruits, thereby causing damage on the fruits, as shown in **FIGURE 54.2**. When thrips grow to large population sizes, they cause devastating damage on these economically important plants by causing deformed flowers and fruits as they suck the sap out of the plants. Such damage makes it harder for farmers to sell the fruit. Researchers noticed that the size of the thrip population seemed to be driven primarily by temperature and rainfall during the spring and preceding fall, rather than other factors that are related to population density. In fact, the researchers observed that the insects did not run out of food in the summer, but instead died from droughts that occurred each summer. When they created a mathematical model that included rainfall and temperature—two density-independent variables—they were able to closely predict changes in the thrip population over 15 years in Australia, as shown in **FIGURE 54.3**. In this figure, the blue bars represent the actual size of the thrip population each year while the dashed red line represents the predicted population size when the researchers incorporated density-independent factors such as temperature and rainfall. Knowing how temperature and rainfall affect the pest populations allowed the ability to predict years of high pest populations, which helped guide the best strategies to control the pests.

Density-dependent Factors

When the growth of a population is increasingly limited as its size increases, we say it is subject to **density-dependent factors.** In a classic example of density-dependent factors, Russian biologist Georgii Gause conducted a lab

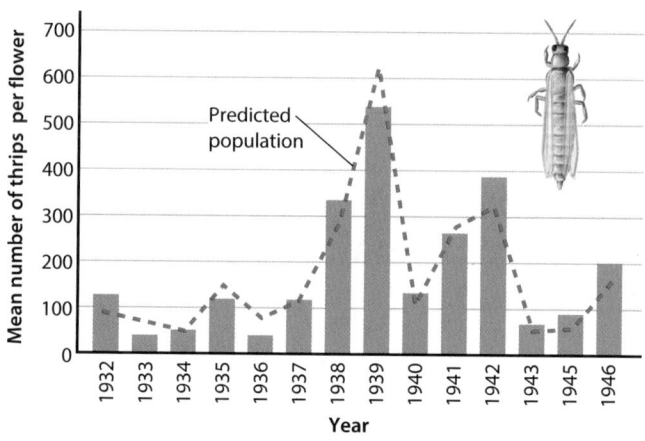

FIGURE 54.3 Density-independent factors in population growth

Populations of apple thrips in Australia exhibited large fluctuations in size over a 15-year period, as shown by the heights of the bars. Large populations occurred when there was abundant rainfall in the spring and preceding fall. The dashed line represents the predicted population size each year when researchers created a mathematical model of population growth that included density-independent factors, such as temperature and rainfall. Data from Davidson J., and H. G. Andrewartha. 1948. "The Influence of Rainfall, Evaporation and Atmospheric Temperature on Fluctuations in the Size of a Natural Population of *Thrips imaginis* (Thysanoptera)." *Journal of Animal Ecology* 17:200–222.

experiment in which he raised two species of single-celled aquatic organisms known as *Paramecium* in test tubes. Each day, he added a constant amount of food and then he tracked the population sizes of the two species for 16 days. As expected from our exponential growth model, both species of *Paramecium* initially experienced exponential growth. Eventually, however, the population sizes began to plateau as the fixed amount of added food began to limit population

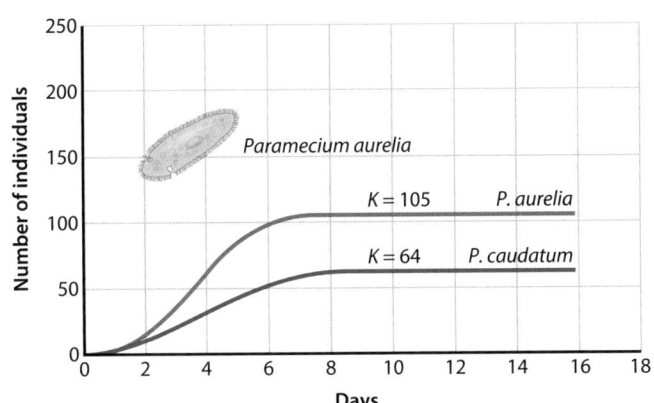

a. Low-food supply

b. High-food supply

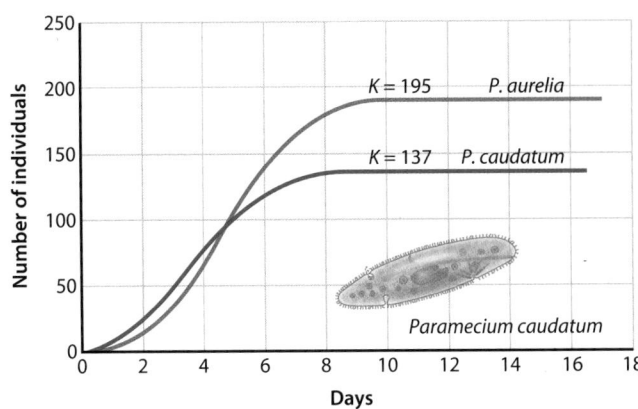

FIGURE 54.4 Density-dependent growth in two species of *Paramecium*

Georgii Gause raised the two species under fixed amounts of food. (a) When the two species were raised under a low-food supply, they initially exhibited exponential growth but then leveled off over time. (b) When the two species were raised under a high-food supply, which contained twice as much food, the populations of the two species once again leveled off over time. However, the leveling off occurred at population sizes that were nearly twice as large as when they were raised under a low-food supply. Data from Gause, G. F. 1932. "Experimental Studies on the Struggle for Existence: 1. Mixed Population of Two Species of Yeast." *Journal of Experimental Biology* 9:389–402.

growth. You can see the results of Gause's experiment in **FIGURE 54.4a**. In this figure, both species achieve a plateau in their sizes, although the population of *P. aurelia* reaches a higher population size than *P. caudatum*.

Gause suspected that the populations reached a plateau because food was a limiting factor. To test this hypothesis, he ran a second experiment in which he doubled the amount of food added to the test tubes each day. When he did this, he found that both *Paramecium* populations once again started with exponential growth but then also leveled off over time. However, this time the populations leveled off at densities that were approximately twice as large as he found in the first experiment, as illustrated in Figure 54.4b.

Gause's experiments are classic because he clearly demonstrated how a limiting resource, in this case food, can limit the size of a population. As the population grows, each individual is competing more intensely for the limited resource. This competition causes birth rates to go down and death rates to go up until the two offset each other and the population

remains stable. As you may recall from the previous module, we refer to this limit on how many individuals can be supported in the environment as the carrying capacity (denoted as K). The carrying capacity is often determined by a limited amount of food, but it can also include other resources, such as a limited number of nesting sites for birds or a limited number of open spots on rocks along an ocean shoreline for barnacles and seaweed to attach. As we will see in the next module, limiting resources combined with predation can cause predator and prey populations to fluctuate in response to changing carrying capacities over time.

✓ Concept Check

1. **Describe** how density-independent factors limit population size and give an example.

2. **Describe** how density-dependent factors limit population size and give an example.

54.2 The logistic growth model includes density-dependent population growth

Now that we understand how density-dependent factors can constrain population growth by the existence of limited resources, we can begin to examine a modified mathematical

model that incorporates the concept of limited resources. To do so, we need to recognize that limited resources set a limit on how many individuals of one species a habitat can support, which is the carrying capacity (K) of the habitat. In this section, we will examine a population growth model that includes the density-dependent effects of the carrying capacity. We will then examine how populations can overshoot their carrying capacity. Finally, we will examine human population growth and whether there are clear capacities.

The Logistic Growth Model

To incorporate carrying capacity into a mathematical model of population growth, we begin with our exponential growth model that we discussed in the previous module:

$$\frac{dN}{dt} = r_{max}N$$

As you recall, this model describes an exponential increase in population sizes over time. To incorporate the idea of a carrying capacity, we need to build on this model by adding some parameters that will slow down population growth as the population increases. We can do this by including the parameter K, which is the carrying capacity, and then determining how large a population is relative to the carrying capacity:

$$\frac{dN}{dt} = r_{max}N\left(\frac{K-N}{K}\right)$$

where:

dt = change in time

N = population size

r_{max} = maximum per capita growth rate of population

K = carrying capacity.

This is the **logistic growth model,** which describes population growth when there is a density-dependent carrying capacity. If we focus on the terms that are in the parentheses, we can see that the value can range from zero to one. When the current population size (N) is much smaller than the carrying capacity (K), the term in the parentheses has a value of approximately one. In this scenario, the few individuals present in the population are experiencing no limits on their population growth, so the model behaves much like our original exponential model. However, when the current population size (N) approaches the carrying capacity (K), the term in the parentheses has a value of approximately zero, which causes the entire right side of the equation to have a value of approximately zero. When $\frac{dN}{dt} = 0$, the population is stable, meaning that it is neither increasing nor decreasing; it is at the carrying capacity of the environment. In this scenario, all individuals are severely limited by the available resources to the point that the birth rate is equivalent to the death rate.

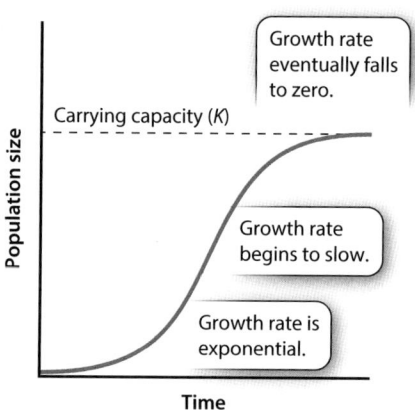

FIGURE 54.5 Logistic growth model

In the logistic growth model, a population initially experiences exponential growth. As the population continues to grow, the carrying capacity begins to limit the growth because resources become more scarce. This causes an increase in the death rate and a decrease in the birth rate. Once the population reaches the carrying capacity of the environment, the population growth rate stabilizes at zero and the population levels off.

When we graph the logistic growth model, we can see the effects of incorporating carrying capacity. As you can see in **FIGURE 54.5**, the growth of the population follows an S-shaped curve. At the beginning the population grows exponentially. Growth then begins to slow as the effects of a carrying capacity limit population growth over time. Eventually, growth falls to zero. This is the same pattern of growth that we observed in Gause's *Paramecium* experiments. In the middle of this S-shaped curve, the curve has its steepest slope, which is when the growth rate of the population is the fastest. To the right of this midway point, population growth begins to slow. You can see how growth rate $\left(\frac{dN}{dt}\right)$ changes by working through the calculations presented in "Analyzing Statistics and Data: The Logistic Growth Model."

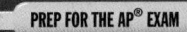
The Logistic Growth Model

To understand how the logistic model of population growth produces an S-shaped curve, we can look at an example using a hypothetical population of rabbits.

PRACTICE THE SKILL

In the population, we will set $r_{max} = 0.5$ and set the carrying capacity at $K = 100$ rabbits. We can then calculate the population's growth rate at several different populations sizes, which tells us how steep the slope of the curve is at different population sizes.

For $N = 10$:

$$\frac{dN}{dt} = r_{max}N\left(\frac{K - N}{K}\right)$$

$$\frac{dN}{dt} = 0.5 \times 10 \times \frac{100 - 10}{100}$$

$$\frac{dN}{dt} = 5 \times 0.9$$

$$\frac{dN}{dt} = 4.5 \text{ rabbits per year}$$

For $N = 20$:

$$\frac{dN}{dt} = r_{max}N\left(\frac{K - N}{K}\right)$$

$$\frac{dN}{dt} = 0.5 \times 20 \times \frac{100 - 20}{100}$$

$$\frac{dN}{dt} = 10 \times 0.8$$

$$\frac{dN}{dt} = 8.0 \text{ rabbits per year}$$

For $N = 50$:

$$\frac{dN}{dt} = r_{max}N\left(\frac{K - N}{K}\right)$$

$$\frac{dN}{dt} = 0.5 \times 50 \times \frac{100 - 50}{100}$$

$$\frac{dN}{dt} = 25 \times 0.5$$

$$\frac{dN}{dt} = 12.5 \text{ rabbits per year}$$

Your Turn

1. Based on the calculations above, determine the population growth rate for $N = 80$, $N = 90$, and $N = 100$.
2. Based on these calculations, what happens to a population's growth rate as it moves from $N = 10$ to $N = 50$ and then from $N = 50$ to $N = 100$?

Population Overshoots and Die-offs

The logistic model suggests that populations approach their carrying capacities very gradually and then become stable. The reality is that populations in nature can often temporarily overshoot their carrying capacity. One reason is that the carrying capacity of the environment is not fixed from month to month or from year to year. Instead, the environment continues to change so the amount of resources such as food varies over time. This becomes even more problematic in large mammals, which breed in the fall when food might be plentiful but do not give birth until spring, when food may be less abundant. These conditions cause the population to dramatically overshoot the environment's carrying capacity, which means the population now has far more individuals to feed than there is food available. When this happens, we commonly observe a large die-off of the population.

A good example of population overshoot and die-off happened in a population of reindeer that was introduced to St. Paul Island in Alaska in 1920. An initial herd of 25 reindeer was introduced to serve as a food source for the island's residents. Since there was plenty of food available to the reindeer and there were no predators, the population exponentially increased, as you can see in the blue line in **FIGURE 54.6** on page 770. If we smooth the trend of the data, as shown in the dashed red line, we see that the population followed a J-shaped curve. By 1938, the herd had grown to nearly 2000 reindeer, which likely far exceeded the island's carrying capacity, and then the population crashed over the next decade to a low of just 8 animals by 1950. Today, hunting is used to keep the herd at about 400 reindeer, which is much more in line with the island's carrying capacity. While this example shows an extreme case of a population

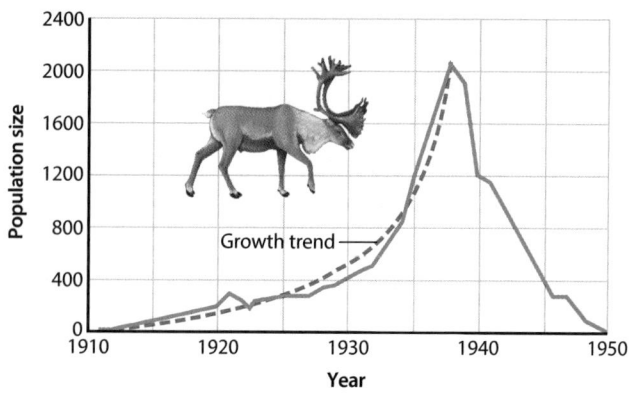

FIGURE 54.6 Overshoot and die-off in reindeer

In 1910, a herd of 25 reindeer were introduced to St. Paul Island. The population experienced exponential growth until 1938, when the reindeer experienced a massive die-off, likely as a result of exceeding their carrying capacity. The solid line shows the raw data while the dashed line shows the general trend of the data in producing a J-shaped curve. Data from Scheffer, V. B. 1951. "The Rise and Fall of a Reindeer Herd." *Scientific Monthly* 73:356–362.

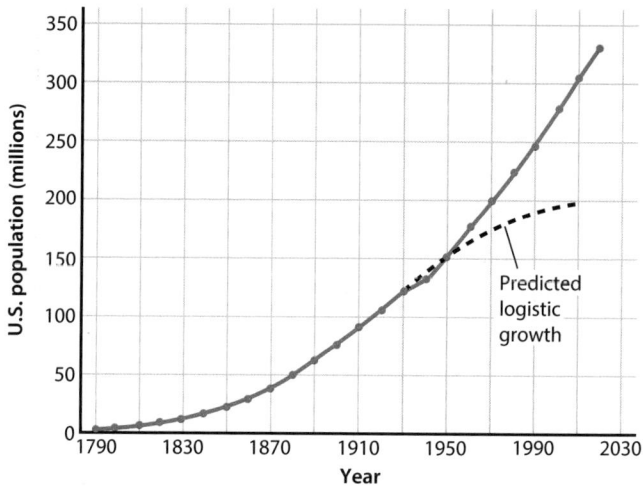

FIGURE 54.7 Human population growth in the United States

Based on U.S. census data from 1790 to 1920, researchers in 1920 used the logistic growth model to predict the carrying capacity of the United States. Their prediction of $K = 197$ million people has been greatly exceeded as a result of increased food production in the United States and increased imports of food from around the world. Data from Pearl, R., and L. J. Reed. 1920. "On the Rate of Growth of the Population of the United States since 1790 and Its Mathematical Representation." *Proceedings of the National Academy of Sciences* 6:275–288; U.S. Census Bureau, http://2010.census.gov/2010census/data/apportionment-pop-text.php.

overshooting its carrying capacity and experiencing a subsequent die-off, many other populations experience smaller overshoots and die-offs, such that the population fluctuates above and below the carrying capacity over time.

Predicting Human Population Growth

Given that the growth of many populations in nature follows the logistic growth model, researchers have been interested in whether human populations also follow the logistic growth model. In the early 1900s, population ecologists were examining U.S. census data from 1790 to 1920. In these census data, the researchers noticed that the U.S. population had exponentially increased in size from 1790 to 1910, but by 1920 the population growth was starting to slow down. Since this pattern closely matched the pattern expected by the logistic growth model, the researchers could estimate that the population in 1910, which was 91 million people, must be about half of the country's carrying capacity. As a result, they used the model to predict that the U.S. carrying capacity was 197 million people, as shown in the black dashed line in **FIGURE 54.7**.

While this carrying capacity was a reasonable prediction in the early 1900s, the U.S. population has far exceeded this prediction, as shown by the red data points in Figure 54.7, with a population of 331 million people in 2020. One of the reasons that the estimated carrying capacity has been far exceeded is that U.S. farmers have more advanced

technologies to produce much more food per hectare than they did 100 years ago. In addition, the researchers were estimating the U.S. carrying capacity based on the assumption that the population could only consume food produced in the United States. However, the modern reality is that the United States imports a great deal of food from other countries, which further increases the number of Americans that can be fed. In addition to these increases in carrying capacity, the rate of population growth has been further increased by medical advances that have dramatically reduced human death rates and by the number of immigrants that have come to the United States.

✓ Concept Check

3. **Describe** how the logistic model of population growth builds on the exponential model.

4. **Describe** the pattern of population growth rate over time in the logistic model.

5. **Identify** what happens when a population overshoots its carrying capacity.

Module 54 Summary

PREP FOR THE AP® EXAM

REVISIT THE BIG IDEAS

SYSTEMS INTERACTIONS: Describe how a population can grow to have a density of individuals that surpasses the system's resource availability.

LG 54.1 Population growth can be limited by density-independent and density-dependent factors.

- When population growth is limited without regard for its population size, we say it is limited by density-independent factors. Page 765

- Common density-independent factors include droughts, extended periods of cold weather, floods, fires, tornadoes, and hurricanes. Page 766

- When population growth is increasingly limited as its population size increases, we say there are density-dependent factors. Page 766

- As the population grows, each individual is competing more intensely for limited resources. This competition causes birth rates to go down and death rates to go up until the two offset each other and the population remains stable $\left(\frac{dN}{dt} = 0 \right)$. Page 767

LG 54.2 The logistic growth model includes density-dependent population growth.

- The logistic growth model describes population growth when there is a density-dependent carrying capacity. Page 768

- The logistic model follows an S-shaped curve, with exponential growth at the beginning followed by the slowing effects of a carrying capacity limiting population growth over time. Page 768

- Populations in nature can often temporarily overshoot their carrying capacity and then experience a die-off. Page 769

- The U.S. population of people has far exceeded its predicted carrying capacity as a result of advances in farming technologies and large amounts of food being imported from other countries. Page 770

Key Terms

Density-independent factor   Density-dependent factor   Logistic growth model

Review Questions

1. Which is a density-dependent factor for populations?

   (A) Hurricane

   (B) Flood

   (C) Fire

   (D) Limited food

2. Which is an expectation of a population exhibiting density-independent growth?

   (A) There is a resource in limited supply.

   (B) Small populations will initially experience rapid growth.

   (C) Large populations can still experience rapid growth.

   (D) There is a carrying capacity that regulates population growth rate.

3. In the logistic growth model, how do birth rates compare to death rates when the population is small versus when the population is at carrying capacity?

   (A) Birth rates = death rates; Birth rates > death rates

   (B) Birth rates > death rates; Birth rates < death rates

   (C) Birth rates = death rates; Birth rates < death rates

   (D) Birth rates > death rates; Birth rates = death rates

4. Which statement is true about population overshoots?
   (A) They can occur because carrying capacity can change over time.
   (B) They can occur because most populations do not have carrying capacities.
   (C) They can occur because most populations do not experience limiting resources.
   (D) They can occur when death rates are exceeding birth rates.

5. Which assumption informed the prediction that the U.S. population would plateau at 197 million people?
   (A) The United States would import large amounts of food.
   (B) The United States had a limited carrying capacity for feeding people.
   (C) The United States would increase its human carrying capacity over time.
   (D) The United States would have substantial immigration.

Module 54
AP® Practice Questions

PREP FOR THE AP® EXAM

Section 1: Multiple-Choice Questions
Choose the best answer for questions 1–4.
Use the following information to answer questions 1 and 2.

Foxes eat a variety of foods, including rabbits and mice. Rabbits eat plants, and mice eat the seeds of plants. Ecologists studying foxes in a prairie ecosystem note that after a particularly rainy spring there was an increase in the prairie grass population and then an increase in the mouse population. During the following summer and fall there was an increased abundance in the fox and rabbit populations.

1. Which of the following statements correctly describes the effect of the grass on the mouse and rabbit populations?
   (A) The increased grass population was a density-independent factor for mice, but a density-dependent factor for rabbits.
   (B) The increased grass population was a density-dependent factor for mice but a density-independent factor for rabbits.
   (C) The increased grass population was a density-dependent factor for mice and rabbits.
   (D) The increased grass population was a density-independent factor for mice and rabbits.

2. Which of the following statements correctly describes the effects of the mouse population on the fox population?
   (A) The increase in mice was a density-dependent factor for the fox population.
   (B) The increase in mice was a density-independent factor for the fox population.
   (C) The increase in mice was neither a density-dependent factor nor a density independent-factor for the fox population.
   (D) The increases in mice had no effect on the fox population.

3. A population of gophers has 1000 individuals with an r_{max} of 0.4 per year and a carrying capacity of 975 gophers. What will be the gopher population size after 2 years? (Use whole numbers in your calculation.)
   (A) The population will be 1000 gophers.
   (B) The population will be 1008 gophers.
   (C) The population will be 984 gophers.
   (D) The population will be 990 gophers.

Use the following figure to answer question 4.

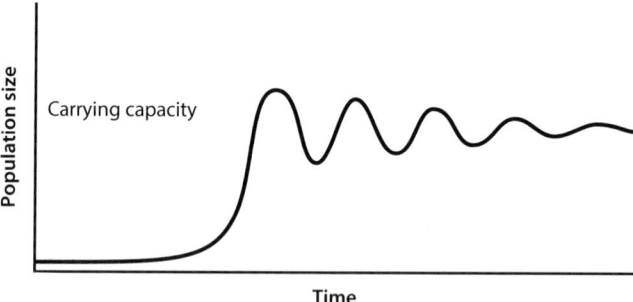

4. The figure depicts the carrying capacity and the population size of an animal species over time. Which of the following best explains why the population size fluctuates less widely with time and reaches the carrying capacity?
   (A) The population runs out of food and stops breeding.
   (B) The population is no longer subjected to density-dependent factors.
   (C) New inputs of resources allow the population to grow at a more uniform rate.
   (D) Over time, the population experiences smaller overshoots and die-offs.

Section 2: Free-Response Question

Write your answer to each part clearly. Support your answers with relevant information and examples. Where calculations are required, show your work.

The data shown in the graph below were collected for the species *Euglena gracilis,* which is a single-celled alga. Cultures of *E. gracilis* were provided with inorganic nutrients, such as nitrogen and phosphate, and the experiment was conducted for 10 days in multiple vessels that could accommodate population sizes 10 times larger than the population sizes observed. To begin the experiment, *E. gracilis* cells were added to the water-based medium containing the nutrients and the vessels were free of competitors, predators, and pathogens. The quantity of light was constant and not growth-limiting. As shown in the graph below, scientists recorded the population size each day and the data from the multiple vessels were averaged.

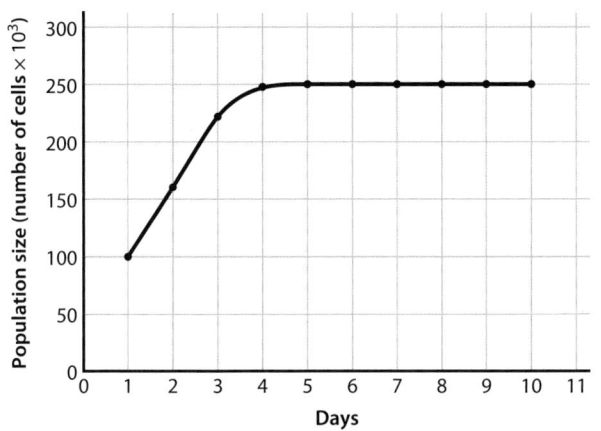

(a) **Predict** how the population size would change by day 20 if, on the 10th day, the population is given two times as many nutrients as was done at the beginning of the experiment. **Justify** your answer.

(b) **Determine** the carrying capacity (*K*) of the *Euglena* for the first 10 days of this experiment.

(c) The scientists conducted the experiment a second time, but this time the amount of light shone on the *E. gracilis* was limiting. The results of this second experiment are shown in the figure below. **Predict** how the carrying capacity of the *Euglena* population would change by day 20 if the researchers reduced the amount of light shining on the growth vessel by 50% on the 11th day of the study. **Justify** your answer.

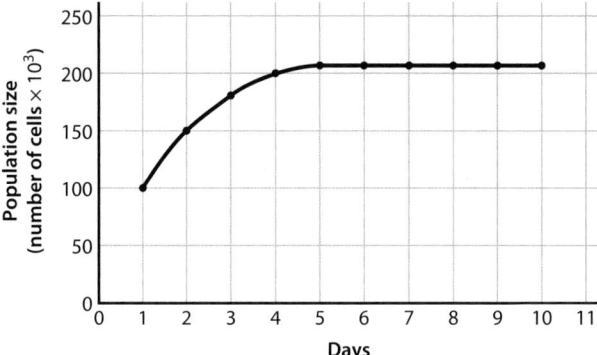

Tutorial 3: Rate and Growth in Population Ecology

As we saw in Module 53, populations fluctuate over time. Biologists interested in understanding population growth might ask questions about why some populations experience rapid or slow growth over time, or why populations have small or large fluctuations in size over time. Analyzing population growth and fluctuations requires the use of two mathematical models that we will review in this tutorial: the exponential growth model and the logistic growth model.

Population Growth

If we assume that a population experiences no immigration or emigration, then population growth is determined by the number of births and deaths occurring in that population. If the number of births is greater than the number of deaths, then that population will increase in size. If the number of deaths is greater than the number of births, the opposite is true and the population will decrease in size. Thus, the change in population size over time is:

$$\frac{dN}{dt} = B - D$$

where:

$\frac{dN}{dt}$ = the change in population size over time

N = population size

dt = change in time

B = birth rate (number of births over time)

D = death rate (number of deaths over time)

Your Turn

A population experienced a birth rate of 7 individuals per month and a death rate of 4 individuals per month. What was the change in population size that year?

Solution

In this example, we see that the birth rate, B, is 7 and the death rate, D, is 4.

$$\frac{dN}{dt} = B - D$$

$$\frac{dN}{dt} = 7 \text{ individuals per month} - 4 \text{ individuals per month}$$

$$\frac{dN}{dt} = 3 \text{ individuals per month}$$

The result tells us that this population increased in size by 3 individuals per month. We know this change was an increase because the birth rate is larger than the death rate and the result is a positive number. If the answer had been −3, that would indicate that the death rate is larger than the birth rate, causing a decrease in population size by 3 individuals.

The question asks us to determine the change in population size that year. To answer this, we can multiply the change in individuals per month by 12, the number of months in a year: $3 \times 12 = 36$. The population increased by 36 individuals that year.

▶ See page 756 for "Analyzing Statistics and Data: Population Growth," which provides an opportunity to practice these concepts in context.

Per Capita Population Growth

While the above simple model of population growth rate considers the birth rate and death rate of a population, it does not tell us how birth and death rates change as the population changes in size. For example, a large population has the potential for many more births, and therefore a higher birth rate, because it has many more breeding individuals. As a result, we need to understand the growth rate of each individual, which we refer to as the per capita growth rate (r). The per capita growth rate is calculated as follows and can be written one of two ways:

$$r = \frac{dN}{dt} \div N \text{ or } r = \frac{B - D}{N}$$

where:

r = per capita growth rate

$\frac{dN}{dt}$ = the change in population size over time or rate of population change

N = population size

As you can see in these equations, per capita growth is the rate of population change divided by the number of individuals in a population. If r is positive, the birth rate exceeds the death rate and the population is increasing. If r is negative, the death rate exceeds the birth rate and the population is decreasing. It is important to take note of the magnitude of the per capita growth rate as well. When r is a large number, the population is increasing or decreasing rapidly. A smaller r value indicates a slower rate of change. A growth rate of 0 indicates that the population is neither increasing nor decreasing. The population size is stable and unchanging because the birth rate offsets the death rate.

The Exponential Growth Model

Knowing the per capita growth rate and the current size of a population allows us to predict future population growth using the exponential growth model:

$$\frac{dN}{dt} = r_{max}N$$

where:

$\frac{dN}{dt}$ = the change in population size over time or rate of population change

r_{max} = the maximum per capita growth rate

N = population size

We call this the exponential growth model because if $r > 0$, the population will grow exponentially. This model assumes that the population does not experience immigration or emigration; that births and deaths occur continuously throughout the year; and that there are no environmental constraints on growth, such as limited food or space. These assumptions are not very realistic in nature, but they provide us with an initial model that we can subsequently modify.

As shown in the graph below, the exponential growth model is represented by a J-shaped curve. At the beginning of the time period, a population grows slowly because there are not many individuals. This is shown by the flat portion of the curve. Over time, as there are more individuals in the population, the population increases rapidly. This increase is represented by a J-shaped curve. Because there are no limitations or constraints acting on the population, individuals will continuously be added to the population at an increasingly fast rate.

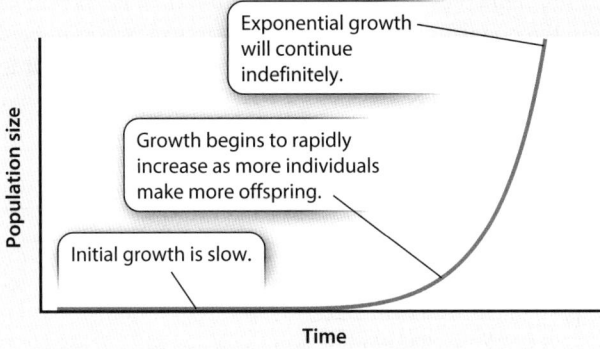

Your Turn

A population of fruit flies is experiencing exponential growth. The population size is 100 flies.

1. What is the size of this population of fruit flies one month from now if $r_{max} = 0.15$?

2. What is the size of this population of fruit flies one month from now if $r_{max} = 0.5$?

Solution

1. We will use the exponential growth model to answer both questions.

$$\frac{dN}{dt} = r_{max}N$$

$$\frac{dN}{dt} = 0.15 \times 100$$

$$\frac{dN}{dt} = 15 \text{ flies per month}$$

Thus, the fly population will increase by 15 additional flies in one month, so the population increases from 100 to 115 flies.

2. In the second example, the population is still 100 flies, but $r_{max} = 0.50$.

$$\frac{dN}{dt} = r_{max}N$$

$$\frac{dN}{dt} = 0.5 \times 100$$

$$\frac{dN}{dt} = 50 \text{ flies per month}$$

In this case, the fly population will increase by 50 additional flies in one month, so the population increases from 100 to 150 flies.

▶ See page 759 for "Analyzing Statistics and Data: Exponential Growth," which provides an opportunity to practice these concepts in context.

The Logistic Growth Model

The exponential growth model is useful for determining changes in size when there are no limitations on population growth. However, this is not typical in nature. Populations may experience a variety of limitations on growth, including limited food and space. When the environment limits population growth, we can modify our exponential growth model to create the logistic growth model. This model includes a term that slows down population growth as the population approaches the carrying capacity (K) of the environment, which is the maximum number of individuals that can be supported in the environment.

The graph on page 776 shows that logistic growth is depicted by an S-shaped curve. Early on, the population experiences exponential growth. Growth is fastest at the middle of the S. However, as the population grows, competition increases as resources become scarcer. This causes population growth to slow and the population experiences a rising death rate and falling birth rate. As the population reaches the carrying capacity, the growth rate stabilizes and birth and death rates offset each other. At this point, the population growth rate is zero.

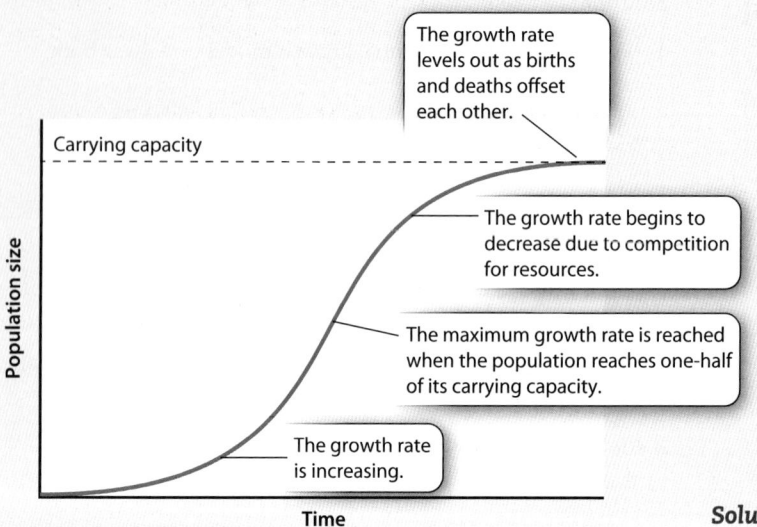

The growth rate levels out as births and deaths offset each other.

The growth rate begins to decrease due to competition for resources.

The maximum growth rate is reached when the population reaches one-half of its carrying capacity.

The growth rate is increasing.

Carrying capacity

Population size

Time

We can understand how the population's growth rate changes with an increasing population size by examining the formula for the logistic growth model:

$$\frac{dN}{dt} = r_{max} N \left(\frac{K - N}{K} \right)$$

where:

$\frac{dN}{dt}$ = the change in population size over time or rate of population change

r_{max} = max per capita growth rate

N = population size

K = carrying capacity

As you can see, the logistic growth model resembles the exponential model, but it contains terms in the parentheses that cause the population to slow its growth as the number of individuals approaches the carrying capacity.

If the population size is much smaller than the carrying capacity, then the term $\left(\frac{K-N}{K} \right)$ has a value of approximately 1. This means that the small number of individuals in the population are not experiencing any limits on their growth and the model behaves like the exponential model. This is seen by plugging a 1 into the logistic growth model for $\frac{K-N}{K}$. The result is the formula for the exponential model:

$$\frac{dN}{dt} = r_{max} N \left(\frac{K - N}{K} \right)$$

$$\frac{dN}{dt} = r_{max} N \times 1$$

$$\frac{dN}{dt} = r_{max} N$$

If the population size is approaching the carrying capacity, then the term $\left(\frac{K-N}{K} \right)$ has a value of approximately 0. This means that the large number of individuals in the population are experiencing harsh limits on their growth due to lessening resource availability.

At this point, the birth rate and the death rate are equal, so the population size is stable at the carrying capacity of the environment.

Your Turn

Georgii Gause's experiments provide a great example of the logistic growth model. We can examine the growth rate of a *Paramecium* population over 10 days at a few different population sizes to see how the growth rate changes.

Calculate the logistic growth rate for population sizes of 25, 100, and 190 individuals, given a maximum per capita growth rate of 1.0 and a carrying capacity of 200. Explain what each growth rate means in terms of the population size.

Solution

To begin, we can see that we have all of the variables needed to use the logistic model formula. We can proceed to plug them in, one population size at a time. It will also be useful to examine each population point on a logistic curve graph to visualize the change in growth rate.

$N = 25$:

$$\frac{dN}{dt} = r_{max} N \left(\frac{K - N}{K} \right)$$

$$\frac{dN}{dt} = 1 \times 25 \times \left(\frac{200 - 25}{200} \right)$$

$$\frac{dN}{dt} = 22 \text{ individuals per day}$$

When the population is at 25 individuals, it is much smaller than the carrying capacity of 200 individuals. As a result, the population is not limited by food. With a growth rate of 22 individuals per day, the population is nearly doubling in size. You can see this growth rate identified at the bottom of the S-curve, where it is just starting to rise.

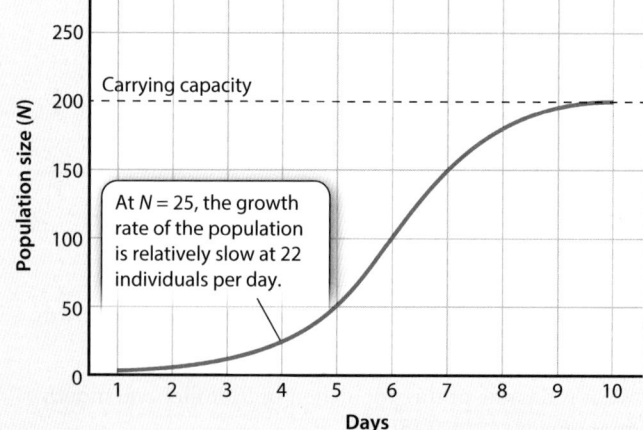

Carrying capacity

At $N = 25$, the growth rate of the population is relatively slow at 22 individuals per day.

Population size (N)

Days

$N = 100$

$$\frac{dN}{dt} = r_{max} N \left(\frac{K - N}{K} \right)$$

$$\frac{dN}{dt} = 1 \times 100 \times \left(\frac{200 - 100}{200}\right)$$

$$\frac{dN}{dt} = 50 \text{ individuals per day}$$

When the population is at 100 individuals, it is halfway to the carrying capacity of 200 individuals. The population is still not limited by food and with a growth rate of 50 individuals per day, the population is rapidly increasing. You can see this growth rate identified near the middle of the S-curve.

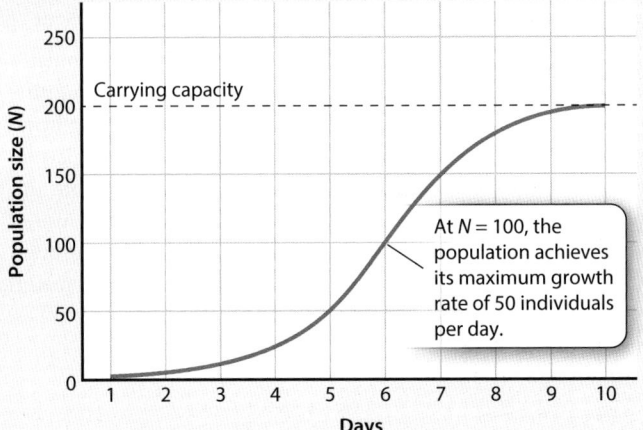

At $N = 100$, the population achieves its maximum growth rate of 50 individuals per day.

$N = 190$

$$\frac{dN}{dt} = r_{max} N \left(\frac{K - N}{K}\right)$$

$$\frac{dN}{dt} = 1 \times 190 \times \left(\frac{200 - 190}{200}\right)$$

$$\frac{dN}{dt} = 10 \text{ individuals per day}$$

When the population is at 190 individuals, it is approaching the carrying capacity of 200 individuals. The population is now limited by food, and the growth rate has slowed to 10 individuals per day. The population is slowly stabilizing to the point where the birth and death rates will be equal. You can see this growth rate identified near the top of the S-curve.

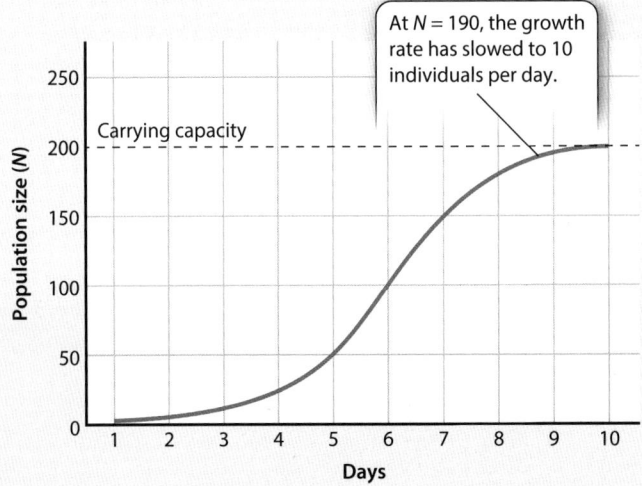

At $N = 190$, the growth rate has slowed to 10 individuals per day.

▶ See page 769 for "Analyzing Statistics and Data: The Logistic Growth Model," which provides an opportunity to practice these concepts in context.

Let's now look at a problem that compares the exponential growth model and the logistic growth model.

Your Turn

A group of scientists is growing a population of bacteria in the laboratory with a fixed amount of food. If r_{max} is 0.9 per hour and the carrying capacity for the bacteria is 500 individuals:

1. Calculate the growth rate for the exponential growth model when N is 64 and when N is 450.

2. Calculate the growth rate for the logistic growth model when N is 64 and when N is 450.

3. Compare and contrast the results given by each model.

4. Which of these models better represents the bacterial colony that the scientists are cultivating and why?

Solution

1. Let's start by calculating the exponential growth rates:

   $N = 64$

   $$\frac{dN}{dt} = r_{max} N$$

   $$\frac{dN}{dt} = 0.9 \times 64$$

   $$\frac{dN}{dt} = 57.6 \text{ bacteria per hour}$$

   $N = 450$

   $$\frac{dN}{dt} = r_{max} N$$

   $$\frac{dN}{dt} = 0.9 \times 450$$

   $$\frac{dN}{dt} = 405 \text{ bacteria per hour}$$

The exponential model shows that at a population size of 64, the number of bacteria added in the subsequent hour is 58. At $N = 450$, that number has increased dramatically to 405 bacteria per hour. With the exponential growth model, the size of the population continues to increase because population growth does not have environmental constraints.

2. Now let's calculate the logistic growth rates:

   $N = 64$

   $$\frac{dN}{dt} = r_{max} N \left(\frac{K - N}{K}\right)$$

   $$\frac{dN}{dt} = 0.9 \times 64 \times \left(\frac{500 - 64}{500}\right)$$

   $$\frac{dN}{dt} = 50 \text{ bacteria per hour}$$

$N = 450$

$$\frac{dN}{dt} = r_{max} N \left(\frac{K - N}{K} \right)$$

$$\frac{dN}{dt} = 0.9 \times 450 \times \left(\frac{500 - 450}{500} \right)$$

$$\frac{dN}{dt} = 41 \text{ bacteria per hour}$$

The logistic model shows that at a population size of 64, the number of bacteria added in the subsequent hour is 50. The population size is still far from reaching the carrying capacity and is experiencing rapid growth. At 450 bacteria, the growth rate slows to 41 bacteria per hour, and will further slow down as the population continues to approach the carrying capacity.

3. Although both models begin with nearly the same rapid growth (58 bacteria per hour compared to 50 bacteria per hour), we can see that as the population continues to increase to 450 individuals, the exponential and logistic rates differ greatly (405 bacteria per hour compared to 41 bacteria per hour). This shows the inherent differences between the models.

When there are no environmental constraints, the population is free to expand and increase indefinitely. However, if there are food or space limitations, the logistic model shows that population growth will eventually slow as it reaches the carrying capacity.

4. Because the scientists are growing the bacteria in laboratory conditions with a fixed amount of food, the scientists are imposing a carrying capacity on the population. As a result, it is likely that the logistic model better reflects the growth rates occurring in the lab.

The two growth models reviewed in this tutorial each have their own specified uses, but all are valuable in determining population growth. The exponential growth model, although it has simplifying assumptions, is useful in introducing us to predicting how populations grow over time, particularly when small populations are far from their carrying capacity. Realistically, populations do not normally undergo indefinite, exponential growth because of the natural constraints that they experience. The logistic growth model accounts for these constraints and offers us a more realistic course of how populations may grow in the future.

Module 55

Community Ecology

LEARNING GOALS ▶**LG 55.1** Community structure includes species diversity and distinct niches.
▶**LG 55.2** Community structure also includes a diverse set of species interactions.
▶**LG 55.3** Species interactions cause direct and indirect effects in communities.
▶**LG 55.4** Communities experience disturbances that cause changes in species abundance and composition.

In the previous module, we focused on populations and how they increase and decrease over time in relation to their carrying capacities. Now it is time for us to move up a level of complexity to think about communities, which we defined as all the interacting species that live in a given area. In doing so, we want to understand how communities are structured in terms of species diversity, considering both the number of species and number of individuals comprising each species. We can also think of community structure in terms of the habitats and niches that species occupy. Communities are also structured in terms of how species interact with each other—including predation, herbivory, and competition—and how this helps us understand the flow of energy throughout the food web. Finally, we can examine how disturbances can dramatically alter the composition and abundance of species in a community and how this composition can change over time.

PREP FOR THE AP® EXAM

FOCUS ON THE BIG IDEAS

ENERGETICS: Consider how the availability of energy resources in the environment affects interactions among populations in a community.

55.1 Community structure includes species diversity and distinct niches

In this section, we begin by discussing how we quantify the diversity of species in a community. We then discuss why a given species lives where it does and why it does not live elsewhere. In doing so, we will distinguish between a habitat and a species' niche.

Species Diversity and Richness

At the species level, biodiversity refers to the variety of genotypes, species, or ecosystems present in an area. When biologists quantify biodiversity, they consider both the number of species, which we refer to as species richness, and the number of individuals representing each species, which we refer to as species evenness. As a result, if two communities both have similar evenness, the one with more species is considered to be more diverse. In addition, if two communities have a similar number of all species, the one with a similar percentage of individuals among all species is considered to be more diverse than a community with one very abundant species. For example, consider two areas with four species of trees, as illustrated in **FIGURE 55.1**. The two areas have the same species richness, but community 1 has high species evenness while community 2 has low species evenness, with one species comprising 70% of all trees. Therefore, community 1 is considered more diverse than community 2.

A common way to quantify the diversity of species in a community is to use an index created by Edward Simpson, which we call **Simpson's diversity index.** Simpson's diversity index incorporates both the number of species in a community and the relative proportion of individuals within each species, which represents a measure of evenness. Using this index, a higher value indicates a more diverse community. To calculate the index, we list the species found in a community and determine the total number of organisms of a particular species (n). By summing

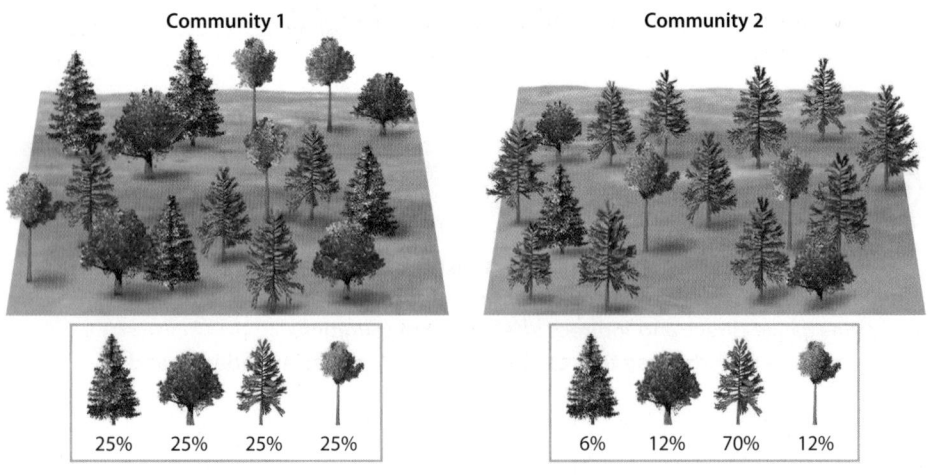

Community 1

25% 25% 25% 25%

Community 2

6% 12% 70% 12%

FIGURE 55.1 Comparing species richness and evenness

The two tree communities both have four species, so they have the same species richness. However, the two communities differ in species evenness, with community 1 having high evenness, with each tree species comprising 25% of the trees, while community 2 has low evenness, with one tree species comprising 70 of the trees.

all individuals across all species, we can determine the total number of organisms of all species (N). With these data, we can calculate Simpson's diversity index (D) using the following formula:

$$\text{Diversity index} = 1 - \Sigma\left(\frac{n}{N}\right)^2$$

where:

  n = the total number of organisms of a particular species
  N = the total number of organisms of all species

In this equation, you can see that the value in the parentheses represents the percentage of individuals that comprise a given species. We then square this percentage and sum the squared percentages for all the species in a community.

As an example, consider **TABLE 55.1,** which contains data on the abundance of mammals in three different communities.

We will begin by calculating Simpson's diversity index for community A:

$$D = 1 - \left[\left(\frac{50}{100}\right)^2 + \left(\frac{30}{100}\right)^2 + \left(\frac{10}{100}\right)^2 + \left(\frac{5}{100}\right)^2 + \left(\frac{5}{100}\right)^2\right]$$

$$D = 1 - [0.25 + 0.09 + 0.01 + 0.0025 + 0.0025]$$

$$D = 1 - [0.355]$$

$$D = 0.645$$

TABLE 55.1 The Abundance of Different Mammal Species in Three Communities

Species	Community A abundance	Community B abundance	Community C abundance
Mouse	50	20	34
Chipmunk	30	20	33
Squirrel	10	20	33
Shrew	5	20	0
Vole	5	20	0
Total abundance	100	100	100

Now we can do the same set of calculations for community B, which has the same number of species as community A, but greater evenness:

$$D = 1 - \left[\left(\frac{20}{100}\right)^2 + \left(\frac{20}{100}\right)^2 + \left(\frac{20}{100}\right)^2 + \left(\frac{20}{100}\right)^2 + \left(\frac{20}{100}\right)^2\right]$$

$$D = 1 - [0.04 + 0.04 + 0.04 + 0.04 + 0.04]$$

$$D = 1 - 0.20$$

$$D = 0.80$$

Finally, we can do the same set of calculations for community C, which has high evenness but a lower species richness:

$$D = 1 - \left[\left(\frac{34}{100}\right)^2 + \left(\frac{33}{100}\right)^2 + \left(\frac{33}{100}\right)^2\right]$$

$$D = 1 - [0.116 + 0.109 + 0.109]$$

$$D = 1 - 0.33$$

$$D = 0.67$$

As you can see in these three examples, Simpson's diversity index produces a higher value for communities that have a higher species richness and for communities that have a higher species evenness. Thus, the index provides us with an objective way to compare the species diversity of different communities, such as communities in different parts of the world or communities that change in their abundance of species over time. For review and practice of Simpson's diversity index, see "Analyzing Statistics and Data: Simpson's Diversity Index."

ANALYZING STATISTICS AND DATA ▶

Simpson's Diversity Index

Biologists often want to measure the biodiversity of species living in a community, such as in a forest, field, or pond. We can calculate the diversity of species in communities using a model known as Simpson's diversity index. This model incorporates both species richness (the number of species) and species evenness (the number of individuals representing each species) to quantify diversity. As we saw in the discussion above, the formula is:

Diversity index $= 1 - \Sigma\left(\dfrac{n}{N}\right)^2$

where:

n = the total number of organisms of a particular species

N = the total number of organisms of all species

A high diversity index value indicates a more diverse community due to either greater species richness or greater species evenness. The diversity index can range from 0 to 1, where 0 is a completely uniform community and 1 is a completely diverse community.

PRACTICE THE SKILL

A team of researchers has conducted a survey of the trees in a plot of land in Yellowstone National Park. Twenty years earlier, this plot was heavily dominated by the lodgepole pine tree and the diversity index was 0.44. The team wishes to know if the diversity in this area has increased, decreased, or remained the same. They collected the following data from their survey of tree species:

Tree species	Abundance
Lodgepole pine	738
Douglas fir	192
Whitebark pine	261
Englemann spruce	313

Calculate the diversity index for this plot of land and compare it to the diversity index from previous years.

To begin, we can assess which variables we have and which we need to find in order to use the diversity index formula. We have the number of each individual tree species surveyed by the researchers, n, but we must find the total number of all tree species, N. This is done simply by adding up each species:

N = 738 + 192 + 261 + 313 = 1504 total trees

With this, we can use Simpson's diversity index formula to calculate the diversity of this plot of land. Remember that the sigma sign indicates that we must sum the abundance of each tree species, divided by the total number of trees.

$$D = 1 - \Sigma\left(\frac{n}{N}\right)^2$$

$$D = 1 - \left[\left(\frac{738}{1504}\right)^2 + \left(\frac{192}{1504}\right)^2 + \left(\frac{261}{1504}\right)^2 + \left(\frac{313}{1504}\right)^2\right]$$

$$D = 1 - (0.241 + 0.016 + 0.030 + 0.043)$$

$$D = 0.67$$

Thus, in the present day, the diversity index of this Yellowstone plot of land is 0.67. Compared to the diversity index of 0.44 in the past, it appears that the forest has become more diverse. The table shows that the lodgepole pine is still the most abundant tree species in this forest, but over time, the abundance of the other species must have increased as well. The higher Simpson's diversity index value indicates that while there has not been an increase in species richness, there has been an increase in species evenness.

Your Turn

Two neighbors living in the northwestern United States decided to begin growing gardens of various wildflowers in their yards in order to attract more pollinators. After several months of growth, they tallied the number of plants of each species to compare the communities. The results are shown in the table. Calculate Simpson's diversity index for each garden to determine which neighbor has a more diverse community of flowers. How do the species richness and evenness of the two gardens compare?

Wildflower species	Neighbor 1 abundance	Neighbor 2 abundance
Scabland penstemon	20	0
Rocky Mountain bee plant	33	86
Bluebell	16	26
White clover	49	67
Bulbous buttercup	61	46
Woods rose	24	54
Common agrimony	28	21
Canada wood sorrel	19	0
Total flowers	250	300

Habitat

A habitat is the physical setting where a species lives. For example, the eastern cottontail rabbit (*Sylvilagus floridanus*) lives in fields throughout eastern North America whereas the moose (*Alces alces*) lives in the deciduous and coniferous forests of Alaska, Canada, the Rocky Mountains, and New England. We typically categorize terrestrial habitats in terms of the dominant plant life that is driven by regional temperature and precipitation, such as tropical grasslands, tropical rainforests, forests, and deserts. In contrast, we categorize aquatic habitats in terms of whether there is salt water or fresh water and whether the water is flowing, such as oceans, lakes, and streams.

Niche

As we discussed in Module 47, a species' niche is a complete description of the role the species plays in its environment and the abiotic and biotic requirements that allow a species to survive, grow, and reproduce. It is helpful to think of the niche in terms of the abiotic conditions a species requires and how interactions with other species may help or hinder a species from persisting in an area.

The abiotic conditions that a species requires are determined by its range of tolerance to different conditions. For example, different species have a specific range of optimal temperatures, as shown in **FIGURE 55.2**. At its optimal temperature range, individuals in the species can survive, grow, and reproduce. Above and below the optimal temperature

range, the species can survive, but not grow and reproduce, because the high or low temperatures are too stressful. Of course, temperature is just one of many abiotic conditions for which a species has an optimal range of tolerance. Other abiotic conditions that affect a species can include pH, nutrients, salinity, oxygen concentrations, and water availability. The full suite of abiotic conditions that allows a species to survive, grow, and reproduce is its fundamental niche.

Although the fundamental niche tells us the abiotic conditions under which a species can live, most species do not

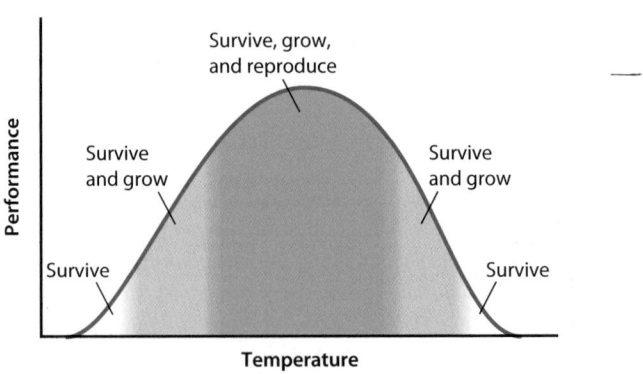

FIGURE 55.2 Range of tolerance

Temperature provides an example of range of tolerance. A species has an optimal temperature range under which it can survive, grow, and reproduce. As temperatures move above and below the optimal temperature, the species loses the ability to reproduce. Under even more extreme temperatures, the species can no longer grow and eventually cannot even survive.

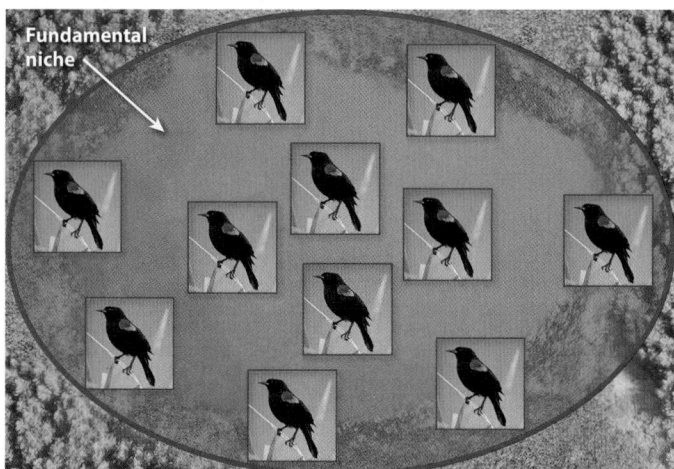

FIGURE 55.3 Fundamental and realized niches

While fundamental niches include the range of abiotic conditions in which a species can live, interactions with other species can further restrict where a species actually lives. (a) When the red-winged blackbird is by itself, it occupies all appropriate nesting sites in the vegetation that emerges from a marsh, from the shallow water along the edge to the deep water in the middle. This represents its fundamental niche. (b) If yellow-headed blackbirds arrive, they drive the red-winged blackbirds out of the nesting sites with the deepest water, so the realized niche of the red-winged blackbird becomes only the shallow-water nesting sites. Photos: (a) Ken Canning/Getty Images; (b) All Canada Photos/Alamy

live everywhere those ideal conditions are found. The reason is because species also have to deal with biotic conditions, such as the presence of predators or competitors, which can prevent the species from persisting in a particular area. The combination of abiotic and biotic conditions under which a species survives, grows, and reproduces is its realized niche.

The red-winged blackbird (*Agelaius phoeniceus*) provides an example of the difference between a fundamental niche and a realized niche. The red-winged blackbird is a common bird that prefers to nest in vegetation that sticks above the water surface of marshes, including both shallow and deep water, as illustrated in **FIGURE 55.3a**. However, if yellow-headed blackbirds (*Xanthocephalus xanthocephalus*) are present, they drive the red-winged blackbirds away from the nesting locations in the middle of the marsh, which have the deepest water. In doing so, they displace the red-winged blackbirds to a realized niche that includes nesting sites around the shallow edges of the marsh, as shown in Figure 55.3b. This example illustrates that the realized niche is often smaller and more restricted than the fundamental niche.

When we compare the niches of different species, we find that some species are considered niche generalists because they can live under a wide range of abiotic and biotic conditions. For example, the gray kangaroo (*Macropus giganteus*) eats a wide range of different plant species, so it is a niche generalist. Other species are considered niche specialists because they only live under a very narrow range of conditions. For example, the panda (*Ailuropoda melanoleuca*) only eats bamboo and not any other plants. Being a niche specialist can be beneficial when a food source is highly reliable throughout the year and year after year. Often the specialist evolves adaptations that make it particularly effective at consuming and digesting one type of food. In contrast, being a niche generalist allows a species to persist as the environment changes over time, such as seasonal changes in the types of food that are available. If some food items are no longer available, the generalist can persist by eating many other food sources. However, the niche specialist faces a major problem if its food sources decline as the environment changes, placing it at an increased risk of future extinction.

✓ Concept Check

1. **Describe** the measures of species diversity that go into Simpson's diversity index.

2. **Describe** the difference between a fundamental niche and realized niche.

3. **Describe** the difference between a niche generalist and niche specialist.

55.2 Community structure also includes a diverse set of species interactions

While a community includes all the species that are interacting within a given area, there are many different ways in which these species can interact. These interactions determine how energy and matter flow in communities among the producers and consumers. Interactions between species can cause negative effects, positive effects, or neutral effects on the populations of these species. As we will see, some of these interactions involve a symbiosis, in which two species live in close proximity.

Predation

Predation is an interaction that involves one species killing another and consuming it. Predation is common in nature, such as lions killing gazelles on an African savanna, grizzly bears catching salmon in a river, and woodpeckers eating insects living in dead trees. Predators can have a large impact on the populations of their prey and their effect sometimes causes predator and prey populations to cycle up and down.

One of the best examples of predator and prey populations cycling involves the Canada lynx (*Lynx canadensis*) as a predator and the snowshoe hare (*Lepus americanus*) as the prey. Beginning in the mid 1800s, both of these species were prized for their pelts and the Hudson Bay Company in Canada would buy the pelts from trappers, so we have a nearly 100-year record of when each species was abundant and when each species was rare. When the ecologist Charles Elton graphed the trapping data, he was struck by the pattern of cycles, which is shown in **FIGURE 55.4**. As you can see in the figure, the lynx and hare populations cycled between high and low numbers about every 10 years. In addition, the cycling of the lynx population lagged behind the cycling of the hare population by 1-2 years.

The hare cycling is caused by changes in the abundance of vegetation that they eat, while the lynx cycling is caused by changes in the size of the hare population. When the hare population is low, the plants they eat have an opportunity to grow more because there are fewer hares to eat them. The increase in the vegetation allows the hare population to increase exponentially over several years. As the hare population grows, more hares are available for the lynx to eat. The abundance of food allows the lynx population to begin increasing its birth rates and decrease its death rates, which

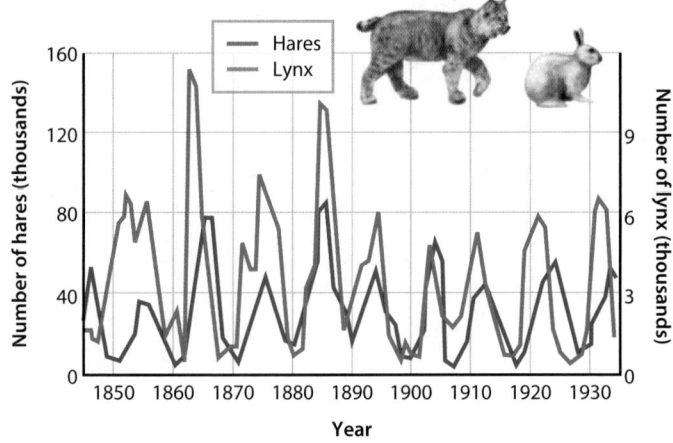

FIGURE 55.4 Population cycles of lynx and hares in Canada

Based on records of pelts purchased from trappers by the Hudson Bay Company, lynx and hare populations cycle up and down about every 10 years. In addition, the lynx population shows a 1- to 2-year lag in its response compared to the hare population. Data from: MacLulich, D. A. 1937. *Fluctuations in the Number of the Varying Hare (Lepus americanus). University of Toronto Studies, Biological Series No. 43.*

causes an exponential increase in the lynx population, albeit with a lag time between the increasing hare population and the resulting increase in the lynx population. Eventually the lynx eat so many hares that the hare population starts to decline. Within 2 years, the decline in hares creates a food scarcity that causes lynx birth rates to go down and death rates to go up, which causes the lynx population to decline. With the decline in lynx predation on the hares and the new increase in plant abundance, the hare population can once again increase, thereby continuing the cycle.

One of the long-term outcomes of predator and prey species interacting is that most species of prey have evolved a variety of antipredator defenses. One of the most common defenses is for prey to hide and become less active to avoid detection by the predator. Some prey species have also evolved camouflage to help blend into to their environment, such as species of insects that resemble twigs and leaves, as shown in **FIGURE 55.5a**. In other species, prey have evolved morphological defenses, such as the spines of porcupines. One of the most fascinating defenses is that of the bombardier beetle (*Stenaptinus insignis*). The beetle has two glands in its abdomen, with each containing a different chemical. When threatened by a predator, the beetle sprays these two chemicals in the predator's direction, as illustrated in Figure 55.5b. The chemicals react to produce a spray that reaches 100°C, which can injure or even kill small predators.

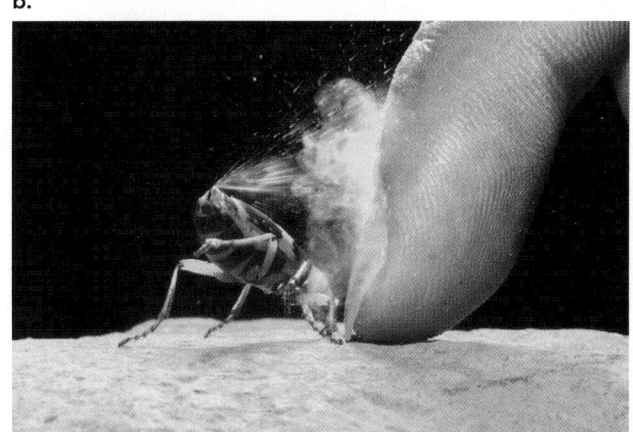

FIGURE 55.5 Prey defenses

Predator–prey interactions have led to the evolution of a suite of prey defenses. (a) Some prey have evolved camouflage, such as katydid species that resemble plant leaves. (b) Other prey have evolved chemical defenses, such as the bombardier beetle that can spray a chemical mixture that produces a nearly 100°C spray into the face of its predator. Photos: (a) Maiapassarak/ Shutterstock; (b) Nature Production/naturepl.com

As discussed in Unit 7, as prey evolve defenses against predators, predators often evolve adaptations to counteract these defenses through the process of coevolution. For example, while most predators cannot successfully attack porcupines due to the abundance of defensive spines, a large species of weasel known as a fisher (*Pekania pennanti*) has coevolved a unique offense. The porcupine has spines all over its body, except for its face and belly. The fisher attacks the porcupine's face to kill it, and then flips the porcupine over and begins consuming the porcupine by attacking the undefended belly and avoiding the spines.

Parasites and Pathogens

Parasitism is an interaction in which one species lives in or on another organism, which we call the host. Parasites generally consume the tissues of their host and reduce the host's growth and reproduction, but they rarely kill the host. Some parasites, such as ticks, fleas, lice, and mites, spend a portion of their life attached to their hosts. Parasitic plants such as mistletoes live on other plants. Other parasites, such as tapeworms, fungi, protists, bacteria, and viruses, live inside their hosts.

Parasites that cause diseases in their hosts are known as pathogens, which we first discussed in Unit 7. Some of the best-known pathogens of humans are those that cause COVID-19, swine flu, bird flu, malaria, and the common cold. Other well known pathogens include a fungus that causes white-nose disease in bats and several species of introduced fungi that cause diseases in crops and wild plants.

When a new pathogen is introduced from another region of the world, it can often have devastating effects on a host species because the host species has not evolved defenses against the pathogen. An excellent example is avian malaria, which is caused by a protist (*Plasmodium relictum*) that was accidentally introduced to the Hawaiian Islands in the early 1900s. The microbe is carried by mosquitoes that feed on the blood of birds. Avian malaria has caused the decline and extinction of several Hawaiian bird species. Similarly, the American elm tree (*Ulmus americana*) was a common species in North America and was a favorite tree to plant along the streets of towns and cities. In the 1930s a fungus that attacks elm trees was accidentally introduced to North America from Europe. Bark beetles that feed on elm trees carry the fungus from one tree to another. Because the North American elm trees have no evolutionary history with the fungus, they have no evolved defenses. As the fungus spread across North America, it killed nearly 95% of the nation's elm trees.

Herbivory

Herbivory is an interaction in which an animal eats an entire plant or part of a plant. Herbivory benefits the animal, known as an herbivore, but harms the plant by reducing its growth and reproduction. Over time, herbivores can have large impacts on plant communities. For example, when researchers fence off areas of the forest from deer, they observe a substantial amount of plant growth, as shown in

FIGURE 55.6 Effects of deer herbivory

When plots of land were fenced to prevent herbivory by deer, the researchers observed a substantial amount of plant growth inside the fenced area. This confirmed the large impact of herbivores on the plant community. Photo: Jean-Louis Martin RGIS/CEFE-CNRS

FIGURE 55.6. Similar to predation, herbivory over long periods of time has favored a variety of anti-herbivore defenses, including spines and distasteful chemicals.

If a new species of herbivore is introduced from a different region of the world, it can have a devastating impact on plants. For example, the emerald ash borer (*Agrilus planipennis*) is native to Asia, and Asian species of ash trees have evolved resistance to the insect. Around 2002, the beetle was accidentally introduced to Detroit, Michigan, on wooden shipping crates. The beetle lays its eggs on ash trees and the larvae feed on the cambium and phloem of the trees, ultimately killing an ash tree within 2–3 years. Once again, because ash trees in North America had no evolutionary history with the beetle, the trees have no evolved defenses. By 2020, the beetle had spread north into Quebec and Ontario and south as far as Georgia, killing hundreds of millions of ash trees along the way. Ash trees are economically important as wood for furniture, hockey sticks, and baseball bats, so the widespread death of ash trees is estimated to have resulted in a $1 billion economic loss.

Competition

Competition is an interaction between species that require the same limited resource. For plants, the limited resource could be water, sunlight, soil nutrients, or physical space. For animals, the limited resource could be food, nesting sites, or physical space. Georgii Gause conducted classic lab experiments on competition with two species of *Paramecium* described in Module 54. In these experiments, the two species were fed yeast and bacteria. In a follow-up experiment, Gause grew the two species of *Paramecium* separately and together to determine if one of the species would outcompete the other for the limited food. As you can see in **FIGURE 55.7**,

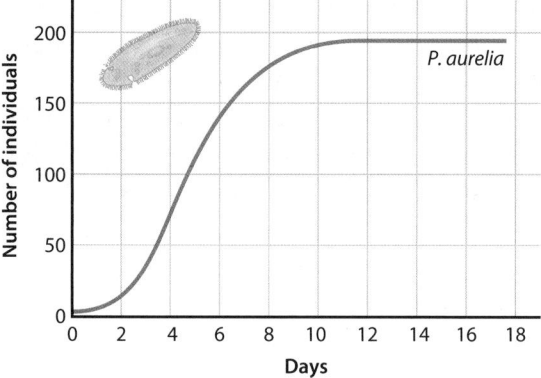

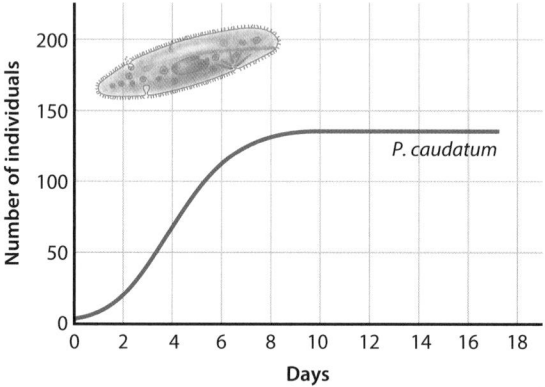

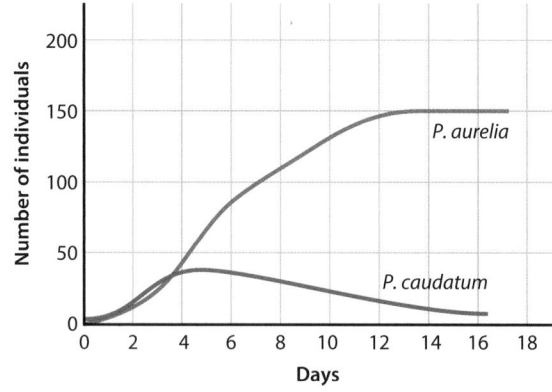

FIGURE 55.7 Competition between *Paramecium* species

When the two species of *Paramecium* were raised separately, shown in parts a and b, both grew well and reached their respective carrying capacities. When the two species were growth together, shown in part c, *P. aurelia* outcompeted *P. caudatum*. Data from G. Gause. 1934. *The Struggle for Existence.* Williams & Wilkins. Baltimore, MD.

each species grew to their respective carrying capacities when raised alone, but *P. aurelia* outcompeted *P. caudatum* when the two were raised together. This led Gause to conclude that two species cannot coexist when they compete for a limited resource.

Competition can occur among individuals within a single species, which we refer to as intraspecific competition, and among individuals of different species, which we refer to as interspecific competition. When competition is interspecific, the competitive outcome can be reversed in the presence of a predator or herbivore if it preferentially consumes the best competitor. For example, abandoned farm fields in New England quickly become dominated by a group of tall wildflowers known as goldenrods. This plant is large enough to outcompete smaller species for sunlight. Occasionally, however, a large population of goldenrod-loving beetles arrive in the field and they eat nearly all of the goldenrod. This allows the other smaller species of wildflowers to gain the resources they need to grow and reproduce.

When two species compete over long periods of time, evolution should favor those individuals with phenotypes that reduce the amount of competition for the limited resource, which we refer to as niche partitioning (also known as resource partitioning). **FIGURE 55.8** visualizes this process. The graph depicts two species of birds that feed on seeds of different sizes, spanning a range from small to large. If the two bird species initially compete for the intermediate-size seeds, evolution will favor individuals of species 1 that feed on smaller seeds and individuals of species 2 that feed on larger seeds. Natural selection will favor individuals of the two species that reduce the amount of overlap in the sizes of seeds that they eat. Over many generations, the two species experience reduced interspecific competition.

Mutualisms

As we discussed in Module 12, some species have evolved to live in a close association with another species, which is known as symbiosis. **Mutualisms** are a special type of symbiosis in which both species benefit in regard to their growth or reproduction. One of the best-known mutualisms is the interaction between flowering plants and their pollinators, which can include insects, birds, and bats. In this case, the plants gain the benefit of achieving breeding through the movement of pollen between flowers, while the pollinators gain a food source from the flower pollen and nectar. Other common mutualisms include lichens and corals, each of which represents two species living in symbiosis with each providing and receiving a benefit. For example, you can often find lichens living on rocks and tree trunks, as shown in **FIGURE 55.9** on page 788. They are comprised of an alga or cyanobacterium and one or two species of fungus; the alga carries out photosynthesis and provides sugars to the fungus, while the fungus provides nutrients to the alga. Similarly, corals live in a mutualism with an alga on coral reefs. The coral is an animal that provides algae with a place to live, while the algae provide valuable sugars to the coral.

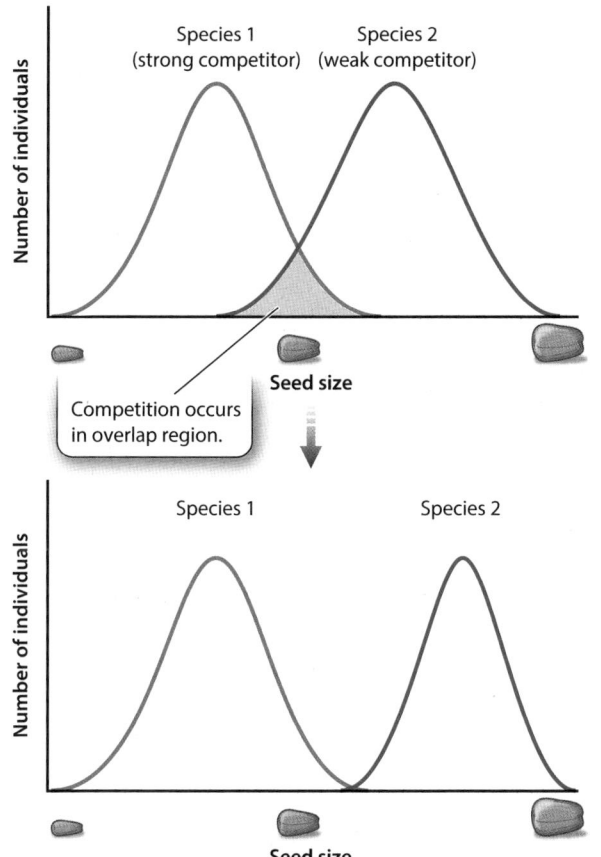

FIGURE 55.8 Resource partitioning

Evolution favors reduced competition between species. (a) Initially, the two bird species show considerable overlap in the seed sizes they eat. (b) Over evolutionary time, the two species evolve to reduce the amount of interspecific competition.

Commensalisms

A **commensalism** is an interaction in which one species receives a benefit while the other species is neither harmed nor helped. For example, gray whales in the ocean often have barnacles attached to them. The barnacles receive the benefit of being carried to plankton-rich water where they can feed on the plankton. However, the whales do not seem

FIGURE 55.9 Lichen mutualism

Lichens are a mutualism between an alga or cyanobacterium and one or two species of fungi. The alga carries out photosynthesis and provides sugars to the fungus while the fungus provides nutrients to the alga. Photo: Ed Reschke/Getty Images

TABLE 55.2 Interactions between Species and Their Effects		
Type of interaction	**Species 1**	**Species 2**
Predation	+	–
Parasitism	+	–
Herbivory	+	–
Competition	–	–
Mutualism	+	+
Commensalism	+	0

to benefit or be impaired by the attached barnacles. Another common commensalism occurs when seeds with tiny hooks become attached the fur or feathers of animals. The plant benefits by having its seeds dispersed to new areas, but the animals are generally not helped or harmed by the attached seeds.

As we have seen, species living in a community can interact in a number of ways that have negative, positive, and neutral effect on the fitness of the two participants. **TABLE 55.2** presents a summary of these interactions.

✓ Concept Check

4. **Describe** why predator and prey populations can cycle over time.

5. **Describe** the difference between parasites and pathogens.

6. **Describe** how predators or herbivores can reverse the outcome of competition.

55.3 Species interactions cause direct and indirect effects in communities

Now that we understand the ways in which pairs of species can interact, we want to consider the full suite of species that live together in communities. In this section, we will examine how a single species can sometimes have a very large effect on the abundance and composition of many other species in the community. We will also consider the interconnectedness of species and how this allows the effect of interactions between two species to have cascading effects through the entire food web.

Keystone Species

While communities can contain a large number of species, sometimes we find that one particular species, known as a **keystone species,** has a disproportionate impact on the other species even though it may not be very abundant. The term "keystone" is a metaphor that comes from the field of architecture, describing an arch built of stones. In such an arch, the keystone is the center stone that carries very little weight of the arch yet is critical in keeping the arch from collapsing. In the same way, a keystone species keeps a community from collapsing into a very different collection of species.

One of the best-known keystone species is the beaver (*Castor canadensis*), which lives in forests throughout North

FIGURE 55.10 A keystone species

Keystone species are those species that have a disproportionate impact on other species even though they may not be very abundant. A common keystone species is the beaver, which builds dams on streams and creates a large pond. When this happens, the beaver completely changes the community of species that are present, including ducks that feed on pond invertebrates and woodpeckers that feed on insects living on the trees killed by the flooding. Photos: (left) Alan & Sandy Carey/Science Source; (right) Larry Lee Photography/Getty Images

America. Beavers construct dams out of branches and mud to convert a flowing stream into a non-flowing pond that floods large areas of the forest, as shown in **FIGURE 55.10**. The water kills many trees that cannot tolerate the flooding. The pond created by the beaver attracts many species with niches that are well adapted to the new habitat, including ducks that feed on pond invertebrates and woodpeckers that feed on insects living under the bark of the dead trees. Thus, despite being not very abundant, a few beavers can completely alter a habitat and the species that live there. Because beavers completely transform the environment in which they live, they are often called ecosystem engineers.

Direct and Indirect Food-Web Effects in Communities

In all the species interactions we have discussed, we have examined how the presence of one species affects the abundance of another species with no intermediate species involved, which we call direct effects. This includes predators reducing the abundance of their prey and pollinators increasing the abundance of a plant species that they pollinate. However, we know that communities can have a large number of species interacting within complex food webs, so we need to realize that the presence of a species not only has direct effects, but its impact can cascade through a food web and affect many other species. When one species affects the abundance of another species by way of one or more intermediate species, we call it an indirect effect.

We saw an example of an indirect effect earlier in our discussion of the cycling lynx and hare populations. When there was a growing population of lynx, they killed more hares and this caused an indirect positive effect on the abundance of the plants that hares eat. Similarly, when wolves (*Canis lupus*) hunt elk (*Cervus canadensis*), the number of elk can decline and then the herbivory of the elk on plants declines. As a result, the presence of the wolf has a negative direct effect on the elk, but a positive indirect effect on the plants. When such indirect effects are initiated by the presence of a predator, we call it a **trophic cascade.**

✓ Concept Check

7. **Describe** the role of beavers as keystone species.

8. **Distinguish** between direct effects and indirect effects of a species on a community.

55.4 Communities experience disturbances that cause changes in species abundance and composition

When we think about the species composition of communities, it is tempting to think that this composition is static. However, the species composition of a community can change dramatically over time as species replace each other in a process known as succession. The process of succession commonly follows a major disturbance.

Succession in Terrestrial Communities

In terrestrial communities, a major disturbance can occur from natural causes such as when volcanoes, fires, or hurricanes destroy most or all the plants and animals in an area. However, a disturbance can also be caused by human activities, such as clearing a forest or field for agriculture, or covering an old landfill with soil. In most cases, the process of succession happens slowly and can take hundreds of years to fully unfold. As a result, biologists have taken advantage of opportunities to follow disturbed areas over long periods of time.

For example, on the island of Krakatau in Indonesia, a volcano erupted in 1883. The eruption blew away three-quarters of the island and destroyed all life by burying it in lava and ash. Researchers immediately began visiting the island to monitor when different species of plants began to appear through the process of succession. When they categorized the plants that arrived on the island based on three modes of seed dispersal, an interesting pattern appeared, which you can see in **FIGURE 55.11**. The first plants to arrive were those with sea-dispersed and wind-dispersed seeds. Only after several plants got a foothold on the ash-covered island did some animal species, such as birds and bats, begin to arrive. When the animals arrived, they inadvertently carried animal-dispersed seeds. This pattern of arrival and change over time is an example of succession.

Another way to observe succession is to study the plants and animals in areas that have been disturbed at different times in the past and then quantify the current composition of species. An excellent example of this happened when Duke University in North Carolina purchased 1900 hectares of what is today known as the Duke Forest. This property

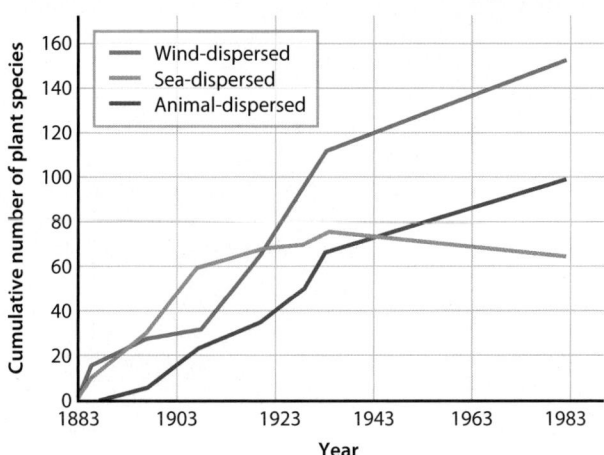

FIGURE 55.11 Succession on Krakatau Island in Indonesia

A volcano on the island that erupted in 1883 blew away three-quarters of the island and covered the remaining land in lava and volcanic ash. Over time, plants began to arrive. The first to arrive were sea-dispersed and wind-dispersed seeds. Later, animals arrived and inadvertently carried with them animal-dispersed seeds. Data from Whittaker, R. J., et al. 1989. "Plant Recolonization and Vegetation Succession on the Krakatau Islands, Indonesia." *Ecological Monographs* 59:59–123.

consisted of a large number of farm fields that had been abandoned at different points in the past. When ecologists visited these sites after the purchase, they noticed that the recently abandoned fields contained grasses and wildflowers, whereas the fields that had been abandoned for more than 100 years were now covered in forests of oak and hickory trees. With records of when each farm was abandoned, the researchers could piece together the process of succession as it transitioned from one composition of plants to another, as shown in **FIGURE 55.12**. Because each animal has a specific niche, the researchers found that the composition of animal species also changed over time as the composition of plant species changed.

Succession in Aquatic Communities

Succession also occurs in aquatic communities. In streams, for example, we can follow the process of succession after major flooding events in which the rushing water destroys all the plants, animals, and algae in the stream. Only rocks and sand remain. In just a matter of days, different groups of producers start to reappear, beginning with a group of algae known as diatoms and

Field	Year 1	Year 2	Years 3-25	Years 25-100	Years 100-200	Years 200+
Crabgrass	Crabgrass, horseweed	Ragweed, heath aster	Broomsedge, perennial flowers, shrubs, pines	Pine forest, hardwood understory	Remnant pines with young oak and hickory trees	Oak-hickory climax forest

FIGURE 55.12 Terrestrial succession in abandoned farm fields

By examining the vegetation present in farm fields that had been abandoned at different time points in the past, researchers in North Carolina could determine how the plant community changed over hundreds of years.

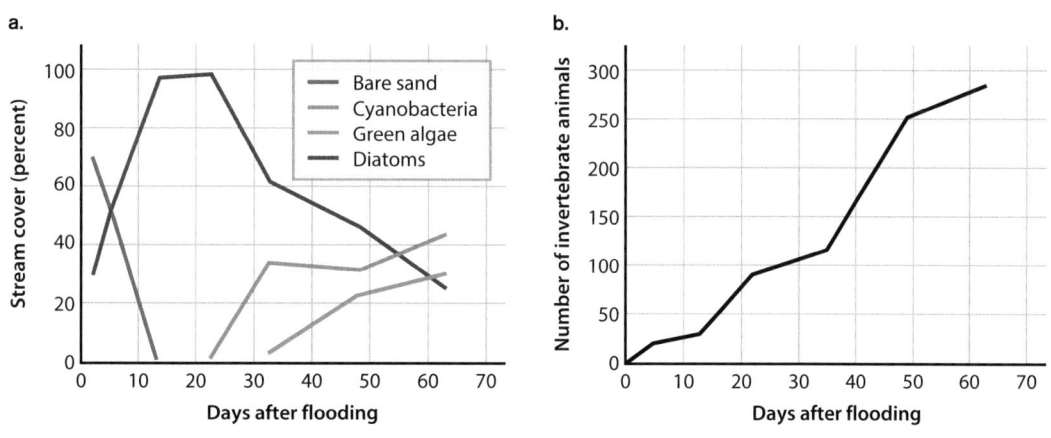

a.

b.

FIGURE 55.13 Stream succession

When the Sycamore Creek in Arizona experienced a major flood, it destroyed all the plants, animals, and algae and left behind bare rocks and sand. (a) Within weeks, different groups of algae began to reappear through the process of succession. (b) Once the algae returned, the number of invertebrate animals rapidly increased over the next 60 days.

followed by cyanobacteria and green algae. You can see these data in **FIGURE 55.13a**. With the return of the various algal groups, the stream invertebrates also started to return over the next 60 days, as shown in Figure 55.13b. In **"Visual Synthesis 8.3: Succession: Ecology in Microcosm"** on page 796 you can see how succession proceeds in ponds and on land, as well as the impacts of human activities on succession.

✓ Concept Check

9. **Describe** how the succession of plants would lead to the succession of animals, using abandoned farms in North Carolina as an example.

10. **Describe** how species richness is expected to change in a stream that experiences a major flood event.

Module 55 Summary

PREP FOR THE AP® EXAM

REVISIT THE BIG IDEA

ENERGETICS: Describe the series of community changes that occurred after the Krakatau volcano eruption, paying particular attention to the energy relationships among the populations in the community.

LG 55.1 Community structure includes species diversity and distinct niches.

- Simpson's diversity index is based on knowing both the number of species in a community and the relative proportion of each species, which represents a measure of evenness. Page 779

- A habitat is the physical setting where a species lives. Page 782

- A species' niche is the range of conditions where it can survive, grow, and reproduce. Page 782

- The full suite of abiotic conditions that allow a species to survive, grow, and reproduce is its fundamental niche. Page 782

- The range of abiotic plus biotic conditions under which a species survives, grows, and reproduces is its realized niche. Page 783

- Some species are categorized as niche generalists because they can live under a wide range of abiotic and biotic conditions whereas other species are considered niche specialists because they only live under a very narrow range of conditions. Page 783

LG 55.2 Community structure also includes a diverse set of species interactions.

- Predation is an interaction that involves one species killing another and consuming it. Page 784

- Predators can have a large impact on the populations of their prey and their effect can cause predator and prey populations to cycle up and down. Page 784

- One of the long-term outcomes of predator and prey species interacting is that most species of prey have evolved a variety of antipredator defenses. Page 785

- Parasitism is an interaction in which one lives in or on another organism, which we call the host. Page 785

- Some parasites cause diseases in their hosts, so we refer to them as pathogens. Page 785

- Herbivory is an interaction in which an animal eats an entire plant or part of a plant. Page 785

- Competition is an interaction between species that require the same limited resource. Page 786

- When competition is interspecific, the competitive outcome can be reversed in the presence of a predator or herbivore that preferentially consumes the best competitor. Page 787

- When two species compete over long periods of time, evolution should favor those individuals with phenotypes that reduce the amount of competition for the limited resource, which we refer to as niche partitioning. Page 787

- Mutualisms are interactions in which both species benefit in regard to their growth or reproduction. Page 787

- A commensalism is an interaction in which one species receives a benefit while the other species is neither harmed nor helped. Page 787

LG 55.3 Species interactions cause direct and indirect effects in communities.

- A keystone species has a disproportionate impact on the other species in its environment, even though it may not be very abundant. Page 788

- When the presence of one species affects the abundance of another species with no intermediate species involved, we call it a direct effect. Page 789

- When one species affects the abundance of another species by way of one or more intermediate species, we call it an indirect effect. Page 789

- When such indirect effects are initiated by the presence of a predator, we call it a trophic cascade. Page 789

LG 55.4 Communities experience disturbances that cause changes in species abundance and composition.

- Succession is a change in the species composition of a community over time. Page 790

- In terrestrial communities, a disturbance can occur from natural causes or from human activities. Page 790

- In streams, we can follow the process of succession after major flooding events in which the rushing water destroys all the plants, animals, and algae in the water. Page 790

Key Terms

Simpson's diversity index
Predation
Parasitism

Competition
Niche partitioning
Mutualism

Commensalism
Keystone species
Trophic cascade

Review Questions

1. Which statement is true about habitats and niches?

   (A) A habitat is the physical setting where a species lives.

   (B) A habitat is the range of conditions where a species can grow but not survive.

   (C) A realized niche is the suite of abiotic conditions that allow a species to survive, grow, and reproduce.

   (D) A fundamental niche is the range of abiotic plus biotic conditions under which a species can survive, grow, and reproduce.

2. Why do some species of predators and prey experience population cycles?

   (A) Predators are able to drive the prey population extinct.

   (B) When prey increase in abundance predators are able to increase in abundance.

   (C) Prey run out of food and the entire prey population dies of starvation.

   (D) When prey decline in abundance the plants they eat also decline in abundance.

3. Which statement is true about competition?

   (A) Natural selection favors those species that increase their niche overlap.

   (B) Competition can be intensified by the presence of predators.

   (C) Natural selection favors individuals that have less resource overlap between two competing species.

   (D) Competition can occur between different species, but not within a species.

4. Which statement is false about Simpson's diversity index?

   (A) It takes into account species richness.

   (B) A high species richness produces a higher index value.

   (C) It includes a measure of species evenness.

   (D) Greater species evenness produces a lower index value.

5. Which statement is true about succession?

   (A) Succession is the gradual replacement of species over time.

   (B) Succession only occurs in terrestrial communities.

   (C) Succession causes a decline in species richness over time.

   (D) Succession only occurs in aquatic communities.

Module 55
AP® Practice Questions

Section 1: Multiple-Choice Questions

Choose the best answer for questions 1–4.

1. A group of students make multiple trips to a local park to record the types and numbers of birds living there. Over the course of a semester, students observe 4 cardinals, 13 robins, and 2 blue jays. What are the species richness and Simpson's diversity index for the birds in this park?

   (A) The species richness is 19 and Simpson's diversity index is 0.377.

   (B) The species richness is 19 and Simpson's diversity index is 0.477.

   (C) The species richness is 3 and Simpson's diversity index is 0.377.

   (D) The species richness is 3 and Simpson's diversity index is 0.477.

2. The rocks of the Pacific Coast of North America are a habitat for a number of marine animal species, including small clam–like animals called mussels. Mussels cling to and live on rocks that are covered with water at high tide and exposed to the air at low tide. Ecologists studying the rocky Pacific Coast observe that mussels proliferate and completely cover the rocks in a particular stretch of coastline, except where sea stars (starfish) are present. On rocks with sea stars, there are few or no mussels. All the rocks in the region have the same exposure to the ocean; sun; and predatory birds, mammals, and fish. Which of the following best describes the observations that the ecologists made?

   (A) The entire rocky Pacific Coast is the mussels' fundamental niche, and the rocks where sea stars are present are the mussels' realized niche.

   (B) The entire rocky Pacific Coast is the mussels' fundamental niche, and the rocks where sea stars are absent are the mussels' fundamental niche.

   (C) The entire rocky Pacific Coast is the mussels' realized niche, and the rocks where the sea stars are present are the mussels' fundamental niche.

   (D) The entire rocky Pacific Coast is the mussels' realized niche, and the rocks where the sea stars are absent are the mussels' realized niche.

Use the following information to answer questions 3 and 4.

A group of researchers study the interaction of two species of small fish in a controlled laboratory environment. The researchers record the population sizes of the two species when they are grown in separate tanks, and when they are grown together in the same tank. They record the population sizes of the two species over 6 months and graph the data collected as in the following graph.

3. Based on the data in the graph, predict what type of interaction the two species exhibit when grown together.

   (A) Parasitism

   (B) Mutualism

   (C) Commensalism

   (D) Predation

4. If the two species were competitors and species A was the better competitor, predict what the curve would look like for species B when living with species A.

   (A) The curve representing species B would have a positive slope but it would always be lower than the curve representing species B living alone.

   (B) The curve representing species B living with species A would be higher than the curve representing species B living alone.

   (C) The curve representing species B living with species A would have the same slope as the curve representing species B living alone.

   (D) The curve representing species B living with species A would have the same slope as it does in the current graph.

Section 2: Free-Response Question

Write your answer to each part clearly. Support your answers with relevant information and examples. Where calculations are required, show your work.

Ecologists are studying a newly discovered valley. They observe that two species of plant, species 1 and species 2, are always present in the fields of the valley. They want to know what type of interaction, if any, the two plants have with one another. To answer this question, the ecologists find three 1–acre plots in a field that are identical in soil type, rain, temperature, the amount of sunshine, and mineral nutrients. The ecologists completely remove the vegetation in each plot and then, in plot A, they plant only species 1. In plot B, they plant only species 2. In plot C, they plant both species. In each plot, the plants are placed an equal distance from each other. After the growing season ends, the ecologists measure the biomass of the plants belonging to species 1 and species 2. The results are listed in the following table.

	Plot A mean dry mass per plant (g)	Plot B mean dry mass per plant (g)	Plot C mean dry mass per plant (g)
Species 1	2022.5	—	3080.7
Species 2	—	1582.1	1579.8

(a) **State** the null and alternative hypotheses for the experiment.

(b) **Identify** the dependent and independent variables of the experiment.

(c) Determine if species 1 and species 2 interact. If the two species do interact, **identify** the type of interaction. **Justify** your answer.

Ecological succession occurs in both aquatic and terrestrial communities. In newly created ponds, the first organisms to arrive are microorganisms and insects, followed by aquatic birds. In some cases, fish can arrive when nearby water bodies overflow into the pond. As the pond ages, it slowly fills in with dead organic material and eventually becomes a terrestrial habitat. The terrestrial

Human impact

Water running off farm fields and home lawns can carry nutrients that cause pond succession to happen more rapidly.

Population growth

When organisms first colonize a pond, their populations increase exponentially until competition for food or space slows further growth. Zooplankton and algae increase exponentially, as do the aquatic insect larvae and tadpoles that feed on them.

Time

Runoff

Excess deposition of N and P

Zooplankton (*Daphnia*)

Water beetles

Fly larvae

Tiny insects

Natural deposition of C, N, and P

Algae

Bacteria

Carrying capacity of environment (*K*)

Population size (*N*)

Time

Colonizing microorganisms, insects, plants, and animals arrive, carried by the wind or by birds. Some colonizers follow cues of light reflecting from water surfaces as well as breezes carrying humid air.

At first, populations grow exponentially, unchecked by competition, predation, or dwindling resources.

Algae, bacteria, and zooplankton

Aquatic insects and other arthropods

Amphibians and fish

Aquatic plants

Grasses and sedges

Herbaceous plants

habitat also undergoes succession. In this example, terrestrial succession moves from grasses to shrubs and then trees. Human activities such as farming and logging speed up or reverse the process of succession.

Species interactions

With increasing diversity comes an increase in the numbers of connections between species as predators, prey, or symbionts. Interactions help to control population sizes.

Succession

As forest communities mature and stabilize, shade-tolerant species may only be disrupted by fire or other disturbances. This can set communities back to earlier successional stages. As soils deepen and store the seeds of sun-loving species, they establish themselves in any openings that are created.

Time

Human impact

Logging, starting fires (or suppressing natural fires), and introducing invasive species can alter the species composition of the pond as well as adjacent fields and forests.

Increased accumulation of dead organic material

As plant and animal succession occurs, death and decomposition leads to the accumulation of organic matter on the pond bottom. When combined with sediments from water running off the land, the pond is gradually converted into a terrestrial ecosystem.

The dominant plants change the physical habitat by shading and spurring organic decay, excluding some animals and plants while permitting others to establish.

Terrestrial animals

Shrubs and trees

Module 56

Biodiversity

LEARNING GOALS ▶LG 56.1 Biodiversity exists at several levels and is driven by multiple processes.

▶LG 56.2 Biodiversity can influence ecosystem organization, productivity, and resilience.

▶LG 56.3 Biodiversity provides benefits to humans and its distribution varies with latitude.

In the previous module, we learned about ecological communities and the many types of interactions that exist among species in a food web. We also learned that we can quantify the diversity of these species in terms of number and evenness by calculating Simpson's diversity index. We also explored the ideas of habitats and niches that species have evolved to fill. We then discussed the various types of species interactions including predation, parasitism, herbivory, competition, and mutualisms. Finally, we explored the importance of keystone species and succession in ecological communities.

In this module, we will explore the concept of biodiversity in more detail by considering biodiversity at several different levels. We will also review the many causes that drive increases in genetic diversity and species diversity. With this understanding, we will explore global patterns in biodiversity and how these patterns have helped to prioritize the protection of biodiversity. Finally, we will discuss the value that biodiversity has for humans and natural ecosystems.

PREP FOR THE AP® EXAM

FOCUS ON THE BIG IDEAS

SYSTEMS INTERACTIONS: Focus on the relationship between ecosystem diversity and the ability of an ecosystem to bounce back after an environmental disturbance.

56.1 Biodiversity exists at several levels and is driven by multiple processes

When we consider biological diversity, also known as biodiversity, we can think about it at multiple levels including genetic diversity, species diversity, and ecosystem diversity. You can see these different levels illustrated in **FIGURE 56.1**. Genetic diversity is the variety of genotypes present within a given population or species. As discussed in the previous module, species diversity is the number of species and the evenness of their abundance. Ecosystem diversity refers to the variety of ecosystems in a region, such as a mix of adjacent aquatic and terrestrial ecosystems. In this section, we will discuss the causes of genetic and species diversity.

Genetic Diversity

As we have discussed in several previous units, including Unit 5 on heredity, Unit 6 on gene expression, and Unit 7 on evolution, genetic diversity arises from multiple processes. At the level of cells, these processes include independent assortment, recombination, and mutation. At the level of populations, these processes include genetic drift, population bottlenecks, founder effects, and natural selection. In all these cases, there is an increase or decrease in the amount of genetic variation over time.

As we have also discussed, high genetic diversity allows populations to respond to changing environmental conditions, so at least a portion of the population can survive the environmental change and continue to persist for many generations to come. For example, the blue moon butterfly (*Hypolimnas bolina*), shown in **FIGURE 56.2**, lives on the South Pacific island of Savaii and more than a century ago the population was parasitized by bacteria. The bacteria kill the male embryos of the butterfly but not the female embryos. As a result, the butterfly population contained 1% males and 99% females. In 2007, however, researchers discovered that the butterfly population rapidly shifted from 1% male to 40% male as a result of a new mutation that is resistant to the parasite. Once the mutation appeared, the mutation favoring male survival spread through 12 generations in a single year, which demonstrated that evolution can occur rapidly once the necessary genetic diversity is present.

a. Genetic diversity

b. Species diversity

c. Ecosystem diversity

FIGURE 56.1 Levels of biodiversity

Biodiversity exists at multiple levels. (a) Genetic diversity is the collection of different genotypes present within a given species. (b) Species diversity is the number of different species and the evenness of their relative abundances. (c) Ecosystem diversity is the number of different ecosystems in a given area.

Another driver of genetic diversity is the process of artificial selection, in which humans purposefully or accidentally favor the survival and reproduction of certain genotypes. As we discussed in Module 41, artificial selection

FIGURE 56.2 The blue moon butterfly

When parasitized by bacteria a century ago, the bacteria killed male embryos, resulting in the species being 1% male and 99% female. When a recent mutation provided resistance to the bacteria, the population rapidly evolved to become 40% male. Photo: Binu Balakrishnan Photography/Getty Images

has produced a tremendous amount of genetic variation in domesticated animals and plants including the diverse breeds of dogs, cattle, sheep, and horses that we see around the world. In many cases, we have breeding records from the past several hundred years, which show how humans have selected for features including size, hair color, meat production, or other heritable features. In plants, humans have also selected for different features. For example, using the wild mustard plant, humans have selected for different stem, leaf, and flower traits as illustrated in **FIGURE 56.3**. This has produced a diversity of genotypes and phenotypes that still make up a single species, including cabbage, cauliflower, kohlrabi, and kale.

In contrast to the intentional breeding of different genetic varieties of animals and plants, sometimes artificial selection is accidental, as we first discussed in Module 41. An excellent example of this is the evolution of antibiotic-resistant bacteria. For instance, tuberculosis is a disease caused by a bacterium (*Mycobacterium tuberculosis*) that infects the lungs and other parts of the body, causing weakness, damaged tissues, and coughing up blood from the lungs. Tuberculosis can be highly lethal, but medical researchers have developed antibiotic medications to kill the bacteria. Although the antibiotic kills nearly all the bacteria, occasionally mutant bacteria have high resistance to the antibiotic, so they survive and proliferate. Over the years, there has been an increase in cases of evolved antibiotic-resistant tuberculosis bacteria and these genotypes require much stronger medications to combat.

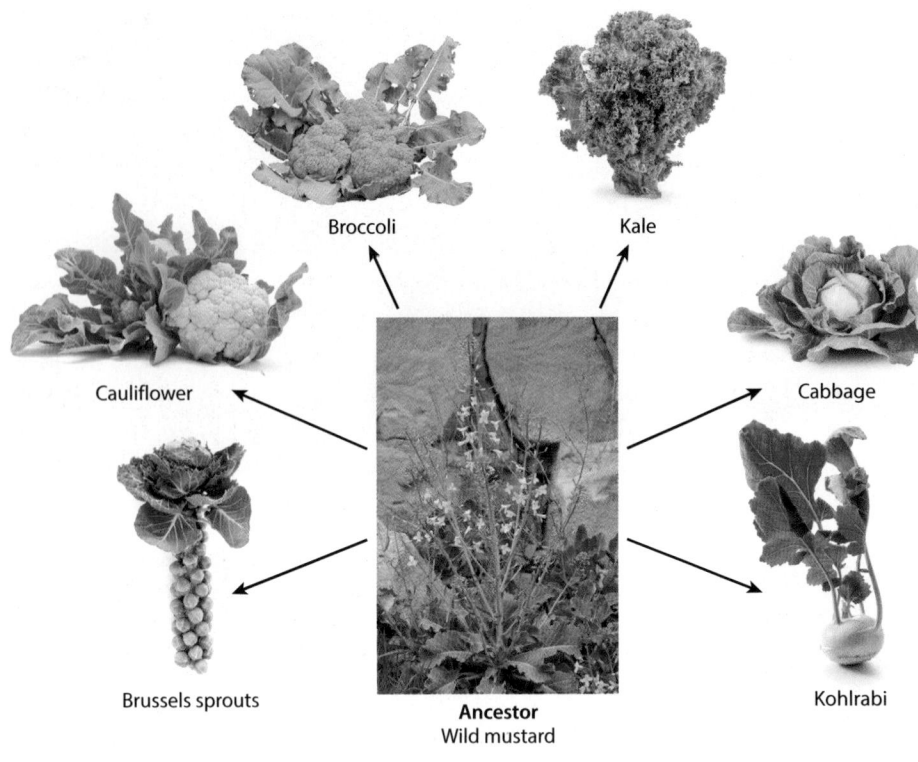

Cauliflower

Broccoli

Kale

Cabbage

Brussels sprouts

Ancestor
Wild mustard

Kohlrabi

FIGURE 56.3 Artificial selection in crop plants

Starting with the wild mustard plant, humans have selected for different stem, leaf, and flower traits to produce a wide diversity of crops that all belong to the same species. Photos: (center) Martin Fowler/Alamy Stock Photo; (clockwise from left) TinasDreamworld/Alamy Stock Photo; white_caty/Getty Images; BoonmeeKim/Shutterstock; nito100/Getty Images; Grygorii Shvets/Dreamstime.com; Melica73/Alamy Stock Photo

This underscores the need for us to be aware of unintentional artificial selection that can alter genetic diversity in ways that benefit the bacteria but harm humans.

Species Diversity

In the previous module, we saw how to quantify species diversity using Simpson's diversity index, which looks at the number of species and the evenness of their relative abundance. We also looked at one reason different places have high or low species diversity: disturbance and ecological succession. These factors produce a change in the composition of species over time, as when a field slowly grows into a forest. A number of other factors also affect species diversity including habitat size, habitat diversity, and keystone species.

For example, if we examine the number of amphibian and reptile species living on islands of different sizes in the Caribbean, we see that there is a positive relationship between the size of the island and the number of species, as shown in **FIGURE 56.4**. One important reason is that larger islands can sustain larger populations of each species because

there are more resources. Larger populations make these species less susceptible to extinction. In addition, larger islands have more distinct habitats, so there are more niches that can accommodate more species on large islands. Finally, larger islands are easier for species to find and colonize.

Another factor that determines species richness is habitat diversity. We mentioned that larger islands have more distinct habitats, but now we want to focus on the diversity of niches within a given habitat. Consider the situation of birds that live in habitats containing only short plants—such as grasses and wildflowers—versus habitats that contains short plants and shrubs and habitats that contain short plants, shrubs, and trees. As we move across this gradient, there are many more ways for bird species to make a living when there is a greater diversity of foliage heights to live in. In a classic study, researchers examined the diversity of foliage height in the United States and Panama by using an index of diversity similar to Simpson's diversity index. They discovered that habitats with a higher diversity of foliage heights also contained a higher

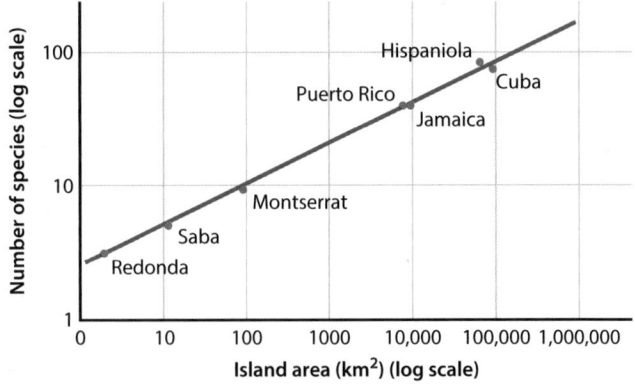

FIGURE 56.4 Effects of island size on species richness

When researchers counted the number of amphibians and reptiles on islands in the Caribbean, they discovered a positive relationship such that larger islands contained more species. Note that the figure uses a log scale for both axes. Data from MacArthur, R. H., and E. O. Wilson. 1967. *The Theory of Island Biogeography.* Princeton, NJ: Princeton University Press.

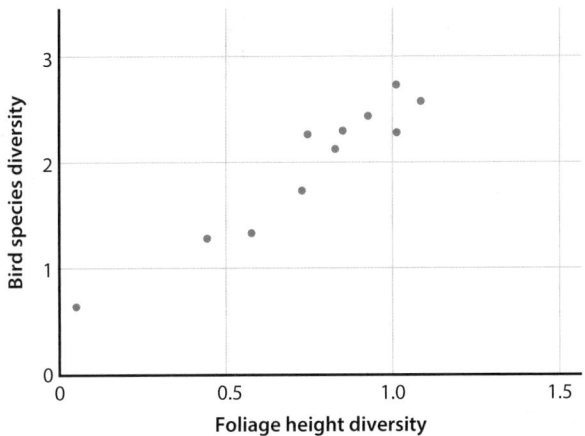

FIGURE 56.5 Effects of habitat diversity on species richness

Researchers examined habitats with a low diversity of foliage height, such as fields with grasses and wildflowers, versus habitats with much greater foliage height diversity, such as a forest containing wildflowers, shrubs, and trees. Habitats with greater foliage height diversity contained more species of birds. Foliage height diversity and bird diversity were quantified using an index of diversity similar to Simpson's diversity index. Data from MacArthur, R. H., and J. W. MacArthur. 1961. "On Bird Species Diversity." *Ecology* 42:594–598.

species richness of birds, as illustrated in **FIGURE 56.5**. Thus, increased habitat diversity helps to drive increases in species richness.

Another important factor in determining species richness is the presence of keystone species, which we introduced in Module 55. As you likely recall, a keystone species is a species that is not very abundant but has a large effect on the species found in a community, such as a beaver that builds a beaver dam. In some cases, the keystone species can have a major effect on species diversity. For example, along the coast of Washington state, we find a community of species that live in the intertidal range of the shoreline, which is the rocky shoreline that is exposed during low tide but submerged in ocean water during high tide. Here we can find about 20 species, including sea stars (*Pisaster ochraceus*), mussels (*Mytilus californianus*), limpets, barnacles, and several species of attached algae, as shown in **FIGURE 56.6a**. However, other coastal areas are dominated by mussels with hardly any other species present, as you can see in Figure 56.6b.

In a classic experiment, researcher Robert Paine suspected that the predatory sea star might be a keystone species in this community. To test this hypothesis, he removed the predatory sea stars from a portion of the intertidal habitat that contained a high diversity species, including the sea star, and then observed how species diversity changed over time compared to a control area where sea stars were not removed. You can see the results of his experiment in **FIGURE 56.7**. Removing the predatory sea stars had a large effect within just a couple of years, reducing species diversity from about 18 species down to just 1 species. The one remaining species was the mussel, which turned out to be such an excellent competitor for rock space that it quickly crowded out all the limpets, barnacles, and attached algae. In the control area that still contained sea stars, there was a small increase in species richness over the decade. The

a.

b.

FIGURE 56.6 Species diversity in the intertidal zone

When Robert Paine observed intertidal communities off the coast of Washington state, he found that (a) some sites contained a high diversity of species, while (b) other sites were dominated by a single species of mussel. Photos: (a) Gary Luhm/DanitaDelimont/Newscom; (b) PePoP Images/Alamy Stock Photo

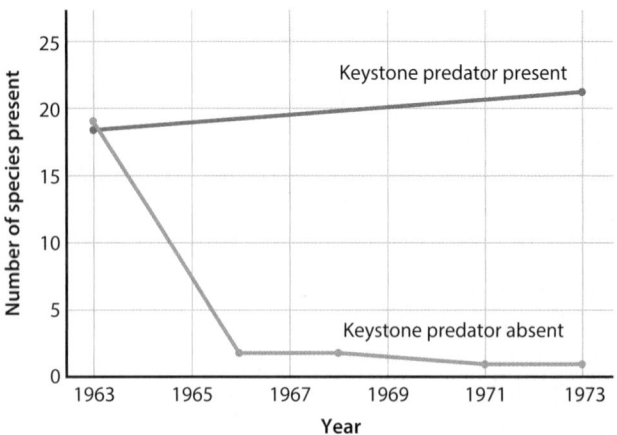

FIGURE 56.7 Effects of a keystone species on species diversity

For intertidal communities off the coast of Washington state, one can often find a high species diversity. If the predatory sea stars are removed, species diversity sharply declines to just one species while the control area does not decline. The one remaining species is the mussel, which is a superior competitor and crowds all other species that are attached to the rocks. When the sea star is present, it preferentially eats the mussels, which allows the other species of attached animals and algae to occupy space on the rocks. Data from Paine, R. T. 1974. "Intertidal Community Structure: Experimental Studies on the Relationship between a Dominant Competitor and Its Principal Predator." *Oecologia* 15:93–120.

reason for this difference is that the sea star prefers to eat mussels, so when the sea star is present it preferentially consumes the best competitor, thereby allowing all the other species to occupy the habitat.

✓ Concept Check

1. **Describe** the difference between genetic diversity and species diversity.

2. **Identify** three processes that can increase the genetic diversity of a species at the level of cells.

3. **Describe** the factors that can affect species diversity in an ecological community.

56.2 Biodiversity can influence ecosystem organization, productivity, and resilience

Biodiversity can have substantial effects on ecosystems. In this section, we will examine how different levels of biodiversity affect the organization of ecosystem, such as when a keystone species is lost from the system. Higher biodiversity can also increase the productivity of an ecosystem by including species with a greater diversity of niches. Finally, increased biodiversity can increase the resilience of ecosystems that experience natural disturbances.

Biodiversity Effects on Ecosystem Organization

The diversity of species can have substantial effects on the organization of ecosystems. For example, a very species-poor ecosystem may only contain producers. As species diversity increases, the ecosystem can include primary consumers, secondary consumers, and tertiary consumers. The organization of an ecosystem can also be strongly impacted when a keystone species is removed from an

ecosystem. As we have discussed, keystone species have a substantial impact on the organization of food webs, so when a keystone species declines in abundance or becomes extinct in an ecosystem, we expect a large change in the ecosystem because no other species in the ecosystem can serve in that role.

For example, sea otters live along the western coast of North America, where they eat sea urchins. The sea urchins consume large amounts of algae known as kelp, which can grow several meters tall. These underwater kelp forests provide protective habitat for many species of young fish. When sea otters were hunted to near extinction during the 1700s and 1800s, the sea urchin population dramatically increased since they were no longer being eaten by the otters. As the sea urchins became abundant, they consumed much of the kelp, thereby removing the protective habitat for the many species of small fish, as you can see in **FIGURE 56.8**. In short, the removal of a single keystone species completely changed the organization of the coastal ecosystem. Today, sea otters are protected from hunting, but new diseases are currently threatening sea otter populations, and these diseases pose the threat of causing another major reorganization of the ecosystem.

a.

b.

c.

FIGURE 56.8 Effects of keystone predators being removed from an ecosystem

Sea otters play a key role in determining the species diversity of coastal ecosystems in western North America. (a) Sea otters are large consumers of sea urchins. (b) When sea otters are present, they keep sea urchin numbers down and the habitat contains an underwater kelp forest. (c) When sea otters became scarce due to overhunting in the 1700s and 1800s, the sea urchins increased in abundance and consumed most of the kelp. This reduction in kelp removed the refuge habitat that many young fish in these coastal areas depend on. Photos: (a) A. Friedlaender/ UWFWS MA186914-2; (b) Brent Durand/Getty Images; (c) Photo by Grant Callegari/Hakai Magazine

Biodiversity Effects on Ecosystem Productivity

Species diversity also affects the function of ecosystems because of its role in ecosystem productivity, which is the amount of biomass accumulated by all species of producers in a year. A given ecosystem can vary widely in how many species of producers are present, from just a few species to hundreds of different species. As a result, we can ask whether having more species allows ecosystems to produce the same amount or a higher amount of total producer biomass compared to having fewer species in a given area.

To answer this question, researchers in Minnesota conducted a large experiment in a field cleared of all the existing plants using herbicide applications and fire, and sectioned off hundreds of large plots within the field. They then manipulated the diversity of grassland species that were allowed to grow in each plot, adding seeds of different species ranging from 0 to 32. After seeding the plots, they spent the summer weeding the plots to remove any additional species that appeared. At the end of the growing season, they collected all the plants in each plot and measured their total biomass. As you can see in **FIGURE 56.9** on page 804, as the number of plant species increased, the total biomass of producers dramatically increased and then began to level off.

Why does species diversity have such a large impact on the productivity of ecosystems? As the number of species increases, we increase the diversity of niches that are present. Some species have niches that allow them to photosynthesize when sunlight is abundant while others can photosynthesize when sunlight is scarce, such as plants that live under tall trees that cast a lot of shade. Similarly, some species of producers do most of their growing early in the spring while others grow more during the summer or fall. In addition, some species of producers are good at obtaining and storing nutrients and water from the shallow region of the soil whereas other species obtain and store nutrients and water from deeper soils. Given the diversity of niches among producer species, an increase in the number of species allows a more complete use of the available sunlight, water, and soil nutrients. This translates into an increase in the productivity of the ecosystem compared to ecosystems with fewer species of producers.

We can see another example of biodiversity affecting ecosystem productivity by considering how soil fungi affect plant growth. For instance, there is a special group of fungi, known as mycorrhizal fungi, that make their living in close connection with the roots of many plants. The fungi are mutualists; they provide soil nutrients to the plants while the plants provide the fungi with the products of photosynthesis. To determine the importance of fungal diversity in a grassland ecosystem, researchers manipulated the number of fungi species in the soil. At the end of the growing

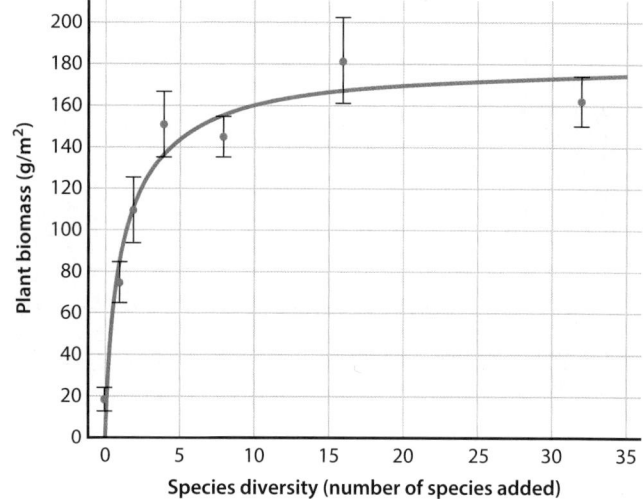

FIGURE 56.9 Effects of plant diversity on ecosystem productivity

In the grasslands of Minnesota, researchers manipulated the number of plant species living in each plot, ranging from 0 to 32. At the end of the growing season, they harvested all the plants in each plot and measured the total plant biomass. The graph shows that as species diversity increased, there was a dramatic increase in the total plant biomass of the field plots. Error bars are calculated as ±1 standard error on either side of the mean. Photo: Jacob Miller. Data from Tilman, D., et al. 1997. "The Influence of Functional Diversity and Composition on Ecosystem Processes." *Science* 277:1300–1302.

season, they measured the biomass of plant roots and shoots as well as the amount of phosphorus that the plants were able to extract from the soil with the assistance of the fungi. As you can see in **FIGURE 56.10**, as the number of fungal species increased in the soil, the plants grew a larger biomass of roots, grew a larger biomass of shoots, and extracted more phosphorus from the soil. This outcome is the result of different fungi having different niches and different abilities to help the plants improve their growth. The more species of soil fungi present, the more the plants can grow and the more productive the ecosystem becomes.

Biodiversity Effects on Ecosystem Resistance and Resilience

We can also consider how biodiversity affects the stability of ecosystems in terms of their resistance and resilience. Resistance is the ecosystem's ability to not be impacted by an environmental disturbance, whereas resilience is the ecosystem's ability to bounce back from an environmental disturbance. A nice example of how species diversity affects the stability of an ecosystem comes from the large grassland plots in a Minnesota field that we discussed earlier, in which researchers manipulated the diversity of plant species. In plots containing 1, 2, 4, 8, or 16 species of plants, the researchers tracked the abundance of the plants and more than 700 species of insects for 11 years to determine how seasonal

and annual variation in environmental disturbances, such as drought, affected the stability of the ecosystem. We can see their data on the stability of herbivore species richness in **FIGURE 56.11a** and the stability of predator and parasitoid species richness in Figure 56.11b. In both figures, we see that increases in plant species richness caused an increase in the stability of the herbivores that depend on these plants and the predators and parasitoids that depend on the herbivores.

We can also see the effects of biodiversity in human-constructed ecosystems, such as agricultural fields used to grow crops. The traditional method of raising a crop is to grow a single genotype, which we often refer to as a crop variety. Different varieties of crops such as wheat, corn, and rice are bred to perform well under certain climates around the world. However, planting a single variety also presents a risk if an herbivore or pathogen arrives and is able to attack the crop. If the herbivore or pathogen grows well on one individual plant, the population can spread rapidly and decimate the crop. However, what would happen if the field contained higher genetic diversity due to planting a mixture of varieties?

Researchers in China answered this question by working with rice farmers to plant mixtures of rice seed varieties. A frequent challenge for rice farmers is a fungal pathogen that attacks the rice and substantially lowers the rice production of a field. To combat the fungus, the farmers have to purchase and apply fungicides. However, when farmers planted

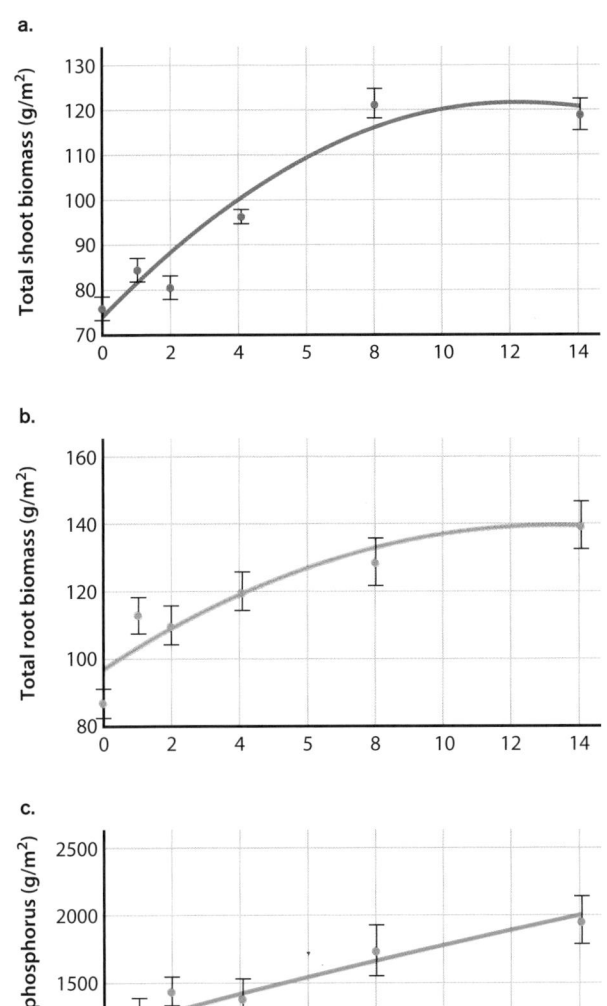

FIGURE 56.10 Effects of fungal species richness on ecosystems

In a grassland ecosystem, researchers manipulated the number of mutualistic fungi in the soil. As the number of fungal species increased, there was an increase in (a) total root biomass, (b) total shoot biomass, and (c) the total amount of phosphorus that the plants were able to extract from the soil. Error bars are calculated as ±1 standard error on either side of the mean. Data from van der Heijden, A. G. A., et al. 1998. "Mycorrhizal Fungal Diversity Determines Plant Biodiversity, Ecosystem Variability, and Productivity." *Nature* 396:69–72.

a diversity of seed varieties, they found that their fungus problem declined dramatically, resulting in an 89% increase in rice production. The reason for this improvement is that the mixture of rice varieties included genotypes that had both high and low resistance to the fungal pathogen. This mix of genotypes made it more difficult for the fungus to spread from one rice plant to another. With such an increase

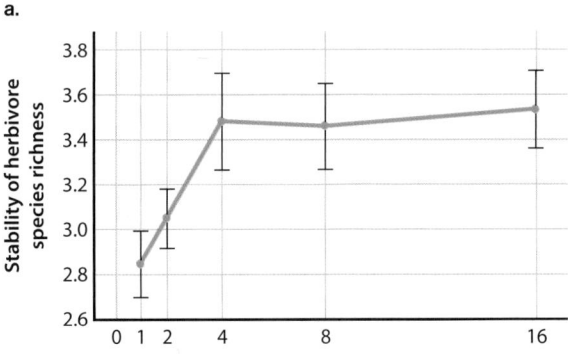

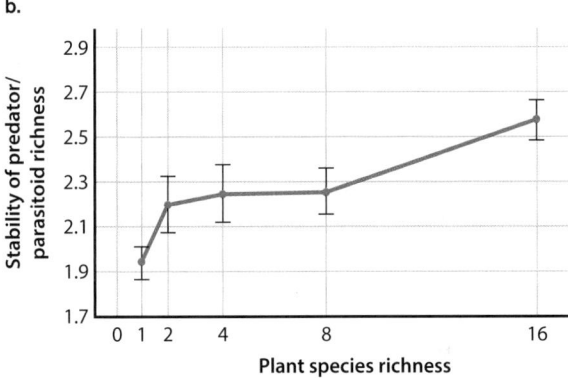

FIGURE 56.11 The effects of plant species richness on the stability of a grassland ecosystem

Researchers manipulated the number of plant species in a grassland in Minnesota. (a) They found that grassland plots containing more plant species caused greater stability in the number of herbivore species over 11 years. (b) They also discovered that grassland plots containing more plant species caused greater stability in the number of predator and parasitoid species. (Stability was quantified as the inverse of the coefficient of variation of species richness, which is the standard deviation divided by the mean.) Error bars are calculated as ±1 standard error on either side of the mean. Data from Haddad, N. M. 2011. "Plant Diversity and the Stability of Foodwebs." *Ecology Letters* 14:42–46.

in productivity, the farmers no longer needed to apply the fungicides, which saved them time and money. Although selection favors the fungus evolving to attack the resistant rice varieties, the mixture of rice varieties is expected to slow the rate of fungal evolution since fungal infections on different rice strains will evolve different strategies.

✓ Concept Check

4. **Describe** how the loss of sea otters as a keystone species affected ecosystem organization.

5. **Describe** how biodiversity affects ecosystem productivity.

6. **Describe** the relationship between biodiversity and the resilience of ecosystems.

56.3 Biodiversity provides benefits to humans and its distribution varies with latitude

Given our discussion of the drivers of biodiversity and the global patterns of its distribution, you might ask why it matters. In this section, we will discuss the values of biodiversity to humans. We will also discuss how biodiversity is distributed around the world and how this distribution, combined with human threats, has led to the identification of biodiversity hotspots that help guide conservation priorities.

Benefits to Humans

The benefits of biodiversity to humans, sometimes referred to as ecosystem services, include both economic values and intrinsic values. Examples of economic values include the natural products that humans use, including lumber, fur, meat, and crops. Medications are a particularly important product from nature. For example, the anticancer drug Taxol comes from the Pacific yew tree (*Taxus brevifolia*), which produces the chemical in its bark to help fend off attacks by harmful fungi. This drug has generated $1.6 billion in annual sales and saved thousands of lives. This is just one of more than 800 pharmaceuticals that come from nature. Beyond these natural products, biodiversity also provides benefits by helping to control floods; filtering our water to make it cleaner; and taking CO_2 out of the air, which helps slow down the rate of global warming caused by higher CO_2 concentrations in the atmosphere. Finally, biodiversity also provides recreational activities such as hiking, fishing, and bird watching as well as the critical service of pollinating wild and domesticated plants.

In addition to these many economic benefits, biodiversity also has an intrinsic value to many people. For example, the effort to bring the bald eagle back from the brink of extinction was motivated not by economics, but by a sense of moral obligation since it is the national bird of the United States.

Latitudinal Diversity Patterns

One of the most striking patterns of species diversity is that species richness is highest near the equator and lowest near the poles. For example, if you were to walk through a forest in the eastern United States, you might see 10 to 20 different species of trees. However, if you took a similar walk in a tropical rainforest in the Amazon, you would see 300 to 400 different species of trees. If we look at a map of species richness for vascular plants, shown in **FIGURE 56.12**, you can see this pattern of more species near the equator and fewer species near the poles. You can also see that the large deserts, such as the Sahara Desert in northern Africa and the many deserts of central Australia, have lower species richness because they receive so little rainfall that there is not a lot of vegetation or habitat diversity to support a larger number of species.

A similar pattern of species richness can be found in the distribution of animal species. For instance, if we look at

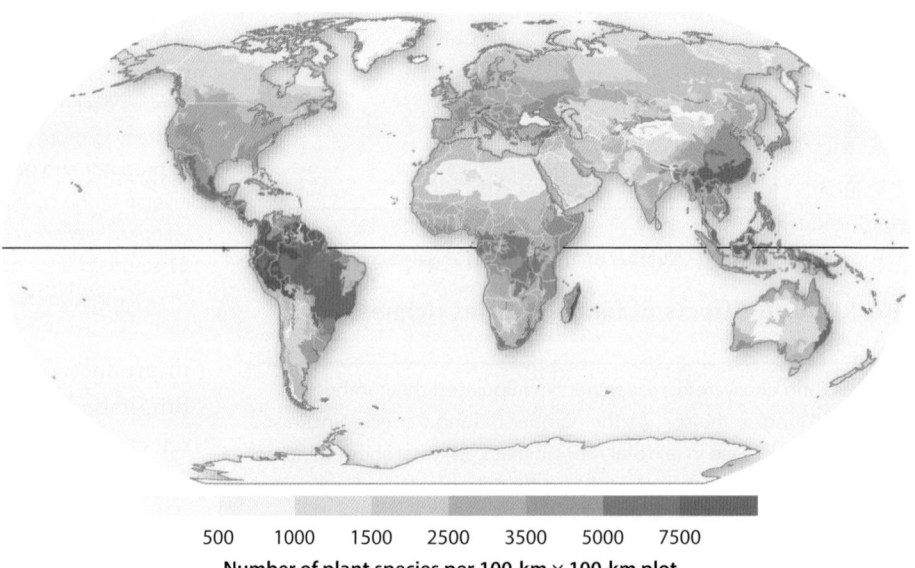

500   1000   1500   2500   3500   5000   7500
Number of plant species per 100-km × 100-km plot

FIGURE 56.12 Species richness patterns in vascular plants

The number of vascular plant species is highest near the equator and lowest near the poles. In addition, species diversity is lower in the large deserts, including northern Africa and central Australia. Data from Mutke, J., and W. Barthlott. 2008. "Biodiversität und ihre Veränderung im Rahmen des Globalen Umweltwandels: Biologische Aspekte" (Biodiversity and Its Change in the Context of Global Environmental Change: Biological Aspects). In *Biodiversität* (Biodiversity) (Ethik in den Biowissenschaften–Sachstandsberichte des DRZE, Bd.5) (Series: Ethics in the Life Sciences: DRZE Expert Reports, vol. 5), edited by D. Lanzerath, J. Mutke, W. Barthlott, S. Baumgärtner, C. Becker, and T. M. Spranger, 63. Freiburg i.B: Alber.

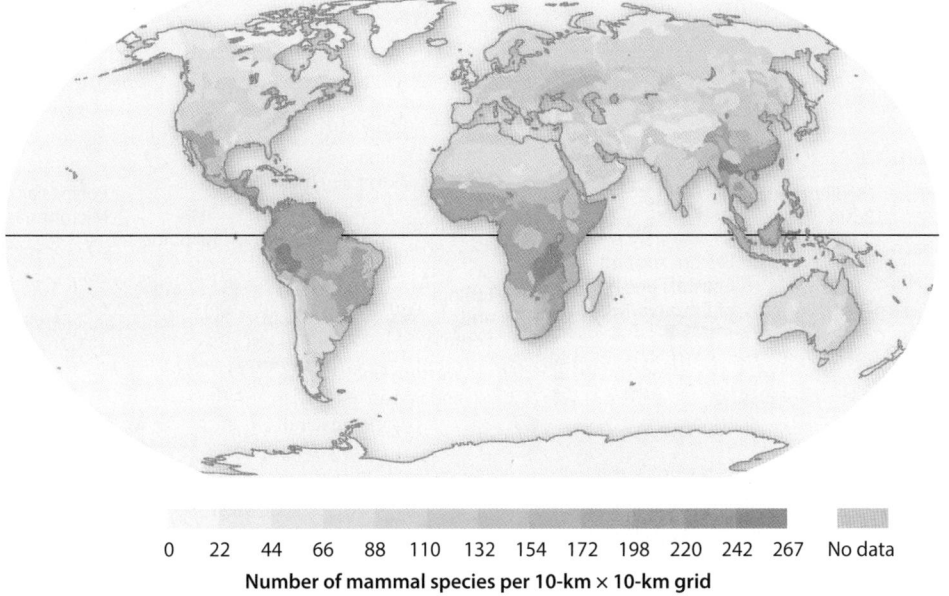

0 22 44 66 88 110 132 154 172 198 220 242 267 No data

Number of mammal species per 10-km × 10-km grid

FIGURE 56.13 Species richness patterns in mammals

Around the world, the highest species richness for mammals occurs near the equator with fewer species near the poles. Data from Kaufman, D. M. 1995. "Diversity of New World Mammals: Universality of the Latitudinal Gradients of Species and Bauplans." *Journal of Mammalogy* 76:322–334.

the number of mammal species around the world, as shown in **FIGURE 56.13**, we see that there are more species near the equator and fewer species near the poles. If we were to look at ocean species, we would see a similar pattern. Understanding these patterns of biodiversity is important because it informs us about where most species have evolved, where most species might be at risk of human impacts, and even where we are most likely to find new pharmaceutical drugs from nature.

Biologists have long wondered why these latitudinal patterns exist. For terrestrial plants and animals, biologists have offered a number of hypotheses. Some hypotheses are related to the fact that the tropical habitats near the equator are more favorable to survival than habitats closer to the poles, which experience cold winters and more variable environmental conditions across the seasons. Other hypotheses emphasize the fact that the tropical locations are much older and contain species that have evolved and diversified into many new species over tens of millions of years. In contrast, habitats closer to the poles, such as northern North America, have experienced multiple instances of glaciers moving southward and eliminating all species. As the glaciers receded, these newly exposed locations had to be colonized by species from more southern latitudes. Thus, regions closer to the poles have had less time for the evolution of new species.

Biodiversity Hotspots

Knowing these latitudinal patterns in species diversity, we can identify places that have a lot of species, including areas of the tropics and islands that have evolved endemic species, which are those species that are found nowhere else on the planet. As human activities continue to negatively impact the populations of many species, an important question is how we best conserve this biodiversity to prevent extinctions. A very practical challenge is that there is a limit to how much habitat can be protected, often because there is a limited amount of conservation funding available to purchase and protect all these habitats.

In response to this challenge, biologists have developed a method of prioritizing conservation efforts by identifying biodiversity hotspots. Biodiversity hotspots are defined as those areas with at least 1500 endemic plant species and experiencing at least 70% vegetation removal as a result of human activities. In short, these are areas with a large number of species that face a large threat of being driven extinct. Based on these criteria, biologists have identified 36 hotspots around the world, as illustrated in **FIGURE 56.14** on page 808. Although plant species are used to define these hotspots, areas with a high diversity of plant species typically also have a high diversity of animal species.

✓ Concept Check

7. **Describe** two economic benefits of biodiversity to humans.

8. **Describe** the relationship between species richness and latitude.

9. **Identify** the two criteria for designating biodiversity hotspots.

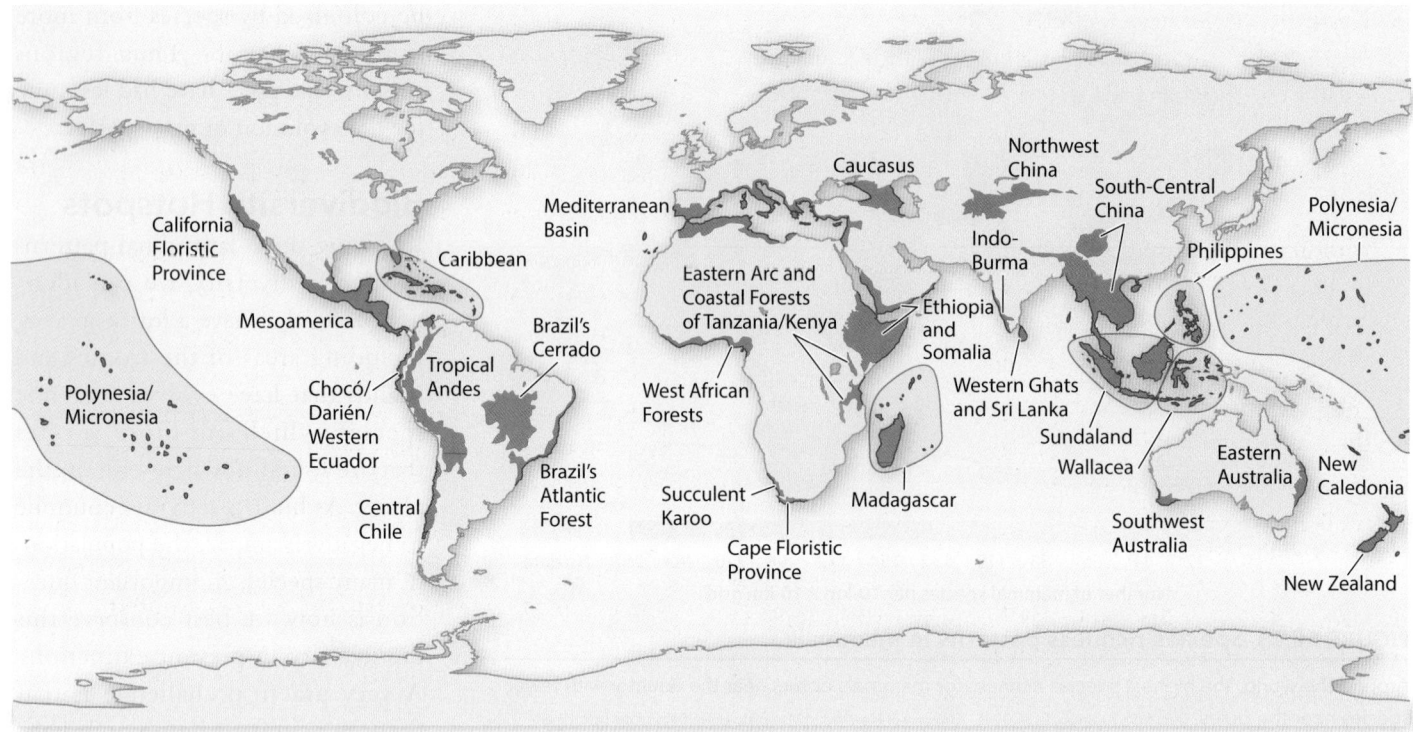

FIGURE 56.14 Biodiversity hotspots

Biologists have identified 36 regions of the world that have at least 1500 endemic plant species and more than 70% vegetation removal. These locations also contain a high diversity of animal species. Data from Meyers, N., et al. 2000. "Biodiversity Hotspots for Conservation Priorities." *Nature* 403:853–858.

Module 56 Summary

PREP FOR THE AP® EXAM

REVIST THE BIG IDEAS

SYSTEMS INTERACTIONS: Draw or **describe** a simple graph that best predicts the relationship between diversity and stability in an ecosystem.

LG 56.1 Biodiversity exists at several levels and is driven by multiple processes.

- Genetic diversity is the variety of genotypes in a given population or species. Page 798

- Species diversity is the number of species and the evenness of their abundance. Page 798

- Ecosystem diversity refers to the variety of ecosystems in a region. Page 798

- At the level of cells, genetic variation is caused by independent assortment, recombination, and mutation. Page 798

- At the level of populations, genetic variation is caused by genetic drift, population bottlenecks, founder effects, natural selection, and artificial selection. Page 798

- In contrast to the intentional breeding of different genetic varieties of animals and plants, sometimes artificial selection is accidental. Page 799

- There is a positive relationship between the size of an island and the number of species. Page 800

- Increased habitat diversity helps to increase species richness. Page 800

- Keystone species can have a major effect on species diversity. Page 801

LG 56.2 Biodiversity can influence ecosystem organization, productivity, and resilience.

- The diversity of species can have substantial effects on the organization of ecosystems. Page 802

- Keystone species have a substantial impact on the organization of food webs, so when a keystone species declines in abundance or is extinct from an ecosystem, we expect a large change in the ecosystem because no other species in the ecosystem can serve in that role. Page 802
- Another way that species diversity can affect the functioning of ecosystems is by altering ecosystem productivity, which is the amount of biomass accumulated by all species of producers in a year. Page 803
- Biodiversity affects the stability of ecosystems in terms of their resistance and resilience. Page 804
- Genetic diversity can improve the productivity and resilience of crops. Page 805

LG 56.3 Biodiversity provides benefits to humans and its distribution varies with latitude.

- The benefits of biodiversity to humans, sometimes referred to as ecosystem services, include both economic values and intrinsic values. Page 806

- Economic benefits of biodiversity include natural products that humans use, such as lumber, fur, meat, crops, and pharmaceuticals. Page 806
- Economic benefits of biodiversity also include flood control, water filtration, recreational opportunities, and pollination of crops. Page 806
- The richness of plant and animal species increases as we move from the poles to the equator. Page 806
- Deserts have lower species richness due to a lack of precipitation. Page 806
- Biologists have developed a method of prioritizing biodiversity conservation by identifying biodiversity hotspots. Page 807
- Biodiversity hotspots are identified as those areas with at least 1500 endemic plant species and experiencing at least 70% vegetation removal as a result of human activities. Page 807

Review Questions

1. Which statement is not true?
   (A) Biodiversity only refers to the number of species.
   (B) Biodiversity includes genetic diversity.
   (C) Biodiversity includes ecosystem diversity.
   (D) Biodiversity includes species diversity.

2. Why does genetic diversity improve resilience in human-created crop ecosystems?
   (A) Genetically diverse crops attract herbivores.
   (B) Genetically diverse crops can make it harder for pathogens to spread.
   (C) Genetically diverse crops attract pathogens.
   (D) Genetically diverse crops repel predators.

3. Why do larger islands contain more species?
   (A) Larger islands support smaller populations of each species.
   (B) Larger islands have species that are more prone to go extinct.
   (C) Larger islands have greater habitat diversity.
   (D) Larger islands are closer to the mainland.

4. What is the pattern of plant and animal richness in relation to latitude?
   (A) More animal species near the equator, but fewer plant species near the equator
   (B) Fewer animal species near the equator, but more plant species near the equator
   (C) Fewer animal species near the equator, and fewer plant species near the equator
   (D) More animal species near the equator, and more plant species near the equator

5. How can increased biodiversity affect ecosystem processes?
   (A) Increases in ecosystem stability
   (B) Reductions in plant growth
   (C) Declines in the movement of phosphorus from the soil
   (D) Reductions in primary productivity

Module 56
AP® Practice Questions

Section 1: Multiple-Choice Questions

Choose the best answer for questions 1–4.

Use the following information to answer questions 1–3.

Ecologists conducted an experiment in a grassy field. They isolated two 3′ × 3′ plots in the same field by digging trenches around each plot. At the start of the experiment, both plots had the same environmental conditions and the same number of species present. In plot 1, they removed one plant species, designated plant species X. In plot 2, they removed a different plant species, designated plant species Y. Both plots were monitored twice a week and, over the 60-month period, plant species X did not regrow in plot 1, and plant species Y did not regrow in plot 2. The ecologists monitored both plots for 60 months to quantify the species richness over time. The data for this experiment are shown in the figure.

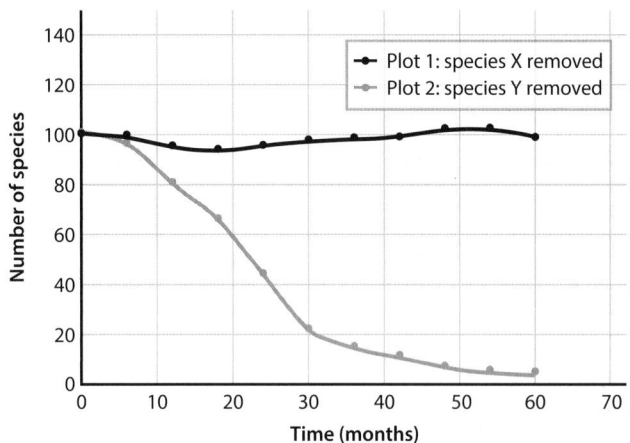

1. Which of the removed species was likely a keystone species?

   (A) Species X

   (B) Species Y

   (C) Both species X and species Y

   (D) Neither species X nor species Y

2. Which of the following best describes the effect of removing species X?

   (A) Removing species X was crucial to maintaining species richness in plot 1.

   (B) Removing species X did not affect species richness in plot 1.

   (C) Removing species X was crucial to maintaining the genetic diversity of the remaining species in plot 1.

   (D) Removing species X was crucial to maintaining the species diversity of plot 2.

3. If both environments contained 100 species at the beginning of the experiment, what percentage of species was lost from plot 2 over the course of the experiment?

   (A) 0%

   (B) 4%

   (C) 90%

   (D) 96%

Use the following information to answer question 4.

Ecologists conducted a 23-year experiment in a prairie community to determine which environmental conditions could increase primary production. They plowed plots of ground in the prairie and then seeded each plot with 1, 2, 4, or 16 different grass species. The ecologists manipulated each plot's amount of nutrients, water, and carbon dioxide, as well as exposure to herbivory, drought, and fire. The ecologists designated one plot as the control plot. The control plot contained 16 species of plants and it did not receive nitrogen-based fertilizer (N), increased water availability, or increased carbon dioxide availability. The control plot was also not exposed to herbivory, drought, or disturbance by fire. The mass of plants that grew in each plot, called biomass, was measured annually for 23 years and compared to the control plot. The results are presented in the figure.

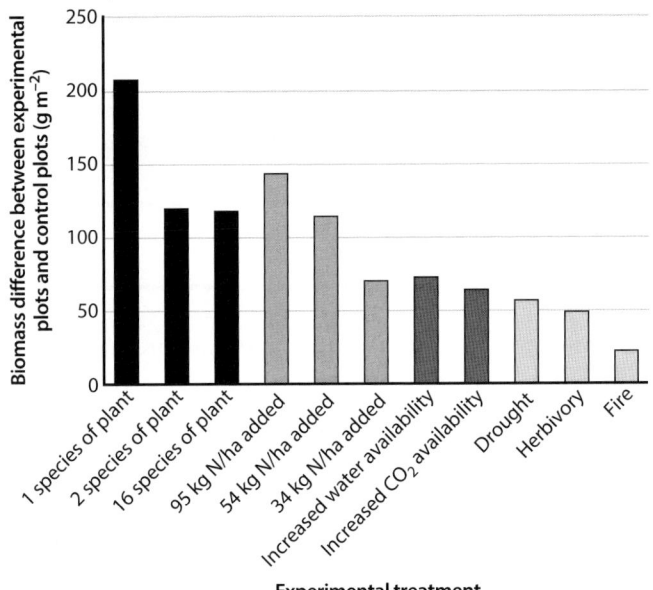

Data from Tilman, D., P. B. Reich, and F. Isbell. 2012. "Biodiversity Impacts Productivity as Much as Resources, Disturbance, or Herbivory." *Proceedings of the National Academy of Sciences USA*, 109:10394–10397.

4. Which manipulation caused the greatest increase in plant biomass?
   (A) Adding nitrogen fertilizer
   (B) Adding water and carbon dioxide together
   (C) Increasing the species diversity
   (D) Adding a disturbance (drought, herbivory, or fire)

Section 2: Free-Response Question

Write your answer to each part clearly. Support your answers with relevant information and examples. Where calculations are required, show your work.

Biologists studied the presence of amphibians and reptiles in seven West Indies islands. The results of their studies are presented in the graph below.

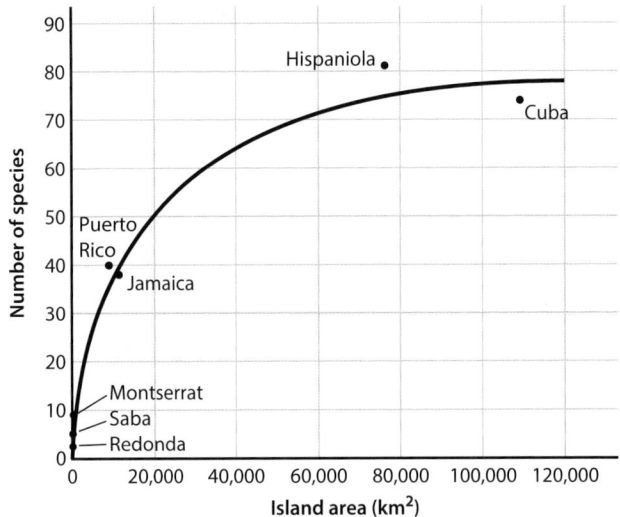

Data from MacArthur, R. H., and E. O. Wilson. 1967. *The Theory of Island Biogeography.* Princeton, NJ: Princeton University Press.

(a) **Describe** the relationship between the number of amphibian and reptile species with respect to island size. **Justify** your answer.

(b) **Explain** why an island the size of Hispaniola has a greater number of amphibian and reptile species than does an island the size of Montserrat.

(c) The West Indies island of Trinidad has an area slightly less than 5000 km². **Determine** the expected number of amphibian and reptile species on Trinidad. **Justify** your answer.

Module 57

Disruptions to Ecosystems

LEARNING GOALS ▶**LG 57.1** Geological and meteorological activity leads to changes in ecosystem structure and dynamics.
▶**LG 57.2** Human impacts accelerate change at local and global levels.
▶**LG 57.3** Human-caused climate change is a global disruption to ecosystems.
▶**LG 57.4** We have solutions to help address species declines and disrupted ecosystems.

In the previous module, we examined the multiple levels of biodiversity and the processes that create this biodiversity. We also looked at patterns in biodiversity across latitudes, how it affects the productivity and resilience of ecosystems, and the many benefits that this biodiversity provides to humans. Given the critical importance of biodiversity to ecosystems, we need to understand how natural and human activities can affect biodiversity on local and global scales. When it comes to human activities, the fundamental question is how we can maintain biodiversity and the benefits it provides as the human population across the globe continues to rapidly grow. In this module, we will begin by examining how geologic factors such as continental drift

have altered the distribution of biodiversity on Earth. We will then assess the current state of the world's major groups of plants and animals and consider the underlying causes that are driving current declines in species diversity across major groups of organisms. Finally, we will discuss some current solutions to help reverse these species declines.

PREP FOR THE AP® EXAM

FOCUS ON THE BIG IDEAS

SYSTEMS INTERACTIONS: Focus on the connections between human population growth and changes or disruptions to ecosystems.

57.1 Geological and meteorological activity leads to changes in ecosystem structure and dynamics

A number of geologic and meteorologic events have affected the distribution of ecosystems and the species that comprise these ecosystems. Important geologic events include the historic drift of continents, ice ages, and historic mass extinctions. Substantial meteorologic events include El Niño events that have major impacts on global climates every 5 to 8 years. In this section, we will examine how these events have caused disruptions to ecosystems and species diversity.

Disruptions Caused by Continental Drift

Looking back 250 million years, Earth possessed a single supercontinent known as Pangaea, as illustrated in **FIGURE 57.1**. By around 100 million years ago, Pangaea had broken up into two large continents, known as Laurasia

and Gondwana, due to a natural geologic process that we call continental drift. Laurasia subsequently broke up into the modern continents of North America, Europe, and Asia, while Gondwana broke up into South America, Africa, Antarctica, and Australia.

Understanding the historic connections among continents helps us understand the modern-day distributions of many species. For example, given that the continents of North America, Europe, and Asia were once all connected, they contain many species that have common ancestors. On all three continents, for instance, we have closely related species of elk, bison, bears, wolves, rabbits, maple trees, and oak trees. In contrast, the unique diversity of marsupial animals in Australia reflects the fact that Australia has been isolated from other continents for a long time. This long-term isolation allowed natural selection to produce new species that are not naturally found outside of Australia and its surrounding islands in southeast Asia, including kangaroos, koalas, and eucalyptus trees.

While the distributions of species were affected by continental drift, so were the ecosystems. As continents moved

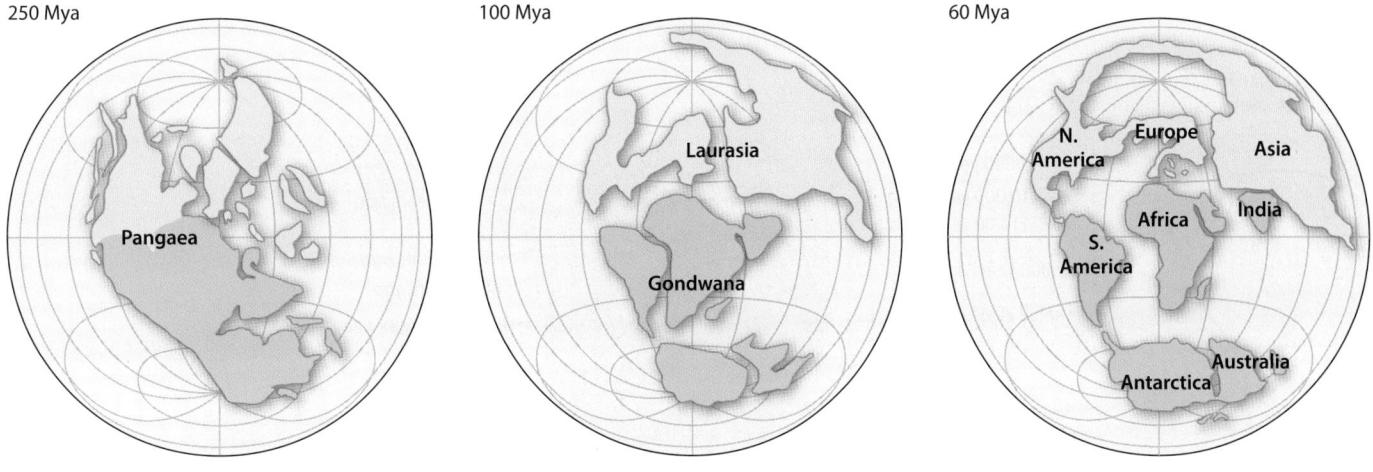

FIGURE 57.1 Disrupted ecosystems due to continental drift

Around 250 million years ago (Mya), modern-day continents were all connected as one supercontinent known as Pangaea.

to different latitudes, the changing climatic conditions caused major changes in the ecosystems that could be supported. For example, prior to the continents breaking apart, the region now known as Antarctica was located near the equator and experienced a warm and rainy environment. As a result, the land supported temperate and tropical forests. Today, Antarctica is located near the South Pole with a much colder climate; 2% of Antarctica has a tundra ecosystem containing small, cold-tolerant plants while the remaining 98% is covered in snow and ice.

Disruptions Caused by Ice Ages

The planet has also experienced multiple ice ages in which the temperature of the planet cooled and glaciers advanced to more southern latitudes, including all of Canada and the northern portion of the United States, from New York City to Chicago. The glaciers eliminated existing plants and animals. As the planet later warmed, the glaciers receded, leaving behind rocks and bare soil that would once again be colonized by species from the south. While some species such as plants with windblown seeds could arrive on bare soil relatively quickly, other species moved northward more slowly. For example, while trees cannot migrate, their seeds can be dispersed slowly northward by animals such as squirrels that collect the seeds and bury them in the ground for winter food. Not all the seeds are eaten, so some germinate into seedlings and this allows the tree populations to slowly expand northward. Thus, the receding of the glaciers allowed ecological succession to proceed, as discussed in the previous module.

We can determine the historic distribution of trees by identifying when pollen grains of each species appear in lake sediments from the surrounding forests. Lake sediments form in layers, with the youngest layers near the sediment surface, and we can date each layer based on the sediment chemistry. In doing so, biologists have discovered that many tree species took thousands of years to move northward after the most recent glaciers receded about 12,000 years ago. For example, when we look at two species of pine trees, we see that the extent of their northern distributions, as shown by the red lines in **FIGURE 57.2a** on page 814, continued to move northward over a period of 5000 years. In addition, the southern limit of the two pine trees also moved northward. In the case of hickory trees, the northern extent of their distribution experienced a northward expansion over 7000 years, as shown in Figure 57.2b. However, the southern limit of the hickory trees did not shift, which allowed the hickory trees to have a much larger modern-day distribution. As climates changed with advancing and receding glaciers, the ecosystems found at a given location in North America also changed; the tundra and conifer forest ecosystems that existed just south of the glaciers slowly moved northward as the glaciers receded and these ecosystems were replaced by a temperate deciduous forest.

Disruptions Caused by Mass Extinction Events

The composition of our modern ecosystems has also been affected by mass extinctions during Earth's history. Mass

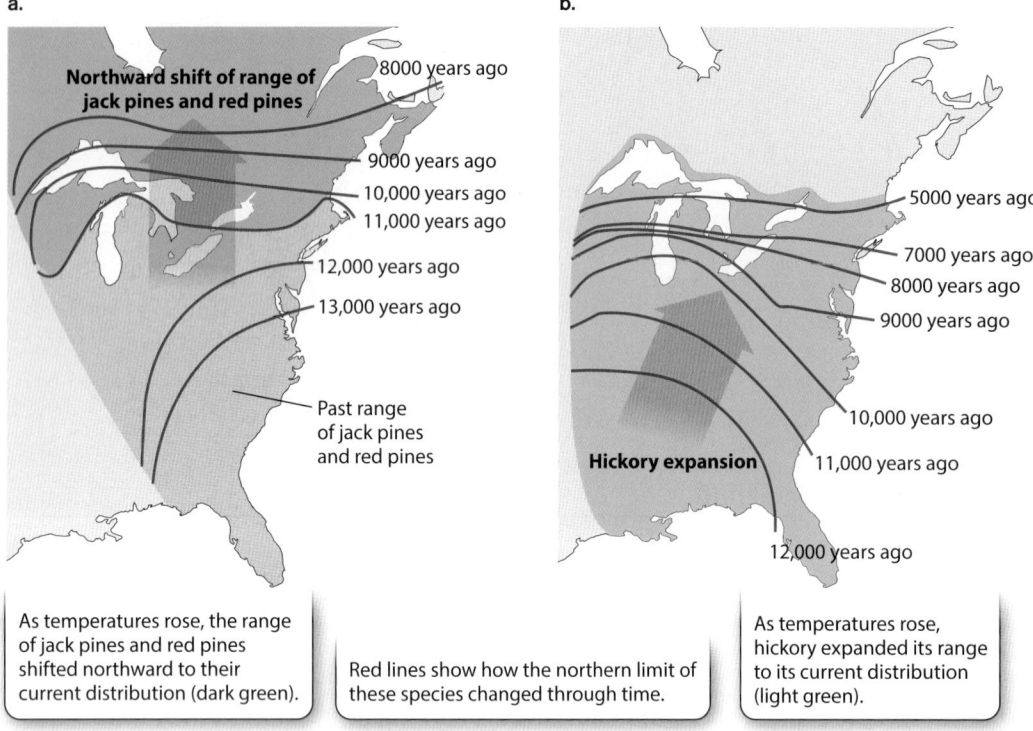

a.

Northward shift of range of jack pines and red pines

8000 years ago
9000 years ago
10,000 years ago
11,000 years ago
12,000 years ago
13,000 years ago

Past range
of jack pines
and red pines

b.

5000 years ago
7000 years ago
8000 years ago
9000 years ago
10,000 years ago
11,000 years ago

Hickory expansion

12,000 years ago

As temperatures rose, the range of jack pines and red pines shifted northward to their current distribution (dark green).

Red lines show how the northern limit of these species changed through time.

As temperatures rose, hickory expanded its range to its current distribution (light green).

FIGURE 57.2 Expanding species distributions after glaciers receded

As the final ice age receded about 12,000 years ago, several tree species slowly began a northward expansion of their distributions, which is an example of ecological succession. (a) In the case of two pine tree species, the northern limit of their distribution expanded northward, as shown by the red lines, as did their southern limit, as shown by the modern-day distribution in green. (b) For hickory trees, their northern limit expanded over thousands of years after the glaciers receded, but the southern limit did not move. Data from Davis, M. B. 1976. "Pleistocene Biogeography of Temperate Deciduous Forests." *Geoscience and Man* 13:13–26.

extinction events are defined as times when at least 75% of the existing species on Earth declined within periods of 2 million years. As you may recall from Unit 7, there have been five mass extinction events in the past. These mass extinctions have several causes in which the environmental conditions of ecosystems around the world have dramatically changed, including ice ages, climate change, volcanic activity, and asteroid collisions. During each event, large numbers of species were unable to adapt to the changing conditions quickly enough. After each event, however, many new species evolved. One interesting result of these mass extinction events is that while there have been an estimated 4 billion species living on Earth during the past 3.5 billion years, approximately 99% of these species have gone extinct.

When there are major disruptions in species composition, the structure and functioning of ecosystems are also affected. For example, many large herbivores and predators that once existed in North America disappeared around 12,000 years ago, such as wooly mammoths, mastodons, native horses, and saber-toothed cats. Given that humans arrived in North America from Asia 15,000 years ago, the extinctions of these large mammals are thought to be a combination of human hunting and natural climate changes that caused the advancing glaciers to recede back north.

Researchers who have studied the fossils of animals and plants during this time have discovered that the decline in large herbivores, caused the landscape to be a mix of forests and open areas, as illustrated in **FIGURE 57.3**, similar to the savannas of Africa today. Mastodons functioned as a keystone species by consuming shrubs and trees, and trampling the landscape. With the extinction of mastodons, more trees were able to grow and regions from California to Alaska became heavily forested. The increase in the forested ecosystem subsequently led to the extinction of many species of small mammals. Thus, the extinction of a keystone species can have large effects on both the food web and the

FIGURE 57.3 Mastodons

Mastodons are thought to have played an important role in North American ecosystems. When the mastodons were abundant, they consumed shrubs and trees and trampled the landscape, causing the landscape to be a mix of forests and open areas. When they were driven to extinction by hunting and climate change, more trees and shrubs were able to grow, resulting in forests growing in these areas from California to Alaska. Photo: North Wind Picture Archives/Alamy Stock Photo

structure of the entire ecosystem for thousands of years. This insight is not only of historical interest, but it also serves as a reminder of why we need to prevent the extinction of species today, such as elephants that are experiencing declining populations in the savanna ecosystem of Africa.

Disruptions Caused by Meteorological Events

Some meteorological events can have a major impact on ecosystems, as we mentioned earlier in this unit in our discussions of ecological succession and density-independent factors. Some of these events, such as tornadoes and hurricanes, cause impact in a matter of hours or days, but their effects on the ecosystems can last for decades. For example, forested islands in the Caribbean can be completely destroyed by hurricane winds and heavy rain. Within hours, the once-forested ecosystem is converted to a more open ecosystem consisting of broken-off trees and newly sprouting

small plants. However, many of the tree species have adapted to hurricanes by rapidly sprouting new stems from the remaining tree stumps of the hurricane-snapped trees, which allows a more rapid return to a forested ecosystem.

Other meteorological events are less frequent but still cause major disruptions to ecosystems. For example, multiple regions of the world experience dramatic changes in their weather every 3 to 7 years due to a phenomenon known as the El Niño–Southern Oscillation (ENSO). These events are initiated by a substantial change in air pressures in the South Pacific, between Peru and Australia, that causes ocean currents near the equator to no longer flow west; instead, the surface water flows east. This change in direction of wind and ocean surface currents explains the name Southern Oscillation.

While ENSO events are initiated in the South Pacific, cascading effects are felt around the world because the altered air and ocean currents impact the movement of warm and cold air around the planet as well as the production of low- and high-precipitation areas. For example, during ENSO years, states in the U.S. Midwest can experience severe droughts while other regions can experience increased rainfall and flooding. These effects can last for 1 to 2 years and have major impacts on natural and human-created ecosystems. ENSO events in recent decades have caused large die-offs of fish off the coast of California, corals in Panama, and kangaroo populations in Australia. They have also caused widespread drought and human starvation in Africa, Australia, and southeast Asia

✓ Concept Check

1. **Describe** how continental drift has caused disruptions to ecosystems.

2. **Describe** how ice ages have altered the structure and function of ecosystems.

3. **Describe** why some species extinctions can have major effects on the structure and functioning of ecosystems.

57.2 Human impacts accelerate change at local and global levels

With our understanding of prehistoric extinctions and disrupted ecosystems, we can now consider these phenomena in more recent times. In this section, we will examine the

current rate of extinctions and how a variety of plant and animal groups are experiencing declines. We then want to ask, what are the causes of these declines? We will explore five major causes: habitat loss, invasive species, overharvesting, pollution, and global climate change. While we will discuss each factor separately, many of these factors are acting together to cause the population declines in many species. We will also see that while some of these factors act locally, others act globally.

A Possible Sixth Mass Extinction

Over the past several centuries, the human population on Earth has grown exponentially, from an estimated 1 billion people in 1800 to more than 7 billion people today. This rapid population growth, as we discussed in Module 48, has resulted in increased human impacts on the environment, including the extinction of many species that have been unable to adapt quickly enough to increasing levels of human activities. Many biologists have suggested that we might be in the midst of a sixth mass extinction that began with the increase in the human population approximately 10,000 years ago. However, species have also gone extinct in the past, so how does the current rate of extinction compare to the historic rate? How can we test this hypothesis, since a mass extinction event is defined as a 75% loss of species over a period of 2 million years?

One way that biologists have started to assess this hypothesis is to compare the current rate of extinctions against a measure of historic rates of extinctions. For example, we can examine species for which we have a very good fossil record, such as mammals. When we compare the rate of extinction for mammals during the most recent 500 years with the rate during previous 500-year periods in the past, we find that the current extinction rate is higher. Across all organisms, the United Nations Convention on Biological Diversity estimates that the rate of species extinctions during the past 50 years may be up to 10,000 times higher than the historic rate. Whether this higher rate will result in 75% of our species going extinct over the next 2 million years is not knowable, but it seems clear that we are currently experiencing a high rate of species extinctions.

The Current Status of Major Plant and Animal Groups

While it is helpful to think about the current versus historic rates of extinctions to assess human impacts on biodiversity, it is also helpful to know how particular groups of plants and animals are faring in regard to whether populations are currently stable or declining. As we discussed in Unit 7, when species face a variety of different human stressors, not all of them can adapt to their changing environments. When this happens, their populations begin to decline.

To determine how many species are in various states of decline, the International Union for Conservation of Nature (IUCN) has quantified the percentage of each group that is threatened with extinction in the future. As you saw in Figure 48.8 on page 690, about one-fourth of all animals are threatened, including 40% percent of amphibian species, 25% of mammal species, and 14% of bird species. Of course, there are many other important groups in the world, such as the flowering plants, but the IUCN does not yet have sufficient data on these groups to have reliable assessments of the different categories.

Habitat Loss

For many species in decline, the primary cause is the loss of habitat, which means these ecosystems are occupying smaller areas than they used to occupy. As we discussed earlier in this unit, each species has a niche that includes the biotic and abiotic conditions required to survive, grow, and reproduce. As we lose habitats due to human activities such as logging, agriculture, home construction, and industries, there is less space for those species that require these habitats, so their populations decline and the structure of the ecosystem changes.

For example, in the lower 48 states of the United States, there were approximately 89 million hectares (ha) of wetlands in the 1600s, including the Everglades wetlands in Florida shown in **FIGURE 57.4**. Over time, many of these wetlands were drained to increase the land available for

FIGURE 57.4 Wetlands

One of the best-known wetlands in the United States is Everglades National Park in Florida. While the Everglades are federally protected, more than half of the wetlands in the United States have been drained to provide more areas for farming and the construction of factories, businesses, and homes. Photo: Jupiterimages/Getty Images

agriculture, factories, and homes. Today, the amount of wetland habitat is less than half of what was present in the 1600s. In California, the amount of wetland habitat has declined by 90%. These wetlands are home to species that are not able to live anywhere else, including many species of amphibians, invertebrates, and aquatic birds. This is just one example of how human activities are causing the loss of ecosystems that are critical for the species that depend on them.

Another high-profile example is the loss of forested habitats. For example, in the 1700s and 1800s, most of the forested land in the United States was cut down to make room for agriculture and buildings and to provide lumber for construction. The good news is that a substantial amount of that cleared land has now grown back into forests as many farms in the eastern United States were abandoned for the more profitable farming opportunities in the Midwest. In fact, much of the modern logging in the United States emphasizes sustainable practices, such as planting trees in logged areas to help speed up ecological succession. However, what are the patterns of deforestation in the rest of the world? If we examine **FIGURE 57.5**, we can see where forested regions of the world are experiencing increases, decreases, or no change in the amount of forested land. As you can see on the map, many areas in the United States are increasing in tree cover, as indicated by the yellow regions. However, other regions of the world, such as Alaska, Russia, and the tropical regions of South

America and central Africa, are experiencing forest loss, as indicated by the blue regions.

Other habitat types are also experiencing declines. For example, according to the most recent Millennium Ecosystem Assessment, the world has lost about 70% of the shrubland habitat that surrounds the Mediterranean Sea, 50% of all grassland habitats, and 30% of all desert habitats. Coral reefs are another ecosystem that is in decline in many parts of the world. While coral reefs are built by living corals, the reefs also serve as critical habitat for many other species of fish and invertebrates. In the coral reefs of the Caribbean Sea, the amount of living coral declined from 50% in the 1970s to just 8% by 2012.

The reasons for the decline in living coral is a combination of pollutants, increased water temperatures, and the collection of coral by people. From an ecosystem perspective, declines in the more sensitive species of coral can favor the growth of the more tolerant species of coral, which in turn can alter the species of fish that occupy the reef. These changes in the fish species result in different amounts and types of reef algae that the fish consume. In short, the decline in sensitive species of coral due to human impacts can have dramatic impacts on the structure of entire coral reef ecosystem.

Invasive Species

With the increasing ability of humans to move people and items around the world, we have also moved many species to new locations around the world. Species that currently live in their historic range are known as native species, whereas species that live outside of their historic range are known as exotic species or alien species. When an exotic species spreads throughout a new area and causes economic or ecological harm, we call it an invasive species. Invasive species are often successful in a new environment because they can exploit unoccupied niches and abundant resources. They usually do not face predators or competition. This allows the invading species to experience rapid population growth to a point where they can alter the structure of ecological communities.

Some exotic species have been intentionally introduced to new areas, such as honeybees (*Apis melifera*), which were introduced to the American

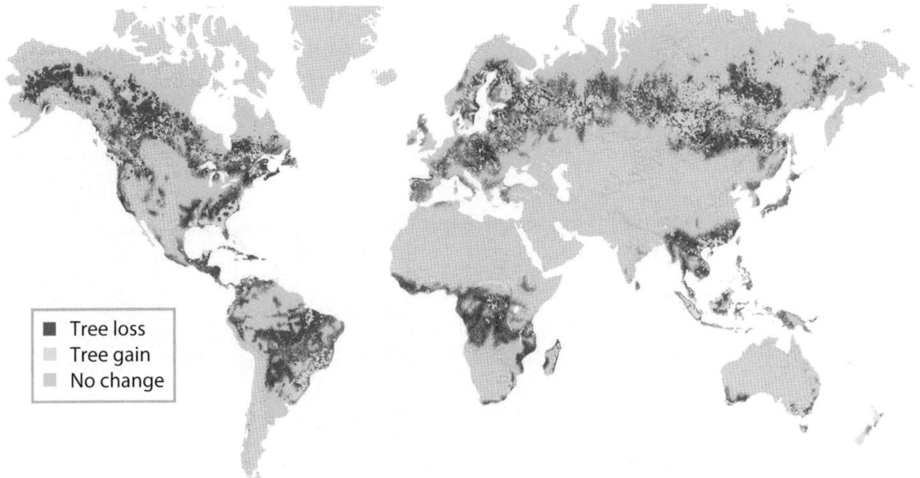

FIGURE 57.5 The gain and loss of forested habitats around the world

Forested habitats occur on six continents. As shown by the three colors, some regions of the world are experiencing increases in forested habitat while other regions are experiencing decreases. Data from Hansen, M. C., P. V. Potapov, R. Moore, M. Hancher, S. A. Turubanova, A. Tyukavina, D. Thau, S. V. Stehman, S. J. Goetz, T. R. Loveland, A. Kommareddy, A. Egorov, L. Chini, C. O. Justice, and J. R. G. Townshend. 2013. "High-Resolution Global Maps of 21st-Century Forest Cover Change." *Science* 342 (15 November): 850–53. Data available on-line from: http:// earthenginepartners.appspot.com/science-2013-global-forest.

Legend:
- ■ Tree loss
- Tree gain
- No change

FIGURE 57.6 Introduced brown trout

The brown trout's historical range is in Europe, and it is a favorite fish for anglers. Given its popularity, the brown trout has been introduced to water bodies throughout North America, South America, Australia, Asia, and Africa. While the introductions provide benefits to anglers, the trout can have devastating effects on ecosystems by consuming large numbers of prey that have no evolved defenses against the predatory trout. Photo: Jess McGlothlin Media/Aurora Photos/Getty Images

colonies in the 1600s from Europe to provide a source of honey. Similarly, the brown trout (*Salmo trutta*), shown in **FIGURE 57.6**, is native to Europe but has been introduced to many water bodies throughout North America, South America, Australia, Asia, and Africa because it is a favorite game fish for many anglers. Although these species provide economic benefits, they can also disrupt the ecosystems. For example, importing honeybees has led to the decline of many native bee species that compete with honeybees for pollen and nectar from flowers. Introducing trout to water bodies that have not historically contained trout results in the trout eating large numbers of invertebrates and amphibians that have no defenses against this predatory fish. As a result, large parts of the ecosystem are disrupted as the trout causes major groups of aquatic prey to decline or go extinct, which dramatically alters the flow of energy and matter through the ecosystem.

Unlike honeybees and brown trout, most invasive species arrive in a new area accidentally. For example, the brown rat (*Rattus norvegicus*) is thought to have a historical range in China but by the time of the Industrial Revolution it had spread across Europe. By the mid-1700s, the rat had been accidentally introduced to North America by hitching a ride on ships from Europe. Today, the rat lives nearly everywhere on the planet where people live. In regions with ground-nesting birds, such as the Hawaiian Islands,

the introduced rats have acted as predators on the eggs of these birds. Given that the island birds have no evolutionary history with rats, many were not able to evolve defenses quickly enough and, as a result, several Hawaiian bird species have been driven extinct by the introduced rats. This means that the functions of these birds in the ecosystem are no longer present, so the loss of the birds causes a disruption in the ecosystem.

Invasive species of plants can also have impacts on ecosystems. In the southeastern United States, for instance, kudzu vine (*Pueraria montana*) was introduced from Asia in the 1930s and promoted as a plant that would help reduce erosion in farm fields and along roadsides. In locations where there is abundant sunlight and no large herbivorous mammals, the vine can dominate the landscape and grow over the tops of trees, as you can see in **FIGURE 57.7**. When this happens, the vines can shade the trees to a point that the trees die and the ecosystem is dominated by the vine. In 2009, however, the kudzu bug (*Megacopta cribraria*) was accidentally introduced from Asia and it favors kudzu. When the kudzu bug finds large areas of kudzu, it can decimate the plant, so an invasive plant is currently being consumed by an invasive insect that evolved to eat kudzu in Asia.

When pathogens arrive in a new region of the world, they can have large impacts because the hosts that they infect have no evolved defenses. For example, in the 1930s a pathogenic fungus that infects elm trees was introduced to North America from Asia. Given that the fungal infections caused a disease that was first described by researchers in Holland, it became known as Dutch elm disease. While species of elm

FIGURE 57.7 The invasive kudzu vine

Widely planted in the 1930s to control erosion on farms and along roadsides, the vine has spread widely and can grow over the tops of buildings and trees. Photo: morgan hill/Alamy Stock Photo

trees in Asia had evolved defenses against the fungus, North American elm tree species had no evolutionary history with the fungus, so nearly 95% of them died from Dutch elm disease. Today, young elm trees continue to sprout, but most are killed within their first decade of life. The small number of trees that have survived appear to have more resistant genotypes, so there is hope that the elm trees will continue to evolve higher resistance over time and be able to coexist with the introduced fungus.

Overharvesting

The harvesting of plants and animals by humans for food, fur, and other uses has gone on for millennia, but there is a limit to how many individuals of a species can be harvested before it declines toward extinction. This is particularly challenging as we try to feed a rapidly growing human population. For example, off the coast of Newfoundland, on the eastern side of Canada, there is a large population of Atlantic cod that has been fished for centuries and supported the local economy. Over the decades, fishing boats have experienced numerous technological improvements for more efficient fishing, including long fishing lines with thousands of fishing hooks and gigantic nets that can surround a 2-km area of fish. With these increases in efficiency, the number of cod caught by commercial fishing boats increased exponentially in the 1960s and 1970s.

However, such a large catch could not be compensated by cod reproduction, so by the 1990s the cod population collapsed to very low numbers, as you can see in **FIGURE 57.8**. In 1992, cod fishing had to be stopped to allow the cod population to rebound. However, this rebound has been slow and a return to sustainable cod fishing off the coast of Newfoundland is not expected to be possible until perhaps 2030. The large decline in cod has meant that the cod's predatory impacts are no longer present, so their prey are beginning to increase. For example, the abundance of Jonah crabs (*Cancer borealis*), a prey of the cod, has increased fourfold; the now abundant crabs can decimate the abundance of sea urchins. As a result, the overharvesting of the cod can have a large impact on the structure of ocean ecosystems.

The story of the Atlantic cod near Newfoundland has been repeated all around the world. For instance, rhino populations in Africa have been illegally killed for their horns, which are sold for their reputed medicinal properties. There are species of rare trees, such as big-leaf mahogany, that are sold for their beautiful lumber, and endangered species of orchids that are dug up in their natural habitats and sold to collectors. The illegal trade in plants and animals is currently

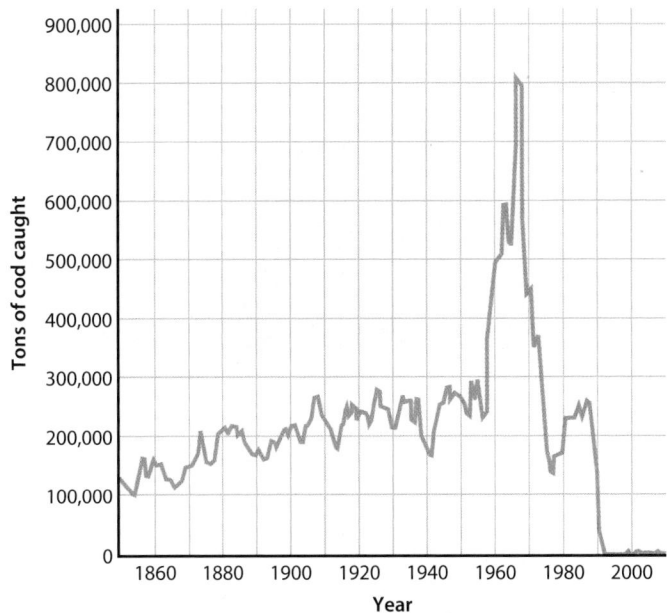

FIGURE 57.8 The overharvesting of cod

As technologies improved the ability of commercial fishing boats to catch cod in the 1960s and 1970s, the number of cod caught off the coast of Newfoundland, Canada, dramatically increased until the population became overharvested. At that point, there was a sharp decline in the cod population and the fishery had to be shut down.

Data from Millennium Ecosystem Assessment. W. Reid, H. Mooney, A. Cropper et. al., 2005: Ecosystems and Human Well-being: Synthesis. Island Press, Washington D.C.

valued at $5 billion to $20 billion annually, and this has led to major issues of overharvesting.

Pollution

Pollution has also contributed to the decline of biodiversity in ecosystems. Pollution can appear in a wide variety of forms, including oil spills, pesticides, heavy metals, and even endocrine disruptors, which inhibit proper reproduction in animals. A high-profile pollution event occurred in the Gulf of Mexico when an oil platform known as the Deepwater Horizon suffered an explosion in 2010 that released a massive amount of oil into the ocean water for months. As the oil continued to pour out of the pipe, it soon became the largest oil spill in U.S. history, totaling 507 million liters of oil.

To break up the oil spill into tiny droplets that could be degraded by bacteria, emergency workers sprayed a chemical onto the oil. However, the chemical is also toxic to many species of aquatic organisms. As the animals were exposed to the oil and the chemical by way of consumption and inhalation, it caused organ damage and reproductive failure. According to a National Oceanic and Atmospheric Association report in 2017, thousands of species were exposed to the

FIGURE 57.9 Pollution effects on biodiversity

During the oil spill from the Deepwater Horizon platform in the Gulf of Mexico, thousands of species were exposed to the oil, including this juvenile Kemp's Ridley sea turtle. Photo: NOAA and Georgia Department of Natural Resources

oil spill. For example, Kemp's Ridley sea turtles (*Lepidochelys kempii*) in the area were exposed to the oil spill, as shown in **FIGURE 57.9**, and 20% of them died. In addition, the bottlenose dolphin (*Tursiops truncatus*) population declined by 50% over the next few years. Multiple species of birds were also killed, including large numbers of pelicans and seagulls.

When it comes to pollution, biologists have observed the evolution of increased tolerance to pollutants over several generations, but only in a limited number of cases. For example, repeated applications of insecticides to kill mosquitoes has resulted in artificial selection for the rare mutants that have a genetic resistance to the insecticide, thereby causing the evolution of insecticide-resistant mosquito populations.

✓ Concept Check

4. **Describe** how biologists can assess whether the current rate of extinctions is higher now than in the past.

5. **Identify** which group of animals has the highest percentage of threatened species.

6. **Identify** the biggest threat to species diversity in ecosystems today.

7. **Distinguish** why birds on islands are commonly susceptible to exotic species that are predators.

57.3 Human-caused climate change is a global disruption to ecosystems

When considering human impacts on disrupted ecosystems, we also have to consider the impacts of global climate change, which is the change in the average weather around the planet over years or decades. An important driver of changing climates is global warming, which is the increase in the temperature of the air, land, and water over years or decades. Over millions of years, the planet has experienced many periods of warming and cooling, as we mentioned in the case of the ice ages that caused glaciers to advance and recede. Of current concern is the role that human activities are having on the temperature of Earth. To understand how global climate change can affect ecosystems, we first need to understand the underlying causes of global warming.

The Greenhouse Effect

Our planet is surrounded by an atmosphere that contains gases, and 99% of those gases are nitrogen (N_2) and oxygen (O_2). While abundant, these two gases do nothing to make the planet warmer. The planet is made warmer by a few scarce gases that we refer to as greenhouse gases, which include water vapor (H_2O), carbon dioxide (CO_2), methane (CH_4), and nitrous oxide (N_2O). These gases have a unique property; they can absorb and release infrared radiation. When sunlight strikes Earth with visible light and ultraviolet radiation, that energy gets converted into infrared radiation at the planet's surface. This infrared radiation gets emitted back toward the atmosphere, where the greenhouse gases absorb the infrared radiation and then reemit it in all directions. Some of this energy goes back to the planet and some goes out into space. The portion of infrared radiation that goes back toward Earth is what makes the planet warmer. You can see these processes illustrated in **FIGURE 57.10**.

It is important to note that greenhouse gases were in the atmosphere long before humans appeared on the planet. Indeed, the natural concentration of greenhouse gases makes the planet about 33°C warmer than it would be otherwise, making life as we know it possible. Thus, these scarce gases have a very large effect on the temperature of our planet. We call this process the greenhouse effect because it resembles

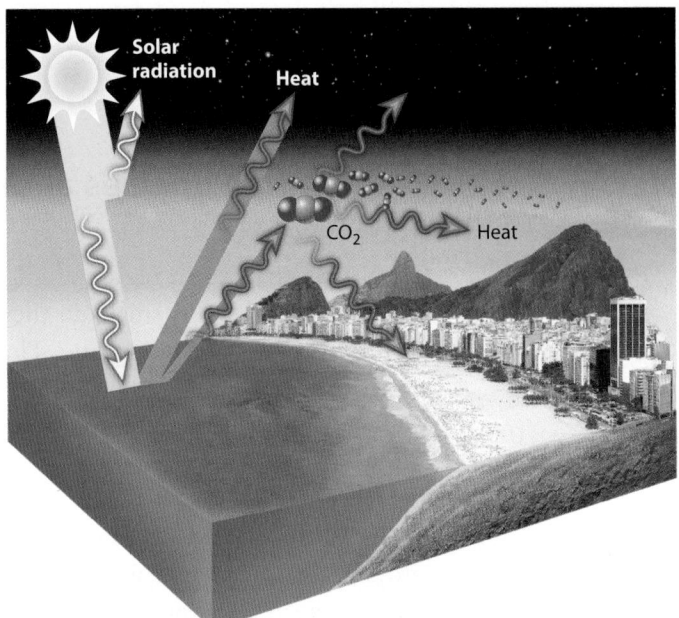

FIGURE 57.10 The greenhouse effect

Solar radiation from the sun passes through the atmosphere and strikes the planet's surface. The solar radiation is converted into infrared radiation, which we can think of as heat. A portion of the heat is emitted from the surface of the planet back toward the atmosphere. In the atmosphere, greenhouse gases absorb the heat and then reemit it in all directions. Some of this heat is reemitted back toward Earth, which increases the planet's temperature, and some is lost to space.

Photo: Fotosearch Stock Images

a gardener's greenhouse, which lets light in through the glass windows and then traps the heat inside. The greenhouse effect operates a bit differently in that the heat is not trapped, but rather the infrared radiation from the Earth's surface is absorbed in the atmosphere and then reemitted in all directions, including back to Earth's surface.

Human Impacts on the Greenhouse Effect

With this knowledge of how the natural greenhouse effect works, we can now examine how humans are altering this process. Humans are increasing the greenhouse effect by increasing the concentrations of carbon dioxide, methane, nitrous oxide, and other chemicals in the atmosphere. The causes of these increasing greenhouse gases include the burning of fossil fuels, raising domesticated animals, deforestation, decomposing landfills, and the production of industrial chemicals.

Beginning in 1958, Dr. Ralph Keeling began measuring the concentration of CO_2 in the atmosphere over several

years on top of a mountain in Hawaii. You can see these data in **FIGURE 57.11**. The first thing that he noticed was that CO_2 concentrations exhibited seasonal cycles, which are caused by plants and algae taking up CO_2 during the spring and summer for photosynthesis. The second thing he noticed was that the amount of CO_2 was increasing each year. More than 60 years later, we see that the amount of CO_2 in the atmosphere has steadily increased. The amount of CO_2 in the atmosphere has increased by 30% since 1958. When we increase the concentration of these greenhouse gases, the atmosphere can absorb and reradiate more infrared radiation back to Earth, so Earth steadily warms.

Based on our knowledge of how greenhouse gases absorb and reradiate infrared energy and based on data showing that greenhouse gas concentrations are increasing, we can hypothesize that the planet should be getting warmer.

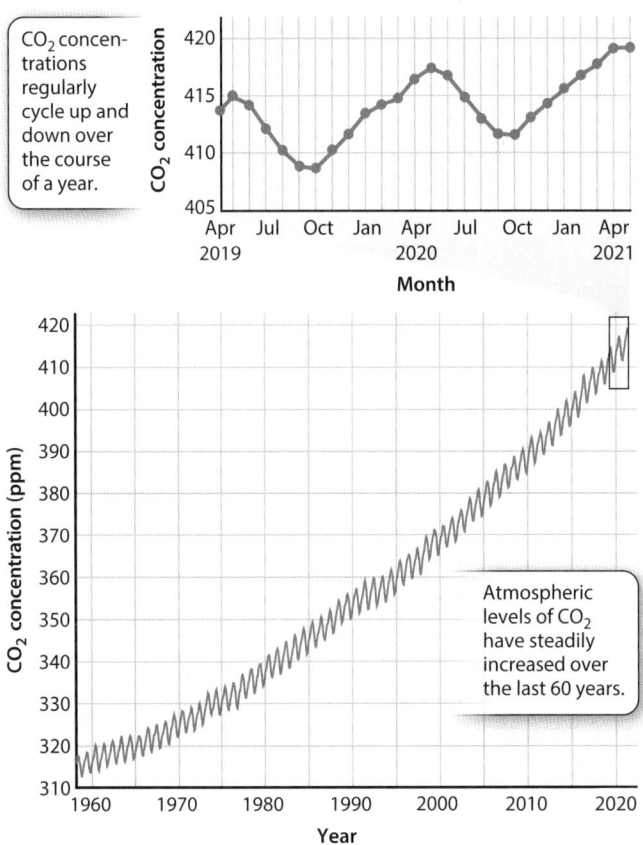

FIGURE 57.11 The Keeling curve of increasing CO_2 in the atmosphere

Based on measurements of CO_2 concentrations made on a mountaintop in Hawaii, we can see seasonal fluctuations in CO_2. Over the past 60 years, we can also see that there is a steadily increasing trend of CO_2 concentrations over time. Data from Dr. Pieter Tans, NOAA/ESRL (www.esrl.noaa.gov/gmd/ccgg/trends/) and Dr. Ralph Keeling, Scripps Institution of Oceanography (scrippsco2.ucsd.edu/).

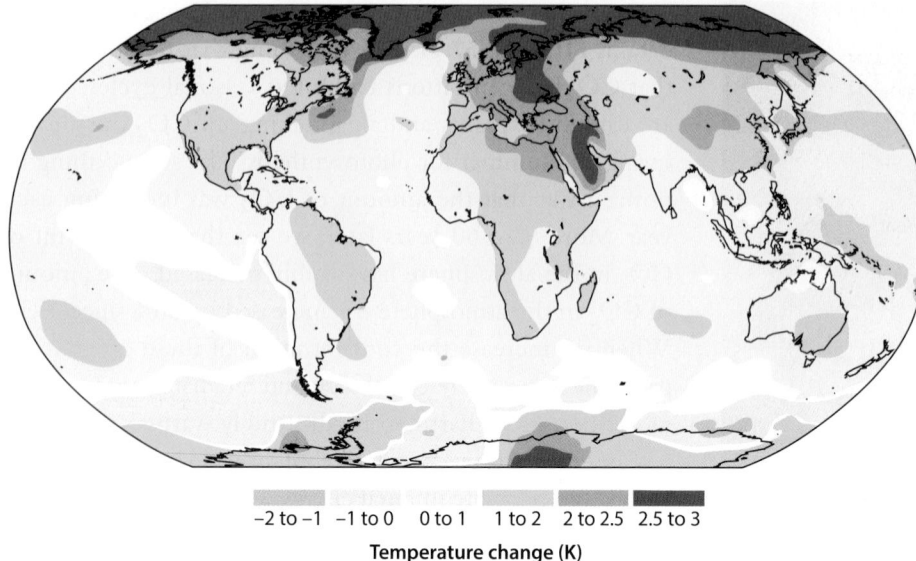

-2 to -1 -1 to 0 0 to 1 1 to 2 2 to 2.5 2.5 to 3
Temperature change (K)

FIGURE 57.12 Global warming around the world

While the mean temperature around the world has increased by about 1°C since the late 1800s, temperature changes vary among different regions. From 1979 to 2019, some areas near Antarctica became 1°C cooler while northern latitudes became up to 3°C warmer. Data from https://climate.nasa.gov/blog/3017/making-sense-of-climate-sensitivity/

However, do temperature data support this hypothesis? Our direct measurements of temperature go back to the 1800s and these measurements tell us that the mean temperature of the planet has increased by nearly 1°C over the past 140 years. While this number seems small, it is important to know that some regions of the world have cooled a bit whereas others have warmed a great deal, such as increases of 4°C (7°F) in Alaska, northern Canada, and northern Europe. In **FIGURE 57.12**, we can see temperature changes in just the past 40 years. While some regions around Antarctica have cooled by 1°C, regions on the northern latitudes of North America and Europe have increased by 3°C.

Impacts of Global Warming on Species and Ecosystems

Given that changes in regional temperatures are relatively small, do they have any effect on biodiversity? The key to understanding the effects is appreciating that small temperature differences can affect important biological processes, such as when flowering plants make their flowers, when birds migrate, when animals breed, and whether species can survive under the warmer temperatures. For example, in the Gulf of Alaska the ocean temperatures rose 2°C to 3°C beginning in 2014 due to an ENSO event. This increase in temperature was enough to cause poor survival in young cod, which led to a ban on cod fishing in 2019. The warmer ocean temperatures also caused many other species of fish to move deeper into cooler waters, which led to the deaths of nearly 1 million sea birds that depend on these fish for their diet. While this small increase in ocean temperature was caused by an ENSO event, it provides a clear example of how a small increase in water temperatures can have a wide range of direct and indirect effects on species and the structure of ecosystems. In "Practicing Science: How do changing CO_2 levels affect coral reefs?" we discuss experiments that examine how increases in CO_2 concentrations can impact coral reef ecosystems.

Practicing Science 57.1

PREP FOR THE AP® EXAM

How do changing CO_2 levels affect coral reefs?

Background We have seen that rising CO_2 concentrations have contributed to global warming. However, they can also affect the pH of the oceans, which can have additional effects on biodiversity. For example, coral reefs are built by animals, known as corals, that live in a mutualistic relationship with algae. The algae receive a home and nutrients from the coral while the coral receives sugars produced by the algae. Together, the coral and algae build a rocky skeleton

that accumulates over thousands of years and becomes a coral reef. A great diversity of other animals and algae depend on coral reefs for habitat, such as the Great Barrier Reef off the coast of Australia, which is home to more than 1500 species of fish and 4000 species of clams and snails.

It turns out that elevated concentrations of CO_2 in the air diffuse into the ocean water, which increases the amount of CO_2 in the oceans. In the oceans, much of this CO_2 combines with water and is rapidly converted into carbonic acid (H_2CO_3), which then loses a hydrogen ion (H^+). All these free hydrogen ions cause the pH of the water

▼

to go down by about 0.1 pH unit, which is a process called ocean acidification. This may not sound like a large change, but you may recall that pH is measured on a log scale, so this 0.1 decline in pH represents a 30% increase in hydrogen ions in the water. A lower pH might make it harder for the coral and algae to access calcium from the water to grow their calcium carbonate skeleton, which is the backbone of the coral reef.

Hypothesis Increasing ocean pH will lead to increased construction of coral reef skeletons while decreasing ocean pH will lead to decreased construction of coral reef skeletons.

Experiment 1 In 2014, American biologist Rebecca Albright and her colleagues identified a lagoon within the Great Barrier Reef where coral reefs were isolated from regional currents. In their first experiment, they set up an apparatus that emitted sodium hydroxide (NaOH) for 3 weeks into a current that flowed in one direction over adjacent reefs. NaOH is a base, so it increased the pH of water flowing over the reef to values that resembled the pH of seawater before the Industrial Revolution. This increase in pH should make it easier to obtain the calcium needed to build the calcium carbonate skeleton. Albright and her colleagues measured the rates of reef building—known as calcification—when NaOH was added to one area and in a control area where NaOH was not added.

Experiment 2 In a second experiment, the researchers added CO_2 to waters flowing over one reef area, in an effort to simulate the projected pH of late-twenty-first-century seawater. Again, they measured rates of reef calcification and compared those to a control area in which no additional CO_2 was added to the water.

Results In part a of the figure to the right, you can see the results of the first experiment in which researchers examined a control area of the reef versus a second area in which they added NaOH to the water to increase pH. Increasing the pH caused an increase in the percent calcification, which is a measure of coral reef growth. Adding NaOH increased the pH to pre–Industrial Revolution levels and calcification increased by 7%. No change in calcification was observed in the control area, as you can see in the figure.

In part b of the figure, you can see the results of the second experiment in which researchers examined a control area of the reef versus a second area in which they added CO_2 to the water to decrease pH. Decreasing the pH reduced calcification by 34%. Again, no change was observed in the control area, as you can see in part b. Error bars are calculated as ±1 standard deviation on either side of the mean.

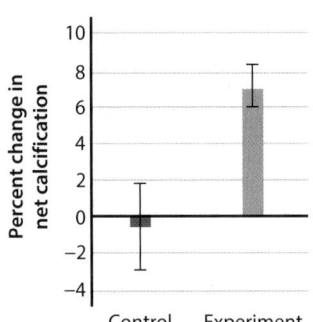
a. Increased pH

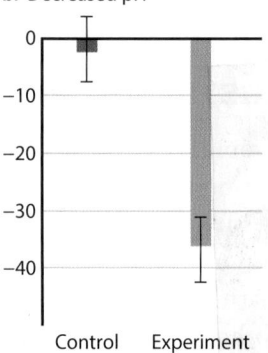
b. Decreased pH

Conclusion Corals and other reef-building organisms in the Great Barrier Reef are sensitive to ocean acidification, showing an increased ability to form calcium carbonate skeletons under pH conditions resembling pre–Industrial Revolution ocean pH conditions and a decreased ability under late-twenty-first-century ocean pH conditions.

Follow-up Work Reef corals show a consistent vulnerability to pH decline, although some populations appear to be more resistant than others. Resistant corals might provide a means to sustain reefs in the future, but their ecology will differ from those we see today.

SOURCES
Albright, R., et al. 2016. "Reversal of Ocean Acidification Enhances Net Coral Reef Calcification." *Nature* 531:362–365; Albright, R., et al. 2018. "Carbon Dioxide Addition to Coral Reef Waters Suppresses Net Community Calcification." *Nature* 555:51–519.

AP® PRACTICE QUESTION

(a) **Identify** the following:
1. The question being investigated by the experimenters
2. The null hypotheses they tested
3. The alternative hypotheses they tested
4. The experimental groups
5. The control group
6. The independent variable
7. The dependent variable

(b) A statistical test was performed comparing the experimental data with the control data, producing a *P*-value less than 0.05. Explain what this value says about the results.

(c) **Explain** the results of the experiments by **identifying** the researchers' claim and **justifying** the claim with the evidence and the reasoning researchers used (CER).

PREP FOR THE AP® EXAM

AP® EXAM TIP

You should understand the causes and effects of global warming, including how it may affect temperatures in various parts of Earth and its role in ocean acidification.

✓ Concept Check

8. **Describe** the reason that the most abundant gases in the atmosphere do not contribute to global warming.

9. **Describe** how global warming has both a natural and human-caused component.

10. **Describe** how a small decline in pH can cause a decline in the growth of coral reefs.

57.4 We have solutions to help address species declines and disrupted ecosystems

From the previous section, it would be understandable if a person felt overwhelmed by the number of threats to biodiversity caused by human activities. However, there is great reason for hope as we look to the future. For example, as we learn about the full range of pollution effects, we learn how to limit these pollutants through new regulations or more limited applications that reduce their harmful effects on ecosystems. In this section, we will discuss the great strides we are making toward additional solutions including habitat protection, species reintroductions, reduced harvesting, and reversing global climate change.

Habitat Protection

As we noted in our discussion of biodiversity hotspots in the previous module, there is a current effort to protect habitats that are home to a large number of endemic species that face threats of extinction. For example, there has been a rapidly growing trend of protecting terrestrial habitats since the 1960s that is now starting to plateau, as shown in **FIGURE 57.13a**. Historically, there has been less emphasis on protecting marine habitats, but the situation began to improve by the 1980s and has dramatically increased since the 2000s, as you can see in Figure 57.13b. These large areas of protected habitats will help many species to avoid extinction.

Controlling Invasive Species and Reintroducing Native Species

It can be exceedingly difficult to rid an area of an invasive species once the population has spread widely. Given that many invasive species have no substantial predators, parasites, or competitors, there are often few natural controls on their populations. In some cases, invasive species can be harvested to reduce their abundance, such as invasive species of aquatic plants that can be harvested by hand or with machines that cut the aquatic plants like a giant lawnmower. In a few cases, researchers have gone to the region of the world where an invasive species originated to find an enemy that can control the invasive species, such as a predator, parasite, or herbivore. We saw an example of this with the kudzu vine. When the kudzu bug was introduced to the United States decades after the kudzu vine was introduced, the bug rapidly increased in

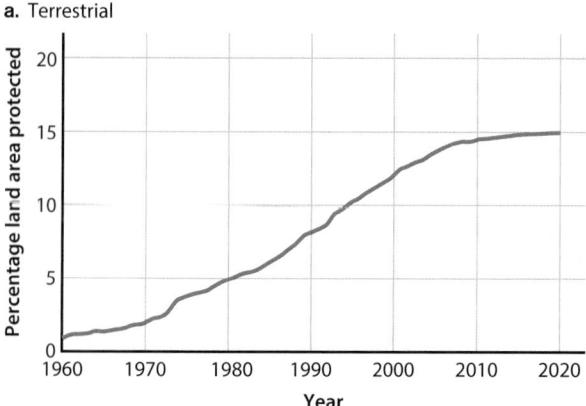

a. Terrestrial

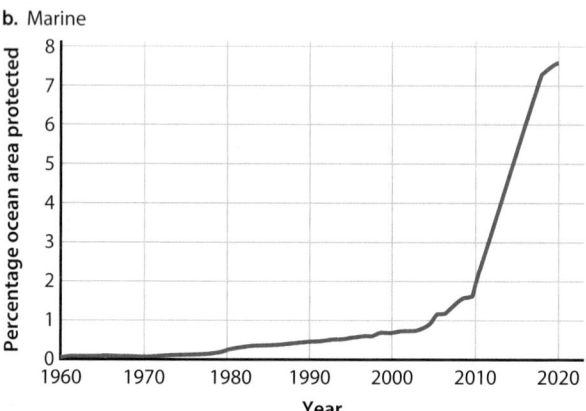

b. Marine

FIGURE 57.13 Protecting habitats

Throughout the world, there has been a growing percentage of protected habitats over time. (a) The protection of terrestrial habitats began to increase rapidly in the 1960s and is currently at a plateau. (b) The protection of marine habitats began much later, but is currently increasing rapidly. Note that the two *y*-axes have different scales. Data from Secretariat of the Convention on Biological Diversity, Global Biodiversity Outlook 4. 2014. https://www.cbd.lnt/gbo/gbo4/publication/gbo4-en-hr.pdf, https://www.cbd.int/gbo2/; https://livereport.protectedplanet.net/chapter-2

abundance and began controlling the kudzu vine's population by eating the invasive plants.

In some cases, we have been able to reintroduce native species into their native habitat, thereby bringing them back from the brink of extinction. For example, the California condor (*Gymnogyps californianus*), shown in **FIGURE 57.14**, is a large vulture that feeds on dead animals. For several decades, its numbers had been declining from a multitude of causes, including illegal shooting, feeding on dead animals that had been killed by bullets and still contained fragments of the lead bullets, and feeding on dead animals that died from eating poisons. By the 1980s, only 22 condors remained. All 22 birds were captured from the wild

FIGURE 57.14 The California condor

The condor is a species of vulture that feeds on dead animals. During much of the twentieth century, condor numbers declined as the result of illegal shooting and inadvertent poisoning. In the 1980s, only 22 individuals remained in the wild. Photo: kojihirano/Shutterstock

and held in captivity to produce offspring. At the same time, steps were taken to reduce the threats in the wild so that individuals that were reintroduced to the wild would have a better probability of surviving. This effort was a tremendous success; today there are more than 500 California condors alive in captivity and in the wild, ranging from California to New Mexico. Similar efforts have been successful with other charismatic species, including black-footed ferrets (*Mustela nigripes*) and gray wolves (*Canis lupus*).

Reduced Harvesting

Reduced harvesting is another way to bring species back from the brink of extinction. One way to do this is to regulate the harvest of plants and animals, particularly those that are commercially harvested, as we saw in the case of Atlantic cod off the coast of Newfoundland. A great example of the power of regulating the harvest of animals is the northern elephant seal (*Mirounga angustirostris*). By the late 1800s, the elephant seal had been hunted so much by humans that

biologists thought it was extinct. However, in the 1890s a few small populations were found that numbered approximately 100 individuals. Soon afterward, the U.S. and Mexican governments protected the elephant seal from any further hunting to allow the population to rebound. Over the subsequent century, the population rebounded to an astounding 150,000 individuals that currently live all along the coasts of California and Mexico.

A similar story occurred with the American alligator (*Alligator mississippiensis*), which was hunted to near extinction in the United States by the 1950s. Once the U.S. government declared the alligator to be endangered in 1967 and stopped all hunting, the alligator rapidly rebounded. By 1987, the alligator was abundant enough to remove it from the endangered species list. Today, there are millions of alligators living in the southeastern United States. The populations are once again hunted, but the number of harvested alligators is now regulated to prevent the species from ever being overharvested again.

Reducing Global Climate Change

The key to reducing global warming and the associated global climate change lies in reducing the concentration of greenhouse gases in the atmosphere. As we have discussed, greenhouse gases have multiple natural and human sources, which means there are multiple solutions that can help. For example, using less fossil fuel results in less CO_2 being released into the atmosphere. Individuals can help by driving less, taking public transportation more often, and using energy-efficient appliances and lights. Many companies are moving to renewable energy using solar panels and wind turbines to also reduce their burning of fossil fuels. In addition, companies are increasingly phasing out the use of human-made chemicals that act as very powerful greenhouse gases. Collectively, these efforts will help slow the current rate of global warming, although the efforts will likely take several decades.

✓ Concept Check

11. **Describe** the trends in global habitat protection over the past several decades.

12. **Describe** how the condor reintroduction program required the mitigation of multiple human impacts.

13. **Describe** how the U.S. government's actions allowed American alligator populations to rebound and remain abundant for the past several decades.

Module 57 Summary

PREP FOR THE AP® EXAM

REVISIT THE BIG IDEAS

SYSTEMS INTERACTIONS: Using one example from the module, **explain** how continued growth of the human population plays a role in reduced species richness and a decrease in biodiversity.

LG 57.1 Geological and meteorological activity leads to changes in ecosystem structure and dynamics.

- Over millions of years, the diversity of species on Earth has been impacted by natural geologic forces, including continental drift. Page 812

- The planet has also experienced multiple ice ages in which the temperature of the planet cooled and glaciers advanced to more southern latitudes, including all of Canada and the northern portion of the United States. Page 813

- We can determine the historic distribution of trees by identifying when each species' pollen grains appear in lake sediments from the surrounding forests. Page 813

- The composition of our modern ecosystems has also been affected by mass extinctions during Earth's history. Page 813

- When there are major disruptions in the species present in an ecosystem, the structure and functioning of ecosystems can be affected. Page 814

- Some meteorological events, such as tornadoes and hurricanes, can have a major impact on ecosystems. Page 815

- Other meteorological events, such as ENSO events, are less frequent but still cause major disruptions to ecosystems. Page 815

LG 57.2 Human impacts accelerate change at local and global levels.

- Rapid human population growth has resulted in increased human impacts on the environment, including the extinction of many species. Page 815

- Many biologists have suggested that we might be in the midst of a sixth mass extinction. Page 816

- For many species in decline, the primary cause is the loss of habitat. Page 816

- Many areas in the United States are increasing in tree cover, but other regions of the world are experiencing forest loss. Page 817

- The world has lost about 70% of the shrubland habitat that surrounds the Mediterranean Sea, 50% of all grassland habitats, and 30% of all desert habitats. Page 817

- With the increasing ability of humans to move people and items around the world, we have also moved many species to new locations around the world. Page 817

- When an exotic species spreads across a new area and causes economic or ecological harm, we call it an invasive species. Page 817

- Most invasive species arrive in a new area accidentally. Page 818

- When pathogens arrive in a new region of the world, they can have large impacts because the hosts that they infect have no evolved defenses. Page 818

- The harvesting of plants and animals by humans for food, fur, and other uses has gone on for millennia, but there is a limit to how many individuals of a species can be harvested before the species declines toward extinction. Page 819

- Overharvesting of a species can have a large impact on the structure of an ecosystem. Page 819

- Pollution has also contributed to the decline of species and disrupted ecosystems. Page 819

LG 57.3 Human-caused climate change is a global disruption to ecosystems.

- Global climate change is the change in the average weather around the planet over years or decades. Page 820

- An important driver of changing climates is global warming. Page 820

- The planet is made warmer by a few scarce gases that we refer to as greenhouse gases, which include water vapor, carbon dioxide, methane, and nitrous oxide. Page 821
- Humans are increasing the greenhouse effect by increasing the concentrations of carbon dioxide, methane, nitrous oxide, and other chemicals in the atmosphere. Page 821
- Some regions of the world have cooled a bit whereas others have warmed a great deal, such as increases of 4°C (7°F) in Alaska, northern Canada, and northern Europe. Page 822
- Small temperature differences can affect important biological processes, such as when flowering plants make their flowers, when birds migrate, when animals breed, and whether species can survive under the warmer temperatures. Page 822
- In addition to rising CO_2 concentrations causing global warming, they also can affect the pH of our oceans, which has additional effects on ecosystems. Page 822

LG 57.4 We have solutions to help address species declines and disrupted ecosystems.

- There has been a rapidly growing trend of protecting terrestrial habitats since the 1960s that is now starting to plateau. Page 824
- There has been less emphasis on protecting marine habitats, but the situation started to improve by the 1980s and has dramatically increased in the 2000s. Page 824
- It can be exceedingly difficult to rid an area of an invasive species once the population has spread widely. Page 824
- In some cases we have been able to reintroduce native species into their native habitat. Page 825
- Reduced harvesting is another way to bring back species from the brink of extinction. Page 825
- The key to reducing global warming and the associated global climate change lies in reducing the concentration of greenhouse gases in the atmosphere. Page 825

Review Questions

1. Which is not true about ice age effects on species diversity?
   (A) Species of pine trees colonized areas where glaciers receded within a few decades.
   (B) Advancing glaciers destroyed all plant and animal diversity in their paths.
   (C) Receding glaciers revealed bare soil for ecological succession to begin.
   (D) Advancing glaciers caused many species to experience altered distributions.

2. Why do many biologists think that we may be experiencing a sixth mass extinction?
   (A) The current rate of extinction is greater than other mass extinction events.
   (B) We have experienced 75% of extinct species during the past 2 million years.
   (C) The current rate of extinction is higher than rates in previous 500-year periods.
   (D) There have been more extinctions during the past century than at any time in Earth's history.

3. Which statement is not true about causes of decline in biodiversity?
   (A) Unregulated commercial harvest has led to declines in many populations.
   (B) Habitat loss is a minor cause.
   (C) Exotic species are often accidentally introduced.
   (D) Pollutants can kill some species, leading to population declines.

4. Which is not a greenhouse gas that contributes to global warming?
   (A) Water vapor
   (B) Carbon dioxide
   (C) Methane
   (D) Nitrogen gas

5. Which statement is true about solutions for reversing the decline in species diversity?

(A) Species reintroductions work even if the human threats are not mitigated.

(B) We are experiencing a decline in the amount of habitat being protected around the world.

(C) Reduced harvesting does not substantially help populations rebound.

(D) It is easy to control populations of exotic species.

Module 57
AP® Practice Questions

Section 1: Multiple-Choice Questions

Choose the best answer for questions 1–4.

Use the following information to answer questions 1–3.

The figure depicts the growth of the human population over the past 200 years and the number of species that went extinct over the same period.

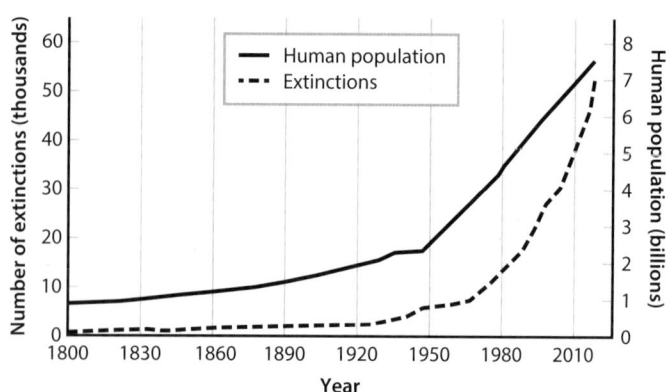

Source: Scott, J. M. 2008. *Threats to Biological Diversity: Global, Continental, Local.* U.S. Geological Survey, Idaho Cooperative Fish and Wildlife, Research Unit, University of Idaho.

1. Which of the following is supported by the data in the figure?

(A) From 1800 to 1920, there was no change in either the human population or the number of extinctions.

(B) From 1800 to 1920, the rate of extinction was lower than was the rate of human population growth.

(C) From 1950 to 2000, the rate of human population growth was inversely related to the rate of increasing extinctions.

(D) From 1950 to 2000, the rate of human population growth and the hunting of species caused the increasing extinctions that occurred during the same period.

2. The figure indicates that the rate of change in the size of the human population was not constant from 1800 to 2010. Estimate the yearly average rate of change of the human population from 1800 to 1950 and from 1950 to 2010.

(A) From 1800 to 1950, the average rate of change of the human population was 1.5 billion people per year while from 1950 to 2010 the average rate of change was 5.5 billion per year.

(B) From 1800 to 1950, the average rate of change of the human population was 1 million people per year while from 1950 to 2010 the average rate of change was 5 million per year.

(C) From 1800 to 1950, the average rate of change of the human population was 10 million people per year while from 1950 to 2010 the average rate of change was 75 million per year.

(D) From 1800 to 1950, the average rate of change of the human population was 100 million people per year while from 1950 to 2010 the average rate of change was 500 million per year.

3. AP® Biology students examining the graph hypothesize that the rate of nonhuman species extinction depends on the size of the human population. Predict which of the following would be correct if the null hypothesis fails to be rejected.

(A) Both the human population and the rate of extinction continue increasing.

(B) The human population declines and the rate of extinctions declines.

(C) The human population stops increasing and the rate of extinctions stops increasing.

(D) The human population stops increasing but the rate of extinctions continues increasing.

4. Ecologists studied the presence of a small mammalian genus that exists across Earth, excluding Antarctica. They observed that a different species within this genus lives on each of the six continents. However, on five of the continents, the species of this genus share many more traits among themselves than they do with the species on the sixth continent. Which of the following best explains these observations?

(A) All species of the genus are niche specialists.

(B) Species of the genus are not capable of migration.

(C) The sixth continent separated from the others earlier than the remaining continents separated from one another.

(D) Species of the genus present on the five continents adapted better to their environments than did the species of the sixth continent.

5. Ecologists studying the evolutionary history of two species of pine trees hypothesize that the two species expanded their ranges by moving northward in North America after the end of the last ice age. If species A expanded its range northward at a faster rate than species B, which of the following would be correct about the pollen grains found in the northern lakes?

(A) Pollen grains of only species A would be found in the lake sediments.

(B) Pollen grains of species A would be found in the same lake sediments as those of species B.

(C) Pollen grains of species A would be found in deeper lake sediments than those of species B.

(D) Pollen grains of species A would be found in shallower lake sediments than those of species B.

Section 2: Free-Response Question

Write your answer to each part clearly. Support your answers with relevant information and examples. Where calculations are required, show your work.

The pied flycatcher (*Ficedula hypoleuca*) is a migratory bird that spends winters in Africa and then migrates to Europe to breed in the spring. The great tit (*Parus major*) is another species of bird that lives in Europe year-round. The two bird species compete for a limited number of nesting sites. Over the past 20 years, global temperature increases have produced warmer winters in Europe, which have resulted in earlier availability of good nesting sites and food sources for both species of birds. Scientists recorded the average December temperature for 5 years and the proportion of high-quality nest sites that were occupied by great tits before pied flycatchers migrated back to Europe each year. The data the scientists obtained is shown in the figure.

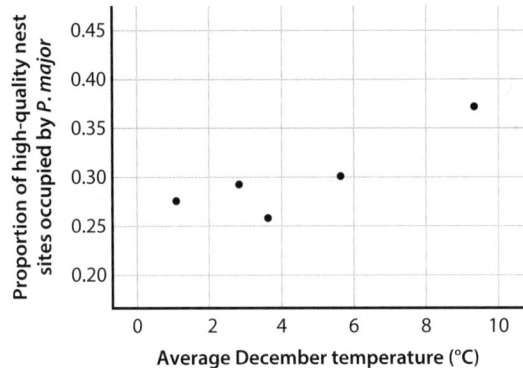

(a) **Identify** the proportion of high-quality nest sites that were occupied by *P. major* when the average December temperature was 2.5°C.

(b) **Calculate** the difference between the maximum and the minimum observed nest site occupancies by *P. major* over the study period.

(c) **Describe** the relationship between the mean December temperature and the proportion of nest sites that are occupied by *P. major*. **Make a claim** to explain this relationship.

(d) Scientists have observed that the timing of pied flycatcher migration from Africa to Europe does not change, regardless of the local weather conditions in Europe. **Predict** the consequence for the pied flycatcher population if European winter temperatures continue to increase.

Unit 8

AP® Practice Questions

Section 1: Multiple-Choice Questions

Choose the best answer for questions 1–15.

1. A population of 180 marine iguanas lives on the Galápagos Islands and has a carrying capacity of 500 individuals. In one year, the population grew to 230. What is the maximum population growth rate (r_{max})?

   (A) 0.16

   (B) 0.29

   (C) 0.40

   (D) 0.46

2. In the 1980s and 1990s, overfishing by humans caused a significant decrease in the cod population in the northwest Atlantic. Predict how the removal of this predator affected the abundance of other species, such as small fish and shrimp, which are consumed by cod, and zooplankton, which are consumed by the small fish.

   (A) Small fish, shrimp, and zooplankton all increased in abundance.

   (B) Small fish and shrimp increased in abundance while zooplankton decreased in abundance.

   (C) There was no effect on the abundance of the other species because other predatory fish species filled the niche of cod.

   (D) There was increased competition between small fish and shrimp for zooplankton, decreasing overall biodiversity.

3. The following table shows the growth of a bacteria population in an experiment.

Hours	Number of bacteria
1	20
2	50
3	120
4	270
5	600

   Which statement is supported by the data in this table?

   (A) The per capita growth rate (r) is greater than zero.

   (B) The max per capita growth rate (r_{max}) is very small in magnitude.

   (C) The carrying capacity is 600 bacteria.

   (D) The population is experiencing growth constraints.

4. Which species interaction correctly depicts a relationship between a heterotroph and an autotroph?

   (A) The parasitism of a parasitoid wasp and a caterpillar

   (B) The competition between tall trees and shorter ferns for light

   (C) The predation of a fish consuming mosquito larvae

   (D) The herbivory of a koala eating eucalyptus leaves

Use the following information to answer question 5.

Feral cats are wild cats that were once domesticated or descended from domesticated cats. In Australia, feral cats are responsible for the extinction of numerous mammal species. One suggestion to control the feral cat population is to introduce dingoes, which are carnivorous mammals that prey on small mammals, to areas infested with feral cats. The hope is that dingoes will prey on feral cats and reduce their population.

5. Which of the following could be a concern about introducing dingoes to control the feral cats?

   (A) Dingoes may prey upon threatened species in the habitats to which they are introduced.

   (B) Dingoes may greatly reduce or eliminate the feral cat population in the habitats to which the dingoes are introduced.

   (C) Dingoes will accelerate the feral cat elimination of native mammal species.

   (D) Dingoes will accelerate the extinction of native plant species.

Use the following information to answer question 6.

Scientists measured the abundance of insects, mammals, birds, and plants at several sites of different temperatures. They were able to calculate the Simpson's diversity index values for each site, as shown in the graph below.

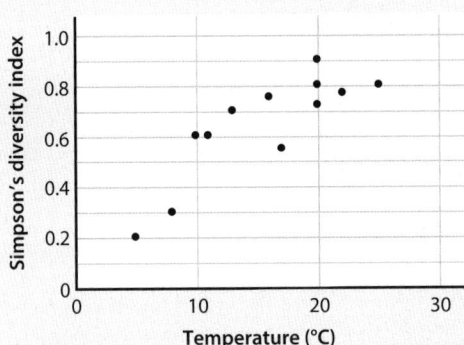

6. What overall conclusion can be drawn from these data about the biodiversity of these sites?

   (A) Biodiversity decreases as temperature increases.

   (B) Mid-range temperatures support the greatest biodiversity.

   (C) Biodiversity increases as temperature increases.

   (D) Extremely cold and extremely warm temperatures have the greatest biodiversity.

Use the following figure to answer question 7.

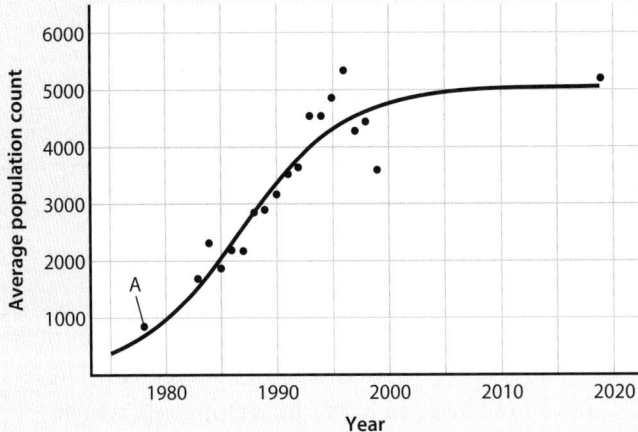

Data from https://www.researchgate.net/publication/341350203_Causes_of_Mortality_in_a_Harbor_Seal_Phoca_vitulina_Population_at_Equilibrium

7. The graph shows a population growth curve for harbor seals in the San Juan Islands. Which statement is true of point A on the curve?

   (A) Competition between seals is at its highest.

   (B) Exponential growth is occurring.

   (C) Linear growth is occurring.

   (D) The death rate is greater than the birth rate.

8. Which is a good indicator that an ecosystem is likely to have high resilience to an environmental disturbance?

   (A) High biodiversity

   (B) High productivity

   (C) High intrinsic value

   (D) High species evenness

9. Which statement is true regarding the difference in metabolic rate between a zebra and a rabbit?

   (A) Larger animals require more food per kilogram of weight because of their higher surface area-to-volume ratio.

   (B) Zebras are physically larger than rabbits and must consume more food to provide energy for everyday activities.

   (C) The metabolic rate of the animals is a negative correlation; the rate decreases as the animals increase in size.

   (D) The difference in metabolic rate between zebras and rabbits is minimal because both are considered endotherms.

10. Large populations with high genetic diversity are more likely to

(A) be resilient to natural or human-caused disturbances.

(B) be outcompeted by invasive species.

(C) suffer great losses due to harmful pathogens or fungi.

(D) experience the effects of a trophic cascade.

11. Robert Paine conducted a famous experiment involving the removal of the starfish *Pisaster* from their environment on the Washington state seacoast. This caused an increase in barnacles and mussels, which crowded out many other herbivores and algae species. Which statement supports these findings?

(A) The removal of *Pisaster* represents a trophic cascade that began from the bottom and worked its way up through the trophic levels.

(B) The number of species present without *Pisaster* is low only because of direct effects.

(C) *Pisaster* is a keystone species responsible for the collapse of diversity in the ecosystem when it is removed.

(D) A disturbance was created by removing *Pisaster* from the ecosystem, but the number of species will return to normal through the process of succession.

Use the following information to answer question 12.

African elephants are a vital member of the grassland ecosystem. They consume saplings, which maintains the grassland habitat and prevents it from becoming a woodland area. This provides a steady supply of grass for herbivores to graze and thus high prey numbers for large predator species. The graph below shows the data collected from an African savanna reserve, counting the number of species within its bounds during a 20-year time span.

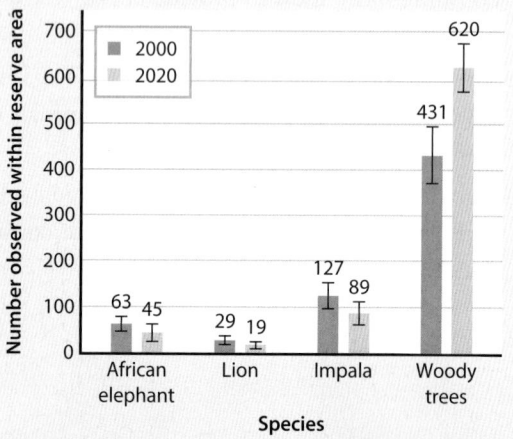

12. Which population has significantly changed in size between 2000 and 2020?

(A) African elephants

(B) Lions

(C) Impala

(D) Woody trees

Use the following information to answer question 13.

Ecologists study three animal species living on a tract of land, species A, B, and C. They track the number of offspring born to each female and the number of the offspring that survive to sexual maturity. The data collected are listed in the table.

Species	Average number of offspring born per female	Number of offspring that survive to sexual maturity
A	44	3
B	3	3
C	36	2

13. Which is a *K*-selected species?

(A) Species A

(B) Species B

(C) Species C

(D) Species A, B, and C

Use the following information to answer question 14.

Researchers examined the resistance of the bacterium *Streptococcus pneumoniae* to the antibiotics co-trimoxazole and erythromycin over a 10-year period. The results are shown in the figure.

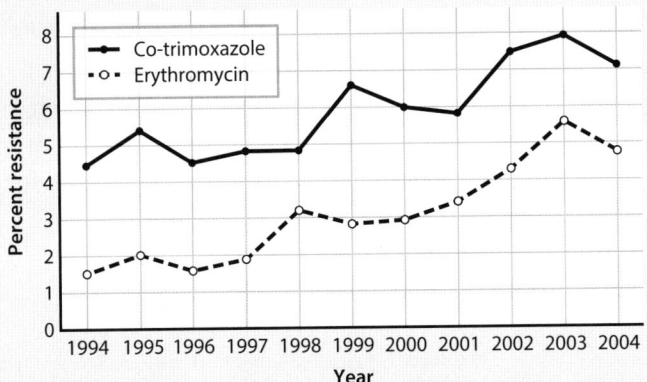

Data from https://ars.els-cdn.com/content/image/1-s2.0-S1473309908700173-gr3_lrg.jpg

14. Which of the following do the data support?

(A) Resistance to either co-trimoxazole or erythromycin was not present in these bacterial populations prior to 1994.

(B) The bacteria were initially more resistant to erythromycin than co-trimoxazole

(C) The percentage of bacteria resistant to erythromycin has declined over the 10-year period but the percentage of bacteria resistant to co-trimoxazole has increased over the same time period.

(D) The percentage of bacteria resistant to co-trimoxazole and erythromycin has increased over the 10-year period.

15. A species of weed has a maximum per capita growth rate of 0.3 per month. If the weed species is experiencing exponential growth and the population size is 20 weed plants, what will the population size be in 3 months?

(A) 26 weed plants
(B) 34 weed plants
(C) 44 weed plants
(D) 78 weed plants

Section 2: Free-Response Questions

Write your answer to each part clearly. Support your answers with relevant information and examples. Where calculations are required, show your work.

1. The Glanville fritillary (*Melitaea cinxia*) is a butterfly that lives in a patchy environment in Finland. Because of frequent environmental change, new patches of habitat regularly appear. These patches are colonized by migrant butterflies, which establish a small local population that grows to the carrying capacity of the patch. Scientists have observed the growth dynamics of Glanville fritillary in one particular patch (patch A) since the habitat appeared. They observed the number of butterflies every year in Glanville fritillary in patch A and obtained the data shown in the figure below.

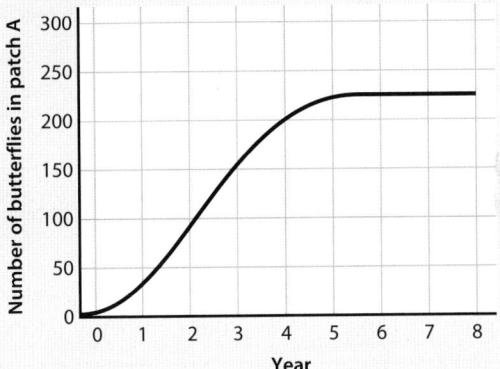

Using these data, they estimated that the maximum growth rate of a Glanville fritillary population is $r_{max} = 0.63$. While the scientists were collecting their data, a second habitat patch appeared (patch B). The second patch was colonized by 26 butterflies during the first year, and the scientists estimated that it has a carrying capacity of 125 butterflies. The scientists assume that the r_{max} value for patch B is the same as the value for patch A.

(a) Based on the data provided, **identify** the carrying capacity of patch A.

(b) Based on the data provided, **predict** the number of Glanville fritillaries that are expected to be observed in patch B one year after it was colonized.

(c) **Explain** why populations grow at slower rates when their population sizes are slightly lower than the carrying capacity than they do when their population sizes are much lower than the carrying capacity.

2. The figure below illustrates a food web showing trophic relationships between species in a field. The table below the food web shows the number of times each species was observed in a 1-km² area of the field.

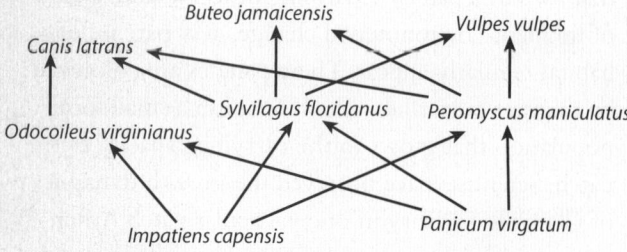

Species	Number of individuals in sampling area
Panicum virgatum	1863
Impatiens capensis	367
Peromyscus maniculatus	295
Sylvilagus floridanus	118
Odocoileus virginianus	26
Vulpes vulpes	18
Canis latrans	10
Buteo jamaicensis	4
Total	2701

(a) **Identify** one species in the food web that is most likely a producer.

(b) **Predict** the effect that removing *Vulpes vulpes* from this ecosystem would have on the abundance of *Panicum virgatum*. **Provide reasoning** to support your prediction.

(c) **Identify** two organisms in the food web that are likely to compete with each other.

(d) Assuming the organisms shown in the food web and the table include all the major species in this ecosystem, **calculate** Simpson's diversity index for this ecosystem.

Cumulative AP® Biology Practice Exam

Section 1: Multiple-Choice Questions

Choose the best answer for questions 1–60.

1. Identify the main source of oxygen gas in Earth's atmosphere.

   (A) Natural geologic processes

   (B) Photosynthesis by plants, algae, and bacteria

   (C) Fermentation by microorganisms

   (D) Decay of organic material

2. Extremophiles are microorganisms that live in environments characterized by extreme temperature, acidity, salt concentrations, or pH. Many extremophiles that live in high-temperature environments without oxygen perform anaerobic cellular respiration with an electron transport chain that uses iron or sulfides as terminal electron acceptors instead of oxygen. Which of the following explains why this form of cellular respiration is observed in extreme environments without oxygen?

   (A) The extremophiles needed to obtain additional energy from glucose so they learned to use alternative electron receptors for respiration.

   (B) The iron and sulfides created a toxic environment so the extremophiles evolved anaerobic cellular respiration to detoxify their surroundings.

   (C) Extremophiles that evolved anerobic cellular respiration had a competitive advantage over other microbes and were able to persist in the environment.

   (D) The extremophiles did not have organelles so they established an electron transport chain in the cell membrane.

Use the following table to answer question 3.

First position (5' end)	Second position				Third position (3' end)
	U	C	A	G	
U	UUU Phe UUC Phe UUA Leu UUG Leu	UCU Ser UCC Ser UCA Ser UCG Ser	UAU Tyr UAC Tyr UAA Stop UAG Stop	UGU Cys UGC Cys UGA Stop UGG Trp	U C A G
C	CUU Leu CUC Leu CUA Leu CUG Leu	CCU Pro CCC Pro CCA Pro CCG Pro	CAU His CAC His CAA Gln CAG Gln	CGU Arg CGC Arg CGA Arg CGG Arg	U C A G
A	AUU Ile AUC Ile AUA Ile AUG Met/Start	ACU Thr ACC Thr ACA Thr ACG Thr	AAU Asn AAC Asn AAA Lys AAG Lys	AGU Ser AGC Ser AGA Arg AGG Arg	U C A G
G	GUU Val GUC Val GUA Val GUG Val	GCU Ala GCC Ala GCA Ala GCG Ala	GAU Asp GAC Asp GAA Glu GAG Glu	GGU Gly GGC Gly GGA Gly GGG Gly	U C A G

3. A portion of a protein is encoded by the following mRNA sequence.

   5'… AUU GCA <u>A</u>GA UCC AGC … 3'

   Which of the following statements correctly predicts the consequence of a mutation that changes the underlined A to a U?

   (A) The mutation would result in a substitution of a different amino acid at the codon position and might change the function of the protein.

   (B) The mutation would change the mRNA sequence but would not change the protein sequence or function.

   (C) The mutation would reduce the length of the polypeptide and could change or eliminate protein function.

   (D) The mutation would change the mRNA expression level and could change protein function.

Use the following information to answer questions 4–6.

As illustrated in the figure below, most plants absorb water from the soil through their roots and then use that water in biochemical reactions. Water vapor is released from plant leaves into the atmosphere through a process called transpiration.

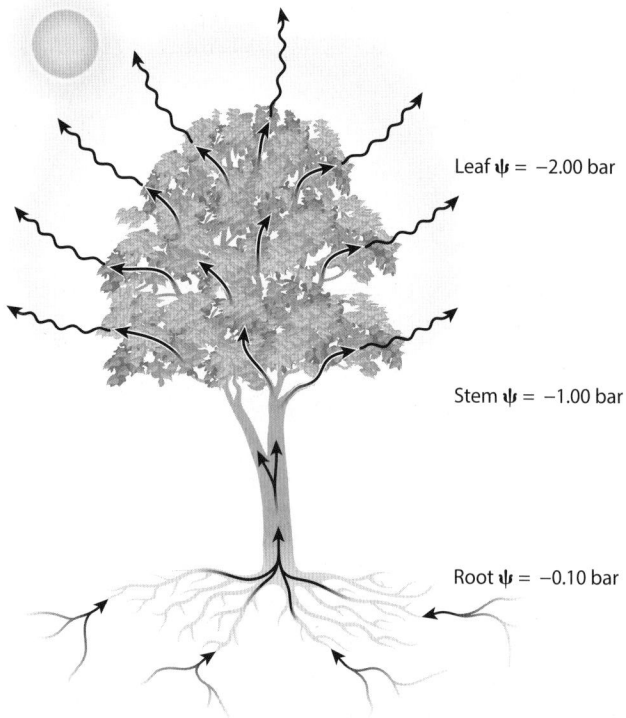

Leaf ψ = −2.00 bar

Stem ψ = −1.00 bar

Root ψ = −0.10 bar

4. Identify a biochemical process in plants that requires water as a reactant.
   (A) Dehydration synthesis to form starch
   (B) Hydrolysis of starch into simple sugars
   (C) Aerobic respiration
   (D) Anaerobic fermentation

5. The driving force for transpiration is the difference in water potential between the soil and the atmosphere surrounding the plant. Which of the following statements best describes the potential for water movement as illustrated in the figure?
   (A) The water potential in the soil is higher than the water potential in the roots, which causes water to move from the soil into root cells.
   (B) The water potential in the roots is higher than the water potential in the soil, which causes water to move from the roots into the stem of the plant.
   (C) The water potential in the stem of the plant is higher than the water potential in the roots, which causes water to distribute into the leaves.
   (D) The water potential in the atmosphere is higher than the water potential in the leaves, which causes water vapor to evaporate from the leaves.

6. Changes in the environment can affect the efficiency of transpiration. Which of the following statements accurately pairs an environmental change with a prediction about water movement?
   (A) Decreasing water concentration in the air is predicted to slow water movement through the plant, thus slowing the rate of transpiration.
   (B) Decreasing the number of leaves on the plant is predicted to decrease the water potential in the leaves and therefore decrease the rate of transpiration.
   (C) Increasing the amount of water in the soil is predicted to increase the rate of water movement into the roots and therefore increase the rate of transpiration.
   (D) Increasing the volume of the plant will decrease the water potential of the stem and is therefore predicted to increase the rate of transpiration.

Use the following figure to answer question 7.

a.

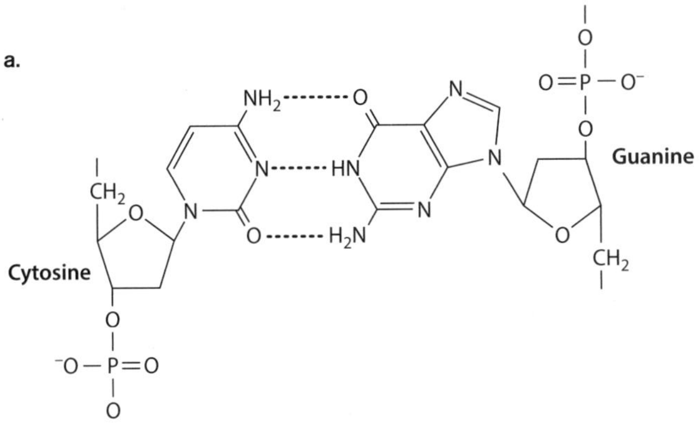

Cytosine ⋯ Guanine

b.

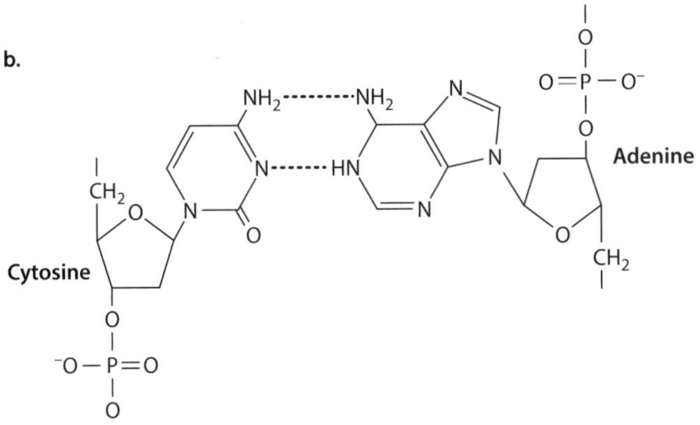

Cytosine ⋯ Adenine

c.

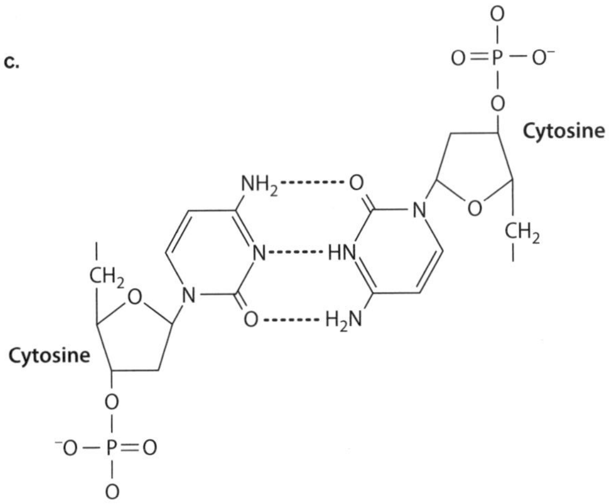

Cytosine ⋯ Cytosine

d.

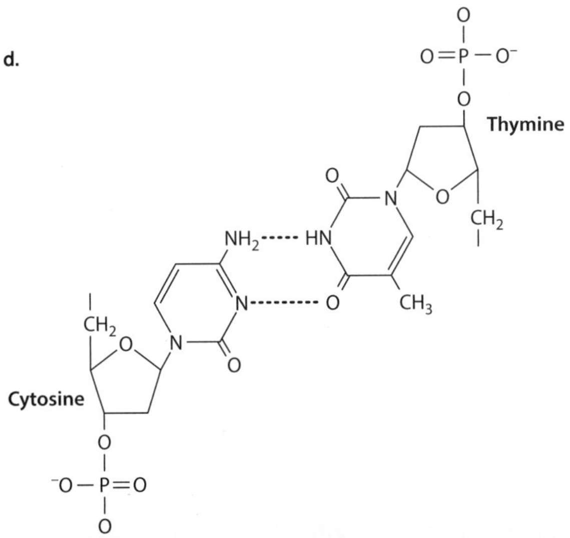

Cytosine ⋯ Thymine

7. Which best illustrates nucleotide base pairing in a strand of DNA?

   (A) a   (B) b   (C) c   (D) d

8. An organism called species X is observed to have a small population size relative to other species in an ecosystem. Scientists experimentally removed species X from the ecosystem and observed very large effects on the population sizes of other species in the system. Which of the following statements best describes species X?

   (A) Species X is a primary producer that supplies most of the energy for other species in the system.

   (B) Species X is an essential species that increases the evolutionary fitness of other species in the system.

   (C) Species X is a keystone species that is critical for ecosystem stability.

   (D) Species X is a top predator that decreases the population size of producers in the system.

Use the following information to answer questions 9 and 10.

In a population of damselflies, there are two alleles of the *Est* gene: Est^F and Est^S. A scientist collected 120 damselflies from a natural population and determined their *Est* genotypes. The scientist obtained the data shown in the table. The scientist used these data to test whether the population was in Hardy–Weinberg equilibrium.

Genotypes Obtained from 120 Damselflies

Genotype	$Est^F Est^F$	$Est^F Est^S$	$Est^S Est^S$
Number of individuals	6	46	68

9. Calculate the expected number of $Est^F Est^S$ individuals in the population if the population is in Hardy–Weinberg equilibrium.

   (A) 13   (B) 46   (C) 44   (D) 55

10. Calculate the chi-square test statistic that describes the goodness of fit of the observed genotype data to Hardy–Weinberg expectations.

   (A) 0.01

   (B) 0.42

   (C) 2.99

   (D) 0.26

11. Mammalian guts contain complex microbiological communities that differ in their biological properties. *Salmonella enterica* is one bacterium that is commonly found in the human gut. *Proteus mirabilis* is also found in the human gut. Some strains of *S. enterica* are highly resistant to multiple antibiotics, including an antibiotic called chloramphenicol. Chloramphenicol is very effective at killing strains of *P. mirabilis* that are found in the environment. However, in an atypical case, scientists isolated both *S. enterica* and *P. mirabilis* from the gut microbial community of a single human patient and discovered that both bacteria were resistant to chloramphenicol. Which of the following claims provides the best explanation for the observed resistance to chloramphenicol by the atypical strain of *P. mirabilis*?

(A) *P. mirabilis* acquired DNA from *S. enterica* in the patient's gut, which made the *P. mirabilis* resistant to chloramphenicol.

(B) *P. mirabilis* absorbed detoxification enzymes from *S. enterica* in the patient's gut, which enabled the *P. mirabilis* to degrade the chloramphenicol.

(C) *S. enterica* degraded the chloramphenicol in the patient's gut so it could no longer kill the *P. mirabilis*.

(D) *S. enterica* activated gene expression by *P. mirabilis* in the patient's gut, which made the *P. mirabilis* resistant to chloramphenicol.

12. Identify which of the following double-stranded DNA molecules would require the greatest energy input to be separated into single strands.

(A) A double-stranded DNA molecule that is 30% cytosine

(B) A double-stranded DNA molecule that is 25% thymine

(C) A double-stranded DNA molecule that is 20% guanine

(D) A double-stranded DNA molecule that is 15% adenine

Use the following figure to answer question 13.

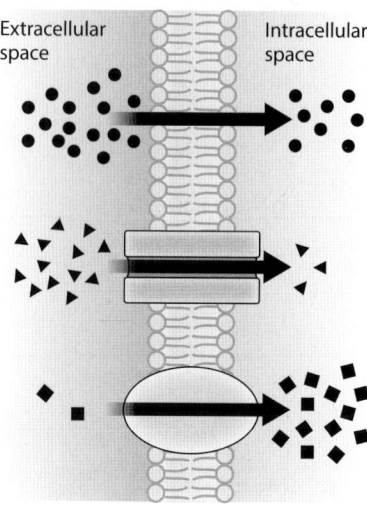

13. The figure shows the transport of three different molecules across a membrane. Each molecule is represented by a different shape. Which of the following statements correctly describes the movement of the molecules across the membrane?

(A) The circular molecule is able to pass through the membrane by simple diffusion.

(B) The triangular molecule is actively transported through a protein channel against its concentration gradient.

(C) The square molecule is passively transported through a protein channel against its concentration gradient.

(D) All three molecules require energy input in order to cross the membrane.

Use the following information to answer questions 14 and 15.

Betta splendens is a popular species of aquarium fish. Body color in *B. splendens* is determined by a gene that has two alleles, *C* and *c*, where the *C* allele is dominant and results in a dark body color. The recessive *c* allele results in a light body color. A different gene with two alleles, *T* and *t*, determines the shape of the fish's tail. The *T* allele is dominant and results in a tail with a single large lobe. The *t* allele is recessive and results in a tail that is split into two lobes. A male fish with the genotype *Cc TT* is crossed to a female fish with genotype *cc Tt*.

14. Identify the phenotype of the parental male fish in the cross.

(A) Dark body and single-lobed tail

(B) Light body and single-lobed tail

(C) Dark body and two-lobed tail

(D) Light body and two-lobed tail

15. Calculate the percentage of progeny from this cross that are expected to have two-lobed tails.

    (A) 0%

    (B) 33%

    (C) 50%

    (D) 100%

Use the following figure to answer question 16.

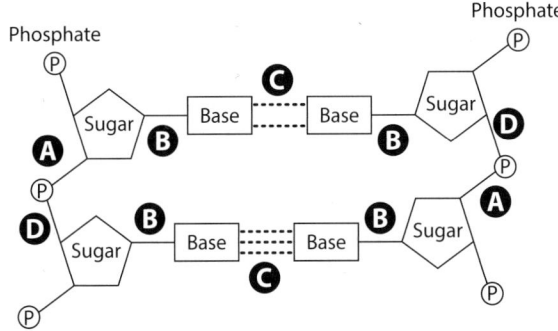

16. A chaotropic agent is a chemical that can disrupt the structure of nucleic acids by interfering with hydrogen bonding. Identify the letter on the figure that corresponds to the bonds that a chaotropic agent is most likely to disrupt.

    (A) A

    (B) B

    (C) C

    (D) D

Use the following information to answer questions 17–19.

The common buzzard (*Buteo buteo*) is a bird of prey whose color is determined by a mutation in the autosomal *mc1r* gene. Buzzards that are homozygous for the *mc1r*⁺ allele have a tan color, buzzards that are homozygous for the *mc1r^mut* allele have a black color, and heterozygous buzzards are brown. In the population, 43% of the birds are tan, 20% of the birds are black, and 36% of the birds are brown. There is no difference between males and females in color frequency. Scientists observed 366 randomly chosen buzzard breeding pairs and recorded the color of the parents. The data they obtained are shown in the table.

		Female parent		
		Tan	Brown	Black
Male parent	Tan	108	30	1
	Brown	29	124	19
	Black	3	19	33

17. Based on the information provided, calculate the frequency of the *mc1r^mut* allele in the population.

    (A) 0.38

    (B) 0.45

    (C) 0.60

    (D) 0.66

18. Based on the information provided, predict the percentage of offspring that are expected to be tan from a breeding pair composed of a tan female and a brown male.

    (A) 0%

    (B) 25%

    (C) 50%

    (D) 75%

19. Based on the information provided, identify the assumption underlying Hardy–Weinberg equilibrium that is most likely to be violated in this system.

    (A) No mutation

    (B) No natural selection

    (C) Random mating

    (D) No migration or gene flow

20. Which of the following statements correctly describes energy flow in an ecosystem?

    (A) All the energy from lower levels of the system is available to higher levels in the system.

    (B) The producer level directly provides energy to each of the higher levels in the system.

    (C) Energy is transferred directly from primary consumers to secondary consumers in the system.

    (D) The highest level of consumer in the system holds the highest amount of energy in the system.

21. Which of the following best describes the flow of free energy within the cells of a eukaryotic organism?

    (A) Catabolic reactions → oxidative phosphorylation → NAD⁺ reduction → anabolic reactions

    (B) Catabolic reactions → NAD⁺ reduction → oxidative phosphorylation → anabolic reactions

    (C) Anabolic reactions → NAD⁺ reduction → oxidative phosphorylation → catabolic reactions

    (D) Anabolic reactions → oxidative phosphorylation → NAD⁺ reduction → catabolic reactions

Use the following figure to answer question 22.

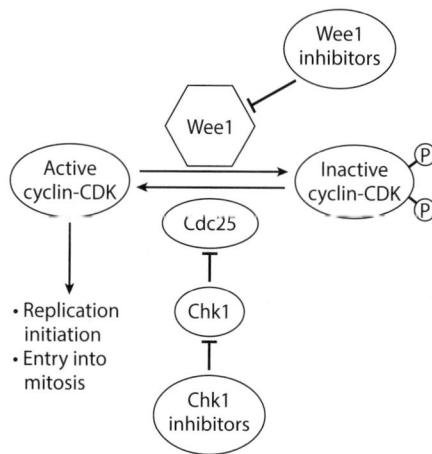

22. The cell cycle is controlled by a complex interaction of molecules, including cyclins and cyclin-dependent kinases (CDKs), that coordinate key events within the cycle. Three major checkpoints, regulated by internal and external signals, determine whether the cell will advance through the cycle. Certain conditions must be met in order to form, activate, and then inactivate the cyclin-dependent kinases. The figure illustrates part of the molecular interactions that determine activation and inhibition of the cycle. Based on the information provided, identify the statement that accurately describes the molecular interactions governing the cell cycle. Note that regular arrowheads indicate activation, and T arrowheads indicate inhibition.

(A) Wee1 inhibitors will block the activity of Cdc25 when they are present in the cell, and will inhibit entry of the cell into mitosis.

(B) The inactive form of CDK cyclin can be activated in the presence of CHK1.

(C) Activated CDK cyclin molecules prevent activation of Cdc25.

(D) The presence of Wee1 molecules blocks the progress of the cell cycle by inhibiting DNA replication.

Use the following information to answer question 23.

A researcher has been studying chemical X, which is hypothesized to change the rate of cell division. To test this hypothesis, the researcher set up two samples of identical cell types. The cells in sample 1 were not exposed to chemical X. The cells in sample 2 were exposed to chemical X. The researcher counted the number of cells in interphase and M phase in a sample of 125 cells from each treatment. The data the researcher obtained are shown in the table. The researcher used these data to perform a chi-square test to determine whether the number of cells undergoing mitosis changes significantly after exposure to chemical X.

	Number of cells in interphase	Number of cells in M phase
Sample 1 (no chemical X)	113	12
Sample 2 (chemical X applied)	79	46

23. Which of the following is the most appropriate null hypothesis for this experiment?

(A) If the cells are treated with chemical X, there will be an increase in the number of cells in mitotic phase.

(B) If the cells are treated with chemical X, there will no change in the number of cells in mitotic phase.

(C) If the cells are not treated with chemical X, there will be a decrease in the number of cells in interphase.

(D) If the cells are not treated with chemical X, there will be no change in the number of cells in interphase.

24. Which of the following correctly describes the role of the enzyme DNA ligase in DNA replication?

(A) Ligase relaxes the structure of the DNA molecule so it can be accessed by DNA polymerase.

(B) Ligase separates the strands of the DNA molecule so DNA polymerase can copy the template strand.

(C) Ligase links the nucleotides bound by DNA polymerase in order to form the sugar–phosphate backbone of the newly copied DNA.

(D) Ligase connects short fragments of copied DNA to form a longer continuous DNA strand.

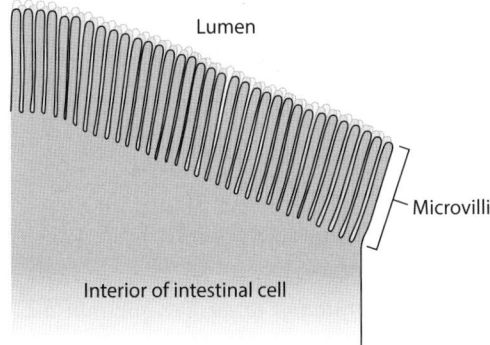

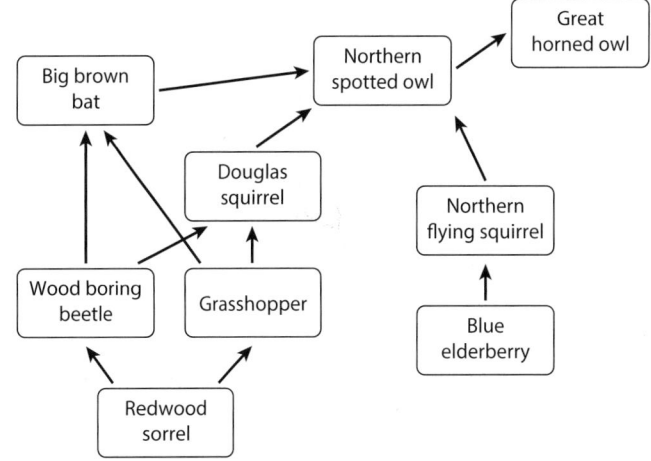

25. In animals, the inner surface of the small intestine absorbs nutrients like sugars and amino acids from digested food. The cells of the inner lining of the intestine have long projections called microvilli that extend into the lumen where the food passes, as shown in the figure. Which of the following statements best explains role of the microvilli?

   (A) The microvilli create a physical structure that slows the passage of food through the intestine in order to increase opportunity for nutrient absorption.

   (B) The microvilli create a barrier that protects the intestinal cells from pathogens and harmful molecules in the food.

   (C) The microvilli create a larger surface area without significantly increasing the volume of the intestinal cells so that nutrients can be absorbed more efficiently.

   (D) The microvilli create membrane barriers that slow the diffusion of digested food in order to increase the absorption of nutrients.

26. Maltase is an enzyme that breaks the sugar maltose into glucose monomers through a chemical reaction called hydrolysis. Which of the following statements best explains the requirement for maltase in the hydrolysis of starch?

   (A) Hydrolysis requires an input of energy, and the maltase enzyme supplies that energy.

   (B) Hydrolysis releases energy, and the maltase enzyme absorbs the released energy.

   (C) Hydrolysis is a spontaneous reaction, and the maltase enzyme makes the reaction happen more quickly.

   (D) Hydrolysis reduces entropy, and the maltase enzyme enables the reaction by stabilizing the products.

27. The figure illustrates part of a food web of species in Redwood National Forest. Based on the relationships shown, predict the most likely result of removing the northern spotted owl.

   (A) There would be an increase in the number of northern flying squirrels and a decrease in the amount of blue elderberry.

   (B) There would be an increase in the number of great horned owls and an increase in the number of Douglas squirrels.

   (C) There would be an increase in the number of big brown bats and an increase in the number of wood boring beetles.

   (D) There would be a decrease in the number of Douglas squirrels and an increase in the number of grasshoppers.

28. Which of the following statements correctly describes photosynthesis?

   (A) Photosynthesis occurs only in green plants and algae.

   (B) Photosynthesis evolved in bacteria and was subsequently acquired by plants.

   (C) Photosynthesis evolved independently in prokaryotes and plants.

   (D) Photosynthesis evolved in plants and was subsequently acquired by bacteria.

29. *Daphnia cucullata* are small aquatic crustaceans that are preyed upon by *Chaoborus flavicans*. *C. flavicans* secretes chemicals known as kairomones into water. When *C. flavicans* kairomones are present, *D. cucullata* develop a large structure known as a helmet. The helmet makes the *D. cucullata* too large to fit into the mouth of *C. flavicans* and therefore protects the *D. cucullata* from predation by *C. flavicans*. The helmet is not produced if there are no *C. flavicans* kairomones in the water. Which of the following statements best explains why *D. cucullata* does not produce a helmet in the absence of *C. flavicans* kairomones?

(A) The helmet is constructed from kairomones, so *D. cucullata* cannot produce it when the kairomones are not present.

(B) Kairomones bind to *D. cucullata* DNA and activate expression of genes required to develop the helmet, so the helmet can only be produced when kairomones are present.

(C) The helmet is costly to produce, so the *D. cucullata* would have lower fitness if they developed it when it is not necessary.

(D) The helmet is a signal to predators that they should not eat the *D. cucullata*, so the *D. cucullata* choose not to make the helmet when *C. flavicans* are absent.

Use the following figure to answer question 30.

$$H_2N-\overset{\overset{\displaystyle H}{|}}{\underset{\underset{\displaystyle OH}{\underset{\displaystyle |}{CH_2}}}{\underset{\displaystyle |}{C}}}-COOH$$

Serine
(Ser, S)

30. Serine is an amino acid that may be found on the hydrophilic interior of a protein channel. The structure of serine is shown in the figure. Identify the atoms in serine that are responsible for its hydrophilic properties.

(A) H_2N

(B) H

(C) COOH

(D) CH_2OH

31. Identify the RNA sequence that would be transcribed from this DNA template strand:

5'-CTC GTA GCC CGA-3'

(A) 5'-CUC GUA GCC CGA-3'

(B) 5'-UCG GGC UAC GAG-3'

(C) 5'-TCG GGC TAC GAG-3'

(D) 5'-GUG CUT CGG GCT-3'

Use the following information to answer question 32.

An investigator is studying dividing cells from the lymph node of a patient. The investigator uses a fluorescent DNA stain to quantify the relative amount of DNA in each cell from the lymph tissue. The figure is a schematic representation of DNA content in one of the cells of the lymph node, where *n* represents the content of DNA that would be found in a haploid gamete.

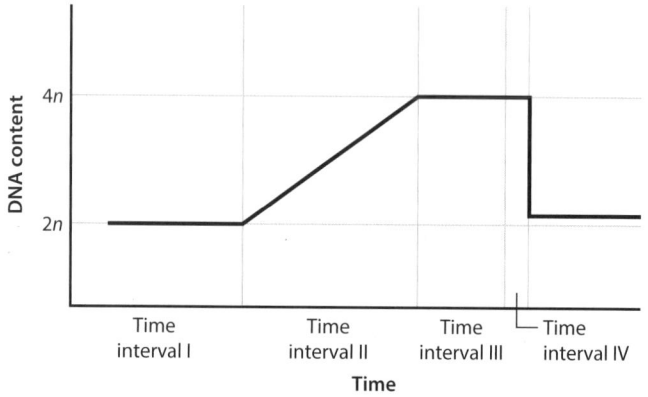

DNA content per cell during the cell cycle

32. Identify the statement that pairs a claim about the graph with correct supporting reasoning.

(A) The cell is in the mitotic phase during time interval I and therefore has half as much DNA as during other time intervals.

(B) Interphase occurs exclusively in time interval III, when the amount of DNA does not change prior to cell division.

(C) The cell is actively replicating its DNA during time interval II so the amount of DNA increases from 2*n* to 4*n*.

(D) The cell represented in the graph is undergoing meiotic division and the change in the amount of DNA between time interval I and time interval IV indicates two cell divisions.

33. Typical eukaryotic cells range in size from 10 to 100 micrometers in diameter. If a spherical cell with a diameter of 50 micrometers divides into two daughter cells, each of which has half the volume of the parent cell, how would the surface area-to-volume ratio of one daughter cell relate to that of the parent cell?

(A) The surface area-to-volume ratio of the daughter cell would be approximately 25% larger than that of the parent cell.

(B) The surface area-to-volume ratio of the daughter cell would be approximately 2.5 times larger than that of the parent cell.

(C) The surface area-to-volume ratio of the daughter cell would be approximately double that of the parent cell.

(D) The surface area-to-volume ratio of the daughter cell would be approximately half that of the parent cell.

34. Which of the following best predicts the consequence of a chemical that disrupts the thylakoid membrane of a chloroplast and makes it permeable to protons (H^+)?

(A) There would be increased accumulation of NADPH.

(B) There would be increased production of O_2.

(C) There would be decreased production of ATP.

(D) There would be decreased production of CO_2.

35. The corn borer (*Ostrinia furncalis*) is a major insect pest of corn crops in the United States. A research team was attempting to produce an insecticide to kill corn borers. The team discovered that chemical Y makes the inner mitochondrial membrane leaky to H^+ ions. Which of the following statements best predicts the outcome of applying chemical Y to corn borers?

(A) Corn borer cells will be unable to couple the oxidation of glucose molecules with the reduction of NAD^+ to NADPH, which will prevent cellular respiration and cause the corn borers to die.

(B) Corn borer cells will be unable to couple the breakdown of high-energy phospholipids in the mitochondrial membrane with regeneration of NADH molecules, which will inhibit cellular respiration and cause the corn borers to die.

(C) Corn borer cells will be unable to couple the hydrolysis of ATP to the creation of H_2O during oxidative phosphorylation, which will inhibit hydrolysis reactions required for cellular metabolism and cause the corn borers to die.

(D) Corn borer cells will be unable to couple the generation of a proton gradient to the production of ATP, so the cells will produce heat energy instead of chemical energy and the corn borers will die.

36. An aquatic ecosystem is susceptible during the summer to massive growths of blue-green algae, known as harmful algal blooms (HABs). HABs pose a threat to drinking water supplies, outdoor recreation, and the health of animals who come into contact with the water. A scientist hypothesized that HABs might be caused by phosphorus input into aquatic ecosystems from agricultural runoff. Which of the following would be the most appropriate dependent variable in an experiment designed to test the researcher's hypothesis?

(A) The health of fish and other aquatic animals in the ecosystem

(B) The abundance of blue-green algae in the aquatic ecosystem

(C) The degree of competition between blue-green algae and native plant species in the ecosystem

(D) The concentration of phosphorus in the water in the ecosystem

Use the following information to answer questions 37–39.

The 10 steps in glycolysis are shown in Figure 1 below. A scientist hypothesized that a chemical known as 3PO might inhibit glycolysis. To test this hypothesis, the scientist measured the levels of glucose, pyruvate, and three intermediate molecules in a flask of cells over a 90-minute period. The scientist added glucose to the flask to begin the experiment. After 30 minutes, the scientist added 3PO to the flask. After the second 30-minute interval, the scientist added additional glucose to the flask. The data the scientist obtained are shown in Figure 2.

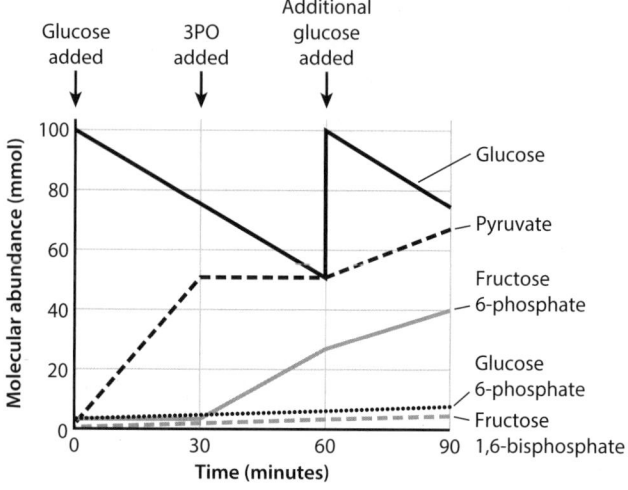

FIGURE 2 Levels of glucose, pyruvate, and three intermediate molecules over a 90-minute time interval

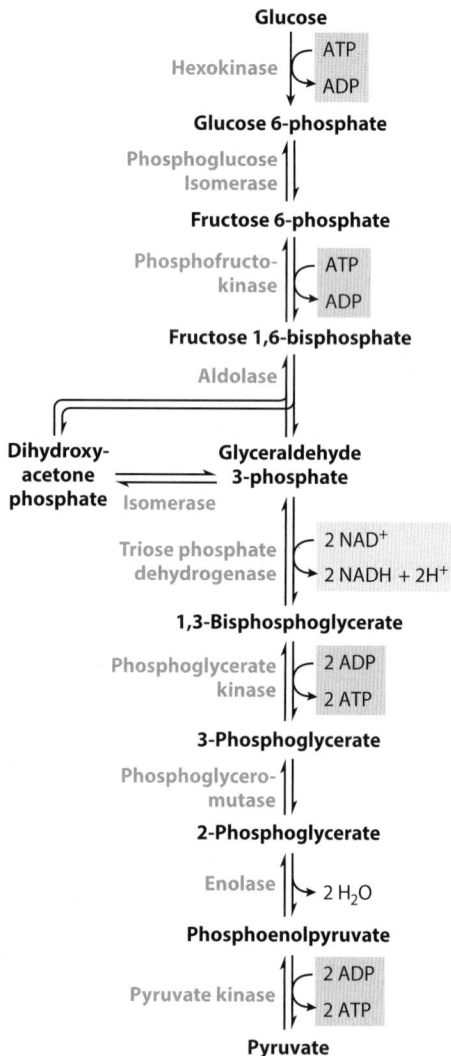

FIGURE 1 Glycolysis

37. Calculate the rate of pyruvate production during the first 30 minutes of the experiment.

(A) 0.7 mmol/minute

(B) 1.7 mmol/minute

(C) 30 mmol/minute

(D) 60 mmol/minute

38. Identify the enzyme that is most likely to be inhibited by 3PO.

(A) Hexokinase

(B) Phosphoglucoisomerase

(C) Phosphofructokinase

(D) Aldolase

39. Identify the type of inhibition exhibited by 3PO.

(A) Competitive

(B) Allosteric

(C) Irreversible

(D) Covalent

40. A scientist determines that a nucleic acid is composed of 17% adenine, 23% uracil, 28% cytosine, and 32% guanine. Identify the structure of the nucleic acid.

(A) Double-stranded DNA

(B) Double-stranded RNA

(C) Single-stranded DNA

(D) Single-stranded RNA

Use the following information to answer questions 41 and 42.

The model shown in the figure illustrates how the human body maintains blood glucose homeostasis. The liver and the pancreas are organs in the body.

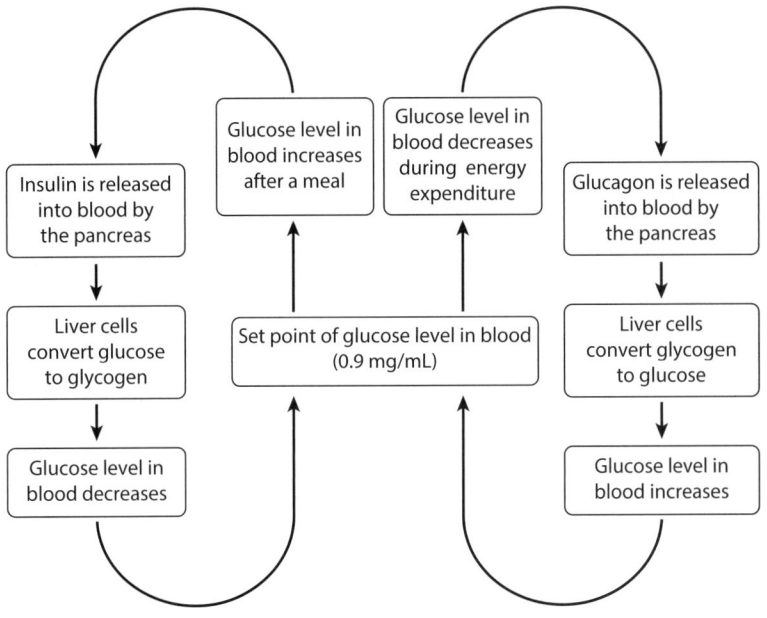

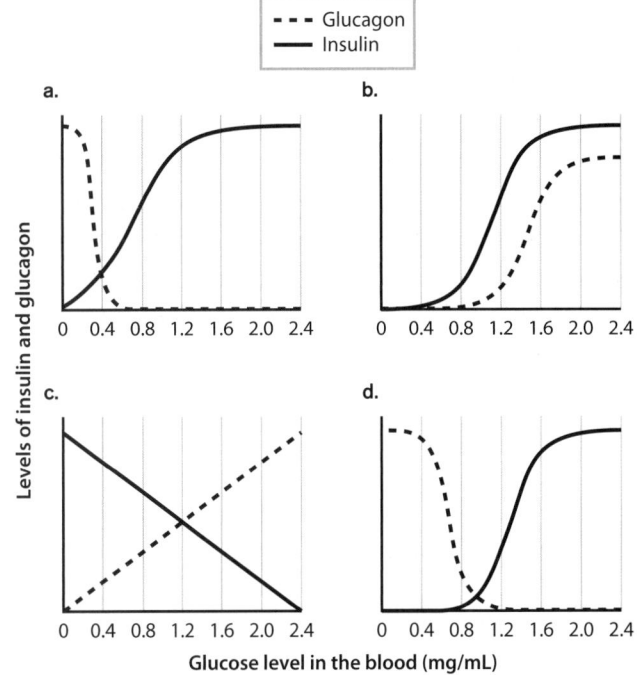

41. Based on the information provided, predict the effect of eating a large meal.

(A) The insulin level in the blood will increase, followed by increased glycogen production by liver cells.

(B) The insulin level in the blood will decrease, followed by an increase in glucose level in the blood.

(C) The glucagon level in the blood will increase, followed by conversion of glycogen to glucose by liver cells.

(D) The glucagon level in the blood will decrease, followed by an increase in glucose level in the blood.

42. Identify the graph above that best illustrates the relationship between the levels of insulin and glucagon produced by the pancreas as a function of glucose level in the blood.

(A) a   (B) b   (C) c   (D) d

43. Identify which organelle uses an electron transport chain to establish a proton gradient across a membrane.

(A) Mitochondrion

(B) Golgi

(C) Ribosome bound to endoplasmic reticulum

(D) Nucleus

44. Photosynthesis is a complex metabolic pathway that utilizes oxidation–reduction reactions as a means of converting light energy to chemical energy. During the light reactions, $NADP^+$ is reduced to NADPH. Which of the following best describes the role of NADPH in the Calvin cycle?

(A) NADPH is used to oxidize CO_2, leading to the release of O_2 into the atmosphere.

(B) Energy from NADPH is used to synthesize ATP, which is then hydrolyzed to provide energy for synthesizing carbohydrates.

(C) NADPH is an essential building block for carbohydrates, and supplies carbon and hydrogen atoms necessary for making sugars.

(D) NADPH reduces organic molecules that are generated from CO_2, and the reduced molecules are then converted to sugar.

45. Which of the following statements best describes how small nonpolar signaling molecules are able to stimulate a cellular response?

(A) Small nonpolar ligands bind to the extracellular domain of cell-surface receptors to initiate a G protein-coupled response.

(B) Small nonpolar ligands diffuse through the cell membrane and interact with receptors found in the cytosol.

(C) Small nonpolar ligands bind to the intracellular domain of cell-surface receptors and initiate a signal transduction cascade.

(D) Small nonpolar ligands bind to phospholipids embedded in the cell membrane and create channels that allow molecules to enter the cell.

Use the following figure to answer question 46.

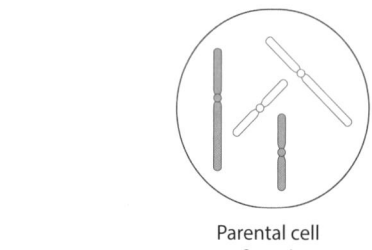

Parental cell
2*n* = 4

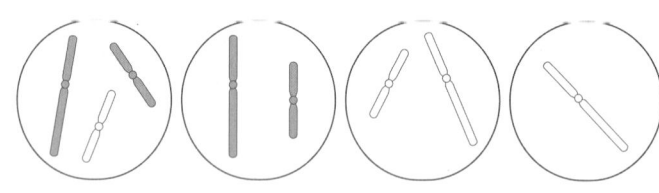

Gametes produced

46. Occasional errors in meiosis result in gametes that contain either too many or too few chromosomes. Embryos that are generated from these gametes typically do not survive. The figure illustrates gametes arising from one possible error in meiosis. Identify the most likely step at which the error in meiosis occurred.

(A) Anaphase I

(B) Telophase I

(C) Anaphase II

(D) Telophase II

47. During periods of intense exercise, the levels of oxygen in a human muscle cell can become depleted and aerobic cellular respiration is suppressed. During these periods, glycolysis may continue to produce ATP through the conversion of glucose to pyruvate. Which of the following statements best explains what happens to pyruvate in this situation?

(A) Pyruvate is converted to lactic acid, which is then converted back to glucose to fuel glycolysis.

(B) Pyruvate is converted to lactic acid, which is coupled with the oxidation of NADH, producing the NAD^+ required for glycolysis.

(C) Pyruvate is converted to lactic acid, which is coupled with the phosphorylation of ADP, producing the ATP required for hydrolysis of glucose.

(D) Pyruvate is converted to lactic acid, which is used as a source of molecular oxygen to restart cellular respiration.

48. Red foxes (*Vulpes vulpes*) typically have red fur. Occasionally they have silver fur. Fur color is coded for by a single gene and silver fur is dominant to red fur. A breeder crossed two foxes with silver fur, each of whom had one parent with red fur. The pair of silver foxes produced 19 offspring, 7 of which were red and 12 of which were silver. Which of the following statements best explains the observed proportion of red foxes in the offspring?

(A) The allele that causes red fur is inherited independently of the allele that causes silver fur.

(B) Red fur is determined by an epigenetic mutation that skips a generation.

(C) The allele that causes silver fur displays its phenotype only when it is homozygous.

(D) The allele that causes silver fur is lethal when it is homozygous.

49. Two closely related species of cricket, *Gryllus pennsyl-vanicus* and *Gryllus firmus*, live in geographic ranges that partially overlap. Male crickets rub their legs together to make sound in a process known as stridulation. The sound is attractive to females, and the males use the sound as courtship songs to attract mates. Different cricket species make slightly different courtship songs. Females can mate with males of the other species, but the offspring that arise from these matings have lower fitness than the offspring of matings between males and females of the same species. Based on the data provided, which of the following statements best predicts the future evolution of courtship songs in these two species?

(A) Mating between members of the two species will cause the species to fuse into a single species that will evolve a shared courtship song.

(B) Females from each species will evolve to recognize the song of the other species and avoid it, and the two species will evolve distinct courtship songs.

(C) Males of each species will evolve to mimic the courtship song of the other species in order to attract as many females as possible.

(D) Females of both species will evolve courtship songs of their own and will begin to attract males.

Use the following figure to answer question 50.

50. Which of the following correctly describes the reaction illustrated in the figure?

(A) Hydrolysis of a polymer through the release of a water molecule

(B) Hydrolysis of a polymer through the addition of a water molecule

(C) Dehydration synthesis of a polymer through the release of a water molecule

(D) Dehydration synthesis of a polymer through the addition of a water molecule

51. Which of the following is common in the generation of eukaryotic mRNA but not in the generation of prokaryotic mRNA?

(A) Uridine (U) is used instead of thymidine (T).

(B) Sequences encoding multiple proteins are produced in a single transcript.

(C) A poly-A cap is added to the 5′ end of the transcript.

(D) Sequences that do not encode protein are removed from the transcript.

Use the following information to answer questions 52 and 53.

Oxidative stress can cause damage to cells. Expression of the *Gpx2* gene is activated in response to oxidative stress. *Gpx2* encodes an antioxidant protein that reduces oxidative stress. Keap1 and Nrf2 are proteins that regulate the expression of *Gpx2* in response to oxidative stress, as shown in the figure. Regular arrowheads indicate activation, and T arrowheads indicate inhibition.

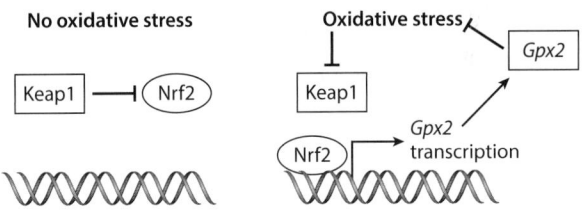

52. Based on the information provided, which of the following statements correctly describes Nrf2?

(A) Nrf2 is a transcriptional activator.

(B) Nrf2 is a negative regulator of gene expression.

(C) Nrf2 is an enzyme in a signaling pathway.

(D) Nrf2 is a biochemical intermediate.

53. Based on the information provided, predict the effect of a mutation that eliminates activity of the Keap1 protein.

(A) The cell will continuously produce Gpx2 protein and therefore will continuously experience oxidative stress.

(B) The cell will be unable to express the *Gpx2* gene and therefore will be unable to respond to oxidative stress.

(C) The cell will be unable to regulate the activity of Nrf2 and therefore will have a high level of protection against oxidative stress.

(D) The cell will be unable to suppress the activity of Keap1 and therefore will be highly sensitive to oxidative stress.

54. A population of 125 pheasants (*Phasianus colchicus torquatus*) was introduced to a small island off the coast of Washington State. After one year, the population had grown to 175 individuals. The estimated carrying capacity of the island is 250 pheasants. Based on the data provided, calculate the maximum population growth rate, r_{max}, for the year.

   (A) 0.29

   (B) 0.40

   (C) 0.57

   (D) 0.80

Use the following figure to answer question 55.

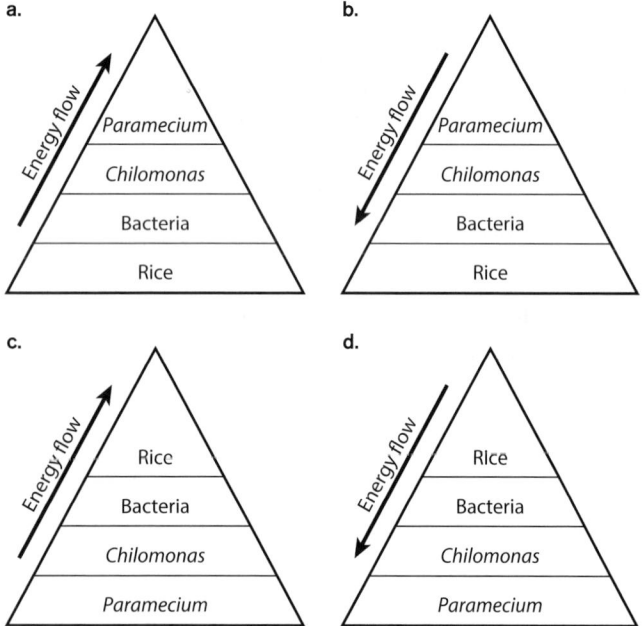

55. A scientist added five grains of rice to 100 mL of sterile water and allowed the mixture to sit at room temperature for 24 hours. She then added 25 mL of pond water in which she had previously observed a small number of microorganisms, covered the dish, and made daily observations for 96 hours. After 24 hours, the scientist observed a rapid growth in the number of bacteria. After 72 hours, she observed a high number of *Chilomonas,* which is a microorganism that feeds on bacteria. After 96 hours, she observed a high number of *Paramecium,* which are a larger microorganism that eats *Chilomonas.* Which diagram correctly illustrates the trophic relationships in this system?

   (A) a   (B) b   (C) c   (D) d

Use the following diagram, which illustrates evolutionary relationships among seven species, labeled J–P, to answer questions 56 and 57.

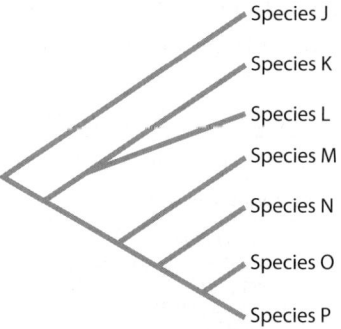

56. Which species is most closely related to species M?

   (A) Species J   (C) Species L

   (B) Species K   (D) Species P

57. Identify the species that is the outgroup on the phylogenetic tree.

   (A) Species J   (C) Species M

   (B) Species L   (D) Species P

58. The brown marmorated stink bug, *Halyomorpha halys,* is native to Asia. In the late 1990s, a small number were found in the northeastern United States. Since their introduction, populations in the United States have grown dramatically and they are now a threat to natural ecosystems and agriculture. One proposed strategy to control brown marmorated stink bug populations in North America is to introduce a species of parasitic wasp that naturally controls them in Asia. However, researchers considering this strategy are concerned that introducing the parasitic wasps might cause additional problems. Which of the following best predicts an environmental problem that might occur if the parasitic wasps are introduced to areas with large populations of marmorated stink bugs?

   (A) The introduced wasps might evolve a preference for native stink bug species and have no effect on the brown marmorated stink bug.

   (B) Brown marmorated stink bugs in the United States might evolve new mechanisms to avoid the parasitic wasps.

   (C) Predators in the United States might prey on the wasps so they would not be effective at controlling brown marmorated stink bugs.

   (D) The wasps may have no natural predators in the Unites States, so their population could become too large and compete with native species.

59. A team of researchers wanted to know the diversity of plants in a small woodland area. They set up a series of quadrats in a sampling area and counted the number of individuals of each species observed within the quadrats. The data obtained from one quadrat is shown in the table.

Species	Abundance
Wood-anemone (*Anemone quinquefolia*)	2
Carolina lupine (*Thermopsis villosa*)	8
Northern maidenhair fern (*Adiantum pedatum*)	1
Dwarf crested iris (*Iris cristata*)	1
Eastern blue star (*Amsonia tabernaemontana*)	3
Total	**15**

Based on the data provided in the table, calculate Simpson's diversity index for this quadrat.

(A) 0.10

(B) 0.35

(C) 0.65

(D) 0.90

60. Frogs of the species *Eleutherodactylus coqui* have brown bodies. Some individuals have light-colored stripes along the length of the body, and the presence of stripes is variable in the population. A scientist examined the striping patterns in a family of *E. coqui* individuals whose pedigree is shown in the figure below. Based on the data presented in the figure, identify the most likely mode of inheritance of the striped phenotype.

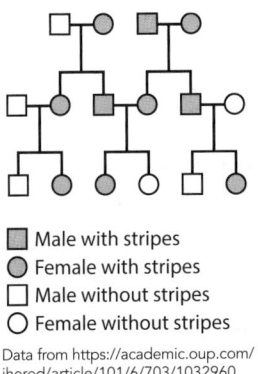

■ Male with stripes
● Female with stripes
□ Male without stripes
○ Female without stripes

Data from https://academic.oup.com/jhered/article/101/6/703/1032960

(A) Stripes are autosomal dominant.

(B) Stripes are autosomal recessive.

(C) Stripes are X-linked dominant.

(D) Stripes are X-linked recessive.

(Continued on the next page.)

Section 2: Free-Response Questions

Write your answer to each part clearly. Support your answers with relevant information and examples. Where calculations are required, show your work.

1. Chorismate is a chemical that can be converted into the amino acids tryptophan, phenylalanine, or tyrosine by the yeast *Saccharomyces cerevisiae*. The metabolic pathway that produces the three amino acids is shown in Figure 1 below. ANTH, PPA, PPY, and HPP are intermediate molecules during the conversion, and Trp3p, Aro7p, Pha2p, and Tyr1p are enzymes in the pathway. Note that regular arrowheads indicate activation and T arrowheads indicate inhibition.

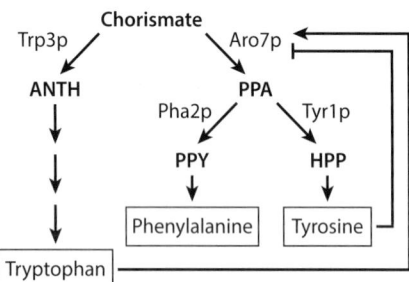

FIGURE 1 The metabolic pathway for production of tryptophan, phenylalanine, and tyrosine by *S. cerevisiae*

A scientist hypothesizes that the amount of tryptophan available to *S. cerevisiae* cells will influence the amount of phenylalanine that the cells produce. The scientist grows five separate cultures of *S. cerevisiae* cells in culture medium. Four cultures are supplemented with different levels of tryptophan and one culture contains no additional tryptophan. The scientist measures the amount of phenylalanine produced by the cells in each culture and obtains the data shown in Figure 2. The amount of phenylalanine is measured in mmol per gram of *S. cerevisiae* cells, and the amount of supplemental tryptophan is measured in μg per mL of culture medium. Error bars are calculated as ±1 standard deviation on either side of the mean.

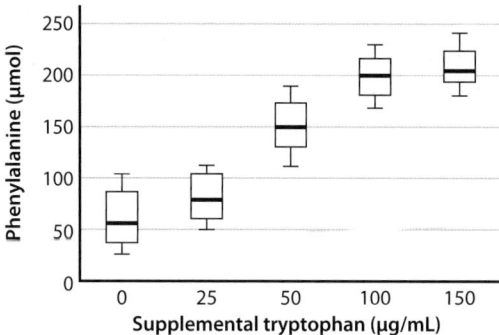

FIGURE 2 The amount of phenylalanine produced by *S. cerevisiae* cells in the presence of different levels of supplemental tryptophan

(a) Using the information provided in Figure 1, **explain** the role of tyrosine in the pathway used by the cell to synthesize phenylalanine.

(b) **State** the null hypothesis for the experiment performed by the scientist. **Identify** the negative control and the independent variable.

(c) Using the information provided in Figure 2, **describe** the effect of adding supplemental tryptophan to the medium on phenylalanine production. **Calculate** the ratio of mean phenylalanine production when the medium is supplemented with 50 μg/mL of tryptophan to mean phenylalanine production when the medium is supplemented with 25 μg/mL of tryptophan. Use the means and confidence intervals in Figure 2 to determine whether there is a statistically significant difference in phenylalanine production when the medium is supplemented with 100 μg/mL of tryptophan compared to when the medium is supplemented with 50 μg/mL of tryptophan, and **provide reasoning** to support your claim.

(d) Assuming that chorismate is present in excess, **predict** how a mutation that eliminates the function of Tryp1 would change the relationship between phenylalanine production and supplemental tryptophan added to the medium. **Provide reasoning** to justify your prediction.

2. The Trinidad guppy *Poecilia reticulata* is a fish that lives in freshwater streams in the Caribbean. Scientists studying *P. reticulata* in nature observed that guppies living in environments with a high density of predators use multiple behaviors to escape predation, but that guppies living in environments with few predators do not display those behaviors. The scientists captured guppies from a region of a stream where there was a particularly high density of a predatory fish, the pike cichlid (*Crenicichla alta*), as well as from a different region of the same stream where pike are not present. The scientists brought the captured guppies from both locations into their laboratory and used them in two experiments.

In experiment 1, the scientists placed five guppies from the high-predation environment and five guppies from the low-predation environment into an aquarium that contained a single predatory pike. The scientists observed the tank until the predator had consumed half of the guppies. The scientists recorded the environment from which each fish that was consumed had originally been collected. The scientists repeated this experiment five times.

In experiment 2, the scientists repeated the experimental procedure using offspring from fish captured in the high-predation environment and offspring from fish captured in the low-predation environment. All offspring were reared in the laboratory.

The data that the scientists collected from both experiments are shown in the table below.

Experiment 1: Fish Captured in the Wild				
	Low-predation environment (mean)	Low-predation environment (standard deviation)	High-predation environment (mean)	High-predation environment (standard deviation)
Proportion preyed upon in the laboratory	0.81	0.12	0.19	0.10
Experiment 2: Offspring of Fish Captured in the Wild				
	Low-predation environment (mean)	Low-predation environment (standard deviation)	High-predation environment (mean)	High-predation environment (standard deviation)
Proportion preyed upon in the laboratory	0.59	0.13	0.41	0.12

(a) Scientists observed that guppies are more abundant than pike in regions of the stream where both species coexist. **Explain** why prey animals tend to have larger population sizes than their predators in most ecosystems.

(b) **Construct** an appropriately labeled bar graph that illustrates the data from both experiments in the table.

(c) The scientists noted the difference in predation between the two populations of guppies in experiment 1 was greater than the difference in predation between the two populations of guppies in experiment 2. **Make a claim** to explain this result. **Provide reasoning** to support your claim.

(d) A scientist claims that wild populations of guppies contain genetic variation for the ability to avoid predation. Based on the information provided, **support** the scientist's claim with one piece of reasoning. Assuming the scientist's claim is correct, **predict** the short-term effect on the natural population of guppies if pike were introduced into the environment where they are currently absent. **Provide reasoning** to support your prediction. Additionally, **predict** the long-term effect on the natural population of guppies if pike were introduced into the environment where they are currently absent and **provide reasoning** to support your prediction.

3. Parasitoid wasps are insects that lay their eggs inside the bodies of other insects, called hosts. The wasp eggs hatch into larvae that eat the host insect from the inside, killing the host. However, the host insect can sometimes kill wasp larvae with a successful immune defense.

A researcher established six populations of the host insect *Drosophila melanogaster,* which is a species of fruit fly. The researcher exposed three of these populations to parasitoid wasps, then collected the surviving *D. melanogaster* and bred them to establish the next generation. The researcher did not expose the other three populations to the parasitoid wasps, and instead chose random individuals to breed for the next generation. The researcher repeated this process for eight generations.

In each generation, the researcher evaluated the ability of *D. melanogaster* from each population to defend themselves against parasitoid wasps. The researcher removed a sample of the *D. melanogaster* from each of the six populations and exposed them to parasitoids, recording the percentage that were able to mount a successful immune defense and survive. The data that the researcher obtained are shown in the figure. After eight generations, the researcher also measured the average number of offspring produced by individual females each population in the absence of parasitoids. The researcher found that females from the populations that were repeatedly exposed to parasitoids over the eight generations produced fewer offspring than females from the populations that were never exposed to parasitoids.

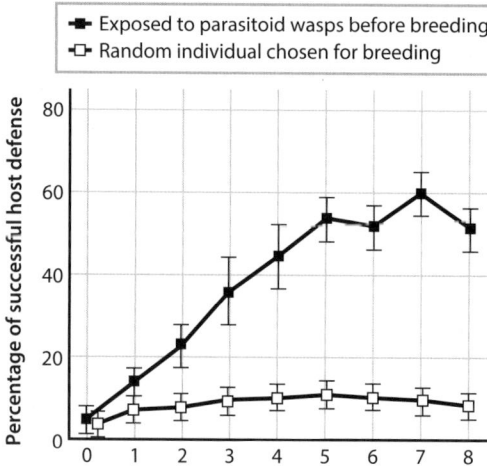

(a) **Describe** the role of genetic variation in evolution by natural selection.

(b) **Justify** the use of the populations that were not exposed to parasitoids in the described experiment.

(c) In the ninth generation, the researcher combined an equal number of individual *Drosophila* from each of the six populations into a single mixed population. Based on the information provided, **predict** how the ability to defend against parasitoids would change in the mixed population after several generations of evolution without exposure to parasitoids.

(d) **Justify** your prediction in part c.

4. Galactokinase is a yeast protein that is encoded by the *gal1* gene. Transcription of *gal1* is activated by the presence of galactose in the environment. Cycloheximide is a chemical that prevents protein translation when it is added to yeast cells.

A scientist measured the level of *gal1* mRNA and galactokinase protein in an actively growing yeast culture for 300 minutes. For the first 60 minutes of the experiment, the growth medium contained no galactose. After 60 minutes, the scientist added a high level of galactose to the medium. After an additional 140 minutes, the scientist replaced the growth medium with fresh medium that contained no galactose. The data that the scientist obtained are shown in the figure.

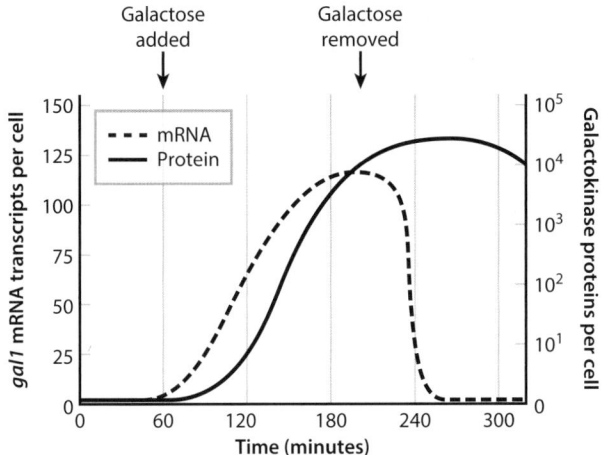

Amount of *gal1* mRNA transcripts and galactokinase proteins per cell

(a) **Identify** the organelle that produces galactokinase protein from *gal1* mRNA transcripts and a location in the cell where that organelle can be found.

(b) **Explain** why there is a difference between galactokinase level and *gal1* mRNA level at 280 minutes in the experiment, after the galactose has been removed.

(c) **Predict** the amount of galactokinase protein that would be observed at minute 200 in the experiment if cycloheximide were added to the culture at minute 180.

(d) **Justify** your prediction from part c.

5. Evening primrose (*Oenothera biennis*) is a small flowering plant that provides habitat for herbivorous and predatory insects. To test the effect of *O. biennis* genetic diversity on the abundance of insects associated with these plants, scientists planted experimental gardens that contained either one *O. biennis* genotype, four *O. biennis* genotypes, or eight *O. biennis* genotypes. Sixteen total plants were planted in each garden. The scientists then counted the mean number of insects associated with each plant in the garden on five different days during the growing season. The data they obtained are shown in the figure. Error bars are calculated as ±1 standard deviation on either side of the mean.

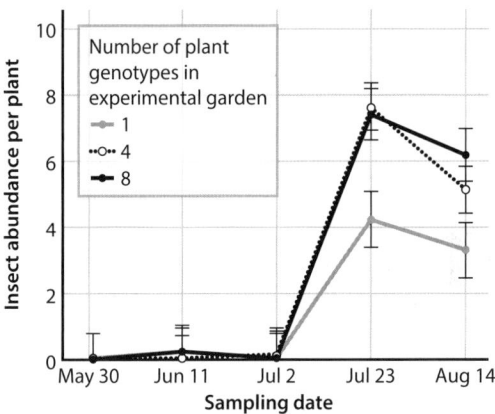

Insect abundance in gardens with plants of varying genetic diversity

(a) Using the data in the figure, **describe** the change in insect abundance on the plants during the growing season.

(b) **Identify** the sampling date that shows the largest difference in insect abundance between gardens planted with four genotypes and gardens planted with eight genotypes.

(c) Before beginning the experiment, the scientists hypothesized that garden plots with genetically diverse *O. biennis* would support larger insect populations. Based on the information provided in the figure, **make a claim** about whether the data support or do not support this hypothesis. **Provide reasoning** to support your claim.

(d) Based on the information provided, **predict** which of the gardens are likely to support the greatest transfer of energy across trophic levels. **Explain** the reasoning behind your answer.

6. The diagram below illustrates the electron transport chain utilized in in the light reactions of photosynthesis.

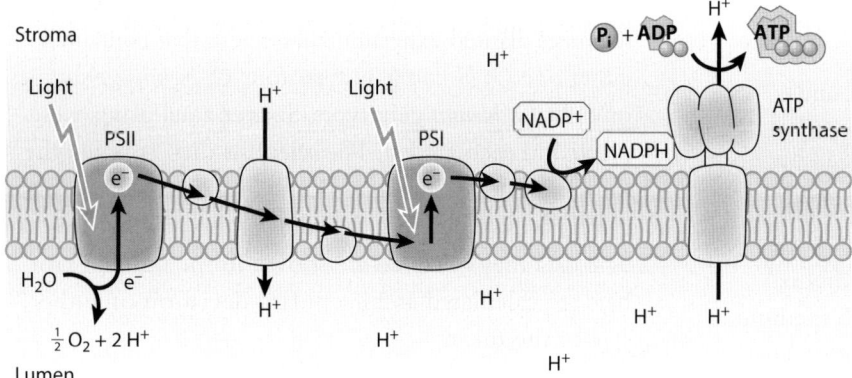

(a) **Describe** the role of light in photosystem II and photosystem I.

(b) **Describe** the reason that the concentration of protons is higher in the thylakoid lumen than in the stroma.

(c) **Identify** a location on the diagram where a protein converts potential energy into a stored chemical energy that can be used by the cell later.

(d) **Explain** how the production of NADPH by the electron transport chain during photosynthesis sup–ports the production of carbohydrates in a plant cell.

Tutorial 4: Graphing

Introduction

Graphs serve as a useful visual tool for scientists to identify and analyze patterns in datasets that they have collected. It is often easier to see these trends in a graph, rather than as data in a table. There are several different kinds of graph, each with its own uses. As you will see, certain types of data are best suited for certain types of graphs.

In this tutorial, we will discuss some of the most commonly used graphs and how to decide when each is appropriate to use. This will serve as a guide to reading and interpreting the results of nearly every graph you encounter throughout this textbook and in your study of AP® Biology.

XY Graphs

The first type of graph that we will examine is the XY graph. Here, we will also cover the parts of a graph and how to construct one. Typically, a graph is created from data that have been organized into a table or chart, such as the one below. It is much easier to see trends visually presented in an XY graph than in a table. Graphs provide a way to quickly see patterns in data.

Example: The table below and its associated graph show the relationship between the average height and diameter of the Douglas fir tree. Diameter measurements were taken 1.37 m from the ground. Total tree height was also measured.

Average diameter (cm)	7.6	12.7	17.8	24.6	27.9
Average height (m)	6.1	9.1	12.8	15.2	17.7

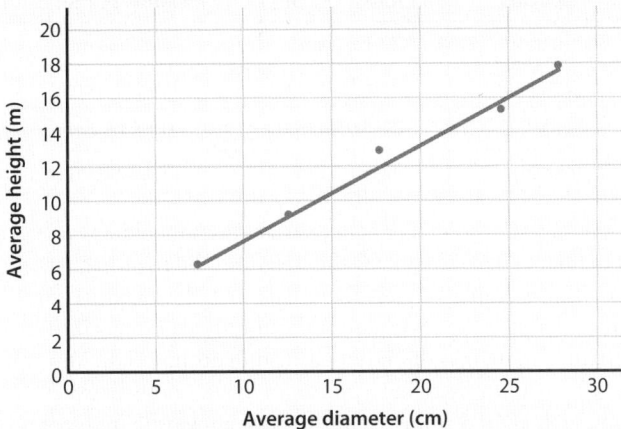

A basic graph, such as this one, is composed of two axes: the horizontal x-axis and the vertical y-axis. Each axis has its own label with units. In this example, the average diameter is on the x-axis and the average height is on the y-axis. Each data point has one corresponding x-axis and y-axis value.

Graphs also feature a scale. The scale is the set of numbers written on each axis. In the graph of tree diameter and height, the scale for the x-axis goes from 0 to 30 cm and increases by increments of 5. The scale of the y-axis goes from 0 to 20 m and increases by increments of 2. Scales depend on the data that are collected. When you are tasked with creating a graph, you must decide what scale to use based on the data you are given. For example, if the data consist of the numbers 2, 3, 6, 8, and 10, it might be good to use a scale from 0 to 10 in increments of 1 or 2. If the data are not as numerically close together, you would create a scale with larger increments.

The orientation of the data is the direction in which the data are plotted along the axes. Generally, data points and values increase as they move up the y-axis and across the x-axis. Sometimes, the 0 point is the same for both axes and is located at the corner where those two axes meet, as in the graph to the left. Other times, the 0 point may be different for each axis. Occasionally, and especially if you are looking at a graph with multiple variables plotted, a graph will contain a key that notes the way that different datasets are depicted.

Each variable plotted on the x- or y-axis is labeled with the appropriate units. These are usually given as part of the axis label. In the graph above, the x-axis diameter is measured in centimeters. The y-axis height is measured in meters. Whenever you are asked to give an answer regarding reading a point on a graph, make sure to include units.

Tables and graphs reveal trends or patterns. For example, in both the table and the graph of tree height and diameter, we can see that as the average diameter of the tree increases, the average height also increases. Sometimes these patterns are represented by a trend line or line of best fit. Trend lines can approximate the data in the graph and provide an even clearer picture of how the data are behaving. They can also be used for predicting the value of certain data points. In the graph of tree height and diameter, the trend line is depicted with a red line.

A trend line is an approximation of the data. Not every point has the line running exactly through it. We are trying to draw a line that best represents the pattern in the data. For this graph, it is straightforward because the data are so linear. The line should go through the middle of your data and have approximately the same number of data points below and above the line. Trend lines can be straight or curved.

Your Turn

During the spring, birds sing at early hours. A student decided to record the number of birds seen or heard between the hours of 4 a.m. and 9 a.m. The results are shown in the table below.

Time of day	4 a.m.	5 a.m.	6 a.m.	7 a.m.	8 a.m.	9 a.m.
Number of birds	6	10	17	14	14	9

Using this table, plot the data on a graph with proper axis labels and scales, and identify trends from this information.

Solution

To begin, let's identify the two types of data from the table. We have data about the time of day and data about the number of birds. We can plot time on the horizontal x-axis and number of birds on the vertical y-axis.

Next, we will determine the scale of the graph. There are only 6 points of data that will be plotted. The scale for the x-axis is simple enough: we can include each individual time as its own data point. The y-axis values range from a minimum of 6 to a maximum of 17. We can begin our y-axis scale at 0. It is a good idea to choose an endpoint to the scale that is close to the maximum value. We could choose 18 or 20 to give ourselves more room to plot the data. Let's choose 20.

The next decision is to pick the increments of the scale. If our y-axis is going to range from 0 to 20, we have a few options. Because the numbers are relatively small, we could choose to increase the scale by increments of 1 (0, 1, 2, 3, and so on), but this can result in a cluttered graph. We want to make the data points as easy to read and interpret as possible. With this in mind, we could choose intervals of 2, 4, or 5. Again, because there is a small range of values in this dataset, any of these options would work. For the sake of having the most uncluttered graph, we will use a y-axis scale that increases by 5. The graph now looks like the figure below.

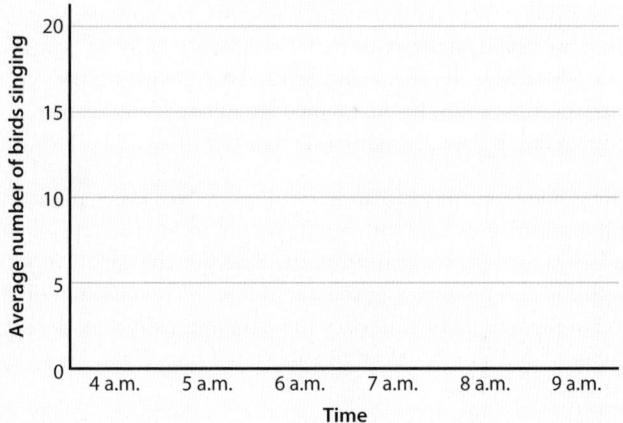

With the axes labeled and scales set, we can plot the data in the table. It is best to plot as you progress across the x-axis. Mark one data point at a time and double-check that you have the correct x value paired with the correct y value. For example, at 4 a.m., there were 6 birds singing. Once you have plotted that point,

move on to the next point. When you've finished, the graph should look like this:

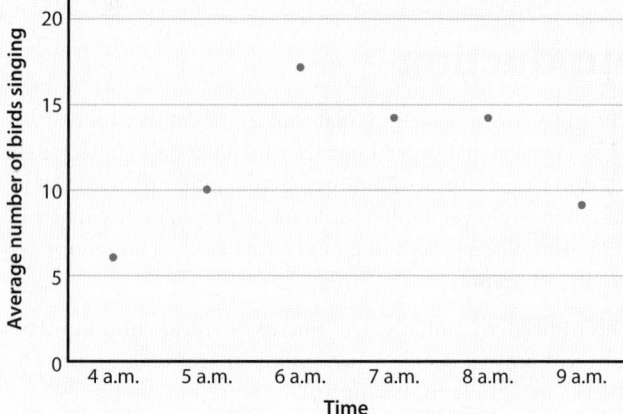

The pattern in this dataset is clear. The points show that the most birds are singing around 6 a.m., while fewer birds are singing earlier and later in the morning.

Bar Graphs

You are probably already familiar with bar graphs. They are a common way for researchers to present data and are often colorful and easy to read. The differences between data points are quickly seen based on the heights of the bars.

Example: Researchers were interested in the ability of Colorado wildflowers to grow in different habitats. Although many of the species are suited to survive in various altitudes, soil types, and shade conditions, they wanted to see if there were any differences between growth in a forest habitat compared to an open meadow habitat. Their data are shown in the table below.

Wildflower species	Maximum height in forest (± 2 SE$_{\bar{x}}$)	Maximum height in meadow (± 2 SE$_{\bar{x}}$)
Beeblossom	20 ± 6.1 cm	48 ± 8.8 cm
False hellebore	95 ± 10.8 cm	110 ± 9.2 cm
Fireweed	79 ± 9.4 cm	91 ± 14.6 cm
Silky lupine	32 ± 7.3 cm	39 ± 4.0 cm

For this dataset, the independent variable is the wildflower species and the dependent variable is the height in different habitats. Typically, bar graphs display discrete data on the x-axis. These are data that have a fixed value and can only be categorized in a certain way, such as the number of students in a class. You can have 19 or 20 students, but you cannot have any in-between values, such as 19.5 students. In the table above, the different categories are species of wildflower. There are beeblossom and fireweed categories, but no beeblossom–fireweed hybrid category. The separate bars in a bar graph reflect the fact that the flowers are distinct categories.

When given discrete data in this format, especially with standard error calculations, bar graphs are a good choice. Having two differently colored bars, one for each habitat type, will make for easy reading and analysis of the graph.

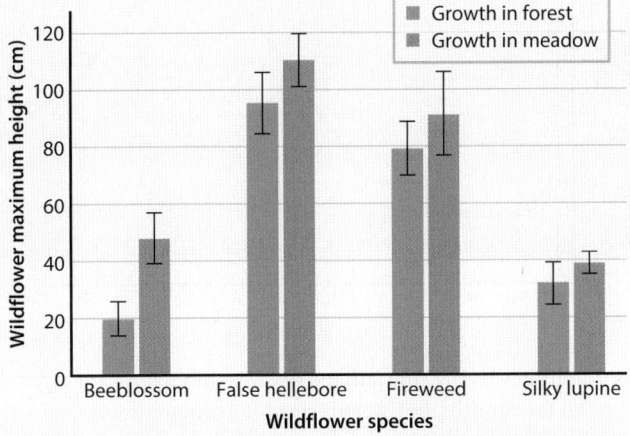

The data in this graph also have error bars. For more on error bars, see the explanations in Module 0 and Tutorial 1.

Bar graphs are versatile and can be presented in many ways. The above graph has single vertical bars, but some graphs have horizontal bars, multiple bars graphed adjacent to each other, or a combination of these features with two *y*-axes. To read the data, you need to study the key carefully. You may also see bar graphs with the different categories listed on the *y*-axis instead of the *x*-axis. Often this arrangement is to accommodate independent variables with longer names and is used to format the graph in a more visually appealing way.

Histogram

Histograms may appear very similar to bar graphs because both include bars of differing heights. However, while bar graphs plot discrete data, histograms present continuous data, where decimal values are possible. Continuous data can take any value within a range. Temperature, time, and length are a few examples of data that can be continuous. Continuous data act as the independent variable and are plotted on the *x*-axis.

Example: The researchers studying wildflowers were also interested in the elevations at which certain species grew. The table in the next column shows data collected for fireweed using a survey

known as the nearest neighbor method. This involves choosing a random sampling point within a set area at these elevations and measuring the distance between a plant and its nearest neighbor.

Elevation (m)	Average nearest neighbor distance (m)
0–610	0.29
610–1219	0.46
1219–1829	1.1
1829–2438	2.8
2438–3048	3.4

The table has given elevations in a range of several values. When data are presented in this way, it is likely that a histogram will be used. Elevation serves as the continuous data and should be plotted on the *x*-axis.

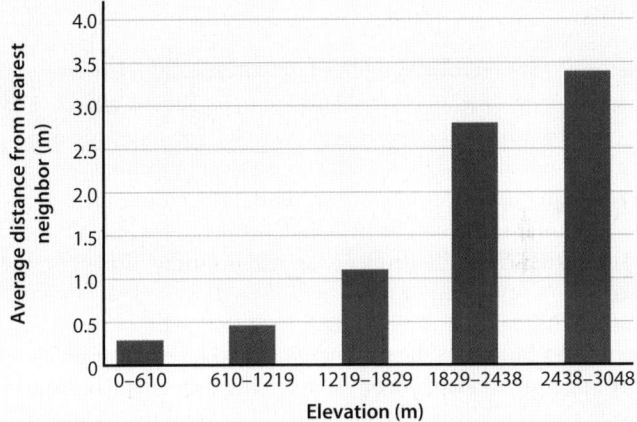

We can conclude that the fireweed flowers grew further from each other as elevation increased.

Your Turn

The graphs below are histograms taken from a study examining the infectiousness of the influenza virus in exhaled breath. The participants in this study were college-aged students, the majority of whom tested positive for influenza. The *x*-axis denotes the severity of the symptoms reported by the students. A higher number indicates more severe symptoms. Describe why these data are presented with a histogram instead of a bar graph.

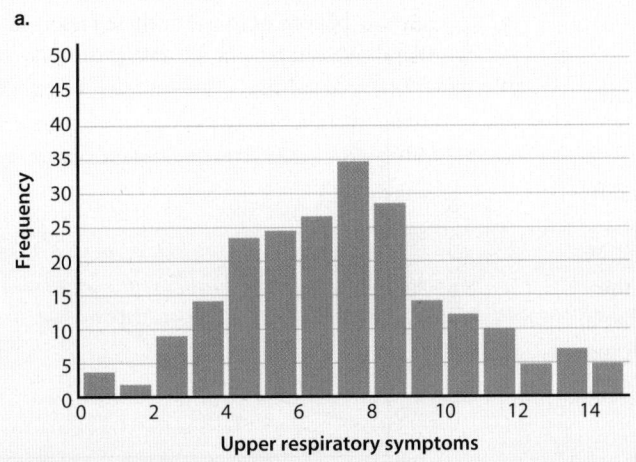

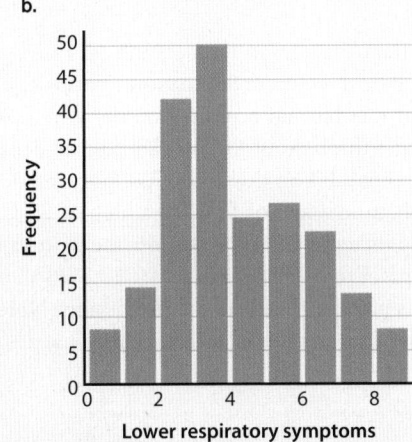

Data from: https://www.pnas.org/content/115/5/1081

Solution

These data are suited for a histogram because the scale on which the students graded the severity of their symptoms is continuous. The x-axes showing the severity values range from 0 to 15 and 0 to 9. However, many of the college students could have given scores that fell between whole numbers. This would not translate to a bar graph because bar graphs use discrete data, where categories are fixed.

Line Graphs

Line graphs are a simple way to track changes in data over time and are useful when plotting variables relative to each other. Each data point is connected with a line. The lines created can be straight, jagged, or curved, as long as they reflect the movement in the data.

Line graphs are useful when plotting two variables relative to each other. In many cases, one variable is the independent variable and the other variable is the dependent variable. The independent variable is not affected by the other variable. For example, if you are collecting data at specific times, time is the independent variable because it elapses on its own, independent of whatever you are measuring. The independent variable is usually plotted on the x-axis.

The second variable is the dependent variable, which varies according to the independent variable. The dependent variable is the effect or result that is being observed or measured. This variable is considered dependent because it is expected to vary based on changes of the independent variable. The dependent variable is usually plotted on the y-axis.

Example: The following table shows the average monthly temperatures for the surface of the northern Atlantic Ocean in 2020.

Month	Jan	Feb	Mar	Apr	May	Jun	Jul	Aug	Sep	Oct	Nov	Dec
Average water temperature (°C)	19	18	18	19	19	21	23	25	25	24	22	21

Because the data track changes in temperature over one year, a line graph would best show these fluctuations. In this case, time is the independent variable and is plotted on the x-axis. Water temperature changes throughout the year, so it is the dependent variable and is plotted on the y-axis. Temperature does not change very much, from a low of 19°C to a high of 25°C, so a scale increasing by increments of 1 will show the pattern well. We can include a range of temperatures from 16°C to 26°C on the y-axis because this range includes all the data points. Note that in this case, the y-axis does not begin at 0, but instead at 16, because there are no lower temperature values.

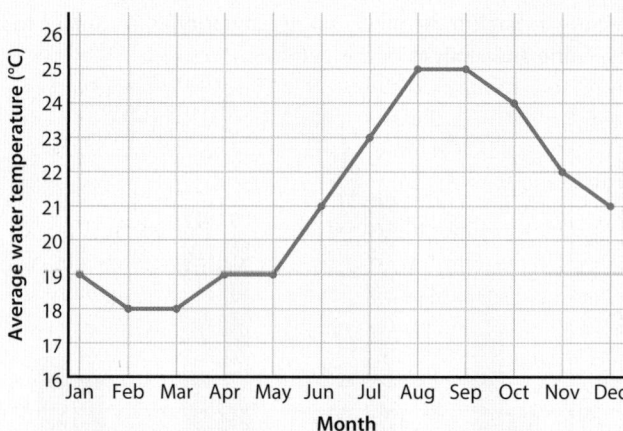

This graph tells us that the surface ocean temperature is colder in the winter months, but warms steadily through the spring and summer, before beginning to fall again.

Dual Y Graphs

There are also graphs that contain two y-axes on a common x-axis. These graphs, which present the findings of multiple dependent variables, are known as dual Y graphs. They are used to show the changes in two variables over time. The variables usually have different units or different ranges in value. These differences are addressed by using one y-axis for each variable, with separate scales that reflect the variable plotted.

Example: The following table shows data on atmospheric CO_2 and CO_2 emissions.

Year	1900	1920	1940	1960	1980	2000	2020
Atmospheric CO_2 (ppm)	296	304	312	319	340	370	413
CO_2 emissions (billions of tons)	2.1	3.7	5.0	9.6	19.0	25.0	36.7

Data from https://www.climate.gov/news-features/understanding-climate/climate-change-atmospheric-carbon-dioxide

As you read this table from left to right, you can see that the value for each variable increases. However, values are increasing at different rates and magnitudes. The variables also have different units: one is in ppm, or parts per million, and the other is in billions of tons. For these reasons, a graph with two y-axes is useful. We can present the data at the same time and compare trends between the two, even with the differences in their scales.

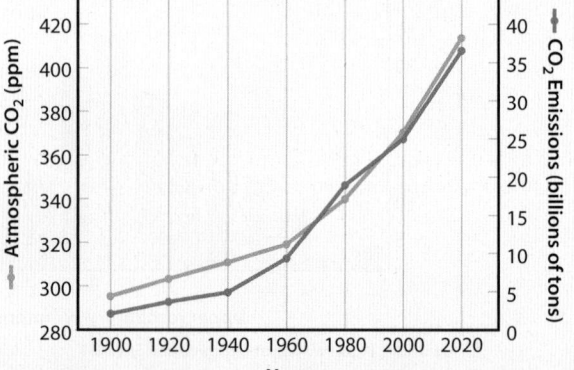

This graph is read in the same way as any other graph with multiple variables. Be sure to pay attention to the color of the line for each variable, so that you read the data from the correct axis. The relationship between these variables is quite clear: both atmospheric CO_2 and CO_2 emissions have increased dramatically over the past century.

Dual-Y graphs are not limited to line graphs. You may also see bar graphs, histograms, and other graph types with two y-axes. To read these graphs, start by differentiating the axis on which each variable is plotted and pay attention to the associated scales and units of each.

Log Scale

So far, we have looked at graphs that use a linear scale. Some graphs use a logarithmic scale on one or more axis. Logarithmic scales are used to display data that have large ranges in value. The examples we have dealt with have all had linear axes because the range between data points has been relatively small. When the largest data are several hundred times larger than the smallest data, a log scale is used to simplify the graph. Smaller log bases are used to convey smaller ranges in the data. Most graphs use a logarithmic scale with log base 10.

Example: A group of researchers is studying the population growth of a type of bacteria at different temperatures. They recorded population numbers every 5°C to 25°C. The results are given in the table.

Temperature (°C)	Number of bacteria
0	5
5	40
10	310
15	2500
20	15,000
25	85,000

There is an extremely large range between the smallest and largest population in this dataset. This is a sign that a logarithmic scale would be useful to present these data. In this example, the temperature is the independent variable and is plotted on the x-axis. Because the temperature data are linear and regular in range, we will use a logarithmic scale for the dependent variable on the y-axis, the number of bacteria.

The difference in presentation of these data with a linear y-axis compared to a logarithmic y-axis is significant. If we were to use a linear scale with increments of 10,000, the graph would look like the one at the top of the next column.

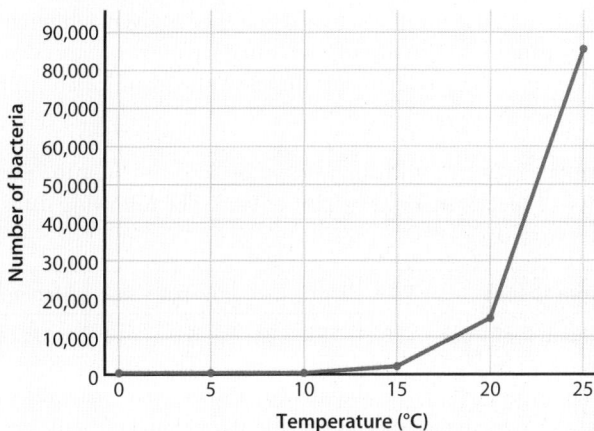

It looks like an exponential growth curve. It is difficult to read the first few data points because the scale is in increments of 10,000. The points at 0, 5, and even 10 could be read as the same number, even though they are different.

With a $\log_{10}$ scale, the minimum value on the y-axis is 1 and the maximum value is 100,000, which is close to the maximum number of bacteria, 85,000. The x-axis remains unchanged. The graph now looks like this:

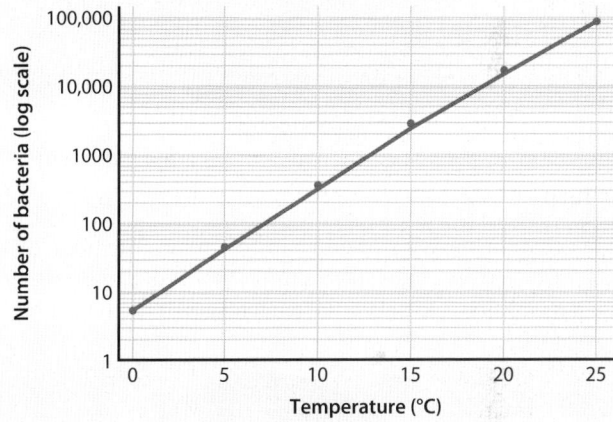

It is possible to read each data point more accurately, without having to refer to the data table. Clearly, as the temperature increases, there is a constant and large increase in the number of bacteria as well. Note that logarithmic gridlines are different from linear gridlines. This changes how the y-axis data are read.

In this example, only one variable (number of bacteria) is plotted logarithmically, so it is considered a semilog ("half"-log) plot. When both variables are plotted logarithmically, it is called a log–log plot.

Pie Charts

Pie charts provide a fast way to visually gather what is happening in certain datasets. Pie charts are circles that represent 100% of the data, with different slices signifying the percentage of a certain category out of the total. Each slice is given a different color or pattern, so the size of each category is easily seen. The largest slice of the pie represents the most abundant variable. The percentages should add up to 100.

Example: A student studying forestry decided to count the number of tree species in a nearby plot of land. The following data were recorded.

Tree species	Pine	Elm	Oak	Walnut	Ash	Maple
Number of trees	38	15	24	9	13	17

This data are showing the abundance of trees. If we wanted to quickly see which tree species was the most abundant, a pie chart would be useful. With one glance at the below graph, we can see immediately that pine trees make up the greatest percentage of this plot of land.

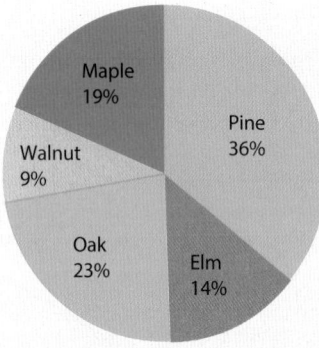

Your Turn

The pie chart below is based on global emissions data from 2010. Which three sectors contributed almost 75% of global greenhouse gas emissions? Estimate the percentages of the remaining three sectors.

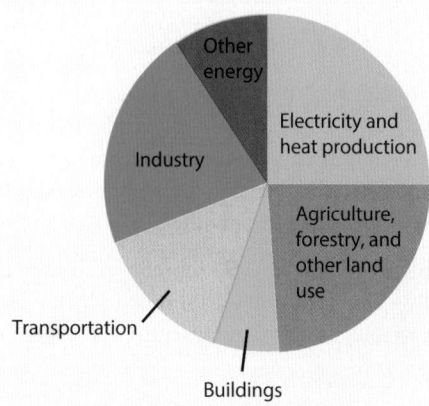

Data from: https://www.epa.gov/ghgemissions/global-greenhouse-gas-emissions-data

Solution

The three sectors that make up nearly 75% of emissions are Electricity and heat production; Agriculture, forestry, and other land use; and Industry. These sectors represent approximately 25%, 25%, and 20% of the pie, respectively. The remaining three sectors are Transportation, Other energy, and Buildings. Their emissions are about 15%, 10%, and 5%, respectively.

Box and Whisker Graphs

A box and whisker graph is a good tool for visualizing the distribution of values in a dataset. These graphs are especially useful for comparing the distributions between several groups or datasets because they are compact. These graphs can have vertical or horizontal boxes, but each has the same components and can be read the same way. Box and whisker plots are so named because of the appearance of the graph. They contain a bar resembling a box and two protruding lines called whiskers. The figure below identifies the major parts of a box and whisker plot, with important numbers listed in red.

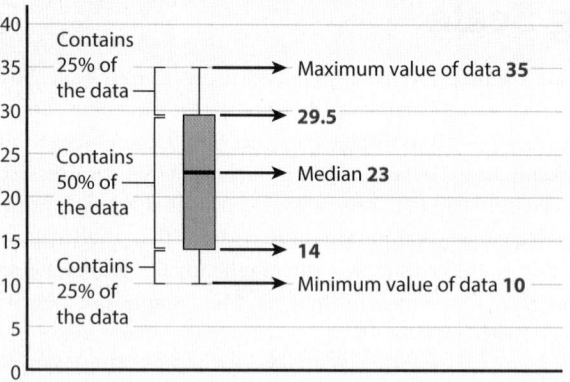

The graph shows a small dataset comprised of these numbers: 35, 23, 18, 10, and 24. In numerical order, the dataset graphed is: 10, 18, 23, 24, 35. Out of these five data points, the median, or middle data point, is 23. This is shown on the graph by the horizontal line in the middle of the box. The median can also be called the 50th percentile. A percentile is a number where a certain percentage of values fall below that number. So, half of the scores in the dataset are found below the median. The ends of each whisker represent the minimum and maximum values in the dataset, 10 and 35.

The line denoting the bottom of the box is the median of the lower half of the data. Twenty-five percent of the dataset falls below this number. The line denoting the top of the box is the median of the upper half of the data. Seventy-five percent of the dataset falls below this number. If we arrange the above dataset to exclude the median, it would look something like this: (10, 18), 23, (24, 25). We can find the median of the lower and upper halves of the data. Because there are an even number of values above and below the median, we have to average the two to calculate the new medians:

Lower box line = (10 + 18) ÷ 2 = 14
Upper box line = (24 + 25) ÷ 2 = 29.5

Looking back at the graph, we can see that the horizontal lines for the box have been drawn at 14 and 29.5, respectively.

Example: The graph on the next page depicts an index of the distribution of rotifer species richness in different habitats. Rotifers are microscopic aquatic invertebrates. Identify the minimum, maximum, and median of rotifers in habitat B, and describe any patterns between the different habitats. Note that this graph contains an outlier, the empty circle above the right-most box. Outliers are data points that are very far from other points in the dataset.

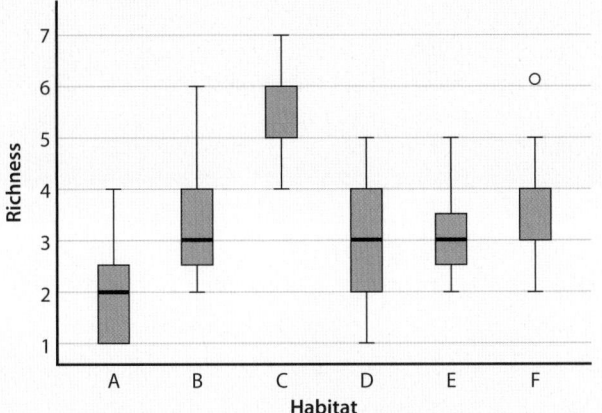

Data from: https://www.researchgate.net/publication/284204454_Testing_the_habitat_selectivity_of_bdelloid_rotifers_in_a_restricted_area

Rotifer richness in habitat B is given by the second bar from the left. The minimum value is 2; the maximum value is 6; and the median is 3.

Habitat C has the largest minimum and largest maximum values of all habitats in the figure. If marked, it would also have the highest median value. We can conclude that it experiences the greatest species richness. Habitat A is the least species rich, having the lowest minimum, maximum, and median values among the habitats examined. The other four habitats are very similar in species richness. Their medians and box values are nearly equal, with differences only in the range of minimum and maximum values.

Area Graphs

Area graphs are related to line graphs. The example at the top of the next column shows the number of COVID-19 cases per county in the San Francisco area. As you can see, the area underneath each line is shaded in. While line graphs measure the changes or differences between data points, area graphs focus on the different volume of data between two or more variables. They are primarily used to show trends in data over time.

In this case, we are seeing the number of COVID-19 cases on the y-axis over time plotted on the x-axis. Each county, such as Napa, Marin, and Sonoma, is graphed using a different color, indicated by the key at the top. Then the data for each county are stacked on top of each other. In this way, we can compare the number of COVID-19 cases in different counties by looking at the area of each color, while also keeping track of the total number of cases, which is indicated by the top of the purple (Santa Clara) line.

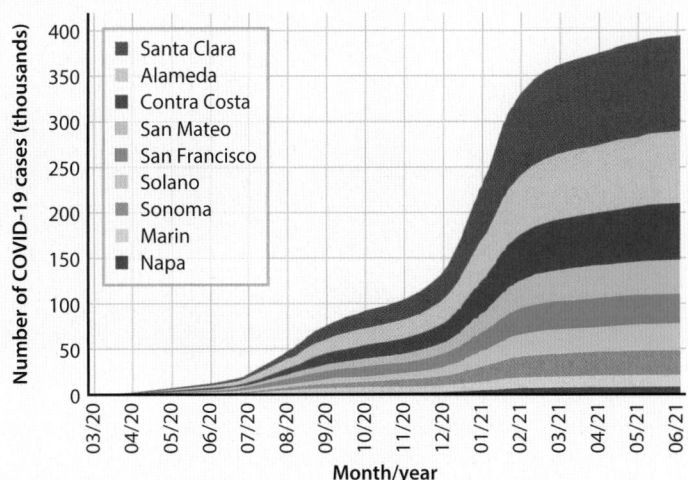

Data from: https://sfist.com/2020/03/26/covid-19-case-numbers-jump-in-san-francisco-and-alameda-counties-young-people/

Scatter Plot Graphs

Sometimes data are presented in a scatter plot. These graphs are used when individual data points are independent of each other and we want to assess if two variables are related. The graph below is a scatter plot showing how the pace of aging, as measured by various functional abilities, changes with a person's actual age.

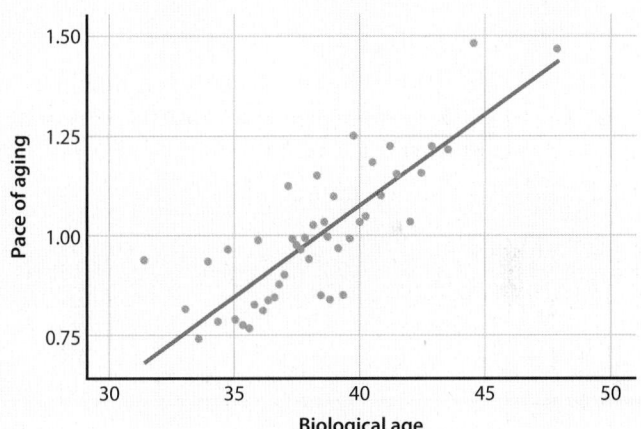

Data from: https://www.pnas.org/content/112/30/E4104

We can use scatter plots to conclude if certain variables have a positive or negative correlation. Oftentimes, these correlations can easily be seen with a line of best fit. If both variables increase or decrease at the same time, they are said to have a positive correlation. If one variable increases while the other decreases, they have a negative correlation. If the scatter plot has no discernable pattern, there is likely no relationship between the two variables.

Glossary/Glosario

GLOSSARY

GLOSARIO

3′ end The end of a DNA or RNA molecule that has a free 3′ hydroxyl group. By convention, the 3′ end is considered the end with the last nucleotide. (p. 80)

extremo 3′ El extremo de una molécula de ADN o ARN que termina en un grupo 3′ hidroxilo libre. Por convención, el extremo 3′ se considera el extremo con el último nucleótido.

5′ end The end of a DNA or RNA molecule that has a free 5′ phosphate group. By convention, the 5′ end is considered the end with the first nucleotide. (p. 80)

extremo 5′ El extremo de una molécula de ADN o ARN que termina en un grupo 5′ fosfato libre. Por convención, el extremo 5′ se considera el extremo con el primer nucleótido.

A

abiotic Pertaining to nonliving things. (p. 4)

abiótico Perteneciente a los seres no vivos.

acid A molecule that donates H⁺ ions to a solution. When the concentration of hydrogen ions is greater than the concentration of hydroxide ions, the pH is acidic. (p. 49)

ácido Molécula que dona iones H⁺ a una solución. Cuando la concentración de iones de hidrógeno es mayor que la concentración de iones de hidróxido, el pH es ácido.

activation energy (E_A) The energy input necessary to reach the transition state. (p. 199)

energía de activación (E_A) El aporte de energía necesario para alcanzar el estado de transición.

activator A compound that increases the activity of an enzyme. (p. 205)

activador Compuesto que aumenta la actividad de una enzima.

active site The portion of an enzyme that binds substrate and converts it to product. (p. 202)

sitio activo La parte de una enzima que une los sustratos y los convierte en productos.

active transport The movement of substances across a cell membrane against a concentration gradient requiring an input of energy. (p. 148)

transporte activo Movimiento de sustancias a través de una membrana celular en dirección opuesta a un gradiente de concentración, que requiere un aporte de energía.

adaptation The close fit between an organism and its environment. (p. 561)

adaptación Proceso de acoplamiento entre un organismo y su entorno.

adaptive radiation A period of unusually rapid evolutionary diversification in which natural selection accelerates the rate of speciation within a group, with new species adapted for specific niches. (p. 674)

radiación adaptativa Periodo de rápida diversificación evolutiva en el que la selección natural acelera el ritmo de especiación dentro de un grupo y emergen varias nuevas especies adaptadas a nichos específicos.

addition rule of probability Rule which states that the probability of two mutually exclusive events happening is the sum of their individual probabilities. (p. 372)

regla de la suma de probabilidades Regla que establece que la probabilidad de que se produzcan dos acontecimientos mutuamente excluyentes es la suma de sus probabilidades individuales.

adenine (A) A purine base. (p. 79)

adenina (A) Una base púrica.

adenosine diphosphate (ADP) A chemical relative of ATP that is composed of the base adenine, the five-carbon sugar ribose, and two phosphate groups. ADP contains less chemical energy than ATP does. (p. 187)

adenosín difosfato (ADP) Pariente químico del ATP compuesto por una base adenina, un azúcar ribosa de cinco carbonos y dos radicales fosfato. El ADP contiene menos energía química que el ATP.

adenosine triphosphate (ATP) The molecule that provides energy in a form that all cells can readily use to perform the work of the cell. (p. 96)

adenosín trifosfato (ATP) Molécula que proporciona energía en una forma que todas las células pueden utilizar fácilmente para realizar su trabajo.

adhesion The tendency of water molecules to stick to other polar or charged molecules. (p. 44)

adhesión La tendencia de las moléculas de agua a adherirse a otras moléculas polares o cargadas.

agonist A ligand that binds a receptor and leads to a response. (p. 305)

agonista Ligando que se une a un receptor celular y provoca una respuesta.

alleles Different forms of a gene, corresponding to different DNA sequences. (p. 369)

alelos Formas alternativas de un gen que corresponden a diferentes secuencias de ADN

allopatric Describes populations that are geographically separated from each other. (p. 668)

alopátrica Describe poblaciones que están separadas geográficamente entre sí.

allosteric site A site other than the active site of an enzyme. (p. 206)

sitio alostérico Parte de la enzima distinta del sitio activo.

alpha helix (α helix) One of the two principal types of secondary structure found in proteins. (p. 70)

hélice alfa (hélice α) Uno de los dos tipos principales de estructura secundaria de las proteínas.

alternative hypothesis A hypothesis that predicts that the independent variable in an experiment will have a real effect, and there will be a difference between the data collected from the control groups and experimental groups. (p. 11)	**hipótesis alternativa** Hipótesis que predice que la variable independiente de un experimento tendrá un efecto real y que habrá una diferencia entre los datos de los grupos de control y los grupos experimentales.
alternative splicing The process by which exons are joined together and introns are removed during RNA processing. A single gene can be spliced in different ways to yield different mRNAs and different protein products. (p. 465)	**empalme alternativo** Proceso por el que se unen los exones y se eliminan los intrones durante el procesamiento del ARN. Un solo gen puede empalmarse de diferentes maneras para generar diferentes ARNm y productos proteicos.
amino acid The subunit of proteins. (p. 38)	**aminoácido** Subunidad básica de las proteínas.
amplification The process in which a small number of signaling molecules can generate a very large response. (p. 293)	**amplificación** Proceso en el que un pequeño número de moléculas señalizadoras puede generar una respuesta muy grande.
anabolism The set of chemical reactions that build molecules from smaller units utilizing an input of energy, usually in the form of ATP. Anabolic reactions result in net energy storage within cells and the organism. (p. 193)	**anabolismo** Conjunto de reacciones químicas que construyen moléculas a partir de unidades más pequeñas utilizando un aporte de energía, generalmente en forma de ATP. Las reacciones anabólicas dan lugar a un almacenamiento neto de energía en las células y el organismo.
anaphase The phase of mitosis in which sister chromatids separate. (p. 328)	**anafase** Fase de la mitosis en la que las cromátidas hermanas separan.
antagonist A ligand that binds a receptor and inhibits a response. (p. 305)	**antagonista** Ligando que se une a un receptor e inhibe una respuesta.
anticodon The sequence of three nucleotides in a tRNA molecule that base pairs with the corresponding codon in an mRNA molecule. (p. 471)	**anticodón** Secuencia de tres nucleótidos en una molécula de ARNt, complementaria al codón correspondiente en una molécula de ARNm.
antiparallel Oriented in opposite directions, like the two strands in a DNA molecule. (p. 82)	**antiparalelo** Orientado en direcciones opuestas, como las dos hebras de una molécula de ADN.
apoptosis The genetically programmed death of a cell. (p. 340)	**apoptosis** Muerte genéticamente programada de una célula.
aquaporin A channel protein that allows water to cross the cell membrane more readily than by diffusing through the lipid bilayer. (p. 147)	**acuaporina** Proteína de canal que permite que el agua atraviese la membrana celular con mayor facilidad que a través de la bicapa lipídica.
archaea One of the three domains of life, consisting of unicellular organisms that differ from bacteria in many aspects of their cell and molecular biology. (p. 98)	**archaea** Uno de los tres dominios de la vida, formado por organismos unicelulares que difieren de las bacterias en muchos aspectos de su biología celular y molecular.
artificial selection A form of directional selection similar to natural selection, but with selection done intentionally by humans, usually with a specific goal in mind, such as increased milk yield in cattle. (p. 580)	**selección artificial** Forma de selección direccional similar a la selección natural pero realizada intencionadamente por el ser humano, generalmente con un objetivo específico, como aumentar la producción de leche en el ganado.
asexual reproduction The reproduction of organisms in which an offspring inherits its DNA from a single parent. (p. 323)	**reproducción asexual** Mecanismo de reproducción de organismos en la que un descendiente hereda su ADN de un solo progenitor.
atom The basic unit of matter. (p. 26)	**átomo** La unidad básica de la materia.
atomic mass The total mass of an atom; consists of the number of protons and neutrons in the atom. (p. 30)	**masa atómica** La masa total de un átomo, compuesta por el número de protones y neutrones en el átomo.
atomic number The number of protons in an element. (p. 30)	**número atómico** El número de protones de un elemento.
ATP synthase An enzyme that couples the movement of protons through the enzyme with the synthesis of ATP. (p. 225)	**ATP sintasa** Enzima que acopla el movimiento de protones a través de la enzima con la síntesis de ATP.
autosome A chromosome that is not a sex chromosome. (p. 352)	**autosoma** Cualquier cromosoma que no sea sexual.
autotroph Organism that obtains energy by photosynthesis or chemosynthesis. Also known as a producer. (p. 741)	**autótrofo** Organismo que obtiene energía por fotosíntesis o quimiosíntesis. También se conoce como productor.

B

bacteria One of the three domains of life, consisting of unicellular organisms that differ from archaea in many aspects of their cell and molecular biology. (p. 98)	**bacteria** Uno de los tres dominios de la vida, formado por organismos unicelulares que difieren de las arqueas en muchos aspectos de su biología celular y molecular.
base A molecule that accepts H^+ ions and removes them from solution. When the concentration of hydrogen ions is less than the concentration of hydroxide ions, the solution is basic or alkaline. (p. 49)	**base** Molécula que acepta iones H^+ y los elimina de la solución. Cuando la concentración de iones de hidrógeno es menor que la concentración de iones de hidróxido, la solución es básica o alcalina.

beta sheet (β sheet) One of the two principal types of secondary structure found in proteins. (p. 71)	**lámina beta (lámina β)** Uno de los dos tipos principales de estructura secundaria que se encuentra en las proteínas.
biodiversity Biological diversity; the total number of species, or, more broadly, the diversity of genetic sequences, cell types, metabolism, life history, phylogenetic groups, communities, and ecosystems. (p. 562)	**biodiversidad** Diversidad biológica; el número total de especies o, en términos más amplios, la diversidad de secuencias genéticas, tipos de células, metabolismos, historias de vida, grupos filogenéticos, comunidades y ecosistemas.
biological species concept (BSC) As described by Ernst Mayr, the concept that "species are groups of actually or potentially interbreeding populations that are reproductively isolated from other such groups." The BSC is the most widely used and accepted definition of a species, but cannot be applied to asexual or extinct organisms. (p. 663)	**concepto de especie biológica (BSC)** Tal y como lo describió Ernst Mayr, el concepto de que "las especies son grupos de poblaciones real o potencialmente entrecruzables que están aisladas reproductivamente de otros grupos afines". El BSC, por sus siglas en inglés, es la definición de especie más utilizada y aceptada, pero no puede aplicarse a organismos asexuados o extintos.
biological system A system made of both biological and physical entities that interact and, as a result, show complex properties. (p. 4)	**sistema biológico** Sistema formado por entidades biológicas y físicas que interactúan y, como resultado, muestran propiedades complejas.
biologist A scientist who studies biology. (p. 1)	**biólogo** Científico que estudia la biología.
biology The science of how life works. (p. 1)	**biología** Ciencia que estudia el funcionamiento de la vida.
biotic Pertaining to life and living organisms. (p. 4)	**biótico** Perteneciente a la vida y a los organismos vivos.
bottleneck An extreme, usually temporary, reduction in population size that may result in marked loss of genetic diversity and, in the process, genetic drift. (p. 603)	**cuello de botella** Reducción extrema y generalmente temporal del tamaño de una población que puede llevar a una marcada pérdida de diversidad genética y acelerar la deriva genética.

C

Calvin cycle The process in which carbon dioxide is reduced to synthesize carbohydrates, with ATP and NADPH as the energy sources. (p. 217)	**ciclo de Calvin** Proceso en el que el dióxido de carbono se reduce para sintetizar carbohidratos, usando ATP y NADPH como fuentes de energía.
cancer A condition of unregulated cell division, often resulting in the development of tumors and disease. (p. 336)	**cáncer** Condición de división celular no regulada que suele resultar en el desarrollo de tumores y enfermedades.
carbohydrate A type of organic molecule. (p. 38)	**carbohidrato** Tipo de molécula orgánica.
carbon-14 (^{14}C) A radioactive isotope of carbon frequently used in radiometric dating. (p. 624)	**carbono-14 (^{14}C)** Isótopo radiactivo del carbono comúnmente utilizado en la datación radiométrica.
carrier protein A transport protein that facilitates movement of molecules across a cell membrane. (p. 147)	**proteína transportadora** Proteína que facilita el movimiento de moléculas a través de una membrana celular.
carrying capacity (K) A limit on how many individuals can be supported in a given environment. (p. 760)	**capacidad de carga (K)** Cantidad de individuos que puede soportar un entorno determinado.
catabolism The set of chemical reactions that break down molecules into smaller units and, in the process, produce ATP to meet the energy needs of the cell. (p. 193)	**catabolismo** Conjunto de reacciones químicas que descomponen las moléculas en unidades más pequeñas y en el proceso producen ATP para satisfacer las necesidades energéticas de la célula.
catalyst A compound that increases the rate of a chemical reaction, but emerges from the process unaltered. (p. 197)	**catalizador** Compuesto que aumenta la velocidad de una reacción química, pero que sale inalterado del proceso.
cell cycle The collective name for the steps that make up the life cycle of a eukaryotic cell, including DNA replication and mitotic cell division. (p. 324)	**ciclo celular** Nombre colectivo de las etapas que componen el ciclo de vida de las células eucariotas, incluyendo la replicación de ADN y la división mitótica.
cell division The process by which a parent cell gives rise to two daughter cells. (p. 322)	**división celular** Proceso por el que una célula madre genera dos células hijas.
cell theory The set of ideas that places the cell at the center of life. It states that the cell is the fundamental unit of life, that organisms are made up of one or more cells, and that cells arise from preexisting cells. (p. 93)	**teoría celular** Conjunto de ideas que sitúan a la célula en el centro de la vida. Afirma que la célula es la unidad fundamental de la vida, que los organismos están formados por una o más células y que las células surgen de células preexistentes.
cell The simplest self-reproducing entity that can exist as an independent unit of life. (p. 1)	**célula** La entidad más sencilla posible que puede reproducirse por sí misma y existir como unidad de vida independiente.
cell membrane The membrane that surrounds the cytoplasm of the cell, separating the inside of the cell from the outside of the cell; also known as **plasma membrane.** (p. 58)	**membrana celular** La membrana que rodea el citoplasma de la célula y separa el interior de la célula del exterior de la célula; también se conoce como **membrana plasmática.**
cellular respiration A series of chemical reactions in which organic molecules are oxidized to carbon dioxide, converting the energy stored in organic molecules to ATP. (p. 110)	**respiración celular** Serie de reacciones químicas en las que las moléculas orgánicas se oxidan y se convierten en dióxido de carbono, transformando la energía almacenada en las moléculas orgánicas en ATP.

central dogma The idea that information in a cell usually flows from DNA to RNA to protein. (p. 84)	**dogma central** Concepto según el cual la información en una célula suele fluir del ADN al ARN y luego a las proteínas.
centromere A region of DNA where sister chromatids can be held together, it is also the place where the mitotic spindle attaches to drive chromosome movement during mitosis. (p. 325)	**centrómero** Región del ADN en la que las cromátidas hermanas se unen. También es el lugar en el que se ancla el huso mitótico para impulsar el movimiento de los cromosomas durante la mitosis.
centrosome A compact structure that is the microtubule organizing center for animal cells. (p. 326)	**centrosoma** Estructura compacta que constituye el centro organizador de los microtúbulos en las células animales.
channel protein A transport protein with a passage that allows the movement of molecules through it. (p. 147)	**canal de proteína** Proteína de transporte con un pasaje que permite el movimiento de moléculas.
character An anatomical, physiological, or molecular feature of an organism that varies among taxa. (p. 651)	**carácter** Atributo anatómico, fisiológico o molecular de un organismo que varía entre taxones.
checkpoint A point of transition between one phase of the cell cycle and the next phase of the cell cycle. (p. 338)	**punto de control** Punto de transición entre dos fases del ciclo celular.
chemical bond A form of attraction between atoms that holds them together. (p. 33)	**enlace químico** Proceso de atracción entre átomos que los mantiene unidos.
chemical energy A form of potential energy held in the chemical bonds between pairs of atoms in a molecule. (p. 185)	**energía química** Forma de energía potencial que se conserva en los enlaces químicos entre pares de átomos en una molécula.
chemical reaction A process by which molecules are transformed into different molecules. (p. 35)	**reacción química** Proceso por el que unas moléculas se transforman en otras moléculas distintas.
chemiosmosis The movement of ions, such as hydrogen ions (protons), from a region of high concentration to a region of low concentration across a selectively permeable membrane, similar to the movement of water by osmosis. In mitochondria and chloroplasts, chemiosmosis powers the synthesis of ATP by ATPase. (p. 251)	**quimiosmosis** Movimiento de iones, como los iones de hidrógeno (protones), desde una región de alta concentración a otra de baja concentración a través de una membrana selectivamente permeable, similar al movimiento del agua por ósmosis. En las mitocondrias y los cloroplastos, la quimiosmosis potencia la síntesis de ATP mediante la ATPasa.
chemosynthesis The biochemical process in which the energy of chemical compounds is used to build sugars. (p. 740)	**quimiosíntesis** Proceso bioquímico en el que la energía de los compuestos químicos se usa para construir azúcares.
Chi-square (χ^2) test A statistical test that determines whether the number of observed events in different categories differs from an expected number of events. (p. 374)	**prueba de Chi-cuadrado (χ^2)** Prueba estadística que determina si el número de eventos observados en diferentes categorías difiere de un número esperado de eventos.
chlorophyll The major photosynthetic pigment contained in the photosynthetic membranes; it plays a key role in the cell's ability to capture energy from sunlight. (p. 112)	**clorofila** El principal pigmento fotosintético contenido en las membranas fotosintéticas; desempeña un papel clave en la capacidad de la célula para captar la energía de la luz solar.
chloroplast An organelle in photosynthetic eukaryotes that converts energy of sunlight into chemical energy by synthesizing simple sugars. (p. 111)	**cloroplasto** Organelo de las células fotosintéticas eucariotas que convierte la energía de la luz solar en energía química a través de la síntesis de azúcares simples.
chromatin Inside the eukaryotic nucleus, chromatin refers to the assemblage of DNA, RNA, and protein that fills the nucleus. (p. 326)	**cromatina** Conjunto de ADN, ARN y proteínas que llena el núcleo de las células eucariotas.
chromosome A cellular structure containing the genetic material in cells, consisting of a single DNA molecule with associated proteins. (p. 324)	**cromosoma** Estructura celular que contiene el material genético de las células, formada por una única molécula de ADN con proteínas asociadas.
cladogram A type of phylogenetic tree that shows clades, or monophyletic groups. (p. 650)	**cladograma** Tipo de árbol filogenético que muestra los clados o grupos monofiléticos.
codon A group of three adjacent nucleotides in mRNA that specifies an amino acid in a protein or that terminates protein synthesis. (p. 471)	**codón** Grupo de tres nucleótidos adyacentes en el ARNm que especifica un aminoácido en una proteína o que termina la síntesis de proteínas.
cohesion Tendency of water molecules to stick to one another due to hydrogen bonding with one another. (p. 44)	**cohesión** Tendencia de las moléculas de agua a mantenerse unidas debido a los enlaces de hidrógeno entre ellas.
commensalism An interaction in which one species receives a benefit while the other species is neither harmed nor helped. (p. 787)	**comensalismo** Interacción en la que una especie recibe un beneficio mientras que la otra no es perjudicada ni beneficiada.
competition An interaction between species that require the same limited resource. (p. 786)	**competencia** Interacción entre especies que requieren el mismo recurso limitado.
complementary Describes the relationship of purine and pyrimidine bases, in which the base A pairs only with T, and G pairs only with C. (p. 83)	**complementario** Describe la relación de las bases púricas con las pirimidínicas, en la que la base A se empareja sólo con la T, y la G sólo con la C.
complex carbohydrate A long, branched chain of monosaccharides. (p. 57)	**carbohidrato complejo** Cadena larga y ramificada de monosacáridos.

concentration gradient A difference in concentration of a substance, with regions of higher and lower concentration. (p. 146)	**gradiente de concentración** Diferencia de concentración de una sustancia con regiones de mayor y menor concentración.
conjugation The direct cell-to-cell transfer of DNA, usually a plasmid; a form of horizontal gene transfer. (p. 520)	**conjugación** Transferencia directa de ADN (por lo general en forma de un plásmido) de una célula a otra; mecanismo de transferencia horizontal de genes.
contractile vacuole An organelle in some protists that actively takes up and expels water that enters the cell by osmosis. (p. 159)	**vacuola contráctil** Orgánulo de algunos protistas que absorbe y expulsa activamente el agua que entra en la célula por ósmosis.
control group A group that is not exposed to the independent variable in an experiment and therefore is not expected to show a change; also known as a **negative control,** or *control* for short. (p. 7)	**grupo de control** Grupo que no está expuesto a la variable independiente en un experimento y, por tanto, no se espera que muestre un cambio; también se conoce como **control negativo,** o simplemente *control*.
controlled experiment An experiment in which there are at least two groups to be tested and the conditions and setup are almost identical for the groups' except that one group (the test or experimental group) deliberately has a variable introduced, whereas the other (control) group does not. (p. 7)	**experimento controlado** Experimento en el que se realiza una prueba con al menos dos grupos. Las condiciones y el montaje son casi idénticos para ambos grupos, salvo que en uno de ellos (el grupo de prueba o experimental) se introduce deliberadamente una variable, mientras que en el otro (el de control) no.
convergent evolution The independent evolution of similar traits in different species as a result of similar environments or selective pressures. (p. 577)	**evolución convergente** Evolución independiente de rasgos similares en diferentes especies como resultado de entornos similares o presiones selectivas.
cooperative behavior A behavior that helps a members of the same species improve another individual's fitness. (p. 731)	**comportamiento cooperativo** Comportamiento que ayuda a los miembros de una misma especie a mejorar la aptitud de otro individuo.
covalent bond A bond in which two atoms share a pair of electrons. (p. 33)	**enlace covalente** Enlace en el que dos átomos comparten un par de electrones.
crossing over A process in prophase I in meiosis when each chromatid can break at the chiasma and then connect to a non-sister chromatid. Also known as **recombination.** (p. 354)	**entrecruzamiento cromosómico** Proceso de la profase I de la meiosis en el que cada cromátida puede romperse en el quiasma y conectarse a una cromátida no hermana. También se conoce como **recombinación.**
cyclic adenosine monophosphate (cAMP) Derived from ATP and serves as an intracellular second messenger in many signal transduction pathways. (p. 293)	**monofosfato de adenosina cíclico (cAMP)** Derivado del ATP que sirve como segundo mensajero intracelular en muchas rutas de transducción de señales.
cyclin-dependent kinase (CDK) An enzyme that is activated by binding to the appropriate cyclin partner, gaining the ability to transfer a phosphate group to a target protein. (p. 337)	**quinasa dependiente de ciclina (CDK)** Enzima que se activa al unirse a la ciclina adecuada, adquiriendo la capacidad de transferir un grupo fosfato a una proteína objetivo.
cyclins A family of several different proteins that are produced in different amounts during different phases of the cell cycle so that they can bind to and regulate cyclin-dependent kinases, thus cueing the progression of the cell cycle. (p. 337)	**ciclinas** Familia de varias proteínas diferentes que se producen en distintas cantidades durante las diferentes fases del ciclo celular, de modo que pueden unirse a las quinasas dependientes de ciclinas y regularlas, indicando así la progresión del ciclo celular.
cytokinesis The stage of M phase in which the cytosol, organelles, and duplicated nuclei from one eukaryotic cell divide into two daughter cells. (p. 324)	**citoquinesis** Etapa de la fase M en la que el citosol, los orgánulos y los núcleos duplicados de una célula eucariota se dividen en dos células hijas.
cytoplasm The region of the cell outside of the nucleus. (p. 98)	**citoplasma** Región de la célula situada fuera del núcleo.
cytosine (C) A pyrimidine base. (p. 79)	**citosina (C)** Una base pirimidínica.
cytoskeleton An internal protein scaffold that helps cells to maintain their shape. (p. 112)	**citoesqueleto** Entramado proteico interno que ayuda a las células a mantener su forma.
cytosol The region of the cell inside the cell membrane but outside the organelles; the jelly-like internal environment that surrounds the organelles. (p. 101)	**citosol** La región de la célula situada dentro de la membrana celular pero fuera de los orgánulos; el medio interno gelatinoso que rodea a los orgánulos.

D

dehydration synthesis reaction A reaction in which a water molecule is released by the two reacting molecules as a new covalent bond is formed between them. Dehydration synthesis reactions build large molecules from smaller subunits. Also called dehydration reaction. (p. 50)	**reacción de síntesis por deshidratación** Reacción en la que dos moléculas forman un nuevo enlace covalente entre ellas y liberan una molécula de agua. Las reacciones de síntesis de deshidratación construyen moléculas grandes a partir de subunidades más pequeñas. También se denomina reacción de deshidratación.
denature Unfold, as a protein. A protein that unfolds loses its function, as the shape of a protein is closely connected to its function. (p. 204)	**desnaturalización** Despliegue de una proteína. Una proteína que se desdobla pierde su función, ya que la forma de una proteína está estrechamente relacionada con su funcionamiento.

density-dependent factor Factors that limit population growth as its size increases. (p. 766)	**factor dependiente de la densidad** Factores que limitan el crecimiento de una población a medida que aumenta su tamaño.
density-independent factor Factors that reduce density regardless of a population's size. (p. 765)	**factor independiente de la densidad** Factores que reducen la densidad de una población independientemente de su tamaño.
deoxyribonucleic acid (DNA) The carrier molecule of genetic information for all organisms. (p. 4)	**ácido desoxirribonucleico (ADN)** La molécula portadora de la información genética de todos los organismos.
deoxyribose The pentose sugar component of DNA. (p. 56)	**desoxirribosa** El componente de azúcar pentosa del ADN.
dependent variable The variable that is being tested, observed, or measured, and that may change due to the independent variable. (p. 8)	**variable dependiente** La variable que se está probando, observando o midiendo, y que puede cambiar debido a la variable independiente.
derived character A character or trait that is newly evolved; an evolutionary innovation. (p. 652)	**carácter derivado** Rasgo que ha evolucionado recientemente; innovación evolutiva.
differentiation The process by which cells become progressively more specialized as a result of gene regulation. (p. 499)	**diferenciación** Proceso por el que las células se vuelven cada vez más especializadas como resultado de la regulación génica.
diffusion The net movement of molecules from areas of higher to lower concentration of the molecules due to random motion. (p. 124)	**difusión** Movimiento neto de moléculas que se produce desde zonas de mayor a menor concentración de moléculas debido al movimiento aleatorio.
dihybrid cross A mating in which hybrid individuals differ in two traits that are coded by two different genes. (p. 378)	**cruzamiento dihíbrido** Apareamiento en el que los individuos híbridos difieren en dos rasgos codificados por dos genes diferentes.
diploid A cell with two sets of chromosomes (denoted as 2n). (p. 351)	**diploide** Célula con dos series de cromosomas (denominados 2n).
directionality An asymmetry such that one end of a structure differs from the other. Also known as **polarity.** (p. 80)	**direccionalidad** Asimetría tal que un extremo de una estructura difiere del otro. También se le conoce como **polaridad.**
disaccharide Two simple sugars joined by a covalent bond; an example is sucrose. (p. 57)	**disacárido** Dos azúcares simples unidos por un enlace covalente; un ejemplo es la sacarosa.
distribution A description which typically takes the form of a list or graph of all of the possible values of a particular variable and their frequencies. The distribution describes the spread of the data and how often each value occurs. (p. 21)	**distribución** Descripción, generalmente presentado como una lista o gráfica, de todos los valores posibles de una variable concreta y sus frecuencias. La distribución describe la dispersión de los datos y la frecuencia con la que ocurre cada valor.
divergent evolution The process by which two groups of organisms become genetically and physically different from each other over time. (p. 668)	**evolución divergente** Proceso por el que dos grupos de organismos se diferencian genética y físicamente entre sí a lo largo del tiempo.
DNA editing Techniques used to "rewrite" a nucleotide sequence so that specific mutations can be introduced into genes to better understand their function, or to correct mutant versions of genes to restore normal function. (p. 536)	**edición del ADN** Técnicas utilizadas para "reescribir" una secuencia de nucleótidos de manera que se puedan introducir mutaciones específicas en ciertos genes para comprender mejor su función, o para corregir versiones mutantes de los genes y restaurar su función normal.
DNA ligase An enzyme that joins two DNA fragments together. (p. 454)	**ADN ligasa** Enzima que une dos fragmentos de ADN.
DNA polymerase An enzyme that is a critical component of a large protein complex that carries out DNA replication. (p. 451)	**ADN polimerasa** Enzima que funciona como componente crítico de un gran complejo proteico.
DNA replication The process of duplicating a DNA molecule, during which the parental strands separate and new partner strands are made. (p. 339)	**replicación del ADN** El proceso de duplicación de una molécula de ADN, durante el cual las hebras parentales se separan y se forma una nueva pareja de hebras.
domain One of the three largest limbs of the tree of life: Eukarya, Bacteria, or Archaea. (p. 98)	**dominio** Una de las tres grandes ramas del árbol de la vida: Eukarya, Bacteria o Arquea.
dominant Describes an allele or trait that is expressed in heterozygotes. Only one dominant allele is needed to express the phenotype. (p. 369)	**dominante** Describe un alelo o carácter que se expresa en heterocigotos. Sólo se necesita un alelo dominante para expresar el fenotipo.
double bond A covalent bond in which two atoms share two pairs of electrons. (p. 33)	**doble enlace** Enlace covalente en el que dos átomos comparten dos pares de electrones.
double helix The structure formed by two strands of complementary nucleotides that coil around each other. (p. 81)	**doble hélice** Estructura formada por dos hebras de nucleótidos complementarios que se enroscan y forman una espiral.
Down syndrome A condition in humans caused by an extra chromosome 21. (p. 425)	**síndrome de Down** Condición en humanos causada por la presencia de una tercera copia del cromosoma 21.

E

ectotherm An animal that obtains most of its heat to warm its body from the environment. (p. 737)	**ectotermo** Animal que obtiene la mayor parte del calor que utiliza para calentar su cuerpo del medio ambiente.

electron An atomic particle that moves around the nucleus and carries a negative charge. (p. 30)

electron shell *See* **energy level.** (p. 31)

electronegativity Difference in the ability of atoms to attract electrons. (p. 34)

element A chemical that cannot be split into other chemicals. (p. 30)

emerging disease An infectious disease that has appeared recently and/or spread rapidly. (p. 547)

endergonic Describes reactions with a positive ΔG that are not spontaneous and so require an input of energy. (p. 191)

energetic coupling The driving of a non-spontaneous reaction by a spontaneous reaction. (p. 192)

endocytosis The process in which a vesicle buds off from the cell membrane, bringing material from outside the cell into that vesicle, which can then fuse with other membranes. (p. 151)

endomembrane system A cellular system present in eukaryotic cells that includes the nuclear envelope, the endoplasmic reticulum, the Golgi apparatus, lysosomes, the cell membrane, and the vesicles that move between them. (p. 106)

endoplasmic reticulum (ER) An organelle composed of a network of membranes that is involved in the synthesis of proteins and lipids. (p. 107)

endosymbiosis A symbiosis in which one partner lives within the other. (p. 173)

endotherm An animal that produces most of its heat as a by-product of metabolic reactions to maintain a warm and steady body temperature. (p. 736)

energetics The science of the properties of energy and how energy is distributed in biological, chemical, and physical processes. (p. 4)

energy level An area in space where electrons circle around the nucleus. Also called an **electron shell.** (p. 31)

energy The ability to do work. (p. 4)

entropy The amount of disorder, or the number of possible positions and motions of molecules, in a system. (p. 188)

enzyme A protein that functions as a catalyst to accelerate the rate of a chemical reaction. (p. 107)

epigenetic Describes effects on gene expression due to differences in DNA packaging and chromosome organization, such as modifications in histones or chromatin structure. (p. 493)

eukarya One of the three domains of life, in which cells have a nucleus. (p. 101)

eukaryotic Describes a cell that has a nucleus; used to refer collectively to animals, plants, fungi, and protists. (p. 98)

evolution A change in the genetic makeup of a population over time. (p. 2)

evolutionarily conserved Characteristics that persist relatively unchanged through diversification of a group of organisms and therefore remain similar in related species. (p. 503)

exergonic Describes reactions with a negative ΔG that proceed spontaneously and release energy. (p. 191)

exocytosis The process in which a vesicle fuses with the cell membrane and empties its contents into the extracellular space or delivers proteins to the cell membrane. (p. 150)

---

electrón Partícula atómica que se desplaza alrededor del núcleo y tiene carga negativa.

capa electrónica *Véase* **nivel energético.**

electronegatividad Diferencia en la capacidad de los átomos para atraer electrones.

elemento Sustancia química que no puede dividirse en otras sustancias químicas.

enfermedad emergente Enfermedad infecciosa que apareció recientemente y/o se ha extendido rápidamente.

endergónico Describe las reacciones con un ΔG positivo que no son espontáneas y que, por tanto, requieren un aporte de energía.

acoplamiento energético La conducción de una reacción no espontánea por una reacción espontánea.

endocitosis Proceso en el que una vesícula se desprende de la membrana celular, introduciendo material del exterior de la célula en esa vesícula, que luego puede fusionarse con otras membranas.

sistema endomembranoso Sistema celular presente en las células eucariotas que incluye la envoltura nuclear, el retículo endoplásmico, el aparato de Golgi, los lisosomas, la membrana celular y las vesículas que se mueven entre ellos.

retículo endoplásmico (RE) Orgánulo compuesto por una red de membranas que participa en la síntesis de proteínas y lípidos.

endosimbiosis Simbiosis en la que un organismo vive dentro del otro.

endotermo Animal que produce la mayor parte de su calor como subproducto de sus reacciones metabólicas y mantiene una temperatura corporal cálida constante.

energética Ciencia que estudia las propiedades de la energía y cómo ésta se distribuye en los procesos biológicos, químicos o físicos.

nivel de energía Área en la que los electrones giran alrededor del núcleo. También llamada **capa electrónica.**

energía La capacidad de trabajar.

entropía La cantidad de desorden, o el número de posiciones y movimientos posibles de moléculas, en un sistema.

enzima Proteína que funciona como catalizador para acelerar la velocidad de una reacción química.

epigenética Describe los efectos en la expresión genética de las diferencias en el empaquetamiento del ADN y la organización de los cromosomas, como modificaciones en las histonas o en la estructura de la cromatina.

eukarya Uno de los tres dominios de la vida, en el que las células tienen un verdadero núcleo.

eucariota Describe una célula que tiene un núcleo; se utiliza para referirse colectivamente a animales, plantas, hongos y protistas.

evolución Cambio en la composición genética de una población a lo largo del tiempo.

evolutivamente conservado Características que persisten relativamente iguales a través de la diversificación de un grupo de organismos y, por lo tanto, siguen siendo similares en las especies relacionadas.

exergónico Describe las reacciones con un ΔG negativo que ocurren espontáneamente y liberan energía.

exocitosis Proceso en el que una vesícula se fusiona con la membrana celular y expulsa su contenido al espacio extracelular o entrega proteínas a la membrana celular.

exon A sequence that is left intact in mRNA after RNA splicing, and is therefore expressed in the protein. (p. 465)

exón Secuencia que permanece intacta en el ARNm después del empalme del ARN y que, por tanto, se expresa en la proteína.

experimental group *See* **Test group.** (p. 7)

grupo experimental *Véase* **Grupo de prueba.**

exponential growth model A model of a population that continues to grow rapidly over time in an exponential manner. (p. 758)

modelo de crecimiento exponencial Modelo en que una población crece rápidamente en el tiempo, de forma exponencial.

extinction The loss of a group of organisms, typically a species. (p. 682)

extinción La pérdida de un grupo de organismos, normalmente una especie.

F

facilitated diffusion Diffusion across a cell membrane through a transmembrane protein, such as a channel or carrier. (p. 147)

difusión facilitada Difusión a través de una membrana celular mediante una proteína transmembranal, como una proteína de canal o transportadora.

fatty acid A long chain of carbons attached to a carboxyl group; three fatty acid chains attached to glycerol form a triacylglycerol, a lipid used for energy storage. (p. 58)

ácido graso Cadena larga de carbonos unida a un grupo carboxilo. Tres cadenas de ácidos grasos unidas al glicerol forman un triacilglicerol, lípido utilizado para el almacenamiento de energía.

fermentation A variety of metabolic pathways that produce ATP from the partial oxidation of organic molecules without oxidative phosphorylation or an electron acceptor, such as oxygen. (p. 254)

fermentación Serie de rutas metabólicas que producen ATP a partir de la oxidación parcial de moléculas orgánicas sin fosforilación oxidativa ni un aceptor de electrones como el oxígeno.

first law of thermodynamics The law of conservation of energy, which states that energy can be neither created nor destroyed; it can only be transformed from one form into another. (p. 188)

primera ley de la termodinámica La ley de conservación de la energía, que establece que la energía no puede crearse ni destruirse, sólo transformarse de una forma a otra.

fitness A measure of the ability of an individual to survive and reproduce in a particular environment; the extent to which the individual's genotype is represented in the next generation. (p. 262)

aptitud Medida de la capacidad de un individuo para sobrevivir y reproducirse en un entorno determinado; según la cual el genotipo del individuo estará representado en la siguiente generación.

fluid mosaic model A model proposing that the lipid bilayer is a dynamic structure that allows molecules to move laterally within the membrane and is a mosaic, or mixture, of several components, including lipids, proteins, and carbohydrates. (p. 141)

modelo de mosaico fluido Modelo que propone que la bicapa lipídica es una estructura dinámica que permite a las moléculas moverse lateralmente dentro de la membrana; y es un mosaico, una mezcla, de varios componentes, incluyendo lípidos, proteínas y carbohidratos.

fossil The remains of a once-living organism. (p. 620)

fósil Los restos de un organismo que estuvo vivo.

founder effect A type of genetic drift that occurs when only a few individuals establish a new population. (p. 603)

efecto fundador Tipo de deriva genética que se produce cuando unos pocos individuos establecen una nueva población.

functional group A group of one or more atoms that has a particular chemical property on its own, regardless of what it is attached to. (p. 55)

grupo funcional Grupo de uno o más átomos que tiene una propiedad química particular por sí mismo, independientemente de a qué esté unido.

G

G protein-coupled receptor A receptor protein in cell membranes that is associated with a G protein that in turn alters activity of adenylyl cyclase and other generators of second messengers in responding cells. (p. 293)

receptor acoplado a proteínas G Proteína receptora en las membranas celulares que se asocia a una proteína G que altera la actividad de la adenilil ciclasa y otros generadores de segundos mensajeros en las células responsivas.

G_0 phase The phase in the cell cycle in which cells pause between M phase and S phase; it may last for periods ranging from days to more than a year. (p. 326)

fase G_0 Fase del ciclo celular en la que las células pausan entre la fase M y la fase S. Puede durar desde unos días hasta más de un año.

G_1 phase The phase during interphase in which the cell synthesizes regulatory proteins controlling the eukaryotic cell cycle. (p. 325)

fase G_1 Fase dentro de la interfase en la que la célula sintetiza proteínas reguladoras que controlan el ciclo celular eucariota.

G_2 phase The phase during interphase after DNA has been replicated, characterized by increases in cell size and protein contents. (p. 325)

fase G_2 Fase dentro de la interfase después de la replicación del ADN, caracterizada por el aumento del tamaño de la célula y del contenido de proteínas.

gamete A reproductive haploid cell; gametes fuse to form a diploid zygote. In many species, there are two types of gametes: eggs in females, sperm in males. (p. 356)

gameto Célula reproductora haploide; los gametos se fusionan para formar un cigoto diploide. En muchas especies hay dos tipos de gametos: los óvulos en las hembras y los espermatozoides en los machos.

gel electrophoresis A technique that is used to separate DNA and RNA based on size, and proteins based on size and charge, using an electric current passed through a jelly-like substance. (p. 529)

electroforesis en gel Técnica utilizada para separar el ADN y el ARN en función de su tamaño, y las proteínas en función de su tamaño y carga, mediante una corriente eléctrica que pasa a través de una sustancia gelatinosa.

gene expression The production of a functional gene product, such as a protein; the "turning on" of a gene. (p. 436)	**expresión génica** La producción de un producto génico funcional, como una proteína; la "activación" de un gen.
gene flow The movement of alleles from one population to another through interbreeding between members of each population. (p. 605)	**flujo de genes** El movimiento de alelos de una población a otra mediante el cruzamiento entre miembros de las distintas poblaciones.
gene regulation The various ways in which cells control gene expression. (p. 440)	**regulación génica** Las diversas formas en que las células controlan la expresión de los genes.
gene The unit of hereditary affecting one or more traits of an organism; the DNA sequence that corresponds to a specific protein or noncoding RNA. (p. 1)	**gen** Secuencia de ADN que corresponde a un producto proteico específico y es la unidad de la herencia.
genetic code The correspondence between codons and amino acids, in which 20 amino acids are specified by 64 codons. (p. 473)	**código genético** Correspondencia entre codones y aminoácidos, en la que 20 aminoácidos son especificados por 64 codones.
genetic drift A random change in the frequency of an allele due to chance. (p. 603)	**deriva genética** Cambio aleatorio en la frecuencia de un alelo debido al azar.
genetic variation The range of different genotypes found among individuals. (p. 351)	**variación genética** Gama de genotipos diferentes que se encuentran entre los individuos.
genetically modified organism (GMO) An organism that has been genetically engineered, such as modified viruses and bacteria, laboratory organisms, agricultural crops, and domestic animals; also called a transgenic organism. (p. 535)	**organismo genéticamente modificado (OGM)** Organismo que ha sido alterado genéticamente, como los virus y bacterias modificados, organismos de laboratorio, cultivos agrícolas y animales domésticos; también se le llama organismo transgénico.
genome All of the genetic information that an individual, or a species, contains. (p. 1)	**genoma** Toda la información genética que contiene un individuo o una especie.
genotype The set of genes that an organism carries. (p. 350)	**genotipo** El conjunto de genes que carga un organismo.
Gibbs free energy (G) The amount of energy available to do work. (p. 191)	**energía libre de Gibbs (G)** La cantidad de energía disponible para realizar trabajo.
glycerol A 3-carbon molecule with hydroxyl groups attached to each carbon; a component of triacylglycerol. (p. 58)	**glicerol** Molécula de 3 carbonos, cada uno de los cuales está unido a un grupo hidroxilo; componente del triacilglicerol.
glycolipid A carbohydrate that is covalently linked to a lipid. (p. 141)	**glucolípido** Carbohidrato unido de modo covalente a un lípido.
glycolysis The breakdown of glucose to pyruvate; the first stage of cellular respiration. (p. 237)	**glucólisis** Descomposición de la glucosa en piruvato; primera etapa de la respiración celular.
glycoprotein A carbohydrate that is covalently linked to a protein. (p. 141)	**glucoproteína** Carbohidrato unido de modo covalente a una proteína.
Golgi apparatus An organelle that modifies proteins and lipids produced by the endoplasmic reticulum and acts as a sorting station as they move to their final destinations. (p. 108)	**aparato de Golgi** Orgánulo que modifica las proteínas y los lípidos producidos por el retículo endoplásmico y que actúa como estación de clasificación cuando se dirigen a su destino final.
gradualism The idea that species change slowly over time. (p. 676)	**gradualismo** La idea de que las especies cambian lentamente con el tiempo.
grana (singular, granum) Interlinked stacks of thylakoids in chloroplasts. (p. 111)	**granas** Apilamientos interconectados de tilacoides en los cloroplastos.
GTP cap The modification of the 5′ end of the primary transcript by the addition of a special nucleotide attached in an unusual chemical linkage; also called a 5′ cap. (p. 464)	**casquete de GTP** La modificación del extremo 5′ de la transcripción primaria mediante la adición de un nucleótido especial unido en un enlace químico inusual; también se denomina caperuza 5′.
guanine (G) A purine base. (p. 79)	**guanina (G)** Una base púrica.

H

half-life The time it takes for an amount of a substance to reach half its original value. Radioactive half-life is the time it takes for half of the atoms in a given sample of a substance to decay. (p. 625)	**vida media** El tiempo que tarda una determinada cantidad de una sustancia en alcanzar la mitad de su valor original. La vida media radiactiva es el tiempo que tarda en descomponerse la mitad de los átomos de una muestra determinada de una sustancia.
haploid A cell with one set of chromosomes (denoted as 1*n*). (p. 352)	**haploide** Célula con una sola serie de cromosomas (denominado 1*n*).
Hardy–Weinberg equilibrium A state in which allele and genotype frequencies do not change over time, implying the absence of evolutionary forces. It also specifies a mathematical relationship between allele frequencies and genotype frequencies. (p. 607)	**equilibrio de Hardy–Weinberg** Estado en el que las frecuencias de alelos y genotipos no cambian con el tiempo, lo que implica la ausencia de fuerzas evolutivas. También especifica una relación matemática entre las frecuencias alélicas y las frecuencias genotípicas.

helicase An enzyme that separates the two DNA strands at the replication fork so that each strand can be copied during DNA replication. (p. 451)

helicasa Enzima que separa las dos cadenas de ADN en la horquilla de replicación para que cada cadena pueda copiarse durante la replicación del ADN.

heterotroph An organism that obtains energy by eating other organisms. Also known as a consumer. (p. 741)

heterótrofo Organismo que obtiene energía al alimentarse de otros organismos. También se conoce como consumidor.

heterozygous A condition in which an individual possesses two different alleles for a given gene. (p. 317)

heterocigoto Condición en la que un individuo posee dos alelos diferentes para un carácter determinado.

homeostasis The active regulation and maintenance of a stable internal physiological state in the face of a changing external environment. (p. 95)

homeostasis La regulación activa y el mantenimiento de un estado fisiológico interno estable frente a un entorno externo cambiante.

homologous chromosomes Two chromosomes that are similar in size and shape, and that carry the same genes. (p. 352)

cromosomas homólogos Dos cromosomas que son similares en tamaño y forma, y que llevan los mismos genes.

homozygous A condition in which an individual possesses two identical alleles for a given gene. (p. 370)

homocigoto Condición en la que un individuo posee dos alelos idénticos para un determinado carácter.

horizontal gene transfer The transfer of genetic material between organisms that are not parent and offspring. (p. 520)

transferencia horizontal de genes Transferencia de material genético entre organismos que no son padres y descendientes.

hormone A signaling molecule that is transported in the circulatory system by endocrine signaling. (p. 279)

hormona Molécula señalizadora que se transporta en el sistema circulatorio por señalización endocrina.

hydrogen bond An interaction between a hydrogen atom with a slight positive charge and an electronegative atom of another molecule. In the case of water, hydrogen bonds form between the hydrogen atom of one water molecule and oxygen atom of another water molecule. (p. 44)

enlace de hidrógeno Una interacción entre un átomo de hidrógeno con una ligera carga positiva y un átomo electronegativo de otra molécula. En el caso del agua, los enlaces de hidrógeno se forman entre el átomo de hidrógeno de una molécula de agua y el átomo de oxígeno de otra molécula de agua.

hydrolysis reaction A reaction that adds water across a covalent bond and breaks the covalent bond. Hydrolysis reactions break down larger molecules into smaller, simpler molecules. (p. 51)

reacción de hidrólisis Reacción que añade agua a través de un enlace covalente y rompe este enlace. Las reacciones de hidrólisis descomponen moléculas grandes en moléculas más pequeñas y sencillas.

hydrophilic "Water loving"; describes polar molecules that readily dissolve in water. (p. 38)

hidrófilo "Amante del agua", describe moléculas polares que se disuelven fácilmente en el agua.

hydrophobic "Water fearing"; describes nonpolar molecules that do not dissolve well in water. (p. 38)

hidrófobo "Temeroso del agua", describe las moléculas no polares que no se disuelven bien en el agua.

hypertonic Describes a solution having a higher solute concentration (lower water potential) than another solution. (p. 159)

hipertónica Describe una solución que tiene una mayor concentración de solutos (menor potencial hídrico) que otra.

hypothesis A tentative explanation for one or more observations that make predictions that can be tested by experimentation or additional observations. (p. 6)

hipótesis Explicación tentativa de una o más observaciones en la cual se hacen predicciones que pueden ser comprobadas o refutadas mediante experimentación u observaciones adicionales.

hypotonic Describes a solution having a lower solute concentration (higher water potential) than another solution. (p. 159)

hipotónica Describe una solución que tiene una concentración de solutos más baja (potencial hídrico más alto) que otra solución.

I

independent variable The variable that is manipulated or added to an experimental group to test a hypothesis. (p. 8)

variable independiente La variable que se manipula o se añade a un grupo experimental para probar una hipótesis.

inhibitor A compound that decreases the activity of an enzyme. (p. 205)

inhibidor Compuesto que disminuye la actividad de una enzima.

innate behavior A behavior that is performed without any previous experience. (p. 728)

comportamiento innato Comportamiento que se manifiesta en un organismo sin necesidad de que exista una experiencia anterior.

integral membrane protein A protein that is permanently associated with the cell membrane and cannot be separated from the membrane experimentally without destroying the membrane itself. (p. 139)

proteína integral de membrana Proteína que está permanentemente asociada a la membrana celular y que no puede separarse de ella sin destruir la membrana misma.

interphase One of two major phases of the eukaryotic cell cycle, in which the cell copies its DNA and synthesizes proteins necessary for mitosis. (p. 324)

interfase Una de las dos fases principales del ciclo celular eucariota, en la que la célula copia su ADN y sintetiza las proteínas necesarias para la mitosis.

intron An intervening sequence that is removed from the primary transcript during RNA splicing. (p. 465)

intrón Secuencia intermedia que se elimina de la transcripción primaria durante el empalme del ARN.

invasive species Non-native species that become established in new ecosystems. (p. 628)	**especies invasoras** Especies no autóctonas que se establecen en nuevos ecosistemas.
ion An electrically charged atom. (p. 31)	**ion** Átomo con carga eléctrica.
ionic bond A bond in which two ions with opposite electrical charges associate with each other. (p. 34)	**enlace iónico** Enlace en el que dos iones con cargas eléctricas opuestas se asocian.
isotonic Describes a solution having the same solute concentration (same water potential) as another solution. (p. 159)	**isotónico** Describe una solución que tiene la misma concentración de soluto (mismo potencial hídrico) que otra.
isotope One of two or more forms of the same element with different numbers of neutrons. (p. 30)	**isótopo** Una de dos o más formas del mismo elemento con diferente cantidad de neutrones.

K

karyotype A visual display of the pairs of chromosomes. (p. 352)	**cariotipo** Representación visual de los pares de cromosomas.
keystone species Pivotal populations that affect other members of the community in ways that are disproportionate to their abundance or biomass. (p. 788)	**especies clave** Poblaciones fundamentales que afectan a otros miembros de la comunidad de forma desproporcionada a su abundancia o biomasa.
kinase An enzyme that catalyzes the transfer of a phosphate group from ATP to another molecule. (p. 293)	**quinasa** Enzima que cataliza la transferencia de un grupo fosfato del ATP a otra molécula.
kinetic energy The energy of motion. (p. 185)	**energía cinética** Energía del movimiento.
kinetochore The protein complex on a chromatid where the mitotic spindle attaches. (p. 328)	**cinetocoro** Complejos proteico de una cromátida donde se une el huso mitótico.
Krebs cycle The third stage of cellular respiration, in which acetyl-CoA is broken down and carbon dioxide is released; also called the citric acid cycle. (p. 237)	**ciclo de Krebs** Tercera etapa de la respiración celular, en la que se descompone el acetil-CoA y se libera dióxido de carbono. También llamado ciclo del ácido cítrico.

L

lagging strand A daughter strand that has its 5′ end pointed toward the replication fork, so as the parental double helix unwinds, a new DNA piece is initiated at intervals, and each new piece is elongated at its 3′ end until it reaches the piece in front of it. (p. 453)	**cadena rezagada** Cadena hija cuyo extremo 5′ apunta hacia la horquilla de replicación, de modo que, según la doble hélice parental se va desenrollando, se inicia un nuevo fragmento de ADN a intervalos, y cada nuevo fragmento se alarga en su extremo 3′ hasta alcanzar al fragmento anterior.
law of independent assortment The principle that segregation of one set of alleles for one gene is independent of the segregation of a second set of alleles for a different gene. (p. 378)	**ley de la segregación independiente** El principio de que la segregación de un conjunto de alelos para un gen es independiente de la segregación de un segundo conjunto de alelos para un gen diferente.
law of segregation The principle that half of the gametes receive one allele of a gene and half receive the other allele. This principle is explained by meiosis, in which the maternal and paternal homologous chromosomes separate during anaphase I. (p. 370)	**ley de segregación** Principio según el cual la mitad de los gametos reciben un alelo de un gen y la otra mitad reciben el otro alelo. Este principio se explica por la meiosis, en la que los cromosomas homólogos maternos y paternos se separan durante la anafase I.
leading strand A daughter strand that has its 3′ end pointed toward the replication fork, so as the parental double helix unwinds, this daughter strand can be synthesized as one long, continuous polymer. (p. 453)	**cadena líder** Cadena hija cuyo extremo 3′ apunta hacia la horquilla de replicación, de modo que cuando la doble hélice parental se desenrolla, esta cadena hija puede sintetizarse como un polímero largo y continuo.
learned behavior A behavior in which the individual's experience in its environment plays a large role. (p. 728)	**comportamiento aprendido** Comportamiento en el que la experiencia del individuo en su entorno juega un papel importante.
ligand-gated channel Receptor proteins in cell membranes that alter membrane permeability to ions after binding to an extracellular hydrophilic signaling molecule. (p. 299)	**canal activado por ligando** Proteínas receptoras de las membranas celulares que alteran la permeabilidad de la membrana a los iones tras unirse a una molécula señalizadora extracelular hidrofílica.
ligand A signaling molecule with a molecular shape and distribution of charge that allows it to match up with and bind to a complementary receptor protein. (p. 277)	**ligando** Molécula señalizadora con una forma molecular y distribución de carga tal que le permite emparejarse y unirse a una proteína receptora complementaria.
light reactions The series of chemical reactions during photosynthesis in which the energy of sunlight is used to synthesize NADPH and ATP. (p. 217)	**reacciones luminosas** La serie de reacciones químicas durante la fotosíntesis en la que la energía de la luz solar se utiliza para sintetizar NADPH y ATP.
linked genes Genes that are physically located near each other on the same chromosome. (p. 398)	**genes ligados** Genes que se encuentran localizados físicamente cerca unos de otros en el mismo cromosoma.

lipid bilayer A two-layered structure of the cell membrane made up of lipids with hydrophilic heads pointing outward toward the aqueous environment and hydrophobic tails oriented inward away from water. (p. 133)	**bicapa lipídica** Estructura de dos capas de la membrana celular formada por lípidos con cabezas hidrofílicas orientadas hacia afuera—hacia el medio acuoso—y colas hidrofóbicas orientadas hacia adentro—en dirección opuesta del agua.
lipid A type of organic molecule that is hydrophobic; lipids make up the barrier between the cell and its environment, store energy, and act as signaling molecules. (p. 38)	**lípido** Tipo de molécula orgánica hidrofóbica: los lípidos almacenan energía, actúan como moléculas señalizadoras y forman una barrera entre la célula y su ambiente.
logistic growth model A growth model that describes population growth when there is a density-dependent carrying capacity. (p. 768)	**modelo de crecimiento logístico** Modelo que describe el crecimiento poblacional cuando hay una capacidad de carga dependiente de la densidad.
lysosome A vesicle derived from the Golgi apparatus that contains enzymes that break down macromolecules such as proteins, nucleic acids, lipids, and carbohydrates. (p. 109)	**lisosoma** Vesícula derivada del aparato de Golgi que contiene enzimas que descomponen macromoléculas como proteínas, ácidos nucleicos, lípidos y carbohidratos.

M

M phase One of two major phases of the eukaryotic cell cycle, consisting of mitosis and cytokinesis. (p. 324)	**fase M** Una de las dos fases principales del ciclo celular eucariota; consiste en la mitosis y la citocinesis.
maternal inheritance A type of inheritance in which the organelles in the offspring cells derive from those in the mother. (p. 401)	**herencia materna** Tipo de herencia en la cual los organelos celulares de la descendencia son derivados de la madre.
matter Material that makes physical objects. (p. 26)	**materia** Material que constituye los objetos físicos.
maximum per capita growth rate (r_{max}) The maximum growth rate possible for a population living without any constraints. (p. 758)	**máximo crecimiento per cápita (r_{max})** La máxima tasa de crecimiento posible para una población que vive sin restricción alguna.
mean A measure of the average of a set of numbers. The mean is determined by adding all the values and dividing by the number of values. (p. 20)	**media** Promedio de un conjunto de números. La media se determina sumando todos los valores y dividiendo por el número de valores.
median The middle number in a set of numbers when ordered from lowest to highest. (p. 20)	**mediana** El número en el medio de un conjunto de números ordenados de menor a mayor.
meiosis A form of cell division in which a parental cell divides in two stages to produce four daughter cells. (p. 352)	**meiosis** Forma de división celular en la que una célula parental se divide en dos etapas para producir cuatro células hijas.
meiosis I The first step in meiosis in which homologous chromosomes are separated from each other. (p. 353)	**meiosis I** La primera etapa de la meiosis, en la que los cromosomas homólogos se separan entre sí.
meiosis II The second step in meiosis in which sister chromatids are separated from each other. (p. 353)	**meiosis II** Segunda etapa de la meiosis, en la que las cromátidas hermanas se separan entre sí.
messenger RNA (mRNA) The RNA molecule that combines with a ribosome to direct protein synthesis; it carries the genetic "message" from the DNA to the ribosome. (p. 463)	**ARN mensajero (ARNm)** Molécula de ARN que se combina con un ribosoma para dirigir la síntesis de proteínas; lleva el "mensaje" genético del ADN al ribosoma.
metabolic pathway A series of chemical reactions that build or break down molecules in cells. (p. 186)	**ruta metabólica** Serie de reacciones químicas que construyen o descomponen moléculas en las células.
metabolic rate The number of calories burned by an organism over time while at rest. (p. 738)	**tasa metabólica** El número de calorías que un organismo quema a través del tiempo cuando descansa.
metabolism The chemical reactions occurring within cells that convert one molecule into another and transfer energy in living organisms. (p. 96)	**metabolismo** Reacciones químicas que se producen en las células y que convierten una molécula en otra y transfieren energía en los organismos vivos.
metaphase The phase of mitosis in which the chromosomes align in the middle of the dividing cell. (p. 328)	**metafase** Fase de la mitosis en la que los cromosomas se alinean en el centro de la célula que se dividirá.
migration The movement of individuals from one population to another or to areas with no prior population. (p. 604)	**migración** El movimiento de individuos de una población a otra, o a regiones que antes no tenían una población.
mitochondria (singular, mitochondrium) Specialized organelles that are the site of cellular respiration in eukaryotic cells, oxidizing chemical compounds such as sugars to carbon dioxide and transferring their chemical energy to ATP. (p. 110)	**mitocondrias** Orgánulos especializados donde sucede la respiración celular en las células eucariotas a través de la oxidación de compuestos químicos como azúcares en dióxido de carbono y la transferencia de su energía química a ATP.
mitochondrial matrix The space enclosed by the inner membrane of the mitochondria. (p. 110)	**matriz mitocondrial** Espacio delimitado por la membrana interna de las mitocondrias.
mitosis The stage of M phase that produces two identical nuclei during the eukaryotic cell cycle. (p. 324)	**mitosis** Etapa de la fase M que produce dos núcleos idénticos durante el ciclo celular eucariota.

mitotic spindle A structure in the cytosol made up predominantly of microtubules that pull the chromosomes into separate daughter cells. (p. 326)

huso mitótico Estructura en el citosol formada predominantemente por microtúbulos que arrastran los cromosomas hacia células hijas separadas.

mode The most common number in a set of numbers. (p. 21)

moda El número más común de un conjunto de números.

molarity The amount of a solute in a volume of solution, expressed as moles of a solute per liter of solution. Also called molar concentration. (p. 156)

molaridad Cantidad de un soluto en un volumen de solución, expresada en moles de un soluto por litro de solución. También se denomina concentración molar.

molecular clock The observation that the extent of genetic divergence between two groups is a reflection of the time since the groups shared a common ancestor. (p. 618)

reloj molecular La observación de que la divergencia genética entre dos grupos refleja el tiempo desde que compartieron un ancestro en común.

molecule A chemical formed of two or more atoms. (p. 26)

molécula Sustancia química formada por dos o más átomos.

monohybrid cross A cross between two individuals that are hybrids for a single gene, meaning they possess two different alleles. (p. 369)

cruzamiento monohíbrido Cruzamiento de dos individuos híbridos para un mismo gen, es decir, que poseen dos alelos distintos.

monomer A single molecule that can bond chemically with similar molecules in a polymer. (p. 37)

monómero Molécula única que puede unirse químicamente a moléculas similares en un polímero.

morphological homology An anatomical structure that is similar in two groups of organisms because it was present in the common ancestor of the two groups and retained over evolutionary time. (p. 616)

homología morfológica Estructura anatómica que es similar en dos grupos de organismos porque estaba presente en el ancestro común de ambos grupos y se conservó a lo largo del tiempo evolutivo.

multiplication rule of probability A rule which states that the probability of two independent events both happening is the product of their individual probabilities. (p. 372)

regla de la multiplicación de la probabilidad Regla que establece que la probabilidad de que dos acontecimientos independientes se produzcan es el producto de sus probabilidades individuales.

mutagen An agent that increases the probability of mutation. (p. 520)

mutágeno Agente que aumenta la probabilidad de mutación.

mutation Any heritable change in the genetic material, usually a change in the nucleotide sequence of a gene. (p. 305)

mutación Cualquier cambio heredable en el material genético, generalmente es un cambio en la secuencia de nucleótidos de un gen.

mutualism A special type of symbiosis in which both species benefit in regard to their growth or reproduction. (p. 787)

mutualismo Tipo especial de simbiosis en el que ambas especies se benefician en cuanto a su crecimiento o reproducción.

N

natural selection A mechanism of evolution that occurs when there is genetic variation in a population of organisms and the variants best suited for survival and reproduction in a particular environment contribute disproportionately to future generations. Of all the evolutionary mechanisms, only natural selection leads to adaptations. (p. 3)

selección natural Mecanismo de evolución que se produce cuando hay variación genética en una población de organismos y las variantes más adecuadas para la supervivencia y la reproducción en un entorno concreto contribuyen desproporcionadamente al material genético de las generaciones futuras. De todos los mecanismos evolutivos, sólo la selección natural lleva a la adaptación.

negative control *See* **control.** (p. 8)

control negativo *Véase* **control.**

negative feedback A process in which the response or output of a system opposes the initial stimulus, resulting in steady conditions, or homeostasis. (p. 285)

retroalimentación negativa Patrón de respuesta en el que un sistema responde a un estímulo original en dirección opuesta a éste, lo que da lugar a condiciones estables u homeostasis.

negative regulatory molecule A molecule, such as a repressor, that binds to the DNA at a site near the gene to prevent transcription. (p. 486)

molécula regulatoria negativa Molécula (por ejemplo, un represor) que se conecta al ADN en un sitio cercano al gen para evitar la transcripción.

neutron An atomic particle of the nucleus that carries no charge. (p. 30)

neutrón Partícula atómica del núcleo que no lleva carga.

niche A complete description of the role a species plays in its environment, and of its requirements, both abiotic and biotic. (p. 665)

nicho Descripción completa del papel que juega una especie en su ambiente y de sus requisitos, tanto bióticos como abióticos.

nicotinamide adenine dinucleotide phosphate (NADPH) An electron carrier in many biochemical reactions; the reducing agent used in the Calvin cycle during photosynthesis. (p. 214)

nicotinamida adenina dinucleótido fosfato (NADPH) Portador de electrones en muchas reacciones bioquímicas; el agente reductor utilizado en el ciclo de Calvin durante la fotosíntesis.

nitrogenous base A nitrogen-containing compound that makes up part of a nucleotide. (p. 78)

base nitrogenada Compuesto que contiene nitrógeno y que forma parte de un nucleótido.

node In phylogenetic trees, the point where a branch splits, representing the common ancestor from which the descendant species diverged. (p. 564)

nodo En los árboles filogenéticos, el punto donde se divide una rama, que representa el ancestro común del que divergen las especies descendientes.

non-sister chromatids Chromatids that are not connected by a centromere. (p. 354)

cromátidas no hermanas Cromátidas que no están conectadas por un centrómero.

nondisjunction The failure of a chromosome to separate during anaphase of cell division. (p. 423)

no disyunción Error en la separación de un cromosoma durante la anafase de la división celular.

nonpolar covalent bond A bond that results when electrons are shared equally, or nearly equally, between two atoms. (p. 34)

enlace covalente apolar Enlace que se produce cuando los electrones se comparten por igual, o casi por igual, entre dos átomos.

nonrandom mating Mate selection that is not random, but instead is based on genotype or relatedness. (p. 606)

apareamiento no aleatorio Selección de pareja que no es aleatoria, sino que se basa en el genotipo o el parentesco.

nontemplate strand The strand of DNA that is not used as a template, or model, for RNA synthesis during transcription. The nontemplate strand is the reverse complement of the template strand. The nontemplate strand is also called the coding, sense, and plus strand. (p. 460)

cadena no molde Cadena de ADN que no se utiliza como plantilla, o modelo, para la síntesis de ARN durante la transcripción. La cadena no molde es el complemento inverso de la cadena molde. También se denomina cadena codificante, cadena de sentido y cadena positiva.

normal distribution A symmetric distribution or bell curve in which most of the values cluster in the middle and fewer values are found at the extremes. In a normal distribution, approximately 68% of the values lie within one standard deviation on either side of the mean, and 95% of the values lie within two standard deviations on either side of the mean. (p. 21)

distribución normal Distribución simétrica o curva de campana en la que la mayoría de los valores se agrupan en el centro, con menos valores en los extremos. En una distribución normal, aproximadamente el 68% de los valores se encuentran dentro de una desviación estándar de la media, y el 95% de los valores se encuentran dentro de dos desviaciones estándar de la media.

nuclear envelope The two membranes, inner and outer, that define the boundary of the nucleus. (p. 106)

envoltura nuclear Las dos membranas, interna y externa, que definen el límite del núcleo.

nucleic acid A type of organic molecule that encodes and transmits genetic information. (p. 38)

ácido nucleico Tipo de molécula orgánica que codifica y transmite la información genética.

nucleotide A subunit of nucleic acids, consisting of a 5-carbon sugar, a nitrogen-containing base, and one or more phosphate groups. (p. 38)

nucleótido Subunidad de los ácidos nucleicos formada por un azúcar pentosa (de 5 carbonos), una base nitrogenada y uno o varios grupos fosfato.

nucleus The dense, central part of an atom. (p. 30)

núcleo La parte densa y central de un átomo.

nucleus (of a cell) The compartment of the cell that houses the DNA in chromosomes. (p. 98)

núcleo (de una célula) El compartimento de la célula que alberga el ADN en los cromosomas.

null hypothesis A hypothesis that predicts that the independent variable in a controlled experiment will not have a real effect, and any difference between the data collected from the control groups and experimental groups is due to chance. (p. 11)

hipótesis nula Hipótesis que predice que la variable independiente en un experimento controlado no tendrá un efecto real, y cualquier diferencia entre los datos recogidos de los grupos de control y los grupos experimentales se debe al azar.

O

observation The act of viewing the world around us. (p. 5)

observación El acto de ver el mundo que nos rodea.

operon A group of functionally related genes located in tandem along DNA and transcribed as a single unit from one promoter in prokaryotes. (p. 478)

operón Grupo de genes funcionalmente relacionados, situados en paralelo a lo largo del ADN y transcritos como una sola unidad a partir de un promotor en células procariotas.

organelle Any one of several compartments in a eukaryotic cell that divides the cell contents into smaller spaces specialized for different functions. (p. 101)

orgánulo Cualquiera de los diversos compartimentos de una célula eucariota que divide el contenido de la célula en espacios más pequeños especializados para diferentes funciones.

organic molecule A molecule made of carbon. (p. 26)

molécula orgánica Compuesto químico que contiene carbono.

organism A living individual entity that displays the properties of life. (p. 1)

organismo Entidad individual viva que presenta las propiedades de la vida.

osmoregulation The regulation of water and solute levels to control osmotic pressure. (p. 163)

osmorregulación La regulación de niveles de agua y solutos para controlar la presión osmótica.

osmosis The net movement of a solvent, such as water, across a selectively permeable membrane toward the side of higher solute concentration. (p. 156)

ósmosis El movimiento neto de un solvente, como el agua, a través de una membrana selectivamente permeable hacia el lado de mayor concentración de soluto.

osmotic pressure The pressure needed to prevent water from moving from one solution into another by osmosis. (p. 157)

presión osmótica Presión necesaria para evitar que el agua pase de una solución a otra por ósmosis.

outgroup A group of organisms that has an older common ancestor than the common ancestor of the group of interest, and therefore serves as a reference for evolutionary relationships within the group of interest. (p. 654)

grupo externo Grupo de organismos que tiene un ancestro común más antiguo que el ancestro común del grupo de interés y que, por tanto, sirve de referencia para las relaciones evolutivas dentro del grupo de interés.

oxidation A chemical reaction in which a molecule loses electrons and energy. (p. 214)	**oxidación** Reacción química en la que una molécula pierde electrones y energía.
oxidative phosphorylation A set of chemical reactions that occurs by passing electrons along an electron transport chain to a final electron acceptor, oxygen, pumping protons across a membrane, and using the proton electrochemical gradient to drive the synthesis of ATP. (p. 236)	**fosforilación oxidativa** Conjunto de reacciones químicas que se producen al pasar electrones a lo largo de una cadena de transporte hasta el aceptor final de electrones —el oxígeno—, bombeando protones a través de una membrana y utilizando el gradiente electroquímico de protones para impulsar la síntesis de ATP.

P

parasitism An interaction in which one species lives in or on another organism, known as the host. (p. 785)	**parasitismo** Interacción en la que una especie vive dentro de o encima de otro organismo, llamado huésped.
passive transport The movement of molecules across a cell membrane by diffusion. (p. 146)	**transporte pasivo** Movimiento de moléculas a través de una membrana celular por difusión.
paternal inheritance A type of inheritance in which the organelles in the offspring cells derive from those in the father. (p. 401)	**herencia paterna** Tipo de herencia en la cual los organelos celulares de la descendencia son derivados del padre.
pedigree A visual mapping of phenotypes using ancestral relationships. (p. 380)	**pedigrí** Representación visual de fenotipos a través de relaciones genealógicas.
peptide bond The bond between the carboxyl end of an amino acid and the amino end of another amino acid that creates a peptide. (p. 69)	**enlace peptídico** Enlace entre el extremo carboxilo de un aminoácido y el extremo amino de otro aminoácido que crea un péptido.
peptide A molecule made by joining amino acids together by peptide bonds. Peptides may range from as few as two to many hundreds of amino acids linked together. (p. 69)	**péptido** Molécula formada por la unión de aminoácidos mediante enlaces peptídicos. Los péptidos pueden estar formados por dos o más aminoácidos, incluso por cientos de ellos.
per capita growth rate (r) The rate of population change divided by the size of the population. (p. 756)	**tasa de crecimiento per cápita (r)** La tasa de cambio de la población dividida por el tamaño de la población.
periodic table of the elements A table in which the elements are indicated by their chemical symbols and arranged in order of increasing atomic number. (p. 31)	**tabla periódica de los elementos** Tabla en la que los elementos están indicados por sus símbolos químicos y ordenados por número atómico.
peripheral membrane protein A protein that is temporarily associated with the lipid bilayer or with integral membrane proteins through weak noncovalent interactions. (p. 139)	**proteína periférica de membrana** Proteína que se asocia temporalmente a la bicapa lipídica o a las proteínas integrales de la membrana mediante interacciones débiles no covalentes.
pH scale A measure of the acidity of a solution. The pH scale ranges from 0 to 14 and is calculated by the negative logarithm of the hydrogen ion concentration. (p. 49)	**escala de pH** Medida de la acidez de una solución. La escala de pH va de 0 a 14 y se calcula mediante el logaritmo negativo de la concentración de iones de hidrógeno.
phenotype The set of traits that an organism expresses, such as hair color or eye color. (p. 350)	**fenotipo** Conjunto de atributos que expresa un organismo, como el color de ojos o de cabello.
phenotypic plasticity The ability of a single genotype to produce different phenotypes in different environments. (p. 406)	**plasticidad fenotípica** La capacidad de un solo genotipo de producir distintos fenotipos en ambientes distintos.
phospholipid A type of lipid and a major component of the cell membrane. (p. 61)	**fosfolípido** Tipo de lípido que actúa como componente principal de la membrana celular.
photophosphorylation The process by which energy from sunlight drives the movement of electrons along an electron transport chain, which leads to the synthesis of ATP. (p. 226)	**fotofosforilación** Proceso por el que la energía de la luz solar impulsa el movimiento de electrones a lo largo de una cadena de transporte de electrones, lo que conduce a la síntesis de ATP.
photosynthesis The biochemical process in which the energy of sunlight is used to synthesize carbohydrates from carbon dioxide and water. (p. 111)	**fotosíntesis** Proceso bioquímico en el que se se utiliza la energía de la luz solar para sintetizar carbohidratos a partir de dióxido de carbono y agua.
photosynthetic electron transport chain A series of redox reactions in which light energy absorbed by chlorophyll is used to power the movement of electrons; in oxygenic photosynthesis, the electrons ultimately come from water, and the terminal electron acceptor is NADP$^+$. (p. 217)	**cadena fotosintética de transporte de electrones** Serie de reacciones redox en las que la energía de la luz absorbida por la clorofila se utiliza para impulsar el movimiento de electrones; en la fotosíntesis oxigénica, los electrones proceden del agua, y el aceptor terminal de electrones es el NADP$^+$.
photosystem A protein–pigment complex that absorbs light energy to drive redox reactions and thereby sets the photosynthetic electron transport chain in motion. (p. 224)	**fotosistema** Complejo proteico-pigmentario que absorbe la energía de la luz para impulsar las reacciones redox y así poner en marcha la cadena fotosintética de transporte de electrones.
phylogenetic tree A tree-like diagram representing a hypothesis about the evolutionary relationships among populations or species. (p. 563)	**árbol filogenético** Diagrama en forma de árbol utilizado para representar visualmente una hipótesis sobre las relaciones evolutivas entre poblaciones o especies.

phylogeny The history of descent with modification and the accumulation of change over time. (p. 645)	**filogenia** La historia de una descendencia con sus modificaciones y la acumulación de cambios a lo largo del tiempo.
plasma membrane *See* **cell membrane.**	**membrana plasmática** *Ver* **membrana celular.**
plasmid In bacteria, a small circular molecule of DNA carrying a small number of genes that replicates independently of the DNA in the bacterium's circular chromosome. (p. 442)	**plásmido** En las bacterias, una pequeña molécula circular de ADN que lleva un pequeño número de genes y que se replica independientemente del ADN del cromosoma circular de la bacteria.
plasmodesmata Connections between two adjacent plant cells that permit molecules and other substances to pass directly from the cytoplasm of one cell to the cytoplasm of another. (p. 282)	**plasmodesmos** Conexiones entre dos células vegetales adyacentes que permiten que las moléculas y otras sustancias pasen directamente del citoplasma de una célula al citoplasma de otra.
polar (molecule) A molecule that has regions of positive and negative charge. (p. 43)	**polar (molécula)** Molécula que tiene regiones de carga positiva y negativa.
polar covalent bond A bond that results when electrons are shared unequally between two atoms. (p. 34)	**enlace covalente polar** Enlace que resulta cuando los electrones se comparten de forma desigual entre dos átomos.
polarity *See* **directionality.** (p. 80)	**polaridad** *Ver* **direccionalidad.**
poly(A) tail The adenine (A) nucleotides added to the 3′ end of the primary transcript. (p. 465)	**cola de poli(A)** Los nucleótidos de adenina (A) que se anexan al extremo 3′ de un transcrito primario.
polymer A large molecule composed of many smaller subunits. (p. 37)	**polímero** Molécula grande compuesta por muchas subunidades más pequeñas.
polymerase chain reaction (PCR) A selective and highly sensitive method for making copies of a piece of DNA, which allows a targeted region of a DNA molecule to be replicated into as many copies as desired. (p. 526)	**reacción en cadena de la polimerasa (PCR)** Método selectivo y altamente sensible para realizar copias de un fragmento de ADN, que permite replicar una región determinada de una molécula de ADN en tantas copias como se desee.
polypeptide A polymer of amino acids connected by peptide bonds. (p. 69)	**polipéptido** Polímero de aminoácidos unidos por enlaces peptídicos.
polyploidy The condition of having more than two complete sets of chromosomes in the genome. (p. 517)	**poliploidía** Condición de tener más de dos series completas de cromosomas en el genoma.
polysaccharide A polymer of simple sugars. Polysaccharides provide long-term energy storage or structural support. (p. 57)	**polisacárido** Polímero de azúcares simples. Los polisacáridos proporcionan almacenamiento de energía a largo plazo o soporte estructural.
population All the individuals of a given species that live and reproduce in a particular place; one of several interbreeding groups of organisms of the same species living in the same geographical area. (p. 276)	**población** Todos los individuos de una especie dada que viven y se reproducen en un lugar en particular; uno de los varios grupos endogámicos de organismos de la misma especie que viven en la misma área geográfica.
positive control A sample or group that receives a treatment or variable with a known effect in a controlled experiment and therefore is expected to show a predictable change. (p. 8)	**control positivo** Muestra o grupo que recibe un tratamiento o variable con un efecto conocido en un experimento controlado y que, por tanto, se espera que muestre un cambio predecible.
positive feedback A pattern of response in which the output or signal of a communication system increases the activity in the same system that produced the signal. (p. 285)	**retroalimentación positiva** Patrón de respuesta en el que un sistema responde a un estímulo original aumentando la actividad en la misma dirección que el estímulo.
positive regulatory molecule A molecule, such as an activator, that binds to DNA at a site near a gene so that transcription can take place. (p. 486)	**molécula reguladora positiva** Cualquier molécula (por ejemplo, un activador) que se une al ADN en un sitio cercano a un gen de tal manera que se pueda producir la transcripción.
postzygotic Describes factors that cause the failure of a fertilized egg to develop into a fertile individual. (p. 666)	**post-cigótico** Describe los factores que provocan que el óvulo fecundado no se convierta en un individuo fértil.
potential energy Stored energy related to an object's structure or position. (p. 185)	**energía potencial** Energía almacenada relacionada con la estructura o la posición de un objeto.
pressure potential The effect of pressure in the movement of water. (p. 160)	**potencial de presión** El efecto de la presión en el movimiento del agua.
prezygotic Describes factors that prevent the fertilization of an egg. (p. 666)	**pre-cigótico** Describe factores que previenen la fertilización de un óvulo.
predation An interaction that involves one species killing and consuming another. (p. 784)	**depredación** Interacción que involucra a una especie que mata a otra y se alimenta de ella.
primary active transport Active transport that uses the energy of ATP directly. (p. 148)	**transporte activo primario** Transporte activo que utiliza directamente la energía del ATP.
primary structure The sequence of amino acids in a peptide or a protein. (p. 70)	**estructura primaria** Secuencia de aminoácidos en un péptido o una proteína.

primary transcript The initial RNA molecule that comes off the template DNA strand. (p. 463)

transcrito primario La molécula inicial de ARN que se desprende de la cadena molde de ADN.

primer A short strand of DNA or RNA; in DNA replication, RNA is used as a primer for DNA synthesis. (p. 451)

cebador Cadena corta de ADN o ARN; en la replicación del ADN, el ARN se utiliza como cebador para la síntesis del ADN. También se conoce como partidor, iniciador, o *primer*.

probability The likelihood of an event happening. (p. 372)

probabilidad Medida de la certidumbre de que se produzca un acontecimiento.

product Atoms or molecules that are made in a chemical reaction. (p. 35)

producto Átomos o moléculas que se producen en una reacción química.

prokaryotic Describes a cell that does not have a nucleus; used to refer collectively to archaeons and bacteria. (p. 98)

procariota Describe una célula que no tiene núcleo; se utiliza para referirse colectivamente a arqueas y bacterias.

prometaphase The phase of mitosis in which the nuclear envelope breaks down and the microtubules of the mitotic spindle attach to chromosomes. (p. 328)

prometafase Fase de la mitosis en la que se rompe la envoltura nuclear y los microtúbulos del huso mitótico se unen a los cromosomas.

promoter A regulatory region where RNA polymerase and associated proteins bind to the DNA molecule in the process of transcription. (p. 461)

promotor Región reguladora donde el RNA polimerasa y otras proteínas asociadas se unen a la molécula de ADN en el proceso de transcripción.

proofreading The process in which a DNA polymerase can immediately correct its own errors by excising and replacing a mismatched base. (p. 455)

corrección de errores Proceso mediante el cual una polimerasa del ADN puede corregir inmediatamente sus errores al eliminar o reemplazar una base que no corresponde.

prophase The phase of mitosis in which the chromosomes condense and become visible through the microscope. (p. 326)

profase Fase de la mitosis en la que los cromosomas se condensan y se hacen visibles al microscopio.

protein A type of organic molecule that provides structural support for the cell and speeds up chemical reactions. (p. 37)

proteína Tipo de molécula orgánica que proporciona soporte estructural a la célula y acelera las reacciones químicas.

proton An atomic particle of the nucleus that carries a positive charge. (p. 30)

protón Partícula atómica del núcleo que lleva una carga positiva.

punctuated equilibrium The idea that species go through periods of relatively little change, which are interrupted by short periods of rapid change and diversification. (p. 676)

equilibrio puntuado La idea de que las especies pasan por períodos de relativamente pocos cambios, que son interrumpidos por períodos cortos de cambio rápido y diversificación.

purine In nucleic acids, either of the bases adenine and guanine, which have a double-ring structure. (p. 78)

purina En los ácidos nucleicos, las bases adenina y guanina, que tienen una estructura de doble anillo.

pyrimidine In nucleic acids, any of the bases thymine, cytosine, and uracil, which have a single-ring structure. (p. 79)

pirimidina En los ácidos nucleicos, las bases timina, citosina y uracilo, que tienen una estructura de anillo único.

Q

quaternary structure The level of protein structure that arises from the interaction of two or more polypeptide chains or subunits, each with its own tertiary structure. (p. 72)

estructura cuaternaria Nivel de la estructura proteica que surge de la interacción de dos o más cadenas o subunidades polipeptídicas, cada una con su propia estructura terciaria.

quorum sensing Density-dependent production of signal molecules in bacteria that leads to population responses, such as bioluminescence or DNA uptake. (p. 278)

percepción de cuórum Producción de moléculas señalizadoras, dependiente de la densidad de población celular en las bacterias, que conduce a respuestas a nivel poblacional, como la bioluminiscencia o la captación de ADN.

R

radiometric dating Dating ancient materials using the decay of radioisotopes as a yardstick, including the decay of radioactive ^{14}C to nitrogen for time intervals up to a few tens of thousands of years, and the decay of radioactive uranium to lead for most of Earth history. (p. 624)

datación radiométrica Datación de materiales antiguos utilizando como criterio la desintegración de radioisótopos. La desintegración del ^{14}C radiactivo en nitrógeno puede medir intervalos de tiempo de hasta unas decenas de miles de años, y la desintegración del uranio radiactivo en plomo se utiliza para medir la mayor parte de la historia de la Tierra.

range A measure of how spread out numbers in a set of numbers are. The range is calculated by determining the difference between the highest and lowest numbers in a set of numbers. (p. 22)

rango Medida de la dispersión de los números en un conjunto de cifras. El rango se calcula determinando la diferencia entre el número más alto y el más bajo de un conjunto de números.

reactant Atoms or molecules that are changed in a chemical reaction. (p. 35)

reactivo Átomos o moléculas que se modifican en una reacción química.

reaction center Specially configured chlorophyll molecules where light energy is converted into electron transport. (p. 223)	**centro de reacción** Moléculas de clorofila especialmente configuradas donde la energía luminosa se convierte en proteínas transportadoras de electrones.
receptor protein A molecule that binds to a signaling molecule and triggers a response in a target cell. (p. 138)	**proteína receptora** Molécula que se une a una molécula de señalización y desencadena una respuesta en una célula objetivo.
receptor-protein kinase A receptor protein in cell membranes that have kinase activity that is activated after binding to an extracellular signaling molecule. (p. 294)	**receptor quinasa** Proteína receptora de las membranas celulares que tiene una actividad enzimática de quinasa que se activa tras la unión a una molécula de señalización extracelular.
recessive Describes an allele or trait that is only expressed in homozygotes, and not expressed in heterozygotes. Two recessive alleles are needed to express the phenotype. (p. 369)	**recesivo** Describe un rasgo o alelo que sólo se expresa en homocigotos, y no en heterocigotos. Se requieren dos alelos recesivos para expresar el fenotipo.
recombination *See* **crossing over.**	**recombinación** *Véase* **entrecruzamiento cromosómico.**
reduction A chemical reaction in which a molecule gains electrons and energy. (p. 214)	**reducción** Reacción química en la que una molécula gana electrones y energía.
replication fork The site where the parental DNA strands separate as the DNA duplex unwinds. (p. 448)	**horquilla de replicación** Lugar en el que se separan las cadenas de ADN parentales al desenrollarse la doble hélice de ADN.
reverse transcriptase An RNA-dependent DNA polymerase that uses a single-stranded RNA as a template to synthesize a DNA strand that is complementary in sequence to the RNA. (p. 544)	**transcriptasa inversa** Enzima de ADN polimerasa dependiente del ARN que utiliza un ARN monocatenario como molde para sintetizar una cadena de ADN que es complementaria en secuencia al ARN.
ribonucleic acid (RNA) A molecule chemically related to DNA that is synthesized from a DNA template and is involved in protein synthesis, among other functions. (p. 38)	**ácido ribonucleico (ARN)** Molécula relacionada químicamente con el ADN que se sintetiza de un molde de ADN y que está involucrada en la síntesis de proteínas, entre otras funciones.
ribose A pentose sugar commonly found in RNA. (p. 56)	**ribosa** Azúcar pentosa que se encuentra en el ARN.
ribosomal RNA (rRNA) Noncoding RNA found in all ribosomes that aid in translation. (p. 465)	**ribosoma ARN (rARN)** ARN no codificante que se encuentra en todos los ribosomas que ayudan en la traslación
ribosome A complex structure of RNA and protein that synthesizes proteins from mRNA. (p. 96)	**ribosoma** Estructura compleja de ARN y proteínas que sintetiza las proteínas a partir del ARNm.
RNA polymerase The enzyme that synthesizes an RNA transcript from a DNA template. (p. 460)	**ARN polimerasa** Enzima que sintetiza un transcrito de ARN a partir de un molde de ADN.
RNA processing A chemical modification that converts the primary transcript into mature mRNA, enabling the RNA molecule to be transported to the cytoplasm and recognized by the translational machinery. (p. 463)	**procesamiento del ARN** Modificación química que convierte el transcrito primario en ARNm maduro, permitiendo que la molécula de ARN sea transportada al citoplasma y reconocida por la maquinaria de traducción.
RNA splicing The process of joining exons and removing introns from the primary transcript. (p. 465)	**empalme del ARN** Proceso de unión de exones y eliminación de intrones del transcrito primario.
RNA world hypothesis The idea that the earliest cells relied on RNA for both information storage and catalysis. (p. 709)	**hipótesis del mundo de ARN** La idea de que las primeras células dependían del ARN tanto para el almacenamiento de información como para la catálisis.
root The base of a phylogenetic tree, representing the common ancestor or group from which all the organisms on the tree evolved. (p. 564)	**raíz** Base de un árbol filogenético, que representa el ancestro o grupo común a partir del cual evolucionaron todos los organismos del árbol.

S

S phase The phase during interphase in which the cell copies its DNA; the S phase follows the G_1 phase but precedes the G_2 phase. (p. 325)	**fase S** Fase dentro de la interfase en la que la célula copia su ADN. La fase S sigue a la fase G_1 pero precede a la fase G_2.
saturated Describes fatty acids that do not contain double bonds; the maximum number of hydrogen atoms is attached to each carbon atom, "saturating" the carbons with hydrogen atoms. (p. 58)	**saturados** Describe los ácidos grasos que no contienen dobles enlaces; el número máximo de átomos de hidrógeno se une a cada átomo de carbono, "saturando" los carbonos con átomos de hidrógeno.
scientific inquiry A deliberate, systematic, careful, and unbiased way of learning about the natural world; the process used to ask questions and seek answers about the natural world in a deliberate and ordered way. (p. 5)	**investigación científica** Forma deliberada, sistemática, cuidadosa e imparcial de aprender sobre el mundo natural; proceso utilizado para formular preguntas y buscar respuestas sobre el mundo natural de forma deliberada y ordenada.
second law of thermodynamics The principle that the transformation of energy is associated with an increase in the degree of disorder in the universe. (p. 189)	**segunda ley de la termodinámica** Principio según el cual la transformación de la energía está asociada a un aumento del grado de desorden en el universo.

second messenger An intermediate signaling molecule that amplifies a response inside a cell. (p. 293)

secondary active transport Active transport that uses the energy of an electrochemical gradient to drive the movement of molecules. (p. 148)

secondary structure The structure that results from hydrogen bonding in the polypeptide backbone. Two kinds of secondary structures are the alpha helix (α helix) and the beta sheet (β sheet). (p. 70)

selective pressure The full set of environmental conditions, both abiotic and biotic, that influence the evolution of a population by natural selection. (p. 576)

semiconservative replication The mechanism of DNA replication in which each strand of a parental DNA molecule serves as a template for the synthesis of a new daughter strand. (p. 448)

sensor In homeostasis, the component that detects a stimulus. (p. 284)

set point In homeostasis, the typical physiological value of a particular parameter, such as body temperature or blood glucose levels, which is actively maintained by the body with very little fluctuation. (p. 284)

sex chromosome A chromosome associated with sex determination. (p. 352)

sex-linked gene A gene that is only located on one of the two sex chromosomes. (p. 394)

sexual reproduction The process of producing offspring that receive genetic material from two parents; in eukaryotes, the process occurs through meiotic cell division and fertilization. (p. 351)

sexual selection A form of selection that promotes traits that increase an individual's ability to find and attract mates. (p. 582)

shared character A character or trait that is present in two or more groups of organisms. (p. 651)

signal transduction The process by which an external signal is converted to an internal response. (p. 290)

signaling cascade A series of chemical reactions inside of a cell that are initiated by a signal and are typically amplified to produce a large cellular response. (p. 294)

signaling molecule A chemical messenger that functions in cell communication by affecting activities of other cells. (p. 277)

Simpson's diversity index An equation that quantifies the diversity of species in a community. (p. 779)

single bond A covalent bond in which two atoms share two electrons. (p. 33)

sister chromatids The two copies of a chromosome resulting from DNA duplication that remain connected at the centromere. (p. 325)

solute A dissolved molecule such as the ions, amino acids, and sugars often found in a solvent such as water. (p. 156)

solute potential The effect of solutes on the movement of water. (p. 160)

solvent A substance that can dissolve another substance. (p. 48)

speciation The process whereby new species are produced. (p. 668)

species A group of individuals that are capable of interbreeding and producing viable, fertile offspring. (p. 2)

segundo mensajero Molécula señalizadora intermedia que amplifica una respuesta dentro de una célula.

transporte activo secundario Transporte activo que utiliza la energía de un gradiente electroquímico para impulsar el movimiento de las moléculas.

estructura secundaria La estructura resultante del enlace de hidrógeno en el esqueleto del polipéptido. Dos tipos de estructuras secundarias son la hélice alfa (hélice α) y la lámina beta (lámina β).

presión selectiva Conjunto completo de condiciones ambientales, bióticas y abióticas, que influyen en la evolución de una población por selección natural.

replicación semiconservativa del ADN Mecanismo de reproducción de ADN en que cada cadena de una molécula parental de ADN sirve como molde para la síntesis de una nueva cadena hija.

sensor En la homeostasis, el componente que detecta un estímulo.

punto de ajuste En la homeostasis, el valor fisiológico típico de un parámetro concreto, como la temperatura corporal o los niveles de glucosa en sangre, que el organismo mantiene activamente con muy poca fluctuación.

cromosoma sexual Cromosoma asociado a la determinación del sexo.

gen ligado al sexo Gen que sólo se encuentra en uno de los dos cromosomas sexuales.

reproducción sexual Proceso de producción de una descendencia que recibe material genético de dos progenitores; en los eucariotas, el proceso se produce mediante la división celular meiótica y la fecundación.

selección sexual Forma de selección que fomenta los rasgos que aumentan la capacidad de un individuo para encontrar y atraer pareja.

carácter compartido Carácter o rasgo que está presente en dos o más grupos de organismos.

transducción de señal Proceso por el que una señal externa se convierte en una respuesta interna.

cascada de señalización Serie de reacciones químicas dentro de una célula que son iniciadas por una señal y suelen ser amplificadas para producir una amplia respuesta celular.

molécula señalizadora Mensajero químico en la comunicación celular que funciona afectando las actividades de otras células.

índice de diversidad de Simpson Ecuación que cuantifica la diversidad de especies en una comunidad.

enlace simple Enlace covalente en el que dos átomos comparten dos electrones.

cromátidas hermanas Las dos copias de un cromosoma resultantes de la duplicación del ADN que permanecen unidas en el centrómero.

soluto Molécula disuelta, como los iones, aminoácidos y azúcares que suelen encontrarse en un solvente como el agua.

potencial de solut El efecto de los solutos en el movimiento del agua.

solvente Sustancia que puede disolver otra sustancia.

especiación Proceso por el que se producen nuevas especies.

especie Grupo de individuos capaces de cruzarse y producir descendencia viable y fértil.

specific heat The amount of heat required per unit mass of a substance to raise the temperature of the mass by 1 degree Celsius. (p. 46)	**calor específico** Cantidad de calor necesaria por unidad de masa de una sustancia para elevar la temperatura de la masa en un grado centígrado.
standard deviation A measure of how spread out numbers in a set of numbers are. The standard deviation is calculated by taking the square root of the variance, or the average of the squared differences from the mean. (p. 22)	**desviación estándar** Medida de la dispersión de los números en un conjunto de cifras. La desviación estándar se calcula tomando la raíz cuadrada de la varianza, o la media del cuadrado de las diferencias con la media.
standard error of the mean A measure of how far the mean of a sample is from the true mean of the whole population. The standard error of the mean is calculated by dividing the standard deviation by the square root of the number of values in a sample. (p. 24)	**error estándar de la media** Medida de la distancia entre la media de una muestra y la media real de toda la población. El error estándar de la media se calcula dividiendo la desviación estándar por la raíz cuadrada del número de valores de una muestra.
steroid A type of lipid; the precursor molecule for cholesterol and steroid hormones. (p. 60)	**esteroide** Un tipo de lípido; la molécula precursora del colesterol y de las hormonas esteroides.
stroma The region of the chloroplast that surrounds the thylakoid, where the Calvin cycle takes place. (p. 218)	**estroma** Región del cloroplasto que rodea al tilacoide, donde tiene lugar el ciclo de Calvin.
substrate-level phosphorylation A way of generating ATP in which a phosphate group is transferred to ADP from an organic molecule, which acts as a phosphate donor or substrate. (p. 235)	**fosforilación a nivel de sustrato** Forma de generar ATP en la que se transfiere un grupo fosfato al ADP desde una molécula orgánica que actúa como donante de fosfato o sustrato.
substrate (S) A molecule acted upon by an enzyme; also called a reactant. (p. 200)	**sustrato (S)** Molécula sobre la que actúa una enzima; también llamada reactante.
surface area The total area of the outside surface of an object. (p. 105)	**área superficial** Área total de la superficie exterior de un objeto.
surface tension A measure of the difficulty of breaking the surface of a liquid. (p. 44)	**tensión superficial** Medida de la dificultad requerida para romper la superficie de un líquido.
symbiosis (plural, symbioses) A close interaction that has evolved between species that live together, often interdependently. (p. 172)	**simbiosis** Interacción estrecha que ha evolucionado entre especies que viven juntas, suelen ser interdependientes.
sympatric Describes populations that are in the same geographic location. (p. 673)	**simpátricas** Describe poblaciones que están en la misma ubicación geográfica.
system A group of entities that function together. A system may be living or nonliving. (p. 4)	**sistema** Conjunto de entidades que funcionan juntas. Un sistema puede o no ser vivo.

T

target cell A cell that has receptor proteins that can bind to a specific signaling molecule. (p. 277)	**célula objetivo** Célula que tiene proteínas receptoras que pueden unirse a una molécula de señalización específica.
telophase The phase of mitosis in which the nuclei of the daughter cells are formed and the chromosomes uncoil to their original state. (p. 328)	**telofase** fase de la mitosis en la que se forman los núcleos de las células hijas y los cromosomas se desenrollan para volver a su estado original.
template strand The strand of DNA that is used as a template, or model, for RNA synthesis during transcription. The template strand is also called the noncoding, antisense, and minus strand. (p. 460)	**cadena molde** Cadena de ADN que se usa como modelo, o molde, para la síntesis de ARN durante la transcripción. La cadena molde se llama también cadena plantilla, cadena no codificante, cadena negativa y cadena antisentido.
tertiary structure The three-dimensional shape of a single polypeptide chain, usually made up of several secondary structure elements. (p. 71)	**estructura terciaria** La forma tridimensional de una cadena polipeptídica única, generalmente formada por varios elementos de estructura secundaria.
test group The group that is exposed to the independent variable in an experiment. Also known as the **experimental group.** (p. 7)	**grupo de prueba** El grupo que se expone a la variable independiente en un experimento. También se le conoce como **grupo experimental.**
theory A general explanation of the world supported by a large body of experiments and observations. (p. 13)	**teoría** Una explicación general del mundo respaldada por un gran número de experimentos y observaciones.
thylakoid A flattened sac within the chloroplast that is bounded by membranes where the light-dependent reactions of photosynthesis occur. (p. 111)	**tilacoide** Saco aplanado dentro del cloroplasto, delimitado por membranas, donde se producen las reacciones de la fotosíntesis que dependen de la luz.
thymine (T) A pyrimidine base. (p. 79)	**timina (T)** Una base pirimidínica.
tonicity A measure of osmotic pressure; the higher the osmotic pressure, the higher the tonicity. (p. 158)	**tonicidad** Medida de la presión osmótica. Cuanto mayor sea la presión osmótica, mayor será la tonicidad.

topoisomerase An enzyme that relieves stress on the DNA double helix that results from overwinding or underwinding during DNA replication and transcription. (p. 451)	**topoisomerasa** Enzima que alivia la tensión en la doble hélice del ADN cuando se desenrolla incorrectamente durante la replicación y transcripción del ADN.
transcription factor A protein that binds to the promoter of a gene and is necessary for transcription to take place. (p. 489)	**factor de transcripción** Proteína que se une al promotor de un gen y es necesaria para que se produzca la transcripción.
transcription The synthesis of RNA from a DNA template. (p. 440)	**transcripción** Síntesis de ARN a partir de un molde de ADN.
transduction The transfer of DNA from cell to cell by means of a virus; a form of horizontal gene transfer. (p. 521)	**transducción** Transferencia de ADN de célula a célula por medio de un virus; forma de transferencia genética horizontal.
transfer RNA (tRNA) Noncoding RNA that carries individual amino acids for use in translation. (p. 465)	**ARN de transferencia (ARNt)** ARN no codificante que transporta aminoácidos individuales para su uso en la traducción.
transformation The conversion of cells from one state to another, as from nonvirulent to virulent, when DNA released to the environment by cell breakdown is taken up by recipient cells. In recombinant DNA technology, the introduction of recombinant DNA into a recipient cell. Transformation is a form of horizontal gene transfer. (p. 521)	**transformación** Conversión de células de un estado a otro, como de no virulento a virulento, cuando el ADN liberado al medio ambiente por la ruptura de la célula es tomado por las células receptoras. En tecnología de ADN recombinante, se utiliza el término para referirse a la introducción de ADN recombinante en una célula receptora. La transformación es una forma de transferencia genética horizontal.
transition state The brief time in a chemical reaction in which chemical bonds in the reactants are broken and new bonds in the product are formed. (p. 198)	**estado de transición** Momento breve de una reacción química en el que se rompen los enlaces químicos de los reactivos y se forman nuevos enlaces en el producto.
translation elongation The process by which successive amino acids are added one by one to a growing polypeptide chain during translation. (p. 475)	**elongación de la traducción** Proceso por el que aminoácidos sucesivos se añaden uno a uno a una cadena polipeptídica en crecimiento durante la traducción.
translation The synthesis of a polypeptide chain corresponding to the coding sequence present in a molecule of messenger RNA. (p. 440)	**traducción** Síntesis de una cadena polipeptídica correspondiente a la secuencia codificadora.
translation initiation The process by which translation begins. During initiation, the initiator AUG codon is recognized, and Met is established as the first amino acid in the new polypeptide chain. (p. 475)	**iniciación de la traducción** Proceso por el que comienza la traducción. Durante la iniciación, se reconoce el codón iniciador AUG y se establece el Met como primer aminoácido de la nueva cadena polipeptídica.
translation termination The process by which the addition of amino acids stops during translation and the completed polypeptide chain is released from the ribosome. (p. 475)	**terminación de la traducción** Proceso por el que se detiene la adición de aminoácidos durante la traducción y la cadena polipeptídica completa se libera del ribosoma.
transmembrane protein A protein that spans the entire lipid bilayer; most integral membrane proteins are transmembrane proteins. (p. 139)	**proteína transmembrana** Proteína que abarca toda la bicapa lipídica. La mayoría de las proteínas integrales de membrana son proteínas transmembrana.
transport protein A membrane protein that moves molecules across the cell membrane. (p. 138)	**proteína transportadora** Proteína de membrana que mueve moléculas a través de la membrana celular.
transposition The process by which a DNA sequence moves from one location to another in a DNA molecule. (p. 515)	**transposición** Proceso por el que una secuencia de ADN se mueve de una ubicación a otra en una molécula de ADN.
triacylglycerol A lipid composed of a glycerol backbone and three fatty acids. (p. 58)	**triacilglicerol** Lípido compuesto por una columna vertebral de glicerol y tres ácidos grasos.
triploidy The condition of having three complete sets of chromosomes in the genome. (p. 517)	**triploidía** Condición de tener tres series completas de cromosomas en el genoma.
trophic cascade When indirect effects are initiated by the presence of a predator. (p. 789)	**cascada trófica** Efectos indirectos que suceden por la presencia de un depredador en un ecosistema.
trophic levels An arrangement of producers and consumers into successive levels that represents the movement of energy. (p. 741)	**niveles tróficos** Disposición de productores y consumidores en niveles sucesivos que representan el movimiento de la energía.
turgor pressure Pressure within a cell resulting from the movement of water into the cell by osmosis. (p. 114)	**turgencia** Presión en el interior de una célula resultante del movimiento del agua en la célula por ósmosis.
Turner syndrome A condition in humans in which females have only one X chromosome. (p. 426)	**síndrome de Turner**. Condición en la que ciertos seres humanos femeninos sólo tienen un cromosoma X.

U

unsaturated Describes fatty acids that contain carbon–carbon double bonds. (p. 59)	**insaturado** Describe los ácidos grasos que contienen dobles enlaces carbono-carbono.
uracil (U) A pyrimidine base in RNA, where it replaces thymine found in DNA. (p. 84)	**uracilo (U)** Una base primidínica del ARN, que sustituye a la timina que se encuentra en el ADN.

V

vacuole A membrane-bound organelle present in some cells, including plant and fungal cells, that contains fluid, ions, and other molecules; in some cases, it absorbs water and contributes to turgor pressure. (p. 114)

vacuola Orgánulo unido a la membrana presente en algunas células, incluidas las vegetales y fúngicas, que contiene líquido, iones y otras moléculas; en algunos casos, absorbe agua y contribuye a la turgencia.

valence electron An electron in the outermost energy level of an atom. (p. 31)

electrón de valencia Electrón del nivel energético más exterior de un átomo.

van der Waals force An interaction of temporarily polarized molecules because of the attraction of opposite charges. (p. 59)

fuerza de Van der Waals Interacción de moléculas temporalmente polarizadas debido a la atracción de cargas opuestas.

vesicle A small membrane-enclosed sac that transports substances within the cell. (p. 105)

vesícula Pequeño saco con membrana que transporta sustancias dentro de la célula.

vestigial structure A structure that has lost its original function over time and is now much reduced in size. (p. 619)

estructura vestigial Estructura que ha perdido su función original con el paso del tiempo y que ahora tiene un tamaño muy reducido.

virus A small infectious agent that contains a nucleic acid genome packaged inside a protein coat called a capsid; some viruses also have a phospholipid envelope. (p. 541)

virus Pequeño agente infeccioso que contiene un genoma de ácido nucleico empaquetado dentro de una cubierta proteica llamada cápside; algunos virus también tienen una envoltura de fosfolípidos.

visible light The portion of the electromagnetic spectrum apparent to our eyes. (p. 221)

luz visible Porción del espectro electromagnético que se percibe con los ojos.

volume The amount of space an object occupies. (p. 120)

volumen La cantidad de espacio que ocupa un objeto.

W

water potential A parameter that combines all the physical and chemical factors that influence the movement of water, such as pressure, osmosis, and gravity; water moves from regions of higher water potential to regions of lower water potential. (p. 157)

potencial hídrico Parámetro que combina todos los factores físicos y químicos que influyen en el movimiento del agua, como la presión, la ósmosis y la gravedad; el agua se desplaza de las regiones de mayor potencial hídrico a las de menor potencial hídrico.

Z

zygote The diploid cell formed by the fusion of two gametes. (p. 360)

cigoto Célula diploide formada por la fusión de dos gametos.

Index

Fatty acids, 58
  cell membranes and, 135–136, 135*f*, 136*f*
  saturated and unsaturated, 58–59, 59*f*
Feedback, 314–318
  See also Negative feedback; Positive feedback
Feathered dinosaur (*Archaeopteryx lithographica*), 622, 622*f*
Feedback loops, 736, 737
Fermentation, 254–256, 263
  in bacteria, 254
  in early cells, 710
  pathways, 254–255, 255*f*
Fertilization, 360, 372, 566, 666
Fetal hemoglobin, 267
Fight-or-flight responses, 279, 279*f*, 294
Fig wasp, 755
Finches. *See* Galápagos finches
Findings, communication of, 13
First law of thermodynamics, 188, 188*f*
Fish
  bony, 656, 656*f*, 657*f*
  cartilaginous, 647*f*, 653*f*
  electrical communication, 728
  fossils and, 623
  gills and, 622, 623
  lobe-finned, 647*f*, 653*f*
  lungfish, 647, 647*f*, 651, 653, 653*f*
  tetrapod link to, 623, 647, 647*f*
Fisher (*Pekania pennanti*), 785
Fish-tetrapod fossil (*Tiktaalik roseae*) 622, 623
Fitness, 262, 575–576, 575*f*, 695
  evolutionary, 725, 731
5′ cap. *See* GTP caps
5′ ends, 80, 80*f*, 82
Flagella, 100*f*
Flavin adenine dinucleotide (FAD, FADH2), 235–236, 247
  electron transport chain and, 249–250, 249*f*
  Krebs cycle and, 247–248, 248*f*
Flies. *See* Drosophila melanogaster (fruit fly)
Flooding, 815
Flowers, 651, 666
  ABC model, 507, 507*f*
  chemical communication and, 727, 728*f*
  development of, 506–507, 506*f*, 507*f*
  diversity of, 507*f*
  fertilization and, 723
  hybridization, 674–675, 675*f*
  mutations and, 506–507, 506*f*
  observation and, 6–7, 7*f*
  pollination of, 7, 667, 728*f*
  visual communication and, 727
Fluid mosaic model, 141
Fluorescent dyes, 442*f*, 443
Fluorescent recovery after photobleaching (FRAP), 139, 140–141
Flu virus, 546–547, 638–639, 657
Foliage height, 800–801, 801*f*
Food chains, 740–742, 742*f*
Food webs, 742–743, 742*f*, 789, 802, 814–815
Foraminifera, 621*f*
Forces
  muscle generation of, 97
  turgor pressure, 114
  van der Waals, 59–60, 59*f*, 135–136
  water movement and, 157–158, 158*f*, 161, 162
Forested habitats, 817, 817*f*

Fossil fuels
  alternatives to, 27, 213, 740, 825
  in carbon cycle, 744*f*
  carbon dioxide from, 639, 640*f*, 743–744, 744*f*, 821, 825
  climate change and, 825
  uses of, 744
Fossilization
  definition of, 620
  molecular fossils and, 621
  trace fossils and, 620–621, 621*f*
Fossil record
  of animals, 712–713
  of Cambrian Period, 621, 712–713, 712*f*
  continuous change in, 632–634
  of dinosaurs, 621*f*, 622, 623, 624
  fossilization and, 620–621, 621*f*
  geologic data and, 624
  human evolution and, 634
  limitations to, 620
  mass extinctions and, 684, 687
  transitional forms, 622–623
  of whales, 633*f*
Fossils, 620–625, 620*f*, 621*f*, 622*f*
  dating, 624–625, 624*f*, 626
Founder effect, 603–604, 604*f*, 698
Four big ideas
  energetics, 4. *See also* Energetics; Energy
  evolution, 2–4. *See also* Evolution
  information storage and transmission, 4, 18*f*, 95–96. *See also* Deoxyribonucleic acid (DNA)
  systems interactions, 4–5, 18*f*
Fox, George, 175
Frameshift mutations, 515*f*
Franklin, Rosalind, 81, 97, 447
Fraternal (dizygotic) twins, 413
Free energy, change in, 237, 238*f*
Frogs, 761, 761*f*
Fructose, 54–55, 55*f*
Fruit fly. *See* Drosophila melanogaster (fruit fly)
Fruits, 38*f*, 122, 318, 566*f*, 579–580, 579*f*, 766, 766*f*
Functional groups, 55, 55*t*, 68, 68*t*
Fundamental niches, 783, 783*f*
Fungi
  agriculture and, 698, 785, 804–805, 805*f*
  earliest, 713
  evolution of, 713
  mutualisms and, 803, 805*f*
  mycorrhizae and, 803–804
  as pathogens, 804–805, 805*f*

G
G. See Gibbs free energy (G); Guanine (G)
G$_0$ phase, 326
G$_1$ phase, 325, 325*f*, 339*f*, 340
G$_2$ phase, 325, 325*f*, 339, 339*f*
Galápagos finches, 627, 627*f*, 635, 635*f*, 670–671, 676*f*
  adaptive radiation and, 674
  phylogenetic tree for, 670*f*, 671*f*
  speciation by hybridization and, 675–676, 676*f*
Galápagos tortoise (*Chelonoidis*), 603, 604*f*
Gametes, 356, 369–370, 371*f*, 401
Gametic isolation, 667*t*
Gap junctions, 282, 282*f*
Gap phases, 325–326

Garden roses, 507*f*
Gause, Georgii, 766–767, 767*f*, 786
Gehring, Walter, 504
Gel electrophoresis, 529–530, 529*f*, 530*f*, 531
  population genetics and, 594, 595–596
Gene expression, 300, 482*f*–483*f*
  epigenetic mechanisms of, 492–494
  mRNA and, 491*f*
  overview of, 482*f*–483*f*
  positive and negative, 486, 486*f*
  regulation of, 484–494, 485*f*, 494*f*
  transcription and, 440–441
Gene flow, 605, 673
Gene pools, 593
General transcription factors, 489
Gene regulation, 440
  epigenetic mechanisms of, 492–494
  in eukaryotes, 488–492, 489*f*, 494*f*
  hypotheses, 500
  in prokaryotes, 485–488
  RNA processing and, 489–491
Genes, 1
  chromosomes and, 484–485, 485*f*, 494*f*
  complex traits and multiple, 390–393
  downstream, 504–505, 505*f*
  health and disease, 412–414
  inheritance of in X chromosomes, 394–396
  linked, 398–400, 399*f*, 400*f*
  X-linked, 399–400, 400*f*
  Y-linked, 396–398, 397*f*
Genetically modified organisms (GMOs), 535–536, 535*f*
  See also Genetic engineering
Genetic code, 473–475, 473*f*
Genetic disorders, 420–423
Genetic diversity, 798–800, 799*f*
  in bacteria, 699, 799–800
  disorders and, 420–423
  sexual reproduction and, 360
  See also Biodiversity; Evolution; Mutations; Natural selection
Genetic drift, 603–604, 603*f*, 604*f*, 605*f*, 669, 680*f*, 698
Genetic engineering
  agriculture and, 535–536, 535*f*
  DNA editing and, 536, 536*f*
  ethics and, 536
  genetically modified organisms (GMOs) and, 535–536, 535*f*
  recombinant DNA and, 534–535, 534*f*
Genetic information
  central dogma of molecular biology, 84, 84*f*, 96, 439–441, 440*f*
  DNA and, 438
  nucleic acids and, 38, 436–437
Genetic mutations. *See* Mutations
Genetic risk factors, 535
Genetics. *See* Mendelian genetics
Genetic tests, 418*f*
Genetic variation, 351
  allele frequencies and, 593–597
  in bacteria, 699
  definition of, 565–566, 694
  environmental adaptation and, 700–701
  extinction and, 694–698
  gel electrophoresis and, 595–596
  high levels of, 698–699

low levels of, 694–698, 699*f*
measurement of, 593–597
migration and, 604–606, 605*f*, 606*f*
mutations and, 566, 577–578, 588, 589–593
phenotypes and, 574
in plants, 695–696, 695*f*, 697–698
population genetics and, 588
recombinants and, 534–535, 534*f*
segregation and, 376–377
sexual reproduction and, 351
sources of, 589
Genome, 1
of bacteria, 568
change in over time, 634–635
comparisons of, 634–635
sequencing of, 530–532, 531*f*
Genome reduction, 500
Genotypes, 350, 518–519
frequency of, 594, 600–602, 607, 608–610, 609*f*, 610*f*
phenotypes, relationship with, 588
Punnett squares and, 371–372
Genus, 648, 649*f*
Geographic isolation, 666–667, 666*f*, 667*t*, 672, 680*f*
Geographic range, of populations, 753–754, 754*f*
Geologic timescale, 625
Germ layers, 652
Germ-line mutations, 592
GH. *See* Growth hormone (GH)
Giant redwoods, 401
Giant viruses, 542
Gibbs free energy (G), 191
Gills, 622, 623
Ginger flower (*Smithatris supraneanae*), 507*f*
Glaciers, 813, 814*f*
Global warming. *See* Climate change
Glucagon, 316–317, 316*f*
Glucose, 54–55
in blood, 285–286, 286*f*
cellular respiration and, 238–239, 239*f*, 244, 245–246
negative feedback and, 315–317, 316*f*
photosynthesis and, 221
regulation of, 286*f*
storage forms of, 238–239, 239*f*
structure of, 55*f*, 56*f*
Glutamic acid, 700
Glycerol, 58
Glycine, 66–67
Glycogen, 238–239, 239*f*, 316
Glycolipids, 141
Glycolysis, 237, 237*f*, 240, 240*f*, 263
biochemistry of, 244–246, 245*f*
in early cells, 710
Glycoproteins, 141
Glycosidic bonds, 57, 57*f*
Glycosylation, 108–109
GMOs (genetically modified organisms), 535–536, 535*f*
See also Genetic engineering
Golden rice, 535, 535*f*
Goldenrod plants, 727
Goldfish (*Carassius auratus*), 412, 412*f*
Golgi apparatus, 106, 106*f*, 108–109, 108*f*, 617, 618*f*
Gondwana, 812

Goodall, Jane, 564*f*
G protein-coupled receptors, 293–294, 293*f*, 296*f*, 301*t*
agonists of, 305–307, 306*f*, 307*f*
antagonists of, 307–309, 307*f*
G proteins, 293, 293*f*
Gradualism, 676
Grana, 111, 218
Grand Canyon, 620, 620*f*
Grant, Peter, 635
Grant, Rosemary, 635
Graphing, TUT-1–TUT-7
area graphs, TUT-7
bar graphs, TUT-2–TUT-4
box and whisker graphs, TUT-6–TUT-7
dual Y graphs, TUT-4–TUT-5
line graphs, TUT-4
log scale, TUT-5
pie charts, TUT-5–TUT-6
scatter plot graphs, TUT-7
XY graphs, TUT-1–TUT-2
Grasslands, 803–804, 804*f*, 805*f*
Graves' disease, 414, 414*t*
Gray kangaroo (*Macropus giganteus*), 783
Gray treefrog (*Hyla versicolor*), 407, 407*f*
Gray whale (*Eschrichtius robustus*), 787–788
Gray wolf (*Canis lupus*), 825
Great Barrier Reef, 822–823
Great Dying. *See* Mass extinctions
Green algae, 111, 791
Greenhouse effect, 820–822, 821*f*
Greenhouse gases, 639, 820, 821*f*, 825
Griffith, Frederick, 437, 438
Griffith, John, 82
Growth factors, 281, 297–298
Growth hormone (GH), 414, 414*t*
GTP caps, 464, 465, 482*f*
Guanine (G), 79, 79*f*, 81, 436–437
Gurdon, John, 500, 501
Gypsy moths, 550, 551*f*

H
Habitat
of communities, 782
conservation of, 807
disturbance of, 812–825
diversity of, 800–801, 801*f*
forested, 817, 817*f*
human activities influencing, 816–817, 817*f*
loss of, 816–817, 817*f*
niches and, 665, 782–783, 783*f*, 803
protection of, 824, 824*f*
wetlands, 816–817, 816*f*
Habituation, 729, 729*f*
Hagfish, 647*f*, 650*f*, 653*f*
Hair cells, 651
Haldane, J. B. S., 123
Half-life, 625, 625*f*, 626
Haliaeetus leucocephalus (bald eagle), 665
Haploids, 352, 355, 356, 358*f*, 359, 360, 361*f*
Hardy, G. H., 607
Hardy–Weinberg equilibrium, 607–611, 609*f*, 610*f*
Harmful (deleterious) mutations, 681*f*
Heat
chemical reactions and, 204
thermodynamics and, 189

Helicase, 451, 452*f*
Hemings, Sally, 655
Hemoglobin, 72
bulk flow and, 128
in cellular respiration, 267–269, 267*f*
oxygen transport by, 72, 513, 700, 700*f*
sickle-cell anemia and, 700, 700*f*
subunits of, 700
variants of, 267
Hemophilia, 396
Herbivores, 741, 804, 805*f*, 814
Herbivory, 785–786, 786*f*, 788*t*
Heterotrophs, 741, 741*f*, 750*f*
Heterozygote advantage, 700
Heterozygous, 371*f*, 393
HEXA gene, 420–421, 421*f*, 512
Hickory trees, 790, 791*f*, 813, 814*f*
Hindenburg (airship), 705, 705*f*
Hierarchy, in ecological systems, 722–724
Histograms, TUT-3
Histone proteins, 444
Histones, 492–493, 492*f*, 618
HIV. *See* Human immunodeficiency virus (HIV)
Holley, Robert W., 474
Home heating system, 284–285, 284*f*
Homeostasis
animals and, 736–738
in biology, 285–286
cell communication and, 284–286
cell membranes and, 95, 132
energy and, 736
membranes and, 710
negative feedback and, 314–317, 315*f*, 316*f*, 737
osmoregulation and, 163–165
overview of, 748*f*–749*f*
Homing pigeons, 726
Homologous chromosomes, 352–353, 354, 354*f*, 355*f*, 356, 358*f*, 359, 360, 398, 399*f*
Homologous traits, 616–618, 617*f*, 651–652, 651*f*
Homology, 651*f*
Homozygous, 370, 371*f*, 393
Honeybee (*Apis melifera*), 731–732, 817–818
Hooke, Robert, 92–93, 94*f*, 114
Horizontal gene transfer, 520–521, 520*f*, 521*f*, 535
Hormones, 279
adrenaline. See Adrenaline
circadian clock and, 726
endocrine signaling, 280–281, 280*f*
glucagon, 316–317, 316*f*
growth, 414, 414*t*, 535
insulin, 107, 285–286, 286*f*, 292, 316–317, 316*f*, 484–485, 485*f*
oxytocin, 317–318, 318*f*
in plants, 725, 725*f*
reproductive system and, 725
as signaling molecules, 128, 279, 292, 300
steroid, 60–61, 108, 300, 300*f*
testosterone, 60, 281, 310
Horses, 392, 392*f*, 653–654, 654*f*
Horticulture, 9
Host cells, 541–542, 544, 547, 547*f*, 548–549
Hosts, co-speciation and. *See* Parasites
Host specificity, 639
House flies, 503*f*

Answers to Concept Check and Review Questions

Module 0

✓ Concept Check 0.1

1. A comparison of a fire ant with an army ant will note the unity of life because both groups have the same general segmented body plan. It will also show the diversity of life because the mandibles, abdomens, and heads of the two groups vary significantly in their structures and sizes.

2. A lack of energy would result in an organism being unable to carry out its cellular functions, maintain itself, grow, and reproduce, which would eventually cause death.

3. Without being able to retrieve information contained in their genes, organisms would be unable to respond to environmental changes and could not perform functions such as growth and reproduction.

4. The Canada lynx, using the systems of its eyes, nose, brain, nerves, bones, and muscles, is able to note the presence of the snowshoe hare, close in on the hare, and give chase that may result in the capture of the hare. This would not be possible without the interaction of these systems.

✓ Concept Check 0.2

5. A scientist uses observation to propose a testable explanation of how, or why, the observation occurs. This is the hypothesis. The scientist investigates the hypothesis either by conducting further observations or through experimental investigation, or both.

6. An experimental group is one in which an independent variable is changed to determine if the dependent variable changes because of it. A control group doesn't change the independent variable, and whether the dependent variable changes or not is also observed. It is important to have both a control and an experimental group if the researcher is to determine that changing the independent variable does change the dependent variable.

7. A guess is a hunch about something, while a hypothesis is a tentative explanation as to why an observation occurred. A theory, in a scientific sense, is a set of hypotheses about a particular natural phenomenon that has gained considerable scientific support via observation or experimental examination.

Review Questions

1. B; 2. C; 3. B; 4. D; 5. A; 6. D; 7. C

Module 1

✓ Concept Check 1.1

1. Matter cycles within living systems and the environment. This means that matter is reused over and over again. By contrast, energy is not reused. Instead, energy flows through living systems and an input of energy is constantly required to sustain life.

2. The energy input for most habitats on Earth comes from the sun. Plants and other organisms convert energy from the sun into chemical energy in the form of carbon-rich organic molecules. Other organisms eat these carbon-rich organic molecules for sources of both matter and energy to survive, grow, reproduce and so on. As a result, the movements of carbon and energy are intimately related.

✓ Concept Check 1.2

3. An atom's components are the protons and neutrons of its nucleus and the electrons that travel around the central nucleus.

4. The periodic table is organized in rows and columns. Rows consist of elements with increasing numbers of protons and electrons in each successive element until the end of the row, where the outermost level is now complete with electrons. Vertical columns define elements with similar chemical properties, and which have the same number of electrons in their outermost level.

5. ^{14}N means that the atomic mass of the element is 14, while ^{15}N means that it has a mass of 15. This is because ^{14}N and ^{15}N have seven protons in their nuclei, but ^{14}N has seven neutrons and ^{15}N has eight neutrons in its nucleus.

6. Four more electrons would bring carbon to eight electrons in its outer level, which would make the level complete.

7. Electrons furthest from the nucleus have the most energy, while those closest have the least amount of energy.

✓ Concept Check 1.3

8. A covalent bond is formed when electrons from two different atoms are shared in the outermost shells of each atom. A polar covalent bond occurs when the shared electrons are not shared equally and one atom more often has the electrons nearer to itself than does the other atom. In nonpolar covalent bonds, electrons are shared equally by the atoms. In an ionic bond, two ions with opposite electrical charges attract each other because of the differences in charge.

9. Due to its polarity, water is attracted to the positive and negative ions of sodium chloride. When water's slightly positive H atoms are close to the Cl^-, and its slightly negative O atom is close to the Na^+ of the NaCl crystal, the water eventually surrounds the ions and dissolves them into the water itself.

10. The reactants are the three H_2 and one N_2 molecules, and the product is the two NH_3 molecules.

✓ Concept Check 1.4

11. The four most common atoms in organic molecules are carbon, hydrogen, oxygen, and nitrogen.

12. Carbon requires four electrons to fill its outer shell, which allows it to form covalent bonds with four other atoms. As a result, carbon-based molecules are very diverse.

13. The four major types of organic molecules are proteins, nucleic acids, carbohydrates, and lipids.

14. A polymer is a molecule built of many small, repeating subunits.

15. Lipids, being hydrophobic, associate with each other and not water, so provide a barrier between the inside and outside of a cell, which is water based.

Review Questions

1. B; 2. C; 3. A; 4. D; 5. C; 6. C; 7. B

Module 2

✓ Concept Check 2.1

1. Water is a polar molecule; its hydrogen atoms are partially positive and its oxygen atom is partially negative. This polarity makes hydrogen bonding possible because the partially negative oxygen atom is attracted to the partially positive hydrogen atom of another water molecule.

2. Water's hydrogen bonding gives water many of its characteristics, such as cohesion, adhesion, surface tension, and high specific heat.

✓ Concept Check 2.2

3. Due to its hydrogen bonds, water can surround an ammonia molecule and separate it from other ammonia molecules. Once the ammonia molecule is completely surrounded by water, it is dissolved.

4. The hydrogen atom of water binds to the nitrogen atom of ammonia, and the oxygen atom of water binds to the hydrogen atom of ammonia.

5. It is slightly basic because it has a higher value than 7.0, which means that it has fewer H^+ than OH^- ions.

6. Because the pH scale is logarithmic and there is one unit of difference between the two solutions, the first solution contains 10 times as many H^+ ions as does the second solution.

✓ Concept Check 2.3

7. A dehydration synthesis reaction removes a water molecule from two reactants, and in the process joins them by a covalent bond.

8. A hydrolysis reaction uses water to break a covalent bond between two units of a molecule to make it into smaller, simpler molecules.

9. Dehydration synthesis reactions build the large molecules and polymers that cells require to survive and function. Hydrolysis reactions break down large molecules so that the simple molecules they release may be taken up and used.

Review Questions

1. A; 2. C; 3. A; 4. D; 5. B; 6. D; 7. D

Module 3

✓ Concept Check 3.1

1. All carbohydrates include the elements carbon, hydrogen, and oxygen.

2. Monosaccharides can take on a linear or ring form.

3. In ring-structured monosaccharides, the atoms and groups attached to the ring are located either above or below the plane of the ring.

✓ Concept Check 3.2

4. A glycosidic bond is the covalent bond that links one monosaccharide to another or to a polymer of carbohydrates. It is formed by a dehydration synthesis reaction.

5. Both starch and cellulose are polysaccharides. Starch is an energy-storage molecule, while cellulose is a structural molecule that makes up the cell wall of plants.

✓ Concept Check 3.3

6. All lipids are hydrophobic and therefore do not dissolve well in water.

7. The components of triacylglycerol are a glycerol molecule and three fatty acids.

8. Unsaturation can introduce a kink or bend in the fatty acid, making it lose its linear shape.

9. Three types of lipids are triacylglycerols, steroids, and phospholipids. Triacylglycerols store energy; steroids are components of cell membranes and act as chemical messengers; and phospholipids are the major components of cell membranes.

Review Questions

1. C; 2. C; 3. D; 4. D; 5. A; 6. B; 7. A

Module 4

✓ Concept Check 4.1

1. The R group or side chain of an amino acid gives a particular amino acid its unique properties because it is what is different among the various amino acids.

2. All amino acids have a central α carbon attached to an amino group, a carboxylic acid group, an H atom, and a variable R group.

3. The groups of amino acids are hydrophobic, hydrophilic, and special amino acids. Hydrophilic amino acids may be polar but uncharged, acidic, or basic.

✓ Concept Check 4.2

4. A peptide bond connects amino acids in a protein.

5. A peptide bond occurs between the amino nitrogen of one amino acid and the carboxyl carbon of a second amino acid.

6. The directionality of a peptide is from the free amino end (the N terminus) to the free carboxyl end (the C terminus) of the peptide.

✓ Concept Check 4.3

7. Secondary structures in proteins include the alpha helix and the beta sheet.

8. Hydrogen bonds form between the carbonyl group in one peptide bond and the amide group in another, thus allowing localized regions of the polypeptide chain to fold into either an alpha helix or a beta sheet.

9. The tertiary structure of a protein is its overall shape. It is determined by the distribution of hydrophilic and hydrophobic R groups along the molecule, as well as chemical bonds and interactions between R groups.

Review Questions

1. C; 2. C; 3. B; 4. D; 5. B; 6. A

Module 5

✓ Concept Check 5.1

1. Adenine and guanine are the purines, and cytosine and thymine are the pyrimidines.

2. Purines are two-ringed nitrogenous bases, while pyrimidines are one-ringed nitrogenous bases.

3. The sugar in DNA is deoxyribose.

✓ Concept Check 5.2

4. Successive nucleotides in DNA are linked by phosphodiester bonds.

5. Genetic information is stored in the sequence, or order, of nucleotides along a DNA molecule.

✓ Concept Check 5.3

6. The sugar–phosphate backbones form the banisters, and the nucleotides or bases make up the steps of the spiral staircase.

7. In DNA, A pairs with T, and C pairs with G.

✓ Concept Check 5.4

8. Characteristics that DNA and RNA share are the nitrogenous bases A, C, and G; a sugar–phosphate backbone; a 5'-phosphate end; a 3'-hydroxyl end; and an ability of their bases to base pair.

9. Characteristics that differ between DNA and RNA are that DNA contains the base T while RNA contains U, DNA contains deoxyribose while RNA contains ribose, DNA tends to be a much longer molecule than RNA, and DNA is a two-stranded double helix while RNA is a single-stranded molecule.

Review Questions

1. D; 2. C; 3. B; 4. C; 5. A; 6. B; 7. A; 8. A

Module 6

✓ Concept Check 6.1

1. The cell theory states that organisms are made up of cells; cells are the fundamental unit of life; and cells arise from other cells.

2. When scientists say that the cell is the fundamental unit of life, they are saying that the cell is the smallest, most basic entity that has all of the features of life.

✓ Concept Check 6.2

3. The cell membrane maintains the internal environment of the cell compatible with life, and different from the environment outside of the cell. The active maintenance of a stable internal environment is called homeostasis and it is a key feature of cells and life.

4. DNA is the genetic material in all cells that stores and transmits information.

5. The central dogma describes the flow of information in a cell, where information is passed from DNA to RNA to protein.

6. Metabolism is the set of chemical reactions that take place in a cell that harness energy and materials from the environment. These metabolic reactions are essential for life.

✓ Concept Check 6.3

7. The close relationship between structure and function can be seen in molecules, cells, organs, and even organisms. Particular structures are necessary to carry out particular functions. The close connection between structure and function is highlighted by the fact that when structure is disrupted, so is function. For

example, when one amino acid is replaced with another in a protein, the resulting form and function may be altered.

8. Neurons are a type of cell that are specialized to transmit signals along their length and convey them to another cell or organ. They have long, thin processes that allow them to carry signals across their length and target specific cells or organs. Red blood cells have a biconcave shape. This shape allows them to deform as they pass through narrow capillaries.

✓ Concept Check 6.4

9. Prokaryotic cells lack a nucleus and extensive internal compartmentalization. They have circular DNA and smaller plasmids that carry additional genes that can be transferred to other bacteria. Prokaryotic cells are small (usually 1–2 micrometers in diameter or smaller). Because of their small size, they have a large surface-area-to-volume ratio, which means they are able to absorb nutrients from the environment to meet their metabolic needs. Eukaryotic cells have a nucleus and specialized internal structures called organelles. They have multiple linear chromosomes. They are 10 times larger in diameter and 1000 times larger in volume than a prokaryotic cell.

10. Prokaryotes include two domains of life: Bacteria and Archaea.

Review Questions

1. A; 2. B; 3. B; 4. C; 5. D; 6. D; 7. D

Module 7

✓ Concept Check 7.1

1. The nucleus stores the cell's genetic information and is the site where the information in DNA is used to synthesize RNA. The endoplasmic reticulum (ER) is the organelle in which proteins and lipids are synthesized. The Golgi apparatus modifies proteins and lipids produced by the ER and acts as a sorting station as those molecules are targeted to their final destinations. Lysosomes contain enzymes that break down macromolecules. Vesicles transport substances from organelle to organelle within a cell and from the inside to the outside of a cell.

2. Spaces within the endomembrane system include the interior of the nucleus, ER, Golgi apparatus, lysosomes, and vesicles. The space outside the endomembrane system makes up the cytosol. One difference between these two spaces is pH. Inside lysosomes, the pH is about 5, while outside lysosomes, the pH is about 7. This difference is significant because enzymes inside lysosomes require an acidic pH of 5, and enzymes outside lysosomes require a neutral pH of 7.

✓ Concept Check 7.2

3. Mitochondria are present in nearly all eukaryotic cells, including both plant and animal cells.

4. Mitochondria and chloroplasts are organelles involved in energy transfer; they are descendants of free-living bacteria; they are surrounded by a double membrane; and they contain circular DNA.

✓ Concept Check 7.3

5. Both the cytoskeleton and cell wall help to maintain the shape and support the structure of the cell.

6. The cell wall is present in archaea, bacteria, fungi, algae, and plants, but not animals.

Review Questions

1. A; 2. D; 3. B; 4. C; 5. B; 6. A; 7. A

Module 8

✓ Concept Check 8.1

1. As an object gets bigger, both the surface area and the volume increase, but the volume increases much more rapidly than the surface area.

2. Prokaryotic cells are smaller than eukaryotic cells, so prokaryotic cells have a higher surface area-to-volume ratio than eukaryotic cells.

3. The surface area of a cube is $6s^2 = 6 \times 5.0$ cm^2 = 6×25.0 cm$^2 = 150$ cm^2. The volume of a cube is $s \times s \times s$, or s^3. In this case, the volume is 5.0 cm$^2 \times 5.0$ cm$^2 \times 5.0$ cm$^2 = 5.0$ cm^3 = 125 cm^3. The surface area-to-volume ratio equals the surface area divided by the volume = $\frac{150}{125} = \frac{6}{5} = 1.2$.

✓ Concept Check 8.2

4. Diffusion is the net movement of molecules from areas of higher concentration to lower concentration due to random motion.

5. Bacterial cells depend on diffusion for taking in nutrients and getting rid of wastes. Diffusion limits the size of prokaryotes because diffusion is effective only over very short distances. In addition, their small size means they have a large amount of surface area relative to the volume of the cell, making it possible for diffusion across the cell membrane to adequately support their functions.

✓ Concept Check 8.3

6. Diffusion is increased by having a large surface area and a short distance.

7. Multicellular organisms have organs with large surface area and thin walls that allow for diffusion. These include the lining of the gut and the lung. In addition, multicellular organisms use bulk flow to distribute oxygen and nutrients around their bodies. Bulk flow does not rely on diffusion, but instead depends on pressure differences, such as those generated by the pumping of the heart to circulate blood.

Review Questions

1. D; 2. B; 3. A; 4. A; 5. A; 6. D; 7. B

Module 9

✓ Concept Check 9.1

1. In an aqueous environment, the polar hydrophilic head group readily interacts with the polar water molecules. In contrast, the nonpolar hydrophobic tail does not readily interact with water, but instead interacts with other nonpolar tail groups or hydrophobic molecules away from water.

2. A micelle forms when the polar head group of a lipid interacts with water, and the hydrophobic tails of the lipids interact with each other, excluding the water. Lipids can also form bilayers and liposomes, depending on the size of the head group relative to the fatty acid tails.

✓ Concept Check 9.2

3. Cell membranes are bilayers made up, in part, of phospholipids. These phospholipids have hydrophobic tails that are oriented inward, away from water. The tails associate loosely with each other by van der Waals interactions, which are individually quite weak, allowing phospholipids to move about laterally. This movement in the plane of the cell membrane is what is meant by membrane fluidity.

4. Membrane fluidity is affected by membrane composition, specifically the length and degree of saturation of the fatty acid tails. The longer the tail, the less fluid the membrane, and the more saturated, the less fluid the membrane. As a result, membranes composed of long saturated fatty acids are less fluid, and membranes composed of short unsaturated fatty acids are more fluid.

✓ Concept Check 9.3

5. Proteins associate with membranes in the following ways. First, integral membrane proteins are permanently associated with the membrane and cannot be removed without destroying the membrane itself. Most integral membrane proteins span the cell membrane, so they have both hydrophilic and hydrophobic regions. Second, peripheral membrane proteins are temporarily associated with the membrane and can easily be experimentally separated. These proteins can be associated with either the internal or external side of the membrane. They are mostly hydrophilic and interact with the polar heads of the lipid bilayer, or the hydrophilic regions of integral membrane proteins.

6. An experiment showing that proteins move in membranes uses the fluorescence recovery

after photobleaching (FRAP) technique. First, the proteins embedded in the cell membrane are labeled with a fluorescent dye. A laser is then used to bleach a small area of the cell surface so that it no longer fluoresces. Eventually, the fluorescently labeled proteins from other parts of the cell move into the bleached area and cause it to fluoresce once again. If the proteins did not move in membranes, that area would stay bleached over time.

Review Questions

1. C; 2. A; 3. B; 4. C; 5. A; 6. B; 7. A

Module 10

✓ Concept Check 10.1

1. In diffusion, molecules move from regions of higher concentration to regions of lower concentration. This movement results from the random motion of molecules.

2. Lipids help maintain the selective permeability of the membrane by preventing charged molecules and ions, as well as most polar molecules, from diffusing freely into the cell. They allow nonpolar molecules such as steroids and small molecules such as gases to diffuse freely through the membrane. Transmembrane proteins help transport specific molecules, such as polar molecules, by acting as channels and carriers that move molecules into and out of the cell.

3. Carbon dioxide and oxygen are two molecules that readily cross the cell membrane by simple diffusion. These are able to cross because they are small, uncharged molecules. Lipids such as steroids are also able to cross the cell membrane by simple diffusion. They are able to cross because they are nonpolar, like the inside of the lipid bilayer.

4. Water crosses the cell membrane by facilitated diffusion through channel proteins called aquaporins. Charged ions such as sodium ions (Na^+) and potassium ions (K^+) cross nerve cell membranes by facilitated diffusion through channel proteins.

✓ Concept Check 10.2

5. Passive transport into or out of cells works by diffusion. When there is a concentration difference of a particular molecule across the cell membrane, the molecule moves from the area of higher concentration to the area of lower concentration. When a molecule cannot move across the cell membrane on its own, it is sometimes able to passively diffuse through channel proteins or carrier proteins in the lipid bilayer by facilitated diffusion. This type of transport does not make use of cellular energy. In contrast, active transport is used to move a molecule into or out of the cell against its concentration gradient. Molecules move

through transport proteins embedded in the cell membrane. This type of transport requires energy, either directly in primary active transport or indirectly in secondary active transport.

6. Both forms of active transport require an input of energy. In primary active transport, energy in the form of ATP is used directly to pump molecules against their concentration gradient. In secondary active transport, energy in the form of ATP is used indirectly to pump molecules against their concentration gradient.

✓ Concept Check 10.3

7. Endocytosis brings macromolecules from the extracellular space into the cell interior by pinching off part of the cell membrane to form a vesicle. Therefore, the macromolecule ends up in a vesicle following endocytosis. By contrast, a macromolecule that passes through the cell membrane by passive or active transport ends up in the cytosol.

8. Endocytosis can transport much larger molecules into the cell than both passive and active transport can.

9. In endocytosis, a region of the cell membrane invaginates to form a vesicle. In exocytosis, a vesicle fuses with the cell membrane, releasing its content to the extracellular space.

Review Questions

1. D; 2. C; 3. B; 4. C; 5. C; 6. A; 7. C

Module 11

✓ Concept Check 11.1

1. Ions, such as sodium ions and potassium ions, are not able to cross the cell membrane on their own. Lipids, such as steroids, are able to diffuse across the cell membrane.

2. Most membranes have channel proteins called aquaporins that allow the rapid diffusion of water across a membrane by facilitated diffusion.

3. Water moves by osmosis from regions of low solute concentration to high solute concentration, or from regions of high water concentration to low water concentration. Note that these both describe the same direction of water movement.

✓ Concept Check 11.2

4. When an animal cell is placed in a hypotonic solution, water moves into the cell by osmosis and the cell swells or bursts.

5. When an animal cell is placed in a hypertonic solution, water leaves the cell by osmosis and the cell shrinks.

6. Differences in solute concentration between two sides of a selectively permeable membrane lead to water movement by osmosis. The cell

wall pushes back and creates pressure called turgor pressure. Gravity also influences the flow of water.

✓ Concept Check 11.3

7. Osmoregulation is the maintenance of constant osmotic pressure inside of cells or organisms.

8. Osmoconformers are animals that regulate their internal osmotic pressure by maintaining their internal solute concentration at a level similar to that of their environment. Osmoregulators actively regulate their internal osmotic pressure, expending considerable energy to maintain an internal solute concentration that is different from that of their environment.

9. Some marine vertebrates are also osmoconformers, including hagfish, lampreys, rays, and coelacanths. Osmoregulators include all freshwater animals, including fishes and amphibians, and all terrestrial animals, including humans. The largest group of marine vertebrates, the teleosts or bony fishes, are also osmoregulators.

Review Questions

1. A; 2. B; 3. C; 4. A; 5. B; 6. C; 7. A

Module 12

✓ Concept Check 12.1

1. Key features of eukaryotic cells are a membrane-bound nucleus that houses DNA, creating separate cellular compartments; membrane-bound organelles that further organize the cell interior and compartmentalize different cellular processes; and dynamic membranes and cytoskeleton that can be remodeled quickly, allowing cells to change shape and transport materials throughout the cell.

2. Eukaryotes can engulf material by endocytosis, and multicellular eukaryotes can ingest whole organisms by predation. Neither of these is possible in prokaryotic organisms, such as bacteria.

✓ Concept Check 12.2

3. Chloroplasts are thought to have originated through endosymbiosis, which is a symbiosis where one partner lives within the other. Chloroplasts closely resemble cyanobacteria and are thought to be descendants of symbiotic cyanobacteria that lived within eukaryotic cells.

4. Mitochondria are thought to have originated through endosymbiosis, which is a symbiosis where one partner lives within the other. Mitochondria closely resemble proteobacteria and are also thought to have evolved as endosymbionts.

5. Scientists hypothesize that archaea are more closely related to eukaryotes than to bacteria.

6. One hypothesis for the origin of eukaryotic cells is that the host for mitochondrion-producing endosymbiosis was itself a true eukaryotic cell with a nucleus, cytoskeleton, and endomembrane system; subsequent engulfment of a proteobacterium led to the evolution of mitochondria. A second hypothesis for the origin of eukaryotic cells argues that the eukaryotic cell as a whole began as a symbiotic association between a proteobacterium and an archaeon, and subsequently evolved a nucleus and endomembrane system.

Review Questions

1. A; 2. D; 3. C; 4. D; 5. C; 6. D; 7. A

Module 13

✓ Concept Check 13.1

1. One form of energy is kinetic energy, or the energy of motion. Examples of kinetic energy are flexing a muscle and throwing a ball. The other form of energy is potential energy, or stored energy. Potential energy depends on the structure of the object or its position relative to its surroundings, and it is released by a change in the object's structure or position. A ball sitting at the top of the stairs has a great deal of potential energy, which is released when the ball rolls down the stairs, at which point the energy is converted into kinetic energy.

2. Chemical energy is a form of potential energy held in chemical bonds (such as covalent bonds) between atoms in a molecule. The stronger the covalent bond, the less chemical energy it contains. The weaker the covalent bond, the more chemical energy it contains. Carbohydrates, lipids, and proteins have many carbon–carbon and carbon–hydrogen bonds. These bonds are relatively weak and are therefore rich sources of chemical energy.

3. ATP is composed of the base adenine and the 5-carbon sugar ribose. The ribose is attached to three phosphate groups. The bonds linking these phosphate groups have a high amount of chemical energy.

✓ Concept Check 13.2

4. The first law of thermodynamics is the law of conservation of energy. It states that energy can be neither created nor destroyed; it can only be transformed from one form into another. The second law of thermodynamics states that the transformation of energy is associated with an increase in the disorder of the universe. The degree of disorder is called entropy.

5. Chemical reactions are subject to the laws of thermodynamics, like everything else. As a result, the total amount of energy remains the same before and after a chemical reaction, but some of the energy is used to increase the entropy of the

system, and only some of the energy is available to do the work of the cell.

✓ Concept Check 13.3

6. The amount of energy available to do work is called Gibbs free energy (G). In a chemical reaction, you compare the free energy of the reactants and products to determine whether there is energy available to do work. This difference is called ΔG.

7. Catabolism and anabolism are both metabolic processes. Catabolism is the set of chemical reactions that breaks down macromolecules into smaller units, releasing energy, usually in the form of ATP. Anabolism is the set of chemical reactions that builds macromolecules from smaller units and requires an input of energy, usually in the form of ATP.

Review Questions

1. C; 2. B; 3. B; 4. C; 5. C; 6. A; 7. C

Module 14

✓ Concept Check 14.1

1. Enzymes increase the rate of chemical reactions. They participate in a reaction but are not themselves consumed in the process. Enzymes do not affect the equilibrium or direction of a reaction.

2. Enzymes increase the rate of chemical reactions by stabilizing the transition state and therefore reducing the activation energy, which is the input of energy necessary to drive all chemical reactions.

✓ Concept Check 14.2

3. Enzymes reduce the activation energy of a chemical reaction by forming a complex with the substrate. As a result, the transition state, an intermediate in all reactions, is stabilized, so has a lower amount of free energy. In this way, enzymes allow the reaction to proceed along a different route than it would otherwise take.

4. An enzyme only acts on those substrates that bind to its active site, which is the portion of the enzyme that binds substrate and converts it to product. An enzyme has to fold into its correct shape for the active site to be the right shape to bind its substrate.

✓ Concept Check 14.3

5. Temperature affects the movement of molecules. As temperatures increase, molecules move around more quickly, increasing their probability of colliding and participating in chemical reactions. As a result, enzyme activity increases with increasing temperature up to a point. Enzymes are proteins that fold into three-dimensional shapes. The shapes of proteins are important for their functions. At high

temperatures, enzymes, like all proteins, can unfold or denature. When an enzyme unfolds, it loses its shape and therefore its function.

6. Inhibitors decrease the activity of enzymes, whereas activators increase this activity.

Review Questions

1. C; 2. D; 3. A; 4. C; 5. D; 6. D; 7. B

Module 15

✓ Concept Check 15.1

1. In words, photosynthesis is a process in which carbon dioxide (CO_2) and water (H_2O) react to form the carbohydrate glucose ($C_6H_{12}O_6$), with oxygen (O_2) given off as a by-product. The overall photosynthetic reaction is $6CO_2 + 6H_2O \rightarrow C_6H_{12}O_6 + 6O_2$.

2. Photosynthesis is carried out by plants, some algae, and photosynthetic bacteria.

✓ Concept Check 15.2

3. In photosynthesis, H_2O is oxidized to O_2.

4. In photosynthesis, CO_2 is reduced to $C_6H_{12}O_6$.

✓ Concept Check 15.3

5. The main products of the light reactions are ATP and NADPH.

6. The main products of the Calvin cycle are carbohydrates. Carbohydrates are synthesized from carbon dioxide using the energy in ATP and NADPH produced by the light reactions.

Review Questions

1. D; 2. C; 3. A; 4. C; 5. B; 6. D; 7. C

Module 16

✓ Concept Check 16.1

1. Antenna chlorophylls transfer absorbed energy from one antenna chlorophyll molecule to another, and ultimately to the reaction center. Reaction center chlorophylls transfer electrons to an electron acceptor, resulting in the oxidation of reaction center chlorophyll molecules.

2. Photosystem II pulls electrons from water and supplies electrons to the photosynthetic electron transport chain. Photosystem I reduces $NADP^+$ to NADPH.

3. The energy from sunlight is absorbed by pigments, which transfer electrons along the photosynthetic electron transport chain. These reactions result in a high concentration of protons in the thylakoid space in two ways: (1) the oxidation of water, and (2) the pumping of protons from the stroma to the thylakoid space. In turn, the high concentration of protons is used to synthesize ATP by ATP synthase.

4. The major products of the light reactions of photosynthesis are ATP and NADPH. Oxygen is produced as a waste product.

✓ Concept Check 16.2

5. The major inputs of the Calvin cycle are CO_2 (from the atmosphere) and ATP and NADPH (from the photosynthetic electron transport chain). The major outputs of the Calvin cycle are carbohydrates (triose phosphates), ADP, and $NADP^+$.

6. The three steps of the Calvin cycle are carbon fixation to produce a 6-carbon molecule, reduction to generate carbohydrates, and regeneration of the 5-carbon molecule.

7. Rubisco catalyzes the first step of the Calvin cycle, in which carbon dioxide is added to a 5-carbon molecule, producing a 6-carbon molecule and therefore incorporating carbon dioxide into an organic molecule.

8. Rubisco faces a fundamental trade-off between selectivity and speed because it can use both CO_2 and O_2 as substrates. High selectivity of CO_2 over O_2 requires that the reaction have a high energy barrier, leading to a lower catalytic rate.

9. Two ways plants limit the formation and effects of reactive oxygen species are antioxidants and xanthophylls. Antioxidants, such as ascorbate and beta-carotene, neutralize reactive oxygen species. Xanthophylls are yellow-orange pigments that slow the formation of reactive oxygen species by reducing excess light energy. They accept absorbed light energy directly from chlorophyll and convert this energy to heat.

Review Questions

1. A; 2. B; 3. B; 4. A; 5. D; 6. A; 7. D

Module 17

✓ Concept Check 17.1

1. In cellular respiration, the reactants are glucose and oxygen, and the products are carbon dioxide and water. ATP is also produced in the reaction.

2. In the overall reaction for cellular respiration, glucose is oxidized and converted to carbon dioxide. Oxygen gas is reduced and converted to water.

3. The reduced molecules are NADH, $FADH_2$, and $C_6H_{12}O_6$, and the oxidized molecules are NAD^+, FAD, and CO_2. The reduced forms have more chemical energy than their corresponding oxidized forms.

4. ATP is generated by substrate-level phosphorylation and oxidative phosphorylation. In substrate-level phosphorylation, a phosphorylated organic molecule directly transfers a phosphate group to ADP. This pathway produces only a small amount of the

total ATP generated in the process of cellular respiration. In contrast, most of the ATP generated in cellular respiration is produced by oxidative phosphorylation. In these reactions, oxidation reactions are coupled to the reduction of electron carriers, which hand off their electrons to the electron transport chain. As electrons are passed along this chain to oxygen, protons are pumped across the inner mitochondrial membrane, providing the potential energy to drive the synthesis of ATP.

✓ Concept Check 17.2

5. The four stages of cellular respiration are glycolysis, pyruvate oxidation, the Krebs cycle, and oxidative phosphorylation.

6. (1) Glycolysis: glucose is partially broken down to pyruvate and a modest amount of energy in the form of ATP and reduced electron carriers is released. (2) Pyruvate oxidation: pyruvate is converted to acetyl-coenzyme A, and carbon dioxide and electron carriers are produced. (3) Krebs cycle: acetyl-CoA is broken down and carbon dioxide, ATP, and reduced electron carriers are produced. (4) Oxidative phosphorylation: electron carriers generated in stages 1–3 donate their electrons to the respiratory electron transport chain. This chain transfers electrons along a series of membrane-associated proteins to a final electron acceptor and, in the process, harnesses the energy of the electrons to produce a large amount of ATP. In aerobic respiration, oxygen is the final electron acceptor, so it is consumed, and water is produced.

✓ Concept Check 17.3

7. Three fuel molecules are carbohydrates such as glucose and sucrose, lipids, and proteins.

8. Plants store glucose as starch, and animals store glucose as glycogen.

Review Questions

1. C; 2. A; 3. A; 4. A; 5. A; 6. C; 7. D

Module 18

✓ Concept Check 18.1

1. The major input of glycolysis is glucose. The major outputs of glycolysis are pyruvate, ATP, and NADH.

2. At the end of glycolysis, the energy in the original glucose molecule is contained in pyruvate, ATP, and NADH.

3. At the end of pyruvate oxidation, the energy in the original glucose molecule is contained in acetyl-CoA and NADH.

4. At the end of the Krebs cycle, the energy in the original glucose molecule is contained in ATP, NADH, and $FADH_2$.

✓ Concept Check 18.2

5. Oxygen gas is consumed in cellular respiration. Specifically, oxygen gas is the final electron acceptor in the electron transport chain. When it accepts an electron, it is converted to water.

6. The movement of electrons along the electron transport chain in the inner mitochondrial membrane is coupled to the transfer of protons through several enzyme complexes and electron carriers. As the electrons pass through the complexes, protons are pumped into the intermembrane space. As a result, the concentration of protons is higher in the intermembrane space compared to the mitochondrial matrix.

7. The proton gradient generated by the electron transport chain provides a source of potential energy that is then used to drive the synthesis of ATP by ATP synthase. The protons accumulated in the intermembrane space cannot passively diffuse across the membrane, so they diffuse through a transport channel called ATP synthase. The movement of protons through the channel allows it to catalyze the synthesis of ATP from ADP and P_i.

✓ Concept Check 18.3

8. Yeast cells are eukaryotes. In bread making, yeast can use sugar as a food source for ethanol fermentation. The carbon dioxide produced in the process causes the bread to rise. The ethanol is removed in the baking process.

9. In the first pathway, pyruvate is converted to acetyl-CoA, which is the starting substrate for the Krebs cycle. During the Krebs cycle, the chemical energy in the bonds of acetyl-CoA is transferred to ATP by substrate-level phosphorylation and to the electron carriers NADH and $FADH_2$. The second pathway is fermentation, a reaction that happens without oxygen. There are many fermentation pathways, but all rely on oxidation of NADH to NAD^+ when pyruvate or a derivative of pyruvate is reduced. Two major fermentation pathways are lactic acid fermentation and ethanol fermentation. In the lactic acid pathway, electrons from NADH are transferred to pyruvate to produce lactic acid and NAD^+. In the ethanol fermentation pathway, pyruvate releases carbon dioxide to form acetaldehyde, and electrons from NADH are transferred to the molecule to produce ethanol and NAD^+.

10. Muscle tissue generates ATP during short-term exercise by converting stored glycogen to glucose. Glucose is rapidly broken down anaerobically to pyruvate, which then feeds into the lactic acid fermentation pathway. During long-term exercise, the liver releases glucose into the blood, which is taken up by muscle cells and oxidized to produce ATP. In addition, adipose tissue releases fatty acids that are also

taken up by muscle cells and broken down. These processes are slower to convert glucose and other molecules to energy; however, the end result is the production of more ATP than the fermentation pathway can produce.

Review Questions

1. B; 2. D; 3. B; 4. B; 5. D; 6. A; 7. B

Module 19

✓ Concept Check 19.1

1. The early atmosphere of Earth had little or no oxygen gas.

2. The earliest cells relied on anaerobic processes, such as fermentation and anaerobic cellular respiration, to meet their energy needs. Neither process requires oxygen, so were likely among the earliest metabolic processes to evolve on Earth.

✓ Concept Check 19.2

3. Cyanobacteria were the first organisms to evolve photosynthesis in which water is oxidized to produce oxygen as a by-product, called oxygenic photosynthesis.

4. Photosynthesis evolved in eukaryotic cells through endosymbiosis, in which a cyanobacterium was incorporated into a eukaryotic cell and eventually evolved into the chloroplast.

✓ Concept Check 19.3

5. Accessory pigments, such as anthocyanins and carotenoids, help to absorb light energy and also protect cells from excess light energy.

6. The various forms of hemoglobin and myoglobin are similar in that they all bind oxygen. They differ in their affinity for oxygen. These differences are important for their function of binding and transporting oxygen necessary for cellular respiration.

Review Questions

1. D; 2. A; 3. D; 4. C; 5. B; 6. D; 7. B

Module 20

✓ Concept Check 20.1

1. The four sequential steps are stimulus, signal release, signal reception, and response. The sequence begins with the arrival of a stimulus to the signaling cell, which causes the signaling cell to release signaling molecules. Target cells are those cells that have receptor proteins that bind signaling molecules. This binding, known as a ligand–receptor interaction, leads to a response by altering the target cell's activity.

2. In quorum sensing, the stimulus is high population density. As population density increases in certain bacteria, a signaling molecule is released by the bacteria. After the signaling molecule binds to receptor proteins on the surface of the bacteria, the bacteria respond. They may respond by bioluminescing or by increasing their uptake of DNA from the environment

✓ Concept Check 20.2

3. Endocrine signals affect distant target cells only after transport in the blood or circulatory system. Paracrine signals affect neighboring target cells. Autocrine signals are released by a signaling cell and bind to receptor proteins on that same cell, such that the signaling cell and the target cell are the same cell.

4. Both gap junctions and plasmodesmata allow materials to move directly from cell to cell by small channels linking adjacent cells.

✓ Concept Check 20.3

5. The thermostat is the sensor, and the heater is the effector. The thermostat might be set at a low temperature or be malfunctioning. If the thermostat is not receiving the stimulus that the house is cold, or if it is not sending signals to the heater, the heater will not heat the house. If the thermostat is working normally, then the heater might be malfunctioning by failing to produce heat.

6. Cold temperature, the stimulus, is sensed by part of the brain called the hypothalamus, the sensor. The hypothalamus sends signals that stimulate involuntary shivering by muscles (the effectors), generating heat (the response) to restore the normal body temperature; increased body temperature terminates the stimulus and signals as the negative feedback in this thermoregulation system.

7. Elevated glucose levels in the blood, the stimulus, are sensed by cells of the pancreas, which in turn secrete insulin, the signal. Cells that respond to insulin, the effectors, increase their rate of glucose uptake. The overall response is to decrease glucose levels in the blood.

Review Questions

1. D; 2. B; 3. A; 4. A; 5. B; 6. C; 7. D

Module 21

✓ Concept Check 21.1

1. A cell will respond to adrenaline only if it has adrenaline receptor proteins. Not all cells have adrenaline receptors. As a result, such cells are not directly affected by adrenaline.

2. The signaling molecule (ligand) binds to the receptor protein, and the binding changes the shape of the receptor protein.

✓ Concept Check 21.2

3. Adrenaline is amplified by a signal transduction cascade. Adrenaline binds to G protein-coupled receptors on a cell. The activated receptor activates G proteins linked to the formation of the second messenger cAMP. Large amounts of cAMP are formed, activating many protein kinases. Each of the kinase enzymes can phosphorylate many target proteins, resulting in a large cellular response.

4. Receptor-protein kinase identifies this protein as having both a receptor function that binds signals and a kinase function that phosphorylates the receptor when the protein has been activated by ligand. These proteins have dual functions: ligand binding and enzyme function.

5. Acetylcholine binds to ligand-gated protein channels on muscles. This binding alters the shape of the receptor protein, opening a small passageway through the membrane-spanning protein, just right for sodium ions to move through and enter the muscle. This sodium entry leads to a relay regulating muscle contraction.

✓ Concept Check 21.3

6. Signals that are water soluble are called hydrophilic signals, whereas signals that are lipid soluble are called hydrophobic signals. Receptors for hydrophilic signals are on the surface of target cells, whereas receptors for hydrophobic signals are inside cells.

7. Because hydrophobic signals are lipid soluble, they diffuse across the cell membrane of target cells, allowing them to bind to intracellular receptor proteins that will trigger a response.

8. Responses to hydrophobic signals usually turn on a gene and require the synthesis of proteins, a slower process than enzyme activation.

Review Questions

1. C; 2. B; 3. B; 4. D; 5. C; 6. A; 7. D

Module 22

✓ Concept Check 22.1

1. Agonists are ligands that activate a receptor and lead to a response in the target cell. Antagonists are ligands that inhibit a receptor and suppress a response of the target cell.

2. Both albuterol and adrenaline act as ligands that bind to and activate the adrenaline receptor in the lungs, leading to a widening of the airways.

3. Coffee contains the chemical caffeine, an antagonist of adenosine receptors, preventing drowsiness that would normally take place as a result of adenosine binding its receptor.

✓ Concept Check 22.2

4. Mutations in the gene encoding leptin or the gene encoding the leptin receptor protein can disrupt leptin signaling, resulting in increased appetite and potentially in obesity.

5. Disrupting the function of the receptor proteins will produce target cells that do not respond to the signaling molecule.

Review Questions

1. A; 2. D; 3. B; 4. B; 5. C; 6. B

Module 23

✓ Concept Check 23.1

1. The penguin entered the huddle to warm up when it was cold. However, the penguins huddle together so closely that they efficiently trap heat to such high levels that a penguin in the middle can get overheated. That is when the initial stimulus, being cold, is gone, and it is when the penguin leaves the huddle.

2. Insulin binds to its receptor to activate a response pathway that increases the rate of glucose uptake by the cell, thus lowering the concentration of glucose in the blood and terminating the stimulus for additional secretion of insulin.

3. Insulin and glucagon have opposite effects on glucose levels in the blood: insulin decreases blood glucose levels and glucagon increases blood glucose levels. Insulin, secreted by pancreatic beta cells in times of high levels of glucose, acts on its target cells to increase uptake of glucose from the blood. In contrast, glucagon, secreted by pancreatic alpha cells in times of low levels of glucose, acts on its target cells to release glucose into the blood.

✓ Concept Check 23.2

4. The baby's head pushes against the birth canal of its mother, which is sensed by cells that cause the brain to secrete oxytocin. Binding to its receptors on the uterus, oxytocin stimulates further uterine contractions, more strongly pushing the baby's head against the birth canal, increasing oxytocin secretion, further increasing uterine contractions, and so on until delivery of the baby.

5. Baby's suckling activity at mother's nipple → nerves communicate sensation to brain → brain secretes oxytocin → oxytocin travels to mother's mammary gland and stimulates release of milk → release of milk from mammary glands.

6. The first ripening apples on a tree produce small amounts of ethylene, which when received by less ripe apples causes them to ripen and produce ethylene, so a small amount of ethylene generates a large response.

Review Questions

1. D; 2. C; 3. C; 4. A; 5. D

Module 24

✓ Concept Check 24.1

1. Cells need two copies of their genetic material, or DNA, so that each daughter cell can receive one copy. They also need to have adequate resources to support two daughter cells. Some cells require a signal in order to divide in two.

2. In binary fission, a cell duplicates its DNA, increases in size, and divides into two daughter cells.

3. Eukaryotic cells produce daughter cells by mitotic cell division. In this form of cell division, the nucleus divides by mitosis and the cytoplasm divides by cytokinesis.

✓ Concept Check 24.2

4. The two main phases of the cell cycle are M phase and interphase.

5. S phase is part of interphase and it is the time when DNA is duplicated so that each daughter cell receives one copy.

✓ Concept Check 24.3

6. Centrosomes organize the mitotic spindle. The mitotic spindle attaches to kinetochores on chromatids to separate and move chromosomes during prometaphase, metaphase, and anaphase.

7. Sister chromatids are two complete copies of DNA attached by a single centromere. The sister chromatids are each a molecule of DNA but are considered a single duplicated chromosome until they split apart in anaphase. When the sister chromatids split apart, they each have their own centromere and so are each considered a chromosome.

8. The centrosome is duplicated during S phase. Because they will eventually move duplicated chromosomes to opposite poles of a cell, lack of duplication means the chromosomes won't be able to be separated and cell division will not proceed.

9. Animal and plant cells both have cell membranes. However, plant cells have an additional covering, the cell wall, that is not present in animal cells. Therefore, plants have to synthesize cell wall plus cell membrane during cytokinesis, while animal cells only synthesize cell membrane.

Review Questions

1. D; 2. B; 3. C; 4. A; 5. B; 6. B

Module 25

✓ Concept Check 25.1

1. Cyclins received this name after researchers observed that these proteins fluctuate in a cyclical pattern in ways that correlated with the cell cycle.

2. CDK is an acronym for cyclin-dependent kinase. The kinase function causes the transfer of phosphates to other regulatory proteins in this process, and the activation of the kinase function is dependent upon the appropriate cyclin protein associating with the CDK protein.

✓ Concept Check 25.2

3. Three checkpoints are the DNA damage checkpoint, the DNA replication checkpoint, and the spindle assembly checkpoint.

4. The DNA damage checkpoint occurs at the G_1/S transition. The DNA replication checkpoint occurs at the G_2/M transition. The spindle assembly checkpoint occurs at the metaphase/anaphase transition.

✓ Concept Check 25.3

5. Cells that have DNA damage can be arrested, or paused, in G_1, allowing the activity of enzymes in the nucleus to repair DNA. Cells that have not completely and accurately copied their DNA can be arrested, or paused, before entering M phase, which provides time for DNA to be completely copied. Similarly, a pause before anaphase at the spindle checkpoint can allow time for microtubules to make the correct connections.

6. If DNA damage has occurred, a DNA-damage-dependent kinase will transfer phosphates from ATP to the p53 protein. The p53 with phosphates blocks the synthesis of enzymes and other materials that are needed for DNA synthesis, resulting in G_1 arrest.

7. If the first error in the DNA is a change that allows the cell to escape regulation by the usual cell cycle regulators, subsequent errors in the DNA might also make it past a defective checkpoint. Furthermore, some of the subsequent changes might cause faster growth and result in the ability of such cells to migrate in the body, or metastasize.

Review Questions

1. D; 2. D; 3. B; 4. A; 5. C; 6. B; 7. B

Module 26

✓ Concept Check 26.1

1. Genes are physical locations on a chromosome that code for particular traits.

2. Asexual reproduction occurs when an individual inherits DNA from a single parent, whereas sexual reproduction occurs when an individual inherits DNA from two parents.

3. Sexual reproduction combines the DNA of two parents, which produces a greater range of different genotypes in the offspring.

✓ Concept Check 26.2

4. During meiosis I, there is synapsis of the maternal and paternal chromosomes and crossing over can occur, causing maternal and paternal homologs to exchange genes.

5. Crossing over and the random segregation of maternal and paternal chromosomes leads to increased genetic variation of the daughter cells.

6. Mitosis produces two daughter cells that are diploid and genetically identical, while meiosis produces four daughter cells that are haploid and each is genetically unique.

✓ Concept Check 26.3

7. In female mammals, there is unequal division of the cytoplasm during meiosis.

8. Fertilization combines genes from two different individuals to produce new genetic combinations.

Review Questions

1. A; 2. C; 3. D; 4. D; 5. B

Module 27

✓ Concept Check 27.1

1. The structures and molecules that are common to all species include a cell membrane, ribosomes, DNA, and RNA.

2. The cell processes that are common to a wide range of distantly related species include cellular respiration, photosynthesis, cell signaling, and cell division.

3. When Mendel crossed two purebred parental pea plants with contrasting traits, he found that only one of the traits appeared in the offspring.

4. We would need to know the phenotype of each parental plant and the phenotypes of the offspring. If the offspring all exhibited only one of the two parental phenotypes, this would be evidence for one allele being dominant.

5. Mendel started with purebred plants because they would carry only a single allele of a given gene.

✓ Concept Check 27.2

6. During meiosis, the maternal and paternal homologs segregate randomly into daughter cells and this causes the alleles on each chromosome to separate from each other.

7. By segregating the alleles, a recessive allele can remain in a population of organisms even if it is not exhibited for several generation, thereby allowing genetic variation to persist.

8. With incomplete dominance, the phenotype is intermediate to that formed by the two alleles.

✓ Concept Check 27.3

9. Mendel found that when he used a dihybrid cross, the ratio of the four phenotypes matched the ratio expected from two monohybrid crosses.

10. During metaphase I, the paternal and maternal homologs line up randomly on either side of the meiotic plate, such that the probability of a given chromosome going into a daughter cell is independent of the probability of another chromosome going into the same daughter cell.

11. 9 dominant, dominant phenotypes
3 dominant, recessive phenotypes
3 recessive, dominant phenotypes
1 recessive, recessive phenotype

✓ Concept Check 27.4

12. Pedigrees can be a useful way to understand allele inheritance in species such as humans in which controlled crosses are not possible.

13. For a rare dominant allele, we expect to see that half of the offspring of an affected parent will exhibit the phenotype and the phenotype will appear in every generation.

14. Incomplete penetrance can complicate interpretations because some individuals who have the genotype that codes for a particular phenotype fail to exhibit that phenotype.

Review Questions

1. A; 2. B; 3. C; 4. B; 5. A

Module 28

✓ Concept Check 28.1

1. Mendel's 9:3:3:1 ratio is based on two genes that act independently of each other, yet when there is epistasis one gene can modify the phenotype produced by another gene.

2. A chicken with the cc genotype has white feathers because it does not produce a dark color pigment; thus, it does not matter whether the chicken also carries the dominant allele (I) that can inhibit pigment production.

3. Palomino horse color is coded by one gene to be a red coat (ee) and then modified by a heterozygous condition at a second gene ($C^{CR}C$).

✓ Concept Check 28.2

4.

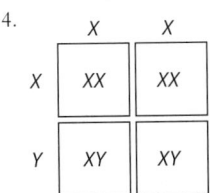

	X	X
X	XX	XX
Y	XY	XY

5. For X-linked genes, females must inherit two copies of the allele to exhibit the trait, whereas males need to inherit only one copy of the allele to exhibit the trait.

6. Fathers transmit the gene to all of their sons and grandsons but not to their daughters. Females cannot inherit or transmit the gene, since females do not possess a Y chromosome.

✓ Concept Check 28.3

7. Linked genes exist when two genes are located in close proximity in a single chromosome.

8. When two genes are linked, they cannot sort independent of each other during meiosis.

9. With crossing over of X-linked genes, the linked genes can undergo recombination, thereby increasing the number of different phenotypic combinations that can be expressed.

✓ Concept Check 28.4

10. Although egg cells and sperm cells both contain mitochondria and the mitochondria are both present at the moment of fertilization, sperm's mitochondria are degraded as the zygote develops.

11. For a mitochondrial disease caused by a mutant mitochondrial gene, we would expect any female with the disease to pass it on to all of her sons and daughters, with all sons and daughters exhibiting the disorder.

Review Questions

1. A; 2. B; 3. D; 4. B; 5. A

Module 29

✓ Concept Check 29.1

1. Phenotypes are the product of genes and the environment.

2. While genes play a role in determining phenotypes, environmental conditions also play a role by activating or inactivating different genes.

3. When no single phenotype is suited to different environments and a given phenotype is better in one environment than another, the evolution of phenotypically plastic traits is favored.

✓ Concept Check 29.2

4. A person's cholesterol level is the product of both genes and diet.

5. Because identical twins are genetically identical whereas nonidentical twins are not, we can look at twins in which at least one carries a disease and then determine how often the other twin also carries the disease. If the disease has a genetic component, the second twin will be more likely to have the disease if he or she is an identical twin compared to a nonidentical twin.

Review Questions

1. B; 2. B; 3. D; 4. D; 5. A

Module 30

✓ Concept Check 30.1

1. Tay-Sachs disease is caused by a mutation in the *HEXA* gene that codes for an enzyme known as beta-hexosaminidase A. Without this enzyme, the lysosomes in nerve cells cannot break down GM2 gangliosides, so they accumulate to toxic levels and destroy the neurons.

2. While Tay-Sachs disease is caused by a recessive allele, Huntington's disease is caused by a dominant allele.

3. Most mutations are harmful because they change the ordering of amino acids in the proteins that are built, which alters the proper functioning of these proteins.

✓ Concept Check 30.2

4. Nondisjunction occurs when chromosomes fail to properly separate during mitosis or meiosis, resulting in some daughter cells having an extra chromosome while other are missing a chromosome.

5. Nondisjunction can cause trisomy 21 when there is a failure of the homologous chromosomes to separate during meiosis I, or a failure of the sister chromatids to separate in meiosis II.

6. In female mammals, only one X chromosome is active, and any additional X chromosomes are inactivated.

Review Questions

1. B; 2. C; 3. D; 4. C; 5. A

Module 31

✓ Concept Check 31.1

1. DNA contains the bases A, G, T, and C. RNA contains the bases A, G, U, and C.

2. DNA stores genetic information in the sequence, or order, of nucleotides along its length.

✓ Concept Check 31.2

3. The usual flow of information in a cell is DNA to RNA to protein.

4. Gene expression refers to the turning on of a gene so that it makes a functional product, such as a protein. Gene regulation refers to the control of gene expression; it is the set of mechanisms that determines when, where, and at what level genes are expressed.

✓ Concept Check 31.3

5. Prokaryotes have a single circular chromosome. Eukaryotes have multiple linear chromosomes.

6. In prokaryotes, the single circular chromosome forms a nucleoid, which is located in the cytoplasm of the cell. In eukaryotes, the linear chromosomes are located in the nucleus.

Review Questions

1. A; 2. A; 3. C; 4. C; 5. D; 6. D; 7. A

Module 32

✓ Concept Check 32.1

1. During DNA replication, each parental strand serves as a template for a new complementary strand of DNA because of the specific pairing of the bases: an A on the template will always specify a T on the new strand, a G on the template will always specify a C on the new strand, and so on. The end result of DNA replication is two DNA molecules that have a sequence that is identical to the original molecule.

2. DNA replicates semiconservatively. In other words, the new DNA molecules are composed of a parental (old) strand and a daughter (new) strand. During DNA replication, the two strands separate ("unzip") from each other and both parental strands are used as a template for the replication of two new daughter strands.

✓ Concept Check 32.2

3. Helicase unwinds the parental strands at the replication fork. Topoisomerase relieves the stress that results from unwinding the double helix. RNA primase synthesizes a short stretch of RNA. DNA polymerase synthesizes new DNA strands. DNA ligase joins DNA (Okazaki) fragments together on the lagging strand.

4. The orientation of the two DNA strands is antiparallel. This means that the 3' hydroxyl end of one strand is opposite the 5' phosphate group of the other strand. When the two strands separate and DNA replication occurs, nucleotides are added to the 3' end of both strands, so DNA replication occurs in the 5'-to-3' direction.

5. Similarities of the resulting daughter strands from DNA replication are that they encode the same genetic information in the form of a nucleotide sequence; they are paired with one of the parental DNA strands; and they are both synthesized in the 5'-to-3' direction.

6. The two daughter strands are synthesized in different ways. The leading strand has the 3' end of its DNA pointed toward the replication fork and thus is synthesized as one long, continuous polymer. The lagging strand has its 5' end pointing toward the replication fork. Because DNA can only be replicated in a 5'-to-3' direction, the lagging strand is synthesized in short, discontinuous pieces. Each new piece, or Okazaki fragment, is elongated at its 3' end until it reaches the piece in front of it.

✓ Concept Check 32.3

7. DNA replication is precise because errors, or mutations, can be harmful or lethal to the cell or organism.

8. DNA polymerase has a proofreading function that allows it to identify a mismatched base and insert the correct one in its place.

Review Questions

1. A, 2. B; 3. B; 4. B; 5. B; 6. B; 7. A

Module 33

✓ Concept Check 33.1

1. The usual flow of genetic information is from DNA to RNA to protein.

2. The RNA transcript is the reverse complement of the template strand, so it has the sequence 5'-AUCGCUGAAAGU-3'.

✓ Concept Check 33.2

3. RNA polymerase is the enzyme that synthesizes RNA from DNA.

4. The incorporation of a nucleotide with a 3'-H atom rather than a 3'-OH group will stop subsequent elongation because the 3'-OH group is necessary to attack the high-energy phosphate bond of the incoming nucleoside triphosphate.

✓ Concept Check 33.3

5. Three modifications of the primary transcript that occur during RNA processing in eukaryotic cells are the addition of the GTP cap, the addition of the poly(A) tail, and RNA splicing.

6. The GTP cap allows the mRNA to be recognized by the ribosome and helps stabilize the mRNA. The poly(A) tail also helps to stabilize the mRNA, and plays a role in the export of the mRNA out of the nucleus and into the cytoplasm.

7. RNA splicing is the process by which exons are joined and introns are removed from a primary transcript. A single transcript with multiple introns may be spliced in different ways to generate different mRNAs and different protein products with different functions. Thus, this alternative splicing contributes to the diversity of the genetic information stored in DNA.

8. Three types of noncoding RNAs are ribosomal RNA (rRNA), transfer RNA (tRNA), and small regulatory noncoding RNA.

Review Questions

1. C; 2. A; 3. C; 4. D; 5. D; 6. D; 7. B

Module 34

✓ Concept Check 34.1

1. Ribosomes are the sites of translation, or protein synthesis. They bring all of the components together and catalyze the chemical reaction in which one amino acid attaches to the next one in line.

2. The tRNAs do the translating between nucleotide sequence and amino acid because each one has an anticodon that base pairs with a specific codon, and it carries an amino acid to the growing polypeptide chain.

3. A codon in mRNA is a group of three nucleotides that codes for a particular amino acid. The sequence of the codons in the mRNA gives rise to the order of the resulting amino acid polypeptide chain.

4. The codons are translated by tRNAs. The sequence of each tRNA includes a group of three nucleotides called an anticodon that is complementary in sequence and, therefore, can recognize and bind to a specific codon in the mRNA. Each tRNA is also bound to a specific amino acid on the 3′ end of the molecule.

✓ Concept Check 34.2

5. There are three reading frames, as follows:

UUU GGG UUU GGG UUU GGG

U UUG GGU UUG GGU UUG GG

UU UGG GUU UGG GUU UGG G

6. UUU codes for phenylalanine (Phe). GGG codes for glycine (Gly).

✓ Concept Check 34.3

7. Initiation factors bind to the GTP cap of the mRNA (in eukaryotic cells) or at an internal sequence (for prokaryotes) and recruit the small subunit of the ribosome and a tRNA bound to methionine. This complex then moves along the mRNA until it finds the start codon AUG, coding for methionine. The large ribosomal subunit then joins the complex and causes the initiation factors to be released. The tRNA attached to Met is then bound to the ribosome.

8. The next tRNA, determined by the codon of the mRNA, binds to the ribosome. The ribosome then catalyzes a reaction in which a peptide bond is formed between the carboxyl group of Met and the amino group of the next amino acid. The tRNA that was originally bound to Met is then released, and the ribosome slides to the next codon on the mRNA.

9. The ribosome continues in this fashion, shifting down the mRNA one codon at a time, adding amino acids to the growing peptide chain.

10. When the ribosome complex comes across a stop codon (UAA, UAG, or UGA), a protein release factor binds to the ribosome and causes the bond between the polypeptide chain and the last tRNA to break. Once the polypeptide chain is released, the ribosomal subunits disassociate from the mRNA and each other, and translation is complete.

Review Questions

1. C; 2. C; 3. C; 4. B; 5. D; 6. C; 7. C

Module 35

✓ Concept Check 35.1

1. Gene expression is the set of processes by which DNA is transcribed into mRNA, and mRNA is translated into protein. Gene regulation controls when, where, and how much gene expression occurs. Gene regulation makes sure that the gene is turned on in the right place, at the right time, and in the right amount.

2. The different levels of gene regulation are the various steps at which the process of gene expression can be controlled, such as transcription, RNA processing, and translation.

✓ Concept Check 35.2

3. An activator is a protein that binds to a sequence of DNA and is required for transcription. So, the gene is usually off, but is turned on in the presence of an activator. An inducer binds to and interferes with the action of a repressor. The repressor binds DNA and turns off transcription. Therefore, in the presence of an inducer, the repressor is not able to bind to DNA and transcription proceeds.

4. Binding with a small molecule results in a conformational change in one part of the protein that alters its structure at other sites in the protein. This is called an allosteric interaction. The changes in structure may affect the ability of the protein to bind to DNA.

✓ Concept Check 35.3

5. One mechanism that produces more than one protein from a single gene is alternative splicing. The mRNA can be spliced in different ways to yield different proteins by either splicing out or leaving in exons. A second mechanism that produces more than one protein from a single gene is RNA editing. RNA editing is a process in which individual bases in mRNA are altered. In some cases, the edit leads to a change in the amino acid sequence of the protein.

6. Small regulatory RNAs are not translated into proteins. As a result, they are often called noncoding RNAs because they do not code for proteins. By contrast, mRNAs are translated into proteins.

7. One way in which gene expression can be influenced after mRNA is processed and leaves the nucleus is through the actions of small regulatory RNAs that can either degrade mRNA or inhibit translation. A second way is to alter the GTP cap of the mRNA, so translation initiation cannot occur. A third way is altering or deleting the poly(A) tail, which inhibits translation. A fourth way is to modify the protein after it is synthesized through post-translational modification.

✓ Concept Check 35.4

8. DNA in a cell is wrapped around a group of histone proteins called a nucleosome. When DNA is coiled tightly around the nucleosome, the DNA is not accessible to transcription factors. However, when the chromatin unravels from the nucleosome through histone modifications, transcription factors can access the DNA and transcribe it.

9. Bases can be modified by attaching a methyl group to cytosine. When several cytosines are methylated near a promoter, that gene is usually repressed or turned off.

Review Questions

1. C; 2. A; 3. B; 4. C; 5. B; 6. D

Module 36

✓ Concept Check 36.1

1. Totipotent cells can give rise to a complete organism. Pluripotent cells are able to give rise to any cell of the body but cannot give rise to an entire organism. Multipotent cells are further along in differentiation than pluripotent cells and can form a limited number of specialized cell types. The terms are appropriate because multipotent cells give rise to multiple cell types, pluripotent cells to more cell types, and totipotent cells to all cell types.

2. A cell from an early embryo has more developmental potential than a cell from a later embryo because it is less differentiated and can give rise to more cell types.

3. Stem cell therapy for an individual might someday consist of taking that person's own cells and reprogramming them into stem cells. These cells could be programmed to replace burned tissue, defective heart muscle, regenerate nerve cells, and so on.

✓ Concept Check 36.2

4. *Pax6* encodes a transcription factor that binds to regulatory sequences in or near eye-specific genes to affect gene expression.

5. Evolutionary conservation is the observation that a gene is the same or very similar among many different organisms, even distantly related ones. From this observation, scientists infer that the gene was present in a common ancestor and

has not changed very much if at all over long periods of time.

✓ Concept Check 36.3

6. Combinatorial control describes the regulation of gene transcription according to the mix of transcription factors in the cell.

7. Although a gene may have binding sites for activators and repressors, it is the combination of transcription factors present in a cell and bound to DNA that determines whether it is expressed. In this case, the gene will be transcribed because of the presence of the activator bound to its regulatory sequence.

Review Questions

1. C; 2. A; 3. D; 4. A; 5. A; 6. C

Module 37

✓ Concept Check 37.1

1. Missense mutations alter a codon so that the result is a single amino acid replacement in a protein; they may affect protein function if the new amino acid causes protein misfolding or impairs interactions with other molecules. Silent mutations produce a codon that encodes the same amino acid as the nonmutant version. Nonsense mutations result in stop codons that terminate polypeptide elongation.

2. An insertion of three nucleotides will add one amino acid to a protein, while a deletion of three nucleotides will delete one amino acid from a protein. In general, the effects of small insertions or deletions in proteins depend on whether their length is a multiple of three. If the length is a multiple of three, then the result is the insertion or deletion of amino acids in the polypeptide chain. Insertions or deletions that are not a multiple of three shift the reading frame so that amino acids attached to the polypeptide chain following the site of the insertion or deletion are incorrect.

✓ Concept Check 37.2

3. A genotype is the genetic composition of a cell or organism, which in this case is SS. A phenotype of a cell or an organism refers to its observable features, which in this case is the disease sickle-cell anemia. The S allele produces an abnormal hemoglobin molecule, which causes it to bind in red blood cells, leading to the symptoms of the disease.

4. Harmful mutations reduce survival and reproduction; beneficial mutations increase survival and reproduction; and neutral mutations have no effect on survival or reproduction.

5. Designating a mutation as simply harmful, beneficial, or neutral does not account for whether it is homozygous or heterozygous

or the influence of the environment on the organism, which may determine how the mutation ultimately affects the individual's ability to survive and reproduce. For example, the S allele is very harmful as a homozygote but not as a heterozygote. Furthermore, the S allele is beneficial as a heterozygote in areas where malaria is widespread, but not in areas where malaria does not occur.

6. Because only a small fraction of the human genome codes for proteins or other functional elements, most mutations are neutral and don't have an observable effect.

✓ Concept Check 37.3

7. Mutagens are agents that increase the probability of mutation. Some common mutagens are: X-rays; ultraviolet light; bleach and hydrogen peroxide; and chemicals that are highly reactive, such as those in cigarette smoke.

8. Conjugation is the direct transfer for DNA from one cell to another; transformation is the uptake of DNA released from cells into other cells; and transduction is the transfer of DNA from one cell to another by a virus.

Review Questions

1. A; 2. B; 3. B; 4. A; 5. D; 6. B; 7. B

Module 38

✓ Concept Check 38.1

1. PCR is used to generate many copies of, or amplify, a piece of DNA.

2. The three steps of PCR are (1) denaturation of the double-stranded DNA into two individual strands, (2) annealing of the two primers to their complementary sequence on the DNA template strands, and (3) extension of the parental DNA strands through elongation (5' to 3') by DNA polymerase (by extending the primers).

3. PCR can be used for a variety of purposes such as matching a person's DNA sequence to evidence found at a crime scene. Paternity tests are performed in a similar way. PCR can be used to identify an organism based on a known conserved region of its DNA. It can also be used to mass produce certain sequences of DNA for DNA-based vaccines.

✓ Concept Check 38.2

4. DNA is negatively charged, so pieces of DNA move through a gel when an electric current is applied, toward the positive electrode and away from the negative electrode.

5. The distance that the DNA pieces move through the gel is based on their size, with smaller fragments moving more slowly and therefore shorter distances, and larger fragments

moving more quickly and therefore longer distances in a given amount of time.

✓ Concept Check 38.3

6. In Sanger sequencing, a DNA template, DNA primer, DNA polymerase, and normal nucleotides are necessary in addition to dideoxynucleotides.

7. The sequence of the template strand is antiparallel to the synthesized strand and inferred from complementary base pairing as 5'-CCTCGGT-3'.

✓ Concept Check 38.4

8. A restriction enzyme cuts DNA at specific, short sites.

9. Gel electrophoresis is a technique that can be used to determine the sizes of DNA fragments following digestion by restriction enzymes.

✓ Concept Check 38.5

10. A recombinant DNA molecule is a single DNA molecule that contains sequences from two or more different species.

11. The mammalian gene can be expressed in a bacterium through its insertion into a vector DNA that can be replicated in the bacterium. Donor DNA containing the mammalian gene and vector DNA are joined in a single plasmid, creating a recombinant DNA molecule. The vector DNA is then inserted into the bacterium by transformation. The plasmid is replicated and expressed in the bacterium just like any other DNA sequence.

Review Questions

1. B; 2. C; 3. B; 4. D; 5. C; 6. C

Module 39

✓ Concept Check 39.1

1. Viruses all have a nucleic acid genome, which can be DNA or RNA. They also all have a protein coat called a capsid.

2. Common shapes of viruses include the head-and-tail structure of some bacteriophages, the helical structure of tobacco mosaic virus, the icosahedral structure of adenovirus, and enveloped viruses such as coronaviruses and influenza virus.

3. Living organisms are made up of one or more cells that regulate the passage of substances into and out of them, store and transmit information, and harness energy from the environment. Viruses do not have these features, so are not considered to be living. They require a host cell to carry out their functions, replicate, and make more viruses.

✓ Concept Check 39.2

4. To infect a cell, a virus binds to a protein on the surface of the cell. Therefore, it is the presence of specific proteins on the cell surface that determines what type of cell a virus is able to infect.

5. Lysis is the process in which a cell bursts open to release progeny viruses. Lysogeny is the process in which the viral DNA is incorporated into the bacterial chromosome.

✓ Concept Check 39.3

6. The influenza virus causes the flu. SARS-CoV-2 causes the disease COVID-19. HIV causes AIDS.

7. Pathogenicity refers to whether an infectious agent, such as a virus, causes disease. A pathogenetic virus causes disease, and a nonpathogenic virus does not cause disease. Virulence describes the degree of severity of the disease. A highly virulent virus causes more severe disease than a less virulent virus does.

✓ Concept Check 39.4

8. Viruses infect all types of organisms, including bacteria, archaea, protists, fungi, plants, and animals.

9. Viruses are able to lyse (break open) phytoplankton, which slows their growth and returns their nutrients to the water.

Review Questions

1. A; 2. D; 3. A; 4. C; 5. B; 6. D; 7. A

Module 40

✓ Concept Check 40.1

1. Species differ in size, shape, color, movement, reproduction, how they harness energy, and many other ways.

2. Species have the same genetic material (DNA), the same amino acids that are used to build proteins, and the same biochemical machinery to build proteins.

3. Evolution is change over time. Traits that organisms share are often inherited from a common ancestor that had that trait. Traits that set organisms apart often arise over time after divergence from a common ancestor. This process was described by Charles Darwin, who called it descent with modification.

✓ Concept Check 40.2

4. Organisms vary because of differences in the environment (environmental variation) and differences in genetic material (genetic variation).

5. Genetic variation ultimately stems from mutations, or changes in the genetic material.

✓ Concept Check 40.3

6. Evolution is change in a population's genetic makeup over time, and natural selection is a mechanism or way by which this change occurs.

7. Natural selection results in antibiotic resistance. Before the application of antibiotics, bacteria with antibiotic resistance exist in low numbers. The application of antibiotics allows these bacteria to grow and reproduce more successfully than those that are susceptible to antibiotics. Over time, the use of antibiotics selects for antibiotic resistance among bacteria.

Review Questions

1. C; 2. B; 3. C; 4. B; 5. A; 6. A; 7. A

Module 41

✓ Concept Check 41.1

1. An adaptation is a close fit between an organism and its environment. Desert plants are adapted for dry conditions. A predator such as a shark or an eagle has sharp teeth and powerful jaws. A bacterium that lives in hot springs has a heat-stable DNA polymerase that functions at high temperatures. Each of these traits allows the organism to survive and reproduce well in the environment in which it lives.

2. The four postulates are: (1) there is variation among individuals; (2) some of this variation is passed on to offspring; (3) there is competition among individuals of a population, with some individuals surviving and reproducing more than others as a result of this variation; and (4) variations that are best suited to a particular environment will be passed on to the next generation more than variations that are less suited.

3. Fitness is the ability to survive and reproduce in a particular environment.

4. Fitness can be measured by the number of individuals that survive, by the number of offspring each individual has, or by the survival of those offspring.

✓ Concept Check 41.2

5. Convergent evolution

6. Natural selection increases the frequency of beneficial alleles, and decreases the frequency of harmful alleles.

7. Stabilizing selection is a form of selection that acts against extreme values of a trait and favors intermediate values of a trait. Human birth weight is an example.

8. Directional selection is a form of selection that favors one extreme and acts against the opposite extreme, so the trait moves in one direction over

time (bigger, faster, redder, etc.). The change in the population from lighter moths to darker moths in the United Kingdom in the late 1800s is an example.

9. Disruptive selection favors two extremes and acts against intermediate values, leading to the evolution of two distinct populations. The evolution of early- and late-feeding apple maggots is an example.

✓ Concept Check 41.3

10. Humans have produced breeds of dogs, cats, and many other animals and plants by artificial selection.

11. In both cases, there is phenotypic variation, and some individuals survive and reproduce more than others. In addition, in both cases the result is a change in the population over time, or evolution.

12. In artificial selection, humans do the selecting, whereas in natural selection, selection is the result of competition in nature. In artificial selection, breeders typically have a goal in mind, for example, cows that produce more milk. In natural selection, there is no goal.

✓ Concept Check 41.4

13. Sexual selection is a form of directional selection.

14. Intrasexual selection typically involves competition between males (such as direct combat), whereas intersexual selection typically involves males advertising to females with bright colors, melodic songs, or elaborate courtship dances.

Review Questions

1. A; 2. D; 3. B; 4. A; 5. B; 6. A

Module 42

✓ Concept Check 42.1

1. A phenotype is a trait or observable characteristic of an organism, whereas a genotype is its genetic makeup.

2. Phenotype variation is the result of differences in the genetic material among organisms as well as differences in the environment.

3. Because mutations are changes in an organism's DNA sequence, the mutation may harm the organism by damaging a protein or process, benefit the organism by improving the function of a protein or process, or have no discernable effect on a protein or process.

✓ Concept Check 42.2

4. Genetic variation is the result of mutation and recombination.

5. Mutation is the ultimate source of all present-day genetic variation.

6. By random, scientists mean that mutations arise without any regard to whether that mutation will be beneficial or harmful to the organism.

7. Point mutations are changes to a single base or nucleotide. Insertions and deletions are additions or removals of one or more bases. Transposable elements can jump into or out of a region of DNA.

✓ Concept Check 42.3

8. Population geneticists typically use allele frequency as a measure of genetic variation in populations.

9. Allele frequency is the proportion of a given allele among all the alleles of a gene in a population.

10. The allele frequency is the number of copies of allele A divided by the total number of alleles of that gene. So, in this case, it is 25 divided by 100, or $\frac{25}{100} = \frac{1}{4} = 0.25$.

Review Questions

1. D; 2. A; 3. B; 4. D; 5. B; 6. C; 7. A

Module 43

✓ Concept Check 43.1

1. Evolution is a change in the genetic makeup of a population over time. Because we can measure the genetic makeup of a population by its allele or genotype frequency, evolution is specifically a change in allele and/or genotype frequency over time.

2. The possible genotypes are CC, Cc, and cc.

3. The frequency of allele C is 30 divided by the total number of alleles, which in this case is $30 + 20 = 50$. So, the frequency of allele C is 30 divided by $50 = \frac{30}{50} = \frac{6}{10} = 0.6$.

✓ Concept Check 43.2

4. The five mechanisms of evolution are natural selection, genetic drift, migration, mutation, and nonrandom mating.

5. Natural selection is the only mechanism of evolution that results in adaptations.

6. Natural selection is an evolutionary mechanism that causes a change in allele or genotype frequencies within a population over time based on the relative fitness of each genotype in a particular environment. Due to competition for limited resources, those individuals with alleles that allow them to

survive and reproduce better than individuals without those alleles are more likely to pass on their genes to the next generation, thereby enriching subsequent generations of the population for these alleles and allowing for adaptation of the population to its environment over time.

7. Nonrandom mating causes the frequency of a genotype in a population to change without a change in the frequency of the alleles. In nonrandom mating, individuals of one genotype may preferentially mate with individuals of the same genotype or a different genotype, maintaining the same frequencies of alleles from generation to generation but configuring them preferentially in certain genotypes.

✓ Concept Check 43.3

8. If a population is at Hardy–Weinberg equilibrium, you can use allele frequencies to calculate genotype frequencies. In this case, p represents the allele frequency of one allele and q represents the frequency of the second allele. The frequency of homozygous dominant individuals is determined by calculating p^2, the frequency of homozygous recessive individuals is determined by calculating q^2, and the frequency of heterozygous individuals is determined by calculating $2pq$.

9. If a population is not in Hardy–Weinberg equilibrium, we can conclude that the population is evolving.

Review Questions

1. C; 2. C; 3. A; 4. C; 5. A; 6. B; 7. A

Module 44

✓ Concept Check 44.1

1. A homologous trait is a trait shared by two or more organisms because it was present in the common ancestor and then retained over time.

2. The amniotic egg in reptiles, birds, and mammals, and four walking legs in tetrapods are examples of homologous traits.

3. Eukaryotes all have a nucleus, mitochondria, membrane-bound organelles, linear chromosomes, and genes with exons and introns.

✓ Concept Check 44.2

4. A vestigial structure is one that was once functional but no longer is.

5. The whale pelvis and human appendix are both examples of vestigial structures.

✓ Concept Check 44.3

6. Fossilization is more likely in areas where sediment accumulates, like the bottom of the sea

floor, and for organisms with hard parts, such as skeletons or shells.

7. Radiometric dating, based on decay of unstable isotopes, can be used to determine the age of a fossil.

✓ Concept Check 44.4

8. The distribution of organisms depends on where they originated and how they are able to spread out or disperse.

9. Invasive species are species that are moved from their native habitat to a new habitat, where their populations grow rapidly, and disrupt the new ecosystem.

Review Questions

1. A; 2. A; 3. C; 4. B; 5. A; 6. A; 7. A

Module 45

✓ Concept Check 45.1

1. In moving from land to water, the hind limbs were lost, the forelimbs became fins, and the tail evolved to be shorter and more powerful.

2. The Grants learned that natural selection could act very quickly, in as little as 2 years, as the average size of the beaks became larger in response to the drought.

✓ Concept Check 45.2

3. Due to mutations, some bacteria are sensitive to antibiotics and some are resistant. In the presence of antibiotics, the resistant bacteria survive and reproduce more than the sensitive bacteria, so the frequency of bacterial resistance to antibiotics increases in the population due to natural selection.

4. Pathogens evolve resistance to the drugs we use to treat them; they sometimes evolve in such a way that they cause more severe disease or spread more easily; and they occasionally alter their host specificity so they can infect new species.

✓ Concept Check 45.3

5. A greenhouse gas allows solar radiation to pass through it in the atmosphere, but traps infrared radiation (heat) from the surface of Earth. In this way, it acts like the panes of glass in a greenhouse and helps to warm Earth.

6. Greenhouse gases include carbon dioxide, water, and methane.

7. Species may adapt to higher temperatures and changes in rainfall patterns; they might migrate to areas to which they are adapted; or they might go extinct.

Review Questions

1. C; 2. B; 3. D; 4. D; 5. A; 6. A

Module 46

✓ Concept Check 46.1

1. The base of the tree is called the root. The lines are the branches. The points where lines split into two are nodes. And the tips of the branches are present-day organisms, called taxa.

2. You can't read a phylogenetic tree by looking at the order of taxa (organisms) along the tips because this order can be changed by rotating the nodes. Rotating nodes does not change the evolutionary relationships, but it does change the order of taxa along the tips. Instead, phylogenetic trees should be read by following the branching order over time, from the past to the present.

3. A clade is a group of organisms that includes a common ancestor and all the descendants of that common ancestor. It is also called a monophyletic group and it reflects the evolutionary history of a particular group.

✓ Concept Check 46.2

4. Two species can share a trait either because the common ancestor of the two species had the trait and it was retained over time, or because the two species evolved the trait independently of each other as an adaptation to similar environments by convergent evolution.

5. Homologous and analogous traits are both shared by two different groups of organisms. However, homologous traits are shared because of common ancestry and analogous traits are shared because of convergent evolution. In addition, homologous traits evolve once, whereas analogous traits evolve more than once.

✓ Concept Check 46.3

6. Evolutionary trees can be built using shared derived characters, or they can be built using distance methods, which rely on the overall similarity between groups of organisms. In both cases, the traits can be morphological or molecular.

7. Molecular traits are useful because they offer an abundance of characters. For example, every base in DNA and every amino acid in a protein is a unique character, so a single molecule provides many independent pieces of data.

✓ Concept Check 46.4

8. Scientists use molecular data to track virus samples and build trees.

9. Building evolutionary trees can be used to track viral and bacterial infections in humans, figure out the source of an invasive species, and place species in their proper taxonomic group which in turn can inform treatment.

Review Questions

1. A; 2. A; 3. A; 4. C; 5. A; 6. A; 7. C

Module 47

✓ Concept Check 47.1

1. Although no one definition perfectly describes the genetic boundaries between different organisms, the biological species concept (BSC) defines a species as a group of organisms that can share genetic material by interbreeding with one another to produce viable, fertile offspring. By this definition, organisms that cannot successfully reproduce with one another belong to different species.

2. The biological species concept defines species based on their ability to interbreed to generate viable, fertile offspring, so organisms that reproduce asexually (such as bacteria) and those that are extinct cannot be easily characterized by this definition. A combination of phylogenetic and ecological species concepts works better to differentiate species in these organisms.

✓ Concept Check 47.2

3. Reproductive barriers that prevent fertilization (pre-zygotic) include reproducing in different places and at different times, differences in mating behaviors such as courtship rituals, mechanical blocks to fertilization such as differences in the genitalia required for mating, and incompatibilities of the gametes that can prevent the fusion of sperm and egg.

4. After fertilization has occurred, post-zygotic barriers include the inability of zygotes formed from gametes of different species to fully develop into adults (hybrid inviability) or the inability to develop into fertile adults (hybrid sterility). These barriers typically result because of differences at the genetic level, such as differences in the number of chromosomes between the two species.

✓ Concept Check 47.3

5. Allopatric speciation occurs after a geographic barrier physically separates two populations from each other. This barrier limits gene flow and promotes genetic divergence of the two populations as they adapt to their different environments. Sympatric speciation is the divergence of one group of organisms into two distinct species in the same geographical area. In sympatric speciation, gene flow between the diverging subpopulations is typically limited by strong selection against hybrid offspring.

6. Allopatric speciation can occur if a subset of a population moves to a new location far from the original population (dispersal)—for example, if a hurricane blew a group of beetles to a distant island off the coast of the mainland. Allopatric speciation can also occur by vicariance, when a geographic barrier divides a population into two distinct groups. An example of allopatric speciation by vicariance would be if a lake receded and began to dry out, leaving only smaller, isolated ponds behind. Fish from a single species might be split among several different ponds, each of which would then undergo genetic divergence. Ultimately, these different pond populations might evolve into new species.

7. Volcanic island chains, such as the Galápagos Islands, offer many examples of adaptive radiation because they are isolated, depend on the chance process of colonization, and have many ecological opportunities available to them.

Review Questions

1. A; 2. C; 3. B; 4. A; 5. B; 6. B; 7. C

Module 48

✓ Concept Check 48.1

1. Extinction is the loss of a group of organisms, such as a species.

2. On a phylogenetic tree, you can illustrate extinction by drawing a branch that does not extend all the way to the present.

3. Speciation and extinction over time determine the total number of species.

✓ Concept Check 48.2

4. Extinction is the loss of a single group of organisms. A mass extinction is a relatively rapid and large loss of many species of organisms.

5. There is strong evidence that a meteorite struck Earth 66 million years ago. As a result, ecosystems were severely disrupted, leading to the extinction of many groups of organisms, including the dinosaurs.

6. All five mass extinctions resulted from global ecological stress, such as climate change, volcanism, and so on. They all led to a large loss of biodiversity. And they all took place over a relatively short period of time.

✓ Concept Check 48.3

7. Adaptive radiation tends to occur after mass extinction events or whenever there are many open habitats. The same process occurs when an organism finds itself on a newly formed volcanic island where there are few colonists.

8. After the extinction of the dinosaurs 66 million years ago, mammals came to dominate many ecosystems. In fact, the next time period is informally called the Age of Mammals.

✓ Concept Check 48.4

9. Humans are affecting other species by destroying habitats, polluting the air and

water, introducing invasive species, hunting, overharvesting, and burning fossil fuels.

10. The dodo, Bali tiger, and northern white rhinoceros are examples of organisms that have gone extinct as a result of human activities.

Review Questions

1. A; 2. A; 3. D; 4. B; 5. B; 6. B

Module 49

✓ Concept Check 49.1

1. Variation is the raw material on which evolution acts. Therefore, populations with little genetic variation may be well adapted to the current environment, but may not be able to adapt if the environment changes.

2. Many of the crops and plants we grow and eat, such as corn, wheat, and bananas, have very little genetic variation and are vulnerable to pathogens. Endangered species, such as sea otters, cheetahs, and Tasmanian devils, also have little genetic variation.

✓ Concept Check 49.2

3. Populations with a lot of genetic variation are able to adapt in many different ways because there are lots of variants. A variant with low fitness in one environment may have high fitness in another.

4. Many strains of bacteria have considerable genetic variation, which enables them to evolve resistance to many kinds of antibiotics.

✓ Concept Check 49.3

5. Individuals that have one A allele and one S allele (AS individuals) have some protection against malaria compared to AA individuals. In addition, AS individuals do not get sickle-cell anemia, like SS individuals do.

6. AA individuals are well adapted to a malaria-free environment because there is no benefit to having the S allele in that environment, but there is a downside to having anemia.

7. The effect of a mutation on fitness (that is, whether it is beneficial, harmful, or neutral) depends on not just the mutation but also the environment. Therefore, a different environment can affect which variant is most fit.

Review Questions

1. C; 2. B; 3. A; 4. B; 5. B; 6. A; 7. D

Module 50

✓ Concept Check 50.1

1. The atmosphere of the early Earth contained carbon dioxide (CO_2), nitrogen (N_2), carbon

monoxide (CO), hydrogen gas (H_2), hydrogen sulfide (H_2S), ammonia (NH_3), and methane (CH_4), but no oxygen gas (O_2). Therefore, it is often described as a reducing atmosphere due to the lack of oxygen gas.

2. The Miller–Urey experiment demonstrated that amino acids can be synthesized from inorganic substances, such as H_2, H_2S, NH_3, and CH_4. Amino acids are the building blocks, or subunits, of proteins. Therefore, organic molecules might have been synthesized under conditions present on ancient Earth, which is a first step in making a cell, and therefore life.

✓ Concept Check 50.2

3. Experiments showing that nucleic acids can be synthesized from nucleotides, and proteins from amino acids, are critical for our understanding of how life might have originated. They demonstrate that organic molecules could be synthesized before life originated.

4. The observation that some RNA molecules can store information in their sequence of nucleotides, copy this information, and catalyze chemical reactions suggests that many of the functions important for life could have been carried out by RNA.

✓ Concept Check 50.3

5. All cells require an archive of information, a membrane to separate inside from outside and maintain the inside compatible with life, and the ability to harness energy from the environment.

6. The fact that phospholipids are able to form a membrane spontaneously, in the absence of an enzyme or a cell, suggests that this might have been a critical step in the formation of the first cell. For example, a liposome might have formed and encapsulated a self-replicating RNA with catalytic properties.

7. Metabolism describes the entire set of chemical reactions that convert molecules into other molecules and transfer energy in living organisms. In essence, these are the chemical reactions that sustain life. The earliest cells might have used fermentation reactions, anaerobic respiration, or photosynthesis to harness energy from the environment.

✓ Concept Check 50.4

8. The first cell was prokaryotic, a cell that lacks a nucleus and other internal membranes.

9. All the oxygen we breathe comes from a process called oxygenic photosynthesis (sometimes simply called photosynthesis). The first organisms to evolve the biochemical machinery for photosynthesis were cyanobacteria, a type of bacteria. They evolved this ability about 2.5 billion years ago.

10. The Cambrian explosion was a rapid diversification of marine animals that took place

about 541 million years ago. Many of the animal groups that are alive today first evolved at this time.

Review Questions

1. A; 2. D; 3. C; 4. C; 5. B; 6. B; 7. D

Module 51

✓ Concept Check 51.1

1. While a population is a group of individuals from the same species, it takes multiple populations to make up a community of interacting species.

2. Communities are comprised of all populations in a given area while ecosystems include both the living and the nonliving factors, such as water, temperature, and sunlight.

3. The laws for the conservation of energy and matter tell us that neither can be created or destroyed, but both can change forms, which means we can follow the movement of matter and energy in ecological systems.

✓ Concept Check 51.2

4. By responding to changes in the environment, organisms can use their behavioral and physiological responses to improve their evolutionary fitness.

5. A decreasing photoperiod indicates that winter is coming, and less food will be available, so the birds could use this environmental cue to migrate south to warmer areas where more food is available.

6. Kinesis is random movement in response to external environmental stimuli, and taxis is movement in a specific direction.

✓ Concept Check 51.3

7. Communication is the transfer of information between two individuals in which the sender is trying to manipulate the behavior of the receiver.

8. If the warning call helps protect the individual's offspring (or other relatives), the calling individuals would experience increased fitness.

9. The predator-specific calls provide unique information regarding the location of the predator, which helps other members of the species know both where to look for the predators and what evasive tactics might work the best against aerial versus ground predators.

✓ Concept Check 51.4

10. If the mice learned through habituation, they would initially show a fear response to the plastic hawk, but over time they would no longer show fear.

11. Innate behaviors are performed correctly with no previous experience, while learned behaviors require past experiences to perform the behavior correctly.

12. Cooperative behavior has several benefits, including finding food, identifying predators, group defense, and prey capture.

Review Questions

1. A; 2. B; 3. D; 4. C; 5. A

Module 52

✓ Concept Check 52.1

1. Endotherms regulate their body temperature to be constant regardless of the external temperature while ectotherms have a body temperature that is more affected by the external environment, although they can behaviorally adjust to thermoregulate.

2. When the air temperature is cold at night, they can burrow to conserve their body heat. When air temperature is cold during the day, they can bask in the sunlight to elevate their body temperature. When air temperature is too hot during the day, they can seek shade and cool soils to lower their body temperatures.

3. Larger animals have a lower per-kg metabolic rate because they lose less body heat due to their small surface area relative to their volume.

✓ Concept Check 52.2

4. Autotrophs, primary consumers, secondary consumers, and tertiary consumers.

5. A food chain is a simple, linear depiction of how energy and matter flow between species. A food web is a more complex depiction of how energy and matter flow between multiple species and includes species that feed on multiple trophic levels.

6. The ecological efficiency between adjacent trophic levels is approximately 10%, so each successive trophic level only contains about 10% of the energy of the trophic level below it.

✓ Concept Check 52.3

7. Pools represent where the carbon resides, while processes indicate how carbon is moved from one pool to another.

8. Human combustion of fossil fuels has moved buried carbon to Earth's surface, and its combustion has further moved carbon into the atmosphere.

Review Questions

1. A; 2. B; 3. C; 4. D; 5. A

Module 53

✓ Concept Check 53.1

1. Given that age structure affects how many individuals are reproductively mature versus immature, age structure affects the number of breeding individuals and therefore the number of offspring produced.

2. $N = \left(\frac{C}{R}\right) \times M = \frac{20}{10} \times 20 = 40$ squirrels

3. A population's geographic range is the entire area occupied by the population whereas the distribution is a measure of how clumped individuals are within their geographic range.

✓ Concept Check 53.2

4. Population growth rate is the birth rate minus the death rate.

5. A low maximum per capita growth rate causes the population to grow more slowly than does a high maximum per capita growth rate.

6. In the exponential growth model, the population growth rate increases slowly at low populations sizes, but increases rapidly over time at high population sizes.

✓ Concept Check 53.3

7. For a Type I survivorship curve, individuals experience high survival throughout most of their life and then their survival declines sharply later in life. For a Type III survivorship curve, individuals experience a sharp drop in survival early in life and then a slower decline throughout the rest of their life, with few individuals reaching adulthood.

8. K-selected species have low maximum per capita growth rates due to slow times to maturity, infrequent reproductive events, and few offspring per reproductive event. These characteristics result in populations that remain near the environment's carrying capacity. Examples include large mammals such as humans, horses, and elephants.

9. Characteristics of r-selected species include a high maximum per capita growth rate due to rapid times to maturity, frequent reproductive events, and many offspring per reproductive event. These characteristics result in populations that exhibit large fluctuations in population sizes. Examples includes mice, insects, and frogs.

Review Questions

1. D; 2. D; 3. A; 4. B; 5. A

Module 54

✓ Concept Check 54.1

1. Density-independent factors cause declines in populations regardless of the size of the population. Examples include fires, droughts, tornadoes, and hurricanes.

2. Density-dependent factors cause stronger limits on population growth as the population increases in size. Examples include limited food, limited nesting sites, and limited open spaces on rocks on an ocean shoreline.

✓ Concept Check 54.2

3. The logistic model adds the parameter of carrying capacity (K), which causes population growth to slow as it approaches the carrying capacity of the environment.

4. For a population that begins small, there is initially a slow rate of growth that rapidly increases as it approaches one-half of the carrying capacity. Beyond one-half carrying capacity, the growth rate slows until it eventually becomes zero, when the population reaches the carrying capacity.

5. When a population exceeds its carrying capacity, death rates increase and birth rates decline, resulting in a rapid die-off of the population.

Review Questions

1. D; 2. C; 3. D; 4. A; 5. B

Module 55

✓ Concept Check 55.1

1. Simpson's diversity index includes both species richness and species evenness.

2. The fundamental niche is the full suite of abiotic conditions that allow a species to survive, grow, and reproduce. The realized niche is the range of abiotic plus biotic conditions under which a species survives, grows, and reproduces. The realized niche is typically smaller than the fundamental niche.

3. A niche generalist lives under a wide range of abiotic and biotic conditions, whereas a niche specialist lives under a narrow range of abiotic and biotic conditions.

✓ Concept Check 55.2

4. When predators become abundant, there is a resulting decline in the prey population and therefore an increase in the food, such as plants, normally consumed by the prey. With a decline in prey, the predator population has less food, so the predator population must decline. With fewer predators and more plants available, the prey population can then rebound, followed by a rebound in the predator population.

5. While parasites are species that live in or on another organism, only a subset of those parasites

cause diseases in their hosts and these are categorized as pathogens.

6. When a predator or herbivore preferentially consumes the superior competitor, the inferior competitors are better able to grow and reproduce.

✓ Concept Check 55.3

7. Beavers are keystone species because it only takes a few beavers to completely convert a flowing stream into a non-flowing pond. This flooding alters the environment by killing trees and favoring a new suite of plants and animals.

8. Direct effects occur between two species without any intermediate species involved. Indirect effects occur between two species with one or more intermediate species involved.

✓ Concept Check 55.4

9. As the plant community transitions from the initial composition of grasses and wildflowers to shrubs and then trees, it favors a change in the animal species that can live there, since each animal has a niche that must align with the plant species present.

10. Immediately after the major flood, we expect few—if any—species to be present. Over the next few weeks, different groups of algae and invertebrate animals will arrive, which will increase species richness over time.

Review Questions

1. A; 2. B; 3. C; 4. D; 5. A

Module 56

✓ Concept Check 56.1

1. Genetic diversity is the variety of genotypes present within a given population or species. Species diversity in the number of species and the evenness of their abundance.

2. Three processes that can affect the genetic diversity of a species at the level of cells include independent assortment, recombination, and mutation.

3. The factors that can affect species diversity in an ecological community include habitat size, habitat diversity, keystone species, and disturbances.

✓ Concept Check 56.2

4. When sea otter populations were sharply reduced by hunting, their sea urchin prey became more abundant. The sea urchins consumed much of the kelp forest, which dramatically altered this important habitat that protects young coastal fish.

5. Increases in biodiversity, such as increases in the number of plant species of soil fungus species, can lead to increases in primary productivity in ecosystems.

6. Ecosystem resilience generally increases with increases in biodiversity.

✓ Concept Check 56.3

7. Economic benefits include natural products that humans use, such as lumber, fur, meat, crops, and medicines. They also include flood control, water filtration, recreational opportunities, and crop pollination.

8. Species richness is highest near the equator and lowest near the poles.

9. Biodiversity hotspots are those areas with at least 1500 endemic plant species and experiencing at least 70% vegetation removal as a result of human activities.

Review Questions

1. A; 2. B; 3. C; 4. D; 5. A

Module 57

✓ Concept Check 57.1

1. Continental drift has caused continents to move across different latitudes, which has caused changes in the climate that any given continent has experienced and, as a result, changes in the ecosystems that can be supported on each continent.

2. With the receding glaciers of the most recent ice age, major shifts in the climate caused the boundaries of ecosystems to shift northward as species of plants and animals shifted their distributions.

3. When an organism functions as a keystone species, such as the mastodon, its extinction can cause widespread changes in the plants and animals that can persist, which can cause a dramatic change from one type of ecosystem (e.g., open habitats) to another (e.g., forested habitats).

✓ Concept Check 57.2

4. Using mammal fossils, we can compare the rate of extinction for mammals during the most recent 500 years with the rate during previous 500-year periods.

5. Amphibians have the highest percentage of threatened species, which is currently 40%.

6. The largest threat to species diversity is habitat loss.

7. The birds on islands have no evolutionary history with the exotic predators, so they often have no defenses against these predators.

✓ Concept Check 57.3

8. The two most common gases, nitrogen and oxygen, comprise 99% of the atmosphere, but they do not absorb and reemit infrared radiation from Earth so they cannot behave as greenhouse gases.

9. Greenhouse gases are naturally in the atmosphere and help to make Earth much warmer. Human activities have increased the concentrations of these gases, causing the planet to become even warmer than would occur naturally.

10. A small decline in pH can make it harder for corals to obtain calcium carbonate from the water to build the skeleton of the coral reef.

✓ Concept Check 57.4

11. Terrestrial habitat protection has been increasing since the 1960s while marine habitat protection has been growing rapidly since the 1980s.

12. The key to the recovery of the condor was a combination of reduced harvesting, reduced poisoning, and a program of captive rearing that allowed reintroductions.

13. Following the decline in the alligator population, the U.S. government first declared the alligator to be endangered, which stopped all hunting. Once the populations rebounded, the government allowed sustainable hunting by controlling how many alligators could be hunted each year.

Review Questions

1. A; 2. C; 3. B; 4. D; 5. A

The Periodic Table of the Elements

Abundance in cells

■ High	■ Low	□ Trace
		■ Rare or none

1 H																	2 He
3 Li	4 Be											5 B	6 C	7 N	8 O	9 F	10 Ne
11 Na	12 Mg											13 Al	14 Si	15 P	16 S	17 Cl	18 Ar
19 K	20 Ca	21 Sc	22 Ti	23 V	24 Cr	25 Mn	26 Fe	27 Co	28 Ni	29 Cu	30 Zn	31 Ga	32 Ge	33 As	34 Se	35 Br	36 Kr
37 Rb	38 Sr	39 Y	40 Zr	41 Nb	42 Mo	43 Tc	44 Ru	45 Rh	46 Pd	47 Ag	48 Cd	49 In	50 Sn	51 Sb	52 Te	53 I	54 Xe
55 Cs	56 Ba	57-71 La-Lu	72 Hf	73 Ta	74 W	75 Re	76 Os	77 Ir	78 Pt	79 Au	80 Hg	81 Tl	82 Pb	83 Bi	84 Po	85 At	86 Rn
87 Fr	88 Ra	89-103 Ac-Lr	104 Rf	105 Db	106 Sg	107 Bh	108 Hs	109 Mt	110 Ds	111 Rg	112 Cn	113 Nh	114 Fl	115 Mc	116 Lv	117 Ts	118 Og

57 La	58 Ce	59 Pr	60 Nd	61 Pm	62 Sm	63 Eu	64 Gd	65 Tb	66 Dy	67 Ho	68 Er	69 Tm	70 Yb	71 Lu
89 Ac	90 Th	91 Pa	92 U	93 Np	94 Pu	95 Am	96 Cm	97 Bk	98 Cf	99 Es	100 Fm	101 Md	102 No	103 Lr